Y0-BXH-222

Organic Electronic Spectral Data
Volume XXVIII 1986

Organic Electronic Spectral Data

Volume XXVIII 1986

JOHN P. PHILLIPS, DALLAS BATES
HENRY FEUER & B.S. THYAGARAJAN

EDITORS

CONTRIBUTORS

Dallas Bates
H. Feuer
L.D. Freedman

C.M. Martini
F.C. Nachod
J.P. Phillips

AN INTERSCIENCE ® PUBLICATION
JOHN WILEY & SONS, INC.
New York • Chichester • Brisbane • Toronto • Singapore

In recognition of the importance of preserving what has been written, it is a policy of John Wiley & Sons, Inc., to have books of enduring value published in the United States printed on acid-free paper, and we exert our best efforts to that end. This text is printed on acid-free paper.

An Interscience ® Publication

Library of Congress Catalog Card Number: 60-16428

ISBN 0-471-58588-2

Printed in the United States of America

10 9 8 7 6 5 4 3 2 1

INTRODUCTION TO THE SERIES

In 1956 a cooperative effort to abstract and publish in formula order all the ultraviolet-visible spectra or organic compounds presented in the journal literature was organized through the enterprise and leadership of M.J. Kamlet and H.E. Ungnade. Organic Electronic Spectral Data was incorporated in 1957 to create a formal structure for the venture, and coverage of the literature from 1946 onward was then carried out by chemists with special interests in spectrophotometry through a page by page search of the major chemical journals. After the first two volumes (covering the literature from 1946 through 1955) were produced, a regular schedule of one volume for each subsequent period of two years was introduced. In 1966 an annual schedule was inaugurated.

Altogether, more than fifty chemists have searched a group of journals totalling more than a hundred titles during the course of this sustained project. Additions and subtractions from both the lists of contributors and of journals have occurred from time to time, and it is estimated that the effort to cover all the literature containing spectra may not be more than 95% successful. However, the total collection is by far the largest ever assembled, amounting to well over a half million spectra in the twenty-eight volumes so far.

Volume XXIX is in preparation.

PREFACE

Processing of the data provided by the contributors to Volume XXVIII as to the last several volumes was performed at the University of Louisville.

John P. Phillips
Dallas Bates
Henry Feuer
B.S. Thyagarajan

ORGANIZATION AND USE OF THE DATA

The data in this volume were abstracted from the journals listed in the reference section at the end. Although a few exceptions were made, the data generally had to satisfy the following requirements: the compound had to be pure enough for satisfactory elemental analysis and for a definite empirical formula; solvent and phase had to be given; and sufficient data to calculate molar absorptivities had to be available. Later it was decided to include spectra even if solvent was not mentioned. Experience has shown that the most probable single solvent in such circumstances is ethanol.

All entries in the compilation are organized according to the molecular formula index system used by Chemical Abstracts. Most of the compound names have been made to conform with the Chemical Abstracts system of nomenclature.

Solvent or phase appears in the second column of the data lists, often abbreviated according to standard practice; there is a key to less obvious abbreviations on the next page. Anion and Cation are used in this column if the spectra are run in relatively basic or acidic conditions respectively but exact specifications cannot be ascertained.

The numerical data in the third column present wavelength values in nanometers (millimicrons) for all maxima, shoulders and inflections, with the logarithms of the corresponding molar absorptivities in parentheses. Shoulders and inflections are marked with a letter s. In spectra with considerable fine structure in the bands a main maximum is listed and labelled with a letter f. Numerical values are given to the nearest nanometer for wavelength and nearest 0.01 unit for the logarithm of the molar absorptivity. Spectra that change with time or other common conditions are labelled "anom." or "changing", and temperatures are indicated if unusual.

The reference column contains the code number of the journal, the initial page number of the paper, and in the last two digits the year (1986). A letter is added for journals with more than one volume or section in a year. The complete list of all articles and authors thereof appears in the References at the end of the book.

Several journals that were abstracted for previous volumes in this series have been omitted, usually for lack of useful data, and several new ones have been added. Most Russian journals have been abstracted in the form of the English translation editions.

ABBREVIATIONS

s	shoulder or inflections
f	fine structure
n.s.g.	no solvent given in original reference
$C_6H_{11}Me$	methylcyclohexane
C_6H_{12}	cyclohexane
DMF	dimethylformamide
DMSO	dimethylsulfoxide
THF	tetrahydrofuran

Other solvent abbreviations generally follow the practice of Chemical Abstracts.

Underlined data were estimated from graphs.

JOURNALS ABSTRACTED

Journal	No.	Journal	No.
Acta Chem. Scand.	1	Tetrahedron Letters	88
Indian J. Chem.	2	Angew. Chem.	89
Anal. Chem.	3	Polyhedron	90
J. Heterocyclic Chem.	4	Chem. Pharm. Bull. Japan	94
Ann. Chem. Liebigs	5	J. Pharm. Soc. Japan	95
Ann. chim. (Rome)	7	The Analyst	96
Applied Spectroscopy	9	Z. Chemie	97
Australian J. Chem.	12	J. Agr. Food Chem.	98
Steroids	13	Theor. Exptl. Chem.	99
Bull. Chem. Soc. Japan	18	J. Natural Products	100
Bull. Polish Acad. Sci.	19	J. Organometallic Chem.	101
Bull. soc. chim. Belges	20	Phytochemistry	102
Bull. soc. chim. France	22	Khim. Geterosikl. Soedin.	103
Can. J. Chem.	23	Zhur. Organ. Khim.	104
Chem. Ber.	24	Khim. Prirodn. Soedin.	105
Chem. and Ind. (London)	25	Die Pharmazie	106
Chimia	27	Synthetic Comm.	107
Compt. rend.	28	Israel J. Chem.	108
Doklady Akad. Nauk S.S.S.R.	30	Doklady Phys. Chem.	109
Gazz. chim. ital.	32	Russian J. Phys. Chem.	110
Helv. Chim. Acta	33	European J. Med. Chem.	111
J. Chem. Eng. Data	34	Spectroscopy Letters	112
J. Am. Chem. Soc.	35	Macromolecules	116
J. Pharm. Sci.	36	Org. Preps. and Procedures	117
J. Chem. Soc., Perkin Trans. II	39B	Synthesis	118
J. Chem. Soc., Perkin Trans. I	39C	J. Macromol. Sci.	121
J. Chim. Phys.	41	Makromol. Chem.	126
J. Indian Chem. Soc.	42	Croatica Chem. Acta	128
J. Org. Chem.	44	Bioorg. Chem.	130
J. Phys. Chem.	46	J. Mol. Structure	131
J. Polymer Sci., Chem. Ed.	47	J. Appl. Spectroscopy U.S.S.R.	135
J. prakt. Chem.	48	Carbohydrate Research	136
Monatsh. Chem.	49	Finnish Chem. Letters	137
Rec. trav. chim.	54	Chemistry Letters	138
Polish. J. Chem.	56	P and S and Related Elements	139
Spectrochim. Acta	59	J. Anal. Chem. U.S.S.R.	140
J. Chem. Soc., Faraday Trans. I	60	Heterocycles	142
Ber. Bunsen Gesell. Phys. Chem.	61	Arzneimittel. Forsch.	145
Z. phys. Chem.	62	Photochem. Photobiol.	149
Biol. Chem. Hoppe-Seyler	63	J. Chem. Research	150
Z. Naturforsch.	64	J. Photochem.	151
Zhur. Obshchei Khim.	65	Nouveau J. Chim.	152
J. Structural Chem.	67	J. Chem. Ecology	154
Biochemistry	69	J. Fluorine Chem.	155
Izvest. Akad. Nauk S.S.S.R.	70	Photobiochem. Photobiophys.	156
Coll. Czech. Chem. Comm.	73	Organometallics	157
Mikrochim. Acta	74	J. Antibiotics	158
J. Chem. Soc., Chem. Comm.	77	J. Carbohydrate Chem.	159
Tetrahedron	78	Anal. Letters	160
Revue Roumaine Chim.	80	Il Farmaco	161
Arch. Pharm.	83	Agr. Biol. Chem.	163
Talanta	86	Pure and Appl. Chem.	164
J. Med. Chem.	87	Acta Phys. Chem. Univ. Seged.	165

Organic Electronic Spectral Data
Volume XXVIII 1986

Compound	Solvent	λ_{max}(log ϵ)	Ref.
CBr_2F_2			
Methane, dibromodifluoro-	gas	229(2.85)	46-6464-86
at 800°K	gas	233(2.8)	46-6464-86
CHN_3S_3			
1,3,5,2,4,6-Trithia(3-S^{IV})triazepine	EtOH	327(3.54)	164-0197-86
CH_2FI			
Methane, fluoroiodo- (at 12°K)	nitrogen	255(2.48)	89-0819-86
CH_2I_2			
Methane, diiodo- (at 12°K) (after irradiation)	nitrogen	370(4.0),545(3.11)	89-0819-86
CH_3NO_2			
Nitrous acid, methyl ester (cross section 14x10^{-19})	gas	220(--)	46-2635-86
CH_3O_2S			
Methylsulfonyl radical	H_2O	332(2.95)	70-1140-86
CN_3O_6			
Methane, trinitro-, ion(1-)	aq buffer	350(4.18)	110-0195-86
C_2HCl_3			
Ethene, trichloro-	benzene	230(3.93)	104-0826-86
$C_2H_2N_2S_3$			
1,3,5,2,4-Trithia(3-S^{IV})diazepine	EtOH	224(3.76),330(3.64)	164-0197-86
$C_2H_3ClN_2O$			
3H-Diazirine, 3-chloro-3-methoxy-	A at 23°K	330(--),345(1.7), 362(--)	35-0099-86
C_2H_3ClO			
Methylene, chloromethoxy-	A at 23°K	325(1.8)(estd.)	35-0099-86
$C_2H_3Cl_3O_2$			
1,1-Ethanediol, 2,2,2-trichloro- (chloral hydrate)	H_2O	198(2.5)	65-2389-86
	pH 13	216(1.0)	65-2389-86
	0.1N H_2SO_4	197(2.4)	65-2389-86
C_2H_3O			
Ethyl, 1-oxo-	pH 1-3	<210(2.64)	39-1003-86B
$C_2H_4Br_2$			
Ethane, 1,2-dibromo-	n.s.g.	200(2.95)	3-2366-86
$C_2H_5NO_2$			
Acetohydroxamic acid, ferric chloride complex	acid	540(2.97)	150-2384-86M
$C_2H_5O_2S$			
Ethylsulfonyl radical	H_2O	332(3.00)	70-1140-86
$C_2H_6N_2S$			
Ethanethioamide, 2-amino-, monohydrochloride	EtOH	264(4.06)	33-1224-86
$C_2N_4S_4$			
1,3,2-Dithiazolo[4,5-e][1,3,2,4,7]-	EtOH	492(3.31)	164-0197-86

Compound	Solvent	λ_{max}(log ε)	Ref.
dithiatriazepine-3,5-S^{IV} (cont.)			164-0197-86
$C_3Cl_2F_4O$			
2-Propanone, 1,3-dichloro-1,1,3,3-tetrafluoro-	n.s.g.	298f(1.8)	46-0607-86
$C_3H_2N_4O$			
1H-1,2,3-Triazole-4-carbonitrile, 5-hydroxy-	H_2O	223(3.76),268(3.95)	103-0740-86
ammonium salt	H_2O	219(3.96),258(3.91)	103-0740-86
sodium salt	H_2O	219(3.86),259(3.89)	103-0740-86
$C_3H_2N_4S$			
1,2,3-Thiadiazole-4-carbonitrile, 5-amino-	EtOH	247(3.66),292s(3.63)	103-0567-86
C_3H_4NP			
1H-1,3-Azaphosphole	MeOH	210(3.48),242(3.34)	88-5699-86
$C_3H_4N_4O$			
Acetamide, 2-cyano-2-hydrazono-	EtOH	230(3.64),286(4.09)	103-0567-86
$C_3H_4N_4OS$			
1,2,3-Thiadiazole-4-carboxamide, 5-amino-	H_2O	260(3.99),291(3.89)	103-0567-86
1H-1,2,3-Triazole-4-carboxamide, 5-mercapto-	H_2O	256(4.03)	103-0567-86
$C_3H_4N_4O_2$			
1H-1,2,3-Triazole-4-carboxamide, 5-hydroxy-	H_2O	223(3.76),268(3.95)	103-0740-86
ammonium salt	H_2O	222(3.78),265(3.98)	103-0740-86
methanamine salt	H_2O	222(3.80),265(4.00)	103-0740-86
$C_3H_4O_3$			
Propanoic acid, 2-oxo-, sodium salt	pH 8	316(1.26)	3-2504-86
acyclic dimer	pH 8	326(1.40)	3-2504-86
cyclic dimer	pH 8	265(3.59)	3-2504-86
$C_3H_5Cl_2NO_2$			
Carbamic acid, dichloro-, ethyl ester	pentane	240(4.39),310(4.32)	104-0826-86
	benzene	310(4.43)	104-0826-86
C_3H_5N			
2-Propyn-1-amine	hexane	210(4.90),272(3.95)	65-2461-86
	MeOH	204(--),280(--)	65-2461-86
$C_3H_5N_3O$			
2-Propanone, 1-amino-3-diazo-, (R*,R*)-(±)-2,3-dihydroxybutanedioate (2:1)	MeOH	210(3.71),237(3.73), 277(4.00)	104-0353-86
$C_3H_5N_3O_2$			
2-Imidazolidinone, 1-nitroso-	neutral	247(3.92)	39-0117-86B
C_3H_5O			
Propyl, 2-oxo-	pH 1-3	220(2.69)	39-1003-86B
$C_3H_5O_2$			
Propyl, 3-hydroxy-2-oxo-	pH 1-3	<210(2.73)	39-1003-86B

Compound	Solvent	$\lambda_{max}(\log \epsilon)$	Ref.
$C_3H_6N_4$			
1H-Pyrazole-1,3-diamine	H_2O	234(3.71)	118-0071-86
1H-Pyrazole-1,5-diamine	H_2O	226(3.84)	118-0071-86
$C_3H_6N_4O$			
1H-Pyrazol-3-amine, 4,5-dihydro-1-nitroso-	EtOH	279(4.14)	104-0947-86
$C_3H_6N_4O_2$			
Propanediamide, 2-hydrazono-	EtOH	275(3.97)	103-0567-86
C_3H_6O			
Acetone	MeOH	270.6(1.20)	59-0771-86
C_3H_7N			
2-Propen-1-amine	hexane	213(4.23),270(3.60)	65-2461-86
	MeOH	207(--),278(--)	65-2461-86
C_3H_7NOS			
Ethanethioamide, N-hydroxy-N-methyl-	MeOH	208s(3.75),269(4.09)	44-5198-86
$C_3H_7N_5$			
1H-1,2,4-Triazole-3,5-diamine, N-methyl-	EtOH	216s(3.72)	4-0401-86
	10% EtOH-HCl	216(3.88)	4-0401-86
$C_3H_7O_2S$			
Propylsulfonyl radical	H_2O	332(3.04)	70-1140-86
C_3H_9N			
Methanamine, N,N-dimethyl-	gas	198(3.6),227(2.9)	35-3907-86
1-Propanamine	hexane	209(4.04)	65-2461-86
	MeOH	203(--)	65-2461-86

Compound	Solvent	λ_{max}(log ϵ)	Ref.
$C_4Cl_6O_3$			
Acetic acid, trichloro-, anhydride	hexane	200s(2.5)	65-2391-86
$C_4H_2ClF_3O$			
3-Buten-2-one, 1-chloro-3,4,4-tri-fluoro-	hexane	227(3.78),316(1.52)	22-0876-86
$C_4H_2Cl_2N_2$			
Pyridazine, 3,6-dichloro-	H_2O	217(3.82),272(3.14)	39-0359-86B
cation	H_0-5.70	217(3.96),275(3.31)	39-0359-86B
dication	H_0-10.02	217(3.87),276(3.35)	39-0359-86B
$C_4H_2N_2$			
2-Butenedinitrile, (E)-	gas	216(--)	46-2670-86
$C_4H_3ClN_2O$			
3(2H)-Pyridazinone, 6-chloro-, cation	H_0-2.95	213(3.79),284(3.31)	39-0359-86B
dication	H_0-10.02	212(3.71),276(3.35)	39-0359-86B
neutral	pH 2.48	213(3.81),225(3.70), 290(3.31)	39-0359-86B
anion	pH 13.98	235(3.89),309(3.41)	39-0359-86B
C_4H_3ClS			
Thiophene, 2-chloro-	hexane	238(3.9)	131-0353-86F
Thiophene, 3-chloro-	hexane	227s(3.7),244(3.7)	131-0353-86F
$C_4H_3F_3O$			
3-Buten-2-one, 3,4,4-trifluoro-	hexane	218(4.00),307(1.20)	22-0876-86
$C_4H_3N_3$			
1H-Imidazole-1-carbonitrile	MeCN	218(3.64)	69-7423-86
$C_4H_3N_3O_3$			
1,2,4-Triazine-6-carboxaldehyde, 2,3,4,5-tetrahydro-3,5-dioxo-	EtOH	266(3.71)	87-0809-86
$C_4H_4Br_2N_2O$			
3H-Pyrrol-3-one, 4,4-dibromo-2,4-dihydro-5-methyl-	EtOH	292(3.12)	150-0920-86M
$C_4H_4ClN_3$			
3-Pyridazinamine, 6-chloro-, cation	H_0-0.58	215(4.16),228(3.93), 300(3.38)	39-0359-86B
dication	H_0-10.02	215(4.16),300(3.39)	39-0359-86B
neutral	pH 5.47	210(4.07),239(4.02), 310(3.30)	39-0359-86B
$C_4H_4N_2$			
Pyrazine	pH 2.22	261(3.76),268(3.69), 302(2.88)	39-0359-86B
cation	H_0-1.0	268(2.85)	39-0359-86B
dication	H_0-7.82	284(4.29)	39-0359-86B
Pyridazine	pH 4.05	242(3.13),247(3.12), 300(2.40)	39-0259-86B
cation	H_0-0.49	215(3.26),238(3.23)	39-0359-86B
dication	H_0-8.14	216(3.36),235(3.26)	39-0359-86B
$C_4H_4N_2O$			
2(1H)-Pyrimidinone	H_2O	300(3.7)	131-0345-86I
	EtOH	307(3.7)	131-0345-86I
	MeCN	312(3.6)	131-0345-86I

Compound	Solvent	λ_{max}(log ϵ)	Ref.
4(1H)-Pyrimidinone	MeCN	222(3.9),298(3.6)	131-0345-86I
$C_4H_4N_2O_2$			
2(1H)-Pyrazinone, 4-oxide (emimycin)	H_2O	332(3.61)	94-3202-86
3,6-Pyridazinediol	pH 3.71	304(3.40)	39-0359-86B
anion	pH 7.11	331(3.36)	39-0359-86B
dianion	pH 14.28	335(3.46)	39-0359-86B
cation	H_0-2.49	213(3.82),287(3.43)	39-0359-86B
dication	H_0-10.02	215(3.81),274(3.18)	39-0359-86B
2,4(1H,3H)-Pyrimidinedione (uracil)	pH 3	204(4.04),257(4.01)	160-0979-86
$C_4H_4N_4O$			
1H-1,2,3-Triazole-4-carbonitrile, 5-hydroxy-1-methyl-	EtOH	219(3.68),261(3.71)	103-0740-86
sodium salt	EtOH	218(3.70),262(3.68)	103-0740-86
$C_4H_4N_6O_3$			
1,2,4-Triazolo[5,1-c][1,2,4]triazine, 4,7-dihydro-7-hydroxy-6-nitro-	H_2O	202(3.5),340(3.84)	103-0543-86
$C_4H_4O_5$			
Butanedioic acid, oxo-, enol with manganese ion present	pH 4.5	282.5(3.2)	69-3752-86
magnesium ion	pH 4.5	263.5(3.1)	69-3752-86
C_4H_4S			
Thiophene	hexane	232(3.8)	131-0353-86F
$C_4H_5BrN_2O$			
3H-Pyrazol-3-one, 4-bromo-2,4-dihydro-5-methyl-	EtOH	246(3.29)	150-0920-86M
$C_4H_5NOS_2$			
2(3H)-Thiazolethione, 3-hydroxy-4-methyl-	EtOH	296(4.13)	39-0039-86C
$C_4H_5N_3O$			
1H-Imidazole-1-carboxamide	H_2O	205(3.59),241(3.38)	69-7423-86
$C_4H_5N_3O_2$			
1H-Imidazole, 1-methyl-4-nitro-	n.s.g.	289(3.84)	112-1049-86
1H-Imidazole, 1-methyl-5-nitro-	n.s.g.	296(3.90)	112-1049-86
C_4H_6			
Butadiene	C_6H_{12}	217(4.32)	89-0578-86
C_4H_6NP			
1H-1,3-Azaphosphole, 2-methyl-	MeOH	209(3.62),247(3.53)	88-5699-86
$C_4H_6N_2O$			
1H-Pyrazole-1-methanol	EtOH	210(3.57)	142-2233-86
1H-Pyrazol-5-ol, 3-methyl-	EtOH	245(3.50)	150-0920-86M
$C_4H_6N_4$			
1,2,4,5-Tetrazine, 3,6-dimethyl-	dioxan	273(3.47),541(2.78)	88-5367-86
	DMSO	537(--)	88-5367-86
$C_4H_6N_4O$			
Acetamide, 2-cyano-2-hydrazono-N-methyl-	H_2O	245s(3.69),286(4.02)	103-0567-86

Compound	Solvent	λ_{max}(log ε)	Ref.
$C_4H_6N_4OS$			
1,2,3-Thiadiazole-4-carboxamide, 5-amino-N-methyl-	H_2O	258(3.35),283(3.78)	103-0567-86
1H-1,2,3-Triazole-4-carboxamide, 5-mercapto-N-methyl-	H_2O	258(4.00)	103-0567-86
1H-1,2,3-Triazole-4-carboxamide, 5-(methylthio)-	EtOH	233s(3.69),263(3.53)	103-0567-86
$C_4H_6N_4O_2$			
1,2,4-Triazine-3,5(2H,4H)-dione, 6-(methylamino)-	pH 1	232(3.90),306(3.67)	4-0033-86
	pH 11	227(4.04),241s(3.89), 295(3.60)	4-0033-86
	MeOH	233(3.88),307(3.68)	4-0033-86
$C_4H_6N_4S_2$			
1,2,4-Triazine-3,5(2H,4H)-dithione, 6-(methylamino)-	pH 1	249s(3.84),301(4.45)	4-0033-86
	pH 11	233(4.01),284(4.30), 327(4.03)	4-0033-86
	MeOH	253s(3.80),304(4.44)	4-0033-86
$C_4H_6O_4$			
Ethanedioic acid, dimethyl ester (dimethyl oxalate)	H_2O	259(1.84)	104-0644-86
$C_4H_7Cl_2O_4P$			
Phosphoric acid , 2,2-dichloroethenyl dimethyl ester	hexane	225(2.58)	65-2389-86
$C_4H_7N_3O$			
2-Butanone, 3-amino-1-diazo-, (R*,R*)-(±)-2,3-dihydroxybutanedioate (2:1)	MeOH	208(3.78),238(3.67), 279(3.75)	104-0353-86
2-Butanone, 4-amino-1-diazo-, (R*,R*)-(±)-2,3-dihydroxybutanedioate (2:1)	MeOH	210(3.83),241(3.71), 278(3.84)	104-0353-86
4(1H)-Pyrimidinone, 2-amino-5,6-dihydro-	H_2O	223(4.28),240(2.72)	142-3023-86
hydrochloride	H_2O	220(3.55),257(3.03)	142-3023-86
$C_4H_7N_5O$			
1H-1,2,3-Triazole-4-carboxamide, 5-amino-1-methyl-	H_2O	223(3.89),257(3.87)	103-0740-86
1H-1,2,3-Triazole-4-carbonitrile, 5-hydroxy-, compd. with methanamine	H_2O	219(4.00),258(3.93)	103-0740-86
C_4H_7O			
Butyl, 2-oxo-	pH 1-3	215(2.95)	39-1003-86B
Ethyl, 1,1-dimethyl-2-oxo-	pH 1-3	235(2.92)	39-1003-86B
Propyl, 1-methyl-2-oxo-	pH 1-3	230(2.82)	39-1003-86B
$C_4H_7O_5P$			
3-Buten-2-one, 1-(phosphonooxy)-, monosodium salt, (E)-	pH 7	215(3.79)(changing)	69-4699-86
$C_4H_8Cl_3O_4P$			
Phosphonic acid, (2,2,2-trichloro-1-hydroxyethyl)-, dimethyl ester	hexane	212(1.98)	65-2389-86
$C_4H_8N_2O$			
Acetic acid, ethylidenehydrazide	pH 4.73	226(4.07)	69-5774-86

Compound	Solvent	λ_{max}(log ε)	Ref.
$C_4H_8N_2S_4$			
Carbamodithioic acid, 1,2-ethanediylbis-, disodium salt	MeOH	212(3.29),261(3.33), 290(3.45),348(1.23)	131-0167-86H
	H_2O	207(--),260(--), 285(--),341(--)	131-0167-86H
	pH 1	240(--),271(--)	131-0167-86H
C_4H_8O			
2-Butanone	H_2O	267.21(1.32)	59-0771-86
	EtOH	273.37(1.26)	59-0771-86
Ethene, ethoxy-	hexane	196(2.00)	70-1677-86
C_4H_8S			
Ethene, (ethylthio)-	hexane	229(2.88),254s(--)	70-1677-86
C_4H_9NOS			
Ethanimidothioic acid, N-methyl-, methyl ester, N-oxide, (Z)-	MeOH	262(4.07)	44-5198-86
hydriodide	MeOH	218(4.22),262(3.93)	44-5198-86
Propanethioamide, N-hydroxy-N-methyl-	MeOH	203s(3.76),269(4.04)	44-5198-86
$C_4H_9N_5O_2$			
1H-1,2,3-Triazole-4-carboxamide, 5-hydroxy-, methanamine salt	H_2O	222(3.80),265(4.00)	103-0740-86
$C_4H_{10}N_4O$			
2-Triazene-1-carboxamide, N,1,3-trimethyl-	MeCN	244(3.96)	44-3751-86
$C_4H_{10}SSi$			
1-Thia-3-silacyclobutane, 3,3-dimethyl-	heptane	215s(3.3),256(2.9)	65-1848-86
$C_4H_{10}Si$			
Silane, ethylidenedimethyl- (at 77°K)	3-Mepentane	255(--)	138-2025-86
$C_4H_{11}N$			
1-Butanamine	hexane	none	23-1491-86
$C_4H_{11}O_2PS_2$			
Phosphorodithioic acid, O,O-diethyl ester, as lead salt	$CHCl_3$	293(3.87)	3-1845-86
$C_4H_{12}SSi$			
Methanethiol, (trimethylsilyl)-	heptane	208(2.32),232s(1.38)	67-0225-86

Compound	Solvent	λ_{max}(log ε)	Ref.
$C_5H_3Cl_2N_3O$			
5-Pyrimidinecarboxaldehyde, 2-amino-4,6-dichloro-	MeOH	281(4.26)	33-1602-86
$C_5H_3F_2N$			
Pyridine, 2,6-difluoro-	C_6H_{12}	253(3.53),254(3.56), 257(3.57),260(3.55), 261(3.53),264(3.43), 266(3.29)	59-1393-86
	EtOH	256(3.58),265(3.32)	59-1393-86
$C_5H_3N_3O_2S$			
1H-Thieno[2,3-d]-1,2,3-triazole-5-carboxylic acid	EtOH	272(3.09)	142-2299-86
C_5H_4BrNO			
2(1H)-Pyridinone, 5-bromo-	dioxan	231(3.90),246s(3.68), 289s(3.32),317(3.43), 328s(3.39),346s(3.04)	59-1289-86
$C_5H_4BrN_3O_2$			
1,2,4-Triazine-3,5(2H,4H)-dione, 6-(2-bromoethenyl)-, (E)-	EtOH	230(4.09),296(4.07)	87-0809-86
$C_5H_4ClIN_2O_2$			
Imidazole-4-carboxylic acid, 2-chloro-5-iodo-, methyl ester	MeOH	267(4.04)	78-1971-86
C_5H_4ClNO			
2-Pyridinol, 5-chloro-	C_6H_{12}	236(3.89),243s(3.80), 294(3.41),288s(3.34), 310(3.43),325(3.30), 340(3.00)	59-1289-86
	dioxan	226(3.82),237s(3.81), 245s(3.70),293(3.35), 317s(3.42),346s(3.04)	59-1289-86
$C_5H_4ClN_5$			
1H-Pyrazolo[3,4-d]pyrimidin-6-amine, 4-chloro-	MeOH	227(4.42),305(3.78)	33-1602-86
$C_5H_4F_3NO_2S$			
1H-Pyrrole, 1-[(trifluoromethyl)sulfonyl]-	CH_2Cl_2	242(1.84)	4-1475-86
$C_5H_4F_3N_5O$			
2-Azetidinone, 1-(1H-tetrazol-5-yl)-4-(trifluoromethyl)-	EtOH	215(3.93)	78-2685-86
$C_5H_4KN_3O_2$			
2-Pyridinamine, N-nitro-, potassium salt	H_2O	210(3.95),265(3.76), 330(3.33)	56-0429-86
$C_5H_4LiN_3O_2$			
2-Pyridinamine, N-nitro-, lithium salt	H_2O	200(3.95),264(3.62), 285s(3.51),325(3.46)	56-0429-86
$C_5H_4N_2O_2$			
Pyrazinecarboxylic acid	pH 0.87	209(3.83),270(3.90)	39-0359-86B
anion	pH 13.98	267(3.86),314(2.81)	39-0359-86B
cation	H_0-5.51	209(3.87),278(3.91)	39-0359-86B

Compound	Solvent	$\lambda_{max}(\log \epsilon)$	Ref.
Pyrazinecarboxylic acid, dication	H_0-8.14	209(2.87),293(3.90)	39-0359-86B
$C_5H_4N_3NaO_2$			
2-Pyridinamine, N-nitro-, sodium salt	H_2O	200(3.57),264(3.30), 286s(3.21),325s(3.10)	56-0429-86
$C_5H_4N_4$			
Pyrazolo[1,5-b][1,2,4]triazine	MeOH	227(4.27),298(3.23)	118-0071-86
C_5H_4OS			
4H-Pyran-4-thione	PF-1,3-DMCH	563(1.04)	46-6314-86
$C_5H_5F_3O$			
1-Penten-3-one, 1,1,2-trifluoro-	hexane	219(4.07),305(1.26)	22-0876-86
C_5H_5N			
Pyridine	1,2-$C_2H_4Cl_2$	257.0(3.23)	46-2589-86
	+ TFA	256.5(3.67)	46-2589-86
C_5H_5NO			
Pyridine N-oxide	EtOH	264(4.1)	94-0007-86
	MeCN	212(4.21),216.5(4.16), 276(4.14)	131-0375-86I
perchlorate	MeCN	215(3.78),258(3.56)	131-0375-86I
semiperchlorate	MeCN	215.5(4.00),220(3.97), 252(4.20)	131-0375-86I
C_5H_5NOS			
2(1H)-Pyridinethione, 1-hydroxy-	C_6H_{12}	292(4.23),370(3.58)	39-0039-86C
	EtOH	251(3.99),295(3.72), 362(3.28)	39-0039-86C
	MeCN	282(4.13),352(3.59)	39-0039-86C
C_5H_5NOSe			
2(1H)-Pyridineselone, 3-hydroxy-	H_2O	215(3.97),245s(--), 270s(--),362(3.77)	39-2075-86C
$C_5H_5NO_2$			
2-Furancarboxaldehyde, oxime, (E)-	EtOH	269(3.89)	94-3202-86
2-Furancarboxaldehyde, oxime, (Z)-	EtOH	264(3.90)	94-3202-86
$C_5H_5N_3O$			
Pyrazinecarboxamide	pH 0.55	210(3.89),270(3.89), 313(2.81)	39-0359-86B
cation	acid	215(3.88),279(3.94)	39-0359-86B
$C_5H_5N_3O_2$			
Formamide, N-(1,4-dihydro-4-oxo-5-pyrimidinyl)-	pH 1	254(3.91),284(3.97)	73-0215-86
	pH 13	245(3.95),285(3.83)	73-0215-86
Pyrazinecarboxylic acid, 3-amino-	pH 3.90	246(3.98),346(3.75)	39-0359-86B
anion	pH 10.91	243(3.98),340(3.76)	39-0359-86B
cation	pH 0.22	242(4.01),356(3.85)	39-0359-86B
dication	H -8.14	240(4.15),393(3.84)	39-0359-86B
2-Pyridinamine, N-nitro-	pH 5.10	205(4.93),238s(4.46), 265(4.79),338(5.20)	56-0429-86 +56-0441-86
	pH 6.45	205(4.78),238(4.58), 265(4.75),295s(4.60), 335(4.76)	56-0429-86
	pH 8.20	205(5.05),238(4.76), 270(4.81),290s(4.76)	56-0429-86

Compound	Solvent	λ_{max}(log ϵ)	Ref.
2-Pyridinamine, N-nitro-, lithium salt	H_2O	200(3.95),264(3.62), 285s(3.51),325(3.46)	56-0429-86
potassium salt	H_2O	210(3.95),265(3.76), 330(3.33)	56-0429-86
sodium salt	H_2O	200(3.57),264(3.30), 286s(3.21),325s(3.10)	56-0429-86
$C_5H_5N_3O_2S$			
Formamide, N-(1,2,3,4-tetrahydro-4-oxo-2-thioxo-5-pyrimidinyl)-	50%MeOH-HCl	222(4.10),287(4.14), 315(4.17)	73-0215-86
	50%MeOH-NaOH	225(4.14),275(4.01), 310(4.00)	73-0215-86
$C_5H_5N_3O_3$			
Formamide, N-(1,2,3,4-tetrahydro-2,4-dioxo-5-pyrimidinyl)-	50%MeOH-HCl	235(4.01),291(3.84)	73-0215-86
	50%MeOH-NaOH	235(3.98),282(3.85)	73-0215-86
$C_5H_5N_5O$			
4H-Imidazo[4,5-d]-1,2,3-triazin-4-one, 3,5-dihydro-3-methyl-	EtOH	249(3.51),260(3.47), 283(3.61)	5-1012-86
$C_5H_5N_5O_2S$			
1H-Pyrazino[2,3-c][1,2,6]thiadiazin-4-amine, 2,2-dioxide (in methanol)	pH 1	246(3.95),336(3.66), 401(3.47)	5-1872-86
	pH 7	259(4.16),381(3.73)	5-1872-86
$C_5H_5N_5S$			
1H-Imidazo[4,5-e]-1,2,4-triazine, 3-(methylthio)-	pH 1	250(4.29)	39-0931-86B
	pH 11	245(4.26),303(3.67), 340(3.47)	39-0931-86B
	MeOH	244(4.31),343(3.61)	39-0931-86B
1H-Pyrazolo[3,4-d]pyrimidine-4(5H)-thione, 6-amino-	pH 1	258(--),326(--)	33-1602-86
	pH 11	277(--),328(--)	33-1602-86
	MeOH	235(4.26),268(3.85), 333(4.29)	33-1602-86
C_5H_5O			
Pyrylium, perchlorate	n.s.g.	269(3.95)	104-0258-86
$C_5H_6ClN_3O$			
2(1H)-Pyrimidinone, 4-amino-5-chloro-hemihydrochloride hemihydrate	pH 8	217(4.11),287(3.86)	4-0505-86
	pH 1	217(4.08),298(4.00)	4-0505-86
$C_6H_6Cl_3N_3O_2$			
2(1H)-Pyrimidinone, 5,5-dichloro-4-(chlorimino)-5,6-dihydro-6-hydroxy-1-methyl-	pH 6	222(3.98)	4-0505-86
C_5H_6N			
Pyridinium (bromochromate)	H_2O	204(3.76),255(3.63), 347(2.65)	2-0228-86
$C_5H_6N_2$			
2-Pyridinamine	pH 0.6	230(4.0),304(3.9)	59-0929-86
	pH 1	300(3.80)	33-1025-86
	EtOH	294(3.60)	33-1025-86
$C_5H_6N_2O$			
1H-Imidazole, 1-acetyl-	C_6H_{12}	207(3.7),241(3.4)	9-0556-86
	EtOH	222(3.4),244(3.4)	9-0556-86
	n.s.g.	216(3.54),243(3.21)	9-0556-86

Compound	Solvent	λ_{max}(log ε)	Ref.
$C_5H_6N_2S_2$			
2-Thiazolecarbothioamide, N-methyl-	MeOH	206(4.06),328(3.78), 408s(2.15)	103-1139-86
$C_5H_6N_2Se$			
1,2,5-Selenadiazole, 3-ethenyl-4-methyl-	MeOH	222(3.85),309(3.99)	150-3060-86M
$C_5H_6N_4O$			
1H-Pyrazole-4-carbonitrile, 3-amino-5-methoxy-	pH 1	221(3.85)	4-1869-86
	pH 7	221(3.88)	4-1869-86
	pH 11	223(3.92)	4-1869-86
$C_5H_6N_4O_2$			
Formamide, N-(2-amino-1,4-dihydro-4-oxo-5-pyrimidinyl)-	pH 1	244(4.03),282(3.92)	73-0215-86
	pH 13	241(3.91),286(3.84)	73-0215-86
$C_5H_6N_4O_3$			
4-Pyridinol, 2,6-diamino-3-nitro-	MeOH	332(3.93),398(4.10)	83-0947-86
Urea, N-(2,5-dihydro-2,5-dioxo-1H-imidazol-4-yl)-N'-methyl-	pH 8	232(4.05),284(3.60)	78-0747-86
$C_5H_6N_4O_3S$			
Imidazo[4,5-c][1,2,6]thiadiazin-4(3H)-one, 1,7-dihydro-6-methyl-, 2,2-dioxide	MeOH	221(3.65),246(3.42), 292(3.74)	142-3451-86
$C_5H_6N_6O_3$			
1,2,4-Triazolo[5,1-c][1,2,4]triazin-4-ol, 1,4-dihydro-7-methyl-3-nitro-	H_2O	203(3.34),341(3.90)	103-0543-86
$C_5H_6O_2$			
4-Pentene-2,3-dione	MeCN	436(1.30)	89-1116-86
$C_5H_6O_3$			
2(5H)-Furanone, 4-(hydroxymethyl)-	EtOH	207.5(4.10)	150-0222-86S
$C_5H_7BrN_2O$			
3H-Pyrazol-3-one, 4-bromo-2,4-dihydro-4,5-dimethyl-	EtOH	272(3.51)	150-0920-86M
C_5H_7NO			
6H-1,2-Oxazine, 3-methyl-	hexane	235(4.15),290(4.14)	108-0033-86
	MeCN	242(3.92),285(3.86)	108-0033-86
$C_5H_7NO_2S$			
L-Proline, 5-thioxo-	EtOH	207(3.48),266(4.00)	5-0269-86
$C_5H_7N_3$			
3-Pyridazinamine, 6-methyl-	H_2O	214(4.05),225s(--), 309(3.30)	39-0359-86B
cation	H_0-1.72	216(4.14),303(3.42)	39-0359-86B
dication	H_0-8.13	216(3.89),270(3.42)	39-0359-86B
$C_5H_7N_3O$			
2(1H)-Pyrimidinone, 4-amino-1-methyl-	pH 2	215(3.9),282(4.0)	35-5109-86
	pH 6	197(4.3),276(3.8)	35-5109-86
$C_5H_7N_3O_2$			
1H-Imidazole, 1,2-dimethyl-4-nitro-	n.s.g.	301(3.84)	112-1049-86
1H-Imidazole, 1,2-dimethyl-5-nitro-	n.s.g.	230s(--),310(3.95)	112-1049-86

Compound	Solvent	$\lambda_{max}(\log \epsilon)$	Ref.
$C_5H_7N_3O_2S$			
1,2,3-Thiadiazole-4-carboxylic acid, 5-amino-, ethyl ester	H_2O	262(3.90),284s(3.78)	103-0567-86
$C_5H_7N_5O$			
2-Azetidinone, 4-methyl-1-(1H-tetrazol-5-yl)-	EtOH	223(3.95)	78-2685-86
C_5H_8BrNO			
4H-1,2-Oxazine, 4-bromo-5,6-dihydro-3-methyl-	hexane	247(3.38)	108-0033-86
	MeCN	242(3.34)	108-0033-86
C_5H_8ClNO			
4H-1,2-Oxazine, 4-chloro-5,6-dihydro-3-methyl-	hexane	239(3.43)	108-0033-86
	MeCN	235(3.38)	108-0033-86
C_5H_8FNO			
4H-1,2-Oxazine, 4-fluoro-5,6-dihydro-3-methyl-	hexane	234(3.13)	108-0033-86
	MeCN	232(3.09)	108-0033-86
C_5H_8INO			
4H-1,2-Oxazine, 5,6-dihydro-4-iodo-3-methyl-	hexane	255(3.55)	108-0033-86
	MeCN	248(3.62)	108-0033-86
C_5H_8NP			
1H-1,3-Azaphosphole, 2-ethyl-	MeOH	211(3.56),253(3.38)	88-5699-86
$C_5H_8N_2O$			
1H-Pyrazol-5-ol, 3,4-dimethyl-	EtOH	247(3.79)	150-0920-86M
$C_5H_8N_2O_3$			
Oxazole, 4,5-dihydro-2-(1-nitroethyl)-	EtOH	338(4.25)	4-0825-86
$C_5H_8N_4OS$			
1H-1,2,3-Triazole-4-carboxamide, 5-(ethylthio)-	H_2O	233s(3.71),262(3.69)	103-0567-86
$C_5H_8N_4O_2$			
1H-Pyrazole-4-carboxamide, 3-amino-5-methoxy-	pH 1	226(3.88)	4-1869-86
	pH 7	230(3.80)	4-1869-86
	pH 11	232(3.88)	4-1869-86
2,4(1H,3H)-Pyrimidinedione, 5-amino-6-(methylamino)-	pH 13	275(3.62)	35-5972-86
$C_5H_8O_2$			
2-Butenoic acid, methyl ester, (E)-	hexane	202(4.15)	35-3016-86
EtAlCl$_2$ complex	CH_2Cl_2	230(3.66)	35-3016-86
2-Butenoic acid, methyl ester, (Z)-	hexane	206(4.15)	35-3016-86
2,4-Pentanedione	50% DMSO	278(3.5)	35-7744-86
anion	50% DMSO	293(4.4)	35-7744-86
$C_5H_8O_2S_2$			
Acetic acid, (methylthio)thioxo-, ethyl ester	MeOH	326(3.50),499(0.91)	118-0968-86
$C_5H_8O_5S$			
Propanoic acid, 3-(methylsulfonyl)-2-oxo-, methyl ester	H_2O	244(1.91)	104-0644-86
$C_5H_8S_2$			
1,3-Dithiolane, 2-ethylidene-	H_2O	235(3.87)	23-1116-86

Compound	Solvent	$\lambda_{max}(\log \epsilon)$	Ref.
C_5H_9NO			
4H-1,2-Oxazine, 5,6-dihydro-3-methyl-	hexane	212(2.95)	108-0033-86
	MeCN	209(2.86)	108-0033-86
C_5H_9NS			
2-Propenethial, 3-(dimethylamino)-	n.s.g.	368(4.74)	70-0458-86
$C_5H_9N_3$			
2,3,4-Triazatricyclo[3.2.1.0^{2,4}]octane	EtOH	200(1.78)(end absorption)	33-2087-86
$C_5H_9N_3O$			
2-Pentanone, 3-amino-1-diazo-, (R*,R*)-(±)-2,3-dihydroxybutanedioate (2:1)	MeOH	208(3.57),240(3.55), 280(3.78)	104-0353-86
4(1H)-Pyrimidinone, 2-amino-5,6-dihydro-1-methyl-	MeOH	205(4.15),240(4.14)	142-3023-86
hydrochloride	MeOH	225(3.75)	142-3023-86
4H-1,2,3-Triazol-4-one, 3,5-dihydro-3,5,5-trimethyl-	hexane	247(3.621),305(2.560)	5-1891-86
$C_5H_9N_3O_2$			
Imidazolidine, 1-methyl-2-(nitromethylene)-, (E)-	EtOH	328(4.42)	24-2208-86
$C_5H_9N_5$			
1H-1,2,4-Triazole-3,5-diamine, N-(2-propenyl)-	EtOH	215s(3.72)	4-0401-86
	10%EtOH-HCl	216s(3.93)	4-0401-86
$C_5H_9N_5O$			
1H-1,2,3-Triazole-4-carbonitrile, 5-hydroxy-, ethanamine salt	H_2O	218(4.08),258(4.06)	103-0740-86
1H-1,2,3-Triazole-4-carboxamide, 1-ethyl-	EtOH	228(3.79),258(3.83)	103-0740-86
$C_5H_9N_5S$			
1,2,4-Triazine-5,6-diamine, N^6-methyl-3-(methylthio)-	pH 1	259(4.37)	4-0033-86
	pH 11	249(4.17),268s(4.10), 322(3.69)	4-0033-86
	MeOH	249(4.17),265s(4.11), 326(3.69)	4-0033-86
C_5H_9O			
Butyl, 1-methyl-3-oxo-	pH 1-3	220(2.85)	39-1003-86B
$C_5H_{10}S$			
Propanethial, 2,2-dimethyl-	n.s.g.	508(1.2)	35-2985-86
C_5H_{11}			
sec-Pentyl radical	pentane	234(3.23)	70-1140-86
	H_2O	234(3.04)	70-1140-86
$C_5H_{11}NOS$			
Butanethioamide, N-hydroxy-N-methyl-	MeOH	208s(3.63),270(4.03)	44-5198-86
Propanimidothioic acid, N-methyl-, methyl ester, N-oxide, hydriodide	MeOH	217(4.28),266(3.85)	44-5198-86
$C_5H_{11}N_3O_2$			
2-Triazene-1-carboxylic acid, 1,3-dimethyl-, ethyl ester	MeCN	232(4.11)	44-3751-86
$C_5H_{11}N_5$			
1H-1,2,3-Triazole-3,5-diamine,	EtOH	216s(3.77)	4-0401-86

Compound	Solvent	$\lambda_{max}(\log \epsilon)$	Ref.
1H-1,2,4-Triazole-3,5-diamine,	EtOH	216s(3.77)	4-0401-86
N-(1-methylethyl)- (cont.)	10%EtOH-HCl	216s(3.95)	4-0401-86
1H-1,2,4-Triazole-3,5-diamine,	EtOH	216(3.79)	4-0401-86
N-propyl-	10%EtOH-HCl	216s(3.90)	4-0401-86
$C_5H_{11}N_5O$			
1H-1,2,4-Triazol-3-amine, 5-[(2-	EtOH	216s(3.98)	4-0401-86
hydroxyethyl)amino]-1-methyl-	10%EtOH-HCl	212(4.14),216s(4.05)	4-0401-86
	10%EtOH-NaOH	221(4.11)	4-0401-86
$C_5H_{12}Cl_2N_2O_2Pt$			
Alanine, 3-amino-, ethyl ester, $PtCl_2$ complex	n.s.g.	272(2.24),297(2.34), 370(1.62)	94-2487-86
$C_5H_{12}N_2O_2$			
Alanine, 3-amino-, ethyl ester, platinum malonate complex	n.s.g.	242s(--),270s(--), 322(1.60)	94-2487-86
platinum oxalate complex	n.s.g.	256s(--),313(2.42)	94-2487-86
platinum sulfate complex	n.s.g.	260(2.36),324(2.00)	94-2487-86
$C_5H_{14}OSi$			
Silane, acetyltrimethyl-	n.s.g.	195(3.62),327f(2.10)	101-0021-86C
$C_5H_{14}O_3SSi$			
Silane, trimethoxy[(methylthio)-methyl]-	MeCN	194(2.42)	67-0062-86
$C_5H_{14}SSi$			
Ethanethiol, 1-(trimethylsilyl)-	heptane	200(2.20),227(1.84)	67-0225-86

Compound	Solvent	$\lambda_{max}(\log \epsilon)$	Ref.
$C_6Cl_4O_2$			
2,5-Cyclohexadiene-1,4-dione, 2,3,5,6-tetrachloro- (chloranil)	hexane	<u>360(2.5)</u>	59-1217-86
	benzene	<u>367(2.5)</u>	59-1217-86
$C_6H_2Cl_3NO$			
2,5-Cyclohexadien-1-one, 2,6-dichloro-4-(chloroimino)-	2M HCl	303(<u>4.3</u>),312(<u>4.3</u>) (changing)	39-0105-86B
	H_2O	303(--),312(--)	39-0105-86B
$C_6H_3Cl_3O$			
Phenol, 2,4,5-trichloro-	acid	289(3.43)	3-0639-86
Phenol, 2,4,6-trichloro-	acid	287(3.35)	3-0639-86
$C_6H_3Cl_3O_3S_2$			
1,3-Oxathiol-1-ium, 4-hydroxy-2-(methylthio)-5-(trichloroacetyl)-, hydroxide, inner salt	ether	268(4.11),280s(4.05), 403.5(3.63)	24-2308-86
	MeCN	266.5(4.08),304.8(3.82), 389(3.76)	24-2308-86
	CH_2Cl_2	267.8(4.11),303(3.82), 399(3.73)	24-2308-86
$C_6H_3N_3O_4$			
Benzofurazan, 5-nitro-, 1-oxide	EtOH	230(3.99),261(4.08), 322s(3.29),338s(3.28), 390(3.69)	12-0089-86
$C_6H_3N_3O_6S$			
Benzenethiol, 2,4,6-trinitro-	acetone	457(3.7)	104-0495-86
	MeCN	445(4.0)	104-0495-86
	DMF	455(4.3)	104-0495-86
	$MeNO_2$	458(3.6)	104-0496-86
	$C_6H_5NO_2$	460(3.4)	104-0495-86
potassium salt	acetone	457(4.3)	104-0495-86
	MeCN	445(4.3)	104-0495-86
	DMF	455(4.3)	104-0495-86
	$MeNO_2$	458(4.3)	104-0495-86
	$C_6H_5NO_2$	460(4.3)	104-0495-86
$C_6H_3N_3O_7$			
Phenol, 2,4,6-trinitro-	acid	355(4.03)	3-0639-86
$C_6H_4Br_2O$			
Phenol, 2,4-dibromo-	acid	285(3.33)	3-0639-86
$C_6H_4Cl_2O$			
Phenol, 2,3-dichloro-	acid	276(3.27)	3-0639-86
Phenol, 2,4-dichloro-	acid	284(3.33)	3-0639-86
Phenol, 3,4-dichloro-	acid	283(3.26)	3-0639-86
$C_6H_4N_2O$			
Benzenediazonium, 4-hydroxy-, hydroxide, inner salt	DMSO	347(4.58),356s(--)	104-0483-86
Benzofurazan	MeOH	<u>208(3.9),270f(3.7)</u>	48-0261-86
$C_6H_4N_2O_4$			
2,3-Pyrazinedicarboxylic acid	pH 0.55	213(3.76),270(3.80)	39-0359-86B
anion	pH 2.50	278(3.77)	39-0359-86B
dianion	pH 7.06	282(3.81)	39-0359-86B
cation	H_0-6.54	215(3.78),283(3.85)	39-0359-86B
dication	H_0-8.14	215(3.81),280(3.86)	39-0359-86B

Compound	Solvent	λ_{max}(log ϵ)	Ref.
$C_6H_4N_2O_5$			
Phenol, 2,4-dinitro-	acid	261(4.08)	3-0639-86
$C_6H_4N_2O_6$			
1-Cyclobutene-1,2-diol, 3,4-bis(nitromethylene)-, disodium salt	H_2O	249(4.0),326(4.1), 435s(4.1),478(4.2)	118-0215-86
$C_6H_4N_2S_3$			
1,3,5,2,4-Benzotrithiadiazepine-3-S^{IV}	EtOH	252(4.25),292(4.02), 365(3.63)	164-0197-86
$C_6H_4N_3O_2$			
Benzenediazonium, 2-nitro-, (E)-	pH 1.5	278(4.2)	65-1397-86
	pH 12.5	247(4.34),320s(3.63)	65-1397-86
Benzenediazonium, 3-nitro-, (E)-	pH 12.5	271(4.23)	65-1397-86
Benzenediazonium, 3-nitro-, (Z)-	pH 12.5	228(3.99),268(3.93)	65-1397-86
Benzenediazonium, 4-nitro-, (E)-	pH 12.5	220(4.10),320(4.13)	65-1397-86
Benzenediazonium, 4-nitro-, (Z)-	pH 12.5	307(3.92)	65-1397-86
Benzenediazonium, 4-nitro-, tetrafluoroborate	H_2O	261(4.301),312(3.447)	104-1114-86
	DMSO	310(3.58)	104-0483-86
$C_6H_4N_4$			
1-Propene-1,1,3-tricarbonitrile, 2-amino-	H_2O	274(4.18)	5-0533-86
anion	pH 13	302(4.44)	5-0533-86
$C_6H_4N_4O_3$			
4-Benzofurazanamine, 7-nitro-	MeOH	213(4.3),222s(4.2), 333(3.8),465(4.1)	48-0261-86
$C_6H_4N_6$			
1H-Imidazo[4,5-c]tetrazolo[1,5-a]pyridine	pH 1	251s(3.81),261(3.92), 270(3.93),280(3.86)	87-0138-86
	pH 13	260(3.76),267(3.77), 291(3.82)	87-0138-86
C_6H_4OS			
Thieno[3,4-b]furan	EtOH	216(3.65),221s(3.62), 259(3.74)	88-3045-86
$C_6H_4O_2$			
2,5-Cyclohexadiene-1,4-dione	pH 7	246(4.34),297(2.54), 424(1.38)	46-6266-86
	CH_2Cl_2	246(4.42),288(2.50), 439(1.35)	154-0561-86
$C_6H_4O_3$			
2,5-Cyclohexadiene-1,4-dione, 2-hydroxy-	acid	256(4.13),362(3.08)	28-1179-86B
anion	n.s.g.	485(3.40)	28-1179-86B
$(C_6H_4S)_n$			
Thiophene, 2-ethynyl-, polymer	1,2-$C_2H_4Cl_2$	290(3.9)	47-2021-86
C_6H_5BrO			
Phenol, 2-bromo-	acid	274(3.32)	3-0639-86
Phenol, 3-bromo-	acid	274(3.26)	3-0639-86
Phenol, 4-bromo-	acid	280(3.15)	3-0639-86
$C_6H_5ClN_2O_2$			
Benzenamine, 2-chloro-4-nitro-	anion	468(4.53)	32-0653-86

Compound	Solvent	λ_{max}(log ε)	Ref.
C_6H_5ClO			
Phenol, 2-chloro-	acid	273(3.28)	3-0639-86
Phenol, 3-chloro-	acid	274(3.22)	3-0639-86
Phenol, 4-chloro-	acid	280(3.15)	3-0639-86
$C_6H_5Cl_2NO$			
Phenol, 4-amino-2,6-dichloro-	pH 1	279(--),285(3.345)	39-0105-86B
	pH 12	244(3.862),324(3.630)	39-0105-86B
	EtOH	239(3.820),312(3.500)	39-0105-86B
$C_6H_5Cl_2NO_2S$			
Benzenesulfonamide, N,N-dichloro-	pentane	260(4.39),300(3.90)	104-0826-86
	benzene	280(4.41),300(4.32)	104-0826-86
$C_6H_5F_2N$			
Benzenamine, 2,5-difluoro-, chromium tricarbonyl complex	$CHCl_3$	216(4.78),318(3.91)	112-0517-86
C_6H_5NO			
Benzene, nitroso-	heptane-EtOH	218(3.81),280(4.01), 300(3.72),760(1.60)	80-0857-86
$C_6H_5NO_3$			
Phenol, 2-nitro-	acid	278(3.81)	3-0639-86
Phenol, 3-nitro-	acid	273(3.84)	3-0639-86
Phenol, 4-nitro-	heptane	285(4.00)	46-2340-86
	benzene	297(2.98)	46-2340-86
	dioxan	304(4.03)	46-2340-86
	THF	308(4.05)	46-2340-86
	CH_2Cl_2	300(4.01)	46-2340-86
	CCl_4	290(4.01)	46-2340-86
	DMF	319(4.04)	46-2340-86
	DMSO	322(4.04)	46-2340-86
	acid	317(4.00)	3-0639-86
	pH 9.5	400(4.3)	18-2032-86
with cyclodextrin	pH 9.5	405(4.3)	18-2032-86
$C_6H_5NO_3Se$			
Benzeneselenenic acid, 2-nitro-	H_2O	425(3.66)	44-0287-86
	NaOH	550(3.69)	44-0287-86
C_6H_5NS			
5H-Thieno[2,3-c]pyrrole	EtOH	230(4.13),285(3.87)	77-0310-86
$C_6H_5N_3O$			
4-Benzofurazanamine	MeOH	232(4.3),270s(2.6), 400(3.3)	48-0261-86
$C_6H_5N_3O_2S$			
1H-Thieno[2,3-d]-1,2,3-triazole-5-carboxylic acid, methyl ester	EtOH	272(3.11)	142-2299-86
$C_6H_5N_3O_3$			
Benzenamine, 4-nitro-N-nitroso-, sodium salt	H_2O	330(4.176)	104-1114-86
$C_6H_5N_3O_4$			
2-Propenoic acid, 3-(2,3,4,5-tetrahydro-3,5-dioxo-1,2,4-triazin-6-yl)-, (E)-	EtOH	225(4.06),299(4.14)	87-0809-86

Compound	Solvent	$\lambda_{max}(\log \epsilon)$	Ref.
$C_6H_5N_3O_5$			
Hydroxylamine, O-(2,4-dinitrophenyl)-	pH 7.5	300(4.03)	69-5602-86
$C_6H_6ClN_3O_2$			
Pyrazinecarboxylic acid, 3-amino-6-chloro-, methyl ester	pH 2.00	256(4.10),372(3.76)	39-0359-86B
cation	H_0-2.22	251(4.07),380(3.80)	39-0359-86B
dication	H_0-10.02	253(4.03),428(3.72)	39-0359-86B
$C_6H_6ClN_5$			
1H-Pyrazolo[3,4-d]pyrimidin-6-amine, 4-chloro-1-methyl-	MeOH	230(4.45),306(3.71)	33-1602-86
2H-Pyrazolo[3,4-d]pyrimidin-6-amine, 4-chloro-2-methyl-	MeOH	225(4.40),281(3.80), 322(3.65)	33-1602-86
$C_6H_6Cl_2O_2$			
2,4-Pentadienoic acid, 5,5-dichloro-, methyl ester, (E)-	n.s.g.	204(3.70),271(4.10)	104-1805-86
$C_6H_6FIN_2O_2$			
2,4(1H,3H)-Pyrimidinedione, 5-fluoro-6-iodo-1,3-dimethyl-	MeCN	271(3.77)	44-5148-86
$C_6H_6I_2N_2O_2$			
2,4(1H,3H)-Pyrimidinedione, 5,6-di-iodo-1,3-dimethyl-	EtOH	288(4.05)	44-5148-86
$C_6H_6N_2$			
1H-Pyrrole-3-carbonitrile, 2-methyl-	EtOH	217(3.85),237(3.79)	35-6739-86
1H-Pyrrole-3-carbonitrile, 5-methyl-	EtOH	216(3.80),238(3.64)	35-6739-86
$C_6H_6N_2O$			
Furo[3,4-d]pyridazine, 1,4-dihydro-	$CHCl_3$	363(c.2.34)	35-8088-86
$C_6H_6N_2O_2$			
Benzenamine, 4-nitro-	hexane	320(4.17)	46-2340-86
	benzene	344(4.15)	46-2340-86
	Bu_2O	346(4.20)	46-2340-86
	CCl_4	324(4.16)	46-2340-86
	DMF	382(4.27)	46-2340-86
3(2H)-Furanone, 5-(diazomethyl)-4-methyl-	EtOH	326(4.50)	103-0225-86
$C_6H_6N_2O_3S$			
Acetic acid, [(1,6-dihydro-6-oxo-2-pyrimidinyl)thio]-	MeOH	228(3.77),287(3.77)	2-0275-86
	MeOH-NaOH	243(3.77),276(3.69)	2-0275-86
$C_6H_6N_2O_4S$			
Diazenesulfonic acid, (4-hydroxy-phenyl)-, monosodium salt (agaradin)	MeOH	237(4.01),329(4.29)	88-0559-86
$C_6H_6N_2S$			
Thieno[3,4-d]pyridazine, 1,4-dihydro-	$CHCl_3$	362(c.2.58)	35-8088-86
2-Thiophenecarbonitrile, 3-amino-5-methyl-	MeOH	220(3.55),257(3.59), 300(3.52)	4-1757-86
$C_6H_6N_4$			
1H-Imidazo[4,5-c]pyridin-4-amine, dihydrochloride	pH 1	260(3.91),278(3.92)	87-0138-86
	pH 13	275(3.86)	87-0138-86
Propanedinitrile, 2-imidazolidin-ylidene-	EtOH	198(4.22),248(4.33)	24-2208-86

Compound	Solvent	λ max (log ε)	Ref.
$C_6H_6N_4O$			
2H-Pyrrolo[2,3-d]pyrimidin-2-one, 4-amino-1,7-dihydro-	M HCl	294(3.89)	39-0525-86B
	pH 13	254(3.83),291(3.85)	39-0525-86B
[1,2,4]Triazolo[1,5-a]pyrimidin-7(4H)-one, 5-methyl-	pH 4.0	212(3.80),271(3.71)	39-0711-86C
	pH 9.9	209(4.04),255(3.77), 280(4.06)	39-0711-86C
$C_6H_6N_4O_2$			
2,4-Pentanedione, 1,5-bis(diazo)-3-methyl-	EtOH	277(4.25)	103-0225-86
2,3-Pyrazinedicarboxamide	pH 2.21	213(3.91),269(3.81)	39-0359-86B
cation	H_0-5.51	241(4.00),268(3.78)	39-0359-86B
dication	H_0-8.14	225(3.89),278(3.82)	39-0359-86B
Pyrazolo[3,4-d]pyrimidin-4-one, 1,5-dihydro-3-methoxy-	pH 1 and 7	218(4.35),243s(3.71)	4-1869-86
	pH 11	223(4.15),265(3.60)	4-1869-86
Pyrazolo[3,4-d]pyrimidin-4-one, 1,5-dihydro-6-methoxy-	MeOH	234(3.93)	5-1213-86
$C_6H_6N_4O_2S$			
1H-Pyrido[2,3-c][1,2,6]thiadiazin-4-amine, 2,2-dioxide	pH -4.0	254(3.92),329(3.57), 364(3.60)	11-0607-86
	pH 2.0	259(4.11),292(3.25), 365(3.77)	11-0607-86
	pH 8.0	252(4.06),355(3.67)	11-0607-86
$C_6H_6N_4O_4$			
Hydrazine, (2,4-dinitrophenyl)-	buffer	349(4.08)	3-0631-86
$C_6H_6N_6O_2$			
Pyrazolo[5,1-d]-1,2,3,5-tetrazine-8-carboxamide, 3,4-dihydro-3-methyl-4-oxo-	$CHCl_3$	310(3.86)	87-1544-86
C_6H_6O			
Methyl, 3,4-furandiylbis- (at <80°K)	n.s.g.	338(--),348(--), 560(3+)	35-8088-86
7-Oxabicyclo[2.2.1]hepta-2,5-diene	MeCN	247(2.09)	24-0589-86
3-Oxatetracyclo[3.2.0.0^{2,7}.0^{4,6}]-heptane	MeOH	220(1.54)(end abs.)	24-0589-86
Phenol	acid	271(3.21)	3-0639-86
$C_6H_6O_2$			
1,2-Benzenediol	acid	275(3.37)	3-0639-86
1,3-Benzenediol	acid	273(3.25)	3-0639-86
1,4-Benzenediol	acid	288(3.42)	3-0639-86
3,8-Dioxatricyclo[3.2.1.0^{2,4}]oct-6-ene	MeCN	220(3.08)(end abs.)	24-0589-86
2(5H)-Furanone, 5-ethylidene, (Z)-	n.s.g.	272(4.1)	102-0649-86
1,5-Hexadiene-3,4-dione	MeCN	434(1.28)	89-1116-86
$C_6H_6O_3$			
1,2,3-Benzenetriol	acid	266(2.86)	3-0639-86
1,3,5-Benzenetriol	acid	267(2.66)	3-0639-86
Ethanone, 1-(2-furanyl)-2-hydroxy-	H_2O	223(3.40),275(4.16)	102-1472-86
2H-Pyran-2-one, 4-hydroxy-6-methyl-	H_2O	283(3.87)	137-0185-86
anion	H_2O	276(3.93)	137-0185-86
2H-Pyran-2-one, 6-hydroxy-4-methyl-	C_6H_{12}	225(3.68)	39-0973-86B
	H_2O	230(--),340(--)	39-0973-86B
	EtOH	240(3.03),340(3.19)	39-0973-86B
	dioxan	242(--)	39-0973-86B
	$CHCl_3$	245(3.44)	39-0973-86B
	80% DMSO	345(3.78)	39-0973-86B

Compound	Solvent	λ_{max}(log ε)	Ref.
4H-Pyran-4-one, 3-hydroxy-2-methyl- (maltol)	MeOH	215(3.92),278(3.72)	94-1015-86
C_6H_6S			
Methyl, 3,4-thiophenediylbis- at <80°K	EtOH	572(4+)	35-8088-86
$C_6H_7BrN_2OS$			
2(1H)-Pyrimidinone, 5-bromo-4-(ethylthio)-	H_2O	278(4.89),312(4.89)	56-0605-86
$C_6H_7ClN_2OS$			
2(1H)-Pyrimidinone, 5-chloro-4-(ethylthio)-	H_2O	278(4.87),312(4.91)	56-0605-86
4(1H)-Pyrimidinone, 5-chloro-2-(ethylthio)-	H_2O	252(4.88),290(4.83)	56-0605-86
$C_6H_7ClN_2O_2S$			
Benzenesulfonic acid, 4-chloro-, hydrazide	pH 1	231(4.21)	140-0140-86
	H_2O	250(4.13)	140-0140-86
$C_6H_7Cl_3O_2$			
2-Pentenoic acid, 5,5,5-trichloro-, methyl ester, (E)-	n.s.g.	205(4.19),266(2.51)	104-1805-86
$C_6H_7Cl_3O_4$			
Propanedioic acid, methyl 2,2,2-trichloroethyl ester	MeOH	214(2.10)	5-1968-86
$C_6H_7F_3O$			
1-Hexen-3-one, 1,1,2-trifluoro-	hexane	220(4.09),307(1.32)	22-0876-86
1-Penten-3-one, 1,1,2-trifluoro-4-methyl-	hexane	221(4.08),314(1.43)	22-0876-86
$C_6H_7IN_2O_2$			
2,4(1H,3H)-Pyrimidinedione, 6-iodo-1,3-dimethyl-	EtOH	273(3.84)	44-5148-86
C_6H_7N			
Aniline	pH 3	228(3.25),272(2.41)	160-0979-86
protonated	98% H_2SO_4	252f(2.3)	112-0265-86
Pyridine, 2-methyl-	pH 3	203(3.54),261(3.89)	160-0979-86
	1,2-$C_2H_4Cl_2$	263.0(3.43)	46-2589-86
	+ TFA	264.0(3.83)	46-2589-86
Pyridine, 3-methyl-	1,2-$C_2H_4Cl_2$	264.5(3.40)	46-2589-86
	+ TFA	264.0(3.74)	46-2589-86
Pyridine, 4-methyl-	1,2-$C_2H_4Cl_2$	257.5(3.23)	46-2589-86
	+ TFA	254.5(3.66)	46-2589-86
C_6H_7NO			
Phenol, 2-amino-	acid	270(3.30)	3-0639-86
Phenol, 3-amino-	acid	270(3.27)	3-0639-86
Phenol, 4-amino-	acid	272(3.22)	3-0639-86
Pyridine, 2-methoxy-	MeOH	269(3.51)	142-2403-86
2(1H)-Pyridinone, 1-methyl-	MeOH	226(3.79),297(3.76)	142-2403-86
$C_6H_7NO_2Se$			
1-Cyclobutene-1-selenol, 2-(dimethylamino)-3,4-dioxo-, sodium salt	H_2O	297(4.0),355(4.4)	24-0182-86

Compound	Solvent	$\lambda_{max}(\log \epsilon)$	Ref.
$C_6H_7NO_3$			
2H-1,3-Oxazine-2,4(3H)-dione, 3,6-dimethyl-	EtOH	232(3.88)	94-1809-86
$C_6H_7NO_3S$			
Benzenesulfonic acid, 4-amino-	pH 3	200(4.47),248(4.21)	160-0979-86
C_6H_7NS			
Benzenethiol, 2-amino-	acid	270(3.83)	41-0589-86
	pH 2.9	293(3.38)	41-0589-86
	pH 8.2	320(3.50)	41-0589-86
	EtOH	212(--),237s(--), 297.5(--)	41-0589-86
2(1H)-Pyridinethione, 1-methyl-	MeOH	281(4.13),351(3.85)	142-2403-86
$C_6H_7N_3OS$			
Propanal, 2-oxo-, 1-(2-thiazolylhydrazone)	50% EtOH	340(4.36)	65-2114-86
$C_6H_7N_3O_2$			
Formamide, N-(1,4-dihydro-2-methyl-4-oxo-5-pyrimidinyl)-	50%MeOH-HCl	256(3.89),285(3.99), 291(3.99)	73-0215-86
	50%MeOH-NaOH	248.5(3.98),292(3.89)	73-0215-86
Hydrazine, (4-nitrophenyl)-	buffer	400(4.02)	3-0631-86
Pyrazinecarboxylic acid, 3-amino-	pH 5.30	246(4.03),344(3.81)	39-0359-86B
cation	H_0-1.00	243(4.03),355(3.85)	39-0359-86B
dication	H_0-8.00	241(4.17),392(3.86)	39-0359-86B
$C_6H_7N_3O_3$			
Formamide, N-(1,4-dihydro-2-methoxy-4-oxo-5-pyrimidinyl)-, dihydrate	50%MeOH-HCl	244(3.96),277(3.85)	73-0215-86
	50%MeOH-NaOH	246(3.94),286(3.88)	73-0215-86
$C_6H_7N_3O_3S$			
Acetic acid, [(6-amino-1,4-dihydro-4-oxo-2-pyrimidinyl)thio]-	MeOH	230s(4.14),273(3.82)	2-0275-86
	MeOH-NaOH	221(4.31),263(4.04)	2-0275-86
$C_6H_7N_3O_4S$			
Benzenesulfonic acid, 3-nitro-, hydrazide	pH 1	253(3.99)	140-0140-86
	H_2O	288(3.89)	140-0140-86
$C_6H_7N_5$			
7H-Purin-6-amine, 7-methyl-	pH 1	272(4.19)	94-2037-86
	H_2O	269(4.08),280s(3.86)	94-2037-86
	pH 13	270(4.05),280s(3.83)	94-2037-86
	EtOH	269(4.06),283s(3.84)	94-2037-86
$C_6H_7N_5O$			
4H-Imidazo[4,5-c]pyridin-4-one, 2,6-diamino-1,5-dihydro-	pH 2	216(4.53),256(4.05), 312(3.98)	87-2034-86
1H-Pyrazolo[3,4-d]pyrimidin-4-amine, 3-methoxy-	pH 1	228(4.34)	4-1869-86
	pH 7	244(3.80),276(3.60)	4-1869-86
	pH 11	276(3.58)	4-1869-86
1H-Pyrazolo[3,4-d]pyrimidin-4-amine, 6-methoxy-	MeOH	263(3.96)	5-1213-86
1H-Pyrazolo[3,4-d]pyrimidin-6-amine, 4-methoxy-	MeOH	244(3.74),276(3.89)	33-1602-86
$C_6H_7N_5OS$			
3H-Pyrazol-3-one, 5-amino-1,2-dihydro-1-methyl-4-(1,3,4-thiadiazol-2-yl)-	MeOH	297(3.17897)	161-0666-86

Compound	Solvent	λ_{max}(log ϵ)	Ref.
$C_6H_7N_5O_2S$			
8H-Pyrazino[2,3-c][1,2,6]thiadiazin-4-amine, 8-methyl-, 2,2-dioxide	pH 7.0	258(3.98),313(3.18), 401(3.76)	5-1872-86
	MeOH	260(4.00),324s(3.23), 410(3.78)	5-1872-86
$C_6H_7N_5S$			
5H-Imidazo[4,5-e]-1,2,4-triazine, 5-methyl-3-(methylthio)-	pH 1	257.5(4.45)	39-0931-86B
	pH 11	236(4.23),252s(4.19), 272s(4.09),323(3.89)	39-0931-86B
	MeOH	247(4.37),344(3.75)	39-0931-86B
7H-Imidazo[4,5-e]-1,2,4-triazine, 7-methyl-3-(methylthio)-	pH 1	256(4.42)	39-0931-86B
	pH 11	266(4.31),287s(4.14), 353(3.71)	39-0931-86B
	MeOH	246(4.38),284(3.62), 356(3.58)	39-0931-86B
2H-Purine-2-thione, 1,3,6,7-tetrahydro-6-imino-1-methyl-, hydrochloride	pH 1	239(4.28),286(4.27)	4-0737-86
	pH 13	237(4.29),287(4.18)	4-0737-86
C_6H_7O			
Pyrylium, 2-methyl-, perchlorate	n.s.g.	279(4.01)	104-0258-86
Pyrylium, 3-methyl-, perchlorate	n.s.g.	263(4.24)	104-0258-86
C_6H_8			
1,3,5-Hexatriene	C_6H_{12}	268(4.54)	89-0578-86
1,4-Pentadiene, 3-methylene-	C_6H_{12}	224(4.41)	89-0578-86
$C_6H_8N_2$			
2(1H)-Pyridinimine, 1-methyl-	EtOH	350(3.50)	33-1025-85
	pH 1	303(--)	33-1025-86
$C_6H_8N_2O_2$			
Isoxazolo[4,3-c]pyridin-3-ol, 4,5,6,7-tetrahydro-	MeOH	257(3.94)	87-0224-86
$C_6H_8N_2O_2S$			
Benzenesulfonic acid hydrazide	pH 1	218(4.10)	140-0140-86
	H_2O	237(3.90)	140-0140-86
1,2,3-Thiadiazole-4-carboxylic acid, 5-ethyl-, methyl ester	C_6H_{12}	224(3.75),254(3.40)	44-4075-86
$C_6H_8N_2O_4$			
2,4,6(1H,3H,5H)-Pyrimidinetrione, 5-(2-hydroxyethyl)-	pH 10.0	268(4.30)	39-1391-86B
$C_6H_8N_3O$			
Pyrazinium, 3-(aminocarbonyl)-1-methyl-, iodide	pH 6.0	226(4.10),280(3.71)	103-1118-86
hydrated form	pH 9.4	229(4.08),267(3.97), 313(3.66)	103-1118-86
$C_6H_8N_4O_3$			
Urea, (1,3-dimethyl-2,5-dioxo-4-imidazolidinylidene)-	pH 8	225(4.15),282(3.53)	78-0747-86
$C_6H_8O_2$			
5-Hexynoic acid	MeOH	206(2.63)	18-3535-86
$C_6H_8O_4$			
Microthecin	dioxan	230(3.64),345(1.37)	102-1472-86

Compound	Solvent	λ_{max}(log ϵ)	Ref.
$C_6H_9NO_2$			
Cyclopropanecarboxamide, N-acetyl-	EtOH	195.7(3.31),200.6(4.01), 201.6(3.97),208.5(4.12)	104-1602-86
$C_6H_9NO_2S$			
L-Proline, 5-thioxo-, methyl ester	EtOH	266(4.21)	5-0269-86
$C_6H_9NO_4$			
Alanine, N-formyl-3-oxo-, ethyl ester, monosodium salt	MeOH	267(3.61)	73-0215-86
$C_6H_9NS_2$			
5(4H)-Thiazolethione, 2,4,4-trimethyl-	EtOH	235s(3.20),314(4.05), 490(1.11)	33-0374-86
$C_6H_9N_3O$			
4-Pyrimidinamine, N-hydroxy-2,6-dimethyl-	n.s.g.	242(4.12),270s(3.79)	103-1232-86
$C_6H_9N_5O_2S$			
Pyrazino[2,3-e]-1,2,4-triazine-6,7-diol, 1,2,6,7-tetrahydro-3-(methylthio)-, trans	pH 1	265(4.40),329s(3.54)	4-0033-86
	H_2O	221(4.04),253.5(4.21), 266s(4.12),330.5(3.76)	4-0033-86
	pH 11	253(4.19),267s(4.08), 330(3.78)	4-0033-86
C_6H_9O			
Cyclohexyl, 2-oxo-	pH 1-3	235(2.86)	39-1003-86B
$C_6H_{10}N_2O$			
1H-Pyrazole-1-methanol, 3,5-dimethyl-	EtOH	214(3.69)	142-2233-86
$C_6H_{10}N_2O_3$			
Oxazole, 4,5-dihydro-4,4-dimethyl-2-(nitromethyl)-	$CHCl_3$	325(4.34)	4-0825-86
$C_6H_{10}N_4O_2$			
1H-Pyrazole-4-carboxamide, 3-amino-5-ethoxy-	pH 1	225(3.97)	4-1869-86
	pH 7	231(3.94)	4-1869-86
	pH 11	233(3.94)	4-1869-86
$C_6H_{10}N_4S_2$			
1,2,4-Triazin-6-amine, N-methyl-3,5-bis(methylthio)-	pH 1	274(4.34),315s(3.95)	4-0033-86
	pH 11	267(4.34),355(3.58)	4-0033-86
	MeOH	267(4.36),362(3.56)	4-0033-86
$C_6H_{10}N_6O$			
[1,3'-Bi-1H-pyrazol]-3-amine, 4,4',5,5'-tetrahydro-1'-nitroso-	EtOH	345(4.46)	104-0947-86
$C_6H_{10}O$			
Cyclohexanone	MeOH	282.1(1.23)	59-0771-86
	EtOH	283.11(1.24)	59-0771-86
$C_6H_{10}O_2S_4$			
Thioperoxydicarbonic acid, diethyl ester	isooctane	240(4.24),286(3.92)	3-0588-86
	EtOH	290(3.91)	3-0588-86
$C_6H_{10}S_2$			
1,3-Dithiolane, 2-(1-methylethylidene)-	H_2O	230(3.89)	23-1116-86

Compound	Solvent	λ_{max}(log ε)	Ref.
C_6H_{11}			
Cyclohexyl	C_6H_{12}	245(3.11)	70-1140-86
$C_6H_{11}NO_2$			
2,3-Butanedione, mono(O-ethyloxime),	C_6H_{12}	234(4.03),318(1.40)	54-0054-86
(E)-	EtOH	238(4.08)	54-0054-86
(Z)-	C_6H_{12}	244(3.51),328(1.78)	54-0054-86
$C_6H_{11}NS$			
2-Propenethial, 3-(dimethylamino)- 2-methyl-	n.s.g.	365(4.79)	70-0458-86
$C_6H_{11}N_3$			
2,3,4-Triazatricyclo[3.2.1.0^{2,4}]octane, 3-methyl-	MeOH	202(1.60)	33-2087-86
$C_6H_{11}N_3O$			
2-Pentanone, 3-amino-1-diazo-4-methyl-, (R*,R*)-(±)-2,3-dihydroxybutanedioate (2:1)	MeOH	245(3.81),282(4.08)	104-0353-86
4(1H)-Pyrimidinone, 5,6-dihydro-1-methyl-2-(methylamino)-	MeOH	220(4.05),238(4.03)	142-3023-86
hydrochloride	MeOH	219(3.88)	142-3023-86
4(1H)-Pyrimidinone, 2-(dimethylamino)-5,6-dihydro-	MeOH	236(4.16)	142-3023-86
hydrochloride	MeOH	219(3.88)	142-3023-86
2,3,4-Triazatricyclo[3.2.1.0^{2,4}]octane-3-methanol	MeOH	202(1.54)	33-2087-86
$C_6H_{11}N_3O_3$			
2-Butenamide, N-methyl-3-(methylamino)-2-nitro-	EtOH	c.243(3.72),350(4.17)	18-3871-86
$C_6H_{11}N_3O_5$			
Formamide, N-(1,4-dihydro-2-methoxy-4-oxo-5-pyrimidinyl)-, dihydrate	50%MeOH-HCl	244(3.96),277(3.85)	73-0215-86
	50%MeOH-NaOH	246(3.94),286(3.88)	73-0215-86
$C_6H_{11}N_5$			
1H-1,2,4-Triazole-3,5-diamine, N^5-cyclopropyl-1-methyl-	EtOH	218(4.01)	4-0401-86
	10%EtOH-HCl	227s(4.14)	4-0401-86
	10%EtOH-NaOH	221(4.15)	4-0401-86
$C_6H_{11}O_2S$			
Cyclohexylsulfonyl	C_6H_{12}	350(2.95)	70-1140-86
$C_6H_{12}N_3$			
3,4-Diaza-2-azoniatricyclo[3.2.1.0^{2,4}]-octane, 2-methyl-, tetrafluoroborate	EtOH	202(1.90)	33-2087-86
$C_6H_{12}N_4S$			
1H-1,2,4-Triazole-1-carbothioamide, 4,5-dihydro-3,5,5-trimethyl-	EtOH	238(4.06),297(4.18)	103-0287-86
$C_6H_{12}O$			
2-Butanone, 3,3-dimethyl-	n.s.g.	186(3.04),278(1.18)	101-0021-86C
C_6H_{13}			
sec-Hexyl radical	hexane	240(3.11)	70-1140-86

Compound	Solvent	$\lambda_{max}(\log \epsilon)$	Ref.
$C_6H_{14}Cl_2N_2O_2Pt$			
Platinum, dichloro(ethyl 2,4-diamino-butanoate-N,N')-	n.s.g.	260(2.26),301(2.06), 357(1.72)	94-2487-86
$C_6H_{14}N_2$			
Diazene, bis(1-methylethyl)-, trans	isooctane	380(2.15)	24-1911-86
$C_6H_{14}N_2O_2$			
Butanoic acid, 2,4-diamino-, ethyl ester, PtSO complex	n.s.g.	250s(--),313(1.79)	94-2487-86
platinum malonate complex	n.s.g.	238(3.5),270s(--), 318(1.68)	94-2487-86
platinum oxalate complex	n.s.g.	248(3.4),312(2.54)	94-2487-86
$C_6H_{15}N_3$			
Triaziridine, bis(1-methylethyl)-, (1α,2β,3α)-	EtOH	202(1.30)	33-2087-86
$C_6H_{16}N_3O_3P$			
Phosphonic acid, (1,3-dimethyl-2-tri-azenyl)-, diethyl ester	MeCN	222(3.90)	44-3751-86
$C_6H_{16}O_3SSi$			
Silane, [(ethylthio)methyl]trimethoxy-	MeCN	194(2.34)	67-0062-86
$C_6H_{16}SSi$			
1-Propanethiol, 3-(trimethylsilyl)-	heptane	200(2.30),227(1.54)	67-0225-86
Silane, [(ethylthio)methyl]trimethyl-	heptane	205s(3.4)	65-1848-86
$C_6H_{16}Si_2$			
Silane, [(dimethylsilylene)methyl]-trimethyl- (at 77°K)	3-Mepentane	265(2.2)	138-2025-86
C_6N_4			
Ethenetetracarbonitrile, N,N-dimethylaniline complex	$CHCl_3$	680(4.17)	35-4920-86
N-ethylcarbazole complex	$CHCl_3$	588(4.23)	35-4920-86
hexamethylbenzene complex	$CHCl_3$	537(4.27)	35-4920-86

Compound	Solvent	λ_{max}(log ϵ)	Ref.
$C_7H_2N_4O_6$			
Benzofurazan, 4-hydroxy-5,7-dinitro-	C_6H_{12}	260(3.2),325(2.95)	65-1020-86
	dioxan	295(2.8),400(3.3)	65-1020-86
	DMSO	430(3.5)	
$C_7H_3Br_3N_2O_5$			
2,5-Cyclohexadien-1-one, 2,3,5-tri-bromo-4-methyl-4,6-dinitro-	$CHCl_3$	262(4.01),306(3.53)	12-0059-86
$C_7H_3Br_3N_2O_6$			
2,5-Cyclohexadien-1-one, 2,3,5-tri-bromo-4-methyl-6-nitro-4-(nitrooxy)-	$CHCl_3$	263(4.16),305(3.66)	12-0059-86
$C_7H_3Br_4NO_4$			
2,5-Cyclohexadien-1-one, 2,3,5,6-tetrabromo-4-methyl-4-(nitrooxy)-	$CHCl_3$	276(4.11),308(3.46)	12-0059-86
$C_7H_3D_2F_3N_2O_2$			
4(3H)-Pyrimidinone, 2,5,6-trifluoro-3-(2-propenyl-1,2-d_2)	C_6H_{12}	226(3.45),263(3.58)	39-0515-86C
$C_7H_3FN_2O_2$			
Benzonitrile, 4-fluoro-2-nitro-	MeOH	255(3.73),293s(3.32)	73-0698-86
$C_7H_3O_2S_4$			
1,3-Dithiole-4-carboxylic acid, 2-(1,3-dithiol-2-ylidene)-, radical cation inner salt	PhCN	400s(3.98),438(4.28), 590(3.80)	104-2131-86
$C_7H_4Br_3NO_4$			
2,5-Cyclohexadien-1-one, 2,3,5-tri-bromo-4-hydroxy-4-methyl-6-nitro-	$CHCl_3$	261(4.14),306(3.60)	12-0059-86
$C_7H_4ClN_3O_3$			
Benzonitrile, 4-chloro-3-(hydroxy-amino)-5-nitro-	MeOH	222(4.42),338(3.30)	4-1091-86
$C_7H_4Cl_3NO_2$			
2,5-Cyclohexadiene-1,4-dione, 2,3,5-trichloro-6-(methylamino)-	CH_2Cl_2	237(4.16),305(4.04), 540(3.24)	106-0832-86
$C_7H_4F_6N_4O$			
Pyrazinecarboxamide, 3-amino-5,6-bis(trifluoromethyl)-	MeOH	203(3.64),257(4.17), 353(3.72)	39-1043-86C
$C_7H_4N_2O_3$			
2-Propenenitrile, 3-(5-nitro-2-furanyl)-	EtOH	344(4.34)	103-0943-86
$C_7H_4N_4O_4$			
Benzoic acid, 2-azido-4-nitro-	EtOH or hexane	250(4.26),330(3.61)	130-0134-86
Benzoic acid, 5-azido-2-nitro-	EtOH or hexane	301(3.98)	130-0134-86
$C_7H_4O_2S_4$			
1,3-Dithiole-4-carboxylic acid, 2-(1,3-dithiol-2-ylidene)- sodium salt	MeCN	290s(4.08),302(4.14), 313(4.15),431(3.30)	104-2131-86
	H_2O	289(4.17),303(4.19), 314(4.19),402(3.47)	104-2131-86

Compound	Solvent	$\lambda_{max}(\log \epsilon)$	Ref.
$C_7H_5ClN_2$			
2-Propenenitrile, 3-(3-chloro-1H-pyrrol-2-yl)-, (E)-	n.s.g.	322(4.46)	150-3601-86M
(Z)-	n.s.g.	324(4.26)	150-3601-86M
$C_7H_5ClN_4O_2$			
2,4(1H,3H)-Pteridinedione, 7-chloro-1-methyl-	pH 1	236(4.12),255s(3.90), 335(4.03)	33-1095-86
	MeOH	203(4.30),237(4.08), 335(3.99)	33-1095-86
2,4(1H,3H)-Pteridinedione, 7-chloro-3-methyl-	MeOH	202(4.24),236(4.12), 333(4.00)	33-1095-86
$C_7H_5FN_2O$			
3H-Diazirine, 3-fluoro-3-phenoxy-	isooctane	325s(--),339(2.26), 350s(--),386(2.30)	44-2168-86
$C_7H_5FN_2O_3$			
Benzamide, 4-fluoro-2-nitro-	MeOH	245(3.74),295s(3.16)	73-0698-86
$C_7H_5FN_2O_4$			
Benzene, 5-fluoro-2-methyl-1,3-dinitro-	MeOH	232(4.2),299(3.6)	35-4115-86
$C_7H_5F_3N_2O$			
4(3H)-Pyrimidinone, 2,5,6-trifluoro-3-(2-propenyl)-	C_6H_{12}	221(3.64),265(3.59)	39-0515-86C
C_7H_5NO			
2-Propenenitrile, 3-(2-furanyl)-	EtOH	300(4.32)	103-0943-86
$C_7H_5NO_2$			
Benzaldehyde, 4-nitroso- (dimer)	EtOH	340(4.08)	23-2076-86
$C_7H_5NO_3$			
Benzaldehyde, 4-nitro-	pH 4.73	268(4.16)	69-5774-86
$C_7H_5N_3O_2$			
Benzoic acid, 3-azido-	EtOH or hexane	250(3.94)	130-0134-86
Benzoic acid, 4-azido-	EtOH or hexane	269(4.26)	130-0134-86
$C_7H_5N_3O_3$			
2H-Benzimidazol-2-one, 1,3-dihydro-5-nitro-	MeOH	204(4.63),223(4.69), 251.3(4.60),341(4.40)	80-0661-86
	MeOH-0.1M HCl	201.4(4.16),222.6(4.22), 251(4.12),340(3.96)	80-0661-86
	MeOH-1M HCl	202.8(4.05),222.8(4.17), 251.7(4.07),341(3.90)	80-0661-86
	MeOH-10M HCl	221.6(4.32),244.3(4.09), 324.4(3.98)	80-0661-86
	MeOH-0.1M NaOH	212(4.14),236.2(4.09), 269.7(4.12),403.7(4.22)	80-0661-86
	MeOH-0.2M NaOH	236.2(3.99),270(4.06), 403.8(4.23)	80-0661-86
	EtOH	200(4.53),224(4.53), 250(4.44),340(4.27)	80-0661-86
	DMF	290(3.32),347(3.57)	80-0661-86
1,2,4-Triazine-3,5(2H,4H)-dione, 6-(3-furanyl)-	EtOH	304(3.89)	87-0809-86

Compound	Solvent	λmax(log ε)	Ref.
$C_7H_5N_3O_4$			
Benzofurazan, 4-methyl-6-nitro-, 1-oxide	EtOH	231(3.99),259(4.04), 289s(3.74),320s(3.40), 336(3.36),393(3.73)	12-0089-86
$C_7H_5N_3O_6$			
Benzene, 2-methyl-1,3,5-trinitro-	aq Na_2SO_3	420(--)	96-0411-86
Formamide, N-(2,4-dinitrophenoxy)-	EtOH	286(4.06)	69-5602-86
$C_7H_5N_5$			
[2,2'-Bi-1H-imidazole]-4-carbonitrile	MeOH	273(4.58)	44-3228-86
$C_7H_6BrNO_2S$			
2,1-Benzisothiazole, 5-bromo-1,3-dihydro-, 2,2-dioxide	MeOH	238(4.04),293(3.18)	4-1645-86
C_7H_6ClNO			
2-Propenal, 3-(3-chloro-1H-pyrrol-2-yl)-, (E)-	EtOH	345(4.35)	150-3601-86M
$C_7H_6ClNO_2S$			
2,1-Benzisothiazole, 5-chloro-1,3-dihydro-, 2,2-dioxide	MeOH	237(4.02),293(2.74)	4-1645-86
$C_7H_6ClN_3O$			
Benzonitrile, 3-amino-4-chloro-5-(hydroxyamino)-	MeOH	209(4.35),231(4.50), 269s(3.56),327(3.57)	4-1091-86
$C_7H_6ClN_3O_2$			
Benzonitrile, 4-chloro-3,5-bis(hydroxyamino)-	EtOH	211(4.36),234(4.47), 270(3.70),324(3.57)	4-1091-86
$C_7H_6Cl_4O_2$			
1,3-Cyclopentadiene, 1,2,3,4-tetrachloro-5,5-dimethoxy-	n.s.g.	308(3.38)	104-1806-86
$C_7H_6F_2N_2O$			
4(3H)-Pyrimidinone, 2,5-difluoro-3-(2-propenyl)-	C_6H_{12}	218(3.45),266(3.59)	39-0515-86C
$C_7H_6F_3N$			
Benzenamine, 2-(trifluoromethyl)-, chromium tricarbonyl complex	$CHCl_3$	214(4.65),260s(4.03), 322(3.99),365(3.66)	112-0517-86
Benzenamine, 3-(trifluoromethyl)-, chromium tricarbonyl complex	$CHCl_3$	215(4.65),322(3.99)	112-0517-86
Benzenamine, 4-(trifluoromethyl)-, chromium tricarbonyl complex	$CHCl_3$	216(4.66),324(3.99)	112-0517-86
$C_7H_6F_3N_5OS$			
3H-Pyrazol-3-one, 5-amino-1,2-dihydro-1-methyl-4-[5-(trifluoromethyl)-1,3,4-thiadiazol-2-yl]-	MeOH	316(4.19865)	161-0666-86
$C_7H_6N_2$			
Imidazo[1,2-a]pyridine	H_2O	215(4.22),277(3.65)	150-0670-86M
	M HCl	211(4.18),277(3.81)	150-0670-86M
	EtOH	221(4.34),277(3.52), 296(3.54)	150-0670-86M
$C_7H_6N_2O$			
2H-Benzimidazol-2-one, 1,3-dihydro-	MeOH	207(5.28),227(4.45),	80-0661-86

Compound	Solvent	λ_{max}(log ϵ)	Ref.
2H-Benzimidazol-2-one, 1,3-dihydro- (cont.)		280.6(4.48)	80-0661-86
	MeOH-0.1M HCl	204(5.05),226s(--), 282(4.09)	80-0661-86
	MeOH-M HCl	204(4.70),226s(--), 282(4.06)	80-0661-86
	MeOH-10M HCl	206(4.32),269(3.88), 276(3.88)	80-0661-86
	MeOH-0.1M NaOH	214(4.22),245s(--), 287(3.79)	80-0661-86
	MeOH-0.2M NaOH	220.8(3.80),248.8(3.38), 287.4(3.65)	80-0661-86
	EtOH	206(4.83),228(4.26), 280(4.31)	80-0661-86
	DMF	288(4.27)	80-0661-86
Imidazo[1,2-a]pyridin-8-ol	H_2O	222(4.23),277(3.88), 298(4.00)	150-0670-86M
	M HCl	216(4.31),282(3.98), 290(3.94)	150-0670-86M
	M NaOH	229(4.27),293(3.96), 304(3.88)	150-0670-86M
	EtOH	223(4.43),282(3.91), 303(3.62)	150-0670-86M
$C_7H_6N_2O_4S$			
2,1-Benzisothiazole, 1,3-dihydro-5-nitro-, 2,2-dioxide	MeOH	207(4.06),318(4.00)	4-1645-86
2,1-Benzisothiazole, 1,3-dihydro-7-nitro-, 2,2-dioxide	MeOH	226(4.09),270(3.84), 355(3.60)	4-1645-86
$C_7H_6N_2S$			
2-Benzothiazolamine	EtOH	263(4.11)	24-1525-86
	MeCN	260(4.13),284s(3.53), 291(3.46)	24-1525-86
$C_7H_6N_4O_2S$			
2,4(1H,3H)-Pteridinedione, 7,8-dihydro-1-methyl-7-thioxo-	pH -1.0	222(4.30),256(4.08), 352(4.18),400(3.65)	33-1095-86
	pH 4.0	223(4.30),234s(4.26), 282s(3.76),300(3.79), 380(4.31)	33-1095-86
dianion	pH 13	228(4.38),256(4.00), 272s(3.92),380(4.33)	33-1095-86
2,4(1H,3H)-Pteridinedione, 7,8-dihydro-3-methyl-7-thioxo-	pH 0	220(4.27),307(3.93), 360(3.94),396(4.12)	33-1095-86
	pH 5	220(4.37),278(3.72), 300(3.82),378(4.28)	33-1095-86
dianion	pH 12	230(4.28),255(4.32), 275(3.90),383(4.26)	33-1095-86
C_7H_6O			
Benzaldehyde	pH 4.73	249(4.06)	69-5774-86
C_7H_6OS			
Benzenecarbothioaldehyde, S-oxide (Z)-	hexane	310(4.28)	77-0964-86
$C_7H_6O_2$			
Benzoic acid	$CHCl_3$-HOAc	276(3.15),282(3.08)	112-0321-86
pressure 0.1 MMPA	heptane	273(3.0),281.2(3.0), 283.0(3.0)	59-0669-86
2-Butynoic acid, 2-propynyl ester	EtOH	212s(3.54)	33-1734-86

Compound	Solvent	λ_{max}(log ϵ)	Ref.
2(3H)-Furanone, dihydro-5-(2-propyn-ylidene)-, (E)-	H_2O	229(4.15)	35-5589-86
	base	220(3.98)	35-5589-86
(Z)-	H_2O	228(4.18)	35-5589-86
	base	220(4.00)	35-5589-86
2(5H)-Furanone, 3-ethynyl-4-methyl-	EtOH	238(4.00)	33-1734-86
2(5H)-Furanone, 4-ethynyl-3-methyl-	EtOH	244(4.16)	33-1734-86
2-Propynoic acid, 2-butynyl ester	EtOH	207s(3.42)	33-1734-86
$C_7H_6O_3$			
2,5-Cyclohexadiene-1,4-dione, 2-meth-oxy-	CH_2Cl_2	254(2.18),358(1.20)	154-0561-86
$C_7H_6O_3S$			
3,5,10-Trioxatricyclo[5.2.1.$0^{2,6}$]dec-8-ene-4-thione, endo	EtOH	237(4.30),300(1.51)	24-0589-86
exo	EtOH	237(4.32),300(1.51)	24-0589-86
$C_7H_7BO_6$			
Boron, [ethanedioato(2-)-O,O'](2,4-pentanedionato-O,O')-, (T-4)-	CH_2Cl_2	290(3.79)	48-0755-86
$C_7H_7BrClNO_2$			
1,3-Cyclohexadiene, 2-bromo-6-chloro-5-methyl-5-nitro-, cis-(±)-	CH_2Cl_2	273(2.54)	23-2382-86
$C_7H_7ClN_2O_4$			
1H-Imidazole-4,5-dicarboxylic acid, 2-chloro-, dimethyl ester	H_2O	258(4.02)	78-1971-86
$C_7H_7ClN_6O_2$			
Pyrazolo[5,1-d]-1,2,3,5-tetrazine-8-carboxamide, 3-(2-chloroethyl)-3,4-dihydro-4-oxo- (pyraoncozine)	$CHCl_3$	314(3.95)	87-1544-86
C_7H_7ClO			
Phenol, 4-chloro-2-methyl-	acid	280(3.26)	3-0639-86
C_7H_7Cs			
Cesium, (phenylmethyl)-	THF	368(4.16)	46-0603-86
C_7H_7DO			
4-Hexen-1-yn-3-one-1-d, 5-methyl-	EtOH	231s(3.77),261(4.08)	33-0560-86
$C_7H_7F_3O_3$			
5-Hexenoic acid, 5,6,6-trifluoro-4-oxo-, methyl ester	hexane	221(4.04),294(1.28)	22-0876-86
C_7H_7Li			
Lithium, (phenylmethyl)-	THF	330(4.13)	46-0603-86
C_7H_7NO			
2-Propenal, 3-(1H-pyrrol-2-yl)-, (E)-	EtOH	348(4.42)	150-3601-86M
2-Propenal, 3-(1H-pyrrol-2-yl)-, (Z)-	EtOH	380(3.92)	150-3601-86M
$C_7H_7NO_2S$			
2,1-Benzisothiazole, 1,3-dihydro-, 2,2-dioxide	MeOH	230(3.83),283(3.19)	4-1645-86
$C_7H_7NO_3$			
Benzene, 1-methoxy-4-nitro-	heptane	291(4.07)	46-2340-86
	benzene	305(4.04)	46-2340-86

Compound	Solvent	λ_{max}(log ε)	Ref.
Benzene, 1-methoxy-4-nitro- (cont.)	ether	299(4.07)	46-2340-86
	EtOAc	305(4.06)	46-2340-86
	THF	305(4.07)	46-2340-86
	CH_2Cl_2	309(4.06)	46-2340-86
	CCl_4	298(4.07)	46-2340-86
	1,2-$C_2H_4Cl_2$	309(4.05)	46-2340-86
	pyridine	313(4.04)	46-2340-86
	DMF	312(4.04)	46-2340-86
	DMSO	315(4.04)	46-2340-86
$C_7H_7NO_3Se$			
Benzeneseleninic acid, 2-nitro-, methyl ester	60% dioxan	419(3.57)	44-0295-86
$C_7H_7NS_2$			
2λ^4-Thieno[3,2-c][1,2]thiazine, 2-methyl-	EtOH	211(3.35),232(3.36), 282(3.29),392(2.73)	39-0483-86C
$C_7H_7N_3O$			
2H-Benzimidazol-2-one, 5-amino-1,3-dihydro-	MeOH	305.8(4.28)	80-0661-86
	MeOH-HCl	286.8(3.86)	80-0661-86
	EtOH	304.4(4.41)	80-0661-86
	DMF	318(3.91)	80-0661-86
$C_7H_7N_3OS$			
4H-Thiazolo[2,3-c][1,2,4]triazin-4-one, 6,7-dihydro-3-methyl-6-methylene-	EtOH	231(3.38),305(3.32)	78-0305-86
7H-Thiazolo[3,2-b][1,2,4]triazin-7-one, 2,3-dihydro-6-methyl-3-methylene-	EtOH	257(4.27)	78-0305-86
$C_7H_7N_3O_2$			
Benzenamine, N-methyl-N,4-dinitroso-	heptane-EtOH	217(4.08),278(3.86), 385(2.34),400(2.36), 606(1.00)	80-0857-86
2-Butenoic acid, 3-amino-2,4-dicyano-, methyl ester	H_2O	277(4.20)	5-0533-86
	pH 13	305(4.50)	5-0533-86
$C_7H_7N_3O_4$			
3-Furanacetic acid, α-[(aminocarbonyl)hydrazono]-	EtOH	268(3.98)	87-0809-86
$C_7H_7N_3O_5$			
Methanamine, N-(2,4-dinitrophenoxy)-	pH 8.5	298(4.06)	69-5602-86
$C_7H_7N_5$			
1H-Imidazole, 1,1'-carbonimidoylbis-	MeCN	201(4.06),235s(--)	69-7423-86
C_7H_8			
1,2,4,6-Heptatetraene, (E)-	hexane	250s(4.49),257(4.60), 268(4.52)	24-1244-86
Toluene	CH_2Cl_2	268.0(2.49)	23-0265-86
NOAlCl$_3$ complex	CH_2Cl_2	338.0(4.04)	23-0265-86
potassium derivative	THF	362(4.19)	46-0603-86
sodium derivative	THF	352(4.20)	46-0603-86
tetracyanoethylene complex	CH_2Cl_2	406.0(3.52)	23-0265-86
$C_7H_8BrN_5O$			
9H-Purine-9-ethanol, 6-amino-8-bromo-	pH 2	267(4.27)	73-0459-86

Compound	Solvent	$\lambda_{max}(\log \epsilon)$	Ref.
$C_7H_8Br_2$			
Cyclopropane, 1,1-dibromo-2-(1,3-butadienyl)-, (E)-	hexane	229.5(4.35)	24-1244-86
C_7H_8ClNO			
2-Propen-1-ol, 3-(3-chloro-1H-pyrrol-2-yl)-, (E)-	EtOH	275(4.36)	150-3601-86M
$C_7H_8ClN_3O_4S_2$			
2H-1,2,4-Benzothiadiazine-7-sulfonamide, 6-chloro-3,4-dihydro-, 1,1-dioxide, product with ethyl acetoacetate	n.s.g.	425(4.50)	86-0170-86
C_7H_8FN			
Benzenamine, 2-fluoro-5-methyl-, chromium tricarbonyl complex	$CHCl_3$	216(4.64),321(3.94)	112-0517-86
Benzenamine, 4-fluoro-N-methyl-, chromium tricarbonyl complex	$CHCl_3$	217(4.69),321(3.89), 365s(3.66)	112-0517-86
$C_7H_8NO_5P$			
Phosphonic acid, [(2-nitrophenyl)-methyl]-	50% EtOH-pH 12	270(3.82)	77-1516-86
Phosphonic acid, [(3-nitrophenyl)-methyl]-	50% EtOH-pH 12	280(3.86)	77-1516-86
Phosphonic acid, [(4-nitrophenyl)-methyl]-	50% EtOH-pH 12	305(4.02)	77-1516-86
$C_7H_8N_2$			
3,5,6-Methenocyclopentapyrazole, 3,3a,4,5,6,6a-hexahydro-	pentane	none above 400 nm	89-0818-86
after irradiation	pentane	450(2.6)	89-0818-86
$C_7H_8N_2O$			
Benzenamine, N-methyl-4-nitroso-	heptane-EtOH	245(3.35),303(2.88), 370(3.90),700(1.30)	80-0857-86
Benzenemethanamine, N-nitroso-	heptane-EtOH	218(4.11),275(3.98), 370(2.00),385(1.99)	80-0857-86
$C_7H_8N_2O_2$			
2-Propenoic acid, 3-(1H-imidazol-4-yl)-, methyl ester, (E)-	CH_2Cl_2	284(4.24)	35-5964-86
$BF_3 \cdot OEt_2$ complex	CH_2Cl_2	266(3.97)	35-5964-86
$EtAlCl_2$ complex	CH_2Cl_2	274(4.20)	35-5964-86
(Z)-	CH_2Cl_2	308(4.25)	35-5964-86
$BF_3 \cdot OEt_2$ complex	CH_2Cl_2	274(4.35)	35-5964-86
1H-Pyrrolizine, 2,3-dihydro-5-nitro-	EtOH	352(4.96)	103-0257-86
$C_7H_8N_2O_2S$			
1,2,3-Thiadiazole-4-carboxylic acid, 5-methyl-, 2-propenyl ester	C_6H_{12}	228(3.56),254(3.35)	44-4075-86
$C_7H_8N_2O_3$			
Acetamide, 2-[(2-furanylmethylene)-amino]oxy]-, (E)-	EtOH	275(3.92)	94-3202-86
3H-Azepine, 2-methoxy-4-nitro-	EtOH	231.5(4.00),347.5(3.69)	18-2317-86
3H-Azepine, 2-methoxy-6-nitro-	EtOH	224.0(3.85),323.0(3.78)	18-2317-86
2-Furancarboxylic acid, 5-[(amino-methylene)amino]-, methyl ester	MeOH	323(3.29)	73-2826-86

Compound	Solvent	λ_{max}(log ε)	Ref.
$C_7H_8N_2O_3S$			
Acetic acid, [(1,4-dihydro-6-methyl-4-oxo-2-pyrimidinyl)thio]-	MeOH	232(3.91),284(3.83)	2-0275-86
	MeOH-NaOH	245(3.83),275(3.70)	2-0275-86
Acetic acid, [(1,6-dihydro-1-methyl-6-oxo-2-pyrimidinyl)thio]-	MeOH	232(3.90),292(3.72)	2-0275-86
	MeOH-NaOH	234(3.92),291(3.75)	2-0275-86
$C_7H_8N_4$			
Propanedinitrile, (1-methyl-2-imidazolidinylidene)-	EtOH	198(4.25),253(4.35)	24-2208-86
Pyrazolo[1,5-b][1,2,4]triazine, 2,3-dimethyl-	EtOH	230(4.38),236(4.32), 299(3.31),329(3.08)	118-0071-86
$C_7H_8N_4O$			
6H-Purin-6-one, 9-ethyl-1,2-dihydro-	pH 1	250.5(4.05)	94-1094-86
	pH 7	251(4.08)	94-1094-86
	pH 13	255.5(4.10)	94-1094-86
7H-Pyrrolo[2,3-d]pyrimidin-4-amine, 2-methoxy-	MeOH	259(3.86),275(3.94)	39-0525-86B
$C_7H_8N_4O_2$			
4H-Pyrazolo[3,4-d]pyrimidin-4-one, 3-ethoxy-1,5-dihydro-	pH 1 and 7	217(4.16),260s(3.40)	4-1869-86
	pH 11	223(3.98),268(3.45)	4-1869-86
$C_7H_8N_4O_2S$			
4H-Pyrazolo[3,4-d]pyrimidin-4-one, 3-ethoxy-1,5,6,7-tetrahydro-6-thioxo-	pH 1	218(3.99),283(4.19)	4-1869-86
	pH 7	243(3.93),284(4.14)	4-1869-86
	pH 11	246(3.97),285(4.10)	4-1869-86
1H-Pyrido[2,3-c][1,2,6]thiadiazin-4-amine, 7-methyl-, 2,2-dioxide	pH -4.0	255(3.86),272s(3.69), 330(3.67),363(3.69)	11-0607-86
	pH 3.0	260(4.04),290s(3.16), 365(3.86)	11-0607-86
	pH 10.0	214(4.32),251(3.99), 285s(3.23),351(3.72)	11-0607-86
8H-Pyrido[2,3-c][1,2,6]thiadiazin-4-amine, 8-methyl-, 2,2-dioxide	pH -4.0	223(4.08),259(3.91), 366(3.82)	11-0607-86
	pH 7.0	259(4.06),292(3.21), 365(3.82)	11-0607-86
	pH 14.0	263(3.98),305(3.44), 369(3.77)	11-0607-86
$C_7H_8N_6$			
1H-Imidazo[4,5-d]-1,2,3-triazin-4-amine, N-2-propenyl-	EtOH	262(4.02),300(3.81)	5-1012-86
1H-1,2,4-Triazole-3,5-diamine, N-3-pyridinyl-	EtOH	260(4.18),302(3.57)	4-0401-86
	EtOH-HCl	213s(3.99),245(4.11), 260(4.06),312(3.53)	4-0401-86
	EtOH-NaOH	264(4.06),308s(3.53)	4-0401-86
C_7H_8O			
2-Cyclopenten-1-one, 3-methyl-5-methylene-	EtOH	239(4.11)	33-0560-86
4,6-Heptadiyn-3-ol	EtOH	217(2.72),228(2.85), 239(2.89),252(2.62)	150-1348-86M
4-Hexen-1-yn-3-one, 5-methyl-	EtOH	232s(3.79),261(4.10)	33-0560-86
7-Oxabicyclo[2.2.1]hept-2-ene, 5-methylene-, (1R)-	isooctane	200(3.88)	157-2500-86
	EtOH	203(3.90)	157-2500-86
Phenol, 2-methyl-	acid	270(3.21)	3-0639-86
Phenol, 3-methyl-	acid	271(3.16)	3-0639-86
Phenol, 4-methyl-	acid	277(3.24)	3-0639-86

Compound	Solvent	$\lambda_{max}(\log \epsilon)$	Ref.
$C_7H_8O_2$			
1-Cyclohexene-1-carboxaldehyde, 3-oxo-	MeOH	231(4.15)	35-4614-86
Phenol, 2-methoxy-	acid	274(3.34)	3-0639-86
Phenol, 3-methoxy-	acid	273(3.25)	3-0639-86
Phenol, 4-methoxy-	acid	287(3.40)	3-0639-86
$C_7H_8O_3$			
7-Oxabicyclo[4.1.0]hept-3-en-2-one, 5-hydroxy-4-methyl-	MeOH	230(3.94)	44-2906-86
$C_7H_8O_4$			
2(5H)-Furanone, 4-(acetoxymethyl)-	EtOH	206.5(4.18)	150-0222-86S
$C_7H_8O_5$			
4H-Pyran-4-one, 3-hydroxy-2,6-bis(hydroxymethyl)-	n.s.g.	272(3.95)	137-0177-86
anion	n.s.g.	322(3.85)	137-0177-86
$C_7H_8S_2$			
1,2-Benzenedithiol, 4-methyl-	MeCN	220(4.38),296(3.23)	35-0936-86
anion	MeCN	240(4.27),294(4.02), 330(3.49)	35-0936-86
dimer	MeCN	238(4.40),294(3.53), 330(3.30)	35-0936-86
$C_7H_9BrN_2OS$			
2(1H)-Pyrimidinone, 5-bromo-4-(propylthio)-	H_2O	278(4.89),312(4.89)	56-0605-86
4(1H)-Pyrimidinone, 5-bromo-2-(propylthio)-	H_2O	252(4.90),290(4.86)	56-0605-86
$C_7H_9BrO_2$			
2,4-Hexadienoic acid, 4-bromo-, methyl ester, (E,Z)-	n.s.g.	268(4.07)	104-2230-86
$C_7H_9ClN_2OS$			
2(1H)-Pyrimidinone, 5-chloro-4-(propylthio)-	H_2O	278(4.84),312(4.89)	56-0605-86
4(1H)-Pyrimidinone, 5-chloro-2-(propylthio)-	H_2O	252(4.95),290(4.89)	56-0605-86
$C_7H_9F_3O$			
1-Penten-3-one, 1,1,2-trifluoro-4,4-dimethyl-	hexane	223(4.06),313(1.54)	22-0876-86
C_7H_9N			
Benzenamine, N-methyl-	heptane-EtOH	247(4.44),294(3.63)	80-0857-86
cation	98% H_2SO_4	252f(2.3)	112-0265-86
Benzenemethanamine	pH 3	206.5(3.90)	160-0979-86
Pyridine, 2,3-dimethyl-	1,2-$C_2H_4Cl_2$	267.0(3.56)	46-2589-86
	+ TFA	268.0(3.90)	46-2589-86
Pyridine, 2,4-dimethyl-	1,2-$C_2H_4Cl_2$	260.5(3.36)	46-2589-86
	+ TFA	260.5(3.76)	46-2589-86
Pyridine, 2,5-dimethyl-	1,2-$C_2H_4Cl_2$	270.5(3.54)	46-2589-86
	+ TFA	272.0(3.83)	46-2589-86
Pyridine, 2,6-dimethyl-	C_6H_{12}	267f(3.7)	64-0223-86B
	pH 3	206(3.58),268(4.05)	160-0979-86
	1,2-$C_2H_4Cl_2$	267.5(3.54)	46-2589-86
	+ TFA	271.5(3.92)	46-2589-86
Pyridine, 3,5-dimethyl-	1,2-$C_2H_4Cl_2$	270.0(3.42)	46-2589-86
	+ TFA	270.0(3.71)	46-2589-86

Compound	Solvent	λ_{max}(log ε)	Ref.
Pyridine, 2-ethyl-	1,2-$C_2H_4Cl_2$	263.0(3.43)	46-2589-86
	+ TFA	265.0(3.83)	46-2589-86
Pyridine, 3-ethyl-	1,2-$C_2H_4Cl_2$	264.5(3.40)	46-2589-86
	+ TFA	264.5(3.75)	46-2589-86
Pyridine, 4-ethyl-	1,2-$C_2H_4Cl_2$	257.5(3.20)	46-2589-86
	+ TFA	254.0(3.65)	46-2589-86
C_9H_7NO			
2-Propen-1-ol, 3-(1H-pyrrol-2-yl)-,	EtOH	275(4.39)	150-3601-86M
(E)-	EtOH	275(4.38)	150-3601-86M
(Z)-	EtOH	271(4.20),277(4.19)	150-3601-86M
$C_7H_9NO_2$			
1H-Pyrrole-3-carboxylic acid, 2-methyl-, methyl ester	EtOH	225(3.86),255(3.81)	35-6739-86
1H-Pyrrole-3-carboxylic acid, 5-methyl-, methyl ester	EtOH	224(3.90),260(3.74)	35-6739-86
$C_7H_9NO_2Se$			
3-Cyclobutene-1,2-dione, 3-(dimethylamino)-4-(methylseleno)-	CH_2Cl_2	278(4.2),330(4.2)	24-0182-86
$C_7H_9NO_3S$			
Benzenesulfonic acid, 2-amino-5-methyl-	pH 3	202.9(4.56),239(3.96), 298(3.47)	160-0979-86
C_7H_9NS			
Benzenamine, 2-(methylthio)-	pH 2.5	250(3.82)	41-0589-86
	pH 4.2	295(3.48)	41-0589-86
$C_7H_9NS_2$			
5(4H)-Thiazolethione, 4-ethenyl-2,4-dimethyl-	EtOH	256s(3.42),315(3.97), 386(3.28),470s(1.54)	33-0374-86
$C_7H_9N_3O_2$			
Acetic acid, cyano-2-imidazolidinylidene-, methyl ester	EtOH	207(4.24),253(4.52)	24-2208-86
$C_7H_9N_3O_2S$			
Formamide, N-[2-(ethylthio)-4-hydroxy-5-pyrimidinyl]-	pH 1	235.5(4.01),296(4.04)	73-0215-86
	pH 13	259(3.93),296(3.95)	73-0215-86
$C_7H_9N_5$			
6H-Purin-6-imine, 3,7-dihydro-3,7-dimethyl-, monohydriodide	pH 1	225(4.38),277(4.21)	94-1821-86
	pH 7	225(4.39),277(4.22)	94-1821-86
	pH 13	225(4.29),282(4.17)	94-1821-86
	EtOH	279(4.21)	94-1821-86
$C_7H_9N_5O$			
4H-Imidazo[4,5-d]-1,2,3-triazin-4-one, 3,5-dihydro-3-propyl-	EtOH	250(3.44),261(3.42), 284(3.56)	5-1012-86
9H-Purine-9-ethanol, 6-amino-	pH 2	261(4.14)	73-0459-86
	pH 7 and 12	263(4.16)	73-0459-86
1H-1,2,4-Triazole-3,5-diamine, N-(2-furanylmethyl)-	EtOH	213(4.10)	4-0401-86
	10%EtOH-HCl	214(4.18)	4-0401-86
$C_7H_9N_5OS$			
3H-Pyrazol-3-one, 5-amino-1,2-dihydro-1-methyl-4-(5-methyl-1,3,4-thiadiazol-2-yl)-	MeOH	289(4.37106)	161-0666-86

Compound	Solvent	λmax(log ε)	Ref.
$C_7H_9N_5O_2S$			
1H-Pyrazino[2,3-c][1,2,6]thiadiazin-4-amine, 6,7-dimethyl-, 2,2-dioxide	pH 1	252(4.02),339(3.74), 411(3.63)	5-1872-86
anion	pH 7	263(4.17),382(3.83)	5-1872-86
$C_7H_9N_5O_2S_2$			
1,2,4-Triazolo[3,4-b][1,3,4]thiadiazolium, 6-[(methoxycarbonyl)amino]-1-methyl-3-(methylthio)-	MeOH	228(4.28),271(4.18)	5-1540-86
$C_7H_9N_5S$			
2H-Purine-2-thione, 1,7-dihydro-7-methyl-6-(methylamino)-	pH 1	245(4.35),287(4.34)	4-0737-86
	pH 13	251(4.30)	4-0737-86
2H-Purine-2-thione, 1,3,6,7-tetrahydro-6-imino-1,7-dimethyl-	pH 1	243(4.34),286(4.30)	4-0737-86
	pH 13	234(4.24),284(4.14)	4-0737-86
C_7H_9O			
Pyrylium, 2,5-dimethyl-, hexachlorostannate (2:1)	n.s.g.	284(4.02)	104-0258-86
$C_7H_{10}Br_2N_2$			
2,3-Diazabicyclo[3.2.0]hept-2-ene, 6,7-dibromo-4,4-dimethyl-, trans	CCl_4	334(2.46)	24-0794-86
$C_7H_{10}N_2O_2$			
4H-Isoxazolo[3,4-c]azepin-3-ol, 5,6,7,8-tetrahydro-	MeOH	262(3.87)	87-0224-86
4H-Isoxazolo[4,3-c]azepin-3-ol, 5,6,7,8-tetrahydro- (zwitterion)	MeOH	248(3.90)	87-0224-86
Pyrazine, trimethyl-, 1,4-dioxide	EtOH	236(4.45),303(4.35)	103-0268-86
$C_7H_{10}N_2O_2S$			
Benzenesulfonic acid, 4-methyl-, hydrazide	pH 1	229(4.15)	140-0140-86
	H_2O	248(4.07)	140-0140-86
$C_7H_{10}N_2O_4$			
2,4,6(1H,3H,5H)-Pyrimidinetrione, 5-(2-hydroxypropyl)-	pH 10.0	270(4.29)	39-1391-86B
$C_7H_{10}N_4$			
1,2,4-Triazine, 3-(1-pyrrolidinyl)-	EtOH	248(4.25),355(3.27)	103-1242-86
$C_7H_{10}N_4O$			
1,2,4-Triazine, 3-(4-morpholinyl)-	EtOH	247(4.24),349(4.19)	103-1242-86
$C_7H_{10}N_4OS$			
Pyrazinecarboxamide, 1,6-dihydro-N-methyl-3-(methylamino)-6-thioxo-, monosodium salt	pH -1.0	266(4.22),400(3.61)	33-0708-86
	pH 3.0	273(4.21),293(3.67)	33-0708-86
	pH 7.0	290(4.22),420(3.51)	33-0708-86
$C_7H_{10}N_4O_3$			
4(1H)-Pyridinone, 6-amino-1-methyl-2-(methylamino)-3-nitro-	MeOH	225(4.28),278(3.69)	83-0947-86
Urea, (1,3-dimethyl-2,5-dioxo-4-imidazolidinylidene)methyl-	pH 8	229(4.23),280(3.62)	78-0747-86
$C_7H_{10}N_5$			
Adeninium, 7,9-dimethyl-, perchlorate	MeOH	212(4.11),272(3.91)	18-2495-86
$C_7H_{10}N_6$			
1H-Imidazo[4,5-d]-1,2,3-triazin-4-am-	EtOH	219(4.13),263(3.98),	5-1012-86

Compound	Solvent	$\lambda_{max}(\log \epsilon)$	Ref.
ine, N-(1-methylethyl)- (cont.)		302(3.78)	5-1012-86
1H-Imidazo[4,5-d]-1,2,3-triazin-4-amine, N-propyl-	EtOH	217(4.12),262(3.97), 302(3.77)	5-1012-86
$C_7H_{10}N_6O$			
8H-Purin-8-one, 6-amino-9-(2-aminoethyl)-7,9-dihydro-	pH 2	269(4.05),280(4.05)	73-0459-86
	pH 7	271(4.11)	73-0459-86
	pH 12	281(4.15)	73-0459-86
$C_7H_{10}O$			
Ethanone, 1-(1-cyclopenten-1-yl)-	MeOH	235(4.03)	107-0957-86
Ethanone, 1-(3-methylbicyclo[1.1.0]-but-1-yl)-	C_6H_{12}	227(3.91),281s(1.69)	54-0386-86
	EtOH	235(3.94)	54-0386-86
Ethanone, 1-(3-methylenecyclobutyl)-	C_6H_{12}	268(1.32),285(1.30)	54-0386-86
	EtOH	272(1.54)	54-0386-86
$C_7H_{10}O_2$			
6-Heptynoic acid	MeOH	206(2.65)	18-3535-86
2,4-Hexadienoic acid, methyl ester	CH_2Cl_2	259(4.42)	35-3016-86
$EtAlCl_2$ complex, (E,E)-	CH_2Cl_2	301(4.36)	35-3016-86
2,4-Hexadienoic acid, methyl ester	CH_2Cl_2	262(4.30)	35-3016-86
$EtAlCl_2$ complex, (Z,E)-	CH_2Cl_2	300s(3.91)	35-3016-86
2,4-Hexadienoic acid, methyl ester	CH_2Cl_2	264(4.23)	35-3016-86
$EtAlCl_2$ complex, (Z,Z)-	CH_2Cl_2	300s(3.58)	35-3016-86
5-Hexynoic acid, methyl ester	MeOH	206(2.18)	18-3535-86
$C_7H_{10}O_3$			
2-Butenoic acid, 3-methyl-4-oxo-, ethyl ester, (E)-	hexane	228(3.35)	44-0256-86 +107-0627-86
$C_7H_{10}O_5$			
1-Cyclohexene-1-carboxylic acid, 3,4,5-trihydroxy- (shikimic acid)	MeOH	211.5(3.70)	95-0989-86
$C_7H_{11}NO_3$			
1,2-Propanedione, 1-(tetrahydro-2-furanyl)-, 1-oxime	EtOH	225(3.95)	118-0473-86
$C_7H_{11}NO_4$			
1,2-Propanedione, 1-(1,4-dioxan-2-yl)-, 1-oxime	EtOH	224(4.00)	118-0473-86
$C_7H_{11}NO_5$			
1,4-Dioxane-2-acetic acid, α-(hydroxyimino)-, methyl ester	EtOH	214(3.90)	118-0473-86
$C_7H_{11}NS$			
2,4-Pentadienethial, 5-(dimethylamino)-	n.s.g.	460(4.54)	70-0458-86
$C_7H_{11}N_3O_4S$			
D-Ribitol, 1-C-(5-amino-1,3,4-thiadiazol-2-yl)-1,4-anhydro-, (R)-	H_2O	261(3.81)	73-1311-86
$C_7H_{11}N_3O_5$			
D-Ribitol, 1-C-(5-amino-1,3,4-oxadiazol-2-yl)-1,4-anhydro-, (R)-	H_2O	225(3.89)	73-1311-86
$C_7H_{11}N_5O_2S$			
Pyrazino[2,3-e][1,2,4]triazine-6,7-diol, 1,5,6,7-tetrahydro-5-methyl-	pH 1	267(4.42),329s(3.59)	4-0033-86
	H_2O	225(4.04),257.5(4.22),	4-0033-86

Compound	Solvent	$\lambda_{max}(\log \epsilon)$	Ref.
3-(methylthio)-, trans (cont.)		266s(4.19),328(3.79)	4-0033-86
	pH 11	257(4.20),264s(4.15), 328(3.85)	4-0033-86
Pyrazino[2,3-e][1,2,4]triazine-6,7-diol, 2,6,7,8-tetrahydro-8-methyl-3-(methylthio)-, trans	pH 1	265(4.40),333s(3.54)	4-0033-86
	pH 11	253.5(4.17),271s(4.05), 332(3.74)	4-0033-86
	MeOH	221(4.04),252(4.18), 270s(4.06),330(3.74)	4-0033-86
Pyrazino[2,3-e][1,2,4]triazin-6-ol, 1,2,6,7-tetrahydro-7-methoxy-3-(methylthio)-, trans	pH 1	263(4.45),327s(3.58)	4-0033-86
	pH 11	255(4.23),269s(4.11), 333(3.82)	4-0033-86
	MeOH	221(4.12),252(4.23), 267s(4.11),331(3.82)	4-0033-86
$C_7H_{11}N_7O$			
9-Purineethanol, 6-amino-8-hydrazino-	pH 2	269.5(4.24)	73-0459-86
	pH 12	267(4.13)	73-0459-86
$C_7H_{12}N_2O$			
2-Propanone, 1-(1-methyl-2-imidazolidinylidene)-, (E)-	EtOH	288(4.43)	24-2208-86
$C_7H_{12}N_2O_2$			
Acetic acid, (1-methyl-2-imidazolidinylidene)-, methyl ester, (E)-	EtOH	271(4.46)	24-2208-86
$C_7H_{12}N_2O_2S$			
6-Oxa-2,4-diazabicyclo[3.3.1]nonane-3-thione, 8-hydroxy-5-methyl-	MeOH	247(2.76)	142-0679-86
$C_7H_{12}N_2O_3$			
Oxazole, 4,5-dihydro-4,4-dimethyl-2-(1-nitroethyl)-	$CHCl_3$	341(4.34)	4-0825-86
$C_7H_{12}N_4O$			
4(1H)-Pteridinone, 2,3,5,6,7,8-hexahydro-6-methyl-	pH 0	258s(2.20)	12-0031-86
	pH 7.3	293(3.02)	12-0031-86
	pH 13.2	286(3.25)	12-0031-86
$C_7H_{12}N_4O_2$			
1H-1,2,3-Triazole-4-carboxylic acid, 5-amino-1-ethyl-, ethyl ester	H_2O	230(3.93),261(4.01)	103-0740-86
$C_7H_{12}N_4O_3$			
2,4(1H,3H)-Pyrimidinedione, 5-amino-6-[(2-hydroxypropyl)amino]-	pH 5.0	200(4.43),224s(4.20), 311(4.18),350(3.00)	44-2461-86
	pH 8.0	238(3.78),311(4.26), 350(3.00),500(2.00)	44-2461-86
	pH 12.0	238(3.60),311(4.26), 350(3.30)	44-2461-86
$C_7H_{12}N_4O_4$			
1H-Pyrazole-4-carboxylic acid, 3,5-diamino-1-(2,3-dihydroxypropyl)-, (±)-	pH 2	239(4.35)	73-1512-86
$C_7H_{12}OS$			
2(3H)-Furanthione, 3-ethyldihydro-3-methyl-	EtOH	250(4.02)	87-1996-86
2(3H)-Furanthione, 4-ethyldihydro-4-methyl-	EtOH	249(4.05)	87-1996-86

Compound	Solvent	$\lambda_{max}(\log \epsilon)$	Ref.
2(3H)-Thiophenone, 3-ethyldihydro-3-methyl-	EtOH	234(3.58)	87-1996-86
2(3H)-Thiophenone, 4-ethyldihydro-4-methyl-	EtOH	236(3.56)	87-1996-86
$C_7H_{12}OS_2$			
Butane(dithioic) acid, 2,2-dimethyl-3-oxo-, methyl ester	MeOH	265(3.58),305(3.70), 314(3.48),478(1.47)	118-0968-86
$C_7H_{12}O_2$			
2-Butenoic acid, 2-methyl-, ethyl ester. (E)-	hexane	216(4.02)	35-3016-86
$EtAlCl_2$ complex	CH_2Cl_2	245(3.52)	35-3016-86
2-Butenoic acid, 2-methyl-, ethyl ester, (Z)-	hexane	213(4.02)	35-3016-86
$EtAlCl_2$ complex	CH_2Cl_2	245s(3.28)	35-3016-86
2,4-Hexadien-1-ol, 2-(hydroxymethyl)-	n.s.g.	232(4.29)	64-0677-86C
$C_7H_{12}S_2$			
2(3H)-Thiophenethione, 3-ethyldihydro-3-methyl-	EtOH	221(3.44),312(4.13)	87-1996-86
2(3H)-Thiophenethione, 4-ethyldihydro-4-methyl-	EtOH	221s(--),312(4.13)	87-1996-86
C_7H_{13}			
Methyl, cyclohexyl-	$C_6H_{11}Me$	245(3.15)	70-1140-86
$C_7H_{13}N_3$			
2-Pyrimidinamine, 1,6-dihydro-4,4,6-trimethyl-	isoPrOH	250(3.36)	142-3023-86
hydrochloride	isoPrOH	240(3.44)	142-3023-86
$C_7H_{13}N_3O_4$			
Butanamide, 4-[[(methyl-aci-nitro)-acetyl]amino]- (YN-0165J-A)	MeOH	252(4.17)	158-0601-86
$C_7H_{13}N_5O$			
1H-1,2,4-Triazol-3-amine, 1-methyl-5-(4-morpholinyl)-	EtOH	203(4.27),229(4.09)	4-0401-86
	10%EtOH-HCl	236(4.06)	4-0401-86
	10%EtOH-NaOH	227(4.20)	4-0401-86
$C_7H_{14}N_2O_2S$			
Glycinethioamide, N-tert-butoxy-carbonyl-	EtOH	266(4.05)	33-1224-86
$C_7H_{14}N_4S$			
1H-1,2,4-Triazole-1-carbothioamide, 3-ethyl-4,5-dihydro-5,5-dimethyl-	EtOH	237(4.07),295(4.18)	103-0287-86
$C_7H_{14}N_6$			
1H-1,2,4-Triazol-3-amine, 5-(4-methyl-1-piperazinyl)-	EtOH	217s(3.89),241s(3.42)	4-0401-86
	10%EtOH-HCl	210(3.97),235(3.74)	4-0401-86
	10%EtOH-NaOH	240s(3.62)	4-0401-86
$C_7H_{14}OSi$			
Silane, (cyclopropylcarbonyl)tri-methyl-	C_6H_{12}	365(2.12)	138-0177-86
$C_7H_{15}NO_2S$			
Ethanethioamide, 2-(2-methoxyethoxy)-N,N-dimethyl-	CH_2Cl_2	276(4.15)	126-2369-86

Compound	Solvent	λ_{max}(log ε)	Ref.
$C_7H_{15}N_3O_3$			
Valinamycin-ammonia adduct	H_2O	223(3.69)	158-0184-86
	acid	220(3.73)	158-0184-86
	base	220s(--)	158-0184-86
$C_7H_{15}N_5$			
1H-1,2,4-Triazole-3,5-diamine, N-(3-methylbutyl)-	EtOH	210s(4.11)	4-0401-86
	10%EtOH-HCl	220(3.98)	4-0401-86
$C_7H_{16}N_4OS$			
Hydrazinecarbothioamide, 2-[1-ethyl-2-(hydroxyamino)-2-methylpropylidene]-	EtOH	274(4.53)	103-0287-86
$C_7H_{16}Si_2$			
1,3-Disilacyclopent-4-ene, 1,1,3,3-tetramethyl-	pentane	190s(4.0),220s(3.05)	44-5051-86
$C_7H_{17}N_3$			
Triaziridine, methylbis(1-methylethyl)-, (1α,2α,3β)-	EtOH	266(1.78)	33-2087-86
(1α,2β,3β)-	EtOH	200(2.28)(end abs.)	33-2087-86
$C_7H_{17}N_3O$			
Triaziridinemethanol, 2,3-bis(1-methylethyl)-, (1α,2α,3β)-	pentane	200(2.79)(end abs.)	33-2087-86
(1α,2β,3β)-	EtOH	240(1.78)	33-2087-86
$C_7H_{18}O_3SSi$			
Silane, trimethoxy[(propylthio)-methyl]-	heptane	190(2.38),210s(--)	67-0062-86
	MeCN	194(2.40)	67-0062-86
$C_7H_{18}SSi$			
Methanethiol, (triethylsilyl)-	heptane	208(2.20),227(1.53)	67-0225-86
Silane, [2-(ethylthio)ethyl]trimethyl-	heptane	214s(3.3)	65-1848-86

Compound	Solvent	λ_{max}(log ε)	Ref.
$C_8Cl_2N_2O_2$			
1,2-Benzenedicarbonitrile, 4,5-dichloro-3,6-dihydroxy-, ion(2-)	EtOH	205(2.85),248(4.37), 398(4.34),452(3.96)	35-4459-86
1,4-Cyclohexadiene-1,2-dicarbonitrile, 4,5-dichloro-3,6-dioxo-	MeCN	209(4.20),216(4.18), 226(4.16),270(4.06), 280(4.09),372(2.94)	35-4459-86
radical anion	MeCN	247(4.21),257s(4.11), 270(3.69),323s(3.28), 347(3.89),432(3.72), 456(3.78),508(3.64), 547(3.76),588(3.80)	35-4459-86
3,6-Cyclohexadiene-1,3-dicarbonitrile, 4,6-dichloro-2,5-dioxo-	MeCN	216s(--),287(4.1)	24-0844-86
$C_8Cl_4N_4$			
Cyanamide, (2,3,5,6-tetrachloro-2,5-cyclohexadiene-1,4-diylidene)bis-	MeCN	327s(4.20),353(4.48), 366(4.52),372s(4.45)	5-0142-86
C_8Cl_6			
Benzene, pentachloro(chloroethynyl)-	C_6H_{12}	227(4.66),233s(4.57), 263(4.19),275(4.40), 306(2.73),318(2.77)	44-1413-86
$C_8F_4N_4$			
Cyanamide, (2,3,5,6-tetrafluoro-2,5-cyclohexadiene-1,4-diylidene)bis-	MeCN	222(3.30),314(4.08), 329(4.25),344s(4.25)	5-0142-86
$C_8F_{14}O_3$			
Butanoic acid, heptafluoro-, anhydride	hexane	245(2.0)	65-1553-86
$C_8HCl_2N_2O_2$			
1,2-Benzenedicarbonitrile, 4,5-dichloro-3,6-dihydroxy-, ion(1-)	EtOH	205(4.38),249(4.38), 397(4.00)	35-4459-86
$C_8HCl_3N_2O$			
1,3-Benzenedicarbonitrile, 2,4,6-trichloro-5-hydroxy-	MeCN	227(4.6),334(3.7), 398(3.1)	24-0844-86
C_8HCl_5			
Benzene, pentachloroethynyl-	C_6H_{12}	224(4.68),231(4.63), 241s(4.28),258(4.10), 267(4.20),293(2.80), 305(2.88),317(2.93)	44-1413-86
$C_8H_2Br_2N_4$			
Cyanamide, (2,5-dibromo-2,5-cyclohexadiene-1,4-diylidene)bis-	MeCN	264s(3.65),331s(4.40), 346s(4.56),348(4.57), 365s(4.43)	5-0142-86
$C_8H_2Cl_2N_2O_2$			
1,2-Benzenedicarbonitrile, 4,5-dichloro-3,6-dihydroxy-	EtOH	217(4.60),230(4.36), 260(3.91),350(3.94), 420(2.30)	35-4459-86
1,3-Benzenedicarbonitrile, 4,6-dichloro-2,5-dihydroxy-	MeCN	216s(--),334(3.9)	24-0844-86
$C_8H_2Cl_2N_4$			
Cyanamide, (2,3-dichloro-2,5-cyclohexadiene-1,4-diylidene)bis-	MeCN	227(3.73),237s(3.64), 316s(4.25),333(4.39), 347(4.35),362s(4.10), 388s(3.55)	5-0142-86

Compound	Solvent	λ_{max}(log ε)	Ref.
Cyanamide, (2,5-dichloro-2,5-cyclohexadiene-1,4-diylidene)bis-	MeCN	254(3.37),347(4.40), 366s(4.25)	5-0142-86
Cyanamide, (2,6-dichloro-2,5-cyclohexadiene-1,4-diylidene)bis-	MeCN	350s(4.40),362(4.47), 390s(3.90)	5-0142-86
$C_8H_2F_6N_4O$			
4(1H)-Pteridinone, 6,7-bis(trifluoromethyl)-	MeOH	251(4.03),355(3.63)	39-1043-86C
$C_8H_2F_6N_4OS$			
4(1H)-Pteridinone, 2,3-dihydro-2-thioxo-6,7-bis(trifluoromethyl)-	MeOH	238(4.19),337(4.47)	39-1043-86C
$C_8H_2F_6N_4O_2$			
2,4(1H,3H)-Pteridinedione, 6,7-bis(trifluoromethyl)-	pH 1	240(4.08),330(3.79)	39-1043-86C
	pH 12	270(4.29),380(3.68)	39-1043-86C
$C_8H_2O_8Os_2$			
Osmium, octacarbonyldihydrodi-	dioxan	231(3.54),286(3.64)	35-0948-86
$C_8H_3ClN_4$			
Cyanamide, (2-chloro-2,5-cyclohexadiene-1,4-diylidene)bis-	MeCN	320s(4.37),335(4.45), 348s(4.40)	5-0142-86
$C_8H_4Cl_2N_4$			
Cyanamide, (2,5-dichloro-1,4-phenylene)bis-	EtOH	278(4.41)(incomplete)	5-0142-86
$C_8H_4F_6O$			
7-Oxabicyclo[2.2.1]hepta-2,5-diene, 2,3-bis(trifluoromethyl)-	EtOH	229s(2.42),258s(2.08)	24-0589-86
7-Oxabicyclo[4.1.0]hepta-2,4-diene, 2,5-bis(trifluoromethyl)-	EtOH	257(3.69)	24-0589-86
3-Oxatetracyclo[3.2.0.0^{2,7}.0^{4,6}]heptane, 1,5-bis(trifluoromethyl)-	EtOH	230(<1.0)(end abs.)	24-0589-86
$C_8H_4N_2$			
1,4-Benzenedicarbonitrile	hexane	288(<u>3.2</u>)	23-1491-86
$C_8H_4N_4$			
Cyanamide, 2,5-cyclohexadiene-1,4-diylidenebis-	MeCN	313s(4.35),330(4.47), 342(4.45),361s(4.24)	5-0142-86
$C_8H_5BrN_3O_2$			
2-Pyrimidineacetic acid, 5-bromo-α-cyano-, methyl ester, ion(1-), salt with piperidinium ion	EtOH	233(4.19),314(4.60)	103-0774-86
$C_8H_5Br_2NO_4$			
2,5-Cyclohexadiene-1,4-dione, 2,3-dibromo-5-ethyl-6-nitro-	$CHCl_3$	287(4.02),378(3.04)	12-0059-86
$C_8H_5Br_3N_2O_5$			
2,5-Cyclohexadien-1-one, 2,3,5-tribromo-4-ethyl-4,6-dinitro-	$CHCl_3$	263(3.96),309(3.48)	12-0059-86
$C_8H_5Br_3N_2O_6$			
2,5-Cyclohexadien-1-one, 2,3,5-tribromo-4-ethyl-6-nitro-4-(nitrooxy)-	$CHCl_3$	263.5(4.21),308(3.69)	12-0059-86

Compound	Solvent	$\lambda_{max}(\log \epsilon)$	Ref.
$C_8H_5Br_3O_2$			
2,5-Cyclohexadiene-1,4-dione, 2,3,5-tribromo-6-ethyl-	$CHCl_3$	304.5(4.38)	12-0059-86
$C_8H_5Br_3O_4$			
2H-Pyran-6-carboxylic acid, 3-bromo-4-(dibromomethyl)-2-oxo-, methyl ester	EtOH	226(4.0),318(4.0)	5-1968-86
$C_8H_5Br_4NO_3$			
2,5-Cyclohexadien-1-one, 2,3,5,6-tetrabromo-4-ethyl-4-nitro-	$CHCl_3$	274(4.18),c.310(3.45)	12-0059-86
$C_8H_5Br_4NO_4$			
2,5-Cyclohexadien-1-one, 2,3,5,6-tetrabromo-4-ethyl-4-(nitrooxy)-	$CHCl_3$	276(4.21),305(3.51)	12-0059-86
$C_8H_5ClN_2O$			
1H-Indazole-3-carboxaldehyde, 4-chloro-	EtOH	242(4.0),248s(3.9), 296s(3.9),301(3.9)	35-4115-86
1H-Indazole-3-carboxaldehyde, 5-chloro-	EtOH	213(4.4),239(4.0), 246(4.0),287(3.9), 305(3.9)	35-4115-86
$C_8H_5ClN_2O_2$			
1H-Indazole-3-carboxylic acid, 5-chloro-	EtOH	213(4.5),268(3.7), 274(3.7),301(3.8), 311s(3.7)	35-4115-86
$C_8H_6BrN_3O_2$			
Acetic acid, (5-bromo-2(1H)-pyrimidinylidene)cyano-, methyl ester, (E)-	EtOH	307(4.61),398(3.46)	103-0774-86
$C_8H_6BrN_3O_7$			
Benzene, 2-(2-bromoethoxy)-1,3,5-trinitro-	n.s.g.	229(4.25)	3-2366-86
$C_8H_6Br_2O_4$			
2H-Pyran-6-carboxylic acid, 3-bromo-4-(bromomethyl)-2-oxo-, methyl ester	EtOH	318(4.0)	5-1968-86
2H-Pyran-6-carboxylic acid, 4-(dibromomethyl)-2-oxo-, methyl ester	EtOH	305(3.8)	5-1968-86
$C_8H_6Br_3NO_4$			
2,5-Cyclohexadien-1-one, 2,3,5-tribromo-4-ethyl-4-hydroxy-6-nitro-	$CHCl_3$	262(4.17),307(3.61)	12-0059-86
$C_8H_6Br_4O_2$			
2,5-Cyclohexadien-1-one, 2,3,5,6-tetrabromo-4-ethyl-4-hydroxy-	$CHCl_3$	272(4.09),309(3.38)	12-0059-86
$C_8H_6ClN_3O_3$			
3H-Indol-3-one, 4-chloro-1,2-dihydro-2-hydroxy-1-nitroso-, oxime (4:1 syn:anti)	EtOH	223s(4.1),262(4.2), 292(3.9)	35-4115-86
$C_8H_6F_6N_4O_2$			
6,7-Pteridinediol, 1,5,6,7-tetrahydro-6,7-bis(trifluoromethyl)-	pH 5.2	250(3.65),295(3.79)	39-1043-86C

Compound	Solvent	λ_{max}(log ϵ)	Ref.
$C_8H_6F_6N_4O_3$			
4(1H)-Pteridinone, 5,6,7,8-tetrahydro-6,7-dihydroxy-6,7-bis(trifluoromethyl)-	MeOH	212(4.28),272(3.92)	39-1043-86C
C_8H_6IN			
1,3,5-Cycloheptatriene-1-carbonitrile, 6-iodo-	dioxan	320(3.65)	89-0720-86
$C_8H_6N_2O$			
Formamide, N-(2-cyanophenyl)-	EtOH	240(4.00),294(3.46)	11-0347-86
$C_8H_6N_2OS$			
4(1H)-Quinazolinone, 2,3-dihydro-2-thioxo-	EtOH	218(4.17),292(4.27)	103-1001-86
	EtOH-KOH	224(4.07),299(4.05)	103-1001-86
$C_8H_6N_2O_3$			
Benzofurazan, 5-acetyl-, 1-oxide	EtOH	302s(3.07),316s(3.20), 333s(3.38),371(3.77), 385s(3.69)	12-0089-86
$C_8H_6N_2S_2$			
1,3-Dithiole-4,5-dicarbonitrile, 2-(1-methylethylidene)-	EtOH	242(4.10),334(3.01), 444(3.34)	33-0419-86
$C_8H_6N_4$			
Cyanamide, 1,4-phenylenebis-	EtOH	246(4.34)	5-0142-86
$C_8H_6O_2S_4$			
1,3-Dithiole-4-carboxylic acid, 2-(1,3-dithiol-2-ylidene)-, methyl ester	MeCN	292s(4.08),302(4.11), 313(4.14),430(3.30)	104-0367-86
$C_8H_6O_3$			
Benzoic acid, 4-formyl-	pH 4.73	258(4.18)	69-5774-86
$C_8H_6O_4$			
3,6-Cycloheptadiene-1,2,5-trione, 3-methoxy-	$CHCl_3$	255(4.15),286(2.74), 356(3.04)	18-0511-86
C_8H_6S			
Benzo[b]thiophene	pet ether	298(3.53)	142-0355-86
Cyclopenta[b]thiapyran (thialene)	C_6H_{12}	233(4.18),268(4.09), 337(3.71),343(3.74), 349(3.77),359(3.59), 368(3.53),535(2.88), 555(2.87),579(2.83), 604(2.71),634(2.59), 661(2.26),702(2.06)	44-4644-86
C_8H_7Br			
Tricyclo[5.1.0.0^{2,8}]octa-3,5-diene, 4-bromo-	pentane	204(3.78),226s(3.51), 282s(3.29),289s(3.34), 294(3.36),299(3.36), 306s(3.29),312s(3.21), 321s(2.93),327s(2.71)	78-1585-86
$C_8H_7BrN_4S$			
1,2,4-Triazole-3-thiol, 4-amino-5-(2-bromophenyl)-	EtOH	251(4.26)	4-1451-86

Compound	Solvent	λ max(log ε)	Ref.
1,2,4-Triazole-3-thiol, 4-amino-5-(3-bromophenyl)-	EtOH	252(4.42),276s(4.18)	4-1451-86
1,2,4-Triazole-3-thiol, 4-amino-5-(4-bromophenyl)-	EtOH	250(4.12),285s(3.96)	4-1451-86
$C_8H_7BrO_3$			
Ethanone, 2-bromo-1-(3,4-dihydroxyphenyl)-	EtOH	236(3.97),287(3.86), 319(3.88)	24-0050-86
$C_8H_7BrO_4$			
Ethanone, 2-bromo-1-(2,3,4-trihydroxyphenyl)-	EtOH	204(4.23),238(3.95), 306(4.11)	24-0050-86
2H-Pyran-6-carboxylic acid, 3-bromo-4-methyl-2-oxo-, methyl ester	EtOH	312(4.0)	5-1968-86
2H-Pyran-6-carboxylic acid, 4-(bromomethyl)-2-oxo-, methyl ester	EtOH	304(3.9)	5-1968-86
$C_8H_7Br_3$			
Bicyclo[4.1.1]octa-2,4-diene, 3,7,8-tribromo- (7-exo,8-syn)	hexane	234(3.36),271s(2.85), 284s(3.16),293(3.29), 304(3.28),317(2.98)	78-1585-86
$C_8H_7ClN_2$			
2-Propenenitrile, 3-(3-chloro-1H-pyrrol-2-yl)-2-methyl-, (E)-	EtOH	317(4.38)	150-3601-86M
(Z)-	EtOH	323(4.22)	150-3601-86M
$C_8H_7ClN_4O_2S$			
2,4(1H,3H)-Pteridinedione, 6-chloro-7-mercapto-1,3-dimethyl-	pH -2.0	226(4.41),256(4.22), 275s(3.90),360(4.19), 410s(2.83)	33-0708-86
	pH 4.0	231(4.41),242s(4.31), 285(3.89),385(4.31)	33-0708-86
$C_8H_7ClN_4O_3$			
2,4(1H,3H)-Pteridinedione, 6-chloro-7-hydroxy-1,3-dimethyl-	pH 7.0	213(4.47),250s(3.80), 279(3.96),336(4.21)	33-0708-86
$C_8H_7ClN_4O_5S$			
7-Pteridinesulfonic acid, 6-chloro-1,2,3,4-tetrahydro-1,3-dimethyl-2,4-dioxo-	pH 7.0	250(4.28),265s(4.08), 352(3.99)	33-0708-86
$C_8H_7ClN_4S$			
1,2,4-Triazole-3-thiol, 4-amino-5-(2-chlorophenyl)-	EtOH	252(4.38)	4-1451-86
1,2,4-Triazole-3-thiol, 4-amino-5-(4-chlorophenyl)-	EtOH	252(4.27),286s(3.98)	4-1451-86
$C_8H_7F_6N_5$			
1,2,4-Triazine-5-carboxaldehyde, 3,6-bis(trifluoromethyl)-, dimethylhydrazone	CH_2Cl_2	239(3.40),387(4.36)	83-0690-86
$C_8H_7IO_3$			
Ethanone, 1-(3,4-dihydroxyphenyl)-2-iodo-	EtOH	216(4.33),243(3.99), 286(3.85),316(3.87)	24-0050-86
$C_8H_7IO_4$			
Ethanone, 2-iodo-1-(2,3,4-trihydroxyphenyl)-	EtOH	214(4.33),236(4.07), 302(4.05)	24-0050-86

Compound	Solvent	$\lambda_{max}(\log \epsilon)$	Ref.
C_8H_7N			
Indole	hexane	266(3.85)	103-1262-86
C_8H_7NO			
Benzoxazole, 2-methyl-	EtOH	231(3.97),264(3.43), 270(3.61),276(3.64)	136-0271-86H
2-Propenenitrile, 3-(2-furanyl)-2-methyl-, (E)-	EtOH	303(4.43)	103-0943-86
2-Propenenitrile, 3-(5-methyl-2-furanyl)-, (E)-	EtOH	312(4.40)	103-0943-86
C_8H_7NOS			
Azirino[1,2-a]thieno[2,3-d]pyridin-7(4H)-one, 6,6a-dihydro-	EtOH	255s(3.84),281(3.99)	39-0729-86C
$C_8H_7NO_2$			
Ethanone, 1-(3-nitrosophenyl)-, (dimer)	20% EtOH	238(4.30),310(4.00)	23-2076-86
Ethanone, 1-(4-nitrosophenyl)-, (dimer)	EtOH	340(4.00)	23-2076-86
$C_8H_7NO_3$			
Acetamide, N-hydroxy-2-oxo-N-phenyl-	phosphate	260(4.8)	142-0013-86
	borate	260s(4.5)	142-0013-86
2H-1,4-Benzoxazin-3(4H)-one, 4-hydroxy-	pH 9	300(4.8)	142-0013-86
$C_8H_7N_3O_3$			
Pyrido[2,3-d]pyrimidine-2,4(1H,3H)-dione, 6-(hydroxymethyl)-	pH 1	220(4.46),249s(3.80), 309(3.95)	87-0709-86
	H_2O	218(4.51),247s(3.83), 307(3.75)	87-0709-86
	pH 13	246(4.13),267(4.04), 336(3.75)	87-0709-86
$C_8H_7N_3O_4$			
[2,4'-Bioxazole]-4-carboxamide, 2'-(hydroxymethyl)-	MeOH	245(3.78)	35-0846-86
$C_8H_7N_3S_2$			
Thiazolium, 5-(dicyanomethyl)-3-methyl-2-(methylthio)-, hydroxide, inner salt	MeCN	286(3.90),425(3.99)	24-2094-86
	toluene	473(--)	24-2094-86
	MeOH	414.5(--)	24-2094-86
	EtOH	422(--)	24-2094-86
	acetone	436(--)	24-2094-86
	EtOAc	451(--)	24-2094-86
C_8H_8			
1,3,7-Octatrien-5-yne, (E)-	EtOH	213(4.08),273(4.33), 283(4.46),299(4.45)	24-1105-86
1,3,7-Octatrien-5-yne, (Z)-	EtOH	217(4.17),273(4.26), 283(4.36),299(4.29)	24-1105-86
$C_8H_8Br_2$			
Bicyclo[4.1.1]octa-2,4-diene, 7,8-dibromo-. (7-exo,8-syn)	hexane	212s(3.65),251s(3.25), 260s(3.36),268(3.46), 279(3.54),291(3.50), 305(3.21)	78-1585-86
C_8H_8ClNO			
3-Buten-2-one, 4-(3-chloro-1H-pyrrol-	EtOH	345(4.35)	150-3601-86M

Compound	Solvent	$\lambda_{max}(\log \epsilon)$	Ref.
2-yl)-, (E)- (cont.)			150-3601-86M
2-Propenal, 3-(3-chloro-1H-pyrrol-2-yl)-2-methyl-, (E)-	EtOH	340(4.46)	150-3601-86
$C_8H_8ClNO_2$			
2-Propenoic acid, 3-(3-chloro-1H-pyrrol-2-yl)-, methyl ester, (E)-	EtOH	327(4.40)	150-3601-86M
(Z)-	EtOH	335(4.22)	150-3601-86M
$C_8H_8ClNO_3$			
Benzene, 1-chloro-5-methoxy-2-methyl-3-nitro-	MeOH	242(3.8),269(3.6), 324(3.5)	35-4115-86
$C_8H_8Cl_2N_4O$			
Methanimidamide, N'-(4,6-dichloro-5-formyl-2-pyrimidinyl)-N,N-dimethyl-	MeOH	335(4.65)	33-1602-86
$C_8H_8F_2N_2O_2$			
4(3H)-Pyrimidinone, 2,5-difluoro-6-methoxy-3-(2-propenyl)-	C_6H_{12}	233(3.54),272(3.59)	39-0515-86C
4(3H)-Pyrimidinone, 5,6-difluoro-2-methoxy-3-(2-propenyl)-	C_6H_{12}	219(3.65),266(3.79)	39-0515-86C
4(3H)-Pyrimidinone, 2,5-difluoro-3-methyl-6-(2-propenyloxy)-	C_6H_{12}	233(3.49),272(3.52)	39-0515-86C
$C_8H_8N_2$			
1H-Benzimidazole, 1-methyl-	C_6H_{12}	211(4.65),254(3.74), 277(3.77),284(3.84), 294(3.67)	39-1917-86B
	pH 2	198(4.55),226(4.26), 266(3.98),274(3.93)	39-1917-86C
	pH 8	202(4.59),252(3.89), 270(3.98),277(3.92)	39-1917-86C
	MeOH	207(4.79),247(4.08), 274(3.93),280(3.96), 289(3.74)	39-1917-86C
	ether	217(4.46),248(4.15), 277(3.97),283(4.07), 294(3.90)	39-1917-86B
	MeCN	207(4.76),247(4.10), 275(3.90),281(3.94), 291(3.73)	39-1917-86B
1H-Benzimidazole, 2-methyl-	pH 2	198(4.52),231(3.55), 266(3.88),274(3.89)	39-1917-86B
	pH 8	202(4.49),241(3.76), 272(3.82)	39-1917-86B
	NaOH	226(3.93),257(3.63), 279(3.89),286(3.74)	39-1917-86B
	MeOH	207(4.36),242(3.82), 272(3.87),279(3.88)	39-1917-86B
	ether	214(4.02),244(3.93), 274(3.90),282(3.93)	39-1917-86B
	MeCN	208(4.42),244(3.80), 274(3.68),281(3.79)	39-1917-86B
	C_6H_{12}	210(--),241(--), 274(--),282(--)	39-1917-86B
$C_8H_8N_2O$			
1H-Benzimidazole-1-methanol	EtOH	244(3.74),272(3.72), 279(3.76)	142-2233-86

Compound	Solvent	$\lambda_{max}(\log \epsilon)$	Ref.
1H-Benzimidazole-2-methanol	pH 3	237(3.49),268(3.83), 275(3.82)	2-1092-86A
	pH 8	243(3.59),265(3.58), 271(3.68),278(3.63)	2-1092-86A
	H_0-8	220(3.75),272(3.93), 278(3.91)	2-1092-86A
	H_0-16	225(3.32),280(3.89), 295(3.80)	2-1092-86A
	MeOH	247(3.69),270(3.63), 276(3.77),283(3.75)	2-1092-86A
	EtOH	247(3.69),270(3.63), 276(3.77),283(3.74)	2-1092-86A
	ether	249(3.63),271(3.49), 277(3.62),284(3.60)	2-1092-86A
	MeCN	249(3.60),271(3.49), 277(3.62),283(3.61)	2-1092-86A
Imidazo[1,2-a]pyridine, 8-methoxy-	H_2O	221(4.42),277(3.88)	150-0670-86M
1H-Indazole-1-methanol	EtOH	251(3.61),287(3.61)	142-2233-86
$C_8H_8N_2O_2$			
Formamide, N-(3-cyano-4,5-dimethyl-2-furanyl)-	EtOH	277(3.88)	11-0347-86
Furo[2,3-d]pyrimidin-4(1H)-one, 5,6-dimethyl-	EtOH	252(3.75),279(3.61)	11-0347-86
$C_8H_8N_2O_5$			
Benzene, 5-methoxy-2-methyl-1,3-dinitro-	MeOH	248(4.0),329(3.6)	35-4115-86
$C_8H_8N_4$			
5H-Cyclopentapyrazine-2-carbonitrile, 3-amino-6,7-dihydro-	MeOH	363(3.92)	33-0793-86
$C_8H_8N_4O_2S$			
2,4(1H,3H)-Pteridinedione, 5,6-dihydro-1,3-dimethyl-6-thioxo-	pH 1.0	241(3.92),276s(3.93), 308(4.23),385s(3.47), 453(3.74)	33-0708-86
	pH 6.0	298(4.35),418(3.71)	33-0708-86
2,4(1H,3H)-Pteridinedione, 7,8-dihydro-1,3-dimethyl-7-thioxo-	pH -1.0	223(4.32),255s(3.87), 354(4.12),412(3.58)	33-0708-86
	pH 5.0	225(4.36),240s(4.14), 281(3.66),300(3.76), 379(4.30)	33-0708-86
$C_8H_8N_4O_3$			
Pyrimido[4,5-d]pyrimidine-2,4,5(1H,3H-6H)-trione, 1,3-dimethyl-	EtOH	232(4.43),267(3.48), 294(3.49)	142-2293-86
$C_8H_8N_4O_4S$			
6-Pteridinesulfinic acid, 1,2,3,4-tetrahydro-1,3-dimethyl-2,4-dioxo-, sodium salt	pH 7.0	243(4.27),260s(4.04), 333(3.89)	33-0708-86
$C_8H_8N_4O_5S$			
6-Pteridinesulfonic acid, 1,2,3,4-tetrahydro-1,3-dimethyl-2,4-dioxo-, potassium salt	pH 6.0	242(4.30),335(3.91)	33-0708-86
7-Pteridinesulfonic acid, 1,2,3,4-tetrahydro-1,3-dimethyl-2,4-dioxo-, sodium salt	pH 6.0	239(4.24),339(3.95)	33-0708-86

Compound	Solvent	$\lambda_{max}(\log \epsilon)$	Ref.
$C_8H_8N_4S$			
3H-1,2,4-Triazole-3-thione, 4-amino-2,4-dihydro-5-phenyl-	EtOH	237(4.16),251(4.30), 276s(3.98)	4-1451-86
$C_8H_8N_6O_2$			
Quinone-semicarbazide product	MeOH	370(4.15)	74-0321-86C
C_8H_8O			
Benzene, (ethenyloxy)-	hexane	226(3.16),270(2.09)	70-1677-86
2,4,6-Cycloheptatrien-1-one, 4-methyl-	EtOH	230(4.40),315(4.16)	39-2006-86C
C_8H_8OS			
Benzenecarbothioaldehyde, 4-methyl-, S-oxide, (Z)-	hexane	316(4.25)	77-0964-86
$C_8H_8O_2$			
Benzoic acid, 2-methyl-	$CHCl_3$-HOAc	282(3.18)	112-0321-86
Benzoic acid, 3-methyl-	$CHCl_3$-HOAc	281(3.15)	112-0321-86
Benzoic acid, 4-methyl-	$CHCl_3$-HOAc	272s(3.00)	112-0321-86
2-Butynoic acid, 2-butynyl ester	EtOH	212s(3.52)	33-1734-86
2(5H)-Furanone, 3-ethyl-4-ethynyl-	EtOH	232(3.79)	33-1734-86
2(5H)-Furanone, 4-ethyl-3-ethynyl-	EtOH	244(4.13)	33-1734-86
1(3H)-Isobenzofuranone, 4,5-dihydro-	EtOH	276(3.46)	33-1734-86
1(3H)-Isobenzofuranone, 6,7-dihydro-	EtOH	284(3.63)	33-1734-86
Spiro[2.5]oct-6-ene-5,8-dione	EtOH	206(3.60),292(2.51), 330s(2.18)	78-6487-86
$C_8H_8O_3$			
Benzoic acid, 2-methoxy-	$CHCl_3$-HOAc	285(3.58)	112-0321-86
Benzoic acid, 3-methoxy-	$CHCl_3$-HOAc	284(3.60)	112-0321-86
Benzoic acid, 4-methoxy-	$CHCl_3$-HOAc	260(4.26)	112-0321-86
2-Propenoic acid, 3-(2-furanyl)-, methyl ester, (E)-	CH_2Cl_2	300(4.38)	35-5964-86
BF_3 complex	CH_2Cl_2	348(4.31)	35-5964-86
BF_3-ether complex	CH_2Cl_2	350s(3.88)	35-5964-86
$EtAlCl_2$ complex	CH_2Cl_2	356(4.37)	35-5964-86
2-Propenoic acid, 3-(2-furanyl)-, methyl ester, (Z)-	CH_2Cl_2	305(4.27)	35-5964-86
BF_3-ether complex	CH_2Cl_2	350s(3.40)	35-5964-86
$EtAlCl_2$ complex	CH_2Cl_2	357(3.96)	35-5964-86
$C_8H_8O_4$			
Benzoic acid, 2,4-dihydroxy-, methyl ester	EtOH	259.0(4.04),295.0(3.74)	33-0734-86
2,5-Cyclohexadiene-1,4-dione, 2,3-dimethoxy-	CH_2Cl_2	253(3.87),389(2.92)	154-0561-86
2,5-Cyclohexadiene-1,4-dione, 2,5-dimethoxy-	CH_2Cl_2	278(4.37),284(4.38), 370(2.48)	154-0561-86
2,5-Cyclohexadiene-1,4-dione, 2,6-dimethoxy-	CH_2Cl_2	285(4.08),375(2.48)	94-1333-86
	CH_2Cl_2	283.5(4.28),377(2.78)	154-0561-86
2H-Pyran-6-carboxylic acid, 4-methyl-2-oxo-, methyl ester	EtOH	301(3.9)	5-1968-86
2H-Pyran-2-one, 5-acetyl-6-hydroxy-3-methyl-	C_6H_{12}	306(--)	39-0973-86B
	H_2O	275(4.02),342(4.42)	39-0973-86B
	EtOH	275(--),345(4.34)	39-0973-86B
	dioxan	310(4.40)	39-0973-86B
	$CHCl_3$	310(4.40)	39-0973-86B
	80% DMSO	275(--),350(4.40)	39-0973-86B
	+ acid	320(3.68)	39-0973-86B
	+ base	275(--),350(4.40)	39-0973-86B

Compound	Solvent	λ_{max}(log ϵ)	Ref.
2H-Pyran-2-one, 5-acetyl-6-hydroxy-4-methyl-	C_6H_{12}	308(--)	39-0973-86B
	H_2O	275(3.89),342(4.44)	39-0973-86B
	EtOH	275(3.88),345(4.34)	39-0973-86B
	dioxan	312(4.18)	39-0973-86B
	$CHCl_3$	312(4.23)	39-0973-86B
	80% DMSO	275(3.83),345(4.30)	39-0973-86B
	+ acid	320(4.01)	39-0973-86B
	+ base	278(--),348(4.35)	39-0973-86B
C_8H_8S			
Benzene, (ethenylthio)-	hexane	248(2.95),267(2.90)	70-1677-86
$C_8H_9BO_6$			
Boron, (2,4-pentanedionato-O,O')[propanedioate(2-)-O,O']-, (T-4)-	CH_2Cl_2	281(3.70),319s(2.86), 400s(1.77)	48-0755-86
C_8H_9BrO			
Phenol, 2-bromo-4,5-dimethyl-	dioxan	220(3.96),282(3.48), 289(3.46)	152-0017-86
C_8H_9CsO			
Cesium, [(2-methoxyphenyl)methyl]-	THF	355(3.92)	46-0603-86
Cesium, [(4-methoxyphenyl)methyl]-	THF	373(3.93)	46-0603-86
C_8H_9F			
Benzene, 1-fluoro-2,6-dimethyl-, chromium tricarbonyl adduct	$CHCl_3$	216(4.43),255s(3.79), 318(4.01)	112-0517-86
Benzene, 1-fluoro-3,4-dimethyl-, chromium tricarbonyl adduct	$CHCl_3$	215(4.41),255s(3.89), 315(4.00)	112-0517-86
C_8H_9LiO			
Lithium, [(2-methoxyphenyl)methyl]-	THF	327(4.04)	46-0603-86
Lithium, [(4-methoxyphenyl)methyl]-	THF	323(4.05)	46-0603-86
C_8H_9KO			
Potassium, [(2-methoxyphenyl)methyl]-	THF	352(4.17)	46-0603-86
Potassium, [(4-methoxyphenyl)methyl]-	THF	368(4.13)	46-0603-86
C_8H_9NO			
3-Buten-2-one, 4-(1H-pyrrol-2-yl)-	EtOH	348(4.36)	150-3601-86M
2-Propenal, 2-methyl-3-(1H-pyrrol-2-yl)-, (E)-	EtOH	345(4.49)	150-3601-86M
(Z)-	EtOH	380(--)	150-3601-86M
$C_8H_9NO_2$			
2,5-Cyclohexadiene-1,4-dione, 2-methyl-5-(methylamino)-	MeOH	483(3.41)	106-0027-86
	CH_2Cl_2	275(4.09),475(3.35)	106-0027-86
2,5-Cyclohexadiene-1,4-dione, 2-methyl-6-(methylamino)-	CH_2Cl_2	278(4.01),474(3.3)	106-0027-86
5(1H)-Indolizinone, 2,3-dihydro-3-hydroxy-	EtOH	231(3.86),305(3.85)	44-2184-86
2-Propenoic acid, 3-(1H-pyrrol-2-yl)-, methyl ester, (E)-	EtOH	331(4.41)	150-3601-86M
(Z)-	EtOH	336(4.25)	150-3601-86M
$C_8H_9NO_3Se$			
Benzeneselenenic acid, 2-nitro-, ethyl ester	60% dioxan	418(3.57)	44-0295-86
$C_8H_9NO_4$			
Acetamide, N,2,2-trihydroxy-N-phenyl-	H_2O	244(4.17)	44-3542-86

Compound	Solvent	λ_{max}(log ϵ)	Ref.
$C_8H_9NO_7$			
2-Furancarboxylic acid, 2-acetoxy-2,5-dihydro-5-nitro-, methyl ester, (E)-	HOAc	285(1.96)	44-3811-86
(Z)-	HOAc	283(1.92)	44-3811-86
$C_8H_9N_3O_2$			
3-Butenoic acid, 3-amino-4,4-dicyano-, ethyl ester	H_2O	272(4.18)	5-0533-86
	pH 13	313(4.60)	5-0533-86
$C_8H_9N_3O_3$			
Imidazo[1,2-a]pyridin-5(1H)-one, 2,3-dihydro-1-methyl-8-nitro-	EtOH	211(3.96),264(3.94), 312(3.85),378(4.11)	24-2208-86
$C_8H_9N_5O$			
4(1H)-Pteridinone, 2-(dimethylamino)-	pH -1.0	220(4.25),245(4.09), 296(3.83),322(3.90), 408(3.43)	5-1705-86
	pH 4.0	227(4.20),284(4.23), 347(3.81),370s(3.81)	5-1705-86
	pH 11.0	230(4.09),269(4.30), 375(3.75)	5-1705-86
Pyrido[2,3-d]pyrimidine-6-methanol, 2,4-diamino-	pH 1	219(4.42),279(3.78), 319(3.79),331s(3.75)	87-0709-86
	H_2O	222(4.41),245(4.23), 272s(3.91),332(3.80)	87-0709-86
	pH 13	247(4.28),269s(3.99), 345(3.83)	87-0709-86
$C_8H_9N_5O_2$			
4,6-Pteridinedione, 2-(dimethyl-amino)-1,5-dihydro-	pH 0.0	244(4.15),263(4.16), 300(3.45),362(3.66)	5-1705-86
	pH 4.0	228(4.19),285(4.14), 396(3.39)	5-1705-86
	pH 10.0	265(4.24),400(3.76)	5-1705-86
$C_8H_9N_5O_5$			
Pyrazolo[5,1-c][1,2,4]triazine-8-carb-oxylic acid, 1,4-dihydro-4-hydroxy-3-nitro-	H_2O	218(3.49),238(3.51), 370(3.43)	103-0543-86
$C_8H_9N_5S$			
1,2,4-Triazole-3-thiol, 4-amino-5-(4-aminophenyl)-	EtOH	260(4.15),288s(4.29)	4-1451-86
C_8H_{10}			
1,5-Hexadiene, 3,4-bis(methylene)-	C_6H_{12}	217(4.56)	89-0578-86
Octatetraene	C_6H_{12}	304(--)	89-0578-86
$C_8H_{10}BrN_5O$			
9H-Purine-9-ethanol, 6-amino-8-bromo-α-methyl-	pH 2	267(4.27)	73-0459-86
9H-Purine-9-propanol, 6-amino-8-bromo-	pH 2	266(4.27)	73-0459-86
$C_8H_{10}BrN_5O_2$			
1,2-Propanediol, 3-(6-amino-8-bromo-9H-purin-9-yl)-	pH 2	267(4.27)	73-0459-86
1,3-Propanediol, 2-(6-amino-8-bromo-9H-purin-9-yl)-	pH 2	266(4.27)	73-0459-86

Compound	Solvent	$\lambda_{max}(\log \epsilon)$	Ref.
$C_8H_{10}ClNO$			
2-Propen-1-ol, 3-(3-chloro-1H-pyrrol-2-yl)-2-methyl-, (E)-	EtOH	270(4.32)	150-3601-86M
(Z)-	EtOH	278(--)	150-3601-86M
$C_8H_{10}NO_2$			
Pyridinium, 1-(carboxymethyl)-4-methyl-, chloride	EtOH	256(3.81)	4-0209-86
$C_8H_{10}N_2O$			
Benzenamine, N,N-dimethyl-4-nitroso-	heptane-EtOH	235(3.85),262(3.74), 270(3.84),303(3.30), 310(3.28),385(4.48), 400(4.54),715(1.75)	80-0857-86
Benzeneacetamide, α-amino-	pH 6.5	254(2.48)	77-1247-86
$C_8H_{10}N_2O_2$			
Benzenamine, N,N-dimethyl-4-nitro-	heptane	362(4.35)	46-2340-86
	benzene	389(4.35)	46-2340-86
	EtOAc	389(4.38)	46-2340-86
	acetone	396(4.38)	46-2340-86
	THF	392(4.38)	46-2340-86
	CH_2Cl_2	401(4.38)	46-2340-86
	CCl_4	374(4.35)	46-2340-86
	1,2-$C_2H_4Cl_2$	399(4.38)	46-2340-86
	DMF	407(4.38)	46-2340-86
	DMSO	411(4.37)	46-2340-86
1H-Pyrrolizine, 2,3-dihydro-2-methyl-5-nitro-	EtOH	353(4.20)	103-0257-86
1H-Pyrrolizine, 2,3-dihydro-3-methyl-5-nitro-	EtOH	354(4.78)	103-0257-86
$C_8H_{10}N_2O_3$			
Benzenamine, 5-methoxy-2-methyl-3-nitro-	MeOH	233(3.9),284(3.0), 345(2.9)	35-4115-86
2-Furanamine, N,N-dimethyl-5-(2-nitroethenyl)-, (E)-	MeOH	520(2.78)	73-0879-86
3-Furancarboxamide, 2-(formylamino)-4,5-dimethyl-	EtOH	288(3.80)	11-0347-86
$C_8H_{10}N_2O_3S$			
Acetic acid, [(1,6-dihydro-1,4-dimethyl-6-oxo-2-pyrimidinyl)thio]-	MeOH	223(3.75),237s(--), 288(3.95)	2-0275-86
	MeOH-NaOH	239(3.55),287(3.88)	2-0275-86
4,7-(Epoxymethano)furo[3,4-d]pyrimidin-5(1H)-one, hexahydro-4-methyl-2-thioxo-	MeOH	252(3.72),302(3.59)	142-0679-86
$C_8H_{10}N_2O_4$			
1H-Pyrazole-4-acetic acid, 3-(methoxycarbonyl)-, methyl ester	EtOH	240s(3.15)	4-0059-86
$C_8H_{10}N_4$			
1H-Imidazo[4,5-c]pyridin-4-amine, N,N-dimethyl-, monohydrochloride	pH 1	270(4.05),290(4.03)	87-0138-86
	pH 13	285(4.00)	87-0138-86
Pyrazinecarbonitrile, 3-amino-5-ethyl-	MeOH	352(3.91)	33-0793-86
$C_8H_{10}N_4O$			
[1,2,4]Triazolo[1,5-a]pyrimidin-7(1H)-one, 2-ethyl-5-methyl-	pH 4.0	216(4.28),271(4.02)	39-0711-86C
	pH 9.9	212(4.47),254(3.78), 280(4.06)	39-0711-86C

Compound	Solvent	λ_{max}(log ε)	Ref.
$C_8H_{10}N_4OS_2$			
1H-Pyrazolo[3,4-d]pyrimidine, 3-methoxy-4,6-bis(methylthio)-	pH 1	255(4.05),295(3.90)	4-1869-86
	pH 7	255(3.92),294(3.76)	4-1869-86
	pH 11	255(4.20),299(3.90)	4-1869-86
$C_8H_{10}N_4O_2$			
Hydrazinecarboximidamide, 2-[(3,4-dihydroxyphenyl)methylene]-	pH 5.0	285(4.3),305(4.3)	74-0027-86A
	pH 9.0	260(4.2),315(4.1)	74-0027-86A
1H-Purine-2,6-dione, 3,7-dihydro-1,3,8-trimethyl-	MeOH	272(4.05)	142-1565-86
$C_8H_{10}N_4O_2S$			
4H-Pyrazolo[3,4-d]pyrimidin-4-one, 3-ethoxy-1,5-dihydro-6-(methylthio)-	pH 1	233(4.14),267(3.95)	4-1869-86
	pH 7	233(4.21),260(4.00)	4-1869-86
	pH 11	236(4.28),259(4.06)	4-1869-86
8H-Pyrido[2,3-c][1,2,6]thiadiazin-4-amine, 1,7-dimethyl-, 2,2-dioxide	pH -3.0	215(4.05),260(4.06), 330(3.98)	11-0607-86
	pH 7.0	217(4.29),247(4.05), 322(3.82)	11-0607-86
	pH 14.0	256(3.96),317(3.84)	11-0607-86
8H-Pyrido[2,3-c][1,2,6]thiadiazin-4-amine, 5,7-dimethyl-, 2,2-dioxide	pH -3.0	227(4.13),256(3.80), 320(3.71),356(3.68)	11-0607-86
	pH 3.0	213(4.25),261(4.02), 292s(3.13),358(3.89)	11-0607-86
	pH 10.0	226(4.30),247s(4.04), 348(3.75)	11-0607-86
8H-Pyrido[2,3-c][1,2,6]thiadiazin-4-amine, 7,8-dimethyl-, 2,2-dioxide	pH -4.0	226s(4.08),262(3.82), 365(3.93)	11-0607-86
	pH 7.0	260(4.02),292s(3.13), 366(3.92)	11-0607-86
	pH 14.0	265(3.96),303(3.42), 368(3.86)	11-0607-86
$C_8H_{10}N_6$			
1H-1,2,4-Triazole-3,5-diamine, N-(2-pyridinylmethyl)-	EtOH	256(3.85),260(3.91)	4-0401-86
	10%EtOH-HCl	260(3.85)	4-0401-86
	10%EtOH-NaOH	255(3.87),259(3.94)	4-0401-86
1H-1,2,4-Triazole-3,5-diamine, N-(3-pyridinylmethyl)-	EtOH	258(3.45)	4-0401-86
	10%EtOH-HCl	258(3.69)	4-0401-86
	10%EtOH-NaOH	257(3.48)	4-0401-86
1H-1,2,4-Triazole-3,5-diamine, N-(4-pyridinylmethyl)-	EtOH	249(3.75),254(3.73)	4-0401-86
	10%EtOH-HCl	252(4.02)	4-0401-86
	10%EtOH-NaOH	248(3.84),254(3.84)	4-0401-86
$C_8H_{10}O$			
Benzene, ethoxy-	hexane	221(2.85),272(2.34)	70-1677-86
Benzenemethanol, α-methyl-, (-)-	MeOH	254(2.18),260(2.26)	39-0635-86B
3-Cyclopenten-1-one, 2-(1-methylethylidene)-	EtOH	228(4.01),289(3.84)	33-0560-86
2-Hepten-5-yn-4-one, 2-methyl-	EtOH	256(4.19)	33-0560-86
4-Hexen-1-yn-3-one, 4,5-dimethyl-	EtOH	221(3.62),276(3.92)	33-0560-86
Phenol, 2,4-dimethyl-	acid	277(3.27)	3-0639-86
Phenol, 2,5-dimethyl-	acid	274(3.25)	3-0639-86
Phenol, 2,6-dimethyl-	acid	269(3.09)	3-0639-86
Phenol, 3,4-dimethyl-	acid	277(3.26)	3-0639-86
	dioxan	217(3.89),279(3.36), 285(3.30)	152-0017-86
Phenol, 3,5-dimethyl-	acid	272(3.08)	3-0639-86
$C_8H_{10}O_2$			
Benzene, 1,4-dimethoxy-	H_2O	285(3.40)	60-0291-86

Compound	Solvent	λ_{max}(log ε)	Ref.
2,4-Cycloheptadien-1-one, 5-methoxy-	EtOH	332(3.67)	77-1480-86
3-Cyclopentene-1,2-dione, 3,5,5-trimethyl-	$CHCl_3$	288(3.9),322(1.3), 486(1.28)	89-1018-86
2-Cyclopenten-1-one, 4-(2-propenyloxy)-	EtOH	215(3.95),313(1.78)	32-0291-86
Ethanol, 2-phenoxy-	H_2O	270(3.11)	60-0291-86
2,4-Octadiyne-1,6-diol, (R)-	EtOH	218(2.67),231(2.58), 243(2.53),257(2.35)	150-1348-86M
2,4-Octadiyne-1,6-diol, (S)-	EtOH	220(2.59),231(2.66), 243(2.64),257(2.42)	150-1348-86M
$C_8H_{10}O_3$			
2,4-Heptadienoic acid, 6-oxo-, methyl ester, (E,E)-	CH_2Cl_2	270(4.39)	70-2409-86
(E,Z)-	CH_2Cl_2	273(4.32)	70-2409-86
(Z,Z)-	CH_2Cl_2	273(4.27)	70-2409-86
2-Oxabicyclo[3.1.0]hex-3-ene-6-carboxylic acid, 3-methyl-, methyl ester	CH_2Cl_2	220(3.89)	70-2409-86
$C_8H_{10}S$			
Benzene, (ethylthio)-	hexane	256(2.91),279(2.18)	70-1677-86
$C_8H_{11}BrN_2OS$			
2(1H)-Pyrimidinone, 5-bromo-4-(butylthio)-	H_2O	280(4.86),312(4.89)	56-0605-86
4(1H)-Pyrimidinone, 5-bromo-2-(butylthio)-	H_2O	252(4.94),290(4.92)	56-0605-86
$C_8H_{11}BrO_2$			
2,4-Hexadienoic acid, 4-bromo-, ethyl ester, (E,E)-	n.s.g.	265(4.37)	104-2230-86
(E,Z)-	n.s.g.	266(4.11)	104-2230-86
$C_8H_{11}ClN_2OS$			
2(1H)-Pyrimidinone, 4-(butylthio)-5-chloro-	H_2O	278(4.86),312(4.90)	56-0605-86
4(1H)-Pyrimidinone, 2-(butylthio)-5-chloro-	H_2O	252(4.90),290(4.84)	56-0605-86
$C_8H_{11}ClO$			
1,4-Hexadien-3-one, 1-chloro-4,5-dimethyl-, (E)-	EtOH	233(4.03),276s(3.57), 341(2.15)	33-0560-86
(Z)-	EtOH	240(3.88),267s(3.67), 341(2.18)	33-0560-86
$C_8H_{11}F_3O$			
1-Octen-3-one, 1,1,2-trifluoro-	hexane	220(4.12),307(1.32)	22-0876-86
$C_8H_{11}N$			
Benzenamine, N,N-dimethyl-	heptane-EtOH	250(4.28),298(3.43)	80-0857-86
	pH 3	201(4.10),241(3.19)	160-0979-86
	MeCN	256(4.3),308(3.5)	39-1115-86B
	98% H_2SO_4	252f(2.3)	112-0265-86
Benzenamine, 2-ethyl-, chromium tricarbonyl complex	$CHCl_3$	218(4.58),319(3.94)	112-0517-86
Benzenamine, 3-ethyl-, chromium tricarbonyl complex	$CHCl_3$	217(4.56),320(3.93)	112-0517-86
Benzenamine, 4-ethyl-, chromium	$CHCl_3$	216(4.71),320(3.92)	112-0517-86
Benzeneethanamine, hydrochloride	pH 3	199(4.10)	160-0979-86
Pyridine, 5-ethyl-2-methyl-	1,2-$C_2H_4Cl_2$	270.0(3.49)	46-2589-86
	+ TFA	271.5(3.83)	46-2589-86

Compound	Solvent	λ_{max}(log ϵ)	Ref.
Pyridine, 2-propyl-	1,2-$C_2H_4Cl_2$	263.5(3.42)	46-2589-86
	+ TFA	265.0(3.84)	46-2589-86
Pyridine, 4-propyl-	1,2-$C_2H_4Cl_2$	258.0(3.20)	46-2589-86
	+ TFA	254.0(3.65)	46-2589-86
Pyridine, 2,4,6-trimethyl-	pH 3	213(3.95),265(4.28)	160-0979-86
$C_8H_{11}NO$			
Benzenamine, 2-ethoxy-, chromium tricarbonyl complex	$CHCl_3$	217(4.71),322(3.86)	112-0517-86
Benzenamine, 3-ethoxy-, chromium tricarbonyl complex	$CHCl_3$	206(4.41),216(4.44), 316(3.90),365s(3.80)	112-0517-86
Benzenamine, 4-ethoxy-, chromium tricarbonyl complex	$CHCl_3$	217(4.69),320(3.86), 365s(3.62)	112-0517-86
Benzenamine, 2-methoxy-5-methyl-, chromium tricarbonyl complex	$CHCl_3$	217(4.71),320(3.88)	112-0517-86
Ethanol, 2-(phenylamino)-	acid	249(2.3),253(2.3)	112-1099-86
2-Propen-1-ol, 2-methyl-3-(1H-pyrrol-2-yl)-, (E)-	EtOH	270(4.35)	150-3601-86M
(Z)-	EtOH	273(4.16)	150-3601-86M
$C_8H_{11}NS_2$			
Thiazole, 2,4-dimethyl-5-(2-propenylthio)-	EtOH	250(3.72),278s(3.40)	33-0374-86
3-Thiophenecarbothioamide, N,2,5-trimethyl-	EtOH	256(4.10),277(4.03), 357(2.62)	78-3683-86
$C_8H_{11}N_3OS$			
Thiourea, N-ethyl-N'-(3-hydroxy-2-pyridinyl)-	benzene	241(4.19)	80-0761-86
	MeOH	260(4.19)	80-0761-86
	EtOH	262(4.21)	80-0761-86
	isoPrOH	261(4.20)	80-0761-86
	BuOH	256(4.40)	80-0761-86
	dioxan	241(4.19)	80-0761-86
	$HOCH_2CH_2OH$	248(4.14)	80-0761-86
	CH_2Cl_2	256(4.18)	80-0761-86
	$CHCl_3$	253(4.18)	80-0761-86
	$C_2H_4Cl_2$	254(4.19)	80-0761-86
	MeCN	260(4.29)	80-0761-86
	DMF	259(4.06)	80-0761-86
	DMSO	254(4.04)	80-0761-86
$C_8H_{11}N_3O_2$			
Acetic acid, cyano(2-imidazolidinylidene)-, ethyl ester	EtOH	207(4.24),253(4.42)	24-2208-86
Acetic acid, cyano(1-methyl-2-imidazolidinylidene)-, methyl ester	EtOH	212(4.23),261(4.37)	24-2208-86
1H-Imidazole, 1-methyl-2-(2-methyl-1-propenyl)-5-nitro-	n.s.g.	235(4.05),340(4.04)	112-1049-86
$C_8H_{11}N_3O_3$			
Imidazo[1,2-a]pyridin-5(1H)-one, 2,3,6,7-tetrahydro-1-methyl-8-nitro-	EtOH	366(4.27)	24-2208-86
$C_8H_{11}N_3O_4$			
4-Oxazoleacetic acid, 2-amino-α-(methoxyimino)-, ethyl ester, (Z)-	MeOH	217(4.28),275(3.53)	158-0121-86
1H-Pyrazole-4-carboxylic acid, 3-amino-1-(1,3-dioxan-5-yl)-	pH 2	265(3.73)	73-1512-86
	pH 7	258(3.75)	73-1512-86

Compound	Solvent	λ_{max}(log ε)	Ref.
$C_8H_{11}N_5$			
6H-Purin-6-imine, 3-ethyl-3,7-dihydro-7-methyl-, monohydriodide	pH 1	225(4.39),277(4.23)	94-1821-86
	pH 7	225(4.39),277(4.23)	94-1821-86
	pH 13	225(4.26),282(4.18)	94-1821-86
	EtOH	280(4.24)	94-1821-86
6H-Purin-6-imine, 7-ethyl-3,7-dihydro-3-methyl-, monohydriodide	pH 1	226(4.41),278(4.23)	94-1821-86
	pH 7	226(4.40),278(4.22)	94-1821-86
	pH 13	225(4.27),282(4.17)	94-1821-86
	EtOH	280(4.21)	94-1821-86
$C_8H_{11}N_5O$			
4H-Imidazo[4,5-d]-1,2,3-triazin-4-one, 3-butyl-3,5-dihydro-	EtOH	250(3.48),260(3.47), 283(3.61)	5-1012-86
9H-Purine-9-propanol, 6-amino-	pH 2	261(4.13)	73-0459-86
8H-Purin-8-one, 3,6,7,9-tetrahydro-3,9-dimethyl-6-(methylimino)-, monohydrochloride (caissarone)	MeOH	228(4.33),302(4.25)	39-2051-86C
4H-Pyrimido[4,5-b][1,4]diazepin-4-one, 2-amino-1,6,7,8-tetrahydro-6-methyl-	pH 1	215(4.22),250(4.10), 267(3.96),325(3.94)	69-4762-86
[1,2,4]Triazolo[1,5-a]pyrimidin-7(1H)-one, 6-amino-2-ethyl-5-methyl-	pH 1.0	208(4.39),235(3.36), 277(3.94)	39-0711-86C
	pH 5.2	221(4.35),258(3.60), 302(3.80)	39-0711-86C
	pH 9.9	223(4.40),263(3.72), 306(3.95)	39-0711-86C
[1,2,4]Triazolo[4,3-a]pyrimidin-5(1H)-one, 6-amino-3-ethyl-7-methyl-	pH 1.1	207(4.19),235?(3.76), 251(3.66),300(3.97), 307(3.95)	39-0711-86C
	pH 5.8	206(4.69),250s(3.82), 325(3.89)	39-0711-86C
	pH 9.9	227(4.21),264(3.71), 340(3.92)	39-0711-86C
$C_8H_{11}N_5O_2S$			
1H-Pyrazino[2,3-c][1,2,6]thiadiazin-4-amine, 1,6,7-trimethyl-, 2,2-dioxide	pH 7.0	253(4.15),348(3.89)	5-1892-86
	MeOH	255(4.16),286s(3.38), 348(3.85)	5-1872-86
$C_8H_{11}N_5O_2S_2$			
1,2,4-Triazolo[3,4-b][1,3,4]thiadiazolium, 6-[(ethoxycarbonyl)amino]-1-methyl-3-(methylthio)-, hydroxide, inner salt	MeOH	228(4.30),272(4.21)	5-1540-86
$C_8H_{11}N_5S$			
2H-Purine-2-thione, 1-ethyl-1,3,6,7-tetrahydro-6-imino-7-methyl-	pH 1	244(4.31),288(4.35)	4-0737-86
	pH 13	235(4.31),286(4.23)	4-0737-86
$C_8H_{12}Br_2N_2O$			
2,3-Diazabicyclo[3.2.0]hept-2-ene, 6,7-dibromo-4-(methoxymethyl)-4-methyl-, (1α,4α,5α,6α,7β)-	EtOH	332(2.36)	24-0794-86
(1α,4β,5α,6α,7β)-	EtOH	331(2.39)	24-0794-86
$C_8H_{12}ClN_5$			
Piperidine, 1-[(5-chloro-1H-1,2,3-triazol-4-yl)iminomethyl]-, monohydrochloride	H_2O	222(3.95)	73-0215-86
	pH 13	233(3.98)	73-0215-86

Compound	Solvent	$\lambda_{max}(\log \epsilon)$	Ref.
$C_8H_{12}F_3NS$			
3-Hexene-2-thione, 4-amino-1,1,1-trifluoro-5,5-dimethyl-	hexane	385(4.34)	140-0239-86
	toluene	400(4.28)	140-0239-86
	pH 12	323(--)	140-0239-86
	5% EtOH	380(4.36)	140-0239-86
$C_8H_{12}N_2$			
1,2-Ethanediamine, N-phenyl-, dication	acid	255(2.3)	112-1099-86
$C_8H_{12}N_2O$			
2,3-Diazabicyclo[3.2.0]hepta-2,6-diene, 4-(methoxymethyl)-4-methyl-, (1α,4α,5α)-	hexane	344(2.45)	24-0794-86
(1α,4β,5α)-	hexane	341(2.40)	24-0794-86
$C_8H_{12}N_2OSe$			
Cyclobutenediylium, 1,3-bis(dimethylamino)-2-hydroxy-4-selenyl-, dihydroxide, bis(inner salt)	CH_2Cl_2	245(4.3),345s(4.2), 381(4.5),435s(3.1)	24-0182-86
2-Cyclobuten-1-one, 2,3-bis(dimethylamino)-4-selenoxo-	CH_2Cl_2	315(3.9),411(4.5), 512(2.7)	24-0182-86
$C_8H_{12}N_2O_2$			
1-Butanone, 4-hydroxy-1-(1-methyl-1H-imidazol-2-yl)-	EtOH	277(3.95)	94-4916-86
2,4(1H,3H)-Pyrimidinedione, 6-methyl-1-(1-methylethyl)-	pH 2	268(3.70)	70-1064-86
	pH 12	269(3.79)	70-1064-86
2,5-Pyrrolidinedione, 3-[(dimethylamino)methylene]-1-methyl-	isoPrOH	305(4.49)	103-1345-86
$C_8H_{12}N_2O_2S$			
4,6(1H,5H)-Pyrimidinedione, 5,5-diethyldihydro-2-thioxo-	pH 9.62	256(4.08),306(4.47)	56-0825-86
	EtOH	238(4.03),289(4.33)	56-0825-86
Pyrimidinium, 2-(ethylthio)-3,6-dihydro-4-hydroxy-1,3-dimethyl-6-oxo-, hydroxide, inner salt	MeCN	222(4.34),250(3.41), 351(3.31)	24-1315-86
$C_8H_{12}N_2O_3$			
2,4(1H,3H)-Pyrimidinedione, 1-(2-hydroxyethyl)-3,6-dimethyl-	EtOH	268(4.02)	94-1809-86
2,4(1H,3H)-Pyrimidinedione, 5-(hydroxymethyl)-1,3,6-trimethyl-	EtOH	273(3.99)	118-0857-86
$C_8H_{12}N_2O_4$			
2,4,6(1H,3H,5H)-Pyrimidinetrione, 5-(2-methoxypropyl)-	pH 10.0	268(4.27)	39-1391-86B
$C_8H_{12}N_2O_6$			
D-glycero-D-gulo-2-Octulose, 3,7-anhydro-1-deoxy-1-diazo-	H_2O	255(3.93),281(4.17), 350(1.62)	136-0143-86G
D-glycero-L-manno-2-Octulose, 3,7-anhydro-1-deoxy-1-diazo-	H_2O	255(3.85),281(4.10), 350(1.63)	136-0143-86G
$C_8H_{12}N_4O$			
4(1H)-Pteridinone, 5,6,7,8-tetrahydro-6,7-dimethyl-, cis	pH 0.0	262(3.83)	12-0031-86
	pH 7.0	292(3.96)	12-0031-86
$C_8H_{12}N_4OS$			
4(1H)-Pteridinone, 3,4,5,6,7,8-hexahydro-6,7-dimethyl-2-thioxo-, cis	pH 0	282(3.73)	12-0031-86

Compound	Solvent	λ_{max}(log ε)	Ref.
$C_8H_{12}N_4O_4$			
1,3,5-Triazin-2-amine, 4-β-D-ribo-furanosyl-	pH 1	248s(3.56)	4-1709-86
	pH 7	256(3.60)	4-1709-86
	pH 11	256(3.56)	4-1709-86
$C_8H_{12}N_4O_4S$			
1,3,5-Triazine-2(1H)-thione, 4-amino-6-β-D-ribofuranosyl-	pH 1	267(4.20)	4-1709-86
	pH 7	279(4.17)	4-1709-86
	pH 11	275(4.09)	4-1709-86
$C_8H_{12}N_4O_5$			
1,3,5-Triazin-2(1H)-one, 4-amino-6-β-D-ribofuranosyl-	pH 1	248(3.90)	4-1709-86
	pH 7	242(3.63)	4-1709-86
	pH 11	251(3.57)	4-1709-86
$C_8H_{12}N_6$			
1H-Imidazo[4,5-d]-1,2,3-triazin-4-amine, N-butyl-	EtOH	218(4.14),263(4.00), 302(3.79)	5-1012-86
$C_8H_{12}N_6O$			
8H-Purin-8-one, 6-amino-9-(2-amino-propyl)-7,9-dihydro-	pH 2	269(4.05),281(4.05)	73-0459-86
	pH 7	271(4.12)	73-0459-86
	pH 12	280(4.15)	73-0459-86
8H-Purin-8-one, 6-amino-9-(3-amino-propyl)-7,9-dihydro-	pH 2	270(4.04),282(4.04)	73-0459-86
	pH 7	271(4.12)	73-0459-86
	pH 12	280(4.15)	73-0459-86
$C_8H_{12}N_6O_2$			
1H-Imidazo[4,5-d]-1,2,3-triazin-4-amine, N-(2,2-dimethoxyethyl)-	EtOH	219(4.12),262(3.92), 299(3.74)	5-1012-86
8H-Purin-8-one, 6-amino-9-(1-amino-3-hydroxy-2-propyl)-7,9-dihydro-	pH 2	268(4.03),281(4.03)	73-0459-86
	pH 7	271(4.11)	73-0459-86
	pH 12	281(4.15)	73-0459-86
8H-Purin-8-one, 6-amino-9-(3-amino-2-hydroxypropyl)-7,9-dihydro-	pH 2	269(4.03),280(4.03)	73-0459-86
	pH 7	272(4.11)	73-0459-86
	pH 12	281(4.15)	73-0459-86
$C_8H_{12}O$			
Ethanone, 1-(1-cyclohexen-1-yl)-	gas	220(--)	131-0119-86N
	hexane	225.0(4.10),332.0(1.54)	131-0119-86N
	MeOH	230(4.0)	107-0957-86
$C_8H_{12}O_2$			
2,4-Hexadienoic acid, 5-methyl-, methyl ester, (E)-	CH_2Cl_2	275(4.45)	35-3016-86
BF_3 complex	CH_2Cl_2	275(4.34)	35-3016-86
2,4-Hexadienoic acid, 5-methyl-, methyl ester, (Z)-	CH_2Cl_2	278(4.24)	35-3016-86
BF_3 complex	CH_2Cl_2	278(4.16)	35-3016-86
5-Hexynoic acid, ethyl ester	MeOH	206(2.23)	18-3535-86
7-Octynoic acid	MeOH	206(2.65)	18-3535-86
4H-Pyran-4-one, 2-ethyl-2,3-dihydro-6-methyl-, (R)-	EtOH	261(4.11)	163-3209-86
$C_8H_{12}O_3$			
2-Cyclobuten-1-one, 3-[(methoxymeth-oxy)methyl]-2-methyl-	isooctane	227(3.88),307(1.61)	35-0806-86
$C_8H_{12}O_4$			
Hexanoic acid, 3,5-dioxo-, ethyl ester	EtOH	272(3.84)	44-4254-86

Compound	Solvent	λmax(log ε)	Ref.
$C_8H_{13}ClOS$			
3-Buten-2-one, 4-(butylthio)-4-chloro-	n.s.g.	297(4.76)	48-0539-86
$C_8H_{13}N$			
1-Buten-3-yn-1-amine, N,N-diethyl-	MeCN	275(4.10)	104-0420-86
$C_8H_{13}NO_2$			
3-Pyridinecarboxylic acid, 1,2,3,6-tetrahydro-1-methyl-, methyl ester	pH 7.2	210(0.24)	87-0125-86
$C_8H_{13}NO_5$			
α-D-erythro-Pentofuranos-3-ulose, 1,2-O-(1-methylethylidene)-, oxime	EtOH	229(2.73)	33-1132-86
$C_8H_{13}NS_2$			
5(4H)-Thiazolethione, 2,4-dimethyl-4-(1-methylethyl)-	EtOH	238(3.36),316(4.00), 484(1.15),500s(1.08)	33-0374-86
$C_8H_{13}N_4$			
1,2,4-Triazinium, 1-methyl-3-(1-pyrrolidinyl)-, iodide	EtOH	222(4.22),258(4.07), 313(2.90),396(2.90), 542(2.16)	103-1242-86
$C_8H_{13}N_4O$			
1,2,4-Triazinium, 1-methyl-3-(4-morpholinyl)-, iodide	EtOH	220(4.30),259(4.20), 307(2.92),392(2.98)	103-1242-86
$C_8H_{13}N_5O$			
4H-Pyrimido[4,5-b][1,4]diazepin-4-one, 2-amino-1,5,6,7,8,9-hexahydro-6-methyl-	pH 1	215(4.20),271(4.17)	69-4762-86
$C_8H_{13}N_5O_2S$			
Pyrazino[2,3-e]-1,2,4-triazin-6-ol, 7-ethoxy-1,2,6,7-tetrahydro-3-(methylthio)-, trans	pH 1	249s(3.84),301(4.45)	4-0033-86
	pH 11	254(4.24),266s(4.17),	4-0033-86
	EtOH	220(4.10),252(4.22), 264s(4.13),330(3.81)	4-0033-86
Pyrazino[2,3-e]-1,2,4-triazin-6-ol, 1,5,6,7-tetrahydro-7-methoxy-5-methyl-3-(methylthio)-, trans	pH 1	267(4.41),327(3.62)	4-0033-86
	pH 11	257(4.16),270s(4.08), 328(3.83)	4-0033-86
	MeOH	224(4.09),255(4.19), 268s(4.11),329(3.80)	4-0033-86
Pyrazino[2,3-e]-1,2,4-triazin-6-ol, 2,6,7,8-tetrahydro-7-methoxy-8-methyl-3-(methylthio)-, trans	pH 1	265(4.42),330s(3.48)	4-0033-86
	pH 11	253(4.22),271s(4.11), 332(3.82)	4-0033-86
	MeOH	220(4.08),251(4.21), 269s(4.08),330(3.78)	4-0033-86
$C_8H_{13}N_5O_4$			
1,3,5-Triazine-2,4-diamine, 6-β-D-ribofuranosyl-	pH 1	248s(3.56)	4-1709-86
	pH 7	256(3.60)	4-1709-86
	pH 11	256(3.56)	4-1709-86
$C_8H_{14}N_2O_2$			
Acetic acid, (1-methyl-2-imidazolidinylidene)-, ethyl ester, (E)-	EtOH	271(4.40)	24-2208-86
$C_8H_{14}N_2O_3$			
Ethanone, 1-(4,5-dihydro-4-hydroxy-4,5,5-trimethyl-1H-imidazol-2-yl)-, N-oxide	EtOH	292(3.54)	103-0856-86

Compound	Solvent	$\lambda_{max}(\log \epsilon)$	Ref.
$C_8H_{14}N_4O_3$			
2,4(1H,3H)-Pyrimidinedione, 5-amino-6-[(4-hydroxybutyl)amino]-	pH 5.0	200(4.42),224s(4.23), 311(4.16),350(2.70)	44-2461-86
	pH 8.0	238(3.74),311(4.27), 350(3.00),500(2.04)	44-2461-86
	pH 12.0	238(3.60),311(4.27), 350(3.40)	44-2461-86
$C_8H_{14}O$			
Cyclohexane, 1-methoxy-3-methylene-	pentane	229(3.15)	35-7575-86
$C_8H_{14}OS$			
2(3H)-Furanthione, dihydro-3,3,4,4-tetramethyl-	EtOH	248(4.07)	87-1996-86
2(3H)-Thiophenone, dihydro-3,3,4,4-tetramethyl-	EtOH	235(3.57)	87-1996-86
$C_8H_{14}OS_2$			
Butane(dithioic) acid, 3,3-dimethyl-2-oxo-, ethyl ester	MeOH	306(3.58),316(3.56), 324(3.55),486(1.42)	118-0968-86
$C_8H_{14}O_2S_4$			
Thioperoxydicarbonic acid, bis(1-methylethyl) ester	isooctane	241(4.25),286(3.95)	3-0588-86
	EtOH	290(3.95)	3-0588-86
$C_8H_{14}O_3$			
Cyclohexanone, 4-hydroxy-4-(2-hydroxyethyl)-	MeOH	287(2.4)	102-2821-86
$C_8H_{14}S_2$			
2(3H)-Thiophenethione, dihydro-3,3,4,4-tetramethyl-	EtOH	221(3.46),313(4.14)	87-1996-86
$C_8H_{15}NO$			
3-Buten-2-one, 4-(diethylamino)-	MeCN	306(4.40)	104-0420-86
$C_8H_{15}NO_2$			
Cyclohexanecarboxylic acid, 2-amino-, methyl ester, cis	EtOH	216(2.46),282(1.83), 305(1.70)	12-0591-86
$C_8H_{15}NO_5$			
α-D-Ribofuranose, 3-deoxy-3-(hydroxyamino)-1,2-O-(1-methylethylidene)-	EtOH	208(2.20)	33-1132-86
$C_8H_{15}N_3$			
2-Pyrimidinamine, 1,4-dihydro-N,4,4,6-tetramethyl-	isoPrOH	266(3.49)	142-3023-86
hydrochloride	isoPrOH	236(3.58)	142-3023-86
$C_8H_{15}N_3O$			
4H-1,2,3-Triazol-4-one, 3-(1,1-dimethylethyl)-3,5-dihydro-5,5-dimethyl-	hexane	250(3.673),309(2.511)	5-1891-86
	10% MeOH	251(3.675),302(2.463)	5-1891-86
$C_8H_{15}N_5O_2$			
1H-1,2,4-Triazole-1-ethanol, 5-amino-3-(4-morpholinyl)-	EtOH	202(4.05),216s(3.93)	4-0401-86
$C_8H_{16}NOS$			
Oxazolidinium, 2-[(ethylthio)methylene]-3,3-dimethyl-, chloride	H_2O	227s(3.76),247(3.80)	70-0994-86

Compound	Solvent	$\lambda_{max}(\log \epsilon)$	Ref.
$C_8H_{16}N_4O_2S$			
Hydrazinecarbothioamide, 2-[2-(acet-oxyamino)-1,2-dimethylpropylidene]-	EtOH	232(4.00),273(4.46)	103-0287-86
$C_8H_{16}N_6O$			
1-Piperazineethanol, 4-(5-amino-1H-1,2,4-triazol-3-yl)-	EtOH	206s(3.90),216s(3.87)	4-0401-86
	10%EtOH-HCl	218s(3.95)	4-0401-86
4(1H)-Pyrimidinone, 2,5-diamino-6-[(3-aminobutyl)amino]-	pH 1	269(4.22)	69-4762-86
$C_8H_{16}O_3S_2$			
Ethane(dithioic) acid, [2-(2-methoxy-ethoxy)ethoxy]-, methyl ester	CH_2Cl_2	312(3.77)	126-2369-86
$C_8H_{17}NO_2S$			
Ethanethioamide, 2-(2-ethoxyethoxy)-N,N-dimethyl-	CH_2Cl_2	274(3.69)	126-2369-86
$C_8H_{18}N_6$			
1H-1,2,4-Triazole-3,5-diamine, N-[2-(diethylamino)ethyl]-	EtOH	203(4.38),217s(3.84)	4-0401-86
$C_8H_{18}O$			
2-Octanol, (+)-	MeOH	none above 220 nm	39-0635-86B
$C_8H_{20}O_3SSi$			
Silane, [(butylthio)methyl]trimethoxy-	heptane	190(2.42),210s(--)	67-0062-86
	MeCN	194(2.40)	67-0062-86
Silane, [[(1,1-dimethylethyl)thio]-methyl]trimethoxy-	heptane	210(2.45)	67-0062-86
	MeCN	194(2.30),210(2.34)	67-0062-86
Silane, trimethoxy[[(2-methylpropyl)-thio]methyl]-	heptane	192(2.60),210s(--)	67-0062-86
	MeCN	194(2.73)	67-0062-86
$C_8H_{20}SSi$			
Ethanethiol, 2-(triethylsilyl)-	heptane	200(2.28),227(1.34)	67-0225-86
Silane, [3-(ethylthio)propyl]tri-methyl-	heptane	211s(3.3)	65-1848-86
$C_8O_2S_6$			
Benzo[1,2-d:4,5-d']bis[1,3]dithiole-4,8-dione, 2,6-dithioxo-, radical anion, salt with N-methylpyridinium	MeCN	215(4.25),237(4.28), 255s(4.10),259(4.12), 266(4.09),278(4.06), 305s(3.70),319s(3.64), 387(4.75),443(4.10), 489(4.05)	77-1489-86

Compound	Solvent	λ_{max}(log ε)	Ref.
$C_9Br_3F_5$			
1H-Indene, 1,1,3-tribromo-2,4,5,6,7-pentafluoro-	heptane	311(3.28),323(3.25)	70-1883-86
$C_9Br_4ClF_3$			
1H-Indene, 1,1,2,3-tetrabromo-5-chloro-4,6,7-trifluoro- (plus isomer)	heptane	332(3.66)	70-1883-86
$C_9Br_4Cl_2F_2$			
1H-Indene, 1,1,2,3-tetrabromo-5,6-dichloro-4,7-difluoro-	heptane	338(3.56)	70-1883-86
$C_9Br_4F_4$			
1H-Indene, 1,1,2,3-tetrabromo-4,5,6,7-tetrafluoro-	heptane	326(3.53)	70-1883-86
$C_9Br_4F_6$			
1H-Indene, 1,1,3,3-tetrabromo-2,2,4,5,6,7-hexafluoro-2,3-dihydro-	heptane	274(3.14),283(3.07)	70-1883-86
$C_9Br_5F_3$			
1H-Indene, 1,1,2,3,5-pentabromo-4,6,7-trifluoro- (plus isomer)	heptane	332(3.72)	70-1883-86
C_9ClF_7			
1H-Indene, 5-chloro-1,1,2,3,4,6,7-heptafluoro- (plus isomer)	heptane	278(3.42),307(3.47)	70-0804-86
$C_9Cl_2F_6$			
1H-Indene, 3,6-dichloro-1,1,2,4,5,7-heptafluoro- (plus isomer)	heptane	312(3.48)	70-0804-86
1H-Indene, 5,6-dichloro-1,1,2,3,4,7-heptafluoro-	heptane	278s(3.39),284(3.43), 303s(3.43),318(3.54)	70-0804-86
$C_9Cl_2F_8$			
1H-Indene, 1,1,3,5-tetrachloro-2,4,6,7-tetrafluoro- (plus isomer)	heptane	272s(3.14),278(3.20)	70-0804-86
$C_9Cl_3F_7$			
1H-Indene, 1,5,6-trichloro-1,2,2,3,3-4,7-heptafluoro-2,3-dihydro-	heptane	281(3.29),290(3.37)	70-0804-86
$C_9Cl_4F_4$			
1H-Indene, 1,1,3,5-tetrachloro-2,4,6,7-tetrafluoro-	heptane	317(3.42)	70-0804-86
$C_9Cl_4F_6$			
1H-Indene, 1,1,5,6-tetrachloro-2,2,3,3,4,7-hexafluoro-2,3-dihydro-	heptane	282(3.35),291(3.42)	70-0804-86
$C_9Cl_5F_3$			
1H-Indene, 1,1,3,5,6-pentachloro-2,4,7-trifluoro-	heptane	324(3.51)	70-0804-86
C_9HClF_6			
1H-Indene, 5-chloro-1,1,2,4,6,7-hexafluoro- (plus isomer)	heptane	284s(3.23),311(3.45)	70-0804-86
$C_9HCl_2F_5$			
1H-Indene, 5,6-dichloro-1,1,2,4,7-	heptane	286(3.30),322(3.55)	70-0804-86

Compound	Solvent	$\lambda_{max}(\log \epsilon)$	Ref.
$C_9H_2BrN_3O$			
Ethenetricarbonitrile, (5-bromo-2-furanyl)-	MeOH	400(3.2)	73-1450-86
$C_9H_3F_5O_2$			
Benzeneacetaldehyde, 2,3,4,5,6-pentafluoro-α-(hydroxymethylene)-	pH 1	248(4.28)	44-3244-86
	pH 13	266(4.47)	44-3244-86
	MeOH	255(4.25)	44-3244-86
$C_9H_3F_6N_3$			
Pyrido[2,3-b]pyrazine, 2,3-bis(trifluoromethyl)-	MeOH	215(4.14),303(3.71), 309(3.71),316(3.82)	39-1043-86C
$C_9H_4Cl_6O$			
Benzene, pentachloro(2-chloro-2-methoxyethenyl)-	C_6H_{12}	213(4.87),228s(4.30), 239s(4.02),277s(2.68), 287s(2.62),292s(2.46)	44-1413-86
$C_9H_5BFe_3O_9$			
Iron, nonacarbonyl-μ-hydro[μ_3-[tetrahydroborato(1-)-H:H':H"]]tri-	hexane	433s(3.3),511(3.1)	35-3304-86
$C_9H_5Br_2NO$			
8-Quinolinol, 5,7-dibromo-, indium chelate	$CHCl_3$	415(3.94)	96-0057-86
$C_9H_5ClN_4$			
Cyanamide, (2-chloro-5-methyl-2,5-cyclohexadiene-1,4-diylidene)bis-	MeCN	328s(4.45),344(4.52), 362s(4.35)	5-0142-86
$C_9H_5N_3O$			
4H-Pyrido[1,2-a]pyrimidine-3-carbonitrile, 4-oxo-	pH 1	247s(3.96),252(3.99), 297s(3.60),306(3.61), 350s(4.17),358(4.21)	44-2988-86
	EtOH	247(3.91),254(3.92), 301s(3.53),311(3.56), 359s(4.19),365(4.20)	44-2988-86
C_9H_6ClNO			
1H-Indole-2-carboxaldehyde, 3-chloro-	EtOH	209(4.16),243(4.09), 322(4.29),357(3.62)	104-2186-86
$C_9H_6ClN_3O_4$			
Acetamide, N-(2-chloro-5-cyano-3-nitrophenyl)-N-hydroxy-	MeOH	221(4.41)	4-1091-86
$C_9H_6F_2$			
Tricyclo[5.2.0.0^{3,5}]nona-1,3(5),6-triene, 4,4-difluoro-	hexane	213(3.74),264(3.52), 269(3.57),275(3.45)	33-1546-86
$C_9H_6N_4$			
Cyanamide, (2-methyl-2,5-cyclohexadiene-1,4-diylidene)bis-	MeCN	220(3.56),332(4.45), 343s(4.42)	5-0142-86
Propanedinitrile, [(2-pyridinylamino)-methylene]- (tautomeric with 4H-pyrido[1,2-a]pyrimidin-4-imine, 3-cyano-)	pH 1	252s(4.00),302(3.73), 310(3.78),352(4.14), 358(4.18)	44-2988-86
	H_2O	261(3.68),269(3.65), 321(3.93),391(4.08)	44-2988-86
1,4,7,9b-Tetraazaphenalene	hexane	245(4.3),375f(4.5), 470(4.2),510(4.3), 580(4.5),622(4.3)	35-0017-86

Compound	Solvent	$\lambda_{max}(\log \epsilon)$	Ref.
$C_9H_6N_4O_4S$			
2(3H)-Thiazolimine, 3-(2,4-dinitro-phenyl)-	n.s.g.	427(3.48)	150-0110-86S
$C_9H_6O_2$			
1H-2-Benzopyran-1-one	10% EtOH-pH 6.3	223s(4.25),228(4.27), 320(3.40)	39-1875-86B
$C_9H_6O_3$			
2H-1-Benzopyran-2-one, 4-hydroxy-	H_2O	270(4.00),280(4.04), 303(3.89)	137-0185-86
	anion	287(4.11)	137-0185-86
2H-1-Benzopyran-2-one, 7-hydroxy-	H_2O	324(4.15)	137-0185-86
	anion	367(4.27)	137-0185-86
$C_9H_7BrN_3O_2$			
2-Pyrimidineacetic acid, 5-bromo-α-cyano-, ethyl ester, ion(1-)-, sodium	EtOH	235(4.08),314(4.46)	103-0774-86
piperidinium salt	EtOH	233(4.12),315(4.54)	103-0774-86
$C_9H_7ClF_3N$			
Benzenecarboximidoyl chloride, N-(2,2,2-trifluoroethyl)-	hexane	246(4.18)	65-2372-86
Ethanamine, 1-chloro-2,2,2-trifluoro-N-(phenylmethylene)-	hexane	250(4.16)	65-2372-86
Ethanimidoyl chloride, 2,2,2-tri-fluoro-N-(phenylmethyl)-	hexane	260(2.54)	65-2372-86
$C_9H_7Cl_2F_3$			
Cyclopropa[cd]pentalene, 1,2-dichloro-1,2,2a,2b,4a,4b-hexahydro-4-(tri-fluoromethyl)-	n.s.g.	254(2.34)	88-5003-86
$C_9H_7Cl_2N$			
Cyclopropa[cd]pentalene-1-carboni-trile, 3,4-dichloro-2a,2b,3,4,4a,4b-hexahydro-	n.s.g.	254(3.73)	88-5003-86
$C_9H_7F_6N_3O_2$			
Pyrido[3,4-b]pyrazine-2,3-diol, 1,2,3,4-tetrahydro-2,3-bis(tri-fluoromethyl)-	MeOH	213(4.17),251(3.37), 291(3.49)	39-1043-86C
C_9H_7NO			
8-Quinolinol	pH 3	249(4.72)	160-0979-86
C_9H_7NOS			
Ethanone, 1-(2-benzothiazolyl)-	EtOH	294(4.05)	94-4916-86
$C_9H_7NOS_2$			
2(3H)-Thiazolethione, 3-hydroxy-4-phenyl-	EtOH	229(3.96),264(4.11), 307(3.90)	39-0039-86C
$C_9H_7NO_2$			
2H-1-Benzopyran-2-one, 3-amino-	C_6H_{12}	210(4.35),241(3.78), 319(4.29)	151-0055-86C
	pH 6	201(4.40),241(3.56), 322(4.03)	151-0055-86C
	H_0-3	198(4.39),282(4.04), 307(3.88)	151-0055-86C

Compound	Solvent	λ_{max}(log ε)	Ref.
2H-1-Benzopyran-2-one, 3-amino- (cont.)	H_0-7	194(4.46),294(4.05), 325s(3.63)	151-0055-86C
	MeOH	204(4.35),244(3.65), 324(4.14)	151-0055-86C
	ether	213(4.16),238(3.77), 321(4.31)	151-0055-86C
	MeCN	203(4.38),242(3.63), 322(4.14)	151-0055-86C
1H-Indole-3-carboxylic acid	EtOH	212(4.11),225s(3.82), 246s(3.55),280(3.62), 287s(3.59)	102-0281-86
$C_9H_7NO_3$			
1,2-Benzisoxazole-7-carboxaldehyde, 6-hydroxy-3-methyl-	EtOH	225(4.1),279(4.1), 321(4.3)	2-0870-86
1H-Indole-2-carboxylic acid, 5-hydroxy-	pH 1	204(4.38),212.5(4.35), 295(4.26),337s(--)	151-0209-86C
	pH 7	212(4.45),282(--), 290(4.19),312(--)	151-0209-86C
	pH 13	224(4.32),286(4.20), 294(4.17),350(4.32)	151-0209-86C
	H_0-5	196(4.54),225(--), 318(4.10)	151-0209-86C
	H_0-10	212(4.32),232(4.20), 331(4.3),350s(--)	151-0209-86C
	MeOH	208(4.67),283(4.31), 292(4.32),312(--)	151-0209-86C
	EtOH	215(4.55),288(4.33), 294(4.33),325(3.72)	151-0209-86C
	ether	225(4.39),283(4.07), 295(4.05),325(3.36)	151-0209-86C
	MeCN	210(4.69),283(4.36), 295(4.37),325(3.79)	151-0209-86C
$C_9H_7NO_3S_2$			
Benzeneethane(dithioic) acid, 2-nitro-α-oxo-, methyl ester	MeOH	238(3.70),351(3.71), 557(1.55)	118-0968-86
Benzeneethane(dithioic) acid, 4-nitro-α-oxo-, methyl ester	MeOH	264(3.86),299(3.81), 484(1.85)	118-0968-86
$C_9H_7N_5O_2$			
1H-1,2,4-Triazol-3-amine, N-[(4-nitrophenyl)methylene]-	dioxan	316(4.32)	80-0273-86
C_9H_7Te			
1-Benzotellurinium perchlorate	CF_3COOH	327(3.99),410(3.31), 560(2.58)	103-1276-86
$C_9H_8BrN_3O_2$			
Acetic acid, (5-bromo-2(1H)-pyrimidinylidene)cyano-, ethyl ester, (E)-	EtOH	303(4.78),390(3.60)	103-0774-86
C_9H_8ClN			
1H-Indole, 2-chloro-3-methyl-	MeOH	223(4.39),274(3.80), 283(3.78),291(3.67)	39-2305-86C
1H-Indole, 3-chloro-2-methyl-	MeOH	224(4.45),282(4.45), 290(4.39)	39-2305-86C
C_9H_8ClNO			
1H-Indole, 4-chloro-6-methoxy-	MeOH	220(4.3),270(3.5), 295(3.5)	35-4115-86

Compound	Solvent	λ_{max}(log ϵ)	Ref.
$C_9H_8ClNO_3S$			
3-Azetidinone, 1-[(4-chlorophenyl)-sulfonyl]-	MeOH	203(4.36),228(4.32)	70-0120-86
$C_9H_8ClN_3O$			
Acetamide, N-(3-amino-2-chloro-5-cyanophenyl)-	MeOH	234(4.58),260s(3.87), 328(3.61)	4-1091-86
$C_9H_8ClN_3O_2$			
Acetic acid, (2-chloro-4(1H)-pyrimidinylidene)cyano-, ethyl ester, (Z)-	EtOH	340(4.25)	103-0774-86
Acetic acid, (4-chloro-2(1H)-pyrimidinylidene)cyano-, ethyl ester, (E)-	EtOH	306(4.46),380(3.38)	103-0774-86
Acetic acid, (6-chloro-4(1H)-pyrimidinylidene)cyano-, ethyl ester, (Z)-	EtOH	310(4.30),350s(3.92)	103-0774-86
sodium salt	EtOH	325(4.15)	103-0774-86
$C_9H_8ClN_3O_3S$			
Benzenesulfonamide, 4-chloro-N-(3-diazo-2-oxopropyl)-	MeOH	204(4.41),244(4.44), 271(4.05)	70-0120-86 +104-0353-86
$C_9H_8ClN_3O_4$			
3H-Indol-3-one, 4-chloro-1,2-dihydro-2-hydroxy-6-methoxy-1-nitroso-, oxime	MeOH	233s(3.5),269(3.5), 302(3.2),348(3.00)	35-4115-86
$C_9H_8ClN_3O_7$			
2-Propanone, 1-chloro-1-(2,4,6-trinitro-2,4-cyclohexadien-1-yl)-, compd. with N,N-diethylethanamine	DMSO-d_6	458(4.15),570(4.45)	150-0024-86S
$C_9H_8F_2$			
Bicyclo[4.1.0]hepta-1,3,5-triene, 7,7-difluoro-3,4-dimethyl-	hexane	200(3.73),211(3.78), 259(3.14),266(3.27), 272(3.20)	33-1546-86
C_9H_8NP			
1H-1,3-Azaphosphole, 2-phenyl-	MeOH	210s(3.80),222(3.85), 243s(3.40),312(3.80)	88-5699-86
$C_9H_8N_2$			
1H-Imidazole, 4-phenyl-	EtOH	207(3.9),260(3.9)	97-0378B-86
$C_9H_8N_2O$			
Cyanamide, (3,5-dimethyl-4-oxo-2,5-cyclohexadien-1-ylidene)-	MeCN	247s(3.65),302(4.29), 322s(4.16),363s(3.77), 378s(3.02)	5-0142-86
Ethanone, 1-(1H-indazol-3-yl)-	EtOH	236(4.0),241s(3.9), 299(4.0)	35-4115-86
3H-Indol-3-one, 2-methyl-, oxime	EtOH	240(4.2),253(4.3), 277s(3.5),313(3.7), 357s(3.4)	35-4115-86
Phenol, 4-(1H-imidazol-1-yl)-	C_6H_{12}	207(3.95),240(3.99), 280(3.34)	2-0513-86A
	pH 1	198(4.23),240(3.84), 278s(--)	2-0513-86A
	pH 8	202(4.18),239(3.94), 277s(--)	2-0513-86A
	pH 12	210(4.12),255(4.11), 295s(--)	2-0513-86A
	H_0-10	198(3.93),226(3.65)	2-0513-86A

Compound	Solvent	$\lambda_{max}(\log \epsilon)$	Ref.
Phenol, 4-(1H-imidazol-1-yl)- (cont.)	MeOH	206(4.11),240(4.03), 277s(--)	2-0513-86A
	EtOH	210(3.94),240(4.00), 277s(--)	2-0513-86A
	dioxan	212(3.92),241(4.06), 283(3.34)	2-0513-86A
	MeCN	204(4.18),240(4.05), 282s(--)	2-0513-86A
$C_9H_8N_2OS$			
4(1H)-Quinazolinone, 2,3-dihydro-1-methyl-2-thioxo-	EtOH	220(4.05),291(4.19)	103-1001-86
	EtOH-KOH	225(4.14),279(4.23)	103-1001-86
4(1H)-Quinazolinone, 2,3-dihydro-3-methyl-2-thioxo-	EtOH	219(4.14),291(4.22)	103-1001-86
	EtOH-KOH	223(4.31),268(3.64)	103-1001-86
4(1H)-Quinazolinone, 2-(methylthio)-	EtOH	232(4.12),275(3.87)	103-1001-86
	EtOH-KOH	240(4.14),285(3.85)	103-1001-86
$C_9H_8N_2O_2$			
1H-Benzimidazole-4,7-dione, 1,2-dimethyl-	pH 7.00	252(4.15),290s(3.34), 400(3.04)	44-0522-86
$C_9H_8N_2O_2S_2$			
[4,4'-Bithiazole]-2-carboxylic acid, 2'-methyl-, methyl ester	MeOH	248(4.08),293(3.41)	35-7089-86
5-Isothiazolecarboxylic acid, 3-(2-methyl-4-thiazolyl)-, methyl ester	MeCN	304(3.66)	88-6385-86
$C_9H_8N_2O_5$			
4H-Furo[3,2-b]pyrrole-5-carboxylic acid, 2-nitro-, ethyl ester	MeOH	384(3.55)	73-0106-86
$C_9H_8N_2S_2$			
2-Benzothiazolecarbothioamide, N-methyl-	MeOH	245s(3.90),250(3.94), 254(4.00),299(4.26), 347s(3.70),417s(2.30)	103-1139-86
$C_9H_8N_4$			
[1,2,4]Triazolo[1,5-a]pyridine-8-carbonitrile, 5,7-dimethyl-	DMF	327(4.13)	118-0860-86
$C_9H_8N_4O$			
Phenol, 2-[(1H-1,2,4-triazol-3-ylimino)methyl]-	dioxan	389(4.26)	80-0273-86
$C_9H_8N_4OS$			
1,2,4-Triazin-5(2H)-one, 6-(2-aminophenyl)-3,4-dihydro-3-thioxo-	EtOH	245(4.14),283(4.26), 384s(3.78)	104-2173-86
$C_9H_8N_4O_5S$			
Benzenesulfonamide, N-(3-diazo-2-oxopropyl)-4-nitro-	MeOH	203(4.20),211s(4.09), 270(4.34)	104-0353-86
$C_9H_8N_5$			
11H-Dipyrimido[1,6-a:6',1'-d][1,3,5]-triazin-10-ium, iodide	H_2O	222(4.297),255(4.239)	142-1893-86
$C_9H_8N_6O$			
1H-Imidazo[4,5-d]-1,2,3-triazin-4-amine, N-(2-furanylmethyl)-	EtOH	218(4.34),262(4.05), 298(3.84)	5-1012-86

Compound	Solvent	$\lambda_{max}(\log \epsilon)$	Ref.
C_9H_8O			
Benzaldehyde, 2-ethenyl-	EtOH	222(--),259(--), 309(--)	47-0133-86
Benzene, (1,2-propadienyloxy)-	hexane	225(3.06),270(2.15)	70-1677-86
Benzene, (1-propynyloxy)-	hexane	223(3.06),268(2.04)	70-1677-86
$(C_9H_8O)_n$			
2-Propen-1-one, 1-phenyl-, homopolymer	$CHCl_3$	246(4.07),280(2.98), 324(1.96)	32-0533-86
$C_9H_8OS_2$			
Benzeneethane(dithioic) acid, α-oxo-, methyl ester	MeOH	274(3.74),306(3.73), 494(1.56)	118-0968-86
$C_9H_8O_2S_4$			
1,3-Dithiole-4-carboxylic acid, 2-(4,5-dimethyl-1,3-dithiol-2-ylidene)-	MeCN	290(4.12),312(4.14), 435(3.34)	104-2131-86
radical cation inner salt	PhCN	418s(3.97),450(4.18), 520s(3.75),625(3.83)	104-2131-86
sodium salt	H_2O	290(4.03),312(4.11), 401(3.25)	104-2131-86
$C_9H_8O_3$			
2,4,6-Cycloheptatrien-1-one, 2-acetoxy-, iron tricarbonyl deriv.	EtOH	230(4.28),270s(3.96), 320s(3.59),450(2.53)	88-3873-86
diastereomer	EtOH	235(4.21),280s(3.89), 320s(3.51),444(2.55)	88-3873-86
Ethanone, 1-(1,3-benzodioxol-5-yl)- (moskachan A)	n.s.g.	229(4.15),273(3.83), 308(3.88)	102-2209-86
2-Propenoic acid, 3-(2-hydroxyphenyl)-	pH 2.0	321.5(3.92)	149-0551-86B
	pH 7.0	313(3.88)	149-0551-86B
	pH 13	359(3.98)	149-0551-86B
	MeOH	318(3.90)	149-0551-86B
	MeOH-NaOAc	310(--)	149-0551-86B
$C_9H_8O_4$			
1,3-Benzodioxole-5-carboxaldehyde, 6-methoxy-	EtOH	241(3.86),277(3.51), 349(3.61)	100-0690-86
$C_9H_8O_6$			
2H-Pyran-2-one, 3,5-diacetyl-4,6-dihydroxy-	C_6H_{12}	268(--),306(4.21)	39-0973-86B
	H_2O	270(4.07),315(4.42)	39-0973-86B
	EtOH	270(4.15),318(4.46)	39-0973-86B
	dioxan	268(--),310(4.31)	39-0973-86B
	$CHCl_3$	268(--),310(4.22)	39-0973-86B
	80% DMSO	270(4.10),320(4.49)	39-0973-86B
C_9H_8S			
Benzene, (1,2-propadienylthio)-	hexane	250(2.99)	70-1677-86
Benzene, (1-propynylthio)-	hexane	254(3.10)	70-1677-86
$C_9H_9BrN_2O_2$			
1H-Benzimidazole-4,7-diol, 2-(bromomethyl)-1-methyl-	H_2O	240(4.30),380(3.72)	44-0522-86
	acid	240(4.30),280(3.56)	44-0522-86
$C_9H_9BrO_2$			
1,3,5-Cycloheptatriene-1-carboxylic acid, 6-bromo-, methyl ester	MeCN	215(4.12),278s(3.55), 305(3.62)	89-0723-86

Compound	Solvent	λ_{max}(log ε)	Ref.
$C_9H_9BrO_3$			
1-Propanone, 3-bromo-1-(2,4-dihydroxy-phenyl)-	EtOH	212(4.16),233(3.88), 279(4.14),313(3.86)	24-0050-86
1-Propanone, 3-bromo-1-(3,4-dihydroxy-phenyl)-	EtOH	232(4.13),279(3.94), 311(3.91)	24-0050-86
$C_9H_9BrO_4$			
1-Propanone, 3-bromo-1-(2,3,4-tri-hydroxyphenyl)-	EtOH	216(4.14),238(4.01), 296(4.17)	24-0050-86
$C_9H_9ClN_2O_2$			
1H-Benzimidazole-4,7-diol, 2-(chloro-methyl)-1-methyl-	H_2O	227(4.20),265(3.65), 290(3.49)	44-0522-86
	acid	235(4.28),276(3.46)	44-0522-86
$C_9H_9ClN_4O_2$			
2,4(1H,3H)-Pteridinedione, 7-chloro-1,3,6-trimethyl-	MeOH	202(4.25),242(4.21), 340(4.01)	33-1095-86
$C_9H_9ClN_4O_4$			
Ethanehydrazonoyl chloride, N-(2,4-dinitrophenyl)-	dioxan	357(4.43)	39-0537-86B
$C_9H_9ClO_2$			
1,3,5-Cycloheptatriene-1-carboxylic acid, 6-chloro-, methyl ester	MeCN	220(3.90),275s(3.53), 302(3.59)	89-0723-86
$C_9H_9ClO_3$			
8-Nonene-4,6-diyne-1,2,3-triol, 9-chloro-, [R*,S*-(E)]-	MeOH	213(4.46),220(4.59), 245(3.73),258(4.05), 273(4.23),289(4.13)	102-1224-86
1-Propanone, 3-chloro-1-(3,4-di-hydroxyphenyl)-	EtOH	231(4.12),278(3.94), 310(3.88)	24-0050-86
$C_9H_9ClO_4$			
1-Propanone, 3-chloro-1-(2,3,4-tri-hydroxyphenyl)-	EtOH	203(4.32),216(4.16), 238(4.01),297(4.17)	24-0050-86
C_9H_9DO			
3,5-Octadien-7-yn-2-one-3-d, 6-methyl-	EtOH	299(4.25)	18-1723-86
$C_9H_9IO_3$			
Ethanone, 1-(4-hydroxy-3-methoxyphen-yl)-2-iodo-	EtOH	231(3.97),290(3.84), 316(3.91)	24-0050-86
1-Propanone, 1-(3,4-dihydroxyphenyl)-3-iodo-	EtOH	223(4.20),284(3.85), 315(3.89)	24-0050-86
$C_9H_9IO_4$			
1-Propanone, 3-iodo-1-(2,3,4-tri-hydroxyphenyl)-	EtOH	217(4.43),235(4.14), 295(4.14)	24-0050-86
C_9H_9N			
Benzenamine, N-2-propynyl-	hexane	205(4.38),241(4.18), 294(3.57)	65-2461-86
	MeOH	203(--),243(--), 291(--)	65-2461-86
1H-Indole, 1-methyl-	MeCN	215(4.5),295(3.8)	18-0569-86
C_9H_9NO			
2-Propenenitrile, 2-methyl-3-(5-meth-yl-2-furanyl)-, (E)-	EtOH	305(4.48)	103-0943-86

Compound	Solvent	λ_{max}(log ϵ)	Ref.
$C_9H_9NO_2$			
Benzamide, N-acetyl-	EtOH	195.5(3.90),203.1(4.45), 206.4(4.35),234.2(4.30)	104-1602-86
$C_9H_9NO_3$			
Benzamide, N-acetyl-4-hydroxy-	EtOH	193.4(3.91),202.7(4.41), 205.7(4.31),265(4.24)	104-1602-86
$C_9H_9NO_4S$			
2-Pyridinecarbo(thioperoxoic) acid, 6-(methoxycarbonyl)-, O-methyl ester	CH_2Cl_2	228(4.02),276(3.71)	78-5969-86
$C_9H_9NO_4S_2$			
2,6-Pyridinedicarbo(thioperoxoic) acid, O,O-dimethyl ester	CH_2Cl_2	228(4.08),287(3.83)	78-5969-86
$C_9H_9NO_5$			
2H-1,4-Benzoxazin-2-one, 3,4-dihydro-3,4-dihydroxy-7-methoxy-	pH 7	258(4.19),285s(--)	142-0335-86
	pH 13	255s(--),300(4.04)	142-0335-86
	EtOH	257(--),287s(--)	142-0335-86
2H-1,4-Benzoxazin-3(4H)-one, 2,4-dihydroxy-7-methoxy-	pH 9	290(<u>5.0</u>)	142-0013-86
$C_9H_9NS_4$			
2H-Pyrrole, 2-[4,5-bis(methylthio)-1,3-dithiol-2-ylidene]-	MeCN	232(3.82),298(3.63), 340s(3.30),446(4.32)	78-0839-86
$C_9H_9N_2$			
Quinoxalinium, 1-methyl-, iodide	pH 4.0	336(3.84)	103-1118-86
	pH 10.5	301(3.33),340(3.13)	103-1118-86
$C_9H_9N_3$			
2-Propenenitrile, 3-(3,6-dimethylpyrazinyl)-, (E)-	EtOH	246(3.96),308s(3.87)	4-1481-86
$C_9H_9N_3O$			
4(3H)-Quinazolinone, 2-amino-6-methyl-	pH 13	228(4.61),268(3.96), 329(3.58)	87-0468-86
$C_9H_9N_3O_2$			
Carbamic acid, 1H-2-benzimidazolyl-, methyl ester	EtOH-pH 7	212(4.37),228(4.16), 245(4.10),250(3.97), 276(4.05),281(4.13), 287(4.15),294(3.73)	103-0057-86
	EtOH-acid	208(4.09),225(4.33), 266(3.95),277(4.16), 283(4.16)	103-0057-86
	EtOH-base	226(4.82),253(4.06), 261(4.07),296(4.26), 304(4.27)	103-0057-86
3H-Indol-3-one, 1,2-dihydro-2-methyl-1-nitroso-	EtOH	219(4.0),260(4.1), 302(4.1)	35-4115-86
$C_9H_9N_3O_2S_2$			
Propanoic acid, 3-[(4-azidophenyl)dithio-	10% MeOH	261(4.20)	47-0241-86
$C_9H_9N_3O_3$			
Benzaldehyde, 4-nitro-, (acetylhydrazone)	pH 4.73	318(4.30)	69-5774-86

Compound	Solvent	λ max(log ε)	Ref.
Pyrido[2,3-d]pyrimidine-2,4(1H,3H)-dione, 6-(hydroxymethyl)-5-methyl-	pH 1	220(4.47),249s(3.76), 309(3.94)	87-0709-86
	H_2O	248(3.94),306(3.84)	87-0709-86
	pH 13	246(4.14),267(4.05), 336(3.74)	87-0709-86
$C_9H_9N_3O_3S$			
Benzenesulfonamide, N-(3-diazo-2-oxopropyl)-	MeOH	203(3.80),225(3.92), 246(3.84),266(3.8), 274(3.8)	104-0353-86
1H-Pyrazole, 1-nitroso-3-(phenylsulfonyl)-4,5-dihydro-	EtOH	314(4.17)	104-0947-86
$C_9H_9N_3O_3S_2$			
Thiazolium, 5-(hexahydro-2,4,6-trioxo-5-pyrimidinyl)-3-methyl-2-(methylthio)-, hydroxide, inner salt	HOAc	270(4.18),361(4.19)	24-2094-86
	DMF	268(4.13),386(4.17)	24-2094-86
$C_9H_9N_3O_4$			
4-Pyrimidinecarbonitrile, 3-(2-acetoxyethyl)-1,2,3,6-tetrahydro-2,6-dioxo-	DMF	292(3.93)	47-0119-86
	DMSO	292(3.90)	47-0119-86
$C_9H_9N_3O_8$			
6,2'-O-Cyclouridine, 5-nitro-	pH 1	290(3.96)	94-3623-86
	H_2O	290(3.97)	94-3623-86
	pH 13	335(4.01)	94-3623-86
$C_9H_9N_5O_3$			
6-Pteridinecarboxylic acid, 2-(dimethylamino)-1,4-dihydro-4-oxo-	pH -2.0	211(4.11),266(4.13), 318(4.02),412(3.23)	5-1705-76
	pH 2.0	223(4.02),266(3.84), 309(4.29),370(3.87)	5-1705-86
	pH 6.0	222(4.18),257(3.90), 301(4.32),354(3.84)	5-1705-86
	pH 12.0	230(4.10),285(4.32), 384(4.02)	5-1705-86
7-Pteridinecarboxylic acid, 2-(dimethylamino)-1,4-dihydro-4-oxo-	pH -1.0	250(4.20),340(3.88)	5-1705-86
	pH 2.0	246(4.08),288(3.98), 342(3.73),402s(3.30)	5-1705-86
	pH 5.0	220(4.14),247(4.05), 287(4.07),354(3.78), 390s(3.46)	5-1705-86
	pH 11.0	275(4.27),395(3.82)	5-1705-86
C_9H_{10}			
1,3,5-Cycloheptatriene, 1-ethenyl-	C_6H_{12}	222.5(3.42),228.5(3.38), 300(2.91)	88-5653-86
1,3,5-Cycloheptatriene, 2-ethenyl-	C_6H_{12}	229(3.41),237s(3.30), 265s(2.55)	88-5653-86
1,3,5-Cycloheptatriene, 3-ethenyl-	C_6H_{12}	230(3.21),237s(3.08), 282.5(3.03)	88-5653-86
$C_9H_{10}BrNO_4$			
2,4-Cyclohexadien-1-ol, 3-bromo-6-methyl-6-nitro-, acetate, (Z)-	CH_2Cl_2	266.4(2.54)	23-2382-86
2,4-Cyclohexadien-1-ol, 5-bromo-2-methyl-6-nitro-, acetate	CH_2Cl_2	279.75(2.86)	23-2382-86
diastereomer	CH_2Cl_2	279.75(2.91)	23-2382-86

Compound	Solvent	$\lambda_{max}(\log \epsilon)$	Ref.
$C_9H_{10}ClNO$			
Phenol, 3-[1-(aminomethyl)-2-chloro-ethenyl]-, (E)-	H_2O	235s(3.77),281(3.29)	130-0103-86
$C_9H_{10}ClNO_2$			
2-Propenoic acid, 3-(3-chloro-1H-pyrrol-2-yl)-2-methyl, methyl ester, (E)-	EtOH	322(4.36)	150-3601-86M
(Z)-	EtOH	335(4.20)	150-3601-86M
$C_9H_{10}ClNO_4$			
2,4-Cyclohexadien-1-ol, 3-chloro-6-methyl-6-nitro-, acetate, (Z)-	MeOH	263(2.59)	23-1764-86
$C_9H_{10}ClN_3O_2$			
Ethanehydrazonoyl chloride, N-methyl-N-(4-nitrophenyl)-	dioxan	360(3.12)	39-0537-86B
$C_9H_{10}ClN_5O_2$			
6H-Purin-6-one, 2-amino-9-(3-chloro-tetrahydro-2-furanyl)-1,9-dihydro-, trans-(±)-	pH 1	255(4.07),278s(3.95)	23-1885-86
	pH 13	257(4.05),265(4.05)	23-1885-86
$C_9H_{10}FNO_4$			
2,4-Cyclohexadien-1-ol, 3-fluoro-6-methyl-6-nitro-, acetate	CH_2Cl_2	256.5(2.60)	23-1764-86
2,5-Cyclohexadien-1-ol, 1-fluoro-4-methyl-4-nitro-, acetate, (E)-	CH_2Cl_2	280(0.90)	23-1764-86
(Z)-	CH_2Cl_2	269(1.45)	23-1764-86
$C_9H_{10}FeO_4$			
Iron, tricarbonyl(η^4-3-methylene-4-penten-2-ol)-	C_6H_{12}	273s(3.36)	101-0061-86D
more polar isomer	C_6H_{12}	275s(3.38)	101-0061-86D
$C_9H_{10}N_2$			
1H-Benzimidazole, 5,6-dimethyl-	pH 2	196(4.43),219(3.86), 274(3.82),281(3.83)	39-1917-86B
	pH 8	202(4.49),247(3.62), 276(3.78),281(3.77), 285(3.77)	39-1817-86B
	NaOH	226(3.94),263(3.61), 285(3.83),289(3.79)	39-1917-86B
	MeOH	207(4.40),247(3.69), 277(3.76),281(3.76), 286(3.79)	39-1917-86B
	ether	214(4.11),247(3.79), 278(3.76),282(3.77), 289(3.81)	39-1917-86B
	MeCN	207(4.44),247(3.72), 278(3.76),282(3.72), 286(3.74)	39-1917-86B
	C_6H_{12}	210(--),245(--), 278(--),282(--), 288(--)	39-1917-86B
Benzonitrile, 4-(dimethylamino)-	hexane	300(3.7)	59-1127-86
	CH_2Cl_2	314(3.8)	59-1127-86
Propanenitrile, 3-(phenylamino)-	acid	255f(2.3)	112-1099-86
$C_9H_{10}N_2O$			
Acetic acid, (phenylmethylene)hydrazide	pH 4.73	282(4.37)	69-5774-86

Compound	Solvent	λ_{max}(log ϵ)	Ref.
1H-Benzimidazole-1-methanol, 2-methyl-	EtOH	243(3.65),274(3.67), 280(3.72)	142-2233-86
3H-Indol-3-one, 1,2-dihydro-2-methyl-, oxime	EtOH	227(4.3),258s(3.9), 358(3.7)	35-4115-86
2-Pyrrolidinone, 5-(1H-pyrrol-2-yl-methylene)-	MeOH	289(4.04)	49-0185-86
$C_9H_{10}N_2O_2$			
1H-Benzimidazole-4,7-diol, 1,2-di-methyl-, hydrobromide	H_2O	218(4.68),256(3.98)	44-0522-86
protonated	acid	218(4.69),270(3.93)	44-0522-86
2-Propenoic acid, 3-(1H-1,2-diazepin-1-yl)-, methyl ester, (E)-	EtOH	224s(3.60),254(3.62), 304(4.15),381s(3.14), 403s(2.98)	39-0595-86C
Pyrano[2,3-c]pyrazol-6(1H)-one, 3,4,5-trimethyl-	EtOH	308(4.05)	150-0920-86M
1H,5H-Pyrazolo[1,2-a]pyrazole-1,5-di-one, 2,3,7-trimethyl-, anti	EtOH	324(4.15)	150-0920-86M
$C_9H_{10}N_2O_3$			
Ethanone, 1-(2-pyridinyl)-, O-acetyl-oxime, N-oxide, (E)-	EtOH	230(4.75),274(4.49)	94-4984-86
(Z)-	EtOH	215(4.69),269(4.48)	94-4984-86
$C_9H_{10}N_2O_5$			
1-Cyclobuten-1-ol, 2-morpholino-3-(nitromethylene)-4-oxo-, sodium salt	H_2O	293(4.1),405(4.1)	118-0216-86
1,3-Diazabicyclo[3.2.0]heptane-2-carb-oxylic acid, 4-(methoxycarbonyl-methylene)-7-oxo-	MeCN	272(4.11)	39-1077-86C
$C_9H_{10}N_4$			
2-Quinoxalinecarbonitrile, 3-amino-5,6,7,8-tetrahydro-	MeOH	368(3.94)	33-0793-86
$C_9H_{10}N_4OS$			
3H-1,2,4-Triazole-3-thione, 4-amino-2,4-dihydro-5-(4-methoxyphenyl)-	EtOH	256(4.54),276s(4.52)	4-1451-86
$C_9H_{10}N_4O_2$			
Carbamic acid, 1H-imidazo[4,5-c]pyri-din-4-yl-, ethyl ester	MeOH	248(3.70),276(3.74)	78-1511-86
1H-Imidazo[4,5-c]pyridinium, 5-[(eth-oxycarbonyl)amino]-, hydroxide, inner salt	MeOH	276(3.78),295(3.77)	78-1511-86
Pyrazinecarboxylic acid, 5-amino-6-cyano-3-methyl-, ethyl ester	MeOH	346(3.95)	33-0793-86
Pyrido[2,3-d]pyrimidin-4(1H)-one, 2-amino-6-(hydroxymethyl)-5-methyl-	pH 1	215(4.35),243s(4.17), 279(4.13),350(3.99)	87-0709-86
	H_2O	218(3.90),271(3.50), 318(3.30)	87-0709-86
	pH 13	216(4.68),241(4.30), 271(3.94),332(3.88)	87-0709-86
$C_9H_{10}N_4O_2S$			
2,4(1H,3H)-Pteridinedione, 5,6-di-hydro-1,3,5-trimethyl-6-thioxo-	MeOH	240s(3.80),306(4.26), 455(3.78)	33-0704-86
2,4(1H,3H)-Pteridinedione, 7,8-di-hydro-1,3,6-trimethyl-7-thioxo-	pH 5	224(4.40),278(3.81), 295s(3.71),380(4.24)	33-1095-86

Compound	Solvent	$\lambda_{max}(\log \epsilon)$	Ref.
2,4(1H,3H)-Pteridinedione, 7,8-dihydro-1,3,6-trimethyl-7-thioxo- (cont.)	pH 0	220(4.32),250s(3.92), 302(3.70),360(4.08), 400(3.83)	33-1095-86
2,4(1H,3H)-Pteridinedione, 1,3-dimethyl-6-(methylthio)-	pH 6.0	230(4.01),279(4.30), 384(3.82)	33-0708-86
$C_9H_{10}N_4O_3$			
Pyrimido[4,5-d]pyrimidine-2,4,5(1H-3H,6H)-trione, 1,3,6-trimethyl-	EtOH	232(4.49),270(3.79), 292(3.64)	142-2293-86
Pyrimido[4,5-d]pyrimidine-2,4,5(1H-3H,6H)-trione, 1,3,7-trimethyl-	EtOH	232(4.47),263(3.54), 293(3.72)	142-2293-86
$C_9H_{10}N_4O_3S$			
2,4(1H,3H)-Pteridinedione, 5,6-dihydro-1,3,5-trimethyl-6-sulfinyl-	MeOH	235(3.96),346(4.09), 575(3.57)	33-0704-86
2,4(1H,3H)-Pteridinedione, 1,3-dimethyl-6-(methylsulfinyl)-	MeOH	246(4.29),265s(4.10), 341(3.85)	33-0708-86
7-Pteridinesulfenic acid, 1,2,3,4-tetrahydro-1,3,6-trimethyl-2,4-dioxo-	pH 2	224(4.27),262(3.75), 360(4.10)	33-1095-86
	pH 9	226(4.19),294(3.16), 416(4.10)	33-1095-86
$C_9H_{10}N_4O_4$			
Acetic acid, cyano(dihydro-4,6-dimethoxy-1,3,5-triazin-2-yl)-, methyl ester	EtOH	309(3.8)	70-1058-86
	DMF	308(4.1)	70-1058-86
	DMSO	308(4.0)	70-1058-86
sodium salt	EtOH	310(4.1)	70-1058-86
Pyrimido[4,5-d]pyrimidine-2,4,5,7-(1H,3H,6H,8H)-tetrone, 1,3,6-trimethyl-	EtOH	227(4.48),274(4.19)	142-2293-86
$C_9H_{10}N_4O_4S$			
2,4(1H,3H)-Pteridinedione, 1,3-dimethyl-6-(methylsulfonyl)-	MeOH	247(4.27),265s(4.00), 333(3.87)	33-0708-86
$C_9H_{10}N_4O_5S$			
7-Pteridinesulfonic acid, 1,2,3,4-tetrahydro-1,3,6-trimethyl-2,4-dioxo-, sodium salt	pH 7	242(4.24),345(3.96)	33-1095-86
$C_9H_{10}N_4S$			
1,2,4-Triazole-3-thiol, 4-amino-5-(2-methylphenyl)-	EtOH	252(4.52),276s(4.21)	4-1451-86
1,2,4-Triazole-3-thiol, 4-amino-5-(3-methylphenyl)-	EtOH	252(4.25),280s(3.91)	4-1451-86
1,2,4-Triazole-3-thiol, 4-amino-5-(4-methylphenyl)-	EtOH	250(4.33),275s(4.06)	4-1451-86
1,2,4-Triazole-3-thiol, 4-amino-5-(phenylmethyl)-	EtOH	251(4.32)	4-1451-86
$C_9H_{10}N_8O_3$			
Uridine, 3',5'-diazido-2',3',5'-trideoxy-	EtOH	260(3.99)	87-0681-86
$C_9H_{10}O_2$			
Benzoic acid, 2,6-dimethyl-	$CHCl_3$-HOAc	274(3.08)	112-0321-86
Benzoic acid, 3,5-dimethyl-	$CHCl_3$-HOAc	284(3.08)	112-0321-86
7-Oxabicyclo[2.2.1]heptane-1-carboxaldehyde, 2,3-bis(methylene)-	isooctane	237s(3.89),242(3.93), 250s(3.79)	33-1287-86
	EtOH	237s(3.94),242(3.96), 251s(3.81)	33-1287-86

Compound	Solvent	λ_{max}(log ε)	Ref.
$C_9H_{10}O_2S_2$			
Propanoic acid, 3-(phenyldithio)-	10% EtOH	238(3.84)	47-0241-86
$C_9H_{10}O_4$			
Benzoic acid, 2,6-dimethoxy-	$CHCl_3$-HOAc	280(3.30)	112-0321-86
Benzoic acid, 3,5-dimethoxy-	$CHCl_3$-HOAc	310(3.45)	112-0321-86
$C_9H_{11}BrO$			
Benzene, 1-bromo-2-methoxy-4,5-dimethyl-	dioxan	221(4.00),281(3.49), 289(3.46)	152-0017-86
$C_9H_{11}BrO_2$			
Benzene, 1-bromo-2-(2-methoxyethoxy)-	EtOH	211(4.14),275(3.32), 283(3.28)	12-1833-86
Bicyclo[2.2.1]heptane-2,3-dione, 1-bromo-7,7-dimethyl-, (S)-	C_6H_{12}	310(1.42),475(1.70)	39-1781-86C
$C_9H_{11}ClN_2$			
Ethanehydrazonoyl chloride, N-methyl-N-phenyl-	dioxan	285(3.75)	39-0537-86B
$C_9H_{11}ClN_4O$			
9H-Purine, 6-chloro-9-(1-ethoxyethyl)-	EtOH	266(4.01)	103-0331-86
$C_9H_{11}FN_2O_3$			
4(3H)-Pyrimidinone, 5-fluoro-2,6-dimethoxy-3-(2-propenyl)-	C_6H_{12}	230(3.56),271(3.81)	39-0515-86C
$C_9H_{11}FO_2$			
Bicyclo[2.2.1]heptane-2,3-dione, 1-fluoro-7,7-dimethyl-, (S)-	C_6H_{12}	281(1.40),486(1.58)	39-1781-86C
$C_9H_{11}IN_2O_6$			
Uridine, 6-iodo-	H_2O	269(3.97)	23-1560-86
$C_9H_{11}N$			
Benzenamine, N-(2-propenyl)-, cation	acid	254f(2.3)	112-1099-86
Benzenamine, N-propylidene-	pH 8.4	330(4.78)	30-0055-86
$C_9H_{11}NO$			
Acetamide, N-(4-methylphenyl)-, chromium tricarbonyl complex	$CHCl_3$	217(4.71),325(4.01)	112-0517-86
5(1H)-Indolizinone, 2,3-dihydro-3-methyl-	EtOH	231(3.85),303(3.81)	44-2184-86
4H-Quinolizin-4-one, 6,7,8,9-tetrahydro-	EtOH	233(3.83),309(3.86)	44-2184-86
$C_9H_{11}NOS$			
Benzenecarboximidothioic acid, N-methyl-, methyl ester, N-oxide	MeOH	273(3.91)	44-5198-86
monohydriodide, (E)-	MeOH	218(4.32),272(3.82)	44-5198-86
monohydriodide, (Z)-	MeOH	217(4.47),272(4.03)	44-5198-86
$C_9H_{11}NO_2$			
Carbamic acid, phenyl-, ethyl ester	C_6H_{12}	275f(3.2)	47-1879-86
2,5-Cyclohexadiene-1,4-dione, 3,5-dimethyl-2-(methylamino)-	CH_2Cl_2	273(4.02),508(3.29)	83-0421-86
2,5-Cyclohexadiene-1,4-dione, 3,6-dimethyl-2-(methylamino)-	CH_2Cl_2	274(3.93),506(3.30)	83-0421-86

Compound	Solvent	λ_{max}(log ε)	Ref.
2,5-Cyclohexadiene-1,4-dione, 5,6-dimethyl-2-(methylamino)-	MeOH	288(--),481(3.34)	83-0421-86
	CH_2Cl_2	288(4.13),475(3.28)	83-0421-86
Ethanone, 1-(2-methoxy-3H-azepin-4-yl)-	EtOH	221.0(4.11),313.0(3.81)	18-2317-86
Ethanone, 1-(2-methoxy-3H-azepin-5-yl)-	EtOH	220.0(4.19),290s(3.29)	18-2317-86
Ethanone, 1-(2-methoxy-3H-azepin-6-yl)-	EtOH	217.0(4.26),288.5(4.14)	18-2317-86
Glycine, N-phenyl-, methyl ester	acid	253f(2.3)	112-1099-86
2-Propenoic acid, 2-methyl-3-(1H-pyrrol-2-yl)-, methyl ester	EtOH	325(4.41)	150-3601-86M
(Z)-	EtOH	335(4.22)	150-3601-86M
$C_9H_{11}NO_2Se$			
1-Cyclobutene-1-selenol, 3,4-dioxo-2-piperidino-, sodium salt	MeOH	304(4.0),359(4.4)	24-0182-86
$C_9H_{11}NO_3$			
Acetamide, N-(3,4-dimethoxyphenyl)-N-methyl-	MeOH	235(3.89),283(3.49)	87-1737-86
3H-Azepine-4-carboxylic acid, 2-methoxy-, methyl ester	EtOH	213.5(3.82),308.0(3.74)	18-2317-86
3H-Azepine-5-carboxylic acid, 2-methoxy-, methyl ester	EtOH	285.0(3.47)	18-2317-86
3H-Azepine-6-carboxylic acid, 2-methoxy-, methyl ester	EtOH	280.0(3.82)	18-2317-86
2,5-Pyrrolidinedione, 3-(tetrahydro-5-methyl-2-furanylidene)-	EtOH	205(4.17),261(4.45)	44-0495-86
Tyrosine	pH 11.5	240(4.04),293(3.36)	35-3186-86
$C_9H_{11}NO_4$			
Benzoic acid, 6-amino-2,3-dimethoxy-	EtOH	240(4.18),284(3.63)	44-2781-86
2-Furancarboxylic acid, 5-[(ethoxymethylene)amino]-, methyl ester	MeOH	300(3.29)	73-2817-86
1H-Pyrrole-2,4-dicarboxylic acid, 1-methyl-, dimethyl ester	EtOH	217(4.34),263(3.98), 268(3.96)	44-3125-86
1H-Pyrrole-3,4-dicarboxylic acid, 2-methyl-, dimethyl ester	EtOH	212(4.08),260(3.89)	35-6739-86
Tyrosine, 3-hydroxy-	pH 7.5	280(3.45)	149-0689-86B
$C_9H_{11}NO_5$			
2-Furancarboxylic acid, 5-[(ethoxycarbonyl)amino]-, methyl ester	MeOH	297(4.20)	73-2817-86
$C_9H_{11}N_3O$			
Imidazo[1,2-a]pyrimidin-5(1H)-one, 7-propyl-	EtOH	219(4.44),248(3.63), 294(4.00)	4-0245-86
$C_9H_{11}N_3O_3$			
7aH-Azeto[1',2':1,5]pyrrolo[2,3-c]pyrazole-7a-carboxylic acid, 3,3a,4,4a,5,6-hexahydro-6-oxo-, methyl ester, (3aRS,4aRS,7aRS)-	EtOH	325(2.60)	39-0973-86C
(3aRS,4aSR,7aRS)-	EtOH	327(2.59)	39-0973-86C
2-Propenamide, N-[2-(3,4-dihydro-2,4-dioxo-1(2H)-pyrimidinyl)ethyl]-	H_2O	268(4.00)	47-3201-86
$C_9H_{11}N_3O_4$			
2,3'-Imino-1-(3-deoxy-β-D-lyxopyranosyl)uracil	MeOH	212(4.45)	44-4417-86

Compound	Solvent	λ_{max}(log ε)	Ref.
$C_9H_{11}N_3O_5$			
1H-Pyrazole-3-carbonitrile, 4-hydroxy-5-β-D-ribofuranosyl-	pH 1	257(3.71)	87-0268-86
	pH 7	296(3.74)	87-0268-86
	pH 11	296(3.90)	87-0268-86
$C_9H_{11}N_3S$			
Pyrrolo[1,2-a]thieno[2,3-d]pyrimidin-4(2H)-imine, 3,6,7,8-tetrahydro-	$CHCl_3$	238(4.30),294(3.78)	24-1070-86
$C_9H_{11}N_5$			
1H-1,2,4-Triazole-3,5-diamine, 1-methyl-N^5-phenyl-	EtOH	259(4.27)	4-0401-86
	10%EtOH-HCl	252(4.14)	4-0401-86
	10%EtOH-NaOH	256(4.20)	4-0401-86
1H-1,2,4-Triazole-3,5-diamine, N-(phenylmethyl)-	EtOH	203(4.38),255(4.15)	4-0401-86
2H-1,2,4-Triazole-3,5-diamine, 2-methyl-N^5-phenyl-	EtOH	202(4.34),259(4.21)	4-0401-86
	EtOH-NaOH	257(4.19)	4-0401-86
$C_9H_{11}N_5O$			
Pyrido[2,3-d]pyrimidine-6-methanol, 2,4-diamino-5-methyl-	pH 1	223(4.45),246s(4.10), 317(3.94),327s(3.91)	87-0709-86
	H_2O	223(4.46),246s(4.09), 272s(3.67),316s(3.98), 328s(3.94)	87-0709-86
	pH 13	226(4.37),247s(4.23), 271(3.88),341(3.84)	87-0709-86
$C_9H_{11}N_5O_2$			
Acetic acid, cyano(6-hydrazino-4(1H)-pyrimidinylidene)-, ethyl ester, (Z)-	EtOH	253(4.11),328(4.59)	103-0774-86
$C_9H_{11}N_5O_4$			
6H-Purin-6-one, 2-amino-1,9-dihydro-9-(tetrahydro-3,4-dihydroxy-2-furanyl)-, (2α,3β,4β)-(±)-	H_2O	253(4.07),274s(3.91)	23-1885-86
Tetrazolo[1,5-c]pyrimidin-5(6H)-one, 6-(2-deoxy-β-D-erythro-pentofuranosyl)-	MeOH	253(3.95)	4-1401-86
$C_9H_{11}N_5O_5$			
1,2,4-Triazolo[3,4-f][1,2,4]triazin-8(5H)-one, 3-β-D-ribofuranosyl-	pH 1	252s(3.36)	87-2231-86
	pH 7	257(3.38)	87-2231-86
	pH 11	255(3.43)	87-2231-86
C_9H_{12}			
1,3,5,7-Octatetraene, 2-methyl- (unstable)	EtOH	255s(--),265(--), 277(--),288(--), 302(--)	39-1203-86C
1,3,5,7-Octatetraene, 3-methyl-	EtOH	260s(--),272(4.26), 284(4.50),296(4.70), 310(4.66)	39-1203-86C
1,3,5,7-Octatetraene, 4-methyl-	EtOH	258s(--),270(4.23), 280(4.52),292(4.69), 306(4.66)	39-1203-86C
$C_9H_{12}BrN_5O_2$			
2,3-Butanediol, 1-(6-amino-8-bromo-9H-purin-9-yl)-, (R*,R*)-(±)-	pH 2	266(4.26)	73-0459-86
$C_9H_{12}BrN_5O_3$			
1,2,3-Butanetriol, 4-(6-amino-8-bromo-	pH 2	267(4.26)	73-0459-86

Compound	Solvent	$\lambda_{max}(\log \epsilon)$	Ref.
9H-purin-9-yl)-, (R*,R*)-(±)-			73-0459-86
$C_9H_{12}DN_3O_4$			
2(1H)-Pyrimidinone, 4-amino-1-(2-deoxy-α-D-arabinofuranosyl-2-C-d)-	pH 7	271(3.96)	78-5427-86
β-	pH 7	271(3.91)	78-5427-86
$C_9H_{12}FN_5O_3$			
6H-Purin-6-one, 2-amino-9-[(2-fluoro-3-hydroxypropoxy)methyl]-, (±)-	pH 1	255(4.10),275s(3.93)	87-0842-86
	pH 7	252(4.12),270s(3.97)	87-0842-86
	pH 13	264(4.08)	87-0842-86
6H-Purin-6-one, 2-amino-9-[(3-fluoro-1-hydroxy-2-propoxy)methyl]-	MeOH	253(4.13),273s(3.96)	87-1384-86
6H-Purin-6-one, 2-amino-9-[(3-fluoro-2-hydroxypropoxy)methyl]-	pH 1	255(4.11),271s(3.94)	87-0842-86
	pH 7	251(4.15),268s(3.99)	87-0842-86
	pH 13	265(4.05)	87-0842-86
$C_9H_{12}N_2O_2$			
Phenylalanine, 4-amino-	H_2O	236.0(3.95),284.3(3.13)	149-0159-86B
	acid	213(--),258(2.23)	149-0159-86B
	MeOH	239.1(--),290.3(--)	149-0159-86B
	BuOH	240.3(--),291.3(--)	149-0159-86B
	dioxan	298.5(--)	149-0159-86B
	CH_2Cl_2	245.3(--),298.5(--)	149-0159-86B
$C_9H_{12}N_2O_5Se$			
4-Selenazolecarboxamide, 2-β-D-ribofuranosyl-	MeOH	215(4.25),259(3.76)	4-0155-86
	EtOH	259(3.80)	4-0155-86
$C_9H_{12}N_2O_6$			
Bicyclo[3.1.1]heptane-6-carboxylic acid, 6,7-dinitro-, methyl ester, 6-endo,7-endo	hexane	281(1.67)	104-0599-86
6-exo,7-endo	hexane	283(1.58)	104-0599-86
Butanedioic acid, (1-aminoethylidene)-(nitromethylene)-, dimethyl ester	EtOH	340(3.81)	18-3871-86
2,4(1H,3H)-Pyrimidinedione, 1-β-D-xylofuranosyl-	H_2O	263(3.97)	87-0203-86
$C_9H_{12}N_3O_8P$			
1H-Pyrazole-3-carboxamide, 4-hydroxy-5-(3,5-O-phosphinico-β-D-ribofuranosyl)-	pH 1	263(3.57)	87-0268-86
	pH 7	268(3.45)	87-0268-86
	pH 11	235(3.46),305(3.76)	87-0268-86
$C_9H_{12}N_4$			
1H-Pyrazole, 3,5-dimethyl-1-(1H-pyrazol-1-ylmethyl)-	EtOH	222(4.09)	142-2233-86
$C_9H_{12}N_4O$			
2-Quinoxalinecarboxamide, 3-amino-5,6,7,8-tetrahydro-	EtOH	362(3.96)	33-0793-86
$C_9H_{12}N_4OS$			
6H-Purine-6-thione, 9-(1-ethoxyethyl)-1,9-dihydro-	EtOH	325(4.29)	103-0331-86
$C_9H_{12}N_4O_2$			
2H-Imidazo[4,5-c]pyridin-2-one, 1,3,4,5-tetrahydro-5-(2-hydroxyethyl)-4-imino-1-methyl-, monohydrochloride	H_2O	234(4.82),304(4.12)	103-0180-86

Compound	Solvent	$\lambda_{max}(\log \epsilon)$	Ref.
1H-Purine-2,6-dione, 8-ethyl-dihydro-1,3-dimethyl-	MeOH	273(4.03)	142-1565-86
6H-Purin-6-one, 9-ethyl-1,9-dihydro-1-(2-hydroxyethyl)-	pH 1	253(4.00)	94-1094-86
	pH 7	253(4.02)	94-1094-86
	pH 13	253(4.01)	94-1094-86
	EtOH	254(3.98)	94-1094-86
$C_9H_{12}N_4O_2S$			
Ethanethioic acid, S-[5-(methylamino)-6-[(methylamino)carbonyl]pyrazinyl] ester	MeOH	272(4.32),373(3.81)	33-0708-86
8H-Pyrido[2,3-c][1,2,6]thiadiazin-4-amine, 1,5,7-trimethyl-, 2,2-dioxide	pH -2.0	226(4.09),261(3.94), 325(3.91)	11-0607-86
	pH 7.0	237(4.31),244s(3.98), 318(3.76)	11-0607-86
	pH 14.0	256(3.86),307(3.76)	11-0607-86
8H-Pyrido[2,3-c][1,2,6]thiadiazin-4-amine, 5,7,8-trimethyl-, 2,2-dioxide	pH -4.0	212(4.23),237(4.21), 262s(3.70),358(3.89)	11-0607-86
	pH 7.0	215(4.22),260(3.95), 358(3.90)	11-0607-86
$C_9H_{12}N_4O_3S$			
6H-Purin-6-one, 1,9-dihydro-9-[(2-hydroxyethoxy)methyl]-2-(methylthio)-	10% EtOH	262(4.18),284s(3.18)	4-0271-86
	+ HCl	268(4.21)	4-0271-86
	+ NaOH	272(4.18)	4-0271-86
$C_9H_{12}N_4O_4$			
1H-Imidazole-4-carbonitrile, 5-amino-1-α-D-xylofuranosyl-	MeOH	249(4.06)	87-0203-86
Uracil, 2,3'-(aminoimino)-1-(3-deoxy-β-D-lyxofuranosyl)-, hydrochloride	MeOH	218(4.32),234s(4.23), 270s(3.56)	44-4417-86
$C_9H_{12}N_5O_5P$			
9H-Purin-6-amine, 9-[[(2-hydroxy-1,3,2-dioxaphosphorinan-5-yl)oxy]-methyl]-, monosodium salt, P-oxide	pH 1	257(4.13)	87-0671-86
	pH 13	260(4.14)	87-0671-86
$C_9H_{12}N_5O_6P$			
6H-Purin-6-one, 2-amino-1,9-dihydro-9-[[(2-hydroxy-1,3,2-dioxaphosphorinan-5-yl)oxy]methyl]-, P-oxide, ammonium salt	H_2O	252(4.12),270s(3.97)	87-0671-86
$C_9H_{12}N_6O_4$			
1,2,4-Triazolo[3,4-f][1,2,4]triazin-8-amine, 3-β-D-ribofuranosyl-	pH 1	226(4.08),272(3.74)	87-2231-86
	pH 7	230(4.05),274(3.66)	87-2231-86
	pH 11	232(4.07),273(3.73)	87-2231-86
$C_9H_{12}N_6O_4S$			
1,2,4-Triazolo[3,4-f][1,2,4]triazin-8(5H)-one, 6-amino-3-β-D-ribofuranosyl-	pH 1	230(4.29),327(4.06)	87-2231-86
	pH 7	230(4.30),327(4.16)	87-2231-86
	pH 11	230(4.30),327(4.16)	87-2231-86
$C_9H_{12}N_6O_5$			
1,2,4-Triazolo[3,4-f][1,2,4]triazin-8(5H)-one, 6-amino-3-β-D-ribofuranosyl-	pH 1	245s(4.13),275s(3.88)	87-2231-86
	pH 7	245(4.08),275s(3.76)	87-2231-86
	pH 11	240(4.13),280(3.87)	87-2231-86
$C_9H_{12}N_8O_3$			
6H-Purin-6-one, 2-amino-9-[(1-azido-3-hydroxy-2-propoxy)methyl]-1,9-dihydro-	pH 1	256(4.05),275s(3.90)	87-1384-86
	pH 13	265(4.02)	87-1384-86

Compound	Solvent	λ max(log ε)	Ref.
$C_9H_{12}O$			
1,3,5-Cycloheptatriene, 3-(methoxymethyl)-	hexane	260(3.45)	24-2889-86
1,3,5-Cycloheptatriene, 7-(methoxymethyl)-	hexane	257(3.46)	24-2889-86
$C_9H_{12}O_2$			
2,4-Nonadiyne-1,6-diol, (R)-	EtOH	220(2.54),231(2.61), 243(2.57),257(2.33)	150-1348-86M
2,4-Nonadiyne-1,6-diol, (S)-	ether	220(2.57),232(2.61), 244(2.57),257(2.38)	150-1348-86M
7-Oxabicyclo[2.2.1]heptane-1-methanol, 2,3-bis(methylene)-	isooctane	237s(3.93),242(3.96), 250s(3.83)	33-1287-86
	EtOH	238s(3.92),242(3.93), 251s(3.81)	33-1287-86
$C_9H_{12}O_2S_2$			
1,3-Dithiole-4-carboxylic acid, 2-(1-methylethylidene)-, ethyl ester	EtOH	254(3.49),296(2.26), 316(2.28),388(2.49)	33-0419-86
3-Thiophenecarboxylic acid, 4,5-dihydro-4,4-dimethyl-5-thioxo-, ethyl ester	EtOH	227(2.74),272(2.48), 345(2.95)	33-0419-86
$C_9H_{12}O_4$			
1-Cyclopentene-1-acetic acid, 3-methoxy-5-oxo-, methyl ester, (±)-	MeOH	214(3.93)	104-1865-86
2-Furancarboxylic acid, 4-(ethoxymethyl)-5-methyl-	MeOH	261(3.09)	73-2186-86
$C_9H_{13}BrO_2$			
2,4-Hexadienoic acid, 4-bromo-, propyl ester, (Z,E)-	n.s.g.	274(4.15)	104-2230-86
$C_9H_{13}N$			
Acetonitrile, (4-methylcyclohexylidene)-, (S)-	C_6H_{12}	213.5(4.20)	44-2863-86
Benzenamine, N-ethyl-N-methyl-	acid	253f(2.3)	112-1099-86
Benzenamine, N-(1-methylethyl)-, sulfate	0.1N H_2SO_4	254(2.30)	65-0179-86
chromium tricarbonyl complex	$CHCl_3$	218(4.51),320(3.92)	112-0517-86
Benzenamine, 4-(1-methylethyl)-, chromium tricarbonyl complex	$CHCl_3$	217(4.66),320(3.90)	112-0517-86
Benzenamine, N-propyl-	hexane	207(4.00),246(5.08), 298(3.18)	65-2461-86
	MeOH	204(--),250(--), 288(--)	65-2461-86
Benzenamine, 2,4,5-trimethyl-, chromium tricarbonyl complex	$CHCl_3$	216(4.55),322(3.86)	112-0517-86
$C_9H_{13}NO$			
Cyclopent[c]azepin-1(2H)-one, 3,4,5,6,7,8-hexahydro-	EtOH	222(4.53)	44-1490-86
5(1H)-Quinolinone, 2,3,4,6,7,8-hexahydro-	EtOH	318(4.45)	44-1490-86
$C_9H_{13}NO_3$			
2,5-Pyrrolidinedione, (4-hydroxypentylidene)-, (E)-	EtOH	230(3.89)	44-0495-86

Compound	Solvent	$\lambda_{max}(\log \epsilon)$	Ref.
$C_9H_{13}NO_3S$			
Benzenamine, N-(1-methylethyl)-, sulfonic acid deriv. sodium salt	NaHCO	239(3.39),269(3.20), 290(3.30)	65-0179-86
$C_9H_{13}NO_5$			
3-Isoxazoleacetic acid, 2,3-dihydro-4-(methoxycarbonyl)-2-methyl-, methyl ester	EtOH	260(4.14)	44-3125-86
$C_9H_{13}NO_6$			
Cyclopentaneacetic acid, 3-hydroxy-2-(nitromethyl)-5-oxo-, methyl ester, trans,cis	MeOH	203(3.70),282(1.72)	104-1865-86
trans,trans	MeOH	204(3.70),282(1.76)	104-1865-86
D-Ribitol, 1-deoxy-1-(2,5-dioxo-3-pyrrolidinylidene)-, (E)-	H_2O	222(3.90)	44-0495-86
$C_9H_{13}N_3$			
Benzaldehyde, 4-(dimethylamino)-, hydrazone	pH 5.0	450(4.36),500(4.75)	30-0055-86
$C_9H_{13}N_3O_2$			
Acetic acid, cyano(1-methyl-2-imidazolidinylidene)-, ethyl ester, (E)-	EtOH	211(4.25),261(4.40)	24-2208-86
$C_9H_{13}N_3O_5$			
2(1H)-Pyrimidinone, 4-amino-1-β-D-xylofuranosyl-	H_2O	273(3.88)	87-0203-86
$C_9H_{13}N_3O_6$			
1H-Pyrazole-4-carboxylic acid, 3-amino-1-β-D-ribofuranosyl-	pH 2	263(3.70)	73-1512-86
	pH 7 and 12	256(3.78)	73-1512-86
$C_9H_{13}N_3S$			
Thiourea, N-ethyl-N'-(4-methyl-2-pyridinyl)-	benzene	280(4.32)	80-0761-86
	MeOH	263(4.29)	80-0761-86
	EtOH	266(4.32)	80-0761-86
	isoPrOH	267(4.35)	80-0761-86
	BuOH	262(4.33)	80-0761-86
	dioxan	241(4.30)	80-0761-86
	HOAc	265(4.28)	80-0761-86
	CH_2Cl_2	267(4.32)	80-0761-86
	$CHCl_3$	269(4.23)	80-0761-86
	$C_2H_4Cl_2$	268(4.24)	80-0761-86
	MeCN	266(4.30)	80-0761-86
	DMF	273(4.44)	80-0761-86
	DMSO	274(4.33)	80-0761-86
Thiourea, N-ethyl-N'-(6-methyl-2-pyridinyl)-	benzene	276(4.20)	80-0761-86
	MeOH	264(4.29)	80-0761-86
	EtOH	267(4.22)	80-0761-86
	isoPrOH	262(4.27)	80-0761-86
	BuOH	258(4.28)	80-0761-86
	dioxan	241(4.25)	80-0761-86
	HOAc	263(4.21)	80-0761-86
	$HOCH_2CH_2OH$	263(4.28)	80-0761-86
	CH_2Cl_2	266(4.23)	80-0761-86
	$CHCl_3$	276(4.22)	80-0761-86
	$C_2H_4Cl_2$	266(4.24)	80-0761-86
	MeCN	266(4.28)	80-0761-86
	DMF	273(4.39)	80-0761-86
	DMSO	271(4.30)	80-0761-86

Compound	Solvent	λ_{max}(log ϵ)	Ref.
$C_9H_{13}N_5$			
6H-Purin-6-imine, 1,9-diethyl-1,9-	pH 1	261(4.10)	94-1094-86
dihydro-, monoperchlorate	pH 7	261(4.10)	94-1094-86
	pH 13	261(4.12)	94-1094-86
	EtOH	261.5(4.12)	94-1094-86
6H-Purin-6-imine, 3,7-diethyl-3,7-	pH 1	225(4.05),277(4.23)	94-1821-86
dihydro-, monoperchlorate	pH 7	225(4.07),277(4.23)	94-1821-86
	pH 13	283(4.17)	94-1821-86
	EtOH	226(4.05),280(4.24)	94-1821-86
$C_9H_{13}N_5O$			
Ethanol, 2-[(9-ethyl-9H-purin-6-yl)-	pH 1	266(4.24)	94-1094-86
amino]-	pH 7	269(4.25)	94-1094-86
	pH 13	269(4.25)	94-1094-86
	EtOH	268.5(4.23)	94-1094-86
1H-Purine-1-ethanol, 9-ethyl-6,9-	pH 1	261(4.10)	94-1094-86
dihydro-6-imino-, hydrobromide	pH 7	261(4.10)	94-1094-86
	pH 13	260.5(4.12)	94-1094-86
	EtOH	260.5(4.10)	94-1094-86
$C_9H_{13}N_5O_2S$			
9H-Purin-6-amine, 9-[(2-hydroxyeth-	10%EtOH-HCl	270(4.22),291s(4.08)	4-0271-86
oxy)methyl]-2-(methylthio)-	10%EtOH-NaOH	276(4.18)	4-0271-86
$C_9H_{13}N_5O_3$			
6H-Purin-6-one, 2-amino-1,9-dihydro-	pH 1	253(4.04),278(3.86)	87-1384-86
9-[3-hydroxy-2-(hydroxymethyl)-	pH 13	254(3.96),268(4.00)	87-1384-86
propyl]-			
6H-Purin-6-one, 2-amino-1,9-dihydro-	pH 1	255(4.10),275s(3.92)	87-0842-86
9-[(2-hydroxypropoxy)methyl]-, (±)-	pH 7	252(4.09),270s(3.94)	87-0842-86
	pH 13	265(4.05)	87-0842-86
$C_9H_{13}N_5O_4$			
6H-Purin-6-one, 2-amino-1,9-dihydro-	pH 1	255(4.08),278s(3.90)	23-1885-86
9-[1-(2-hydroxyethoxy)-2-hydroxy-	pH 13	254(4.03),265(4.03)	23-1885-86
ethyl]-			
$C_9H_{13}N_6O_5P$			
9H-Purine-2,6-dione, 9-[[(2-hydroxy-	pH 1	252(4.02),292(3.95)	87-0671-86
1,3,2-dioxaphosphorinan-5-yl)oxy]-	pH 13	256(3.97),280(4.00)	87-0671-86
methyl]-, P-oxide			
$C_9H_{13}N_7O_4$			
1,2,4-Triazolo[3,4-f][1,2,4]triazine-	pH 1	232(4.06),276(3.75)	87-2231-86
6,8-diamine, 3-β-D-ribofuranosyl-	pH 7	230(4.08),280(3.69)	87-2231-86
	pH 11	231(4.08),278(3.70)	87-2231-86
$C_9H_{14}Cl_2N_4O_3Pt$			
Platinum, dichloro(3',5'-diamino-2',3',5'-trideoxyuridine-$N^{3'}$,$N^{5'}$)-, (SP-4-3)-	EtOH	261(4.01)	87-0681-86
$C_9H_{14}N$			
Benzenaminium, N,N,N-trimethyl-	98% H_2SO_4	247(2.3),252(2.3), 258(2.3),263(2.2)	112-0265-86
$C_9H_{14}N_2O_2$			
2,6-Piperidinedione, 3-[(dimethyl-amino)methylene]-1-methyl-	isoPrOH	217(4.33),320(4.60)	103-1345-86
2,4(1H,3H)-Pyrimidinedione, 6-methyl-	pH 1	267(3.70)	70-1064-86
1-(2-methylpropyl)-	pH 12	269(4.15)	70-1064-86

Compound	Solvent	λ$_{max}$(log ε)	Ref.
2,4(1H,3H)-Pyrimidinedione, 6-methyl-3-(2-methylpropyl)-	pH 7	260(3.50)	70-1064-86
	pH 12	281(3.76)	70-1064-86
$C_9H_{14}N_2O_2S$			
Pyrimidinium, 2-(ethylthio)-3,6-dihydro-4-hydroxy-1,3,5-trimethyl-6-oxo-, hydroxide, inner salt	MeCN	226(4.41),262(3.54), 360(3.38)	24-1315-86
$C_9H_{14}N_2O_4$			
2,4,6(1H,3H,5H)-Pyrimidinetrione, 5-(2-ethoxypropyl)-	pH 10.0	269(4.21)	39-1391-86B
$C_9H_{14}N_2O_4S$			
6-Oxa-2,4-diazabicyclo[3.3.1]nonane-9-carboxylic acid, 8-hydroxy-5-methyl-3-thioxo-, methyl ester	MeOH	250.6(3.37)	142-0679-86
$C_9H_{14}N_3O_9P$			
1H-Pyrazole-5-carboxamide, 4-hydroxy-3-(5-O-phosphono-β-D-ribofuranosyl)-	pH 1	257(3.63)	87-0268-86
	pH 7	257(3.57)	87-0268-86
	pH 11	303(3.73)	87-0268-86
$C_9H_{14}N_4O$			
4(3H)-Pteridinone, 5,6,7,8-tetrahydro-2,6,7-trimethyl-, cis	HCl	263(3.92)	12-0031-86
$C_9H_{14}N_4OS$			
4(3H)-Pteridinone, 5,6,7,8-tetrahydro-2-(methylthio)-6,7-dimethyl-, cis	1.5M HCl	274(3.83)	12-0031-86
$C_9H_{14}N_4O_3$			
5-Pyrimidinepentanamide , 2-amino-1,4-dihydro-6-hydroxy-4-oxo-	50%MeOH-HCl	266(4.02)	73-1140-86
	50%MeOH-NaOH	270(4.06)	73-1140-86
$C_9H_{14}N_4O_4S$			
1,3,5-Triazin-2-amine, 4-(methylthio)-6-β-D-ribofuranosyl-	pH 1	228(4.16),238s(4.11)	4-1709-86
	pH 7	223(4.24),267(3.54)	4-1709-86
	pH 11	250(3.65)	4-1709-86
$C_9H_{14}N_4O_5$			
1H-Pyrazole-3-methanimidamide, 4-hydroxy-5-β-D-ribofuranosyl-	pH 1	227s(3.94),276(3.64)	87-0268-86
	pH 7	326(3.64)	87-0268-86
	pH 11	325(3.85)	87-0268-86
$C_9H_{14}N_4O_5S$			
1,3,5-Triazin-2-amine, 4-(methylsulfinyl)-6-β-D-ribofuranosyl-	pH 1	228(3.44),265s(2.96)	4-1709-86
	pH 7	228(3.34),260s(3.10)	4-1709-86
	pH 11	218(3.37),258(3.03)	4-1709-86
$C_9H_{14}N_4O_6$			
1H-Pyrazole-4-carboxylic acid, 3,5-diamino-1-β-D-ribofuranosyl-	pH 2	248(4.35)	73-1512-86
$C_9H_{14}N_5O$			
7H-Purinium, 6-amino-9-ethyl-7-(2-hydroxyethyl)-, bromide	pH 1	270(4.06)	94-1094-86
	pH 7	270(4.07)	94-1094-86
	EtOH	273(4.04)	94-1094-86
$C_9H_{14}N_5O_4P$			
Phosphonic acid, [1-[(6,7-dihydro-6-oxo-1H-purin-2-yl)amino]-1-	pH 1	253(4.17),280s(3.83)	94-0471-86
	H_2O	251.5(4.12),275s(3.93)	94-0471-86

Compound	Solvent	λ_{max}(log ε)	Ref.
methylpropyl]- (cont.)	pH 13	261(4.01),280s(3.95)	94-0471-86
$C_9H_{14}N_5O_7P$			
6H-Purin-6-one, 2-amino-1,9-dihydro-9-[[1-(hydroxymethyl)-2-(phosphono-oxy)ethoxy]methyl]-, diammonium salt	H_2O	252(4.03),272(3.88)	87-0671-86
$C_9H_{14}N_6O_2$			
8H-Purin-8-one, 6-amino-9-(3-amino-2-hydroxybutyl)-7,9-dihydro-	pH 2	269(4.04),282s(--)	73-0459-86
	pH 7	270(4.12)	73-0459-86
	pH 12	280(4.15)	73-0459-86
$C_9H_{14}N_6O_3$			
6H-Purin-6-one, 2-amino-9-[[2-amino-1-(hydroxymethyl)ethoxy]methyl]-1,9-dihydro-, hydrochloride	pH 1	257(4.08),276s(3.92)	87-1384-86
	pH 13	266(4.03)	87-1384-86
8H-Purin-8-one, 6-amino-9-(3-amino-2,4-dihydroxybutyl)-7,9-dihydro-	pH 2	269(4.17),284(4.17)	73-0459-86
	pH 7	270(4.11)	73-0459-86
	pH 12	281(4.15)	73-0459-86
$C_9H_{14}O$			
Ethanone, 1-(1,2-dimethyl-2-cyclo-penten-1-yl)-	gas	199(3.6),300f(1.9)	35-5527-86
$C_9H_{14}O_2$			
1,4-Dioxaspiro[4.5]decane, 7-methylene-	pentane	234(3.03)	35-7575-86
2,4-Heptadienoic acid, 6,6-dimethyl-, (Z,Z)-	EtOH	262(4.22)	39-1809-86C
6-Heptynoic acid, ethyl ester	MeOH	206(2.23)	18-3535-86
5-Hexynoic acid, 1-methylethyl ester	MeOH	205(2.18)	18-3535-86
8-Nonynoic acid	MeOH	206(2.69)	18-3535-86
6-Oxabicyclo[3.1.0]hexan-3-one, 2,2,4,4-tetramethyl-	pentane	295(1.38),304(1.34), 316(1.11)	35-4556-86
$C_9H_{14}O_3$			
1,3-Dioxolane-2-butanal, α-ethyli-dene-, (E)-	EtOH	229(4.19)	70-1851-86
$C_9H_{14}O_3S$			
2-Pentenoic acid, 2-(butylthio)-4-oxo-	MeOH	275(3.90),315s(2.85)	48-0539-86
$C_9H_{14}S$			
3-Cyclopentene-1-thione, 2,2,5,5-tetramethyl-	hexane	209(3.93),225(4.09), 260(3.61),491(0.90)	78-1693-86
$C_9H_{14}S_2$			
2-Cyclohexene-1-thione, 5,5-dimethyl-3-(methylthio)-	pH 2.93	370(3.4)	65-0365-86
	pH 13	338(2.9)	65-0365-86
$C_9H_{15}NO$			
Acetaldehyde, (4-methylcyclohexyli-dene)-, oxime, anti	C_6H_{12}	237(4.33)	44-2863-86
syn	C_6H_{12}	234(4.31)	44-2863-86
6H-1,2-Oxazine, 3-pentyl-	hexane	237(3.23),289(3.08)	108-0033-86
$C_9H_{15}NO_2$			
1,2-Cyclohexanedione, 3,3-dimethyl-, 1-(O-methyloxime), (E)-	C_6H_{12}	235(3.77)	54-0054-86
	EtOH	245(3.76),330s(2.11)	54-0054-86

Compound	Solvent	λ$_{max}$(log ε)	Ref.
1,2-Cyclohexanedione, 3,3-dimethyl-, 1-(O-methyloxime)	C_6H_{12}	245(3.36),315s(2.08)	54-0054-86
	EtOH	245(3.40),315(2.10)	54-0054-86
Propanamide, N-cyclohexyl-2-oxo-	MeOH	244(3.24),345(1.23)	33-0469-86
1,2-Propanedione, 1-cyclohexyl-, 1-oxime	EtOH	221(3.91)	118-0473-86
3-Pyridinecarboxylic acid, 1,2,3,6-tetrahydro-1,3-dimethyl-, methyl ester, (±)-	pH 7.2	214(0.13)	87-0125-86
$C_9H_{15}NO_2S$			
Proline, 5-thioxo-, 1,1-dimethylethyl ester	MeOH	215(3.60),271(4.04)	5-0269-86
$C_9H_{15}NO_3$			
Cyclohexaneacetic acid, α-(hydroxyimino)-, methyl ester	EtOH	215(3.88)	118-0473-86
$C_9H_{15}NO_4$			
Propanedioic acid, [(2-methylpropyl)-imino]-, dimethyl ester	MeOH	296(2.06)	5-1990-86
$C_9H_{15}N_3O_4$			
Propanedioic acid, azido(2-methylpropyl)-, dimethyl ester	MeOH	222(2.96),270(1.76)	5-1990-86
$C_9H_{15}N_4O$			
1,2,4-Triazinium, 1-ethyl-3-morpholino-, tetrafluoroborate	EtOH	224(3.81),259(4.21), 313(2.88),393(2.94)	103-1242-86
$C_9H_{15}N_5O_2S$			
Pyrazino[2,3-e]-1,2,4-triazin-6-ol, 7-ethoxy-5,6,7,8-tetrahydro-5-methyl-3-(methylthio)-, trans	pH 1	267(4.44),328s(3.59)	4-0033-86
	pH 11	257(4.21),273(4.11), 329(3.85)	4-0033-86
	EtOH	223.5(4.14),254(4.24), 268s(4.15),328(3.83)	4-0033-86
Pyrazino[2,3-e]-1,2,4-triazin-6-ol, 7-ethoxy-5,6,7,8-tetrahydro-8-methyl-3-(methylthio)-, trans	pH 1	266(4.43),332s(3.43)	4-0033-86
	pH 11	254(4.21),270s(4.11), 332(3.80)	4-0033-86
	EtOH	221(4.09),251(4.22), 269s(4.09),330(3.80)	4-0033-86
$C_9H_{15}N_5O_3$			
5-Pyrimidinepentanoic acid, 2-amino-1,4-dihydro-6-hydroxy-4-oxo-, hydrazide	50%MeOH-HCl	267(4.20)	73-1140-86
	50%MeOH-NaOH	271(4.14)	73-1140-86
$C_9H_{15}N_5O_{10}P_2$			
6H-Purin-6-one, 2-amino-1,9-dihydro-9-[[2-(phosphonooxy)-1-[(phosphonooxy)methyl]ethoxy]methyl]-, tetrasodium salt	H_2O	251(4.11),270s(3.96)	87-0671-86
$C_9H_{16}BrNO$			
4H-1,2-Oxazine, 4-bromo-5,6-dihydro-3-pentyl-	hexane	216(3.43),251(3.35)	108-0033-86
$C_9H_{16}FNO$			
4H-1,2-Oxazine, 4-fluoro-5,6-dihydro-3-pentyl-	hexane	236(4.155)	108-0033-86

Compound	Solvent	λ_{max}(log ϵ)	Ref.
$C_9H_{16}N_4O_3$			
2,4(1H,3H)-Pyrimidinedione, 5-amino-6-[(5-hydroxypentyl)amino]-	pH 5.0	220(4.44),224s(4.26), 311(4.11),350(2.70)	44-2461-86
	pH 8.0	238(3.74),311(4.23), 350(3.00),500(2.06)	44-2461-86
	pH 12.0	238(3.60),311(4.23), 350(3.30)	44-2461-86
$C_9H_{16}O_2$			
Cyclohexane, 1,1-dimethoxy-3-methylene-	pentane	241(3.04)	35-7575-86
2-Pentenoic acid, 4-methyl-3-(1-methylethyl)-	EtOH	219(4.08)	33-0560-86
$C_9H_{16}O_5S_2$			
Ethane(dithioic) acid, 2-methoxyethoxyethoxy-, carboxymethyl ester	CH_2Cl_2	310(3.59)	126-2369-86
$C_9H_{17}NO$			
4H-1,2-Oxazine, 5,6-dihydro-3-pentyl-	hexane	223(3.24)	108-0033-86
$C_9H_{17}NO_2S_2$			
Propane(dithioic) acid, 2-[[(1,1-dimethylethoxy)carbonyl]amino]-, methyl ester, (S)-	ether	300(4.00)	78-6555-86
$C_9H_{17}N_2$			
Methanaminium, N-[5-(dimethylamino)-2,4-pentadienylidene]-N-methyl-, perchlorate	MeOH	410(5.1)	18-0843-86
$C_9H_{17}N_3$			
2-Pyrimidinamine, 1,6-dihydro-N,N,4,6,6-pentamethyl-	isoPrOH	269(3.63)	142-3023-86
2-Pyrimidinamine, N-ethyl-1,6-dihydro-4,6,6-trimethyl-	isoPrOH	260(3.56)	142-3023-86
hydrochloride	isoPrOH	240(3.59)	142-3023-86
$C_9H_{17}N_3O$			
4H-1,2,3-Triazol-4-one, 3-(2,2-dimethylpropyl)-3,5-dihydro-5,5-dimethyl-	hexane	250(3.720),305(2.627)	5-1891-86
$C_9H_{18}NOS$			
Oxazolium, tetrahydro-N,N-dimethyl-2-[(propylthio)methylene]-, chloride	H_2O	227s(3.76),247(3.82)	70-0994-86
$C_9H_{18}NO_2S$			
Oxazolium, 2-[(ethylthio)methylene]-tetrahydro-3-(2-hydroxyethyl)-3-methyl-, chloride	H_2O	232s(3.51),248(3.56)	70-0994-86
$C_9H_{18}N_4O_2S$			
Hydrazinecarbothioamide, 2-[2-(acetoxyamino)-1-ethyl-2-methylpropylidene]-	EtOH	233s(3.98),276(4.44)	103-0287-86
$C_9H_{18}OSi$			
Methanone, (2,2-dimethylcyclopropyl)-(trimethylsilyl)-	C_6H_{12}	371(2.09)	138-0177-86

Compound	Solvent	$\lambda_{max}(\log \epsilon)$	Ref.
Methanone, (2,3-dimethylcyclopropyl)-(trimethylsilyl)-, (1α,2α,3β)-	C_6H_{12}	366(2.03)	138-0177-86
(1α,2β,3β)-	C_6H_{12}	371(2.00)	138-0177-86
$C_9H_{19}NO_3S$			
Ethanethioamide, 2-[2-(2-methoxyethoxy)ethoxy]-N,N-dimethyl-	CH_2Cl_2	276(4.16)	126-2369-86
$(C_9H_{20}Si_2)_n$			
Silane, [(dimethyl-1-propynylsilyl)-methyl]trimethyl-, polymer	C_6H_{12}	273(2.14)	47-1839-86
$C_9H_{22}O_3SSi$			
Silane, triethoxy[(ethylthio)methyl]-	heptane	192(2.54),210s(--)	67-0062-86
	MeCN	194(2.54)	67-0062-86
Silane, triethoxy[2-(methylthio)-ethyl]-	heptane	196(2.54),210s(--)	67-0062-86
	MeCN	189(2.60),194(2.58), 210s(--)	67-0062-86
$C_9H_{22}SSi$			
1-Propanethiol, 3-(triethylsilyl)-	heptane	200(2.34),227(1.26)	67-0225-86

Compound	Solvent	$\lambda_{max}(\log \epsilon)$	Ref.
$C_{10}Br_4F_8$			
Naphthalene, 1,1,4,4-tetrabromo-2,2,3,3,5,6,7,8-octafluoro-1,2,3,4-tetrahydro-	heptane	288(3.41),293(3.39)	70-1883-86
$C_{10}Cl_6O_3$			
2,5-Furandione, 3-chloro-4-(pentachlorophenyl)-	C_6H_{12}	213(4.91),229s(4.54),241s(4.32),288s(2.78),300s(2.70)	44-1413-86
$C_{10}Cl_{10}$			
Benzene, pentachloro(2,3,3,4,4-pentachloro-1-cyclobuten-1-yl)-	C_6H_{12}	216(4.75),243s(4.18),275(3.09),296(2.92),306(3.00)	44-1413-86
Benzene, pentachloro[2,3,3-trichloro-1-(dichloromethylene)-2-propenyl]-	C_6H_{12}	216(4.81),250s(4.30),302s(3.00)	44-1413-86
$C_{10}HCl_9$			
Benzene, pentachloro[2,3-dichloro-1-(dichloromethylene)-2-propenyl]-	C_6H_{12}	215(4.86),248s(4.28),300s(2.98)	44-1413-86
Benzene, pentachloro(2,3,4,4-tetrachloro-1-cyclobuten-1-yl)-	C_6H_{12}	215(4.73),245s(4.11),277s(2.90),297(2.92),307(3.00)	44-1413-86
$C_{10}H_2Cl_2F_6N_2$			
Quinoxaline, 6,7-dichloro-2,3-bis(trifluoromethyl)-	MeOH	211(4.53),246(4.72),330(3.79),342(3.81)	39-1043-86C
$C_{10}H_2Cl_6O_4$			
2-Butenedioic acid, 2-chloro-3-(pentachlorophenyl)-, (E)-	EtOH	213(4.84),228s(4.40),241s(4.24),287s(2.75),298(2.66)	44-1413-86
$C_{10}H_2N_8S$			
Cyanamide, [thiobis[3-(cyanoamino)-2-cycloprop-2-yl-1-ylidene]]bis-, dipotassium	H_2O	238(4.63),296(4.25)	24-2104-86
$C_{10}H_4F_6N_2$			
Quinoxaline, 2,3-bis(trifluoromethyl)-	MeOH	237(4.68),315(3.67)	39-1043-86C
$C_{10}H_4S_8$			
1,3-Dithiolo[4,5-b][1,4]dithiin, 2-(1,3-dithiolo[4,5-b][1,4]dithiin-2-ylidene)-	dioxan	313(4.15),340(4.16),480(2.55)	103-1186-86
$C_{10}H_5ClN_2$			
Propanedinitrile, [(3-chlorophenyl)methylene]-	H_2O	300(4.26)	35-2372-86
	50% DMSO	300(4.24)	35-2372-86
	60% DMSO	300(4.24)	35-2372-86
$C_{10}H_5ClOS$			
Thieno[3,2-b]benzofuran, 3-chloro-	MeOH	283(3.07),290(3.06),302(3.85)	73-1685-86
$C_{10}H_5NO_3$			
Furo[3,4-f]-1,3-benzodioxole-5-carbonitrile	MeOH	316(3.33)	44-3973-86

Compound	Solvent	$\lambda_{max}(\log \epsilon)$	Ref.
$C_{10}H_5N_3O_2$			
Propanedinitrile, [(4-nitrophenyl)-methylene]-	H_2O	307(4.33)	35-2372-86
	50% DMSO	308(4.32)	35-2372-86
	60% DMSO	308(4.28)	35-2372-86
$C_{10}H_6BrNO_3$			
1H-Indole-2,3-dione, 1-acetyl-5-bromo-	heptane	241.5(4.55),292(3.66), 400(2.85)	104-2163-86
$C_{10}H_6Cl_2N_4$			
Cyanamide, (2,5-dichloro-3,6-dimethyl-2,5-cyclohexadiene-1,4-diylidene)-bis-	MeCN	346s(4.54),352(4.55), 366s(4.49)	5-0142-86
$C_{10}H_6F_6N_4O_2$			
2,4(1H,3H)-Pteridinedione, 1,3-dimethyl-6,7-bis(trifluoromethyl)-	pH 5.2	250(4.21),335(3.77)	39-1043-86C
$C_{10}H_6N_2$			
Propanedinitrile, (phenylmethylene)-	H_2O	308(4.33)	35-2372-86
	50% DMSO	308(4.31)	35-2372-86
	60% DMSO	309(4.32)	35-2372-86
$C_{10}H_6N_2O_3S$			
2-Furancarboxamide, N-furo[2,3-d]thiazol-2-yl-	MeOH	254(2.98),325(2.98)	73-1678-86
$C_{10}H_6N_2O_5S$			
Phenol, 2,4-dinitro-6-(2-thienyl)-	n.s.g.	212(4.12),242(4.11), 280(4.21),397(3.77)	44-3453-86
$C_{10}H_6N_2O_6$			
Phenol, 4,6-dinitro-2-(2-furanyl)-	n.s.g.	212(4.05),240(4.05), 278(4.28)	44-3453-86
$C_{10}H_6N_4O_2S$			
Thiazolo[4,5-b]quinoxaline, 2-methyl-7-nitro-	n.s.g.	405(4.1703)	18-1245-86
$C_{10}H_6N_4O_3$			
3H-Imidazo[4,5-g]quinazoline-4,8,9(5H)-trione, 3-methyl-	neutral	237(4.24),322(3.96), 410(2.86)	44-4784-86
	anion	222(4.18),254(4.26), 330(3.81),43?(3.04)	44-4784-86
$C_{10}H_6O_3$			
1,4-Naphthalenedione, 5-hydroxy-	C_6H_{12}	245(3.87),320(3.14), 406s(3.58),423(3.59)	42-0035-86
	pH 5	208(4.22),248(4.13), 310(2.98),420(3.56)	42-0035-86
	pH 10.7	215(4.26),258(3.97), 350(3.17),515(3.61)	42-0035-86
	MeOH	248(4.11),340s(3.05), 407s(3.55),422(3.56)	42-0035-86
	CCl_4	328(3.07),412s(3.57), 423(3.58)	42-0035-86
	DMF	329(3.31),416(4.87)	111-0385-86
manganese chelate	DMF	346(4.02),430(3.82), 540(4.06),618s(3.76)	111-0385-86

Compound	Solvent	$\lambda_{max}(\log \epsilon)$	Ref.
$C_{10}H_6O_4$			
6H-1,3-Dioxolo[4,5-g][1]benzopyran-6-one (ayapin)	MeOH	233(4.04),260s(3.56), 293(3.46),344(3.90)	102-0077-86
Furo[2,3-f]-1,3-benzodioxole-6-carboxaldehyde	EtOH	255(3.43),266(3.45), 293(3.57),344(3.69)	4-1715-86
1,4-Naphthalenedione, 5,8-dihydroxy-(naphthazarin)	C_6H_{12}	264(4.01),330(3.1), 486(3.87),516s(3.90), 524(3.92),544(3.74), 562(3.80)	42-0035-86
	pH 5	209(4.54),266(3.89), 340(3.03),482(3.72), 513(3.73),556(3.47)	42-0035-86
	pH 10.7	211(4.76),292(3.73), 528s(3.64),573(3.81), 610(3.76)	42-0035-86
	pH 13	308(3.73),575(3.94), 613(4.06)	42-0035-86
	MeOH	265(3.86),330(3.10), 483(3.71),511(3.73), 552s(3.49)	42-0035-86
	CCl_4	334(3.05),489(3.78), 524(3.83),546(3.63), 564(3.65)	42-0035-86
$C_{10}H_6S_2$			
Benzo[1,2-b:4,5-b']dithiophene	EtOH	235s(4.31),239s(4.36), 247.5(4.60),256(4.74), 291(3.75),301(3.81), 321(3.84),326(3.79), 334(4.05)	24-3198-86
$C_{10}H_7BO_6$			
Boron, [ethanedionato(2-)-O,O'][1-(2-hydroxyphenyl)ethanonato-O,O']-, (T-4)-	CH_2Cl_2	272(3.64),276s(3.64), 323(4.44)	48-0755-86
$C_{10}H_7ClN_2O_2$			
2,4-Imidazolidinedione, 5-[(4-chlorophenyl)methylene]-, (Z)-	EtOH	320(4.351)	39-1941-86B
4H-Pyrido[1,2-a]pyrimidine-3-carboxaldehyde, 2-chloro-6-methyl-4-oxo-	EtOH	266s(4.11),272(3.58), 398(4.15)	4-1295-86
$C_{10}H_7ClN_6$			
1H-Imidazo[4,5-d]-1,2,3-triazin-4-amine, N-(4-chlorophenyl)-	EtOH	238(3.84),291(4.07), 313s(3.91)	5-1012-86
$C_{10}H_7Cl_2NO_2$			
Cycloprop[a]indene, 1,1-dichloro-1,1a,6,6a-tetrahydro-2-nitro-	MeCN	217(4.00),230s(--), 278(3.76)	44-2550-86
Cycloprop[a]indene, 1,1-dichloro-1,1a,6,6a-tetrahydro-3-nitro-	MeCN	218(3.98),269(3.72), 313(3.00)	44-2550-86
Cycloprop[a]indene, 1,1-dichloro-1,1a,6,6a-tetrahydro-4-nitro-	MeCN	224s(--),285?(3.97)	44-2550-86
$C_{10}H_7IN_2O$			
1H-Imidazole, 1-(4-iodobenzoyl)-	C_6H_{12}	263(4.19)	9-0556-86
$C_{10}H_7N$			
Pyrrolo[2,1,5-cd]indolizine	EtOH	255(4.5),308(3.8), 390s(3.5),415(3.7)	142-3071-86

Compound	Solvent	$\lambda_{max}(\log \epsilon)$	Ref.
$C_{10}H_7NO_2$			
1,4-Naphthalenedione, 2-amino-	EtOH	233(3.98),266(4.19), 333(3.20),445(3.44)	42-0448-86
$C_{10}H_7NO_2S_2$			
Thiophene, 2-nitro-5-[2-(2-thienyl)-ethenyl]-, (Z)-	MeOH	288(3.02),426(3.20)	73-1127-86
$C_{10}H_7NO_3$			
1H-Indole-2,3-dione, 1-acetyl-	EtOH	232(4.34),263.5(3.94), 269.5(3.93),330(3.51)	104-2163-86
	$CHCl_3$	232(4.35),240(4.26), 247.5(4.12),293(3.76), 374(2.79)	104-2163-86
$C_{10}H_7N_3O$			
4H-Pyrido[1,2-a]pyrimidine-3-carbonitrile, 2-methyl-4-oxo-	pH 1	247s(3.91),253(3.97), 287s(3.72),295(3.77), 331(4.04),353s(3.60)	44-2988-86
	EtOH	251(4.00),258(4.03), 299s(3.86),312(3.91), 356(4.18)	44-2988-86
$C_{10}H_7N_3O_3$			
1H-Imidazole, 1-(4-nitrobenzoyl)-	EtOH	258(4.18)	9-0556-86
$C_{10}H_7N_3O_3S$			
Spiro[benzothiazole-2(3H),5'(2'H)-pyrimidine]-2',4',6'(1'H,3'H)-trione	MeCN	222(4.30),302(3.45)	94-0664-86
$C_{10}H_7N_3S$			
Thiazolo[4,5-b]quinoxaline, 2-methyl-	n.s.g.	384(4.1535)	18-1245-86
$C_{10}H_7N_3S_2$			
Benzenamine, N-(3H-thiazolo[4,3-c]-[1,2,4]thiadiazol-3-ylidene)-	$CHCl_3$	240(3.80),319(4.02)	118-1027-86
$C_{10}H_7N_7O_2$			
1H-Imidazo[4,5-d]-1,2,3-triazin-4-amine, N-(3-nitrophenyl)-	EtOH	235(3.90),285(4.01)	5-1012-86
$C_{10}H_8$			
Bicyclo[6.2.0]deca-1,3,5,7,9-pentaene	C_6H_{12}	275(4.3),280(4.3), 400f(2.6)	164-0007-86
1,3-Cyclopentadiene, 5-(2,4-cyclopentadien-1-ylidene)-	hexane	265s(3.51),278s(3.89), 287(4.24),299(4.55), 313(4.64),342s(2.98), 416(2.45)	33-1644-86
$C_{10}H_8BrF$			
Bullvalene, bromofluoro-	ether	277s(2.32)	24-2339-86
$C_{10}H_8BrN$			
1-Naphthalenamine, 4-bromo-, diazonium salt	acid	382(3.94)	61-0891-86
$C_{10}H_8BrN_5S$			
Acetonitrile, [[4-amino-5-(4-bromophenyl)-4H-1,2,4-triazol-3-yl]thio]-	EtOH	266(4.44)	4-1451-86

Compound	Solvent	$\lambda_{max}(\log \epsilon)$	Ref.
$C_{10}H_8Br_2O_2$			
Dicyclopropa[cd,gh]pentalene-1,2-dione, 1a,2a-dibromodihydro-2c,2d-dimethyl-	EtOH	291(1.79)	33-0071-86
$C_{10}H_8Br_2O_4$			
1,4-Pentalenedione, 3,6-dibromo-3a,6a-dihydro-2,5-dihydroxy-3a,6a-dimethyl-, cis	EtOH	264(3.24),321(3.00), 339(3.09),346(3.08), 370s(2.81)	33-0071-86
1,6-Pentalenedione, 3,4-dibromo-3a,6a-dihydro-2,5-dihydroxy-3a,6a-dimethyl-, cis	EtOH	274(4.19),346(3.05)	33-0071-86
$C_{10}H_8ClN$			
1-Naphthalenamine, 4-chloro-, diazonium salt	acid	375(3.93)	61-0891-86
$C_{10}H_8ClNO_4$			
2-Propenoic acid, 3-chloro-3-(4-nitrophenyl)-, methyl ester E:Z 6:4	EtOH	285(3.94)	44-4112-86
$C_{10}H_8ClN_5S$			
Acetonitrile, [[4-amino-5-(4-chlorophenyl)-4H-1,2,4-triazol-3-yl]thio]-	EtOH	263(4.07)	4-1451-86
$C_{10}H_8FN_3OS_2$			
1,2,4-Triazine-3,5(2H,4H)-dione, 6-[[(2-fluorophenyl)methyl]thio]-	pH 1	318(3.95)	80-0881-86
	pH 13	352(3.73)	80-0881-86
	EtOH	318(3.90)	80-0881-86
1,2,4-Triazine-3,5(2H,4H)-dione, 6-[[(3-fluorophenyl)methyl]thio]-	pH 1	318(3.90)	80-0881-86
	pH 13	352(3.72)	80-0881-86
	EtOH	318(3.88)	80-0881-86
1,2,4-Triazine-3,5(2H,4H)-dione, 6-[[(4-fluorophenyl)methyl]thio]-	pH 1	318(3.91)	80-0881-86
	pH 13	352(3.79)	80-0881-86
	EtOH	318(3.90)	80-0881-86
$C_{10}H_8FN_3S_3$			
1,2,4-Triazine-3,5(2H,4H)-dithione, 6-[[(2-fluorophenyl)methyl]thio]-	pH 1	305(4.43)	80-0881-86
	pH 13	290(4.28)	80-0881-86
	EtOH	304(4.46)	80-0881-86
1,2,4-Triazine-3,5(2H,4H)-dithione, 6-[[(3-fluorophenyl)methyl]thio]-	pH 1	305(4.41)	80-0881-86
	pH 13	290(4.28)	80-0881-86
	EtOH	304(4.44)	80-0881-86
1,2,4-Triazine-3,5(2H,4H)-dithione, 6-[[(4-fluorophenyl)methyl]thio]-	pH 1	305(4.40)	80-0881-86
	pH 13	290(4.21)	80-0881-86
	EtOH	304(4.38)	80-0881-86
$C_{10}H_8F_2$			
Cycloprop[f]indene, 1,1-difluoro-1,3,4,5-tetrahydro-	hexane	199(3.65),214(3.61), 271(3.37),277(3.29)	33-1546-86
$C_{10}H_8F_5N$			
Ethenamine, N,N-dimethyl-2-(pentafluorophenyl)-, (E)-	MeOH	224(3.64),311(4.41)	44-3244-86
$C_{10}H_8F_6N_4O_4$			
2,4(1H,3H)-Pteridinedione, 7,8-dihydro-7-hydroxy-8-(2-hydroxyethyl)-6,7-bis(trifluoromethyl)-	pH 1	225(4.00),267(4.12), 340(3.81)	39-1051-86C
	pH 13	250(4.09),275(3.87), 345(3.85)	39-1051-86C

Compound	Solvent	λ_{max}(log ϵ)	Ref.
$C_{10}H_8MgN_6O_4$			
Magnesium, bis(N-nitro-2-pyridinaminato-N^1,O^2)-	H_2O	200s(--),265(4.10), 288s(3.98),325s(3.85)	56-0429-86
$C_{10}H_8N_2$			
1,2-Diazabenzo[a]cyclopropa[cd]pentalene, 2a,2b,6b,6c-tetrahydro-	MeCN	337(2.45)	35-4111-86
$C_{10}H_8N_2O$			
3H,5H-Dipyrrolo[1,2-c:2',1'-f]pyrimidin-3-one	THF	397(4.22)	54-0360-86
1H-Imidazole, 1-benzoyl-	EtOH?	225(4.0),267(2.87), 272(2.95),278(2.86)	9-0556-86
1H-Imidazole-4-carboxaldehyde, 2-phenyl-	MeOH	286(4.28)	87-0261-86
$C_{10}H_8N_2OS$			
Propanenitrile, 3-(2-benzoxazolylthio)-	EtOH	280(0.04),287(4.04)	103-0456-86
$C_{10}H_8N_2O_2$			
2,4-Imidazolidinedione, 5-(phenylmethylene)-, (E)-	EtOH	329(4.193)	39-1941-86B
(Z)-	EtOH	315(4.092)	39-1941-86B
$C_{10}H_8N_2O_2S_2$			
Pyridine, 2,2'-dithiobis-, 1,1'-dioxide	EtOH	243(4.58),270(4.18)	39-0039-86C
	MeCN	247(4.54),281(4.07)	39-0039-86C
$C_{10}H_8N_2O_2Se_2$			
3-Pyridinol, 2,2'-diselenobis-	MeOH	215(4.24),235s(--), 300(4.04)	39-2075-86C
$C_{10}H_8N_2O_7$			
2-Benzofuranol, 2,3-dihydro-5,7-dinitro-, acetate	n.s.g.	204(3.95),238(3.99), 283(4.21)	44-3453-86
$C_{10}H_8N_2S_2$			
Propanenitrile, 3-(2-benzothiazolylthio)-	EtOH	276(4.08)	103-0456-86
Pyridine, 2,2'-dithiobis-	MeOH	236(4.16),280(3.96)	142-2403-86
2-Thiazolecarbothioamide, N-phenyl-	MeOH	315(4.15),353s(3.88), 441s(2.32)	103-1139-86
$C_{10}H_8N_4$			
Cyanamide, (2,3-dimethyl-2,5-cyclohexadiene-1,4-diylidene)bis-	MeCN	320s(4.35),330(4.39), 343s(4.35),398s(3.75)	5-0142-86
Cyanamide, (2,5-dimethyl-2,5-cyclohexadiene-1,4-diylidene)bis-	MeCN	330s(4.45),344(4.49), 363s(4.31)	5-0142-86
Cyanamide, (2,6-dimethyl-2,5-cyclohexadiene-1,4-diylidene)bis-	MeCN	322s(4.40),338(4.51), 351(4.47),380s(3.45)	5-0142-86
Dipyrido[1,2-b:1',2'-e][1,2,4,5]-tetrazine	CH_2Cl_2	270(4.4),310f(4.0), 425f(3.3),640f(2.08)	33-1521-86
Propanedinitrile, [[(6-methyl-2-pyridinyl)amino]methylene]- (or tautomeric ring structure)	pH 1	229(3.75),266s(3.62), 271(3.70),312s(4.09), 319(4.10)	44-2988-86
Propanedinitrile, [1-(2-pyridinylamino)ethylidene]-	EtOH	266(3.78),274.5(3.76), 328(3.88),394(4.07)	44-2988-86
	pH 1	253s(4.04),260(4.11), 297(3.65),309(3.70), 347s(4.07),355(4.11)	44-2988-86

Compound	Solvent	λ_{max}(log ϵ)	Ref.
Pyrazine, 2,2'-(1,2-ethenediyl)bis-, cis	hexane	240s(3.9),295(4.2)	35-1006-86
trans	hexane	260(4.1),320(4.2)	35-1006-86
$C_{10}H_8N_4O_2$			
Cyanamide, (2,5-dimethoxy-2,5-cyclohexadiene-1,4-diylidene)bis-	MeCN	260(3.69),302s(3.57), 360(4.50),382s(4.33)	5-0142-86
$C_{10}H_8N_4O_4S$			
2-Thiazolamine, 4-methyl-N-(2,4-dinitrophenyl)-	n.s.g.	382(3.99)	150-0110-86S
2-Thiazolimine, dihydro-3-methyl-N-(2,4-dinitrophenyl)-	n.s.g.	434(4.27)	150-0110-86S
$C_{10}H_8N_4S$			
Propanedinitrile, (5,6-dihydro-2-methylthieno[2,3-d]pyrimidin-4-yl)-	$CHCl_3$	284(3.96),317(3.93), 377(4.11)	94-0516-86
Thiazolo[4,5-b]quinoxalin-7-amine, 2-methyl-	n.s.g.	376(4.05)	18-1245-86
$C_{10}H_8N_6$			
1H-Imidazo[4,5-d]-1,2,3-triazin-4-amine, N-phenyl-	EtOH	233(3.81),288(3.99), 310s(3.81)	5-1012-86
$C_{10}H_8N_6O_3$			
[1,2,4]Triazolo[5,1-c][1,2,4]triazin-4-ol, 1,4-dihydro-3-nitro-7-phenyl-	H_2O	256(3.00),340(3.92)	103-0543-86
$C_{10}H_8O$			
1H-Inden-1-one, 7-methyl-	EtOH	233(4.50),239s(4.47), 328(3.38),344s(3.31)	33-0560-86
2-Propyn-1-one, 1-(2-methylphenyl)-	EtOH	207(4.08),266(4.08)	33-0560-86
$C_{10}H_8O_2$			
Furan, 2,2'-(1,2-ethenediyl)bis-	EtOH	318(3.67)	142-1291-86
2(5H)-Furanone, 4-phenyl-	CH_2Cl_2	270(4.25)	35-3005-86
	+ BF_3	273(4.42)	35-3005-86
	+$EtAlCl_2$	278(4.33)	35-3005-86
$C_{10}H_8O_3$			
2-Benzofurancarboxaldehyde, 6-methoxy-	EtOH	247(3.10),325(3.45)	4-1715-86
$C_{10}H_8O_3S$			
2-Naphthalenesulfonic acid	pH 3	225(5.76),273(4.67)	160-0979-86
$C_{10}H_8O_4$			
2-Benzofurancarboxylic acid, 6-methoxy-	EtOH	236(3.64),272(3.92), 300(4.13),305(4.15)	4-1347-86
3(2H)-Furanone, 2-[1-(3-hydroxy-2-furanyl)ethylidene]-, (E)-	EtOH	230(3.82),257(3.56)	136-0107-86A
4,5-Isobenzofurandione, 6-methoxy-3-methyl- (albidin)	MeOH	240(4.02),350(3.27), 434(3.50)	39-1145-86C
$C_{10}H_8S_2$			
Thiophene, 2,2'-(1,2-ethenediyl)bis-, (E)-	$CHCl_3$	350(4.4)	142-2639-86
$C_{10}H_9BrClNO_2$			
Butanamide, 2-bromo-N-(4-chlorophenyl)-3-oxo-	benzene	205(3.30),260(4.15), 286(3.89)	80-0869-86
	MeOH	204(3.20),253(4.10),	80-0869-86

Compound	Solvent	$\lambda_{max}(\log \epsilon)$	Ref.
(cont.)		278(3.85)	80-0869-86
	EtOH	206(3.21),254(4.10), 281(3.86)	80-0869-86
	PrOH	206(3.25),255(4.11), 283(3.89)	80-0869-86
	BuOH	205(3.27),256(4.12), 284(3.88)	80-0869-86
	dioxan	205(3.30),260(4.15), 286(3.89)	80-0869-86
	HOAc	200(3.21),252(4.07), 282(3.86)	80-0869-86
	CH_2Cl_2	201(3.24),250(4.10), 281(3.87)	80-0869-86
	$CHCl_3$	202(3.24),256(4.12), 284(3.86)	80-0869-86
	CCl_4	203(3.26),259(4.12), 285(3.88)	80-0869-86
	MeCN	205(3.31),258(4.14), 284(3.90)	80-0869-86
	DMF	202(3.28),265(4.10), 288(3.87)	80-0869-86
	pyridine	212(3.09),305(3.77)	80-0869-86
	DMSO	201(3.24),259(4.06), 288(3.88)	80-0869-86
Butanamide, 4-bromo-N-(4-chlorophen-yl)-3-oxo-	benzene	204(3.26),255(4.11), 282(3.84)	80-0869-86
	MeOH	201(3.20),251(4.07), 278(3.82)	80-0869-86
	EtOH	203(3.20),252(4.08), 279(3.83)	80-0869-86
	PrOH	204(3.24),253(4.09), 280(3.84)	80-0869-86
	BuOH	202(3.26),254(4.09), 281(3.84)	80-0869-86
	dioxan	206(3.28),256(4.12), 291(3.85)	80-0869-86
	HOAc	200(3.20),251(4.23), 278(3.82)	80-0869-86
	CH_2Cl_2	201(3.22),252(4.08), 281(3.83)	80-0869-86
	$CHCl_3$	202(3.23),256(4.08), 283(3.83)	80-0869-86
	CCl_4	203(3.25),258(4.10), 284(3.83)	80-0869-86
	MeCN	206(3.30),254(4.11), 285(3.86)	80-0869-86
	DMF	201(3.24),265(4.03), 191(3.83)	80-0869-86
	pyridine	209(3.04),309(3.74)	80-0869-86
	DMSO	202(3.21),260(4.02), 287(3.84)	80-0869-86
$C_{10}H_9BrN_4O$			
1H-Tetrazolium, 1-[2-(4-bromophenyl)-2-oxoethyl]-3-methyl-, hydroxide, inner salt	H_2O	373(4.20)	39-1157-86C
1H-Tetrazolium, 1-methyl-, 2-(4-bromo-phenyl)-2-oxoethylide	H_2O	378(4.39)	39-1157-86C
$C_{10}H_9BrN_4OS$			
Acetamide, N-[3-(4-bromophenyl)-1,5-	EtOH	253(4.29),288s(4.03)	4-1451-86

Compound	Solvent	$\lambda_{max}(\log \epsilon)$	Ref.
dihydro-5-thioxo-4H-1,2,4-triazol-4-yl]- (cont.)			4-1451-86
$C_{10}H_9BrO_2$			
Tricyclo[5.3.0.0^{2,5}]dec-8-ene-3,10-dione, 9-bromo-	EtOH	199(3.38),245(3.70)	78-1903-86
$C_{10}H_9Br_3OTe$			
Tellurium, tribromo(1-methyl-3-oxo-3-phenyl-1-propenyl-C,O)-	CH_2Cl_2	305(4.32)	44-1692-86
$C_{10}H_9ClN_2$			
Quinazoline, 2-(chloromethyl)-4-methyl-	EtOH	242(3.20),272(3.00), 308(2.90),318(2.80)	4-1263-86
$C_{10}H_9ClOS$			
3-Buten-2-one, 4-chloro-4-(phenylthio)-	n.s.g.	267(4.51),290(4.80)	48-0539-86
$C_{10}H_9ClO_2$			
2-Propenoic acid, 3-(4-chlorophenyl)-, methyl ester, (E)-	CH_2Cl_2	285(4.39)	35-3005-86
	+ BF_3	317(4.45)	35-3005-86
(Z)-	CH_2Cl_2	276(4.10)	35-3005-86
	+ BF_3	305(4.23)	35-3005-86
$C_{10}H_9Cl_3OTe$			
Tellurium, trichloro(1-methyl-3-oxo-3-phenyl-1-propenyl-C,O)-	CH_2Cl_2	308(4.26)	44-1692-86
$C_{10}H_9FN_4OS$			
1,2,4-Triazin-3(2H)-one, 5-amino-6-[[(2-fluorophenyl)methyl]thio]-	pH 1	308(3.60)	80-0881-86
	pH 13	314(3.78)	80-0881-86
	EtOH	308(3.58)	80-0881-86
1,2,4-Triazin-3(2H)-one, 5-amino-6-[[(3-fluorophenyl)methyl]thio]-	pH 1	308(3.61)	80-0881-86
	pH 13	314(3.77)	80-0881-86
	EtOH	308(3.59)	80-0881-86
1,2,4-Triazin-3(2H)-one, 5-amino-6-[[(4-fluorophenyl)methyl]thio]-	pH 1	308(3.62)	80-0881-86
	pH 13	314(3.76)	80-0881-86
	EtOH	308(3.59)	80-0881-86
$C_{10}H_9N$			
Benzeneacetonitrile, α-ethylidene-, (E)-	C_6H_{12}	243.2(4.04)	88-6225-86
(Z)-	C_6H_{12}	255.1(4.02)	88-6225-86
Quinoline, 2-methyl-	aq dioxan	314f(3.8)	116-2932-86
	acid	314(4.0)	116-2932-86
$C_{10}H_9NO$			
2-Naphthalenol, 5-amino-, diazonium salt	acid	430(3.83)	61-0891-86
6H-1,2-Oxazine, 3-phenyl-	hexane	233(4.25),308(3.25)	108-0033-86
1H-Pyrrole, 2-[2-(2-furanyl)ethenyl]-, (E)-	EtOH	203(3.89),330(4.52), 342(4.45)	118-0620-86
(Z)-	EtOH	226(4.14),329(4.39), 345(4.32)	118-0620-86
$C_{10}H_9NO_2$			
2H-1-Benzopyran-2-one, 7-amino-4-methyl-	MeOH	231(4.11),271(2.82), 293(3.43),348(4.23)	151-0069-86A
	$CHCl_3$	333(4.13)	151-0069-86A

Compound	Solvent	λ_{max}(log ε)	Ref.
6H-1,4-Dioxino[2,3-f]indole, 2,3-dihydro-	EtOH	220(4.25),280(3.74), 308(3.12)	103-1311-86
7H-1,4-Dioxino[2,3-e]indole, 2,3-dihydro-	EtOH	222(4.39),280(3.65), 294(3.87),305(3.76)	103-1311-86
Ethanone, 1-(3-hydroxy-1H-indol-2-yl)-	EtOH	208(4.19),239(4.16), 262(3.98),321(4.25), 358(3.88)	104-2186-86
1H-Indene-1,2(3H)-dione, 7-methyl-, 2-oxime	EtOH	279(4.26)	54-0188-86
	EtOH-NaOH	272(3.92),325(4.25)	54-0188-86
$C_{10}H_9NO_3$			
2-Benzofurancarboxamide, 6-methoxy-	EtOH	220(4.15),270(4.03), 277(4.08),300(4.27), 307(4.27)	4-1347-86
1H-Indole-2-carboxylic acid, 5-hydroxy-, methyl ester	pH 1	202(4.83),212(4.81), 295(4.75)	151-0209-86C
	pH 7	212(4.97),287(4.78), 294(--),348(4.21)	151-0209-86C
	H_0-5	192(4.7),200(--), 262(3.92),330(--)	151-0209-86C
	H_0-10	200(4.75),236(4.33), 331(4.55),362(--)	151-0209-96C
	MeOH	208(4.28),215(4.25), 287(--),296(4.21), 329(3.59)	151-0209-86C
	MeCN	209(4.35),215(4.34), 187(4.24),295(4.28), 329(3.78)	151-0209-86C
	C_6H_{12}	190(--),256(--), 267(--),287(--), 302(--),317(--)	151-0209-86C
	ether	202(--),260(--), 271(--),307.5(--)	151-0209-86C
$C_{10}H_9NO_3S$			
Propanoic acid, 3-(2-benzoxazolylthio)-	EtOH	250(3.90),280(4.0), 287(4.20)	103-0456-86
$C_{10}H_9NO_3S_2$			
2-Propenoic acid, 2-cyano-3-[2-methoxy-5-(methylthio)-3-thienyl]-	EtOH	223(3.90),277(4.19), 365(4.20)	70-1470-86
compd. with 2-aminophenol	EtOH	204(4.71),228(4.06), 283(4.31),365(4.23)	70-1470-86
$C_{10}H_9NO_4$			
Formamide, N-[2-(1,3-benzodioxol-5-yl)-2-oxoethyl]-	MeOH	229(4.27),272(3.85), 307(3.91)	44-3374-86
4H-Furo[3,2-b]pyrrole-5-carboxylic acid, 2-formyl-, ethyl ester	MeOH	343(3.55)	73-0106-86
$C_{10}H_9NO_5$			
Benzenepropanoic acid, 4-nitro-β-oxo-, methyl ester	EtOH	243(4.21),311(3.89)	44-4112-86
$C_{10}H_9NS$			
1H-Pyrrole, 2-[2-(2-thienyl)ethenyl]-, (E)-	EtOH	248(3.68),344(4.38)	118-0620-86
(Z)-	EtOH	248(3.86),342(4.08)	118-0620-86
$C_{10}H_9N_3O$			
Methanone, (1-methyl-1H-imidazol-	EtOH	236(3.77),306(4.06)	94-4916-86

Compound	Solvent	$\lambda_{max}(\log \epsilon)$	Ref.
2-yl)-4-pyridinyl- (cont.)			94-4916-86
Propanedinitrile, [[5-(dimethyl-amino)-2-furanyl]methylene]-	MeOH	234(3.17),466(3.83)	73-0879-86
4-Pyrimidinamine, N-hydroxy-6-phenyl-	n.s.g.	248(4.54),300(3.66)	103-1232-86
$C_{10}H_9N_3O_4$			
Pyrido[2,3-d]pyrimidine-2,4(1H,3H)-dione, 6-(acetoxymethyl)-	H_2O	212(4.59),246(4.04), 309(3.84)	87-0709-86
$C_{10}H_9N_5O_3$			
1H-Tetrazolium, 1-methyl-, 2-(4-nitro-phenyl)-2-oxoethylide	H_2O	393(4.30)	39-1157-86C
1H-Tetrazolium, 3-methyl-1-[2-(4-nitrophenyl)-2-oxoethyl]-, hydroxide, inner salt	H_2O	390(4.18)	39-1157-86C
$C_{10}H_9N_5O_5$			
1,2,4-Triazine-3,5(2H,4H)-dione, 2-[3-(3,4-dihydro-2,4-dioxo-1(2H)-pyrimidinyl)-2-oxopropyl]-	H_2O	264(4.18)	107-1177-86
$C_{10}H_9N_5S$			
Acetonitrile, [(4-amino-5-phenyl-4H-4H-1,2,4-triazol-3-yl)thio]-	EtOH	263(4.13)	4-1451-86
$C_{10}H_9N_7O_2S$			
1,2,4-Triazino[5,6-g]pteridine-6,8(7H,10H)-dione, 7,10-di-methyl-3-(methylthio)-	MeCN	365(3.88)(changing to)	150-0382-86S
		345(4.14),510(3.52)	150-0382-86S
reduced	MeCN	358(4.23),420s(--)	150-0382-86S
$C_{10}H_{10}$			
1,3-Cyclopentadiene, 5-(2-cyclopen-ten-1-ylidene)-	hexane	290s(--),302s(4.27), 312(4.33),322s(4.25), 338s(3.89),386(2.46)	33-1644-86
$C_{10}H_{10}BrNO_2$			
Butanamide, 2-bromo-3-oxo-N-phenyl-	benzene	206(3.31),255(4.13), 286(3.86)	80-0869-86
	MeOH	199(3.23),250(4.08), 278(3.84)	80-0869-86
	EtOH	200(3.24),252(4.09), 281(3.85)	80-0869-86
	PrOH	201(3.26),254(4.10), 283(3.85)	80-0869-86
	BuOH	202(3.28),256(4.12), 285(3.86)	80-0869-86
	dioxan	205(3.33),256(4.14), 285(3.86)	80-0869-86
	HOAc	201(3.21),251(4.01), 280(3.82)	80-0869-86
	CH_2Cl_2	202(3.26),250(4.08), 282(3.84)	80-0869-86
	$CHCl_3$	202(3.28),252(4.10), 281(3.85)	80-0869-86
	CCl_4	203(3.29),254(4.11), 282(3.85)	80-0869-86
	MeCN	207(3.32),257(4.13), 285(3.86)	80-0869-86
	pyridine	210(3.11),306(3.76)	80-0869-86

Compound	Solvent	$\lambda_{max}(\log \epsilon)$	Ref.
Butanamide, 2-bromo-3-oxo-N-phenyl- (cont.)	DMF	206(3.28),266(4.03), 287(3.81)	80-0869-86
	DMSO	204(3.24),262(4.10), 285(3.75)	80-0869-86
Butanamide, 4-bromo-3-oxo-N-phenyl-	benzene	204(3.30),252(4.09), 283(3.84)	80-0869-86
	MeOH	204(3.21),248(4.06), 280(3.81)	80-0869-86
	EtOH	203(3.23),249(4.07), 281(3.82)	80-0869-86
	PrOH	202(3.25),250(4.09), 283(3.82)	80-0869-86
	BuOH	203(3.28),251(4.11), 284(3.83)	80-0869-86
	dioxan	203(3.33),253(4.11), 285(3.83)	80-0869-86
	HOAc	200(3.21),249(4.00), 280(3.80)	80-0869-86
	CH_2Cl_2	200(3.24),249(4.06), 281(3.82)	80-0869-86
	$CHCl_3$	201(3.26),251(4.07), 282(3.83)	80-0869-86
	CCl_4	202(3.28),252(4.08), 283(3.83)	80-0869-86
	$C_2H_4Cl_2$	201(3.23),251(4.04), 282(3.81)	80-0869-86
	MeCN	201(3.29),254(4.10), 276(3.83)	80-0869-86
	pyridine	210(3.08),306(3.73)	80-0869-86
	DMF	202(3.24),264(3.97), 289(3.71)	80-0869-86
	DMSO	204(3.22),258(4.06), 282(3.71)	80-0869-86
$C_{10}H_{10}BrN_3O_3S$			
Benzenesulfonamide, 4-bromo-N-(4-diazo-3-oxobutyl)-	MeOH	203(4.31),235(4.28), 271(3.94)	104-0353-86
$C_{10}H_{10}BrN_4O$			
1H-Tetrazolium, 1-[2-(4-bromophenyl)-2-oxoethyl]-3-methyl-, bromide	EtOH-HCl	263(4.24)	39-1157-86C
1H-Tetrazolium, 3-[2-(4-bromophenyl)-2-oxoethyl]-1-methyl-, bromide	EtOH-HCl	263(4.26)	39-1157-86C
$C_{10}H_{10}ClNO_2$			
2,4-Pentadienoic acid, 5-(3-chloro-1H-pyrrol-2-yl)-, methyl ester, (E,E)-	EtOH	365(4.46)	150-3601-86M
$C_{10}H_{10}ClNO_3$			
DL-Phenylalanine, β-(chloromethylene)-3-hydroxy-, (E)-	H_2O	237s(3.59),279(3.27)	130-0103-86
$C_{10}H_{10}ClNO_3S$			
3-Pyrrolidinone, N-[(4-chlorophenyl)-sulfonyl]-	MeOH	203(4.16),235(4.12)	70-0120-86
$C_{10}H_{10}ClN_3O_2$			
Acetic acid, (6-chloro-1-methyl-4(1H)-pyrimidinylidene)cyano-, ethyl ester	EtOH	354(4.49)	103-0774-86

Compound	Solvent	$\lambda_{max}(\log \epsilon)$	Ref.
$C_{10}H_{10}ClN_3O_3S$			
Benzenesulfonamide, 4-chloro-N-(4-diazo-3-oxobutyl)-	MeOH	204(4.37),239(4.39), 271(3.94)	70-0120-86 +104-0353-86
$C_{10}H_{10}FNO$			
1H-Indole-3-ethanol, 6-fluoro-	MeOH	219(4.36),282(3.63)	158-1495-86
4H-1,2-Oxazine, 4-fluoro-5,6-dihydro-3-phenyl-	hexane	207(4.22),255(3.99)	108-0033-86
$C_{10}H_{10}FNO_3$			
DL-Phenylalanine, β-(fluoromethylene)-3-hydroxy-, (E)-	H_2O	239s(3.66),279(3.28)	130-0103-86
(Z)-	H_2O	239s(3.64),280(3.25)	130-0103-86
DL-Tyrosine, β-(fluoromethylene)-, (E)-	H_2O	228s(3.92),280(3.06)	130-0103-86
$C_{10}H_{10}FeO_5$			
Iron, tricarbonyl[(η^4-2-methylene-3-butenyl)acetate]-	MeCN	282s(2.40)	101-0061-86D
$C_{10}H_{10}MnNO_3$			
Manganese, [(1,2,3,4,5-η)-1-(1-aminoethyl)-2,4-cyclopentadien-1-yl]tricarbonyl-	EtOH	330(3.05)	101-0173-86D
$C_{10}H_{10}NO_2S$			
Benzothiazolium, 3-(carboxymethyl)-2-methyl-, chloride	EtOH	278(3.82)	4-0209-86
$C_{10}H_{10}N_2O$			
1H-Cyclopenta[e]pyrrolo[1,2-c]pyrimidin-1-one, 2,3,4,5-tetrahydro-	THF	235(4.18),269(3.86), 316(3.70)	54-0360-86
Dipyrrolo[1,2-c:2',1'-f]pyrimidin-3-one, 1,2-dihydro-	THF	302(4.11)	54-0360-86
$C_{10}H_{10}N_2OS$			
Formamide, N-(3-cyano-4,5,6,7-tetrahydrobenzo[b]thien-2-yl)-	EtOH	217(4.11),253(3.74), 261(3.72),305(3.90)	11-0347-86
4(1H)-Quinazolinone, 2,3-dihydro-1,3-dimethyl-2-thioxo-	EtOH	219(4.55),292(3.74)	103-1001-86
4(1H)-Quinazolinone, 1-methyl-2-(methylthio)-	EtOH	203(4.12),254(4.16)	103-1001-86
4(3H)-Quinazolinone, 3-methyl-2-(methylthio)-	EtOH	232(4.17),280(3.86)	103-1001-86
$C_{10}H_{10}N_2OS_2$			
3(2H)-Benzothiazolepropanamide, 2-thioxo-	EtOH	326(4.40)	103-0564-86
Propanamide, 3-(2-benzothiazolylthio)-	EtOH	230(4.35),280(4.14)	103-0456-86
$C_{10}H_{10}N_2O_2$			
Benzofuro[2,3-d]pyrimidin-4(3H)-one, 5,6,7,8-tetrahydro-	EtOH	253(3.96),290(3.86)	11-0347-86
Formamide, N-(3-cyano-4,5,6,7-tetrahydro-2-benzofuranyl)-	EtOH	279(3.92)	11-0347-86
$C_{10}H_{10}N_2O_2S$			
Propanamide, 3-(2-benzoxazolylthio)-	EtOH	250(4.10),279(4.12), 287(4.09)	103-0456-86

Compound	Solvent	λ_{max}(log ε)	Ref.
$C_{10}H_{10}N_2O_3$			
Benzoic acid, 4-[(acetylhydrazono)-methyl]-	pH 4.73	296(4.44)	69-5774-86
$C_{10}H_{10}N_2O_4$			
2,3-Quinoxalinedimethanol, 1,4-dioxide	H_2O	240(4.37),260(4.54), 375(4.08)	70-0733-86
	EtOH	235(4.11),265(4.18), 380(3.76)	70-0733-86
	$CHCl_3$	267(4.13),387(3.81)	70-0733-86
$C_{10}H_{10}N_2O_5$			
4H-Furo[3,2-b]pyrrole-5-carboxylic acid, 4-methyl-2-nitro-, ethyl ester	MeOH	382(3.56)	73-0106-86
$C_{10}H_{10}N_2S_3Ti$			
Titanium, bis[η^5-2,4-cyclopentadien-1-yl)(thiosulfeno)sulfur diimidato(2-)]-	CH_2Cl_2	324(3.70),378(3.70), 488(3.30)	157-1395-86
$C_{10}H_{10}N_4$			
Cyanamide, (2,5-dimethyl-1,4-phenylene)bis-	EtOH	250(4.27),274(3.47), 294(3.53)	5-0142-86
[1,2,4]Triazolo[1,5-a]pyridine-8-carbonitrile, 2,5,7-trimethyl-	DMF	338(4.00)	118-0860-86
$C_{10}H_{10}N_4O$			
Benzo[g]pteridin-4(1H)-one, 6,7,8,9-tetrahydro-	pH 13	338(3.91)	33-0793-86
1H-Tetrazolium, 1-methyl-, 3-(2-oxo-2-phenylethylide)	H_2O	376(4.36)	39-1157-86C
	MeOH	386(--)	39-1157-86C
	EtOH	387(--)	39-1157-86C
	isoPrOH	388(--)	39-1157-86C
	acetone	395(--)	39-1157-86C
	CH_2Cl_2	400(--)	39-1157-86C
	$CHCl_3$	400(--)	39-1157-86C
1H-Tetrazolium, 3-methyl-1-(2-oxo-2-phenylethyl)-, hydroxide, inner salt	H_2O	371(4.15)	39-1157-86C
	MeOH	387(--)	39-1157-86C
	EtOH	390(--)	39-1157-86C
	isoPrOH	396(--)	39-1157-86C
	acetone	416(--)	39-1157-86C
	CH_2Cl_2	420(--)	39-1157-86C
	$CHCl_3$	420(--)	39-1157-86C
1H-1,2,4-Triazol-3-amine, N-[(4-methoxyphenyl)methylene]-	dioxan	319(4.23)	80-0273-86
$C_{10}H_{10}N_4OS$			
Acetamide, N-(1,5-dihydro-3-phenyl-5-thioxo-4H-1,2,4-triazol-4-yl)-	EtOH	251(4.32)	4-1451-86
$C_{10}H_{10}N_4O_2$			
Benzo[g]pteridine-2,4(1H,3H)-dione, 6,7,8,9-tetrahydro-	H_2O	245(3.85),325(4.01)	33-0793-86
$C_{10}H_{10}N_4O_5S$			
Benzenesulfonamide, N-(4-diazo-3-oxobutyl)-4-nitro-	MeOH	208(3.95),215s(3.89), 270(4.17)	104-0353-86
$C_{10}H_{10}N_4O_6$			
Acetic acid, [(2,4-dinitrophenyl)-	EtOH	350(4.41)	18-0073-86

Compound	Solvent	$\lambda_{max}(\log \epsilon)$	Ref.
hydrazono]-, (E)- (cont.)			18-0073-86
(Z)-	EtOH	349(4.41)	18-0073-86
$C_{10}H_{10}N_4S_3Ti$			
Titanium, bis(η^5-2,4-cyclopentadien-1-yl)[[N,N"-thiobis[sulfur diimidato]](2-)-N',N"]-	CH_2Cl_2	324(4.00),458(3.60), 584(3.30)	157-1395-86
$C_{10}H_{10}N_5O_3$			
1H-Tetrazolium, 1-methyl-3-[2-(4-nitrophenyl)-2-oxoethyl]-, bromide	EtOH-HCl	263(4.17)	39-1157-86C
1H-Tetrazolium, 3-methyl-1-[2-(4-nitrophenyl)-2-oxoethyl]-, bromide	EtOH-HCl	263(4.16)	39-1157-86C
$C_{10}H_{10}O$			
6(1H)-Azulenone, 2,3-dihydro-	MeOH	232(4.01),324(4.26)	88-6051-86
Cyclopenta[cd]pentalen-1(2H)-one, 2a,4a,5,6b-tetrahydro-	C_6H_{12}	241(3.78)	78-1831-86
Cycloprop[cd]azulen-2(1H)-one, 2a,2b,6a,6b-tetrahydro-	$CDCl_3$	258(3.62)	89-0906-86
1H-Inden-1-one, 2,3-dihydro-7-methyl-	C_6H_{12}	243(4.07),251(4.00), 286(3.26),297(3.30)	54-0188-86
	EtOH	247(4.11),297(3.38)	54-0188-86
Tricyclo[5.3.0.0^{2,8}]deca-3,5-dien-9-one	MeCN	275(3.59)	89-0906-86
$C_{10}H_{10}O_2$			
Dispiro[2.0.2.4]dec-8-ene-7,10-dione	EtOH	224(4.12),275(2.48), 335(2.04),380s(1.28)	78-6487-86
4,7-Methano-1H-indene-5,6-dione, 3a,4,7,7a-tetrahydro-	EtOH	263(2.15),475(1.38)	18-2811-86
2-Propenoic acid, bicyclo[2.2.1]hepta-2,5-dien-7-yl ester	CH_2Cl_2	229(2.64)	18-1501-86
polymer	CH_2Cl_2	228(2.74)	18-1501-86
2-Propenoic acid, 3-phenyl-, methyl ester, (E)-	CH_2Cl_2	277(4.38)	35-5964-86
with BF_3	CH_2Cl_2	313(4.44)	35-3005-86
with BF_3	CH_2Cl_2	313(4.52)	35-3005-86
2-Propenoic acid, 3-phenyl-, methyl	CH_2Cl_2	269(4.03)	35-3005-86
ester, (Z)-	CH_2Cl_2	267(4.03)	35-5964-86
with BF_3	CH_2Cl_2	276(4.24)	35-3005-86
with BF_3	CH_2Cl_2	303(4.08)	35-5964-86
Tetracyclo[3.3.0.0^{2,8}.0^{4,6}]octane-3,7-dione, 1,5-dimethyl-	C_6H_{12}	219(2.88),278s(1.32), 289s(1.44),295s(1.46), 311(1.59),319s(1.57)	33-0071-86
	EtOH	228s(3.00),272s(1.61), 282(1.62),313(1.84)	33-0071-86
Tricyclo[5.3.0.0^{2,5}]dec-8-ene-3,10-dione	EtOH	221(3.85)	78-1903-86
$C_{10}H_{10}O_2S_2$			
Benzeneethane(dithioic) acid, 4-methoxy-α-oxo-, methyl ester	MeOH	297(4.55),496(1.67)	118-0968-86
$C_{10}H_{10}O_3$			
Benzenepropanoic acid, β-oxo-, methyl ester	EtOH	243(4.33),284(3.89)	44-4112-86
1(3H)-Isobenzofuranone, 5-hydroxy-3,6-dimethyl-	EtOH	254(4.09),281(3.79), 288(3.79)	44-4840-86
	EtOH-NaOH	228s(--),307(--)	44-4840-86

Compound	Solvent	λ_{max}(log ε)	Ref.
2-Propenoic acid, 3-(2-hydroxy-phenyl)-, methyl ester	pH 2	322(3.95)	149-0551-86B
	pH 11	380(3.97)	149-0551-86B
	MeOH	317(3.97)	149-0551-86B
2-Propenoic acid, 3-(2-methoxyphenyl)-	pH 2.0	323(3.92)	149-0551-86B
	pH 11	311(3.88)	149-0551-86B
	MeOH	315(3.91)	149-0551-86B
$C_{10}H_{10}O_5$			
2H-Pyran-2-one, 3,5-diacetyl-6-hy-droxy-4-methyl-	C_6H_{12}	320(4.18)	39-0973-86B
	H_2O	283(3.89),355(4.38)	39-0973-86B
	EtOH	280(4.00),360(4.43)	39-0873-86B
	dioxan	322(4.20)	39-0973-86B
	$CHCl_3$	322(4.20)	39-0973-86B
	80% DMSO	280(3.96),361(4.41)	39-0973-86B
	+ acid	330(--)	39-0973-86B
	+ base	282(--),361(--)	39-0973-86B
$C_{10}H_{10}O_6$			
1,3-Benzodioxole-4-carboxylic acid, 6,7-dimethoxy-	EtOH	220(4.01),288(3.86), 322(3.71)	102-1427-86
$C_{10}H_{10}S_2$			
1,3-Benzodithiole, 2-(1-methyleth-ylidene)-	EtOH	240(4.43),257s(3.99), 316(3.50)	33-0419-86
1,3-Dithiolane, 2-(phenylmethylene)-	H_2O	304(4.28)	23-1116-86
$C_{10}H_{10}S_4Se_2$			
1,3-Dithiolo[4,5-b][1,4]dithiin, 2-(4,5-dimethyl-1,3-diselenol-2-ylidene)-5,6-dihydro-	MeCN	208(4.29),313(4.18), 340s(3.78),490(2.81)	77-1472-86
$C_{10}H_{11}BrO_2$			
Phenol, 2-bromo-4,5-dimethyl-, acetate	dioxan	216s(4.04),226s(3.80), 270(2.83),278(2.88)	152-0017-86
Tricyclo[3.3.0.0^{2,8}]octane-3,7-dione, 4-bromo-1,5-dimethyl-, exo	EtOH	292(1.93)	33-0071-86
$C_{10}H_{11}BrO_3$			
1-Butanone, 4-bromo-1-(2,4-dihydroxy-phenyl)-	EtOH	211(4.21),230(3.90), 276(4.13),313(3.85)	24-0050-86
1-Butanone, 4-bromo-1-(3,4-dihydroxy-phenyl)-	EtOH	231(4.15),275(3.96), 307(3.89)	24-0050-86
1-Propanone, 2-bromo-1-(3,4-dihy-droxyphenyl)-2-methyl-	EtOH	200(4.43),232(4.06), 282(3.83),313(3.81)	24-0050-86
$C_{10}H_{11}BrO_4$			
1-Butanone, 4-bromo-1-(2,3,4-tri-hydroxyphenyl)-	EtOH	215(4.18),237(4.04), 293(4.19),343(3.30)	24-0050-86
$C_{10}H_{11}ClN_3O$			
3H-1,2,4-Triazolium, 1-(3-chloro-phenyl)-4,5-dihydro-3,3-dimethyl-5-oxo-, tetrafluoroborate	MeCN	336(3.87)	89-1132-86
3H-1,2,4-Triazolium, 1-(4-chloro-phenyl)-4,5-dihydro-3,3-dimethyl-5-oxo-, tetrafluoroborate	MeCN	364(3.98)	89-1132-86
$C_{10}H_{11}ClO_3$			
1-Butanone, 4-chloro-1-(2,4-di-hydroxyphenyl)-	EtOH	212(4.21),231(3.91), 277(4.14),313(3.86)	24-0050-86

Compound	Solvent	λ max(log ε)	Ref.
1-Butanone, 4-chloro-1-(3,4-di-hydroxyphenyl)-	EtOH	232(4.15),277(3.98), 309(3.89)	24-0050-86
$C_{10}H_{11}ClO_4$			
Benzoic acid, 3-chloro-4-hydroxy-6-methoxy-2-methyl-, methyl ester	MeOH	207(4.38),242(3.64), 287(3.48)	100-0111-86
Benzoic acid, 3-chloro-6-hydroxy-4-methoxy-2-methyl-, methyl ester	MeOH	215(4.38),253(3.94), 305(3.56)	100-0111-86
1-Butanone, 4-chloro-1-(2,3,4-tri-hydroxyphenyl)-	EtOH	215(4.16),238(3.99), 294(4.16)	24-0050-86
$C_{10}H_{11}Cl_2N_5O$			
1H-1,2,4-Triazole-3,5-diamine, N-[2-(2,6-dichlorophenoxy)ethyl]-	EtOH	217s(3.99),270(4.00)	4-0401-86
	10%EtOH-HCl	216s(4.13),270(3.28)	4-0401-86
	10%EtOH-NaOH	266(4.18)	4-0401-86
$C_{10}H_{11}FN_2O_4$			
2-Propenoic acid, 2-methyl-, 2-(5-fluoro-3,4-dihydro-2,4-dioxo-1(2H)-pyrimidinyl)ethyl ester	EtOH	272.5(3.87)	121-1233-86
$C_{10}H_{11}IO_3$			
1-Butanone, 1-(3,4-dihydroxyphenyl)-4-iodo-	EtOH	233(4.35),276(3.98), 308(3.89)	24-0050-86
Ethanone, 1-(2,5-dimethoxyphenyl)-2-iodo-	EtOH	223(4.18),248(3.77), 349(3.55)	24-0050-86
1-Propanone, 1-(4-hydroxy-3-methoxy-phenyl)-3-iodo-	EtOH	227(4.22),279(3.97), 307(3.99)	24-0050-86
$C_{10}H_{11}IO_4$			
1-Butanone, 4-iodo-1-(2,3,4-tri-hydroxyphenyl)-	EtOH	217(4.44),236(4.11), 293(4.19)	24-0050-86
$C_{10}H_{11}NO$			
Acetamide, N-(2-phenylethenyl)-	EtOH	221(4.14),287(4.33)	39-0727-86B
Oxazole, 2,5-dihydro-4-methyl-5-phenyl-	EtOH	250(2.95),258(2.93)	35-6739-86
Oxazole, 4,5-dihydro-2-methyl-5-phenyl-	EtOH	250(2.04),257(2.18)	35-6739-86
Pyridine, 4-(1-cyclopenten-1-yl)-, 1-oxide	EtOH	303(4.3)	94-0007-86
$C_{10}H_{11}NO_2$			
Benzamide, N-(1-oxopropyl)-	EtOH	195.5(3.90),203.1(4.45), 206.4(4.35),236.7(4.30)	104-1602-86
Benzene, 1-(2-methyl-1-propenyl)-4-nitro-	C_6H_{12}	265(3.86)	12-0281-86
Phenol, 2-[3-(methylimino)-1-propen-yl]-, N-oxide	EtOH	312(4.11),357(4.12)	104-1302-86
	dioxan	312(3.64),358(3.94)	104-1302-86
	DMSO	312(3.97),360(4.00)	104-1302-86
$C_{10}H_{11}NO_2S_2$			
1H-Thieno[2,3-e]-1,2-thiazine-3-carb-oxylic acid, 1-methyl-, ethyl ester	EtOH	245(3.93),400(3.81)	39-0483-86C
2H-Thieno[3,2-c]-1,2-thiazine-3-carb-oxylic acid, 2-methyl-, ethyl ester	EtOH	217(3.94),268(3.53), 321(3.88),457(3.64)	39-0483-86C
$C_{10}H_{11}NO_3$			
DL-Phenylalanine, 3-hydroxy-β-methyl-ene-	H_2O	239s(3.80),283(3.26)	130-0103-86

Compound	Solvent	$\lambda_{max}(\log \epsilon)$	Ref.
$C_{10}H_{11}NO_6$			
Benzoic acid, 4-hydroxy-3-nitro-, 2-methoxyethyl ester	MeCN	240(4.59)	96-0209-86
$C_{10}H_{11}N_2$			
Quinoxalinium, 1-ethyl-, iodide	pH 4.0	338(3.90)	103-1118-86
hydrated	pH 10.5	309(3.34),347(3.35)	103-1118-86
$C_{10}H_{11}N_2O_7S$			
2-Propanone, 1-[5-(methylsulfonyl)-2,4-dinitro-2,4-cyclohexadien-1-yl]-, ion(1-), potassium	acetone	370(4.02),539(4.36)	104-1138-86
$C_{10}H_{11}N_3O$			
2-Butanone, 3-amino-1-diazo-4-phenyl-, (R*,R*)-(±)-2,3-dihydroxybutanedioate (2:1)	MeOH	205s(4.32),250(3.98), 280(4.31)	104-0353-86
$C_{10}H_{11}N_3O_2$			
2-Propenamide, 2-cyano-3-[5-(dimethylamino)-2-furanyl]-, (E)-	MeOH	230(3.10),460(3.35)	73-0879-86
$C_{10}H_{11}N_3O_3$			
2-Butenamide, 3-amino-2-nitro-N-phenyl-	EtOH	252(4.15),338(4.32)	18-3871-86
$C_{10}H_{11}N_3O_3S$			
Benzenesulfonamide, N-(4-diazo-3-oxobutyl)-	MeOH	203(4.03),224(4.06), 246(3.97),266(3.94), 272(3.94)	104-0353-86
Benzenesulfonamide, N-(3-diazo-2-oxopropyl)-4-methyl-	MeOH	203(4.29),232(4.21), 267s(3.91)	104-0353-86
$C_{10}H_{11}N_3O_4S$			
Benzenesulfonamide, N-(3-diazo-2-oxopropyl)-4-methoxy-	MeOH	204(4.41),244(4.44), 271(4.05)	104-0353-86
$C_{10}H_{11}N_3O_5$			
Imidazo[1,2-a]pyridine-7-carboxylic acid, 1,2,3,5-tetrahydro-1-methyl-8-nitro-5-oxo-, methyl ester	EtOH	228(4.05),267(3.86), 317(3.80),380(4.30)	24-2208-86
$C_{10}H_{11}N_3O_7$			
2,4(1H,3H)-Pyrimidinedione, 1-(5,6-dideoxy-6-nitro-β-D-ribohex-5-enofuranosyl)-, (E)-	n.s.g.	258(4.07)	111-0111-86
$C_{10}H_{11}N_4O$			
1H-Tetrazolium, 1-methyl-3-(2-oxo-2-phenylethyl)-, bromide	EtOH-HCl	248(4.17)	39-1157-86C
1H-Tetrazolium, 3-methyl-1-(2-oxo-2-phenylethyl)-, bromide	EtOH-HCl	248(4.16)	39-1157-86C
$C_{10}H_{11}N_5O$			
Benzo[g]pteridin-4(1H)-one, 2-amino-6,7,8,9-tetrahydro-	pH 13	249(4.23),358(3.84)	33-0793-86
$C_{10}H_{11}N_5O_2$			
2',3'-Dideoxy-2',3'-secoadenosine-1',3'-diene	H_2O	206(4.23),259(4.00)	87-2445-86
	pH 13	217(3.94),259(3.87)	87-2445-86

Compound	Solvent	$\lambda_{max}(\log \epsilon)$	Ref.
$C_{10}H_{12}$			
Bicyclo[2.2.1]heptane, 2,3,5-tris(methylene)-	isooctane	228s(3.66),242s(3.72), 258s(3.59)	33-1310-86
Tricyclo[5.3.0.0^{2,8}]deca-3,5-diene	hexane	253(2.85),263(3.06), 272(3.26),283(3.39), 295(3.38),309(3.11)	33-1872-86
$C_{10}H_{12}BrNO_2$			
Benzene, 1-(2-bromo-2-methylpropyl)-4-nitro-	C_6H_{12}	262(4.06)	12-0281-86
$C_{10}H_{12}BrN_3O_5$			
1,2,4-Triazine-3,5(2H,4H)-dione, 6-(2-bromoethenyl)-2-(2-deoxy-α-D-erythro-pentofuranosyl)-, (E)-	EtOH	228(4.13),300(4.09)	87-0809-86
	EtOH-NaOH	292(--)	87-0809-86
β-	EtOH	230(4.09),300(4.06)	87-0809-86
	EtOH-NaOH	291(--)	87-0809-86
$C_{10}H_{12}Br_2O_2$			
2,5(1H,3H)-Pentalenedione, 1,4-dibromotetrahydro-3a,6a-dimethyl-, (1α,3aα,4α,6aα)-	EtOH	218(3.07),298(1.84)	33-0071-86
$C_{10}H_{12}ClNO$			
1-Penten-3-one, 1-(3-chloro-1H-pyrrol-2-yl)-4-methyl-, (E)-	EtOH	346(4.35)	150-3601-86M
$C_{10}H_{12}ClNO_2$			
1-Penten-3-one, 1-(3-chloro-1H-pyrrol-2-yl)-4-hydroxy-4-methyl-, (E)-	EtOH	355(4.36)	150-3601-86M
$C_{10}H_{12}ClN_3O_5$			
1,2,4-Triazine-3,5(2H,4H)-dione, 6-(2-chloroethenyl)-2-(2-deoxy-β-D-erythro-pentofuranosyl)-, (E)-	EtOH	224(4.11),296(4.01)	87-0809-86
$C_{10}H_{12}ClN_5$			
1H-1,2,4-Triazole-3,5-diamine, N-[1-(4-chlorophenyl)ethyl]-	EtOH	218(4.17)	4-0401-86
	10%EtOH-HCl	219(4.24)	4-0401-86
$C_{10}H_{12}DN_5O_3$			
9H-Purin-6-amine, 9-(2-deoxy-β-D-erythro-pentofuranosyl-2-C-d)-	EtOH	259(4.14)	78-5427-86
threo	EtOH	259(4.13)	118-0196-86
$C_{10}H_{12}FNO_4S_2$			
4-Thia-1-azabicyclo[3.2.0]hept-2-ene-2-carboxylic acid, 3-[(2-fluoroethyl)thio]-6-(1-hydroxyethyl)-7-oxo-, sodium salt, [5R-[5α,6α(R*)]]-	H_2O	250.7(3.77),320.2(3.88)	158-1551-86
$C_{10}H_{12}IN_5O_3$			
9H-Purin-6-amine, 9-(3-deoxy-3-iodo-xylofuranosyl)-	EtOH	260(4.20)	11-0013-86
$C_{10}H_{12}IN_5O_4$			
Guanosine, 5'-deoxy-5'-iodo-	pH 1	258(4.03),278(3.86)	23-1885-86
	pH 13	260(4.06),266(4.07)	23-1885-86
$C_{10}H_{12}N_2$			
1H-Benzimidazole, 1-ethyl-2-methyl-	C_6H_{12}	214(4.19),253(3.84),	39-1917-86B

Compound	Solvent	λmax(log ε)	Ref.
(cont.)		278(3.69),284(3.76)	39-1917-86B
	pH 2	203(4.27),240(3.60), 268(3.89),276(3.92)	39-1917-86B
	pH 8	207(4.28),250(3.73), 274(3.74),280(3.63)	39-1917-86B
	MeOH	212(4.13),252(3.77), 275(3.71),281(3.74)	39-1917-86B
	ether	216(4.01),253(3.90), 277(3.77),284(3.83)	39-1917-86B
	MeCN	212(4.26),253(3.77), 276(3.68),283(3.69)	39-1917-86B
1H-Benzimidazole, 2,5,6-trimethyl-	pH 2	197(4.53),219(3.97), 273(3.94),282(3.97)	39-1917-86B
	pH 8	201(4.40),245(3.42), 277(3.68),282(3.68), 286(3.67)	39-1917-86B
	NaOH	227(4.06),263(3.78), 286(3.95),289(3.99)	39-1917-86B
	MeOH	208(4.40),246(3.67), 277(3.77),282(3.78), 288(3.79)	39-1917-86B
	ether	216(4.28),246(4.04), 278(4.02),284(4.04), 290(4.04)	39-1917-86B
	MeCN	209(4.54),246(3.89), 278(3.89),284(3.91), 289(3.90)	39-1917-86B
	C_6H_{12}	210(--),246(--), 278(--),285(--), 290(--)	39-1917-86B
Diazene, methyl(2-phenylcyclopropyl)-, cis	C_6D_6	346(1.73)	44-4792-86
trans	$CDCl_3$	342(--)	44-4792-86
Pyrazole, 4,5-dihydro-1-methyl-5-phenyl-	hexane	245(3.60)	44-4792-86
3-Pyridinepentanenitrile	C_6H_{12}	229s(3.30),235s(3.16), 251s(3.22),257(3.29), 262(3.29),267s(3.21), 320.5(2.36),318.5(2.27), 324(2.26)[sic]	150-2049-86M
$C_{10}H_{12}N_2O$			
2-Pyrrolidinone, 5-[1-methyl-1H-pyrrol-2-yl)methylene]-, (E)-	MeOH	292(4.18)	49-0185-86
(Z)-	MeOH	288(3.42)	49-0185-86
$C_{10}H_{12}N_2O_2$			
Imidazo[1,2-a]pyridin-5(1H)-one, 8-acetyl-2,3-dihydro-1-methyl-	EtOH	233(4.08),327(4.23)	24-2208-86
1H,5H-Pyrazolo[1,2-a]pyrazole-1,5-dione, 2,3,6,7-tetramethyl-	H_2O	314.5(4.26)	150-0920-86M
	dioxan	321(4.16)	150-0920-86M
1H,7H-Pyrazolo[1,2-a]pyrazole-1,7-dione, 2,3,5,6-tetramethyl-	H_2O	225(4.15),251(4.04), 385(3.64)	150-0920-86M
	dioxan	229(4.12),356(3.69)	150-0920-86M
$C_{10}H_{12}N_2O_3$			
3-Benzofurancarboxamide, 2-(formylamino)-4,5,6,7-tetrahydro-	EtOH	291(3.83)	11-0347-86
Imidazo[1,2-a]pyridine-8-carboxylic acid, 1,2,3,5-tetrahydro-1-methyl-5-oxo-, methyl ester	EtOH	285(4.18),322(4.07)	24-2208-86

Compound	Solvent	λ_{max}(log ε)	Ref.
Propanoic acid, 3-[(6-methyl-2-pyridinyl)amino]-3-oxo-, methyl ester	EtOH	237(4.08),278(4.00)	4-1295-86
1H-Pyrrole-3-carbonitrile, 1-(2-deoxy-β-D-erythro-pentofuranosyl)-	EtOH	224(4.01)	78-5869-86
$C_{10}H_{12}N_2O_4S$			
2,1-Benzisothiazole-1(3H)-carboxylic acid, 5-amino-, ethyl ester, 2,2-dioxide, monohydrochloride	MeOH	255(4.21),310(3.09)	4-1645-86
$C_{10}H_{12}N_2O_5$			
1,3-Diazabicyclo[4.2.0]octane-2-carboxylic acid, 4-[(methoxycarbonyl)-methylene]-8-oxo-, (2RS,6SR)-	EtOH	282(4.13)	39-1077-86C
$C_{10}H_{12}N_2O_6$			
6,3'-Methanouridine	H_2O	266(4.04)	94-0423-86
$C_{10}H_{12}N_3O$			
3H-1,2,4-Triazolium, 4,5-dihydro-3,3-dimethyl-5-oxo-1-phenyl-, tetrafluoroborate	MeCN	346(3.97)	89-1132-86
$C_{10}H_{12}N_3S$			
S-(1-Methyl-3-indolyl)isothiouronium iodide	dioxan	220(4.33),286(3.89), 360(3.37)	83-0120-86
$C_{10}H_{12}N_4$			
5H-Cycloheptapyrazine-2-carbonitrile, 3-amino-6,7,8,9-tetrahydro-	MeOH	352(3.94)	33-0793-86
2-Quinoxalinecarbonitrile, 3-amino-4,6,7,8-tetrahydro-4-methyl-	EtOH	381(3.52)	33-1025-86
$C_{10}H_{12}N_4O$			
2H-Diimidazo[1,2-a:4',5'-c]pyridin-2-one, 1,3,7,8-tetrahydro-1,3-dimethyl-	H_2O	238(4.35),264(3.36), 315(3.57)	103-0180-86
$C_{10}H_{12}N_4OS$			
Pyridine, 2,6-diacetyl-, mono(thiosemicarbazone)	DMF	323(4.34)	59-0701-86
$C_{10}H_{12}N_4O_2$			
5H-Imidazo[4,5-c]pyridine-5-acetaldehyde, 1,2,3,4-tetrahydro-4-imino-1,3-dimethyl-2-oxo-, monohydrobromide	H_2O	229(4.38),254(3.74), 293(3.95)	103-0180-86
$C_{10}H_{12}N_4O_2S$			
2,4(1H,3H)-Pteridinedione, 1,3,6-trimethyl-7-(methylthio)-	MeOH	223(4.36),263(3.97), 275s(3.90),352(4.28)	33-1095-86
$C_{10}H_{12}N_4O_3$			
2,4,7(1H,3H,8H)-Pteridinetrione, 1,3,6,8-tetramethyl-	EtOH	233(4.49),267(3.75), 295(3.72)	142-2293-86
$C_{10}H_{12}N_4O_3S$			
2,4(1H,3H)-Pteridinedione, 1,3,6-trimethyl-7-(methylsulfinyl)-	MeOH	245(4.18),351(3.95)	33-1095-86

Compound	Solvent	λ_{max}(log ε)	Ref.
$C_{10}H_{12}N_4O_4S$			
2,4(1H,3H)-Pteridinedione, 1,3,6-trimethyl-7-(methylsulfonyl)-	MeOH	246(4.23),349(3.93)	33-1095-86
$C_{10}H_{12}N_4O_5$			
1H-Pyrazolo[3,4-d]pyrimidine-4,6(5H,7H)-dione, 1-(2-deoxy-β-D-erythro-pentofuranosyl)-5,7-dihydro-	MeOH	232(3.82),249(3.87)	5-1213-86
$C_{10}H_{12}N_6O$			
2-Propenamide, N-[2-(6-amino-9H-purin-9-yl)-	H_2O	262(4.09)	47-3201-86
$C_{10}H_{12}N_6O_4$			
2,6-Diamino-8,2'-anhydro-8-hydroxy-9-β-D-arabinofuranosyl-	pH 1	250(4.08),301(4.03)	94-2609-86
	pH 7	249(4.03),284(3.93)	94-2609-86
	pH 13	251(4.00),285(3.93)	94-2609-86
$C_{10}H_{12}O$			
Benzaldehyde, 2,3,4-trimethyl-	EtOH	210(2.17),237(1.99), 285s(--)	102-0505-86
2,4,6-Cycloheptatrien-1-one, 4-(1-methylethyl)-	EtOH	230(4.47),313(4.17)	39-2005-86C
$C_{10}H_{12}O_2$			
3-Cyclopentene-1,2-dione, 4-ethenyl-3,5,5-trimethyl-	$CHCl_3$	325(4.0),496(1.45)	89-1018-86
2(3H)-Furanone, dihydro-3-(1-methylethyl)-5-(2-propynylidene)-	H_2O	232(4.09)	35-5589-86
	base	221(3.91)	35-5589-86
2(5H)-Furanone, 4-(4-methyl-1,3-pentadienyl)-, (E)- (scobinolide)	ether	305(4.18)	150-0102-86S
(Z)-	ether	303(4.09)	150-0102-86S
Tricyclo[3.3.0.0^{2,8}]octane-3,7-dione, 1,5-dimethyl-	C_6H_{12}	209(3.18),293s(1.51), 300(1.55),311(1.49), 322(1.16)	33-0071-86
	EtOH	216(3.29),290(2.01)	33-0071-86
$C_{10}H_{12}O_3$			
1,2-Benzenediol, 3,6-dimethyl-, 2-acetate	MeOH	272(3.16),276(3.16)	33-0469-86
1,3-Benzenediol, 2,4-dimethyl-, 1-acetate	MeCN	271.5(3.12),276.5(3.11)	33-0469-86
1,4-Benzenediol, 2,5-dimethyl-, 4-acetate	MeOH	280(3.42)	33-0469-86
Bicyclo[3.1.0]hex-3-en-2-one, 6-acetoxy-3,6-dimethyl-, (1α,5α,6α)-(±)-	MeOH	224(3.74),260(3.51), 331(2.48)	33-0469-86
	+ CF_3COOH	223(3.78),263(3.51), 323(2.59)	33-0469-86
2-Cyclopenten-1-one, 5-(acetoxyethylidene)-2-methyl-, (Z)-	MeOH	254(4.09)	33-0469-86
Ethanone, 1-(5-hydroxy-4-methoxy-2-methylphenyl)-	EtOH	231(4.29),271(3.94), 312(3.74)	22-0659-86
2-Oxatetracyclo[4.3.1.1^{4,6}.0^{1,4}]undecan-3-one	pentane	218(1.99)	78-1581-86
Phenol, 4-(3-hydroxy-1-propenyl)-2-methoxy-	EtOH	214(4.11),260(3.98), 292(3.55)	102-1701-86
$C_{10}H_{12}O_4$			
Benzoic acid, 2-hydroxy-4-methoxy-6-methyl-, methyl ester	MeOH	215(4.30),260(4.13), 298(3.68)	100-0111-86

Compound	Solvent	λ_{max}(log ε)	Ref.
2,5-Cyclohexadiene-1,4-dione, 2,5-dimethoxy-3,6-dimethyl-	EtOH	285(4.15)	94-1299-86
$C_{10}H_{12}O_4S_2$			
1,3-Dithiole-4,5-dicarboxylic acid, 2-(1-methylethylidene)-, dimethyl ester	EtOH	242(3.43),330s(2.23), 405(2.51)	33-0419-86
2,3-Thiophenedicarboxylic acid, 4,5-dihydro-4,4-dimethyl-5-thioxo-, dimethyl ester	EtOH	228(3.69),280(3.55), 346(4.00)	33-0419-86
$C_{10}H_{13}BrO_2$			
2,5(1H,3H)-Pentalenedione, 1-bromo-tetrahydro-3a,6a-dimethyl-, (1α,3aα,6aα)-	EtOH	210s(2.81),298(1.73), 310s(1.64)	33-0071-86
$C_{10}H_{13}DN_2O_5$			
2,4(1H,3H)-Pyrimidinedione, 1-(2-deoxy-β-D-erythro-pentofuranosyl-2-C-d)-5-methyl-	EtOH	267(3.94)	78-5427-86
$C_{10}H_{13}DO$			
Adamantan-2-one-4-d, (+)-(1R,3S)-4R	EPA	287(1.36)	35-4484-86
	MI	293(1.26)	35-4484-86
4S	EPA	283(1.32)	35-4484-86
	MI	293(1.26)	35-4484-86
$C_{10}H_{13}F_3O_3$			
4-Heptenoic acid, 2-methyl-6-oxo-, 2,2,2-trifluoroethyl ester, (E)-(±)-	MeOH	220.5(4.18)	33-0469-86
$C_{10}H_{13}N$			
Benzenamine, N-methyl-N-2-propenyl-	acid	255f(2.3)	112-1099-86
$C_{10}H_{13}NO$			
Benzenamine, 2-(3-butenyloxy)-	EtOH	236(3.86),286(3.46)	33-0927-86
Benzenamine, 2-[(2-methyl-2-propenyl)oxy]-	EtOH	235(3.89),286(3.47)	33-0927-86
2-Butanone, 1-amino-4-phenyl-, hydrochloride	MeOH	242(3.15),247(2.83), 251(2.28),260(2.35), 262(2.28),266(2.23)	44-3374-86
Isoquinoline, 1,2,3,4-tetrahydro-6-methoxy- (longimammatine)	EtOH-HCl	220s(3.81),226(3.85), 277(3.23),285(3.20)	100-0745-86
Isoquinoline, 1,2,3,4-tetrahydro-7-methoxy- (weberidine)	EtOH-HCl	214(3.8),280(3.4), 288(3.4)	100-0745-86
6-Isoquinolinol, 1,2,3,4-tetrahydro-2-methyl- (longimammosine)	EtOH-HCl	221(3.82),228(3.79), 286(3.23)	100-0745-86
8-Isoquinolinol, 1,2,3,4-tetrahydro-2-methyl- (longimammidine)	pH 13	242(3.94),288(3.52)	100-0745-86
	EtOH	217(3.81),279(3.32)	100-0745-86
	isoPrOH	274(3.26),280(3.26)	100-0745-86
	+ HCl	218(3.80),275(3.32), 280s(3.31)	100-0745-86
Morpholine, N-phenyl-	acid	254f(2.3)	112-1099-86
1-Penten-3-one, 4-methyl-1-(1H-pyrrol-2-yl)-, (E)-	EtOH	350(4.36)	150-3601-86M
2(1H)-Quinolinone, 5,6,7,8-tetrahydro-5-methyl-	EtOH	230.5(3.94),314(3.88)	44-2184-86
$C_{10}H_{13}NOS$			
Benzenecarboximidothioic acid, N-methyl-, ethyl ester, N-oxide, (Z)-	MeOH	273(3.92)	44-5198-86

Compound	Solvent	$\lambda_{max}(\log \epsilon)$	Ref.
Pyridine, 2-[(tetrahydro-2H-pyran-2-yl)thio]-	MeOH	245(3.96),287(3.63)	142-2403-86
$C_{10}H_{13}NO_2$			
3-Buten-2-one, 4-[5-(dimethylamino)-2-furanyl]-, (E)-	MeOH	242(2.88),437(3.41)	73-0879-86
2,5-Cyclohexadiene-1,4-dione, 3,5,6-trimethyl-2-(methylamino)-	CH_2Cl_2	272s(3.90),278(3.98), 504(3.15)	83-0421-86
Formamide, N-(2-hydroxy-1-methyl-2-phenylethyl)-, [R-(R*,S*)]-	MeOH	208(3.8),252(2.1), 258(2.2),264(2.1)	102-2241-86
5(1H)-Indolizinone, 3-ethoxy-2,3-dihydro-	EtOH	231(3.82),307(3.80)	44-2184-86
4,8-Isoquinolinediol, 1,2,3,4-tetrahydro-2-methyl-	EtOH-HCl	216(3.61),279(3.23)	100-0745-86
6,7-Isoquinolinediol, 1,2,3,4-tetrahydro-1-methyl- (salsolinol)	EtOH	225s(3.81),288(3.59)	100-0745-86
1-Penten-3-one, 4-hydroxy-4-methyl-1-(1H-pyrrol-2-yl)-	EtOH	360(4.37)	150-3601-86M
2-Propenoic acid, 2-methyl-3-(1H-pyrrol-2-yl)-, ethyl ester, (E)-	EtOH	320(4.42)	150-3601-86M
(Z)-	EtOH	336(4.20)	150-3601-86M
2(1H)-Pyridinone, 1-(tetrahydro-2H-pyran-2-yl)-	MeOH	228(3.78),302(3.72)	142-2403-86
$C_{10}H_{13}NO_2Se$			
3-Cyclobutene-1,2-dione, 3-(methylseleno)-4-piperidino-	CH_2Cl_2	278(4.2),331(4.2)	24-0182-86
$C_{10}H_{13}NO_3$			
Acetamide, N-(4-hydroxy-3-methoxyphenyl)-N-methyl-	MeOH	235(3.90),283(3.51)	87-1737-86
7,8-Isoquinolinediol, 1,2,3,4-tetrahydro-6-methoxy-	EtOH-HCl	206(4.58),227s(3.84), 271(2.96),280s(2.86)	100-0745-86
$C_{10}H_{13}NO_4$			
2-Furancarboxylic acid, 5-[(1-ethoxyethylidene)amino]-, methyl ester	MeOH	300(3.54)	73-2817-86
$C_{10}H_{13}NO_5$			
2,4-Cyclohexadien-1-ol, 3-methoxy-6-methyl-6-nitro-, acetate, cis	CH_2Cl_2	271.3(2.46)	23-1764-86
$C_{10}H_{13}NO_6Se$			
4-Selenazolecarboxylic acid, 2-β-D-ribofuranosyl-, methyl ester	MeOH	214(4.16),260(3.72)	4-0155-86
$C_{10}H_{13}N_2O_4$			
Pyrazinium, 1-ethyl-2,3-bis(methoxycarbonyl)-, tetrafluoroborate hydrated	pH 1.5	287(3.85)	103-1118-86
	pH 7.0	265(3.83),312(3.85)	103-1118-86
$C_{10}H_{13}N_2O_8P$			
Uridine, 2'-O-(phosphonomethyl)-, intramol. 2',3'-ester	EtOH	223(3.83),253(3.83)	142-0617-86
$C_{10}H_{13}N_3O$			
Furo[2,3-d]pyrrolo[1,2-a]pyrimidin-4(2H)-imine, 3,6,7,8-tetrahydro-2-methyl-	$CHCl_3$	253(3.88),296(3.37)	24-1070-86
5H-Pyrano[2,3-d]pyrrolo[1,2-a]pyrimidin-5-imine, 2,3,4,7,8,9-hexahydro-	$CHCl_3$	253(3.92),263(3.64)	24-1070-86

Compound	Solvent	$\lambda_{max}(\log \epsilon)$	Ref.
$C_{10}H_{13}N_3O_2$			
2-Quinoxalinecarboxylic acid, 3-amino-5,6,7,8-tetrahydro-, methyl ester	EtOH	364(3.94)	33-0793-86
$C_{10}H_{13}N_3O_2S$			
4H-Pyrrolo[2,3-d]pyrimidin-4-one, 3,7-dihydro-3-(methoxymethyl)-5-methyl-2-(methylthio)-	MeOH	280s(3.96),302(4.04)	78-0199-86
$C_{10}H_{13}N_3O_3$			
Benzenamine, N-(2-methyl-2-nitropropyl)-4-nitroso-	heptane-EtOH	280(4.04),303s(--), 390(4.15),715(1.72)	80-0857-86
2-Propenamide, N-[2-(3,4-dihydro-2,4-dioxo-1(2H)-pyrimidinyl)ethyl]-2-methyl-	H_2O	269(3.98)	47-3201-86
polymer	DMSO	268.5(3.93)	47-3201-86
$C_{10}H_{13}N_3O_4$			
2,5-Methano-5H-pyrimido[1,2-c][1,3,5]-oxadiazepin-9(1H)-one, 2,3-dihydro-11-hydroxy-3-(hydroxymethyl)-1-methyl-	MeOH	217.5(4.38),228(4.30), 258s(3.73)	44-4417-86
2,6-Methano-2H,6H-pyrimido[1,2-c]-[1,3,5]oxadiazocin-10(1H)-one, 3,4-dihydro-3,12-dihydroxy-1-methyl-	MeOH	217.5(4.43)	44-4417-86
5-Pyrimidinecarbonitrile, 1,2,3,4-tetrahydro-1,3-bis(methoxymethyl)-6-methyl-2,4-dioxo-	H_2O	217(4.03),276(4.09)	87-0709-86
$C_{10}H_{13}N_3O_8$			
2,4(1H,3H)-Pyrimidinedione, 5-nitro-6-C-[(β-D-ribofuranosyl)methyl]-	pH 1	254s(3.76),276(3.79), 335(3.26)	44-1058-86
	pH 11	268(3.68),273s(3.67), 352(3.78)	44-1058-86
	MeOH	274(3.79),335(3.38)	44-1058-86
$C_{10}H_{13}N_3S$			
4H-Pyrido[1,2-a]thieno[2,3-d]pyrimidin-4-imine, 2,3,6,7,8,9-hexahydro-	$CHCl_3$	248(4.36),310(3.75)	24-1070-86
$C_{10}H_{13}N_3S_4$			
4H-Pyrazol-3-amine, 4-[4,5-bis(methylthio)-1,3-dithiol-2-ylidene)-N,N-dimethyl-	MeCN	300s(3.92),416(4.11)	88-0159-86
tetrafluoroborate	MeCN	232(4.11),250s(3.98), 297(4.01),464(4.25)	88-0159-86
$C_{10}H_{13}N_4$			
Pyrido[2,3-b]pyrazinium, 6-(dimethylamino)-4-methyl-, iodide	pH 6.3	227(4.23),251(4.28), 439(4.20)	103-1118-86
hydrated	pH 13.9	358(4.26)	103-1118-86
$C_{10}H_{13}N_5$			
1H-1,2,4-Triazole-3,5-diamine, 1-methyl-N^5-(2-methylphenyl)-	EtOH	203(4.37),253(4.07)	4-0401-86
	10%EtOH-NaOH	218(4.30),249(4.32)	4-0401-86
1H-1,2,4-Triazole-3,5-diamine, 1-methyl-N^5-(4-methylphenyl)-	EtOH	202(4.29),261(4.29)	4-0401-86
	10%EtOH-NaOH	257(4.24)	4-0401-86
1H-1,2,4-Triazole-3,5-diamine, 1-methyl-N^5-(phenylmethyl)-	EtOH	203(4.12),226s(3.78)	4-0401-86
	10%EtOH-HCl	225s(3.90)	4-0401-86
	10%EtOH-NaOH	219(3.90),225s(3.74)	4-0401-86

Compound	Solvent	λ max(log ε)	Ref.
$C_{10}H_{13}N_5O_3$			
Adenine, 9-[6(R)-(hydroxymethyl)-1,4-dioxacyclohexan-2(R)-yl]-	pH 1	257(4.16)	87-2445-86
	pH 13	259(4.16)	87-2445-86
2,3-Oxetanedimethanol, 4-(6-amino-9H-purin-9-yl)- (oxetanocin)	pH 1	257(4.17)	158-1623-86
	H_2O	259(4.18)	158-1623-86
	pH 13	259(4.18)	158-1623-86
$C_{10}H_{13}N_5O_3S$			
1H-Pyrazolo[3,4-d]pyrimidine-4(5H)-thione, 6-amino-1-(2-deoxy-β-D-erythro-pentofuranosyl)-	MeOH	238(4.20),272(3.83), 336(4.33)	33-1602-86
2H-Pyrazolo[3,4-d]pyrimidine-4(5H)-thione, 6-amino-2-(2-deoxy-β-D-erythro-pentofuranosyl)-	MeOH	240(4.43),282(3.36), 298(3.30),346(3.91)	33-1602-86
$C_{10}H_{13}N_5O_4$			
Guanine, 9-[(2-acetoxyethoxy)methyl]-	MeOH	254(4.15),275s(4.00)	87-1384-86
Guanine, 9-[6(R)-(hydroxymethyl)-1,4-dioxacyclohexan-2(R)-yl]-	pH 1	256(4.11)	87-2445-86
	pH 13	264(4.07)	87-2445-86
Guanosine, 5'-deoxy-	pH 1	257(4.08),278s(3.91)	23-1885-86
	pH 13	257(4.00),267(4.01)	23-1885-86
9H-Purin-6-amine, 9-α-D-xylofuranosyl)-	pH 1	258(--)	87-0203-86
	H_2O	260(4.34)	87-0203-86
	pH 13	260(--)	87-0203-86
β-	EtOH	258(4.16)	87-0203-86
1H-Pyrazolo[3,4-d]pyrimidin-4(5H)-one, 6-amino-1-(2-deoxy-β-D-erythro-pentofuranosyl)-	MeOH	354(4.15)	33-1602-86
2H-Pyrazolo[3,4-d]pyrimidin-4(5H)-one, 6-amino-2-(2-deoxy-β-D-erythro-pentofuranosyl)-	MeOH	224(4.44),240s(3.84), 279(3.82)	33-1602-86
6H-Tetrazolo[1,5-c]pyrimidin-5-one, 6-(2-deoxy-β-D-erythro-pentofuranosyl)-8-methyl-	MeOH	252(3.90)	4-1401-86
$C_{10}H_{13}N_5O_4S$			
6H-Purine-6-thione, 2-amino-1,7-dihydro-7-β-D-xylofuranosyl-	H_2O	223(4.17),352(4.31)	87-0203-86
6H-Purine-6-thione, 2-amino-1,9-dihydro-9-β-D-xylofuranosyl-	H_2O	225(4.11),260(3.89), 341(4.40)	87-0203-86
$C_{10}H_{13}N_5O_5$			
Guanine, 7-hydroxy-, deoxyriboside	pH 1	254(4.04),272s(--)	158-1288-86
	H_2O	234(4.28),267(3.99), 282s(--)	158-1288-86
	pH 13	234(4.31),278(4.01)	158-1288-86
6H-Purin-6-one, 2-amino-1,7-dihydro-7-β-D-xylofuranosyl-	H_2O	244s(--),285(3.82)	87-0203-86
6H-Purin-6-one, 2-amino-1,9-dihydro-9-α-D-xylofuranosyl-	pH 1	255(4.06),278s(3.90)	87-0203-86
	H_2O	252(--),272s(--)	87-0203-86
	pH 13	266(--)	
β-	H_2O	253(4.12),275s(--)	87-0203-86
$C_{10}H_{13}N_5O_6$			
Guanosine 7-oxide	pH 1	260(4.04),280s(3.89)	163-2697-86
	pH 1	260(4.03),280s(--)	158-1288-86
	pH 7	236(4.29),270(3.99)	163-2697-86
	H_2O	236(4.26),268(3.97)	158-1288-86
	pH 13	234(4.23),276(3.94)	158-1288-86
	pH 13	231(4.27),279(3.96)	163-2697-86

Compound	Solvent	$\lambda_{max}(\log \epsilon)$	Ref.
$C_{10}H_{13}N_5O_6S$			
1H-Pyrazino[2,3-c][1,2,6]thiadiazin-4-amine, 1-β-D-ribofuranosyl-, 2,2-dioxide	MeOH	203(3.81),246(3.96), 338(3.48)	5-1872-86
$C_{10}H_{13}O_{10}P$			
1-Cyclohexene-1-carboxylic acid, 5-[(1-carboxyethenyl)oxy]-4-hydroxy-3-(phosphonooxy)-, tetrasodium salt	pH 6.5	240(3.41)	44-0075-86
$C_{10}H_{14}$			
1,3,5,8-Nonatetraene, 5-methyl-	EtOH	246s(--),255(4.51), 264(4.63),275(4.54)	39-1203-86C
1,3,7-Octatriene, 2-methyl-6-methylene-, (E)-	EtOH	224(4.56)	88-2591-86
$C_{10}H_{14}BrN_3O_3$			
2(1H)-Pyrimidinone, 4-amino-5-bromo-1-[3-hydroxy-4-(hydroxymethyl)-cyclopentyl]-, (1α,3β,4α)-(±)-	pH 1	217(4.05),303(4.04)	87-1720-86
	pH 7	217s(--),290(3.88)	87-1720-86
	pH 13	290(3.87)	87-1720-86
$C_{10}H_{14}BrN_3O_4$			
2(1H)-Pyrimidinone, 4-amino-5-bromo-1-[2,3-dihydroxy-4-(hydroxymethyl)-cyclopentyl]-, (1α,2β,3β,4α)-(±)-	pH 1	217(4.05),303(4.05)	87-1720-86
	pH 7	217s(--),291(3.89)	87-1720-86
	pH 13	292(3.89)	87-1720-86
$C_{10}H_{14}ClN_3O_3$			
2(1H)-Pyrimidinone, 4-amino-5-chloro-1-[3-hydroxy-4-(hydroxymethyl)-cyclopentyl]-, (1α,3β,4α)-(±)-	pH 1	218(4.05),302(4.08)	87-1720-86
	pH 7	218(4.08),290(3.90)	87-1720-86
	pH 13	290(3.90)	87-1720-86
$C_{10}H_{14}ClN_3O_4$			
2(1H)-Pyrimidinone, 4-amino-5-chloro-1-[2,3-dihydroxy-4-(hydroxymethyl)-cyclopentyl]-, (1α,2β,3β,4α)-(±)-	pH 1	218(4.01),301(3.98)	87-1720-86
	pH 7	218(4.02),290(3.79)	87-1720-86
	pH 13	290(3.83)	87-1720-86
$C_{10}H_{14}ClN_5O$			
4(1H)-Pteridinone, 2-(chloroamino)-6,7-dihydro-6,6,7,7-tetramethyl-	CH_2Cl_2	246(3.86),318(3.60), 368(3.68)	130-0017-86
$C_{10}H_{14}IN_3O_3$			
2(1H)-Pyrimidinone, 4-amino-1-[2-hydroxy-4-(hydroxymethyl)cyclopentyl]-5-iodo-, (1α,2β,4α)-(±)-	pH 1	223(4.13),313(3.99)	87-1720-86
	pH 7	222(4.19),298(3.81)	87-1720-86
	pH 13	297(3.82)	87-1720-86
2(1H)-Pyrimidinone, 4-amino-1-[3-hydroxy-4-(hydroxymethyl)cyclopentyl]-5-iodo-, (1α,3β,4α)-	pH 1	223(4.14),312(3.99)	87-1720-86
	pH 7	221(4.19),296(3.81)	87-1720-86
	pH 13	297(3.83)	87-1720-86
$C_{10}H_{14}IN_3O_4$			
2(1H)-Pyrimidinone, 4-amino-1-[2,3-dihydroxy-4-(hydroxymethyl)cyclopentyl]-5-iodo-, (1α,2α,3β,4α)-	pH 1	223(4.14),312(4.01)	87-1720-86
	pH 7	222(4.21),296(3.85)	87-1720-86
	pH 13	223(4.19),297(3.85)	87-1720-86
(1α,2β,3β,4α)-	pH 1	223(4.13),313(3.97)	87-1720-86
	pH 7	222(4.18),297(3.81)	87-1720-86
	pH 13	297(3.81)	87-1720-86
$C_{10}H_{14}NO_2$			
Pyridinium, 3-(methoxycarbonyl)-1,2,5-trimethyl-, iodide	H_2O	225(4.25),278(3.76)	5-0765-86

Compound	Solvent	$\lambda_{max}(\log \epsilon)$	Ref.
$C_{10}H_{14}NO_6P$			
Phosphoric acid, diethyl 4-nitrophenyl ester	MeCN	274(4.30)	78-1315-86
$C_{10}H_{14}N_2$			
Piperazine, N-phenyl-, dication	acid	253f(2.4)	112-1099-86
$C_{10}H_{14}N_2O$			
Benzenamine, 4-morpholino-	MeOH	250(4.15),305(3.40)	73-0937-86
$C_{10}H_{14}N_2O_2$			
Phenylalanine, 4-amino-, methyl ester	H_2O	238.0(3.95),285.1(3.13)	149-0159-86B
$C_{10}H_{14}N_2O_2Se$			
Cyclobutenediylium, 1-(dimethylamino)-2-hydroxy-3-morpholino-4-selenyl-, dihydroxide, bis(inner salt)	CH_2Cl_2	247(4.3),381(4.5), 435s(3.0)	24-0182-86
$C_{10}H_{14}N_2O_3$			
2-Furancarboxylic acid, 5-[[[(1-methylethyl)amino]methylene]amino]-, methyl ester	MeOH	330(3.34)	73-2826-86
$C_{10}H_{14}N_2O_3S$			
1H-Pyrrole-3-thiocarboxamide, 1-(2-deoxy-β-D-erythro-pentofuranosyl)-	pH 1	257(3.95),304(3.94)	78-5869-86
	pH 7	257(3.93),304(3.93)	78-5869-86
	pH 11	257(3.95),302(3.86)	78-5869-86
$C_{10}H_{14}N_2O_4$			
2-Pentenedioic acid, 3-amino-2-cyano-, diethyl ester	H_2O	277(4.21)	5-0533-86
	pH 13	312(--)	5-0533-86
2,4(1H,3H)-Pyrimidinedione, 5-(acetoxymethyl)-1,3,6-trimethyl-	EtOH	272(3.98)	118-0857-86
1H-Pyrrole-3-carboxamide, 1-(2-deoxy-β-D-erythro-pentofuranosyl)-	pH 1	232(4.03)	78-5869-86
	pH 7	231(4.03)	78-5869-86
	pH 11	230(4.06)	78-5869-86
$C_{10}H_{14}N_2O_6$			
Butanedioic acid, [1-(methylamino)ethylidene](nitromethylene)-, dimethyl ester	EtOH	322(3.81)	18-3871-86
1-Cyclobutene-1,2-dicarboxylic acid, 3-methyl-3-(methylamino)-4-nitro-, dimethyl ester	EtOH	c.238(c.3.77),354(3.83)	18-3871-86
2,4(1H,3H)-Pyrimidinedione, 5-methyl-1-α-D-xylofuranosyl-	pH 1	268(--)	87-0203-86
	H_2O	269(4.06)	87-0203-86
	pH 13	269(--)	87-0203-86
β-	H_2O	268(3.99)	87-0203-86
$C_{10}H_{14}N_2S$			
2-Pyridinecarbothioamide, N-butyl-	hexane	230(4.04),273(3.88), 323(3.84)	65-2088-86
	CCl_4	274(3.91),325(3.87)	65-2088-86
	$CHCl_3$	275(3.93),319(3.86)	65-2088-86
	12M HCl	269(4.08),329(3.57)	65-2088-86
$C_{10}H_{14}N_4OS$			
9H-Purine, 9-(1-ethoxyethyl)-6-(methylthio)-	EtOH	285(4.23)	103-0331-86

Compound	Solvent	$\lambda_{max}(\log \epsilon)$	Ref.
$C_{10}H_{14}N_4O_2$			
[4,4'-Bi-3H-pyrazole]-3,3'-dione, 2,2',4,4'-tetrahydro-4,4',5,5'-tetramethyl-	EtOH	228(4.17)	150-0920-86M
2H-Imidazo[4,5-c]pyridin-2-one, 1,3,4,5-tetrahydro-5-(2-hydroxyethyl)-4-imino-1,3-dimethyl-, monohydrochloride	H_2O	235(4.74),260(3.61), 304(3.92)	103-0180-86
1H-Purine-2,6-dione, 3,7-dihydro-1,3-dimethyl-8-propyl-	MeOH	273(4.06)	142-1565-86
6H-Purin-6-one, 9-ethyl-1,9-dihydro-1-(3-hydroxypropyl)-	pH 1	253(4.00)	94-1094-86
	pH 7	253(4.00)	94-1094-86
	pH 13	253(4.01)	94-1094-86
	EtOH	254(3.97)	94-1094-86
$C_{10}H_{14}N_4O_3$			
1H-Purine-2,6-dione, 3,7-dihydro-8-(2-hydroxypropyl)-1,3-dimethyl-	MeOH	273(4.06)	142-1565-86
$C_{10}H_{14}N_4O_3S$			
Imidazo[4,5-c]-1,2,6-thiadiazin-4(3H)-one, 7-cyclohexyl-1,7-dihydro-, 2,2-dioxide	MeOH	218(3.81),248(3.32), 293(3.68)	142-3451-86
$C_{10}H_{14}N_4O_4$			
2,5-Methano-5H-pyrimido[1,2-c][1,3,5]-oxadiazepin-9(1H)-one, 2,3-dihydro-11-hydroxy-3-(hydroxymethyl)-1-(methylamino)-	MeOH	223(4.35),270s(3.53)	44-4417-86
2,6-Methano-2H,6H-pyrimido[1,2-c]-[1,3,5]oxadiazocin-10(1H)-one, 3,4-dihydro-3,12-dihydroxy-1-(methylamino)-	MeOH	221.5(4.33)	44-4417-86
$C_{10}H_{14}N_4O_5$			
Histidine, N-acetyl-1-methyl-5-nitro-, methyl ester	H_2O	311(3.85),312(3.89)	4-0921-86
$C_{10}H_{14}N_5O_8P$			
2'-Guanylic acid, disodium salt	D_2O	253(4.13)	46-0328-86
5'-Guanylic acid, disodium salt	D_2O	252(4.14)	46-0328-86
$C_{10}H_{14}N_6$			
1H-Imidazo[4,5-d]-1,2,3-triazin-4-amine, N-cyclohexyl-	EtOH	263(4.02),303(3.79)	5-1012-86
$C_{10}H_{14}N_6O$			
1,2,4-Triazolo[4,3-a]pyrimidin-6-amine, 5-methyl-7-morpholino-	pH 2.0	203(4.36),302s(3.98), 319(4.04)	39-0711-86C
	pH 7.2	235(4.32),288s(4.31), 304(3.99)	39-0711-86C
$C_{10}H_{14}N_6O_4$			
9H-Purine-2,6-diamine, 9-β-D-arabinofuranosyl-	pH 1	251.8(4.06),290(4.01)	94-2609-86
	pH 7.0	255(3.99),279(4.03)	94-2609-86
	pH 13	255(3.99),279(4.03)	94-2609-86
$C_{10}H_{14}O$			
2,7-Cyclodecadien-1-one	pentane	213(3.53),315(1.53)	33-1872-86
3,7-Cyclodecadien-1-one	MeCN	288(1.53)	33-1872-86

Compound	Solvent	λ_{max}(log ϵ)	Ref.
2-Cyclopenten-1-one, 4,4-dimethyl-2-(2-propenyl)-	EtOH	228(3.89)	94-4939-86
1(3aH)-Pentalenone, 4,5,6,6a-tetrahydro-3,6a-dimethyl-	EtOH	229(4.10)	33-0659-86
Phenol, 4-(1,1-dimethylethyl)-	acid	274(3.20)	3-0639-86
Phenol, 2-methyl-5-(1-methylethyl)-(carvacrol)	acid	275(3.20)	3-0639-86
Phenol, 5-methyl-2-(1-methylethyl)-(thymol)	acid	274(3.30)	3-0639-86
$C_{10}H_{14}O_2$			
Benzene, 1,4-dimethoxy-2,5-dimethyl-	dioxan	230(3.8),294(3.3)	88-5367-86
1,3,5-Cycloheptatriene, 3-(dimethoxymethyl)-	hexane	264(3.61)	24-2889-86
2-Cyclohexene-1,4-dione, 5,5,6,6-tetramethyl-	EtOH	224(3.78),332(1.26)	78-6487-86
2-Cyclopenten-1-one, 4-[(2-methyl-2-butenyl)oxy]-, (E)-	EtOH	210(4.03),314(1.60)	32-0291-86
(Z)-	EtOH	209(3.95),318(1.48)	32-0291-86
2-Cyclopenten-1-one, 4-[(3-methyl-2-butenyl)oxy]-	EtOH	213(3.95),314(1.81)	32-0291-86
2-Cyclopenten-1-one, 2-(2-oxopropyl)-4,4-dimethyl-, (6S)-	EtOH	227(3.88)	94-4947-86
2,4-Decadiyne-1,6-diol, (6S)-	ether	219(2.59),231(2.62), 243(2.59),257(2.37)	150-1348-86M
2(3H)-Furanone, dihydro-3-methyl-5-(3-methyl-1,3-butadienyl)-	MeOH	230(4.5)	163-0813-86
3,5-Octadiene-2,7-dione, 4,5-dimethyl-, (E,E)-	MeCN	277(4.27)	89-1131-86
7-Oxabicyclo[2.2.1]heptane, 1-(methoxymethyl)-2,3,5,6-tetrakis(methylene)-	isooctane	216s(4.09),222(4.22), 229(4.24),239s(3.94), 249s(3.83),263s(3.38)	33-1287-86
	EtOH	215s(4.12),222(4.24), 228(4.24),241s(3.95), 249s(3.85),263s(3.36)	33-1287-86
2,5(1H,3H)-Pentalenedione, tetrahydro-3a,6a-dimethyl-	C_6H_{12}	302f(1.59)	33-0071-86
	EtOH	291(1.61)	33-0071-86
2H-Pyran-2-one, 6-pentyl-	EtOH	222(3.55),299(3.73)	163-2943-86
$C_{10}H_{14}O_3$			
Bicyclo[2.2.1]heptane-2,3-dione, 1-(hydroxymethyl)-7,7-dimethyl-, (1R)-	C_6H_{12}-dioxan	280s(1.56),474(1.61)	39-1781-86C
1-Cyclopentene-1-acetic acid, 3,3-dimethyl-5-oxo-, methyl ester	EtOH	225(4.00)	94-4947-86
4,6-Decadiyne-1,3,8-triol	MeOH	220(3.33),228(3.32), 243(3.24),256(3.03)	94-3465-86
1,7-Dioxaspiro[5.5]undec-2-en-4-one, 2-methyl-	MeOH	260(4.04)	44-4254-86
Ethanone, 1-[4-(1-hydroperoxy-1-methylethyl)-1,3-cyclopentadien-1-yl]-	EtOH	222(3.83),300(3.33)	78-3311-86
$C_{10}H_{14}O_4$			
Ethanol, 2,2'-[1,3-phenylenebis(oxy)]-bis-	H_2O	273(3.28)	60-0291-86
Ethanol, 2,2'-[1,4-phenylenebis(oxy)]-bis-	H_2O	285(3.36)	60-0291-86
2(5H)-Furanone, 3-[2-(1,1-dimethylethoxy)ethenyl]-4-hydroxy-, (Z)-	MeOH	270(3.83),280(3.93)	4-0963-86
$C_{10}H_{14}O_5$			
8-Oxabicyclo[3.2.1]oct-3-en-2-one,	MeOH	219(3.78)	18-0511-86

Compound	Solvent	λ_{max}(log ϵ)	Ref.
1,5,7-trimethoxy- (cont.)			18-0511-86
$C_{10}H_{14}SSi$			
Methanethione, phenyl(trimethyl-silyl)-	ether	678(1.57)	39-0381-86C
$C_{10}H_{15}BrO_2$			
2,4-Hexadienoic acid, 4-bromo-, butyl ester, (Z,E)-	n.s.g.	273(4.08)	104-2230-86
$C_{10}H_{15}N$			
Benzenamine, 4-(1,1-dimethylethyl)-, chromium tricarbonyl complex	$CHCl_3$	216(4.77),318(3.90)	112-0517-86
Benzenamine, N,N-diethyl-	acid	255f(2.3)	112-1099-86
sulfate	0.1N H_2SO_4	254(2.34)	65-0179-86
Benzenamine, 2-methyl-6-(1-methyl-ethyl)-, chromium tricarbonyl complex	$CHCl_3$	217(4.53),320(3.91)	112-0517-86
Benzenamine, N,N,3,5-tetramethyl-, chromium tricarbonyl complex	$CHCl_3$	220(4.70),255s(4.21), 319(3.92)	112-0517-86
Benzenebutanamine	pH 6	260f(2.3)	112-0033-86
	MeOH	260f(2.3)	112-0033-86
$C_{10}H_{15}NO$			
6H-1-Benzazepin-6-one, 1,2,3,4,5,7,8,9-octahydro-	EtOH	330(4.49)	44-1490-86
9H-1-Benzazepin-9-one, 1,2,3,4,5,6,7,8-octahydro-	EtOH	289(4.54)	44-1490-86
$C_{10}H_{15}NO_2$			
Ethanol, 2,2'-(phenylimino)di-	acid	248(2.2),254(2.4), 259(2.3),263(2.1)	112-1099-86
2-Propanol, 1-(2-aminophenoxy)-2-methyl-	EtOH	236(3.88),286(3.47)	33-0927-86
3-Pyridinecarboxylic acid, 1,4-dihy-dro-1,2,5-trimethyl-, methyl ester	MeOH	354(3.59)	5-0765-86
	5% MeOH	363(3.59)	5-0765-86
3-Pyridinecarboxylic acid, 1,2,5,6-tetrahydro-1-methyl-3-(1-methyl-ethylidene)-	pH 7.2	210(0.94),276(0.22)	87-0125-86
$C_{10}H_{15}NO_3$			
2-Azabicyclo[4.2.0]oct-4-en-3-one, 1,5-dimethoxy-2-methyl-	MeOH	215(4.06),243(3.76)	94-3658-86
$C_{10}H_{15}NO_3S$			
Benzenamine, N,N-diethyl-, sulfonic acid derivative	$NaHCO_3$	233(3.16),274(3.72), 292(2.85)	65-0179-86
$C_{10}H_{15}NO_6$			
Cyclopentaneacetic acid, 3-methoxy-2-(nitromethyl)-5-oxo-, methyl ester, trans,cis	MeOH	205(3.67),283(1.68)	104-1865-86
trans,trans	MeOH	204(3.71),283(1.75)	104-1865-86
$C_{10}H_{15}N_2OS$			
1H-Imidazol-1-yloxy, 2,5-dihydro-2,2,5,5-tetramethyl-4-[(methyl-thio)ethynyl]-	n.s.g.	223(4.15),263(4.15)	88-1625-86
$C_{10}H_{15}N_3O_2$			
Acetic acid, cyano(2-imidazolidinyli-	EtOH	208(4.25),254(4.40)	24-2208-86

Compound	Solvent	λ_{max}(log ϵ)	Ref.
dene)-, butyl ester, (E)- (cont.)			24-2208-86
$C_{10}H_{15}N_3O_3$			
1H-Pyrrole-3-carboximidamide, 1-(2-deoxy-β-D-erythro-pentofuranosyl)-	pH 1	237(4.04)	78-5869-86
	pH 7	236(4.03)	78-5869-86
	pH 11	237(4.06)	78-5869-86
$C_{10}H_{15}N_3O_4$			
1H-Pyrazole-4-carboxylic acid, 5-amino-1-(1,3-dioxan-5-yl)-, ethyl ester	MeOH	228(3.93),256(3.90)	73-1512-86
4-Pyrimidinamine, 6-(α-D-ribofuranosylmethyl)-	MeOH	237(3.91),277(3.33)	5-0957-86
β-	MeOH	236(3.91),267(3.33)	5-0957-86
5-Pyrimidinepentanoic acid, 2-amino-1,4-dihydro-6-hydroxy-4-oxo-, methyl ester	50%MeOH-HCl	266.5(4.14)	73-1140-86
	50%MeOH-NaOH	271(4.12)	73-1140-86
1H-Pyrrole-3-carboximidamide, 1-(2-deoxy-β-D-erythro-pentofuranosyl)-N-hydroxy-	pH 1	235(4.09)	78-5869-86
	pH 7	245(4.03)	78-5869-86
	pH 11	244(3.93)	78-5869-86
$C_{10}H_{15}N_3O_4S$			
1H-Imidazole-4-carbothioamide, 5-methyl-1-β-D-ribofuranosyl-	H_2O	266(3.28),303(4.31)	4-0679-86
$C_{10}H_{15}N_3O_6$			
1H-Pyrazole-3-carboxamide, 4-methoxy-5-β-D-ribofuranosyl-	pH 1 and 7	224(4.16)	87-0268-86
	pH 11	224(4.18)	87-0268-86
1H-Pyrazole-5-carboxamide, 1-methyl-3-β-D-ribofuranosyl-4-hydroxy-	pH 1	230(3.91),267(3.75)	87-0268-86
	pH 7	232(3.81),308(3.72)	87-0268-86
	pH 11	237(3.78),309(3.91)	87-0268-86
1H-Pyrazole-4-carboxylic acid, 3-amino-1-β-D-ribofuranosyl-, methyl ester	pH 2	263(3.76)	73-1512-86
	pH 7 and 12	263(3.78)	73-1512-86
$C_{10}H_{15}N_3O_7$			
2,4,6(1H,3H,5H)-Pyrimidinetrione, 5-(2-amino-2-deoxy-β-D-glucopyranosyl)-	H_2O	220(3.79),250(4.26)	136-0053-86M
sodium salt	H_2O	220(3.79),250(4.26)	136-0053-86M
$C_{10}H_{15}N_3O_8$			
2,4(1H,3H)-Pyrimidinedione, 6-amino-5-β-D-glucopyranosyloxy-	pH 7	261(4.02),300(4.02)	1-0806-86
	pH 2	262(--),298(--)	1-0806-86
	pH 13	274(--)	1-0806-86
$C_{10}H_{15}N_5$			
9H-Purin-6-amine, 9-ethyl-N-propyl-	pH 1	265.5(4.26)	94-1094-86
	pH 7	270(4.24)	94-1094-86
	pH 13	270(4.24)	94-1094-86
	EtOH	269(4.23)	94-1094-86
9H-Purin-6-imine, 9-ethyl-1,9-dihydro-1-propyl-, monohydriodide	pH 1 and 7	261(4.11)	94-1094-86
	pH 13	261.5(4.13)	94-1094-86
	EtOH	261(4.10)	94-1094-86
monoperchlorate	pH 1 and 7	261.5(4.10)	94-1094-86
	pH 13	261(4.13)	94-1094-86
	EtOH	261(4.11)	94-1094-86
$C_{10}H_{15}N_5O$			
1-Propanol, 3-[(9-ethyl-9H-purin-6-yl)amino]-	pH 1	266(4.25)	94-1094-86
	pH 7	269.5(4.25)	94-1094-86

Compound	Solvent	λmax(log ε)	Ref.
(cont.)	pH 13	269.5(4.24)	94-1094-86
	EtOH	269(4.21)	94-1094-86
1H-Purine-1-propanol, 6,9-dihydro-6-imino-, monohydrobromide	pH 1	261(4.10)	94-1094-86
	pH 7	261.5(4.09)	94-1094-86
	pH 13	261(4.12)	94-1094-86
	EtOH	261(4.11)	94-1094-86
$C_{10}H_{15}N_5O_4$			
2',3'-Secoguanosine, 5'-deoxy-	pH 1	256(4.08),278s(3.91)	23-1885-86
	pH 13	257(4.02),266(4.03)	23-1885-86
$C_{10}H_{15}N_5O_5$			
6H-Purin-6-one, 2-amino-1,9-dihydro-9-[2-hydroxy-1-[2-hydroxy-1-(hydroxymethyl)ethoxy]ethyl]-, (R)-(2',3'-secoguanosine)	pH 1	255(4.09),277s(3.91)	23-1885-86
	pH 13	257(4.04),266(4.05)	23-1885-86
6H-Purin-6-one, 2-amino-9-[[1,3-dihydroxy-2-(hydroxymethyl)-2-propoxy]methyl]-1,9-dihydro-	pH 1	256(4.08),276s(3.93)	87-1384-86
	pH 13	256(4.05),266(4.05)	87-1384-86
	MeOH	253(4.12),273s(3.97)	87-1384-86
1,2,3-Triazolo[4,5-d][1,3]diazepin-8-ol, 3,4,7,8-tetrahydro-3-β-D-ribofuranosyl-	pH 1	260(3.92)	44-1050-86
	pH 11	275(4.08)	44-1050-86
	MeOH	277(4.00)	44-1050-86
$C_{10}H_{15}N_5S_2$			
Purine-2-thione, 7-(2-ethylthioethyl)-6-(methylamino)-	pH 1	247(4.10),289(4.16)	4-0737-86
	pH 13	253(4.24)	4-0737-86
$C_{10}H_{16}Cl_2N_4O_3Pt$			
Platinum, dichloro(3',5'-diamino-3',5'-dideoxythymidine-$N^{3'}$,$N^{5'}$)-, (SP-4-3)-	pH 2	266(4.02)	87-0681-86
	pH 12	267(3.92)	87-0681-86
$C_{10}H_{16}N_2$			
1,4-Benzenediamine, N,N,N',N'-tetramethyl-, radical cation	MeCN	575(4.20),620(--)	65-1167-86
1,4:6,9-Dimethanopyridazino[1,2-a]-pyridazine, octahydro-, (1α,4α,5β-6α,9α,10β)-, radical cation, nitrate	MeCN	321(2.85)	35-7926-86
$C_{10}H_{16}N_2O$			
1-Pentanone, 2-methyl-1-(1-methyl-1H-imidazol-2-yl)-	EtOH	279(4.20)	94-4916-86
$C_{10}H_{16}N_2O_3$			
Butanoic acid, 2-(1-methyl-2-imidazolidinylidene)-3-oxo-, ethyl ester, (Z)-	EtOH	249(4.24)	24-2208-86
Ethanone, 1-(3a,4,5,6,7,7a-hexahydro-3a-hydroxy-7a-methyl-1H-benzimidazol-2-yl)-, N-oxide	EtOH	294(3.46)	103-0856-86
$C_{10}H_{16}N_2O_4$			
2,4,6(1H,3H,5H)-Pyrimidinetrione, 5-[2-(1-methylethoxy)propyl]-	pH 10.0	269(4.27)	39-1391-86B
2,4,6(1H,3H,5H)-Pyrimidinetrione, 5-(2-propoxypropyl)-	pH 10.0	268(4.26)	39-1391-86B
$C_{10}H_{16}N_2O_4S$			
5-Pyrimidinecarboxylic acid, 4-(1,2-dihydroxyethyl)-1,4-dihydro-6-methyl-2-(methylthio)-, methyl ester	MeOH	226(3.79),307(3.48)	142-0679-86

Compound	Solvent	$\lambda_{max}(\log \epsilon)$	Ref.
$C_{10}H_{16}N_3O$			
Pyrylium, 2-amino-3-cyano-6-phenyl-4-pyrrolidino-, perchlorate	MeCN	222(3.79),257(4.50), 311(4.32)	48-0314-86
$C_{10}H_{16}N_4O_4$			
1H-Imidazole-4-carboximidamide, 5-methyl-1-β-D-ribofuranosyl-, monohydrochloride	H_2O	250(3.46)	4-0679-86
1H-Pyrazole-4-carboxylic acid, 3,5-dinitro-1-(1,3-dioxan-5-yl)-, ethyl ester	MeOH	246(4.17)	73-1512-86
$C_{10}H_{16}N_4O_5$			
1H-Imidazole-4-carboximidamide, N-hydroxy-5-methyl-1-β-D-ribofuranosyl-	H_2O	227(3.62)	4-0679-86
$C_{10}H_{16}N_5O_6P$			
Phosphonic acid, [3-[(2-amino-1,6-dihydro-6-oxo-9H-purin-9-yl)methoxy]-4-hydroxybutyl]-	pH 1	256(4.06),280s(3.90)	87-0671-86
	pH 13	267(3.90)	87-0671-86
$C_{10}H_{16}O$			
Cyclopropanemethanol, 2,2-dimethyl-α-(1-methylene-2-propenyl)-, (R*,R*)-	EtOH	224(4.20)	73-1316-86
(R*,S*)-	EtOH	223(4.21)	73-1316-86
4,7-Octadien-2-ol, 2-methyl-6-methylene-	EtOH	215(4.02),238(3.95)	73-1316-86
2-Propanone, 1-(4-methylcyclohexylidene)-	C_6H_{12}	236(4.13)	35-2691-86
(S)-	C_6H_{12}	236(4.13),329(1.75)	44-2863-86
$C_{10}H_{16}OS_2$			
Cyclohexaneethane(dithioic) acid, α-methyl-3-oxo-, methyl ester	C_6H_{12}	304(3.94),456(1.30)	22-0817-86
$C_{10}H_{16}OSi$			
4-Hexen-1-yn-3-one, 5-methyl-1-(trimethylsilyl)-	EtOH	263(4.16)	33-0560-86
$C_{10}H_{16}O_2$			
Acetic acid, (4-methylcyclohexylidene)-, methyl ester	C_6H_{12}	218(4.22)	35-2691-86
9-Decynoic acid	MeOH	206(2.71)	18-3535-86
6-Heptynoic acid, 1-methylethyl ester	MeOH	205(2.23)	18-3535-86
7-Octynoic acid, ethyl ester	MeOH	206(2.31)	18-3535-86
$C_{10}H_{16}O_6$			
D-arabino-Hept-2-enonic acid, 2,3-dideoxy-4,5-O-(1-methylethylidene)-, (E)-	MeOH	227(4.08)	136-0283-86G
$C_{10}H_{17}ClOS$			
1-Penten-3-one, 1-(butylthio)-1-chloro-4-methyl-	n.s.g.	250(4.16),297(4.77)	48-0539-86
$C_{10}H_{17}NO_2$			
1,2-Cyclopentanedione, 3,3,5,5-tetramethyl-, mono(O-methyloxime), (E)-	C_6H_{12}	247(3.95),357(1.40)	54-0054-86
	EtOH	252(3.91),347(1.60)	54-0054-86
(Z)-	C_6H_{12}	269(3.83),404(1.98), 425(1.90)	54-0054-86
	EtOH	268(3.82),400(2.00)	54-0054-86

Compound	Solvent	λ_{max}(log ε)	Ref.
$C_{10}H_{17}NO_2$			
1,2-Propanedione, 1-cycloheptyl-, 1-oxime	EtOH	226(3.96)	118-0473-86
$C_{10}H_{17}NO_4$			
Butanoic acid, 2-[(ethoxycarbonyl)-imino]-3-methyl-, ethyl ester	MeOH	238(2.88),300(1.94)	5-1990-86
Propanedioic acid, [(1-methylethyl)-imino]-, diethyl ester	MeOH	289(2.13)	5-1990-86
$C_{10}H_{17}N_3O_4$			
Propanedioic acid, azido(1-methyl-ethyl)-, diethyl ester	EtOH	228(3.05),270(1.70)	5-1990-86
$C_{10}H_{17}N_6O_{12}P_3$			
Adenosine 5-(tetrahydrogen triphos-phate), monoanhydride with phos-phoramidic acid, trisodium salt	pH 1	258(4.17)	87-0318-86
	pH 6.8	260(4.18)	87-0318-86
$C_{10}H_{18}N_2O_7$			
α-D-Allofuranose, 3-C-[amino(hydroxy-imino)methyl]-1,2-O-(1-methylethyl-idene)-, monohydrochloride	EtOH	207(3.86)	159-0631-86
$C_{10}H_{18}N_4O_3$			
2,4(1H,3H)-Pyrimidinedione, 5-amino-6-[(6-hydroxyhexyl)amino]-	pH 5.0	200(4.46),224s(4.22), 311(4.08),350(3.00)	44-2461-86
	pH 8.0	238(3.74),311(4.22), 350(3.00),500(2.11)	44-2461-86
	pH 12.0	238(3.60),311(4.22), 350(3.30)	44-2461-86
$C_{10}H_{18}N_6O_9$			
2-Propanone, 1-amino-3-diazo-, (R*,R*)-(±)-2,3-dihydroxybut-anedioate (2:1)	MeOH	210(3.71),237(3.73), 277(4.00)	104-0353-86
$C_{10}H_{18}O$			
4,6-Heptadien-2-ol, 2,3,6-trimethyl-, (E)-	EtOH	230(4.35)	73-1316-86
$C_{10}H_{18}O_2S_4$			
Thioperoxydicarbonic acid, bis(2-methylpropyl) ester	isooctane	242(4.25),286(3.94)	3-0588-86
	EtOH	290(3.85)	3-0588-86
$C_{10}H_{19}N_3O_2$			
2-Propenoic acid, 3-(dimethylamino)-2-[(dimethylamino)methylene]-amino]-, ethyl ester, (Z)-	MeOH	211(4.07),281(4.23)	5-1749-86
$C_{10}H_{19}N_3O_3S$			
Glycinethioamide, N-(tert-butoxy-carbonyl)-L-alanyl-	EtOH	267(4.06)	33-1224-86
$C_{10}H_{20}NOS$			
Oxazolidinium, 2-[[(1,1-dimethyleth-yl)thio]methylene]-3,3-dimethyl-, chloride	H_2O	217(3.74),249(3.83)	70-0944-86
Oxazolidinium, 3,3-dimethyl-2-[(but-ylthio)methylene]-, chloride	H_2O	227s(3.77),247(3.83)	70-0994-86

Compound	Solvent	$\lambda_{max}(\log \epsilon)$	Ref.
$C_{10}H_{20}NO_2S$			
Oxazolidinium, 3-(2-hydroxyethyl)-3-methyl-2-[(propylthio)methylene]-, chloride	H_2O	227s(3.60),247(3.54)	70-0994-86
$C_{10}H_{20}N_4O$			
1,2,4-Triazolidinium, 1,1-dimethyl-4-(1-methylethyl)-5-[(1-methylethyl)imino]-3-oxo-, hydroxide, inner salt	CH_2Cl_2	254(2.56)	44-1719-86
1H-1,2,4-Triazolium, 5-(dimethylamino)-2,3-dihydro-1,4-bis(1-methylethyl)-3-oxo-, hydroxide, inner salt	CH_2Cl_2	252(3.68)	44-1719-86
$C_{10}H_{20}O_3Si$			
2-Hexenoic acid, 5-(trimethylsilyloxy)-, methyl ester, (E)-	MeOH	213(4.04)	48-0777-86
$C_{10}H_{21}$			
sec-Decyl	decane	245(3.11)	70-1140-86
$(C_{10}H_{22}Si_2)_n$			
Silane, [[dimethyl(1-methylethyl)-silyl]ethynyl]trimethyl-, polymer	C_6H_{12}	276(2.24)	47-1839-86
$C_{10}H_{24}O_3SSi$			
Silane, triethoxy[(propylthio)methyl]-	heptane	196(2.57),210s(--)	67-0062-86
	MeCN	194(2.43),210s(--)	67-0062-86
$C_{10}H_{26}N_2NiP_2S_2$			
Nickel, bis(P,P-diethyl-N-methylphosphinothioic amidate-N,S)-	toluene	545(1.75),682(1.74), 775(1.69)	24-1569-86

Compound	Solvent	λmax(log ε)	Ref.
$C_{11}H_4Cl_2O_2S$			
Thieno[3,2-b]benzofuran-2-carbonyl chloride, 3-chloro-	MeOH	323(3.47)	73-1685-86
$C_{11}H_4F_6N_2O_2$			
6-Quinoxalinecarboxylic acid, 2,3-bis(trifluoromethyl)-	EtOH	257(4.08),340(3.67)	39-1043-86C
$C_{11}H_5Br_2ClO_3$			
1,4-Naphthalenedione, 2,3-dibromo-5-chloro-8-hydroxy-6-methyl-	$CHCl_3$	240(4.01),262s(3.90), 349(4.05),360(3.06), 420s(3.56),450(3.73), 468s(3.72),500s(3.46)	64-0377-86B
$C_{11}H_5ClO_3S$			
Thieno[3,2-b]benzofuran-2-carboxylic acid, 3-chloro-	MeOH	313(3.33)	73-1685-86
$C_{11}H_5Cl_2N_5O_4$			
1H-Purine-2,6,8(3H)-trione, 7-[(3,5-dichloro-4-oxo-2,5-cyclohexadien-1-ylidene)amino]-7,9-dihydro-, disodium salt	pH 9.2	436(4.20)	83-0081-86
$C_{11}H_5F_6N_3O$			
Pyridinium 4,5-dihydro-4-oxo-2,6-bis(trifluoromethyl)-5(4H)-pyrimidinylide	EtOH	229(4.23),259(3.91), 299.5(3.58)	39-1769-86C
$C_{11}H_6BrNOSe$			
[1,4]Benzoxaselenino[3,2-b]pyridine, 7-bromo-	MeOH	216(4.19),242f(4.21), 305(3.73)	39-2075-86C
$C_{11}H_6Br_2O_4$			
1,4-Naphthalenedione, 2,6-dibromo-5,8-dihydroxy-3-methyl-	$CHCl_3$	312(3.95),495(3.85), 524(3.90),563(3.69)	5-1655-86
1,4-Naphthalenedione, 3,6-dibromo-5,8-dihydroxy-2-methyl-	$CHCl_3$	312(3.95),495(3.86), 524(3.91),563(3.70)	5-1655-86
$C_{11}H_6ClF_8N$			
1H-Inden-5-amine, 1-chloro-1,2,2,3,3-4,6,7-octafluoro-2,3-dihydro-N,N-dimethyl-	heptane	284(4.16)	70-0804-86
$C_{11}H_6ClNOSe$			
[1,4]Benzoxaselenino[3,2-b]pyridine, 7-chloro-	MeOH	225s(--),240f(4.29), 305(3.81)	39-2075-86C
$C_{11}H_6ClN_3O_5$			
Pyridinium, 2-chloro-1-(2-hydroxy-3,5-dinitrophenyl)-, hydroxide, inner salt	EtOH	360(4.08)	5-0947-86
$C_{11}H_6Cl_2F_7N$			
1H-Inden-5-amine, 1,6-dichloro-1,2,2,3,3,4,7-heptafluoro-2,3-dihydro-N,N-dimethyl- (plus isomer)	heptane	243(4.06),290(3.94), 305s(3.82)	70-0804-86
$C_{11}H_6Cl_5NO_2S$			
L-Proline, 5-thioxo-, pentachlorophenyl ester	CH_2Cl_2	230(4.29),269(4.29)	5-0269-86

Compound	Solvent	λ_{max}(log ϵ)	Ref.
$C_{11}H_6F_3NO_2$			
5(4H)-Oxazolone, 4-(phenylmethylene)-2-(trifluoromethyl)-, (E)-	dioxan	235(3.79),240(3.76), 340(4.32)	70-2267-86
$C_{11}H_6N_2O_3Se$			
[1,4]Benzoxaselenino[3,2-b]pyridine, 7-nitro-	MeOH	212(4.33),230s(--), 264s(--),311(3.79), 361(3.71)	39-2075-86C
[1,4]Benzoxaselenino[3,2-b]pyridine, 9-nitro-	MeOH	225(4.38),305(3.91), 415(3.45)	39-2075-86C
$C_{11}H_6O_4$			
Psoralen, 5-hydroxy-	saline	315(4.11)	149-0221-86A
Psoralen, 8-hydroxy-	saline	322(4.00)	149-0391-86A
$C_{11}H_7BrO_4$			
1,4-Naphthalenedione, 2-bromo-5,8-dihydroxy-3-methyl-	$CHCl_3$	297(3.91),460s(3.58), 490(3.79),522(3.85), 562(3.65)	5-1655-86
$C_{11}H_7ClN_4$			
1H-Pyrazolo[3,4-d]pyridazine, 3-(4-chlorophenyl)-	MeOH	240(4.23),274(3.93), 290s(3.87)	39-0169-86C
$C_{11}H_7ClN_4O_2$			
Benzo[g]pteridine-2,4(1H,3H)-dione, 7-chloro-8-methyl-	DMF	324(3.80),383(4.00), 399s(--)	35-0490-86
$C_{11}H_7ClO$			
6-Azulenecarboxaldehyde, 1-chloro-	hexane	294(4.76),341(3.56), 357(3.67),371(2.91), 630s(--),682(2.54), 720(2.47),760(2.47), 805s(--),862(2.09)	5-1222-86
$C_{11}H_7ClO_3$			
1,4-Naphthalenedione, 2-chloro-3-methoxy-	MeOH	245(4.37),250(4.35), 326(3.87),500(3.05)	78-2771-86
$C_{11}H_7Cl_4NO_2$			
2,4-Cyclohexadien-1-one, 3,4,5,6-tetrachloro-2-hydroxy-6-(1-methyl-1H-pyrrol-2-yl)-	EtOH	340(3.92)	142-0579-86
$C_{11}H_7N$			
4-Azulenecarbonitrile	hexane	250(4.36),254(4.52), 282(4.61),287(4.62), 298(4.10),311(3.64), 326(3.10),335(2.20), 341(3.37),351(3.27), 358(3.49),376(2.96), 390s(1.57),517s(1.93), 534s(2.09),557s(2.26), 578(2.39),603(2.46), 628(2.55),657(2.48), 691(2.53),727(2.23), 772(2.23)	5-1222-86
$C_{11}H_7NOS$			
Pyrano[3,2-b]indole-4(5H)-thione	MeOH	204(4.42),221(4.26), 305(3.89),386s(4.34),	83-0108-86

Compound	Solvent	λ_{max}(log ϵ)	Ref.
(cont.)		396(4.37)	83-0108-86
Thiopyrano[2,3-b]indol-4(9H)-one	dioxan	229(4.24),270(4.28), 304(3.90)	83-0120-86
Thiopyrano[3,2-b]indol-4(5H)-one	dioxan	222(4.63),260(4.13), 270(4.16),310(4.35), 350(4.19)	83-0120-86
$C_{11}H_7NOSe$			
[1,4]Benzoxaselenino[3,2-b]pyridine	MeOH	235(4.21),315(3.64)	39-2075-86C
$C_{11}H_7NO_2$			
Pyrano[3,2-b]indol-4(5H)-one	MeOH	208(4.35),242(3.99), 298(4.28)	83-0108-86
$C_{11}H_7NO_2S_2$			
4H-[1]Benzothieno[2,3-e]-1,3-thiazin-4-one, 2-methoxy-	EtOH	215(3.01),255(2.88), 305(2.83)	73-2002-86
$C_{11}H_7NO_3S$			
Thiopyrano[2,3-b]indol-4(9H)-one, 1,1-dioxide	dioxan	237(4.10),275(3.47), 348(3.67)	83-0120-86
Thiopyrano[3,2-b]indol-4(5H)-one, 1,1-dioxide	dioxan	241(3.98),350(3.64)	83-0120-86
$C_{11}H_7NO_4$			
Furo[2,3-g]-1,2-benzisoxazole-7-carboxylic acid, 3-methyl-	EtOH	220(4.3),256(4.5), 310(3.8)	2-0870-86
$C_{11}H_7NS_2$			
Thiopyrano[2,3-b]indol-4(9H)-one	dioxan	228(4.26),279(4.20), 391(4.12)	83-0120-86
Thiopyrano[3,2-b]indol-4(5H)-one	dioxan	238(4.22),324(3.67), 419(4.39),435(4.38)	83-0120-86
$C_{11}H_7N_3OS$			
1,2-Benzisothiazol-3(2H)-one, 2-pyrazinyl-	MeOH	235(3.60),345(4.27)	18-1601-86
$C_{11}H_7N_3O_5$			
Pyridinium, 1-(2-hydroxy-3,5-dinitrophenyl)-, hydroxide, inner salt	EtOH	360(4.11)	5-0947-86
$C_{11}H_8BrN$			
Bullvalenecarbonitrile, bromo-	ether	225s(3.98),237s(3.81), 270s(3.40)	24-2339-86
$C_{11}H_8Br_2O_5$			
1,3-Dioxane-4,6-dione, 5-[(4,5-dibromo-2-furanyl)methylene]-2,2-dimethyl-	EtOH	215(4.15),262(4.42), 380(4.33)	103-1072-86
	acid	380(4.48)	103-1072-86
	base	262(4.66)	103-1072-86
$C_{11}H_8ClF_3N_2O_5$			
Acetamide, N-acetoxy-N-[2-chloro-3-nitro-5-(trifluoromethyl)phenyl]-	EtOH	240s(3.72)	4-1091-86
$C_{11}H_8ClNO_2$			
1H-Indole-2-carboxaldehyde, 1-acetyl-3-chloro-	EtOH	210(4.20),244(3.98), 312(4.24)	104-2186-86

Compound	Solvent	λmax(log ε)	Ref.
$C_{11}H_8ClNO_2S_2$			
Carbamothioic acid, [(3-chloro-2-benzo[b]thien-2-yl)carbonyl]-, O-methyl ester	EtOH	211(3.18),310(3.06)	73-2002-86
$C_{11}H_8ClNO_5$			
2-Propenoic acid, 3-chloro-2-formyl-3-(4-nitrophenyl)-, methyl ester, (E)-	EtOH	250s(3.70),284(3.88)	44-4112-86
(Z)-	EtOH	244(3.84),275(4.22), 335s(3.20)	44-4112-86
$C_{11}H_8ClNO_6$			
Propanedioic acid, [chloro(4-nitrophenyl)methylene]-, monomethyl ester, (Z)-	EtOH	288(3.91)	44-4112-86
Pyrano[3,4-b]pyrrole-2,7-dicarboxylic acid, 3-chloro-1,5-dihydro-5-oxo-, dimethyl ester	EtOH	259(4.5),265(4.6), 293(4.0),325(3.5)	5-1968-86
	dioxan	258(4.8),266(4.8), 297(4.2),325(3.9), 343(3.2)	5-1968-86
$C_{11}H_8ClN_3O_5$			
Acetamide, N-acetoxy-N-(2-chloro-5-cyano-3-nitrophenyl)-	MeOH	223(4.34),300s(3.04)	4-1091-86
$C_{11}H_8F_5NO$			
Benzeneacetaldehyde, α-[(dimethylamino)methylene]-2,3,4,5,6-pentafluoro-	MeOH	283(4.53)	44-3244-86
$C_{11}H_8F_5NO_2$			
2-Propenal, 3-(dimethylamino)-2-(pentafluorophenoxy)-	MeOH	298(4.55)	44-3244-86
$C_{11}H_8N_2$			
1-Naphthalenecarbonitrile, 4-amino-, diazonium salt	acid	362(3.74)	61-0891-86
1H-Perimidine	DMF	321s(4.03),334(4.15), 348(3.91),410s(2.88), 433s(2.49),468s(1.52)	48-0812-86
$C_{11}H_8N_2O$			
Propanedinitrile, [(4-methoxyphenyl)-methylene]-	H_2O	350(4.45)	35-2372-86
	50% DMSO	351(4.44)	35-2372-86
	60% DMSO	351(4.43)	35-2372-86
$C_{11}H_8N_2O_6$			
Phenol, 2-(5-methyl-2-furanyl)-4,6-dinitro-	n.s.g.	210(3.98),238(3.95), 283(4.25)	44-3453-86
$C_{11}H_8N_2S$			
2-Thiophenecarbonitrile, 3-amino-5-phenyl-	MeOH	285(4.22),305(4.05), 335(3.89)	4-1757-86
$C_{11}H_8N_4$			
Pyrazinecarbonitrile, 3-amino-5-phenyl-	EtOH	368(4.11)	33-0793-86
$C_{11}H_8N_4O_2$			
Benzo[g]pteridine-2,4(1H,3H)-dione,	DMF	335(3.92),373(3.99),	35-0490-86

Compound	Solvent	λ_{max}(log ϵ)	Ref.
8-methyl- (cont.)		392s(--)	35-0490-86
$C_{11}H_8N_4O_4$			
3H-Imidazo[4,5-g]quinazoline-4,6,8,9(5H,7H)-tetrone, 2,3-dimethyl-	neutral	245(4.13),312(4.05)	44-4784-86
	anion	232(4.26),262(4.00), 325(4.03),450(3.00)	44-4784-86
$C_{11}H_8N_4S$			
Benzenamine, N-3H-[1,2,4]thiadiazolo-[4,3-a]pyrazin-3-ylidene-	$CHCl_3$	252(4.04),287(4.00), 415(3.48)	118-1027-86
$C_{11}H_8O$			
4-Azulenecarboxaldehyde	hexane	238(4.29),244(4.34), 260(4.16),266(4.18), 288(4.41),295(4.45), 317(3.91),343(3.18), 354(3.29),370(3.29), 390(2.85),592(2.42), 616(2.49),642(2.55), 667(2.52),703(2.51), 738(2.35),787(2.14), 844(1.70)	5-1222-86
6-Azulenecarboxaldehyde	hexane	218(3.98),248(4.07), 286(4.84),324(3.25), 328(3.31),336(3.55), 342(3.49),351(3.76), 356(3.49),365(3.71), 380(1.96),393(1.32), 402(1.15),414(0.78), 581(2.30),604(2.41), 627(2.47),651(2.56), 678(2.50),719(2.53), 756(2.23),810(2.19), 867(1.04)	5-1222-86
$C_{11}H_8O_2$			
1,4-Naphthalenedione, 2-methyl-	EtOH	245(4.27),251(4.26), 264s(4.16),330(3.34)	94-2810-86
$C_{11}H_8O_3$			
1,4-Naphthalenedione, 5-hydroxy-2-methyl- (plumbagin)	EtOH	252s(3.47),268(3.50), 420(3.05)	94-2810-86
Pentacyclo[$6.3.0.0^{2,6}.0^{3,10}.0^{5,9}$]undecane-4,7,11-trione	MeCN	265(1.78),300(2.45), 309(2.44)	78-1797-86
$C_{11}H_8O_4$			
4H-1-Benzopyran-2-carboxylic acid, 4-oxo-, methyl ester	EtOH	250(3.91),315(3.52)	42-0600-86
2H-1-Benzopyran-2-one, 7-acetoxy-	EtOH	263(3.7),325(4.15)	100-0692-86
Cyclopenta[c]pyran-4-carboxylic acid, 7-formyl- (cerbinal)	MeOH	249(4.03),277(3.63), 288(3.63),326(3.42), 428(3.38)	163-2655-86
1,4-Naphthalenedione, 6-hydroxy-2-methoxy-	MeOH	268(4.21),290(4.19), 332(3.45)	100-0122-86
1,4-Naphthalenedione, 7-hydroxy-2-methoxy-	MeOH	263(4.29),289(4.04), 336(3.34)	100-0122-86
2H-Pyran-2,5(6H)-dione, 3-hydroxy-4-phenyl-	dioxan	228.4(4.04),298.8(3.98)	5-0177-86
$C_{11}H_8S_2$			
2,2'-Bithiophene, 5-(1-propynyl)-	EtOH	239(3.75),326(4.29),	163-0565-86

Compound	Solvent	λ max (log ε)	Ref.
(cont.)		333(4.30)	163-0565-86
$C_{11}H_9Br$			
Azulene, 6-bromo-2-methyl-	CH_2Cl_2	282(4.86),291.5(4.95), 311.5(3.73),338.5(3.59), 354(3.72),369.5(3.40), 565.5(2.42),604.5(2.34), 666.5(1.88)	24-2272-86
$C_{11}H_9BrO$			
Bullvalenecarboxaldehyde, bromo-	ether	225(3.93),266(3.58)	24-2339-86
$C_{11}H_9BrO_5$			
1,3-Dioxane-4,6-dione, 5-[(5-bromo-2-furanyl)methylene]-2,2-dimethyl	EtOH	215(3.75),262(2.95), 379(4.43)	103-1072-86
	acid	380(4.44)	103-1072-86
	base	261(4.23)	103-1072-86
$C_{11}H_9ClN_2O_2$			
2,4-Imidazolidinedione, 5-[(4-chlorophenyl)methylene]-1-methyl-, (E)-	EtOH	336(4.304)	39-1941-86B
(Z)-	EtOH	312(4.228)	39-1941-86B
2,4-Imidazolidinedione, 5-[(4-chlorophenyl)methylene]-3-methyl-, (Z)-	EtOH	320(4.228)	39-1941-86B
$C_{11}H_9ClN_2O_3$			
1,3-Benzenedicarbonitrile, 5-chloro-2,4,6-trimethoxy-	MeCN	226(4.8),310(3.5)	24-0844-86
$C_{11}H_9ClN_2S$			
3(2H)-Pyridazinethione, 6-(4-chlorophenyl)-4-methyl-	EtOH	329(3.81)	48-0522-86
$C_{11}H_9ClN_4O$			
Methanone, (5-amino-4-pyridazinyl)(4-chlorophenyl)-, oxime	MeOH	254(4.49)	39-0169-86C
$C_{11}H_9ClO_3$			
2-Propenoic acid, 3-chloro-2-formyl-3-phenyl-, methyl ester E:Z 35:65	EtOH	279(3.96)	44-4112-86
$C_{11}H_9ClO_5$			
1,3-Dioxane-4,6-dione, 5-[(5-chloro-2-furanyl)methylene]-2,2-dimethyl-	EtOH	217(3.90),261(3.88), 370(4.25)	103-1072-86
	acid	370(4.54)	103-1072-86
	base	261(4.20)	103-1072-86
$C_{11}H_9Cl_2NO_2S$			
1H-Pyrrole, 3,4-dichloro-1-[(4-methylphenyl)sulfonyl]-	EtOH	222(4.09),227(4.10), 245(4.11),275s(3.18)	24-0616-86
$C_{11}H_9F_3N_2O_2$			
Imidazo[1,2-a]pyridine-3-carboxylic acid, 2-(trifluoromethyl)-, ethyl ester	EtOH	234(4.43),240.5(4.41), 289(3.99),300(3.98)	39-1769-86C
$C_{11}H_9F_3O_2$			
2-Propenoic acid, 3-[4-(trifluoromethyl)phenyl]-, methyl ester, (E)-	CH_2Cl_2	270(4.35)	35-3005-86
with BF_3	CH_2Cl_2	295(4.46)	35-3006-86
(Z)-	CH_2Cl_2	261(4.03)	35-3006-86

Compound	Solvent	λ max (log ε)	Ref.
with BF_3 (cont.)	CH_2Cl_2	268(4.15)	35-3005-86
$C_{11}H_9NO$			
4-Azulenecarboxaldehyde, oxime	dioxan	230(4.15),283(4.51), 293(4.54),301(4.48), 354(3.46),369(3.31), 386(2.75),586(2.57), 606(2.62),654(2.55), 722(2.11)	5-1222-86
1,4-Oxazepine, 5-phenyl-	EtOH	253(4.08)	77-1188-86
$C_{11}H_9NO_2$			
2-Benzofurancarbonitrile, 6-ethoxy-	EtOH	222(3.18),236(3.15), 269(3.36),275(3.43), 298(3.60),307(3.63)	4-1347-86
1,4-Naphthalenedione, 2-(methylamino)-	CH_2Cl_2	240s(4.0),263s(4.2), 271(4.22),290s(3.9), 330s(3.3),441(3.31)	83-0607-86
$C_{11}H_9NO_2S$			
2-Propenoic acid, 3-(1H-indol-3-ylthio)-, (E)-	dioxan	215(4.73),267(4.26)	83-0120-86
$C_{11}H_9NO_2S_2$			
Carbamothioic acid, benzo[b]thien-2-ylcarbonyl-, O-methyl ester	EtOH	211(3.11),312(3.13)	73-2002-86
$C_{11}H_9NO_3$			
2-Benzofurancarbonitrile, 4,6-dimethoxy-	EtOH	235(4.31),240(4.30), 290(4.43)	4-1347-86
1H-Indole-3-carboxaldehyde, 5,6-(ethylenedioxy)-	EtOH	208(4.30),260(4.06), 291(4.14)	103-1311-86
$C_{11}H_9NO_7$			
1,3-Dioxane-4,6-dione, 2,2-dimethyl-5-[(5-nitro-2-furanyl)methylene]-	EtOH	257(4.32),366(4.09)	103-1072-86
	acid	366(4.46)	103-1072-86
	base	260(4.33)	103-1072-86
$C_{11}H_9N_3$			
8H-Benzo[cd]triazirino[a]indazole, 8-methyl- (85°K) (relative abs. given)	MTHF	289s(0.69),295s(0.80), 301(0.94),309(0.99), 313(1.00),321s(0.63), 325(0.76),328.5(0.76)	89-0828-86
1H-Naphtho[1,8-de]-1,2,3-triazinium, 2-methyl-, hydroxide, inner salt	MTHF	231(4.59),235s(4.56), 251s(3.59),286s(3.65), 300s(3.68),315s(3.74), 329s(3.85),344s(4.06), 351s(4.15),355.5(4.20), 363s(3.95),367s(3.83), 375.5(3.55),384(3.48), 450s(2.38),485s(2.69), 524s(2.88),555(2.97), 598.5(2.98),654(2.82), 720(2.37)	89-0828-86
$C_{11}H_9N_3O_2S$			
4(1H)-Pyrimidinethione, 6-methyl-5-nitro-2-phenyl-	EtOH	c.270(c.4.00),305(4.08)	18-3871-86
$C_{11}H_9N_3O_3S$			
Spiro[benzothiazole-2(3H),5'(2'H)-	MeCN	220(4.30),302(3.30)	94-0664-86

Compound	Solvent	λ_{max}(log ε)	Ref.
pyrimidine]-2',4',6'(1'H,3'H)-trione, 1'-methyl- (cont.)			94-0664-86
$C_{11}H_9N_5O_3$			
3-Pyridinol, 2-amino-4-[(4-nitrophenyl)azo]- (also silver complex)	DMF	500(4.8)	74-0097-86A
$C_{11}H_9N_5S$			
2H-Purine-2-thione, 1,3,6,7-tetrahydro-6-imino-1-phenyl-	pH 1	243(3.20),287(3.49)	4-0737-86
	pH 13	240(4.25),272(4.36), 314(4.20)	4-0737-86
$C_{11}H_{10}BrNO_2$			
1H-Indole-3-propanoic acid, 2-bromo-, sodium salt	H_2O	222(4.48),280(3.85)	35-2023-86
$C_{11}H_{10}ClFN_4$			
4,5-Pyrimidinediamine, 6-chloro-N^4-[(2-fluorophenyl)methyl]-	10%EtOH-HCl	305(4.11)	4-1189-86
	10%EtOH-pH7	263(3.98),268(3.96), 292(3.97)	4-1189-86
	10%EtOH-NaOH	263(3.98),268(3.96), 292(3.97)	4-1189-86
$C_{11}H_{10}ClNO$			
Cyclohepta[b]pyrrol-2(1H)-one, 6-chloro-1,3-dimethyl-	MeOH	269(--),272(--), 276(4.45),279(--), 301s(--),402(3.81), 420(3.80)	44-4087-86
$C_{11}H_{10}ClNO_2$			
1H-Indole-3-propanoic acid, 2-chloro-, sodium salt	H_2O	220(4.49),272(3.84)	35-2023-86
$C_{11}H_{10}ClN_3O_3$			
Acetamide, N-acetoxy-N-(3-amino-2-chloro-5-cyanophenyl)-	EtOH	226(4.46),259(3.85), 335(3.66)	4-1091-86
$C_{11}H_{10}FN_3OS_2$			
1,2,4-Triazin-3(2H)-one, 6-[[(2-fluorophenyl)methyl]thio]-5-(methylthio)-	pH 1	293(3.87)	80-0881-86
	pH 13	335(3.85)	80-0881-86
	EtOH	293(3.85)	80-0881-86
1,2,4-Triazin-3(2H)-one, 6-[[(3-fluorophenyl)methyl]thio]-5-(methylthio)-	pH 1	293(3.83)	80-0881-86
	pH 13	335(3.75)	80-0881-86
	EtOH	293(3.81)	80-0881-86
1,2,4-Triazin-3(2H)-one, 6-[[(4-fluorophenyl)methyl]thio]-5-(methylthio)-	pH 1	293(3.88)	80-0881-86
	pH 13	335(3.78)	80-0881-86
	EtOH	293(3.87)	80-0881-86
$C_{11}H_{10}F_2$			
1H-Cyclopropa[b]naphthalene, 1,1-difluoro-3,4,5,6-tetrahydro-	hexane	198(3.78),218(3.54), 266(3.14),271(3.23), 278(3.13)	33-1546-86
$C_{11}H_{10}FeO_5$			
Iron, tricarbonyl(η^4-methyl 4-methylene-2,5-hexadienoate)-	MeCN	247(4.05)	101-0061-86D
$C_{11}H_{10}MnNO_4$			
Manganese, tricarbonyl[(1,2,3,4,5-η)-1-[1-(formylamino)ethyl]-2,4-cyclopentadien-1-yl]-	EtOH	336(2.91)	101-0173-86D

Compound	Solvent	$\lambda_{max}(\log \epsilon)$	Ref.
$C_{11}H_{10}N_2O$			
Ethanone, 1-(2-quinolinyl)-, oxime, (E)-	EtOH	247.2(4.53)	94-0564-86
1H-Imidazole, 1-(4-methylbenzoyl)-	C_6H_{12}	215(4.1),238(4.16), 280s(3.1)	9-0556-86
1H-Imidazole-4-carboxaldehyde, 1-methyl-2-phenyl-	MeOH	266(4.18)	87-0261-86
1H-Imidazole-5-carboxaldehyde, 1-methyl-2-phenyl-	MeOH	265(4.18)	87-0261-86
2(1H)-Pyrazinone, 6-methyl-5-phenyl-	EtOH	257(4.42),342(4.04)	103-0417-86
$C_{11}H_{10}N_2O_2$			
Benzoic acid, 4-(3-methyl-1H-pyrazol-1-yl)-	toluene	294.5(4.71)	24-1627-86
	EtOH	284(4.22)	24-1627-86
	DMSO	289(4.41)	24-1627-86
Ethanone, 1-(1-isoquinolinyl)-, oxime, N-oxide, (Z)-	EtOH	258.4(4.43)	94-0564-86
Ethanone, 1-(2-quinolinyl)-, oxime, N-oxide, (E)-	EtOH	234.2(4.39)	94-0564-86
Ethanone, 1-(4-quinolinyl)-, oxime, N-oxide, (E)-	EtOH	233.6(4.47)	94-0564-86
(Z)-	EtOH	232.5(4.49)	94-0564-86
1H-Imidazole, 1-(4-methoxybenzoyl)-	C_6H_{12}	214(4.17),262(4.13)	9-0556-86
2,4-Imidazolidinedione, 1-methyl-5-(phenylmethylene)-, (E)-	EtOH	338(4.097)	39-1941-86B
(Z)-	EtOH	308(4.085)	39-1941-86B
2,4-Imidazolidinedione, 3-methyl-5-(phenylmethylene)-, (Z)-	EtOH	316(4.247)	39-1941-86B
2,4-Imidazolidinedione, 5-[(4-methylphenyl)methylene]-, (Z)-	EtOH	320(3.314)	39-1941-86B
1H-Pyrazole-4-carboxaldehyde, 5-hydroxy-3-methyl-1-phenyl-, sodium salt	H_2O	262(4.28)	104-0582-86
2(1H)-Quinoxalinone, 3-(2-oxopropyl)-	MeOH	374(4.18),395(4.31), 417(4.14)	2-0525-86
$C_{11}H_{10}N_2O_2S_2$			
5(4H)-Thiazolethione, 4,4-dimethyl-2-(4-nitrophenyl)-	EtOH	274(4.26),293(4.20), 315s(4.13),480(0.90)	33-0374-86
$C_{11}H_{10}N_2O_3$			
2,4-Imidazolidinedione, 5-[(4-methoxyphenyl)methylene]-, (Z)-	EtOH	331(3.877)	39-1941-86B
2-Quinazolinecarboxylic acid, 1,4-dihydro-4-oxo-, ethyl ester	MeOH	204(4.43),230(4.29), 297(3.97)	33-1017-86
$C_{11}H_{10}N_2S$			
Pyridazine, 3-(methylthio)-6-phenyl-	EtOH	277(3.97)	48-0522-86
$C_{11}H_{10}N_4$			
Cyanamide, (2,3,5-trimethyl-2,5-cyclohexadiene-1,4-diylidene)bis-	MeCN	324s(4.36),341(4.46), 355s(4.40)	5-0142-86
Pyridine, 2-(3-phenyltriazenyl)-	C_6H_{12}	303(3.13),338(3.29), ?s(1.95)	56-0767-86
Pyridine, 3-(3-phenyltriazenyl)-	C_6H_{12}	259(2.53),294(3.00), 310(2.94),353(3.29)	56-0767-86
Pyridine, 4-(3-phenyltriazenyl)-	C_6H_{12}	258(1.90),279s(2.46), 308s(3.09),342(3.28)	56-0767-86
$C_{11}H_{10}N_4O$			
5H,10H-[1,2,4]Triazolo[5,1-b]quinazo-	MeOH	235(4.38),263(3.85),	83-0188-86

Compound	Solvent	λmax(log ε)	Ref.
lin-5-one, 2,10-dimethyl- (cont.)		327(3.84)	83-0188-86
$C_{11}H_{10}N_4O_2S$			
1H-Indole-2,3-dione, 1-acetyl-, 3-[(aminothioxomethyl)hydrazone]	$CHCl_3$	329s(3.74),394(4.28)	104-2173-86
	DMF	330s(3.86),398(4.27)	104-2173-86
1,2,3-Thiadiazole-4-carboxamide, 5-(acetylimino)-2,5-dihydro-2-phenyl-	DMF	274(3.99),380(4.33)	48-0741-86
$C_{11}H_{10}N_4O_2S_2$			
1,2,3-Thiadiazole-4-carboxylic acid, 2,5-dihydro-5-[[(methylamino)thioxomethyl]imino]-2-phenyl-	DMF	422(4.19)	48-0741-86
$C_{11}H_{10}N_4O_3S$			
Imidazo[4,5-c][1,2,6]thiadiazin-4(3H)-one, 1,7-dihydro-7-(phenylmethyl)-, 2,2-dioxide	MeOH	228(3.61),250(3.30), 293(3.62)	142-3451-86
1H-[1,2,6]Thiadiazino[3,4-b]quinoxalin-4(3H)-one, 7,8-dimethyl-, 2,2-dioxide	MeOH	220(4.4),259(4.3), 343(3.59),435(3.56)	83-0079-86
$C_{11}H_{10}N_4O_4S$			
Benzenamine, 2,4-dinitro-N-(3,4-dimethyl-2H-thiazol-2-ylidene)-	n.s.g.	438(4.29)	150-0110-86S
$C_{11}H_{10}N_4S$			
Propanedinitrile, (5,6-dihydro-2,6-dimethylthieno[2,3-d]pyrimidin-4-yl)-	$CHCl_3$	284(4.15),317(4.11), 377(4.30)	94-0516-86
$C_{11}H_{10}N_4S_2$			
Benzenamine, N-(6-ethyl-3H-[1,3,4]thiadiazolo[2,3-c][1,2,4]thiadiazol-3-ylidene)-	$CHCl_3$	240(3.92),317(4.00)	118-1027-86
$C_{11}H_{10}N_6$			
1H-Imidazo[4,5-d]-1,2,3-triazin-4-amine, N-(4-methylphenyl)-	EtOH	233(3.84),290(3.98), 314s(3.81)	5-1012-86
1H-Imidazo[4,5-d]-1,2,3-triazin-4-amine, N-(phenylmethyl)-	EtOH	263(4.02),301(3.81)	5-1012-86
$C_{11}H_{10}O$			
Benzene, 1-ethynyl-2-(2-propenyloxy)-	EtOH	237(4.32),287(3.80), 298(3.76)	44-3125-86
Bicyclo[6.3.0]undeca-2,4,6,11-tetraen-10-one	C_6H_{12}	246(4.25),317(3.77)	164-0007-86
2-Butyn-1-one, 1-(2-methylphenyl)-	EtOH	217s(4.05),240s(3.89), 262(4.14),298s(3.37)	33-0560-86
1H-Inden-1-one, 2,3-dihydro-4-methyl-2-methylene-	EtOH	209s(4.10),230(4.01), 266(4.09),308s(3.48)	33-0560-86
1H-Inden-1-one, 2,3-dihydro-7-methyl-2-methylene-	EtOH	209(4.10),226(3.94), 265(4.14),305s(3.44)	33-0560-86
1H-Inden-1-one, 2,7-dimethyl-	EtOH	235(4.54),241(4.54), 327(3.23),340s(3.18)	33-0560-86
1H-Inden-1-one, 6,7-dimethyl-	EtOH	238(4.47),328(3.29), 396(2.89)	33-0560-86
1H-Inden-1-one, 2-ethylidene-2,3-dihydro-	EtOH	266(4.23),302s(3.47)	33-0560-86
1-Naphthalenol, 3-methyl-	EtOH	236(3.83),300(2.97), 313s(2.84),328(2.69)	94-2810-86

Compound	Solvent	$\lambda_{max}(\log \epsilon)$	Ref.
1-Naphthalenol, 5-methyl-	EtOH	221(4.59),229s(4.54), 297(3.83),309(3.71), 324(3.53)	33-0560-86
1-Naphthalenol, 6-methyl-	EtOH	214(4.58),235(4.51), 296(3.66),311(3.54), 317s(3.40),325(3.41)	33-0560-86
2-Propyn-1-one, 1-(2,3-dimethyl-phenyl)-	EtOH	209(4.22),271(4.06), 313s(3.24)	33-0560-86
2-Propyn-1-one, 1-(2,6-dimethyl-phenyl)-	EtOH	209(4.15),275(3.42)	33-0560-86
Tetracyclo[5.3.0.0^{2,4}.0^{3,5}]deca-6,8,10-triene, 6-methoxy-	hexane	300(4.08),367(2.84)	78-1745-86
Tricyclo[5.3.0.0^{2,5}]deca-3,6,8,10-tetraene, 6-methoxy-	hexane	289(4.08),357(2.98)	78-1745-86
$C_{11}H_{10}OS$			
Thiophene, 2-(4-methoxyphenyl)-	EtOH	267(4.60),292s(4.28), 310s(3.98)	18-0083-86
$C_{11}H_{10}O_2$			
2(3H)-Furanone, dihydro-3-(phenyl-methylene)-, (E)-	CH_2Cl_2	284(4.41)	35-3005-86
with BF_3	CH_2Cl_2	284(4.52),325s(3.68)	35-3005-86
with $EtAlCl_2$	CH_2Cl_2	282(4.43)	35-3005-86
(Z)-		285(4.18)	35-3005-86
with $BtAlCl_2$	CH_2Cl_2	285(4.23)	35-3005-86
1H-Indene-2-carboxylic acid, methyl ester	CH_2Cl_2	285(4.25)	35-3005-86
with BF_3	CH_2Cl_2	287(4.77)	35-3005-86
with $EtAlCl_2$	CH_2Cl_2	289(4.35)	35-3005-86
$C_{11}H_{10}O_2S$			
1-Benzothiepin-5(2H)-one, 3,4-dihydro-4-(hydroxymethylene)-	EtOH	240(4.15),269s(3.68), 327(3.75)	4-0449-86
$C_{11}H_{10}O_3$			
2-Benzofurancarboxaldehyde, 6-ethoxy-	EtOH	247(3.55),328(3.91)	4-1715-86
2H-1-Benzopyran-2-one, 7-ethoxy-	H_2O	324(<u>4.1</u>)	18-0961-86
$C_{11}H_{10}O_3S$			
2,5-Furandione, dihydro-3,3-dimethyl-4-(2-thienylmethylene)-, (E)-	EtOH	290s(3.68),338(4.19)	34-0369-86
2-Pentenoic acid, 4-oxo-2-(phenyl-thio)-	MeOH	249(4.15),270(4.08)	48-0539-86
$C_{11}H_{10}O_4$			
2-Benzofurancarboxaldehyde, 4,6-di-methoxy-	EtOH	249(3.57),330(3.78)	4-1715-86
2-Benzofurancarboxylic acid, 6-eth-oxy-	EtOH	235(3.56),264(3.90), 271(3.94),299(4.11), 304(4.12)	4-1347-86
1,4-Naphthalenedione, 2,3-dihydro-2,8-dihydroxy-3-methyl-	EtOH	212(4.42),260(4.03), 338(3.71)	25-0823-86
2-Propenal, 3-(6-methoxy-1,3-benzo-dioxol-5-yl)-	EtOH	248(4.06),300(3.06), 376(4.23)	100-0690-86
$C_{11}H_{10}O_5$			
1H-2-Benzopyran-5-carboxylic acid, 3,4-dihydro-8-hydroxy-3-methyl-1-oxo-, (R)-	EtOH	228(4.28),248(3.89), 315(3.45)	163-0997-86

Compound	Solvent	$\lambda_{max}(\log \epsilon)$	Ref.
1,3-Dioxane-4,6-dione, 5-(2-furanyl-methylene)-2,2-dimethyl-	EtOH	216(3.78),262(3.04), 364(4.48)	103-1072-86
	acid	364(4.52)	103-1072-86
	base	262(4.30)	103-1072-86
$C_{11}H_{10}S_4Se_2$			
1,3-Dithiolo[4,5-b][1,4]dithiin, 2-(5,6-dihydro-4H-cyclopenta-1,3-diselenol-2-ylidene)-5,6-dihydro-	MeCN	209(4.29),315(4.15), 335s(3.80),480(2.48)	77-1472-86
$C_{11}H_{11}BO_4$			
3-Penten-2-one, 4-(1,3,2-benzodioxaborol-2-yloxy)-, (Z)-	CH_2Cl_2	288(4.15),342s(2.35), 368s(2.33),389s(2.24)	48-0755-86
$C_{11}H_{11}Br$			
Bullvalene, x-bromo-y-methyl-	ether	230s(3.71)	24-2339-86
$C_{11}H_{11}BrN_2O_2$			
L-Tryptophan, 2-bromo-	H_2O	218(4.51),280(3.85), 288(3.76)	35-2023-86
$C_{11}H_{11}BrN_2O_3$			
4-Isoxazolecarboxamide, 3-(4-bromophenyl)-4,5-dihydro-5-(hydroxymethyl)-	MeOH	271(3.19)	73-2167-86
$C_{11}H_{11}BrO$			
Bullvalenemethanol, bromo-	ether	end absorption	24-2339-86
$C_{11}H_{11}BrO_2$			
Tricyclo[5.2.1.0^{2,6}]deca-4,8-dien-3-one, 4-bromo-5-methoxy-	EtOH	264(4.11)	78-1903-86
$C_{11}H_{11}BrO_4S$			
Tricyclo[5.2.1.0^{2,6}]deca-4,8-dien-3-one, 4-bromo-5-[(methylsulfonyl)oxy]-	EtOH	275(3.91)	78-1903-86
$C_{11}H_{11}ClN_2O_2$			
L-Tryptophan, 2-chloro-	H_2O	218(4.54),271(3.88), 288(3.74)	35-2023-86
$C_{11}H_{11}ClN_2O_3$			
4-Isoxazolecarboxamide, 3-(2-chlorophenyl)-4,5-dihydro-5-(hydroxymethyl)-	MeOH	248(3.80)	73-2167-86
4-Isoxazolecarboxamide, 3-(4-chlorophenyl)-4,5-dihydro-5-(hydroxymethyl)-	MeOH	269(4.13)	73-2167-86 +118-0565-86
2-Propenamide, 3-(4-chlorophenyl)-2-formyl-3-[(hydroxymethyl)amino]-, (Z)-	MeOH	244(3.71),288(3.58)	73-2167-86 +118-0565-86
$C_{11}H_{11}FN_2O_3$			
4-Isoxazolecarboxamide, 3-(4-fluorophenyl)-4,5-dihydro-5-(hydroxymethyl)-	MeOH	264(4.08)	73-2167-86 +118-0565-86
2-Propenamide, 3-(4-fluorophenyl)-2-formyl-3-[(hydroxymethyl)amino]-, (Z)-	MeOH	244(3.89),288(3.88)	73-2167-86 +118-0565-86

Compound	Solvent	λ_{max}(log ε)	Ref.
$C_{11}H_{11}FN_4O$			
1H-Pyrazole-4-carboxamide, 5-amino-N-(4-fluorophenyl)-1-methyl-	EtOH	274(3.40)	11-0331-86
$C_{11}H_{11}N$			
Benzeneacetonitrile, α-propylidene-, (E)-	C_6H_{12}	244.2(4.02)	88-6225-86
(Z)-	C_6H_{12}	256.0(4.05)	88-6225-86
1H-Indole, 3-(2-propenyl)-	hexane	265(3.75)	44-2343-86
Quinoline, 2,6-dimethyl-	75% H_2SO_4	245(4.7),318(4.0)	94-1468-86
$C_{11}H_{11}NO$			
Ethanone, 1-[(2-propynylamino)phenyl]-	hexane	209(4.0),232(3.78), 326(4.20)	65-2461-86
	MeOH	207(--),236(--), 323(--)	65-2461-86
1H-Pyrrole, 2-[2-(2-furanyl)ethenyl]-1-methyl-, (E)-	EtOH	332(3.92),350s(3.74)	118-0620-86
2-Pyrrolidinone, 5-(phenylmethylene)-, (E)-	EtOH	283(4.36)	49-0185-86
(Z)-	MeOH	269(4.12)	49-0185-86
$C_{11}H_{11}NO_2$			
1H-Indene-1,2(3H)-dione, 7-methyl-, 2-(methyloxime), (E)-	C_6H_{12}	273(4.27),279(4.27), 370(1.60),395s(1.54), 415s(1.30)	54-0188-86
	EtOH	283(4.27)	54-0188-86
(Z)-	C_6H_{12}	285(4.19),408(1.70), 428(1.81),450(1.60)	54-0188-86
1-Propanone, 1-(3-hydroxy-1H-indol-2-yl)-	EtOH	210(4.03),242(4.05), 258(3.83),323(4.19), 357(3.69)	104-2186-86
$C_{11}H_{11}NO_2S_2$			
3(2H)-Benzothiazolepropanoic acid, 2-thioxo-, methyl ester	EtOH	328(4.33)	103-0564-86
Propanoic acid, 3-(2-benzothiazolylthio)-, methyl ester	EtOH	276(4.06)	103-0456-86
$C_{11}H_{11}NO_3$			
2-Benzofurancarboxamide, 6-ethoxy-	EtOH	270(3.93),278(3.98), 302(4.19),308(4.19)	4-1347-86
Ethanone, 1-(3-ethyl-6-hydroxy-1,2-benzisoxazol-7-yl)-	EtOH	223(4.1),280(4.1), 328(4.1)	2-0870-86
$C_{11}H_{11}NO_3S$			
Propanoic acid, 3-(2-benzoxazolylthio)-, methyl ester	EtOH	250(4.06),280(4.08), 284(4.08)	103-0456-86
$C_{11}H_{11}NO_3S_2$			
Benzeneethane(dithioic) acid, 4-nitro-α-oxo-, 1-methylethyl ester	MeOH	264(3.70),305(3.67), 500(1.69)	118-0968-86
$C_{11}H_{11}NO_4$			
2-Benzofurancarboxamide, 4,6-dimethoxy-	EtOH	234(4.21),297(4.39)	4-1347-86
4H-Furo[3,2-b]pyrrole-5-carboxylic acid, 2-formyl-4-methyl-, ethyl ester	MeOH	344(3.55)	73-0106-86
1H-Indole-2-carboxylic acid, 4,6-dimethoxy-	n.s.g.	239(4.38),298(4.33)	12-0015-86

Compound	Solvent	λ max(log ε)	Ref.
$C_{11}H_{11}NO_4S$			
Benzenesulfonic acid, 3-[3-(ethylimino)-1-oxo-2-propenyl]-	pH 8.0	310(4.18)	35-5543-86
$C_{11}H_{11}NS$			
1H-Pyrrole, 1-methyl-2-[2-(2-thienyl)ethenyl]-, (E)-	EtOH	250(3.81),336(3.87)	118-0620-86
$C_{11}H_{11}N_3O$			
4-Pyrimidinamine, N-hydroxy-2-methyl-6-phenyl-	n.s.g.	246(4.39),294s(3.68)	103-1232-86
4-Pyrimidinamine, N-hydroxy-6-methyl-2-phenyl-	n.s.g.	246(4.44),275(3.87)	103-1232-86
$C_{11}H_{11}N_3OS_2$			
Morpholinium, 4-[4-(dicyanomethyl)-5-methyl-1,3-dithiol-2-ylidene]-, hydroxide, inner salt	MeCN	220(4.34),293(4.35), 451(3.59)	24-2094-86
$C_{11}H_{11}N_3O_3$			
Butanamide, N-(2,3-dihydro-2-oxo-1H-benzimidazol-5-yl)-3-oxo-	MeOH	216.7(5.27),260(4.65), 295.6(4.69)	80-0661-86
	MeOH-HCl	214.9(4.37),260(4.00), 269.9(4.02)	80-0661-86
	EtOH	215.2(5.11),259.5(4.50), 294.9(4.58)	80-0661-86
	DMF	268.1(4.21),304.4(4.32)	80-0661-86
$C_{11}H_{11}N_3O_3S$			
Benzamide, N-(3-amino-2-nitro-1-thioxo-2-butenyl)-	EtOH	253(4.04),304(4.23), 357(3.90)	18-3871-86
$C_{11}H_{11}N_3O_4$			
Pyrido[2,3-d]pyrimidine-2,4(1H,3H)-dione, 6-(acetoxymethyl)-5-methyl-	H_2O	218(4.60),246(4.00), 306(3.80)	87-0709-86
$(C_{11}H_{11}N_3O_4)_n$			
Poly[1-(2-methacryloyloxy)ethyl]-6-cyanouracil	DMF	292(3.88)	47-0119-86
	DMSO	292(3.86)	47-0119-86
$C_{11}H_{11}N_3O_5$			
β-D-ribo-Hexofuranurononitrile, 1,5-dideoxy-1-(3,4-dihydro-2,4-dioxo-1(2H)-pyrimidinyl)-5-methylene-	n.s.g.	260(3.91)	111-0111-86
1-Imidazolidinecarboxylic acid, 3-methyl-2-oxo-, 4-nitrophenyl ester	MeCN	272(4.16)	78-1315-86
1H-Pyrrolo[3,4-c]pyridine-7-carboxylic acid, 3-amino-2,4,5,6-tetrahydro-2,5-dimethyl-1,4,6-trioxo-, methyl ester	pH 1	268(4.23),332(3.81), 437(3.87)	44-0149-86
	pH 7	268(4.23),327(3.37), 436(3.89)	44-0149-86
$C_{11}H_{11}N_5OS$			
1,2,4-Triazol-4-amine, 3-(cyanomethylthio)-5-(4-methoxyphenyl)-	EtOH	261(4.25)	4-1451-86
$C_{11}H_{11}N_5O_5$			
1,2,4-Triazine-3,5(2H,4H)-dione, 2-[3-(3,4-dihydro-5-methyl-2,4-dioxo-1(2H)-pyrimidinyl)-2-oxopropyl]-	H_2O	269(4.16)	107-1177-86

Compound	Solvent	λmax(log ε)	Ref.
1,2,4-Triazine-3,5(2H,4H)-dione, 2-[3-(3,4-dihydro-2,4-dioxo-1(2H)-pyrimidinyl)-2-oxopropyl]-6-methyl-	H_2O	265(4.17)	107-1177-86
$C_{11}H_{11}N_5S$			
Acetonitrile, [[4-amino-5-(4-methylphenyl)-4H-1,2,4-triazol-3-yl]thio]-	EtOH	256(4.23)	4-1451-86
Acetonitrile, [[4-amino-5-(phenylmethyl)-4H-1,2,4-triazol-3-yl]thio]-	EtOH	264(4.12)	4-1451-86
$C_{11}H_{12}BrNO_2$			
Butanamide, 2-bromo-N-(2-methylphenyl)-3-oxo-	benzene	206(3.28),258(4.10), 286(3.85)	80-0869-86
	MeOH	203(3.15),252(4.05), 281(3.82)	80-0869-86
	EtOH	204(3.17),253(4.07), 282(3.83)	80-0869-86
	PrOH	204(3.20),254(4.08), 283(3.83)	80-0869-86
	BuOH	205(3.22),256(4.09), 282(3.85)	80-0869-86
	dioxan	206(3.28),258(4.10), 286(3.85)	80-0869-86
	HOAc	205(3.14),252(3.99), 281(3.79)	80-0869-86
	CH_2Cl_2	202(3.18),255(4.05), 285(3.82)	80-0869-86
	$CHCl_3$	203(3.20),254(4.07), 285(3.83)	80-0869-86
	CCl_4	204(3.23),256(4.08), 285(3.83)	80-0869-86
	$C_2H_4Cl_2$	200(3.17),256(4.04), 283(3.81)	80-0869-86
	MeCN	206(3.31),256(4.08), 286(3.84)	80-0869-86
	pyridine	211(2.98),306(3.74)	80-0869-86
	DMF	205(3.23),265(3.98), 288(3.81)	80-0869-86
	DMSO	204(3.20),261(4.07), 284(3.73)	80-0869-86
Butanamide, 4-bromo-N-(2-methylphenyl)-3-oxo-	benzene	205(3.22),254(4.06), 281(3.81)	80-0869-86
	MeOH	206(3.13),249(4.01), 278(3.77)	80-0869-86
	EtOH	205(3.17),250(4.02), 280(3.78)	80-0869-86
	PrOH	204(3.18),251(4.04), 282(3.79)	80-0869-86
	BuOH	203(3.21),250(4.05), 283(3.80)	80-0869-86
	dioxan	204(3.26),255(4.07), 283(3.82)	80-0869-86
	HOAc	200(3.13),248(3.96), 279(3.75)	80-0869-86
	CH_2Cl_2	201(3.17),252(4.02), 279(3.79)	80-0869-86
	$CHCl_3$	203(3.21),252(4.04), 281(3.80)	80-0869-86
	CCl_4	204(3.22),253(4.05), 282(3.82)	80-0869-86

Compound	Solvent	λ max(log ε)	Ref.
(cont.)	MeCN	206(3.27),252(4.06), 278(3.81)	80-0869-86
	pyridine	209(2.93),307(3.70)	80-0869-86
	DMF	202(3.20),266(3.96), 286(3.69)	80-0869-86
	DMSO	201(3.18),260(4.05), 280(3.70)	80-0869-86
$C_{11}H_{12}BrNO_3$			
Butanamide, 2-bromo-N-(2-methoxyphenyl)-3-oxo-	benzene	205(3.36),256(4.14), 287(3.91)	80-0869-86
	MeOH	200(3.32),252(4.11), 282(3.87)	80-0869-86
	EtOH	202(3.34),253(4.12), 283(3.88)	80-0869-86
	PrOH	204(3.36),254(4.13), 284(3.89)	80-0869-86
	BuOH	206(3.37),255(4.14), 286(3.89)	80-0869-86
	dioxan	205(3.37),256(4.16), 288(3.91)	80-0869-86
	HOAc	203(3.32),253(4.10), 282(3.86)	80-0869-86
	CH_2Cl_2	204(3.36),253(4.10), 283(3.89)	80-0869-86
	$CHCl_3$	203(3.33),255(4.12), 284(3.89)	80-0869-86
	CCl_4	202(3.34),257(4.13), 285(3.90)	80-0869-86
	MeCN	206(3.38),258(4.14), 285(3.90)	80-0869-86
	pyridine	209(3.06),304(3.75)	80-0869-86
	DMF	205(3.35),265(4.07), 284(3.86)	80-0869-86
	DMSO	205(3.33),258(4.09), 282(3.86)	80-0869-86
Butanamide, 2-bromo-N-(4-methoxyphenyl)-3-oxo-	benzene	203(3.28),259(4.14), 285(3.89)	80-0869-86
	MeOH	202(3.23),251(4.10), 280(3.86)	80-0869-86
	EtOH	204(3.24),253(4.11), 287(3.86)	80-0869-86
	PrOH	205(3.26),254(4.12), 284(3.87)	80-0869-86
	BuOH	206(3.28),254(4.14), 285(3.88)	80-0869-86
	dioxan	204(4.30),260(4.15), 284(3.89)	80-0869-86
	HOAc	205(3.19),255(4.05), 283(3.86)	80-0869-86
	CH_2Cl_2	199(4.28),251(4.11), 283(3.88)	80-0869-86
	$CHCl_3$	202(4.25),257(4.11), 284(3.88)	80-0869-86
	CCl_4	202(3.27),258(4.12), 285(3.88)	80-0869-86
	MeCN	204(3.31),259(4.14), 286(3.88)	80-0869-86
	pyridine	210(3.08),305(3.75)	80-0869-86
	DMF	202(3.25),265(4.07), 285(3.87)	80-0869-86

Compound	Solvent	$\lambda_{max}(\log \epsilon)$	Ref.
Butanamide, 2-bromo-N-(4-methoxy-phenyl)-3-oxo- (cont.)	DMSO	201(3.27),258(4.09), 288(3.87)	80-0869-86
Butanamide, 4-bromo-N-(2-methoxy-phenyl)-3-oxo-	benzene	203(3.33),254(4.12), 285(3.88)	80-0869-86
	MeOH	201(3.29),250(4.09), 279(3.84)	80-0869-86
	EtOH	202(3.30),251(4.10), 280(3.85)	80-0869-86
	PrOH	203(3.31),252(4.11), 281(3.86)	80-0869-86
	BuOH	204(3.34),254(4.50), 282(3.87)	80-0869-86
	dioxan	204(3.34),255(4.13), 286(3.88)	80-0869-86
	HOAc	202(3.28),257(4.05), 277(3.85)	80-0869-86
	CH_2Cl_2	204(3.32),249(4.09), 278(3.87)	80-0869-86
	$CHCl_3$	204(3.30),252(4.11), 283(3.87)	80-0869-86
	CCl_4	202(3.32),252(4.11), 284(3.87)	80-0869-86
	MeCN	204(3.34),254(4.11), 276(3.87)	80-0869-86
	pyridine	210(3.01),310(3.72)	80-0869-86
	DMF	205(3.30),265(4.01), 289(3.81)	80-0869-86
	DMSO	202(3.27),258(4.05), 286(3.82)	80-0869-86
Butanamide, 4-bromo-N-(4-methoxy-phenyl)-3-oxo-	benzene	201(3.27),256(4.11), 283(3.86)	80-0869-86
	MeOH	201(3.22),253(4.07), 279(3.83)	80-0869-86
	EtOH	203(3.23),254(4.09), 280(3.84)	80-0869-86
	PrOH	204(3.24),255(4.10), 281(3.85)	80-0869-86
	BuOH	201(3.27),253(4.11), 282(3.86)	80-0869-86
	dioxan	202(3.29),257(4.13), 285(3.87)	80-0869-86
	HOAc	202(3.17),255(4.04), 281(3.83)	80-0869-86
	CH_2Cl_2	200(3.26),254(4.08), 284(3.83)	80-0869-86
	$CHCl_3$	201(3.24),255(4.09), 282(3.84)	80-0869-86
	CCl_4	202(3.26),257(4.10), 284(3.85)	80-0869-86
	MeCN	205(3.30),256(4.12), 278(3.86)	80-0869-86
	pyridine	210(3.02),306(3.73)	80-0869-86
	DMF	202(3.22),265(4.02), 290(3.82)	80-0869-86
	DMSO	204(3.24),260(4.06), 285(3.83)	80-0869-86
$C_{11}H_{12}BrN_3O_3$			
4-Isoxazolecarboxylic acid, 3-(4-bromophenyl)-4,5-dihydro-5-(hydroxy-methyl)-, hydrazide, cis	MeOH	272(3.22)	73-2167-86

Compound	Solvent	$\lambda_{max}(\log \epsilon)$	Ref.
$C_{11}H_{12}ClNO_2$			
2,4-Pentadienoic acid, 5-(3-chloro-1H-pyrrol-2-yl)-4-methyl-, methyl ester. (E,E)-	EtOH	360(4.31)	150-3601-86M
$C_{11}H_{12}ClNO_3$			
Acetamide, 2-chloro-N-hydroxy-N-(1-methyl-2-oxo-2-phenylethyl)-	EtOH	244(4.09)	103-0417-86
$C_{11}H_{12}ClN_3O_3$			
1H-Imidazo[4,5-c]pyridine, 4-chloro-1-(2-deoxy-β-D-ribofuranosyl)-	pH 1	266(3.78),273(3.69)	87-0138-86
	pH 13	255(3.79),265(3.76), 273(3.65)	87-0138-86
$C_{11}H_{12}ClN_5O_3$			
Acetamide, N-[7-(3-chlorotetrahydro-2-furanyl)-6,7-dihydro-6-oxo-1H-purin-2-yl]-, trans	MeOH	264(4.13)	23-1885-86
Acetamide, N-[9-(3-chlorotetrahydro-2-furanyl)-6,9-dihydro-6-oxo-1H-purin-2-yl]-, trans	MeOH	259(4.13),278s(3.99)	23-1885-86
2,4(1H,3H)-Pteridinedione, 6-chloro-7-[1-(hydroxyimino)propyl]-1,3-dimethyl-	MeOH	248(4.20),353(4.00)	108-0081-86
$C_{11}H_{12}Cl_3NO_2$			
2,5-Cyclohexadiene-1,4-dione, 2-(butylmethylamino)-3,5,6-trichloro-	CH_2Cl_2	251(4.07),290s(3.80), 315s(3.60),570(3.40)	106-0832-86
$C_{11}H_{12}FNO_3$			
Phenylalanine, β-(fluoromethylene)-3-hydroxy-4-methyl-, (E)-	H_2O	236s(3.74),280(3.34)	130-0103-86
$C_{11}H_{12}FNO_4$			
Phenylalanine, β-(fluoromethylene)-3-hydroxy-4-methoxy-, (E)-	H_2O	243s(3.78),282(3.52)	130-0103-86
$C_{11}H_{12}FNO_4S$			
2-Oxazolidinone, 4-(fluoromethyl)-5-[4-(methylsulfonyl)phenyl]-	MeOH	223(4.09),266(2.87), 273(2.82)	32-0485-86
$C_{11}H_{12}FN_3O_3$			
4-Isoxazolecarboxylic acid, 3-(2-fluorophenyl)-4,5-dihydro-5-(hydroxymethyl)-, hydrazide, cis	MeOH	272(3.14)	73-2167-86
$C_{11}H_{12}N_2$			
1H-Pyrrole, 1-methyl-2-[2-(1H-pyrrol-2-yl)ethenyl]-, (E)-	EtOH	219(3.83),338(4.39)	118-0620-86
(Z)-	EtOH	200(4.08),318(4.09)	118-0620-86
$C_{11}H_{12}N_2O$			
1H-Imidazole-4-methanol, 1-methyl-2-phenyl-	MeOH	260(4.04)	87-0261-86
4-Quinolinol, 2-(dimethylamino)-	MeOH	225(4.44),245(4.45), 303(4.18)	83-0347-86
$C_{11}H_{12}N_2OS_2$			
[1]Benzothieno[2,3-d]pyrimidine-4(3H)-thione, 5,6,7,8-tetrahydro-3-hydroxy-2-methyl-	EtOH-dioxan	250(4.32),277(4.14), 316(4.09),343s(3.92)	106-0096-86

Compound	Solvent	λ_{max}(log ε)	Ref.
$C_{11}H_{12}N_2O_2$			
1,5-Methano-2H-pyrido[1,2-a][1,5]diazocine-4,8(1H,3H)-dione, 5,6-dihydro-	EtOH	234(3.40),304(3.45)	102-2000-86
Tryptophan	pH 11.5	220(4.56),280(3.74)	35-3186-86
$C_{11}H_{12}N_2O_2S$			
β-Alanine, N-2-benzothiazolyl-N-methyl-	pH 1	215(4.47),262(4.10), 280(4.04),290(4.08)	103-0795-86
	EtOH	227(4.50),274(4.23), 305(3.65)	103-0795-86
3(2H)-Benzothiazolepropanoic acid, 2-(methylimino)-	pH 1	223(4.33),256(4.04), 281(3.98),290(4.02)	103-0795-86
	EtOH	223(4.44),264(3.97), 281(3.70),306(3.59)	103-0795-86
Pyrazine, trimethyl-2-thienyl-, 1,4-dioxide	EtOH	241(4.25),267(4.09), 298(4.39)	103-0268-86
$C_{11}H_{12}N_2O_3$			
4-Isoxazolecarboxamide, 4,5-dihydro-5-(hydroxymethyl)-3-phenyl-	MeOH	264(3.99)	118-0565-86
2-Propenamide, 2-formyl-3-[(hydroxymethyl)amino]-3-phenyl-, (Z)-	MeOH	240(3.97),284(4.11)	73-2167-86 +118-0565-86
2-Propenoic acid, 2-cyano-3-[5-(dimethylamino)-2-furanyl]-, methyl ester	MeOH	230(3.38),461(3.98)	73-0879-86
Pyrazine, 2-furanyltrimethyl-, 1,4-dioxide	EtOH	239(4.16),267(4.21), 294(4.45)	103-0268-86
$C_{11}H_{12}N_2O_4$			
Phenylalanine, N-acetyl-N-nitroso-, dicyclohexylamine salt	MeOH	390(2.24),405(2.37), 424(2.36)	39-0727-86B
2-Butenedioic acid, 2-(1H-1,2-diazepin-1-yl)-, dimethyl ester, (E)-	EtOH	223s(3.81),255s(3.71), 305(4.19),374s(3.04)	39-0595-86C
2H-1,2-Oxazine, tetrahydro-2-(4-nitrobenzoyl)-	MeCN	215(3.68),261(3.66)	4-1423-86
$C_{11}H_{12}N_2O_5$			
Uridine, 2'-deoxy-6-ethynyl-	H_2O	285.5(4.03)	23-1560-86
$C_{11}H_{12}N_2O_6$			
Uridine, 6-ethynyl-	H_2O	285(4.00)	23-1560-86
$C_{11}H_{12}N_2S_2$			
[1]Benzothieno[2,3-d]pyrimidine-4(3H)-thione, 5,6,7,8-tetrahydro-2-methyl-	EtOH and Meglycol	231(4.46),289(4.24), 321(4.26),357(4.11)	106-0096-86
$C_{11}H_{12}N_4OS$			
1,2,4-Triazole-3-thiol, 4-(acetylamino)-5-(phenylmethyl)-	EtOH	227(4.05),252(4.26), 278s(4.10)	4-1451-86
$C_{11}H_{12}N_4O_2$			
2,4(1H,3H)-Pyrimidinedione, 5-amino-6-(methylphenylamino)-	pH 13	238(4.19),322(4.10)	35-5972-86
$C_{11}H_{12}N_4O_3S$			
Benzeneacetic acid, 2-(acetylamino)-α-oxo-, 2-(aminothioxomethyl)hydrazide	EtOH	272s(3.86),342(3.51)	104-2173-86

Compound	Solvent	λmax(log ε)	Ref.
$C_{11}H_{12}N_4O_4$			
Butanoic acid, 2-(hydroxyimino)-3-oxo-, 2-methyl-2-nitroso-1-phenylhydrazide	pH 7.5	288(4.06),371(3.03)	94-3866-86
	$CHCl_3$	365(2.15)	94-3866-86
$C_{11}H_{12}N_4O_4S$			
Acetamide, N-[4-[[(3-diazo-2-oxopropyl)amino]sulfonyl]phenyl]-	MeOH	206(4.30),266(4.42)	104-0353-86
Acetic acid, [(1,2,3,4-tetrahydro-1,3-dimethyl-2,4-dioxo-6-pteridinyl)thio]-, methyl ester	MeOH	252(4.05),277(4.29), 376(3.83)	108-0081-86
Acetic acid, [(1,2,3,4-tetrahydro-1,3-dimethyl-2,4-dioxo-7-pteridinyl)thio]-, methyl ester	MeOH	226(4.38),263(3.95), 355(4.30)	108-0081-86
$C_{11}H_{12}N_4O_5S$			
Benzenesulfonamide, N-(3-diazo-1-methyl-2-oxopropyl)-N-methyl-4-nitro-	MeOH	204(4.13),212s(3.98), 255(4.22),280(4.20)	104-0353-86
$C_{11}H_{12}N_4O_6$			
Propanoic acid, 2-[(2,4-dinitrophenyl)hydrazono]-, ethyl ester, (E)-	EtOH	348(4.38)	18-0073-86
(Z)-	EtOH	360(4.42)	18-0073-86
$C_{11}H_{12}N_6OS_2$			
1,2,3-Thiadiazole-4-carboxylic acid, 2,5-dihydro-5-[[(methylamino)thioxomethyl]imino]-2-phenyl-, hydrazide	DMF	284(3.99),414(4.18)	48-0741-86
$C_{11}H_{12}N_6O_2$			
1H-Pyrazolo[3,4-g]pteridine-6,8(5H,7H)-dione, 3-ethyl-5,7-dimethyl-	pH -4.3	239(4.46),290(3.78), 400(3.96),410s(3.92)	108-0081-86
	pH 5.0	242(4.44),320(3.89), 330s(3.85),397(3.83)	108-0081-86
	pH 12.0	254(4.35),346(4.15), 440(3.37)	108-0081-86
1H-Pyrazolo[4,3-g]pteridine-5,7(1H,8H)-dione, 3-ethyl-6,8-dimethyl-	pH -4.0	220(4.39),244(4.29), 316s(3.82),357(4.21)	108-0081-86
	pH 5.0	232(4.50),352(4.26)	108-0081-86
	pH 12.0	240(4.43),250s(4.38), 360(4.19)	108-0081-86
$C_{11}H_{12}O$			
4,8-Methanoazulen-5(1H)-one, 3a,4,8,8a-tetrahydro-	EtOH	230(3.69)	18-2811-86
4,8-Methanoazulen-5(3H)-one, 3a,4,8,8a-tetrahydro-	EtOH	236(3.76)	18-2811-86
Tetracyclo[3.3.0.0^{2,8}.0^{2,6}]octan-3-one, 1,5-dimethyl-7-methylene-	C_6H_{12}	227(3.10),277(1.88)	33-0071-86
	EtOH	236(3.48),279(2.28)	33-0071-86
$C_{11}H_{12}OS_2$			
Benzenecarbodithioic acid, 4-methoxy-, 1-propenyl ester	CH_2Cl_2	347(3.72),500(1.71)	18-1403-86
$C_{11}H_{12}O_2$			
Ethanone, 1-[(2-propenyloxy)phenyl]-	EtOH	247(3.91),305(3.56)	44-3125-86
7-Oxabicyclo[2.2.1]heptane-1-methanol, 2,3,5,6-tetrakis(methylene)-	isooctane	212(4.09),221(4.21), 228(4.23),238s(3.93), 250s(3.81),263s(3.36)	33-1287-86

Compound	Solvent	λ_{max}(log ε)	Ref.
(cont.)	EtOH	216(4.07),221(4.19), 226(4.19),240s(3.90), 249s(3.79),262s(3.40)	33-1287-86
1-Propanone, 3-oxiranyl-1-phenyl-	EtOH	241.5(4.08),279(3.0)	35-1617-86
2-Propenoic acid, 2-methyl-, 7-bi- cyclohepta-2,5-dienyl ester	EtOH	212(3.91)	18-1501-86
	CH_2Cl_2	229(3.04)	18-1501-86
polymer	CH_2Cl_2	228(2.60)	18-1501-86
2-Propenoic acid, 2-methyl-3-phenyl-, methyl ester, (E)-	CH_2Cl_2	268(4.15)	35-3005-86
(Z)-	CH_2Cl_2	258(4.21)	35-3005-86
2-Propenoic acid, 3-(4-methylphen- yl)-, methyl ester, (E)-	CH_2Cl_2	285(4.29)	35-3005-86
with BF_3	CH_2Cl_2	327(4.46)	35-3005-86
(Z)-	CH_2Cl_2	280(4.09)	35-3005-86
with BF_3	CH_2Cl_2	322(4.18)	35-3005-86
2-Propenoic acid, 3-phenyl-, ethyl ester, (E)-	CH_2Cl_2	276(4.24)	35-3005-86
with BF_3	CH_2Cl_2	311(4.33)	35-3005-86
(Z)-	CH_2Cl_2	269(4.03)	35-3005-86
Tricyclo[5.2.1.$0^{2,6}$]deca-4,8-dien- 3-one, 5-methoxy-	EtOH	241(3.79)	78-1903-86
$C_{11}H_{12}O_2S$			
Benzene, [(1-methyl-1,3-butadienyl)- sulfonyl]-	MeOH	249(4.43)	78-5321-86
$C_{11}H_{12}O_3$			
1H-2-Benzopyran-1-one, 3,4-dihydro- 8-hydroxy-3,5-dimethyl-, (-)-(3R)-	EtOH	218(4.28),247(3.74), 322(3.54)	163-0997-86
Epiisoshinanolone (plus epimer)	n.s.g.	217(4.16),259(3.87), 335(3.48)	102-0764-86
$C_{11}H_{12}O_4$			
1H-2-Benzopyran-1-one, 3,4-dihydro- 4,8-dihydroxy-3,5-dimethyl-, (+)-(3R,4S)-	EtOH	216(4.25),245(3.42), 324(3.37)	163-0997-86
(-)-(3R,4R)-	EtOH	216(4.11),245(3.41), 322(3.35)	163-0997-86
1H-2-Benzopyran-1-one, 3,4-dihydro- 8-hydroxy-5-(hydroxymethyl)-3- methyl-, (-)-(3R)-	EtOH	216(4.19),246(3.65), 319(4.47)	163-0997-86
$C_{11}H_{12}O_4S_2$			
1,3-Dithiole-4,5-dicarboxylic acid, 2-(1-methyl-2-propenylidene)-, dimethyl ester	EtOH	229(4.07),306(4.25), 416(3.15)	33-0419-86
$C_{11}H_{12}O_6$			
7-Oxabicyclo[2.2.1]hepta-2,5-diene- 2,3-dicarboxylic acid, 1-(hydroxy- methyl)-, dimethyl ester	MeCN	284(3.08)	24-0589-86
$C_{11}H_{12}S_4Se_2$			
5H-1,3-Dithiolo[4,5-b][1,4]dithiepin, 2-(4,5-dimethyl-1,3-diselenol-2- ylidene)-6,7-dihydro-	MeCN	212(4.30),300(4.17), 336(3.88),380(3.17)	77-1472-86
$C_{11}H_{13}BrN_2O_6$			
Uridine, 5-(2-bromoethenyl)-, (E)-	EtOH	251(4.11),294(4.01)	87-0213-86

Compound	Solvent	λ_{max}(log ϵ)	Ref.
$C_{11}H_{13}ClHgO$			
Mercury, chloro[(3,4-dihydro-2-methyl-2H-1-benzopyran-2-yl)methyl]-	EtOH	281s(--)	32-0441-86
$C_{11}H_{13}ClHgO_3$			
Mercury, chloro[(3,4-dihydro-2,7-dimethyl-5-oxo-2H,5H-pyrano[4,3-b]pyran-2-yl)methyl]-	EtOH	284(3.8)	32-0441-86
$C_{11}H_{13}ClN_2O_2$			
1H-Benzimidazole, 2-(chloromethyl)-4,7-dimethoxy-1-methyl-	H_2O	226(4.29),264(3.62)	44-0522-86
	acid	231(4.15),274(3.45)	44-0522-86
Propanoic acid, 2-[(4-chlorophenyl)-hydrazono]-, ethyl ester, (E)-	EtOH	329(4.36)	18-0073-86
(Z)-	EtOH	349(4.29)	18-0073-86
$C_{11}H_{13}F_3N_2O_2$			
Butanoic acid, 3-[[2-amino-4-(trifluoromethyl)phenyl]amino]-	n.s.g.	219(4.46),264(4.02), 303(3.68)	150-2471-86M
$C_{11}H_{13}F_3O_2$			
2,5-Cyclohexadien-1-one, 2,4,6-trimethyl-4-(2,2,2-trifluoroethoxy)-	MeOH	233(4.08),265s(3.05)	33-0469-86
Phenol, 2,6-dimethyl-4-[(2,2,2-trifluoroethoxy)methyl]-	MeOH	223(3.85),227s(3.83), 274(3.12),280(3.13)	33-0469-86
$C_{11}H_{13}F_3O_4$			
3,5-Heptadienoic acid, 6-acetoxy-, 2,2,2-trifluoroethyl ester, (E,Z)-	MeOH	234(4.33)	33-0469-86
$C_{11}H_{13}IO_3$			
1-Propanone, 1-(3,4-dimethoxyphenyl)-3-iodo-	EtOH	226(4.29),275(4.01), 306(3.95)	24-0050-86
$C_{11}H_{13}NO$			
2-Butenal, 3-[(4-methylphenyl)amino]-	63.1% H_2SO_4	295(4.2)	94-1468-86
3-Buten-2-one, 4-phenyl-, O-methyloxime, (E,E)-	EtOH	286(4.50)	39-1691-86C
(E,Z)-	EtOH	231(3.92),264(4.14)	39-1691-86C
(Z,E)-	EtOH	291(4.31)	39-1691-86C
(Z,Z)-	EtOH	234(3.83),259(4.09)	39-1691-86C
1H-Inden-1-one, 2,3-dihydro-7-methyl-, O-methyloxime, (E)-	C_6H_{12}	261(4.13),287(3.52), 299(3.48)	54-0188-86
(Z)-	C_6H_{12}	259(3.94)	54-0188-86
Pyridine, 4-cyclohexenyl-, 1-oxide	EtOH	269(5.2)	94-0007-86
$C_{11}H_{13}NOS_2$			
Morpholine, 4-[(phenylthioxomethyl)-thio]-	C_6H_{12}	286(4.17),323(3.87), 532(1.89)	88-4595-86
$C_{11}H_{13}NO_2$			
2,4,6-Cycloheptatrien-1-one, 4-morpholino-	MeOH	226(4.18),275(3.43), 368(4.18)	88-3005-86
1H-Indole, 4,6-dimethoxy-2-methyl-	n.s.g.	220(4.55),264(3.96)	12-0015-86
Isopycnarrhine	EtOH	213(3.63),225s(3.51), 271(3.54),278(3.54), 320(3.21),413(4.14)	100-1028-86
	EtOH-HCl	208(4.13),253(3.74), 314(3.31)	100-1028-86

Compound	Solvent	λ_{max}(log ε)	Ref.
$C_{11}H_{13}NO_2S_2$			
2λ^4-Thieno[3,2-c][1,2]thiazine-3-carboxylic acid, 2-ethyl-, ethyl ester	EtOH	220(4.05),270(3.96), 323(3.96),468(3.72)	39-0483-86C
$C_{11}H_{13}NO_3$			
2,5-Cyclohexadiene-1,4-dione, 2-(1-aziridinyl)-5-methoxy-3,6-dimethyl-	EtOH	303(4.28)	94-1299-86
$C_{11}H_{13}NO_4$			
1(4H)-Pyridineacetic acid, 3,5-di-formyl-4-methyl-, methyl ester	H_2O	236(4.28),262(3.90), 384(3.94)	35-8283-86
3,5-Pyridinedicarboxylic acid, 2,6-di-methyl-, dimethyl ester	MeOH	233(--),270(4.42), 279(--)	87-1596-86
Serine, N-benzoyl-, methyl ester	EtOH	226(4.11)	35-7609-86
$C_{11}H_{13}NO_4S$			
2-Pyridinecarbo(thioperoxoic) acid, 6-(ethoxycarbonyl)-, O-ethyl ester	CH_2Cl_2	233(4.00),278(3.73)	78-5969-86
$C_{11}H_{13}NO_4S_2$			
2,6-Pyridinedicarbo(thioperoxoic) acid, SO,SO-diethyl ester	CH_2Cl_2	228(4.07),287(3.84)	78-5969-86
$C_{11}H_{13}NO_5S$			
Benzenesulfonic acid, 3-[3-(ethyl-amino)-1,3-dioxopropyl]-	pH 7.0	245(4.06),283.5(3.04)	35-5543-86
conjugate base		313(4.13)	35-5543-86
2-Oxazolidinone, 4-(hydroxymethyl)-5-[4-(methylsulfonyl)phenyl]-, (4R,5R)-	MeOH	223(4.14),266(2.94), 273(2.89)	32-0485-86
$C_{11}H_{13}NO_9$			
Methanediol, (2-acetoxy-2,5-dihydro-5-nitro-2-furanyl)-, diacetate, (E)-	THF	287(1.87)	44-3811-86
(Z)-	THF	295(2.11)	44-3811-86
$C_{11}H_{13}NS_2$			
3H-Pyrrole, 3-[2,3-bis(methylthio)-2-cyclopropen-1-ylidene]-2,5-di-methyl-, tetrafluoroborate	MeCN	207(4.24),248s(3.87), 326(4.29)	78-0839-86
$C_{11}H_{13}NS_4$			
3H-Pyrrole, 3-[4,5-bis(methylthio)-1,3-dithiol-2-ylidene]-2,5-di-methyl-, tetrafluoroborate	MeCN	206(4.09),238(3.98), 305(4.01),460(4.37)	78-0839-86
$C_{11}H_{13}N_2$			
Quinoxalinium, 1,5,7-trimethyl-, iodide	pH 6.3	225(4.25),255(4.52), 347(3.92)	103-1118-86
hydrated	pH 12.6	229(4.40),290(3.60), 346(3.51)	103-1118-86
$C_{11}H_{13}N_3$			
Propenenitrile, 3-(3,6-diethyl-2-pyrazinyl)-, (E)-	EtOH	247(4.26),309s(4.16), 315(4.17)	4-1481-86
(Z)-	EtOH	246(4.06),308(3.91), 316s(3.89)	4-1481-86
2,4-Quinolinediamine, N^2,N^2-dimethyl-	MeOH	229(4.17),257(4.27), 306(3.73)	83-0347-86

Compound	Solvent	$\lambda_{max}(\log \epsilon)$	Ref.
$C_{11}H_{13}N_3O$			
Sydnone imine, 3-(1-methyl-2-phenyl-ethyl)-	benzene	351(--)	65-2109-86
	H_2O	330(--)	65-2109-86
protonated	benzene	306(3.8)	65-2109-86
	H_2O	294(3.8)	65-2109-86
$C_{11}H_{13}N_3O_2S$			
2-Butenethioamide, 3-(methylamino)-2-nitro-N-phenyl-	EtOH	316(4.15),366(4.06)	18-3871-86
$C_{11}H_{13}N_3O_3$			
2-Butenamide, 3-(methylamino)-2-nitro-N-phenyl-	EtOH	249(4.16),350(4.14)	18-3871-86
$C_{11}H_{13}N_3O_3S$			
Benzenesulfonamide, N-(3-diazo-1-methyl-2-oxopropyl)-4-methyl-	MeOH	203(4.22),233(4.22), 246s(4.12),267s(3.94)	104-0353-86
Benzenesulfonamide, N-(4-diazo-1-oxobutyl)-4-methyl-	MeOH	203(4.24),233(4.18), 270s(3.91)	104-0353-86
$C_{11}H_{13}N_3O_3S_2$			
Thiazolium, 5-(hexahydro-1,3-dimethyl-2,4,6-trioxo-5-pyrimidinyl)-3-methyl-2-(methylthio)-, hydroxide, inner salt	MeCN	267(4.25),381.5(4.32)	24-2094-86
	toluene	409(--)	24-2094-86
	MeOH	369(--)	24-2094-86
	EtOH	372.5(--)	24-2094-86
	EtOAc	394(--)	24-2094-86
	CH_2Cl_2	392.5(--)	24-2094-86
$C_{11}H_{13}N_3O_4$			
1H-Imidazo[4,5-c]pyridine, 1-β-D-ribofuranosyl-	pH 1	255(3.58),265(3.60)	87-0138-86
	pH 13	250(3.65),263s(3.57), 270s(3.43)	87-0138-86
Propanoic acid, 2-[(4-nitrophenyl)hydrazono]-, ethyl ester, (E)-	EtOH	373(4.47)	18-0073-86
(Z)-	EtOH	381(4.50)	18-0073-86
$C_{11}H_{13}N_3O_4S$			
Benzenesulfonamide, N-(4-diazo-3-oxobutyl)-4-methoxy-	MeOH	204(4.41),243(4.43), 272(4.04)	104-0353-86
$C_{11}H_{13}N_3O_5$			
Pyrazolo[1,5-a]pyrimidin-7(4H)-one, 3-β-D-ribopyranosyl-	pH 1	231(4.07),268(3.65)	4-0349-86
	pH 7	231(4.07),268(3.65)	4-0349-86
	pH 12	243s(3.25),338(4.25)	4-0349-86
$C_{11}H_{13}N_3O_7$			
2-Propenoic acid, 3-[2-(2-deoxy-β-D-erythro-pentofuranosyl)-2,3,4,5-tetrahydro-3,5-dioxo-1,2,4-triazin-6-yl]-, (E)-	EtOH	227(4.09),304(4.04)	87-0809-86
$C_{11}H_{13}N_3S_2$			
[1]Benzothieno[2,3-d]pyrimidine-4(3H)-thione, 3-amino-5,6,7,8-tetrahydro-2-methyl-	EtOH-dioxan	251(4.26),284(4.03), 318(4.09),350s(3.91)	106-0096-86
Ethanaminium, N-[4-(dicyanomethyl)-5-methyl-1,3-dithiol-2-ylidene]-N-ethyl-, hydroxide, inner salt	MeCN	217(4.28),290(4.26), 437(3.51)	24-2094-86
	MeOH	421(--)	24-2094-86
	EtOH	432(--)	24-2094-86
	acetone	453(--)	24-2094-86
	dioxan	488(--)	24-2094-86

Compound	Solvent	$\lambda_{max}(\log \epsilon)$	Ref.
$C_{11}H_{13}N_5$			
1H-1,2,4-Triazol-3-amine, N-[[4-(dimethylamino)phenyl]-	dioxan	371(4.51)	80-0273-86
	+ CF_3COOH	435(--)	80-0273-86
$C_{11}H_{13}N_5O$			
4H-Imidazo[4,5-h]pyrrolo[2,3-f]indolizin-8(1H)-one, 2-amino-5,6,9,11b-tetrahydro-, hydrochloride	MeOH	215(3.80),275(3.91)	88-3177-86
$C_{11}H_{13}N_5O_2$			
2-Propenamide, N-[2-(1,6-dihydro-6-oxo-9H-purin-9-yl)ethyl]-2-methyl-	H_2O	256.6(3.99)	47-3201-86
polymer	DMSO	256.0(3.95)	47-3201-86
$C_{11}H_{13}N_5O_3$			
Adenosine, 2'-deoxy-2'-methylene-	H_2O	259.5(4.16)	94-1518-86
$C_{11}H_{13}N_5O_3S$			
Adenosine, 4',5'-didehydro-5'-S-methyl-5'-thio-, (Z)-	MeOH	258(4.28)	44-1258-86
$C_{11}H_{13}N_5O_4$			
8,2'-Methanoadenosine	H_2O	207.7(4.39),262(4.22)	94-1518-86
2,4,6(1H,3H,5H)-Pteridinetrione, 7-[1-(hydroxyimino)propyl]-1,3-dimethyl-	MeOH	217(4.30),256(4.10), 284s(3.85),384(4.11)	108-0081-86
2,4,7(1H,3H,8H)-Pteridinetrione, 6-[1-(hydroxyimino)propyl]-1,3-dimethyl-	MeOH	210(4.44),240s(4.00), 285(3.91),338(4.21)	108-0081-86
$C_{11}H_{13}N_5O_5$			
8,2'-Methanoguanosine	pH 1	252(4.14),278(3.95)	94-1961-86
	H_2O	250(4.17),277(3.97)	94-1961-86
	pH 11	214(4.26),255(4.06), 270(4.05)	94-1961-86
$C_{11}H_{14}$			
Tricyclo[5.3.1]undeca-2,4-diene	EtOH	245(3.46)	78-1823-86
$C_{11}H_{14}Br_2O$			
2(3H)-Naphthalenone, 1,8-dibromo-4,4a,5,6,7,8-hexahydro-4a-methyl-, cis	EtOH	273(4.01)	107-0195-86 +118-0461-86
$C_{11}H_{14}ClN_5O$			
2-Pteridinamine, 7-chloro-4-(pentyloxy)-	MeOH	242(4.32),268(3.96), 364(3.98)	33-0708-86
$C_{11}H_{14}FeO_4$			
Iron, tricarbonyl(η^4-2-methyl-4-methylene-5-hexen-3-ol)-	C_6H_{12}	270s(3.40)	101-0061-86D
stereoisomer	C_6H_{12}	275s(3.54)	101-0061-86D
$C_{11}H_{14}N_2O$			
2H-Pyrrol-2-one, 1,5-dihydro-3,4-dimethyl-5-[(1H-pyrrol-2-yl)methyl]-	MeOH	220(3.85)	49-0185-86
$C_{11}H_{14}N_2O_2$			
1H-Benzimidazole, 4,7-dimethoxy-1,2-dimethyl-	H_2O	216(4.70),256(4.11), 280s(--)	44-0522-86

Compound	Solvent	$\lambda_{max}(\log \epsilon)$	Ref.
(cont.)	acid	270(3.96)	44-0522-86
2-Butenoic acid, 3-[(6-methyl-2-pyridinyl)amino]-, methyl ester	EtOH	289s(3.97),318(4.37)	4-1295-86
Propanoic acid, 2-(phenylhydrazono)-, ethyl ester, (E)-	EtOH	325(4.33)	18-0073-86
(Z)-	EtOH	349(4.26)	18-0073-86
Propenoic acid, 3-(3,6-dimethylpyrazin-2-yl)-, ethyl ester, (E)-	EtOH	248(3.76),313(3.73)	4-1481-86
$C_{11}H_{14}N_2O_2S$			
Benzenecarbimidothioic acid, N-(1-methylethyl)-4-nitro-, methyl ester	MeCN	268(4.0)	64-0265-86B
$C_{11}H_{14}N_2O_3$			
1H-Benzimidazole-2-methanol, 4,7-dimethoxy-1-methyl-	H_2O	218(4.72),258(4.07), 280s(3.91)	44-0522-86
	acid	226(4.45),270(3.89)	44-0522-86
Imidazo[1,2-a]pyridine-8-carboxylic acid, 1,2,3,5-tetrahydro-1-methyl-5-oxo-, ethyl ester	EtOH	285(4.18),320(4.08)	24-2208-86
Propanoic acid, 3-methoxy-3-[(6-methyl-2-pyridinyl)imino]-, methyl ester	EtOH	218(3.81),280(3.90), 301s(3.61)	4-1295-86
2-Propenoic acid, 3-(3,6-dimethylpyrazinyl)-, ethyl ester, N-oxide, (E)-	EtOH	254(4.04),282s(3.43), 332(3.30)	4-1481-86
$C_{11}H_{14}N_2O_3S$			
Ethanone, 1-(2,5-dihydro-1-hydroxy-2,5-dimethyl-4-(2-thienyl)-1H-imidazol-2-yl)-, N-oxide	EtOH	224(4.03),320(4.20)	103-0268-86
$C_{11}H_{14}N_2O_3S_2$			
Glycine, N-[(methylthio)[(2-thienylcarbonyl)amino]methylene]-, ethyl ester	EtOH	260(4.07),266s(4.02), 303(4.24)	118-0484-86
$C_{11}H_{14}N_2O_4$			
Ethanone, 1-[4-(2-furanyl)-2,5-dihydro-1-hydroxy-2,5-dimethyl-1H-imidazol-2-yl]-, N-oxide, cis	EtOH	307(4.28)	103-0268-86
$C_{11}H_{14}N_2O_4S$			
Glycine, N-[[(2-furanylcarbonyl)amino](methylthio)methy;ene]-, ethyl ester	EtOH	258s(3.96),266(3.97), 298(4.41)	118-0484-86
$C_{11}H_{14}N_2O_8$			
Uridine, 5-acetoxy-	H_2O	265.5(3.98)	94-4585-86
$C_{11}H_{14}N_3O_2$			
3H-1,2,4-Triazolium, 1-(4-methoxyphenyl)-3,3-dimethyl-5-oxo-, tetrafluoroborate	MeCN	423(4.15)	89-1132-86
$C_{11}H_{14}N_4$			
2-Cyclooctapyrazinecarbonitrile, 3-amino-5,6,7,8,9,10-hexahydro-	MeOH	354(3.95)	33-0793-86
$C_{11}H_{14}N_4O$			
Benzenamine, 4-(4,5-dihydro-1-nitroso-1H-pyrazol-3-yl)-N,N-dimethyl-	EtOH	444(4.52)	104-0947-86

Compound	Solvent	$\lambda_{max}(\log \epsilon)$	Ref.
1H-1,2,4-Triazole-1-carboxamide, 4,5-dihydro-5,5-dimethyl-3-phenyl-	EtOH	232(4.24),324(3.90)	103-0287-86
$C_{11}H_{14}N_4O_2$			
Acetic acid, cyano[6-(dimethylamino)-4(1H)-pyrimidinylidene]-, ethyl ester, (Z)-	EtOH	253(4.24),275(4.20), 333(4.68)	103-0774-86
2H-Imidazo[4,5-c]pyridin-2-one, 1,3,4,5-tetrahydro-4-imino-1,3-dimethyl-5-(2-oxopropyl)-, monohydrobromide	H_2O	229(4.36),253(3.74), 290(3.86)	103-0180-86
$C_{11}H_{14}N_4O_2S_2$			
2,4(1H,3H)-Pteridinedione, 7-(ethyldithio)-1,3,6-trimethyl-	MeOH	220(4.29),254(4.03), 357(4.24)	33-1095-86
$C_{11}H_{14}N_4O_3$			
1H-Imidazo[4,5-c]pyridin-4-amine,	pH 1	262(4.03)	87-0138-86
1-(2-deoxy-β-D-ribofuranosyl)-	pH 13	266(4.03)	87-0138-86
1H-Imidazo[4,5-c]pyridin-4-amine,	pH 1	263(3.96)	87-0138-86
1-(5-deoxy-β-D-ribofuranosyl)-	pH 13	267(3.92)	87-0138-86
2,4(3H,8H)-Pteridinedione, 8-(2-hydroxypropyl)-6,7-dimethyl-	pH 5.0	228(3.81),256(4.20), 278(4.06),313(3.30), 407(4.11)	44-2461-86
	pH 8.0	228(4.02),256(4.20), 278(4.10),313(3.54), 407(4.02)	44-2461-86
	pH 12.0	228(4.51),278(4.16), 313(4.08),407(3.00)	44-2461-86
$C_{11}H_{14}N_4O_4$			
1H-Imidazo[4,5-c]pyridin-4-amine,	pH 1	261(4.02)	87-0138-86
1-β-D-ribofuranosyl-	pH 13	266(4.02)	87-0138-86
1H-Pyrazole-4-carboxamide, 3-(cyano-	pH 1	270(3.32)	4-0059-86
methyl)-1-(2-deoxy-β-D-erythro-	pH 7	274(3.52)	4-0059-86
pentofuranosyl)-	pH 11	282(3.46)	4-0059-86
1H-Pyrazole-4-carboxamide, 5-(cyano-	pH 1	278(3.65)	4-0059-86
methyl)-1-(2-deoxy-β-D-erythro-	pH 7	280(3.79)	4-0059-86
pentofuranosyl)-	pH 11	282(3.75)	4-0059-86
4H-Pyrazolo[4,3-c]pyridin-4-one,	pH 1	285(3.86)	4-0059-86
6-amino-1-(2-deoxy-β-D-erythro-	pH 7	228(4.12)	4-0059-86
pentofuranosyl)-1,5-dihydro-	pH 11	229(4.16),282(3.72)	4-0059-86
4H-Pyrazolo[4,3-c]pyridin-4-one,	pH 1	245(4.02)	4-0059-86
6-amino-2-(2-deoxy-β-D-erythro-pentofuranosyl)-2,5-dihydro-	pH 7	228(4.30),267(3.86), 308s(3.46)	4-0059-86
	pH 11	229(3.39),265(3.69), 314(3.57)	4-0059-86
2H-Pyrrolo[2,3-d]pyrimidin-2-one, 4-amino-7-(2-deoxy-β-D-erythro-pentofuranosyl)-3,7-dihydro-	MeOH	255(3.88),305(3.86)	39-0525-86B
$C_{11}H_{14}N_4O_5$			
1H-Pyrazole-4-carboxamide, 3-(cyano-	pH 1	270(3.77)	4-0059-86
methyl)-1-β-D-ribofuranosyl-	pH 7	274(3.88)	4-0059-86
	pH 11	282(3.81)	4-0059-86
4H-Pyrazolo[4,3-c]pyridin-4-one,	pH 1	245(4.09)	4-0059-86
6-amino-2,5-dihydro-2-β-D-ribofuranosyl-	pH 7	228(3.36),267(3.93), 309s(3.57)	4-0059-86
	pH 11	266(3.95),312s(3.65)	4-0059-86

Compound	Solvent	λ_{max}(log ε)	Ref.
1H-Pyrazolo[3,4-d]pyrimidin-4(5H)-one, 1-(2-deoxy-β-D-erythro-pentofuranosyl)-6-methoxy-	MeOH	246(4.03)	5-1213-86
$C_{11}H_{14}N_4O_6$			
4H-Pyrazolo[3,4-d]pyrimidin-4-one, 1,5-dihydro-3-methoxy-1-β-D-ribofuranosyl-	pH 1	219(4.39),260s(3.54)	4-1869-86
	pH 7	219(4.32),260s(3.52)	4-1869-86
	pH 11	273(3.67)	4-1869-86
$C_{11}H_{14}N_4S$			
1H-1,2,4-Triazole-1-carbothioamide, 4,5-dihydro-5,5-dimethyl-3-phenyl-	EtOH	235(4.23),342(4.11)	103-0287-86
$C_{11}H_{14}N_6O$			
2-Propenamide, N-[2-(6-amino-9H-purin-9-yl)ethyl]-2-methyl-	H_2O	263(4.11)	47-3201-86
polymer	DMSO	264.0(4.08)	47-3201-86
$C_{11}H_{14}O$			
Bicyclo[3.2.0]hepta-3,6-dien-2-one, 6-butyl-, cis	hexane	350(1.81)	39-2005-86C
Bicyclo[5.3.1]undeca-3,5,7-trien-1-ol	EtOH	283(3.64)	78-1823-86
Tricyclo[3.3.0.0^{2,8}]octan-3-one, 1,5-dimethyl-7-methylene-	C_6H_{12}	207(4.03),285(1.59)	33-0071-86
	EtOH	202(4.07),275(1.89)	33-0071-86
Tricyclo[5.4.0.0^{7,11}]undec-9-en-2-one, cis	MeCN	282(1.26)	35-4881-86
trans	C_6H_{12}	288(2.00)	35-4881-86
$C_{11}H_{14}O_2$			
Bicyclo[2.2.2]oct-7-ene-3,6-dione, 2,2,5-trimethyl-, (-)-	hexane	228(3.34),289(2.20), 299(2.20),309(2.19), 321(2.05)	35-4149-86
Bicyclo[5.3.1]undeca-1,3-dien-5-one, 7-hydroxy-	EtOH	298(3.76)	78-1823-86
2-Cyclopenten-1-one, 4-(1-cyclopenten-1-ylmethoxy)-, (±)-	EtOH	215(3.88),320(1.79)	32-0291-86
Ethanone, 1-(3',3'-dimethylspiro[bicyclo[3.1.0]hex-3-ene-2,2'-oxiran]-5-yl)-, (1α,2β,5α)-	EtOH	260(3.46)	78-3311-86
2-Oxatetracyclo[4.4.1^{4,9}.0^{1,4}]dodecan-3-one	pentane	217(1.68)	78-1581-86
$C_{11}H_{14}O_3$			
Bicyclo[3.1.0]hex-3-en-2-one, 6-acetoxy-3,5,6-trimethyl-, (1α,5α,6α)-(±)-	MeOH	225.5(3.78),259(3.42), 333(2.40)	33-0469-86
2,5-Cyclohexadien-1-one, 4-acetoxy-2,4,5-trimethyl-, (±)-	MeOH	243.5(4.13),237[sic] (1.78),355(1.40)	33-0469-86
1-Propanone, 1-(2-hydroxy-3-methoxy-5-methylphenyl)-	EtOH	224(4.22),266(3.90)	22-0659-86
1-Propanone, 1-(3-hydroxy-2-methoxy-5-methylphenyl)-	EtOH	215(4.23),255(3.73), 307(2.30)	22-0659-86
1-Propanone, 1-(4-hydroxy-5-methoxy-2-methylphenyl)-	EtOH	229(4.20),274(3.94), 306(3.78)	22-0659-86
1-Propanone, 1-(5-hydroxy-4-methoxy-2-methylphenyl)-	EtOH	231(4.26),272(3.91), 312(3.73)	22-0659-86
2-Propen-1-ol, 3-(3,4-dimethoxyphenyl)-	EtOH	205(4.56),260(4.34)	100-0677-86
$C_{11}H_{14}O_4S$			
Acetic acid, [[[3-(5-oxo-1,3-pentadienyl)oxiranyl]methyl]thio]-, methyl	EtOH	278(4.44)	104-1598-86 +104-1875-86

Compound	Solvent	$\lambda_{max}(\log \epsilon)$	Ref.
ester, (E)- (cont.)			104-1875-86
$C_{11}H_{14}O_6$			
2H-Pyran-4-acetic acid, 3-formyl-3,4-dihydro-5-(methoxycarbonyl)-2-methyl- (elenolic acid)	MeOH	237(3.94)	102-0865-86
$C_{11}H_{15}ClN_2O_6$			
1H-Imidazole-4-carboxylic acid, 2-chloro-5-methyl-1-β-D-ribofuranosyl-, methyl ester	H_2O	241(4.07)	78-1971-86
$C_{11}H_{15}ClN_4O$			
9H-Purine, 9-(1-butoxyethyl)-6-chloro-	EtOH	266(4.02)	103-0331-86
$C_{11}H_{15}N$			
2-Butenenitrile, 4-(4-methylcyclohexylidene)-, [S-(E)]-	C_6H_{12}	266(4.53)	44-2863-86
[S-(Z)]-	C_6H_{12}	268(4.46)	44-2863-86
Piperidine, N-phenyl-	acid	253f(2.3)	112-1099-86
Pyridine, 4-cyclohexyl-	n.s.g.	255f(3.2)	94-0007-86
$C_{11}H_{15}NO$			
Benzenamine, 2-(4-pentenyloxy)-	EtOH	236(3.71),286(3.30)	33-0927-86
Benzenamine, 3-(4-pentenyloxy)-	EtOH	236.5(3.82),286(3.32)	33-0927-86
2-Butanone, 3,3-dimethyl-1-(2-pyridinyl)-	CF_3COOH	263(3.88)	116-0291-86
copolymers	CF_3COOH	263(3.89)	116-0291-86
$C_{11}H_{15}NO_2$			
Isoquinoline, 1,2,3,4-tetrahydro-6,7-dimethoxy- (heliamine)	aq HCl	203(4.5),220(3.8), 284(3.6),288s(3.6)	100-0745-86
Isoquinoline, 1,2,3,4-tetrahydro-7,8-dimethoxy-	aq HCl	223s(3.9),278(3.4), 286(3.4),293s(3.3)	100-0745-86
6-Isoquinolinol, 1,2,3,4-tetrahydro-7-methoxy-1-methyl- (salsoline)	MeOH	210(4.14),225(3.81), 286(3.56),290(3.51)	100-0745-86
(+)-	isoPrOH-HCl	204(4.60),227(3.77), 284(3.55),286(3.55)	100-0745-86
6-Isoquinolinol, 1,2,3,4-tetrahydro-7-methoxy-2-methyl- (isocorypalline)	EtOH	227(3.55),286(3.44)	100-0745-86
	EtOH-NaOH	247(3.69),301(3.55)	100-0745-86
7-Isoquinolinol, 1,2,3,4-tetrahydro-6-methoxy-1-methyl-, (+)-	isoPrOH-HCl	204(4.62),227s(3.89), 286(3.56)	100-0745-86
7-Isoquinolinol, 1,2,3,4-tetrahydro-6-methoxy-2-methyl- (corypalline)	MeOH	202(4.43),225(3.68), 285(3.56)	100-0745-86
	MeOH-NaOH	245(3.97),293(3.92)	100-0745-86
8-Isoquinolinol, 1,2,3,4-tetrahydro-7-methoxy-2-methyl-	isoPrOH	233(3.75),283(3.44)	100-0745-86
	pH 13	247(3.85),292(3.60)	100-0745-86
1-Penten-3-one, 1-[5-(dimethylamino)-2-furanyl]-, (E)-	MeOH	243(2.95),435(3.38)	73-0879-86
2-Propenoic acid, 2-methyl-3-(1H-pyrrol-2-yl)-, 1-methylethyl ester, (E)-	EtOH	323(4.39)	150-3601-86M
$C_{11}H_{15}NO_3$			
6-Isoquinolinol, 1,2,3,4-tetrahydro-7,8-dimethoxy- (isoanhalamine)	isoPrOH	230s(3.90),273s(3.20), 282(3.26)	100-0745-86
	acid	223s(3.92),273s(3.11), 279(3.15)	100-0745-86
	base	240s(3.89),295(3.56)	100-0745-86
8-Isoquinolinol, 1,2,3,4-tetrahydro-6,7-dimethoxy- (anhalamine)	isoPrOH	227s(3.98),272(2.90), 280s(2.84)	100-0745-86

Compound	Solvent	λ_{max}(log ε)	Ref.
Anhalamine (cont.)	pH 1	225s(3.92),270(2.88), 280s(2.70)	100-0745-86
	pH 13	240s(3.90),285(3.38)	100-0745-86
$C_{11}H_{15}NO_4$			
α-D-xylo-Hexofuranurononitrile, 5-deoxy-3-O-methyl-5-methylene-1,2-O-(1-methylethylidene)-	n.s.g.	220(3.00)	111-0111-86
3,5-Pyridinedicarboxylic acid, 1,4-dihydro-2,6-dimethyl-, dimethyl ester	MeOH	226(--),258(--), 271(--),368(3.94)	87-1596-86
$C_{11}H_{15}NO_5$			
2-Furancarboxylic acid, 5-[(diethoxymethylene)amino]-, methyl ester	MeOH	305(4.49)	73-2817-86
4H-Pyran-4-one, 3-hydroxy-6-(hydroxymethyl)-2-(4-morpholinomethyl)-	cation	274(4.01)	137-0119-86
	neutral	274(3.96)	137-0119-86
	anion	324(3.89)	137-0119-86
$C_{11}H_{15}N_2O_8P$			
Phosphonic acid, [[3,3a-dihydro-3-hydroxy-2-(hydroxymethyl)-6-oxo-6H-furo[2',3':4,5]oxazolo[3,2-a]pyrimidin-9a(2H)-yl]methyl]-, monomethyl ester	EtOH	222(3.87),253(3.82)	142-0617-86
$C_{11}H_{15}N_3$			
Cyclopenta[c]pyrrole-1,3-diamine, N,N,N',N'-tetramethyl-	hexane	534(2.76)	24-3213-86
	dioxan	504(2.74)	24-3213-86
	CH_2Cl_2	489(2.73)	24-3213-86
$C_{11}H_{15}N_3O$			
4H-Furo[2,3-d]pyrido[1,2-a]pyrimidin-4-imine, 2,3,6,7,8,9-hexahydro-2-methyl-	$CHCl_3$	253(4.17),293(3.58)	24-1070-86
2H,5H-Pyrano[2,3-d]pyrido[1,2-a]pyrimidin-5-imine, 3,4,7,8,9,10-hexahydro-	$CHCl_3$	252(3.94),272s(3.56)	24-1070-86
Triacsin B	70% MeCN	363(4.59)	158-1211-86
$C_{11}H_{15}N_3O_2$			
2-Quinoxalinecarboxylic acid, 3-amino-4,6,7,8-tetrahydro-4-methyl-, methyl ester	EtOH	360(3.68)	33-1025-86
$C_{11}H_{15}N_3O_4$			
2,5-Methano-5H-pyrimido[1,2-c][1,3,5]-oxadiazepin-9(1H)-one, 1-ethyl-2,3-dihydro-11-hydroxy-3-(hydroxymethyl)-	MeOH	220(4.36),227(4.32), 258s(3.76)	44-4417-86
2,6-Methano-2H,6H-pyrimido[1,2-c]-[1,3,5]oxadiazocin-10(1H)-one, 1-ethyl-3,4-dihydro-3,12-dihydroxy-	MeOH	218.5(4.46)	44-4417-86
$C_{11}H_{15}N_3O_4Se$			
1,2,4-Triazine-5,6-diacetic acid, 2,3-dihydro-3-selenoxo-, diethyl ester	MeOH	283(4.3),309(4.4)	56-0541-86
	EtOH	282(4.1),309(4.2)	56-0541-86
$C_{11}H_{15}N_3O_5$			
Pyrrole-3,4-dicarboxamide, 1-(2-deoxy-β-D-erythro-pentofuranosyl)-	pH 1	256(4.04)	78-5869-86
	pH 7	252(4.03)	78-5869-86
	pH 11	251(4.03)	78-5869-86

Compound	Solvent	$\lambda_{max}(\log \epsilon)$	Ref.
$C_{11}H_{15}N_3O_6$			
1H-Pyrrole-3,4-dicarboxamide, 1-β-D-arabinofuranosyl-	pH 1	256(3.99)	78-5869-86
	pH 7	250(3.90)	78-5869-86
	pH 11	249(3.94)	78-5869-86
$C_{11}H_{15}N_3S$			
Thieno[2',3':4,5]pyrimido[1,2-a]azepin-4(6H)-imine, 2,3,7,8,9,10-hexahydro-	$CHCl_3$	248(4.42),318(3.74)	24-1070-86
$C_{11}H_{15}N_4O_8P$			
4H-Pyrazolo[4,3-c]pyridin-4-one, 6-amino-2,5-dihydro-2-(5-O-phosphono-β-D-ribofuranosyl)-	pH 1	246(4.27)	4-0059-86
	pH 7	267(4.24),309s(3.88)	4-0059-86
	pH 11	266(4.02),317s(4.05)	4-0059-86
$C_{11}H_{15}N_5$			
7H-Purin-6-amine, N-methyl-7-(3-methyl-2-butenyl)-	MeOH	272s(4.09),277(4.10)	18-2495-86
9H-Purin-6-amine, N-methyl-9-(3-methyl-2-butenyl)-	MeOH	267(4.21)	18-2495-86
$C_{11}H_{15}N_5O$			
9H-Purin-6-amine, N-(4-hydroxy-1,3-dimethyl-2-buten-1-yl)-, trans	pH 7	269(4.07)	102-0525-86
	pH 10.0	275(4.08)	102-0525-86
	EtOH	270(4.06)	102-0525-86
$C_{11}H_{15}N_5OS$			
7(1H)-Pteridinethione, 2-amino-4-(pentyloxy)-	pH -2.0	220(4.29),314(3.87), 368(4.15),396(4.17)	33-0708-86
	pH 4.0	234(4.44),298(3.82), 412(4.31)	33-0708-86
	pH 10.0	233(4.36),253(4.28), 286(3.82),387(4.28)	33-0708-86
$C_{11}H_{15}N_5O_2$			
Pyrido[2,3-d]pyrimidine-2,4-diamine, 6-[(methoxymethoxy)methyl]-5-methyl-	pH 1	223(4.54),245s(4.21), 271s(3.77),316(4.02), 326(3.99)	87-0709-86
	H_2O	223(4.53),245s(4.29), 272s(3.92),318s(3.98), 327.5(4.00)	87-0709-86
	pH 13	221(4.52),247(4.37), 272(4.03),340(4.96)	87-0709-86
$C_{11}H_{15}N_5O_3$			
Adenosine, 2'-deoxy-N-methyl-	H_2O	265(4.19)	33-1034-86
Adenosine, 3'-deoxy-3'-C-methyl-	EtOH	259(4.10)	11-0013-86
1,2-Cyclopentanediol, 3-(6-amino-9H-purin-9-yl)-5-(hydroxymethyl)- (±-aristeromycin)	pH 1	260(4.13)	44-1287-86
	pH 13	262(4.14)	44-1287-86
2,4(1H,3H)-Pyrimidinedione, 1-[3-azido-4-(hydroxymethyl)cyclopentyl]-5-methyl-, (1α,3α,4α)-(±)-	pH 1	273(4.01)	87-0483-86
	pH 7	273(4.02)	87-0483-86
	pH 13	271(3.90)	87-0483-86
(1α,3β,4α)-(±)-	pH 1	211(3.99),273(4.04)	87-0483-86
	pH 7	210(4.00),273(4.04)	87-0483-86
	pH 13	271(3.92)	87-0483-86
$C_{11}H_{15}N_5O_4$			
Ethanehydrazonamide, N'-(2,4-dinitrophenyl)-N,N,N'-trimethyl-, (E)-	dioxan	401(4.38)	39-0537-86B
	NaOH	413(--)	39-0537-86B

Compound	Solvent	λ_{max}(log ε)	Ref.
Ethanehydrazonamide, N'-(2,4-dinitro-phenyl)-N,N,N'-trimethyl-, (Z)-	dioxan	406(4.23)	39-0537-86B
	NaOH	420(--)	39-0537-86B
4(1H)-Pteridinone, 2-(dimethylamino)-6-(D-erythro-1,2,3-trihydroxy-propyl)-	pH -1.0	214(4.18),248(4.23), 328(3.88),415(2.86)	5-1705-86
	pH 4.0	220(4.18),246s(3.97), 287(4.28),350(3.74), 370s(3.74)	5-1705-86
	pH 11.0	227(4.06),272(4.38), 384(3.90)	5-1705-86
threo-	pH -1.0	215(4.16),248(4.23), 328(3.87),415(2.75)	5-1705-86
	pH 4.0	224(4.11),245s(4.01), 288(4.28),350(3.75), 370s(3.65)	5-1705-86
	pH 11.0	229(4.06(4.06),273(4.38), 384(3.88)	5-1705-86
L-erythro-	pH -1.0	214(4.17),248(4.23), 328(3.84),415(2.87)	5-1705-86
	pH 4.0	221(4.16),244s(4.01), 287(4.28),350(3.75), 370s(3.67)	5-1705-86
	pH 11.0	228(4.07),272(4.37), 384(3.89)	5-1705-86
threo-	pH -1.0	215(4.20),248(4.27), 329(3.91),415(2.93)	5-1705-86
	pH 4.0	226(4.14),244s(4.05), 287(4.30),350(3.77), 370s(3.67)	5-1705-86
	pH 11	229(4.08),273(4.39), 384(3.90)	5-1705-86
1H-Pyrazolo[3,4-d]pyrimidin-4-amine, 1-(2-deoxy-β-D-erythro-pento-furanosyl)-6-methoxy-	MeOH	264(4.03)	5-1213-86
1H-Pyrazolo[3,4-d]pyrimidin-6-amine, 1-(2-deoxy-α-D-erythro-pento-furanosyl)-4-methoxy-	MeOH	223(4.36),253(3.87), 277(3.91)	33-1602-86
β-	MeOH	223(4.37),253(3.86), 277(3.94)	33-1602-86
2H-Pyrazolo[3,4-d]pyrimidin-6-amine, 2-(2-deoxy-β-D-erythro-pento-furanosyl)-4-methoxy-	MeOH	221(4.47),266(3.83), 295(3.87)	33-1602-86
$C_{11}H_{15}N_5O_4S$			
7-Pteridinesulfonic acid, 2-amino-4-(pentyloxy)-	pH 7.0	243(4.40),265s(4.02), 372(3.96)	33-0708-86
$C_{11}H_{15}N_5O_5$			
1H-Pyrazolo[3,4-d]pyrimidin-4-amine, 3-methoxy-1-β-D-ribofuranosyl)-	pH 1	230(4.30)	4-1869-86
	pH 7 and 11	248(3.63),278(3.56)	4-1869-86
$C_{11}H_{15}N_5O_6$			
Pyrazolo[3,4-d]pyrimidin-4-one, 6-am-ino-1,5-dihydro-3-methoxy-1-β-D-ribofuranosyl-	pH 1	253(3.89)	4-1869-86
	pH 7	256(3.93)	4-1869-86
	pH 11	255(3.95)	4-1869-86
$C_{11}H_{15}N_7S_2$			
Hydrazinecarbothioamide, 2,2'-(2,6-pyridinediyldiethylidyne)bis-	DMF	330(4.63)	59-0701-86
$C_{11}H_{16}$			
Bicyclo[4.1.0]hept-2-ene, 3,7,7-tri-	C_6H_{12}	257(3.91)	5-1104-86

Compound	Solvent	λ_{max}(log ε)	Ref.
methyl-4-methylene-, (1R,6R)-(+)-			5-1104-86
$C_{11}H_{16}BrN_3O_6S$			
Thymidine, 5'-[[(bromomethyl)sulfonyl]amino]-5'-deoxy-	pH 1	265(3.99)	87-1052-86
	pH 7	266(3.99)	87-1052-86
	pH 13	266(3.88)	87-1052-86
$C_{11}H_{16}N_2$			
1,4-Ethano-6,9-methanopyridazino[1,2-a]pyridazine, 1,4,6,7,8,9-hexahydro-, radical cation, nitrate	MeCN	280(3.08),304s(--)	35-7926-86
$C_{11}H_{16}N_2O$			
Methanone, cyclohexyl(1-methyl-1H-imidazol-2-yl)-	EtOH	278(4.13)	94-4916-86
$C_{11}H_{16}N_2O_2$			
1H-Pyrrolizine, 3-(1,1-dimethylethyl)-2,3-dihydro-5-nitro-	EtOH	355(4.29)	103-0257-86
$C_{11}H_{16}N_2O_3$			
Benzenamine, N-butyl-2-methoxy-4-nitro-	MeOH	230(2.79),266(2.66), 402(3.13)	73-1665-86
Benzenamine, N-butyl-2-methoxy-5-nitro-	MeOH	233(2.57),263(2.72), 307(2.18),400(2.00)	73-1665-86
2-Furancarboxylic acid, 5-[[1-[(1-methylethyl)amino]ethylidene]amino]-, methyl ester	MeOH	320(3.15)	73-2826-86
1H-Indazole-4-carboxylic acid, 4,5,6,7-tetrahydro-3-hydroxy-1-methyl-, ethyl ester	EtOH	207(3.95),230(3.93), 257s(3.28)	78-2377-86
2H-Indazole-4-carboxylic acid, 4,5,6,7-tetrahydro-3-hydroxy-2-methyl-, ethyl ester	EtOH	206(3.84),251(3.94)	78-2377-86
$C_{11}H_{16}N_2O_4$			
Isoxazolo[4,3-c]pyridine-5-carboxylic acid, 2,3,6,7-tetrahydro-3-oxo-, 1,1-dimethylethyl ester	MeOH	261(3.91)	87-0224-86
2,4(1H,3H)-Pyrimidinedione, 1-[3-hydroxy-4-(hydroxymethyl)cyclopentyl]-5-methyl-	pH 1	273(4.03)	87-0483-86
	pH 7	273(4.03)	87-0483-86
	pH 13	271(3.90)	87-0483-86
$C_{11}H_{16}N_2O_7$			
2,4,6(1H,3H,5H)-Pyrimidinetrione, 1,3-dimethyl-5-α-D-arabinopyranosyl-	H_2O	259(4.14)	136-0053-86M
2,4,6(1H,3H,5H)-Pyrimidinetrione, 1,3-dimethyl-5-β-D-xylopyranosyl-	H_2O	259(4.17)	136-0053-86M
$C_{11}H_{16}N_4O$			
2H-Imidazo[4,5-c]pyridin-2-one, 1,3,4,5-tetrahydro-4-imino-1,3-dimethyl-5-propyl-, HI salt	H_2O	232(4.70),259(3.62), 302(3.76)	103-0180-86
$C_{11}H_{16}N_4OS$			
Hydrazinecarbothioamide, 2-[2-(hydroxyamino)-2-methyl-1-phenylpropylidene]-	EtOH	234s(4.10),278(4.40)	103-0287-86

Compound	Solvent	λ_{max}(log ε)	Ref.
$C_{11}H_{16}N_4O_2$			
Hydrazinecarboxamide, 2-[2-(hydroxy-amino)-2-methyl-1-phenylpropyli-dene]-	EtOH	234s(4.09)	103-0287-86
$C_{11}H_{16}N_4O_5$			
Glycine, N-[5-(2-amino-2,4-dihydro-6-hydroxy-4-oxo-5-pyrimidinyl)-1-oxopentyl]-	50%MeOH-HCl	266(4.13)	73-1140-86
	50%MeOH-NaOH	270(4.10)	73-1140-86
Pyrazolo[3,4-d][1,3]diazepin-4-ol, 1,4,5,6-tetrahydro-1-β-D-ribo-furanosyl-	pH 1	235(3.77),263(3.82)	44-1050-86
	pH 11	238(3.66),277(3.97)	44-1050-86
	MeOH	241(3.60),278(3.95)	44-1050-86
$C_{11}H_{16}N_5O_3$			
1-Adenosinium, 2'-deoxy-1-methyl-, iodide	H_2O	257(4.16)	33-1034-86
$C_{11}H_{16}N_5O_4P$			
Phosphonic acid, [1-[(6,7-dihydro-6-oxo-1H-purin-2-yl)amino]cyclo-hexyl]-	pH 1	254.5(4.18),280s(3.85)	94-0471-86
	H_2O	252(4.15),275s(3.95)	94-0471-86
	pH 13	262(4.04),280s(3.97)	94-0471-86
$C_{11}H_{16}N_5O_7P$			
4(1H)-Pteridinone, 6-[1,2-dihydroxy-3-(phosphonooxy)propyl]-2-(dimeth-ylamino)-, barium salt	pH -1.0	213(4.19),247(4.24), 327(3.89),415(2.89)	5-1705-86
	pH 4.0	221(4.17),286(4.27), 350(3.76)	5-1705-86
	pH 11.0	225(4.07),272(4.38), 384(3.90)	5-1705-86
$C_{11}H_{16}O$			
2-Butenal, 4-(4-methylcyclohexyli-dene)-, [S-(E)]-	C_6H_{12}	278(4.54)	44-2863-86
[S-(Z)]-	C_6H_{12}	287(4.35)	44-2863-86
4-Hepten-1-yn-3-one, 6-methyl-5-(1-methylethyl)-	EtOH	232(3.74),268(4.08)	33-0560-86
1-Naphthalenol, 1,2,3,4,6,8a-hexa-hydro-1-methyl-	hexane	210(3.31)	35-4676-86
2(3H)-Naphthalenone, 4,4a,5,6,7,8-hexahydro-4a-methyl-	EtOH	237(4.20),312(1.75)	35-4561-86
2(1H)-Pentalenone, hexahydro-3a,6a-dimethyl-5-methylene-, cis	C_6H_{12}	295(1.32)	33-0071-86
	EtOH	290(1.36)	33-0071-86
Phenol, 4-(1,1-dimethylpropyl)-	acid	275(3.19)	3-0639-86
Tricyclo[3.3.1.1^{3,7}]decanone, 1-meth-yl-, axial	C_6H_{11}Me-iso-pentane	290(1.23)	35-4484-86
	EFA	287(1.32)	35-4484-86
equatorial	C_6H_{11}Me-iso-pentane	293(1.28)	35-4484-86
	EFA	290(1.34)	35-4484-86
$C_{11}H_{16}O_2$			
Bicyclo[5.3.1]undeca-1,3-diene-5,7-diol	EtOH	243(3.75)	78-1823-86
1-Butanol, 2-(phenylmethoxy)-	MeOH	213(3.58)	32-0025-86
2,4-Undecadiyne-1,6-diol, (S)-	ether	220(2.59),232(2.63), 244(2.59),257(2.34)	150-1348-86
$C_{11}H_{16}O_3$			
Cyclopentanecarboxylic acid, 1-(3-butenyl)-2-oxo-, methyl ester	MeCN	215(2.37),280(2.06)	44-4196-86

Compound	Solvent	λ_{max}(log ε)	Ref.
2,9-Dioxabicyclo[3.3.1]non-7-en-6-one, 3-ethyl-1,8-dimethyl-, (1R,3S,5R)-	EtOH	226(4.04)	78-5281-86
1,7-Dioxaspiro[5.5]undec-2-en-4-one, 2,3-dimethyl-	MeOH	272(3.84)	44-4254-86
7-Oxabicyclo[2.2.1]heptane, 1-(dimethoxymethyl)-2,3-bis(methylene)-	isooctane	237s(3.85),243(3.98), 251s(3.94)	33-1287-86
	EtOH	237s(3.85),242(3.98), 250s(3.95)	33-1287-86
$C_{11}H_{16}O_4$			
2-Furancarboxylic acid, 4-(butoxymethyl)-5-methyl-	MeOH	264(3.26)	73-2186-86
2,4-Heptadienoic acid, 6-acetoxy-2-methyl-, methyl ester, (2E,4E,6RS)-	MeOH	259.5(4.46)	33-0469-86
3,5-Heptadienoic acid, 6-acetoxy-2-methyl-, methyl ester, (2RS,3E,5Z)-	MeOH	235(4.37)	33-0469-86
(2RS,3Z,5E)-	MeOH	237.5(4.35)	33-0469-86
3,5-Heptadienoic acid, 6-acetoxy-5-methyl-, methyl ester, (3Z,5E)-	MeOH	218(3.79)	33-0469-86
$C_{11}H_{16}O_5$			
2H-Pyran-2-one, 6-(1,2-dihydroxypentyl)-4-methoxy-, [S-(R*,S*)]-	MeOH	225s(3.54),280(3.92)	163-1649-86
$C_{11}H_{16}O_6$			
2-Butenoic acid, 4-acetoxy-3-(acetoxymethyl)-, ethyl ester	EtOH	209(4.14)	150-0222-86S
$C_{11}H_{17}BrO_2$			
2,4-Hexadienoic acid, 4-bromo-, pentyl ester	n.s.g.	268(4.24)	104-2230-86
$C_{11}H_{17}ClN_2O_2$			
2,4(1H,3H)-Pyrimidinedione, 3-(6-chlorohexyl)-6-methyl-	pH 2	263(3.96)	70-1064-86
	pH 12	282(4.22)	70-1064-86
$C_{11}H_{17}NO_2$			
3-Pyridinecarboxylic acid, 1,4-dihydro-1,4,4,5-tetramethyl-, methyl ester	MeOH	240(3.45),280(3.11), 356(3.71)	118-0190-86
3-Pyridinecarboxylic acid, 1,2,3,4-tetrahydro-1-methyl-3-(2-propenyl)-, (3RS)-, hydrochloride	pH 7.2	210(0.10)	87-0125-86
$C_{11}H_{17}N_2O_9P$			
2,4(1H,3H)-Pyrimidinedione, 1-[1-deoxy-1-(hydroxymethoxyphosphinyl)-β-D-fructofuranosyl]-	EtOH	263(4.04)	142-2133-86
$C_{11}H_{17}N_3O_2$			
Acetic acid, cyano(1-methyl-2-imidazolidinylidene)-, butyl ester, (E)-	EtOH	210(4.31),260(4.46)	24-2208-86
$C_{11}H_{17}N_3O_3$			
2,4(1H,3H)-Pyrimidinedione, 1-[3-amino-4-(hydroxymethyl)cyclopentyl]-5-methyl-, (1α,3α,4α)-(±)-	pH 1	272(4.01)	87-0483-86
	pH 7	271(4.01)	87-0483-86
	pH 13	272(3.91)	87-0483-86
(1α,3β,4α)-(±)-	pH 1	272(4.00)	87-0483-86
	pH 7	273(4.01)	87-0483-86
	pH 13	271(3.90)	87-0483-86

Compound	Solvent	λmax(log ε)	Ref.
2,4(1H,3H)-Pyrimidinedione, 1-[3-(aminomethyl)-4-hydroxycyclopentyl]-5-methyl-, (1α,3β,4α)-(±)-	pH 1	211(3.94),273(4.00)	87-0483-86
	pH 7	209(3.95),273(4.02)	87-0483-86
	pH 13	272(3.91)	87-0483-86
$C_{11}H_{17}N_3O_4$			
Thymidine, 5'-deoxy-5'-(methylamino)- (spectra in methanol)	pH 1	265(3.99)	87-1052-86
	pH 7	265(3.99)	87-1052-86
	pH 13	266(3.89)	87-1052-86
$C_{11}H_{17}N_3O_5$			
5-Pyrimidinepentanoic acid, 2-amino-1,4-dihydro-6-hydroxy-4-oxo-, 2-hydroxyethyl ester	50%MeOH-HCl	266(4.35)	73-1140-86
	50%MeOH-NaOH	269(4.10)	73-1140-86
$C_{11}H_{17}N_3O_6$			
1H-Pyrazole-5-carboxamide, 4-methoxy-1-methyl-3-β-D-ribofuranosyl-	pH 1 and 7	228(3.79)	87-0268-86
	pH 11	228(3.78)	87-0268-86
1H-Pyrazole-4-carboxylic acid, 3-amino-1-β-D-ribofuranosyl-, ethyl ester	pH 2	263(3.76)	73-1512-86
	pH 7 and 12	263(3.80)	73-1512-86
$C_{11}H_{17}N_5$			
6H-Purin-6-amine, N-butyl-9-ethyl-, monoperchlorate	pH 1	266(4.27)	94-1094-86
	pH 7 and 13	270(4.25)	94-1094-86
	EtOH	270(4.23)	94-1094-86
$C_{11}H_{18}$			
1,3,7-Nonatriene, 4,8-dimethyl-, (E)-	EtOH	235(4.60)	88-2111-86
(Z)-	EtOH	236(4.44)	88-2111-86
1,4,6-Undecatriene, (Z,Z)-	EtOH	234(4.26)	39-1809-86C
$C_{11}H_{18}N_2$			
1,4-Ethano-6,9-methanopyridazino[1,2-a]pyridazine, octahydro-, radical cation, nitrate	MeCN	283(3.26)	35-7926-86
$C_{11}H_{18}N_2O$			
1-Heptanone, 1-(1-methyl-1H-imidazol-2-yl)-	EtOH	278(4.12)	94-4916-86
$C_{11}H_{18}N_2OS$			
1H-Imidazole, 2,5-dihydro-1-methoxy-2,2,5,5-tetramethyl-4-[(methylthio)ethynyl]-	n.s.g.	220(3.96),267(4.0)	88-1625-86
$C_{11}H_{18}N_2O_4$			
Propanedioic acid, (1-methyl-2-imidazolidinylidene)-, diethyl ester	EtOH	270(4.11)	24-2208-86
2,4,6(1H,3H,5H)-Pyrimidinetrione, 5-(2-butoxypropyl)-	pH 10.0	268(4.11)	39-1391-86B
$C_{11}H_{18}N_3O_9P$			
1H-Pyrazole-3-carboxamide, 5-[5-O-(dimethoxyphosphinyl)-β-D-ribofuranosyl]-4-hydroxy-	pH 1	262.5(3.81)	87-0268-86
	pH 7	265.5(3.67),305(3.46)	87-0268-86
	pH 11	307(3.95)	87-0268-86
1H-Pyrazole-3-carboxamide, 5-[5-O-(ethoxyhydroxyphosphinyl)-β-D-ribofuranosyl-4-hydroxy-	pH 1	218(3.82),263(3.74)	87-0268-86
	pH 7	263(3.70),307s(3.20)	87-0268-86
	pH 11	234s(3.72),307(3.93)	87-0268-86
$C_{11}H_{18}N_4O_6$			
1H-Pyrazole-4-carboxylic acid, 3,5-diamino-1-β-D-ribofuranosyl-, ethyl ester	pH 2	218(4.11),242(4.43)	73-1512-86
	pH 7 and 12	244(4.16)	73-1512-86

Compound	Solvent	λmax(log ε)	Ref.
$C_{11}H_{18}N_5O_{13}P_3$			
Adenosine 5'-(tetrahydrogen triphosphate, 2'-O-methyl-, tetrasodium salt	pH 6.8	260(4.18)	87-0318-86
Adenosine 5'-(tetrahydrogen triphosphate, 3'-O-methyl-, tetrasodium salt	pH 6.8	260(4.17)	87-0318-86
$C_{11}H_{18}N_5O_{13}P_3S$			
Adenosine 5'-(tetrahydrogen triphosphate), 2-(methylthio)-, tetrasodium salt	pH 1	270(4.20)	87-0318-86
	pH 6.8	273(4.19)	87-0318-86
$C_{11}H_{18}O$			
2-Buten-1-ol, 4-(4-methylcyclohexylidene)-, [S-(E)]-	C_6H_{12}	243(4.48)	44-2863-86
[S-(Z)]-	C_6H_{12}	244.5(4.46)	44-2863-86
$C_{11}H_{18}OSi$			
4-Hexen-1-yn-3-one, 4,5-dimethyl-1-(trimethylsilyl)-	EtOH	241(3.80),282(3.91)	33-0560-86
$C_{11}H_{18}O_2$			
2-Cyclopenten-1-one, 2-hexyl-4-hydroxy-	MeOH	221(4.00)	2-0675-86
Ethanone, 1-(3,4-dihydro-2,2,3,6-tetramethyl-2H-pyran-5-yl)-	MeOH	263(4.13)	35-1234-86
Ethanone, 1-[5-(1,1-dimethylethyl)-4,5-dihydro-2-methyl-3-furanyl]-	MeOH	260(4.10)	35-1234-86
8-Nonynoic acid, ethyl ester	MeOH	206(2.43)	18-3535-86
7-Octynoic acid, 1-methylethyl ester	MeOH	205(2.28)	18-3535-86
10-Undecynoic acid	MeOH	206(2.73)	18-3535-86
$C_{11}H_{18}O_3S$			
2-Hexenoic acid, 2-(butylthio)-5-methyl-4-oxo-	MeOH	275(4.09),310(3.95)	48-0539-86
$C_{11}H_{18}O_6$			
D-arabino-Hept-2-enonic acid, 2,3-dideoxy-4,5-O-(1-methylethylidene)-, methyl ester, (E)-	MeOH	222(4.03)	136-0283-86G
$C_{11}H_{19}ClOSi$			
1,4-Hexadien-3-one, 1-chloro-4,5-dimethyl-1-(trimethylsilyl)-	EtOH	255(3.85)	33-0560-86
$C_{11}H_{19}NO_2$			
1,2-Cyclohexanedione, 3,3,6,6-tetramethyl-, 1-(O-methyloxime)-, (Z)-	C_6H_{12}	230(3.45),240(3.45), 315(1.74)	54-0054-86
	EtOH	242(3.36),315(1.78)	54-0054-86
1,2-Propanedione, 1-cyclooctyl-, 1-oxime	EtOH	225(4.00)	118-0473-86
$C_{11}H_{19}NO_3$			
5-Heptenoic acid, 6-(butylamino)-4-oxo-, cis	H_2O	312(4.36)	35-5589-86
3-Pyridinecarboxylic acid, 1,2,5,6-tetrahydro-5-(1-hydroxy-1-methylethyl)-1-methyl-, methyl ester, (±)-	pH 7.2	218(1.03)	87-0125-86

Compound	Solvent	λ_{max}(log ϵ)	Ref.
$C_{11}H_{19}N_3$			
2-Quinazolinamine, 1,4,5,6,7,8-hexahydro-4,4,7-trimethyl-, (R)-	EtOH	204(3.76),254(3.53)	4-0705-86
$C_{11}H_{19}N_3O$			
Triacsin A	70% MeCN	300(4.60)	158-1211-86
$C_{11}H_{19}N_5O$			
Methanimidamide, N'-[2-(dimethylamino)-1-(3-ethyl-1,2,4-oxadiazol-5-yl)ethenyl]-N,N-dimethyl-	MeOH	205(4.22),330(4.21)	5-1749-86
$C_{11}H_{19}N_5O_6$			
1,2-Hydrazinedicarboxylic acid, 1-[(1-methyl-2-imidazolidinylidene)nitromethyl]-, diethyl ester, (E)-	EtOH	330(4.27)	24-2208-86
$C_{11}H_{19}N_6O_{13}P_3$			
Adenosine (tetrahydrogen triphosphate), 8-(methylamino)-, tetrasodium salt	pH 1	277(4.17)	87-0318-86
	pH 6.8	279(4.24)	87-0318-86
$C_{11}H_{20}N_2O$			
8a(1H)-Quinazolinol, 4,4a,5,6,7,8-hexahydro-4,4,7-trimethyl-, [4aS-(4aα,7α,8aβ)]-	EtOH	208(3.63)	4-0705-86
$C_{11}H_{20}OSi$			
Silane, (bicyclo[4.1.0]hept-7-ylcarbonyl)trimethyl-	C_6H_{12}	365(1.83)	138-0177-86
$C_{11}H_{20}O_2$			
2,4-Pentanedione, 3-(3,3-dimethylbutyl)-	MeOH	275(3.86)	35-1234-86
	MeOH-NaOH	287(3.72)	35-1234-86
$C_{11}H_{21}NO$			
Piperidine, 4-(ethenyloxy)-2,2,6,6-tetramethyl-	EtOH	200.2(4.05)	70-2522-86
$C_{11}H_{21}NO_2$			
3,4-Hexanedione, 2,2,5,5-tetramethyl-, mono(O-methyloxime)-, (E)-	C_6H_{12}	231(3.42),311(1.74)	54-0054-86
	EtOH	234(3.30),308(1.70)	54-0054-86 +54-0188-86
(Z)-	C_6H_{12}	227(3.19),318(1.81)	54-0054-86
	EtOH	228(3.10),315(1.85)	54-0054-86 +54-0188-86
Nonanamide, N-acetyl-	EtOH	199.2(2.87),207.3(4.13), 210(4.05),212(4.01)	104-1602-86
Octanamide, N-(1-oxopropyl)-	EtOH	194.0(3.03),200.2(4.05), 202(4.13),205.5(4.18)	104-1602-86
3,4-Octanedione, 2,7-dimethyl-, 4-(O-ethyloxime), (E)-	C_6H_{12}	238(4.04),328(1.51)	54-0054-86
(Z)-	C_6H_{12}	242(3.32),322(1.66)	54-0054-86
1-Propanone, 2,2-dimethyl-1-(2,4,4-trimethyloxazolidin-2-yl)-	MeCN	282(4.32)	116-0291-86
	CF_3CH_2OH	287(4.05)	116-0291-86
	CF_3COOH	264(3.23)	116-0291-86
$C_{11}H_{22}O_3Si$			
2-Heptenoic acid, 6-(trimethylsilyloxy)-, (E)- , methyl ester	MeOH	212(4.10)	48-0777-86

Compound	Solvent	$\lambda_{max}(\log \epsilon)$	Ref.
$C_{11}H_{26}O_3SSi$			
Silane, triethoxy[3-(ethylthio)-	heptane	198(2.52),210s(--)	67-0062-86
propyl]-	MeCN	194(2.53),210s(--)	67-0062-86

Compound	Solvent	λ max(log ε)	Ref.
$C_{12}Br_2F_8N_2O$			
Diazene, bis(4-bromo-2,3,5,6-tetrafluorophenyl)-, 1-oxide	n.s.g.	239(4.36),262(4.23), 299(3.98)	70-1391-86
$C_{12}H_2F_8N_2O$			
Diazene, bis(2,3,4,5-tetrafluorophenyl)-, 1-oxide	n.s.g.	230(3.97),253(3.90), 316(4.13)	70-1391-86
Diazene, bis(2,3,5,6-tetrafluorophenyl)-, 1-oxide	n.s.g.	234(4.00),288(3.78)	70-1391-86
$C_{12}H_4Cl_2N_4$			
Cyanamide, (2,5-dichloro-1,4-naphthalenediylidene)bis-	MeCN	213(4.38),254(3.90), 277s(3.80),339s(4.30), 346(4.45),360s(4.37)	5-0142-86
$C_{12}H_4Cl_4O$			
1(2H)-Acenaphthylenone, 2,2,5,6-tetrachloro-	isoPrOH	232(4.69),298(3.64), 328(3.79),356(3.74)	104-1979-86
$C_{12}H_4Cl_6$			
Acenaphthylene, 1,1,2,2,5,6-hexachloro-1,2-dihydro-	isoPrOH	236(3.91),315(3.91), 329(3.78),335(3.75)	104-1979-86
$C_{12}H_5ClF_3NO$			
10H-Phenoxazine, 3-chloro-1,2,4-trifluoro-	EtOH	242(4.65),317(3.89)	70-1685-86
$C_{12}H_5ClN_2O_3S$			
3H-Phenothiazin-3-one, 1-chloro-4-nitro-	MeOH	232(4.35),253s(4.23), 259s(4.21),282(4.21), 369(4.06),390s(4.04), 494(4.03)	80-0137-86
3H-Phenothiazin-3-one, 2-chloro-4-nitro-	MeOH	233(4.32),257(4.14), 265(4.12),280(4.20), 403(4.01),490(3.97)	80-0137-86
3H-Phenothiazin-3-one, 2-chloro-7-nitro-	MeOH	230(4.25),278(4.46), 337(4.11),358(4.09), 376(4.16),390(4.09), 510(3.84)	80-0137-86
3H-Phenothiazin-3-one, 6-chloro-4-nitro-	MeOH	208(4.20),255(4.22), 262(4.20),277(4.73), 395(4.04),476(4.00)	80-0137-86
3H-Phenothiazin-3-one, 7-chloro-4-nitro-	MeOH	230(4.39),211(4.23), 260(4.33),267(4.34), 393(3.99),476(4.08)	80-0137-86
3H-Phenothiazin-3-one, 8-chloro-4-nitro-	MeOH	207(4.25),235(4.33), 256(4.29),264(4.30), 276(4.29),363(4.10), 490(4.05)	80-0137-86
3H-Phenothiazin-3-one, 9-chloro-4-nitro-	MeOH	235(4.36),258(4.20), 266(4.20),279(4.25), 360(3.94),410(4.08), 482(4.00)	80-0137-86
$C_{12}H_5Cl_3O$			
1(2H)-Acenaphthylenone, 2,5,6-trichloro-	isoPrOH	229(4.49),255(4.20), 335(3.78),355(3.81)	104-1979-86
	+ base	322(3.93),537(3.20)	104-1979-86
$C_{12}H_5FeN_6$			
Iron, benzonitrilepentacyano-	n.s.g.	339(3.3),476(3.7)	27-0074-86

Compound	Solvent	$\lambda_{max}(\log \epsilon)$	Ref.
$C_{12}H_5N_7O_{12}$			
Benzenamine, 2,4,6-trinitro-N-(2,4,6-trinitrophenyl)-	MeCN	275(--),380(4.18)	104-1547-86
sodium salt	MeCN	430(4.54)	104-1547-86
$C_{12}H_6BrNO_3$			
5H-[1]Benzopyrano[4,3-c]pyridin-5-one, 1-bromo-4-hydroxy-	DMF	350(4.29)	2-0137-86
$C_{12}H_6ClNO_2$			
1H-Pyrrolo[1,2-a]indole-2-carboxaldehyde, 9-chloro-1-oxo-	MeOH	245(4.21),277(4.56), 286(4.56),314(3.73), 329(3.85)	83-0108-86
$C_{12}H_6ClNO_3S$			
Thieno[3,2-b]furan, 6-chloro-2-(2-nitrophenyl)-	MeOH	239(3.10)	73-1685-86
$C_{12}H_6Cl_2O$			
Acenaphth[1,2-b]oxirene, 3,4-dichloro-6b,7a-dihydro-	isoPrOH	235(4.68),295(3.87), 308(4.02),321(3.88)	104-2329-86
1(2H)-Acenaphthylenone, 5,6-dichloro-	isoPrOH	222(4.48),253(4.42)	104-1979-86
	+ base	319(3.93),534(3.28)	104-1979-86
$C_{12}H_6Cl_2OS$			
Thieno[3,2-b]furan, 3,6-dichloro-2-phenyl-	MeOH	307(3.28)	73-1685-86
$C_{12}H_6Cl_6O_4$			
2-Butenedioic acid, 2-chloro-3-(pentachlorophenyl)-, dimethyl ester, (E)-	C_6H_{12}	212(4.88),227s(4.48), 238s(4.28),288s(2.83), 300(2.76)	44-1413-86
$C_{12}H_6F_2N_2O_4$			
1,1'-Biphenyl, 4,4'-difluoro-2,2'-dinitro-	MeOH	210(4.32),216(4.32), 252.5(4.02),303(3.57)	73-0698-86
$C_{12}H_6F_3NO$			
10H-Phenoxazine, 1,2,4-trifluoro-	EtOH	239(4.75),315(3.83)	70-1685-86
$C_{12}H_6F_3NOSe$			
[1,4]Benzoxaselenino[3,2-b]pyridine, 7-(trifluoromethyl)-	MeOH	220s(4.19),235(4.19), 247(4.14),308(3.80)	39-2075-86C
$C_{12}H_6F_3N_3O_2$			
Pyrido[1,2-a]benzimidazole, 9-nitro-7-(trifluoromethyl)-	EtOH	213(4.43),233(4.67), 263(4.16),327(3.59), 376(3.55)	4-1091-86
$C_{12}H_6F_4N_2O$			
Diazene, bis(2,4-difluorophenyl)-, 1-oxide	n.s.g.	230(3.95),316(4.03)	70-1391-86
Diazene, bis(2,5-difluorophenyl)-, 1-oxide	n.s.g.	228(3.87),248(3.83)	70-1391-86
$C_{12}H_6I_2S_3$			
2,2':5',2"-Terthiophene, 5,5"-diiodo-	MeOH	368(4.51)	149-0441-86B
$C_{12}H_6N_2O_3S$			
3H-Phenothiazin-3-one, 4-nitro-	MeOH	230(4.37),254(4.27), 260(4.27),275(4.25),	80-0137-86

Compound	Solvent	$\lambda_{max}(\log \epsilon)$	Ref.
(cont.)		369(4.04),385(4.04), 483(4.09)	80-0137-86
3H-Phenothiazin-3-one, 7-nitro-	MeOH	225(1.16),277(4.40), 322s(4.06),331(4.08), 345(4.04),367(4.00), 385(3.85),501(3.83)	80-0137-86
$C_{12}H_6N_4$			
Cyanamide, 1,4-naphthalenediylidene-bis-	MeCN	263s(4.01),269(4.15), 278(4.14),329(4.30), 344s(4.20),399(4.00)	5-0142-86
$C_{12}H_6N_4O_2$			
Pyrido[1,2-a]benzimidazole-7-carbonitrile, 9-nitro-	MeOH	237(4.74),268(4.25), 340(3.61),378(3.57)	4-1091-86
Pyrido[1,2-a]benzimidazole-8-carbonitrile, 6-nitro-	EtOH	219(4.47),226(4.46), 240(4.39),247(4.40), 267(4.48),305s(3.51), 398(4.00)	4-1091-86
$C_{12}H_7Br$			
Acenaphthylene, 1-bromo-	isoPrOH	310s(3.70),317(3.84), 324(3.89),330s(3.79), 339(3.57),346(3.56)	104-0563-86
$C_{12}H_7Cl_2FN_4$			
9H-Purine, 6,8-dichloro-9-[(2-fluorophenyl)methyl]-	EtOH-HCl	269(4.16)	4-1189-86
	EtOH-pH 7	269(4.16)	4-1189-86
	pH 13	269(4.12)	4-1189-86
$C_{12}H_7Cl_2NO_4S$			
7H-Thieno[2',3':4,5]furo[3,2-b]pyrrole-6-carboxylic acid, 3-chloro-2-(chlorocarbonyl)-, ethyl ester	MeOH	250(2.98)	73-1685-86
$C_{12}H_7Cl_3O$			
1-Acenaphthylenol, 2,5,6-trichloro-1,2-dihydro-	isoPrOH	235(4.84),298(3.92), 310(4.05),324(3.90)	104-2329-86
$C_{12}H_7F_2N_3O_4$			
Benzenamine, 4-fluoro-N-(4-fluoro-2-nitrophenyl)-2-nitro-	MeOH	222(4.16),242(4.25), 257s(4.16),287s(3.87), 394s(3.89),430(3.98)	73-0698-86
$C_{12}H_7F_3N_2$			
1H-Perimidine, 2-(trifluoromethyl)-	DMF	324s(3.95),339(4.03), 348s(3.86),412s(2.84), 438s(2.73),468s(2.41)	48-0812-86
$C_{12}H_7IS_3$			
2,2':5',2"-Terthiophene, 5-iodo-	MeOH	362(4.45)	149-0441-86B
$C_{12}H_7NOS$			
3H-Phenothiazin-3-one	MeOH	237(4.39),272(4.26), 287(4.25),287(4.24), 367(4.02),380(3.99), 504(3.91)	80-0137-86
$C_{12}H_7NO_2$			
1H-Carbazole-1,4(9H)-dione	EtOH	225(5.27),251s(5.02), 265(5.05),267s(4.86),	77-1019-86

Compound	Solvent	λ_{max}(log ε)	Ref.
(cont.)		304(4.13),399(5.02)	77-1019-86
$C_{12}H_7NO_3$			
2H-1-Benzopyran-4-carbonitrile, 3-acetyl-2-oxo-	MeOH	365(2.78)	2-0137-86
3H-Phenoxazin-3-one, 2-hydroxy-	EtOH	223(4.20),400(4.09)	4-1003-86
$C_{12}H_7NO_3S$			
Thiopyrano[2,3-b]indole-2-carboxylic acid, 4,9-dihydro-4-oxo-	dioxan	236(4.34),255(4.19), 277(4.24),358(3.74)	83-0332-86
Thiopyrano[3,2-b]indole-2-carboxylic acid, 4,5-dihydro-4-oxo-	dioxan	226(4.38),283(4.04), 320(3.80),380(3.68)	83-0332-86
$C_{12}H_7N_5$			
Pyrimido[1',2':1,2]imidazo[4,5-b]-quinoxaline	DMF	337(4.04)	2-1057-86
$C_{12}H_7N_5O_4$			
Benzonitrile, 3,5-dinitro-4-(2-pyridinylamino)-	MeOH	221(4.34),239s(4.25), 276(4.22),368(3.90)	4-1091-86
$C_{12}H_8BrClN_2O$			
Phenol, 4-bromo-2-[[(5-chloro-2-pyridinyl)imino]methyl]-	$CHCl_3$	360(3.42)	2-0127-86A
$C_{12}H_8BrN_3$			
Imidazo[1,2-a]pyridine, 1-bromo-3-(2-pyridinyl)-	MeOH	211(3.9),225(3.9), 245(3.8),258(3.2), 351(3.7)	83-0043-86
$C_{12}H_8BrN_3O_4$			
Benzenamine, 2-bromo-4-nitro-N-(3-nitrophenyl)-	EtOH	257(4.35),372(4.31)	104-0720-86M
$C_{12}H_8Br_2$			
Acenaphthylene, 1,2-dibromo-1,2-dihydro-, (E)-	isoPrOH	288s(3.89),300(3.95), 312s(3.79)	104-0563-86
(Z)-	isoPrOH	288s(3.84),297(3.90), 303s(3.84),320s(3.35), 326(3.11)	104-0563-86
$C_{12}H_8Br_2N_2$			
Tricyclo[3.3.0.0^{2,8}]octa-3,6-diene-3,7-dicarbonitrile, 2,4-dibromo-1,5-dimethyl-	MeCN	220(4.009),264(3.717)	24-1801-86
Tricyclo[3.3.0.0^{2,8}]octa-3,6-diene-3,7-dicarbonitrile, 2,6-dibromo-1,5-dimethyl-	MeCN	223(4.014),257(3.713)	24-1801-86
$C_{12}H_8ClFN_4$			
9H-Purine, 6-chloro-9-[(2-fluorophenyl)methyl]-	10% EtOH-pH 7	266s(4.01)	4-1189-86
$C_{12}H_8ClNO_3S$			
Phenol, 2-chloro-4-[2-(5-nitro-2-thienyl)ethenyl]-, (E)-	MeOH	287(3.10),427(3.37)	73-1127-86
Phenol, 3-chloro-4-[2-(5-nitro-2-thienyl)ethenyl]-, (E)-	MeOH	287(1.94),421(3.18)	73-1127-86
Phenol, 4-chloro-2-[2-(5-nitro-2-thienyl)ethenyl]-, (E)-	MeOH	272(3.03),303(2.94), 420(3.35)	73-1127-86

Compound	Solvent	λmax(log ε)	Ref.
$C_{12}H_8ClN_3O_4$			
Benzenamine, 2-chloro-4-nitro-N-(3-nitrophenyl)-	EtOH	256(4.39),372(4.38)	104-0720-86
$C_{12}H_8Cl_2N_2O$			
Phenol, 4-chloro-2-[[(5-chloro-2-pyridinyl)imino]methyl]-	$CHCl_3$	360(4.18)	2-0127-86A
$C_{12}H_8Cl_2O$			
1-Acenaphthylenol, 5,6-dichloro-1,2-dihydro-	isoPrOH	234(4.88),298(3.88), 309(3.99),324(3.84), 334(3.48)	104-1979-86
$C_{12}H_8Cl_2S_2$			
Disulfide, bis(4-chlorophenyl)	C_6H_{12}	246(4.0)	65-0758-86
with BBr_3	C_6H_{12}	255(5.2)	65-0758-86
with $GaCl_3$	C_6H_{12}	235s(3.7)	65-0758-86
$C_{12}H_8F_2N_2O$			
Diazene, bis(3-fluorophenyl)-, 1-oxide	n.s.g.	226(3.96),260(3.89), 322(4.21)	70-1391-86
$C_{12}H_8F_6N_2$			
Quinoxaline, 6,7-dimethyl-2,3-bis(trifluoromethyl)-	EtOH	247(4.65),325(3.75)	39-1043-86C
$C_{12}H_8F_6O_2$			
7-Oxabicyclo[2.2.1]hepta-2,5-diene, 1-[(2-propynyloxy)methyl]-2,3-bis(trifluoromethyl)-	MeCN	260(2.18)	24-0589-86
$C_{12}H_8IN_3O_4$			
Benzenamine, 2-iodo-4-nitro-N-(3-nitrophenyl)-	EtOH	257(4.34),272(4.21)	104-0720-86
$C_{12}H_8N_2OS$			
1,2-Benzisothiazol-3(2H)-one, 2-(2-pyridinyl)-	MeOH	255(3.98),335(3.81)	18-1601-86
1,2-Benzisothiazol-3(2H)-one, 2-(3-pyridinyl)-	MeOH	254(4.11),330(3.73)	18-1601-86
$C_{12}H_8N_2O_2$			
3H-Phenoxazin-3-one, 2-amino-	EtOH	237(4.17),268(3.90), 421-436(4.10)	4-1003-86
$C_{12}H_8N_2O_2S$			
Benzamide, N-furo[2,3-d]thiazol-2-yl-	MeOH	232(3.21),323(3.21)	73-1678-86
$C_{12}H_8N_4$			
Dibenzo[b,e]-1,3a,6,6a-tetraazapentalene	EtOH	234(4.54),271(3.77), 280(3.92),343(4.51), 356(4.60)	20-1107-86
Dibenzo[b,f]-1,3a,4,6a-tetraazapentalene	EtOH	255(4.80),364(3.89), 382(4.37),402(4.58)	20-1107-86
9H-Indeno[1,2-b]pyrazine-3-carbonitrile, 2-amino-	MeOH	384(4.18)	33-0793-86
Pyrido[1",2":1',2']imidazo[4',5':4,5]-imidazo[1,2-a]pyridine (picrate)	MeOH	255(4.34),262(4.34), 286(3.95),299(3.95)	35-8002-86
Pyrido[2",1":2',3']imidazo[4',5':4,5]-imidazo[1,2-a]pyridine	pH 1	215(4.20),252(4.51), 354(4.18)	35-8002-86
	pH 11	219(4.15),256(4.60),	35-8002-86

Compound	Solvent	λ_{max}(log ε)	Ref.
(cont.)		356(4.15),372(4.20), 392(4.0)	35-8002-86
	MeOH	221(4.32),258(4.61), 358(4.11),375(4.18), 396(4.00)	35-8002-86
picrate	MeOH	256(4.30),343(4.00), 357(4.00),373(4.00), 395(3.85)	35-8002-86
7H-1,2,4-Triazolo[4,3-a]perimidine	dioxan	215(4.13),250(3.52), 287(3.11),297(3.17), 345(3.40),357(3.46)	48-0237-86
$C_{12}H_8N_4OS$			
Imidazo[5,1-a]phthalazine-10b(1H)-carbonitrile, 2,3-dihydro-2-methyl-3-oxo-1-thioxo-	n.s.g.	270(3.65),316(3.35)	150-1901-86M
$C_{12}H_8O$			
Azuleno[1,2-b]furan	MeOH	296(4.72),354(3.52), 371(3.69),389(3.79), 619(2.41),652(2.30), 677(2.38),721(1.96), 750(2.07)	138-1021-86
$C_{12}H_8OS_2$			
[2,2'-Bithiophene]-5-carboxaldehyde, 5'-(1-propynyl)- (arctinal)	EtOH	220(3.87),262(4.10), 280(3.74),372(4.45)	163-0263-86
$C_{12}H_8O_2S_2$			
[2,2'-Bithiophene]-5-carboxylic acid, 5'-(1-propynyl)-	EtOH	218(4.03),254(3.93), 355(4.46)	163-0565-86
$C_{12}H_8O_4$			
Psoralen, 5-methoxy-	saline	313(4.20)	149-0221-86A
Psoralen, 8-methoxy-	1% DMSO	303(4.11)	149-0391-86A
$C_{12}H_8S_3$			
2,2':3',2"-Terthiophene	MeOH	205(4.27),254(4.13), 296(3.97)	142-2261-86
2,2':3',3"-Terthiophene	MeOH	203(4.27),250(4.10), 295(3.90)	142-2261-86
2,2':4',3"-Terthiophene	MeOH	210(4.46),263(4.37), 310(3.99)	142-2901-86
2,2':5',2"-Terthiophene	MeOH	352(4.35)	149-0441-86B
2,3':2',3"-Terthiophene	MeOH	207(4.16),244(4.17), 292(4.04)	142-2901-86
2,3':4',3"-Terthiophene	MeOH	210(4.24),242(4.17)	142-2901-86
2,3':5',3"-Terthiophene	MeOH	205(3.99),272(4.17)	142-2901-86
3,2':3',3"-Terthiophene	MeOH	210(4.31),255(4.05), 275(4.03)	142-2261-86
3,2':4',3"-Terthiophene	MeOH	222(4.38),262(4.35)	142-2261-86
3,3':4',3"-Terthiophene	MeOH	210(4.51),250(4.20)	142-2261-86
$C_{12}H_9BO_6$			
Boron, [ethanedioato(2-)-O,O'](1-phenyl-1,3-butanedionato-O,O')-, (T-4)-	CH_2Cl_2	271(3.72),339(4.45), 344s(4.13)	48-0755-86
$C_{12}H_9BrN_2$			
Tricyclo[3.3.0.0^{2,8}]octa-3,6-diene-3,7-dicarbonitrile, 4(2)-bromo-1,5-dimethyl-	MeCN	222(3.984),255s(3.737)	24-1801-86

Compound	Solvent	λ_{max}(log ε)	Ref.
$C_{12}H_9BrN_2O$			
Benzo[f]quinoxalin-6(2H)-one, 5-bromo-3,4-dihydro-	CH_2Cl_2	234(4.08),273(4.23), 420(3.54)	83-0052-86
Phenol, 4-bromo-2-[(2-pyridinylimino)methyl]-	$CHCl_3$	360(3.63)	2-0127-86A
$C_{12}H_9ClN_2O$			
Phenol, 2-[[(5-chloro-2-pyridinyl)-imino]methyl]-	$CHCl_3$	356(4.08)	2-0127-86A
Phenol, 4-chloro-2-[(2-pyridinylimino)methyl]-	$CHCl_3$	362(3.96)	2-0127-86A
$C_{12}H_9ClN_4$			
1H-Pyrazolo[3,4-d]pyridazine, 3-(4-chlorophenyl)-1-methyl-	MeOH	241(4.30),276(4.01), 305s(3.92)	39-0169-86C
$C_{12}H_9ClN_4O_2$			
Benzo[g]pteridine-2,4(3H,10H)-dione, 8-chloro-7,10-dimethyl-	DMF	272(4.56),335(3.90), 420s(--),443(4.08), 470s(--)	35-0490-86
$C_{12}H_9ClO_4$			
1,4-Naphthalenedione, 2-chloro-5,8-dimethoxy-	$CHCl_3$	283(3.74),466(3.64)	150-3137-86M
1,4-Naphthalenedione, 6-chloro-5,8-dimethoxy-	$CHCl_3$	409(3.57),420(3.56)	150-3137-86M
$C_{12}H_9Cl_2N_3$			
5H-Pyrimido[5,4-d][1]benzazepine, 4,9-dichloro-6,7-dihydro-	EtOH	221s(--),245(4.31), 264(4.06),288(3.88), 385(3.77)	142-0143-86
$C_{12}H_9N$			
1H-Azuleno[1,2-b]pyrrole	MeOH	279(4.35),312(4.69), 357s(3.29),376(3.46), 394(3.57),415(3.52), 590(2.30),642(2.46), 708(2.39),794(2.02)	138-1021-86
9H-Carbazole	C_6H_{12}	215(4.81),229(4.80), 245s(4.54),255s(4.32), 262s(4.16),284s(4.40), 290(4.43),317(4.08), 330(4.04)	20-0631-86
	EtOH	210(4.65),231(4.74), 242s(4.54),254(4.47), 290(4.40),320(4.08), 334(4.06)	20-0631-86
	PrOH	215(4.11),237(4.54), 244s(4.51),256(4.43), 294(4.36),323(4.09), 337(4.06)	20-0631-86
	MeCN	209(4.65),230(4.74), 243s(4.55),253(4.44), 289(4.41),318(4.11), 333(4.10)	20-0631-86
	20% MeCN	210(4.64),231(4.73), 242s(4.56),255(4.46), 290(4.39),320(4.11), 335(4.08)	20-0631-86

Compound	Solvent	λmax(log ε)	Ref.
$C_{12}H_9NO$			
9H-Carbazol-2-ol	pH 6	210(4.43),234(4.63), 256(4.31),300(4.14), 312s(--)	59-0793-86
	pH 12	208(4.35),240(4.61), 259(4.32),326(4.24)	59-0793-86
	MeOH	211(4.41),235(4.64), 253(4.31),257(4.33), 301(4.15),312s(--)	59-0793-86
	ether	221(4.69),235(4.56), 254(4.26),257(4.29), 301(4.06),312(3.90), 324s(--)	59-0793-86
	MeCN	210(4.41),235(4.65), 253(4.31),256(4.32), 300(4.15),312(3.98), 323s(--)	59-0793-86
	C_6H_{12}	210(--),233(--), 250(--),255(--), 297(--),311(--), 322(--),334(--)	59-0793-86
dianion	KOH	250(4.42),259(4.55), 284(4.54),338(4.09)	59-0763-86
monocation	H_2SO_4	199(4.46),238(4.71), 263(4.35),274(4.17), 394(3.20)	59-0763-86
dication	H_2SO_4	220(--),235(--), 276(--),293(--), 328(--),378(--)	59-0763-86
10H-Phenoxazine	EtOH	239(4.60),318(3.90)	4-1003-86
$C_{12}H_9NOS$			
Thiopyrano[2,3-b]indol-4(9H)-one, 5-methyl-	dioxan	231(4.17),272(4.31), 305(3.87)	83-0120-86
Thiopyrano[3,2-b]indol-4(5H)-one, 5-methyl-	dioxan	226(4.37),273(3.91), 307(4.13),360(3.86)	83-0120-86
$C_{12}H_9NO_2S$			
Benzoxazole, 2-(5-methoxy-3-thienyl)-	EtOH	216(4.30),224(4.31), 229(4.30),237(4.25), 285(4.19),295(4.36), 302(4.36)	70-1470-86
$C_{12}H_9NO_2S_2$			
4H-[1]Benzothieno[2,3-e]-1,3-thiazin-4-one, 2-ethoxy-	EtOH	215(3.05),255(2.19), 305(2.87)	73-2002-86
$C_{12}H_9NO_2Se$			
[1,4]Benzoxaselenino[3,2-b]pyridine, 7-methoxy-	MeOH	216(4.39),235s(--), 296(3.82),325s(--)	39-2075-86C
$C_{12}H_9NO_3$			
Furo[2,3-b]quinolin-4(9H)-one, 7-methoxy-	n.s.g.	252(3.67),320(3.08)	100-0048-86
$C_{12}H_9NO_3S$			
Phenol, 4-[2-(5-nitro-2-thienyl)ethenyl]-, (Z)-	MeOH	293(3.03),444(3.27)	73-1127-86
$C_{12}H_9NO_4$			
Furo[2,3-b]quinoline-3,4(2H,9H)-dione,	$CHCl_3$	254(3.25),266(3.22),	100-0048-86

Compound	Solvent	λ_{max}(log ε)	Ref.
7-methoxy- (cont.)		318(3.22)	100-0048-86
$C_{12}H_9NO_4S$			
2-Butenedioic acid, 2-(1H-indol-3-ylthio)-, (Z)-	dioxan	216(4.56),280(4.15)	83-0332-86
$C_{12}H_9NS$			
Azuleno[1,2-d]thiazole, 2-methyl-	C_6H_{12}	260(4.0),305f(5.0), 380f(3.9),610f(2.2)	18-3320-86
10H-Phenothiazine	MeOH	254(4.665),317.5(3.670)	106-0571-86
$C_{12}H_9N_3$			
Imidazo[1,5-a]pyridine, 3-(2-pyridinyl)-	MeOH	211(3.9),221(3.9), 245(3.8),351(3.7)	83-0043-86
$C_{12}H_9N_3O$			
Imidazo[1,2-a]pyrimidin-5(1H)-one, 7-phenyl-	EtOH	247(4.55),273(4.20), 288s(4.00),321(3.88)	4-0245-86
$C_{12}H_9N_3O_2$			
Benzenamine, N,4-dinitroso-N-phenyl-	heptane-EtOH	240(3.39),286(3.42), 330(3.42),408(2.53), 715(1.97)	80-0857-86
Diazene, (4-nitrophenyl)phenyl-, trans	n.s.g.	330(4.34)	46-1179-86
$C_{12}H_9N_3O_3$			
Phenol, 2-nitro-6-[(2-pyridinylimino)methyl]-	$CHCl_3$	446(3.02)	2-0127-86A
$C_{12}H_9N_3O_9S_4$			
Phenothiazin-5-ium, 3,7-diamino-2,4,6-trisulfo-, hydroxide, inner salt disodium salt	0.1N H_2SO_4	286(4.45),570(4.72)	150-1555-86M
$C_{12}H_9N_3S$			
[3,4'-Bipyridine]-3'-carbonitrile, 1',2'-dihydro-6'-methyl-2'-thioxo-	EtOH	253(4.19),270(4.19), 316(4.46),424(3.62)	104-1762-86
[4,4'-Bipyridine]-3-carbonitrile, 1,2-dihydro-6-methyl-2-thioxo-	DMF	320(4.45),431(3.69)	104-1762-86
$C_{12}H_{10}BrNO_6$			
Propanedioic acid, [bromo(4-nitrophenyl)methylene]-, dimethyl ester	EtOH	279(4.09)	44-4107-86
$C_{12}H_{10}BrN_3O_3$			
4(1H)-Pyridinone, 2-[(4-bromophenyl)azo]-3-hydroxy-6-(hydroxymethyl)-	MeOH	246(4.17),282s(4.00), 374(4.02),450s(3.89)	4-0333-86
$C_{12}H_{10}BrN_5OS$			
Ethanone, 1-[5-(4-bromophenyl)-1-(2-thiazolyl)formazanyl]-	50% EtOH	470(4.23)	65-2114-86
$C_{12}H_{10}ClNO_2S_2$			
Carbamothioic acid, [(3-chlorobenzo[b]thien-2-yl)carbonyl]-, O-ethyl ester	EtOH	211(3.16),307(3.07)	73-2002-86
$C_{12}H_{10}ClN_3$			
5H-Pyrimido[5,4-d][1]benzazepine, 4-chloro-6,7-dihydro-	EtOH	242(4.26),262(4.05), 280(3.89),390(3.69)	142-0143-86

Compound	Solvent	$\lambda_{max}(\log \epsilon)$	Ref.
$C_{12}H_{10}ClN_3O$			
4H-Pyrimido[5,4-d][1]benzazepin-4-one, 9-chloro-1,5,6,7-tetrahydro-	EtOH	237(4.40),258s(--), 284(3.77),296(3.72), 309s(--),374(3.66)	142-0143-86
$C_{12}H_{10}ClN_3O_3$			
4(1H)-Pyridinone, 2-[(3-chlorophenyl)-azo]-3-hydroxy-6-(hydroxymethyl)-	MeOH	244(4.03),455(3.90)	4-0333-86
4(1H)-Pyridinone, 2-[(4-chlorophenyl)-azo]-3-hydroxy-6-(hydroxymethyl)-	MeOH	245(3.87),458(3.77)	4-0333-86
$C_{12}H_{10}FN_5$			
9H-Purin-6-amine, 9-[(2-fluorophenyl)-methyl]-	pH 7.4	262(4.22)	4-1189-86
$C_{12}H_{10}F_3N_5O$			
2-Azetidinone, 1-[(2-phenylmethyl)-2H-tetrazol-5-yl]-4-(trifluoro-methyl)-	EtOH	231s(3.95)	78-2685-86
$C_{12}H_{10}FeN_2O_5$			
Iron, tricarbonyl[methyl 1(or 3)-[(4,5,6,7-η)-1H-1,2-diazepin-1-yl]-2-propenoate]-, cis	EtOH	304(4.20),341s(4.01), 419s(3.38)	39-0595-86C
trans	EtOH	307(4.08),346(4.08)	39-0595-86C
$C_{12}H_{10}N_2$			
Bicyclo[2.2.2]oct-2-ene-2,3-dicarbo-nitrile, 5,6-bis(methylene)-	EtOH	239(4.28),244s(4.24), 253s(4.06),272s(3.08), 299(2.77)	138-0969-86
	CH_2Cl_2	241(4.29),246s(4.26), 254s(4.11),275s(3.05), 303(2.78)	138-0969-86
Propanedinitrile, [5,6-bis(methylene)-bicyclo[2.2.1]hept-2-ylidene]-	isooctane	236(4.35),284(3.51)	33-1310-86
	EtOH	236(4.30),284(3.41)	33-1310-86
9H-Pyrido[3,4-b]indole, 1-methyl-(harman)	C_6H_{12}	229(4.61),235(4.51), 244(4.26),278(3.95), 283(4.18),313(3.31), 324(3.65),338(3.64)	2-0178-86A
	pH 2	204(4.26),246(4.44), 297(4.18),365(3.66)	39-1247-86B
	pH 8	235(4.57),248(4.38), 280s(--),286(4.17), 334(3.66),347(3.66)	2-0178-86A
	MeOH	233(4.57),235(4.56), 247(4.37),280s(--), 286(4.21),319s(--), 334(3.67),346(3.68)	2-0178-86A
	dioxan	231(4.50),235(4.52), 245(4.34),280(3.93), 284(4.13),315s(--), 329(3.65),342(3.67)	2-0178-86A
	MeCN	232(4.53),235(4.53), 245(4.36),280s(--), 285(4.12),317s(--), 329(3.63),343(3.65)	2-0178-86A
dication	H_0 -8.5	202(4.15),250(4.60), 297(4.30),360(3.71)	39-1247-86B
Tricyclo[3.3.0.0^{2,8}]octa-3,6-diene-3,7-dicarbonitrile, 1,5-dimethyl-	MeCN	216(4.015),253(3.788)	24-1801-86

Compound	Solvent	$\lambda_{max}(\log \epsilon)$	Ref.
$C_{12}H_{10}N_2O$			
Benzenamine, N-nitroso-N-phenyl-	heptane-EtOH	238(4.12),294(3.82), 400(2.38)	80-0857-86
Benzenamine, 4-nitroso-N-phenyl-	heptane-EtOH	260(4.11),417(4.41), 715(1.78)	80-0857-86
Diazene, diphenyl-, 1-oxide	benzene	326(4.217)	44-0088-86
	MeOH	322(4.151)	44-0088-86
Phenol, 2-[(2-pyridinylimino)methyl]-	CHCl	332(4.06)	2-0127-86A
9H-Pyrido[3,4-b]indol-7-ol, 1-methyl- (harmol)	pH 2	204(4.31),247(4.53), 320(4.26),360(3.82)	39-1247-86B
dication	H_0 -8.5	203(4.26),252(4.55), 315(4.20),357(3.92)	39-1247-86B
$C_{12}H_{10}N_2O_2$			
1H-Pyrido[3,4-b]indole-1,3(2H)-dione, 4,9-dihydro-2-methyl-	isoPrOH	219(4.39),239(4.22), 312(4.18)	103-1345-86
$C_{12}H_{10}N_2O_2S$			
Benzenamine, 4-[(4-nitrophenyl)thio]-	MeOH	280f(4.3),350(4.2)	46-5654-86
4(1H)-Pyrimidinone, 2,3-dihydro-5-[2-(3-hydroxyphenyl)ethenyl]-2-thioxo-	MeOH	354(4.32)	56-0599-86
4(1H)-Pyrimidinone, 2,3-dihydro-5-[2-(4-hydroxyphenyl)ethenyl]-2-thioxo-, (E)-	MeOH	362(4.56)	56-0599-86
4(1H)-Pyrimidinone, 2,3-dihydro-6-[2-(4-hydroxyphenyl)ethenyl]-2-thioxo-, (E)-	MeOH	337(4.10)	56-0599-86
$C_{12}H_{10}N_2O_3$			
Methanone, (1,3-benzodioxol-5-yl)(1-methyl-1H-imidazol-2-yl)-	EtOH	295(4.08),330(4.16)	94-4916-86
2,4(1H,3H)-Pyrimidinedione, 5-[2-(3-hydroxyphenyl)ethenyl]-, (E)-	MeOH	322(4.10)	56-0599-86
2,4(1H,3H)-Pyrimidinedione, 5-[2-(4-hydroxyphenyl)ethenyl]-, (E)-	MeOH	328(3.89)	56-0599-86
2,4(1H,3H)-Pyrimidinedione, 6-[2-(4-hydroxyphenyl)ethenyl]-, (E)-	MeOH	342(3.72)	56-0599-86
$C_{12}H_{10}N_2O_6S$			
Thiophene, 2,5-dihydro-3-methyl-4-nitro-2-[(4-nitrophenyl)methylene]-, 1,1-dioxide, (Z)-	$CHCl_3$	350(4.314)	104-0379-86
$C_{12}H_{10}N_2S$			
2-Pyridinecarbothioamide, N-phenyl-	benzene	282(3.89),355(3.58)	65-2088-86
	hexane	236(4.32),283(4.26), 350(4.04)	65-2088-86
	$CHCl_3$	288(4.00),348(3.81)	65-2088-86
	CCl_4	284(3.99),351(3.79)	65-2088-86
$C_{12}H_{10}N_3O_6S_3$			
Phenothiazin-5-ium, 3,7-diamino-2,6-disulfo-	pH 1	582(4.72)	46-6068-86
Phenothiazin-5-ium, 3,7-diamino-4,6-disulfo-	pH 1	571(4.75)	46-6068-86
$C_{12}H_{10}N_3S$			
Phenothiazin-5-ium, 3,7-diamino- (thionine)	pH 1	599(4.71)	46-6068-86

Compound	Solvent	$\lambda_{max}(\log \epsilon)$	Ref.
$C_{12}H_{10}N_4$			
Imidazo[1,2-a]pyridin-3-amine, N-2-pyridinyl-	pH 1	206(4.18),244(4.00), 282(3.90)	35-8002-86
	pH 11	230(4.34),290(3.90)	35-8002-86
	MeOH	210(4.30),231(4.36), 280(3.85),291(3.85)	35-8002-86
Pyrazinecarbonitrile, 3-amino-6-methyl-5-phenyl-	pH 1	378(4.00)	33-1025-86
	EtOH	366(4.08)	33-0793-86
$C_{12}H_{10}N_4OS$			
Acetamide, N-(2-methylthiazolo[4,5-b]-quinoxalin-7-yl)-	n.s.g.	368(4.0351)	18-1245-86
$C_{12}H_{10}N_4O_2$			
Benzenamine, 4-[(4-nitrophenyl)azo]-	base	505(4.54)	74-0417-86C
Benzo[g]pteridine-2,4(3H,10H)-dione, 7,10-dimethyl-	DMF	269(4.52),330(3.85), 395s(--),420s(--), 445(3.97),476s(--)	35-0490-86
$C_{12}H_{10}N_6O_2$			
1,4-Naphthalenedione, disemicarbazone	MeOH	380(4.47)	74-0321-86C
$C_{12}H_{10}N_6O_3S$			
Ethanone, 1-[5-(4-nitrophenyl)-1-(2-thiazolyl)formazanyl]-	50% EtOH	480(4.39)	65-2114-86
$C_{12}H_{10}OS_2$			
[2,2'-Bithiophene]-5-methanol, 5'-(1-propynyl)-	EtOH	244(3.77),262(3.54), 330(4.30),338(4.32)	163-0263-86
$C_{12}H_{10}O_2$			
[1,1'-Biphenyl]-4,4'-diol	H_2O	260(4.30)	60-0291-86
1,4-Naphthalenedione, 2,3-dimethyl-	H_2SO_4	257(4.01),307(3.85), 414(3.39)	88-0255-86
$C_{12}H_{10}O_2S_2$			
Benzo[1,2-b:4,5-b']dithiophene, 4,8-dimethoxy-	EtOH	226(4.44),246(4.51), 253(4.66),275s(3.57), 285(3.78),298(3.79), 318s(3.50),333(3.93), 347.5(4.10)	24-3198-86
$C_{12}H_{10}O_2S_4$			
1,3-Benzenediethane(dithioic) acid, α,α'-dioxo-, dimethyl ester	MeOH	309(3.81),497(2.00)	118-0968-86
1,4-Benzenediethane(dithioic) acid, α,α'-dioxo-, dimethyl ester	MeOH	271(3.87),305(3.87), 494(2.03)	118-0968-86
$C_{12}H_{10}O_3$			
2(5H)-Furanone, 5-hydroxy-4-(2-phenylethenyl)-, (E)-	EtOH	313(4.53)	94-4346-86
1,4-Naphthalenediol, monoacetate	MeOH	237(4.31),252s(3.89), 300(3.58),311s(3.55), 326s(3.48)	12-2067-86
1,2-Naphthalenedione, 8-methoxy-3-methyl-	EtOH	240(3.82),430(3.25)	94-2810-86
1,2-Naphthalenedione, 8-methoxy-6-methyl-	MeOH	245(4.21),416(3.79)	39-0675-86C
1,4-Naphthalenedione, 5-methoxy-2-methyl-	EtOH	252(3.22),270s(3.12), 398(2.63)	94-2810-86

Compound	Solvent	λ_{max}(log ϵ)	Ref.
$C_{12}H_{10}O_4$			
4H-1-Benzopyran-2-carboxylic acid, 6-methyl-4-oxo-, methyl ester	EtOH	241(4.26),320(3.79)	42-0600-86
4H-1-Benzopyran-2-carboxylic acid, 8-methyl-4-oxo-, methyl ester	EtOH	235(4.18),309(3.87)	42-0600-86
1,4-Naphthalenedione, 2,6-dimethoxy-	MeOH	265(4.27),290(4.23), 328(3.54)	100-0122-86
1,4-Naphthalenedione, 2,7-dimethoxy-	MeOH	261(4.39),288(4.12), 333(3.47)	100-0122-86
1,4-Naphthalenedione, 6-hydroxy-5-methoxy-2-methyl-	MeOH	262(4.18),399(3.38), 530s(2.59)	39-0675-86C
$C_{12}H_{10}O_5$			
4H-1-Benzopyran-2-carboxylic acid, 7-methoxy-, methyl ester	EtOH	236(4.24),307(3.99)	42-0600-86
3(2H)-Furanone, 2-[1-(3-acetoxy-2-furanyl)ethylidene]-, (E)-	n.s.g.	230(3.82),257(3.46)	136-0107-86A
2H-Pyran-2,5(6H)-dione, 3-hydroxy-4-(4-methoxyphenyl)-	MeOH	245(4.10),329(3.82)	5-0177-86
$C_{12}H_{10}O_6$			
2-Benzofurancarboxylic acid, 5-formyl-4,6-dimethoxy-, monohydrate	EtOH	222(4.03),231(4.02), 238(4.01),273(4.34), 295(4.32),329(4.08)	4-1347-86
8H-1,3-Dioxolo[4,5-h][1]benzopyran-8-one, 4,5-dimethoxy- (isosabandin)	EtOH	218(4.58),254s(4.07), 259(4.09),320(4.29)	5-2142-86
$C_{12}H_{10}S_2$			
Benzo[1,2-b:4,5-b']dithiophene, 4,8-dimethyl-	EtOH	226.5(4.38),246(4.55), 253(4.68),277s(3.64), 287(3.91),298.5(4.02), 315s(3.54),328.5(3.98), 343(4.19)	24-3198-86
Disulfide, diphenyl	C_6H_{12}	200(4.7),241(4.3)	65-0758-86
with $GaCl_3$	C_6H_{12}	200(5.0),285(2.5)	65-0758-86
Naphtho[1,2-d][1,2]dithiin, 1,4-dihydro-	C_6H_{12}	230(4.84),264(3.61), 275(3.73),285(3.76), 295(3.63)	56-0315-86
Naphtho[2,3-d][1,2]dithiin, 1,4-dihydro-	C_6H_{12}	229(4.88),256(3.66), 266(3.67),276(3.71), 287(3.71),299(3.54)	56-0315-86
$C_{12}H_{11}BO_5$			
Boron, [2-hydroxybenzoato(2-)-O^1,O^2]-(2,4-pentanedionato-O,O')-, (T-4)-	CH_2Cl_2	237(3.80),291(4.11)	48-0755-86
$C_{12}H_{11}BrO$			
Azulene, 6-bromo-2-ethoxy-	CH_2Cl_2	285.5(4.84),295(4.98), 327.5(3.85),351.5(3.69), 367(3.92),385(4.00), 518.5(2.25)	24-2272-86
$C_{12}H_{11}BrO_3$			
Tricyclo[5.2.1.$0^{2,6}$]deca-4,8-dien-3-one, 5-acetoxy-4-bromo-	EtOH	252(3.91)	78-1903-86
$C_{12}H_{11}ClN_2OS_2$			
Benzo[b]thiophene-2-carboxamide, 3-chloro-N-[(ethylamino)thioxomethyl]-	EtOH	207(3.35),251(3.30), 305(3.30)	73-2839-86

Compound	Solvent	$\lambda_{max}(\log \epsilon)$	Ref.
$C_{12}H_{11}ClN_4O_4$			
Pyrido[2,3-d]pyrimidine-6-carbonitrile, 7-chloro-1,2,3,4-tetrahydro-1,3-bis(methoxymethyl)-2,4-dioxo-	H_2O	221(4.70),272(4.22), 311(3.99)	87-0709-86
$C_{12}H_{11}ClO_3$			
2-Propenoic acid, 3-chloro-2-formyl-3-phenyl-, ethyl ester 4:1 E:Z	EtOH	258(3.74)	44-4112-86
$C_{12}H_{11}FO$			
2-Cyclopenten-1-one, 3-(4-fluorophenyl)-5-methyl-	EtOH	283(3.8)	83-0465-86
$C_{12}H_{11}N$			
Benzenamine, N-phenyl-	C_6H_{12}	202(4.42),284(4.14)	20-0631-86
	EtOH	203(4.54),285(4.39)	20-0631-86
	PrOH	220s(3.75),234(3.81), 286(4.38)	20-0631-86
	MeCN	282(4.45)	20-0631-86
	20% MeCN	282(4.27)	20-0631-86
	heptane-EtOH	241s(--),283(4.37)	80-0857-86
cation	acid	254f(2.3)	112-1099-86
Bullvalenecarbonitrile, methyl-	ether	254s(3.53)	24-2339-86
1H-Pyrrole, 2-(2-phenylethenyl)-, (E)-	EtOH	237(3.75),330(4.38)	118-0620-86
(Z)-	EtOH	228(4.05),315(3.95)	118-0620-86
$C_{12}H_{11}NO$			
Phenol, 4-(3-pyridinylmethyl)-	EtOH	226(3.96),258s(3.59), 264(3.64)	87-1461-86
2H-Pyrrol-2-one, 1,5-dihydro-5-[(4-methylphenyl)methylene]-, (Z)-	MeOH	341(4.36)	49-0185-86
$C_{12}H_{11}NO_2S$			
Phenol, 2-[[(5-methoxy-3-thienyl)-methylene]amino]-	EtOH	204(4.45),226(4.21), 245(4.14),287(3.84), 338(3.74)	70-1470-86
2-Propenoic acid, 3-[(1-methyl-1H-indol-3-yl)thio]-, (E)-	dioxan	222(4.48),280(4.15)	83-0120-86
(Z)-	dioxan	222(4.48),280(4.15)	83-0120-86
$C_{12}H_{11}NO_3$			
1H-Indole-2,3-dione, 1-(1-oxobutyl)-	EtOH	236(4.40),267s(3.99), 271(3.99),334(3.60)	104-2164-86
	$CHCl_3$	233(4.34),241s(4.25), 248(4.11),294(3.69), 376(2.75)	104-2163-86
3H-Indol-3-one, 1,2-diacetyl-1,2-dihydro-	EtOH	206(3.98),239(4.51), 262(4.14),350(3.81)	104-2186-86
$C_{12}H_{11}NO_4$			
1H-Indole-5,6-diol, diacetate	MeOH	254(4.10),300(3.50)	32-0407-86
Spiro[1,3-dioxolane-2,3'-[3H]indol-2'(1'H)-one, 1'-acetyl-	heptane	219(4.33)	104-2163-86
	EtOH	217.5(4.26),279(2.70)	104-2163-86
$C_{12}H_{11}NO_4S$			
Thiophene, 2,5-dihydro-3-methyl-4-nitro-2-(phenylmethylene)-, 1,1-dioxide, (Z)-	$CHCl_3$	250(3.544),365(4.338)	104-0379-86
$C_{12}H_{11}NO_5$			
2-Benzofurancarboxamide, N-formyl-4,6-	EtOH	288(4.07),315(4.03),	4-1347-86

Compound	Solvent	λ max(log ε)	Ref.
dimethoxy- (cont.)		333(4.03)	4-1347-86
2-Benzofurancarboxamide, 7-formyl-4,6-dimethoxy-	EtOH	221(3.42),277(3.73), 298(3.67),332(3.42)	4-1347-86
4H-Furo[3,2-b]pyrrole-5-carboxylic acid, 2-(2-carboxyethenyl)-, 5-ethyl ester	MeOH	351(3.67)	73-1685-86
$C_{12}H_{11}NO_6$			
1H-Indole-2,3-dicarboxylic acid, 4,6-dimethoxy-	n.s.g.	255(3.91),310(3.83)	12-0015-86
Propanedioic acid, [(4-nitrophenyl)-methylene]-, dimethyl ester	EtOH	294(4.47)	44-4107-86
$C_{12}H_{11}NS_2$			
5(4H)-Thiazolethione, 4-ethenyl-4-methyl-2-phenyl-	EtOH	220(4.09),262(4.08), 318(3.97),480(0.30)	33-0374-86
$C_{12}H_{11}N_3$			
Benzenamine, 4-(phenylazo)-	buffer	384(4.40)	3-0631-86
Propanedinitrile, [[4-(dimethylamino)-phenyl]methylene]-	50% DMSO	442(4.72)	35-2372-86
5H-Pyrido[4,3-b]indol-3-amine, 1-methyl-	pH 7.4	264(4.75)	94-0944-86
$C_{12}H_{11}N_3O$			
Pyridine, 3-[(4-methoxyphenyl)azo]-	EtOH	351(4.44)	23-2115-86
monoprotonated	n.s.g.	367(--)	23-2115-86
diprotonated	n.s.g.	465(--)	23-2115-86
4H-Pyrimido[5,4-d][1]benzazepin-4-one, 1,5,6,7-tetrahydro-	EtOH	235(4.38),259s(--), 286(3.71),296(3.70), 309s(--),376(3.49)	142-0143-86
$C_{12}H_{11}N_3OS$			
Thiourea, N-(3-hydroxy-2-pyridinyl)-N'-phenyl-	benzene	291(4.28)	80-0761-86
	MeOH	252(4.23)	80-0761-86
	EtOH	255(4.25)	80-0761-86
	isoPrOH	256(4.27)	80-0761-86
	BuOH	254(4.43)	80-0761-86
	dioxan	255(4.30)	80-0761-86
	HOAc	255(4.18)	80-0761-86
	$CHCl_3$	252(4.24)	80-0761-86
	CH_2Cl_2	253(4.30)	80-0761-86
	MeCN	256(4.25)	80-0761-86
	DMF	260(4.13)	80-0761-86
	DMSO	262(4.14)	80-0761-86
$C_{12}H_{11}N_3O_2$			
Acetic acid, [(2-amino-2-phenylethenyl)imino]cyano-, methyl ester, (Z,?)-	EtOH	451(4.59)	33-0793-86
1,3-Benzenediol, 4-[(4-methyl-2-pyridinyl)azo]-	EtOH-acid	388(4.35)	18-1481-86
	EtOH-Et_3N	406(4.54)	18-1481-86
Pyrazinecarboxylic acid, 3-amino-5-phenyl-, methyl ester	EtOH	373(4.16)	33-0793-86
$C_{12}H_{11}N_3O_2S$			
4(1H)-Pyrimidinethione, 1,6-dimethyl-5-nitro-2-phenyl-	EtOH	251(3.90),335(4.26)	18-3871-86
$C_{12}H_{11}N_3O_3$			
4(1H)-Pyridinone, 3-hydroxy-6-(hydroxymethyl)-2-(phenylazo)-	MeOH	240(4.13),371(3.91), 435(3.97)	4-0333-86

Compound	Solvent	λ_{max}(log ε)	Ref.
$C_{12}H_{11}N_3O_5$			
2,4,6(1H,3H,5H)-Pyrimidinetrione, 5-ethyl-5-(4-nitrophenyl)-	EtOH	254(3.73)	117-0209-86
Quinoxaline, 5,6,7,8-tetrahydro-2-(5-nitro-2-furanyl)-, 1,4-dioxide	EtOH	236(4.20),268(4.11), 318(4.40),344(4.31)	103-0268-86
$C_{12}H_{11}N_5$			
7H-Purin-6-amine, 7-(phenylmethyl)-	pH 1	273.5(4.18)	94-2037-86
	H_2O	271(4.06),280s(3.90)	94-2037-86
	pH 13	270(4.07),280s(3.91)	94-2037-86
	EtOH	272(4.05),283s(3.86)	94-2037-86
9H-Purin-6-amine, 9-(phenylmethyl)-	pH 1	260(4.16)	94-2037-86
	H_2O	261(4.17)	94-2037-86
	pH 13	261(4.17)	94-2037-86
	EtOH	261(4.18)	94-2037-86
$C_{12}H_{11}N_5OS$			
Ethanone, 1-[5-phenyl-1-(2-thiazolyl)formazanyl]-	50% EtOH	480(4.36)	65-2114-86
3H-Pyrazol-3-one, 5-amino-1,2-dihydro-1-methyl-4-(5-phenyl-1,3,4-thiadiazol-2-yl)-	MeOH	331(4.35024)	161-0666-86
$C_{12}H_{11}N_5OS_2$			
1,2,4-Triazolo[3,4-b][1,3,4]thiadiazolium, 6-(benzoylamino)-1-methyl-3-(methylthio)-, hydroxide, inner salt	MeOH	241(4.21),304(4.13)	5-1540-86
$C_{12}H_{11}N_5S$			
2H-Purine-2-thione, 1,3,6,7-tetrahydro-6-imino-7-methyl-1-phenyl-	pH 1	249(4.19),288(4.31)	4-0737-86
	pH 13	237(4.22),274(4.26)	4-0737-86
$C_{12}H_{12}$			
Tricyclo[5.5.0.0^{2,8}]dodeca-3,5,9,11-tetraene	C_6H_{12}	218(3.95),318(3.48)	35-0512-86
$C_{12}H_{12}BrNO_2$			
1H-Indole-3-propanoic acid, 2-bromo-, methyl ester	MeOH	221(4.56),282(3.94), 290(3.86)	35-2023-86
$C_{12}H_{12}BrN_5O_3$			
7H-Pyrrolo[2,3-d]pyrimidine-5-carbonitrile, 4-amino-6-bromo-7-(2-deoxy-β-D-erythro-pentofuranosyl)-	pH 1	233(4.19),282(4.19)	78-5869-86
	pH 7	218(4.25),286(4.16)	78-5869-86
	pH 11	286(4.21)	78-5869-86
$C_{12}H_{12}BrN_2$			
2,3'-Bipyridinium, 1'-(2-bromoethyl)-, bromide	pH 8.2	209(4.17),230(4.16), 276(4.04)	64-0239-86B
$C_{12}H_{12}ClN$			
1H-Carbazole, 5(and 7)-chloro-2,3,4,9-tetrahydro-	$HOCH_2CH_2OH$	238(4.50),287(3.74), 296s(3.70)	103-0713-86
1H-Carbazole, 6-chloro-2,3,4,9-tetrahydro-	$HOCH_2CH_2OH$	238(4.58),292(3.97), 299(3.93)	103-0713-86
1H-Carbazole, 7-chloro-2,3,4,9-tetrahydro-	$HOCH_2CH_2OH$	238(4.48),288(3.78), 295s(3.70)	103-0713-86
$C_{12}H_{12}ClNO_2$			
2,4,6-Heptatrienoic acid, 7-(3-chloro-1H-pyrrol-2-yl)-, methyl ester, (E)-	EtOH	385(4.62)	150-3601-86M

Compound	Solvent	λ_{max}(log ϵ)	Ref.
1H-Indole-3-propanoic acid, 2-chloro-, methyl ester	MeOH	219(4.61),272(3.99), 289(3.75)	35-2023-86
$C_{12}H_{12}FN_3S_3$			
1,2,4-Triazine, 6-[[(2-fluorophenyl)-methyl]thio]-3,5-bis(methylthio)-	EtOH	355(3.59)	80-0881-86
1,2,4-Triazine, 6-[[(3-fluorophenyl)-methyl]thio]-3,5-bis(methylthio)-	EtOH	355(3.59)	80-0881-86
1,2,4-Triazine, 6-[[(4-fluorophenyl)-methyl]thio]-3,5-bis(methylthio)-	EtOH	355(3.59)	80-0881-86
$C_{12}H_{12}F_6O_4$			
3,5-Heptadienoic acid, acid, 6-acet-oxy-, 2,2,2-trifluoro-1-(tri-fluoromethyl)ethyl ester, (E,Z)-	MeOH	228(4.28)	33-0469-86
$C_{12}H_{12}N_2$			
Bicyclo[2.2.1]hepta-2,5-diene-2,3-di-carbonitrile, 1,5,6-trimethyl-	MeCN	337(2.3)	35-1589-86
	benzene	335(--)	35-1589-86
	C_6H_{12}	325(--)	35-1589-86
2,2'-Bipyridine, 4,4'-dimethyl-	MeCN	250(4.14),290(4.21), 350s(1.85),380(1.90)	152-0165-86
Dipyrido[1,2-a:2',1'-c]pyrazinediium, 6,7-dihydro- (radical cation)	pH 8	760(3.43)	18-1709-86
2,5-Pentalenedicarbonitrile, 1,3a,4,6a-tetrahydro-3a,6a-dimethyl- (contains isomer)	MeCN	207(4.207)	24-1801-86
$C_{12}H_{12}N_2O$			
Pyrazine, 2,5-dimethyl-3-phenyl-, 1-oxide	EtOH	224(4.29),251(4.24), 264s(4.13),311(3.63)	142-0785-86
Pyrazine, 2,5-dimethyl-3-phenyl-, 4-oxide	EtOH	225(4.28),265(4.01), 305(3.65)	142-0785-86
3-Pyridinecarboxamide, 1,4-dihydro-1-phenyl-	EtOH	272(4.14),350(3.89)	150-0901-86M
9H-Pyrido[3,4-b]indol-7-ol, 3,4-di-hydro-1-methyl- (harmalol)	pH 2-8	203(4.26),215(4.27), 258(3.73),369(4.20)	149-0571-86B
	pH 11	205(4.16),219(4.20), 271(3.76),363(3.91), 431(4.22)	149-0571-86B
anion	pH 13	217(4.33),269(3.86), 363(4.27)	149-0571-86B
dianion	n.s.g.	230(4.22),269(4.12), 329(3.96),388(3.74)	149-0571-86B
	MeOH	204(4.49),214(4.43), 259(3.78),373(4.23)	149-0571-86B
	MeCN	226(4.43),261(4.01), 330(4.24)	149-0571-86B
	+ acid	204(2.41),214(4.43), 259(3.78),373(4.23)	149-0571-86B
cation	pH 2	203(4.26),215(4.21), 258(3.73),368(4.19)	39-1247-86B
dication	H_0 -8.5	197(4.16),215(4.21), 276(3.87),367(4.20)	39-1247-86B
$C_{12}H_{12}N_2O_2$			
2,4-Imidazolidinedione, 1-methyl-5-[(4-methylphenyl)methylene]-, (E)-	EtOH	338(4.025)	39-1941-86B
(Z)-	EtOH	312(4.126)	39-1941-86B
2,4-Imidazolidinedione, 3-methyl-5-[(4-methylphenyl)methylene]-, (Z)-	EtOH	320(4.257)	39-1941-86B

Compound	Solvent	λ_{max}(log ϵ)	Ref.
Pyrazine, 2,3-dimethyl-5-phenyl-, 1,4-dioxide	EtOH	263(4.56),314(4.21)	103-0268-86
Pyrazine, 2,5-dimethyl-3-phenyl-, 1,4-dioxide	EtOH	237(4.22),247s(4.04), 309(4.27)	103-0268-86
2(1H)-Quinoxalinone, 1-methyl-3-(2-oxopropyl)-	MeOH	372(4.33),394(4.38), 416(4.21)	2-0525-86
2(1H)-Quinoxalinone, 3-(2-oxobutyl)-	MeOH	372(4.21),392(4.34), 414(4.19)	2-0525-86
$C_{12}H_{12}N_2O_3$			
1H-Imidazole-4-carboxaldehyde, 2-(3,4-dimethoxyphenyl)-	MeOH	253(4.03),307(4.26)	87-0261-86
2,4-Imidazolidinedione, 5-[(4-methoxyphenyl)methylene]-1-methyl-, (E)-	EtOH	348(4.385)	39-1941-86B
(Z)-	EtOH	323(4.302)	39-1941-86B
2,4-Imidazolidinedione, 5-[(4-methoxyphenyl)methylene]-3-methyl-, (Z)-	EtOH	332(4.491)	39-1941-86B
1,6-Naphthyridine-3-carboxylic acid, 5,6-dihydro-2-methyl-5-oxo-, ethyl ester	EtOH	250(3.97),320(4.12)	39-0753-86C
$C_{12}H_{12}N_2O_4$			
Benzamide, 2-nitro-N-(1-oxo-4-pentenyl)-	MeOH	209(4.29),259(3.76)	142-0777-86
$C_{12}H_{12}N_2O_5$			
Dibenzofuran, 1,2,3,4,4a,9b-hexahydro-6,8-dinitro-	n.s.g.	218(4.20),252(4.16), 312(4.05)	44-3453-86
Phenol, 2-(1-cyclohexenyl)-4,6-dinitro-	n.s.g.	218(4.08),250(3.95)	44-3453-86
$C_{12}H_{12}N_2S$			
3(2H)-Pyridazinethione, 4-methyl-6-(4-methylphenyl)-	EtOH	320(4.28)	48-0522-86
Thieno[2.3-c:5,4-c']dipyridinium, 2,7-dimethyl-, diperchlorate	pH 5	290(4.1),395(3.7)	77-1645-86
radical cation	n.s.g.	370(4.2),420(3.6), 622(3.78)	77-1645-86
$C_{12}H_{12}N_4$			
Cyanamide, (2,6-diethyl-2,5-cyclohexadiene-1,4-diylidene)bis-	MeCN	321s(4.40),339(4.53), 352(4.49),380s(3.85)	5-0142-86
Cyanamide, 2,3,5,6-tetramethyl-2,5-cyclohexadiene-1,4-diylidene)bis-	MeCN	331s(3.87),346(4.37), 368s(4.15)	5-0142-86
1,4,7,9b-Tetraazaphenalene, 2,5,8-trimethyl-	pentane	246(4.39),333(4.21), 347(4.31),359(4.22), 365(4.32),371s(4.04), 426(1.83),454(2.09), 490(2.30),532(2.39), 582(2.26)	35-0017-86
$C_{12}H_{12}N_4O$			
Pyrazinecarboxamide, 3-amino-6-methyl-5-phenyl-	EtOH	370(4.00)	33-0793-86
$C_{12}H_{12}N_4O_4$			
Cyanamide, (2,3,5,6-tetramethoxy-2,5-cyclohexadiene-1,4-diylidene)bis-	MeCN	356s(4.48),368(4.54)	5-0142-86
Pyrido[2,3-d]pyrimidine-6-carbonitrile, 1,2,3,4-tetrahydro-1,3-bis(methoxymethyl)-2,4-dioxo-	H_2O	217(4.58),263(4.27), 307(3.84),320s(3.68)	87-0709-86

Compound	Solvent	$\lambda_{max}(\log \epsilon)$	Ref.
$C_{12}H_{12}N_4O_5$			
Pyrido[2,3-d]pyrimidine-6-carbonitrile, 1,2,3,4,7,8-hexahydro-1,3-bis(methoxymethyl)-2,4,7-trioxo-	H_2O	227(4.78),282(4.37), 323(4.51)	87-0709-86
$C_{12}H_{12}N_6$			
Hydrazinecarboximidamide, 2-(di-2-pyridinylmethylene)-	pH 1.0	260(4.2),325(4.3)	74-0251-86C
	pH 4.0	297(4.3)	74-0251-86C
	pH 8.0	305(4.3)	74-0251-86C
	pH 12.0	325(4.3)	74-0251-86C
$C_{12}H_{12}N_6OS$			
Ethanone, 1-[5-phenyl-1-(2-thiazolyl)formazanyl]-, oxime	50% EtOH	460(4.27)	65-2114-86
$C_{12}H_{12}N_6S$			
Cyclopropenylium, [[2-(cyanoamino)-3-(cyanoimino)-1-cyclopropen-1-yl]thio]bis(dimethylamino)-, hydroxide, inner salt	CH_2Cl_2	237(4.45),311(3.87)	24-2104-86
$C_{12}H_{12}O_2$			
1H-Cyclopenta[a]pentalene-1,4(3aH)-dione, 3b,6a,7,7a-tetrahydro-3-methyl-	MeOH	223(4.17),315(3.15)	39-0291-86C
1H-Cyclopenta[a]pentalene-3,4(2H,3bH)-dione, 6a,7-dihydro-6a-methyl-	MeOH	214(4.21),244(3.73)	35-3443-86
1H-Cyclopenta[a]pentalene-3,4(3aH)-dione, 3b,6a,7,7a-tetrahydro-1-methyl-	MeOH	217(4.04),222(4.09)	35-3443-86
1H-Cyclopenta[a]pentalene-3,4(3aH)-dione, 3b,6a,7,7a-tetrahydro-6a-methyl-	MeOH	216(4.07)	35-3443-86
1H-Inden-1-one, 3-acetyl-2,3-dihydro-2-methyl-	MeOH	226s(3.98),245(3.95), 284(3.17)	39-1789-86C
4,8-Methanoazulene-1,5-dione, 3a,4,8,8a-tetrahydro-3-methyl-, (3aα,4α,8α,8aα)-	MeOH	221(4.29)	39-0291-86C
1-Naphthalenol, 8-methoxy-3-methyl-	EtOH	290s(3.44),306(3.51), 320(3.50),335(3.57)	94-2810-86
2,4-Pentadienoic acid, 5-phenyl-, methyl ester, (E,E)-	CH_2Cl_2	309(4.50)	35-3016-86
BF_3 complex	CH_2Cl_2	352(4.51)	35-3016-86
2,4-Pentanedione, 3-(phenylmethylene)-	50% DMSO	287(4.3)	35-7744-86
$C_{12}H_{12}O_2S$			
Thiophene, 2-(2,4-dimethoxyphenyl)-	MeOH	226s(4.07),283(4.09), 310(4.05)	18-0083-86
$C_{12}H_{12}O_3$			
2H-1-Benzopyran-2-one, 7-ethoxy-4-methyl-	acid	321(4.1)	18-0961-86
1,3,5-Hexanetrione, 1-phenyl-	dioxan	326(4.06),350(4.65)	83-0242-86
2-Naphthalenol, 1,8-dimethoxy-	MeOH	247.5(4.32),278s(3.84), 287(4.00),298(3.98), 336(3.69),348(3.75)	39-0675-86C
$C_{12}H_{12}O_3S$			
Furan, 2-[[(4-methylphenyl)sulfonyl]-methyl]-	EtOH	226(4.22)	142-1291-86

Compound	Solvent	λ_{max}(log ε)	Ref.
$C_{12}H_{12}O_4$			
1H-2-Benzopyran-1-one, 8-hydroxy-6-methoxy-3,4-dimethyl- (polygonolide)	EtOH	242(4.11),248(4.20), 260(3.48),280(3.28), 298(3.04),330(3.26)	102-0517-86
1(3H)-Isobenzofuranone, 4-(2,5-dihydro-2-oxo-3-furanyl)-3a,4,5,6-tetrahydro-, trans-(±)- (differolide)	EtOH	214(4.19)	33-1833-86
$C_{12}H_{12}O_5$			
1,3-Dioxane-4,6-dione, 2,2-dimethyl-5-[(5-methyl-2-furanyl)methylene]-	EtOH	219(3.66),262(3.30), 396(4.27)	103-1072-86
	acid	389(4.38)	103-1072-86
	base	262(4.09)	103-1072-86
5-Oxatetracyclo[4.3.0.0^{2,4}.0^{3,7}]non-8-ene-8,9-dicarboxylic acid, dimethyl ester	MeCN	235(3.88)	24-0589-86
9-Oxatricyclo[4.2.1.0^{2,5}]nona-3,7-diene-7,8-dicarboxylic acid, dimethyl ester, (1α,2α,5α,6α)-	EtOH	242(3.52)	24-0589-86
$C_{12}H_{12}O_6$			
Benzoic acid, 2,4-diacetoxy-, methyl ester	EtOH	233(4.14),270s(3.31), 280s(3.23)	33-0734-86
4H-1-Benzopyran-4-one, 6,7-dihydroxy-5,8-dimethoxy-2-methyl-	EtOH	212(4.38),229s(4.28), 252(4.11),300(3.85), 323(3.87)	44-3116-86
$C_{12}H_{12}S$			
Azulene, 4-[(methylthio)methyl]-	hexane	240(4.27),280(4.57), 320s(--),330(3.54), 343(3.60),356(3.15), 545s(--),568(2.60), 587(2.60),615(2.57), 635s(--),675(2.17)	5-1222-86
Thiophene, 2-(2,4-dimethylphenyl)-	EtOH	229(4.03),268(4.03)	18-0083-86
$C_{12}H_{12}S_2$			
1,2-Naphthalenedimethanethiol	C_6H_{12}	231(4.86),279(3.83), 289(3.88),299(3.75), 321(3.61)	56-0315-86
2,3-Naphthalenedimethanethiol	C_6H_{12}	232(4.91),270(3.77), 279(3.74),291(3.51)	56-0315-86
$C_{12}H_{12}S_4Se_2$			
5H-1,3-Dithiolo[4,5-b][1,4]dithiepin, 2-(5,6-dihydro-4H-cyclopenta-1,3-diselenol-2-ylidene)-6,7-dihydro-	MeCN	208(4.33),303(4.20), 332(3.89),378(3.17)	77-1472-86
$C_{12}H_{13}Cl$			
5,9-Methano-5H-benzocycloheptene, 7-chloro-6,7,8,9-tetrahydro-, endo	MeOH	264(--),270(2.87), 276(--)	44-4681-86
exo	MeOH	263(--),268(2.75), 274(--)	44-4681-86
$C_{12}H_{13}ClN_2O_2$			
Quinazoline, 2-(chloromethyl)-6,7-dimethoxy-4-methyl-	EtOH	250(4.54),318(3.83), 331(3.93)	4-1263-86
$C_{12}H_{13}ClOS$			
1-Penten-3-one, 1-chloro-4-methyl-1-(phenylthio)-	n.s.g.	256(4.07),291(4.52)	48-0539-86

Compound	Solvent	$\lambda_{max}(\log \epsilon)$	Ref.
$C_{12}H_{13}FN_2O_3$			
4-Isoxazolecarboxamide, 3-(4-fluorophenyl)-4,5-dihydro-5-(hydroxymethyl)-N-methyl-	MeOH	262(2.96)	73-2167-86
$C_{12}H_{13}N$			
Benzeneacetonitrile, α-(2-methylpropylidene)-, (E)-	C_6H_{12}	243.3(4.00)	88-6225-86
(Z)-	C_6H_{12}	256.6(4.06)	88-6225-86
1H-Carbazole, 2,3,4,9-tetrahydro-	$HOCH_2CH_2OH$	230(4.61),285(4.13), 292s(4.08)	103-0713-86
$C_{12}H_{13}NO$			
Cyclohepta[b]pyrrol-2(1H)-one, 1,3,6-trimethyl-	MeOH	269(--),273(--), 275(4.36),295s(--), 400(3.79),418(3.81)	44-4087-86
2H-Cyclopent[cd]isoindol-2-one, 1,2a,7a,7b-tetrahydro-7a,7b-dimethyl-	EtOH	305(3.69)	39-0145-86C
$C_{12}H_{13}NOS$			
1H-Pyrrole, 1-(methoxymethyl)-2-[2-(2-thienyl)ethenyl]-, (E)-	EtOH	245(3.88),345(4.40)	118-0620-86
$C_{12}H_{13}NOS_2$			
5(4H)-Thiazolethione, 2-(4-methoxyphenyl)-4,4-dimethyl-	EtOH	280(4.21),290s(4.24), 300(4.28),480(1.20)	33-0374-86
$C_{12}H_{13}NO_2$			
2,4,6-Heptatrienoic acid, 7-(1H-pyrrol-2-yl)-, methyl ester	EtOH	394(4.63)	150-3601-86M
A-Norsecurinan-11-one, (±)-	n.s.g.	256(4.13)	100-0614-86
1H-Pyrrole, 2-[2-(2-furanyl)ethenyl]-1-(methoxymethyl)-, (E)-	EtOH	221(3.81),335(4.39)	118-0620-86
$C_{12}H_{13}NO_3$			
[1,3]Dioxepino[5,6-d]isoxazole, 3a,4,8,8a-tetrahydro-3-phenyl-, cis	MeOH	260(2.97)	73-2158-86
2,5-Furandione, dihydro-3,3-dimethyl-4-[(1-methyl-1H-pyrrol-2-yl)methylene]-, (E)-	EtOH	227(3.50),337(4.11)	34-0369-86
4,8a-Methano-8aH-furo[3,2-g]oxazolo-[2,3,4-cd]indolizin-7(4H)-one, hexahydro- (nirurine)	MeOH	243(4.17)	39-1551-86C
$C_{12}H_{13}NO_4$			
Benzeneacetic acid, 2-(acetylamino)-α-oxo-, ethyl ester	EtOH	236(4.379),265.5(3.913), 272.5(3.904),340(3.496)	104-2163-86
1,6-Dioxecino[2,3,4-gh]pyrrolizine-2,5-dione, 7,9,11,12,12a,12b-hexahydro-	EtOH	<204(>4.26)	39-0585-86C
1H-Indole-2-carboxylic acid, 4,6-dimethoxy-, methyl ester	n.s.g.	243(4.28),308(4.31)	12-0015-86
$C_{12}H_{13}NO_5$			
D-Ribitol, 1,4-anhydro-1-C-2-benzoxazolyl-, (R)-	EtOH	236(4.13),265(3.70), 273(3.82),280(3.75)	136-0271-86H
$C_{12}H_{13}N_3$			
Cyclohepta[b]pyrrol-2(1H)-one, (1-methylethylidene)hydrazone	EtOH	280(4.51),288(4.46), 325(4.18),386(4.02),	18-3326-86

Compound	Solvent	λ_{max}(log ε)	Ref.
(cont.)		400(4.01),444(3.84), 495s(3.32)	18-3326-86
$C_{12}H_{13}N_3O_2S$			
Thieno[2,3-d]pyrimidine-4-acetic acid, α-cyano-5,6-dihydro-2-methyl-, ethyl ester	$CHCl_3$	279(4.07),315(3.99), 371(4.32)	94-0516-86
$C_{12}H_{13}N_3O_3$			
2,4,6(1H,3H,5H)-Pyrimidinetrione, 5-(4-aminophenyl)-5-ethyl-	EtOH	261(3.67)	117-0209-86
$C_{12}H_{13}N_3O_3S$			
Benzamide, N-[3-(methylamino)-2-nitro-1-thioxo-2-butenyl]-	EtOH	248(4.00),310(4.19), 377(4.03)	18-3871-86
$C_{12}H_{13}N_3O_4S$			
4-Pyrimidinecarboxylic acid, 1,4,5,6-tetrahydro-5-oxo-2-[(2-thienylcarbonyl)amino]-, ethyl ester	EtOH	235s(3.89),255s(4.02), 265(4.07),292(4.28)	118-0484-86
$C_{12}H_{13}N_3O_5$			
4-Pyrimidinecarboxylic acid, 2-[(2-furanylcarbonyl)amino]-1,4,5,6-tetrahydro-5-oxo-, ethyl ester	EtOH	228(3.94),287(4.45)	118-0484-86
1H-Pyrrolo[3,4-c]pyridine-7-carboxylic acid, 3-amino-2,4,5,6-tetrahydro-2,5-dimethyl-1,4,6-trioxo-, ethyl ester	pH 7	275(3.56),327(2.94), 438(3.31)	44-0149-86
$C_{12}H_{13}N_3O_6$			
1,2,4-Triazine-3,5(2H,4H)-dione, 2-(2-deoxy-α-D-erythro-pentofuranosyl)-6-(3-furanyl)-	EtOH	309(3.88)	87-0809-86
β-	pH 7	308(3.87)	87-0809-86
	pH 13	298(3.85)	87-0809-86
$C_{12}H_{13}N_3O_7$			
1,2,4-Triazine-3,5(2H,4H)-dione, 6-(3-furanyl)-2-β-D-ribofuranosyl-	EtOH	308(3.92)	87-0809-86
$C_{12}H_{13}N_3S$			
2-Thiazolamine, N-[[4-(dimethylamino)phenyl]methylene]-	dioxan	391(4.53)	80-0273-86
	+ CF_3COOH	470(--)	80-0273-86
$C_{12}H_{13}N_5O$			
2-Azetidinone, 4-methyl-1-[2-(phenylmethyl)-2H-tetrazol-5-yl]-, (±)-	EtOH	236s(3.97)	78-2685-86
$C_{12}H_{13}N_5O_3$			
Benzoic acid, 2-[[(3-diazo-1-methyl-2-oxopropyl)amino]carbonyl]-, hydrazide	MeOH	242(4.20),272(4.15)	104-0353-86
$C_{12}H_{13}N_5O_4$			
Pyrido[2,3-d]pyrimidine-6-carbonitrile, 7-amino-1,2,3,4-tetrahydro-1,3-bis(methoxymethyl)-2,4-dioxo-	H_2O	229(4.59),284(4.18), 329(4.25)	87-0709-86

Compound	Solvent	λ_{max}(log ε)	Ref.
$C_{12}H_{13}N_5O_5$			
1,2,4-Triazine-3,5(2H,4H)-dione, 2-[3-(3,4-dihydro-5-methyl-2,4-dioxo-1(2H)-pyrimidinyl)-2-oxopropyl]-6-methyl-	H_2O	269(4.17)	107-1177-86
$C_{12}H_{14}$			
Bicyclo[6.2.2]dodeca-4,8,10,12-tetraene, (Z)-	hexane	227(3.73),260(3.59), 310(2.53)	89-0369-86
Bicyclo[3.2.1]octane, 2,3,6,7-tetrakis(methylene)-	isooctane	223(4.00),239(4.00), 246s(3.96),259s(3.75)	44-2385-86
	EtOH	224(3.98),238(3.98), 249s(3.92),259s(3.73)	44-2385-86
Bullvalene, dimethyl-	ether	222s(3.65),235s(3.34)	24-2339-86
Tetracyclo[3.3.0.0^{2,8}.0^{4,6}]octane, 1,5-dimethyl-3,7-bis(methylene)-	C_6H_{12}	228(4.26)	33-0071-86
$C_{12}H_{14}BrNO_2$			
Butanamide, 2-bromo-N-(2,4-dimethylphenyl)-3-oxo-	benzene	205(3.27),257(4.11), 284(3.85)	80-0869-86
	MeOH	202(3.19),253(4.06), 278(3.83)	80-0869-86
	EtOH	202(3.21),255(4.07), 279(3.83)	80-0869-86
	PrOH	203(3.23),256(4.09), 280(3.84)	80-0869-86
	BuOH	199(3.26),257(4.10), 282(3.85)	80-0869-86
	dioxan	205(3.29),257(4.11), 284(3.85)	80-0869-86
	HOAc	201(3.18),252(4.00), 280(3.81)	80-0869-86
	CH_2Cl_2	200(3.26),253(4.04), 279(3.83)	80-0869-86
	$CHCl_3$	202(4.25),253(4.08), 281(3.84)	80-0869-86
	CCl_4	204(3.23),256(4.09), 282(3.84)	80-0869-86
	$C_2H_4Cl_2$	201(3.25),252(4.04), 278(3.83)	80-0869-86
	MeCN	204(3.32),257(4.10), 284(3.85)	80-0869-86
	pyridine	212(3.00),302(3.75)	80-0869-86
	DMF	203(3.26),264(3.99), 282(3.81)	80-0869-86
	DMSO	205(3.21),264(4.07), 282(3.73)	80-0869-86
Butanamide, 4-bromo-N-(2,4-dimethylphenyl)-3-oxo-	benzene	203(3.26),255(4.07), 280(3.82)	80-0869-86
	MeOH	201(3.18),251(4.04), 276(3.81)	80-0869-86
	EtOH	202(3.20),252(4.05), 278(3.79)	80-0869-86
	PrOH	200(3.22),252(4.07), 279(3.80)	80-0869-86
	BuOH	204(3.26),253(4.08), 280(3.84)	80-0869-86
	dioxan	204(3.28),257(4.08), 282(3.83)	80-0869-86
	HOAc	205(3.17),250(3.98), 278(3.80)	80-0869-86

Compound	Solvent	λ_{max}(log ε)	Ref.
(cont.)	CH_2Cl_2	200(3.24),248(4.01), 276(3.80)	80-0869-86
	$CHCl_3$	202(3.23),253(4.05), 280(3.80)	80-0869-86
	CCl_4	202(3.25),254(4.06), 282(3.81)	80-0869-86
	MeCN	206(3.28),253(4.06)	80-0869-86
	pyridine	212(2.98),308(3.68)	80-0869-86
	DMF	204(3.22),263(3.95), 288(3.70)	80-0869-86
	DMSO	226(3.20),260(4.03), 283(3.71)	80-0869-86
$C_{12}H_{14}BrN_5O_4$			
7H-Pyrrolo[2,3-d]pyrimidine-5-carbox- amide, 4-amino-6-bromo-7-(2-deoxy- β-D-erythro-pentofuranosyl)-	pH 1	231(4.44),276(4.39)	78-5869-86
	pH 7	231(4.29),280(4.44)	78-5869-86
	pH 11	228(4.32),280(4.45)	78-5869-86
$C_{12}H_{14}Br_2N_2O_2$			
1H,7H-Pyrazolo[1,2-a]pyrazole-1,7- dione, 3,5-bis(1-bromoethyl)- 2,6-dimethyl-, 75% dl	pH 7.29	257(4.12),399(3.65)	35-4532-86
65% meso	pH 7.29	258(4.13),403(3.65)	35-4532-86
$C_{12}H_{14}Br_2O_2$			
Cyclopropanemethanol, 2,2-dibromo- 1-(4-ethoxyphenyl)-	C_6H_{12}	229(3.90)	12-0271-86
$C_{12}H_{14}ClNO_3$			
Acetamide, 2-chloro-N-(1,1-dimethyl- 2-oxo-2-phenylethyl)-	EtOH	244(4.12)	103-0417-86
1-Propanone, 2-[[(2-chloroacetyl)oxy]- amino]-2-methyl-1-phenyl-	EtOH	246(4.17)	103-0417-86
$C_{12}H_{14}Cl_2O_2$			
Cyclopropanemethanol, 2,2-dichloro- 1-(4-ethoxyphenyl)-	C_6H_{12}	231(4.26)	12-0271-86
$C_{12}H_{14}Cl_2O_4$			
2-Pentanone, 5,5-dichloro-4-hydroxy- 1-(3-hydroxy-5-methoxyphenyl)- (citreovirone)	MeOH	209(4.15),222s(3.85), 280(3.00)	138-1129-86
$C_{12}H_{14}F_2O_2$			
Cyclopropanemethanol, 1-(4-ethoxy- phenyl)-2,2-difluoro-	C_6H_{12}	228(4.02)	12-0271-86
$C_{12}H_{14}NO_2$			
Isoquinolinium, 6,7-dimethoxy- 2-methyl-, chloride	MeOH	253(4.91),310(3.95)	100-0745-86
$C_{12}H_{14}N_2$			
2,3'-Bipyridinium, 1,1'-dimethyl-, diperchlorate	pH 8.2	202(4.18),270(4.10)	64-0239-86B
4,4'-Bipyridinium, 1,1'-dimethyl-	n.s.g.	255(4.3)	23-0845-86
Dipyrido[1,2-a:2',1'-c]pyrazine, 6,7,12a,12b-tetrahydro-	MTHF	275s(3.6),295(3.6), 340s(3.4)	138-2139-86
dl-	MTHF	330(3.6)	138-2139-86
1H-Pyrrole, 2,2'-(1,2-ethenediyl)- bis[1-methyl-, (E)-	EtOH	231(3.76),344(4.23)	118-0620-86
(Z)-	EtOH	214(4.08),328(4.19)	118-0620-86

Compound	Solvent	$\lambda_{max}(\log \epsilon)$	Ref.
$C_{12}H_{14}N_2O$			
1H-Pyrrole, 1-(methoxymethyl)-2-[2-(1H-pyrrol-2-yl)ethenyl]-, (E)-	EtOH	205(2.81),338(4.40)	118-0620-86
(Z)-	EtOH	207(4.22),321(4.30)	118-0620-86
$C_{12}H_{14}N_2OS_2$			
Carbamodithioic acid, [(1-methoxy-1H-indol-3-yl)methyl]-, methyl ester	MeOH	218(4.57),267(4.20), 287s(3.98),297s(3.76)	77-1077-86
$C_{12}H_{14}N_2O_2$			
2,5-Cyclohexadiene-1,4-dione, 2,5-bis(1-aziridinyl)-3,6-dimethyl-	EtOH	328(4.20)	94-1299-86
2(1H)-Pyrazinone, 3,6-dihydro-1-hydroxy-6,6-dimethyl-5-phenyl-	EtOH	235(3.60)	103-0417-86
$C_{12}H_{14}N_2O_2S$			
β-Alanine, N-2-benzothiazolyl-N-ethyl-	pH 1	220(4.43),265(4.12), 282(4.08),291(4.10)	103-0795-86
	EtOH	226(4.59),275(4.31), 302(4.01)	103-0795-86
3(2H)-Benzothiazolepropanoic acid, 2-(ethylimino)-	pH 1	223(4.19),256(3.82), 280(3.74),289(3.78)	103-0795-86
	EtOH	223(4.26),264(3.80), 281(3.57),290(3.52), 306(3.39)	103-0795-86
6-Thia-2a,7b-diazacyclopent[cd]indene-2,3(5H,7H)-dione, 1,4,5,7-tetramethyl-, cis	MeCN	232(4.22),349(3.70)	35-4532-86
	3% MeCN	234(4.17),255s(3.77), 365(3.62)	35-4532-86
	dioxan-MeCN	232(4.18),347.5(3.73)	35-4532-86
trans	MeCN	233.5(4.17),353(3.72)	35-4532-86
	3% MeCN	234(4.20),255s(3.78), 367.5(3.64)	35-4532-86
	dioxan-MeCN	232(4.14),351(3.74)	35-4532-86
$C_{12}H_{14}N_2O_3$			
Ethanone, 1-(2,5-dihydro-1-hydroxy-2-methyl-4-phenyl-1H-imidazol-2-yl)-, N-oxide	EtOH	226(4.03),294(--)	103-0268-86
1H-Imidazole-4-methanol, 2-(3,4-dimethoxyphenyl)-	MeOH	278(4.26)	87-0261-86
4-Isoxazolecarboxamide, 4,5-dihydro-5-(hydroxymethyl)-3-(4-methylphenyl)-	MeOH	270(4.06)	73-2167-86 +118-0565-86
Methanone, (4,5-dihydro-4-hydroxy-5,5-dimethyl-1H-imidazol-2-yl)-phenyl-, N-oxide	EtOH	256(4.03),321s(3.34)	103-0856-86
Oxazole, 4,5-dihydro-4,4-dimethyl-2-(nitrophenylmethyl)-	EtOH	340(4.16)	4-0825-86
Propanal, 2-methyl-2-[(2-oxo-2-phenylethylidene)amino]-, 1-oxime, N-oxide	EtOH	259(3.90),308(4.17)	103-0856-86
2-Propenamide, 2-formyl-3-[(hydroxymethyl)amino]-3-(4-methylphenyl)-, (Z)-	MeOH	244(4.01),284(4.06)	73-2167-86 +118-0565-86
$C_{12}H_{14}N_2O_4$			
Imidazo[1,2-a]pyridine-7-carboxylic acid, 8-acetyl-1,2,3,5-tetrahydro-1-methyl-5-oxo-, methyl ester	EtOH	302(3.77),346(3.85)	24-2208-86
4-Isoxazolecarboxamide, 4,5-dihydro-5-(hydroxymethyl)-3-(2-methoxyphenyl)-	MeOH	254(3.71),286(4.17)	73-2167-86

Compound	Solvent	λ_{max}(log ε)	Ref.
4-Isoxazolecarboxamide, 4,5-dihydro-5-(hydroxymethyl)-3-(4-methoxyphenyl)-	MeOH	276(4.11)	73-2167-86 +118-0565-86
2-Propenamide, 2-formyl-3-[(hydroxymethyl)amino]-3-(4-methoxyphenyl)-, (Z)-	MeOH	244(3.88),291(4.12)	73-2167-86 +118-0565-86
2,4(1H,3H)-Pyrimidinedione, 5-ethynyl-1-[3-hydroxy-4-(hydroxymethyl)cyclopentyl]-, (±)-	pH 1	230(4.00),292(4.08)	87-0079-86
	pH 7	230(4.01),292(4.08)	87-0079-86
	pH 13	231(4.05),288(3.98)	87-0079-86
$C_{12}H_{14}N_2O_5$			
Benzofuran, 2,3-dihydro-2,2,3,3-tetramethyl-5,7-dinitro-	n.s.g.	218(4.20),251(4.18), 312(4.10)	44-3453-86
$C_{12}H_{14}N_2O_5S$			
2,1-Benzisothiazole-1(3H)-carboxylic acid, 5-(acetylamino)-, ethyl ester, 2,2-dioxide	MeOH	258(4.35)	4-1645-86
$C_{12}H_{14}N_2O_6$			
Benzofuran, 2,3-dihydro-5,7-dinitro-2-(2-methylpropoxy)-	n.s.g.	213(4.01),250(4.00), 302(3.81)	44-3453-86
$C_{12}H_{14}N_2O_7$			
Uridine, 6-(3-hydroxy-1-propynyl)-	H_2O	287(4.04)	23-1560-86
$C_{12}H_{14}N_2O_8$			
2-Propenoic acid, 3-(1,2,3,4-tetrahydro-2,4-dioxo-1-β-D-ribofuranosyl-5-pyrimidinyl)-, (E)-	EtOH	260(4.01),300(4.15)	87-0213-86
$C_{12}H_{18}N_2S_2$			
[1]Benzothieno[2,3-d]pyrimidine-4(3H)-thione, 5,6,7,8-tetrahydro-2,3-dimethyl-	EtOH-1% dioxan	221(4.44),284(4.10), 320(4.10),348s(3.94)	106-0096-86
$C_{12}H_{14}N_4O$			
1,3,5-Triazino[1,2-a]benzimidazol-4(1H)-one, 2,3-dihydro-2,2,3-trimethyl-	EtOH	282(3.88),287(3.89)	98-1005-86
$C_{12}H_{14}N_4O_2$			
2,4(1H,3H)-Pyrimidinedione, 5-amino-6-(ethylphenylamino)-	pH 13	239(4.14),324(4.06)	35-5972-86
2,4(1H,3H)-Pyrimidinedione, 5-amino-3-methyl-6-(methylphenylamino)-	pH 13	238(4.21),321(4.14)	35-5972-86
$C_{12}H_{14}N_4O_3$			
Acetamide, 2-azido-N-(1,1-dimethyl-2-oxo-2-phenylethyl)-N-hydroxy-	EtOH	245(4.17)	103-0417-86
$C_{12}H_{14}N_4O_4S$			
Acetamide, N-[4-[[(3-diazo-1-methyl-2-oxopropyl)amino]sulfonyl]phenyl]-	MeOH	206(4.26),265(4.41), 278s(4.30)	104-0353-86
Acetamide, N-[4-[[(4-diazo-3-oxobutyl)amino]sulfonyl]phenyl]-	MeOH	207(4.28),268(4.41)	104-0353-86
$C_{12}H_{14}N_4O_7$			
1H-Pyrrolo[2,3-d]pyrimidine-5-carboxylic acid, 2-amino-4,7-dihydro-4-oxo-7-β-D-arabinofuranosyl-	H_2O	231(4.21),268(3.85), 295(3.84)	18-1915-86

Compound	Solvent	λ_{max}(log ε)	Ref.
$C_{12}H_{14}N_6O_2$			
1H-Pyrazolo[3,4-g]pteridine-6,8(5H,7H)-dione, 3-ethyl-1,5,7-trimethyl-	pH -4.3	242(4.48),300(3.76), 408(3.92),420s(3.87)	108-0081-86
	pH 4.0	245(4.48),330(3.96), 412(3.76)	108-0081-86
1H-Pyrazolo[4,3-g]pteridine-5,7(6H,8H)-dione, 3-ethyl-1,6,8-trimethyl-	pH -4.3	224(4.45),236s(4.39), 357(4.24)	108-0081-86
	pH 5.0	236(4.55),356(4.28)	108-0081-86
$C_{12}H_{14}N_6O_5$			
Acetamide, N-[4-amino-6a,7,8,9a-tetrahydro-7-hydroxy-8-(hydroxymethyl)-furo[2',3':4,5]oxazolo[3,2-e]purin-2-yl]-, [6aS-(6aα,7α,8β,9aα)]-	pH 1	268(4.27),296.5(3.99)	94-2609-86
	pH 7	218(4.38),286(4.19)	94-2609-86
	pH 13	272(4.15)	94-2609-86
$C_{12}H_{14}O$			
1,4-Cyclohexadiene, 1,1'-oxybis-	MeOH	206(3.79)	44-3468-86
4,8-Methanoazulen-5(1H)-one, 3a,4,8,8a-tetrahydro-7-methyl-, (3aα,4α,8α,8aα)-	EtOH	237(3.86)	18-2811-86
4,8-Methanoazulen-5(3H)-one, 3a,4,8,8a-tetrahydro-7-methyl- (3aα,4α,8α,8aα)-	EtOH	238(3.90)	18-2811-86
2-Naphthalenol, 1,2-dihydro-1,1-dimethyl-	MeOH	224s(4.16),264(3.88)	39-1789-86C
$C_{12}H_{14}OS_2$			
Benzenecarbodithioic acid, 4-methoxy-, 1-butenyl ester	CH_2Cl_2	345(4.13),498(2.14)	18-1403-86
$C_{12}H_{14}O_2$			
1H-Cyclopenta[a]pentalene-3,4(2H-3bH)-dione, 5,6,6a,7-tetrahydro-6a-methyl-, cis	MeOH	242(3.88)	35-3443-86
1,4-Ethano-2,3-benzodioxin, 1,4,5,6-7,8-hexahydro-2,3-bis(methylene)-	MeCN	245(3.85)	33-0761-86
4,8-Methanoazulen-5(1H)-one, 3a,4,8,8a-tetrahydro-7-methoxy-, (3aα,4α,8α,8aα)-	EtOH	254(4.11)	18-2811-86
4,8-Methanoazulen-5(3H)-one, 3a,4,8,8a-tetrahydro-7-methoxy-, (3aα,4α,8α,8aα)-	EtOH	254(3.97)	18-2811-86
$C_{12}H_{14}O_3$			
3,10-Dioxatricyclo[5.2.1.$0^{1,5}$]decane, 2-methoxy-6,8,9-tris(methylene)-	isooctane	240(3.70),247(3.79), 253s(3.70)	33-1287-86
	EtOH	240(3.74),247(3.81), 253s(3.72)	33-1287-86
2-Propenoic acid, 2-(4-methoxyphenyl)-, ethyl ester	EtOH	200(2.51)(end abs.)	33-2087-86
$C_{12}H_{14}O_3S$			
2-Propenoic acid, 3-(methylthio)-, (2-methoxyphenyl)methyl ester, (E)-	EtOH	207(3.87),271(3.83), 275(4.03)	102-2543-86
$C_{12}H_{14}O_4$			
1,3-Benzodioxole, 4,5-dimethoxy-7-(1-propenyl)-	EtOH	218(3.38),264(3.95), 305(2.52)	102-1427-86
1,3-Benzodioxole, 4,5-dimethoxy-7-(2-propenyl)-	EtOH	208(3.52),285(2.47)	102-1427-86
Benzoic acid, 2-acetyl-4-hydroxy-	EtOH	264(3.87)	44-4840-86

Compound	Solvent	λmax(log ε)	Ref.
5-methyl-, ethyl ester (cont.)	EtOH-NaOH	236(--),310(--)	44-4840-86
2,8-Decadiynedioic acid, dimethyl ester	EtOH	215(3.98)	44-3125-86
Ethanone, 1-(5-acetoxy-4-methoxy-2-methylphenyl)-	EtOH	220(4.23),266(4.12)	22-0659-86
2-Propenal, 3-(2,4,5-trimethoxyphenyl)-	EtOH	205(4.18),244(4.09), 298(4.18),366(4.29)	2-0981-86
$C_{12}H_{14}O_4S$			
Butanedioic acid, dimethyl(2-thienylmethylene)-, 4-methyl ester, (E)-	EtOH	228(4.07)	34-0369-86
$C_{12}H_{14}O_6$			
7-Oxabicyclo[2.2.1]hepta-2,5-diene-2,3-dicarboxylic acid, 1-(2-hydroxyethyl)-, dimethyl ester	MeCN	284(3.15)	24-0589-86
4H-Pyran-3-carboxylic acid, 4-(1,3-dioxolan-2-ylmethyl)-5-formyl-, methyl ester, (S)-	hexane	209(3.57),289(3.46)	5-1413-86
$C_{12}H_{15}$			
Cyclopropenylium, tricyclopropyl-, (chloride)	pH 3.7	210(4.36)	35-0134-86
$C_{12}H_{15}BrN_2O_6$			
Thymidine, 5'-(bromoacetate)	pH 1	266(3.98)	87-1052-86
(all spectra in ethanol)	pH 7	266(3.98)	87-1052-86
	pH 13	266(3.86)	87-1052-86
$C_{12}H_{15}BrO_2$			
Cyclopropanemethanol, 2-bromo-1-(4-ethoxyphenyl)-, isomer A	C_6H_{12}	232(4.08)	12-0271-86
isomer B	C_6H_{12}	229(4.10)	12-0271-86
$C_{12}H_{15}BrO_3$			
1-Butanone, 4-bromo-1-(3,4-dimethoxyphenyl)-	EtOH	227(4.20),272(4.02), 302(3.85)	24-0050-86
$C_{12}H_{15}ClFN_5O_6$			
Uridine, 3'-[[[(2-chloroethyl)nitrosoamino]carbonyl]amino]-2',3'-dideoxy-5-fluoro-	MeOH	269(4.02)	87-0862-86
$C_{12}H_{15}ClHgO$			
Mercury, chloro[(2,3,4,5-tetrahydro-2-methyl-1-benzoxepin-2-yl)methyl]-	EtOH	272(3.0)	32-0441-86
$C_{12}H_{15}ClN_2$			
Piperidine, 1-[[(3-chlorophenyl)imino]methyl]-	$HOCH_2CH_2OH$	278(4.09)	103-0713-86
Piperidine, 1-[[(4-chlorophenyl)imino]methyl]-	$HOCH_2CH_2OH$	280(4.34)	103-0713-86
$C_{12}H_{15}ClN_2O_7$			
1H-Imidazole-4-carboxylic acid, 2-chloro-1-[2,3-O-(methoxymethylene)-β-D-ribofuranosyl]-, methyl ester	MeOH	234(4.03)	78-1971-86
$C_{12}H_{15}ClN_2O_8$			
1H-Imidazole-4,5-dicarboxylic acid, 2-chloro-1-β-D-ribofuranosyl-, dimethyl ester	H_2O	246(3.93)	78-1971-86

Compound	Solvent	$\lambda_{max}(\log \epsilon)$	Ref.
$C_{12}H_{15}ClO_2$			
Cyclopropanemethanol, 2-chloro-1-(4-ethoxyphenyl)-, isomer A	C_6H_{12}	230(4.11)	12-0271-86
isomer B	C_6H_{12}	229(4.15)	12-0271-86
$C_{12}H_{15}ClO_4$			
Benzoic acid, 3-chloro-6-hydroxy-2-methyl-4-(1-methylethoxy)-, methyl ester	MeOH	215(4.41),262(4.01), 307(3.64)	100-0111-86
$C_{12}H_{15}Cl_3O$			
2-Cyclohexen-1-one, 5,6-dichloro-4-(chloromethylene)-2-(1,1-dimethylethyl)-6-methyl-, (4E,5α,6β)-(±)-	$CHCl_3$	239(2.78),294(4.15)	12-2121-86
(4Z,5α,6β)-(±)-	$CHCl_3$	291(4.20)	12-2121-86
$C_{12}H_{15}FO_2$			
Cyclopropanemethanol, 1-(4-ethoxyphenyl)-2-fluoro-	C_6H_{12}	229(3.90)	12-0271-86
geometric isomer	C_6H_{12}	228(4.16)	12-0271-86
$C_{12}H_{15}F_3NO_3P$			
Benzenemethanamine, α-methyl-N-[2,2,2-trifluoro-1-(dimethoxyphosphinyl)ethylidene]-	hexane	232(3.23)	65-2372-86
$C_{12}H_{15}F_3O_4$			
3,5-Heptadienoic acid, 6-acetoxy-2-methyl-, 2,2,2-trifluoroethyl ester, (E,E)-(±)-	MeOH	235(4.36)	33-0469-86
(E,Z)-(±)-	MeOH	238(4.34)	33-0469-86
(Z,E)-(±)-	MeOH	217.5(3.78)	33-0469-86
$C_{12}H_{15}IO_3$			
1-Butanone, 1-(3,4-dimethoxyphenyl)-4-iodo-	EtOH	224(4.45),272(4.05), 303(3.90)	24-0050-86
$C_{12}H_{15}N$			
Benzenamine, N-cyclohexylidene-	pH 8.4	330(4.76)	30-0055-86
$C_{12}H_{15}NO_2$			
3H-[1]Benzopyrano[4,3-c]isoxazole, 3a,4-dihydro-1-methyl-9b(1H)-methyl-	EtOH	273(4.33),282(3.28)	44-3125-86
Formamide, N-(1-benzoyl-2-methylpropyl)-	MeOH	245(4.09)	44-3374-86
Formamide, N-[2-(1-methylethoxy)-2-phenylethenyl]-	MeOH	222(3.98),283(4.31)	44-3374-86
1H-Indole, 4,6-dimethoxy-2,3-dimethyl-	n.s.g.	224(4.56),268(3.93)	12-0015-86
2,4-Pentadienoic acid, 5-(1H-pyrrol-2-yl)-, 1-methylethyl ester, (E)-	EtOH	365(4.39)	150-3601-86M
$C_{12}H_{15}NO_3$			
1,3-Dioxolo[4,5-g]isoquinoline, 5,6,7,8-tetrahydro-4-methoxy-6-methyl- (hydrocotarnine)	n.s.g.	287(3.23)	100-0745-86
1,3-Dioxolo[4,5-h]isoquinoline, 6,7,8,9-tetrahydro-4-methoxy-9-methyl-, (±)-	EtOH	214(4.68),250s(3.43), 277(2.99),286s(2.94)	100-0745-86

Compound	Solvent	λ_{max}(log ε)	Ref.
(S)- (cont.)	EtOH	277(2.96),285s(2.92)	100-0745-86
Isoquinolinium, 3,4-dihydro-8-hydroxy-6,7-dimethoxy-2-methyl-, hydroxide, inner salt	EtOH	341(4.26),425(3.90)	100-0745-86
	EtOH-HCl	337(--)	100-0745-86
	EtOH-NaOH	340(--),420(--)	100-0745-86
8-Isoquinolinol, 3,4-dihydro-6,7-dimethoxy-1-methyl-	EtOH	324(--),408(--)	100-0745-86
	EtOH-HCl	321.5(--)	100-0745-86
	EtOH-NaOH	280s(--),320(--),404(--)	100-0745-86
	dioxan	268(--)	100-0745-86
	$CHCl_3$	270(--)	100-0745-86
$C_{12}H_{15}NO_4$			
Benzeneacetic acid, 4-morpholino-, N-oxide	EtOH	211.8(4.38)	104-1968-86
Butanedioic acid, dimethyl[(1-methyl-1H-pyrrol-2-yl)methylene]-, (E)-	EtOH	227(3.81),300(3.06),372(3.57)	34-0369-86
1H-Pyrrole-2-carboxaldehyde, 5-(hydroxymethyl)-1-(tetrahydro-4,5-dimethyl-2-oxo-3-furanyl)-, [3S-(3α,4β,5α)]- (funebral)	MeOH	257(3.88),291.5(4.13)	100-0695-86
3(1H)-Quinolinecarboxylic acid, hexahydro-4-hydroxy-2-oxo-, ethyl ester	EtOH	226(3.92),258(3.23),319(3.72)	44-1374-86
$C_{12}H_{15}NO_5$			
2H-Pyran, tetrahydro-2-(2-methoxy-4-nitrophenoxy)-	MeOH	237(3.98),298(3.82),334(3.88)	87-1737-86
$C_{12}H_{15}N_3O$			
4H-Pyrido[1,2-a]pyrimidin-4-one, 2-(butylamino)-	EtOH	232(3.98),258(4.42),328(3.56)	4-1295-86
$C_{12}H_{15}N_3OS$			
3-Pyridinepropanethioamide, α-cyano-1,4-dihydro-β,1,5-trimethyl-4-oxo-	EtOH	208(4.03),277(4.09),333(4.13)	70-1251-86
$C_{12}H_{15}N_3O_2S_3$			
Ethanaminium, N-ethyl-N-[4-(hexahydro-4,6-dioxo-2-thioxo-5-pyrimidinyl)-5-methyl-1,3-dithiol-2-ylidene)-, hydroxide, inner salt	HOAc	296(4.54)	24-2094-86
	DMF	307(4.41),370s(3.54)	24-2094-86
$C_{12}H_{15}N_3O_3$			
Acetic acid, cyano(6-ethoxy-1-methyl-4(1H)-pyrimidinylidene)-, ethyl ester	EtOH	348(4.52)	103-0774-86
$C_{12}H_{15}N_3O_3S$			
Benzenesulfonamide, N-[1-(diazoacetyl)propyl]-4-methyl-	MeOH	203(4.24),234(4.24),248s(4.12),270s(3.94)	104-0353-86
Benzenesulfonamide, N-(5-diazo-4-oxopentyl)-4-methyl-	MeOH	203(4.28),232(4.23),268s(3.94)	104-0353-86
Glycine, N-[(methylthio)[(3-pyridinylcarbonyl)amino]methylene]-, ethyl ester	EtOH	232(3.98),286(4.29)	118-0484-86
1H-Imidazo[4,5-c]pyridine, 1-(2-deoxy-β-D-ribofuranosyl)-4-(methylthio)-	pH 1	223(4.18),238s(4.02),307(4.23)	87-0138-86
	pH 13	285(4.12)	87-0138-86
$C_{12}H_{15}N_3O_3S_2$			
Ethanaminium, N-ethyl-N-[4-(hexahy-	HOAc	258(4.52),317(4.02)	24-2094-86

Compound	Solvent	$\lambda_{max}(\log \epsilon)$	Ref.
dro-2,4,6-trioxo-5-pyrimidinyl)-5-methyl-1,3-dithiol-2-ylidene]-, hydroxide, inner salt (cont.)	DMF	288.5(4.24),378(3.44)	24-2094-86
$C_{12}H_{15}N_3O_4$			
4-Isoxazolecarboxylic acid, 4,5-dihydro-5-(hydroxymethyl)-3-(4-methoxyphenyl)-, hydrazide, cis	MeOH	277(4.11)	73-2167-86
$C_{12}H_{15}N_3O_5$			
1H-Pyrazole-4-carboxylic acid, 3-(cyanomethyl)-1-(2-deoxy-β-D-erythro-pentofuranosyl)-, methyl ester	pH 1	222(4.03)	4-0059-86
	pH 7	222(4.00)	4-0059-86
	pH 11	222(3.97)	4-0059-86
1H-Pyrazole-4-carboxylic acid, 5-(cyanomethyl)-1-(2-deoxy-β-D-erythro-pentofuranosyl)-, methyl ester	pH 1	218(4.08)	4-0059-86
	pH 7	218(4.05)	4-0059-86
	pH 11	224(4.01)	4-0059-86
$C_{12}H_{15}N_3O_6$			
1H-Pyrazole-4-carboxylic acid, 3-(cyanomethyl)-1-β-D-ribofuranosyl-, methyl ester	pH 1	222(3.90)	4-0059-86
	pH 7	222(3.93)	4-0059-86
$C_{12}H_{15}N_3O_7$			
2-Propenoic acid, 3-[2-(2-deoxy-β-D-erythro-pentofuranosyl)-2,3,4,5-tetrahydro-3,5-dioxo-1,2,4-triazin-6-yl]-, methyl ester, (E)-	EtOH	221(4.16),303(4.12)	87-0809-86
$C_{12}H_{15}N_3S_2$			
Ethanamine, N,N-dimethyl-2-[[4-(2-thienyl)-2-pyrimidinyl]thio]-	n.s.g.	324(4.16)	87-1311-86
$C_{12}H_{15}N_5OS$			
1-Propanone, 1-[4-(dimethylamino)-2-(methylthio)-7-pteridinyl]-	MeOH	244(4.41),296(4.15), 394(3.95)	128-0183-86
$C_{12}H_{15}N_5O_3$			
8,2'-Ethanoadenosine, 2'-deoxy-	H_2O	207(4.32),262(4.22)	94-0015-86
$C_{12}H_{15}N_5O_3S$			
2,4(1H,3H)-Pteridinedione, 6-[1-(hydroxyimino)propyl]-1,3-dimethyl-7-(methylthio)-	MeOH	237(4.45),298(4.06), 366(4.33)	108-0081-86
7H-Pyrrolo[2,3-d]pyrimidine-5-carbothioamide, 4-amino-7-(2-deoxy-β-D-erythro-pentofuranosyl)-	pH 1	241(4.20),286(4.08)	78-5869-86
	pH 7	246(4.05),263(4.06)	78-5869-86
	pH 11	235(4.10),279(4.18)	78-5869-86
$C_{12}H_{15}N_5O_4$			
8,2'-Ethanoadenosine	H_2O	262(4.22)	94-1518-86
Glycine, N-(1,2,3,4-tetrahydro-1,3-dimethyl-2,4-dioxo-7-pteridinyl)-, ethyl ester	MeOH	279(4.05),338(4.22)	108-0081-86
$C_{12}H_{15}N_5O_4S$			
Propanamide, 3-(methylsulfonyloxy)-N-[2-(phenylmethyl)-2H-tetrazol-5-yl]-	EtOH	230s(3.90)	78-2685-86
$C_{12}H_{15}N_5O_5$			
7H-Pyrrolo[2,3-d]pyrimidine-5-carboxamide, 4-amino-7-β-D-arabinofurano-	pH 1	234(4.17),275(4.18)	78-5869-86
	pH 7	233(3.97),245(3.91),	78-5869-86

Compound	Solvent	λmax(log ε)	Ref.
syl- (cont.)		279(4.16)	78-5869-86
	pH 11	231(4.00),280(4.16)	78-5869-86
$C_{12}H_{15}N_5O_7$			
Carbamic acid, [[(4-cyano-1-β-D-ribo-furanosyl-1H-imidazol-5-yl)amino]-carbonyl]-, methyl ester	pH 1	228(3.96)	4-0153-86 +44-1065-86
	pH 11	272(3.00)	4-0153-86
	MeOH	227(3.95)	4-0153-86 +44-1065-86
$C_{12}H_{16}$			
Bicyclo[6.2.2]dodeca-8,10,11-triene	hexane	214(4.23),251(3.81), 299(2.66)	89-0369-86
Tricyclo[3.3.0.0^{2,8}]octane, 1,5-di-methyl-3,7-bis(methylene)-	C_6H_{12}	210(4.12)	33-0071-86
	EtOH	208(4.11)	33-0071-86
$C_{12}H_{16}ClFN_4O_5$			
Uridine, 3'-[[[(2-chloroethyl)amino]-carbonyl]amino]-2',3'-dideoxy-5-fluoro-	MeOH	270(3.91)	87-0862-86
$C_{12}H_{16}ClN_5O_6$			
Uridine, 3'-[[[(2-chloroethyl)nitroso-amino]carbonyl]amino]-2',3'-di-deoxy-	EtOH	260(--)	87-0862-86
$C_{12}H_{16}Cl_2N_2O_2$			
2,5-Cyclohexadiene-1,4-dione, 3,6-di-chloro-2-(butylmethylamino)-5-(meth-ylamino)-	CH_2Cl_2	250s(4.00),383(4.12), 565(2.73)	106-0832-86
$C_{12}H_{16}Cl_2O$			
2-Cyclohexen-1-one, 5,6-dichloro-2-(1,1-dimethylethyl)-6-methyl-4-methylene-	$CHCl_3$	282(4.09)	12-2121-86
$C_{22}H_{16}NOS$			
Oxazolidinium, 3,3-dimethyl-2-[(phen-ylthio)methylene]-, chloride	H_2O	249(3.97),266(3.91)	70-0944-86
$C_{12}H_{16}N_2$			
Cyclohexanone, phenylhydrazone	$HOCH_2CH_2OH$	275(3.89)	103-0713-86
1,3-Pentalenediamine, N,N,N',N'-tetramethyl-	CH_2Cl_2	618(2.56)	24-3213-86
$C_{12}H_{16}N_2O$			
2H-Pyrrol-2-one, 1,5-dihydro-3,4-di-methyl-5-[(1-methyl-1H-pyrrol-2-yl)-methyl]-	MeOH	209(4.30)	49-0185-86
$C_{12}H_{16}N_2OSe$			
Cyclobutanediylium, 1-hydroxy-2,4-di-1-pyrrolidinyl-3-selenyl-, dihy-droxide bis(inner salt)	CH_2Cl_2	252(4.3),386(4.4), 435s(3.0)	24-0182-86
$C_{12}H_{16}N_2O_2$			
Ethanone, 1-[2-(4-morpholinyl)-3H-azepin-5-yl]-	EtOH	233.0(3.89),274.5(3.80)	18-2317-86
Ethanone, 1-[2-(4-morpholinyl)-3H-azepin-6-yl]-	EtOH	231.5(3.78),330.5(3.96)	18-2317-86

Compound	Solvent	$\lambda_{max}(\log \epsilon)$	Ref.
Propanoic acid, 2-[(4-methylphenyl)hydrazono]-, ethyl ester, (E)-	EtOH	332(4.33)	18-0073-86
(Z)-	EtOH	355(4.26)	18-0073-86
2H-Pyrrol-2-one, 1,5-dihydro-5-hydroxy-3,4-dimethyl-5-[(1-methyl-1H-pyrrol-2-yl)methyl]-	MeOH	207(c.4.18)	49-0185-86
2H-Pyrrol-2-one, 1,5-dihydro-5-methoxy-3,4-dimethyl-5-(1H-pyrrol-2-ylmethyl)-	MeOH	208(4.23)	49-0185-86
$C_{12}H_{16}N_2O_2S$			
Benzenecarboximidothioic acid, N-(1-methylethyl)-4-nitro-, ethyl ester	MeCN	268(4.0)	64-0265-86B
$C_{12}H_{16}N_2O_2S_2$			
Ethanaminium, N-[4-(1-cyano-2-methoxy-2-oxoethyl)-5-methyl-1,3-dithiol-2-ylidene]-N-ethyl-, hydroxide, inner salt	MeCN	224(4.12),244(4.11), 300(4.29),424(3.63)	24-2094-86
$C_{12}H_{16}N_2O_3$			
1H,7H-Pyrazolo[1,2-a]pyrazole-1,7-dione, 3-ethyl-5-(1-hydroxyethyl)-2,6-dimethyl-, (±)-	MeCN	232.5(4.18),250s(3.86), 375(3.75)	35-4532-86
	3% MeCN	231(4.13),258(3.40), 392(3.65)	35-4532-86
2H-Pyrrol-2-one, 1,3-dihydro-3,4-dimethyl-3-(methyldioxy)-5-(1H-pyrrol-2-ylmethyl)-	MeOH	210(4.30)	49-0185-86
$C_{12}H_{16}N_2O_3Se$			
Cyclobutenediylium, 1-hydroxy-2,4-di-4-morpholinyl-3-selenyl-, dihydroxide, bis(innner salt)	CH_2Cl_2	248(4.3),259(4.3), 355s(4.3),379(4.5), 441s(3.0)	24-0182-86
2-Cyclobuten-1-one, 2,3-di-4-morpholinyl-4-selenoxo-	CH_2Cl_2	275(4.0),320(4.2), 411(4.8),520(2.7)	24-0182-86
$C_{12}H_{16}N_2O_4S$			
Benzenesulfonic acid, 3-[3-(ethylamino)-3-(methylamino)-1-oxo-2-propenyl]-	H_2O	246.5(4.08),285s(3.07)	35-5543-86
	pH 10.55	319(4.23)	35-5543-86
$C_{12}H_{16}N_2O_6S$			
Tyrosine, 3-[(2-amino-2-carboxyethyl)-thio]-5-hydroxy-	pH 7.5	293(3.47)(abs. at 290 nm)	149-0689-86B
$C_{12}H_{16}N_2O_8$			
5-Pyrimidineacetic acid, 1,2,3,4-tetrahydro-2,4-dioxo-1-β-D-ribofuranosyl-, methyl ester	H_2O	265(4.00)	94-4585-86
$C_{12}H_{16}N_4O$			
Propanedinitrile, [[1-methyl-2-(4-morpholinyl)-1-butenyl]imino]-	EtOH	444(4.62)	33-0793-86
2-Quinoxalinecarbonitrile, 3-amino-4,6,7,8-tetrahydro-4-(2-methoxyethyl)-	EtOH	386(3.58)	33-1025-86
$C_{12}H_{16}N_4O_2$			
Acetic acid, cyano[6-(dimethylamino)-1-methyl-4(1H)-pyrimidinylidene]-, ethyl ester	EtOH	275(3.83),366(4.55)	103-0774-86

Compound	Solvent	λ_{max}(log ϵ)	Ref.
$C_{12}H_{16}N_4O_2S$			
2,4(1H,3H)-Pteridinedione, 1,3,6-trimethyl-7-[(1-methylethyl)thio]-	MeOH	223(4.53),263(4.12), 275s(4.04),353(4.46)	33-1095-86
$C_{12}H_{16}N_4O_3$			
2,4(3H,8H)-Pteridinedione, 8-(4-hydroxybutyl)-6,7-dimethyl-	pH 5.0	228(3.78),256(4.18), 278(4.08),313(2.78), 350(4.10)	44-2461-86
	pH 8.0	228(3.98),256(4.23), 278(4.13),313(3.30), 407(4.08)	44-2461-86
	pH 12.0	228(4.46),245(4.20), 265s(4.00),313(4.23), 370(3.60)	44-2461-86
$C_{12}H_{16}N_4O_3S$			
2,4(1H,3H)-Pteridinedione, 1,3,6-trimethyl-7-[(1-methylethyl)sulfinyl]-	MeOH	245(4.16),353(4.00)	33-1095-86
$C_{12}H_{16}N_4O_4$			
3-Cyclopentene-1,2-diol, 5-(7,8-dihydro-8-hydroxyimidazo[4,5-d][1,3]-diazepin-3(4H)-yl)-3-(hydroxymethyl)-(adecpenol)	H_2O	279(3.90)	158-0309-86
1H-Pyrazole-4-carbonitrile, 5-amino-1-[2,3-O-(1-methylethylidene)-β-D-ribofuranosyl]-	pH 1	233(4.03)	44-1050-86
	pH 11	235(4.05)	44-1050-86
	MeOH	236(4.02),288(2.75)	44-1050-86
7H-Pyrrolo[2,3-d]pyrimidin-4-amine, 7-(2-deoxy-β-D-erythro-pentofuranosyl)-2-methoxy-	MeOH	261(3.98),271(3.98)	39-0525-86B
$C_{12}H_{16}N_4O_5$			
1H-Pyrazolo[3,4-d]pyrimidine, 1-(2-deoxy-β-D-erythro-pentofuranosyl)-4,6-dimethoxy-	MeOH	221(4.31),257(3.91)	5-1213-86
$C_{12}H_{16}N_4O_6$			
Inosine, 1-(2-hydroxyethyl)-	pH 1	250.5(4.00)	94-1094-86
	pH 7	250(4.01)	94-1094-86
	pH 13	250(4.00)	94-1094-86
	EtOH	251(3.95)	94-1094-86
1H-Purine-2,6-dione, 3,7-dihydro-1,3-dimethyl-8-β-D-ribofuranosyl-	EtOH	277(4.67)	136-0271-86H
4H-Pyrazolo[3,4-d]pyrimidin-4-one, 3-ethoxy-1,5-dihydro-1-β-D-ribofuranosyl-	pH 1 and 7	219(4.33),260(3.57)	4-1869-86
	pH 11	274(3.68)	4-1869-86
4H-Pyrazolo[3,4-d]pyrimidin-4-one, 3-ethoxy-2,5-dihydro-2-β-D-ribofuranosyl-	pH 1 and 7	219(4.26),247s(3.81)	4-1869-86
	pH 11	227(4.06),250(3.79), 287(3.77)	4-1869-86
$C_{12}H_{16}N_4S$			
1,2-Ethanediamine, N,N-dimethyl-N'-[4-(2-thienyl)-2-pyrimidinyl]-	n.s.g.	333(4.03)	87-1311-86
$C_{12}H_{16}N_5O$			
Pteridinium, 8-ethyl-4-(4-morpholinyl)-, tetrafluoroborate	pH 3.0	247(4.18),272(3.85), 445(3.97)	103-1118-86
hydrated	pH 9.0	262(4.35),311(3.81)	103-1118-86
$C_{12}H_{16}N_6O_3$			
L-Alanine, N-(N-1H-purin-6-yl-β-ala-	pH 2.85	272(4.12)	63-0757-86

Compound	Solvent	λ_{max}(log ϵ)	Ref.
nyl)-, methyl ester (cont.)	pH 7.0	268(4.14)	63-0757-86
	pH 10.5	272(4.02)	63-0757-86
7H-Pyrrolo[2,3-d]pyrimidine-5-carboximidamide, 4-amino-7-(2-deoxy-β-D-erythro-pentofuranosyl)-, hydrochloride	pH 1	283(3.83)	78-5869-86
	pH 7	284(3.84)	78-5869-86
	pH 11	275(3.62)	78-5869-86
$C_{12}H_{16}N_6O_4$			
7H-Pyrrolo[2,3-d]pyrimidine-5-carboximidamide, 4-amino-7-(2-deoxy-β-D-erythro-pentofuranosyl)-N-hydroxy-	pH 1	223(4.47),275(4.31)	78-5869-86
	pH 7	277(4.34)	78-5869-86
	pH 11	277(4.34)	78-5869-86
$C_{12}H_{16}O$			
Bicyclo[5.3.1]undeca-1,3,5-triene, 7-methoxy-	EtOH	245(3.38)	78-1823-86
2-Cyclohexen-1-one, 2-(2-methyl-1,3-pentadienyl)-, (Z,E)-	C_6H_{12}	233(4.27),290(3.65)	35-4881-86
2(3H)-Pentalenone, 4-ethenyl-3a,4,5,6-tetrahydro-5,5-dimethyl-, (3aR,4R*)-	EtOH	233(4.15)	94-4947-86
$C_{12}H_{16}O_2$			
Bicyclo[2.2.2]oct-7-ene-2,5-dione, 3,3,6,6-tetramethyl-, (±)-	hexane	223(3.30),291(2.13), 303(2.24),314(2.25), 325(2.17)	35-4149-86
2-Cyclopenten-1-one, 4-(1-cyclohexen-1-ylmethoxy)-	EtOH	215(3.90),314(1.67)	32-0291-86
Cyclopropanemethanol, 1-(4-ethoxyphenyl)-	C_6H_{12}	227(3.83)	12-0271-86
2(3H)-Furanone, dihydro-5-(2-octynylidene)-, (E)-	H_2O	236(4.12)	35-5589-86
	base	226(c.3.93)	35-5589-86
$C_{12}H_{16}O_3$			
Bicyclo[3.1.0]hex-3-en-2-one, 6-acetoxy-3,4,5,6-tetramethyl-, (1α,5α,6α)-(±)-	MeOH	221(3.84),266(3.62), 323(2.63)	33-0469-86
2,5-Cyclohexadien-1-one, 4-acetoxy-2,3,4,5-tetramethyl-, (±)-	MeOH	245(4.13),323s(2.26), 276s(3.59)[sic]	33-0469-86
$C_{12}H_{16}O_4$			
Benzoic acid, 4-hydroxy-2-(4-hydroxybutyl)-, methyl ester	EtOH	255(4.00)	44-4254-86
Bicyclo[2.2.1]heptane-2,3-dione, 1-(acetoxymethyl)-7,7-dimethyl-, (1R)-	C_6H_{12}	280(1.34),480(1.54)	39-1781-86C
1,3-Dioxane-4,6-dione, 5-cyclohexylidene-2,2-dimethyl-	EtOH	245(3.95)	39-1507-86C
2-Propen-1-ol, 3-(3,4,5-trimethoxyphenyl)-	EtOH	214(4.36),260(4.03)	100-0677-86
$C_{12}H_{16}O_8$			
4H-Pyran-4-one, 5-[(6-deoxy-α-L-mannopyranosyl)oxy]-3-hydroxy-2-methyl-	MeOH	220(4.29),287(4.19)	95-0989-86
4H-Pyran-4-one, 3-(β-D-glucopyranosyloxy)-2-methyl-	MeOH	258(3.98)	94-1015-86
$C_{12}H_{17}BO_6$			
Boron, (diethylpropanedioato(2-)-O,O'](2,4-pentanedionato-O,O')-(T-4)-	CH_2Cl_2	292(4.10)	48-0755-86

Compound	Solvent	$\lambda_{max}(\log \epsilon)$	Ref.
$C_{12}H_{17}ClHgO_2$			
Mercury, chloro[(2,3,4,5,6,7-hexahydro-2,6,6-trimethyl-4-oxo-2-benzofuranyl)methyl]-	EtOH	271(4.1)	32-0441-86
$C_{12}H_{17}ClN_4O_5$			
5-Pyrimidinepentanoic acid, 2-[[[(2-chloroethyl)amino]carbonyl]amino]-1,4-dihydro-6-hydroxy-4-oxo-	50%MeOH-HCl	242(4.05),266(4.07)	73-1140-86
	50%MeOH-NaOH	279(4.10)	73-1140-86
Uridine, 3'-[[[(2-chloroethyl)nitrosoamino]carbonyl]amino]-2',3'-dideoxy-	MeOH	260(4.00)	87-0862-86
$C_{12}H_{17}Cl_3O$			
2-Cyclohexen-1-one, 2-(1,1-dimethylethyl)-4,6-dimethyl-, (4α,5α,6β)-	$CHCl_3$	241(3.86)	12-2121-86
(4α,5β,6α)-(±)-	$CHCl_3$	241(3.88)	12-2121-86
$C_{12}H_{17}N$			
Benzenamine, 2-(2-butenyl)-N,N-dimethyl-	MeOH	246(3.67),278s(3.06)	33-0184-86
	+ H_2SO_4	257s(2.62),260(2.68), 267(2.58)	33-0184-86
Benzenamine, N,N-dimethyl-2-(2-methylcyclopropyl)-, (1R-cis)	EtOH	249.1(3.77),280.6s(3.22)	33-1936-86
(1S-trans)	EtOH	251.6(3.79),281.9s(3.21)	33-1936-86
Benzenamine, N,N-dimethyl-2-(1-methyl-2-propenyl)-, (S)-	EtOH	249.2(3.58),280.7s(2.98)	33-1936-86
	MeCN	251.0(3.65),278.7s(3.13)	33-1936-86
(±)-	MeOH	247.5(3.74),282s(3.14)	33-0184-86
$C_{12}H_{17}NO_2$			
2-Butanone, 4-(4-hydroxy-3,5-dimethylphenyl)-, oxime	EtOH	281(3.76)	103-0052-86
Isoquinoline, 1,2,3,4-tetrahydro-6,7-dimethoxy-1-methyl-, (salsolidine)	EtOH	212(4.08),232(3.99), 285(3.84)	100-0745-86
Isoquinoline, 1,2,3,4-tetrahydro-6,7-dimethoxy-2-methyl- (N-methylheliamine)	aq HCl	210(4.4),217(3.6), 282(3.4),288(3.3)	100-0745-86
Isoquinoline, 1,2,3,4-tetrahydro-7,8-dimethoxy-2-methyl-	n.s.g.	280(3.26)	100-0745-86
7-Isoquinolinol, 1,2,3,4-tetrahydro-7-hydroxy-6-methoxy-1,2-dimethyl-	isoPrOH-HCl	203(4.70),226(3.88), 285(3.59)	100-0745-86
$C_{12}H_{17}NO_2S$			
Benzenesulfonamide, N-(3-butenyl)-N,4-dimethyl-	EtOH	230(4.05)	24-0813-86
$C_{12}H_{17}NO_3$			
1H,5H-Cyclobuta[ij]quinolizin-5-one, 2,3,7a,8,8a,8b-hexahydro-7,8b-dimethoxy-	MeOH	217(4.00),250(3.71)	94-3658-86
2-Furancarboxylic acid, 5-methyl-4-(1-pyrrolidinylmethyl)-, methyl ester	MeOH	266(3.13)	73-2186-86
Isoquinoline, 1,2,3,4-tetrahydro-5,6,7-trimethoxy- (nortehaunine)	H_2O	203(4.4),223s(3.7), 281(3.2),292s(3.0)	100-0745-86
8-Isoquinolinol, 1,2,3,4-tetrahydro-6,7-dimethoxy-1-methyl- (analonidine)	pH 1	270(2.79),278s(2.72)	100-0745-86
	pH 13	245s(3.84),286(3.38)	100-0745-86
	isoPrOH	270(2.87),278s(2.81)	100-0745-86
6-Isoquinolinol, 1,2,3,4-tetrahydro-7,8-dimethoxy-1-methyl- (isoanhalonidine)	pH 1	225s(3.95),274s(3.11), 280(3.15)	100-0745-86
	pH 13	237s(3.95),295(3.54)	100-0745-86

Compound	Solvent	λ_{max}(log ε)	Ref.
(cont.)	isoPrOH-HBr	230s(3.95),275s(3.11), 282(3.15)	100-0745-86
8-Isoquinolinol, 1,2,3,4-tetrahydro-6,7-dimethoxy-2-methyl-	pH 1	225s(4.03),270(2.95), 277s(2.84)	100-0745-86
	pH 13	240s(3.97),285(3.47)	100-0745-86
	isoPrOH	227s(4.06),272(2.98), 280s(2.84)	100-0745-86
$C_{12}H_{17}NO_3Se$			
2-Cyclobuten-1-one, 3-butoxy-2-(4-morpholinyl)-4-selenoxo-	CH_2Cl_2	243(3.9),314(3.6), 390(4.7)	24-0182-86
$C_{12}H_{17}NO_4$			
2-Furancarboxylic acid, 5-methyl-4-(4-morpholinylmethyl)-, methyl ester	MeOH	265(3.19)	73-2186-86
4H-Pyran-4-one, 3-hydroxy-6-(hydroxymethyl)-2-(1-piperidinylmethyl)-	neutral	320(3.85)	137-0119-86
	cation	274(4.01)	137-0119-86
	anion	324(3.89)	137-0119-86
$C_{12}H_{17}NO_6$			
D-Ribitol, 1-deoxy-1-(2,5-dioxo-3-pyrrolidinylidene)-2,3-O-(1-methylethylidene)-, (E)-	EtOH	243(3.63)	44-0495-86
$C_{12}H_{17}N_2O_8P$			
Phosphonic acid, [[3,3a-dihydro-3-hydroxy-2-(hydroxymethyl)-6-oxo-6H-furo[2',3':4,5]oxazolo[3,2-a]pyrimidin-9a(2H)-yl]methyl]-, dimethyl ester	EtOH	224(3.99),249(3.87)	142-0617-86
$C_{12}H_{17}N_3O$			
Furo[2',3':4,5]pyrimido[1,2-a]azepin-4(2H)-imine, 3,6,7,8,9,10-hexahydro-2-methyl-	$CHCl_3$	248(4.12),303(3.70)	24-1070-86
5H-Pyrano[2',3':4,5]pyrimido[1,2-a]-azepin-5-imine, 2,3,4,7,8,9,10,11-octahydro-	$CHCl_3$	248(3.83),252(3.84), 310(3.54)	24-1070-86
$C_{12}H_{17}N_3O_4$			
5-Pyrimidinepentanoic acid, 2-amino-1,4-dihydro-6-hydroxy-4-oxo-, 2-propenyl ester	50%MeOH-HCl	267(4.16)	73-1140-86
	50%MeOH-NaOH	270(4.13)	73-1140-86
$C_{12}H_{17}N_3O_5$			
1H-Pyrazole-4-carboxaldehyde, 5-amino-1-[2,3-O-(1-methylethylidene)-β-D-ribofuranosyl]-	pH 1	236(3.86),284(3.93)	44-1050-86
	pH 11	222(3.87),237(3.92), 284(3.93)	44-1050-86
	MeOH	284(3.95)	44-1050-86
$C_{12}H_{17}N_3S$			
4H-Thieno[2',3':4,5]pyrimido[1,2-a]-azocin-4-imine, 2,3,6,7,8,9,10,11-octahydro-	$CHCl_3$	250(4.32),314(3.60)	24-1070-86
$C_{12}H_{17}N_5O$			
Guanidine, [3-(1,2-dihydro-2-methyl-1-oxopyrrolo[1,2-a]pyrazin-3-yl)-propyl]- (peramine)	MeOH	237(--),285(--)	77-0935-86
diacetate	MeOH	232(4.59),257(4.29),	77-0935-86

Compound	Solvent	$\lambda_{max}(\log \epsilon)$	Ref.
(cont.)		290s(3.91)	77-0935-86
1H-1,2,4-Triazole-3,5-diamine, N-[2-(2,6-dimethylphenoxy)ethyl]-	EtOH	212s(4.20)	4-0401-86
	10%EtOH-HCl	211s(4.26)	4-0401-86
$C_{12}H_{17}N_5O_3S_2$			
Adenosine, 5'-S-methyl-5'-C-(methylthio)-5'-thio-	MeOH	259(4.14)	44-1258-86
$C_{12}H_{17}N_5O_5$			
Adenosine, N-(2-hydroxyethyl)-	pH 1	263(4.26)	94-1054-86
	pH 7	266(4.26)	94-1054-86
	pH 13	266(4.26)	94-1054-86
	EtOH	266.5(4.25)	94-1094-86
$C_{12}H_{17}N_5O_6$			
4H-Pyrazolo[3,4-d]pyrimidin-4-one, 6-amino-3-ethoxy-1,5-dihydro-1-β-D-ribofuranosyl-	pH 1	252(4.20)	4-1869-86
	pH 7	254(4.41)	4-1869-86
	pH 11	254(4.21)	4-1869-86
$C_{12}H_{17}N_6$			
1H-1,2,4-Triazolium, 5-[[4-(dimethylamino)phenyl]azo]-1,4-dimethyl-, bromide	H_2O	540(4.79)	140-0439-86
isomer	H_2O	522(4.76)	140-0439-86
$C_{12}H_{18}$			
Benzene, hexamethyl-	CH_2Cl_2	271.0(3.61)	23-0265-86
dinitroso hexafluorogermanate complex	CH_2Cl_2	338.0(3.85)	23-0265-86
$NOAlCl_4$ complex	CH_2Cl_2	325.0(3.86)	23-0265-86
$NOBF_4$ complex	CH_2Cl_2	334.0(4.00)	23-0265-86
TCNE complex	CH_2Cl_2	545.0(3.64)	23-0265-86
Cyclohexane, 1-methyl-4-(2,4-pentadienylidene)-, [S-(E)]-	C_6H_{12}	265(4.58),275(4.73), 282(4.63)	44-2863-86
[S-(Z)]-	C_6H_{12}	265(4.50),275(4.59), 286(4.45)	44-2863-86
Pentalene, octahydro-3a,6a-dimethyl-2,5-bis(methylene)-, cis	C_6H_{12}	197(4.24)	33-0071-86
	EtOH	198(4.26)	33-0071-86
$C_{12}H_{18}ClNO_6$			
α-D-Mannofuranose, 1-C-chloro-1-deoxy-2,3:5,6-bis-O-(1-methylethylidene)-1-nitroso-	EtOH	204(3.09),654(1.18)	33-1137-86
$C_{12}H_{18}NO_2$			
Isoquinolinium, 1,2,3,4-tetrahydro-7-hydroxy-6-methoxy-2,2-dimethyl-, iodide	EtOH	287(4.14)	100-0745-86
$C_{12}H_{18}N_2$			
1,4:6,9-Diethanopyridazino[1,2-a]-pyridazine, 1,2,3,4,6,9-hexahydro-, radical cation, hexafluorophosphate	MeCN	221(3.28),272(3.18), 286(3.15)	35-7926-86
nitrate	MeCN	272(3.18),286(3.15)	35-7926-86
$C_{12}H_{18}N_2O_4$			
3H-Isoxazolo[3,4-c]azepin-7(8H)-carboxylic acid, 3a,4,5,6-tetrahydro-3-oxo-, 1,1-dimethylethyl ester	MeOH	257(3.86)	87-0224-86
3H-Isoxazolo[4,3-c]azepin-5(6H)-carboxylic acid, 3a,4,7,8-tetrahydro-3-	MeOH	261(3.73)	87-0224-86

Compound	Solvent	λ_{max}(log ϵ)	Ref.
oxo-, 1,1-dimethylethyl ester (cont.)			87-0224-86
Piperazinetetrone, bis(2-methylpropyl)-	MeCN	242(4.15),330s(1.90)	44-0247-86
2,4(1H,3H)-Pyrimidinedione, 5-ethyl-	pH 1	274(4.01)	87-0079-86
1-[3-hydroxy-4-(hydroxymethyl)-	pH 7	273(4.02)	87-0079-86
cyclopentyl]-, (1α,3β,4α)-(±)-	pH 13	271(3.90)	87-0079-86
$C_{12}H_{18}N_2O_4S$			
4,12-Epoxy-5H,10H-1,3-dioxolo[4,5-e]-pyrimido[6,1-b][1,3]oxazocine-10-thione, octahydro-2,2-dimethyl-, [3aR-(3aα,4β,6aβ,12β,12aα)]-	n.s.g.	250(4.14)	103-1336-86
5-Pyrimidinecarboxylic acid, 4-(2,2-dimethyl-1,3-dioxolan-4-yl)-1,2,3,4-tetrahydro-6-methyl-2-thioxo-, methyl ester	MeOH	305(4.19)	142-0679-86
$C_{12}H_{18}N_2O_8$			
2,4,6(1H,3H,5H)-Pyrimidinetrione, 5-β-D-galactopyranosyl-	H_2O	225(3.76),255(4.19)	136-0053-86M
sodium salt	H_2O	225(3.75),255(4.21)	136-0053-86M
2,4,6(1H,3H,5H)-Pyrimidinetrione, 5-β-D-glucopyranosyl-, sodium salt	H_2O	225(3.70),253(4.16)	136-0053-86M
2,4,6(1H,3H,5H)-Pyrimidinetrione, 5-β-D-mannopyranosyl-, sodium salt	H_2O	225(3.69),256(4.15)	136-0053-86M
$C_{12}H_{18}N_4OS$			
9H-Purine, 9-(1-butoxyethyl)-6-(methylthio)-	EtOH	285(4.25)	103-0331-86
$C_{12}H_{18}N_4S_4$			
4H-Pyrazole-3,5-diamine, 4-[4,5-bis(methylthio)-1,3-dithiol-2-ylidene]-, perchlorate	MeCN	300s(3.63),396(4.11)	88-0159-86
tetrafluoroborate	MeCN	220s(4.22),280(3.79), 300(3.76),425(4.22)	88-0159-86
$C_{12}H_{18}O$			
1,3-Dodecadiyn-5-ol, (R)-	EtOH	215(2.82),224(2.85), 237(2.83),252(2.62)	150-1348-86M
3-Penten-2-one, 5-(4-methylcyclohexylidene)-, [S-(E)]-	C_6H_{12}	281(4.48),335s(2.18)	44-2863-86
[S-(Z)]-	C_6H_{12}	291(4.29),340s(2.26)	44-2863-86
Tricyclo[3.3.1.1^{3,7}]decanone, 4-ethyl-, axial	C_6H_{11}Me-isopentane	295(1.23)	35-4484-86
	EPA	288(1.34)	35-4484-86
equatorial	C_6H_{11}Me-isopentane	295(1.34)	35-4484-86
	EPA	292(1.40)	35-4484-86
$C_{12}H_{18}O_2$			
2-Cyclohepten-1-one, 6-methyl-7-(3-oxobutyl)-, trans-(±)-	C_6H_{12}	227(2.99)	12-0433-86
2,4-Pentanedione, 3-bicyclo[2.2.1]-hept-2-yl-	MeOH	282(2.72),306(2.63)	35-1234-86
	MeOH-NaOH	295(2.98)	35-1234-86
2,4,6-Undecatrienoic acid, methyl ester, (E,Z,Z)-	EtOH	304(4.26)	39-1809-86C
3,5,7-Undecatrienoic acid, methyl ester, (E,Z,Z)-	EtOH	261(4.42)	39-1809-86C
$C_{12}H_{18}O_3$			
Cyclohexane, 2-(3-acetoxy-1-methyl-	EtOH	240(3.78)	94-3097-86

Compound	Solvent	λ_{max}(log ε)	Ref.
propylidene)-5-methyl- (cont.)			94-3097-86
1-Cyclopentene-1-heptanoic acid, 5-oxo-	MeOH	223(4.00)	2-0675-86
2-Cyclopenten-1-one, 4,4-dimethyl-2-[(2-methyl-1,3-dioxolan-2-yl)-methyl]-	EtOH	230(3.90)	94-4947-86
Oxacyclododec-3-ene-2,5-dione, 12-methoxy- (patulolide B)	MeOH	207(3.91)	158-0629-86
$C_{12}H_{18}O_4$			
2,4-Heptadienoic acid, 6-acetoxy-2,4-dimethyl-, methyl ester, (Z,Z)-(±)-	MeOH	218s(3.83),238s(3.69)	33-0469-86
3,5-Heptadienoic acid, 6-acetoxy-2,4-dimethyl-, methyl ester, (E,E)-(±)-	MeOH	234(4.09)	33-0469-86
(E,Z)-(±)-	MeOH	236(4.20)	33-0469-86
(Z,E)-(±)-	MeOH	229s(3.84)	33-0469-86
$C_{12}H_{19}BrO_2$			
2,4-Hexadienoic acid, 4-bromo-, hexyl ester	n.s.g.	264(4.11)	104-2230-86
$C_{12}H_{19}N$			
Benzene, 2,6-bis(1-methylethyl)-, chromium tricarbonyl complex	$CHCl_3$	218(4.51),320(3.91)	112-0517-86
Benzenamine, 4-(1,1-dimethylethyl)-N,N-dimethyl-, chromium tricarbonyl complex	$CHCl_3$	219(4.81),255s(4.33), 322(3.90)	112-0517-86
$C_{12}H_{19}NO_2$			
Bicyclo[2.2.1]heptane-2,3-dione, 1,7,7-trimethyl-, 3-(O-ethyloxime), (E)-	C_6H_{12}	246(4.03)	54-0054-86
(Z)-	C_6H_{12}	265(3.89),379(2.00), 396(2.03)	54-0054-86
$C_{12}H_{19}NO_3$			
2(1H)-Pyridinone, 6-(3,3-diethoxy-propyl)-	EtOH	228(3.88),305(3.89)	44-2184-86
$C_{12}H_{19}NO_9$			
β-D-Fructopyranose, 1-deoxy-1-[[3-methoxy-2-(methoxycarbonyl)-3-oxo-1-propenyl]amino]-	EtOH	222(3.96),278(4.37)	136-0329-86D
$C_{12}H_{19}NSSi$			
Methanamine, [1-(phenylthio)ethylidene]-1-(trimethylsilyl)-	EtOH	221(4.25),255(3.74)	35-6739-86
$C_{12}H_{19}N_2O_9P$			
2,4(1H,3H)-Pyrimidinedione, 1-[1-deoxy-1-(dimethoxyphosphinyl)-β-D-fructofuranosyl]-	EtOH	263(4.02)	142-2133-86
$C_{12}H_{19}N_3O_4$			
1H-Pyrazole-4-carboxylic acid, 3-amino-1-[(2,2-dimethyl-1,3-dioxolan-4-yl)methyl]-, ethyl ester	MeOH	263(3.80)	73-1512-86
1H-Pyrazole-4-carboxylic acid, 5-amino-1-[(2,2-dimethyl-1,3-dioxolan-4-	MeOH	227(3.78),254(3.96)	73-1512-86

Compound	Solvent	λ_{max}(log ε)	Ref.
yl)methyl]-, ethyl ester (cont.)			73-1512-86
5-Pyrimidinepentanoic acid, 2-amino-1,4-dihydro-6-hydroxy-4-oxo-, propyl ester	50%MeOH-HCl	266.5(4.19)	73-1140-86
	50%MeOH-NaOH	270(4.15)	73-1140-86
5-Pyrimidinepentanoic acid, 2-amino-1,6-dihydro-4-methoxy-1-methyl-6-oxo-, methyl ester	50%MeOH-HCl	238(3.85),280(4.07)	73-1140-86
	50%MeOH-NaOH	238(3.90),281(4.12)	73-1140-86
5-Pyrimidinepentanoic acid, 2-amino-4,6-dimethoxy-, methyl ester	50%MeOH-HCl	236(4.07),281(4.03)	73-1140-86
	50%MeOH-NaOH	238(4.13),268(3.89)	73-1140-86
$C_{12}H_{19}N_3O_5$			
5-Pyrimidinepentanoic acid, 2-amino-1,4-dihydro-6-hydroxy-4-oxo-, 2-methoxyethyl ester	50%MeOH-HCl	266.5(4.18)	73-1140-86
	50%MeOH-NaOH	271(4.14)	73-1140-86
$C_{12}H_{20}N_2$			
1,4:6,9-Diethanopyridazino[1,2-a]pyridazine, octahydro-, radical cation nitrate	MeCN	204(3.98),266(3.20)	35-7926-86
bis(hexafluorophosphate)	MeCN	218(3.26),244(3.23), 266(3.26)	35-7926-86
bis(tetrafluoroborate)	MeCN	243(3.18),264(3.20)	35-7926-86
Quinazoline, 1,4,5,6,7,8-hexahydro-2,4,4,7-tetramethyl-, (R)-	EtOH	204(3.54),260(3.75)	4-0705-86
$C_{12}H_{20}N_2O_3S$			
5(4H)-Thiazolone, 2-[1-(tert-butoxycarbonylamino)ethyl]-4,4-dimethyl-, (S)-	EtOH	241(3.40)	78-6555-86
$C_{12}H_{20}N_4O_4$			
1H-Pyrazole-4-carboxylic acid, 3,5-diamino-1-[(2,2-dimethyl-1,3-dioxolan-4-yl)methyl]-, ethyl ester	MeOH	245(4.18)	73-1512-86
$C_{12}H_{20}N_5P$			
Pyrrolo[3,4-d]-1,3,2-diazaphosphole-4,6-diamine, N,N,N',N'-tetraethyl-	hexane	275(4.6),305(4.6), 520(3.02)	24-3213-86
	dioxan	492(2.47)	24-3213-86
	THF	285(4.2),305(4.2), 490(2.99)	24-3213-86
	CH_2Cl_2	310(4.2),480(2.56)	24-3213-86
	MeCN	320(4.0),460(2.61)	
$C_{12}H_{20}O_2$			
Acetaldehyde, (4-hydroxy-2,2,6,6-tetramethylcyclohexylidene)-, (S)-(+)-	C_6H_{12}	242(4.14)	35-2691-86
9-Decynoic acid, ethyl ester	MeOH	206(2.67)	18-3535-86
11-Dodecynoic acid	MeOH	206(2.89)	18-3535-86
3-Furanhexanol, α,4-dimethyl-	MeCN	215(3.58)	100-0593-86
3-Furanpentanol, α-ethyl-4-methyl-	MeCN	215(3.58)	100-0593-86
8-Nonynoic acid, 1-methylethyl ester	MeOH	205(2.28)	18-3535-86
$C_{12}H_{20}O_3$			
2(5H)-Furanone, 4-(5-hydroxyheptyl)-3-methyl- (isoseiridin)	MeCN	216(4.24)	100-0593-86
2(5H)-Furanone, 4-(6-hydroxyheptyl)-3-methyl- (seiridin)	MeCN	215(4.21)	100-0593-86
Oxacyclododec-3-en-2-one, 5-hydroxy-12-methyl- (patulolide C)	MeOH	212(3.96)	158-0629-86

Compound	Solvent	λ_{max}(log ϵ)	Ref.
$C_{12}H_{20}O_4$			
2-Dodecenedioic acid	MeOH	213(4.11)	2-0675-86
$C_{12}H_{21}NO_2$			
1,2-Cycloheptanedione, 3,3,7,7-tetra-	C_6H_{12}	240s(2.93),305(1.65)	54-0054-86
methyl-, mono(O-methyloxime), (Z)-	EtOH	240s(2.96),302(1.74)	54-0054-86
$C_{12}H_{21}NO_6$			
α-D-Allofuranose, 3-deoxy-3-(hydroxy-amino)-1,2:5,6-bis-O-(1-methyleth-ylidene)-	EtOH	213(3.30)	33-1132-86
$C_{12}H_{21}NO_6Si$			
Cyclopentaneacetic acid, 2-(nitrometh-yl)-5-oxo-3-[(trimethylsilyl)oxy]-, methyl ester, trans,cis	MeOH	204(3.78),280(1.83)	104-1865-86
trans,trans	MeOH	204(3.70),280(1.79)	104-1865-86
$C_{12}H_{21}N_3O_8$			
D-Glucitol, 2-amino-2-deoxy-1-C-(hexa-hydro-1,3-dimethyl-2,4,6-trioxo-5-pyrimidinyl)-, (S)-	H_2O	225(3.81),253(4.23)	136-0053-86M
sodium salt	H_2O	229(3.76),258(4.30)	136-0053-86M
$C_{12}H_{22}$			
5,7-Dodecadiene, (E,E)-	hexane	230.4(4.41)	44-4934-86
5,7-Dodecadiene, (E,Z)-	hexane	232.4(4.32)	44-4934-86
$C_{12}H_{22}N_2O$			
8a(1H)-Quinazolinol, 4,4a,5,6,7,8-hexahydro-2,4,4,7-tetramethyl-	EtOH	208(3.92)	4-0705-86
$C_{12}H_{22}N_2O_4S$			
Alanine, N-[N-[(1,1-dimethylethoxy)-carbonyl]thio-L-alanyl]-2-methyl-	EtOH	200(3.60),267(3.90)	78-6555-86
$C_{12}H_{22}O$			
8,10-Dodecadien-1-ol, (E,E)-	EtOH	229.9(4.45)	44-4934-86
$C_{12}H_{22}O_2$			
1,3-Butadiene, 1,4-bis(1,1-dimethyl-ethoxy)-, (E,E)-	heptane	245(4.48)	44-1440-86
(E,Z)-	heptane	251(4.40)	44-1440-86
(Z,Z)-	heptane	254(4.49)	44-1440-86
$C_{12}H_{22}O_2S_4$			
Thioperoxydicarbonic acid, dipentyl ester	isooctane	242(4.26),286(3.93)	3-0588-86
$C_{12}H_{22}O_3$			
2(3H)-Furanone, dihydro-4-(5-hydroxy-heptyl)-3-methyl-	MeCN	213(2.40)	100-0593-86
2(3H)-Furanone, dihydro-4-(6-hydroxy-heptyl)-3-methyl-	MeCN	213(2.30)	100-0593-86
$C_{12}H_{22}O_3Si$			
2-Propenoic acid, 3-[2-(trimethylsil-yloxy)-1-cyclopentyl]-, methyl ester, (E)-	MeOH	217(4.02)	48-0777-86

Compound	Solvent	$\lambda_{max}(\log \epsilon)$	Ref.
$C_{12}H_{23}NO$			
Piperidine, 4-(ethenyloxy)-1,2,2,6,6-pentamethyl-	EtOH	201.0(4.25)	70-2522-86
hydriodide	EtOH	196.3(4.23),218.9(4.17)	70-2522-86
$C_{12}H_{23}N_3O_3S$			
Ethanimidothioic acid, 2-[[2-[[(1,1-dimethylethoxy)carbonyl]amino]-1-oxopropyl]amino]-, ethyl ester, (S)-	MeCN	215(3.62),227(3.66)	33-1224-86
$C_{12}H_{24}NOS$			
Oxazolidinium, 2-[(butylthio)methylene]-3,3-diethyl-, chloride	H_2O	232s(3.34),250(3.36)	70-0994-86
$C_{12}H_{24}NO_2S$			
Oxazolidinium, 3-butyl-2-[(ethylthio)-methylene]-3-(2-hydroxyethyl)-, chloride	H_2O	227s(3.68),248(3.75)	70-0994-86
$C_{12}H_{24}N_2$			
1,3-Butadiene-1,3-diamine, N,N,N',N'-tetraethyl-	MeCN	325(4.25)	104-0420-86
$C_{12}H_{28}O_3SSi$			
Silane, [2-[(1,1-dimethylethyl)thio]-ethyl]triethoxy-	heptane	210(2.32)	67-0062-86
	MeCN	194(2.62),210(2.43)	67-0062-86
$C_{12}H_{30}CuN_6$			
Copper, (6,13-dimethyl-1,4,8,11-tetraazacyclotetradecane-6,13-diamine-N^1,N^4,N^8,N^{11})-, (SP-4-1)-	H_2O	249(3.90),516(1.85)	12-1101-86
$C_{12}H_{33}GeN_3Si_3$			
3H-1,2,3,4-Triazagermole, 4,5-dihydro-4,4-dimethyl-3,5,5-tris(trimethylsilyl)-	C_6H_{12}	251(4.60)	24-2980-86
$C_{12}H_{36}Si_6$			
Cyclopentasilane, nonamethyl(trimethylsilyl)-	C_6H_{12}	280(3.0)	157-0128-86

Compound	Solvent	λ_{max}(log ϵ)	Ref.
$C_{13}H_3Cl_9$			
Benzene, pentachloro[chloro(2,4,6-trichlorophenyl)methyl]-	C_6H_{12}	218(4.84),240(4.36), 287s(2.76),294(2.82), 305(2.81)	118-0064-86
$C_{13}H_4Cl_4O_2S$			
Thieno[3,2-b]furan-5-carbonyl chloride, 3,6-dichloro-2-(4-chlorophenyl)-	MeOH	340(3.46)	73-1685-86
$C_{13}H_5Cl_2NO_4S$			
Thieno[3,2-b]furan-5-carbonyl chloride, 6-chloro-2-(2-nitrophenyl)-	MeOH	319(3.04)	73-1685-86
$C_{13}H_5Cl_3O_2S$			
Thieno[3,2-b]furan-5-carbonyl chloride, 3,6-dichloro-2-phenyl-	MeOH	336(3.36)	73-1685-86
$C_{13}H_5Cl_3O_3S$			
Thieno[3,2-b]furan-5-carboxylic acid, 3,6-dichloro-2-(4-chlorophenyl)-	MeOH	341(3.38)	73-1685-86
$C_{13}H_5F_3N_6O_{12}S$			
Benzenamine, N-[2,6-dinitro-4-[(trifluoromethyl)sulfonyl]phenyl]-2,4,6-trinitro-	MeCN	220(--),400(4.23)	104-1547-86
	+ H_2SO_4	368(4.3)	104-1547-86
sodium salt	MeCN	220(--),400(4.26)	104-1547-86
$C_{13}H_5F_{10}N_3O$			
Pyridinium, 2-(heptafluoropropyl)-4-oxo-6-(trifluoromethyl)-5(4H)-pyrimidinylide	EtOH	230(4.14),257(3.91), 304(3.60)	39-1769-86C
$C_{13}H_6ClNO_5S$			
Thieno[3,2-b]furan-5-carboxylic acid, 6-chloro-2-(2-nitrophenyl)-	MeOH	313(3.16)	73-1685-86
$C_{13}H_6Cl_2O_3S$			
Thieno[3,2-b]furan-5-carboxylic acid, 3,6-dichloro-2-phenyl-	MeOH	338(3.12)	73-1685-86
$C_{13}H_6N_2O_4$			
2H,8H-Benzo[1,2-b:3,4-b']dipyran-2,8-dione, 4-(diazomethyl)-	EtOH	252.5(4.24),285(4.41), 305(4.38)	94-0390-86
2H,5H-Pyrano[3,2-c][1]benzopyran-2,5-dione, 4-(diazomethyl)-	EtOH	227(4.24),265(3.83), 297.5(4.41),320.5(4.31), 335(4.21)	94-0390-86
$C_{13}H_7BF_2O_2$			
Boron, difluoro(1H-phenalene-1,9(9aH)-dionato-O,O')-, (T-4)-	MeCN	231(4.27),253(3.67), 335s(3.70),369(4.32), 396s(3.46),410s(3.54), 421(3.70),435s(3.72), 446(3.85)	78-6293-86
$C_{13}H_7ClN_4$			
Propanedinitrile, (2-chloro-6-phenyl-4(1H)-pyrimidinylidene)-	EtOH	260(4.24),315(4.37), 367(4.09)	103-0774-86
$C_{13}H_7N_3O_5S$			
Benzenamine, N-(5,7-dinitro-1,3-benzoxathiol-2-ylidene)-	EtOH	233(4.30),258(4.25), 356(3.61)	5-0947-86

Compound	Solvent	λ_{max}(log ε)	Ref.
$C_{13}H_8$			
Benzene, 1,3,5-heptatriynyl-	MeOH	237(4.88),249(5.11), 273(4.16),289(4.44), 309(4.54),330(4.41)	138-1075-86
	MeOH	310(4.29)	149-0441-86B
$C_{13}H_8Br_3NO$			
2,5-Cyclohexadien-1-one, 2,3,6-tribromo-4-[(4-methylphenyl)imino]-	heptane	312(4.13),504(3.66)	104-0894-86
$C_{13}H_8Br_3NO_3S$			
Benzenesulfonamide, 4-methyl-N-(2,3,5-tribromo-4-oxo-2,5-cyclohexadien-1-ylidene)-	heptane	322(4.34),401(3.18)	104-0894-86
$C_{13}H_8Cl_2N_4$			
Pyridazine, 3,4-dichloro-6-[4-(1H-imidazol-1-yl)-	MeOH	280(4.35)	4-1515-86
$C_{13}H_8INO$			
Phenanthridone, 2-iodo-	hexane	240f(4.9),330(4.1), 342(4.1)	110-1666-86
$C_{13}H_8N_2O_3S$			
4(1H)-Pyrimidinone, 2,3-dihydro-5-(2-oxo-2H-1-benzopyran-3-yl)-2-thioxo-	MeOH	356(4.61)	56-0599-86
$C_{13}H_8N_2O_4$			
2H,8H-Benzo[1,2-b:5,4-b']dipyran-2,8-dione, 6-(diazomethyl)-3,4-dihydro-	EtOH	250(4.10),258(4.09), 291(3.95),323(4.21)	94-0390-86
$C_{13}H_8N_2S_2$			
$2\lambda^4$-Thieno[3,2-c][1,2]thiazine-3-carbonitrile, 2-phenyl-	EtOH	220(4.18),286(3.86), 330(3.88),457(3.43)	39-0483-86C
$C_{13}H_8N_4$			
Cyanamide, (2-methyl-1,4-naphthalenediylidene)bis-	MeCN	205(4.45),259s(3.95), 268(4.00),276s(3.90), 324s(4.37),338(4.40), 351s(4.36),400s(3.77)	5-0142-86
Pyrido[1',2':1,2]imidazo[4,5-b]quinoxaline	DMF	345(4.07)	2-1057-86
$C_{13}H_8O_2$			
2H-Naphth[1,8-bc]oxepin-2-one	$CHCl_3$	263(5.59),320(4.86), 335(4.86),370(4.90)	104-1342-86
$C_{13}H_8O_3$			
1H-Naphtho[2,3-c]pyran-5,10-dione	MeOH	260(4.16),445(3.38)	94-1505-86
4H-Naphtho[2,3-b]pyran-5,10-dione	MeOH	253(4.36),294(3.59)	94-1505-86
$C_{13}H_8O_6$			
Acetic acid, [(7-oxo-7H-furo[3,2-g]-[1]benzopyran-9-yl)oxy]-	1% DMSO	305.5(4.11)	149-0391-86A
$C_{13}H_9BrN_2$			
Imidazo[1,5-a]pyridine, 1-bromo-3-phenyl-	MeOH	209(4.5),221(4.4), 318(4.1)	83-0043-86
$C_{13}H_9Br_2NO$			
2,5-Cyclohexadien-1-one, 3-bromo-	heptane	267(4.32),483(3.28)	104-0894-86

Compound	Solvent	λ_{max}(log ε)	Ref.
4-[(2-bromo-4-methylphenyl)imino]- (cont.)			104-0894-86
2,5-Cyclohexadien-1-one, 2,5-dibromo- 4-[(4-methylphenyl)imino]-	heptane	306(4.18),493(3.66)	104-0894-86
2,5-Cyclohexadien-1-one, 2,6-dibromo- 4-[(4-methylphenyl)imino]-	heptane	316(4.22),482(3.75)	104-0894-86
$C_{13}H_9Br_2NO_3S$			
Benzenesulfonamide, N-(3,5-dibromo- 4-oxo-2,5-cyclohexadien-1-ylidene)- 4-methyl-	heptane	317(4.35),392(3.20)	104-0894-86
$C_{13}H_9ClN_2$			
Benzeneacetonitrile, 4-chloro-α-[(1H- pyrrol-2-yl)methylene]-, (Z)-	EtOH	249(4.04),373(4.54)	4-1747-86
Imidazo[1,2-a]pyridine, 8-(2-chloro- phenyl)-	MeOH	214(4.5),250(3.7), 304(3.8)	83-0043-86
$C_{13}H_9ClN_4O$			
1,2,4-Benzotriazin-3(2H)-one, 6-amino- 2-(4-chlorophenyl)-	EtOH	234(4.44),270(4.37), 390(4.25)	97-0134B-86
3(2H)-Pyridazinone, 4-chloro-6-[4-(1H- imidazol-1-yl)phenyl]-	MeOH	272(4.45)	4-1515-86
$C_{13}H_9ClOS$			
3-Butyn-2-ol, 4-[5-(1,3-pentadiynyl)- 2-thienyl]-1-chloro-	ether	210(4.23),236(--), 250(3.9),322(4.34), 342(4.30)	154-1205-86
$C_{13}H_9ClO_3S$			
Thieno[3,2-b]benzofuran-2-carboxylic acid, 3-chloro-, ethyl ester	MeOH	322(3.33)	73-1685-86
$C_{13}H_9N$			
Cyclobuta[a]naphthalene-2-carbo- nitrile, 2a,8b-dihydro-	C_6H_{12}	262s(3.82),268(3.88), 278s(3.82),291s(3.51), 302s(3.81)	23-0237-86
Cyclobuta[a]naphthalene-2a(8bH)-carbo- nitrile	C_6H_{12}	258s(3.76),266(3.89), 276(3.88),285s(3.67), 287s(3.56),293(3.00)	23-0237-86
1,2,7-Metheno-1H-cyclopropa[b]naph- thalene-1-carbonitrile, 1a,2,7,7a- tetrahydro-	C_6H_{12}	249s(2.08),254s(2.30), 260(2.51),265s(2.64), 266(2.68),273(2.75)	23-0237-86
Tetracyclo[5.5.0.0^{2,4}.0^{3,5}]dodeca- 6,8,10,12-tetraene-6-carbonitrile	C_6H_{12}	232(4.04),357(4.01), 371(4.03),392(3.80), 430(2.66),461(2.66), 499(2.57),544(2.41), 597(2.00),660(1.37)	35-2773-86
Tricyclo[5.5.0.0^{2,5}]dodeca-3,6,8,10,12- pentaene-6-carbonitrile	C_6H_{12}	230(4.13),270s(3.80), 347(4.11),361(4.19), 380(3.98),436(2.64), 466(2.67),504(2.61), 546(2.43),598(2.08), 650(1.30)	78-1745-86
$C_{13}H_9NO$			
Benzoxazole, 2-phenyl-	heptane	286(4.18)	103-1262-86
2-Propenenitrile, 3-(5-phenyl- 2-furanyl)-, (E)-	EtOH	237(4.50)	103-0943-86

Compound	Solvent	λ_{max}(log ε)	Ref.
$C_{13}H_9NOS_2$			
2λ⁴-Thieno[3,2-c][1,2]thiazine-3-carboxaldehyde, 2-phenyl-	EtOH	219(4.02),334(3.78), 476(3.48)	39-0483-86C
$C_{13}H_9NOSe$			
1,2-Benzisoselenazol-3(2H)-one, 2-phenyl-	MeOH	260(3.37),325(3.30)	18-2179-86
$C_{13}H_9NO_2$			
4H-Furo[3,2-b]pyrrole-5-carboxaldehyde, 2-phenyl-	MeOH	358(3.72),368(3.73)	73-0106-86
1H-Indeno[1,2-b]pyridine-2,5-dione, 1-methyl- (dielsine)	EtOH	243s(4.15),256(4.18), 262s(4.17),272s(4.04), 282(3.56),341(3.50)	102-1691-86
	EtOH-NaOH	238(4.00),260s(4.12), 270(4.17),393(3.87), 489(3.77)	102-1691-86
$C_{13}H_9NO_3$			
5H-Indeno[1,2-b]pyridine-2,5-dione, 1-(hydroxymethyl)- (dielsinol)	EtOH	253(4.26),269s(4.11), 283(4.03),333(3.77)	102-1691-86
	EtOH-NaOH	230(4.09),237s(4.09), 260s(4.18),278(4.24), 380(3.81)	102-1691-86
3H-Phenoxazin-3-one, 2-methoxy-	$CHCl_3$	376(3.28)	4-1003-86
$C_{13}H_9NO_3S$			
Thiopyrano[2,3-b]indole-2-carboxylic acid, 4,9-dihydro-9-methyl-4-oxo-	dioxan	236(4.47),257(4.22), 276(4.28),360(3.80)	83-0332-86
Thiopyrano[2,3-b]indole-2-carboxylic acid, 4,9-dihydro-4-oxo-, methyl ester	dioxan	208(4.09),235(4.11), 258(3.9), 360(3.56)	83-0332-86
Thiopyrano[3,2-b]indole-2-carboxylic acid, 4,5-dihydro-5-methyl-4-oxo-	dioxan	232(4.35),286(3.89), 322(3.99),400(3.62)	83-0332-86
Thiopyrano[3,2-b]indole-2-carboxylic acid, 4,5-dihydro-4-oxo-, methyl ester	dioxan	228(4.21),283(3.84), 323(3.75),? (3.58)	83-0332-86
$C_{13}H_9N_3$			
Cyclopropa[cd]pentalene-1,2,3-tricarbonitrile, 2a,2b,4a,4b-tetrahydro-4a,4b-dimethyl- (or isomer)	MeCN	224(4.041),272s(3.692), 307s(3.383)	24-1801-86
$C_{13}H_9N_3O_2$			
Benzeneacetonitrile, 4-nitro-α-(1H-pyrrol-2-ylmethylene)-, (Z)-	EtOH	250s(3.78),414(4.49)	4-1747-86
Imidazo[1,5-a]pyridine, 3-(2-nitrophenyl)-	MeOH	221(4.5),259(4.1)	83-0043-86
Imidazo[1,5-a]pyridine, 3-(4-nitrophenyl)-	MeOH	239(4.0),396(4.0)	83-0043-86
$C_{13}H_9N_3O_3$			
Acetic acid, cyano(1,5-dihydro-3-methyl-5-oxo-1-phenyl-4H-pyrazol-4-ylidene)-	DMAA	288(3.84),368(3.25)	104-0582-86
potassium salt	DMAA	288(3.82),370(3.26)	104-0582-86
$C_{13}H_{10}BrF_3N_2O_3$			
L-Tryptophan, 2-bromo-N-(trifluoroacetyl)-, sodium salt	H_2O	220(4.53),280(3.89), 288(3.81)	35-2023-86

Compound	Solvent	λ_{max}(log ϵ)	Ref.
$C_{13}H_{10}BrN$			
5H-Benzocycloheptene-5-carbonitrile, 6-bromo-9-methyl-	EtOH	206(4.29),238(4.03), 294(3.89)	39-1479-86C
5H-Benzocycloheptene-5-carbonitrile, 8-bromo-5-methyl-	EtOH	203(4.30),228s(3.81), 282(3.85)	39-1479-86C
5H-Benzocycloheptene-9-carbonitrile, 8-bromo-5-methyl-	EtOH	203(3.91),219s(3.80), 278(3.47)	39-1479-86C
$C_{13}H_{10}BrNO$			
5H-Benzocycloheptene-5-carbonitrile, 8-bromo-6,9-dihydro-5-methyl-9-oxo-	EtOH	209(3.79),260(3.74)	39-1479-86C
2,5-Cyclohexadien-1-one, 2-bromo-4-[(4-methylphenyl)imino]-	heptane	265(4.17),301(4.09), 470(3.61)	104-0894-86
2,5-Cyclohexadien-1-one, 3-bromo-4-[(4-methylphenyl)imino]-	heptane	263(4.26),301(4.11), 479(3.57)	104-0894-86
2,5-Cyclohexadien-1-one, 4-[(2-bromo-4-methylphenyl)imino]-	heptane	262(4.44),466(3.38)	104-0894-86
$C_{13}H_{10}BrNO_2$			
Benzaldehyde, 3-bromo-2-hydroxy-5-(3-pyridinylmethyl)-	EtOH	221(4.33),258.5(4.06), 263(4.09),342(3.48), 400(3.24)(anom.)	87-1461-86
$C_{13}H_{10}BrNO_3S$			
Benzenesulfonamide, N-(3-bromo-4-oxo-2,5-cyclohexadien-1-ylidene)-4-methyl-	heptane	287(4.31),357(3.66)	104-0894-86
$C_{13}H_{10}Br_2O_4$			
1,4-Naphthalenedione, 2,6-dibromo-5,8-dimethoxy-3-methyl-	$CHCl_3$	256(3.95),291(4.02), 415(3.70)	5-1655-86
1,4-Naphthalenedione, 3,6-dibromo-5,8-dimethoxy-2-methyl-	$CHCl_3$	257(3.93),292(3.99), 417(3.71)	5-1655-86
$C_{13}H_{10}ClF_3N_2O_3$			
L-Tryptophan, 2-chloro-N-(trifluoroacetyl)-, sodium salt	H_2O	219(4.59),272(3.95), 288(3.83)	35-2023-86
$C_{13}H_{10}ClNO$			
Methanone, (2-amino-5-chlorophenyl)-phenyl-	20% MeOH	236(4.37),258s(4.08), 385(3.68)	96-1051-86
cation	20% MeOH	258(4.12)	96-1051-86
$C_{13}H_{10}ClNO_3S$			
Phenol, 4-chloro-5-methyl-2-[2-(5-nitro-2-thienyl)ethenyl]-, (E)-	MeOH	279(3.06),308(2.97), 431(3.38)	73-1127-86
$C_{13}H_{10}ClNO_7$			
Pyrano[3,4-b]pyrrole-2,7-dicarboxylic acid, 1-acetyl-3-chloro-1,5-dihydro-5-oxo-	EtOH	256(4.1),317(4.0), 416(2.6)	5-1968-86
	dioxan	259(4.1),319(4.1), 424(1.3)	5-1968-86
$C_{13}H_{10}FN_5O$			
Formamide, N-[9-[(2-fluorophenyl)-methyl]-9H-purin-6-yl]- (spectra in 9.5% ethanol)	pH 7.00	273(4.27)	4-1189-86
	HCl	267(4.18)	4-1189-86
	NaOH	261(4.18)	4-1189-86
$C_{13}H_{10}F_3NO_7$			
2H-Pyran-6-carboxylic acid, 4-[3-methoxy-3-oxo-2-[(trifluoroacetyl)amino]-	EtOH	261(4.0),304(3.9)	5-1968-86

Compound	Solvent	λ_{max}(log ε)	Ref.
1-propenyl]-2-oxo-, methyl ester, (E)- (cont.)			5-1968-86
(Z)-	EtOH	255(4.1),307(4.0)	5-1968-86
$C_{13}H_{10}N_2$			
9-Acridinamine	pH 7.4	260(4.89)	94-0944-86
Imidazo[1,5-a]pyridine, 3-phenyl-	MeOH	208(3.8),220(3.7), 311(3.3)	83-0043-86
$C_{13}H_{10}N_2O$			
1H-Imidazo[1,2-a]pyridin-4-ium, 3-hydroxy-2-phenyl-, hydroxide, inner salt	MeOH	207(4.23),216(4.17), 273(4.16),400(4.10)	88-1627-86
	MeOH-HCl	208(4.30),245(4.20), 323(4.10)	88-1627-86
4H-Pyrimido[2,1-a]isoquinolin-4-one, 2-methyl-	CH_2Cl_2	367(4.0)	39-1561-86B
$C_{13}H_{10}N_2O_2$			
Benzoic acid, 4-(phenylazo)-, trans	$CF_3CHOHCF_3$	230(4.0),310(4.3), 410(3.3)	116-2472-86
irradiated to cis form	$CF_3CHOHCF_3$	260(4.0),305(4.0), 415(3.5)	116-2472-86
$C_{13}H_{10}N_2S_2$			
1H-Perimidine-2-carbodithioic acid, methyl ester	DMF	295(4.12),346(4.11), 356s(4.05),607(2.99)	48-0812-86
	C_6H_{12}	231s(--),236(--), 294(--),324(--), 344(--),353(--), 501s(--),548s(--), 593(--),605(--), 700s(--)	48-0812-86
Thieno[3,2-d]pyrimidine-4(1H)-thione, 2-methyl-6-phenyl-	MeOH	266(4.39),310(4.25), 325(4.10),370(4.15)	4-1757-86
Thieno[3,2-d]pyrimidine-4(1H)-thione, 6-methyl-2-phenyl-	MeOH	248(4.49),254(3.99), 295(4.11)	4-1757-86
$C_{13}H_{10}N_4$			
Benzo[f]quinoxaline-2-carbonitrile, 3-amino-5,6-dihydro-	EtOH	381(4.03)	33-0793-86
Benzo[f]quinoxaline-3-carbonitrile, 2-amino-5,6-dihydro-	MeOH	387(4.21)	33-0793-86
Pyridine, 3-(5-phenyl-1H-1,2,4-triazol-3-yl)-	EtOH	210(4.15),232(4.20), 255(4.22)	128-0027-86
7H-1,2,4-Triazolo[4,3-a]perimidine, 10-methyl-	dioxan	233(3.90),252(3.81), 286(3.72),297(3.79), 353(3.70)	48-0237-86
$C_{13}H_{10}N_4O$			
1,2,4-Benzotriazin-3(2H)-one, 6-amino-2-phenyl-	EtOH	235(4.49),267(4.39), 386(4.25)	97-0134B-86
4(1H)-Pteridinone, 6-methyl-7-phenyl-	pH 13	346(3.99)	33-0793-86
$C_{13}H_{10}N_4O_5$			
Pyrido[2,1-i]purine-9,10-dicarboxylic acid, 1,7-dihydro-7-oxo-, dimethyl ester	C_6H_{12}	340(3.9),400(4.0), 425s(3.9),450s(3.7)	39-1561-86B
	EtOH	332(3.9),336(3.9), 425(4.0),442s(3.9)	39-1561-86B
$C_{13}H_{10}N_4O_8$			
Benzene, 1-nitro-2-[(2,4,6-trinitro-	EtOH	472(4.73),536(4.29)	109-1028-86

Compound	Solvent	λ max (log ε)	Ref.
2,4-cyclohexadien-1-yl)methyl]- (cont.)			109-1028-86
$C_{13}H_{10}N_6O_2$			
1,2,4-Triazolo[1,5-a]pyrimidine, 1,7-dihydro-7-(1H-indol-3-yl)-	EtOH	222(3.99),276(3.13), 363(4.00)	103-1250-86
$C_{13}H_{10}O$			
Benzophenone	MeCN	342(2.1)	35-0128-86
	n.s.g.	337(2.18)	89-0999-86
$C_{13}H_{10}OS_2$			
Ethanone, 1-[5'-(1-propynyl)[2,2'-bithiophene]-5-yl]-	EtOH	222(3.86),262(4.05), 267(4.04),370(4.41)	163-0565-86
$C_{13}H_{10}O_2$			
Benzophenone O-oxide	Ar/O_2 at 40°K	422(--)	89-0255-86
2(3H)-Furanone, dihydro-5-(3-phenyl-2-propynylidene)-, (E)-	H_2O	278(4.40)	35-5589-86
	base	242(4.39)	35-5589-86
Naphtho[1,8-bc]pyran-3(2H)-one, 6-methyl-	EtOH	210(4.55),257(4.36), 350(3.59),369(3.63), 385s(3.52)	12-0635-86
3-Oxatricyclo[7.3.2^{1,9}.0^{5,13}]tetradeca-1(13),5,7,9,11-pentaen-2-one	MeOH	265(4.49),310(3.80)	33-1263-86
$C_{13}H_{10}O_2S_2$			
[2,2'-Bithiophene]-5-carboxylic acid, 5'-(1-propynyl)-, methyl ester	EtOH	257(3.94),270(3.81), 358(4.51)	163-0263-86
	EtOH	220(4.02),257(3.95), 358(4.50)	163-0565-86
Ethanone, 2-hydroxy-1-[5'-(1-propynyl)[2,2'-bithiophene]-5-yl]-	EtOH	224(3.90),263(4.10), 370(4.46)	163-0263-86 +163-0565-86
$C_{13}H_{10}O_4$			
1,4-Naphthalenedione, 6-acetyl-5-hydroxy-7-methyl-	EtOH	216(4.38),227s(4.27), 251(4.18),427(3.63)	64-0377-86B
	$CHCl_3$	252(4.17),340(3.33), 390s(3.49),413(3.64), 432(3.66),456s(3.46)	64-0377-86B
Psoralen, 5-ethoxy-	saline	313(4.20)	149-0221-86A
$C_{13}H_{10}O_5$			
7H-Furo[3,2-g][1]benzopyran-7-one, 9-(2-hydroxyethoxy)-	1% DMSO	303(4.11)	149-0391-86A
1,4-Naphthalenedione, 6-acetoxy-2-methoxy-	MeOH	251(4.27),280(4.22), 328(3.51)	100-0122-86
1,4-Naphthalenedione, 7-acetoxy-2-methoxy-	MeOH	249(4.33),272s(4.13), 279(4.15),328(3.49)	100-0122-86
$C_{13}H_{11}BrN_2O$			
Benzo[f]quinoxalin-6(2H)-one, 5-bromo-3,4-dihydro-2-methyl-	CH_2Cl_2	238(4.08),276(4.26), 422(3.53)	83-0052-86
Benzo[f]quinoxalin-6(2H)-one, 5-bromo-3,4-dihydro-3-methyl-	CH_2Cl_2	236(4.10),276(4.26), 422(3.54)	83-0052-86
Phenol, 4-bromo-2-[[(5-methyl-2-pyridinyl)imino]methyl]-	$CHCl_3$	360(3.35)	2-0127-86A
$C_{13}H_{11}BrO$			
5H-Benzocycloheptene-5-carboxaldehyde, 6-bromo-9-methyl-	EtOH	224(4.03),281(3.72)	39-1479-86C

Compound	Solvent	$\lambda_{max}(\log \epsilon)$	Ref.
5H-Benzocycloheptene-5-carboxaldehyde, 8-bromo-5-methyl-	EtOH	204(4.27),282(3.88)	39-1479-86C
$C_{13}H_{11}Br_2NO_7$			
2H-Pyran-6-carboxylic acid, 4-[2-(acetylamino)-1-bromo-3-methoxy-3-oxo-1-propenyl]-3-bromo-2-oxo-, methyl ester, (Z)-	EtOH	228(4.2),320(4.0)	5-1968-86
$C_{13}H_{11}ClFN_5$			
9H-Purin-6-amine, 8-chloro-9-[(2-fluorophenyl)methyl]-N-methyl- (spectra in 9.5% ethanol)	pH 7.00	268(4.28)	4-1189-86
	HCl	267(4.34)	4-1189-86
	NaOH	268(4.28)	4-1189-86
$C_{13}H_{11}ClN_2$			
5H-Benzo[6,7]cyclohepta[1,2-d]pyrimidine, 4-chloro-6,7-dihydro-	EtOH	274(4.05)	4-1685-86
$C_{13}H_{11}ClN_2O$			
Benzo[f]quinoxalin-6(2H)-one, 5-chloro-3,4-dihydro-2-methyl-	CH_2Cl_2	238(4.15),275(4.31), 432(3.60)	83-0052-86
Benzo[f]quinoxalin-6(2H)-one, 5-chloro-3,4-dihydro-3-methyl-	CH_2Cl_2	238(4.13),274(4.30), 432(3.51)	83-0052-86
Phenol, 4-chloro-2-[[(5-methyl-2-pyridinyl)imino]methyl]-	$CHCl_3$	360(4.03)	2-0127-86A
Phenol, 2-[[(5-chloro-2-pyridinyl)-imino]methyl]-4-methyl-	$CHCl_3$	360(4.18)	2-0127-86A
Phenol, 2-[[(5-chloro-2-pyridinyl)-imino]methyl]-5-methyl-	$CHCl_3$	340(4.10)	2-0127-86A
$C_{13}H_{11}ClN_2OS_2$			
Benzo[b]thiophene-2-carboxamide, 3-chloro-N-[(2-propenylamino)-thioxomethyl]-	EtOH	208(3.03),250(2.97), 306(2.98)	73-2839-86
$C_{13}H_{11}ClN_2O_2$			
Phenol, 2-[[(5-chloro-2-pyridinyl)-imino]methyl]-6-methoxy-	$CHCl_3$	470(2.49)	2-0127-86A
$C_{13}H_{11}ClO_3$			
2,5-Furandione, 4-[(2-chlorophenyl)-methylene]dihydro-3,3-dimethyl-	EtOH	263(4.53),293(3.55)	34-0369-86
2,5-Furandione, 4-[(3-chlorophenyl)-methylene]dihydro-3,3-dimethyl-	EtOH	225(3.97),292(4.07)	34-0369-86
2,5-Furandione, 4-[(4-chlorophenyl)-methylene]dihydro-3,3-dimethyl-	EtOH	228(3.87),306(4.14)	34-0369-86
1H-Indene-2-acetic acid, 4-chloro-α,α-dimethyl-1-oxo-	EtOH	241(4.67),248(4.68), 328(2.96)	34-0369-86
1H-Indene-2-acetic acid, 5-chloro-α,α-dimethyl-1-oxo-	EtOH	242(4.28),245(4.29), 325(2.46)	34-0369-86
1H-Indene-2-acetic acid, 6-chloro-α,α-dimethyl-1-oxo-	EtOH	241(4.60),248(4.64), 280(3.07)	34-0369-86
2H-Indeno[1,2-b]furan-2,4(3H)-dione, 6-chloro-3a,8b-dihydro-3,3-dimethyl-	EtOH	246(3.96),293(3.15), 300(3.14)	34-0369-86
2H-Indeno[1,2-b]furan-2,4(3H)-dione, 8-chloro-3a,8b-dihydro-3,3-dimethyl-	EtOH	246(3.59),252(3.64), 276(3.26)	34-0369-86
$C_{13}H_{11}ClO_5$			
1,4-Naphthalenedione, 2-chloro-5,6,8-trimethoxy-	MeOH	221(4.64),237(4.02), 243(4.06),248(4.15), 254(4.24),260(4.27),	78-3767-86

Compound	Solvent	λ_{max}(log ε)	Ref.
(cont.)		270(4.26),436(3.74)	78-3767-86
1,4-Naphthalenedione, 2-chloro-5,7,8-trimethoxy-	MeOH	220(4.58),242s(3.91), 248s(3.99),254s(4.08), 260s(4.15),271(4.21), 438(3.67)	78-3767-86
$C_{13}H_{11}Cl_2NO_2S$			
7-Azabicyclo[2.2.1]hepta-2,5-diene, 2,3-dichloro-7-[(4-methylphenyl)-sulfonyl]-	EtOH	230(4.26),266(3.31), 276s(3.20)	24-0616-86
1H-Azepine, 3,6-dichloro-1-[(4-methyl-phenyl)sulfonyl]-	MeCN	219(4.35),251s(3.83), 269s(3.49),276s(3.38), 334s(2.63)	24-0616-86
$C_{13}H_{11}F_3N_4S$			
2,4-Cyclohexadiene-1,1,3-tricarbo-nitrile, 2-amino-6,6-dimethyl-4-[[(trifluoromethyl)thio]methyl]-	EtOH	210.5(4.15),312.5(4.27)	155-0201-86
$C_{13}H_{11}IN_4O_5S$			
Iodine, (2-hydroxy-3,5-dinitrophenyl)-phenyl(thioureato-S)-	EtOH	358(4.20)	5-0947-86
$C_{13}H_{11}N$			
4-Azulenecarbonitrile, 6,8-dimethyl-	hexane	253(4.36),258(4.40), 289(4.62),294(4.66), 304s(4.21),319(3.93), 351(4.43),369(3.31), 387(2.71),504s(2.46), 584s(2.53),602(2.59), 625(2.53),659(2.51), 685(2.22),731(2.09)	5-1222-86
Benzenamine, N-(phenylmethylene)-	pH 5.0	330(3.60)	30-0055-86
5H-Benzocycloheptene-5-carbonitrile, 5-methyl-	EtOH	205(4.29),274(3.90)	39-1479-86C
Cyclobuta[a]naphthalene-2a(2H)-carbo-nitrile, 1,8b-dihydro-	C_6H_{12}	257s(3.80),265(3.90), 269s(3.90),273s(3.87), 281(3.69),286s(3.55)	23-0237-86
Cyclopent[a]indene-2-carbonitrile, 3,3a,8,8a-tetrahydro-	C_6H_{12}	248s(1.30),253s(2.53), 259(2.78),263s(2.87), 266(2.99),272(3.08)	23-0237-86
9H-Fluoren-2-amine	pH 7.4	282(4.38)	94-0944-86
$C_{13}H_{11}NO$			
9H-Carbazole, 2-methoxy-	C_6H_{12}	210(4.16),233(4.49), 250(4.10),255(4.06), 297(3.98),311(3.56), 321s(--)	59-0793-86
	pH 6	208(4.21),233(4.23), 255(4.09),299(3.98), 312(--)	59-0793-86
	MeOH	209(4.23),234(4.51), 252(4.14),256(4.15), 300(4.01),312s(--)	59-0793-86
	ether	235(4.52),253(4.13), 257(4.14),300(4.03), 312(3.78),324s(--)	59-0793-86
	MeCN	209(4.25),234(4.52), 251(4.16),256(4.15), 299(4.02),312(3.78)	59-0793-86

Compound	Solvent	$\lambda_{max}(\log \epsilon)$	Ref.
9H-Carbazole, 2-methoxy-, anion	KOH	225(4.19),267(4.18), 302(3.98)	59-0793-86
cation	H_2SO_4	198(4.42),236(3.74), 272(4.38),399(3.29)	59-0793-86
dication	H_2SO_4	200(--),221(--), 270(--),277(--), 340(--)	59-0793-86
9H-Carbazol-2-ol, 3-methyl-	EtOH	235(4.65),254(4.25), 258(4.26),304(4.19), 332(3.66)	25-0246-86
4,5,6-Metheno-4H-cyclopent[d]isoxazole, 3a,5,6,6a-tetrahydro-3-phenyl-, cis	C_6H_{12}	265(4.05),271s(4.03), 280s(3.82),285s(3.63), 291s(2.96)	24-0950-86
$C_{13}H_{11}NO_2$			
Benzamide, N-hydroxy-N-phenyl-, Se(IV) complex	$CHCl_3$	345(5.18)	160-0025-86
Benzo[f]quinoline-7,8-diol, 7,8-dihydro-, trans	EtOH-THF	202(4.50),229(4.47), 244(4.53),306s(3.72), 318(3.78),325(3.75), 338(3.60)	44-3407-86
Benzo[f]quinoline-9,10-diol, 9,10-dihydro-, trans	EtOH-THF	204(4.37),223(4.22), 253(4.47),293s(3.51), 302(3.58),314(3.61), 330(3.60),345(3.56)	44-3407-86
2(1H)-Pyridinone, 3-benzoyl-6-methyl-	EtOH	254(4.0),335(3.9)	103-1097-86
	0.01M H_2SO_4	260(4.0),345(3.9)	103-1097-86
	pH 13	230(4.2),365(3.7)	103-1097-86
$C_{13}H_{11}NO_2S_2$			
4H-[1]Benzothieno[2,3-e]-1,3-thiazin-4-one, 2-(1-methylethoxy)-	EtOH	215(3.10),252(2.89), 303(3.01)	73-2002-86
4H-[1]Benzothieno[2,3-e]-1,3-thiazin-4-one, 2-propoxy-	EtOH	216(3.07),250(2.96), 304(2.80)	73-2002-86
Benzoxazole, 2-[2-methoxy-5-(methylthio)-3-thienyl]-	EtOH	217(4.24),260(4.25), 310(4.31)	70-1470-86
Benzoxazole, 2-[5-methoxy-2-(methylthio)-3-thienyl]-	EtOH	205(4.00),233(4.03), 256(4.05),288(4.06), 336(3.95)	70-1470-86
$C_{13}H_{11}NO_3$			
Oxireno[3,4]benzo[1,2-f]quinoline-8,9-diol, 1a,8,9,9a-tetrahydro-, (1aα,8α,9β,9aα)-(±)-	THF	240(3.76),280(3.43), 308(3.30),314(3.15), 320(3.29)	44-3407-86
Oxireno[5,6]benzo[1,2-f]quinoline-2,3-diol, 1a,2,3,9c-tetrahydro-, (1aα,2α,3β,9cα)-(±)-	THF	240(3.59),284(3.58), 308(3.42),320(3.38)	44-3407-86
$C_{13}H_{11}NO_3S$			
Thiophene, 2-[2-(4-methoxyphenyl)ethenyl]-5-nitro-, (Z)-	MeOH	286(3.23),430(3.56)	73-1127-86
$C_{13}H_{11}NO_3S_2$			
1,3-Dithiolo[4,5-g]quinoline-7-carboxylic acid, 5-ethyl-5,8-dihydro-8-oxo-	aq Et_3N	237s(4.13),286(4.58)	49-1339-86
$C_{13}H_{11}NO_4$			
7H-Furo[3,2-g][1]benzopyran-7-one, 9-(2-aminoethoxy)-, hydrobromide	1% DMSO	302.5(4.08)	149-0391-86A

Compound	Solvent	$\lambda_{max}(\log \epsilon)$	Ref.
Oxireno[5,6]benzo[1,2-f]quinoline-2,3-diol, 1a,2,3,9c-tetrahydro-, 6-oxide, (1aα,2β,3α,9cα)-(±)-	THF	246(4.13),254(4.08), 338s(3.70),353(3.76), 368s(3.61)	44-3407-86
2-Quinolizinecarboxylic acid, 1-acetyl-4-oxo-, methyl ester	CH_2Cl_2	420(4.0)	39-1561-86B
$C_{13}H_{11}NO_4S$			
2-Butenedioic acid, 2-[(1-methyl-1H-indol-3-yl)thio]-, (Z)-	dioxan	220(4.55),283(4.19)	83-0332-86
$C_{13}H_{11}NO_5$			
4H-1-Benzopyran-2-carboxylic acid, 7-(acetylamino)-4-oxo-, methyl ester	EtOH	254(4.29),313(4.13)	42-0600-86
$C_{13}H_{11}NO_6$			
2-Benzofurancarboxamide, N,7-diformyl-4,6-dimethoxy-	EtOH	223(4.12),290(4.16), 315(4.04),332(4.06)	4-1347-86
$C_{13}H_{11}NO_7$			
Pyrano[3,4-b]pyrrole-2,7-dicarboxylic acid, 1-acetyl-1,5-dihydro-5-oxo-, dimethyl ester	EtOH	262(4.2),269(4.1), 305(4.1),316(4.1)	5-1968-86
	dioxan	263(4.4),271(4.4), 306(4.3),316(4.3)	5-1968-86
$C_{13}H_{11}NS$			
Benzenecarbothioamide, N-phenyl-	hexane	241(4.11),319(3.86)	65-2088-86
	$CHCl_3$	320(3.56),410(2.57)	65-2088-86
	CCl_4	320s(3.45),430(2.45)	65-2088-86
10H-Phenothiazine, 10-methyl-	80% MeCN	252(4.60),306(3.71)	46-2469-86
$C_{13}H_{11}NS_2$			
3H-Indole, 3-[2,3-bis(methylthio)-2-cyclopropen-1-ylidene]-, tetrafluoroborate	MeCN	213(4.55),250(4.11), 275s(4.13),286(3.95), 342(4.45)	78-0839-86
$C_{13}H_{11}NS_4$			
3H-Indole, 3-[4,5-bis(methylthio)-1,3-dithiol-2-ylidene]-	MeCN	211(4.40),254(4.00), 270(4.03),310(3.38), 442(4.40)	78-0839-86
hydrobromide	MeCN	250(4.00),282(4.02), 354(3.40),475(4.40)	78-0849-86
tetrafluoroborate	MeCN	208(4.40),250(4.04), 283(4.06),350(3.50), 477(4.37)	78-0839-86
$C_{13}H_{11}N_3$			
3,6-Acridinediamine, protonated	n.s.g.	445(4.60),490(--)	60-0281-86
diprotonated	n.s.g.	345(--),360(--), 455(4.30)	60-0281-86
dimer	n.s.g.	428(4.48),473(--)	60-0281-86
$C_{13}H_{11}N_3O$			
Imidazo[1,2-a]pyrimidin-5(1H)-one, 1-methyl-7-phenyl-	EtOH	249(4.51),276(4.20), 290s(4.00),325(3.90)	4-0245-86
Imidazo[1,2-a]pyrimidin-5(8H)-one, 8-methyl-7-phenyl-	EtOH	212s(4.28),233(4.40), 259s(3.90),269s(3.79), 307(3.85)	4-0245-86
$C_{13}H_{11}N_3O_2S$			
7H-Thiazolo[3,2-a]pyrimidinium, 5-ami-	MeCN	226(3.94),273(4.05),	2-0275-86

Compound	Solvent	$\lambda_{max}(\log \epsilon)$	Ref.
no-3-hydroxy-8-methyl-7-oxo-2-phenyl-, hydroxide, inner salt (cont.)		323(3.95),410(3.67)	2-0275-86
	benzene	426(--)	2-0275-86
$C_{13}H_{11}N_3S$			
1H-Imidazo[4,5-c]pyridine, 4-[(phenylmethyl)thio]-	pH 1	310(4.03)	87-0138-86
	pH 13	293(4.04),301s(4.02)	87-0138-86
$C_{13}H_{11}N_5O$			
3(2H)-Pyridazinone, 4-amino-6-[4-(1H-imidazol-1-yl)phenyl]-	MeOH	268(4.68)	4-1515-86
$C_{13}H_{12}$			
1H-Cyclopent[e]azulene, 2,3-dihydro-	hexane	246(4.34),275s(4.56), 280(4.61),283s(4.58), 316(3.27),323(3.37), 328(3.46),338(3.58), 343(3.13),354(3.15), 362(2.32),484s(1.85), 532s(2.38),554(2.52), 578(2.63),600(2.60), 630(2.63),656(2.42), 694(2.34)	44-2961-86
$C_{13}H_{12}BrN$			
Pyridine, 2-(4-bromophenyl)-5-ethyl-	n.s.g.	207(4.58),257(4.53), 285(4.49)	104-1375-86
$C_{13}H_{12}BrNS_2$			
3-Thiophenecarbothioamide, N-(4-bromophenyl)-2,5-dimethyl-	EtOH	262(4.11),315(4.13), 401s(2.72)	78-3683-86
$C_{13}H_{12}BrNS_4$			
3H-Indole, 3-[4,5-bis(methylthio)-1,3-dithiol-2-ylidene]-, hydrobromide	MeCN	250(4.00),282(4.02), 354(3.40),475(4.40)	78-0849-86
$C_{13}H_{12}BrN_3O_3$			
4(1H)-Pyridinone, 2-[(4-bromophenyl)-azo]-3-hydroxy-6-(hydroxymethyl)-1-methyl-	MeOH	227(4.32),338(4.03), 422(3.99)	4-0333-86
$C_{13}H_{12}BrO$			
Pyrylium, 2-(4-bromophenyl)-5-ethyl-, tetrafluoroborate	EtOH	242(3.70),270(4.39), 385(4.76)	104-1375-86
$C_{13}H_{12}ClNO_2S_2$			
Carbamothioic acid, [(3-chlorobenzo-[b]thien-2-yl)carbonyl]-, O-(1-methylethyl) ester	EtOH	211(3.12),305(3.04)	73-2002-86
Carbamothioic acid, [(3-chlorobenzo-[b]thien-2-yl)carbonyl]-, O-propyl ester	EtOH	213(3.10),307(3.03)	73-2002-86
$C_{13}H_{12}ClNO_7$			
2H-Pyran-6-carboxylic acid, 4-[2-(acetylamino)-1-chloro-3-methoxy-3-oxo-1-propenyl]-2-oxo-, methyl ester, (E)-	EtOH	262(4.0),316(4.0)	5-1968-86
(Z)-	EtOH	261(4.0),316(4.0)	5-1968-86
$C_{13}H_{12}ClNS_2$			
3-Thiophenecarbothioamide, N-(4-chlorophenyl)-2,5-dimethyl-	EtOH	260(4.13),314(4.15), 400(2.72)	78-3683-86

Compound	Solvent	λ_{max}(log ε)	Ref.
$C_{13}H_{12}ClN_3O$			
4H-Indazol-4-one, 3-[(4-chlorophenyl)-amino]-1,5,6,7-tetrahydro-	dioxan	244(4.240),268(4.322), 305(4.139)	48-0635-86
$C_{13}H_{12}ClN_3O_3$			
4(1H)-Pyridinone, 2-[(3-chlorophenyl)-azo]-3-hydroxy-6-(hydroxymethyl)-1-methyl-	MeOH	286(3.78),324(3.81), 422(3.81)	4-0333-86
4(1H)-Pyridinone, 2-[(4-chlorophenyl)-azo]-3-hydroxy-6-(hydroxymethyl)-1-methyl-	MeOH	227(4.32),324(3.81), 420(3.81)	4-0333-86
$C_{13}H_{12}ClN_5S$			
2H-Purine-2-thione, 7-[(4-chlorophenyl)methyl]-1,7-dihydro-6-(methylamino)-	pH 1	246(4.13),288(4.17)	4-0737-86
	pH 13	253(4.52)	4-0737-86
$C_{13}H_{12}FN_5$			
9H-Purin-6-amine, 9-[(2-fluorophenyl)-methyl]-N-methyl-	10%EtOH-HCl	263(4.22)	4-1189-86
	10%EtOH-NaOH	268(4.18)	4-1189-86
6H-Purin-6-imine, 9-[(2-fluorophenyl)-methyl]-1,9-dihydro-1-methyl-, monohydriodide	pH 7.4	259(4.16)	4-1189-86
$C_{13}H_{12}F_3NO_7$			
2H-Pyran-4-propanoic acid, 6-(methoxycarbonyl)-2-oxo-α-[(trifluoroacetyl)amino]-, methyl ester	EtOH	302(3.8)	5-1968-86
$C_{13}H_{12}F_6N_4$			
1,2,4-Triazin-4(5H)-amine, N,N-dimethyl-5-phenyl-3,6-bis(trifluoromethyl)-	CH_2Cl_2	300(3.57)	83-0690-86
$C_{13}H_{12}IN_3$			
Imidazo[1,5-a]pyridine, 3-(2-pyridinyl)-, methiodide	MeOH	233(3.8),305(3.8)	83-0043-86
$C_{13}H_{12}N_2O$			
Benzo[6,7]cyclohepta[1,2-d]pyrimidin-4-one, 3,5,6,7-tetrahydro-	EtOH	249(4.23),287(3.83)	4-1685-86
Phenol, 2-[[(3-methyl-2-pyridinyl)-imino]methyl]-	$CHCl_3$	352(4.14)	2-0127-86A
Phenol, 2-[[(4-methyl-2-pyridinyl)-imino]methyl]-	$CHCl_3$	350(4.04)	2-0127-86A
Phenol, 2-[[(5-methyl-2-pyridinyl)-imino]methyl]-	$CHCl_3$	350(4.10)	2-0127-86A
Phenol, 2-[[(6-methyl-2-pyridinyl)-imino]methyl]-	$CHCl_3$	350(4.10)	2-0127-86A
Phenol, 4-methyl-2-[(2-pyridinyl-imino)methyl]-	$CHCl_3$	362(4.03)	2-0127-86A
Phenol, 5-methyl-2-[(2-pyridinyl-imino)methyl]-	$CHCl_3$	354(4.16)	2-0127-86A
2-Propen-1-one, 1-(1-methyl-1H-imidazol-2-yl)-3-phenyl-	EtOH	324(4.44)	94-4916-86
9H-Pyrido[3,4-b]indole, 7-methoxy-1-methyl- (harmine)	pH 2	204(4.33),246(4.56), 319(4.29),361(3.90)	39-1247-86B
dication	H_0 -8.5	203(4.26),253(4.60), 318(4.29),359(3.94)	39-1247-86B
9H-Pyrido[3,4-b]indole-1-ethanol	EtOH	214(4.32),235(4.57), 240s(4.58),250(4.40),	100-0303-86

Compound	Solvent	λ_{max}(log ε)	Ref.
(cont.)		282s(4.0),288(4.21), 336(3.67),350(3.66)	100-0303-86
$C_{13}H_{12}N_2OS$			
2-Pyridinecarbothioamide, N-(2-methoxyphenyl)-	hexane	239(4.18),287(4.08), 363(4.00)	65-2088-86
	CCl_4	277(4.04),367(3.92)	65-2088-86
	$CHCl_3$	291(4.04),365(3.92)	65-2088-86
	12M HCl	264(3.94),355s(3.58)	65-2088-86
$C_{13}H_{12}N_2O_2$			
Azepino[3,4-b]indole-1,3(2H,4H)-dione, 5,10-dihydro-2-methyl-	isoPrOH	241(4.22),316(4.29)	103-1345-86
Benzenamine, 5-methoxy-2-nitroso-N-phenyl-	EtOH	334(4.27),441(3.81)	104-1407-86
Diazene, (4-methoxyphenyl)phenyl-, 1-oxide	benzene	339(4.306)	44-0088-86
	MeOH	334(4.265)	44-0088-86
Diazene, (4-methoxyphenyl)phenyl-, 2-oxide	benzene	350(4.297)	44-0088-86
	MeOH	349(4.293)	44-0088-86
Phenol, 2-methoxy-6-[(2-pyridinylimino)methyl]-	$CHCl_3$	480(2.87)	2-0127-86A
1H-Pyrido[3,4-b]indole-1,3(2H)-dione, 4,9-dihydro-2,9-dimethyl-	isoPrOH	209(4.28),241(4.17), 309(4.14),336s(3.71)	103-1345-86
$C_{13}H_{12}N_2O_3$			
Benzenamine, 5-methoxy-2'-methyl-2-nitro-N-phenyl-	EtOH	408(4.00)	104-1407-86
2-Furancarboxylic acid, 5-[[(phenylamino)methylene]amino]-, methyl ester	MeOH	260(2.89),350(3.36)	73-2826-86
$C_{13}H_{12}N_2O_3S$			
Benzenesulfonamide, 2-(formylamino)-N-phenyl-	MeOH	246(4.34),280(3.72)	44-1967-86
$C_{13}H_{12}N_2O_5$			
Phenol, 2-bicyclo[2.2.1]hept-2-en-2-yl-4,6-dinitro-	n.s.g.	205(3.81),262(4.31)	44-3453-86
$C_{13}H_{12}N_2O_6$			
2-Furancarboxylic acid, 5-[[[[5-(methoxycarbonyl)-2-furanyl]amino]methylene]amino]-, methyl ester	MeOH	277(3.01),360(3.57)	73-2817-86 +73-2826-86
$C_{13}H_{12}N_2S$			
2-Pyridinecarbothioamide, N-(2-methylphenyl)-	hexane	235(4.11),281(3.98), 348(3.73)	65-2088-86
	benzene	285(3.96),343(3.72)	65-2088-86
	$CHCl_3$	283(4.04),336(3.76), 440s(2.15)	65-2088-86
	CCl_4	283(3.89),343(3.72), 450s(2.04)	65-2088-86
	hexanol	281(3.86),337(3.61)	65-2088-86
	12M HCl	260(3.98),350(3.59)	65-2088-86
2-Pyridinecarbothioamide, N-(phenylmethyl)-	hexane	227(3.96),273(3.92), 323(3.66)	65-2088-86
	$CHCl_3$	276(3.95),317(3.85), 401(2.04)	65-2088-86
	CCl_4	272(3.83),323(3.83), 425(1.93)	65-2088-86
	12M HCl	269(3.99),327(3.49)	65-2088-86

Compound	Solvent	λ$_{max}$(log ε)	Ref.
$C_{13}H_{12}N_3$			
Imidazo[1,5-a]pyridine, 3-(2-pyridinyl)-, methiodide	MeOH	233(3.8),305(3.8)	83-0043-86
$C_{13}H_{12}N_4$			
1H-Imidazo[4,5-b]pyridin-2-amine, 1-methyl-6-phenyl-	MeOH	225(4.46),273(4.00), 316(4.46)	142-1815-86
1H-Imidazo[4,5-b]pyridin-2-amine, 3-methyl-6-phenyl-	MeOH	240(4.31),312(4.14)	142-1815-86
1H-Imidazo[4,5-c]pyridin-4-amine, N-(phenylmethyl)-	pH 1	263s(4.08),277(4.11)	87-0138-86
	pH 13	278(4.08)	87-0138-86
Pyrazinecarbonitrile, 3,4-dihydro-3-imino-4,6-dimethyl-5-phenyl-	pH 1	378(4.13)	33-1025-86
	EtOH	437(3.83)	33-1025-86
Pyrazolo[1,5-b][1,2,4]triazine, 2,3-dimethyl-7-phenyl-	EtOH	262(4.4),287(3.5), 336(3.74)	118-0071-86
$C_{13}H_{12}N_4O$			
4H-Imidazol-4-one, 5-(dimethylamino)-3-(1H-indol-3-yl)-	EtOH	213(4.18),261(3.85), 284(3.78),342(3.71), 430(3.73)	88-2621-86
Purine, 6-(2-phenylethoxy)-	pH 1	254(4.08)	163-2243-86
	pH 7	252(4.08)	163-2243-86
	pH 13	261(4.04)	163-2243-86
$C_{13}H_{12}N_4O_2$			
Benzo[g]pteridine-2,4(3H,10H)-dione, 10-ethyl-3-methyl-	MeCN	436(3.99)	39-1825-86C
with $KClO_4$	MeCN	436(3.99)	39-1825-86C
with $Mg(ClO_4)_2$	MeCN	436(3.98)(anom.)	39-1825-86C
$C_{13}H_{12}N_4O_2S$			
1H-Pyrido[2,3-c][1,2,6]thiadiazin-4-amine, 5-methyl-7-phenyl-, 2,2-dioxide	pH -4.0	228s(4.20),271(4.02), 354(4.20)	11-0607-86
	pH 2.0	265(4.23),294s(4.06), 380(4.14)	11-0607-86
	pH 10.0	229(4.25),257(4.37), 288s(4.03),366(3.97)	11-0607-86
1H-Pyrido[2,3-c][1,2,6]thiadiazin-4-amine, 7-methyl-5-phenyl-, 2,2-dioxide	pH -4.0	254(4.19),324(3.97), 368s(3.75)	11-0607-86
	pH 2.0	260(4.35),294(3.89), 372(3.88)	11-0607-86
	pH 10.0	236(4.38),288s(4.31), 358(3.61)	11-0607-86
$C_{13}H_{12}N_4O_3S$			
2H-1,3,5-Thiadiazin-2-one, 6-(4-nitrophenyl)-4-(1-pyrrolidinyl)-	MeCN	275(4.17)	97-0132-86
$C_{13}H_{12}N_4S$			
3H-1,2,4-Thiadiazolo[4,3-c]pyrimidine, 5,7-dimethyl-3-(phenylimino)-	$CHCl_3$	238(3.97),295(3.94), 373(3.50)	118-1027-86
$C_{13}H_{12}N_6O_2S$			
Carbonothioic dihydrazide, bis(3-hydroxy-2-pyridinyl)methylene-, cobalt chelate	20% DMF-pH 4.60	462(4.37)	131-0509-86I
copper chelate	"	455(3.96)	131-0509-86I
ferrous chelate	"	393(4.39)	131-0509-86I
$C_{13}H_{12}O$			
4-Azulenecarboxaldehyde, 6,8-dimethyl-	hexane	215(4.20),244s(4.37),	5-1222-86

Compound	Solvent	$\lambda_{max}(\log \epsilon)$	Ref.
(cont.)		250(4.45),269(4.20), 295s(4.41),303(4.51), 326(4.08),353s(3.32), 363(3.37),382(3.27), 403(2.79),599s(2.59), 617(2.64),637s(2.61), 673(2.56),702(2.35), 748(2.14),795s(1.40)	5-1222-86
5H-Benzocycloheptene-5-carboxaldehyde, 5-methyl-	EtOH	204(4.31),276(3.86)	39-1479-86C
$C_{13}H_{12}O_2$			
3-Butyn-2-one, 4-[2-(2-propenyloxy)-phenyl]-	EtOH	280(3.76),321(3.66)	44-3125-86
1H-Dibenzo[b,d]pyran-4(6H)-one, 2,3-dihydro-	ether	235(4.10),242(4.08), 320(4.15)	44-4424-86
$C_{13}H_{12}O_2S_2$			
1,2-Ethanediol, 1-[5'-(1-propynyl)-[2,2'-bithiophene]-5-yl]-	EtOH	243(3.76),260s(3.58), 338(4.31)	163-0263-86
$C_{13}H_{12}O_3$			
2H-1-Benzopyran-2-one, 4-hydroxy-3-(2-methyl-2-propenyl)-	EtOH	240(3.89),312(4.08)	2-1167-86
2H-1-Benzopyran-2-one, 4-[(2-methyl-2-propenyl)oxy]-	EtOH	214(4.36),266(4.02), 276(4.01),303(3.82)	2-1167-86
Cycloprop[a]inden-6(1H)-one, 1-acetoxy-1a,6a-dihydro-6a-methyl-, endo	MeOH	222(4.39),290(3.36), 321s(3.09)	39-1789-86C
Ethanone, 1-(2,4-dihydroxy-3-methyl-1-naphthalenyl)-	MeOH	232(4.44),325(3.81), 352(3.85),358s(3.82)	39-1789-86C
2,5-Furandione, dihydro-3,3-dimethyl-4-(phenylmethylene)-	EtOH	225(3.54),226(3.54), 302(4.11)	34-0369-86
2(5H)-Furanone, 5-hydroxy-4-[2-(2-methylphenyl)ethenyl]-, (E)-	EtOH	316(4.46)	94-4346-86
2(5H)-Furanone, 5-hydroxy-4-[2-(4-methylphenyl)ethenyl]-, (E)-	EtOH	322(4.55)	94-4346-86
1H-Indene-2-acetic acid, α,α-dimethyl-1-oxo-	EtOH	238(4.54),244(4.60), 315(2.67)	34-0369-86
2H-Indeno[1,2-b]furan-2,4(3H)-dione, 3a,8b-dihydro-3,3-dimethyl-	EtOH	241(4.04),278(3.26)	34-0369-86
1,3-Naphthalenediol, 2-methyl-, 3-acetate	MeOH	231(4.55),299(3.63), 324s(3.42)	39-1789-86C
1,4-Naphthalenediol, 2-methyl-, 4-acetate	MeOH	225(3.46),266s(3.65), 302(3.64)	12-2067-86
	MeOH-KOH	388(3.78)	12-2067-86
1,4-Naphthalenedione, 5-methoxy-2,7-dimethyl-	EtOH	255(3.27),274s(3.13), 405(2.66)	94-2810-86
1(2H)-Naphthalenone, 2-acetoxy-2-methyl-	EtOH	233(4.52),265(3.61), 326(3.27)	12-2067-86
$C_{13}H_{12}O_4$			
4H-1-Benzopyran-4-one, 7-hydroxy-2-methyl-5-(2-oxopropyl)-	EtOH	243(4.13),251(4.16), 293(3.97)	102-1727-86
	EtOH-NaOH	257(4.30),333.5(4.04)	102-1727-86
2(5H)-Furanone, 5-hydroxy-4-[2-(2-methoxyphenyl)ethenyl]-, (E)-	EtOH	302(4.25),342(4.30)	94-4346-86
2(5H)-Furanone, 5-hydroxy-4-[2-(4-methoxyphenyl)ethenyl]-, (E)-	EtOH	341(4.53)	94-4346-86
2(5H)-Furanone, 4-methoxy-5-(methoxyphenylmethylene)-, (E)-	MeOH	232(4.20),307(4.57)	163-3053-86
(Z)-	MeOH	234(3.91),307(4.43)	163-3053-86

Compound	Solvent	λ_{max}(log ϵ)	Ref.
1,2-Naphthalenedione, 5,8-dimethoxy-3-methyl-	EtOH	228(4.43),260s(4.09),510(3.85)	94-2810-86
1,4-Naphthalenedione, 5,6-dimethoxy-2-methyl-	EtOH	262(4.32),278s(4.16),392(3.60)	94-2810-86
1,4-Naphthalenedione, 5,7-dimethoxy-2-methyl-	EtOH	267(4.28),410(3.60)	94-2810-86
1,4-Naphthalenedione, 5,8-dimethoxy-2-methyl-	EtOH	264(3.69),450(3.22)	94-2810-86
1,4-Naphthalenedione, 5,8-dimethoxy-6-methyl-	$CHCl_3$	256(4.49),327(3.54),420(3.99)	5-1655-86
1,4-Naphthalenedione, 5-(methoxymethoxy)-2-methyl-	EtOH	250(4.20),272s(4.07),382(3.53)	94-2810-86
$C_{13}H_{12}O_5$			
2,5-Furandione, 3-[(3,4-dimethoxyphenyl)methylene]dihydro-	EtOH-$CHCl_3$ (1:1)	345(4.28)	150-0735-86M
2-Furanpropanoic acid, 2,5-dihydro-4-hydroxy-5-oxo-3-phenyl-	MeOH	285(4.25)	88-2015-86
	MeOH-NaOH	320(4.17)	88-2015-86
$C_{13}H_{12}O_6$			
4H-1-Benzopyran-6-carboxaldehyde, 7-hydroxy-5,8-dimethoxy-2-methyl-4-oxo-	EtOH	217(4.10),219s(4.09),261(4.48),272(4.47),300s(3.44),375s(3.08)	44-3116-86
8H-1,3-Dioxolo[4,5-g][1]benzopyran-8-one, 4,9-dimethoxy-6-methyl-	EtOH	213(4.38),239(4.29),253(4.17),286(3.86),313s(3.73)	44-3116-86
2H-Pyran-2,5(6H)-dione, 4-(2,5-dimethoxyphenyl)-3-hydroxy-	dioxan	277.8(4.09),327s(3.32)	5-0177-86
2H-Pyran-2,5(6H)-dione, 4-(3,4-dimethoxyphenyl)-3-hydroxy-	dioxan	250(4.02),271.3(4.01),328.7(3.87)	5-0177-86
$C_{13}H_{13}BO_5$			
Boron, [α-hydroxybenzeneacetato(2-)]-(2,4-pentanedionato-O,O')-, (T-4)-	CH_2Cl_2	290(3.79)	48-0755-86
$C_{13}H_{13}BrO$			
Azulene, 6-bromo-2-propoxy-	CH_2Cl_2	285.5(4.90),295(5.04),327.5(3.87),367(3.95),385(4.04),519(2.31),555.5(2.19),604(1.70)	24-2272-86
5H-Benzocycloheptene-5-methanol, 8-bromo-5-methyl-	EtOH	205(4.24),288(3.90)	39-1479-86C
$C_{13}H_{13}BrO_4$			
Propanedioic acid, [bromo(4-methylphenyl)methylene]-, dimethyl ester	EtOH	286(4.01)	44-4107-86
$C_{13}H_{13}BrO_8$			
Propanedioic acid, [[3-bromo-6-(methoxycarbonyl)-2-oxo-2H-pyran-4-yl]-methyl]-, dimethyl ester	MeOH	313(3.95)	5-1968-86
$C_{13}H_{13}ClN_4O$			
Phenol, 2-[(5-chloro-2-pyridinyl)-azo]-5-(dimethylamino)-	pH 9.2 in 50% EtOH	440(4.6)	86-0375-86
$C_{13}H_{13}ClN_4O_4$			
Pyrido[2,3-d]pyrimidine-6-carbonitrile, 7-chloro-1,2,3,4-tetrahydro-1,3-bis(methoxymethyl)-5-methyl-2,4-dioxo-	H_2O	230(4.62),273(4.11),310(3.94)	87-0709-86

Compound	Solvent	$\lambda_{max}(\log \epsilon)$	Ref.
$C_{13}H_{13}ClO_4$			
Butanedioic acid, [(2-chlorophenyl)-methylene]dimethyl-, (E)-	EtOH	254(3.92)	34-0369-86
Butanedioic acid, [(3-chlorophenyl)-methylene]dimethyl-, (E)-	EtOH	256(3.69)	34-0369-86
Butanedioic acid, [(4-chlorophenyl)-methylene]dimethyl-, (E)-	EtOH	265(4.37)	34-0369-86
Propanedioic acid, [chloro(4-methyl-phenyl)methylene]-, dimethyl ester	EtOH	284(3.90)	44-4107-86
$C_{13}H_{13}F_3N_2O_2$			
Imidazo[1,2-a]pyridine-3-carboxylic acid, 2-(trifluoromethyl)-, 1,1-dimethylethyl ester	EtOH	234.5(4.42),241(4.41), 289(4.00),300(3.98)	39-1769-86C
$C_{13}H_{13}N$			
Cyclobuta[a]naphthalene-2-carbo-nitrile, 1,2,2a,3,4,8b-hexahy-dro-, (2α,2aβ,8bβ)-	C_6H_{12}	247s(2.02),253s(2.28), 260s(2.46),265s(2.56), 266(2.58),272(2.58)	23-0237-86
1H-Indole, 3-(1,1-dimethyl-2-prop-ynyl)-	MeOH	216(4.55),261(4.29), 298(4.03)	44-2343-86
1H-Pyrrole, 1-methyl-2-(2-phenyl-ethenyl)-, (E)-	EtOH	237(3.95),338(4.27)	118-0620-86
(Z)-	EtOH	233(4.14),322(4.01)	118-0620-86
Quinoline, 3-(2-methyl-1-propenyl)-	MeOH	228(4.53),285(3.72), 323(3.62)	44-2343-86
$C_{13}H_{13}NO$			
4-Azulenecarboxaldehyde, 6,8-dimeth-yl-, oxime	dioxan	246(4.25),292(4.60), 339(3.48),351(3.56), 571(2.66),592(2.63), 651(2.29)	5-1222-86
5H-Benzocycloheptene-5-carboxalde-hyde, 5-methyl-, oxime	EtOH	205(4.32),279(3.82)	39-1479-86C
Bicyclo[4.4.1]undeca-3,6,8,10-tetra-en-2-one, 5-[(methylamino)methyl-ene]-, (±)-	CH_2Cl_2	296(3.90),407(4.30)	33-0315-86
2-Buten-1-one, 1-(1H-indol-3-yl)-3-methyl-	MeOH	246(4.06),262(3.96), 312(4.03)	44-2343-86
1H-Indole, 1-(3-methyl-1-oxo-2-but-enyl)-	MeOH	248(4.39),300(3.91)	44-2343-86
1,3-Oxazepine, 6,7-dimethyl-4-phenyl-	EtOH	223(4.18),262(4.15), 312(4.00)	77-1188-86
2,4-Pentadienal, 5-(2,3-dihydro-1H-indol-1-yl)-	MeOH	403(4.84)	33-1588-86
$C_{13}H_{13}NO_2$			
2H-1-Benzopyran-2-one, 4-methyl-7-(2-propenylamino)-	MeOH	237(4.17),293(3.72), 365(4.30)	151-0069-86A
	$CHCl_3$	310(3.68),345(4.27)	151-0069-86A
1-Naphthalenecarboxamide, 3-methoxy-5-methyl-	MeOH	225(4.71),280(3.69), 292(3.69),322(3.45), 335(3.53)	94-4554-86
Pyrano[2,3-e]indol-2(7H)-one, 8,9-di-hydro-4,8-dimethyl-	MeOH	238(4.04),270(3.63), 367(4.12)	151-0069-86A
	$CHCl_3$	261(3.55),325(4.10), 338(4.20)	151-0069-86A
1H-Pyrrole-3-carboxylic acid, 1-(phen-ylmethyl)-, methyl ester	EtOH	232(4.06),249(3.85)	24-0813-86

Compound	Solvent	$\lambda_{max}(\log \epsilon)$	Ref.
$C_{13}H_{13}NO_2S$			
7-Azabicyclo[2.2.1]hepta-2,5-diene, 7-[(4-methylphenyl)sulfonyl]-	MeCN	230(4.14),265(2.93), 271s(2.70),276s(2.55)	24-0616-86
3-Azatetracyclo[3.2.0.0^{2,7}.0^{4,6}]heptane, 3-[(4-methylphenyl)sulfonyl]-	MeCN	228(4.04),263s(3.01), 268s(2.90),274(2.77)	24-0616-86
1H-Azepine, 1-[(4-methylphenyl)-sulfonyl]-	MeCN	222(4.10),230s(4.03), 264s(3.52),275s(4.43)	24-0616-86
2-Propen-1-one, 3-[5-(dimethylamino)-2-furanyl]-1-(2-thienyl)-, (E)-	MeOH	284(3.18),499(3.48)	73-0879-86
$C_{13}H_{13}NO_2S_2$			
Phenol, 2-[[[2-methoxy-5-(methylthio)-3-thienyl]methylene]amino]-	EtOH	203(4.77),234(4.60), 278(4.20)	70-1470-86
Phenol, 2-[[[5-methoxy-2-(methylthio)-3-thienyl]methylene]amino]-	EtOH	201(4.71),260(4.42), 357(4.25)	70-1470-86
$C_{13}H_{13}NO_3$			
Benzeneacetamide, N-methyl-N-(2-methyl-1-oxo-2-propenyl)-α-oxo-	C_6H_{12}	218(4.02),256(4.00), 350(1.95)	39-1759-86C
3H-Indol-3-one, 1-acetyl-1,2-dihydro-2-(1-oxopropyl)-	EtOH	206(4.12),239(4.68), 262(4.27),345(3.82)	104-2186-86
6-Oxa-3-azabicyclo[3.1.1]heptane-2,4-dione, 1,3-dimethyl-5-phenyl-	C_6H_{12}	225(3.75),251(2.69), 257(2.70),264(2.60)	39-1759-86C
2-Propen-1-one, 3-[5-(dimethylamino)-2-furanyl]-1-(2-furanyl)-, (E)-	MeOH	285(3.20),499(3.52)	73-0879-86
$C_{13}H_{13}NO_3S$			
3-Oxa-8-azatricyclo[3.2.1.0^{2,4}]oct-6-ene, 8-[(4-methylphenyl)sulfonyl]-, (1α,2α,4α,5α)-	EtOH	230(4.06),257s(2.82), 264(2.75),270s(2.60), 275s(2.46)	24-0616-86
$C_{13}H_{13}NO_3Se$			
Glycine, N-[(2-selenyl-3-benzofuranyl)methylene]-, ethyl ester	EtOH	217(4.36),266(4.24), 292(4.55),435(4.10)	103-0608-86
$C_{13}H_{13}NO_4S$			
3,7-Dioxa-9-azatetracyclo[3.3.1.0^{2,4}-0^{6,8}]nonane, 9-[(4-methylphenyl)-sulfonyl]-, (1α,2α,4α,5α,6α,8α)-	EtOH	230(4.05)	24-0616-86
$C_{13}H_{13}NO_5S$			
Thiophene, 2,5-dihydro-2-[(4-methoxyphenyl)methylene]-3-methyl-4-nitro-, 1,1-dioxide	$CHCl_3$	263(3.699),410(4.399)	104-0379-86
$C_{13}H_{13}NO_6S$			
Phenol, 2-methoxy-4-[(3-methyl-4-nitro-2(5H)-thienylidene)methyl]-, S,S-dioxide	$CHCl_3$	270(3.613),425(4.255)	104-0379-86
$C_{13}H_{13}NO_7$			
2H-Pyran-6-carboxylic acid, 4-[2-(acetylamino)-3-methoxy-3-oxo-1-propenyl]-2-oxo-, methyl ester, (E)-	EtOH	255(4.1),318(4.1)	5-1968-86
(Z)-	EtOH	252(4.1),317(4.0)	5-1968-86
Pyrano[3,4-b]pyrrole-2,5-dicarboxylic acid, 1-acetyl-1,2,3,7-tetrahydro-7-oxo-, dimethyl ester	MeOH	202(3.95),230(3.91), 268(3.35),348(4.15)	5-1968-86
Pyrano[3,4-b]pyrrole-2,7-dicarboxylic acid, 1-acetyl-1,2,3,5-tetrahydro-5-oxo-, dimethyl ester	MeOH	302(4.2)	5-1968-86

Compound	Solvent	$\lambda_{max}(\log \epsilon)$	Ref.
$C_{13}H_{13}NO_8$			
2H-Pyran-6-carboxylic acid, 4-[3-methoxy-2-[(methoxycarbonyl)amino]-3-oxo-1-propenyl]-2-oxo-, methyl ester, (Z)-	EtOH	257(4.0),316(4.1)	5-1990-86
Pyrano[3,4-b]pyrrole-2,2,5(1H)-tricarboxylic acid, 3,7-dihydro-7-oxo-, trimethyl ester	MeOH	237(--),287(3.63), 360(4.14)	5-1968-86
$C_{13}H_{13}NS_2$			
Thiazole, 4-methyl-2-phenyl-5-(2-propenylthio)-	EtOH	214s(3.94),312(3.98)	33-0374-86
3-Thiophenecarbothioamide, N,2-dimethyl-5-phenyl-	EtOH	268(4.30),298(4.22), 393s(2.57)	78-3683-86
3-Thiophenecarbothioamide, N,5-dimethyl-2-phenyl-	EtOH	259(4.19),270s(4.18)	78-3683-86
3-Thiophenecarbothioamide, 2,5-dimethyl-N-phenyl-	EtOH	262(4.07),312(4.11), 400(2.70)	78-3683-86
$C_{13}H_{13}N_3$			
5H-Benzo[6,7]cyclohepta[1,2-d]pyrimidin-4-amine, 6,7-dihydro-	EtOH	241(4.38),288(3.75)	4-1685-86
$C_{13}H_{13}N_3O$			
4H-Indazol-4-one, 1,5,6,7-tetrahydro-3-(phenylamino)-	dioxan	242(4.349),264(4.279), 301(4.075)	48-0635-86
2-Pyrrolidinone, 3,4-dimethyl-5-[(2-pyridinyl)cyanomethylene]-, (E)-	MeOH	265(4.16),314(4.27)	49-0631-86
	CHCl	267(4.19),317(4.29)	49-0631-86
$C_{13}H_{13}N_3OS$			
2H-1,3,5-Thiadiazin-2-one, 6-phenyl-4-(1-pyrrolidinyl)-	MeCN	237(4.18),276(4.27)	97-0132-86
Thiourea, N-(3-hydroxy-2-pyridinyl)-N'-(2-methylphenyl)-	benzene	270(4.29)	80-0761-86
	MeOH	251(4.25)	80-0761-86
	EtOH	252(4.25)	80-0761-86
	isoPrOH	266(4.27)	80-0761-86
	BuOH	256(4.44)	80-0761-86
	dioxan	251(4.29)	80-0761-86
	HOAc	250(4.22)	80-0761-86
	CH_2Cl_2	252(4.29)	80-0761-86
	$CHCl_3$	252(4.28)	80-0761-86
	MeCN	250(4.28)	80-0761-86
	DMF	264(4.17)	80-0761-86
	DMSO	261(4.13)	80-0761-86
Thiourea, N-(3-hydroxy-2-pyridinyl)-N'-(4-methylphenyl)-	benzene	275(4.29)	80-0761-86
	MeOH	260(4.25)	80-0761-86
	EtOH	255(4.25)	80-0761-86
	isoPrOH	263(4.26)	80-0761-86
	BuOH	256(4.44)	80-0761-86
	CH_2Cl_2	255(4.30)	80-0761-86
	$CHCl_3$	257(4.29)	80-0761-86
	MeCN	253(4.24)	80-0761-86
	DMF	263(4.16)	80-0761-86
	DMSO	261(4.14)	80-0761-86
	dioxan	256(4.30)	80-0761-86
	HOAc	258(4.22)	80-0761-86
$C_{13}H_{13}N_3OS_2$			
2H-1,3,5-Thiadiazine-2-thione, 4-(4-morpholinyl)-6-phenyl-	$CHCl_3$	294s(4.48),303(4.51), 390s(3.43)	97-0132-86

Compound	Solvent	λ_{max}(log ε)	Ref.
$C_{13}H_{13}N_3O_2$			
Acetic acid, [(2-amino-1-methyl-2-phenylethenyl)imino]cyano-, methyl ester, (Z,?)-	EtOH	455(4.50)	33-0793-86 +33-1025-86
1H-Imidazole, 1-methyl-2-(2-methyl-2-phenylethenyl)-5-nitro-	n.s.g.	270(4.05),355(4.17)	112-1049-86
Pyrazinecarboxylic acid, 3-amino-6-methyl-5-phenyl-, methyl ester	EtOH	372(4.08)	33-0793-86
$C_{13}H_{13}N_3O_3$			
4(1H)-Pyridinone, 3-hydroxy-6-(hydroxymethyl)-1-methyl-2-(phenylazo)-	MeOH	286(3.78),290s(3.82), 411(3.71)	4-0333-86
$C_{13}H_{13}N_3O_8$			
Propanedioic acid, azido[6-(methoxycarbonyl)-2-oxo-2H-pyran-4-ylmethyl]-, dimethyl ester	EtOH	301(3.7)	5-1990-86
Propanedioic acid, [[3-azido-6-(methoxycarbonyl)-2-oxo-2H-pyran-4-yl]-methyl]-, dimethyl ester	MeOH	225(3.95),332(4.22)	5-1968-86
$C_{13}H_{13}N_3S$			
Thiourea, N-(4-methyl-2-pyridinyl)-N'-phenyl-	benzene	273(4.29)	80-0761-86
	MeOH	260(4.33)	80-0761-86
	EtOH	267(4.35)	80-0761-86
	isoPrOH	266(4.38)	80-0761-86
	BuOH	256(4.43)	80-0761-86
	dioxan	252(4.26)	80-0761-86
	HOAc	263(4.35)	80-0761-86
	CH_2Cl_2	262(4.37)	80-0761-86
	$CHCl_3$	261(4.30)	80-0761-86
	$C_2H_4Cl_2$	263(4.35)	80-0761-86
	MeCN	262(4.36)	80-0761-86
	DMF	268(4.45)	80-0761-86
	DMSO	270(4.37)	80-0761-86
Thiourea, N-(6-methyl-2-pyridinyl)-	benzene	272(4.27)	80-0761-86
	MeOH	263(4.35)	80-0761-86
	EtOH	267(4.35)	80-0761-86
	isoPrOH	265(4.32)	80-0761-86
	BuOH	260(4.41)	80-0761-86
	dioxan	252(4.30)	80-0761-86
	HOAc	264(4.34)	80-0761-86
	$HOCH_2CH_2OH$	258(4.31)	80-0761-86
	CH_2Cl_2	262(4.38)	80-0761-86
	$CHCl_3$	263(4.29)	80-0761-86
	$C_2H_4Cl_2$	264(4.33)	80-0761-86
	MeCN	260(4.31)	80-0761-86
	DMF	266(4.42)	80-0761-86
	DMSO	268(4.40)	80-0761-86
$C_{13}H_{13}N_3S_2$			
2H-1,3,5-Thiadiazine-2-thione, 6-phenyl-4-pyrrolidino-	$CHCl_3$	293s(4.53),303(4.56), 390s(3.35)	97-0132-86
$C_{13}H_{13}N_5$			
6H-Purin-6-imine, 3,7-dihydro-3-methyl-7-(phenylmethyl)-, monohydrobromide	pH 1	224s(4.12),278(4.21)	94-1821-86
	pH 7	224s(4.11),278(4.20)	94-1821-86
	pH 13	285(4.15)	94-1821-86
	EtOH	281(4.20)	94-1821-86
6H-Purin-6-imine, 3,7-dihydro-7-methyl-3-(phenylmethyl)-, monohydriodide	pH 1	223.5s(4.10),278(4.25)	94-1821-86
	pH 7	224s(4.12),278(4.24)	94-1821-86

Compound	Solvent	λ_{max}(log ε)	Ref.
(cont.)	pH 13	280(4.21)	94-1821-86
	EtOH	226s(4.07),281.5(4.26)	94-1821-86
6H-Purin-6-imine, 3,9-dihydro-3-methyl-9-(phenylmethyl)-, hydrobromide	pH 1	271(4.24)	94-1821-86
	pH 7	271(4.24)	94-1821-86
	EtOH	272.5(4.23)	94-1821-86
6H-Purin-6-imine, 3,9-dihydro-9-methyl-3-(phenylmethyl)-, hydriodide	pH 1	271(4.20)	94-1821-86
	pH 7	271(4.20)	94-1821-86
	EtOH	273(4.20)	94-1821-86
$C_{13}H_{13}N_5OS$			
Ethanone, 1-[5-(4-methylphenyl)-1-(2-thiazolyl)formazanyl]-	50% EtOH	485(4.20)	65-2114-86
$C_{13}H_{13}N_5OS_2$			
1,2,4-Triazolo[3,4-b][1,3,4]thiadiazolium, 1-methyl-6-[(4-methylbenzoyl)amino]-3-(methylthio)-, hydroxide, inner salt	MeOH	250(4.18),306(4.40)	5-1540-86
$C_{13}H_{13}N_5O_2S$			
Ethanone, 1-[5-(4-methoxyphenyl)-1-(2-thiazolyl)formazanyl]-	50% EtOH	455(4.16)	65-2114-86
$C_{13}H_{13}N_5O_4$			
Acetic acid, cyano(1,3,4,5-tetrahydro-1,3,5-trimethyl-2,4-dioxo-6(2H)-ylidene)-, methyl ester	MeOH	217(4.18),328(4.24), 487(3.97)	33-0704-86
$C_{13}H_{13}N_5S$			
2H-Purine-2-thione, 1,7-dihydro-6-(methylamino)-7-(phenylmethyl)-	pH 1	246(4.27),287(4.32)	4-0737-86
	pH 13	253(4.43)	4-0737-86
$C_{13}H_{14}$			
Bicyclo[2.2.2]octane, 2,3,5,6,7-pentakis(methylene)-	isooctane	211(3.95),239s(4.09), 244(4.10),256s(3.98)	33-1310-86
	EtOH	209(3.99),243(4.10), 256s(3.97)	33-1310-86
Bicyclo[2.2.2]oct-2-ene, 2-methyl-5,6,7,8-tetrakis(methylene)-	isooctane	220s(4.03),228(4.06), 237(4.05),251(4.03)	33-1310-86
	EtOH	204(4.10),221s(4.02), 228(4.04),237(4.04), 252(4.01)	33-1310-86
$C_{13}H_{14}BrNO_3S$			
2(1H)-Azocinone, 5-bromo-5,6,7,8-tetrahydro-3-(phenylsulfonyl)-	MeOH	212(4.08),234(4.10), 255(3.49),261(3.28)	12-0687-86
$C_{13}H_{14}BrNO_7$			
2H-Pyran-4-propanoic acid, α-(acetylamino)-3-bromo-6-(methoxycarbonyl)-2-oxo-, methyl ester	EtOH	314(4.0)	5-1968-86
$C_{13}H_{14}BrN_2$			
2,3'-Bipyridinium, 1'-(3-bromopropyl)-, bromide	pH 8.2	210(4.18),229(4.17), 276(4.04)	64-0239-86B
$C_{13}H_{14}ClN$			
1H-Carbazole, 6-chloro-2,3,4,9-tetrahydro-9-methyl-	$HOCH_2CH_2OH$	239(4.61),294(3.83), 300s(3.82)	103-0713-86

Compound	Solvent	$\lambda_{max}(\log \epsilon)$	Ref.
$C_{13}H_{14}ClNO$			
1-Butanone, 3-chloro-1-(1H-indol-3-yl)-3-methyl-	MeOH	238(4.07),255(3.94), 296(4.05)	44-2343-86
$C_{13}H_{14}ClNOS$			
2(1H)-Azocinone, 4-chloro-5,6,7,8-tetrahydro-3-(phenylthio)-	MeOH	258(3.91)	12-0687-86
$C_{13}H_{14}ClNO_2$			
2,4,6-Heptatrienoic acid, 7-(3-chloro-1H-pyrrol-2-yl)-6-methyl-, methyl ester, (E,E,E)-	EtOH	380(4.56)	150-3601-86M
$C_{13}H_{14}ClNO_3S$			
2(1H)-Azocinone, 1-chloro-5,6,7,8-tetrahydro-3-(phenylsulfonyl)-, (E)-	MeOH	211(4.09),230(4.17), 258(3.43),265(3.38), 273(3.23)	12-0687-86
2(1H)-Azocinone, 5-chloro-5,6,7,8-tetrahydro-3-(phenylsulfonyl)-, (E)-	MeOH	212(4.06),232(3.08), 265(3.40),273(3.20)	12-0687-86
$C_{13}H_{14}ClNS_2$			
2(3H)-Thiazolethione, 3-(3-chlorophenyl)-4-(1,1-dimethylethyl)-	EtOH	321.8(4.09)	152-0399-86
$C_{13}H_{14}Cl_2O_6$			
2-Butenoic acid, 4-[(2,5-dichloro-1-hydroxy-4-oxo-2,5-cyclohexadien-1-yl)-3,4-dimethoxy-, methyl ester	MeOH	228(4.21),296(3.60)	78-3767-86
$C_{13}H_{14}FeO_3$			
Iron, tricarbonyl(η^4-7-methyl-3-methylene-1,4,6-octatriene)-, (E)-	C_6H_{12}	275(4.42)	101-0061-86D
(Z)-	C_6H_{12}	272(4.32)	101-0061-86D
$C_{13}H_{14}N_2$			
1H-Cyclopent[2,3]azirino[1,2-b]pyrazole, 4a,5,6,7-tetrahydro-2-phenyl-	EtOH	253.5(4.09)	35-1617-86
$C_{13}H_{14}N_2O$			
Benzenepropanenitrile, α-(1-iminopropyl)-α-methyl-β-oxo-	MeCN	273(4.59)	2-1133-86
2-Benzofurancarbonitrile, 6-(diethylamino)-	EtOH	230(3.75),302(3.65), 346(3.79)	4-1347-86
3-Pyridinecarboxamide, 1,4-dihydro-1-(4-methylphenyl)-	EtOH	272(4.23),354(3.92)	150-0088-86S +150-0901-86M
3-Pyridinecarboxamide, 1,4-dihydro-1-(phenylmethyl)-	EtOH	351(3.82)	150-0088-86S +150-0901-86M
9H-Pyrido[3,4-b]indole, 3,4-dihydro-7-methoxy-1-methyl- (harmaline)	pH 2	203(4.38),213(4.36), 260(3.86),373(4.35)	39-1247-86B
	pH 2-8	204(4.37),214(4.35), 260(3.83),374(4.35)	149-0571-86B
	pH 11	218(4.44),258(4.00), 328(4.20)	149-0571-86B
	dioxan	226(4.34),258(3.95), 327(4.14)	149-0571-86B
	MeCN	226(4.37),257(3.89), 328(4.11)	149-0571-86B
	+ acid	204(4.23),216(4.28), 261(3.78),374(4.15)	149-0571-86
anion	n.s.g.	228(4.51),266(3.95), 330(4.05),382(3.70)	149-0571-86B

Compound	Solvent	$\lambda_{max}(\log \epsilon)$	Ref.
Harmaline, cation	MeOH	204(4.23),215(4.29), 261(3.84),376(4.26)	149-0571-86B
dication	H_0 -8.5	196(4.26),213(4.30), 275(4.00),317(4.34)	39-1247-86B
$C_{13}H_{14}N_2O_2$			
Pyrazine, trimethylphenyl-, 1,4-di-oxide	EtOH	241(4.37),250s(4.27), 308(4.37)	103-0268-86
$C_{13}H_{14}N_2O_2S_2$			
[1]Benzothieno[2,3-d]pyrimidine-3(4H)-acetic acid, 5,6,7,8-tetra-hydro-2-methyl-4-thioxo-	EtOH	234(4.43),287(4.23), 320(4.19),355(3.98)	106-0096-86
$C_{13}H_{14}N_2O_3$			
1H-Imidazole-4-carboxaldehyde, 2-(3,4-dimethoxyphenyl)-1-methyl-	MeOH	262(4.19)	87-0261-86
1H-Imidazole-5-carboxaldehyde, 2-(3,4-dimethoxyphenyl)-1-methyl-	MeOH	262(4.19)	87-0261-86
$C_{13}H_{14}N_2O_4S$			
2(1H)-Cycloheptimidazolethione, 1-β-D-ribofuranosyl-	n.s.g.	230(4.16),257(4.44), 310(3.62),442(4.25)	150-2625-86M
$C_{13}H_{14}N_2O_4Se_2$			
3-Cyclobutene-1,2-dione, 3,3'-[meth-ylenebis(seleno)]bis[4-(dimethyl-amino)-	CH_2Cl_2	279(4.4),341(4.4)	24-0182-86
$C_{13}H_{14}N_2O_5$			
2(1H)-Cycloheptimidazolone, 1-β-D-ribofuranosyl-	n.s.g.	220s(4.05),253(4.42), 345(3.95),369(3.89), 374(3.89),388s(3.83)	150-2625-86M
2-Pyrrolidinecarboxaldehyde, 5-meth-oxy-1-(2-nitrobenzoyl)-	MeOH	215(4.00),256(3.80)	142-0777-86
$C_{13}H_{14}N_2O_7$			
2-Pyridinecarboxylic acid, 3-[5-carb-oxy-4-(carboxymethyl)-3-pyrrolidin-yl]-1,6-dihydro-6-oxo-, (3α,4α,5β)-, (acromelic acid B)	pH 2	241(3.67),312(3.47)	138-1399-86
	pH 7	239(3.71),311(3.51)	138-1399-86
	pH 12	236(3.80),302(3.47)	138-1399-86
2-Pyridinecarboxylic acid, 5-[5-carb-oxy-4-(carboxymethyl)-3-pyrrolidin-yl]-1,6-dihydro-6-oxo-, (3α,4α,5β)-, (acromelic acid A)	pH 2	242(3.64),317(3.95)	88-0607-86
	pH 7	240(3.71),313(3.98)	88-0607-86
	pH 12	242(3.71),313(3.96)	88-0607-86
$C_{13}H_{14}N_2S$			
3(2H)-Pyridazinethione, 2,4-dimethyl-6-(4-methylphenyl)-	EtOH	314(3.85)	48-0522-86
$C_{13}H_{14}N_3$			
6H-Dipyrido[1,2-a:2',1'-d][1,3,5]tri-azin-5-ium, 1,11-dimethyl-, iodide	H_2O	205(4.429),223s(4.285), 273(3.887),316(3.702), 406(4.318)	142-1893-86
6H-Dipyrido[1,2-a:2',1'-d][1,3,5]tri-azin-5-ium, 2,10-dimethyl-, iodide	H_2O	275(4.082),305(3.854), 388(4.431)	142-1893-86
6H-Dipyrido[1,2-a:2',1'-d][1,3,5]tri-azin-5-ium, 3,9-dimethyl-, iodide	H_2O	208(4.531),265(4.020), 310(3.670),406(4.204)	142-1893-86
$C_{13}H_{14}N_4$			
Benzenamine, N,N-dimethyl-4-(3-pyri-	acetone	420(4.46)	7-0473-86

Compound	Solvent	λ_{max}(log ε)	Ref.
dinylazo)- (cont.)			7-0473-86
$C_{13}H_{14}N_4O$			
Ethanol, 2-[[4-(3-pyridinylazo)phenyl]amino]-	acetone	410(4.47)	7-0473-86
$C_{13}H_{14}N_4O_2S$			
1H-Indole-2,3-dione, 1-(1-oxobutyl)-, 3-[(aminothioxomethyl)hydrazone]-, (Z)-	$CHCl_3$	330s(3.49),390(4.09)	104-2173-86
	DMF	331s(3.76),349(4.16)	104-2173-86
Spiro[benzothiazole-2(3H),5'(2'H)-pyrimidine]-2',4'(3'H)-dione, 6'-(propylamino)-	MeCN	223(4.34),266(4.11)	94-0664-86
$C_{13}H_{14}N_4O_2S_2$			
1,2,3-Thiadiazole-4-carboxylic acid, 2,5-dihydro-5-[[(methylamino)thioxomethyl]imino]-2-phenyl-, ethyl ester	DMF	289(3.91),413(4.11)	48-0741-86
$C_{13}H_{14}N_4O_3S$			
1H-[1,2,6]Thiadiazino[3,4-b]quinoxalin-4(3H)-one, 1,3,7,8-tetramethyl-, 2,2-dioxide	MeOH	218(4.34),253(4.58), 330(3.74),338(3.74), 381s(3.44)	83-0079-86
10H-[1,2,6]Thiadiazino[3,4-b]quinoxalin-4(3H)-one, 3,7,8,10-tetramethyl-, 2,2-dioxide	MeOH	222(4.63),262(4.57), 345(3.88),424(3.85)	83-0079-86
$C_{13}H_{14}N_4O_4$			
Pyrido[2,3-d]pyrimidine-6-carbonitrile, 1,2,3,4-tetrahydro-1,3-bis(methoxymethyl)-5-methyl-2,4-dioxo-	H_2O	223(4.56),265(4.21), 304(3.71),315s(3.63)	87-0709-86
4-Pyrimidinecarboxylic acid, 1,4,5,6-tetrahydro-5-oxo-2-[(3-pyridinylcarbonyl)amino]-, ethyl ester	EtOH	233(4.16),249(3.87), 273(4.34)	118-0484-86
$C_{13}H_{14}N_4O_4S$			
2-Butanamine, N-[3-(2,4-dinitrophenyl)-2(3H)-thiazolylidene]-	n.s.g.	433(3.50)	150-0110-86S
$C_{13}H_{14}N_4O_5$			
Pyrido[2,3-d]pyrimidine-6-carbonitrile, 1,2,3,4,7,8-hexahydro-1,3-bis(methoxymethyl)-5-methyl-2,4,7-trioxo-	H_2O	228(4.78),282(4.37), 322(4.50)	87-0709-86
$C_{13}H_{14}N_4O_5S$			
2-Propenoic acid, 3-[(1,2,3,4-tetrahydro-1,3,6-trimethyl-2,4-dioxo-7-pteridinyl)sulfinyl]-, methyl ester	CH_2Cl_2	245(4.31),353(4.02), 365s(3.96)	33-1095-86
$C_{13}H_{14}N_6O$			
Pyrazinecarboxamide, 3-amino-N-(aminoiminomethyl)-6-methyl-5-phenyl-	EtOH	374(4.05)	33-0793-86
$C_{13}H_{14}N_6O_3$			
Acetamide, N,N'-[7-(1-oxopropyl)-2,4-pteridinediyl]bis-	pH -1.0	214(4.41),254(4.09), 288(3.90),345(4.05)	128-0183-86
	pH 4.0	218(4.37),254(4.38), 356(4.03)	128-0183-86

Compound	Solvent	λ_{max}(log ε)	Ref.
(cont.)	MeOH	219(4.37),257(4.39), 357(3.99)	128-0183-86
$C_{13}H_{14}N_8$			
2(1H)-Pyrimidinone, 4-(3,5-diamino-1H-pyrazol-4-yl)-6-phenyl-, hydrazone	EtOH	252(4.36),350(4.22)	103-0774-86
$C_{13}H_{14}O$			
5H-Benzocycloheptenemethanol, 5-methyl-	EtOH	206(4.21),285(3.83)	39-1479-86C
Cycloprop[a]inden-6(6aH)-one, 1,1a-dihydro-1,1,1a-trimethyl-	MeOH	226(4.20),298(3.25), 310s(3.14),350(2.59)	39-2011-86C
1(2H)-Naphthalenone, 2,2,3-trimethyl-	MeOH	235(4.54),265(3.66), 272(3.64),282s(3.42), 340(3.25)	39-1789-86C
1(4H)-Naphthalenone, 3,4,4-trimethyl-	MeOH	254(4.10),291s(3.49)	39-2011-86C
$C_{13}H_{14}O_2$			
Bullvalenecarboxylic acid, methyl-, methyl ester	ether	218s(3.88),258(3.51)	24-2339-86
2-Cyclohexen-1-one, 2-methoxy-3-phenyl-	MeOH	220s(3.9),282(4.2)	44-4424-86
	ether	220(3.86),280(4.22)	44-4424-86
1H-Cyclopenta[a]pentalene-1,4(3aH)-dione, 5,6,7,7a-tetrahydro-3,7a-dimethyl-	MeOH	216(4.20),223(4.31)	35-3443-86
1H-Cyclopenta[a]pentalene-3,4-dione, tetrahydro-2,3b-dimethyl-	MeOH	228(4.01)	35-3443-86
3H-Cyclopenta[a]pentalene-3,4(3aH)-dione, tetrahydro-2,5-dimethyl-	MeOH	216(3.90),241(3.60)	39-0291-86C
Naphthalene, 1,8-dimethoxy-3-methyl-	EtOH	287s(3.28),300(3.34), 318(3.28),333(3.32)	94-2810-86
1-Naphthalenol, 8-methoxy-3,6-dimethyl-	EtOH	290s(3.12),305(3.21), 320(3.21),334(3.26)	94-2810-86
1,2,4-[1]Propanyl[3]ylidenepentalene-5,8(1H)-dione, hexahydro-6-methyl-9-methylene-	MeOH	239(3.72)	39-0291-86C
Tricyclo[6.3.0.0^{2,6}]undeca-2(6),9(10)-diene-3,11-dione, 1,4-dimethyl-	MeOH	218(4.09),245(3.68)	35-3443-86
$C_{13}H_{14}O_3$			
Ethanone, 1-(2,5-dihydro-8-hydroxy-3-methyl-1-benzoxepin-9-yl)-	EtOH	221(4.10),260(3.93), 341(3.40)	18-3983-86
Ethanone, 1-[2,3-dihydro-5-hydroxy-2-(1-methylethenyl)-4-benzofuranyl]-	EtOH	233(4.21),261s(3.90), 368(3.65)	18-3983-86
Ethanone, 1-[2,3-dihydro-6-hydroxy-2-(1-methylethenyl)-7-benzofuranyl]-	EtOH	230(4.10),268(4.13), 356(3.66)	18-3983-86
Ethanone, 1-(2-ethenyl-2,3-dihydro-5-hydroxy-2-methyl-4-benzofuranyl)-	EtOH	232(4.31),261s(4.08), 366(3.80)	18-3983-86
2(3H)-Furanone, 4-(3,4-dimethoxybenzoyl)dihydro-	EtOH	259(3.90)	56-0143-86 +80-0541-86
1-Naphthalenol, 5,8-dimethoxy-3-methyl-	EtOH	294s(3.22),310(3.32), 328(3.34),342(3.38)	94-2810-86
1-Naphthalenol, 6,8-dimethoxy-3-methyl-	EtOH	296s(3.29),302(3.41), 322(3.26),335(3.29)	94-2810-86
1-Naphthalenol, 7,8-dimethoxy-3-methyl-	EtOH	280s(3.64),295(3.79), 306(3.80),336(3.60), 347s(3.59)	94-2810-86
1-Naphthalenol, 8-(methoxymethoxy)-3-methyl-	EtOH	265(2.69),292s(2.61), 305(2.67),320(2.59), 335(2.59)	94-2810-86

Compound	Solvent	$\lambda_{max}(\log \epsilon)$	Ref.
$C_{13}H_{14}O_3S$			
2-Hexenoic acid, 5-methyl-4-oxo-2-(phenylthio)-	MeOH	251(4.18),270(4.20)	48-0539-86
$C_{13}H_{14}O_4$			
Butanedioic acid, dimethyl(phenylmethylene)-, (E)-	EtOH	258(4.08)	34-0369-86
$C_{13}H_{14}O_5$			
Samin	EtOH	236(4.06),288(4.04)	142-0923-86
$C_{13}H_{14}O_6$			
Bicyclo[2.2.1]hepta-2,5-diene-2,3-dicarboxylic acid, 7-acetoxy-, dimethyl ester, anti	EtOH	227(3.61),268s(3.20)	39-0157-86C
syn	EtOH	230(3.67),267(3.57)	39-0157-86C
Tetracyclo[3.2.0.0^{2,7}.0^{4,6}]heptane-1,5-dicarboxylic acid, 3-acetoxy-, dimethyl ester, anti	EtOH	270(3.08)	39-0157-86C
$C_{13}H_{14}O_8$			
Propanedioic acid, [[6-(methoxycarbonyl)-2-oxo-2H-pyran-4-yl]methyl]-, dimethyl ester	EtOH	301(3.9)	5-1990-86
$C_{13}H_{15}BrO_2$			
Benzenepentanal, 4-bromo-α-ethyl-δ-oxo-	EtOH	257(4.11)	104-1375-86
$C_{13}H_{15}ClN_2O_3S$			
Glycine, N-[[(2-chlorobenzoyl)amino]-(methylthio)methylene]-, ethyl ester	EtOH	250s(3.99),278(4.20)	118-0484-86
Glycine, N-[[(4-chlorobenzoyl)amino]-(methylthio)methylene]-, ethyl ester	EtOH	255(4.21),280s(4.36), 284(4.37),291s(4.34)	118-0484-86
$C_{13}H_{15}ClN_2O_6$			
2(1H)-Pyrimidinone, 4-chloro-1-(3,5-di-O-acetyl-2-deoxy-β-D-erythro-pentofuranosyl)-	$CHCl_3$	308(3.65)	4-1401-86
$C_{13}H_{15}ClOS$			
2-Propen-1-one, 3-(butylthio)-3-chloro-1-phenyl-	n.s.g.	257(4.70),332(4.80)	48-0539-86
$C_{13}H_{15}N$			
Benzeneacetonitrile, α-(2,2-dimethylpropylidene)-, cis	C_6H_{12}	255.9(4.03)	88-6225-86
trans	C_6H_{12}	238s(3.79)	88-6225-86
1H-Carbazole, 2,3,4,9-tetrahydro-5(and 7)-methyl-	$HOCH_2CH_2OH$	233(4.40),282(3.93), 295s(3.90)	103-0713-86
1H-Carbazole, 2,3,4,9-tetrahydro-6-methyl-	$HOCH_2CH_2OH$	233(4.50),287(3.98), 298s(3.90)	103-0713-86
1H-Indole, 3-(1,1-dimethyl-2-propenyl)-	hexane	265(3.68)	44-2343-86
1H-Indole, 2-(3-methyl-2-butenyl)-	MeOH	218(4.40),268(3.99), 284(3.86)	44-2343-86
1H-Indole, 3-(3-methyl-2-butenyl)-	MeOH	220(4.40),273(3.69), 279(3.71),287(3.65)	44-2343-86
$C_{13}H_{15}NO$			
1-Butanone, 1-(1H-indol-3-yl)-3-meth-	MeOH	237(4.05),252(3.92),	44-2343-86

Compound	Solvent	$\lambda_{max}(\log \epsilon)$	Ref.
yl- (cont.)		292(4.05)	44-2343-86
4(1H)-Quinolinone, 1-methyl-2-propyl- (leptomerine)	EtOH	213(3.63),230.5(3.61), 285.5(3.31),294(3.40)	105-0078-86
$C_{13}H_{15}NOS$			
1-Benzothiepin-5(2H)-one, 4-[(dimethylamino)methylene]-3,4-dihydro-, (E)-	EtOH	248(4.07),350(4.24)	4-0449-86
$C_{13}H_{15}NO_2$			
2-Benzofurancarboxaldehyde, 6-(diethylamino)-	EtOH	245(4.21),300(3.54), 311(3.51),395(4.37)	4-1715-86
$C_{13}H_{15}NO_2S$			
2(1H)-Azocinone, 5,6,7,8-tetrahydro-3-(phenylsulfinyl)-, (E)-	MeOH	220(4.19),262(3.53), 267(3.40),275(3.20)	12-0687-86
$C_{13}H_{15}NO_3$			
[1,3]Dioxepino[5,6-d]isoxazole, 3a,4,8,8a-tetrahydro-6-methyl-3-phenyl-, endo	MeOH	265(2.97)	73-2158-86
exo	MeOH	261(3.00)	73-2158-86
$C_{13}H_{15}NO_3S$			
1-Azabicyclo[4.2.0]octan-8-one, 7-(phenylsulfonyl)-, trans	MeOH	217(3.46),252(2.18), 257(2.30),264(2.48), 271(2.30)	12-0687-86
2(1H)-Azocinone, 3,4,5,6-tetrahydro-3-(phenylsulfonyl)-, (Z)-	MeOH	222(4.16),257(3.30), 264(3.26),271(3.15)	12-0687-86
2(1H)-Azocinone, 3,6,7,8-tetrahydro-5-(phenylsulfonyl)-, (E)-	MeOH	224(3.32),257(3.60), 264(3.51),271(3.32)	12-0687-86
2(1H)-Azocinone, 5,6,7,8-tetrahydro-3-(phenylsulfonyl)-, (E)-	MeOH	230(4.18),266(3.23), 274(3.04)	12-0687-86
$C_{13}H_{15}NO_4S$			
9-Oxa-3-azabicyclo[6.1.0]nonan-2-one, 1-(phenylsulfonyl)-	MeOH	220(3.99),253(2.90), 259(3.00),266(3.11), 273(3.04)	12-0687-86
$C_{13}H_{15}NO_5$			
1,3-Dioxane-4,6-dione, 5-[[5-(dimethylamino)-2-furanyl]methylene]-2,2-dimethyl-	EtOH	236(3.90),271(4.00), 456(4.58)	103-1072-86
$C_{13}H_{15}NO_6$			
1H-Azepine-1,2,5-tricarboxylic acid, 1-ethyl 2,5-dimethyl ester	MeOH	230(4.38),350(3.28)	23-1969-86
$C_{13}H_{15}NO_7$			
2H-Pyran-4-propanoic acid, α-(acetylamino)-6-(methoxycarbonyl)-2-oxo-, methyl ester	EtOH	225(3.7),302(3.9)	5-1968-86
$C_{13}H_{15}NS_2$			
5(4H)-Thiazolethione, 4-methyl-4-(1-methylethyl)-2-phenyl-	EtOH	238(4.08),260(4.14), 318(4.02),485(1.23), 500s(1.18)	33-0374-86
$C_{13}H_{15}NS_3$			
1-Piperidine(dithioic) acid, thiobenzoyl ester	CH_2Cl_2	293(4.24),522(2.08)	88-4595-86

Compound	Solvent	$\lambda_{max}(\log \epsilon)$	Ref.
$C_{13}H_{15}N_3$			
Cyclohepta[b]pyrrol-2(1H)-one, (1-methylpropylidene)hydrazone	EtOH	280(4.48),288(4.44), 322(4.18),332(4.17), 382(4.02),400(4.00), 447(3.81),464s(3.73), 496s(3.31)	18-3326-86
$C_{13}H_{15}N_3O$			
3H-Pyrazol-3-one, 4-[(dimethylamino)-methylene]-2,4-dihydro-5-methyl-2-phenyl-	DMAA	445(2.29),560(1.12)	104-0582-86
4H-Pyrido[1,2-a]pyrimidin-4-one, 2-(1-piperidinyl)-	EtOH	235(3.96),270(4.58), 330(3.51)	4-1295-86
1H-Pyrrole-3-carboxamide, 2-amino-4,5-dimethyl-1-phenyl-	EtOH	240(4.09),281(3.86)	11-0347-86
$C_{13}H_{15}N_3O_2$			
4H-Pyrido[1,2-a]pyrimidine-3-carbox-aldehyde, 2-(butylamino)-4-oxo-	EtOH	253(4.35),273s(4.08), 347(4.01),358s(3.98)	4-1295-86
$C_{13}H_{15}N_3O_2S$			
Ethanone, 1,1'-[3,4-diamino-6-(2-prop-enyl)-6H-thieno[2,3-b]pyrrole-2,5-diyl]bis- (hydrate)	EtOH	278(4.20),317(4.53), 341(4.25)	48-0459-86
Thieno[2,3-d]pyrimidine-4-acetic acid, α-cyano-5,6-dihydro-2,6-dimethyl-, ethyl ester	$CHCl_3$	279(4.05),315(3.98), 370(4.30)	94-0516-86
$C_{13}H_{15}N_3O_3S$			
Piperidine, 1-[diazo(phenylsulfonyl)-acetyl]-	hexane	228(4.13)	12-0687-86
$C_{13}H_{15}N_3O_4S$			
6H-Thieno[2,3-b]pyrrole-2,5-dicarbox-ylic acid, 3,4-diamino-6-(2-propen-yl)-, dimethyl ester	EtOH	260(4.34),293(4.54), 320(4.09)	48-0459-86
$C_{13}H_{15}N_3O_5$			
1H-Pyrrolo[3,4-c]pyridine-7-carbox-ylic acid, 3-(dimethylamino)-2,4,5,6-tetrahydro-2,5-dimethyl-1,4,6-trioxo-, methyl ester	DMF	283(4.35),447(4.00)	44-0149-86
$C_{13}H_{15}N_3O_5S$			
Glycine, N-[(methylthio)[(2-nitrobenz-oyl)amino]methylene]-, ethyl ester	EtOH	262(4.12)	118-0484-86
Glycine, N-[(methylthio)[(3-nitrobenz-oyl)amino]methylene]-, ethyl ester	EtOH	228(4.33),283(4.31)	118-0484-86
Glycine, N-[(methylthio)[(4-nitrobenz-oyl)amino]methylene]-, ethyl ester	EtOH	241(4.08),275s(4.20), 296(4.30)	118-0484-86
$C_{13}H_{15}N_3O_7$			
2,4(1H,3H)-Pyrimidinedione, 1-[5,6-dideoxy-2,3-O-(1-methylethylidene)-6-nitro-β-D-ribo-hex-5-enofurano-syl)-, (E)-	n.s.g.	252(4.00)	111-0111-86
$C_{13}H_{15}N_5O_3$			
Benzoic acid, 2-[[(3-diazo-1-ethyl-2-oxopropyl)amino]carbonyl]-, hydrazide	MeOH	238(4.12),277(4.06)	104-0353-86

Compound	Solvent	λ_{max}(log ε)	Ref.
$C_{13}H_{15}N_5O_4$			
Pyrido[2,3-d]pyrimidine-6-carbonitrile, 7-amino-1,2,3,4-tetrahydro-1,3-bis(methoxymethyl)-5-methyl-2,4-dioxo-	H_2O	232(4.39),284(3.86), 325(3.95)	87-0709-86
$C_{13}H_{15}N_5O_6$			
Tetrazolo[1,5-c]pyrimidin-5(6H)-one, 6-(3,5-di-O-acetyl-2-deoxy-β-D-erythro-pentofuranosyl)-	$CHCl_3$	253(4.03)	4-1401-86
$C_{13}H_{15}O_3P$			
Phosphonic acid, (4-azulenylmethyl)-, dimethyl ester	CH_2Cl_2	243(4.39),280(4.61), 286(4.60),312s(--), 328(3.48),343(3.60), 355(3.06),540s(--), 572(2.58),612(2.52), 675s(--)	5-1222-86
Phosphonic acid, (6-azulenylmethyl)-, dimethyl ester	CH_2Cl_2	234(4.20),278(4.75), 284(4.75),325s(--), 333(3.56),340(3.52), 347(3.69),535s(--), 550s(--),575(2.44), 590s(--),621(2.38), 685(1.93)	5-1222-86
$C_{13}H_{16}ClN$			
1H-Indole, 3-(3-chloro-3-methylbutyl)-	MeOH	220(4.43),271(3.77), 279(3.81),286(3.74)	44-2343-86
$C_{13}H_{16}ClNO_2S$			
Piperidine, 1-[2-[(4-chlorophenyl)sulfonyl]ethenyl]-, (E)-	MeCN	226(3.99),287(4.16)	104-2085-86
$C_{13}H_{16}Cl_2NO_3P$			
Phosphonic acid, [2,2-dichloro-1-[2,2-dichloro-1-[(phenylmethylene)amino]ethenyl]-, diethyl ester	hexane	229(4.05),250s(4.10), 263(4.15)	65-2372-86
$C_{13}H_{16}F_3N_2O_2$			
Pyridinium, 1-[2-amino-1-[(1,1-dimethylethoxy)carbonyl]-3,3,3-trifluoro-1-propenyl]-, perchlorate	EtOH	263(4.24)	39-1769-86C
$C_{13}H_{16}FeO_3$			
Iron, tricarbonyl(η^4-7-methyl-3-methylene-1,4-octadiene)-	C_6H_{12}	278(3.40)	101-0061-86D
$C_{13}H_{16}N_2$			
2-Cyclohexen-1-one, 3-methyl-, phenylhydrazone	EtOH	303(4.27)	126-1573-86
$C_{13}H_{16}N_2O$			
Acetamide, N-[2-(1H-indol-3-yl)-ethyl]-N-methyl-	MeOH	221(4.41),273s(3.61), 281(3.64),290(3.59)	100-0901-86
$C_{13}H_{16}N_2OS_2$			
[1]Benzothieno[2,3-d]pyrimidine-4(3H)-thione, 5,6,7,8-tetrahydro-3-(2-hydroxyethyl)-2-methyl-	EtOH	235(4.27),287(4.04), 322(4.03),357(3.88)	106-0096-86

Compound	Solvent	λ_{max}(log ϵ)	Ref.
$C_{13}H_{16}N_2O_2$			
2-Benzofurancarboxamide, 6-(diethyl-amino)-	EtOH	228(3.99),302(3.66), 352(4.16)	4-1347-86
3(2H)-Pyridazinone, 6-(3,4-dimethyl-phenyl)-4,5-dihydro-5-(hydroxy-methyl)-	EtOH	285(4.38)	56-0143-86 +80-0541-86
$C_{13}H_{16}N_2O_2S$			
β-Alanine, N-2-benzothiazolyl-N-propyl-	pH 1	220(4.30),265(4.00), 282(3.98),291(4.00)	103-0795-86
	EtOH	227(4.39),275(4.13), 302(3.83)	103-0695-86
3(2H)-Benzothiazolepropanoic acid, 2-(propylimino)-	pH 1	219(4.34),258(3.91), 281(3.89),290(3.91)	103-0795-86
	EtOH	222(4.39),263(3.94), 281(3.79),290(3.81), 306(3.37)	103-0795-86
$C_{13}H_{16}N_2O_3$			
2-Butanone, 3-methyl-3-[(2-oxo-2-phen-ylethylidene)amino]-, 2-oxime, N-oxide	EtOH	259(3.94),312(4.15)	103-0856-86
Ethanone, 1-(4,5-dihydro-4-hydroxy-5,5-dimethyl-4-phenyl-1H-imidazol-2-yl)-, N-oxide	EtOH	291(3.52)	103-0856-86
1H-Imidazole-4-methanol, 2-(3,4-di-methoxyphenyl)-1-methyl-	MeOH	264(4.14)	87-0261-86
Methanone, (2,5-dihydro-1-hydroxy-4,5,5-trimethyl-1H-imidazol-2-yl)phenyl-, N-oxide	EtOH	245(4.35)	103-0856-86
Methanone, (4,5-dihydro-4-hydroxy-4,5,5-trimethyl-1H-imidazol-2-yl)phenyl-, N-oxide	EtOH	256(4.02),321s(3.31)	103-0856-86
$C_{13}H_{16}N_2O_3S$			
Glycine, N-[(benzamido)(methylthio)-methylene]-, ethyl ester	EtOH	247(4.10),279(4.30), 292s(4.18)	118-0484-86
$C_{13}H_{16}N_2O_4$			
Benzamide, N-(1-methoxy-4-pentenyl)-2-nitro-	MeOH	213(4.06),251(3.75)	142-0777-86
$C_{13}H_{16}N_2O_4S$			
Piperidine, 1-[2-[(4-nitrophenyl)sul-fonyl]ethenyl]-, (E)-	MeCN	256(4.46),340(3.88)	104-2085-86
$C_{13}H_{16}N_2O_5$			
Benzamide, N-(1-methoxy-3-oxiranyl-propyl)-2-nitro-	MeOH	206(4.23),250(3.81)	142-0777-86
2-Pyrrolidinemethanol, 5-methoxy-1-(2-nitrobenzoyl)-	MeOH	219(4.01),257(3.83)	142-0777-86
$C_{13}H_{16}N_2O_8$			
2-Propenoic acid, 3-(1,2,3,4-tetra-hydro-2,4-dioxo-1-β-D-ribofurano-syl-5-pyrimidinyl)-, methyl ester, (E)-	EtOH	269s(4.03),302(4.23)	87-0213-86
$C_{13}H_{16}N_2S_2$			
[1]Benzothieno[2,3-d]pyrimidine-4(1H)-thione, 5,6,7,8-tetrahydro-	EtOH	230(4.36),287(4.14), 321(4.16),355(4.00)	106-0096-86

Compound	Solvent	λ_{max}(log ε)	Ref.
2-(1-methylethyl)- (cont.)			106-0096-86
$C_{13}H_{16}N_4O_3S$			
Benzeneacetic acid, 2-(acetylamino)-α-oxo-, 2-[(ethylamino)thioxomethyl]hydrazide	EtOH	234(4.41),272s(4.09), 330s(3.64)	104-2173-86
	$CHCl_3$	279(3.88),356s(3.61)	104-2173-86
$C_{13}H_{16}N_4O_4S$			
Acetamide, N-[4-[[(3-diazo-1-ethyl-2-oxopropyl)amino]sulfonyl]phenyl]-	MeOH	206(4.26),265(4.41), 278s(4.30)	104-0353-86
Acetamide, N-[4-[[(5-diazo-4-oxopentyl)amino]sulfonyl]phenyl]-	MeOH	207(4.32),266(4.43), 278s(4.31)	104-0353-86
$C_{13}H_{16}N_4O_5$			
4,5-Pyrimidinedicarboxylic acid, 2-cyano-1,6-dihydro-1-(4-morpholinyl)-, dimethyl ester	CH_2Cl_2	382(3.72)	83-0798-86
$C_{13}H_{16}N_6O_4$			
Adenosine, 8-cyano-N,N-dimethyl-	MeOH	233(4.29),270s(3.72), 279(3.78),315(4.16)	94-3635-86
$C_{13}H_{16}O$			
3H-Cyclopenta[a]pentalen-3-one, 1,2,3b,4,5,6,6a,7-octahydro-6a-methyl-4-methylene-, cis-(±)-	MeOH	236(4.04)	35-3443-86
$C_{13}H_{16}OSi$			
2-Propyn-1-one, 1-(2-methylphenyl)-3-(trimethylsilyl)-	EtOH	222(3.88),244s(3.91), 267(4.14)	33-0560-86
$C_{13}H_{16}O_2$			
Bicyclo[6.2.2]dodeca-8,10,11-triene-9-carboxylic acid	EtOH	231(4.30),260(3.70), 328(3.10)	78-1851-86
2-Propenoic acid, 3-phenyl-, 1,1-dimethylethyl ester, (E)-	n.s.g.	275(4.43)	35-3005-86
	+ BF_3	310(4.63)	35-3005-86
$(C_{13}H_{16}O_2)_n$			
2-Propen-1-one, 1-[5-(1,1-dimethylethyl)-2-hydroxyphenyl]-, polymer	$CHCl_3$	262(--),345(3.54)	32-0533-86
$C_{13}H_{16}O_3$			
3aH-Cyclopenta[3,4]cyclobuta[1,2]benzen-3a-ol, 1,2,3,7b-tetrahydro-3,3-dimethoxy-	MeOH	260(2.72),267(2.86), 273(2.86)	44-1419-86
2-Hexanone, 6-(1,3-benzodioxol-5-yl)-(moskachan B)	n.s.g.	231(3.78),287(3.75)	102-2209-86
7-Oxabicyclo[2.2.1]heptane, 1-(dimethoxymethyl)-2,3,5,6-tetrakis(methylene)-	isooctane	216s(4.15),222(4.25), 229(4.26),240s(3.97), 250s(3.88),265s(3.42)	33-1287-86
	EtOH	216s(4.13),222(4.18), 228(4.19),239s(3.90), 249s(3.81),263s(3.28)	33-1287-86
$C_{13}H_{16}O_6$			
2-Pentenedioic acid, 2-(2-hydroxy-4,4-dimethyl-6-oxo-1-cyclohexen-1-yl)-	EtOH	261(4.14)	2-0347-86
$C_{13}H_{16}O_7$			
6,9-Epoxy-1H-pyrano[3,4-d]oxepin-	ether	233(4.08)	5-1413-86

Compound	Solvent	λ max (log ε)	Ref.
1-acetoxy-4a,5,6,8,9,9a-hexahydro-, methyl ester, [1S-(1α,4aα,6β,9β-9aα)]- (cont.)			5-1413-86
$C_{13}H_{17}BrN_4O_2$			
2-Pyrimidineacetic acid, 5-bromo-α-cyano-, methyl ester, ion(1-), piperidinium	EtOH	233(4.19),314(4.60)	103-0774-86
$C_{13}H_{17}ClN_2$			
Cyclohexanone, (4-chlorophenyl)methylhydrazone	$HOCH_2CH_2OH$	260(4.17),290(3.72)	103-0713-86
$C_{13}H_{17}F_3NO_3P$			
Phosphonic acid, [2,2,2-trifluoro-1-[(phenylmethyl)imino]ethyl]-, diethyl ester	hexane	246(3.72)	65-2372-86
Phosphonic acid, [2,2,2-trifluoro-1-(phenylmethylene)amino]ethyl]-, diethyl ester	hexane	251(4.27)	65-2372-86
$C_{13}H_{17}F_3O_4$			
3,5-Heptadienoic acid, 6-acetoxy-2,4-dimethyl-, 2,2,2-trifluoroethyl ester, (E,Z)-(±)-	MeOH	219.5(3.80)	33-0469-86
$C_{13}H_{17}NO$			
3-Buten-2-one, 4-[2-[(dimethylamino)-methyl]phenyl]-	n.s.g.	280(4.20)	88-2169-86
	+ HCl	270(3.00)	88-2169-86
Pyridine, 4-(cyclooctenyl)-, 1-oxide	EtOH	264(4.1),300s(3.8)	94-0007-86
$C_{13}H_{17}NOS$			
Piperidine, 1-[(phenylthio)acetyl]-	MeOH	214(4.11),250(3.70)	12-0687-86
$C_{13}H_{17}NO_2$			
1,3-Butanediol, 2-(1H-indol-3-yl)-3-methyl-, (+)- (tanakamine)	MeOH	221(4.31),277s(3.37), 281(3.56),290(3.49)	100-0901-86
2,3-Butanediol, 1-(1H-indol-3-yl)-3-methyl-, (+)- (tanakine)	MeOH	222(4.31),273s(3.70), 281(3.71),290(3.66)	100-0901-86
$C_{13}H_{17}NO_2S$			
Acetic acid, cyano(2,2,4,4-tetramethyl-3-thioxocyclobutylidene)-, ethyl ester	C_6H_{12}	229(3.37),301(2.58), 496(0.90)	151-0331-86A
Piperidine, 1-[2-(phenylsulfonyl)ethenyl]-	MeCN	225(4.00),287(4.17)	104-2085-86
$C_{13}H_{17}NO_3$			
Benzeneacetic acid, 4-(4-morpholinyl)-, methyl ester	EtOH	205(4.15),252(4.02)	104-1968-86
1,3-Dioxolo[4,5-h]isoquinoline, 6,7,8,9-tetrahydro-4-methoxy-8,9-dimethyl-, (-)-,(lophophorine)	EtOH	276(2.95),284s(2.91)	100-0745-86
(±)-	EtOH-HBr	211(4.63),250s(3.47), 278(3.03),286s(2.98)	100-0745-86
$C_{13}H_{17}NO_3S$			
Piperidine, 1-[(phenylsulfonyl)-acetyl]-	MeOH	220(4.11),257(2.78), 263(3.00)	12-0687-86
3-Thiopheneacetamide, 5-ethyl-2,5-dihydro-4-hydroxy-5-(2-methyl-1,3-	pH 2	239(4.50)	158-0026-86
	pH 10	238(4.49),303(4.12)	158-0026-86

Compound	Solvent	$\lambda_{max}(\log \epsilon)$	Ref.
butadienyl)-2-oxo- (U-68,204) (cont.)			158-0026-86
$C_{13}H_{17}NO_4$			
Butanedioic acid, dimethyl[(1-methyl-1H-pyrrol-2-yl)methylene]-, 4-methyl ester, (E)-	EtOH	298(3.60),378(4.38)	34-0369-86
Butanoic acid, 2-[[5-(dimethylamino)-2-furanyl]methylene]-3-oxo-, ethyl ester, (Z)-	MeOH	235(2.88),443(3.44)	73-0879-86
$C_{13}H_{17}NO_5S_2$			
Benzenesulfonic acid, 3-[3-(ethylamino)-3-[(2-hydroxyethyl)thio]-1-oxo-2-propenyl]-	H_2O	349(4.38)	35-5543-86
$C_{13}H_{17}NO_6$			
1,2-Cyclopentanediacetic acid, α^1-cyano-5-hydroxy-3-oxo-, α^1-ethyl α^2-methyl ester	EtOH	205(3.19),208s(3.08), 286(1.42)	104-1865-86
$C_{13}H_{17}N_3$			
2-Propenenitrile, 3-[3,6-bis(1-methylethyl)pyrazinyl]-, (E)-	EtOH	246(4.19),309(4.09), 314(4.08)	4-1481-86
(Z)-	EtOH	247(4.14),309s(4.00), 315(4.02)	4-1481-86
$C_{13}H_{17}N_3$			
1H-Pyrrol-1-amine, N,N-dimethyl-2-[2-(1-methyl-1H-pyrrol-2-yl)ethenyl]-, (E)-	EtOH	225(3.97),340.5(4.14)	118-0620-86
$C_{13}H_{17}N_3O$			
4H-Pyrido[1,2-a]pyrimidin-4-one, 2-(butylamino)-6-methyl-	EtOH	240s(4.20),257(4.31), 346(3.67)	4-1295-86
$C_{13}H_{17}N_3O_6$			
7H-Pyrrolo[3,2-d]pyrimidine, 7-C-(α-D-ribofuranosyl)-2,4-dimethoxy-	pH 1	278(4.11)	44-1058-86
	pH 11	265(3.90),283s(3.72)	44-1058-86
	MeOH	266(3.88),283s(3.71)	44-1058-86
β-	pH 1	277(4.15)	44-1058-86
	pH 11	266(3.93),283s(3.76)	44-1058-86
	MeOH	265(3.90),283s(3.74)	44-1058-86
$C_{13}H_{17}N_3O_8$			
2,4(1H,3H)-Pyrimidinedione, 1-[6-deoxy-2,3-O-(1-methylethylidene)-6-methoxy-β-D-allofuranosyl]-	n.s.g.	258(3.93)	111-0111-86
$C_{13}H_{17}N_4$			
Pyrido[2,3-b]pyrazinium, 4-methyl-6-(1-piperidinyl)-, iodide hydrated	pH 6.0	225(4.42),446(4.21)	103-1118-86
	pH 13.9	358(4.23)	103-1118-86
$C_{13}H_{17}N_4O$			
Methanaminium, N-[4-(dimethylamino)-6-phenyl-2H-1,3,5-oxadiazin-2-ylidene]-N-methyl-, perchlorate	MeCN	237(4.33),283(4.32)	97-0094-86
$C_{13}H_{17}N_5$			
Guanidine, N''-(6,6-dicyano-1,3,5-	benzene	482(4.85),506(4.75)	24-3276-86

Compound	Solvent	λ_{max}(log ε)	Ref.
hexatrienyl)-N,N,N',N'-tetramethyl-, (E,E)- (cont.)	EtOH	484(4.85),506(4.93)	24-3276-86
	acetone	486s(--),511(5.00)	24-3276-86
	$CHCl_3$	490(4.81),510(4.81)	24-3276-86
$C_{13}H_{17}N_5O$			
1H-1,2,4-Triazol-3-amine, 5-(4-morpholinyl)-1-(phenylmethyl)-	EtOH	230(3.76)	4-0401-86
	10%EtOH-HCl	239(3.89)	4-0401-86
	10%EtOH-NaOH	228(3.79)	4-0401-86
2H-1,2,4-Triazol-3-amine, 5-(4-morpholinyl)-2-(phenylmethyl)-	EtOH	215s(4.05),225s(3.86)	4-0401-86
	10%EtOH-HCl	207(4.27),223s(4.07)	4-0401-86
	10%EtOH-NaOH	216(4.24),225s(4.07)	4-0401-86
$C_{13}H_{17}N_5O_4$			
Glycine, N-methyl-N-(1,2,3,4-tetrahydro-1,3-dimethyl-2,4-dioxo-7-pteridinyl)-, ethyl ester	MeOH	222(4.46),281(4.09), 349(4.29)	108-0081-86
$C_{13}H_{17}N_5O_6$			
9H-Purine-8-carboxylic acid, 6-(dimethylamino)-9-β-D-ribofuranosyl-, monosodium salt	H_2O	238(4.27),274(3.75), 309(4.23)	94-3635-86
$C_{13}H_{17}N_5O_7$			
Carbamic acid, (6,9-dihydro-6-oxo-9-β-D-ribofuranosyl-1H-purin-2-yl)-, ethyl ester	pH 1	257(4.11)	44-1277-86
	pH 11	263(4.08)	44-1277-86
	MeOH	256(4.04)	44-1277-86
Carbamic acid, (6-methoxy-9-β-D-ribofuranosyl-9H-purin-2-yl)-, methyl ester	pH 1	270(4.04)	44-1065-86
	pH 11	266(4.08)	44-1065-86
	MeOH	267(4.15)	44-1065-86
$C_{13}H_{18}BrN_3O_5$			
Thymidine, 5'-[(bromoacetyl)methylamino]-5'-deoxy- (in methanol)	pH 1 and 7	265(4.00)	87-1052-86
	pH 13	265(3.88)	87-1052-86
Thymidine, 5'-[(2-bromo-1-oxopropyl)-amino]-5'-deoxy-	pH 1	267(3.98)	87-1052-86
	pH 7	267(4.00)	87-1052-86
	pH 13	267(3.89)	87-1052-86
Thymidine, 5'-[(3-bromo-1-oxopropyl)-amino]-5'-deoxy-	pH 1 and 7	266(4.01)	87-1052-86
	pH 13	265(3.91)	87-1052-86
$C_{13}H_{18}ClN_3O_5$			
Thymidine, 5'-[(chloroacetyl)methylamino]-5'-deoxy- (in methanol)	pH 1 and 7	265(3.98)	87-1052-86
	pH 13	266(3.86)	87-1052-86
$C_{13}H_{18}Cl_4$			
Bicyclo[2.2.2]octane, 2,3,5,6-tetrakis(chloromethyl)-7-methylene-, (1α,2α,3β,4α,5α,6β)-	EtOH	255(<2.0)(end abs.)	33-1310-86
$C_{13}H_{18}FeO_4$			
Iron, tricarbonyl(η^4-7-methyl-3-methylene-1-octen-4-ol)-	C_6H_{12}	276s(3.38)	101-0061-86D
minor isomer	C_6H_{12}	276s(3.36)	101-0061-86D
$C_{13}H_{18}N_2$			
Cyclohexanone, (3-methylphenyl)hydrazone	$HOCH_2CH_2OH$	278(4.10)	103-0713-86
Cyclohexanone, (4-methylphenyl)hydrazone	$HOCH_2CH_2OH$	278(4.02)	103-0713-86
$C_{13}H_{18}N_2O$			
Ethanone, 1-[2-(1-piperidinyl)-3H-	EtOH	230s(4.19),325.0(3.64)	18-2317-86

Compound	Solvent	$\lambda_{max}(\log \epsilon)$	Ref.
azepin-5-yl]- (cont.)			18-2317-86
Ethanone, 1-[2-(1-piperidinyl)-3H-azepin-6-yl]-	EtOH	231.5(4.13),335(4.36)	18-2317-86
$C_{13}H_{18}N_2O_2$			
2-Propenoic acid, 3-(3,6-di et hyl-pyrazinyl)-, ethyl ester, (E)-	EtOH	248(4.21),312(4.17)	4-1481-86
2H-Pyrrol-2-one, 1,5-dihydro-5-methoxy-3,4-dimethyl-5-[(1-methyl-1H-pyrrol-2-yl)methyl]-	MeOH	210(c.4.15)	49-0185-86
$C_{13}H_{18}N_2O_3$			
2,5-Cyclohexadiene-1,4-dione, 2-(1-aziridinyl)-5-[(2-methoxyethyl)-amino]-3,6-dimethyl-	EtOH	339(4.33)	94-1299-86
2-Propenoic acid, 3-(3,6-diethyl-pyrazinyl)-, ethyl ester, N-oxide, (E)-	EtOH	256(4.62),287s(4.03), 333(3.85),344s(3.79)	4-1481-86
$C_{13}H_{18}N_2O_6$			
Uridine, 2'-deoxy-5-ethyl-, 3'-acetate	EtOH	265(3.96)	83-0154-86
$C_{13}H_{18}N_4O_2S$			
Hydrazinecarbothioamide, 2-[2-(acet-oxyamino)-2-methyl-1-phenylpropyl-idene]-	EtOH	236(4.08),279(4.36)	103-0287-86
$C_{13}H_{18}N_4O_3$			
Hydrazinecarboxamide, 2-[2-(acetoxy-amino)-2-methyl-1-phenylpropyli-dene]-	EtOH	225s(4.15)	103-0287-86
3,5,7-Nonanetrione, 2,2,8,8-tetra-methyl-4,6-bis(diazo)-	CH_2Cl_2	238(4.37),290s(3.90), 381s(2.27)	89-0999-86
2,4(3H,8H)-Pteridinedione, 8-(5-hy-droxypentyl)-6,7-dimethyl-	pH 5.0	228(3.85),256(4.23), 278(4.11),313(2.74), 407(4.15)	44-2461-86
	pH 8.0	228(4.00),256(4.24), 278(4.15),313(3.30), 407(4.11)	44-2461-86
	pH 12.0	245(4.34),265s(4.04), 313(4.33),370(3.81)	44-2461-86
$C_{13}H_{18}N_4O_4$			
Quinoxaline, 1,2,3,4-tetrahydro-1-methyl-2,3-bis(1-nitroethyl)-	EtOH	222(4.54),260(3.78), 310(3.71)	103-0318-86
$C_{13}H_{18}N_4O_5S_2$			
1H-Pyrazolo[3,4-d]pyrimidine, 3-meth-oxy-4,6-bis(methylthio)-1-β-D-ribo-furanosyl-	pH 1	258(4.58),297(4.19)	4-1869-86
	pH 7	258(4.61),296(4.20)	4-1869-86
	pH 11	257(4.61),296(4.21)	4-1869-86
$C_{13}H_{18}N_4O_6S$			
4H-Pyrazolo[3,4-d]pyrimidin-4-one, 3-ethoxy-1,5-dihydro-6-(methyl-thio)-1-β-D-ribofuranosyl-	pH 1	234(4.17),275(3.92)	4-1869-86
	pH 7	234(4.20),274(3.95)	4-1869-86
	pH 11	238(4.18),261(3.97)	4-1869-86
$C_{13}H_{18}N_5O$			
Pteridinium, 8-ethyl-4-methyl-2-(4-morpholinyl)-, tetrafluoroborate hydrated	pH 3.5	249(4.36),286(4.03)	103-1118-86
	pH 9.4	234(4.30),323(4.18)	103-1118-86

Compound	Solvent	λ_{max}(log ε)	Ref.
$C_{13}H_{18}N_6O$			
1-Propanone, 1-[4-amino-2-(butyl-amino)-7-pteridinyl]-	MeOH	223(4.27),280(4.24), 406(3.77)	128-0183-86
$C_{13}H_{18}N_6O_3$			
L-Alanine, N-[1-oxo-4-(1H-purin-6-yl-amino)butyl]-, methyl ester	pH 2.85	272(4.12)	63-0757-86
	pH 7.0	268(4.14)	63-0757-86
	pH 10.5	272(4.12)	63-0757-86
$C_{13}H_{18}N_6O_4S$			
Adenosine, 8-(aminothioxomethyl)-N,N-dimethyl-	MeOH	238(4.30),274s(3.96), 283(3.97),325(4.06)	94-3635-86
$C_{13}H_{18}N_6O_5$			
Adenosine, 8-(aminocarbonyl)-N,N-di-methyl-	MeOH	233(4.32),270s(3.78), 279s(3.88),310(4.11)	94-3635-86
$C_{13}H_{18}N_6O_6$			
1,2,4-Triazine-5,6-dicarboxylic acid, 3-cyanotetrahydro-3-hydroxy-5-[(4-morpholinylimino)methyl]-, dimethyl ester	CH_2Cl_2	314(3.51)	83-0798-86
$C_{13}H_{18}OSi$			
Silane, trimethyl[(2-phenylcyclo-propyl)carbonyl]-, trans	C_6H_{12}	396(2.03)	138-0177-86
$C_{13}H_{18}O_2$			
1H-Cyclopenta[a]pentalen-1-one, 3a,3b,4,5,6,6a,7,7a-octahydro-4-hydroxy-3,7a-dimethyl-	MeOH	232(4.07)	35-3443-86
2H-Pyran-2-one, tetrahydro-6-(2-oct-ynylidene)-	H_2O	240(4.14)	35-5589-86
	base	225(3.99)	35-5589-86
$C_{13}H_{18}O_3$			
Bicyclo[3.1.0]hex-3-en-2-one, 6-acet-oxy-1,3,4,5,6-pentamethyl-, (1α,5α,6α)-(±)-	MeOH	228(3.78),268(3.54), 328(2.58)	33-0469-86
Moskachan C (α-methyl-1,3-benzodi-oxole-5-pentanol)	n.s.g.	234(3.78),286(3.76)	102-2209-86
Spiro[1,3-dioxolane-2,1'(2'H)-naph-thalen]-6'(7'H)-one, 3',4',8',8'a-tetrahydro-8'a-methyl-	MeOH	238(4.33)	44-0773-86
$C_{13}H_{18}O_5$			
3,4,5,6,7-Nonanepentone, 2,2,8,8-tetramethyl-	CH_2Cl_2	352s(2.29),436(2.00), 559(1.98)	89-0999-86
1,7-Dioxaspiro[5.5]undec-2-ene-2-carboxylic acid, 8,9-dimethyl-4-oxo-, methyl ester, [6S-(6α,8β,9α)]-	EtOH	252(4.07)	44-4840-86
$C_{13}H_{19}ClHgO_2$			
Mercury, chloro[(3,4,5,6,7,8-hexahy-dro-2,7,7-trimethyl-5-oxo-2H-1-benzopyran-2-yl)methyl]-	EtOH	262(4.2)	32-0441-86
$C_{13}H_{19}N$			
Benzenamine, N,N-dimethyl-2-(3-methyl-2-butenyl)-	C_6H_{12}	211(4.24),249(3.76)	33-0184-86

Compound	Solvent	$\lambda_{max}(\log \epsilon)$	Ref.
$C_{13}H_{19}NO$			
Pyridine, 4-(1-methylheptyl)-, didehydro deriv., 1-oxide	EtOH	264(4.1),292(3.8)	94-0007-86
$C_{13}H_{19}NO_2S$			
4-Piperidinone, 3-hydroxy-2,2,6,6-tetramethyl-3-(2-thienyl)-	EtOH	236(3.90)	70-2511-86
$C_{13}H_{19}NO_3$			
Anhalonidine, O-methyl-, (+)-	EtOH	207(4.61),228s(3.97), 273(3.09),280(3.11)	100-0745-86
2-Furancarboxylic acid, 5-methyl-4-(1-piperidinylmethyl)-, methyl ester	MeOH	266(3.21)	73-2186-86
Isoquinoline, 1,2,3,4-tetrahydro-5,6,7-trimethoxy-2-methyl- (tehaunine)	aq HCl	203(4.35),211s(3.68), 280(2.92),289s(2.61)	100-0745-86
4-Isoquinolinol, 1,2,3,4-tetrahydro-6,7-dimethoxy-1,2-dimethyl-, cis-(±)-	MeOH	228s(3.99),283(3.55), 298s(3.51)	100-0745-86
trans-(±)-	MeOH	230(3.95),282(3.53), 288s(3.48)	100-0745-86
8-Isoquinolinol, 1,2,3,4-tetrahydro-6,7-dimethoxy-1,2-dimethyl- (±-pellotine)	EtOH-HCl	230s(3.99),271(2.96), 281s(2.85)	100-0745-86
	EtOH-KOH	281(3.22),286s(3.21)	100-0745-86
$C_{13}H_{19}NO_4$			
1-Isoquinolinemethanol, 1,2,3,4-tetrahydro-5-hydroxy-6,7-dimethoxy-2-methyl- (deglucopterocereine)	aq HCl	215s(3.15),268(2.16)	100-0745-86
$C_{13}H_{19}NO_9$			
D-Fructose, 1-deoxy-1-[[(2,2-dimethyl-4,6-dioxo-1,3-dioxan-5-ylidene)methyl]amino]-	H_2O	228(4.02),280(4.37)	136-0329-86D
$C_{13}H_{19}NS_2$			
3-Thiophenecarbothioamide, N-cyclohexyl-2,5-dimethyl-	EtOH	255(4.05),281(4.01), 361(2.63)	78-3683-86
$C_{13}H_{19}NS_4$			
3,5-Pyridinedicarbothioic acid, 1,4-dihydro-2,6-dimethyl-, diethyl ester	EtOH	207(--),309(--), 500(--)	103-0401-86
	anion	705(4.44)	103-0401-86
$C_{13}H_{19}N_3O$			
4H-Furo[2',3':4,5]pyrimido[1,2-a]azocin-4-imine, 2,3,6,7,8,9,10,11-octahydro-2-methyl-	$CHCl_3$	252s(3.92),257(3.92), 312s(3.42)	24-1070-86
4H,5H-Pyrano[2',3':4,5]pyrimido[1,2-a]azocin-5-imine, 2,3,7,8,9,10,11,12-octahydro-	$CHCl_3$	248(3.87),280(3.49)	24-1070-86
$C_{13}H_{19}N_3O_3$			
2-Quinoxalinecarboxylic acid, 3-amino-4,6,7,8-tetrahydro-4-(2-methoxyethyl)-	EtOH	371(3.63)	33-1025-86
$C_{13}H_{19}N_3O_3S$			
Pyrrolidinium, 1-[(7-amino-2-carboxy-8-oxo-5-thia-1-azabicyclo[4.2.0]oct-2-en-3-yl)methyl]-1-methyl-, hydroxide, inner salt, (6R-trans)-	pH 7	267(3.95)	158-1092-86

Compound	Solvent	λ_{max}(log ϵ)	Ref.
$C_{13}H_{19}N_3O_4$			
1H-Imidazole-1-acetic acid, α-[(dimethylamino)methylene]-4-(ethoxycarbonyl)-, ethyl ester, (Z)-	MeOH	231(4.04),277(4.39)	5-1749-86
$C_{13}H_{19}N_3O_8S$			
Thymidine, 5'-deoxy-5'-[[[(methylsulfonyl)oxy]acetyl]amino]- (in methanol)	pH 1 and 7	266(4.00)	87-1052-86
	pH 13	266(3.87)	87-1052-86
$C_{13}H_{19}N_3S$			
4,13-Imino-2H-thieno[2,3-b][1,5]diazacyclotridecine, 3,6,7,8,9,10,11,12-octahydro-	$CHCl_3$	237(4.38),286(3.94)	24-1070-86
Thieno[2',3':4,5]pyrimido[1,2-a]azonin-4(2H)-imine, 3,6,7,8,9,10,11,12-octahydro-	$CHCl_3$	251(4.40),314(3.75)	24-1070-86
$C_{13}H_{19}N_5O$			
1H-1,2,4-Triazole-3,5-diamine, N-[3-(2,6-dimethylphenoxy)propyl]-	EtOH	212s(4.20)	4-0401-86
	10%EtOH-HCl	211s(4.27)	4-0401-86
$C_{13}H_{19}N_5O_3$			
6H-Purin-6-one, 2-amino-1,9-dihydro-9-[[2-(1-methylethyl)-1,3-dioxan-5-yl]methyl]-	pH 1	254(4.04),281(3.89)	87-1384-86
	pH 13	254s(3.98),268(4.01)	87-1384-86
$C_{13}H_{19}N_5O_6S$			
Adenosine, N,N-dimethyl-8-methylsulfonyl-	MeOH	226(4.23),274s(4.20), 299(4.32)	94-3635-86
$C_{13}H_{19}O_2PS_4$			
Disulfide, [bis(1-methylethoxy)phosphinothioyl](phenylthioxomethyl)	C_6H_{12}	303(4.06),524(1.96)	88-4595-86
$C_{13}H_{20}$			
1,3-Cyclohexadiene, 1,2,3,4,5,5-hexamethyl-6-methylene-	C_6H_{12}	313(3.81)	5-1098-86
$C_{13}H_{20}NO_3$			
Anhalotine (anhalidine methiodide)	HCl	272(2.96),282s(2.86)	100-0745-86
	KOH	291(3.48)	100-0745-86
	MeOH	272(2.95),282s(2.94)	100-0745-86
$C_{13}H_{20}N_2O_4S$			
4,12-Epoxy-5H,10H-1,3-dioxolo[4,5-e]-pyrimido[6,1-b][1,3]oxazocine-10-thione, 2,2,8-trimethyl-	n.s.g.	251(4.17)	103-1336-86
5-Pyrimidinecarboxylic acid, 4-(2,2-dimethyl-1,3-dioxolan-4-yl)-1,4-dihydro-6-methyl-2-(methylthio)-, methyl ester	MeOH	310(3.78)	142-0679-86
hydriodide	MeOH	220(4.14),307(3.72)	142-0679-86
$C_{13}H_{20}N_2O_9$			
2,4(1H,3H)-Pyrimidinedione, 1-β-D-glucopyranosyl-6-methoxy-5-(methoxymethyl)-	$CHCl_3$	261(3.95)	97-0437-86
$C_{13}H_{20}N_6O_7$			
Carbamic acid, [amino[[4-(aminocarbonyl)-1-β-D-ribofuranosyl-1H-imida-	pH 1	237s(4.00)	44-1277-86
	pH 11	261(3.95)	44-1277-86

Compound	Solvent	$\lambda_{max}(\log \epsilon)$	Ref.
zol-5-yl]amino]methylene]-, ethyl ester (cont.)	MeOH	263(4.00)	44-1277-86
$C_{13}H_{20}N_8$			
1-Propanone, 1-(2,4-diamino-7-pteridinyl)-, (1,1-dimethylethyl)hydrazone	pH 8	220(4.23),255(4.27), 392(4.13)	128-0183-86
$C_{13}H_{20}O$			
Cyclopentanone, 3-(1,3-octadien-1-yl)-, (Z,Z)-	EtOH	233(4.26)	39-1809-86C
Tricyclo[3.3.1.1^{3,7}]decanone, 4-(1-methylethyl)-, axial	MI	290(1.30)	35-4484-86
	EPA	292(1.30)	35-4484-86
equatorial	MI	290(1.34)	35-4484-86
	EPA	290(1.43)	35-4484-86
$C_{13}H_{20}O_2$			
Bicyclo[5.1.0]octan-2-one, 8-acetyl-3,3,7-trimethyl-	pentane	285(2.19)	33-0555-86
2-Butenoic acid, 4-[(4-methylcyclohexylidene)-, ethyl ester, [S-(E)]-	C_6H_{12}	271(4.49)	44-2863-86
[S-(Z)]-	C_6H_{12}	275.1(4.44)	44-2863-86
2,4-Tridecadiyne-1,6-diol, (R)-	EtOH	220(2.49),231(2.62), 243(2.63),257(2.41)	150-1348-86M
$C_{13}H_{20}O_3$			
1-Cyclopentene-1-heptanoic acid, 5-oxo-, methyl ester	MeOH	228(3.96)	2-0675-86
	EtOH	227(4.04)	118-0212-86
$C_{13}H_{20}O_4$			
1-Cyclopentene-1-heptanoic acid, 3-hydroxy-5-oxo-, methyl ester	MeOH	222(3.97)	2-0675-86
$C_{13}H_{20}O_6$			
3,4,6,7-Nonanetetrone, 5,5-dihydroxy-2,2,8,8-tetramethyl-	CH_2Cl_2	402(2.06)	89-0999-86
$C_{13}H_{20}O_7$			
D-arabino-Hept-2-enonic acid, 2,3-dideoxy-4,5-O-(1-methylethylidene)-, methyl ester, 7-acetate, (E)-	MeOH	225(3.91)	136-0283-86G
Heptonic acid, 3,6-anhydro-2-deoxy-4,5-O-(1-methylethylidene)-, methyl ester, acetate	MeOH	240(3.65)	136-0283-86G
$C_{13}H_{21}BrO_2$			
2,4-Hexadienoic acid, 4-bromo-, heptyl ester	n.s.g.	272(4.08)	104-2230-86
$C_{13}H_{21}ClN_2O_6Si$			
1H-Imidazole-4-carboxylic acid, 2-chloro-1-β-D-ribofuranosyl-5-(trimethylsilyl)-, methyl ester	MeOH	233(4.06)	78-1971-86
$C_{13}H_{21}NO_9$			
α-D-Fructofuranoside, methyl 1-deoxy-1-[[3-methoxy-2-(methoxycarbonyl)-3-oxo-1-propenyl]amino]-	EtOH	223(4.94),278(4.36)	136-0329-86D
$C_{13}H_{21}N_3O_4$			
5-Pyrimidinepentanoic acid, 2-amino-	50%MeOH-HCl	267(4.17)	73-1140-86

Compound	Solvent	λ max (log ε)	Ref.
1,4-dihydro-6-hydroxy-4-oxo-, butyl ester (cont.)	50%MeOH-NaOH	271(4.15)	73-1140-86
5-Pyrimidinepentanoic acid, 2-amino-1,4-dihydro-6-hydroxy-4-oxo-, 2-methylpropyl ester	50%MeOH-HCl	266(4.19)	73-1140-86
	50%MeOH-NaOH	270.5(4.15)	73-1140-86
$C_{13}H_{22}Ge$			
Germane, triethyl(phenylmethyl)-	n.s.g.	270(2.62)	18-3169-86
$C_{13}H_{22}N_2O_7$			
α-D-Allofuranose, 3-C-[amino(hydroxyimino)methyl]-1,2:5,6-bis-O-(1-methylethylidene)-, (Z)-	EtOH	204(3.70),272(3.91)	159-0631-86
α-D-Glucofuranose, 3-C-[amino(hydroxyimino)methyl]-1,2:5,6-bis-O-(1-methylethylidene)-	EtOH	205(3.82),258(2.68)	159-0631-86
$C_{13}H_{22}N_4O_3$			
5-Pyrimidinepentanamide, 2-amino-N-butyl-1,4-dihydro-6-hydroxy-4-oxo-	50%MeOH-HCl	268(4.25)	73-1140-86
	+ NaOH	270(4.18)	73-1140-86
$C_{13}H_{22}N_4O_5$			
1,2-Hydrazinedicarboxylic acid, 1-[1-(1-methyl-2-imidazolidinylidene)-2-oxopropyl]-, diethyl ester, (Z)-	EtOH	340(4.20)	24-2208-86
$C_{13}H_{22}N_4O_6$			
1,2-Hydrazinedicarboxylic acid, 1-[2-methoxy-1-(1-methyl-2-imidazolidinylidene)-2-oxoethyl]-, diethyl ester, (Z)-	EtOH	273(4.35)	24-2208-86
$C_{13}H_{22}O$			
Acetaldehyde, (2,2,4,6,6-pentamethylcyclohexylidene)-, (S)-(+)-	C_6H_{12}	242(4.06)	35-2691-86
2-Butanone, 3,3-dimethyl-1-(4-methylcyclohexylidene)-, (S)-	C_6H_{12}	236(4.15),333(1.86)	44-2863-86
Cyclohexanol, 3,3,5,5-tetramethyl-4-(2-propenylidene)-, (S)-(+)-	C_6H_{12}	241(4.31)	35-2691-86
3-Penten-2-ol, 2-methyl-5-(4-methylcyclohexylidene)-	C_6H_{12}	241.5(4.46)	44-2863-86
isomer	C_6H_{12}	244(4.36)	44-2863-86
$C_{13}H_{22}O_2$			
Acetic acid, (2,2,4,6,6-pentamethylcyclohexylidene)-, (R)-(-)-	C_6H_{12}	227(3.82)	35-2691-86
(S)-(+)-	C_6H_{12}	211s(3.70),226(3.78)	35-2691-86
2-Cyclohexene-1-methanol, 2-(3-hydroxy-1-butenyl)-1,3-dimethyl-, (±)-	EtOH	231(3.56)	163-1589-86
9-Decynoic acid, 1-methylethyl ester	MeOH	205(2.92)	18-3535-86
2,4-Tridecadienoic acid, (Z,Z)-	EtOH	256(4.15)	39-1809-86C
12-Tridecynoic acid	MeOH	206(2.93)	18-3535-86
10-Undecynoic acid, ethyl ester	MeOH	206(2.67)	18-3535-86
$C_{13}H_{22}O_3$			
Acetic acid, (4-hydroxy-2,2,6,6-tetramethylcyclohexylidene)-, methyl ester	C_6H_{12}	218(3.84)	35-2691-86

Compound	Solvent	$\lambda_{max}(\log \epsilon)$	Ref.
$C_{13}H_{22}S_2$			
1,3-Dithiane, 2-(2,6-dimethyl-1,5-heptadienyl)-, (E)-	pentane	235(3.13),244s(3.10)	33-1378-86
(Z)-	pentane	235(3.05),246(2.98)	33-1378-86
$C_{13}H_{23}N_3O_4S$			
L-Alanine, N-[N-(N-acetyl-L-alanyl)-2-methylthioalanyl]-, methyl ester	MeOH	207(3.60),266(3.70)	78-6555-86
L-Alanine, N-[2-methyl-N-[N-(1-thioxoethyl)-L-alanyl]alanyl]-, methyl ester	MeOH	255(3.06)	78-6555-86
$C_{13}H_{23}N_3O_5$			
L-Alanine, N-[N-(N-acetyl-L-alanyl)-2-methylalanyl]-, methyl ester	MeOH	201(3.00)	78-6555-86
$C_{13}H_{24}N_2O_2S$			
Carbamic acid, [1-methyl-2-(1-piperidinyl)-2-thioxoethyl]-, 1,1-dimethylethyl ester, (S)-	ether	280(4.10)	78-6555-86
$C_{13}H_{24}N_2O_5$			
L-Alanine, N-[N-(1,1-dimethylethoxy)-carbonyl]-2-methylalanyl]-, methyl ester	EtOH	201(3.02)	78-6555-86
$C_{13}H_{24}O$			
Ethanol, 2-(2,2,4,6,6-pentamethylcyclohexylidene)-, (S)-(+)-	C_6H_{12}	200(3.98)(end abs.?)	35-2691-86
$C_{13}H_{25}NO_2$			
Decanamide, N-(1-oxopropyl)-	EtOH	194.5(2.78),200.1(3.98), 201(3.93),206.0(4.04)	104-1602-86
$C_{13}H_{27}$			
Undecyl, 1,1-dimethyl- (radical)	tridecane	245(3.11)	70-1140-86
$C_{13}H_{30}O_3SSi$			
Silane, [3-[(1,1-dimethylethyl)thio]-propyl]triethoxy-	heptane	212(2.53)	67-0062-86
	MeCN	194(2.48),210(2.40)	67-0062-86
$C_{13}H_{33}GeN_3Si_2$			
3H-1,2,3,4-Triazagermole, 3-(1,1-dimethylethyl)-4,5-dihydro-4,4-dimethyl-5,5-bis(trimethylsilyl)-	C_6H_{12}	258(4.59)	24-2980-86

Compound	Solvent	$\lambda_{max}(\log \epsilon)$	Ref.
$C_{14}Cl_{10}$			
Benzene, 1,1'-(1,2-ethynediyl)-bis[2,3,4,5,6-pentachloro-	$CHCl_3$	245(4.60),254(4.55), 295s(4.33),303s(4.41), 314(4.61),320s(4.49), 335(4.60)	44-1100-86
$C_{14}Cl_{10}O_2$			
Ethanedione, bis(pentachlorophenyl)-	C_6H_{12}	212(4.00),238(4.41), 290(3.31),298(3.30), 315s(3.15)	44-1100-86
$C_{14}Cl_{10}O_4S$			
1,3,2-Dioxathiole, 4,5-bis(pentachlorophenyl)-, 2,2-dioxide	C_6H_{12}	224(4.69),243s(4.46), 284(4.00),304s(3.64)	44-1100-86
$C_{14}F_{14}N_2O$			
Diazene, bis[2,3,5,6-tetrafluoro-4-(trifluoromethyl)phenyl]-, 1-oxide	n.s.g.	234(3.83),282(3.64)	70-1391-86
$C_{14}H_5F_6N_5O_{12}S_2$			
Benzenamine, N-[2,6-dinitro-4-(trifluoromethyl)sulfonyl]phenyl]-	MeCN	240(--),315(4.36), 380(4.26)	104-1547-86
2,6-dinitro-4-[(trifluoromethyl)-sulfonyl]-	$MeCN-H_2SO_4$	260s(4.52),365(4.26)	104-1547-86
sodium salt	MeCN	240(--),315(4.48), 380(4.28)	104-1547-86
$C_{14}H_6BrN_3O_2$			
11-Aza-5H-pyrido[2,3-a]phenoxazin-5-one, 6-bromo-	MeOH	241(4.37),282(4.08), 336(4.05),447(4.07)	4-1697-86
11-Aza-5H-pyrido[3,2-a]phenoxazin-5-one, 6-bromo-	MeOH	243(4.25),283(3.91), 337(3.92),443(3.97)	4-1697-86
$C_{14}H_6ClNO_5$			
1,4-Anthracenedione, 9-chloro-10-hydroxy-5-nitro-	benzene	467(3.84)	104-0129-86
1,4-Anthracenedione, 10-chloro-9-hydroxy-5-nitro-	benzene	467(3.84)	104-0129-86
$C_{14}H_6FNO_2$			
6H-Anthra[1,9-cd]isoxazol-6-one, 5-fluoro-	EtOH	446(4.07)	104-1192-86
$C_{14}H_6F_6N_2$			
Benzo[g]quinoxaline, 2,3-bis(trifluoromethyl)-	EtOH	227(4.24),276(4.53), 365(2.99)	39-1043-86C
$C_{14}H_6F_8N_2O$			
Diazene, bis(2,3,5,6-tetrafluoro-4-methylphenyl)-, 1-oxide	n.s.g.	236(4.09),300(3.86)	70-1391-86
$C_{14}H_6F_8N_2O_3$			
Diazene, bis(2,3,5,6-tetrafluoro-4-methoxyphenyl)-, 1-oxide	n.s.g.	232(3.53),314(3.28)	70-1391-86
$C_{14}H_6N_4O_2$			
6H-Anthra[1,9-cd]isoxazol-6-one, 3-azido-	EtOH	440(4.13),463(4.25)	104-1192-86
6H-Anthra[1,9-cd]isoxazol-6-one, 5-azido-	EtOH	448(3.94),468(4.03)	104-1192-86

Compound	Solvent	$\lambda_{max}(\log \epsilon)$	Ref.
$C_{14}H_7ClN_4OS$			
5H-Naphtho[2,1-b]pyrimido[5,4-e][1,4]-thiazin-5-one, 11-amino-6-chloro-	MeOH	234(4.21),279(4.24), 329(3.98),476(3.99)	78-2771-86
$C_{14}H_7ClO_4$			
9,10-Anthracenedione, 2-chloro-1,4-dihydroxy-	EtOH	235(4.59),252(4.28), 293(3.92),480(3.98), 492(3.99),511s(3.87), 524(3.75)	150-3162-86M
$C_{14}H_7NO_2$			
6H-Anthra[1,9-cd]isoxazol-6-one	heptane	430(4.1),470(4.2)	104-2121-86
$C_{14}H_7NO_3$			
6H-Anthra[1,9-cd]isoxazol-6-one, 5-hydroxy-	heptane	440(4.0)	104-2121-86
$C_{14}H_8Cl_2OS$			
Dibenzo[b,f]thiepin-10(11H)-one, 2,6-dichloro-	MeOH	240(4.30),262s(4.62), 268(3.98),273(3.87)	73-0156-86
Dibenzo[b,f]thiepin-10(11H)-one, 2,7-dichloro-	MeOH	249(4.35),278s(3.94), 327(3.57)	73-0156-86
$C_{14}H_8Cl_2S$			
Dibenzo[b,f]thiepin, 2,6-dichloro-	MeOH	267.5(4.34),305(3.70)	73-0156-86
$C_{14}H_8Cl_3NO_3S$			
3H-Indolium, 3-[4-hydroxy-5-(trichloroacetyl)-1,3-oxathiol-2-ylidene]-1-methyl-, hydroxide, inner salt	MeCN	270(4.12),274(4.11), 319(3.98),337s(3.85), 444.6(4.22)	24-2308-86
	CH_2Cl_2	270(4.08),274.5(4.07), 319(3.99),338s(3.87), 453.6(4.17)	24-2308-86
	ether	270s(--),275(--), 464.8(--)	24-2308-86
$C_{14}H_8N_2O$			
6H-Indolo[3,2,1-de][1,5]naphthyridin-6-one (6-canthinone)	EtOH	257(4.03),266(4.00), 295(3.84),360(4.06), 378(4.01)	100-0428-86
$C_{14}H_8N_2O_2$			
[1,4]Benzoxazino[3,2-b][1,4]benzoxazine	isooctane	224(4.63),230(4.71), 265(3.62),308(3.59), 324(3.82),341(4.11), 359(4.34),380(4.42), 404(4.22)	24-3316-86
6H-Canthin-6-one, 2-hydroxy-	MeOH	256(3.43),266(3.55), 274(3.56),315(3.41), 342(3.50),358(3.73), 375(3.70)	102-1010-86
	MeOH-HCl	256(3.45),270(3.54), 342(3.52),358(3.68), 375(3.70)	102-1010-86
	MeOH-NaOH	256(3.55),266(3.61), 274(3.63),315(3.45), 342(3.52),358(3.73)	102-1010-86
6H-Canthin-6-one, 11-hydroxy-	EtOH	326(3.91),383(3.92)	100-0428-86
2H-Naphtho[1,2-b]pyran-2-one, 4-(diazomethyl)-	EtOH	224.5(4.69),262.5(4.58), 321(4.28),349(4.08)	94-0390-86

Compound	Solvent	λ_{max}(log ε)	Ref.
2H-Naphtho[2,3-b]pyran-2-one, 4-(diazomethyl)-	EtOH	227(4.75),276(3.99), 298(4.03),348(4.18)	94-0390-86
3H-Naphtho[2,1-b]pyran-3-one, 1-(diazomethyl)-	EtOH	259(4.47),296(4.21), 308.5(4.43),331(4.12), 345.5(4.21)	94-0390-86
$C_{14}H_8N_2O_4$			
Formamide, N-(4'-cyano[2,2':3',2"-terfuran]-5'-yl)-	EtOH	238(3.88),260(3.86), 315(3.96)	11-0347-86
$C_{14}H_8N_2O_5$			
Benzofuran, 5,7-dinitro-2-phenyl-	n.s.g.	203(3.65),255(3.43)	44-3453-86
$C_{14}H_8N_4$			
4-Cyclohexene-1,1,2,2-tetracarbonitrile, 3-(3-buten-1-ynyl)-	EtOH	222(4.13),233(4.07)	24-1105-86
$C_{14}H_8N_4S_2$			
[1,2,4,5]Tetrazino[3,4-b:6,1-b']bisbenzothiazole	CH_2Cl_2	230(4.5),250s(4.3), 330(4.2)	33-1521-86
$C_{14}H_8N_6O_{14}$			
Ethylene dipicrate	n.s.g.	229(4.59)	3-2366-86
$C_{14}H_8O$			
Acenaphtho[5,4-b]furan	n.s.g.	208(4.42),242(4.53), 320(3.75),336(4.00), 352(3.85),370(3.79)	4-1551-86
$C_{14}H_8O_2$			
9,10-Anthracenedione	benzene-C_6H_{12}	326.1(3.69)	60-1307-86
physically adsorbed	n.s.g.	370(3.80)	60-1307-86
$C_{14}H_8O_4$			
9,10-Anthracenedione, 1,4-dihydroxy-	dioxan	328(3.59),475(3.95)	99-0219-86
in PMMA block	solid	474(3.93)	99-0219-86
6H-Dibenzo[b,d]pyran-2-carboxaldehyde, 3-hydroxy-6-oxo-	n.s.g.	260(4.32),338(4.0)	2-0619-86
$C_{14}H_8S_2$			
1,6-Dithiapyrene	toluene	413(4.09),454(3.84)	35-3460-86
$C_{14}H_8S_{12}$			
Bis(ethylenedithio)tetrathiafulvalene	dioxan	321(4.22),346(4.06), 490(2.60)	103-1186-86
$C_{14}H_9BO_6$			
Boron, [ethanedioato(2-)-O,O'][1-(1-hydroxy-2-naphthalenyl)ethanonato-O,O']-, (T-4)-	DMF	265(4.49),281s(3.67), 297(3.69),308s(3.75), 323(3.99),377(3.44), 428(3.44)	48-0755-86
$C_{14}H_9BO_7$			
Boron, [ethanedioato(2-)-O,O'](methyl 1-hydroxy-2-naphthalenecarboxylato-$O^1,O^{2'}$)-, (T-4)-	CH_2Cl_2	354(3.61),383s(3.09)	48-0755-86
Boron, [ethanedioato(2-)-O,O'](methyl 3-hydroxy-2-naphthalenecarboxylato-$O^{2'},O^3$)-, (T-4)-	CH_2Cl_2	286(3.75),366(3.26), 398(3.59)	48-0755-86

Compound	Solvent	λmax(log ε)	Ref.
$C_{14}H_9BrCl_2O_2$			
1-Acenaphthylenol, 2-bromo-5,6-dichloro-1,2-dihydro-	isoPrOH	235(4.80),300(3.94), 313(4.08),326(3.93)	104-2329-86
$C_{14}H_9BrN_2S_2$			
2-Benzothiazolecarbothioamide, N-(4-bromophenyl)-	MeOH	230s(4.40),328(4.35), 457s(2.51)	103-1139-86
$C_{14}H_9BrO_3$			
3,6-Cycloheptadiene-1,2,5-trione, 3-[(4-bromophenyl)methyl]-	$CHCl_3$	253(4.28),258(4.30), 265(4.28),271(4.22)	18-1125-86
$C_{14}H_9ClNO$			
Cyclohept[d]isoxazol-1-ium, 3-(4-chlorophenyl)-, tetrafluoroborate	EtOH	237(4.28),278(3.83)	138-1925-86
	MeCN	230s(4.56),239(4.63), 283(3.98),303s(3.77), 355(3.28)	138-1925-86
$C_{14}H_9ClN_2O$			
Pyridine, 4-[5-(4-chlorophenyl)-2-oxazolyl]-	EtOH	322(4.57)	135-0244-86
	5% HOAc	378(4.20)	135-0244-86
$C_{14}H_9ClN_2S_2$			
2-Benzothiazolecarbothioamide, N-(4-chlorophenyl)-	MeOH	230s(4.39),326(4.33), 455s(2.48)	103-1139-86
$C_{14}H_9Cl_2N_3O_2$			
Quinoxaline, 5,7(and 6,8)-dichloro-1,2-dihydro-3-(4-nitrophenyl)-63:37 ratio of isomers	MeOH	286(4.20),344(3.88), 438(3.52)	103-0918-86
$C_{14}H_9Cl_3N_2$			
Quinoxaline, 5,7(and 6,8)-dichloro-3-(4-chlorophenyl)-1,2-dihydro-	MeOH	265(4.37),302(3.94), 410(3.59)	103-0918-86
	toluene	411(--)	103-0918-86
$C_{14}H_9Cl_3O_2$			
1-Acenaphthylenol, 2,5,6-trichloro-1,2-dihydro-, acetate. trans	isoPrOH	235(4.88),297(3.94), 310(4.08),324(3.94)	104-2329-86
$C_{14}H_9N$			
Benzo[g]cycl[3.2.2]azine	EtOH	255(4.5),285(4.4), 330f(3.9),430(3.4)	142-3071-86
Indolizino[3,4,5-ab]isoindole	EtOH	231(4.40),260(4.70), 279(4.16),291(4.16), 314(3.73),326(3.66), 424(4.02),450(4.06)	142-3071-86
$C_{14}H_9NO_2$			
9,10-Anthracenedione, 1-amino-	dioxan	305(3.70),468(3.80)	99-0219-86
9,10-Anthracenedione, 2-amino-	dioxan	325(3.80),422(3.40)	99-0219-86
in PMMA block	solid	423(3.60)	99-0219-86
[1,3]Dioxolo[4,5-j]phenanthridine (trisphaeridine)	n.s.g.	252(4.58),265s(--), 278(4.13),308(3.77), 335(3.51),350(3.45)	102-2399-86
$C_{14}H_9NO_3$			
9,10-Anthracenedione, 1-amino-4-hydroxy-	EtOH	577(4.18),617(4.19)	99-0219-86
	DMF	285(4.19),492s(2.89), 526(3.05),565s(2.99)	111-0385-86

Compound	Solvent	λ_{max}(log ϵ)	Ref.
$C_{14}H_9N_3O_4$			
1,3,4-Oxadiazole, 2-(2-furanyl)-5-[2-(3-nitrophenyl)ethenyl]-	dioxan	320(4.34)	103-1262-86
$C_{14}H_9N_3S_2$			
Benzenamine, N-3H-[1,2,4]thiadiazolo-[3,4-b]benzothiazol-3-ylidene-	$CHCl_3$	239(4.17),311(4.02)	118-1027-86
$C_{14}H_9N_5O$			
Propanedinitrile, 2,2'-[[5-(dimethyl-amino)-2,4-furandiyl]dimethyli-dyne]bis-	MeOH	257(2.87),382(2.88), 498(3.42),642(3.34)	73-0879-86
4-Pyridazinecarbonitrile, 1,6-dihydro-3-[4-(1H-imidazol-1-yl)phenyl]-6-oxo-, monoperchlorate (hydrate)	MeOH	280(4.34),360(3.20)	4-1515-86
$C_{14}H_{10}$			
Anthracene	C_6H_{12}	308(3.14)	41-0311-86
triplet	C_6H_{12}	422(4.81)	41-0311-86
Bicyclo[4.1.0]hepta-1,3,5-triene, 7-(phenylmethylene)-	C_6H_{12}	236s(3.84),242(3.89), 248(3.94),357s(3.92), 362s(3.96),370(4.04), 380s(3.89),394(3.90)	35-5949-86
	MeCN	236(3.86),246(3.85), 355s(3.83),366(3.90), 387(3.73)	35-5949-86
$C_{14}H_{10}BrClN_2$			
Quinoxaline, 3-(4-bromophenyl)-6(and 7)-chloro-1,2-dihydro-	MeOH	268(4.43),307s(--), 408(3.61)	103-0918-86
65:35 ratio of isomers	toluene	405(--)	103-0918-86
$C_{14}H_{10}BrNO_2S$			
1H-Indole, 3-bromo-1-(phenylsulfonyl)-	MeOH	252(4.08),280(3.69), 287(3.68)	44-2343-86
$C_{14}H_{10}ClFOS$			
Ethanone, 1-[5-chloro-2-[(3-fluoro-phenyl)thio]phenyl]-	MeOH	231(4.30),234.5(4.30), 265(3.89),285s(3.77), 340(3.63)	73-2598-86
Ethanone, 1-[5-chloro-2-[(4-fluoro-phenyl)thio]phenyl]-	MeOH	233.7(4.35),271(3.94), 348(3.65)	73-2598-86
$C_{14}H_{10}ClN$			
1H-Indole, 3-chloro-2-phenyl-	MeOH	246(4.32),306(4.27)	39-2305-86C
$C_{14}H_{10}ClNO$			
6H-Cyclohept[d]isoxazole, 3-(4-chloro-phenyl)-	EtOH	237(4.09),281(3.50)	138-1925-86
$C_{14}H_{10}ClNO_3S$			
2,1-Benzisothiazole, 1-(4-chlorobenz-oyl)-1,3-dihydro-, 2,2-dioxide	MeOH	209(4.26),247(4.17)	4-1645-86
$C_{14}H_{10}ClN_3O_2$			
Quinoxaline, 6(and 7)-chloro-1,2-di-hydro-3-(4-nitrophenyl)- 55:45 ratio of isomers	MeOH	286(4.29),305(4.03), 455(3.67)	103-0918-86
$C_{14}H_{10}Cl_2OS$			
Ethanone, 1-[5-chloro-2-[(2-chloro-	MeOH	234(4.38),266(3.92),	73-0156-86

Compound	Solvent	$\lambda_{max}(\log \epsilon)$	Ref.
phenyl)thio]phenyl]- (cont.)		284s(3.75),341(3.62)	73-0156-86
Ethanone, 1-[5-chloro-2-[(3-chloro-phenyl)thio]phenyl]-	MeOH	233(4.37),263(3.90), 289s(3.69),341(3.58)	73-0156-86
$C_{14}H_{10}F_6N_4S_2$			
2,4-Cyclohexadiene-1,1,3-tricarbo-nitrile, 2-amino-6,6-dimethyl-5-[(trifluoromethyl)thio]-4-[[(trifluoromethyl)thio]methyl]-	EtOH	210.5(4.15),362.3(4.51)	155-0201-86
$C_{14}H_{10}NO$			
Cyclohept[d]isoxazol-1-ium, 3-phenyl-, tetrafluoroborate	EtOH	228s(4.54),279(4.10)	138-1925-86
	MeCN	235(4.37),280(3.85), 302s(3.52),353(2.81)	138-1925-86
$C_{14}H_{10}N_2$			
Cyclopent[a]indene-1,1(2H)-dicarbo-nitrile, 3,3a-dihydro-	C_6H_{12}	221(4.06),227(4.16), 267(3.95)	88-6221-86
Cyclopent[a]indene-1,1(2H)-dicarbo-nitrile, 3,8-dihydro-	C_6H_{12}	212(4.05),218(4.04), 225(3.92),258(3.78)	88-6221-86
Propanedinitrile, (1,2,3,4-tetrahydro-1,4-methanonaphthalen-9-ylidene)-	C_6H_{12}	219(4.27),283(3.77)	88-6221-86
$C_{14}H_{10}N_2O$			
6H-Indolo[3,2,1-de][1,5]naphthyridin-6-one, 4,5-dihydro-	CH_2Cl_2	260(4.35),270s(4.25), 282(4.25),300s(3.78), 314(3.99),326(4.06)	102-1010-86
3H-Indol-3-one, 2-phenyl-, oxime	EtOH	265(4.6),335(3.6), 387(3.5)	35-4115-86
Pyridine, 4-(5-phenyl-2-oxazolyl)-	EtOH	320(4.37)	135-0244-86
	5% HOAc	368(4.31)	135-0244-86
$C_{14}H_{10}N_2O_2$			
2-Anthracenamine, 3-nitro-	EtOH	223(3.00),240s(3.20), 267(3.70),292(3.30), 336(3.10),381(2.70), 388(2.70)	78-3943-86
9,10-Anthracenedione, 1,4-diamino-	dioxan	545(4.04),585(4.02)	99-0219-86
[1,4]Benzoxazino[3,2-b][1,4]benzoxa-zine, 5a,6,11a,12-tetrahydro-	$C_6H_{11}Me$	290(3.85)	24-3316-86
Phenol, 2-[(2-benzoxazolylmethylene)-amino]-	C_6H_{12}	254(3.61),312(4.21), 325(4.22),341(4.15), 381(4.23)	24-3316-86
$C_{14}H_{10}N_2O_3$			
Acetamide, N-(3-oxo-3H-phenoxazin-2-yl)-	$CHCl_3$	403(4.24)	4-1003-86
Benzaldehyde, 3,3'-azoxybis-	EtOH	235(4.34),320(3.99)	23-2076-86
$C_{14}H_{10}N_2O_5$			
Phenol, 2,4-dinitro-6-(2-phenyleth-enyl)- (cis-trans mixture)	n.s.g.	241(4.46)	44-3453-86
[2,2':3',2"-Terfuran]-4'-carboxamide, 5'-(formylamino)-	EtOH	236(4.24),305(4.30)	11-0347-86
$C_{14}H_{10}N_2S_2$			
2-Benzothiazolecarbothioamide, N-phenyl-	MeOH	250s(3.98),323(4.29), 454s(2.42)	103-1139-86
2-Thiazolecarbothioamide, N-1-naph-thalenyl-	MeOH	289(4.20),317s(4.02)	103-1139-86

Compound	Solvent	λ_{max}(log ε)	Ref.
$C_{14}H_{10}N_4O$			
Naphtho[2,1-g]pteridin-8(5H)-one, 6,9-dihydro-	pH 13	382(3.95)	33-0793-86
$C_{14}H_{10}N_4O_2$			
Cyanamide, (5,8-dimethoxy-1,4-naphthalenediylidene)bis-, (E,E)-	MeCN	218(4.30),304s(3.97), 331(4.05),366s(3.97), 544(3.70)	5-0142-86
8-Quinolinol, 5-(4-pyridinylazo)-,	DMF	335(3.62),525(4.23)	140-0502-86
N-oxide	80% DMF-pH 5.6	335(3.59),530(4.20)	140-0502-86
in 60% DMF	pH 2.4	310(3.79),435(4.13)	140-0502-86
	pH 4.8	515(4.13)	140-0502-86
	pH 5.6	330(3.54),530(4.19)	140-0502-86
in 40% DMF	pH 5.6	330(3.57),525(4.18)	140-0502-86
in 20% dioxan	pH 1	308(3.71),415(3.89)	140-0502-86
	pH 2.4	310(3.91),420(3.92)	140-0502-86
	pH 5.1	308(3.92),455(3.99)	140-0502-86
	pH 8.7	308(3.78),545(4.11)	140-0502-86
	pH 9.9	310(3.79),550(4.25)	140-0502-86
8-Quinolinol, 7-(4-pyridinylazo)-,	pH 1	308(3.75),423(4.21)	140-0502-86
N-oxide (in 20% DMF)	pH 2.4	320(3.93),425(4.19)	140-0502-86
	pH 4.2	305(3.96),465(4.13)	140-0502-86
	pH 6.8	305(3.86),540(4.21)	140-0502-86
	pH 9.9	315(3.79),370(3.62), 555(4.46)	140-0502-86
$C_{14}H_{10}N_6O_4$			
Pyrazolo[1,5-a]pyrimidine, 4,7-dihydro-7-(1H-indol-3-yl)-3,6-dinitro-	EtOH	218(4.59),276(4.02), 387(4.16)	103-1250-86
$C_{14}H_{10}N_8O_4S_2$			
2,4(1H,3H)-Pteridinedione, 7,7'-dithiobis[1-methyl-	MeCN	254(4.23),350s(4.40), 360(4.44)	33-1095-86
2,4(1H,3H)-Pteridinedione, 7,7'-dithiobis[3-methyl-	MeCN	249(4.22),345s(4.40), 355(4.44)	33-1095-86
$C_{14}H_{10}O$			
Acenaphtho[5,4-b]furan, 2,3-dihydro-	n.s.g.	206(4.49),240(4.57), 312(3.94),326(4.09), 348(4.06),364(4.11)	4-1551-86
9(10H)-Anthracenone	EtOH	270(4.26)	104-0158-86
$C_{14}H_{10}O_2$			
9(10H)-Anthracenone, 10-hydroxy-	EtOH	276(4.19)	104-0158-86
3,6-Cycloheptadiene-1,2-dione, 5-(2,4,6-cycloheptatrien-1-ylidene)-	MeCN	236(4.30),280s(3.64), 294(3.64),343(3.86), 545(4.30)	88-5515-86
Dibenz[c,e]oxepin-6(7H)-one	MeOH	248(4.13),302(3.46)	73-2848-86
Ethanedione, diphenyl-	n.s.g.	380(1.88)	89-0999-86
1H-Phenalen-1-one, 6-methoxy-	hexane	415(4.0)	135-0967-86
	EtOH	445(4.0)	135-0967-86
$C_{14}H_{10}O_2S_3$			
Naphtho[2,3-b]thiophene-4,9-dione, 2,3-bis(methylthio)-	$CHCl_3$	274(4.31),472(3.69)	24-2859-86
$C_{14}H_{10}O_2S_4$			
Naphtho[2,3-b]thiophene-4,9-dione, 2-(methyldithio)-3-(methylthio)-	$CHCl_3$	270(4.44),449(3.64)	24-2859-86

Compound	Solvent	λmax(log ε)	Ref.
$C_{14}H_{10}O_3$			
Acenaphtho[1,2-c]furan-9(7H)-one, 6b,9a-dihydro-9a-hydroxy-	$CHCl_3$	292(4.10)	39-1965-86C
Azuleno[1,2-b]furan-4-carboxylic acid, methyl ester	MeOH	216(4.37),265(4.10), 315(4.84),327(4.88), 368(3.90),385(4.07), 403(3.70),577(2.58), 609(2.50),629(2.51), 667(2.17),693(2.03)	138-1021-86
3,6-Cycloheptadiene-1,2,5-trione, 3-(phenylmethyl)-	$CHCl_3$	252(4.10),272(3.88)	18-1125-86
2H-Naphth[1,8-bc]oxepin-2-one, 7-methoxy-	$CHCl_3$	263(5.17),322(4.92), 338(4.98),377(4.95)	104-1342-86
$C_{14}H_{10}O_3S_2$			
[2,2'-Bithiophene]-5-acetic acid, α-oxo-5'-(1-propynyl)-, methyl ester	EtOH	225(3.87),280(3.89), 402(4.42)	163-0565-86
$C_{14}H_{10}O_4$			
1,4,9,10-Anthracenetetrol	11.1% EtOH	415(3.9)	96-0429-86
	base	535(2.7)	96-0429-86
1,3-Benzenediol, 4-(6-hydroxy-2-benzofuranyl)-	MeOH	280s(3.86),320(4.49), 335(4.47)	2-0481-86
$C_{14}H_{10}S$			
Benzo[b]thiophene, 2-phenyl-	hexane	318s(4.17)	142-0355-86
Benzo[b]thiophene, 3-phenyl-	hexane	293(3.82),302(3.82)	142-0355-86
9-Phenanthrenethiol	MeOH	223(4.33),251(4.58), 258.5(4.58),300(4.00), 312s(3.93)	73-2848-86
Phenanthro[9,10-b]thiirene, 1a,9b-dihydro-	MeOH	252(4.37),257(4.36), 305(3.69),316(3.58)	73-2848-86
$C_{14}H_{10}S_2$			
Thiophene, 2,2'-(1,3-phenylene)bis-	MeOH	284(4.43)	149-0441-86B
$C_{14}H_{11}BO_4$			
Boron, [1,2-benzenediolato(2-)-O,O']-[1-(2-hydroxyphenyl)ethanonato-O,O']-	CH_2Cl_2	274(3.20),333(4.37), 405s(2.36)	48-0755-86
$C_{14}H_{11}BrN_2$			
Quinoxaline, 3-(4-bromophenyl)-1,2-dihydro-	MeOH	270(4.42),296(4.01), 405(3.69)	103-0918-86
	toluene	400(--)	103-0918-86
$C_{14}H_{11}BrO$			
Ethanone, 2-bromo-1-(1,2-dihydro-5-acenaphthylenyl)-	$CHCl_3$	258(4.351),352(3.979)	65-0301-86
Phenol, 2-[2-(4-bromophenyl)ethenyl]-, (E)-	EtOH	326(4.39),333(4.38)	150-0433-86S 150-3514-86M
$C_{14}H_{11}BrO_3$			
2,4,6-Cycloheptatrien-1-one, 3-[(4-bromophenyl)methyl]-2,5-dihydroxy-	MeOH	230s(4.38),237(4.38), 336(3.91),379(3.86), 394(3.87)	18-1125-86
$C_{14}H_{11}Br_3$			
5H-Benzocycloheptene, 8-bromo-5-(2,2-dibromoethenyl)-5-methyl-	EtOH	205(4.40),283(3.91)	39-1479-86C

Compound	Solvent	λ_{max}(log ϵ)	Ref.
Bicyclo[4.1.1]octa-2,4-diene, 4,7,8-tribromo-2-phenyl-, (7-exo,8-syn)	hexane	222(3.94),228(3.96), 237s(3.91),316(3.91)	78-1585-86
$C_{14}H_{11}ClN_2$			
Benzeneacetonitrile, 4-chloro-α-[(1-methyl-1H-pyrrol-2-yl)methylene]-, (Z)-	EtOH	244(4.08),280s(3.45), 375(4.55)	4-1747-86
Quinoxaline, 3-(4-chlorophenyl)-1,2-dihydro-	MeOH	265(4.38),294s(--), 405(3.65)	103-0918-86
	toluene	400(--)	103-0918-86
$C_{14}H_{11}ClN_2O_2$			
Imidazo[1,2-a]pyridin-5(1H)-one, 8-(4-chlorobenzoyl)-2,3-dihydro-	EtOH	244(4.21),310(3.95), 341(4.24)	142-2247-86
$C_{14}H_{11}ClO$			
Phenol, 2-[2-(4-chlorophenyl)ethenyl]-, (E)-	EtOH	323(4.40),332(4.39)	150-3514-86M
$C_{14}H_{11}FO$			
Phenol, 2-[2-(4-fluorophenyl)ethenyl]-, (E)-	EtOH	320(4.29),328(4.29)	150-3514-86M
$C_{14}H_{11}IO$			
Phenol, 2-[2-(4-iodophenyl)ethenyl]-, (E)-	EtOH	328(4.42),336(4.42)	150-0433-86S +150-3514-86M
$C_{14}H_{11}NO$			
6H-Cyclohept[d]isoxazole, 3-phenyl-	EtOH	226(4.24),282(3.63)	138-1925-86
2-Propenenitrile, 2-methyl-3-(5-phenyl-2-furanyl)-, (E)-	EtOH	230(4.46),351(4.80)	103-0943-86
$C_{14}H_{11}NOS_2$			
Ethanone, 1-(2-phenyl-2λ^4-thieno[3,2-c][1,2]thiazin-3-yl)-	EtOH	221(4.20),332(3.96), 478(3.66)	39-0483-86C
$C_{14}H_{11}NOSe$			
1,2-Benzisoselenazol-3(2H)-one, 2-(4-methylphenyl)-	MeOH	265(3.61),325(3.25)	18-2179-86
$C_{14}H_{11}NO_2$			
4H-Furo[3,2-b]pyrrole, 4-acetyl-2-phenyl-	MeOH	326(3.22)	73-1455-86
4H-Furo[3,2-b]pyrrole-5-carboxaldehyde, 2-(4-methylphenyl)-	MeOH	361(3.75),371(3.76)	73-0106-86
5H-Indeno[1,2-b]pyridin-5-one, 8-methoxy-1-methyl-	EtOH	252(4.22),280(3.79), 292(3.75),312s(3.43)	102-1691-86
	EtOH-HCl	250(4.15),282s(3.77), 297(3.77)	102-1691-86
$C_{14}H_{11}NO_2Se$			
1,2-Benzisoselenazol-3(2H)-one, 2-(4-methoxyphenyl)-	MeOH	265(4.57),325(4.19)	18-2179-86
$C_{14}H_{11}NO_3$			
9(10H)-Acridinone, 1,3-dihydroxy-10-methyl-	EtOH	242s(4.89),249(5.02), 255(5.11),262(5.19), 268(5.14),295(4.18), 318(3.87),394(3.81)	64-0187-86C
Phenol, 2-[2-(4-nitrophenyl)ethenyl]-, (E)-	EtOH	371(4.38)	150-0433-86S

Compound	Solvent	$\lambda_{max}(\log \epsilon)$	Ref.
$C_{14}H_{11}NO_3S$			
Thiopyrano[2,3-b]indole-2-carboxylic acid, 4,9-dihydro-9-methyl-4-oxo-, methyl ester	dioxan	210(4.07),236(4.49), 257(4.23),277(4.28), 360(3.81)	83-0332-86
Thiopyrano[3,2-b]indole-2-carboxylic acid, 4,5-dihydro-5-methyl-4-oxo-, methyl ester	dioxan	231(4.49),276(4.12), 287(4.18),321(4.18), 392(3.78)	83-0332-86
$C_{14}H_{11}N_3O$			
Formamide, N-(3-cyano-6-methyl-4-phenyl-2-pyridinyl)-	EtOH	217(4.26),249(4.44), 307(3.69)	11-0347-86
$C_{14}H_{11}N_3OS$			
[3,4'-Bipyridine]-3'-carbonitrile, 5'-acetyl-1',2'-dihydro-6'-methyl-2'-thioxo-	DMF	318(4.40),415(3.85)	104-1762-86
$C_{14}H_{11}N_3O_2$			
2-Propenenitrile, 3-(1-methyl-1H-pyrrol-2-yl)-2-(4-nitrophenyl)-, (Z)-	EtOH	253(3.75),325(3.59), 416(4.34)	4-1747-86
Quinoxaline, 1,2-dihydro-3-(4-nitrophenyl)-	MeOH	282(4.23),304(4.05), 448(3.69)	103-0918-86
$C_{14}H_{11}N_3O_3$			
Ethanone, 1-[5-(2H-benzotriazin-2-yl)-2,4-dihydroxyphenyl]-	$CHCl_3$	243(4.13),283(4.45), 315(4.42)	49-0805-86
$C_{14}H_{11}O_2$			
Cycloheptatrienylium, (4-hydroxy-5-oxo-1,3,6-cycloheptatrien-1-yl)-	MeCN	282(4.02),350s(3.92), 366(3.96),440(4.11)	88-5515-86
$C_{14}H_{12}$			
9aH-Benz[cd]azulene, 9a-methyl-	hexane	211(4.12),257(4.34), 265(4.38),375(3.61)	89-0632-86
9bH-Benz[cd]azulene, 9b-methyl-	hexane	262(--),286(--), 351(--),368(--), 387s(--),491s(--), 559s(--),567(--)	89-0632-86
Benzene, 1,1'-(1,2-ethanediyl)-, trans	$C_6H_{11}Me$	294(4.42),307(4.38), 320s(4.15)	35-0841-86
radical anion	THF	500(4.79)	46-1921-86
5H-Benzocycloheptene, 5-ethynyl-5-methyl-	EtOH	203(4.29),279(3.87)	39-1479-86C
5H-Benzocycloheptene, 5-ethynyl-9-methyl-	EtOH	203(4.13),266(3.59)	39-1479-86C
Cyclohepta[de]naphthalene, 3,10a-dihydro-	C_6H_{12}	208.5(4.33),282.5(3.90), 374.5(2.15)	35-3739-86
$C_{14}H_{12}BrCl_3O_8$			
Propanedioic acid, [[3-bromo-6-(methoxycarbonyl)-2-oxo-2H-pyran-4-yl]-methyl]-, methyl 2,2,2-trichloroethyl ester	MeOH	314(4.12)	5-1968-86
$C_{14}H_{12}BrF_3N_2O_3$			
DL-Tryptophan, 2-bromo-1-methyl-N-(trifluoroacetyl)-, sodium salt	H_2O	223(4.52),283(3.89), 291(3.86)	35-2023-86
L-Tryptophan, 2-bromo-N-(trifluoroacetyl)-, methyl ester	EtOH	219(4.61),282(3.98), 289(3.91)	35-2023-86

Compound	Solvent	λ_{max}(log ϵ)	Ref.
$C_{14}H_{12}BrN_5O_4$			
Benzenamine, 4-[(2-bromo-4,6-dinitrophenyl)azo]-N,N-dimethyl-	$CHCl_3$	291(3.95),367(3.69), 538(4.39)	48-0497-86
$C_{14}H_{12}Br_2$			
5H-Benzocycloheptene, 5-(2,2-dibromoethenyl)-5-methyl-	EtOH	205(4.42),276(3.82)	39-1479-86C
$C_{14}H_{12}Br_2N_4O_2$			
Isoxazolo[4,5-b]pyridine, 5-(6,7-dibromo-6,7-dihydro-3aH-[1,2,5]oxadiazolo[2,3-a]pyridin-3a-yl)-3-methyl-	EtOH	293(4.38)	94-4984-86
$C_{14}H_{12}Br_4N_4O_2$			
Isoxazolo[4,5-b]pyridine, 3-methyl-5-(4,5,6,7-tetrabromo-4,5,6,7-tetrahydro-3-methyl-3aH-[1,2,5]-oxadiazolo[2,3-a]pyridin-3a-yl)-	EtOH	230(4.52),294(4.35)	94-4984-86
$C_{14}H_{12}ClF_3N_2O_3$			
L-Tryptophan, 2-chloro-N-(trifluoroacetyl)-, methyl ester	EtOH	218(4.64),272(4.00), 289(3.89)	35-2023-86
$C_{14}H_{12}Cl_3N_3O_8$			
Propanedioic acid, [[3-azido-6-(methoxycarbonyl)-2-oxo-2H-pyran-4-yl]-methyl]-, methyl 2,2,2-trichloroethyl ester	MeOH	224(3.93),332(4.19)	5-1968-86
$C_{14}H_{12}FeN_2O_7$			
Iron, tricarbonyl[dimethyl 2-[(4,5,6,7-η)-1H-1,2-diazepin-1-yl]-2-butenedioate]-, cis	EtOH	305(4.11),345(4.08)	39-0595-86C
trans	EtOH	242s(4.15),281s(3.82), 370(3.84)	39-0595-86C
$C_{14}H_{12}FeO_4$			
Iron, tricarbonyl[α-(η^4-1-methylene-2-propenyl)benzenemethanol]-	C_6H_{12}	275s(3.58)	101-0061-86D
more polar isomer	C_6H_{12}	278s(3.57)	101-0061-86D
$C_{14}H_{12}N_2$			
1H-Benzimidazole, 1-methyl-2-phenyl-	C_6H_{12}	211(4.39),232(4.01), 292(4.20)	39-1917-86B
	pH 2	197(4.59),237(4.25), 285(4.27)	39-1917-86B
	pH 8	203(4.58),233(4.11), 287(4.21)	39-1917-86B
	MeOH	209(4.43),233(4.07), 288(4.19)	39-1917-86B
	ether	218(4.31),232(4.04), 291(4.27)	39-1917-86B
	MeCN	208(4.43),234(4.09), 290(4.27)	39-1917-86B
Imidazo[1,5-a]pyridine, 3-(2-methylphenyl)-	MeOH	213(4.6),285(3.9)	83-0043-86
1H-Indol-3-amine, 2-phenyl-	EtOH	247(4.3),312(4.1), 334(4.1)	35-4115-86
8H-Pyrrolo[1,2-a]perimidine, 9,10-dihydro-	MeOH	235(4.51),317s(4.01), 328(4.19),346s(3.95)	48-0906-86

Compound	Solvent	λ_{max}(log ϵ)	Ref.
Quinoxaline, 1,2-dihydro-3-phenyl-	MeOH	258(4.30),292s(--), 392(3.56)	103-0918-86
	toluene	392(--)	103-0918-86
$C_{14}H_{12}N_2O$			
Benzeneacetonitrile, 4-methoxy-α-(1H-pyrrol-2-ylmethylene)-, (Z)-	EtOH	242(4.03),307s(3.66), 369(4.48)	4-1747-86
Imidazo[1,2-a]pyridine, 3-methoxy-2-phenyl-	MeOH	207(4.36),244(4.47), 331(3.91)	88-1627-86
	MeOH-HCl	208(4.37),224(4.25), 240(4.30),311(4.16)	88-1627-86
8H-Pyrrolo[1,2-a]perimidin-8-ol, 9,10-dihydro-	$CHCl_3$	236(4.58),318s(4.13), 335(4.30),350s(4.07)	48-0906-86
$C_{14}H_{12}N_2O_2$			
Benzaldehyde, 2-hydroxy-5-methyl-3-[(2-pyridinylimino)methyl]-	$CHCl_3$	488(2.90)	2-0127-86A
Benzenamine, 2-methyl-N-[(2-nitrophenyl)methylene]-	THF	285(4.15),359(3.95)	56-0831-86
Benzenamine, 3-methyl-N-[(2-nitrophenyl)methylene]-	THF	294(4.16),351(4.06)	56-0831-86
Benzenamine, 4-methyl-N-[(4-nitrophenyl)methylene]-	MeOH	238(4.27),267(4.02), 355(3.92)	165-0017-86
5H-[1]Benzopyrano[2,3-b]pyridin-5-one, 2-(dimethylamino)-	MeOH	283(3.54),354(4.06)	83-0347-86
Benzo[h]quinazoline-2,4(1H,3H)-dione, 1,3-dimethyl-	$CHCl_3$	288(--),298(--), 310(c.4.0),360(--)	151-0369-86A
Imidazo[1,2-a]pyridin-5(1H)-one, 8-benzoyl-2,3-dihydro-	EtOH	237(4.09),310(3.93), 349(4.31)	142-2247-86
1H-Perimidine-2-carboxylic acid, ethyl ester	DMF	338s(4.06),349(4.09), 465(2.97)	48-0812-86
Phenazine, 8-methoxy-1-methyl-, 5-oxide	MeOH	384(4.13),422(3.82), 440(3.73)	104-1407-86
$C_{14}H_{12}N_2O_2S$			
2H-1,2,4-Benzothiadiazine, 3-methyl-2-phenyl-, 1,1-dioxide	MeOH	260(4.15),270s(4.08), 297s(3.94)	44-1967-86
5H-Dibenzo[b,g][1,4,6]thiadiazocine, 6-methyl-, 12,12-dioxide	EtOH	254(4.23),300(3.83), 350(2.40)	44-1967-86
7H-Thiazolo[3,2-a]pyrimidinium, 3-hydroxy-5,8-dimethyl-7-oxo-2-phenyl-, hydroxide, inner salt	MeCN	226(4.09),283(4.15), 370(3.95),440s(3.59)	2-0275-86
	benzene	465(--)	2-0275-86
$C_{14}H_{12}N_2O_3$			
Benzenamine, 4-methoxy-N-[(4-nitrophenyl)methylene]-	MeOH	240(4.22),263(4.10), 374(3.98)	165-0017-86
$C_{14}H_{12}N_2O_4S$			
1H-Indole, 3-[(3-methyl-4-nitro-2(5H)-thienylidene)methyl]-, (Z)-	$CHCl_3$	280(3.663),470(4.318)	104-0379-86
$C_{14}H_{12}N_2O_9$			
2,3-Benzofurandicarboxylic acid, 5,7-dinitro-, diethyl ester	n.s.g.	208(4.34),254(4.48)	44-3453-86
$C_{14}H_{12}N_4$			
Benzo[f]quinoxaline-3-carbonitrile, 1,2,5,6-tetrahydro-2-imino-1-methyl-	EtOH	478(4.07)	33-1025-86
Imidazo[1,2-a]pyrazine-8-carbonitrile, 2,3-dihydro-6-methyl-5-phenyl-	pH 1	418(3.91)	33-1025-86
	EtOH	495(3.50)	33-1025-86

Compound	Solvent	λ_{max}(log ε)	Ref.
Pyrido[3,4-d]pyrimidin-4-amine, 2-methyl-N-phenyl-	pH 1	285(3.78),340(4.06)	163-2243-86
	pH 7	289(3.96),344(3.99)	163-2243-86
	pH 12	289(3.96),344(3.99)	163-2243-86
7H-[1,2,4]Triazolo[4,3-a]perimidine, 10-ethyl-	dioxan	230(4.28),252(4.01), 285(3.79),296(3.80), 348(3.90),380(2.75)	48-0237-86
$C_{14}H_{12}N_4O$			
1,2,4-Benzotriazin-3(2H)-one, 6-amino-2-(4-methylphenyl)-	EtOH	237(4.47),262(4.34), 283(4.25)	97-0134B-86
3(2H)-Pyridazinone, 6-(1-methyl-2-phenyl-1H-imidazol-4-yl)-	MeOH	277(4.47)	87-0261-86
3(2H)-Pyridazinone, 6-(1-methyl-2-phenyl-1H-imidazol-5-yl)-	MeOH	269(4.44)	87-0261-86
$C_{14}H_{12}N_4O_2$			
1,2,4-Benzotriazin-3(2H)-one, 6-amino-2-(4-methoxyphenyl)-	EtOH	236(4.51),267(4.40), 408(4.21)	97-0134B-86
Isoxazolo[4,5-b]pyridine, 3-methyl-5-(3-methyl-3aH-[1,2,5]oxadiazolo-[2,3-a]pyridin-3a-yl)-	EtOH	336(4.92)	94-4984-86
Pyrazino[2,3-g]quinoxaline-5,10-dione, 2,3,7,8-tetramethyl-	MeCN	226s(4.38),228(4.38), 281(4.31),303(4.36)	138-0715-86
$C_{14}H_{12}N_4O_2S$			
Pyrimido[4,5-d]pyrimidine-2,4(1H,3H)-dione, 5,6-dihydro-1,3-dimethyl-6-phenyl-5-thioxo-	EtOH	222(4.19),263(4.20), 318(4.19)	142-2293-86
$C_{14}H_{12}N_4O_3$			
Pyrimido[4,5-d]pyrimidine-2,4,5(1H,3H-6H)-trione, 1,3-dimethyl-6-phenyl-	EtOH	234(4.53),274(3.95)	142-2293-86
Pyrimido[4,5-d]pyrimidine-2,4,5(1H,3H-6H)-trione, 1,3-dimethyl-7-phenyl-	EtOH	238(4.26),278(3.92), 314(3.93)	142-2293-86
$C_{14}H_{12}N_4O_4$			
Pyrimido[4,5-d]pyrimidine-2,4,5,7-(1H,3H,6H,8H)-tetrone, 1,3-dimethyl-6-phenyl-	EtOH	228(4.58),276(4.18)	142-2293-86
$C_{14}H_{12}N_4O_5S$			
Carbamimidothioic acid, (phenylmethyl)-, 2-hydroxy-3,5-dinitrophenyl ester	EtOH	358(4.23)	5-0947-86
$C_{14}H_{12}N_4O_8$			
Benzene, 1-[(5-methyl-2,4,6-trinitro-2,4-cyclohexadien-1-yl)methyl]-2-nitro-	EtOH	463(4.52),520s(4.26)	109-1028-86
$C_{14}H_{12}N_6O_2$			
1,2,4-Triazolo[1,5-a]pyrimidine, 1,7-dihydro-7-(1H-indol-3-yl)-2-methyl-6-nitro-	EtOH	218(4.03),273(3.83), 366(4.00)	103-1250-86
isomer	EtOH	219(3.06),276(3.97), 369(3.91)	103-1250-86
1,2,4-Triazolo[1,5-a]pyrimidine, 4,7-dihydro-7-(2-methyl-1H-indol-3-yl)-6-nitro-	EtOH	223(3.37),278(3.20), 364(3.95)	103-1250-86

Compound	Solvent	$\lambda_{max}(\log \epsilon)$	Ref.
$C_{14}H_{12}O$			
Acenaphtho[5,4-b]furan, 4,5,7,8-tetra-hydro-	n.s.g.	240(4.64),274(3.56), 284(3.62),296(3.58), 324(3.18),338(3.38)	4-1551-86
Benzaldehyde, 2-(phenylmethyl)-	n.s.g.	207(4.48),249(4.04), 293(3.35)	33-0560-86
6,11-Epoxybenzocyclodecene, 5,6-di-hydro-	C_6H_{12}	206s(4.16),241.5(4.27), 329(3.79)	35-3731-86
Phenol, 2-(2-phenylethenyl)-, (E)-	EtOH	320(4.28),329(4.28)	150-3514-86M
Tetracyclo[4.4.4.0^{1,6}.0^{5,8}]tetradeca-2,9,11,13-tetraen-4-one	C_6H_{12}	221s(3.80),260(3.62), 358(1.76)	35-3731-86
$C_{14}H_{12}OS_2$			
Benzenecarbodithioic acid, 4-methoxy-, phenyl ester	CH_2Cl_2	361(4.23),511(2.27)	18-1403-86
$C_{14}H_{12}O_2$			
1,4-Anthracenediol, 9,10-dihydro-	MeOH	217(3.87),290(3.65)	5-0839-86
3,8-Heptalenedione, 2,4-dimethyl-	n.s.g.	267(--),276(--), 295(--),368(4.01)	131-0055-86C
1,4-Naphthalenedione, 2-methyl-3-(1-methylethenyl)-	EtOH	245s(4.26),251(4.28), 259(4.22),329(3.41)	39-2217-86C
1(4H)-Naphthalenone, 4-(2-butenyli-dene)-2-hydroxy-	$CHCl_3$	310(4.12),406(3.53)	39-1627-86C
Phenol, 2-[2-(2-hydroxyphenyl)ethen-yl]-, (E)-	EtOH	332(4.32)	150-3514-86M
Phenol, 2-[2-(4-hydroxyphenyl)ethen-yl]-, (E)-	EtOH	324(4.29),333(4.29)	150-0433-86S +150-3514-86M
2H-Pyran-2-one, tetrahydro-6-(3-phen-yl-2-propynylidene)-, (E)-	H_2O	281(4.35)	35-5589-86
	base	240(4.31)	35-5589-86
$C_{14}H_{12}O_3$			
2,4,6-Cycloheptatrien-1-one, 2,5-di-hydroxy-3-(phenylmethyl)-	MeOH	239(4.33),270s(3.59), 342(3.94),380s(3.86), 393(3.89),422s(3.42)	18-1125-86
7H-Furo[3,2-g][1]benzopyran-7-one, 2,5,9-trimethyl- (trioxsalen)	1% EtOH	342.5(3.51)	149-0391-86A
	1% DMSO	342.5(3.82)	149-0391-86A
Naphth[2,3-b]oxirene-2,7-dione, 1a,7a-dihydro-1a-methyl-7a-(1-methyleth-enyl)-	EtOH	230(4.44),303(3.30), 337(2.47)	39-2217-86C
$C_{14}H_{12}O_3S_2$			
[2,2'-Bithiophene]-5-acetic acid, α-hydroxy-5'-(1-propynyl)-, methyl ester	EtOH	241(3.84),330(4.30), 337(4.30)	163-0263-86
$C_{14}H_{12}O_4$			
2,4,6-Cycloheptatrien-1-one, 2,5-di-hydroxy-3-[(4-hydroxyphenyl)methyl]-	MeOH	229(4.32),239(4.32), 280s(3.59),344(3.93), 395(3.83),422s(3.49)	18-1125-86
Naphtho[1,2-b]furan-4,5-dione, 2,3-dihydro-3-(hydroxymethyl)-9-methyl-	EtOH	265(4.05),338(3.50), 436(3.40)	88-3935-86
Naphtho[2,3-b]furan-4,9-dione, 2,3-dihydro-3-(hydroxymethyl)-5-methyl-	EtOH	253(4.16),289(4.01), 355(3.43)	88-3935-86
Spiro[furan-3(2H),5'(3'H)-isobenzo-furan]-1',2(3'H)-dione, 4,5,6',7'-tetrahydro-4,6',7'-tris(methylene)-	EtOH	210(4.00)	33-1734-86
Spiro[furan-3,5'-isobenzofuran]-1'(3H),2(3'H)-dione, 4,4',5,5',6',7'-hexahydro-4-methylene-4-ethenylidene-	EtOH	255(4.18)	33-1734-86

Compound	Solvent	$\lambda_{max}(\log \epsilon)$	Ref.
$C_{14}H_{12}O_5$			
Cycloprop[a]inden-6(6aH)-one, 1,6a-diacetoxy-1,1a-dihydro-, (1α,4aβ,6aβ)-	MeOH	256(3.87),297(3.28), 327(2.95)	39-1789-86C
1(2H)-Naphthalenone, 2,2-diacetoxy-	MeOH	226(4.18),249s(3.90)	12-2067-86
$C_{14}H_{12}O_6$			
5H-Furo[3,2-g][1]benzopyran-2,5(3H)-dione, 4,9-dimethoxy-7-methyl-	$CHCl_3$	249(4.30),304(3.73)	44-3116-86
$C_{14}H_{12}S$			
Acenaphtho[5,4-b]thiophene, 4,5,7,8-tetrahydro-	n.s.g.	220(4.78),234(4.72), 260(5.04),284(4.23), 294(4.35),306(4.25)	4-1551-86
$C_{14}H_{12}S_8$			
1,3-Dithiolo[4,5-b][1,4]dithiin, 2-(5,6-dimethyl-1,3-dithiolo[4,5-b]-[1,4]dithiin-2-ylidene)-5,6-dimethyl-	dioxan	312(4.12),344(4.15), 490(2.63)	103-1186-86
	THF	314(3.50),344(3.52)	138-0781-86
$C_{14}H_{13}BrO$			
2-Butanone, 3-(3-bromo-1-naphthalenyl)-	EtOH	228(4.68),272s(3.67), 281(3.70)	39-1479-86C
Ethanone, 1-(8-bromo-5-methyl-5H-benzocyclohepten-5-yl)-	EtOH	202(4.33),279(3.94)	39-1479-86C
$C_{14}H_{13}BrO_2$			
5H-Benzocycloheptene-5-carboxylic acid, 8-bromo-5-methyl-, methyl ester	EtOH	204(4.31),236s(3.90), 275(3.91)	39-1479-86C
$C_{14}H_{13}BrO_3$			
5H-Benzocycloheptene-5-carboxylic acid, 8-bromo-6,9-dihydro-5-methyl-9-oxo-, methyl ester	EtOH	207(3.64),259(3.69)	39-1479-86C
$C_{14}H_{13}ClN_2$			
Pyrazine, 5-chloro-3,6-dimethyl-2-(2-phenylethenyl)-, (E)-	EtOH	229(3.96),232(3.94), 282(4.23),343(4.30)	4-1481-86
$C_{14}H_{13}ClN_4$			
1H-Imidazo[1',2':1,6]pyrimido[5,4-d]-[1]benzazepine, 8-chloro-2,4,5,6-tetrahydro-	EtOH	245(4.57),268s(--), 274(4.22),289s(--), 325s(--),338s(--), 357(3.78),376s(--)	142-0143-86
$C_{14}H_{13}ClO_3$			
2-Naphthalenecarboxylic acid, 7-chloro-3,4-dihydro-3,3-dimethyl-4-oxo-, methyl ester	EtOH	244(4.75),325(3.37)	34-0369-86
2-Naphthalenecarboxylic acid, 8-chloro-3,4-dihydro-3,3-dimethyl-4-oxo-, methyl ester	EtOH	241(4.20),247(4.27), 300(3.23)	34-0369-86
$C_{14}H_{13}Cl_6NO_6S_2$			
4,5-Dithia-1-azabicyclo[4.2.0]oct-2-ene-2-carboxylic acid, 3-methyl-8-oxo-7-[1-[[(2,2,2-trichloroethoxy)-carbonyl]oxy]ethyl]-, 2,2,2-trichloroethyl ester, [6R-[6α,7α(R*)]]-	EtOH	281(3.80),330(3.59)	44-3413-86

Compound	Solvent	$\lambda_{max}(\log \epsilon)$	Ref.
$C_{14}H_{13}Cl_6NO_7S_2$			
4,5-Dithia-1-azabicyclo[4.2.0]oct-2-ene-2-carboxylic acid, 3-methyl-8-oxo-7-[1-[[(2,2,2-trichloroethoxy)-carbonyl]oxy]ethyl]-, 2,2,2-trichloroethyl ester, 5-oxide, [6R-[6α,7α(R*)]]-	EtOH	299(3.76)	44-3413-86
$C_{14}H_{13}Cl_6NO_8S_2$			
4,5-Dithia-1-azabicyclo[4.2.0]oct-2-ene-2-carboxylic acid, 3-methyl-8-oxo-7-[1-[[(2,2,2-trichloroethoxy)-carbonyl]oxy]ethyl]-, 2,2,2-trichloroethyl ester, 5,5-dioxide, [6R-[6α,7α(R*)]]-	EtOH	284(3.63)	44-3413-86
$C_{14}H_{13}FN_4S_2$			
9H-Purine, 9-[(2-fluorophenyl)methyl]-2,6-bis(methylthio)- (spectra in 9.5% ethanol)	0.1M HCl	262(4.30),312(4.04)	4-1189-86
	pH 7	261(4.32),308(4.10)	4-1189-86
	0.1M NaOH	260(4.32),309(4.10)	4-1189-86
$C_{14}H_{13}IN_2$			
Imidazo[1,5-a]pyridinium, 2-methyl-3-phenyl-, iodide	MeOH	227(4.1),301(4.0)	83-0043-86
$C_{14}H_{13}IN_4O_5S$			
Iodine, (2-hydroxy-3,5-dinitrophenyl)-(methylthioureato-S)phenyl-	EtOH	359(4.17)	5-0947-86
$C_{14}H_{13}N$			
Benzenamine, N-[(4-methylphenyl)methylene]-	C_6H_{12}	227(4.16),268(4.27), 319(3.91)	165-0017-86
Benzenamine, 2-methyl-N-(phenylmethylene)-	C_6H_{12}	212(4.26),263(4.23), 330(3.77)	165-0017-86
Benzenamine, 4-methyl-N-(phenylmethylene)-	MeOH	224(4.12),265(4.18), 321(4.00)	165-0017-86
Carbazole, N-ethyl-	MeCN	240(4.7),260(4.4), 295(4.2),330(3.7), 345(3.7)	35-6579-86
1H-Indeno[2,1-b]pyridine, 1,9-dimethyl-	EtOH	485s(2.71),585(3.36)	103-0980-86
$C_{14}H_{13}NO$			
Benzenamine, N-[(4-methoxyphenyl)-methylene]-	C_6H_{12}	222(4.27),280(4.29), 315(4.11)	165-0017-86
Benzenamine, 2-methoxy-N-(phenylmethylene)-	EtOH	246(4.07),263s(--), 330(3.59)	165-0017-86
Benzenamine, 4-methoxy-N-(phenylmethylene)-	EtOH	229(4.11),264(4.12), 330(4.12)	165-0017-86
Phenol, 2-[2-(4-aminophenyl)ethenyl]-, (E)-	n.s.g.	339(4.45)	150-0433-86S
$C_{14}H_{13}NO_2$			
Bullvalenecarbonitrile, (acetoxymethyl)-	ether	262(3.51)	24-2339-86
Furo[3,2-c]quinolin-4(2H)-one, 3,5-dihydro-3,3-dimethyl-2-methylene-	MeOH	236(4.30),282(3.75), 294(3.78),318(3.80), 331(3.71)	83-0720-86
1H-Indene-1,3(2H)-dione, 2-[3-(dimethylamino)-2-propenylidene]-	hexane	438(4.66)	70-1446-86
	EtOH	440(4.71)	70-1446-86
	$CHCl_3$	443(4.78)	70-1446-86

Compound	Solvent	$\lambda_{max}(\log \epsilon)$	Ref.
Methanone, (2-methoxy-3H-azepin-4-yl)phenyl-	EtOH	240s(3.87),318.0(3.73)	18-2317-86
Methanone, (2-methoxy-3H-azepin-5-yl)phenyl-	EtOH	251.0(4.18)	18-2317-86
Methanone, (2-methoxy-3H-azepin-6-yl)phenyl-	EtOH	245.0(3.97),296.0(3.98)	18-2317-86
1-Naphthalenol, 4-[3-(methylimino)-1-propenyl]-, N-oxide	EtOH	388(4.18),520(2.78)	104-1302-86
	dioxan	365(3.62)	104-1302-86
	acetone	370(4.21)	104-1302-86
	DMSO	380(4.10)	104-1302-86
	+ NaOMe	480(4.29)	104-1302-86
2-Naphthalenol, 1-[3-(methylimino)-1-propenyl]-, N-oxide	dioxan	313(3.69),350(3.76), 378(3.84)	104-1302-86
	acetone	340(3.92),380(4.06)	104-1302-86
	DMSO	340(4.12),385(4.30)	104-1302-86
	+ NaOMe	355(3.96),475(4.22)	104-1302-86
5H-Pyrano[3,2-c]quinolin-5-one, 2,6-dihydro-2,2-dimethyl- (flindersine)	MeOH	236(4.30),261(3.88), 332(3.84),346(3.93)	83-0720-86
$C_{14}H_{13}NO_3$			
Furo[2,3-b]quinolin-4(9H)-one, 9-ethyl-7-methoxy-	MeOH	255(4.37),322(4.05)	100-0048-86
2H-Pyran-2,6(3H)-dione, 4-methyl-3-[1-(phenylamino)ethylidene]-	EtOH	352(4.39)	39-0973-86B
	dioxan	355(4.50)	39-0973-86B
	$CHCl_3$	355(4.50)	39-0973-86B
	80% DMSO	356(4.45)	39-0973-86B
	+ acid	355(--)	39-0973-86B
	+ base	355(--)	39-0973-86B
$C_{14}H_{13}NO_4$			
Furo[2,3-b]quinoline, 4-ethoxy-7-methoxy-	$CHCl_3$	264(3.76),330(4.09)	100-0048-86
Furo[2,3-b]quinolin-4(9H)-one, 9-ethyl-7-methoxy-	$CHCl_3$	255(3.64),276(3.68), 318(3.30)	100-0048-86
Indole, 1,3-diacetyl-5,6-(ethylenedioxy)-	EtOH	222(4.45),285(3.90), 312(3.56)	103-1311-86
2(1H)-Pyridinone, 3-acetyl-4-hydroxy-5-(4-methoxyphenyl)-	EtOH	246(4.28),342(3.65)	44-1374-86
Pyrrolo[2,1-a]isoquinoline-2,3-dione, 5,6-dihydro-8,9-dimethoxy-	EtOH	260(3.83),310(3.85), 376(3.89),434s(3.81), 470s(3.72)	44-2781-86
$C_{14}H_{13}NO_4S$			
3-Pyridinecarboxylic acid, 1,6-dihydro-4-hydroxy-2-mercapto-6-oxo-1-phenyl-, ethyl ester	EtOH	285(3.76),344(4.03)	5-1109-86
Thiophene, 2,5-dihydro-3-methyl-4-nitro-2-(3-phenyl-2-propenylidene)-, 1,1-dioxide, (Z,Z)-	$CHCl_3$	292(3.556),410(4.524)	104-0379-86
$C_{14}H_{13}NO_6$			
[1,3]Dioxolo[4,5-j]phenanthridin-6(2H)-one, 3,4,4a,5-tetrahydro-2,3,4-trihydroxy- (7-deoxynarciclasine)	MeOH	233(4.14),248(4.15), 302(3.75)	102-0995-86
$C_{14}H_{13}N_2$			
Benzo[g]quinoxalinium, 1-ethyl-, tetrafluoroborate	pH 3.4	227(--),285(--), 386(--),479(--)	103-1118-86
hydrated	pH 9.1	234(--),354(--)	103-1118-86

Compound	Solvent	$\lambda_{max}(\log \epsilon)$	Ref.
Imidazo[1,5-a]pyridinium, 2-methyl-3-phenyl-, iodide	MeOH	227(4.1),301(4.0)	83-0043-86
$C_{14}H_{13}N_3O$			
Formamide, N-(3-cyano-4,5-dimethyl-1-phenyl-1H-pyrrol-2-yl)-	EtOH	227(4.25),268(3.72)	11-0347-86
Isoxazole, 3-methyl-5-(1-methyl-3-phenyl-1H-pyrazol-4-yl)-	dioxan	238(4.55),263(4.45)	83-0242-86
$C_{14}H_{13}N_3OS$			
[3,4'-Bipyridine]-3'-carbonitrile, 5'-acetyl-1',4'-dihydro-2'-mercapto-6'-methyl-	DMF	282(3.84),335(4.11)	104-1762-86
[3,4'-Bipyridine]-3'-carbonitrile, 5'-acetyl-1',2',3',4'-tetrahydro-6'-methyl-2'-thioxo-, compd. with hexahydro-1H-azepine	EtOH	235(4.08),294(4.07), 385s(2.44)	104-1762-86
compd. with morpholine	EtOH	235(4.03),287(4.08), 385s(2.90)	104-1762-86
compd. with piperidine	EtOH	234(4.03),303(4.05), 390(2.95)	104-1762-86
[4,4'-Bipyridine]-3-carbonitrile, 5-acetyl-1,2,3,4-tetrahydro-6-methyl-2-thioxo-, compd. with piperidine	EtOH	237(4.22),297(4.21), 388s(2.46)	104-1762-86
$C_{14}H_{13}N_3S_2$			
4H-Pyrazol-3-amine, N,N-dimethyl-4-(4-phenyl-1,3-dithiol-2-ylidene)-	MeCN	235(4.25),324(3.53), 412(4.29)	88-0159-86
monoperchlorate	MeCN	234(4.29),282s(3.76), 360s(3.68)	88-0159-86
$C_{14}H_{13}N_4$			
Pyridinium, 1-methyl-3-(5-phenyl-1H-1,2,4-triazol-3-yl)-, iodide	EtOH	212(4.43),222(4.39), 254(4.31)	128-0027-86
Pyrido[1",2":1',2']imidazo[4',5':4,5]-imidazo[1,2-a]pyridinium, ethyl-, bromide	MeOH	233(4.49),243(4.40), 278(4.08),287(4.08), 334(4.30)	35-8002-86
$C_{14}H_{13}N_5O_3$			
2-Propenamide, 3,3'-[5-(dimethyl-amino)-2,4-furandiyl]bis[2-cyano-	MeOH	258(2.30),370(2.40), 472(2.73)	73-0879-86
$C_{14}H_{14}$			
3H-Benz[cd]azulene, 4,9b-dihydro-9b-methyl-	hexane	262(4.25),266(4.25), 289s(3.23),304s(3.04), 350s(3.39),370s(3.53), 384(3.59),402s(3.45)	89-0632-86
5H-Benzocycloheptene, 5-ethenyl-5-methyl-	EtOH	204(4.25),277(3.80)	39-1479-86C
Bi-1,3,5-cycloheptatrien-1-yl	hexane	221(4.26),265(2.54), 340(4.04)	89-0367-86
4a,8a-[2]Butenonaphthalene	C_6H_{12}	250(4.00)	35-3739-86
Cyclopenta[ef]heptalene, 3,4,5,6-tetrahydro-	hexane	245(4.30),269s(4.27), 274s(4.47),279s(4.58), 283(4.66),288s(4.64), 293s(4.45),302(4.02), 322s(3.25),327s(3.36), 335(3.53),347s(3.64), 350(3.70),363(3.22), 393s(1.18),502s(1.94),	89-0633-86

Compound	Solvent	λ_{max}(log ϵ)	Ref.
(cont.)		520s(2.14),540s(2.30), 563s(2.44),582(2.52), 604(2.60),631(2.54), 660(2.56),693(2.22), 732(2.19)	89-0633-86
Heptalene, 1,6-dimethyl-	hexane	251(4.34),332(3.59)	88-1669-86
Heptalene, 1,10-dimethyl-	hexane	250(4.33),333(3.59)	88-1669-86
$C_{14}H_{14}BrN_3O_3$			
2-Pyridinemethanol, 6-[(4-bromophenyl)azo]-4,5-dimethoxy-	MeOH	232(4.23),337(4.05), 450s(3.03)	4-0333-86
$C_{14}H_{14}Br_4$			
Benzocyclobutacyclooctene, 3,4,6a,10b-tetrabromooctahydro-	C_6H_{12}	261(4.23),287(4.49), 297(4.50),382(3.31), 402(3.30),426(3.13), 454(2.77)	164-0007-86
$C_{14}H_{14}ClHgNO_2$			
Mercury, chloro[(3,4,5,6-tetrahydro-2-methyl-5-oxo-2H-pyrano[3,2-c]quinolin-2-yl)methyl]-	EtOH	226(4.6),270(3.8), 280(3.8),312(3.8)	32-0441-86
$C_{14}H_{14}ClN_3O_3$			
2-Pyridinemethanol, 6-[(3-chlorophenyl)azo]-4,5-dimethoxy-	MeOH	233s(4.22),335(3.92), 446s(2.93)	4-0333-86
2-Pyridinemethanol, 6-[(4-chlorophenyl)azo]-4,5-dimethoxy-	MeOH	232(4.26),335(4.04), 450s(3.02)	4-0333-86
$C_{14}H_{14}ClN_3O_4$			
4-Pyrimidinecarboxylic acid, 2-[(2-chlorobenzoyl)amino]-1,4,5,6-tetrahydro-5-oxo-, ethyl ester	EtOH	230s(4.15),263(4.33)	118-0484-86
4-Pyrimidinecarboxtlic acid, 2-[(4-chlorobenzoyl)amino]-1,4,5,6-tetrahydro-5-oxo-, ethyl ester	EtOH	235s(4.02),271(4.52), 280s(4.32),287s(4.15)	118-0484-86
$C_{14}H_{14}ClN_5S_2$			
2H-Purine-2-thione, 7-[2-[(4-chlorophenyl)thio]ethyl]-1,7-dihydro-6-(methylamino)-	pH 1	254(4.18),289(4.29)	4-0737-86
	pH 13	254(4.16)	4-0737-86
$C_{14}H_{14}FN_5$			
9H-Purin-6-amine, 9-[(2-fluorophenyl)methyl]-N,N-dimethyl- (in 10% ethanol)	0.1M HCl	269(4.26)	4-1189-86
	0.1M NaOH	277(4.25)	4-1189-86
$C_{14}H_{14}F_3NS_2$			
2(3H)-Thiazolethione, 4-(1,1-dimethylethyl)-3-[3-(trifluoromethyl)-phenyl]-	EtOH	321.8(3.96)	152-0399-86
$C_{14}H_{14}N_2$			
[2.2](2,5)Pyridinophane, pseudo-geminal-	isooctane	215(4.0),240s(3.5), 286(3.5),309(2.6)	46-1541-86
pseudo-ortho-	isooctane	215(4.0),240s(3.3), 288(3.3),309(3.3)	46-1541-86
pseudo-meta-	isooctane	215(4.0),240s(3.3), 288(3.3),309(3.3)	46-1541-86
pseudo-para	isooctane	215(4.0),240s(3.5), 283(3.5),306s(2.5)	46-1541-86

Compound	Solvent	λ_{max}(log ε)	Ref.
$C_{14}H_{14}N_2O$			
Phenol, 2-[[(4,6-dimethyl-2-pyridinyl)imino]methyl]-	$CHCl_3$	350(4.11)	2-0127-86A
Phenol, 4-methyl-2-[[(4-methyl-2-pyridinyl)imino]methyl]-	$CHCl_3$	360(3.42)	2-0127-86A
Phenol, 4-methyl-2-[[(5-methyl-2-pyridinyl)imino]methyl]-	$CHCl_3$	360(4.10)	2-0127-86A
Phenol, 5-methyl-2-[[(5-methyl-2-pyridinyl)imino]methyl]-	$CHCl_3$	340(4.04)	2-0127-86A
Pyrazine, 2,5-dimethyl-3-(2-phenylethenyl)-, 1-oxide, (E)-	EtOH	228(3.77),236s(3.74), 273(4.10),301s(3.72), 347(3.79),358s(3.75)	4-1481-86
$C_{14}H_{14}N_2O_2$			
Azepino[3,4-b]indole-1,3(2H,4H)-dione, 5,10-dihydro-2,10-dimethyl-	isoPrOH	241(4.26),316(4.27)	103-1345-86
Benzenamine, 5-methoxy-N-(2-methylphenyl)-2-nitroso-	EtOH	333(4.33),436(3.98)	104-1407-86
3H-Benz[f]indazole-4,9-dione, 3a,9a-dihydro-3,3,9a-trimethyl-, cis-(±)-	EtOH	229(4.49),255(4.05), 302(3.27),311(3.25)	39-2217-86C
Imidazo[1,2-a]pyridin-5(1H)-one, 8-benzoyl-2,3,6,7-tetrahydro-	EtOH	236(4.13),334(4.23)	142-2247-86
Phenol, 2-methoxy-6-[[(3-methyl-2-pyridinyl)imino]methyl]-	$CHCl_3$	480(2.61)	2-0127-86A
Phenol, 2-methoxy-6-[[(4-methyl-2-pyridinyl)imino]methyl]-	$CHCl_3$	480(2.91)	2-0127-86A
Phenol, 2-methoxy-6-[[(5-methyl-2-pyridinyl)imino]methyl]-	$CHCl_3$	480(2.88)	2-0127-86A
Phenol, 2-methoxy-6-[[(6-methyl-2-pyridinyl)imino]methyl]-	$CHCl_3$	480(2.95)	2-0127-86A
1,3-Propanediol, 1-(1H-perimidin-2-yl)-	MeOH	234(4.32),330(4.23)	48-0906-86
2,4(1H,3H)-Pyrimidinedione, 1,3-dimethyl-5-(2-phenylethenyl)-, (E)-	$CHCl_3$	316(4.39)	18-3257-86
(Z)-	$CHCl_3$	255(--),305(3.70)	18-3257-86
5,8-Quinolinedione, 6-piperidino-	$CHCl_3$	467(3.7)	138-1059-86
$C_{14}H_{14}N_2O_3$			
1,3-Diazabicyclo[3.2.0]hept-3-ene-2-carboxylic acid, 4-methyl-7-oxo-, phenylmethyl ester, (2RS,5RS)-	EtOH	209(4.00)	39-1077-86C
(2RS,5SR)-	EtOH	209(3.99)	39-1077-86C
2-Furancarboxylic acid, 5-[[1-(phenylamino)ethylidene]amino]-, methyl ester	MeOH	206(3.89),292(3.31)	73-2826-86
1H-Imidazole-4-butanoic acid, 1-methyl-γ-oxo-2-phenyl-	MeOH	275(4.30)	87-0261-86
$C_{14}H_{14}N_2O_3S$			
Benzeneacetic acid, α-[(1,4-dihydro-4-oxo-2-pyrimidinyl)thio]-, ethyl ester	MeOH	220(4.10),283(3.79)	2-0275-86
$C_{14}H_{14}N_2O_3S_2$			
Methanone, [2,5-dihydro-1-hydroxy-2,5-dimethyl-4-(2-thienyl)-1H-imidazol-2-yl]-2-thienyl-, N-oxide	EtOH	222(4.06),270(4.11), 314(4.18)	103-0268-86
$C_{14}H_{14}N_2O_4S$			
2,1-Benzisothiazole-1(3H)-carboxylic acid, 5-(1H-pyrrol-1-yl)-, ethyl ester, 2,2-dioxide	MeOH	265(4.34)	4-1645-86

Compound	Solvent	λ_{max}(log ε)	Ref.
Methanone, [4-(2-furanyl)-2,5-dihydro-1-hydroxy-2,5-dimethyl-1H-imidazol-2-yl]-2-thienyl-, N-oxide	EtOH	309(4.34)	103-0268-86
$C_{14}H_{14}N_2O_5$			
Methanone, 2-furanyl[4-(2-furanyl)-2,5-dihydro-1-hydroxy-2,5-dimethyl-1H-imidazol-2-yl]-, N-oxide	EtOH	304(4.44)	103-0268-86
3(4H)-Quinazolineacetic acid, 2-(ethoxycarbonyl)-4-oxo-, methyl ester	MeOH	204(4.32),228(4.29), 285(3.81)	33-1017-86
$C_{14}H_{14}N_2O_6$			
Butanedioic acid, (aminophenylmethylene)(nitromethylene)-, dimethyl ester	EtOH	240(3.93),350(4.00)	18-3871-86
1-Cyclobutene-1,2-dicarboxylic acid, 3-amino-4-nitro-3-phenyl-, dimethyl ester	EtOH	240(3.93),310(3.72), 355(4.07)	18-3871-86
3-Furancarboxylic acid, 5-[4-(aminocarbonyl)-4-cyano-3,4-dihydro-3-hydroxy-2H-pyran-6-yl]-2-methyl-, methyl ester, (3S-trans)-	MeOH	261(4.18)	136-0237-86B
2-Furancarboxylic acid, 5-[[1-[[5-(methoxycarbonyl)-2-furanyl]amino]-ethylidene]amino]-, methyl ester	MeOH	290(3.11),347(3.51)	73-2826-86
$C_{14}H_{14}N_2S$			
2-Pyridinecarbothioamide, 5-ethyl-N-phenyl-	hexane	235(4.11),296(4.15), 326(3.88)	65-2088-86
	CCl_4	299(4.18),330s(--), 350s(--)	65-2088-86
	$CHCl_3$	300(4.18),330s(--), 350s(--)	65-2088-86
	12M HCl	282(3.85),345(3.87)	65-2088-86
$C_{14}H_{14}N_3S_2$			
1H-Pyrazolium, 3-(dimethylamino)-1-(5-phenyl-3H-1,2-dithiol-3-ylidene)-, perchlorate	MeCN	231s(4.03),328(4.19), 486(4.41)	88-0159-86
Thiazolo[3,4-b][1,2,4]triazin-5-ium, 2,3-dimethyl-6-(methylthio)-8-phenyl-, perchlorate	n.s.g.	324(3.94),425(3.78)	103-1373-86
$C_{14}H_{14}N_4$			
1H-Imidazo[1',2':1,6]pyrimido[5,4-d]-[1]benzazepine, 2,4,5,6-tetrahydro-	EtOH	242(4.54),268(4.25), 289s(--),325s(--), 339s(--),358(3.75), 386s(--)	142-0143-86
monohydrochloride (hydrate)	EtOH	241(4.39),268s(--), 289s(--),308s(--), 322s(--),335s(--), 364s(--),392(3.50)	142-0143-86
$C_{14}H_{14}N_4O$			
8aH-Oxazolo[3,2-a]pyrazine-8-carbonitrile, 8a-amino-2,3-dihydro-6-methyl-5-phenyl-	pH 1	379(4.12)	33-1025-86
	EtOH	333(--)	33-1025-86
	acetone	330(3.73)	33-1025-86
1H-Purine, 6-(3-phenylpropoxy)-	pH 1	254(3.99)	163-2243-86
	pH 7	253(3.95)	163-2243-86
	pH 13	261(4.03)	163-2243-86

Compound	Solvent	$\lambda_{max}(\log \epsilon)$	Ref.
3(2H)-Pyridazinone, 4,5-dihydro-6-(1-methyl-2-phenyl-1H-imidazol-4-yl)-	MeOH	303(4.42)	87-0261-86
3(2H)-Pyridazinone, 4,5-dihydro-6-(1-methyl-2-phenyl-1H-imidazol-5-yl)-	MeOH	290(4.34)	87-0261-86
$C_{14}H_{14}N_4O_3S$			
Benzo[g]pteridine-2,4(3H,10H)-dione, 8-[(2-hydroxyethyl)thio]-7,10-dimethyl-	DMF	272(4.47),388s(--), 425s(--),453(4.40), 477(4.34)	35-0490-86
formate	DMF	272(4.48),390s(--), 426s(--),452(4.41), 476(4.34)	35-0490-86
$C_{14}H_{14}N_4O_4$			
Azirino[2',3':3,4]pyrrolo[1,2-a]indole-4,7-dione, 6-amino-8-[[(aminocarbonyl)oxy]methyl]-1,1a,2,8b-tetrahydro-5-methyl-, (1aS)-	MeOH	245(4.26),305(4.06), 250s(3.86),518(3.00)	87-1864-86
Pyrimido[4,5-b]quinoline-3(2H)-acetic acid, 8-amino-4,6,7,10-tetrahydro-10-methyl-2,4-dioxo-	pH 8.0	229(4.14),247(4.18), 294(3.95),409s(--), 426(4.50)	77-1385-86
$C_{14}H_{14}N_4O_4S$			
Thieno[2,3-g]pteridine-7-carboxylic acid, 8-ethyl-1,2,3,4-tetrahydro-1,3-dimethyl-2,4-dioxo-, methyl ester	MeOH	239(4.43),258(4.32), 347(4.16),388(3.85)	108-0081-86
Thieno[3,2-g]pteridine-7-carboxylic acid, 6-ethyl-1,2,3,4-tetrahydro-1,3-dimethyl-2,4-dioxo-, methyl ester	MeOH	262(4.43),288(4.29), 369(4.08)	108-0081-86
$C_{14}H_{14}N_4O_6$			
4-Pyrimidinecarboxylic acid, 1,4,5,6-tetrahydro-2-[(2-nitrobenzoyl)-amino]-5-oxo-, ethyl ester	EtOH	238(4.36),261s(4.30)	118-0484-86
4-Pyrimidinecarboxylic acid, 1,4,5,6-tetrahydro-2-[(3-nitrobenzoyl)-amino]-5-oxo-, ethyl ester	EtOH	228(4.45),267(4.46)	118-0484-86
4-Pyrimidinecarboxylic acid, 1,4,5,6-tetrahydro-2-[(4-nitrobenzoyl)-amino]-5-oxo-, ethyl ester	EtOH	229(4.12),283(4.21)	118-0484-86
$C_{14}H_{14}O$			
2a,6a-[1,3]Butadienonaphth[2,3-b]oxirene, 1a,2,7,7a-tetrahydro-	C_6H_{12}	247.5(3.82)	35-3739-86
Ethanone, 1-(1,2,3,4-tetrahydro-2(and 3)-methylene-1,4-methanonaphthalen-6-yl)-	isooctane	254(4.16),284(3.26), 294(3.11)	33-1310-86
1(2H)-Phenanthrenone, 3,4,9,10-tetrahydro-	MeOH	229(4.16),235(4.16), 297(4.24),394(4.18)	44-4424-86
$C_{14}H_{14}O_2$			
1H-Cyclopenta[a]pentalene-1,4(3aH)-dione, 3b,6a,7,7a-tetrahydro-3-(2-propenyl)-, (3aα,3bα,6aα,7aα)-	MeOH	222(4.19),328(2.00)	39-0291-86C
1H-Dibenzo[b,d]pyran-4(6H)-one, 2,3-dihydro-6-methyl-	ether	235(3.89),242(3.86), 320(3.93)	44-4424-86
4,5-Dioxatricyclo[6.2.2.0^{2,7}]dodec-2(7)-ene, 9,10,11,12-tetrakis(methylene)-	EtOH	220(4.02),228(4.01), 237(3.95),257(3.95)	33-0761-86

Compound	Solvent	λ_{max}(log ε)	Ref.
Ethanone, 1-(6-methoxy-3-methyl-1-azulenyl)-	MeOH	270(3.90),329(4.48), 406(3.72),502(2.81)	138-2045-86
4,8-Methanoazulene-1,5-dione, 3a,4,8,8a-tetrahydro-3-(1-propenyl)-, (3aα,4α,8α,8aα)-	MeOH	230(4.11),281(4.31)	39-0291-96C
1,4-Methanonaphthalene-6-carboxylic acid, 1,2,3,4-tetrahydro-2(and 3)-methylene-, methyl ester	isooctane	245(4.11),276(3.20), 286(3.15)	33-1310-86
Methanone, (4,5-dihydro-3-furanyl)(2,3-dihydro-1H-inden-1-yl)-	n.s.g.	206(4.11),270(4.21)	4-1551-86
$C_{14}H_{14}O_2S$			
Benzene, 1-methoxy-2-[(4-methylphenyl)sulfinyl]-, (S)-	EtOH	239(4.26),283(3.66)	12-1833-86
Benzene, 1,1'-sulfonylbis[4-methyl-	MeOH	245(4.30)	73-2034-86
$C_{14}H_{14}O_2S_2$			
1,3-Dithiole-4-carboxylic acid, 2-(1-phenylethylidene)-, ethyl ester	EtOH	224(3.76),309(3.47), 334s(3.36),392(2.93)	33-0419-86
$C_{14}H_{14}O_2S_4$			
1,3-Benzenediethane(dithioic) acid, α,α'-dioxo-, diethyl ester	MeOH	311(3.95),495(1.97)	118-0968-86
1,4-Benzenediethane(dithioic) acid, α,α'-dioxo-, diethyl ester	MeOH	271(4.06),308(4.05), 498(2.12)	118-0968-86
$C_{14}H_{14}O_3$			
Cycloprop[a]inden-6(1H)-one, 1-acetoxy-1a,6a-dihydro-1,6-dimethyl-, (1α,4aβ,6aβ)-	MeOH	235(4.17),256s(3.72), 297(3.31),326s(2.91)	39-1789-86C
1H-Dibenzo[b,d]pyran-4(6H)-one, 2,3-dihydro-8-methoxy-	ether	235(4.11),330(4.26)	44-4424-86
1-Naphthalenecarboxylic acid, 3-methoxy-5-methyl-, methyl ester	MeOH	218(4.85),245s(--), 298(3.80),340(3.83)	94-4554-86
1(2H)-Naphthalenone, 2-acetoxy-2,3-dimethyl-	MeOH	236(4.38),260(3.81), 269s(3.79),299s(3.44), 327s(3.21)	12-2067-86
$C_{14}H_{14}O_4$			
2,5-Furandione, dihydro-4-[(2-methoxyphenyl)methylene]-3,3-dimethyl-	EtOH	239(3.76),300(3.81), 338(3.77)	34-0369-86
2,5-Furandione, dihydro-4-[(4-methoxyphenyl)methylene]-3,3-dimethyl-	EtOH	233(4.10),335(4.39)	34-0369-86
1H-Indene-2-acetic acid, 6-methoxy-α,α-dimethyl-1-oxo-	EtOH	251(4.48),290(2.99)	34-0369-86
2H-Indeno[1,2-b]furan-2,4(3H)-dione, 3a,8b-dihydro-6-methoxy-3,3-dimethyl-	EtOH	223(4.19),252(3.53), 315(3.12)	34-0369-86
2H,5H-Pyrano[2,3-b][1]benzopyran-5-one, 3,4-dihydro-2-(hydroxymethyl)-2-methyl-	EtOH	206(4.34),225(4.36), 276(4.09),288(4.88)	32-0491-86
2H,5H-Pyrano[3,2-c][1]benzopyran-5-one, 3,4-dihydro-2-(hydroxymethyl)-2-methyl-	EtOH	270(4.01),281(4.06), 304(3.99),317(3.82)	32-0501-86
3H,10H-Pyrano[4,3-b][1]benzopyran-10-one, 3-ethoxy-4,4a-dihydro-	EtOH	218(4.290),272s(4.009), 284.5(4.045),330(3.664)	39-0789-86C
$C_{14}H_{14}O_5$			
3,11a-Epoxy-11aH-oxepino[2,3-b][1]-benzopyran-6(2H)-one, 3,4,5,5a-tetrahydro-5a-hydroxy-3-methyl-,	EtOH	251(4.04),311(3.52)	32-0285-86

Compound	Solvent	λ_{max}(log ϵ)	Ref.
(3α,5aα,11aα)- (cont.)			32-0285-86
(3α,5aβ,11aα)-	EtOH	252(4.11),313(3.56)	32-0285-86
2(5H)-Furanone, 4-[2-(2,3-dimethoxy-phenyl)ethenyl]-5-hydroxy-, (E)-	EtOH	317(4.51)	94-4346-86
2(5H)-Furanone, 4-[2-(2,5-dimethoxy-phenyl)ethenyl]-5-hydroxy-, (E)-	EtOH	306(4.43),370(4.21)	94-4346-86
2(5H)-Furanone, 4-[2-(3,4-dimethoxy-phenyl)ethenyl]-5-hydroxy-, (E)-	EtOH	352(4.44)	94-4346-86
$C_{14}H_{14}O_6$			
3,12-Dioxatetracyclo[5.4.1.0^{1,5}.0^{8,11}]-dodeca-5,9-diene-9,10-dicarboxylic acid, dimethyl ester, (1α,7α,8α,11α)-	MeCN	250(3.45)	24-0589-86
1,4-Naphthalenedione, 2,5,6,8-tetra-methoxy-	MeOH	276(4.18),288s(4.16), 422(3.72)	78-3767-86
1,4-Naphthalenedione, 2,5,7,8-tetra-methoxy-	MeOH	218(4.54),243s(3.97), 249s(4.05),254s(4.14), 260s(4.22),265(4.24), 290(4.03),428(3.68)	78-3767-86
$C_{14}H_{14}O_7$			
5H-Furo[3,2-g]benzopyran-5-one, 2,3-dihydro-2,3-dihydroxy-4,9-dimeth-oxy-7-methyl-, cis	EtOH	206(4.18),231(4.23), 250s(4.25),254(4.26), 292(3.74)	44-3116-86
$C_{14}H_{14}S_2$			
1,3-Dithiane, 2-(4-azulenyl)-	hexane	242(4.44),281(4.60), 332(3.47),345(3.57), 360(3.06),542s(--), 563(2.62),584(2.70), 605(2.67),635(2.67), 660(2.39),700(2.29)	5-1222-86
1,3-Dithiane, 2-(6-azulenyl)-	hexane	240(4.15),279(4.91), 330(3.63),338(3.59), 345(3.76),543s(--), 564(2.42),586(2.50), 609(2.45),640(2.46), 670(2.15),709(2.10)	5-1222-86
$C_{14}H_{15}BrN_4O_3$			
Piperazine, 1-(5-bromo-1,6-dihydro-1-methyl-6-oxo-4-pyridazinyl)-4-(2-furanylcarbonyl)-	EtOH	244(4.3),307(3.9)	83-0060-86
$C_{14}H_{15}ClN_2O_3$			
1,3-Benzenedicarbonitrile, 5-chloro-2,4,6-triethoxy-	MeCN	229(4.8),288(3.8)	24-0844-86
$C_{14}H_{15}ClN_4O$			
Ethanol, 2-[(9-chloro-6,7-dihydro-5H-pyrimido[5,4-d][1]benzazepin-4-yl)amino]-	EtOH	239(4.54),259s(--), 303(3.67),310s(--), 348(3.57)	142-0143-86
$C_{14}H_{15}ClN_4OS$			
5-Thiazolecarboxaldehyde, 4-chloro-2-[[4-(diethylamino)phenyl]azo]-	CH_2Cl_2	574(4.71)	77-1639-86
	DMSO	592(--)	77-1639-86
$C_{14}H_{15}ClN_4O_3$			
Piperazine, 1-(5-chloro-1,6-dihydro-1-methyl-6-oxo-4-pyridazinyl)-4-(2-furanylcarbonyl)-	EtOH	242(4.4),305(3.9)	83-0060-86

Compound	Solvent	$\lambda_{max}(\log \epsilon)$	Ref.
$C_{14}H_{15}ClO_4$			
Butanedioic acid, [(2-chlorophenyl)-methylene]dimethyl-, 4-methyl ester, (E)-	EtOH	250(3.95)	34-0369-86
Butanedioic acid, [(3-chlorophenyl)-methylene]dimethyl-, 4-methyl ester, (E)-	EtOH	256(4.13)	34-0369-86
Butanedioic acid, [(4-chlorophenyl)-methylene]dimethyl-, 4-methyl ester, (E)-	EtOH	259(4.22)	34-0369-86
$C_{14}H_{15}NO$			
1H-Pyrrole, 1-(methoxymethyl)-2-(2-phenylethenyl)-, (E)-	EtOH	237(3.94),335(4.05)	118-0620-86
2H-Pyrrol-2-one, 1,5-dihydro-3,4-di-methyl-5-[(2-methylphenyl)methyl-ene]-, (E)-	EtOH	281(4.18)	49-0631-86
(Z)-	EtOH	317(4.22)	49-0631-86
2H-Pyrrol-2-one, 1,5-dihydro-3,4-di-methyl-5-[(4-methylphenyl)methyl-ene]-, (E)-	EtOH	313(4.07)	49-0631-86
(Z)-	EtOH	336(4.40)	49-0631-86
$C_{14}H_{15}NOS$			
Sulfoximine, N-methyl-S-(4-methyl-phenyl)-S-phenyl-, (R)-	EtOH	216(4.04),241(4.16), 268s(3.54),275s(3.34)	12-1655-86
$C_{14}H_{15}NOS_2$			
3-Thiophenecarbothioamide, N-(4-meth-oxyphenyl)-2,5-dimethyl-	EtOH	259s(4.09),304(4.06), 313(4.06),400s(2.86)	78-3683-86
$C_{14}H_{15}NO_2$			
Methanamine, N-[1-methyl-3-[2-(2-pro-penyloxy)phenyl]-2-propynylidene]-, N-oxide	MeOH	234(4.04),300(4.28), 328(4.34),339(4.32)	44-3125-86
$C_{14}H_{15}NO_2S$			
Sulfoximine, S-(2-methoxyphenyl)-S-(4-methylphenyl)-, (S)-	EtOH	241(4.00),289(3.57)	12-1833-86
$C_{14}H_{15}NO_2S_2$			
Benzenamine, 4-methoxy-N-[[2-methoxy-5-(methylthio)-3-thienyl]methylene]-	EtOH	234(4.40),254(4.30), 287(4.24),338(4.37)	70-1470-86
$C_{14}H_{15}NO_3$			
3-Azabicyclo[4.1.0]hept-4-ene-3-carb-oxylic acid, 2-hydroxy-7-phenyl-, methyl ester, (1α,2α,6α,7α)-	EtOH	242(3.69)	142-1831-86
2-Furancarboxaldehyde, 5-(dimethyl-amino)-4-(hydroxyphenylamino)-	MeOH	367(3.37)	73-0573-86
Methacrylamide, N-benzoylformyl-N-ethyl-	C_6H_{12}	216(3.92),258(3.88), 355(1.78)	39-1759-86C
6-Oxa-3-azabicyclo[3.1.1]heptane-2,4-dione, 3-ethyl-1-methyl-5-phenyl-	C_6H_{12}	218(3.92),252(2.50), 258(2.59),264(2.51)	39-1759-86C
$C_{14}H_{15}NO_3S_2$			
[1]Benzothiepino[4,5-e]-1,2-oxathiin-4-amine, 5,6-dihydro-N,N-dimethyl-, 2,2-dioxide	EtOH	226(4.13),233s(4.08), 251s(3.79),277s(3.92), 297(4.02)	4-0455-86

Compound	Solvent	λ_{max}(log ϵ)	Ref.
$C_{14}H_{15}NO_4$			
Furo[2,3-b]quinolin-4(2H)-one, 9-ethyl-3,9-dihydro-3-hydroxy-7-methoxy-	$CHCl_3$	252(3.99),310(3.68)	100-0048-86
$C_{14}H_{15}NO_5$			
Cyclohepta[b]pyrrol-2(1H)-one, 1-α-D-arabinofuranosyl-	n.s.g.	263(4.53),286s(4.22), 388(3.95),408(3.94), 428s(3.61),452s(3.22), 488s(2.62)	150-2625-86M
Cyclohepta[b]pyrrol-2(1H)-one, 1-β-D-arabinofuranosyl-	n.s.g.	293s(4.05),391(3.81), 406s(3.85),411(3.86), 454s(3.15),486(2.52)	150-2625-86M
Cyclohepta[b]pyrrol-2(1H)-one, 3-α-D-arabinofuranosyl-	n.s.g.	231s(3.53),264(3.92), 288s(3.69),393s(3.40), 411(3.46),463s(2.70)	150-2625-86M
β-	n.s.g.	230s(3.80),264(4.18), 283s(3.96),395s(3.66), 411(3.72),461s(2.99)	150-2625-86M
$C_{14}H_{15}NO_6$			
1H-Indole-2,3-dicarboxylic acid, 4,6-dimethoxy-, dimethyl ester	n.s.g.	246(4.52),312(4.37)	12-0015-86
$C_{14}H_{15}NO_8$			
[1,3]Dioxolo[4,5-j]phenanthridin-6(2H)-one, 1,3,4,4a,5,11-hexahydro-1,2,3,4,7-pentahydroxy- (pancratistatin)	MeOH	209s(--),219s(--), 233(4.32),278(3.91), 308s(--)	102-0995-86
$C_{14}H_{15}N_3$			
Benzenamine, N,N-dimethyl-4-(phenylazo)-	benzene	410(4.45)	60-1307-86
	MeOH	228.7(4.03),256.9(3.94), 410.5(4.40)	56-0797-86
1.47M H_2SO_4	MeOH	523.0(5.03)	56-0797-86
14.7M H_2SO_4	MeOH	420.9(4.45)	56-0797-86
0.02M p-toluenesulfonic acid	MeOH	229.1(5.43),262.3(4.25), 413.9(4.11),518.7(4.39)	56-0797-86
	dioxan	275.0(3.89),310.5(3.71), 412.5(4.45)	56-0797-86
0.01M p-toluenesulfonic acid	dioxan	411.1(3.96),525(4.07)	56-0797-86
	DMF	268.5(3.97),420.5(4.25)	56-0797-86
0.02M p-toluenesulfonic acid	DMF	271.1(4.11),423.7(4.36)	56-0797-86
2.8M p-toluenesulfonic acid	DMF	527.4(4.57)	56-0797-86
$C_{14}H_{15}N_3O$			
Benzenamine, N,N-dimethyl-4-(phenylazo)-, N-oxide	benzene	321(3.86),440(2.77)	39-1439-86B
	MeOH	228(4.05),318(4.26), 436(2.64)	39-1439-86B
Benzenamine, N,N-dimethyl-4-(phenyl-NNO-azoxy)-	benzene	391(4.397)	44-0088-86
	MeOH	393(4.354)	44-0088-86
Benzenamine, N,N-dimethyl-4-(phenyl-ONN-azoxy)-	benzene	410(4.452)	44-0088-86
	MeOH	416(4.452)	44-0088-86
$C_{14}H_{15}N_3O_2$			
Benzenamine, N,N-dimethyl-4-(phenylazo)-, $N^4,N^{4'}$-dioxide	benzene	324(4.08)	39-1439-86C
	MeOH	223(4.04),318(4.13)	39-1439-86C
1H-Pyrrole-3-carboxamide, 2-(formylamino)-4,5-dimethyl-1-phenyl-	EtOH	236(4.18)	11-0347-86
$C_{14}H_{15}N_3O_3$			
Ethanone, 1-[1,4-dihydro-2,6-dimethyl-	EtOH	255(3.98),308(3.81)	39-0083-86C

Compound	Solvent	λmax(log ε)	Ref.
4-(3-nitrophenyl)-5-pyrimidinyl]-(cont.)			39-0083-86C
Formamide, N-butyl-N-(3-formyl-4-oxo-4H-pyrido[1,2-a]pyrimidin-2-yl)-	EtOH	253(4.17),279(4.06), 374s(4.17),383(4.19)	4-1295-86
Glycine, N-[2-hydroxy-5-[(3-pyridinyl-methyl)amino]phenyl]-, sodium salt	DMAA	275(3.68),320(3.14)	104-0582-86
2-Pyridinemethanol, 4,5-dimethoxy-6-(phenylazo)-	MeOH	229(4.20),330(3.94), 440s(2.96)	4-0333-86
$C_{14}H_{15}N_3O_3S$			
Benzeneacetic acid, α-[(6-amino-1,4-dihydro-4-oxo-2-pyrimidinyl)thio]-, ethyl ester	MeOH	220(4.20),270(3.80)	2-0275-86
$C_{14}H_{15}N_3O_4$			
4-Pyrimidinecarboxylic acid, 2-(benz-oylamino)-1,4,5,6-tetrahydro-5-oxo-, ethyl ester	EtOH	240s(4.10),263s(4.37), 267(4.39)	118-0484-86
$C_{14}H_{15}N_3O_4S$			
Benzenesulfonic acid, 3-[3-(ethyl-amino)-3-(1H-imidazol-1-yl)-1-oxo-2-propenyl]-	H_2O	342(4.28)	35-5543-86
conjugate base	H_2O	348.5(4.37)	35-5543-86
$C_{14}H_{15}N_3O_5S$			
2(1H)-Pyrimidinone, 4-(2-pyridinyl-thio)-1-β-D-ribofuranosyl-	pH 7	277(--),306(4.07)	1-0806-86
	pH 2	263(--),312(--)	1-0806-86
	pH 13	275s(--),307(--)	1-0806-86
$C_{14}H_{15}N_3O_6$			
2(1H)-Pyrimidinone, 4-(2-oxo-1(2H)-pyridinyl)-1-β-D-ribofuranosyl-	pH 7	314(3.88)	1-0806-86
	pH 2	314(--)	1-0806-86
	pH 13	314(--)	1-0806-86
2(1H)-Pyrimidinone, 4-(3-pyridinyl-oxy)-1-β-D-ribofuranosyl-	pH 7	270(3.89),280(3.92)	1-0806-86
	pH 2	277(--)	1-0806-86
	pH 13	284(--)	1-0806-86
$C_{14}H_{15}N_3S$			
Thiourea, N-(2-methylphenyl)-N'-(4-methyl-2-pyridinyl)-	benzene	272(4.30)	80-0761-86
	MeOH	264(4.35)	80-0761-86
	EtOH	267(4.36)	80-0761-86
	isoPrOH	265(4.34)	80-0761-86
	BuOH	262(4.41)	80-0761-86
	dioxan	253(4.27)	80-0761-86
	HOAc	265(4.37)	80-0761-86
	CH_2Cl_2	259(4.38)	80-0761-86
	$CHCl_3$	266(4.32)	80-0761-86
	$C_2H_4Cl_2$	264(4.38)	80-0761-86
	MeCN	260(4.39)	80-0761-86
	DMF	267(4.45)	80-0761-86
	DMSO	270(4.39)	80-0761-86
Thiourea, N-(2-methylphenyl)-N'-(6-methyl-2-pyridinyl)-	benzene	271(4.29)	80-0761-86
	MeOH	262(4.37)	80-0761-86
	EtOH	267(4.29)	80-0761-86
	isoPrOH	265(4.28)	80-0761-86
	BuOH	256(4.43)	80-0761-86
	dioxan	253(4.30)	80-0761-86
	HOAc	264(4.35)	80-0761-86
	$HOCH_2CH_2OH$	259(4.32)	80-0761-86
	CH_2Cl_2	262(4.37)	80-0761-86

Compound	Solvent	$\lambda_{max}(\log \epsilon)$	Ref.
(cont.)	$CHCl_3$	263(4.30)	80-0761-86
	$C_2H_4Cl_2$	264(4.34)	80-0761-86
	MeCN	262(4.38)	80-0761-86
	DMF	265(4.46)	80-0761-86
	DMSO	269(4.37)	80-0761-86
Thiourea, N-(4-methylphenyl)-N'-(4-methyl-2-pyridinyl)-	benzene	275(4.31)	80-0761-86
	MeOH	263(4.35)	80-0761-86
	EtOH	267(4.35)	80-0761-86
	isoPrOH	267(4.35)	80-0761-86
	BuOH	257(4.44)	80-0761-86
	dioxan	255(4.27)	80-0761-86
	HOAc	263(4.36)	80-0761-86
	CH_2Cl_2	261(4.39)	80-0761-86
	$CHCl_3$	264(4.32)	80-0761-86
	$C_2H_4Cl_2$	266(4.38)	80-0761-86
	MeCN	262(4.39)	80-0761-86
	DMF	266(4.44)	80-0761-86
	DMSO	269(4.40)	80-0761-86
Thiourea, N-(4-methylphenyl)-N'-(6-methyl-2-pyridinyl)-	benzene	271(4.29)	80-0761-86
	MeOH	262(4.38)	80-0761-86
	EtOH	267(4.29)	80-0761-86
	isoPrOH	264(4.30)	80-0761-86
	BuOH	258(4.42)	80-0761-86
	dioxan	244(4.30)	80-0761-86
	HOAc	261(4.35)	80-0761-86
	$HOCH_2CH_2OH$	262(4.34)	80-0761-86
	CH_2Cl_2	258(4.34)	80-0761-86
	$CHCl_3$	262(4.29)	80-0761-86
	$C_2H_4Cl_2$	262(4.34)	80-0761-86
	MeCN	259(4.36)	80-0761-86
	DMF	266(4.43)	80-0761-86
	DMSO	267(4.36)	80-0761-86
$C_{14}H_{15}N_3S_2$			
2H-1,3,5-Thiadiazine-2-thione, 6-phenyl-4-(1-piperidinyl)-	$CHCl_3$	290s(4.42),303(4.46), 390s(3.37)	97-0132-86
$C_{14}H_{15}N_5$			
Imidazo[1,2-a]pyrazine-8-carbonitrile, 8a-amino-1,2,3,8a-tetrahydro-6-methyl-5-phenyl-	pH 1	378(4.05)	33-1025-86
	EtOH	341(3.62)	33-1025-86
6H-Purin-6-imine, 3-ethyl-3,7-dihydro-7-(phenylmethyl)-, monohydrobromide	pH 1	222.5s(4.10),277(4.18)	94-1821-86
	pH 7	222.5s(4.10),277(4.18)	94-1821-86
	pH 13	283(4.12)	94-1821-86
	EtOH	223.5s(4.08),280(4.20)	94-1821-86
6H-Purin-6-imine, 3-ethyl-3,9-dihydro-9-(phenylmethyl)-, monohydrobromide	pH 1	271(4.23)	94-1821-86
	pH 7	271(4.23)	94-1821-86
	EtOH	273(4.22)	94-1821-86
6H-Purin-6-imine, 7-ethyl-3,7-dihydro-3-(phenylmethyl)-, monohydrochloride	pH 1	223.5s(4.16),278.5(4.24)	94-1821-86
	pH 7	223s(4.18),278(4.23)	94-1821-86
	pH 13	282(4.14)	94-1821-86
	EtOH	226s(4.17),282(4.27)	94-1821-86
$C_{14}H_{16}$			
4a,8a-[2]Butenonaphthalene, 1,4-dihydro-	C_6H_{12}	267(3.42)	35-3731-86
iron tricarbonyl deriv.	C_6H_{12}	280s(3.54)	35-3731-86
$C_{14}H_{16}AgFeS_3$			
Silver(1+), [1,1'-[thiobis(2,1-ethanediylthio)]ferrocene]-, tetrafluoroborate	MeCN	450(2.39)	18-3611-86

Compound	Solvent	λ_{max}(log ε)	Ref.
$C_{14}H_{16}BrN_2$			
2,3'-Bipyridinium, 1'-(4-bromobutyl)-, bromide	pH 8.2	210(4.20),229(4.12), 276(4.02)	64-0239-86B
$C_{14}H_{16}BrN_7O$			
2,4-Pyrimidinediamine, 5-[(4-bromo-phenyl)azo]-6-morpholino-	EtOH	204(4.29),234(4.15), 258(4.04),272(4.03), 392(4.26)	42-0420-86
$C_{14}H_{16}Br_2O_3$			
Cyclopropanecarboxylic acid, 2,2-di-bromo-1-(4-ethoxyphenyl)-, ethyl ester	C_6H_{12}	228(4.08)	12-0271-86
$C_{14}H_{16}ClFN_2O_2$			
4-Isoxazolecarboxamide, 5-(chlorometh-yl)-3-(4-fluorophenyl)-4,5-dihydro-N-propyl-	MeOH	262(3.03)	73-2167-86
$C_{14}H_{16}ClNO_2$			
2,4,6-Heptatrienoic acid, 7-(3-chloro-1H-pyrrol-2-yl)-6-methyl-, ethyl ester	EtOH	382(4.55)	150-3601-86M
$C_{14}H_{16}Cl_2FePdS_3$			
Palladium, dichloro[1,1'-[thiobis(2,1-ethanediylthio)ferrocene]-, (SP-4-2)-	MeCN	410(2.92),558(2.61)	18-3611-86
$C_{14}H_{16}Cl_2N_2O_2$			
4-Isoxazolecarboxamide, 5-(chlorometh-yl)-3-(4-chlorophenyl)-4,5-dihydro-N-propyl-	MeOH	269(3.01)	73-2169-86
$C_{14}H_{16}Cl_2O_3$			
Cyclopropanecarboxylic acid, 2,2-di-chloro-1-(4-ethoxyphenyl)-, ethyl ester, (±)-	C_6H_{12}	234(4.15)	12-0271-86
$C_{14}H_{16}F_2O_3$			
Cyclopropanecarboxylic acid, 1-(4-eth-oxyphenyl)-2,2-difluoro-, ethyl ester	C_6H_{12}	230(4.04)	12-0271-86
$C_{14}H_{16}FeO_5$			
Iron, tricarbonyl[methyl 2,2-dimethyl-3-(η^4-1-methylene-2-propenyl)cyclo-propanecarboxylate]-	C_6H_{12}	280s(3.41)	101-0061-86D
minor isomer	C_6H_{12}	280s(3.33)	101-0061-86D
$C_{14}H_{16}FeS_3$			
Ferrocene, 1,1'-[thiobis(2,1-ethane-diylthio)]-	MeCN	462(2.54)	18-3611-86
$C_{14}H_{16}N_2$			
4,9-Azo-5,8-methano-1H-benz[f]indene, 3a,4,4a,5,8,8a,9,9a-octahydro-	hexane	396(2.19)	89-0187-86
1H,6H-Benz[2,3]azirino[1,2-b]pyra-zole, 4a,5,7,8-tetrahydro-2-phenyl-	EtOH	254.5(4.10)	35-1617-86
Pyrazine, 2,5-diethyl-3-phenyl-	EtOH	211(3.94),227(3.90), 285(3.97)	142-0785-86

Compound	Solvent	λ_{max}(log ϵ)	Ref.
1H-Pyrrol-1-amine, N,N-dimethyl-2-(2-phenylethenyl)-, (E)-	EtOH	235(4.03),331(4.27)	118-0620-86
$C_{14}H_{16}N_2O$			
Ethanone, 1-(1,4-dihydro-2,6-dimethyl-4-phenyl-5-pyrimidinyl)-	EtOH	228(3.94),322(3.83)	39-0083-86C
Pyrazine, 2,5-diethyl-3-phenyl-, 1-oxide	EtOH	225(4.31),249(4.16), 266(4.10),307(3.74)	142-0785-86
Pyrazine, 2,5-diethyl-3-phenyl-, 4-oxide	EtOH	227.5(4.22),269(3.92), 302s(3.57)	142-0785-86
$C_{14}H_{16}N_2O_2$			
1-Butanone, 4-hydroxy-1-(1-methyl-1H-imidazol-2-yl)-4-phenyl-	EtOH	278(3.99)	94-4916-86
$C_{14}H_{16}N_2O_2S$			
Pyrimidinium, 2-(ethylthio)-3,6-dihydro-4-hydroxy-1,3-dimethyl-6-oxo-5-phenyl-, hydroxide, inner salt	MeCN	229(4.37),257(3.93), 369(3.66)	24-1315-86
$C_{14}H_{16}N_2O_3$			
2-Benzofurancarboxamide, 6-(diethylamino)-N-formyl-	EtOH	246(3.82),300(2.90), 311(2.90),387(4.07)	4-1347-86
$C_{14}H_{16}N_2O_4S$			
Benzenamine, N,N-dimethyl-4-[(3-methyl-4-nitro-2(5H)-thienylidene)methyl]-, S,S-dioxide	$CHCl_3$	325(3.643),520(4.509)	104-0379-86
$C_{14}H_{16}N_2O_5$			
2,4(1H,3H)-Pyrimidinedione, 1-[(2-hydroxyethoxy)methyl]-5-[(3-hydroxyphenyl)methyl]-	pH 1	267(4.06)	4-1651-86
	pH 11	265(3.84)	4-1651-86
$C_{14}H_{16}N_2S_2$			
Benzenamine, N,N'-dithiobis[N-methyl-	EtOH	200(4.54),230(4.08), 268(3.85),287(3.94), 315(3.98)	44-1866-86
2-Benzothiazolecarbothioamide, N-cyclohexyl-	MeOH	246s(3.81),250(3.83), 355(3.86),308(4.26), 322s(4.14),350s(3.66), 420s(2.30)	103-1139-86
$C_{14}H_{16}N_4$			
Benzenamine, 4-[(4-nitrophenyl)azo]-N,N-dimethyl-	EtOH	410(4.48),440(4.46)	48-0113-86
Benzenamine, N-ethyl-3-methyl-4-(3-pyridinylazo)-	acetone	418(4.42)	7-0473-86
6,7,17,18-Tetraazatricyclo[10.2.2.2^{5,8}]-octadeca-5,7,12,14,15,17-hexaene	dioxan	215(4.09),260(3.45), 328(2.62),335s(2.57), 522(2.67)	88-5367-86
	DMSO	518(--)	88-5367-86
$C_{14}H_{16}N_4O$			
5H-Pyrimido[5,4-d][1]benzazepin-4-amine, N-(2-hydroxyethyl)-6,7-dihydro-	EtOH	235(4.65),249s(--), 268s(--),300(3.93), 308s(--),345s(--)	142-0143-86
$C_{14}H_{16}N_4O_2S$			
Spiro[benzothiazole-2(3H),5'(2'H)-	MeCN	225(4.40),276(3.89),	94-0664-86

Compound	Solvent	λ_{max}(log ε)	Ref.
pyrimidine]-2',4'(3'H)-dione, 3'-methyl-6'-(propylamino)- (cont.)		361(2.11)	94-0664-86
$C_{14}H_{16}O$			
8a,4a-[1]Butenonaphthalen-11-ol, 1,4-dihydro-	C_6H_{12}	271(3.39)	35-3731-86
isomer	C_6H_{12}	270(3.43)	35-3731-86
2a,6a-[2]Butenonaphth[2,3-b]oxirene, 1a,2,7,7a-tetrahydro-	EtOH	263(3.45)	35-3731-86
2-Cyclohexen-1-one, 2-ethyl-3-phenyl-	MeOH	249(4.12),325(1.79)	44-4424-86
1(2H)-Naphthalenone, 2,2,5,7-tetramethyl-	MeOH	233(4.30),242(4.31), 314(3.12),345(3.06)	39-2011-86C
$C_{14}H_{16}O_2$			
Bicyclo[6.2.2]dodeca-4,8,10,11-tetraene-9-carboxylic acid, methyl ester, anti	EtOH	225(3.8),247(3.7), 270s(3.6),345(3.0)	138-1217-86
syn	EtOH	220(3.8),343(3.0)	138-1217-86
2a,5a-[1,3]Butadienonaphtho[2,3-b:6,7-b']bisoxirene, 1a,2,3,3a,4a,5,6,6a-octahydro-	C_6H_{12}	264.5(3.54)	35-3731-86
2-Cyclohexen-1-one, 2-ethoxy-3-phenyl-	ether	215(3.88),280(4.11)	44-4424-86
2-Naphthalenol, 1,2-dihydro-1,1-dimethyl-, acetate	MeOH	233(4.32),264(4.02)	39-1789-86C
4H-Pyran, 4-(2,6-dimethyl-4H-pyran-4-ylidene)-2,6-dimethyl-	MeCN	326(4.46),341(4.52)	104-1814-86
$C_{14}H_{16}O_3$			
2-Cyclohexen-1-one, 2-methoxy-3-(4-methoxyphenyl)-	ether	222(3.89),302(4.20)	44-4424-86
2(3H)-Furanone, 4-(3,4-dimethylbenzoyl)dihydro-5-methyl-	EtOH	264(4.20)	56-0143-86 +80-0541-86
Naphthalene, 1,2,8-trimethoxy-6-methyl- (macassar III)	EtOH	232(3.80),278s(3.71), 290(3.87),302(3.86), 334(3.56),348(3.53)	94-2810-86
$C_{14}H_{16}O_3S$			
2-Butenoic acid, 2-(butylthio)-4-oxo-4-phenyl-	MeOH	249(4.15)	48-0539-86
$C_{14}H_{16}O_4$			
Butanedioic acid, dimethyl(phenylmethylene)-, 4-methyl ester, (E)-	EtOH	255(4.06)	34-0369-86
4H-Pyran-4-one, 2,3-dihydro-6-[(3-hydroxy-5-methoxyphenyl)methyl]-2-methyl- (citreovirenone)	MeOH	209(4.30),222s(3.85), 267(4.08)	138-1129-86
$C_{14}H_{16}O_4S$			
7-Oxabicyclo[2.2.1]heptane-2-carboxylic acid, 3-(phenylsulfinyl)-, methyl ester, endo,endo-	EtOH	250(3.42)	12-0575-86
isomer	EtOH	252(3.42)	12-0575-86
exo,exo-	EtOH	254(3.42)	12-0575-86
isomer	EtOH	249(3.26)	12-0575-86
$C_{14}H_{16}O_5$			
Butanedioic acid, [(2-methoxyphenyl)methylene]dimethyl-, (E)-	EtOH	257(3.82),295(3.57)	34-0369-86
Butanedioic acid, [(4-methoxyphenyl)methylene]dimethyl-, (E)-	EtOH	270(4.23)	34-0369-86

Compound	Solvent	λmax(log ε)	Ref.
$C_{14}H_{16}S$			
Azulene, 4,6-dimethyl-8-[(methyl-thio)methyl]-	hexane	245(4.42),290(4.65), 332s(3.48),337(3.54), 349s(3.61),352(3.64), 363s(3.02),550(2.69), 563(2.69),587(2.66), 610s(2.51),640s(2.28)	5-1222-86
$C_{14}H_{17}BrN_2O_3$			
4-Isoxazolecarboxamide, 3-(4-bromo-phenyl)-4,5-dihydro-5-(hydroxy-methyl)-N-propyl-	MeOH	272(3.10)	73-2107-86
$C_{14}H_{17}BrO_3$			
Cyclopropanecarboxylic acid, 2-bromo-1-(4-ethoxyphenyl)-, ethyl ester	C_6H_{12}	235(4.04)	12-0271-86
isomer B	C_6H_{12}	230(4.12)	12-0271-86
$C_{14}H_{17}ClN_2O_3$			
4-Isoxazolecarboxamide, 3-(4-chloro-phenyl)-4,5-dihydro-5-(hydroxy-methyl)-N-propyl-	MeOH	270(3.12)	73-2167-86
$C_{14}H_{17}ClN_2O_4$			
Alanine, N-[N-(4-chlorobenzoyl)gly-cyl]-2-methyl-, methyl ester	MeOH	236(4.24)	23-0277-86
$C_{14}H_{17}ClO_3$			
Cyclopropanecarboxylic acid, 2-chloro-1-(4-ethoxyphenyl)-, ethyl ester	C_6H_{12}	233(4.16)	12-0271-86
isomer B	C_6H_{12}	230(3.97)	12-0271-86
$C_{14}H_{17}Cl_3O_3S_2$			
1,4-Epoxy-3H-cycloocta[c]thiopyran-3-one, decahydro-1-(methylthio)-4-(trichloroacetyl)-	hexane	211(3.77),231(3.77)	24-2317-86
$C_{14}H_{17}FN_2O_3$			
4-Isoxazolecarboxamide, 3-(4-fluoro-phenyl)-4,5-dihydro-5-(hydroxy-methyl)-N-propyl-	MeOH	267(3.03)	73-2167-86
$C_{14}H_{17}FO_3$			
Cyclopropanecarboxylic acid, 1-(4-eth-oxyphenyl)-2-fluoro-, ethyl ester	C_6H_{12}	230(4.18)	12-0271-86
isomer B	C_6H_{12}	228(4.04)	12-0271-86
$C_{14}H_{17}N$			
1H-Carbazole, 2,3,4,9-tetrahydro-6,9-dimethyl-	$HOCH_2CH_2OH$	235(4.52),293(3.86), 298s(3.83)	103-0713-86
$C_{14}H_{17}NO$			
2H-Pyrrol-2-one, 1,5-dihydro-3,4-di-methyl-5-[(4-methylphenyl)methyl]-	isooctane	211(4.23),219(4.19)	49-0185-86
$C_{14}H_{17}NOS_2$			
2(3H)-Thiazolethione, 4-(1,1-dimeth-ylethyl)-3-(3-methoxyphenyl)-	EtOH	320.8(4.10)	152-0399-86
$C_{14}H_{17}NO_2$			
2H-1-Benzopyran-2-one, 7-(diethyl-	EtOH	374(4.37)	2-0509-86A

Compound	Solvent	λ_{max}(log ε)	Ref.
amino)-4-methyl- (cont.)	DMF	376(4.41)	2-0496-86
Ethanedione, cyclohexylphenyl-, 1-oxime	EtOH	257(3.99)	118-0473-86
2,4,6-Heptatrienoic acid, 6-methyl-7-(1H-pyrrol-2-yl)-, ethyl ester	EtOH	395(4.61)	150-3601-86M
3-Pyridinecarboxylic acid, 1,2,3,6-tetrahydro-1-methyl-3-phenyl-, methyl ester, (±)-, hydrochloride	pH 7.2	210(0.82)	87-0125-86
2H-Pyrrol-2-one, 1,5-dihydro-5-hydroxy-3,4-dimethyl-5-[(2-methylphenyl)methyl]-	MeOH	none above 210 nm	49-0185-86
2H-Pyrrol-2-one, 1,5-dihydro-5-hydroxy-3,4-dimethyl-5-[(4-methylphenyl)methyl]-	MeOH	212(4.13)	49-0185-86
$C_{14}H_{17}NO_2S$			
3,4-Piperidinedione, 2,2,6,6-tetramethyl-5-(2-thienylmethylene)-	EtOH	200(3.83),230(3.74)	70-2511-86
$C_{14}H_{17}NO_3$			
Acetamide, N-spiro[1,3-benzodioxole-2,1'-cyclohexan]-5-yl-	MeOH	259(3.89),297(3.70)	87-1737-86
4H-1,3,5-Dioxazocine-7-carboxaldehyde, 2-ethyl-5,8-dihydro-6-phenyl-	MeOH	236(2.93),299(3.06)	73-2158-86
2-Propanone, 1-(3a,4-dihydro-1-methyl-3H-[1]benzopyrano[4,3-c]isoxazol-9b(1H)-yl)-	EtOH	216s(3.88),271(3.33), 283(3.29)	44-3125-86
3,5,10-Trioxa-9-azabicyclo[5.3.0]dec-8-ene, 4-ethyl-8-phenyl-, endo	MeOH	263(3.06)	73-2158-86
exo	MeOH	258(2.99)	73-2158-86
$C_{14}H_{17}NS$			
Cyclobutanethione, 2,2,4,4-tetramethyl-3-(phenylimino)-	C_6H_{12}	226(3.60),312(2.49), 518(1.11)	151-0331-86A
$C_{14}H_{17}NS_2$			
2(3H)-Thiazolethione, 4-(1,1-dimethylethyl)-3-(3-methylphenyl)-, (±)-	EtOH	321.6(4.14)	152-0399-86
$C_{14}H_{17}N_3O$			
Acetonitrile, (4-oxo-2,3-di-1-pyrrolidinyl-2-cyclobuten-1-ylidene)-, (E)-	CH_2Cl_2	290s(4.4),366(4.6)	118-0216-86
4H-Pyrido[1,2-a]pyrimidin-4-one, 6-methyl-2-piperidino-	EtOH	241(4.09),269(4.60), 350(3.60),366s(3.72)	4-1295-86
$C_{14}H_{17}N_3O_2$			
3-Pyridinecarboxylic acid, 1,2,5,6-tetrahydro-1-methyl-5-(phenylhydrazono)-, methyl ester	pH 7.2	208(1.06),250(1.07), 366(2.20)	87-0125-86
4H-Pyrido[1,2-a]pyrimidine-3-carboxaldehyde, 2-(butylamino)-6-methyl-4-oxo-	EtOH	248(4.39),276s(4.02), 374(4.08),392(3.97)	4-1295-86
$C_{14}H_{17}N_3O_3S$			
6H-Thieno[2,3-b]pyrrole-5-carboxylic acid, 2-acetyl-3,4-diamino-6-(2-propenyl)-	EtOH	278(4.33),305(4.55), 365(3.99)	48-0459-86
$C_{14}H_{17}N_3O_4$			
Ethanone, 1-[1,4,5,6-tetrahydro-4-hydroxy-2,4-dimethyl-6-(3-nitrophenyl)-5-pyrimidinyl]-	EtOH	202(4.27),254(3.85)	39-0083-86C

Compound	Solvent	λ_{max}(log ε)	Ref.
$C_{14}H_{17}N_3O_5$			
4-Isoxazolecarboxamide, 4,5-dihydro-5-(hydroxymethyl)-3-(3-nitrophenyl)-N-propyl-	MeOH	260(3.08)	73-2167-86
4-Isoxazolecarboxamide, 4,5-dihydro-5-(hydroxymethyl)-3-(4-nitrophenyl)-N-propyl-	MeOH	308(3.02)	73-2167-86
$C_{14}H_{17}N_3O_6$			
Pyrido[2,3-d]pyrimidine-2,4(1H,3H)-dione, 6-(acetoxymethyl)-1,3-bis(methoxymethyl)-	H_2O	215(4.61),246(4.00), 307(3.82),318s(3.72)	87-0709-86
$C_{14}H_{17}N_5O_3$			
Benzoic acid, 2-[[[3-diazo-1-(1-methylethyl)-2-oxopropyl]amino]carbonyl]-, hydrazide	MeOH	240(4.01),276(3.97)	104-0353-86
$C_{14}H_{17}N_5O_3S$			
7H-Pyrrolo[2,3-d]pyrimidine-5-carbonitrile, 4-amino-7-(2-deoxy-β-D-erythro-pentofuranosyl)-6-(ethylthio)-	pH 1	237(4.03),292(3.95)	78-5869-86
	pH 7	221s(4.13),239s(3.91), 297(4.01)	78-5869-86
	pH 11	239(3.97),297(4.03)	78-5869-86
$C_{14}H_{17}N_5O_4$			
Propanamide, N-[6,7-dihydro-6-(hydroxymethyl)-11-oxo-4,7-methano-4H,11H-[1,3,5]oxadiazepino[5,4,3-cd]purin-9-yl]-2-methyl-. [4R-(4α,6β,7α)]-	MeOH	279(4.36)	44-0755-86
$C_{14}H_{17}N_5O_6$			
Acetamide, N-nitroso-N-[[1,2,3,4-tetrahydro-1,3-bis(methoxymethyl)-2,4-dioxopyrido[2,3-d]pyrimidin-6-yl]methyl]-	H_2O	215(4.60),243(4.22), 309(3.79)	87-0709-86
Tetrazolo[1,5-c]pyrimidin-5(6H)-one, 6-(3,5-di-O-acetyl-2-deoxy-β-D-erythro-pentofuranosyl)-8-methyl-	$CHCl_3$	252(3.93)	4-1401-86
$C_{14}H_{17}N_5O_7$			
1H-1,2,3-Triazole-4-carbonitrile, 5-amino-1-(2,3,5-tri-O-acetyl-β-D-ribofuranosyl)-	pH 1	228(3.90),255(3.70)	44-1050-86
	pH 11	233(3.89),254(3.80)	44-1050-86
	MeOH	230(3.94),255(3.75)	44-1050-86
$C_{14}H_{17}N_7O_4S$			
Benzenesulfonic acid, 4-[[2,4-diamino-6-(4-morpholinyl)-5-pyrimidinyl]azo]-	EtOH	204(5.04),232(4.82), 255(4.64),285(4.45), 390(4.10)	42-0420-86
$C_{14}H_{18}$			
Naphthalene, 2,6-dihydro-2,2,6,6-tetramethyl-	C_6H_{12}	263(4.37)	89-0578-86
$C_{14}H_{18}BrIN_2O_7$			
Uridine, 5-bromo-6-iodo-5'-O-(methoxymethyl)-2',3'-O-(1-methylethylidene)-	MeOH	278(3.91)	23-1560-86
$C_{14}H_{18}Cl_6O_2Sn$			
Pyrylium, 2,5-dimethyl-, hexachlorostannate (2:1)	n.s.g.	284(4.02)	104-0258-86

Compound	Solvent	$\lambda_{max}(\log \epsilon)$	Ref.
$C_{14}H_{18}N_2O_2$			
Ethanone, 1-(1,4,5,6-tetrahydro-4-hydroxy-2,4-dimethyl-6-phenyl-5-pyrimidinyl)-	EtOH	202(4.26),252(2.85), 258(2.88)	39-0083-86C
3(2H)-Pyridazinone, 6-(3,4-dimethylphenyl)-4,5-dihydro-5-(1-hydroxyethyl)-	EtOH	289(4.45)	56-0143-86 +80-0541-86
$C_{14}H_{18}N_2O_2S$			
β-Alanine, N-2-benzothiazolyl-N-butyl-	pH 1	220(4.14),265(3.84), 283(3.81),290(3.85)	103-0795-86
	EtOH	227(4.27),275(4.00), 302(3.52)	103-0795-86
3(2H)-Benzothiazolepropanoic acid, 2-(butylimino)-	pH 1	215(4.30),258(3.89), 280(3.86),290(3.89)	103-0795-86
	EtOH	222(4.47),265(4.01), 283(3.84),290(3.82), 305(3.65)	103-0795-86
$C_{14}H_{18}N_2O_3S$			
Glycine, N-[[(2-methylbenzoyl)amino]-(methylthio)methylene]-, ethyl ester	EtOH	251(3.99),279(4.15)	118-0484-86
Glycine, N-[[(3-methylbenzoyl)amino]-(methylthio)methylene]-, ethyl ester	EtOH	253(4.03),282(4.25)	118-0484-86
Glycine, N-[[(4-methylbenzoyl)amino]-(methylthio)methylene]-, ethyl ester	EtOH	257(4.12),284(4.35), 292s(4.29)	118-0484-86
$C_{14}H_{18}N_2O_4S$			
Butanamide, 2-[(dimethylamino)methylene]-3-oxo-N-[(4-methylphenyl)sulfonyl]-	EtOH	228(4.16),263(4.06), 304(4.10)	161-0539-86
Glycine, N-[[(2-methoxybenzoyl)amino]-(methylthio)methylene]-, ethyl ester	EtOH	250(4.06),277(4.17), 302s(3.93)	118-0484-86
Glycine, N-[[(4-methoxybenzoyl)amino]-(methylthio)methylene]-, ethyl ester	EtOH	222(4.05),285s(4.32), 296(4.38)	118-0484-86
$C_{14}H_{18}N_2O_5$			
β-D-Alanine, N-γ-L-glutamyl-3-phenyl-	H_2O	241s(2.05),246s(2.15), 268s(2.04),262(2.23), 256(2.34),251(2.27)	102-0527-86
$C_{14}H_{18}N_2O_5S$			
5H-Pyrano[3,2-d]thiazole-6,7-diol, 3a,6,7,7a-tetrahydro-5-(hydroxymethyl)-2-[(4-methoxyphenyl)-amino]-, monohydrobromide	EtOH	226(3.84),252(3.73)	136-0049-86I
$C_{14}H_{18}N_2Pd$			
Palladium, (2,2'-bi-N,N'-pyridinyl)-diethyl-, (SP-4-2)- (approx. abs.)	THF	248(4.07),280(4.14), 367(3.3),496(2.5)	24-2531-86
$C_{14}H_{18}N_4O_2$			
Acetic acid, cyano[2-(1-piperidinyl)-4(1H)-pyrimidinylidene]-, ethyl ester, (Z)-	EtOH	276(4.04),367(4.39)	103-0774-86
Acetic acid, cyano[6-(1-piperidinyl)-4(1H)-Pyrimidinylidene]-, ethyl ester, (Z)-	EtOH	266(4.23),278(4.19), 336(4.61)	103-0774-86
$C_{14}H_{18}N_4O_4$			
4,5-Pyrimidinedicarboxylic acid,	CH_2Cl_2	386(3.73)	83-0798-86

Compound	Solvent	$\lambda_{max}(\log \epsilon)$	Ref.
2-cyano-1,6-dihydro-1-piperidino-, dimethyl ester (cont.)			83-0798-86
$C_{14}H_{18}N_4O_5$			
Acetamide, N-[[1,2,3,4-tetrahydro-1,3-bis(methoxymethyl)-2,4-dioxo-pyrido[2,3-d]pyrimidin-6-yl]-methyl]-	H_2O	214(4.60),244s(4.03), 310(3.85)	87-0709-86
$C_{14}H_{18}N_4O_7$			
Imidazo[4,5-d][1,2]diazepine-6(1H)-carboxylic acid, 2,3-dihydro-2-oxo-1-β-D-ribofuranosyl-, ethyl ester	MeOH	264(3.79),370(3.00)	78-1511-86
Imidazo[4,5-d][1,2]diazepine-6(1H)-carboxylic acid, 2,3-dihydro-2-oxo-3-β-D-ribofuranosyl-, ethyl ester	MeOH	265(3.81),348(3.26)	78-1511-86
$C_{14}H_{18}N_6O$			
1H-Imidazole, 2,2',2"-(methoxymethylidyne)tris[1-methyl-	MeOH	225(4.3)	35-2479-86
$C_{14}H_{18}N_6O_4$			
1,2,4-Triazine-5,6-dicarboxylic acid, 3-cyano-2,5-dihydro-5-[(1-piperidinylimino)methyl]-, dimethyl ester	CH_2Cl_2	238(3.98),313(3.51)	83-0798-86
$C_{14}H_{18}N_6O_5$			
Adenosine, N-(5-oxo-2-pyrrolidinyl)-	EtOH	266(4.23)	142-0625-86
$C_{14}H_{18}N_8O_2S_2$			
Pyrazinecarboxamide, 6,6'-dithiobis[N-methyl-3-(methylamino)-	MeOH	270(4.34),295s(4.23), 389(4.08)	33-0708-86
$C_{14}H_{18}O$			
1(2H)-Naphthalenone, 3,4-dihydro-2,2,5,7-tetramethyl-	MeOH	232(4.72),257(4.01), 303(3.32)	39-2011-86C
Spiro[3H-cyclopenta[a]pentalene-3,1'-cyclopropan]-4(2H)-one, 1,3a,5,6,7,7a-hexahydro-7a-methyl-, cis-(±)-	MeOH	240(3.94)	35-3443-86
Tricyclo[6.3.0.0^{1,5}]undeca-2,6-dien-4-one, 5,6,8-trimethyl-	EtOH	231(3.81)	33-0659-86
$C_{14}H_{18}OS$			
Spiro[2H-thiopyran-2,2'-tricyclo-[3.3.1.1^{3,7}]decan]-4(3H)-one	hexane	303(3.96)	44-0314-86
$C_{14}H_{18}OSi$			
2-Propyn-1-one, 1-(2,3-dimethylphenyl)-3-(trimethylsilyl)-	EtOH	207(4.25),229(3.89), 243(3.89),273(4.10)	33-0560-86
2-Propyn-1-one, 1-(2,6-dimethylphenyl)-3-(trimethylsilyl)-	EtOH	212s(4.13),231(3.88), 239(3.87),275(3.47)	33-0560-86
$C_{14}H_{18}O_2$			
1,8(2H,5H)-Benzocyclooctenedione, 3,6,7,10a-tetrahydro-4,10a-dimethyl-	MeOH	218(3.46)	78-2757-86
Bicyclo[6.2.2]dodeca-8,10,11-triene-9-carboxylic acid, methyl ester	EtOH	231(4.30),261(3.70), 329(3.10)	78-1851-86

Compound	Solvent	λ_{max}(log ε)	Ref.
(cont.)	EtOH	230(4.3),260s(3.8), 330(3.1)	138-1217-86
Bullvalene, bis(methoxymethyl)-	ether	235s(3.40)	24-2339-86
1,6-Heptalenedimethanol, 5,5a,10,10a-tetrahydro-, cis	MeOH	236(4.11),243(4.12), 248(4.11),254(4.11)	88-1669-86
2-Naphthalenecarboxylic acid, 5,6,7,8-tetrahydro-5,5-dimethyl-, methyl ester	EtOH	242(4.14)	88-0703-86
2(3H)-Naphthalenone, 5-ethynyl-4,4a,5,6,7,8-hexahydro-5-hydroxy-1,4a-dimethyl-, (4aS-cis)-	MeOH	250(4.14)	78-2757-86
Spiro[cyclohexan-1,1'-[1H]indene]-5',7'-diol, 2',3'-dihydro-	MeOH	210(4.50),222(4.21), 281(3.37)	102-1992-86
$(C_{14}H_{18}O_2)_n$			
2-Propen-1-one, 1-[5-(1,1-dimethylethyl)-2-methoxyphenyl]-, polymer	$CHCl_3$	250(3.81),314(3.47)	32-0533-86
$C_{14}H_{18}O_2S_4$			
Benzo[1,2-b:4,5-b']bis[1,4]dithiinium, 1,6-diethyl-2,3,5,7,8,10-hexahydro-5,10-dioxo-, bis(tetrafluoroborate)	H_2SO_4	263(4.05),387(4.11), 493(3.07)	88-0691-86
$C_{14}H_{18}O_3$			
Bicyclo[4.2.0]octa-1,3,5-trien-7-ol, 8,8-dimethyl-7-(2-methyl-1,3-dioxolan-2-yl)-	MeOH	260(3.08),266(3.24), 272.5(3.23)	44-1419-86
Bicyclo[4.2.0]octa-1,3,5-trien-7-ol, 8-ethyl-7-(2-methyl-1,3-dioxolan-2-yl)-	MeOH	260.5(3.17),266.5(3.32), 272(3.3)	44-1419-86
4a(2H)-Biphenylenol, 1,3,4,8b-tetrahydro-4,4-dimethoxy-, cis	MeOH	260(3.00),267(3.22), 274(3.39)	44-1419-86
2-Cyclohexen-1-one, 3-hydroxy-5,5-dimethyl-2-(1-oxo-2,4-hexadienyl)-,	MeOH	249(4.01),316s(4.20), 343(4.35)	78-3157-86
(E,E)-	MeOH-NaOH	272(4.51),330(3.95)	78-3157-86
Cyclopropanecarboxylic acid, 1-(4-ethoxyphenyl)-, ethyl ester	C_6H_{12}	228(4.08)	12-0271-86
2,4-Hexadienoic acid, 5,5-dimethyl-3-oxo-1-cyclohexen-1-yl ester, (E,E)-	MeOH	230(4.00),272(4.51)	78-3157-86
$C_{14}H_{18}O_4$			
1-Propanone, 1-[4-methoxy-2-methyl-5-(1-oxopropoxy)phenyl]-	EtOH	221(4.22),266(4.09)	22-0659-86
4,5,11,12-Tetraoxatetracyclo[6.6.2-$0^{2,7}.0^{9,14}$]hexadeca-2(7),9(14)-diene, 15,16-bis(methylene)-	EtOH	212(3.90),237(3.94), 244(3.96)	33-0761-86
$C_{14}H_{18}O_4S_2$			
1,3-Dioxane-4,6-dione, 5-(1,4-dithiaspiro[4.5]dec-8-ylidene)-2,2-dimethyl-	EtOH	230(4.24),252s(4.14)	39-1507-86C
$C_{14}H_{18}O_6$			
1,3-Dioxane-4,6-dione, 5-(1,4-dioxaspiro[4.5]dec-8-ylidene)-2,2-dimethyl-	EtOH	238(3.96)	39-1507-86C
$C_{14}H_{18}O_7$			
1,4,7,10,13-Benzopentaoxacyclopentadecin-14,17-dione, 2,3,5,6,8,9,11,12-octahydro-	MeCN	251(4.10),399(3.14)	77-0268-86

Compound	Solvent	$\lambda_{max}(\log \epsilon)$	Ref.
$C_{14}H_{18}S_6$			
1H,3H-Thieno[3,4-c]thiophene-1,3-dithione, 4,6-bis(1,1-dimethylethyl)thio]-	hexane	260s(4.12),280(4.41), 326s(3.70),350(4.06), 385(4.12),408(4.10), 458s(3.69),493(3.81), 578(2.33)	77-0019-86
$C_{14}H_{18}S_8$			
1,3-Dithiole, 4,5-bis(methylthio)-2-[2,3,4,5-tetrakis(methylthio)-2,4-cyclopentadien-1-ylidene]-	MeCN	208(4.47),270s(4.05), 335s(4.05),490(4.47)	78-0839-86
$C_{14}H_{19}BrN_2O_7$			
Uridine, 5-bromo-5'-O-(methoxymethyl)-2',3'-O-(1-methylethylidene)-	n.s.g.	276(3.97)	23-1560-86
$C_{14}H_{19}ClHgO$			
Mercury, chloro[[2-(1,1-dimethylethyl)-3,4-dihydro-2H-1-benzopyran-2-yl]methyl]-	EtOH	278(3.4)	32-0441-86
$C_{14}H_{19}F_3NO_3P$			
Phosphonic acid, [2,2,2-trifluoro-1-[(1-phenylethyl)imino]ethyl]-, diethyl ester	hexane	230(3.22)	65-2372-86
$C_{14}H_{19}N$			
1H-Indole, 2-(1,1-dimethylethyl)-1,3-dimethyl-	n.s.g.	230(4.33),286(3.86)	150-0514-86M
$C_{14}H_{19}NO$			
1-Propanone, 2,2-dimethyl-3-[(1-methylethyl)imino]-1-phenyl-	EtOH	243(4.0),280(3.1)	39-0091-86C
$C_{14}H_{19}NOS$			
4-Piperidinone, 2,2,6,6-tetramethyl-3-(2-thienylmethylene)-	EtOH	202(3.92),225(3.70), 255(3.48),340(3.94)	70-2511-86
$C_{14}H_{19}NO_2S$			
3,2,1-Benzoxathiazine, 1,4,4a,5,6,7-8,8a-octahydro-1-(phenylmethyl)-, cis-fused, 2-oxide	EtOH	213(3.85)	12-0591-86
diastereomer	EtOH	212(3.86)	12-0591-86
trans-fused, 2-oxide	EtOH	208(3.94)	12-0591-86
diastereomer	EtOH	220(3.52)	12-0591-86
Piperidine, 1-[2-[(4-methylphenyl)-sulfonyl]ethenyl]-, (E)-	MeCN	224(3.98),279(4.17)	104-2085-86
$C_{14}H_{19}NO_4$			
Acetamide, N-[3-methoxy-4-[(tetrahydro-2H-pyran-2-yl)oxy]phenyl]-	MeOH	251(4.13),286(3.62)	87-1737-86
$C_{14}H_{19}NO_5$			
Propanedioic acid, [[5-(dimethylamino)-2-furanyl]methylene]-, diethyl ester	MeOH	231(3.02),422(3.49)	73-0879-86
$C_{14}H_{19}NO_8$			
3-Pyrrolidineacetic acid, 1-acetyl-2-(1-hydroxy-2-methoxy-2-oxoethylidene)-5-(methoxycarbonyl)-, methyl ester	EtOH	230(3.9)	5-1968-86

Compound	Solvent	$\lambda_{max}(\log \epsilon)$	Ref.
$C_{14}H_{19}N_3O_3S_2$			
Ethanaminium, N-ethyl-N-[4-(hexahydro-1,3-dimethyl-2,4,6-trioxo-5-pyrimidinyl)-5-methyl-1,3-dithiol-2-ylidene]-, hydroxide, inner salt	MeCN	257(4.20),287(4.33), 372(3.58)	24-2094-86
	toluene	419(--)	24-2094-86
	EtOH	340s(--)	24-2094-86
	EtOAc	395(--)	24-2094-86
	acetone	379(--)	24-2094-86
	CH_2Cl_2	391(--)	24-2094-86
$C_{14}H_{19}N_3O_7$			
Thymidine, 5'-deoxy-5'-[(ethoxyoxoacetyl)amino]-	pH 1 and 7	267(3.97)	87-1052-86
	pH 13	267(3.85)	87-1052-86
$C_{14}H_{19}N_5O_3S$			
Adenosine, 4',5'-didehydro-5'-(S-(2-methylpropyl)-5'-thio-, (Z)-	MeOH	258(4.29)	44-1258-86
$C_{14}H_{19}N_5O_6$			
Acetamide, N-(6-ethoxy-9-β-D-ribofuranosyl)-9H-purin-2-yl)-	H_2O	218(4.31),260(4.21)	94-1961-86
9H-Purine-8-carboxylic acid, 6-(dimethylamino)-9-β-D-ribofuranosyl-, methyl ester	MeOH	234(4.33),270s(3.85), 278(3.89),315(4.18)	94-3635-86
$C_{14}H_{19}N_5O_8$			
1H-1,2,3-Triazole-4-carboxamide, 5-amino-1-(2,3,5-tri-O-acetyl-β-D-ribofuranosyl)-	pH 1	233(4.11),261(4.02)	44-1050-86
	pH 11	236(3.89),252(3.93)	44-1050-86
	MeOH	236(4.02),259(3.96)	44-1050-86
$C_{14}H_{19}S_8$			
1,3-Dithiol-1-ium, 4,5-bis(methylthio)-2-[2,3,4,5-tetrakis(methylthio)-1,3-cyclopentadien-1-yl]-, perchlorate	$MeCN-HClO_4$	234(4.19),357(3.77), 400(3.80),586(4.33)	78-0839-86
$C_{14}H_{20}$			
5-Decene-3,7-diyne, 2,2,9,9-tetramethyl-, (E)-	hexane	261(4.49),275(4.44)	35-4685-86
Naphthalene, 2,3,4,6-tetrahydro-2,2,6,6-tetramethyl-	C_6H_{12}	253(4.46)	89-0578-86
$C_{14}H_{20}NO_3$			
Lophotine (iodide)	MeOH	276(3.04),285s(2.95)	100-0745-86
	HCl	276(3.06),287s(2.95)	100-0745-86
	KOH	276(3.15),287s(3.09)	100-0745-86
$C_{14}H_{20}N_2$			
Cyclohexanone, methyl(4-methylphenyl)hydrazone	$HOCH_2CH_2OH$	253(4.13),295(3.59)	103-0713-86
$C_{14}H_{20}N_2OSe$			
2-Cyclobuten-1-one, 2,3-di-1-piperidinyl-4-selenoxo-	MeOH	324(3.9),412(4.6), 520(2.9)	24-0182-86
$C_{14}H_{20}N_2O_4$			
Butanamide, 3-methyl-N-(5a,6,7,8-tetrahydro-4-methyl-1,5-dioxo-1H,5H-pyrrolo[1,2-c][1,3]oxazepin-3-yl)-	acid and neutral	274(3.99)	88-1161-86
	basic	282(4.02),328(4.08)	88-1161-86

Compound	Solvent	$\lambda_{max}(\log \epsilon)$	Ref.
$C_{14}H_{20}N_2O_4S$			
1-Azabicyclo[3.2.0]hept-2-ene-2-carboxylic acid, 3-[(3-aminocyclopentyl)thio]-6-(1-hydroxyethyl)-7-oxo-	H_2O	300(3.87)	23-2184-86
isomer B	H_2O	300(3.83)	23-2184-86
$C_{14}H_{20}N_2O_5S$			
1-Azabicyclo[3.2.0]hept-2-ene-2-carboxylic acid, 3-[[2-(acetylamino)ethyl]thio]-6-ethyl-6-methoxy-7-oxo-, cis, sodium salt	H_2O	305.5(3.90)	88-4335-86
$C_{14}H_{20}N_4O$			
2-Cyclooctapyrazinecarbonitrile, 3,4,5,6,7,8,9,10-octahydro-3-imino-4-(2-methoxyethyl)-	EtOH	437(3.82)	33-1025-86
$C_{14}H_{20}N_4O_2$			
2H-Imidazo[4,5-c]pyridin-2-one, 5-(3,3-dimethyl-2-oxobutyl)-1,3,4,5-tetrahydro-4-imino-1,3-dimethyl-, monohydrobromide	H_2O	225(4.71),252(3.73), 292(3.95)	103-0180-86
Pyrazino[2,3-g]quinoxaline-5,10-dione, 1,2,3,4,6,7,8,9-octahydro-1,4,6,9-tetramethyl-	CH_2Cl_2	232(4.10),235s(4.02), 394(4.05),627(2.28)	88-0691-86
radical ion, salt with trifluoroethanesulfonic acid (1:2)	MeCN	270(4.10),347(3.92), 514(4.13),589(4.39)	88-0691-86
bis(tetrafluoroborate)	CF_3COOH	236(4.25),331(4.00), 390(3.65),509(4.29), 590(4.60)	88-0691-86
$C_{14}H_{20}N_4O_3$			
2,4(3H,8H)-Pteridinedione, 8-(6-hydroxyhexyl)-6,7-dimethyl-	pH 5.0	228(3.85),256(4.15), 278(4.04),313(3.00), 407(4.04)	44-2461-86
	pH 8.0	228(4.00),256(4.20), 278(4.11),313(3.30), 407(4.04)	44-2461-86
	pH 12.0	245(4.28),265s(4.00), 313(4.27),370(3.78)	44-2461-86
$C_{14}H_{20}N_4O_6$			
1H-1,2-Diazepine-1-carboxylic acid, 4,5-bis[(ethoxycarbonyl)amino]-, ethyl ester	MeOH	262(4.11),351(3.08)	78-1511-86
1H-1,2-Diazepine-1-carboxylic acid, 5,6-bis[(ethoxycarbonyl)amino]-, ethyl ester	MeOH	257(4.04),340(2.74)	78-1511-86
Pyridinium, 1,3,4-tris[(ethoxycarbonyl)amino]-, hydroxide, inner salt	MeOH	241(4.12),268(4.03), 313(4.05)	78-1511-86
$C_{14}H_{20}N_6O_3$			
L-Alanine, N-[1-oxo-5-(1H-purin-6-ylamino)pentyl]-, methyl ester	pH 2.85	272(4.16)	63-0757-86
	pH 7.0	268(4.17)	63-0757-86
	pH 10.5	272(4.16)	63-0757-86
$C_{14}H_{20}N_6O_4$			
β-D-Ribofuranuronamide, 1-deoxy-1-[6-(dimethylamino)-9H-purin-9-yl]-N-ethyl-	n.s.g.	275(4.28)	87-1683-86

Compound	Solvent	λ_{max}(log ε)	Ref.
$C_{14}H_{20}N_6O_4S$			
7H-Pyrrolo[2,3-d]pyrimidine-5-carboximidamide, 4-amino-7-(2-deoxy-β-D-erythro-pentofuranosyl)-6-(ethylthio)-N-hydroxy-	pH 1 pH 7 pH 11	286(4.10) 292(4.15) 292(4.16)	78-5869-86 78-5869-86 78-5869-86
$C_{14}H_{20}N_6O_5$			
9H-Purine-8-carboximidic acid, 6-(dimethylamino)-9-β-D-ribofuranosyl-, methyl ester	MeOH	226(4.38),271s(3.85), 278(3.90),312(4.25)	94-3635-86
1,2,4-Triazine-5,6-dicarboxylic acid, 3-cyano-1,2,5,6-tetrahydro-6-hydroxy-5-[(1-piperidinylimino)-methyl]-, dimethyl ester	CH_2Cl_2	316(3.53)	83-0798-86
$C_{14}H_{20}O$			
4-Penten-2-yn-1-ol, 5-(2,6,6-trimethyl-1-cyclohexen-1-yl)-, (E)-	EtOH	213s(3.77),232(3.89), 268(4.06)	5-1398-86
$C_{14}H_{20}O_2$			
1,8(2H,5H)-Benzocyclooctenedione, 3,4,4a,6,7,10a-hexahydro-4,10a-dimethyl-	MeOH	224(2.88)	78-2757-86
2,7(3H,8H)-Benzocyclooctenedione, 4,4a,5,6,9,10-hexahydro-1,4a-dimethyl-	MeOH	254(4.21)	78-2757-86
2-Cyclopenten-1-one, 4-[(3,3-dimethyl-1-cyclohexen-1-yl)methoxy]-, (±)-	EtOH	216(3.91),315(1.65)	32-0291-86
2H-Cyclopent[a]pentalen-2-one, 3,3a,3b,4,5,6,6a,7-octahydro-5-hydroxy-3a,4,5a-trimethyl-	EtOH	233(3.81)	33-0659-86
Naphthalene, 5,5-(ethylenedioxy)-2,3,4,4a,5,6,7,8-octahydro-4aβ-methyl-2-methylene-	MeOH	236(4.46)	44 0773-86
2-Naphthalenecarboxylic acid, 1,5,6,7,8,8a-hexahydro-5,5-dimethyl-, methyl ester	EtOH	306(4.07)	88-0699-86
2(3H)-Naphthalenone, 5-ethenyl-4,4a,5,6,7,8-hexahydro-5-hydroxy-1,4a-dimethyl-, (4aS-cis)-	MeOH	253(4.13)	78-2757-86
$C_{14}H_{20}O_3$			
2H-1-Benzopyran-6(5H)-one, 3,4-dihydro-5-hydroxy-2,2,5,7,8-pentamethyl-	MeCN	340s(3.16),376(3.20)	44-1435-86
5H-1-Benzopyran-5-one, 2,3,4,6-tetrahydro-6-hydroxy-2,2,6,7,8-pentamethyl-	MeCN	249(3.74),327(3.45)	44-1435-86
$C_{14}H_{20}O_5$			
laH,6H-Oxireno[e][l]benzopyran-2(4aH)-one, 7,8-dihydro-4a-hydroperoxy-la,3,4,6,6-pentamethyl-, (lα,4aα,8aS*)-	MeCN	246(3.80)	77-1076-86
$C_{14}H_{21}ClNO_2S$			
Ethenaminium, 2-[(4-chlorophenyl)sulfonyl]-N,N,N-triethyl-, chloride, (E)-	MeCN	247(4.13)	104-2085-86

Compound	Solvent	$\lambda_{max}(\log \epsilon)$	Ref.
$C_{14}H_{21}NO$			
Cyclohexanemethanol, 2-[(phenylmeth-yl)amino]-, cis	EtOH	212(3.74),260(2.27)	12-0591-86
trans	EtOH	216(3.60)	12-0591-86
$C_{14}H_{21}NO_2$			
3-Pyridinecarboxylic acid, 5-cyclo-hexylidene-1,2,5,6-tetrahydro-1-methyl-, methyl ester, hydrochloride	pH 7.2	292(2.22)	87-0125-86
$C_{14}H_{21}NO_3$			
2-Furancarboxylic acid, 4-[(cyclohex-ylamino)methyl]-5-methyl-, methyl ester	MeOH	252(3.28),266s(--)	73-2186-86
Pyridine, 2-methoxy-6-[3-[(tetrahydro-2H-pyran-2-yl)oxy]propyl]-	EtOH	217(3.89),274(3.77)	44-2184-86
$C_{14}H_{21}NO_4$			
Isoquinoline, 1,2,3,4-tetrahydro-5,6,7,8-tetramethoxy-2-methyl-(weberine)	aq HCl	205(4),223s(3.5), 282(2.9),292s(2.8)	100-0745-86
3,4-Pyridinedicarboxylic acid, 1-(1,1-dimethylethyl)-1,4-dihydro-4-meth-yl-, dimethyl ester	MeCN	244(3.63),340(3.76)	118-0190-86
3,5-Pyridinedicarboxylic acid, 1,4-dihydro-2,6-dimethyl-4-(1-methyl-ethyl)-, dimethyl ester	MeOH	232(--),258(--), 272(--),338(3.93)	87-1596-86
$C_{14}H_{21}N_2O_4S$			
Ethenaminium, N,N,N-triethyl-2-[(4-nitrophenyl)sulfonyl]-, chloride, (E)-	MeCN	250(4.17)	104-2085-86
$C_{14}H_{21}N_2O_8P$			
Phosphonic acid, [[3,3a-dihydro-3-hy-droxy-2-(hydroxymethyl)-6-oxo-6H-furo[2',3':4,5]oxazolo[3,2-a]pyrimi-din-9a(2H)-yl]methyl]-, diethyl ester	EtOH	255(3.75)	142-0617-86
$C_{14}H_{21}N_3O$			
13-Oxa-2,11,18-triazatricyclo-[8.7.1.0^{12,17}]octadeca-1(18),10,12(17)-triene	$CHCl_3$	250(3.98)	24-1070-86
13-Oxa-2,11,17-triazatricyclo-[8.6.1.0^{12,16}]heptadeca-1(17),10-12(16)-triene, 14-methyl-	$CHCl_3$	250(4.27),313(3.48)	24-1070-86
4H-1,2,3-Triazol-4-one, 3-(1-adaman-tyl)-3,5-dihydro-5,5-dimethyl-	hexane	251(3.663),309(2.492)	5-1891-86
$C_{14}H_{21}N_3O_4$			
5-Pyrimidinepentanoic acid, 2-amino-1,4-dihydro-6-hydroxy-4-oxo-, cyclopentyl ester	50%MeOH-HCl	265(4.16)	73-1140-86
	50%MeOH-NaOH	268(4.10)	73-1140-86
$C_{14}H_{21}N_5S_2$			
2H-Purine-2-thione, 7-[(2-cyclohexyl-thio)ethyl]-3,7-dihydro-6-(methyl-amino)-	pH 1	245(4.13),287(4.18)	4-0737-86
	pH 13	253(4.32)	4-0737-86

Compound	Solvent	λ_{max}(log ϵ)	Ref.
$C_{14}H_{21}N_6$			
1H-1,2,4-Triazolium, 5-[[4-(diethylamino)phenyl]azo]-1,4-dimethyl-, bromide	H_2O	542(4.81)	140-0439-86
isomer II	H_2O	524(4.77)	140-0439-86
$C_{14}H_{22}$			
Naphthalene, 2,3,4,6,7,8-hexahydro-2,2,6,6-tetramethyl-	C_6H_{12}	243(4.21)	89-0578-86
Spiro[2.5]octa-4,6-diene, 4,5,6,7,8,8-hexamethyl-	C_6H_{12}	275(3.76)	5-1098-86
$C_{14}H_{22}ClN_2Ru$			
Ruthenium(1+), (η^6-benzene)chloro-[N,N'-1,2-ethenediylidenebis[2-propanamine]-N,N']-, tetrafluoroborate	MeCN	358(3.23),418(3.32), 492s(2.51)	157-1449-86
$C_{14}H_{22}Cl_2N_4O_2$			
2,5-Cyclohexadiene-1,4-dione, 2,5-dichloro-3,6-bis[[2-(dimethylamino)ethyl]amino]-, dihydrochloride	MeOH	226(4.30),354(4.39)	87-1792-86
$C_{14}H_{22}NO_2S$			
Ethenaminium, N,N,N-triethyl-2-(phenylsulfonyl)-, chloride, (E)-	MeCN	239(3.77)	104-2085-86
$C_{14}H_{22}NO_3$			
Peyotine (iodide)	MeOH	271(2.92),282s(2.80)	100-0745-86
	HCl	271(2.94),282s(2.80)	100-0745-86
	KOH	292(3.52)	100-0745-86
$C_{14}H_{22}N_2$			
Diazene, bis(dicyclopropylmethyl)-, cis	hexane	386(1.87)	44-2969-86
	C_6H_{12}	384(1.79)	24-1911-86
	mesitylene	383(1.88)	24-1911-86
trans	hexane	358(1.31)	44-2969-86
	C_6H_{12}	361(1.28)	24-1911-86
$C_{14}H_{22}N_2O_4$			
2,5-Cyclohexadiene-1,4-dione, 2,5-bis[(2-methoxyethyl)amino]-3,6-dimethyl-	EtOH	347(4.41)	94-1299-86
$C_{14}H_{22}N_2O_4S$			
4,12-Epoxy-5H,10H-1,3-dioxolo[4,5-e]-pyrimido[6,1-b][1,3]oxazocine-10-thione, octahydro-2,2,8,8-tetramethyl-	n.s.g.	249(4.19)	103-1336-86
$C_{14}H_{22}N_2O_9$			
2,4(1H,3H)-Pyrimidinedione, 6-ethoxy-1-β-D-glucopyranosyl-5-(methoxymethyl)-	$CHCl_3$	260(3.95)	97-0437-86
$C_{14}H_{22}N_2Ru$			
Ruthenium, (η^6-benzene)[N,N'-1,2-ethanediylidenebis[2-propanamino]-N,N']-	CH_2Cl_2	412(4.20),488s(3.08)	157-1449-86

Compound	Solvent	$\lambda_{max}(\log \epsilon)$	Ref.
$C_{14}H_{22}N_3O_4$			
Ethanaminium, N,N,N-trimethyl-2-[[[[1-(2-nitrophenyl)ethyl]amino]-carbonyl]oxy]-, iodide	pH 7.0	265(3.1)	69-1799-86
$C_{14}H_{22}N_5O_8P$			
Guanosine, N-(1-methyl-1-phosphono-propyl)-, barium salt	pH 1	262(4.20),285s(3.91)	94-0471-86
	H_2O	259(4.26),275s(4.11)	94-0471-86
	pH 13	261(4.21),275s(4.10)	94-0471-86
$C_{14}H_{22}N_6S_2$			
Thiazole, 2,2'-(2-tetrazene-1,4-diyli-dene)bis[3-ethyl-2,3-dihydro-4,5-di-methyl-	$CHCl_3$	405.9(4.04)	142-3097-86
$C_{14}H_{22}O$			
2-Adamantanone, 4-(1,1-dimethyl-ethyl)-, (-)-(1S,3R)-4(S)(a)-	MI	290(1.40)	35-4484-86
	EPA	288(1.43)	35-4484-86
(+)-(1S,3R)-4(R)(e)-	MI	288(1.42)	35-4484-86
	EPA	284(1.43)	35-4484-86
2-Adamantanone, 4-methyl-4-(1-methyl-ethyl)-, (+)-(1S,3R)-4(S)(a)-	MI	295(1.30)	35-4484-86
	EPA	285(1.38)	35-4484-86
isomer	MI	290(1.36)	35-4484-86
	EPA	286(1.40)	35-4484-86
$C_{14}H_{22}O_3$			
1,5-Dioxaspiro[5.5]undec-3-en-2-one, 4,10-dimethyl-7-(1-methylethyl)-	hexane	244(3.78)	89-1117-86
isomer	hexane	241(3.76)	89-1117-86
$C_{14}H_{22}O_3S_2$			
2,4-Nonadienoic acid, 8-(1,3-dithian-2-yl)-7-hydroxy-6-methyl-, [6S-(2E,4E,6R*,7R*,8S*)]-	EtOH	254(4.28)	88-4741-86
$C_{14}H_{22}O_4$			
2(5H)-Furanone, 4-(5-acetoxyheptyl)-3-methyl-	MeCN	215(4.04)	100-0593-86
2(5H)-Furanone, 4-(6-acetoxyheptyl)-3-methyl-	MeCN	215(4.17)	100-0593-86
3,5-Heptadienoic acid, 6-acetoxy-2-methyl-, 1,1-dimethylethyl ester, (E,Z)-(±)-	MeOH	239(4.34)	33-0469-86
$C_{14}H_{23}BrO_2$			
2,4-Hexadienoic acid, 4-bromo-, octyl ester	n.s.g.	270(3.93)	104-2230-86
$C_{14}H_{23}NOS$			
1,4-Thiazepine, 2,7-bis(1,1-dimethyl-ethyl)-5-methoxy-	C_6H_{12}	244(3.77),340(2.36)	89-0635-86
$C_{14}H_{23}NO_3$			
3-Pyridinecarboxylic acid, 1,2,5,6-tetrahydro-5-(1-hydroxycyclohexyl)-1-methyl-, methyl ester, (±)-	pH 7.2	220(1.05)	87-0125-86
$C_{14}H_{23}N_2$			
Pyrrolidinium, 1-[1-methyl-5-(1-pyrro-lidinyl)-2,4-pentadienylidene]-, perchlorate	EtOH	429(5.09)	33-1588-86

Compound	Solvent	$\lambda_{max}(\log \epsilon)$	Ref.
Pyrrolidinium, 1-[3-methyl-5-(1-pyrrolidinyl)-2,4-pentadienylidene]-, iodide	EtOH	422s(4.51),444(4.91)	33-1588-86
perchlorate	EtOH	422s(4.69),444(5.08)	33-1588-86
tetrafluoroborate	EtOH	422s(4.72),444(5.11)	33-1588-86
$C_{14}H_{23}N_2O_9P$			
2,4(1H,3H)-Pyrimidinedione, 1-[1-deoxy-1-(diethoxyphosphinyl)methyl]-arabinofuranosyl]-	EtOH	263(4.06)	142-2133-86
$C_{14}H_{23}N_6O_{15}P_3S$			
L-Homocysteine, S-[6-amino-9-[5-O-[hydroxy[[hydroxy(phosphonooxy)phosphinyl]oxy]phosphinyl]-β-D-ribofuranosyl]-9H-purin-8-yl]-, tetrasodium salt	pH 1	281(4.28)	87-0318-86
	pH 6.8	281(4.27)	87-0318-86
$C_{14}H_{24}$			
Cyclohexane, 1,1,3,3,5-pentamethyl-2-(2-propenylidene)-, (S)-(+)-	C_6H_{12}	244(4.30)	35-2691-86
$C_{14}H_{24}N_2O_2$			
1(2H)-Pyridinecarboxylic acid, 2-butyl-4-(dimethylamino)-, ethyl ester	hexane	253(3.92),299(3.37)	78-0835-86
$C_{14}H_{24}N_4$			
1,4,8,11-Tetraazacyclotetradeca-4,6,11,13-tetraene, 5,7,12,14-tetramethyl-	DMF	305(4.53)	138-0907-86
copper chelate	DMF	341(4.40),598(2.41)	138-0907-86
$C_{14}H_{24}N_4Ni$			
Nickel(2+), (2,3,9,10-tetramethyl-1,4,8,11-tetraazacyclotetradeca-1,3,8,10-tetraene-N^1,N^4,N^8,N^{11})-	n.s.g.	392(3.36)	77-1616-86
$C_{14}H_{24}N_4O_2$			
2,5-Cyclohexadiene-1,4-dione, 2,5-bis[[2-(dimethylamino)ethyl]-amino]-	MeOH	210(4.27),245(3.51), 340(4.22)	87-1792-86
$C_{14}H_{24}N_4O_6$			
1,2-Hydrazinedicarboxylic acid, 1-[2-ethoxy-1-(1-methyl-2-imidazolidinylidene)-2-oxoethyl]-, diethyl ester, (Z)-	EtOH	273(4.39)	24-2208-86
$C_{14}H_{24}O$			
2-Propanone, 1-(2,2,4,6,6-pentamethylcyclohexylidene)-, (S)-	C_6H_{12}	240(3.76)	35-2691-86
$C_{14}H_{24}OSi$			
4-Hepten-1-yn-3-one, 6-methyl-5-(1-methylethyl)-1-(trimethylsilyl)-	EtOH	251s(4.02),269(4.14)	33-0560-86
$C_{14}H_{24}O_2$			
Acetic acid, (2,2,4,6,6-pentamethylcyclohexylidene)-, methyl ester, (S)-(+)-	C_6H_{12}	210(3.75),226(3.79)	35-2691-86
11-Dodecynoic acid, ethyl ester	MeOH	206(2.92)	18-3535-86

Compound	Solvent	λ max(log ε)	Ref.
10-Undecynoic acid, 1-methylethyl ester	MeOH	205(3.15)	18-3535-86
$C_{14}H_{24}O_3$			
Ethanone, 1-(1-methoxy-2,2,6-trimethyl-9-oxabicyclo[4.2.1]non-7-yl)-	pentane	287(1.60)	33-0555-86
$C_{14}H_{26}N_4$			
1,2,4,5-Benzenetetramine, N,N,N',N',-N",N",N"',N"'-octamethyl-	MeCN	242(4.32),278(3.85), 320(3.68)	89-1023-86
Methanaminium, N,N'-[4,5-bis(dimethylamino)-3,5-cyclohexadiene-1,2-diylidene]bis[N-methyl-, iodide, (triiodide)	MeCN	290(4.74),368(4.46), 573(3.38)	89-1023-86
$C_{14}H_{26}N_6O_9$			
2-Pentanone, 3-amino-1-diazo-, tartrate (2:1)	MeOH	208(3.57),240(3.55), 280(3.78)	104-0353-86
$C_{14}H_{26}O_3$			
2-Butanone, 4-(3-methoxy-3,7,7-trimethyl-2-oxa-1-cycloheptylidene)-	pentane	279(1.96)	33-0555-86
$C_{14}H_{27}N_2$			
[5-(Diethylamino)-3-methyl-2,4-pentadienylidene]diethylammonium perchlorate	EtOH	414s(4.67),437(5.09)	33-1588-86
$C_{14}H_{30}N_2$			
Diazene, bis[2-methyl-1-(1-methylethyl)propyl]-	hexane	377(1.28)	44-2969-86
$C_{14}H_{32}N_4O_2Os$			
Osmium, dioxo(1,4,8,11-tetramethyl-1,4,8,11-tetraazacyclotetradecane-N^1,N^4,N^8,N^{11})-	MeCN	270(3.2),305f(3.2)	35-4644-86
$C_{14}H_{32}N_8O_4$			
1,2-Ethanediaminium, N,N'-bis[2-[[3-(methylnitrosoamino)carbonyl]amino]ethyl]-N,N,N',N'-tetramethyl-, dibromide	H_2O	198(4.30),235(4.04), 395(2.23)	70-0555-86
$C_{14}H_{32}P_3Re$			
Rhenium, (η^5-2,4-cyclopentadien-1-yl)tris(trimethylphosphine)-	hexane	234(4.20)	35-4856-86
$C_{14}H_{42}Si_7$			
Cyclopentasilane, 1,1,2,2,3,3,4,4-octamethyl-5,5-bis(trimethylsilyl)-	C_6H_{12}	241s(3.9),281(2.9)	157-0128-86
$C_{14}K_2N_8S$			
Dipotassium 2,3-bis(cyanimino)-2',3'-bis(dicyanomethylene)-1,1'-thiodicyclopropanide	H_2O	236(4.41),286(4.56), 327(4.17)	24-2104-86
$C_{14}N_4O_2S_4$			
Benzo[1,2-b:4,5-b']bis[1,4]dithiin-2,3,7,8-tetracarbonitrile, radical anion, salt with N-methylpyridinium	MeCN	216(4.46),269s(4.69), 281(4.74),296(4.70), 357s(3.79),405(3.83),	77-1489-86

Compound	Solvent	$\lambda_{max}(\log \epsilon)$	Ref.
(cont.)		430(3.81),736(3.56)	77-1489-86
Propanedinitrile, 2,2'-(4,8-dihydro-4,8-dioxobenzo[1,2-d:4,5-d']bis-[1,3]dithiole-2,6-diylidene)bis-	MeCN	220s(4.25),232(4.35), 261(3.99),267(4.02), 277(4.10),308(3.89), 369s(4.56),384(4.78), 444s(4.09),465(4.14), 476(4.16)	77-1489-86

Compound	Solvent	$\lambda_{max}(\log \epsilon)$	Ref.
$C_{15}H_5F_{14}N_3O$			
Pyridinium 2,6-bis(heptafluoropropyl)-4-oxo-5(4H)-pyrimidinylide	EtOH	232.5(4.14),259(3.92),307.5(3.62)	39-1769-86C
$C_{15}H_6ClN_3O$			
Ethenetricarbonitrile, [5-(4-chlorophenyl)-2-furanyl]-	acetone	461(3.48)	73-1450-86
$C_{15}H_6ClN_3O_4$			
5H-Naphtho[2,1-b]pyrido[2,3-e][1,4]-oxazin-5-one, 6-chloro-1-nitro-	$CHCl_3$	246(4.36),259(4.25),297(4.08),309s(4.08),336(4.12),457(4.16)	4-1697-86
5H-Naphtho[2,1-b]pyrido[2,3-e][1,4]-oxazin-5-one, 6-chloro-4-nitro-	$CHCl_3$	245(4.35),259(4.24),295(4.01),342(4.05),457(4.18)	4-1697-86
$C_{15}H_6Cl_2N_2O$			
Cyanamide, (1,5-dichloro-10-oxo-9(10H)-anthracenylidene)-	MeCN	216(4.28),261(4.40),286(4.05),345(3.80)	5-0142-86
$C_{15}H_6Cl_2N_2OS$			
5H-Naphtho[2,1-b]pyrido[3,2-e][1,4]-thiazin-5-one, 6,9-dichloro-	MeOH	245(4.25),251(4.26),328(3.78),462(3.75)	78-2771-86
$C_{15}H_6F_5N_5O$			
Pyrimido[1,2-a]purin-10(1H)-one, 1-methyl-7-(pentafluorophenyl)-	MeCN	233(4.39),269(4.39),330(3.76)	44-3244-86
$C_{15}H_7BrN_2O_2$			
Naphtho[2,1-b]pyrido[2,3-e][1,4]-oxazin-5-one, 6-bromo-	MeOH	234(4.30),242(4.26),256(4.15),263(4.14),350(4.09),442(4.10)	4-1697-86
$C_{15}H_7ClN_2OS$			
5H-Naphtho[2,1-b]pyrido[3,2-e][1,4]-thiazin-5-one, 6-chloro-	MeOH	245(4.22),250(4.22),325(3.64),498(3.38)	78-2771-86
$C_{15}H_7N_3O$			
Ethenetricarbonitrile, (5-phenyl-2-furanyl)-	MeOH	467(3.23)	73-1450-86
$C_{15}H_8ClF_3N_2$			
Quinoxaline, 2-chloro-3-phenyl-6-(trifluoromethyl)-	EtOH	206(4.56),243(4.50),266s(4.00),332(3.98)	150-0277-86M
Quinoxaline, 3-chloro-2-phenyl-6-(trifluoromethyl)-	EtOH	205(4.59),242(4.56),272s(3.95),332s(4.04)	150-0277-86M
$C_{15}H_8ClNO_5$			
1,4-Anthracenedione, 10-chloro-9-methoxy-5-nitro-	benzene	407(3.73)	104-0129-86
$C_{15}H_8ClN_3O_2$			
5H-Naphtho[2,1-b]pyrido[2,3-e][1,4]-oxazin-5-one, 1-amino-5-chloro-	$CHCl_3$	266(4.11),357(4.00),375s(3.97),430(4.09),524(4.03)	4-1697-86
5H-Naphtho[2,1-b]pyrido[2,3-e][1,4]-oxazin-5-one, 4-amino-6-chloro-	$CHCl_3$	264(4.14),296(4.01),356(4.12),402s(3.90),421(3.93),446(3.85),538(4.07)	4-1697-86

Compound	Solvent	λ_{max}(log ε)	Ref.
$C_{15}H_8Cl_4N_2$			
4-Quinolinamine, 2,3,6-trichloro-N-(4-chlorophenyl)-	EtOH	239(4.62),353(3.91)	104-0973-86
$C_{15}H_8Cl_4N_8$			
1H-Tetrazole, 5,5'-(dichloromethylene)bis[1-(4-chlorophenyl)-	EtOH	220(4.31)	104-0177-86
$C_{15}H_8CrO_4S$			
Chromium, tricarbonyl[(1,2,3,4,4a,10a-η)-phenoxathiin]-	EtOH	222(4.62),330(3.88)	104-0138-86
$C_{15}H_8CrO_5$			
Chromium, tricarbonyl[(1,2,3,4,4a,10a-η)-dibenzo[b,e][1,4]dioxin]-	EtOH	215(4.77),327(3.98)	104-0138-86
$C_{15}H_8F_6N_4O_2$			
2,4(1H,3H)-Pteridinedione, 1-(phenylmethyl)-6,7-bis(trifluoromethyl)-	MeOH	255(4.12),335(3.71)	39-1051-86C
$C_{15}H_8N_2O_2$			
5H-Naphtho[2,1-b]pyrido[2,3-e][1,4]-oxazin-5-one	MeOH	232(4.15),239(4.13), 256(4.25),345(4.05), 424(4.16)	4-1697-86
$C_{15}H_8O_3$			
6H-Benzofuro[3,2-c][1]benzopyran-6-one	EtOH	249(4.49),270(4.47), 286(4.31),338(4.11), 365(3.43)	149-0001-86B
$C_{15}H_8O_4$			
2-Anthracenecarboxaldehyde, 9,10-dihydro-1-hydroxy-9,10-dioxo-	n.s.g.	255(4.44),285(4.06), 339(3.40),412(3.78)	12-2075-86
$C_{15}H_8O_5$			
6H-Benzofuro[3,2-c][1]benzopyran-6-one, 1,8-dihydroxy-	MeOH	236(4.38),288(4.20), 345(4.29)	42-1060-86
6H-Benzofuro[3,2-c][1]benzopyran-6-one, 2,8-dihydroxy-	MeOH	240(4.43),273(4.06), 348(4.22)	42-1060-86
6H-Benzofuro[3,2-c][1]benzopyran-6-one, 3,8-dihydroxy-	MeOH	235(4.45),275(4.09), 345(4.21)	42-1060-86
$C_{15}H_8O_6$			
6H-Benzofuro[3,2-c][1]benzopyran-6-one, 1,3,8-trihydroxy-	MeOH	240(4.41),280(4.12), 348(4.25)	42-1060-86
6H-Benzofuro[3,2-c][1]benzopyran-6-one, 1,8,9-trihydroxy-	MeOH	235(4.41),285(4.10), 340(4.27)	42-1060-86
6H-Benzofuro[3,2-c][1]benzopyran-6-one, 2,8,9-trihydroxy-	MeOH	235(4.42),278(4.09), 335(4.20)	42-1060-86
6H-Benzofuro[3,2-c][1]benzopyran-6-one, 3,8,9-trihydroxy-	MeOH	240(4.38),270(4.08), 345(4.30)	42-1060-86
$C_{15}H_8O_7$			
6H-Benzofuro[3,2-c][1]benzopyran-6-one, 1,3,8,9-tetrahydroxy-	MeOH	245(4.40),275(4.10), 330(4.28)	42-1060-86
$C_{15}H_9BrN_4$			
1H-Pyrazolo[3,4-b]quinoxaline, 1-(4-bromophenyl)-	MeOH	271(5.03),333(4.35)	2-0215-86

Compound	Solvent	$\lambda_{max}(\log \epsilon)$	Ref.
$C_{15}H_9BrO_2$			
4H-1-Benzopyran-4-one, 2-(4-bromo-phenyl)-	EtOH	256(4.49),306(4.37)	104-1658-86
$C_{15}H_9ClN_4$			
1H-Pyrazolo[3,4-b]quinoxaline, 1-(4-chlorophenyl)-	MeOH	269.8(5.18),331.6(4.54)	2-0901-86
1H-Pyrazolo[3,4-b]quinoxaline, 3-chloro-1-phenyl-	MeOH	267.8(4.89),334.2(4.28)	2-0960-86
$C_{15}H_9ClO_2S$			
Benzo[b]thiophene-6-carboxylic acid, 4-(4-chlorophenyl)-	EtOH	231(4.50),248(4.52), 279(4.13)	5-1003-86
$C_{15}H_9ClO_3$			
1,4-Anthracenedione, 9-chloro-10-meth-oxy-	benzene	414(3.76)	104-0129-86
$C_{15}H_9Cl_2NO_5S$			
Benzenesulfonamide, N-(4,7-dichloro-1,3-dihydro-1,3-dioxo-5-isobenzo-furanyl)-4-methyl-	MeCN	231(4.54),250s(4.32), 285(3.85),335(3.75)	24-0616-86
$C_{15}H_9CrNO_4$			
Chromium, tricarbonyl[(1,2,3,4,4a,10a-η)-10H-phenoxazine]-	EtOH	223(4.74),318(4.08)	104-0138-86
$C_{15}H_9F_3N_2$			
Quinoxaline, 2-phenyl-6-(trifluoro-methyl)-	EtOH	208(4.55),242(4.33), 250(4.34),257(4.33), 264s(4.30),292(3.89), 334(4.12)	150-0277-86M
Quinoxaline, 2-phenyl-7-(trifluoro-methyl)-	EtOH	207(4.57),244s(4.33), 251(4.35),259s(4.34), 263s(4.32),335(4.08), 352(3.90)	150-0277-86M
$C_{15}H_9F_3N_2O$			
Quinoxaline, 2-phenyl-6-(trifluoro-methyl)-, 4-oxide	EtOH	210(4.30),246(4.27), 265s(4.31),273(4.38), 292(4.19),301s(4.14), 332(4.02),344s(3.99), 364s(3.76)	150-0277-86M
Quinoxaline, 2-phenyl-7-(trifluoro-methyl)-, 4-oxide	EtOH	206(4.32),251(4.31), 263(4.29),272(4.32), 291(4.23),295s(4.18), 328(3.98),348(3.86), 364(3.70)	150-0277-86M
2(1H)-Quinoxalinone, 3-phenyl-6-(tri-fluoromethyl)-	EtOH	205(4.46),225(4.39), 302(4.00),352(4.07)	150-0277-86M
2(1H)-Quinoxalinone, 3-phenyl-7-(tri-fluoromethyl)-	EtOH	205(4.54),220(4.49), 340(4.14),360(4.10)	150-0277-86M
$C_{15}H_9N$			
1-Azapyrene	EtOH	237(4.59),261(4.17), 271(4.28),304(3.62), 323(4.10),335(4.26), 351(3.94),361(3.58), 369(4.08)	4-0747-86
2-Azapyrene	EtOH	239(4.66),261(4.26), 271(4.29),294(3.61),	4-0747-86

Compound	Solvent	$\lambda_{max}(\log \epsilon)$	Ref.
(cont.)		304(3.99),317(4.38), 332(4.58),356(3.56), 370(3.20),375(3.79)	4-0747-86
$C_{15}H_9NO$			
1-Azapyren-2-ol	EtOH	238(4.55),250(4.54), 258(4.60),283(3.95), 320(3.51),334(3.97), 350(4.24),369(3.48), 390(3.85),409(4.06), 432(4.01)	4-0747-86
2-Azapyren-1-ol	EtOH	236(4.51),250(4.43), 274(4.29),285(4.44), 346(4.09),381(3.82), 403(3.96),426(3.92)	4-0747-86
$C_{15}H_9NO_6$			
9,10-Anthracenedione, 1-hydroxy-8-methoxy-2-nitro-	$CHCl_3$	254(4.29),268(4.18), 280(4.08),416(3.90), 432(3.85)	78-3303-86
9,10-Anthracenedione, 4-hydroxy-5-methoxy-1-nitro-	$CHCl_3$	262(4.31),280(4.08), 416(3.92),432(3.88)	78-3303-86
$C_{15}H_9N_3O_2$			
5H-Naphtho[2,3-b]pyrido[2,3-e][1,4]-oxazin-5-one, 1-amino-	$CHCl_3$	263s(4.06),354(3.96), 368(3.97),415(4.09), 517(3.99)	4-1697-86
5H-Naphtho[2,3-b]pyrido[2,3-e][1,4]-oxazin-5-one, 4-amino-	$CHCl_3$	260s(4.15),295(4.01), 354(4.09),392s(3.97), 410(3.98),437s(3.86), 525(4.03)	4-1697-86
$C_{15}H_9N_3O_5$			
Isoquinolinium, 2-(2-hydroxy-3,5-di-nitrophenyl)-, hydroxide, inner salt	EtOH	357(4.09)	5-0947-86
$C_{15}H_9N_5$			
5H-Benzimidazo[1',2':1,2]imidazo[4,5-b]quinoxaline	DMF	359(4.28)	2-1057-86
$C_{15}H_{10}BrClN_4S$			
6H-Thieno[3,2-f][1,2,4]triazolo[4,3-a][1,4]diazepine, 2-bromo-4-(2-chlorophenyl)-9-methyl-	pH 6.7	209(--),243(4.31)	160-1853-86
	pH <2.3	209(--),263(--), 290s(--)	160-1853-86
$C_{15}H_{10}ClNO_3S$			
9H-Thieno[2',3':4,5]furo[3,2-b]indole-2-carboxylic acid, 3-chloro-, ethyl ester	MeOH	370(4.77)	73-1685-86
$C_{15}H_{10}ClNO_5S$			
Thieno[3,2-b]furan-5-carboxylic acid, 6-chloro-2-(2-nitrophenyl)-, ethyl ester	MeOH	319(3.26)	73-1685-86
$C_{15}H_{10}ClN_3O_2$			
1H-Pyrazole, 3-(4-chlorophenyl)-5-(4-nitrophenyl)-	PrOH	262(4.51),317(4.44)	30-0176-86

Compound	Solvent	λ_{max}(log ε)	Ref.
$C_{15}H_{10}Cl_2$			
Anthracene, 1,8-dichloro-9-methyl-	C_6H_{12}	263(5.05),335s(3.11), 350(3.42),369(3.72), 389(3.91),412(3.82)	44-0921-86
$C_{15}H_{10}Cl_2N_2$			
4-Quinolinamine, 2,3-dichloro- 4-phenyl-	EtOH	235(4.64)	104-0973-86
$C_{15}H_{10}Cl_2N_8$			
1H-Tetrazole, 5,5'-(dichloromethyl- ene)bis[1-phenyl-	EtOH	248(4.28)	104-0177-86
1H-Tetrazole, 5,5'-methylenebis[1-(4- chlorophenyl)-	EtOH	229(4.19)	104-0177-86
$C_{15}H_{10}Cl_2O$			
Anthracene, 1,8-dichloro-9-methoxy-	C_6H_{12}	260(5.09),347(3.49), 365(3.82),385(4.00), 402s(3.78),407(3.93)	44-0921-86
$C_{15}H_{10}Cl_2O_3S$			
Thieno[3,2-b]furan-5-carboxylic acid, 3,6-dichloro-2-phenyl-, ethyl ester	MeOH	336(3.46)	73-1685-86
$C_{15}H_{10}Cl_4N_2$			
Propanediimidoyl dichloride, 2,2-di- chloro-N,N'-diphenyl-	EtOH	300(3.74)	104-0177-86
$C_{15}H_{10}CrO_3$			
Chromium, [(1,2,3,4,4a,10a-η)-benzo- cyclooctene]tricarbonyl-	C_6H_{12}	240s(4.35),280s(3.94), 305(3.85),340(3.83)	77-1518-86
Chromium, [(5,6,7,8,9,10-η)-benzo- cyclooctene]tricarbonyl-	C_6H_{12}	250(3.96),335(4.03), 410(3.32)	77-1518-86
$C_{15}H_{10}F_6N_4O_3$			
2,4(1H,3H)-Pteridinedione, 7,8-dihydro- 7-hydroxy-8-(phenylmethyl)-6,7- bis(trifluoromethyl)-	pH 1	267(4.19),340(3.78)	39-1051-86C
	pH 12	250(4.19),275(4.01), 345(3.88)	39-1051-86C
$C_{15}H_{10}N_2$			
5H-Cyclohepta[4,5]pyrrolo[2,3-b]- indole	MeOH	240(4.30),284(4.33), 296(4.32),325(4.57), 472(3.17),495(3.16), 545s(2.84)	18-3326-86
	MeOH-HCl	263(4.30),282(4.31), 335(4.25),415s(3.94)	18-3326-86
6H-Indeno[1,2-b][1,8]naphthyridine	EtOH	254(4.43),331(4.44), 345(4.45)	4-0689-86
5H-Pyrido[3',2':4,5]cyclopenta[1,2-b]- quinoline	EtOH	220(4.13),246(4.15), 270(4.05),312(4.03), 318(4.06),326(4.10), 332(4.12),341(4.14)	4-0689-86
$C_{15}H_{10}N_2O_2$			
2H-Benz[f]isoindole-2-acetic acid, 1-cyano-, sodium salt	MeOH	252(4.67),400s(3.50), 418(3.70),442(3.68)	44-3978-86
$C_{15}H_{10}N_4O$			
3H-Pyrazolo[3,4-b]quinoxalin-3-one, 1,2-dihydro-1-phenyl-	MeOH	283.8(5.16)	2-0960-86
	MeOH-NaOH	300(5.18)	2-0960-86

Compound	Solvent	λ max(log ε)	Ref.
$C_{15}H_{10}N_4O_4S$			
2-Thiazolimine, dihydro-3-(2,4-dinitrophenyl)-N-phenyl-	n.s.g.	420(3.41)	150-0110-86S
$C_{15}H_{10}N_4O_5S$			
2-Benzothiazolamine, N-[(2,4-dinitrophenyl)methylene]-6-methoxy-	dioxan	423(4.11)	80-0273-86
	+ CF_3COOH	423(--)	80-0273-86
$C_{15}H_{10}N_4S$			
Propanedinitrile, (5,6-dihydro-2-phenylthieno[2,3-d]pyrimidin-4-yl)-	$CHCl_3$	267(4.40),342(4.29), 388(4.01)	94-0516-86
$C_{15}H_{10}O_2$			
4H-1-Benzopyran-4-one, 2-phenyl- (flavone)	EtOH	253(4.31),293(4.41)	104-1658-86
$C_{15}H_{10}O_2S$			
Benzo[b]thiophene-6-carboxylic acid, 4-phenyl-	EtOH	225(4.51),244(4.53), 278(4.08),317(3.85)	5-1003-86
$C_{15}H_{10}O_3$			
Cyclohepta[de]naphthalene-7,8-dione, 1-methoxy-	$CHCl_3$	254(4.30),316(3.67), 396(4.11)	39-1965-86C
Cyclohepta[de]naphthalene-7,8-dione, 3-methoxy-	$CHCl_3$	252(4.35),260(4.37), 370(4.23)	39-1965-86C
Cyclohepta[de]naphthalene-7,8-dione, 6-methoxy-	$CHCl_3$	266(4.36),363(4.10), 368(3.74),390s(3.98)	39-1965-86C
Propanetrione, diphenyl-	n.s.g.	450(1.65)	89-0999-86
$C_{15}H_{10}O_4$			
1,2-Anthracenedione, 7,8-dihydroxy-3-methyl-	MeOH	265(4.53),316(4.39), 570(3.77)	100-0145-86
6H-Dibenzo[b,d]pyran-6-one, 2-acetyl-3-hydroxy-	n.s.g.	253(4.29),345(--)	2-0619-86
$C_{15}H_{10}O_5$			
9,10-Anthracenedione, 1,3-dihydroxy-4-methoxy-	MeOH	248(4.44),263s(4.16), 285(4.14),430(3.83)	78-3767-86
4H-1-Benzopyran-4-one, 5,7-dihydroxy-2-(2-hydroxyphenyl)-	MeOH	267(4.35),335(3.83)	94-1667-86
	MeOH-NaOMe	273(4.35),399(3.94)	94-1667-86
4H-1-Benzopyran-4-one, 5,7-dihydroxy-2-(4-hydroxyphenyl)- (apigenin)	MeOH	268(4.11),333(4.15)	94-1667-86
	MeOH-NaOMe	275(4.19),388(4.49)	94-1667-86
4H-1-Benzopyran-4-one, 5,7-dihydroxy-3-(4-hydroxyphenyl)-	EtOH	263(4.42),329s(3.81)	105-0107-86
4H-1-Benzopyran-4-one, 3,5,7-trihydroxy-2-phenyl- (galangin)	EtOH	268(4.20),362(4.01)	105-0107-86
6H-Dibenzo[b,d]pyran-6-one, 2-acetyl-1,3-dihydroxy-	n.s.g.	264(4.30),340(4.0)	2-0619-86
$C_{15}H_{10}O_6$			
9,10-Anthracenedione, 1,3,5-trihydroxy-2-methoxy-	MeOH	253(4.19),281(4.43), 420(4.05)	78-3767-86
9,10-Anthracenedione, 1,3,5-trihydroxy-4-methoxy-	MeOH	232(4.39),249(4.30), 270s(4.10),290(4.11), 335(3.40),442(4.04)	78-3767-86
2H,8H-Benzo[1,2-b:3,4-b']dipyran-2,8-dione, 4-(acetoxymethyl)-	EtOH	291(4.33)	94-0390-86
4H-1-Benzopyran-4-one, 2-(2,6-dihydroxyphenyl)-5,7-dihydroxy-	MeOH	259(4.23),308(4.00), 330s(3.75)	94-1667-86
	MeOH-NaOMe	266(4.29),335(4.00)	94-1667-86

Compound	Solvent	$\lambda_{max}(\log \epsilon)$	Ref.
4H-1-Benzopyran-4-one, 3,5,7-trihydroxy-2-(4-hydroxyphenyl)-	EtOH	266(3.99),370(4.16)	105-0601-86
Pyrano[2,3-b]benzobyran-2,5-dione, 4-(acetoxymethyl)-	EtOH	237.5(3.91),257(4.04), 266.5(4.03),333(4.09), 345(4.14)	94-0390-86
$C_{15}H_{11}BrN_4OS$			
Benzamide, N-[3-(3-bromophenyl)-1,5-dihydro-5-thioxo-4H-1,2,4-triazol-4-yl]-	EtOH	265(4.09)	4-1451-86
$C_{15}H_{11}ClN_2O$			
2H-1,4-Benzodiazepin-2-one, 7-chloro-1,3-dihydro-5-phenyl- (in 20% MeOH)	neutral	230(4.54),250s(4.23), 310(3.30)	96-1051-86
	cation	236(4.48),282(4.13), 330(3.60)	96-1051-86
$C_{15}H_{11}ClN_2OS$			
2-Benzothiazolamine, N-[(4-chlorophenyl)methylene]-6-methoxy-	dioxan	376(4.28)	80-0273-86
	+ CF_3COOH	376(--)	80-0273-86
$C_{15}H_{11}ClOS$			
Dibenzo[b,f]thiepin, 2-chloro-6-methoxy-	MeOH	263(4.24),313(3.89)	73-0141-86
Dibenzo[b,f]thiepin-10(11H)-one, 2-chloro-6-methyl-	MeOH	235(4.33),260(4.00), 335(3.58)	73-0141-86
2-Propen-1-one, 3-chloro-1-phenyl-3-(phenylthio)-	n.s.g.	263(4.22),325(4.55)	48-0539-86
$C_{15}H_{11}ClO_2$			
4H-1-Benzopyran-4-one, 6-chloro-2,3-dihydro-2-phenyl-	C_6H_{12}	245(3.90),324(3.54), 368s(1.00)	39-1217-86B
	EtOH	250(3.81),328(3.51)	39-1217-86B
$C_{15}H_{11}ClO_5$			
9H-Xanthen-9-one, 4-chloro-1,7-dihydroxy-3-methoxy-5-methyl-	MeOH	262(4.57),307(3.98), 382(3.77)	94-0858-86
$C_{15}H_{11}ClS$			
Dibenzo[b,f]thiepin, 2-chloro-6-methyl-	MeOH	224(4.51),267(4.30), 303(3.71)	73-0141-86
$C_{15}H_{11}Cl_2NO_6S$			
1,2-Benzenedicarboxylic acid, 3,6-dichloro-4-[[(4-methylphenyl)sulfonyl]amino]-	MeCN	219(4.56),228s(4.51), 253(4.05),269s(3.85), 275s(3.68),292(3.34), 300s(3.30)	24-0616-86
$C_{15}H_{11}Cl_3OTe$			
Tellurium, trichloro(3-oxo-1,3-diphenyl-1-propenyl-C,O)-	CH_2Cl_2	311(4.28)	44-1692-86
$C_{15}H_{11}FeIN_2$			
Ferrocene, [2,2-dicyano-1-(iodomethyl)ethenyl]-	CH_2Cl_2	343(4.029),568(3.535)	70-2105-86
$C_{15}H_{11}NO$			
Benzonitrile, 4-[2-(2-hydroxyphenyl)ethenyl]-, (E)-	n.s.g.	342(4.42)	150-0433-86S
Benzoxazole, 2-(2-phenylethenyl)-	hexane	318(4.56)	103-1262-86

Compound	Solvent	$\lambda_{max}(\log \epsilon)$	Ref.
Oxazole, 2,5-diphenyl-	toluene	307(4.41)	103-1026-86
	EtOH	304(4.43)	135-0244-86
2(1H)-Quinolinone, 4-phenyl-	CH_2Cl_2	225(4.57),230s(4.53), 275s(3.86),279(3.87), 322s(3.675),332(3.74), 345s(3.595)	24-3109-86
$C_{15}H_{11}NOS$			
1,5-Benzothiazepin-4(5H)-one, 2-phenyl-	MeOH	230(4.20),258.5(4.22), 295s(3.68),345s(3.06), 370s(2.83),395s(2.29)	24-3109-86
	CH_2Cl_2	232(4.20),260(4.22), 290s(3.73),298s(3.69), 328s(3.56),368s(2.99), 395s(2.37)	24-3109-86
2H-1,4-Benzothiazin-3(4H)-one, 2-(phenylmethylene)-, (E)-	CH_2Cl_2	237(4.32),255s(4.17), 265s(4.13),283s(3.96), 315s(3.76),343(3.90), 365s(3.79)	24-3109-86
(Z)-	80% MeOH	235s(4.29),260(4.19), 264(4.19),285s(3.88), 297s(3.71),316s(3.64), 335s(3.86),343s(3.95), 356(4.01),370s(3.95), 390s(3.65)	24-3109-86
$C_{15}H_{11}NOSe$			
2-Benzofuranselenol, 3-[(phenylimino)methyl]-	EtOH	263(4.37),297(4.56), 333(3.96),478(4.30)	103-0608-86
$C_{15}H_{11}NO_2$			
Methanone, (3-hydroxy-1H-indol-2-yl)-phenyl-	EtOH	208(4.21),222(4.17), 256(4.19),334(4.22), 400(3.81)	104-2186-86
$C_{15}H_{11}NO_2S_2$			
Benzenecarbodithioic acid, 2-(4-nitrophenyl)ethenyl ester, (Z)-	CH_2Cl_2	328(4.42),512(2.38)	18-1403-86
$C_{15}H_{11}NO_3$			
2-Benzofurancarboxylic acid, 5-(3-pyridinylmethyl)-, sodium salt	EtOH	211(4.45),261s(4.30), 268(4.35),276s(4.16), 288s(3.89),299(3.76)	87-1461-86
$C_{15}H_{11}NO_4$			
9,10-Anthracenedione, 1-amino-4-hydroxy-5-methoxy-	$CHCl_3$	248(4.31),284(3.91), 280(3.45),506(3.91), 523(3.98),564(3.83)	78-3303-86
9,10-Anthracenedione, 2-amino-1-hydroxy-8-methoxy-	$CHCl_3$	240(4.32),256(4.42), 268(4.22),288(4.00), 308(3.89),394(3.73), 472(3.80)	78-3303-86
Dielsiquinone	EtOH	247s(4.07),274(4.27), 291(4.25),322s(4.04)	102-1691-86
	EtOH-NaOH	251(4.23),270(4.22), 332(4.19),363s(4.03)	102-1691-86
	EtOH-NaOAc	250(4.18),273(4.18), 324(4.17),361s(4.03)	102-1691-86
$C_{15}H_{11}NO_4S$			
1H-Indene-2,3-dione, 1-[(4-methyl-	$CHCl_3$	277(3.69),299(3.78),	104-2173-86

Compound	Solvent	$\lambda_{max}(\log \epsilon)$	Ref.
phenyl)sulfonyl]- (cont.)		392(2.83)	104-2173-86
$C_{15}H_{11}NS$			
Benzothiazole, 2-(2-phenylethenyl)-, (E)-	80% MeOH	255(3.75),263(3.79), 315s(4.45),334s(4.58), 340(4.59),356s(4.40)	24-3109-86
	CH_2Cl_2	230s(--),238s(--), 257(--),264(--), 305s(--),315s(--), 332s(--),339.5(--), 355s(--)	24-3109-86
(Z)-	80% MeOH	246s(3.99),257(3.94), 264(3.94),290s(4.09), 301(4.13),330s(3.87)	24-3109-86
$C_{15}H_{11}N_3O_2$			
1H-Benzo[h]pyrimido[1,2-c]quinazolin-1-one, 5,6-dihydro-3-hydroxy-	EtOH	230(4.10),239(4.05), 254(4.20),264(4.25), 289(3.78),320(3.76), 330(3.90),345(4.01)	142-1119-86
2-Quinoxalinecarboxylic acid, 3-(phenylamino)-, sodium salt	MeOH	283.6(4.99),398(4.26)	2-0215-86
$C_{15}H_{11}N_3O_3$			
1,2,4-Benzenetriol, 5-(2-quinolinylazo)-	n.s.g.	420(4.3)	112-0669-86
$C_{15}H_{11}N_3O_3S$			
2-Benzothiazolamine, 6-methoxy-N-[(2-nitrophenyl)methylene]-	dioxan	390(4.08)	80-0273-86
	+ CF_3COOH	390(--)	80-0273-86
2-Benzothiazolamine, 6-methoxy-N-[(3-nitrophenyl)methylene]-	dioxan	380(4.23)	80-0273-86
	+ CF_3COOH	380(--)	80-0273-86
2-Benzothiazolamine, 6-methoxy-N-[(4-nitrophenyl)methylene]-	dioxan	402(4.23)	80-0273-86
	+ CF_3COOH	402(--)	80-0273-86
$C_{15}H_{11}N_4O_{11}S$			
2-Propanone, 1-[5-[(2,4-dinitrophenyl)sulfonyl]-2,4-dinitro-2,4-cyclohexadien-1-yl]-, ion(1-)-, potassium	acetone	366(3.98),536(4.37)	104-1138-86
$C_{15}H_{11}N_5$			
1H-Pyrazolo[3,4-b]quinoxalin-3-amine, 1-phenyl-	MeOH	290.6(5.13),335s(--)	2-0960-86
$C_{15}H_{11}N_5O$			
1,1,2,2-Ethanetetracarbonitrile, 1-(4,5,6,7-tetrahydro-3-methyl-4-oxo-1H-indol-5-yl)-	MeOH	202(4.22),234(4.30), 440(4.38)	103-0172-86
Pyrimido[1,2-a]purin-10(1H)-one, 7-phenyl-	MeCN	240(4.49),274(4.45), 326(3.81),377s(3.41)	44-3244-86
$C_{15}H_{11}N_5O_2$			
Spiro[3H-indole-3,4'(1'H)-pyrano[2,3-c]pyrazole]-5'-carbonitrile, 6'-amino-1,2-dihydro-3'-methyl-2-oxo-	EtOH	253(4.18)	18-1235-86
$C_{15}H_{11}OTe$			
1,2-Oxatellrol-1-ium, 3,5-diphenyl-, tribromide	CH_2Cl_2	310(4.23)	44-1692-86

Compound	Solvent	λ_{max}(log ε)	Ref.
$C_{15}H_{12}$			
Bicyclo[4.1.0]hepta-1,3,5-triene, 7-(1-phenylethylidene)-	C_6H_{12}	242(4.04),249(4.09), 256s(3.95),?s(4.15), 371(4.26),394(4.12)	35-5949-86
$C_{15}H_{12}BrN_3O$			
9H-Pyrido[3,4-b]indol-8-ol, 6-bromo-1-(3,4-dihydro-2H-pyrrol-5-yl)- (eudistomidin A)	MeOH	223(4.52),254(4.23), 371(3.74)	88-1191-86
$C_{15}H_{12}ClNO_5$			
2-Propenoic acid, 2-chloro-3-[5-(2-nitrophenyl)-2-furanyl]-, ethyl ester	MeOH	399(3.55)	73-1685-86
2-Propenoic acid, 2-chloro-3-[5-(3-nitrophenyl)-2-furanyl]-, ethyl ester	MeOH	350(3.54)	73-1685-86
2-Propenoic acid, 2-chloro-3-[5-(4-nitrophenyl)-2-furanyl]-, ethyl ester	MeOH	382(3.57)	73-1685-86
$C_{15}H_{12}ClN_3O_2$			
Acetic acid, (2-chloro-6-phenyl-4(1H)-pyrimidinylidene)cyano-, ethyl ester, (Z)-	EtOH	234(4.08),300(4.27), 376(4.18)	103-0774-86
sodium salt	EtOH	236(4.27),263(4.11), 310(4.17),364(4.08)	103-0774-86
$C_{15}H_{12}ClN_3O_3S$			
Ethanediamide, [6-chloro-2-[(2-oxo-2-phenylethyl)thio]-3-pyridinyl]-	EtOH	208(4.18),245(4.05), 283(3.84)	103-0800-86
$C_{15}H_{12}Cl_2N_2$			
Benzenamine, 4-chloro-N-[3-(4-chlorophenyl)amino]-2-propenylidene]-	pH 1	387(4.84)	94-1794-86
	pH 7	389(4.76)	94-1794-86
	pH 13	364(--)(unstable)	94-1794-86
	MeOH	364(4.67)	94-1794-86
	EtOH	364(4.68)	94-1794-86
Quinoxaline, 5,7(and 6,8)-dichloro-1,2-dihydro-3-(4-methylphenyl)-	MeOH	274(4.38),305(3.97), 392(3.82)	103-0918-86
	toluene	394(--)	103-0918-86
$C_{15}H_{12}Cl_2N_2O$			
Quinoxaline, 5,7(and 6,8)-dichloro-1,2-dihydro-3-(4-methoxyphenyl)-	MeOH	286(4.34),312(4.16), 385(3.96)	103-0918-86
	toluene	388(--)	103-0918-86
$C_{15}H_{12}Cl_2O_2$			
Acenaphtho[1,2-d][1,3]dioxole, 3,4-dichloro-6b,9a-dihydro-8,8-dimethyl-	isoPrOH	234(4.89),294(3.90), 306(4.03),314(3.84), 321(3.89),332(3.23)	104-2329-86
$C_{15}H_{12}Cl_3NO_3S_2$			
1,4-Epoxythiopyrano[4,3-b]indol-3(1H)-one, 4,4a,5,9b-tetrahydro-5-methyl-1-(methylthio)-4-(trichloroacetyl)-, (1α,4α,4aα,9bα)-	CH_2Cl_2	238(3.94),251(3.93), 310.8(3.49)	24-2317-86
$C_{15}H_{12}NO_2$			
Cyclohept[d]isoxazol-1-ium, 3-(4-methoxyphenyl)-, tetrafluoroborate	EtOH	245(4.23),278s(3.99), 284s(3.97)	138-1925-86

Compound	Solvent	λ_{max}(log ϵ)	Ref.
(cont.)	MeCN	233(4.51),246(4.46), 294(4.00),416(3.09)	138-1925-86
$C_{15}H_{12}N_2$			
1H-Imidazole, 2,4-diphenyl-	EtOH	207(3.8),221(3.8), 301(3.9)	97-0378B-86
1H-Pyrazole, 3,5-diphenyl-	EtOH	260(4.55)	30-0176-86
Pyrrolo[3,2-a]carbazole, 3,10-dihydro-1-methyl-	MeOH	202(4.22),229(4.07), 255(4.28),295(4.36), 328(4.38)	103-0172-86
$C_{15}H_{12}N_2O$			
4H-Pyrido[1,2-a]pyrimidin-4-one, 6-methyl-2-phenyl-	EtOH	275(4.40),363(3.89), 378(3.87)	4-1295-86
$C_{15}H_{12}N_2OS_2$			
2-Benzothiazolecarbothioamide, N-(4-methoxyphenyl)-	MeOH	253s(3.90),329(4.24), 383s(3.85)	103-1139-86
$C_{15}H_{12}N_2O_2$			
6H-[1]Benzopyrano[2,3-b]pyrrolo[2,3-e]pyridin-6-one, 3,4-dihydro-2-methyl-	MeOH	292(3.8),358(4.37)	83-0347-86
1H-Indole, 3-methyl-7-nitro-2-phenyl-	EtOH	250(4.15),267(4.09), 285(4.12),300(4.00), 323(3.47),350(3.79), 371(3.92),400(3.75)	64-0768-86B
Pyridine, 4-[5-(4-methoxyphenyl)-2-oxazolyl]-	EtOH	338(4.43)	135-0244-86
	5% HOAc	390(4.32)	135-0244-86
$C_{15}H_{12}N_2O_2S$			
Phenol, 2-[[(6-methoxy-2-benzothiazolyl)imino]methyl]-	dioxan	384(4.30)	80-0273-86
	+ CF_3COOH	384(--)	80-0273-86
$C_{15}H_{12}N_2O_3$			
1-Phenazinecarboxylic acid, 6-(1-hydroxyethyl)-	MeOH	253(4.92),367(4.20)	158-0624-86
$C_{15}H_{12}N_2O_3S$			
7H-Thiazolo[3,2-a]pyrimidinium, 2-benzoyl-3-hydroxy-5,8-dimethyl-7-oxo-, hydroxide, inner salt	MeCN	245(4.23),370(4.14)	2-0275-86
$C_{15}H_{12}N_2O_4$			
Benzamide, N-(2-acetyl-6-nitrophenyl)-	EtOH	240(4.41)	64-0768-86B
$C_{15}H_{12}N_2O_5$			
4H-Furo[3,2-b]pyrrole-5-carboxylic acid, 6-nitro-2-phenyl-, ethyl ester	MeOH	311(3.52)	73-0106-86
$C_{15}H_{12}N_2S_2$			
2-Benzothiazolecarbothioamide, N-(4-methylphenyl)-	MeOH	253s(3.95),325(4.31), 385s(3.81),459s(2.59)	103-1139-86
$C_{15}H_{12}N_3O_9S$			
2-Propanone, 1-[2,4-dinitro-5-[(4-nitrophenyl)sulfonyl]-2,4-cyclohexadien-1-yl]-, ion(1-), potassium	acetone	374(4.00),538(4.39)	104-1138-86
$C_{15}H_{12}N_4$			
1H-Indol-6-amine, 2-(1H-benzimidazol-	EtOH	332(4.18)	2-0509-86A

Compound	Solvent	λ_{max}(log ϵ)	Ref.
2-yl)- (cont.)			2-0509-86A
[1,2,4]Triazolo[1,5-a]pyridine-8-carbonitrile, 5,7-dimethyl-2-phenyl-	DMF	347(4.04)	118-0860-86
$C_{15}H_{12}N_4O$			
Benzaldehyde, 2-hydroxy-, 1-phthalazinylhydrazone	H_2O	356(4.16)	86-0627-86
	MeOH	370(4.43)	86-0627-86
	EtOH	368(4.36)	86-0627-86
	$PhCH_2OH$	359(4.31)	86-0627-86
	acetone	362(4.37)	86-0627-86
	$CHCl_3$	376(4.18)	86-0627-86
	DMF	380(4.37)	86-0627-86
2-Quinoxalinecarboxamide, 3-(phenylamino)-	MeOH	220.4(5.00),288.2(5.08)	2-0215-86
2-Quinoxalinecarboxylic acid 2-phenylhydrazide	MeOH	242.6(5.04)	2-0960-86
$C_{15}H_{12}N_4OS$			
Benzamide, N-(1,5-dihydro-3-phenyl-5-thioxo-4H-1,2,4-triazol-4-yl)-	EtOH	223(4.33),255(4.29)	4-1451-86
$C_{15}H_{12}N_4O_3$			
Benzenamine, N-methyl-2-[5-(4-nitrophenyl)-1,3,4-oxadiazol-2-yl]-	MeOH	240(4.65),295(4.82), 391(4.76)	18-1575-86
$C_{15}H_{12}N_4O_4$			
1H-Pyrazolo[3,4-b]pyrazine-5,6-dicarboxylic acid, 1-phenyl-, dimethyl ester	MeOH	268.6(5.72)	2-0215-86
$C_{15}H_{12}N_4O_7$			
Benzamide, 3,5-dinitro-N-[1-(4-nitrophenyl)ethyl]-, (+)-	EtOH	259(4.36),340(2.72)	104-1943-86
(-)-	EtOH	259(4.36),340(2.72)	104-1943-86
$C_{15}H_{12}N_6O_4$			
Benzonitrile, 2-[[4-(dimethylamino)-phenyl]azo]-3,5-dinitro-	$CHCl_3$	299(3.80),348s(3.37), 586(4.37)	48-0497-86
Pyrazolo[1,5-a]pyrimidine, 4,7-dihydro-7-(1-methyl-1H-indol-3-yl)-3,6-dinitro-	EtOH	222(4.09),279(4.13), 390(4.24)	103-1250-86
Pyrazolo[1,5-a]pyrimidine, 4,7-dihydro-7-(2-methyl-1H-indol-3-yl)-3,6-dinitro-	EtOH	222(4.09),279(3.96), 390(4.20)	103-1250-86
$C_{15}H_{12}O$			
Bicyclo[4.1.0]hepta-1,3,5-triene, 7-[(4-methoxyphenyl)methylene]-	C_6H_{12}	238s(3.88),258(3.94), 361s(3.93),377(4.09), 387s(3.95),401(4.00)	35-5949-86
	MeCN	257(4.08),356s(4.11), 372(4.24),393(4.12)	35-5949-86
Cyclopenta[ef]heptalen-2-one, 10b-methyl-	CH_2Cl_2	260s(4.36),269(4.43), 343(3.47),475(3.14)	35-4105-86
4a,9a:4b,8a-Dimethano-9H-fluoren-9-one, (4aα,4bβ,8aβ,9aα)-	hexane	195(4.11),230s(3.71), 270(3.66)	89-0367-86
2-Propen-1-one, 1,3-diphenyl-	benzene-C_6H_{12}	309(4.35)	60-1307-86
adsorbed form		405(4.50)	60-1307-86
$(C_{15}H_{12}O)_n$			
Methanone, (4-ethenylphenyl)phenyl-,	$CHCl_3$	260(4.25),336(2.34)	32-0533-86

Compound	Solvent	λ_{max}(log ϵ)	Ref.
polymer (cont.)			32-0533-86
$C_{15}H_{12}OS$			
Benzo[b]thiophene, 4-(4-methoxy-phenyl)-	EtOH	229(4.38),276(4.10)	5-1003-86
Dibenzo[b,f]thiepin-10(11H)-one, 6-methyl-	MeOH	241(4.29),265s(3.97), 332(3.59)	73-0141-86
1(2H)-Naphthalenone, 3,4-dihydro-2-(2-thienylmethylene)-, (E)-	EtOH	269(4.07),352(4.30)	4-0135-86
Thiophene, 2-(4-methoxy-1-naphthalenyl)-	MeOH	218(4.57),238s(4.37), 310(4.02)	18-0083-86
$C_{15}H_{12}O_2$			
4H-1-Benzopyran-4-one, 2,3-dihydro-2-phenyl-	C_6H_{12}	245(3.94),312(3.56), 347s(1.40),365s(1.18)	39-1217-86B
	EtOH	250(3.98),317(3.56)	39-1217-86B
1(2H)-Naphthalenone, 2-(2-furanyl-methylene)-3,4-dihydro-, (E)-	EtOH	260(3.89),353(4.25)	4-0135-86
$C_{15}H_{12}O_3$			
4H-1-Benzopyran-4-one, 2,3-dihydro-7-hydroxy-3-phenyl-	MeOH	218(4.43),228s(4.32), 273(4.45),312(4.23)	39-0215-86C
Cyclohepta[de]naphthalene-7,8-dione, 9,10-dihydro-1-methoxy-	$CHCl_3$	258(4.42),346(3.86)	39-1965-86C
Cyclohepta[de]naphthalene-7,8-dione, 9,10-dihydro-3-methoxy-	$CHCl_3$	275(3.34),280s(3.36), 284s(3.39),289s(3.41), 323(4.01),339(3.98)	39-1965-86C
Cyclohepta[de]naphthalene-7,8-dione, 9,10-dihydro-6-methoxy-	$CHCl_3$	247(4.39),313(3.65), 346(3.54)	39-1965-86C
Naphtho[1,8-bc]pyran-7,8-dione, 3,6,9-trimethyl-	EtOH	232(4.48),336(3.64), 549(3.67)	12-0647-86
$C_{15}H_{12}O_3S_2$			
Ethanone, 2-acetoxy-1-[5'-(1-propyn-yl)[2,2'-bithiophen]-5-yl]-	EtOH	224(3.89),263(4.10), 372(4.45)	163-0263-86
$C_{15}H_{12}O_4$			
4H-1-Benzopyran-4-one, 2,3-dihydro-5,7-dihydroxy-3-phenyl-	MeOH	212(4.40),290(4.39), 322(4.31)	39-0215-86C
$C_{15}H_{12}O_5$			
4H-1-Benzopyran-4-one, 2,3-dihydro-5,7-dihydroxy-2-(4-hydroxyphenyl)-	EtOH	290(4.19),327s(3.62)	105-0107-86
6H-Dibenz[b,d]oxocin-7(8H)-one, 3,10,11-trihydroxy- (protosappanin A)	MeOH	260(4.00),284(3.84)	94-0001-86 +142-0601-86
9H-Xanthen-9-one, 3-hydroxy-2,4-dimethoxy-	MeOH	239(4.47),280s(3.84), 316(4.04),355s(3.95)	100-0095-86
	MeOH-NaOMe and MeOH-NaOAc	232(--),275(--), 383(--)	100-0095-86
$C_{15}H_{12}O_5S$			
2H-Naphtho[1,2-b]pyran-2-one, 4-[[(methylsulfonyl)oxy]methyl]-	EtOH	266(4.41),275.5(4.48), 292(3.82),304(3.90), 314(3.88),355(3.71)	94-0390-86
$C_{15}H_{12}O_6$			
Acetic acid, [(7-oxo-7H-furo[3,2-g]-[1]benzopyran-9-yl)oxy]-, ethyl ester	1% DMSO	302(4.08)	149-0391-86A

Compound	Solvent	λmax(log ε)	Ref.
2H,8H-Benzo[1,2-b:5,4-b']dipyran-2,8-dione, 6-(acetoxymethyl)-3,4-dihydro-	EtOH	283(3.95),320(4.02)	94-0390-86
2H,6H-Benzo[1,2-b:5,4-b']dipyran-2,6-dione, 5,10-dimethoxy-8-methyl-	EtOH	234(4.21),263(4.35), 275(4.37),283s(4.34), 318(4.04),330(3.98)	44-3116-86
Violaceic acid	MeOH	232(4.41),253(4.24), 276(4.18),283(4.14), 310(3.85)	39-0109-86C
9H-Xanthen-9-one, 1,3-dihydroxy-5,6-dimethoxy-	EtOH	246(4.45),285(3.88), 318(4.18)	102-0503-86
	EtOH-NaOH	246(4.32),264(4.21), 284s(3.80),374(4.17)	102-0503-86
	EtOH-$AlCl_3$	238(4.22),253(4.27), 265s(4.18),285s(3.86), 329(4.18)	102-0503-86
	EtOH-NaOAc-H_3BO_3	246(4.45),285(3.88), 318(4.18)	102-0503-86
9H-Xanthen-9-one, 1,6-dihydroxy-5,7-dimethoxy-	MeOH	219(4.19),240s(4.22), 255(4.40),313(3.93), 373(3.68)	100-0095-86
	MeOH-NaOMe	248(--),268(--), 280s(--),350(--), 408s(--)	100-0095-86
	MeOH-NaOAc	248(--),268(--), 280s(--),345(--), 388s(--)	100-0095-86
9H-Xanthen-9-one, 3,7-dihydroxy-1,8-dimethoxy-	EtOH and EtOH-$AlCl_3$	243(4.43),256(4.46), 309(4.13),356(3.79)	102-2681-86
	EtOH-NaOH	250(4.42),270(4.46), 299s(3.93),343(4.24)	102-2681-86
	EtOH-NaOAc	236(4.42),254(4.33), 267s(4.22),289(3.86), 344(4.16)	102-2681-86
	EtOH-NaOAc-H_3BO_3	243(4.43),256(4.46), 309(4.13),356(3.79)	102-2681-86
$C_{15}H_{12}S$			
Benzo[b]thiophene, 4-(4-methylphenyl)-	EtOH	229(4.70),275(4.28)	5-1003-86
Dibenzo[b,f]thiepin, 4-methyl-	MeOH	223(4.62),264(4.47), 302(3.88)	73-0141-86
1H-Indene-3-thiol, 2-phenyl-	MeOH or MeCN	237(4.17),360(4.31)	104-0513-86
	DMF or DMSO	305(--),407(--)	104-0513-86
$C_{15}H_{12}S_4$			
Disulfide, [(4-methylphenyl)thioxomethyl](phenylthioxomethyl)-	CH_2Cl_2	319(4.42),522(2.42)	88-4595-86
$C_{15}H_{13}BO_2$			
Boron, dimethyl(1H-phenalene-1,9(9aH)-dionato-O,O')-, (T-4)-	MeCN	235(4.44),265(3.86), 335s(3.87),352(4.22), 370(4.20),414(3.79), 437(3.87),464(3.54)	78-6293-86
$C_{15}H_{13}BO_4$			
Boron, [1,8-naphthalenediolato(2-)-O,O'](2,4-pentanedionato-O,O')-, (T-4)-	CH_2Cl_2	326(4.10),341(4.47), 378s(4.20)	48-0755-86
$C_{15}H_{13}BrN_2O_2$			
1H-Benz[de]isoquinoline-1,3(2H)-dione,	pH 7.4	432(4.10)	4-0849-86

Compound	Solvent	λ_{max}(log ϵ)	Ref.
6-amino-2-(3-bromopropyl)- (cont.)			4-0849-86
$C_{15}H_{13}BrO_3$			
2,4,6-Cycloheptatrien-1-one, 2-[(4-bromophenyl)methoxy]-5-methoxy-	MeOH	227(4.22),252s(3.61),327(3.77),332(3.77),341(3.75),362s(3.68),376s(3.53)	18-1125-86
2,4,6-Cycloheptatrien-1-one, 3-[(4-bromophenyl)methoxy]-2-hydroxy-5-methoxy-	MeOH	228(4.32),241(4.32),271s(3.83),330(3.88),370(3.73),384s(3.68)	18-1125-86
$C_{15}H_{13}ClN_2$			
Quinoxaline, 6(and 7)-chloro-1,2-dihydro-3-(4-methylphenyl)- (60:40 ratio of 6:7 isomers)	MeOH	270(4.36),304(3.92),391(3.67)	103-0918-86
	toluene	397(--)	103-0918-86
$C_{15}H_{13}ClN_2O$			
Quinoxaline, 6(and 7)-chloro-1,2-dihydro-3-(4-methoxyphenyl)- (60:40 ratio of 6:7 isomers)	MeOH	286(4.43),308(4.17),390(3.79)	103-0918-86
	toluene	389(--)	103-0918-86
$C_{15}H_{13}ClOS$			
Ethanone, 1-[5-chloro-2-[(2-methylphenyl)thio]phenyl]-	MeOH	273(3.95),347(3.56)	73-0141-86
$C_{15}H_{13}Cl_2NO_6S$			
7-Azabicyclo[2.2.1]hept-5-ene-2,3-dicarboxylic acid, 5,6-dichloro-7-[(4-methylphenyl)sulfonyl]-, (endo,exo)-	EtOH	231(4.16),265(2.85),270s(2.77),276(2.65)	24-0616-86
$C_{15}H_{13}Cl_2N_3O$			
Benzenamine, 2,4-dichloro-N-(1,5,6-trimethylfuro[2,3-d]pyrimidin-4(1H)-ylidene)-, monohydriodide	EtOH	306(4.24)	11-0337-86
$C_{15}H_{13}F_7N_2O_2$			
Imidazo[1,2-a]pyridine-3-carboxylic acid, 2-(heptafluoropropyl)-, 1,1-dimethylethyl ester	EtOH	235(4.40),241(4.35),288.5(3.96)	39-1769-86C
$C_{15}H_{13}N$			
Benzenamine, N-phenyl-N-2-propenyl-	hexane	206(4.23),242(4.88),289(3.40)	65-2461-86
	MeOH	205(--),244(--),285(--)	65-2461-86
$C_{15}H_{13}NO$			
Acetamide, N-9H-fluoren-2-yl-	pH 7.4	277(4.34)	94-0944-86
Azacyclotetradeca-3,5,11,13-tetraene-7,9-diyn-2-one, 6,11-dimethyl-	THF	253s(4.20),305(4.54),312s(4.53),365s(3.72),435(3.49)	39-0933-86C
2H-1-Azapyren-2-one, 3,3a,4,5-tetrahydro-	$CHCl_3$	232(4.33),252(4.61),260(4.47),282(3.71),294(3.81),306(3.78),324(3.39),339(3.27)	4-0747-86
2H-2-Azapyren-1-one, 3,3a,4,5-tetrahydro-	EtOH	242(4.79),278(3.79),289(3.85),301(3.72),327(3.46),341(3.56)	4-0747-86
Benzo[f]quinolin-3(4H)-one, 1,4-dimethyl-	EtOH	243(4.86),283(4.09),302(3.89),314(3.93),	4-1207-86

Compound	Solvent	λmax(log ε)	Ref.
(cont.)		354(3.89),371(3.88)	4-1207-86
Benzo[g]quinolin-2(1H)-one, 1,4-dimethyl-	EtOH	241(4.70),256(4.55), 266(4.61),277(4.58), 319(3.98),328(4.09)	4-1207-86
Benzo[h]quinolin-2(1H)-one, 1,4-dimethyl-	EtOH	236(4.37),272(4.37), 282(4.48),307(3.91), 359(3.92)	4-1207-86
1H-Cyclopenta[cd]phenalen-1-one, 2,2a,3,4-tetrahydro-, oxime	MeOH	247(4.61),253(4.63), 280(3.97),296(4.03), 307(3.93),330(3.59), 346(3.67)	4-0747-86
2,4,10,12-Cyclotridecatetraene-6,8-diyn-1-one, 5,10-dimethyl-, oxime, (E,E,Z,Z)-	THF	215(4.43),273(4.57), 349s(3.99)	39-0933-86C
1H-Phenalen-1-one, 6-(dimethylamino)-	hexane	435(4.0)	135-0967-86
	EtOH	526(4.0)	135-0967-86
$C_{15}H_{13}NOS$			
1,5-Benzothiazepin-4(5H)-one, 2,3-dihydro-2-phenyl-	CH_2Cl_2	242.5(4.14),273(3.70)	24-3109-86
9H-Thioxanthen-9-one, 2-(dimethylamino)-	MeOH	268(4.58),297.5(4.37), 450(3.61)	73-0937-86
$C_{15}H_{13}NO_2$			
1H-Azuleno[1,2-b]pyrrole-1-carboxylic acid, ethyl ester	MeOH	235s(4.08),277s(4.46), 302(4.71),310(4.71), 340s(3.49),355(3.64), 372(3.67),390(3.65), 560s(2.57),608(2.88), 661(2.83),730(3.42)	138-1021-86
6H-Cyclohept[d]isoxazole, 3-(4-methoxyphenyl)-	EtOH	244(4.29),277s(3.91), 285s(3.88)	138-1925-86
9H-Carbazol-2-ol, 3-methyl-, acetate	EtOH	236(4.40),262(4.05), 296(4.05),330(3.30), 342(3.30)	25-0246-86
4H-Furo[3,2-b]pyrrole, 4-acetyl-2-(4-methylphenyl)-	MeOH	323(3.20)	73-1455-86
1,2-Naphthalenediol, 3-(1-methyl-1H-pyrrol-2-yl)-	EtOH	268(4.19)	142-0579-86
Phenol, 2-[3-(phenylimino)-1-propenyl]-, N-oxide	EtOH	315(3.93),370(3.71)	104-1302-86
	dioxan	310(3.93),350(3.04)	104-1302-86
	DMSO	320(3.93),370(3.83)	104-1302-86
	+ NaOMe	350(3.83),460(3.79)	104-1302-86
$C_{15}H_{13}NO_2S$			
Azuleno[1,2-d]thiazole-9-carboxylic acid, 2-methyl-, ethyl ester	C_6H_{12}	240(4.2),315f(4.8), 400f(3.9),580f(2.3)	18-3320-86
$C_{15}H_{13}NO_2S_2$			
2λ⁴-Thieno[3,2-c][1,2]thiazine-3-carboxylic acid, 2-phenyl-, ethyl ester	EtOH	220(4.15),262s(3.64), 326(3.93),459(3.63)	39-0483-86C
2λ⁴-Thieno[3,4-c][1,2]thiazine-3-carboxylic acid, 2-phenyl-, ethyl ester	EtOH	228(4.06),266(3.99), 320(3.89)	39-0483-86C
4aH-Thieno[3,2-c][1,2]thiazine-3-carboxylic acid, 4a-phenyl-, ethyl ester	EtOH	260s(3.89),267(3.89), 272s(3.87),296(3.78), 330(3.82)	39-0491-86C
$C_{15}H_{13}NO_3$			
9(10H)-Acridinone, 1-hydroxy-3-methoxy-N-methyl-	EtOH	242s(4.18),249(5.02), 255(5.11),262(5.19), 268(5.14),295(4.18),	64-0187-86C

Compound	Solvent	$\lambda_{max}(\log \epsilon)$	Ref.
(cont.)		318(3.87),394(3.81)	64-0187-86C
9H-Carbazole-1-carboxaldehyde, 4-hydroxy-3-methoxy-2-methyl-	MeOH	214(4.39),227(4.37), 263(4.09),295(4.21), 320(3.71),372(3.92)	158-0727-86
$C_{15}H_{13}NO_4$			
1,3-Dioxane-4,6-dione, 5-(1H-indol-3-ylmethylene)-2,2-dimethyl-	EtOH	213(4.58),270(4.20), 411(4.79)	39-1651-86C
$C_{15}H_{13}NO_5$			
2-Propenoic acid, 3-[1-(methoxycarbonyl)-4-oxoquinolizin-3-yl]-, methyl ester	CH_2Cl_2	421(4.3)	39-1561-86B
$C_{15}H_{13}NO_7S$			
3-Oxa-8-azatricyclo[3.2.1.0^{2,4}]octane-6,7-dicarboxylic acid, 8-[(4-methylphenyl)sulfonyl]-, (1α,2α,4α,5α-6α,7β)-	MeCN	230(4.10),263(2.85), 268s(2.74),274(2.63)	24-0616-86
$C_{15}H_{13}NS$			
1H-1-Azapyrene-2-thione, 3,3a,4,5-tetrahydro-	EtOH	207(4.34),227(4.39), 243(3.77),259(3.67), 269(3.98),281(4.31), 291(4.45),340(4.21)	4-0747-86
2H-2-Azapyrene-1-thione, 3,3a,4,5-tetrahydro-	EtOH	213(4.33),247(4.66), 255(4.69),283(3.97), 344(3.54)	4-0747-86
$C_{15}H_{13}N_2$			
Quinoxalinium, 1-methyl-3-phenyl-, iodide	pH 6.0	259(4.50),381(3.99)	103-1118-86
hydrated form	pH 10.4	230(4.48),372(3.89)	103-1118-86
$C_{15}H_{13}N_2O_7S$			
2-Propanone, 1-[2,4-dinitro-5-(phenylsulfonyl)-2,4-cyclohexadien-1-yl]-, ion(1-), potassium	acetone	378(3.99),540(4.38)	104-1138-86
$C_{15}H_{13}N_3O$			
Benzenamine, N-methyl-2-(5-phenyl-1,3,4-oxadiazol-2-yl)-	MeOH	253(4.62),272(4.71), 280(4.78),293(4.83)	18-1575-86
1H-Benzo[h]pyrimido[1,2-c]quinazolin-1-one, 2,3,5,6-tetrahydro-	EtOH	264(4.05),290(3.72), 312(3.41),330(3.63), 345(3.85),362(3.92), 379(3.72)	142-1119-86
3H-Benzo[h]pyrimido[1,2-c]quinazolin-3-one, 1,2,5,6-tetrahydro-	EtOH	251(3.54),274(3.43), 302(3.33),323(3.35), 338(3.25)	142-1119-86
3H-Benzo[h]pyrimido[1,2-c]quinazolin-3-one, 1,2,5,6-tetrahydro-, monohydrochloride	EtOH	228(3.47),260(3.59), 268(3.64),280(3.50), 314(3.46),345(3.68), 370(3.35)	142-1119-86
$C_{15}H_{13}N_3O_4$			
Ethanone, 1-[2,4-dihydroxy-5-(5-methoxy-2H-benzotriazol-2-yl)phenyl]-	$CHCl_3$	248(4.25),285(4.39), 334(4.34)	49-0805-86
$C_{15}H_{13}N_3O_5$			
Benzamide, 3,5-dinitro-N-(1-phenylethyl)-, (+)-	EtOH	230(3.47),340(1.58)	104-1943-86

Compound	Solvent	λ_{max}(log ε)	Ref.
$C_{15}H_{13}N_3S$			
3-Quinolinecarbonitrile, 1,2,5,6,7,8-hexahydro-4-(3-pyridinyl)-2-thioxo-	EtOH	251(4.02),267(3.88), 314(4.33),423(3.63)	104-1762-86
3-Quinolinecarbonitrile, 1,2,5,6,7,8-hexahydro-4-(4-pyridinyl)-2-thioxo-	EtOH	244(4.15),314(4.39), 420(3.70)	104-1762-86
$C_{15}H_{14}$			
Cyclopenta[ef]heptalene, 1,10b-dihydro-10b-methyl-	hexane	245(4.23),253(4.28), 270(4.30),413(3.49)	35-4105-86
$C_{15}H_{14}BrF_3N_2O_3$			
DL-Tryptophan, 2-bromo-1-methyl-N-(trifluoroacetyl)-, methyl ester	MeOH	221(4.60),283(4.04), 292(3.96)	35-2023-86
$C_{15}H_{14}BrNO_3S_2$			
8H-Thieno[3,2-c]azepin-8-one, 7-bromo-4,5,6,7-tetrahydro-5-[(4-methylphenyl)sulfonyl]-	EtOH	229(4.14),283(3.95)	39-0729-86C
$C_{15}H_{14}ClNO_2S$			
2H-[1]Benzothiepino[5,4-b]pyran-2-one, 3-chloro-4-(dimethylamino)-5,6-dihydro-	EtOH	232s(3.97),256(4.07), 263s(4.05),320(4.04)	4-0449-86
$C_{15}H_{14}ClN_3O$			
Acetamide, 2-[[(2-amino-5-chlorophenyl)phenylmethylene]amino]-, (E)-	MeOH	204(4.46),232(3.46), 260s(3.85),365(3.79)	96-1051-86
	ether	365(3.82)	96-1051-86
	CH_2Cl_2	365(3.82)	96-1051-86
	80% MeOH	360(3.74)	96-1051-86
	50% MeOH	356(3.70)	96-1051-86
	20% MeOH	352(3.63)	96-1051-86
cation	20% MeOH	240(4.27),270s(4.03), 420(3.44)	96-1051-86
(Z)-	MeOH	206(4.59),248(3.38), 300s(3.35)	96-1051-86
cation	MeOH	240(4.33),275(4.00), 420(3.44)	96-1051-86
Benzoic acid, 2-(methylamino)-, [(4-chlorophenyl)methylene]hydrazide	MeOH	220(4.22),252(4.36), 297(4.38),360(4.43)	18-1575-86
$C_{15}H_{14}F_3NO_9$			
Propanedioic acid, [[6-(methoxycarbonyl)-2-oxo-2H-pyran-4-yl]methyl]-[(trifluoroacetyl)amino]-, dimethyl ester	EtOH	303(3.8)	5-1968-86
$C_{15}H_{14}FeN_2O_7$			
Iron, tricarbonyl[dimethyl 2-[(4,5,6,7-η)-3-methyl-1H-1,2-diazepin-1-yl)-2-butenedioate]-, cis	EtOH	306(4.16),352(4.13)	39-0595-86C
trans	EtOH	251s(4.17),315s(3.77), 373(3.78)	39-0595-86C
Iron, tricarbonyl[dimethyl 2-[(4,5,6,7-η)-5-methyl-1H-1,2-diazepin-1-yl)-2-butenedioate], cis	EtOH	305(4.19),346(4.18)	39-0595-86C
trans	EtOH	242s(4.09),281s(3.73), 370(3.77)	39-0595-86C
$C_{15}H_{14}FeO_3$			
Iron, tricarbonyl[η^4-(2,3,6,7-tetra-	EtOH	218(4.27),307s(3.15)	44-2385-86

Compound	Solvent	λ_{max}(log ε)	Ref.
kis(methylene)bicyclo[3.2.1]octane] (cont.)			44-2385-86
$C_{15}H_{14}NO$			
Methanaminium, N-methyl-N-2H-naphtho[1,2-b]pyran-2-ylidene-, perchlorate	n.s.g.	210(5.2),270(5.2), 275(5.2),310(5.1), 365(5.1),440(4.4)	104-1342-86
Methanaminium, N-methyl-N-2H-naphtho[1,8-bc]oxepin-2-ylidene-, perchlorate	MeCN	227(5.30),288(5.25), 412(4.94)	104-1342-86
$C_{15}H_{14}N_2$			
Benzenamine, N-[3-(phenylamino)-2-propenylidene]-	pH 1	381(4.72)	94-1794-86
	pH 7	381(4.69)	94-1794-86
	pH 13	357(--)(unstable)	94-1794-86
	MeOH	365(4.58)	94-1794-86
	EtOH	358(4.56)	94-1794-86
1H-Cyclohepta[4,5]pyrrolo[2,3-b]indole, 2,3,4,5-tetrahydro-	MeOH	233(4.16),258(4.34), 312(4.63),340(4.22), 417(3.75),522(2.69), 560s(2.63)	18-3326-86
	MeOH-HCl	230(4.31),259(4.38), 265(4.35),322(4.19), 350s(4.35)	18-3326-86
Quinoxaline, 1,2-dihydro-3-(4-methylphenyl)-	MeOH	267(4.37),292s(--), 389(3.67)	103-0918-86
	toluene	388(--)	103-0918-86
$C_{15}H_{14}N_2O$			
Benzeneacetonitrile, α-(1,5-dihydro-3,4-dimethyl-5-oxo-2H-pyrrol-2-ylidene)-2-methyl-, (E)-	EtOH	293(4.17)	49-0631-86
	$CHCl_3$	299(4.17)	49-0631-86
(Z)-	EtOH	289(4.10)	49-0631-86
	$CHCl_3$	291(4.09)	49-0631-86
Benzeneacetonitrile, α-(1,5-dihydro-3,4-dimethyl-5-oxo-2H-pyrrol-2-ylidene)-4-methyl-, (E)-	EtOH	333(4.08)	49-0631-86
	$CHCl_3$	334(4.11)	49-0631-86
(Z)-	EtOH	302(4.16)	49-0631-86
	$CHCl_3$	305(4.14)	49-0631-86
Benzeneacetonitrile, α-[[5-(dimethylamino)-2-furanyl]methylene]-, (E)-	MeOH	242(3.02),463(3.43)	73-0879-86
Benzeneacetonitrile, 4-methoxy-α-[(1-methyl-1H-pyrrol-2-yl)methylene]-, (Z)-	EtOH	245(4.05),290(3.60), 378(4.47)	4-1747-86
Quinoxaline, 1,2-dihydro-3-(4-methoxyphenyl)-	MeOH	277(4.32),296(4.14), 388(3.79)	103-0918-86
	toluene	385(--)	103-0918-86
$C_{15}H_{14}N_2O_2$			
Benzaldehyde, 2-hydroxy-5-methyl-3-[[(3-methyl-2-pyridinyl)imino]methyl]-	$CHCl_3$	488(3.08)	2-0127-86A
Benzaldehyde, 2-hydroxy-5-methyl-3-[[(4-methyl-2-pyridinyl)imino]methyl]-	$CHCl_3$	488(3.00)	2-0127-86A
Benzaldehyde, 2-hydroxy-5-methyl-3-[[(5-methyl-2-pyridinyl)imino]methyl]-	$CHCl_3$	488(3.00)	2-0127-86A
Benzaldehyde, 2-hydroxy-5-methyl-3-[[(6-methyl-2-pyridinyl)imino]methyl]-	$CHCl_3$	488(3.10)	2-0127-86A

Compound	Solvent	λmax(log ε)	Ref.
Benzenamine, 2,3-dimethyl-N-[(4-nitrophenyl)methylene]-	THF	292.5(4.16),362.5(3.96)	56-0831-86
Benzenamine, 2,5-dimethyl-N-[(4-nitrophenyl)methylene]-	THF	289s(4.16),366.5(3.96)	56-0831-86
6H-Pyrido[1,2-a]pyrimidin-6-one, 9-benzoyl-1,2,3,4-tetrahydro-	EtOH	239(4.16),308(4.03), 341(4.33)	142-2247-86
$C_{15}H_{14}N_2O_2S$			
2H-1,2,4-Benzothiadiazine, 3-methyl-2-(4-methylphenyl)-, 1,1-dioxide	MeOH	260(4.18),275s(4.04), 300(3.93)	44-1967-86
5H-Dibenzo[b,g][1,4,6]thiadiazocine, 2,6-dimethyl-, 12,12-dioxide	MeOH	250(4.60),300(4.00), 350(2.95)	44-1967-86
5H-Dibenzo[b,g][1,4,6]thiadiazocine, 4,6-dimethyl-, 12,12-dioxide	EtOH	254(4.24),301(3.83), 350(2.30)	44-1967-86
$C_{15}H_{14}N_2O_3$			
Imidazo[1,2-a]pyridin-5(1H)-one, 2,3-dihydro-8-(4-methoxybenzoyl)-	EtOH	229(4.17),312(4.15), 341(4.32)	142-2247-86
$C_{15}H_{14}N_2O_3S$			
2H-1,2,4-Benzothiadiazine, 2-(4-methoxyphenyl)-3-methyl-, 1,1-dioxide	MeOH	251(4.17),266(4.18), 295(4.97)	44-1967-86
5H-Dibenzo[b,g][1,4,6]thiadiazocine, 2-methoxy-6-methyl-, 12,12-dioxide	MeOH	257(4.18),300(4.10), 354(3.08)	44-1967-86
$C_{15}H_{14}N_2O_4S$			
1H-Indole, 1-methyl-3-[(3-methyl-4-nitro-2(5H)-thienylidene)methyl]-, S,S-dioxide, (Z)-	$CHCl_3$	290(3.699),485(4.430)	104-0379-86
$C_{15}H_{14}N_2S$			
2H-Benz[g]indazole, 3,3a,4,5-tetrahydro-3-(2-thienyl)-	EtOH	237(4.10),300(4.19)	4-0135-86
$C_{15}H_{14}N_2S_2$			
Thieno[2,3-d]pyrimidine-4(3H)-thione, 3,5,6-trimethyl-2-phenyl-	EtOH	238(4.52),288(4.16), 328(4.23),366s(3.98)	106-0096-86
$C_{15}H_{14}N_4$			
2H-Pyrazino[1,2-a]pyrimidine-9-carbonitrile, 3,4-dihydro-7-methyl-6-phenyl-	EtOH	454(3.71)	33-1025-86
Pyrido[3,4-d]pyrimidin-4-amine, 2-methyl-N-(phenylmethyl)-	pH 1	275(3.88),324(4.06), 338(3.94)	163-2243-86
	pH 7	282.5(3.97),330.5(3.89)	163-2243-86
	pH 13	282.5(3.98),331(3.88)	163-2243-86
$C_{15}H_{14}N_4O_2S$			
Pyrimido[4,5-d]pyrimidine-2,4(1H,3H)-dione, 5,6-dihydro-1,3,7-trimethyl-6-phenyl-5-thioxo-	EtOH	226(4.13),257(4.09), 318(4.08)	142-2293-86
$C_{15}H_{14}N_4O_3$			
Benzoic acid, 2-(methylamino)-, [(4-nitrophenyl)methylene]hydrazide	MeOH	224(4.21),253(4.31), 310(4.29),372(4.38)	18-1575-86
Pyrimido[4,5-d]pyrimidine-2,4,5(1H,3H-6H)-trione, 1,3,6-trimethyl-7-phenyl-	EtOH	235(4.49),271(3.87), 306(3.90)	142-2293-86
Pyrimido[4,5-d]pyrimidine-2,4,5(1H,3H-6H)-trione, 1,3,7-trimethyl-6-phenyl-	EtOH	234(4.56),266(3.82), 297(3.66)	142-2293-86

Compound	Solvent	λ_{max}(log ε)	Ref.
$C_{15}H_{14}N_4O_5$			
3H-Azeto[1',2':1,5]pyrrolo[2,3-c]-pyrazole-7a-carboxylic acid, 3a,4,4a,5,6,7a-hexahydro-6-oxo-, (4-nitrophenyl)methyl ester, (3aα,4aα,7aα)-(±)-	EtOH	263(4.11),326s(2.69)	39-0973-86C
(3aα,4aβ,7aα)-(±)-	EtOH	263(4.10),327(2.63)	39-0973-86C
$C_{15}H_{14}N_6O_2$			
[1,2,4]Triazolo[1,5-a]pyrimidine, 1,7-dihydro-2-methyl-7-(2-methyl-1H-indol-3-yl)-6-nitro-	EtOH	220(3.92),272(3.84), 356(4.06)	103-1250-86
$C_{15}H_{14}O$			
1H-Cyclopenta[cd]phenalen-1-ol, 2,2a,3,4-tetrahydro-	$CHCl_3$	239(4.14),268(3.39), 280(3.61),290(3.66), 295(3.59),309(3.10), 324(3.11)	4-0747-86
Phenol, 2-[2-(4-methylphenyl)ethenyl]-, (E)-	EtOH	321(4.30),330(4.30)	150-0433-86S +150-3514-86M
$C_{15}H_{14}O_2$			
2(3H)-Furanone, 5-(2-butynylidene)-dihydro-3-(phenylmethyl)-, (E)-	H_2O	234(4.21)	35-5589-86
	+ base	228s(c.3.95)	35-5589-86
1(4H)-Naphthalenone, 2-hydroxy-4-(3-methyl-2-butenylidene)-	n.s.g.	315(4.11),420(4.59)	39-1627-86C
Phenol, 2-[2-(4-methoxyphenyl)ethenyl]-, (E)-	EtOH	324(4.43),333(4.44)	150-0433-86S +150-3514-86C
$C_{15}H_{14}O_2S$			
Propanoic acid, 2-(4-biphenylylthio)-, 1-	MeCN	214(4.28),257(4.26), 280(4.27)	56-0065-86
$C_{15}H_{14}O_3$			
6H-Benzofuro[3,2-c][1]benzopyran-6-one, 6b,7,8,9,10,10a-hexahydro-	EtOH	218(4.06),272(3.89), 283(3.99),295(3.85), 306(3.98)	2-1167-86
11H-Benzofuro[2,3-b][1]benzopyran-11-one, 6a,7,8,9,10,10a-hexahydro-	EtOH	229(4.05),278(4.01), 288(3.99)	2-1167-86
2H-1-Benzopyran-2-one, 3-(2-cyclohexen-1-yl)-4-hydroxy-	EtOH	211(4.48),236(3.93), 314(4.11)	2-1167-86
2H-1-Benzopyran-2-one, 4-(2-cyclohexen-1-yloxy)-	EtOH	267(4.07),278(4.05), 303(3.87)	2-1167-86
2,4,6-Cycloheptatrien-1-one, 2-hydroxy-5-methoxy-3-(phenylmethyl)-	MeOH	227(4.08),253(3.88), 322(3.53),372(3.46), 386(3.45)	18-1125-86
2,4,6-Cycloheptatrien-1-one, 5-methoxy-2-(phenylmethoxy)-	MeOH	227(4.38),252s(3.88), 328(4.05),347(4.01), 361s(3.97),376s(3.81)	18-1125-86
Naphtho[1,8-bc]pyran-7,8-dione, 2,3-dihydro-3,6,9-trimethyl-, (±- mansonone E)	EtOH	222(4.16),265(4.51), 373(2.95),445(3.14)	12-0647-86
Naphtho[1,8-bc]pyran-3(2H)-one, 7-methoxy-6,9-dimethyl-	EtOH	214(4.58),270(4.46), 332(3.43),346(3.49), 410(3.65)	12-0647-86
$C_{15}H_{14}O_4$			
Ethanone, 1-[2,3,4-trihydroxy-5-(phenylmethyl)phenyl]-	MeOH	226(4.6),301(4.1)	2-0259-86
1,4-Naphthalenediol, 2-methyl-, diacetate	MeOH	226(4.70),267(3.73)	12-2067-86

Compound	Solvent	λ_{max}(log ε)	Ref.
Naphtho[1,2-b]furan-4,5-dione, 2,3-dihydro-6-hydroxy-2,2,3-trimethyl-	MeOH	239s(4.26),255(4.31), 290(3.86),414(3.78)	100-0122-86
Naphtho[1,2-b]furan-4,5-dione, 2,3-dihydro-6-hydroxy-2,3,3-trimethyl-	MeOH	238s(4.19),259(4.30), 291(3.76),412(3.72)	100-0122-86
Naphtho[1,2-b]furan-4,5-dione, 2,3-dihydro-7-hydroxy-2,2,3-trimethyl-	MeOH	269(4.38),277(4.41), 298(3.80),496(3.18)	100-0122-86
Naphtho[1,2-b]furan-4,5-dione, 2,3-dihydro-7-hydroxy-2,3,3-trimethyl-	MeOH	270(4.47),277(4.50), 303(3.79),496(3.31)	100-0122-86
Naphtho[1,8-bc]pyran-7,8-dione, 2,3-dihydro-3-hydroxy-3,6,9-trimethyl-	EtOH	217(4.62),262(4.71), 370(3.59),440(3.79)	12-0647-86
Phebalosin	MeOH	217(4.53),322(4.13)	100-0180-86
$C_{15}H_{14}O_4S_2$			
1,3-Dithiole-4,5-dicarboxylic acid, 2-(1-phenylethylidene)-, dimethyl ester	EtOH	228(4.06),304(3.87), 408(3.18)	33-0419-86
$C_{15}H_{14}O_5$			
Ethanone, 1-(5-acetoxy-1,8-dihydroxy-3-methyl-2-naphthalenyl)-	$CHCl_3$	240s(3.77),271(4.29), 300s(3.38),310s(3.25)	64-0377-86B
2-Furancarboxylic acid, 4-[(2-formyl-phenoxy)methyl]-5-methyl-, methyl ester	MeOH	265(3.32),318(2.65)	73-2186-86
$C_{15}H_{15}ClN_2OS$			
Ethanimidamide, N'-[3-(2-chlorobenz-oyl)-2-thienyl]-N,N-dimethyl-	$CHCl_3$	243(4.28),340(3.78)	83-0347-86
$C_{15}H_{15}ClN_2O_3$			
2-Butenoic acid, 4-[(4-chlorobenz-oyl)-2-imidazolidinylidene]-, methyl ester, (E)-	EtOH	240(4.25),325(4.41)	142-2247-86
$C_{15}H_{15}ClN_4O_2$			
Carbamic acid, [5-amino-2-[(4-chloro-phenyl)azo]phenyl]-, ethyl ester	EtOH	223(4.32),286(4.04), 419(4.45)	97-0134B-86
$C_{15}H_{15}N$			
1-Azapyrene, 1,2,3,3a,4,5-hexahydro-	$CHCl_3$	240(4.07),266(3.59), 278(3.72),286(3.76), 295(3.66),309(3.13), 324(3.03)	4-0747-86
2-Azapyrene, 1,2,3,3a,4,5-hexahydro-	$CHCl_3$	237(4.07),277(3.70), 286(3.74),294(3.64), 306(3.03),323(2.86)	4-0747-86
Benzenamine, N-[(2,6-dimethylphenyl)-methylene]-	MeOH	212(4.35),267(4.32), 313(3.47)	165-0017-86
Benzenamine, 2,6-dimethyl-N-(phenyl-methylene)-	C_6H_{12}	213(4.33),253(4.35), 340(3.26)	165-0017-86
Benzenamine, N-phenyl-N-2-propenyl-	hexane	204(4.28),244(4.78), 290(3.32)	65-2461-86
	MeOH	202(--),247(--), 287(--)	65-2461-86
[2]Paracyclo[2](2,5)pyridinophane	isooctane	215(4.1),240s(3.4), 287(3.2),308(2.9)	46-1541-86
$C_{15}H_{15}NO$			
Benzenamine, N-[(4-methoxyphenyl)-methylene]-4-methyl-	C_6H_{12}	221(4.26),280(4.31), 322(4.16)	165-0017-86

Compound	Solvent	$\lambda_{max}(\log \epsilon)$	Ref.
$C_{15}H_{15}NOS$			
Benzenecarboximidothioic acid, N-(phenylmethyl)-, methyl ester, N-oxide, (Z)-	MeOH	277(4.04)	44-5198-86
4H-Thieno[2,3-c]azepin-4-one, 5,6,7,8-tetrahydro-7-(phenylmethyl)-	EtOH	253(4.07)	39-0729-86C
8H-Thieno[3,2-c]azepin-8-one, 4,5,6,7-tetrahydro-5-(phenylmethyl)-	EtOH	271(3.99)	39-0729-86C
$C_{15}H_{15}NO_2$			
1H-Indene-1,3(2H)-dione, 2-[3-(dimethylamino)-2-methyl-2-propenylidene]-	EtOH	440(4.91)	70-1446-86
	$CHCl_3$	440(4.83),430(4.54), 460(4.43)	70-1446-86
1-Naphthalenol, 4-methyl-2-[3-(methylimino)-1-propenyl]-, N-oxide	dioxan	313(3.84),335(3.85), 370s(3.99),380(3.94)	104-1302-86
	acetone	370s(3.97),380(3.96)	104-1302-86
	DMSO	295(4.06),330(4.07), 385(4.01),510(3.82)	104-1302-86
	+ NaOMe	330(4.21),355(3.96), 500(4.23)	104-1302-86
2-Propen-1-one, 3-[5-(dimethylamino)-2-furanyl]-1-phenyl-, (E)-	MeOH	272(2.94),492(3.46)	73-0879-86
$C_{15}H_{15}NO_2S$			
Benzoic acid, 2-[[4-(dimethylamino)-phenyl]thio]-	MeOH	278(4.39),310s(3.93)	73-0937-86
$C_{15}H_{15}NO_3$			
Furo[3,2-c]quinolin-4(5H)-one, 6-hydroxy-5-methyl-2-(1-methylethyl)-	n.s.g.	217(4.48),228.8(4.46), 248.5(4.47),254(4.49), 279(3.91),292(3.90), 326(3.48)	25-0669-86
$C_{15}H_{15}NO_3S_2$			
10H-Phenothiazine-10-propanoic acid, sodium salt	80% MeCN	253(4.57),305(3.65)	46-2469-86
8H-Thieno[3,2-c]azepin-8-one, 4,5,6,7-tetrahydro-5-[(4-methylphenyl)sulfonyl]-	EtOH	271(4.04)	39-0729-86C
$C_{15}H_{15}NO_4$			
3H-Cyclopent[d][1,4]oxazino[3,4-b]-[1,3]benzoxazepin-3-one, 1,2,5,6-8,8a-hexahydro-12-hydroxy-	EtOH	305(3.95),315(3.94), 365(3.98)	24-3487-86
1H-Pyrrole-3,4-dicarboxylic acid, 1-(phenylmethyl)-, dimethyl ester	EtOH	253(3.88)	24-0813-86
Pyrrolo[2,1-a]isoquinoline-2,3-dione, 5,6-dihydro-8,9-dimethoxy-1-methyl-	MeOH	264(3.91),312(3.80), 370(3.85),446s(3.66), 480s(3.64)	44-2781-86
$C_{15}H_{15}NO_4S$			
3-Pyridinecarboxylic acid, 1,6-dihydro-4-hydroxy-2-(methylthio)-6-oxo-1-phenyl-, ethyl ester	EtOH	312(3.73)	5-1109-86
$C_{15}H_{15}NO_7$			
Pyrano[3,4-b]pyrrole-2,7-dicarboxylic acid, 1-acetyl-1,5-dihydro-5-oxo-, diethyl ester	EtOH	262(4.2),269(4.1), 306(4.1),316(4.1)	5-1968-86

Compound	Solvent	$\lambda_{max}(\log \epsilon)$	Ref.
$C_{15}H_{15}N_3$			
Benz[h]imidazo[1,2-c]quinazoline, 1,2,4,5-tetrahydro-1-methyl-	EtOH	262(4.14),271(4.11), 328(3.36),340(3.37)	4-0685-86
Benz[h]imidazo[1,2-c]quinazoline, 1,2,4,5-tetrahydro-2-methyl-	EtOH	235(3.68),242(3.84), 262(4.13),271(4.11), 325(4.28),340(3.37)	4-0685-86
4H-Benzo[3,4]cyclohept[1,2-e]imidazo-[1,2-c]pyrimidine, 1,2,5,6-tetra-hydro-, monohydrochloride	EtOH	253(4.24),315(3.52)	4-1685-86
1H-Benzo[h]pyrimido[1,2-c]quinazoline, 2,3,5,6-tetrahydro-	EtOH	258(4.22),303(3.73), 318(3.77),330(4.83), 345(3.71)	142-1119-86
hydrochloride	EtOH	220(3.78),256(4.02), 303(3.63),318(3.68), 329(3.73),343(3.53)	142-1119-86
$C_{15}H_{15}N_3O$			
Benzoic acid, 2-(methylamino)-, (phenylmethylene)hydrazide	MeOH	220(4.29),257(4.27), 362(4.36)	18-1575-86
$C_{15}H_{15}N_3OS$			
3-Pyridinecarbonitrile, 1,4-dihydro-2-mercapto-6-methyl-5-(1-oxopropyl)-4-(3-pyridinyl)-, morpholine salt	EtOH	232(4.10),295(4.12), 383s(2.45)	104-1762-86
$C_{15}H_{15}N_3O_2$			
Benzenamine, N,N-dimethyl-4-[[(4-nitrophenyl)imino]methyl]-	MeOH	246(4.02),302(3.89), 386(4.47)	165-0017-86
1,4-Benzenediamine, N,N-dimethyl-N'-[(4-nitrophenyl)methylene]-	MeOH	277(4.31),436(--)	165-0017-86
	EtOH	276(4.28),442(4.24)	56-0831-86
	THF	279(4.30),446.5(4.30)	56-0831-86
Benzo[g]phthalazin-1(2H)-one, 4-(3-hydroxypropylamino)-	EtOH	222(4.31),245(4.78), 253(4.85)	111-0143-86
Benzo[f]quinoxaline-3-carboxylic acid, 1,2,5,6-tetrahydro-2-imino-1-methyl-, methyl ester	EtOH	479(3.92)	33-1025-86
Hydrazinecarboxamide, 2-(2-hydroxy-1,2-diphenylethylidene)-, (E)-	MeOH	238(4.09)	48-0181-86
(Z)-	MeOH	275(4.11)	48-0181-86
$C_{15}H_{15}N_3O_3$			
Benzamide, N-[(2,3-dihydro-6-methyl-7-oxo-7H-oxazolo[3,2-a]pyrimidin-2-yl)methyl]-, (±)-	n.s.g.	229(3.97),261s(3.68)	128-0297-86
2,4(1H,3H)-Pyrimidinedione, 1-[(4,5-dihydro-2-phenyl-5-oxazolyl)methyl]-5-methyl-, (±)-	EtOH	220s(4.16),252(4.14)	128-0297-86
$C_{15}H_{15}N_3O_4$			
2,5-Methano-5H-pyrimido[1,2-c][1,3,5]-oxadiazepin-9(1H)-one, 2,3-dihydro-11-hydroxy-3-(hydroxymethyl)-1-phenyl-	MeOH	214(4.25),233.5(4.25)	44-4417-86
2,6-Methano-2H,6H-pyrimido[1,2-c]-[1,3,5]oxadiazocin-10(1H)-one, 3,4-dihydro-3,12-dihydroxy-1-phenyl-	MeOH	208(4.38),228(4.28)	44-4417-86
$C_{15}H_{15}N_5O_2$			
Acetic acid, cyano(2-hydrazino-6-phenyl-4(1H)-pyrimidinylidene)-, ethyl ester, (Z)-	EtOH	282(4.54),394(4.45)	103-0774-86

Compound	Solvent	$\lambda_{max}(\log \epsilon)$	Ref.
2,4(1H,3H)-Pteridinedione, 5,6-dihydro-1,3,5-trimethyl-6-(phenylimino)-	pH 3	278(4.29),415(3.82)	33-0704-86
	pH 8	282(4.23),455(3.72)	33-0704-86
$C_{15}H_{15}N_5O_6$			
6-Acetoxy-1,3-dimethyllumazin-7-yl ethyl O-acetylketoxime	MeOH	220(4.23),252(4.23), 357(4.02)	108-0081-86
$C_{15}H_{15}N_7O$			
Propanedinitrile, [[3-amino-2-cyano-4,6,7,8-tetrahydro-4-(2-methoxyethyl)-5-quinoxalinyl]imino]-	EtOH	584(4.67)	33-1025-86
$C_{15}H_{16}BrNO$			
2H-Pyrrol-2-one, 5-[bromo(4-methylphenyl)methylene]-1,5-dihydro-1,3,4-trimethyl-, (Z)-	EtOH	297(4.27)	49-0631-86
$C_{15}H_{16}BrNO_6$			
Propanedioic acid, [bromo(4-nitrophenyl)methylene]-, 1,1-dimethylethyl methyl ester, (E)-	EtOH	277(4.18)	44-4112-86
(Z)-	EtOH	278(4.12)	44-4112-86
$C_{15}H_{16}BrNO_9$			
Propanedioic acid, (acetylamino)[[3-bromo-6-(methoxycarbonyl)-2-oxo-2H-pyran-4-yl]methyl]-, dimethyl ester	EtOH	315(4.0)	5-1968-86
$C_{15}H_{16}ClHgNO_2$			
Mercury, chloro[(3,4,5,6-tetrahydro-2,6-dimethyl-5-oxo-2H-pyrano[3,2-c]quinolin-2-yl)methyl]-	EtOH	228(4.6),275(3.8), 285(3.8),314(3.8)	32-0441-86
$C_{15}H_{16}ClHgNO_3$			
Mercury, chloro[(3,4,5,6-tetrahydro-9-methoxy-2-methyl-5-oxo-2H-pyrano[3,2-c]quinolin-2-yl)methyl]-	EtOH	274(3.7),283(3.7), 333(3.7)	32-0441-86
$C_{15}H_{16}ClN_3O$			
4H-Indazol-4-one, 3-[(4-chlorophenyl)-amino]-1,5,6,7-tetrahydro-6,6-dimethyl-	dioxan	245(4.264),268(4.356), 306(4.149)	48-0635-86
$C_{15}H_{16}FN_3O_5S_2$			
1,2,4-Triazin-3(2H)-one, 6-[[(2-fluorophenyl)methyl]thio]-4,5-dihydro-2-β-D-ribofuranosyl-5-thioxo-	pH 1	326(4.00)	80-0881-86
	pH 13	334(4.03)	80-0881-86
	EtOH	326(3.97)	80-0881-86
1,2,4-Triazin-3(2H)-one, 6-[[(3-fluorophenyl)methyl]thio]-4,5-dihydro-2-β-D-ribofuranosyl-5-thioxo-	pH 1	326(3.97)	80-0881-86
	pH 13	334(3.96)	80-0881-86
	EtOH	326(4.00)	80-0881-86
1,2,4-Triazin-3(2H)-one, 6-[[(4-fluorophenyl)methyl]thio]-4,5-dihydro-2-β-D-ribofuranosyl-5-thioxo-	pH 1	326(3.93)	80-0881-86
	pH 13	334(3.91)	80-0881-86
	EtOH	326(3.96)	80-0881-86
$C_{15}H_{16}F_3NO_7$			
2H-Pyran-4-propanoic acid, 6-(ethoxycarbonyl)-2-oxo-α-[(trifluoroacetyl)amino]-, ethyl ester	EtOH	302(3.9)	5-1968-86

Compound	Solvent	λ_{max}(log ε)	Ref.
$C_{15}H_{16}IN_3O$			
Furo[2,3-d]pyrimidin-4-amine, 1,5,6-trimethyl-N-phenyl-, iodide	EtOH	316(4.22)	11-0337-86
$C_{15}H_{16}NO$			
Pyrylium, 2-[2-[4-(dimethylamino)phenyl]ethenyl]-, hexachlorostannate, (2:1)	CH_2Cl_2	620(4.21)	104-0258-86
Pyrylium, 4-[2-[4-(dimethylamino)phenyl]ethenyl]-, perchlorate	CH_2Cl_2	620(4.13)	104-0258-86
$C_{15}H_{16}N_2$			
Benzenamine, N,N-dimethyl-4-[(phenylimino)methyl]-	C_6H_{12}	238(4.18),356(4.59)	165-0017-86
	pH 5.0	450(4.56)	30-0055-86
1,4-Benzenediamine, N,N-dimethyl-N'-(phenylmethylene)-	MeOH	252(4.30),374(4.26)	165-0017-86
$C_{15}H_{16}N_2O$			
2H-Pyrrolidin-2-one, 3,4-dimethyl-5-[(2-methylphenyl)cyanomethylene]-, (E)-	MeOH	261(4.36)	49-0631-86
	$CHCl_3$	260(4.38)	49-0631-86
(Z)-	MeOH	262(4.26)	49-0631-86
	$CHCl_3$	262(4.30)	49-0631-86
$C_{15}H_{16}N_2O_2$			
3H-Benz[f]indazole-4,9-dione, 9a-(1,1-dimethylethyl)-3a,9a-dihydro-, cis-(±)-	EtOH	227(4.50),251s(4.04), 302(3.26)	39-2217-86C
Phenol, 2-[[(4,6-dimethyl-2-pyridinyl)-imino]methyl]-6-methoxy-	$CHCl_3$	480(2.85)	2-0127-86A
2-Propen-1-one, 3-(dimethylamino)-2-(3-methyl-5-isoxazolyl)-1-phenyl-	dioxan	233(4.35),317(4.41)	83-0242-86
6H-Pyrido[1,2-a]pyrimidin-6-one, 9-benzoyl-1,2,3,4,7,8-hexahydro-	EtOH	230(4.08),340(4.17)	142-2247-86
$C_{15}H_{16}N_2O_3$			
6H-1,3-Oxazine-5-carboxaldehyde, 2-(diethylamino)-6-oxo-4-phenyl-	EtOH	204(4.21),256(4.28), 280s(3.99),342(4.28)	44-0945-86
2-Pentenoic acid, 4-(2-imidazolidinylidene)-5-oxo-5-phenyl-, methyl ester, (E)-	EtOH	236(4.16),322(4.29)	142-2247-86
$C_{15}H_{16}N_2O_3S$			
Benzeneacetic acid, α-[(1,4-dihydro-6-methyl-4-oxo-2-pyrimidinyl)-thio]-, ethyl ester	MeOH	220(4.23),280(3.90)	2-0275-86
$C_{15}H_{16}N_2O_5$			
L-Glutamic acid, N-[4-(2-propynylamino)benzoyl]-	pH 13	281.5(4.29)	87-0468-86
L-Tryptophan, N-acetyl-1-(2-hydroxyacetyl)-	MeOH	240(4.20),292(3.79), 301(3.80)	163-2315-86
$C_{15}H_{16}N_2O_6$			
Butanedioic acid, [(methylamino)phenylmethylene](nitromethylene)-, dimethyl ester	EtOH	323(3.82)	18-3871-86
Butanedioic acid, (nitromethylene)-[1-(phenylamino)ethylidene]-, dimethyl ester	EtOH	356(3.82)	18-3871-86

Compound	Solvent	λ_{max}(log ε)	Ref.
1-Cyclobutene-1,2-dicarboxylic acid, 3-(methylamino)-4-nitro-3-phenyl-, dimethyl ester	EtOH	240(3.86),362(4.08)	18-3871-86
1-Cyclobutene-1,2-dicarboxylic acid, 3-methyl-4-nitro-3-(phenylamino)-, dimethyl ester	EtOH	230(4.06),362(4.09)	18-3871-86
3-Furancarboxylic acid, 5-[4-(aminocarbonyl)-4-cyano-3,4-dihydro-3-hydroxy-2H-pyran-6-yl]-2-methyl-, ethyl ester, (3S-trans)-	MeOH	263(4.31)	136-0237-86B
$C_{15}H_{16}N_2O_6S$			
Uridine, 5-(phenylthio)-	MeOH	246(4.11),268s(4.01), 303(3.66)	78-4187-86
$C_{15}H_{16}N_3$			
Benz[h]imidazo[1,2-c]quinazolinium, 1,2,4,5-tetrahydro-2-methyl-, chloride	EtOH	258(3.98),285(3.76), 328(3.55),340(3.56)	4-0685-86
Benz[h]imidazo[1,2-c]quinazolinium, 1,2,4,5-tetrahydro-3-methyl-, chloride	EtOH	259(3.95),273(3.79), 290(3.69),326(3.49), 338(3.52)	4-0685-86
$C_{15}H_{16}N_3O$			
Furo[2,3-d]pyrimidin-4-amine, 1,5,6-trimethyl-N-phenyl-, iodide	EtOH	316(4.22)	11-0337-86
$C_{15}H_{16}N_3S_3$			
1H-Pyrazolium, 3-(dimethylamino)-4-(methylthio)-1-(4-phenyl-1,3-dithiol-2-ylidene)-, perchlorate	MeCN	230(4.28),247s(4.22), 309(3.64),462(4.53), 484s(4.44)	88-0159-86
$C_{15}H_{16}N_4$			
Pyrazinecarbonitrile, 5-ethyl-3,4-dihydro-3-imino-6-methyl-4-(phenylmethyl)-	EtOH	440(3.81)	33-1025-86
$C_{15}H_{16}N_4O$			
Benzamide, 4-[[4-(dimethylamino)phenyl]azo]-	MeOH	230.4(3.56),277.2(3.99), 435.5(4.45)	56-0797-86
with 0.02M p-toluenesulfonic acid		230.4(5.29),265.0(4.19)	56-0797-86
	DMF	270.0(3.80),437.1(4.39)	56-0797-86
with 0.02M p-toluenesulfonic acid	DMF	271.1(4.22),438.6(4.44)	56-0797-86
1H-Purine, 6-(4-phenylbutoxy)-	pH 1	254(3.99)	163-2243-86
	pH 7	252(3.97)	163-2243-86
	pH 13	254(3.99)	163-2243-86
Pyrazinecarbonitrile, 3,4-dihydro-4-(3-hydroxypropyl)-3-imino-6-methyl-5-phenyl-	pH 1	378(4.11)	33-1025-86
	EtOH	438(2.82)	33-1025-86
$C_{15}H_{16}N_4O_2$			
Carbamic acid, [5-amino-2-(phenylazo)phenyl]-, ethyl ester, (E)-	EtOH	225(4.27),285(4.03), 407(4.14)	97-0134B-86
$C_{15}H_{16}O$			
9(2H)-Anthracenone, 1,3,4,9a-tetrahydro-9a-methyl-	MeOH	237(4.34),275s(3.44), 285s(3.25),344(3.06)	39-1789-86C
1(2H)-Phenanthrenone, 3,4,9,10-tetrahydro-9-methyl-	MeOH	228(4.09),235(4.07), 294(4.10),306(4.02)	44-4424-86
2(3H)-Phenanthrenone, 4,4a,9,10-tetrahydro-4a-methyl-	isoPrOH	234(4.27),315(1.79)	35-4561-86

Compound	Solvent	λ_{max}(log ϵ)	Ref.
$C_{15}H_{16}O_2$			
Curzeone	EtOH	290(4.5),323(4.3)	102-1351-86
1,2-Heptalenedimethanol, 10-methyl-	dioxan	255(4.34),317(3.53)	88-1669-86
$C_{15}H_{16}O_2S_2$			
1,3-Dithiole-4-carboxylic acid, 2-(1-methylethylidene)-5-phenyl-, ethyl ester	EtOH	248(4.15),340s(3.00), 400(3.37)	33-0419-86
$C_{15}H_{16}O_3$			
1H-Dibenzo[b,d]pyran-4(6H)-one, 2,3-dihydro-8-methoxy-6-methyl-	ether	236(3.82),244(3.74), 330(4.04)	44-4424-86
1,2-Naphthalenedione, 3-(1-hydroxy-1-methylethyl)-5,8-dimethyl-	EtOH	218.4(4.51),254.6(4.62), 432(3.86)	2-1247-86
1,4-Naphthalenedione, 2-hydroxy-3-(3-methylbutyl)-	EtOH	251(4.25),281(3.92), 339(3.26),383(3.00)	39-0659-86C
Naphtho[1,2-b]furan-2,8(3H,4H)-dione, 3a,5,5a,9b-tetrahydro-5a,9-dimethyl-3-methylene-	EtOH	205(4.21),234(3.96)	36-0784-86
[7]Paracyclophane-9,10-dicarboxylic acid anhydride	MeCN	237(4.43),275s(3.39), 344(3.59)	24-2698-86
$C_{15}H_{16}O_4$			
Butanoic acid, 4-hydroxy-4-(6-methoxy-1-oxo-1,2,3,4-tetrahydro-2-naphthalenyl)-, γ-lactone	EtOH	227(4.01),275(4.15)	150-2492-86M
Phenol, 4,4'-methylenebis[2-methoxy-	EtOH	286(3.90)	104-0158-86
Pleurotellic acid	MeOH	226(3.70)	78-3587-86
2-Propenoic acid, 3-(3,4-dihydro-2,2-dimethyl-4-oxo-2H-1-benzopyran-6-yl)-, methyl ester, (E)-	EtOH	249(4.44),292(4.47)	163-3083-86
2-Propenoic acid, 3-[4-hydroxy-3-(3-methyl-1-oxo-2-butenyl)phenyl]-, methyl ester, (E)-	EtOH	280(4.47),300s(4.44), 355s(3.71)	163-3083-86
2H,5H-Pyrano[2,3-b][1]benzopyran-5-one, 3,4-dihydro-2-(hydroxymethyl)-2,4-dimethyl-	EtOH	207(--),225(4.32), 275(4.05),288(4.01)	32-0491-86
2H,5H-Pyrano[3,2-c][1]benzopyran-5-one, 3,4-dihydro-2-(hydroxymethyl)-2,4-dimethyl-	EtOH	270(3.98),282(4.02), 304(3.95),316s(--)	32-0501-86
3H,10H-Pyrano[4,3-b][1]benzopyran-10-one, 3-ethoxy-4,4a-dihydro-1-1-methyl-	EtOH	219.5(4.305),266(3.930), 290(4.086),329(3.790)	39-0789-86C
$C_{15}H_{16}O_5$			
2H-1-Benzopyran-2-one, 7-[(3,3-dimethyloxiranyl)methoxy]-8-methoxy- (lacinartinepoxide)	EtOH	218(4.12),246(3.60), 255(3.61),305(4.08)	5-2142-86
3,11a-Epoxy-11aH-oxepino[2,3-b][1]-benzopyran-6(2H)-one, 3,4,5,5a-tetrahydro-5a-hydroxy-3,5-dimethyl-, (3α,5α,5aα,11aα)-	EtOH	250(3.78),310(3.26)	32-0285-86
(3α,5α,5aβ,11aα)-	EtOH	251(3.96),312(3.45)	32-0285-86
(3α,5β,5aβ,11aα)-	EtOH	250(4.13),310(3.61)	32-0285-86
$C_{15}H_{16}O_7$			
4H-1-Benzopyran-6-acetic acid, 7-hydroxy-5,8-dimethoxy-2-methyl-4-oxo-, methyl ester	EtOH	208(4.28),228(4.32), 248(4.31),254(4.33), 268s(3.89),293(3.91), 342s(3.32)	44-3116-86

Compound	Solvent	λmax(log ε)	Ref.
$C_{15}H_{17}BrN_4O$			
Phenol, 2-[(5-bromo-2-pyridinyl)azo]-5-(diethylamino)-	pH 4	445(4.5)	160-0713-86
$C_{15}H_{17}BrO_2$			
Spiro[5.5]undeca-2,9-diene-1,8-dione, 4-bromo-3,11,11-trimethyl-7-methylene-	EtOH	237(3.94)	33-0091-86
Tricyclo[6.3.1.0^{1,6}]dodeca-3,9-diene-5,11-dione, 8-bromo-2,2,9-trimethyl-	EtOH	233(3.94)	33-0091-86
$C_{15}H_{17}N$			
Benzenamine, N-phenyl-N-propyl-	hexane	213(4.11),246(5.18), 302(3.45)	65-2461-86
	MeOH	208(--),249(--), 299(--)	65-2461-86
$C_{15}H_{17}NO$			
2-Butanone, 3,3-dimethyl-1-(2-quinolinyl)-	CF_3COOH	319(4.02)	116-0291-86
1-Propanone, 1-[6-(dimethylamino)-2-naphthalenyl]-	C_6H_{12}	230(4.5),260(4.5), 282(4.3),344(4.3)	131-0013-86E
	$HOCH_2CH_2OH$	388(4.3)	131-0013-86E
	MeCN	354(4.4)	131-0013-86E
2H-Pyrrol-2-one, 5-[(2,6-dimethylphenyl)methylene]-1,5-dihydro-3,4-dimethyl-, (E)-	EtOH	270(4.25)	49-0631-86
(Z)-	EtOH	290(4.20)	49-0631-86
$C_{15}H_{17}NOS$			
1-Benzothiepin-5(2H)-one, 3,4-dihydro-4-(1-pyrrolidinylmethylene)-, (E)-	EtOH	249(3.96),355(4.23)	4-0449-86
$C_{15}H_{17}NO_2$			
1,6-Octadiene-3,5-dione, 7-(methylamino)-1-phenyl-	n.s.g.	334(4.06),406(4.56)	4-1721-86
$C_{15}H_{17}NO_2S$			
1-Benzothiepin-5(2H)-one, 3,4-dihydro-4-(4-morpholinylmethylene)-, (E)-	EtOH	249(4.03),301(3.80), 351(4.07)	4-0449-86
$C_{15}H_{17}NO_3$			
3-Azabicyclo[4.1.0]hept-4-ene-3-carboxylic acid, 2-hydroxy-7-phenyl-, ethyl ester, (1α,2α,6α,7α)-	EtOH	240(3.60)	142-1831-86
Benzeneacetamide, N-(1-methylethyl)-N-(2-methyl-1-oxo-2-propenyl)-α-oxo-	C_6H_{12}	219(4.12),258(4.17), 360(2.15)	39-1759-86C
Benzeneacetamide, N-(1-methylethyl)-α-oxo-N-(1-oxo-2-butenyl)-	C_6H_{12}	229(4.16),253(4.16)	39-1759-86C
6-Oxa-3-azabicyclo[3.1.1]heptane-2,4-dione, 1-methyl-3-(1-methylethyl)-5-phenyl-	C_6H_{12}	218(4.02),252(2.61), 257(2.65),264(2.60)	39-1759-86C
6-Oxa-3-azabicyclo[3.1.1]heptane-2,4-dione, 7-methyl-3-(1-methylethyl)-1-phenyl-	C_6H_{12}	218(4.27),252(2.95), 258(2.93),263(2.93)	39-1759-86C
$C_{15}H_{17}NO_3S$			
2-Thiophenecarboxaldehyde, 5-[(2,2,6,6-tetramethyl-4,5-dioxo-3-piperidin-	EtOH	221(3.94),263(3.85), 368(4.45)	70-2511-86

Compound	Solvent	λmax(log ε)	Ref.
ylidene)methyl]- (cont.)			70-2511-86
higher melting isomer	EtOH	221(3.93),370(4.34)	70-2511-86
$C_{15}H_{17}NO_4$			
2-Cyclopenten-1-one, 3-(2,5-dihydroxyphenyl)-2-(4-morpholinyl)-	EtOH	285(4.12),356(3.76)	24-3487-86
$C_{15}H_{17}NO_6$			
1H-Indole-3-acetic acid, 1-β-D-ribofuranosyl-	pH 1	220(4.40),270(3.66), 278s(3.66),288s(3.54)	4-1777-86
	pH 7	222(4.39),272s(3.61), 280(3.64),289s(3.57)	4-1777-86
	pH 13	223(4.39),272s(3.59), 281(3.65),289s(3.57)	4-1177-86
Propanedioic acid, [(4-nitrophenyl)-methylene]-, 1,1-dimethylethyl methyl ester (Z)-	EtOH	296(4.27)	44-4112-86
$C_{15}H_{17}NO_7$			
2H-Pyran-6-carboxylic acid, 4-[2-(acetylamino)-3-ethoxy-3-oxo-1-propenyl]-2-oxo-, ethyl ester, (Z)-	EtOH	253(4.0),317(4.0)	5-1968-86
$C_{15}H_{17}NO_9$			
Propanedioic acid, (acetylamino)[[6-(methoxycarbonyl)-2-oxo-2H-pyran-4-yl]methyl]-, dimethyl ester	EtOH	304(3.8)	5-1968-86
$C_{15}H_{17}N_2$			
1H-Indolium, 1-[(2,3-dihydro-4(1H)-pyridinylidene)ethylidene]-2,3-dihydro-, perchlorate	MeOH	265(3.97),302(3.38), 311(3.32),494(4.86)	33-1588-86
	MeOH-NaOH	241(3.96),265(3.91), 315(3.79),398(4.61)	33-1588-86
$C_{15}H_{17}N_2O$			
1H-Imidazol-1-yloxy, 2,5-dihydro-2,2,5,5-tetramethyl-4-(phenylethynyl)-	n.s.g.	274(4.33)	88-1625-86
$C_{15}H_{17}N_3$			
Benzenamine, N,N-dimethyl-4-[(2-methylphenyl)azo]-	C_6H_{12}	395(4.47)	56-0163-86
	C_6H_5Cl	405(--)	56-0163-86
Benzenamine, N,N,3-trimethyl-4-(phenylazo)-	C_6H_{12}	405(4.48)	56-0163-86
Cyclohepta[b]pyrrol-2(1H)-one, cyclohexylidenehydrazone	MeOH	279(4.43),288(4.40), 316(4.22),435(4.07), 490s(3.52)	18-3326-86
$C_{15}H_{17}N_3O$			
Ethanol, 2-[6,7-dihydro-5H-benzo-[6,7]cyclohepta[1,2-d]pyrimidin-4-yl)amino]-	EtOH	247(4.31),297(3.60)	4-1685-86
4H-Indazol-4-one, 1,5,6,7-tetrahydro-6,6-dimethyl-3-(phenylamino)-	dioxan	243(4.329),264(4.278), 305(4.064)	48-0635-86
$C_{15}H_{17}N_3O_2$			
Benzenamine, 4-[(4-methoxyphenyl)-azo]-N,N-dimethyl-, N-oxide	benzene	324(3.93),436(2.95)	39-1439-86B
	MeOH	241(4.00),350(4.25), 434(3.15)	39-1439-86B
Benzenamine, 4-[(4-methoxyphenyl)-NNO-azoxy]-N,N-dimethyl-	benzene	410(4.475)	44-0088-86
	MeOH	415(4.472)	44-0088-86

Compound	Solvent	$\lambda_{max}(\log \epsilon)$	Ref.
Benzenamine, 4-[(4-methoxyphenyl)-	benzene	393(4.509)	44-0088-86
ONN-azoxy]-N,N-dimethyl-	MeOH	395(4.510)	44-0088-86
4H-Pyrido[1,2-a]pyrimidine-3-carbox-aldehyde, 6-methyl-4-oxo-2-(1-piperidinyl)-	EtOH	250s(4.20),270(4.37), 386(4.06)	4-1295-86
$C_{15}H_{17}N_3O_3$			
Benzenamine, 4-[(4-methoxyphenyl)-	benzene	349(4.19)	39-1439-86B
ONN-azoxy]-N,N-dimethyl-, N-oxide	MeOH	241(3.98),349(4.23)	39-1439-86B
Formamide, N-butyl-N-(3-formyl-6-methyl-4-oxo-4H-pyrido[1,2-a]-pyrimidin-2-yl)-	EtOH	252(4.22),289(4.06), 394s(4.18),405(4.21)	4-1295-86
$C_{15}H_{17}N_3O_4$			
4-Pyrimidinecarboxylic acid, 1,4,5,6-tetrahydro-2-[(2-methylbenzoyl)-amino]-5-oxo-, ethyl ester	EtOH	227s(4.12),266(4.46)	118-0484-86
4-Pyrimidinecarboxylic acid, 1,4,5,6-tetrahydro-2-[(3-methylbenzoyl)-amino]-5-oxo-, ethyl ester	EtOH	234s(3.82),269(4.26)	118-0484-86
4-Pyrimidinecarboxylic acid, 1,4,5,6-tetrahydro-2-[(4-methylbenzoyl)-amino]-5-oxo-, ethyl ester	EtOH	233s(3.85),270(4.39)	118-0484-86
$C_{15}H_{17}N_3O_5$			
4-Pyrimidinecarboxylic acid, 1,4,5,6-tetrahydro-2-[(2-methoxybenzoyl)-amino]-5-oxo-, ethyl ester	EtOH	213(4.32),265(4.29), 305(3.80)	118-0484-86
4-Pyrimidinecarboxylic acid, 1,4,5,6-tetrahydro-2-[(4-methoxybenzoyl)-amino]-5-oxo-, ethyl ester	EtOH	220(4.25),277s(4.41), 287(4.46)	118-0484-86
2,4(1H,3H)-Pyrimidinedione, 1-[3-deoxy-3-(phenylamino)-β-D-lyxofuranosyl]-	MeOH	202.5(4.44),246.5(4.26), 262.5s(4.12),295s(3.26)	44-4417-86
$C_{15}H_{17}N_3O_6$			
2(1H)-Pyrimidinone, 4-[(6-methyl-2-	pH 7	270(3.90),282s(--)	1-0806-86
pyridinyl)oxy]-1-β-D-ribofuranosyl-	pH 2	270(--),282s(--)	1-0806-86
	pH 13	272(--),284s(--)	1-0806-86
2(1H)-Pyrimidinone, 4-[(6-methyl-3-	pH 7	278(3.93)	1-0806-86
pyridinyl)oxy]-1-β-D-ribofuranosyl-	pH 2	281(--)	1-0806-86
	pH 13	278(--)	1-0806-86
$C_{15}H_{17}N_5$			
Imidazo[1,2-a]pyrazine-8-carbonitrile,	pH 1	378(4.12)	33-1025-86
8a-amino-1,2,3,8a-tetrahydro-1,6-dimethyl-5-phenyl-	EtOH	343(3.70)	33-1025-86
9H-Purin-6-amine, 9-ethyl-N-[(2-meth-	pH 1	268(4.30)	94-1094-86
ylphenyl)methyl]-	pH 7	271.5(4.30)	94-1094-86
	pH 13	271.5(4.30)	94-1094-86
	EtOH	271(4.30)	94-1094-86
9H-Purin-6-imine, 9-ethyl-1,9-di-	pH 1	262(4.14)	94-1094-86
hydro-1-[(2-methylphenyl)methyl]-,	pH 7	262(4.14)	94-1094-86
monohydrobromide	pH 13	261(4.14)	94-1094-86
	EtOH	261.5(4.13)	94-1094-86
Pyrazinecarbonitrile, 4-(3-amino-	pH 1	378(4.11)	33-1025-86
phenyl)-3,4-dihydro-3-imino-6-methyl-5-phenyl-	EtOH	349(3.72)	33-1025-86
$C_{15}H_{17}N_5O$			
Benzenemethanol, 2-[(9-ethyl-6,9-di-	pH 1	261.5(4.12)	94-1094-86

Compound	Solvent	$\lambda_{max}(\log \epsilon)$	Ref.
hydro-6-imino-1H-purin-1-yl)methyl]-, monohydrobromide (cont.)	pH 7	261(4.12)	94-1094-86
	pH 13	260.5(4.13)	94-1094-86
	EtOH	262(4.13)	94-1094-86
Benzenemethanol, 2-[[(9-ethyl-9H-purin-6-yl)amino]methyl]-, monohydrobromide	pH 1	268(4.30)	94-1094-86
	pH 7	271(4.31)	94-1094-86
	pH 13	271(4.31)	94-1094-86
	EtOH	271(4.28)	94-1094-86
Pyrazinecarboxylic acid, 3-amino-6-methyl-5-phenyl-, (1-methylethylidene)hydrazide	EtOH	381(4.09)	33-0793-86
$C_{15}H_{18}BrClO_2$			
Spiro[5.5]undec-3-ene-2,7-dione, 10-bromo-9-chloro-5,5,9-trimethyl-1-methylene-, [6R-(6α,9α,10β)]-	EtOH	203(3.78),225(3.69)	33-0091-86
$C_{15}H_{18}BrN_2$			
2,3'-Bipyridinium, 1'-(5-bromopentyl)-, bromide	pH 8.2	210(4.22),229(4.15), 276(4.05)	64-0239-86B
$C_{15}H_{18}BrN_7$			
2,4-Pyrimidinediamine, 5-[(4-bromophenyl)azo]-6-piperidino-	EtOH	202(4.69),213(4.63), 229(4.61),271(4.58), 396(4.73)	42-0420-86
$C_{15}H_{18}ClNO_7$			
2H-Pyran-4-propanoic acid, α-(acetylamino)-3-chloro-6-(ethoxycarbonyl)-2-oxo-, ethyl ester	EtOH	310(3.9)	5-1968-86
2H-Pyran-4-propanoic acid, α-(acetylchloroamino)-6-(ethoxycarbonyl)-2-oxo-, ethyl ester	EtOH	303(3.8)	5-1968-86
$C_{15}H_{18}Cl_2O_4$			
Butanoic acid, 4-chloro-, 5(and 4)-(4-chloro-1-oxobutyl)-2-methoxyphenyl ester	EtOH	218(4.18),267(4.18)	24-0050-86
$C_{15}H_{18}N_2$			
4,4'-Bipyridinium, 1-ethenyl-1'-propyl-, dibromide	H_2O	283(4.22)	77-0574-86
1H-Cyclohept[2,3]azirino[1,2-b]pyrazole, 4a,5,6,7,8,9-hexahydro-2-phenyl-	EtOH	254(4.13)	35-1617-86
$C_{15}H_{18}N_2O$			
Indolo[2,3-a]quinolizin-1-ol, 1,2,3,4,6,7,12,12b-octahydro-, cis	MeOH	224(4.57),274s(3.86), 282(3.87),290(3.79)	94-3713-86
5,9-Methanocycloocta[b]pyridin-2(1H)-one, 5-amino-11-ethylidene-5,6,9,10-tetrahydro-7-methyl-, [5R-(5α,9β,11E)- (huperzine A)	EtOH	231(4.01),313(3.89)	23-0837-86
$C_{15}H_{18}N_2OS$			
1-Piperidinamine, N-[(3-methoxybenzo[b]thien-2-yl)methylene]-	PrOH	335(4.43)	104-2108-86
$C_{15}H_{18}N_2O_2$			
2-Furancarboxaldehyde, 5-(dimethylamino)-4-[(methylamino)phenylmethyl]-	MeOH	368(3.29)	73-0573-86

Compound	Solvent	λ max (log ε)	Ref.
5-Pyrimidinecarboxylic acid, 1,4-dihydro-2,6-dimethyl-1-phenyl-, ethyl ester	EtOH	310(3.66)	150-0088-86S +150-0901-86M
5-Pyrimidinecarboxylic acid, 1,4-dihydro-1,6-dimethyl-4-phenyl-, ethyl ester	EtOH	227s(4.07),323(3.76)	103-0990-86
$C_{15}H_{18}N_2O_2S$			
5-Pyrimidinecarboxylic acid, 1,2,3,4-tetrahydro-1,6-dimethyl-4-phenyl-2-thioxo-, ethyl ester	EtOH	305(4.24)	103-0990-86
$C_{15}H_{18}N_2O_3$			
Methanone, (3a,4,5,6,7,7a-hexahydro-3a-hydroxy-7a-methyl-1H-benzimidazol-2-yl)phenyl-, N-oxide	EtOH	256(4.00),324(3.30)	103-0856-86
1H-Pyrido[3,4-b]indole-1-carboxylic acid, 2,3,4,9-tetrahydro-8-methoxy-1-methyl-, methyl ester	MeOH	218(4.49),266(3.84), 288s(3.55)	78-6719-86
$C_{15}H_{18}N_2O_4$			
2,4(1H,3H)-Pyrimidinedione, 1-[1-(2-hydroxyethoxy)ethyl]-5-(phenylmethyl)-	pH 1	269(3.76)	4-1651-86
	pH 11	267(3.61)	4-1651-86
$C_{15}H_{18}N_2O_5$			
1H-Cyclopent[cd]indazole-1,2-dicarboxylic acid, 7a,7b-dihydro-2a-methoxy-7b-methyl-, dimethyl ester	EtOH	296(3.81)	39-0145-86C
$C_{15}H_{18}N_4$			
Benzenamine, N,N-diethyl-4-(3-pyridinylazo)-	acetone	428(4.50)	7-0473-86
2(1H)-Pyrimidinone, 4,6-dimethyl-, [1-(2-methylphenyl)ethylidene]-hydrazone, (E)-	MeOH	273(4.38)	163-2427-86
(Z)-	MeOH	264(4.40)	163-2427-86
$C_{15}H_{18}N_4O$			
Azepino[2,1-b]benzo[g]pteridin-13(2H)-one, 1,3,4,7,8,9,10,11-octahydro-	H_2O	323(3.93)	33-0793-86
Ethanol, 2-[ethyl[4-(3-pyridinylazo)-phenyl]amino]-	acetone	426(4.50)	7-0473-86
$C_{15}H_{18}N_4O_2$			
Ethanol, 2,2'-[[4-(3-pyridinylazo)-phenyl]imino]bis-	acetone	425(4.47)	7-0473-86
$C_{15}H_{18}N_4O_2S$			
1H-Indole-2,3-dione, 1-(1-oxobutyl)-3-[[(ethylamino)thioxomethyl]hydrazone	EtOH	226(4.36),338(4.08), 372(4.16)	104-2173-86
	$CHCl_3$	333s(4.01),373(4.41)	104-2173-86
	DMF	285(3.76),342(3.78), 463(4.42)	104-2173-86
geometric isomer	$CHCl_3$	281(3.99),336s(4.15), 394(4.49)	104-2173-86
	DMF	331s(3.82),392(4.12)	104-2173-86
$C_{15}H_{18}N_5$			
7H-Purinium, 6-amino-9-ethyl-7-[(2-	pH 1	271(4.07)	94-1094-86

Compound	Solvent	λ_{max}(log ε)	Ref.
methylphenyl)methyl]-, bromide	pH 7	272(4.07)	94-1094-86
(cont.)	EtOH	274(4.04)	94-1094-86
$C_{15}H_{18}N_6S$			
1H-Purine-2-thione, 7-[2-(dimethyl-amino)ethyl]-1,3,6,7-tetrahydro-6-imino-1-phenyl-, hydrochloride	pH 1	250(4.17),284(4.33)	4-0737-86
	pH 13	238(4.29),275(4.23)	4-0737-86
$C_{15}H_{18}O$			
2-Cyclohexen-1-one, 3-phenyl-2-propyl-	MeOH	255(4.02)	44-4424-86
Naphthalene, 1-methoxy-3,4,5,7-tetra-methyl-	MeOH	231(4.59),240s(4.57), 303(3.80),322(3.55), 337(3.47)	39-2011-86C
Naphtho[2,3-b]furan, 4,4a,8a,9-tetra-hydro-3,5,8a-trimethyl- (tubipo-furan)	EtOH	216(3.73),263(3.59)	138-1789-86
$C_{15}H_{18}O_2$			
2-Cyclohexen-1-one, 2-ethyl-3-(4-meth-oxyphenyl)-	MeOH	200(4.08),226(4.11), 282(4.10)	44-4424-86
4,8-Heptanofuro[3,4-d]oxepin	EtOH	219(4.29),241s(3.59)	24-2698-86
4,7-Heptanoisobenzofuran-1(3H)-one	EtOH	222(4.09),248(3.51), 308(3.02)	24-2698-86
Naphtho[1,2-b]furan-2(4H)-one, 5,5a,6,7-tetrahydro-3,5a,9-trimethyl-, (R)-	n.s.g.	330(4.30)	2-0233-86
$C_{15}H_{18}O_3$			
1-Cyclohexen-1-one, 2-ethoxy-3-(4-methoxyphenyl)-	ether	225(3.79),302(4.09)	44-4424-86
4,8-Heptanofuro[3,4-d]oxepin-1(3H)-one	EtOH	213(4.37),280(3.45)	24-2698-86
Pleurotellol	MeOH	226s(3.70)	78-3587-86
α-Santonin	EtOH	240(3.77)	105-0608-86
Spiro[5H-benzo[3,4]cyclobuta[1,2]cyclo-heptene-5,2'-[1,3]dioxolan]-4b(6H)-ol, 7,8,9,9a-tetrahydro-, cis	MeOH	260(3.07),266(3.25), 273(3.23)	44-1419-86
isomer	MeOH	261(3.28),267(3.40), 273(3.39)	44-1419-86
Zedoarol	EtOH	256(3.7)	102-1351-86
$C_{15}H_{18}O_4$			
8-Epiartemisin	EtOH	241(4.06),266s(--)	94-1319-86
Ethanone, 1-[1-(dimethoxymethyl)-1,2,3,4-tetrahydro-1,4-epoxy-naphthalen-6-yl]-	isooctane	205(4.28),252(4.07), 256s(4.02),281(3.15), 292(3.08)	33-1287-86
	EtOH	203(4.35),255(4.12), 280s(3.32)	33-1287-86
Ethanone, 1-[4-(dimethoxymethyl)-1,2,3,4-tetrahydro-1,4-epoxy-naphthalen-6-yl]-	isooctane	209(4.41),250(4.12), 256s(4.06),280(3.15), 289(3.04)	33-1287-86
	EtOH	206(4.38),254(4.11)	33-1287-86
Naphtho[2,3-b]furan-2,6(3H,4H)-dione, 3a,8a,9,9a-tetrahydro-4-hydroxy-3,5,8a-trimethyl-	EtOH	240(4.00),263s(--)	94-1319-86
$C_{15}H_{18}O_5$			
Artelin	EtOH	268(3.42),300(3.31)	105-0112-86
Butanedioic acid, [(2-methoxyphenyl)-methylene]dimethyl-, 4-methyl ester, (E)-	EtOH	258(3.88),296(3.67)	34-0369-86

Compound	Solvent	$\lambda_{max}(\log \epsilon)$	Ref.
Butanedioic acid, [(4-methoxyphenyl)-methylene]dimethyl-, 4-methyl ester, (E)-	EtOH	269(4.05)	34-0369-86
$C_{15}H_{18}O_6$			
2H-1-Benzopyran-2-one, 7-(2,3-dihydroxy-3-methylbutoxy)-8-methoxy-, (+)-	EtOH	217(4.24),247(3.66), 256(3.68),309(4.10)	5-2142-86
$C_{15}H_{18}S_2$			
Azulene, 4-[bis(ethylthio)methyl]-	hexane	242(4.39),281(4.61), 284(4.62),295(4.18), 321(3.56),329(3.59), 333(3.61),344(3.62), 348(3.66),361(3.24), 556(2.64),574(2.76), 592(2.70),622(2.68), 642(2.52),682(2.29)	5-1222-86
Azulene, 6-[bis(ethylthio)methyl]-	hexane	237(4.11),277(4.76), 280(4.79),285(4.79), 299(4.19),324(3.68), 333(3.70),341(3.63), 349(3.80),364(3.16), 509(2.08),527(2.23), 547(2.35),564(2.42), 586(2.51),610(2.45), 641(2.46),668(2.15), 708(2.10)	5-1222-86
Azulene, 4-[bis(methylthio)methyl]-6,8-dimethyl-	hexane	244(4.41),289(4.64), 325s(3.50),335s(3.51), 340(3.56),351 s(3.60), 354(3.63),366s(2.96), 555(2.76),567s(2.76), 590(2.72),645s(2.33)	5-1222-86
$C_{15}H_{19}BrO$			
Phenol, 4-bromo-5-(1,2-dimethylbicyclo[3.1.0]hex-2-yl)-2-methyl-, [1S-(1α,2β,3α)]- (cyclolaurenol)	EtOH	206(4.08),269(2.76), 278s(2.75)	44-3364-86
Phenol, 4-bromo-2-methyl-5-(1,2,2-trimethyl-3-cyclopenten-1-yl)-, (R)- (cupalaurenol)	EtOH	206(4.15),224s(3.82), 282(3.18)	44-3364-86
Spiro[5.5]undeca-1,7-dien-3-one, 9-(bromomethylene)-1,2,5-trimethyl-, (Z)-	EtOH	253(4.43)	33-0091-86
$C_{15}H_{19}ClN_2OS$			
1,8-Naphthalenediamine, 4-[(chloromethyl)sulfinyl]-N,N,N',N'-tetramethyl-	MeOH	205(3.26),245(3.38), 353(2.99)	88-0167-86
$C_{15}H_{19}ClN_2O_2$			
4-Isoxazolecarboxamide, 5-(chloromethyl)-4,5-dihydro-3-(4-methylphenyl)-N-propyl-, cis	MeOH	279(3.15)	73-2167-86
$C_{15}H_{19}ClO_4S$			
Sulfonium, [(6,8-dimethyl-4-azulenyl)-methyl]dimethyl-, perchlorate	MeCN	248(4.40),290(4.63), 293(4.62),309s(3.93), 341(3.51),355(3.55), 370s(2.79),568(2.70),	5-1222-86

Compound	Solvent	$\lambda_{max}(\log \epsilon)$	Ref.
(cont.)		598s(2.65),661s(2.21),	5-1222-86
$C_{15}H_{19}N$			
1H-Carbazole, 2,3,4,9-tetrahydro-1,1,9-trimethyl-	n.s.g.	231(4.27),285(3.88)	150-0514-86M
Cyclohept[b]indole, 5,6,7,8,9,10-hexahydro-5,6-dimethyl-	n.s.g.	229(4.40),286(3.85)	150-2782-86M
Trikentrin A, cis	MeOH	241(4.05),271(4.05) (unstable)	78-6545-86
Trikentrin A, trans	MeOH	222(4.84),271(4.10)	78-6545-86
$C_{15}H_{19}NO$			
Phenol, 4-(1,2,3,5,8,8a-hexahydro-7-methyl-6-indolizinyl)-, (±)-(ipalbidine)	EtOH	238(4.01),278(3.24)	33-2048-86
	EtOH-NaOH	261?(4.09)	33-2048-86
$C_{15}H_{19}NOS$			
Azocine, 8-ethoxy-2,3,4,5-tetrahydro-7-(phenylthio)-, (E)-	MeOH	214(4.14),257(3.85)	12-0687-86
1-Benzothiepin-5(2H)-one, 4-[(diethylamino)methylene]-3,4-dihydro-, (E)-	EtOH	248(4.09),351(4.26)	4-0449-86
$C_{15}H_{19}NO_2$			
Cyclohexanecarboxylic acid, 2-[(phenylmethylene)amino]-, methyl ester, cis	EtOH	209(4.32),249(4.26)	12-0591-86
trans	EtOH	212(4.18),250(4.20)	12-0591-86
2,4,6-Heptatrienoic acid, 2-methyl-7-(1H-pyrrol-2-yl)-, 1-methylethyl ester, (E,E,E)-	EtOH	392(4.61)	150-3601-86M
3-Pyridinecarboxylic acid, 1,2,3,6-tetrahydro-1-methyl-3-(phenylmethyl)-, (±)-, hydrochloride	pH 7.2	210(0.74)	87-0125-86
2H-Pyrrol-2-one, 1,5-dihydro-5-methoxy-3,4-dimethyl-5-[(2-methylphenyl)methyl]-	MeOH	207(4.23)	49-0185-86
2H-Pyrrol-2-one, 1,5-dihydro-5-methoxy-3,4-dimethyl-5-[(4-methylphenyl)methyl]-	MeOH	207(4.24)	49-0185-86
$C_{15}H_{19}NO_2S$			
2-Thiophenecarboxaldehyde, 5-[(2,2,6,6-tetramethyl-4-oxo-3-piperidinylidene)methyl]-	EtOH	227(3.96),277(3.76), 349(3.43)	70-2511-86
$C_{15}H_{19}NO_3$			
1,3,5-Dioxazocine-7-carboxaldehyde, 5,8-dihydro-2-(1-methylethyl)-6-phenyl-	MeOH	235(3.07),299(3.21)	73-2158-86
1,3,5-Dioxazocine-7-carboxaldehyde, 5,8-dihydro-6-phenyl-2-propyl-	MeOH	235(3.03),298(3.16)	73-2158-86
[1,3]Dioxepino[5,6-d]isoxazole, 3a,4,8,8a-tetrahydro-6-(1-methylethyl)-3-phenyl-, endo	MeOH	240(2.98),265(2.89)	73-2158-86
exo	MeOH	257(3.00)	73-2158-86
[1,3]Dioxepino[5,6-d]isoxazole, 3a,4,8,8a-tetrahydro-3-phenyl-6-propyl-, endo	MeOH	263(3.03)	73-2158-86
exo	MeOH	258(3.03)	73-2158-86
Spiro[1,3-benzodioxole-2,1'-cyclohexane], 5-amino-N-acetyl-N-methyl-	MeOH	244(3.68),290(3.70)	87-1737-86

Compound	Solvent	λ_{max}(log ε)	Ref.
Spiro[cyclohexan-1,3'(4'H)-isoquinolin]-4'-one, 1',2'-dihydro-6'-hydroxy-7'-methoxy-	EtOH	232(4.25),276(4.06), 319(3.75)	39-1329-86C
	EtOH-KOH	255(4.39),290(3.90), 367(3.61)	39-1329-86C
Spiro[cyclohexan-1,3'(4'H)-isoquinolin]-4'-one, 1',2'-dihydro-7'-hydroxy-6'-methoxy-	EtOH	232(4.15),278(3.97), 314(3.89)	39-1329-86C
	EtOH-KOH	255(3.95),298(3.61), 351(4.07)	39-1329-86C
isomer?	EtOH	232(4.21),278(4.06), 314(3.89)	39-1329-86C
	EtOH-KOH	260(3.87),285(3.72), 350(4.30)	39-1329-86C
$C_{15}H_{19}NO_3S$			
Azocine, 8-ethoxy-2,3,4,5-tetrahydro-3-(phenylsulfonyl)-, (E)-	MeOH	228(4.22),258(3.23), 266(3.20),274(3.04)	12-0687-86
2-Thiophenecarboxylic acid, 5-[(2,2,6-6-tetramethyl-4-oxo-3-piperidinylidene)methyl]-	EtOH	212(3.88),305(4.07)	70-2511-86
$C_{15}H_{19}NO_4$			
2-Pentenedioic acid, 3-[(1-phenylethyl)amino]-, dimethyl ester, (R)-	MeOH	274(4.36)(changing to 282(4.23))	44-1498-86
$C_{15}H_{19}NO_6$			
α-D-Ribofuranose, 3-deoxy-3-(hydroxyamino)-1,2-O-(1-methylethylidene)-, 5-benzoate	EtOH	229(4.04),265(3.90), 272(3.97),278(2.89)	33-1132-86
$C_{15}H_{19}NO_7$			
2H-Pyran-4-propanoic acid, α-(acetylamino)-6-(ethoxycarbonyl)-2-oxo-, ethyl ester	EtOH	301(3.7)	5-1968-86
$C_{15}H_{19}NO_9S$			
α-D-Glucopyranose, 2-deoxy-2-isothiocyanato-, 1,3,4,6-tetraacetate	EtOH	259(4.24)	136-0049-86I
$C_{15}H_{19}N_3O_2S$			
2-Butenethioamide, 2-nitro-N-phenyl-3-piperidino-	EtOH	334(4.20),355(4.19)	18-3871-86
$C_{15}H_{19}N_3O_5$			
2,5-Cyclohexadiene-1,4-dione, 2-[2-[(aminocarbonyl)oxy]-1-methoxyethyl]-3,6-bis(1-aziridinyl)-5-methyl-	EtOH	333(4.21)	94-1299-86
$C_{15}H_{19}N_3O_6$			
Pyrido[2,3-d]pyrimidine-2,4(1H,3H)-dione, 6-(acetoxymethyl)-1,3-bis(methoxymethyl)-5-methyl-	H_2O	220(4.60),250s(3.95), 303(3.81)	87-0709-86
$C_{15}H_{19}N_3O_8$			
1H-Pyrazole-4-carboxaldehyde, 5-amino-1-(2,3,5-tri-O-acetyl-β-D-ribofuranosyl)-	pH 1	236(3.87),285(3.94)	44-1050-86
	pH 11	236(3.87),284(3.89)	44-1050-86
	MeOH	264(2.45),286(3.94)	44-1050-86
$C_{15}H_{19}N_3O_9$			
α-Cymarofuranoside, methyl, (3,5-dinitrophenyl)carbamate	EtOH	226(4.51),247(4.24), 343(3.45)	94-3130-86

Compound	Solvent	λ_{max}(log ε)	Ref.
β-Cymarofuranoside, methyl, (3,5-dinitrophenyl)carbamate	EtOH	226(4.62),248(4.34), 334(3.52)	94-3130-86
α-Cymaropyranoside, methyl, (3,5-dinitrophenyl)carbamate	EtOH	227(4.64),248(4.37), 343(3.54)	94-3130-86
β-Cymaropyranoside, methyl, (3,5-dinitrophenyl)carbamate	EtOH	226(4.55),247(4.26), 343(3.41)	94-3130-86
$C_{15}H_{19}N_5O_3$			
Benzoic acid, 2-[[[1-(diazoacetyl)-2-methylbutyl]amino]carbonyl]-, hydrazide	MeOH	238(4.09),278(4.02)	104-0353-86
Benzoic acid, 2-[[[1-(diazoacetyl)-3-methylbutyl]amino]carbonyl]-, hydrazide	MeOH	236(4.07),275(3.99)	104-0353-86
$C_{15}H_{19}N_5O_5$			
Glycine, N-[1,2,3,4-tetrahydro-1,3-dimethyl-2,4-dioxo-6-(1-oxopropyl)-7-pteridinyl]-, ethyl ester	MeOH	233(4.57),297(4.22), 369(4.24)	108-0081-86
$C_{15}H_{19}N_5O_6$			
Acetamide, N-nitroso-N-[[1,2,3,4-1,3-bis(methoxymethyl)-5-methyl-2,4-dioxopyrido[2,3-d]pyrimidin-6-yl]methyl]-	H_2O	221(4.60),249s(4.02), 305(3.73)	87-0709-86
$C_{15}H_{19}N_7O$			
2,4-Pyrimidinediamine, 5-[(2-methylphenyl)azo]-6-morpholino-	EtOH	204(4.45),272(3.19), 392(4.40)	42-0420-86
$C_{15}H_{19}N_7O_2$			
2,4-Pyrimidinediamine, 5-[(2-methoxyphenyl)azo]-6-morpholino-	EtOH	204(4.47),275(4.34), 400(4.48)	42-0420-86
2,4-Pyrimidinediamine, 5-[(3-methoxyphenyl)azo]-6-morpholino-	EtOH	210(4.19),275(3.99), 394(4.05)	42-0420-86
$C_{15}H_{19}N_7O_3S$			
Benzenesulfonic acid, 4-[[2,4-diamino-6-(1-piperidinyl)-5-pyrimidinyl]-azo]-	EtOH	202(4.93),224(4.87), 270(4.62),394(4.53)	42-0420-86
$C_{15}H_{19}O_3P$			
Phosphonic acid, (4-azulenylmethyl)-, diethyl ester	CH_2Cl_2	245(4.42),281(4.64), 287(4.64),332(3.49), 345(3.63),358(3.06), 550s(--),573(2.62), 613(2.57),675s(--)	5-1222-86
Phosphonic acid, (6-azulenylmethyl)-, diethyl ester	CH_2Cl_2	237(4.22),280(4.79), 286(4.79),325(3.49), 332(3.60),339(3.58), 347(3.74),550s(--), 574(2.56),618(2.49), 680(2.10)	5-1222-86
$C_{15}H_{19}S$			
Sulfonium, [(6,8-dimethyl-4-azulenyl)-methyl]dimethyl-	MeOH	248(4.37),290(4.62), 309s(3.92),342(3.56), 356(3.58),371s(3.00), 570(2.71),602s(2.66), 666s(2.18)	5-1222-86

Compound	Solvent	$\lambda_{max}(\log \epsilon)$	Ref.
Sulfonium, [(6,8-dimethyl-4-azulenyl)-methyl]dimethyl-, methyl sulfate	MeCN	247(4.36),290(4.61), 309s(3.93),342(3.54), 355(3.57),370s(3.01), 568(2.76),602s(2.71), 662s(2.47)	5-1222-86
perchlorate	MeCN	248(4.40),290(4.63), 293(4.62),309s(3.93), 341(3.51),355(3.55), 370s(2.79),568(2.70), 598s(2.65),661s(2.21)	5-1222-86
tetrafluoroborate	MeOH	248(4.39),289(4.65), 342(3.50),356(3.54), 370s(2.74),572(2.69), 597s(2.64),665s(2.20)	5-1222-86
$C_{15}H_{20}$			
Laurencenyne	MeOH	224(4.14),231s(4.06)	78-3781-86
trans	MeOH	223(4.13)	78-3781-86
$C_{15}H_{20}BrClO$			
Spiro[5.5]undec-3-en-2-one, 8-bromo-9-chloro-5,5,9-trimethyl-1-methylene-	EtOH	235(3.86)	33-0091-86
$C_{15}H_{20}Br_2O_2$			
2,7-Dioxabicyclo[4.2.1]nonane, 5-bromo-3-(1-bromopropyl)-8-(2-penten-4-ynyl)-, (Z)- (isoprelaurefucin)	EtOH	214s(4.11),222(4.16), 231s(4.04)	18-2953-86
Pyrano[3,2-b]pyran, 3,7-dibromo-2-ethyloctahydro-6-(2-penten-4-ynyl)-, [2S-[2α,3α,4aβ,6α(Z),7α,8aβ]]-	EtOH	224(4.10),233s(3.99)	12-1401-86
$C_{15}H_{20}N_2$			
6-Quinolinecarbonitrile, 1,2,3,4-tetrahydro-1,3,3,4,4-pentamethyl-	n.s.g.	304(4.08)	88-5967-86
$C_{15}H_{20}N_2O$			
1-Heptanone, 1-(1-methyl-1H-benzimidazol-2-yl)-	EtOH	308(4.22)	94-4916-86
$C_{15}H_{20}N_2O_3$			
3H-3-Benzazonine-3-carbonitrile, 1,2,4,5,6,7-hexahydro-7-hydroxy-9,10-dimethoxy-, (±)-	MeOH	230(3.95),282(3.49), 285s(3.46)	12-0893-86
4-Isoxazolecarboxamide, 4,5-dihydro-5-(hydroxymethyl)-3-(4-methylphenyl)-N-propyl-, cis	MeOH	268(3.13)	73-2167-86
$C_{15}H_{20}N_2O_4$			
4H-1-Benzopyran-4-one, 5-amino-6-(3-amino-4-hydroxy-1-oxobutyl)-2,3-dihydro-2,2-dimethyl- (fusarochromanone)	MeOH	248(4.34),277(3.94), 383(4.07)	23-1308-86
6H-3,6-Benzoxazecine-6-carbonitrile, 1,2,4,5,7,8-hexahydro-1-hydroxy-10,11-dimethoxy-, (±)-	MeOH	232(3.95),282(3.45), 287s(3.42)	12-0893-86
2,5-Benzoxazonine-5(1H)-carbonitrile, 3,4,6,7-tetrahydro-1,9,10-trimethoxy-, (±)-	MeOH	234(3.98),281(3.45), 285s(3.45)	12-0893-86
4-Isoxazolecarboxamide, 4,5-dihydro-5-(hydroxymethyl)-3-(4-methoxy-	MeOH	279(3.17)	73-2167-86

Compound	Solvent	λ_{max}(log ε)	Ref.
phenyl)-N-propyl-, cis (cont.)			73-2167-86
$C_{15}H_{20}N_2O_5$			
2,4(1H,3H)-Pyrimidinedione, 1-[2,6-dideoxy-2-C-methylene-3,4-O-(1-methylethylidene)-β-L-lyxo-hexopyranosyl]-5-methyl-	n.s.g.	264(4.06)	39-1297-86C
$C_{15}H_{20}N_2O_5S$			
1-Azabicyclo[3.2.0]hept-2-ene-2-carboxylic acid, 3-[[3-(aminocarbonyl)-cyclopentyl]thio]-6-(1-hydroxy-ethyl)-7-oxo-, sodium salt	H_2O	300(3.92)	23-2184-86
diastereomer B	H_2O	300(3.90)	23-2184-86
$C_{15}H_{20}N_2O_6$			
2,4(1H,3H)-Pyrimidinedione, 1-[2,2[1]-anhydro-6-deoxy-1-C-(hydroxymethyl)-3,4-O-(1-methylethylidene)-β-L-galacto-hexopyranosyl]-5-methyl-	MeOH	264(4.12)	39-1297-86C
2,4(1H,3H)-Pyrimidinedione, 1-[2,2[1]-anhydro-6-deoxy-1-C-(hydroxymethyl)-3,4-O-(1-methylethylidene)-β-L-talo-hexopyranosyl]-5-methyl-	MeOH	264(4.07)	39-1297-86C
Uridine, 6-(1-hexynyl)-	H_2O	287(4.08)	23-1560-86
$C_{15}H_{20}N_2O_8$			
5-Pyrimidineacetic acid, 1,2,3,4-tetrahydro-1-[2,3-O-(1-methylethylidene)-β-D-ribofuranosyl]-2,4-dioxo-, methyl ester	MeOH	263(4.00)	94-4585-86
$C_{15}H_{20}N_2S_2$			
2(3H)-Thiazolethione, 3-[3-(dimethylamino)phenyl]-4-(1,1-dimethylethyl)-	EtOH	257(4.07),319.2(4.15)	152-0399-86
$C_{15}H_{20}N_4O_2$			
Acetic acid, cyano[1-methyl-6-(1-piperidinyl)-4(1H)-pyrimidinylidene]-, ethyl ester	EtOH	276(3.86),368(4.66)	103-0774-86
$C_{15}H_{20}N_4O_4$			
1H-Imidazo[4,5-c]pyridinium, 1-[(2,2-dimethyl-1-oxopropoxy)methyl]-5-[(ethoxycarbonyl)amino]-, hydroxide, inner salt	MeOH	296(3.85)	78-1511-86
Propanoic acid, 2,2-dimethyl-, [4-[(ethoxycarbonyl)amino]-1H-imidazo[4,5-c]pyridin-1-yl]methyl ester	MeOH	263(4.16)	78-1511-86
4,5-Pyrimidinedicarboxylic acid, 2-cyano-1-(hexahydro-1H-azepin-1-yl)-1,6-dihydro-, dimethyl ester	CH_2Cl_2	386(3.72)	83-0798-86
2,4(1H,3H)-Pyrimidinedione, 5,5'-methylenebis[1,3,6-trimethyl-	EtOH	277(4.24)	118-0857-86
$C_{15}H_{20}N_4O_5$			
Acetamide, N-[[1,2,3,4-tetrahydro-1,3-bis(methoxymethyl)-5-methyl-2,4-dioxopyrido[2,3-d]pyrimidin-6-yl]-methyl]-	H_2O	224(4.62),244s(3.96), 306(3.72)	87-0709-86

Compound	Solvent	$\lambda_{max}(\log \epsilon)$	Ref.
$C_{15}H_{20}N_6O_4$			
1,2,4-Triazine-5,6-dicarboxylic acid, 3-cyano-5-[[(hexahydro-1H-azepin-1-yl)imino]methyl]-2,5-dihydro-, dimethyl ester, (E)-(±)-	CH_2Cl_2	253(4.01),312(3.44), 369(3.34)	83-0798-86
$C_{15}H_{20}O$			
Naphtho[1,2-c]furan, 4,5,5a,6,7,8,9,9a-octahydro-7,9a-dimethyl-6-methylene-, (5aα,7β,9aβ)-(±)-	MeOH	225(3.41)	44-0773-86
$C_{15}H_{20}O_2$			
Benzaldehyde, 4-hydroxy-3-(1,2,2-trimethylcyclopentyl)-, (S)-	EtOH	292(4.04)	39-0701-86C
Bullvalenecarboxaldehyde diethyl acetal	ether	218(3.65)	24-2889-86
1H-Cyclopenta[a]pentalene-3,5(2H,3aH)-dione, 3b,4,7,7a-tetrahydro-2,2,3b,4-tetramethyl-	MeOH	226(4.08)	35-3443-86
epimer	MeOH	228(4.00)	35-3443-86
1(2H)-Naphthalenone, 3,4-dihydro-4-methoxy-3,3,5,7-tetramethyl-	MeOH	262(3.97),306(3.28)	39-2011-86C
2(1H)-Naphthalenone, 3,4-dihydro-6-hydroxy-4,7-dimethyl-1-(1-methylethyl)-, (1R-trans)-	EtOH	222(3.86),283(3.44)	12-0103-86
Periplanone A	hexane	220(4.18)	88-6189-86
Spiro[cyclohexan-1,1'-[1H]inden]-5'-ol, 2',3'-dihydro-7'-methoxy-	MeOH	210(4.44),222(3.97), 280(3.37)	102-1992-86
Spiro[cyclohexan-1,1'-[1H]inden]-7'-ol, 2',3'-dihydro-5'-methoxy-	MeOH	210(4.46),222(4.03), 280(3.35)	102-1992-86
$C_{15}H_{20}O_2S$			
1,3-Nonadiene, 2-benzenesulfonyl-, (E,E)-	MeOH	221(4.08),260(4.38), 265(4.34)	78-5321-86
(E,Z)-	MeOH	222(4.18),262(4.26), 265(4.23)	78-5321-86
$C_{15}H_{20}O_3$			
Artesovin	EtOH	239(3.86)	105-0609-86
Bicyclo[4.2.0]octa-1,3,5-trien-7-ol, 7-(2-methyl-1,3-dioxolan-2-yl)-8-propyl-	MeOH	261(2.87),267(3.05), 273(3.01)	44-1419-86
10βH-Eremophil-7(11)-en-8α,12-olide, 1-oxo-	EtOH	218(4.16),283(1.8)	102-2412-86
3-Furancarboxylic acid, 5-(2,6-dimethyl-5,7-octadienyl)-	EtOH	230(3.88)	100-0729-86
Hypnophilin	MeOH	234(3.65)	78-3587-86
Merulidial	MeOH	268(4.20)	78-3579-86
1,2-Naphthalenedicarboxaldehyde, 1,4,4a,5,6,7,8,8a-octahydro-1-hydroxy-6,8a-dimethyl-5-methylene- (muzigadial)	MeOH	223.4(3.73)	44-0773-86
Naphtho[1,2-b]furan-2(3aH)-one, 4,5,5a,6,7,9a-hexahydro-3a-hydroxy-1,5a,8-trimethyl-	n.s.g.	211(3.94)	2-0233-86
Naphtho[1,2-b]furan-2(3H)-one, 3a,4,5,5a,6,9b-hexahydro-4-hydroxy-3,5a,9-trimethyl-, [3S-(3α,3aα-4α,5aβ,9bβ)]-	EtOH	264(4.00)	78-3655-86
2-Octanone, 8-(1,3-benzodioxol-5-yl)- (moskachan D)	n.s.g.	235(3.74),288(3.71)	102-2209-86

Compound	Solvent	λ_{max}(log ε)	Ref.
$C_{15}H_{20}O_4$			
Bicyclo[5.3.1]undeca-1,3-diene-5,7-diol, diacetate	EtOH	241(3.77)	78-1823-86
Bisoxireno[4,5:8,9]cyclodeca[1,2-b]-furan-4(laH)-one, 2,2a,6,6a,7a,8-9,9a-octahydro-la,5,7a-trimethyl-	MeOH	220(4.2)	100-0336-86
2-Naphthaleneacetic acid, 1,2,3,4,4a,7-hexahydro-1-hydroxy- ,4a,8-tri-methyl-7-oxo-	EtOH	253(3.91),263(3.81)	102-1757-86
Naphtho[1,2-b]furan-2,6(3H,4H)-dione, hexahydro-4-hydroxy-3,5,8a-tri-methyl-	EtOH	248(4.17)	94-1319-86
Naphtho[1,2-b]furan-2,8(3H,4H)-dione, 3a,5,5a,6,7,9b-hexahydro-4-hydroxy-3,5a,9-trimethyl-, [3S-(3α,3aα,4α-5aβ,9bβ)]-	EtOH	242(4.23)	78-3655-86
2H-Oxireno[8,8a]naphtho[1,2-b]furan-7(6H)-one, octahydro-5a-hydroxy-3a,9a-dimethyl-6-methylene-	n.s.g.	214(3.83)	2-0233-86
$C_{15}H_{20}O_5$			
Bisoxireno[4,5:8,9]cyclodeca[1,2-b]-furan-4(laH)-one, 2,2a,6,6a,7a,8-9,9a-octahydro-2a-hydroxy-la,5,7a-trimethyl-	MeOH	220(4.2)	100-0336-86
$C_{15}H_{21}Cl_3O$			
2-Cyclohexen-1-one, 5,6-dichloro-4-(chloromethylene)-2,6-bis(1,1-dimethylethyl)-, (E)-cis	$CHCl_3$	254(3.69),296(4.07)	12-2121-86
(Z)-cis	$CHCl_3$	238(3.76),290(4.19)	12-2121-86
(Z)-trans	$CHCl_3$	239(3.52),294(3.85)	12-2121-86
$C_{15}H_{21}NO$			
Propanimine, 2-benzoyl-2-methyl-N-(1,1-dimethylethyl)-	EtOH	242(3.98),280(3.03)	39-0091-86C
$C_{15}H_{21}NO_3$			
2,8-Decadienoic acid, 2-cyano-4,5-epoxy-5,9-dimethyl-, ethyl ester	EtOH	245(3.78)	103-0370-86
$C_{15}H_{21}NO_4$			
Acetamide, N-[3-methoxy-4-[(tetrahy-dro-2H-pyran-2-yl)oxy]phenyl]-N-methyl-	MeOH	233(3.97),280(3.46)	87-1737-86
$C_{15}H_{21}NO_6$			
2(3H)-Furanone, 3,3',3"-[nitrilo-tris(methylene)]tris[dihydro-	EtOH	207(3.20),275(2.42)	102-0972-86
3-Pyrrolidineacetic acid, 2-carboxy-4-(5-carboxy-1-methylene-4-hexen-yl)- (isodomoic acid C)	n.s.g.	213(3.70)	94-4892-86
3-Pyrrolidineacetic acid, 2-carboxy-4-(5-carboxy-1-methyl-1,3-hexadi-enyl)-, (E,E)-	n.s.g.	211(3.94)	94-4892-86
(E,Z)-	n.s.g.	214(3.99)	94-4892-86
$C_{15}H_{21}NO_8$			
1H-Pyrrole-3,4-dicarboxylic acid, 1-β-D-arabinofuranosyl-, diethyl ester	EtOH	249(3.95)	78-5869-86

Compound	Solvent	$\lambda_{max}(\log \epsilon)$	Ref.
$C_{15}H_{21}N_3$			
2-Propenenitrile, 3-[3,6-bis(1-methylpropyl)pyrazinyl]-, (E)-	EtOH	247(4.14),309s(4.01), 315(4.02)	4-1481-86
(Z)-	EtOH	247(4.13),308(3.92), 313s(3.88)	4-1481-86
$C_{15}H_{21}N_3O_4S$			
1-Azabicyclo[3.2.0]hept-2-ene-2-carboxylic acid, 6-(1-hydroxyethyl)-3-[[3-[(iminomethyl)amino]cyclopentyl]thio]-7-oxo-	H_2O	303(3.98)	23-2184-86
diastereomer B	H_2O	302(3.92)	23-2184-86
$C_{15}H_{21}N_3O_8S_2$			
Tyrosine, 2,5-bis[(2-amino-2-carboxyethyl)thio]-3-hydroxy-, [R-(R*,R*)]-	pH 7.5	303(3.53)	149-0689-86B
$C_{15}H_{21}N_3O_{10}$			
D-glycero-α-D-galacto-2-Nonulopyranosonic acid, 5-(acetylamino)-2,3,5-trideoxy-2-(3,4-dihydro-2,4-dioxo-1(2H)-pyrimidinyl)-	H_2O	263(3.86)	94-1479-86
β-	H_2O	263(3.89)	94-1479-86
$C_{15}H_{21}N_3S_2$			
[1]Benzothieno[2,3-d]pyrimidine-4(3H)-thione, 3-[2-(dimethylamino)ethyl]-5,6,7,8-tetrahydro-2-methyl-, hydrochloride	EtOH	234(4.28),287(4.10), 322(4.04),357(3.86)	106-0096-86
$C_{15}H_{21}N_5O_6S$			
Adenosine, N-methyl-2',3'-O-(1-methylethylidene)-8-(methylsulfonyl)-	MeOH	222(4.35),266s(4.13), 292(4.31)	94-3635-86
$C_{15}H_{22}$			
1,3-Cyclopentadiene, 1,3-bis(1,1-dimethylethyl)-5-ethenylidene-	hexane	250(4.13),354(2.80)	89-0466-86
Neolaurencenyne	MeOH	224(4.10),231s(4.03)	78-3781-86
trans	MeOH	223(4.18)	78-3781-86
$C_{15}H_{22}BrNO_2$			
2(3H)-Naphthalenone, 1-bromo-4,4a,5,6-7,8-hexahydro-4a-methyl-3-morpholino-	EtOH	261.5(4.15)	118-0461-86
$C_{15}H_{22}BrN_4Zn$			
Zinc(1+), bromo(2,12-dimethyl-3,7,11,17-tetraazabicyclo[11.3.1]heptadeca-1(17),2,11,13,15-pentaene-N^3,N^7,N^{11},N^{17})-, perchlorate	MeCN	219(4.33),297(3.52)	35-2388-86
$C_{15}H_{22}Cl_2N_2O_2$			
2,5-Cyclohexadiene-1,4-dione, 2-(butylamino)-5-(butylmethylamino)-3,6-dichloro-	CH_2Cl_2	252s(4.20),388(4.26), 558(2.70)	106-0832-86
$C_{15}H_{22}Cl_2O$			
2-Cyclohexen-1-one, 5,6-dichloro-2,6-bis(1,1-dimethylethyl)-4-methylene-, cis-(±)-	$CHCl_3$	283(4.12)	12-2121-86
trans	$CHCl_3$	285(4.20)	12-2121-86

Compound	Solvent	λ_{max}(log ε)	Ref.
2,5-Cyclohexadien-1-one, 4-chloro-4-(chloromethyl)-2,6-bis(1,1-dimethylethyl)-	$CHCl_3$	239(3.86),294(2.96)	12-2121-86
$C_{15}H_{22}Cl_4O$			
2-Cyclohexen-1-one, 4,5,6-trichloro-4-(chloromethyl)-2,6-bis(1,1-dimethylethyl)-	$CHCl_3$	240(3.90)	12-2121-86
$C_{15}H_{22}N_2O_2$			
2-Propenoic acid, 3-[3,6-bis(1-methylethyl)pyrazinyl]-, ethyl ester, (E)-	EtOH	247(4.03),308(3.92), 315s(3.95)	4-1481-86
$C_{15}H_{22}N_2O_3$			
2-Propenoic acid, 3-[3,6-bis(1-methylethyl)pyrazinyl]-, ethyl ester, N-oxide, (E)-	EtOH	258(4.53),291s(3.94), 333(3.71)	4-1481-86
$C_{15}H_{22}N_2O_4S$			
1-Azabicyclo[3.2.0]hept-2-ene-2-carboxylic acid, 3-[(4-aminocyclohexyl)thio]-6-(1-hydroxyethyl)-7-oxo-	H_2O	303(3.83)	23-2184-86
$C_{15}H_{22}N_2O_6$			
2,4(1H,3H)-Pyrimidinedione, 1-[6-deoxy-2-C-methyl-3,4-O-(1-methylethylidene)-β-L-talopyranosyl]-5-methyl-	MeOH	265(4.08)	39-1297-86C
Uridine, 2'-deoxy-5-ethyl-, 5'-butanoate	EtOH	268(3.97)	83-0154-86
Uridine, 2'-deoxy-5-ethyl-, 5'-(2-methylpropanoate)	EtOH	268(3.95)	83-0154-86
$C_{15}H_{22}N_2O_7$			
α-D-Allofuranose, 1,2:5,6-bis-O-(1-methylethylidene)-3-C-(5-methyl-1,2,4-oxadiazol-3-yl)-	EtOH	203(3.62),234(2.93)	159-0631-86
α-D-Glucofuranose, 1,2:5,6-bis-O-(1-methylethylidene)-3-C-(5-methyl-1,2,4-oxadiazol-3-yl)-	EtOH	202(3.09)	159-0631-86
$C_{15}H_{22}N_4O_4$			
Glycine, N-[N-(4-amino-L-phenylalanyl)glycyl]-, ethyl ester	H_2O	237.6(3.95),285.7(3.13)	149-0159-86B
$C_{15}H_{22}N_6O_3$			
L-Leucine, N-(N-1H-purin-6-yl-L-alanyl)-, methyl ester	pH 2.85	274(4.24)	63-0757-86
	pH 7	266(4.24)	63-0757-86
	pH 10.5	271(4.20)	63-0757-86
$C_{15}H_{22}N_6O_5$			
1,2,4-Triazine-5,6-dicarboxylic acid, 3-cyano-5-[[(hexahydro-1H-azepin-1-yl)imino]methyl]tetrahydro-3(or 6)-hydroxy-, dimethyl ester	CH_2Cl_2	320(3.51)	83-0798-86
$C_{15}H_{22}O$			
Phenol, 2-methyl-4-(1,2,2-trimethylcyclopentyl)-, (S)-	EtOH	225(3.71),277(3.18), 284(3.14)	39-0701-86C
Phenol, 4-methyl-2-(1,2,2-trimethylcyclopentyl)-, (S)- (α-herbertenol)	EtOH	220(3.72),283(3.41), 289(3.36)	39-0701-86C

Compound	Solvent	$\lambda_{max}(\log \epsilon)$	Ref.
$C_{15}H_{22}O_2$			
1,2-Benzenediol, 5-methyl-3-(1,2,2-trimethylcyclopentyl)-, (S)- (herbertenediol)	EtOH	285(3.49)	39-0701-86C
Bicyclo[7.2.2]trideca-9,11,12-triene-10,11-dimethanol	EtOH	215(4.63),235(3.89), 293(2.81)	24-2698-86
3,7-Cyclodecadien-1-one, 10-(2-hydroxy-1-methylethylidene)-3,7-dimethyl-, (E,E,E)-	EtOH	240(3.4)	102-1351-86
2,4-Hexadienoic acid, 6-(4-methylcyclohexylidene)-, ethyl ester, [S-(E,E)]-	C_6H_{12}	308(4.69),319(4.69)	44-2863-86
[S-(E,Z)]-	C_6H_{12}	310(4.59)	44-2863-86
[S-(Z,E)]-	C_6H_{12}	314(4.60)	44-2863-86
[S-(Z,Z)]-	C_6H_{12}	314(4.52)	44-2863-86
2,6-Naphthalenediol, 1,2,3,4-tetrahydro-4,7-dimethyl-1-(1-methylethyl)-	EtOH	211(3.58),217(3.51), 282(3.05)	12-0103-86
3,7-calamenediol, (1S,3R,4S)-	EtOH	225(3.63),282(3.31)	12-0103-86
	EtOH-base	240(3.96),291(3.59)	12-0103-86
$C_{15}H_{22}O_3$			
3-Buten-2-one, 4-[6-(acetoxymethyl)-2,6-dimethyl-1-cyclohexen-1-yl)-, (E)-(±)-	EtOH	222(3.81),289(3.91)	163-1589-86
1,2,6-Naphthalenetriol, 1,2,3,4-tetrahydro-4,7-dimethyl-1-(1-methylethyl)- (1S,3R,4R-calamene-3,4,7-triol)	EtOH	208(3.51),221(3.27), 280(2.73)	12-0103-86
	EtOH-base	219(3.98),248(3.97), 296(3.51)	12-0103-86
(1S,3R,4S)-	EtOH	225(3.57),280(3.06)	12-0103-86
	EtOH-base	246(3.94),294(3.52)	12-0103-86
Naphtho[1,2-c]furan-1(3H)-one, 4,5,5a,6,7,8,9,9a-octahydro-7-hydroxy-6,6,9a-trimethyl-	MeOH	217(4.05)	102-0253-86
11-Oxabicyclo[7.3.2]tetradeca-9,12,13-triene-13,14-dimethanol	EtOH	209(4.28),254(3.50)	24-2698-86
2(1H)-Pentalenone, 5,5-dimethyl-6-[2-(1,3-dioxolan-2-yl)ethyl]-4,5,6,6a-tetrahydro-, cis-(±)-	EtOH	232(4.10)	94-4939-86
1,2-Propanediol, 3-(1,3,5,7,9-dodecapentaenyloxy)-, (all-E)-	CH_2Cl_2	323(4.68),338(4.87), 357(4.83)	35-1338-86
$C_{15}H_{22}O_4$			
2,4-Pentadienoic acid, 5-(1,4-dihydroxy-2,6,6-trimethyl-2-cyclohexen-1-yl)-3-methyl-, [1α,1(2Z,4E),4β]-(±)-	EtOH	262(4.30)	102-1865-86
$C_{15}H_{22}O_5$			
Benzoic acid, 2,4-dihydroxy-6-(4-hydroxy-3-methylpentyl)-, ethyl ester, (R*,S*)-(±)-	EtOH	266(3.83),300(3.48)	44-4254-86
4H-1-Benzopyran-8-acetic acid, 3-acetyl-4a,5,6,7,8,8a-hexahydro-8a-hydroxy- ,5-dimethyl-	MeOH	259(4.11)	3-0289-86
1,7-Dioxaspiro[5.5]undec-2-ene-2-acetic acid, 8,9-dimethyl-4-oxo-, ethyl ester, (6α,8β,9α)-(±)-	EtOH	264(4.01)	44-4254-86
Oxireno[4,5]cyclodeca[1,2-b]furan-8(1aH)-one, 2,3,4,5,6,6a,10,10a-octahydro-4,5-dihydroxy-1a,5,9-	MeOH	216(4.2)	100-0336-86

Compound	Solvent	λ_{max}(log ε)	Ref.
trimethyl- (cont.)			100-0336-86
Qinghaosu	MeOH	203(2.57)	3-0289-86
$C_{15}H_{22}O_8$			
D-arabino-Hept-2-enonic acid, 2,3-dideoxy-4,5-O-(1-methylethylidene)-, methyl ester, diacetate, (E)-	MeOH	225(3.23)	136-0283-86G
$C_{15}H_{22}Si_2$			
Disilane, pentamethyl(1-naphthalenyl)-	THF	285(3.7)	47-1943-86
	THF	283(3.7)	116-0196-86
$C_{15}H_{23}Cl$			
1,3-Cyclopentadiene, 5-(1-chloroethylidene)-1,3-bis(1,1-dimethylethyl)-	hexane	260s(4.19),266(4.27), 272(4.27),281s(4.06), 376(2.66)	89-0466-86
$C_{15}H_{23}ClO$			
2,5-Cyclohexadien-1-one, 4-chloro-2,6-bis(1,1-dimethylethyl)-4-methyl-	$CHCl_3$	242(3.66)	12-2121-86
$C_{15}H_{23}NO_3$			
7-Azabicyclo[4.1.0]heptane-7-carboxylic acid, 2,2,6-trimethyl-1-(3-oxo-1-butenyl)-, methyl ester, (E)-	pentane	227(3.80)	33-0692-86
2-Furancarboxylic acid, 5-methyl-4-[[(2-methylcyclohexyl)amino]-methyl]-, methyl ester	MeOH	266(3.06)	73-2186-86
3,5-Heptadienamide, 6-acetoxy-N-cyclohexyl-, (E,Z)-	MeOH	237(4.38)	33-0469-86
1H-Pyrrole-3-carboxylic acid, 5-(1,1-dimethyl-5-oxohexyl)-2-methyl-, methyl ester	pentane	226(3.81),265(3.66)	33-0692-86
$C_{15}H_{23}NO_7$			
D-Fructose, 1-deoxy-1-[[(4,4-dimethyl-2,6-dioxocyclohexylidene)methyl]-amino]-	H_2O	253(4.21),301(4.39)	136-0329-86D
$C_{15}H_{23}N_5O_6$			
Glycine, N-[N-[5-(2-amino-1,4-dihydro-6-hydroxy-4-oxo-5-pyrimidinyl)-1-oxopentyl]glycyl]-, ethyl ester (trihydrate)	50% MeOH-HCl	266(4.21)	73-1140-86
	50%MeOH-NaOH	271(4.18)	73-1140-86
$C_{15}H_{24}$			
Azulene, 1,2,3,4,5,6-hexahydro-1,4-dimethyl-7-(1-methylethyl)-	$CHCl_3$	268(3.87)	44-5134-86
$C_{15}H_{24}NO_2S$			
Ethenaminium, N,N,N-triethyl-2-[(4-methylphenyl)sulfonyl]-, chloride, (E)-	MeCN	249(3.93)	104-2085-86
$C_{15}H_{24}N_2O_2$			
Tetracaine, hydrochloride	pH 7.0	310(4.37)	69-7118-86
	EtOH	308(4.46)	69-7118-86
$C_{15}H_{24}N_2O_8$			
α-D-Allofuranose, 3-C-[(acetoxyimino)-aminomethyl]-1,2:5,6-bis-O-(1-	EtOH	203(3.95),217(3.99)	159-0631-86

Compound	Solvent	λ_{max}(log ϵ)	Ref.
methylethylidene)- (cont.)			159-0631-86
α-D-Glucofuranose, 3-C-[(acetoxy-imino)aminomethyl]-1,2:5,6-bis-O-(1-methylethylidene)-	EtOH	215(3.83)	159-0631-86
$C_{15}H_{24}N_4O_3$			
5-Pyrimidinepentanamide, 2-amino-N-cyclohexyl-1,4-dihydro-6-hydroxy-4-oxo-	50%MeOH-HCl	266(4.35)	73-1140-86
	50%MeOH-NaOH	268(4.27)	73-1140-86
$C_{15}H_{24}N_5NiO_3$			
Nickel, [4-nitro-2-(1,4,7,10-tetraaza-cyclotridec-11-yl)phenolato-$N^2,N^{2'},N^{2''},N^{2'''},O^1$]-, (SP-5-14)-	H_2O	386(4.36)	77-1110-86
$C_{15}H_{24}O$			
Bicyclo[8.1.0]undeca-2,6-diene-11-methanol, 3,7,11-trimethyl- (de-acetylcoralloidin B)	C_6H_{12}	226(3.58)	100-0608-86
4-Hexen-3-one, 2,2-dimethyl-6-(4-meth-ylcyclohexylidene)-, [S-(E)]-	C_6H_{12}	287(4.42),335s(2.26)	44-2863-86
1H-Inden-1-one, 3a,4,5,6,7,7a-hexa-hydro-2,3,3a,4,4,7a-hexamethyl- (6-thapsen-8-one)	hexane	244(4.30)	102-0703-86
2-Naphthalenol, 1,2,3,4,6,7,8,8a-octa-hydro-5,8a-dimethyl-3-(1-methyleth-ylidene)- (+-deacetylcoralloidin A)	C_6H_{12}	248(4.20)	100-0608-86
Zonarene, 12-hydroxy-	EtOH	248(4.23)	12-0103-86
$C_{15}H_{24}O_2$			
2-Cyclohexen-1-one, 6-(1-hydroxy-1,5-dimethyl-4-hexenyl)-3-methyl-, (+)- (hernandulcin)	EtOH	236(4.12)	88-0981-86
2H-Cyclopenta[a]pentalen-2-one, deca-hydro-4-hydroxy-3,3a,5,5-tetra-methyl-	MeOH	225(4.20)	35-3443-86
$C_{15}H_{24}O_3S_2$			
2,4-Nonadienoic acid, 8-(1,3-dithian-2-yl)-7-hydroxy-6-methyl-, methyl ester	EtOH	261(4.39)	88-4741-86
$C_{15}H_{24}O_7$			
D-gluco-Oct-2-enonic acid, 2,3-dide-oxy-4,5:7,8-bis-O-(1-methylethyli-dene)-, methyl ester, (E)-	MeOH	215(3.78)	136-0283-86G
$C_{15}H_{25}BrN_2O_2$			
2,4(1H,3H)-Pyrimidinedione, 1-(10-bromodecyl)-6-methyl-	pH 7	268(3.95)	70-1064-86
	pH 12	267(4.04)	70-1064-86
2,4(1H,3H)-Pyrimidinedione, 3-(10-bromodecyl)-6-methyl-	pH 7	267(3.89)	70-1064-86
	pH 12	282(4.04)	70-1064-86
$C_{15}H_{25}BrO$			
2H-Oxocin, 3-(1-bromopropyl)dihydro-8-pentyl-	EtOH	225(3.69)	18-2953-86
$C_{15}H_{25}N_2$			
Piperidinium, 1-[5-(1-piperidinyl)-2,4-pentadienylidene]-, perchlor-ate	H_2O	413(5.09)	33-1588-86
	EtOH	415(5.07)	33-1588-86

Compound	Solvent	λ_{max}(log ϵ)	Ref.
$C_{15}H_{25}N_2Ru$			
Ruthenium(1+), (η^6-benzene)[N,N'-1,2-ethanediylidenebis[2-propanamino]-N,N']methyl-, iodide	CH_2Cl_2	300s(--),428(3.88)	157-1449-86
$C_{15}H_{25}N_4NiO$			
Nickel(1+), [2-(1,4,7,10-tetraaza-cyclotridec-11-yl)phenolato-N,N',N",N"',O]-, (SP-5-14)-, ClO_4^-	pH 3.5	272(3.30),424(2.11)	138-1137-86
	pH 10+	291(3.53),559(0.78)	138-1137-86
Nickel(2+), [2-(1,4,7,10-tetraaza-cyclotridec-11-yl)phenolato-N,N',N",N"',O]-	H_2O	413(3.63)	77-1110-86
$C_{15}H_{25}N_5S$			
2H-Purine-2-thione, 1,7-dihydro-6-(methylamino)-7-nonyl-	pH 1	246(3.53),287(3.54)	4-0737-86
	pH 13	252(4.35)	4-0737-86
$C_{15}H_{25}N_6O_{14}P_3S$			
9H-Purin-6-amine, 9-[5-S-(3-amino-3-carboxypropyl)-6-deoxy-6-[hydroxy-[[hydroxy(phosphonooxy)phosphinyl]-oxy]phosphinyl]-5-thio-β-D-allo-furanosyl]-, (S)-, tetrasodium salt	pH 2	257(4.18)	87-1030-86
	pH 11	259(4.19)	87-1030-86
epimer	pH 2	257(4.18)	87-1030-86
	pH 11	259(4.18)	87-1030-86
$C_{15}H_{26}N_2O_3$			
2,4(1H,3H)-Pyrimidinedione, 3-(10-hydroxydecyl)-6-methyl-	pH 7	257(3.69)	70-1064-86
	pH 12	283(3.64)	70-1064-86
$C_{15}H_{26}N_2O_7$			
Acmimycin	H_2O	end absorption	163-0239-86
$C_{15}H_{26}N_3O_4$			
1-Piperidinyloxy, 4-[[2-[(3-carboxy-1-oxo-2-propenyl)amino]ethyl]-amino]-2,2,6,6-tetramethyl-, (Z)-	H_2O	202(3.91),332s(3.75)	70-0363-86
$C_{15}H_{26}O$			
1-Naphthalenemethanol, 1,4,4a,5,6,7-8,8a-octahydro-2,5,5,8a-tetra-methyl-, [1S-(1α,4aβ,8aα)]-(drimenol)	EtOH	226(4.3)	102-0253-86
$C_{15}H_{26}O_2$			
11-Dodecynoic acid, 1-methylethyl ester	MeOH	205(3.23)	18-3535-86
12-Tridecynoic acid, ethyl ester	MeOH	206(2.98)	18-3535-86
$C_{15}H_{27}F_3O_3S_2Si_2$			
Sulfonium, bis[(trimethylsilyl)-phenyl-, triflate	EtOH	220(3.92),259(3.72), 266(3.79),273(3.78)	35-6739-86
$C_{15}H_{27}N$			
Azete, tris(1,1-dimethylethyl)-	hexane	320(2.16)	89-0842-86
$C_{15}H_{27}NO_2$			
1,2-Propanedione, 1-cyclododecyl-, 1-oxime	EtOH	227(3.95)	118-0473-86

Compound	Solvent	$\lambda_{max}(\log \epsilon)$	Ref.
$C_{15}H_{27}N_3O_3$			
2-Butenoic acid, 4-oxo-4-[[2-(2,2,6,6-tetramethyl-4-piperidinyl)amino]-ethyl]amino-, (Z)-	H_2O	206(3.88),228s(3.72)	70-0363-86
$C_{15}H_{27}N_5Ni$			
Nickel(2+), [5-(2-pyridinyl)-1,4,8,11-tetraazacyclotetradecene-N^1,N^4,N^5-N^8,N^{11}]-, (SP-5-15)-, diperchlorate	H_2O	523(0.90)	77-1322-86
$C_{15}H_{30}$			
3-Hexene, 3-(1,1-dimethylethyl)-2,2,4,5,5-pentamethyl-	C_6H_{12}	234(3.70)	78-1693-86
$C_{15}H_{36}N_3PSi_2$			
Piperidine, 1-[[bis(trimethylsilyl)-hydrazono]phosphino]-2,2,6,6-tetramethyl-, (E)-	hexane	304(3.78),398(2.86)	77-1086-86
$C_{15}H_{39}GeN_3Si_3$			
3H-1,2,3,4-Triazagermole, 3-[(1,1-dimethylethyl)dimethylsilyl]-4,5-dihydro-4,4-dimethyl-5,5-bis(trimethylsilyl)-	C_6H_{12}	283(4.60)	24-2980-86

Compound	Solvent	λ_{max}(log ε)	Ref.
$C_{16}Cl_{12}$			
Benzene, 1,1'-(1,2-dichloro-1-buten-3-yne-1,4-diyl)bis[2,3,4,5,6-pentachloro-	C_6H_{12}	216(4.87),225s(4.79), 240s(4.61),253s(4.40), 293s(4.29),305(4.36), 310s(4.32),325(4.37)	44-1413-86
Cyclobuta[l]phenanthrene, 1,1,2,2,3,4-5,6,7,8,9,10-dodecachloro-1,2-dihydro-	C_6H_{12}	212(4.48),242(4.40), 261(4.31),290(4.68), 318(4.43),329s(4.27), 342(4.19),357(4.18), 382s(3.09),404(2.72)	44-1100-86
$C_{16}H_5F_{12}N_3O_{12}S_4$			
Benzenamine, 2-nitro-N-[2-nitro-4,6-bis[(trifluoromethyl)sulfonyl]phenyl]-4,6-bis[(trifluoromethyl)sulfonyl]-	MeCN	230(--),280(--), 320(4.43),370(4.36)	104-1547-86
	+ H_2SO_4	263(4.57),278(4.71), 339(4.26)	104-1547-86
sodium salt	MeCN	230(--),280(--), 320(4.45),370(4.40)	104-1547-86
$C_{16}H_6BrN_3O_7$			
2H-1-Benzopyran-3-carbonitrile, 8-[(dromodimethylphenyl)azo]-7-hydroxy-2-oxo-	n.s.g.	485(4.47)	2-0638-86
$C_{16}H_6Cl_2N_4$			
Cyanamide, (1,5-dichloro-9,10-anthracenediylidene)bis-	MeCN	273(4.94),306(4.17), 363s(3.35)	5-0142-86
$C_{16}H_6ClN_2O_3S$			
5H-Benzo[a]phenothiazin-5-one, 6-chloro-1-nitro-	$CHCl_3$	240(4.37),261(4.37), 323(4.27),368s(3.99), 387(3.96),495(4.08)	4-0589-86
5H-Benzo[a]phenothiazin-5-one, 6-chloro-4-nitro-	$CHCl_3$	242(4.37),260(4.36), 322(4.16),352(4.03), 371(4.01),387(3.99), 497(4.09)	4-0589-86
$C_{16}H_7ClN_4O_5$			
2H-1-Benzopyran-3-carbonitrile, 8-[(chloronitrophenyl)azo]-7-hydroxy-2-oxo-	n.s.g.	575(4.76)	2-0638-86
$C_{16}H_7Cl_2N_3O_3$			
2H-1-Benzopyran-3-carbonitrile, 8-[(dichlorophenyl)azo]-7-hydroxy-2-oxo-	n.s.g.	410(4.35)	2-0638-86
$C_{16}H_7Cl_2N_3O_7$			
2H-1-Benzopyran-3-carboxylic acid, 8-[(dichloronitrophenyl)azo]-7-hydroxy-2-oxo-	n.s.g.	420(4.39)	2-0638-86
$C_{16}H_7F_{12}NO_8S_4$			
Benzenamine, N-[2,4-bis[(trifluoromethyl)sulfonyl]phenyl]-2,4-bis[(trifluoromethyl)sulfonyl]-	MeCN	360(4.42)	104-1547-86
sodium salt	MeCN	420(4.23)	104-1547-86
$C_{16}H_7N_5O_7$			
2H-1-Benzopyran-3-carbonitrile, 7-hydroxy-8-[(dinitrophenyl)azo]-2-oxo-	n.s.g.	530(4.52)	2-0638-86

Compound	Solvent	λ_{max}(log ϵ)	Ref.
$C_{16}H_8$			
Dicyclopenta[a,e]dicyclopropa[c,g]-cyclooctene	C_6H_{12}	244(4.63),332(4.42), 384(4.20),480(2.81)	35-7032-86
	CH_2Cl_2	241(4.63),328(4.51), 390(4.20),470s(2.81)	35-7032-86
$C_{16}H_8ClN_3O_7$			
2H-1-Benzopyran-3-carboxylic acid, 8-(chloronitrophenyl)azo]-7-hydroxy-2-oxo-	n.s.g.	560(4.74)	2-0638-86
$C_{16}H_8N_2O_4$			
Benzo[f]pyrido[1,2-a]indole-6,11-dione, 12-nitro-	DMF	362(3.8062),489(3.6800)	2-1126-86
Fluoranthene, 1,2-dinitro-	EtOH	250(4.52)	35-4126-86
$C_{16}H_8N_4$			
Cyanamide, (1,4-anthracenediylidene)-bis-	MeCN	243(4.50),302s(4.21), 316(4.31),348s(4.00), 384s(3.82),462(3.74), 488(3.77)	5-0142-86
Cyanamide, (9,10-anthracenediylidene)-bis-	MeCN	251(4.54),322(4.36)	5-0142-86
$C_{16}H_8N_4O_5$			
2H-1-Benzopyran-3-carbonitrile, 7-hydroxy-8-[(nitrophenyl)azo]-2-oxo-	n.s.g.	430(4.35)	2-0638-86
$C_{16}H_8N_4O_9$			
2H-1-Benzopyran-3-carboxylic acid, 7-hydroxy-8-[(dinitrophenyl)-azo]-2-oxo-	n.s.g.	520(4.37)	2-0638-86
$C_{16}H_8O_3$			
4H-Cyclopenta[def]phenanthrene-3-carboxylic acid, 4-oxo-	EtOH	236(4.81),289(4.06), 301(3.91),375(2.72)	24-3521-86
Phenanthro[3,4-c]furan-1,3-dione	dioxan	301(4.30),347(3.65), 366(3.60),385(3.56)	24-3521-86
$C_{16}H_8O_4$			
5H,9H-[2]Benzopyrano[4,3-g][1]benzopyran-5,9-dione	n.s.g.	270(4.40),330(4.32)	2-0619-86
$C_{16}H_8O_5$			
5H,9H-[2]Benzopyrano[4,3-g][1]benzopyran-5,9-dione, 11-hydroxy-	n.s.g.	265(4.38),345(4.31)	2-0619-86
6H,11H-[2]Benzopyrano[4,3-c][1]-benzopyran-6,11-dione, 2-hydroxy-	MeOH	260(4.1),325(3.9), 342(4.2),360(4.2)	2-0619-86
6H,11H-[2]Benzopyrano[4,3-c][1]-benzopyran-6,11-dione, 3-hydroxy-	MeOH	245(4.1),310(4.18), 335(4.20),360(4.29)	2-0619-86
6H,11H-[2]Benzopyrano[4,3-c][1]-benzopyran-6,11-dione, 4-hydroxy-	MeOH	258(4.32),310(4.18), 355(4.20),360(4.29)	2-0619-86
$C_{16}H_8O_6$			
5H,9H-[2]Benzopyrano[4,3-g][1]benzopyran-5,9-dione, 11,12-dihydroxy-	n.s.g.	272(4.36),340(4.29)	2-0619-86
6H,11H-[2]Benzopyrano[4,3-c][1]benzopyran-6,11-dione, 2,4-dihydroxy-	MeOH	261(3.91),312(4.24), 343(4.13),358(4.24)	2-0619-86
$C_{16}H_8S_2$			
Fluorantheno[3,4-cd]-1,2-dithiole	CS_2	438(4.29),461(4.29),	88-2011-86

Compound	Solvent	$\lambda_{max}(\log \epsilon)$	Ref.
(cont.)		500s(--)	88-2011-86
$C_{16}H_8Se_2$			
Fluorantheno[3,4-cd]-1,2-diselenole	CS_2	442(4.29),466(4.45), 540s(--)	88-2011-86
$C_{16}H_8Te_2$			
Fluorantheno[3,4-cd]-1,2-ditellurole	CS_2	503(4.54),719(2.43)	88-2011-86
$C_{16}H_9Cl$			
Dicyclopenta[ef,kl]heptalene, 1(and 4)-chloro- (3:2 ratio)	hexane	266(4.90),287(4.42), 301(4.35),312(4.35), 336(4.05),346(4.21), 359(3.54),410(3.07), 444(3.15),460(4.20), 486(3.82),492(3.86)	44-2961-86
$C_{16}H_9ClN_2O$			
Benzonitrile, 4-[5-(4-chlorophenyl)-2-oxazolyl]-	toluene	337(4.39)	103-1026-86
$C_{16}H_9ClN_2OS$			
5H-Benzo[a]phenothiazin-5-one, 1-amino-6-chloro-	$CHCl_3$	246(4.57),305(4.23), 335(3.96),363(3.96), 379(3.92),484(4.14), 515s(4.12)	4-0589-86
5H-Benzo[a]phenothiazin-5-one, 4-amino-6-chloro-	$CHCl_3$	250(4.56),281(4.13), 327(4.03),343s(4.03), 360(3.99),380(3.99), 442s(3.61),470s(3.73), 542(4.05)	4-0589-86
$C_{16}H_9ClN_2O_2S$			
5H-Naphtho[2,1-b]pyrido[3,2-e][1,4]-thiazin-5-one, 6-chloro-9-methoxy-	MeOH	256(3.97),335(3.56), 480(3.75)	78-2771-86
$C_{16}H_9Cl_3N_4$			
1H-Pyrazolo[3,4-b]quinoxaline, 1-phenyl-3-(trichloromethyl)-	MeOH	270(5.32),340(4.63)	2-0901-86
$C_{16}H_9Cl_4NO_2$			
2,4-Cyclohexadien-1-one, 3,4,5,6-tetrachloro-2-hydroxy-6-(1-phenyl-1H-pyrrol-2-yl)-	EtOH	345(3.60)	142-0579-86
$C_{16}H_9NO_2$			
Fluoranthene, 2-nitro-	EtOH	258(4.59)	35-4126-86
$C_{16}H_9NO_2S$			
5H-Benzo[a]phenothiazin-5-one, 6-hydroxy-	MeOH	253(3.55),273s(3.44), 329(3.27),353(3.27), 510(3.01)	39-2233-86C
$C_{16}H_9NO_3$			
3H-Naphtho[2,1-b]pyran-1-carbonitrile, 2-acetyl-3-oxo-	MeOH	401(3.53)	2-0137-86
$C_{16}H_9N_3O$			
Ethenetricarbonitrile, [5-(4-methylphenyl)-2-furanyl]-	MeOH	479(3.50)	73-1450-86

Compound	Solvent	λ_{max}(log ϵ)	Ref.
$C_{16}H_{10}$			
Pyrene	n.s.g.	241.3(4.9),273.3(4.7), 335.4(4.7)	1-0665-86
$C_{16}H_{10}Br_2O_6$			
9H-Xanthene-1-carboxylic acid, 4,5-dibromo-2,8-dihydroxy-6-methyl-9-oxo-, methyl ester	MeOH	243(4.56),254s(4.55), 293(3.89),319(3.80), 373(3.96)	158-0164-86
$C_{16}H_{10}ClNO_2$			
Benzenepropanenitrile, α-benzoyl-4-chloro-β-oxo-	DMSO-d_6	315(4.61)	2-1133-86
Benzoyl chloride, 4-(5-phenyl-2-oxazolyl)-	toluene	354(4.35)	103-1022-86 +103-1026-86
$C_{16}H_{10}ClNO_3$			
Benzoic acid, 4-[5-(4-chlorophenyl)-2-oxazolyl]-	toluene	338(4.42)	103-1022-86
$C_{16}H_{10}Cl_2O_4$			
9,10-Anthracenedione, 1,4-dichloro-5,8-dimethoxy-	MeOH	228(4.38),246(4.44), 420(3.72)	4-1491-86
$C_{16}H_{10}N_2O$			
Benzonitrile, 4-(5-phenyl-2-oxazolyl)-	toluene	337(4.54)	103-1026-86
$C_{16}H_{10}N_2OS$			
5H-Benzo[a]phenothiazin-5-one, 1-amino-	$CHCl_3$	246(4.60),277(4.14), 303(4.25),325(4.00), 363(3.99),378(3.96), 490(4.16)	4-0589-86
5H-Benzo[a]phenothiazin-5-one, 4-amino-	$CHCl_3$	248(4.61),278(4.21), 322(4.12),359(4.02), 378(3.95),533(4.08)	4-0589-86
$C_{16}H_{10}N_2O_2$			
4H-1-Benzopyran-4-one, 2-(1H-benzimid-azol-2-yl)-	EtOH	230(4.36),284(3.95), 333(3.92)	42-0600-86
Benzo[f]pyrido[1,2-a]indole-6,11-di-one, 12-amino-	DMF	382(4.1106),454(3.6180), 601(3.7818)	2-1126-86
$C_{16}H_{10}N_4O_3S$			
Thiazolo[4,5-b]quinoxaline, 2-(4-meth-oxyphenyl)-7-nitro-	n.s.g.	412(4.3102)	18-1245-86
$C_{16}H_{10}O_2$			
1,2-Naphthalenedione, 3-phenyl-	EtOH	262(3.95),340(3.43), 445(2.95)	94-2810-86
1,4-Naphthalenedione, 2-phenyl-	EtOH	250(4.45),255s(4.43), 268s(4.03),292s(3.87), 300(3.94),340(3.75), 356s(3.71)	94-2810-86
$C_{16}H_{10}O_3$			
2H-Benzofuro[3,2-g]-1-benzopyran-2-one, 4-methyl-	EtOH	249(4.50),268(4.45), 283(4.31),337(4.17), 365(3.20)	149-0001-86B
2H-Benzofuro[3,2-g]-1-benzopyran-2-one, 11-methyl-	EtOH	249(4.46),270(4.44), 288(4.40),337(4.04), 365(3.46)	149-0001-86B

Compound	Solvent	λmax(log ε)	Ref.
1,4-Epoxyanthracene-5,8-dione, 1,2,3,4-tetrahydro-2,3-bis(methylene)-	dioxan	256(4.44),337(3.52)	44-4160-86
$C_{16}H_{10}O_4$			
Butanetetrone, diphenyl-	n.s.g.	515(2.30)	89-0999-86
Tricyclo[8.2.2.2^{4,7}]hexadeca-2,4(16),6,8,10,12,13-heptaene-5,15-dione, 11,13-dihydroxy-	EtOH	330(3.43),517(2.14)	89-0171-86
$C_{16}H_{10}O_6$			
8H-1,3-Dioxolo[4,5-g][1]benzopyran-8-one, 9-hydroxy-7-(2-hydroxyphenyl)-	EtOH	217(4.51),243(4.24), 269(4.27),290s(4.17), 340s(3.59)	102-0281-86
	EtOH-$AlCl_3$	220(--),242s(--), 282(--),317(--)	102-0281-86
$C_{16}H_{11}BO_6$			
Boron, [ethanedioato(2-)-O,O'][1-(2-naphthalenyl)-1,3-butanedionato-O,O']-, (T-4)-	CH_2Cl_2	470(3.92)	48-0755-86
$C_{16}H_{11}BrN_5O_2$			
Tetrazolo[1,5-b]isoquinolinium, 5-bromo-10-methyl-3-(4-nitrophenyl)-, bromide	EtOH	270(4.37),394(3.58)	78-5415-86
$C_{16}H_{11}BrO_2$			
2(5H)-Furanone, 4-(4-bromophenyl)-3-phenyl-	$CHCl_3$	244(5.00),297(4.26)	2-0760-86
$C_{16}H_{11}BrO_6$			
9H-Xanthene-1-carboxylic acid, 4-bromo-2,8-dihydroxy-6-methyl-9-oxo-, methyl ester	MeOH	238(4.42),263(4.50), 292(4.04),324s(3.73), 385(3.81)	158-0164-86
$C_{16}H_{11}BrO_7$			
1H-Oxepino[4,3-b][1]benzopyran-1-carboxylic acid, 5-bromo-3,11-dihydro-10-hydroxy-8-methyl-3,11-dioxo-, methyl ester	MeOH	279(4.59)	158-0164-86
$C_{16}H_{11}Br_4NO_4$			
[2,7]Benzodioxecino[3,4,5-gh]pyrrolizine-8,13-dione, 9,10,11,12-tetrabromo-1,2,4,6,14a,14b-hexahydro-, (14aR-cis)-	EtOH	230(4.43),300s(--), 308(2.98)	39-0585-86C
$C_{16}H_{11}ClN_2O$			
Anthra[1,9-cd]pyrazol-6(1H)-one, 5-chloro-1-ethyl-	MeOH	280(4.18),304(4.06), 348(4.73),433(3.99)	4-1491-86
Anthra[1,9-cd]pyrazol-6(2H)-one, 5-chloro-2-ethyl-	MeOH	247(4.22),259(4.15), 268(4.18),299(4.05), 340(3.58),415(3.92)	4-1491-86
2-Naphthalenol, 1-[(5-chloro-2-pyridinyl)imino]methyl]-	$CHCl_3$	464(4.21)	2-0127-86A
$C_{16}H_{11}ClN_2OS_2$			
Benzo[b]thiophene-2-carboxamide, 3-chloro-N-[(phenylamino)thioxomethyl]-	EtOH	210(3.47),258(3.31), 311(3.44)	73-2839-86

Compound	Solvent	λ_{max}(log ε)	Ref.
$C_{16}H_{11}ClN_2O_2$			
Benzamide, 4-[5-(4-chlorophenyl)-2-oxazolyl]-	toluene	333(4.45)	103-1022-86
$C_{16}H_{11}ClO_2S$			
Benzo[b]thiophene-6-carboxylic acid, 4-(4-chlorophenyl)-, methyl ester	EtOH	230(4.49),250(4.57), 280(4.09),325(3.93)	5-1003-86
$C_{16}H_{11}ClO_3$			
Cyclobut[a]acenaphthylen-7(6bH)-one, 8a-acetoxy-8-chloro-8,8a-dihydro-, (6bα,8α,8aα)-	$CHCl_3$	287s(3.85),297.5(3.92), 311s(3.71),331s(2.91)	39-1965-86C
$C_{16}H_{11}ClO_5$			
9,10-Anthracenedione, 2(and 3)-chloro-6-hydroxy-1,4-dimethoxy-	EtOH	274(5.41),388(4.84)	150-3137-86M
$C_{16}H_{11}ClO_6$			
9H-Xanthene-1-carboxylic acid, 4-chloro-2,8-dihydroxy-6-methyl-9-oxo-, methyl ester	MeOH	237(4.57),263(4.65), 294(4.11),324s(3.76), 386(3.94)	158-0164-86
$C_{16}H_{11}Cl_4NO_4$			
[2,7]Benzodioxecino[3,4,5-gh]pyrrolizine-8,13-dione, 9,10,11,12-tetrachloro-1,2,4,6,14a,14b-hexahydro-, (14aR-cis)-	EtOH	227(4.51),298(3.04), 308(3.06)	39-0585-86C
$C_{16}H_{11}F_2NO_3S$			
Oxazole, 2-[4-(difluoromethyl)sulfonyl]phenyl]-5-phenyl-	toluene	340(4.44)	103-1026-86
$C_{16}H_{11}F_3N_2$			
Quinoxaline, 3-methyl-2-phenyl-6-(trifluoromethyl)-	EtOH	205(4.61),240(4.49), 260s(4.0),323(3.95)	150-0277-86M
	EtOH	205(4.60),237(4.50), 268s(3.91),324(3.99)	150-0277-86M
$C_{16}H_{11}F_3N_2O$			
Quinoxaline, 2-methyl-3-phenyl-6-(trifluoromethyl)-, 1-oxide	EtOH	203(4.32),243(4.51), 264s(4.16),280s(4.03), 325(4.01)	150-0277-86M
Quinoxaline, 3-methyl-2-phenyl-6-(trifluoromethyl)-, 4-oxide	EtOH	205(4.44),249(4.29), 296(3.92),304s(3.91)	150-0277-86M
	EtOH-$HClO_4$	203(4.47),232(4.34)	150-0277-86M
$C_{16}H_{11}N$			
5H-Benzo[b]carbazole	MeOH	228(4.45),270(4.59), 280(4.52),290(4.34), 314(3.84),329(3.93), 370(3.48),390(3.49)	116-2397-86
7H-Benzo[c]carbazole	MeOH	215(4.38),230(4.27), 265(4.67+),285(4.00), 325(4.16),342(3.72), 360(3.63)	116-2397-86
11H-Indeno[1,2-b]quinoline	EtOH	222(4.16),260(4.18), 312(4.00),318(4.03), 326(4.14),334(4.10), 341(4.16)	4-0689-86

Compound	Solvent	$\lambda_{max}(\log \epsilon)$	Ref.
$C_{16}H_{11}NO$			
Benzeneacetonitrile, α-(2-oxo-2-phenylethylidene)-, (E)-	n.s.g.	263(4.11)	150-0196-86S
(Z)-	n.s.g.	311(4.21)	150-0196-86S
$C_{16}H_{11}NO_2$			
Anthracene, 9-(2-nitroethenyl)-, (E)-	benzene	425(3.9)	44-3223-86
	C_6H_{12}	415(3.9)	44-3223-86
	EtOH	420(3.9)	44-3223-86
	CH_2Cl_2	430(3.9)	44-3223-86
(Z)-	C_6H_{12}	370f(3.8)	44-3223-86
Benzaldehyde, 4-(5-phenyl-2-oxazolyl)-	toluene	346(4.40)	103-1022-86
$C_{16}H_{11}NO_3$			
Benzoic acid, 4-(5-phenyl-2-oxazolyl)-	toluene	335(4.68)	103-1022-86
[1,3]Dioxolo[4,5-j]pyrrolo[3,2,1-de]phenanthridinium, 4,5-dihydro-2-hydroxy-, hydroxide, inner salt (ungeremine)	MeOH	262(4.28),285s(4.07), 300s(3.74),418(3.88)	102-1097-86
	MeOH	263(4.48),285s(3.98), 325(3.82),415(3.48)	150-0112-86S
	MeOH-HCl	261(--),275(--), 368(--),380(--)	102-1097-86
	MeOH-NaOH	265(--),275(--), 323(--),422(--)	102-1097-86
$C_{16}H_{11}NO_3S$			
5H-Benzo[b]carbazole-6-sulfonic acid, sodium salt	MeOH	232(4.47),268(4.71), 278(4.89),285s(4.61), 305(3.69),320(3.94), 322(4.12),376(3.65), 395(3.75)	116-2397-86
$C_{16}H_{11}NO_5S$			
9H-Thioxanthene-1-carboxylic acid, 3-nitro-9-oxo-, ethyl ester	CH_2Cl_2	407(3.63)	151-0353-86D
$C_{16}H_{11}N_3O$			
4H-Pyrazolo[3,4-c]quinolin-4-one, 2,5-dihydro-2-phenyl-	EtOH	251s(4.47),257(4.33), 274(4.43),297s(4.15)	142-3181-86
7-Quinolinamine, 3-(2-benzoxazolyl)-	EtOH	340(4.32)	2-0509-86A
$C_{16}H_{11}N_3OS$			
Thiazolo[4,5-b]quinoxaline, 2-(4-methoxyphenyl)-	n.s.g.	385(4.3038)	18-1245-86
$C_{16}H_{11}N_3O_2$			
Pyrimidine, 2-nitro-4,6-diphenyl-	EtOH	278(4.38),310(4.27)	70-0871B-86
Pyrimidine, 4-nitro-2,6-diphenyl-	EtOH	268(4.46),324s(3.74)	70-0871B-86
$C_{16}H_{11}N_3O_4$			
3,5-Pyrazolidinedione, 4-[(4-nitrophenyl)methylene]-1-phenyl-	EtOH-acid	250(4.14),319(4.10)	103-0176-86
	EtOH-base	270(4.30),320(4.05)	103-0176-86
$C_{16}H_{11}N_5$			
1-Phthalazinamine, N-1-phthalazinyl-	$CHCl_3$	258(4.13),280(4.15), 288(4.17),393(4.37), 813s(3.84),848(4.07)	94-1840-86
$C_{16}H_{12}$			
10H-4b,10-Ethenobenz[a]azulene	EtOH	236(3.77),275(3.56), 280(3.58),336(3.45), 351(3.46),370(3.24)	78-6713-86

Compound	Solvent	$\lambda_{max}(\log \epsilon)$	Ref.
Tricyclo[8.2.2.2^{4,7}]hexadeca-2,4,6-8,10,12,13,15-octaene, $Cr(CO)_3$ complex	CH_2Cl_2	342(3.97)	88-2353-86
$C_{16}H_{12}BrN_4$			
Tetrazolo[1,5-b]isoquinolinium, 3-(4-bromophenyl)-10-methyl-, bromide	EtOH	277(4.18),358(3.98)	78-5415-86
$C_{16}H_{12}BrN_5O_2$			
Isoquinoline, 1-bromo-4-methyl-3-[3-(4-nitrophenyl)-1-triazenyl]-	EtOH	408(4.49)	78-5415-86
$C_{16}H_{12}Br_2$			
Tricyclo[8.2.2.2^{4,7}]hexadeca-2,4,6-10,12,13,15-heptaene, 2,3-dibromo-	MeCN	220(4.32)	24-1836-86
$C_{16}H_{12}ClNOS$			
Benzo[b]thiophene-2-carboxamide, 3-chloro-N-(phenylmethyl)-	EtOH	211(3.14),237(3.03), 287(2.89)	73-2002-86
$C_{16}H_{12}ClN_3O_2$			
6H-1,4-Dioxino[2,3-f]indole, 8-[(4-chlorophenyl)azo]-2,3-dihydro-	EtOH	218(4.25),298(4.08)	103-1311-86
$C_{16}H_{12}ClN_4$			
Tetrazolo[1,5-b]isoquinolinium, 3-(4-chlorophenyl)-10-methyl-, bromide	EtOH	275(4.36),356(4.12)	78-5415-86
$C_{16}H_{12}Cl_2$			
Tricyclo[8.2.2.2^{4,7}]hexadeca-2,4,6-10,12,13,15-heptaene, 2,3-dichloro-	MeCN	220(4.18)	24-1836-86
$C_{16}H_{12}Cl_2N_8O_4S_2$			
2,4(1H,3H)-Pteridinedione, 7,7'-dithiobis[6-chloro-1,3-dimethyl-	MeOH	222(4.48),256(4.40), 373(4.41)	33-0708-86
$C_{16}H_{12}FN_4$			
Tetrazolo[1,5-b]isoquinolinium, 3-(4-fluorophenyl)-10-methyl-, bromide	EtOH	263(4.16),354(4.45)	78-5415-86
$C_{16}H_{12}INO$			
Acetamide, N-(3-iodo-2-anthracenyl)-	EtOH	237s(4.30),266(4.90), 330(3.40),346(3.60), 364(3.60),383(3.50)	78-3943-86
$C_{16}H_{12}N_2$			
[Bi-1,3,5-cycloheptatrien-1-yl]-6,6'-dicarbonitrile	dioxan	237(4.74),294(3.56), 377(4.06)	89-0720-86
Naphtho[1,2-b][1,8]naphthyridine, 5,6-dihydro-	EtOH	256(4.24),335(4.16), 349(4.20)	4-0689-86
$C_{16}H_{12}N_2O$			
2-Naphthalenol, 1-[(2-pyridinylimino)methyl]-	$CHCl_3$	460(4.19)	2-0127-86A
$C_{16}H_{12}N_2O_2$			
Benzamide, 4-(5-phenyl-2-oxazolyl)-	toluene	327(4.52)	103-1022-86 +103-1026-86
2H-Benz[f]isoindole-2-acetic acid, 1-	MeOH	252(4.81),400s(3.64),	44-3978-86

Compound	Solvent	λ_{max}(log ε)	Ref.
cyano-α-methyl-, (±)-, sodium salt		420(3.85),442(3.81)	44-3978-86
1,4-Methanonaphthalene-6-carboxylic acid, 2(and 3)-(dicyanomethylene)-1,2,3,4-tetrahydro-, methyl ester	isooctane	231(4.18),244(4.16), 264(4.14),282s(3.76), 296s(2.60)	33-1310-86
	MeCN	236(4.18),246(4.19), 264(4.13),294s(3.58)	33-1310-86
3,5-Pyrazolidinedione, 1-phenyl-4-(phenylmethylene)-	EtOH-acid	249(4.15),325(4.39)	103-0176-86
	EtOH-base	232(3.95),270(4.1), 312(4.39)	103-0176-86
4H-Pyrido[1,2-a]pyrimidine-3-carboxaldehyde, 6-methyl-4-oxo-2-phenyl-	EtOH	250(4.28),283(4.14), 390s(4.22),400(4.26)	4-1295-86
2(1H)-Quinoxalinone, 3-(2-oxo-2-phenylethyl)-	MeOH	392(4.19),414(4.34), 438(4.28)	2-0525-86
Unidentified quinoxaline from o-phenylenediamine and albidin	MeOH	262(4.55),270(4.55), 436(3.96)	39-1145-86C
$C_{16}H_{12}N_2O_3$			
4H-1-Benzopyran-2-carboxamide, N-(2-aminophenyl)-4-oxo-	EtOH	231(4.34),276(3.96), 436(4.54)	42-0600-86
2-Butyn-1-ol, 4-[(4-nitrophenyl)-imino]-4-phenyl-	PrOH	222(4.18),305(4.67)	30-0176-86
6H-Indolo[3,2,1-de][1,5]naphthyridin-6-one, 1,11-dimethoxy-	EtOH	253(3.89),271(3.93), 280(3.93),355(4.21)	100-0428-86
$C_{16}H_{12}N_2O_5S$			
3H-Phenothiazine-1,9-dicarboxylic acid, 2-amino-4,6-dimethyl-3-oxo-	DMF	470(4.43)	104-0207-86
$C_{16}H_{12}N_2O_6$			
1-Phenazinecarboxylic acid, 6-formyl-4,7,9-trihydroxy-8-methyl-, methyl ester	MeOH	226(4.20),271(4.42), 385(3.78)	158-0800-86
	MeOH-KOH	262(4.32),294(4.38), 351(4.30),415(3.71), 440(3.67)	158-0800-86
$C_{16}H_{12}N_4$			
1,4-Methanonaphthalene-6,6,7,7-tetracarbonitrile, 1,2,3,4,5,8-hexahydro-2-methylene-	MeCN	210(3.87)	33-1310-86
1H-Pyrazolo[3,4-b]quinoxaline, 1-(4-methylphenyl)-	MeOH	266(4.63),332.8(4.02)	2-0901-86
$C_{16}H_{12}N_4OS$			
Thiazolo[4,5-b]quinoxalin-7-amine, 2-(4-methoxyphenyl)-	n.s.g.	380(4.2945)	18-1245-86
$C_{16}H_{12}N_4O_2S_2$			
1,2,3-Thiadiazole-4-carboxylic acid, 2,5-dihydro-2-phenyl-5-[[(phenylamino)thioxomethyl]imino]-	EtOH	232(4.34),277(4.09), 419(4.11)	48-0741-86
$C_{16}H_{12}N_4O_3$			
Butanoic acid, 2,4-dicyano-4-(1,5-dihydro-3-methyl-5-oxo-1-phenyl-4H-pyrazol-4-ylidene)-	DMAA	362(3.54)	104-0582-86
$C_{16}H_{12}N_4O_4$			
6H-1,4-Dioxino[2,3-f]indole, 2,3-dihydro-8-[(4-nitrophenyl)azo]-	EtOH	211(4.23),310(3.61)	103-1311-86

Compound	Solvent	$\lambda_{max}(\log \epsilon)$	Ref.
$C_{16}H_{12}N_4S$			
Propanedinitrile, (5,6-dihydro-2-methyl-5-phenylthieno[2,3-d]pyrimidin-4-yl)-	$CHCl_3$	267(4.18),341(4.09), 388(3.82)	94-0516-86
Propanedinitrile, (5,6-dihydro-6-methyl-2-phenylthieno[2,3-d]pyrimidin-4-yl)-	$CHCl_3$	285(4.04),317(4.05), 375(4.15)	94-0516-86
$C_{16}H_{12}N_5O$			
Tetrazolo[1,5-b]isoquinolinium, 10-methyl-3-(4-nitrophenyl)-, bromide	EtOH	233(4.63),362(4.19)	78-5415-86
$C_{16}H_{12}N_6NiO_2$			
Nickel, [2-[[1-(1H-benzimidazol-2-ylazo)ethyl]azo]benzoato(2-)]-, (SP-4-2)-	isoPrOH	535(4.43),1060(3.90)	65-0727-86
$C_{16}H_{12}N_6O_2$			
1,3-Benzenedicarbonitrile, 2-[[4-(dimethylamino)phenyl]azo]-5-nitro-	$CHCl_3$	301(3.91),312s(3.76), 333s(3.53),590(4.68)	48-0497-86
$C_{16}H_{12}N_6S$			
Cyclopropanylium, [[2-(dicyanomethyl)-3-(dicyanomethylene)-1-cyclopropen-1-yl]thio]bis(dimethylamino)-, hydroxide, inner salt	CH_2Cl_2	257s(4.14),288(4.54), 346(3.92)	24-2104-86
$C_{16}H_{12}N_8O_4S_2$			
[1,4]Dithiino[2,3-g:5,6-g']dipteridine-2,4,9,11(1H,3H,8H,10H)-tetrone, 1,3,8,10-tetramethyl-	MeOH	256(4.67),272s(4.52), 402(4.35)	33-0708-86
[1,4]Dithiino[2,3-g:6,5-g']dipteridine-2,4,8,10(1H,3H,9H,11H)-tetrone, 1,3,9,11-tetramethyl-	MeOH	240(4.50),275(4.46), 334(3.92),410(4.05)	33-0708-86
$C_{16}H_{12}O$			
1-Naphthalenol, 3-phenyl-	EtOH	258(4.05),305(3.28), 344s(2.92)	94-2810-86
1-Naphthalenol, 4-phenyl-	EtOH	212(4.57),219s(4.52), 238(4.47),261(3.70), 306(3.97),326(2.86)	33-0560-86
2-Propyn-1-one, 1-[(2-phenylmethyl)-phenyl]-	EtOH	216s(4.26),266(4.04), 299s(3.42)	33-0560-86
$C_{16}H_{12}OS_2$			
1,2,5-Oxadithiepin, 4,7-diphenyl-	EtOAc	253(4.27),317(3.78), 390s(3.18)	2-0319-86
$C_{16}H_{12}O_2$			
Propynoic acid, 3-phenyl-, 4-methylphenyl ester	hexane	259(4.2)	44-4432-86
2-Propyn-1-one, 1-(2-hydroxy-5-methylphenyl)-3-phenyl-	hexane	294(4.2),306(4.3), 370(3.7)	44-4432-86
Tricyclo[8.4.1.1^{3,8}]hexadeca-3,5,7-10,12,14-hexaene-2,9-dione, anti	dioxan	250(4.42),282(3.97), 362s(2.48)	89-1000-86
syn	dioxan	238(4.31),273(4.44), 303s(4.09),371(3.04)	89-1000-86
$C_{16}H_{12}O_2S$			
Benzo[b]thiophene-6-carboxylic acid, 4-(4-methylphenyl)-	EtOH	224(4.52),244(4.56), 277(4.11),318(3.88)	5-1003-86

Compound	Solvent	λmax(log ε)	Ref.
Benzo[b]thiophene-6-carboxylic acid, 4-phenyl-, methyl ester	EtOH	225(4.47),249(4.55), 281(4.07),326.5(3.83)	5-1003-86
$C_{16}H_{12}O_3$			
Methanone, (6-methoxy-2-benzofuranyl)-phenyl-	EtOH	233(4.19),254(4.22), 341(4.54)	142-0771-86
2-Propenoic acid, 4-benzoylphenyl ester	$CHCl_3$	256.5(4.24),337(2.23)	32-0533-86
2-Propynoic acid, 3-phenyl-, 4-methoxyphenyl ester	hexane	258(4.4)	44-4432-86
2-Propyn-1-one, 1-(2-hydroxy-5-methoxyphenyl)-3-phenyl-	hexane	288(4.3),306(4.4), 397(3.8)	44-4432-86
$C_{16}H_{12}O_3S$			
Benzo[b]thiophene-6-carboxylic acid, 4-(4-methoxyphenyl)-	EtOH	228(4.51),248(4.54), 277(4.19),318(3.90)	5-1003-86
2-Butenoic acid, 4-oxo-4-phenyl-2-(phenylthio)-	MeOH	257(4.04),331(4.11)	48-0539-86
9H-Thioxanthene-1-carboxylic acid, 9-oxo-, ethyl ester	CH_2Cl_2	383(3.81)	151-0353-86D
9H-Thioxanthene-2-carboxylic acid, 9-oxo-, ethyl ester	CH_2Cl_2	377(3.76)	151-0353-86D
9H-Thioxanthene-3-carboxylic acid, 9-oxo-, ethyl ester	CH_2Cl_2	397(3.78)	151-0353-86D
9H-Thioxanthene-4-carboxylic acid, 9-oxo-, ethyl ester	CH_2Cl_2	386(3.89)	151-0353-86D
$C_{16}H_{12}O_4$			
1,2-Anthracenedione, 8-hydroxy-7-methoxy-3-methyl-	MeOH	259(4.51),300(4.34), 309(4.44),327(4.09), 540(3.87)	100-0145-86
	MeOH-NaOH	286(--),312(--), 370(--),710(--)	100-0145-86
1,4-Anthracenedione, 9,10-dimethoxy-	CH_2Cl_2	239(4.81),287s(3.97), 300(4.00),421(3.80)	5-0839-86
Formononetin	EtOH	239s(4.28),251(4.32), 261s(4.28),304(4.01)	105-0603-86
2H-Naphtho[1,2-b]pyran-2-one, 4-(acetoxymethyl)-	EtOH	265(4.42),275(4.50), 291.5(3.80),303.5(3.89), 316(3.88),353(3.72)	94-0390-86
2H-Naphtho[2,3-b]pyran-2-one, 4-(acetoxymethyl)-	EtOH	226(4.71),263(4.46), 273(4.47),321(4.24)	94-0390-86
3H-Naphtho[2,1-b]pyran-3-one, 1-(acetoxymethyl)-	EtOH	230(4.75),248(4.17), 317(3.97),348(4.01)	94-0390-86
$C_{16}H_{12}O_4S_8$			
Cyclobuta[1,2-d:3,4-d']bis[1,3]dithiole-3a,6a(3bH,6bH)-dicarboxylic acid, 2,5-di-1,3-dithiol-2-ylidene-, dimethyl ester	dioxan	294(4.32),312s(4.22), 410(3.60)	104-0367-86
$C_{16}H_{12}O_5$			
9,10-Anthracenedione, 2-hydroxy-1,4-dimethoxy-	MeOH	229s(4.39),243(4.47), 280(4.31),410(3.83)	78-3767-86
9,10-Anthracenedione, 6-hydroxy-1,4-dimethoxy-	EtOH	271(4.31),454(3.75)	150-3137-86M
4H-1-Benzopyran-4-one, 3-[(3,4-dihydroxyphenyl)methylene]-2,3-dihydro-7-hydroxy-	MeOH	208(4.76),258(4.11), 310(4.14),370(4.35)	94-2506-86
4H-1-Benzopyran-4-one, 5-hydroxy-2-(4-hydroxyphenyl)-7-methoxy-	MeOH	270(4.30),300s(4.15), 338(4.04)	102-1255-86

Compound	Solvent	$\lambda_{max}(\log \epsilon)$	Ref.
(cont.)	MeOH-NaOH	275(4.40),301(4.30), 359s(4.41)	102-1255-86
	MeOH-NaOAc	270(4.25),301(4.03), 359s(4.23)	102-1255-86
Biochanin A	EtOH	263(4.48),335s(3.73)	105-0603-86
Inermin	EtOH	282(3.53),287(3.59), 311(3.78)	105-0603-86
9H-Xanthene-1-carboxylic acid, 8-hydroxy-6-methyl-9-oxo-, methyl ester	MeOH	232(4.40),255(4.38), 290(4.04),299(4.04), 360(3.62)	163-1669-86
$C_{16}H_{12}O_6$			
9,10-Anthracenedione, 1,3-dihydroxy-2,5-dimethoxy-	MeOH	244(4.18),279(4.44), 400(4.00)	78-3767-86
9,10-Anthracenedione, 1,4-dihydroxy-5-(2-hydroxyethoxy)-	EtOH	253(3.96),270(3.80), 283(3.70),476(3.75), 493(3.73),513(3.56), 528(3.42)	150-3162-86M
9,10-Anthracenedione, 1,4-dihydroxy-6-(2-hydroxyethoxy)-	EtOH	225(4.47),273(4.44), 294s(3.76),477(3.96), 482(3.94),500(3.81), 515(3.68)	150-3162-86M
Pratensein	EtOH	264(4.39),283s(4.05)	105-0603-86
$C_{16}H_{12}O_7$			
9H-Xanthene-1-carboxylic acid, 6,8-dihydroxy-2-methoxy-3-methyl-9-oxo-	MeOH	240(4.4),275(4.1), 338(3.75),396(3.52)	2-0860-86
$C_{16}H_{12}O_8$			
4H-1-Benzopyran-4-one, 2-(3,4-dihydroxyphenyl)-3,5,7-trihydroxy-6-methoxy-	MeOH	256(4.28),270s(4.11), 293(3.81),370(4.32)	102-0231-86
	MeOH-NaOMe	240s(4.12),334(4.48)	102-0231-86
	MeOH-NaOAc	261(4.23),270s(4.22), 322(4.02),388(4.29)	102-0231-86
	MeOH-$AlCl_3$	274(4.33),340s(3.63), 454(4.48)	102-0231-86
	MeOH-$AlCl_3$-HCl	268(4.33),305(3.66), 370s(3.98),430(4.39)	102-0231-86
	MeOH-NaOAc-H_3BO_3	261(4.34),386(4.37)	102-0231-86
$C_{16}H_{12}S$			
Thiophene, 2,5-diphenyl-	C_6H_{12}	229.5(3.98),261.5s(3.80), 322(4.37)	24-3158-86
$C_{16}H_{13}BO_4$			
Boron, [1,2-benzenediolato(2-)-O,O'](1-phenyl-1,3-butanedionato-O,O')-, (T-4)-	CH_2Cl_2	334(4.33),375(3.61), 410s(3.29)	48-0755-86
$C_{16}H_{13}BO_5$			
Boron, [1-hydroxy-2-naphthalenecarboxylato(2-)-O^1,O^2](2,4-pentanedionato-O,O')-	CH_2Cl_2	335(3.62),348(3.66)	48-0755-86
Boron, [3-hydroxy-2-naphthalenecarboxylato(2-)-O^2,O^3](2,4-pentanedionato-O,O')-	DMF	239(4.75),253s(4.45), 276s(4.25),288(4.32), 296s(4.25),358(3.38), 369s(3.32)	48-0755-86
$C_{16}H_{13}Br$			
Tricyclo[8.2.2.$2^{4,7}$]hexadeca-	MeCN	217(4.27)	24-1836-86

Compound	Solvent	$\lambda_{max}(\log \epsilon)$	Ref.
2,4,6,10,12,13,15-heptaene, 2-bromo- (cont.)			24-1836-86
$C_{16}H_{13}BrIN_7$			
2,4,6-Pyrimidinetriamine, 5-[(4-bromophenyl)azo]-N^4-(4-iodophenyl)-	EtOH	205(5.05),256(4.78), 284(4.81),428(4.56)	42-0420-86
$C_{16}H_{13}BrN_4$			
Isoquinoline, 3-[3-(4-bromophenyl)-1-triazenyl]-4-methyl-	EtOH	226(4.42),354(4.37), 382(4.38)	78-5415-86
$C_{16}H_{13}BrO_3$			
Benzeneacetic acid, 2-(4-bromophenyl)-2-oxoethyl ester	MeCN	212(4.17),225(4.29)	2-0760-86
$C_{16}H_{13}Br_3$			
Tricyclo[8.2.2.2^{4,7}]hexadeca-4,6,10-12,13,15-hexaene, 2,2,3-tribromo-	MeCN	202(4.32),233(4.18)	24-1836-86
$C_{16}H_{13}Cl$			
Tricyclo[8.2.2.2^{4,7}]hexadeca-2,4,6,10-12,13,15-heptaene, 2-chloro-	MeCN	221(4.12)	24-1836-86
$C_{16}H_{13}ClN_4$			
Isoquinoline, 3-[3-(4-chlorophenyl)-1-triazenyl]-4-methyl-	EtOH	228(4.50),356(4.40), 380(4.41)	78-5415-86
$C_{16}H_{13}ClO_5$			
Spiro[benzofuran-2(3H),1'-[2,5]cyclohexadiene]-3,4'-dione, 7-chloro-4,6-dimethoxy-2'-methyl-	MeOH	214(4.51),294(4.43)	94-0858-86
$C_{16}H_{13}ClO_6$			
Spiro[benzofuran-2(3H),1'-[2,5]cyclohexadiene]-3,4'-dione, 7-chloro-4-hydroxy-2',6-dimethoxy-6'-methyl-	MeOH	216(4.46),290(4.51)	94-0858-86
$C_{16}H_{13}Cl_2NO_4$			
[2,7]Benzodioxecino[3,4,5-gh]pyrrolizine-8,13-dione, 10,11-dichloro-1,2,4,6,14a,14b-hexahydro-, (14aR-cis)-	EtOH	275(2.88),287(2.87), 296s(--)	39-0585-86C
$C_{16}H_{13}Cl_3$			
Tricyclo[8.2.2.2^{4,7}]hexadeca-4,6,10-12,13,15-hexaene, 2,2,3-trichloro-	MeCN	203(4.08),230(4.22)	24-1836-86
$C_{16}H_{13}FN_4$			
Isoquinoline, 3-[(4-fluorophenyl)-1-triazenyl]-4-methyl-	EtOH	226(4.49),349(4.37), 375(4.33)	78-5415-86
$C_{16}H_{13}N$			
Indeno[1,2-b]indole, 5,10-dihydro-5-methyl-	n.s.g.	215(4.15),245(4.32), 318(4.35)	150-0514-86M
$C_{16}H_{13}NO$			
Oxiranecarbonitrile, 3-(2-ethynyl-1-cyclopenten-1-yl)-2-phenyl-, cis	MeCN	220(4.07),248(4.15)	78-2221-86
2(1H)-Quinolinone, 1-methyl-4-phenyl-	CH_2Cl_2	231(4.58),275s(3.80), 280.5(3.82),325s(3.68), 336(3.76),348s(3.62)	24-3109-86

Compound	Solvent	λ_{max}(log ε)	Ref.
$C_{16}H_{13}NOS$			
1,5-Benzothiazepine, 4-methoxy-2-phenyl-	CH_2Cl_2	241(4.30),255s(4.12),275s(3.63),377(4.49)	24-3109-86
1,5-Benzothiazepin-4(5H)-one, 5-methyl-2-phenyl-	CH_2Cl_2	240(4.22),255(4.17),295s(3.72)	24-3109-86
2H-1,4-Benzothiazin-3(4H)-one, 4-methyl-2-(phenylmethylene)-, (E)-	CH_2Cl_2	244(4.39),265s(4.16),286s(3.96),325(3.94),365s(3.55)	24-3109-86
(Z)-	CH_2Cl_2	239(4.34),259(4.24),285s(3.90),348(4.05)	24-3109-86
$C_{16}H_{13}NOSe$			
Benzenamine, N-[[3-(methylseleno)-2-benzofuranyl]methylene]-	EtOH	342(4.28)	103-0608-86
2-Benzofuranselenol, 3-[[(2-methylphenyl)imino]methyl]-	EtOH	264(4.46),297(4.62),337(3.88),479(4.24)	103-0608-86
2-Benzofuranselenol, 3-[[(4-methylphenyl)imino]methyl]-	EtOH	266(4.36),299(4.70),337(3.97),484(4.37)	103-0608-86
$C_{16}H_{13}NO_2$			
2H-1-Benzopyran-2-one, 7-amino-4-methyl-3-phenyl-	MeOH	240(4.09),355(4.38)	151-0069-86A
	$CHCl_3$	339(4.20)	151-0069-86A
$C_{16}H_{13}NO_2S_2$			
Benzenecarbodithioic acid, 4-methyl-, 2-(4-nitrophenyl)ethenyl ester, (Z)-	CH_2Cl_2	332(4.43),507(2.34)	18-1403-86
1H-Pyrrole, 1-(phenylsulfonyl)-2-[2-(2-thienyl)ethenyl]-, (E)-	EtOH	230(4.15),321(3.97)	118-0620-86
$C_{16}H_{13}NO_2Se$			
2-Benzofuranselenol, 3-[[(4-methoxyphenyl)imino]methyl]-	EtOH	270(4.39),299(4.70),340(4.10),478(4.40)	103-0608-86
$C_{16}H_{13}NO_3$			
Pyrrolo[3,2,1-de]phenanthridinium, 4,5-dihydro-2,9-dihydroxy-10-methoxy-, hydroxide, inner salt	MeOH	262(4.70),278s(4.59),285s(4.55),408(3.78)	102-1975-86
	MeOH-NaOMe	265(--),280s(--),300s(--),458(--)	102-1975-86
	MeOH-HCl	263(--),278(--),283(--),378(--)	102-1975-86
$C_{16}H_{13}NO_3S$			
1H-Indole, 3-acetyl-1-(phenylsulfonyl)-	MeOH	222(4.25),287(3.93)	5-2065-86
9H-Thioxanthene-1-carboxylic acid, 3-amino-9-oxo-, ethyl ester	CH_2Cl_2	333(4.04)	151-0353-86D
$C_{16}H_{13}NO_3S_2$			
Benzenecarbodithioic acid, 4-methoxy-, 2-(4-nitrophenyl)ethenyl ester	CH_2Cl_2	367(4.58),504(2.57)	18-1403-86
$C_{16}H_{13}NO_4$			
4H-Furo[3,2-b]pyrrole-5-carboxylic acid, 4-formyl-2-phenyl-, ethyl ester	MeOH	353(3.70),377(3.75)	73-0106-86
4H-Furo[3,2-b]pyrrole-5-carboxylic acid, 6-formyl-2-phenyl-, ethyl ester	MeOH	360(3.36)	73-0106-86
$C_{16}H_{13}NO_4S$			
2(1H)-Quinolinone, 1-[[(4-methyl-	EtOH	230(4.61)	94-0564-86

Compound	Solvent	$\lambda_{max}(\log \epsilon)$	Ref.
phenyl)sulfonyl]oxy]- (cont.)			94-0564-86
$C_{16}H_{13}N_3O$			
4-Pyrimidinamine, N-hydroxy-2,6-diphenyl-	n.s.g.	255(4.66),305s(3.69)	103-1232-86
$C_{16}H_{13}N_3O_2$			
Benzoic acid, 4-(4-methyl-5-phenyl-2H-1,2,3-triazol-2-yl)-	toluene	314(4.49)	24-1627-86
	EtOH	308(3.89)	24-1627-86
	DMSO	309(4.21)	24-1627-86
$C_{16}H_{13}N_3O_3$			
6-Quinoxalinecarboxylic acid, 5-hydroxy-8-(phenylamino)-, methyl ester	MeOH	252(4.30),284(4.50), 388(3.40),480(3.40)	4-1641-86
6-Quinoxalinecarboxylic acid, 8-hydroxy-5-(phenylamino)-, methyl ester	MeOH	252(4.20),284(4.40), 376(3.30),460(3.40)	4-1641-86
$C_{16}H_{13}N_3O_6S_2$			
1,4-Cyclohexadiene-1-carboxylic acid, 6-[(3-ethyl-6-sulfo-2(3H)-benzothiazolylidene)hydrazono]-3-oxo-, disodium salt	H_2O	194(4.46),210(4.44), 295(3.95),514(4.58)	163-1355-86
1,5-Cyclohexadiene-1-carboxylic acid, 3-[(3-ethyl-6-sulfo-2(3H)-benzothiazolylidene)hydrazono]-4-oxo-, disodium salt	H_2O	196(4.44),216(4.39), 293(3.85),587(4.24)	163-1355-86
$C_{16}H_{13}N_5O_2$			
Isoquinoline, 4-methyl-3-[(4-nitrophenyl)-1-triazenyl]-	EtOH	221(4.54),402(4.57)	78-5415-86
$C_{16}H_{14}$			
Anthracene, 4a,10-dihydro-4a-methyl-10-methylene-	hexane	238(4.47),247(4.46), 253s(4.35),268s(4.21), 277(4.33),289(4.22), 377(4.21),390s(3.96)	77-1257-86
Anthracene, 1,4-dimethyl-	dioxan	224(4.08),238s(4.35), 257(5.16),317(3.10), 331(3.43),347(3.73), 365(3.90),385(3.85)	64-0223-86B
1,6:7,12-Dimethano[14]annulene, anti	C_6H_{12}	243(4.43),275s(4.08), 345(3.73)	89-0723-86
syn	C_6H_{12}	245(3.92),300(4.67), 367(3.87),450s(2.53)	89-0720-86
Tricyclo[6.5.2^{13,14}.0^{7,15}]pentadeca-1,3,5,7,9,11,13-heptaene, 15-methyl-	CH_2Cl_2	307(4.46),380(3.69)	35-4105-86
$C_{16}H_{14}BrN_3O_2$			
Pyrrolo[1,2-a]pyrazine-1,4-dione, 3-[(6-bromo-1H-indol-3-yl)methylene]-hexahydro-, (-)- (barettin)	MeOH	235(4.11),294s(3.72), 340(3.99)	88-3283-86
$C_{16}H_{14}Br_2$			
Tricyclo[8.2.2.2^{4,7}]hexadecahexaene, 2,3-dibromo-	MeCN	206(4.18),225(4.18)	24-1836-86
Tricyclo[8.4.1.1^{3,8}]hexadeca-3,5,7-10,12,14-hexaene, 2,9-dibromo-	dioxan	237s(4.37),253(4.41), 348s(2.99)	89-1000-86

Compound	Solvent	λ_{max}(log ε)	Ref.
$C_{16}H_{14}ClN_2O$			
Pyridinium, 4-[5-(4-chlorophenyl)-2-oxazolyl]-1-ethyl-, salt with 4-methylbenzenesulfonic acid	EtOH	255(4.19),380(4.39)	135-0244-86
	H_2O	374(4.06)	135-0244-86
$C_{16}H_{14}Cl_2$			
Tricyclo[8.2.2.2^{4,7}]hexadeca-4,6,10-12,13,15-hexaene, 2,3-dichloro-	MeCN	201(4.11),227(4.29)	24-1836-86
$C_{16}H_{14}IN_7O_3S$			
Benzenesulfonic acid, 4-[[2,4-diamino-6-[(4-iodophenyl)amino]-5-pyrimidinyl]azo]-	EtOH	254(4.25),288(4.40), 428(4.28)	42-0420-86
$C_{16}H_{14}NO_3$			
Isoquinolinium, 2-[[5-(methoxycarbonyl)-2-furanyl]methyl]-, chloride	MeOH	232(3.45),340(2.53)	73-0412-86
$C_{16}H_{14}N_2$			
2H-Benz[f]isoindole-1-carbonitrile, 2-propyl-	EtOH	400s(3.68),417(3.89), 441(3.88)	44-3978-86
2,3'-Bi-1H-indole, 2,3-dihydro-	EtOH	223(4.74),282(4.01)	78-5019-86
3H-Pyrazolo[1,5-a]indole, 3a,4-dihydro-2-phenyl-	EtOH	227(4.16),242.5s(4.04), 300(3.97),340s(3.51)	33-0927-86
	EtOH-HCl	251(4.20)	33-0927-86
1H-Pyrrolo[3,2-c]pyridine, 1-methyl-4-(2-phenylethenyl)-	EtOH	238(4.36),277(4.16), 340(4.25)	150-0343-86M
$C_{16}H_{14}N_2O$			
Ethanone, 1-(9,10-dihydro-8H-pyrrolo-[1,2-a]perimidin-8-yl)-	MeOH	235(4.03),330(3.64)	48-0906-86
Ethanone, 1-[3-(phenylamino)-1H-indol-2-yl]-	MeCN	240(4.28),260(4.08), 298(4.14),390(3.81)	24-2289-86
$C_{16}H_{14}N_2O_2S$			
Benzenesulfonamide, 4-methyl-N-8-quinolinyl-	neutral	232(4.66)	11-0325-86
2-Benzothiazolamine, 6-methoxy-N-[(4-methoxyphenyl)methylene]-	dioxan	373(4.39)	80-0273-86
	+ CF_3COOH	373(--)	80-0273-86
Pyrazine, 2,6-dimethyl-3-phenyl-5-(2-thienyl)-, 1,4-dioxide	EtOH	241(4.32),303(4.43)	103-0268-86
$C_{16}H_{14}N_2O_3$			
Ethanone, 1,1'-(azoxydi-3,1-phenylene)bis-	EtOH	227(4.36),315(4.00)	23-2076-86
Pyrazine, 2-(2-furanyl)-3,5-dimethyl-6-phenyl-, 1,4-dioxide	EtOH	238(4.18),275s(4.24), 300(4.41)	103-0268-86
$C_{16}H_{14}N_2O_3S$			
1H-Indole, 3-[1-(hydroxyimino)ethyl]-1-(phenylsulfonyl)-	MeOH	222(4.42),266(4.15), 294(3.94)	5-2065-86
$C_{16}H_{14}N_2O_4$			
Imidazo[1,2-a]pyridine-7-carboxylic acid, 8-benzoyl-1,2,3,5-tetrahydro-5-oxo-, methyl ester	EtOH	238(4.16),362(4.05)	142-2247-86
$C_{16}H_{14}N_2O_4S$			
Benzoic acid, 4-(3-methyl-2H-1,2,4-benzothiadiazin-2-yl)-, methyl ester, S,S-dioxide	MeOH	258(4.27),263(4.34), 296(4.07)	44-1967-86

Compound	Solvent	$\lambda_{max}(\log \epsilon)$	Ref.
5H-Dibenzo[b,g][1,4,6]thiadiazocine-2-carboxylic acid, 6-methyl-, methyl ester, 12,12-dioxide	MeOH	225(4.11),252(4.26), 300(3.88),361(3.28)	44-1967-86
$C_{16}H_{14}N_2O_5$			
4H-Furo[3,2-b]pyrrole-5-carboxylic acid, 2-(4-methylphenyl)-6-nitro-, ethyl ester	MeOH	302(3.46)	73-0106-86
$C_{16}H_{14}N_2O_7$			
1-Azabicyclo[3.2.0]hept-2-ene-2-carboxylic acid, 4-acetoxy-7-oxo-, (4-nitrophenyl)methyl ester	EtOH	270(4.12)	88-6001-86
$C_{16}H_{14}N_3O$			
Pyridinium, 1-[2-[1-(hydroxyimino)-ethyl]-4-quinolinyl]-, chloride, (E)-	EtOH	249(4.79)	94-0564-86
$C_{16}H_{14}N_4$			
2,3:7,8-Dipyridazino[4,5]tetracyclo-[7.2.1.0^{4,10}.0^{6,10}]dodeca-2,7-diene, acs-	EtOH	240(3.19),247(3.18), 302(2.97)	24-3442-86
4,5:9,10-Dipyridazino[4,5]tetracyclo-[6.2.1.1^{3,6}.0^{2,7}]dodeca-4,9-diene	EtOH	242(3.27),248(3.27), 313(2.99)	24-3442-86
$C_{16}H_{14}N_4OS$			
Benzamide, N-[1,5-dihydro-3-(phenylmethyl)-5-thioxo-4H-1,2,4-triazol-4-yl]-	EtOH	227(4.06),253(4.16)	4-1451-86
$C_{16}H_{14}N_4O_2$			
1H-Isoindole-1,3(2H)-dione, 5-[[4-(dimethylamino)phenyl]azo]-	DMF	272.9(4.05),480.8(4.44)	56-0797-86
with 0.02M p-toluenesulfonic acid	DMF	272.5(4.24),480.8(4.44)	56-0797-86
$C_{16}H_{14}N_4O_3S_2$			
Hydrazinecarbothioamide, 2-[2,3-dihydro-2-oxo-1-1-[(4-methylphenyl)-sulfonyl]-1H-indol-3-ylidene]-	EtOH	239(4.38),275(3.90), 392(4.20)	104-2173-86
$C_{16}H_{14}N_4O_6$			
Benzeneacetic acid, α-[(2,4-dinitrophenyl)hydrazono]-, ethyl ester, (E)-	EtOH	354(4.42)	18-0073-86
(Z)-	EtOH	389(4.46)	18-0073-86
Ethanedioic acid, bis[[(2,4-dihydroxyphenyl)methylene]hydrazide]	DMF	346(4.62)	131-0513-86I
	CCl_4	351(4.59)	131-0513-86I
	benzene	420(--)	131-0513-86I
	40% DMF	347(--)	131-0513-86I
	+ acid	278(--)	131-0513-86I
$C_{16}H_{14}N_6O_2$			
Benzoic acid, 2-[5-(1H-benzimidazol-2-yl)-3-methyl-1-formazano]-	EtOH	480(4.65)	65-0727-86
$C_{16}H_{14}N_6O_4$			
[1,2,4]Triazolo[1,5-a]pyrimidine-2-carboxylic acid, 1,7-dihydro-7-(1H-indol-3-yl)-6-nitro-, ethyl ester	EtOH	221(4.59),272(4.92), 360(4.98)	103-1250-86

Compound	Solvent	$\lambda_{max}(\log \epsilon)$	Ref.
$C_{16}H_{14}N_8O_4S_2$			
2,4(1H,3H)-Pteridinedione, 6,6'-dithiobis[1,3-dimethyl-	MeCN	249(4.40),263(4.38), 360(4.13)	33-0708-86
2,4(1H,3H)-Pteridinedione, 7,7'-dithiobis[1,3-dimethyl-	MeOH	222(4.59),245(4.33), 365(4.41)	33-0708-86
$C_{16}H_{14}OS$			
Dibenzo[b,f]thiepin-10(11H)-one, 6-ethyl-	MeOH	241(4.31),260s(4.04), 330(3.60)	73-0141-86
Thiophene, 2-(2-ethoxy-1-naphthalenyl)-	MeOH	232(4.71),283(3.80), 294(3.81),337(3.60)	18-0083-86
$C_{16}H_{14}OS_2$			
Benzenecarbodithioic acid, 4-methoxy-, 2-phenylethenyl ester	CH_2Cl_2	356(4.47),503(2.53)	18-1403-86
$C_{16}H_{14}O_2$			
Anthracene, 1,4-dimethoxy-	MeCN	239(4.95),259(4.67), 345(3.49),354s(3.58), 362(3.74)	24-1016-86
4H-1-Benzopyran-4-one, 2,3-dihydro-6-methyl-2-phenyl-	C_6H_{12}	247(3.95),320(3.57), 365s(1.48)	39-1217-86B
	EtOH	250(3.89),330(3.53)	39-1217-86B
[Bi-1,3,5-cycloheptatrien-1-yl]-6,6'-dicarboxaldehyde	dioxan	245(4.69),305(3.53), 410(3.97)	89-0720-86
4,5-Pyrenediol, 4,5,9,10-tetrahydro-, (4S-trans)-	n.s.g.	274s(--),283(4.16), 294s(--)	44-1773-86
$C_{16}H_{14}O_3$			
2,4-Benzofurandione, 5,6-dihydro-6,6-dimethyl-3-phenyl-	EtOH	227(4.09),350(4.25)	2-0347-86
Benzoic acid, 4-[2-(2-hydroxyphenyl)-ethenyl]-, methyl ester, (E)-	n.s.g.	340(4.49)	150-0433-86S
4H-1-Benzopyran-4-one, 2,3-dihydro-6-methoxy-2-phenyl-	C_6H_{12}	248(3.87),343(3.64), 357s(1.00)	39-1217-86B
	EtOH	254(3.87),350(3.60)	39-1217-86B
4H-1-Benzopyran-4-one, 2,3-dihydro-7-methoxy-3-phenyl-	MeOH	232s(4.17),240(4.21), 282(4.29),320(4.05)	39-0215-86C
Cyclohepta[de]naphthalen-7(8H)-one, 8,8-dimethoxy-	$CHCl_3$	322(3.77),340s(3.71)	39-1965-86C
1,4-Naphthalenedione, 2-(1-cyclohexen-1-yl)-3-hydroxy-	EtOH	252(4.26),274(4.26), 327(3.46),407(3.11)	39-0659-86C
2H-Naphtho[2,3-b]pyran-5,10-dione, 3-ethyl-2-methyl-	EtOH	265(4.27),280(4.26), 332(3.33),459(3.23)	39-0659-86C
2H-Naphtho[1,2-b]pyran-2-one, 4-(ethoxymethyl)-	EtOH	265(4.44),275(4.52), 291.5(3.81),303.5(3.89), 316.5(3.86),351.5(3.70)	94-0390-86
9-Oxatricyclo[4.3.0.0^{1,4}]non-6-ene-5,8-dione, 3,3-dimethyl-7-phenyl-	n.s.g.	313(3.50)	2-0347-86
Phenol, 2-[2-(4-acetoxyphenyl)ethenyl]-, (E)-	n.s.g.	322(4.29),329(4.28)	150-0433-86S +150-3514-86M
4-Phenanthrenol, 5,7-dimethoxy- (thebaol)	MeOH	212(4.26),246(4.58), 301(4.06),311(4.08)	83-0694-86
$C_{16}H_{14}O_4$			
1H-Benz[e]indene-3-carboxylic acid, 2,3-dihydro-7-methoxy-1-oxo-, methyl ester	EtOH	247(4.55)	2-0015-86
4H-1-Benzopyran-4-one, 2,3-dihydro-5-hydroxy-7-methoxy-2-phenyl-	EtOH	226(4.23),287(4.26), 328(3.56)	102-1427-86

Compound	Solvent	λ_{max}(log ε)	Ref.
4H-1-Benzopyran-4-one, 2,3-dihydro-7-hydroxy-3-(4-methoxyphenyl)-	MeOH	219(4.65),273(4.49), 312(4.33)	39-0215-86C
[1,1'-Biphenyl]-3,3'-dicarboxaldehyde, 6,6'-dimethoxy-	EtOH	234(4.18),278(4.29)	100-1018-86
Hexacyclo[$5.4.1.0^{2,6}.0^{3,10}.0^{4,8}.0^{9,12}$]-dodecane-7,8,9,12-tetracarboxaldehyde	MeCN	289(1.95)	24-3442-86
$C_{16}H_{14}O_5$			
4H-1-Benzopyran-4-one, 3-[(2,4-dihydroxyphenyl)methyl]-2,3-dihydro-7-hydroxy-	MeOH	204(4.65),231(4.26), 276(4.25),312(3.98)	94-2506-86
$C_{16}H_{14}O_6$			
Benzoic acid, 2-(2,6-dihydroxy-4-methylbenzoyl)-3-hydroxy-, methyl ester (moniliphenone)	MeOH	282(4.11),354(3.36)	163-1669-86
4H-1-Benzopyran-4-one, 2-(3,4-dihydroxyphenyl)-2,3-dihydro-5-hydroxy-7-methoxy-	MeOH	290(4.07),328s(3.28)	95-0989-86
4H-1-Benzopyran-4-one, 3-[(3,4-dihydroxyphenyl)methyl]-2,3-dihydro-3,7-dihydroxy-	MeOH	205(4.61),231(4.18), 277(4.18),311(3.89)	94-2506-86
1,3,5-Cyclohexanetrione, 2-hydroxy-4-[3-(4-hydroxyphenyl)-1-oxo-2-propenyl]-2-methyl-	EtOH	405(4.33)	138-0495-86
Dicyclohepta[b,d]furan-1,4,9(5H)-trione, 5a,11b-dihydro-10,11b-dimethoxy-, cis	MeOH	234(4.36),336(3.94), 359(3.88),377(3.90), 395(3.84)	18-0511-86
5H-Furo[3,2-g][1]benzopyran-5-one, 2-acetyl-4,9-dimethoxy-7-methyl-	EtOH	228s(4.21),245(4.33), 282(4.48),337(4.02)	44-3116-86
1,2,3-Naphthalenetriol, triacetate	MeOH	225(4.89),260s(3.57), 267(3.71),279(3.74), 288(3.57)	39-1789-86C
9H-Xanthen-9-one, 1-hydroxy-3,5,6-trimethoxy-	EtOH	245(4.67),280s(4.02), 315(4.38),340(3.97)	102-0503-86
$C_{16}H_{14}O_7$			
5H-Furo[3,2-g][1]benzopyran-7-carboxylic acid, 4,9-dimethoxy-5-oxo-, ethyl ester	EtOH	242(4.40),271(4.53), 328(4.43)	44-3116-86
$C_{16}H_{14}S$			
Dibenzo[b,f]thiepin, 4-ethyl-	MeOH	261.5(4.31),300(3.74)	73-0141-86
$C_{16}H_{14}S_2$			
Benzenecarbodithioic acid, 4-methyl-, 2-phenylethenyl ester	CH_2Cl_2	328(4.32),503(2.40)	18-1403-86
$C_{16}H_{15}BrN_2O$			
Benzo[a]phenazin-5(7H)-one, 6-bromo-7a,8,9,10,11,11a-hexahydro-, trans	CH_2Cl_2	236(4.11),277(4.27), 422(3.60)	83-0052-86
$C_{16}H_{15}BrN_6O_5$			
Acetamide, N-[2-[(2-bromo-4,6-dinitrophenyl)azo]-5-(dimethylamino)phenyl]-	$CHCl_3$	294(3.93),563(4.67)	48-0497-86
$C_{16}H_{15}Br_2N_5O_3$			
Acetamide, N-[2-[(2,6-dibromo-4-nitrophenyl)azo]-5-(dimethylamino)phenyl]-	$CHCl_3$	301(4.00),376s(3.66), 498(4.43)	48-0497-86

Compound	Solvent	$\lambda_{max}(\log \epsilon)$	Ref.
$C_{16}H_{15}ClHgO$			
Mercury, chloro[(3,4-dihydro-2-phenyl-2H-1-benzopyran-2-yl)methyl]-	EtOH	274(3.3)	32-0441-86
$C_{16}H_{15}ClN_4$			
1H-Pyrazolo[3,4-b]quinolin-4-amine, N-(2-chlorophenyl)-5,6,7,8-tetrahydro-	EtOH	216s(--),253(4.24), 273(4.05),314(4.36)	5-1728-86
1H-Pyrazolo[3,4-b]quinolin-4-amine, N-(4-chlorophenyl)-5,6,7,8-tetrahydro-	EtOH	219(4.33),248(3.98), 324(4.15)	5-1728-86
$C_{16}H_{15}ClO_3$			
1-Propanone, 3-chloro-1-(2-hydroxy-4-methoxyphenyl)-2-phenyl-	MeOH	217(4.39),227s(4.18), 273(4.28),317(4.02)	39-0215-86C
$C_{16}H_{15}ClO_6$			
Methanone, (3-chloro-2,6-dihydroxy-4-methoxyphenyl)(4-hydroxy-2-methoxy-6-methylphenyl)-	MeOH	208(4.60),294(4.27), 342(3.73)	94-0858-86
$C_{16}H_{15}FN_4$			
1H-Pyrazolo[3,4-b]quinolin-4-amine, N-(4-fluorophenyl)-5,6,7,8-tetrahydro-	EtOH	218(4.88),239s(--), 271(4.05),315(4.65)	5-1728-86
$C_{16}H_{15}FO_2S$			
Benzoic acid, 4-fluoro-2-[[4-(1-methylethyl)phenyl]thio]-	MeOH	257(4.00),306(3.60)	73-0698-86
$C_{16}H_{15}IO_2$			
Benzene, 1-iodo-4-[2-[2-(methoxymethoxy)phenyl]ethenyl]-, (E)-	EtOH	321(4.46)	150-3514-86M
$C_{16}H_{15}N$			
1H-Indole, 1,3-dimethyl-2-phenyl-	n.s.g.	227(4.44),297(3.60)	150-0514-86M
1H-Indole, 1-(1-phenylethyl)-	MeOH	219(4.58),264s(3.81), 274(3.83),281(3.84), 291.5(3.73)	1-0625-86
$C_{16}H_{15}NO$			
Azacyclotetradeca-3,5,11,13-tetraene-7,9-diyn-2-one, 6,11,14-trimethyl-, (E,E,Z,Z)-	THF	298s(4.57),306(4.59), 365s(3.73),436(3.52)	18-1791-86
2,4,10,12-Cyclotridecatetraene-6,8-diyn-1-one, 2,5,10-trimethyl-, oxime, (E,E,E,Z,Z)-	THF	213(4.48),275(4.49), 350s(3.89)	18-1791-86
1(2H)-Naphthalenone, 3,4-dihydro-2-[(1-methyl-1H-pyrrol-2-yl)methylene]-, (E)-	EtOH	225(3.85),265(4.00), 394(4.44)	4-0135-86
$C_{16}H_{15}NOS$			
1,5-Benzothiazepin-4(5H)-one, 2,3-dihydro-5-methyl-2-phenyl-	CH_2Cl_2	248(4.08),271(3.69)	24-3109-86
$C_{16}H_{15}NO_2$			
Formamide, N-[2-oxo-2-phenyl-1-(phenylmethyl)ethyl]-	MeOH	245(4.11),280(4.12)	44-3374-86
Galanthan, 1,2,3,4-tetradehydro-9,10-[methylenebis(oxy)]-	EtOH	235(3.85),278(3.73), 290(3.73)	102-2739-86

Compound	Solvent	λ_{max}(log ε)	Ref.
1H-Indene-1,3(2H)-dione, 2-[5-(dimethylamino)-2,4-pentadienylidene]-	EtOH	250(4.30),540(4.16)	70-1446-86
Phenol, 2-[3-[(4-methylphenyl)imino]-1-propenyl]-, N-oxide	EtOH	280(4.05),315(4.03), 380(3.52)	104-1302-86
	dioxan	270(4.08),312(3.94)	104-1302-86
	DMSO	282(4.18),315(4.17), 370(3.57)	104-1302-86
	DMSO-NaOMe	380(3.94),480(3.64)	104-1302-86
$C_{16}H_{15}NO_3$			
Crinan-3-one, 1,2-didehydro-	n.s.g.	223(4.25),292.5(3.64)	102-2399-86
Galanthan, 3,12-didehydro-1,2-epoxy-9,10-[methylenebis(oxy)]-, (1α,2α)-	EtOH	235s(--),290(3.74)	102-2739-86
$C_{16}H_{15}NO_3S$			
1H-Cyclopenta[cd]phenalen-1-one, 2,2a,3,4-tetrahydro-, O-(methylsulfonyl)oxime	$CHCl_3$	229(4.07),244s(4.34), 252(4.51),260(4.57), 283(3.76),293(3.80), 302(3.62),332(3.48), 348(3.63)	4-0747-86
Dibenzo[c,e]thiepin-5-carboxamide, 5,7-dihydro-N-methyl-, 6,6-dioxide	MeOH	243(4.08)	73-2846-86
[1,3]Dioxepino[5,6-d]isoxazole, 3a,4,8,8a-tetrahydro-3-phenyl-6-(2-thienyl)-, endo	MeOH	236(3.12),264(3.05)	73-2158-86
exo	MeOH	248(3.10),260(3.00)	73-2158-86
$C_{16}H_{15}NO_4$			
9H-Carbazole-1-carboxaldehyde, 4-hydroxy-3,6-dimethoxy-2-methyl-	MeOH	215(4.43),227(4.39), 245(4.11),268(4.11), 310(4.23),382(3.93)	158-0727-86
1,3-Dioxane-4,6-dione, 2,2-dimethyl-5-[(1-methyl-1H-indol-2-yl)methylene]-	EtOH	210(4.36),259(3.81), 421(4.45)	39-1651-86C
1,3-Dioxane-4,6-dione, 2,2-dimethyl-5-[(1-methyl-1H-indol-3-yl)methylene]-	EtOH	217.5(4.39),244.5(3.84), 275(4.06),281.5(3.95), 416(4.54)	39-1651-86C
1,3-Dioxane-4,6-dione, 2,2-dimethyl-5-[(2-methyl-1H-indol-3-yl)methylene]-	EtOH	218s(5.00),260(4.46), 417(4.66)	39-1651-86C
[1,3]Dioxepino[5,6-d]isoxazole, 6-(2-furanyl)-3a,4,8,8a-tetrahydro-3-phenyl-, endo	MeOH	265(3.03)	73-2158-86
exo	MeOH	260(3.00)	73-2158-86
2H-Pyran-2,6(3H)-dione, 5-acetyl-4-methyl-3-[1-(phenylamino)ethylidene]-, (Z)-	EtOH	280(3.96),355(4.32)	39-0973-86B
	dioxan	376(4.45)	39-0973-86B
	$CHCl_3$	375(4.45)	39-0973-86B
	C_6H_{12}	375(--)	39-0973-86B
$C_{16}H_{15}NO_5$			
3,4-Pyridinedicarboxylic acid, 1,6-dihydro-5-methyl-6-oxo-1-phenyl-, dimethyl ester	MeCN	201(4.36),268(4.16), 304(3.75)	24-1315-86
$C_{16}H_{15}N_2O$			
Pyridinium, 1-ethyl-4-(5-phenyl-2-oxazolyl)-, salt with 4-methylbenzenesulfonic acid	H_2O	375(4.37)	135-0244-86
	EtOH	250(4.25),375(4.50)	135-0244-86
$C_{16}H_{15}N_2O_2$			
Pyridinium, 4-[5-(4-methoxyphenyl)-	H_2O	394(4.25)	135-0244-86

Compound	Solvent	λ_{max}(log ε)	Ref.
2-oxazolyl]-1-methyl-, salt with 4-methylbenzenesulfonic acid (cont.)	EtOH	260(4.23),410(4.35)	135-0244-86
$C_{16}H_{15}N_3$			
2-Quinolinamine, N,N-dimethyl-4-(2-pyridinyl)-	MeOH	214(4.47),253(4.49), 365(3.78)	83-0338-86
$C_{16}H_{15}N_3O$			
1H-Benzo[h]pyrimido[1,2-c]quinazolin-1-one, 2,3,5,6-tetrahydro-3-methyl-	EtOH	228(3.79),234(3.75), 268(4.23),280(4.02), 310(3.65),325(3.82), 342(4.03),354(4.12), 370(3.97)	142-1119-86
2-Benzoxazolamine, N-[[4-(dimethyl-amino)phenyl]methylene]-	dioxan	405(4.65)	80-0273-86
	+ CF_3COOH	469(--)	80-0273-86
$C_{16}H_{15}N_3OS$			
Morpholine, 4-(1H-perimidin-2-yl-thioxomethyl)-	DMF	295(4.11),336(4.14), 349s(4.07),391s(3.16), 411s(3.05),437s(2.87)	48-0812-86
	C_6H_{12}	231s(--),236(--), 253s(--),327s(--), 337(--),352s(--), 412(--),468s(--)	48-0812-86
$C_{16}H_{15}N_3O_2$			
Benzenamine, 2-[5-(4-methoxyphenyl)-1,3,4-oxadiazol-2-yl]-N-methyl-	MeOH	212(4.56),235(4.67), 260(4.89),280(4.68), 288(4.77),300(4.76)	18-1575-86
Benzoic acid, 4-[2-(6-amino-5-cyano-3-pyridinyl)ethyl]-, methyl ester	EtOH	248(4.42),336(3.75)	4-0001-86
$C_{16}H_{15}N_3O_3$			
2-Butenamide, 2-nitro-N-phenyl-3-(phenylamino)-	EtOH	248(4.16),355(4.10)	18-3871-86
$C_{16}H_{15}N_3O_3Se$			
Furo[3',4':5,6]selenino[2,3-c]pyra-zole-5,7-dione, 1,4,4a,7a-tetra-hydro-1,3-dimethyl-4-(phenylamino)-	EtOH	246s(--)	104-2126-86
$C_{16}H_{15}N_3O_4S$			
Pyrrolidine, 1-[4-[(2,4-dinitrophen-yl)thio]phenyl]-	MeOH	280(4.45),317s(4.05), 342(4.06)	73-0937-86
$C_{16}H_{15}N_3O_5$			
2-Propenoic acid, 3,3'-[5-(dimethyl-amino)-2,4-furandiyl]bis[2-cyano-, dimethyl ester	MeOH	258(2.90),380(3.09), 482(3.23)	73-0879-86
$C_{16}H_{15}N_3S$			
2-Benzothiazolamine, N-[[4-(dimethyl-amino)phenyl]methylene]-	dioxan	404(4.61)	80-0273-86
	+CF_3COOH	480(--)	80-0273-86
Pyrrolidine, 1-(1H-perimidin-2-yl-thioxomethyl)-	DMF	291(4.00),328(4.11), 351s(4.08),389s(3.15), 409s(3.05),437s(2.90)	48-0812-86
	C_6H_{12}	231s(--),236(--), 254s(--),277s(--), 328s(--),336(--), 353(--),424(--), 470s(--),492(--),	48-0812-86

Compound	Solvent	λ_{max}(log ϵ)	Ref.
(cont.)		522s(--),570s(--)	48-0812-86
$C_{16}H_{15}N_5O_2$			
Benzoic acid, 4-[2-(2,4-diaminopyrido[2,3-d]pyrimidin-6-yl)ethyl]-	pH 13	245(4.41),344(3.76)	4-0001-86
$C_{16}H_{16}$			
Cyclopenta[ef]heptalene, 2a,10b-dihydro-2a,10b-dimethyl-	hexane	234(4.22),272.5(4.16), 282.5(4.19),438(3.31)	35-4105-86
Cyclopenta[ef]heptalene, 6,10b-dihydro-6,10b-dimethyl-	hexane	247(4.09),255(4.11), 276(4.17),420(3.37)	35-4105-86
Cyclopenta[ef]heptalene, 6a,10b-dihydro-6a,10b-dimethyl-	hexane	232(3.93),290(3.92), 300(3.87),475(3.31)	35-4105-86
1H-Cyclopropa[b]naphthalene, 1-(2,2-dimethylpropylidene)-	C_6H_{12}	231(4.50),273(4.17), 335s(3.73),351(4.02), 361(3.95),371(4.27)	35-5949-86
	MeCN	230(4.59),272(4.24), 332s(3.82),349(4.09), 359(4.04),369(4.32)	35-5949-86
1,6:7,12-Dimethano[14]annulene, dihydro-, anti	C_6H_{12}	243(4.00),285(3.85)	89-0723-86
syn	C_6H_{12}	258(4.17),352(3.91)	89-0723-86
Tricyclo[8.2.2.2^{4,7}]hexadeca-4,6,10-12,13,15-hexaene ([2.2]paracyclophane)	isooctane	225(4.3),244s(3.5), 286(2.4),305s(2.1)	46-1541-86
	5N H_2SO_4	220s(4.0),268(3.7)	46-1541-86
$C_{16}H_{16}BrN_5O_4$			
Benzenamine, 4-[(2-bromo-4,6-dinitrophenyl)azo]-N,N-diethyl-	$CHCl_3$	296(4.00),374(3.76), 552(4.44)	48-0497-86
$C_{16}H_{16}BrN_5O_5$			
7H-Pyrrolo[2,3-d]pyrimidine-5-carbonitrile, 4-amino-6-bromo-7-(3,4-di-O-acetyl-2-deoxy-β-D-erythro-pentofuranosyl)-	EtOH	220(4.21),286(4.18)	78-5869-86
$C_{16}H_{16}Br_2$			
Bi-1,3,5-cycloheptatrien-1-yl, 6,6'-bis(bromomethyl)-	dioxan	238(4.59),282(3.75), 360(4.08)	89-0723-86
$C_{16}H_{16}ClNO_3$			
Galanthan-1-ol, 2-chloro-3,12-didehydro-9,10-[methylenebis(oxy)]-, (1α,2β)-	EtOH	280(3.60),290(3.54)	102-2739-86
$C_{16}H_{16}ClN_3O$			
Morpholine, 4-[4-[(4-chlorophenyl)-azo]phenyl]-	EtOH	396(4.40)	39-0123-86B
	EtOH-HCl	530(4.23)	39-0123-86B
$C_{16}H_{16}ClN_3O_2S$			
Thiomorpholine, 4-[4-[(4-chlorophenyl)azo]phenyl]-, 1,1-dioxide	EtOH	388(4.40)	39-0123-86B
	EtOH-HCl	535(4.82)	39-0123-86B
$C_{16}H_{16}ClN_3S$			
Thiomorpholine, 4-[4-[(4-chlorophenyl)azo]phenyl]-	EtOH	407(4.40)	39-0123-86B
	EtOH-HCl	534(4.53)	39-0123-86B
$C_{16}H_{16}NO_2$			
Methanaminium, N-(7-methoxy-2H-naphth[1,8-bc]oxepin-2-ylidene)-N-methyl-, perchlorate	MeCN	238(4.59),295(4.31), 450(4.15)	104-1342-86

Compound	Solvent	$\lambda_{max}(\log \epsilon)$	Ref.
$C_{16}H_{16}N_2O$			
Quinoxaline, 1,2-dihydro-3-(4-ethoxyphenyl)-	MeOH	279(4.35),306(4.19), 388(3.81)	103-0918-86
	toluene	385(--)	103-0918-86
$C_{16}H_{16}N_2O_2$			
Benzaldehyde, 3-[[(4,6-dimethyl-2-pyridinyl)imino]methyl]-2-hydroxy-5-methyl-	$CHCl_3$	488(2.94)	2-0127-86A
$C_{16}H_{16}N_2O_3$			
Acetamide, N-acetyl-N-[1-(1-acetyl-1H-indol-3-yl)ethenyl]-	MeOH	220(4.19),240(4.17), 282(3.90),297(3.99), 306(3.99)	5-2065-86
2H-1-Benzopyran-4-carbonitrile, 3-acetyl-7-(diethylamino)-2-oxo-	MeOH	490(3.95)	2-0137-86
$C_{16}H_{16}N_2O_3S$			
Methanone, (2,5-dihydro-1-hydroxy-2,5-dimethyl-4-phenyl-1H-imidazol-2-yl)-2-thienyl-, N-oxide	EtOH	227(4.05),300(4.43)	103-0268-86
Methanone, [2,5-dihydro-1-hydroxy-2,5-dimethyl-4-(2-thienyl)-1H-imidazol-2-yl]phenyl-, N-oxide	EtOH	225(4.89),250(4.96), 320(5.02)	103-0268-86
$C_{16}H_{16}N_2O_4$			
Methanone, (2,5-dihydro-1-hydroxy-2,5-dimethyl-4-phenyl-1H-imidazol-2-yl)-2-furanyl-, N-oxide	EtOH	226(4.05),292(4.39)	103-0268-86
Methanone, [4-(2-furanyl)-2,5-dihydro-1-hydroxy-2,5-dimethyl-1H-imidazol-2-yl]phenyl-, N-oxide	EtOH	247(4.14),308(4.34)	103-0268-86
$C_{16}H_{16}N_2O_4S$			
4-Thiazolidinecarboxylic acid, 2-[2,5-dihydro-5-oxo-2-(phenylmethylene)-4-oxazolyl]-5,5-dimethyl-, (2R-trans)-	MeOH	242(3.85),359(4.33), 377(4.28)	44-3232-86
$C_{16}H_{16}N_2O_7$			
[1,1'-Biphenyl]-2,3'-diacetic acid, α,α'-diamino-4,6,6'-trihydroxy- (actinoidinic acid)	n.s.g.	220s(3.37),287(2.70)	44-4272-86
$C_{16}H_{16}N_2S$			
Acetonitrile, [(phenylmethyl)[(phenylthio)methyl]amino]-	EtOH	251(3.89)	24-0813-86
2H-Benz[g]indazole, 3,3a,4,5-tetrahydro-2-methyl-3-(2-thienyl)-	EtOH	237(3.99),311(4.21)	4-0135-86
$C_{16}H_{16}N_2S_4$			
4H-Imidazole, 2,5-bis(ethylthio)-4-(4-phenyl-1,3-dithiol-2-ylidene)-	MeCN	220(4.21),338(4.19), 376(4.03),535(4.25)	78-0839-86
$C_{16}H_{16}N_3$			
Pyridinium, 1-[1-cyano-2-[4-(dimethylamino)phenyl]ethenyl]-, (E)-, tetrafluoroborate	MeOH	262(4.18),442(4.47)	78-0699-86
$C_{16}H_{16}N_3O_2$			
Pyrylium, 2-amino-3-cyano-4-morpho-	MeCN	225(3.91),259(4.50),	48-0314-86

Compound	Solvent	$\lambda_{max}(\log \epsilon)$	Ref.
lino-6-phenyl-, perchlorate (cont.)		318(4.43)	48-0314-86
$C_{16}H_{16}N_4$			
1H-Benzimidazol-2-amine, N-[[4-(dimethylamino)phenyl]methylene]-	dioxan	392(4.64)	80-0273-86
	+ CF_3COOH	392(--)	80-0273-86
1H-Pyrazolo[3,4-b]quinolin-4-amine, 5,6,7,8-tetrahydro-N-phenyl-	EtOH	214s(--),242s(--), 261(3.71),318(4.26)	5-1728-86
Pyrido[3,4-d]pyrimidin-4-amine, 2-methyl-N-(2-phenylethyl)-	pH 1	284(3.86),324.5(4.04), 338(3.96)	163-2243-86
	pH 7	284.5(3.96),332(3.90)	163-2243-86
	pH 13	284.5(3.96),332(3.90)	163-2243-86
$C_{16}H_{16}N_4O$			
Benzo[f]quinoxaline-2-carbonitrile, 3-amino-4,6-dihydro-4-(2-methoxyethoxy)-	EtOH	443(3.80)	33-1025-86
Propanedinitrile, [[1-methyl-2-(4-morpholinyl)-2-phenylethenyl]imino]-	EtOH	460(4.53)	33-1025-86
$C_{16}H_{16}N_4O_2S$			
Thiomorpholine, 4-[4-[(4-nitrophenyl)azo]phenyl]-	EtOH	456(4.45)	39-0123-86B
	EtOH-HCl	526(4.83)	39-0123-86B
$C_{16}H_{16}N_4O_3$			
1H-Imidazo[4,5-c]pyridinium, 5-[(ethoxycarbonyl)amino]-2,3-dihydro-2-oxo-1-(phenylmethyl)-, hydroxide, inner salt	MeOH	304(3.98)	78-1511-86
Morpholine, 4-[4-[(4-nitrophenyl)azo]phenyl]-	EtOH	437(4.44)	39-0123-86B
	EtOH-HCl	520(4.68)	39-0123-86B
3(2H)-Pyridazinone, 6-[2-(3,4-dimethoxyphenyl)-1-methyl-1H-imidazol-4-yl]-	MeOH	208(4.52),271(4.44)	87-0261-86
$C_{16}H_{16}N_4O_4$			
Pyrido[2,1-i]purine-9-carboxylic acid, 3,7-dihydro-7-oxo-3-(tetrahydro-2H-pyran-2-yl)-, methyl ester	CH_2Cl_2	266(3.9),335(3.4), 354(3.4),432(3.9), 457(3.9)	39-1561-86B
$C_{16}H_{16}N_4O_4S$			
Thiomorpholine, 4-[4-[(4-nitrophenyl)azo]phenyl]-, 1,1-dioxide	EtOH	422(4.41)	39-0123-86B
	EtOH-HCl	547(4.86)	39-0123-86B
$C_{16}H_{16}N_4O_4S_2$			
Benzeneacetic acid, 2-[[(4-methylphenyl)sulfonyl]amino]-α-oxo-, 2-(nitrothioxomethyl)hydrazide	EtOH	267s(4.19),340(3.63)	104-2173-86
$C_{16}H_{16}N_4O_5$			
2,3'-(Aminoimino)-1-(5'-O-benzoyl-3'-deoxy-β-D-lyxofuranosyl)uracil	MeOH	201(4.24),224(4.50), 262s(3.66)	44-4417-86
$C_{16}H_{16}N_6O_2S$			
Cyclopropenylium, [[2-(cyanoamino)-3-(cyanoimino)-1-cyclopropen-1-yl]thio]di-4-morpholinyl-, hydroxide, inner salt	H_2O	234(4.46),294(4.10)	24-2104-86
$C_{16}H_{16}O$			
Methanone, [4-(1-methylethyl)phenyl]-	$CHCl_3$	260(4.37),336(2.33)	32-0533-86

Compound	Solvent	$\lambda_{max}(\log \epsilon)$	Ref.
phenyl- (cont.)			32-0533-86
Tetracyclo[9.2.1.0^{3,10}.0^{4,9}]tetradec-3-en-2-one, 15,16-dimethyl-	EtOH	253(3.84)	78-3491-86
$C_{16}H_{16}O_2$			
2-Naphthalenecarboxylic acid, 3-methyl-2-butenyl ester	MeCN	236(4.79)	130-0046-86
Naphtho[1,8-bc]pyran, 7-methoxy-3,6,9-trimethyl-	EtOH	206(4.43),239(4.40), 253s(4.21),297(3.69), 310(3.75),353(3.97), 363(3.97)	12-0647-86
Pyrene-4,5-diol, 1,2,3,3a,4,5-hexahydro-, trans	n.s.g.	232(1.00),287(0.07)(rel. abs.)	44-1773-86
Tricyclo[9.3.1.1^{4,8}]hexadeca-1(15),4-6,8(16),11,13-hexaene-15,16-diol	dioxan	234(4.47),255s(4.07), 315s(2.76)	89-1000-86
$C_{16}H_{16}O_3$			
1(2H)-Acenaphthylenone, 2,2-diethoxy-	$CHCl_3$	316(3.69),336(3.67)	39-1965-86C
9,10-Anthracenediol, 1,2,3,4-tetrahydro-, 9-acetate	MeOH	238(4.58),270(3.62), 303(3.57),313s(3.52), 328s(3.38)	12-2067-86
	MeOH-KOH	253(4.38),273(3.81), 342(3.70)	12-2067-86
9(2H)-Anthracenone, 9a-acetoxy-1,3,4,9a-tetrahydro-	MeOH	237(4.43),268(3.80), 304s(3.53),328s(3.28)	12-2067-86
2H-1-Benzopyran-7-ol, 3,4-dihydro-2-(4-methoxyphenyl)-, (S)-	MeOH	227s(4.08),277(3.04)	102-1097-86
2H-1-Benzopyran-7-ol, 3,4-dihydro-5-methoxy-2-phenyl-, (S)-	MeOH	265s(2.85),271(2.90), 285.5s(2.76)	163-1655-86
	base	279.5(--),285s(--)	163-1655-86
9a,10-Cycloanthracen-9(9aH)-one, 4a-acetoxy-1,2,3,4,4a,10-hexahydro-, endo	MeOH	237(4.17),259s(3.75), 298(3.32),328s(2.99), 348s(2.56)	39-1789-86C
Methanone, (4-ethoxyphenyl)(4-methoxyphenyl)-	EtOH	225(4.18),295(4.34)	13-0073-86B
Phenol, 4-[[2-(methoxymethoxy)phenyl]-ethenyl]-, (E)-	EtOH	323(4.47)	150-3514-86M
1-Propanone, 3-hydroxy-2-methoxy-1,2-diphenyl-	THF	249(4.00),343(2.18)	121-0355-86
$C_{16}H_{16}O_4$			
2,4,6-Cycloheptatrien-1-one, 2-hydroxy-5-methoxy-3-[(4-methoxyphenyl)methyl]-	MeOH	227(4.39),240s(4.33), 251s(4.17),276(3.62), 330(3.86),371(3.79), 387(3.79)	18-1125-86
2,4,6-Cycloheptatrien-1-one, 5-methoxy-2-[(4-methoxyphenyl)methoxy]-	MeOH	229(4.18),252s(3.62), 274(3.11),282(3.18), 327(3.73),332(3.73), 348(3.69),362s(3.62), 378s(3.49)	18-1125-86
1H-Dibenzo[b,d]pyran-6-carboxylic acid, 2,3,4,6-tetrahydro-4-oxo-, ethyl ester	ether	243(3.85),250(3.88), 320(3.85)	44-4424-86
1-Propanone, 1-(2,6-dihydroxy-4-methoxyphenyl)-3-phenyl-	EtOH	208(3.30),226(3.16), 285(3.25)	102-1427-86
1-Propanone, 3-hydroxy-1-(2-hydroxy-4-methoxyphenyl)-2-phenyl-	MeOH	216(4.25),224s(4.10), 272(4.18),312(4.60)	39-0215-86C
$C_{16}H_{16}O_5$			
Benzoic acid, 2-[[3-(5-oxo-1,3-pentadienyl)oxiranyl]methoxy]-, methyl ester	n.s.g.	274(4.43)	104-1597-86

Compound	Solvent	λ_{max}(log ε)	Ref.
2,5-Furandione, 3-[(2,5-dimethoxyphenyl)methylene]dihydro-4-(1-methylethylidene)-, (E)-	EtOH	316(3.90),386(3.86)	39-1599-86C
(Z)-	EtOH	328(4.11),392(4.08)	39-1599-86C
Hongconin	EtOH	221(3.487),267(3.480), 430(2.633)	94-2743-86
1H-Naphtho[2,3-c]pyran-5,10-dione, 3,4-dihydro-9-hydroxy-3-(2-hydroxyethyl)-1-methyl-, (1S-trans)-	MeOH	248(3.70),273(3.80), 422(3.48)	158-1343-86
$C_{16}H_{16}O_6$			
2H-1-Benzopyran-3,4,7-triol, 3-[(3,4-dihydroxyphenyl)methyl]-3,4-dihydro-	MeOH	205(4.80),219(4.26), 280(3.81),285(3.79)	94-2506-86
6H-Dibenz[b,d]oxocin-3,7,10,11-tetrol, 7,8-dihydro-7-(hydroxymethyl)-(protosappanin B)	EtOH	255(4.08),288(3.86)	142-0601-86
$C_{16}H_{16}O_9$			
7-Oxabicyclo[2.2.1]hepta-2,5-diene-2,3-dicarboxylic acid, 1-[[(4-methoxy-1,4-dioxo-2-butenyl)oxy]methyl]-, dimethyl ester, (E)-	MeCN	281(3.15)	24-0589-86
$C_{16}H_{17}BrN_2O$			
Benzo[f]quinoxalin-6-one, 5-bromo-2,3,4,6-tetrahydro-2,2,3,3-tetramethyl-	CH_2Cl_2	238(4.19),279(4.27), 425(3.56)	83-0052-86
$C_{16}H_{17}ClN_2OS_2$			
Benzo[b]thiophene-2-carboxamide, 3-chloro-N-[(cyclohexylamino)thioxomethyl]-	EtOH	207(3.08),250(3.07), 306(3.08)	73-2839-86
$C_{16}H_{17}ClN_2O_6S$			
1H-Imidazole-4-carboxylic acid, 2-chloro-5-(phenylthio)-1-β-D-ribofuranosyl-, methyl ester	H_2O	243(4.14),297(3.00)	78-1971-86
$C_{16}H_{17}ClN_3$			
Benz[h]imidazo[1,2-c]quinazolinium, 3-(2-chloroethyl)-1,2,4,5-tetrahydro-, chloride	EtOH	262(4.01),285(3.86), 347(3.53)	4-0685-86
Phenazinium, 5-(2-chloroethyl)-3-(dimethylamino)-, chloride	EtOH	238(4.56),297(4.50), 385(4.00),398(4.02), 560(4.20)	103-1342-86
$C_{16}H_{17}N$			
Benzenamine, N-[(2,4,6-trimethylphenyl)methylene]-	C_6H_{12}	210(4.32),273(4.24), 326(3.80)	165-0017-86
Benzenamine, 2,4,6-trimethyl-N-(phenylmethylene)-	MeOH	213(4.29),258(4.31), 322(3.57)	165-0017-86
$C_{16}H_{17}NO$			
Phenol, 2-[2-[4-(dimethylamino)phenyl]ethenyl]-, (E)-	n.s.g.	350(4.57)	150-0433-86S
$C_{16}H_{17}NOS$			
Sulfoximine, S-(4-methylphenyl)-S-phenyl-N-2-propenyl-, (R)-	EtOH	245(4.01)	12-1655-86

Compound	Solvent	$\lambda_{max}(\log \epsilon)$	Ref.
$C_{16}H_{17}NO_2$			
Lycorene	EtOH	237s(--),290(3.67)	102-2739-86
$C_{16}H_{17}NO_3$			
3H,6H-5,10b-Ethanophenanthridin-3-one, 4,4a-dihydro-8-hydroxy-9-methoxy- (noroxomaritidine)	MeOH	240s(3.92),290(3.98), 300s(3.68)	150-0028-86S
Galanthan-1-ol, 3,12-didehydro- 9,10-[methylenebis(oxy)]- (caranine)	EtOH	237(3.49),280(3.49), 291(3.44)	102-2739-86
Phenol, 5-[[[2-(4-hydroxyphenyl)- ethyl]imino]methyl]-2-methoxy- (craugsodine)	MeOH	227(4.30),278(3.92), 309(3.68)	150-0028-86S
$C_{16}H_{17}NO_3S_2$			
[1]Benzothiepino[4,5-e]-1,2-oxathiin, 5,6-dihydro-4-(1-pyrrolidinyl)-, 2,2-dioxide	EtOH	226(4.12),235s(4.07), 275s(3.95),295(4.00)	4-0455-86
$C_{16}H_{17}NO_4$			
Flexinine	n.s.g.	203(4.66),242(3.89), 295(3.93)	102-2399-86
$C_{16}H_{17}NO_4S_2$			
[1]Benzothiepino[4,5-e]-1,2-oxathiin, 5,6-dihydro-4-(4-morpholinyl)-, 2,2-dioxide	EtOH	226(4.06),233s(4.01), 254s(3.74),278s(3.90), 300(4.03)	4-0455-86
$C_{16}H_{17}NO_5$			
Galanthan-1,2,3-triol, 4,12-didehydro- 9,10-[methylenebis(oxy)]- (pan-crassidine)	MeOH	228(3.96),284(3.53)	102-1097-86
$C_{16}H_{17}N_3$			
Benz[h]imidazo[1,2-c]quinazoline, 1-ethyl-1,2,4,5-tetrahydro-	EtOH	235(3.70),262(4.15), 271(4.14),325(3.17), 340(3.26)	4-0685-86
Benz[h]imidazo[1,2-c]quinazoline, 2-ethyl-1,2,4,5-tetrahydro-	EtOH	263(4.26),270(4.26), 340(3.27)	4-0685-86
2H-Benz[g]indazole, 3,3a,4,5-tetra-hydro-3-(1-methyl-1H-pyrrol-2-yl)-	EtOH	225(3.80),295(3.70)	4-0135-86
5,10-Ethanophenazin-2-amine, N,N-di-methyl-	EtOH	213(4.36),272(4.15)	103-1342-86
1H-1,2,4-Triazole, 4,5-dihydro-5,5- dimethyl-1,3-diphenyl-	EtOH	228(4.29),263s(3.94), 350(4.08)	103-0287-86
$C_{16}H_{17}N_3O$			
Benzoic acid, 2-(methylamino)-, [(4-methylphenyl)methylene]hy-drazide	MeOH	217(4.25),258(4.31), 349(4.19)	18-1575-86
Morpholine, 4-[4-(phenylazo)phenyl]-	EtOH	386(4.38)	39-0123-86B
	EtOH-HCl	525(4.37)	39-0123-86B
$C_{16}H_{17}N_3O_2$			
1,4-Benzenediamine, N,N-dimethyl- N'-[(3-methyl-4-nitrophenyl)- methylene]-	THF	272(4.25),434(4.27)	56-0831-86
1,4-Benzenediamine, N^4,N^4,2-trimethyl- N^1-[(4-nitrophenyl)methylene]-	THF	277(4.27),401(4.12)	56-0831-86
Benzoic acid, 2-(methylamino)-, [(4- methoxyphenyl)methylene]hydrazide	MeOH	212(4.21),224(4.21), 263(4.39),358(4.42)	18-1575-86

Compound	Solvent	λ_{max}(log ε)	Ref.
Benzo[g]phthalazin-1(2H)-one, 4-[(3-methoxypropyl)amino]-	EtOH	245(4.68),253(4.96)	111-0143-86
$C_{16}H_{17}N_3O_3$			
1H-Imidazole-4-butanenitrile, 2-(3,4-dimethoxyphenyl)-1-methyl-γ-oxo-	MeOH	263(4.08)	87-0261-86
$C_{16}H_{17}N_3O_4$			
1H-Benz[de]isoquinoline-1,3(2H)-dione, 5-amino-2-(2-hydroxyethyl)-6-[(2-hydroxyethyl)amino]-	pH 7.4	361(3.80),440(3.87)	4-0849-86
$C_{16}H_{17}N_3O_4S$			
5-Thia-1-azabicyclo[4.2.0]oct-2-ene-2-carboxylic acid, 7-[(aminophenylacetyl)amino]-3-methyl-8-oxo-, [6R-[6α,7β(R*)]-	H_2O	261(3.89)	107-1469-86
$C_{16}H_{17}N_3O_5$			
2,3'-[(4-Methoxyphenyl)imino]-1-(3'-deoxy-β-D-lyxopyranosyl)-	MeOH	214.5(4.24),227(4.21)	44-4417-86
$C_{16}H_{17}N_3S$			
Thiomorpholine, 4-[4-(phenylazo)-phenyl]-	EtOH	395(4.37)	39-0123-86B
	EtOH-HCl	531(4.53)	39-0123-86B
$C_{16}H_{17}N_5O_3S$			
Ethanol, 2-[[6-amino-2-(methylthio)-9H-purin-9-yl]methoxy]-, benzoate	10% EtOH	276(4.13)	4-0271-86
	+ HCl	270(4.18),300s(3.85)	4-0271-86
	+ NaOH	276(4.17)	4-0271-86
$C_{16}H_{17}N_5O_5$			
7H-Pyrrolo[2,3-d]pyrimidine-5-carbonitrile, 4-amino-7-(3,5-di-O-acetyl-2-deoxy-β-D-erythro-pentofuranosyl)-	EtOH	229s(4.18),278(4.29)	78-5869-86
$C_{16}H_{17}N_5O_6S$			
Uridine, 3'-azido-2',3'-dideoxy-, 5'-[(4-methylphenyl)sulfonate]	EtOH	261(4.01)	87-0681-86
$C_{16}H_{17}N_5O_6S_2$			
Epialthiomycin	MeOH	223(4.54),240(4.42), 286(3.93)	18-2185-86
$C_{16}H_{17}N_5O_7$			
D-Glucofuranuronic acid, 1-deoxy-1-[6-(methylamino)-9H-purin-9-yl]-, γ-lactone, 2,5-diacetate	MeOH	265(4.24)	103-0085-86
$C_{16}H_{17}N_5O_8S$			
5-Thia-1-azabicyclo[4.2.0]oct-2-ene-2-carboxylic acid, 3-(acetoxymethyl)-7-[[(2-amino-4-oxazolyl)(methoxyimino)acetyl]amino]-8-oxo-, (Z)-	MeOH	217(4.28),265(4.01)	158-0121-86
$C_{16}H_{18}$			
1,4-p-Benzenonaphthalene, 1,2,3,4,9-10,11,12-octahydro-, (4RS,12RS)-	EtOH	204(4.42),261(2.56), 267(2.70),275(2.66)	39-0885-86C
1,4-p-Benzenonaphthalene, 1,4,9,10,11-12,13,14-octahydro-, (4RS,12RS)-	EtOH	207(4.28),267(2.73), 273(2.67)	39-0885-86C

Compound	Solvent	$\lambda_{max}(\log \epsilon)$	Ref.
Cyclopenta[ef]heptalene, 3,4,5,6-tetrahydro-8,10-dimethyl-	hexane	216(4.09),247(4.40), 290(4.69),295(4.67), 307s(4.08),327s(3.33), 341(3.57),351s(3.62), 357(3.71),368s(3.00), 522s(2.37),545s(2.52), 567s(2.61),582(2.65), 603(2.61),631(2.58), 663s(2.30),695(2.14)	89-0633-86
Heptalene, 1,3,5,6-tetramethyl-	hexane	225(4.36),317s(3.57)	88-1669-86
Heptalene, 1,5,6,10-tetramethyl-	hexane	251(4.34),295s(3.47)	88-1669-86
$C_{16}H_{18}BrN$			
Pyridine, 2-(4-bromophenyl)-5-pentyl-	EtOH	207(4.58),257(4.53), 285(4.49)	104-0957-86
$C_{16}H_{18}ClN_3$			
Benzo[h]quinolizin-4-amine, N-[1-(chloromethyl)propyl]-5,6-dihydro-, monohydrochloride, (±)-	EtOH	249(4.01),298(3.71), 312(3.77),322(3.77)	4-0685-86
$C_{16}H_{18}DN_5O_7$			
Acetamide, N-[9-(3,5-di-O-acetyl-2-deoxy-β-D-arabinofuranosyl-2-C-d)-6,9-dihydro-6-oxo-1H-purin-2-yl]-(S)-	EtOH-pH 7	256(4.15)	78-5427-86
(R)-	EtOH-pH 7	256(4.20)	78-5427-86
$C_{16}H_{18}FN_3O_5S_2$			
1,2,4-Triazin-3(2H)-one, 6-[[(2-fluorophenyl)methyl]thio]-5-(methylthio)-2-β-D-ribofuranosyl-	EtOH	312(3.89)	80-0881-86
1,2,4-Triazin-3(2H)-one, 6-[[(3-fluorophenyl)methyl]thio]-5-(methylthio)-2-β-D-ribofuranosyl-	EtOH	312(3.87)	80-0881-86
1,2,4-Triazin-3(2H)-one, 6-[[(4-fluorophenyl)methyl]thio]-5-(methylthio)-2-β-D-ribofuranosyl-	EtOH	312(3.89)	80-0881-86
$C_{16}H_{18}INO$			
Bicyclo[4.4.1]undeca-1,3,5,7,9-pentaene-2-carboxamide, N,N-diethyl-9-iodo-	MeOH	260(4.63),320(3.92)	33-1851-86
Bicyclo[4.4.1]undeca-1,3,5,7,9-pentaene-2-carboxamide, N,N-diethyl-10-iodo-	MeOH	255(4.50),320(3.88), 390(2.93)	33-1263-86
$C_{16}H_{18}N_2$			
7,10-Methanophenanthridine-2-carbonitrile, 5,6,6a,7,8,9,10,10a-octahydro-5-methyl-	n.s.g.	306(4.00)	88-5967-86
Pyrazine, 2,5-diethyl-3-(2-phenylethenyl)-, (E)-	EtOH	231(3.97),237(3.95), 276(4.17),337(4.28)	4-1481-86
$C_{16}H_{18}N_2O$			
Pyrazine, 2,5-diethyl-3-(2-phenylethenyl)-, 1-oxide, (E)-	EtOH	231(4.24),237s(4.23), 275(4.59),300s(4.29), 347(4.36),359s(4.32)	4-1481-86
$C_{16}H_{18}N_2O_2S$			
Acetamide, N-[(3-oxobenzo[b]thien-	dodecane	305(4.15),425(3.85)	104-2108-86

Compound	Solvent	$\lambda_{max}(\log \epsilon)$	Ref.
2(3H)-ylidene)methyl]-N-(1-piperi-dinyl)- (cont.)			104-2108-86
Benzo[b]thiophene-3-ol, 2-[(1-piperi-dinylimino)methyl]-, acetate	dodecane	330(4.31)	104-2108-86
$C_{16}H_{18}N_2O_3$			
2-Pentenoic acid, 5-oxo-5-phenyl-4-(tetrahydro-2(1H)-pyrimidin-ylidene)-, methyl ester, (E)-	EtOH	236(4.11),342(4.20)	142-2247-86
$C_{16}H_{18}N_2O_4$			
3-Pyridinecarboxylic acid, 5-(2,3-di-hydro-3-hydroxy-2-oxo-1H-indol-3-yl)-1,2,5,6-tetrahydro-1-methyl-, methyl ester, (R*,R*)-(±)-	pH 7.2	216(1.34)	87-0125-86
$C_{16}H_{18}N_2O_5$			
1H-Imidazole-4-butanoic acid, 2-(3,4-dimethoxyphenyl)-1-methyl-γ-oxo-	MeOH	246(4.08),294(4.26)	87-0261-86
2-Quinazolinecarboxylic acid, 4-(2-ethoxy-1-methyl-2-oxoethoxy)-, ethyl ester	MeOH	205(4.36),228(4.47), 283(3.86)	33-1017-86
L-Tryptophan, N-acetyl-1-(1-hydroxy-2-oxopropyl)-	MeOH	274(3.77),280(3.77)	33-1017-86
$C_{16}H_{18}N_2O_6$			
Uridine, 3-(phenylmethyl)-	pH 1 and 6	261(3.88)	18-2947-86
	pH 11	261(3.88)	18-2947-86
$C_{16}H_{18}N_2O_7$			
Uridine, 2'-deoxy-5-ethyl-, 5'-(2-furancarboxylate)	EtOH	258(4.29)	83-0154-86
$C_{16}H_{18}N_2S$			
3-Pyridinecarbonitrile, 1,2-dihydro-2-thioxo-6-tricyclo[3.3.1.1^{3,7}]dec-1-yl-	EtOH	244(3.66),309(4.03), 408(3.52)	70-0131-86
$C_{16}H_{18}N_2Se$			
3-Pyridinecarbonitrile, 1,2-dihydro-2-selenoxo-6-tricyclo[3.3.1.1^{3,7}]-dec-1-yl-	EtOH	222(4.11),335(3.75), 438(3.08)	70-0376-86
$C_{16}H_{18}N_3O$			
Phenazinium, 3-(dimethylamino)-5-(2-hydroxyethyl)-, chloride	EtOH	239(4.49),299(4.52), 385(3.98),398(4.02), 560(4.20)	103-1342-86
$C_{16}H_{18}N_4$			
1H-Pyrazole, 3,5-dimethyl-1-[(4-meth-yl-5-phenyl-1H-pyrazol-1-yl)methyl]-	EtOH	251(4.04)	142-2233-86
$C_{16}H_{18}N_4O$			
Acetamide, N-[4-[[4-(dimethylamino)-phenyl]azo]phenyl]-	DMF	268.9(4.03),428.3(4.44)	56-0797-86
with 0.02M p-toluenesulfonic acid	DMF	269.4(4.20),426.8(4.45)	56-0797-86
with 2.8M p-toluenesulfonic acid	DMF	555.5(4.45)	56-0797-86
Bicyclo[4.4.1]undeca-1,3,5,7,9-penta-ene-2-carboxamide, 10-azido-N,N-di-ethyl-	MeOH	210(4.19),240(4.05), 320(3.30)	33-1263-86

Compound	Solvent	λ max(log ε)	Ref.
Pyrazinecarbonitrile, 3,4-dihydro-4-(4-hydroxybutyl)-3-imino-6-methyl-5-phenyl-	EtOH	439(3.87)	33-1025-86
$C_{16}H_{18}N_4O_2$			
Carbamic acid, [5-amino-2-[(4-methylphenyl)azo]phenyl]-, ethyl ester, (E)-	EtOH	222(4.28),285(4.05), 406(4.43)	97-0134B-86
$C_{16}H_{18}N_4O_3$			
Carbamic acid, [5-amino-2-[(4-methoxyphenyl)azo]phenyl]-, ethyl ester, (E)-	EtOH	223(4.30),286(4.05), 405(4.44)	97-0134B-86
3(2H)-Pyridazinone, 6-[2-(3,4-dimethoxyphenyl)-1-methyl-1H-imidazol-4-yl]-4,5-dihydro-	MeOH	238(4.05),298(4.38)	87-0261-86
$C_{16}H_{18}N_4O_4$			
Pyrimido[4,5-b]quinoline-3(2H)-acetic acid, 8-(dimethylamino)-4,6,7,10-tetrahydro-10-methyl-2,4-dioxo-	pH 8.0	231(4.16),251(4.26), 303(3.83),419s(--), 438(4.67)	77-1385-86
$C_{16}H_{18}N_4S_2$			
4H-Pyrazole-3,5-diamine, N,N,N',N'-tetramethyl-4-(4-phenyl-1,3-dithiol-2-ylidene)-	MeCN	230(4.31),290(3.59), 398(4.37)	88-0159-86
perchlorate	MeCN	290(3.62),405(4.39)	88-0159-86
$C_{16}H_{18}O_2$			
Phenol, 2-[1-(2-hydroxyphenyl)ethyl]-4,6-dimethyl-, (-)-	EtOH	279(3.63)	116-0509-86
Tricyclo[9.2.1.0^{3,8}]tetradecane-2,10-dione, 15,16-dimethyl-	EtOH	220(2.39),300(1.75)	78-3491-86
$C_{16}H_{18}O_2S_4$			
1,3-Benzenediethane(dithioic) acid, 5-(1,1-dimethylethyl)-α,α'-dioxo-, dimethyl ester	MeOH	259(4.38),309(4.28), 496(1.85)	118-0968-86
$C_{16}H_{18}O_3$			
9-Anthracenol, 1,2,3,4-tetrahydro-5,8-dimethoxy-	EtOH	298s(3.29),312(3.42), 327(3.44),342(3.47)	94-2810-86
1,4-Benzenediol, 2-[4-(4-methyl-3-pentenyl)-2-furanyl]-	MeOH	207(4.18),266s(4.02), 272(4.10),284(4.39), 324(3.94)	94-2290-86
9-Phenanthrenol, 1,2,3,4-tetrahydro-5,8-dimethoxy-	EtOH	298(3.38),310(3.44), 338(3.44),353(3.54)	94-2810-86
$C_{16}H_{18}O_3S$			
Benzene, 1-(2-methoxyethoxy)-2-[(4-methylphenyl)sulfinyl]-, (S)-	EtOH	211(4.66),238(4.11), 284(3.32)	12-1833-86
$C_{16}H_{18}O_4$			
Acetic acid, [(6-oxo-2-phenyl-1-cyclohexen-1-yl)oxy]-, ethyl ester	ether	220(3.82),285(4.19)	44-4424-86
Bullvalene, bis(acetoxymethyl)-	ether	237s(3.48)	24-2339-86
Butanoic acid, 4-(3,4-dihydro-6-methoxy-1-oxo-2(1H)-naphthalenylidene)-, methyl ester, (E)-	EtOH	212(4.02),237(4.01), 308(4.24)	150-2492-86M
Spiro[cyclopropane-1,1'(4'H)-naphtha-	EtOH	244(3.80)	117-0007-86

Compound	Solvent	λ_{max}(log ε)	Ref.
lene]-5',8'-dione, 4'-acetoxy-4'a,8'a-dihydro-4'a,6'-dimethyl- (cont.)			117-0007-86
$C_{16}H_{18}O_5$			
2H-1-Benzopyran-2-one, 7,8-dimethoxy-5-[(3-methyl-2-butenyl)oxy]- (neo-artanin)	EtOH	217(4.27),255(3.90), 258(3.91),317(4.07)	5-2142-86
$C_{16}H_{18}O_6$			
2H-1-Benzopyran-2-one, 5-[(2,3-dimethyloxiranyl)methoxy]-7,8-dimethoxy-	EtOH	218(4.19),254s(3.86), 260(3.88),317(4.09)	5-2142-86
$C_{16}H_{18}O_7$			
5H-Furo[3,2-g][1]benzopyran-5-one, 2,3-dihydro-2,3,4,9-tetramethoxy-7-methyl-, trans	EtOH	231(4.41),249s(4.37), 253(4.38),288(3.84)	44-3116-86
$C_{16}H_{18}O_9$			
4H-1-Benzopyran-4-one, 7-(β-D-glucopyranosyloxy)-5-hydroxy-2-methyl-	MeOH	229(4.26),248(4.27), 257(4.28),285(3.91), 322(3.78)	56-0837-86
Scopolin	MeOH	205(4.4),227(4.1), 289(3.6),340(3.8)	94-4012-86
	MeOH	230(3.85),282(3.27), 340(2.79)	105-0228-86
$C_{16}H_{19}ClN_2O_2$			
2(1H)-Pyrimidinone, 1-(4-chlorophenyl)-4-(3-ethoxypropyl)-6-methyl-	EtOH	216(4.17),307(3.88)	39-1147-86C
$C_{16}H_{19}Cl_2N_3Pd$			
Palladium, dichloro[4,5,6,7-tetrahydro-7,8,8-trimethyl-2-(2-pyridinyl)-4,7-methano-2H-indazole-N^1,N^2]-, [SP-4-3-(4S)]-	MeOH	265.7(4.30),395.1(4.01), 317.7(4.04),373s(2.70)	12-1525-86
$C_{16}H_{19}NO$			
Bicyclo[4.4.1]undeca-1,3,5,7,9-pentaene-2-carboxamide, N,N-diethyl-	MeOH	255(4.68),295(3.80)	33-1263-86
2H-Pyrrol-2-one, 1,5-dihydro-3,4-dimethyl-5-[(2,4,6-trimethylphenyl)-methylene]-, (E)-	EtOH	266(4.16)	49-0631-86
(Z)-	EtOH	297(4.16)	49-0631-86
$C_{16}H_{19}NOS$			
1-Benzothiepin-5(2H)-one, 3,4-dihydro-4-(1-piperidinylmethylene)-, (E)-	EtOH	248(3.97),353(4.14)	4-0449-86
Sulfoximine, N-(1-methylethyl)-S-(4-methylphenyl)-S-phenyl-, (R)-	EtOH	241(4.26)	12-1655-86
Sulfoximine, S-(4-methylphenyl)-1-1-phenyl-N-propyl-	EtOH	221s(4.00),241(4.09), 267s(3.46),276s(3.23)	12-1655-86
$C_{16}H_{19}NO_2$			
Cyclopenta[3,4]cyclobuta[1,2]cyclohepten-6(1H)-one, 2,3,3a,8b-tetrahydro-3a-morpholino-	MeOH	232(4.27),324(4.04)	88-3005-86
$C_{16}H_{19}NO_2S$			
2(1H)-Pyridinethione, 1-[(tricyclo-[3.3.1.1^{3,7}]dec-1-ylcarbonyl)oxy]-	EtOH	286(3.15),364(3.64)	39-0039-86C

Compound	Solvent	$\lambda_{max}(\log \epsilon)$	Ref.
$C_{16}H_{19}NO_3$			
2-Furancarboxylic acid, 4-[(ethylphenylamino)methyl]-5-methyl-, methyl ester	MeOH	256(2.71)	73-2186-86
1,6-Octadiene-3,5-dione, 1-(4-methoxyphenyl)-7-(methylamino)-	n.s.g.	328(4.34),412(4.48)	4-1721-86
$C_{16}H_{19}NO_3S$			
Sulfoximine, S-[2-(2-methoxyethoxy)-phenyl]-S-(4-methylphenyl)-, (S)-	EtOH	227(4.14),289(3.53)	12-1833-86
$C_{16}H_{19}NO_3S_2$			
[1]Benzothiepino[4,5-e]-1,2-oxathiin-4-amine, N,N-diethyl-5,6-dihydro-, 2,2-dioxide	EtOH	227(4.07),234s(4.03), 250s(3.74),276s(3.84), 299(3.98)	4-0455-86
$C_{16}H_{19}NO_4$			
Pseudolycorine	EtOH	212(4.22),288(3.61)	102-1453-86
	EtOH-NaOH	252(3.85),302(3.73)	102-1453-86
Retronecine, 7,9-O,O-(3,4,5,6-tetrahydrophthaloyl)-	EtOH	248(3.29),288(2.99)	39-0585-86C
$C_{16}H_{19}N_3$			
Benzenamine, N,N,3-trimethyl-4-[(2-methylphenyl)azo]-	C_6H_{12}	400(4.46)	56-0163-86
Cyclohepta[b]pyrrol-2(1H)-one, cycloheptylidenehydrazone	EtOH	280(4.43),288(4.40), 324(4.18),385(4.03), 398(4.01),444(3.83), 463s(3.75),495s(3.33)	18-3326-86
$C_{16}H_{19}N_3O$			
Benzenamine, N,N-diethyl-4-(phenylazo)-, N-oxide	benzene	334(4.14),440s(--)	39-1439-86B
	MeOH	228(3.99),317(4.19), 441(2.82)	39-1439-86B
4H-Indazol-4-one, 1,5,6,7-tetrahydro-1,6,6-trimethyl-3-(phenylamino)-	dioxan	242s(4.309),256s(4.16), 265s(4.091),296(4.085)	48-0635-86
1-Propanone, 2-(hydroxyamino)-2-methyl-1-phenyl-, phenylhydrazone	EtOH	269(4.26),294(4.00)	103-0287-86
$C_{16}H_{19}N_3OS$			
4(1H)-Pyrimidinone, 5-[2-[4-(diethylamino)phenyl]ethenyl]-2,3-dihydro-2-thioxo-	MeOH	400(4.62)	56-0599-86
4(1H)-Pyrimidinone, 6-[2-[4-(diethylamino)phenyl]ethenyl]-2,3-dihydro-2-thioxo-	MeOH	422(4.31)	56-0599-86
$C_{16}H_{19}N_3O_2$			
Benzenamine, N,N-diethyl-4-(phenyl-ONN-azoxy)-, N-oxide	benzene	330(4.09)	39-1439-86B
	MeOH	231(3.95),319(4.22)	39-1439-86B
2-Propenoic acid, 2-[[(dimethylamino)-methylene]amino]-3-(1H-indol-3-yl)-, ethyl ester, (Z)-	MeOH	228(4.43),268(4.05), 348(4.35)	5-1749-86
2,4(1H,3H)-Pyrimidinedione, 5-[2-[4-(diethylamino)phenyl]ethenyl]-, (E)-	MeOH	367(4.09)	56-0599-86
2,4(1H,3H)-Pyrimidinedione, 6-[2-[4-(diethylamino)phenyl]ethenyl]-, (E)-	MeOH	413(4.41)	56-0599-86

Compound	Solvent	λ_{max}(log ϵ)	Ref.
$C_{16}H_{19}N_3O_3$			
2,5-Cyclohexadien-1-one, 3-amino-4-[[4-[bis(2-hydroxyethyl)amino]-phenyl]imino]-	pH 8.0	580(4.16)	39-0065-86B
$C_{16}H_{19}N_3O_3S$			
Benzamide, N-[2-nitro-3-(1-piperidin-yl)-1-thioxo-2-butenyl]-	EtOH	242(4.00),312(4.17), 397(4.09)	18-3871-86
$C_{16}H_{19}N_3O_5S$			
1H-Pyrazole-3-carbothioamide, 4-(phen-ylmethoxy)-5-β-D-ribofuranosyl-	pH 1	253(3.82),297(3.98)	87-0268-86
	pH 7	253(3.88),297(4.04)	87-0268-86
	pH 11	265(3.94),305(4.08)	87-0268-86
$C_{16}H_{19}N_3O_6$			
1H-Pyrazole-3-carboxamide, 4-(phenyl-methoxy)-5-β-D-ribofuranosyl-	pH 1	245s(3.51)	87-0268-86
	pH 7	245s(3.62)	87-0268-86
	pH 11	243(3.80)	87-0268-86
2,4(1H,3H)-Pyrimidinedione, 1-[3-de-oxy-3-[(4-methoxyphenyl)amino]-β-D-lyxofuranosyl]-	MeOH	201(4.42),247(4.21), 305(3.20)	44-4417-86
$C_{16}H_{19}N_3O_7$			
1H-Imidazole-4-carbonitrile, 5-methyl-1-(2,3,5-tri-O-acetyl-β-D-ribofuran-osyl)-	EtOH	223(3.82)	4-0679-86
1H-Imidazole-5-carbonitrile, 4-methyl-1-(2,3,5-tri-O-acetyl-β-D-ribofuran-osyl)-	EtOH	230(3.65)	4-0679-86
$C_{16}H_{19}N_4S_2$			
1H-Pyrazolium, 3,5-bis(dimethylamino)-1-(5-phenyl-3H-1,2-dithiol-3-yli-dene)-, perchlorate	MeCN	234s(4.13),334(4.25), 492(4.21)	88-0159-86
$C_{16}H_{19}N_5$			
Pyrazinecarbonitrile, 4-(4-aminobut-yl)-3,4-dihydro-3-imino-6-methyl-5-phenyl-	MeOH	434(3.97)	33-1025-86
Pyrazino[1,2-a][1,3]diazepine-10-carbo-nitrile, 2-amino-1,2,3,4,5,10a-hexa-hydro-8-methyl-7-phenyl-	pH 1	378(4.09)	33-1025-86
	MeOH	345(--)	33-1025-86
	acetone	346(3.72)	33-1025-86
$C_{16}H_{19}N_5O_4$			
1H-Pyrrolo[3,2-g]pteridine-7-carbox-ylic acid, 6-ethyl-2,3,4,8-tetra-hydro-1,3,8-trimethyl-2,4-dioxo-, ethyl ester	MeOH	252(4.58),275s(4.24), 363(4.32)	108-0081-86
$C_{16}H_{19}N_5O_5$			
8,6'-Cyclo-9-(5-O-acetyl-6-deoxy-2,3-O-isopropylidene-β-D-allofurano-syl)adenine	H_2O	260.5(4.17)	94-1573-86
8,6'-Cyclo-9-(5-O-acetyl-6-deoxy-2,3-O-isopropylidene-β-L-talofurano-syl)adenine	H_2O	261.5(4.22)	94-1573-86
$C_{16}H_{19}N_5O_6$			
8,3'-Ethanoadenosine, 3'-deoxy-, 2',5'-diacetate	H_2O	262.5(4.22)	94-0015-86

Compound	Solvent	$\lambda_{max}(\log \epsilon)$	Ref.
$C_{16}H_{19}N_5O_9S$			
1H-Pyrazino[2,3-c][1,2,6]thiadiazin-4-amine, 1-(2,3,5-tri-O-acetyl-β-D-ribofuranosyl)-, 2,2-dioxide	MeOH	217s(4.03),246(4.38), 334(3.96)	5-1872-86
8H-Pyrazino[2,3-c][1,2,6]thiadiazin-4-amine, 8-(2,3,5-tri-O-acetyl-D-ribofuranosyl)-, 2,2-dioxide	MeOH	222(4.09),258(4.16), 332s(3.63),410(3.92)	5-1872-86
$C_{16}H_{20}BrN_7$			
2,4,6-Pyrimidinetriamine, 5-[(4-bromophenyl)azo]-N^4-cyclohexyl-	EtOH	202(4.25),258(4.12), 388(4.24)	42-0420-86
$C_{16}H_{20}ClN_3O_4S$			
4-Morpholineacetamide, N-[6-chloro-2-[(2-oxopentyl)thio]pyridinyl]-α-oxo-	EtOH	258(4.06),302(3.90)	103-0800-86
$C_{16}H_{20}Cl_2FeHg$			
Mercury, dichloro[1,1'-[thiobis(3,1-propanediylthio)]ferrocene]-	MeCN	449(2.23)	18-3611-86
$C_{16}H_{20}Cl_2FePdS_3$			
Palladium, dichloro[1,1'-[thiobis(3,1-propanediylthio)]ferrocene]-, (SP-4-3)-	MeCN	388(3.34),552(2.90)	18-3611-86
$C_{16}H_{20}FeS_3$			
Ferrocene, 1,1'-[thiobis(3,1-propanediyjthio)]-	MeCN	446(2.32)	18-3611-86
$C_{16}H_{20}FeS_4$			
Ferrocene, 1,1'-[1,2-ethanediylbis(thio-2,1-ethanediylthio)]-	MeCN	442(2.35)	18-1515-86
copper perchlorate salt	MeCN	440(2.14)	18-1515-86
mercuric chloride salt	MeCN	442(2.15)	18-1515-86
silver perchlorate salt	MeCN	438(2.29)	18-1515-86
$C_{16}H_{20}N_2$			
[1,1'-Biphenyl]-4,4'-diamine, N,N,N',N'-tetramethyl-	H_2O?	565(4.10),615(4.08)	60-0359-86
2-Cyclohexen-1-one, 2-methyl-5-(1-methylethenyl)-, phenylhydrazone	EtOH	302(4.35),320(4.38)	126-1573-86
Pyrazine, 2,5-bis(1-methylethyl)-3-phenyl-	EtOH	227(4.04),284(4.11)	142-0785-86
$C_{16}H_{20}N_2O$			
Bicyclo[4.4.1]undeca-1,3,5,7,9-pentaene-2-carboxamide, 10-amino-N,N-diethyl-	MeOH	250(4.38),280(4.14), 360(3.65),420(3.18)	33-1263-86
Hyperzine B	MeOH	231(3.95),312(3.85)	23-0837-86
Pyrazine, 2,5-bis(1-methylethyl)-3-phenyl-, 1-oxide	EtOH	227(4.40),251(4.15), 269(4.12),305(3.78)	142-0785-86
Pyrazine, 2,5-bis(1-methylethyl)-3-phenyl-, 4-oxide	EtOH	218.5(4.24),270(3.91), 306s(3.49)	142-0785-86
$C_{16}H_{20}N_2OS$			
1-Benzothiepin-5(2H)-one, 3,4-dihydro-4-[(4-methyl-1-piperazinyl)methylene]-, (E)-	EtOH	249(4.03),301(3.80), 351(4.07)	4-0449-86

Compound	Solvent	λ_{max}(log ε)	Ref.
$C_{16}H_{20}N_2O_2$			
2-Furancarboxaldehyde, 5-(dimethylamino)-4-[(dimethylamino)phenylmethyl]-	MeOH	369(3.31)	73-0573-86
5-Pyrimidinecarboxylic acid, 1,4-dihydro-2,6-dimethyl-1-(4-methylphenyl)-, ethyl ester	EtOH	309(3.72)	150-0088-86S +150-0901-86M
5-Pyrimidinecarboxylic acid, 1,4-dihydro-2,6-dimethyl-1-(phenylmethyl)-, ethyl ester	EtOH	297(3.68)	150-0088-86S +150-0901-86M
2(1H)-Pyrimidinone, 4-(3-ethoxypropyl)-6-methyl-1-phenyl-	EtOH	205(4.18),237s(3.83), 307(3.71)	39-1147-86C
$C_{16}H_{20}N_2O_2S_4$			
1,2-Ethanediamine, N,N'-bis[[2-methoxy-5-(methylthio)-3-thienyl]methylene]-	EtOH	234(4.71),250(4.54), 283(4.36)	70-1470-86
$C_{16}H_{20}N_2O_4S$			
1-Azabicyclo[3.2.0]hept-2-ene-2-carboxylic acid, 3-[(4-cyanocyclohexyl)thio]-6-(1-hydroxyethyl)-7-oxo-, sodium salt	H_2O	305(3.90)	23-2184-86
$C_{16}H_{20}N_2O_8$			
Pyrrolo[3,2-b]pyrrole-1,3,4,6-tetracarboxylic acid, 3a,6a-dihydro-, 1,4-diethyl 3,6-dimethyl ester, cis	MeOH	260(3.55)	23-1969-86
$C_{16}H_{20}N_3O_9P$			
1H-Pyrazole-3-carboxamide, 4-(phenylmethoxy)-5-(5-O-phosphono-β-D-ribofuranosyl)-, disodium salt	pH 1 and 7	240s(3.60)	87-0268-86
	pH 11	240s(3.73)	87-0268-86
$C_{16}H_{20}N_4$			
Benzenamine, 4,4'-azobis[N,N-dimethyl-	MeOH	422(4.30),457(4.34)	48-0113-86
	EtOH	421(4.48),459(4.49)	48-0113-86
Cyanamide, [2,5-bis(1,1-dimethylethyl)-2,5-cyclohexadiene-1,4-diylidene]bis-, (E,E)-	MeCN	343(4.46),367s(4.25)	5-0142-86
$C_{16}H_{20}N_4O_2$			
Ethanol, 2,2'-[[4-[(2-amino-4-imino-2,5-cyclohexadien-1-ylidene)amino]phenyl]imino]bis-	neutral	525(4.02)	39-0065-86B
	cation	633(4.38)	39-0065-86B
Ethanol, 2,2'-[[3-methyl-4-(3-pyridinyl)azo]phenyl]imino]bis-	acetone	429(4.45)	7-0473-86
6,7,17,18-Tetraazatricyclo[10.2.2.2^{5,8}]octadeca-5,7,12,14,15,17-hexaene, 13,15-dimethoxy-	dioxan	228(4.02),263(3.47), 280(3.44),302(3.23), 405(2.49),518(2.76)	88-5367-86
	DMSO	516(--)	88-5367-86
	2-Mebutane	523(--)	88-5367-86
$C_{16}H_{20}N_4O_6$			
1H-Pyrazole-3-methanimidamide, N-hydroxy-4-(phenylmethoxy)-5-β-D-ribofuranosyl-	pH 1	235(3.86)	87-0268-86
	pH 7	235(3.81)	87-0268-86
	pH 11	235(3.84)	87-0268-86
$C_{16}H_{20}N_6O_4$			
Adenosine, 8-cyano-N,N-dimethyl-2',3'-O-(1-methylethylidene)-	MeOH	234(4.30),270s(3.73), 312(4.16)	94-3635-86

Compound	Solvent	$\lambda_{max}(\log \epsilon)$	Ref.
$C_{16}H_{20}O$			
2-Cyclohexen-1-one, 2-butyl-3-phenyl-	MeOH	255(3.95)	44-4424-86
2-Cyclohexen-1-one, 2-(2-methyl-propyl)-3-phenyl-	MeOH	255(3.90)	44-4424-86
$C_{16}H_{20}O_3$			
Naphtho[1,2-b]furan-8-ol, 6,7,8,9-tetrahydro-3-(methoxymethyl)-5,9-dimethyl-, (8R-trans)-	$CHCl_3$	248(3.98),278(3.02), 288(3.02)	78-4493-86
Spiro[benzo[3,4]cyclobuta[1,2]cyclo-octene-5(4bH),2'-[1,3]dioxolan]-4b-ol, 6,7,8,9,10,10a-hexahydro-, cis	MeOH	260.5(3.11),267(3.26), 273(3.25)	44-1419-86
$C_{16}H_{20}O_4$			
4a(2H)-Phenanthrenecarboxylic acid, 3,4,4b,5,6,7,8,10-octahydro-1-hy-droxy-2-oxo-, methyl ester, (cis-(±)-	EtOH	278(3.68)	32-0115-86
4a(4H)-Phenanthrenecarboxylic acid, 1,4b,5,6,7,8,10,10a-octahydro-2-hydroxy-1-oxo-, methyl ester	EtOH	272(3.78)	32-0115-86
Propanedioic acid, [(4-methylphenyl)-methylene]-, 1,1-dimethylethyl methyl ester, (Z)-	EtOH	223(4.13),227s(4.05), 288(4.39)	44-4112-86
$C_{16}H_{20}O_7$			
2H-1-Benzopyran-2-one, 5-(2,3-di-hydroxy-3-methylbutoxy)-7,8-di-methoxy-, (+)-	EtOH	223s(4.13),237s(3.82), 256s(3.96),262(3.99), 323(4.06)	5-2142-86
$C_{16}H_{21}BrN_2$			
Pyrrolo[2,3-b]indole, 6-bromo-3a-(1,1-dimethyl-2-propenyl)-1,2,3,3a,8,8a-hexahydro-1-methyl-, cis-(-)- (dihydroflustramine C)	MeOH	213(4.30),250(3.75), 309(3.53)	23-1312-86
$C_{16}H_{21}BrN_2O$			
Pyrrolo[2,3-b]indole, 6-bromo-3a-(1,1-dimethyl-2-propenyl)-1,2,3,3a,8,8a-hexahydro-1-methyl-, 1-oxide, (3aα,8aα)-	MeOH	226(3.58),283(3.30)	23-1312-86
$C_{16}H_{21}ClHgO_3$			
Mercury, [(6-acetoxy-3,4-dihydro-2,5,7,8-tetramethyl-2H-1-benzo-pyran-2-yl)methyl]chloro-	EtOH	284(3.1)	32-0441-86
$C_{16}H_{21}ClN_6O_7$			
Adenosine, N-(3-chloro-4,4-dimethoxy-5-oxo-2-pyrrolidinyl)-	EtOH	265(4.29)	142-0625-86
$C_{16}H_{21}FN_2O_9$			
α-D-Glucofuranose, 1,2-O-(1-methyleth-ylidene)-3-(5-fluoro-3,4-dihydro-2,4-dioxo-1(2H)-pyrimidinepropan-oate)	MeOH	270.5(3.90)	47-2059-86
$C_{16}H_{21}NO$			
2H-Pyrrol-2-one, 1,5-dihydro-3,4-di-methyl-5-[(2,4,6-trimethylphenyl)-methyl]-	isooctane	217.5(4.14)	5-1241-86

Compound	Solvent	$\lambda_{max}(\log \epsilon)$	Ref.
$C_{16}H_{21}NO_2$			
3-Pyridinecarboxylic acid, 1,2,3,6-tetrahydro-1-methyl-3-(1-phenylethyl)-, methyl ester, hydrochloride	pH 7.2	210(0.80)	87-0125-86
3-Pyridinecarboxylic acid, 1,2,3,6-tetrahydro-1-methyl-3-(2-phenylethyl)-, methyl ester, hydrochloride, (±)-	pH 7.2	210(0.73)	87-0125-86
$C_{16}H_{21}NO_3$			
4H-1,3,5-Dioxazocine-7-carboxaldehyde, 2-butyl-5,8-dihydro-6-phenyl-	MeOH	235(3.05),299(3.21)	73-2158-86
4H-1,3,5-Dioxazocine-7-carboxaldehyde, 5,8-dihydro-2-(2-methylpropyl)-6-phenyl-	MeOH	235(2.92),289(3.13)	73-2158-86
4H-1,3,5-Dioxazocine-7-carboxaldehyde, 2-(1,1-dimethylethyl)-5,8-dihydro-6-phenyl-	MeOH	236(3.04),300(3.14)	73-2158-86
[1,3]Dioxepino[5,6-d]isoxazole, 6-butyl-3a,4,8,8a-tetrahydro-3-phenyl-, exo	MeOH	258(2.99)	73-2158-86
[1,3]Dioxepino[5,6-d]isoxazole, 6-(1,1-dimethylethyl)-3a,4,8,8a-tetrahydro-3-phenyl-, endo	MeOH	265(2.95)	73-2158-86
exo	MeOH	256(3.00)	73-2158-86
[1,3]Dioxepino[5,6-d]isoxazole, 6-(2-methylpropyl)-3a,4,8,8a-tetrahydro-3-phenyl-, exo	MeOH	268(3.03)	73-2158-86
3-Pyridinecarboxylic acid, 1,2,5,6-tetrahydro-5-(1-hydroxy-1-phenylethyl)-1-methyl-, methyl ester, (5RS,1'RS)-	pH 7.2	218(1.38)	87-0125-86
(5RS,1'SR)-	pH 7.2	216(1.31)	87-0125-86
Spiro[cyclohexan-1,3'(4'H)-isoquinolin]-4'-one, 6'-ethoxy-1',2'-dihydro-7'-hydroxy-	EtOH	232(4.19),279(4.04), 313(3.93)	39-1329-86C
	EtOH-KOH	252(3.99),295(3.61), 348(4.36)	39-1329-86C
$C_{16}H_{21}N_3O_4$			
1,3-Dioxane-4,6-dione, 5-[5-[(3,4-dimethyl-2-imidazolidinylidene)amino]-2,4-pentadienylidene]-2,2-dimethyl-, (E,E)-	$CHCl_3$	521(4.79)	24-3276-86
$C_{16}H_{21}N_3O_6$			
[1,4,7,10,13]Pentaoxacyclopentadecino[2,3-g]quinoxalin-2(1H)-one, 7,8,10,11,13,14,16,17-octahydro-	EtOH	218(4.79),255(4.84), 380(4.29)	24-3870-86
$C_{16}H_{21}N_3O_8$			
Cytidine, N-methyl-, 2',3',5'-triacetate	pH 7	260(4.05),299(4.05)	1-0806-86
	pH 2	261(--),299(--)	1-0806-86
	pH 13	274(--)	1-0806-86
$C_{16}H_{21}N_5O_3$			
Peramine diacetate	MeOH	232(4.59),257(4.29), 290s(3.91)	77-0935-86
$C_{16}H_{21}N_5O_6$			
9H-Purine-8-carboxylic acid, 6-(dimethylamino)-9-[2,3-O-(1-methyl-	MeOH	240(4.28),274(3.78), 309(4.23)	94-3635-86

Compound	Solvent	λ_{max}(log ϵ)	Ref.
ethylidene)-β-D-ribofuranosyl]-, monosodium salt (cont.)			94-3635-86
$C_{16}H_{21}N_5O_8$			
1,2,3-Triazolo[4,5-d][1,3]diazepin-8-ol, 3,4,7,8-tetrahydro-3-(2,3,5-tri-O-acetyl-β-D-ribofuranosyl)-	pH 1	261(3.91)	44-1050-86
	pH 11	277(4.09)	44-1050-86
	MeOH	279(4.02)	44-1050-86
$C_{16}H_{21}N_7$			
2,4-Pyrimidinediamine, 5-[(2-methylphenyl)azo]-6-(1-piperidinyl)-	EtOH	240(4.28),282(4.34), 395(4.36)	42-0420-86
$C_{16}H_{21}N_7O$			
2,4-Pyrimidinediamine, 5-[(2-methoxyphenyl)azo]-6-(1-piperidinyl)-	EtOH	245(3.97),285(4.21), 325(3.74),402(4.32)	42-0420-86
2,4-Pyrimidinediamine, 5-[(3-methoxyphenyl)azo]-6-(1-piperidinyl)-	EtOH	205(4.91),270(4.71), 394(4.81)	42-0420-86
$C_{16}H_{21}N_7O_3S$			
Benzenesulfonic acid, 4-[[2,4-diamino-6-(cyclohexylamino)-5-pyrimidinyl]-azo]-	EtOH	211(4.50),264(4.28), 394(4.39)	42-0420-86
$C_{16}H_{22}N_2$			
2,6-Octadienal, 3,7-dimethyl-, phenylhydrazone	EtOH	297(3.98)	126-1573-86
$C_{16}H_{22}N_2OS$			
1-Benzothiepin-5(2H)-one, 4-[[[2-(dimethylamino)ethyl]methylamino]-methylene]-3,4-dihydro-, (E)-	EtOH	247(3.71),348(3.82)	4-0449-86
Benzo[b]thiophen-3(2H)-one, 2-[[2,2-dimethyl-1-(3-methylbutyl)hydrazino]methylene]-	CCl_4	318(4.19),437(4.15)	104-2108-86
$C_{16}H_{22}N_2O_3$			
3-Benzazecine-3(2H)-carbonitrile, 1,4,5,6,7,8-hexahydro-8-hydroxy-10,11-dimethoxy-, (±)-	MeOH	231(3.95),282(3.45), 287s(3.40)	12-0893-86
3H-3-Benzazonine-3-carbonitrile, 1,2,4,5,6,7-hexahydro-7,9,10-trimethoxy-, (±)-	MeOH	232(3.93),282(3.50), 286s(3.47)	12-0893-86
1H-Pyrrole-2-carboxylic acid, 5-[(2,5-dihydro-3,4-dimethyl-5-oxo-1H-pyrrol-2-yl)methyl]-3,4-dimethyl-, ethyl ester	MeOH	281.5(4.24)	5-1241-86
$C_{16}H_{22}N_2O_4$			
1H-2,6-Benzoxazecine-6(3H)-carbonitrile, 4,5,7,8-tetrahydro-1,10,11-trimethoxy-	MeOH	229(3.96),280(3.40), 285s(3.36)	12-0893-86
3,5-Pyridinedicarboxylic acid, 2-[2-(dimethylamino)ethenyl]-6-methyl-, diethyl ester	EtOH	270(4.01),286s(--), 390(4.44)	39-0753-86C
Pyrrolidinium, 2-carboxy-1-[5-(2-carboxy-1-pyrrolidinyl)-3-methyl-2,4-pentadienylidene]-, hydroxide, inner salt, [S-(R*,R*)]-	MeOH	251(3.76),280(3.58), 428s(4.65),449(4.99)	33-1588-86
$C_{16}H_{22}N_2O_5S$			
1-Azabicyclo[3.2.0]hept-2-ene-2-carb-	H_2O	303(3.94)	23-2184-86

Compound	Solvent	$\lambda_{max}(\log \epsilon)$	Ref.
oxylic acid, 3-[[4-(aminocarbonyl)-cyclohexyl]thio]-6-(1-hydroxyethyl)-7-oxo-, sodium salt (cont.)			23-2184-86
$C_{16}H_{22}N_4O_7$			
Acetamide, N-[7-(β-D-arabinofuranosyl)-4,7-dihydro-3-(methoxymethyl)-5-methyl-4-oxo-3H-pyrrolo[2,3-d]-pyrimidin-2-yl]-	MeCN	305(3.88)	18-1915-86
$C_{16}H_{22}N_4O_9$			
2H-Imidazo[4,5-c]pyridin-2-one, 4-amino-1,3-dihydro-1,3-di-β-D-ribofuranosyl-	MeOH	229(3.79),251(3.48), 292(3.34)	78-1511-86
$C_{16}H_{22}N_6O_4S$			
Adenosine, 8-(aminothioxomethyl)-N,N-dimethyl-2',3'-O-(1-methylethylidene)-	MeOH	242(4.25),274s(3.87), 285(3.89),333(4.06)	94-3635-86
$C_{16}H_{22}N_6O_7$			
Adenosine, N-(4,4-dimethoxy-5-oxo-2-pyrrolidinyl)-	EtOH	265(4.13)	142-0625-86
$C_{16}H_{22}O$			
1-Butanone, 4-cyclohexyl-1-phenyl-	hexane	269(2.54),324(1.66)	35-5264-86
Ethanone, 2-cyclohexyl-1-(4-ethyl-phenyl)-	hexane	322(1.83)	35-5264-86
Naphthalene, 1,2-dihydro-7-methoxy-1,6-dimethyl-4-(1-methylethyl)-, (S)-	EtOH	215(4.47),273(4.11), 305(3.58)	12-0103-86
$C_{16}H_{22}O_2$			
1,4-Benzenediol, 2-(3,7-dimethyl-2,6-octadienyl)-, (E)-	MeOH	208(4.05),293(3.53)	94-2290-86
Cyclodeca[b]furan, 4,7,8,11-tetra-hydro-8-methoxy-3,6,10-trimethyl-, [R-(E,E)]-	isooctane	206(4.05)	32-0303-86
2(1H)-Naphthalenone, 3,4-dihydro-6-methoxy-4,7-dimethyl-1-(1-methyl-ethyl)-, (1R-trans)-	EtOH	208(3.79),282(3.13)	12-0103-86
$C_{16}H_{22}O_3$			
Acetic acid, 1,3,4-trimethyl-4-(1-oxo-2-propynyl)bicyclo[3.3.0]oct-2-yl ester	EtOH	212(3.65),218s(3.56)	33-0659-86
Acetic acid, 5,7,8-trimethyl-9-oxo-tricyclo[6.3.0.0^{1,5}]undec-10-en-6-yl ester	EtOH	229(3.90)	33-0659-86
Benzo[1,2]cyclobuta[3,4]cyclooctene, 1,2,3,4,5,6,6a,10b-octahydro-10b-hydroxy-1,1-dimethoxy-	MeOH	261(3.11),267.5(3.27), 274(3.25)	44-1419-86
Cyclopenta[c]pentalene-2-carboxylic acid, 3,3a,4,5,5a,6,7,8-octahydro-1,3a,6-trimethyl-3-oxo-, methyl ester	EtOH	239(4.21)	100-0845-86
2-Naphthaleneacetic acid, 1,2,3,4,4a,7-hexahydro- ,4a,8-trimethyl-7-oxo-, methyl ester, [2R-[2α(S*),4aα]]-	EtOH	245(4.15)	2-0180-86
$C_{16}H_{22}O_4$			
Butanoic acid, 3-methyl-, 3-(3,4-di-	EtOH	212(3.20),266(3.15)	100-0677-86

Compound	Solvent	λ_{max}(log ε)	Ref.
methoxyphenyl)-2-propenyl ester, (E)- (cont.)			100-0677-86
Cyclopenta[6,6a]pentaleno[1,2-b]oxirene-6a(7aH)-carboxylic acid, octahydro-3,5a,7a-trimethyl-6-oxo-, methyl ester	n.s.g.	205(3.45)	100-0845-86
$C_{16}H_{22}O_4S$			
Phenanthro[1,2-b]thiophen-7(2H)-one, 1,3a,3b,4,5,8,9,9a,9b,10,11,11a-dodecahydro-1-hydroxy-, 3,3-dioxide	EtOH	238(4.31),299(1.93)	104-0107-86
Phenanthro[2,1-b]thiophen-7(3bH)-one, 2,3,3a,4,5,8,9,9a,9b,10,11,11a-dodecahydro-3-hydroxy-, 1,1-dioxide	EtOH	238(4.28),298(1.96)	104-0107-86
$C_{16}H_{22}O_5$			
2,4-Pentadienoic acid, 5-(8-hydroxy-1,5-dimethyl-3-oxo-6-oxabicyclo-[3.2.1]oct-8-yl)-3-methyl-, methyl ester (methyl phaseate)	EtOH	265(4.23)	163-1589-86
2E isomer	EtOH	265(4.34)	163-1589-86
methyl epiphaseate	EtOH	264(4.29)	163-1589-86
2E isomer	EtOH	265(4.36)	163-1589-86
2,4-Pentadienoic acid, 5-[1-hydroxy-2-(hydroxymethyl)-6,6-dimethyl-4-oxo-2-cyclohexen-1-yl]-3-methyl-, methyl ester	MeOH	266(4.33)	102-1103-86
$C_{16}H_{22}O_6$			
α-D-ribo-Hexofuranos-3-ulose, 1,2:5,6-di-O-cyclopentylidene-	EtOH	203(2.17)	33-1132-86
$C_{16}H_{22}O_8$			
2-Butanone, 4-[3-(β-D-glucopyranosyl-oxy)-4-hydroxyphenyl]-	MeOH	222(3.68),278(3.26)	100-0318-86
	MeOH-NaOMe	240(--),294(--)	100-0318-86
acetylation product	MeOH	270(3.20),276(3.19)	100-0318-86
$C_{16}H_{22}O_9$			
7-Deoxygardoside	n.s.g.	234(3.9)	32-0067-86
2,8-Dioxabicyclo[3.3.1]non-6-en-3-one, 9-ethenyl-6-[(β-D-glucopyranosyl-oxy)methyl]- (isosweroside)	MeOH	204(3.79)	33-1113-86
$C_{16}H_{23}CuN_5$			
Copper(2+), [2-methyl-N^1,N^3-bis(2-pyridinylmethyl)-1,2,3-propanetriamine-N^1,$N^{1'}$:N^3,$N^{3'}$]-, (SP-4-2)-, diperchlorate	H_2O	256(4.18),600(1.98)	12-1101-86
$C_{16}H_{23}NO_6$			
α-D-ribo-Hexofuranos-3-ulose, 1,2:5,6-di-O-cyclopentylidene-, oxime	EtOH	205(3.84)	33-1132-86
3,4,4(1H)-Pyridinetricarboxylic acid, 1-(1,1-dimethylethyl)-5-methyl-, trimethyl ester	MeCN	245(3.56),326(3.80)	118-0190-86
$C_{16}H_{23}N_3O_3$			
Piperidine, 1,1'-[(5-nitro-2-furanyl)-ethenylidene]bis-	MeCN	538(4.44)	104-1371-86

Compound	Solvent	$\lambda_{max}(\log \epsilon)$	Ref.
$C_{16}H_{23}N_3O_3S$			
Glycinamide, N-[(1,1-dimethylethoxy)-carbonyl]-L-phenylalanylthio-, (+)-	EtOH	265(4.08)	33-1224-86
$C_{16}H_{23}N_3O_4$			
Guanidine, N"-[5-(2,2-dimethyl-4,6-dioxo-1,3-dioxan-5-ylidene)-1,3-pentadienyl]-N,N,N',N'-tetra-methyl-, (E,E)-	benzene	495(4.74),525(4.94)	24-3276-86
	EtOH	471(5.25)	24-3276-86
	acetone	521(5.16)	24-3276-86
	$CHCl_3$	527(5.07)	24-3276-86
$C_{16}H_{23}N_3O_4S$			
1-Azabicyclo[3.2.0]hept-2-ene-2-carb-oxylic acid, 6-(1-hydroxyethyl)-3-[[4-(iminomethylamino)cyclohexyl]-thio]-7-oxo-	H_2O	307(3.95)	23-2184-86
$C_{16}H_{23}N_3O_6$			
2,5-Cyclohexadiene-1,4-dione, 2-[2-[(aminocarbonyl)oxy]-1-methoxyeth-yl]-6-(1-aziridinyl)-3-[(2-methoxy-ethyl)amino]-5-methyl-	EtOH	342(4.34)	94-1299-86
$C_{16}H_{23}N_5O_3$			
Guanidine, N,N,N',N'-tetramethyl-N"-[5-(tetrahydro-1,3-dimethyl-2,4,6-trioxo-5(2H)-pyrimidinylidene)-1,3-pentadienyl]-	benzene	515(4.92),541(5.00)	24-3276-86
	EtOH	535(4.90)	24-3276-86
	acetone	539(5.15)	24-3276-86
	$CHCl_3$	546(5.06)	24-3276-86
$C_{16}H_{23}N_5O_4S$			
Adenosine, N,N-dimethyl-2',3'-O-(1-methylethylidene)-8-(methylthio)-	MeOH	229(4.25),288(4.29)	94-3635-86
$C_{16}H_{23}N_5O_6S$			
Adenosine, N,N-dimethyl-2',3'-O-(1-methylethylidene)-8-(methylsulfonyl)-	MeOH	226(4.27),274s(4.19), 299(4.34)	94-3635-86
$C_{16}H_{23}N_6O_2$			
Pteridinium, 2,4-dimorpholino-8-ethyl-, tetrafluoroborate	pH 3.5	242(4.24),299(3.95), 444(4.04)	103-1118-86
hydrated	pH 9.5	227(4.18),265(4.42), 328(3.98)	103-1118-86
$C_{16}H_{23}O_2$			
2H-1-Benzopyran-6-yloxy, 7-(1,1-di-methylethyl)-3,4-dihydro-2,2,5-trimethyl-	EtOH	338(3.57),398(3.38), 419(3.59)	18-2899-86
$C_{16}H_{24}Cl_2N_2O_2$			
2,5-Cyclohexadiene-1,4-dione, 2,5-bis(butylmethylamino)-3,6-dichloro-	CH_2Cl_2	242(4.22),260s(4.00), 433(4.03),570s(2.70)	106-0832-86
$C_{16}H_{24}Cl_3NO$			
2,4-Cyclohexadien-1-one, 6-chloro-4-[1-(dichloroamino)ethyl]-2,6-bis(1,1-dimethylethyl)-	hexane	322(3.52)	70-0446-86
$C_{16}H_{24}N_2O$			
2H-Pyrrol-2-one, 4-ethyl-5-[(4-ethyl-3,5-dimethyl-1H-pyrrol-2-yl)methyl]-1,5-dihydro-3-methyl-	MeOH	210(4.27)	5-1241-86

Compound	Solvent	$\lambda_{max}(\log \epsilon)$	Ref.
$C_{16}H_{24}N_2O_2S$			
Benzenesulfonamide, N-[3-(dimethylamino)-1-(1,1-dimethylethyl)-2-propenylidene]-4-methyl-, (?,E)-	EtOH	222s(4.12),265s(3.32), 351(4.47)	161-0539-86
$C_{16}H_{24}N_2O_4S$			
1-Azabicyclo[3.2.0]hept-2-ene-2-carboxylic acid, 3-[[4-(aminomethyl)cyclohexyl]thio]-6-(1-hydroxyethyl)-7-oxo-	H_2O	303(3.73)	23-2184-86
$C_{16}H_{24}N_2O_6$			
Uridine, 2'-deoxy-5-ethyl-, 3'-(2,2-dimethylpropanoate)	MeOH	268(3.68)	83-0154-86
Uridine, 2'-deoxy-5-ethyl-, 5'-(2,2-dimethylpropanoate)	MeOH	267(3.99)	83-0154-86
Uridine, 2'-deoxy-5-ethyl-, 5'-(3-methylbutanoate)	MeOH	267(3.95)	83-0154-86
Uridine, 2'-deoxy-5-ethyl-, 5'-pentanoate	EtOH	267(3.95)	83-0154-86
$C_{16}H_{24}N_2S_3$			
Benzo[b]thiophene-3-carbothioamide, 4,5,6,7-tetrahydro-N-(1-methylethyl)-2-[(2-methyl-1-thioxopropyl)-amino]-	EtOH	245(4.13),286(4.11), 357(3.93),379s(3.90)	106-0096-86
$C_{16}H_{24}N_6O_4$			
β-D-Ribofuranuronamide, 1-deoxy-N-methyl-1-[6-(pentylamino)-9H-purin-9-yl]-	n.s.g.	269(4.27)	87-1683-86
$C_{16}H_{24}O$			
Tricyclo[9.3.1.0^{3,8}]pentadec-11-en-2-one, 8-methyl-	MeOH	218(4.20)	39-1303-86C
$C_{16}H_{24}O_2$			
2-Naphthaleneacetic acid, 1,2,3,4,4a,5-hexahydro-α,4a,8-trimethyl-, methyl ester	EtOH	269(3.81)	2-0180-86
1-Naphthalenol, 1,2,3,4-tetrahydro-6-methoxy-4,7-dimethyl-1-(1-methylethyl)-, (1R,4S)-	EtOH	207(3.78),275(3.30)	12-0103-86
(1S,4S)-	EtOH	210(3.85),276(3.30)	12-0103-86
2-Naphthalenol, 1,2,3,4-tetrahydro-6-methoxy-4,7-dimethyl-1-(1-methylethyl)-, [1R-(1α,2β,4α)]-	EtOH	223(3.70),280(3.20), 286(3.18)	12-0103-86
$C_{16}H_{24}O_3$			
1-Cyclopentene-1-carboxylic acid, 3-oxo-, 5-methyl-2-(1-methylethyl)-cyclohexyl ester, [1R-(1α,2β,5α)]-	$CHCl_3$	242(3.63),346(1.42)	78-3547-86
$C_{16}H_{24}O_4$			
1,2,3,4-Cyclododecanetetrone, 5,5,12,12-tetramethyl-	CH_2Cl_2	368(1.99),507(1.63)	89-0449-86
2-Naphthaleneacetic acid, 1,2,3,4,4a-5,6,8a-octahydro-8a-hydroxy-α,4a,8-trimethyl-6-oxo-, methyl ester	EtOH	240(4.08)	2-0180-86

Compound	Solvent	λ_{max}(log ε)	Ref.
$C_{16}H_{24}O_5$			
Butanoic acid, 2-[[3-[2-(3,3-dimethyloxiranyl)ethyl]-3-methyloxiranyl]methylene]-3-oxo-, ethyl ester	EtOH	220(3.59),250(3.52)	103-0370-86
$C_{16}H_{24}O_6$			
α-D-Glucofuranose, 1,2:5,6-di-O-cyclopentylidene-	EtOH	201(1.04)	33-1132-86
$C_{16}H_{24}O_8$			
Cyclopenta[c]pyran-4-carboxaldehyde, 1-(β-D-glucopyranosyloxy)-1,4,5,6-7,7a-hexahydro-7-methyl- (boschnaloside-11-d)	EtOH	249(4.16)	102-2309-86
$C_{16}H_{25}N$			
Cyclohexane, 4-(1,5-dimethyl-1,3-hexadienyl)-1-isocyano-1-methyl-	MeOH	203(3.81),232s(3.04), 239(4.06)	44-5136-86
Cyclohexene, 4-(1-isocyano-1,5-dimethyl-4-hexenyl)-1-methyl-	hexane	195(4.16)	44-5136-86
$C_{16}H_{25}NO$			
Cyclohexene, 4-(1-isocyanato-1,5-dimethyl-4-hexenyl)-1-methyl-	hexane	205(3.70),248(2.70)	44-5136-86
2,4,8,10-Dodecatetraenamide, N-(2-methylpropyl)-, (E,E,E,Z)-	EtOH	260(4.48)	102-2289-86
$C_{16}H_{25}NO_3$			
2,4-Heptadienamide, 6-acetoxy-N-cyclohexyl-5-methyl-, (Z,Z)-	hexane	261(4.42)	33-0469-86
3,5-Heptadienamide, 6-acetoxy-N-cyclohexyl-2-methyl-, (E,Z)-	EtOH	240(4.38)	33-0469-86
3,5-Heptadienamide, 6-acetoxy-N-cyclohexyl-5-methyl-, (E,Z)-	EtOH	216s(3.84)	33-0469-86
$C_{16}H_{25}NO_4$			
1(4H)-Pyridinepentanoic acid, 3-(methoxycarbonyl)-4,4,5-trimethyl-, methyl ester	MeCN	287(3.17),350(3.83)	118-0190-86
$C_{16}H_{25}NO_6$			
α-D-Allofuranose, 1,2:5,6-di-O-cyclopentylidene-3-deoxy-3-(hydroxyamino)-	EtOH	202(2.61)	33-1132-86
$C_{16}H_{26}$			
1,3,7,11-Tridecatetraene, 4,8,12-trimethyl-, (E,E)-	EtOH	236(4.16)	88-2111-86
$C_{16}H_{26}O$			
Tricyclo[3.3.1.1^{3,7}]decanone, 4-(3,3-dimethylbutyl)-, axial	EPA	286(1.33)	35-4484-86
	MI	289(1.26)	35-4484-86
equatorial	EPA	285(1.47)	35-4484-86
	MI	285(1.34)	35-4484-86
$C_{16}H_{26}O_3$			
2-Cyclopenten-1-one, 2-hexyl-4-[(tetrahydro-2H-pyran-2-yl)oxy]-	MeOH	219(4.08)	2-0675-86
$C_{16}H_{26}O_4$			
2(1H)-Pentalenone, 4,5,6,6a-tetrahy-	EtOH	229(4.00)	94-4947-86

Compound	Solvent	λ_{max}(log ϵ)	Ref.
dro-6-[2-[(2-methoxyethoxy)methoxy]ethyl]-5,5-dimethyl- (cont.)			94-4947-86
$C_{16}H_{26}O_7$			
Cyclohexanone, 2-[2-(β-D-glucopyranosyloxy)-1-methylethylidene]-5-methyl-, [S-(E)]- (schizonepetoside C)	EtOH	247(3.61)	94-3097-86
$C_{16}H_{26}Si$			
Silane, bicyclo[4.1.0]hepta-1,3,5-trien-7-yltris(1-methylethyl)-	CH_2Cl_2	280(3.51)	33-1623-86
$C_{16}H_{27}NO_4SSi$			
4-Thia-1-azabicyclo[3.2.0]hept-2-ene-2-carboxylic acid, 6-[1-[[(1,1-dimethylethyl)dimethylsilyl]oxy]ethyl]-3-methyl-7-oxo-, methyl ester	EtOH	269(3.57),313(3.82)	44-3413-86
$C_{16}H_{27}NO_4S_2Si$			
4,5-Dithia-1-azabicyclo[4.2.0]oct-2-ene-2-carboxylic acid, 7-[1-[[(1,1-dimethylethyl)dimethylsilyl]oxy]ethyl]-3-methyl-8-oxo-, methyl ester, [6R-[6α,7α(R*)]]-	EtOH	277(3.78),331(3.49)	44-3413-86
[6R-[6α,7β(R*)]]-	EtOH	278(3.83),327(3.49)	44-3413-86
$C_{16}H_{27}NO_5S_2Si$			
4,5-Dithia-1-azabicyclo[4.2.0]oct-2-ene-2-carboxylic acid, 7-[1-[[(1,1-dimethylethyl)dimethylsilyl]oxy]ethyl]-3-methyl-8-oxo-, methyl ester, 4α-oxide	hexane	272(3.62),305s(--)	44-3413-86
4β-oxide	hexane	273(3.69),309s(3.44)	44-3413-86
5-oxide	hexane	276(3.73)	44-3413-86
$C_{16}H_{27}NO_6S_2Si$			
4,5-Dithia-1-azabicyclo[4.2.0]oct-2-ene-2-carboxylic acid, 7-[1-[[(1,1-dimethylethyl)dimethylsilyl]oxy]ethyl]-3-methyl-8-oxo-, methyl ester, 4,4-dioxide	hexane	276(3.78),297s(3.65)	44-3413-86
$C_{16}H_{27}N_2$			
Piperidinium, 1-[1-methyl-5-(1-piperidinyl)-2,4-pentadienylidene]-, perchlorate	EtOH	430(5.09)	33-1588-86
Piperidinium, 1-[3-methyl-5-(1-piperidinyl)-2,4-pentadienylidene]-, bromide	EtOH	418s(4.72),440(5.08)	33-1588-86
iodide	EtOH	418s(4.69),440(5.09)	33-1588-86
perchlorate	EtOH	418s(4.72),440(5.07)	33-1588-86
$C_{16}H_{27}N_3O_2S$			
2,6-Diazabicyclo[2.2.2]oct-7-ene-3,5-dione, 8-(diethylamino)-1-(ethylthio)-2,4,6,7-tetramethyl-	MeCN	219(4.51),261(4.11), 320(3.65)	24-1315-86
$C_{16}H_{27}N_3O_4$			
5-Pyrimidinepentanoic acid, 2-amino-1,4-dihydro-6-hydroxy-4-oxo-, heptyl ester	50%MeOH-HCl	266.5(4.17)	73-1140-86
	50%MeOH-NaOH	271(4.14)	73-1140-86

Compound	Solvent	$\lambda_{max}(\log \epsilon)$	Ref.
$C_{16}H_{28}N_4O_4$			
2,5-Cyclohexadiene-1,4-dione, 2,5-bis-[[2-(dimethylamino)ethyl]amino]-3,6-dimethoxy-, dihydrochloride	MeOH	228(4.39),360(4.34)	87-1792-86
$C_{16}H_{28}O_2$			
4-Octene-3,6-dione, 4-(1,1-dimethyl-ethyl)-2,2,7,7-tetramethyl-, (Z)-	ether	233(3.91)	78-1345-86
12-Tridecynoic acid, 1-methyethyl ester	MeOH	205(3.23)	18-3535-86
$C_{16}H_{29}N_3O_4S_2$			
L-Alanine, N-[N-[N-[(1,1-dimethyleth-oxy)carbonyl]thio-L-alanyl]-2-methylthioalanyl]-, methyl ester	MeOH	216(3.30),268(3.40)	78-6555-86
$C_{16}H_{29}N_3O_5S$			
L-Alanine, N-[N-[N-[(1,1-dimethyleth-oxy)carbonyl]-L-alanyl]-2-methyl-thioalanyl]-, methyl ester	MeOH	270(3.70)	78-6555-86
L-Alanine, N-[N-[N-[(1,1-dimethyleth-oxy)carbonyl]thio-DL-alanyl]-2-methylalanyl]-, methyl ester	MeOH	202(3.90)	78-6555-86
$C_{16}H_{29}N_3O_6$			
L-Alanine, N-[N-[N-[(1,1-dimethyleth-oxy)carbonyl]-L-alanyl]-2-methyl-alanyl]-, methyl ester	MeOH	202(3.60)	78-6555-86
$C_{16}H_{30}N_4O_4S_2$			
6,14,17,22-Tetraoxa-1,3,9,11-tetraaza-bicyclo[9.8.5]tetracosane-2,10-di-thione	EtOH	210.6(4.32),249.3(4.42)	104-1589-86
$C_{16}H_{30}N_6O_9$			
2-Pentanone, 3-amino-1-diazo-4-meth-yl-, 2:1 tartrate	MeOH	245(3.81),282(4.08)	104-0353-86
$C_{16}H_{30}O$			
11,13-Hexadecadien-1-ol, (E,E)-	EtOH	230(4.36)	44-4934-86
$C_{16}H_{30}OSi$			
Ethanone, 1-(1,1-dimethylethyl)dimeth-ylsilyl-2-[1-methyl-2-(2-methyl-1-propenyl)cyclopropyl]-	pentane	349s(1.78),364(2.00), 378(2.15),395(2.06)	33-1378-86
isomer B	pentane	350s(1.78),366(2.00), 381(2.15),398(2.11)	33-1378-86
1,6-Octadiene, 1-[(1,1-dimethylethyl)-dimethylsilyloxy]-7-methyl-3-meth-ylene-, (E)-	pentane	240(4.31)	33-1378-86
2,6-Octadien-1-one, 1-[(1,1-dimethyl-ethyl)dimethylsilyl]-3,7-dimethyl-, (E)-	pentane	254(4.14),441(2.16), 461(2.13),490(1.78)	33-1378-86
(Z)-	pentane	256(3.91),443(1.98), 461(1.95),490(1.65)	33-1378-86
3,6-Octadien-1-one, 1-[(1,1-dimethyl-ethyl)dimethylsilyl]-3,7-dimethyl-, (E)-	pentane	346(1.90),361(2.06), 374(2.18),388(2.18)	33-1378-86
1,3,6-Octatriene, 1-[(1,1-dimethyl-ethyl)dimethylsilyloxy]-3,7-di-methyl-, (E,E)-	pentane	242(4.27),282(3.38)	33-1378-86

Compound	Solvent	$\lambda_{max}(\log \epsilon)$	Ref.
$C_{16}H_{30}S$			
Thiepin, 4-(1,1-dimethylethyl)-5-ethyl-2,3,6,7-tetrahydro-3,3,6,6-tetramethyl-	C_6H_{12}	210(3.65),225(3.53), 251(3.58)	78-1693-86
$C_{16}H_{31}NO_2$			
Tetradecanamide, N-acetyl-	EtOH	195.3(2.91),200.2(4.06), 202.2(4.16),205.7(4.22)	104-1602-86
$C_{16}H_{32}$			
1-Butene, 1,1,2-tris(1,1-dimethylethyl)-	C_6H_{12}	245(3.57)	78-1693-86
$C_{16}H_{34}Cl_2N_8O_4$			
1,2-Ethanediamine, N,N'-bis[2-[[[(2-chloroethyl)nitrosoamino]carbonyl]amino]ethyl]-N,N,N',N'-tetramethyl-, dibromide	H_2O	198(4.30),235(3.96), 398(2.20)	70-0555-86
$C_{16}H_{48}Si_8$			
Cyclopentasilane, heptamethyl-1,1,3-tris(trimethylsilyl)-	C_6H_{12}	237(4.21),275s(3.3)	157-0128-86

Compound	Solvent	$\lambda_{max}(\log \epsilon)$	Ref.
$C_{17}H_7N_3O_4$			
Benzo[f]pyrido[1,2-a]indole-2-carbonitrile, 6,11-dihydro-12-nitro-6,11-dioxo-	DMF	474(3.8195)	2-1126-86
$C_{17}H_7O_2S_8$			
Tetrathiofulvalene complex with tetrathiofulvalenecarboxylic acid radical cation	C_6H_5CN	400s(--),438(4.28), 590(3.80)	104-2131-86
$C_{17}H_8Cl_2O$			
Benzanthrone, 6,8-dichloro-	benzene	345s(3.63),361(3.91), 386(4.02),402(3.99)	104-0566-86
$C_{17}H_8F_6N_2O$			
Methanone, [2,3-bis(trifluoromethyl)-6-quinoxalinyl]phenyl-	MeOH	251(4.43)	39-1043-86C
$C_{17}H_8N_2O_2$			
Benzo[f]pyrido[1,2-a]indole-12-carbonitrile, 6,11-dihydro-6,11-dioxo-	DMF	352(3.9031),476(3.8325)	2-1126-86
$C_{17}H_8N_4O_7$			
2H-1-Benzopyran-3-carboxylic acid, 8-[(cyanonitrophenyl)azo]-7-hydroxy-2-oxo-	n.s.g.	570(4.46)	2-0638-86
$C_{17}H_9ClO_3$			
1,6-Pyrenedione, 8-chloro-3-methoxy-	C_6H_5Cl	396(3.09),430(3.08), 452(3.02)	104-1159-86
$C_{17}H_9FN_4OS$			
Imidazo[5,1-a]phthalazine-10b(1H)-carbonitrile, 2-(4-fluorophenyl)-2,3-dihydro-3-oxo-1-thioxo-	EtOH	210(4.65),278(4.35), 316(3.97)	150-1901-86M
$C_{17}H_9NO_4$			
1,3-Isobenzofurandione, 5-(5-phenyl-2-oxazolyl)-	toluene	364(4.27)	103-1026-86
$C_{17}H_{10}ClN_3O_3$			
Acetamide, N-(6-chloro-5-oxo-5H-naphtho[2,1-b]pyrido[2,3-e][1,4]oxazin-1-yl)-	$CHCl_3$	267(4.16),354(4.01), 370s(3.97),456(4.15)	4-1697-86
Acetamide, N-(6-chloro-5-oxo-5H-naphtho[2,1-b]pyrido[2,3-e][1,4]oxazin-4-yl)-	$CHCl_3$	265(4.10),285s(3.99), 297s(3.96),353(4.01), 485(4.16)	4-1697-86
$C_{17}H_{10}Cl_2O_2S$			
Ethanone, 2,2-dichloro-1-[9-(methylthio)acenaphtho[1,2-c]furan-7-yl]-	CH_2Cl_2	240(4.30),266.7(3.97), 280(3.97),315(4.15), 343s(3.82),356s(3.77), 405(3.86)	24-2317-86
$C_{17}H_{10}N_4$			
Cyanamide, (2-methyl-9,10-anthracenediylidene)bis-	MeCN	214(4.39),275(4.56), 36[sic](4.38)	5-0142-86
$C_{17}H_{10}N_4O$			
Azepino[2,1-b]pteridin-11-one, 6,7,8-9,10,10a-hexahydro-3-phenyl-	MeOH	338(4.17)	33-0793-86

Compound	Solvent	λ_{max}(log ϵ)	Ref.
$C_{17}H_{10}O_4$			
Anthra[1,2-b]furan-6,11-dione, 2-(hydroxymethyl)-	n.s.g.	234(4.25),244(4.24), 275(4.34),358(3.71)	12-2075-86
$C_{17}H_{10}O_5$			
Pentanepentone, diphenyl-	CH_2Cl_2	437(2.19),547(2.11)	89-0999-86
$C_{17}H_{11}BO_6$			
Boron, (1,3-diphenyl-1,3-propanedionato-O,O')[ethanedioato(2-)-O,O']-, (T-4)-	CH_2Cl_2	276(3.81),294(3.95), 376(4.60),390(4.63)	48-0755-86
	CH_2Cl_2	385(4.33)	97-0399-86
$C_{17}H_{11}ClO_2$			
1,2-Naphthalenedione, 4-[(4-chlorophenyl)methyl]-	$CHCl_3$	335(4.44),399(4.54), 519(1.82)	39-1627-86C
1(4H)-Naphthalenone, 4-[(4-chlorophenyl)methylene]-2-hydroxy-	$CHCl_3$	311(4.00),396(4.38)	39-1627-86C
$C_{17}H_{11}F_3N_2O_7S_2$			
2,7-Naphthalenedisulfonic acid, 6-hydroxy-5-[[3-(trifluoromethyl)phenyl]azo]-, disodium salt	pH 7.1	320(4.0),400s(3.9), 483(4.3),500s(4.3)	60-3141-86
$C_{17}H_{11}N_3O$			
Ethenetricarbonitrile, [5-(2,4-dimethylphenyl)-2-furanyl]-	MeOH	417(3.15),475(3.18)	73-1450-86
Ethenetricarbonitrile, [5-(2,5-dimethylphenyl)-2-furanyl]-	acetone	468(3.39)	73-1450-86
Ethenetricarbonitrile, [5-(3,4-dimethylphenyl)-2-furanyl]-	MeOH	486(3.16)	73-1450-86
$C_{17}H_{11}N_3O_2$			
Benzoic acid, 4-(2H-naphtho[1,2-d]triazol-2-yl)-	toluene	332(4.17)	24-1627-86
	EtOH	339(4.18)	24-1627-86
	DMSO	343.5(4.31)	24-1627-86
$C_{17}H_{11}N_3O_3$			
Acetamide, N-(5-oxo-5H-naphtho[2,1-b]pyrido[2,3-e][1,4]oxazin-1-yl)-	$CHCl_3$	262(4.28),352(4.08), 365s(4.07),437(4.25)	4-1697-86
Acetamide, N-(5-oxo-5H-naphtho[2,1-b]pyrido[2,3-e][1,4]oxazin-4-yl)-	$CHCl_3$	265(4.24),292(4.09), 352(4.06),472(4.23)	4-1697-86
2H-1-Benzopyran-3-carbonitrile, 7-hydroxy-8-[(methylphenyl)azo]-2-oxo-	n.s.g.	435(4.65)	2-0638-86
$C_{17}H_{11}N_3O_9S$			
2H-1-Benzopyran-3-carboxylic acid, 7-hydroxy-8-[(methylsulfonyl)nitrophenyl]-2-oxo-	n.s.g.	560(4.40)	2-0638-86
$C_{17}H_{12}$			
Benzo[a]fluorene, cesium salt	THF	433(3.99)	35-7016-86
lithium salt	THF	454(3.93)	35-7016-86
Benzo[b]fluorene, cesium salt	THF	415(4.34)	35-7016-86
lithium salt	THF	430(4.41)	35-7016-86
Benzo[c]fluorene, cesium salt	THF	379(3.88),394(3.72)	35-7016-86
lithium salt	THF	387(3.89),410(3.82)	35-7016-86
3H-Cyclonona[def]biphenylene, (Z,Z)-	C H	215(4.53),257(4.38), 268(4.36),280(4.36), 368(3.65),388(3.68), 404(3.50)	35-7693-86

Compound	Solvent	λ_{max}(log ϵ)	Ref.
Dicyclopenta[ef,kl]heptalene, 1-methyl-	hexane	254(4.30),267(4.55), 287(4.14),302(3.90), 312(3.88),334(3.56), 347(3.68),359(3.14), 410(2.55),442(2.71), 452(2.78),460(2.68), 472(3.01),484(3.60)	44-2961-86
$C_{17}H_{12}ClNO_3$			
Benzoic acid, 4-[5-(4-chlorophenyl)-2-oxazolyl]-, methyl ester	toluene	338(4.45)	103-1022-86
$C_{17}H_{12}ClNO_3$			
3H-Indol-3-one, 1-acetyl-2-(4-chlorobenzoyl)-1,2-dihydro-	EtOH	207(4.08),243(4.33), 263(4.28),345(3.40)	104-2186-86
$C_{17}H_{12}Cl_2N_2O_2$			
Methanimidamide, N'-(2,4-dichloro-9,10-dihydro-9,10-dioxo-1-anthracenyl)-N,N-dimethyl-	EtOH	260(4.2),270(4.2), 325(3.8),400(3.7)	104-1566-86
hydrochloride	EtOH	320(3.7)	104-1566-86
$C_{17}H_{12}N_2O$			
Benzonitrile, 4-[5-(4-methylphenyl)-2-oxazolyl]-	toluene	339(4.34)	103-1026-86
5H-Cyclohepta[4,5]pyrrolo[2,3-b]indole, 5-acetyl-	MeOH	241(4.27),284(4.28), 297(4.29),325(4.56), 480(3.12),500(3.11), 545s(2.83)	18-3326-86
$C_{17}H_{12}N_2O_2$			
Benzonitrile, 4-[5-(4-methoxyphenyl)-2-oxazolyl]-	toluene	350(4.39)	103-1026-86
4H-1-Benzopyran-4-one, 2-(1H-benzimidazol-2-yl)-6-methyl-	EtOH	231(4.34),280(3.92), 337(3.95)	42-0600-86
4H-1-Benzopyran-4-one, 2-(1H-benzimidazol-2-yl)-8-methyl-	EtOH	230(4.40),284(3.94), 338(4.22)	42-0600-86
$C_{17}H_{12}N_2O_2S$			
4H-Thiopyran, 4-diazo-2,6-diphenyl-, 1,1-dioxide	CH_2Cl_2	328(5.35)	44-3551-86
$C_{17}H_{12}N_2O_3$			
1H-Anthra[1,2-c]pyrazolium, 2,3,6,11-tetrahydro-2,2-dimethyl-3,6,11-trioxo-, hydroxide, inner salt	toluene	658(3.92)	104-2117-86
4H-1-Benzopyran-4-one, 2-(1H-benzimidazol-2-yl)-7-methoxy-	EtOH	229(4.48),257(4.22), 279(4.06),336(4.07)	42-0600-86
$C_{17}H_{12}N_2O_4$			
9H-Pyrido[3,4-b]indole-3-carboxylic acid, 1-[5-(hydroxymethyl)-2-furanyl]- (flazin)	MeOH	263(4.43),290(4.36), 355(4.07),370(4.09)	88-3399-86
$C_{17}H_{12}N_2O_5S$			
3-Pyrrolidinesulfonic acid, 1-(9-acridinyl)-2,5-dioxo-	pH 6.0	251(5.19),362(4.09)	163-1139-86
$C_{17}H_{12}N_4$			
Pyrazolo[1,5-b][1,2,4]triazine, 2,3-diphenyl-	MeOH	244(4.45),276(4.25), 337(2.20)	118-0071-86

Compound	Solvent	$\lambda_{max}(\log \epsilon)$	Ref.
$C_{17}H_{12}N_4O_3$			
2-Anthracenecarboxamide, 1-azido-9,10-dihydro-N,N-dimethyl-9,10-dioxo-	toluene	342(3.45),362(3.43)	104-2117-86
$C_{17}H_{12}O$			
Cyclopenta[a]phenalene, 7-methoxy-	MeCN	259s(4.21),265(4.22), 278s(4.15),294s(4.06), 305s(4.03),352s(3.57), 369s(3.63),402(3.80), 425(3.86),451(3.72), 519(3.46)	77-1130-86
$C_{17}H_{12}O_2$			
1,2-Naphthalenedione, 4-(phenylmethyl)-	$CHCl_3$	337(3.41),402(3.39), 518(1.78)	39-1627-86C
$C_{17}H_{12}O_3$			
2H-Benzofuro[3,2-g]-1-benzopyran-2-one, 4,11-dimethyl-	EtOH	251(4.49),268(4.43), 286(4.38),338(4.09), 365(3.34)	149-0001-86B
1,4-Naphthalenedione, 5-methoxy-2-phenyl-	EtOH	235s(4.43),300(3.92), 408(3.81)	94-2810-86
$C_{17}H_{12}O_5$			
9,10-Anthracenedione, 1-hydroxy-2-(1-hydroxy-2-oxopropyl)-	n.s.g.	227(4.30),255(4.46), 280s(3.53),407(3.79)	12-2075-86
$C_{17}H_{12}O_6$			
Anthra[2,3-c]furan-5,10-dione, 1,3-dihydro-4,11-dihydroxy-1-methoxy-	n.s.g.	228(3.82),250(4.09), 285(3.56),436(3.49)	12-2075-86
1,2,4,5-Pentanetetrone, 3,3-dihydroxy-1,5-diphenyl-	CH_2Cl_2	284(4.28),403(2.10)	89-0999-86
$C_{17}H_{13}Br_2ClO_6$			
Ethanone, 1-(5,8-diacetoxy-6,7-dibromo-4-chloro-1-hydroxy-3-methyl-2-naphthalenyl]-	$CHCl_3$	244(4.59),249(4.59), 277(4.37),307(3.77), 320(3.75),336(3.67), 350(3.58),397(3.69)	64-0377-86B
$C_{17}H_{13}ClN_2O$			
[1]Benzopyrano[4,3,2-de]quinolin-2-amine, 6-chloro-N,N-dimethyl-	MeOH	219(4.6),257(4.65), 310(4.09),338(3.98), 358(4.07),392(3.62), 416(3.44)	83-0347-86
$C_{17}H_{13}ClN_2OS_2$			
Benzo[b]thiophene-2-carboxamide, 3-chloro-N-[[(phenylmethyl)amino]-thioxomethyl]-	EtOH	209(3.39),251(3.28), 306(3.27)	73-2839-86
$C_{17}H_{13}ClN_4$			
1H-Pyrazolo[3,4-b]quinolin-4-amine, N-(4-chlorophenyl)-1-methyl-	EtOH	218(4.43),250(4.75), 382s(4.05),395(4.11)	11-0331-86
$C_{17}H_{13}ClO$			
1(2H)-Naphthalenone, 2-[(3-chlorophenyl)methylene]-3,4-dihydro-	EtOH	227(4.39),295(4.53), 321(4.39)	4-0135-86
1(2H)-Naphthalenone, 2-[(4-chlorophenyl)methylene]-3,4-dihydro-	EtOH	240(3.73),272(3.92), 310(4.16)	4-0135-86

Compound	Solvent	$\lambda_{max}(\log \epsilon)$	Ref.
$C_{17}H_{13}CoN_2S_2$			
Cobalt, (η^5-2,4-cyclopentadien-1-yl)-[1,2-di-2-pyridinyl-1,2-ethenedithiolato(2-)-S,S']-	MeOH	293(4.78),577(4.14)	138-1537-86
$C_{17}H_{13}CoN_2Se_2$			
Cobalt, (η^5-2,4-cyclopentadien-1-yl)-[1,2-di-2-pyridinyl-1,2-ethenediselenolato(2-)-Se,Se']-	MeOH	304(--),590(--), 802(--)	138-1537-86
$C_{17}H_{13}DO$			
2,4,6,12,14-Cyclopentadecapentaene-8,10-diyn-1-one-15-d, 7,12-dimethyl-, (E,E,Z,Z,E)-	ether	245(4.06),257(4.12), 302(4.55),382(3.77)	18-1723-86
$C_{17}H_{13}FN_4$			
1H-Pyrazolo[3,4-b]quinolin-4-amine, N-(4-fluorophenyl)-1-methyl-	EtOH	213(4.27),251(4.72), 380(3.90),396(3.92)	11-0331-86
$C_{17}H_{13}N$			
Benz[c]acridine, 5,6-dihydro-	EtOH	224(4.14),263(4.16), 298(3.88),315(3.91), 330(4.04),345(4.07)	4-0689-86
Quinoline, 3-(2-phenylethenyl)-, trans	C_6H_{12}	280(4.40),315(4.46), 360s(3.75)	151-0177-86D
	EtOH	283(--)	151-0177-86D
cation	C_6H_{12}	390(3.9)	151-0177-86D
	H_2O-EtOH	380(3.8)	151-0177-86D
Tricyclo[8.2.2.2^{4,7}]hexadeca-2,4,6-10,12,13,15-heptaene-2-carbonitrile	MeCN	218(4.22)	24-1836-86
$C_{17}H_{13}NO_2$			
1H-Pyrrole-2-carboxaldehyde, 3,4-diphenyl-	MeOH	206(4.35),236(4.16), 342(4.15)	83-0749-86
$C_{17}H_{13}NO_2S$			
1,5-Benzothiazepin-4(5H)-one, 5-acetyl-2-phenyl-	CH_2Cl_2	263.5(4.18),275s(4.10), 285s(4.02),323(3.89)	24-3109-86
$C_{17}H_{13}NO_2Se$			
Ethanone, 1-phenyl-2-[[(2-selenyl-3-benzofuranyl)methylene]amino]-	EtOH	245(4.02),266s(3.90), 293(4.11),435(3.95)	103-0608-86
$C_{17}H_{13}NO_3$			
Benzoic acid, 4-[5-(4-methylphenyl)-2-oxazolyl]-	toluene	341(4.36)	103-1022-86
Benzoic acid, 4-(5-phenyl-2-oxazolyl)-, methyl ester	toluene	335(4.50)	103-1022-86 103-1026-86
1H-Indol-3-ol, 1-acetyl-, benzoate	EtOH	206(4.20),242(4.36), 264(3.97),294(3.86), 305(3.89)	104-2186-86
3H-Indol-3-one, 1-acetyl-2-benzoyl-	EtOH	206(4.04),243(4.28), 265(3.78),345(3.52)	104-2186-86
1(2H)-Naphthalenone, 3,4-dihydro-2-[(4-nitrophenyl)methylene]-, (E)-	EtOH	273(4.20),310(4.55)	4-0135-86
$C_{17}H_{13}NO_3S$			
1,3-Propanedione, 2-(3-methyl-5-isoxazolyl)-1-phenyl-3-(2-thienyl)-	dioxan	254(4.13),355(4.22)	83-0242-86

Compound	Solvent	λ_{max}(log ϵ)	Ref.
$C_{17}H_{13}NO_4$			
[1,3]Dioxolo[4,5-j]pyrrolo[3,2,1-de]-phenanthridinium, 4,5-dihydro-2-hydroxy-8-methoxy-, hydroxide, inner salt (zeflabetaine)	MeOH	222s(4.02),267(4.68), 278s(4.15),288s(4.08), 292(3.90),305(3.98), 415(3.55)	102-1975-86
	MeOH-HCl	267(--),275(--), 282(--),292(--), 300s(--),368(--)	102-1975-86
	MeOH-NaOMe	255s(--),267(--), 280s(--),435(--)	102-1975-86
Enterocarpam II	MeOH	246(4.8),255(4.47), 293.5(4.4),405(4.18)	102-0965-86
	NaOH	255(4.80),323(4.05), 450(4.00)	102-0965-86
Graveoline, 3'-hydroxy-	MeOH	224(4.5),270s(--), 330s(--),338(4.5)	102-2692-86
$C_{17}H_{13}N_2$			
Di(3-indolyl)carbenium tetrafluoroborate	MeOH	481(5.26)	5-1621-86
$C_{17}H_{13}N_3$			
1H-Indene-4,6-dicarbonitrile, 5-amino-2,3-dihydro-7-phenyl-	EtOH	363(3.89)	64-1471-86B
2H-Naphtho[1,2-d]triazole, 2-(4-methylphenyl)-	EtOH	337(4.38)	61-0439-86
$C_{17}H_{13}N_3O$			
4H-Pyrazolo[3,4-c]quinolin-4-one, 2,5-dihydro-1-methyl-2-phenyl-	EtOH	246(4.53),253s(4.46), 268s(4.15),295(4.00)	142-3181-86
4H-Pyrazolo[3,4-c]quinolin-4-one, 2,5-dihydro-5-methyl-2-phenyl-	EtOH	253s(4.41),259(4.44), 280(4.36),298s(4.13)	142-3181-86
$C_{17}H_{13}N_3O_2S$			
4(1H)-Pyrimidinethione, 1-methyl-5-nitro-2,6-diphenyl-	EtOH	245(4.04),338(4.28)	18-3871-86
4(1H)-Pyrimidinethione, 6-methyl-5-nitro-1,2-diphenyl-	EtOH	243(4.15),308(4.23), 386(4.12)	18-3871-86
$C_{17}H_{13}N_3O_3$			
Glycine, N-[(1-cyano-2H-benz[f]isoindol-2-yl)acetyl]-, monosodium salt	MeOH	251(4.79),400s(3.62), 418(3.82),442(3.78)	44-3978-86
$C_{17}H_{13}N_3O_3S$			
2,4,6(1H,3H,5H)-Pyrimidinetrione, 1,3-dimethyl-5-thiopyrano[3,2-b]indol-4(5H)-ylidene-	dioxan	228(4.23),245(4.27), 361(3.70),502(4.53)	83-0120-86
$C_{17}H_{13}N_3O_4$			
3,5-Pyrazolidinedione, 1-methyl-4-[(4-nitrophenyl)methylene]-2-phenyl-	EtOH	260(4.38),324(4.15)	103-0176-86
2,4,6(1H,3H,5H)-Pyrimidinetrione, 1,3-dimethyl-5-pyrano[3,2-b]indol-4(5H)-ylidene-	MeOH	213(4.19),238(4.26), 327(3.84),452(4.49)	83-0108-86
$C_{17}H_{13}N_5$			
1,2,4-Triazolo[3,4-f][1,6]naphthyridine-6-carbonitrile, 5,7-dihydro-7-methyl-3-phenyl-	EtOH	218(4.35),243(4.40), 335(4.01),348(3.96), 450(3.36)	128-0027-86

Compound	Solvent	λ_{max}(log ε)	Ref.
[1,2,4]Triazolo[5,1-a][2,7]naphthyridine-6-carbonitrile, 5,9-dihydro-9-methyl-2-phenyl-	EtOH	210(4.39),250(4.41), 352(4.29),368(4.28)	128-0027-86
$C_{17}H_{13}N_5NiO_2S$			
Nickel, [2-[[1-(2-benzothiazolylazo)-propyl]azo]benzoato(2-)]-, (SP-4-2)-	isoPrOH	540(4.26),1080(3.26)	65-0727-86
$C_{17}H_{13}N_5O_2S$			
1H-Pyrazino[2,3-c][1,2,6]thiadiazin-4-amine, 6,7-diphenyl-, 2,2-dioxide anion	pH 1.0	276(4.26),369(3.97), 441(3.45)	5-1872-86
	pH 7.0	285(4.33),405(3.98)	5-1872-86
$C_{17}H_{13}N_5O_2SZn$			
Zinc, [2-[[1-(2-benzothiazolylazo)-propyl]azo]benzoato(2-)]-, (T-4)-	isoPrOH	630(3.34)	65-0727-86
$C_{17}H_{13}O$			
Pyrylium, 2,6-diphenyl-, tetrafluoroborate	THF	285(4.0),410(4.0)	88-4489-86
$C_{17}H_{14}$			
Azulene, 6-(4-methylphenyl)-	CH_2Cl_2	241(4.16),306.5(4.73), 357(3.92),372.5(3.95), 580.5(2.54),624.5(2.45), 689.5(1.97)	24-2272-86
$C_{17}H_{14}BrN_3Se$			
3H-Pyrazole-3-selone, 4-[[(4-bromophenyl)amino]methylene]-2,4-dihydro-5-methyl-2-phenyl-	EtOH	350(4.2),450(3.7)	104-2126-86
$C_{17}H_{14}ClFN_2O_2S$			
1H-1,4-Benzodiazepine, 7-chloro-5-(2-fluorophenyl)-2,3-dihydro-2-[(methylsulfonyl)methylene]-	isoPrOH	263(4.35),287(3.38), 343(3.40)	4-1303-86
$C_{17}H_{14}ClFN_2O_3S$			
1H-1,4-Benzodiazepine, 7-chloro-5-(2-fluorophenyl)-2,3-dihydro-2-[(methylsulfonyl)methylene]-, 4-oxide	isoPrOH	220s(4.40),240s(4.33), 282(4.58),355s(3.34)	4-1303-86
$C_{17}H_{14}ClN_3Se$			
3H-Pyrazole-3-selone, 4-[[(4-chlorophenyl)amino]methylene]-2,4-dihydro-5-methyl-2-phenyl-	EtOH	347(4.2),450(3.6)	104-2126-86
$C_{17}H_{14}Cl_2N_2$			
2-Quinolinamine, 6-chloro-4-(2-chlorophenyl)-N,N-dimethyl-	MeOH	219(4.64),255(4.54), 369s(3.86)	83-0338-86
4-Quinolinamine, 2,3-dichloro-6-methyl-N-(4-methylphenyl)-	EtOH	235(4.66)	104-0973-86
$C_{17}H_{14}Cl_2N_2O_2$			
4-Quinolinamine, 2,3-dichloro-6-methoxy-N-(4-methoxyphenyl)-	EtOH	244(4.63),351(3.92)	104-0973-86
$C_{17}H_{14}Cl_2N_8$			
1H-Tetrazole, 5,5'-(dichloromethylene)bis[1-(4-methylphenyl)-	EtOH	250(3.7)	104-0177-86

Compound	Solvent	λ_{max}(log ε)	Ref.
$C_{17}H_{14}Cl_2N_8O_2$			
1H-Tetrazole, 5,5'-(dichloromethylene)bis[1-(4-methoxyphenyl)-	EtOH	226(4.32),267(3.82)	104-0177-86
$C_{17}H_{14}Cl_4N_2$			
Propanediimidoyl dichloride, 2,2-dichloro-N,N'-bis(4-methylphenyl)-	EtOH	300(3.82)	104-0177-86
$C_{17}H_{14}Cl_4N_2O_2$			
Benzaldehyde, 2,2'-(1,3-propanediyldiimino)bis[3,6-dichloro-	EtOH	218(4.42),253(4.42), 402(3.91)	12-1231-86
$C_{17}H_{14}Cl_4N_2O_4$			
Benzoic acid, 2,2'-(1,3-propanediyldiimino)bis[3,6-dichloro-	EtOH	220(4.56),258(4.20), 312(3.69)	12-1231-86
$C_{17}H_{14}N_2$			
5H-Benzo[6,7]cyclohepta[1,2-b][1,8]-naphthyridine, 6,7-dihydro-	EtOH	232(4.18),327(4.05)	4-0689-86
Pyridine, 2,2'-bicyclo[2.2.1]hepta-2,5-diene-2,3-diylbis-	EtOH	223(4.06),316(3.93)	138-1279-86
Pyridine, 3,3'-bicyclo[2.2.1]hepta-2,5-diene-2,3-diylbis-	EtOH	231(4.08),252s(3.96), 306(3.92)	138-1279-86
Pyridine, 4,4'-bicyclo[2.2.1]hepta-2,5-diene-2,3-diylbis-	EtOH	243(4.06),315(3.83)	138-1279-86
5H-Pyrido[3',2':6,7]cyclohepta[1,2-b]-quinoline, 6,7-dihydro-	EtOH	240(4.15),269(4.09), 310(3.74),324(3.72)	4-0689-86
$C_{17}H_{14}N_2O$			
1H-Imidazole-4-carboxaldehyde, 2-phenyl-1-(phenylmethyl)-	MeOH	263(4.18)	87-0261-86
2-Naphthalenol, 1-[[(3-methyl-2-pyridinyl)imino]methyl]-	$CHCl_3$	462(4.26)	2-0127-86A
2-Naphthalenol, 1-[[(4-methyl-2-pyridinyl)imino]methyl]-	$CHCl_3$	460(4.25)	2-0127-86A
2-Naphthalenol, 1-[[(5-methyl-2-pyridinyl)imino]methyl]-	$CHCl_3$	462(4.30)	2-0127-86A
2-Naphthalenol, 1-[[(6-methyl-2-pyridinyl)imino]methyl]-	$CHCl_3$	462(4.31)	2-0127-86A
Pyrazolo[5,1-d][1,5]benzoxazepine, 4,5-dihydro-2-phenyl-	EtOH	271(4.34),275s(4.34), 284s(4.29),291s(4.26)	33-0927-86
$C_{17}H_{14}N_2O_2$			
Benzamide, 4-[5-(4-methylphenyl)-2-oxazolyl]-	toluene	334(4.35)	103-1022-86
1,5-Methano-1,2,3-oxadiazolo[3,2-a]-cinnolin-2(1H)-one, 5,6-dihydro-1-phenyl-	EtOH	230s(3.71),255.5s(2.96), 258s(2.93),262(2.86), 264s(2.83),268(2.73), 274s(2.45)	33-0927-86
2-Naphthalenol, 1-[(2-methoxyphenyl)-azo]-	hexane	490(4.5),510s(4.5)	48-0089-86
	EtOH	510(4.5)	48-0089-86
Nordragabine	EtOH	228(4.36),264s(3.92), 297(3.70)	77-1481-86
	EtOH-HCl	228(4.29),318(3.66), 375(3.32)	77-1481-86
3,5-Pyrazolidinedione, 1-methyl-2-phenyl-4-(phenylmethylene)-	EtOH	235(4.01),244(3.95), 330(4.24)	103-0176-86
3H-Pyrazol-3-one, 4-benzoyl-2,4-dihydro-5-methyl-2-phenyl-	pH 2.3	250(4.3)	86-0288-86
	pH 10.3	270(4.2)	86-0288-86
Sydnone, 4-phenyl-3-[2-(2-propenyl)-phenyl]-	EtOH	245(3.88),260s(3.71), 332(3.95)(changing)	33-0927-86

Compound	Solvent	$\lambda_{max}(\log \epsilon)$	Ref.
$C_{17}H_{14}N_2O_3$			
2-Anthracenecarboxamide, 1-amino-9,10-dihydro-N,N-dimethyl-9,10-dioxo-	toluene	474(4.03)	104-2117-86
Benzamide, 4-[5-(4-methoxyphenyl)-2-oxazolyl]-	toluene	345(4.35)	103-1022-86
3,5-Pyrazolidinedione, 4-[(4-methoxyphenyl)methylene]-1-phenyl-	EtOH-acid	252(4.33),372(4.54)	103-0176-86
	EtOH-base	246(4.33),348(4.37)	103-0176-86
6-Quinolinecarboxylic acid, 5-hydroxy-8-(phenylamino)-, methyl ester	MeOH	240(4.30),270(4.40), 372(3.60),418(3.30)	4-1641-86
6-Quinolinecarboxylic acid, 8-hydroxy-5-(phenylamino)-, methyl ester	MeOH	244(4.40),260(4.20), 348(3.40),392(3.50)	4-1641-86
7-Quinolinecarboxylic acid, 5-hydroxy-8-(phenylamino)-, methyl ester	MeOH	244(4.40),272(4.40), 364(3.60),408(3.70)	4-1641-86
Sydnone, 4-phenyl-3-[2-(2-propenyl)-phenyl]-	EtOH	244(3.97),265s(3.74), 331.5(4.03)	33-0927-86
$C_{17}H_{14}N_2O_4$			
2-Anthracenecarboxylic acid, 1-(2,2-dimethylhydrazino)-9,10-dihydro-9,10-dioxo-	n.s.g.	275(4.02),313(3.73), 488(3.76)	104-2117-86
6,11-Methanocyclodeca[d]pyridazine-1,4-dicarboxylic acid, dimethyl ester	MeOH	244(4.27),268(4.36), 288(4.40),390(3.55)	24-3862-86
$C_{17}H_{14}N_2O_5$			
1-Phenazinecarboxylic acid, 6-[1-[(hydroxyacetyl)oxy]ethyl]-	MeOH	251(4.90),363(4.15)	158-0624-86
$C_{17}H_{14}N_2S$			
1-Isoquinolinecarbothioamide, 3-methyl-N-phenyl-	EtOH	218(4.43),292(4.13), 325(4.14)	150-1901-86M
$C_{17}H_{14}N_4$			
[1,2,4]Triazolo[1,5-a]pyridine-8-carbonitrile, 5,7-dimethyl-2-(2-phenylethenyl)-	DMF	344(4.28)	118-0860-86
$C_{17}H_{14}N_4O_2S$			
3H-Pyrazole-3-thione, 2,4-dihydro-5-methyl-4-[[(3-nitrophenyl)amino]-methylene]-2-phenyl-	toluene	355(4.4),445(3.8)	103-0503-86
3H-Pyrazole-3-thione, 2,4-dihydro-5-methyl-4-[[(4-nitrophenyl)amino]-methylene]-2-phenyl-	toluene	375(4.5),460(4.0)	103-0503-86
Spiro[benzothiazole-2(3H),5'(2'H)-pyrimidine]-2',4'(3'H)-dione, 6'-[(phenylmethyl)amino]-	MeCN	221(4.34),266(3.95)	94-0664-86
$C_{17}H_{14}N_4O_2Se$			
3H-Pyrazole-3-selone, 2,4-dihydro-5-methyl-4-[[(3-nitrophenyl)amino]-methylene]-2-phenyl-	EtOH	345(4.0),460(3.3)	104-2126-86
$C_{17}H_{14}O$			
1(2H)-Naphthalenone, 3,4-dihydro-2-(phenylmethylene)-	EtOH	225(4.00),275(4.01), 305(4.16)	4-0135-86
1,4-Pentadien-3-one, 1,5-diphenyl-	C_6H_{12}	229(4.3),320(4.5), 410f(1.8)	39-1147-86B
	EtOH	230(4.2),326(4.4)	39-1147-86B

Compound	Solvent	$\lambda_{max}(\log \epsilon)$	Ref.
Tricyclo[8.2.2.2^{4,7}]hexadeca-2,4,6-10,12,13,15-heptaene-2-carboxaldehyde	MeCN	208(4.08),221(4.20)	24-1836-86
$C_{17}H_{14}OS$			
2H-Thiopyran-3(6H)-one, 6,6-diphenyl-	hexane	228(4.03)	44-0314-86
$C_{17}H_{14}O_2$			
3(2H)-Benzofuranone, 5,7-dimethyl-2-(phenylmethylene)-, (Z)-	hexane	290(4.1),308(4.4), 323(4.4),380(4.2)	44-4432-86
4H-1-Benzopyran-4-one, 6,8-dimethyl-2-phenyl-	hexane	211(4.2),263(4.3), 300(4.2)	44-4432-86
1H-Cyclopenta[a]pentalene-1,4(3aH)-dione, 3b,6a,7,7a-tetrahydro-3-phenyl-	MeOH	220(4.19),296(4.24)	39-0291-86C
Furan, 2-methoxy-4,5-diphenyl-	n.s.g.	305(4.03)	104-2213-86
1-Naphthalenol, 8-methoxy-3-phenyl-	EtOH	256(4.15),302(3.31), 345(3.29)	94-2810-86
2-Propynoic acid, 3-phenyl-, 2,4-dimethylphenyl ester	hexane	260(4.4)	44-4432-86
2-Propyn-1-one, 1-(2-hydroxy-3,5-dimethylphenyl)-3-phenyl-	hexane	295(4.2),309(4.3), 380(3.7)	44-4432-86
$C_{17}H_{14}O_2S$			
Benzo[b]thiophene-6-carboxylic acid, 4-(4-methylphenyl)-, methyl ester	EtOH	244(4.51),249(4.61), 279(4.15),324(3.98)	5-1003-86
$C_{17}H_{14}O_3$			
7H-Benz[6,7]indeno[1,2-b]furan-7,9(8H)-dione, 7a,10a-dihydro-8,8-dimethyl-	EtOH	251(4.58),277s(3.74), 285(3.79),299s(3.64)	34-0369-86
2H-1-Benzopyran-2-one, 7-hydroxy-4-methyl-6-(phenylmethyl)-	MeOH	212(4.11),327(4.09)	2-0862-86
2H-1-Benzopyran-2-one, 4-methyl-7-(phenylmethoxy)-	MeOH	208(4.12),318(3.96)	2-0862-86
Cyclohepta[de]naphthalene-7,8-dione, 3-(1-methylethoxy)-	$CHCl_3$	364(3.95)	39-1965-86C
Cyclohepta[de]naphthalene-7,8-dione, 3-propoxy-	$CHCl_3$	252(4.36),260(4.40), 368(4.24)	39-1965-86C
2,5-Furandione, dihydro-3,3-dimethyl-4-(1-naphthalenylmethylene)-, (E)-	EtOH	221(4.80),325(3.78)	34-0369-86
Methanone, (6-ethoxy-2-benzofuranyl)-phenyl-	EtOH	232(3.94),254(3.93), 342(4.21)	142-0771-86
$C_{17}H_{14}O_3S$			
Benzo[b]thiophene-6-carboxylic acid, 4-(4-methoxyphenyl)-, methyl ester	EtOH	229.5(4.51),251.5(4.57), 276.5(4.17),328(3.91)	5-1003-86
9H-Thioxanthene-3-carboxylic acid, 7-methyl-9-oxo-, ethyl ester	CH_2Cl_2	403(3.78)	151-0353-86D
$C_{17}H_{14}O_4$			
9,10-Anthracenedione, 1,5-dihydroxy-2-propyl-	$CHCl_3$	256(4.17),280(3.80), 290(3.82),404(3.65), 426(3.81),436(3.82)	78-3303-86
7H-1-Benzopyran-7-one, 5-methoxy-2-(4-methoxyphenyl)- (bifloridin)	MeOH	240(4.0),265(3.82), 300s(3.08),315(3.36), 332(3.24),388(3.02)	102-1097-86
2(3H)-Furanone, dihydro-4-(4-phenoxybenzoyl)-	EtOH	262(4.05)	56-0143-86 +80-0541-86
Methanone, (4,6-dimethoxy-2-benzofuranyl)phenyl-	EtOH	251(4.35),349(4.42)	142-0771-86

Compound	Solvent	$\lambda_{max}(\log \epsilon)$	Ref.
$C_{17}H_{14}O_4S$			
9H-Thioxanthene-3-carboxylic acid, 7-methoxy-9-oxo-, ethyl ester	CH_2Cl_2	416(3.80)	151-0353-86D
$C_{17}H_{14}O_6$			
9,10-Anthracenedione, 2,5-dihydroxy-1,4-dimethoxy-7-methyl-	MeOH	225(4.63),250s(4.34), 254(4.37),260(4.34), 290(4.31),440(4.05)	78-3767-86
1,3-Azulenedicarboxylic acid, 2,6-diformyl-, 1-ethyl 3-methyl ester	CH_2Cl_2	249.5(4.39),278(4.17), 316(4.72),352(3.96), 581.5(2.80)	24-2272-86
4H-1-Benzopyran-4-one, 5-hydroxy-2-(4-hydroxyphenyl)-6,7-dimethoxy-	MeOH	275(4.23),336(4.49)	102-1255-86
	MeOH-NaOH	280(4.03),289s(3.74), 375(4.25)	102-1255-86
	MeOH-NaOAc	276(4.16),345s(4.17), 388(4.24)	102-1255-86
	MeOH-NaOAc-H_3BO_3	275(4.23),336(4.49)	102-1255-86
	MeOH-$AlCl_3$	265s(3.97),286s(4.23), 303(4.40),355(4.46)	102-1255-86
Sophorocarpan B	MeOH	282s(3.72),287(3.76), 310(3.89)	94-3067-86
9H-Xanthene-1-carboxylic acid, 8-hydroxy-2-methoxy-3,6-dimethyl-9-oxo-	MeOH	238(4.45),260(4.32), 296(4.00),346(3.80)	2-0860-86
$C_{17}H_{14}O_8$			
4H-1-Benzopyran-4-one, 3,5,7-trihydroxy-2-(4-hydroxy-3-methoxyphenyl)-6-methoxy-	MeOH	255(4.29),270s(4.13), 295(3.86),340s(4.22), 367(4.35)	102-0231-86
	MeOH-NaOMe	240s(4.22),333(4.32)	102-0231-86
	MeOH-NaOAc	270(4.25),275s(4.25), 320(4.04),390(4.32)	102-0231-86
	MeOH-NaOAc-H_3BO_3	255(4.28),270s(4.14), 370(4.33)	102-0231-86
	MeOH-$AlCl_3$	266(4.35),300s(3.78), 380s(4.17),430(4.41)	102-0231-86
	MeOH-$AlCl_3$-HCl	266(4.36),305(3.75), 370s(4.10),430(4.34)	102-0231-86
$C_{17}H_{15}BrN_2O_3$			
Acetamide, N-[2-(3-bromopropyl)-2,3-dihydro-1,3-dioxo-1H-benz[de]isoquinolin-6-yl]-	pH 7.4	354(4.15)	4-0849-86
$C_{17}H_{15}ClN_2$			
2H-Benz[g]indazole, 3-(3-chlorophenyl)-3,3a,4,5-tetrahydro-	EtOH	243(3.73)	4-0135-86
2H-Benz[g]indazole, 3-(4-chlorophenyl)-3,3a,4,5-tetrahydro-	EtOH	223(3.90),240(3.64), 300(4.06)	4-0135-86
$C_{17}H_{15}ClN_2O_4$			
2H-Pyrido[1,2-a]pyrimidine-8-carboxylic acid, 9-(4-chlorobenzoyl)-1,3,4,6-tetrahydro-6-oxo-, methyl ester	EtOH	242(4.20),360(3.86)	142-2247-86
$C_{17}H_{15}ClN_6S$			
Propanedinitrile, [[4-chloro-2-[[4-(diethylamino)phenyl]azo]-5-thiazolyl]methylene]-	CH_2Cl_2	645(4.79)	77-1639-86
	DMSO	668(--)	77-1639-86

Compound	Solvent	λ_{max}(log ϵ)	Ref.
$C_{17}H_{15}ClO_6$			
Ethanone, 1-(5,8-diacetoxy-4-chloro-1-hydroxy-3-methyl-2-naphthalenyl)-	$CHCl_3$	242(4.25),268(4.33), 300(3.67),313(3.71), 325(3.69),391(3.70)	64-0377-86B
Spiro[benzofuran-2(3H),1'-[2,5]cyclohexadiene]-3,4'-dione, 7-chloro-2',4,6-trimethoxy-6'-methyl-	MeOH	214(4.48),292(4.48)	94-0858-86
$C_{17}H_{15}Cl_2NO_6S$			
1,2-Benzenedicarboxylic acid, 3,6-dichloro-4-[[(4-methylphenyl)sulfonyl]amino]-, dimethyl ester	MeCN	220(4.56),229s(4.47), 256(4.05),269s(3.89), 275s(3.72),295(3.32), 300(3.32)	24-0616-86
7-Azabicyclo[2.2.1]hepta-2,5-diene-2,3-dicarboxylic acid, 5,6-dichloro-7-[(4-methylphenyl)sulfonyl]-, dimethyl ester	EtOH	228(4.31),274s(4.07), 310s(2.53)	24-0616-86
$C_{17}H_{15}DO$			
3,5,8,10,12-Pentadecapentaene-1,14-diyn-7-one-6-d, 3,13-dimethyl-	ether	230s(3.90),246s(4.08), 360(4.51)	18-1723-86
$C_{17}H_{15}FS$			
Dibenzo[b,f]thiepin, 7-fluoro-2-(1-methylethyl)-	MeOH	263.5(4.33),293s(3.58)	73-0698-86
$C_{17}H_{15}F_3N_4S$			
Spiro[cyclopentane-1,4'-[4H]indene]-5',5',7'(1'H)-tricarbonitrile, 6'-amino-2',3'-dihydro-1'-[(trifluoromethyl)thio]-	EtOH	208.3(4.16),317.5(4.20)	155-0201-86
$C_{17}H_{15}N$			
5H-Benzo[a]carbazole, 10,11-dihydro-5-methyl-	n.s.g.	217(4.22),245(4.35), 320(4.30)	150-0514-86M
Tricyclo[8.2.2.2^{4,7}]hexadeca-4,6,10-12,13,15-hexaene-2-carbonitrile	MeCN	203(3.95),224(4.19)	24-1836-86
$C_{17}H_{15}NO$			
Azacyclohexadeca-3,5,11,13,15-pentaene-7,9-diyn-2-one, 6,11-dimethyl-, (Z,E,Z,Z,E)-	THF	253s(4.14),266s(4.27), 292(4.48),310s(4.32), 342s(3.75),463(2.89)	39-0933-86C
2,4,6,12,14-Cyclopentadecapentaene-8,10-diyn-1-one, 7,12-dimethyl-, oxime	THF	262s(4.43),276(4.55), 302(4.63),394s(3.55)	39-0933-86C
Oxiranecarbonitrile, 3-(2-ethynyl-1-cyclohexen-1-yl)-2-phenyl-, cis	MeCN	218(4.12),243(4.04)	78-2221-86
4(1H)-Quinolinone, 1-methyl-2-(4-methylphenyl)-	EtOH	250(4.33)	78-1139-86
$C_{17}H_{15}NOS$			
9H-Thioxanthen-9-one, 2-pyrrolidino-	MeOH	269(4.59),300(4.37), 335s(3.63)	73-0937-86
$C_{17}H_{15}NO_2$			
2H-1-Benzopyran-2-one, 4-methyl-7-[(phenylmethyl)amino]-	MeOH	240(4.08),310(3.70), 364(4.21)	151-0069-86A
	$CHCl_3$	310(3.51),355(4.09)	151-0069-86A
$C_{17}H_{15}NO_2S$			
1,5-Benzothiazepin-4(5H)-one, 5-acetyl-	CH_2Cl_2	271.5(3.61)	24-3109-86

Compound	Solvent	$\lambda_{max}(\log \epsilon)$	Ref.
2,3-dihydro-2-phenyl- (cont.)			24-3109-86
Benzo[c][1λ^4,2]thiazine-3-carboxylic acid, 2-phenyl-, ethyl ester	EtOH	213(4.33),251(4.33), 310(3.85)	39-0483-86C
9H-Thioxanthen-9-one, 2-morpholino-	MeOH	263(4.57),288s(4.40), 416(3.63)	73-0937-86
$C_{17}H_{15}NO_3$			
Phenol, 2-[3-(phenylimino)-1-propen-yl]-, acetate, N-oxide	dioxan	268(4.19),350(3.40)	104-1302-86
Pyrrolo[3,2,1-de]phenanthridinium, 4,5-dihydro-2-hydroxy-9,10-dimethoxy-, hydroxide, inner salt (criasbetaine)	MeOH	267(4.35),288s(4.14), 418(3.88)	150-0112-86S
7H-Pyrrolo[3,2,1-de]phenanthridin-7-one, 4,5-dihydro-9,10-dimethoxy- (oxoassoanine)	EtOH	252(4.36),272(4.27), 326(3.65),342(3.67)	102-2637-86
$C_{17}H_{15}NO_4$			
9,10-Anthracenedione, 4-amino-1,5-dihydroxy-2-propyl-	$CHCl_3$	244(4.45),252(4.37), 306(3.80),396(3.20), 512(3.91),540(4.10), 580(4.06)	78-3303-86
9,10-Anthracenedione, 1,5-dihydroxy-4-(propylamino)-	$CHCl_3$	240(4.23),252(4.09), 290(3.59),388(2.95), 534(3.68),592(3.94), 614(3.95)	78-3303-86
1H-Azuleno[1,2-b]pyrrole-1,4-dicarboxylic acid, 1-ethyl 4-methyl ester	MeOH	213(4.46),283(4.28), 287s(4.24),320(4.66), 331(4.73),370s(3.80), 387(3.95),405s(3.61), 560(2.62),598(2.59), 670s(2.07)	138-1021-86
4H-Furo[3,2-b]pyrrole-5-carboxylic acid, 4-acetyl-2-phenyl-, ethyl ester	MeOH	352(3.27)	73-1455-86
4H-Furo[3,2-b]pyrrole-5-carboxylic acid, 4-formyl-2-(4-methylphenyl)-, ethyl ester	MeOH	340(3.73),355(3.68)	73-0106-86
4H-Furo[3,2-b]pyrrole-5-carboxylic acid, 6-formyl-2-phenyl-4-methyl-, ethyl ester	MeOH	302(3.45),367(3.27)	73-0106-86
Unnamed pyrindenotropolone pigment	EtOH	213(4.00),252(4.19), 305(3.74),318(3.67), 399(3.62),426(3.67), 530(3.32)	88-6049-86
$C_{17}H_{15}NO_4S$			
Benzenesulfonamide, 4-methyl-N-(4-methyl-2-oxo-2H-1-benzopyran-7-yl)-	MeOH	232(4.21),292(3.62), 345(4.25)	151-0069-86A
	$CHCl_3$	294(3.82),332(4.33)	151-0069-86A
$C_{17}H_{15}NO_5S$			
L-Cysteine, S-(9,10-dihydro-4,5-dihydroxy-10-oxo-9-anthracenyl)-	EtOH	375(4.04)	118-0430-86
$C_{17}H_{15}N_3$			
Benzeneacetaldehyde, cyclohepta[b]-pyrrol-2-ylhydrazone	EtOH	278(4.47),287(4.47), 317(4.22),380(4.01), 398(3.95),440(3.83), 462s(3.76),495s(3.31)	18-3326-86

Compound	Solvent	$\lambda_{max}(\log \epsilon)$	Ref.
$C_{17}H_{15}N_3O_2$			
2H-Benz[g]indazole, 3,3a,4,5-tetra-hydro-3-(4-nitrophenyl)-	EtOH	285(4.18)	4-0135-86
$C_{17}H_{15}N_3O_2S$			
Acetic acid, cyano(5,6-dihydro-2-phenylthieno[2,3-d]pyrimidin-4-yl)-, ethyl ester	$CHCl_3$	270(4.32),343(4.26), 376s(3.99)	94-0516-86
$C_{17}H_{15}N_3O_3S$			
Benzamide, N-[3-(methylamino)-2-nitro-3-phenyl-1-thioxo-2-propenyl]-	EtOH	247(4.10),309(4.04), 378(4.00),400(3.95)	18-3871-86
Benzamide, N-[2-nitro-3-(phenylamino)-1-thioxo-2-butenyl]-	EtOH	243(4.15),308(4.23), 386(4.12)	18-3871-86
$C_{17}H_{15}N_3S$			
3H-Pyrazole-3-thione, 2,4-dihydro-5-methyl-2-phenyl-4-[(phenylamino)methylene]-	toluene	355(4.3),440(3.9)	103-0503-86
$C_{17}H_{15}N_3Se$			
3H-Pyrazole-3-selone, 2,4-dihydro-5-methyl-2-phenyl-4-[(phenylamino)-methylene]-	EtOH	344(4.4),448(3.8)	104-2126-86
$C_{17}H_{15}N_5O_2S$			
Benzoic acid, 2-[5-(2-benzothiazolyl)-3-ethyl-1-formazano]-	EtOH	500(4.64)	65-0727-86
nickel complex	EtOH	550(4.14)	65-0727-86
zinc complex	EtOH	630(4.41)	65-0727-86
$C_{17}H_{15}N_5O_4$			
Pyrazolo[1,5-a]pyrimidine-3-carboxylic acid, 4,7-dihydro-7-(1H-indol-3-yl)-6-nitro-, ethyl ester	EtOH	220(3.84),281(3.47), 370(3.36)	103-1250-86
$C_{17}H_{15}N_7O_5$			
Acetamide, N-[2-[(2-cyano-4,6-dinitrophenyl)azo]-5-(dimethylamino)-phenyl]-	$CHCl_3$	286(3.81),308s(3.76), 578s(4.67),614(4.77)	48-0497-86
$C_{17}H_{16}$			
Benzo[a]fluorene, 1,2,3,4-tetrahydro-	EtOH	259(4.06)	39-1243-86C
Stilbene, 2-(2-propenyl)-, cis	MeOH	204(4.36),275(3.96)	151-0105-86A
trans	MeOH	225(4.00),297(4.16), 310s(4.10),325s(3.79)	151-0105-86A
Tricyclo[8.2.2.2^{4,7}]hexadeca-2,4,6-10,12,13,15-heptaene, 2-methyl-	MeCN	219(4.10)	24-1836-86
$C_{17}H_{16}BrN_7O$			
2,4,6-Pyrimidinetriamine, 5-[(4-bromophenyl)azo]-N^4-(4-methoxyphenyl)-	EtOH	204(3.95),232(3.67), 274(3.67),430(3.26)	42-0420-86
$C_{17}H_{16}ClNO_2S$			
2H-[1]Benzothiepino[5,4-b]pyran-2-one, 3-chloro-5,6-dihydro-4-(1-pyrrolidinyl)-	EtOH	231s(3.97),257s(4.07), 266(4.11),318(4.00)	4-0449-86
$C_{17}H_{16}ClNO_3S$			
2H-[1]Benzothiepino[5,4-b]pyran-2-one, 3-chloro-5,6-dihydro-4-(4-morpholinyl)-	EtOH	235s(3.99),253(4.05), 264s(3.97),323(4.08)	4-0449-86

Compound	Solvent	$\lambda_{max}(\log \epsilon)$	Ref.
$C_{17}H_{16}ClN_3O_3S$			
Ethanediamide, N-[6-chloro-2-[(2-oxo-propyl)thio]-3-pyridinyl]-N'-(phen-ylmethyl)-	EtOH	205(4.28),230(4.36), 284(3.98)	103-0800-86
$C_{17}H_{16}ClO_2S_8$			
Chloride of tetramethyltetrathioful-valene-thiofulvalenecarboxylic acid complex	MeCN	289(3.86),302(3.88), 313(3.91),391s(--), 427s(--),459(3.96), 529?(3.34),598s(--), 650(3.58)	104-2131-86
$C_{17}H_{16}Cl_2N_2O_2$			
Benzaldehyde, 2,2'-(1,3-propanediyldi-imino)bis[5-chloro-	EtOH	231(4.65),274(4.16), 396(4.04)	12-1231-86
$C_{17}H_{16}Cl_2N_2O_4$			
Benzoic acid, 2,2'-(1,3-propanediyldi-imino)bis[5-chloro-	EtOH	224(4.52),265(4.42), 366(3.99)	12-1231-86
$C_{17}H_{16}F_2N_2O_2$			
Benzaldehyde, 2,2'-(1,3-propanediyldi-imino)bis[5-fluoro-	EtOH	218(4.48),235(4.46), 261(3.94),272s(3.81), 398(4.07)	12-1231-86
$C_{17}H_{16}F_2N_2O_4$			
Benzoic acid, 2,2'-(1,3-propanediyldi-imino)bis[5-fluoro-	EtOH	223(4.47),256(4.16), 366(4.00)	12-1231-86
$C_{17}H_{16}F_3N_3O$			
Morpholine, 4-[4-[[4-(trifluorometh-yl)phenyl]azo]phenyl]-	EtOH	404(4.36)	39-0123-86B
	EtOH-HCl	514(4.48)	39-0123-86B
$C_{17}H_{16}F_3N_3O_2S$			
Thiomorpholine, 4-[4-[[4-(trifluoro-methyl)phenyl]azo]phenyl]-, 1,1-dioxide	EtOH	394(4.33)	39-0123-86B
	EtOH-HCl	535(4.82)	39-0123-86B
$C_{17}H_{16}F_3N_3S$			
Thiomorpholine, 4-[4-[[4-(trifluoro-methyl)phenyl]azo]phenyl]-	EtOH	417(4.40)	39-0123-86B
	EtOH-HCl	517(4.66)	39-0123-86B
$C_{17}H_{16}IN_7$			
2,4,6-Pyrimidinetriamine, N^4-(4-iodo-phenyl)-5-[(2-methylphenyl)azo]-	EtOH	202(4.76),254(4.41), 290(4.62),400(4.51)	42-0420-86
$C_{17}H_{16}IN_7O$			
2,4,6-Pyrimidinetriamine, N^4-(4-iodo-phenyl)-5-[(2-methoxyphenyl)azo]-	EtOH	208(4.72),292(4.49), 397(4.22)	42-0420-86
2,4,6-Pyrimidinetriamine, N^4-(4-iodo-phenyl)-5-[(3-methoxyphenyl)azo]-	EtOH	202(4.49),255(4.17), 290(4.48),400(4.37), 420(4.39)	42-0420-86
$C_{17}H_{16}NO_3$			
Isoquinolinium, 2-[[5-(methoxycarbo-nyl)-2-methyl-3-furanyl]methyl]-, bromide	MeOH	233(3.67),262(3.28), 338(2.64)	73-0412-86
Quinolinium, 1-[[5-(methoxycarbonyl)-2-methyl-3-furanyl]methyl]-, bromide	MeOH	239(3.60),259(3.16), 316(2.90)	73-0412-86

Compound	Solvent	λ_{max}(log ε)	Ref.
$C_{17}H_{16}N_2$			
2H-Benz[g]indazole, 3,3a,4,5-tetra-hydro-3-phenyl-	EtOH	240(4.08),295(3.97)	4-0135-86
2H-Benz[f]isoindole-1-carbonitrile, 2-butyl-	EtOH	400s(3.56),416(3.78), 439(3.75)	44-3978-86
2H-Benz[f]isoindole-1-carbonitrile, 2-(1,1-dimethylethyl)-	EtOH	405s(3.54),423(3.73), 446(3.69)	44-3978-86
Pyridine, 4,4'-bicyclo[2.2.1]hept-2-ene-1,3-diylbis-	EtOH	227(4.06),239(3.97), 288(3.98)	138-1279-86
$C_{17}H_{16}N_2O$			
5H-Cyclohepta[4,5]pyrrolo[2,3-b]in-dole, 5-acetyl-1,2,3,4-tetrahydro-	MeOH	232(4.14),258(4.32), 313(4.59),341(4.17), 417(3.71),510(2.61), 550s(2.55)	18-3326-86
1H-Imidazole-4-methanol, 2-phenyl-1-(phenylmethyl)-	MeOH	258(3.96)	87-0261-86
Indolo[2,3-a]quinolizin-4(6H)-one, 3-ethyl-7,12-dihydro-	MeOH	252(--),261(--), 278(--),291(--), 352s(--),367(--), 386(--)	94-0077-86
Pyrazolo[5,1-d][1,5]benzoxazepine, 3,3a,4,5-tetrahydro-2-phenyl-	EtOH	229(4.16),253s(3.87), 259s(3.77),292s(3.86), 332(4.25)	33-0927-86
3H-Pyrazolo[5,1-c][1,4]benzoxazine, 3a,4-dihydro-3a-methyl-2-phenyl-	EtOH	228(4.18),244s(3.99), 327(4.11)	33-0927-86
$C_{17}H_{16}N_2OS$			
2H-Benz[g]indazole, 2-acetyl-3,3a,4,5-tetrahydro-3-(2-thienyl)-	EtOH	285(4.02)	4-0135-86
$C_{17}H_{16}N_2O_2$			
Benzenepropanenitrile, α-[[5-(dimeth-ylamino)-2-furanyl]methylene]-4-methyl-β-oxo-, (E)-	MeOH	286(2.22),499(3.60)	73-0879-86
1,3,4-Oxadiazole, 3-benzoyl-2,3-di-hydro-2,2-dimethyl-5-phenyl-	MeOH	303(4.29)	24-2963-86
$C_{17}H_{16}N_2O_2S$			
1H-Pyrrole, 1-methyl-2-[2-[1-(phenyl-sulfonyl)-1H-pyrrol-2-yl]ethenyl]-, (E)-	EtOH	220(4.07),354(3.80)	118-0620-86
$C_{17}H_{16}N_2O_3$			
3(2H)-Pyridazinone, 4,5-dihydro-5-(hy-droxymethyl)-6-(4-phenoxyphenyl)-	EtOH	296(4.37)	56-0143-86 +80-0541-86
$C_{17}H_{16}N_2O_4$			
6,11-Methanocyclodeca[d]pyridazine-1,4-dicarboxylic acid, 7,10-dihy-dro-, dimethyl ester	MeOH	268(4.32),390(3.07)	24-3862-86
6H-Pyrido[1,2-a]pyrimidine-8-carbox-ylic acid, 9-benzoyl-1,2,3,4-tetra-hydro-6-oxo-, methyl ester	EtOH	236(4.22),359(3.92)	142-2247-86
$C_{17}H_{16}N_2O_5$			
1,3-Diazabicyclo[4.2.0]oct-2-ene-2-carboxylic acid, 4-(methoxycarbonyl-methylene)-8-oxo-, phenylmethyl ester	EtOH	315(4.16)	39-1077-86C

Compound	Solvent	λ_{max}(log ϵ)	Ref.
$C_{17}H_{16}N_2O_6$			
Uridine, 6-(phenylethynyl)-	MeOH	308(4.28)	23-1560-86
$C_{17}H_{16}N_2O_7$			
16H-Dibenzo[h,k][1,4,7]trioxacyclododecin, 6,7,9,10-tetrahydro-2,14-dinitro-	$CHCl_3$	245(3.96),313(4.15)	78-0137-86
$C_{17}H_{16}N_2S_2$			
[1]Benzothieno[2,3-d]pyrimidine-4(3H)-thione, 5,6,7,8-tetrahydro-3-methyl-2-phenyl-	EtOH	238(4.46),288(4.10), 329(4.16),366s(3.94)	106-0096-86
$C_{17}H_{16}N_4$			
1,1,2,2-Propanetetracarbonitrile, 3-(2,5,5-trimethyl-1,3,6-cycloheptatrien-1-yl)-	MeOH	270(3.46)	5-1104-86
$C_{17}H_{16}N_4O$			
Benzonitrile, 4-[[4-(4-morpholinyl)-phenyl]azo]-	EtOH	420(4.45)	39-0123-86B
	EtOH-HCl	520(4.58)	39-0123-86B
$C_{17}H_{16}N_4O_2S$			
Benzonitrile, 4-[[4-(4-thiomorpholinyl)phenyl]azo]-, S,S-dioxide	EtOH	408(4.43)	39-0123-86B
	EtOH-HCl	542(4.89)	39-0123-86B
$C_{17}H_{16}N_4O_4$			
Spiro[3H-indole-3,4'(1'H)-pyrano[2,3-c]pyrazole]-5'-carboxylic acid, 6'-amino-1,2-dihydro-3'-methyl-2-oxo-, ethyl ester	EtOH	248.5(4.23)	18-1235-86
$C_{17}H_{16}N_4S$			
Benzonitrile, 4-[[4-(4-thiomorpholinyl)phenyl]azo]-	EtOH	434(4.45)	39-0123-86B
	EtOH-HCl	522(4.75)	39-0123-86B
$C_{17}H_{16}N_6$			
Propanenitrile, 3,3'-[[4-(3-pyridinylazo)phenyl]imino]bis-	acetone	400(4.24)	7-0473-86
$C_{17}H_{16}N_6O_4$			
Benzonitrile, 2-[[4-(diethylamino)-phenyl]azo]-3,5-dinitro-	$CHCl_3$	301(3.94),352s(3.61), 590(4.69)	48-0497-86
[1,2,4]Triazolo[1,5-a]pyrimidine-2-carboxylic acid, 1,7-dihydro-7-(2-methyl-1H-indol-3-yl)-6-nitro-, ethyl ester	EtOH	222(4.65),272(5.01), 380(5.14)	103-1250-86
$C_{17}H_{16}N_8O_2$			
1H-Tetrazole, 5,5'-methylenebis[1-(4-methoxyphenyl)-	EtOH	236(4.23)	104-0177-86
$C_{17}H_{16}O$			
Tricyclo[8.2.2.2^{4,7}]hexadeca-4,6,10-12,13,15-hexaene-5-carboxaldehyde	MeCN	204(4.02),224(4.18)	24-1836-86
Tricyclo[5.3.1.0^{2,6}]undeca-3,9-dien-8-one, 10-phenyl-	EtOH	222(3.88),290(4.14)	18-2811-86
Tricyclo[5.3.1.0^{2,6}]undeca-4,9-dien-8-one, 10-phenyl-	EtOH	222(3.98),290(4.20)	18-2811-86

Compound	Solvent	$\lambda_{max}(\log \epsilon)$	Ref.
$C_{17}H_{16}O_2$			
1-Propanone, 1-phenyl-3-(3-phenyloxiranyl)-	EtOH	219(4.30),241(4.29)	35-1617-86
$(C_{17}H_{16}O_3)_n$			
2-Propenoic acid, 2-methyl-, (4-benzoylphenyl) ester, polymer	$CHCl_3$	258(4.31),336(2.32)	32-0533-86
$C_{17}H_{16}O_4$			
Benzofuran, 2-(2,4-dimethoxyphenyl)-6-methoxy-	MeOH	282s(3.89),320(4.50), 335(4.45)	2-0481-86
4H-1-Benzopyran-4-one, 2,3-dihydro-5,7-dimethoxy-3-phenyl-	MeOH	208(4.47),230(4.30), 281(4.29)	39-0215-86C
4H-1-Benzopyran-4-one, 2,3-dihydro-7-methoxy-3-(4-methoxyphenyl)-	MeOH	222(4.57),226(4.46), 310(4.16)	39-0215-86C
Butanedioic acid, dimethyl(naphthalenylmethylene)-	EtOH	225(5.06),290(4.03)	34-0369-86
1,2-Heptalenedicarboxylic acid, 10-methyl-, dimethyl ester	hexane	205(4.40),264(4.21), 323s(3.52),405s(2.66)	88-1669-86
$C_{17}H_{16}O_5$			
6H-Benzofuro[3,2-c][1]benzopyran-3-ol, 6a,11a-dihydro-6,9-dimethoxy- (sophoracarpan A)	MeOH	283s(3.89),287(3.92)	94-3067-86
2-Propen-1-one, 3-(2,4-dihydroxy-6-methoxyphenyl)-1-(4-methoxyphenyl)-	MeOH	388(4.11)	102-1097-86
$C_{17}H_{16}O_6$			
Ethanone, 1-(6-hydroxy-4-methoxy-1,3-benzodioxol-5-yl)-2-(2-methoxyphenyl)-	EtOH	242(4.68),282(4.74), 348(4.30)	102-0281-86
9H-Xanthen-9-one, 1,2,6,8-tetramethoxy-	EtOH	242(4.53),252(4.56), 303(4.21),351(3.69)	102-2681-86
9H-Xanthen-9-one, 1,3,5,6-tetramethoxy-	EtOH	245(4.58),285s(3.97), 305(4.23)	102-0503-86
$C_{17}H_{17}BrN_6O_5$			
Propanamide, N-[2-[(2-bromo-4,6-dinitrophenyl)azo]-5-(dimethylamino)phenyl]-	$CHCl_3$	293(3.93),396(3.45), 561(4.65)	48-0497-86
$C_{17}H_{17}BrO_2$			
1-Azulenecarboxylic acid, 6-bromo-2-methyl-3-(1-methylethenyl)-, ethyl ester	CH_2Cl_2	310.5(4.72),316.5(4.77), 356(3.74),396.5(3.75), 390.5(3.42),550.5(2.55)	24-2272-86
$C_{17}H_{17}BrO_4$			
1,3-Azulenedicarboxylic acid, 6-bromo-2-methyl-, diethyl ester	CH_2Cl_2	236.5(4.46),270(4.26), 316.7(4.85),353.5(3.90), 361.5(3.90),380.5(3.57), 507(2.57)	24-2272-86
$C_{17}H_{17}Cl$			
Heptalene, 1,2-dihydro-2-(chloromethylene)-1-methylene-6,8,10-trimethyl-	dioxan	223(4.37),286(3.82), 389(4.01)	88-1669-86
$C_{17}H_{17}ClHgO_3$			
Mercury, chloro[[3,4-dihydro-7-methyl-2-(4-methylphenyl)-5-oxo-2H,5H-pyrano[4,3-b]pyran-2-yl]methyl]-	EtOH	285(3.8)	32-0441-86

Compound	Solvent	$\lambda_{max}(\log \epsilon)$	Ref.
$C_{17}H_{17}ClHgO_4$			
Mercury, chloro[[3,4-dihydro-2-(4-methoxyphenyl)-7-methyl-5-oxo-2H,5H-pyrano[4,3-b]pyran-2-yl]-methyl]-	EtOH	282(3.9)	32-0441-86
$C_{17}H_{17}ClN_2O$			
1H-Indazole, 1-(3-chloropropyl)-5-(phenylmethoxy)-	EtOH	208(4.63),255(3.81), 262(3.74),310(3.67)	95-1002-86
2H-Indazole, 2-(3-chloropropyl)-5-(phenylmethoxy)-	EtOH	212(4.69),263(3.83), 272(3.77),309(3.71)	95-1002-86
$C_{17}H_{17}ClN_2O_5$			
2-Butenedioic acid, 2-[2-[(4-chlorophenyl)-1-(2-imidazolidinylidene)-2-oxoethyl]-, dimethyl ester, (Z)-	EtOH	248(4.23),364(4.11)	142-2247-86
$C_{17}H_{17}ClN_2O_7$			
Benzeneacetic acid, α-amino-5-[5-[carboxy(methylamino)methyl]-2-hydroxyphenoxy]-2-chloro-3-hydroxy-	pH 2.0	283(3.60)	44-4272-86
	pH 11.0	300(3.74)	44-4272-86
$C_{17}H_{17}ClN_4$			
1H-Pyrazolo[3,4-b]quinolin-4-amine, N-(4-chlorophenyl)-5,6,7,8-tetrahydro-1-methyl-	EtOH	221(5.89),249(5.53), 326(5.65)	11-0331-86
$C_{17}H_{17}ClN_6O_2$			
2,4(1H,3H)-Pteridinedione, 6-chloro-1,3-dimethyl-7-[1-(phenylhydrazono)propyl]-	MeOH	256(4.50),320(3.80), 437(4.41)	108-0081-86
$C_{17}H_{17}Cl_2NO_6S$			
7-Azabicyclo[2.2.1]hept-5-ene-2,3-dicarboxylic acid, 5,6-dichloro-7-[(4-methylphenyl)sulfonyl]-, dimethyl ester, (endo,endo)-	EtOH	231(4.20),265(2.85), 276s(2.60)	24-0616-86
$C_{17}H_{17}FN_4$			
1H-Pyrazolo[3,4-b]quinolin-4-amine, N-(4-fluorophenyl)-5,6,7,8-tetrahydro-1-methyl-	EtOH	218(4.22),242(3.87), 272(3.51),321(3.97)	11-0331-86
$C_{17}H_{17}N$			
1H-Indole, 1-methyl-3-(1-phenylethyl)-	MeOH	225.0(4.56),272s(3.69), 279s(3.73),290(3.79), 296.5s(3.73)	1-0625-86
1H-Indole, 2-methyl-1-(1-phenylethyl)-	MeOH	221.5(4.55),269.5s(3.86), 274.5(3.89),280.0(3.90), 290.5(3.81)	1-0625-86
$C_{17}H_{17}NO$			
Azacyclotetradeca-1,3,5,11,13-pentaene-7,9-diyne, 2-ethoxy-6,11-dimethyl-, (E,E,E,Z,Z)-	THF	285s(4.34),308s(4.57), 317(4.60),374(3.84), 403s(3.66),459(3.31), 494(3.30),540(3.14)	39-0933-86C
Azacyclotetradeca-3,5,11,13-tetraene-7,9-diyn-2-one, 14-ethyl-6,11-dimethyl-, (E,E,Z,Z)-	THF	218(4.14),299s(4.45), 307(4.48),441s(3.40)	18-1791-86
Cyclotrideca-2,4,10,12-tetraene-6,8-diyn-1-one, 2-ethyl-5,10-dimethyl-,	THF	217(4.30),274(4.49), 351s(3.86)	18-1791-86

Compound	Solvent	λ_{max}(log ε)	Ref.
oxime, (E,E,E,Z,Z)- (cont.)			18-1791-86
Isoxazole, 2,3-dihydro-2,3-dimethyl-3,5-diphenyl-	EtOH	218(4.18),225(4.16), 274(3.92)	44-3125-86
1-Propanone, 2,2-dimethyl-1-phenyl-3-(phenylimino)-, (E)-	EtOH	244(4.2),280(3.6)	39-0091-86C
$C_{17}H_{17}NO_2$			
Benzeneacetic acid, α-[[2-(2-propenyl)phenyl]amino]-	EtOH	242(4.12),291(3.40)	33-0927-86
7H-Pyrrolo[3,2,1-de]phenanthridine, 4,5-dihydro-9,10-dimethoxy-(assoanine)	EtOH	251(4.10),254(3.96), 325(3.70),347(3.68)	102-2637-86
$C_{17}H_{17}NO_3$			
Benzeneacetic acid, α-[[2-(2-propenyloxy)phenyl]amino]-	EtOH	244(4.08),290(3.52)	33-0927-86
Furo[2,3-b]quinolin-7-ol, 4-methoxy-8-(3-methyl-2-butenyl)-	MeOH	252(4.54),306s(3.77), 318(3.85),330(3.84), 342(3.79)	100-1091-86
1-Naphthalenol, 2-[3-(methylimino)-2-propenyl]-4-methyl-, N-oxide, acetate	EtOH	385(4.09)	104-1302-86
	acetone	385(3.92)	104-1302-86
	CCl_4	385(3.80)	104-1302-86
	DMSO	390(4.15)	104-1302-86
$C_{17}H_{17}NO_3S$			
Benzoic acid, 2-[[4-(4-morpholinyl)-phenyl]thio]-	MeOH	273(4.29),320s(3.78)	73-0937-86
$C_{17}H_{17}NO_4$			
1(2H)-Anthracenone, 3,4-dihydro-5,9,10-trihydroxy-4-(propylimino)-	$CHCl_3$	260(4.53),276(4.49), 420(4.06),445(4.41), 474(4.55)	78-3303-86
1,3-Dioxane-4,6-dione, 5-[(1,2-dimethyl-1H-indol-3-yl)methylene]-2,2-dimethyl-	EtOH	221.5(4.19),261(3.74), 273(3.67),286s(3.52), 422(4.45)	39-1651-86C
1,3-Dioxane-4,6-dione, 5-[(1,3-dimethyl-1H-indol-2-yl)methylene]-2,2-dimethyl-	EtOH	223(4.45),266(3.96), 427(4.19)	39-1651-86C
3,5-Pyridinedicarboxylic acid, 2,6-dimethyl-4-phenyl-, dimethyl ester	MeOH	268(4.45),277(--)	87-1596-86
$C_{17}H_{17}NO_5$			
3H-Benzo[f]cyclopent[d][1,4]oxazino-[3,4-b][1,3]oxazepin-3-one, 12-acetoxy-1,2,5,6,8,8a-hexahydro-	EtOH	309(3.79),345(3.84), 362(3.78)	24-3487-86
$C_{17}H_{17}NO_6$			
9(10H)-Acridinone, 1,6-dihydroxy-2,3,5-trimethoxy-10-methyl-	MeOH	220(3.54),258(3.66), 267s(3.64),330(3.12)	102-0429-86
	MeOH-NaOMe	224(3.55),258(3.63), 272s(3.44),336(3.09)	102-0429-86
	MeOH-$AlCl_3$	222(3.64),258(3.54), 267(3.57),333(3.09)	102-0429-86
$C_{17}H_{17}NO_7S$			
3-Oxa-8-azatricyclo[3.2.1.0^{2,4}]oct-6-ene-6,7-dicarboxylic acid, 8-[(4-methylphenyl)sulfonyl]-, dimethyl ester, (1α,2α,4α,5α)-	EtOH	227(4.21),225s(3.59), 263s(3.52),274s(3.34)	24-0616-86

Compound	Solvent	$\lambda_{max}(\log \epsilon)$	Ref.
$C_{17}H_{17}NO_{10}S$			
7-Azabicyclo[2.2.1]heptane-2,3,5,6-tetracarboxylic acid, 7-[(4-methylphenyl)sulfonyl]-, tetramethyl ester, (1α,2α,3β,4α,5α,6β)-	MeCN	230(4.09),256(2.85), 263(2.79),268s(2.68), 274(2.59)	24-0616-86
$C_{17}H_{17}N_2$			
1H-Pyrazolo[1,2-a]indazol-4-ium, 2,3-dihydro-7-methyl-9-phenyl-, bromide	EtOH	210(4.50),244(4.15), 323(4.06)	95-1002-86
$C_{17}H_{17}N_2O_2$			
Pyridinium, 1-ethyl-4-[5-(4-methoxyphenyl)-2-oxazolyl]-, salt with 4-methylbenzenesulfonic acid	H_2O	400(4.34)	135-0244-86
	EtOH	260(4.18),410(4.28)	135-0244-86
$C_{17}H_{17}N_3OS$			
2-Benzothiazolamine, N-[[4-(dimethylamino)phenyl]methylene]-6-methoxy-	dioxan	409(4.61)	80-0273-86
	+ CF_3COOH	481(--)	80-0273-86
$C_{17}H_{17}N_3O_3$			
Benzo[f]quinoxaline-2-carboxylic acid, 3,4-dihydro-3-imino-4-(2-methoxyethyl)-, methyl ester	EtOH	440(3.75)	33-1025-86
2-Butenamide, 2-nitro-N-phenyl-3-[(phenylmethyl)amino]-	EtOH	249(4.19),352(4.19)	18-3871-86
$C_{17}H_{17}N_3O_3S$			
Benzenesulfonamide, N-[3-diazo-2-oxo-1-(phenylmethyl)propyl]-4-methyl-	MeOH	204(4.34),235(4.19), 248s(4.13),274s(3.91)	104-0353-86
6H-Thieno[2,3-b]pyrrole-5-carboxylic acid, 2-acetyl-3,4-diamino-6-phenyl-, ethyl ester	EtOH	277(4.35),304(4.45), 362(3.94)	48-0459-86
$C_{17}H_{17}N_3O_4S$			
Piperidine, 1-[4-[(2,4-dinitrophenyl)thio]phenyl]-	MeOH	274(4.40),324s(4.15)	73-0937-86
$C_{17}H_{17}N_3O_5$			
2,3'-(Methylimino)-1-(5'-O-benzoyl-3'deoxy-β-D-lyxofuranosyl)uracil	MeOH	200(4.12),223(4.37), 260s(3.69)	44-4417-86
$C_{17}H_{17}N_3O_7$			
1,3-Diazabicyclo[3.2.0]heptane-2-carboxylic acid, 4-(2-methoxy-2-oxoethylidene)-7-oxo-6-[(phenoxyacetyl)amino]-	EtOH	276(4.22)	39-1077-86C
$C_{17}H_{17}N_3S$			
Piperidine, 1-(1H-perimidin-2-ylthioxomethyl)-	DMF	292(4.09),337(4.14), 349(4.07),390s(3.22), 411s(3.09),437s(2.87)	48-0812-86
	C_6H_{12}	231s(--),236(--), 326s(--),335(--), 353(--),407(--), 476s(--)	48-0812-86
$C_{17}H_{17}N_3S_2$			
4H-Cyclohepta[4,5]thieno[2,3-d]pyrimidine-4-thione, 3-amino-3,5,6,7,8,9-hexahydro-2-phenyl-	EtOH	259(4.30),332(4.10), 368s(3.81)	106-0096-86

Compound	Solvent	$\lambda_{max}(\log \epsilon)$	Ref.
$C_{17}H_{17}N_5O_4S$			
2,4(1H,3H)-Pteridinedione, 1,3,6-tri-methyl-7-[[2-(4-nitrophenyl)ethyl]-thio]-	MeOH	254(4.26),353(4.31)	33-1095-86
$C_{17}H_{17}N_5O_5S$			
2,4(1H,3H)-Pteridinedione, 1,3,6-tri-methyl-7-[[2-(4-nitrophenyl)ethyl]-sulfinyl]-	MeOH	248(4.29),353(4.00)	33-1095-86
$C_{17}H_{17}N_7O_4S$			
Benzenesulfonic acid, 4-[[2,4-diamino-6-[(4-methoxyphenyl)amino]-5-pyrimi-dinyl]azo]-	EtOH	202(4.76),252(4.40), 276(4.46),428(4 .25)	42-0420-86
$C_{17}H_{17}O_6P$			
Phosphoric acid, 1-benzoyl-2-oxo-2-phenylethyl dimethyl ester	MeOH	205(4.26),255(4.49), 335(2.98)	70-1690-86
	ether	205(4.26),255(4.49), 335(2.98)	70-1690-86
$C_{17}H_{18}$			
Tricyclo[8.2.2.2^{4,7}]hexadeca-4,6,10-12,13,15-hexaene, 2-methyl-	MeCN	207(4.00),225(4.18)	24-1836-86
$C_{17}H_{18}BrN$			
Isoquinoline, 4-[(2-bromophenyl)meth-yl]-1,2,3,4-tetrahydro-2-methyl-	MeOH	206(4.47),265(2.78), 272(2.66)	83-0735-86
$C_{17}H_{18}Br_2N_8$			
1,3,5,7-Tetraazabicyclo[3.3.1]nonane, 3,7-bis[(4-bromophenyl)azo]-	EtOH	287(4.46)	23-1567-86
$C_{17}H_{18}ClN$			
Isoquinoline, 4-[(2-chlorophenyl)meth-yl]-1,2,3,4-tetrahydro-2-methyl-	MeOH	207(4.38),264(3.18), 272(3.12)	83-0735-86
$C_{17}H_{18}ClNO_2S$			
2H-[1]Benzothiepino[5,4-b]pyran-2-one, 3-chloro-4-(diethylamino)-5,6-di-hydro-	EtOH	235s(4.08),254(4.08), 266s(3.99),327(4.17)	4-0449-86
$C_{17}H_{18}ClNO_3$			
1H-3-Benzazepin-7-ol, 6-chloro-2,3,4,5-tetrahydro-1-(4-hydroxy-phenyl)-8-methoxy-, hydrochloride	MeOH	206(4.57),227s(4.14), 283(3.54)	4-1805-86
	EtOH	208(4.67),228s(4.24), 282(3.60)	4-1805-86
1H-3-Benzazepin-7-ol, 9-chloro-2,3,4,5-tetrahydro-5-(4-hydroxy-phenyl)-8-methoxy-, hydrochloride	MeOH	207(4.65),224s(4.25), 282(3.55)	4-1805-86
	EtOH	209(4.68),227s(4.26), 282(3.59)	4-1805-86
$C_{17}H_{18}ClN_3$			
Piperidine, 1-[4-(4-chlorophenyl)-azo]phenyl]-	EtOH	412(4.45)	39-0123-86B
	EtOH-HCl	534(3.42)	39-0123-86B
$C_{17}H_{18}ClN_3OS$			
Hydrazinecarbothioamide, 2-[1-(4-chlorophenyl)-1,3-dihydro-3,3-di-methyl-1-isobenzofuranyl]-	EtOH	244(4.24),270(3.32), 314(2.49)	104-0647-86

Compound	Solvent	λ_{max}(log ε)	Ref.
$C_{17}H_{18}Cl_2N_2S$			
10H-Phenothiazine-10-propanamine, 2,4-dichloro-N,N-dimethyl-, hydrochloride (dichloropromazine)	MeOH	240.5(4.286),262(4.62), 312.5(3.622)	106-0571-86
$C_{17}H_{18}Cl_4N_2O_2$			
Benzenemethanol, 2,2'-(1,3-propanediyldiimino)bis[3,6-dichloro-	EtOH	222(4.58),257(4.10), 299(3.58)	12-1231-86
$C_{17}H_{18}FN$			
Isoquinoline, 4-[(2-fluorophenyl)methyl]-1,2,3,4-tetrahydro-2-methyl-	MeOH	211(4.08),262(3.15), 268(3.09)	83-0735-86
$C_{17}H_{18}FN_3OS$			
Hydrazinecarbothioamide, 2-[1-(3-fluorophenyl)-1,3-dihydro-3,3-dimethyl-1-isobenzofuranyl]-	EtOH	245(4.12),263s(3.58), 270(3.46),311(3.06)	104-0647-86
$C_{17}H_{18}FN_3O_7S$			
Thymidine, 5'-deoxy-5'-[[3-(fluorosulfonyl)benzoyl]amino]- (in EtOH)	pH 1	267(4.00)	87-1052-86
	pH 7	266(3.98)	87-1052-86
	pH 13	266(3.89)	87-1052-86
Thymidine, 5'-deoxy-5'-[[4-(fluorosulfonyl)benzoyl]amino]- (in EtOH)	pH 1	222(4.25),260(4.09)	87-1052-86
	pH 7	222(4.25),260(4.09)	87-1052-86
	pH 13	227(4.36),265s(3.99)	87-1052-86
$C_{17}H_{18}N_2$			
1,8a,3,4a-Azino-4,9-etheno-1H-benz[f]ind e, 2,3,3a,4,9,9a-hexahydro-2,2-dimethyl-	hexane	363(2.57)	89-0187-86
isomer	hexane	364(2.81)	89-0187-86
Benzenamine, 4-methyl-N-[3-[(4-methylphenyl)amino]-2-propenylidene]-	pH 1	389(4.71)	94-1794-86
	pH 7	389(4.70)	94-1794-86
	pH 13	361(--)(unstable)	94-1794-86
	MeOH	378(4.63)	94-1794-86
	EtOH	367(4.60)	94-1794-86
Benzonitrile, 4-(5-pentyl-2-pyridinyl)-	EtOH	208(4.34),265(4.26), 290(4.31)	104-0957-86
$C_{17}H_{18}N_2O$			
Benzeneacetonitrile, α-(1,5-dihydro-3,4-dimethyl-5-oxo-2H-pyrrol-2-ylidene)-2,4,6-trimethyl-, (E)-	EtOH	282(4.15)	49-0631-86
	$CHCl_3$	285(4.15)	49-0631-86
(Z)-	EtOH	285(4.19)	49-0631-86
	$CHCl_3$	285(4.18)	49-0631-86
Indolo[2,3-a]quinolizin-4-one, 3-ethyl-1,4,6,7,12,12b-hexahydro-	MeOH	223(4.63),273(3.96), 282s(3.93),290(3.81)	94-0077-86
$C_{17}H_{18}N_2O_2$			
Benzenamine, 4-methoxy-N-[3-[(4-methoxyphenyl)amino]-2-propenylidene]-	pH 1	393(4.70)	94-1794-86
	pH 7	395(4.68)	94-1794-86
	pH 13	368(--)	94-1794-86
	MeOH	392(4.62)	94-1794-86
	EtOH	375(4.58)	94-1794-86
1,3-Dioxolo[4,5-g]isoquinolin-5-amine, 5,6,7,8-tetrahydro-6-methyl-N-phenyl-	MeOH	247(4.34),305(3.82), 362(3.93)	83-0910-86
Methanone, [2-(4-morpholinyl)-3H-azepin-6-yl]phenyl-	EtOH	237.0(4.09),340.5(4.18)	18-2317-86
3H-Naphth[1,8-de]azocine-3-carbonitrile, 1,2,4,5-tetrahydro-7,8-dimethoxy-	MeOH	217(4.36),238(4.74), 291(3.75),302(3.75), 328(3.40),340(3.42)	12-0001-86

Compound	Solvent	λ_{max}(log ϵ)	Ref.
Pyridinium, 1-(carboxymethyl)-4-[2-[4-(dimethylamino)phenyl]ethenyl]-, hydroxide, inner salt	EtOH	472(4.61)	4-0209-86
$C_{17}H_{18}N_2O_3S$			
Glycine, N-[(methylthio)[(1-naphthalenylcarbonyl)amino]methylene]-, ethyl ester	EtOH	217(4.51),242(4.16), 273s(3.90),312(3.97)	118-0484-86
$C_{17}H_{18}N_2O_5$			
1,3-Diazabicyclo[4.2.0]octane-2-carboxylic acid, 4-(2-methoxy-2-oxoethylidene)-8-oxo-, phenylmethyl ester, (2α,4Z,6β)-(±)-	EtOH	283(4.27)	39-1077-86C
$C_{17}H_{18}N_2O_6S$			
2(1H)-Pyrimidinone, 4-(phenylthio)-1-β-D-ribofuranosyl-, 5'-acetate	pH 7	301(4.09)	1-0806-86
	pH 2	301(--)	1-0806-86
	pH 13	272(--),309(--)	1-0806-86
$C_{17}H_{18}N_3O$			
Pyrylium, 2-amino-3-cyano-6-phenyl-4-piperidino-, perchlorate	MeCN	258.5(4.54),314(4.39)	48-0314-86
$C_{17}H_{18}N_4$			
1H-Pyrazolo[3,4-b]quinolin-4-amine, 5,6,7,8-tetrahydro-N-(4-methylphenyl)-	EtOH	215(4.34),242s(--), 271(3.69),316(4.10)	5-1728-86
1H-Pyrazolo[3,4-b]quinolin-4-amine, 5,6,7,8-tetrahydro-1-methyl-N-phenyl-	EtOH	217(4.51),243(5.19), 272(3.75),322(4.27)	11-0331-86
Pyrido[3,4-d]pyrimidin-4-amine, 2-methyl-N-(3-phenylpropyl)-	pH 1	276(3.86),324.5(3.90), 338(3.96)	163-2243-86
	pH 7	285(3.96),332.5(3.90)	163-2243-86
	pH 13	285(3.96),332.5(3.90)	163-2243-86
$C_{17}H_{18}N_4O_2$			
Acetic acid, cyano[2-(dimethylamino)-6-phenyl-4(1H)-pyrimidinylidene]-, ethyl ester, (Z)-	EtOH	280(4.48),395(4.35)	103-0774-86
Piperidine, 1-[4-[(4-nitrophenyl)-azo]phenyl]-	EtOH	470(4.44)	39-0123-86B
	EtOH-HCl	522(4.12)	39-0123-86B
$C_{17}H_{18}N_4O_4S$			
Piperazine, 1-[4-[(2,4-dinitrophenyl)-thio]phenyl]-4-methyl-	MeOH	274(4.38),323s(3.03)	73-0937-86
$C_{17}H_{18}N_4O_5$			
2,5-Methano-5H-pyrimido[1,2-c][1,3,5]-oxadiazepin-9(1H)-one, 3-benzoyloxy-2,3-dihydro-11-hydroxy-1-(methylamino)-, [2R-(2α,3β,5α,11S*)]-	MeOH	200.5(4.23),226(4.54), 263s(3.73)	44-4417-86
$C_{17}H_{18}N_4O_6$			
2,4,6(1H,3H,5H)-Pyrimidinetrione, 5-[5-(hexahydro-1,3-dimethyl-2,4,6-trioxo-5-pyrimidinyl)-2,4-pentadienylidene]-1,3-dimethyl-, compd. with pyridine	aq pyridine	251(--),361(--), 588(5.20)	3-3184-86

Compound	Solvent	$\lambda_{max}(\log \epsilon)$	Ref.
$C_{17}H_{18}N_{10}O_4$			
1,3,5,7-Tetraazabicyclo[3.3.1]nonane, 3,7-bis[(4-nitrophenyl)azo]-	DMSO	348(4.47)	23-1567-86
$C_{17}H_{18}O$			
2-Heptalenecarboxaldehyde, 1,5,6,10-tetramethyl-	dioxan	253(4.27),271(4.30), 330(3.65),394s(2.84)	88-1669-86
2-Heptalenecarboxaldehyde, 1,6,8,10-tetramethyl-	dioxan	253(4.25),271(4.29), 328(3.63),396s(2.85)	88-1669-86
$C_{17}H_{18}OSi$			
2-Silacyclohexanone, 1,1-diphenyl-	n.s.g.	383f(2.45)	101-0021-86C
$C_{17}H_{18}O_3$			
1,3-Dioxane, 2-phenyl-5-(phenylmethoxy)-	MeOH	213(4.05)	32-0025-86
$C_{17}H_{18}O_4$			
2H,6H-Benzo[1,2-b:5,4-b']dipyran-6-one, 10-acetyl-3,4-dihydro-2,2,8-trimethyl-	MeOH	224(4.02),253(3.82), 305(3.90)	2-0377-86
2H,10H-Benzo[1,2-b:3,4-b']dipyran-10-one, 6-acetyl-3,4-dihydro-2,2,8-trimethyl-	MeOH	240(3.82),261(3.74), 312(3.82)	2-0377-86
2H-1-Benzopyran-7-ol, 3,4-dihydro-5-methoxy-2-(4-methoxyphenyl)-	MeOH	232s(4.30),278(3.24), 282(3.20)	102-1097-86
2H-1-Benzopyran-2-one, 4-hydroxy-3-[(3-methyl-2-oxocyclohexyl)-methyl]-	MeOH	269(4.09),281(4.14), 308(4.09)	2-0102-86
Ethanone, 1-[2-hydroxy-3,4-dimethoxy-5-(phenylmethyl)phenyl]-	MeOH	228(4.9),269(4.6)	2-0259-86
Ethanone, 1-[1-(dimethoxymethyl)-1,2,3,4-tetrahydro-2,3-bis(methylene)-1,4-epoxynaphthalen-6-yl]-, (±)-	EtOH	205(4.37),209s(4.36), 228(4.23),256(4.21)	33-1287-86
Ethanone, 1-[4-(dimethoxymethyl)-1,2,3,4-tetrahydro-2,3-bis(methylene)-1,4-epoxynaphthalen-6-yl]-, (±)-	isooctane	204(4.39),208s(4.37), 230(4.23),254(4.21), 260s(4.09),286(3.11), 296(3.04)	33-1287-86
	EtOH	206(4.31),230(4.18), 256(4.14),293s(3.11)	33-1287-86
1,2-Naphthalenedione, 3-(1-acetoxy-1-methylethyl)-5,8-dimethyl-	EtOH	220(4.52),255(4.66), 425(3.88)	2-1247-86
1,2-Naphthalenedione, 3-(2,2-dimethyl-1,3-dioxan-5-yl)-5-methyl-	EtOH	237(4.13),256(4.21), 351(3.41),422(3.32)	88-3935-86
Phenol, 4-(3,4-dihydro-5,7-dimethoxy-2H-1-benzopyran-2-yl)-, (S)-	MeOH	274(3.34)	102-2395-86
	MeOH-base	276(--),298s(--)	102-2395-86
(±)-	EtOH	230s(4.51),277(3.46)	102-2395-86
	EtOH-NaOEt	247(4.43),292(3.54)	102-2395-86
$C_{17}H_{18}O_4S$			
Phenanthro[1,2-b]thiophen-1(2H)-one, 3a,4,5,10,11,11a-hexahydro-7-methoxy-, 3,3-dioxide, cis	EtOH	212(4.14),220(4.09), 266(4.02),299(3.64)	104-0107-86
	EtOH-HCl	220(4.23),227(4.05), 265(4.13),299(3.76)	104-0107-86
	dioxan	220(4.32),276.5(4.26)	104-0107-86
Phenanthro[2,1-b]thiophen-2(1H)-one, 3a,4,5,10,11,11a-hexahydro-7-methoxy-, S,S-dioxide	EtOH	236(4.01),245(4.01), 268(4.01),295(4.03), 306(4.09),320(4.15), 332(4.03)	104-0107-86

Compound	Solvent	λ max (log ε)	Ref.
(cont.)	EtOH-HCl	238(3.94),246(3.93), 285(4.02),295(4.06), 306(4.15),320(4.20), 332(4.05)	104-0104-86
	dioxan	222(4.28),269(4.22)	104-0104-86
$C_{17}H_{18}O_5$			
2H,6H-Benzo[1,2-b:5,4-b']dipyran-6-one, 10-acetyl-3,4-dihydro-5-hydroxy-2,2,8-trimethyl-	MeOH	206(4.4),229(4.16), 266(4.28),280(4.16)	2-0377-86
2H,10H-Benzo[1,2-b:3,4-b']dipyran-10-one, 6-acetyl-3,4-dihydro-5-hydroxy-2,2,8-trimethyl-	MeOH	204(3.75),234(3.77), 258(3.75),295(4.40)	2-0377-86
2,5-Furandione, 3-[1-(2,5-dimethoxyphenyl)ethylidene]dihydro-4-(1-methylethylidene)-, (E)-	EtOH	295(4.01)	39-1599-86C
(Z)-	EtOH	281(4.02)	39-1599-86C
Naphtho[2,3-c]furan-1,3-dione, 4,9-dihydro-4,6-dimethoxy-4,9,9-trimethyl-	EtOH	245(3.28),278(3.30), 286(3.25)	39-1599-86C
1-Propanone, 3-hydroxy-1-(2-hydroxy-4,6-dimethoxyphenyl)-2-phenyl-	MeOH	207(4.49),226s(4.32), 283(4.34)	39-0215-86C
2H,5H-Pyrano[2,3-b][1]benzopyran-5-one, 2-(acetoxymethyl)-3,4-dihydro-2,4-dimethyl-	EtOH	206(4.31),226(4.35), 274(4.02),288(3.97), 298s(--),338s(--)	32-0491-86
2H,5H-Pyrano[3,2-c][1]benzopyran-5-one, 2-(acetoxymethyl)-3,4-dihydro-2,4-dimethyl-	EtOH	271(3.99),282(3.06), 306(3.95),318s(--)	32-0501-86
$C_{17}H_{18}O_6$			
2H-1-Benzopyran-3,7-diol, 3-[(3,4-dihydroxyphenyl)methyl]-3,4-dihydro-4-methoxy-	MeOH	205(4.74),219(4.20), 280(3.72),285(3.70)	94-2506-86
stereoisomer 6	MeOH	207(4.86),220(4.39), 280(4.01),285(3.99)	94-2506-86
4H-1-Benzopyran-4-one, 5,6,7,8-tetrahydro-5,6,7,8-tetrahydroxy-2-(2-phenylethyl)-	EtOH	208(4.20),253(4.08)	94-2766-86
isomer (isoaguratetrol)	EtOH	208(4.24),252(4.13)	94-2766-86
2,5-Furandione, dihydro-3-(1-methylethylidene)-4-[(3,4,5-trimethoxyphenyl)methylene]-	toluene	348(4.10)	39-0315-86C
7H-Furo[3,2-g][1]benzopyran-7-one, 9-(2,2-diethoxyethoxy)-	1% DMSO	302(4.04)	149-0391-86A
Naphtho[2,3-c]furan-1,3-dione, 4,9-dihydro-4-hydroxy-5,8-dimethoxy-4,9,9-trimethyl-	EtOH	291(3.58)	39-1599-86C
$C_{17}H_{18}O_7$			
4H-1-Benzopyran-4-one, 5,6,7,8-tetrahydro-5,6,7,8-tetrahydroxy-2-[2-(2-hydroxyphenyl)ethyl]-, (5α,6β,7α,8β)-	EtOH	202(4.35),214.8(4.31), 253(4.18)	94-3033-86
$C_{17}H_{18}O_8$			
Acetic acid, [(6-formyl-5,8-dimethoxy-2-methyl-4-oxo-4H-1-benzopyran-7-yl)oxy]-, ethyl ester	EtOH	236(4.36),254(4.39), 295s(3.71)	44-3116-86
$C_{17}H_{18}O_9$			
7-Oxabicyclo[2.2.1]hepta-2,5-diene-	MeCN	282(3.15)	24-0589-86

Compound	Solvent	λ_{max}(log ε)	Ref.
2,3-dicarboxylic acid, 1-[2-[(4-methoxy-1,4-dioxo-2-butenyl)oxy]-ethyl]-, dimethyl ester, (E)- (cont.)			24-0589-86
$C_{17}H_{19}ClHgO_2$			
Mercury, chloro[(2,3,4,5,6,7-hexahydro-6,6-dimethyl-4-oxo-2-phenyl-2-benzofuranyl)methyl]-	EtOH	271(4.1)	32-0441-86
$C_{17}H_{19}ClN_2S$			
Chlorpromazine hydrochloride	MeOH	256.5(4.536),309.5(3.602)	106-0571-86
$C_{17}H_{19}ClN_4O_7S$			
9H-Purine, 6-chloro-2-(methylthio)-9-(2,3,5-tri-O-acetyl-β-D-ribofuranosyl)-	EtOH	234.5(4.18),263(4.04), 304.5(3.85)	118-0450-86
$C_{17}H_{19}NO$			
3-Pentanone, 2-amino-1,5-diphenyl-, hydrobromide	MeOH	220(3.57),248(2.36), 253(2.49),258(2.60), 264(2.52),267(2.40)	44-3374-86
$C_{17}H_{19}NO_2$			
Bicyclo[4.4.1]undeca-1,3,5,7,9-pentaene-2-carboxamide, N,N-diethyl-10-formyl-	MeOH	235(4.24),255(4.33), 345(3.91)	33-1263-86
$C_{17}H_{19}NO_2S$			
1-Propanone, 3-[1,2-dihydro-1-(1-methylethyl)-2-thioxo-3-pyridinyl]-3-hydroxy-1-phenyl-	EtOH	241(4.14),273(4.05), 362(3.81)	4-0567-86
$C_{17}H_{19}NO_3S$			
Sulfoximine, N-(2-ethoxy-2-oxoethyl)-S-(4-methylphenyl)-S-phenyl-, (R)-	EtOH	220(3.94),243(4.10), 267s(3.46)	12-1655-86
$C_{17}H_{19}NO_3S_2$			
[1]Benzothiepino[4,5-e]-1,2-oxathiin, 5,6-dihydro-4-(1-piperidinyl)-, 2,2-dioxide	EtOH	227(4.13),233s(4.10), 252s(3.78),278s(3.91), 300(4.05)	4-0455-86
$C_{17}H_{19}NO_4$			
α-D-xylo-Hexofuranurononitrile, 5-deoxy-5-methylene-1,2-O-(1-methylethylidene)-3-O-(phenylmethyl)-	n.s.g.	225(2.72)	111-0111-86
β-L-arabino-Hexofuranurononitrile, 5-deoxy-5-methylene-1,2-O-(1-methylethylidene)-3-O-(phenylmethyl)-	n.s.g.	257(2.38)	111-0111-86
3,5-Pyridinedicarboxylic acid, 1,4-dihydro-2,6-dimethyl-4-phenyl-, dimethyl ester	MeOH	237(--),270(--), 348(3.92)	87-1596-86
$C_{17}H_{19}NO_5$			
4,5-Isoxazoledicarboxylic acid, 3-(2-ethenylphenyl)-2,3-dihydro-2,3-dimethyl-, dimethyl ester	EtOH	282s(4.03)	44-3125-86
Lutessine, deacetyl-	EtOH	233(3.78),290(3.71)	100-0090-86
$C_{17}H_{19}NO_7$			
Furo[4,3,2-ef][2]benzazepine-2a(2H)-	n.s.g.	258(4.05),297(3.75)	94-4050-86

Compound	Solvent	$\lambda_{max}(\log \epsilon)$	Ref.
propanoic acid, 2-carboxy-3,4,5,6-tetrahydro-9-methoxy-5-methyl-6-oxo-, (2R-cis)- (cont.)			94-4050-86
$C_{17}H_{19}NO_7S$			
3-Oxa-8-azatricyclo[3.2.1.0^{2,4}]octane-6,7-dicarboxylic acid, 8-[(4-methylphenyl)sulfonyl]-, dimethyl ester, (1α,2β,4β,5α,6α,7α)-	EtOH	228(4.44),273s(3.65)	24-0616-86
$C_{17}H_{19}N_2S_2$			
Pyridinium, 1-[2-[4-(dimethylamino)-phenyl]-1-[(methylthio)thioxomethyl]ethenyl]-, iodide, (Z)-	MeOH	269(4.61),492(4.63)	78-0699-86
$C_{17}H_{19}N_3$			
2H-Benz[g]indazole, 3,3a,4,5-tetrahydro-2-methyl-3-(1-methyl-1H-pyrrol-2-yl)-	EtOH	225(4.16),312(4.00)	4-0135-86
1H-Benzo[h]pyrimido[1,2-c]quinazoline, 2,3,5,6-tetrahydro-2,2-dimethyl-	EtOH	230(3.22),258(3.70), 290(3.27),304(3.28), 318(3.31),331(3.38), 345(3.22)	142-1119-86
monohydrochloride	EtOH	258(3.63),304(3.19), 318(3.23),350(3.12)	142-1119-86
Piperidine, 1-[4-(phenylazo)phenyl]-	EtOH	400(4.40)	39-0123-86B
	EtOH-HCl	528(3.54)	39-0123-86B
$C_{17}H_{19}N_3OS$			
Hydrazinecarbothioamide, 2-(1,3-dihydro-3,3-dimethyl-1-phenyl-1-isobenzofuranyl)-	EtOH	244(4.22),270(3.34), 310(3.38)	104-0647-86
Thiomorpholine, 4-[4-[(4-methoxyphenyl)azo]phenyl]-	EtOH	393(4.40)	39-0123-86B
	EtOH-HCl	569(4.18)	39-0123-86B
$C_{17}H_{19}N_3O_2$			
1,4-Benzenediamine, N,N-dimethyl-N'-[(3,5-dimethyl-4-nitrophenyl)methylene]-	EtOH	259(4.27),402(4.23)	56-0831-86
	THF	259.5(4.27),406.5(4.27)	56-0831-86
1,4-Benzenediamine, N^4,N^4,2,6-tetramethyl-N^1-[(4-nitrophenyl)methylene]-	EtOH	276(4.20),376(3.97)	56-0831-86
1,3-Benzodioxole-5-carboxaldehyde, 6-[2-(methylamino)ethyl]-, phenylhydrazone	MeOH	248(4.42),305(3.90), 365(3.96)	83-0910-86
Morpholine, 4-[4-[(4-nitrophenyl)-azo]phenyl]-	EtOH	386(4.44)	39-0123-86B
	EtOH-HCl	566(4.02)	39-0123-86B
$C_{17}H_{19}N_3O_3$			
Acetamide, N-[1,4-dihydro-3-(4-methyl-1-piperazinyl)-1,4-dioxo-2-naphthalenyl]-	heptane	235s(4.12),240(4.13), 246(4.05),287(4.24), 305s(4.16),346s(3.09), 373(2.79),392(2.82), 412s(2.87),515(3.57)	104-1560-86
$C_{17}H_{19}N_3O_3S$			
Thiomorpholine, 4-[4-[(4-methoxyphenyl)azo]phenyl]-, 1,1-dioxide	EtOH	381(4.48)	39-0123-86B
	EtOH-HCl	558(4.78)	39-0123-86B
$C_{17}H_{19}N_3O_6$			
Thymidine, 5'-deoxy-5'-[(phenoxy-	pH 1	266(3.98)	87-1052-86

Compound	Solvent	λ max(log ε)	Ref.
carbonyl)amino]- (cont.)	pH 7	266(3.98)	87-1052-86
spectra in ethanol	pH 13	232(4.23),267(3.93)	87-1052-86
$C_{17}H_{19}N_5O_3$			
Benzo[g]pteridine-2,4(3H,10H)-dione,	benzene	452(4.48),479(4.59)	130-0119-86
10-ethyl-3-methyl-8-(4-morpholinyl)-	H_2O	496(4.57)	130-0119-86
	MeOH	485(4.61)	130-0119-86
	30% MeOH	492(4.65)	130-0119-86
	dioxan	450s(--),473(4.58)	130-0119-86
	EtOAc	455s(--),478(4.63)	130-0119-86
	THF	457s(--),478(4.62)	130-0119-86
	MeCN	480(4.66)	130-0119-86
	C_6H_{12}-benzene	450(4.48),477(4.56)	130-0119-86
$C_{17}H_{19}N_7O_6$			
Riboflavin, 6-azido-	pH 6.0	262(4.5),423(4.3)	69-3282-86
	20% $HClO_4$	250(4.4),415(4.2)	69-3282-86
$C_{17}H_{20}$			
Heptalene, 1,3,5,6,10-pentamethyl-	hexane	252(4.36),299s(3.49)	88-1669-86
$C_{17}H_{20}ClHgNO_4$			
Mercury, chloro[(3,4,5,6-tetrahydro-8,9-dimethoxy-2,6-dimethyl-5-oxo-2H-pyrano[3,2-c]quinolin-2-yl)-methyl]-	EtOH	277(3.7),286(3.8), 329(4.1),342(4.1)	32-0441-86
$C_{17}H_{20}ClN_4O$			
1,3,5-Oxadiazin-1-ium, 2-(2-chloro-phenyl)-4,6-di-1-pyrrolidinyl-, perchlorate	MeCN	218(4.37),280(4.19)	97-0094-86
$C_{17}H_{20}Cl_2N_2O_2$			
Benzenemethanol, 2,2'-(1,3-propane-diyldiimino)bis[5-chloro-	EtOH	213(4.17),258(4.45), 308(3.69)	12-1231-86
$C_{17}H_{20}F_2N_2O_2$			
Benzenemethanol, 2,2'-(1,3-propane-diyldiimino)bis[5-fluoro-	EtOH	210(4.21),246(4.21), 307(3.72)	12-1231-86
$C_{17}H_{20}N_2$			
1,8a,3,4a-Azino-4,9-ethano-1H-benz-[f]indene, 2,3,3a,4,9,9a-hexahy-dro-2,2-dimethyl-	hexane	381(2.19)	89-0187-86
$C_{17}H_{20}N_2O$			
Bicyclo[4.4.1]undeca-3,6,8,10-tetra-en-2-one, 5-(tetrahydro-1,3-dimeth-yl-2(1H)-pyrimidinylidene)-, (±)-	MeOH	260(4.41),334(3.66), 379(3.59)	33-0315-86
	CH_2Cl_2	338(3.59),458(4.32)	33-0315-86
	MeCN	225(4.27),262(4.10), 338(3.48),450(4.11)	33-0315-86
	benzene	461(--)	33-0315-86
2-Pyrrolidinone, 3,4-diethyl-5-[(2-methylphenyl)cyanomethylene]-, (E)-	$CHCl_3$	261(4.28)	49-0631-86
(Z)-	$CHCl_3$	263(4.28)	49-0631-86
2-Pyrrolidinone, 3,4-dimethyl-5-[(2,4,6-trimethylphenyl)cyano-methylene]-, (E)-	EtOH	258(4.41)	49-0631-86
(Z)-	EtOH	263(4.36)	49-0631-86

Compound	Solvent	λ_{max}(log ε)	Ref.
$C_{17}H_{20}N_2O_2$			
3H-Benz[f]indazole-4,9-dione, 9a-(1,1-dimethylethyl)-3a,9a-dihydro-3,3-dimethyl-, cis-(±)-	EtOH	229(4.43),254(3.93), 299(3.16),307s(3.15)	39-2217-86C
Indolo[2,3-a]quinolizin-1-ol, 1,2,3-4,6,7,12,12b-octahydro-, acetate, cis-(±)-	MeOH	225(4.57),274s(3.86), 281(3.88),290(3.80)	94-3713-86
Spiro[9H-benz[f]indazole-9,2'-oxiran]-4(3H)-one, 3a,9a-dihydro-3,3,3',3'-9a-pentamethyl-, (3aα,9α,9aα)-(±)-	EtOH	212(4.29),254(3.96), 292(3.20),333(2.54)	39-2217-86C
$C_{17}H_{20}N_2O_3$			
2,5-Cyclohexadien-1-one, 2-methyl-4-[[[4-bis(2-hydroxyethyl)amino]-phenyl]imino]-	pH 8.0	616(4.29)	39-0065-86B
$C_{17}H_{20}N_2O_4$			
5a,9a-Methanobenzo[g]phthalazine-1,4-dicarboxylic acid, 2,4a,5,6,9,10-hexahydro-, dimethyl ester	CH_2Cl_2	270(3.87),360(3.5)	24-3862-86
2-Pentenoic acid, 5-(4-methoxyphenyl)-5-oxo-4-(tetrahydro-2(1H)-pyrimidinylidene)-, methyl ester, (E)-	EtOH	242(4.06),348(4.33)	142-2247-86
$C_{17}H_{20}N_2O_5$			
2-Quinazolinecarboxylic acid, 4-[1-(ethoxycarbonyl)propoxy]-, ethyl ester	MeOH	205(4.33),227(4.48), 284(3.83)	33-1017-86
$C_{17}H_{20}N_2O_6$			
Uridine, 2'-O-methyl-3-(phenylmethyl)-	pH 1 and 6	260.5(4.03)	18-2947-86
	pH 11	262(4.02)	18-2947-86
$C_{17}H_{20}N_2S$			
Methanethione, bis[4-(dimethylamino)-phenyl]-	pH 9	440(4.3)	140-0653-86
	40% DMF-pH 2-8	470(--)	140-0653-86
silver complex	n.s.g.	530(5.15)	140-0653-86
Promazine (phosphate)	MeOH	254(4.426),304(3.637)	106-0571-86
Promethazine, hydrochloride	MeOH	252(4.491),301.5(3.558)	106-0571-86
3-Pyridinecarbonitrile, 2-(methylthio)-6-tricyclo[3.3.1.1^{3,7}]dec-1-yl-	EtOH	221(4.36),267(4.15), 315(3.75)	70-0131-86
$C_{17}H_{20}N_2Se$			
3-Pyridinecarboniteile, 2-(methylseleno)-6-tricyclo[3.3.1.1^{3,7}]dec-1-yl-	EtOH	226(4.29),280(4.06), 323(3.57)	70-0376-86
$C_{17}H_{20}N_3O$			
Furo[2,3-d]pyrimidine-4-amine, N-(3,5-dimethylphenyl)-2,3-dimethyl-, 7-methiodide	EtOH	317(4.24)	11-0337-86
$C_{17}H_{20}N_4O$			
Benzoic acid, 2-(methylamino)-, [[4-(dimethylamino)phenyl]-methylene]hydrazide	MeOH	215(4.19),226(4.33), 264(4.27),368(4.43)	18-1575-86
Benzo[g]phthalazin-1(2H)-one, 4-[[3-(dimethylamino)propyl]amino]-	EtOH	245(4.65),253(4.71)	111-0143-86

Compound	Solvent	λ_{max}(log ε)	Ref.
$C_{17}H_{20}N_4O_4$			
4(1H)-Azulenone, 2,3,3a,5,6,7-hexahydro-3a-methyl-, 2,4-dinitrophenylhydrazone	$CHCl_3$	366(4.31)	44-2408-86
$C_{17}H_{20}N_4O_5S$			
Riboflavin, 2-thio-, neutral semiquinone	pH 5	360(4.0),500(4.0), 620s(3.7)	23-0067-86
radical anion	pH 10	400(4.2),510(3.9)	23-0067-86
$C_{17}H_{20}N_4O_6S$			
Riboflavin, 6-mercapto-	n.s.g.	440(4.29)	69-3282-86
$C_{17}H_{20}N_4O_7S$			
9H-Purine, 6-(methylthio)-9-(2,3,5-tri-O-acetyl-β-D-ribofuranosyl)-	EtOH	215(4.00),282(4.18), 290(4.18)	118-0450-86
$C_{17}H_{20}N_5O_8P$			
Guanosine, N-(phenylphosphonomethyl)-, barium salt (1:1)	pH 1	259.5(4.19),280s(3.97)	94-0471-86
	H_2O	255(4.20),277s(4.03)	94-0471-86
	pH 13	260(4.12),275s(4.03)	94-0471-86
$C_{17}H_{20}O$			
9a-Carbamorphinan-10-one	EtOH	255(4.16)	39-1243-86C
14α-	EtOH	254(4.08)	39-1243-86C
2(3H)-Naphthalenone, 4,4a,5,6,7,8-hexahydro-1-(phenylmethyl)-	EtOH	248(4.02)	39-1243-86C
$C_{17}H_{20}O_2$			
1,2-Heptalenedimethanol, 5,6,10-trimethyl-	dioxan	253(4.32),300s(3.49)	88-1669-86
1,2-Heptalenedimethanol, 6,8,10-trimethyl-	dioxan	246s(4.22),258(4.33), 316(3.58)	88-1669-86
$C_{17}H_{20}O_3$			
Naphtho[2,3-b]furan-5-methanol, 4,4a,8a,9-tetrahydro-3,8a-dimethyl-, acetate, (4aR-cis)-	EtOH	215(3.74),262(3.63)	138-1789-86
$C_{17}H_{20}O_4$			
Benzenebutanoic acid, α-oxo-γ-(2-oxocyclohexyl)-, methyl ester	MeOH	206(3.94),260s(--)	83-0694-86
Butanoic acid, 4-(3,4-dihydro-6-methoxy-1-oxo-2(1H)-naphthalenylidene)-, ethyl ester, (E)-	EtOH	214(3.86),234(3.97), 305(4.25)	150-2492-86M
1,7-Dioxaspiro[5.5]undec-2-en-4-one, 5-(hydroxyphenylmethyl)-2-methyl-, [5α(S*),6β]-(±)-	EtOH	262(4.05)	44-4254-86
1,7-Dioxaspiro[5.5]undec-2-en-4-one, 2-methyl-5-(phenoxymethyl)-	EtOH	260(4.05)	44-4254-86
geometric isomer	EtOH	262(4.06)	44-4254-86
1,7-Dioxaspiro[5.5]undec-2-en-4-one, 2-[(phenylmethoxy)methyl]-, (±)-	EtOH	248(4.03)	44-4254-86
2H,5H-Pyrano[2,3-b][1]benzopyran-5-one, 2-(1,1-dimethylethyl)-3,4-dihydro-2-(hydroxymethyl)-	EtOH	206(4.29),226(4.31), 278(4.09)	32-0491-86
2H,5H-Pyrano[3,2-c][1]benzopyran-5-one, 3,4-dihydro-2-(hydroxymethyl)-2,4,8,10-tetramethyl-	EtOH	283s(--),293(4.16), 308(4.06),321s(--)	32-0501-86
2H,5H-Pyrano[3,2-c][1]benzopyran-5-one, 2-(1,1-dimethylethyl)-3,4-	EtOH	270(3.93),281(4.01), 304(3.98),317(3.83)	32-0501-86

Compound	Solvent	λmax(log ε)	Ref.
dihydro-2-(hydroxymethyl)- (cont.)			32-0501-86
$C_{17}H_{20}O_5$			
3,11a-Epoxy-11aH-oxepino[2,3-b][1]-benzopyran-6(2H)-one, 3-(1,1-dimethylethyl)-3,4,5,5a-tetrahydro-8a-hydroxy-, (3α,5aα,11aα)-	EtOH	251(4.02),312(3.49)	32-0285-86
(3α,5aβ,11aα)-	EtOH	251(3.97),314(3.54)	32-0285-86
$C_{17}H_{20}O_6$			
4-Hexenoic acid, 6-(1,3-dihydro-4-hydroxy-6-methoxy-7-methyl-3-oxo-5-isobenzofuranyl)-4-methyl-, (E)-	EtOH	249(4.01),305(3.72)	35-0806-86
$C_{17}H_{20}O_6S$			
Benzo[b]thiophene-2,3-dicarboxylic acid, 4,5-dimethoxy-7-methyl-, diethyl ester	EtOH	208(4.39),245(4.30), 296(4.13),356(3.52)	142-2749-86
$C_{17}H_{20}O_{10}$			
Fraxidin, 8-O-β-D-glucopyranoside	MeOH	228(3.76),293(3.58), 338(3.43)	105-0228-86
Isofraxidin, 7-O-β-D-glucopyranoside (calycanthoside)	MeOH	230(3.48),308s(3.60), 342(3.79)	105-0228-86
$C_{17}H_{21}BrO_2$			
Cupalaurenol acetate	EtOH	205(4.32),225s(4.00), 280(3.48)	44-3364-86
Phenol, 4-bromo-5-(1,2-dimethylbicyclo-[3.1.0]hex-2-yl)-2-methyl-, acetate (cyclolaurenol acetate)	MeOH	206(4.23),270(2.85), 278(2.85)	44-3364-86
$C_{17}H_{21}Cl_2N_3Pd$			
Palladium, dichloro[4,5,6,7-tetrahydro-7,8,8-trimethyl-1-(2-pyridinylmethyl)-4,7-methano-1H-indazol-NN^1,N^2]-	MeOH	374.5(2.56)	12-1525-86
Palladium, dichloro[4,5,6,7-tetrahydro-7,8,8-trimethyl-2-(2-pyridinylmethyl)-4,7-methano-2H-indazol-NN^1,N^2]-	MeOH	255s(3.95),361.8(2.51)	12-1525-86
$C_{17}H_{21}FeN_2O_2$			
1H-Imidazol-1-yloxy, 2-ferrocenyl-4,5-dihydro-4,4,5,5-tetramethyl-, 3-oxide	n.s.g.	271(3.94),295(3.97), 303(3.95),358(3.73), 463(2.75),685(2.85)	64-1587-86B
$C_{17}H_{21}N$			
1H-Benzo[a]carbazole, 2,3,4,4a,5,6-11,11b-octahydro-11-methyl-	n.s.g.	?(4.51),284(3.90)	150-2782-86M
Trikentrin B, trans	MeOH	249(4.38),275(3.94)	78-6545-86
$C_{17}H_{21}NO$			
Bicyclo[4.4.1]undeca-1,3,5,7,9-pentaene-2-carboxamide, N,N-diethyl-10-methyl-	MeOH	260(4.65),310(3.81)	33-1263-86
Methanamine, N-[[5-(1,1-dimethylethoxy)bicyclo[4.4.1]undeca-2,4,6,8,10-pentaen-1-yl]methylene]-, (±)-	CH_2Cl_2	276(4.54),344(4.08)	33-0315-86

Compound	Solvent	$\lambda_{max}(\log \epsilon)$	Ref.
$C_{17}H_{21}NO_2$			
7H-Benzo[3,4]cyclobuta[1,2]cyclohepten-7-one, 1,2,3,4,4a,9b-hexahydro-4a-(4-morpholinyl)-	MeOH	232(4.26),324(4.07)	88-3005-86
Bicyclo[4.4.1]undeca-1,3,5,7,9-pentaene-2-carboxamide, N,N-diethyl-10-(hydroxymethyl)-	MeOH	260(4.72),310(3.85)	33-1263-86
2-Dibenzofuranol, 1-[[(1,1-dimethylethyl)imino]methyl]-6,7,8,9-tetrahydro-, (E)-	toluene	308(4.29),345(3.85)	104-0503-86
	MeOH	310(3.93),345(4.01), 437(3.58)	104-0503-86
	DMSO	308(4.25),345(3.89), 437(2.90)	104-0503-86
1,6-Octadiene-3,5-dione, 7-[(1-methylethyl)amino]-1-phenyl-	n.s.g.	334(4.00),404(4.57)	4-1721-86
$C_{17}H_{21}NO_3$			
2(1H)-Quinolinone, 4-ethoxy-8-hydroxy-1-methyl-3-(3-methyl-2-butenyl)-(homoglycosolone)	n.s.g.	222.0(4.56),237.2(4.28), 252.5(4.34),260(4.45), 287(3.87),295(3.88), 333(3.48),345(3.44)	25-0669-86
	base	218(4.17),361(3.48)	25-0669-86
$C_{17}H_{21}NO_4$			
Phenol, 5-[[[2-(4-hydroxyphenyl)ethyl]amino]methyl]-2,4-dimethoxy-	MeOH	228(4.27),279(3.83)	150-0028-86S
Pyrrolidine, 1-(3,5-dimethoxy-1,4-dioxo-5-phenyl-2-pentenyl)-, (Z)-	MeOH	220(4.23),275(2.11)	163-3053-86
$C_{17}H_{21}NO_5$			
Morpholine, 4-(3,5-dimethoxy-1,4-dioxo-5-phenyl-2-pentenyl)-, (Z)-	MeOH and MeOH- base	225(4.26),280(2.00)	163-3053-86
2-Pentenedioic acid, 2-acetyl-3-[(1-phenylethyl)amino]-, dimethyl ester, [R-(E)]-	MeOH	244(3.88),309(4.21)	44-1498-86
$C_{17}H_{21}N_2O_7P$			
Benzo[1,2-b:4,3-b']dipyrrole-2-carboxylic acid, 5-[(diethoxyphosphinyl)oxy]-3,6-dihydro-4-methoxy-, methyl ester	MeOH	208(4.35),247(4.06), 316(4.24)	77-1162-86
$C_{17}H_{21}N_3$			
4,7-Methano-1H-indazole, 4,5,6,7-tetrahydro-7,8,8-trimethyl-1-(2-pyridinylmethyl)-, (4S-cis)-	MeOH	231.4(--),253.7(--), 259.3(--),265s(--)	12-1525-86
4,7-Methano-2H-indazole, 4,5,6,7-tetrahydro-7,8,8-trimethyl-2-(2-pyridinylmethyl)-, (4S-cis)-	MeOH	235.6(3.92),254.7(3.68), 261.2(3.69),266s(3.56)	12-1525-86
$C_{17}H_{21}N_3O$			
Benzo[h]quinazolin-4-amine, 5,6-dihydro-N-[1-(methoxymethyl)propyl]-	EtOH	250(4.44),292(3.76), 315(3.91)	4-0685-86
$C_{17}H_{21}N_3O_3$			
Benzenamine, N,N-diethyl-4-[(4-methoxyphenyl)-ONN-azoxy]-, N-oxide	benzene	349(4.10)	39-1439-86B
	MeOH	239(4.02),347(4.18)	39-1439-86B
$C_{17}H_{21}N_3S$			
3(2H)-Pyridazinethione, 6-(4-methylphenyl)-2-(1-piperidinylmethyl)-	EtOH	314(4.17)	48-0522-86

Compound	Solvent	$\lambda_{max}(\log \epsilon)$	Ref.
$C_{17}H_{21}N_4O$			
1,3,5-Oxadiazin-1-ium, 2-phenyl-4,6-di-1-pyrrolidinyl-, perchlorate	MeCN	239(4.39),279(4.24)	97-0094-86
$C_{17}H_{21}N_4O_3$			
1,3,5-Oxadiazin-1-ium, 2,4-di-4-morpholinyl-6-phenyl-, perchlorate	MeCN	246(4.54),284(4.40)	97-0094-86
$C_{17}H_{21}N_4S_3$			
1H-Pyrazolium, 3,5-bis(dimethylamino)-4-(methylthio)-1-(5-phenyl-3H-1,2-dithiol-3-ylidene)-, perchlorate	MeCN	215s(4.65),342(4.59), 490(4.51)	88-0159-86
$C_{17}H_{21}N_5O_6$			
Riboflavin, 6-amino-	pH 7	428(4.26)	69-8095-86
$C_{17}H_{21}N_5O_7$			
2-Amino-6-ethoxy-8,2'-methano-9-(3,5-di-O-acetyl-α-D-arabinofuranosyl)-purine	MeOH	247(4.07),284(3.98)	94-1961-86
8,2'-Methanoguanosine, 3',5'-di-O-acetyl-O^6-ethyl-	MeOH	246(4.06),284(3.98)	94-1961-86
$C_{17}H_{21}N_5O_7S$			
9H-Purin-2-amine, 6-(methylthio)-9-(2,3,5-tri-O-acetyl-β-D-ribofuranosyl)-	EtOH	216(4.15),244(4.15), 308(4.00)	118-0450-86
9H-Purin-6-amine, 2-(methylthio)-9-(2,3,5-tri-O-acetyl-β-D-ribofuranosyl)-	EtOH	236(4.28),273(4.04)	118-0450-86
$C_{17}H_{22}BrClO_3$			
Spiro[5.5]undec-3-en-2-one, 7-acetoxy-10-bromo-9-chloro-5,5,9-trimethyl-1-methylene-	EtOH	235(3.82)	33-0091-86
$C_{17}H_{22}Br_2N_4O_4$			
2,4(1H,3H)-Pyrimidinedione, 5,5'-methylenebis[1-(2-bromoethyl)-3,6-dimethyl-	EtOH	295(4.27)	94-1809-86
$C_{17}H_{22}FeS_4$			
Ferrocene, 1,1'-[1,3-propanediylbis(thio-2,1-ethanediylthio)]-	MeCN	443(2.29)	18-1515-86
copper perchlorate salt	MeCN	434(2.11)	18-1515-86
mercuric chloride salt	MeCN	447(2.15)	18-1515-86
silver perchlorate salt	MeCN	437(2.27)	18-1515-86
$C_{17}H_{22}N_2$			
Quinazoline, 1,4,5,6,7,8-hexahydro-4,4,7-trimethyl-2-phenyl-, (R)-	EtOH	204(4.15),230(4.04), 319(3.51)	4-0705-86
$C_{17}H_{22}N_2O_2$			
2(1H)-Pyrimidinone, 4-(3-ethoxypropyl)-6-methyl-1-(4-methylphenyl)-	EtOH	307(3.86)	39-1147-86C
$C_{17}H_{22}N_2O_3$			
2(1H)-Pyrimidinone, 4-(3-ethoxypropyl)-1-(4-methoxyphenyl)-6-methyl-	EtOH	233(4.16),307(3.92)	39-1147-86C

Compound	Solvent	$\lambda_{max}(\log \epsilon)$	Ref.
$C_{17}H_{22}N_2O_8$			
Uridine, 6-(3-hydroxy-1-propynyl)-5'-O-(methoxymethyl)-2',3'-O-(1-methylethylidene)-	MeOH	287(4.03)	23-1560-86
$C_{17}H_{22}N_2O_{10}$			
Propanedioic acid, [1,2,3,4-tetrahydro-1-[2,3-O-(1-methylethylidene)-β-D-ribofuranosyl]-2,4-dioxo-5-pyrimidinyl]-, dimethyl ester	MeOH	264(4.00)	94-4585-86
2,4,6(1H,3H,5H)-Pyrimidinetrione, 1,3-dimethyl-5-(2,3,4-tri-O-acetyl-β-D-ribopyranosyl)-	MeOH	226(3.80),259(3.00)	136-0053-86M
$C_{17}H_{22}N_4O_2$			
Ethanol, 2,2'-[[4-[(2-amino-4-imino-5-methyl-2,5-cyclohexadien-1-ylidene)amino]phenyl]imino]bis-	neutral	523(3.92)	39-0065-86B
	cation	636(4.25)	39-0065-86B
$C_{17}H_{22}N_4O_6$			
Piperazine, 1-[5-(2,3-dihydroxypropoxy)-1,6-dihydro-1-methyl-6-oxo-4-pyridazinyl]-4-(2-furanylcarbonyl)-	EtOH	239(4.2)	83-0060-86
1H-Purine-2,6-dione, 9-(2,2'-anhydro-6-deoxy-2-C-(hydroxymethyl)-3,4-O-(1-methylethylidene)-α-L-galactopyranosyl]-3,9-dihydro-1,3-dimethyl-	MeOH	275(4.03)	39-1297-86C
β-	MeOH	275(4.17)	39-1297-86C
1H-Purine-2,6-dione, 9-[2,2'-anhydro-6-deoxy-2-C-(hydroxymethyl)-3,4-O-(1-methylethylidene)-β-L-talopyranosyl)-3,9-dihydro-1,3-dimethyl-	MeOH	275(3.82)	39-1297-86C
$C_{17}H_{22}N_4O_7$			
7H-Pyrrolo[2,3-d]pyrimidin-4-one, 2-amino-4,7-dihydro-3-(methoxymethyl)-7-[2,3-O-(1-methylethylidene)-β-D-ribofuranosyl]-5-formyl-	MeOH	249(4.33),290(3.72), 317(3.82)	78-0207-86
$C_{17}H_{22}N_4O_8$			
Pyrazolo[3,4-d][1,3]diazepin-4-ol, 1,4,5,6-tetrahydro-1-(2,3,5-tri-O-acetyl-β-D-ribofuranosyl)-	pH 1	234(3.74),264(3.87)	44-1050-86
	pH 11	235(3.83),277(4.07)	44-1050-86
	MeOH	279(3.99)	44-1050-86
$C_{17}H_{22}N_6O_6$			
Pentanediamide, N,N'-bis[2-(3,4-dihydro-2,4-dioxo-1(2H)-pyrimidinyl)-ethyl]-	H_2O	268(4.33)	47-3201-86
polymer	DMSO	268.0(3.94)	47-3201-86
$C_{17}H_{22}O$			
7-Octene-3,5-diyn-2-ol, 8-(2,6,6-trimethyl-1-cyclohexen-1-yl)-, (E)-	EtOH	213(4.21),228s(3.85), 243(3.79),258s(3.76), 270s(4.01),287s(4.21), 298(4.23)	5-1398-86
$C_{17}H_{22}O_3$			
Benzoic acid, 3-(3,7-dimethyl-2,6-octadienyl)-4-hydroxy-	MeOH	207(4.12),254(3.92)	94-2290-86
Cyclodeca[b]furan-8-ol, 4,7,8,11-tetrahydro-3,6,10-trimethyl-,	isooctane	206(4.03)	32-0303-86

Compound	Solvent	λ_{max}(log ε)	Ref.
acetate, [R-(E,E)]- (cont.)			32-0303-86
2(5H)-Furanone, 5-hydroxy-4-[4-(2,6,6-trimethyl-1-cyclohexen-1-yl)-1,3-butadienyl]-, (E/Z,E)-	EtOH	318(4.37)	94-4346-86
2(1H)-Naphthalenone, 6-acetoxy-3,4-dihydro-4,7-dimethyl-1-(1-methylethyl)-, (1R-trans)- (7-acetoxycalamenen-3-one)	EtOH	208(3.92),270(2.93)	12-0103-86
Spiro[5H-benzo[3,4]cyclobuta[1,2]-cyclononene-5,2'-[1,3]dioxolan-4b(6H)-ol, 7,8,9,10,11,11a-hexahydro-, (4bR*,11aR*)-	MeOH	261(3.14),267(3.31), 273(3.29)	44-1419-86
$C_{17}H_{22}O_4$			
Bicyclo[7.2.2]trideca-9,11,13-triene-10,11-dicarboxylic acid, dimethyl ester	EtOH	222(4.49),308(3.38)	24-2698-86
Ethanone, 1-[1-(dimethoxymethyl)-1,2,3,4,5,6,7,8-octahydro-2,3-bis(methylene)-1,4-epoxynaphthalen-6-yl]-	isooctane	224s(3.96),232(3.88), 242s(3.78)	33-1287-86
	EtOH	210(4.10),223s(4.00), 232s(3.91),241s(3.81)	33-1287-86
Kushequinone A	MeOH	205(4.22),291(4.15), 420(2.52)	95-0022-86
Naphtho[2,3-b]furan-2(3H)-one, 6-acetoxy-3a,7,8,8a,9,9a-hexahydro-3,5,8a-trimethyl-. [3S-(3α,3aβ,8aα,9aβ)]-	EtOH	238(4.33)	94-1319-86
4a(4H)-Phenanthrenecarboxylic acid, 1,4b,5,6,7,8,10,10a-octahydro-2-methoxy-1-oxo-, methyl ester, (4aα,4bα,10aβ)-(±)-	EtOH	262(4.18)	32-0115-86
$C_{17}H_{22}O_5$			
Artegallin	EtOH	246(4.03)	102-1757-86
1(3H)-Isobenzofuranone, 7-hydroxy-6-(6-hydroxy-3-methyl-2-hexenyl)-5-methoxy-4-methyl-, (E)-	isooctane	249(3.79),305(3.57)	35-0806-86
11-Oxabicyclo[7.3.2]tetradeca-9,12,13-triene-13,14-dicarboxylic acid, dimethyl ester	EtOH	210s(4.19),279(3.58)	24-2698-86
Oxireno[5,6]naphtho[2,3-b]furan-6(2H)-one, 1a-acetoxy-1a,3,3a,4-4a,7,7a,8b-octahydro-3a,7,8b-trimethyl-, [1aS-(1aα,3aα,4aβ,7α,7aβ,-8bα)]-	EtOH	213(3.32),291(1.90)	94-1319-86
$C_{17}H_{23}NOS$			
1-Benzothiepin-5(2H)-one, 4-[[bis(1-methylethyl)amino]methylene]-3,4-dihydro-, (E)-	EtOH	248(3.99),355(4.21)	4-0449-86
$C_{17}H_{23}NO_4$			
3,5-Pyridinedicarboxylic acid, 4-(3-cyclohexen-1-yl)-1,4-dihydro-2,6-dimethyl-, dimethyl ester	MeOH	235(--),274(--), 340(3.99)	87-1596-86
$C_{17}H_{23}N_3OS$			
1-Cyclohexene-1-carbothioamide, 2-[(2-aminoethyl)amino]-4,4-dimethyl-6-oxo-N-phenyl-	dioxan	281(4.303),312(4.122)	48-0635-86

Compound	Solvent	λ_{max}(log ϵ)	Ref.
$C_{17}H_{23}N_3O_8$			
Cytidine, N,N-dimethyl-, 2',3',5'-tri-acetate	pH 7	275(4.07)	1-0806-86
	pH 2	285(--)	1-0806-86
	pH 13	282(--)	1-0806-86
$C_{17}H_{23}N_5$			
1,3-Propanediamine, N'-(2,3-dimethyl-2H-pyrazolo[3,4-b]quinolin-4-yl)-N,N-dimethyl-, dihydrochloride	EtOH	213(4.21),248(4.70), 300(4.15)	73-1692-86
$C_{17}H_{23}N_5O_6$			
9H-Purine-8-carboxylic acid, 6-(di-methylamino)-9-[2,3-O-(1-methyl-ethylidene)-β-D-ribofuranosyl]-, methyl ester	MeOH	235(4.40),270s(3.88), 278(3.92),318(4.25)	94-3635-86
$C_{17}H_{23}N_5O_7$			
3H-Pyrrolo[2,3-d]pyrimidine-5-carbox-aldehyde, 2-amino-4,7-dihydro-3-(methoxymethyl)-7-[2,3-O-(1-methylethylidene)-β-D-ribofurano-syl]-4-oxo-, 5-oxime	MeOH	245(4.38),284(3.89), 306(3.97)	78-0207-86
$C_{17}H_{23}N_5O_8$			
Methanimidamide, N'-[4-formyl-1-(2,3,5-tri-O-acetyl-β-D-ribofuranosyl)-1H-1,2,3-triazol-5-yl]-N,N-dimethyl-	pH 1	263(3.93),312(3.79)	44-1050-86
	pH 11	229(4.21),264(3.94), 311(3.81)	44-1050-86
	MeOH	224(4.21),261(3.92), 315(3.86)	44-1050-86
$C_{17}H_{23}N_7$			
2,4,6-Pyrimidinetriamine, N[4]-cyclo-hexyl-5-[(2-methylphenyl)azo]-	EtOH	204(4.00),258(4.19), 392(4.29)	42-0420-86
$C_{17}H_{23}N_7O$			
2,4,6-Pyrimidinetriamine, N[4]-cyclo-hexyl-5-[(2-methoxyphenyl)azo]-	EtOH	204(4.10),258(4.38), 388(4.42)	42-0420-86
$C_{17}H_{23}N_7O_8$			
Azotomycin, (-)-	H_2O	245(4.19),275(4.39)	44-1282-86
$C_{17}H_{24}ClN_4O$			
1,3,5-Oxadiazin-1-ium, 2-(4-chloro-phenyl)-4,6-bis(diethylamino)-, perchlorate	MeCN	222(4.34),241(4.44)	97-0094-86
$C_{17}H_{24}Cl_2O_4$			
Butanoic acid, 2-[[3-[2-(2,2-dichloro-3,3-dimethylcyclopropyl)ethyl]-3-methyloxiranyl]methylene]-3-oxo-, ethyl ester	EtOH	207(4.32),265(4.30)	103-0370-86
$C_{17}H_{24}N_2O_3$			
3-Benzazecine-3(2H)-carbonitrile, 1,4,5,6,7,8-hexahydro-8,10,11-trimethoxy-, (±)-	MeOH	232(3.90),282(3.42), 287s(3.39)	12-0893-86
$C_{17}H_{24}N_2O_4$			
3,4-Pyridinedicarboxylic acid, 4-cyano-1,4-dihydro-1-(1,1-dimethyl-ethyl)-5-(1-methylethyl)-, dimethyl ester	MeCN	254(3.40),342(3.76)	118-0190-86

Compound	Solvent	λ max (log ε)	Ref.
$C_{17}H_{24}N_2O_5S_2$			
Uridine, 2',3'-O-cyclohexylidene-5'-S-methyl-5'-C-(methylthio)-5'-thio-	MeOH	258(4.00)	44-1258-86
$C_{17}H_{24}N_4O_4S_2$			
2,4(1H,3H)-Pyrimidinedione, 5,5'-methylenebis[1-(2-mercaptoethyl)-3,6-dimethyl-	EtOH	276(4.16)	94-1809-86
$C_{17}H_{24}N_4O_6$			
1H-Purine-2,6-dione, 9-[6-deoxy-2-C-methyl-α-L-galactopyranosyl-3,4-O-(1-methylethylidene)]-3,9-dihydro-1,3-dimethyl-	MeOH	275(3.92)	39-1297-86C
1H-Purine-2,6-dione, 9-[6-deoxy-2-C-methyl-β-L-talopyranosyl-3,4-O-(1-methylethylidene)]-3,9-dihydro-1,3-dimethyl-	MeOH	275(3.93)	39-1297-86C
2,4(1H,3H)-Pyrimidinedione, 5,5'-methylenebis[1-(2-hydroxyethyl)-3,6-dimethyl-	EtOH	278(4.30)	94-1809-86 +118-0857-86
$C_{17}H_{24}N_6O_5$			
9H-Purine-8-carboximidic acid, 6-(dimethylamino)-9-[2,3-O-(1-methylethylidene)-β-D-ribofuranosyl]-, methyl ester	MeOH	226(4.35),271s(3.83), 278(3.90),312(4.24)	94-3635-86
$C_{17}H_{24}O$			
2-Cyclohexen-1-one, 3-[2-(2,6,6-trimethyl-1-cyclohexen-1-yl)ethenyl]-, (E)-	$CHCl_3$	325(4.23)	35-4614-86
1,9-Heptadecadiene-4,6-diyn-3-ol, [R-(Z)]- (falcarinol)	hexane	230(2.99),242(2.87), 256(2.65)	102-0529-86
2(1H)-Pentalenone, 4-[6-(5-methyl-3,5-hexadienyl)-5,5-dimethyl-4,5,6,6a-tetrahydro-, (E)-	EtOH	230(4.48)	94-4939-86
(Z)-	EtOH	232(4.32)	94-4939-86
1-Pentanone, 5-cyclohexyl-1-phenyl-	hexane	264(2.60),324(1.68)	35-5264-86
$C_{17}H_{24}O_4$			
2-Cyclohexene-1-methanol, 2-(3-acetoxy-1-butynyl)-1,3-dimethyl-, acetate	EtOH	230(3.81)	163-1589-86
Naphtho[1,2-c]furan-1(3H)-one, 7-acetoxy-4,5,5a,6,7,8,9,9a-octahydro-6,6,9a-trimethyl-	MeOH	216(4.02)	102-0253-86
Trichodermin	EtOH	204(3.69)	163-2667-86
$C_{17}H_{24}O_4S$			
15-Thiaestr-4-en-3-one, 17-hydroxy-, 15,15-dioxide, (14β,17β)	EtOH	240(4.30),305(1.99)	104-0107-86
$C_{17}H_{24}O_5$			
Bisoxireno[4,5:8,9]cyclodeca[1,2-b]-furan-4(1aH)-one, 2a-ethoxy-2,2a-6,6a,7a,8,9,9a-octahydro-1a,5,7a-trimethyl-, [1aR-(1aR*,2aS*.6aR*-7aR*)]-	MeOH	218(4.3)	100-0336-86
Butanoic acid, 3-methyl-, 3-(3,4,5-	EtOH	218(3.66),266(3.52)	100-0677-86

Compound	Solvent	$\lambda_{max}(\log \epsilon)$	Ref.
trimethoxyphenyl)-2-propenyl ester, (E)- (cont.)			100-0677-86
2αH-Cedr-8-ene-12,15-dioic acid, 13-hydroxy-, dimethyl ester	MeOH	224(4.26)	2-0799-86
$C_{17}H_{24}O_6$			
Oxireno[4,5]cyclodeca[1,2-b]furan-4,8(1aH,5H)-dione, 6a-ethoxy-2,3,5,6,10,10a-hexahydro-5-hydroxy-1a,5,9-trimethyl-	MeOH	220(4.3)	100-0336-86
$C_{17}H_{24}O_{10}$			
7-Dehydrologanin	MeOH	234(4.01)	33-0156-86
$C_{17}H_{25}NO_4$			
4,4a(2H)-Isoquinolinedicarboxylic acid, 2-(1,1-dimethylethyl)-5,6,7,8-tetrahydro-, dimethyl ester	MeCN	240(3.35),347(3.48)	118-0190-86
$C_{17}H_{25}NO_4S$			
1-Piperidinecarboxylic acid, 2-[(phenylsulfonyl)methyl]-, 1,1-dimethylethyl ester	MeOH	217(3.98),258(2.85), 264(3.00),271(2.90)	12-0687-86
$C_{17}H_{25}N_4O$			
1,3,5-Oxadiazin-1-ium, 2,4-bis(diethylamino)-6-phenyl-, perchlorate	MeCN	241(4.39),282(4.25)	97-0094-86
$C_{17}H_{25}N_7O_4$			
L-Leucine, N-[N-(N-1H-purin-6-ylglycyl)-L-alanyl]-, methyl ester	pH 2.95	272(4.15)	63-0757-86
	pH 7.0	265(4.16)	63-0757-86
	pH 10.5	271(4.11)	63-0757-86
$C_{17}H_{25}O_2$			
2H-1-Benzopyran-6-yloxy, 3,4-dihydro-2,2-dimethyl-5,7-bis(1-methylethyl)-	EtOH	338(3.61),401(3.40), 420(3.60)	18-2899-86
$C_{17}H_{26}N_2O_2$			
Phenol, 4-[2-(2,5-dihydro-2,4,5,5-tetramethyl-1H-imidazol-2-yl)ethyl]-2,6-dimethyl-, N-oxide	EtOH	224.2(4.23),277.8(3.27)	103-0052-86
2-Propenoic acid, 3-[3,6-bis(2-methylpropyl)pyrazinyl]-, ethyl ester, (E)-	EtOH	250(4.24),313(4.17)	4-1481-86
$C_{17}H_{26}N_2O_3$			
Acetic acid, diazo-, 9-(3,3-dimethyloxiranyl)-3,7-dimethyl-2,6-nonadienyl ester, (E,E)-	hexane	244(4.18)	20-0895-86
2-Propenoic acid, 3-[3,6-bis(2-methylpropyl)pyrazinyl]-, ethyl ester, N-oxide, (E)-	EtOH	257(4.36),290s(3.78), 335(3.58),346s(3.52)	4-1481-86
$C_{17}H_{26}N_2O_9$			
α-D-Allofuranose, 3-C-[(acetoxyimino)aminomethyl]-1,2:5,6-bis-O-(1-methylethylidene)-, 3-acetate, (Z)-	EtOH	216(3.92)	159-0631-86
$C_{17}H_{26}N_3O_4$			
1-Piperidinyloxy, 4-[acetyl[2-(2,5-di-	H_2O	208(4.12),216s(4.09),	70-0363-86

Compound	Solvent	λ_{max}(log ϵ)	Ref.
hydro-2,5-dioxo-1H-pyrrol-1-yl)-ethyl]amino]-2,2,6,6-tetramethyl- (cont.)		223s(3.99),232s(3.73), 256s(3.16),316s(2.56), 508(1.70)	70-0363-86
$C_{17}H_{26}N_6O_4$			
β-D-Ribofuranuronamide, 1-deoxy-N-ethyl-1-[6-[(1-ethylpropyl)amino]-9H-purin-9-yl]-	n.s.g.	269(4.26)	87-1683-86
β-D-Ribofuranuronamide, 1-deoxy-1-[6-[(1-ethylpropyl)amino]-9H-purin-9-yl]-N,N-dimethyl-	n.s.g.	269(4.27)	87-1683-86
$C_{17}H_{26}O_2$			
1-Naphthaleneethanol, 2,3,4,4a,5,6-hexahydro-β,4,7-trimethyl-, acetate (12-acetoxyzonarene)	EtOH	245(4.17)	12-0103-86
2-Naphthalenol, 1,2,3,5,6,7,8,8a-octahydro-5,8a-dimethyl-3-(1-methylethylidene)-, acetate, (2α,5β,8aα)-(+)-coralloidin A	C_6H_{12}	249(4.35)	100-0608-86
(-)-coralloidin B	C_6H_{12}	227(3.59)	100-0608-86
$C_{17}H_{26}O_3$			
1-Cyclohexene-1-carboxylic acid, 3-oxo-, 5-methyl-2-1-(methylethyl)cyclohexyl ester, [1R-(1α,2β,5α)]-	$CHCl_3$	232(3.99),360(1.42)	78-3547-86
2(1H)-Pentalenone, 4,5,6,6a-tetrahydro-5,5-dimethyl-6-[2-[(tetrahydro-2H-pyran-2-yl)oxy]ethyl]-, (6α,6aα)-(±)-	EtOH	232(4.05)	94-4947-86
$C_{17}H_{26}O_4$			
3,6,11-Dodecatrien-5-one, 10-acetoxy-2-hydroxy-2,6,10-trimethyl-, [S-(E,E)]-	EtOH	239(3.96)	102-0185-86
$C_{17}H_{26}O_5$			
1-Cyclopentene-1-heptanoic acid, 5-acetoxy-3-oxo-, 1-methylethyl ester	MeOH	245(4.08)	2-0675-86
$C_{17}H_{26}O_{10}$			
Adoxoside	MeOH	236(3.93)	102-1227-86
$C_{17}H_{26}O_{11}$			
Morroniside	MeOH	238(4.17)	33-1113-86
$C_{17}H_{26}Si_2$			
Silane, bicyclo[4.4.1]undeca-1,3,5-7,9-pentaene-2,7-diylbis[trimethyl-	MeOH	265(4.78),310(3.94)	33-1851-86
$C_{17}H_{27}NO_3$			
3,5-Heptadienamide, 6-acetoxy-N-cyclohexyl-2,4-dimethyl-, (E,Z)-(±)-	EtOH	222s(3.82)	33-0469-86
$C_{17}H_{27}NO_4$			
2H-2,6-Benzoxazecine, 6-ethyl-5,6,7,8-tetrahydro-1,10,11-trimethyl-, (±)-	MeOH	232(4.23),283(3.77), 287s(3.73)	12-0893-86

Compound	Solvent	λ_{max}(log ϵ)	Ref.
$C_{17}H_{27}N_3O$			
4H-1,2,3-Triazol-4-one, 3-(1-adamantyl)-5-(1,1-dimethylethyl)-3,5-dihydro-5-methyl-	hexane	253(3.951),312(2.718)	5-1891-86
$C_{17}H_{27}N_3O_7$			
2,5-Cyclohexadiene-1,4-dione, 2-[2-[(aminocarbonyl)oxy]-1-methoxyethyl]-3,6-bis(2-methoxyethyl)amino]-5-methyl-	EtOH	349(4.43)	94-1299-86
$C_{17}H_{27}N_3S$			
4,17-Imino-2H-thieno[2,3-b][1,5]diazacycloheptadecine, 3,6,7,8,9,10,11,12-13,14,15,16-dodecahydro-	$CHCl_3$	237(4.30),286(3.86)	24-1070-86
Thieno[2',3':4,5]pyrimido[1,2-a]azacyclotridecin-4(2H)-imine, 3,6,7,8-9,10,11,12,13,14,15,16-dodecahydro-	$CHCl_3$	249(4.41),313(3.81)	24-1070-86
$C_{17}H_{28}Br_2NTe$			
4H-Tellurin, 1,1-dibromo-2,6-bis(1,1-dimethylethyl)-4-[(methyliminio)-ethylidene]-1,1-dihydro-, tetrafluoroborate	CH_2Cl_2	388(4.48)	157-2250-86
$C_{17}H_{18}NTe$			
Tellurinium, 4-[2-(dimethylamino)ethenyl]-2,6-bis(1,1-dimethylethyl)-, tetrafluoroborate	CH_2Cl_2	508(4.94)	157-2250-86
$C_{17}H_{28}N_4O_5$			
L-Leucine, N-[5-(2-amino-1,4-dihydro-6-hydroxy-4-oxo-5-pyrimidinyl)-1-oxopentyl]-, ethyl ester	50% MeOH-HCl	266(4.15)	73-1140-86
	50% MeOH-NaOH	270(4.13)	73-1140-86
$C_{17}H_{29}NO$			
Acetamide, N-[1,5-dimethyl-1-(4-methyl-3-cyclohexen-1-yl)-4-hexenyl]-(7-acetamido-7,8-dihydro-α-bisabolene)-	hexane	201(4.00)	44-5136-86
Pyridine, 4-dodecyl-, N-oxide	EtOH	268(4.3)	94-0007-86
$C_{17}H_{29}NO_6$			
α-D-xylo-Hex-5-enofuranose, 5,6-dideoxy-1,2-O-(1-methylethylidene)-6-nitro-3-O-octyl-	n.s.g.	233(3.76)	111-0111-86
$C_{17}H_{30}$			
Cyclohexane, 2-(2,3-dimethyl-2-butenylidene)-1,1,3,3,5-pentamethyl-, (±)-	C_6H_{12}	198(4.17)	44-2361-86
$C_{17}H_{32}O_3S_2$			
4-Thiepinethanol, 5-(1,1-dimethylethyl)-2,3,6,7-tetrahydro-3,3,6,6-tetramethyl-, methanesulfonate	C_6H_{12}	237(3.42),251(3.42)	78-1693-86
$C_{17}H_{34}OSi$			
Silane, (1,1-dimethylethyl)[[1-(2,2-dimethylpropyl)-2-methylenecyclopentyl]oxy]dimethyl-	MeOH	237(4.05)	23-0180-86

Compound	Solvent	$\lambda_{max}(\log \epsilon)$	Ref.
$C_{18}H_5F_{18}NO_{12}S_6$			
Benzenamine, 2,4,6-tris[(trifluoromethyl)sulfonyl]-N-[2,4,6-tris-[(trifluoromethyl)sulfonyl]phenyl]-	MeCN	315(4.48),370(4.40)	104-1547-86
	CF_3COOH	350(4.23)	104-1547-86
sodium salt	MeCN	315(4.48),370(4.42)	104-1547-86
$C_{18}H_6$			
Benzene, hexaethynyl-	DME	267s(4.77),279(5.02), 289s(4.69),298(4.76)	89-0268-86
$C_{18}H_6F_6N_2O_2$			
Naphtho[2,3-f]quinoxaline-7,12-dione, 2,3-bis(trifluoromethyl)-	MeCN	206(4.50),247(4.77)	39-1043-86C
$C_{18}H_6N_6$			
Propanedinitrile, 2,2'-benzo[g]quinoxaline-5,10-diylidenebis-	MeCN	205(4.05),350(4.36)	138-0715-86
$C_{18}H_7Cl_2NO_6$			
[2,3'-Bi-4H-1-benzopyran]-4,4'-dione, 6,6'-dichloro-3-nitro-	n.s.g.	303(4.42)	2-0212-86
$C_{18}H_8BrCl_2NO_6$			
[2,3'-Bi-4H-1-benzopyran]-4,4'-dione, 3-bromo-6,6'-dichloro-2,3-dihydro-3-nitro-	n.s.g.	342(3.82)	2-0212-86
$C_{18}H_8F_6N_2$			
Dibenzo[f,h]quinoxaline, 2,3-bis(trifluoromethyl)-	dioxan	258(4.77),296(4.19)	39-1043-86C
$C_{18}H_9Cl_2NO_6$			
[2,3'-Bi-4H-1-benzopyran]-4,4'-dione, 6,6'-dichloro-2,3-dihydro-3-nitro-	n.s.g.	316(4.16)	2-0212-86
$C_{18}H_9N_3O_2$			
Naphtho[2,3-h]quinoline-3-carbonitrile, 2-amino-7,12-dihydro-7,12-dioxo-	DMSO	218(4.349),427(3.699)	2-0616-86
$C_{18}H_9N_3O_3$			
19,20,21-Triazatetracyclo[$13.3.1.1^{3,7}.1^{9,13}$]heneicosa-1(19),3,5,7(21),9-11,13(20),15,17-nonaene-2,8,14-trione	MeOH	201(4.43),215s(4.36), 245s(4.11),270s(4.04)	35-6074-86
	MeCN	227(4.43),250s(4.32)	35-6074-86
$C_{18}H_{10}$			
Benzene, 1,1'-(1,3,5-hexatriyne-1,6-diyl)bis-	MeOH	254(4.98),267(4.93), 283(4.86),311(4.53), 332(4.64),358(4.47)	138-1075-86
$C_{18}H_{10}N_2$			
7,10-Ethenocyclohepta[de]naphthalene-8,9-dicarbonitrile, 7,10-dihydro-	EtOH	205.5(4.62),209(4.62), 221.5(4.80),262(3.73), 275s(3.71),286s(3.76), 295s(3.70),322s(3.34), 326s(3.37),344(3.50)	138-0969-86
	CH_2Cl_2	262.5(3.74),276s(3.72), 287s(3.77),296s(3.71), 322s(3.30),326s(3.34), 348(3.50)	138-0969-86

Compound	Solvent	λ_{max}(log ϵ)	Ref.
$C_{18}H_{10}N_2O_2$			
Triphenodioxazine	$CHCl_3$	471(4.61),505(4.74)	4-1003-86
$C_{18}H_{10}N_2S_2$			
1,3-Dithiole-4,5-dicarbonitrile, 2-(diphenylmethylene)-	EtOH	241(4.25),308(3.94), 450(3.26)	33-0419-86
$C_{18}H_{10}N_6$			
1,4-Methanonaphthalene-6,6,7,7-tetracarbonitrile, 2-(dicyanomethylene)-1,2,3,4,5,8-hexahydro-	MeCN	236(4.09),266s(3.77)	33-1310-86
$C_{18}H_{10}O_2$			
Benzaldehyde, 2,2'-(1,3-butadiyne-1,4-diyl)bis-	THF	227s(4.76),235(4.79), 244s(4.76),261s(4.55), 270s(4.44),286s(4.15), 308(4.25),334(4.37), 355(4.30)	18-2209-86
1,2-Chrysenedione	EtOH	261(4.51),305(4.34), 319(4.39)	44-1407-86
1,4-Naphthacenedione	$CHCl_3$	329(4.4),345(4.5), 386(3.8),410(3.8), 480(4.1)	44-3762-86
5,12-Naphthacenedione	hexane	250(4.7),300(4.3), 400f(3.7),416s(2.5)	110-0379-86
$C_{18}H_{10}O_3$			
7H-Benzo[c]fluorene-6-carboxylic acid, 7-oxo-	EtOH	238(4.26),290(4.17), 334(3.60),436(2.88)	24-3521-86
$C_{18}H_{10}O_7$			
Anthra[2,3-c]furan-1,5,10(3H)-trione, 11-acetoxy-4-hydroxy-	n.s.g.	228(4.23),260(4.42), 282s(3.92),402(3.75), 420s(3.66)	12-2075-86
2H-1-Benzopyran-2-one, 6,7-dihydroxy-3-[(2-oxo-2H-1-benzopyran-7-yl)oxy]- (demethyldaphnoretin)	EtOH	210(4.8),268(3.7), 323(4.2)	100-0692-86
1H-Pyrano[3,4-b]benzofuran-3,4-dione, 1-[(3,4-dihydroxyphenyl)methylene]-6-hydroxy- (anhydrogrevillin D)	dioxan	246.2(4.13),277.8(3.92), 345s(3.67),439.4(4.35)	5-0177-86
	+ NaOH	473.1(3.69),580.5(3.75)	5-0177-86
	MeOH	250(--),282(--), 456(--)	5-0177-86
$C_{18}H_{10}S_2$			
3,10-Dithiaperylene	CH_2Cl_2	273(4.02),297(4.00), 341(3.94),399s(--), 417(4.32)	35-3460-86
$C_{18}H_{11}BrN_2$			
1,1(8aH)-Azulenedicarbonitrile, 2-(4-bromophenyl)-	MeCN	270(4.1),358(4.1)	24-2631-86
Propanedinitrile, [1-(4-bromophenyl)-2-(2,4,6-cycloheptatrien-1-ylidene)ethylidene]-	MeCN	474(4.3)	24-2631-86
$C_{18}H_{11}BrN_4O_9$			
2H-1-Benzopyran-3-carboxylic acid, 7-hydroxy-8-[(bromodinitrophenyl)-azo]-, ethyl ester	n.s.g.	430(4.39)	2-0638-86
$C_{18}H_{11}ClN_2O_2S$			
Acetamide, N-(6-chloro-5-oxo-5H-	$CHCl_3$	244(4.58),276(4.14),	4-0589-86

Compound	Solvent	$\lambda_{max}(\log \epsilon)$	Ref.
benzo[a]phenothiazin-1-yl)- (cont.)		297(4.15),323(4.19), 363(3.95),379(3.96), 486(4.14)	4-0589-86
Acetamide, N-(6-chloro-5-oxo-5H-benzo[a]phenothiazin-4-yl)-	$CHCl_3$	247(4.78),330(4.27), 379(4.04),512(4.33)	4-0589-86
$C_{18}H_{11}Cl_2N_3S$			
1H-Perimidine-2-carbothioamide, N-(2,4-dichlorophenyl)-	DMF	342(4.24),354(4.24), 515(3.00)	48-0812-86
$C_{18}H_{11}NO_5$			
Acetamide, N-(1,3-dihydro-1,3-dioxo-7-phenylbenzo[1,2-b:3,4-c']difuran-4-yl)-	dioxan	288(3.62),367(3.42), 406(3.28)	73-1455-86
$C_{18}H_{11}N_3O_2$			
1,1(8aH)-Azulenedicarbonitrile, 2-(4-nitrophenyl)-	MeCN	235(4.0),270s(3.9), 385(4.2)	24-2631-86
Propanedinitrile, [2-(2,4,6-cyclohep-tatrien-1-ylidene)-1-(4-nitrophenyl)-	MeCN	483(4.45)	24-2631-86
$C_{18}H_{12}$			
Benzanthrene, lithium salt	THF	447(4.54),661(3.90)	35-7016-86
Bicyclo[4.1.0]hepta-1,3,5-triene, 7-(7H-benzocyclohepten-7-ylidene)-	C_6H_{12}	394(4.34),419(4.69), 439(4.55),451.5(4.79)	88-5159-86
	MeCN	391(4.33),414(4.61), 433(4.48),444(4.64)	88-5159-86
1H-Cyclopropa[b]naphthalene, 1-(phen-ylmethylene)-	C_6H_{12}	229(4.60),245(4.19), 284(4.29),376s(4.28), 394(4.61),406s(4.42), ?(4.73)	35-5949-86
	MeCN	228(4.59),252s(4.20), 283(4.34),373s(4.29), 391(4.60),417(4.69)	35-5949-86
1H-Indene, 1-(1H-inden-1-ylidene)-	C_6H_{12}	248(4.27),284(4.30), 291s(4.26),320s(3.51), 362s(4.19),379(4.43), 400(4.47)	33-1644-86
Pyrene, 1-ethenyl-	THF	342(4.5),362(4.6)	116-2390-86
$C_{18}H_{12}BrN_3S$			
1H-Perimidine-2-carbothioamide, N-(4-bromophenyl)-	DMF	282(4.23),342(4.31), 356(4.31),521(3.04)	48-0812-86
$C_{18}H_{12}Br_3N$			
Benzenamine, 4-bromo-N,N-bis(4-bromo-phenyl)-, radical cation, tetra-fluoroborate	MeCN	365(4.36),705(4.52) (changing)	1-0210-86
$C_{18}H_{12}ClN_3O_7$			
2H-1-Benzopyran-3-carboxylic acid, 8-[(chloronitrophenyl)azo]-7-hy-droxy-, ethyl ester	n.s.g.	435(4.53)	2-0638-86
$C_{18}H_{12}ClN_3S$			
1H-Perimidine-2-carbothioamide, N-(4-chlorophenyl)-	DMF	282(4.23),342(4.31), 356(4.31),356(4.32), 522(3.04)	48-0812-86
$C_{18}H_{12}N_2$			
1,1(8aH)-Azulenedicarbonitrile,	MeCN	237s(--),265(4.2),	24-2631-86

Compound	Solvent	$\lambda_{max}(\log \epsilon)$	Ref.
2-phenyl- (cont.)		354(4.1)	24-2631-86
Benzo[f]cinnoline, 2-phenyl-	EtOH	214(4.47),233(4.32), 271(4.72),273.5(4.72), 355(3.42),371(3.37)	35-1617-86
7,10-Ethanocyclohepta[de]naphthalene-8,9-dicarbonitrile, 7,10-dihydro-	EtOH	223(5.00),257.5(3.88), 276s(3.70),286s(3.57), 309s(3.24),324s(3.44), 342(3.56)	138-0969-86
	CH_2Cl_2	259(3.90),277s(3.71), 288s(3.58),310s(3.19), 325s(3.40),346(3.56)	138-0969-86
Propanedinitrile, [2-(2,4,6-cyclo-heptatrien-1-ylidene)-1-phenyl-ethylidene]-	MeCN	254s(--),298s(--), 471(4.4)	24-2631-86
$C_{18}H_{12}N_2O$			
Quinoline, 2-(5-phenyl-2-oxazolyl)-	hexane	341(4.37),357(4.32)	110-1000-86
	EtOH	353(4.39)	110-1000-86
	EtOH-HOAc	353(4.41),401(3.0)	110-1000-86
Quinoline, 4-(5-phenyl-2-oxazolyl)-	hexane	341(4.16),353s(--)	110-1000-86
	EtOH	235(4.48),350(4.20)	103-0229-86
	EtOH	353(4.16)	110-1000-86
	+ 60% HOAc	353(4.03),401(4.03)	110-1000-86
	+ 20% H_2SO_4	409(4.26)	110-1000-86
$C_{18}H_{12}N_2O_2$			
1H-Benz[de]isoquinoline-1,3(2H)-dione, 6-amino-2-phenyl-	EtOH	430(4.16)	65-1105-86
Formamide, N-(3-cyano-4,5-diphenyl-2-furanyl)-	EtOH	284(4.25)	11-0347-86
Furo[2,3-d]pyrimidin-4(1H)-one, 5,6-diphenyl-	EtOH	224(4.07),311(4.29)	11-0347-86
$C_{18}H_{12}N_2O_2S$			
Acetamide, N-(5-oxo-5H-benzo[a]pheno-thiazin-1-yl)-	$CHCl_3$	242(4.52),274(4.30), 294(4.27),317(4.30), 361(4.05),377(4.06), 479(4.22)	4-0589-86
Acetamide, N-(5-oxo-5H-benzo[a]pheno-thiazin-4-yl)-	$CHCl_3$	246(4.76),322(4.34), 362(4.05),378(4.05), 503(4.33)	4-0589-86
$C_{18}H_{12}N_2O_3$			
Acetamide, N-(6,11-dihydro-6,11-dioxo-benzo[f]pyrido[1,2-a]indol-12-yl)-	DMF	488(3.6946)	2-1126-86
$C_{18}H_{12}N_2O_{12}$			
Ethanedioic acid, bis[4-(methoxy-carbonyl)-2-nitrophenyl] ester	MeCN	236(4.63),340(3.73)	96-0209-86
$C_{18}H_{12}N_4$			
Cyanamide, (2,3-dimethyl-9,10-anthra-cenediylidene)bis-	MeCN	210(4.39),277(4.55), 339(4.36)	5-0142-86
[1,2,4,5]Tetrazino[1,6-a:4,3-a']di-isoquinoline	CH_2Cl_2	255(4.6),420(3.9), 610f(2.35)	33-1521-86
[1,2,4,5]Tetrazino[1,6-a:4,3-a']di-quinoline	CH_2Cl_2	299(4.6),380(4.0), 425f(3.3),660(2.33)	33-1521-86
7H-1,2,4-Triazolo[4,3-a]perimidine, 10-phenyl-	dioxan	228(4.58),253(4.11), 303(3.92),353(4.00)	48-0237-86

Compound	Solvent	$\lambda_{max}(\log \epsilon)$	Ref.
$C_{18}H_{12}N_4OS$			
2(1H)-Pteridinone, 3,4-dihydro-6,7-diphenyl-4-thioxo-	MeOH	225(4.49),293(4.12), 407(4.21)	128-0199-86
$C_{18}H_{12}N_4O_2$			
Cyanamide, (1,4-dimethoxy-9,10-anthracenediylidene)bis-	MeCN	220(4.29),252s(4.21), 271(4.31),311(4.14), 460(3.64)	5-0142-86
2,4(1H,3H)-Pteridinedione, 6,7-diphenyl-	pH 5	220(4.43),272(4.16), 361(4.17)	128-0199-86
	pH 11	220s(4.41),288(4.30), 376(4.06)	128-0199-86
[1,2,4]Triazolo[1,5-a]pyridine-4-carbonitrile, 5,7-dimethyl-2-(2-oxo-1H-benzopyran-3-yl)-	DMF	352(4.25)	118-0860-86
$C_{18}H_{12}N_4O_5$			
1,3,4-Oxadiazole, 2,5-bis[2-(3-nitrophenyl)ethenyl]-	dioxan	330(4.46)	103-1262-86
$C_{18}H_{12}N_4O_9$			
2H-1-Benzopyran-3-carboxylic acid, 7-hydroxy-8-[(dinitrophenyl)azo]-	n.s.g.	440(4.52)	2-0638-86
$C_{18}H_{12}N_5O_6$			
Hydrazyl, 2,2-diphenyl-1-(2,4,6-trinitrophenyl)-	EtOH-pH 6	528(4.02)	41-0577-86
$C_{18}H_{12}O$			
1-Pyrenecarboxaldehyde, 2-methyl-	MeOH	237(4.63),292(4.35), 364(4.27),397(4.05)	35-4498-86
$C_{18}H_{12}O_3$			
2H-1-Benzopyran-2-one, 4-(2-benzofuranyl)-6-methyl-	n.s.g.	335(4.26)	2-0779-86
1,6-Pyrenedione, 3-ethoxy-	C_6H_5Cl	395(3.05),427(3.08), 450(3.04)	104-1159-86
1,8-Pyrenedione, 3-ethoxy-	C_6H_5Cl	377(2.99),458(2.94)	104-1159-86
$C_{18}H_{12}O_4$			
2,5-Cyclohexadiene-1,4-dione, 2,5-dihydroxy-3,6-diphenyl-	dioxan	259.2(4.39),331(3.84), 425(2.52)	5-0195-86
1,8-Pyrenedione, 3,6-dimethoxy-	C_6H_5Cl	362(3.07),430(2.92), 452(2.94)	104-1159-86
$C_{18}H_{12}O_5$			
2,5-Cyclohexadiene-1,4-dione, 2,5-dihydroxy-3-(4-hydroxyphenyl)-6-phenyl- (ascocorynin)	dioxan	265(4.39),352(3.64), 475(2.39)	5-0195-86
$C_{18}H_{12}O_6$			
Benzeneacetic acid, 4-hydroxy-α-(3-hydroxy-5-oxo-4-phenyl-2(5H)-furanylidene)-	EtOH	263(4.16),367(3.98)	39-2127-86C
Benzeneacetic acid, α-[3-hydroxy-4-(4-hydroxyphenyl)-5-oxo-2(5H)-furanylidene]-	EtOH	256(4.13),397(3.84)	39-2127-86C
2H-Pyran-2,5(6H)-dione, 6-[(3,4-dihydroxyphenyl)methylene]-3-hydroxy-4-phenyl-	MeOH	284(4.21),409(4.21)	5-0177-86

Compound	Solvent	λ_{max}(log ϵ)	Ref.
2H-Pyran-2,5(6H)-dione, 3-hydroxy-4-(4-hydroxyphenyl)-6-[(4-hydroxyphenyl)methylene]- (grevillin A)	MeOH	268(4.27),389(4.23)	5-0177-86
	H_2SO_4	530(--)	5-0177-86
$C_{18}H_{12}O_7$			
Anthra[2,3-c]furan-5,10-dione, 11-acetoxy-1,3-dihydro-1,4-dihydroxy-	n.s.g.	226(4.27),256(4.49), 282(4.01),332(3.41), 404(3.81)	12-2075-86
2,5-Cyclohexadiene-1,4-dione, 2-(3,4-dihydroxyphenyl)-3,6-dihydroxy-5-(4-hydroxyphenyl)- (leucomelon)	pH 13	257.1(4.12),326.3(4.22), 384.6(4.11),682.6(3.48)	5-0195-86
2H-Pyran-2,5(6H)-dione, 6-[(3,4-dihydroxyphenyl)methylene]-3-hydroxy-4-(4-hydroxyphenyl)- (grevillin B)	MeOH	280(4.20),396(4.20)	5-0177-86
$C_{18}H_{12}O_8$			
2,5-Cyclohexadiene-1,4-dione, 2-(2,5-dihydroxyphenyl)-5-(3,4-dihydroxyphenyl)-3,6-dihydroxy-	pH 13	318.6(4.26),430(3.78)	5-0195-86
	MeOH	262.7(4.21),286.1(4.31)	5-0195-86
2H-Pyran-2,5(6H)-dione, 4-(2,5-dihydroxyphenyl)-6-[(3,4-dihydroxyphenyl)methylene]-3-hydroxy-	MeOH	287.6(4.14),408.9(4.18)	5-0177-86
2H-Pyran-2,5(6H)-dione, 4-(3,4-dihydroxyphenyl)-6-[(3,4-dihydroxyphenyl)methylene]-3-hydroxy- (grevillin C)	MeOH	287.3(4.24),405.6(4.23)	5-0177-86
	H_2SO_4	555(--)	5-0177-86
$C_{18}H_{12}O_9$			
Benzeneacetic acid, 4-hydroxy-2-[3-hydroxy-5-oxo-4-(3,4,5-trihydroxyphenyl)-2(5H)-furanylidene]-, (E)- (gomphidic acid)	EtOH	260(4.14),398(3.81)	88-0403-86
	EtOH-NaOH	304(4.00),402(4.13)	39-2127-86C
Benzeneacetic acid, 3,4,5-trihydroxy-α-[3-hydroxy-4-(4-hydroxyphenyl)-5-oxo-2(5H)-furanylidene]-, (E)- (isogomphidic acid)	EtOH	263(4.30),397(3.89)	39-2127-86C +88-0403-86
	EtOH-NaOH	269(4.24),403(3.62)	39-2127-86C
$C_{18}H_{12}S$			
Benzo[b]thiophene, 2-(1-naphthalenyl)-	hexane	303(4.19)	142-0355-86
Benzo[b]thiophene, 3-(1-naphthalenyl)-	hexane	230(5.0),304(4.04)	142-0355-86
$C_{18}H_{12}S_4$			
2,2'-Bithiophene, 5,5"-(1,2-ethenediyl)bis-, (E)-	$CHCl_3$	423(4.65)	142-2639-86
$C_{18}H_{13}BO_4$			
Boron, [1,2-benzenediolato(2-)-O,O']-[(1-hydroxy-2-naphthalenyl)ethanonato-O,O']-, (T-4)-	DMF	266(4.48),281s(3.89), 296(3.75),307s(3.50), 314s(3.41),368(3.68), 409s(2.86)	48-0755-86
$C_{18}H_{13}BO_6$			
Boron, (1,3-diphenyl-1,3-propanedionato-O,O')[propanedioato(2-)-O,O']-, (T-4)-	MeCN	282(3.80),369(4.46), 385s(4.40)	48-0755-86
$C_{18}H_{13}Br_2N_3O_2$			
Discorhabdin C	MeOH	245(4.46),351(4.00), 545(2.70)	44-5476-86
	MeOH-KOH	337(4.11),481(3.18)	44-5476-86

Compound	Solvent	λ_{max}(log ϵ)	Ref.
$C_{18}H_{13}ClN_2O$			
1H-[1]Benzopyrano[4,3,2-de]pyrrolo-[2,3-b]quinoline, 7-chloro-2,3-dihydro-3-methyl-	MeOH	256(4.68),308(4.03), 357(4.08)	83-0347-86
$C_{18}H_{13}ClN_4O_2$			
1H,5H-Pyrazolo[1,2-a]pyrazol-1-one, 7-amino-2-(4-chlorophenyl)-3-hydroxy-5-imino-6-phenyl-	MeCN	280(4.68)	118-0239-86
$C_{18}H_{13}ClO_3$			
9,10-Anthracenedione, 2-(2-chloro-2-propenyl)-1-methoxy-	n.s.g.	225(4.05),255(4.27), 276s(3.92),332(3.41), 400(3.03)	12-2075-86
$C_{18}H_{13}ClO_4$			
9,10-Anthracenedione, 2-(2-chloro-3-hydroxy-1-propenyl)-1-methoxy-	n.s.g.	223(4.29),258(4.49), 289s(4.10),361(3.76)	12-2075-86
$C_{18}H_{13}N$			
2-Chrysenamine	C_6H_{12}	217(4.40),228.6(4.51), 243.2(4.33),274.2(4.81), 338(4.09),382(3.35)	2-0517-86A
	pH 6	225(--),270.2(4.39)	2-0517-86A
	MeOH	218(4.49),228.6(4.52), 242.8(4.38),273.4(4.80), 339.8(4.07),388(3.57)	2-0517-86A
	EtOH	218(4.48),227.6(4.49), 242.5(4.38),274.6(4.79), 340(4.09),388(3.68)	2-0517-86A
	ether	218.2(4.53),229.2(4.60), 242.8(4.48),274.6(4.80), 342.8(4.19),387(3.50)	2-0517-86A
	dioxan	220(4.38),230(4.44), 242.8(4.32),278.6(4.68), 346.7(4.01),385(3.47)	2-0517-86A
	MeCN	218(4.49),228.6(4.55), 242.8(4.42),274.6(4.79), 343(4.12),388(3.51)	2-0517-86A
	KOH	229(4.17),287.5(4.01), 356(3.50)	2-0517-86A
in sulfuric acid	H_0 -6	215(4.58),257.6(4.86), 267(5.05),296(4.07), 308(4.09),322(4.08)	2-0517-86A
in sulfuric acid	H_0 -10	220(4.41),267.5(4.89), 304(4.20)	2-0517-86A
$C_{18}H_{13}NO$			
4-Benzazacyclotetradecin-3(4H)-one, 8-methyl-9,10,11,12-tetradehydro-, (E,E,Z)-	THF	226(4.56),258s(4.43), 273s(4.59),284(4.63), 300s(4.39),388(3.90)	39-0933-86C
7H-Benzocyclotridecen-7-one, 11-methyl-12,13,14,15-tetradehydro-, oxime, (Z,E,E,Z)-	THF	227(4.42),265s(4.51), 276(4.54),346(3.83), 371s(3.72)	39-0933-86C
5H-Indeno[1,2-b]pyridin-5-ol, 5-phenyl-	$CHCl_3$	272s(3.73),282(3.95), 305(3.96),319(4.14)	104-1352-86
9H-Indeno[2,1-b]pyridin-9-ol, 9-phenyl-	$CHCl_3$	270(4.06),283(4.09), 292(3.90),303(3.83), 316(3.88),330s(2.06)	104-1352-86

Compound	Solvent	$\lambda_{max}(\log \epsilon)$	Ref.
$C_{18}H_{13}NOS$			
Bicyclo[4.4.1]undeca-1,3,5,7,9-pentadecaen-2-ol, 5-(2-benzothiazolyl)-, (±)-	MeCN	224(4.33),261s(4.17), 284(4.27),374(4.15), 429s(3.55)	33-0315-86
Bicyclo[4.4.1]undeca-3,6,8,10-tetraen-2-one, 5-(2(3H)-benzothiazolylidene)-, (Z)-(±)-	MeCN	220(4.28),283(4.08), 368(3.84),489(3.68)	33-0315-86
$C_{18}H_{13}NO_3$			
7H-Dibenzo[de,h]quinolin-7-one, 5,9-dimethoxy- (bianfugecine)	EtOH	254(4.54),268s(4.34), 288(4.26),291s(4.25), 318(4.25),331s(4.27), 382s(4.32),420(4.37)	142-0437-86
Methanone, (3,8-dimethylfuro[2,3-g]-1,2-benzisoxazol-7-yl)phenyl-	EtOH	220(4.3),255(4.8), 320(3.9)	2-0870-86
$C_{18}H_{13}NO_4$			
Indolizino[3,4,5-ab]isoindole-1,2-dicarboxylic acid, dimethyl ester	EtOH	223(4.69),241(4.53), 256(4.46),271(4.58), 304(4.14),314(4.17), 428(4.06),454(4.04)	142-3071-86
$C_{18}H_{13}N_3$			
1,1(8aH)-Azulenedicarbonitrile, 2-(4-aminophenyl)-	MeCN	228(--),284(--), 388(4.4)	24-2631-86
Propanedinitrile, [1-(4-aminophenyl)-2-(2,4,6-cycloheptatrien-1-ylidene)ethylidene]-	MeCN	258(--),456(4.5)	24-2631-86
$C_{18}H_{13}N_3O$			
Ethenetricarbonitrile, [5-(2,4,6-trimethylphenyl)-2-furanyl]-	MeOH	364(3.02),422(3.02)	73-1450-86
4H-Furo[3,2-b]pyrrole, 2-phenyl-5-(phenylazo)-	MeOH	475(3.67)	73-0106-86
Pyrazolo[1,5-c]pyrimidin-7(6H)-one, 2,5-diphenyl-	isoPrOH	245(4.6),305(4.4)	103-1322-86
$C_{18}H_{13}N_3O_3$			
Acetamide, N-[2-(1H-benzimidazol-2-yl)-4-oxo-4H-1-benzopyran-7-yl]-	EtOH	230(4.46),279(3.72), 342(4.32)	42-0600-86
$C_{18}H_{13}N_3O_4$			
Acetamide, N-(1,2,3,4-tetrahydro-1,4-dioxo-8-phenylfuro[3,2-f]phthalazin-5-yl)-	dioxan	342(3.28),368(3.34), 383(3.33)	73-1455-86
$C_{18}H_{13}N_3O_5S$			
10H-Thiazolo[5",4":5',6']pyrano-[4',3':3,4]furo[2,3-b]indole-2-acetic acid, 10a,11-dihydro-11-imino-10-oxo-, ethyl ester	EtOH	262(5.98),279(5.396), 414(5.76)	18-1235-86
$C_{18}H_{13}N_3S$			
1H-Perimidine-2-carbothioamide, N-phenyl-	DMF	277(4.16),341(4.27), 355(4.28),467s(2.98), 511(3.01)	48-0812-86
	C_6H_{12}	232s(--),236(--), 281(--),303s(--), 339(--),353(--), 367s(--),384s(--), 424s(--),456s(--),	48-0812-86

Compound	Solvent	λ_{max}(log ε)	Ref.
(cont.)		488(--),524(--), 566(--),624(--)	48-0812-86
Pyrazolo[1,5-c]pyrimidine-7(6H)-thione, 2,5-diphenyl-	isoPrOH	260(4.5),288(4.4), 325(4.2)	103-1322-86
$C_{18}H_{14}$			
7H-Benzo[c]fluorene, 9-methyl-	EtOH	233(4.54),254(4.16), 283(3.70),313(4.04), 323(4.03),330(3.94), 338(4.08)	2-0887-86
1,1'-Bi-1H-indene	C_6H_{12}	219(4.58),227s(4.48), 256(4.24),283s(3.17), 289(2.88)	33-1644-86
Chrysene, 4b,10b-dihydro-	MeOH	218.5(4.55),225(4.54), 256(4.20),266.5s(4.13), 298.5(3.23)	80-0503-86
Naphthalene, 2-(2-phenylethenyl)-, trans	n.s.g.	226(4.45),282(4.51), 320(4.61)	109-0607-86
conformer B	n.s.g.	282(4.51),330(4.61)	109-0607-86
$C_{18}H_{14}BrNS_2$			
3-Thiophenecarbothioamide, N-(4-bromophenyl)-2-methyl-5-phenyl-	EtOH	269(4.51),295s(4.38), 400(2.77)	78-3683-86
3-Thiophenecarbothioamide, N-(4-bromophenyl)-5-methyl-2-phenyl-	EtOH	265(4.33),320(4.21), 400s(2.82)	78-3683-86
$C_{18}H_{14}ClN_7S$			
Ethenetricarbonitrile, [4-chloro-2-[[4-(diethylamino)phenyl]azo]-5-thiazolyl]-	CH_2Cl_2	725(4.88)	77-1639-86
	DMSO	745(--)	77-1639-86
$C_{18}H_{14}Fe_2O_6$			
Iron, hexacarbonyl[μ-[η^4:η^4-2,3,6,7-tetrakis(methylene)bicyclo[3.2.1]octane]]di-	isooctane	285s(3.77)	44-2385-86
$C_{18}H_{14}NO_6P$			
Phosphoric acid, 4-nitrophenyl diphenyl ester	MeCN	256(4.20)	78-1315-86
$C_{18}H_{14}N_2$			
Benzo[f]cinnoline, 5,6-dihydro-2-phenyl-	EtOH	262(4.36)	35-1617-86
1H-Pyrrolo[2,3-b]quinoline, 1-methyl-4-phenyl-	MeOH	253(4.6),332(3.77), 367(3.68)	83-0338-86
$C_{18}H_{14}N_2OS$			
2H-Thiopyran-5-carbonitrile, 3,4-dihydro-4-oxo-2-phenyl-6-(phenylamino)-	EtOH	315(3.52)	5-1109-86
$C_{18}H_{14}N_2O_2$			
1H-Pyrido[3,4-b]indole-1,3(2H)-dione, 4,9-dihydro-2-methyl-9-phenyl-	isoPrOH	212(4.56),239(4.42), 308(4.34),334s(3.95)	103-1345-86
$C_{18}H_{14}N_2O_2S$			
5H-Benzo[b]naphtho[1,2-g][1,4,6]thiadiazocine, 6-methyl-, 12,12-dioxide	MeOH	256(4.44),262(4.53), 307(3.79),333(3.63)	44-1967-86
2H-1,2,4-Benzothiadiazine, 3-methyl-2-(1-naphthalenyl)-, 1,1-dioxide	MeOH	252(4.21),262(4.20), 271(4.20)	44-1967-86

Compound	Solvent	λ_{max}(log ε)	Ref.
$C_{18}H_{14}N_2O_3$			
3-Furancarboxamide, 2-(formylamino)-4,5-diphenyl-	EtOH	221(4.35),302(4.27)	11-0347-86
$C_{18}H_{14}N_3S_2$			
Thiazolo[3,4-b][1,2,4]triazin-5-ium, 6-(methylthio)-2,3-diphenyl-, perchlorate	n.s.g.	317(4.36),420s(--)	103-1373-86
Thiazolo[3,4-b][1,2,4]triazin-5-ium, 6-(methylthio)-2,8-diphenyl-, perchlorate	n.s.g.	333(4.37),468(3.68)	103-1373-86
Thiazolo[3,4-b][1,2,4]triazin-5-ium, 6-(methylthio)-3,8-diphenyl-, perchlorate	n.s.g.	303(4.40),457(3.57)	103-1373-86
$C_{18}H_{14}N_4$			
Tricyclo[7.2.1.0^{2,7}]dodec-2(7)-ene-4,4,5,5-tetracarbonitrile, 10,11-bis(methylene)-	isooctane	221(3.78),226(3.78), 238(3.81),244(3.81)	44-2385-86
	EtOH	239(3.85)	44-2385-86
$C_{18}H_{14}N_4O$			
Benzoic acid, (di-2-pyridinylmethylene)hydrazide	H_2O	305(4.20)	9-1058-86
	EtOH	320(4.21)	9-1058-86
	$CHCl_3$	325(4.18)	9-1058-86
	DMF	320(4.15)	9-1058-86
$C_{18}H_{14}N_4O_2$			
Benzoic acid, 2-hydroxy-, (di-2-pyridinylmethylene)hydrazide	H_2O	300(4.22)	9-1058-86
	EtOH	325(4.37)	9-1058-86
	DMF	318(4.37)	9-1058-86
1H,5H-Pyrazolo[1,2-a]pyrazol-1-one, 7-amino-3-hydroxy-5-imino-2,6-diphenyl-	MeCN	277(4.65)	118-0239-86
$C_{18}H_{14}N_4O_2S$			
Acetamide, N-[2-(4-methoxyphenyl)-thiazolo[4,5-b]quinoxalin-7-yl]-	n.s.g.	359(4.2630)	18-1245-86
1H-Pyrido[2,3-c][1,2,6]thiadiazin-4-amine, 5,7-diphenyl-, 2,2-dioxide	pH -4.0	265(4.46),304(4.12), 368(4.32)	11-0607-86
	pH 2.0	267(4.50),296s(4.17), 396(4.22)	11-0607-86
	pH 10.0	233(4.48),260(4.55), 292s(4.22),378(4.12)	11-0607-86
$C_{18}H_{14}N_4O_4$			
Imidazo[5,1,2-cd]indolizine-3,4-dicarboxylic acid, 2-(2-pyridinylamino)-, dimethyl ester	MeOH	208(4.26),244(4.56), 286(4.43),340(3.95), 394(4.43)	88-4129-86
$C_{18}H_{14}N_4O_4S$			
10H-Thiazolo[5",4":5',6']pyrano-[4',3':3,4]furo[2,3-b]indole-2-acetic acid, 10a,11-dihydro-10,11-diimino-, ethyl ester	EtOH	258(3.76),270(4.58), 425(4.91)	18-1235-86
$C_{18}H_{14}O$			
Oxepin, 2,7-diphenyl-	hexane	253(4.23),259(4.41), 266(4.49),358(3.74)	44-2784-86
	EtOH	205(4.48),263(4.55), 352(3.82)	44-2784-86

Compound	Solvent	λ_{max}(log ε)	Ref.
$C_{18}H_{14}OS_2$			
Benzenecarbodithioic acid, 4-methoxy-, 4-phenyl-1-buten-3-ynyl ester, cis	CH_2Cl_2	294(4.19),362(4.33), 505(2.41)	18-1403-86
trans	CH_2Cl_2	293(4.33),361(4.41), 502(2.51)	18-1403-86
$C_{18}H_{14}O_2$			
Bicyclo[2.2.1]hepta-2,5-diene-7-carboxylic acid, 1-naphthalenyl ester	MeCN	280(3.96)	32-0281-86
Bicyclo[2.2.1]hepta-2,5-diene-7-carboxylic acid, 2-naphthalenyl ester	MeCN	265(4.00)	32-0281-86
1,2-Chrysenediol, 1,2-dihydro-, trans	EtOH	223(4.87),248(4.57), 272(4.61),284(4.48)	44-1407-86
5,6-Methanodibenzo[a,c]cyclopropa[e]-cycloheptene-11-carboxylic acid, 4b,5,5a,6-tetrahydro-	MeOH	209(4.38),263(3.92), 299s(2.88)	80-1019-86
$C_{18}H_{14}O_2Te$			
2-Propen-1-one, 3,3'-tellurobis[1-phenyl-	CH_2Cl_2	425(4.77)	44-1692-86
$C_{18}H_{14}O_3$			
2H-1-Benzopyran-2-one, 4-[(3-phenyl-2-propenyl)oxy]-	EtOH	211(4.56),255(4.41), 293(3.78),303(3.79)	2-1167-86
1,2-Chrysenediol, 3,4-epoxy-1,2,3,4-tetrahydro-, trans-anti	EtOH	216(4.58),259(4.74), 301(4.05)	44-1407-86
syn	EtOH	215(4.22),260(4.45), 280(3.92),300(3.77)	44-1407-86
4H-Furo[3,2-c][1]benzopyran-4-one, 2,3-dihydro-3-methyl-2-phenyl-	EtOH	278(3.84),289(3.98), 313(4.01),327(3.87)	2-1167-86
1,2-Naphthalenedione, 4-[(4-methoxyphenyl)methyl]-	$CHCl_3$	344(4.45),405(4.44), 516(1.82)	39-1627-86C
1(4H)-Naphthalenone, 2-hydroxy-4-[(4-methoxyphenyl)methylene]-	$CHCl_3$	292(3.92),419(4.45)	39-1627-86C
$C_{18}H_{14}O_4$			
9,10-Anthracenedione, 1-methoxy-2-(2-oxopropyl)-	n.s.g.	225(4.18),256(4.57), 344(3.69)	12-2075-86
$C_{18}H_{14}O_5$			
7-Benzofurancarboxaldehyde, 2-benzoyl-4,6-dimethoxy-	EtOH	231(3.52),264(3.39), 293(3.64),344(3.72)	142-0771-86
4,5-Benzofurandicarboxylic acid, 2-phenyl-, dimethyl ester	dioxan	260(3.26),290(3.19), 301(3.19),331(3.27)	73-1455-86
Pyrano[3,2-c]xanthen-12(3H)-one, 5,8-dihydroxy-3,3-dimethyl-(hypericanarin)	MeOH	240(4.08),260(3.97), 306(3.71),382(3.44)	100-0095-86
	MeOH-NaOMe	260(--),268s(--), 296(--),343(--), 380(--)	100-0095-86
	MeOH-NaOAc	229(--),239(--), 300(--),315(--), 375(--)	100-0095-86
$C_{18}H_{15}BO_5$			
Boron, [α-hydroxybenzeneacetato(2-)]-(1-phenyl-1,3-butanedionato-O,O')-, (T-4)-	CH_2Cl_2	267(3.80),334(4.43), 345s(4.30)	48-0755-86
$C_{18}H_{15}ClFNO_2S_2$			
Morpholine, 4-[2-[5-chloro-2-[(4-fluorophenyl)thio]phenyl]-2-oxo-1-thi-	MeOH	237.5(4.35),272.5(4.32), 356(3.78)	73-2598-86

Compound	Solvent	$\lambda_{max}(\log \epsilon)$	Ref.
oxoethyl]- (cont.)			73-2598-86
$C_{18}H_{15}ClN_2OS_2$			
Benzo[b]thiophene-2-carboxamide, 3-chloro-N-[[(2-phenylethyl)amino]-thioxomethyl]-	EtOH	208(3.43),251(3.29), 306(3.30)	73-2839-86
$C_{18}H_{15}ClO_6$			
9H-Xanthen-9-one, 6-acetoxy-4-chloro-1,3-dimethoxy-8-methyl-	MeOH	242(4.56),302(4.22), 334(3.89)	94-0858-86
9H-Xanthen-9-one, 7-acetoxy-4-chloro-1,3-dimethoxy-5-methyl-	MeOH	240(4.60),306(4.18), 348(3.87)	94-0858-86
$C_{18}H_{15}F_3O$			
Anthracene, 9-ethoxy-9,10-dihydro-10-(2,2,2-trifluoroethylidene)-	pentane	254(4.19)	44-1324-86
$C_{18}H_{15}N$			
5H-Benzo[6,7]cyclohepta[1,2-b]quinoline, 6,7-dihydro-	EtOH	236(4.16),252(4.16), 283(3.82),298(3.80), 312(3.86),326(3.90)	4-0689-86
$C_{18}H_{15}NO_2$			
5(4H)-Oxazolone, 4-[(2,4-dimethylphenyl)methylene]-2-phenyl-, (Z)-	MeCN	376(4.56)	23-2064-86
5(4H)-Oxazolone, 4-[(2,5-dimethylphenyl)methylene]-2-phenyl-, (Z)-	MeCN	372(4.55)	23-2064-86
$C_{18}H_{15}NO_2S$			
1H-Pyrrole, 2-(2-phenylethenyl)-1-(phenylsulfonyl)-, (E)-	EtOH	225(4.16),318(4.11)	118-0620-86
$C_{18}H_{15}NO_3$			
2-Anthracenecarboxylic acid, 3-(acetylamino)-, methyl ester	EtOH	209(4.00),234(4.20), 260(4.60),278(4.90), 328(3.60),344(3.70), 362(3.20),384(3.20), 412(3.30)	78-3943-86
Benzoic acid, 4-[5-(4-methylphenyl)-2-oxazolyl]-, methyl ester	toluene	340(4.41)	103-1022-86
$C_{18}H_{15}NO_3S$			
4H-Carbazol-4-one, 1,2,3,9-tetrahydro-9-(phenylsulfonyl)-	MeOH	226(4.37),269s(3.98), 277s(4.01),287(4.03)	5-2065-86
$C_{18}H_{15}NO_3S_2$			
Benzenecarbodithioic acid, 4-(2-nitrophenyl)-1,3-butadienyl ester	CH_2Cl_2	361(4.52),491(2.72)	18-1403-86
Benzenecarbodithioic acid, 4-(4-nitrophenyl)-1,3-butadienyl ester	CH_2Cl_2	365(4.55),513(2.96)	18-1403-86
$C_{18}H_{15}NO_4$			
Benzoic acid, 4-[5-(4-methoxyphenyl)-2-oxazolyl]-, methyl ester	toluene	347(4.36)	103-1022-86
$C_{18}H_{15}NO_5$			
4,5-Benzofurandicarboxylic acid, 6-amino-2-phenyl-, dimethyl ester	dioxan	264(3.37),307(3.25), 381(3.25)	73-1455-86
2-Butenedioic acid, 2-(2-phenyl-4H-furo[3,2-b]pyrrol-5-yl)-, dimethyl ester	MeOH	432(3.42)	73-1455-86

Compound	Solvent	$\lambda_{max}(\log \epsilon)$	Ref.
1-Phenanthrenecarboxylic acid, 10-amino-3-hydroxy-4,8,9-trimethoxy-, lactam	MeOH	239(4.12),253(4.36), 268(4.06),298(3.80), 397(3.66)	102-0965-86
	MeOH-NaOH	261(4.42),319(3.62), 448(3.66)	102-0965-86
$C_{18}H_{15}NS_2$			
3-Thiophenecarbothioamide, 2-methyl-N,5-diphenyl-	EtOH	269(4.47),298(4.39), 394s(2.56)	78-3683-86
3-Thiophenecarbothioamide, 5-methyl-N,2-diphenyl-	EtOH	265(4.26),315(4.16), 400s(2.79)	78-3683-86
$C_{18}H_{15}N_3$			
Benzenamine, N-phenyl-4-(phenylazo)-	benzene	404.6(4.49)	60-1307-86
2H-Pyrazolo[3,4-c]quinoline, 1,4-dimethyl-2-phenyl-	EtOH	235s(4.49),242s(4.56), 247(4.59),263(4.44), 298s(3.95),322s(3.70), 337s(3.54)	142-3181-86
5-Pyrimidinecarbonitrile, 1,4-dihydro-6-methyl-1,4-diphenyl-	EtOH	247(3.98),292(3.74)	103-0990-86
$C_{18}H_{15}N_3O$			
4H-Pyrazolo[3,4-c]quinolin-4-one, 2,5-dihydro-1,5-dimethyl-2-phenyl-	EtOH	248(4.54),255s(4.46), 272s(4.14),299(3.94)	142-3181-86
$C_{18}H_{15}N_3O_2S$			
4(1H)-Pyrimidinethione, 6-methyl-5-nitro-2-phenyl-1-(phenylmethyl)-	EtOH	250(3.85),338(4.27)	18-3871-86
$C_{18}H_{15}N_3O_3$			
1H-Benzo[h]pyrimido[1,2-c]quinazoline-2-carboxylic acid, 5,6-dihydro-1-oxo-, ethyl ester	EtOH	235(3.54),242(3.57), 250(3.51),293(3.35), 382(3.99),398(4.02)	142-1119-86
$C_{18}H_{15}N_3S$			
5-Pyrimidinecarbonitrile, 1,2,3,4-tetrahydro-6-methyl-1,4-diphenyl-2-thioxo-	EtOH	303(4.26)	103-0990-86
$C_{18}H_{15}N_4O$			
1,2,4-Triazolo[4,3-c]pyrimidinium, 5,6-dihydro-6-methyl-5-oxo-1,3-diphenyl-, chloride	EtOH	268(4.12),314(4.07)	103-0532-86
$C_{18}H_{15}N_7O_3$			
Acetamide, N-[2-[(2,6-dicyano-4-nitrophenyl)azo]-5-(dimethylamino)phenyl]-	$CHCl_3$	304(3.84),314s(3.72), 355s(3.40),601(4.61)	48-0497-86
$C_{18}H_{16}$			
1,10:3,6-Ethanediylidene[14]annulene, 15,16-dimethyl-, trans	CH_2Cl_2	335(4.87),356(4.64), 390(3.85),418s(3.47), 440s(3.35),505(2.26), 544(2.62),570(2.63), 599(2.96)	35-4105-86
$C_{18}H_{16}Br_2$			
Cyclohexene, 4,5-dibromo-1,2-diphenyl-, trans	hexane	256(3.93)	44-2784-86

Compound	Solvent	$\lambda_{max}(\log \epsilon)$	Ref.
$C_{18}H_{16}Br_2O$			
7-Oxabicyclo[4.1.0]heptane, 3,4-dibromo-1,6-diphenyl-, trans	hexane	201(4.35),245(2.64), 253(2.67),258(2.69), 263(2.54)	44-2784-86
$C_{18}H_{16}ClN_3O_2$			
Anthra[1,9-cd]pyrazol-6(2H)-one, 5-chloro-2-[2-[(2-hydroxyethyl)amino]-ethyl]-	n.s.g.	247(4.23),258(4.12), 267(4.15),299(4.04), 341(3.57),410(3.91)	4-1491-86
Anthra[1,9-cd]pyrazol-6(2H)-one, 7-chloro-2-[2-[(2-hydroxyethyl)amino]-ethyl]-, hydrochloride	n.s.g.	235(4.16),277(4.03), 307(3.80),408(4.04)	4-1491-86
$C_{18}H_{16}ClN_3O_4$			
Anthra[1,9-cd]pyrazol-6(2H)-one, 5-chloro-7,10-dihydroxy-2-[2-(2-hydroxyethyl)amino]ethyl-, hydrochloride	n.s.g.	210(4.56),244(4.24), 283(3.96),490(3.93)	4-1491-86
$C_{18}H_{16}N_2$			
2,3'-Bi-1H-indole, 1,1'-dimethyl-	EtOH	220(4.45),283(3.97)	78-5019-86
3,3'-Bi-1H-indole, 2,2'-dimethyl-	EtOH	285(4.25),291(4.24)	39-2305-86C
2,4-Hexadiyne-1,6-diamine, N,N'-diphenyl-	hexane	215(4.64),240(4.20), 299(3.51)	65-2461-86
	MeOH	213(--),243(--), 293(--)	65-2461-86
Pyrido[3',2':7,8]cycloocta[1,2-b]quinoline, 5,6,7,8-tetrahydro-	EtOH	267(3.98),308(3.63), 321(3.65)	4-0685-86
$C_{18}H_{16}N_2O$			
Acetamide, N-(2,5-diphenyl-1H-pyrrol-3-yl)-	EtOH	205(4.38),232(4.36), 325(4.40)	103-0499-86
2-Naphthalenol, 1-[[(4,6-dimethyl-2-pyridinyl)imino]methyl]-	$CHCl_3$	460(4.28)	2-0127-86A
$C_{18}H_{16}N_2O_2$			
Benzo[d]-1,3-dioxolo[4,5-g]pyrido-[4,3,2-jk][2]benzazepine, 5,6,7,7a-tetrahydro-7-methyl- (dragabine)	EtOH	228(4.43),260s(4.02), 300(3.78)	77-1481-86
	EtOH-HCl	232s(4.35),256s(4.13), 316(3.81),372(3.65)	77-1481-86
Methanone, (4,4-dimethyl-5-phenyl-4H-imidazol-2-yl)phenyl-, N-oxide	EtOH	241s(4.16),261(4.24), 313s(4.00),339(4.06)	103-0856-86
Pyrazine, 2,6-dimethyl-3,5-diphenyl-, 1,4-dioxide	EtOH	238(4.32),260(4.33), 316(4.34)	103-0268-86
5-Pyrimidinecarboxylic acid, 1,4-dihydro-6-methyl-1,4-diphenyl-	EtOH	313(3.58)	103-0990-86
$C_{18}H_{16}N_2O_2S$			
Acetic acid, 2-methyl-1-[(3-oxobenzo-[b]thien-2(3H)-ylidene)methyl]-2-phenylhydrazide	dodecane	300(4.16),415(3.85)	104-2108-86
Benzo[b]thiophene-2-carboxaldehyde, 3-acetoxy-, 2-(methylphenylhydrazone)	dodecane	360(4.34)	104-2108-86
$C_{18}H_{16}N_2O_3$			
Benzoic acid, 4-[5-[4-(dimethylamino)-phenyl]-2-oxazolyl]-	toluene	355(4.25)	103-1022-86
3,5-Pyrazolidinedione, 4-[(4-methoxyphenyl)methylene]-1-methyl-2-phenyl-	EtOH	235s(4.35),247(4.40), 374(4.66),379(4.60)	103-0176-86

Compound	Solvent	λ_{max}(log ϵ)	Ref.
6-Quinolinecarboxylic acid, 8-methoxy-5-(phenylamino)-, methyl ester	MeOH	245(4.50),369(4.20), 360(3.50),398(3.60)	4-1641-86
7-Quinolinecarboxylic acid, 5-methoxy-8-(phenylamino)-, methyl ester	MeOH	244(4.40),271(4.50), 368(3.50),408(3.70)	4-1641-86
Sydnone, 3-[2-(3-butenyloxy)phenyl]-4-phenyl-	EtOH	244(3.97),263s(3.77), 332(4.05)	33-0927-86
Sydnone, 3-[2-[(2-methyl-2-propenyl)-oxy]phenyl]-4-phenyl-	EtOH	244(3.97),265s(3.75), 278s(3.68),330.5(4.03)	33-0927-86
$C_{18}H_{16}N_2O_3S$			
Acetamide, N-[1-(phenylsulfonyl)-1H-indol-3-yl]ethenyl]-	MeOH	219(4.46),259s(4.04), 289s(3.81)	5-2065-86
4H-Carbazol-4-one, 1,2,3,9-tetrahydro-9-(phenylsulfonyl)-, oxime	MeOH	223(--),252s(4.18), 265s(4.15)	5-2065-86
Ethanone, 1-(2-quinolinyl)-, O-[(4-methylphenyl)sulfonyl]oxime, (E)-	EtOH	245(4.60)	94-0564-86
$C_{18}H_{16}N_2O_4$			
3,5-Pyridinedicarboxylic acid, 4-(3-cyanophenyl)-2,6-dimethyl-, dimethyl ester	MeOH	221(--),230(--), 270(4.43)	87-1596-86
$C_{18}H_{16}N_2O_4S$			
Ethanone, 1-(1-isoquinolinyl)-, O-[(4-methylphenyl)sulfonyl]oxime, N-oxide, (Z)-	EtOH	223.7(4.55)	94-0564-86
Ethanone, 1-(4-quinolinyl)-, O-[(4-methylphenyl)sulfonyl]oxime, N-oxide, (Z)-	EtOH	232.2(4.61)	94-0564-86
1H-Indole, 3-[1-(acetoxyimino)ethyl]-1-(phenylsulfonyl)-	MeOH	220(4.40),275(4.16)	5-2065-86
$C_{18}H_{16}N_2O_5S$			
3H-Phenothiazine-1,9-dicarboxylic acid, 2-amino-4,6-dimethyl-3-oxo-, dimethyl ester	EtOH	260(4.48),465(4.39)	104-0207-86
$C_{18}H_{16}N_2S$			
[1]Benzothiopyrano[4,3,2-de]quinolin-2-amine, N,N,6-trimethyl-	MeOH	204(4.62),235(4.61), 263(4.61),377(4.02), 410(3.63)	83-0347-86
$C_{18}H_{16}N_4$			
1H-Pyrazolo[3,4-b]quinolin-4-amine, 1-methyl-N-(4-methylphenyl)-	EtOH	245(4.55),279(4.47), 383(4.01),397(4.06)	11-0331-86
$C_{18}H_{16}N_4OS_3$			
4-Thiazolidinone, 5-(2,3-dimethyl-8-phenyl-6H-thiazolo[3,4-b][1,2,4]-triazin-6-ylidene)-3-ethyl-2-thioxo-	n.s.g.	446(4.44),560(3.72)	103-1373-86
$C_{18}H_{16}N_4O_2$			
Benzoic acid, 2-(1,2-dihydro-1-methyl-2-oxo-4-pyrimidinyl)-2-phenylhydrazide	pH 2	294(3.94)	103-0532-86
	pH 10	252(4.06),320(4.01)	103-0532-86
	EtOH	289(4.0)	103-0532-86
2,4(1H,3H)-Pyrimidinedione, 1-methyl-3-[phenyl(phenylhydrazono)methyl]-	EtOH	240(4.29),282(4.13), 340(4.39)	103-0532-86
[1,2,4]Triazolo[5,1-a][2,7]naphthyridine-6-carboxylic acid, 5,9-dihydro-9-methyl-2-phenyl-, methyl ester	EtOH	212(4.43),248(4.48), 372(4.39)	128-0027-86

Compound	Solvent	$\lambda_{max}(\log \epsilon)$	Ref.
$C_{18}H_{16}N_4O_2S$			
1H-Indole-2,3-dione, 1-acetyl-, 3-[[[(phenylmethyl)amino]thioxomethyl]hydrazone], (Z)-	DMF	333s(3.87),390(4.22)	104-2173-86
	$CHCl_3$	274.5(4.45),331s(3.86), 394(4.23)	104-2173-86
	DMSO	282s(3.87),340s(3.81), 459(4.36)	104-2173-86
Spiro[benzothiazole-2(3H),5'(2'H)-pyrimidine]-2',4'(3'H)-dione, 3'-methyl-6'-[(phenylmethyl)amino]-	MeCN	221(4.38),275(3.89), 361(1.96)	94-0664-86
$C_{18}H_{16}N_4O_2S_2$			
1,2,3-Thiadiazole-4-carboxylic acid, 2,5-dihydro-2-phenyl-5-[[(phenylamino)thioxomethyl]imino]-, ethyl ester	DMF	278(4.37),421(4.30)	48-0741-86
$C_{18}H_{16}N_6NiO_2$			
Nickel, [2-[[1-(1H-benzimidazol-2-ylazo)-2-methylpropyl]azo]benzoato-(20)]-, (SP-4-2)-	isoPrOH	540(4.11),1060(3.89)	65-0727-86
$C_{18}H_{16}N_6O_2$			
1,3-Benzenedicarbonitrile, 2-[[4-(diethylamino)phenyl]azo]-5-nitro-	$CHCl_3$	304(3.84),314s(3.72), 355s(3.40),601(4.61)	48-0497-86
$C_{18}H_{16}N_8O_2S$			
1,2,4-Thiadiazole-3,5-diacetamide, α,α'-bis(phenylhydrazono)-	DMF	292(3.80),300s(3.79), 395(4.40)	48-0741-86
$C_{18}H_{16}O$			
1(2H)-Naphthalenone, 3,4-dihydro-2-[(4-methylphenyl)methylene]-, (E)-	EtOH	242(4.05),270(4.19), 326(4.59)	4-0135-86
1(2H)-Naphthalenone, 2-methyl-2-(4-methylphenyl)-	MeOH	238(4.65)	39-2011-86C
$C_{18}H_{16}OS_2$			
Benzenecarbodithioic acid, 4-methoxy-, 4-phenyl-1,3-butadienyl ester	CH_2Cl_2	296(4.44),350(4.39), 400(4.31)	18-1403-86
11H-5,8-Epoxy-1,12-metheno-2H-cyclohepta[d][1,8]dithiacyclotetradecin, 4,9-dihydro-	CH_2Cl_2	295(4.54),379(3.73), 630(2.47)	88-1929-86
12H-6,9-Epoxy-2,13-metheno-3H-cyclopenta[1][1,8]dithiacyclohexadecin, 5,10-dihydro-	CH_2Cl_2	288(4.72),368(3.87), 600(2.65)	88-1929-86
$C_{18}H_{16}O_2$			
4H-1-Benzopyran-4-one, 2-[4-(1-methylethyl)phenyl]-	EtOH	253(4.18),301(4.38)	104-1658-86
2-Naphthalenol, 1-[1-(2-hydroxyphenyl)ethyl]-, (S)-	EtOH	279(3.71)	116-0509-86
1(2H)-Naphthalenone, 3,4-dihydro-2-[(4-methoxyphenyl)methylene]-	EtOH	250(3.79),335(3.79)	4-0135-86
Tricyclo[$8.2.2.2^{4,7}$]hexadecaoctaene, 5,15-dimethoxy-, complex with tricarbonylchromium	CH_2Cl_2	334(3.83),410(3.57)	88-2353-86
$C_{18}H_{16}O_3$			
1H-Cyclopenta[a]pentalene-1,4(3aH)-dione, 3b,6a,7,7a-tetrahydro-3-(4-methoxyphenyl)-	MeOH	222(4.19),325(4.35)	39-0291-86C

Compound	Solvent	λ_{max}(log ε)	Ref.
1-Naphthalenol, 5,8-dimethoxy-3-phenyl-	EtOH	260(4.67),317s(3.82), 355(3.97)	94-2810-86
1-Naphthalenol, 6,8-dimethoxy-3-phenyl-	EtOH	262(4.74),296s(3.92), 349(3.64)	94-2810-86
$C_{18}H_{16}O_4$			
4H-1-Benzopyran-4-one, 5,7-dihydroxy-2,3-dimethyl-6-(phenylmethyl)-	MeOH	205(4.20),260(4.29), 296(3.63)	2-0862-86
4H-1-Benzopyran-4-one, 5,7-dihydroxy-2,3-dimethyl-8-(phenylmethyl)-	MeOH	212(3.93),252(4.00), 294(3.60)	2-0862-86
2(3H)-Furanone, dihydro-5-methyl-4-(4-phenoxybenzoyl)-	EtOH	266(4.10)	56-0143-86 +80-0541-86
2-Propenoic acid, 2-formyl-3-(4-methylphenoxy)-3-phenyl-, methyl ester	EtOH	243(4.43),279(3.69)	44-4112-86
$C_{18}H_{16}O_4S$			
9H-Thioxanthene-1-carboxylic acid, 3-ethoxy-9-oxo-, ethyl ester	CH_2Cl_2	368(3.79)	151-0353-86D
$C_{18}H_{16}O_5$			
Acenaphtho[5,4-b]furan-4,5-dione, 7,8-dihydro-3,6-dihydroxy-1,7,7,8-tetramethyl- (scleroderodione)	MeOH	219(4.4),231s(4.36), 255(4.82),286(4.13), 333(3.91),343(3.89), 360s(3.74)	44-4813-86
4H-1-Benzopyran-4-one, 5,7,8-trimethoxy-2-phenyl-	MeOH	270(4.46),331(3.92)	102-1255-86
$C_{18}H_{16}O_6$			
9,10-Anthracenedione, 5-hydroxy-1,2,4-trimethoxy-7-methyl-	MeOH	227(4.65),254(4.34), 260(4.29),290(4.26), 432(4.06)	78-3767-86
4H-1-Benzopyran-4-one, 7-hydroxy-5,8-dimethoxy-2-(2-methoxyphenyl)-	MeOH	225s(4.35),270(4.42), 331(4.11)	94-0406-86
	MeOH-NaOMe	230s(4.37),280(4.50), 320(4.02),370(4.02)	94-0406-86
	MeOH-NaOAc	279(4.47),320(3.99), 370(4.00)	94-0406-86
	MeOH-NaOAc-H_3BO_3	273(4.42),327(4.06)	94-0406-86
	MeOH-$AlCl_3$	225s(4.41),270(4.47), 331(4.19)	94-0406-86
	MeOH-$AlCl_3$-HCl	225s(4.40),270(4.45), 299s(4.15),331(4.15), 355s(4.16)	94-0406-86
4H-1-Benzopyran-4-one, 5-hydroxy-3,6,7-trimethoxy-2-phenyl-	MeOH	249(4.29),272(4.54), 321(4.33)	102-1255-86
	MeOH-NaOH	227(4.54),289(4.52), 385(3.70)	102-1255-86
	MeOH-NaOAc	249(4.29),272(4.54), 321(4.33)	102-1255-86
	MeOH-$AlCl_3$	253(4.23),285(4.55), 338(4.34),390s(3.90)	102-1255-86
	MeOH-$AlCl_3$-HCl	256(4.22),288(4.54), 341(4.33),392s(3.89)	102-1255-86
4H-1-Benzopyran-4-one, 7-hydroxy-3,5,8-trimethoxy-2-phenyl-	MeOH	270(4.51),342(4.04)	102-1255-86
	MeOH-NaOH	240(4.21),280(4.57), 375(4.01)	102-1255-86
	MeOH-NaOAc	240(4.20),280(4.60), 375(3.99)	102-1255-86
	MeOH-$AlCl_3$	270(4.51),342(4.04)	102-1255-86

Compound	Solvent	$\lambda_{max}(\log \epsilon)$	Ref.
8H-1,3-Dioxolo[4,5-g][1]benzopyran-8-one, 6,7-dihydro-9-methoxy-6-(2-methoxyphenyl)-, (S)-	MeOH	220(4.28),245(4.21), 282(3.97),340(3.51)	102-2223-86
4H,6H-Furo[3',2':3,4]naphtho[1,8-cd]-pyran-4,6-dione, 8,9-dihydro-3,7-dihydroxy-1,8,8,9-tetramethyl-	EtOH	215(4.45),256(4.51), 297(3.97),310(3.91), 347(4.13),361(4.14), 378(4.09)	44-4813-86
	EtOH-NaOH	234(4.57),253s(4.38), 308(4.26),320s(4.19), 375s(4.17),390(4.31)	44-4813-86
$C_{18}H_{16}O_7$			
4H-1-Benzopyran-4-one, 3,5-dihydroxy-6,7,8-trimethoxy-2-phenyl-	EtOH	278(4.36),325(4.15), 378(4.01)	105-0230-86
4H-1-Benzopyran-4-one, 2-(2,6-dimethoxyphenyl)-5,7-dihydroxy-8-methoxy-	MeOH	267(4.42),310s(3.92), 350s(3.72)	94-0406-86
	MeOH-NaOMe	277(4.48),335s(3.86), 360(3.90)	94-0406-86
	MeOH-NaOAc	277(4.45),330s(3.84), 360(3.88)	94-0406-86
	MeOH-NaOAc-H_3BO_3	270(4.37),341(3.78)	94-0406-86
	MeOH-$AlCl_3$	277(4.40),300s(4.10), 326(3.93),395(3.72)	94-0406-86
	MeOH-$AlCl_3$-HCl	278(4.41),300s(4.11), 324(3.91),395(3.72)	94-0406-86
4H-1-Benzopyran-4-one, 5-hydroxy-2-(4-hydroxy-3-methoxyphenyl)-6,7-dimethoxy- (fastigenin)	MeOH	275(4.07),340(4.18)	105-0732-86
	MeOH-NaOAc	274(--),338(--)	105-0732-86
	MeOH-$AlCl_3$	263(--),286(--), 370(--)	105-0732-86
$C_{18}H_{16}O_9$			
1,3-Dioxane-4,6-dione, 5,5'-(2,5-furandiyldimethylidyne)bis[2,2-dimethyl-	EtOH	258(4.26),404(4.56)	103-1072-86
$C_{18}H_{16}S_3$			
11H-5,8-Epithio-1,12-metheno-2H-cyclohepta[d][1,8]dithiacyclotetradecin, 4,9-dihydro-	CH_2Cl_2	303(4.43),384(3.62), 628(2.31)	88-1929-86
12H-6,9-Epithio-2,13-metheno-3H-cyclopenta[l][1,8]dithiacyclohexadecin, 5,10-dihydro-	CH_2Cl_2	289(4.67),352(3.75), 608(2.60)	88-1929-86
$C_{18}H_{17}ClFNOS_2$			
Morpholine, 4-[2-[5-chloro-2-[(4-fluorophenyl)thio]phenyl]-1-thioxoethyl]-	MeOH	280(4.32)	73-2598-86
$C_{18}H_{17}ClN_2$			
2H-Benz[g]indazole, 3-(3-chlorophenyl)-3,3a,4,5-tetrahydro-2-methyl-	EtOH	215(3.86),239(3.18), 312(4.22)	4-0135-86
2H-Benz[g]indazole, 3-(4-chlorophenyl)-3,3a,4,5-tetrahydro-2-methyl-	EtOH	315(3.85)	4-0135-86
$C_{18}H_{17}N$			
Benzo[6,7]cyclohept[1,2-b]indole, 5,6,7,12-tetrahydro-12-methyl-	n.s.g.	218(4.29),240(4.33), 307(4.26)	150-0514-86M
2-Naphthaleneethanamine, N-phenyl-	hexane	276(3.83)	18-1626-86
$C_{18}H_{17}NO$			
Azacyclohexadeca-3,5,11,13,15-penta-	THF	265s(4.37),293(4.58)	18-1801-86

Compound	Solvent	$\lambda_{max}(\log \epsilon)$	Ref.
ene-7,9-diyn-2-one, 3,6,11-trimethyl-, (Z,E,Z,Z,E)- (cont.)		310s(4.47),341s(3.97), 464(3.02)	18-1801-86
Azacyclohexadeca-3,5,11,13,15-pentaene-7,9-diyn-2-one, 6,11,16-trimethyl-, (Z,E,Z,Z,E)-	THF	291(4.43),342s(3.80), 459(3.05)	18-1801-86
2,4,10,12,14-Cyclopentadecapentaene-6,8-diyn-1-one, 2,5,10-trimethyl-, oxime	THF	279(4.44),309(4.59), 395(3.53)	18-1801-86
2,4,10,12,14-Cyclopentadecapentaene-6,8-diyn-1-one, 5,10,15-trimethyl-, oxime	THF	278s(4.56),300(4.70), 385s(4.31)	18-1801-86
Oxiranecarbonitrile, 3-(2-ethynyl-1-cyclohepten-1-yl)-2-phenyl-, (Z)-	MeCN	220(4.10),250(4.03)	78-2221-86
$C_{18}H_{17}NOS$			
1-Benzothiepin-5(2H)-one, 3,4-dihydro-4-[(methylphenylamino)methylene]-, (E)-	EtOH	248(4.06),358(4.20)	4-0449-86
9H-Thioxanthen-9-one, 2-(1-piperidinyl)-	MeOH	266(4.56),292(4.39), 421(3.58)	73-0937-86
$C_{18}H_{17}NO_2$			
1H-Indene-1,3(2H)-dione, 2-[7-(dimethylamino)-2,4,6-heptatrienylidene]-	EtOH	630(5.01)	70-1446-86
1H-Indole-2-carboxylic acid, 1-(1-phenylethyl)-, methyl ester	MeOH	204.5(4.48),228.0(4.32), 293.5(4.28),308s(4.02)	1-0625-86
1H-Indole-2-carboxylic acid, 3-(1-phenylethyl)-, methyl ester, (±)-	MeOH	203.5(4.50),230(4.46), 296(4.31),310s(4.10)	1-0625-86
2-Propen-1-one, 1-(2-hydroxyphenyl)-3-phenyl-3-(2-propenylamino)-, (Z)-	EtOH	245s(4.02),367(4.52)	104-1658-86
$C_{18}H_{17}NO_3$			
[1,3]Dioxepino[5,6-d]isoxazole, 3a,4,8,8a-tetrahydro-3,6-diphenyl-, endo	MeCN	266(3.03)	73-2158-86
exo	MeCN	260(2.99)	73-2158-86
5H-9,12a-Methanocycloocta[c]quinoline-6,8,13-trione, 9,10,11,12-tetrahydro-5,9-dimethyl-	n.s.g.	211(4.15),234(4.26), 359(3.54)	150-2782-86M
Norlaureline, (-)-	EtOH	218(4.28),232s(4.24), 265(4.03),275(4.06), 305(3.81)	100-1078-86
$C_{18}H_{17}NO_4$			
9H-Carbazole-3,4-dicarboxylic acid, 1,9-dimethyl-, dimethyl ester	n.s.g.	233(4.44),245(4.32), 269(4.40),280(4.60), 313(3.90)	150-0514-86M
Elmerrillicine, (-)-	EtOH	224(4.41),240s(4.23), 268s(4.16),276(4.20), 298(4.02)	100-1078-86
4H-Furo[3,2-b]pyrrole-5-carboxylic acid, 4-acetyl-2-(4-methylphenyl)-, ethyl ester	MeOH	354(3.44)	73-1455-86
4H-Furo[3,2-b]pyrrole-5-carboxylic acid, 6-formyl-4-methyl-2-(4-methylphenyl)-, ethyl ester	MeOH	302(3.44),371(3.29)	73-0106-86
$C_{18}H_{17}NS_3$			
6H-1,3-Thiazine, 6,6-bis(methylthio)-2,4-diphenyl-	MeOH	268(4.11),312(3.95)	118-0916-86

Compound	Solvent	λ_{max}(log ε)	Ref.
$C_{18}H_{17}N_3O$			
2-Naphthalenol, 1-[[4-(dimethylamino)-phenyl]azo]-	hexane	470(4.5),490(4.5)	48-0089-86
	EtOH	498(4.5)	48-0089-86
5-Pyrimidinecarboxamide, 1,4-dihydro-6-methyl-1,4-diphenyl-	EtOH	264(4.27)	103-0990-86
$C_{18}H_{17}N_3OS$			
3H-Pyrazole-3-thione, 2,4-dihydro-4-[[(3-methoxyphenyl)amino]methylene]-5-methyl-2-phenyl-	toluene	355(4.4),435(4.0)	103-0503-86
3H-Pyrazole-3-thione, 2,4-dihydro-4-[[(4-methoxyphenyl)amino]methylene]-5-methyl-2-phenyl-	toluene	360(4.5),435(4.2)	103-0503-86
5-Pyrimidinecarboxamide, 1,2,3,4-tetrahydro-6-methyl-1,4-diphenyl-2-thioxo-	EtOH	291(4.13)	103-0990-86
$C_{18}H_{17}N_3OSe$			
3H-Pyrazole-3-selone, 2,4-dihydro-4-[[(4-methoxyphenyl)amino]methylene]-5-methyl-2-phenyl-	EtOH	356(4.3),444(4.0)	104-2126-86
$C_{18}H_{17}N_3O_2$			
2H-Benz[g]indazole, 3,3a,4,5-tetrahydro-2-methyl-3-(4-nitrophenyl)-	EtOH	300(4.14)	4-0135-86
3-Pyrrolidinecarboxaldehyde, 1-methyl-2,5-dioxo-, 3-(diphenylhydrazone)	isoPrOH	270s(4.17),286(4.22)	103-1345-86
enehydrazine form	isoPrOH	288(4.35)	103-1345-86
$C_{10}H_{17}N_3O_2S$			
Thieno[2,3-d]pyrimidine-4-acetic acid, α-cyano-5,6-dihydro-6-methyl-2-phenyl-	$CHCl_3$	270(4.43),342(4.34), 375s(4.07)	94-0516-86
$C_{18}H_{17}N_3O_3S$			
6H-Thieno[2,3-b]pyrrole-2-carboxylic acid, 3,4-diamino-5-benzoyl-6-(2-propenyl)-, methyl ester	$CHCl_3$	260(4.29),318(4.43), 383(3.77)	48-0459-86
$C_{18}H_{17}N_3O_4$			
4-Pyrimidinecarboxylic acid, 1,4,5,6-tetrahydro-2-[(1-naphthalenylcarbonyl)amino]-5-oxo-, ethyl ester	EtOH	217(4.54),237s(4.28), 259s(4.06),310(4.01)	118-0484-86
$C_{18}H_{17}N_3S$			
3H-Pyrazole-3-thione, 2,4-dihydro-5-methyl-4-[[(3-methylphenyl)-amino]methylene]-2-phenyl-	toluene	353(4.3),435(3.9)	103-0503-86
3H-Pyrazole-3-thione, 2,4-dihydro-5-methyl-4-[[(4-methylphenyl)-amino]methylene]-2-phenyl-	toluene	355(4.3),435(3.9)	103-0503-86
$C_{18}H_{17}N_3Se$			
3H-Pyrazole-3-selone, 2,4-dihydro-5-methyl-4-[[(3-methylphenyl)-amino]methylene]-2-phenyl-	toluene	346(4.3),446(3.8)	104-2126-86
3H-Pyrazole-3-selone, 2,4-dihydro-5-methyl-4-[[(4-methylphenyl)-amino]methylene]-2-phenyl-	toluene	348(4.3),444(3.8)	104-2126-86

Compound	Solvent	λ_{max}(log ε)	Ref.
$C_{18}H_{17}N_5$			
Propanedinitrile, [6-phenyl-2-(1-piperidinyl)-4(1H)-pyrimidinylidene]-	EtOH	247(4.43),288(4.23), 357(4.35)	103-0774-86
1H-Pyrazolo[3,4-b]quinoxalin-7-amine, N,N,3-trimethyl-1-phenyl-	BuOAc	410(3.98),440(3.90)	48-0342-86
$C_{18}H_{17}N_5O_3$			
Benzoic acid, 2-[[(4-diazo-3-oxo-1-phenylbutyl)amino]carbonyl]-, hydrazide	MeOH	240(4.19),274(4.04)	104-0353-86
$C_{18}H_{17}N_5O_4$			
Pyrazolo[1,5-a]pyrimidine-3-carboxylic acid, 4,7-dihydro-7-(1-methyl-1H-indol-3-yl)-6-nitro-	EtOH	221(4.13),270(3.52), 380(4.18)	103-1250-86
Pyrazolo[1,5-a]pyrimidine-3-carboxylic acid, 4,7-dihydro-7-(2-methyl-1H-indol-3-yl)-6-nitro-	EtOH	219(4.03),270(3.34), 358(4.01)	103-1250-86
$C_{18}H_{17}N_7O_5$			
Propanamide, N-[2-[(2-cyano-4,6-dinitrophenyl)azo]-5-(dimethylamino)-phenyl]-	$CHCl_3$	287(3.80),306s(3.73), 575s(4.64),612(4.73)	48-0497-86
$C_{18}H_{18}$			
Stilbene, 2-(2-buten-2-yl)-	MeOH	202(4.34),224(4.21), 308(4.31)	151-0105-86A
Tetracyclo[$6.6.2.1^{3,13}.1^{6,10}$]octadeca-1,3(17),6,8,10(18),13-hexaene	hexane	257(3.0)	88-2907-86
Tricyclo[$8.2.2.2^{4,7}$]hexadeca-4,6,10,12,13,15-hexaene, 2-ethenyl-	MeCN	206(4.04),225(4.22)	24-1836-86
$C_{18}H_{18}BrNO_6$			
1,1-Cyclohexanedicarboxylic acid, 4-benzoyloxy-3-bromo-3-cyano-, dimethyl ester, trans	n.s.g.	229(4.17)	128-0491-86
$C_{18}H_{18}ClNO_2S$			
2H-[1]Benzothiepino[5,4-b]pyran-2-one, 3-chloro-5,6-dihydro-4-(1-piperidinyl)-	EtOH	233(3.95),255(4.01), 264s(3.99),323(4.06)	4-0449-86
$C_{18}H_{18}ClNO_4$			
Furo[2,3-b]quinoline, 7-[(4-chloro-3-methyl-2-butenyl)oxy]-4,8-dimethoxy- (haplobine)	EtOH	251(4.73),319(3.60), 330(3.67)	105-0684-86
$C_{18}H_{18}ClN_2$			
2-Quinolinaminium, 6-chloro-N,N,N-trimethyl-4-phenyl-, iodide	MeOH	236(4.41),299(3.51)	83-0338-86
$C_{18}H_{18}Cl_2N_2O_6$			
Uridine, 2'-deoxy-5-ethyl-, 5'-(2,4-dichlorobenzoate)	octanol	245(4.11)	83-0154-86
Uridine, 2'-deoxy-5-ethyl-, 5'-(2,6-dichlorobenzoate)	octanol	267(3.96)	83-0154-86
$C_{18}H_{18}F_3N_3$			
Piperidine, 1-[4-[[4-(trifluoromethyl)-phenyl]azo]phenyl]-	EtOH	423(4.43)	39-0123-86B
	EtOH-HCl	515(3.81)	39-0123-86B

Compound	Solvent	$\lambda_{max}(\log \epsilon)$	Ref.
$C_{18}H_{18}N_2$			
4,11-Azo-5,10-methano-1H-cyclopent[b]-anthracene, 3a,4,4a,5,10,10a,11,11a-octahydro-	hexane	387(2.27)	89-0187-86
2H-Benz[g]indazole, 3,3a,4,5-tetra-hydro-2-methyl-3-phenyl-, cis	EtOH	310(3.72)	4-0135-86
2H-Benz[g]indazole, 3,3a,4,5-tetra-hydro-3-(4-methylphenyl)-, cis	EtOH	360(3.87)	4-0135-86
Cyclopentapyrazole, 3,3a,4,5,6,6a-hexahydro-3,6a-diphenyl-	hexane	254(2.64),260(2.68), 265.5(2.48),333(2.36)	35-1617-86
$C_{18}H_{18}N_2O$			
2H-Benz[g]indazole, 3,3a,4,5-tetra-hydro-3-(4-methoxyphenyl)-	EtOH	224(4.31),294(4.12)	4-0135-86
3H-Pyrazolo[5,1-e][1,6]benzoxazocine, 3a,4,5,6-tetrahydro-2-phenyl-	EtOH	241s(4.03),254(4.04), 260s(4.02),281s(3.70), 292s(3.70),303(3.73), 368(4.13)	33-0927-86
$C_{18}H_{18}N_2OS$			
9H-Thioxanthen-9-one, 2-(4-methyl-1-piperazinyl)-	MeOH	263(4.57),282s(4.41), 416(3.56)	73-0937-86
$C_{18}H_{18}N_2OS_2$			
Thieno[2,3-d]pyrimidine-4(3H)-thione, 2-(4-methoxyphenyl)-5,6-dimethyl-3-(2-propenyl)-	EtOH	233(4.35),286(4.19), 333(3.80),386(4.12)	106-0096-86
$C_{18}H_{18}N_2O_2S$			
Benzonitrile, 4-[3-[1,2-dihydro-1-(1-methylethyl)-2-thioxo-3-pyridinyl]-3-hydroxy-1-oxopropyl]-	EtOH	248(4.33),284(4.08), 364(3.83)	4-0567-86
$C_{18}H_{18}N_2O_3$			
Methanone, (2,5-dihydro-1-hydroxy-2,5-dimethyl-4-phenyl-1H-imidazol-2-yl)phenyl-, N-oxide	EtOH	235(4.13),248(4.18)	103-0268-86
Methanone, (4,5-dihydro-4-hydroxy-5,5-dimethyl-4-phenyl-1H-imidazol-2-yl)phenyl-, N-oxide	EtOH	257(4.02),322s(3.32)	103-0856-86
3(2H)-Pyridazinone, 4,5-dihydro-5-(1-hydroxyethyl)-6-(4-phenoxyphenyl)-	EtOH	298(4.47)	56-0143-86 +80-0541-86
9H-Pyrido[3,4-b]indole-3-carboxylic acid, 1-(tetrahydro-5-methyl-2-furanyl)-, methyl ester	MeOH	235(4.32),269(4.50)	88-3399-86
$C_{18}H_{18}N_2O_4$			
[2.2](2,5)-Pyridinophane-8,15-dicarb-oxylic acid, dimethyl ester, pseudo-gem.	MeCN	228(4.05),295(3.58), 325s(2.87)	5-0751-86
pseudo-meta-	MeCN	252(3.77),297(3.66), 320s(3.02)	5-0751-86
[2.2](2,5)-Pyridinophane-8,16-dicarb-oxylic acid, dimethyl ester, pseudo-ortho-	MeCN	251(3.72),294(3.62), 318s(3.06)	5-0751-86
pseudo-para-	MeCN	227(4.09),292(3.58), 317s(2.94)	5-0751-86
$C_{18}H_{18}N_2O_5S$			
1-Azabicyclo[3.2.0]hept-2-ene-2-carb-	MeCN	269(4.07),300(4.07)	39-0421-86C

Compound	Solvent	$\lambda_{max}(\log \epsilon)$	Ref.
oxylic acid, 6-ethylidene-3-(ethylthio)-7-oxo-, (4-nitrophenyl)methyl ester, [R-(Z)]- (cont.)			39-0421-86C
$C_{18}H_{18}N_2O_6$			
3,5-Pyridinedicarboxylic acid, 2,6-dimethyl-4-(3-nitrophenyl)-, ethyl methyl ester	MeOH	268(4.42),277(--)	87-1596-86
$C_{18}H_{18}N_2O_{10}S$			
Paulinone, 11-O-pauloyl-	MeOH	231(4.19),266(4.26), 438(3.23)	44-2493-86
$C_{18}H_{18}N_4$			
Benzonitrile, 4-[[4-(1-piperidinyl)-phenyl]azo]-	EtOH	442(4.46)	39-0123-86B
	EtOH-HCl	521(4.00)	39-0123-86B
$C_{18}H_{18}N_4O$			
Azepino[2,1-b]pteridine-12(6H)-one, 7,8,9,10-tetrahydro-2-methyl-3-phenyl-	MeOH	331(4.06)	33-0793-86
$C_{18}H_{18}N_4O_3S$			
Benzeneacetic acid, 2-(acetylamino)-α-oxo-, 2-[[(phenylmethyl)amino]-thioxomethyl]hydrazide	EtOH	236(4.48),272s(4.04), 335(3.63)	104-2173-86
	$CHCl_3$	277.5(4.00),358(3.74)	104-2173-86
$C_{18}H_{18}N_4O_4$			
2-Propenoic acid, 3-[3,7-dihydro-7-oxo-3-(tetrahydro-2H-pyran-2-yl)-pyrido[2,1-i]purin-10-yl]-, methyl ester	CH_2Cl_2	380(4.3)	39-1561-86B
$C_{18}H_{18}N_4O_6$			
Pyrido[2,1-i]purine-9,10-dicarboxylic acid, 3,7-dihydro-7-oxo-3-(tetrahydro-2H-pyran-2-yl)-, dimethyl ester	C_6H_{12}	330(4.2),355(4.2), 430(4.5),455s(4.4)	39-1561-86B
	EtOH	330(3.8),345(3.8), 423(4.1),448s(4.0)	39-1561-86B
	CH_2Cl_2	268(4.0),335(3.7), 350(3.7),428(4.1), 451(4.0)	39-1561-86B
$C_{18}H_{18}N_4S_4$			
Pyrrolo[1,2-a]pyrazine, 8,8'-dithiobis[7-methyl-6-(methylthio)-	EtOH	242(4.20),302(3.60), 312(3.60),350(3.56), 372(3.51)	78-0409-86
$C_{18}H_{18}N_6$			
Propanenitrile, 3,3'-[[3-methyl-4-(3-pyridinylazo)phenyl]imino]bis-	acetone	406(4.40)	7-0473-86
5H,12H-[1,2,4,5]Tetrazino[1,6-a:4,3-a']bisbenzimidazole, 5,12-diethyl-	CH_2Cl_2	240(4.3),380(4.0)	33-1521-86
$C_{18}H_{18}N_6O_2$			
Benzoic acid, 2-[5-(1H-benzimidazol-2-yl)-3-(1-methylethyl)-1-formazano]-	EtOH	450(4.26)	65-0727-86
1:1 nickel complex	EtOH	530(4.11)	65-0727-86
1:1 zinc complex	EtOH	630(4.36)	65-0727-86

Compound	Solvent	λ_{max}(log ϵ)	Ref.
$C_{18}H_{18}N_8O_4S_2$			
2,4(1H,3H)-Pteridinedione, 7,7'-dithiobis[1,3,6-trimethyl-	CH_2Cl_2	252(4.42),352(4.43), 369(4.48)	33-1095-86
$C_{18}H_{18}O_2$			
Benzo[c]phenanthrene-5,6-diol, 5,6,9,10,11,12-hexahydro-, (5S-trans)	n.s.g.	269(4.12)	44-1773-86
5,6-Chrysenediol, 1,2,3,4,5,6-hexahydro-, (5R-trans)	n.s.g.	277(4.26)	44-1773-86
Equilenin	pH 1.57	325(3.4),338(3.4)	69-1186-86
	pH 12.1	298(3.7),358(3.5)	69-1186-86
$C_{18}H_{18}O_3$			
2H-1-Benzopyran-7-ol, 3-(4-methoxyphenyl)-2,4-dimethyl-	EtOH	212(3.65),230(3.63), 296(3.61),312(3.62)	4-1781-86
Cyclohepta[de]naphthalen-7(8H)-one, 8,8-diethoxy-	$CHCl_3$	325(3.94),339s(3.90), 357s(3.74)	39-1965-86C
$C_{18}H_{18}O_3S$			
9(10H)-Anthracenone, 10-(butylthio)-1,8-dihydroxy-	EtOH	375(3.89)	118-0430-86
$C_{18}H_{18}O_4$			
Anthracene, 1,4,5,8-tetramethoxy-	MeCN	234(4.19),263(4.64), 350(3.37),368(3.57), 392(3.35),415(3.22)	24-1016-86
4H-1-Benzopyran-4-one, 7-(ethoxymethoxy)-2,3-dihydro-3-phenyl-	MeOH	204(4.40),268(4.51), 308(4.16)	39-0215-86C
[Bi-1,3,5-cycloheptatrien-1-yl]-6,6'-dicarboxylic acid, dimethyl ester	dioxan	240(4.71),295(3.61), 385(4.04)	89-0723-86
Butanedioic acid, dimethyl(1-naphthalenylmethylene)-, 4-methyl ester, (E)-	EtOH	225(4.72),293(3.96)	34-0369-86
Phenol, 4-[2-[2-(methoxymethoxy)phenyl]ethenyl]-, acetate, (E)-	EtOH	314(4.33)	150-3514-86M
1-Propanone, 3-acetoxy-2-methoxy-1,2-diphenyl-	THF	252(4.18),340(2.29)	121-0355-86
$C_{18}H_{18}O_5$			
4H-1-Benzopyran-4-one, 2,3-dihydro-5,7-dimethoxy-3-(4-methoxyphenyl)-	MeOH	223(4.60),271(4.49)	39-0215-86C
1H-Indene-5,6-diol, 2-[2-(3,4-dihydroxyphenyl)-1-(hydroxymethyl)ethyl]-	MeOH	208.5(4.19),287.5(4.01)	39-1181-86C
1-Propanone, 3-acetoxy-1-(2-hydroxy-4-methoxyphenyl)-2-phenyl-	MeOH	219(4.35),230s(4.24), 282(4.30),316(4.06)	39-0215-86C
$C_{18}H_{18}O_6$			
1(3H)-Isobenzofuranone, 3-[(3,4-dimethoxyphenyl)methyl]-7-hydroxy-5-methoxy- (balantiolide)	EtOH	204.5(4.79),214(4.85), 252(4.40),295s(3.95), 281(4.08)	102-2543-86
	EtOH-$AlCl_3$	218(5.08)	102-2543-86
1,3-Propanedione, 1-(2,4-dimethoxyphenyl)-3-(2-hydroxy-4-methoxyphenyl)-	MeOH	308(3.95),380(4.13), 395(4.08)	2-0304-86
$C_{18}H_{18}O_8$			
[Bi-1,5-cycloheptadien-1-yl]-4,4',7,7'-tetrone, 3,3,3',3'-tetramethoxy-	MeOH	229(4.25),380(2.95)	18-0511-86

Compound	Solvent	$\lambda_{max}(\log \epsilon)$	Ref.
$C_{18}H_{18}S_6$			
1,3-Dithiole, 4-phenyl-2-[2,3,4,5-tetrakis(methylthio)-2,4-cyclopentadien-1-ylidene]-	MeCN	202(4.64),235(4.45), 343(3.46),490(4.53)	78-0839-86
3H-1,2-Dithiole, 5-phenyl-3-[2,3,4,5-tetrakis(methylthio)-2,4-cyclopentadien-1-ylidene]-	MeCN	267(3.93),304s(3.42), 320s(3.82),338(4.00), 542(4.37)	78-0839-86
$C_{18}H_{19}BrN_6O_5$			
Acetamide, N-[2-[(2-bromo-4,6-dinitrophenyl)azo]-5-(diethylamino)phenyl]-	$CHCl_3$	297(3.90),377(3.62), 572(4.71)	48-0497-86
$C_{18}H_{19}Br_2N_5O_3$			
Acetamide, N-[2-[(2,6-dibromo-4-nitrophenyl)azo]-5-(diethylamino)phenyl]-	$CHCl_3$	304(3.94),392(3.66), 507(4.40)	48-0497-86
$C_{18}H_{19}ClN_2O_2S$			
2H-[1]Benzothiepino[5,4-b]pyran-2-one, 3-chloro-5,6-dihydro-4-(4-methyl-1-piperazinyl)-	EtOH	236s(4.00),254(4.07), 264s(4.00),321(4.06)	4-0449-86
$C_{18}H_{19}ClN_3O$			
Phenazinium, 5-(2-chloroethyl)-3-(4-morpholinyl)-, chloride	EtOH	239(4.44),298(4.46), 390(3.99),405(4.00), 555(4.19)	103-1342-86
$C_{18}H_{19}ClN_4$			
1H-Pyrazolo[3,4-b]quinolin-4-amine, N-(4-chloro-2-methylphenyl)-5,6,7,8-tetrahydro-1-methyl-	EtOH	219(4.46),250(3.79), 271(3.64),320(4.16)	11-0331-86
$C_{18}H_{19}ClN_4O$			
Piperazine, 1-acetyl-4-[4-[(4-chlorophenyl)azo]phenyl]-	pH 1	532(4.80)	39-0123-86B
	EtOH	396(4.44)	39-0123-86B
$C_{18}H_{19}ClO_4S$			
2-Cyclohexene-1-carboxylic acid, 6-acetyl-1-[(4-chlorophenyl)thio]-3-methyl-4-oxo-, ethyl ester, trans-(±)-	EtOH	234s(4.15)	44-4840-86
$C_{18}H_{19}Cl_2N_5O_3$			
Acetamide, N-[2-[(2,6-dichloro-4-nitrophenyl)azo]-5-(diethylamino)phenyl]-	$CHCl_3$	300(3.95),377(3.65), 512(4.40)	48-0497-86
$C_{18}H_{19}F_3N_2S$			
Fluopromazine hydrochloride	MeOH	259(4.518),308.5(3.557)	106-0571-86
$C_{18}H_{19}I_2N_5O_3$			
Acetamide, N-[5-(diethylamino)-2-[(2,6-diiodo-4-nitrophenyl)azo]phenyl]-	$CHCl_3$	294s(3.92),373(3.65), 510(4.45)	48-0497-86
$C_{18}H_{19}N$			
1H-Indole, 1-butyl-2-phenyl-	hexane	297(4.30)	103-1262-86
1H-Indole, 1,2-dimethyl-3-(1-phenylethyl)-, (±)-	MeOH	228.5(4.54),274s(3.74), 280s(3.79),286(3.85), 293.5(3.83)	1-0625-86
$C_{18}H_{19}NO$			
Azacyclotetradeca-1,3,5,11,13-pentaene-7,9-diyne, 2-ethoxy-6,11,14-	THF	277s(4.28),307(4.42), 386s(3.60),470s(3.25),	18-1791-86

Compound	Solvent	$\lambda_{max}(\log \epsilon)$	Ref.
trimethyl- (cont.)		511s(3.07),553s(2.72)	18-1791-86
1-Propanone, 2,2-dimethyl-1-phenyl-3-[(phenylmethyl)imino]-, (E)-	EtOH	245(4.0),280(3.1)	39-0091-86C
$C_{18}H_{19}NO_2$			
2-Furancarboxamide, N-[2-(phenylmethylene)cyclohexyl]-, (E)-	EtOH	246(4.41)	44-5002-86
(Z)-	EtOH	248(4.39)	44-5002-86
1H-Indole, 1,3-bis(3-methyl-1-oxo-2-butenyl)-	MeOH	236(4.47),247(4.45), 313(4.21)	44-2343-86
2-Propen-1-one, 1-(2-hydroxyphenyl)-3-[(1-methylethyl)amino]-3-phenyl-	EtOH	246s(4.01),365(4.48)	104-1658-86
$C_{18}H_{19}NO_2S$			
Benzoic acid, 2-[[4-(1-piperidinyl)-phenyl]thio]-	MeOH	279(4.31),313s(3.92)	73-0937-86
$C_{18}H_{19}NO_3$			
Benzeneacetic acid, α-[[2-(3-butenyl-oxy)phenyl]amino]-	EtOH	245(4.07),290(3.52)	33-0927-86
Benzeneacetic acid, α-[[2-[(2-methyl-2-propenyl)oxy]phenyl]amino]-	EtOH	243.5(4.08),289(3.54)	33-0927-86
1,3-Dioxolo[4,5-g]isoquinoline, 5,6,7,8-tetrahydro-6-[(4-methoxy-phenyl)methyl]- (viguine)(same spectra in acid or base)	n.s.g.	224(--),294(--)	142-3359-86
$C_{18}H_{19}NO_4$			
Crinine, O-acetyl-	n.s.g.	203(4.35),240(3.30), 293(3.86)	102-2399-86
6H-Dibenzo[a,g]quinolizine-3,10,11-triol, 5,8,13,13a-tetrahydro-2-methoxy- (artavenustine)	MeOH	228(4.09),288(3.86)	100-0602-86
3,5-Pyridinedicarboxylic acid, 2,6-dimethyl-4-(4-methylphenyl)-, dimethyl ester	MeOH	230(--),268(4.51)	87-1596-86
$C_{18}H_{19}NO_5$			
1-Naphthalenecarboxylic acid, 3-methoxy-5-methyl-, 2-amino-1-(2-methyl-oxiranyl)-2-oxoethyl ester	MeOH	217(4.67),245s(--), 303(3.52),343(3.69)	94-4554-86
Stephanaberrine	EtOH	240.5(3.52),293(3.71)	100-0588-86
$C_{18}H_{19}NO_6$			
9(10H)-Acridinone, 1-hydroxy-2,3,5,6-tetramethoxy-10-methyl-	MeOH	220(3.99),258(4.29), 267(4.28),332(3.84), 396(3.32)	102-0429-86
	MeOH-NaOMe	222(3.97),258(4.27), 268(4.24),333(3.81), 405(3.32)	102-0429-86
	MeOH-AlCl$_3$	240(4.11),265s(4.13), 279(4.24),356(4.02)	102-0429-86
Benzo[f]quinolin-3(4H)-one, 2-acetyl-5,6-dihydro-1-hydroxy-7,8,9-tri-methoxy-	EtOH	226(3.86),278(3.95), 364(4.22)	44-1374-86
$C_{18}H_{19}NO_9$			
7-Azatricyclo[4.3.0.0^{2,5}]nona-3,8-diene-1,2,3,9-tetracarboxylic acid, 7-acetyl-, tetramethyl ester, (1α,2β,5β,6α)-	EtOH	210(4.06),278(4.12)	24-0616-86

Compound	Solvent	λ_{max}(log ϵ)	Ref.
9-Azatricyclo[4.2.1.0^{2,5}]nona-3,7-diene-3,4,7,8-tetracarboxylic acid, 9-acetyl-, tetramethyl ester, (1α,2β,5β,6α)-	EtOH	220(4.15)	24-0616-86
$C_{18}H_{19}N_3O$			
2H-Benz[g]indazole, 2-acetyl-3,3a,4,5-tetrahydro-3-(1-methyl-1H-pyrrol-2-yl)-	EtOH	220(3.86),290(3.82)	4-0135-86
5,10-Ethanophenazine, 2-(4-morpholinyl)-	EtOH	215(4.26),260(3.76)	103-1342-86
$C_{18}H_{19}N_3O_2$			
Acetic acid, cyano[[3,4-dihydro-2-(1-pyrrolidinyl)-1-naphthalenyl]imino]-, methyl ester	CH_2Cl_2	467(4.45)	33-1025-86
Propanedinitrile, [3-ethoxy-1-(4-morpholinyl)-3-phenyl-2-propenylidene]-	EtOH	300(4.20)	48-0314-86
$C_{18}H_{19}N_3O_3$			
Benzoic acid, 4-(5-butoxy-6-methyl-2H-benzotriazol-2-yl)-	toluene	341(4.29)	24-1627-86
	EtOH	336(4.35)	24-1627-86
	DMSO	339(4.61)	24-1627-86
$C_{18}H_{19}N_3O_4S$			
6H-Thieno[2,3-b]pyrrole-2,5-dicarboxylic acid, 3,4-diamino-6-phenyl-, diethyl ester	EtOH	262(4.39),292(4.54), 331(4.13)	48-0459-86
$C_{18}H_{19}N_3O_5$			
2H-Imidazo[4,5-c]pyridin-2-one, 1,3-dihydro-1-(phenylmethyl)-3-β-D-ribofuranosyl-	MeOH	274(3.48)	78-1511-86
2,5-Methano-5H-pyrimido[1,2-c][1,3,5]-oxadiazepin-9(1H)-one, 3-(benzoyloxymethyl)-1-ethyl-2,3-dihydro-11-hydroxy-	MeOH	200(4.23),224.5(4.51), 266s(3.49)	44-4417-86
$C_{18}H_{19}N_3O_7$			
1,3-Diazabicyclo[4.2.0]octane-2-carboxylic acid, 4-(2-methoxy-2-oxoethylidene)-8-oxo-7-[(phenoxyacetyl)-amino]-	EtOH	285(4.19)	39-1077-86C
$C_{18}H_{19}N_3S$			
3-Pyridinecarbonitrile, 2-[(cyanomethyl)thio]-6-adamantyl-	EtOH	218(4.41),261(4.12), 304(3.89)	70-0131-86
Thieno[2,3-b]pyridine-2-carbonitrile, 3-amino-6-adamantyl-	EtOH	202(4.38),228(4.16), 279(4.50),355(3.71)	70-0131-86
$C_{18}H_{19}N_3Se$			
3-Pyridinecarbonitrile, 2-[(cyanomethyl)seleno]-6-adamantyl-	EtOH	204(4.25),227(4.20), 284(4.26),365(3.48)	70-0376-86
$C_{18}H_{19}N_5O_3$			
Piperazine, 1-acetyl-4-[4-[(4-nitrophenyl)azo]phenyl]-	pH 1	546(4.89)	39-0123-86B
	EtOH	437(4.41)	39-0123-86B
$C_{18}H_{19}N_5O_6S$			
Riboflavin, 6-thiocyanato-	pH 7.0	441(4.13)	69-8103-86

Compound	Solvent	λ_{max}(log ε)	Ref.
$C_{18}H_{19}O_3$			
1H-Isobenzofurylium, 3-(3,4-dimethoxy-phenyl)-1,1-dimethyl-, perchlorate	HOAc	289(3.82),370s(4.19), 414(4.37)	104-0647-86
$C_{18}H_{20}$			
[2.2]Metacyclophane, 5,13-dimethyl-, anti	EtOH at 0°	273(2.8)	88-2907-86
syn	EtOH at 0°	258(3.28)	88-2907-86
[2.2]Paracyclophane, 1-ethyl-	MeCN	224(4.16)	24-1836-86
$C_{18}H_{20}ClNO_2S$			
1-Propanone, 1-(4-chlorophenyl)-3-[1,2-dihydro-1-(1-methylethyl)-2-thioxo-3-pyridinyl]-3-hydroxy-2-methyl-, (R*,R*)-(±)-	EtOH	252(4.26),283(3.99), 362(3.82)	4-0567-86
$C_{18}H_{20}Cl_2$			
Heptalene, 1,2-bis(chloromethyl)-5,6,8,10-tetramethyl-	hexane	211(4.34),247s(4.21), 263(4.33),305s(3.53), 364s(2.78)	88-1665-86
stereoisomer	hexane	246s(4.26),257(4.32), 300s(3.55)	88-1665-86
$C_{18}H_{20}Cl_2N_3$			
Methanaminium, N-[(4-chlorophenyl)-[[(4-chlorophenyl)(dimethylamino)-methylene]amino]methylene]-N-methyl-, iodide	EtOH	270(4.15),298(4.18)	39-0619-86C
$C_{18}H_{20}FN_5O_4$			
Acetamide, N-[7-[[1-(fluoromethyl)-2-(phenylmethoxy)ethoxy]methyl]-6,7-dihydro-6-oxo-1H-purin-2-yl]-, (±)-	MeOH	264(4.14)	87-1384-86
Acetamide, N-[9-[[1-(fluoromethyl)-2-(phenylmethoxy)ethoxy]methyl]-6,7-dihydro-6-oxo-1H-purin-2-yl]-	MeOH	258(4.20),279(4.06)	87-1384-86
$C_{18}H_{20}N_2O$			
Indolo[2,3-a]quinolizine, 2-acetyl-hexahydro-1-methyl-	MeOH	221(3.99),303(4.44)	44-2995-86
Methanone, phenyl[2-(1-piperidinyl)-3H-azepin-6-yl]-	EtOH	235.5(3.97),347(4.09)	18-2317-86
$C_{18}H_{20}N_2O_2$			
2-Propen-1-one, 2-(3-methyl-5-isoxa-zolyl)-1-phenyl-3-(1-piperidinyl)-	dioxan	235(4.31),322(4.39)	83-0242-86
$C_{18}H_{20}N_2O_2S$			
Benzoic acid, 2-[[4-(4-methyl-1-piper-azinyl)phenyl]thio]-	MeOH	270(3.89)	73-0937-86
$C_{18}H_{20}N_2O_3$			
4H-Furo[3,2-b]pyrrole-5-carboxylic acid, 6-[(dimethylamino)methyl]-2-phenyl-, ethyl ester	MeOH	336(3.68),351(3.63)	73-0106-86
3-Pyridinecarboxylic acid, 3-(1-acet-yl-1H-indol-3-yl)-1,2,3,6-tetrahy-dro-1-methyl-, methyl ester, hydro-chloride, (±)-	pH 7.2	240(1.58),296(0.62)	87-0125-86

Compound	Solvent	$\lambda_{max}(\log \epsilon)$	Ref.
$C_{18}H_{20}N_2O_4Se$			
3-Cyclobutene-1,2-dione, 3,3'-seleno-bis[4-(1-piperidinyl)-	CH_2Cl_2	262(4.3),320s(3.9), 337(4.0),410(3.5)	24-0182-86
$C_{18}H_{20}N_2O_6$			
L-Tyrosine, O-[5-(2-amino-2-carboxy-ethyl)-2-hydroxyphenyl]-	H_2O	274(3.47)	158-1685-86
	acid	274(3.47)	158-1685-86
	0.05M NaOH	297(3.59)	158-1685-86
$C_{18}H_{20}N_2O_6S$			
2(1H)-Pyrimidinone, 4-[(4-methylphen-yl)thio]-1-(5-O-acetyl-β-D-ribo-furanosyl)-	pH 7	275s(--),302(4.11)	1-0806-86
	pH 2	275s(--),302(--)	1-0806-86
	pH 13	273s(--),307(--)	1-0806-86
$C_{18}H_{20}N_2O_7$			
Butanedioic acid, mono[2-[[3,4-dihydro-2,4-dioxo-5-(phenylmethyl)-1(2H)-pyrimidinyl]methoxy]ethyl] ester	pH 1	266(4.09)	4-1651-86
	pH 11	263(3.78)	4-1651-86
$C_{18}H_{20}N_3O$			
Phenazinium, 5-(2-hydroxyethyl)-3-(4-morpholinyl)-, chloride	EtOH	238(4.53),299(4.54), 388(4.02),405(4.03), 555(4.22)	103-1342-86
$C_{18}H_{20}N_4$			
1H-Pyrazolo[3,4-b]quinolin-4-amine, N-(3,5-dimethylphenyl)-5,6,7,8-tetrahydro-	EtOH	217(4.73),242s(--), 270(3.78),317(4.42)	5-1728-86
1H-Pyrazolo[3,4-b]quinolin-4-amine, 5,6,7,8-tetrahydro-1-methyl-N-(3-methylphenyl)-	EtOH	216(5.87),243(5.51), 272(5.05),322(5.59)	11-0331-86
1H-Pyrazolo[3,4-b]quinolin-4-amine, 5,6,7,8-tetrahydro-1-methyl-N-(4-methylphenyl)-	EtOH	216(5.95),244(5.56), 272(5.16),321(5.68)	11-0331-86
$C_{18}H_{20}N_4O$			
Piperazine, 1-acetyl-4-[4-(phenylazo)-phenyl]-	EtOH	387(4.39)	39-0123-86B
	EtOH-HCl	535(4.77)	39-0123-86B
$C_{18}H_{20}N_4O_3$			
Hydrazinecarboxamide, 2-[[6-[2-(meth-ylamino)ethyl]-1,3-benzodioxol-5-yl]methylene]-N-phenyl-	MeOH	243(4.47),306(3.81), 364(3.92)	83-0910-86
$C_{18}H_{20}N_4O_4$			
1H-Imidazo[4,5-c]pyridin-4-amine, N-(phenylmethyl)-1-β-D-ribo-furanosyl-	MeOH	274(4.22)	87-0138-86
$C_{18}H_{20}N_5O_4$			
Methanaminium, N-[[[(dimethylamino)-(4-nitrophenyl)methylene]amino]-(4-nitrophenyl)methylene]-N-methyl-, iodide	EtOH	255(4.40),322(3.92)	39-0619-86C
$C_{18}H_{20}N_6O_2S$			
2,4(1H,3H)-Pteridinedione, 1,3-dimeth-yl-7-(methylthio)-6-[1-(phenylhy-drazono)propyl]-	MeOH	250(4.11),286(3.86), 308(3.84),364(4.05), 410s(3.63)	108-0081-86

Compound	Solvent	$\lambda_{max}(\log \epsilon)$	Ref.
$C_{18}H_{20}N_6O_4$			
1,2-Hydrazinedicarboxylic acid, 1-[3-(2-pyridinylamino)imidazo[1,2-a]pyridin-2-yl]-, diethyl ester	pH 1	213(4.48),283(4.15)	35-8002-86
	pH 11	215(4.43),233(4.42), 290(3.90)	35-8002-86
	MeOH	229(4.42),289(3.90)	35-8002-86
$C_{18}H_{20}N_6S$			
Cyclopropenylium, [[2-(cyanamino)-3-(cyanimino)-1-cyclopropen-1-yl]-thio]di-1-piperidinyl-, hydroxide, inner salt	CH_2Cl_2	325(3.89)	24-2104-86
$C_{18}H_{20}N_6Se$			
Cyclopropenylium, [[2-(cyanoamino)-3-(cyanimino)-1-cyclopropen-1-yl]-seleno]di-1-piperidinyl-, hydroxide, inner salt	n.s.g.	237(4.49),312(3.67)	89-0183-86
$C_{18}H_{20}O$			
2-Heptalenecarboxaldehyde, 1,5,6,8,10-pentamethyl-	dioxan	218(4.26),247s(4.19), 256s(4.23),271(4.28), 372s(2.92)	88-1669-86
Phenol, 2-[2-[4-(1,1-dimethylethyl)-phenyl]ethenyl]-, (E)-	EtOH	320(4.41),330(4.36)	150-0433-86S +150-3514-86M
$C_{18}H_{20}O_2$			
Estra-1,3,5,7,9-pentaene-3,17-diol (dihydroequilenin)	pH 1.57	324(3.4),337(3.4)	69-1186-86
	pH 12	355(3.3)	69-1186-86
$C_{18}H_{20}O_3$			
1H-Cyclopropa[3,4]cyclohept[1,2-e]in-dene-4,7-dione, 1a,5,6,8,9,9a-hexa-hydro-3-hydroxy-1,1,2-trimethyl-	EtOH	201(4.07),255(4.27), 327(3.53)	35-3040-86
	EtOH-base	206(4.66),267(4.32), 333(4.00)	35-3040-86
1H-Cyclopropa[3,4]cyclohept[1,2-e]in-dene-6,7-dione, 1a,4,5,8,9,9a-hexa-hydro-3-hydroxy-1,1,2-trimethyl-	EtOH	229(4.32),276(3.90), 317(3.59)	35-3040-86
	EtOH-base	213(4.65),258(4.25), 372(3.83)	35-3040-86
Gibba-1,3,4a(10a)-triene-10-carbox-ylic acid, 1,7-dimethyl-8-oxo- (gibberic acid)	MeOH	266f(2.51),274.5(2.44)	59-0405-86
(epigibberic acid)	MeOH	266f(2.51),275(2.45)	59-0405-86
Gibba-1,3,4a(10a)-triene-10-carbox-ylic acid, 7-hydroxy-1-methyl-8-methylene- (allogibberic acid)	MeOH	261(2.46),265(2.51), 274.5(2.38)	59-0405-86
(epiallogibberic acid)	MeOH	261(2.45),266.5(2.51), 270(2.45),273.5(2.36)	59-0405-86
1,4-Naphthalenedione, 2-hydroxy-3-(1-octenyl)-, (E)-	EtOH	270(4.34),284s(--), 308s(--),335s(--), 448(3.17)	39-0659-86C
$C_{18}H_{20}O_4$			
1,2-Azulenedicarboxylic acid, 3,4,6,8-tetramethyl-, dimethyl ester	dioxan	229s(4.18),253(4.44), 302s(4.64),309(4.69), 347s(3.77),358(3.82), 370s(3.60),581(2.88)	88-1673-86
1,3-Azulenedicarboxylic acid, 2,6-di-methyl-, diethyl ester	CH_2Cl_2	234.5(4.48),274.5(4.38), 313(4.77),348(3.88), 376(3.75),487(2.68)	24-2272-86
3-Cyclopentene-1,2-dione, 4,4'-(1,2-ethenediyl)bis[3,5,5-trimethyl-	$CHCl_3$	383(4.45),406(4.35), 520(2.26)	89-1018-86

Compound	Solvent	$\lambda_{max}(\log \epsilon)$	Ref.
$C_{18}H_{20}O_4S$			
15-Thiaestra-1,3,5(10),8-tetraen-17-one, 3-methoxy-, S,S-dioxide, (14β)-	EtOH	212(4.28),276(4.36)	104-0107-86
	EtOH-HCl	275(4.05),362(3.10)	104-0107-86
$C_{18}H_{20}O_5$			
Phenol, 2-[3-(1,3-benzodioxol-5-yl)-propyl]-3,5-dimethoxy-	EtOH	240(4.30),285(3.80)	102-2395-86
	EtOH-NaOEt	234(4.32),290(3.88)	102-2395-86
Phenol, 6-[2-(3,4-dimethoxyphenyl)-ethenyl]-2,3-dimethoxy-, (E)-	$CHCl_3$	251(4.23),299(4.24), 340(4.26)	142-1943-86
1-Propanone, 1-[4-(ethoxymethoxy)-2-hydroxyphenyl]-3-hydroxy-2-phenyl-	MeOH	212(4.33),221(4.28), 264(4.34),308(4.05)	39-0215-86C
$C_{18}H_{20}O_6$			
2,5-Furandione, dihydro-3-(1-methyl-ethylidene)-4-[1-(3,4,5-trimethoxy-phenyl)ethylidene]-, (E)-	toluene	341(3.81)	39-0315-86C
	CH_2Cl_2	340(4.01)	39-0315-86C
	CCl_4	340(3.90)	39-0315-86C
	C_6H_5Cl	342(3.86)	39-0315-86C
(Z)-	toluene	338(4.01)	39-0315-86C
Naphtho[2,3-c]furan-1,3-dione, 3a,4-dihydro-5,6,7-trimethoxy-4,4,9-trimethyl-	toluene	322(4.12)	39-0315-86C
Naphtho[2,3-c]furan-1,3-dione, 4,9-dihydro-5,6,7-trimethoxy-4,4,9-trimethyl-, (E)-	hexane	248(3.58),268(3.42), 310(3.06)	39-0315-86C
1-Propanone, 3-hydroxy-1-(2-hydroxy-4,6-dimethoxyphenyl)-2-(4-methoxy-phenyl)-	MeOH	220(4.32),228(4.39), 292(4.29)	39-0215-86C
$C_{18}H_{20}O_7$			
4H-1-Benzopyran-4-one, 5,6,7,8-tetra-hydro-5,6,7,8-tetrahydroxy-2-[2-(4-methoxyphenyl)ethyl]-, [5S-(5α,6β-7α,8β)]-	EtOH	202(4.28),223(4.26), 253(4.17)	94-3033-86
$C_{18}H_{20}O_{10}$			
2H-1-Benzopyran-2-one, 7-[(2-O-acetyl-β-D-glucopyranosyl)oxy]-6-methoxy-(2'-O-acetylscopolin)	MeOH	204(4.4),224(4.1), 287(3.6),342(3.5)	94-4012-86
2H-1-Benzopyran-2-one, 7-[(6-O-acetyl-β-D-glucopyranosyl)oxy]-6-methoxy-(6'-O-acetylscopolin)	MeOH	205(4.2),227(3.8), 287(3.2),338(3.5)	94-4012-86
$C_{18}H_{21}ClN_2O_2S$			
2H-[1]Benzothiepino[5,4-b]pyran-2-one, 3-chloro-4-[[2-(dimethylamino)ethyl]-methylamino]-5,6-dihydro-	EtOH	234(4.02),255(4.06), 266s(3.99),325(4.07)	4-0449-86
$C_{18}H_{21}NO$			
2-Furanmethanamine, N-[2-(phenylmeth-ylene)cyclohexyl]-, (E)-, (Z)-2-butenedioate	EtOH	241(4.21)	44-5002-86
$C_{18}H_{21}NO_2$			
Benzenamine, 4-[2-[2-(methoxymeth-oxy)phenyl]ethenyl]-N,N-dimethyl-, (E)-	EtOH	349(4.52)	150-3514-86M
Bicyclo[4.4.1]undeca-1,3,5,7,9-penta-ene-2-carboxamide, 10-acetyl-N,N-diethyl-, (±)-	MeOH	260(4.47),330(3.88)	33-1263-86
$C_{18}H_{21}NO_2S$			
1-Propanone, 3-[1,2-dihydro-1-(1-	EtOH	241(4.13),284(4.03),	4-0567-86

Compound	Solvent	$\lambda_{max}(\log \epsilon)$	Ref.
methylethyl)-2-thioxo-3-pyridinyl]-3-hydroxy-2-methyl-1-phenyl- (cont.)		367(3.85)	4-0567-86
1-Propanone, 3-[1-(1,1-dimethylethyl)-2-thioxo-3-pyridinyl]-3-hydroxy-1-phenyl-	EtOH	241(4.04),289(3.90), 371(3.76)	4-0567-86
$C_{18}H_{21}NO_2S_2$			
Benzenesulfonamide, N-(3-butenyl)-4-methyl-N-[(phenylthio)methyl]-	EtOH	228(4.14)	24-0813-86
$C_{18}H_{21}NO_3$			
Bicyclo[4.4.1]undeca-1,3,5,7,9-pentaene-2-carboxylic acid, 10-[(diethylamino)carbonyl]-, methyl ester, (±)-	MeOH	265(4.46),330(3.87)	33-1263-86
$C_{18}H_{21}NO_3S$			
Sulfoximine, N-(2-ethoxy-1-methyl-2-oxoethyl)-S-(4-methylphenyl)-S-phenyl-, (-)-	EtOH	206(4.26),237(4.18)	12-1833-86
$C_{18}H_{21}NO_3S_2$			
10H-Phenothiazine-10-hexanesulfonic acid, sodium salt	80% MeCN	253(4.52),306(3.61)	46-2469-86
$C_{18}H_{21}NO_4$			
Cyclopent[b]indol-3a(1H)-acetic acid, 2,3,4,8b-tetrahydro-2,2,4,8-tetramethyl-1,3-dioxo-, methyl ester	n.s.g.	217(4.08),276(3.51), 344(3.00)	150-2782-86
3,5-Pyridinedicarboxylic acid, 1,4-dihydro-2,6-dimethyl-4-(phenylmethyl)-, dimethyl ester	MeOH	233(--),276(--), 348(3.99)	87-1596-86
$C_{18}H_{21}NO_4S$			
Sulfoximine, N-(2-ethoxy-2-oxoethyl)-S-(2-methoxyphenyl)-S-(4-methylphenyl)-, (S)-	EtOH	245(4.01),289(3.65)	12-1833-86
$C_{18}H_{21}NO_5$			
1H-Benz[f]isoindole-1,3(2H)-dione, 3a,4-dihydro-5,6,7-trimethoxy-4,4,9-trimethyl-	toluene	314(4.14)	39-0315-86C
1H-Carbazole-4a,9a(2H,9H)-diacetic acid, 3,4-dihydro-α^{4a}-oxo-, dimethyl ester	EtOH	237(3.89),298(3.27)	44-3125-86
Pseudolycorine, 1-O-acetyl-	EtOH	212(4.26),286(3.64)	102-1453-86
	EtOH-NaOH	254(3.84),304(3.74)	102-1453-86
Pseudolycorine, 2-O-acetyl-	EtOH	210(4.26),286(3.62)	102-1453-86
	EtOH-NaOH	254(3.84),298(3.73)	102-1453-86
2,5-Pyrrolidinedione, 3-(1-methylethylidene)-4-[1-(3,4,5-trimethoxyphenyl)ethylidene]-, (E)-	toluene	328(3.93)	39-0315-86C
$C_{18}H_{21}NO_5S$			
Sulfoximine, N-(carboxymethyl)-S-[2-(2-methoxyethoxy)phenyl]-S-(4-methylphenyl)-, (S)-	EtOH	223(4.23),236(4.16), 288(3.65)	12-1833-86
$C_{18}H_{21}NO_6$			
Crinafoline	MeOH	240(3.29),262s(3.08), 292(3.61),318s(2.98)	150-0312-86S

Compound	Solvent	$\lambda_{max}(\log \epsilon)$	Ref.
$C_{18}H_{21}NS$			
Benzenemethanamine, N-3-butenyl-N-[(phenylthio)methyl]-	EtOH	240(3.84)	24-0813-86
$C_{18}H_{21}N_2O_3$			
Pyrylium, 2-amino-3-(ethoxycarbonyl)-6-phenyl-4-(1-pyrrolidinyl)-, perchlorate	MeCN	220s(4.09),255.5(4.56), 310(4.34)	48-0314-86
$C_{18}H_{21}N_2O_4$			
Pyrylium, 2-amino-3-(ethoxycarbonyl)-4-(4-morpholinyl)-6-phenyl-, perchlorate	MeCN	225(4.10),257.5(4.50), 320(4.40)	48-0314-86
$C_{18}H_{21}N_3O$			
Piperidine, 1-[4-[(4-methoxyphenyl)-azo]phenyl]-	EtOH	398(4.45)	39-0123-86B
	EtOH-HCl	558(3.08)	39-0123-86B
$C_{18}H_{21}N_3OS$			
Acetamide, 2-[(3-cyano-6-tricyclo-[3.3.1.1^{3,7}]dec-1-yl-2-pyridinyl)thio]-	EtOH	222(4.40),265(4.17), 310(3.83)	70-0131-86
Hydrazinecarbothioamide, 2-[1,3-dihydro-3,3-dimethyl-1-(2-methylphenyl)-1-isobenzofuranyl]-	EtOH	245(4.21),269(3.39), 304(3.40)	104-0647-86
Hydrazinecarbothioamide, 2-[1,3-dihydro-3,3-dimethyl-1-(3-methylphenyl)-1-isobenzofuranyl]-	EtOH	245(4.12),269(3.33), 272s(3.18),311(3.15)	104-0647-86
Thieno[2,3-b]pyridine-2-carboxamide, 3-amino-6-tricyclo[3.3.1.1^{3,7}]dec-1-yl-	EtOH	204(4.32),233(4.03), 285(4.50),365(3.70)	70-0131-86
$C_{18}H_{21}N_3OSe$			
Acetamide, 2-[(3-cyano-6-tricyclo-[3.3.1.1^{3,7}]dec-1-yl-2-pyridinyl)seleno]-	EtOH	227(3.94),275(3.62), 313(3.27)	70-0376-86
Selenolo[2,3-b]pyridine-2-carboxamide, 3-amino-6-tricyclo[3.3.1.1^{3,7}]dec-1-yl-	EtOH	207(4.51),234(4.27), 291(4.64),375(3.90)	70-0376-86
$C_{18}H_{21}N_3O_2$			
1-Propanone, 2-(acetoxyamino)-2-methyl-1-phenyl-, 1-(phenylhydrazone)	EtOH	271(4.22),295(4.03)	103-0287-86
$C_{18}H_{21}N_3O_2S$			
Hydrazinecarbothioamide, 2-[1,3-dihydro-1-(2-methoxyphenyl)-3,3-dimethyl-1-isobenzofuranyl]-	EtOH	244(4.24),269(3.63), 280(3.52),306s(2.74)	104-0647-86
Hydrazinecarbothioamide, 2-[1,3-dihydro-1-(3-methoxyphenyl)-3,3-dimethyl-1-isobenzofuranyl]-	EtOH	245(4.28),269(3.60), 275(3.56),282(3.54), 311(3.31)	104-0647-86
Hydrazinecarbothioamide, 2-[1,3-dihydro-1-(4-methoxyphenyl)-3,3-dimethyl-1-isobenzofuranyl]-	EtOH	238(4.36),269(3.53), 274s(3.45),281(3.44), 319(3.69)	104-0647-86
$C_{18}H_{21}N_4O_6$			
Pyridinium, 4-[2-[[(1,1-dimethylethoxy)carbonyl]amino]ethyl]-1-(2,4-dinitrophenyl)-, chloride	MeOH	230(4.36),260s(4.16), 295s(3.56)	33-1588-86

Compound	Solvent	λ_{max}(log ε)	Ref.
$C_{18}H_{21}O_3P$			
1,1'(3H,3'H)-Spirobi[2,1-benzoxaphosphole], 1-hydroxy-3,3,3',3'-tetramethyl-	pH 4	267(3.2),272(3.2)	35-2416-86
	pH 13	267(3.3),274(3.3)	35-2416-86
$C_{18}H_{22}$			
Heptalene, 1,2,5,6,8,10-hexamethyl-	hexane	253(4.37),300s(3.54)	88-1665-86
stereoisomer	hexane	252(4.37),297s(3.49)	88-1665-86
5,9-Tetradecadiene-3,7,11-triyne, 2,2,13,13-tetramethyl-	hexane	312(4.65),333(4.70)	35-4685-86
$C_{18}H_{22}N_2$			
Pyrazine, 2,5-bis(1-methylethyl)-3-(2-phenylethenyl)-, (E)-	EtOH	231(4.03),236(3.99), 276(4.23),333(4.32)	4-1481-86
$C_{18}H_{22}N_2O$			
Pyrazine, 2,5-bis(1-methylethyl)-3-(2-phenylethenyl)-, 1-oxide, (E)-	EtOH	231(4.13),237s(4.10), 277(4.47),302s(4.16), 347(4.17),359s(4.10)	4-1481-86
$C_{18}H_{22}N_2OS$			
10H-Phenothiazine-10-propanamine, 2-methoxy-N,N-dimethyl-, (Z)-2-butenedioate	MeOH	253(4.436),307(3.692)	106-0571-86
$C_{18}H_{22}N_2O_2$			
Bicyclo[4.4.1]undeca-1,3,5,7,9-pentaene-2-carboxamide, 10-(acetylamino)-N,N-diethyl-, (±)-	MeOH	265(3.64),450(2.82)	33-1263-86
Brafovedine	EtOH	220(4.02),230s(3.93), 300(3.65),313s(3.60)	100-0428-86
2-Furancarboxaldehyde, 5-(dimethylamino)-4-(phenyl-1-pyrrolidinylmethyl)-	MeOH	370(3.26)	73-0573-86
Isobrafovedine	EtOH	227(4.18),275s(3.50), 285(3.52),292(3.51)	100-0428-86
$C_{18}H_{22}N_2O_3$			
2-Furancarboxaldehyde, 5-(dimethylamino)-4-[(4-morpholinyl)phenylmethyl]-	MeOH	368(3.32)	73-0573-86
2-Furancarboxylic acid, 5-methyl-4-[(4-phenyl-1-piperazinyl)methyl]-, methyl ester	MeOH	255(3.28)	73-2186-86
$C_{18}H_{22}N_2O_3S$			
2,5-Piperazinedione, 1,4-dimethyl-3-[[4-[(3-methyl-2-butenyl)oxy]phenyl]methyl]-6-thioxo-, (-)-(silvathione)	MeOH	228(4.43),274(4.30), 284s(3.93),300s(3.84)	77-1495-86
$C_{18}H_{22}N_2O_6S$			
2-Indolizinecarboxylic acid, 3-[(2-methyl-1-[(2-propenyloxy)carbonyl]-2-sulfinopropyl]amino]-, 2-methyl ester, monosodium salt	MeOH	243(4.41)	88-3449-86
$C_{18}H_{22}N_2S$			
10H-Phenothiazine-10-propanamine, N,N,β-trimethyl-, tartrate (2:1)	MeOH	255(4.643),306(3.753)	106-0571-86

Compound	Solvent	$\lambda_{max}(\log \epsilon)$	Ref.
$C_{18}H_{22}N_3$			
Acridinium, 3,6-bis(dimethylamino)-10-methyl-, chloride	pH 6.9	495(4.7)(anom.)	18-3393-86
Methanaminium, N-[[[(dimethylamino)-phenylmethylene]amino]phenylmethylene]-N-methyl-, iodide	EtOH	260(4.00),295(4.19)	39-0619-86C
$C_{18}H_{22}N_4$			
Acetonitrile, 2,2'-(3,4-di-1-piperidinyl-3-cyclobutene-1,2-diylidene)bis-, (E,Z)-	CH_2Cl_2	268(4.3),364(4.4), 381(4.4)	118-0216-86
(Z,Z)-	CH_2Cl_2	268(4.2),362(4.2), 380(4.3)	118-0216-86
$C_{18}H_{22}N_4O_2$			
1-Propanol, 3,3'-(benzo[g]phthalazine-1,4-diyldiimino)bis-	EtOH	236(4.80),254(4.57)	111-0143-86
$C_{18}H_{22}N_4O_2S$			
3-Pyridinecarbonitrile, 5-acetyl-6-methyl-4-(3-pyridinyl)-2-mercapto-, morpholine salt	EtOH	235(4.03),287(4.08), 385s(2.90)	104-1762-86
$C_{18}H_{22}N_4O_3$			
Benzo[g]pteridine-2,4(3H,10H)-dione, 3,10-dibutyl-, 5-oxide	MeCN	212(3.96),269(4.43), 339(3.90),451(3.78), 475(3.70)	35-6039-86
$C_{18}H_{22}N_4O_3S$			
Pyridinium, N-[[2m3m4-triazatricyclo-[3.2.1.0^{2,4}]oct-3-yl]methyl]-, 4-methylbenzenesulfonate	EtOH	219(4.14),260(3.65), 266s(3.57)	33-2087-86
$C_{18}H_{22}N_4O_7S_2$			
9H-Purine, 2,6-bis(methylthio)-9-(2,3,5-tri-O-acetyl-β-D-ribofuranosyl)-	EtOH	226(3.90),260(4.18), 306(3.86)	118-0450-86
$C_{18}H_{22}O_2$			
7H-Cyclopropa[3,4]cyclohept[1,2-e]inden-7-one, 1,1a,4,5,6,8,9,9a-octahydro-3-hydroxy-1,1,2-trimethyl-, (1aR-cis)-	EtOH	204(4.31),225(4.14), 279(3.85)	35-3040-86
	EtOH-base	215(4.87),261(3.98), 374(4.15)	35-3040-86
Estra-3,5,7-trien-17-one, 3-hydroxy-	MeOH	320(4.11)	35-3145-86
1,2-Heptalenedimethanol, 5,6,8,10-tetramethyl-	dioxan	257(4.31),300s(3.53)	88-1665-86
stereoisomer	dioxan	248s(4.33),254(4.37), 300s(3.53)	88-1665-86
$C_{18}H_{22}O_3$			
11-Oxaandrosta-1,4-diene-3,17-dione	EtOH	242(4.1)	13-0381-86A
$C_{18}H_{22}O_4$			
Naphtho[1,2-b]furan-8-ol, 6,7,8,9-tetrahydro-3-(methoxymethyl)-5,9-dimethyl-, acetate, (8R-trans)-	$CHCl_3$	248(3.98),278(3.02), 288(3.02)	78-4493-86
$C_{18}H_{22}O_5$			
1H-2-Benzoxacyclotetradecin-1,7(8H)-dione, 3,4,5,6,9,10-hexahydro-14,16-dihydroxy-3-methyl-	MeOH	234.0(4.40),273.0(4.04), 314.5(3.69)	33-0734-86

Compound	Solvent	$\lambda_{max}(\log \epsilon)$	Ref.
Bicyclo[6.2.2]dodeca-8,10,11-triene-9,10-dicarboxylic acid, 4-oxo-, diethyl ester	EtOH	228(4.50),262s(3.93), 324(3.49)	24-0297-86
Phenol, 2-[3-(4-hydroxy-3-methoxyphenyl)propyl]-3,5-dimethoxy-	EtOH	234(4.31),284(3.71)	102-2395-86
	EtOH-NaOEt	240(4.28),292(3.74)	102-2395-86
$C_{18}H_{22}O_6$			
10-Oxabicyclo[6.3.2]trideca-8,11,12-triene-12,13-dicarboxylic acid, 4-oxo-, diethyl ester	EtOH	203s(4.12),277(3.56)	24-0297-86
$C_{18}H_{22}O_{10}$			
Bicyclo[2.2.2]oct-7-ene-2,3,5,6,7-pentacarboxylic acid, pentamethyl ester	MeCN	215(3.96)	33-1310-86
$C_{18}H_{23}Br$			
Azulene, 6-bromo-2-octyl-	CH_2Cl_2	235.5(4.09),283.5(4.91), 293.5(5.01),312(3.78), 339.5(3.66),354.4(3.79), 370.5(3.59),565.5(2.43), 605.5(2.35),668.5(1.91)	24-2272-86
$C_{18}H_{23}NO_2$			
7H-Benzo[3,4]cyclobuta[1,2]cyclohepten-7-one, 1,2,3,4,4a,9b-hexahydro-1-methyl-4a-(4-morpholinyl)-	MeOH	232(4.18),325(3.98)	88-3005-86
7H-Benzo[3,4]cyclobuta[1,2]cyclohepten-7-one, 1,2,3,4,4a,9b-hexahydro-1-methyl-9b-(4-morpholinyl)-	MeOH	232(4.27),326(4.07)	88-3005-86
Cyclobuta[1,2:3,4]dicyclohepten-3(5bH)-one, 6,7,8,9,10,10a-hexahydro-5b-(4-morpholinyl)-	MeOH	233(4.21),325(4.04)	88-3005-86
9H-Indeno[2,1-b]pyridine-3-carboxylic acid, 1,2,3,9a-tetrahydro-1,9,9-trimethyl-, ethyl ester	$EtOH-HClO_4$	251(3.92),286(3.49), 295(3.41)	94-2786-86
3-Pyridinecarboxylic acid, 1,4-dihydro-4,4,5-trimethyl-1-(1-phenylethyl)-, methyl ester	ether	278(3.54),348(3.65)	118-0190-86
$C_{18}H_{23}NO_3$			
1,6-Octadiene-3,5-dione, 1-(4-methoxyphenyl)-7-[(1-methylethyl)amino]-	n.s.g.	330(4.28),412(4.56)	4-1721-86
$C_{18}H_{23}NO_4$			
3H,6H-5,10b-Ethanophenanthridin-6-ol, 4,4a-dihydro-3,8,9-trimethoxy-	MeOH	232(3.86),283(3.42)	142-2089-86
6H-1,3-Oxazin-6-one, 4,5-bis(1,1-dimethylethoxy)-2-phenyl-	C_6H_{12}	202(4.41),240(4.22), 338(3.97)	39-0961-86B
Piperidine, 1-(3,5-dimethoxy-1,4-dioxo-5-phenyl-2-pentenyl)-, (Z)-	MeOH and MeOH-base	208(4.20),270(2.03)	163-3053-86
$C_{18}H_{23}N_3O_4$			
L-Lysine, N^2-acetyl-N^6-(1H-indol-3-ylacetyl)-	MeOH	220(4.33),272(3.58), 279(3.61),288(3.54)	102-0125-86
$C_{18}H_{23}N_3O_{11}$			
2,4,6(1H,3H,5H)-Pyrimidinetrione, 5-[3,4,6-tri-O-acetyl-2-(acetylamino)-2-deoxy-β-D-glucopyranosyl]-	MeOH	253(4.28)	136-0053-86M

Compound	Solvent	$\lambda_{max}(\log \epsilon)$	Ref.
$C_{18}H_{23}N_4O_3$			
1,3,5-Oxadiazin-1-ium, 2-(4-methyl-phenyl)-4,6-di-4-morpholinyl-, perchlorate	MeCN	244(4.54),297(4.59)	97-0094-86
$C_{18}H_{23}N_4S_4$			
1H-Pyrazolium, 1-[4,5-bis(methylthio)-1,3-dithiol-2-ylidene]-3,5-bis(di-methylamino)-4-phenyl-, perchlorate	MeCN	305s(3.84),466(4.37)	88-0159-86
$C_{18}H_{24}F_2S_4$			
Ferrocene, 1,1'-[1,2-ethanediyl-bis(thio-3,1-propanediylthio)]-	MeCN	443(2.28)	18-1515-86
copper perchlorate salt	MeCN	430(2.09)	18-1515-86
$C_{18}H_{24}Ge$			
Germane, diethyl-bis(phenylmethyl)-	n.s.g.	267(2.92)	18-3169-86
$C_{18}H_{24}N_2$			
Cyclododeca[b]quinoxaline, 6,7,8,9-10,11,12,13,14,15-decahydro-	n.s.g.	241(4.48),321(3.98)	44-3257-86
Pyrazine, 2,5-bis(2-methylpropyl)-3-phenyl-	EtOH	222s(3.92),286(4.00), 305s(3.65)	142-0785-86
$C_{18}H_{24}N_2O$			
Pyrazine, 2,5-bis(2-methylpropyl)-3-phenyl-, 1-oxide	EtOH	228(4.38),250s(4.19), 266s(4.14),312(3.80)	142-0785-86
Pyrazine, 2,5-bis(2-methylpropyl)-3-phenyl-, 4-oxide	EtOH	229.5(4.32),270.5(4.00), 303s(3.64)	142-0785-86
$C_{18}H_{24}N_2O_2$			
2-Furancarboxaldehyde, 4-[(diethyl-amino)phenylmethyl]-5-(dimethyl-amino)-	MeOH	370(3.25)	73-0573-86
7,11a-(Iminoethano)-11aH-dibenz[c,e]-azepin-5(6H)-one, 7,7a,8,9,10,11-hexahydro-2-methoxy-14-methyl-, [7S-(7α,7aα,11aα)]-	EtOH	212(4.46),258(4.08)	4-1897-86
Morphinan-10-one, 3-methoxy-17-meth-yl-, oxime	EtOH	268(4.18),294s(3.76), 305s(3.54)	4-1897-86
3a(1H)-Pentaleneacetic acid, hexahy-dro-6a-(phenylazo)-, ethyl ester	EtOH	215(4.68),265(4.60), 413(2.24)	78-5081-86
$C_{18}H_{24}N_2S_2$			
Propanethioamide, 2-(4,4-dimethyl-2-phenyl-5(4H)-thiazolylidene)-N,N-diethyl-, (E)-	EtOH	242(4.38),280(4.09), 372s(2.68)	33-0174-86
(Z)-	EtOH	240(4.37),278(4.09), 384(2.65)	33-0174-86
2-Propanethione, 1-(diethylamino)-1-(4,4-dimethyl-2-phenyl-5(4H)-thiazolylidene)-	EtOH	230s(3.81),316s(3.97), 357(4.18),450(4.01)	33-0174-86
$C_{18}H_{24}N_3O_9P$			
1H-Pyrazole-3-carboxamide, 5-[5-O-(dimethoxyphosphinyl)-β-D-ribo-furanosyl]-4-(phenylmethoxy)-	pH 1	246(3.80)	87-0268-86
	pH 7	246(3.87)	87-0268-86
	pH 11	241(4.19)	87-0268-86
1H-Pyrazole-3-carboxamide, 5-[5-O-(ethoxyhydroxyphosphinyl)-β-D-ribofuranosyl]-4-(phenylmethoxy)-	pH 1	246(3.81)	87-0268-86
	pH 7	246(3.81)	87-0268-86
	pH 11	243(3.89)	87-0268-86

Compound	Solvent	λ_{max}(log ε)	Ref.
$C_{18}H_{24}N_4O_8$			
Methanimidamide, N'-[4-formyl-1-(2,3,5-tri-O-acetyl-β-D-ribofuranosyl)-1H-pyrazol-5-yl]-N,N-dimethyl-	pH 1	217(4.45)	44-1050-86
	pH 11	232(4.79)	44-1050-86
	MeOH	234(4.57),322(4.03)	44-1050-86
$C_{18}H_{24}N_4O_{10}$			
1H-Imidazole-4-carboxamide, 5-amino-1-(2,3,4,6-tetra-O-acetyl-D-glucopyranosyl)-	MeOH	265(4.01)	142-3451-86
$C_{18}H_{24}N_6O_6$			
1H-Imidazole-4-carboxamide, 5-[[imino[[(4-methoxyphenyl)methyl]amino]-methyl]amino]-1-β-D-ribofuranosyl-	pH 1	264s(3.40)	44-1277-86
	pH 11	263(3.83)	44-1277-86
	MeOH	264(3.86)	44-1277-86
$C_{18}H_{24}O$			
2-Cyclopenten-1-one, 4-[1-methyl-3-(2,6,6-trimethyl-1-cyclohexen-1-yl)-2-propenylidene]-, cis	EtOH	346(4.16)	39-0905-86C
trans	EtOH	350(4.24)	39-0905-86C
Phenanthrene, 1,2,3,4,4a,10a-hexahydro-7-methoxy-1,1,4a-trimethyl-	MeOH	200(4.15),275(3.96), 315(3.20)	78-6615-86
$C_{18}H_{24}O_2$			
2-Cyclopenten-1-one, 3,3'-(1,2-ethenediyl)bis[2,4,4-trimethyl-, (E)-	$CHCl_3$	336(4.35),355s(4.18)	89-1018-86
$C_{18}H_{24}O_3$			
Benzoic acid, 4-methoxy-3,5-bis(3-methyl-2-butenyl)-	EtOH	208(3.84),242(4.42)	102-1427-86
	MeOH-KOH	208(3.84),242(4.42)	102-1427-86
11-Oxa-5α-androst-1-ene-3,17-dione	EtOH	227(3.9)	13-0381-86A
11-Oxaandrost-4-ene-3,17-dione	EtOH	238(4.14)	13-0381-86A
$C_{18}H_{24}O_5$			
1H-2-Benzoxacyclotetradecin-1,7(8H)-dione, 3,4,5,6,9,10,11,12-octahydro-14,16-dihydroxy-3-methyl-	MeOH	217.5(4.38),264(4.11), 302.5(3.72)	33-0734-86
Bicyclo[6.2.2]dodeca-8,10,11-triene-9,10-dicarboxylic acid, 4-hydroxy-, diethyl ester	EtOH	233(4.09),262s(3.46), 328(3.07)	24-0297-86
stereoisomer	EtOH	233(4.34),263s(3.77), 328(3.34)	24-0297-86
$C_{18}H_{24}O_6$			
10-Oxabicyclo[6.3.2]trideca-1(11),8,12-triene-12,13-dicarboxylic acid, 4-hydroxy-, diethyl ester	EtOH	210s(3.90),275(3.28)	24-0297-86
stereoisomer	EtOH	210s(4.05),275(3.46)	24-0297-86
13-Oxatricyclo[8.2.1.0^{2,9}]trideca-2(9),11-diene-11,12-dicarboxylic acid, 5-hydroxy-, diethyl ester	EtOH	233s(3.86),280(2.92), 315(2.75)	24-0297-86
$C_{18}H_{24}O_{10}$			
Bicyclo[2.2.2]octane-2,3,5,6,7-pentacarboxylic acid, pentamethyl ester	MeCN	212(3.00)	33-1310-86
$C_{18}H_{25}ClO_{11}$			
β-D-Mannopyranoside, 4-chlorophenyl 2-O-β-D-mannopyranosyl-	H_2O	275.4(2.91)	94-5140-86

Compound	Solvent	λ_{max}(log ε)	Ref.
β-D-Mannopyranoside, 4-chlorophenyl 6-O-β-D-mannopyranosyl-	H_2O	275.6(2.91)	94-5140-86
$C_{18}H_{25}F_3N_2O_3$			
Propanoic acid, 2-diazo-3,3,3-trifluoro-, 9-(3,3-dimethyloxiranyl)-3,7-dimethyl-2,6-nonadienyl ester, (E,E)-	hexane	226(4.00)	20-0895-86
$C_{18}H_{25}NO_{12}$			
β-D-Fructofuranose, 1-deoxy-1-[[3-methoxy-2-(methoxycarbonyl)-3-oxo-1-propenyl]amino]-, 3,4,6-triacetate	$CHCl_3$	278(4.31)	136-0329-86D
β-D-Fructopyranose, 1-deoxy-1-[[3-methoxy-2-(methoxycarbonyl)-3-oxo-1-propenyl]amino]-, 3,4,5-triacetate	$CHCl_3$	279(4.38)	136-0329-86D
$C_{18}H_{25}NO_{13}$			
β-D-Mannopyranoside, 4-nitrophenyl 2-O-β-D-mannopyranosyl-	H_2O	303(3.98)	94-5140-86
β-D-Mannopyranoside, 4-nitrophenyl 6-O-β-D-mannopyranosyl-	H_2O	303(4.00)	94-5140-86
$C_{18}H_{25}N_5O_2$			
Pyrazinecarbonitrile, 3-amino-3,4-dihydro-4-(2-methoxyethyl)-3-[(2-methoxyethyl)amino]-6-methyl-5-phenyl-	EtOH pH 1	328(4.00) 379(3.94)	33-1025-86 33-1025-86
$C_{18}H_{26}Cl_2N_6O_8Pt$			
Platinum, bis(3'-amino-2',3'-dideoxyuridine-$N^{3'}$)dichloro-	H_2O	263(4.22)	87-0681-86
Platinum, bis(5'-amino-2',5'-dideoxyuridine-$N^{5'}$)dichloro-	H_2O	261(4.25)	87-0681-86
$C_{18}H_{26}N_2O_3$			
1H-Pyrrole-3-propanoic acid, 5-[(3-ethyl-2,5-dihydro-4-methyl-5-oxo-1H-pyrrol-2-yl)methyl]-2,4-dimethyl-, methyl ester	MeOH	210(4.28)	5-1241-86
$C_{18}H_{26}N_4O_4$			
Pyrazino[2,3-g]quinoxaline-5,10-diol, 1,2,3,4,6,7,8,9-octahydro-1,4,6,9-tetramethyl-, diacetate, dication bis(tetrafluoroborate)	MeCN	543(--)	88-0691-86
$C_{18}H_{26}N_6O_4$			
β-D-Ribofuranuronamide, 1-[6-(cyclohexylamino)-9H-purin-9-yl]-1-deoxy-N-ethyl-	n.s.g.	270(4.27)	87-1683-86
β-D-Ribofuranuronamide, N-cyclopropyl-1-deoxy-1-[6-(1-ethylpropyl)amino]-9H-purin-9-yl]-	n.s.g.	269(4.30)	87-1683-86
β-D-Ribofuranuronamide, 1-deoxy-1-[6-[(1-ethylpropyl)amino]-9H-purin-9-yl]-N-2-propenyl-	n.s.g.	269(4.28)	87-1683-86
$C_{18}H_{26}O_3$			
2-Naphthalenol, 1,2,3,4-tetrahydro-6-methoxy-4,7-dimethyl-1-(1-methylethyl)-, acetate, [1R-(1α,2β,4β)]-	EtOH	226(3.55),279(3.12), 286(3.11)	12-0103-86

Compound	Solvent	λmax(log ε)	Ref.
11-Oxaandrost-4-en-3-one, 17β-hydroxy-	EtOH	238(4.18)	13-0047-86B
$C_{18}H_{26}O_4$			
2-Oxaandrost-4-en-3-one, 1α,17β-dihydroxy-	MeOH	226(4.14)	78-5693-86
$C_{18}H_{26}O_5$			
1H-2-Benzoxacyclotetradecin-1-one, 3,4,5,6,7,8,9,10,11,12-decahydro-7,14,16-trihydroxy-3-methyl-, [3S-(3R*,7R*)]-	MeOH	218.0(4.47),263.0(4.10), 302.5(3.76)	33-0734-86
[3S-(3R*,7S*)]-	MeOH	214.5(4.32),262(3.90), 237.0(3.78)[sic]	33-0734-86
$C_{18}H_{26}O_{11}$			
2H-Pyran-4-acetic acid, 3-ethylidene-2-(β-D-glucopyranosyloxy)-3,4-dihydro-5-(methoxycarbonyl)-, methyl ester, [2S-(2α,3E,4β)]- (oleoside)	EtOH	233(3.89)	102-0865-86
$C_{18}H_{27}NO_6$			
3,4,4(1H)-Pyridinetricarboxylic acid, 1-(1,1-dimethylethyl)-5-(1-methylethyl)-, trimethyl ester	MeCN	247(3.60),331(3.84)	118-0190-86
$C_{18}H_{27}N_3O_2$			
Guanidine, N"-[5-(4,4-dimethyl-2,6-dioxocyclohexylidene)-1,3-pentadienyl]-N,N,N',N'-tetramethyl-	benzene	512(4.87),544(4.84)	24-3276-86
	EtOH	541(4.90)	24-3276-86
	acetone	512s(--),546(5.01)	24-3276-86
	$CHCl_3$	511s(--),552(4.94)	24-3276-86
$C_{18}H_{27}N_4O$			
1,3,5-Oxadiazin-1-ium, 2,4-bis(diethylamino)-6-(4-methylphenyl)-, perchlorate	MeCN	223(4.54),295(4.67)	97-0094-86
$C_{18}H_{27}N_4O_2$			
1,3,5-Oxadiazin-1-ium, 2,4-bis(diethylamino)-6-(4-methoxyphenyl)-, perchlorate	MeCN	230(4.53),317(4.57)	97-0094-86
$C_{18}H_{27}N_7O_4$			
L-Leucine, N-[N-(N-1H-purin-6-yl)-β-alanyl-L-alanyl-, methyl ester	pH 2.95	272(4.18)	63-0757-86
	pH 7.0	267(4.20)	63-0757-86
	pH 10.5	272(4.17)	63-0757-86
$C_{18}H_{27}O_2$			
2H-1-Benzopyran-6-yloxy, 7-(1,1-dimethylethyl)-3,4-dihydro-2,2-dimethyl-5-(1-methylethyl)-	EtOH	341(3.60),400(3.40), 419(3.60)	18-2899-86
$C_{18}H_{28}$			
Cyclopentene, 3,3,5,5-tetramethyl-4-(2,2,5,5-tetramethyl-3-cyclopenten-1-ylidene)-	C_6H_{12}	220(3.53)	78-1693-86
$C_{18}H_{28}N_2S$			
6-Thia-12,13-diazadispiro[4.1.4.2]trideca-2,9,12-triene, 1,1,4,4,8,8,11,11-octamethyl-	hexane	205(3.32),283(3.01), 333(2.49)	78-1693-86

Compound	Solvent	λ_{max}(log ε)	Ref.
4-Thiepinacetonitrile, 2,3,6,7-tetra-hydro-α,α,3,3,6,6-hexamethyl-5-[(2-methyl-1-propenylidene)amino]-	C_6H_{12}	214(4.03)	78-1693-86
3H-Thiepino[4,5-b]pyrrole-2-acetoni-trile, 4,5,7,8-tetrahydro-α,α,3,3-4,4,8,8-octamethyl-	C_6H_{12}	230(3.09),244(3.04), 276(3.10)	78-1693-86
$C_{18}H_{28}N_6O_4$			
β-D-Ribofuranuronamide, 1-deoxy-1-[6-(1-ethylpropyl)amino]-9H-purin-9-yl]-N-(1-methylethyl)-	n.s.g.	269(4.26)	87-1683-86
$C_{18}H_{28}O$			
Spiro[4.5]deca-6,9-dien-8-one, 7,9-bis(1,1-dimethylethyl)-	EtOH	247(4.12)	104-1079-86
$C_{18}H_{28}O_2$			
4H-1-Benzopyran-6-ol, 2,3-dihydro-2,2-dimethyl-7-(1,1-dimethylethyl)-5-(1-methylethyl)-	EtOH	267(4.29),295(3.56)	18-2899-86
$C_{18}H_{28}O_4$			
1,2,3,4-Cyclotetradecanetetrone, 5,5,14,14-tetramethyl-	CH_2Cl_2	385(1.97),533(1.94)	89-0449-86
3-Hexenedial, 3,4-bis(1,1-dimethyl-2-oxoethyl)-2,2,5,5-tetramethyl-	C_6H_{12}	207(3.33),241(3.18)	78-1693-86
$C_{18}H_{28}O_5$			
1-Cyclopentene-1-heptanoic acid, 5-oxo-3-[(tetrahydro-2H-pyran-2-yl)-oxy]-, methyl ester	MeOH	223(4.00)	2-0675-86
$C_{18}H_{28}O_8$			
1,2,9,10-Tetraoxacyclodocosane-3,8,11,12-tetrone	C_6H_{12}	215(2.90)	78-1285-86
$C_{18}H_{28}S$			
11-Thiadispiro[4.0.4.1]undeca-2,8-di-ene, 1,1,4,4,7,7,10,10-octamethyl-	hexane	200(3.41),265(2.08)	78-1693-86
$C_{18}H_{29}ClN_2O_7Si$			
1H-Imidazole-4-carboxylic acid, 2-chloro-1-[5-O-[(1,1-dimethylethyl)-dimethylsilyl]-2,3-O-(methoxymeth-ylene)-β-D-ribofuranosyl]-, methyl ester	MeOH	233(4.17)	78-1971-86
$C_{18}H_{29}ClO$			
Benzene, 2-(4-chlorobutoxy)-1,3-bis(1,1-dimethylethyl)-	EtOH	220(3.95),267(2.53)	104-1079-86
2,4-Cyclohexadien-1-one, 6-chloro-2,4,6-tris(1,1-dimethylethyl)-	hexane	260(3.75),320(3.20)	70-0446-86
2,5-Cyclohexadien-1-one, 4-chloro-2,4,6-tris(1,1-dimethylethyl)-	hexane	242(3.95),325(3.27)	70-0446-86
$C_{18}H_{29}NO_2$			
2-Butanone, 4-[3,5-bis(1,1-dimethyl-ethyl)-4-hydroxyphenyl]-, oxime	EtOH	278(3.29)	103-0052-86
$C_{18}H_{29}NO_3$			
3,5-Heptadienamide, 6-acetoxy-N-cyclo-	MeOH	219s(3.82)	33-0469-86

Compound	Solvent	$\lambda_{max}(\log \epsilon)$	Ref.
hexyl-2,3,4-trimethyl-, (E,Z)-(±)-			33-0469-86
$C_{18}H_{29}NO_4$			
α-D-xylo-Hexofuranurononitrile, 5-deoxy-5-methylene-1,2-O-(1-methylethylidene)-3-O-octyl-	n.s.g.	217(3.04)	111-0111-86
$C_{18}H_{29}N_3O$			
4,17-Imino-2H-furo[2,3-b][1,5]diazacycloheptadecine, 3,6,7,8,9,10,11-12,13,14,15,16-dodecahydro-2-methyl-	$CHCl_3$	254(4.11)	24-1070-86
5,18-Iminopyrano[2,3-b][1,5]diazacycloheptadecine, 2,3,4,7,8,9,10,11-12,13,14,15,16,17-tetradecahydro-	$CHCl_3$	248(3.97)	24-1070-86
$C_{18}H_{29}N_5O_3S_2$			
Adenosine, 5'-S-(2-methylpropyl)-5'-C-[(2-methylpropyl)thio]-5'-thio-	MeOH	259(4.15)	44-1258-86
$C_{18}H_{30}O_3$			
2H-Pyran-2-one, 3,5-dibutyl-4-hydroxy-6-pentyl-	MeOH	293(4.95)	44-0268-86
Sterebin D	MeOH	228(3.87)	78-6443-86
$C_{18}H_{30}O_4$			
Sterebin A	MeOH	228(4.08)	78-6443-86
$C_{18}H_{31}N$			
Benzenamine, 2,4,6-tris(1,1-dimethylethyl)-, chromium tricarbonyl complex	$CHCl_3$	217(4.61),322(3.87)	112-0517-86
$C_{18}H_{32}N$			
Pyridinium, 1-dodecyl-4-methyl-, chloride	HOAc	256(3.61)	4-0209-86
$C_{18}H_{32}N_4$			
1,5,10,14-Tetraazacyclooctadecatetraene, 2,4,11,13-tetramethyl-	DMF	308(4.47)	138-0907-86
copper chelate	DMF	354(4.11),494(3.23), 1167(2.05)	138-0907-86
$C_{18}H_{32}OSi$			
2-Propen-1-one, 1-[(1,1-dimethylethyl)-dimethylsilyl]-3-(2,6,6-trimethyl-2-cyclohexenyl)-, (E)-	pentane	225(4.11),412(2.08)	33-1378-86
2-Sila-4,6-heptadien-2-ol, 7-(2,6,6-trimethyl-2-cyclohexenyl)-2,3,3-trimethyl-, (4E,6E)-	pentane	250(4.45)	33-1378-86
Silane, [(1-cyclohexyl-2-methylene-3-cyclopenten-1-yl)oxy](1,1-dimethylethyl)dimethyl-	MeOH	231(4.12)	23-0180-86
$C_{18}H_{34}N_4O_5S_2$			
6,9,17,20,25-Pentaoxa-1,3,12,14-tetraazabicyclo[12.8.5]heptacosane-2,13-dithione	EtOH	211.7(4.26),249.0(4.44)	104-1589-86
6,14,17,22,25-Pentaoxa-1,3,9,11-tetraazabicyclo[9.8.8]heptacosane-2,10-dithione	EtOH	210.2(4.31),249.1(4.38)	104-1589-86

Compound	Solvent	$\lambda_{max}(\log \epsilon)$	Ref.
$C_{18}H_{34}O_2Si$			
Acetaldehyde, [4-[[(1,1-dimethylethyl)dimethylsilyl]oxy]-2,2,6,6-tetramethylcyclohexylidene]-, (S)-	C_6H_{12}	241(4.12)	35-2691-86
$C_{18}H_{36}NiS_6$			
Nickel(2+), (1,5,9,13,17,21-hexathiacyclotetracosane-$S^1,S^5,S^9,S^{13},S^{17}-S^{21}$)-, (OC-6-11)-, bis(tetrafluoroborate)	$MeNO_2$	590(1.85),905(2.00)	77-1083-86
$C_{18}H_{37}NO_4SSi_2$			
Furo[2,3-d]oxazole-2(3H)-thione, 6-[[(1,1-dimethylethyl)dimethylsilyl]oxy]-5-[[[(1,1-dimethylethyl)-dimethylsilyl]oxy]methyl]tetrahydro-, [3aS-(3aα,5α,6α,6aα)]-	MeOH	244(4.27)	87-0203-86
$C_{18}H_{38}N_2$			
Diazene, bis[1-(1,1-dimethylethyl)-2,2-dimethylpropyl]-, cis	toluene	411(--)	24-1911-86
trans	toluene	384.5(1.413)	24-1911-86
$C_{18}H_{42}N_3O_{12}P_3S_3$			
Phosphonic acid, 1,3,5,2,4,6-trithiatriazine-2,4,6-triyltris-, hexakis(1-methylethyl) ester, S,S',S"-trioxide	hexane or ether or MeCN	216(3.86),233(4.04), 249(3.80),273(3.58), 312(3.53)	30-0387-86
$C_{18}H_{45}GeN_3Si_3$			
3H-1,2,3,4-Triazagermole, 3-[bis(1,1-dimethylethyl)methylsilyl]-4,5-dihydro-4,4-dimethyl-5,5-bis(trimethylsilyl)-	C_6H_{12}	253(4.60)	24-2980-86
$C_{18}H_{54}Si_9$			
Cyclopentasilane, hexamethyl-1,1,3,3-tetrakis(trimethylsilyl)-	C_6H_{12}	240(4.3),280s(3.2)	157-0128-86
$C_{18}N_8S$			
Propanedinitrile, 2,2'-[thiobis[3-(dicyanomethyl)-2-cyclopropen-2-yl-1-ylidene]]bis-, ion(2-), dipotassium	H_2O	288(4.82),347(4.26)	24-2104-86
$C_{18}N_8Se$			
Propanedinitrile, 2,2'-[selenobis-[3-(dicyanomethyl)-2-cyclopropen-2-yl-1-ylidene]]bis-, ion(2-)-, disodium	n.s.g.	287(4.78),343(4.10)	89-0183-86

Compound	Solvent	λ_{max}(log ε)	Ref.
$C_{19}H_2Cl_{14}$			
Benzene, 1,1'-[(2,3,4,5-tetrachlorophenyl)methylene]bis[2,3,4,5,6-pentachloro-	C_6H_{12}	220(5.11),292(3.02), 302s(2.98)	118-0064-86
$C_{19}H_3Cl_{13}$			
Benzene, [bis(2,3,5,6-tetrachlorophenyl)methyl]pentachloro-	C_6H_{12}	220(5.05),283s(3.05), 292(3.30),302(3.35)	118-0064-86
Benzene, 1,1'-[(2,4,6-trichlorophenyl)methylene]bis[2,3,4,5,6-pentachloro-	C_6H_{12}	218(5.06),280(2.85), 291(2.90),303(2.85)	118-0064-86
$C_{19}H_4Cl_{12}$			
Benzene, 1,1',1"-methylidynetris-[2,3,5,6-tetrachloro-	C_6H_{12}	217(5.04),241s(4.72), 284s(3.22),291(3.39), 300(3.45)	118-0064-86
$C_{19}H_6Cl_{10}$			
Benzene, 1,1'-(phenylmethylene)-bis[2,3,4,5,6-pentachloro-	C_6H_{12}	213(4.98),291(2.79), 301(2.76)	118-0064-86
$C_{19}H_7Cl_9$			
Benzene, 1,1',1"-methylidynetris-[2,4,6-trichloro-	C_6H_{12}	213(4.97),230s(4.65), 243(4.56),270(2.77), 278(2.86),286(2.75)	118-0064-86
$C_{19}H_8Br_4O_3$			
3H-Xanthen-3-one, 2,4,5,7-tetrabromo-6-hydroxy-9-phenyl-	EtOH-NaOH	254(4.53),310(4.25), 341s(3.93),516(4.68)	39-0671-86C
$C_{19}H_9N_3O$			
Ethenetricarbonitrile, [5-(1-naphthalenyl)-2-furanyl]-	acetone	486(3.34)	73-1450-86
$C_{19}H_{10}ClNO_6$			
[2,3'-Bi-4H-1-benzopyran]-4,4'-dione, 6'-chloro-6-methyl-3-nitro-	n.s.g.	301(4.51)	2-0212-86
$C_{19}H_{10}ClNO_7$			
[2,3'-Bi-4H-1-benzopyran]-4,4'-dione, 6'-chloro-7-methoxy-3-nitro-	n.s.g.	301(4.52)	2-0212-86
$C_{19}H_{10}Cl_4N_2O_4$			
1H-Indole-2,3-dione, 1,1'-(1,3-propanediyl)bis[4,7-dichloro-	EtOH	225(4.51),250(4.29), 312(3.40)	12-1231-86
$C_{19}H_{10}N_2O_4$			
6H-[1,4]Diazepino[1,2-a:4,3-a']diindole-6,8,14,15(17H)-tetrone	benzene	292(4.40),350(3.52), 435(3.85)	138-1393-86
$C_{19}H_{10}O_4$			
1H-Indene-1,3(2H)-dione, 2-[(1,3-dihydro-1,3-dioxo-2H-inden-2-ylidene)methyl]-	EtOH	240(4.94),312(4.78), 470(5.06)	70-1446-86
$C_{19}H_{11}BrClNO_7$			
[2,3'-Bi-4H-1-benzopyran]-4,4'-dione, 3-bromo-6'-chloro-2,3-dihydro-7-methoxy-3-nitro-	n.s.g.	344(3.08)	2-0212-86

Compound	Solvent	$\lambda_{max}(\log \epsilon)$	Ref.
$C_{19}H_{11}F_3$			
1H-Cyclopropa[b]naphthalene, 1-(1,2,2-trifluoro-1-phenylethylidene)-	C_6H_{12}	227(4.63),236s(4.35), 253(4.29),263(4.24), 281s(4.07),396(4.50), 420(4.36)	35-5949-86
5λ,6ε	MeCN	226(4.63),252(4.39), 265(4.38),?s(4.25), 393(4.55),414s(4.40)	35-5949-86
$C_{19}H_{11}N_3O_2$			
1H-Naphtho[2,3-h]pyrazolo[3,4-b]quinoline-7,12-dione, 3-methyl-	DMSO	230(4.455),455(4.146)	2-0616-86
$C_{19}H_{11}N_3S$			
Propanedinitrile, (4,6-diphenyl-2H-1,3-thiazin-2-ylidene)-	MeOH	261(4.38),322(4.55), 448(4.17)	118-0916-86
$C_{19}H_{12}BClO_6$			
Boron, [5-(4-chlorophenyl)-1-phenyl-4-pentene-1,3-dionato-O,O'](ethanedioato(2-)-O,O']-	CH_2Cl_2	440(4.82)	97-0399-86
$C_{19}H_{12}ClNO_6$			
[2,3'-Bi-4H-1-benzopyran]-4,4'-dione, 6'-chloro-2,3-dihydro-6-methyl-3-nitro-	n.s.g.	314(4.13)	2-0212-86
$C_{19}H_{12}ClNO_7$			
[2,3'-Bi-4H-1-benzopyran]-4,4'-dione, 6'-chloro-2,3-dihydro-6-methoxy-3-nitro-	n.s.g.	305(4.14)	2-0212-86
$C_{19}H_{12}Cl_2N_2O_4$			
1H-Indole-2,3-dione, 1,1'-(1,3-propanediyl)bis[5-chloro-	EtOH	220(4.35),252(4.42), 258s(4.38),275s(3.93), 301(3.43)	12-1231-86
$C_{19}H_{12}Cl_2O_5$			
2,5-Cyclohexadiene-1,4-dione, 2-(2,4-dichlorophenyl)-3,6-dihydroxy-5-(4-methoxyphenyl)-	dioxan	262(4.47),340(3.67)	5-0195-86
2H-Pyran-2,5(6H)-dione, 6-[(2,4-dichlorophenyl)methylene]-3-hydroxy-4-(4-methoxyphenyl)-	dioxan	270(4.18),349(4.13)	5-0195-86
$C_{19}H_{12}CrO_3$			
Chromium, tricarbonyl[(1,10,11,12,13-14-η)-tricyclo[8.2.2.2^{4,7}]hexadeca-2,4,6,8,10,12,13,15-octaene]-	CH_2Cl_2	342(3.97)	88-2353-86
$C_{19}H_{12}FN_3OS$			
Imidazo[5,1-a]isoquinoline-10b(1H)-carbonitrile, 2-(4-fluorophenyl)-2,3-dihydro-5-methyl-3-oxo-1-thioxo-	$CHCl_3$	252(3.79),279(3.97)	150-1901-86M
$C_{19}H_{12}F_2$			
Bicyclo[4.1.0]hepta-1,3,5-triene, 7,7-difluoro-3,4-diphenyl-	hexane	201(4.15),213(4.42), 238(4.21),277(3.66)	33-1546-86
$C_{19}H_{12}F_2N_2O_4$			
1H-Indole-2,3-dione, 1,1'-(1,3-prop-	EtOH	213(4.32),247(4.40),	12-1231-86

Compound	Solvent	λ_{max}(log ϵ)	Ref.
anediyl)bis[5-fluoro- (cont.)		255s(4.32),270s(3.76), 296(3.48)	12-1231-86
$C_{19}H_{12}N_2O_6$			
Benzo[f]pyrido[1,2-a]indole-2-carboxylic acid, 6,11-dihydro-12-nitro-6,11-dioxo-, ethyl ester	DMF	476(3.8751)	2-1126-86
$C_{19}H_{12}N_4O_4$			
Benzenepropanenitrile, α-(1,5-dihydro-3-methyl-5-oxo-1-phenyl-4H-pyrazol-4-ylidene)-4-nitro-β-oxo-	DMAA	360(3.45)	104-0582-86
$C_{19}H_{12}N_4S$			
1H-Perimidine-2-carbothioamide, N-(4-cyanophenyl)-	DMF	346(4.29),354s(4.28), 466s(2.97),532(3.07)	48-0812-86
	C_6H_{12}	230s(--),236(--), 256(--),288(--), 299s(--),340(--), 351s(--),355(--), 365s(--),384s(--), 442s(--),476s(--), 508(--),548(--), 595(--),650(--)	48-0812-86
$C_{19}H_{12}O_2$			
1,2-Chrysenedione, 11-methyl-	EtOH	261(4.54),305(4.38), 325(4.38)	44-1407-86
1,4-Triphenylenedione, 2-methyl-	EtOH	249(5.17),273s(4.65), 295(4.46),307(4.30), 418(3.59)	94-2810-86
$C_{19}H_{12}O_3$			
1H-Naphtho[2,3-c]pyran-5,10-dione, 3-phenyl-	MeCN	264(4.57),487(3.93)	88-0255-86
	CH_2Cl_2-CF_3COOH	310(4.31),529(3.92)	88-0255-86
	H_2SO_4	310(3.94),352(3.83), 390(3.74),660(3.73)	88-0255-86
oxidation by air	H_2SO_4	234(4.14),408(3.78)	88-0255-86
3H-Xanthen-3-one, 6-hydroxy-9-phenyl-	neutral	455(4.5),480(4.5)	110-0857-86
	cation	435(4.7)	110-0857-86
$C_{19}H_{12}O_4$			
Anthra[1,2-b]furan-6,11-dione, 3-acetyl-2-methyl-	n.s.g.	230(4.43),251(4.45), 275(4.61),348(3.91)	12-2075-86
6H-Fluoren-6-one, 2,3,7-trihydroxy-9-phenyl-	10% MeOH	510(4.82)	140-0957-86
	10% EtOH	510(4.87)	140-0957-86
	10% MeOH-pH acid	460(4.59)	140-0957-86
	pH 4-5.5	460(--),490(--)	140-0957-86
	pH basic	555(4.72)	140-0957-86
$C_{19}H_{12}O_6$			
Anthra[1,2-b]furan-4-acetic acid, 6,11-dihydro-5-hydroxy-6,11-dioxo-, methyl ester	n.s.g.	223(4.18),255(4.47), 274(4.26),433(3.87)	12-0821-86
$C_{19}H_{13}BO_6$			
Boron, (1,5-diphenyl-4-pentene-1,3-dionato-O,O'][ethanedioato(2-)-O,O']-, [T-4-(E)]-	CH_2Cl_2	430(4.73)	97-0399-86

Compound	Solvent	$\lambda_{max}(\log \epsilon)$	Ref.
$C_{19}H_{13}IN_4O_5S$			
Iodine, (1,3-dihydro-2H-benzimidazole-2-thionato-S)(2-hydroxy-3,5-dinitrophenyl)phenyl-	EtOH	360(3.97)	5-0947-86
$C_{19}H_{13}NO$			
1H-Pyrrolizin-1-one, 6,7-diphenyl-	MeOH	204(4.39),250(4.11), 286s(--),452(3.28)	83-0749-86
3H-Pyrrolizin-3-one, 6,7-diphenyl-	$CHCl_3$	255(4.29),354(4.06), 446(3.06)	83-0749-86
$C_{19}H_{13}NO_5$			
Acetamide, N-[1,3-dihydro-7-(4-methylphenyl)-1,3-dioxobenzo[1,2-b:3,4-c']-difuran-4-yl]-	dioxan	291(3.51),321(3.42), 408(3.27)	73-1455-86
8H-Benzo[g]-1,3-benzodioxolo[6,5,4-de]quinolin-8-one, 10,11-dimethoxy-(dicentrinone)	EtOH	211(4.40),249(4.26), 271(4.18),308(3.76), 349(3.77),388(3.66), 433(3.57)	102-1999-86
$C_{19}H_{13}NO_7$			
2,5-Cyclohexadiene-1,4-dione, 2,5-dihydroxy-3-(4-methoxyphenyl)-6-(4-nitrophenyl)-	dioxan	251(4.33),297(4.28)	5-0195-86
2H-Pyran-2,5(6H)-dione, 3-hydroxy-4-(4-methoxyphenyl)-6-[(4-nitrophenyl)-methylene]-	dioxan	252(4.18),290s(4.26), 348(4.39)	5-0177-86
$C_{19}H_{13}N_3$			
Benzaldehyde, 5H-indeno[1,2-b]pyridin-5-ylidenehydrazone	EtOH	207(4.52),237(4.54), 243(4.55),246s(4.51), 264s(4.35),293(4.34), 304(4.38),315(4.40), 335(4.44),345(4.47), 357s(4.34)	103-0200-86
	$EtOH-H_2SO_4$	206(4.58),241(4.54), 293s(4.22),303(4.32), 330(4.36),345(4.52)	103-0200-86
$C_{19}H_{13}N_3O$			
Benzaldehyde, 2-hydroxy-, 5H-indeno-[1,2-b]pyridin-5-ylidenehydrazone	EtOH	217(4.53),227(4.56), 240(4.48),267(4.32), 310(4.30),340(4.25), 375(4.28)	103-0200-86
	$EtOH-H_2SO_4$	217(4.48),236(4.34), 253(4.37),329(4.39), 390(4.27)	103-0200-86
$C_{19}H_{13}N_3OS$			
Imidazo[5,1-a]isoquinoline-10b(1H)-carbonitrile, 2,3-dihydro-5-methyl-3-oxo-2-phenyl-1-thioxo-	EtOH	225(4.73),277(4.69), 435(3.09)	150-1901-86M
$C_{19}H_{13}N_3O_3$			
9,10-Anthracenedione, 1-amino-2-[(1,5-dihydro-3-methyl-5-oxo-4H-pyrazol-4-ylidene)methyl]-	DMSO	330(4.491),450(4.455)	2-0616-86
Methanone, [5-(2H-benzotriazol-2-yl)-2,4-dihydroxyphenyl]phenyl-	$CHCl_3$	248(4.19),302(4.47)	49-0805-86

Compound	Solvent	λ_{max}(log ε)	Ref.
$C_{19}H_{13}O$			
Diindeno[1,2-b:2',1'-e]pyrylium, 10,12-dihydro-, perchlorate	MeCN	266(4.29),277(4.16), 432(4.30)	22-0600-86
$C_{19}H_{14}$			
Bicyclo[5.4.1]dodecapentaene, 4-(2,4,6-cycloheptatrien-1-ylidene)-	EtOH	264(4.38),336(4.45), 390(4.36)	88-0485-86
	isooctane	390(--)	88-0485-86
1H-Cyclopropa[b]naphthalene, 1-(1-phenylethylidene)-	C_6H_{12}	230(4.70),242s(4.38), 253s(4.24),289(4.38), 377s(4.32),394(4.62), 412s(4.49),422(4.72)	35-5949-86
	MeCN	229(4.62),241s(4.31), 252s(4.22),288(4.43), 373s(4.27),391(4.56), 417(4.66)	35-5949-86
9H-Fluorene, 9-phenyl-, lithium salt	THF	411(4.40)	35-7016-86
cesium salt	THF	397(4.38)	35-7016-86
$C_{19}H_{14}Br_2OSe_2$			
Benzene, 1,1'-[(phenoxymethylene)-bis(seleno)]bis[4-bromo-	hexane	230(4.39),242s(4.36), 300(3.73)	104-0092-86
$C_{19}H_{14}ClNO_3$			
1,3-Propanedione, 1-(4-chlorophenyl)-2-(3-methyl-5-isoxazolyl)-3-phenyl-	dioxan	256(4.25),325(4)	83-0242-86
$C_{19}H_{14}Cl_2N_2$			
Quinoline, 4,7-dichloro-3-(3,4-dihydro-3-methyl-1-isoquinolinyl)-	EtOH	235(4.77)	2-1072-86
$C_1\ H_{14}Cl_2OSe_2$			
Benzene, 1,1'-[(phenoxymethylene)-bis(seleno)]bis[4-chloro-	hexane	227(4.42),237s(4.39), 298(3.78)	104-0092-86
$C_{19}H_{14}Fe_2O_6$			
Iron, hexacarbonyl[μ-[η⁴:η⁴-pentakis(methylene)bicyclo[2.2.2]octane]]di-	isooctane	285s(3.70)	33-0865-86
$C_{19}H_{14}N_2$			
1,1(8aH)-Azulenedicarbonitrile, 2-(4-methylphenyl)-	MeCN	212(4.2),237(4.0), 268(4.3),355(4.3)	24-2631-86
Propanedinitrile, [2-(2,4,6-cycloheptatrien-1-ylidene)-1-(4-methylphenyl)ethylidene]-	MeCN	302(3.8),344(3.7), 469(4.4)	24-2631-86
$C_{19}H_{14}N_2O$			
1,1(8aH)-Azulenedicarbonitrile, 2-(4-methoxyphenyl)-	MeCN	220(4.2),275(4.2), 365(4.4)	24-2631-86
Propanedinitrile, [2-(2,4,6-cycloheptatrien-1-ylidene)-1-(4-methoxyphenyl)ethylidene]-	MeCN	237(4.2),467(4.5)	24-2631-86
$C_{19}H_{14}N_2OS$			
Methanone, [3-(phenylamino)-1H-indol-2-yl]-2-thienyl-	MeCN	230(4.21),262(4.32), 336(4.26),440(4.09)	24-2289-86
$C_{19}H_{14}N_2O_4$			
1-Azulenecarboxylic acid, 1-cyano-1,8a-dihydro-2-(4-nitrophenyl)-,	MeCN	236(4.2),390(4.3)	64-1151-86B

Compound	Solvent	$\lambda_{max}(\log \epsilon)$	Ref.
methyl ester, cis-(±)- (cont.)			64-1151-86B
Spiro[2H-anthra[1,2-c]pyrazole-2,4'-morpholinium], 1,3,6,11-tetrahydro-3,6,11-trioxo-, hydroxide, inner salt	toluene	658(3.94)	104-2117-86
$C_{19}H_{14}N_4$			
5,9-Ethano-1H-benzocycloheptene-2,2,3,3-tetracarbonitrile, 4,5,6,9-tetrahydro-10,11-bis(methylene)-	EtOH	243s(3.91),248s(3.89), 260s(3.69)	33-1310-86
	dioxan	224s(3.95),230s(3.92), 243(3.91),249s(3.90), 260s(3.70)	33-1310-86
1,4-Ethanonaphthalene-6,6,7,7-tetracarbonitrile, 1,4,5,8-tetrahydro-2-methyl-9,10-bis(methylene)-	MeCN	250s(3.84),256(3.90), 262(3.92),271(3.89)	33-1310-86
$C_{19}H_{14}N_4OS$			
2(1H)-Pteridinone, 3,4-dihydro-1-methyl-6,7-diphenyl-4-thioxo-	pH 3.9	226(4.53),292(4.20), 403(4.28)	128-0199-86
	pH 10.8	227s(4.53),301(4.37), 435(4.17)	128-0199-86
	MeOH	226(4.54),292(4.20), 402(4.25)	128-0199-86
2(1H)-Pteridinone, 3,4-dihydro-3-methyl-6,7-diphenyl-4-thioxo-	pH 1	225(4.53),291(4.20), 403(4.28)	128-0199-86
	pH 10	229s(4.51),302(4.37), 435(4.19)	128-0199-86
	MeOH	226(4.57),291(4.23), 401(4.28)	128-0199-86
2(1H)-Pteridinone, 4-(methylthio)-6,7-diphenyl-	MeOH	222(4.53),282(4.24), 388(4.26)	128-0199-86
$C_{19}H_{14}N_4O_2$			
2,4(1H,3H)-Pteridinedione, 1-methyl-6,7-diphenyl-	pH 6	222(4.44),276(4.21), 365(4.18)	128-0199-86
	pH 11	220s(4.32),266(4.25), 367(4.16)	128-0199-86
	MeOH	225(4.40),278(4.20), 365(4.14)	128-0199-86
2,4(1H,3H)-Pteridinedione, 3-methyl-6,7-diphenyl-	pH 2	223(4.44),270(4.16), 361(4.19)	128-0199-86
	pH 11	240s(4.41),288(4.19), 384(4.18)	128-0199-86
	MeOH	226(4.46),271(4.20), 362(4.19)	128-0199-86
$C_{19}H_{14}N_4O_4$			
Morpholine, 4-[(1-azido-9,10-dihydro-9,10-dioxo-2-anthracenyl)carbonyl]-	toluene	342(3.46),362(3.43)	104-2117-86
$C_{19}H_{14}O$			
1H-Cyclopropa[b]naphthalene, 1-[(4-methoxyphenyl)methylene]-	C_6H_{12}	232(4.72),250(4.43), 259s(4.42),287(4.30), 300s(4.17),386s(4.33), 403(4.71),423s(4.52), 432(4.86)	35-5949-86
	MeCN	231(4.75),248s(4.45), 283(4.36),379s(4.41), 400(4.75),427(4.84)	35-5949-86
Pyrene, 1-(2-methoxyethenyl)-, cis	MeOH	243(4.50),276(4.39),	35-4498-86

Compound	Solvent	λ_{max}(log ε)	Ref.
(cont.)		287(4.48),352(5.52), 367(4.48)	35-4498-86
Pyrene, 1-(2-methoxyethenyl)-, trans	MeOH	243(4.45),284(4.37), 352(4.44)	35-4498-86
1-Triphenylenol, 3-methyl-	EtOH	254s(4.46),262(4.62), 288s(3.87),300(3.57), 324s(3.25),340(3.17), 356(3.11)	94-2810-86
$C_{19}H_{14}O_5$			
2,5-Cyclohexadiene-1,4-dione, 2,5-dihydroxy-3-(4-methoxyphenyl)-6-phenyl-	dioxan	264(4.48),353(3.69), 465(2.90)	5-0195-86
2H-Pyran-2,5(6H)-dione, 3-hydroxy-4-(4-methoxyphenyl)-6-(phenylmethylene)-	dioxan	264(4.35),349(4.28)	5-0195-86
$C_{19}H_{14}O_6$			
9,10-Anthracenedione, 2-(1-acetoxy-2-oxopropyl)-1-hydroxy-	n.s.g.	227(4.29),254(4.45), 277s(4.07),334(3.52), 405(3.82)	12-2075-86
Benz[a]anthracene-1,7,12(2H)-trione, 3,4-dihydro-2,3,8-trihydroxy-3-methyl- (PD 116,779)	MeOH	266(4.48),398(3.70)	158-0469-86
	MeOH-KOH	257(4.40),312(3.88), 408(3.38),488(3.69)	158-0469-86
2H-Pyran-2,5(6H)-dione, 3-hydroxy-6-[(3-hydroxyphenyl)methylene]-4-(4-methoxyphenyl)-	MeOH	280(4.28),358(4.20)	5-0177-86
2H-Pyran-2,5(6H)-dione, 3-hydroxy-6-[(4-hydroxyphenyl)methylene]-4-(4-methoxyphenyl)-	MeOH	263(4.08),386(4.05)	5-0177-86
$C_{19}H_{14}O_7$			
9,10-Anthracenedione, 1,4-dihydroxy-2-(tetrahydro-4-hydroxy-4-methyl-5-oxo-2-furanyl)-	MeOH	211(4.15),229(4.34), 250(4.55),255(4.49), 284(3.99),310(3.43), 468(3.94),480(3.95), 509(3.74),566(3.07)	78-6635-86
Anthra[2,3-c]furan-5,10-dione, 11-acetoxy-1,3-dihydro-4-hydroxy-1-methoxy-	n.s.g.	213(4.08),226(4.18), 256(4.39),340(3.34), 406(3.72)	12-2075-86
2H-Pyran-2,5(6H)-dione, 6-[(3,4-dihydroxyphenyl)methylene]-3-hydroxy-4-(4-methoxyphenyl)- (4'-O-methylgrevillin B)	MeOH	276(4.28),398(4.23)	5-0177-86
$C_{19}H_{15}BO_6$			
Boron, [1,3-bis(4-methylphenyl)-1,3-propanedionato-O,O'][ethanedioato-(2-)-O,O']-	CH_2Cl_2	327(3.97),406s(4.66), 417(4.74)	48-0755-86
$C_{19}H_{15}BO_8$			
Boron, [1,3-bis(4-methoxyphenyl)-1,3-propanedionato][ethanedioato(2-)-O,O']-	CH_2Cl_2	415(4.64),442(4.76)	48-0755-86
	CH_2Cl_2	431(4.78)	97-0399-86
$C_{19}H_{15}BrN_2O$			
1,1(2H)-Azulenedicarbonitrile, 2-(4-bromophenyl)-3,8a-dihydro-3-methoxy-, (2α,3α,8aβ)-	MeCN	225(4.2),273(3.5)	64-1151-86B

Compound	Solvent	λ_{max}(log ϵ)	Ref.
$C_{19}H_{15}ClN_2O$			
4-Quinolinol, 7-chloro-3-(3,4-dihydro-3-methyl-1-isoquinolinyl)-	EtOH	245(4.22),280(4.31), 360(3.92),395(4.06)	2-1072-86
$C_{19}H_{15}ClN_2O_6$			
2H-1-Benzopyran-3-carboxylic acid, 8-[(chloromethoxyphenyl)azo]-7-hydroxy-2-oxo-, ethyl ester	n.s.g.	465(4.72)	2-0638-86
$C_{19}H_{15}ClN_4O_3$			
1H,5H-Pyrazolo[1,2-a]pyrazol-1-one, 7-amino-2-(4-chlorophenyl)-3-hydroxy-5-imino-6-(4-methoxyphenyl)-	MeCN	281(4.70)	118-0239-86
$C_{19}H_{15}ClO_7$			
9H-Xanthen-9-one, 1,7-diacetoxy-4-chloro-3-methoxy-5-methyl-	MeOH	246(4.69),300(4.13), 342(3.76)	94-0858-86
$C_{19}H_{15}F_6N_3O_2$			
1,2,4-Triazine, 4,5-dihydro-4,5-bis(4-methoxyphenyl)-3,6-bis(trifluoromethyl)-	CH_2Cl_2	277(3.89),282(3.90), 314(3.90)	83-0690-86
$C_{19}H_{15}N$			
1H-Indeno[2,1-b]pyridine, 1-methyl-2-phenyl-	EtOH	308(4.54),350(4.0), 480-570(3.1),608(3.23)	103-0980-86
1H-Indeno[2,1-b]pyridine, 1-methyl-4-phenyl-	EtOH	300(4.3),335s(3.90), 480s(3.14),590(3.50)	103-0980-86
$C_{19}H_{15}NO$			
4-Benzazacyclotetradecin-3(4H)-one, 9,10,11,12-tetradehydro-2,8-dimethyl-, (E,E,Z)-	THF	257s(4.32),272(4.46), 286(4.49),326s(3.67), 388(3.84),422s(3.67)	18-1791-86
4-Benzazacyclotetradecin-3(4H)-one, 9,10,11,12-tetradehydro-5,8-dimethyl-, (E,E,Z)-	THF	227(4.46),268s(4.42), 281(4.51),295s(4.42), 300s(4.38),379(3.86)	18-1791-86
7H-Benzocyclotridecen-7-one, 12,13-14,15-tetradehydro-6,11-dimethyl-, oxime, (Z,E,E,Z)-	THF	225(4.53),253s(4.40), 266(4.58),279(4.67), 328s(3.77),350s(3.88), 371(3.98),392s(3.88)	18-1791-86
7H-Benzocyclotridecen-7-one, 12,13-14,15-tetradehydro-8,11-dimethyl-, oxime, (E,E,E,Z)-	THF	225(4.48),272(4.56), 281s(4.55),380(3.74), 412s(3.62)	18-1791-86
5H-Indeno[1,2-b]pyridin-5-ol, 7-methyl-5-phenyl-	$CHCl_3$	282s(4.05),292(4.10), 310(4.09),322(4.21)	104-1352-86
5H-Indeno[1,2-b]pyridin-5-ol, 5-(phenylmethyl)-	$CHCl_3$	286(4.02),282s(3.70), 301(4.02),313(4.16)	104-1352-86
9H-Indeno[2,1-b]pyridin-9-ol, 9-(phenylmethyl)-	$CHCl_3$	270(4.16),281(4.27), 300(3.94),312(3.98), 335s(2.54)	104-1352-86
9H-Indeno[2,1-c]pyridin-9-ol, 3-methyl-9-phenyl-	$CHCl_3$	277(4.14),298(3.77), 311(3.79)	104-1352-86
$C_{19}H_{15}NOS$			
Bicyclo[4.4.1]undeca-3,6,8,10-tetraen-2-one, 5-(3-methyl-2(3H)-benzothiazolylidene)-	benzene	474(4.57)	33-0315-86
	MeOH	329(3.94),512(4.70)	33-0315-86
	CH_2Cl_2	328(3.86),489(4.60)	33-0315-86
	MeCN	323(3.81),488(4.61)	33-0315-86
$C_{19}H_{15}NO_2$			
1-Naphthalenol, 4-(3-(phenylimino)-1-	EtOH	275(3.79),340(3.74),	104-1302-86

Compound	Solvent	λ_{max}(log ε)	Ref.
propenyl]-, N-oxide (cont.)		415(4.14)	104-1302-86
	dioxan	400(4.03)	104-1302-86
	DMSO	350(3.62),415(4.17)	104-1302-86
	+ NaOMe	365(4.01),550(3.36)	104-1302-86
2-Naphthalenol, 1-[3-(phenylimino)-1-propenyl]-, N-oxide	acetone	343s(3.94),357(3.98), 400(4.18)	104-1302-86
	dioxan	278s(4.00),335s(3.88), 347(3.86),415(4.25)	104-1302-86
	DMSO	280s(4.22),305s(3.92), 350(3.97),410(4.17)	104-1302-86
	+ NaOMe	280s(4.30),305(4.03), 353(3.91),400(3.84), 535(4.08)	104-1302-86
$C_{19}H_{15}NO_3$			
7H-Benzo[g]-1,3-benzodioxolo[6,5,4-de]quinoline-7-carboxaldehyde, 5,6-dihydro-8-methyl- (trichoguattine)	EtOH	234(4.19),249(4.21), 267(4.04),279(4.01), 292s(3.74),304s(3.67), 340(3.40),430(3.68), 450s(3.62)	100-1078-86
$C_{19}H_{15}NO_3Se_2$			
Benzene, 1-[bis(phenylseleno)methoxy]-4-nitro-	hexane	211(4.56),298(4.16)	104-0092-86
$C_{19}H_{15}NO_4$			
7H-Dibenzo[de,g]quinolin-7-one, 1,2,3-trimethoxy-	MeOH	233(4.31),238(4.27), 274(4.34),315(3.67), 436.5(3.83)	102-1509-86
$C_{19}H_{15}NO_5$			
4H-Dibenzo[de,g]quinoline-4,5(6H)-dione, 1,2,3-trimethoxy-	EtOH	228(4.21),256(4.50), 270s(4.26),488(3.80)	100-0878-86
$C_{19}H_{15}N_3O$			
4H-Furo[3,2-b]pyrrole, 2-(4-methylphenyl)-5-(phenylazo)-	MeOH	475(3.72)	73-0106-86
Isoxazole, 5-(1,3-diphenyl-1H-pyrazol-4-yl)-3-methyl-	dioxan	239(4.42),276(4.45)	83-0242-86
Pyrazolo[5,1-c]pyrimidin-7(6H)-one, 6-methyl-2,5-diphenyl-	isoPrOH	247(4.5),300(4.3)	103-1322-86
$C_{19}H_{15}N_3OS$			
1H-Perimidine-2-carbothioamide, N-(4-methoxyphenyl)-	DMF	287(4.21),342(4.27), 355(4.34),378s(3.90), 494s(3.09)	48-0812-86
	C_6H_{12}	230s(--),235(--), 287(--),301s(--), 333(--),353(--), 369s(--),385s(--), 408s(--),445s(--), 480(--),517(--), 558(--),608(--)	48-0812-86
$C_{19}H_{15}N_3O_3$			
1-Azulenecarboximidic acid, 1-cyano-1,8a-dihydro-2-(4-nitrophenyl)-, methyl ester	MeCN	237(4.0),267s(3.9), 305s(3.9),382(4.1)	64-1151-86B
$C_{19}H_{15}N_3O_4$			
Acetamide, N-[1,2,3,4-tetrahydro-	dioxan	343(3.36),363(3.39),	73-1455-86

Compound	Solvent	$\lambda_{max}(\log \epsilon)$	Ref.
8-(4-methylphenyl)-1,4-dioxofuro-[3,2-f]phthalazin-5-yl]- (cont.)		384(3.23)	73-1455-86
$C_{19}H_{15}N_3O_4S$			
Thiopyrano[2,3-b]indole, 9-acetyl-4,9-dihydro-4-(tetrahydro-1,3-dimethyl-2,4,6-trioxo-5(2H)-pyrimidinylidene)-	dioxan	240(4.06),287(3.57), 300(3.59),331(3.47), 478(4.01)	83-0120-86
$C_{19}H_{15}N_3O_7$			
2H-1-Benzopyran-3-carboxylic acid, 7-hydroxy-8-[(methylnitrophenyl)-azo]-2-oxo-	n.s.g.	415(4.69)	2-0638-86
$C_{19}H_{15}N_3S$			
1H-Perimidine-2-carbothioamide, N-(4-methylphenyl)-	DMF	281(4.18),342(4.26), 356(4.29),470s(3.04), 499(3.04)	48-0812-86
Pyrazolo[1,5-c]pyrimidine, 7-(methylthio)-2,5-diphenyl-	isoPrOH	270(4.7),320(4.2)	103-1322-86
$C_{19}H_{15}N_5O$			
2(1H)-Pteridinone, 4-amino-1-methyl-6,7-diphenyl-	pH 0	222(4.46),282(4.23), 393(4.18)	128-0199-86
	pH 7	222(4.36),278(4.30), 374(4.20)	128-0199-86
2(1H)-Pteridinone, 3,4-dihydro-4-imino-3-methyl-6,7-diphenyl-	pH 0	220(4.49),278(4.20), 377(4.22)	128-0199-86
	pH 14	223s(4.44),292(4.19), 394(4.06)	128-0199-86
	MeOH	224(4.45),277(4.16), 366(4.18)	128-0199-86
$C_{19}H_{16}$			
7H-Benzo[c]fluorene, 3,9-dimethyl-	EtOH	232(4.38),255(3.99), 270(3.41),312(3.95), 327(3.84),333(3.9), 342(3.98)	2-0887-86
1H-Cyclopenta[l]phenanthrene, 1,3-dimethyl-	EtOH	243(4.26),260s(4.31), 323(3.93),333(3.89)	39-1471-86C
Pyrene, 1-(1-methylethyl)-	n.s.g.	243.3(4.8),275.9(4.7), 342.8(4.6)	1-0665-86
Pyrene, 2-(1-methylethyl)-	n.s.g.	245.7(4.9),275.8(4.6), 337.1(4.6)	1-0665-86
Pyrene, 4-(1-methylethyl)-	n.s.g.	243.1(4.8),275.8(4.6), 338.8(4.5)	1-0665-86
$C_{19}H_{16}ClNO_2S$			
4H-Thiopyran-4-one, 5-acetyl-6-[(4-chlorophenyl)amino]-2,3-dihydro-2-phenyl-	EtOH	255(3.21),314(3.26)	5-1109-86
$C_{19}H_{16}ClNO_3S$			
2H-Thiopyran-5-carboxylic acid, 6-[(4-chlorophenyl)amino]-3,4-dihydro-4-oxo-2-phenyl-, methyl ester	EtOH	244(3.15),308(3.32)	5-1109-86
$C_{19}H_{16}ClN_3$			
4-Quinolinamine, 7-chloro-3-(3,4-dihydro-3-methyl-1-isoquinolinyl)-	EtOH	255(4.63),280(4.28)	2-1072-86

Compound	Solvent	$\lambda_{max}(\log \epsilon)$	Ref.
$C_{19}H_{16}CrO_2S$			
Chromium, (carbonothioyl)dicarbonyl-[(1,2,3,4,4a,10a-η)-9,10-dihydro-4,5-dimethylphenanthrene]-	$CHCl_3$	263(4.27),343(3.87), 440s(3.04)	49-1423-86
$C_{19}H_{16}CrO_3$			
Chromium, tricarbonyl[(1,2,3,4,4a,10a-η)-9,10-dihydro-4,5-dimethylphenanthrene]-	EtOH	246(4.22),328(3.92), 415s(3.19)	49-1423-86
$C_{19}H_{16}F_6N_4$			
Benzenamine, 4-[4,5-dihydro-4-phenyl-3,6-bis(trifluoromethyl)-1,2,4-triazin-5-yl]-N,N-dimethyl-	CH_2Cl_2	263(4.23),313(4.00)	83-0690-86
$C_{19}H_{16}Fe_2O_6$			
Iron, hexacarbonyl[μ-[η^4:η^4-7-methyl-2,3,5,6-tetrakis(methylene)bicyclo-[2.2.2]octane]]di-	isooctane	215s(4.71),295s(3.78)	33-0865-86
$C_{19}H_{16}N_2$			
Benzenecarboximidamide, N,N'-diphenyl-	benzene	400(4.53)	3-0192-86
NbO(SCN)$_3$ complex	$CHCl_3$	398(4.52)	3-0192-86
Benzo[de]cinnoline, 1-ethyl-3-phenyl-	EtOH	225(4.60),266(4.48), 336(3.60),348(3.90)	32-0405-86
Pyrazine, 3-methyl-5-phenyl-2-(2-phenylethenyl)-, (E)-	EtOH	233(4.08),238(4.08), 295(4.20),355(4.29)	4-1481-86
$C_{19}H_{16}N_2O$			
Benzofuro[3,2-b]pyridin-2-amine, N,N-dimethyl-4-phenyl-	MeOH	234(4.63),256(4.49), 381(3.99)	83-0347-86
$C_{19}H_{16}N_2O_2$			
Azepino[3,4-b]indole-1,3(2H,4H)-dione, 5,10-dihydro-2-methyl-10-phenyl-	isoPrOH	243(4.05),288(4.27)	103-1345-86
2-Propen-1-one, 2-(3-methyl-5-isoxazolyl)-1-phenyl-3-(phenylamino)-	dioxan	242(4.52),293(4.15), 349(4.26)	83-0242-86
1H-Pyrido[3,4-b]indole-1,3(2H)-dione, 4,9-dihydro-2-methyl-9-(phenylmethyl)-	isoPrOH	241(4.33),310(4.28), 338s(3.86),380s(2.40)	103-1345-86
$C_{19}H_{16}N_2O_4$			
2-Dibenzofuranol, 6,7,8,9-tetrahydro-1-[[(3-nitrophenyl)imino]methyl]-, (E)-	toluene	355(4.03)	104-0503-86
	PrOH	348(4.01)	104-0503-86
Morpholine, 4-[(1-amino-9,10-dihydro-9,10-dioxo-2-anthracenyl)carbonyl]-	toluene	476(3.97)	104-2117-86
$C_{19}H_{16}N_2O_5$			
2-Anthracenecarboxylic acid, 9,10-dihydro-1-(4-morpholinylamino)-9,10-dioxo-	n.s.g.	275(4.03),313(3.72), 493(3.79)	104-2117-86
6,7-Quinolinedicarboxylic acid, 5-hydroxy-8-(phenylamino)-, dimethyl ester	MeOH	260(4.10),356(3.20)	4-1641-86
$C_{19}H_{16}N_2S$			
1H-[1]Benzothiopyrano[4,3,2-de]pyrrolo[2,3-b]quinoline, 2,3-dihydro-3,7-dimethyl-	$CHCl_3$	236(4.42),266(4.54), 335(3.8),377(3.95)	83-0347-86

Compound	Solvent	$\lambda_{max}(\log \epsilon)$	Ref.
$C_{19}H_{16}N_3S_2$			
Thiazolo[3,4-b][1,2,4]triazin-5-ium, 2-methyl-6-(methylthio)-3,8-diphenyl-, perchlorate	n.s.g.	329(4.27),446(3.74)	103-1373-86
Thiazolo[3,4-b][1,2,4]triazin-5-ium, 3-methyl-6-(methylthio)-2,8-diphenyl-, perchlorate	n.s.g.	329(4.24),447(3.71)	103-1373-86
$C_{19}H_{16}N_4O_3$			
Benzoic acid, 4-[[(2-amino-1,4-dihydro-4-oxo-6-quinazolinyl)methyl]-2-propynylamino]-	pH 13	228.5(4.63),279(4.39), 300s(4.27)	87-1114-86
1,4-Epoxynaphthalene-6,6,7,7-tetracarbonitrile, 1-(dimethoxymethyl)-1,2,3,4,5,8-hexahydro-2,3-bis(methylene)-	MeCN	246(3.89)	33-1287-86
1H,5H-Pyrazolo[1,2-a]pyrazol-1-one, 7-amino-3-hydroxy-5-imino-2-(4-methoxyphenyl)-6-phenyl-	MeCN	273(4.68)	118-0239-86
1H,5H-Pyrazolo[1,2-a]pyrazol-1-one, 7-amino-3-hydroxy-5-imino-6-(4-methoxyphenyl)-2-phenyl-	MeCN	278(4.59)	118-0239-86
$C_{19}H_{16}N_6OS$			
Carbonothioic dihydrazide, (di-2-pyridinylmethylene)[(4-hydroxyphenyl)-methylene]-	benzene	350(4.51)	131-0517-86I
	H_2O	335(4.51)	131-0517-86I
	MeOH	342(4.53)	131-0517-86I
	EtOH	344(4.52)	131-0517-86I
	dioxan	349(4.56)	131-0517-86I
	ether	350(4.57)	131-0517-86I
	acetone	345(4.57)	131-0517-86I
	$CHCl_3$	349(4.54)	131-0517-86I
	CCl_4	350(4.51)	131-0517-86I
	DMF	348(4.53)	131-0517-86I
	DMSO	350(4.51)	131-0517-86I
$C_{19}H_{16}O$			
Methanone, (5-methyl-5H-benzocyclohepten-5-yl)phenyl-	EtOH	217(4.09),240s(4.00), 270s(3.83)	39-1479-86C
1-Propanone, 2-(1-naphthalenyl)-1-phenyl-	EtOH	227(4.30),239(4.08), 271s(3.87),281(3.89)	39-1479-86C
$C_{19}H_{16}OSe_2$			
Benzene, 1,1'-[(phenoxymethylene)-bis(seleno)]bis-	hexane	225(4.20),295(3.52)	104-0092-86
$C_{19}H_{16}O_2$			
7,8-Chrysenediol, 7,8-dihydro-5-methyl-, trans	EtOH	223(4.68),248(4.51), 270(4.59),316(4.13)	44-1407-86
Dibenzo[c,e]tricyclo[5.2.0.0^{2,9}]nona-3,5-diene-8-carboxylic acid, methyl ester	MeOH	213(4.29),263(3.89), 296s(2.88)	80-1019-86
$C_{19}H_{16}O_2S_2$			
1,3-Dithiole-4-carboxylic acid, 2-(diphenylmethylene)-, ethyl ester	EtOH	224s(3.91),324(3.67), 390s(2.71)	33-0419-86
$C_{19}H_{16}O_3$			
1,2-Chrysenediol, 3,4-epoxy-1,2,3,4-tetrahydro-5-methyl-, trans	EtOH	194(4.05),206(4.06), 209(4.09),220(4.24), 260(4.56)	44-1407-86

Compound	Solvent	λmax(log ε)	Ref.
7,8-Chrysenediol, 9,10-epoxy-7,8,9,10-tetrahydro-5-methyl-, trans-anti	EtOH	214(4.46),259(4.82), 294(4.08),305(4.08)	44-1407-86
syn	EtOH	213(4.24),256(4.65), 304(3.87)	44-1407-86
$C_{19}H_{16}O_4$			
2H-1-Benzopyran-2-one, 6-[2-(2,3-dimethoxyphenyl)ethenyl]-, cis	MeOH	228.5(4.42),279(4.49)	94-3727-86
2H-1-Benzopyran-2-one, 6-[2-(3,4-dimethoxyphenyl)ethenyl]-, cis	MeOH	233(3.99),283.5(3.89), 320(3.77)	94-3727-86
trans	MeOH	234(4.18),287.5(4.25), 320(4.21),324.5(4.26)	94-3727-86
5,12-Naphthacenedione, 7,8,9,10-tetrahydro-1,9-dihydroxy-9-methyl-	MeOH	224(4.20),259(4.11), 286(3.85),431(3.66), 450(3.62)	89-0339-86
2H,5H-Pyrano[2,3-b][1]benzopyran-5-one, 3,4-dihydro-2-(hydroxymethyl)-2-phenyl-	EtOH	205(4.47),224(4.36), 275(4.08)	32-0491-86
2H,5H-Pyrano[3,2-c][1]benzopyran-5-one, 3,4-dihydro-2-(hydroxymethyl)-2-phenyl-	EtOH	270(3.96),282(4.02), 305(3.95),318(3.78)	32-0501-86
$C_{19}H_{16}O_5$			
3,11a-Epoxy-11aH-oxepino[2,3-b][1]-benzopyran-6(2H)-one, 3,4,5,5a-tetrahydro-5a-hydroxy-3-phenyl-, (3α,5aα,11aα)-	EtOH	251(3.93),311(3.39)	32-0285-86
(3α,5aβ,11aα)-	EtOH	251(3.94),312(3.40)	32-0285-86
9,10-Phenanthrenedione, 1,8-dihydroxy-2-methyl-3-(1-oxobutyl)-	MeOH	242(4.26),288(4.14), 401(3.50),524(3.65)	88-0143-86
$C_{19}H_{16}O_7$			
4H-1-Benzopyran-4-one, 2-(1,3-benzodioxol-5-yl)-5,6,7-trimethoxy-	MeOH	267(4.17),330(4.25)	102-2625-86
$C_{19}H_{16}O_8$			
4H-1-Benzopyran-4-one, 2 - [4-hydroxy-3 - (isobutyroxy)phenyl]-3,5,7-trihydroxy- (plus isomer)	MeOH	257(3.82),272s(3.60), 304s(3.43),372(3.95)	23-0514-86
	MeOH-$AlCl_3$	228(3.79),271(3.90), ?(3.28),458(4.12)	23-0514-86
	MeOH-$AlCl_3$-HCl	266(3.87),340(3.32), 360(3.53),428(4.00)	23-0514-86
	MeOH-NaOMe	246(3.57),332(4.00)	23-0514-86
	MeOH-NaOAc	275(3.70),329(3.71), 396(3.77)	23-0514-86
	MeOH-NaOAc-H_3BO_3	260(3.77),296(3.60), 386(3.82)	23-0514-86
Quercetin 3-O-isobutyrate	MeOH	253(4.10),262(4.09), 347(4.04)	23-0514-86
	MeOH-$AlCl_3$	275(4.22),304(3.69), 332(3.57),435(4.24)	23-0514-86
	MeOH-$AlCl_3$-HCl	268(4.16),298(3.86), 358(4.04),393(4.04)	23-0514-86
	MeOH-NaOMe	269(4.21),328s(3.81), 397(4.19)	23-0514-86
	MeOH-NaOAc	259(4.23),260s(3.67), 366(4.13)	23-0514-86
$C_{19}H_{17}BO_5$			
Boron, [α-hydroxy-α-phenylbenzeneacetate(2-)](2,4-pentanedionato-O,O')-	CH_2Cl_2	254s(3.22),259s(3.36), 291(4.11)	48-0755-86

Compound	Solvent	λ_{max}(log ε)	Ref.
$C_{19}H_{17}Br_4NOS_2$			
4-Piperidinone, 3,5-bis[(4,5-dibromo-2-thienyl)methylene]-2,2,6,6-tetramethyl-	EtOH	247(4.35),296(4.33), 385(4.23),416(4.26)	70-2511-86
$C_{19}H_{17}ClN_2O$			
2H-Benz[g]indazole, 2-acetyl-3-(3-chlorophenyl)-3,3a,4,5-tetrahydro-	EtOH	295(3.50)	4-0135-86
2H-Benz[g]indazole, 2-acetyl-3-(4-chlorophenyl)-3,3a,4,5-tetrahydro-	EtOH	290(3.59)	4-0135-86
4-Quinolinol, 7-chloro-3-(1,2,3,4-tetrahydro-3-methyl-1-isoquinolinyl)-	EtOH	255(4.58),290(3.71), 325(4.05)	2-1072-86
$C_{19}H_{17}ClN_2O_2$			
3-Quinolinecarboxamide, 7-chloro-4-hydroxy-N-(1-methyl-2-phenylethyl)-	EtOH	260(4.41),315(4.01)	2-1072-86
$C_{19}H_{17}ClN_4$			
Quinoline, 7-chloro-3-(3,4-dihydro-3-methyl-1-isoquinolinyl)-4-hydrazino-	EtOH	245(4.55),295(3.90)	2-1072-86
$C_{19}H_{17}ClN_4O_2$			
Acetamide, N-[[4-(2-benzoyl-4-chlorophenyl)-5-methyl-4H-1,2,4-triazol-3-yl]methyl]-	H_2O	260(3.96)	87-1346-86
$C_{19}H_{17}ClN_4O_2S$			
5H-Pyrimido[5,4-d][1]benzazepin-4-amine, 9-chloro-6,7-dihydro-7-[(4-methylphenyl)sulfonyl]-	EtOH	241(4.33),268s(--), 294(3.42),300s(--)	142-0143-86
$C_{19}H_{17}N$			
Dicyclopenta[ef,kl]heptalene-1-methanamine, N,N-dimethyl-	hexane	254(4.40),267(4.59), 287(4.22),302(4.00), 312(3.98),336(3.68), 347(3.74),359(3.26), 408(2.70),442(2.85), 452(2.90),460(2.78), 472(3.15),484(3.69)	44-2961-86
$C_{19}H_{17}NO$			
Azacyclooctadeca-3,5,7,13,15,17-hexaene-9,11-diyn-2-one, 8,13-dimethyl-, (Z,E,Z,Z,E,E)-	THF	267s(4.09),296s(4.34), 333(4.63),403s(3.80)	39-0933-86C
Bicyclo[4.4.1]undeca-3,6,8,10-tetraen-2-one, 5-[(1-methyl-2(1H)-pyridinylidene)ethylidene]-	benzene	366(3.45),533(4.45)	33-0315-86
	MeOH	234(4.17),598(4.67)	33-0315-86
	CH_2Cl_2	368(3.71),567(4.58)	33-0315-86
	MeCN	380(3.73),584(4.58)	33-0315-86
Bicyclo[4.4.1]undeca-3,6,8,10-tetraen-2-one, 5-[(1-methyl-4(1H)-pyridinylidene)ethylidene]-, (E)-(±)-	benzene	349(--),552(--)	33-0315-86
	MeOH	247(4.22),324(3.80), 481(4.05),624(4.40)	33-0315-86
	CH_2Cl_2	366(3.50),584s(4.50), 603(4.51)	33-0315-86
	MeCN	365(3.40),609(4.31)	33-0315-86
2,4,6,12,14,16-Cycloheptadecahexaene-8,10-diyn-1-one, 7,12-dimethyl-, oxime, (E,E,Z,Z,E,E)-	THF	216(4.31),301(4.68), 384s(3.93)	39-0933-86C
1H-Pyrrolizin-1-one, 2,3,5,6-tetrahydro-6,7-diphenyl-	MeOH	204(4.40),224(4.29), 265(4.02),287(4.03)	83-0749-86

Compound	Solvent	$\lambda_{max}(\log \epsilon)$	Ref.
$C_{19}H_{17}NO_2$			
Benz[c]acridin-5-ol, 5,6-dihydro-6-methoxy-7-methyl-, trans	MeOH	259.9(3.32),267.2(3.36), 294(2.82),311.5(2.73), 326.0(2.71),341.2(2.69)	39-0741-86C
Benz[c]acridin-6-ol, 5,6-dihydro-5-methoxy-7-methyl-, trans	MeOH	257.9(3.28),267.7(3.31), 293.3(2.78),312.4(2.71), 326.4(2.69),341.5(2.66)	39-0741-86C
2-Dibenzofuranol, 6,7,8,9-tetrahydro-1-[(phenylimino)methyl]-	toluene	350(4.03)	104-0503-86
	EtOH	347(4.04)	104-0503-86
	MeCN	344(4.05)	104-0503-86
5(4H)-Oxazolone, 2-phenyl-4-[(2,4,6-trimethylphenyl)methylene]-, (Z)-	MeCN	348(4.29)	23-2064-86
3-Pyridinecarboxylic acid, 5-(diphenylmethylene)-1,2,5,6-tetrahydro-	pH 7.2	210(1.41),240(1.34), 308(1.92)	87-0125-86
$C_{19}H_{17}NO_2S$			
4H-Thiopyran-4-one, 5-acetyl-2,3-dihydro-2-phenyl-6-(phenylamino)-	EtOH	254(3.18),313(3.23)	5-1109-86
$C_{19}H_{17}NO_3$			
Benzeneacetamide, N-(2-methyl-1-oxo-2-propenyl)-N-(4-methylphenyl)-α-oxo-	C_6H_{12}	213(4.24),253(4.22)	39-1759-86C
Benzeneacetamide, N-(2-methyl-1-oxo-2-propenyl)-α-oxo-N-(phenylmethyl)-	C_6H_{12}	214(4.18),257(4.09), 360(1.48)	39-1759-86C
Benzeneacetamide, N-(4-methylphenyl)-α-oxo-N-(1-oxo-2-butenyl)-	C_6H_{12}	240(4.40)	39-1759-86C
Dehydrostephalagine	EtOH	264(4.72),330(4.09)	100-1078-86
6-Oxa-3-azabicyclo[3.1.1]heptane-2,4-dione, 1-methyl-3-(4-methylphenyl)-5-phenyl-	C_6H_{12}	222(3.99),257(3.04), 264(2.91)	39-1759-86C
6-Oxa-3-azabicyclo[3.1.1]heptane-2,4-dione, 1-methyl-5-phenyl-3-(phenylmethyl)-	C_6H_{12}	215(4.19),252(2.85), 258(2.88),264(2.81)	39-1759-86C
Pyrano[2,3-a]carbazole-8-carboxylic acid, 2,11-dihydro-2,2-dimethyl-, methyl ester	n.s.g.	195(3.38),255(3.12), 310(3.40)	142-2195-86
Pyrano[2,3-a]carbazole-9-carboxylic acid, 2,11-dihydro-2,2-dimethyl-, methyl ester	n.s.g.	196(3.25),248(3.52), 326(3.34)	142-2195-86
Pyrano[2,3-a]carbazole-10-carboxylic acid, 2,11-dihydro-2,2-dimethyl-, methyl ester	n.s.g.	198(3.52),247(3.64), 317(3.20)	142-2195-86
2-Pyrrolidinone, 1-(1,3-dioxo-2,3-diphenylpropyl)-	n.s.g.	207(4.53),245s(--)	83-0231-86
$C_{19}H_{17}NO_3S$			
1H-Indole, 2-(3-methyl-1-oxo-2-butenyl)-1-(phenylsulfonyl)-	MeOH	247(4.25),259(4.19), 269(4.16),295(4.23)	44-2343-86
1H-Indole, 3-(3-methyl-1-oxo-2-butenyl)-1-(phenylsulfonyl)-	MeOH	224(4.20),242(4.21), 296(4.08)	44-2343-86
2H-Thiopyran-5-carboxylic acid, 3,4-dihydro-4-oxo-2-phenyl-6-(phenylamino)-	EtOH	248(3.16),307(3.32)	5-1109-86
$C_{19}H_{17}NO_4$			
3H-Indol-3-one, 1-acetyl-2-(acetoxyphenylmethyl)-1,2-dihydro-	EtOH	208(3.86),239(4.08), 262(3.67),350(3.24)	104-2186-86
$C_{19}H_{17}NO_5$			
4,5-Benzo[b]furandicarboxylic acid,	dioxan	267(3.48),310(3.41),	73-1455-86

Compound	Solvent	λ_{max}(log ε)	Ref.
6-amino-2-(4-methylphenyl)-, dimethyl ester (cont.)		383(3.34)	73-1455-86
2-Butenedioic acid, (4-methyl-2-phenyl-4H-furo[3,2-b]pyrrol-5-yl)-, dimethyl ester	MeOH	415(3.53)	73-1455-86
Eschscholtzine N-oxide	MeOH	230s(3.74),293(3.90)	100-0922-86
$C_{19}H_{17}NO_6$			
D-Ribitol, 1,4-anhydro-1-C-2-benzoxazolyl-, 5-benzoate, (R)-	n.s.g.	232(4.40),273(3.87), 281(3.81)	136-0271-86H
$C_{19}H_{17}N_2$			
3H-Indolium, 1-methyl-3-[(1-methyl-1H-indol-3-yl)methylene]-, tetrafluoroborate	MeOH	523(5.96)	5-1621-86
$C_{19}H_{17}N_3O$			
Evodiamine	MeCN	268(4.12),281(4.09), 291(3.98),346(3.19)	36-0612-86
$C_{19}H_{17}N_3O_2Se$			
Benzoic acid, 3-[[(1,5-dihydro-3-methyl-1-phenyl-5-selenoxo-4H-pyrazol-4-ylidene)methyl]amino]-, methyl ester	EtOH	346(4.2),450(3.6)	104-2126-86
$C_{19}H_{17}N_3O_3$			
L-Alanine, N-[2-(1-cyano-2H-benz[f]-isoindol-2-yl)-1-oxopropyl]-, monosodium salt, (S)-	H_2O	400s(3.63),420(3.83), 444(3.78)	44-3978-86
2H-Benz[g]indazole, 2-acetyl-3,3a,4,5-tetrahydro-3-(4-nitrophenyl)-	EtOH	289(4.29)	4-0135-86
Ethanone, 1-[1,4-dihydro-6-methyl-4-(3-nitrophenyl)-2-phenyl-5-pyrimidinyl]-	EtOH	245(4.35),324(3.70)	39-0083-86C
$C_{19}H_{17}N_5$			
6H-Purin-6-imine, 3,7-dihydro-3,7-bis(phenylmethyl)-, monohydrobromide	pH 1	225s(4.10),279(4.24)	94-1821-86
	pH 7	225s(4.10),279(4.24)	94-1821-86
	pH 13	282(4.18)	94-1821-86
	EtOH	226s(4.08),282(4.27)	94-1821-86
1H-Pyrazolo[3,4-b]quinoxaline, 1-phenyl-3-(1-pyrrolidinyl)-	MeOH	300(5.30),340s(--)	2-0960-86
$C_{19}H_{17}N_5O_4$			
2-Propenoic acid, 3-[10-cyano-3,7-dihydro-7-oxo-3-(tetrahydro-2H-pyran-2-yl)pyrido[2,1-i]purin-8-yl]-, methyl ester	CH_2Cl_2	267(4.4),417(4.4), 432(4.4),440(4.5)	39-1561-86B
$C_{19}H_{17}N_7O_3$			
Propanamide, N-[2-[(2,6-dicyano-4-nitrophenyl)azo]-5-(dimethylamino)-phenyl]-	$CHCl_3$	268s(3.90),301(3.78), 418(3.36),577(4.64), 618(4.83)	48-0497-86
$C_{19}H_{18}$			
Benzene, 1-(1-cyclopenten-1-yl)-2-(2-phenylethenyl)-, (E)-	MeOH	218(4.07),254(3.96), 286(3.89),326(3.32)	151-0105-86A
$C_{19}H_{18}ClNO_3S$			
1H-Indole, 2-(3-chloro-3-methyl-1-oxobutyl)-1-(phenylsulfonyl)-	MeOH	236(3.98),287(4.14)	44-2343-86

Compound	Solvent	$\lambda_{max}(\log \epsilon)$	Ref.
$C_{19}H_{18}ClN_3O_4S$			
4-Morpholineacetamide, N-[6-chloro-2-[(2-oxo-2-phenylethyl)thio]-3-pyridinyl]-α-oxo-	EtOH	248(4.33),304(3.91)	103-0800-86
$C_{19}H_{18}Cl_3N_2O_3P$			
Phosphonic acid, [3,6-dichloro-4-[(4-chlorophenyl)amino]-2-quinolinyl]-, diethyl ester	EtOH	238(4.72),366(3.90)	104-0973-86
$C_{19}H_{18}F_3N_3OS$			
Spiro[cyclohexane-1,1'(2'H)-naphthalene]-2',2',4'-tricarbonitrile, 3',5',6',7',8',8'a-hexahydro-3'-oxo-5'-[(trifluoromethyl)thio]-	EtOH	252.5(4.06)	155-0201-86
$C_{19}H_{18}N_2O$			
Acetamide, N-(1-methyl-2,5-diphenyl-1H-pyrrol-3-yl)-	EtOH	209(4.32),305(4.22)	103-0499-86
Ethanone, 1-(1,4-dihydro-6-methyl-2,4-diphenyl-5-pyrimidinyl)-	EtOH	248(4.39),329(3.59)	39-0083-86C
$C_{19}H_{18}N_2O_2$			
Benzene, 4-nitro-N-(1-phenyl-2-heptynylidene)-	PrOH	223(4.34),299(4.53)	30-0176-86
Ethanone, 1-[4-[[3-[(4-acetylphenyl)-amino]-2-propenylidene]amino]phenyl]-	MeOH	393(4.74)	94-1794-86
	EtOH	394(4.73)	94-1794-86
$C_{19}H_{18}N_2O_2S$			
Benzothiazolium, 3-(carboxymethyl)-2-[2-[4-(dimethylamino)phenyl]ethenyl]-, hydroxide, inner salt	H_2O	521(4.72)	4-0209-86
$C_{19}H_{18}N_2O_3$			
Benzoic acid, 4-[5-[4-(dimethylamino)-phenyl]-2-oxazolyl]-, methyl ester	toluene	353(4.11)	103-1022-86
7H-1-Benzopyrano[5,6-b][1,8]naphthyridin-7-one, 3,12-dihydro-6-methoxy-3,3,12-trimethyl-	MeOH	216(4.21),231(4.23), 268s(4.44),284(4.59), 293(4.58),308s(4.13), 340(2.56),395(3.79)	83-0025-86
3,5-Pyrazolidinedione, 1-ethyl-4-[(4-methoxyphenyl)methylene]-2-phenyl-	EtOH	252(4.46),374(4.46)	103-0176-86
Sydnone, 3-[2-(4-pentenyloxy)phenyl]-4-phenyl-	EtOH	244(4.00),260s(3.79), 279s(3.68),332.5(4.06)	33-0927-86
Sydnone, 3-[3-(4-pentenyloxy)phenyl]-4-phenyl-	EtOH	241(4.05),332.5(4.01)	33-0927-86
$C_{19}H_{18}N_2O_4$			
2-Anthracenecarboxylic acid, 1-(2,2-dimethylhydrazino)-9,10-dihydro-9,10-dioxo-, ethyl ester	n.s.g.	273(4.04),313(3.72), 483(3.76)	104-2117-86
$C_{19}H_{18}N_4OS_2$			
Propanamide, N-2-benzothiazolyl-3-(2-benzothiazolylmethylamino)-N-methyl-	pH 1	210(4.55),259(4.18), 281(4.12),289(4.13)	103-0795-86
	EtOH	223(--),273(--), 300(--)	103-0795-86
$C_{19}H_{18}N_4O_2S$			
5H-Pyrimido[5,4-d][1]benzazepin-4-amine, 6,7-dihydro-7-tosyl-	EtOH	236(4.21),293(3.34), 329s(--)	142-0143-86

Compound	Solvent	$\lambda_{max}(\log \epsilon)$	Ref.
$C_{19}H_{18}N_{10}$			
Benzonitrile, 4,4'-[1,3,5,7-tetraaza-bicyclo[3.3.1]nonane-3,7-diyl-bis(azo)]bis-	EtOH-DMSO	302(4.54)	23-1567-86
$C_{19}H_{18}O$			
Cyclobutanone, 3,3-diphenyl-2-(1-prop-enyl)-, (E)-	EtOH	250(2.55),256(2.77), 261(2.76),272(2.38), 290(1.81),310(1.30)	35-6276-86
2-Cyclohexen-1-one, 4-methyl-5,5-di-phenyl-	EtOH	250(3.70),272(3.46), 290(2.70),320(1.45)	35-6276-86
$C_{19}H_{18}O_2$			
Azulene, 6-ethoxy-2-(4-methoxyphenyl)-	CH_2Cl_2	235.5(4.08),309(4.80), 316.5(4.87),396(4.29), 415.5(4.33),521(2.54)	24-2272-86
Cycloprop[a]inden-6(1H)-one, 1a,6a-di-hydro-1-(4-methoxyphenyl)-1-methyl-, endo	MeOH	225(4.42),257s(3.94), 308s(3.18)	39-2011-86C
exo	MeOH	231(4.49),255s(3.89), 304s(3.28)	39-2011-86C
1(2H)-Naphthalenone, 2-(4-methoxyphen-yl)-2,4-dimethyl-	MeOH	238(4.51),258(3.76), 265(3.75),273(3.60)	39-2011-86C
1(4H)-Naphthalenone, 4-(4-methoxyphen-yl)-3,4-dimethyl-	MeOH	237(4.34),269s(4.08)	39-2011-86C
$C_{19}H_{18}O_4$			
2H-1-Benzopyran-2-one, 6-[2-(2,3-di-methoxyphenyl)ethyl]-	MeOH	217.5(4.62),277.5(4.20), 324(3.75)	94-3727-86
3,9-Phenanthrenedione, 4-hydroxy-6-methoxy-8-methyl-2-(1-methylethyl)-	MeOH	210(4.44),230(4.63), 300(4.57),340(4.23), 400(3.63)	88-5459-86
2-Propenoic acid, 2-methoxy-3-oxo-2,3-diphenylpropyl ester	THF	250(--),340(2.35)	121-0355-86
2-Propenoic acid, 2-formyl-3-(4-meth-ylphenoxy)-3-phenyl-, ethyl ester 40:60 E:Z	EtOH	244(4.37),278(4.20)	44-4112-86
$C_{19}H_{18}O_5$			
2H,8H,12H-Benzo[1,2-b:3,4-b':5,6-b"]-tripyran-8,12-dione, 3,4-dihydro-2,2,6,10-tetramethyl-	MeOH	203(3.76),231(3.55), 263(3.84),304(3.25)	2-0377-86
$C_{19}H_{18}O_6$			
4H-1-Benzopyran-4-one, 3-(3,4-dimeth-oxyphenyl)-7,8-dimethoxy-	$CHCl_3$	256(4.49),290(4.17)	142-1943-86
4H-1-Benzopyran-4-one, 3,5,7,8-tetra-methoxy-2-phenyl-	MeOH	231(4.13),270(4.46), 290s(4.14),345(4.04)	102-1255-86
4H-1-Benzopyran-4-one, 5,7,8-trimeth-oxy-2-(2-methoxyphenyl)-	MeOH	238s(4.21),269(4.50), 336(4.19)	94-0406-86
$C_{19}H_{18}O_7$			
4H-1-Benzopyran-4-one, 2-(4-hydroxy-3-methoxyphenyl)-5,6,7-trimethoxy-(ageconyflavone B)	MeOH	267(4.23),332(4.32)	102-2625-86
	MeOH-NaOMe	267(--),418(--)	102-2625-86
	MeOH-NaOAc	267(--),415(--)	102-2625-86
	+ H_3BO_3	270(--),335(--)	102-2625-86
	MeOH-$AlCl_3$	267(--),332(--)	102-2625-86
4H-1-Benzopyran-4-one, 5-hydroxy-3,6,7,8-tetramethoxy-	EtOH	282(4.02),317(3.91), 366(3.66)	105-0230-86
4H-1-Benzopyran-4-one, 5-hydroxy-6,7,8-trimethoxy-2-(2-methoxyphenyl)-	MeOH	266(4.48),338(3.84)	94-0406-86
	+ NaOMe	267(4.47),365(3.72)	94-0406-86

Compound	Solvent	λ_{max}(log ε)	Ref.
(cont.)	MeOH-NaOAc	266(4.47),338(3.83)	94-0406-86
	MeOH-AlCl$_3$	276(4.48),300s(4.16), 323(3.98),400(3.83)	94-0406-86
	MeOH-AlCl$_3$-HCl	276(4.48),300s(4.17), 320(3.97),400(3.84)	94-0406-86
	MeOH-NaOAc-H$_3$BO$_3$	266(4.52),338(3.90)	94-0406-86
4H,6H-Furo[3',2':3,4]naphtho[1,8-cd]-pyran-4,6-dione, 8,9-dihydro-2,3-dihydroxy-7-methoxy-1,8,8,9-tetra-methyl-, (S)-	MeOH	211(4.45),234(4.24), 257(4.43),309(3.51), 356(3.92),396s(4.03), 409(4.04)	23-1585-86
$C_{19}H_{18}O_9$			
4H-1-Benzopyran-4-one, 2-(2,4-dihy-droxy-5-methoxyphenyl)-5-hydroxy-3,7,8-trimethoxy-	MeOH	263(4.0),364(3.7)	102-1257-86
	MeOH-NaOMe	268(--),416(--)	102-1257-86
	MeOH-AlCl$_3$	277(--),370s(--), 425(--)	102-1257-86
$C_{19}H_{19}BrN_6O_3$			
Acetamide, N-[2-[(2-bromo-6-cyano-4-nitrophenyl)azo]-5-(diethylamino)-phenyl]-	CHCl$_3$	310(3.86),361(3.49), 372(3.54),588s(3.70), 613(4.73)	48-0497-86
$C_{19}H_{19}ClN_6O_3$			
Acetamide, N-[2-[(2-chloro-6-cyano-4-nitrophenyl)azo]-5-(diethylamino)-phenyl]-	CHCl$_3$	310(3.87),353s(3.52), 604(4.71)	48-0497-86
$C_{19}H_{19}F_3N_2O_4$			
Butanoic acid, 3-[[2-amino-4-(triflu-oromethyl)phenyl](1-hydroxy-2-oxo-2-phenylethyl)amino]-	n.s.g.	204(4.33),252(4.45), 284s(4.02),292s(4.00), 364(3.91)	150-2471-86M
$C_{19}H_{19}F_3N_4O$			
Piperazine, 1-acetyl-4-[4-[[4-(tri-fluoromethyl)phenyl]azo]phenyl]-	pH 1	537(4.80)	39-0123-86B
	EtOH	405(4.39)	39-0123-86B
$C_{19}H_{19}F_3N_4S$			
Spiro[cyclohexane-1,1'(2'H)-naphtha-lene]-2',2',4'-tricarbonitrile, 3'-amino-5',6',7',8'-tetrahydro-5'-[(trifluoromethyl)thio]-	EtOH	222.2(4.11),328.9(3.69)	155-0201-86
$C_{19}H_{19}N$			
1H-Indole, 2-(1-methylethenyl)-1-(1-phenylethyl)-	MeOH	205(4.44),215(4.41), 221s(4.38),287s(4.02), 291.5(4.04)	1-0625-86
$C_{19}H_{19}NO$			
Azacyclohexadeca-1,3,5,11,13,15-hexa-ene-7,9-diyne, 2-ethoxy-6,11-di-methyl-, (E,E,E,Z,Z,E)-	THF	250s(4.37),287(4.67), 345s(3.77),479(2.55)	39-0933-86C
2-Oxiranecarbonitrile, 3-(2-ethynyl-1-cyclooctenyl)-2-phenyl-, cis	MeCN	220(4.12),247(4.04)	78-2221-86
1H-Pyrrolizin-1-one, hexahydro-6,7-diphenyl-	MeOH	204(4.39),258(3.02)	83-0749-86
$C_{19}H_{19}NOS$			
1-Benzothiepin-5(2H)-one, 3,4-dihydro-4-[[methyl(phenylmethyl)amino]-methylene]-, (E)-	EtOH	248(4.02),350(4.20)	4-0449-86

Compound	Solvent	λ_{max}(log ϵ)	Ref.
$C_{19}H_{19}NO_2$			
1H-Indole-2-carboxylic acid, 1-methyl-3-(1-phenylethyl)-, methyl ester	MeOH	206(4.45),230(4.47), 296(4.33),311s(4.08)	1-0625-86
Methanone, [6-(diethylamino)-2-benzofuranyl]phenyl-	EtOH	260(3.93),344(3.79), 414(3.90)	142-0771-86
$C_{19}H_{19}NO_2S$			
1H-Indole, 2-(3-methyl-2-butenyl)-1-(phenylsulfonyl)-	MeOH	248(4.15)	44-2343-86
1H-Indole, 3-(3-methyl-2-butenyl)-1-(phenylsulfonyl)-	MeOH	249(4.02),279(3.53), 286(3.51)	44-2343-86
$C_{19}H_{19}NO_3$			
6H-Benzo[g]-1,3-benzodioxolo[5,6-a]quinolizine, 5,8,13,13a-tetrahydro-10-methoxy-, (±)-	MeOH	208(4.46),286.5(3.94)	2-0256-86
4H-Dibenzo[de,g]quinoline, 5,6-dihydro-1,2,3-trimethoxy- (O-methyldehydroisopiline)	EtOH	214(4.48),260(4.69), 327(3.96)	100-0878-86
6H-Dibenzo[de,g]quinoline-6-carboxaldehyde, 4,5,6a,7-tetrahydro-1,2-dimethoxy-	EtOH	212(4.48),226s(4.32), 269(4.26),312s(3.74)	100-0878-86
9,13a-Methano-13aH-cyclonona[c]quinoline-6,8,14(5H,9H)-trione, 5,9-dimethyl-10,11,12,13-tetrahydro-	n.s.g.	213(4.08),235(4.24), 342(3.42)	150-2782-86M
2-Propen-1-one, 1-(4-hydroxyphenyl)-3-(4-morpholinyl)-3-phenyl-	EtOH	243(3.85),355(4.51)	104-1658-86
$C_{19}H_{19}NO_3S$			
1H-Indole, 3-(3-methyl-1-oxobutyl)-1-(phenylsulfonyl)-	MeOH	217(4.38),265(3.96), 273(4.01),285(4.04)	44-2343-86
1H-Indole-3-methanol, α-(2-methyl-1-propenyl)-1-(phenylsulfonyl)-	MeOH	249(4.11),276(3.74), 283(3.71)	44-2343-86
$C_{19}H_{19}NO_4$			
5H-Benzo[g]-1,3-benzodioxolo[6,5,4-de]quinolin-12-ol, 6,7,7a,8-tetrahydro-4-methoxy-7-methyl-, (R)- (N-methylelmerrilicine)	EtOH	222(4.46),240s(4.23), 268s(4.12),276(4.15), 298(3.98)	100-1078-86
3H-[1]Benzopyrano[4,3-c]isoxazole-9b(1H)-acetic acid, 3a,4-dihydro-1-phenyl-, methyl ester	EtOH	248(3.68),277(3.56)	44-3125-86
Stepharine, N-formyl-, (-)-	MeOH	213(4.58),232(4.35), 282(3.50)	94-1148-86
$C_{19}H_{19}NO_6S$			
5-Azatetracyclo[4.3.0.0^{2,4}.0^{3,7}]non-8-ene-8,9-dicarboxylic acid, 5-(4-methylphenylsulfonyl)-, dimethyl ester	MeCN	231(4.24),263s(3.70), 274s(3.43)	24-0616-86
9-Azatricyclo[4.2.1.0^{2,5}]nona-3,7-diene-7,8-dicarboxylic acid, 9-[(4-methylphenyl)sulfonyl]-, dimethyl ester, (1α,2β,5β,6α)-	MeCN	227(4.19),265s(3.37)	24-0616-86
$C_{19}H_{19}NS_3$			
6H-1,3-Thiazine, 6-(ethylthio)-6-(methylthio)-2,4-diphenyl-	$CHCl_3$	261(4.32),314(3.75)	118-0916-86
$C_{19}H_{19}N_3O_4$			
Ethanone, 1-[1,4,5,6-tetrahydro-4-hy-	EtOH	210(4.35),250(4.13)	39-0083-86C

Compound	Solvent	λ_{max}(log ε)	Ref.
droxy-4-methyl-6-(3-nitrophenyl)-2-phenyl-5-pyrimidinyl]- (cont.)			39-0083-86C
$C_{19}H_{19}N_3O_5$			
1H-Benz[de]isoquinoline-1,3(2H)-dione, 2-[3-(4-morpholinyl)propyl]-5-nitro-	pH 7.4	335(3.92)	4-0849-86
3,5-Pyrazolidinedione, 4-[methoxy(4-nitrophenyl)methyl]-1,4-dimethyl-2-phenyl-	EtOH	240(3.88)	103-0176-86
$C_{19}H_{19}N_5O$			
Piperazine, 1-acetyl-4-[4-[(4-cyanophenyl)azo]phenyl]-	pH 1	540(4.84)	39-0123-86B
	EtOH	421(4.45)	39-0123-86B
$C_{19}H_{19}N_7O_5$			
Acetamide, N-[2-[(2-cyano-4,6-dinitrophenyl)azo]-5-(diethylamino)phenyl]-	$CHCl_3$	290s(3.67),349(3.35), 442s(3.37),585s(4.58), 621(4.72)	48-0497-86
$C_{19}H_{20}$			
Bicyclo[4.1.0]heptane, 1,7-diphenyl-	hexane	260(2.79),266.5(2.75), 277(2.53)	35-1617-86
$C_{19}H_{20}BrFN_2O_{11}$			
2,4(1H,3H)-Pyrimidinedione, 5-(bromomethyl)-6-fluoro-1-(2,3,4,6-tetra-O-acetyl-β-D-glucopyranosyl)-	$CHCl_3$	253(3.85)	97-0437-86
$C_{19}H_{20}ClN_2O_3P$			
Phosphonic acid, [3-chloro-4-(phenylamino)-2-quinolinyl]-, diethyl ester	EtOH	235(3.88)	104-0973-86
$C_{19}H_{20}ClN_3O_3S$			
Ethanediamide, N-[6-chloro-2-[(2-oxopentyl)thio]-3-pyridinyl]-N'-(phenylmethyl)-	EtOH	230(4.19),282(3.97), 308(3.92)	103-0800-86
$C_{19}H_{20}N_2$			
3H-Indazole, 3a,4,5,6,7,7a-hexahydro-3,7a-diphenyl-	hexane	252(2.87),259(2.80), 265(2.59),340(2.36)	35-1617-86
Pyridinium, 2,2'-bicyclo[2.2.1]hepta-2,5-diene-2,3-diylbis[1-methyl-, bis(tetrafluoroborate)	EtOH	284(3.99),328s(3.64)	138-1279-86
Pyridinium, 3,3'-bicyclo[2.2.1]hepta-2,5-diene-2,3-diylbis[1-methyl-, bis(tetrafluoroborate)	EtOH	245(4.03),280s(3.84), 314(3.79)	138-1279-86
Pyridinium, 4,4'-bicyclo[2.2.1]hepta-2,5-diene-2,3-diylbis[1-methyl-, bis(tetrafluoroborate)	EtOH	233(4.03),255(3.98), 286(4.00),364(3.80)	138-1279-86
$C_{19}H_{20}N_2O$			
2H-Benz[g]indazole, 3,3a,4,5-tetrahydro-3-(4-methoxyphenyl)-2-methyl-	EtOH	227(4.08),313(4.26)	4-0135-86
$C_{19}H_{20}N_2OS_2$			
[1]Benzothieno[2,3-d]pyrimidine-4(3H)-thione, 5,6,7,8-tetrahydro-3-(2-hydroxyethyl)-2-(4-methylphenyl)-	EtOH	241(4.39),291(4.07), 329(4.12),362s(3.93)	106-0096-86

Compound	Solvent	$\lambda_{max}(\log \epsilon)$	Ref.
$C_{19}H_{20}N_2O_2$			
Ethanone, 1-(1,4,5,6-tetrahydro-4-hydroxy-4-methyl-2,6-diphenyl-5-pyrimidinyl)-	EtOH	203(4.27),223(4.20)	39-0083-86C
Pyrrolo[3,4-b]indol-3(2H)-one, 1,3a,4,8b-tetrahydro-3a-methoxy-2-methyl-8b-(phenylmethyl)-	EtOH	239(3.73),292.5(3.34)	118-0731-86
$C_{19}H_{20}N_2O_3$			
10(9H)-Acridineacetic acid, 9-oxo-, 2-(dimethylamino)ethyl ester, hydrochloride	H_2O	254(4.69),383(3.96), 398(3.97)	56-0615-86
1H-Benz[de]isoquinoline-1,3(2H)-dione, 2-[3-(4-morpholinyl)propyl]-	pH 7.4	344(4.17)	4-0849-86
Cyanamide, [2-(2-benzoyl-4,5-dimethoxyphenyl)ethyl]methyl-	MeOH	243(4.12),286(3.59), 308(3.52)	12-0893-86
3,5-Pyrazolidinedione, 4-(methoxyphenylmethyl)-1,4-dimethyl-2-phenyl-	EtOH	235(4.04)	103-0176-86
$C_{19}H_{20}N_2O_4$			
1H-Benz[de]isoquinoline-1,3(2H)-dione, 5-hydroxy-2-[3-(4-morpholinyl)-propyl]-	pH 7.4	341(3.96),383(3.79)	4-0849-86
$C_{19}H_{20}N_2O_4S$			
Acetamide, N-[1-methoxy-1-[1-(phenylsulfonyl)-1H-indol-3-yl]-	MeOH	217(4.36),253(4.08), 281s(3.63),288s(3.59)	5-2065-86
$C_{19}H_{20}N_2O_5$			
Acetic acid, [7-(diethylamino)-1,4-dihydro-1,4-dioxo[1]benzopyrano[3,4-c]pyrrol-3(2H)-ylidene]-, ethyl ester	MeOH	510(4.49)	2-0137-86
$C_{19}H_{20}N_2O_8$			
19H-Dibenzo[k,n][1,4,7,10]tetraoxacyclopentadecin, 6,7,9,10,12,13-hexahydro-2,17-dinitro-	$CHCl_3$	245(3.93),315(4.30)	78-0137-86
$(C_{19}H_{20}N_4O_2)_n$			
Poly[N^{ε}-(phenylazobenzoyl)-L-lysine]	HFIP	230(4.01),318(4.30), 425(3.03)	116-2472-86
irradiated at 360 nm	HFIP	252(3.17),318(3.98), 425(3.24)	116-2472-86
$C_{19}H_{20}N_4O_5S$			
Spiro[1-azabicyclo[3.2.0]hept-2-ene-6,3'-[3H]pyrazole]-2-carboxylic acid, 3-(ethylthio)-4',5'-dihydro-4'-methyl-7-oxo-, (4-nitrophenyl)-methyl ester, [5R-[5α,6α(R*)]]-	MeCN	271(4.02),317(4.05), 326s(--)	39-0421-86C
[5R-[5α,6β(S*)]]-	MeCN	266(--),319(--)	39-0421-86C
5-Thia-1-azabicyclo[4.2.0]oct-2-ene-2-carboxylic acid, 7-(2,2-dimethyl-3-nitroso-5-oxo-4-phenyl-1-imidazolidinyl)-3-methyl-8-oxo-, [6R-(6α,7β(R*)]]-	EtOH	235s(4.08),375(1.88)	107-1469-86
[6R-(6α,7β(S*)]]-	EtOH	235s(4.06),375(1.87)	107-1469-86
$C_{19}H_{20}N_4O_7$			
1H-Purine-2,6-dione, 8-(5-O-benzoyl-	EtOH	214(4.07),230(4.00),	136-0271-86H

Compound	Solvent	λ_{max}(log ε)	Ref.
β-D-ribofuranosyl)-3,7-dihydro-1,3-dimethyl- (cont.)		277(3.91)	136-0271-86H
$C_{19}H_{20}N_4S$			
3H-Pyrazole-3-thione, 4-[[4-(dimethylamino)phenyl]amino]methylene]-2,4-dihydro-5-methyl-2-phenyl-	toluene	385(4.3),460(4.3)	103-0503-86
$C_{19}H_{20}N_4Se$			
3H-Pyrazole-3-selone, 4-[[[4-(dimethylamino)phenyl]amino]methylene]-2,4-dihydro-5-methyl-2-phenyl-	EtOH	390(4.2),470(4.3)	104-2126-86
$C_{19}H_{20}OSi$			
2-Propyn-1-one, 1-[2-(phenylmethyl)-phenyl]-3-(trimethylsilyl)-	EtOH	248s(3.93),268(4.07), 309s(3.34)	33-0560-86
$C_{19}H_{20}O_3$			
2H-1-Benzopyran-7-ol, 2-ethyl-3-(4-methoxyphenyl)-4-methyl-	EtOH	215(3.64),231(3.63), 293(3.61),316(3.62)	4-1781-86
2H-1-Benzopyran-7-ol, 4-ethyl-3-(4-methoxyphenyl)-2-methyl-	EtOH	212(3.65),228(3.63), 285(3.61),312(3.61)	4-1781-86
1H-Cyclopenta[a]phenanthrene-17-carboxylic acid, 2,3,6,7,10,15,16,17-octahydro-10-methyl-3-oxo-, cis	EtOH	232(4.26)	2-0015-86
$C_{19}H_{20}O_4$			
Benzo[a]heptalen-10(5H)-one, 6,7-dihydro-1,2,3-trimethoxy-	EtOH	232(4.46),321(4.14)	35-6713-86
1,2-Heptalenedicarboxylic acid, 5,6,10-trimethyl-, dimethyl ester	dioxan	262(4.19),273s(4.17), 326s(3.38)	88-1669-86
1,2-Heptalenedicarboxylic acid, 6,8,10-trimethyl-, dimethyl ester	dioxan	268(4.19),323s(3.57), 393s(2.79)	88-1669-86
1,3-Heptalenedicarboxylic acid, 6,8,10-trimethyl-, dimethyl ester	dioxan	224s(4.31),279(4.25), 324s(3.58),434(3.13)	88-1673-86
$C_{19}H_{20}O_5$			
Benzo[a]heptalen-10(5H)-one, 6,7-dihydro-11-hydroxy-1,2,3-trimethoxy-	EtOH-HCl	241(--),310(--), 324s(--),368(--), 382(--)	35-6713-86
	EtOH-NaOH	244(--),288(--), 300s(--),345(--), 412(--)	35-6713-86
2-Butenoic acid, 2-methyl-, 1-(8,9-dihydro-2-oxo-2H-furo[3,2-h]-1-benzopyran-8-yl)-1-methylethyl ester, [S-(Z)]- (spregelianin)	n.s.g.	205(2.32),216(2.26), 248(1.62),255(1.58), 297(1.76),332(2.12)	102-0505-86
Propanedioic acid, [(6-methoxy-2-naphthalenyl)methylene]-, diethyl ester	EtOH	257(4.34),278(4.40), 332(4.40)	2-0015-86
$C_{19}H_{20}O_6$			
4H-1-Benzopyran-4-one, 2,3-dihydro-5,6,7-trimethoxy-2-(4-methoxyphenyl)-, (S)-	MeOH	225(4.29),275(3.44), 325(3.68)	102-2223-86
Phenol, 4,4'-methylenebis[2-methoxy-, diacetate	EtOH	278(3.78)	104-0158-86
1-Propanone, 3-acetoxy-1-(2-hydroxy-4,6-dimethoxyphenyl)-2-phenyl-	MeOH	220(4.43),276(4.90), 316(4.31)	39-0215-86C
$C_{19}H_{20}O_{10}$			
4H-1-Benzopyran-4-one, 7-acetoxy-	EtOH	209s(4.20),229(4.50),	44-3116-86

Compound	Solvent	λ_{max}(log ϵ)	Ref.
6-(diacetoxymethyl)-5,8-dimethoxy-2-methyl- (cont.)		249s(4.21),304(3.76)	44-3116-86
$C_{19}H_{21}BrN_6O_5$			
Propanamide, N-[2-[(2-ethyl-4,6-dinitrophenyl)azo]-5-(diethylamino)-phenyl]-	$CHCl_3$	279(3.89),386(3.60), 575(4.68)	48-0497-86
$C_{19}H_{21}BrO_5$			
Benzene, 1-[2-(2-bromo-4,5-dimethoxy-phenyl)ethenyl]-2,3,4-trimethoxy-	$CHCl_3$	298(4.25),322(4.26)	142-1943-86
$C_{19}H_{21}ClN_2O_2$			
1H,5H-Benz[b]indolizino[8,1-hi]indolizine-5,12(13H)-dione, 6-chloro-14a-ethyl-2,3,6,6a,14,14a-hexahydro- (α-chlorodiazaspiroleuconolam)	EtOH	205(4.44),238s(3.92), 278s(3.39)	88-2501-86
β-chlorodiazaspiroleuconolam	EtOH	205(4.42),244s(3.82), 280s(3.16)	88-2501-86
$C_{19}H_{21}Cl_2N_2O_6P$			
Phosphonic acid, [acetoxy[5-chloro-3-[(4-chlorophenyl)azo]-2-hydroxy-phenyl]methyl]-, diethyl ester	MeOH	330(4.23),392(3.93)	12-1747-86
$C_{19}H_{21}N$			
5H-Indolo[3,2,1-de]acridine, 6,7,7a,8-9,10,11,12-octahydro-	MeOH	312(4.24)	103-0038-86
19-Norretinoic acid nitrile, 9,10,11,12-tetradehydro-, all-trans	EtOH	241(4.23),250s(4.19), 269(4.12),292s(4.17), 311s(4.22),322(4.23), 330(4.23)	5-1398-86
13-cis	EtOH	234(4.17),242(4.17), 267(4.08),292s(4.11), 322(4.17),332(4.17)	5-1398-86
$C_{19}H_{21}NO$			
Azacyclotetradeca-1,3,5,11,13-pentaene-7,9-diyne, 2-ethoxy-14-ethyl-6,11-dimethyl-	THF	219(4.23),298s(4.44), 308(4.49),383s(3.62), 422s(3.29)	18-1791-86
Cyclopent[b]indole, 8b-ethoxy-1,2,3,3a,4,8b-hexahydro-3a-phenyl-	n.s.g.	249(3.97),312(3.46)	25-0107-86
1H-Indole, 5-methoxy-2,3-dimethyl-1-(1-phenylethyl)-, (S)-	MeOH	228.5(4.41),284.5(3.92), 298(3.86),303s(3.82), 309s(3.72)	1-0625-86
1H-Indole-2-methanol, α,α-dimethyl-1-(1-phenylethyl)-, (S)-	MeOH	222(4.56),269s(3.90), 275(3.92),282.5(3.94), 292.0(3.83)	1-0625-86
$C_{19}H_{21}NOS_2$			
4-Piperidinone, 2,2,6,6-tetramethyl-3,5-bis(2-thienylmethylene)-	EtOH	236(4.04),278(4.16), 375(4.11)	70-2511-86
$C_{19}H_{21}NO_2$			
Benzeneacetic acid, α-[[2-(2-propenyl)phenyl]amino]-, ethyl ester	C_6H_{12}	242(4.14),292(3.50)	33-0927-86
$C_{19}H_{21}NO_3$			
Benzeneacetic acid, α-[[2-(4-pentenyloxy)phenyl]amino]-	EtOH	239(4.09),290(3.59)	33-0927-86

Compound	Solvent	$\lambda_{max}(\log \epsilon)$	Ref.
Benzeneacetic acid, α-[[2-(2-propenyloxy)phenyl]amino]-, ethyl ester	EtOH	243(4.08),289(3.49)	33-0927-86
4H-Dibenzo[de,g]quinolin-1-ol, 5,6,6a,7-tetrahydro-2,3-dimethoxy-6-methyl-, (R)- (N-methylisopiline)	EtOH	214(4.52),271(4.29), 301s(3.91)	100-0878-86
	EtOH-NaOH	233(4.31),253s(4.23), 279s(4.10),341(3.91)	100-0878-86
Nuciferidine, (-)-	EtOH	230s(4.29),272(4.16), 305s(3.84)	100-1078-86
$C_{19}H_{21}NO_4$			
Benzo[a]heptalen-10(5H)-one, 9-amino-6,7-dihydro-1,2,3-trimethoxy-	EtOH	247(4.40),354(4.18), 372(4.11),402(3.87)	35-6713-86
Clavizepine	MeOH	207(4.40),250(3.44), 295(3.62)	88-4077-86
	MeOH-NaOH	209(4.61),262(3.83), 306(3.81)	88-4077-86
4H-Dibenzo[de,g]quinolin-9-ol, 5,6,6a,7-tetrahydro-1,2,3-trimethoxy-, (R)- (oureguattine)	EtOH	207(4.24),222s(4.17), 281(3.78)	100-0878-86
4H-Dibenzo[de,g]quinolin-10-ol, 5,6,6a,7-tetrahydro-1,2,11-trimethoxy- (glaufinine)	n.s.g.	222(4.35),270(3.82), 309(3.43)	105-0236-86
Sinoacutine, (-)-	MeOH	240(4.24),276(3.76)	89-1025-86
$C_{19}H_{21}NO_5$			
Prostephanaberrine	EtOH	273.5(3.86)	100-0588-86
Wilsonirine, 4-hydroxy-	EtOH	222(4.48),273s(3.99), 281(4.09),304(4.12), 313s(4.09)	100-1028-86
$C_{19}H_{21}NO_6$			
Lutessine	EtOH	235(3.54),290(3.52)	100-0090-86
$C_{19}H_{21}N_3$			
2H-Benz[f]isoindole-1-carbonitrile, 5-(dimethylamino)-2-(1,1-dimethylethyl)-	EtOH	410(3.90),430(4.10), 453(4.06)	44-3978-86
$C_{19}H_{21}N_3O_3$			
1H-Benz[de]isoquinoline-1,3(2H)-dione, 5-amino-2-[3-(4-morpholinyl)propyl]-	pH 7.4	345(3.93),405(3.65)	4-0849-86
1H-Benz[de]isoquinoline-1,3(2H)-dione, 6-amino-2-[3-(4-morpholinyl)propyl]-	pH 7.4	432(4.08)	4-0849-86
$C_{19}H_{21}N_3O_3S_2$			
Ethanaminium, N-ethyl-N-[4-(hexahydro-1,3-dimethyl-2,4,6-trioxo-5-pyrimidinyl)-5-phenyl-1,3-dithiol-2-ylidene]-, hydroxide, inner salt	MeCN	230(4.33),250s(3.98), 280s(3.80),362(3.87)	24-2094-86
$C_{19}H_{21}N_3O_4S$			
5-Thia-1-azabicyclo[4.2.0]oct-2-ene-2-carboxylic acid, 7-(2,2-dimethyl-5-oxo-4-phenyl-1-imidazoldinyl)-3-methyl-8-oxo-, sodium salt, [6R-[6α,7β(R*)]]-	H_2O	259.5(3.90)	107-1469-86
$C_{19}H_{21}N_5O_4S_2$			
Adenosine, N-benzoyl-5'-S-methyl-5'-C-(methylthio)-5'-thio-	MeOH	230(4.15),280(4.31)	44-1258-86

Compound	Solvent	$\lambda_{max}(\log \epsilon)$	Ref.
$C_{19}H_{21}N_5O_8$			
β-D-Glucofuranuronic acid, 1-deoxy-1-[6-(4-morpholinyl)-9H-purin-9-yl]-, γ-lactone, 2,5-diacetate	MeOH	277(4.28)	103-0085-86
$C_{19}H_{21}N_7O_6$			
Benzamide, 2-[[2-(acetylamino)-4-(diethylamino)phenyl]azo]-3,5-dinitro-	$CHCl_3$	305(3.88),424(3.54), 588(4.77)	48-0497-86
$C_{19}H_{22}ClNO_2S$			
2H-[1]Benzothiepino[5,4-b]pyran-2-one, 4-[bis(1-methylethyl)amino]-3-chloro-5,6-dihydro-	EtOH	236(4.05),255s(3.84)	4-0449-86
$C_{19}H_{22}FNO_4S$			
Benzamide, N-[1-(fluoromethyl)-2-hydroxy-2-[4-(methylsulfonyl)phenyl]-ethyl]-2,3-dimethyl-, [R-(R*,S*)]-	MeOH	205(4.46)	32-0485-86
$C_{19}H_{22}N_2$			
Pyridinium, 4,4'-bicyclo[2.2.1]hept-2-ene-2,3-diylbis[1-methyl-, bis(tetrafluoroborate)	EtOH	238(4.08),280(3.96), 341(4.04)	138-1279-86
$C_{19}H_{22}N_2O$			
Strempeliopine, (±)-	MeOH	255(4.10),281(3.69), 291(3.61)	73-1731-86
$C_{19}H_{22}N_2OS$			
Acepromazine, (Z)-2-butenedioate (1:1)	MeOH	243.5(4.434),278(4.327), 370(3.328)	106-0571-86
$C_{19}H_{22}N_2O_2$			
Benzaldehyde, 2,2'-(1,3-propanediyl-dimino)bis[5-methyl-	EtOH	265(4.05),396(4.05)	12-1231-86
Spiro[3H-indole-3,1'(5'H)-indolizine]-2,5'(1H)-dione, 2',3',6',7',8',8'a-hexahydro-8'-(1-methyl-2-propenyl)-	EtOH	221(4.12),234(3.83), 266s(3.62),286(3.03)	78-6047-86
$C_{19}H_{22}N_2O_2S$			
Acetic acid, [(3-cyano-6-tricyclo-[3.3.1.1^{3,7}]dec-1-yl-2-pyridinyl)-thio]-, methyl ester	EtOH	220(4.50),264(4.23), 310(3.91)	70-0131-86
Thieno[2,3-b]pyridine-2-carboxylic acid, 3-amino-6-tricyclo[3.3.1.1^{3,7}]-dec-1-yl-, methyl ester	EtOH	202(4.34),236(4.04), 283(4.60),368(3.73)	70-0131-86
$C_{19}H_{22}N_2O_2Se$			
Acetic acid, [(3-cyano-6-tricyclo-[3.3.1.1^{3,7}]dec-1-yl-2-pyridinyl)-seleno]-, methyl ester	EtOH	225(4.47),275(4.18), 315(3.82)	70-0376-86
Selenolo[2,3-b]pyridine-2-carboxylic acid, 3-amino-6-tricyclo[3.3.1.1^{3,7}]-dec-1-yl-, methyl ester	EtOH	206(4.31),236(3.96), 291(4.46),380(3.71)	70-0376-86
$C_{19}H_{22}N_2O_3$			
1,2-Butanedione, 1-[3-(2,2-dimethyl-1-oxopropyl)-2-quinoxalinyl]-3,3-dimethyl-	CH_2Cl_2	315(3.86),425(2.00)	89-0999-86
Epileuconolam	EtOH	203(4.11),251(4.22), 342(3.55)	88-2501-86

Compound	Solvent	$\lambda_{max}(\log \epsilon)$	Ref.
Meloscine, 14,15,18,19-tetrahydro-21-hydroxy-5-oxo-	EtOH	212(4.48),251(3.99),258(3.92),284(3.42),293(3.34)	88-2501-86
$C_{19}H_{22}N_2O_4$			
Benzaldehyde, 2,2'-(1,3-propanediyl-diimino)bis[5-methoxy-	EtOH	227(4.54),242s(4.46),267(3.91),416(4.06)	12-1231-86
Benzoic acid, 2,2'-(1,3-propanediyl-diimino)bis[5-methyl-	EtOH	226(4.51),261(4.30),360(4.03)	12-1231-86
$C_{19}H_{22}N_2O_6$			
Benzoic acid, 2,2'-(1,3-propanediyl-diimino)bis[5-methoxy-	EtOH	224(4.47),262(4.22),376(4.01)	12-1231-86
Butanedioic acid, (nitromethylene)-(phenyl-1-piperidinylmethylene)-, dimethyl ester	EtOH	287(3.92),404(4.10)	18-3871-86
L-Tyrosine, 3-[4-(2-amino-2-carboxy-ethyl)phenoxy]-O-methyl-, (S)-	H_2O	273(3.52)	158-1685-86
	HCl	273(3.52)	158-1685-86
	NaOH	273(3.54)	158-1685-86
$C_{19}H_{22}N_2O_9$			
1,5-Diazabicyclo[4.4.1]undeca-2,7,9-triene-7,8,9,10-tetracarboxylic acid, 2,5-dimethyl-4-oxo-, tetra-methyl ester	MeOH	224(4.23),312(4.27)	142-0009-86
1,5-Diazecine-6,7,8,9-tetracarboxylic acid, 1,2-dihydro-1,4,10-trimethyl-2-oxo-, tetramethyl ester	MeOH	247(4.23),285s(3.79),380(3.64)	142-0009-86
$C_{19}H_{22}N_2S$			
Pecazine (lactate)	MeOH	254.5(4.548),305.5(3.608)	106-0571-86
$C_{19}H_{22}N_4$			
1H-Pyrazolo[3,4-b]quinolin-4-amine, N-(3,5-dimethylphenyl)-5,6,7,8-tetrahydro-1-methyl-	EtOH	218(4.47),243s(4.09),273(3.58),323(4.17)	11-0331-86
1H-Pyrazolo[3,4-b]quinolin-4-amine, N-(4-ethylphenyl)-5,6,7,8-tetra-hydro-1-methyl-	EtOH	216(4.88),244(4.30),272(3.88),321(4.42)	11-0331-86
$C_{19}H_{22}N_4O_2$			
Piperazine, 1-acetyl-4-[4-[(4-methoxy-phenyl)azo]phenyl]-	EtOH	387(4.42)	39-0123-86B
	EtOH-HCl	565(4.68)	39-0123-86B
$C_{19}H_{22}N_4O_3$			
1H-Benz[de]isoquinoline-1,3(2H)-dione, 5,8-diamino-2-[3-(4-morpholinyl)-propyl]-	pH 7.4	405(3.94)	4-0849-86
$C_{19}H_{22}N_4O_9$			
Acetamide, N-[4,5-dihydro-4-oxo-1-(2,3,5-tri-O-acetyl-β-D-ribo-furanosyl)-1H-imidazo[4,5-c]-pyridin-6-yl]-	H_2O	269(4.17),275(4.15),305(4.15)	78-1971-86
$C_{19}H_{22}N_6O_5$			
β-D-Ribofuranuronamide, 1-deoxy-N-ethyl-1-[6-[(4-methoxyphenyl)-amino]-9H-purin-9-yl]-	n.s.g.	287(4.29)	87-1683-86

Compound	Solvent	λ_{max}(log ϵ)	Ref.
$C_{19}H_{22}O$			
19-Norretinal, 9,10,11,12-tetradehydro-, all-trans	CH_2Cl_2	240s(4.12),253(4.16), 282s(4.14),320(4.28), 344s(4.24)	5-1398-86
13-cis	CH_2Cl_2	231(4.10),253(4.09), 264(4.08),320(4.19), 343s(4.16)	5-1398-86
$C_{19}H_{22}OSe_2$			
Benzene, 1,1'-[[(cyclohexyloxy)methylene]bis(seleno)]bis-	hexane	220(4.40),290(3.68)	104-0092-86
$C_{19}H_{22}OSi$			
Benzenemethanol, 2-(phenylmethyl)-α-[(trimethylsilyl)ethynyl]-	EtOH	219s(4.21),262(3.27), 282s(3.08)	33-0560-86
$C_{19}H_{22}O_3$			
1H-Cyclopropa[3,4]cyclohept[1,2-e]indene-6,7-dione, 1a,4,5,8,9,9a-hexahydro-3-methoxy-1,1,2-trimethyl-, (1aR-cis)-	EtOH	212(4.19),256(3.72), 295(3.09)	35-3040-86
3,4-Phenanthrenedione, 5,6,7,8-tetrahydro-5-hydroxy-8,8-dimethyl-2-(1-methylethyl)-	MeOH	226(4.37),264(4.24), 280s(4.04),356(3.20), 422(3.38)	102-1935-86
	MeOH-NaOMe	220(4.20),257(4.38), 275s(4.00),335(3.14), 430(2.60)	102-1935-86
4(1H)-Phenanthrenone, 2,3-dihydro-5,6-dihydroxy-1,1-dimethyl-7-(1-methylethyl)- (arucadiol)	MeOH	228(4.58),242s(4.39), 270(4.25),278(4.24), 362(3.54),410(3.65)	102-1935-86
	MeOH-NaOMe	225(4.49),244(4.40), 252s(4.39),260s(4.35), 325(3.47)	102-1935-86
	MeOH-$AlCl_3$	230(4.57),269(4.23), 280(4.21),458(3.47)	102-1935-86
	MeOH-H_3BO_3	228(4.58),270s(4.29), 279(4.30),360(3.47), 416(3.60)	102-1935-86
$C_{19}H_{22}O_5$			
1,3-Azulenedicarboxylic acid, 6-ethoxy-2-methyl-, diethyl ester	CH_2Cl_2	231(4.36),276.5(4.29), 323(4.73),364.5(4.01), 453.5(2.91)	24-2272-86
Bicyclo[3.2.1]octan-3-one, 7-(1,3-benzodioxol-5-yl)-2,8-dihydroxy-6-methyl-5-(2-propenyl)-, (2-exo-6-endo,7-exo,8-anti)-(-)-	EtOH	237(3.58),289(3.51)	102-1953-86
Naphtho[1,2-b]furan-6-carboxylic acid, 2,3-dihydro-7-hydroxy-4-methoxy-2,3,3,9-tetramethyl-, methyl ester	MeOH	221(4.64),254(4.33), 269(4.32),277(4.36), 341(3.84),361s(3.75)	44-4813-86
$C_{19}H_{22}O_6$			
Armexifolin, 8α-methacryloyloxy-	EtOH	247(3.57)	102-1365-86
Butanoic acid, 3-methyl-, 9,10-dihydro-10-hydroxy-8,8-dimethyl-2-oxo-2H,8H-benzo[1,2-b:3,4-b']dipyran-9-yl ester, (9R-trans)- (junosmarin)	EtOH	221s(3.73),260(3.66), 278(3.71),328(3.61)	142-2777-86
1-Propanone, 1-[4-(ethoxymethoxy)-2-hydroxyphenyl]-3-hydroxy-2-(4-methoxyphenyl)-	MeOH	214(4.53),224(4.53), 290(4.58),326s(4.07)	39-0215-86C

Compound	Solvent	$\lambda_{max}(\log \epsilon)$	Ref.
$C_{19}H_{22}O_8$			
2H-Pyran-4-acetic acid, 3-formyl-3,4-dihydro-5-(methoxycarbonyl)-2-methyl-, 2-(3,4-dihydroxyphenyl)ethyl ester	MeOH	230(4.1),280(3.5)	102-0865-86
$C_{19}H_{23}FN_2O_{10}$			
α-D-Glucofuranose, 1,2-O-(1-methylethylidene)-3-(5-fluoro-3,4-dihydro-1(2H)-pyrimidinepropanoate)-6-(2-propenoate)	MeOH	270.5(3.92)	47-2059-86
$C_{19}H_{23}FN_2O_{11}$			
2,4(1H,3H)-Pyrimidinedione, 6-fluoro-5-methyl-1-(2,3,4,6-tetra-O-acetyl-β-D-glucopyranosyl)-	$CHCl_3$	251(3.67)	97-0437-86
$C_{19}H_{23}FN_2O_{12}$			
2,4(1H,3H)-Pyrimidinedione, 6-fluoro-5-(hydroxymethyl)-1-(2,3,4,6-tetra-O-acetyl-β-D-glucopyranosyl)-	$CHCl_3$	249(3.90)	97-0437-86
$C_{19}H_{23}F_3N_4S$			
2,4-Cyclohexadiene-1,1,3-tricarbonitrile, 6-methyl-6-(2-methylpropyl)-4-[2-methyl-1-[(trifluoromethyl)-thio]propyl]-	EtOH	217.4(4.13),333.3(3.68)	155-0201-86
$C_{19}H_{23}N$			
6-Azulenecarbonitrile, 2-octyl-	CH_2Cl_2	282(4.87),290(5.02), 341.5(3.58),357.5(3.84), 375(3.93),418.5(1.66), 616.5(2.52),670.5(2.44), 745.5(1.99)	24-2272-86
$C_{19}H_{23}NO$			
2-Furanmethanamine, N-methyl-N-[2-(phenylmethylene)cyclohexyl]-, (E)-(±)-	EtOH	242(4.10)	44-5002-86
$C_{19}H_{23}NO_4$			
Acetic acid, [8a-(1,1-dimethylethyl)-8,8a-dihydro-3a,8-dimethyl-2-oxo-2H-furo[2,3-b]indol-3(3aH)-ylidene]-, methyl ester	MeOH	234(3.99),287(3.11), 318(3.00)	150-0514-86M
Acetic acid, [4-(2,2-dimethyl-1-oxopropyl)-1,4-dihydro-1,4-dimethyl-2-oxo-3(2H)-quinolinylidene]-, methyl ester, (E)-	n.s.g.	214(4.18),223s(4.06), 283(3.59),312(3.63)	150-0514-86M
α-D-xylo-Hexofuranurononitrile, 5-deoxy-5-methylene-1,2-O-(1-methylethylidene)-3-O-(3-phenylpropyl)-	n.s.g.	217(3.40)	111-0111-86
3,5-Pyridinedicarboxylic acid, 1,4-dihydro-2,6-dimethyl-1-phenyl-, diethyl ester	EtOH	231(4.26),359(3.76)	150-0088-86S +150-0901-86M
3,5-Pyridinedicarboxylic acid, 1,4-dihydro-2,6-dimethyl-4-phenyl-, diethyl ester	MeOH	268(4.45),277(--)	87-1596-86
$C_{19}H_{23}NO_4S$			
Sulfoximine, N-(2-ethoxy-1-methyl-	EtOH	204(4.45),225(4.23),	12-1833-86

Compound	Solvent	λ_{max}(log ε)	Ref.
2-oxoethyl)-S-(2-methoxyphenyl)-S-(4-methylphenyl)-, (+)- (cont.)		238s(4.17),289(3.67)	12-1833-86
$C_{19}H_{23}NO_5$			
2,5-Pyrrolidinedione, 1-methyl-3-(1-methylethylidene)-4-[1-(3,4,5-trimethoxyphenyl)ethylidene]-, (E)-	toluene	323(3.92)	39-0315-86C
$C_{19}H_{23}NO_6$			
Crinafolidine	MeOH	238s(3.08),288s(3.12), 322(3.28)	150-0312-86S
$C_{19}H_{23}N_2O_3$			
Pyrylium, 2-amino-3-(ethoxycarbonyl)-6-phenyl-4-(1-piperidinyl)-, perchlorate	MeCN	225.5(4.07),258.5(4.50), 320(4.36)	48-0314-86
$C_{19}H_{23}N_3O_2$			
1,4-Benzenediamine, N^4-(3,5-dimethyl-4-nitrophenyl)methylene]-N^1,N^1,2,6-tetramethyl-	EtOH	260(4.21),355(3.94)	56-0831-86
$C_{19}H_{23}N_3O_2S$			
Hydrazinecarbothioamide, 2-[1-(4-ethoxyphenyl)-1,3-dihydro-3,3-dimethyl-1-isobenzofuranyl]-	EtOH	239(4.35),269(3.48), 274s(3.38),282(3.36), 320(3.61)	104-0647-86
$C_{19}H_{23}N_3O_3$			
Methanimidic acid, N-[3-[[(ethoxymethylene)amino]carbonyl]-4,5-dimethyl-1-phenyl-1H-pyrrol-2-yl]-, ethyl ester	EtOH	235(4.27),292s(3.86), 303(3.90)	11-0347-86
$C_{19}H_{23}N_3O_3S$			
Hydrazinecarbothioamide, 2-[1-(3,4-dimethoxyphenyl)-1,3-dihydro-3,3-dimethyl-1-isobenzofuranyl]-	EtOH	242(4.30),270(3.56), 277(3.54),283(3.52), 319(2.96)	104-0647-86
$C_{19}H_{23}N_3O_{11}$			
1H-Pyrazole-5-carboxamide, 4-acetoxy-1-acetyl-3-(2,3,5-tri-O-acetyl-β-D-ribofuranosyl)-	MeOH	252(4.46)	87-0268-86
$C_{19}H_{23}N_5O_4$			
Adenosine, N-(3-phenylpropyl)-	pH 1	263(4.17)	163-2243-86
	pH 7	267(4.20)	163-2243-86
	pH 13	267(4.17)	163-2243-86
$C_{19}H_{23}N_5O_7$			
β-D-Glucofuranuronic acid, 1-[6-(butylamino)-9H-purin-9-yl)-1-deoxy-, γ-lactone, 2,5-diacetate	MeOH	267(4.21)	103-0085-86
$C_{19}H_{24}ClN_3O_2$			
1,4-Naphthalenedione, 5-amino-2-chloro-3-[(2,2,6,6-tetramethyl-4-piperidinyl)amino]-	PrOH	258(4.86),302(3.73), 462(3.77)	104-1560-86
$C_{19}H_{24}ClN_4O$			
Piperidinium, 1-[6-(4-chlorophenyl)-4-(1-piperidinyl)-2H-1,3,5-oxadia-	MeCN	243(4.50),290(4.46)	97-0094-86

Compound	Solvent	$\lambda_{max}(\log \epsilon)$	Ref.
zin-2-ylidene]- (cont.)			97-0094-86
$C_{19}H_{24}ClO$			
Pyrylium, 4-(3-chlorophenyl)-2,6-bis(1,1-dimethylethyl)-, perchlorate	MeOH	309(4.38)	44-4385-86
Pyrylium, 4-(4-chlorophenyl)-2,6-bis(1,1-dimethylethyl)-, perchlorate	MeOH	307(4.24),342(4.44)	44-4385-86
$C_{19}H_{24}NO_3$			
Isoquinolinium, 1,2,3,4-tetrahydro-6,7-dihydroxy-1-[(4-methoxyphenyl)-methyl]-2,2-dimethyl-, chloride (luxandrine chloride)	EtOH	210(4.47),231(4.22),283(3.75)	102-2693-86
	EtOH-NaOH	251(4.10),310(3.63)	102-2693-86
Pyrylium, 2,6-bis(1,1-dimethylethyl)-4-(4-nitrophenyl)-, trifluoromethane sulfonate	MeOH	310(4.48)	44-4385-86
$C_{19}H_{24}N_2O$			
Aspidospermidine, 1,2-didehydro-, N-oxide	MeOH	206(--),222(--),260(--)	164-0663-86
Indolo[2,3-a]quinolizine-2-ethanol, β-ethenyl-1,2,3,4,6,7,12,12b-octahydro- (3-epiantirhine)	EtOH	227(3.82),275(3.27),286(3.29),291(3.25)	102-1130-86
Rhazinilam, 5,21-dihydro-	EtOH	213(4.26),227(3.65),261(3.43)	88-2501-86
$C_{19}H_{24}N_2OS$			
Levomepromazine, (Z)-2-butenedioate (1:1)	MeOH	253(4.425),306(3.663)	106-0571-86
$C_{19}H_{24}N_2O_2$			
2-Furancarboxaldehyde, 5-(dimethylamino)-4-(phenyl-1-piperidinylmethyl)-	MeOH	369(3.63)	73-0573-86
2H-3,13-Methanooxireno[9,10]azacycloundecino[5,4-b]indol-4-ol, 13-ethyl-1a,4,5,10,11,12,13,13a-octahydro-(stapfinine)	MeOH	222(4.58),275(3.84),292(3.83)	102-1781-86 +164-0663-86
$C_{19}H_{24}N_2O_6S$			
2-Indolizinecarboxylic acid, 3-[[2-methyl-2-(methylsulfonyl)-1-[(2-propenyloxy)carbonyl]propyl]-amino]-, methyl ester	MeOH	240(4.39)	88-3449-86
$C_{19}H_{24}N_2O_{11}$			
D-Arabinitol, 1-deoxy-1-(hexahydro-1,3-dimethyl-2,4,6-trioxo-5(2H)-pyrimidinylidene)-, 2,3,4,5-tetraacetate	MeOH	239(4.15)	136-0053-86M
D-Xylitol, 1-deoxy-1-(hexahydro-1,3-dimethyl-2,4,6-trioxo-6(2H)-pyrimidinylidene)-, 2,3,4,5-tetraacetate	EtOH	239(4.10)	136-0053-86M
$C_{19}H_{24}N_2S$			
Profenamine, monohydrochloride	MeOH	252(4.489),300.5(3.56)	106-0571-86
$C_{19}H_{24}N_4OS$			
[3,4'-Bipyridine]-3'-carbonitrile, 5'-acetyl-1',2',3',4'-tetrahydro-6'-methyl-2'-thioxo-, compd. with piperidine	EtOH	234(4.03),303(4.05),390(2.95)	104-1762-86

Compound	Solvent	λ max (log ε)	Ref.
[4,4'-Bipyridine]-3-carbonitrile, 5-acetyl-1,2,3,4-tetrahydro-6-methyl-2-thioxo-, compd. with piperidine	EtOH	237(4.22),297(4.21), 388s(2.46)	104-1762-86
Hydrazinecarbothioamide, 2-[1-[4-(dimethylamino)phenyl]-1,3-dihydro-3,3-dimethyl-1-isobenzofuranyl]-	EtOH	244(4.32),265(4.32), 310s(3.49),355(3.72)	104-0647-86
$C_{19}H_{24}N_4O_3$			
1-Propanol, 3,3'-[(6-methoxybenzo[g]phthalazine-1,4-diyl)diimino]bis-	EtOH	225(4.59),237s(4.45), 263(4.63),269(4.65), 360(3.68)	111-0143-86
$C_{19}H_{24}N_4O_7$			
Hexanoic acid, 6-[[5,6-dihydro-5-oxo-6-(1,2,3,6-tetrahydro-1,3-dimethyl-2,6-dioxo-9H-purin-9-yl)-2H-pyran-2-yl]methoxy]-, (2S-cis)-	$CHCl_3$	278(3.90)	136-0256-86K
$C_{19}H_{24}N_6O_5S_2$			
Pyrrolidinium, 1-[[7-[[(2-amino-4-thiazolyl)(methoxyimino)acetyl]amino]-2-carboxy-8-oxo-5-thia-1-azabicyclo[4.2.0]oct-2-en-3-yl]methyl]-1-methyl-, hydroxide, inner salt	pH 7	235(4.22),257(4.21)	158-1092-86
sulfate	pH 7	236(4.24),258(4.23)	158-1092-86
$C_{19}H_{24}N_{12}O_2$			
Pentanediamide, N,N'-bis[2-(6-amino-9H-purin-9-yl)ethyl]-	H_2O	265(4.43)	47-3201-86
polymer	DMSO	264.0(4.11)	47-3201-86
$C_{19}H_{24}O$			
Benzenepentanol, α-(2-phenylethyl)-, (R)-	EtOH	243(2.29),248(2.44), 255(2.57),261(2.64), 268(2.51)	18-1181-86
1H-Cyclopropa[3,4]cyclohept[1,2-e]inden-3-ol, 1a,4,5,6,7,8,9,9a-octahydro-1,1,2-trimethyl-7-methylene-, (1aR-cis)-	EtOH	210(4.35)	35-3040-86
	EtOH-base	214(4.63)	35-3040-86
10αH-5β-Gona-8,11,13-trien-17-one, 5,12-dimethyl-	EtOH	263(4.27)	2-1200-86
19-Norretinal, 9,10-didehydro-, all-trans	CH_2Cl_2	246(3.85),357(4.37)	5-1398-86
13-cis	CH_2Cl_2	250(3.97),353(4.27)	5-1398-86
2-Propen-1-one, 1-(4b,5,6,7,8,8a,9,10-octahydro-3,8a-dimethyl-2-phenanthrenyl)-	EtOH	263(3.98)	2-1200-86
$C_{19}H_{24}O_2$			
Androsta-4,6-diene-3,17-dione	EtOH	283(4.47)	56-0311-86
7H-Cyclopropa[3,4]cyclohept[1,2-e]inden-7-one, 1,1a,4,5,6,8,9,9a-octahydro-3-methoxy-1,1,2-trimethyl-, (1aR-cis)-	EtOH	208(4.16),223(4.15), 262(3.65)	35-3040-86
3,5-Heptanediol, 1,7-diphenyl-, [S-(R*,R*)]-	EtOH	243(2.16),249(2.39), 255(2.55),261(2.63), 268(2.50)	18-1181-86
6,10-Methanobenzocyclodecen-5(6H)-one, 7,8,11,12-tetrahydro-4-methoxy-9,13,13-trimethyl-, endo	n.s.g.	280(3.74)	35-4953-86
exo	n.s.g.	280(3.81)	35-4953-86

Compound	Solvent	λ_{max}(log ε)	Ref.
Phenol, 2-(1,1-dimethylethyl)-6-[1-(2-hydroxyphenyl)ethyl]-4-methyl-, (-)-	n.s.g.	280(3.64)	116-0509-86
$C_{19}H_{24}O_3$			
1(3H)-Isobenzofuranone, 4-acetyl-5-(1,3,3-trimethylcyclohexyl)-, (S)-	MeOH	242(3.99)	44-2601-86
1-Phenanthrenecarboxylic acid, 7-ethenyl-4b,5,6,7,8,8a,9,10-octahydro-3-hydroxy-4b,7-dimethyl- (ar-maximic acid)	MeOH	291(3.47)	94-1015-86
2-Propenoic acid, 3-(4-hydroxyphenyl)-, 1,7,7-trimethylbicyclo[2.2.1]hept-2-yl ester, (1S-endo)-	MeOH	213(4.00),228(4.02), 313(4.32)	102-1130-86
$C_{19}H_{24}O_5$			
1H-Indene-1,5(4H)-dione, hexahydro-7-hydroxy-4-[1-hydroxy-2-(5-hydroxy-2-methylphenyl)ethyl]-7a-methyl-	MeOH	217(4.39),280(4.00)	13-0439-86B
	MeOH-base	217(--),241(--), 300(--)	13-0439-86B
1H-Xanthene-9-carboxylic acid, 2,3,4-5,6,7,8,9-octahydro-3,3,6,6,9-pentamethyl-1,8-dioxo-	EtOH	229(4.21),297(3.64)	2-0347-86
$C_{19}H_{24}O_6$			
Armexifolin, 8α-isobutyryloxy-	EtOH	242(4.35)	102-1365-86
1H-Indene-1,5(4H)-dione, 4-[2-(2,3-dihydroxy-6-methylphenyl)-1-hydroxyethyl]hexahydro-7-hydroxy-7a-methyl-	MeOH	216(3.93),283(3.36)	13-0439-86B
	MeOH-base	217(--),290(--)	13-0439-86B
$C_{19}H_{24}O_7$			
15,18-Methanonaphth[2,3-b]-1,4,7,10,13-pentaoxacyclopentadecin-14,19-dione, 2,3,5,6,8,9,11,12,14a,15,18,18a-dodecahydro-, (14aα,15α,18α,18aα)-	MeCN	294(3.97)	77-0268-86
$C_{19}H_{25}ClN_2S$			
Profenamine monohydrochloride	MeOH	252(4.489),300.5(3.56)	106-0571-86
$C_{19}H_{25}NO_2$			
Benzenemethanol, α-[2-[(2-furanylmethyl)methylamino]cyclohexyl]-, cis	EtOH	212(4.51),216s(4.47), 247s(2.71),252(2.68), 257(2.69),261s(2.58), 264(2.57),267(2.40)	44-5002-86
6a,9-Epoxy-6aH-benzo[3,4]cyclohepta-[1,2-c]quinolin-7-ol, 1,2,3,4,4a,5-6,7,8,9,13b,13c-dodecahydro-5-methyl-	EtOH	215s(3.94),258s(2.47), 264(2.55),272(2.51), 290s(1.79),300s(1.75)	44-5002-86
stereoisomer	EtOH	259s(2.47),265(2.59), 272(2.54)	44-5002-86
9H-Indeno[2,1-b]pyridine-3-carboxylic acid, 1,2,3,9a-tetrahydro-1,3,9,9-tetramethyl-, ethyl ester, cis-(±)-	EtOH	250(4.14),286(3.60), 295(3.53)	94-2786-86
$C_{19}H_{25}N_3O_2S_2$			
Dimethothiazine	MeOH	206.5(4.37),236(4.275), 267(4.531),314(3.537)	106-0571-86
$C_{19}H_{25}N_3O_6S$			
L-Histidine, N-[(1,1-dimethylethoxy)-carbonyl]-N-methyl-1-[(4-methylphenyl)sulfonyl]-	EtOH	225(4.06),228(4.08), 223[sic](4.06),275.5 (2.88)	87-2088-86

Compound	Solvent	$\lambda_{max}(\log \epsilon)$	Ref.
$C_{19}H_{25}N_3O_9$			
2(1H)-Pyrimidinone, 4-(4-morpholinyl)-1-(2,3,5-tri-O-acetyl-β-D-ribofuranosyl)-	pH 7	278(4.11)	1-0806-86
	pH 2	285(--)	1-0806-86
	pH 13	284(--)	1-0806-86
$C_{19}H_{25}N_3O_{10}$			
Glycine, N-[1,2-dihydro-2-oxo-1-(2,3-5-tri-O-acetyl-β-D-ribofuranosyl)-4-pyrimidinyl]-, ethyl ester	pH 7	264(3.98)	1-0806-86
	pH 2	279(--)	1-0806-86
	pH 13	275(--)	1-0806-86
$C_{19}H_{25}N_3O_{11}$			
L-Serine, N-[1,2-dihydro-2-oxo-1-(2,3,5-tri-O-acetyl-β-D-ribofuranosyl)-4-pyrimidinyl]-, methyl ester	pH 7	268(3.97)	1-0806-86
	pH 2	280(--)	1-0806-86
	pH 13	248s(--),276(--)	1-0806-86
$C_{19}H_{25}N_3S$			
Aminopromazine, (E)-2-butenedioate (1:1)	MeOH	254(4.892),305(3.921)	106-0571-86
$C_{19}H_{25}N_4O$			
Piperidinium, 1-[6-phenyl-4-(1-piperidinyl)-2H-1,3,5-oxadiazin-2-ylidene]-, perchlorate	MeCN	242(4.17),282(3.96)	97-0094-86
$C_{19}H_{25}N_5O_9$			
4(1H)-Pyrimidinone, 2-amino-6-[[3-[4-(β-D-galactopyranosyloxy)phenoxy]propyl]amino]-5-nitroso-	MeOH	223(4.44),290s(3.99), 323(4.26)	4-0629-86
$C_{19}H_{25}O$			
Pyrylium, 2,6-bis(1,1-dimethylethyl)-4-phenyl-, perchlorate	MeOH	306(4.24),332(4.31)	44-4385-86
$C_{19}H_{26}$			
Azulene, 6-methyl-2-octyl-	CH_2Cl_2	233.5(4.16),281(4.81), 290.5(4.91),307(3.76), 335(3.59),351(3.70), 364.5(3.18),551(2.43), 582.5(2.38)	24-2272-86
$C_{19}H_{26}Br_2O_5$			
11-Oxatricyclo[7.3.2.0^{10,12}]tetradec-13-ene-13,14-dimethanol, 1,9-dibromo-, diacetate	EtOH	227(4.07)	24-2698-86
$C_{19}H_{26}FeN_2O_3$			
Cyclopenta[cd]pentalene-2a,4a-diamine, N,N,N',N'-tetramethyl-, tricarbonyl iron deriv.	THF	238s(3.98),313(4.01), 316(3.81)	78-1721-86
$C_{19}H_{26}FeS_4$			
Ferrocene, 1,1'-[1,3-propanediylbis(thio-3,1-propanediylthio)]-	MeCN	443(2.28)	18-1515-86
copper perchlorate complex	MeCN	434(2.14)	18-1515-86
mercuric chloride complex (1:2)	MeCN	445(2.23)	18-1515-86
$C_{19}H_{26}N_2$			
3-Buten-2-one, 4-(2,6,6-trimethyl-1-cyclohexen-1-yl)-, phenylhydrazone	EtOH	329(4.15)	126-1573-86

Compound	Solvent	$\lambda_{max}(\log \epsilon)$	Ref.
Vallesamidine, 1-demethyl-, (±)-	MeOH	246(3.86),298(3.39)	73-1731-86
$C_{19}H_{26}N_2O_2$			
Benzenemethanol, 2,2'-(1,3-propanediyldiimino)bis[5-methyl-	EtOH	215(4.18),253(4.33), 305(3.66)	12-1231-86
$C_{19}H_{26}N_2O_2S$			
Benzenesulfonamide, 4-methyl-N-[2-(1-piperidinylmethylene)cyclohexylidene]-	EtOH	221s(4.08),380(4.12)	161-0539-86
$C_{19}H_{26}N_2O_4$			
Benzenemethanol, 2,2'-(1,3-propanediyldiimino)bis[5-methoxy-	EtOH	214(4.14),249(4.28), 313(3.67)	12-1231-86
1H-Pyrrole-2-carboxylic acid, 5,5'-methylenebis[3,4-dimethyl-, diethyl ester	EtOH	250(3.90),275(4.17), 290(4.24)	49-0631-86
$C_{19}H_{26}N_2O_5S$			
Uridine, 2',3'-O-cyclohexylidene-4',5'-didehydro-5'-S-(2-methylpropyl)-5'-thio-	MeOH	256(4.16)	44-1258-86
$C_{19}H_{26}N_2O_6Si$			
2,4(1H,3H)-Pyrimidinedione, 1-[3-acetoxy-4-(acetoxymethyl)cyclopentyl]-5-[(trimethylsilyl)ethynyl]-, (1α,3β,4α)-(±)-	pH 1	237(4.09),298(4.16)	87-0079-86
	pH 7	236(4.09),297(4.17)	87-0079-86
	pH 13	232(4.09),288(4.00)	87-0079-86
$C_{19}H_{26}N_2O_9$			
1,5-Diazabicyclo[4.4.1]undec-7-ene-7,8,9,10-tetracarboxylic acid, 2,5-dimethyl-4-oxo-, tetramethyl ester	MeOH	301(2.49)	142-0009-86
1,5-Diazecine-6,7,8,9-tetracarboxylic acid, 1,2,5,6,7,10-hexahydro-1,4,10-trimethyl-2-oxo-, tetramethyl ester	MeOH	228(3.72),288(3.75), 334(3.11)	142-0009-86
$C_{19}H_{26}N_4O_6$			
1H-Purine-2,6-dione, 9-[3,6-dihydro-6-[[(6-hydroxyhexyl)oxy]methyl]-3-oxo-2H-pyran-2-yl]-3,9-dihydro-1,3-dimethyl-, (2R-cis)-	$CHCl_3$	278(4.04)	136-0256-86K
$C_{19}H_{26}N_4O_7$			
Acetamide, N-[4,7-dihydro-3-(methoxymethyl)-5-methyl-7-[2,3-O-(1-methylethylidene)-α-D-ribofuranosyl]-4-oxo-3H-pyrrolo[2,3-d]pyrimidin-2-yl]-	MeOH	304(3.98)	78-0199-86
β-	MeOH	302(3.99)	78-0199-86
$C_{19}H_{26}N_8O_{15}P_2$			
Adenosine 5'-(trihydrogen phosphate), 5'→5'-ester with 4-hydroxy-5-β-D-ribofuranosyl-1H-pyrazole-3-carboxamide	pH 1	259(4.23)	87-0268-86
	pH 7	259(4.26)	87-0268-86
	pH 11	259(4.19),306(3.94)	87-0268-86
$C_{19}H_{26}O$			
Androsta-3,5-dien-17-one	MeOH	233(4.13)	150-0718-86M

Compound	Solvent	$\lambda_{max}(\log \epsilon)$	Ref.
2-Cyclopenten-1-one, 3-[2-methyl-4-(2,6,6-trimethyl-1-cyclohexen-1-yl)-1,3-butadienyl]-, all-trans	CH_2Cl_2	339(4.23)	35-4614-86
(EZ)-	CH_2Cl_2	338(4.08)	35-4614-86
2,4,6,8-Nonatetraenal, 7-methyl-9-(2,6,6-trimethyl-1-cyclohexen-1-yl)-, all-trans	heptane	367(4.68)	5-0479-86
11Z	heptane	244(--),368(--)	5-0479-86
$C_{19}H_{26}O_2$			
Androst-4-ene-6,17-dione	MeOH	241(3.70)	150-0718-86M
Estr-5(10)-ene-3,17-dione, 7α-methyl-	EtOH	240(1.42)	54-0111-86
D-Homo-18-norandrosta-5,13(17a)-dien-17-one, 3β-hydroxy-	EtOH	241(4.20)	39-0261-86C
1-Phenanthrenemethanol, 7-ethenyl-4b,5,6,7,8,8a,9,10-octahydro-3-hydroxy-4b,7-dimethyl-, [4bS-(4bα,7α,8aβ)]- (ar-maximol)	MeOH	221(3.95),285(3.47)	94-1015-86
$C_{19}H_{26}O_3$			
4-Octene-3,6-dione, 4-(4-methoxyphenyl)-2,2,7,7-tetramethyl-, (Z)-	ether	227(4.19),310(4.10)	78-1345-86
Spiro[5H-benzo[3,4]cyclobuta[1,2]-cycloundecene-5,2'-[1,3]dioxolan]-4b(6H)-ol, 7,8,9,10,11,12,13,13a-octahydro-, (4bR*,13aR*)-	MeOH	260(3.16),266(3.31), 272.5(3.29)	44-1419-86
$C_{19}H_{26}O_4$			
Bicyclo[7.2.2]trideca-9,11,12-triene-10,11-dimethanol, diacetate	EtOH	215(4.61),236s(3.83), 297(3.25)	24-2698-86
$C_{19}H_{26}O_5$			
11-Oxabicyclo[7.3.2]tetradeca-9,12,13-triene-13,14-dimethanol, diacetate	EtOH	209(4.04),257(3.21)	24-2698-86
$C_{19}H_{26}O_6$			
Leucanthanolide	MeOH	213(4.01)	100-0313-86
$C_{19}H_{26}O_6S$			
Estra-1,3,5(10)-triene-3,17β-diol, 4-methoxy-, 17-(hydrogen sulfate), monopotassium salt	EtOH	208(4.24),222s(3.88), 279(3.20),283(3.17)	94-2231-86
Estra-1,3,5(10)-triene-4,17β-diol, 3-methoxy-, 17-(hydrogen sulfate), monopotassium salt	EtOH	210(4.48),222s(3.89), 278(3.25),282(3.22)	94-2231-86
$C_{19}H_{27}NOSi$			
Bicyclo[4.4.1]undeca-1,3,5,7,9-pentaene-2-carboxamide, N,N-diethyl-9-(trimethylsilyl)-	MeOH	305(3.82),365(4.69)	33-1851-86
Bicyclo[4.4.1]undeca-1,3,5,7,9-pentaene-2-carboxamide, N,N-diethyl-10-(trimethylsilyl)-	MeOH	260(4.60),310(3.89)	33-1263-86
$C_{19}H_{27}NO_3S$			
2-Thiophenecarboxaldehyde, 5-[1-(2,2,6,6-tetramethyl-4,5-dioxo-3-piperidinyl)pentyl]-	EtOH	266(3.98),302(4.06)	70-2511-86
$C_{19}H_{27}NO_5$			
2H-Benzo[a]quinolizine-2-acetic acid,	pH 1	273(3.02),279s(3.01)	142-0345-86

Compound	Solvent	λ_{max}(log ϵ)	Ref.
3-ethyl-1,3,4,6,7,11b-hexahydro-8-hydroxy-9,10-dimethoxy-, [2R-(2α,3β,11bβ)]- (-)-(alancine)	pH 13	287(3.43)	142-0345-86
	MeOH	230s(3.97),273(3.01), 280s(2.99)	142-0345-86
hydrochloride	pH 1	273(3.02),279s(3.01)	142-0345-86
	pH 13	287(3.42)	142-0345-86
	MeOH	230s(3.98),273(3.04), 280s(3.01)	142-0345-86
$C_{19}H_{27}NO_{12}$			
α-D-Fructofuranoside, methyl 1-deoxy-1-[[(3-methoxy-2-(methoxycarbonyl)-3-oxo-1-propenyl]amino]-, 3,4,6-triacetate	$CHCl_3$	278(4.83)	136-0329-86D
β-D-Fructopyranoside, methyl 1-deoxy-1-[[(3-methoxy-2-(methoxycarbonyl)-3-oxo-1-propenyl]amino]-, 3,4,5-triacetate	$CHCl_3$	278(4.42)	136-0329-86D
$C_{19}H_{27}N_3O_2S_2$			
Morpholine, 4-[(4,5,6,7-tetrahydro-2-[[1-(4-morpholinyl)ethylidene]-amino]benzo[b]thien-3-yl]thioxo-methyl]-	EtOH	232(4.22),287(4.19), 397(3.02)	106-0096-86
$C_{19}H_{28}N_4O_6$			
1H-Purine-2,6-dione, 9-[3,4-dideoxy-6-O-(6-hydroxyhexyl)-β-D-erythro-hex-3-enopyranosyl)-3,9-dihydro-1,3-dimethyl-	$CHCl_3$	279(3.92)	136-0256-86K
$C_{19}H_{28}N_8O_4S_2$			
Carbamimidothioic acid, methylene-bis[(3,4-dihydro-3,6-dimethyl-2,4-dioxo-5,1(2H)-pyrimidinediyl)-2,1-ethenediyl] ester, dihydrobromide	EtOH	275(4.29)	94-1809-86
$C_{19}H_{28}O_2$			
5β-Androst-15-en-17-one, 3-hydroxy-	EtOH	229(3.91)	19-0313-86
$C_{19}H_{28}O_{11}$			
2H-Pyran-5-carboxylic acid, 4-(1,3-dioxolan-2-ylmethyl)-3-ethenyl-2-(β-D-glucopyranosyloxy)-3,4-dihydro-, methyl ester, [2S-(2α,3β,4β)]-	MeOH	332(4.01)	5-1413-86
$C_{19}H_{29}NO_4$			
Ankorine, (+)-	pH 13	287(3.42)	94-0669-86
	EtOH	273(2.95)	94-0669-86
Pterocereine, (-)-	H_2O	214s(3.29),268(2.60)	100-0745-86
$C_{19}H_{29}N_7O$			
L-Leucine, N-[N-[1-oxo-4-(1H-purin-6-ylamino)butyl]-L-alanyl]-, methyl ester	pH 2.95	271(4.07)	63-0757-86
	pH 7.0	268(4.10)	63-0757-86
	pH 10.5	272(4.07)	63-0757-86
$C_{19}H_{30}N_2O_3$			
2(3H)-Naphthalenone, 4,4a,5,6,7,8-hexahydro-4a-methyl-3,8-di-4-morpholinyl-	EtOH	236(4.08)	107-0195-86

Compound	Solvent	$\lambda_{max}(\log \epsilon)$	Ref.
$C_{19}H_{30}N_2O_9S$			
α-D-Glucopyranose, 2-deoxy-2-[[(diethylamino)thioxomethyl]amino]-, 1,3,4,6-tetraacetate	EtOH	239(4.35)	136-0049-86I
$C_{19}H_{30}N_3O_4$			
1-Piperidinyloxy, 4-[acetyl[4-(2,5-dihydro-2,5-dioxo-1H-pyrrol-1-yl)butyl]amino]-2,2,6,6-tetramethyl-	H_2O	207(4.18),217s(4.14), 223s(4.05),232s(3.76), 254s(3.20),309s(2.69), 451(1.43)	70-0363-86
$C_{19}H_{30}N_4O_8$			
1H-Purine-2,6-dione, 3,9-dihydro-9-[6-O-(6-hydroxyhexyl)-β-D-galactopyranosyl]-1,3-dimethyl-	MeOH	275(3.94)	136-0256-86K
$C_{19}H_{30}N_6O_4$			
β-D-Ribofuranuronamide, 1-deoxy-N,N-diethyl-1-[6-[(1-ethylrpopyl)amino]-9H-purin-9-yl]-	n.s.g.	269(4.28)	87-1683-86
$C_{19}H_{30}O$			
Spiro[5.5]undeca-1,4-dien-3-one, 2,4-bis(1,1-dimethylethyl)-	EtOH	246(3.8)	104-1079-86
$C_{19}H_{30}O_2$			
Gracilin F	MeOH	224(3.83)	78-5369-86
$C_{19}H_{30}O_3$			
2-Naphthalenecarboxylic acid, 1,4,4a-5,6,7,8,8a-octahydro-5,5,8a-trimethyl-1-(3-oxobutyl)-, methyl ester, [1R-(1α,4aβ,8aα)]-	EtOH	215(3.65)	102-0711-86
$C_{19}H_{31}ClN_4O_5$			
5-Pyrimidinepentanoic acid, 2-[[[(2-chloroethyl)amino]carbonyl]amino]-1,4-dihydro-6-hydroxy-4-oxo-, heptyl ester	50% MeOH	247(4.01),276(4.25)	73-1140-86
$C_{19}H_{31}IO_3$			
2,4,12-Tridecatrienoic acid, 11-hydroxy-13-iodo-3,7,11-trimethyl-, 1-methylethyl ester, (E,E)-	hexane	259(4.31)	44-5447-86
$C_{19}H_{31}N_3O_3$			
Acetamide, N-[4-(2,5-dihydro-2,5-dioxo-1H-pyrrol-1-yl)butyl]-N-(2,2-6,6-tetramethyl-4-piperidinyl)-	H_2O	212s(4.23),216s(4.20), 224s(4.09),231s(3.79), 301(3.61)	70-0363-86
$C_{19}H_{32}N_3O_6$			
1-Piperidinyloxy, 4-[[4-[(3-carboxy-1-oxo-2-propenyl)amino]butyl](methoxycarbonyl)amino]-2,2,6,6-tetramethyl-, (Z)-	H_2O	206(4.22),243s(3.75), 428(1.48)	70-0363-86
$C_{19}H_{32}O_3$			
9-Octadecynoic acid, 8-oxo-, methyl ester	MeOH	222(3.70)	102-0449-86
11-Octadecynoic acid, 10-oxo-, methyl ester	MeOH	222(3.70)	102-0449-86

Compound	Solvent	λ_{max}(log ε)	Ref.
4H-Pyran-4-one, 3,5-dibutyl-2-methoxy-6-pentyl-	MeOH	256(3.95)	44-0268-86
$C_{19}H_{32}O_8$			
Rehmaionoside C	MeOH	232(4.03)	94-2294-86
$C_{19}H_{33}N_3O_4$			
2-Butenoic acid, 4-[[4-[acetyl(2,2,6,6-tetramethyl-4-piperidinyl)amino]-butyl]amino]-4-oxo-, (Z)-	H_2O	203(4.25),228(3.70)	70-0363-86
$C_{19}H_{34}OSi$			
Cyclohexene, 3-[3-(1,1-dimethylethyl)-dimethylsilyloxy-2-methyl-2-propen-ylidene]-2,4,4-trimethyl-	pentane	221(3.97),267(3.98)	33-1378-86
2-Propen-1-one, 1-[(1,1-dimethylethyl)-dimethylsilyl]-2-methyl-3-(2,6,6-trimethyl-2-cyclohexenyl)-, (E)-	pentane	244(4.15),418(2.11)	33-1378-86
(Z)-	pentane	263(3.71),402(2.11), 418(2.11)	33-1378-86
$C_{19}H_{34}O_3$			
Heptadecanoic acid, 9-methylene-8-oxo-, methyl ester	MeOH	228(3.68)	102-0449-86
$C_{19}H_{35}NS_2$			
Thiazole, 4-methyl-2-(pentadecyl-thio)-	EtOH	283(3.66)	39-0039-86C
$C_{19}H_{36}OSi$			
Silane, [(1-cycloheptyl-2-methylene-cyclopentyl)oxy](1,1-dimethyl-ethyl)dimethyl-	MeOH	233(3.95)	23-0180-86
Silane, (1,1-dimethylethyl)dimethyl-[[3,3,5,5-tetramethyl-4-(2-propen-ylidene)cyclohexyl]oxy]-, (S)-(+)-	C_6H_{12}	236(4.32),242(4.35)	35-2691-86
$C_{19}H_{36}O_2Si$			
2-Propanone, 1-[4-[[(1,1-dimethyleth-yl)dimethylsilyl]oxy]-2,2,6,6-tetra-methylcyclohexylidene]-, (S)-	C_6H_{12}	240(3.82)	35-2691-86
$C_{19}H_{36}S_2Si$			
Silane, (1,1-dimethylethyl)[2-(2,6-dimethyl-1,5-heptadienyl)-1,3-di-thian-2-yl]dimethyl-, (E)-	pentane	257(2.86)	33-1378-86
(Z)-	pentane	257s(2.85)	33-1378-86

Compound	Solvent	λ_{max}(log ε)	Ref.
$C_{20}Cl_{16}$			
Bicyclo[4.2.0]octa-2,4,7-triene, 1,2,3,5,6,?-hexachloro-4,?-bis(pentachlorophenyl)-	C_6H_{12}	216(4.94),250s(4.51), 294(4.10)	44-1413-86
$C_{20}H_2BrCl_{14}$			
Methyl, [4-(bromomethyl)-2,3,5,6-tetrachlorophenyl]bis(pentachlorophenyl)-	C_6H_{12}	223(4.90),290(3.81), 340s(3.76),370s(4.27), 388(4.56),482s(3.02), 508(3.04),566(3.01)	44-2472-86
$C_{20}H_2Cl_{14}O$			
Benzaldehyde, 4-[bis(pentachlorophenyl)methyl]-2,3,5,6-tetrachloro-	$CHCl_3$	273s(4.85),306(3.29), 332(3.21)	44-2472-86
$C_{20}H_2Cl_{15}$			
Methyl, bis(pentachlorophenyl)-[2,3,5,6-tetrachloro-4-(chloromethyl)phenyl]-	C_6H_{12}	222(5.00),290(3.82), 338s(3.77),370s(3.28), 385(4.58),482s(3.02), 515(3.03),565(3.00)	44-2472-86
$C_{20}H_3BrCl_{14}$			
Benzene, 1,1'-[[4-(bromomethyl)-2,3,5,6-tetrachlorophenyl]methylene]bis[2,3,4,5,6-pentachloro-	C_6H_{12}	223(5.02),303(3.21)	44-2472-86
$C_{20}H_3Cl_{14}I$			
Benzene, 1,1'-[[2,3,5,6-tetrachloro-4-(iodomethyl)phenyl]methylene]-bis[2,3,4,5,6-pentachloro-	C_6H_{12}	222(4.97),291s(3.57), 302s(3.53)	44-2472-86
$C_{20}H_3Cl_{14}O$			
Methyl, bis(pentachlorophenyl)-[2,3,5,6-tetrachloro-4-(hydroxymethyl)phenyl]-	C_6H_{12}	220(4.91),289(3.83), 369s(4.24),386(4.56), 509(3.04),565(3.01)	44-2472-86
$C_{20}H_3Cl_{15}$			
Benzene, 1,1'-[[2,3,5,6-tetrachloro-4-(chloromethyl)phenyl]methylene]-bis[2,3,4,5,6-pentachloro-	C_6H_{12}	221(5.05),250s(3.52), 297(3.22),305(3.25)	44-2472-86
$C_{20}H_4Cl_{14}O$			
Benzenemethanol, 4-[bis(pentachlorophenyl)methyl]-2,3,5,6-tetrachloro-	$CHCl_3$	294(3.20),304(3.25)	44-2472-86
$C_{20}H_6F_{12}N_4$			
6,6'-Biquinoxaline, 2,2',3,3'-tetrakis(trifluoromethyl)-	EtOH	270(4.65),345(4.28)	39-1043-86C
$C_{20}H_8I_4O_5$			
Erythrosine, disodium salt	H_2O	526(4.96)	48-0081-86
	H_2O	527(4.89)	48-0081-86
	EtOH	532(5.03)	48-0081-86
$C_{20}H_8S_3$			
Thieno[2',3',4',5'-6,7]perylo[1,12-cde]-1,2-dithiin	KOH	254(4.41),248(4.44), 254(4.45),258(4.39), 275(4.39),275(4.52), 291(4.66),368(4.09), 389(4.33),413(4.43) [sic]	104-0943-86

Compound	Solvent	λ_{max}(log ϵ)	Ref.
$C_{20}H_{10}Br_2N_2S_2$			
Benzothiazole, 2,2'-(1,4-phenylene)-bis[6-bromo-	C_6H_5Cl	360(4.81)	103-1262-86
$C_{20}H_{10}N_2O_4$			
Benz[5,6]indolo[2,1-a]isoquinoline-8,13-dione, 14-nitro-	DMF	467(3.8633)	2-1126-86
Benz[5,6]indolo[1,2-a]quinoline-8,13-dione, 7-nitro-	DMF	464(3.9845)	2-1126-86
$C_{20}H_{10}N_4O_3$			
6H-Anthra[1,9-cd]isoxazol-6-one, 3-azido-5-phenoxy-	EtOH	460(4.19)	104-1192-86
$C_{20}H_{10}N_6$			
Propanedinitrile, 2,2'-(2,3-dimethyl-benzo[g]quinoxaline-5,10-diylidene)bis-	MeCN	218(4.34),296(4.29), 352(4.40)	138-0715-86
$C_{20}H_{11}N_3O$			
Ethenetricarbonitrile, [5-(4-methyl-1-naphthalenyl)-2-furanyl]-	acetone	495(3.18)	73-1450-86
$C_{20}H_{12}$			
Benzo[a]fluoranthene	n.s.g.	250f(4.6),400(3.9)	131-0275-86B
Benzo[b]fluoranthene	n.s.g.	250f(4.2),310(4.2), 360f(3.9)	131-0275-86B
Benzo[j]fluoranthene	n.s.g.	232(4.4),320f(4.3), 415f(4.1)	131-0275-86B
Benzo[k]fluoranthene	n.s.g.	250f(4.4),310f(4.4), 400(4.1)	131-0275-86B
9H-Fluorene, 9-bicyclo[4.1.0]hepta-1,3,5-trien-7-ylidene)-	C_6H_{12}	392(4.40),411(4.57), 438.5(4.63)	88-5159-86
	MeCN	388(4.40),407.5(4.56), 432(4.51)	88-5159-86
$C_{20}H_{12}Br_2N_2O$			
1,3,4-Oxadiazole, 2-(4'-bromo[1,1'-biphenyl]-4'-yl)-5-(4-bromophenyl)-	dioxan	312(4.73)	103-1262-86
$C_{20}H_{12}N_2O$			
Benzonitrile, 4-[5-(1-naphthalenyl)-2-oxazolyl]-	toluene	340(4.28)	103-1026-86
$C_{20}H_{12}N_2O_2$			
Benzoxazole, 2,2'-(1,4-phenylene)bis-	C_6H_5Cl	343(4.83)	103-1262-86
$C_{20}H_{12}N_2O_4$			
1,4-Benzenediol, 2,5-bis(2-benzoxazolyl)-	2-MeTHF	312(4.5),333(4.6), 380(4.3)	46-1455-86
$C_{20}H_{12}N_2O_5$			
Benzofuran, 5,7-dinitro-2,3-diphenyl-	n.s.g.	278(3.48)	44-3453-86
$C_{20}H_{12}N_2S_2$			
Benzothiazole, 2,2'-(1,4-phenylene)-bis-	dioxan	352(4.66)	103-1262-86
$C_{20}H_{12}N_4$			
Cyanamide, (2,5-diphenyl-2,5-cyclohexadiene-1,4-diylidene)bis-	MeCN	237(4.31),340(4.30), 387s(3.95)	5-0142-86

Compound	Solvent	$\lambda_{max}(\log \epsilon)$	Ref.
$C_{20}H_{12}N_8$			
Propanedinitrile, 2,2'-(2,3,7,8-tetramethylpyrazino[2,3-g]quinoxaline-5,10-diylidene)bis-	MeCN	238(4.27),262s(3.81), 307s(4.25),318(4.44), 382s(4.53),398(4.60)	138-0715-86
radical anion, sodium salt	MeCN	238(4.38),322(4.29), 366(4.33),380s(4.32), 402(4.36),604s(3.81), 614(3.92),672(4.52)	138-0715-86
$C_{20}H_{12}O_3$			
1,4-Epoxynaphthacene-6,11-dione, 1,2,3,4-tetrahydro-2,3-bis(methylene)-	dioxan	262(4.59),321(3.72)	44-4160-86
$C_{20}H_{12}O_5$			
Fluorescein, cation	H_2O	440(4.7)	110-0857-86
	MeOH	440(4.78)	131-0197-86H
	EtOH	437(4.71)	131-0197-86H
	isoPrOH	445(4.72)	131-0197-86H
	$HOCH_2CH_2OH$	443(3.67)	131-0197-86H
dianion	MeOH	493(4.96)	131-0197-86H
	EtOH	500(5.01)	131-0197-86H
	isoPrOH	500(5.01)	131-0197-86H
	$HOCH_2CH_2OH$	498(3.86)	131-0197-86H
$C_{20}H_{12}O_6$			
Anthra[1,2-b]furan-6,11-dione, 4-(1,2-dioxopropyl)-5-hydroxy-2-methyl-	n.s.g.	224(3.98),257(4.32), 290(3.71),487(3.72), 522s(3.49)	12-2075-86
Perylo[1,2-b:7,8-b']bisoxirene-5,10-dione, 1a,1b,5a,6a,6b,10a-hexahydro-4,9-dihydroxy- (altertoxin III)	MeOH	210(4.29),265(4.16), 352(3.72)	100-0866-86
$NaBH_4$ reduction product	MeOH	234(4.15),287(3.72)	100-0866-86
Perylo[1,2-b]oxirane-7,11-dione, 7a,8a,8b,8c-tetrahydro-1,6,8c-trihydroxy-, (7aα,8aα,8bβ,8cα)-(±)- (stemphyltoxin III)	EtOH	268(4.46),274(4.42), 287s(4.31),374(3.68)	39-0525-86C
$C_{20}H_{12}O_7$			
Perylo[1,2-b:11,12-b']bisoxirane-7,10-dione, 7a,8a,8b,8c,8d,9a-hexahydro-1,6,8b-trihydroxy- (stemphyltoxin IV)	EtOH	268(4.49),300s(4.12), 372(3.74)	39-0525-86C
$C_{20}H_{13}BO_6$			
Boron, [1-(9-anthracenyl)-1,3-butanedionato-O,O'][ethanedioato(2-)-O,O']-	CH_2Cl_2	474(3.90)	48-0755-86
$C_{20}H_{13}BrN_2O$			
1,3,4-Oxadiazole, 2-[1,1'-biphenyl]-4-yl-5-(4-bromophenyl)-	dioxan	306(4.64)	103-1262-86
$C_{20}H_{13}F_3N_2O_5$			
4-Isoxazolecarboxylic acid, 5-(dibenzoylamino)-3-(trifluoromethyl)-, methyl ester	EtOH	249(4.49)	4-1535-86
$C_{20}H_{13}NO_2$			
1H-Pyrrolizine-5-carboxaldehyde, 1-oxo-6,7-diphenyl-	$CHCl_3$	274(4.27),332(4.03), 461(3.50)	83-0749-86

Compound	Solvent	λmax(log ε)	Ref.
$C_{20}H_{13}NO_3$			
Benzoic acid, 4-[5-(1-naphthalenyl)-2-oxazolyl]-	toluene	340(4.28)	103-1022-86
$C_{20}H_{13}NO_6$			
[2,3'-Bi-4H-1-benzopyran]-4,4'-dione, 6,6'-dimethyl-3-nitro-	n.s.g.	300(4.42)	2-0212-86
$C_{20}H_{13}N_3O$			
5H-Indolo[2,3-a]pyrrolo[3,4-c]carbazol-5-one, 6,7,12,13-tetrahydro-(K-252c)	MeOH	290(4.9),330(4.3), 360(4.0)	158-1066-86
	MeOH	230(4.57),238s(4.53), 246s(4.45),257s(4.46), 287(4.93),320s(4.20), 331(4.30),341(4.20), 358(4.04)	158-1072-86
$C_{20}H_{14}$			
Anthracene, 1-phenyl-	n.s.g.	258(5.14),305(3.08), 333(3.45),349(3.74), 366(3.92),387(3.84)	97-0209-86
Anthracene, 2-phenyl-	n.s.g.	261(4.72),281(4.85), 320(3.15),334(3.45), 350(3.70),368(3.82), 389(3.67)	97-0209-86
Bicyclo[4.1.0]hepta-1,3,5-triene, 7-(diphenylmethylene)-	C_6H_{12}	249(4.34),265s(4.25), 385(4.41),405s(4.32)	35-5949-86
	MeCN	246(4.34),262s(4.43), 398s(4.31)	35-5949-86
Phenanthrene, 9-phenyl-	85% MeOH	250s(4.73),257(4.80), 270s(4.33),278s(4.14), 290s(4.01),299(4.12), 326s(2.60),335(2.56), 343(2.42),351(2.48)	24-1525-86
$C_{20}H_{14}BrNO_6$			
[2,3'-Bi-4H-1-benzopyran]-4,4'-dione, 3-bromo-2,3-dihydro-6,6'-dimethyl-3-nitro-	n.s.g.	338(3.84)	2-0212-86
$C_{20}H_{14}ClN$			
3H-Indole, 3-chloro-2,3-diphenyl-	CH_2Cl_2	223(4.01),326(4.06)	39-2305-86C
$C_{20}H_{14}Cl_2N_2$			
Quinoxaline, 3-[1,1'-biphenyl]-4-yl-5,7-dichloro-1,2-dihydro-	MeOH	287(4.44),312(4.33), 398(3.88)	103-0918-86
	toluene	397(--)	103-0918-86
$C_{20}H_{14}Cl_2O_4$			
9,10-Anthracenedione, 1,4-dichloro-5,8-bis(2-propenyloxy)-	MeOH	228(4.41),246(4.47), 410(3.70)	4-1491-86
4a,9a-[2]Butenoanthracene-1,4,9,10-tetrone, 2,3-dichloro-12,13-dimethyl-	EtOH	228(4.51),262(4.24), 322(3.39)	39-1923-86C
$C_{20}H_{14}HgO_4$			
Mercury, bis[2-oxo-6-(3-penten-1-ynyl)-2H-pyran-5-yl]-, (Z,Z)-	MeOH	210(4.45),263(4.25), 348(4.44)	106-0703-86
$C_{20}H_{14}N_2$			
Pyrrolo[3,2-a]carbazole, 3,10-di-	MeOH	202(4.40),229(4.27),	103-0172-86

Compound	Solvent	$\lambda_{max}(\log \epsilon)$	Ref.
hydro-2-phenyl- (cont.)		256(4.41),295(4.45), 328(4.26)	103-0172-86
$C_{20}H_{14}N_2O$			
Ethanone, 1-[1,1'-biphenyl]-2-yl-2-diazo-2-phenyl-	MeOH	242(4.4),310(3.7)	64-0772-86B
Pyridine, 4-(5-[1,1'-biphenyl]-4-yl-2-oxazolyl)-	EtOH	335(4.64)	135-0244-86
	60% HOAc	388(4.46)	135-0244-86
$C_{20}H_{14}N_2O_2$			
Benzamide, 4-[5-(1-naphthalenyl)-2-oxazolyl]-	toluene	336(4.31)	103-1022-86
Benzoic acid, 4-[2,2-dicyano-1-(2,4,6-cycloheptatrien-1-ylidenemethyl)ethenyl]-, methyl ester	MeCN	235(--),476(4.4)	24-2631-86
Benzoic acid, 4-(1,1-dicyano-1,8a-dihydro-2-azulenyl)-, methyl ester	MeCN	273(4.2),365(4.2)	24-2631-86
$C_{20}H_{14}N_2O_4$			
3H-Indol-3-one, 1-acetyl-2-(1-acetyl-1,3-dihydro-3-oxo-2H-indol-2-ylidene)-1,2-dihydro-	benzene	430(3.59),560(3.81)	44-2529-86
$C_{20}H_{14}N_2O_4S$			
1-Naphthalenesulfonic acid, 4-[(2-hydroxy-1-naphthalenyl)azo-, sodium salt (rocellin)	pH 11-12	505(4.3)(anom.)	60-2333-86
$C_{20}H_{14}N_2O_5$			
Benzofuran, 2,3-dihydro-5,7-dinitro-2,3-diphenyl-	n.s.g.	206(4.31),252(3.98), 305(3.87)	44-3453-86
$C_{20}H_{14}N_2O_7$			
Pyrrolo[1,2-a]quinoline-2,3-dicarboxylic acid, 1-(5-nitro-2-furanyl)-, dimethyl ester	MeOH	232(3.52),269s(3.33), 337(3.23),390s(2.73)	73-0412-86
$C_{20}H_{14}N_4$			
Porphycene	benzene	358(5.14),370(5.03), 558(4.53),596(4.48), 630(4.71)	89-0257-86
	toluene	362(5.03),373(4.94), 562(4.45),600(4.41), 633(4.63)	149-0555-86B
	MeOH	356(--),368(--), 556(--),593(--), 627(--)	149-0555-86B
	CH_2Cl_2	360(--),372(--), 559(--),597(--), 629(--)	149-0555-86B
$C_{20}H_{14}N_6O_3$			
Ethanone, 1-[3,5-bis(2H-benzotriazol-2-yl)-2,4-dihydroxyphenyl]-	$CHCl_3$	252(4.39),273(4.51), 322(4.53),342s(4.42)	49-0805-86
$C_{20}H_{14}O$			
Benz[a]azulen-10(9aH)-one, 9a-phenyl-	MeOH	232(4.3),270(3.8), 322(3.8),366(3.5)	64-0772-86B
$C_{20}H_{14}O_2$			
Azuleno[2,1-a]azulene-2-carboxylic	CH_2Cl_2	255(4.13),360(4.87),	138-2039-86

Compound	Solvent	λ_{max}(log ε)	Ref.
acid, methyl ester (cont.)		480(3.91),507(4.02), 1110(2.48),1200(2.35)	138-2039-86
2,3:7,8-Dibenzotetracyclo[7.2.1.0^{4,11}-0^{6,10}]dodeca-2,7-diene-5,12-dione, acs-	MeCN	261(2.96),268(3.13), 275(3.05),307s(2.90), 316(2.92),330s(2.72)	24-3442-86
$C_{20}H_{14}O_3$			
Benzo[a]pyrene-7,8-diol, 9,10-epoxy-7,8,9,10-tetrahydro-, syn-trans	pH 9.5	313(--),328(--), 344(4.5)	69-3290-86
Cyclohepta[de]naphthalen-7(8H)-one, 8,8-bis(2-propynyloxy)-	$CHCl_3$	322(3.92),336(3.90), 354(3.83)	39-1965-86C
$C_{20}H_{14}O_3S$			
9(10H)-Anthracenone, 1,8-dihydroxy-10-(phenylthio)-	EtOH	375(3.90)	118-0430-86
$C_{20}H_{14}O_4$			
5,12-Naphthacenedione, 6,11-dihydroxy-2,3-dimethyl- (or isomer)	dioxan	308(4.50),429s(3.04), 456(3.38),484(3.63), 518(3.46)	39-1923-86C
Phenolphthalein, dianion	MeOH	560(3.1)	140-1223-86
	EtOH	565(3.9)	140-1223-86
	8% EtOH	560(4.47)	140-1223-86
$C_{20}H_{14}O_5$			
2,5-Furandione, 3-[2-(1,3-benzodioxol-5-yl)ethylidene]dihydro-4-(phenylmethylene)-	hexane	406(3.98)	46-2651-86
$C_{20}H_{14}O_6$			
Anthra[1,2-b]furan-6,11-dione, 5-hydroxy-4-(1-hydroxy-2-oxopropyl)-2-methyl-	n.s.g.	215(4.21),257(4.46), 289(4.30),430(4.00)	12-2075-86
Perylo[1,2-b]oxirane-7,11-dione, 7a,8a,8b,8c,9,10-hexahydro-1,6,8c-trihydroxy- (altertoxin II)	MeOH	215(4.43),258(4.50), 286s(4.23),297s(4.13), 358(3.72)	100-0866-86
stemphyltoxin II	EtOH	260(4.47),285s(4.29), 364(3.79)	39-0525-86C
$C_{20}H_{14}O_7$			
Naphtho[2,3-c]furan-1(3H)-one, 9-(1,3-benzodioxol-5-yl)-4,7-dihydroxy-6-methoxy- (haplomyrtin)	MeOH	228(4.32),268(4.48), 309(3.81),322(3.79), 361(3.55)	102-1949-86
	MeOH-base	211(4.78),242(4.34), 264(4.29),290(4.30), 348(3.72)	102-1949-86
Perylo[1,2-b]oxirane-7,11-dione, 7a,8a,8b,8c,9,10-hexahydro-1,6,8c,9-tetrahydroxy-	EtOH	257(4.38),283s(4.18), 295s(4.08),364(3.71)	39-0525-86C
$C_{20}H_{14}S$			
Fluorantheno[8,9-c]thiophene, 8,10-dimethyl-	EtOH	236s(4.47),261(4.32), 270s(4.29),300s(4.65), 308(4.75),347(3.64), 358s(3.72),365(3.82), 388(3.87),411(3.92), 438(3.74)	39-0415-86C
$C_{20}H_{15}BO_7$			
Boron, [ethanedioato(2-)-O,O'][5-(4-methoxyphenyl)-1-phenyl-4-pentene-	CH_2Cl_2	473(4.68)	97-0399-86

Compound	Solvent	λ_{max}(log ε)	Ref.
1,3-dionato-O[1],O[3]]- (cont.)			97-0399-86
$C_{20}H_{15}BrN_2O_3$			
1H-Pyrazole, 5-(3-bromophenyl)-1-(2-furancarbonyl)-4,5-dihydro-5-hy-hydroxy-3-phenyl-	hexane	258(4.15),305(4.20)	4-0727-86
	EtOH	258(4.26),305(4.34)	4-0727-86
$C_{20}H_{15}ClFNO_2S$			
1-Propanone, 3-[1-(4-chlorophenyl)-2-thioxo-3-pyridinyl]-1-(4-fluoro-phenyl)-3-hydroxy-	EtOH	245(4.23),290(3.92), 375(3.71)	4-0567-86
$C_{20}H_{15}ClN_2$			
Quinoxaline, 3-[1,1'-biphenyl]-4-yl-6(and 7)-chloro-1,2-dihydro-	MeOH	286(4.43),309(4.33), 406(3.77)	103-0918-86
	toluene	405(--)	103-0918-86
$C_{20}H_{15}ClO_4$			
4a,9a-[2]Butenoanthracene-1,4,9,10-tetrone, 2-chloro-12,13-dimethyl-	EtOH	228(4.67),254(4.42), 307(2.51)	39-1923-86C
$C_{20}H_{15}FO$			
Phenol, 4-[1-(4-fluorophenyl)-2-phen-ylethenyl]-	EtOH	246(4.01),307(4.08)	13-0073-86B
$C_{20}H_{15}NO_6$			
[2,3'-Bi-4H-1-benzopyran]-4,4'-dione, 2,3-dihydro-6,6'-dimethyl-3-nitro-	n.s.g.	314(4.08)	2-0212-86
$C_{20}H_{15}N_3$			
Benzaldehyde, (3-methyl-9H-indeno[2,1-c]pyridin-9-ylidene)hydrazone	EtOH	208(4.70),220s(4.52), 258(4.69),264(4.68), 340(4.55),354(4.57), 388s(4.10)	103-0200-86
	EtOH-H_2SO_4	215(4.48),242(4.53), 270(4.53),295(4.14), 310(4.29),368(4.40), 380(4.10)	103-0200-86
$C_{20}H_{15}N_3O$			
Benzaldehyde, 2-hydroxy-, (3-methyl-9H-indeno[2,1-c]pyridin-9-ylidene)-hydrazone	EtOH	212(4.61),220(4.58), 260(4.65),340(4.30), 378(4.34)	103-0200-86
	EtOH-H_2SO_4	217(4.37),242(4.37), 263(4.50),297(4.16), 315(4.11),324(4.32), 400(4.15)	103-0200-86
Benzaldehyde, 4-methoxy-, 5H-indeno-[1,2-b]pyridin-5-ylidenehydrazone	EtOH	213(4.50),246(4.52), 240(4.53),247(4.52), 300(4.16),315(4.14), 367(4.50)[sic]	103-0200-86
	EtOH-H_2SO_4	212(4.53),220(4.67), 234(4.50),253(4.48), 305(4.27),317(4.80), 387(4.45)	103-0200-86
$C_{20}H_{15}N_3O_2$			
9,10-Anthracenedione, 1,5-diamino-4-(phenylamino)-	EtOH	519(4.14),631(4.15)	104-0547-86
Benzaldehyde, 4-hydroxy-3-methoxy-,	EtOH	212(4.47),227(4.53),	103-0200-86

Compound	Solvent	λmax(log ε)	Ref.
5H-indeno[1,2-b]pyridin-5-ylidene-hydrazone (cont.)		250(4.38),287(4.16), 384(4.34),500(3.30)	103-0200-86
	EtOH-H_2SO_4	212(4.29),232(4.52), 280(4.22),323(4.46), 412(4.15)	103-0200-86
$C_{20}H_{15}N_3O_4$			
Methanone, [2,4-dihydroxy-5-(5-methoxy-2H-benzotriazol-2-yl)phenyl]-phenyl-	$CHCl_3$	248(4.26),294s(4.35), 332(4.44),340s(4.42)	49-0805-86
$C_{20}H_{15}N_5O$			
4-Pyrimidinamine, 6-phenyl-N-[(6-phenyl-4-pyrimidinyl)oxy]-	n.s.g.	266(4.58),340s(3.81)	103-1232-86
$C_{20}H_{15}O$			
Diindeno[1,2-b:1',2'-d]pyrylium, 7,12-dihydro-6-methyl-, perchlorate	MeCN	252(4.18),270s(3.96), 330(4.32),385(4.62)	22-0600-86
$C_{20}H_{16}$			
Cyclobuta[1,2:3,4]dicyclopentene, 1,6-bis(2,4-cyclopentadien-1-ylidene)-1,3a,3b,6,6a,6b-hexahydro-	hexane	293(4.70),303(4.70), 330s(4.27),400(2.77)	33-1644-86
1,3,9,11-Cyclotridecatetraene-5,7-diyne, 13-(2,4-cyclopentadien-1-ylidene)-4,9-dimethyl-, (E,E,Z,Z)-	THF	248s(4.19),261(4.37), 310(4.28),376s(4.40), 397(4.53),417s(4.45)	18-1723-86
relative absorbance given	MeCN	247s(0.58),259(0.79), 272(0.69),308(0.60), 373s(0.76),392(1.00), 413s(0.84)	18-1723-86
9H-Fluorene, 9-(phenylmethyl)-, cesium salt	THF	371(4.12)	35-7016-86
lithium salt	THF	381(4.19)	35-7016-86
$C_{20}H_{16}ClNO_2S$			
2H-[1]Benzothiepino[5,4-b]pyran-2-one, 3-chloro-5,6-dihydro-4-(methylphenylamino)-	EtOH	241(4.25),273s(3.83), 340(4.07)	4-0449-86
$C_{20}H_{16}ClNO_3S$			
1-Propanone, 1-(4-chloro-2-hydroxyphenyl)-3-(1,2-dihydro-1-phenyl-2-thioxo-3-pyridinyl)-3-hydroxy-	EtOH	261(4.22),290(4.06), 330(3.84),367(3.82)	4-0567-86
$C_{20}H_{16}Cl_2N_2O$			
Diazene, (3,4-dichlorophenyl)(methoxydiphenylmethyl)-	CH_2Cl_2	279.0(4.19),421.0(2.26)	107-0741-86
$C_{20}H_{16}F_3N_2Se_2$			
Benzoselenazolium, 3-methyl-2-[3,3,3-trifluoro-2-[(3-methyl-2(3H)-benzoselenazolylidene(methyl]-1-propenyl]-3-methyl-, perchlorate	n.s.g.	606(4.50)	104-0144-86
$C_{20}H_{16}N_2$			
Quinoxaline, 3-[1,1'-biphenyl]-4-yl-1,2-dihydro-	MeOH	259(4.43),301(4.36), 427(3.93)	103-0918-86
	toluene	400(--)	103-0918-86
$C_{20}H_{16}N_2O$			
1H-Benzofuro[3,2-b]pyrrolo[3,2-e]pyri-	MeOH	328(4.67),380(4.42)	83-0347-86

Compound	Solvent	λ_{max}(log ε)	Ref.
dine, 2,3-dihydro-1-methyl-4-phen-yl- (cont.)			83-0347-86
$C_{20}H_{16}N_2O_2$			
5H-[1]Benzopyrano[2,3-b]pyridin-5-one, 2-(dimethylamino)-4-phenyl-	MeOH	232(4.6),317(4.15), 351(4.48)	83-0347-86
$C_{20}H_{16}N_2O_2S$			
9H-Acridin-9-amine, N-[4-(methyl-sulfonyl)phenyl]-	MeOH	203(4.59),258(4.89), 357(4.26)	150-3425-86M
$C_{20}H_{16}N_2O_3$			
Propanenitrile, 3,3'-[(7-oxocyclohep-ta[de]naphthalen-8(7H)-ylidene)-bis(oxy)]bis-	$CHCl_3$	320(3.92),336(3.91), 354(3.83)	39-1965-86C
Spiro[2H-anthra[1,2-c]pyrazole-2,1'-piperidinium], 1,3,6,11-tetrahydro-3,6,11-trioxo-, hydroxide, inner salt	toluene	658(3.81)	104-2117-86
$C_{20}H_{16}N_2O_8$			
2-Propenoic acid, 3-(3-nitrophenyl)-, 1,2-ethanediyl ester, (E,E)-	MeCN	261(4.76)	18-1379-86
2-Propenoic acid, 3-(4-nitrophenyl)-, 1,2-ethanediyl ester, (E,E)-	MeCN	301(4.62)	18-1379-86
$C_{20}H_{16}N_4$			
[1,2,4,5]Tetrazino[1,6-a:4,3-a']di-quinoline, 5,13-dimethyl-	n.s.g.	299(4.7),310(4.6), 375(4.0),610f(2.40)	33-1521-86
$C_{20}H_{16}N_4O$			
3(2H)-Pyridazinone, 6-[2-phenyl-1-(phenylmethyl)-1H-indazol-4-yl]-	MeOH	266(4.45)	87-0261-86
$C_{20}H_{16}N_4OS$			
2(1H)-Pteridinone, 3,4-dihydro-1,3-di-methyl-6,7-diphenyl-4-thioxo-	pH 1-10	226(4.54),294(4.23), 406(4.27)	128-0199-86
2(1H)-Pteridinone, 1-methyl-4-(methyl-thio)-6,7-diphenyl-	MeOH	223(4.67),284(4.40), 393(4.38)	128-0199-86
$C_{20}H_{16}N_4O_2$			
1H-Benz[de]isoquinoline-1,3(2H)-dione, 6-[[4-(dimethylamino)phenyl]azo]-	DMF	282.2(3.98),523(4.42)	56-0797-86
with 0.02M p-toluenesulfonic acid	DMF	271.1(4.35),523(4.41)	56-0797-86
2,4(1H,3H)-Pteridinedione, 1,3-dimeth-yl-6,7-diphenyl-	pH 7	226(4.41),276(4.19), 365(4.18)	128-0199-86
$C_{20}H_{16}N_4O_2S$			
2(1H)-Pteridinone, 3,4-dihydro-1,3-di-methyl-6,7-diphenyl-4-sulfinyl-	$CHCl_3$	260s(4.26),348(4.03), 457(4.18)	128-0199-86
$C_{20}H_{16}N_4O_3$			
Piperidine, 1-[(1-azido-9,10-dihydro-9,10-dioxo-2-anthracenyl)carbonyl]-	toluene	342(3.45),362(3.38)	104-2117-86
$C_{20}H_{16}N_4O_4$			
11,7-Metheno-7H-dibenzo[b,m][1,15-4,5,11,12]dioxatetraazacyclohepta-decine-8,21-diol, 19,20-dihydro-	EtOH	416(4.38)	70-1741-86

Compound	Solvent	$\lambda_{max}(\log \epsilon)$	Ref.
$C_{20}H_{16}O$			
1H-Cyclopenta[l]phenanthrene-1-carboxaldehyde, 1,3-dimethyl-	EtOH	245(4.39),254(4.34), 266(4.38),281s(4.24), 323(3.85)	39-1471-86C
1H-Cyclopenta[l]phenanthrene-2-carboxaldehyde, 1,3-dimethyl-	EtOH	264(4.35),300s(4.02), 360.5(4.11)	39-1471-86C
2-Cyclopenten-1-one, 3-phenyl-5-(3-phenyl-2-propynyl)-	EtOH	250(4.20),287(4.45)	39-1235-86C
9H-Fluoren-9-ol, 2-methyl-9-phenyl-	$CHCl_3$	271(4.41),278(4.45), 290(4.43),303s(3.90), 312s(3.58)	104-1352-86
9H-Fluoren-9-ol, 3-methyl-9-phenyl-	$CHCl_3$	280(4.02),288(3.99), 305(3.78),315(3.74)	104-1352-86
Phenol, 2-(2-[1,1'-biphenyl]-4-ylethenyl)-, (E)-	n.s.g.	335(4.60)	150-0433-86S
Pyrene, 1-(2-methoxyethenyl)-2-methyl-	EtOH	245(4.65),280(4.39), 343(4.35)	35-4498-86
$C_{20}H_{16}O_2$			
Naphth[2,3-a]azulene, 5,12-dimethoxy-	$CHCl_3$	215(4.4),274(4.4), 283(4.5),292(4.5), 313(4.7),326(4.6), 377(4.0),401(3.9), 425(3.9),452(3.8), 549(2.8),590(2.9), 647(2.8),715(2.6)	118-0686-86
$C_{20}H_{16}O_3$			
1,4-Epoxynaphthacene-6,11-dione, 1,2,3,4,5,5a,11a,12-octahydro-2,3-bis(methylene)-	dioxan	222(4.71),242(4.28), 292(3.34),304(3.32)	44-4160-86
$C_{20}H_{16}O_4$			
1,3-Benzodioxol-5-ol, 6-[(2-hydroxyphenyl)phenylmethyl]-, (S)-	EtOH	302(3.78)	116-0509-86
Cyclohepta[ef]heptalene-1,2-dicarboxylic acid, dimethyl ester	hexane	233(4.44),257s(4.25), 308(4.09),372s(3.35)	89-0633-86
1,4-Naphthalenedione, 2-[(2,5-dimethyl-3,6-dioxo-1,4-cyclohexadien-1-yl)methyl]-3-methyl-	EtOH	251(4.42),260(4.41), 332(3.50)	78-3663-86
1,6-Pyrenedione, 3,8-diethoxy-	C_6H_5Cl	376(3.02),448(3.93)	104-1159-86
1,8-Pyrenedione, 3,6-diethoxy-	C_6H_5Cl	362(3.06),435(2.88), 456(2.89)	104-1159-86
$C_{20}H_{16}O_4S_2$			
1,3-Dithiole-4,5-dicarboxylic acid, 2-(diphenylmethylene)-, dimethyl ester	EtOH	230s(4.03),322(3.78), 408(2.88)	33-0419-86
$C_{20}H_{16}O_5$			
Methanone, (4,8-dimethoxy-3-methylbenzo[1,2-b:5,4-b']difuran-2-yl)phenyl-	EtOH	220s(4.36),225(4.37), 230s(4.35),270(4.34), 326(4.24),365(3.84)	44-3116-86
$C_{20}H_{16}O_6$			
2,5-Cyclohexadiene-1,4-dione, 2-(2,5-dimethoxyphenyl)-3,6-dihydroxy-5-phenyl-	dioxan	253.8(4.32),295(4.14), 321.9(3.89),395(2.84)	5-0195-86
3,9-Perylenedione, 1,2,6b,7,8,12b-hexahydro-1,4,7,10-tetrahydroxy-, (1α,6bα,7α,12bα)-	EtOH	212(4.46),255(4.41), 335(3.83)	39-0525-86C

Compound	Solvent	λ_{max}(log ε)	Ref.
3,10-Perylenedione, 1,2,11,12,12a,12b-hexahydro-1,4,9,12b-tetrahydroxy- (altertoxin I)	MeOH	215(4.41),256(4.54), 285(4.21),296s(4.12), 356(3.78)	100-0866-86
$NaBH_4$ reduction product	MeOH	214(4.49),299(4.38)	100-0866-86
2H-Pyran-2,5(6H)-dione, 4-(2,5-dimethoxyphenyl)-3-hydroxy-6-(phenylmethylene)-	MeOH	247(4.17),336(4.30)	5-0177-86
2H-Pyran-2,5(6H)-dione, 3-hydroxy-4-(4-methoxyphenyl)-6-[(4-methoxyphenyl)methylene]- (4',4"-di-O-methylgrevillin A)	MeOH	266(4.11),379(4.06)	5-0177-86
$C_{20}H_{16}O_7$			
9,10-Anthracenedione, 1,4-diacetoxy-2-methoxy-7-methyl-	MeOH	265(4.53),281s(4.28), 337s(3.69),363(3.70)	78-3767-86
Anthra[1,2-b]furan-4-acetic acid, 2,3,6,11-tetrahydro-5-hydroxy-2-methoxy-6,11-dioxo-, methyl ester	n.s.g.	234s(4.24),252(4.46), 280(3.97),456(3.87)	12-0821-86
1,3-Benzenediol, 4-(6-acetoxy-2-benzofuranyl)-, diacetate	MeOH	279s(4.1),304(4.53), 316(4.45)	2-0481-86
$C_{20}H_{16}O_8$			
2H-Pyran-2,5(6H)-dione, 6-[(3,4-dihydroxyphenyl)methylene]-4-(2,5-dimethoxyphenyl)-3-hydroxy-	dioxan	284(4.19),399(4.19)	5-0177-86
$C_{20}H_{16}O_8S_8$			
Cyclobuta[1,2-d:3,4-d']bis[1,3]dithiole-3a,6a(3bH,6bH)-dicarboxylic acid, 2,5-bis[4-(methoxycarbonyl)-1,3-dithiol-2-ylidene]-, dimethyl ester. isomer A	dioxan	297(4.47),312(4.39), 390(3.86)	104-0367-86
isomer B	dioxan	300(4.50),310(4.54), 386(3.79)	104-0367-86
$C_{20}H_{17}BrN_2O$			
Diazene, (4-bromophenyl)(methoxydiphenylmethyl)-	CH_2Cl_2	258.0(4.33),281s(4.26), 421.0(2.32)	107-0741-86
$C_{20}H_{17}ClN_2O$			
Diazene, (4-chlorophenyl)(methoxydiphenylmethyl)-	CH_2Cl_2	280.5(4.19),420.0(2.30)	107-0741-86
$C_{20}H_{17}ClO_4$			
2H,5H-Pyrano[3,2-c][1]benzopyran-5-one, 4-(4-chlorophenyl)-3,4-dihydro-2-(hydroxymethyl)-2-methyl-	EtOH	270(3.84),281(3.87), 303(3.77),318s(--)	32-0501-86
isomer 14a	EtOH	270(4.05),281(4.08), 304(3.99),319s(--)	32-0501-86
$C_{20}H_{17}Cl_2NO_7$			
Azuleno[6,5-d]isoxazole-3,6,8-tricarboxylic acid, 4,7-dichloro-, triethyl ester	CH_2Cl_2	260(4.41),313(4.73), 327.5(4.65),361(3.88), 368.7(3.87),553.5(3.17)	24-2956-86
$C_{20}H_{17}NO$			
4-Benzazacyclotetradecine, 9,10,11,12-tetradehydro-3-ethoxy-8-methyl-	THF	229(4.52),290s(4.54), 300(4.58),321s(4.40), 398(3.85),424s(3.76), 458s(3.40),491s(3.00)	39-0933-86C

Compound	Solvent	λ_{max}(log ε)	Ref.
5H-Indeno[1,2-b]pyridin-5-ol, 7-methyl-5-(phenylmethyl)-	$CHCl_3$	274s(4.29),288(4.45) 308(4.44),318(4.52)	104-1352-86
9H-Indeno[2,1-c]pyridin-9-ol, 3-methyl-9-(phenylmethyl)-	$CHCl_3$	277(4.20),294(3.88), 308(3.83)	104-1352-86
$C_{20}H_{17}NOS_2$			
Ethanone, 1-[1,2-dihydro-4-(methylthio)-1,6-diphenyl-2-thioxo-3-pyridinyl]-	EtOH	245(3.72),290(4.01), 395(3.44)	118-0952-86
Methanone, [1,2-dihydro-6-methyl-4-(methylthio)-1-phenyl-2-thioxo-3-pyridinyl]phenyl-	EtOH	249(4.30),285(4.50), 387(3.90)	118-0952-86
$C_{20}H_{17}NO_2$			
1-Naphthalenol, 4-[3-[(4-methylphenyl)-imino]-1-propenyl]-, N-oxide	EtOH	415(3.94)	104-1302-86
	dioxan	395(4.08)	104-1302-86
	DMSO	410(4.08)	104-1302-86
	+ NaOMe	355(3.80),530(4.26)	104-1302-86
1-Naphthalenol, 4-methyl-2-[3-(phenylimino)-1-propenyl]-, N-oxide	EtOH	280(3.78),330(3.88), 415(4.03),550(3.60)	104-1302-86
	dioxan	280(4.32),315(4.00), 350s(3.76),380(3.38)	104-1302-86
	acetone	335(3.74),407(4.08)	104-1302-86
	DMSO	330s(3.91),365s(3.97), 418(4.18),585(3.87)	104-1302-86
	+ NaOMe	355(3.86),570(4.30)	104-1302-86
2-Naphthalenol, 1-[3-[(4-methylphenyl)imino]-1-propenyl]-, N-oxide	dioxan	300s(3.79),360(3.88), 405(4.02)	104-1302-86
	DMSO	355(3.90),415(4.30)	104-1302-86
	+ NaOMe	345(3.90),355(3.91), 415(4.31),540(3.30)	104-1302-86
Pyrene, 2-(1,1-dimethylethyl)-1-nitro-	hexane	234(4.7),244(5.0), 253s(4.1),264(4.3), 275(4.5),310(4.0), 322(4.3),337(4.7)	54-0156-86
Pyrene, 7-(1,1-dimethylethyl)-1-nitro-	hexane	248(4.7),278(4.7), 340(4.0),370(4.0)	54-0156-86
$C_{20}H_{17}NO_2S$			
1-Propanone, 3-(1,2-dihydro-1-phenyl-2-thioxo-3-pyridinyl)-3-hydroxy-1-phenyl-	EtOH	243(4.25),289(3.98), 374(3.74)	4-0567-86
$C_{20}H_{17}NO_3S$			
1-Propanone, 3-(1,2-dihydro-1-phenyl-2-thioxo-3-pyridinyl)-3-hydroxy-1-(2-hydroxyphenyl)-	EtOH	250(4.18),289(4.01), 341(3.79),372(3.80)	4-0567-86
$C_{20}H_{17}NO_4$			
3H-Benzo[f]cyclopent[d][1,3]oxazepin-3-one, 9-acetoxy-1,2,4,5-tetrahydro-5-phenyl-	EtOH	252(4.12),352(4.17)	24-3487-86
1H-Indene-1,3(2H)-dione, 2-(5,6,7,8-tetrahydro-6-methyl-1,3-dioxolo-[4,5-g]isoquinolin-5-yl)-	MeOH	253(4.53),262(4.47), 295(3.70),372(3.30)	83-1018-86
8H,15H-Naphtho[1',8':8,9,10][1,6]-diazacycloundecino[2,3,4-gh]pyrrolizine-8,15-dione, 1,2,4,6,16a-16b-hexahydro-	EtOH	226(4.14),286(3.45)	39-0585-86C
Naphtho[2',3':8,9][1,6]dioxecino-	EtOH	221(4.63),324(3.08),	39-0585-86C

Compound	Solvent	λ_{max}(log ε)	Ref.
[2,3,4-gh]pyrrolizine-8,15-dione, 1,2,4,6,16a,16b-hexahydro- (cont.)		337(3.22),357(2.22)	39-0585-86C
$C_{20}H_{17}NO_5$			
4H-Dibenzo[de,g]quinoline-4,5(6H)-dione, 1,2,3-trimethoxy-6-methyl-	MeOH	241(4.49),271(4.15), 303(3.97),316(4.06), 416(3.97)	102-1509-86
$C_{20}H_{17}NO_6$			
Africanine, (+)-	EtOH	206(4.29),233(4.42), 259s(4.00),290(3.62)	142-2781-86
	EtOH-base	310(--),350(--)	142-2781-86
4,5-Benzofurandicarboxylic acid, 6-(acetylamino)-2-phenyl-, dimethyl ester	MeOH	256(3.42),302(3.25), 432(3.36)	73-1455-86
1H-[1]Benzoxepino[2,3,4-ij]isoquinoline-2,3-dione, 12,12a-dihydro-6,9,10-trimethoxy-1-methyl- (dioxocularine)(same spectra in acid or base)	EtOH	224(4.40),260(4.15), 304(3.96),340(4.06), 452(3.75)	142-3359-86
$C_{20}H_{17}N_2$			
Acridinium, 9-(4-aminophenyl)-10-methyl-, iodide	MeOH	205.5(4.54),259(4.83), 360(4.17)	150-3425-86M
$C_{20}H_{17}N_3$			
1,1(8aH)-Azulenedicarbonitrile, 2-[4-(dimethylamino)phenyl]-	MeCN	234(--),290s(--), 406(4.5)	24-2631-86
Propanedinitrile, [2-(2,4,6-cycloheptatrien-1-ylidene)-1-[4-(dimethylamino)phenyl]ethylidene]-	MeCN	284(--),460(4.5)	24-2631-86
$C_{20}H_{17}N_3O$			
1H-Imidazole-4-butanenitrile, γ-oxo-2-phenyl-1-(phenylmethyl)-	MeOH	260(3.91)	87-0261-86
$C_{20}H_{17}N_3O_3$			
Diazene, (methoxydiphenylmethyl)(4-nitrophenyl)-	CH_2Cl_2	285.0(4.30),431.5(2.48)	107-0741-86
4H-Imidazo[1,2-a]pyrrolo[3,4-b]indole-3,5,6(2H)-trione, 1,11b-dihydro-2-methyl-11b-(phenylmethyl)-	EtOH	264(3.88),286s(3.75), 306s(3.50)	118-0731-86
2-Propenoic acid, 2-cyano-3-[4-(2-cyano-2-phenylethenyl)-5-(dimethylamino)-2-furanyl]-, methyl ester	MeOH	262(3.03),381(3.16), 468(3.68)	73-0879-86
$C_{20}H_{17}N_4$			
1,2,4,5-Tetrazinium, 5,6-dihydro-1,3,5-triphenyl-, iodide	MeCN	540(4.06)	65-0391-86
1,2,4,5-Tetrazin-1(2H)-yl, 3,4-dihydro-2,4,6-triphenyl-	CCl_4	720(3.63)	65-0391-86
	$C_2H_4Cl_2$	720(3.66)	65-0391-86
	o-$C_6H_4Cl_2$	720(3.63)	65-0391-86
triiodide	MeCN	540(4.10)	65-0391-86
$C_{20}H_{17}N_5O$			
2(1H)-Pteridinone, 3,4-dihydro-4-imino-1,3-dimethyl-6,7-diphenyl-	pH 3	225(4.49),282(4.22), 382(4.20)	128-0199-86
	pH 8	224(4.47),284(4.18), 370(4.21)	128-0199-86
	MeOH	224(4.49),283(4.20), 369(4.22)	128-0199-86

Compound	Solvent	$\lambda_{max}(\log \epsilon)$	Ref.
2(1H)-Pteridinone, 3,4-dihydro-1-methyl-4-(methylimino)-6,7-diphenyl-	pH 0	223(4.46),274(4.25),383(4.23)	128-0199-86
	pH 5	222(4.47),276(4.29),377(4.23)	128-0199-86
	MeOH	222(4.46),277(4.32),377(4.24)	128-0199-86
2(1H)-Pteridinone, 3,4-dihydro-3-methyl-4-(methylimino)-6,7-diphenyl-	pH 1	222(4.45),273(4.25),379(4.19)	128-0199-86
	pH 7.2	228(4.48),297(4.27),373(4.02)	128-0199-86
	pH 13	233(4.44),312(4.19),400(4.08)	128-0199-86
	MeOH	227(4.47),286(4.14),367(4.13)	128-0199-86
$C_{20}H_{17}N_5O_5$			
3H-Azeto[1',2':1,5]pyrrolo[2,3-d]-1,2,3-triazole-3a-carboxylic acid, 1a,5,6,6a,7,7a-hexahydro-5-oxo-1-phenyl-, (4-nitrophenyl)methyl ester, (3aα,6aα,7aα)-(±)-	EtOH	268(4.21),287s(4.11)	39-0973-86C
(3aα,6aβ,7aα)-(±)-	EtOH	266(4.23),290s(4.03)	39-0973-86C
$C_{20}H_{18}$			
Azulene, 2-methyl-6-[2-(4-methylphenyl)ethenyl]-, (E)-	CH_2Cl_2	264.5(3.93),323(4.66),400(4.52),507(2.35),593.5(2.59)	24-2272-86
5H-Benzocycloheptene, 5-methyl-5-(2-phenylethenyl)-, (E)-	EtOH	204(4.44),252(4.18)	39-1479-86C
(Z)-	EtOH	206(4.43),278s(3.79)	39-1479-86C
7H-Benzo[c]fluorene, 1,3,9-trimethyl-	EtOH	260(3.99),301(3.26),330(3.54)	2-0887-86
Dibenzo[a,e]cyclooctene, 5,6-didehydro-1,4,7,10-tetramethyl-	hexane	272(4.91),282(4.94),358(3.55),369s(3.54),388s(3.31)	78-0655-86
Pyrene, 2-(1,1-dimethylethyl)-	hexane	235(4.8),241s(4.9),244(5.0),253s(4.3),263(4.4),274(4.6),305(4.0),319(4.5),335(4.7)	54-0156-86
$C_{20}H_{18}ClNO_2S$			
5-Hexene-2,4-dione, 3-[[(4-chlorophenyl)amino](methylthio)methylene]-6-phenyl-	EtOH	234(3.15),352(3.34)	5-1109-86
4(1H)-Pyridinone, 5-acetyl-1-(4-chlorophenyl)-2,3-dihydro-6-(methylthio)-2-phenyl-	EtOH	265(3.02),342(3.16)	5-1109-86
$C_{20}H_{18}ClNO_3S$			
4-Pentenoic acid, 2-[[(4-chlorophenyl)amino](methylthio)methylene]-3-oxo-5-phenyl-, methyl ester, (E)-	EtOH	333(3.43)	5-1109-86
$C_{20}H_{18}ClNO_7$			
Azuleno[6,5-d]isoxazole-3,6,8-tricarboxylic acid, 7-chloro-, triethyl ester	CH_2Cl_2	258(4.48),308.5(4.64),323.5(4.60),367.5(3.94),542.5(3.15)	24-2956-86
$C_{20}H_{18}ClN_2O_3S$			
1H-Benzimidazolium, 2-[5-(4-chloro-	EtOH	362(4.61)	61-0439-86

Compound	Solvent	$\lambda_{max}(\log \epsilon)$	Ref.
phenyl)-2-furanyl]-1,3-dimethyl-5-(methylsulfonyl)-, trifluoromethanesulfonate (cont.)			61-0439-86
$C_{20}H_{18}NO_6P$			
Phosphoric acid, 4-nitrophenyl bis(phenylmethyl) ester	MeCN	276(4.98)	78-1315-86
$C_{20}H_{18}N_2O$			
Diazene, (methoxydiphenylmethyl)phenyl-	CH_2Cl_2	271.5(4.04),418.5(2.21)	107-0741-86
Ethanone, 2-hydroxy-1,2-diphenyl-, phenylhydrazone, (E)-	MeOH	303s(4.04)	48-0181-86
(Z)-	MeOH	300s(4.0),335(4.28)	48-0181-86
$C_{20}H_{18}N_2O_2$			
Azepino[3,4-b]indole-1,3(2H,4H)-dione, 5,10-dihydro-2-methyl-10-(phenylmethyl)-	isoPrOH	240(4.29),316(4.30)	103-1345-86
$C_{20}H_{18}N_2O_2S$			
1-Propanone, 1-(4-aminophenyl)-3-(1,2-dihydro-1-phenyl-2-thioxo-3-pyridinyl)-3-hydroxy-	EtOH	294(3.95),321(3.96), 373(3.75)	4-0567-86
$C_{20}H_{18}N_2O_3$			
5H-Naphtho[1',2':5,6]pyrano[4,3-c]pyridin-5-one, 1-(butylamino)-4-hydroxy-	MeOH	450(4.39),480(4.47)	2-0137-86
Piperidine, 1-[(1-amino-9,10-dihydro-9,10-dioxo-2-anthracenyl)carbonyl]-	toluene	476(4.06)	104-2117-86
2-Propen-1-one, 3-[(4-methoxyphenyl)-amino]-2-(3-methyl-5-isoxazolyl)-1-phenyl-	dioxan	241(4.28),295(4.24), 362(4.43)	83-0242-86
$C_{20}H_{18}N_2O_3S$			
Acetamide, N-[2,9-dihydro-9-(phenylsulfonyl)-1H-carbazol-4-yl]-	MeOH	225(4.49),273s(3.93), 300s(3.75)	5-2065-86
$C_{20}H_{18}N_2O_4$			
2-Anthracenecarboxylic acid, 9,10-dihydro-9,10-dioxo-1-(1-piperidinylamino)-	n.s.g.	275(3.85),313(3.65), 493(3.64)	104-2117-86
$C_{20}H_{18}N_2O_4S$			
Acetamide, N-acetyl-N-[1-[1-(phenylsulfonyl)-1H-indol-3-yl]ethenyl]-	MeOH	219(4.31),267(3.96)	5-2065-86
4H-Carbazol-4-one, 1,2,3,9-tetrahydro-9-(phenylsulfonyl)-, O-acetyloxime	MeOH	222(4.37),255s(4.09), 270s(4.09),275(4.10), 281s(4.09)	5-2065-86
$C_{20}H_{18}N_2O_5$			
6,7-Quinolinedicarboxylic acid, 5-methoxy-8-(phenylamino)-, dimethyl ester	MeOH	267(4.30),388(3.70)	4-1641-86
$C_{20}H_{18}N_2O_6$			
Butanedioic acid, (nitromethylene)-(phenylamino)methylene-, dimethyl ester	EtOH	233(4.11),378(4.04)	18-3871-86
1-Cyclobutene-1,2-dicarboxylic acid, 4-nitro-3-phenyl-3-(phenylamino)-, dimethyl ester	EtOH	231(4.13),310(3.88), 378(4.20)	18-3871-86

Compound	Solvent	λ_{max}(log ε)	Ref.
$C_{20}H_{18}N_3S_2$			
1H-Pyrazolium, 3-(dimethylamino)-4-phenyl-1-(5-phenyl-3H-1,2-dithiol-3-ylidene)-, perchlorate	MeCN	217s(4.30),240(4.17), 335(4.21),487(4.44)	88-0159-86
$C_{20}H_{18}N_4O$			
3(2H)-Pyridazinone, 4,5-dihydro-6-[2-phenyl-1-(phenylmethyl)-1H-imidazol-4-yl]-, monohydrochloride	MeOH	292(4.34)	87-0261-86
$C_{20}H_{18}N_4O_3S$			
1,2,4-Thiadiazole-5-acetic acid, 3-(7-oxo-2-phenylethyl)-α-(phenylhydrazono)-, ethyl ester	DMF	295(4.17),408(4.38)	48-0741-86
2-Thiazoleacetic acid, 4-amino-5-benzoyl-α-(phenylhydrazono)-, ethyl ester	DMF	422(4.19)	48-0741-86
$C_{20}H_{18}N_4O_4$			
Imidazo[5,1,2-cd]indolizine-3,4-dicarboxylic acid, 6-methyl-2-[(4-methyl-2-pyridinyl)amino]-, dimethyl ester	MeOH	208(4.53),250(4.60), 264(4.53),349(3.95), 400(4.42)	88-4129-86
1H,5H-Pyrazolo[1,2-a]pyrazol-1-one, 7-amino-3-hydroxy-5-imino-2,6-bis(4-methoxyphenyl)-	MeCN	273(4.59)	118-0239-86
$C_{20}H_{18}N_4S$			
1H-Perimidine-2-carbothioamide, N-[4-(dimethylamino)phenyl]-	DMF	326(4.14),341(4.16), 356(4.22),430(4.00)	48-0812-86
	C_6H_{12}	230s(--),235(--), 259(--),310(--), 321(--),337(--), 353(--),424(--), 511(--),552(--), 602(--)	48-0812-86
$C_{20}H_{18}O$			
1H-Cyclopenta[l]phenanthrene-1-methanol, 1,3-dimethyl-	EtOH	242(4.54),262(4.55), 322(4.01),334(3.96)	39-1471-86C
2-Cyclopenten-1-one, 3-phenyl-5-(3-phenyl-2-propenyl)-, (Z)-	EtOH	251(4.12),285(4.34)	39-1235-86C
$C_{20}H_{18}OS_2$			
Benzenecarbodithioic acid, 4-methoxy-, 6-phenyl-1,3,5-hexatrienyl ester	CH_2Cl_2	334(4.67),418(4.43)	18-1403-86
$C_{20}H_{18}O_2$			
9,10-Anthracenedione, 2-methyl-1-(3-methyl-2-butenyl)-	$CHCl_3$	260(4.71),337(3.90)	2-0360-86
Benz[a]anthracene-3,4-diol, 3,4-dihydro-7,12-dimethyl-, trans	THF	273.1(4.94)	44-3502-86
Phenol, 2-[(2-hydroxyphenyl)(3-methylphenyl)methyl]-, (-)-	EtOH	275(3.67)	116-0509-86
$C_{20}H_{18}O_3$			
Benz[a]anthracene-3,4-diol, 1,2-epoxy-1,2,3,4-tetrahydro-7,12-dimethyl-, trans-syn	THF	269.2(4.62)	44-3502-86
4H,8H-Benzo[1,2-b:3,4-b']dipyran-4-one, 2,3-dihydro-8,8-dimethyl-2-phenyl-	MeOH	273(4.41),331(4.48)	2-0481-86

Compound	Solvent	λ_{max}(log ϵ)	Ref.
Cyclohepta[de]naphthalen-7(8H)-one, 8,8-bis(2-propenyloxy)-	$CHCl_3$	317(3.91),336s(3.86)	39-1965-86C
$C_{20}H_{18}O_4$			
5H-Benzo[c]fluorene-5,8,11-trione, 7,7a,11a,11b-tetrahydro-11b-hydroxy-6,7a,10-trimethyl-	EtOH	244(4.15),274(3.79)	78-3663-86
1,4-Naphthalenedione, 2-[(1,4-dimethyl-2,5-dioxo-3-cyclohexen-1-yl)-methyl]-3-methyl-	EtOH	231(4.42),256(3.98), 301(3.27)	78-3663-86
2H,5H-Pyrano[2,3-b][1]benzopyran-5-one, 3,4-dihydro-2-(hydroxymethyl)-2-methyl-4-phenyl-	EtOH	207(--),224(4.44), 275(4.05)	32-0491-86
isomer	EtOH	207(--),223(4.44), 274(4.02)	32-0491-86
2H,5H-Pyrano[2,3-b][1]benzopyran-5-one, 3,4-dihydro-2-(hydroxymethyl)-2-(4-methylphenyl)-	EtOH	221(4.41),276(4.03)	32-0491-86
2H,5H-Pyrano[3,2-c][1]benzopyran-5-one, 3,4-dihydro-2-(hydroxymethyl)-2-methyl-4-phenyl-	EtOH	271(4.02),282(4.06), 304(3.97),318s(--)	32-0501-86
isomer 14j	EtOH	270(4.07),282(4.11), 304(4.02),318s(--)	32-0501-86
2H,5H-Pyrano[3,2-c][1]benzopyran-5-one, 3,4-dihydro-2-(hydroxymethyl)-4-methyl-2-phenyl-	EtOH	271(3.96),282(4.01), 305(3.94),318s(--)	32-0501-86
isomer 14h	EtOH	271(3.99),282(4.03), 305(3.96),318s(--)	32-0501-86
2H,5H-Pyrano[3,2-c][1]benzopyran-5-one, 3,4-dihydro-2-(hydroxymethyl)-2-(4-methylphenyl)-	EtOH	270(3.95),282(4.00), 304(3.93),312(3.76)	32-0501-86
3H,10H-Pyrano[4,3-b][1]benzopyran-10-one, 3-ethoxy-4,4a-dihydro-1-phenyl-	EtOH	219(4.253),237s(4.090), 267(3.962),313s(3.96), 334(3.978)	39-0789-86C
$C_{20}H_{18}O_5$			
Androsta-5,8-dieno[6,5,4-bc]furan-3,7,17-trione, 1,2-epoxy-, (1α,2α)- (dehydrowortmannolone)	MeOH	231(4.01),261(3.96), 300(3.81)	39-1317-86C
9,10-Anthracenedione, 1,4-dihydroxy-2-(2-hydroxy-2-methyl-4-pentenyl)-, (R)-	MeOH	210(4.14),228(4.27), 235(4.32),250(4.59), 255(4.55),288(3.97), 312(3.51),457(3.92), 475(3.96),485(4.00), 498(3.90),519(3.81), 563(2.91)	78-6635-86
$C_{20}H_{18}O_6$			
1,4-Anthracenediol, 9,10-dimethoxy-, diacetate	$CHCl_3$	260(5.04),350s(3.57), 368(4.82),386(3.98)	5-0839-86
9,10-Anthracenedione, 2-(1,2-dihydroxy-2-methyl-4-pentenyl)-1,4-dihydroxy-	MeOH	210(4.16),230(4.29), 236(4.32),250(4.60), 255(4.55),288(3.97), 312(3.52),460(3.96), 476(3.98),485(4.01), 498(3.92),512(3.82), 564(2.01)	78-6635-86
2,6'-Bi[2H-1-benzopyran]-4(3H)-one, 5,7,8'-trihydroxy-2',2'-dimethyl-, (S)- (sigmoidin C)	EtOH	285(4.45)	39-0033-86C

Compound	Solvent	$\lambda_{max}(\log \epsilon)$	Ref.
$C_{20}H_{18}O_7$			
2-Anthracenebutanoic acid, 9,10-di-hydro-α,1,4-trihydroxy-α-methyl-9,10-dioxo-, methyl ester, (R)-	MeOH	210(4.12),233(4.32),250(4.59),255(4.53),285(4.00),310(3.47),458(3.94),475(3.97),483(3.99),497(3.88)	78-6635-86
$C_{20}H_{18}O_8$			
2-Anthraceneacetic acid, 9,10-dihydro-1,4-dihydroxy-3-(dimethoxymethyl)-9,10-dioxo-, methyl ester	n.s.g.	252(4.61),289(4.04),324s(3.51),491(3.68),525s(3.75)	12-0487-86
Arborone	MeOH	206(4.44),234(4.17),285(4.01)	100-1061-86
$C_{20}H_{19}BrN_2O_2$			
1H-Pyrazole-1-carboxylic acid, 5-(3-bromophenyl)-3-phenyl-, 1,1-dimethyl-ethyl ester	EtOH	262(4.46)	4-0727-86
$C_{20}H_{19}BrN_2O_2S$			
5-Pyrimidinecarboxylic acid, 4-(4-bromophenyl)-1,2,3,4-tetrahydro-6-methyl-1-phenyl-2-thioxo-, ethyl ester	EtOH	308(4.23)	103-0990-86
$C_{20}H_{19}BrN_2O_4$			
Pyrimidine, 5-(2-bromo-4,5-dimethoxy-phenyl)-4-(3,4-dimethoxyphenyl)-	$CHCl_3$	240(4.27),292(4.11),310(4.07)	142-1943-86
$C_{20}H_{19}BrN_6O$			
Acetamide, N-[2-[(4-bromo-2,6-dicyano-phenyl)azo]-5-(diethylamino)phenyl]-	$CHCl_3$	309s(3.81),315(3.81),409(3.62),548s(4.67),575(4.71)	48-0497-86
$C_{20}H_{19}BrO_3$			
Benz[a]anthracene-1,3,4-triol, 2-bro-mo-1,2,3,4-tetrahydro-7,12-dimethyl-, (1α,2β,3β,4α)-	dioxan	273(4.63)	44-3502-86
$C_{20}H_{19}Cl_3N_2O_2$			
17-Norcorynan-18-carboxylic acid, 16,16,16-trichloro-18,19,20,21-tetradehydro-, methyl ester, (3β)-(±)-	MeOH	279(3.94),286(3.91),345(4.65)	44-2995-86
2-Propenoic acid, 3-[1,6-dihydro-1-[2-(1H-indol-3-yl)ethyl]-6-(tri-chloromethyl)-3-pyridinyl]-, methyl ester, (±)-	MeOH	277(3.90),287(3.97),334(4.36),354(4.34)	44-2995-86
$C_{20}H_{19}Cl_3O_6$			
1,3-Azulenedicarboxylic acid, 2-chlo-ro-6-(1,1-dichloro-2-ethoxy-2-oxo-ethyl)-, diethyl ester	CH_2Cl_2	246.5(4.43),268.7(4.21),304.5(4.71),314.7(4.81),360(3.96),526.8(2.77)	24-2956-86
$C_{20}H_{19}F_3N_2O_3$			
6H-1,3-Oxazin-6-one, 3-[2-amino-4-(trifluoromethyl)phenyl]-2-benzoyltetrahydro-5,5-dimethyl-	n.s.g.	206(4.41),254(4.42),274s(4.02),360(3.87)	150-2471-86M
$C_{20}H_{19}NO$			
Azacyclooctadeca-3,5,7,13,15,17-hexa-	THF	327(4.72),403s(3.80)	18-1801-86

Compound	Solvent	$\lambda_{max}(\log \epsilon)$	Ref.
ene-9,11-diyn-2-one, (E,E,Z,Z,E,E)-, 8,13,18-trimethyl- (cont.)			18-1801-86
2,4,6,12,14,16-Cycloheptadecahexaene-8,10-diyn-1-one, 2,7,12-trimethyl-, oxime, (E,E,E,Z,Z,E,E)-	THF	300(4.69),307(4.69), 370s(3.98),398(3.86)	18-1801-86
1H-Pyrrolizin-1-one, 5,6,7,7a-tetrahydro-7a-methyl-2,3-diphenyl-	MeCN	238(4.15),296(3.91), 369(3.73)	118-0899-86
$C_{20}H_{19}NOS$			
Sulfoximine, S-(4-methylphenyl)-S-phenyl-N-(phenylmethyl)-, (R)-	EtOH	222(4.09),241(4.13), 268s(3.53),276s(3.40)	12-1655-86
$C_{20}H_{19}NO_2$			
2H-1-Benzopyran-2-one, 4-methyl-7-[(phenylmethyl)-2-propenylamino]-	MeOH	241(4.11),310(3.57), 364(4.32)	151-0069-86A
	$CHCl_3$	313(3.29),361(4.23)	151-0069-86A
2-Dibenzofuranol, 6,7,8,9-tetrahydro-1-[(phenylmethyl)imino]methyl]-, (E)-	toluene	312(4.05),343(3.62)	104-0503-86
	PrOH	309(3.94),343(3.58), 440(2.60)	104-0503-86
1H-Indene-1,3(2H)-dione, 2-[9-(dimethylamino)-2,4,6,8-nonatetraenylidene]-	EtOH-DMSO	745(4.67)	70-1446-86
3-Pyridinecarboxylic acid, 5-(diphenylmethylene)-1,2,5,6-tetrahydro-1-methyl-	pH 7.2	210(1.39),240(1.34), 308(1.94)	87-0125-86
$C_{20}H_{19}NO_2S$			
4(1H)-Pyridinone, 5-acetyl-2,3-dihydro-6-(methylthio)-1,2-diphenyl-	EtOH	265(3.10),342(3.11)	5-1109-86
$C_{20}H_{19}NO_3$			
Benzeneacetamide, N-(2,6-dimethylphenyl)-N-(2-methyl-1-oxo-2-propenyl)-α-oxo-	C_6H_{12}	225(4.15),253(4.22)	39-1759-86C
Benzeneacetamide, N-(1-methylethyl)-α-oxo-N-(1-oxo-3-phenyl-2-propenyl)-	C_6H_{12}	211(4.07),222(4.08), 228(4.09)	39-1759-86C
2-Dibenzofuranol, 6,7,8,9-tetrahydro-1-[[(4-methoxyphenyl)imino]methyl]-	toluene	360(4.09)	104-0503-86
	PrOH	357(4.05)	104-0503-86
	MeCN	357(4.11)	104-0503-86
11,14a-Methano-4H,14aH-benzo[ij]cycloocta[b]quinolizine-8,10,15-trione, 5,6,11,12,13,14-hexahydro-11-methyl-	n.s.g.	213(4.18),238(4.23), 371(3.54)	150-2782-86M
6-Oxa-3-azabicyclo[3.1.1]heptane-2,4-dione, 3-(2,6-dimethylphenyl)-1-methyl-5-phenyl-	C_6H_{12}	217(4.19),258(2.91), 264(2.93),272(2.79)	39-1759-86C
6-Oxa-3-azabicyclo[3.1.1]heptane-2,4-dione, 3-(1-methylethyl)-1,7-diphenyl-	C_6H_{12}	218(4.30),253(3.15), 259(3.20),263(3.16)	39-1759-86C
$C_{20}H_{19}NO_3S$			
3-Pyridinecarboxylic acid, 1,4,5,6-tetrahydro-2-(methylthio)-4-oxo-1,6-diphenyl-	EtOH	335(3.17)	5-1109-86
4H-Thiopyran-4-one, 5-acetyl-2,3-dihydro-6-[(4-methoxyphenyl)amino]-2-phenyl-	EtOH	249(3.19),315(3.27)	5-1109-86
$C_{20}H_{19}NO_4$			
6H-Dibenzo[de,g]quinoline-6-carboxaldehyde, 4,5-dihydro-1,2,3-trimethoxy- (dehydroformouregine)	EtOH	221(3.92),251s(4.29), 263(4.32),310(3.67)	100-0878-86

Compound	Solvent	$\lambda_{max}(\log \epsilon)$	Ref.
$C_{20}H_{19}NO_4S$			
Benzenesulfonamide, 4-methyl-N-(4-methyl-2-oxo-2H-1-benzopyran-7-yl)-N-2-propenyl-	MeOH	218(4.25),280(3.92), 344(3.92)	151-0069-86A
	$CHCl_3$	285(3.97),318(4.08)	151-0069-86A
Pyrano[2,3-e]indol-2(7H)-one, 8,9-dihydro-4,8-dimethyl-7-[(4-methylphenyl)sulfonyl]-	MeOH	219(4.35),250(3.94), 326(4.19)	151-0069-86A
	$CHCl_3$	328(4.31)	151-0069-86A
2H-Thiopyran-5-carboxylic acid, 3,4-dihydro-6-[(4-methoxyphenyl)amino]-4-oxo-2-phenyl-, methyl ester	EtOH	306(3.25)	5-1109-86
$C_{20}H_{19}NO_5$			
Isoparfumine, (+)-	EtOH	210(4.30),234(4.35), 260(4.60),293s(--), 350(3.47)	142-2781-86
	EtOH-base	300(--),350(--)	142-2781-86
Pyrano[3,2-b]acridin-6-one, 2,11-dihydro-5,9-dihydroxy-10-methoxy-2,2,11-trimethyl- (honyumine)	EtOH	224(3.98),262(4.15), 290s(4.64),301(4.75), 327(4.05),350s(3.55), 393(3.59)	142-0041-86
$C_{20}H_{19}NO_6$			
Corpaine, (-)-	MeOH	236(4.41),291(4.05), 313(3.97)	102-2245-86
Corysolidine, (±)-	MeOH	236(4.32),290(3.90), 310(3.84)	102-2245-86
$C_{20}H_{19}NO_7$			
Azuleno[6,5-d]isoxazole-3,6,8-tricarboxylic acid, triethyl ester	CH_2Cl_2	256.5(4.56),303.5(4.52), 321(4.66),352.5(3.86), 554(3.22)	24-2956-86
$C_{20}H_{19}NO_9$			
[1,3]Dioxolo[4,5-j]phenanthridin-6(2H)-one, 2,3,4-triacetoxy-3,4,4a,5-tetrahydro- (7-deoxynarciclasine triacetate)	MeOH	231(4.20),251(4.21), 305(3.93)	102-0995-86
$C_{20}H_{19}N_3O_2$			
Acetamide, N,N'-(2,5-diphenyl-1H-pyrrole-3,4-diyl)bis-	n.s.g.	208(4.37),220s(4.18), 313(4.36)	103-0499-86
1H-1,4-Diazepino[6,5-c]quinoline-2,5-dione, 3,4,7,11b-tetrahydro-4-methyl-7-(phenylmethyl)-, (±)-	CH_2Cl_2	329(4.0)	5-1127-86
$C_{20}H_{19}N_3O_2S$			
Benzoic acid, 3-[[(1,5-dihydro-3-methyl-1-phenyl-5-thioxo-4H-pyrazol-4-ylidene)methyl]amino]-, ethyl ester	toluene	362(4.4),445(3.9)	103-0503-86
Benzoic acid, 4-[[(1,5-dihydro-3-methyl-1-phenyl-5-thioxo-4H-pyrazol-4-ylidene)methyl]amino]-, ethyl ester	toluene	353(4.2),440(3.7)	103-0503-86
$C_{20}H_{19}N_3O_4S$			
5-Pyrimidinecarboxylic acid, 1,2,3,4-tetrahydro-6-methyl-4-(4-nitrophenyl)-1-phenyl-2-thioxo-, ethyl ester	EtOH	298(4.16)	103-0990-86
$C_{20}H_{19}N_3O_8$			
1,3-Azulenedicarboxylic acid, 2-diazo-	CH_2Cl_2	289(4.47),316(4.33),	24-2956-86

Compound	Solvent	λ_{max}(log ε)	Ref.
6-(2-ethoxy-1-nitro-2-oxoethylidene)-2,6-dihydro-, diethyl ester (cont.)		359.5(4.16),478(4.45), 507(4.33)	24-2956-86
$C_{20}H_{19}N_5$			
1H-Pyrazolo[3,4-b]quinoxaline, 1-phenyl-3-(1-pyrrolidinylmethyl)-	MeOH	267.6(4.78),336.6(4.15)	2-0960-86
$C_{20}H_{19}N_5O_5$			
Benzamide, 4-cyano-N-[4-cyano-1-[2,3-O-(1-methylethylidene)-β-D-ribofuranosyl]-1H-imidazol-5-yl]-	MeOH	235(4.40),285s(3.61)	23-2109-86
$C_{20}H_{19}N_5O_8$			
β-D-Glucofuranuronic acid, 1-deoxy-1-[6-[(2-furanylmethyl)amino]-9H-purin-9-yl]-, γ-lactone, 2,5-diacetate	MeOH	266(4.26)	103-0085-86
$C_{20}H_{19}N_6O_3$			
Tetrazolo[1,5-b]isoquinolinium, 10-methyl-5-(4-morpholinyl)-3-(4-nitrophenyl)-, bromide	EtOH	275(4.10),365(4.03)	78-5415-86
$C_{20}H_{19}N_7O_3$			
Acetamide, N-[2-[(2,6-dicyano-4-nitrophenyl)azo]-5-(diethylamino)phenyl]-	$CHCl_3$	266s(3.96),299(3.79), 420(3.54),586s(4.72), 625(4.93)	48-0497-86
$C_{20}H_{20}$			
Benzene, 1-(1-cyclohexen-1-yl)-2-(2-phenylethenyl)-, (E)-	MeOH	220(4.10),252(3.80), 290(4.05),310s(3.97)	151-0105-86A
Bicyclo[2.2.1]heptane, 2-(diphenylmethylene)-	hexane	250(4.10)	78-6175-86
Bicyclo[2.2.1]hept-2-ene, 5-(diphenylmethyl)-, endo	hexane	258(2.71)	78-6175-86
exo	hexane	258(2.59)	78-6175-86
Cyclobuta[1]phenanthrene, 1,2-dihydro-1,1,2,2-tetramethyl-	MeCN	210(4.45),221(4.28), 246s(4.63),254(4.71), 270(4.15),278(3.96), 289(3.95),302(4.08), 324(2.80),339(2.98), 356(3.08)	78-6195-86
Tricyclo[12.2.2.2.$2^{4,7}$]eicosa-4,6,8,12,14,16,17,19-octaene, (E,E)-	isooctane	204(4.35),244s(4.27), 253(4.33),274s(3.83), 303(3.33),313(3.28)	47-1726-86
isomer	C_6H_{12}	252(4.2),287(3.3), 295(3.0)	88-5923-86
$C_{20}H_{20}ClNO_7$			
1,3-Azulenedicarboxylic acid, 2-chloro-6-[2-ethoxy-1-(hydroxyimino)-2-oxoethyl]-, diethyl ester, anti	CH_2Cl_2	241.5(4.51),264.5(4.28), 314(4.82),358(4.03), 520(2.74)	24-2956-86
syn	CH_2Cl_2	241.5(4.45),264.5(4.16), 323(4.80),366.5(4.11), 384.5(3.88),535.5(2.75)	24-2956-86
1,3-Azulenedicarboxylic acid, 2-chloro-6-(2-ethoxy-1-nitroso-2-oxoethyl)-, diethyl ester	CH_2Cl_2	245(4.49),304(4.70), 313.5(4.80),359.5(3.94), 527.5(2.69)	24-2956-86

Compound	Solvent	$\lambda_{max}(\log \epsilon)$	Ref.
$C_{20}H_{20}ClN_3O$			
Anthra[1,9-cd]pyrazol-6(2H)-one, 5-chloro-2-[2-(diethylamino)ethyl]-, monohydrochloride	n.s.g.	246(4.25),266(4.15), 298(4.07),340(3.61), 407(3.92)	4-1491-86
$C_{20}H_{20}ClN_3O_3$			
Anthra[1,9-cd]pyrazol-6(2H)-one, 5-chloro-2-[2-(diethylamino)ethyl]-7,10-dihydroxy-, hydrochloride	n.s.g.	246(4.22),284(3.99), 485(4.00)	4-1491-86
$C_{20}H_{20}ClN_5O$			
1H-1,2,4-Triazol-3-amine, N- [(4-chlorophenyl)methylene]-5-(4-morpholinyl)-1-(phenylmethyl)-	n.s.g.	222s(4.16),272(4.16), 308(4.13)	4-0401-86
1H-1,2,4-Triazol-5-amine, N- [(4-chlorophenyl)methylene]-3-(4-morpholinyl)-1-(phenylmethyl)-	n.s.g.	285(4.27),361(3.98)	4-0401-86
$C_{20}H_{20}Cl_6N_2O_6S_2$			
4-Thia-1-azabicyclo[3.2.0]heptane-2-carboxylic acid, 6-[3,3-dimethyl-7-oxo-2-[(2,2,2-trichloroethoxy)-carbonyl]-4-thia-1-azabicyclo[3.2.0]-hept-6-ylidene]-3,3-dimethyl-7-oxo-, 2,2,2-trichloroethyl ester, (E)-	n.s.g.	231(4.10),293(3.65)	39-2213-86C
(Z)-	n.s.g.	233(4.20),285(3.50)	39-2213-86C
$C_{20}H_{20}NO_5$			
4H-Bis[1,3]benzodioxolo[5,6-a:4',5'-g]quinolizinium, 6,7,12b,13-tetrahydro-13-hydroxy-5-methyl-, iodide	MeOH	222(4.24),243s(4.00), 292(3.97)	73-2232-86
$C_{20}H_{20}N_2O$			
2H-Benz[g]indazole, 2-acetyl-3,3a,4,5-tetrahydro-3-(4-methylphenyl)-	EtOH	291(4.07)	4-0135-86
$C_{20}H_{20}N_2O_2$			
2H-Benz[g]indazole, 2-acetyl-3,3a,4,5-tetrahydro-3-(4-methoxyphenyl)-	EtOH	225(4.03),285(4.00)	4-0135-86
4-Oxa-1,2-diazaspiro[4.5]dec-2-ene, 1-benzoyl-3-phenyl-	MeOH	304(4.25)	24-2963-86
Phenol, 2,2'-[1,2-ethanediylbis(nitrilomethylidyne)]bis[4-ethenyl-	dioxan-pH 3.45	347(4.0)	2-0923-86A
	dioxan-pH 11.45	400(4.3)	2-0923-86A
5-Pyrimidinecarboxylic acid, 1,4-dihydro-6-methyl-1,4-diphenyl-, ethyl ester	EtOH	247(3.94),315(3.74)	103-0990-86
$C_{20}H_{20}N_2O_2S$			
5-Pyrimidinecarboxylic acid, 1,2,3,4-tetrahydro-6-methyl-1,4-diphenyl-2-thioxo-, ethyl ester	EtOH	305(4.17)	103-0990-86
$C_{20}H_{20}N_2O_3$			
1H-Indole-3-propanoic acid, 2-[(methylamino)carbonyl]-1-(phenylmethyl)-	isoPrOH	214(4.46),292(4.13), 308s(3.94)	103-1345-86
$C_{20}H_{20}N_2O_3S$			
Acetamide, N-[2,3,4,9-tetrahydro-9-(phenylsulfonyl)-1H-carbazol-4-yl]-	MeOH	220(4.34),254(4.16), 288s(3.53)	5-2065-86

Compound	Solvent	λ_{max}(log ε)	Ref.
Benzeneacetic acid, α-[(1,4,5,6-tetrahydro-6-oxo-1-phenyl-2-pyrimidinyl)thio]-, ethyl ester	MeOH	220(4.29)	2-0275-86
$C_{20}H_{20}N_2O_4$			
Pyrimidine, 4,5-bis(3,4-dimethoxyphenyl)-	$CHCl_3$	240(4.40),292(4.16), 310(4.07)	142-1943-86
$C_{20}H_{20}N_2O_5$			
Phenol, 6-[5-(3,4-dimethoxyphenyl)-4-pyrimidinyl]-2,3-dimethoxy-	$CHCl_3$	240(4.22),307(4.13), 335(4.08)	142-1943-86
$C_{20}H_{20}N_4$			
Propanedinitrile, [bis[4-(dimethylamino)phenyl]methylene]-	acetone	429(4.73)	24-3276-86
	$CHCl_3$	266(4.30),433(4.70)	24-3276-86
1H-Pyrrolo[2,3-b]pyridine-5-carbonitrile, 4-methyl-1-(phenylmethyl)-6-(1-pyrrolidinyl)-	$CHCl_3$	244(1.05),270(1.39), 358(0.36)	103-0076-86
$C_{20}H_{20}N_4O$			
1(4H)-Naphthalenone, 4-[(2-aminoethyl)imino]-, (4-acetylphenyl)hydrazone	H_2O	550(4.75)	96-0983-86
	MeOH	555(4.74)	96-0983-86
	EtOH	554(4.69)	96-0983-86
	dioxan	555(4.70)	96-0983-86
	acetone	545(4.62)	96-0983-86
	DMF	531(4.52)	96-0983-86
	DMSO	532(4.52)	96-0983-86
1H-Pyrrolo[2,3-b]pyridine-5-carbonitrile, 4-methyl-6-(4-morpholinyl)-1-(phenylmethyl)-	$CHCl_3$	248(1.27),266(1.20), 334(0.26)	103-0076-86
$C_{20}H_{20}N_4O_2$			
Benzoic acid, 1,2-cyclohexanediylidenedihydrazide	MeOH	266(4.08),329(4.05)	86-0209-86
	EtOH	265(4.06),328(4.05)	86-0209-86
	3-MeBuOH	335(4.05)	86-0209-86
	pentanol	273(4.06),334(4.06)	86-0209-86
	acetone	335(4.01)	86-0209-86
	DMF	277(3.94),335(4.03)	86-0209-86
Ti(IV) complex	n.s.g.	477(4.02)	86-0209-86
$C_{20}H_{20}N_4O_2S$			
1H-Indole-2,3-dione, 1-(1-oxobutyl)-, 3-[[[(phenylmethyl)amino]thioxomethyl]hydrazone]	dioxan	390(4.27)	104-2173-86
	$CHCl_3$	336s(4.15),392(4.23)	104-2173-86
	DMF	336s(3.96),391(4.25)	104-2173-86
isomer	dioxan	330s(4.03),380(4.21)	104-2173-86
	EtOH	337s(4.38),376(4.46)	104-2173-86
	$CHCl_3$	338s(4.10),372(4.17)	104-2173-86
	DMF	281s(3.85),340(3.69), 462(4.53)	104-2173-86
	DMSO	282s(3.87),459(4.36), 340s(3.81)	104-2173-86
$C_{20}H_{20}N_4O_6$			
2-Propenoic acid, 3-[10-(methoxycarbonyl)-1-(tetrahydro-2H-pyran-2-yl)-7-oxopyrido[1,2-e]purin-8-yl]-, methyl ester	CH_2Cl_2	277(4.3),425(4.5), 442(4.5)	39-1561-86C
$C_{20}H_{20}N_6O_2$			
1H-Pyrazolo[3,4-b]quinoxalin-7-amine, N,N-diethyl-3-methyl-1-(4-nitrophenyl)-	BuOAc	344(4.11),465(3.95)	48-0342-86

Compound	Solvent	$\lambda_{max}(\log \epsilon)$	Ref.
$C_{20}H_{20}O$			
Azulene, 6-(4-methylphenyl)-2-propoxy-	CH_2Cl_2	237.5(4.12),300.5(4.86), 377(4.06),391(4.23), 521.5(2.36)	24-2272-86
Bicyclo[2.2.1]hept-5-ene-2-methanol, α,α-diphenyl-, endo	hexane	257(2.60)	78-6175-86
exo	hexane	257(2.78)	78-6175-86
$C_{20}H_{20}O_2$			
Cyclohepta[ef]heptalene-1,2-dimethanol, 10,12-dimethyl-	hexane	240(4.48),258s(4.29), 287s(3.94),318s(3.67), 363(3.34)	89-0633-86
Cyclohexanone, 2-[[2-(phenylmethoxy)-phenyl]methylene]-, (E)-	EtOH	204(4.44),227(4.05), 282(4.02),314(3.86)	78-5637-86
2-Naphthalenol, 1-[1-(2-hydroxyphen-yl)-2-methylpropyl]-, (S)-	EtOH	280(3.84)	116-0509-86
4-Octene-3,6-dione, 4,5-diphenyl-, (E)-	ether	260(3.89)	78-1345-86
(Z)-	ether	228(4.11),283(3.90)	78-1345-86
$C_{20}H_{20}O_3$			
Naphtho[1,8-bc]pyran-7,8-dione, 6,9-dimethyl-3-(4-methyl-3-pentenyl)- (biflorin)	EtOH	204(4.26),234(4.48), 254s(4.15),338(3.69), 552(3.72)	12-0647-86
2H-Naphtho[1,2-b]pyran-2-one, 4-[(cy-clohexyloxy)methyl]-	EtOH	265(4.46),275(4.56), 291.5(3.83),303.5(3.92), 316.5(3.91),351(3.75)	94-0390-86
$C_{20}H_{20}O_4$			
4H-1-Benzopyran-4-one, 2,3-dihydro-5,7-dihydroxy-6-(3-methyl-2-buten-yl)-2-phenyl-	EtOH	297(3.95),335s(3.00)	105-0107-86
4H-1-Benzopyran-4-one, 2,3-dihydro-5,7-dihydroxy-8-(3-methyl-2-buten-yl)-2-phenyl- (8-prenylpinocembrin)	MeOH	294(3.98),330s(--)	102-1505-86
	MeOH-NaOAc	332(--)	102-1505-86
	MeOH-$AlCl_3$	317(--)	102-1505-86
2,3-Benzotetracyclo[7.2.1.0^{4,11}.0^{6,10}]-dodeca-2,7-diene-5,12-dione, bis-(ethyleneacetal), acs	MeCN	264s(2.81),271(2.99), 278(2.96)	24-3442-86
Cyclohepta[ef]heptalene-1,2-dicarbox-ylic acid, 5,6,7,8-tetrahydro-, dimethyl ester	hexane	214(4.35),272(4.32), 329(3.56)	89-0633-86
Cyclohepta[ef]heptalene-2,3-dicarbox-ylic acid, 5,6,7,8-tetrahydro-, dimethyl ester	hexane	253s(4.23),266(4.29), 341(3.79)	89-0633-86
1H-Dibenzo[a,d]cycloheptene-1,4(10H)-dione, 3-hydroxy-7-methoxy-9-methyl-2-(1-methylethyl)- (fruticulin A)	MeOH	215(4.51),250(4.00), 278(4.06),325(4.22), 420(3.62)	88-5459-86
3,6-Phenanthrnedione , 4,4a-dihydro-5,8-dihydroxy-1,2,4a-trimethyl-7-(2-propenyl)-, (S)- (coleon E)	ether	217s(4.06),279.5(4.17), 436s(4.13)	33-1395-86
$C_{20}H_{20}O_5$			
Androsta-5,8-dieno[6,5,4-bc]furan-7,17-dione, 1,2-epoxy-3-hydroxy-, (1α,2α,3β)- (wortmannolone)	MeOH	264(3.92),308(3.84)	39-1317-86C
6-Benzofuranol, 2-(1,3-benzodioxol-5-yl)-2,3-dihydro-5-methoxy-3-methyl-7-(2-propenyl)-, (2S-trans)-	MeOH	224s (4.41),293(4.05), 308s(3.86),325(3.69), 345s(3.45)	102-1953-86
	MeOH-NaOMe	223(--),294(--), 309s(--),344(--)	102-1953-86
Bicyclo[3.2.1]oct-3-ene-2,8-dione,	MeOH	226(3.60),289(3.46)	102-1953-86

Compound	Solvent	λ_{max}(log ϵ)	Ref.
7-(1,3-benzodioxol-5-yl)-3-methoxy-6-methyl-5-(2-propenyl)- (cont.)			102-1953-86
Linearifoline	MeOH	208(4.11),300(3.99)	102-2381-86
Phenol, 4-[4-(1,3-benzodioxol-5-yl)-tetrahydro-1H,3H-furo[3,4-c]furan-6-yl]-2-methoxy-	EtOH	222(3.98),284(3.77)	142-0923-86
Plectrinone B	ether	252s(4.10),278s(4.24), 289(4.25),305s(4.07), c.398(3.54)	33-0972-86
	ether	253(3.96),279s(4.22), 290(4.30),306s(4.16), 395s(3.57)	33-1395-86
1,3-Propanedione, 1-[2,4-dihydroxy-5-(3-methyl-2-butenyl)phenyl]-3-(4-hydroxyphenyl)- (5'-prenyl-licodione)	MeOH	283(4.13),382(4.27)	102-2803-86
Salviarin, 1(10)-dehydro-	MeOH	208(4.20),263(4.18), 272(4.15),281s(--)	102-2381-86
Scleroderodione dimethyl ether	MeOH	221(4.49),260(4.88), 286(4.12),333(3.86), 352(3.88),370s(3.75)	44-4813-86
$C_{20}H_{20}O_6$			
4H-1-Benzopyran-4-one, 2-[3,4-dihy-droxy-5-(3-methyl-2-butenyl)phen-yl]-2,3-dihydro-5,7-dihydroxy-(sigmoidin B)	MeOH	288(4.11)	39-0033-86C
	MeOH-NaOMe	325(4.34)	39-0033-86C
	MeOH-NaOAc	323(4.29)	39-0033-86C
	MeOH-$AlCl_3$	309(4.20)	39-0033-86C
	MeOH-$AlCl_3$-HCl	309(4.20)	39-0033-86C
[2,6'-Bi-4H-1-benzopyran]-4-one, 2,2',3,3'-tetrahydro-5,7,8'-tri-hydroxy-2',2'-dimethyl-	MeOH	288(4.08)	39-0033-86C
1α,10α-Epoxysalviarin	MeOH	210(4.27)	102-2381-86
Fibleucin	EtOH	208(4.02),230(3.87)	102-0905-86
4H,6H-Furo[3',2':3,4]naphtho[1,8-cd]-pyran-4,6-dione, 8,9-dihydro-3,7-dimethoxy-1,8,8,9-tetramethyl-, (±)-	MeOH	263(4.53),344(4.03), 379(4.06),393s(4.00)	44-4813-86
$C_{20}H_{20}O_7$			
Balantiolide acetate	EtOH	208(4.34),217(4.22), 254(3.92)	102-2543-86
Epoxysalviacoccin	MeOH	217(3.87),225(3.63), 230(3.40),235(3.20), 240(3.00),250(2.70)	102-0272-86
Fibraurin	EtOH	209(3.97),230(3.81)	102-0905-86
Sigmoidin D, (-)-	EtOH	290(4.40)	100-0932-86
$C_{20}H_{20}O_8$			
4H-1-Benzopyran-4-one, 2-(3,4-dimeth-oxyphenyl)-3-hydroxy-5,6,7-tri-methoxy-	MeOH	252(4.35),356(4.39)	102-0231-86
	MeOH-NaOAc	252(4.35),356(4.39)	102-0231-86
	MeOH-$AlCl_3$	265(4.39),421(4.49)	102-0231-86
	+ HCl	265(4.39),421(4.49)	102-0231-86
4H-1-Benzopyran-4-one, 4-hydroxy-2-(3-methoxyphenyl)-5,6,7,8-tetramethoxy-	MeOH	270(4.21),337(4.21)	102-2625-86
	MeOH-NaOMe	266(--),418(--)	102-2625-86
	MeOH-NaOAc	268(--),415(--)	102-2625-86
	+ H_3BO_3	270(--),337(--)	102-2625-86
	MeOH-$AlCl_3$	273(--),338(--)	102-2625-86
4H-1-Benzopyran-4-one, 2-(4-hydroxy-3,5-dimethoxyphenyl)-6,7-dimethoxy-(ageconyflavone C)	MeOH	265(4.27),323(4.30)	102-2625-86
	MeOH-NaOMe	265(--),394(--)	102-2625-86
	MeOH-NaOAc	264(--),388(--)	102-2625-86
	+ H_3BO_3	268(--),330(--)	102-2625-86

Compound	Solvent	λ_{max}(log ε)	Ref.
$C_{20}H_{20}O_9$			
4H-1-Benzopyran-4-one, 2-(2,4-dihydroxy-5-methoxyphenyl)-3,5,6,7-tetramethoxy-	MeOH	253(4.3),324(4.1)	94-1656-86
	MeOH-NaOMe	263s(--),290(--),313s(--),401(--)	94-1656-86
	MeOH-$AlCl_3$	253(--),324(--)	94-1656-86
	+ HCl	253(--),324(--)	94-1656-86
4H-1-Benzopyran-4-one, 2-(2,5-dihydroxy-4-methoxyphenyl)-3,5,6,7-tetramethoxy-	MeOH	252(4.5),307(4.2)	94-1656-86
	MeOH-NaOMe	246(--),290(--),310s(--)(decomposes)	94-1656-86
	MeOH-$AlCl_3$	252(--),307(--)	94-1656-86
	+ HCl	252(--),307(--)	94-1656-86
4H-1-Benzopyran-4-one, 3-hydroxy-2-(2-hydroxy-4,5-dimethoxyphenyl)-5,6,7-trimethoxy-	MeOH	255(4.5),304s(4.1),353(4.3)	94-1656-86
	MeOH-NaOMe	262(--),310(--),334(--),399(--)	94-1656-86
	MeOH-$AlCl_3$	270(--),335(--),425(--)	94-1656-86
	+ HCl	268(--),332(--),422(--)	94-1656-86
4H-1-Benzopyran-4-one, 3-hydroxy-2-(4-hydroxy-2,5-dimethoxyphenyl)-5,6,7-trimethoxy-	MeOH	253(4.2),303s(3.9),332(3.9)	94-1656-86
	MeOH-NaOMe	270s(--),295s(--),395(--)(decomposes)	94-1656-86
	MeOH-$AlCl_3$	266(--),324(--),410(--)	94-1656-86
	+ HCl	265(--),321(--),408(--)	94-1656-86
4H-1-Benzopyran-4-one, 5-hydroxy-2-(4-hydroxy-2,5-dimethoxyphenyl)-3,6,7-trimethoxy-	MeOH	260(4.5),303(3.1),345(4.2)	94-1656-86
	MeOH-NaOMe	267(--),301s(--),396(--)	94-1656-86
	MeOH-$AlCl_3$	275(--),323(--),396(--)	94-1656-86
	+ HCl	274(--),317(--),383(--)	94-1656-86
4H-1-Benzopyran-4-one, 2-(2-hydroxy-3,4,6-trimethoxyphenyl)-6,7-dimethoxy-	MeOH	264(4.2),312(4.1)	94-1656-86
	MeOH-NaOMe	262(--),291s(--),340(--)	94-1656-86
	MeOH-$AlCl_3$	274(--),299(--),361(--)	94-1656-86
	+ HCl	276(--),295s(--),344(--)	94-1656-86
4H-1-Benzopyran-4-one, 5-hydroxy-2-(3-hydroxy-2,4,6-trimethoxyphenyl)-6,7-dimethoxy-	MeOH	263(4.3),300(4.0),331s(4.0)	94-1656-86
	MeOH-NaOMe	269(--),360(--)	94-1656-86
	MeOH-$AlCl_3$	276(--),326(--),379(--)	94-1656-86
	+ HCl	276(--),322(--),371(--)	94-1656-86
4H-1-Benzopyran-4-one, 5-hydroxy-2-(4-hydroxy-2,3,6-trimethoxyphenyl)-6,7-dimethoxy-	MeOH	263(4.2),323(4.0)	94-1656-86
	MeOH-NaOMe	269(--),372(--)	94-1656-86
	MeOH-$AlCl_3$	274(--),290(--),364(--)	94-1656-86
	+ HCl	275(--),295s(--),351(--)	94-1656-86
4H-1-Benzopyran-4-one, 5-hydroxy-2-(6-hydroxy-2,3,4-trimethoxyphenyl)-6,7-	MeOH	262(4.3),326(4.1)	94-1656-86
	MeOH-NaOMe	264(--),380(--)	94-1656-86
	MeOH-$AlCl_3$	275(--),293s(--),368(--)	94-1656-86
	+ HCl	275(--),296s(--),356(--)	94-1656-86

Compound	Solvent	λ_{max}(log ϵ)	Ref.
4H-1-Benzopyran-4-one, 2-(2,4,5-tri-methoxyphenyl)-3,5-dihydroxy-6,7-dimethoxy-	MeOH	259(4.3),304s(3.8), 356(4.1)	94-1656-86
	MeOH-NaOMe	266(--),394(--)	94-1656-86
	MeOH-AlCl3	269(--),327(--), 414(--)	94-1656-86
	+ HCl	272(--),327s(--), 410(--)	94-1656-86
$C_{20}H_{20}O_{10}$			
4H-1-Benzopyran-4-one, 3,7-dihydroxy-2-(2-hydroxy-4,5-dimethoxyphenyl)-5,6,8-trimethoxy-	MeOH	255(4.5),358(4.3)	94-2228-86
	MeOH-AlCl3	271(--),332s(--), 430(--)	94-2228-86
	+ HCl	270(--),332s(--), 427(--)	94-2228-86
	MeOH-NaOMe	274(--),335(--), 400(--)	94-2228-86
	MeOH-NaOAc	272(--),335(--), 404(--)	94-2228-86
	+ H3BO3	262(--),378(--)	94-2228-86
4H-1-Benzopyran-4-one, 3,7-dihydroxy-2-(4-hydroxy-2,5-dimethoxyphenyl)-5,6,8-trimethoxy-	MeOH	254(4.5),305s(4.2), 350(4.4)	94-2228-86
	MeOH-AlCl3	267(--),293(--), 325(--),416(--)	94-2228-86
	+ HCl	266(--),323(--), 412(--)	94-2228-86
	MeOH-NaOMe	270(--),330(--), 390(--)	94-2228-86
	MeOH-NaOAc	270(--),330s(--), 368(--)	94-2228-86
	+ H3BO3	260(--),280s(--), 345(--)	94-2228-86
4H-1-Benzopyran-4-one, 3,7-dihydroxy-2-(5-hydroxy-2,4-dimethoxyphenyl)-5,6,8-trimethoxy-	MeOH	256(4.6),308(4.2), 352(4.4)	94-2228-86
	MeOH-AlCl3	269(--),293s(--), 328(--),419(--)	94-2228-86
	+ HCl	276(--),290s(--), 325(--),415(--)	94-2228-86
	MeOH-NaOMe	272(--),335s(--), 378(--)	94-2228-86
	MeOH-NaOAc	270(--),335s(--), 370(--)	94-2228-86
	+ H3BO3	256(--),276s(--), 352(--)	94-2228-86
4H-1-Benzopyran-4-one, 5,7-dihydroxy-2-(2-hydroxy-4,5-dimethoxyphenyl)-3,6,8-trimethoxy-	MeOH	266(4.3),305s(3.9), 355(4.0)	94-2228-86
	MeOH-AlCl3	275(--),329(--), 390(--)	94-2228-86
	+ HCl	277(--),329(--), 371(--)	94-2228-86
	MeOH-NaOMe	277(--),340s(--), 392(--)	94-2228-86
	MeOH-NaOAc	271(--),364(--)	94-2228-86
	+ H3BO3	276(--),357(--)	94-2228-86
4H-1-Benzopyran-4-one, 5,7-dihydroxy-2-(4-hydroxy-2,5-dimethoxyphenyl)-3,6,8-trimethoxy-	MeOH	257(4.8),312s(4.4), 350(4.5)	94-2228-86
	MeOH-AlCl3	278(--),328(--), 400(--)	94-2228-86
	+ HCl	276(--),325s(--), 364(--),410s(--)	94-2228-86
	MeOH-NaOMe	275(--),335s(--), 398(--)	94-2228-86

Compound	Solvent	λ_{max}(log ε)	Ref.
(cont.)	MeOH-NaOAc	270(--),360(--)	94-2228-86
	+ H_3BO_3	256(--),352(--)	94-2228-86
4H-1-Benzopyran-4-one, 5,7-dihydroxy-2-(5-hydroxy-2,4-dimethoxyphenyl)-3,6,8-trimethoxy-	MeOH	266(4.4),306(3.9), 354(4.1)	94-2228-86
	MeOH-$AlCl_3$	270(--),335s(--), 370(--)	94-2228-86
	MeOH-$AlCl_3$-HCl	277(--),330s(--), 366(--)	94-2228-86
	MeOH-NaOMe	278(--),374(--)	94-2228-86
	MeOH-NaOAc	276(--),366(--)	94-2228-86
	MeOH-NaOAc-H_3BO_3	267(--),315s(--), 358(--)	94-2228-86
$C_{20}H_{21}BrN_2O_3$			
1H-Pyrazole-1-carboxylic acid, 5-(3-bromophenyl)-4,5-dihydro-5-hydroxy-3-phenyl-, 1,1-dimethylethyl ester	hexane	284(4.41)	4-0727-86
	EtOH	282(4.40)	4-0727-86
$C_{20}H_{21}Cl_3N_2O_2$			
Ethanone, 2,2,2-trichloro-1-[3-(diethylamino)-4-methyl-5-(1-methyl-1H-indol-3-yl)-2-furanyl]-	MeCN	228(4.57),253s(4.18), 302s(3.84),390.4(4.50)	24-2317-86
$C_{20}H_{21}N$			
1H-Indole, 1-methyl-2-(1-methylethenyl)-3-(1-phenylethyl)-	MeOH	227.5(4.48),291.2(3.92)	1-0625-86
$C_{20}H_{21}NO$			
Azacyclohexadeca-1,3,5,11,13,15-hexaene-7,9-diyne, 2-ethoxy-3,6,11-trimethyl-, (E,E,Z,Z,E)-	THF	237(3.72),292(4.13), 340s(3.34),455(2.42)	18-1801-86
Azacyclohexadeca-1,3,5,11,13,15-hexaene-7,9-diyne, 2-ethoxy-6,11,16-trimethyl-	THF	266s(4.22),278(4.27), 304(4.17),321s(4.08), 380s(3.24)	18-1801-86
Oxiranecarbonitrile, 3-[2-(1-hexynyl)-1-cyclopenten-1-yl]-2-phenyl-, cis	MeCN	223(4.10),255(4.21)	78-2221-86
Oxiranecarbonitrile, 3-[2-(3-methyl-1-butynyl)-1-cyclohexen-1-yl]-2-phenyl-, cis	MeCN	220(4.20),251(4.06)	78-2221-86
3-Pyridinemethanol, 5-(diphenylmethylene)-1,2,5,6-tetrahydro-1-methyl-	pH 7.2	214(1.46),288(1.86)	87-0125-86
$C_{20}H_{21}NOS$			
1-Benzothiepin-5(2H)-one, 3,4-dihydro-4-[[methyl(2-phenylethyl)amino]-methylene]-, (E)-	EtOH	248(4.02),353(4.18)	4-0449-86
$C_{20}H_{21}NO_3$			
1H-Indole-3-carboxylic acid, 5-methoxy-2-methyl-1-(1-phenylethyl)-, methyl ester	MeOH	217.0(4.36),242.5(4.35), 286.5(4.17),294s(4.16), 304s(3.99)	1-0625-86
$C_{20}H_{21}NO_4$			
9,10-Anthracenedione, 1,5-dihydroxy-2-propyl-4-(propylamino)-	$CHCl_3$	240(4.53),252(4.40), 308(3.82),392(3.40), 536(3.96),570(4.21), 614(4.25)	78-3303-86
1H-[1]Benzoxepino[2,3,4-ij]isoquinoline, 2,3-dihydro-6,9,10-trimethoxy-1-methyl- (1,α-dehydrocularine)	EtOH	228(3.56),280(2.98), 362(2.89)	88-5535-86
	EtOH-HCl	216(3.37),242s(3.09), 262(2.87),386(2.57)	88-5535-86

Compound	Solvent	λ_{max}(log ϵ)	Ref.
Berberine, tetrahydro-, cyanoborane complex	MeOH	212(4.21),228s(--), 286(3.65)	83-0694-86
6H-Dibenzo[de,g]quinoline-6-carboxaldehyde, 4,5,6a,7-tetrahydro-1,2,3-trimethoxy-	EtOH	222(4.27),233s(4.12), 274(4.05)	100-0878-86
Isoquinoline, 1-[(3,4-dimethoxyphenyl)methyl]-6,7-dimethoxy- (papaverine)	n.s.g.	313(3.6),326(3.7)	140-0584-86
$C_{20}H_{21}NO_5$			
5H-Benzo[g]-1,3-benzodioxolo[6,5,4-de]quinolin-8-ol, 6,7,7a,8-tetrahydro-10,11-dimethoxy-7-methyl- (dasymachaline)	EtOH	221(4.37),285(4.12), 297(4.07)	102-1999-86
7H-Pyrano[2,3-c]acridin-7-one, 1,2,3,12-tetrahydro-1,2-dihydroxy-6-methoxy-3,3,12-trimethyl-, cis-(-)-	MeOH	230(3.93),269s(4.22), 276(4.30),298(3.74), 320(3.57),384(3.52)	100-1091-86
trans	MeOH	230(3.94),269s(4.20), 276(4.29),298(3.72), 320(3.57),384(3.50)	100-1091-86
$C_{20}H_{21}N_3O_2$			
Acetamide, 2-[[[1,4-dihydro-1-(phenylmethyl)-3-quinolinyl]carbonyl]methylamino]-	CH_2Cl_2	336(3.8)	5-1127-86
$C_{20}H_{21}N_3O_4$			
5-Pyrimidinecarboxylic acid, 1,2,3,4-tetrahydro-6-methyl-4-(4-nitrophenyl)-1-phenyl-, ethyl ester	EtOH	238s(4.06),307(4.09)	103-0990-86
$C_{20}H_{21}N_3O_5$			
3,5-Pyrazolidinedione, 4-[ethoxy(4-nitrophenyl)methyl]-1,4-dimethyl-2-phenyl-	EtOH	237(4.15)	103-0176-86
$C_{20}H_{21}N_3O_8S$			
2(1H)-Pyrimidinone, 4-(2-pyridinylthio)-1-(2,3,5-tri-O-acetyl-β-D-ribofuranosyl)-	pH 7	287(4.11)	1-0806-86
	pH 2	287(--)	1-0806-86
	pH 13	273(--),310(--)	1-0806-86
$C_{20}H_{21}N_3O_9$			
2(1H)-Pyrimidinone, 4-(2-oxo-1(2H)-pyridinyl)-1-(2,3,5-tri-O-acetyl-β-D-ribofuranosyl)-	pH 7	313(3.91)	1-0806-86
	pH 2	313(--)	1-0806-86
	pH 13	315(--)	1-0806-86
2(1H)-Pyrimidinone, 4-(4-oxo-1(4H)-pyridinyl)-1-(2,3,5-tri-O-acetyl-β-D-ribofuranosyl)-	pH 7	228s(--),318(4.51)	1-0806-86
	pH 2	318(--),328s(--)	1-0806-86
	pH 13	241s(--),270(--)	1-0806-86
2(1H)-Pyrimidinone, 4-(3-pyridinyloxy)-1-(2,3,5-tri-O-acetyl-β-D-ribofuranosyl)-	pH 7	270(3.86),284s(--)	1-0806-86
	pH 2	270(--),284s(--)	1-0806-86
	pH 13	260(--),289(--)	1-0806-86
$C_{20}H_{21}N_5$			
1,4-Benzenediamine, N,N-diethyl-N'-(2-methyl-3H-pyrazolo[1,5-a]benzimidazol-3-ylidene)-	BuOAc	290(4.00),375(3.59), 536(4.60)	48-0342-86
1H-Pyrazolo[3,4-b]quinoxalin-7-amine, N,N-diethyl-3-methyl-1-phenyl-	BuOAc	414(4.05),454(4.07)	48-0342-86
$C_{20}H_{21}N_5O_4$			
Pyrimido[2,1-f]purine-2,4,8(1H,3H,9H)-	MeOH	261(4.55),289s(3.97)	87-1099-86

Compound	Solvent	$\lambda_{max}(\log \epsilon)$	Ref.
trione, 6-hydroxy-1,3-dimethyl-9-(phenylmethyl)-7-propyl- (cont.)	MeOH-base	261(4.66),290(3.97), 300(3.95)	87-1099-86
$C_{20}H_{21}N_7O_5$			
L-5-Deazaaminopterin	pH 1	220(4.59),245s(4.33), 278s(3.96),315(4.00), 329s(3.90)	87-0709-86
	H_2O	218(4.54),279s(4.28), 297(4.31)	87-0709-86
	pH 13	249(4.42),279(4.45), 298s(4.42),342s(3.00)	87-0709-86
Propanamide, N-[2-[(2-cyano-4,6-dinitrophenyl)azo]-5-(diethylamino)-phenyl]-	$CHCl_3$	285(3.90),308(3.82), 585s(4.69),623(4.81)	48-0497-86
$C_{20}H_{22}$			
Bicyclo[2.2.1]heptane, 2-(diphenylmethyl)-, endo	hexane	259(2.55)	78-6175-86
exo	hexane	260(2.69)	78-6175-86
$C_{20}H_{22}BrNO_2$			
1-Propanone, 1-(4-bromophenyl)-2-hydroxy-3-phenyl-3-(1-piperidinyl)-	n.s.g.	250(4.60)	11-0379-86
1-Propanone, 3-(4-bromophenyl)-2-hydroxy-1-phenyl-3-(1-piperidinyl)-	n.s.g.	250(3.38)	11-0379-86
$C_{20}H_{22}BrNO_6$			
1,1-Cyclohexanedicarboxylic acid, 4-(benzoyloxy)-3-bromo-3-cyano-, diethyl ester, trans	n.s.g.	231.5(4.22)	128-0491-86
$C_{20}H_{22}ClNO_2$			
1-Propanone, 3-(4-chlorophenyl)-2-hydroxy-1-phenyl-3-(1-piperidinyl)-	n.s.g.	250(3.38)	11-0379-86
$C_{20}H_{22}N_2O$			
Schizozygane, 10-formyl-18,19-didehydro-, (±)-	MeOH	244(3.65),298(3.59), 351(3.98)	73-1731-86
$C_{20}H_{22}N_2OS_2$			
4H-Cyclohepta[4,5]thieno[2,3-d]pyrimidine-4-thione, 3,5,6,7,8,9-hexahydro-3-(3-hydroxypropyl)-2-phenyl-	EtOH	238(4.25),268(4.17), 385(4.02)	106-0096-86
$C_{20}H_{22}N_2O_2$			
17-Norcorynan-18-carboxylic acid, 18,19,20,21-tetradehydro-, methyl ester, (3β)-(±)-	MeOH	235(2.63),283(0.66), 299(0.55),360(4.56)	44-2995-86
$C_{20}H_{22}N_2O_3$			
3-Pyridinecarboxylic acid, 1,2,5,6-tetrahydro-5-[(hydroxymethyl)-2-pyridinylmethyl]-1-methyl-, methyl ester	pH 7.2	210(1.12),266(0.59)	87-0125-86
diastereoisomer	pH 7.2	210(1.19),266(0.56)	87-0125-86
3-Pyridinecarboxylic acid, 1,2,5,6-tetrahydro-5-[(hydroxymethyl)-3-pyridinylmethyl]-1-methyl-, methyl ester	pH 7.2	222(1.68)	87-0125-86
diastereoisomer	pH 7.2	222(1.72)	87-0125-86

Compound	Solvent	λ_{max}(log ε)	Ref.
3-Pyridinecarboxylic acid, 1,2,5,6-tetrahydro-5-[(hydroxyphenyl)-2-pyridinylmethyl]-1-methyl-, methyl ester	pH 7.2	222(1.64)	87-0125-86
diastereoisomer	pH 7.2	222(1.74)	87-0125-86
$C_{20}H_{22}N_2O_4$			
1H-Benz[de]isoquinoline-1,3(2H)-dione, 5-methoxy-2-[3-(4-morpholinyl)propyl]-	pH 7.4	340(4.00),380(3.84)	4-0849-86
4-Isoxazolecarboxamide, 4,5-dihydro-5-(hydroxymethyl)-3-(4-methoxyphenyl)-N-(2-phenylethyl)-, cis	MeOH	278(3.16)	73-2167-86
$C_{20}H_{22}N_2O_5$			
1H-Carbazole-1-carboxylic acid, 9-acetyl-4-(diacetylamino)-2,3,9,9a-tetrahydro-, methyl ester	MeOH	214(4.08),243(4.10), 273(3.93),317(3.57)	5-2065-86
$C_{20}H_{22}N_2O_7$			
1,3-Azulenedicarboxylic acid, 2-amino-6-[2-ethoxy-1-(hydroxyimino)-2-oxoethyl]-, diethyl ester, (Z)-	CH_2Cl_2	249(4.56),333(4.78), 418.5(4.07)	24-2956-86
$C_{20}H_{22}N_4$			
1H-Pyrrolo[2,3-b]pyridine-5-carbonitrile, 6-(butylamino)-4-methyl-1-(phenylmethyl)-	$CHCl_3$	242(1.08),262(1.36), 349(0.41)	103-0076-86
1H-Pyrrolo[2,3-b]pyridine-5-carbonitrile, 6-(diethylamino)-4-methyl-1-(phenylmethyl)-	$CHCl_3$	244(1.07),272(1.28), 353(0.29)	103-0076-86
1H-Pyrrolo[2,3-b]pyridine-5-carbonitrile, 2,3-dihydro-4-methyl-1-(phenylmethyl)-6-(1-pyrrolidinyl)-	$CHCl_3$	240(0.73),244(0.74), 296(0.51),347(0.81)	103-0076-86
$C_{20}H_{22}N_4O$			
3H-Pyrazol-3-one, 4-[[4-(diethylamino)phenyl]imino]-2,4-dihydro-5-methyl-2-phenyl-	MeOH	248(4.35),440(4.07), 528(4.59)	48-0149-86
	BuOAc	444(4.23),512(4.59)	48-0149-86
1H-Pyrrolo[2,3-b]pyridine-5-carbonitrile, 2,3-dihydro-4-methyl-6-(4-morpholinyl)-1-(phenylmethyl)-	$CHCl_3$	205(0.79),255(0.74), 335(0.8)	103-0076-86
$C_{20}H_{22}N_4O_2$			
Acetic acid, cyano[6-phenyl-2-(1-piperidinyl)-4(1H)-pyrimidinylidene]-, ethyl ester, (Z)-	EtOH	281(4.20),397(4.30)	103-0774-86
3H-Pyrazol-3-one, 4-[[4-[ethyl(2-hydroxyethyl)amino]phenyl]imino]-2,4-dihydro-5-methyl-2-phenyl-	MeOH	248(4.37),438(4.07), 548?(4.55)	48-0149-86
	BuOAc	442(4.20),512(4.56)	48-0149-86
$C_{20}H_{22}N_6O_5S$			
Acetamide, N-[2-[[2-cyano-6-(methylsulfonyl)-4-nitrophenyl]azo]-5-(diethylamino)phenyl]-	$CHCl_3$	299(3.76),309(3.76), 402(3.39),580s(4.62), 622(4.86)	48-0497-86
$C_{20}H_{22}O_2$			
Cyclohexanol, 2-[[2-(phenylmethoxy)phenyl]methylene]-, (E)-	MeOH	208(4.58),243(4.15), 287(3.71)	78-5637-86

Compound	Solvent	λ_{max}(log ϵ)	Ref.
$C_{20}H_{22}O_3$			
2H-1-Benzopyran-7-ol, 2,4-diethyl-3-(4-methoxyphenyl)-	EtOH	209(3.64),228(3.63), 290(3.61),312(3.61)	4-1781-86
Cyclohepta[de]naphthalen-7(8H)-one, 8,8-bis(1-methylethoxy)-	$CHCl_3$	316(3.94),336(3.84)	39-1965-86C
Cyclohepta[de]naphthalen-7(8H)-one, 8,8-dipropoxy-	$CHCl_3$	320(3.92),336s(3.86), 360s(3.73)	39-1965-86C
Cyclopenta[b]naphtho[2,3-e]pyran-5,10-dione, 1,2,3,3a,11,11a-hexahydro-1-methyl-3a-(1-methylethyl)-	EtOH	251(4.35),281(4.04), 338(3.36),383(3.08)	39-0659-86C
1,2-Naphthalenedione, 7-methyl-3-(1-methylethyl)-8-(4-methyl-1-oxo-4-pentenyl)-	MeOH and MeOH-NaOMe	261.5(4.20),330(2.69), 428(2.77)	102-1935-86
1,4-Naphthalenedione, 2-(3,7-dimethyl-2,6-octadienyl)-, (Z)-	EtOH	270(4.47),335s(--), 444(3.40)	39-0659-86C
3-Oxogona-4,8,11,13-tetraene-17β-carboxylic acid, 10-methyl-, methyl ester	EtOH	232(4.24)	2-0015-86
$C_{20}H_{22}O_4$			
2H-1-Benzopyran-2-one, 4-hydroxy-3-[(4,6,6-trimethyl-3-oxobicyclo-[3.1.1]hept-2-yl)methyl]-	MeOH	270(4.10),283(4.15), 308(4.10)	2-0102-86
1,3-Heptalenedicarboxylic acid, 4,6,8,10-tetramethyl-, dimethyl ester	dioxan	278(4.33),325s(3.68), 412s(3.11)	88-1673-86
1,3-Heptalenedicarboxylic acid, 5,6,8,10-tetramethyl-, dimethyl ester	hexane	208(4.39),227(4.31), 279(4.31),318s(3.59), 413(3.06)	88-1673-86
Naphtho[1,8-bc]pyran-7,8-dione, 2,3-dihydro-3-hydroxy-6,9-dimethyl-3-(4-methyl-3-pentenyl)-	EtOH	263(4.42),375(3.31), 446(3.51)	12-0647-86
$C_{20}H_{22}O_5$			
6(2H)-Benzofuranone, 2-(1,3-benzodioxol-5-yl)-3,3a,4,5-tetrahydro-5-methoxy-3-methyl-3a-(2-propenyl)-, (2α,3α,3aβ,5β)-	MeOH	260(4.15),290s(3.71)	102-2613-86
(2α,3β,3aβ,5α)-	MeOH	260(4.07)	102-2613-86
1(10)-Dehydrosalviarin, dihydro-	MeOH	204(4.26)	102-2381-86
Dehydrowortmannolone, $NaBH_4$ product diol (hydrate)	MeOH	263(3.96),307(3.85)	39-1317-86C
isomer	MeOH	265(3.91),309(3.82)	39-1317-86C
$C_{20}H_{22}O_6$			
4H-1-Benzopyran-4-one, 5,6,7,8-tetrahydro-5,6,7,8-tetrahydroxy-2-(2-phenylethyl)-, monoacetonide	EtOH	208(4.23),250(4.08)	94-2766-86
isomer	EtOH	208(4.11),252(3.97)	94-2766-86
1H,10H-Furo[3',4':4a,5]naphtho[2,1-c]-pyran-1,8(4bH)-dione, 3-(3-furanyl)-3,4,4a,5,6,11,12,12a-octahydro-12-hydroxy-4a-methyl-	MeOH	208(3.59)	102-1677-86
Methanone, (4-hydroxy-3-methoxyphenyl)-[4-[(4-hydroxy-3-methoxyphenyl)methyl]tetrahydro-3-furanyl]-	MeOH	229(--),287.0(4.02), 311.0(3.98)	39-1181-86C
3,9-Phenanthrenedione, 4,4a-dihydro-5,6,8-trihydroxy-7-(2-hydroxypropyl)-1,2,4a-trimethyl-, [R-(R*,S*)]-	ether	253s(3.96),293(4.27), 414(3.61)	33-0972-86
[S-(R*,R*)]- (plectrinone A)	ether	254(3.98),279s(4.23),	33-1395-86

Compound	Solvent	λmax(log ε)	Ref.
(cont.)		291.5(4.34),310s(4.16), 416s(3.58)	33-1395-86
$C_{20}H_{22}O_{11}$			
2H-1-Benzopyran-2-one, 7-[(2,6-di-O-acetyl-β-D-glucopyranosyl)oxy]- (2',6'-di-O-acetylscopolin)	MeOH	205.3(4.4),227.5(4.2), 250(3.7),257.5(3.6), 287(3.8),339(4.0)	94-4012-86
2H-1-Benzopyran-2-one, 7-[(3,6-di-O-acetyl-β-D-glucopyranosyl)oxy]-	MeOH	204(4.5),226(4.2), 287(3.7),340(4.9)	94-4012-86
$C_{20}H_{22}S_2$			
Thiophene, 2,2'-(1,3-butadiyne-1,4-diyl)bis[5-butyl-	EtOH	344(4.22)	142-0365-86
$C_{20}H_{23}BrN_6O_5$			
Acetamide, N-[2-[(2-bromo-4,6-dinitrophenyl)azo]-5-(dipropylamino)phenyl]-	$CHCl_3$	295(3.92),382(3.63), 575(4.74)	48-0497-86
$C_{20}H_{23}BrO_5$			
1,3-Azulenedicarboxylic acid, 6-bromo-2-butoxy-, diethyl ester	CH_2Cl_2	239(4.29),266.5(4.19), 318.5(4.81),352.5(3.91), 363.5(3.90),477.5(2.49)	24-2272-86
$C_{20}H_{23}N$			
5H-Indolo[3,2,1-de]acridine, 6,7,7a-8,9,10,11,12-octahydro-8-methyl-	MeOH	314(3.90)	103-0038-86
$C_{20}H_{23}NO$			
1H-Indole-2-methanol, α,α,1-trimethyl-3-(1-phenylethyl)-	MeOH	227(4.48),270(3.71), 280s(3.80),290(3.83), 296.5s(3.78)	1-0625-86
$C_{20}H_{23}NOS$			
1H-Phenothiazin-1-one, 2,4-bis(1,1-dimethylethyl)-	$CHCl_3$	247(3.98),312(3.93), 351s(3.84),630(3.53)	39-2233-86C
$C_{20}H_{23}NO_2$			
3-Heptanone, 6-(2-oxaziridinyl)-4,4-diphenyl-	MeOH	208(2.23),255(2.68), 265(2.74),296(2.70)	78-0245-86
1-Propanone, 2-hydroxy-1,3-diphenyl-3-(1-piperidinyl)-	n.s.g.	245(4.13)	11-0379-86
$C_{20}H_{23}NO_3$			
Benzeneacetic acid, α-[[2-(3-butenyloxy)phenyl]amino]-, ethyl ester	EtOH	243.5(4.09),285(3.55)	33-0927-86
Benzeneacetic acid, α-[[2-[(2-methyl-2-propenyl)oxy]phenyl]amino]-, ethyl ester	EtOH	242(4.09),288(3.58)	33-0927-86
Bicyclo[4.4.1]undeca-1,3,5,7,9-pentaene-2-carboxamide, 10-(1-acetoxyethenyl)-N,N-diethyl-	CH_2Cl_2	265(4.57),330(3.99)	33-1263-86
6H-Dibenzo[a,g]quinolizine, 5,8,13,13a-tetrahydro-2,3,10-trimethoxy-	MeOH	232(4.09),279.8(3.87)	2-0256-86
$C_{20}H_{23}NO_4$			
Acetic acid, [1,2,3,4-tetrahydro-1,1,9-trimethyl-11-oxo-9a,4a-(epoxyethano)-9H-carbazol-12-ylidene]-, methyl ester, (E)-	MeOH	235(4.16),289(3.32)	150-0514-86M
(Z)-	MeOH	237(4.09),290(3.23)	150-0514-86M

Compound	Solvent	λ_{max}(log ϵ)	Ref.
Bicyclo[4.4.1]undeca-1,3,5,7,9-pentaene-2-acetic acid, 10-[(diethylamino)carbonyl]-α-oxo-, ethyl ester	CH_2Cl_2	245(4.47),280(4.47), 360(3.92)	33-1263-86
Boldinemethine (5λ,4ε)	MeOH	263(4.48),282s(--), 305(4.11),318(3.75), 348(3.07)	100-0398-86
Flavinantine, O-methyl-	EtOH	231(3.4),280(3.0)	102-2155-86
Laurotetanine, N-methyl-	MeOH	220(4.46),283(4.08), 305(4.03),316s(3.96)	73-1743-86
Norsecocularine	EtOH	220(4.23),296s(3.68), 316(3.76)	142-3359-86
$C_{20}H_{23}NO_6$			
1,3-Azulenedicarboxylic acid, 2-amino-6-(2-ethoxy-2-oxoethyl)-, diethyl ester	CH_2Cl_2	246.5(4.53),265.5(4.33), 319.5(4.72),330.5(4.83), 360(3.91),369.5(3.88), 455.5(3.43)	24-2956-86
$C_{20}H_{23}N_2O_3$			
Ethanaminium, N,N,N-trimethyl-2-[[(9-oxo-10(9H)-acridinyl)acetyl]oxy]-, chloride	H_2O	253(4.66),384(3.91), 400(3.92)	56-0615-86
$C_{20}H_{23}N_3OS$			
Hydrazinecarbothioamide, 2-[2,5-dihydro-5,5-dimethyl-2-(4-methylphenyl)-4-phenyl-2-furanyl]-	EtOH	243(4.38),312s(1.74)	104-0647-86
$C_{20}H_{23}N_3O_4$			
1H-Benz[de]isoquinoline-1,3(2H)-dione, 6-amino-5-methoxy-2-[3-(4-morpholinyl)propyl]-	pH 7.4	458(4.04)	4-0849-86
Butanamide, 2-[[4-bis(2-hydroxyethyl)-amino]phenylimino]-3-oxo-N-phenyl-	H_2O	452(4.12)	39-0065-86B
1,3-Dioxane-4,6-dione, 5-[3-[2-[(1,3-dimethyl-2-imidazolidinylidene)amino]phenyl]-2-propenylidene]-2,2-dimethyl-	benzene	363(4.22),472(4.12)	24-3276-86
	acetone	362(4.24),480(4.16)	24-3276-86
	$CHCl_3$	361(4.18),472(4.13)	24-3276-86
	MeCN	364(4.15),476(4.06)	24-3276-86
	pyridine	367(4.19),486(4.14)	24-3276-86
$C_{20}H_{23}N_3O_5$			
K-252d	MeOH	223(4.60),242s(4.56), 248s(4.49),260s(4.53), 268s(4.58),280s(4.78), 290(4.98),322(4.23), 335(4.36),347(4.20), 364(4.11)	158-1072-86
$C_{20}H_{23}N_5$			
1,4-Benzenediamine, N'-(1,5-dihydro-5-imino-3-methyl-1-phenyl-4H-pyrazol-4-ylidene)-N,N-diethyl-	BuOAc	564(3.99)	48-0342-86
$C_{20}H_{23}N_5O_3$			
2,4,6(1H,3H,5H)-Pyrimidinetrione, 5-[3-[2-[(1,3-dimethyl-2-imidazolidinylidene)amino]phenyl]-2-propenylidene]-1,3-dimethyl-	benzene	367(4.22),475(4.26)	24-3276-86
	MeOH	362(3.99),445(3.87)	24-3276-86
	acetone	362(4.10),474(4.07)	24-3276-86
	$CHCl_3$	377(4.23),485(4.27)	24-3276-86
	MeCN	368(4.16),475(4.11)	24-3276-86
	pyridine	374(4.20),494(4.21)	24-3276-86

Compound	Solvent	λ_{max}(log ε)	Ref.
$C_{20}H_{23}N_5O_4S$			
9H-Purin-6-amine, 9-[2,3-O-(1-methylethylidene)-6-S-phenyl-6-thio-β-D-allo (and α-L-talo)furanosyl]-	H_2O	254.5(4.24)	94-1573-86
	0.05M HCl	254(4.25)	94-1573-86
$C_{20}H_{23}N_5O_8S$			
1H,4aH-[1,2,4]Triazolo[1',2':1,2][1,2]-diazeto[3,4-d]azepine-4a,9-dicarboxylic acid, 8-amino-2,3,5,6,7,9a-hexahydro-2-methyl-7-[(4-methylphenyl)sulfonyl]-1,3-dioxo-, dimethyl ester	$CHCl_3$	234(4.10),288(4.28)	24-2114-86
$C_{20}H_{24}F_2O_2$			
18-Norandrosta-1,4,13(17)-trien-3-one, 6,9-difluoro-11-hydroxy-16,17-dimethyl-, (6α,11β,16α)-	MeOH	234(4.20)	44-2315-86
$C_{20}H_{24}N$			
2H-Pyrrolium, 5-ethyl-3,4-dihydro-1,2-dimethyl-4,4-diphenyl-, perchlorate, (R)-	MeOH	237(3.62),266(3.32)	4-0369-86
(S)-	MeOH	237(3.64),266(3.34)	4-0369-86
$C_{20}H_{24}NO_2P$			
2H-3,1,2-Benzoxazaphosphorine, 1,4,4a-5,6,7,8,8a-octahydro-2-phenyl-1-(phenylmethyl)-, cis-fused	EtOH	262(2.86),265(2.89), 270(2.64),272(2.68)	12-0591-86
diastereoisomer	EtOH	258(2.92),265(2.89), 272(2.68)	12-0591-86
trans-fused	EtOH	258(2.92),264(2.92), 271(2.76)	12-0591-86
diastereoisomer	EtOH	262(2.83),265(2.87), 270(2.61),272(2.64)	12-0591-86
$C_{20}H_{24}N_2O_2$			
16-Epiaffinine (same in acid or base)	EtOH	209(4.35),238(4.19), 318(4.26)	102-0965-86
Mauiensine, 12-hydroxy-	EtOH	255(3.88),295(3.36)	102-1783-86
Quinine	pH 8	331(3.80)	3-0455-86
	base	331(3.75)	3-0455-86
$C_{20}H_{24}N_2O_2S$			
Acetic acid, [(3-cyano-6-tricyclo-[3.3.1.1^{3,7}]dec-1-yl-2-pyridinyl)-thio]-, ethyl ester	EtOH	220(4.35),273(3.60), 309(3.77)	70-0131-86
Thieno[2,3-b]pyridine-2-carboxylic acid, 3-amino-6-tricyclo[3.3.1.1^{3,7}]-dec-1-yl-, ethyl ester	EtOH	203(4.37),237(4.00), 285(4.49),368(3.74)	70-0131-86
$C_{20}H_{24}N_2O_2Se$			
Acetic acid, [(3-cyano-6-tricyclo-[3.3.1.1^{3,7}]dec-1-yl-2-pyridinyl)-seleno]-, ethyl ester	EtOH	226(4.29),275(4.00), 314(3.66)	70-0376-86
Selenolo[2,3-b]pyridine-2-carboxylic acid, 3-amino-6-tricyclo[3.3.1.1^{3,7}]-dec-1-yl-, ethyl ester	EtOH	235(4.04),293(4.50), 377(3.76)	70-0376-86
$C_{20}H_{24}N_2O_3$			
Rankinidine	MeOH	217(4.02),256(3.75)	100-0806-86

Compound	Solvent	$\lambda_{max}(\log \epsilon)$	Ref.
$C_{20}H_{24}N_2O_4$			
Cholinium 9-oxo-10-acridineacetate	H_2O	255(4.76),391(3.92), 407(3.91)	56-0615-86
2,4-Pentadienoic acid, 2-cyano-5-ethyl-5-(1-naphthalenyl)-5-phenyl-, ethyl ester	EtOH	311(4.21)	48-0314-86
[2.2](2,5)-Pyridiniophane, 7,15-bis(methoxycarbonyl)-5,13-dimethyl-, pseudoortho-, diiodide	H_2O	225(4.57),302(3.59)	5-0765-86
[2.2](2,5)-Pyridiniophane, 7,15-bis(methoxycarbonyl)-5,13-dimethyl-, pseudopara-, diiodide	H_2O	225(4.57),293(3.80), 312s(3.07)	5-0765-86
[2.2](2,5)-Pyridiniophane, 8,15-bis(methoxycarbonyl)-4,13-dimethyl-, pseudogem., diiodide	H_2O	225(4.57),297(3.76), 318s(2.83)	5-0765-86
[2.2](2,5)-Pyridiniophane, 8,15-bis(methoxycarbonyl)-4,13-dimethyl-, pseudometa, diiodide	H_2O	225(4.57),307(3.45)	5-0765-86
$C_{20}H_{24}N_2O_6S$			
2-Thiazolebutanoic acid, 4-carboxy-γ-[[(1,1-dimethylethoxy)carbonyl]amino]-, α-phenylmethyl ester, (S)-	n.s.g.	233(3.81)	44-4590-86
$C_{20}H_{24}N_4$			
1H-Pyrrolo[2,3-b]pyridine-5-carbonitrile, 6-(butylamino)-2,3-dihydro-4-methyl-1-(phenylmethyl)-	$CHCl_3$	247(0.61),292(0.47), 347(1.03)	103-0076-86
1H-Pyrrolo[2,3-b]pyridine-5-carbonitrile, 6-(diethylamino)-2,3-dihydro-4-methyl-1-(phenylmethyl)-	$CHCl_3$	240(0.77),258(0.80), 297(0.57),346(0.81)	103-0076-86
$C_{20}H_{24}N_6O_4$			
β-D-ribo-Heptofuranuronamide, 1-(6-amino-9H-purin-9-yl)-1,5,6-trideoxy-N-(2-phenylethyl)-	pH 6.8	260(4.1)	94-1871-86
$C_{20}H_{24}O_2$			
Androsta-4,6-diene-3,17-dione, 2-methylene-	EtOH	302(4.16)	56-0311-86
Circusone A	EtOH	206(3.986),257(4.010)	88-2439-86
Circusone B	EtOH	206(3.924),255(3.936)	88-2439-86
1H-Inden-5-ol, 2,3-dihydro-1-(4-hydroxy-3-methylphenyl)-1,3,3,6-tetramethyl-	EtOH	283.5(2.78)	12-0817-86
1,2-Naphthalenedione, 7-methyl-3-(1-methylethyl)-8-(4-methyl-3-pentenyl)-	MeOH	253(4.20),341(3.44), 438(3.48)	102-0755-86
$C_{20}H_{24}O_3$			
Benz[e]azulene-1,4-dione, 2,3,6a,7,8-9,10,10a-octahydro-2-hydroxy-2,5-dimethyl-10-methylene-7-(1-methylethenyl)- (circusone C)	EtOH	208(4.015),258(3.961)	88-2439-86
circusone D (epimer)	EtOH	204(4.051),260(3.941)	88-2439-86
$C_{20}H_{24}O_4$			
Ethanone, 1,1'-(7',7'a-dihydro-3,3-3",3"-tetramethyldispiro[oxirane-2,1'-[4,7]methano(1H-indene-8',2"-oxirane]-3'a,5'(4'H)-diyl)bis-,	EtOH	215(3.97),235s(3.87)	78-3311-86

Compound	Solvent	λmax(log ε)	Ref.
(1'α,3'aβ,4'β,7'β,7'aβ,8'S*)- (cont.)			78-3311-86
Ethanone, 1,1'-(7',7'a-dihydro-3,3-3",3"-tetramethyldispiro[oxirane-2,1'-[4,7]methano[1H]indene-8',2"-oxirane]-3'a,6'(4'H)-diyl)bis-, (1'α,3'aβ,4'β,7'β,7'aβ,8'S*)-	EtOH	215(4.00),235s(3.87)	78-3311-86
1H-Naphtho[1,8a-c]furan-3,9(5H,10H)-dione, 7-[2-(3-furanyl)ethyl]-6,6a,7,8-tetrahydro-7,8-dimethyl-, [6aR-(6aα,7α,8β)]- (bacchasmacranone)	EtOH	209(4.09)	102-2175-86
1,4-Phenanthrenedione, 4b,5,6,7,8,8a-hexahydro-3,7-dihydroxy-4b,8,8-trimethyl-2-(2-propenyl)- (plectranthone F)	ether	212(1.0),245(0.45), 324(0.44),450s(0.06) (relative abs.)	33-1395-86
$C_{20}H_{24}O_4S$			
Thiophene, 2,4-bis(2,4-dimethoxyphenyl)tetrahydro-	MeOH	231(4.29),277(3.78)	18-0083-86
$C_{20}H_{24}O_5$			
Bacchasmacranone, 2β-hydroxy-	EtOH	240(4.19)	102-2175-86
7-Benzofuranol, 2-(1,3-benzodioxol-5-yl)-2,3,3a,4,7,7a-hexahydro-7a-methoxy-3-methyl-3a-(2-propenyl)-	MeOH	235(4.45),280(4.30)	102-2613-86
Naphtho[1,2-b]furan-6-carboxylic acid, 2,3-dihydro-4,7-dimethoxy-2,3,3,9-tetramethyl-, methyl ester, (±)-	MeOH	222(4.52),240s(4.48), 254(4.75),319(3.69), 337s(3.58)	44-4813-86
1H-Naphtho[1,8a-c]furan-3,6(3aH,6aH)-dione, 7-[2-(2,5-dihydro-5-oxo-3-furanyl)ethyl]-7,8,9,10-tetrahydro-7,8-dimethyl-	n.s.g.	230(3.79)	102-1393-86
$C_{20}H_{24}O_6$			
1,3-Azulenedicarboxylic acid, 2,6-diethoxy-, diethyl ester	CH_2Cl_2	230(4.37),273(4.35), 323.5(4.81),365(4.08), 435.5(3.02)	24-2272-86
3-Furanmethanol, tetrahydro-α-(4-hydroxy-3-methoxyphenyl)-4-[(4-hydroxy-3-methoxyphenyl)methyl]-	MeOH	232.5(4.04)	39-1181-86C
1H-Indene-1,5-diol, 2,3-dihydro-2-[2-(4-hydroxy-3-methoxyphenyl)-1-(hydroxymethyl)ethyl]-6-methoxy-(3λ,2ε)	MeOH	211.0(--),230.0(4.18), 284.0(4.04)	39-1181-86C
Sculponeatin A	EtOH	230(3.85)	142-0001-86
Sculponeatin C	EtOH	231(3.87)	142-0001-86
$C_{20}H_{24}O_7$			
Budlein A, 17,18-dihydro-	MeOH	267s(3.88)	64-0695-86C
Ketone from teugnaphthalodin oxidn.	MeOH	254(3.48)	102-0171-86
$C_{20}H_{24}O_9$			
4H-1-Benzopyran-4-one, 8-β-D-glucopyranosyl-7-methoxy-5-methyl-2-(2-oxopropyl)- (7-O-methylaloesin)	MeOH	227(4.33),244(4.30), 252(4.29),296(4.09)	102-2219-86
$C_{20}H_{24}S_3$			
2,2':5',2"-Terthiophene, 5,5"-dibutyl-	EtOH	360(4.42)	142-0365-86
$C_{20}H_{25}FN_2O_{12}$			
2,4(1H,3H)-Pyrimidinedione, 6-fluoro-5-(methoxymethyl)-1-(2,3,4,6-tetra-	$CHCl_3$	251(3.92)	97-0437-86

Compound	Solvent	λ_{max}(log ε)	Ref.
O-acetyl-β-D-glucopyranosyl)- (cont.)			97-0437-86
$C_{20}H_{25}FO_2$			
18-Norandrosta-1,4,13(17)-trien-3-one, 9-fluoro-11-hydroxy-16,17-dimethyl-, (11β,16β)-	MeOH	236(4.30)	44-2315-86
$C_{20}H_{25}N$			
6-Azuleneacetonitrile, 2-octyl-	CH_2Cl_2	239.5(4.25),280(4.88), 289.5(4.99),307.5(3.71), 335.5(3.58),351(3.71), 365(3.02),564.5(2.43), 601(2.36),658.5(1.93)	24-2272-86
$C_{20}H_{25}NO_4$			
3,5-Pyridinedicarboxylic acid, 1,4-dihydro-2,6-dimethyl-1-(4-methylphenyl)-, diethyl ester	EtOH	231(4.21),320(3.69), 355(3.67)	150-0088-86S +150-0901-86M
3,5-Pyridinedicarboxylic acid, 1,4-dihydro-2,6-dimethyl-1-(phenylmethyl)-, diethyl ester	EtOH	262(3.84),317(3.44), 347(3.43)	150-0088-86S +150-0901-86M
$C_{20}H_{25}NO_5$			
4,6-Isoquinolinediol, 1-[(3,4-dimethoxyphenyl)methyl]-1,2,3,4-tetrahydro-7-methoxy-2-methyl-, (1S-cis)- (roemecarine)	MeOH	229s(4.03),282(3.66)	142-1227-86
	MeOH-base	237s(3.98),253s(3.83), 286(3.64),305s(3.53)	142-1227-86
$C_{20}H_{25}NO_5S$			
Sulfoximine, N-(2-ethoxy-2-oxoethyl)-S-[2-(2-methoxyethoxy)phenyl]-S-(4-methylphenyl)-, (S)-	EtOH	210(4.24),225(4.16), 240(4.08),291(3.61)	12-1833-86
$C_{20}H_{25}NO_6$			
Roemecarine 2α-oxide, (+)-	MeOH	231(4.07),282(3.70)	142-1227-86
$C_{20}H_{25}NO_7$			
Butanoic acid, 2,2'-[[5-(dimethylamino)-2,4-furandiyl]dimethylidyne]-bis[3-oxo-, diethyl ester	MeOH	253(3.06),380(2.82), 451(3.21)	73-0879-86
$C_{20}H_{25}N_3OS$			
Hydrazinecarbothioamide, 2-[3,3-diethyl-1,3-dihydro-1-(2-methylphenyl)-1-isobenzofuranyl]-	EtOH	246(4.15),270(3.29), 300(2.95)	104-0647-86
Hydrazinecarbothioamide, 2- [3,3-diethyl-1,3-dihydro-1-(3-methylphenyl)-1-isobenzofuranyl]-	EtOH	245(4.14),270(3.28), 272s(3.07),308(2.62)	104-0647-86
Hydrazinecarbothioamide, 2-[[2-(1-hydroxy-1-methylethyl)phenyl](2,4,6-trimethylphenyl)methylene]-	EtOH	237s(4.22),305(4.35)	104-0647-86
$C_{20}H_{25}N_3O_2S$			
Hydrazinecarbothioamide, 2-[3,3-diethyl-1,3-dihydro-1-(2-methoxyphenyl)-1-isobenzofuranyl]-	EtOH	245(4.13),262(3.56), 274s(3.50),280(3.43), 309s(2.27)	104-0647-86
Hydrazinecarbothioamide, 2-[3,3-diethyl-1,3-dihydro-1-(4-methoxyphenyl)-1-isobenzofuranyl]-	EtOH	240(4.37),270(3.43), 281(3.42),316(2.56)	104-0647-86

Compound	Solvent	$\lambda_{max}(\log \epsilon)$	Ref.
$C_{20}H_{25}N_3S$			
Perazine (as hydrogen malonate)	MeOH	255.5(4.548),309(3.669)	106-0571-86
$C_{20}H_{25}N_5O_4$			
Adenosine, N-(4-phenylbutyl)-	pH 1	263(4.21)	163-2243-86
	pH 7	266(4.18)	163-2243-86
	pH 13	267(4.18)	163-2243-86
$C_{20}H_{26}ClN_7S$			
2H-Purine-2-thione, 7-[3-[4-[(4-chlorophenyl)methyl]piperazinyl]propyl]-1,7-dihydro-6-(methylamino)-	pH 1	247(4.52),288(4.54)	4-0737-86
	pH 13	254(4.52)	4-0737-86
$C_{20}H_{26}N_2$			
Diazene, (4-pentylphenyl)(4-propylphenyl)-, (E)-	hexane	330(5.74),444(2.72)	19-0183-86
Pyrazine, 2,5-bis(2-methylpropyl)-3-(2-phenylethenyl)-, (E)-	EtOH	231(4.11),237(4.00), 278(4.26),338(4.33)	4-1481-86
$C_{20}H_{26}N_2O$			
Diazene, (4-pentylphenyl)(4-propylphenyl-, monooxide, (Z)-	hexane	333(5.67)	19-0183-86
Pyrazine, 2,5-bis(2-methylpropyl)-3-(2-phenylethenyl)-, 1-oxide, (E)-	EtOH	230(3.92),238s(3.87), 276(4.19),303s(3.86), 348(3.90),362s(3.84)	4-1481-86
Vallesamidin-22-al, (±)-	MeOH	253(4.01),281(3.56), 288(3.50)	73-1731-86
$C_{20}H_{26}N_2O_3$			
Erichsonine (same spectra in acid or base)	EtOH	237(4.05),318(4.19)	102-0969-86
$C_{20}H_{26}N_2O_{12}$			
2,4(1H,3H)-Pyrimidinedione, 6-methoxy-5-methyl-1-(2,3,4,6-tetra-O-acetyl-β-D-glucopyranosyl)-	$CHCl_3$	264(3.92)	97-0437-86
2,4,6(1H,3H,5H)-Pyrimidinetrione, 1,3-dimethyl-5-(2,3,4,6-tetra-O-acetyl-β-D-glucopyranosyl)-	MeOH	226(3.93),256(3.65)	136-0053-86M
$C_{20}H_{26}N_3O_2$			
Methanaminium, N-[[[(dimethylamino)(4-methoxyphenyl)methylene]amino](4-methoxyphenyl)methylene]-N-methyl-, iodide	EtOH	248(4.17),283(4.40)	39-0619-86C
$C_{20}H_{26}N_4OS$			
3-Pyridinecarbonitrile, 5-acetyl-1,4-dihydro-2-mercapto-4-(3-pyridinyl)-6-methyl-, azepinium salt	EtOH	235(4.08),294(4.07), 385s(2.44)	104-1762-86
$C_{20}H_{26}N_4O_2$			
Benzo[g]phthalazine-1,4-diamine, N,N'-bis(3-methoxypropyl)-	EtOH	236(4.77),251s(4.54), 257(4.58),299(3.77), 307(3.73)	111-0143-86
2H-Imidazo[4,5-c]pyridin-2-one, 1,3,4,5-tetrahydro-4-imino-1,3-dimethyl-5-(2-oxo-2-tricyclo[3.3.1-$1^{3,7}$]dec-2-ylethyl]-, monohydrobromide	H_2O	229(4.72),272(3.90), 291(4.09)	103-0180-86

Compound	Solvent	λ_{max}(log ε)	Ref.
$C_{20}H_{26}N_4O_6$			
Piperazine, 1-[5-[(2,2-dimethyl-1,3-dioxolan-4-yl)methoxy]-1,6-dihydro-1-methyl-6-oxo-4-pyridazinyl]-4-(2-furanylcarbonyl)-	EtOH	243(4.5),312(3.1)	83-0060-86
$C_{20}H_{26}N_6O_7$			
Carbamic acid, [[[4-(aminocarbonyl)-1-β-D-ribofuranosyl-1H-imidazol-5-yl]amino][(phenylmethyl)amino]-methylene]-, ethyl ester	pH 1	250(3.90)	44-1277-86
	pH 11	268(3.90)	44-1277-86
	MeOH	278(3.91)	44-1277-86
$C_{20}H_{26}O$			
Acetaldehyde, [4-[1-methyl-3-(2,6,6-trimethyl-1-cyclohexen-1-yl)-2-propenylidene]-2-cyclopenten-1-ylidene]-, "9-cis,11-cis"	EtOH	402(4.06)	39-0905-86C
"11-cis"	EtOH	263(4.12),405(4.27)	39-0905-86C
"11-cis,13-cis"	EtOH	392(4.00)	39-0905-86C
4-Heptanone, 2-(3,8-dimethyl-5-azulenyl)-6-methyl-	MeOH	286(4.46),292(4.49), 305(4.15),334(3.66), 352(3.66),368(3.51), 603(2.54),655(2.48)	94-4641-86
$C_{20}H_{26}O_2$			
1H-Phenalene-1,2(4H)-dione, 5,6,6a-7,8,9-hexahydro-3,6,9-trimethyl-4-(2-methyl-1-propenyl)-	EtOH	237(3.95),260(3.77), 285(3.52),338(3.45), 450s(2.77)	44-5140-86
4(1H)-Phenanthrenone, 4a,9,10,10a-tetrahydro-6-hydroxy-1,1,4a-trimethyl-7-(1-methylethyl)-, (4aS-trans)- (15-deoxyfuerstione)	MeOH	252(3.73),262s(3.63), 340s(3.58),440(4.05)	102-0755-86
	MeOH-$AlCl_3$	237s(4.04),260s(3.72), 333(3.59),430(3.99), 550(3.81)	102-0755-86
	MeOH-$AlCl_3$-HCl	240s(3.99),253s(3.84), 260s(3.76),337(3.51), 424(3.92),552(3.70)	102-0755-86
$C_{20}H_{26}O_2S_4$			
1,3-Benzenebisethane(dithioic) acid, 5-(1,1-dimethylethyl)-α,α'-dioxo-, bis(1-methylethyl) ester	MeOH	312(3.99),500(2.02)	118-0968-86
$C_{20}H_{26}O_3$			
Androsta-4,6-diene-3,17-dione, 19-hydroxy-7-methyl-	EtOH	283(3.42)	54-0111-86
Androsta-4,6-diene-3,17-dione, 6-methoxy-	MeOH	304(4.22)	150-0718-86M
Androst-4-en-19-al, 7α-methyl-3,17-dioxo-	EtOH	248(3.04)	54-0111-86
7β-	EtOH	247(3.00)	54-0111-86
11-Oxaandrost-4-en-3-one, 17-ethynyl-17β-hydroxy-	EtOH	237(4.1)	13-0381-86A
1-Phenanthrenecarboxylic acid, 7-ethenyl-4b,5,6,7,8,8a,9,10-octahydro-3-hydroxy-4b,7-dimethyl-, methyl ester	MeOH	240(3.94),305(3.79)	94-1015-86
$C_{20}H_{26}O_4$			
2-Butenoic acid, 3-methyl-, 2,3,3a,4-5,8,9,11a-octahydro-6,10-dimethyl-3-methylene-2-oxocyclodeca[b]furan-	MeOH	216(4.30)	100-0360-86

Compound	Solvent	λ_{max}(log ϵ)	Ref.
4-yl ester, [3aR-(3aR*,4S*,6E,10E-11aR*)]- (cont.)			100-0360-86
11-Oxaandrosta-1,4-dien-3-one, 17β-acetoxy-	EtOH	242(4.19)	13-0047-86B
1H-Oxireno[1,10a]phenanthro[3,2-b]furan-9(7aH)-one, 2,3,4,4a,5,6,11a,11b-octahydro-8-(hydroxymethyl)-4,4,11b-trimethyl-, [4aR-(4aα,6aS*,7aβ,11aα-11bβ)]-	MeOH	285(4.228)	102-1411-86
Torilolide	EtOH	221(4.35)	94-4682-86
$C_{20}H_{26}O_5$			
1H-Naphtho[1,8a-c]furan-3(5H)-one, 7-[2-(2,5-dihydro-5-oxo-3-furanyl)-ethyl]-6,6a,7,8,9,10-hexahydro-6-hydroxy-7,8-dimethyl-	n.s.g.	216(4.21)	102-1393-86
Oxytorilolide	EtOH	218(4.36)	94-4682-86
$C_{20}H_{26}O_6$			
Leptocephalide, 8β-angeloyloxy-	MeOH	end absorption	102-0467-86
Montafrusin B (several formulas)	MeOH	211(4.35)	102-0695-86
Semiatrin	MeOH	208(4.58)	102-1484-86
$C_{20}H_{26}O_7$			
Deacetylcalein A	MeOH	end absorption	102-0467-86
10-Oxabicyclo[6.3.2]trideca-1(11),8,12-triene-12,13-dicarboxylic acid, 4-acetoxy-, diethyl ester	EtOH	210s(4.02),275(3.41)	24-0297-86
diastereoisomer	EtOH	210s(4.03),275(3.44)	24-0297-86
$C_{20}H_{26}O_9$			
4H-1-Benzopyran-4-one, 8-β-D-glucopyranosyl-2-(2-hydroxypropyl)-7-methoxy-5-methyl-, (R)-	MeOH	228(4.36),244(4.29), 252(4.30),294(4.12)	102-2219-86
Shinjulactone M	EtOH	238(4.19)	18-1638-86
$C_{20}H_{27}ClO_2$			
2H-Pyran, 4-(3-chlorophenyl)-2,6-bis(1,1-dimethylethyl)-2-methoxy-	MeOH	238(4.31),292(3.49)	44-4385-86
2H-Pyran, 4-(4-chlorophenyl)-2,6-bis(1,1-dimethylethyl)-2-methoxy-	MeOH	245(4.46),292(3.55)	44-4385-86
$C_{20}H_{27}NO_4$			
2H-Pyran, 2,6-bis(1,1-dimethylethyl)-2-methoxy-4-(4-nitrophenyl)-	MeOH	283(4.19),346(3.49)	44-4385-86
$C_{20}H_{27}NO_{13}$			
β-D-Fructofuranose, 1-deoxy-1-[(3-methoxy-2-(methoxycarbonyl)-3-oxo-1-propenyl]amino]-, 2,3,4,6-tetraacetate	CHCl	278(4.24)	136-0329-86D
$C_{20}H_{27}N_3O_8$			
2(1H)-Pyrimidinone, 4-(1-piperidinyl)-1-(2,3,5-tri-O-acetyl-β-D-ribofuranosyl)-	pH 7	278(4.17)	1-0806-86
	pH 2	287(--)	1-0806-86
	pH 13	284(--)	1-0806-86
$C_{20}H_{27}N_3O_{11}$			
1H-Imidazole-4-carboxylic acid, 5-amino-1-(2,3,4,6-tetra-O-acetyl-β-D-glucopyranosyl)-, ethyl ester	MeOH	266(4.18)	142-3451-86

Compound	Solvent	λ_{max}(log ε)	Ref.
2,4,6(1H,3H,5H)-Pyrimidinetrione, 1,3-dimethyl-5-[3,4,6-tri-O-acetyl-2-(acetylamino)-2-deoxy-β-D-glucopyranosyl]-	MeOH	224(3.67),255(3.34)	136-0053-86M
$C_{20}H_{27}N_4O$			
Piperidinium, 1-[2-(4-methylphenyl)-6-(1-piperidinyl)-4H-1,3,5-oxadiazin-4-ylidene]-, perchlorate	MeCN	239(4.34),287(4.10)	97-0094-86
$C_{20}H_{27}N_4O_2$			
Piperidinium, 1-[2-(4-methoxyphenyl)-6-(1-piperidinyl)-4H-1,3,5-oxadiazin-4-ylidene]-, perchlorate	MeCN	295(3.95)	97-0094-86
$C_{20}H_{27}O$			
Pyrylium, 2,6-bis(1,1-dimethylethyl)-4-(4-methylphenyl)-, perchlorate	MeOH	304(4.17),353(4.49)	44-4385-86
$C_{20}H_{27}O_2$			
Pyrylium, 2,6-bis(1,1-dimethylethyl)-4-(4-methoxyphenyl)-, perchlorate	MeOH	299(4.12),390(4.62)	44-4385-86
$C_{20}H_{28}N_2O$			
1H-Indeno[2,1-b]pyridine-3-carboxamide, N,N-diethyl-2,3,9,9a-tetrahydro-1,9,9-trimethyl-	EtOH	252(4.16),286(3.65), 295(3.57)	94-2786-86
$C_{20}H_{28}N_2O_3S_2$			
Sulfoximine, N,N'-(oxydi-2,1-ethanediyl)bis[S-methyl-S-(4-methylphenyl)-, [R-(R*,R*)]-	EtOH	224(4.31)	12-1655-86
$C_{20}H_{28}N_2S_2$			
Butanethioamide, 2-(4,4-dimethyl-2-phenyl-5(4H)-thiazolylidene)-N,N-diethyl-3-methyl-, (E)-	EtOH	244(4.38),282(4.07), 328s(3.08),392s(2.18)	33-0174-86
2-Butanethione, 1-(diethylamino)-1-(4,4-dimethyl-2-phenyl-5(4H)-thiazolylidene)-3-methyl-	EtOH	236(4.06),308(4.07), 366(4.24),456(4.14)	33-0174-86
$C_{20}H_{28}N_6O_8$			
4-Imidazolecarboxamide, 5-[3-(ethoxycarbonyl)-3'-(4-methoxybenzyl)-1-guanidino]-1-β-D-ribofuranosyl-	pH 1	250(4.00)	44-1277-86
	pH 11	273(4.00)	44-1277-86
	MeOH	276(4.04)	44-1277-86
$C_{20}H_{28}N_6S$			
Cyclopropenylium, bis[bis(1-methylethyl)amino][[2-(cyanoamino)-3-(cyanoimino)-1-cyclopropen-1-yl]thio]-, hydroxide, inner salt	MeCN	233(4.40),267s(4.10)	24-2104-86
$C_{20}H_{28}O$			
Azulene, 6-ethoxy-2-octyl-	CH_2Cl_2	290(4.84),297.5(4.90), 346.5(3.62),358.5(3.70), 376.5(3.74),514.5(2.33)	24-2272-86
2-Cyclohexen-1-one, 3-[2-methyl-4-(2,6,6-trimethyl-1-cyclohexen-1-yl)-1,3-butadienyl]-, all-trans	$CHCl_3$	342(4.30)	35-4614-86
Cyclopenta[g]cyclopropa[a]naphthalene-3-carboxaldehyde, 7a,8b-dimethyl-	C_6H_{12}	195(4.04),247(3.93), 323(1.70)	100-0236-86

Compound	Solvent	$\lambda_{max}(\log \epsilon)$	Ref.
5-(1-methylethenyl)-1,1a,3a,4,4a,5-6,7,7a,8,8a,8b-dodecahydro- (cont.)			100-0236-86
Retinal, 13-demethyl-14-methyl-, all-E-	heptane	369(4.70)	5-0479-86
11Z-	heptane	246(--),369(4.60)	5-0479-86
$C_{20}H_{28}O_2$			
2,9(1H,3H)-Phenanthrenedione, 7-ethenyl-4,4a,4b,5,6,7,10,10a-octahydro-1,1,4a,7-tetramethyl-	MeOH	247(3.1)	102-0909-86
9(1H)-Phenanthrenone, 2,3,4,4a,10,10a-hexahydro-7-(1-hydroxy-1-methylethyl)-1,1,4a-trimethyl-, (4aS-trans)-	MeOH	205(4.69),249(4.27), 295(3.54)	102-1931-86
2H-Pyran, 2,6-bis(1,1-dimethylethyl)-2-methoxy-4-phenyl-	MeOH	235(4.35),292(3.59)	44-4385-86
Retinoic acid, all-trans	MeOH	348(4.63)	39-0635-86B
$C_{20}H_{28}O_3$			
Androst-4-ene-3,17-dione, 19-hydroxy-7β-methyl-	EtOH	243(3.17)	54-0111-86
Oxireno[9,10]cyclotetradeca[1,2-b]-furan-9(1aH)-one, 2,3,6,7,7a,8,10a-13,14,14a-decahydro-1a,5,12-trimethyl-8-methylene- (7,8-epoxycembra-3,11,15-trien-16,2-olide, 1R,2R,3E-7R,8R,11E)-	EtOH	218(3.98)	12-0123-86
(1R,2S,3E,7S,8S,11E)-	EtOH	215(3.90)	12-0123-86
Spiro[benzo[3,4]cyclobuta[1,2]cyclododecene-5(4bH),2'-[1,3]dioxolan]-4b-ol, 6,7,8,9,10,11,12,13,14,14a-decahydro-, (4bR*,14aR*)-	MeOH	261(3.09),267(3.27), 273.5(3.24)	44-1419-86
$C_{20}H_{28}O_4$			
Curculathyrane A	EtOH	272(3.6285)	88-5675-86
Curculathyrane B	EtOH	272(3.5625)	88-5675-86
1,4-Ethanonaphthalene-6,10(4H)-dione, 4,4a,5,8a-tetrahydro-5,9-dihydroxy-5,9-dimethyl-2,8-bis(1-methylethyl)-	EtOH	234(3.98)	12-1843-86
11-Oxaandrost-4-en-3-one, 17 -acetoxy-	EtOH	238(4.18)	13-0047-86B
$C_{20}H_{28}O_5$			
2-Butenoic acid, 2-methyl-, 1a,2,3,6-7,7a,8,9,10a,10b-decahydro-1a,5,8-trimethyl-9-oxooxireno[9,10]cyclodeca[1,2-b]furan-7-yl ester	EtOH	218(3.6)	39-1363-86C
8(14),15-Isopimaradiene-20,6-γ-lactone, 6α,7β,9α-trihydroxy-	MeOH	206(3.90)	163-2003-86
$C_{20}H_{29}NO$			
Retinal oxime, 13-cis, anti	hexane	356(4.73)	149-0075-86B
	EtOH	356(4.77)	149-0075-86B
	hexane-ether-EtOH	359(4.76)	149-0075-86B
syn	hexane	352(4.74)	149-0075-86B
	EtOH	353(4.78)	149-0075-86B
	hexane-ether-EtOH	355(4.76)	149-0075-86B
$C_{20}H_{29}NOSi$			
Bicyclo[4.4.1]undeca-1,3,5,7,9-penta-	MeOH	270(4.45),330(3.87)	33-1851-86

Compound	Solvent	$\lambda_{max}(\log \epsilon)$	Ref.
ene-2-carboxamide, N,N-diethyl-3-methyl-9-(trimethylsilyl)- (cont.)			33-1851-86
$C_{20}H_{30}Cl_2N_6O_8Pt$			
Platinum, bis(3'-amino-3'-deoxythymidine-$N^{3'}$)dichloro-	pH 2	268(4.25)	87-0681-86
	pH 12	268(4.15)	87-0681-86
Platinum, bis(5'-amino-5'-deoxythymidine-$N^{5'}$)dichloro-	pH 2	268(4.23)	87-0681-86
	pH 12	268(4.17)	87-0681-86
$C_{20}H_{30}N_2O_6$			
1,8-Naphthyridine-3,4a,6(4H)-tricarboxylic acid, 1,5,8,8a-tetrahydro-2,7,8a-trimethyl-, triethyl ester, cis	EtOH	270(4.48),285s(--)	39-0753-86C
$C_{20}H_{30}N_2S_2Si$			
Ethanethioamide, 2-(4,4-dimethyl-2-phenyl-5(4H)-thiazolylidene)-N,N-diethyl-2-(trimethylsilyl)-	EtOH	242(4.37),264s(4.24), 384s(2.48)	33-0174-86
$C_{20}H_{30}N_8O_4$			
β-D-Ribofuranuronamide, 1-deoxy-N-(1-piperidinyl)-1-[6-(1-piperidinylamino)-9H-purin-9-yl]-	n.s.g.	281(4.34)	87-1683-86
$C_{20}H_{30}N_8O_5$			
D-α-Glutamine, N^2-[N-(N-1H-purin-6-yl)-β-alanyl]-L-alanyl]-, 1,1-dimethylethyl ester	pH 2.9	272(4.17)	63-0757-86
	pH 7.0	267(4.18)	63-0757-86
	pH 10.5	271(4.19)	63-0757-86
L-α-Glutamine, N^2-[N-(N-1H-purin-6-yl)-β-alanyl]-L-alanyl]-, 1,1-dimethylethyl ester	pH 2.9	272(4.06)	63-0757-86
	pH 7.0	265(4.09)	63-0757-86
	pH 10.5	271(4.05)	63-0757-86
$C_{20}H_{30}O$			
Benz[f]azulen-8-ol, 2,3,3a,5,6,7,8,8a-9,10-decahydro-3a,8a-dimethyl-5-methylene-1-(1-methylethyl)- (amijitrienol)	EtOH	230(4.23)	18-0661-86
Cyclopenta[g]cyclopropa[a]naphthalene-3-methanol, 1,1a,3a,4,4a,5,6,7,7a,8-8a,8b-dodecahydro-7a,8b-dimethyl-5-(1-methylethenyl)-	C_6H_{12}	194(4.02),320(1.68)	100-0236-86
Furan, 3-(4,8,12-trimethyl-3,7,11-trienyl)-, (E,E)-	EtOH	241(3.19)	12-1629-86
2-Hepten-4-one, 2-methyl-6-(octahydro-7-methyl-3-methylene-1H-cyclopenta-[1,3]cyclopropa[1,2]benzen-4-yl)-, [3aR-(3aα,3bβ,4β(R*),7α,7aS*]]-	EtOH	236(4.00)	20-0815-86
$C_{20}H_{30}O_2$			
Androst-15-en-17-one, 3-methoxy-, (3β,5α)-	EtOH	229(3.90)	19-0313-86
Dolasta-1,3-dien-6-one, 9-hydroxy-	MeOH	204(3.20),261(3.54)	44-2736-86
Linifoliol	MeOH	247(3.4)	102-0909-86
2-Phenanthrenemethanol, 4b,5,6,7,8-8a,9,10-octahydro-10-hydroxy-α,α,4b,8,8-pentamethyl- (7β,15-dihydroxyabietatriene)	MeOH	203(4.37),266(4.56), 274(4.48)	102-1931-86
Phenanthro[3,2-b]furan-4a(2H)-ol, 1,3,4,5,6,6a,7,11,11a,11b-decahydro-4,4,7,11b-tetramethyl-	EtOH	220(3.98)	102-0167-86

Compound	Solvent	λ_{max}(log ϵ)	Ref.
[4aR-(4aα,6aβ,7α,11aα,11bβ)]- (voua-capen-5α-ol) (cont.)			102-0167-86
$C_{20}H_{30}O_3$			
Androst-4-en-3-one, 17,19-dihydroxy-7-methyl-, (7α,17β)- (98-99%)	EtOH	244(3.20)	54-0111-86
1,2-Butanedione, 1-[3,5-bis(1,1-dimethylethyl)-4-hydroxyphenyl]-3,3-dimethyl-	C_6H_{12}	205(4.15),225(4.04), 282(4.08)	44-2257-86
1,2-Hexanedione, 1-[3,5-bis(1,1-dimethylethyl)-4-hydroxyphenyl]-	C_6H_{12}	203(4.14),228(3.90), 295(3.99)	44-2257-86
7-Oxabicyclo[4.1.0]hept-4-en-3-one, 2,4-bis(1,1-dimethylethyl)-6-(1-hexynyl)-2-hydroxy-	C_6H_{12}	205(3.72),253(3.64)	44-2257-86
7-Oxabicyclo[4.1.0]hept-4-en-3-one, 2,4-bis(1,1-dimethylethyl)-2-hydroxy-6-(3,3-dimethyl-1-butynyl)-	C_6H_{12}	205(3.74),252(3.66)	44-2257-86
Pisiferdiol	MeOH	217(3.86),280(3.32)	78-0529-86
$C_{20}H_{30}O_3S$			
Acetic acid, [[[3-(1,3,5,8-tetradecatetraenyl)oxiranyl]methyl]thio]-, methyl ester, [2α,3β(1E,3E,5Z,8Z)]-(±)-	EtOH	274(4.47),280(4.54), 292(4.40)	104-1875-86
$C_{20}H_{30}O_4$			
2-Butenoic acid, 2-methyl-, 6,10-dimethyl-3-(1-methylethyl)-2-oxo-11-oxabicyclo[8.1.0]undec-6-en-4-yl ester	EtOH	218(3.9)	39-1363-86C
Vaginatin	EtOH	218(3.9)	39-1363-86C
$C_{20}H_{30}O_5$			
2-Butenoic acid, 2-methyl-, decahydro-5a-hydroxy-2a,7a-dimethyl-5-(1-methylethyl)-3-oxoazuleno[5,6-b]oxiren-2-yl ester	EtOH	218(3.8)	39-1363-86C
1H-Cyclopenta[c]furan-1-one, hexahydro-4-(3-oxo-1-octenyl)-5-[(tetrahydro-2H-pyran-2-yl)oxy]-	MeOH	227(4.13)	5-1179-86
Portulenone	MeOH	232(3.76)	138-1585-86
$C_{20}H_{30}P_2$			
Diphosphene, bis(1,2,3,4,5-pentamethyl-2,4-cyclopentadien-1-yl)-	pentane	236(4.2),275(4.1), 407(3.3)	89-0919-86
$C_{20}H_{31}NOSi$			
Bicyclo[4.4.1]undeca-4,6,8,10-tetraene-2-carboxamide, N,N-diethyl-3-methyl-10-(trimethylsilyl)-	MeOH	270(4.07),320(3.29)	33-1597-86
$C_{20}H_{31}NO_5$			
Cyclobuta[d]pyrrolo[2,1-b]oxazole-2a(2H)-acetic acid, 6-[2-(1,1-dimethylethoxy)-2-oxoethylidene]-hexahydro-, 1,1-dimethylethyl ester	EtOH	272(3.15)	78-2575-86
$C_{20}H_{31}N_7O_4$			
L-Leucine, N-[N-[1-oxo-5-(1H-purin-6-ylamino)pentyl]-L-alanyl-, methyl ester	pH 2.9	271(4.19)	63-0757-86
	pH 7.0	267(4.20)	63-0757-86
	pH 10.5	272(4.21)	63-0757-86

Compound	Solvent	$\lambda_{max}(\log \epsilon)$	Ref.
$C_{20}H_{32}$			
Bicyclo[8.8.2]eicosa-1(19),10(20),19-triene	n.s.g.	268(4.2)	35-0343-86
Cyclohexane, 1,1'-(2,3-dimethyl-2-butene-1,4-diylidene)bis[4-methyl-, trans	C H	230(3.99)	44-2361-86
$C_{20}H_{32}NS$			
Benzothiazolium, 3-dodecyl-2-methyl-, iodide	EtOH	277(3.80)	4-0209-86
$C_{20}H_{32}N_6O_4$			
β-D-Ribofuranuronamide, N-butyl-1-deoxy-1-[6-[(1-ethylpropyl)amino]-9H-purin-9-yl]-N-methyl-	n.s.g.	270(4.28)	87-1683-86
β-D-Ribofuranuronamide, 1-deoxy-N-(1-ethylpropyl)-1-[6-[(1-ethylpropyl)-amino]-9H-purin-9-yl]-	n.s.g.	270(4.25)	87-1683-86
$C_{20}H_{32}O$			
Cyclohexanemethanol, 4-ethenyl-α,4-dimethyl-3-(1-methylethenyl)-α-(4-methyl-1,3-pentadienyl)- (isofuscol)	EtOH	237s(--),239(4.26), 247s(--),275(3.62), 286(3.56)	20-0815-86
2,4-Cyclopentadien-1-one, 2,5-bis(1,1-dimethylethyl)-3-[2-methyl-1-(1-methylethyl)-1-propenyl]-	C_6H_{12}	412(4.65)?	44-2257-86
3,5-Heptadien-2-ol, 2-(4,8-dimethyl-3,7-cyclodecadien-1-yl)-6-methyl-	EtOH	238(4.30)	20-0815-86
3,5-Heptadien-2-ol, 6-(4,8-dimethyl-3,7-cyclodecadien-1-yl)-2-methyl-	EtOH	240(4.31)	20-0815-86
3,5-Heptadien-2-ol, 6-[4-ethenyl-4-methyl-3-(1-methylethenyl)cyclohexyl]-2-methyl- (fuscol)			
4-Heptanone, 2-(4,8-dimethyl-1,3,7-cyclodecatrien-1-yl)-6-methyl-, [S-(E,E,E)]- (germacrexeniolone)	MeOH	254(4.00)	94-4641-86
$C_{20}H_{32}O_2$			
4-Cyclodecen-1-ol, 7-(5-hydroxy-1,5-dimethyl-1,3-hexadienyl)-4-methyl-10-methylene-	EtOH	241(4.27)	20-0815-86
Dolabella-3,6-dien-9-one, 12-hydroxy-	MeOH	213(3.28)	44-2736-86
Dolabella-3,7-dien-9-one, 12-hydroxy-, (E)-	MeOH	203(3.83),235(3.86)	44-2736-86
(Z)-	MeOH	207(3.53),237(3.53)	44-2736-86
1-Naphthalenol, decahydro-6-(5-hydroxy-1,5-dimethyl-1,3-hexadienyl)-8a-methyl-4-methylene-	EtOH	239(4.30)	20-0815-86
1-Naphthalenol, 1,2,4a,5,6,7,8,8a-octahydro-6-(5-hydroxy-1,5-dimethyl-1,3-hexadienyl)-4,8a-dimethyl-	EtOH	240(4.26)	20-0815-86
$C_{20}H_{32}O_3$			
Dolabella-2,7-dien-9-one, 4,12-dihydroxy-	MeOH	250(3.30)	44-2736-86
Dolabella-3-en-9-one, 7,8-epoxy-12-hydroxy-	MeOH	205(3.51),221s(3.20)	44-2736-86
Dolabella-7-en-9-one, 3,4-epoxy-12-hydroxy-	MeOH	231(3.70)	44-2736-86

Compound	Solvent	λ_{max}(log ε)	Ref.
$C_{20}H_{32}O_4$			
2-Butenoic acid, 2-methyl-, 2-hydroxy-6,10-dimethyl-3-(1-methylethyl)-11-oxabicyclo[8.1.0]undec-6-en-4-yl ester	EtOH	218(3.9)	39-1363-86C
2-Butenoic acid, 2-methyl-, 1,2,3,5,6-7,8,8a-octahydro-1,8-dihydroxy-1,4-dimethyl-7-(1-methylethyl)-6-azulenyl ester (cycloshiromodiol 8-O-angelate)	EtOH	218(3.9)	39-1363-86C
LTB$_4$	MeOH	260(4.68),270.5(4.72), 281(4.62)	44-1253-86
Shiromodiol 8-O-angelate	EtOH	218(3.8)	39-1363-86C
$C_{20}H_{32}O_5$			
Neorustmicin C	MeOH	245(3.65)	158-1016-86
Neorustmicin D	MeOH	220(3.60)	158-1016-86
Sterebin B	MeOH	228(4.02)	78-6443-86
Sterebin C	MeOH	228(3.97)	78-6443-86
$C_{20}H_{32}Si_4$			
Dibenzo[c,g][1,2,5,6]tetrasilocin, 5,6,11,12-tetrahydro-5,5,6,6,11-11,12,12-octamethyl-	hexane	230s(4.3)	138-1781-86
2,3,8,9-Tetrasilatricyclo[8.2.2.2^{4,7}]-hexadeca-4,6,10,12,13,15-hexaene, 2,2,3,3,8,8,9,9-octamethyl-	hexane	223(4.28),263(4.35)	138-1781-86
2,3,9,10-Tetrasilatricyclo[9.3.1.1^{4,8}]-hexadeca-1(15),4,6,8(16),11,13-hexaene, 2,2,3,3,9,9,10,10-octamethyl-	hexane	235(4.3)	138-1781-86
$C_{20}H_{33}IO_3$			
2,4,12-Tridecatrienoic acid, 13-iodo-11-methoxy-3,7,11-trimethyl-, 1-methylethyl ester, (E,E)-	hexane	259(4.33)	44-5447-86
$C_{20}H_{33}N_3O_2$			
2(3H)-Naphthalenone, 4,4a,5,6,7,8-hexahydro-4a-methyl-8-(4-methyl-1-piperazinyl)-3-(4-morpholinyl)-, (3α,4aβ,8β)-	EtOH	236.0(4.16)	118-0461-86
$C_{20}H_{34}O$			
2-Naphthalenol, decahydro-2,5,5,8a-tetramethyl-1-(3-methylene-4-pentenyl)- (isoabienol)	EtOH	225(4.08)	163-3179-86
$C_{20}H_{35}N_3O_2$			
Hydrazinium, 1-(2-hydroxydodecyl)-1,1-dimethyl-2-(2-pyridinylcarbonyl)-, hydroxide, inner salt	pH 7.0	263(3.93)	138-1519-86
with zinc ions	pH 7.0	270(4.08)	138-1519-86
$C_{20}H_{36}OSi$			
Silane, [[1-(3,3-dimethylcyclohexyl)-2-methylene-3-cyclopenten-1-yl]oxy](1,1-dimethylethyl)dimethyl-	MeOH	232(4.07)	23-0180-86
$C_{20}H_{36}O_3$			
Octadecanoic acid, 10-methylene-	MeOH	228(3.66)	102-0449-86

Compound	Solvent	$\lambda_{max}(\log \epsilon)$	Ref.
9-oxo-, methyl ester (cont.)			102-0449-86
$C_{20}H_{38}N_4O_6S_2$			
6,9,12,20,23,28-Hexaoxa-1,3,15,17-tetraazabicyclo[15.8.5]triacontane-2,16-dithione	EtOH	210.1(4.30),249.0(4.40)	104-1589-86
6,9,17,20,25,28-Hexaoxa-1,3,12,14-tetraazabicyclo[12.8.8]triacontane-2,13-dithione	EtOH	209.5(4.33),249.0(4.40)	104-1589-86
$C_{20}H_{42}OSi_4$			
Trisilane, 1,1,1,3,3,3-hexamethyl-2-(tricyclo[3.3.1.1^{3,7}]dec-1-yl-[(trimethylsilyloxy)methylene]-	n.s.g.	340(3.87)	101-0021-86C

Compound	Solvent	$\lambda_{max}(\log \epsilon)$	Ref.
$C_{21}H_3Cl_{14}N$			
Benzeneacetonitrile, 4-[bis(pentachlorophenyl)methyl]-2,3,5,6-tetrachloro-	$CHCl_3$	293(3.20),303(3.25)	44-2472-86
$C_{21}H_4Cl_{14}O_2$			
Benzeneacetic acid, 4-[bis(pentachlorophenyl)methyl]-2,3,5,6-tetrachloro-	$CHCl_3$	296(3.10),305(3.10)	44-2472-86
$C_{21}H_{11}NO_2S$			
Naphtho[2,3-h]quinoline-7,12-dione, 2-(2-thienyl)-	DMSO	312(4.255),460(4.580)	2-0616-86
$C_{21}H_{12}ClN_3O_4$			
Benzenamine, N-[1-(4-chlorophenyl)-3-(4-nitrophenyl)-2-propynylidene]-4-nitro-	PrOH	218s(4.76),292(4.93)	30-0176-86
$C_{21}H_{12}N_4O_3$			
6H-Anthra[1,9-cd]isoxazol-6-one, 3-azido-5-(4-methoxyphenoxy)-	EtOH	454(4.22),471(4.21)	104-1192-86
6H-Anthra[1,9-cd]isoxazol-6-one, 3-azido-5-(phenylmethoxy)-	EtOH	460(4.20),478(4.18)	104-1192-86
$C_{21}H_{12}O_4$			
1H-Indene-1,3(2H)-dione, 2-[3-(1,3-dihydro-1,3-dioxo-2H-inden-2-ylidene)-1-propenyl]-	EtOH	250(4.54),560(4.94)	70-1446-86
potassium salt	EtOH	250(4.60),560(4.94)	70-1446-86
salt with N,N,N',N'-tetramethylpropenediamine	EtOH	248(4.79),312(4.86), 559(5.13)	70-1446-86
	$CHCl_3$	248(4.60),312(4.61), 575(5.13)	70-1446-86
$C_{21}H_{13}BCl_2O_4$			
Boron, [1,2-benzenediolato(2-)-O,O']-[1,3-bis(4-chlorophenyl)-1,3-propanedionato-O,O']-, (T-4)-	CH_2Cl_2	379(4.58),396s(4.51), 446s(2.55)	48-0755-86
$C_{21}H_{13}BrN_2O_2$			
Methanimidamide, N-(4-bromophenyl)-N'-(9,10-dihydro-9,10-dioxo-1-anthracenyl)-	EtOH	230(4.246),240(4.230), 265(4.034),470(3.58)	104-1566-86
$C_{21}H_{13}BrN_4$			
1H-Pyrazolo[3,4-b]quinoxaline, 1-(4-bromophenyl)-3-phenyl-	MeOH	261.2(5.43),289.4(5.49), 336.6(4.87)	2-0215-86
$C_{21}H_{13}ClN_2O_2$			
Benzenamine, N-[1-(4-chlorophenyl)-3-phenyl-2-propynylidene]-4-nitro-	PrOH	228(4.61),299(4.53)	30-0176-86
$C_{21}H_{13}ClN_4$			
1H-Pyrazolo[3,4-b]quinoxaline, 1-(4-chlorophenyl)-3-phenyl-	MeOH	270(5.32),340(4.63)	2-0901-86
$C_{21}H_{13}Cl_2N_3O$			
4(3H)-Quinazolinone, 2-(2-chlorophenyl)-3-[(2-chlorophenyl)methylene]amino]-	MeOH	212(3.51),258(3.63), 282(3.61),313(3.74)	18-1575-86

Compound	Solvent	λ max(log ε)	Ref.
4(3H)-Quinazolinone, 2-(4-chlorophenyl)-3-[[(4-chlorophenyl)methylene]amino]-	MeOH	228(3.50),255(3.59), 280(3.71),342(3.69)	18-1575-86
$C_{21}H_{13}NO$			
Dibenzo[d,j]azacyclotetradecin-8(7H)-one, 15,16,17,18-tetradehydro-, (E,E)-	THF	228(4.44),282(4.56), 326s(4.15),340s(4.06), 372s(3.61)	39-0933-86C
15H-Dibenzo[a,g]cyclotridecen-15-one, 5,6,7,8-tetradehydro-, oxime, (E,E)-	THF	220s(4.53),229(4.55), 282(4.61),295s(4.54), 340(3.90)	39-0933-86C
$C_{21}H_{13}NO_4$			
6H-Anthra[1,9-cd]isoxazol-6-one, 5-hydroxy-3-(4-methylphenoxy)-	heptane	410(4.1),430s(4.0)	104-2121-86
$C_{21}H_{13}NO_5S$			
1,2-Benzisothiazol-3(2H)-one, 2-[(2-oxo-2H-naphtho[1,2-b]pyran-4-yl)methyl]-, 1,1-dioxide	EtOH	264.5(4.38),274.5(4.47), 290.5(3.85),303(3.85), 315.5(3.81),350(3.61)	94-0390-86
$C_{21}H_{13}NS$			
Benzothiazole, 2-(9-phenanthrenyl)-	CH_2Cl_2	251(4.02),265s(3.92), 275s(3.85),287s(3.68), 315(3.64),360s(3.00)	24-1525-86
$C_{21}H_{13}N_5O_2$			
6H-Anthra[1,9-cd]isoxazol-6-one, 3-azido-5-[(4-methylphenyl)amino]-	EtOH	498(4.19),529(4.21)	104-1192-86
$C_{21}H_{13}N_5O_5$			
4(3H)-Quinazolinone, 2-(4-nitrophenyl)-3-[[(4-nitrophenyl)methylene]amino]-	MeOH	235(3.53),250(3.58), 273(3.73),317(3.66)	18-1575-86
$C_{21}H_{14}$			
2,3:6,7-Dibenzofluorene, cesium salt	THF	480(4.73)	35-7016-86
2,3:6,7-Dibenzofluorene, lithium salt	THF	497(4.75)	35-7016-86
13H-Indeno[1,2-c]phenanthrene	EtOH	238(4.60),256(4.66), 263(4.73),285(3.98), 296(4.24),310(4.48), 325(4.46),343(3.49), 352(3.08),360(3.53)	44-2428-86
$C_{21}H_{14}Br_4O_3$			
9H-Xanthene, 2,4,5,7-tetrabromo-3,6-dimethoxy-9-phenyl-	EtOH	272(4.18),302s(3.55)	39-0671-86C
$C_{21}H_{14}N_2O_2$			
6H-Anthra[1,9-cd]isoxazol-6-one, 5-[(2-methylphenyl)amino]-	EtOH	495(4.24),521(4.29)	103-1363-86
Benzenamine, N-(1,3-diphenyl-2-propynylidene)-4-nitro-	PrOH	214s(4.32),225s(4.25), 290(4.41)	30-0176-86
$C_{21}H_{14}N_2O_4$			
Phenol, 2,5-bis(2-benzoxazolyl)-4-methoxy-	2-MeTHF	312(4.5),378(4.4)	46-1455-86
$C_{21}H_{14}N_4$			
2,4-Cyclohexadiene-1,1,3-tricarbonitrile, 2-amino-4,6-diphenyl-	n.s.g.	259(4.28),339(4.30)	104-0230-86

Compound	Solvent	λ_{max}(log ε)	Ref.
1,3,5-Hexatriene-1,1,3-tricarbonitrile, 2-amino-4,6-diphenyl-	n.s.g.	245(4.04),270(4.27), 340(4.42)	104-0230-86
Propanedinitrile, (3-cyano-5,6-dihydro-4,6-diphenyl-2(1H)-pyridinylidene)-	n.s.g.	249(4.01),328(4.06), 370(4.03)	104-0230-86
1H-Pyrazolo[3,4-b]quinoxaline, 1,3-diphenyl-	MeOH	262.4(5.31),337.2(4.59)	2-0901-86
$C_{21}H_{14}N_4O_2$			
Naphtho[2,1-b]pyrido[2,3-e][1,4]oxazin-5-one, 4-amino-6-(phenylamino)-	$CHCl_3$	288(4.20),308s(4.18), 482(3.92),588(3.97)	4-1697-86
$C_{21}H_{14}N_4S$			
Propanedinitrile, (5,6-dihydro-2,5-diphenylthieno[2,3-d]pyrimidin-4-yl)-	$CHCl_3$	268(4.12),341(4.03), 386(3.70)	94-0516-86
$C_{21}H_{14}N_6NiO_2$			
Nickel, [2-[[(1H-benzimidazol-2-ylazo)phenylmethyl]azo]benzoate(2-)]-, (SP-4-2)-	isoPrOH	545(--),1040(3.40)	65-0727-86
$C_{21}H_{14}O_4$			
2H-Naphtho[1,2-b]pyran-2-one, 4-[(benzoyloxy)methyl]-	EtOH	265.5(4.40),275.5(4.48), 291.5(3.78),304(3.86), 317(3.85),354(3.70)	94-0390-86
$C_{21}H_{14}O_8$			
2H-1-Benzopyran-2-one, 7-acetoxy-6-methoxy-3-[(2-oxo-2H-1-benzopyran-7-yl)oxy]- (7-acetoxydaphnoretin)	EtOH	212(4.25),294(3.90), 325(4.01)	102-0557-86
$C_{21}H_{15}BO_4$			
Boron, [1,2-benzenediolato(2-)-O,O']-(1,3-diphenyl-1,3-propanedionato-O,O')-, (T-4)-	CH_2Cl_2	370(4.56),397s(4.50), 433s(2.76)	48-0755-86
$C_{21}H_{15}BO_5$			
Boron, [1-hydroxy-2-naphthalenecarboxylate(2-)-O^1,O^2](1-phenyl-1,3-butanedionato-O,O')-, (T-4)-	CH_2Cl_2	239s(3.84),268(3.83), 336(4.43)	48-0755-86
Boron, [2-hydroxy-3-naphthalenecarboxylate(2-)-O,O'](1-phenyl-1,3-butanedionato-O,O')-	CH_2Cl_2	334(4.43),346s(4.35)	48-0755-86
$C_{21}H_{15}BO_6$			
Boron, 1,7-diphenyl-1,6-heptadiene-3,5-dionato-O,O')[ethanedioato(2-)-O,O']-, (T-4)-	CH_2Cl_2	467(4.90)	97-0399-86
$C_{21}H_{15}Br$			
5,11[1',2']-Benzeno-5H-cyclohepta[b]-naphthalene, 5-bromo-5a,11-dihydro-	EtOH	273(3.84)	18-1095-86
$C_{21}H_{15}Br_3O_3$			
5H-Tribenzo[a,d,g]cyclononene-2,7,12-triol, 3,8,13-tribromo-10,15-dihydro-, (±)-	dioxan	210(4.74),232s(4.42), 287s(3.96),296(4.01)	152-0017-86
$C_{21}H_{15}Cl$			
5,11[1',2']-Benzeno-5H-cyclohepta[b]-naphthalene, 5-chloro-5a,11-dihydro-	EtOH	275(3.91)	18-1095-86

Compound	Solvent	$\lambda_{max}(\log \epsilon)$	Ref.
$C_{21}H_{15}ClFN_3O$			
1H-1,4-Benzodiazepine, 7-chloro-5-(2-fluorophenyl)-2,3-dihydro-2-(2-pyridinylmethylene)-, 4-oxide	isoPrOH	227(4.36),236s(4.34), 283(4.41),384(4.43), 405s(3.49)	4-1303-86
$C_{21}H_{15}ClN_2$			
1H-Pyrazole, 3-(4-chlorophenyl)-1,5-diphenyl-	n.s.g.	258(4.51)	78-3807-86
1H-Pyrazole, 5-(4-chlorophenyl)-1,3-diphenyl-	n.s.g.	255(4.58)	78-3807-86
$C_{21}H_{15}Cl_2NO$			
Spiro[aziridine-2,9'-[9H]fluorene], 3,3-dichloro-1-(4-methoxyphenyl)-	$CHCl_3$	208(4.46),240(4.50), 278(4.04),338s(2.60)	103-0710-86
$C_{21}H_{15}Cl_2N_3O$			
Benzoic acid, 2-[[(2-chlorophenyl)-methylene]amino]-, [(2-chlorophenyl)methylene]hydrazide	MeOH	206(3.18),226(3.32), 260(3.41),305(3.41)	18-1575-86
Benzoic acid, 2-[[(4-chlorophenyl)-methylene]amino]-, [(4-chlorophenyl)methylene]hydrazide	MeOH	224(3.19),253(3.33), 295(3.42),308(3.43)	18-1575-86
$C_{21}H_{15}NO$			
Benzonitrile, 4-[3-oxo-4-(3-phenyl-2-propynyl)-1-cyclopenten-1-yl]-	EtOH	249(4.41),289(4.5)	39-1235-86C
$C_{21}H_{15}NO_2$			
5,11[1',2']-Benzeno-5H-cyclohepta[b]-naphthalene, 5a,11-dihydro-5-nitro-	EtOH	273(3.52)	18-1095-86
Dibenz[a,h]acridine-1,2-diol, 1,2-dihydro-, trans	EtOH-THF	225(4.33),243(4.21), 282s(4.56),289(4.72), 282s(4.56),289(4.72), 301s(4.57),348(3.68), 361(3.76),366(3.75), 300(3.92),411(3.95)	44-2445-86
Dibenz[a,h]acridine-3,4-diol, 3,4-dihydro-, trans	EtOH-THF	222s(4.39),228(4.48), 280s(4.70),289(4.78), 358(3.70),369(3.79), 393(3.77)	44-2445-86
Dibenz[a,h]acridine-8,9-diol, 8,9-dihydro-, trans	EtOH-THF	222(4.47),241(4.38), 288(4.76),291(4.76), 346(3.76),367(3.82), 378(3.95),398(3.99)	44-2445-86
Dibenz[c,h]acridine-5,6-diol, 5,6-dihydro-, trans	55% THF	295(4.36),365(4.20)	44-1773-86
Naphth[2,3-c]acridine-5,14-dione, 9,10,11,12-tetrahydro-	DMSO	230(3.732),450(3.544)	2-0616-86
Spiro[9H-fluorene-9,2'-[2H]indol]-3'(1'H)-one, 5'-methoxy-	CHCl	214(4.78),233(4.44), 266(4.55),297(3.92), 308(4.00)	103-0710-86
$C_{21}H_{15}NO_3$			
Benzoic acid, 4-[5-(1-naphthalenyl)-2-oxazolyl]-, methyl ester	toluene	339(4.28)	103-1022-86
$C_{21}H_{15}NO_5$			
2,5-Cyclohexadiene-1,4-dione, 3-(1H-indol-2-yl)-6-(4-methoxyphenyl)-2,5-dihydroxy-	dioxan	268(5.09),400(4.23)	5-0195-86

Compound	Solvent	$\lambda_{max}(\log \epsilon)$	Ref.
2H-Pyran-2,5(6H)-dione, 3-hydroxy-6-(1H-indol-3-ylmethylene)-4-(4-methoxyphenyl)-	dioxan	276(4.37),428(4.21)	5-0177-86
$C_{21}H_{15}NS_2$			
5(4H)-Thiazolethione, 2,4,4-triphenyl-	EtOH	244(4.16),256s(4.10), 268s(4.00),328(3.72), 511(1.42),538(1.36)	33-0374-86
$C_{21}H_{15}N_2O_3$			
3H-Indolizinium, 1-hydroxy-3-[(4-nitrophenyl)methylene]-2-phenyl-, chloride	80% MeOH + HCl	580(3.70) 455(3.68)	4-1315-86 4-1315-86
$C_{21}H_{15}N_3O_4$			
9,10-Anthracenedione, 1-(methylamino)-5-nitro-4-(phenylamino)-	EtOH	616(4.21),661(4.26)	104-0547-86
9,10-Anthracenedione, 4-(methylamino)-5-nitro-1-(phenylamino)-	EtOH	616(4.12),661(4.17)	104-0547-86
$C_{21}H_{15}N_5O_2$			
Spiro[3H-indole-3,4'(1'H)-pyrano[2,3-c]pyrazole]-5'-carbonitrile, 6'-amino-1,2-dihydro-3'-methyl-2-oxo-1'-phenyl-	EtOH	246(4.94)	18-1235-86
$C_{21}H_{15}N_5O_5$			
Benzoic acid, 2-[[(4-nitrophenyl)methylene]amino]-, [(4-nitrophenyl)methlene]hydrazide	MeOH	229(3.20),258(3.27), 297(3.34),310(3.41)	18-1575-86
$C_{21}H_{15}N_5S_4$			
2H-Pyrrole, 2-[4,5-bis(methylthio)-1,3-dithiol-2-ylidene]-, conjugate monoacid salt with 2,2'-(2,5-cyclohexadiene-1,4-diylidene)bis(propanedinitrile)	MeCN	233(4.17),250(4.11), 278(3.84),305(3.88), 395(4.44),406(4.47), 420(4.51),436(4.47), 472(4.44),665(3.84), 680(3.01),696(3.90), 727(4.17),743(4.37), 760(4.30),820(4.46), 843(4.62)	78-0849-86
$C_{21}H_{16}$			
1H-Cyclopenta[l]phenanthrene, 1-ethynyl-1,3-dimethyl-	EtOH	239(4.54),263(4.46), 271s(4.38),278s(4.23), 323(3.99),334(3.95)	39-1471-86C
$C_{21}H_{16}BrNO_2$			
2-Propen-1-one, 1-(5-bromo-2-hydroxyphenyl)-3-phenyl-3-(phenylamino)-, (Z)-	EtOH	270(3.81),394(4.49)	104-1658-86
$C_{21}H_{16}BrN_3O_3Se$			
Furo[3',4':5,6]selenino[2,3-c]pyrazole-5,7-dione, 4-[(4-bromophenyl)amino]-1,4,4a,7a-tetrahydro-3-methyl-1-phenyl-	EtOH	259(3.9)	104-2126-86
$C_{21}H_{16}Br_2$			
1H-Cyclopenta[l]phenanthrene, 1-(2,2-dibromoethenyl)-1,3-dimethyl-	EtOH	227(4.48),241(4.34), 262(4.21),273(4.12)	39-1471-86C

Compound	Solvent	λ_{max}(log ε)	Ref.
$C_{21}H_{16}Br_2O_3$			
2,4,6-Cycloheptatrien-1-one, 2,5-bis[(4-bromophenyl)methoxy]-	MeOH	228(4.59),254s(3.98), 327(4.07),346s(4.01), 360s(3.95),376s(3.77)	18-1125-86
2,4,6-Cycloheptatrien-1-one, 5-[(4-bromophenyl)methoxy]-3-[(4-bromophenyl)methyl]-2-hydroxy-	MeOH	227(4.47),241s(4.38), 250s(4.29),332(3.92), 370(3.76),387(3.75), 422(3.15)	18-1125-86
$C_{21}H_{16}ClNO_2$			
4,1-Benzoxazepin-2(3H)-one, 7-chloro-1,5-dihydro-5,5-diphenyl-	EtOH	254(4.20),287s(3.34), 294s(3.20)	44-5001-86
$C_{21}H_{16}ClN_2O$			
Pyridinium, 4-[5-(4-chlorophenyl)-2-oxazolyl]-1-(phenylmethyl)-, 4-methylbenzenesulfonate	EtOH	255(4.18),385(4.34)	135-0244-86
	5% HOAc	378(4.40)	135-0244-86
$C_{21}H_{16}N_2$			
1H-Benzo[f]cyclopenta[c]quinoline, 2,3-dihydro-4-(4-pyridinyl)-	EtOH	269(4.53),293(4.19), 344(3.49),358(3.51)	103-0759-86
$C_{21}H_{16}N_2O$			
Methanone, phenyl[3-(phenylamino)-1H-indol-2-yl]-	MeCN	258(4.37),326(4.20), 426(3.99)	24-2289-86
$C_{21}H_{16}N_2O_2$			
9,10-Anthracenedione, 1-(methylamino)-4-(phenylamino)-	EtOH	601(4.13),644(4.16)	104-0547-86
2-Butene-1,4-dione, 2-(1H-1,2-diazepin-1-yl)-1,4-diphenyl-, cis	EtOH	259(4.10),347(4.00)	39-0595-86C
trans	EtOH	256(4.35),362(4.32)	39-0595-86C
Spiro[9H-fluorene-9,3'(3'aH)-pyrazolo[1,5-a]pyridine]-2'-carboxylic acid, methyl ester	CH_2Cl_2	401(3.95)	89-0572-86
$C_{21}H_{16}N_2O_2S$			
Benzonitrile, 4-[3-(1,2-dihydro-1-phenyl-2-thioxo-3-pyridinyl)-3-hydroxy-1-oxopropyl]-	EtOH	247(4.42),289(4.04), 373(3.78)	4-0567-86
$C_{21}H_{16}N_2O_5$			
1H-Pyrazol-5-ol, 5-(1,3-benzodioxol-5-yl)-1-(2-furanylcarbonyl)-4,5-dihydro-3-phenyl-	EtOH	257(4.21),297(4.42), 304(4.41)	4-0727-86
	hexane	257(4.34),295(4.41), 306(4.90)	4-0727-86
$C_{21}H_{16}N_4O_5S$			
Furo[3',4':5,6]thiopyrano[2,3-c]pyrazole-5,7-dione, 1,4,4a,7a-tetrahydro-3-methyl-4-[(4-nitrophenyl)amino]-1-phenyl-	EtOH	330(3.9)	103-0503-86
$C_{21}H_{16}N_4O_5Se$			
Furo[3',4':5,6]selenino[2,3-c]pyrazole-5,7-dione, 1,4,4a,7a-tetrahydro-3-methyl-4-[(3-nitrophenyl)amino]-1-phenyl-	EtOH	257(3.9)	104-2126-86
$C_{21}H_{16}N_6O_2$			
Benzoic acid, 2-[5-(1H-benzimidazol-	EtOH	500(4.22)	65-0727-86

Compound	Solvent	λ_{max}(log ε)	Ref.
2-yl)-3-phenyl-1-formazano]- (cont.)			65-0727-86
nickel complex	EtOH	540(4.08)	65-0727-86
zinc complex	EtOH	630(4.34)	65-0727-86
$C_{21}H_{16}O$			
2-Propyn-1-one, 3-(5-methyl-5H-benzo-cyclohepten-5-yl)-1-phenyl-	EtOH	202(4.45),262(4.24)	39-1479-86C
2-Propyn-1-one, 3-(5-methyl-5H-benzo-cyclohepten-9-yl)-1-phenyl-	EtOH	202(4.36),266(4.11), 334(3.96)	39-1479-86C
2-Propyn-1-one, 3-(9-methyl-5H-benzo-cyclohepten-5-yl)-1-phenyl-	EtOH	203(4.42),261(4.22)	39-1479-86C
$C_{21}H_{16}O_2$			
Naphth[2,3-a]azulen-5(12H)-one, 12-hy-droxy-12-(2-propenyl)-	$CHCl_3$	240(4.16),275(4.37), 313s(4.37),325(4.56), 387(3.13),406(4.20), 485s(2.76),518(2.85), 556(2.74),610(2.03)	118-0686-86
$C_{21}H_{16}O_3$			
2H-Naphtho[1,2-b]pyran-2-one, 4-(phen-ylmethoxy)methyl]-	EtOH	265(4.45),275(4.54), 291.5(3.82),303.5(3.92), 316(3.90),351.5(3.74)	94-0390-86
$C_{21}H_{16}O_3S$			
9(10H)-Anthracenone, 1,8-dihydroxy-10-[(phenylmethyl)thio]-	EtOH	375(3.92)	118-0430-86
$C_{21}H_{16}O_5$			
2,5-Cyclohexadiene-1,4-dione, 2,5-di-hydroxy-3-(4-methoxyphenyl)-6-(2-phenylethenyl)-	dioxan	308(4.49),385(3.75), 505(3.13)	5-0195-86
2H-Pyran-2,5(6H)-dione, 3-hydroxy-4-(4-methoxyphenyl)-6-(3-phenyl-2-propenylidene)-	MeOH	282(4.55),306(4.13), 384(4.31)	5-0177-86
$C_{21}H_{16}O_5S$			
2H-Naphtho[1,2-b]pyran-2-one, 4-[[[(4-methylphenyl)sulfonyl]-oxy]methyl]-	EtOH	265.5(4.38),275.5(4.44), 292.5(3.81),304.5(3.88), 317(3.85),355(3.68)	94-0390-86
$C_{21}H_{16}O_6$			
4H,8H-Benzo[1,2-b:3,4-b']dipyran-4-one, 3-(1,3-benzodioxol-5-yl)-5-hydroxy-8,8-dimethyl-	MeOH	270(1.40)	102-2425-86
5H-Furo[3,2-g][1]benzopyran-5-one, 2-benzoyl-4,9-dimethoxy-7-methyl-	EtOH	244(4.38),264s(4.29), 292(4.45),350(4.06)	44-3116-86
$C_{21}H_{16}S_2$			
1,3-Dithiane, 2-(1-pyrenylmethylene)-	MeOH	236(4.71),267(4.29), 277(4.38),363(4.44)	35-4498-86
$C_{21}H_{17}BrN_6O_5$			
Benzamide, N-[2-[(2-bromo-4,6-dinitro-phenyl)azo]-5-(dimethylamino)phenyl]-	$CHCl_3$	310s(3.99),415(3.72), 555(4.61)	48-0497-86
$C_{21}H_{17}BrO$			
Methanone, (3-bromo-1,1a,4,8b-tetra-hydro-8b-methyl-1,4-methanobenzo-[a]cyclopropa[c]cyclohepten-9-yl)-phenyl-	EtOH	202(4.46),239(4.03)	39-1479-86C

Compound	Solvent	λ_{max}(log ε)	Ref.
2-Propen-1-one, 3-(8-bromo-5-methyl-5H-benzocyclohepten-5-yl)-1-phenyl-, (E)-	EtOH	203(4.43),259(4.20)	39-1479-86C
$C_{21}H_{17}ClN_2$			
1H-Pyrazole, 3-(4-chlorophenyl)-4,5-dihydro-1,5-diphenyl-	n.s.g.	246(4.29)	78-3807-86
1H-Pyrazole, 5-(4-chlorophenyl)-4,5-dihydro-1,3-diphenyl-	n.s.g.	225(4.32),354(4.27)	78-3807-86
$C_{21}H_{17}ClN_2O_2$			
Benzoic acid, 4-chloro-, (2-hydroxy-1,2-diphenylethylidene)hydrazide, (E)-	MeOH	260(4.23)	48-0181-86
(Z)-	MeOH	295(4.30)	48-0181-86
$C_{21}H_{17}Cl_2NO_2$			
Acetamide, 2-chloro-N-[4-chloro-2-(hydroxydiphenylmethyl)phenyl]-	EtOH	252(4.16),289s(3.02)	44-5001-86
$C_{21}H_{17}FO$			
Benzene, 1-fluoro-4-[1-(4-methoxyphenyl)-2-phenylethenyl]-	EtOH	230(4.41),244s(4.33), 303(4.40)	13-0073-86B
$C_{21}H_{17}F_6NO_6S$			
9-Azatricyclo[4.2.1.0^{2,5}]nona-3,7-diene-3,4-dicarboxylic acid, 9-[(4-methylphenyl)sulfonyl]-7,8-bis(trifluoromethyl)-, dimethyl ester, (1α,2β,5β,6α)-	EtOH	224(4.25),254s(3.57), 268s(3.23),273s(3.08)	24-0616-86
9-Azatricyclo[4.2.1.0^{2,5}]nona-3,7-diene-7,8-dicarboxylic acid, 9-[(4-methylphenyl)sulfonyl]-3,4-bis(trifluoromethyl)-, dimethyl ester, (1α,2β,5β,6α)-	EtOH	226(4.17),253s(3.52), 262s(3.41),273s(3.22)	24-0616-86
$C_{21}H_{17}N$			
4,9[1',2']-Benzenobenzo[a]cyclobuta[e]cyclooctene-4(2H)-carbonitrile, 1,3,9,10-tetrahydro-	EtOH	263(2.8),270(2.8)	88-5621-86
$C_{21}H_{17}NO$			
Benzonitrile, 4-[3-oxo-4-(3-phenyl-2-propenyl]-1-cyclopenten-1-yl]-	EtOH	245(4.20),288(4.41)	39-1235-86C
$C_{21}H_{17}NOS$			
1,5-Benzothiazepin-4(5H)-one, 2,3-dihydro-2,3-diphenyl-, trans	CH_2Cl_2	244(4.11),278(3.70)	24-1525-86
$C_{21}H_{17}NO_2$			
2H-Naphtho[1,2-b]pyran-2-one, 4-(dimethylamino)-3-phenyl-	EtOH	218(4.60),270s(4.45), 275(4.48),298(4.07), 306(4.06),347(4.04)	4-1067-86
2-Propen-1-one, 1-(2-hydroxyphenyl)-3-phenyl-3-(phenylamino)-, (Z)-	EtOH	260(4.01),390(4.40)	104-1658-86
2-Propen-1-one, 1-(4-hydroxyphenyl)-3-phenyl-3-(phenylamino)-	EtOH	295(4.00),357(4.56)	104-1658-86
$C_{21}H_{17}NO_3$			
1-Naphthalenol, 4-[3-(phenylimino)-1-propenyl]-, acetate, N-oxide	dioxan	385(4.33)	104-1302-86

Compound	Solvent	λmax(log ε)	Ref.
2-Naphthalenol, 1-[3-(phenylimino)-1-propenyl]-, acetate, N-oxide	CCl_4	385(4.18)	104-1302-86
	DMSO	385(4.50)	104-1302-86
$C_{21}H_{17}NO_4$			
Acetic acid, (5-methyl-12-oxo-5H,10H-4b,9b-(epoxyethano)indeno[1,2-b]indol-11-ylidene)-, methyl ester	MeOH	237(4.01),274(3.66)	150-0514-86M
1(2H)-Anthracenone, 3,4-dihydro-5,9,10-trihydroxy-4-[(phenylmethyl)imino]-	$CHCl_3$	258(4.44),278(4.42), 400(3.72),420(4.03), 444(4.37),474(4.51)	78-3303-86
$C_{21}H_{17}NO_5$			
Oxychelerythrine	MeOH	240(4.60),280(4.65), 289(4.74),323(4.19), 341(4.16)	39-2253-86C
$C_{21}H_{17}NO_6$			
Enterocarpam-II diacetate	MeOH	220(4.66),251(4.66), 279(4.38),289(4.39), 320(3.92),375(3.70), 393(3.79)	102-0965-86
$C_{21}H_{17}N_2O$			
Pyridinium, 1-(phenylmethyl)-4-(5-phenyl-2-oxazolyl)-, salt with 4-methylbenzenesulfonic acid	H_2O	380(4.30)	135-0244-86
	EtOH	250(4.21),385(4.36)	135-0244-86
$C_{21}H_{17}N_2O_3$			
Xanthylium, 3,6-diamino-9-[2-(methoxycarbonyl)phenyl]-, chloride	neutral	499.7(4.88)	156-0139-86A
$C_{21}H_{17}N_3O$			
Benzaldehyde, 4-methoxy-, (3-methyl-9H-indeno[2,1-c]pyridin-9-ylidene)-hydrazone	EtOH	213(5.17),248(4.54), 367(4.52)	103-0200-86
	EtOH-H_2SO_4	223(4.59),237(4.35), 273(4.59),295s(4.43), 330(4.20),403(4.31)	103-0200-86
$C_{21}H_{17}N_3OS_2$			
Benzenecarbothioic acid, 4-[(1H-perimidin-2-ylthioxomethyl)amino]-, S-ethyl ester	DMF	348s(4.34),357(4.36), 470s(3.04),536(3.07)	48-0812-86
	C_6H_{12}	231s(--),236(--), 272(--),290(--), 302s(--),341(--), 356(--),391s(--), 396s(--),439s(--), 473s(--),503(--), 542(--),586(--), 641(--)	48-0812-86
$C_{21}H_{17}N_3O_2$			
Benzaldehyde, 3,4-dimethoxy-, 5H-indeno[1,2-b]pyridin-5-ylidenehydrazone	EtOH	214(4.56),228(4.60), 252(4.47),267(4.38), 287(4.29),313(4.19), 387(4.43)	103-0200-86
	EtOH-H_2SO_4	212(4.66),230(4.68), 277(4.44),324(4.54), 395(4.36)	103-0200-86
Benzaldehyde, 4-hydroxy-3-methoxy-, (3-methyl-9H-indeno[2,1-c]pyridin-9-ylidene)hydrazone	EtOH	216(4.58),226(4.55), 260(4.64),294(3.04), 390(4.43)	103-0200-86

Compound	Solvent	λ_{max}(log ϵ)	Ref.
(cont.)	EtOH-H_2SO_4	212(4.21),233(4.49), 270(4.43),298(4.26), 312(4.25),364(3.61), 424(4.35)	103-0200-86
$C_{21}H_{17}N_3O_2S$			
Benzoic acid, 4-[(1H-perimidin-2-yl-thioxomethyl)amino]-, ethyl ester	DMF	345(4.31),356(4.31), 463s(3.01),525(3.05)	48-0812-86
	C_6H_{12}	231s(--),236(--), 261(--),289(--), 341(--),356s(--), 387s(--),435s(--), 471s(--),500(--), 539(--),582(--), 633(--)	48-0812-86
$C_{21}H_{17}N_3O_3S$			
Furo[3',4':5,6]thiopyrano[2,3-c]pyrazole-5,7-dione, 1,4,4a,7a-tetrahydro-3-methyl-1-phenyl-4-(phenylamino)-	toluene	325(3.8)	103-0503-86
$C_{21}H_{17}N_3O_3Se$			
Furo[3',4':5,6]selenino[2,3-c]pyrazole-5,7-dione, 1,4,4a,7a-tetrahydro-3-methyl-1-phenyl-4-(phenylamino)-	EtOH	256(3.9)	104-2126-86
$C_{21}H_{17}O$			
Dibenzo[c,h]xanthylium, 5,6,8,9-tetrahydro-, perchlorate	MeCN	228(4.20),290(4.11), 445(4.45)	22-0600-86
$C_{21}H_{18}$			
Tricyclo[3.3.1.0^{2,8}]nona-3,6-diene,	hexane	218s(4.30),271(4.24)	78-1805-86
2,6-diphenyl-	decalin	273(4.21)	78-1805-86
at 350° K	decalin	273(4.19)	78-1805-86
at 400° K	decalin	272(4.19)	78-1805-86
at 450° K	decalin	270(4.21)	78-1805-86
$C_{21}H_{18}BNO_6$			
Boron, [5-[4-(dimethylamino)phenyl]-1-phenyl-4-pentene-1,3-dionato-O,O'](ethanedioato(2-)-O,O')-	CH_2Cl_2	575(4.91)	97-0399-86
$C_{21}H_{18}Br_2O_{12}$			
Propanedioic acid, bis[[3-bromo-6-(methoxycarbonyl)-2-oxo-2H-pyran-4-yl]methyl]-, dimethyl ester	MeOH	217(4.38),314(4.34)	5-1968-86
$C_{21}H_{18}ClNO_2S$			
2H-[1]Benzothiepino[5,4-b]pyran-2-one, 3-chloro-5,6-dihydro-4-[methyl(phenylmethyl)amino]-	EtOH	235s(3.98),255(4.02), 265s(3.98),323(4.04)	4-0449-86
1-Propanone, 1-(4-chlorophenyl)-3-(1,2-dihydro-1-phenyl-2-thioxo-3-pyridinyl)-3-hydroxy-2-methyl-	EtOH	249(4.25),289(3.90), 374(3.73)	4-0567-86
$C_{21}H_{18}Cl_2O_5Si$			
4a,9a-(2-Buteno)anthracene-1,4,9,10-tetrone, 2,3-dichloro-11-[(trimethylsilyl)oxy]-	EtOH	228(4.27),262(3.97), 314s(3.10)	39-1923-86C

Compound	Solvent	λ$_{max}$(log ε)	Ref.
$C_{21}H_{18}N_2$			
1H-Benzo[f]cyclopenta[c]quinoline, 2,3,4,5-tetrahydro-4-(4-pyridinyl)-	EtOH	257(4.64),303(3.87), 315(3.78),370(3.63)	103-0759-86
1H-Pyrrolizine-5-acetonitrile, 2,3-dihydro-6,7-diphenyl-	MeOH	206(4.48),239(4.26), 267s(--)	83-0354-86
$C_{21}H_{18}N_2O_2$			
Benzoic acid, (2-hydroxy-1,2-diphenylethylidene)hydrazide, (E)-	MeOH	262(4.15)	48-0181-86
(Z)-	MeOH	295(4.26)	48-0181-86
1-Naphthalenecarboxylic acid, (1-methyl-3-oxo-3-phenylpropylidene)hydrazide	DMF	284s(3.86),331(4.26) (changing)	90-0983-86
after 24 hours	DMF	282(3.98),292(3.96), 306(3.86)	90-0983-86
$C_{21}H_{18}N_2O_2S$			
Methanesulfonamide, N-[4-(9-acridinyl)phenyl]-N-methyl-	MeOH	206(4.56),254(5.00), 358(4.23)	150-3425-86M
$C_{21}H_{18}N_2O_3$			
2,4-Cyclohexadien-1-one, 4-nitro-6-[(5,7,8,9-tetrahydro-5-methyl-10-phenanthridinyl)methylene]-	EtOH	245(4.60),300(4.22), 330(4.27),370(4.24), 405(4.33),495(4.11)	135-1208-86
1H-Pyrazol-5-ol, 1-(2-furanylcarbonyl)-4,5-dihydro-5-(4-methylphenyl)-3-phenyl-	hexane	256(4.30),306(4.38)	4-0727-86
	EtOH	257(4.18),306(4.25)	4-0727-86
$C_{21}H_{18}N_2O_4$			
1H-Indole-2,3-dione, 1,1'-(1,3-propanediyl)bis[5-methyl-	EtOH	219(4.33),249(4.56), 256s(4.47),302(3.61)	12-1231-86
$C_{21}H_{18}N_2O_5$			
Benzeneacetic acid, α-acetoxy-3-[(2-hydroxy-1-naphthalenyl)azo]-, methyl ester	EtOH	207(4.55),228(4.56), 306(3.89),430s(4.06), 479(4.21)	5-1956-86
$C_{21}H_{18}N_2O_6$			
1H-Indole-2,3-dione, 1,1'-(1,3-propanediyl)bis[5-methoxy-	EtOH	219(4.26),256(4.48), 303(3.43)	12-1231-86
$C_{21}H_{18}N_2O_8$			
2-Propenoic acid, 3-(3-nitrophenyl)-, 1,3-propanediyl ester, (E,E)-	MeCN	261(4.78)	18-1379-86
2-Propenoic acid, 3-(4-nitrophenyl)-, 1,3-propanediyl ester, (E,E)-	MeCN	301(4.64)	18-1379-86
$C_{21}H_{18}N_2S$			
2H-Benz[g]indazole, 3,3a,4,5-tetrahydro-2-phenyl-3-(2-thienyl)-	EtOH	224(4.19),339(4.25)	4-0135-86
2-Benzothiazolamine, N-(1,2-diphenylethyl)-	MeCN	269(4.20),285s(3.96), 295s(3.64)	24-1525-86
$C_{21}H_{18}N_4O_3$			
Spiro[1,3-dioxolane-2,4'(1'H)-pteridin]-2'(3'H)-one, 3'-methyl-6',7'-diphenyl-	MeOH	227(4.45),276(4.11), 338(4.16)	128-0199-86
	pH 14	302(4.22),366(4.08)	128-0199-86
$C_{21}H_{18}N_4O_4$			
1,2-Ethanediol, 1-(1-phenyl-1H-pyra-	MeOH	233s(4.4),267(4.7),	159-0049-86

Compound	Solvent	λ$_{max}$(log ε)	Ref.
zolo[3,4-b]quinoxalin-3-yl)-, D-, diacetate (cont.)		334(4.1),408(3.6)	159-0049-86
L-	MeOH	233s(4.3),267(4.6), 334(4.0),408(3.5)	159-0049-86
$C_{21}H_{18}N_4S_2$			
1H-Perimidine-2-carbothioamide, N-[4-[(dimethylamino)thioxomethyl]-phenyl]-	DMF	342(4.27),356(4.30), 500s(3.06)	48-0812-86
$C_{21}H_{18}O$			
Benz[a]anthracene, 3-methoxy-7,12-dimethyl-	THF	226(4.32),243(4.31), 292(4.76),297(4.78), 367(3.85)	44-3502-86
9H-Fluoren-9-ol, 2-methyl-9-(phenylmethyl)-	$CHCl_3$	262(3.93),270(4.18), 283(4.23),291(4.18), 300s(3.85),312s(3.60)	104-1352-86
9H-Fluoren-9-ol, 3-methyl-9-(phenylmethyl)-	$CHCl_3$	276(4.01),282(4.05), 288(4.02),300(3.74), 313(3.64)	104-1352-86
Naphthacene, 2-methoxy-6,11-dimethyl-	THF	288(5.14),290(5.14)	44-3502-86
1,3,6,8-Nonatetraen-5-one, 1,9-diphenyl-	benzene	369(4.72)	60-1307-86
$C_{21}H_{18}O_2$			
2-Cyclopenten-1-one, 3-(4-methoxyphenyl)-5-(3-phenyl-2-propynyl)-	EtOH	250(4.22),315(4.33)	39-1235-86C
$C_{21}H_{18}O_3$			
2,4,6-Cycloheptatrien-1-one, 2,5-bis(phenylmethoxy)-	MeOH	227(4.12),253s(3.72), 327(3.80),332(3.80), 344(3.76),360s(3.69), 370s(3.53)	18-1125-86
2,4,6-Cycloheptatrien-1-one, 2-hydroxy-5-(phenylmethoxy)-3-(phenylmethyl)-	MeOH	237(4.35),250s(4.27), 332(3.93),370(3.79), 387(3.77)	18-1125-86
Ethanone, 1-[3-(diphenylmethyl)-2,4-dihydroxyphenyl]-	MeOH	210(4.58),280(4.24)	2-0755-86
Ethanone, 1-[5-(diphenylmethyl)-2,4-dihydroxyphenyl]-	MeOH	202(4.67),276(4.21)	2-0755-86
$C_{21}H_{18}O_5S$			
Cresol Red	pH 4-5	434(3.7)	110-0205-86
with cetylpyridinium bromide	5% H_2SO_4	520(4.4)	110-0205-86
$C_{21}H_{18}O_6$			
Dispiro[furan-3,5'-benzo[g]isobenzofuran-7',3"-furan]-1'(3'H),2(3H)-2"(3"H)-trione, 4,4',4",5,5',5"-6',7',8',9'-decahydro-4,4",8'-tris(methylene)-	EtOH	264s(3.49)	33-1734-86
5,12-Naphthacenedione, 2-acetyl-1,2,3,4-tetrahydro-2,11-dihydroxy-7-methoxy-	MeOH	242.5(4.74),275s(4.21), 284s(4.15),324.5(3.57), 342s(3.40),480.5(3.96)	35-3088-86
2H-Pyran-2,5(6H)-dione, 3-methoxy-4-(4-methoxyphenyl)-6-[(4-methoxyphenyl)methylene]-	MeOH	228(4.28),331(4.37), 390(3.58)	5-0177-86
$C_{21}H_{18}O_7$			
4H-1-Benzopyran-6-carboxaldehyde, 5,8-dimethoxy-2-methyl-4-oxo-7-(2-oxo-2-	EtOH	252(4.56),290s(3.86), 305s(3.73)	44-3116-86

Compound	Solvent	λ_{max}(log ε)	Ref.
phenylethoxy)- (cont.)			44-3116-86
2,5-Cyclohexadiene-1,4-dione, 2-(3,4-dimethoxyphenyl)-3,6-dihydroxy-5-(4-methoxyphenyl)-	dioxan	267.6(4.50),388.2(3.43)	5-0195-86
Daunomycinone, 4-demethoxy-7-O-methyl-, (±)-	n.s.g.	251(4.32),287(3.78), 327s(3.18),465s(3.75), 486(3.78),520s(3.60)	12-0487-86
Epidaunomycinone, 4-demethoxy-7-O-methyl-, (±)-	n.s.g.	252(3.68),258(3.68), 284(3.08),330(2.60), 485(3.08),518s(2.85)	12-0487-86
5,12-Naphthacenedione, 8-acetyl-7,8,9,10-tetrahydro-6,8,11-trihydroxy-1-methoxy-, (8R)-	MeOH	235.0(4.50),252.5(4.48), 289.0(3.90),493(4.08), 528.0(3.86)	35-3088-86
2H-Pyran-2,5(6H)-dione, 6-[(3,4-dimethoxyphenyl)methylene]-3-hydroxy-4-(4-methoxyphenyl)-	MeOH	273(4.19),395(4.11)	5-0177-86
$C_{21}H_{18}O_8$			
2-Anthraceneacetic acid, 9,10-dihydro-1,2,15-trihydroxy-9,10-dioxo-3-(3-oxobutyl)-, methyl ester	n.s.g.	207(4.21),235(4.51), 252(4.41),293(3.94), 472(3.99),485s(--), 493(4.05),510s(--), 528(3.88),568s(--)	5-1506-86
2-Anthraceneacetic acid, 3-(1,3-dioxolan-2-ylmethyl)-9,10-dihydro-1,4-dihydroxy-9,10-dioxo-, methyl ester	n.s.g.	228s(4.24),257(4.66), 292(4.02),489(4.06), 521s(3.85)	12-0821-86
1-Naphthacenecarboxylic acid, 1,2,3,4,6,11-hexahydro-2,5,7,12-tetrahydroxy-2-methyl-6,11-dioxo-, methyl ester	n.s.g.	218(4.50),235(4.76), 253(4.72),293(4.12), 463(4.27),470(4.30), 489(4.40),510(4.26), 525(4.28),573(3.60)	5-1506-86
Protosappanin A, triacetate	MeOH	248(4.01),281(3.52)	94-0001-86
$C_{21}H_{18}O_9$			
δ-Rhodomycinone, (±)-	n.s.g.	210(4.19),234(4.62), 253(4.42),291(3.94), 463(4.08),479(4.14), 491(4.18),508(4.06), 526(4.00)	5-1506-86
$C_{21}H_{19}ClO_5Si$			
4a,9a-(But-2-eno)anthracene-1,4,9,10-tetrone, 2-chloro-4a,9a-dihydro-11-(trimethylsilyloxy)-	EtOH	228(4.20),253s(3.92), 312(3.04)	39-1923-86C
$C_{21}H_{19}NO$			
Azacycloeicosa-3,5,7,13,15,17,19-heptaene-9,11-diyn-2-one, 8,13-dimethyl-, (Z,E,Z,Z,E,E,E)-	THF	244(4.20),298s(4.64), 328(4.82),346s(4.68), 446(3.37)	39-0933-86C
4-Benzazacyclotetradecine, 9,10,11,12-tetradehydro-3-ethoxy-2,8-dimethyl-, (E,E,E,Z)-	THF	260s(4.25),274(4.33), 288(4.34),364s(3.67), 387(3.73),418s(3.60)	18-1791-86
4-Benzazacyclotetradecine, 9,10,11,12-tetradehydro-3-ethoxy-5,8-dimethyl-, (E,E,E,Z)-	THF	228(4.47),245s(4.38), 271s(4.49),281(4.53), 322s(3.99),386(3.58)	18-1791-86
2,4,6,8,14,16,18-Cyclononadecaheptaene-10,12-diyn-1-one, 9,14-dimethyl-, oxime	THF	300s(4.51),330(4.70), 440s(3.52)	39-0933-86C
1H-Pyrrolizine-5-acetaldehyde, 2,3-dihydro-6,7-diphenyl-	MeOH	207(4.47),243(4.24), 274(4.08)	83-0500-86

Compound	Solvent	λ_{max}(log ε)	Ref.
$C_{21}H_{19}NO_2$			
1-Naphthalenol, 4-methyl-2-[3-[(4-methylphenyl)imino]-1-propenyl]-, N-oxide	dioxan	295(4.13),320(3.92), 410(4.31)	104-1302-86
	acetone	335(3.95),370s(3.98), 400(4.02)	104-1302-86
	DMSO	325(4.02),420(4.32)	104-1302-86
	+ NaOMe	315(3.93),355(4.04), 375(3.98),560(4.37)	104-1302-86
$C_{21}H_{19}NO_2S$			
1-Propanone, 3-(1,2-dihydro-1-phenyl-2-thioxo-3-pyridinyl)-3-hydroxy-2-methyl-1-phenyl-, (R*,R*)-(±)-	EtOH	242(4.21),289(3.98), 374(3.78)	4-0567-86
$C_{21}H_{19}NO_3Se_2$			
Benzene, 1,1'-[[(4-nitrophenoxy)methylene]bis(seleno)]bis[4-methyl-	hexane	213(4.38),227(4.37), 300(3.93)	104-0092-86
$C_{21}H_{19}NO_4$			
Acetic acid, (8,8a-dihydro-3a,8-dimethyl-2-oxo-8a-phenyl-2H-furo[2,3-b]indol-3(3aH)-ylidene)-, methyl ester, (E)-	MeOH	231(4.23),285(3.40), 340(3.00)	150-0514-86M
(Z)-	MeOH	233(4.13),288(3.30), 326(3.18)	150-0514-86M
Chelerythrine, 12,13-dihydro-	MeOH	228(4.53),283(4.67), 319(4.19),349(3.52)	39-2253-86C
$C_{21}H_{19}NO_5$			
2H-Pyran-2,5(6H)-dione, 6-[[4-(dimethylamino)phenyl]methylene]-3-hydroxy-4-(4-methoxyphenyl)-	MeOH	256(4.20),320(4.17), 417(4.15)	5-0177-86
$C_{21}H_{19}NO_6$			
Azuleno[2,1-c]pyridine-5,10-dione, 6,8-diacetoxy-3-methyl-1-propyl-	EtOH	212(4.16),250(4.55), 316(4.03),362(3.92)	88-6049-86
4,5-Benzofurandicarboxylic acid, 6-(acetylamino)-2-(4-methylphenyl)-, dimethyl ester	MeOH	260(3.23),303(3.07), 436(3.20)	73-1455-86
$C_{21}H_{19}N_3O_3$			
2-Propenoic acid, 2-cyano-3-[4-(2-cyano-2-phenylethenyl)-5-(dimethylamino)-2-furanyl]-, ethyl ester	MeOH	260(3.03),378(3.04), 464(3.67)	73-0879-86
$C_{21}H_{19}N_3O_3S$			
Acetamide, N-[1-cyano-2,3,9,9a-tetrahydro-9-(phenylsulfonyl)-1H-carbazol-4-yl]-	MeOH	218(4.35),267(4.14), 272s(4.12),312(3.89)	5-2065-86
$C_{21}H_{19}N_5O$			
2(1H)-Pteridinone, 3,4-dihydro-1,3-dimethyl-4-(methylimino)-6,7-diphenyl-	pH 1	226(4.45),282(4.27), 386(4.20)	128-0199-86
	pH 7	228(4.48),292(4.15), 370(4.17)	128-0199-86
	MeOH	228(4.47),292(4.14), 370(4.16)	128-0199-86
2(1H)-Pteridinone, 4-(ethylimino)-3,4-dihydro-3-methyl-6,7-diphenyl-	MeOH	226(4.48),284(4.15), 366(4.13)	128-0199-86

Compound	Solvent	λ_{max}(log ε)	Ref.
$C_{21}H_{19}N_5O_4$			
Glycine, N-[4-[[(2-amino-1,4-dihydro-4-oxo-6-quinazolinyl)methyl]-2-propynylamino]benzoyl]-	pH 13	228(4.66),281s(4.36), 300(4.39)	87-1114-86
$C_{21}H_{19}N_5O_5$			
Acetic acid, [[5-(acetoxymethyl)-1-(4-nitrophenyl)-1H-pyrazol-3-yl]methylene]phenylhydrazide	MeOH	276(4.35),313(4.32)	136-0316-86G
2-Pentenedioic acid, 3-amino-4-[(2-amino-3-cyano-4-quinolinyl)carbonyl]-2-cyano-, diethyl ester	EtOH	235(5.30),285(5.32), 360(4.88)	18-1235-86
$C_{21}H_{19}N_7O$			
Acetamide, N-[5-(diethylamino)-2-[(2,4,6-tricyanophenyl)azo]phenyl]-	$CHCl_3$	299(3.76),317(3.74), 333s(3.22),417(3.59), 572s(4.73),604(4.87)	48-0497-86
$C_{21}H_{20}$			
Bicyclo[3.3.1]nona-2,6-diene, 2,6-diphenyl-	hexane	200(4.61),220s(4.21), 246(4.34),283s(3.95)	78-1805-86
Pyrene, 1-pentyl-	THF	265(4.5),275(4.8), 328(4.5),342(4.7)	116-2390-86
$C_{21}H_{20}Cu_2N_2O_4$			
Copper(2+), [μ-[[3,3'-propane-1,3-diylbis(nitrilomethylidyne)]bis[2-hydroxy-5-methylbenzaldehydato]](2-)-$N^3,N^{3'},O^2,O^{2'}:O^1,O^{1'},O^2,O^{2'}$]]di-, diperchlorate	MeOH	210(4.8),255(4.8), 377(4.0),717(2.0)	126-0763-86
$C_{21}H_{20}N_2O$			
Acetaldehyde, (2,6-diphenyl-4H-pyran-4-ylidene)-, dimethylhydrazone	n.s.g.	406(4.53)	44-4477-86
Diazene, (methoxydiphenylmethyl)(4-methylphenyl)-	CH_2Cl_2	256.5(4.39),277s(4.26), 416.0(2.31)	107-0741-86
1H-Indole, 2,3-dihydro-2-(1H-indol-3-yl)-1-(3-methyl-1-oxo-2-butenyl)-	MeOH	220(4.68),268(4.05), 273(4.04),285(3.94)	44-2343-86
$C_{21}H_{20}N_2O_2S$			
Acridinium, 9-(N-methyl-4-methanesulfonamidophenyl)-10-methyl-, picrate	MeOH	207(4.68),262(4.98), 362(4.53)	150-3425-86M
$C_{21}H_{20}N_2O_4$			
1H-Pyrazole-1-carboxylic acid, 5-(1,3-benzodioxol-5-yl)-3-phenyl-, 1,1-dimethylethyl ester	EtOH	268(4.51)	4-0727-86
$C_{21}H_{20}N_2O_4S$			
Acetamide, N-[1-formyl-2,3,9,9a-tetrahydro-9-(phenylsulfonyl)-1H-carbazol-4-yl]-, trans	MeOH	218(4.31),267(4.10), 272s(4.08),311(3.89)	5-2065-86
$C_{21}H_{20}N_2O_5$			
2-Anthracenecarboxylic acid, 9,10-dihydro-1-(4-morpholinylamino)-9,10-dioxo-, ethyl ester	n.s.g.	273(4.07),313(3.79), 485(3.79)	104-2117-86
$C_{21}H_{20}N_2O_6$			
Butanedioic acid, (nitromethylene)-	EtOH	327(3.79)	18-3871-86

Compound	Solvent	$\lambda_{max}(\log \epsilon)$	Ref.
[phenyl[(phenylmethyl)amino]methylene]-, dimethyl ester (cont.)			18-3871-86
$C_{21}H_{20}N_4O_3$			
2(1H)-Pteridinone, 3,4-dihydro-4,4-dimethoxy-3-methyl-6,7-diphenyl-	pH 8	228(4.41),268(4.14), 340(4.17)	128-0199-86
	pH 14	240s(4.28),291(4.30), 370(4.07)	128-0199-86
	MeOH	228(4.43),275(4.14), 340(4.17)	128-0199-86
$C_{21}H_{20}N_6O$			
2,4(1H,3H)-Pteridinedione, 1,3-dimethyl-6,7-diphenyl-, 4-(methylhydrazone)	pH 0.7	226(4.43),286(4.26), 376(4.18)	128-0199-86
	pH 6.8	229(4.45),246s(4.36), 323(3.87),418(3.94)	128-0199-86
	MeOH	228(4.52),247s(4.44), 323(3.91),426(4.02)	128-0199-86
$C_{21}H_{20}O$			
Azulene, 6-ethoxy-2-[2-(4-methylphenyl)ethenyl]-	CH_2Cl_2	324.5(4.89),414.5(4.53), 438(4.46),525.5(2.70)	24-2272-86
$C_{21}H_{20}O_2$			
2-Cyclopenten-1-one, 3-(4-methoxyphenyl)-5-(3-phenyl-2-propenyl)-, (E)-	EtOH	248(4.18),316(4.32)	39-1235-86C
(Z)-	EtOH	234(4.15),316(4.40)	39-1235-86C
$C_{21}H_{20}O_3$			
2H-Furo[2,3-h]-1-benzopyran, 8,9-dihydro-5-methoxy-8-(1-methylethenyl)-2-phenyl- (abbottin)	MeOH	205(3.49),243(3.09), 294(2.87)	25-0827-86
Hildgardtene	EtOH	246(--),294(--), 320s(--)	102-1711-86
$C_{21}H_{20}O_4$			
4H,8H-Benzo[1,2-b:3,4-b']dipyran-4-one, 2,3-dihydro-3-(4-methoxyphenyl)-8,8-dimethyl-, (±)-	MeOH	267(4.23),321(4.2)	2-0481-86
$C_{21}H_{20}O_5$			
2H,5H-Pyrano[3,2-c][1]benzopyran-5-one, 3,4-dihydro-2-(hydroxymethyl-8-methoxy-2-methyl-4-phenyl-	EtOH	273s(--),284(4.02), 311(4.31),323s(--)	32-0501-86
isomer	EtOH	273s(--),284(4.03), 311(4.32),323s(--)	32-0501-86
$C_{21}H_{20}O_6$			
9,10-Anthracenedione, 1,4-dihydroxy-2-[3-(2-methyl-1,3-dioxolan-2-yl)-propyl]-	n.s.g.	209(4.16),235(4.31), 249(4.59),254(4.54), 285(4.00),311(3.56), 461(3.96),483(4.01), 498(3.89),514(3.82)	5-1506-86
5(8H)-Naphthacenone, 8-acetyl-7,9,10,12-tetrahydro-6,8,11-trihydroxy-1-methoxy-, (R)-	MeOH	237.0(4.16),269.5(4.21), 297.5(4.05),396.0(3.60)	35-3088-86
$C_{21}H_{20}O_7$			
9,10-Anthracenedione, 1,4-dihydroxy-8-methoxy-2-[2-(2-methyl-1,3-dioxolan-2-yl)ethyl]-	n.s.g.	231(4.88),249(4.66), 282(4.25),390s(--), 477(4.35),493(4.36),	5-1506-86

Compound	Solvent	$\lambda_{max}(\log \epsilon)$	Ref.
(cont.)		521(4.11),560s(--)	5-1506-86
5,12-Naphthacenedione, 2-acetyl-1,2,3,4,4a,12a-hexahydro-2,6,11-trihydroxy-7-methoxy-	MeOH	241.0(4.45),266.5(4.37), 376.5(3.63),397.5(3.94), 417.5(4.18),442(4.20)	35-3088-86
$C_{21}H_{20}O_8$			
2-Anthraceneacetic acid, 3-(2,2-dimethoxyethyl)-9,10-dihydro-1,4-dihydroxy-9,10-dioxo-, methyl ester	n.s.g.	233s(4.19),251(4.45), 287(3.90),490(3.87), 523s(3.62)	12-0821-86
Butanedioic acid, bis[(4-hydroxy-3-methoxyphenyl)methylene]-, monomethyl ester, (E,E)-	EtOH	220(4.23),245(4.26), 289(3.90),310s(4.00)	150-0735-86M
$C_{21}H_{20}O_9$			
Emodin 1-O-α-L-rhamnoside	MeOH	225(4.32),290(4.38), 420(3.96)	100-0343-86
Sophoroflavone B	MeOH	256(3.97),316(4.30)	100-0645-86
	MeOH-NaOMe	264(4.69),360(4.16)	100-0645-86
	MeOH-AlCl$_3$	256(3.97),316(4.30)	100-0645-86
	MeOH-NaOAc	266(4.36),306(4.16), 320s(4.07),359(4.14)	100-0645-86
$C_{21}H_{20}O_{10}$			
Apigenin 7-O-β-D-glucoside	MeOH	269(4.22),335(4.28)	94-3097-86
Isovitexin	MeOH	272(4.23),335(4.25)	95-0982-86
Vitexin	MeOH	271(4.20),332(4.21)	95-0982-86
$C_{21}H_{20}O_{11}$			
Astragalin	MeOH	268(4.31),352(4.24)	95-0989-86
Glucoluteolin	MeOH	253(4.43),265(4.40), 348(4.45)	94-0082-86
	MeOH-AlCl$_3$	272(--),425(--)	94-0082-86
Kaempferol 3-O-α-D-galactopyranoside	MeOH	265(4.36),350(4.26)	95-0982-86
	MeOH-NaOMe	275.5(--),326.5(--), 401(--)	95-0982-86
β-	MeOH	267(4.18),297s(3.94), 352(4.10)	95-0982-86
$C_{21}H_{20}O_{12}$			
Propanedioic acid, bis[[6-(methoxycarbonyl)-2-oxo-2H-pyran-4-yl]methyl]-, dimethyl ester	EtOH	302(4.1)	5-1990-86
$C_{21}H_{21}BrN_2O_5$			
Pyrimidine, 5-(2-bromo-4,5-dimethoxyphenyl)-4-(2,3,4-trimethoxyphenyl)-	CHCl$_3$	243(4.25),275(4.30), 292(4.29)	142-1943-86
$C_{21}H_{21}ClFN_3O_3$			
1H-1,4-Benzodiazepine-2-acetic acid, 7-chloro-5-(2-fluorophenyl)-2,3-dihydro-α-(hydroxyimino)-, 1,1-dimethylethyl ester	isoPrOH	226(4.42),244s(4.38), 372(3.48)	4-1303-86
$C_{21}H_{21}F_6N_5$			
Benzenamine, 4,4'-[3,6-bis(trifluoromethyl)-1,2,4-triazine-4,5(5H)-diyl]bis[N,N-dimethyl-	CH$_2$Cl$_2$	263(4.48),316(4.05), 354(4.07)	83-0690-86
$C_{21}H_{21}NO$			
Azacyclooctadeca-1,3,5,7,13,15,17-heptaene-9,11-diyne, 2-ethoxy-	THF	346(4.77),385s(4.02), 412s(3.86),535s(3.08),	39-0933-86C

Compound	Solvent	$\lambda_{max}(\log \epsilon)$	Ref.
8,13-dimethyl- (cont.)		587s(2.89),654s(2.38)	39-0933-86C
1(5H)-Indolizinone, 6,7,8,8a-tetrahydro-8a-methyl-2,3-diphenyl-	MeCN	227(4.07),276(4.07), 348(3.98)	118-0899-86
$C_{21}H_{21}NO_2$			
1,2-Ethanediol, 1-(2,3-dihydro-6,7-diphenyl-1H-pyrrolizin-5-yl)-	MeOH	206(4.51),242(4.31), 268(4.12)	83-0500-86
3-Pyridinecarboxylic acid, 5-(diphenylmethylene)-1,2,5,6-tetrahydro-1-methyl-, methyl ester	pH 7.2	208(0.92),246(0.63), 320(0.98)	87-0125-86
$C_{21}H_{21}NO_3$			
Benzenamine, 4-methoxy-N,N-bis(4-methoxyphenyl)-, radical ion(1+)-, tetrafluoroborate	MeCN	375(4.18),720(4.46)	1-0210-86
5H,9H-Methanobenzo[5,6]cycloocta[1,2-c]quinoline-6,8,15-trione, 10,11,12-12a.13.14-hexahydro-5-methyl-	n.s.g.	210(4.20),234(4.26), 363(3.56)	150-2782-86
$C_{21}H_{21}NO_3S$			
4(1H)-Pyridinone, 5-acetyl-2,3-dihydro-1-(4-methoxyphenyl)-6-(methylthio)-2-phenyl-	EtOH	342(3.11)	5-1109-86
$C_{21}H_{21}NO_4S$			
3-Pyridinecarboxylic acid, 1,4,5,6-tetrahydro-1-(4-methoxyphenyl)-2-(methylthio)-4-oxo-6-phenyl-, methyl ester	EtOH	335(3.10)	5-1109-86
$C_{21}H_{21}NO_4S_2$			
Sulfoximine, S-(2-methoxyphenyl)-S-(4-methylphenyl)-N-[(4-methylphenyl)-sulfonyl]-, (S)-	EtOH	244(4.21),295(3.62)	12-1833-86
$C_{21}H_{21}NO_6$			
1(2H)-Isoquinolinone, 4-[6-(2-hydroxyethyl)-1,3-benzodioxol-5-yl]-7,8-dimethoxy-2-methyl-	MeOH	289(4.28),299(4.33), 354(3.90)	39-2253-86C
$C_{21}H_{21}NO_7S$			
7-Azabicyclo[2.2.1]hepta-2,5-diene-2,3-dicarboxylic acid, 7-[(4-methylphenyl)sulfonyl]-1-[(2-propynyloxy)methyl]-, dimethyl ester	MeCN	270s(3.15),288s(3.08)	24-0616-86
3-Oxa-12-azatetracyclo[5.4.1.0^{1,5}-0^{8,11}]dodeca-5,9-diene-9,10-dicarboxylic acid, 12-[(4-methylphenyl)-sulfonyl]-, dimethyl ester	MeCN	270s(3.30)	24-0616-86
$C_{21}H_{21}NS$			
Cycloprop[h,i]indolizine-2(1H)-thione, hexahydro-6-methyl-1,6b-diphenyl-, (1α,6aα,6bα,6cα)-	MeCN	269s(--),293(3.93), 331s(--)	118-0899-86
$C_{21}H_{21}N_2$			
1H-Indolium, 1-[5-(2,3-dihydro-1H-indol-1-yl)-2,4-pentadienylidene]-2,3-dihydro-, perchlorate	MeOH	500s(5.01),525(5.14)	33-1588-86
3H-Indolium, 3-[(1,2-dimethyl-1H-	MeOH	488(6.49)	5-1621-86

Compound	Solvent	$\lambda_{max}(\log \epsilon)$	Ref.
indol-3-yl)methylene]-1,2-dimethyl-, tetrafluoroborate (cont.)			5-1621-86
$C_{21}H_{21}N_3O_2$			
L-Phenylalaninamide, glycyl-N-2-naphthalenyl-, monohydrochloride	H_2O	239(4.56),283(3.94)	104-1554-86
4H,5H-Pyrrolo[1",2":4',5']pyrazino-[1',2':1,6]pyrido[3,4-b]indole-5,14-dione, 1,2,3,11,12,14a-hexahydro-12-(2-methyl-1-propenyl)-, (12S-trans)-	EtOH	234(4.45),262s(4.18), 282s(4.14),369(4.21)	88-6217-86
$C_{21}H_{21}N_3O_3S$			
Benzamide, N-[2-nitro-3-phenyl-3-(1-piperidinyl)-1-thioxo-2-propenyl]-	EtOH	242(4.19),308(4.11), 368(4.05),c.400(c.4.00)	18-3871-86
$C_{21}H_{21}N_3O_7$			
2,4(1H,3H)-Pyrimidinedione, 1-[(2-hydroxyethoxy)methyl]-5-[[3-[(4-nitrophenyl)methoxy]phenyl]methyl]-	pH 1	270(4.14)	4-1651-86
	pH 11	272(4.20)	4-1651-86
$C_{21}H_{21}N_5O_6S$			
Spiro[1-azabicyclo[3.2.0]hept-2-ene-6,3'-[3H]pyrazole]-2-carboxylic acid, 3-[[2-(acetylamino)ethenyl]-thio]-4',5'-dihydro-4'-methyl-7-oxo-, (4-nitrophenyl)methyl ester	MeCN	263(4.25),322(4.23)	39-0421-86C
isomer	MeCN	264(4.27),322(4.21)	39-0421-86C
$C_{21}H_{21}N_7O_3$			
Propanamide, N-[2-[(2,6-dicyano-4-nitrophenyl)azo]-5-(diethylamino)-phenyl]-	$CHCl_3$	295(3.85),419(3.59), 588s(4.71),626(4.91)	48-0497-86
$C_{21}H_{21}N_7O_4$			
Acetamide, N-[2-[(2,6-dicyano-4-nitrophenyl)azo]-5-(diethylamino)-4-methoxyphenyl]-	$CHCl_3$	279(4.02),292s(3.97), 318s(3.58),392s(3.50), 462s(3.67),610s(4.67), 656(5.00)	48-0497-86
$C_{21}H_{21}N_7O_5$			
Propanamide, N-[5-[bis(2-hydroxyethyl)amino]-2-[(2,6-dicyano-4-nitrophenyl)azo]phenyl]-	MeCN	422s(3.29),581(4.43), 614(4.52)	48-0497-86
$C_{21}H_{22}$			
Cycloheptene, 1-[2-(2-phenylethenyl)-phenyl]-, (E)-	MeOH	220(4.27),252(4.00), 286(4.09),328(3.54)	151-0105-86A
$C_{21}H_{22}ClNO_7$			
1,3-Azulenedicarboxylic acid, 2-chloro-6-[2-ethoxy-1-(methoxyimino)-2-oxoethyl]-, diethyl ester, anti	CH_2Cl_2	242.5(4.51),263(4.25), 314(4.77),358(3.99), 513.5(2.74)	24-2956-86
syn	CH_2Cl_2	243.5(4.48),329(4.80), 366(4.15),388.5(4.04), 534(2.75)	24-2956-86
$C_{21}H_{22}F_2O_2$			
Androsta-1,4,9(11),16-tetraene-17-carbonyl fluoride, 6α-fluoro-16-methyl-3-oxo-	MeOH	237(4.37)	44-2315-86

Compound	Solvent	λ_{max}(log ε)	Ref.
$C_{21}H_{22}N_2O_2$			
1H-Pyrazole-1-carboxylic acid, 5-(4-methylphenyl)-3-phenyl-, 1,1-dimethylethyl ester	EtOH	258(4.42)	4-0727-86
$C_{21}H_{22}N_2O_2S$			
5-Pyrimidinecarboxylic acid, 1,4-dihydro-6-methyl-2-(methylthio)-1,4-diphenyl-, ethyl ester	EtOH	248s(4.15),314(3.49)	103-0990-86
$C_{21}H_{22}N_2O_3$			
8-Oxotabersonine, (±)-	EtOH	202(4.20),292(3.98), 326(4.11)	77-1756-86
5-Pyrimidinecarboxylic acid, 1,4-dihydro-4-(4-methoxyphenyl)-6-methyl-1-phenyl-, ethyl ester	EtOH	223s(4.04),252s(3.79), 310(3.68)	103-0990-86
$C_{21}H_{22}N_2O_3S$			
5-Pyrimidinecarboxylic acid, 1,2,3,4-tetrahydro-4-(4-methoxyphenyl)-6-methyl-1-phenyl-2-thioxo-, ethyl ester	EtOH	307(4.06)	103-0990-86
$C_{21}H_{22}N_2O_4$			
21-Oxogelseverine	MeOH	210(4.38),256(3.77)	100-0483-86
$C_{21}H_{22}N_2O_5$			
1H-Pyrazole-1-carboxylic acid, 5-(1,3-benzodioxol-5-yl)-4,5-dihydro-5-hydroxy-3-phenyl-, 1,1-dimethylethyl ester	hexane	263(4.44),280(4.40)	4-0727-86
	EtOH	266(4.30),286(4.21)	4-0727-86
$C_{21}H_{22}N_2O_7$			
Butanedioic acid, 2-hydroxy-3-[[4-(phenylazo)benzoyl]oxy]-, diethyl ester	EtOH	228(4.07),325(4.39), 448(2.88)	5-1956-86
$C_{21}H_{22}N_2S_2$			
Propanethioamide, 2-(4,4-dimethyl-2-phenyl-5(4H)-thiazolylidene)-N-methyl-N-phenyl-, (E)-	EtOH	238(4.02),284(4.09), 388s(2.65)	33-0174-86
$C_{21}H_{22}N_4$			
1H-Pyrrolo[2,3-b]pyridine-5-carbonitrile, 4-methyl-1-(phenylmethyl)-6-(1-piperidinyl)-	$CHCl_3$	248(1.10),270(1.11), 338(0.25)	103-0076-86
$C_{21}H_{22}N_4OS_2$			
Propanamide, N-2-benzothiazolyl-3-(2-benzothiazolylethylamino)-N-ethyl-	pH 1	210(4.53),259(4.27), 281(4.10),289(4.10)	103-0795-86
	EtOH	224(4.56),278(4.36), 301(4.12)	103-0795-86
$C_{21}H_{22}N_4O_{11}$			
3,6-Pyridazinedicarboxylic acid, 4-[1-(3,5-di-O-acetyl-β-D-erythro-pentofuranosyl)-1,2,3,4-tetrahydro-2,4-dioxo-5-pyrimidinyl]-, dimethyl ester	pH 1	293(4.02)	44-0950-86
	H_2O	295(4.00)	44-0950-86
	pH 13	293(3.91)	44-0950-86

Compound	Solvent	λ_{max} (log ϵ)	Ref.
$C_{21}H_{22}N_6O_5$			
5,10-Dideazaaminopterin	pH 13	247(4.52),346(3.82)	4-0001-86
$C_{21}H_{22}N_6O_6$			
L-Glutamic acid, N-[4-[[(2-amino-1,4-dihydro-5-methyl-4-oxopyrido[2,3-d]pyrimidin-6-yl)methyl]amino]benzoyl]-	pH 1	220(4.52),280(4.42), 295s(4.37),334s(4.02)	87-1080-86
	pH 7	223(4.60),280(4.41), 290s(4.41)	87-1080-86
	pH 13	221(4.53),283(4.40), 295s(4.39)	87-1080-86
$C_{21}H_{22}O_3$			
2H-1-Benzopyran, 7-[2-(2,4-dimethoxyphenyl)ethenyl]-2,2-dimethyl-, (E)-	MeOH	232(1.30),274(1.24), 318(1.40)	102-2425-86
$C_{21}H_{22}O_3Se$			
Naphtho[1,2-b]furan-2,8(3H,4H)-dione, 3a,5,5a,9b-tetrahydro-3,5a,9-trimethyl-3-(phenylseleno)-, [3R-(3α,3aβ,5aα,9bα)]-	EtOH	202(4.34),224(4.28)	36-0784-86
$C_{21}H_{22}O_4$			
2H,8H-Benzo[1,2-b:3,4-b']dipyran-4-ol, 3,4-dihydro-5-methoxy-8,8-dimethyl-2-phenyl- (tephrobbottin)	MeOH	203(4.43),234(4.31), 241(4.40),287(3.94)	25-0827-86
4H-1-Benzopyran-4-one, 2,3-dihydro-8-(3-hydroxy-3-methyl-1-butenyl)-7-methoxy-2-phenyl- (falciformin)	MeOH	260(4.11),290(3.81)	102-1767-86
Hildgardtol A	EtOH	280(4.13)	102-1711-86
$C_{21}H_{22}O_5$			
6H-Benzofuro[3,2-c][1]benzopyran-6a,9(11aH)-diol, 3-methoxy-2-(3-methyl-2-butenyl)-, (6aS-cis)-(erythrabissin-1)	EtOH	213(4.48),280(3.90), 286(3.90)	102-0757-86
$C_{21}H_{22}O_7$			
Teucvidin, 2β-hydroxy-, acetate	EtOH	224.5(4.01)	102-2857-86
$C_{21}H_{22}O_8$			
4H-1-Benzopyran-4-one, 5,8-diacetoxy-5,6,7,8-tetrahydro-6,7-dihydroxy-2-(2-phenylethyl)-, [5S-(5α,6β-7α,8β)]-	EtOH	208(4.22),249(4.05)	94-2766-86
isomer	MeOH	207(4.24),250(4.09)	94-2766-86
$C_{21}H_{22}O_9$			
4H-1-Benzopyran-4-one, 8-hydroxy-3,6,7-trimethoxy-2-(2,4,5-trimethoxyphenyl)-	MeOH	252(4.4),305s(4.0), 388(4.1)	94-1656-86
	MeOH-NaOMe	265(--),378(--)	94-1656-86
	MeOH-$AlCl_3$	265(--),325(--), 406(--)	94-1656-86
4H-1-Benzopyran-4-one, 2-(4-hydroxy-3,5-dimethoxyphenyl)-5,6,7,8-tetramethoxy-	MeOH	272(4.26),310(4.24)	102-2625-86
	MeOH-NaOMe	268(--),355(--)	102-2625-86
	MeOH-NaOAc	268(--),370(--)	102-2625-86
	MeOH-NaOAc-H_3BO_3	268(--),308(--)	102-2625-86
	MeOH-$AlCl_3$	268(--),310(--)	102-2625-86
$C_{21}H_{23}ClFN_3OS$			
1-Piperazinepropanamide, 4-(2-chloro-	MeOH	242s(3.64),267s(3.58)	73-2598-86

Compound	Solvent	$\lambda_{max}(\log \epsilon)$	Ref.
7-fluoro-10,11-dihydrodibenzo[b,f]-thiepin-10-yl)-, monomethanesulfonate (cont.)			73-2598-86
$C_{21}H_{23}Cl_2FO_3$			
Androsta-1,4,16-triene-17-carboxylic acid, 9,11-dichloro-6-fluoro-16-methyl-3-oxo-, (6α,11β)-	MeOH	234(4.32)	44-2315-86
$C_{21}H_{23}FO_3$			
Androsta-1,4,9(11),16-tetraene-17-carboxylic acid, 6α-fluoro-16-methyl-3-oxo-	MeOH	235(4.32)	44-2315-86
$C_{21}H_{23}NO_2$			
2H-Pyrrolo[2,1-b][1,3]oxazine, 6,7,8,8a-tetrahydro-2-methoxy-8a-methyl-3,4-diphenyl-	MeCN	226(4.07),306(3.91)	118-0899-86
$C_{21}H_{23}NO_3$			
3-Pyridinecarboxylic acid, 1,2,5,6-tetrahydro-5-(hydroxydiphenylmethyl)-1-methyl-, (±)-	pH 7.2	222(1.65)	87-0125-86
$C_{21}H_{23}NO_7$			
Lutessine, 2-acetoxy-	EtOH	225(3.95),286(3.69)	100-0090-86
Noyaine	EtOH	220(4.20),254(3.98), 312(3.73)	88-5535-86
$C_{21}H_{23}NO_{10}S$			
7-Azabicyclo[2.2.1]hept-2-ene-2,3,5,6-tetracarboxylic acid, 7-[(4-methylphenyl)sulfonyl]-, tetramethyl ester, (exo,exo)-	MeCN	227(4.26)	24-0616-86
$C_{21}H_{23}NO_{13}$			
Paulinone, 9,10,11,13-tetra-O-acetyl-	MeOH	209(4.30),259(4.08), 446(3.28)	44-2493-86
$C_{21}H_{23}N_3OS$			
Propericiazine	MeOH	232.5(4.319),271.5(4.503)	106-0571-86
$C_{21}H_{23}N_3O_2$			
Benzoic acid, 4-[5,5-dicyano-3-(1-piperidinylmethylene)-4-pentenyl]-, methyl ester	EtOH	236(4.18),380(4.71)	4-0001-86
$C_{21}H_{23}N_3O_4$			
Acetamide, N-[2,3-dihydro-2-[3-(4-morpholinyl)propyl]-1,3-dioxo-1H-benz[de]isoquinolin-5-yl]-, hydrochloride	pH 7.4	343(3.98)	4-0849-86
$C_{21}H_{23}N_3O_9$			
2(1H)-Pyrimidinone, 4-[(6-methyl-2-pyridinyl)oxy]-1-(2,3,5-tri-O-acetyl-β-D-ribofuranosyl)-	pH 7	271(3.86),284s(--)	1-0806-86
	pH 2	271(--),284s(--)	1-0806-86
	pH 13	273(--),285s(--)	1-0806-86
2(1H)-Pyrimidinone, 4-[(6-methyl-3-pyridinyl)oxy]-1-(2,3,5-tri-O-acetyl-β-D-ribofuranosyl)-	pH 7	270s(--),276(3.93)	1-0806-86
	pH 2	278(--)	1-0806-86
	pH 13	276(--)	1-0806-86

Compound	Solvent	λ_{max}(log ε)	Ref.
$C_{21}H_{23}N_5O_5$			
Adenosine, N-(4-methylbenzoyl)-2',3'-O-(1-methylethylidene)-	MeOH	255(4.20),279(4.34)	23-2109-86
$C_{21}H_{23}N_7O_5$			
Acetamide, N-[2-[(2-cyano-4,6-dinitrophenyl)azo]-5-(dipropylamino)phenyl]-	$CHCl_3$	285s(3.83),311s(3.75), 349(4.07),588s(4.73), 625(4.83)	48-0497-86
L-Glutamic acid, N-[4-[[(2,4-diamino-5-methylpyrido[2,3-d]pyrimidin-6-yl)methyl]amino]benzoyl]-	pH 1	222(4.59),245s(4.24), 279s(4.24),301(3.96)	87-0709-86
	pH 1	224(4.60),298(4.33)	87-1080-86
	H_2O	222(4.49),276s(4.18), 298(4.25),332s(3.88)	87-0709-86
	pH 7	222(4.56),282(4.43)	87-1080-86
	pH 13	248s(4.34),286(4.39), 295s(4.38),340s(3.97)	87-0709-86
	pH 13	223(4.38),283(4.43), 340s(3.94)	87-1080-86
$C_{21}H_{24}ClN_2O_5P$			
Phosphonic acid, [3-chloro-6-methoxy-4-[(4-methoxyphenyl)amino]-2-quinolinyl]-, diethyl ester	EtOH	250(4.59),355(3.84)	104-0973-86
$C_{21}H_{24}ClN_5O$			
1H-Pyrazolo[3,4-b]quinolin-4-amine, N-(4-chlorophenyl)-5,6,7,8-tetrahydro-1-(4-morpholinylmethyl)-	EtOH	220(4.51),249(4.15), 325(4.23)	5-1728-86
$C_{21}H_{24}FN_5$			
1H-Pyrazolo[3,4-b]quinolin-4-amine, N-(4-fluorophenyl)-5,6,7,8-tetrahydro-1-(1-pyrrolidinylmethyl)-	EtOH	218(4.39),241(4.03), 272(3.71),316(4.17)	5-1728-86
$C_{21}H_{24}FN_5O$			
1H-Pyrazolo[3,4-b]quinolin-4-amine, N-(4-fluorophenyl)-5,6,7,8-tetrahydro-1-(4-morpholinylmethyl)-	EtOH	219(4.27),243(3.90), 273(3.54),318(4.01)	5-1728-86
$C_{21}H_{24}F_2O_4$			
18-Norandrosta-1,4-diene-17-carboxylic acid, 6,9-difluoro-11,13-dihydroxy-16,17-dimethyl-, β-lactone, (6α,11β,13α,16α,17α)-	MeOH	236(4.20)	44-2315-86
$C_{21}H_{24}F_3N_3S$			
Trifluoperazine, dihydrochloride	MeOH	260.5(4.604),312(3.642), 316(3.447)	106-0571-86
$C_{21}H_{24}NO_5$			
6H-Benzo[g]-1,3-benzodioxolo[5,6-a]quinolizinium, 5,8,13,13a-tetrahydro-9,10-dimethoxy-7-methyl-, iodide (N-methylophiocarpinium iodide)	MeOH	202(4.43),211s(4.32), 275(3.75)	73-2232-86
$C_{21}H_{24}N_2O$			
3-Quinolinecarboxamide, 1,4-dihydro-1,2,4-trimethyl-N-(1-phenylethyl)-	MeCN	313(4.02)	78-0961-86

Compound	Solvent	λ_{max}(log ε)	Ref.
$C_{21}H_{24}N_2O_2$			
7,20(2H,19H)-Cyclovobasan, 1,2,18,19-tetradehydro-3,17-epoxy-11-methoxy-(11-methoxykoumine)	MeOH	232(4.40),276(3.50), 301s(3.27)	88-4585-86
$C_{21}H_{24}N_2O_3$			
Gelsevirine	MeOH	210(4.26),256(3.67)	100-0483-86
1H-Pyrazole-1-carboxylic acid, 4,5-dihydro-5-hydroxy-5-(4-methylphenyl)-3-phenyl-, 1,1-dimethylethyl ester	hexane	283(4.31)	4-0727-86
	EtOH	283(4.39)	4-0727-86
Rhazine	n.s.g.	227(4.52),282(3.87), 291(3.78)	42-0449-86
	+ acid	222(4.56),274(3.93), 290(3.81)	42-0449-86
Scholaricine	MeOH	210(--),235(--), 285(--),335(--)	164-0663-86
$C_{21}H_{24}N_2O_3S$			
2,4,6-Heptatrienoic acid, 6-formyl-7-[[2-(1H-indol-3-yl)ethyl]amino]-2-(methylthio)-, ethyl ester	MeOH	221(4.58),291(4.10), 378(4.30)	44-2995-86
$C_{21}H_{24}N_2O_4$			
5-Pyrimidinecarboxylic acid, 1,2,3,4-tetrahydro-2-hydroxy-4-(4-methoxyphenyl)-6-methyl-1-phenyl-, ethyl ester	EtOH	223(4.08),309(4.21)	103-0990-86
$C_{21}H_{24}N_2O_4S$			
Pentanamide, 4-methyl-2-[(methylphenylamino)methylene]-N-[(4-methylphenyl)sulfonyl]-3-oxo-	EtOH	232(4.31),252(4.31), 274s(4.21)	161-0539-86
$C_{21}H_{24}N_2O_7$			
1,3-Azulenedicarboxylic acid, 2-amino-6-[2-ethoxy-1-(methoxyimino)-2-oxoethyl]-, diethyl ester	CH_2Cl_2	249(4.62),337(4.85), 420(4.24),435.5(4.26)	24-2956-86
$C_{21}H_{24}N_2O_9$			
22H-Dibenzo[n,q][1,4,7,10,13]pentaoxacyclooctadecin, 6,7,9,10,12,13,15,16-octahydro-2,20-dinitro-	$CHCl_3$	245(3.88),315(4.30)	78-0137-86
$C_{21}H_{24}N_4$			
1H-Pyrrolo[2,3-b]pyridine-5-carbonitrile, 2,3-dihydro-4-methyl-1-(phenylmethyl)-6-(1-piperidinyl)-	$CHCl_3$	257(0.85),312(0.54), 339(0.75)	103-0076-86
$C_{21}H_{24}N_4O$			
3H-Pyrazol-3-one, 4-[[1-(diethylamino)-2-methylphenyl]imino]-2,4-dihydro-5-methyl-1-phenyl-	MeOH	246(4.47),443(4.24), 538(4.66)	48-0149-86
	BuOAc	448(4.21),519(4.55)	48-0149-86
$C_{21}H_{24}N_4O_2$			
3H-Pyrazol-3-one, 4-[[4-[ethyl(2-hydroxyethyl)amino]-2-methylphenyl]imino]-2,4-dihydro-5-methyl-1-phenyl-	MeOH	248(4.32),442(4.15), 536(4.64)	48-0149-86
	BuOAc	448(4.22),521(4.56)	48-0149-86
$C_{21}H_{24}N_8O_2$			
Ethanone, 1,1'-[1,3,5,7-tetraaza-	EtOH	306(4.39)	23-1567-86

Compound	Solvent	$\lambda_{max}(\log \epsilon)$	Ref.
bicyclo[3.3.1]nonaene-3,7-diyl-bis(azo-4,1-phenylene)]bis- (cont.)			23-1567-86
$C_{21}H_{24}O$			
Bicyclo[2.2.1]heptane, 2-(diphenyl-methyl)-2-methoxy-, endo	hexane	259(2.75)	78-6175-86
exo	hexane	259(2.76)	78-6175-86
$C_{21}H_{24}O_2$			
2-Naphthalenecarboxylic acid, 3,7-di-methyl-2,6-octadienyl ester	MeCN	236(4.77)	130-0046-86
$C_{21}H_{24}O_3$			
2H-1-Benzopyran, 6-[2-(2,3-dimethoxy-phenyl)ethyl]-2,2-dimethyl-	MeOH	271.5(3.39),279.5(3.44), 290.5(3.40)	94-3727-86
2H-1-Benzopyran, 6-[2-(3,4-dimethoxy-phenyl)ethyl]-2,2-dimethyl-	MeOH	224(4.55),283(4.01)	94-3727-86
$C_{21}H_{24}O_4$			
1(4H)-Anthracenone, 7-ethyl-8,9-di-hydroxy-3-methoxy-2,4,4,6-tetra-methyl- (garveatin B)	MeOH	210(4.09),238(4.21), 265(3.97),287(3.93), 320s(3.59),426(3.58)	44-0057-86
1,4-Heptanonaphthalene-6,7-dicarbox-ylic acid, dimethyl ester	EtOH	239s(4.54),261(4.88), 311(4.33),345s(4.32), 356(4.35)	24-2698-86
Racemosol	EtOH	224(4.43),276(3.77), 430(2.28)	78-2417-86
$C_{21}H_{24}O_5$			
2-Butenoic acid, 3-methyl-, 1-[5-(2,5-dihydroxyphenyl)-3-furanyl]-4-meth-yl-3-pentenyl ester (shikonofuran E)	MeOH	214(4.27),264s(3.99), 269(4.08),281(3.99), 323(3.96)	94-2290-86
Ethanone, 1,1'-[10-(dimethoxymethyl)-1,2,3,4,9,10-hexahydro-9,10-epoxy-anthracene-2,7-diyl]bis-	MeCN	201(4.29),234(4.25), 276(3.57),301(3.46)	33-1287-86
$C_{21}H_{24}O_6$			
1,3-Azulenedicarboxylic acid, 6-(2-ethoxy-2-oxoethyl)-2-methyl-, diethyl ester	CH_2Cl_2	238.5(4.47),272(4.32), 312(4.77),348(3.88), 376.5(3.75),500.5(2.63)	24-2272-86
1,5-Heptano-3-benzoxepin-7,8-dicarbox-ylic acid, 6,9-dihydro-6,9-epoxy-, dimethyl ester	EtOH	209(4.25),234s(3.89), 266s(3.57)	24-2698-86
1,4-Heptanonaphthalene-6,7-dicarbox-ylic acid, 1,2-dihydro-8-hydroxy-2-oxo-, dimethyl ester	MeCN	219(4.60),258s(3.90), 313(4.26)	24-2698-86
$C_{21}H_{24}O_9$			
4H-1-Benzopyran-4-one, 2,3-dihydro-5-hydroxy-6,7-dimethoxy-2-(2,3,4,5-tetramethoxyphenyl)-	MeOH	280(4.28),330s(3.86)	102-2223-86
	$MeOH-AlCl_3$	315(4.27),380s(--)	102-2223-86
	+ HCl	310(4.83),380s(--)	102-2223-86
$C_{21}H_{24}O_{10}$			
Trifolirhizin	EtOH	278(3.57),285(3.58), 311(3.81)	105-0603-86
$C_{21}H_{25}BrN_2O_9S$			
α-D-Glucopyranose, 2-[[[(4-bromophen-yl)amino]thioxomethyl]amino]-2-de-oxy-, 1,3,4,6-tetraacetate	EtOH	271(3.91),281(3.94), 292(3.83)	136-0049-86I

Compound	Solvent	λ_{max}(log ϵ)	Ref.
$C_{21}H_{25}BrN_4O_{10}$			
3H-Pyrrolo[2,3-d]pyrimidine-5-carboxylic acid, 2-(acetylamino)-7-[5-O-acetyl-2,3-O-(1-methylethylidene)-β-D-ribofuranosyl]-6-bromo-4,7-dihydro-3-(methoxymethyl)-4-oxo-	MeOH	223(4.26),287(4.09), 305s(4.00)	78-0199-86
$C_{21}H_{25}Cl_2FO_4$			
Androsta-1,4-diene-17-carboxylic acid, 9,11-dichloro-6-fluoro-17-hydroxy-16-methyl-3-oxo-, (6α,11β,16β,17α)-	MeOH	236(4.18)	44-2315-86
$C_{21}H_{25}FN_2OS$			
Piperazine, 1-[3-fluoro-10,11-dihydro-8-(1-methylethyl)dibenzo[b,f]thiepin-10-yl]-, S-oxide	MeOH	226(4.13),273(3.40), 280(3.32)	73-0698-86
$C_{21}H_{25}FO_4$			
Androsta-1,4,9(11)-triene-17-carboxylic acid, 6-fluoro-17-hydroxy-16-methyl-3-oxo-, (6α,16β,17α)-	MeOH	238(4.21)	44-2315-86
18-Norandrosta-1,4-diene-17-carboxylic acid, 9-fluoro-11,13-dihydroxy-16,17-dimethyl-3-oxo-, β-lactone, (11β,13α,16β,17α)-	MeOH	237(4.18)	44-2315-86
$C_{21}H_{25}N$			
3-Azahepta-3,6-diene, 2,5,5-trimethyl-7,7-diphenyl-	EtOH	249(4.14)	150-0631-86M
$C_{21}H_{25}NO_2$			
1-Propanone, 2-hydroxy-1-(4-methylphenyl)-3-phenyl-3-(1-piperidinyl)-	n.s.g.	255(3.95)	11-0379-86
1-Propanone, 2-hydroxy-3-(4-methylphenyl)-1-phenyl-3-(1-piperidinyl)-	n.s.g.	250(3.80)	11-0379-86
$C_{21}H_{25}NO_3$			
Benzeneacetic acid, α-[[2-(4-pentenyloxy)phenyl]amino]-, ethyl ester	EtOH	243(4.09),289(3.55)	33-0927-86
Benzeneacetic acid, α-[[3-(4-pentenyloxy)phenyl]amino]-, ethyl ester	EtOH	245(4.02),289(3.45)	33-0927-86
Dehydropipernonaline	EtOH	267(3.3)	36-1188-86
1-Propanone, 2-hydroxy-1-(4-methoxyphenyl)-3-phenyl-3-(1-piperidinyl)-	n.s.g.	270(4.02)	11-0379-86
1-Propanone, 2-hydroxy-3-(4-methoxyphenyl)-1-phenyl-3-(1-piperidinyl)-	n.s.g.	245(3.99)	11-0379-86
$C_{21}H_{25}NO_4$			
4-Phenanthrenol, 9-(dimethylamino)-8-ethenyl-9,10-dihydro-3,5,6-trimethoxy-	MeOH	229(4.66),235(4.65), 247(4.64),310(3.86)	100-0159-86
	MeOH-base	255(4.68),311(4.02)	100-0159-86
4-Phenanthrenol, 10-(dimethylamino)-1-ethenyl-9,10-dihydro-3,5,6-trimethoxy-	MeOH	227(4.45),259(4.32), 276s(4.18),310(3.77)	100-0159-86
	MeOH-base	261(4.27),312(3.95)	100-0159-86
4-Phenanthrenol, 1-[2-(dimethylamino)-ethyl]-3,5,6-trimethoxy- (5λ,4ε) (corydinemethine)	MeOH	244(4.32),259(3.94), 319(3.92),328(3.35), 366(3.33),383(?)	100-0159-86
	MeOH-base	259(4.42),317(3.94), 328(3.94),364(3.59), 381(3.55)	100-0159-86

Compound	Solvent	$\lambda_{max}(\log \epsilon)$	Ref.
$C_{21}H_{25}NO_5$			
Thaicanine, (-)-	EtOH	208(4.34),221s(3.91), 280(3.11)	100-0253-86
	EtOH-base	216(4.51),280(3.51)	100-0253-86
$C_{21}H_{25}NO_6$			
1,3-Dioxolo[4,5-g]isoquinolin-4-ol, 5,6,7,8-tetrahydro-5-[[2-(hydroxy-methyl)-3,4-dimethoxyphenyl]meth-yl]-6-methyl-, (S)- (narcotolinol)	MeOH	232(4.15),283(3.56)	102-2403-86
$C_{21}H_{25}NO_{10}S$			
7-Azabicyclo[2.2.1]heptane-2,3,5,6-tetracarboxylic acid, 7-[(4-methyl-phenyl)sulfonyl]-, tetramethyl ester, (1α,2α,3α,4α,5β,6β)-	MeCN	230(4.13),256s(2.87), 263(2.83),268s(2.73), 274(2.65)	24-0616-86
(1α,2α,3β,4β,5β,6β)-	MeCN	232(4.10)	24-0616-86
$C_{21}H_{25}N_3O_2$			
1H-Naphtho[2,3-d]imidazole-4,9-dione, 2-methyl-1-(2,2,6,6-tetramethyl-4-piperidinyl)-	PrOH	225s(3.88),249(4.61), 274s(4.13),283(4.15), 334(3.34),379(2.56)	104-1560-86
$C_{21}H_{25}N_5O$			
1H-Pyrazolo[3,4-b]quinolin-4-amine, 5,6,7,8-tetrahydro-1-(4-morphol-inylmethyl)-N-phenyl-	EtOH	217(4.36),244s(--), 273(3.61),326(4.11)	5-1728-86
$C_{21}H_{25}N_5O_3$			
9H-Purin-6-amine, 9-[dihydro-2,2-di-methyl-6-[(phenylmethoxy)methyl]-4H-cyclopenta-1,3-dioxol-5-yl]-, (3aα,4α,6α,6aα)-(±)-	MeOH	261(4.12)	44-1287-86
$C_{21}H_{25}N_5O_4$			
4H-Cyclopenta-1,3-dioxol-5-ol, 4-(6-amino-9H-purin-9-yl)dihydro-2,2-dimethyl-6-[(phenylmethoxy)methyl]-	MeOH	260(4.15)	44-1287-86
$C_{21}H_{25}N_5O_7$			
β-D-Glucofuranuronic acid, 1-[6-(cy-clohexylamino)-9H-purin-9-yl)-1-deoxy-, γ-lactone, 2,5-diacetate	MeOH	267(4.30)	103-0085-86
$C_{21}H_{26}BrN_7O_8$			
7H-Pyrrolo[2,3-d]pyrimidin-4-one, 2-(acetylamino)-5-(azidomethyl)-6-bromo-3,4-dihydro-3-(methoxy-methyl)-7-[5-O-acetyl-2,3-O-(1-methylethylidene)-β-D-ribofuranosyl]-	MeOH	272(3.97),305(3.99)	78-0207-86
$C_{21}H_{26}ClN_3OS$			
Perphenazine	MeOH	213(4.286),257(4.527), 314(3.615)	106-0571-86
$C_{21}H_{26}ClN_3O_3$			
Acetamide, N-[6-chloro-5,8-dihydro-5,8-dioxo-7-[(2,2,6,6-tetramethyl-4-piperidinyl)amino]-1-naphthal-enyl]-	isoPrOH	211(4.40),246(4.33), 287(4.31),425(3.78), 508(3.35)	104-1560-86

Compound	Solvent	λ max(log ε)	Ref.
Acetamide, N-[7-chloro-5,8-dihydro-5,8-dioxo-6-[(2,2,6,6-tetramethyl-4-piperidinylamino)-1-naphthalenyl]-	heptane	243(4.19),273s(4.37), 283(4.40),296s(4.17), 309s(3.62),411s(3.55), 437(3.63),479(3.67)	104-1560-86
	benzene	318s(3.59),415s(3.57), 441(3.64),487(3.71)	104-1560-86
	PrOH	245(4.24),254s(4.20), 278s(4.35),283(4.37), 411s(2.57),481(3.72)	104-1560-86
	DMSO	285(4.39),411s(3.54), 488(3.75)	104-1560-86
$C_{21}H_{26}F_2O_5$			
Androsta-1,4-diene-17-carboxylic acid, 6,9-difluoro-11,17-dihydroxy-16-methyl-3-oxo-, (6α,11β,16α,17α)-	MeOH	238(4.21)	44-2315-86
$C_{21}H_{26}N_2O$			
Vallesamidine, N-formyl-N-demethyl-18-methylene-, (±)-	MeOH	252(4.00),279(3.58), 286(3.53)	73-1731-86
$C_{21}H_{26}N_2OS_2$			
Mesoridazine (benzenesulfonate)	MeOH	216.5(4.305),240.5(4.168), 263(4.513),312(3.571)	106-0571-86
$C_{21}H_{26}N_2O_2$			
7,20(2H,19H)-Cyclovobasan, 18,19-didehydro-3,17-epoxy-11-methoxy-, (3R,7α,20α)-	EtOH	210(4.56),240s(3.74), 297(3.69)	88-4585-86
$C_{21}H_{26}N_2O_3$			
Bhimberine	MeOH	222(4.20),268s(3.61), 273s(3.61),282(3.61), 290(3.55)	142-0703-86
$C_{21}H_{26}N_2O_4$			
Humantenirine	MeOH	226(4.24),261(3.59)	100-0806-86
Vincamine N-oxide, as dimer	EtOH-HCl	203(4.52),224(4.82), 276(4.19)	88-2775-86
$C_{21}H_{26}N_2O_7S$			
Uridine, 5'-O-(methoxymethyl)-5-methyl-2',3'-O-(1-methylethylidene)-6-(phenylthio)-	MeOH	243(3.99),275(3.96), 313s(3.56)	78-4187-86
$C_{21}H_{26}N_2S_2$			
Thioridazine, hydrochloride	MeOH	208(4.263),230(4.215), 264(4.577),315(3.669)	106-0571-86
$C_{21}H_{26}N_4O_4$			
Guanidine, N''-[5-(2,2-dimethyl-4,6-dioxo-1,3-dioxolan-5-ylidene)-3-(4-pyridinyl)-1,3-pentadienyl]-N,N,N',N'-tetramethyl-	$CHCl_3$	514s(--),552(4.63)	24-3276-86
	benzene	516(--),544(--)	24-3276-86
	acetone	545(5.09)	24-3276-86
$C_{21}H_{26}N_4O_{10}$			
7H-Pyrrolo[2,3-d]pyrimidine-5-carboxylic acid, 2-(acetylamino)-3,4-dihydro-3-(methoxymethyl)-4-oxo-7-[5-O-acetyl-2,3-O-(1-methylethylidene)-β-D-ribofuranosyl]-	MeOH	226(4.16),281(3.95), 298(3.91)	78-0199-86

Compound	Solvent	λ_{max} (log ε)	Ref.
$C_{21}H_{26}N_4O_{12}$			
1H-Pyrazole-5-methanimidamide, N,4-di-acetoxy-1-acetyl-3-(2,3,5-tri-O-acetyl-β-D-ribofuranosyl)-	pH 1	230(3.96)	87-0268-86
	pH 7	230(3.94)	87-0268-86
	pH 11	300(3.78)	87-0268-86
$C_{21}H_{26}N_6O_3$			
Guanidine, N,N,N',N'-tetramethyl-N"-[3-(4-pyridinyl)-5-(tetrahydro-1,3-dimethyl-2,4,6-trioxo-5(2H)-pyrimidinylidene)-1,3-pentadienyl]-	acetone	533s(--),561(5.13)	24-3276-86
	$CHCl_3$	536(4.69),569(4.92)	24-3276-86
$C_{21}H_{26}N_6O_4$			
β-D-Ribofuranuronamide, 1-deoxy-N-ethyl-1-[6-[(1-methyl-2-phenylethyl)amino]-9H-purin-9-yl]-, (R)-	n.s.g.	270(4.26)	87-1683-86
(S)-	n.s.g.	270(4.25)	87-1683-86
$C_{21}H_{26}N_6O_5$			
Adenosine, 2-[[4-(1,1-dimethylethyl)-benzoyl]amino]-	pH 7	248(4.32),280(4.29)	1-0806-86
	pH 2	253(--),291(--)	1-0806-86
	pH 13	248(--),280(--)	1-0806-86
$C_{21}H_{26}O$			
18-Norpregna-4,6,8(14),13(17)-tetraen-3-one, 20-methyl-	EtOH	412(4.30)	94-1561-86
	90% H_2SO_4	372(4.5)	94-1561-86
after dilution with ethanol		600(4.7)	94-1561-86
$C_{21}H_{26}OSe_2$			
Benzene, 1,1'-[[(cyclohexyloxy)methylene]bis(seleno)]bis[4-methyl-	hexane	210(4.26),235(4.31), 300(3.50)	104-0092-86
$C_{21}H_{26}O_2$			
Gestodene	MeOH	240(4.24)	145-0784-86
D(17a)-Homo-18-norpregna-5,13,15,17-tetraen-20-one, 3β-hydroxy-	EtOH	265(4.26)	39-0261-86C
$C_{21}H_{26}O_5$			
1(3H)-Isobenzofuranone, 3-acetoxy-4-acetyl-5-(1,3,3-trimethylcyclohexyl)- (macfarlandin A)	MeOH	216(4.24),246(4.20)	44-2601-86
Macfarlandin B	MeOH	209(4.27),245(4.08)	44-2601-86
$C_{21}H_{26}O_7$			
Bicyclo[3.2.1]octan-3-one, 2,8-dihydroxy-1-methoxy-7-(7-methoxy-1,3-benzodioxol-5-yl)-6-methyl-5-(2-propenyl)-	MeOH	238s(3.45),270(3.06)	102-0213-86
3-Furanmethanol, tetrahydro-2-(4-hydroxy-3,5-dimethoxyphenyl)-4-[(4-hydroxy-3-methoxyphenyl)methyl]-, (2α,3β,4β)-(±)- (5'-methoxylaciresinol)	MeOH	230(4.71),281(4.19)	100-0706-86
1H-Xanthene-9-propanoic acid, 9-carboxy-2,3,4,5,6,7,8,9-octahydro-3,3,6,6-tetramethyl-1,8-dioxo-	EtOH	230(4.24),302(3.62)	2-0347-86
$C_{21}H_{26}O_{13}$			
Lariside	n.s.g.	229(3.96),250s(3.49), 260s(3.38),290(3.62), 340(3.83)	105-0267-86

Compound	Solvent	λ_{max}(log ε)	Ref.
$C_{21}H_{27}BrN_2O$			
Pyrrolo[2,3-b]indol-2(1H)-one, 6-bromo-3,3a,8,8a-tetrahydro-1-methyl-3a,8-bis(3-methyl-2-butenyl)-, cis (flustramide B)	EtOH	216(2.36),260(3.89), 310(3.55)	1-0555-86
$C_{21}H_{27}BrO_4$			
Androst-4-ene-3,17-dione, 2α-(bromoacetoxy)-	MeOH	239(4.19)	13-0347-86B
Androst-4-ene-3,17-dione, 6α-(bromoacetoxy)-	MeOH	234(4.10)	13-0347-86B
6β-	MeOH	232(4.16)	13-0347-86B
$C_{21}H_{27}FO_5$			
Androsta-1,4-diene-17-carboxylic acid, 9-fluoro-11,17-dihydroxy-16-methyl-3-oxo-, (11β,16β,17α)-	MeOH	239(4.33)	44-2315-86
$C_{21}H_{27}NO_3$			
6a,9-Epoxy-6aH-benzo[3,4]cyclohepta-[1,2-c]quinolin-7-ol, 1,2,3,4,4a,5-6,7,8,9,13b,13c-dodecahydro-5-methyl-, acetate, (4aα,6aβ,7α,9β,13bβ-13cα)-	EtOH	264s(2.47),272(2.37)	44-5002-86
$C_{21}H_{27}NO_5S$			
Sulfoximine, N-(2-ethoxy-1-methyl-2-oxoethyl)-S-[2-(2-methoxyethoxy)-phenyl]-S-(4-methylphenyl)-, [S-(R*,S*)]-	EtOH	209(4.39),226(4.21), 290(3.57)	12-1833-86
$C_{21}H_{27}N_2O$			
Pyronine B, tetrachloroferrate salt	pH 5.70	533(3.9),553(4.00)	60-2123-86
$C_{21}H_{27}N_3OS$			
Hydrazinecarbothioamide, 2-(1,3-dihydro-3,3-bis(1-methylethyl)-1-phenyl-1-isobenzofuranyl]-	EtOH	243(4.30),270(3.21), 309s(1.86)	104-0647-86
$C_{21}H_{27}N_3O_3$			
Acetamide, N-[1,4-dihydro-1,4-dioxo-3-[(2,2,6,6-tetramethyl-4-piperidinyl)amino]-2-naphthalenyl]-	heptane	233(4.14),241(4.10), 280(4.34),301s(4.16), 344s(3.10),501(3.45)	104-1560-86
	benzene	486(3.48)	104-1560-86
	isoPrOH	458(3.53)	104-1560-86
	DMSO	474(3.52)	104-1560-86
$C_{21}H_{27}N_3O_6$			
Proline, N^{α}-furylacryloylphenylalanylalanyl-	MeOH	305(4.40)	65-0613-86
$C_{21}H_{27}O_2$			
[1,1'-Biphenyl]-4-yloxy, 3,5-bis(1,1-dimethylethyl)-4'-methoxy-	EtOH	376(4.25),580(3.63)	18-2899-86
$C_{21}H_{28}Br_2O_3$			
Androst-4-en-3-one, 17-acetoxy-4,6-dibromo-, (6β,17β)-	EtOH	273(4.11)	107-0195-86 +118-0461-86
$C_{21}H_{28}NO_3$			
Isoquinolinium, 1,2,3,4-tetrahydro-	EtOH	221(4.52),281(3.90)	102-2693-86

Compound	Solvent	λ_{max}(log ϵ)	Ref.
6,7-dimethoxy-1-[(4-methoxyphenyl)-methyl]-2,2-dimethyl-, iodide (O,O-dimethylluxandrine iodide)	EtOH-NaOH	221(4.52),281(3.90)	102-2693-86
$C_{21}H_{28}N_2O_2$			
Bicyclo[4.4.1]undeca-1,3,5,7,9-pentaene-2,10-dicarboxamide, N,N,N',N'-tetraethyl-	MeOH	265(4.49),315(3.96)	33-1263-86
$C_{21}H_{28}N_2O_3$			
Aspidospermidin-21-oic acid, 17-methoxy-, methyl ester, (±)-	MeOH	212(4.40),242(3.67), 286(3.27)	78-6719-86
$C_{21}H_{28}N_2O_8$			
1,2-Pyrrolidinedicarboxylic acid, 4-[2-(methoxycarbonyl)-3-pyridinyl]-3-(2-methoxy-2-oxoethyl)-, 1-(1,1-dimethylethyl) 2-methyl ester	EtOH	269(3.71)	88-0607-86
$C_{21}H_{28}N_2O_9$			
1,2-Pyrrolidinedicarboxylic acid, 4-[2-(methoxycarbonyl)-1,6-dihydro-6-oxo-3-pyridinyl]-3-(2-methoxy-2-oxoethyl)-, 1-(1,1-dimethylethyl 2-methyl ester	EtOH	231(4.22),241(3.90)	88-0607-86
$C_{21}H_{28}N_4O_3$			
Benzo[g]phthalazine-1,4-diamine, 6-methoxy-N,N'-bis(3-methoxypropyl)-	EtOH	226(4.61),239(4.58), 263(4.65),269(4.67)	111-0143-86
$C_{21}H_{28}N_4O_5S$			
Benzenesulfonic acid, 2,4,6-trimethyl-, salt with 5-amino-1-[(2,2-dimethyl-1-oxopropoxy)methyl]-1H-imidazo[4,5-c]pyridinium cation	MeOH	274(3.79)	78-1511-86
$C_{21}H_{28}O$			
Acetaldehyde, [3-[2-methyl-4-(2,6,6-trimethyl-1-cyclohexen-1-yl)-1,3-butadienyl]-2-cyclopenten-1-ylidene]-, (E,E,E)-	hexane	362(4.34)	35-4614-86
(Z,E,E)-	hexane	361(3.90)	35-4614-86
$C_{21}H_{28}O_2$			
18-Norpregna-4,13-diene-3,20-dione, 17β-methyl-	EtOH	237(4.23)	94-1561-86
19-Norpregn-5(10)-en-20-yn-3-one, 17-hydroxy-7-methyl-, (7α,17α)-	EtOH	240(1.42)	54-0111-86
Phenol, 2-(1,1-dimethylethyl)-6-[1-(2-hydroxyphenyl)-2-methylpropyl]-4-methyl-, (S)-	EtOH	284(3.76)	116-0509-86
Pregna-1,4-diene-3,20-dione	EtOH	244(4.17)	39-1797-86C
$C_{21}H_{28}O_2S$			
Androsta-2,4-diene-6,17-dione, 3-(ethylthio)-	MeOH	277(3.77)	150-0718-86M
$C_{21}H_{28}O_3$			
1H-Dibenzo[b,d]pyran-1,2(6H)-dione, 6a,7,10,10a-tetrahydro-6,6,9-trimethyl-3-pentyl-, (6aR-trans)-	EtOH	269(4.06),362(2.85), 400(2.99)	142-1973-86

Compound	Solvent	λ_{max}(log ε)	Ref.
18-Norpregn-4-ene-3,20-dione, 13,21-epoxy-17-methyl-, (13α,17α)-	EtOH	241(4.02)	70-1720-86
19-Norpregn-4-en-20-yn-3-one, 10β,17β-dihydroxy-7α-methyl-	EtOH	236.5(3.19)	54-0111-86
Pregn-4-ene-3β,20-dione, 17α,21-epoxy-	EtOH	241(4.02)	70-1720-86
Sapranthin	EtOH	227(3.59),238(3.82), 250(4.10),264(4.21), 281(4.14)	102-1903-86
3-Undecen-6-one, 1,11-di-3-furanyl-4,8-dimethyl-, (3E,8S)-(-)-	$CHCl_3$	220(3.68)	33-0726-86
4-Undecen-6-one, 1,11-di-3-furanyl-4,8-dimethyl-, [S-(E)]- (caco-spongienone B)	$CHCl_3$	242(4.01)	33-0726-86
[S-(Z)]- (cacospongienone A)	$CHCl_3$	242(4.03)	33-0726-86
$C_{21}H_{28}O_4$			
19-Norpregn-4-en-20-yn-3-one, 10-hy-droperoxy-17-hydroxy-7-methyl-, (7α,17α)-	EtOH	236.5(3.19)	54-0111-86
1-Phenanthrenecarboxylic acid, 7-eth-enyl-1,2,3,4,4a,5,6,7,8,9,10,10a-dodecahydro-1,4a,7-trimethyl-5,9-dioxo-, methyl ester	MeOH	263(3.96)	102-1240-86
1-Phenanthrenecarboxylic acid, 1,2,3,4-4a,9,10,10a-octahydro-7-(1-hydroxy-1-methylethyl)-1,4a-dimethyl-9-oxo-, methyl ester	EtOH	228(3.13),262(2.83), 330(2.30)	105-0536-86
Pregna-1,4-diene-3,20-dione, 2,17-di-hydroxy-	MeOH	253(4.16)	78-5693-86
$C_{21}H_{29}BrN_2$			
Pyrrolo[2,3-b]indole, 6-bromo-3a-(1,1-dimethyl-2-propenyl)-1,2,3,3a,8,8a-hexahydro-1-methyl-5-(3-methyl-2-butenyl)-, cis-(-)- (flustramine D)	MeOH	231(3.85),292(3.49)	23-1312-86
$C_{21}H_{29}BrN_2O$			
1,2-Oxazino[6,5-b]indole, 7-bromo-2,3,4,4a,9,9a-hexahydro-2-methyl-4a,9-bis(3-methyl-2-butenyl)-(flustrarine B)	EtOH	216(4.34),255(3.59), 304(3.28)	1-0555-86
Pyrrolo[2,3-b]indole , 6-bromo-3a-(1,1-dimethyl-2-propenyl)-1,2,3,3a,8,8a-hexahydro-1-methyl-5-(3-methyl-2-butenyl)-, 1-oxide (flustramine D N-oxide)	MeOH	224(3.85),285(3.49)	23-1312-86
$C_{21}H_{29}BrO_2$			
Pregn-4-ene-3,20-dione, 6β-bromo-	MeOH	248(4.22)	13-0431-86B
$C_{21}H_{29}BrO_4$			
Androst-4-en-3-one, 17β-(bromoacet-oxy)-2α-hydroxy-	MeOH	241(4.09)	13-0347-86B
Prosta-5,7,10,14-tetraen-1-oic acid, 10-bromo-12-hydroxy-9-oxo-, methyl ester, (5Z,7E,14Z)- (bromovulone I)	n.s.g.	247(4.08),312(4.08)	77-0981-86
$C_{21}H_{29}IO_4$			
Iodovulone I	n.s.g.	240(4.34),313(4.29)	77-0981-86

Compound	Solvent	$\lambda_{max}(\log \epsilon)$	Ref.
$C_{21}H_{29}NO_3$			
3-Pyridinecarboxylic acid, 5-(cyclohexylhydroxyphenylmethyl)-1,2,5,6-tetrahydro-1-methyl-, methyl ester, (R*,R*)-(±)-	pH 7.2	216(1.31)	87-0125-86
(R*,S*)-(±)-	pH 7.2	214(1.38)	87-0125-86
$C_{21}H_{29}N_2S$			
10H-Phenothiazine-10-propanaminium, N,N,N-triethyl-, chloride	80% MeCN	251(4.51),301(3.61)	46-2469-86
$C_{21}H_{29}N_3O_2S$			
2,6-Diazabicyclo[2.2.2]oct-7-ene-3,5-dione, 8-(diethylamino)-1-(ethylthio)-2,6,7-trimethyl-4-phenyl-	MeCN	222(4.47),258(4.16), 340(3.88)	24-1315-86
$C_{21}H_{30}Br_2NTe$			
Methanaminium, N-[4-[1,1-dibromo-2,6-bis(1,1-dimethylethyl)-1,1-dihydro-4H-tellurin-4-ylidene]-2,5-cyclohexadien-1-ylidene]-N-methyl-, tribromide	CH_2Cl_2	496(4.48)	157-2250-86
$C_{21}H_{30}NO$			
Pyrylium, 4-[4-(dimethylamino)phenyl]-2,6-bis(1,1-dimethylethyl)-, tetrafluoroborate	MeOH	288(4.12),499(4.83)	44-4385-86
$C_{21}H_{30}NTe$			
Tellurinium, 4-[4-(dimethylamino)phenyl]-2,6-bis(1,1-dimethylethyl)-, tetrafluoroborate	CH_2Cl_2	611(4.82)	157-2250-86
$C_{21}H_{30}O$			
Retinal, 13-demethyl-13-ethyl-, all-E-	heptane	372(4.69)	5-0479-86
11Z	EtOH	253(--),379(--)	5-0479-86
13Z	EtOH	258(--),377(--)	5-0479-86
Retinal, 14-methyl-, all-E-	heptane	374(4.69)	5-0479-86
$C_{21}H_{30}O_2$			
1H-Phenalen-4-ol, 2,3,7,8,9,9a-hexahydro-5-methoxy-3,6,9-trimethyl-7-(2-methyl-1-propenyl)-, [3S-(3α,7β,9α,9aα)]-	MeOH	226(3.86),274(3.20), 283(3.18)	44-5140-86
2H-Pyran, 2,6-bis(1,1-dimethylethyl)-2-methoxy-4-(4-methylphenyl)-	MeOH	242(4.37),288(3.62)	44-4385-86
$C_{21}H_{30}O_3$			
5β-Androst-15-en-17-one, 3β-acetoxy-	EtOH	231(3.86)	19-0313-86
Benzaldehyde, 2-[2-(2,2-dimethyl-6-oxocyclohexyl)ethyl]-5-methoxy-4-(1-methylethyl)-, (S)-	MeOH	227(4.40),267(4.11), 323(3.68)	78-0529-86
1-Phenanthrenecarboxylic acid, 7-ethenyl-1,2,3,4,4a,5,6,7,8,9,10,10a-dodecahydro-1,4a,7-trimethyl-6-oxo-, methyl ester	MeOH	244(3.93)	102-1240-86
1-Phenanthrenecarboxylic acid, 7-ethenyl-1,2,3,4,4a,6,7,8,8a,9,10,10a-dodecahydro-1,4a,7-trimethyl-6-oxo-, methyl ester	MeOH	245(4.05)	102-1240-86

Compound	Solvent	λ_{max}(log ε)	Ref.
Pregn-4-ene-3,20-dione, 21-hydroxy-	97% H_2SO_4	288(4.28),372(--), 436(--)	94-1561-86
diluted with ethanol		600(4.3)	94-1561-86
2H-Pyran, 2,6-bis(1,1-dimethylethyl)-2-methoxy-4-(4-methoxyphenyl)-	MeOH	248(4.37)	44-4385-86
$C_{21}H_{30}O_4$			
Androst-15-en-17-one, 3-acetoxy-14-hydroxy-, (3β,5β,15β)-	EtOH	208(3.98)	19-0313-86
Cyclopenta[b][1]benzopyran-8(3H)-one, 3a,5,6,7,9,9a-hexahydro-5-hydroxy-1-(6-hydroxy-1,5-dimethyl-4-hexenyl)-3a-methyl- (toxin E)	MeOH	262(3.91)	163-1597-86
Cyclopenta[b][1]benzopyran-8(3H)-one, 3a,5,6,7,9,9a-hexahydro-7-hydroxy-1-(6-hydroxy-1,5-dimethyl-4-hexenyl)-3a-methyl- (toxin D)	MeOH	263(4.01)	163-1597-86
$C_{21}H_{30}O_5$			
2-Heptenal, 6-[5-[(2,5-dihydroxy-6-oxo-1-cyclohexen-1-yl)methyl]-4-hydroxy-4-methyl-1-cyclopenten-1-yl]-2-methyl- (toxin F)	MeOH	231(3.93),263(3.79), 288(3.74)	163-1597-86
	MeOH-HCl	223(--),261(--)	163-1597-86
Spongionellin	MeOH	234(4.05)	78-5369-86
$C_{21}H_{30}O_6$			
6,7a(2H)-Benzofurandiol, 2-(3,4-dimethoxyphenyl)hexahydro-5-methoxy-3-methyl-3a-(2-propenyl)-, (2α,3α,3aβ,5α,6α,7aα)-	MeOH	230(4.90),275(4.08)	102-2613-86
(2α,3α,3aβ,5α,6β,7aα)-	MeOH	220(4.43),287(3.90)	102-2613-86
$C_{21}H_{30}O_7$			
1,4-Phenanthrenedione, 4b,5,6,7,8,8a-9,10-octahydro-3,9,10-trihydroxy-8-(hydroxymethyl)-2-(2-methoxypropyl)-4b,8-dimethyl-	ether	271(3.96),400s(2.78)	33-0456-86
epimer	ether	271(3.94),400s(2.76)	33-0456-86
$C_{21}H_{31}N_3O_2S$			
1,3-Benzenediol, 4-dodecyl-6-(2-thiazolylazo)-	neutral	455(4.30)	138-0225-86
anion	pH 9.4+	495(4.5)	138-0225-86
$C_{21}H_{32}N_2O_9S$			
α-D-Glucopyranose, 2-[[(cyclohexylamino)thioxomethyl]amino]-2-deoxy-, 1,3,4,6-tetraacetate	EtOH	244(4.21)	136-0049-86I
$C_{21}H_{32}N_8O_5$			
L-Isoglutamine, N^2-[N-1-oxo-4-(1H-purin-6-ylamino)butyl]-L-alanyl-, 1,1-dimethylethyl ester	pH 2.9	272(4.13)	63-0757-86
	pH 7.0	268(4.14)	63-0757-86
	pH 10.5	271(4.13)	63-0757-86
$C_{21}H_{32}O_3$			
4aH-Dibenzo[a,d]cycloheptene-4a,5-diol, 1,2,3,4,5,10,11,11a-octahydro-7-methoxy-1,1-dimethyl-8-(1-methylethyl)-, [4aR-(4aα,5α,11aβ)]-	MeOH	221(3.93),278(3.48)	78-0529-86
Gracilin E	MeOH	224(3.82)	78-5369-86

Compound	Solvent	λ_{max}(log ϵ)	Ref.
$C_{21}H_{32}O_4$			
2-Cyclohexen-1-one, 2-[[5-(1,5-dimethyl-1,4-hexadienyl)-2-hydroxy-2-methylcyclopentyl]methyl]-3,6-dihydroxy- (toxin A)	MeOH	292(3.92)	163-0801-86
	MeOH-HCl	263(--)	163-0801-86 +163-1597-86
Toxin B	MeOH	295(4.03)	163-0801-86
	MeOH-HCl	263(--)	163-1597-86
$C_{21}H_{32}O_5$			
Citricolic acid	MeOH	217(4.00)	20-0699-86
2-Cyclohexen-1-one, 2-[[5-(2,6-dimethyl-1,5-heptadienyl)tetrahydro-3-hydroxy-3-methyl-2-furanyl]methyl]-3,6-dihydroxy- (toxin C)	MeOH	287(3.89)	163-0801-86
	MeOH-HCl	261(--)	163-0801-86
$C_{21}H_{32}O_6$			
Dendrillol-4	hexane	230(2.48)	12-1643-86
Oxacyclotetradeca-7,12-diene-2,4-dione, 14-ethyl-5-hydroxy-5-(hydroxymethyl)-7-methoxy-3,9,13-trimethyl-11-methylene- (galbonolide A)	MeOH	220s(3.7)	158-1760-86
	MeOH-HCl	230(3.9)	158-1760-86
	MeOH-NaOH	259(3.9)	158-1760-86
$C_{21}H_{32}O_{10}$			
Ebuloside	MeOH	204(3.79),279(1.39)	33-0156-86
$C_{21}H_{33}PS_3Si_3$			
Phosphine, tris[5-(trimethylsilyl)-2-thienyl]-	CH_2Cl_2	252(4.28),285(4.16)	70-1919-86
$C_{21}H_{34}N_4O_4$			
Amphibine-I, (-)-	MeOH	282(3.49)	100-0745-86
$C_{21}H_{34}O$			
2-Cyclohexen-1-one, 3-(4'-propyl-[1,1'-bicyclohexyl]-4-yl)-, [trans(trans)]-	MeOH	239(4.24)	24-0387-86
$C_{21}H_{34}O_3$			
Benzenemethanol, 2-[2-(6-hydroxy-2,2-dimethylcyclohexyl)ethyl]-5-methoxy-4-(1-methylethyl)-	MeOH	227(3.91),278(3.34)	78-0529-86
$C_{21}H_{34}O_5$			
3-Furancarboxylic acid, 2,5-dihydro-4-methyl-5-oxo-2-(14-oxopentadecyl)- (isomuronic acid)	MeOH	229(4.22)	102-0453-86
3-Furancarboxylic acid, tetrahydro-4-methylene-5-oxo-2-(14-oxopentadecyl)-, (2S-cis)- (allopertusaric acid)	MeOH	211(4.0)	102-0453-86
$C_{21}H_{34}O_7$			
Neorustmicin B	MeOH	216(3.28)	158-1016-86
$C_{21}H_{36}N_4O$			
2(3H)-Naphthalenone, 4,4a,5,6,7,8-hexahydro-4a-methyl-3,8-bis(4-methyl-1-piperazinyl)-, (3α,4aβ,8β)-	EtOH	236(4.12)	107-0195-86
$C_{21}H_{36}N_6O_8Si$			
1,2,3-Triazole, 5-[[(dimethylamino)-	MeOH	265(3.82)	44-1050-86

Compound	Solvent	λ_{max}(log ε)	Ref.
methylene]amino]-4-[cyano(trimethyl-silyloxy)methyl]-1-(2,3,5-tri-O-acetyl-β-D-ribofuranosyl)- (cont.)			44-1050-86
$C_{21}H_{36}OS_2$			
Benzene, 1-methyl-4-[[1-(methylthio)-tridecyl]sulfinyl]-	dioxan	263(3.7)	88-6381-86
$C_{21}H_{36}O_4$			
Cyclohexanehexanoic acid, 2-(2,3-di-oxobutyl)-β,1,3,3-tetramethyl-, methyl ester, [1R-[1α(S*),2β]]-	EtOH	265(3.87)	78-4563-86
5-Nonanone, 1-[3-hydroxy-6-(2-hydroxy-ethylidene)-2-methyl-2-oxepanyl]-4,8-dimethyl-7-methylene- (tomentol)	MeOH	210.5(3.70)	102-2567-86
1-Nonen-5-one, 9-[3-hydroxy-6-(2-hy-droxyethylidene)-2-methyl-2-oxepan-yl]-2,3,6-trimethyl- (tomentanol)	MeOH	210(3.78)	102-2567-86
1,2-Propanediol, 3-[(18-hydroxy-1,5-octadecadien-3-ynyl)oxy]-, (Z,Z)-(+)- (raspailyne A)	MeOH	261s(4.15),273(4.26), 288(4.18)	77-0077-86
$C_{21}H_{36}O_5$			
3-Furancarboxylic acid, tetrahydro-4-methyl-5-oxo-2-(14-oxopenta-decyl)- (dihydropertusaric acid)	MeOH	207(2.74)	102-0453-86
$C_{21}H_{38}O_2Si$			
1-Penten-3-one, 5-[2-[1-[[(1,1-dimeth-ylethyl)dimethylsilyl]oxy]-2-propen-yl]-1-methylcyclohexyl]-, [1α,2β(S*))]-(±)-	MeOH	210(3.93)	39-1303-86C
$C_{21}H_{40}N_2O_5Si_2$			
Uridine, 2'-deoxy-3',5'-bis-O-(1,1-dimethylethyl)dimethylsilyl]-	EtOH	262(3.94)	152-0643-86
$C_{21}H_{42}N_4O_5Si_2$			
1H-Imidazole-4-carboxamide, 5-amino-1-[3,5-bis-O-[(1,1-dimethylethyl)-dimethylsilyl]-α-D-xylofuranosyl]-	MeOH	266(4.21)	87-0203-86
$C_{21}H_{46}O_7Si_4$			
2-Propenoic acid, 3-[2,3,4,5-tetrakis-(trimethylsilyloxy)pyran-6-yl]-, methyl ester, (E)-	MeOH	205(4.13)	48-0777-86

Compound	Solvent	λ_{max}(log ε)	Ref.
$C_{22}H_4Cl_{15}O$			
Methyl, bis(pentachlorophenyl)-[2,3,5,6-tetrachloro-4-(3-chloro-3-oxopropyl)phenyl]-	$CHCl_3$	283(3.79),340s(3.84), 368s(4.30),383(4.58), 480s(3.03),507(3.06), 560(3.05)	44-2472-86
$C_{22}H_5Cl_{14}O_2$			
Methyl, [4-(acetoxymethyl)-2,3,5,6-tetrachlorophenyl]bis(pentachlorophenyl)-	C_6H_{12}	223(4.99),287(3.85), 366s(4.28),384(4.59), 506(3.06),562(3.03)	44-2472-86
Methyl, [4-(2-carboxyethyl)-2,3,5,6-tetrachlorophenyl]bis(pentachlorophenyl)-	$CHCl_3$	288(3.84),367s(4.23), 384(4.55),510(3.02), 563(3.01)	44-2472-86
$C_{22}H_6Cl_{14}O_2$			
Benzenemethanol, 4-[bis(pentachlorophenyl)methyl]-2,3,5,6-tetrachloro-, acetate	C_6H_{12}	221(5.06),294(3.13), 304(3.18)	44-2472-86
Benzenepropanoic acid, 4-[bis(pentachlorophenyl)methyl]-2,3,5,6-tetrachloro-	$CHCl_3$	294(3.09),303(3.09)	44-2472-86
$C_{22}H_9ClF_{18}O_6S_2$			
1-Butanesulfonic acid, 1,1,2,2,3,3-4,4,4-nonafluoro-1-(4-chlorophenyl)-2-phenyl-1,2-ethenediyl ester, (E)-	C_6H_{12}	264(4.33)	24-2220-86
(Z)-	C_6H_{12}	227(4.33),272(4.13)	24-2220-86
$C_{22}H_{10}$			
Benzo[7,8]cyclodeca[1,2,3,4-def]biphenylene, 5,6,13,14-tetradehydro-	C_6H_{12}	235(4.17),263(4.75), 268(4.76),278(4.88), 292(4.51),301(4.31), 309(4.42),353(3.48), 372(3.66),392(3.79)	44-1088-86
$C_{22}H_{10}Br_4N_2O_2$			
2,5-Cyclohexadiene-1,4-dione, 2,5-dibromo-3,6-bis(5-bromo-1H-indol-3-yl)-	MeCN	230(4.66),285(4.37), 491(3.74)	5-1765-86
2,5-Cyclohexadiene-1,4-dione, 2,6-dibromo-3,5-bis(5-bromo-1H-indol-3-yl)-	MeCN	230(4.68),280(4.34), 500(3.66)	5-1765-86
$C_{22}H_{10}F_{18}O_6S_2$			
1-Butanesulfonic acid, 1,1,2,2,3,3-4,4,4-nonafluoro-1,2-diphenyl-1,2-ethenediyl ester, (E)-	MeCN	255(4.25)	24-2220-86
(Z)-	C_6H_{12}	218(4.28),265(4.07)	24-2220-86
$C_{22}H_{10}N_4O_8$			
Benzo[f]pyrido[1,2-a]indole-6,11-dione, 12-(2,4,6-trinitrophenyl)-	DMF	495(3.8722)	2-1126-86
$C_{22}H_{10}O_5$			
6,13-Epoxypentacene-1,4,8,11-tetrone, 6,13-dihydro-	dioxan	250(4.27),262(4.38), 270(4.44),334(3.40)	44-4160-86
$C_{22}H_{11}N_3O_2S_2$			
Benzo[a][1,4]benzothiazino[3,2-c]phenothiazine, 6-nitro-	$CHCl_3$	277(4.68),351(4.27), 375(4.23),557(4.26),	4-0589-86

Compound	Solvent	λ_{max}(log ε)	Ref.
(cont.)		590s(4.21)	4-0589-86
$C_{22}H_{11}N_3O_6$			
Benzo[f]pyrido[1,2-a]indole-6,11-dione, 12-(2,4-dinitrophenyl)-	DMF	414(3.663),497(3.884)	2-1126-86
$C_{22}H_{12}Cl_{10}O_6$			
Benzene, 1,4-dichloro-2,3-dimethoxy-5,6-bis(2,3,4,5-tetrachloro-6-methoxyphenoxy)-	EtOH	295(3.59)	23-0950-86
$C_{22}H_{12}N_2O_2$			
Naphtho[2,3-h]quinoline-7,12-dione, 2-(2-pyridinyl)-	DMSO	400s(4.591),475(4.532), 540(4.591)	2-0616-86
Naphtho[2,3-h]quinoline-7,12-dione, 2-(3-pyridinyl)-	DMSO	410s(3.732),485(3.681), 540(3.663)	2-0616-86
Naphtho[2,3-h]quinoline-7,12-dione, 2-(4-pyridinyl)-	DMSO	320s(4.146),440(4.114)	2-0616-86
$C_{22}H_{12}N_4$			
Dibenzo[cd,c'd'][1,2,4,5]tetrazino-[1,6-a:4,3-a']diindole	CH_2Cl_2	295(4.4),490(4.2)	33-1521-86
$C_{22}H_{12}O$			
Indeno[1,2,3-cd]pyren-1-ol	EtOH	242(4.80),247(4.86), 296(4.47),303(4.49), 313(4.42),340(3.95), 358(4.12),378(4.15), 386(4.02)	44-2428-86
Indeno[1,2,3-cd]pyren-2-ol	EtOH	252(4.80),295(4.38), 306(4.54),318(4.46), 338(4.01),355(4.14), 365(4.03),374(3.96)	44-2428-86
Indeno[1,2,3-cd]pyren-6-ol	EtOH	249(4.80),270(4.47), 281(4.48),292(4.44), 303(4.37),314(4.22), 344(3.88),360(3.91), 378(4.02)	44-2428-86
Indeno[1,2,3-cd]pyren-7-ol	EtOH	223(4.74),239(4.83), 250(4.80),270(4.44), 286s(4.44),297(4.54), 313(4.22),325s(3.92), 350s(3.85),370s(4.19), 388s(4.32),399(4.33)	44-2428-86
Indeno[1,2,3-cd]pyren-8-ol	EtOH	252(4.69),274(4.40), 282(4.41),312(4.59), 324(4.61),356(4.08), 367(4.01),375(4.05)	44-2428-86
Indeno[1',2',3':1,10]pyreno[4,5-b]oxirene, la,llb-dihydro-	EtOH	223(4.60),251(4.56), 258(4.53),267(4.45), 283(4.49),288(4.50), 293(4.53),298(4.50), 305(4.58),348(4.03), 357(4.05),373(3.85)	44-2428-86
$C_{22}H_{13}Br_2N_3$			
Pyrimido[1,2-a]benzimidazole, 2,4-bis(4-bromophenyl)-	n.s.g.	345(4.28),390(3.75)	103-0927-86
$C_{22}H_{13}ClN_2O_2$			
1H-Oxazolo[3,4-a]quinazolin-1-one,	CH_2Cl_2	269(4.42),299(4.42),	39-2295-86C

Compound	Solvent	$\lambda_{max}(\log \epsilon)$	Ref.
3-(4-chlorophenyl)-5-phenyl- (cont.)		505(3.96)	39-2295-86C
$C_{22}H_{13}Cl_2N_3$			
Pyrimido[1,2-a]benzimidazole, 2,4-bis(4-chlorophenyl)-	n.s.g.	346(4.26),390(3.75)	103-0927-86
$C_{22}H_{13}N_3O_4$			
Benzo[f]pyrido[1,2-a]indole-6,11-dione, 12-(4-amino-2-nitrophenyl)-	DMF	538(3.7853)	2-1126-86
$C_{22}H_{13}N_3S_2$			
Benzo[a][1,4]benzothiazino[3.2-c]phenothiazin-6-amine	$CHCl_3$	266(4.73),323(4.10), 377(4.10),530(4.33), 575s(4.20)	4-0589-86
$C_{22}H_{13}N_5O_2S$			
Benzo[de][1,2,3]triazolo[4,5-g]isoquinoline-4,6(5H,9H)-dione, 5-(4-methylphenyl)-9-(2-thiazolyl)-	DMF	383(3.947)	2-0496-86
$C_{22}H_{14}$			
Bicyclo[4.1.0]hepta-1,3,5-triene, 7-(5H-dibenzo[a,d]cyclohepten-5-ylidene)-	C_6H_{12}	308(3.82),383(4.31)	88-5159-86
	MeCN	299(3.89),376(4.30)	88-5159-86
1H-Cyclopropa[b]naphthalene, 1-(7H-benzocyclohepten-7-ylidene)-	C_6H_{12}	407(4.09),438(4.54), 458(4.35),475(4.92)	88-5159-86
	MeCN	404(4.17),434(4.63), 454(4.45),470(4.93)	88-5159-86
Dibenzo[a,g]cyclotetradecene, 15,16-17,18-tetradehydro-	isooctane	210s(4.81),236s(4.32), 244(4.38),289(4.91), 299(4.92),307(4.93), 337(4.31),372(4.14), 414s(2.79)	18-2209-86
$C_{22}H_{14}Br_2N_4$			
2,4-Cyclohexadiene-1,1,3-tricarbonitrile, 2-amino-4,6-bis(4-bromophenyl)-6-methyl-	n.s.g.	212(4.44),228(4.44), 258(4.58),320(3.90)	104-0230-86
1,3,5-Heptatriene-1,1,3-tricarbonitrile, 2-amino-4,6-bis(4-bromophenyl)-	n.s.g.	228(4.47),264(4.56), 320(4.20)	104-0230-86
Propanedinitrile, [4,6-bis(4-bromophenyl)-3-cyano-5,6-dihydro-6-methyl-2(1H)-pyridinylidene]-	n.s.g.	220(4.34),252(4.42), 336(4.28)	104-0230-86
$C_{22}H_{14}Br_4$			
1,1'-Binaphthalene, 2,2'-bis(dibromomethyl)-	EtOH	228(4.9),280(4.2)	44-0589-86
$C_{22}H_{14}ClFN_4$			
4H-Imidazo[1,5-a][1,4]benzodiazepine, 8-chloro-6-(2-fluorophenyl)-3-(2-pyridinyl)-	isoPrOH	217(4.58),247(4.46), 259s(4.43),292(4.27)	4-1303-86
$C_{22}H_{14}Cl_2N_4$			
2,4-Cyclohexadiene-1,1,3-tricarbonitrile, 2-amino-4,6-bis(4-chlorophenyl)-6-methyl-	n.s.g.	212(4.41),224(4.45), 258(4.59),320(3.95)	104-0230-86
1,3,5-Heptatriene-1,1,3-tricarbonitrile, 2-amino-4,6-bis(4-chlorophenyl)-	n.s.g.	264(4.59),320(4.28)	104-0230-86

Compound	Solvent	λ_{max}(log ε)	Ref.
Propanedinitrile, [4,6-bis(4-chlorophenyl)-3-cyano-5,6-dihydro-6-methyl-2(1H)-pyridinylidene]-	n.s.g.	222(4.19),254(4.27), 340(4.28)	104-0230-86
$C_{22}H_{14}NO$			
7H-Isoindolo[2,1-b]isoquinolinium, 7-oxo-12-phenyl-, bromide	H_2SO_4	235(4.35),270(4.56), 305s(4.09),350s(3.73), 463(3.48)	142-0969-86
$C_{22}H_{14}N_2O$			
Benzonitrile, 4-(5-[1,1'-biphenyl]-4-yl-2-oxazolyl)-	toluene	346(4.43)	103-1026-86
12H-Isoquino[2,3-a]quinazolin-12-one, 5-phenyl-	EtOH	236(4.69),246s(4.68), 270s(4.43),303(4.41), 400(4.42)	39-2295-86C
$C_{22}H_{14}N_2O_2$			
1H-Oxazolo[3,4-a]quinazolin-1-one, 3,5-diphenyl-	CH_2Cl_2	268(4.24),297(4.29), 453(3.85)	39-2295-86C
$C_{22}H_{14}O$			
Benzo[b]naphtho[2,1-d]furan, 5-phenyl-	EtOH	205(4.35),248(4.73), 255(4.78),278(3.90), 288(3.90),296(3.70), 326(3.09),341(2.94)	4-1625-86
$C_{22}H_{14}O_2$			
Indeno[1,2,3-cd]pyrene-1,2-diol, 1,2-dihydro-, cis	EtOH	225(4.57),245(4.53), 258(4.55),265(4.47), 286(4.42),293(4.42), 304(4.43),345(4.02), 355(4.02),375(3.74)	44-2428-86
trans	EtOH	225(4.58),245(4.56), 258(4.56),265(4.51), 284(4.40),293(4.39), 305(4.39),345(4.09), 355(4.02),376(3.74)	44-2428-86
2-Propenal, 3,3'-(1,3-butadiyne-1,4-diyldi-2,1-phenylene)bis-, (E,E)-	THF	224s(4.27),254(4.51), 277(4.45),294s(4.38), 330s(4.10),367s(3.92)	18-2209-86
$C_{22}H_{14}O_4$			
Ethanedione, 1,1'-(1,4-phenylene)-bis[2-phenyl-	dioxan	274(5.12),402(2.78)	104-1117-86
1H-Indene-1,3(2H)-dione, 2-[3-(1,3-dihydro-1,3-dioxo-2H-inden-2-ylidene)-2-methyl-1-propenyl]-	EtOH	248(4.47),561(4.89)	70-1446-86
$C_{22}H_{14}O_5$			
6,13-Epoxypentacene-1,4,8,11-tetrone, 4a,5,6,13,14,14a-hexahydro-	dioxan	257(4.41),330(3.54)	44-4160-86
$C_{22}H_{14}O_6$			
[1,2'-Binaphthalene]-1',4',5,6-tetrone, 4,5'-dihydroxy-2,7'-dimethyl-	EtOH	252.5(4.30),445(3.77), 580s(3.08)	39-0675-86C
$C_{22}H_{14}O_7$			
[1',2-Binaphthalene]-1,4,5',8'-tetrone, 4',5,8-trihydroxy-2',3-dimethyl-	EtOH	217(4.52),253(4.22), 295s(3.87),462s(3.83), 480(3.86),511(3.83), 549(3.61)	5-1669-86

Compound	Solvent	$\lambda_{max}(\log \epsilon)$	Ref.
(cont.)	MeOH-NaOH	573(--),612(--)	5-1669-86
[1',2-Binaphthalene]-5,5',8,8'-tetrone, 1,4,4'-trihydroxy-6,7'-dimethyl-	$CHCl_3$	261(4.33),442s(3.79), 467s(3.84),492(3.91), 524(3.93),565(3.72)	5-1669-86
	EtOH	437s(--),465s(--), 487(--),517(--), 557(--)	5-1669-86
	EtOH-NaOH	578(--),615(--)	5-1669-86
[2,2'-Binaphthalene]-1,1',4,4'-tetrone, 5,8,8'-trihydroxy-3,3'-dimethyl-	$CHCl_3$	276(4.16),492(3.87), 525(3.80),566(3.59)	5-1669-86
	EtOH	442s(--),462s(--), 488(--),521(--), 560(--)	5-1669-86
	EtOH-NaOH	586(--),624(--)	5-1669-86
[2,2'-Binaphthalene]-1,4,5',8'-tetrone, 1',5,8-trihydroxy-3,3'-dimethyl-	$CHCl_3$	256(4.29),292s(3.97), 420s(3.76),442s(3.84), 470s(3.89),490(3.93), 522(3.92),562(3.70)	5-1669-86
	EtOH	442s(--),465s(--), 486(--),517(--), 556(--)	5-1669-86
	EtOH-NaOH	586(--),629(--)	5-1669-86
[2,2'-Binaphthalene]-1',4',5,8-tetrone, 1,4,5'-trihydroxy-7,7'-dimethyl- (euclanone)	$CHCl_3$	257(4.25),280(4.19), 497(3.96),522(3.95), 557s(3.74)	5-1669-86
[2,2'-Binaphthalene]-1',4',5,8-tetrone, 1,4,8'-trihydroxy-6,6'-dimethyl-	$CHCl_3$	280(4.22),495(3.92), 520(3.91),557s(3.69)	5-1669-86
	EtOH	490(--),517(--), 552s(--)	5-1669-86
	EtOH-NaOH	595(--),625(--)	5-1669-86
[2,2'-Binaphthalene]-5,5',8,8'-tetrone, 1,1',4-trihydroxy-6,6'-dimethyl-	$CHCl_3$	266(4.33),497(3.94), 525(3.95),562s(3.72)	5-1669-86
	EtOH	492(--),521(--), 557s(--)	5-1669-86
	EtOH-NaOH	602s(--),627(--)	5-1669-86
$C_{22}H_{14}O_8$			
[2,2'-Binaphthalene]-5,5',8,8'-tetrone, 5,5',8,8'-tetrahydroxy-3,3'-dimethyl-	$CHCl_3$	282(4.28),496(4.09), 526(4.18),566(4.02)	5-1669-86
	EtOH	492(--),521(--), 560(--)	5-1669-86
	EtOH-NaOH	587(--),635(--)	5-1669-86
[2,2'-Binaphthalene]-5,5',8,8'-tetrone, 1,1',4,4'-tetrahydroxy-6,6'-dimethyl-	$CHCl_3$	280(4.23),502s(4.04), 527(4.11),562s(3.97)	5-1669-86
	EtOH	495(--),521(--), 557(--)	5-1669-86
	EtOH-NaOH	592(--),632(--)	5-1669-86
[2,2'-Binaphthalene]-5,5',8,8'-tetrone, 1,1',4,4'-tetrahydroxy-7,7'-dimethyl-	$CHCl_3$	282(4.09),507s(3.96), 529(4.03),560s(3.91)	5-1669-86
	EtOH	502(--),522(--), 555(--)	5-1669-86
	EtOH-NaOH	592(--),627(--)	5-1669-86
$C_{22}H_{15}BO_5$			
Boron, (1,3-diphenyl-1,3-propanedionato-O,O')[2-hydroxybenzoato-(2-)-O^1,O^2]-, (T-4)-	CH_2Cl_2	277(4.13),288(4.15), 305s(4.05)	48-0755-86
$C_{22}H_{15}BrN_2O_2$			
6H-Anthra[1,9-cd]isoxazol-6-one, 3-bromo-5-[(3,5-dimethylphenyl)amino]-	EtOH	505(4.24),530(4.29)	103-1363-86

Compound	Solvent	$\lambda_{max}(\log \epsilon)$	Ref.
Methanimidamide, N'-(4-bromophenyl)-N-(9,10-dihydro-9,10-dioxo-1-anthracenyl)-N-methyl-	EtOH	202(4.556),245(4.77), 405(3.462)	104-1566-86
$C_{22}H_{15}ClN_2O$			
1,4-Naphthalenedione, 2-chloro-, 4-(diphenylhydrazone)	EtOH	243(4.25),286(4.12), 339s(3.73),474(4.43)	104-0124-86
$C_{22}H_{15}ClN_2O_2$			
6H-Anthra[1,9-cd]isoxazol-6-one, 3-chloro-5-[(3,5-dimethylphenyl)-amino]-	EtOH	505(4.30),530(4.35)	103-1363-86
1,2-Naphthalenedione, 3-chloro-4-hydroxy-, 2-(diphenylhydrazone)	hexane	225s(4.28),242s(4.24), 247s(4.22),286(4.34), 314(4.21),330(4.12), 379(3.45),417(3.30), 480(3.58)	104-0124-86
	$CHCl_3$	245(4.23),251s(4.22), 289(4.31),321(4.21), 386(3.44),511(3.54)	104-0124-86
1,4-Naphthalenedione, 2-chloro-3-[[4-(phenylamino)phenyl]amino]-	EtOH	260(4.29),294(4.55), 407s(3.26),546(3.64)	104-0124-86
	$CHCl_3$	253(4.25),295(4.54), 405s(3.16),530(3.60)	104-0124-86
$C_{22}H_{15}ClN_4$			
1H-Pyrazolo[3,4-b]quinolin-4-amine, N-(4-chlorophenyl)-1-phenyl-	EtOH	245(4.53),270(4.50), 385s(4.01),397(4.07)	11-0331-86
$C_{22}H_{15}FN_4$			
1H-Pyrazolo[3,4-b]quinolin-4-amine, N-(4-fluorophenyl)-1-phenyl-	EtOH	245(4.44),277(4.42), 388(3.89),397(3.95)	11-0331-86
$C_{22}H_{15}N$			
5,11[1',2']-Benzeno-5H-cyclohepta[b]-naphthalene-5-carbonitrile, 5a,11-dihydro-	EtOH	273(4.16)	18-1095-86
1-Cyclopropene-1-carbonitrile, 2,3,3-triphenyl-	MeOH	293(4.27)	104-1389-86
1H-Indene-1-carbonitrile, 2,3-diphenyl-	MeOH	260(4.43)	104-1389-86
1H-Indene-2-carbonitrile, 1,3-diphenyl-	MeOH	237(4.40),290(4.06)	104-1389-86
$C_{22}H_{15}NO_2$			
1,2-Naphthalenedione, 3-phenyl-4-(phenylamino)-	$CHCl_3$	378(3.48),479(3.58)	39-2233-86C
$C_{22}H_{15}N_2O_2$			
3H-Indolizinium, 3-(1,2-dihydro-2-oxo-3H-indol-3-ylidene)-1-hydroxy-2-phenyl-, chloride, (Z)-	80% MeOH	475(3.30),670(3.18)	4-1315-86
	MeOH-HCl	670(3.15)	4-1315-86
	MeOH-NaOH	460(3.87)	4-1315-86
$C_{22}H_{15}N_3$			
1H-Pyrazole-4-carbonitrile, 1,3,5-triphenyl-	MeOH	249(4.48)	104-1389-86
3H-Pyrazole-4-carbonitrile, 3,3,5-triphenyl-	MeOH	245(4.28),295(3.85), 320(3.85)	104-1389-86
4H-Pyrazole-4-carbonitrile, 3,4,5-triphenyl-	MeOH	331(4.21)	104-1389-86

Compound	Solvent	λ_{max}(log ϵ)	Ref.
Pyrimido[1,2-a]benzimidazole, 2,4-diphenyl-	n.s.g.	343(4.23),390(3.74)	103-0927-86
$C_{22}H_{15}N_3O$			
4H-Pyrazolo[3,4-c]quinolin-4-one, 2,5-dihydro-1,2-diphenyl-	EtOH	247(4.41),256s(4.38), 270s(4.24),300(4.04)	142-3181-86
$C_{22}H_{15}N_3O_2$			
Spiro[9H-fluorene-9,3'(3'aH)-pyrazolo-[1,5-a]pyridine]-2'-carboxylic acid, 5'-cyano-, methyl ester	CH_2Cl_2	406(3.96)	89-0572-86
$C_{22}H_{15}N_3O_2S$			
4(1H)-Pyrimidinethione, 5-nitro-1,2,6-triphenyl-	EtOH	256(4.01),342(4.09)	18-3871-86
$C_{22}H_{15}N_3O_7S$			
4-Thia-1-azabicyclo[3.2.0]hept-2-ene-2-carboxylic acid, 6-(1,3-dihydro-1,3-dioxo-2H-isoindol-2-yl)-3-methyl-7-oxo-, (4-nitrophenyl)methyl ester	$CHCl_3$	270(4.20),306(4.12)	44-3413-86
$C_{22}H_{15}N_3O_8S_2$			
4,5-Dithia-1-azabicyclo[4.2.0]oct-2-ene-2-carboxylic acid, 7-(1,3-dihydro-1,3-dioxo-2H-isoindol-2-yl)-3-methyl-8-oxo-, (4-nitrophenyl)methyl ester, 5-oxide	EtOH	268(4.18)	44-3413-86
$C_{22}H_{15}N_3O_9S_2$			
4,5-Dithia-1-azabicyclo[4.2.0]oct-2-ene-2-carboxylic acid, 7-(1,3-dihydro-1,3-dioxo-2H-isoindol-2-yl)-3-methyl-8-oxo-, (4-nitrophenyl)methyl ester, 4,4-dioxide	EtOH	271(4.17)	44-3413-86
$C_{22}H_{15}N_5O_2$			
6H-Anthra[1,9-cd]isoxazol-6-one, 3-azido-5-[(3,5-dimethylphenyl)amino]-	EtOH	495(4.22),527(4.27)	104-1192-86
$C_{22}H_{15}N_5O_3$			
6H-Anthra[1,9-cd]isoxazol-6-one, 3-azido-5-[(4-ethoxyphenyl)amino]-	EtOH	500(3.97),520(4.02)	104-1192-86
$C_{22}H_{16}$			
9H-Benzo[def]fluorene, 9-(phenylmethyl)-, lithium salt	THF	535(3.83)	35-7016-86
Phenanthrene, 1-(2-phenylethenyl)-, cis	$C_6H_{11}Me$	301(4.155)	32-0705-86
trans	$C_6H_{11}Me$	320(4.39),370s(3.079)	32-0705-86
	MeOH	224(4.66),230(4.70), 264(4.55),319(4.40), 370(3.16)	151-0296-86B
Phenanthrene, 2-(2-phenylethenyl)-, cis	$C_6H_{11}Me$	280(4.633),294(4.536), 310(4.269)	32-0705-86
trans	$C_6H_{11}Me$	281(4.713),294(4.701), 312s(4.589),325(4.699), 340s(4.548),372(3.146)	32-0705-86
	MeOH	220(4.43),238(4.54), 247(4.23),271s(4.52),	151-0296-86B

Compound	Solvent	λmax(log ε)	Ref.
(cont.)		280(4.68),293(4.65), 311(4.52),324(4.63), 338s(4.49),370(3.05)	151-0296-86B
Phenanthrene, 3-(2-phenylethenyl)-, cis	$C_6H_{11}Me$	318(4.318)	32-0705-86
trans	$C_6H_{11}Me$	285s(4.322),320s(4.544), 330(4.617),345s(4.477), 368(3.602)	32-0705-86
	MeOH	241(4.38),249(4.49), 260s(4.31),270(4.40), 278(4.40),287s(4.13), 320s(4.43),330(4.52), 343(4.40),367(3.57)	151-0296-86B
Phenanthrene, 4-(2-phenylethenyl)-, cis	$C_6H_{11}Me$	263(4.477),284s(4.354)	32-0705-86
trans	$C_6H_{11}Me$	275(4.415),283(4.425), 294(4.408),310s(4.238)	32-0705-86
Phenanthrene, 9-(2-phenylethenyl)-, cis	$C_6H_{11}Me$	305(4.041)	32-0705-86
trans	$C_6H_{11}Me$	321(4.342)	32-0705-86
$C_{22}H_{16}Br_2$			
1,1'-Binaphthalene, 2,2'-bis(bromo-methyl)-	EtOH	230(4.9),285(4.2)	44-0589-86
$C_{22}H_{16}Br_2N_2O$			
Ethanone, 1-(4-bromophenyl)-2-[3-(4-bromophenyl)-1,2-dihydro-2-quinox-alinyl]-	toluene	403(3.67)	103-0538-86
	MeOH	403(--)	103-0538-86
$C_{22}H_{16}Br_2N_2S_4$			
6H-1,3-Thiazine, 4-(4-bromophenyl)-6-[4-(4-bromophenyl)-2-(methylthio)-6H-1,3-thiazin-6-ylidene]-2-(methyl-thio)-	$CHCl_3$	287(4.43),334(4.18), 505(3.91)	118-0916-86
$C_{22}H_{16}ClN_3$			
1H-Pyrrole, 3-[(4-chlorophenyl)azo]-2,5-diphenyl-	EtOH	205(4.32),232(4.18), 260(4.18),316(4.55), 430(4.16)	103-0499-86
$C_{22}H_{16}Cl_2N_2O$			
Ethanone, 1-(4-chlorophenyl)-2-[3-(4-chlorophenyl)-1,2-dihydro-2-quinox-alinyl]-	toluene	403(3.70)	103-0538-86
	MeOH	402(--)	103-0538-86
$C_{22}H_{16}Fe_2N_2O_2$			
Iron, dicarbonyl(η^5-2,4-cyclopenta-dien-1-yl)(3,3-dicyano-2-ferrocen-yl-2-propenyl)-, iodide triiodide	CH_2Cl_2	294(4.676),369(4.606), 612(3.640)	70-2105-86
$C_{22}H_{16}N_2O$			
1,4-Naphthalenedione, mono(diphenyl-hydrazone)	hexane	242(4.22),248s(4.18), 287(4.21),337(3.42), 444(4.25)	104-0124-86
	EtOH	240(4.21),288(4.16), 338s(3.52),465(4.38)	104-0124-86
1(4H)-Naphthalenone, 4-[[4-(phenyl-amino)phenyl]imino]-	hexane	245(4.24),250s(4.20), 301(4.47),332s(3.68), 470(3.97)	104-0124-86

Compound	Solvent	λ_{max}(log ϵ)	Ref.
(cont.)	EtOH	247(4.23),252(4.21), 314(4.44),531(3.94)	104-0124-86
Pyrazine, triphenyl-, 1-oxide	EtOH	270(4.26),339(3.52)	142-0785-86
Pyrazine, triphenyl-, 4-oxide	EtOH	271(4.19),341(3.44)	142-0785-86
$C_{22}H_{16}N_2O_2$			
6H-Anthra[1,9-cd]isoxazol-6-one, 5-[(3,4-dimethylphenyl)amino]-	EtOH	500(4.29),526(4.33)	103-1363-86
6H-Anthra[1,9-cd]isoxazol-6-one, 5-[(3,5-dimethylphenyl)amino]-	EtOH	500(4.31),526(4.35)	103-1363-86
1,4-Naphthalenedione, 2-hydroxy-, 4-(diphenylhydrazone)	EtOH	243(4.40),289(4.19), 342s(3.66),480(4.26)	104-0124-86
1,4-Naphthalenedione, 5-hydroxy-, 1-(diphenylhydrazone)	hexane	251s(4.07),286(4.14), 392(3.80),498(4.18)	104-0124-86
	EtOH	263s(4.05),286(4.14), 392(3.84),513(4.29)	104-0124-86
$C_{22}H_{16}N_2O_4$			
Benzoxazole, 2,2'-(2,3-dimethoxy-1,4-phenylene)bis-	2-MeTHF	310(4.3),368(4.4)	46-1455-86
$C_{22}H_{16}N_4$			
2,4-Cyclohexadiene-1,1,3-tricarbonitrile, 2-amino-6-methyl-4,6-diphenyl-	n.s.g.	224(4.48),254(4.60), 320(4.11)	104-0230-86
1,3,5-Heptatriene-1,1,3-tricarbonitrile, 2-amino-4,6-diphenyl-	n.s.g.	257(4.21),274(4.17), 328(4.19)	104-0230-86
1H-Pyrazolo[3,4-b]quinoxaline, 1-(4-methylphenyl)-3-phenyl-	MeOH	261.8(4.77),285s(--), 336.2(4.24)	2-0901-86
1H-Pyrazolo[3,4-b]quinoxaline, 3-(4-methylphenyl)-1-phenyl-	MeOH	266.2(5.54),337(4.84)	2-0901-86
$C_{22}H_{16}N_4OS_3$			
4-Thiazolidinone, 5-(2,3-diphenyl-6H-thiazolo[3,4-b][1,2,4]triazin-6-ylidene)-3-ethyl-2-thioxo-	n.s.g.	447(4.54),590(3.45)	103-1373-86
4-Thiazolidinone, 5-(2,8-diphenyl-6H-thiazolo[3,4-b][1,2,4]triazin-6-ylidene)-3-ethyl-2-thioxo-	n.s.g.	460(4.63),650(3.60)	103-1373-86
4-Thiazolidinone, 5-(3,8-diphenyl-6H-thiazolo[3,4-b][1,2,4]triazin-6-ylidene)-3-ethyl-2-thioxo-	n.s.g.	451(4.47),640(3.54)	103-1373-86
$C_{22}H_{16}N_4O_2$			
Propanenitrile, 3,3'-[(6,11-dihydro-6,11-dioxobenzo[f]pyrido[1,2-a]indol-12-yl)imino]bis-	DMF	493(3.8808)	2-1126-86
$C_{22}H_{16}N_4O_5$			
Ethanone, 2-[1,2-dihydro-3-(4-nitrophenyl)-2-quinoxalinyl]-1-(4-nitrophenyl)-	toluene	443(3.80)	103-0538-86
	MeOH	446(--)	103-0538-86
$C_{22}H_{16}N_6O_2S_2$			
Imidazo[5,1-a]phthalazin-3(2H)-one, 1,1'-dithiobis[2-methyl-	EtOH	271(4.47),346(3.91), 394(3.84),424(3.83)	150-1901-86M
$C_{22}H_{16}O_3$			
4H-1-Benzopyran-4-one, 7-hydroxy-3-phenyl-6-(phenylmethyl)-	MeOH	234(4.6),302(4.0), 410(3.1)	2-0646-86

Compound	Solvent	λ_{max}(log ε)	Ref.
4H-1-Benzopyran-4-one, 7-hydroxy-3-phenyl-8-(phenylmethyl)-	MeOH	232(4.51),277(4.05), 330(3.58)	2-0646-86
$C_{22}H_{16}O_4$			
Azuleno[2,1-a]azulene-2,4-dicarboxylic acid, dimethyl ester	CH_2Cl_2	370(4.79),508(4.02), 541(4.11),1200(2.20)	138-2039-86
4H-1-Benzopyran-4-one, 5,7-dihydroxy-3-phenyl-6-(phenylmethyl)-	MeOH	208(4.29),258(4.37), 300(3.74)	2-0166-86
4H-1-Benzopyran-4-one, 5,7-dihydroxy-3-phenyl-8-(phenylmethyl)-	MeOH	210(4.30),259(4.39), 299(3.77)	2-0166-86
$C_{22}H_{16}O_5S_2$			
9H-Thioxanthene-1-carboxylic acid, 9-oxo-3-(phenylsulfonyl)-, ethyl ester	CH_2Cl_2	398(3.74)	151-0353-86D
$C_{22}H_{16}O_7$			
Anthra[1,2-b]furan-6,11-dione, 4-(1-acetoxy-2-oxopropyl)-5-hydroxy-2-methyl-	n.s.g.	221(4.17),257(4.39), 289(4.26),427(3.93)	12-2075-86
$C_{22}H_{16}S_8$			
1,3-Dithiolo[4,5-c][2,5]benzodithiocin, 2-(5,10-dihydro-1,3-dithiolo-[4,5-c][2,5]benzodithiocin-2-ylidene)-5,10-dihydro-	1,1,2-Cl_3-ethane	275(4.04),340(4.09), 380(3.70),408(3.11)	103-0680-86
$C_{22}H_{17}BrN_2$			
Benzo[a]-4,7-phenanthroline, 8-(4-bromophenyl)-9,10,11,12-tetrahydro-	EtOH	242(4.86),278(4.88), 334(3.42),350(3.20)	30-0105-86
$C_{22}H_{17}BrN_2O$			
Ethanone, 1-(4-bromophenyl)-2-(1,2-dihydro-3-phenyl-2-quinoxalinyl)-	toluene	396(3.77)	103-0538-86
	MeOH	394(--)	103-0538-86
$C_{22}H_{17}BrO_7$			
Furo[3',4':6,7]naphtho[2,3-d]-1,3-dioxol-6(8H)-one, 5-(2-bromo-3,4,5-trimethoxyphenyl)-	MeOH	207(4.71),250(4.78), 257(4.79),310(4.00), 330(3.74),350(3.77)	94-2056-86
Furo[3',4':6,7]naphtho[2,3-d]-1,3-dioxol-6(8H)-one, 10-bromo-5-(3,4,5-trimethoxyphenyl)-	MeOH	205(4.61),260(4.77), 315(4.07),352(3.77)	94-2056-86
$C_{22}H_{17}Br_2Cl_3O_{12}$			
Propanedioic acid, bis[[3-bromo-6-(methoxycarbonyl)-2-oxo-2H-pyran-4-yl]methyl]-, methyl 2,2,2-trichloroethyl ester	MeOH	314(4.26)	5-1968-86
$C_{22}H_{17}ClN_2$			
Benzo[a]-4,7-phenanthroline, 8-(4-chlorophenyl)-9,10,11,12-tetrahydro-	EtOH	244(4.87),280(4.92), 333(3.18),348(2.86)	30-0105-86
$C_{22}H_{17}ClN_2O$			
Ethanone, 2-(5-chloro-1,2-dihydro-3-phenyl-2-quinoxalinyl)-1-phenyl-	toluene	397(3.86)	103-0538-86
	MeOH	397(--)	103-0538-86
Ethanone, 2-(7-chloro-1,2-dihydro-3-phenyl-2-quinoxalinyl)-1-phenyl-	toluene	397(3.83)	103-0538-86
	MeOH	400(--)	103-0538-86
Ethanone, 1-(4-chlorophenyl)-2-(1,2-dihydro-3-phenyl-2-quinoxalinyl)-	toluene	395(3.63)	103-0538-86
	MeOH	392(--)	103-0538-86

Compound	Solvent	λ_{max}(log ε)	Ref.
$C_{22}H_{17}ClN_2O_5$			
3,5-Pyridinedicarboxylic acid, 4-(9-chloro-1-oxo-1H-pyrrolo[1,2-a]indol-2-yl)-2,6-dimethyl-, dimethyl ester	MeOH	204(4.32),244(4.44), 276(4.24),285(4.29), 330(4.09)	83-0108-86
$C_{22}H_{17}FN_2$			
Benzo[a]-4,7-phenanthroline, 8-(4-fluorophenyl)-9,10,11,12-tetrahydro-	EtOH	242(4.78),280(4.79), 336(3.43),348(3.25)	30-0105-86
$C_{22}H_{17}NO$			
Oxiranecarbonitrile, 2-phenyl-3-[2-(phenylethynyl)-1-cyclopenten-1-yl]-, cis	MeCN	220(4.07),239s(4.11), 289(4.29)	78-2221-86
$C_{22}H_{17}NO_7$			
Pyrrolo[2,1-a]isoquinoline-1,2-dicarboxylic acid, 3-[5-(methoxycarbonyl)-2-furanyl]-, dimethyl ester	MeOH	213(3.48),274(3.36), 320(3.30)	73-0412-86
$C_{22}H_{17}NS$			
Benzenamine, N-(3,4-diphenyl-2H-thiet-2-ylidene)-4-methyl-	EtOH	246(4.41),250(4.40), 282s(4.27),350(3.86), 425s(3.69)	24-0162-86
$C_{22}H_{17}N_3$			
1H-Pyrrole, 2,5-diphenyl-3-(phenylazo)-	EtOH	205(4.58),225(4.30), 315(4.77),415(4.24)	103-0499-86
$C_{22}H_{17}N_3O_3$			
Ethanone, 2-(1,2-dihydro-3-phenyl-2-quinoxalinyl)-1-(4-nitrophenyl)-	toluene	390(3.96)	103-0538-86
	MeOH	390(--)	103-0538-86
$C_{22}H_{18}$			
5,11[1',2']-Benzeno-5H-cyclohepta[b]-naphthalene, 5a,11-dihydro-5-methyl-	EtOH	273(3.70)	18-1095-86
6,6'-Biazulene, 2,2'-dimethyl-	CH_2Cl_2	252.5(4.39),315.7(5.09), 393(4.28),578(2.78)	24-2272-86
Bullvalene, diphenyl-	ether	218s(4.29),253(4.25)	24-2339-86
1,3,5,11,13-Cyclopentadecapentaene-7,9-diyne, 15-(2,4-cyclopentadien-1-ylidene)-6,11-dimethyl-, (E,E,Z-Z,E)-	THF	268s(4.29),279(4.37), 343(4.46),432(4.26)	18-1723-86
	MeCN	260s(0.66),277(0.88), 338(1.00),427(0.60) (rel. absorbances)	18-1723-86
$C_{22}H_{18}BrNO_2$			
2-Propen-1-one, 1-(5-bromo-2-hydroxyphenyl)-3-phenyl-3-[(phenylmethyl)amino]-, (Z)-	EtOH	252(4.15),373(4.44)	104-1658-86
$C_{22}H_{18}Br_2O_4$			
1,3-Azulenedicarboxylic acid, 6-bromo-2-(4-bromophenyl)-, diethyl ester	CH_2Cl_2	234.5(4.51),271.5(4.27), 315.5(4.77),515(2.73)	24-2272-86
$C_{22}H_{18}ClNO$			
1H-Indole-2-carboxaldehyde, 5-chloro-2,3-dihydro-1-methyl-3,3-diphenyl-	EtOH	258(3.97),319(3.45)	44-5001-86
$C_{22}H_{18}ClNOS_2$			
2-Propen-1-one, 1-[1-(4-chlorophenyl)-	EtOH	289(4.29),362(3.76)	4-0567-86

Compound	Solvent	$\lambda_{max}(\log \epsilon)$	Ref.
1,2-dihydro-6-methyl-4-(methylthio)-2-thioxo-3-pyridinyl]-3-phenyl-, (E)- (cont.)			4-0567-86
$C_{22}H_{18}ClNO_2$			
4,1-Benzoxazepin-2(3H)-one, 7-chloro-1,5-dihydro-1-methyl-5,5-diphenyl-	EtOH	253(4.11),258s(4.06), 264s(3.92),271s(3.69), 287s(3.02),291s(2.90)	44-5001-86
$C_{22}H_{18}N_2$			
Benzeneacetonitrile, α-[(methylphenyl-amino)phenylmethylene]-	EtOH	213(4.14),254(4.24), 369(4.11)	22-0781-86
Benzo[a]phenanthridine, 1,2,3,4-tetra-hydro-5-(4-pyridinyl)-	EtOH	258(4.39),322(3.12), 335(3.30),350(3.34)	103-0759-86
2,4,10,12-Cyclotridecatetraene-6,8-di-yn-1-one, 5,10-dimethyl-, (2,4,6-cycloheptatrien-1-ylidene)hydrazone	THF	276(4.51),442(4.45)	18-2509-86
rel. absorbances given	CF_3COOH	261s(0.66),284(0.85), 417(1.00),526(0.21)	18-2509-86
1H-Pyrazole, 3-(4-methylphenyl)-1,5-diphenyl-	n.s.g.	255(4.53)	78-3807-86
1H-Pyrazole, 5-(4-methylphenyl)-1,3-diphenyl-	n.s.g.	254(4.54)	78-3807-86
$C_{22}H_{18}N_2O$			
Ethanone, 2-(1,2-dihydro-3-phenyl-2-quinoxalinyl)-1-phenyl-	toluene	395(3.78)	103-0538-86
	MeOH	395(--)	103-0538-86
1H-Pyrazole, 3-(4-methoxyphenyl)-1,5-diphenyl-	n.s.g.	259(4.54)	78-3807-86
1H-Pyrazole, 5-(4-methoxyphenyl)-1,3-diphenyl-	n.s.g.	256(4.58)	78-3807-86
1H-Pyrrolizine, 2,3-dihydro-5-(5-oxa-zolyl)-6,7-diphenyl-	MeOH	207(4.52),247(4.46), 304(4.30)	83-0354-86
$C_{22}H_{18}N_2O_2$			
Spiro[9H-fluorene-9,3'(3'aH)-pyrazolo-[1,5-a]pyridine]-2'-carboxylic acid, 5'-methyl-, methyl ester	CH_2Cl_2	399(3.96)	89-0572-86
$C_{22}H_{18}N_2S_2$			
6H-1,3-Thiazine, 4-methyl-6-(4-methyl-2-phenyl-6H-1,3-thiazin-6-ylidene)-2-phenyl-	$CHCl_3$	283(4.39),326(4.28), 524(3.58)	118-0916-86
$C_{22}H_{18}N_2S_4$			
6H-1,3-Thiazine, 2-(methylthio)-6-[2-(methylthio)-4-phenyl-6H-1,3-thiazin-6-ylidene]-4-phenyl-	$CHCl_3$	276(4.52),335(4.27), 508(3.93)	118-0916-86
$C_{22}H_{18}N_4O_2$			
1H-Purine-2,6-dione, 3,7-dihydro-1,3,7-trimethyl-8-(9-phenan-threnyl)-	CH_2Cl_2	242s(4.61),250s(4.73), 256(4.81),273s(4.47), 286s(4.12),299(4.29), 305s(4.25),309s(4.21), 314s(4.15),325s(3.96), 337s(3.59)	24-1525-86
$C_{22}H_{18}N_6O$			
2-Propenamide, 2-azido-3-phenyl-3-(phenylazo)-N-(phenylmethyl)-	$CHCl_3$	252(3.90),380(4.55), 470(2.71)	97-0165-86

Compound	Solvent	λ_{max}(log ε)	Ref.
$C_{22}H_{18}N_6O_5$			
Ethanone, 1-[2,4-dihydroxy-3,5-bis(5-methoxy-2H-benzotriazol-2-yl)-phenyl]-	$CHCl_3$	248(4.48),278(4.45), 335(4.51),350s(4.44)	49-0805-86
$C_{22}H_{18}O_2$			
[1,1'-Binaphthalene]-4,4'-diol, 3,3'-dimethyl-	MeOH	222(3.86),307(4.11), 315s(4.09),329s(4.00)	39-1789-86C
	MeOH-KOH	347(4.13)	39-1789-86C
$C_{22}H_{18}O_3$			
Benzo[g]chrysene-3,6-dione, 4,4a,4b-5,9,10-hexahydro-12-hydroxy-, trans	EtOH	344(4.27),424(4.11)	149-0121-86A
2-Naphthalenol, 7,8-dihydro-5,6-bis(4-hydroxyphenyl)-	EtOH	308(4.25)	149-0121-86A
$C_{22}H_{18}O_4$			
9(10H)-Anthracenone, 10-hydroxy-10-[(4-hydroxy-3-methoxyphenyl)methyl]-	EtOH	274(4.13)	104-0158-86
$C_{22}H_{18}O_7$			
2H,8H-Benzo[1,2-b:3,4-b']dipyran-2-one, 3-(1,3-benzodioxol-5-yl)-4-hydroxy-5-methoxy-8,8-dimethyl-	MeOH	232(4.38),278(4.19), 339(4.07)	102-2425-86
Furo[3',4':6,7]naphtho[2,3-d]-1,3-dioxol-6(8H)-one, 5-(3,4,5-trimethoxyphenyl)-	MeOH	205(4.69),256(4.80), 310(4.00),350(3.69)	94-2056-86
Naphtho[2,3-c]furan-1,3-dione, 4-(3,4-dimethoxyphenyl)-6,7-dimethoxy-	EtOH-$CHCl_3$ (1:1)	290(4.60)	150-0735-86M
$C_{22}H_{19}ClN_2O_5$			
3,5-Pyridinedicarboxylic acid, 4-(9-chloro-1-oxo-1H-pyrrolo[1,2-a]indol-2-yl)-1,4-dihydro-2,6-dimethyl-	MeOH	221(4.14),244(4.20), 267(4.15),288(4.17), 332(3.92)	83-0108-86
$C_{22}H_{19}ClN_4$			
1H-Pyrazolo[3,4-b]quinolin-4-amine, N-(4-chlorophenyl)-5,6,7,8-tetrahydro-1-phenyl-	EtOH	220s(4.36),255(4.42), 334(4.23)	11-0331-86
$C_{22}H_{19}ClN_4S$			
Thieno[2,3-b]pyridine, 4-(2-chlorophenyl)-6-[4-(2-pyridinyl)-1-piperazinyl]-	$CDCl_3$	244(4.38),292(4.12), 335(3.78)	83-0347-86
$C_{22}H_{19}FN_4$			
1H-Pyrazolo[3,4-b]quinolin-4-amine, N-(4-fluorophenyl)-5,6,7,8-tetrahydro-1-phenyl-	EtOH	225s(4.36),256(4.38), 328(4.21)	11-0331-86
$C_{22}H_{19}N$			
5,10[1',2']-Benzeno-5H-benzo[a]cyclopenta[e]cyclooctene-5-carbonitrile, 1,2,3,4,10,11-hexahydro-	EtOH	263(2.9),270(2.9)	88-5621-86
1H-Indole, 1-methyl-2-phenyl-3-(phenylmethyl)-	n.s.g.	226(4.43),296(4.15)	150-0514-86M
$C_{22}H_{19}NOS$			
1H-Phenothiazin-1-one, 2-(1,1-dimethylethyl)-4-phenyl-	$CHCl_3$	310(4.15),645(3.58)	39-2233-86C

Compound	Solvent	$\lambda_{max}(\log \epsilon)$	Ref.
2-Propen-1-one, 1-(1,2-dihydro-4,6-dimethyl-1-phenyl-2-thioxo-3-pyridinyl)-3-phenyl-	EtOH	287(4.28),370(3.87)	4-0567-86
$C_{22}H_{19}NO_2$			
Acetamide, N-[4-[1-(4-hydroxyphenyl)-2-phenylethenyl]phenyl]-	EtOH	229(4.18),265(4.23), 318(4.21)	13-0073-86B
1H-Indole, 4,6-dimethoxy-2,3-diphenyl-	n.s.g.	240(4.58),325(4.49)	12-0015-86
2-Propen-1-one, 1-(2-hydroxyphenyl)-3-phenyl-3-[(phenylmethyl)amino]-, (Z)-	EtOH	254(4.24),369(4.60)	104-1658-86
$C_{22}H_{19}NO_2S$			
2-Propen-1-one, 1-(1,2-dihydro-4-methoxy-6-methyl-1-phenyl-2-thioxo-3-pyridinyl)-3-phenyl-, (E)-	EtOH	229(4.49),287(4.43), 355(3.99)	4-0567-86
$C_{22}H_{19}NO_4$			
Acetic acid, (5,6-dihydro-11-methyl-13-oxo-11a,6a-(epoxyethano)-11H-benzo[a]carbazol-14-ylidene)-, methyl ester, (E)-	MeOH	237(4.25),295(3.40)	150-0514-86M
(Z)-	MeOH	241(4.05),270(3.57)	150-0514-86M
Dibenzo[8,9:10,11][1,6]dioxacyclododecino[2,3,4-gh]pyrrolizine-7,16-dione, 2,4,5,5a,18,18b-hexahydro-, [5aR-(5aR*,18bR*)]-	EtOH	278(3.42)	39-0585-86C
$C_{22}H_{19}NO_7$			
Dibenz[cd,f]indol-4(5H)-one, 5-acetyl-2-acetoxy-1,6,7-trimethoxy-	MeOH	249(4.68),293(4.36), 340(3.96),358(3.86), 380(3.96)	102-0965-86
$C_{22}H_{19}N_2O$			
Pyrylium, 3-cyano-2,6-diphenyl-4-(1-pyrrolidinyl)-, perchlorate	MeCN	285s(4.52),303(4.53)	48-0314-86
$C_{22}H_{19}N_2O_2$			
Pyridinium, 4-[5-(4-methoxyphenyl)-2-oxazolyl]-1-(phenylmethyl)-, chloride	H_2O	403(4.27)	135-0244-86
	EtOH	265(4.27),420(4.35)	135-0244-86
$C_{22}H_{19}N_3O_2$			
9,10-Anthracenedione, 1,5-bis(methylamino)-4-(phenylamino)-	EtOH	612(4.20),659(4.26)	104-0547-86
Benzaldehyde, 3,4-dimethoxy-, (3-methyl-9H-indeno[2,1-c]pyridin-9-ylidene)hydrazone	EtOH	216(4.59),225(4.57), 260(4.65),287(4.07), 380(4.44)	103-0200-86
	$EtOH-H_2SO_4$	207(4.61),233(4.63), 272(4.61),313(4.45), 410(4.38)	103-0200-86
$C_{22}H_{19}N_3O_3S$			
6H-Thieno[2,3-b]pyrrole-2-carboxylic acid, 3,4-diamino-5-benzoyl-6-phenyl-, ethyl ester	EtOH	257(4.36),315(3.43), 388(3.94)	48-0459-86
$C_{22}H_{19}N_3O_3Se$			
Furo[3',4':5,6]selenino[2,3-c]pyrazole-5,7-dione, 1,4,4a,7a-tetrahydro-3-methyl-4-[(3-methylphenyl)amino]-1-phenyl-	EtOH	256(3.9)	104-2126-86

Compound	Solvent	$\lambda_{max}(\log \epsilon)$	Ref.
Furo[3',4':5,6]selenino[2,3-c]pyrazole-5,7-dione, 1,4,4a,7a-tetrahydro-3-methyl-4-[(4-methylphenyl)amino]-1-phenyl-	EtOH	258(3.9)	104-2126-86
$C_{22}H_{19}N_3O_4Se$			
Furo[3',4':5,6]selenino[2,3-c]pyrazole-5,7-dione, 1,4,4a,7a-tetrahydro-4-[(4-methoxyphenyl)amino]-3-methyl-1-phenyl-	EtOH	262(3.9)	104-2126-86
$C_{22}H_{19}N_3O_5$			
2,5-Methano-5H-pyrimido[1,2-c][1,3,5]-oxadiazepin-9(1H)-one, 3-(benzoyloxymethyl)-2,3-dihydro-11-hydroxy-1-phenyl-, [2R-(2α,3β,5α,11S*)]-	MeOH	200(4.45),227(4.52), 245.5(4.27)	44-4417-86
$C_{22}H_{19}N_5O$			
4-Pyrimidinamine, 2-methyl-N-[(2-methyl-6-phenyl-4-pyrimidinyl)oxy]-6-phenyl-	n.s.g.	268(4.58),340s(3.87)	103-1232-86
$C_{22}H_{19}O$			
Dinaphtho[1,2-b:1',2'-d]pyrylium, 7,8,13,14-tetrahydro-6-methyl-, perchlorate	MeCN	222(4.38),256(4.15), 285(4.17),330(3.88), 438(4.30)	22-0600-86
$C_{22}H_{20}$			
4,14:7,11-Diethenodicyclopenta[a,g]-cyclododecene, 1,5,6,8,12,13-hexahydro-	EtOH	221s(4.33),230(4.35), 257(4.15),265(4.14), 277(3.69)	78-1655-86
Naphthalene, 1,2-dihydro-2-[2-(2-naphthalenyl)ethyl]-	hexane	266(4.14)	18-1626-86
Pyrene, 1-(3,3-dimethyl-1-butenyl)-, (E)-	MeOH	254(4.59),258(4.58), 283(4.47),343(4.50), 359(4.52)	35-4498-86
$C_{22}H_{20}BrN_2O$			
Pyridinium, 1-[1-(4-bromobenzoyl)-2-[4-(dimethylamino)phenyl]ethenyl]-, iodide, (Z)-	MeOH	264(4.37),425(4.60)	78-0699-86
$C_{22}H_{20}ClNO$			
1H-Indole-2-methanol, 5-chloro-2,3-dihydro-1-methyl-3,3-diphenyl-	EtOH	262(3.98),266s(3.97), 272s(3.89),318(3.43)	44-5001-86
$C_{22}H_{20}ClNO_2S$			
2H-[1]Benzothiepino[5,4-b]pyran-2-one, 3-chloro-5,6-dihydro-4-[methyl(2-phenylethyl)amino]-	EtOH	232s(4.02),255(4.06), 265s(4.01),324(4.08)	4-0449-86
$C_{22}H_{20}ClN_2O$			
Pyridinium, 1-[1-(4-chlorobenzoyl)-2-[4-(dimethylamino)phenyl]ethenyl]-, iodide, (Z)-	MeOH	267(4.44),431(4.64)	78-0699-86
$C_{22}H_{20}Cl_2N_2O_2$			
2,5-Cyclohexadiene-1,4-dione, 3,6-bis[4-(dimethylamino)phenyl]-	MeOH	265(4.18),295(4.22), 570(3.61)	87-1792-86

Compound	Solvent	$\lambda_{max}(\log \epsilon)$	Ref.
$C_{22}H_{20}F_6Pt$			
Platinum, [(1,2,5,6-η)-1,5-cycloocta-diene]bis[4-(trifluoromethyl)-phenyl]-	CH_2Cl_2	266(3.81),297(3.48)	101-0027-86N
$C_{22}H_{20}N_2$			
Benzo[a]phenanthridine, 1,2,3,4,5,6-hexahydro-5-(4-pyridinyl)-	EtOH	257(4.70),298(3.96), 310(3.88),370(3.65)	103-0759-86
Pyrazine, 2,5-dimethyl-3,6-bis(2-phen-ylethenyl)-, (E,E)-	EtOH	226(4.11),237s(4.05), 244s(3.95),294(4.34), 302s(4.32)	4-1481-86
1H-Pyrazole, 4,5-dihydro-3-(4-methyl-phenyl)-1,5-diphenyl-	n.s.g.	244(4.25),355(4.30)	78-3807-86
1H-Pyrazole, 4,5-dihydro-5-(4-methyl-phenyl)-1,3-diphenyl-	n.s.g.	240(4.23),356(4.30)	78-3807-86
$C_{22}H_{20}N_2O$			
1H-Pyrazole, 4,5-dihydro-3-(4-methoxy-phenyl)-1,5-diphenyl-	n.s.g.	249(4.22),351(4.32)	78-3807-86
1H-Pyrazole, 4,5-dihydro-5-(4-methoxy-phenyl)-1,3-diphenyl-	n.s.g.	230(4.38),356(4.31)	78-3807-86
$C_{22}H_{20}N_2O_3$			
2,4-Cyclohexadien-1-one, 6-[(5-ethyl-5,7,8,9-tetrahydro-10-phenanthri-dinyl)methylene]-4-nitro-	EtOH	240(4.59),300(4.13), 325(4.18),405(4.19), 490(4.00)	135-1208-86
$C_{22}H_{20}N_2O_4S$			
Acetamide, N-acetyl-N-[2,9-dihydro-9-(phenylsulfonyl)-1H-carbazol-4-yl]-	MeOH	220(4.46),260s(3.93), 300s(3.74)	5-2065-86
$C_{22}H_{20}N_2O_8$			
2-Propenoic acid, 3-(3-nitrophenyl)-, 1,4-butanediyl ester, (E,E)-	MeCN	261(4.81)	18-1379-86
2-Propenoic acid, 3-(4-nitrophenyl)-, 1,4-butanediyl ester, (E,E)-	MeCN	302(4.63)	18-1379-86
$C_{22}H_{20}N_2O_{14}$			
Ethanedioic acid, bis[2-[(2-methoxy-ethoxy)carbonyl]-4-nitrophenyl] ester	MeCN	222(4.58),296(4.16)	96-0209-86
Ethanedioic acid, bis[4-[(2-methoxy-ethoxy)carbonyl]-2-nitrophenyl] ester	MeCN	233(4.59),340(3.61)	96-0209-86
$C_{22}H_{20}N_4$			
2,3'-Bipyridinium, 1',1"'-(1,2-ethane-diyl)bis-, dibromide	pH 8.2	209(4.38),232(4.37), 278(4.26)	64-0239-86B
Dibenzo[f,lmn][2,9]phenanthroline-2,7-diamine, N,N,N',N'-tetramethyl-	dioxan	250(4.22),282(4.72), 373(3.89),482(3.49)	83-0338-86
1H-Pyrazolo[3,4-b]quinolin-4-amine, N,1-diphenyl-5,6,7,8-tetrahydro-	EtOH	225s(4.48),256(4.51), 331(4.54)	11-0331-86
$C_{22}H_{20}N_4O_2$			
1H-Purine-2,6-dione, 8-(1,2-diphenyl-ethenyl)-3,7-dihydro-1,3,7-tri-methyl-, (E)-	70% MeOH	283s(3.91),322(4.16)	24-1525-86
(Z)-	70% MeOH	278(4.37),299s(4.11), 322s(3.85)	24-1525-86

Compound	Solvent	λ max (log ε)	Ref.
$C_{22}H_{20}N_4O_3$			
Spiro[1,3-dioxolane-2,4'(1'H)-pteridin]-2'(3'H)-one, 1',3'-dimethyl-6',7'-diphenyl-	MeOH	229(4.43),280(4.16), 339(4.19)	128-0199-86
$C_{22}H_{20}N_4O_5$			
11,7-Metheno-7H-dibenzo[b,m][1,15,18-4,5,11,12]trioxatetraazacycloeicosine-8,25-diol, 19,20,22,23-tetrahydro-	EtOH	430.4(4.49)	70-1741-86
$C_{22}H_{20}N_6S$			
Cyclopropenylium, [[2-(dicyanomethyl)-3-(dicyanomethylene)-1-cyclopropen-1-yl]thio]di-1-piperidinyl-, hydroxide, inner salt	CH_2Cl_2	289(4.56),344(3.93)	24-2104-86
$C_{22}H_{20}N_6Se$			
Cyclopropenylium, [[2-(dicyanomethyl)-3-(dicyanomethylene)-1-cyclopropen-1-yl]seleno]di-1-piperidinyl-, hydroxide, inner salt	n.s.g.	287(4.54),332s(3.93)	89-0183-86
$C_{22}H_{20}O_2$			
Cyclohexanone, 2-[(4-phenyl-1-naphthalenyl)oxy]-	EtOH	214(4.31),236(4.18), 302(3.69)	4-1625-86
1H-Cyclopenta[l]phenanthrene-1-carboxylic acid, 1,3-dimethyl-, ethyl ester	EtOH	241(4.71),263(4.62), 272s(4.52),323(4.12), 335s(4.08)	39-1471-86C
4,14:7,11-Diethenodicyclopenta[a,g]cyclododecene-1,8-dione, 2,3,5,6-9,10,12,13-octahydro-	EtOH	221(4.48),276(3.85), 334(3.28)	78-1655-86
4,14:7,11-Diethenodicyclopenta[a,g]cyclododecene-1,10-dione, 2,3,5,6-8,9,12,13-octahydro-	EtOH	219(4.58),261(4.10), 317(3.52)	78-1655-86
Naphth[2,3-a]azulene, 5,12-diethoxy-	$CHCl_3$	319(4.69),329(4.64), 380(3.99),404(3.89), 428(3.92),455(3.76), 497(2.68),542(2.88), 590(2.99),598(2.97), 722(2.77)	118-0686-86
$C_{22}H_{20}O_3$			
Ethanone, 1-[2-hydroxy-4-(phenylmethoxy)-3-(phenylmethyl)phenyl]-	MeOH	220(3.9),270(3.8)	2-0259-86
Ethanone, 1-[2-hydroxy-4-(phenylmethoxy)-5-(phenylmethyl)phenyl]-	MeOH	229(4.4),280(4.0)	2-0259-86
1-Propanone, 1-[2,4-dihydroxy-3-(phenylmethyl)phenyl]-3-phenyl-	MeOH	228(4.9),269(4.8)	2-0259-86
1-Propanone, 1-[2,4-dihydroxy-5-(phenylmethyl)phenyl]-3-phenyl-	MeOH	219(4.7),280(4.5)	2-0259-86
$C_{22}H_{20}O_4$			
2H-1-Benzopyran-2-one, 4-hydroxy-3-[(2-oxocyclohexyl)phenylmethyl]-	MeOH	271(4.10),281(4.14), 304(4.10)	2-0102-86
Cyclohepta[ef]heptalene-1,2-dicarboxylic acid, 10,12-dimethyl-, dimethyl ester	hexane	222(4.43),233(4.44), 260s(4.26),318(3.94)	89-0633-86
1,4:8,11-Ethanediylidene[14]annulene-2,3-dicarboxylic acid, 15,16-dimethyl-, dimethyl ester, trans	CH_2Cl_2	360(4.96),408(3.80), 430(3.79),453(3.68), 518(2.47),562(3.02),	35-4105-86

Compound	Solvent	λ_{max}(log ϵ)	Ref.
(cont.)		582(2.71),620(3.10)	35-4105-86
Naphth[2,3-a]azulene, 5,6-bis(methoxy-methoxy)-	$CHCl_3$	278(4.25),297(4.34), 319(4.57),330(4.52), 360(3.93),380(3.91), 404(3.82),428(3.86), 454(3.70)	118-0686-86
1-Propanone, 1-[3-(diphenylmethyl)-2,4,6-trihydroxyphenyl]-	MeOH	258(4.59),281(4.48)	2-0478-86
1-Propanone, 3-phenyl-1-[2,3,4-tri-hydroxy-5-(phenylmethyl)phenyl]-	MeOH	216(4.4),274(4.1)	2-0259-86
$C_{22}H_{20}O_5$			
2,5-Furandione, 3-[(2,5-dimethoxyphen-yl)phenylmethylene]dihydro-4-(1-methylethylidene)-, (E)-	n.s.g.	345(4.02)	39-1599-86C
(Z)-	n.s.g.	338(3.92)	39-1599-86C
Naphtho[2,3-c]furan-1,3-dione, 4,9-di-hydro-4,6-dimethoxy-9,9-dimethyl-4-phenyl-	EtOH	281(3.60),288(3.57)	39-1599-86C
$C_{22}H_{20}O_6$			
Naphtho[2,3-c]furan-1,3-dione, 4,9-di-hydro-4-hydroxy-5,8-dimethoxy-9,9-dimethyl-4-phenyl-	EtOH	292(3.66)	39-1599-86C
$C_{22}H_{20}O_7$			
2-Anthraceneacetic acid, 9,10-dihydro-1,4-dihydroxy-9,10-dioxo-3-(3-oxo-pentyl)-, methyl ester	n.s.g.	208(4.19),250(4.63), 256(4.61),284(4.15), 315s(--),475(3.97), 483(3.99),517(3.78), 560s(--)	5-1506-86
2-Anthraceneacetic acid, 1,4-dihydro-9-hydroxy-5-methoxy-1,4-dioxo-3-(3-oxobutyl)-, methyl ester	n.s.g.	207(4.15),244(4.70), 269(4.05),325s(--), 492(3.97),603s(--)	5-1506-86
2-Anthraceneacetic acid, 1,4-dihydro-10-hydroxy-5-methoxy-1,4-dioxo-3-(3-oxobutyl)-, methyl ester	n.s.g.	214(3.88),245(4.52), 280(3.80),326s(--), 510(3.76)	5-1506-86
1,7,8-Anthracenetriol, 2-methoxy-6-methyl-, triacetate	MeOH	262(--),322(--), 339(--),358(--), 389(--),410(--)	100-0145-86
1-Naphthacenecarboxylic acid, 1,2,3,4-5,12-hexahydro-2,11-dihydroxy-7-methoxy-2-methyl-5,12-dioxo-, methyl ester	n.s.g.	209(3.87),244(4.52), 275(3.89),285(3.89), 322s(--),485(3.74)	5-1506-86
2H-Pyran-2,5(6H)-dione, 6-[(3,4-di-methoxyphenyl)methylene]-3-methoxy-4-(4-methoxyphenyl)-	MeOH	228(4.28),332(4.37), 400(3.25)	5-0177-86
ξ-Rhodomycinone, 4-deoxy-, (±)-	n.s.g.	209(4.24),228(4.23), 251(4.66),254(4.65), 288(3.98),325s(--), 455(3.99),480(4.05), 495(3.92),514(3.87)	5-1506-86
$C_{22}H_{20}O_8$			
2-Anthraceneacetic acid, 9,10-dihydro-1,4-dihydroxy-5-methoxy-9,10-dioxo-3-(3-oxobutyl)-, methyl ester	n.s.g.	220(4.33),234(4.54), 246(4.47),288(3.96), 425s(--),455s(--), 479(4.04),497(4.04), 531(3.80)	5-1506-86
2-Anthraceneacetic acid, 9,10-dihydro-1,4,5-trihydroxy-9,10-dioxo-3-(3-	n.s.g.	207(4.21),235(4.51), 252(4.41),293(3.94),	5-1506-86

Compound	Solvent	λ_{max}(log ϵ)	Ref.
oxopentyl)-, methyl ester (cont.)		472(3.99),485s(--), 493(4.05),510s(--), 528(3.88),568s(--)	5-1506-86
2,5-Cyclohexadiene-1,4-dione, 2-(2,5-dimethoxyphenyl)-5-(3,4-dimethoxyphenyl)-3,6-dihydroxy-	MeOH	261.5(4.29),284.3(4.36)	5-0195-86
Edulon A, 11,14-di-O-dehydro- (relative absorbances)	ether	250(1.00),276s(0.74), 283s(0.69),310(0.57), 324s(0.47),366s(0.26), 396(0.28)	33-1513-86
1-Naphthacenecarboxylic acid, 1,2,3,4-6,11-hexahydro-2,5,12-trihydroxy-7-methoxy-2-methyl-6,11-dioxo-, methyl ester	n.s.g.	218(4.32),234(4.50), 251(4.44),289(3.91), 378s(--),475(4.05), 492(4.07),528(3.86), 575s(--)	5-1506-86
2H-Pyran-2,5(6H)-dione, 4-(2,5-dimethoxyphenyl)-6-[(3,4-dimethoxyphenyl)-methylene]-3-hydroxy- (grevillin C)	MeOH	277(4.23),397(4.21)	5-0177-86
2H-Pyran-2,5(6H)-dione, 4-(3,4-dimethoxyphenyl)-6-[(3,4-dimethoxyphenyl)-methylene]-3-hydroxy-	MeOH	299(4.27),386(4.29)	5-0177-86
	H_2SO_4	565(--)	5-0177-86
ε-Rhodomycinone, 4-deoxy-, (±)-	n.s.g.	209(4.16),238(4.34), 250(4.59),254(4.54), 285(3.96),327s(--), 456(3.94),481(3.98), 500(3.85),511(3.77)	5-1506-86
ξ-Rhodomycinone, 10-epi-, (±)-	n.s.g.	218(4.50),235(4.76), 253(4.72),293(4.12), 463(4.27),470(4.30), 489(4.40),510(4.26), 525(4.28),573(3.60)	5-1506-86
$C_{22}H_{20}O_9$			
ε-Rhodomycinone, (±)-	n.s.g.	212(4.19),233(4.60), 253(4.40),291(3.91), 414s(--),464(4.06), 478(4.11),492(4.16), 508(4.04),526(3.98)	5-1506-86
$C_{22}H_{21}BrN_2O$			
Cyclohexanone, 2-[(4-bromophenyl)(6-quinolinylamino)methyl]-	EtOH	255(4.56),293(3.82), 369(3.60)	30-0105-86
$C_{22}H_{21}ClN_2O$			
Cyclohexanone, 2-[(4-chlorophenyl)(6-quinolinylamino)methyl]-	EtOH	254(4.59),295(3.85), 369(3.60)	30-0105-86
$C_{22}H_{21}ClN_2O_4$			
Phenylalanine, N-[(7-chloro-4-hydroxy-3-quinolinyl)carbonyl]-α-methyl-, ethyl ester	EtOH	260(4.38),315(4.04)	2-1072-86
$C_{22}H_{21}ClN_4O_2$			
4,5-Ethano-7H-purine-2,6(1H,3H)-dione, 8-chloro-1,3,7-trimethyl-10,11-diphenyl-, (10-anti,11-syn)	CH_2Cl_2	252s(3.25),260(3.21), 265s(3.14),270s(3.03)	24-1525-86
(10-syn,11-anti)	MeOH	254s(--),260s(--), 266s(--),270s(--)	24-1525-86
1H-Purine-2,6-dione, 8-(2-chloro-1,2-diphenylethyl)-3,7-dihydro-1,3,7-trimethyl-, erythro-	CH_2Cl_2	281(4.13),289s(4.03)	24-1525-86

Compound	Solvent	λ_{max}(log ε)	Ref.
1H-Purine-2,6-dione, 8-(2-chloro-1,2-diphenylethyl)-3,7-dihydro-1,3,7-trimethyl-, threo-	CH_2Cl_2	280(4.13),288s(4.03)	24-1525-86
$C_{22}H_{21}Cl_3O_2Si$			
Silane, tris(4-chlorophenyl)[(1,1-dimethylethyl)dioxy]-	MeCN	259(4.01),265.5(4.06), 270(3.94)	116-0968-86
$C_{22}H_{21}FN_2O$			
Cyclohexanone, 2-[(4-fluorophenyl)(6-quinolinylamino)methyl]-	EtOH	253(4.54),292(3.85), 368(3.60)	30-0105-86
$C_{22}H_{21}NO$			
Benzenemethanamine, α-(methoxyphenylmethylene)-N-methyl-N-phenyl-	EtOH	213(4.24),260(4.35), 348(3.96)	22-0781-86
$C_{22}H_{21}NOS$			
Benzothiazole, 2-[5-(1,1-dimethylethoxy)bicyclo[4.4.1]undeca-1,3,5,7,9-pentaen-2-yl]-	CH_2Cl_2	280(4.48),367(4.39)	33-0315-86
$C_{22}H_{21}NO_2$			
1H-Pyrrolizine-5-propanoic acid, 2,3-dihydro-6,7-diphenyl-	$CHCl_3$	245(4.07),329(3.38), 355(3.49)	83-0234-86
$C_{22}H_{21}NO_3$			
Ethanone, 2-[(3,5-dimethoxyphenyl)-amino]-1,2-diphenyl-	n.s.g.	250(4.34),325(3.76)	12-0015-86
$C_{22}H_{21}NO_5$			
[1,3]Benzodioxolo[5,6-c]phenanthridine, 12,13-dihydro-2,3,4-trimethoxy-12-methyl-	MeOH	228(4.51),284(4.65), 320(4.18)	39-2253-86C
Butanedioic acid, (5,6,11,12-tetrahydro-5-methyl-6-oxodibenzo[b,f]-azocin-12-ylidene)-, dimethyl ester	MeOH	209(4.36),218(4.37)	150-0514-86M
Spiro[2H-indene-2,3'-[3H]indole]-2'-acetic acid, 1,1',2',3-tetrahydro-2'-(methoxycarbonyl)-1'-methyl-1-oxo-, methyl ester	MeOH	242(4.31),288(3.80)	150-0514-86M
$C_{22}H_{21}N_2O$			
Pyridinium, 1-[1-benzoyl-2-[4-(dimethylamino)phenyl]ethenyl]-, bromide, (Z)-	MeOH	261(4.30),427(4.58)	78-0699-86
$C_{22}H_{21}N_2O_6P$			
Phosphoric acid, 4-(6-amino-1,3-dioxo-1H-benz[de]isoquinolin-2(3H)-yl)-phenyl diethyl ester	EtOH	430(4.15)	65-1105-86
$C_{22}H_{21}N_3$			
2H-Benz[g]indazole, 3,3a,4,5-tetrahydro-3-(1-methyl-1H-pyrrol-2-yl)-2-phenyl-	EtOH	230(4.13),252(4.07), 300s(3.78),351(4.11)	4-0135-86
$C_{22}H_{21}N_3O_2$			
7H-Pyrrolo[1',2':1,2][1,4]diazepino-[6,5-c]quinoline-7,12(5H)-dione, 9,10,11,11a,13,13a-hexahydro-5-(phenylmethyl)-, (11aS-cis)-	CH_2Cl_2	329(3.9)	5-1127-86

Compound	Solvent	λ_{max}(log ε)	Ref.
$C_{22}H_{21}N_3O_3$			
7H-Pyrrolo[1',2':1,2][1,4]diazepino-[6,5-c]quinoline-7,12(5H)-dione, 9,10,11,11a,13,13a-hexahydro-10-hydroxy-5-(phenylmethyl)-, [10R-(10α,11aα,13aα)]-	CH_2Cl_2	329(3.9)	5-1127-86
$C_{22}H_{21}N_5O_2$			
2,4(1H,3H)-Pteridinedione, 1,3-dimethyl-6,7-diphenyl-, 4-(O-ethyloxime)	MeOH	230(4.56),298(4.06), 376(4.25)	128-0199-86
$C_{22}H_{21}N_5O_4$			
L-Alanine, N-[4-[N-[(2-amino-1,4-dihydro-4-oxo-6-quinazolinyl)methyl]-2-propynylamino]benzoyl]-	pH 13	231(4.71),280.5(4.41), 303(4.42)	87-1114-86
$C_{22}H_{21}N_5O_6S$			
1H-Pyrazino[2,3-c][1,2,6]thiadiazin-4-amine, 6,7-diphenyl-1-β-D-ribofuranosyl-, 2,2-dioxide	MeOH	215(4.42),232s(4.23), 290(4.17),370(3.92)	5-1872-86
$C_{22}H_{21}N_5O_7$			
β-D-Glucofuranuronic acid, 1-deoxy-1-[6-[(phenylmethyl)amino]-9H-purin-9-yl]-, γ-lactone, 2,5-diacetate	MeOH	266(4.27)	103-0085-86
$C_{22}H_{21}O_3$			
Methylium, tris(4-methoxyphenyl)-	CH_2Cl_2-TFA	487(5.01)	39-1987-86B
$C_{22}H_{22}$			
Pyrene, 1,3-bis(1-methylethyl)-	n.s.g.	245(4.8),279(4.7), 281.5(4.7)	1-0665-86
Pyrene, 1,6-bis(1-methylethyl)-	n.s.g.	245.1(4.8),278.4(4.7), 348.7(4.6)	1-0665-86
Pyrene, 1,8-bis(1-methylethyl)-	n.s.g.	244.8(4.8),278.3(4.7), 348.0(4.6)	1-0665-86
$C_{22}H_{22}ClN_2Ru$			
Ruthenium(1+), (η^6-benzene)chloro-[N,N'-(1,2-dimethyl-1,2-ethanediylidene)bis[benzenamine]-N,N']-, tetrafluoroborate	MeCN	344(3.54),406(3.45), 520(2.38)	157-1449-86
$C_{22}H_{22}Cl_2N_4O_2$			
2,5-Cyclohexadiene-1,4-dione, 2,5-bis[[4-(dimethylamino)phenyl]-amino]-3,6-dichloro-	$CHCl_3$	270(4.50),340(4.27), 530(4.03)	87-1792-86
$C_{22}H_{22}N_2O$			
Bicyclo[4.4.1]undeca-3,6,8,10-tetraen-2-one, 5-(1,3-dihydro-1,3,5,6-tetramethyl-2H-benzimidazol-2-ylidene)-	benzene	402(3.82),495(4.61)	33-0315-86
	MeOH	280(4.23),412s(4.04), 454(4.17)	33-0315-86
	MeCN	276(4.22),358(3.88), 422s(3.94),481(4.35)	33-0315-86
	CH_2Cl_2	418(3.89),493(4.61)	33-0315-86
Cyclohexanone, 2-[phenyl(6-quinolinylamino)methyl]-	EtOH	254(4.54),292(3.80), 369(3.57)	30-0105-86

Compound	Solvent	λ_{max}(log ε)	Ref.
$C_{22}H_{22}N_2O_2$			
Cyclohexanone, 2-[(2-hydroxyphenyl)(6-quinolinylamino)methyl]-	EtOH	256(4.46),299(3.78), 372(3.52)	30-0105-86
Indolo[2,3-a]quinolizin-1-ol, 1,2,3,4,6,7,12,12b-octahydro-, benzoate, cis	MeOH	225(4.71),274s(3.98), 280(3.98),290(3.87)	94-3713-86
$C_{22}H_{22}N_2O_2S$			
1-Propanone, 3-[1-[4-(dimethylamino)-phenyl]-1,2-dihydro-2-thioxo-3-pyridinyl]-3-hydroxy-1-phenyl-	EtOH	245(4.34),287(4.12), 374(3.74)	4-0567-86
$C_{22}H_{22}N_2O_4$			
2-Anthracenecarboxylic acid, 9,10-dihydro-9,10-dioxo-1-(1-piperidinyl-amino)-, ethyl ester	n.s.g.	277(4.07),313(3.81), 490(3.79)	104-2117-86
2,5-Cyclohexadiene-1,4-dione, 2,5-bis[4-(dimethylamino)phenyl]-3,6-dihydroxy-	MeOH	292(4.55),460(3.35)	87-1792-86
2-Propenoic acid, 3-[1,2,6,7,12,12b-hexahydro-2-(2-methoxy-2-oxoethyl-idene)indolo[2,3-a]quinolizin-3-yl]-, methyl ester, (±)-	MeOH	222(4.07),283(3.35), 291(3.28),360(3.93)	44-2995-86
$C_{22}H_{22}N_2O_5$			
2,4(1H,3H)-Pyrimidinedione, 1-[1-[2-(benzoyloxy)ethoxy]ethyl]-5-(phenylmethyl)-	EtOH	267(4.03)	4-1651-86
$C_{22}H_{22}N_2O_5S$			
1H-Carbazole-1-carboxylic acid, 4-(acetylamino)-2,3,9,9a-tetra-hydro-9-(phenylsulfonyl)-, methyl ester	MeOH	218(4.35),267(4.13), 273s(4.12),312(3.94)	5-2065-86
$C_{22}H_{22}N_2O_6$			
2(1H)-Pyrimidinone, 1-[(2-benzoyloxy-ethoxy)methyl]-5-[(3-hydroxyphenyl)-methyl]-4-methoxy-	EtOH	281(3.82)	4-1651-86
$C_{22}H_{22}N_2O_9$			
2-Naphthacenecarboxamide, 4-(dimethyl-amino)-1,4,4a,5,6,11,12,12a-octahy-dro-6-hydroperoxy-3,10,12a-trihy-droxy-6-methyl-1,11,12-trioxo-	$CHCl_3$	220(4.56),265(4.63), 345(3.83)	35-4237-86
$C_{22}H_{22}N_4O$			
Acetamidine, N'-[2-(dimethylamino)-7-oxo-1,4-dihydronaphtho[1,2,3-de]-quinolin-6-yl]-N,N-dimethyl-	dioxan	227(4.36),272(4.44), 342(3.86),358(3.77), 502(3.8)	83-0338-86
$C_{22}H_{22}N_4O_3$			
Benzenebutanoic acid, 4-[[(2-amino-1,4-dihydro-4-oxo-6-quinazolinyl)-methyl]-2-propynylamino]-	pH 13	230.5(4.67),267s(4.21), 274s(4.20),325(3.59)	87-1114-86
2(1H)-Pteridinone, 3,4-dihydro-4,4-di-methoxy-1,3-dimethyl-6,7-diphenyl-	MeOH	227(4.45),276(4.11), 338(4.16)	128-0199-86
$C_{22}H_{22}N_4O_7S$			
Spiro[1-azabicyclo[3.2.0]hept-2-ene-	MeCN	263(4.24),322(4.20)	39-0421-86C

Compound	Solvent	$\lambda_{max}(\log \epsilon)$	Ref.
6,5'(4H)-isoxazole]-2-carboxylic acid, 3-[[2-(acetylamino)ethenyl]-thio]-3',4'-dimethyl-7-oxo-, (4-nitrophenyl)methyl ester, [5R-[3(E),5α,6α(R*)]-			39-0421-86C
[5R-[3(E),5α,6β(S*)]-	MeCN	265(4.25),327(4.20)	39-0421-86C
isomer	MeCN	264(4.27),324(4.23)	39-0421-86C
$C_{22}H_{22}N_6O_4S$			
1,2,4-Thiadiazole-3,5-diacetic acid, α,α'-bis(phenylhydrazono)-, diethyl ester	DMF	303(3.77),395(4.50)	48-0741-86
$C_{22}H_{22}O$			
4,14:7,11-Diethenodicyclopenta[a,g]-cyclododecen-1(2H)-one, 3,5,6,8-9,10,12,13-octahydro-	EtOH	217(4.43),244s(3.92), 290(3.52),325s(3.23)	78-1655-86
3,10-Ethanodiindeno[1,7-bc:7',1'-gh]-oxonin, 1,2,6,7,11,12,12a,13a-octahydro-	EtOH	210(2.98),232(4.30), 248s(3.49),277(2.90)	78-1655-86
$C_{22}H_{22}O_2$			
Cyclohexanone, 2-(7,8,9,10-tetrahydro-benzo[b]naphtho[2,1-d]furan-5-yl)-	EtOH	210(4.41),252(4.67), 259(4.70),282(3.82), 292(3.83),300(3.64), 328(3.15),345(3.04)	4-1625-86
$C_{22}H_{22}O_4$			
1-Azulenecarboxylic acid, 6-ethoxy-2-(4-methoxyphenyl)-, ethyl ester	CH_2Cl_2	262.5(4.04),328(4.78), 399.5(4.08),491(2.91)	24-2272-86
1,4-Pentalenediol, octahydro-, di-benzoate, (1α,3aβ,4α,6aβ)-(±)-	EtOH	228.5(4.365),267(3.168), 272(3.227),279(3.140)	65-1891-86
2H,5H-Pyrano[2,3-b][1]benzopyran-5-one, 3,4-dihydro-2-(hydroxymethyl)-2,6,8-trimethyl-4-phenyl-	EtOH	211(4.44),230(4.52), 275(4.06),290(3.91)	32-0491-86
2H,5H-Pyrano[3,2-c][1]benzopyran-5-one, 3,4-dihydro-2-(hydroxymethyl)-2,8,10-trimethyl-4-phenyl-	EtOH	283s(--),293(4.19), 307s(--),321s(--)	32-0501-86
$C_{22}H_{22}O_5$			
4H,8H-Benzo[1,2-b:3,4-b']dipyran-4-one, 3-(2,4-dimethoxyphenyl)-2,3-dihydro-8,8-dimethyl-, (±)-	MeOH	275(4.47),311(4.2)	2-0481-86
$C_{22}H_{22}O_6$			
4H-1-Benzopyran-4-one, 7-hydroxy-3-[2-hydroxy-4,6-dimethoxy-3-(3-methyl-2-butenyl)phenyl]-	EtOH	248.5(4.34),264(4.20), 297s(4.01),305s(3.97)	94-2369-86
	EtOH-NaOAc	261(4.40),289s(4.06), 341(3.96)	94-2369-86
4H-1-Benzopyran-4-one, 7-hydroxy-3-[6-hydroxy-2,4-dimethoxy-3-(3-methyl-2-butenyl)phenyl]- (lico-ricone)	EtOH	240(4.39),248(4.38), 285(4.11),305s(4.04)	94-2369-86
	EtOH-NaOAc	249.5(4.38),256s(4.37), 307s(3.95),337.4(3.93)	94-2369-86
Sigmoidin C, dimethyl ether	EtOH	286(4.24)	39-0033-86C
$C_{22}H_{22}O_7$			
9,10-Anthracenedione, 1,4-dihydroxy-2-[2-(2-ethyl-1,3-dioxolan-2-yl)-ethyl]-1,4-dihydroxy-8-methoxy-	n.s.g.	231(4.88),249(4.66), 282(4.25),390s(--), 477(4.35),493(4.36), 521(4.11),560s(--)	5-1506-86

Compound	Solvent	$\lambda_{max}(\log \epsilon)$	Ref.
2(3H)-Furanone, 4-(1,3-benzodioxol-5-ylmethyl)dihydro-3-[(3,4,5-trimethoxyphenyl)methylene]-, (±)-(anhydropodorhizol)	EtOH	234(4.30),298s(4.24),313(4.26)	102-2093-86
4H-Phenaleno[1,2-b]furan-4,6(5H)-dione, 8,9-dihydro-3,5,7-trihydroxy-1,8,8,9-tetramethyl-5-(2-oxopropyl)-	MeOH	221(4.52),230s(4.43),252s(4.48),260(4.61),341(4.18),364s(4.04)	23-1585-86
$C_{22}H_{22}O_8$			
Butanedioic acid, bis[(3,4-dimethoxyphenyl)methylene]-, (E,E)-	$EtOH-CHCl_3$ (1:1)	326(4.20)	150-0735-86M
Butanedioic acid, bis[(4-hydroxy-3-methoxyphenyl)methylene]-, dimethyl ester, (E,E)-	EtOH	219(4.46),246(4.34),287(3.95),310s(4.04)	150-0735-86M
$C_{22}H_{22}O_9$			
Ononin	EtOH	251s(4.01),262(4.03),302(3.55)	105-0603-86
$C_{22}H_{22}O_{12}$			
Pedaliin	EtOH	258(4.16),272(4.17),351(4.32)	94-0030-86
	EtOH-NaOAc	270(4.18),370s(4.10),412(4.19)	94-0030-86
	$EtOH-AlCl_3$	279(4.19),378(4.28)	94-0030-86
$C_{22}H_{23}BrO_3$			
Benzoic acid, 4-bromo-, 5,6,7,8-tetrahydro-3,8-dimethyl-5-(1-methylethyl)-6-oxo-2-naphthalenyl ester, (5R-trans)-	EtOH	210(4.34),247(4.15),276(4.34)	12-0103-86
$C_{22}H_{23}ClO_3$			
Benzoic acid, 4-chloro-, 4,7,8,11-tetrahydro-3,6,10-trimethylcyclodeca[b]furan-8-yl ester, [R-(E,E)]-	isooctane	206(4.36),242(4.37)	32-0303-86
$C_{22}H_{23}NO$			
Azacyclooctadeca-1,3,5,7,13,15,17-heptaene-9,11-diyne, 2-ethoxy-8,13,18-trimethyl-, (E,E,E,Z,Z,E,E)-	THF	291(4.43),333(4.74),412s(3.75),510s(3.10)	18-1801-86
1(5H)-Indolizinone, 8a-ethyl-6,7,8,8a-tetrahydro-2,3-diphenyl-	MeCN	226(4.07),276(4.06),348(3.96)	118-0899-86
$C_{22}H_{23}NO_2$			
2-Dibenzofuranol, 6,7,8,9-tetrahydro-1-[[(2,4,6-trimethylphenyl)imino]methyl]-, (E)-	toluene	315(3.89),347(3.88)	104-0503-86
	PrOH	315(3.89),345(3.86)	104-0503-86
$C_{22}H_{23}NO_3$			
1,6-Octadiene-3,5-dione, 1-(4-methoxyphenyl)-7-[(phenylmethyl)amino]-	n.s.g.	330(4.24),412(4.56)	4-1721-86
$C_{22}H_{25}NO_5$			
8H-Dibenzo[a,g]quinolizin-8-one, 5,6-dihydro-2,3,9,10-tetramethoxy-13-methyl-	MeOH	254s(4.25),330(4.41),360s(4.12),380s(3.97)	44-2781-86
$C_{22}H_{23}NO_6S$			
Propanedioic acid, [[(4-methylphenyl)thio](4-nitrophenyl)methylene]-,	EtOH	255s(4.21),276(4.32)	44-4112-86

Compound	Solvent	$\lambda_{max}(\log \epsilon)$	Ref.
1,1-dimethylethyl methyl ester, (E)- (cont.)			44-4112-86
$C_{22}H_{23}NO_7$			
Narcotine	n.s.g.	290(3.56),308(3.68)	105-0234-86
Propanedioic acid, [(4-methylphenoxy)-(4-nitrophenyl)methylene]-, 1,1-dimethylethyl methyl ester, (E)-	EtOH	235s(3.91),285(3.89)	44-4112-86
$C_{22}H_{23}NS$			
Cycloprop[h,i]indolizine-2(1H)-thione, 6c-ethylhexahydro-1,6b-diphenyl-, (1α,6aα,6bα,6cα)-	MeCN	277s(--),293(4.05), 334s(--)	118-0899-86
$C_{22}H_{23}N_2$			
3H-Indolium, 1-[5-(2,3-dihydro-1H-indol-1-yl)-3-methyl-2,4-pentadienylidene]-2,3-dihydro-, perchlorate	MeOH	519s(4.98),550(5.20)	33-1588-86
$C_{22}H_{23}N_3O_3$			
5H,14H-Pyrrolo[1",2":4',5']pyrazino-[1',2':1,6]pyrido[3,4-b]indole-5,14-dione, 1,2,3,11,12,14a-hexahydro-9-methoxy-12-(3-methyl-1-propenyl)-, (12S-trans)-	EtOH	234(4.41),268.5(4.21), 299(4.15),377.5(4.13)	88-6217-86
$C_{22}H_{23}N_4S_2$			
1H-Pyrazolium, 3,5-bis(dimethylamino)-4-phenyl-1-(4-phenyl-1,3-dithiol-2-ylidene)-, perchlorate	MeCN	225(4.32),290s(3.58), 448(4.42)	88-0159-86
1H-Pyrazolium, 3,5-bis(dimethylamino)-4-phenyl-1-(5-phenyl-3H-1,2-dithiol-3-ylidene)-, perchlorate	MeCN	219s(4.36),242s(4.20), 349(4.29),492(4.15)	88-0159-86
$C_{22}H_{23}N_7O_3$			
Acetamide, N-[2-[(2,6-dicyano-4-nitrophenyl)azo]-5-(dipropylamino)phenyl]-	$CHCl_3$	271(3.99),290(3.87), 420(3.55),597s(4.74), 628(4.93)	48-0497-86
Butanamide, N-[2-[(2,6-dicyano-4-nitrophenyl)azo]-5-(diethylamino)-phenyl]-	$CHCl_3$	266s(3.97),298s(3.81), 340s(3.37),420(3.56), 588s(4.75),629(4.96)	48-0497-86
$C_{22}H_{24}$			
5,9,13-Octadecatriene-3,7,11,15-tetrayne, 2,2,17,17-tetramethyl-, (E,E,E)-	hexane	347(4.78),374(4.74)	35-4685-86
$C_{22}H_{24}Cl_2O_6$			
1,3-Azulenedicarboxylic acid, 2-chloro-6-[1-chloro-1-(ethoxycarbonyl)-propyl]-, diethyl ester	CH_2Cl_2	241(4.59),269(4.41), 304.5(4.82),313.3(4.91), 356.5(4.16),507(2.77)	24-2956-86
$C_{22}H_{24}FN_3O_2$			
2H-Benzimidazol-2-one, 1-[1-[4-(4-fluorophenyl)-4-oxobutyl]-4-piperidinyl]-1,3-dihydro- (benperidol)	hexane	283f(3.8)	96-0923-86
	isoPrOH	282(3.9)	96-0923-86
	ether	288(3.9)	96-0923-86
	heptane	286(--)	96-0923-86
	H_2SO_4	232(--),249(--), 279(--)	96-0923-86
	NaOH	248(--),288(--)	96-0923-86

Compound	Solvent	$\lambda_{max}(\log \epsilon)$	Ref.
$C_{22}H_{24}NS$			
1H-Cycloprop[hi]indolizinium, 4,5,6,6a,6b,6c-hexahydro-6c-methyl-2-(methylthio)-1,6b-diphenyl-, iodide, (1α,6aα,6bα,6cα)-	MeCN	246(4.30),291s(--)	118-0899-86
$C_{22}H_{24}N_2$			
Pyrazine, 2,5-bis(1,1-dimethylethyl)-3,6-diphenyl-	EtOH	244(4.02),293(4.04)	142-0785-86
1H-Pyrrolizine-5-methanamine, 2,3-dihydro-N,N-dimethyl-6,7-diphenyl-	EtOH	208(4.50),242(4.31), 270(4.13)	83-0065-86
$C_{22}H_{24}N_2O_5$			
2,4(1H,3H)-Pyrimidinedione, 1-[1-(2-hydroxyethoxy)ethyl]-5-[[3-(phenylmethoxy)phenyl]methyl]-	pH 1	269(4.15)	4-1651-86
	pH 11	269(3.75)	4-1651-86
$C_{22}H_{24}N_2O_6$			
5,5'-Bi-1,3-dioxolo[4,5-g]isoquinoline, 5,5',6,6',7,7',8,8'-octahydro-4,4'-dimethoxy-, dihydrochloride	MeCN	229(4.1),260(3.2), 285(3.6)	83-1122-86
isomer b	MeCN	223(4.2),260(3.2), 285(3.6)	83-1122-86
$C_{22}H_{24}N_2O_8$			
Tetracycline	MeOH-HCl	268(4.60),363(4.45)	35-4237-86
$C_{22}H_{24}N_4O_2$			
2,5-Cyclohexadiene-1,4-dione, 2,5-bis[[4-(dimethylamino)phenyl]amino]-	$CHCl_3$	270(4.49),318(4.40), 485(4.15)	87-1792-86
$C_{22}H_{24}N_4O_3S$			
Acetic acid, 1-[[[(8)-6-methylergolin-8-yl]methyl]-5-oxo-2-thioxo-4-imidazolidinylidene]-, methyl ester, (Z)-	EtOH	224(4.55),286(4.00), 293(4.01),394(4.30)	83-0266-86
$C_{22}H_{24}N_6O_6$			
L-Glutamic acid, N-[4-[[(2-amino-1,4-dihydro-5-methyl-4-oxopyrido[2,3-d]pyrimidin-6-yl)methyl]methylamino]-benzoyl]- (5,10-dimethyl-5-deazafolic acid)	pH 1	223(4.51),278(4.33), 306(4.40)	87-1080-86
	pH 7	224(4.61),307(4.48), 275s(4.25)	87-1080-86
	pH 13	226(4.54),309(4.45)	87-1080-86
$C_{22}H_{24}O$			
Benzo[b]naphtho[2,1-d]furan, 5-cyclohexyl-7,8,9,10-tetrahydro-	EtOH	208(4.30),253(4.67), 260(4.71),284(3.78), 293(3.79),302(3.63), 329(3.13),344(3.03)	4-1625-86
$C_{22}H_{24}O_2$			
Cyclohexanol, 2-(7,8,9,10-tetrahydrobenzo[b]naphtho[2,1-d]furan-5-yl)-	EtOH	209(4.18),253(4.57), 261(4.62),284(3.67), 293(3.67),304(3.49), 328(3.04),345(2.93)	4-1625-86
$C_{22}H_{24}O_3$			
2,5,9-Trioxabicyclo[4.2.1]non-7-ene, 1,6-diethyl-7,8-diphenyl-	ether	227(4.15),247(4.11)	78-1345-86

Compound	Solvent	$\lambda_{max}(\log \epsilon)$	Ref.
$C_{22}H_{24}O_4$			
2H,8H-Benzo[1,2-b:3,4-b']dipyran, 3,4-dihydro-4,5-dimethoxy-8,8-dimethyl-2-phenyl-, (2S-trans)- (methylhildgardtol B)	EtOH	248(3.98),290(3.96)	102-1711-86
Cyclohepta[ef]heptalene-1,2-dicarboxylic acid, 5,6,7,8-tetrahydro-10,12-dimethyl-, dimethyl ester	hexane	274(4.24),324s(3.56), 386s(2.91)	89-0633-86
Cyclohepta[ef]heptalene-2,3-dicarboxylic acid, 5,6,7,8-tetrahydro-10,12-dimethyl-, dimethyl ester	hexane	225s(4.27),239s(4.42), 275(4.30),340(3.80), 404s(3.01)	89-0633-86
3-Cyclopentene-1,2-dione, 4,4'-(1,3,5-hexatriene-1,6-diyl)bis[3,5,5-trimethyl-	$CHCl_3$	465(4.89),546s(2.53)	89-1018-86
Pregna-5,8-dieno[6,5,4-bc]furan-3,7,20-trione (virone)	MeOH	230(4.00),264(3.92), 300(3.83)	39-1317-86C
$C_{22}H_{24}O_6$			
1(2H)-Anthracenone, 3-acetoxy-3,4-dihydro-8,9-dihydroxy-3-methyl-6-[(3-methyl-2-butenyl)oxy]- (vismione H)	$CHCl_3$	276(4.74),347s(4.01), 397(4.26)	102-1217-86
4H-Phenaleno[1,2-b]furan-4-one, 8,9-dihydro-6-hydroxy-3,5,7-trimethoxy-1,8,8,9-tetramethyl-, (±) (atrovenetin trimethyl ether)	MeOH	215(4.55),274(4.51), 365(4.09),416(4.16)	44-4813-86
$C_{22}H_{24}O_7$			
Chrysophyllon I-A	MeOH	269(3.94)	102-2609-86
Chrysophyllon II-B	MeOH	334(3.86)	102-2609-86
Edulon A, 16-O-deacetyl-11,12-di-O-methyl- (relative absorbances)	ether	287(0.69),298(0.68), 315s(0.70),345(1.00), 354s(0.97),398s(0.62)	33-1513-86
2(3H)-Furanone, 3-(1,3-benzodioxol-5-ylmethyl)-4-[(3,4,5-trimethoxyphenyl)methyl]dihydro- (isoyatein)	MeOH	232s(3.90),286(3.32)	102-0487-86
2(3H)-Furanone, 4-(1,3-benzodioxol-5-ylmethyl)-3-[(3,4,5-trimethoxyphenyl)methyl]dihydro-	EtOH	285(3.65)	102-2093-86
1,4,6(5H)-Phenanthrenetrione, 9-acetoxy-4b,8a,9,10-tetrahydro-3,10-dihydroxy-4b,7,8-trimethyl-2-(2-propenyl)- (allylroyleanone)	ether	247(4.17),269s(4.02), 404(2.89)	33-0972-86
2-Propenal, 3-[2,3-dihydro-8-methoxy-2-methyl-3-(3,4,5-trimethoxyphenyl)-1,4-benzodioxin-6-yl]-, [2α,3β,6(E)]-	MeOH	230(3.95),280(3.23), 328(3.53)	102-0213-86
$C_{22}H_{24}O_8$			
2(3H)-Furanone, 4-(1,3-benzodioxol-5-ylmethyl)dihydro-3-[hydroxy(3,4,5-trimethoxyphenyl)methyl]-, (±)- (podorhizol)	EtOH	285(3.59)	102-2093-86
epipodorhizol	EtOH	285(3.63)	102-2093-86
$C_{22}H_{24}O_9$			
Strobopinin 7-O-β-D-glucoside, (2S)-	MeOH	229s(4.12),290(4.17), 344(3.64)	95-0982-86
	MeOH-NaOAc	290(--),345(--)	95-0982-86
$C_{22}H_{24}O_{11}$			
Eriodictyol, 7-O-methyl ether, 3'-O-β-D-glucoside	MeOH	289(4.29),326s(3.56)	95-0989-86

Compound	Solvent	$\lambda_{max}(\log \epsilon)$	Ref.
$C_{22}H_{25}BrN_4O_{12}$			
7H-Pyrrolo[2,3-d]pyrimidine-5-carboxylic acid, 2-(acetylamino)-6-bromo-3,4-dihydro-3-(methoxymethyl)-4-oxo-7-(2,3,5-tri-O-acetyl-β-D-arabinofuranosyl)-	MeCN	287(4.07)	18-1915-86
$C_{22}H_{25}ClN_2O_3S$			
1-Piperazinecarboxylic acid, 4-(2-chloro-10,11-dihydro-6-methoxydibenzo[b,f]thiepin-10-yl)-, ethyl ester	MeOH	280(3.80),294(3.86), 300(3.86)	73-0141-86
$C_{22}H_{25}FO_2S$			
Androsta-1,4,9(11),16-tetraene-17-carbothioic acid, 6-fluoro-16-methyl-3-oxo-, S-methyl ester, (6α)-	MeOH	240(4.26)	44-2315-86
$C_{22}H_{25}NO_2$			
2H-Pyrrolo[2,1-b][1,3]oxazine, 2-ethoxy-6,7,8,8a-tetrahydro-8a-methyl-3,4-diphenyl-	MeCN	226(4.10),308(3.84)	118-0899-86
$C_{22}H_{25}NO_6$			
Benzaldehyde, 2,3-dimethoxy-6-[(5,6,7-8-tetrahydro-4-methoxy-6-methyl-1,3-dioxolo[4,5-g]isoquinolin-5-yl)-methyl-, (S)- (macrantaldehyde)	MeOH	241(3.98),283(3.59)	102-2403-86
$C_{22}H_{25}NO_7$			
1-Isobenzofuranol, 1,3-dihydro-6,7-dimethoxy-3-(5,6,7,8-tetrahydro-4-methoxy-6-methyl-1,3-dioxolo[4,5-g]-isoquinolin-5-yl)-	MeOH	237(3.97),282(3.55)	102-2403-86
$C_{22}H_{25}N_3$			
Pyridinium, 1-[[4-(dimethylamino)phenyl][4-(dimethyliminio)-2,5-cyclohexadien-1-ylidene]methyl]-, salt with trifluoromethanesulfonic acid	MeOH	204(4.58),252(4.40), 367(3.88),425(3.51), 660(4.46)	24-3276-86
	acetone	428(3.85),662(4.77)	24-3276-86
$C_{22}H_{25}N_3O_3S$			
FR 900452	MeOH	246(4.13),347(4.16)	158-0198-86
$C_{22}H_{25}N_3O_7S$			
1-Azabicyclo[3.2.0]hept-2-ene-2-carboxylic acid, 3-[[3-(aminocarbonyl)-cyclopentyl]thio]-6-(1-hydroxyethyl)-7-oxo-, (4-nitrophenyl)methyl ester	EtOH	267(4.05),322(4.12)	23-2184-86
Benzenesulfonic acid, 3-[3-[4-[2-(acetylamino)-3-amino-3-oxopropyl]phenoxy]-3-(ethylamino)-1-oxo-2-propenyl]-, (S)-	H_2O	329(4.28)	35-5543-86
$C_{22}H_{25}N_3O_{10}$			
L-Aspartic acid, N-[3-[4-(2-amino-2-carboxyethyl)phenoxy]-L-tyrosyl]-3-hydroxy-, [erythro-(S)]-	H_2O and H_2O-HCl	273(3.50)	158-1685-86
	0.05M NaOH	296(3.53)	158-1685-86

Compound	Solvent	λ_{max}(log ε)	Ref.
$C_{22}H_{25}N_5O_4S_2$			
Adenosine, N-benzoyl-5'-S-methyl-2',3'-O-(1-methylethylidene)-5'-C-(methylthio)-5'-thio-	n.s.g.	280(4.29)	44-1258-86
$C_{22}H_{25}N_7O_4$			
β-D-ribo-Heptofuranuronamide, 1-(6-amino-9H-purin-9-yl)-1,5,6-trideoxy-N-[2-(1H-indol-3-yl)ethyl]-	pH 6.8	266(4.1)	94-1871-86
$C_{22}H_{25}N_7O_5$			
L-Glutamic acid, N-[4-[[(2,4-diamino-5-methylpyrido[2,3-d]pyrimidin-6-yl)methyl]methylamino]benzoyl]-	pH 1	227(4.60),310(4.35)	87-1080-86
	pH 7	227(4.57),307(4.46)	87-1080-86
	pH 13	228(4.56),306(4.45)	87-1080-86
$C_{22}H_{26}Br_2O_3$			
2,19-Dioxabicyclo[16.3.1]docosa-6,9,18(22),21-tetraen-12-yn-20-one, 4,21-dibromo-3-ethyl-	MeOH	304(3.87)	88-5189-86
$C_{22}H_{26}ClN_5$			
1H-Pyrazolo[3,4-b]quinolin-4-amine, N-(4-chlorophenyl)-5,6,7,8-tetrahydro-1-(1-piperidinylmethyl)-	EtOH	220(4.44),249(4.07), 273(3.61),325(4.19)	5-1728-86
$C_{22}H_{26}FN_5$			
1H-Pyrazolo[3,4-b]quinolin-4-amine, N-(4-fluorophenyl)-5,6,7,8-tetrahydro-1-(1-piperidinylmethyl)-	EtOH	218(4.39),242s(--), 272(3.69),318(4.15)	5-1728-86
$C_{22}H_{26}F_3N_3OS$			
Fluphenazine, dihydrochloride	MeOH	260(4.458),311(3.519)	106-0571-86
$C_{22}H_{26}N_2OS$			
1-Propanamine, N,N-dimethyl-3-[2-(4-morpholinyl)-9H-thioxanthen-9-ylidene]-	MeOH	234(4.52),279(4.27), 342(3.49)	73-0937-86
$C_{22}H_{26}N_2O_3$			
16-Epiaffinine, O-acetyl- (same in acid or base)	EtOH	212(4.27),238(4.17), 318(4.32)	102-0965-86
Vincamajine	MeOH	249(3.93),293(3.50)	105-0618-86
$C_{22}H_{26}N_2O_3S$			
2,4-Octadienoic acid, 6-[[2-(1H-indol-3-yl)ethyl]amino]methylene]-2-(methylthio)-7-oxo-, ethyl ester	MeOH	264(3.90),289(3.05), 366(4.27),380(4.28)	44-2995-86
$C_{22}H_{26}N_2O_4S$			
18-Norcorynan-17,19-dioic acid, 20,21-didehydro-16-(methylthio)-, 17-ethyl 19-methyl ester	MeOH	220(4.40),287(4.43)	44-2995-86
isomer m. 170-172°	MeOH	220(4.40),289(4.43)	44-2995-86
$C_{22}H_{26}N_2O_9S$			
Acetamide, N-[6,7-diacetoxy-5-(acetoxymethyl)-3a,6,7,7a-tetrahydro-5H-pyrano[3,2-d]thiazol-2-yl]-N-(4-methoxyphenyl)-	EtOH	230(4.18),252(4.15)	136-0049-86I

Compound	Solvent	λ_{max}(log ϵ)	Ref.
$C_{22}H_{26}N_3O_2$			
Morpholinium, 4-[[[(4-morpholinyl)-phenylmethylene]amino]phenyl-methylene]-, iodide	EtOH	260(4.06),305(4.20)	39-0619-86C
$C_{22}H_{26}N_4O_{12}$			
7H-Pyrrolo[2,3-d]pyrimidine-5-carbox-ylic acid, 2-(acetylamino)-3,4-di-hydro-3-(methoxymethyl)-4-oxo-7-(2,3,5-tri-O-acetyl-β-D-ara-binofuranosyl)-	MeCN	281(3.96),297s(3.95)	18-1915-86
$C_{22}H_{26}O_2$			
2,5-Cyclohexadien-1-one, 2,6-bis(1,1-dimethylethyl)-4-(2-oxo-2-phenyl-ethylidene)-	C_6H_{12}	331(4.52)	44-2257-86
$C_{22}H_{26}O_3$			
Cyclohepta[de]naphthalen-7(8H)-one, 8,8-dibutoxy-	$CHCl_3$	320(3.92),330s(3.80)	39-1965-86C
2,19-Dioxabicyclo[16.3.1]docosa-3,6,9,18(22),21-pentaen-12-yn-20-one, 3-ethyl-	MeOH	282(3.88)	88-5189-86
2,22-Dioxabicyclo[16.3.1]docosa-3,6,9,18,21-pentaen-12-yn-20-one, 3-ethyl-	MeOH	240(4.00)	88-5189-86
Ethanedione, [3,5-bis(1,1-dimethyl-ethyl)-4-hydroxyphenyl]phenyl-	C_6H_{12}	210(4.28),253(4.12), 292(4.10)	44-2257-86
Furan, 2,5-diethyl-2,5-dihydro-2,5-di-methoxy-3,4-diphenyl-	ether	225(4.31),270(4.14)	78-1345-86
7-Oxabicyclo[4.1.0]hept-4-en-3-one, 2,4-bis(1,1-dimethylethyl)-2-hy-droxy-6-(phenylethynyl)-	C_6H_{12}	205(4.26),246(4.31)	44-2257-86
$C_{22}H_{26}O_4$			
2,22-Dioxabicyclo[16.3.1]docosa-3,5,9,18,21-pentaen-12-yn-20-one, 3-ethyl-7-hydroxy-	MeOH	231(4.26)	88-5189-86
1,4-Heptanonaphthalene-6,7-dicarbox-ylic acid, 5-methyl-, dimethyl ester	EtOH	233s(4.28),261(4.67), 310(3.67),347s(3.53), 358(3.56)	24-2698-86
$C_{22}H_{26}O_5$			
1,4-Epoxy-5,13-etheno-1H-benzocyclo-undecene-2,3-dicarboxylic acid, 4,6,7,8,9,10,11,12-octahydro-1-methyl-, dimethyl ester	EtOH	213(4.46),234s(3.97), 255s(3.44),320(2.97)	24-2698-86
$C_{22}H_{26}O_6$			
Naphtho[1,2-b]furan-6-carboxylic acid, 5-acetyl-2,3-dihydro-4,7-dimethoxy-2,3,3,9-tetramethyl-, methyl ester, (±)-	MeOH	217(4.37),253(4.65), 312s(3.68),324(3.77), 347(3.81)	44-4813-86
$C_{22}H_{26}O_7$			
1,4-Benzodioxin-2-ol, 2,3-dihydro-5-methoxy-3-methyl-7-(2-propen-yl)-2-(3,4,5-trimethoxyphenyl)-, trans	MeOH	235s(--),255(3.18)	102-0213-86
6(2H)-Benzofuranone, 3,3a,4,5-tetra-hydro-5,7-dimethoxy-2-(7-methoxy-	MeOH	213(3.94),258(3.79)	102-0213-86

Compound	Solvent	λ_{max}(log ε)	Ref.
1,3-benzodioxol-5-yl)-3-methyl-3a-(2-propenyl)-, [3S-(2α,3α-3aβ,5α)]- (cont.)			102-0213-86
1H-2-Benzoxacyclotetradecin-1,7(8H)-dione, 14,16-diacetoxy-3,4,5,6,9,10-hexahydro-3-methyl-, [S-(E)]-	MeOH	205.5(4.46)	33-0734-86
Bicyclo[3.2.1]oct-3-en-2-one, 8-hydroxy-1,3-dimethoxy-7-(7-methoxy-benzodioxol-5-yl)-6-methyl-5-(2-propenyl)-, [1S-(6-endo,7-exo,8-syn)]-	MeOH	231s(3.94),268(3.64)	102-0213-86
$C_{22}H_{26}O_8$			
Calyculatolide, 8β-angeloyloxy-9α-acetoxy-	MeOH	212(4.29),258(4.02)	102-0467-86
Jamaicolide A	MeOH	end absorption	102-0877-86
Methanone, (3,4-dimethoxyphenyl)-[5-(3,4-dimethoxyphenyl)tetrahydro-4-hydroxy-4-(hydroxymethyl)-3-furanyl]- (7-oxodihydrogmelinol)	MeOH	231(4.06),276(3.8), 310(3.47)	100-1061-86
$C_{22}H_{26}O_9$			
4H-1-Benzopyran-4-one, 2,3-dihydro-5,6,7-trimethoxy-2-(2,3,4,5-tetra-methoxyphenyl)-, (S)-	n.s.g.	230s(--),280(4.18), 322(3.65)	102-2223-86
Jamaicolide B	MeOH	end absorption	102-0877-86
$C_{22}H_{27}Cl_2FO_3S$			
Androsta-1,4-diene-17-carbothioic acid, 9,11-dichloro-6-fluoro-17-hydroxy-3-oxo-16-methyl-, S-methyl ester, (6α,11β,16β,17α)-	MeOH	237(4.30)	44-2315-86
$C_{22}H_{27}Cl_2FO_4$			
Pregna-1,4-diene-3,20-dione, 9,11-dichloro-17,21-dihydroxy-6-fluoro-16-methyl-, (6α,11β,16β)-	MeOH	236(4.18)	44-2315-86
$C_{22}H_{27}FO_4$			
Pregna-1,4,9(11)-triene-3,20-dione, 6-fluoro-17,21-dihydroxy-16-methyl-, (6α,16β)-	MeOH	238(4.21)	44-2315-86
$C_{22}H_{27}F_3O_5$			
Benzeneacetic acid, α-methoxy-α-(tri-fluoromethyl)-, 5-(2,5-dihydro-4-methyl-5-oxo-3-furanyl)-1-ethyl-pentyl ester	MeCN	209(4.18),260(2.34)	100-0593-86
$C_{22}H_{27}NO_4$			
6H-Dibenzo[a,g]quinolizine, 5,8,13,13a-tetrahydro-2,3,9,10-tetramethoxy-8-methyl-, cis-(±)-	MeOH	238(3.59),280(3.58)	78-2075-86
trans	MeOH	238(3.60),281(3.59)	78-2075-86
$C_{22}H_{27}NO_5$			
6H-Dibenzo[a,g]quinolizine, 5,8,13,13a-tetrahydro-2,3,4,9,10-pentamethoxy-(O-methylthaicanine), (-)-	EtOH	208(4.42),229s(4.18), 278(3.23)	100-0253-86
$C_{22}H_{27}NO_7$			
Narcotinediol	MeOH	236(4.02),281(3.52)	102-2403-86

Compound	Solvent	$\lambda_{max}(\log \epsilon)$	Ref.
$C_{22}H_{27}N_5$			
1H-Pyrazolo[3,4-b]quinolin-4-amine, 5,6,7,8-tetrahydro-N-phenyl-1-(1-piperidinylmethyl)-	EtOH	215(4.45),242s(--), 272(3.71),318(4.22)	5-1728-86
$C_{22}H_{27}N_5O$			
1H-Pyrazolo[3,4-b]quinolin-4-amine, 5,6,7,8-tetrahydro-N-(4-methylphenyl)-1-(4-morpholinylmethyl)-	EtOH	216(4.42),243s(--), 272(3.66),319(4.15)	5-1728-86
$C_{22}H_{28}$			
Benzene, 1,1'-[1,2-bis(1,1-dimethylethyl)-1,2-ethenediyl]bis-, (Z)-	MeOH	207(4.04),240s(3.42), 262s(2.58)	88-5339-86
$C_{22}H_{28}F_2O_4S$			
Androsta-1,4-diene-17-carbothioic acid, 6,9-difluoro-11,17-dihydroxy-16-methyl-3-oxo-, S-methyl ester, (6α,11β,16α,17α)-	MeOH	239(4.29)	44-2315-86
$C_{22}H_{28}N_2O_2$			
Ethanedione, bis[4-(diethylamino)-phenyl]-	EtOH	245(3.91),353s(--), 374(4.62)	48-0253-86
$C_{22}H_{28}N_2O_4$			
Androst-4-ene-3,11-dione, 17-(2,3-dihydro-2-oxo-1H-imidazol-4-yl)-17 -hydroxy-	MeOH	233(4.26)	150-2731-86M
Aspidospermidin-21-oic acid, 1-formyl-17-methoxy-, methyl ester, (±)-	MeOH	215(4.34),254(3.99)	78-6719-86
$C_{22}H_{28}N_2O_7S$			
Uridine, 5-ethyl-5'-(methoxymethyl)-2',3'-O-(1-methylethylidene)-6-(phenylthio)-	MeOH	243(3.99),276(3.98), 313s(3.57)	78-4187-86
$C_{22}H_{28}N_2O_9S$			
α-D-Glucopyranose, 2-deoxy-2-[[[(phenylmethyl)amino]thioxomethyl]amino]-, 1,3,4,6-tetraacetate	EtOH	245(4.21)	136-0049-86I
$C_{22}H_{28}N_2O_{10}S$			
α-D-Glucopyranose, 2-deoxy-2-[[[(4-methoxyphenyl)amino]thioxomethyl]-amino]-, 1,3,4,6-tetraacetate	EtOH	245(4.25)	136-0049-86I
$C_{22}H_{28}N_2O_{13}$			
D-Galactitol, 1-deoxy-1-(tetrahydro-1,3-dimethyl-2,4,6-trioxo-5(2H)-pyrimidinylidene)-, pentaacetate	MeOH	235(4.16)	136-0053-86M
D-Glucitol, 1-deoxy-1-(tetrahydro-1,3-dimethyl-2,4,6-trioxo-5(2H)-pyrimidinylidene)-, 2,3,4,5,6-pentaacetate	MeOH	235(4.19)	136-0053-86M
D-Mannitol, 1-deoxy-1-(tetrahydro-1,3-dimethyl-2,4,6-trioxo-5(2H)-pyrimidinylidene)-, 2,3,4,5,6-pentaacetate	MeOH	238(4.13)	136-0053-86M
2,4(1H,3H)-Pyrimidinedione, 6-acetoxy-1,3-dimethyl-5-(2,3,4,6-tetra-O-acetyl-β-D-mannopyranosyl)-	EtOH	266(4.06)	136-0053-86M

Compound	Solvent	λ_{max}(log ε)	Ref.
$C_{22}H_{28}N_6O_4$			
β-D-Ribofuranuronamide, 1-deoxy-N-ethyl-1-[6-[[1-(phenylmethyl)propyl]amino]-9H-purin-9-yl]-, (R)-	n.s.g.	271(4.26)	87-1683-86
(S)-	n.s.g.	270(4.28)	87-1683-86
β-D-Ribofuranuronamide, 1-deoxy-1-[6-[(1-ethylpropyl)amino]-9H-purin-9-yl]-N-(phenylmethyl)-	n.s.g.	269(4.24)	87-1683-86
$C_{22}H_{28}N_{10}O_2S_2$			
2-Pteridinamine, 7,7'-dithiobis-[4-(pentyloxy)-	MeOH	233(4.70),266(4.40), 382(4.42)	33-0708-86
$C_{22}H_{28}O_2$			
2,5-Cyclohexadien-1-one, 2,6-bis(1,1-dimethylethyl)-4-(2-hydroxy-2-phenylethylidene)-	C_6H_{12}	306(4.48)	44-2257-86
2-Cyclopenten-1-one, 3,3'-(1,3,5-hexatriene-1,6-diyl)bis[2,4,4-trimethyl-, (E,E,E)-	$CHCl_3$	412(4.69)	89-1018-86
$C_{22}H_{28}O_3$			
7-Oxabicyclo[4.1.0]hept-3-en-2-one, 3,5-bis(1,1-dimethylethyl)-5-hydroxy-1-(1-phenylethenyl)-	C_6H_{12}	207(4.14),230(4.10)	44-2257-86
7-Oxabicyclo[4.1.0]hept-3-en-2-one, 3,5-bis(1,1-dimethylethyl)-5-hydroxy-1-(2-phenylethenyl)-	C_6H_{12}	208(4.24),253(4.29)	44-2257-86
Pregna-4,14,16-triene-21-carboxylic acid, 3-oxo-	n.s.g.	242(4.22)	88-5463-86
$C_{22}H_{28}O_7$			
1H-2-Benzoxacyclotetradecine-1,7(8H)-dione, 14,16-diacetoxy-3,4,5,6,9,10-11,12-octahydro-3-methyl-, (S)-	MeOH	218.0(4.42),251.5(4.19)	33-0734-86
Jamaicolide D	MeOH	end absorption	102-0877-86
1,4-Phenanthrenedione, 9-acetoxy-4b,5,6,7,8,8a,9,10-octahydro-3,7,10-trihydroxy-4b,8,8-trimethyl-2-(2-propenyl)- (plectanthone I)	ether	270(4.10),400s(2.89)	33-1395-86
$C_{22}H_{28}O_8$			
Coleon C, 2α-acetoxy-	ether	266(4.20),288s(3.98), 326(3.78),396s(3.92)	33-1395-86
Coleon H, (15S)-	ether	267(4.09),288(3.86), 329(3.74),400s(3.9)	33-1395-86
Eupafortunin	MeOH	213(4.25)	94-5157-86
Gracilin C	MeOH	292(4.29)	100-0823-86
$C_{22}H_{28}Si_2$			
Tricyclo[8.2.2.2^{4,7}]hexadeca-2,4,6,8-10,12,13,15-octaene, 2,9-bis(trimethylsilyl)-, chromium tricarbonyl complex	CH_2Cl_2	345(3.97)	88-2353-86
$C_{22}H_{29}FO_4S$			
Androsta-1,4-diene-17-carbothioic acid, 9-fluoro-11,17-dihydroxy-16-methyl-3-oxo-, S-methyl ester, (11β,16α,17α)-	MeOH	239(4.29)	44-2315-86

Compound	Solvent	λ_{max}(log ε)	Ref.
$C_{22}H_{29}NO_9$			
Propanedioic acid, 2,2'-[[5-(dimethylamino)-2,4-furandiyl]dimethylidyne]bis-, tetraethyl ester	MeOH	249(2.73),365(2.87), 454(3.14)	73-0879-86
$C_{22}H_{29}NO_{14}$			
D-arabino-Hex-1-enitol, 1-deoxy-1-[[3-methoxy-2-(methoxycarbonyl)-3-oxo-1-propenyl]amino]-, 2,3,4,5,6-pentaacetate	$CHCl_3$	310(4.44),370s(3.77)	136-0329-86D
$C_{22}H_{29}N_3O_3$			
Acetamide, N-[1,4-dihydro-1,4-dioxo-3-[(2,2,6,6-tetramethyl-4-piperidinyl)amino]-2-naphthalenyl]-N-methyl-	heptane	233(4.13),242(4.08), 262s(4.19),273(4.31), 297s(3.85),341s(3.28), 453(3.35)	104-1560-86
$C_{22}H_{29}N_3S_2$			
Thiethylperazine bis(hydrogen maleate)	MeOH	216(4.652),265(4.806), 317(3.675)	106-0571-86
$C_{22}H_{29}N_5$			
1H-Pyrazolo[3,4-b]quinoline-1-methanamine, N,N-diethyl-5,6,7,8-tetrahydro-4-[(4-methylphenyl)amino]-	EtOH	214(4.44),242(4.08), 271(3.71),317(4.20)	5-1728-86
$C_{22}H_{30}$			
4,4'-Bitricyclo[4.3.1.1^{3,8}]undec-4-ene	C_6H_{12}	200(3.795),244(3.852)	35-5503-86
Tricyclo[3.3.1.1^{3,7}]decane, 2,2'-(1,2-ethanediylidene)bis-	hexane	239(4.518),247(4.577), 256(4.403)	35-5503-86
$C_{22}H_{30}F_6O_2$			
Bicyclo[2.2.0]hexa-2,5-diene-1-carboxylic acid, 2,6-bis(trifluoromethyl)-3,4,5-tris(1,1-dimethylethyl)-, methyl ester	hexane	222(3.41)	78-5341-86
Bicyclo[2.2.0]hexa-2,5-diene-2-carboxylic acid, 1,6-bis(trifluoromethyl)-3,4,5-tris(1,1-dimethylethyl)-, methyl ester	hexane	229(3.41),254(3.08)	78-5341-86
$C_{22}H_{30}N_2O_2$			
Ethanone, 1,2-bis[4-(diethylamino)-phenyl]-2-hydroxy-	EtOH	267(4.13),348(4.37)	48-0253-86
$C_{22}H_{30}N_2O_6$			
Uridine, 2'-deoxy-5-ethyl-, 3'-tricyclo[3.3.1.1^{3,7}]decane-1-carboxylate	octanol	268(4.20)	83-0154-86
Uridine, 2'-deoxy-5-ethyl-, 5'-tricyclo[3.3.1.1^{3,7}]decane-1-carboxylate	EtOH	270(4.28)	83-0154-86
$C_{22}H_{30}N_4O_2S_2$			
Thioproperazine (dimethanesulfonate)	MeOH	236.5(4.293),267(4.611), 317(3.539)	106-0571-86
$C_{22}H_{30}O$			
2-Butenal, 3-methyl-4-[3-[2-(2,6,6-trimethyl-1-cyclohexen-1-yl)ethenyl]-2-cyclohexen-1-ylidene]-, (E,E,E)-	hexane	365(4.45)	35-4614-86

Compound	Solvent	λ_{max}(log ε)	Ref.
(E,E,Z)- (cont.)	hexane	355(3.98)	35-6440-86
(2-cis)-	hexane	320(4.20)	35-4614-86
	CH_2Cl_2	306(4.11),369(4.13)	35-4614-86
(2,4-dicis)-	CH_2Cl_2	384(3.70)	35-4614-86
$C_{22}H_{30}O_2$			
6-Azuleneacetic acid, 2-octyl-, ethyl ester	CH_2Cl_2	238.5(4.13),280.5(4.85), 290.5(4.95),306(3.76), 335(3.62),351(3.74), 364.5(3.22),561.5(2.45), 597.5(2.39),663.5(1.94)	24-2272-86
Phenol, 2,4-bis(1,1-dimethylethyl)-6-[1-(2-hydroxyphenyl)ethyl]-, (S)-	EtOH	279(3.66)	116-0509-86
$C_{22}H_{30}O_4$			
Benzaldehyde, 2,3,6-trihydroxy-5-[(1,4,4a,5,6,7,8,8a-octahydro-2,5,5,8a-tetramethyl-1-naphthalenyl)methyl]- (siphonodictyal C)	MeOH	209(4.07),228s(3.90), 276(3.76),368(3.20)	44-4568-86
	MeOH-KOH	212(4.21),244(3.89), 291(3.65),406(3.26)	44-4568-86
Benzoic acid, 3-[(decahydro-1,2,4a-trimethyl-5-methylene-1-naphthalenyl)methyl]-4,5-dihydroxy-	MeOH	216(3.92),245(3.52), 294(3.22)	44-4568-86
	MeOH-KOH	216(3.94),244s(3.57), 295(3.22)	44-4568-86
2,5-Cyclohexadiene-1,4-dione, 2-hydroxy-5-methoxy-3-(8,11,14-pentadecatrienyl)-, (Z,Z)-	CH_2Cl_2	283(4.22),288(4.30), 412(2.68)	35-7858-86
$C_{22}H_{30}O_6$			
6,10-Dodecadienoic acid, 12-(3-formyl-2,4,5-trihydroxyphenyl)-2,6,10-trimethyl-, (E,E)- (siphonodictyal E)	MeOH	276(3.74),366(3.20)	44-4568-86
$C_{22}H_{30}O_7S$			
Benzaldehyde, 2,3-dihydroxy-5-[(1,2,3-5,6,7,8,8a-octahydro-1,2,5,5-tetramethyl-1-naphthalenyl)methyl]-6-(sulfooxy)-, (1α,2α,8aα)- (siphonodictyal D)	MeOH	214(4.15),228s(4.09), 276(3.97),367(3.43)	44-4568-86
	MeOH-KOH	212(4.24),244(4.12), 288(3.82),401(3.47)	44-4568-86
$C_{22}H_{31}BrN_2O$			
Urea, N-(4-bromophenyl)-N'-[1,5-dimethyl-1-(4-methyl-3-cyclohexen-1-yl)-4-hexenyl]-, [R-(R*,R*)]-	MeOH	204.2(4.67),249.7(4.46)	44-5136-86
$C_{22}H_{31}ClN_4O_6$			
L-Valine, N-[N-[N^2-(4-chlorobenzoyl)-L-glutaminyl]-2-methylalanyl]-, methyl ester	MeOH	238(4.18)	23-0277-86
$C_{22}H_{32}NO_6P$			
1H-Indeno[2,1-b]pyridine-3-carboxylic acid, 4-[(diethylphosphinyl)oxy]-2,3,9,9a-tetrahydro-1,9,9-trimethyl-, ethyl ester	EtOH	252(4.20),284(3.62), 294(3.57)	94-2786-86
isomer	EtOH	252(4.16),284(3.66), 293(3.60)	94-2786-86
$C_{22}H_{32}N_2O_2S_2$			
Sulfoximine, N,N'-1,6-hexanediyl-bis[S-methyl-S-(4-methylphenyl)-, [R-(R*,R*)]-	EtOH	224(4.33)	12-1655-86

Compound	Solvent	λ max(log ε)	Ref.
$C_{22}H_{32}N_6$			
1,3-Propanediamine, N,N"-benzo[g]-phthalazine-1,4-diylbis[N',N'-dimethyl-	EtOH	213s(4.19),222s(4.32), 236(4.60),256(4.47), 300(3.54),309(3.48)	111-0143-86
$C_{22}H_{32}O$			
Retinal, 13-demethyl-13-propyl-, all-E-	heptane	372(4.65)	5-0479-86
11Z-	EtOH	253(--),379(--)	5-0479-86
$C_{22}H_{32}O_3$			
6-Cyclopentacycloundecenecarboxaldehyde, 11-acetoxy-3,3a,4,7,8,11,12-12a-octahydro-3a,10-dimethyl-1-(1-methylethyl)- (9-acetoxydolabella-3,7,12-trien-16-al)	MeOH	229(3.77)	44-2736-86
$C_{22}H_{32}O_4$			
Benzoic acid, 2-[2-(2,2-dimethyl-6-oxocyclohexyl)ethyl]-5-methoxy-4-(1-methylethyl)-, methyl ester, (S)-	MeOH	213(4.49),245(3.95), 297(3.54)	78-0529-86
6-Cyclopentacycloundecenecarboxylic acid, 11-acetoxy-3,3a,4,7,8,11,12-12a-octahydro-3a,10-dimethyl-1-(1-methylethyl)-, (3aR*,5E,9E,11R*-12aR*)-(+)-	MeOH	207(3.74),222(3.76)	44-2736-86
2(5H)-Furanone, 4-[2-(4-acetoxy-1,2,3,4,4a,7,8,8a-octahydro-1,2,4a,5-tetramethyl-1-naphthalenyl)ethyl]-, [1R-(1α,2α,4β-4aα,8aα)]-	EtOH	204(4.26)	78-3461-86
1,4-Phenanthrenedione, 10-ethoxy-4b,5,6,7,8,8a,9,10-octahydro-3-hydroxy-4b,8,8-trimethyl-2-(1-methylethyl)-, [4bS-(4bα,8aβ,10β)]-(7α-ethoxyroyleanone)	MeOH	271.5(4.04),410(2.48)	102-0266-86
	MeOH-NaOMe	273(3.88),514(3.08)	102-0266-86
	MeOH-AlCl$_3$	289(3.92),520(2.60)	102-0266-86
	MeOH-AlCl$_3$-HCl	277(3.94),295s(3.70), 366(2.60)	102-0266-86
$C_{22}H_{32}O_5$			
2(5H)-Furanone, 4-[2-(7-acetoxydecahydro-4,5,7a,7b-tetramethylnaphth-[1,2-b]oxiren-4-yl)ethyl]-, [1aR-(1aα,3aβ,4β,5β,7α,7aβ,7bα)]-	EtOH	204(4.21)	78-3461-86
Furo[4,3,2-ij][2]benzopyran-7(2H)-one, 2-acetoxy-2a,3,4,5,8a,8b-hexahydro-6-methyl-5-(1,3,3-trimethylcyclohexyl)-, [(1R*,1'S*,1"R*,3R*,8S*,9S*)]-	hexane	205s(3.69),235(3.93)	12-1643-86
Furo[4,3,2-ij][2]benzopyran-7(2H)-one, 2-acetoxy-2a,3,4,6,8a,8b-hexahydro-6-methyl-5-(1,3,3-trimethylcyclohexyl)-	hexane	206(3.90)	12-1643-86
isomer	hexane	205(3.90)	12-1643-86
$C_{22}H_{32}O_6S$			
1,2-Benzenediol, 5-(hydroxymethyl)-3-[(3,4,4a,5,6,7,8,8a-octahydro-2,5,5,8a-tetramethyl-1-naphthalenyl)methyl]-, 2-(hydrogen sulfate), trans, sodium salt (siphonodictyol H)	MeOH	213(3.96),274(3.18), 280(3.19)	44-4568-86
	MeOH-KOH	218(4.04),244(3.49), 282(3.22),292(3.20)	44-4568-86
$C_{22}H_{32}O_7$			
1,4-Phenanthrenedione, 4b,5,6,7,8,8a-	ether	270(3.98),400s(2.82)	33-0456-86

Compound	Solvent	λmax(log ε)	Ref.
9,10-octahydro-3,9-dihydroxy-8-(hydroxymethyl)-10-methoxy-2-(2-methoxypropyl)-4b,8-dimethyl- (cont.)			33-0456-86
epimer	ether	270(3.98),400s(2.79)	33-0456-86
$C_{22}H_{33}NO_2$			
Benzenamine, 4-[2,6-bis(1,1-dimethylethyl)-2-methoxy-2H-pyran-4-yl]-N,N-dimethyl-	MeOH	290(4.28)	44-4385-86
$C_{22}H_{33}N_5O_8Si$			
Methanimidamide, N'-[4-[cyano(trimethylsilyl)oxy]methyl]-1-(2,3,5-tri-O-acetyl-β-D-ribofuranosyl)-1H-pyrazol-5-yl]-N,N-dimethyl-	MeOH	273(4.04)	44-1050-86
$C_{22}H_{34}N_4O_8$			
1H-Purine-2,6-dione, 3,9-dihydro-9-[6-O-(6-hydroxyhexyl)-3,4-O-(1-methylethylidene)-β-D-galactopyranosyl]-1,3-dimethyl-	$CHCl_3$	279(4.05)	136-0256-86K
$C_{22}H_{34}O_3Si$			
5,7-Octadien-4-ol, 1-[[(1,1-dimethylethyl)dimethylsilyl]oxy]-7-methyl-, benzoate, [R-(Z)]-	EtOH	230(4.26)	163-0187-86
$C_{22}H_{34}O_4$			
5(1H)-Cyclopentacycloundecenone, 12-acetoxy-2,3,3a,4,8,9,12,12a-octahydro-3-hydroxy-6,10,12a-trimethyl-3-(1-methylethyl)-	MeOH	205(3.65),237(3.83)	44-2736-86
16-Oxospongian-7-yl acetate	EtOH	230(2.48)	12-1629-86
$C_{22}H_{34}O_5$			
2-Butenoic acid, 2-methyl-, 8-acetoxy-1,2,3,5,6,7,8,8a-octahydro-1-hydroxy-1,4-dimethyl-7-(1-methylethyl)-6-azulenyl ester (cycloshiromodiol 6-O-acetate 8-O-angelate)	EtOH	214(3.9)	39-1363-86C
Shiromodiol 6-O-acetate 8-O-angelate	EtOH	218(3.9)	39-1363-86C
$C_{22}H_{36}N_3O_4$			
1-Piperidinyloxy, 4-[acetyl[7-(2,5-dihydro-2,5-dioxo-1H-pyrrol-1-yl)heptyl]amino]-2,2,6,6-tetramethyl-	H_2O	207(4.18),217s(4.14), 233s(3.73),245s(3.35), 448(1.18)	70-0363-86
$C_{22}H_{36}O_3$			
1,11-Cyclopentacycloundecenediol, 1,2,3,3a,4,7,8,11,12,12a-decahydro-3,6,10-trimethyl-1-(1-methylethyl)-, 11-acetate (9-acetoxydolabella-3,7-dien-12-ol)	MeOH	215(3.15)	44-2736-86
1,3-Cyclopentadiene-1-carboxylic acid, 2,3,4-tris(1,1-dimethylethyl)-5-oxo-, 1,1-dimethylethyl ester	hexane	218(4.03),417(2.20)	24-2159-86
1,4-Cyclopentadiene-1-carboxylic acid, 2,4,5-tris(1,1-dimethylethyl)-3-oxo-, 1,1-dimethylethyl ester	hexane	213s(3.79),410(2.21)	24-2159-86

Compound	Solvent	$\lambda_{max}(\log \epsilon)$	Ref.
$C_{22}H_{36}O_4$			
Cyclopenta[5,6]cycloundec[1,2-b]oxirene-9,11-diol, 1a,2,3,6,6a,7,8,9,9a-10,11,11a-dodecahydro-4,6a,11a-trimethyl-9-(1-methylethyl)-, 11-acetate	MeOH	207(3.40)	44-2736-86
$C_{22}H_{36}O_5$			
3-Furancarboxylic acid, 2,5-dihydro-4-methyl-5-oxo-2-(14-oxopentadecyl)-, methyl ester, (S)- (methyl isomuronate)	MeOH	230(4.14)	102-0453-86
$C_{22}H_{38}N_3O_6$			
1-Piperidinyloxy, 4-[[7-[(3-carboxy-1-oxo-2-propenyl)amino]heptyl]-(methoxycarbonyl)amino]-2,2,6,6-tetramethyl-, (Z)-	H_2O	204(4.23),239s(3.72), 428(1.26)	70-0363-86
$C_{22}H_{38}O$			
1,3-Docosadiyn-5-ol, (R)-	EtOH	228(2.80),239(2.82), 252(2.65)	150-1348-86M
$C_{22}H_{38}Si_5$			
1,2,3,10,11-Pentasila[3.2]-p-cyclophane, 1,1,2,2,3,3,10,10,11,11-decamethyl-	n.s.g.	227(4.5),248(4.5)	138-1781-86
$C_{22}H_{40}N_4$			
1,5,12,16-Tetraazacyclodocosa-1,3,12,14-tetraene, 2,4,13,15-tetramethyl-	DMF	315(4.47)	138-0907-86
copper chelate	DMF	349(4.31),462(2.46), 568s(2.18)	138-0907-86
$C_{22}H_{40}OSi$			
Silane, (1,1-dimethylethyl)dimethyl-[[1-[2-methyl-2-(3-methylene-4-pentenyl)cyclohexyl]-2-propenyl]oxy]-	hexane	224.5(4.26)	39-1303-86C
$C_{22}H_{42}N_4O_7S_2$			
6,9,12,20,23,28,31-Heptaoxa-1,3,15,17-tetraazabicyclo[15.8.8]tritriacontane-2,16-dithione	EtOH	212.2(4.30),249.3(4.42)	104-1589-86
$C_{22}H_{42}OSi$			
Silane, [[[4-(2,3-dimethyl-2-butenylidene)-3,3,5,5-tetramethylcyclohexyl]-oxy](1,1-dimethylethyl)dimethyl-, (±)-	C_6H_{12}	196(4.18)	44-2361-86
$C_{22}H_{66}Si_{12}$			
1,1'-Bicyclohexasilane, 1,1,1',1',2,2-2',2',3,3,3',3',4,4,4',4',5,5,5',5'-6,6,6',6'-docosamethyl-	n.s.g.	237(3.67),257(3.51), 276(3.54)	101-0001-86Q

Compound	Solvent	λ_{max}(log ε)	Ref.
$C_{23}H_{11}N_3O_4$			
Benzonitrile, 2-(6,11-dihydro-6,11-dioxobenzo[f]pyrido[1,2-a]indol-12-yl)-5-nitro-	DMF	500(3.872)	2-1126-86
$C_{23}H_{12}F_{18}O_6S_2$			
1-Butanesulfonic acid, 1,1,2,2,3,3,4-4,4-nonafluoro-1-(4-methylphenyl)-2-phenyl-1,2-ethenediyl ester, (E)-	EtOH	263(4.29)	24-2220-86
(Z)-	C_6H_{12}	224(4.32),273(4.09)	24-2220-86
$C_{23}H_{12}F_{18}O_7S_2$			
1-Butanesulfonic acid, 1,1,2,2,3,3,4-4,4-nonafluoro-1-(4-methoxyphenyl)-2-phenyl-1,2-ethenediyl ester, (E)-	C_6H_{12}	278.5(4.29)	24-2220-86
$C_{23}H_{12}N_2O_2$			
5H-Indolo[2,3-b]naphtho[2,3-h]isoquinoline-5,15(13H)-dione	DMSO	214(4.431)	2-0616-86
$C_{23}H_{12}N_4O_4$			
Benzonitrile, 2-(2-amino-6,11-dihydro-6,11-dioxobenzo[f]pyrido[1,2-a]indol-12-yl)-5-nitro-	DMF	440(3.5911),557(3.9294)	2-1126-86
$C_{23}H_{12}O$			
1H-Benzo[cd]perylen-1-one	CF_3COOH	390(4.0),495(3.7), 580(4.6),620(5.0)	44-0251-86
$C_{23}H_{13}$			
1H-Benzo[cd]perylenylium tetrafluoroborate	CF_3COOH	400(4.1),572f(5.1)	44-0251-86
$C_{23}H_{13}ClO_2$			
7H-Benz[de]anthracen-7-one, 8-chloro-6-phenoxy-	benzene	342s(3.73),360(4.04), 392(4.07),405s(3.73)	104-0566-86
7H-Benz[de]anthracen-7-one, 11-chloro-6-phenoxy-	benzene	343s(3.75),360(4.02), 393(4.03)	104-0566-86
$C_{23}H_{13}Cl_2N_3$			
5-Pyrimidinecarbonitrile, 4,6-bis(4-chlorophenyl)-2-phenyl-	EtOH	215(3.09),245(2.96), 320(3.05)	2-1133-86
$C_{23}H_{13}N_3O_2$			
Indolo[2,1-b]quinazolin-12(6H)-one, 6-(1,3-dihydro-3-oxo-2H-indol-2-ylidene)- (candidin)	$CHCl_3$	252(4.87),283(4.73), 538(4.49),573(4.63)	5-1847-86
$C_{23}H_{14}ClNO$			
7H-Benz[de]anthracen-7-one, 8-chloro-6-(phenylamino)-	benzene	366(4.07),453s(4.16), 474(4.27)	104-0566-86
$C_{23}H_{14}Cl_2N_6O_6$			
4a,8a-Methanonaphthalene-1,4,5,8-tetrone, 1,8-bis(4-chloro-2-nitrophenyl)hydrazone	acetone	342(4.24),439(4.08)	64-0093-86B
	MeOH	247(4.59),346(4.36), 426(4.27)	64-0093-86B
	CH_2Cl_2	248(4.65),348(4.43), 432(4.34)	64-0093-86B
$C_{23}H_{14}N_2O_3$			
9,10-Anthracenedione, 1-amino-2-	DMSO	340(4.519),440(3.431)	2-0616-86

Compound	Solvent	λ_{max}(log ϵ)	Ref.
[(1,2-dihydro-2-oxo-3H-indol-3-ylidene)methyl]- (cont.)			2-0616-86
Ethanedione, (3-benzoyl-2-quinoxalinyl)phenyl-	CH_2Cl_2	261(4.64),320s(3.96), 402s(2.09)	89-0999-86
$C_{23}H_{14}O_4$			
1H-Indene-1,3(2H)-dione, 2-[5-(1,3-dihydro-1,3-dioxo-2H-inden-2-ylidene)-1,3-pentadienyl]-	EtOH-1% DMSO	555s(4.15),665(4.65)	70-1446-86
$C_{23}H_{14}O_5$			
Furo[3,2-b]furan-2,5-dione, 3-(4-methoxyphenyl)-6-(2-naphthalenyl)-	dioxan	238(4.78),417(4.64)	5-0195-86
$C_{23}H_{15}Br_2S$			
Thiopyrylium, 2,6-bis(4-bromophenyl)-4-phenyl-, perchlorate	MeCN	256(4.31),277(4.32), 372(4.46),410s(4.40)	48-0373-86
$C_{23}H_{15}Cl_2S$			
Thiopyrylium, 2,6-bis(4-chlorophenyl)-4-phenyl-, perchlorate	MeCN	252(4.29),275(4.31), 371(4.48),406s(4.41)	48-0373-86
$C_{23}H_{15}N$			
[2](1,4)Anthraceno[2](2,6)pyridinophane-1,13-diene	C_6H_{12}	263(5.07),303s(3.75), 370(3.64)	64-0223-86B
$C_{23}H_{15}NO_3$			
9,10-Anthracenedione, 1-amino-2-(3-oxo-3-phenyl-1-propenyl)-	DMSO	213(4.431),440(4.146)	2-0616-86
$C_{23}H_{15}N_3$			
5-Pyrimidinecarbonitrile, 2,4,6-triphenyl-	EtOH	243(3.91),321(4.08)	2-1133-86
$C_{23}H_{15}N_3O_8$			
Pyrrolo[1,2-a]quinoline-3-carboxylic acid, 1,2-bis(5-nitro-2-furanyl)-, ethyl ester	MeOH	228(3.42),270s(3.51), 421(2.58)	73-0412-86
$C_{23}H_{16}$			
7H-Benzocyclotridecene, 7-(2,4-cyclopentadien-1-ylidene)-12,13,14,15-tetradehydro-11-methyl-, (E,E,Z)-	THF	274(4.41),285(4.43), 318(4.30),392(4.39), 413s(4.28)	18-1723-86
relative absorbances	MeCN	269s(0.98),283(1.00), 316(0.73),387(0.75), 411s(0.59)	18-1723-86
$C_{23}H_{16}BrS$			
Thiopyrylium, 4-(4-bromophenyl)-2,6-diphenyl-, perchlorate	MeCN	251(4.29),275s(4.17), 381(4.51)	48-0373-86
$C_{23}H_{16}ClS$			
Thiopyrylium, 4-(4-chlorophenyl)-2,6-diphenyl-, perchlorate	MeCN	249(4.27),274s(4.19), 375(4.48)	48-0373-86
$C_{23}H_{16}N_2O_2$			
2-Butene-1,4-dione, 2-(1H-benzimidazol-1-yl)-1,4-diphenyl-, (E)-	MeOH	218(4.35),251(4.45), 314(3.91)	44-3420-86
(Z)-	MeOH	220(4.38),260(4.47), 323(3.94)	44-3420-86

Compound	Solvent	$\lambda_{max}(\log \epsilon)$	Ref.
1H-Oxazolo[3,4-a]quinazolin-1-one, 3-(4-methylphenyl)-5-phenyl-	CH_2Cl_2	271(3.96),301(4.35), 458(3.83)	39-2295-86C
$C_{23}H_{16}N_2O_3$			
1H-Oxazolo[3,4-a]quinazolin-1-one, 3-(4-methoxyphenyl)-5-phenyl-	CH_2Cl_2	269(4.47),297(4.54), 449(4.04)	39-2295-86C
$C_{23}H_{16}N_6O$			
Propanedinitrile, (2,3-dihydro-1,3-dimethyl-2-oxo-6,7-diphenyl-4(1H)-pteridinylidene)-	MeOH	231(4.58),265(4.17), 303(4.27),421(4.34), 434s(4.30)	128-0199-86
$C_{23}H_{16}O_2S$			
1,2-Naphthalenedione, 4-[(4-methylphenyl)thio]-3-phenyl-	$CHCl_3$	339(3.55),424(3.62), 479s(3.54)	39-2233-86C
$C_{23}H_{16}O_4$			
2-Propenoic acid, 3-phenyl-, (2-oxo-2H-naphtho[1,2-b]pyran-4-yl)methyl ester	EtOH	266(4.61),275.5(4.69), 316(3.96),353(3.76)	94-0390-86
$C_{23}H_{16}O_5$			
2,5-Cyclohexadiene-1,4-dione, 2,5-dihydroxy-3-(4-methoxyphenyl)-6-(2-naphthalenyl)-	dioxan	263(4.70),357(3.75)	5-0195-86
2H-Pyran-2,5(6H)-dione, 3-hydroxy-4-(4-methoxyphenyl)-6-(2-naphthalenylmethylene)-	dioxan	256(4.62),282s(4.46), 360(4.39)	5-0177-86
$C_{23}H_{16}O_{10}$			
1,3-Naphthacenedicarboxylic acid, 6,11-dihydro-2,5,12-trihydroxy-7-methoxy-6,11-dioxo-, dimethyl ester	MeOH	227(4.33),253(4.31), 269(3.96),408(3.71), 492(3.88),520(3.93), 552(3.83)	118-0942-86
$C_{23}H_{17}BO_5$			
Boron, (1,3-diphenyl-1,3-propanedionato-O,O')[α-hydroxybenzeneacetato-(2-)]-, (T-4)-	MeCN	272(3.98),368(4.40), 381s(4.30)	48-0755-86
$C_{23}H_{17}BrN_4O$			
1H-Imidazo[4,5-c]isoquinolin-1-amine, N-(4-bromophenyl)-2-(4-methoxyphenyl)-	EtOH	300(4.21),353(4.22)	78-5415-86
$C_{23}H_{17}NO$			
Dibenzo[d,j]azacyclotetradecine, 15,16,17,18-tetradehydro-8-ethoxy-, (E,E,E)-	THF	222(4.43),264s(4.19), 280s(4.37),292(4.45), 313s(4.31),330(4.27), 355s(3.85),394s(3.58)	39-0933-86C
$C_{23}H_{17}NOS$			
Benzo[h]quinolin-2(1H)-one, 5,6-dihydro-3-phenyl-4-(2-thienyl)-	EtOH	235(3.65),261(3.42), 370(3.54)	4-1789-86
$C_{23}H_{17}NO_2$			
Benzo[h]quinolin-2(1H)-one, 4-(2-furanyl)-5,6-dihydro-3-phenyl-	EtOH	235(3.89),259(3.66), 368(3.70)	4-1789-86
1,3-Indolizinedicarboxaldehyde, 6-methyl-2,7-diphenyl-	$CHCl_3$	246(4.37),278(4.60), 294(4.42),302(4.40), 362(4.42)	103-0844-86

Compound	Solvent	λ_{max}(log ε)	Ref.
Isoindolo[2,1-b]isoquinolin-7(5H)-one, 5-methoxy-12-phenyl-	MeOH	222(4.38),232(4.35), 243s(4.06),298(3.70), 311(3.65),362(4.12), 381s(3.94)	142-0969-86
$C_{23}H_{17}NO_3$			
Benzoic acid, 4-(5-[1,1'-biphenyl]-4-yl-2-oxazolyl)-, methyl ester	toluene	347(4.51)	103-1022-86
$C_{23}H_{17}N_3$			
2H-Pyrazolo[3,4-c]quinoline, 4-methyl-1,2-diphenyl-	EtOH	231(4.48),269(4.49), 299s(4.06),322s(3.84), 338s(3.62)	142-3181-86
$C_{23}H_{17}N_3O_2$			
1H-Indole, 3,3'-[(4-nitrophenyl)methylene]bis-	EtOH	223(4.79),275(4.35), 290(4.26)	2-1204-86
	EtOH-$HClO_4$	203(4.41),221(4.30), 261(4.31),308(4.06)	2-1204-86
$C_{23}H_{17}N_3O_2S$			
4(1H)-Pyrimidinethione, 5-nitro-2,6-diphenyl-1-(phenylmethyl)-	EtOH	248(4.04),340(4.20)	18-3871-86
$C_{23}H_{17}N_7O_3$			
Benzamide, N-[2-[(2,6-dicyano-4-nitrophenyl)azo]-5-(dimethylamino)phenyl]-	$CHCl_3$	257(4.35),318s(3.79), 471(3.53),574s(4.62)	48-0497-86
$C_{23}H_{18}BrN$			
Benzo[f]quinoline, 3-(4-bromophenyl)-1-(3-butenyl)-	EtOH	282(4.74),294(4.58), 346(3.90),363(3.93)	103-1229-86
$C_{23}H_{18}ClNOS$			
Benzo[h]quinolin-2(1H)-one, 3-(4-chlorophenyl)-3,4,5,6-tetrahydro-4-(2-thienyl)-	EtOH	225(3.87),295(3.25)	4-1789-86
$C_{23}H_{18}ClNO_2$			
Benzo[h]quinolin-2(1H)-one, 3-(4-chlorophenyl)-4-(2-furanyl)-3,4,5,6-tetrahydro-	EtOH	224(4.02),295(3.20)	4-1789-86
$C_{23}H_{18}ClN_3$			
1H-Pyrrole, 3-[(4-chlorophenyl)azo]-1-methyl-2,5-diphenyl-	EtOH	206(4.12),250(3.91), 300(4.07),390(3.89)	103-0499-86
$C_{23}H_{18}FN$			
Benzo[f]quinoline, 1-(3-butenyl)-3-(4-fluorophenyl)-	EtOH	279(4.67),294(4.47), 345(3.85),362(3.88)	103-1229-86
$C_{23}H_{18}N$			
Pyridinium, 1,2,4-triphenyl-, cation	DMSO	306(4.35)	44-2481-86
Pyridinium, 1,2,6-triphenyl-, cation	DMSO	300(4.03)	44-2481-86
$C_{23}H_{18}N_2O$			
Phenol, 2-(di-1H-indol-3-ylmethyl)-	EtOH	226(4.74),282(4.10), 291(4.03)	2-1204-86
	EtOH-$HClO_4$	205(4.48),256(4.26), 262(4.26),321(4.29)	2-1204-86

Compound	Solvent	λ_{max}(log ϵ)	Ref.
$C_{23}H_{18}N_2O_2$			
6H-Anthra[1,9-cd]isoxazol-6-one, 5-[(3,5-dimethylphenyl)amino]-3-methyl-	EtOH	495(4.30),521(4.36)	103-1363-86
Benzo[f]quinoline, 1-(3-butenyl)-3-(4-nitrophenyl)-	EtOH	287(4.44),297(4.43), 367(4.08)	103-1229-86
$C_{23}H_{18}N_2O_2S$			
Acetic acid, 1-[(3-oxobenzo[b]thien-2(3H)-ylidene)methyl]-2,2-diphenylhydrazide	dodecane	295(4.37),420(4.00)	104-2108-86
Benzo[b]thiophene-2-carboxaldehyde, 3-acetoxy-, 2-(diphenylhydrazone)	dodecane	360(4.38)	104-2108-86
$C_{23}H_{18}N_2O_3$			
1H-Benzimidazole-1-acetic acid, α-(2-phenoxy-2-phenylethenyl)-, (Z)-	MeOH	256(4.59)	44-3420-86
$C_{23}H_{18}N_4OS_3$			
4-Thiazolidinone, 3-ethyl-5-(2-methyl-3,8-diphenyl-6H-thiazolo[3,4-b]-[1,2,4]triazin-6-ylidene)-2-thioxo-	n.s.g.	453(4.60),604(3.44)	103-1373-86
4-Thiazolidinone, 3-ethyl-5-(3-methyl-2,8-diphenyl-6H-thiazolo[3,4-b]-[1,2,4]triazin-6-ylidene)-2-thioxo-	n.s.g.	452(4.60),595(3.43)	103-1373-86
$C_{23}H_{18}O$			
Ethanone, 1-(5a,11-dihydro-5,11[1',2']-benzeno-5H-cyclohepta[b]naphthalen-5-yl)-	EtOH	246(4.41)	18-1095-86
$C_{23}H_{18}O_2$			
2-Naphthalenol, 1-[(2-hydroxyphenyl)-phenylmethyl]-	EtOH	280(3.90)	116-0509-86
2-Pentene-1,4-dione, 1,2,3-triphenyl-, (E)-	ether	254(4.29)	78-1345-86
(Z)-	ether	235(4.22),280(3.95)	78-1345-86
Spiro[1H-cyclohepta[c]furan-1,1'-[2,4,10,12]cyclotridecatetraene-[6,8]diyn]-3(8aH)-one, 5',10'-dimethyl-	EtOH	258(4.38),316(3.90)	18-1713-86
$C_{23}H_{18}O_3$			
2H-1-Benzopyran-2-one, 3-(diphenylmethyl)-7-hydroxy-4-methyl-	MeOH	210(4.22),327(3.91)	2-0623-86
2H-1-Benzopyran-2-one, 8-(diphenylmethyl)-7-hydroxy-4-methyl-	MeOH	212(4.29),323(3.80)	2-0623-86
4H-1-Benzopyran-4-one, 6-(diphenylmethyl)-7-hydroxy-2-methyl-	MeOH	195(4.68),248(4.25), 292(4.09)	2-0755-86
4H-1-Benzopyran-4-one, 8-(diphenylmethyl)-7-hydroxy-2-methyl-	MeOH	195(4.48),246(4.39), 292(4.21)	2-0755-86
$C_{23}H_{18}O_5$			
4H-1-Benzopyran-4-one, 5,7-dihydroxy-3-methoxy-6-(phenylmethyl)-2-phenyl-	MeOH	218(4.01),276(4.52), 322(3.92),358(3.92)	2-0166-86
4H-1-Benzopyran-4-one, 5,7-dihydroxy-3-methoxy-8-(phenylmethyl)-2-phenyl-	MeOH	277(4.31),322(3.95), 361(3.69)	2-0166-86
$C_{23}H_{18}O_6$			
[1,2'-Binaphthalene]-1',4'-dione, 4,5',6-trihydroxy-5-methoxy-2,7'-	MeOH	232(4.72),240s(4.66), 294(3.92),304(3.91),	39-0675-86C

Compound	Solvent	λ_{max}(log ε)	Ref.
dimethyl- (cont.)		335(3.80),348(3.87), 426(3.60),545s(2.86)	39-0675-86C
[2,2'-Binaphthalene]-1,4-dione, 1',5,5'-trihydroxy-4'-methoxy-7,7'-dimethyl-	$CHCl_3$	327s(4.05),342(4.07), 454(3.72),645(3.30)	39-0675-86C
$C_{23}H_{18}O_{10}$			
1,3-Naphthacenedicarboxylic acid, 3,4,6,11-tetrahydro-2,5,12-trihydroxy-7-methoxy-6,11-dioxo-, dimethyl ester	MeOH	231(4.42),253(4.38), 280(4.06),477(3.96), 491(3.96),525(3.83), 545(3.59)	118-0942-86
$C_{23}H_{18}S_2$			
Azulene, 6-[bis(phenylthio)methyl]-	CH_2Cl_2	286(4.76),334(3.75), 350(3.78),364(3.34), 550s(--),585(2.49), 631(2.41),696(1.99)	5-1222-86
$C_{23}H_{19}BO_4$			
Boron, [1,2-benzenediolato(2-)-O,O']-[1,3-bis(4-methylphenyl)-1,3-propanedionato-O,O']-, (T-4)-	DMF	396(4.70),466s(2.49)	48-0755-86
$C_{23}H_{19}BO_6$			
Boron, [1,2-benzenediolato(2-)-O,O']-[1,3-bis(4-methoxyphenyl)-1,3-propanedionato-O,O']-, (T-4)-	CH_2Cl_2	327(3.97),406s(4.66), 417(4.74)	48-0755-86
$C_{23}H_{19}BO_8$			
Boron, [1,7-bis(4-methoxyphenyl)-1,6-heptadiene-3,5-dionato-O^3,O^5](ethanedioato(2-)-O,O')-	CH_2Cl_2	516(4.96)	97-0399-86
$C_{23}H_{18}BrN_2O$			
Ethanone, 1-(4-bromophenyl)-2-[1,2-dihydro-3-(4-methylphenyl)-2-quinoxalinyl]-	toluene	397(3.84)	103-0538-86
	MeOH	393(--)	103-0538-86
$C_{23}H_{19}ClN_2$			
2H-Benz[g]indazole, 3-(3-chlorophenyl)-3,3a,4,5-tetrahydro-2-phenyl-	EtOH	343(3.97)	4-0135-86
2H-Benz[g]indazole, 3-(4-chlorophenyl)-3,3a,4,5-tetrahydro-2-phenyl-	EtOH	246(3.95),340(4.21)	4-0135-86
$C_{23}H_{19}ClN_2O$			
Ethanone, 1-(4-chlorophenyl)-2-[1,2-dihydro-3-(4-methylphenyl)-2-quinoxalinyl]-	toluene	397(3.60)	103-0538-86
	MeOH	397(--)	103-0538-86
$C_{23}H_{19}Cl_3N_2O_2$			
2,5-Cyclohexadiene-1,4-dione, 2,3,5-trichloro-6-[[3-(10,11-dihydro-5H-dibenz[b,f]azepin-5-yl)propyl]amino]-	CH_2Cl_2	240s(4.30),303(4.12), 542(3.31)	106-0832-86
$C_{23}H_{19}N$			
[2](1,4)Anthraceno[2](2,6)pyridinophane	dioxan	229(4.09),262(4.88), 311s(3.19),364(3.55), 382(3.64),398(3.53), 402s(3.50)	64-0223-86B

Compound	Solvent	λ$_{max}$(log ε)	Ref.
5,12[1',2']-Benzenodibenzo[a,e]cyclo-octene-5(6H)-carbonitrile, 7,10,11,12-tetrahydro-	EtOH	267(3.0),275(3.0)	88-5621-86
Benzo[f]quinoline, 1-(3-butenyl)-3-phenyl-	EtOH	280(4.59),294(4.37), 346(3.67),361(3.67)	103-1229-86
$C_{23}H_{19}NO$			
Phenol, 2-[1-(3-butenyl)benzo[f]quin-olin-3-yl]-	EtOH	267(4.60),282(4.63), 294(4.50),353(4.16), 370(4.22)	103-1229-86
$C_{23}H_{19}NOS$			
Benzo[g]quinolin-2(1H)-one, 3,4,5,6-tetrahydro-3-phenyl-4-(2-thienyl)-	EtOH	230(4.39),296(3.69)	4-1789-86
1-Benzothiepin-5(2H)-one, 4-[(diphen-ylamino)methylene]-3,4-dihydro-, (E)-	EtOH	251(4.12),372(4.26)	4-0449-86
$C_{23}H_{19}NO_2$			
Benzo[h]quinolin-2(1H)-one, 4-(2-fur-anyl)-3,4,5,6-tetrahydro-3-phenyl-	EtOH	224(4.23),292(3.61), 370(3.01)	4-1789-86
$C_{23}H_{19}NO_2S$			
3,4-Isothiazoledimethanol, α,α',5-tri-phenyl-	C_6H_{12}	220(4.22),265(4.07)	5-1796-86
$C_{23}H_{19}NO_3$			
2H-Naphtho[1,2-b]pyran-2-one, 4-(4-morpholinyl)-3-phenyl-	EtOH	265(4.36),274(4.39), 304(4.06),337s(4.09), 346(4.10)	4-1067-86
$C_{23}H_{19}NO_4S$			
Benzenesulfonamide, 4-methyl-N-(4-meth-yl-2-oxo-3-phenyl-2H-1-benzopyran-7-yl)-	MeOH	235(4.07),294(3.81), 342(4.27)	151-0069-86A
	$CHCl_3$	297(3.86),323(4.30)	151-0069-86A
$C_{23}H_{19}NS_3$			
6H-1,3-Thiazine, 5-[bis(phenylthio)-methyl]-2-phenyl-	$CHCl_3$	263(4.47),370(3.95)	118-0916-86
$C_{23}H_{19}N_3O$			
4(3H)-Quinazolinone, 2-(4-methylphen-yl)-3-[[(4-methylphenyl)methylene]-amino]-	MeOH	225(3.51),253(3.62), 284(3.70),324(3.74)	18-1575-86
$C_{23}H_{19}N_3O_2$			
2H-Benz[g]indazole, 3,3a,4,5-tetra-hydro-3-(4-nitrophenyl)-2-phenyl-	EtOH	295(4.26),330(4.25)	4-0135-86
$C_{23}H_{19}N_3O_2S$			
Thieno[2,3-d]pyrimidine-4-acetic acid, α-cyano-5,6-dihydro-2,5-diphenyl-, ethyl ester	$CHCl_3$	277(4.36),343(4.28), 377s(3.95)	94-0516-86
$C_{23}H_{19}N_3O_3$			
4(3H)-Quinazolinone, 2-(4-methoxyphen-yl)-3-[[(4-methoxyphenyl)methylene]-amino]-	MeOH	228(3.54),248(3.59), 281(3.62),326(3.74)	18-1575-86
$C_{23}H_{19}N_3O_3S$			
Benzamide, N-[2-nitro-3-phenyl-	EtOH	247(4.18),309(4.15),	18-3871-86

Compound	Solvent	$\lambda_{max}(\log \epsilon)$	Ref.
3-(phenylmethyl)amino]-1-thioxo-2-propenyl]- (cont.)	EtOH	381(4.07),400(4.00)	18-3871-86
$C_{23}H_{19}N_3O_4$			
1H-Imidazole, 4,5-bis(4-methoxyphenyl)-2-(4-nitrophenyl)-	benzene	412(4.24)	39-1623-86B
	EtOH	406(4.24)	39-1623-86B
	dioxan	410(4.29)	39-1623-86B
	HOAc	372(4.10)	39-1623-86B
	$CHCl_3$	415(4.23)	39-1623-86B
	DMF	422(4.26)	39-1623-86B
$C_{23}H_{19}N_3O_5Se$			
Benzoic acid, 3-[(1,4,4a,5,7,7a-hexahydro-3-methyl-5,7-dioxo-1-phenylfuro[3',4':5,6]selenino[2,3-c]pyrazol-4-yl)amino]-, methyl ester	EtOH	258(3.9)	104-2126-86
$C_{23}H_{19}N_5O_2$			
Quino[8,7-g]pteridine-9,11(7H,10H)-dione, 2,4,7-trimethyl-10-(phenylmethyl)-	MeCN	444(4.03)	39-1825-86C
with potassium perchlorate	MeCN	444(4.03)	39-1825-86C
with silver nitrate	MeCN	455(3.98)	39-1825-86C
$C_{23}H_{19}N_{11}O_2$			
Methanone, bis[1-(2-methylphenyl)-1H-tetrazol-5-yl]-, (4-nitrophenyl)hydrazone	EtOH	389(4.47)	104-0177-86
Methanone, bis[1-(4-methylphenyl)-1H-tetrazol-5-yl]-, (4-nitrophenyl)hydrazone	EtOH	380(4.46)	104-0177-86
$C_{23}H_{19}N_{11}O_4$			
Methanone, bis[1-(4-methoxyphenyl)-1H-tetrazol-5-yl]-, (4-nitrophenyl)hydrazone	EtOH	229(4.42),380(4.99)	104-0177-86
$C_{23}H_{20}BrN_3O_2$			
Ethanone, 1-(4-bromophenyl)-2-[5,6-dimethyl-4-[(4-methylphenyl)imino]furo[2,3-d]pyrimidin-1(4H)-yl]-, monohydrobromide	EtOH	263(4.42),319(4.30)	11-0337-86
$C_{23}H_{20}ClN_2O$			
1H-Indolizinium, 3-(4-chlorophenyl)-2-[[4-(dimethylamino)phenyl]methylene]-2,3-dihydro-1-oxo-, perchlorate	80% MeOH	545(4.34)	4-1315-86
	MeOH-HCl	540(2.94)	4-1315-86
$C_{23}H_{20}N_2$			
2H-Benz[g]indazole, 3,3a,4,5-tetrahydro-2,3-diphenyl-	EtOH	345(4.33)	4-0135-86
9H-Benzo[f]cyclohepta[c]quinoline, 10,11,12,13-tetrahydro-8-(4-pyridinyl)-	EtOH	267(4.61),325(3.36), 340(3.57),355(3.62)	103-0759-86
$C_{23}H_{20}N_2O$			
Dipyrrolo[1,2-a:2',1'-c]pyrazin-1(10bH)-one, 5,6-dihydro-10b-methyl-2,3-diphenyl-	MeCN	244(4.20),316(3.83), 354(3.81)	118-0899-86

Compound	Solvent	$\lambda_{max}(\log \epsilon)$	Ref.
Ethanone, 2-[1,2-dihydro-3-(4-methylphenyl)-2-quinoxalinyl]-1-phenyl-	toluene	392(3.81)	103-0538-86
	MeOH	391(--)	103-0538-86
$C_{23}H_{20}N_2OS_2$			
1-Triphenodithiazinol, 4-(1,1-dimethylethyl)-2-methyl-	MeOH-20% $CHCl_3$	285(4.42),325(4.31), 530s(4.37),565(4.47)	39-2233-86C
	+ HCl	281(4.37),336(4.42), 614(4.17),672(4.34), 736(4.25)	39-2233-86C
$C_{23}H_{20}N_2O_2$			
1H-Imidazole, 4,5-bis(4-methoxyphenyl)-2-phenyl-	EtOH	235(4.37),304(4.45)	39-1623-86B
$C_{23}H_{20}N_2O_3$			
2,4-Cyclohexadien-1-one, 4-nitro-6-[(5,6,10,11-tetrahydro-4H,9H-pyrido[3,2,1-de]phenanthridin-12-yl)methylene]-	EtOH	240(4.74),310(4.27), 335(4.24),370(4.22), 410(4.27),490(3.95)	135-1208-86
Methanone, [7-methoxy-3-[(2-methoxyphenyl)amino]-1H-indol-2-yl]phenyl-	MeCN	260(4.51),316(4.17), 426(4.06)	24-2289-86
$C_{23}H_{20}N_2O_8$			
Dipyrido[1,2-a:2',1'-b]benzimidazole-1,2,3,4-tetracarboxylic acid, tetramethyl ester	EtOH	210(4.40),275(4.05), 320(4.00),430(3.80), 570s(3.20)	103-0192-86
$C_{23}H_{20}N_4$			
1H-Pyrrolo[2,3-b]pyridine-5-carbonitrile, 4-methyl-1-(phenylmethyl)-6-[(phenylmethyl)amino]-	$CHCl_3$	265(1.01),350(0.38)	103-0076-86
$C_{23}H_{20}N_4O_4$			
Spiro[3H-indole-3,4'(1'H)-pyrano[2,3-c]pyrazole]-5'-carboxylic acid, 6'-amino-1,2-dihydro-3'-methyl-2-oxo-1'-phenyl-, ethyl ester	EtOH	248(4.91)	18-1235-86
$C_{23}H_{20}O_2$			
1-Propanone, 3-(3,3-diphenyloxiranyl)-1-phenyl-	EtOH	242(4.10)	35-1617-86
$C_{23}H_{20}O_3$			
2-Propen-1-one, 1-[2-hydroxy-4-methoxy-3-(phenylmethyl)phenyl]-3-phenyl-	MeOH	208(4.6),270(4.2), 322(4.3)	2-0259-86
2-Propen-1-one, 1-[2-hydroxy-4-methoxy-5-(phenylmethyl)phenyl]-3-phenyl-	MeOH	220(4.2),256(3.8), 340(4.1)	2-0259-86
$C_{23}H_{20}O_7$			
2H,8H-Benzo[1,2-b:3,4-b']dipyran-2-one, 3-(1,3-benzodioxol-5-yl)-4,5-dimethoxy-8,8-dimethyl-	MeOH	235(4.50),289(4.25), 349(4.16)	102-2425-86
$C_{23}H_{20}O_{10}$			
2-Pentenedioic acid, 4-(9,10-dihydro-1,4-dihydroxy-8-methoxy-9,10-dioxo-2-anthracenyl)methyl]-3-hydroxy-, dimethyl ester	MeOH	230(4.54),248(4.36), 264(4.20),290(3.89), 400(3.64),419(3.82), 495(3.98),475(3.95), 492(3.92),525(3.62)	118-0942-86

Compound	Solvent	$\lambda_{max}(\log \epsilon)$	Ref.
$C_{23}H_{21}BrN_6O_5$			
Benzamide, N-[2-[(2-bromo-4,6-dinitrophenyl)azo]-5-(diethylamino)phenyl]-	$CHCl_3$	255(4.26),412(3.73), 565(4.65)	48-0497-86
$C_{23}H_{21}BrO_5$			
1,3-Azulenedicarboxylic acid, 6-bromo-2-(4-methoxyphenyl)-, diethyl ester	CH_2Cl_2	237.5(4.43),270.5(4.31), 307(4.67),341.5(4.43), 400(3.80),517(2.68)	24-2272-86
$C_{23}H_{21}NOS$			
2-Propen-1-one, 3-(1,2-dihydro-1-phenyl-2-thioxo-3-pyridinyl)-1-(2,4,6-trimethylphenyl)-	EtOH	257(4.35),339(4.03), 427(3.74)	4-0567-86
$C_{23}H_{21}NO_2$			
Acetamide, N-[4-[1-(4-methoxyphenyl)-2-phenylethenyl]phenyl]-	EtOH	228(4.26),265(4.34), 315(4.29)	13-0073-86B
Formamide, N-[2-oxo-2-phenyl-1,1-bis(phenylmethyl)ethyl]-	MeOH	219(4.01),250(4.00)	44-3374-86
2H-Naphtho[1,2-b]pyran-2-one, 4-(diethylamino)-3-phenyl-	EtOH	269(4.41),276(4.46), 294(4.02),308(4.09), 349(4.01)	4-1067-86
$C_{23}H_{21}NO_4$			
Acetic acid, (6,7-dihydro-12-methyl-14-oxo-5H,12H-12a,7a-(epoxyethano)-benzo[6,7]cyclohepta[1,2-b]indol-15-ylidene)-, methyl ester	n.s.g.	243(4.15),294(3.30), 333(3.20)	150-0514-86M
Propanedioic acid, [(2,3-dihydro-6,7-diphenyl-1H-pyrrolizin-5-yl)methyl]-	$CHCl_3$	243(4.43),354s(--)	83-0234-86
$C_{23}H_{21}NO_4S$			
Benzoic acid, 4-[3-(1-hydroxy-3-oxo-3-phenylpropyl)-2-thioxo-1(2H)-pyridinyl]-, ethyl ester	EtOH	224(4.39),237(4.37), 285(4.08),374(3.76)	4-0567-86
$C_{23}H_{21}N_2O$			
3H-Indolizinium, 3-[[4-(dimethylamino)phenyl]methylene]-1-hydroxy-2-phenyl-, chloride	80% MeOH	610(3.34)	4-1315-86
	MeOH-HCl	530(2.95)	4-1315-86
	MeOH-NaOH	480(3.53)	4-1315-86
$C_{23}H_{21}N_2O_6P$			
1H-Benz[de]isoquinoline-1,3(2H)-dione, 6-amino-2-[4-[(5,5-dimethyl-1,3,2-dioxaphosphorinan-2-yl)oxy]-phenyl]-, P-oxide	EtOH	436(4.08)	65-1105-86
$C_{23}H_{21}N_3$			
Benzenamine, 4-(1,3-diphenyl-1H-pyrazol-5-yl)-N,N-dimethyl-	n.s.g.	273(4.57)	78-3807-86
$C_{23}H_{21}N_3O$			
Benzoic acid, 2-[[(4-methylphenyl)-methylene]amino]-, [(4-methylphenyl)methylene]hydrazide	MeOH	226(3.18),257(3.34), 305(3.41)	18-1575-86
$C_{23}H_{21}N_3O_2$			
[3]Benzazocino[7,6-ab]carbazole-6(5H)-carbonitrile, 4,7,8,14-tetrahydro-1,2-dimethoxy-	MeOH	260(4.80),283(4.74), 301(4.45),335(3.97), 354(3.89),372(3.93)	12-0001-86

Compound	Solvent	λ_{max}(log ε)	Ref.
$C_{23}H_{21}N_3O_3$			
Benzoic acid, 2-[[(4-methoxyphenyl)-methylene]amino]-, [(4-methoxy-phenyl)methylene]hydrazide	MeOH	224(3.17),261(3.26), 303(3.33)	18-1575-86
$C_{23}H_{21}N_4S_2$			
Thiazolo[3,4-b][1,2,4]triazin-5-ium, 6-[(3-ethyl-2(3H)-benzothiazolylidene)methyl]-2,3-dimethyl-8-phenyl-, perchlorate	n.s.g.	438(4.50),505(4.19)	103-1373-86
$C_{23}H_{21}N_5O_2$			
2(1H)-Pteridinone, 1-methyl-4-(4-morpholinyl)-6,7-diphenyl-	pH -1	224(4.45),287(4.30), 396(4.24)	128-0199-86
	pH 5	222(4.47),276(4.29), 377(4.23)	128-0199-86
	MeOH	222(4.46),277(4.32), 377(4.24)	128-0199-86
$C_{23}H_{21}N_5O_6$			
L-Aspartic acid, N-[4-[[[(2-amino-1,4-dihydro-4-oxo-6-quinazolinyl)methyl]-prop-2-ynylamino]benzoyl]-	pH 13	228.5(4.70),279.5(4.38), 301.5(4.42)	87-1114-86
$C_{23}H_{21}O$			
Bisbenzo[6,7]cyclohepta[1,2-b:2',1'-e]pyrylium, 5,6,7,9,10,11-hexahydro-, perchlorate	MeCN	232(4.13),276(4.05), 306(3.61),388(4.41)	22-0600-86
$C_{23}H_{21}O_3$			
Dibenzo[c,h]xanthenylium, 5,6,8,9-tetrahydro-3,11-dimethoxy-	MeCN	234(4.24),268(4.23), 310(4.25),350s(3.85), 496(4.68)	22-0600-86
$C_{23}H_{22}BrN_3O$			
Benzenamine, N-[1-[(4-bromophenyl)-methyl]-5,6-dimethylfuro[2,3-d]-pyrimidin-4(1H)-ylidene]-3,5-dimethyl-, monohydrobromide	EtOH	320(4.26)	11-0337-86
$C_{23}H_{22}ClN_3O$			
Quinoline, 7-chloro-3-(3,4-dihydro-3-methyl-1-isoquinolinyl)-4-morpholino-	pH 1	225(4.52),245(4.26), 265(4.24),310(4.27), 400(3.86)	2-1072-86
	EtOH	235(3.71),260(3.41)	2-1072-86
$C_{23}H_{22}FNO$			
6-Hepten-3-one, 1-(4-fluorophenyl)-1-(2-naphthalenylamino)-	EtOH	246(4.64),275(3.89), 284(3.96),294(3.91)	103-1229-86
$C_{23}H_{22}N_2O_4S$			
Benzoic acid, 4-[3-[3-(4-aminophenyl)-1-hydroxy-3-oxopropyl]-2-thioxo-1(2H)-pyridinyl]-, ethyl ester	EtOH	235(4.03),294(4.24), 322(4.26),370(3.76)	4-0567-86
$C_{23}H_{22}N_2O_5S$			
Acetamide, N-acetyl-N-[1-formyl-2,3,9,9a-tetrahydro-9-(phenylsulfonyl)-1H-carbazol-4-yl]-	MeOH	217(4.71),260(4.41), 314(4.10)	5-2065-86

Compound	Solvent	λ_{max}(log ε)	Ref.
$C_{23}H_{22}N_2O_8$			
2-Propenoic acid, 3-(4-nitrophenyl)-, 1,5-pentanediyl ester, (E,E)-	MeCN	301(4.61)	18-1379-86
$C_{23}H_{22}N_4$			
2,3'-Bipyridinium, 1',1"'-(1,3-propanediyl)bis-, dibromide	pH 8.2	209(4.42),230(4.39), 277(4.28)	64-0239-86B
1H-Pyrazolo[3,4-b]quinolin-4-amine, 5,6,7,8-tetrahydro-N-(4-methylphenyl)-1-phenyl-	EtOH	225s(4.46),256(4.45), 330(4.29)	11-0331-86
$C_{23}H_{22}O$			
3,8[1',4']-Benzeno-2H-naphth[2,3-b]-oxete, 2a,3,8,8a,9,10,11,12-octahydro-2-phenyl-	EtOH	203(4.54),253(2.86), 260(2.85),266(2.83), 274(2.70)	39-0885-86C
$C_{23}H_{22}O_3$			
1-Propanone, 1-[2-hydroxy-4-methoxy-3-(phenylmethyl)phenyl]-3-phenyl-	MeOH	220(4.5),280(3.9)	2-0259-86
1-Propanone, 1-[2-hydroxy-4-methoxy-5-(phenylmethyl)phenyl]-3-phenyl-	MeOH	212(4.5),276(4.1)	2-0259-86
$C_{23}H_{22}O_4$			
2H-1-Benzopyran-2-one, 4-hydroxy-3-[(3-methyl-2-oxocyclohexyl)-phenylmethyl]-	MeOH	271(4.09),281(4.14), 305(4.08)	2-0102-86
$C_{23}H_{22}O_5$			
2,4,6-Cycloheptatrien-1-one, 2,5-bis[(4-methoxyphenyl)methoxy]-	MeOH	229(4.62),253s(4.03), 330(4.09),382s(3.79)	18-1125-86
Dictyoceratin B	EtOH	221(4.53),275(4.10)	78-4197-86
2H,6H-Pyrano[3,2-b]xanthen-6-one, 5,8-dihydroxy-2,2-dimethyl-7-(3-methyl-2-butenyl)- (calothwaitesixanthone)	MeOH	238(4.52),246(4.53), 286(4.70),319(4.41), 391(3.88)	102-1957-86
$C_{23}H_{22}O_6$			
2-Propenoic acid, 3-phenyl-, [2-[4-formyl-2-(2-methyl-1-oxopropoxy)-phenyl]oxiranyl]methyl ester	EtOH	215(4.35),254(4.14), 280(4.17)	102-1743-86
2H,6H-Pyrano[3,2-b]xanthone, 5,9,10-trihydroxy-2,2-dimethyl-12-(3-methyl-2-butenyl)-	MeOH	240(4.31),285(4.66), 337(4.31),380s(3.76)	102-1217-86
	MeOH-NaOAc	284(--),368(--)	102-1217-86
	MeOH-AlCl$_3$	260s(--),307(--), 400(--)	102-1217-86
$C_{23}H_{22}O_7$			
2-Anthraceneacetic acid, 1,4-dihydro-9-hydroxy-5-methoxy-1,4-dioxo-3-(3-oxopentyl)-, methyl ester	n.s.g.	207(4.15),244(4.70), 269(4.05),325s(--), 492(3.97),603s(--)	5-1506-86
2-Anthraceneacetic acid, 1,4-dihydro-10-hydroxy-5-methoxy-1,4-dioxo-3-(3-oxopentyl)-, methyl ester	n.s.g.	214(3.88),245(4.52), 280(3.80),326s(--), 510(3.76)	5-1506-86
2,5-Cyclohexadiene-1,4-dione, 2-(3,4-dimethoxyphenyl)-3,6-dimethoxy-5-(4-methoxyphenyl)-	dioxan	256.8(4.38),275.4(4.32), 396.4(3.69)	5-0195-86
$C_{23}H_{22}O_8$			
2-Anthraceneacetic acid, 9,10-dihydro-1,4-dihydroxy-5-methoxy-9,10-dioxo-3-(3-oxopentyl)-, methyl ester	n.s.g.	220(4.33),234(4.54), 246(4.47),288(3.96), 425s(--),455s(--),	5-1506-86

Compound	Solvent	λ_{max}(log ϵ)	Ref.
(cont.)		479(4.04),497(4.04), 531(3.80)	5-1506-86
1-Naphthacenecarboxylic acid, 2-ethyl-1,2,3,4,6,11-hexahydro-2,5,12-trihydroxy-7-methoxy-6,11-dioxo-, methyl ester	n.s.g.	218(4.32),234(4.50), 251(4.44),289(3.91), 378s(--),475(4.05), 492(4.07),528(3.86), 575s(--)	5-1506-86
2H-Pyran-2,5(6H)-dione, 4-(2,5-dimethoxyphenyl)-6-[(3,4-dimethoxyphenyl)-methylene]-3-methoxy-	dioxan	284(4.26),407(4.23)	5-0177-86
2H-Pyran-2,5(6H)-dione, 4-(3,4-dimethoxyphenyl)-6-[(3,4-dimethoxyphenyl)-methylene]-3-methoxy-	dioxan	284(4.36),411(4.28)	5-0177-86
$C_{23}H_{22}O_9$			
ε-Rhodomycinone, 4-O-methyl-, (±)-	n.s.g.	209(4.22),233(4.57), 250(4.36),263s(--), 288(3.91),377(3.93), 475(4.06),494(4.05), 530(3.77)	5-1506-86
$C_{23}H_{23}NO$			
Azacycloeicosa-1,3,5,7,13,15,17,19-octaene-9,11-diyne, 2-ethoxy-8,13-dimethyl-, (E,E,E,Z,Z,E,E,E)-	THF	244(4.10),298s(4.62), 326(4.80),480(2.90)	39-0933-86C
6-Hepten-3-one, 1-(2-naphthalenyl-amino)-1-phenyl-	EtOH	246(4.69),273(3.85), 283(3.97),294(3.94)	103-1229-86
$C_{23}H_{23}NO_2$			
Bicyclo[4.4.1]undeca-1,3,5,7,9-pentaene-2-carboxamide, 10-benzoyl-N,N-diethyl-, (±)-	MeOH	265(4.30),320(3.89)	33-1263-86
6-Hepten-3-one, 1-(2-hydroxyphenyl)-1-(2-naphthalenylamino)-	EtOH	245(4.88),275(4.17), 284(4.28),294(4.15)	103-1229-86
1H-Pyrrolizine, 5-(1,3-dioxolan-2-ylmethyl)-2,3-dihydro-6,7-diphenyl-	MeOH	205(4.59),242(4.32), 274(4.16)	83-0500-86
$C_{23}H_{23}NO_5$			
2,5-Pyrrolidinedione, 3-(1-methylethylidene)-1-phenyl-4-[(3,4,5-trimethoxyphenyl)methylene]-, (E)-	toluene	352(4.15)	39-0315-86C
$C_{23}H_{23}NO_6S$			
Benzo[b]thiophen-3(2H)-one, 4-methoxy-7-(4,5,6,7-tetrahydro-10-methoxy-5-methyl-3H-furo[4,3,2-fg][1,3]-benzazocin-6-yl)-, 1,1-dioxide, (R)-	EtOH	247(4.08),255(3.95), 283(3.72),290(3.70), 415(2.04)	78-0591-86
$C_{23}H_{23}NO_{10}S$			
9-Azatricyclo[4.2.1.0^{2,5}]nona-3,7-diene-3,4,7,8-tetracarboxylic acid, 9-[(4-methylphenyl)sulfonyl]-, tetramethyl ester	EtOH	225(4.37),248s(3.90), 254s(3.75),272s(3.43), 283s(3.13)	24-0616-86
$C_{23}H_{23}N_2O$			
Pyridinium, 1-[2-[4-(dimethylamino)-phenyl]-1-(4-methylbenzoyl)ethenyl]-, iodide, (Z)-	MeOH	266(4.41),425(4.63)	78-0699-86
$C_{23}H_{23}N_3$			
Benzenamine, 4-(4,5-dihydro-1,3-di-	n.s.g.	255(4.44),358(4.26)	78-3807-86

Compound	Solvent	$\lambda_{max}(\log \epsilon)$	Ref.
phenyl-1H-pyrazol-5-yl)-N,N-dimethyl- (cont.)			78-3807-86
$C_{23}H_{23}N_5O$			
2(1H)-Pteridinone, 4-(butylimino)-3,4-dihydro-3-methyl-6,7-diphenyl-	pH 1.6	222(4.47),274(4.28), 382(4.21)	128-0199-86
	pH 14	292s(4.15),312(4.17), 401(4.06)	128-0199-86
	MeOH	227(4.47),286(4.13), 368(4.14)	128-0199-86
$C_{23}H_{23}N_5O_7$			
5,8-Dideazafolic acid, 10-acetyl-	pH 7.5	229(4.67),252s(4.26)	87-2117-86
$C_{23}H_{24}N_2OS$			
Indolo[2,3-a]quinolizin-4(1H)-one, 3-ethyl-2,3,6,7,12,12b-hexahydro-3-(phenylthio)-	MeOH	225(4.71),274(3.98), 280s(3.97),290(3.85)	94-0077-86
more polar form	MeOH	224(4.71),274(3.97), 281s(3.96),290(3.83)	94-0077-86
$C_{23}H_{24}N_2O_4S$			
1H-Cyclopent[g]indolo[2,3-a]quinolizine-1,2-dicarboxylic acid, 6,7,12-12b,13,13a-hexahydro-1-(methylthio)-, dimethyl ester	MeOH	222(4.33),266(3.83), 290(3.69),374(4.24)	44-2995-86
$C_{23}H_{24}N_2O_7$			
Benzamide, N-[[[5-O-benzoyl-2,3-O-(1-methylethylidene)-β-D-ribofuranosyl]amino]carbonyl]-	MeOH	230(4.45),270s(3.36)	94-4585-86
$C_{23}H_{24}N_2O_8$			
1H-7a,11b-Diazacyclohepta[jk]fluorene-4,5,6,7-tetracarboxylic acid, 2,3,6,7-tetrahydro-, tetramethyl ester	EtOH	219(4.30),260(3.97), 307(3.62),460(4.80)	103-0192-86
Dipyrido[1,2-a:2',1'-b]benzimidazole-1,2,3,4-tetracarboxylic acid, 5,6,7,8-tetrahydro-, tetramethyl ester	EtOH	219(4.28),260(4.10), 327(3.80),465(3.80)	103-0192-86
$C_{23}H_{24}N_6O_2S$			
Benzenecarbothioic acid, 4-[[2-(acetylamino)-4-(diethylamino)phenyl]azo]-3,5-dicyano-, S-ethyl ester	$CHCl_3$	304(3.87),314s(3.83), 406(3.53),574s(4.68), 607(4.77)	48-0497-86
$C_{23}H_{24}N_6O_3$			
Benzoic acid, 4-[[2-(acetylamino)-4-(diethylamino)phenyl]azo]-3,5-dicyano-, ethyl ester	$CHCl_3$	302(3.82),316(3.90), 401(3.51),563s(4.70), 594(4.77)	48-0497-86
$C_{23}H_{24}O_3$			
Cyclohexanol, 2-(7,8,9,10-tetrahydrobenzo[b]naphtho[2,1-d]furan-5-yl)-, formate	EtOH	206(4.35),253(4.83), 260(4.86),284(3.89), 293(3.89),304(3.70), 330(3.27),344(3.13)	4-1625-86
$C_{23}H_{24}O_4$			
5,7-Octadiene-1,4-diol, 7-methyl-, dibenzoate, [R-(E)]-	EtOH	230(4.60)	163-0187-86

Compound	Solvent	$\lambda_{max}(\log \epsilon)$	Ref.
5,7-Octadiene-1,4-diol, 7-methyl-, dibenzoate, [R-(Z)]-	EtOH	230(4.54)	163-0187-86
$C_{23}H_{24}O_5$			
6-Deoxy-γ-mangostin	MeOH	240(4.60),264(4.59), 314(4.35),367(3.80)	102-1957-86
$C_{23}H_{24}O_6$			
9H-Xanthen-9-one, 1,3,5,6-tetrahydroxy-2,4-bis(3-methyl-2-butenyl)-	MeOH	253(4.14),330(3.91)	102-1217-86
	MeOH-NaOAc	261(--),372(--)	102-1217-86
	MeOH-$AlCl_3$	278(--),401(--)	102-1217-86
$C_{23}H_{25}BrN_2O_9S$			
β-D-Galactopyranosylamine, N-[5-(4-bromophenyl)-2-thiazolyl]-, 2,3,4,6-tetraacetate	CH_2Cl_2	305(4.26)	136-0318-86H
β-D-Glucopyranosylamine, N-[5-(4-bromophenyl)-2-thiazolyl]-, 2,3,4,6-tetraacetate	CH_2Cl_2	310(4.31)	136-0318-86H
$C_{23}H_{25}BrN_6O_{10}$			
Acetamide, N-[5-[bis(2-acetoxyethyl)-amino]-2-[(2-bromo-4,6-dinitrophenyl)azo]-4-methoxyphenyl]-	$CHCl_3$	284(4.03),594(4.54)	48-0497-86
$C_{23}H_{25}FeNO_4$			
Ferrocene, [3,5-bis(ethoxycarbonyl)-2,6-dimethyl-4-pyridinyl]-	MeCN	207(4.1),284(4.0), 361(3.1),473(2.7)	103-0886-86
$C_{23}H_{25}NO_5S_2$			
Sulfoximine, S-[2-(2-methoxyethoxy)-phenyl]-S-(4-methylphenyl)-N-[(4-methylphenyl)sulfonyl]-, (S)-	EtOH	230s(4.37),294(3.84)	12-1833-86
$C_{23}H_{25}NO_6$			
2-Furancarboxylic acid, 4,4'-[[(4-methylphenyl)imino]bis(methylene)]-bis[5-methyl-, dimethyl ester	MeOH	264(3.53)	73-2186-86
$C_{23}H_{25}NO_6S$			
4a,8-Etheno-4,13-methano-4aH-benzofuro[3,2-e]isoquinolin-7(6H)-one, 1,2,3,4,4b,7a,8,8a-octahydro-8,10-dimethoxy-3-methyl-, 5,5-dioxide	EtOH	205(4.43),287(3.30)	78-0591-86
$C_{23}H_{25}NO_7$			
1(2H)-Isoquinolinone, 3-[6-(2,2-dimethoxyethyl)-1,3-benzodioxol-5-yl]-7,8-dimethoxy-2-methyl-	MeOH	222(4.41),290(4.21), 299(4.25),355(4.02)	39-2253-86C
$C_{23}H_{25}N_3O_6S$			
1-Azabicyclo[3.2.0]hept-2-ene-2-carboxylic acid, 3-[(4-cyanocyclohexyl)-thio]-6-(1-hydroxyethyl)-7-oxo-, (4-nitrophenyl)methyl ester	EtOH	267(4.04),322(4.11)	23-2184-86
$C_{23}H_{25}N_7O_3$			
Pentanamide, N-[2-[(2,6-dicyano-4-nitrophenyl)azo]-5-(diethylamino)-phenyl]-	$CHCl_3$	267s(4.00),297s(3.86), 347(3.44),421(3.63), 588s(4.77),625(4.97)	48-0497-86

Compound	Solvent	$\lambda_{max}(\log \epsilon)$	Ref.
$C_{23}H_{26}NO$			
Pyridinium, 4-[2-[5-(1,1-dimethylethoxy)bicyclo[4.4.1]undeca-2,4,6,8,10-pentaen-2-yl)ethenyl]-1-methyl-, iodide	CH_2Cl_2	256(4.39),316(3.80), 471(3.94)	33-0315-86
$C_{23}H_{26}N_2O$			
1H-Pyrrolizine-5-methanamine, 2,3-dihydro-6-(4-methoxyphenyl)-N,N-dimethyl-7-phenyl-	n.s.g.	204(4.42),240(4.34), 363s(3.30)	83-0065-86
$C_{23}H_{26}N_2OS_3$			
Morpholine, 4-[6-(ethylthio)-4-(4-methylphenyl)-6-(phenylthio)-6H-1,3-thiazin-2-yl]-	MeOH	256(4.02),323(3.90)	118-0916-86
$C_{23}H_{26}N_2O_3$			
20,21-Dinoraspidospermidine-3-carboxylic acid, 2,3-didehydro-5-(1-methyl-2-propenyl)-8-oxo-, methyl ester	EtOH	228(4.06),299(4.12), 330(4.24)	78-6047-86
$C_{23}H_{26}N_2O_4S$			
Corynan-18-carboxylic acid, 18,19,20-21-tetradehydro-17-methoxy-16-(methylthio)-17-oxo-, methyl ester	MeOH	222(4.32),270(3.77), 290(3.64),356(4.42)	44-2995-86
$C_{23}H_{26}N_2O_9S$			
β-D-Galactopyranosylamine, N-(5-phenyl-2-thiazolyl)-, 2,3,4,6-tetraacetate	CH_2Cl_2	300(4.25)	136-0318-86H
β-D-Glucopyranosylamine, N-(5-phenyl-2-thiazolyl)-, 2,3,4,6-tetraacetate	CH_2Cl_2	302(4.41)	136-0318-86H
$C_{23}H_{26}N_2O_{10}S$			
1H-Indole-3a,5,7,7a-tetracarboxylic acid, 4-amino-2,3-dihydro-1-[(4-methylphenyl)sulfonyl]-, tetramethyl ester	$CHCl_3$	275(4.12),357(4.08)	24-2114-86
$C_{23}H_{26}N_2S_2$			
Benzeneethanethioamide, α-(4,4-dimethyl-2-phenyl-5(4H)-thiazolylidene)-N,N-diethyl-, (E)-	EtOH	241(4.29),280s(4.15), 380s(2.70)	33-0174-86
(Z)-	EtOH	236(4.42),282s(4.05), 360s(2.85),448s(2.36)	33-0174-86
Propanethioamide, N,N-diethyl-2-(4-methyl-2,4-diphenyl-5(4H)-thiazolylidene)-, (E)-	EtOH	246(4.31),280s(3.98), 330s(3.30),380s(2.88)	33-0174-86
2-Propanethione, 1-(diethylamino)-1-(4-methyl-2,4-diphenyl-5(4H)-thiazolylidene)-	EtOH	236s(3.98),296(3.96), 356(3.91),446(3.81)	33-0174-86
$C_{23}H_{26}N_2S_3$			
Ethanethioamide, 2-(4,4-dimethyl-2-phenyl-5(4H)-thiazolylidene)-N,N-diethyl-2-(phenylthio)-, (E)-	EtOH	242(4.25),270s(4.05), 298s(3.86),392s(2.64)	33-0174-86
6H-1,3-Thiazine, 6-(ethylthio)-4-(4-methylphenyl)-6-(phenylthio)-2-(1-pyrrolidinyl)-	$CHCl_3$	247(3.97),322(3.92)	118-0916-86
$C_{23}H_{26}N_4OS_2$			
Propanamide, N-2-benzothiazolyl-3-(2-	pH 1	212(4.54),260(4.22),	103-0795-86

Compound	Solvent	$\lambda_{max}(\log \epsilon)$	Ref.
benzothiazolylpropylamino)-N-propyl-		282(4.14),290(4.12)	103-0795-86
(cont.)	EtOH	227(4.53),279(4.29), 302(4.11)	103-0795-86
$C_{23}H_{26}O_2$			
Bicyclo[2.1.0]pentane-5-carboxylic acid, 2,2,3,3-tetramethyl-1,4-diphenyl-, methyl ester	n.s.g.	260(2.10)	104-2213-86
$C_{23}H_{26}O_3$			
3,5,7,9,11,13,15,17,19-Docosanonaenoic acid, 21-oxo-, methyl ester	MeOH	276(4.26),430(4.82)	5-0741-86
1H-Xanthene-1,8(2H)-dione, 3,4,5,6,7-9-hexahydro-3,3,6,6-tetramethyl-9-phenyl-	EtOH	234(4.17),300(3.69)	2-0347-86
$C_{23}H_{26}O_4S$			
Propanedioic acid, [(4-methylphenyl)-[(4-methylphenyl)thio]methylene]-, 1,1-dimethylethyl methyl ester (9:1 E:Z)	EtOH	262(3.68),293(3.82)	44-4112-86
$C_{23}H_{26}O_5$			
Propanedioic acid, [(4-methoxyphenyl)-(4-methylphenyl)methylene]-, 1,1-dimethylethyl methyl ester (85:15 E:Z)	EtOH	278(4.19)	44-4112-86
$C_{23}H_{26}O_6$			
Fercolide	MeOH	261(4.03)	102-1673-86
Sigmoidin B trimethyl ether	MeOH	287(4.03)	39-0033-86C
$C_{23}H_{26}O_7$			
Spiro[1,3-dioxan-2,1'-[1H-1b,10](epoxymethano)benzo[a]cyclopropa[3,4]-benzo[1,2-c]cyclohepten]-11'-one, 1's,2',3',4',10',10'a-hexahydro-6',7',8'-trimethoxy-, exo	MeCN	229(4.11),270(3.96)	35-6713-86
$C_{23}H_{26}O_8$			
2(3H)-Furanone, dihydro-3-[(7-methoxy-1,3-benzodioxol-5-yl)methyl]-4-[(3,4,5-trimethoxyphenyl)methyl]-(cubebinone)-(-)-	MeOH	275(3.73)	102-0487-86
$C_{23}H_{26}O_{11}$			
β-D-Glucopyranoside, 2-(3,4-dihydroxyphenyl)ethyl 4-[3-(3,4-dihydroxyphenyl)-2-propenoate, (E)- (calceolarioside A)	MeOH	290(4.05),334(4.13)	32-0431-86
	MeOH-NaOH	310(3.93),383(4.14)	32-0431-86
	MeOH-NaOAc-H_3BO_3	297(4.03),349(4.10)	32-0431-86
β-D-Glucopyranoside, 2-(3,4-dihydroxyphenyl)ethyl 6-[3-(3,4-dihydroxyphenyl)-2-propenoate, (E)- (calceo-	MeOH	290(4.05),331(4.13)	32-0431-86
	MeOH-NaOH	310(3.93),373(4.14)	32-0431-86
	MeOH-NaOAc-H_3BO_3	297(4.03),352(4.10)	32-0431-86
$C_{23}H_{27}BrN_2O_{10}S$			
Thiourea, N-[2-(4-bromophenyl)-2-oxoethyl]-N'-(2,3,4,6-tetra-O-acetyl-β-D-galactopyranosyl)-	CH_2Cl_2	253(4.45)	136-0318-86H
Thiourea, N-[2-(4-bromophenyl)-2-oxoethyl]-N'-(2,3,4,6-tetra-O-acetyl-β-D-glucopyranosyl)-	CH_2Cl_2	253(4.42)	136-0318-86H

Compound	Solvent	λ_{max}(log ε)	Ref.
$C_{23}H_{27}BrN_4O_{10}$			
7H-Pyrrolo[2,3-d]oyrimidin-4-one, 6-bromo-2-(diacetylamino)-5-formyl-3,4-dihydro-3-(methoxymethyl)-7-[5-O-acetyl-2,3-O-(1-methylethylidene)-β-D-ribofuranosyl]-	MeOH	239(4.18),296(4.14)	78-0199-86
$C_{23}H_{27}F_3O_5$			
Androsta-1,4-diene-17β-carboxylic acid, 6α,9α-difluoro-11β,17α-dihydroxy-16α-methyl-3-oxo-, 17,20-(1-fluoro-1-methyl)methylene ketal, (R)-	MeOH	236(4.19)	44-2315-86
(S)-	MeOH	237(4.20)	44-2315-86
$C_{23}H_{27}FeNO_4$			
Ferrocene, [3,5-bis(ethoxycarbonyl)-1,4-dihydro-2,6-dimethyl-4-pyridinyl]-	MeCN	211(4.52),240(4.36), 280(4.0),337(3.9), 456(2.5)	103-0886-86
$C_{23}H_{27}NO$			
Oxiranecarbonitrile, 3-(2-ethynyl-1-cyclododecen-1-yl)-2-phenyl-, (Z)-	MeCN	217(4.06),243(4.06)	78-2221-86
$C_{23}H_{27}NO_2$			
2H,6H-Pyrido[2,1-b][1,3]oxazine, 7,8,9,9a-tetrahydro-2-ethoxy-9a-methyl-3,4-diphenyl-	MeCN	227(4.07),290(3.93)	118-0899-86
2H,6H-Pyrido[2,1-b][1,3]oxazine, 7,8,9,9a-tetrahydro-9a-ethyl-2-methoxy-3,4-diphenyl-	MeCN	230(4.08),292(3.91)	118-0899-86
$C_{23}H_{27}NO_6S$			
4a,8-Etheno-4,13-methano-4aH-benzofuro[3,2-e]thieno[3,2-h]isoquinolin-7-ol, 1,2,3,4,4b,6,7,7a,8,8a-decahydro-8,10-dimethoxy-3-methyl-, 5,5-dioxide	EtOH	215(4.30),238(3.96), 288(3.12),366(2.10), 420(1.85)	78-0591-86
$C_{23}H_{27}N_3O_4S$			
8aH-Carbazole-8a-acetic acid, 4b,5,6-7,8,9-hexahydro-4b-[[[(4-methylphenyl)sulfonyl]hydrazono]methyl]-, methyl ester	EtOH	230(4.40),296(4.52)	44-3125-86
$C_{23}H_{27}N_3O_7S$			
1-Azabicyclo[3.2.0]hept-2-ene-2-carboxylic acid, 3-[[4-(aminocarbonyl)-cyclohexyl]thio]-6-(1-hydroxyethyl)-7-oxo-, (4-nitrophenyl)methyl ester	EtOH	267(4.03),324(4.11)	23-2184-86
$C_{23}H_{27}N_3O_{10}$			
L-Aspartic acid, N-[3-[4-(2-amino-2-carboxyethyl)phenoxy]-O-methyl-L-tyrosyl]-3-hydroxy-, [erythro-(S)]-	H_2O	273(3.53)	158-1685-86
	HCl	273(3.53)	158-1685-86
	NaOH	273(3.55)	158-1685-86
$C_{23}H_{27}N_3O_{11}$			
2,3,6-Pyridinetricarboxylic acid, 4-[1-(2-deoxy-β-D-erythro-pento-	H_2O	275(4.01)	44-0950-86

Compound	Solvent	$\lambda_{max}(\log \epsilon)$	Ref.
furanosyl)-1,2,3,4-tetrahydro-2,4-dioxo-5-pyrimidinyl]-, triethyl ester (cont.)			44-0950-86
$C_{23}H_{27}N_5O_9$			
7H-Pyrrolo[2,3-d]pyrimidin-4-one, 2-(diacetylamino)-5-cyano-3,4-dihydro-3-(methoxymethyl)-7-[5-O-acetyl-2,3-O-(1-methylethylidene)-β-D-ribofuranosyl]-	MeOH	221(4.26),263(3.97), 294(3.85)	78-0207-86
$C_{23}H_{27}N_7O_5$			
L-Glutamic acid, N-[4-[[(2,4-diamino-5-methylpyrido[2,3-d]pyrimidin-6-yl)methyl]ethylamino]benzoyl]-	pH 1	228(4.61),313(4.23)	87-1080-86
	pH 7	227(4.59),308(4.48)	87-1080-86
	pH 13	228(4.56),308(4.46)	87-1080-86
$C_{23}H_{28}ClN_3O_2S$			
Thiopropazate (dihydrochloride)	MeOH	213(4.341),257(4.556), 310(3.634)	106-0571-86
$C_{23}H_{28}F_2O_5$			
Androsta-1,4-diene-17β-carboxylic acid, 9α-fluoro-17-(1-fluoro-1-hydroxyethoxy)-11β-hydroxy-16β-methyl-3-oxo-, γ-lactone	MeOH	238(4.20)	44-2315-86
isomeric ketal	MeOH	238(4.18)	44-2315-86
$C_{23}H_{28}F_2O_6$			
Androsta-1,4-diene-17-carboxylic acid, 17-acetoxy-6,9-difluoro-11-hydroxy-16-methyl-3-oxo-, (6α,11β,16α,17α)-	MeOH	238(4.20)	44-2315-86
Androsta-1,4-diene-17-carboxylic acid, 6,9-difluoro-11-hydroxy-16,17-[(1-nethylethylidene)bis(oxy)]-3-oxo-, (6α,11β,16α,17α)-	MeOH	237(4.20)	44-2315-86
$C_{23}H_{28}NO_5$			
Isoquinolinium, 6,7-diacetoxy-1,2,3,4-tetrahydro-1-[(4-methoxyphenyl)methyl]-2,2-dimethyl-, chloride, (R)-	EtOH and EtOH-NaOH	224(4.37),270(3.18)	102-2693-86
$C_{23}H_{28}N_2O_3S$			
1H-Benz[de]isoquinoline-1,3(2H)-dione, 6-(butylthio)-2-[3-(4-morpholinyl)propyl]-, monohydrochloride	pH 7.4	402(4.08)	4-0849-86
$C_{23}H_{28}N_2O_4$			
Aspidospermidine-3-carboxylic acid, 2,3-didehydro-21-ethoxy-21-oxo-, methyl ester, (±)-	MeOH	224(3.98),296(4.01), 324(4.15)	78-6719-86
Spiro[3H-indole-3,1'(5'H)-indolizine]-2',3',6',7',8',8'a-hexahydro-2-methoxy-5'-oxo-8"-(2-propenyl)-, methyl ester	EtOH	220(4.28),261(3.69), 272s(3.49),285s(3.10)	78-6047-86
$C_{23}H_{28}N_2O_4S$			
1H-Benz[de]isoquinoline-1,3(2H)-dione, 6-(butylsulfinyl)-2-[3-(4-morpholinyl)propyl]-	pH 7.4	347(4.21)	4-0849-86

Compound	Solvent	λ max (log ε)	Ref.
$C_{23}H_{28}N_2O_5S$			
Methotrimeprazine, (Z)-2-butenedioate	MeOH	253(4.425),306(3.663)	106-0571-86
$C_{23}H_{28}N_2S$			
1-Propanamine, N,N-dimethyl-3-[2-(1-piperidinyl)-9H-thioxanthen-9-ylidene)-, (Z)-	MeOH	233(4.51),282(4.27), 342(3.45)	73-0937-86
$C_{23}H_{28}N_8O_4$			
Benzoic acid, 4,4'-[1,3,5,7-tetraazabicyclo[3.3.1]nonane-3,7-diylbis-(azo)]bis-, diethyl ester	EtOH	304(4.52)	23-1567-86
$C_{23}H_{28}O_2$			
Oxirane, 2-(4-pentenyl)-3-(10-phenoxy-1,3,5,8-decatetraenyl)-, [2α,3β(1E-3Z,5Z,8Z)]-(±)-	EtOH	268(4.38),276(4.48), 289(4.33)	104-1457-86
$C_{23}H_{28}O_3$			
3'H-Cyclopropa[15,16]pregna-4,6,15-triene-21-carboxylic acid, 15,16-dihydro-17-hydroxy-3-oxo-, γ-lactone	n.s.g.	284(4.43)	88-5463-86
$C_{23}H_{28}O_4$			
3a,7-Ethano-3aH-indene-4-carboxylic acid, 2-(benzoyloxy)octahydro-8,8-dimethyl-3-methylene-, methyl ester, [2R-(2α,3aα,4β,7α,7aβ)]-	EtOH	229.5(4.13)	158-0149-86
Oxiranebutanoic acid, 3-(10-phenoxy-1,3,5,8-decatetraenyl)-, methyl ester, [2α,3β(1E,3E,5Z,8Z)]-(±)-	EtOH	267(4.33),275(4.41), 286(4.31)	104-1457-86
$C_{23}H_{28}O_7$			
Chrysophyllon I-B	MeOH	269(3.94)	102-2609-86
Chrysophyllon II-A	MeOH	334(3.86)	102-2609-86
Chrysophyllon III-B	MeOH	294(4.00)	102-2609-86
2(3H)-Furanone, 4-[(3,4-dimethoxyphenyl)methyl]dihydro-3-[(3,4,5-trimethoxyphenyl)methyl]-	MeOH	228(4.15),277(3.51)	102-0487-86
Naphtho[1,2-b]furan-6-carboxylic acid, 2,3-dihydro-4,7-dimethoxy-5-(methoxyacetyl)-2,3,3,9-tetramethyl-, methyl ester, (±)-	MeOH	218(--),253(--), 311s(--),323(--), 346(--)	44-4813-86
$C_{23}H_{29}BrN_4O_9$			
7H-Pyrrolo[2,3-d]pyrimidin-4-one, 6-bromo-2-(diacetylamino)-3,4-dihydro-3-(methoxymethyl)-5-methyl-7-[5-O-acetyl-2,3-O-(1-methylethylidene)-β-D-ribofuranosyl]-	MeOH	271(3.83),310(4.03)	78-0199-86
$C_{23}H_{29}BrN_4O_{10}$			
7H-Pyrrolo[2,3-d]pyrimidin-4-one, 6-bromo-2-(diacetylamino)-3,4-dihydro-5-(hydroxymethyl)-3-(methoxymethyl)-7-[5-O-acetyl-2,3-O-(1-methylethylidene)-β-D-ribofuranosyl]-	MeOH	269(3.92),306(4.02)	78-0199-86
$C_{23}H_{29}FO_6$			
Androsta-1,4-diene-17-carboxylic acid,	MeOH	239(4.25)	44-2315-86

Compound	Solvent	$\lambda_{max}(\log \epsilon)$	Ref.
17-acetoxy-9-fluoro-11-hydroxy-16-methyl-3-oxo-, (11β,16β,17α)- (cont.)			44-2315-86
$C_{23}H_{29}NO_4$			
Pyridinium, 1-[[1-(2-hydroxy-4,4-dimethyl-6-oxo-1-cyclohexen-1-yl)-4,4-dimethyl-2,6-dioxocyclohexyl]methyl]-2-methyl-, hydroxide, inner salt	MeCN	289(4.1),375(3.0)	39-0043-86B
$C_{23}H_{29}NO_{12}$			
Hygromycin A	H_2O	217(4.28),272(4.16), 302s(3.99)	163-0143-86
$C_{23}H_{29}N_2O_4$			
3-Dehydromitragynine	MeOH	205(4.2),247(4.0), 317s(3.75),342(3.85), 395s(3.70)	102-2910-86
	MeOH-NaOH	205(4.03),230s(4.1), 280(3.85),302(3.82), 318s(3.79)	102-2910-86
$C_{23}H_{29}N_3O_2$			
1H-Naphtho[2,3-d]imidazole-4,9-dione, 2-propyl-1-(2,2,6,6-tetramethyl-4-piperidinyl)-	PrOH	226(3.88),250(4.65), 275s(4.13),283(4.14), 335(3.33),380(2.64)	104-1560-86
$C_{23}H_{29}N_3O_3$			
Aspidospermidin-21-oic acid, 2-cyano-17-methoxy-, ethyl ester	MeOH	210(4.51),241(3.81), 287(3.41)	78-6719-86
$C_{23}H_{29}N_3O_3S$			
1H-Benz[de]isoquinoline-1,3(2H)-dione, 5-amino-6-(butylthio)-2-[3-(4-morpholinyl)propyl]-	pH 7.4	351(3.89),425(3.74)	4-0849-86
$C_{23}H_{29}N_5$			
1H-Pyrazolo[3,4-b]quinolin-4-amine, 5,6,7,8-tetrahydro-N-(4-methylphenyl)-1-(piperidinomethyl)-	EtOH	215(4.44),243s(--), 271(3.70),316(4.20)	5-1728-86
$C_{23}H_{29}N_5O$			
1H-Pyrazolo[3,4-b]quinolin-4-amine, N-(3,5-dimethylphenyl)-5,6,7,8-tetrahydro-1-(4-morpholinylmethyl)-	EtOH	218(4.52),244s(--), 271(3.66),321(4.20)	5-1728-86
$C_{23}H_{30}$			
Benzene, 1,3,5-tris(1-methylethyl)-2-(2-phenylethenyl)-	$C_6H_{11}Me$	262(4.26)	35-0841-86
$C_{23}H_{30}ClN_3O$			
Quinacrine	pH 7.4	280(4.69)	94-0944-86
$C_{23}H_{30}F_2O_4S$			
Androsta-1,4-diene-17-carbothioic acid, 6,9-difluoro-11,17-dihydroxy-16-methyl-3-oxo-, S-ethyl ester, (6α,11β,16α,17α)-	MeOH	238(4.30)	44-2315-86
$C_{23}H_{30}N_2S$			
Androst-16-eno[17,16-b]pyridine-5'-carbonitrile, 1',6'-dihydro-6'-	EtOH	217(4.09),245(3.76), 315(4.21),416(3.71)	70-1504-86

Compound	Solvent	$\lambda_{max}(\log \epsilon)$	Ref.
thioxo-, (5α)- (cont.)			70-1504-86
$C_{23}H_{30}N_4$			
7H-1,2,4-Triazolo[4,3-a]perimidine, 10-undecyl-	dioxan	223(4.37),253(3.96), 257(3.66),259(3.72), 350(3.92),364(3.72)	48-0237-86
$C_{23}H_{30}N_4O_9$			
7H-Pyrrolo[2,3-d]pyrimidin-4-one, 2-(diacetylamino)-3,4-dihydro-3-(methoxymethyl)-5-methyl-7-[5-O-acetyl-2,3-O-(1-methylethylidene)-β-D-ribofuranosyl]-	MeOH	304(3.93)	78-0199-86
$C_{23}H_{30}N_4O_{10}$			
7H-Pyrrolo[2,3-d]pyrimidin-4-one, 2-(diacetylamino)-3,4-dihydro-5-(hydroxymethyl)-3-(methoxymethyl)-7-[5-O-acetyl-2,3-O-(1-methylethylidene)-β-D-ribofuranosyl]-	MeOH	265(3.73),300(3.94)	78-0207-86
$C_{23}H_{30}N_6O_7$			
β-D-Ribofuranuronamide, 1-deoxy-N-ethyl-1-[6-[[2-(3,4,5-trimethoxyphenyl)ethyl]amino]-9H-purin-9-yl]-	n.s.g.	271(4.20)	87-1683-86
$C_{23}H_{30}O_3Se$			
Benzoic acid, 4-[[2,4,6-tris(1-methylethyl)phenyl]seleninyl]-, methyl ester, (-)-	MeOH	214(4.36),237(4.31), 275(3.97)	138-0161-86
$C_{23}H_{30}O_4$			
7H-Furo[3,2-g][1]benzopyran-7-one, 9-(dodecyloxy)-	1% DMSO	308(4.04)	149-0391-86A
1-Phenanthrenemethanol, 3-acetoxy-7-ethenyl-4b,5,6,7,8,8a,9,10-octahydro-4b,7-dimethyl-, acetate, [4bS-(4bα,7α,8aβ)]-	MeOH	220(3.89),272(3.12)	94-1015-86
$C_{23}H_{30}O_4S$			
Acetic acid, [[[3-[7-(2-butoxyphenyl)-1,3,5-heptatrienyl]oxiranyl]methyl]-thio]-, methyl ester, [2α,3β(1E-3E,5Z)]-(±)-	EtOH	271(4.40),282(4.49), 296(4.37)	104-1598-86 +104-1875-86
$C_{23}H_{30}O_5$			
Fercomin	MeOH	261(4.04)	102-1673-86
$C_{23}H_{30}O_8$			
Gracilin D	MeOH	296(4.36)	100-0823-86
Picrasine B, 6α-acetoxy-	EtOH	253(3.86)	100-0303-86
$C_{23}H_{31}NO_{12}$			
neo-Inositol, 2-deoxy-2-[[3-[4-(6-deoxy-β-D-arabino-hexofuranos-5-ulos-1-yl)oxy]-3-hydroxyphenyl]-2-methyl-1-oxo-2-propenyl]- (methoxyhygromycin)	H_2O	217(4.22),272(4.13), 300s(3.95)	163-0143-86
$C_{23}H_{31}N_3O_2S$			
Pagisulfine	EtOH	224(4.36),284(3.89), 291(3.92),298(3.86)	142-1567-86

Compound	Solvent	λ_{max}(log ϵ)	Ref.
$C_{23}H_{31}N_3O_3$			
Butanamide, N-[[1,4-dihydro-1,4-dioxo-3-(2,2,6,6-tetramethyl-4-piperidinyl)amino]-2-naphthalenyl]-	PrOH	234(4.11),242s(4.09), 273(4.33),335s(2.92), 460(3.32)	104-1560-86
$C_{23}H_{31}N_5O_6$			
Benzo[g]pteridine-2,4(3H,10H)-dione, 10-ethyl-3-methyl-8-(1,4,7,10-tetraoxa-13-azacyclopentadec-13-yl)-	benzene	457(4.50),486(4.73)	130-0119-86
	H_2O	498(4.67)	130-0119-86
	MeOH	495(4.71)	130-0119-86
	EtOAc	458s(--),485(4.74)	130-0119-86
	dioxan	455s(--),483(4.71)	130-0119-86
	THF	455s(--),486(4.74)	130-0119-86
	30% MeOH	498(4.71)	130-0119-86
	MeCN	487(4.71)	130-0119-86
	C_6H_{12}-benzene	455(4.51),483(4.77)	130-0199-86
$C_{23}H_{31}N_5O_9$			
7H-Pyrrolo[2,3-d]pyrimidin-4-one, 5-(N-acetylaminomethyl)-2-(diacetylamino)-3,4-dihydro-3-(methoxymethyl)-7-[5-O-acetyl-2,3-O-(1-methylethylidene)-β-D-ribofuranosyl]-	MeOH	265(3.74),301(3.93)	78-0207-86
$C_{23}H_{31}N_5O_{12}$			
Capuramycin	MeOH	214(4.21),257s(3.99)	158-1047-86
$C_{23}H_{32}Cl_2N_4Pd$			
Palladium, dichloro[2,2'-methylenebis[4,5,6,7-tetrahydro-7,8,8-trimethyl-4,7-methano-2H-indazole-N^1,$N^{1'}$]-	MeCN	305s(3.00),358(2.63)	12-1525-86
$C_{23}H_{32}N_2O_7SSi$			
Uridine, 5'-(methoxymethyl)-2',3'-O-(1-methylethylidene)-6-(phenylthio)-5-(trimethylsilyl)-	MeOH	245(3.98),277(4.00), 313s(3.58)	78-4187-86
$C_{23}H_{32}N_4$			
4,7-Methano-1H-indazole, 1,1'-methylenebis[4,5,6,7-tetrahydro-7,8,8-trimethyl-	MeOH	233.6(3.72)	12-1525-86
4,7-Methano-2H-indazole, 1,1'-methylenebis[4,5,6,7-tetrahydro-7,8,8-trimethyl-	MeOH	238.9(4.27)	12-1525-86
4,7-Methano-1H-indazole, 4,5,6,7-tetrahydro-7,8,8-trimethyl-1-[(4,5,6,7-tetrahydro-7,8,8-trimethyl-4,7-methano-2H-indazol-2-yl)methyl]-, (4S,6R)-	MeOH	234.2(4.17)	12-1525-86
$C_{23}H_{32}O_4$			
Dictyoceratin A	EtOH	218(4.40),268(4.00), 300(3.64)	78-4197-86
$C_{23}H_{32}O_5$			
Benzenepentanol, α-[2-(3,4-dimethoxyphenyl)ethyl]-3,4-dimethoxy-, (R)-	EtOH	229(4.13),280(3.70)	18-1181-86
$C_{23}H_{32}O_{12}$			
1(3H)-Isobenzofuranone, 3-[3-[[6-O-(6-	MeOH	225(3.88),230s(3.86),	100-0514-86

Compound	Solvent	$\lambda_{max}(\log \epsilon)$	Ref.
deoxy-α-L-mannopyranosyl)-β-D-gluco-pyranosyl]oxy]propyl]- (pedirutino-side) (cont.)		265s(3.01),272(3.17), 278(3.16)	100-0514-86
$C_{23}H_{33}BrN_2O$			
Urea, N-[(4-bromophenyl)methyl]-N'-[1,5-dimethyl-1-(4-methyl-3-cyclo-hexen-1-yl)-4-hexenyl]-, [R-(R*,R*)]-	MeOH	204(4.56),229(4.18)	44-5136-86
$C_{23}H_{33}N_3O_{14}$			
Uridine, 5'-O-(N-acetyl-α-D-neuramino-syl)-2',3'-O-(1-methylethylidene)-	MeOH	260(3.98)	136-0037-86M
β-	MeOH	260(3.96)	136-0037-86M
$C_{23}H_{34}N_6O$			
1,3-Propanediamine, N,N"-(6-methoxy-benzo[g]phthalazine-1,4-diyl)-bis[N',N'-dimethyl-	EtOH	226(4.68),239s(4.53), 264s(4.72),268(4.73), 360(3.87)	111-0143-86
$C_{23}H_{34}N_6O_4S_4$			
Carbamodithioic acid, dimethyl-, meth-ylenebis[3,4-dihydro-3,6-dimethyl-2,4-dioxo-5,1(2H)-pyrimidinediyl)-2,1-ethanediyl] ester	EtOH	252(4.36),277(4.6)	94-1809-86
$C_{23}H_{34}O$			
24-Norchola-1,4-dien-3-one	EtOH	244(4.13)	39-1805-86C
$C_{23}H_{34}O_3$			
Phellinic acid	EtOH	236(3.25),243(3.26), 296(3.01)	39-0551-86C
$C_{23}H_{34}O_4$			
1,4-Phenanthrenedione, 10-ethoxy-4b,5,6,7,8,8a,9,10-octahydro-3-methoxy-4b,8,8-trimethyl-2-(1-methylethyl)-, [4bS-(4bα,8aβ,10β)]-, (7α-ethoxy-12-O-methylroyleanone)	EtOH	269(4.00),364(2.70)	102-0266-86
$C_{23}H_{35}ClO_5$			
Prosta-5,7,14-trien-1-oic acid, 12-acetoxy-10-chloro-9-hydroxy-, methyl ester, (5E,7Z,9α,10β,14Z)-	EtOH	247(3.65)	88-0223-86
9β-	EtOH	248(3.86)	88-0223-86
$C_{23}H_{35}NOSi$			
Bicyclo[4.4.1]undeca-1,3,5,7,9-penta-ene-2-carboxamide, 3-butyl-N,N-di-ethyl-10-(trimethylsilyl)-	MeOH	270(4.44),320(3.96)	33-1851-86
$C_{23}H_{36}$			
Cyclohexane, 1,1'-(1,2-propadiene-1,3-diylidene)bis[2,2,6,6-tetra-methyl-	C_6H_{12}	234(4.93),256(4.51), 302(2.68),342(1.97)	89-0340-86
$C_{23}H_{36}N_2O_5S_2$			
Uridine, 2',3'-O-cyclohexylidene-5'-S-(2-methylpropyl)-5'-C-[(2-meth-ylpropyl)thio]-5'-thio-	MeOH	258(3.99)	44-1258-86

Compound	Solvent	λ_{max}(log ε)	Ref.
$C_{23}H_{36}N_4O_3$			
Piperazine, 1,1'-(1,2,3,4,4a,5,6,7-octahydro-4a-methyl-7-oxo-1,6-naphthalenediyl)bis[4-acetyl-, (1α,4aα,6β)-	EtOH	235(4.07)	107-0195-86
$C_{23}H_{37}NOSi$			
Bicyclo[4.4.1]undeca-4,6,8,10-tetraene-2-carboxamide, 3-butyl-N,N-diethyl-10-(trimethylsilyl)-	MeOH	270(4.20)	33-1597-86
$C_{23}H_{37}NO_3Si$			
Bicyclo[4.4.1]undeca-4,6,8,10-tetraene-2-carboxamide, 3-butyl-N,N-diethyl-2-hydroperoxy-10-(trimethylsilyl)-	MeOH	280(3.88)	33-1597-86
Bicyclo[4.4.1]undeca-4,6,8,10-tetraene-2-carboxamide, N,N-diethyl-3-(1-methylpropyl)-2-hydroperoxy-10-(trimethylsilyl)-	MeOH	270(3.96)	33-1597-86
$C_{23}H_{37}N_3O_3$			
Phenol, 4-[2-(2,5-dihydro-2,4,5,5-tetramethyl-1-nitroso-1H-imidazol-2-yl)ethyl]-2,6-bis(1,1-dimethylethyl)-, N-oxide	EtOH	229(4.32),278(3.41), 304(3.00)	103-0052-86
$C_{23}H_{38}N_2O_2$			
Phenol, 4-[2-(2,5-dihydro-2,4,5,5-tetramethyl-1H-imidazol-2-yl)ethyl]-2,6-bis(1,1-dimethylethyl)-, N-oxide	EtOH	228(4.20),278(3.30)	103-0052-86
$C_{23}H_{38}N_2O_3$			
2,5-Cyclohexadien-1-one, 4-[2-(2,5-dihydro-2,4,5,5-tetramethyl-1H-imidazol-2-yl)ethyl]-2,6-bis(1,1-dimethylethyl)-4-hydroxy-, N-oxide	EtOH	233(4.38)	103-0052-86
$C_{23}H_{38}N_2O_6$			
Uridine, 2'-deoxy-5-ethyl-, 5'-dodecanoate	octanol	267(4.04)	83-0154-86
$C_{23}H_{38}N_8O_5$			
L-α-Glutamine, N -[N-[(1-oxo-6-(1H-purin-6-yl)amino]hexyl]-L-alanyl-, 1,1-dimethylethyl ester	pH 2.95	271(4.12)	63-0757-86
	pH 7.0	268(4.13)	63-0757-86
	pH 10.5	272(4.14)	63-0757-86
$C_{23}H_{38}O_2$			
1,3-Benzenediol, 5-(12-heptadecenyl)-, (Z)-	EtOH	233(3.28),274(2.96), 280(2.93)	102-1093-86
	EtOH-NaOH	244(3.15),292(2.96)	102-1093-86
$C_{23}H_{38}O_3$			
[1,1':4',1"-Tercyclohexane]-4-carboxylic acid, 3-oxo-4"-propyl-, methyl ester, [1 [trans(trans)],4α]-	MeOH	257(3.80)	24-0387-86
$C_{23}H_{38}O_5$			
Muamvatin	CH_2Cl_2	236(4.01)	35-6680-86

Compound	Solvent	λ_{max}(log ϵ)	Ref.
$C_{23}H_{39}N_5O_5Si_2$			
8,2'-Methanoadenosine, 3',5'-O-[tetrakis(1-methylethyl)disiloxane-1,3-diyl]-	MeOH	261(4.22)	94-1518-86
$C_{23}H_{40}N_4O_3$			
2(3H)-Naphthalenone, 4,4a,5,6,7,8-hexahydro-3,8-bis[4-(2-hydroxyethyl)-1-piperazinyl]-4a-methyl-, (3α,4aβ,8β)-	EtOH	236(4.08)	107-0195-86
$C_{23}H_{40}N_6O_6Si_2$			
2(1H)-Pyrimidinone, 1-[2-deoxy-3,5-bis-O-[(1,1-dimethylethyl)dimethylsilyl]-β-D-erythro-pentofuranosyl]-4-(3-nitro-1H-1,2,4-triazol-1-yl)-	EtOH	248(3.08)	152-0643-86
$C_{23}H_{40}O_2$			
2,4-Tricosadiyne-1,6-diol, (R)-	EtOH	220(2.48),231(2.67), 243(2.62),257(2.38)	150-1348-86M
$C_{23}H_{42}N_4O_5Si_2$			
Acetamide, N-[1-[3,5-bis-O-[(1,1-dimethylethyl)dimethylsilyl]-α-D-xylofuranosyl]-4-cyano-1H-imidazol-5-yl]-	MeOH	239s(3.98),270(4.03)	87-0203-86
1H-Imidazole-4-carbonitrile, 1-[2-O-acetyl-3,5-bis-O-[(1,1-dimethylethyl)dimethylsilyl]-α-D-xylofuranosyl]-5-amino-	MeOH	246(4.07)	87-0203-86
$C_{23}H_{44}N_2O_6Si_2$			
2(1H)-Pyrimidinone, 1-[2-deoxy-3,5-bis-O-[(1,1-dimethylethyl)dimethylsilyl]-β-D-erythro-pentofuranosyl]-4-(2-hydroxyethoxy)-	EtOH	275(3.74)	152-0643-86
Uridine, 2'-deoxy-3',5'-O-[(1,1-dimethylethyl)dimethylsilyl]-3-(2-hydroxyethyl)-	EtOH	262(3.97)	152-0643-86
$C_{23}H_{44}N_4O_6Si_2$			
1H-Imidazole-4-carboxamide, 1-[2-O-acetyl-3,5-bis-O-[(1,1-dimethylethyl)dimethylsilyl]-α-D-xylofuranosyl]-5-amino-	MeOH	267(4.43)	87-0203-86
$C_{23}H_{45}N_3O_5Si_2$			
Cytidine, 2'-deoxy-3',5'-bis-O-[(1,1-dimethylethyl)dimethylsilyl]-N-(2-hydroxyethyl)-	EtOH	273(3.98)	152-0643-86
Cytidine, N,4-didehydro-2'-deoxy-3',5'-bis-O-[(1,1-dimethylethyl)dimethylsilyl]-3,4-dihydro-3-(2-hydroxyethyl)-	EtOH	268(3.94)	152-0643-86

Compound	Solvent	λ_{max}(log ε)	Ref.
$C_{24}Cl_{18}$			
Compd., m. 319-321°	C_6H_{12}	212(5.11),224s(4.96), 294(3.54)	44-1413-86
$C_{24}H_{12}$			
Bisbenzo[3,4]cyclobuta[1,2-a:1',2'-c]-biphenylene	hexane	227s(4.22),234(4.39), 243(4.56),269(4.38), 282(4.61),297(4.85), 324s(3.91),336(4.02), 352(3.90),363s(3.7), 379s(3.4)	35-3150-86
$C_{24}H_{12}Br_4N_2O_2$			
Oxazole, 2,2'-(1,4-phenylene)bis[4-bromo-5-(4-bromophenyl)-	dioxan	372(4.72)	103-1262-86
$C_{24}H_{12}Br_4N_2S_2$			
Thiazole, 2,2'-(1,4-phenylene)bis[4-bromo-5-(4-bromophenyl)-	C_6H_5Cl	380(4.71)	103-1262-86
$C_{24}H_{12}N_4O_2S_2$			
9H,19H-Bisnaphtho[1',2':4,5]thiazolo-[3,2-a:3',2'-e][1,3,5,7]tetraazocine-9,19-dione	MeOH	267(4.00),340(3.66), 375(3.76)	4-1265-86
$C_{24}H_{12}O_3$			
Dibenzo[b,b']furo[2,3-f:5,4-f']bis-benzofuran	EtOH	354(--)	46-2666-86
	dioxan	362(4.94)	46-2666-86
$C_{24}H_{13}NO_2$			
Naphtho[1,2,3,4-def]chrysene, 8-nitro-	benzene	294(4.39),307(4.42), 351(3.73),368(3.85), 386(3.80)	24-1683-86
$C_{24}H_{13}N_3S_3$			
17H-Bis[1,4]benzothiazino[2,3-a:3',2'-c]phenothiazine	$CHCl_3$	267(4.42),375(4.37), 537s(4.16),576(4.31)	39-2233-86C
$C_{24}H_{14}$			
9H-Fluorene, 9-(1H-cyclopropa[b]naph-thalen-1-ylidene)-	C_6H_{12}	435.5(4.79),470.5(5.00)	88-5159-86
	MeCN	433.5(4.73),466.5(4.87)	88-5159-86
$C_{24}H_{14}Br_2N_2O_2$			
Oxazole, 2,2'-(1,4-phenylene)bis[5-(4-bromophenyl)-	dioxan	365(4.76)	103-1262-86
$C_{24}H_{14}Br_2N_2S_2$			
Thiazole, 2,2'-(1,4-phenylene)-bis[5-(4-bromophenyl)-	dioxan	380(4.77)	103-1262-86
$C_{24}H_{14}Cl_2F_8N_4Pt$			
Platinum, [N,N'-bis(4-chloro-2,3,5,6-tetrafluorophenyl)-1,2-ethanediami-nato(2-)-N,N']bis(pyridine)-, (SP-4-2)-	EtOH	365(3.97)	12-2013-86
$C_{24}H_{14}F_8I_2N_4Pt$			
Platinum, [N,N'-bis(2,3,5,6-tetra-fluoro-4-iodophenyl)-1,2-ethane-diaminato(2-)-N,N']bis(pyridine)-, (SP-4-2)-	EtOH	369(3.98)	12-2013-86

Compound	Solvent	λ_{max}(log ε)	Ref.
$C_{24}H_{14}F_{10}N_4Pt$			
Platinum, [N,N'-bis(pentafluorophenyl)-1,2-ethanediaminato(2-)-N,N']-bis(pyridine)-, (SP-4-2)-	EtOH	355(3.70)	12-2013-86
$C_{24}H_{14}F_{18}O_8S_2$			
1-Butanesulfonic acid, 1,1,2,2,3,3,4-4,4-nonafluoro-1,2-bis(4-methoxyphenyl)-1,2-ethenediyl ester, (E)-	EtOH	222(4.21),284(4.37)	24-2220-86
(Z)-	C_6H_{12}	237(4.18),296(4.14)	24-2220-86
$C_{24}H_{14}N_4O_7$			
Acetamide, N-[12-(2,4-dinitrophenyl)-6,11-dihydro-6,11-dioxobenzo[f]-pyrido[1,2-a]indol-2-yl]-	DMF	537(3.7160)	2-1126-86
$C_{24}H_{14}N_8$			
5,10-Methano-1H-dibenzo[a,d]cyclohep-tene-2,2,3,3,7,7,8,8-octacarbonitrile, 4,5,6,9,10,11-hexahydro-	EtOH	210(3.85)(end abs.)	44-2385-86
$C_{24}H_{14}O_2$			
3,3'-Bidibenzofuran	EtOH	319(4.77)	46-2666-86
	dioxan	323(4.74)	46-2666-86
$C_{24}H_{15}Cl_2N_3$			
1H-Indene-4,6-dicarbonitrile, 5-amino-7-(4-chlorophenyl)-3-[(4-chlorophenyl)methylene]-2,3-dihydro-	EtOH	408(4.18)	64-1471-86B
$C_{24}H_{15}N_3$			
5H-Diindolo[3,2-a:3',2'-c]carbazole, 10,15-dihydro-	EtOH	224(4.48),280(3.84)	78-5019-86
$C_{24}H_{15}N_3O_5$			
Acetamide, N-[4-(6,11-dihydro-6,11-dioxobenzo[f]pyrido[1,2-a]indol-12-yl)-3-nitrophenyl]-	DMF	424(3.5378),498(3.7559)	2-1126-86
$C_{24}H_{16}$			
1H-Cyclopropa[b]naphthalene, 1-(diphenylmethylene)-	C_6H_{12}	230(4.82),250(4.54), 258s(4.52),269s(4.48), 291(4.32),412(4.69), 438(4.73)	35-5949-86
	MeCN	228(4.73),249(4.48), 287(4.31),409(4.62), 433(4.65)	35-5949-86
Dibenzo[a,g]cyclohexadecene, 17,18,19,20-tetradehydro-, (E,E,E,Z)-	isooctane	298(4.89),308s(4.92), 319(4.97),390(4.70), 433s(4.40),462s(4.22)	18-2209-86
$C_{24}H_{16}D_4O$			
2-Cyclohexen-1-one-2,4,6,6-d_4, 4,5,5-triphenyl-	EtOH	252(3.73),270(3.52), 290(2.90),320(1.48)	35-6276-86
$C_{24}H_{16}F_8N_4Pt$			
Platinum, [N,N'-bis(2,3,5,6-tetrafluorophenyl)-1,2-ethanediaminato(2-)-N,N']bis(pyridine)-, (SP-4-2)-	EtOH	360(3.73)	12-2013-86

Compound	Solvent	$\lambda_{max}(\log \epsilon)$	Ref.
$C_{24}H_{16}FeN_2O_5$			
Iron, tricarbonyl[2-[(4,5,6,7-η)-1H-1,2-diazepin-1-yl]-1,4-diphenyl-2-butene-1,4-dione]-	EtOH	256(4.40),405(4.32)	39-0595-86C
$C_{24}H_{16}N_2$			
Benzo[c]benzo[3,4]cinnolino[1,2-a]-cinnoline	dioxan	262(4.2),312(3.8)	89-0374-86
3-Pyridinecarbonitrile, 2,4,6-tri-phenyl-	MeCN	320(3.98)	73-1061-86
$C_{24}H_{16}N_2O_2$			
Oxazole, 2,2'-(1,4-phenylene)bis[5-phenyl-	benzene	363(4.69)	151-0341-86A
Pyridazino[1,2-b]phthalazine-6,11-di-one, 1,4-diphenyl-	EtOH	262(4.76),402(4.17)	44-3123-86
$C_{24}H_{16}N_8$			
Phthalazine, 1,1',1"-(1-hydrazinyl-2-ylidene)tris-	$CHCl_3$	252(5.04),272s(4.84), 378(4.72),812s(4.05), 836(4.14)	94-1840-86
$C_{24}H_{16}O$			
Dibenzofuran, 3,7-diphenyl-	EtOH	321(4.68)	46-2666-86
	dioxan	323(4.71)	46-2666-86
$C_{24}H_{16}O_2$			
2,4-Pentadienal, 5-[2-[4-[2-(3-oxo-1-propenyl)phenyl]-1,3-butadiynyl]-phenyl]-, (E,E,E)-	THF	235s(4.51),261(4.61), 271(4.58),285(4.66), 301(4.59),313(4.61), 338s(4.51)	18-2209-86
$C_{24}H_{16}O_8$			
4H-Anthra[1,2-b]pyran-4,7,12-trione, 11-acetoxy-2-[2,2'-bioxiran]-2-yl-5-methyl-	$CHCl_3$	372(4.01)	158-0780-86
$C_{24}H_{17}ClO_7$			
[2,2'-Binaphthalene]-1,4,5',8'-tetr-one, 4'-chloro-1'-hydroxy-5,8-di-methoxy-3,3'-dimethyl-	$CHCl_3$	262(4.47),445(4.01), 462s(3.99)	5-1669-86
$C_{24}H_{17}N_3$			
Benzenamine, 2-(5,12-dihydroindolo-[3,2-a]carbazol-6-yl)-	EtOH	242(4.46),272(4.38), 285(4.40),340(3.85), 358(3.73)	78-5019-86
1H-Indene-4,6-dicarbonitrile, 5-amino-2,3-dihydro-7-phenyl-3-(phenylmeth-ylene)-	EtOH	408(4.21)	64-1471-86B
2,3':2',3"-Ter-1H-indole	EtOH	297(4.36)	78-5019-86
$C_{24}H_{17}N_5O_4$			
Propanoic acid, 3-cyano-2,3-bis(1,5-dihydro-3-methyl-5-oxo-1-phenyl-4H-pyrazol-4-ylidene)-	EtOH	255(4.40),450(2.89)	104-0682-86
$C_{24}H_{17}N_5S$			
1H-Perimidine-2-carbothioamide, N-[4-(phenylazo)phenyl]-	DMF	324(4.32),357(4.37), 380s(4.29),556s(3.17)	48-0812-86

Compound	Solvent	λ_{max}(log ε)	Ref.
$C_{24}H_{18}Br_2OS$			
2H-Thiopyran, 2,6-bis(4-bromophenyl)-2-methoxy-4-phenyl-	dioxan	259(4.40),333(4.03)	48-0373-86
$C_{24}H_{18}Cl_2OS$			
2H-Thiopyran, 2,6-bis(4-chlorophenyl)-2-methoxy-4-phenyl-	dioxan	257(4.41),332(4.01)	48-0373-86
$C_{24}H_{18}N_2$			
2-Pyridinecarbonitrile, 1,2-dihydro-1,4,6-triphenyl-	DMSO	323(4.10)	44-2481-86
4-Pyridinecarbonitrile, 1,4-dihydro-1,2,6-triphenyl-	DMSO	319(3.95)	44-2481-86
$C_{24}H_{18}N_2O_6$			
4,5-Benzofurandicarboxylic acid, 6-[(4-hydroxyphenyl)azo]-2-phenyl-, dimethyl ester	dioxan	257(3.42),345(3.54), 410(3.53)	73-1455-86
$C_{24}H_{18}N_3S_2$			
Thiazolo[3,4-b][1,2,4]triazin-5-ium, 6-(methylthio)-2,3,8-triphenyl-, perchlorate	n.s.g.	339(4.38),470(3.69)	103-1373-86
$C_{24}H_{18}N_4O_2S_2$			
Imidazo[5,1-a]isoquinolin-3(2H)-one, 1,1'-dithiobis[2-methyl-	EtOH	265(4.04),356(3.75), 480(3.50)	150-1901-86M
$C_{24}H_{18}N_4O_3$			
6H-Anthra[1,9-cd]isoxazol-6-one, 3-azido-5-[4-(1,1-dimethylethyl)phenoxy]-	EtOH	455(4.20),472(4.22)	104-1192-86
$C_{24}H_{18}N_4O_4$			
2,4-Imidazolidinedione, 5,5'-(1,4-phenylene)bis[5-phenyl-	dioxan	320(4.92)	104-1117-86
$C_{24}H_{18}O_3$			
Naphtho[2,3-b]oxirene-2,7-dione, 1a,7a-dihydro-3,6-dimethyl-4,5-diphenyl-	MeCN	216(4.43),249(4.49), 325(3.77)	12-1537-86
$C_{24}H_{18}O_4$			
4H-1-Benzopyran-4-one, 7-acetoxy-3-phenyl-8-(phenylmethyl)-	MeOH	240(4.5),300(4.0), 390(3.1)	2-0646-86
$C_{24}H_{18}O_6$			
6,11-Naphthacenedione, 5,12-diacetoxy-2,3-dimethyl-	dioxan	285(4.35),293(4.36), 396(3.65),422s(3.30), 464s(3.19)	39-1923-86C
$C_{24}H_{18}O_7$			
[2,2'-Binaphthalene]-1,4,5',8'-tetrone, 1'-hydroxy-5,8-dimethoxy-3,3'-dimethyl-	$CHCl_3$	259(4.42),442(3.99)	5-1669-86
$C_{24}H_{18}O_8$			
[2,2'-Binaphthalene]-1,1',4,4'-tetrone, 5,5',8,8'-tetramethoxy-	$CHCl_3$	267(4.38),462(3.99)	5-1669-86

Compound	Solvent	λ_{max}(log ε)	Ref.
$C_{24}H_{18}O_9$			
2H-Pyran-2,5(6H)-dione, 3-acetoxy-4-(4-acetoxyphenyl)-6-[(4-acetoxyphenyl)methylene]- (tri-O-acetylgrevillin A)	MeOH	221(4.21),264(3.85), 360(3.64)	5-0177-86
$C_{24}H_{19}BO_5$			
Boron, [α-hydroxy-α-phenylbenzeneacetato(2-)(1-phenyl-1,3-butanedionato-O,O')]-, (T-4)-	CH_2Cl_2	266(3.83),334(4.40), 346s(4.27)	48-0755-86
$C_{24}H_{19}BrOS$			
2H-Thiopyran, 4-(4-bromophenyl)-2-methoxy-2,6-diphenyl-	dioxan	258(4.43),332(3.94)	48-0373-86
$C_{24}H_{19}BrO_4$			
1,3-Azulenedicarboxylic acid, 6-bromo-2-(2-phenylethynyl)-, diethyl ester	CH_2Cl_2	233.5(4.33),329.5(4.78), 344(4.72),420(4.34), 545(2.83)	24-2272-86
$C_{24}H_{19}ClN_2$			
Quinoline, 6-chloro-2-(3,4-dihydro-2(1H)-isoquinolinyl)-4-phenyl-	$CHCl_3$	262(4.56)	83-0338-86
$C_{24}H_{19}ClN_2O$			
Benzo[h]quinolin-2(1H)-one, 3-(4-chlorophenyl)-3,4,5,6-tetrahydro-4-(3-pyridinyl)-	EtOH	224(4.16),252s(3.73), 295(3.46)	4-1789-86
Benzo[h]quinolin-2(1H)-one, 3-(4-chlorophenyl)-3,4,5,6-tetrahydro-4-(4-pyridinyl)-	EtOH	225(4.34),295(3.64)	4-1789-86
$C_{24}H_{19}ClOS$			
2H-Thiopyran, 4-(4-chlorophenyl)-2-methoxy-2,6-diphenyl-	dioxan	257(4.41),331(3.93)	48-0373-86
$C_{24}H_{19}NO$			
11H-Indeno[1,2-b]quinolin-11-ol, 2-methyl-11-(phenylmethyl)-	$CHCl_3$	266(4.62),274(4.65), 290s(4.22),313(4.04), 320(4.10),330(4.10), 335(4.37),343(4.23), 353(4.52)	104-1352-86
11H-Indeno[1,2-b]quinolin-11-ol, 6-methyl-11-(phenylmethyl)-	$CHCl_3$	268(4.63),276(4.63), 305(4.02),320(4.08), 335(4.24),344(4.06), 350(4.29)	104-1352-86
Pyrrolo[2,1-a]isoquinolin-1-ol, 5,6-dihydro-2,3-diphenyl-	MeCN	255(4.11),279(4.12), 320(4.17)	118-0899-86
Pyrrolo[2,1-a]isoquinolin-1(5H)-one, 6,10b-dihydro-2,3-diphenyl-	MeCN	240(4.16),260(4.14), 375(3.69)	118-0899-86
$C_{24}H_{19}NO_2$			
Isoindolo[2,1-b]isoquinolin-7(5H)-one, 5-ethoxy-5,7-dihydro-12-phenyl-	MeOH	223(4.35),232(4.33), 244(4.02),300(3.69), 312(3.65),364(4.16), 381s(3.92)	142-0969-86
$C_{24}H_{19}NO_3$			
Benzeneacetamide, N-(4-methylphenyl)-α-oxo-N-(1-oxo-3-phenyl-2-propenyl)-	C_6H_{12}	230(4.36),250(4.29), 305(4.45)	39-1759-86C

Compound	Solvent	$\lambda_{max}(\log \epsilon)$	Ref.
$C_{24}H_{19}NO_3$			
2(1H)-Pyridinone, 3-(1-hydroxy-3-oxo-3-phenylpropyl)-1-(2-naphthalenyl)-	EtOH	221(4.87),278(3.94), 287(3.96),312(3.94)	4-0567-86
$C_{24}H_{19}N_3O$			
Benzeneacetonitrile, α,α'-[[5-(dimethylamino)-2,4-furandiyl]dimethylidyne]bis-	MeOH	264(3.03),388(2.88), 462(3.27)	73-0879-86
$C_{24}H_{19}N_3O_2$			
Benzonitrile, 4-[4,5-bis(4-methoxyphenyl)-1H-imidazol-2-yl]-	EtOH	240(4.29),253s(--), 280(4.08),359(4.31)	39-1623-86B
$C_{24}H_{20}$			
1,3,5,11,13,15-Cycloheptadecahexaene-7,9-diyne, 17-(2,4-cyclopentadien-1-ylidene)-6,11-dimethyl-, (E,E,Z,Z,E,E)-	THF	276s(4.36),288(4.54), 304(4.46),355(4.41), 428(4.42),469s(4.18)	18-1723-86
(relative absorbances)	MeCN	275s(0.74),286(1.00), 302(0.82),354(0.74), 419(0.75),468s(0.51)	18-1723-86
$C_{24}H_{20}F_3N_2$			
Quinolinium, 4-[3,3,3-trifluoro-2-[(1-methyl-4(1H)-quinolinylidene)methyl]-1-propenyl]-1-methyl-, perchlorate	n.s.g.	698(4.56)	104-0144-86
$C_{24}H_{20}N$			
Pyridinium, 1-methyl-2,4,6-triphenyl-, iodide	$CHCl_3$	315(4.3),395(2.5)	62-1107-86
perchlorate	$CHCl_3$	320(4.2)	62-1107-86
$C_{24}H_{20}N_2$			
2,4,6,12,14-Cyclopentadecapentaene-8,10-diyn-1-one, 7,12-dimethyl-, 2,4,6-cycloheptatrien-1-ylidenehydrazone	THF	293(4.50),375(4.32), 444(4.42)	18-2509-86
relative absorbances given	CF_3COOH	258(0.47),367(0.98), 378(1.00),461s(0.14), 491s(0.20),542(0.36), 579(0.66)	18-2509-86
$C_{24}H_{20}N_2O$			
Benzo[h]quinolin-2(1H)-one, 3,4,5,6-tetrahydro-3-phenyl-4-(3-pyridinyl)-	EtOH	226(3.70),251s(3.45), 290(3.20),298(3.20)	4-1789-86
Benzo[h]quinolin-2(1H)-one, 3,4,5,6-tetrahydro-3-phenyl-4-(4-pyridinyl)-	EtOH	237s(3.82),300(3.25)	4-1789-86
$C_{24}H_{20}N_2O_2$			
Pyridine, 1,2-dihydro-2-(nitromethyl)-1,4,6-triphenyl-	DMSO	328(4.15)	44-2481-86
Pyridine, 1,4-dihydro-4-(nitromethyl)-1,2,6-triphenyl-	DMSO	326(4.07)(changing)	44-2481-86
$C_{24}H_{20}N_2O_2S$			
Pyrimidinium, 2-(ethylthio)-3,6-dihydro-4-hydroxy-6-oxo-1,3,5-triphenyl-, hydroxide, inner salt	MeCN	207(4.40),224(4.49), 376(3.67)	24-1315-86
$C_{24}H_{20}N_2O_3$			
1H-Benzimidazole-1-acetic acid, α-(2-	MeOH	246(4.68)	44-3420-86

Compound	Solvent	$\lambda_{max}(\log \epsilon)$	Ref.
phenoxy-2-phenylethenyl)-, methyl ester, (Z)- (cont.)			44-3420-86
$C_{24}H_{20}N_4$			
2,4-Cyclohexadiene-1,1,3-tricarbonitrile, 2-amino-6-methyl-4,6-bis(4-methylphenyl)-	n.s.g.	264(4.46),328(3.78)	104-0230-86
1,3,5-Heptatriene-1,1,3-tricarbonitrile, 2-amino-4,6-bis(4-methylphenyl)-	n.s.g.	228(4.54),260(4.62), 332(4.34)	104-0230-86
Propanedinitrile, [3-cyano-5,6-dihydro-6-methyl-4,6-bis(4-methylphenyl)-2(1H)-pyridinylidene)-	n.s.g.	226(4.18),254(4.26), 350(4.33)	104-0230-86
$C_{24}H_{20}N_4O_8$			
Demethylstreptonigrin	MeOH	245(4.52),385(4.09)	158-1013-86
$C_{24}H_{20}N_6NiO_2$			
Nickel, [2-[[1-[[1-(phenylmethyl)-1H-benzimidazol-2-yl]azo]propyl]azo]-benzoato(2-)]-, (SP-4-2)-	isoPrOH	535(4.30),1040(3.80)	65-0727-86
$C_{24}H_{20}N_6O_2S_2$			
Imidazo[5,1-a]phthalazin-3(2H)-one, 1,1'-dithiobis[2-ethyl-	EtOH	271(4.49),350(3.92), 386(3.89),446(3.80)	150-1901-86M
$C_{24}H_{20}N_8O_8$			
1H-Inden-1-one, 3-[1-[(2,4-dinitrophenyl)hydrazono]ethyl]-2,3-dihydro-2-methyl-, 2,4-dinitrophenylhydrazone, trans	MeOH	223s(4.20),374(4.35)	39-1789-86C
$C_{24}H_{20}O$			
Bicyclo[3.1.0]hexan-2-one, 4,4,6-triphenyl-, exo	EtOH	270(3.97),280(2.79), 313(2.45)	35-6276-86
Cyclobutanone, 3,3-diphenyl-2-(2-phenylethenyl)-, cis-(±)-	EtOH	258(3.66),288(3.04), 296(2.85)	35-6276-86
trans-(±)-	EtOH	260(3.69),290(3.08), 296(2.85)	35-6276-86
2-Cyclohexen-1-one, 3,5,5-triphenyl-	EtOH	270(4.07),290(4.23), 310(4.04)	35-6276-86
2-Cyclohexen-1-one, 4,5,5-triphenyl-	EtOH	252(3.79),270(3.51), 290(2.93),320(1.56)	35-6276-86
$C_{24}H_{20}OS$			
2H-Thiopyran, 2-methoxy-2,4,6-triphenyl-	dioxan	253(4.35),330(3.92)	48-0373-86
$C_{24}H_{20}O_2$			
Naphth[2,3-a]azulene, 5,12-bis(2-propenyloxy)-	$CHCl_3$	258(4.36),291(4.45), 317(4.67),329(4.63), 361(4.01),390(3.98), 404(3.87),428(3.92), 455(3.76),541(2.70), 588(2.78),696(2.76), 741(2.57)	118-0686-86
$C_{24}H_{20}O_3$			
2H-1-Benzopyran-2-one, 7-hydroxy-4-methyl-3,8-bis(phenylmethyl)-	MeOH	218(4.36),321(4.23)	2-0862-86

Compound	Solvent	$\lambda_{max}(\log \epsilon)$	Ref.
2H-1-Benzopyran-2-one, 7-hydroxy-4-methyl-6,8-bis(phenylmethyl)-	MeOH	219(4.29),328(4.12)	2-0862-86
1,4-Naphthalenedione, 2,3-dihydro-2-hydroxy-5,8-dimethyl-6,7-diphenyl-	MeCN	239(4.54),320(3.67)	12-1537-86
$C_{24}H_{20}O_3S$			
9,10-Ethanoanthracene-11-carboxylic acid, 9,10-dihydro-12-(phenylsulfinyl)-, [11α,12α(R*)]-, methyl ester	n.s.g.	257(3.64),273s(3.46)	12-0575-86
[11α,12α(S*)]-	n.s.g.	252(3.42),266s(3.57), 273s(3.73)	12-0575-86
(11α,12β)-	n.s.g.	238(4.40),255(3.49), 260(3.51)	12-0575-86
$C_{24}H_{20}O_4$			
4H-1-Benzopyran-4-one, 6-(diphenylmethyl)-5,7-dihydroxy-2,3-dimethyl-	MeOH	262(4.7),299(4.33), 324(4.23)	2-0478-86
4H-1-Benzopyran-4-one, 8-(diphenylmethyl)-5,7-dihydroxy-2,3-dimethyl-	MeOH	297(4.38),312(4.18)	2-0478-86
$C_{24}H_{20}O_4S$			
9,10-Ethanoanthracene-11-carboxylic acid, 9,10-dihydro-12-(phenylsulfonyl)-, methyl ester, trans	EtOH	260(3.04),266(3.18), 273(3.15)	12-0575-86
$C_{24}H_{20}O_6$			
1(2H)-Naphthalenone, 5-hydroxy-2-(5-hydroxy-4-methoxy-7-methyl-1-oxo-2(1H)-naphthalenylidene)-4-methoxy-7-methyl-	$CHCl_3$	296(4.47),330s(4.27), 678s(4.53),698(4.55)	39-0675-86C
$C_{24}H_{20}O_8$			
7H-1,4-Dioxino[2,3-c]xanthen-7-one, 2,3-dihydro-3-(4-hydroxy-3-methoxyphenyl)-2-(hydroxymethyl)-5-methoxy-, trans-(±)- (kielcorin)	MeOH	239(4.29),253(4.25), 287(3.72),318(3.85), 362s(3.44)	100-0095-86
	MeOH-NaOMe	257(--),290(--), 319(--),362s(--)	100-0095-86
$C_{24}H_{21}Br_3N_2O_2$			
2,5-Cyclohexadiene-1,4-dione, 2,3,5-tribromo-6-[[3-(10,11-dihydro-5H-dibenzo[b,f]azepin-5-yl)propyl]methylamino]-	CH_2Cl_2	255(4.18),278(4.14), 335s(3.36),570(3.33)	106-0832-86
$C_{24}H_{21}Br_3O_3$			
5H-Tribenzo[a,d,g]cyclononene, 2,7,12-tribromo-10,15-dihydro-3,8,13-trimethoxy-	dioxan	211(4.68),231(4.48), 287(3.95),296(3.99)	152-0017-86
$C_{24}H_{21}ClN_2O$			
Ethanone, 2-[5-chloro-1,2-dihydro-3-(4-methylphenyl)-2-quinoxalinyl]-1-(4-methylphenyl)-	toluene	397(3.87)	103-0538-86
	MeOH	397(--)	103-0538-86
Ethanone, 2-[7-chloro-1,2-dihydro-3-(4-methylphenyl)-2-quinoxalinyl]-1-(4-methylphenyl)-	toluene	390(3.96)	103-0538-86
	MeOH	392(--)	103-0538-86
$C_{24}H_{21}ClN_2O_5$			
3,5-Pyridinedicarboxylic acid, 4-(9-chloro-1-oxo-1H-pyrrolo[1,2-a]indol-2-yl)-2,6-dimethyl-, diethyl ester	MeOH	220(4.37),241(4.32), 276(4.16),286(4.12), 329s(3.59)	83-0108-86

Compound	Solvent	λ_{max}(log ϵ)	Ref.
$C_{24}H_{21}Cl_2N_5O_6$			
L-Glutamic acid, N-[4-[[(2-amino-1,4-dihydro-4-oxo-6-quinazolinyl)methyl]-2-propynylamino]-3,5-dichlorobenzoyl]-	pH 13	228.5(4.74),276(4.18), 309(3.94)	87-0468-86
$C_{24}H_{21}Cl_3N_2O_2$			
2,5-Cyclohexadiene-1,4-dione, 2,3,5-trichloro-6-[[3-(10,11-dihydro-5H-dibenzo[b,f]azepin-5-yl)propyl]methylamino]-	CH_2Cl_2	254(4.29),330s(3.60), 572(3.41)	106-0832-86
$C_{24}H_{21}NO$			
Benzo[f]quinoline, 1-(3-butenyl)-3-(4-methoxyphenyl)-	EtOH	285(4.64),294(4.58), 348(3.95),365(3.98)	103-1229-86
Pyridine, 1,2-dihydro-2-methoxy-1,4,6-triphenyl-	DMSO	336(4.14)	44-2481-86
Pyridine, 1,4-dihydro-4-methoxy-1,2,6-triphenyl-	DMSO	323(4.00)	44-2481-86
$C_{24}H_{21}NO_2$			
2H-Naphtho[1,2-b]pyran-2-one, 3-phenyl-4-(1-piperidinyl)-	EtOH	268(4.40),295(3.99), 306(4.01),347(3.96)	4-1067-86
$C_{24}H_{21}NO_4$			
9,10-Anthracenedione, 1,5-dihydroxy-4-[(phenylmethyl)amino]-2-propyl-	$CHCl_3$	240(4.54),252(4.43), 304(3.82),528(3.94), 562(4.20),605(4.21)	78-3303-86
$C_{24}H_{21}NS_3$			
6H-1,3-Thiazine, 6-(ethylthio)-2,4-diphenyl-6-(phenylthio)-	$CHCl_3$	272(4.39),367(4.02)	118-0916-86
$C_{24}H_{21}N_3$			
Benzenamine, 2-(2,2-di-1H-indol-3-ylethyl)-	EtOH	220(4.63),282(3.94)	78-5019-86
Benzenamine, 2-[2-(1H-indol-2-yl)-2-(1H-indol-3-yl)ethyl]-	EtOH	220(4.63),282(3.91)	78-5019-86
$C_{24}H_{21}N_3O_4$			
1H-Imidazole, 4,5-bis(4-methoxyphenyl)-1-methyl-2-(4-nitrophenyl)-	EtOH	239(4.40),267(4.40), 386(4.12)	39-1623-86B
$C_{24}H_{21}N_7O_5$			
Benzamide, N-[2-[(2-cyano-4,6-dinitrophenyl)azo]-5-(diethylamino)phenyl]-	$CHCl_3$	405s(3.52),612(4.68)	48-0497-86
$C_{24}H_{22}Br_2N_2O_2$			
2,5-Cyclohexadiene-1,4-dione, 2,3-dibromo-5-[[3-(10,11-dihydro-5H-dibenzo[b,f]azepin-5-yl)propyl]methylamino]-	CH_2Cl_2	247s(4.30),322(3.96), 530(3.41)	106-0832-86
$C_{24}H_{22}ClN_5O_6$			
L-Glutamic acid, N-[4-[[(2-amino-1,4-dihydro-4-oxoquinazolinyl)methyl]-2-propynylamino]-2-chlorobenzoyl]-	pH 13	227.5(4.71),275.5(4.46)	87-0468-86
L-Glutamic acid, N-[4-[[(2-amino-1,4-dihydro-4-oxoquinazolinyl)methyl]-2-propynylamino]-3-chlorobenzoyl]-	pH 13	229(4.72),277(4.37)	87-0468-86

Compound	Solvent	$\lambda_{max}(\log \epsilon)$	Ref.
$C_{24}H_{22}IN$			
1H-Indole, 1-butyl-2-(4'-iodo-[1,1'-biphenyl]-4-yl)-	dioxan	316(4.46)	103-1262-86
$C_{24}H_{22}N_2$			
2H-Benz[g]indazole, 3,3a,4,5-tetrahydro-3-(3-methylphenyl)-2-phenyl-	EtOH	247(4.45),347(4.62)	4-0135-86
$C_{24}H_{22}N_2O$			
2H-Benz[g]indazole, 3,3a,4,5-tetrahydro-3-(4-methoxyphenyl)-2-phenyl-	EtOH	246(3.81),345(4.24)	4-0135-86
Ethanone, 2-[1,2-dihydro-3-(4-methylphenyl)-2-quinoxalinyl]-1-(4-methylphenyl)-	toluene	392(3.84)	103-0538-86
	MeOH	391(--)	103-0538-86
1,3,4-Oxadiazole, 2-[1,1'-biphenyl]-4-yl-5-[4-(1,1-dimethylethyl)-phenyl]-	EtOH	305(4.60)	61-0439-86
$C_{24}H_{22}N_2O_2$			
1,2-Ethenediamine, 1,2-di-2-furanyl-N,N-dimethyl-N,N'-diphenyl-, (E)-	EtOH	249(4.39),309(4.31), 383(3.93)	22-0781-86
(Z)-	EtOH	254(4.22),298(4.14), 376(4.10)	22-0781-86
$C_{24}H_{22}N_2O_4$			
5H-1,2-Benzoxazin-5-one, 6,7-dihydro-7,7-dimethyl-8-(2-nitro-1-phenylethyl)-4-phenyl-	MeOH	207(4.44),235(4.23), 358(4.14)	150-0116-86S
$C_{24}H_{22}N_2O_8$			
Bis-1,3-dioxolo[4,5-g:4',5'-g']pyrazino[2,1-a:3,4-a']diisoquinoline-4,5-dione, 1,2,7,8,13b,13c-hexahydro-13,14-dimethoxy-	MeOH	216(4.6),269(3.5), 279(3.5)	83-1122-86
isomer b	MeOH	218(4.4),265(3.5), 279(3.6)	83-1122-86
1H-Indolium, 2-carboxy-1-[5-(2-carboxy-2,3-dihydro-5,6-dihydroxy-1H-indol-1-yl)-3-methyl-2,4-pentadienylidene]-2,3-dihydro-5,6-dihydroxy-, hydroxide, inner salt, [S-(R*,R*)]-	MeOH-HCl	220(4.12),258(4.18), 289(4.12),566s(4.94), 598(4.99)	33-1588-86
1,3(2H,4H)-Isoquinolinedione, 4-(2,3-dihydro-5,7-dimethoxy-1,3-dioxo-4(1H)-isoquinolinylidene)-5,6-dimethoxy-2-methyl-, (E)-	$CHCl_3$	417(4.08)	139-0337-86C
$C_{24}H_{22}N_2S_4$			
6H-1,3-Thiazine, 2-(ethylthio)-6-[2-(ethylthio)-4-phenyl-6H-1,3-thiazin-6-ylidene]-4-phenyl-	$CHCl_3$	278(4.51),335(4.22), 504(3.96)	118-0916-86
$C_{24}H_{22}N_3O_2P$			
2,4(1H,3H)-Pyrimidinedione, 1,3-dimethyl-6-[(triphenylphosphoranylidene)amino]-	$CHCl_3$	268(4.17),274(4.20)	44-0149-86
$C_{24}H_{22}N_4$			
2,2':6',2":6",2"'-Quaterpyridine, 3",5,5',5"'-tetramethyl- (approx. spectrum)	MeCN	250(4.42),290(4.64), 375s(1.76)	152-0165-86

Compound	Solvent	$\lambda_{max}(\log \epsilon)$	Ref.
$C_{24}H_{22}N_6O_2$			
Benzoic acid, 2-[3-ethyl-5-[1-(phenylmethyl)-1H-benzimidazol-2-yl]-1-formazano]-	EtOH	500(4.52)	65-0727-86
1:1 nickel complex	EtOH	535(4.18)	65-0727-86
1:1 zinc complex	EtOH	640(4.38)	65-0727-86
$C_{24}H_{22}O_2$			
[1,1'-Binaphthalene]-4,4'-diol, 2,2',3,3'-tetramethyl-	MeOH	252(4.71),342(4.08)	39-1789-86C
2,4,6,8,14,16,18,20-Docosaoctaene-10,12-diynedial, 9,14-dimethyl-, (E,E,Z,Z,E,E,E,E)-	THF	254s(4.21),264(4.26), 290s(4.27),302s(4.35), 315s(4.43),347s(4.70), 364(4.78),408(4.72), 457s(4.48)	18-2209-86
2-Propenoic acid, 3-(1,3-dimethyl-1H-cyclopenta[l]phenanthren-1-yl)-, ethyl ester, (E)-	EtOH	241(4.53),262.5(4.48), 271(4.38),282s(4.30), 322(3.96)	39-1471-86C
2-Propenoic acid, 3-(1,3-dimethyl-1H-cyclopenta[l]phenanthren-2-yl)-, ethyl ester	EtOH	240.5(4.37),261(4.49), 282(4.39),294s(4.36), 306(4.22),338(4.32)	39-1471-86C
$C_{24}H_{22}O_4$			
2-Propen-1-one, 1-[2-hydroxy-3,4-dimethoxy-5-(phenylmethyl)phenyl]-3-phenyl-	MeOH	216(4.8),268(4.3), 329(4.6)	2-0259-86
$C_{24}H_{22}O_4S$			
15-Thia-18-norestra-1,3,5(10),8(9)-tetraen-17-one, 3-methoxy-16-(phenylmethylene)-, S,S-dioxide	EtOH	272(4.32),325(3.46), 333(3.33)	104-0107-86
	EtOH-HCl	272(4.32),335(3.53)	104-0107-86
17-Thia-18-norestra-1,3,5(10),8(9)-tetraen-15-one, 3-methoxy-16-(phenylmethylene)-, S,S-dioxide	EtOH	257(3.88),335(3.95), 412(4.32)	104-0107-86
	EtOH-HCl	256(3.63),345(3.91), 400(4.05)	104-0107-86
$C_{24}H_{22}O_6$			
[2,2'-Binaphthalene]-1,1',5,5'-tetrol, 4,4'-dimethoxy-7,7'-dimethyl-	$CHCl_3$	265s(4.43),321(4.05), 340s(4.10),353(4.22)	39-0675-86C
$C_{24}H_{22}O_8$			
2,5-Cyclohexadiene-1,4-dione, 2-[1-(3,4-dimethoxyphenyl)-2-(5-methoxy-3,6-dioxo-1,4-cyclohexadien-1-yl)-ethyl]-5-methoxy-	MeOH	206.5(4.69),237(4.32), 265(4.54),404(3.24)	94-4418-86
$C_{24}H_{22}O_{10}$			
Arborone diacetate	MeOH	234(4.44),284(4.17), 313(4.11)	100-1061-86
$C_{24}H_{23}BrN_4O_5$			
Acetic acid, [[5-(1,2-diacetoxyethyl)-1-(4-bromophenyl)-1H-pyrazol-3-yl]-methylene]phenylhydrazide, (±)-	MeOH	282(4.67)	136-0316-86G
$C_{24}H_{23}BrO_4$			
1,3-Azulenedicarboxylic acid, 6-bromo-2-(2-phenylethyl)-, diethyl ester	CH_2Cl_2	234.5(4.81),270(4.40), 319.5(4.89),364.5(4.08), 381(3.51),508(2.61)	24-2272-86

Compound	Solvent	λ_{max}(log ϵ)	Ref.
$C_{24}H_{23}BrO_5$			
1,3-Azulenedicarboxylic acid, 6-bromo-2-(2-phenylethoxy)-, diethyl ester	CH_2Cl_2	236.5(4.45),267(4.22), 318.5(4.84),352(3.93), 363.5(3.93),479(2.53)	24-2272-86
$C_{24}H_{23}ClN_2O_5$			
3,5-Pyridinedicarboxylic acid, 4-(9-chloro-1-oxo-1H-pyrrolo[1,2-a]indol-2-yl)-1,4-dihydro-2,6-dimethyl-, diethyl ester	MeOH	220(4.32),242(4.24), 276(4.13),287(4.09), 330s(3.65)	83-0108-86
$C_{24}H_{23}N$			
Benzenamine, N-(2,2-dimethyl-4,4-diphenyl-3-butenylidene)-	tert-BuOH	251(4.42)	150-0631-86M
$C_{24}H_{23}NO$			
Azacyclodocosa-3,5,7,9,15,17,19,21-octaene-11,13-diyn-2-one, 10,15,22-trimethyl-	THF	238(4.19),273(4.23), 350s(4.90),358(4.91), 433s(4.60)	39-0933-86C
2,4,6,8,14,16,18,20-Cycloheneicosa-octaene-10,12-diyn-1-one, 2,9,14-trimethyl-, oxime	THF	335(3.96),357s(4.02), 310s(4.66),333s(4.79), 341(4.80),432s(3.85)	39-0933-86C
$C_{24}H_{23}NO_2$			
2-Propen-1-one, 1-(2-hydroxyphenyl)-3-[4-(1-methylethyl)phenyl]-3-(phenylamino)-	EtOH	260(4.12),391(4.50)	104-1658-86
$C_{24}H_{23}N_3O_2$			
1H-[3]Benzazecino[9,8,7-abc]carbazole-4-carbonitrile, 2,3,4,5,6,10-hexahydro-8,9-dimethoxy-	MeOH	262(4.73),283(4.73), 298(4.48),333(3.90), 353(3.74),371(3.74)	12-0001-86
$C_{24}H_{23}N_3O_7$			
1,3-Diazabicyclo[3.2.0]heptane-2-carboxylic acid, 4-(2-methoxy-2-oxoethylidene)-7-oxo-6-[(phenoxyacetyl)-amino]-, phenylmethyl ester, (2RS,5RS,6RS)-	EtOH	275(4.22)	39-1077-86C
(2RS,5SR,6SR)-	EtOH	274(4.25)	39-1077-86C
$C_{24}H_{23}N_5O_6$			
L-Glutamic acid, N-[4-[[(2-amino-1,4-dihydro-4-oxo-6-quinazolinyl)methyl]-2-propynylamino]benzoyl]-	pH 13	229(4.70),279(4.38), 301.5(4.42)	87-0468-86
$C_{24}H_{23}N_5O_7$			
Acetic acid, [[5-(1,2-diacetoxyethyl)-1-(4-nitrophenyl)-1H-pyrazol-3-yl]-methylene]phenylhydrazide, (S)-	MeOH	276(4.38),307s(4.26)	136-0316-86G
$C_{24}H_{23}O$			
Bisbenzo[6,7]cyclohepta[1,2-b:1',2'-d]pyrylium, 7,8,9,14,15,16-hexahydro-6-methyl-, perchlorate	MeCN	230(4.55),272(4.37), 368(4.36)	22-0600-86
$C_{24}H_{23}O_3$			
Dinaphtho[1,2-b:1',2'-d]pyrylium, 7,8,13,14-tetrahydro-2,10-dimethoxy-6-methyl-, perchlorate	MeCN	220s(4.42),264(4.32), 308(4.15),470(4.31)	22-0600-86

Compound	Solvent	λ$_{max}$(log ε)	Ref.
$C_{24}H_{24}$			
Tricyclo[12.2.2.2^{6,9}]eicosa-2,6,8,10-14,16,17,19-octaene, 5,13-diethenyl-	C_6H_{12}	256(4.6)	88-5923-86
$C_{24}H_{24}FeN$			
Pyridinium, 1-(4-ferrocenylphenyl)-2,4,6-trimethyl-, perchlorate	MeCN	209(4.7),250s(4.6), 281(4.8),352s(3.6), 461(3.1)	103-0886-86
$C_{24}H_{24}N_2$			
4a,4'a-Bi-4aH-carbazole, 1,1',2,2'-3,3',4,4'-octahydro-, dl-	EtOH	227(4.04),262(3.70)	142-0687-86
meso-	EtOH	225(3.99),262(3.73)	142-0687-86
4a,9'(2H,2'H)-Bi-1H-carbazole, 3,3',4,4'-tetrahydro-	EtOH	230(4.25),272(3.95)	142-0687-86
Pyrazine, 2,5-diethyl-3,6-bis(2-phenylethenyl)-, (E,E)-	EtOH	226(3.96),237s(3.88), 244s(3.78),294(4.20), 302s(4.18)	4-1481-86
$C_{24}H_{24}N_2OS$			
Methanone, (3-amino-6-adamantylthieno[2,3-b]pyridin-2-yl)phenyl-	EtOH	281(4.31),314(4.34)	70-0131-86
3-Pyridinecarbonitrile, 2-[(2-oxo-2-phenylethyl)thio]-6-adamantyl]-	EtOH	202(4.57),222(4.41), 261(4.28),312(3.80)	70-0131-86
2H,11bH-Pyrrolo[2',1':3,4]pyrazino-[2,1-b][1,3]thiazine, 6,7-dihydro-2-methoxy-11b-methyl-3,4-diphenyl-	MeCN	223(4.33),286(3.91)	118-0899-86
$C_{24}H_{24}N_2OSe$			
Methanone, (3-amino-6-adamantylselenolo[2,3-b]pyridin-2-yl)phenyl-	EtOH	319(4.51),424(4.35)	70-0376-86
3-Pyridinecarbonitrile, 2-[(2-oxo-2-phenylethyl)seleno]-6-adamantyl-	EtOH	227(4.22),240(4.19), 275(3.99),366(3.55)	70-0376-86
$C_{24}H_{24}N_2O_2$			
2,5-Cyclohexadiene-1,4-dione, 2-[[3-(10,11-dihydro-5H-dibenz[b,f]azepin-5-yl)propyl]amino]-5-methyl-	CH_2Cl_2	258s(4.20),274(4.28), 482(3.41)	106-0027-86
$C_{24}H_{24}N_2O_4$			
1,7,11-Ethenylylideneazacyclotridecino[6,5-b]indole-4(3H)-carboxylic acid, 2,5,6,12-tetrahydro-9,10-dimethoxy-, methyl ester	MeOH	261s(4.37),267(4.41), 292(4.46),303s(4.28), 336s(3.76),363s(3.55), 382(3.44)	12-0001-86
$C_{24}H_{24}N_2O_5S$			
6,8a-Ethano-8aH-carbazole-8-carboxylic acid, 5-(acetylamino)-6,7,8,9-tetrahydro-9-(phenylsulfonyl)-, methyl ester, (6α,8β,8aα)-	MeOH	218(4.33),238(4.18), 274(4.15),318(4.03), 328s(4.01)	5-2065-86
$C_{24}H_{24}N_2O_8$			
[4,4'-Biisoquinoline]-1,1',3,3'(2H-2'H,4H,4'H)-tetrone, 5,5',7,7'-tetramethoxy-2,2'-dimethyl-, (R*,R*)-(±)-	$CHCl_3$	327(3.40)	139-0337-86C
2-Propenoic acid, 3-(3-nitrophenyl)-, 1,6-hexanediyl ester, (E,E)-	MeCN	261(4.78)	18-1379-86
2-Propenoic acid, 3-(4-nitrophenyl)-, 1,6-hexanediyl ester, (E,E)-	MeCN	303(4.58)	18-1379-86

Compound	Solvent	$\lambda_{max}(\log \epsilon)$	Ref.
$C_{24}H_{24}N_2O_{13}S$			
Paulinone, 9,10,11-tri-O-acetyl-11-O-pauloyl-	MeOH	227(4.19),265(4.14), 437(3.18)	44-2493-86
$C_{24}H_{24}N_4$			
2,3'-Bipyridinium, 1',1'''-(1,4-butanediyl)bis-, dibromide	pH 8.2	210(4.34),230(4.39), 277(4.29)	64-0239-86B
1H-Pyrazolo[3,4-b]quinolin-4-amine, N-(3,5-dimethylphenyl)-5,6,7,8-tetrahydro-1-phenyl-	EtOH	224s(4.42),256(4.40), 331(4.32)	11-0331-86
$C_{24}H_{24}N_4O_6$			
11,7-Metheno-7H-dibenzo[b,m][1,15,18-21,4,5,11,12]tetraoxatetraazacyclotricosine-8,28-diol, 19,20,22,23-25,26-hexahydro-	EtOH	417.5(4.55)	70-1741-86
$C_{24}H_{24}N_4S_2$			
6H-1,3-Thiazin-2-amine, 6-[2-(dimethylamino)-4-phenyl-6H-1,3-thiazin-6-ylidene]-N,N-dimethyl-4-phenyl-	$CHCl_3$	267(4.53),320s(3.93), 510(4.04)	118-0916-86
$C_{24}H_{24}O$			
2,4,10,12-Cyclopentadecatetraene-6,8-diyn-1-one, 14-(2,4-cyclopentadien-1-yl)-2-ethyl-5,10-dimethyl-, (E,Z,Z,E)-	THF	239s(4.07),250s(4.16), 276s(4.55),287(4.64), 381(3.80),423s(3.63)	18-1723-86
$C_{24}H_{24}O_2$			
Naphth[2,3-a]azulene, 5,12-bis(1-methylethoxy)-	$CHCl_3$	318(4.68),330(4.42), 365(4.01),382(4.01), 407(3.89),430(3.89), 458(3.72),554(2.44), 590(2.43),664(2.31)	118-0686-86
$C_{24}H_{24}O_4$			
1-Propanone, 1-[2-hydroxy-3,4-dimethoxy-5-(phenylmethyl)phenyl]-3-phenyl-	MeOH	219(4.9),280(3.9)	2-0259-86
$C_{24}H_{24}O_7$			
[3,8'-Bi-4H-1-benzopyran]-4-one, 7-acetoxy-2',3'-dihydro-5',7'-dimethoxy-2',2'-dimethyl-	EtOH	256s(4.21),292(3.82), 301.5s(3.79)	94-2369-86
2H,6H-Pyrano[3,2-b]xanthen-6-one, 5,9,10-trihydroxy-8-methoxy-2,2-dimethyl-12-(3-methyl-2-butenyl)-	MeOH	289(4.49),339(4.10)	102-1217-86
	MeOH-NaOAc	287(--),387(--)	102-1217-86
	MeOH-$AlCl_3$	299(--),408(--)	102-1217-86
$C_{24}H_{24}O_9$			
Benzeneacetic acid, 4-methoxy-α-[3-methoxy-5-oxo-4-(3,4,5-trimethoxyphenyl)-2(5H)-furanylidene]-, methyl ester, (E)-	n.s.g.	363(4.34)	39-2127-86C
(Z)-	n.s.g.	356(--)	39-2127-86C
Benzeneacetic acid, 3,4,5-trimethoxy-α-[3-methoxy-4-(4-methoxyphenyl)-5-oxo-2(5H)-furanylidene]-, methyl ester, (E)-	$CHCl_3$	364(4.30)	39-2127-86C
$C_{24}H_{24}S_2$			
Dicyclopenta[a,e]dicyclopropa[c,g]-	C_6H_{12}	246(4.43),341(4.63),	35-7032-86

Compound	Solvent	λ_{max}(log ε)	Ref.
cyclooctene, 1,5-bis[(1,1-dimethylethyl)thio]- (cont.)		380(4.35),530(2.74)	35-7032-86
	CH_2Cl_2	?(4.51),266(4.49), 337(4.74),372(4.43), 413(4.39),525(2.74)	35-7032-86
$C_{24}H_{25}Cl_4N_5O_4$			
Benzeneacetic acid, α-[amino[(4-chloro-2,6-di-4-morpholinylphenyl)azo]methylene]-2,4,6-trichloro-, methyl ester, (E,E)-	MeOH	444(3.15)	88-4281-86
$C_{24}H_{25}F_6N_3O_6S_2$			
Pyridinium, 1-[[4-(dimethylamino)phenyl][4-(dimethylimino)-2,5-cyclohexadien-1-ylidene]methyl]-, salt with trifluoromethanesulfonic acid	MeOH	204(4.58),252(4.40), 367(3.88),425(3.51), 660(4.46)	24-3276-86
	acetone	428(3.85),662(4.77)	24-3276-86
$C_{24}H_{25}NO$			
1H-Pyrrolo[2,1-j]quinolin-1-one, 5,6,7,7a,8,9,10,11-octahydro-2,3-diphenyl-	MeCN	227(4.11),270(4.08), 353(3.89)	118-0899-86
$C_{24}H_{25}NO_2$			
6-Hepten-3-one, 1-(4-methoxyphenyl)-1-(2-naphthalenylamino)-	EtOH	248(4.73),273(4.04), 284(4.13),294(4.09)	103-1229-86
$C_{24}H_{25}NO_5$			
2,5-Pyrrolidinedione, 3-(1-methylethylidene)-1-phenyl-4-[1-(3,4,5-trimethoxyphenyl)ethylidene]-, (E)-	toluene	285(4.11),326(3.97)	39-0315-86C
$C_{24}H_{25}NO_8$			
2-Butenedioic acid, 2-[5,6,8,9,10,10a-hexahydro-7-(methoxycarbonyl)-5,9,10a-trimethyl-6,10-dioxo-9-phenanthridinyl]-, dimethyl ester	n.s.g.	221(4.31),276(3.61), 306(3.54)	150-0514-86M
$C_{24}H_{25}NO_{13}$			
[1,3]Dioxolo[4,5-j]phenanthridin-6(2H)-one, 1,2,3,4,7-pentaacetoxy-1,3,4,4a,5,11b-hexahydro- (pancratistatin pentaacetate)	MeOH	227(4.31),247s(--), 271s(--),299(3.75)	102-0995-86
$C_{24}H_{25}N_5O$			
2(1H)-Pteridinone, 4-(butylimino)-3,4-dihydro-1,3-dimethyl-6,7-diphenyl-	MeOH	228(4.48),292(4.14), 370(4.16)	128-0199-86
$C_{24}H_{25}N_7O_{10}$			
Acetamide, N-[5-[bis(2-acetoxyethyl)amino]-2-[(2-cyano-4,6-dinitrophenyl)azo]-4-methoxyphenyl]-	$CHCl_3$	282(4.00),482s(3.75), 360(3.53),602s(4.60), 644(4.79)	48-0497-86
$C_{24}H_{26}Br_2N_4O_2$			
2,5-Cyclohexadiene-1,4-dione, 2,5-dibromo-3,6-bis[[4-[(dimethylamino)methyl]phenyl]amino]-	MeOH	252(4.26),395(4.17)	87-1792-86
$C_{24}H_{26}Cl_2N_4O_2$			
2,5-Cyclohexadiene-1,4-dione, 2,5-di-	MeOH	270(4.32),395(4.23)	87-1792-86

Compound	Solvent	$\lambda_{max}(\log \epsilon)$	Ref.
chloro-3,6-bis[[4-[(dimethylamino)-methyl]phenyl]amino]- (cont.)			87-1792-86
$C_{24}H_{26}N_2O$			
1H-Benzimidazole, 2-[5-(1,1-dimethyl-ethoxy)bicyclo[4.4.1]undeca-2,4,6-8,10-pentaen-2-yl]-5,6-dimethyl-, (±)-	CH_2Cl_2	266(4.46),285(4.39), 364(4.31)	33-0315-86
$C_{24}H_{26}N_2O_4$			
2,5-Cyclohexadiene-1,4-dione, 2,5-bis[[4-(dimethylamino)methyl]phenyl]-3,6-dihydroxy-	MeOH	308(3.32),530(3.18)	87-1792-86
$C_{24}H_{26}N_2O_6$			
3-Isoxazolidineacetic acid, 5-(hydroxyphenylamino)-2-phenyl-4-[(2-propenyloxy)carbonyl]-, 2-propenyl ester	EtOH	224(4.09),318(4.42)	44-3125-86
$C_{24}H_{26}N_4$			
2,3'-Bipyridinium, 1',1'''-(1,2-ethane-diyl)bis[1-methyl-, tetraperchlorate	pH 2.3	208(4.40),273(4.4)	64-0239-86B
4,4'-Bipyridinium, 1-methyl-1'-[2-[4-(1-methyl-4(1H)-pyridinylidene)-1(4H)-pyridinyl]ethyl]-	n.s.g.	264(4.62)	35-3380-86
diradical dication	n.s.g.	600(4.34)	35-3380-86
radical trication	n.s.g.	560s(3.86),600(4.04)	35-3380-86
$C_{24}H_{26}N_4O_3$			
4,5-Ethano-7H-purine-2,6(1H,3H)-dione, 8-ethoxy-1,3,7-trimethyl-10,11-di-phenyl-, (10R*,11R*)-	CH_2Cl_2	259s(3.22),265s(3.12), 274s(2.95)	24-1525-86
(10S*,11S*)-	70% MeOH	255s(--),260s(--), 266s(--),270s(--), 280s(--)	24-1525-86
$C_{24}H_{26}N_6$			
1H-Indole, 3,3'-[1,2,5,6-tetrahydro-1-methyl-3-(1-pyrrolidinyl)-1,2,4-triazine-5,6-diyl]bis-, mono-hydriodide	EtOH	220(4.92),274(5.02), 282(5.04),290(5.00), 317(2.76),375(2.65)	103-1242-86
$C_{24}H_{26}N_6O$			
1H-Indole, 3,3'-[1,4,5,6-tetrahydro-1-methyl-3-(4-morpholinyl)-1,2,4-triazine-5,6-diyl]bis-, mono-hydriodide	EtOH	221(5.23),282(4.09), 290(3.95),421(3.20)	103-1242-86
$C_{24}H_{26}N_6O_4$			
Benzoic acid, 4-[[2-(acetylamino)-4-(diethylamino)-5-methoxyphenyl]-azo]-3,5-dicyano-, ethyl ester	$CHCl_3$	303(3.97),318(3.79), 364s(3.34),466s(3.58), 588s(4.66),630(4.89)	48-0497-86
$C_{24}H_{26}N_6S_2$			
Thiazole, 2,2'-(2-tetrazene-1,4-diyli-dene)bis[2,3-dihydro-4,5-dimethyl-3-(phenylmethyl)-	$CHCl_3$	408(3.04)	142-3097-86
$C_{24}H_{26}O_2$			
Phenol, 2-(1,1-dimethylethyl)-6-[(2-	EtOH	279(3.73)	116-0509-86

Compound	Solvent	λ max (log ε)	Ref.
hydroxyphenyl)phenylmethyl]-4-methyl-, (S)- (cont.)			116-0509-86
$C_{24}H_{26}O_3$			
Phenol, 2-[1-(2-hydroxy-3,5-dimethylphenyl)ethyl]-6-[1-(2-hydroxyphenyl)ethyl]-	EtOH	279(3.82)	116-0509-86
$C_{24}H_{26}O_4$			
5,7-Octadiene-1,4-diol, 2,7-dimethyl-, dibenzoate, [R-[R*,R*-(E)]]-[S-[R*,S*-(E)]]-	EtOH	230(4.70)	163-0187-86
$C_{24}H_{26}O_5$			
9H-Xanthen-9-one, 1,3-dihydroxy-7-methoxy-2,8-bis(3-methyl-2-butenyl)-	MeOH	240(4.78),262(4.74), 314(4.56),365(3.97)	102-1957-86
$C_{24}H_{26}O_6$			
4H-1-Benzopyran-4-one, 7-methoxy-3-[2,4,6-trimethoxy-3-(3-methyl-2-butenyl)phenyl]-	EtOH	247(4.41),283.5(4.12), 294s(4.08),303.5(4.03)	94-2369-86
$C_{24}H_{26}O_7$			
9H-Xanthen-9-one, 1,3,5,6-tetrahydroxy-7-methoxy-2,4-bis(3,3-dimethyl-2-propenyl)-	MeOH	267(3.82),283s(3.78), 332(3.57)	102-1217-86
	MeOH-NaOAc	267(--),385(--)	102-1217-86
changing	MeOH-AlCl₃	280(--),373(--)	102-1217-86
$C_{24}H_{26}O_8$			
1H-Benz[4,5]indeno[7,1-bc]pyran-4,6(2H,10bH)-dione, 8-(2-acetoxy-1-methylethyl)-7-hydroxy-9,10-dimethoxy-3,10b-dimethyl-, [S-(R*,R*)]-(11,21-di-O-methyledulon A)	ether	249s(3.72),288s(3.75), 298.4(3.77),311(3.79), 347.2(4.00),357.3(4.00), 393.6(3.92)	33-1513-86
8H-1,3-Dioxolo[4,5-g][1]benzopyran-8-one, 4,9-diacetoxy-3a,4,9,9a-tetrahydro-2,2-dimethyl-6-(2-phenylethyl)-	MeOH	207(4.23),251(4.02)	94-2766-86
isomer	EtOH	208(4.26),250(4.06)	94-2766-86
$C_{24}H_{26}O_9$			
Mulberroside C	EtOH	218(4.45),257s(3.80), 288s(4.05),298s(4.07), 309s(4.22),322(4.41), 337(4.36)	100-0218-86
$C_{24}H_{27}BrN_4O_{12}$			
Acetamide, N-acetyl-N-[6-bromo-5-formyl-4,7-dihydro-3-(methoxymethyl)-4-oxo-7-(2,3,5-tri-O-acetyl-β-D-arabinofuranosyl)-3H-pyrrolo[2,3-d]pyrimidin-2-yl]-	MeCN	239(4.16),296(4.10)	18-1915-86
$C_{24}H_{27}Fe$			
Iron(1+), [(1,2,3,4,4a,9a-η)-9,10-dihydroanthracene][(1,2,3,4,5-η)-1,2,3,4,5-pentamethyl-2,5-cyclopentadien-1-yl]-, hexafluorophosphate	n.s.g.	253(3.53),259(3.59), 268(3.70),275(3.82), 451(2.12)	101-0097-860

Compound	Solvent	λ_{max}(log ϵ)	Ref.
$C_{24}H_{27}NO_2S$			
1H-Indole, 2-[4-methyl-1-(2-methyl-1-propenyl]-1-(phenylsulfonyl)-	MeOH	250(4.30)	44-2343-86
$C_{24}H_{27}NO_8$			
Papaveroxine, (-)-	MeOH	238(3.95),282(3.54)	102-2403-86
$C_{24}H_{27}NO_9$			
Lactoquinomycin B	MeOH	239(4.18),287(3.54), 369(3.72)	158-0001-86
	MeOH-HCl	240(4.22),285(3.55), 366(3.75)	158-0001-86
	MeOH-NaOH	223s(4.30),287(3.81), 442(3.80)	158-0001-86
$C_{24}H_{27}N_5O_8S_2$			
2',3'-Secoadenosine, 2',3'-di-O-[(4-methylphenyl)sulfonyl]-	EtOH	261(4.15)	87-2445-86
$C_{24}H_{27}N_7O_5$			
Acetamide, N-[5-[bis(2-ethoxyethyl)-amino]-2-[(2,6-dicyano-4-nitro-phenyl)azo]phenyl]-	$CHCl_3$	267s(3.94),298s(3.79), 315s(3.68),393s(3.45), 427(3.54),588s(4.71), 628(4.89)	48-0497-86
$C_{24}H_{28}FeNO_4$			
Pyridinium, 3,5-bis(ethoxycarbonyl)-4-ferrocenyl-1,2,6-trimethyl-, perchlorate	MeCN	210(4.5),288(4.3), 316(4.2),438(3.7), 568(3.5)	103-0886-86
$C_{24}H_{28}N_2$			
Pyrazine, 2,5-bis(2-methylpropyl)-3,6-diphenyl-	EtOH	248(3.99),297(4.01), 310s(3.93)	142-0785-86
$C_{24}H_{28}N_2O_2$			
1H-Pyrrolizine-5-methanamine, 6-(2,5-dimethoxyphenyl)-2,3-dihydro-N,N-dimethyl-7-phenyl-, hydrochloride	n.s.g.	205(4.56),221s(4.46), 263s(4.07)	83-0065-86
1H-Pyrrolizine-5-methanamine, 6-(3,4-dimethoxyphenyl)-2,3-dihydro-N,N-dimethyl-7-phenyl-, hydrochloride	n.s.g.	203(4.65),244(4.33), 275s(4.13)	83-0065-86
Pyrrolo[3,4-b]indol-3(2H)-one, 2-cyclo-hexyl-1,3a,4,8b-tetrahydro-3a-meth-oxy-8b-(phenylmethyl)-	EtOH	240(3.75),292.5(3.34)	118-0731-86
$C_{24}H_{28}N_2O_4S$			
Corynan-18-carboxylic acid, 18,19,20-21-tetradehydro-17-ethoxy-16-(meth-ylthio)-17-oxo-, methyl ester	MeOH	222(4.32),270(3.77), 290(3.64),356(4.42)	44-2995-86
$C_{24}H_{28}N_2O_6$			
Bishydrocotarnine	EtOH	213(4.6),260(3.2), 285(3.5)	83-1122-86
form b	EtOH	212(4.6),260(3.1), 285(3.4)	83-1122-86
Spiro[3H-indole-3,1'(5'H)-indolizine]-8'-propanoic acid, 8'-(3-acetoxy-1-propenyl)-1,2,2',3',6',7',8',8'a-octahydro-2,5'-dioxo-, methyl ester	EtOH	219(4.12),255(3.78), 265s(3.65),285(3.09)	78-6047-86

Compound	Solvent	$\lambda_{max}(\log \epsilon)$	Ref.
$C_{24}H_{28}N_2O_9S$			
β-D-Galactopyranosylamine, N-[5-(4-methylphenyl)-2-thiazolyl]-, 2,3,4,6-tetraacetate	CH_2Cl_2	300(4.34)	136-0318-86H
β-D-Glucopyranosylamine, N-[5-(4-methylphenyl)-2-thiazolyl]-, 2,3,4,6-tetraacetate	CH_2Cl_2	302(4.33)	136-0318-86H
$C_{24}H_{28}N_2O_{10}S$			
β-D-Glucopyranosylamine, N-[5-(4-methoxyphenyl)-2-thiazolyl]-, 2,3,4,6-tetraacetate	CH_2Cl_2	303(4.32)	136-0318-86H
$C_{24}H_{28}N_4O_2$			
2,5-Cyclohexadiene-1,4-dione, 2,5-bis[[4-[(dimethylamino)methyl]phenyl]amino]-	MeOH	273(4.26),388(4.28)	87-1792-86
$C_{24}H_{28}N_4O_7$			
1H-Imidazole-4-carboxamide, 5-[[bis(4-methoxyphenyl)methyl]amino]-1-β-D-ribofuranosyl-	pH 1	270(4.23)	44-1277-86
	pH 11	272(4.20)	44-1277-86
	MeOH	273(4.20)	44-1277-86
$C_{24}H_{28}N_5O_6P$			
Phosphonic acid, [3-[(2-amino-1,6-dihydro-6-oxo-9H-purin-9-yl)methoxy]-4-hydroxybutyl]-, bis(phenylmethyl) ester	MeOH	255(4.15),270s(4.00)	87-0671-86
$C_{24}H_{28}O_4$			
1,2-Benzenediol, 3-[2-[2-(5-hydroxy-4-methyl-3-pentenyl)-2-methyl-2H-1-benzopyran-6-yl]ethyl]-, [S-(E)]- (ternatin)	MeOH	227(4.62),268(4.06), 317(3.88)	94-3727-86
2H,8H-Benzo[1,2-b:5,4-b']dipyran-2-one, 10-(1,1-dimethyl-2-propenyl)-5-[(1,1-dimethyl-2-propenyl)oxy]-8,8-dimethyl- (ponfolin)	MeOH	206(4.29),230(4.29), 273(4.41),329(4.05)	94-3922-86
$C_{24}H_{28}O_9$			
Ethanone, 1-[3-[(8-acetyl-3,4-dihydro-3,5,7-trihydroxy-2,2-dimethyl-2H-1-benzopyran-6-yl)methyl]-2,6-dihydroxy-4-methoxy-5-methylphenyl]-, (±)-	EtOH	288(4.08),323(3.97)	100-0298-86
Gmelinol, 7-oxodihydro-, acetate	MeOH	231(4.47),281(4.28), 308(4.11)	100-1061-86
$C_{24}H_{29}BrN_4O_{11}$			
Acetamide, N-acetyl-N-[6-bromo-4,7-dihydro-3-(methoxymethyl)-5-methyl-4-oxo-7-(2,3,5-tri-O-acetyl-β-D-arabinofuranosyl)-3H-pyrrolo[2,3-d]pyrimidin-2-yl]-	MeCN	269(3.73),310(3.92)	18-1915-86
$C_{24}H_{29}BrN_4O_{12}$			
Acetamide, N-acetyl-N-[6-bromo-4,7-dihydro-5-(hydroxymethyl)-3-(methoxymethyl)-4-oxo-7-(2,3,5-tri-O-acetyl-β-D-arabinofuranosyl)-3H-pyrrolo[2,3-d]pyrimidin-2-yl]-	MeCN	269(3.77),306(3.86)	18-1915-86

Compound	Solvent	λ_{max}(log ε)	Ref.
$C_{24}H_{29}ClO_9$			
Benzo[4,5]cyclodeca[1,2-b]furan-2,11(1H,3aH)-dione, 8,13-diacetoxy-4-chloro-4,5,8,8a,12,12a,13,13a-octahydro-12,13a-dihydroxy-1,8a,12-trimethyl-5-methylene- (minabein-1)	MeOH	218(4.08)	20-0835-86
Ptilosarcenone, 11-hydroxy-	EtOH	219(4.07)	78-6565-86
$C_{24}H_{29}Cl_2FO_4S$			
Androsta-1,4-diene-17-carbothioic acid, 17-acetoxy-9,11-dichloro-6-fluoro-16-methyl-3-oxo-, S-methyl ester, (6α,11β,16β,17α)-	MeOH	235(4.29)	44-2315-86
$C_{24}H_{29}Cl_2FO_5$			
Androsta-1,4-diene-17-carboxylic acid, 9,11-dichloro-6-fluoro-16-methyl-3-oxo-17-(1-oxopropoxy)-, (6α,11β,16β-17α)-	MeOH	236(4.16)	44-2315-86
$C_{24}H_{29}Fe$			
Iron(1+), [(1,2,3,4,5-η)-1,2,3,4,5-pentamethyl-2,4-cyclopentadien-1-yl][(5,6,7,8,8a,10a-η)-1,2,3,4-tetrahydroanthracene]-, hexafluorophosphate	n.s.g.	253(3.64),259(3.69), 268(3.78),285(3.98), 445(2.43),483(2.45)	101-0097-860
$C_{24}H_{29}FeNO_4$			
Ferrocene, [3,5-bis(ethoxycarbonyl)-1,4-dihydro-1,2,6-trimethyl-4-pyridinyl]-	MeCN	209(4.6),241(4.3), 267(4.1),338(3.8), 458(2.7)	103-0886-86
$C_{24}H_{29}NO_2$			
2H,6H-Pyrido[2,1-b][1,3]oxazine, 2-ethoxy-9a-ethyl-7,8,9,9a-tetrahydro-3,4-diphenyl-	MeCN	228(4.07),293(3.91)	118-0899-86
$C_{24}H_{29}NO_5$			
Ethanone, 1-[(6α,18R,19S)-7,8-didehydro-4,6-epoxy-3,6,19-trimethoxy-17-methyl-5,14-ethanomorphinan-18-yl]-	EtOH	245(4.28),287(3.31), 305(1.76)	78-0591-86
$C_{24}H_{29}NO_8$			
1,2-Benzenedimethanol, 3,4-dimethoxy-α^1-(5,6,7,8-tetrahydro-4-methoxy-6-methyl-1,3-dioxolo[4,5-g]isoquinolin-5-yl)-, α^1-acetate (papaveroxinoline)	MeOH	234(4.18),281(3.56)	102-2403-86
$C_{24}H_{30}F_2O_5S$			
Androsta-1,4-diene-17-carbothioic acid, 17-acetoxy-6,9-difluoro-11-hydroxy-16-methyl-3-oxo-, S-methyl ester, (6α,11β,16α,17α)-	MeOH	238(4.30)	44-2315-86
Androsta-1,4-diene-17-carbothioic acid, 6,9-difluoro-11-hydroxy-16,17-[(1-methylethylidene)bis(oxy)]-3-oxo-, S-methyl ester, (6α,11β,16α,17α)-	MeOH	238(4.31)	44-2315-86
Androsta-1,4-diene-17-carboxylic acid, 6,9-difluoro-11-hydroxy-17-[1-hydroxy-1-(methylthio)ethoxy]-16-	MeOH	237(4.22)	44-2315-86

Compound	Solvent	$\lambda_{max}(\log \epsilon)$	Ref.
methyl-3-oxo-, γ-lactone, (6α,11β,16α,17α)- (cont.)			44-2315-86
$C_{24}H_{30}F_2O_6$			
Androsta-1,4-diene-17-carboxylic acid, 6,9-difluoro-11-hydroxy-17-(1-hydroxy-1-methoxyethoxy)-16-methyl-3-oxo-, γ-lactone, [6α,11β,16α-17α(R)]-	MeOH	238(4.21)	44-2315-86
Androsta-1,4-diene-17-carboxylic acid, 6,9-difluoro-11-hydroxy-16-methyl-3-oxo-17-(1-oxopropoxy)-, (6α,11β,16α-17α)-	MeOH	238(4.23)	44-2315-86
$C_{24}H_{30}N_2O_5$			
Aspidospermidine-3-carboxylic acid, 2,3-didehydro-21-ethoxy-17-methoxy-21-oxo-, methyl ester, (±)-	EtOH	206(4.77),227(4.65), 288(4.36),331(4.68)	78-6719-86
Erichsonine, di-O-acetyl-	EtOH	238(4.11),319(4.25)	102-0969-86
$C_{24}H_{30}N_4$			
1H-Pyrrolo[2,3-b]pyridine-5-carbonitrile, 6-(dibutylamino)-4-methyl-1-(phenylmethyl)-	$CHCl_3$	244(0.93),272(1.05), 354(0.28)	103-0076-86
$C_{24}H_{30}N_4O_{11}$			
Acetamide, N-acetyl-N-[4,7-dihydro-3-(methoxymethyl)-5-methyl-4-oxo-7-(2,3,5-tri-O-acetyl-β-D-arabinofuranosyl)-3H-pyrrolo[2,3-d]pyrimidin-2-yl]-	MeCN	305(3.94)	18-1915-86
$C_{24}H_{30}O_3$			
Cyclohepta[de]naphthalen-7(8H)-one, 8,8-bis(3-methylbutoxy)-	$CHCl_3$	320(3.92),336s(3.84), 356s(3.68)	39-1965-86C
$C_{24}H_{30}O_4$			
1-Cyclohexene-1-carboxylic acid, 3-oxo-4,7,7-trimethyl-3-(phenylmethoxy)bicyclo[2.2.1]hept-2-yl ester, [1S-(exo,exo)]-	hexane	206(4.13),232(4.09), 355(1.45)	78-3547-86
$C_{24}H_{30}O_7$			
Phenanthro[3,2-b]furan-1,2,3,4a(2H)-tetrol, 1,3,4,5,6,11b-hexahydro-4,4,7,11b-tetramethyl-, 1,2-diacetate	EtOH	252(3.96),282(3.38), 293(3.36)	100-0913-86
$C_{24}H_{30}O_8$			
Montafrusin B diacetate	MeOH	212(4.43)	102-0695-86
Semiatrin diacetate	MeOH	208(4.65)	102-1484-86
$C_{24}H_{30}O_{13}$			
Isosweroside tetraacetate	MeOH	205(3.33)	33-1113-86
$C_{24}H_{31}BrO_4$			
1,3-Azulenedicarboxylic acid, 6-bromo-2-octyl-, diethyl ester	CH_2Cl_2	236.5(4.48),270.5(4.23), 318.5(4.85),363.5(3.95), 380.5(3.44),508(2.58)	24-2272-86

Compound	Solvent	$\lambda_{max}(\log \epsilon)$	Ref.
$C_{24}H_{31}BrO_5$			
1,3-Azulenedicarboxylic acid, 6-bromo-2-(octyloxy)-, diethyl ester	CH_2Cl_2	236.5(4.41),267(4.18), 319(4.82),351.5(3.91), 363.5(3.90),477(2.51)	24-2272-86
$C_{24}H_{31}ClO_6$			
Benzoic acid, 3-chloro-6-hydroxy-4-methoxy-2-methyl-, 2,4,4a,5,6,7-7a,7b-octahydro-4-hydroxy-3-(hydroxymethyl)-6,6,7b-trimethyl-1H-cyclobut[e]inden-2-yl ester (arnamiol)	MeOH	217(4.36),259(4.16), 298(3.93)	100-0111-86
$C_{24}H_{31}ClO_8$			
12-Ptilosarcenol	EtOH	218(3.69)	78-6565-86
$C_{24}H_{31}ClO_9$			
Minabein-2	MeOH	218(3.76)	20-0835-86
Minabein-5	MeOH	220(3.70)	20-0835-86
$C_{24}H_{31}FO_5S$			
Androsta-1,4-diene-17-carbothioic acid, 17-acetoxy-9-fluoro-11-hydroxy-16-methyl-3-oxo-, S-methyl ester, (11β,16β,17α)-	MeOH	293(4.29)	44-2315-86
Androsta-1,4-diene-17-carboxylic acid, 9-fluoro-11-hydroxy-17-[1-hydroxy-1-(methylthio)ethoxy]-16-methyl-3-oxo-, γ-lactone, (11β,16β,17α)-	MeOH	?(4.20)(no wavelength)	44-2315-86
$C_{24}H_{31}FO_6$			
Androsta-1,4-diene-17-carboxylic acid, 9-fluoro-11-hydroxy-17-(1-hydroxy-1-methoxyethoxy)-16-methyl-3-oxo-, γ-lactone, [11β,16β,17α(R)]-	MeOH	238(4.19)	44-2315-86
Androsta-1,4-diene-17-carboxylic acid, 9-fluoro-11-hydroxy-16-methyl-3-oxo-17-(1-oxopropoxy)-, (11β,16α,17α)-	MeOH	239(4.15)	44-2315-86
$C_{24}H_{31}NO_5Si$			
Carbamic acid, 2-[3,8-dihydro-5-methoxy-3-oxophenanthro[4,5-bcd]furan-9b(3aH)-ylethyl]methyl]-, 2-(trimethylsilyl)ethyl ester	EtOH	281(3.94)	44-2594-86
$C_{24}H_{31}N_3OS$			
Butyrylperazine (hydrogen maleate)	MeOH	244.5(4.421),279(4.342), 370(3.263)	106-0571-86
$C_{24}H_{31}N_3O_2S$			
Proketazine (hydrogen maleate)	MeOH	244(4.422),278.5(4.346), 368s(3.286)	106-0571-86
$C_{24}H_{31}N_5$			
1H-Pyrazolo[3,4-b]quinolin-4-amine, N-(3,5-dimethylphenyl)-5,6,7,8-tetrahydro-1-(piperidinylmethyl)-	EtOH	217(4.45),243s(--), 271(3.60),319(4.14)	5-1728-86
$C_{24}H_{32}N_2O_7S$			
Uridine, 5-butyl-5'-O-(methoxymethyl)-	MeOH	243(3.97),276(3.98),	78-4187-86

Compound	Solvent	λ_{max}(log ϵ)	Ref.
2',3'-O-(1-methylethylidene)-6-(phenylthio)- (cont.)		312s(3.60)	78-4187-86
$C_{24}H_{32}N_4$			
1H-Pyrrolo[2,3-b]pyridine-5-carbonitrile, 6-(dibutylamino)-2,3-dihydro-4-methyl-1-(phenylmethyl)-	$CHCl_3$	259(0.76),298(0.51), 348(0.75)	103-0076-86
$C_{24}H_{32}O_3$			
Androst-4-en-3-one, 17β-(3-ethoxy-1-oxo-2-propynyl)-	EtOH	238(4.39)	69-7288-86
7-Oxabicyclo[4.1.0]hept-3-en-2-one, 3,5-bis(1,1-dimethylethyl)-5-hydroxy-1-(2-methyl-1-phenyl-1-propenyl)-	C_6H_{12}	207(4.12),230(3.87)	44-2257-86
$C_{24}H_{32}O_4$			
Oxiranebutanoic acid, 3-[7-(2-butoxyphenyl)-1,3,5-heptatrienyl]-, methyl ester, [2α,3β(1E,3E,5Z)]-(±)-	EtOH	272(4.41),281(4.48), 292(4.35)	104-1457-86
4H-Phenalene-1,2-diol, 5,6,6a,7,8,9-hexahydro-3,6,9-trimethyl-4-(2-methyl-1-propenyl)-, diacetate	MeOH	222(4.08),230(3.87), 260s(3.46)	44-5140-86
$C_{24}H_{32}O_6$			
Benzoic acid, 2-hydroxy-4-methoxy-6-methyl-, 2,4,4a,5,6,7,7a,7b-octahydro-4-hydroxy-3-(hydroxymethyl)-6,6,7b-trimethyl-1H-cyclobut[e]inden-2-yl ester (armillyl everninate)	MeOH	211(4.33),261(4.04), 298(3.78)	100-0111-86
$C_{24}H_{32}O_7$			
Melleolide B	EtOH	264(4.16),303(3.79)	102-0471-86
$C_{24}H_{32}O_8$			
Dibenz[b,n][1,4,7,10,13,16,19,22]octaoxacyclotetracosin, 6,7,9,10,12,13-20,21,23,24,26,27-dodecahydro-	MeOH	276(3.64)	90-0877-86
Melleolide C	EtOH	264(4.17),303(3.80)	102-0471-86
$C_{24}H_{32}O_9$			
Benzo[4,5]cyclodeca[1,2-b]furan-2,11(1H,3aH)-dione, 8,13-diacetoxy-6,7,8,8a,12,12a,13,13a-octahydro-12,13a-dihydroxy-1,5,8a,12-tetramethyl- (minabein-8)	MeOH	218(3.81)	20-0835-86
Renillafoulin A	MeOH	203(4.18),220(3.98)	44-4450-86
$C_{24}H_{32}O_{11}$			
Vernojalcanolide 8-O-methacrylate	EtOH	215(4.20)	64-1119-86C
$C_{24}H_{33}NO_6$			
Carbamic acid, [2-[2-[2-(3,4-dimethoxyphenyl)ethyl]-4,5-dimethoxyphenyl]ethyl]methyl-, ethyl ester	MeOH	207(4.50),227(4.27), 279(3.85)	83-0694-86
2,4(1H,3H)-Pyridinedione, 1-benzoyl-3,5,6-tris(1,1-dimethylethoxy)-	C_6H_{12}	208(4.43),240(4.14), 337(4.12)	39-0961-86B
$C_{24}H_{33}N_3O_2S$			
Dixyrazine	MeOH	255.5(4.808),308(3.656)	106-0571-86

Compound	Solvent	$\lambda_{max}(\log \epsilon)$	Ref.
$C_{24}H_{33}PS_3$			
Phosphine, tris[5-(1,1-dimethylethyl)-2-thienyl]-	CH_2Cl_2	246(4.10),285(4.00)	70-1919-86
$C_{24}H_{34}O_3$			
Androst-4-en-3-one, 17-[(3-ethoxy-1-hydroxy-2-propynyl]-, [17β(R)]-	EtOH	241.4(4.17)	69-7288-86
[17β(S)]-	EtOH	240.6(4.20)	69-7288-86
Chola-4,6-dien-24-oic acid, 3-oxo-	EtOH	284(4.48)	94-1929-86
$C_{24}H_{34}O_4$			
4H-Pyran-4-one, 2-(10-hydroxy-3,7,9,11-tetramethyl-2,5,7,11-tridecatetraenyl)-6-methoxy-3-methyl- (actinopyrone B)	MeOH	239(4.59)	158-0038-86
$C_{24}H_{34}O_5$			
4aH-Dibenzo[a,d]cycloheptene-4a,5,7-triol, 1,2,3,4,5,10,11,11a-octahydro-1,1-dimethyl-8-(1-methylethyl)-, 5,7-diacetate, [4aR-(4aα,5α,11aβ)]-	MeOH	215(4.06),265(3.11)	78-0529-86
$C_{24}H_{34}O_6$			
Gitoxigenin 3β-formate	MeOH	220(4.17)	87-0997-86
Gitoxigenin 16β-formate	MeOH	218(4.17)	87-0997-86
$C_{24}H_{34}O_8$			
Dendrillol-2	hexane	220(2.54)	12-1643-86
$C_{24}H_{34}O_{11}$			
Schizonepetoside C tetraacetate	EtOH	246(3.81)	94-3097-86
$C_{24}H_{35}N_2S$			
10H-Phenothiazine-10-hexanaminium, N,N,N-triethyl-, chloride	80% MeCN	253(4.57),306(3.67)	46-2469-86
$C_{24}H_{36}F_2N_2O$			
Chol-5-en-3-ol, 24-azi-23,23-difluoro-, (3β)-	EtOH	306(2.2),314(2.2), 319(--)	94-0931-86
$C_{24}H_{36}N_2O_2$			
Cyanamide, [(3β)-3-(methoxymethoxy)-pregn-5-en-20-ylidene]-	EtOH	223(3.93)	118-1055-86
$C_{24}H_{36}N_4O_2S$			
2,6-Diazabicyclo[2.2.2]oct-7-ene-3,5-dione, 7,8-bis(diethylamino)-1-(ethylthio)-2,6-dimethyl-4-phenyl-	MeCN	224(4.48),253s(4.15), 393(3.82)	24-1315-86
$C_{24}H_{36}O_3$			
Dibenzo[1,7:3,4]cyclohepta[1,2-d]-1,3-dioxole, 4,5,6,7,7a,8,9,13b-octahydro-12-methoxy-2,2,7,7-tetramethyl-11-(1-methylethyl)-, [7aS-(3aS*,7aα-13bα)]-	MeOH	228(3.94),278(3.48)	78-0529-86
$C_{24}H_{36}O_4$			
Chol-4-en-24-oic acid, 7α-hydroxy-3-oxo-	EtOH	242(4.20)	94-1929-86
21-Norchol-20(22)-en-24-oic acid,	EtOH	201(3.953)	48-0695-86

Compound	Solvent	λ_{max}(log ε)	Ref.
3,14-dihydroxy-22-(hydroxymethyl)-, γ-lactone, (3β,5β,14β)-			48-0695-86
21-Norchol-23-en-24-oic acid, 3,14-dihydroxy-22-(hydroxymethyl)-, γ-lactone, (3β,5β,14β)-	EtOH	211.5(4.239)	48-0695-86
$C_{24}H_{36}O_5$			
5α-Pregn-8-ene-11,20-dione, 3,7,15-trihydroxy-4,4,14-trimethyl-, (3β,7β,15α)- (lucidone C)	MeOH	255(3.89)	163-0809-86
$C_{24}H_{36}O_6$			
1,2,3-Benzenetricarboxylic acid, 4,5,6-tris(1,1-dimethylethyl)-, trimethyl ester	C_6H_{12}	253(3.85),302(3.43)	24-1111-86
Bicyclo[2.2.0]hexa-2,5-diene-1,2,3-tricarboxylic acid, 4,5,6-tris(1,1-dimethylethyl)-, trimethyl ester	C_6H_{12}	212(4.15),240s(3.74)	24-1111-86
18-Nor-16-oxaandrostane-8-carboxaldehyde, 15-hydroxy-4,4-dimethyl-17-oxo-7-(1-oxobutoxy)-, cyclic 8,15-hemiacetal, [5α,7α,8R),15β]- (aplyroseol-1)	EtOH	226(2.48)	12-1629-86
18-Nor-16-oxaandrostan-17-one, 7-acetoxy-8-(acetoxymethyl)-4,4-dimethyl-, (5α,7α,13α)-	EtOH	230(2.48)	12-1629-86
$C_{24}H_{36}O_7$			
18-Nor-16-oxaandrostane-8-carboxaldehyde, 6,15-dihydroxy-4,4-dimethyl-17-oxo-7-(1-oxobutoxy)-, cyclic 8,15-hemiacetal (aplyroseol-3)	EtOH	235(2.48)	12-1629-86
18-Nor-16-oxaandrostane-8-carboxaldehyde, 7,15-dihydroxy-4,4-dimethyl-17-oxo-6-(1-oxobutoxy)-, cyclic 8,15-hemiacetal, [5α,6α,7α,8(R),13α-15β)]- (aplyroseol-5)	EtOH	222(2.30)	12-1629-86
$C_{24}H_{37}NS_2$			
Thiazole, 2-(pentadecylthio)-4-phenyl-	EtOH	228(3.99),246(4.15), 273(3.95)	39-0039-86C
$C_{24}H_{38}N_2$			
4,4'-Bipyridinium, 1,1'-diheptyl-, dibromide	pH 9.5	<u>265(4.4)</u>	18-2032-86
compd. with cyclodextrin	pH 9.5	<u>266(4.4)</u>	18-2032-86
$C_{24}H_{39}NOSi$			
Bicyclo[4.4.1]undeca-4,6,8,10-tetraene-2-carboxamide, 3-(1,1-dimethylethyl)-N,N-diethyl-2-methyl-10-(trimethylsilyl)-	MeOH	260(4.67),318(3.99)	33-1597-86
$C_{24}H_{39}N_3O_8Si_2$			
1,2,4-Triazine-3,5(2H,4H)-dione, 6-(3-furanyl)-2-[3,5-O-[1,1,3,3-tetrakis(1-methylethyl)disiloxane-1,3-diyl]-β-D-ribofuranosyl]-	EtOH	308(3.88)	87-0809-86
$C_{24}H_{39}N_5O_5Si_2$			
8,2'-Ethenoadenosine, 3',5'-O-(tetra-	MeOH	233(4.00),282s(--),	94-1518-86

Compound	Solvent	λ_{max}(log ε)	Ref.
isopropyldisiloxane-1,3-diyl)- (cont.)		304(4.26)	94-1518-86
$C_{24}H_{40}NO$			
Benzoxazolium, 3-hexadecyl-2-methyl-, iodide	isoPrOH	276(3.61)	4-0209-86
$C_{24}H_{40}NS$			
Benzothiazolium, 3-hexadecyl-2-methyl-, iodide	EtOH	277(3.80)	4-0209-86
$C_{24}H_{40}O_3$			
[1,1':4',1"-Tercyclohexane]-4-carboxylic acid, 4-methyl-3-oxo-4"-propyl-, methyl ester, [trans[trans-(trans)]]-	MeOH	259(3.96)	24-0387-86
$C_{24}H_{40}O_4$			
2H-Pyran-2-one, 3,5-dibutyl-6-(1-butyl-2-oxoheptyl)-4-hydroxy- (elasnin)	EtOH	291(3.89)	44-0268-86
$C_{24}H_{41}BrO_5Si$			
Phenol, 2-bromo-6-[6-[[(1,1-dimethylethyl)dimethylsilyl]oxy]-3-methyl-2-hexenyl]-5-methoxy-3-[(methoxymethoxy)methyl]-4-methyl-, (E)-	isooctane	288(3.50)	35-0806-86
$C_{24}H_{41}N_5O_4Si_2$			
8,2'-Ethanoadenosine, 3',5'-O-(tetraisopropyldisiloxane-1,3-diyl)-2'-deoxy-	MeOH	261.8(4.23)	94-0015-86
$C_{24}H_{41}N_5O_6Si_2$			
8,2'-Ethanoadenosine, 3',5'-di-O-(tetraisopropyldisiloxane-1,3-diyl)-6'-hydroxy-	MeOH	261.3(4.24)	94-1518-86
$C_{24}H_{42}O_4$			
2H-Pyran-2-one, 3,5-dibutyl-6-(1-butyl-2-hydroxyheptyl)-4-hydroxy- (dihydroelasnin)	MeOH	293(3.91)	44-0268-86
$C_{24}H_{42}O_5Si$			
Phenol, 2-[6-[[(1,1-dimethylethyl)-dimethylsilyl]oxy]-3-methyl-2-hexenyl]-3-methoxy-5-[(methoxymethoxy)-methyl]-4-methyl-	isooctane	281(3.26)	35-0806-86
$C_{24}H_{43}N_2O_9PSi_2$			
Phosphonic acid, [[6,6a,13a,13b-tetrahydro-2,2,4,4-tetrakis(1-methylethyl)-11-oxo-7aH,11H-1,3,5,2,4-trioxadisilocino[6",7":4',5']furo-[2',3':4,5]oxazolo[3,2-a]pyrimidin-7a-yl]methyl]-, dimethyl ester	EtOH	226(3.95),248(3.89)	142-0617-86
$C_{24}H_{44}Si_6$			
1,2,3,10,11,12-Hexasila[3.3]paracyclophane, 1,1,2,2,3,3,10,10,11,11,12,12-dodecamethyl-	hexane	223(4.4),247(4.4)	138-1781-86

Compound	Solvent	λ_{max}(log ε)	Ref.
$C_{24}H_{46}N_4O_8S_2$			
6,9,12,15,23,26,31,34-Octaoxa-1,3,18,20-tetraazabicyclo[18.8.8]-hexatriacontane-2,19-dithione	EtOH	200(4.55),246.3(4.56)	104-1589-86
$C_{24}H_{54}Si_2$			
Disilane, hexakis(1,1-dimethylethyl)-	n.s.g.	198(c.3.9)	89-0079-86
$C_{24}H_{56}Si_4$			
Cyclotetrasilane, octakis(1-methylethyl)-	C_6H_{12}	290(2.30)	138-1643-86
$C_{24}H_{72}Si_{12}$			
Cyclotetrasilane, octakis(trimethylsilyl)-	C_6H_{12}	300s(3.11)	138-1643-86
$C_{24}H_{72}Si_{13}$			
Cyclohexasilane, 1,1'-(dimethylsilylene)bis[1,2,2,3,3,4,4,5,5,6,6-undecamethyl-	n.s.g.	250(3.34),279(3.68)	101-0001-86Q

Compound	Solvent	$\lambda_{max}(\log \epsilon)$	Ref.
$C_{25}H_{14}N_4O_5$			
Acetamide, N-[12-(2-cyano-4-nitrophenyl)-6,11-dihydro-6,11-dioxobenzo[f]pyrido[1,2-a]indol-2-yl]-	DMF	528(3.6532)	2-1126-86
$C_{24}H_{14}N_5S_2$			
2H-Pyrrole, 2-(4-phenyl-1,3-dithiol-2-ylidene)-, conjugate monoacid salt with 2,2'-(2,5-cyclohexadiene-1,4-diylidene)bis(propanedinitrile)-	MeCN	240(4.22),373(4.49), 394(4.71),422(4.48), 442(4.55),664(3.56), 680(3.63),725(3.92), 743(4.12),760(4.05), 820(4.17),845(4.37)	78-0849-86
2H-Pyrrole, 2-(5-phenyl-3H-1,2-dithiol-3-ylidene)-, conjugate monoacid salt with 2,2'-(2,5-cyclohexadiene-1,4-diylidene)bis(propanedinitrile)	MeCN	233(4.06),272(3.82), 376(4.47),394(4.62), 420(4.39),438(4.39), 455(4.32),495(4.06), 530(3.85),662(3.52), 680(3.61),725(3.89), 742(4.05),761(3.97), 820(4.12),845(4.28)	78-0849-86
$C_{25}H_{14}N_8$			
9,10-Ethenoanthracene-2,2,3,3,6,6,7,7-(1H,4H)-octacarbonitrile, 5,8,9,10-tetrahydro-11-methyl-	MeCN	236(3.65)	33-1310-86
10,9-Propenoanthracene-2,2,3,3,6,6,7,7-(1H,4H)-octacarbonitrile, 5,8,9,10-tetrahydro-	EtOH	203(3.85)(end abs.)	33-1310-86
$C_{25}H_{15}BCl_2O_4$			
Boron, [1,3-bis(4-chlorophenyl)-1,3-propanedionato-O,O'][1,8-naphthalenediolato(2-)-O,O']-, (T-4)-	DMF	313s(4.16),318s(4.25), 327s(4.36),332s(4.44), 342(4.51),354s(4.41), 377s(4.20),475s(1.97)	48-0755-86
Boron, [1,3-bis(4-chlorophenyl)-1,3-propanedionato-O,O'][2,3-naphthalenediolato(2-)-O,O']-, (T-4)-	CH_2Cl_2	362(4.36),376s(4.26), 398s(3.87),446(2.25)	48-0755-86
$C_{25}H_{15}N_3$			
3,5-Pyridinedicarbonitrile, 2,4,6-triphenyl-	MeCN	274(4.60)	73-1061-86
$C_{25}H_{15}N_3O_2$			
1H-Naphtho[2,3-h]pyrazolo[3,4-b]quinoline-7,12-dione, 3-methyl-1-phenyl-	DMSO	220(4.447),450(4.176)	2-0616-86
$C_{25}H_{16}ClNO_4S$			
2H,5H-[1]Benzothiopyrano[3,4-e]-1,3-oxazin-5-one, 9-chloro-2,4-di-2-furanyl-3,4-dihydro-3-phenyl-	MeOH	238(4.2),331(3.9)	42-0323-86
$C_{25}H_{16}F_{10}N_4Pt$			
Platinum, [N,N'-bis(pentafluorophenyl)-1,3-propanediaminato(2-)-N,N']-bis(pyridine)-, (SP-4-2)-	EtOH	350(3.90)	12-2013-86
$C_{25}H_{16}N_4$			
Bicyclo[3.2.2]nona-2,8-diene-6,6,7,7-tetracarbonitrile, 4-bicyclo[5.4.1]dodeca-2,5,7,9,11-pentaen-4-ylidene, anti	EtOH	231(4.44),320(4.45), 378(4.05)	88-0485-86

Compound	Solvent	$\lambda_{max}(\log \epsilon)$	Ref.
syn (cont.)	EtOH	231(4.43),318(4.41), 372(4.00)	88-0485-86
9,14-Methanocyclohepta[3,4]benzo[1,2]-cycloundecene-6,6,7,7-tetracarbonitrile, 5a,7a-dihydro-	EtOH	265(4.23),324(4.17), 390(3.99)	88-0485-86
$C_{25}H_{16}N_5S_4$			
3H-Indole, 3-[4,5-bis(methylthio)-1,3-dithiol-2-ylidene)-, conjugate monoacid salt with 2,2'-(2,5-cyclohexadiene-1,4-diylidene)bis(propanedinitrile)	MeCN	216(4.75),252(4.27), 281(4.27),394(4.52), 406(4.53),421(4.54), 436(4.50),476(4.56), 665(3.07),679(3.94), 695(3.92),726(4.20), 742(4.41),760(4.34), 820(4.66),845(4.66)	78-0849-86
$C_{25}H_{16}N_6O_3$			
Methanone, [3,5-bis(2H-benzotriazol-2-yl)-2,4-dihydroxyphenyl]phenyl-	$CHCl_3$	246(4.33),285(4.55), 327(4.47),350s(4.26)	49-0805-86
$C_{25}H_{16}O_2$			
13,15-Methanodibenzo[d,d']benzo[1,2-a:5,4-a']dicycloheptene-6,17(13aH)-dione, 14,14a-dihydro-	EtOH	317(4.60)	77-0444-86
$C_{25}H_{17}BO_4$			
Boron, (1,3-diphenyl-1,3-propanedionato-O,O')[1,8-naphthalenediolato(2-)-O,O']-, (T-4)-	DMF	317s(4.28),328s(4.37), 333s(4.44),343(4.47), 354s(4.36),369s(4.32), 454s(2.25)	48-0755-86
Boron, (1,3-diphenyl-1,3-propanedionato-O,O')[2,3-naphthalenediolato(2-)-O,O']-, (T-4)-	CH_2Cl_2	372(4.42),389(4.34), 440s(2.74)	48-0755-86
$C_{25}H_{17}N$			
Benz[g]isoquinoline, 5,10-diphenyl-	$CHCl_3$	267.0(3.87),389.7(3.81), 408.7(3.81)	83-0886-86
$C_{25}H_{17}NO$			
4H-Benzo[a]quinolizin-4-one, 2,3-diphenyl-	MeCN	307(4.05),320(3.95), 403(4.16)	118-0908-86
$C_{25}H_{17}NO_4S$			
2H,5H-[1]Benzothiopyrano[3,4-e]-1,3-oxazin-5-one, 2,4-di-2-furanyl-3,4-dihydro-N-phenyl-	MeOH	239(4.2),325(3.6)	42-0323-86
$C_{25}H_{17}N_3O_3$			
9,10-Anthracenedione, 1-amino-2-[(1,5-dihydro-3-methyl-5-oxo-1-phenyl-4H-pyrazol-4-ylidene)methyl]-	DMSO	360(4.634),470(4.462)	2-0616-86
$C_{25}H_{18}ClNO$			
Benzo[h]quinolin-2(1H)-one, 4-(3-chlorophenyl)-5,6-dihydro-3-phenyl-	EtOH	235(4.90),260(3.68), 366(3.81)	4-1789-86
$C_{25}H_{18}ClNO_2S$			
2H-[1]Benzothiepino[5,4-b]pyran-2-one, 3-chloro-4-(diphenylamino)-5,6-dihydro-	EtOH	239(4.18),277(4.22), 343(4.08)	4-0449-86

Compound	Solvent	$\lambda_{max}(\log \epsilon)$	Ref.
$C_{25}H_{18}ClOP_2RhS_6$			
Rhodium, carbonylchlorobis[tri-(2-thienyl)phosphine-P]-	CH_2Cl_2	247(4.77),281s(4.61), 366(3.58)	70-1919-86
Rhodium, carbonylchlorobis[tri-(3-thienyl)phosphine-P]-	CH_2Cl_2	245(4.59),278s(4.24), 365(3.42)	70-1919-86
$C_{25}H_{18}F_8N_4Pt$			
Platinum, [N,N'-bis(2,3,5,6-tetrafluorophenyl)propane-1,3-diaminato(2-)-N,N']bis-, (SP-4-2)-	EtOH	347(3.88)	12-2013-86
$C_{25}H_{18}N_2$			
3-Pyridinecarbonitrile, 6-[1,1'-biphenyl]-4-yl-2-methyl-4-phenyl-	MeCN	322(4.60)	73-1061-86
$C_{25}H_{18}O_5$			
2,5-Cyclohexadiene-1,4-dione, 2,5-dihydroxy-3-phenyl-6-[(4-phenylmethoxy)phenyl]-	dioxan	263(4.46),353(3.67), 480(2.88)	5-0195-86
2H-Pyran-2,5(6H)-dione, 3-hydroxy-4-phenyl-6-[[(4-phenylmethoxy)phenyl]methylene]-	dioxan	257(4.32),379(4.23)	5-0195-86
$C_{25}H_{19}Cl_2NO$			
Benzo[h]quinolin-2(1H)-one, 4-(3-chlorophenyl)-3-(4-chlorophenyl)-3,4,5,6-tetrahydro-	EtOH	227(4.08),295(3.23)	4-1789-86
Benzo[h]quinolin-2(1H)-one, 3,4-bis-(4-chlorophenyl)-3,4,5,6-tetrahydro-	EtOH	222(3.90),297(3.65)	4-1789-86
$C_{25}H_{19}NO$			
Pyrrolo[2,1-a]isoquinolin-1(10bH)-one, 10b-methyl-2,3-diphenyl-	MeCN	262(4.25),273(4.28), 414(3.94)	118-0899-86
$C_{25}H_{19}NO_2$			
Isoindolo[2,1-b]isoquinolin-7(5H)-one, 5-(2-oxopropyl)-12-phenyl-	MeOH	221(4.33),233(4.33), 248(4.03),371(4.15)	142-0969-86
$C_{25}H_{19}NO_2S$			
Benzo[h]quinolin-2-ol, 5,6-dihydro-3-phenyl-4-(2-thienyl)-, acetate	EtOH	265(4.16),292(3.92), 319(4.15),370(3.29)	4-1789-86
$C_{25}H_{19}NO_3$			
Benzo[h]quinolin-2-ol, 4-(2-furanyl)-5,6-dihydro-3-phenyl-, acetate	EtOH	267(4.25),293(3.99), 321(4.19),372(3.29)	4-1789-86
$C_{25}H_{19}N_3$			
Benzenamine, 2-(5,12-dihydro-7-methylindolo[3,2-a]carbazol-6-yl)-	EtOH	242(4.45),268(4.39), 283(4.34),338(3.52), 353(3.72)	78-5019-86
$C_{25}H_{19}N_3O_2$			
5-Pyrimidinecarbonitrile, 4,6-bis(4-methoxyphenyl)-2-phenyl-	EtOH	223(4.00),328(4.23)	2-1133-86
$C_{25}H_{20}ClNO$			
Benzo[h]quinolin-2(1H)-one, 4-(3-chlorophenyl)-3,4,5,6-tetrahydro-3-phenyl-	EtOH	226(4.16),297(3.59)	4-1789-86
Benzo[h]quinolin-2(1H)-one, 4-(4-	EtOH	227(4.42),298(3.68)	4-1789-86

Compound	Solvent	$\lambda_{max}(\log \epsilon)$	Ref.
chlorophenyl)-3,4,5,6-tetrahydro-3-phenyl- (cont.)			4-1789-86
$C_{25}H_{20}ClNS$			
1H-Pyrrolizine, 6-(4-chlorophenyl)-2,3-dihydro-7-phenyl-5-(phenylthio)-	n.s.g.	207(4.56),242(4.53), 275s(--)	83-0065-86
$C_{25}H_{20}F_5N_2$			
Quinolinium, 4-[3,3,4,4,4-pentafluoro-2-[(1-methyl-4(1H)-quinolinylidene)-methyl]-1-butenyl]-1-methyl-, perchlorate	n.s.g.	696(4.53)	104-0144-86
$C_{25}H_{20}N_2O_5$			
2-Furancarboxylic acid, 4-(2,3-dihydro-1-nitro-2-phenylpyrrolo[2,1-a]isoquinolin-3-yl)-5-methyl-, methyl ester	MeOH	215(3.15),242(3.27), 263s(2.90),347(2.20), 480(2.98)	73-0412-86
$C_{25}H_{20}N_6O$			
2,4(1H,3H)-Pteridinedione, 1-methyl-6,7-diphenyl-, 4-(phenylhydrazone)	pH 0	225(4.46),282(4.13), 393(4.11)	128-0199-86
	pH 5.7	226(4.28),264(4.35), 346(3.76),464(4.02)	128-0199-86
	pH 13	222(4.33),268(4.27), 337(3.80),492(3.97)	128-0199-86
2,4(1H,3H)-Pteridinedione, 3-methyl-6,7-diphenyl-, 4-(phenylhydrazone)	pH 5	262s(4.36),345(3.85), 464(4.06),535(3.63)	128-0199-86
	pH 14	275(4.56),352(3.87), 505(4.17)	128-0199-86
	MeOH	222(--),264(--), 341(--),485(--)	128-0199-86
$C_{25}H_{20}OSi$			
Silane, benzoyltriphenyl-	n.s.g.	257(4.21),424f(2.47)	101-0021-86C
$C_{25}H_{20}O_3$			
2-Oxapentacyclo[$6.4.0.0^{1,4}.0^{3,7}.0^{5,9}$]-dodec-11-ene-6,10-dione, 5,7-dimethyl-3,4-diphenyl-	$CHCl_3$	261(3.84)	12-1537-86
Pentacyclo[$5.4.0.0^{2,6}.0^{3,10}.0^{5,9}$]undecane-4,8,11-trione	MeCN	216(4.11)	12-1537-86
Tricyclo[$6.2.1.0^{2,7}$]undeca-4,9-diene-3,6,11-trione, 1,8-dimethyl-9,10-diphenyl-, endo	MeCN	225s(4.31),263(3.81)	12-1537-86
$C_{25}H_{20}O_4$			
2,12-Dioxahexacyclo[$6.5.0.0^{1,4}.0^{3,7}.0^{5,9}.0^{11,13}$]tridecane-6,10-dione, 5,7-dimethyl-3,4-diphenyl-	MeCN	221(4.06),258(3.92)	12-1537-86
5-Oxatetracyclo[$7.2.1.0^{2,8}.0^{4,6}$]dodec-10-ene-3,7,12-trione, 1,9-dimethyl-10,11-diphenyl-, endo,anti	MeCN	218(4.34),238(3.86)	12-1537-86
9-Oxatricyclo[$6.2.2.0^{2,7}$]dodeca-4,11-diene-3,6,10-trione, 1,8-dimethyl-11,12-diphenyl-, exo	MeCN	220(4.32)	12-1537-86
2H,5H-Pyrano[2,3-b][1]benzopyran-5-one, 3,4-dihydro-2-(hydroxymethyl)-2,4-diphenyl-	EtOH	269(4.06),280(4.09), 303(4.00)	32-0491-86
2H,5H-Pyrano[3,2-c][1]benzopyran-5-one, 3,4-dihydro-2-(hydroxy-	EtOH	269(4.06),280(4.09), 303(4.00),317s(--)	32-0501-86

Compound	Solvent	λ_{max}(log ϵ)	Ref.
methyl)-2,4-diphenyl- (cont.)			32-0501-86
$C_{25}H_{20}O_6$			
1,4-Naphthalenedione, 2,2'-(3-methylbutylidene)bis[3-hydroxy-	EtOH	255(4.54),273(4.52), 329(3.67),40?(3.41)	39-0659-86C
$C_{25}H_{20}O_7$			
4H-1-Benzopyran-8-propanoic acid, 5,7-dihydroxy-2-(4-hydroxyphenyl)-4-oxo-β-phenyl-, methyl ester	MeOH	273(4.2),305s(4.0), 327(4.1)	64-0681-86C
	MeOH-AlCl$_3$	280(--),308(--), 349(--),390(--)	64-0681-86C
$C_{25}H_{20}SSi$			
Methanethione, phenyl(triphenylsilyl)-	CCl$_4$	692(1.72)	39-0381-86C
$C_{25}H_{21}NO$			
Benzo[h]quinolin-2(1H)-one, 3,4,5,6-tetrahydro-3,4-diphenyl-	EtOH	225(4.09),295(3.49)	4-1789-86
Pyrrolo[2,1-a]isoquinolin-1(5H)-one, 6,10b-dihydro-10b-methyl-2,3-diphenyl-	MeCN	234(4.15),273(4.07), 354(3.85)	118-0899-86
$C_{25}H_{21}NO_3$			
8H-Dibenzo[a,g]quinolizin-8-one, 5,6-dihydro-2,3-dimethoxy-13-phenyl-(8-oxoprotoberberine)	EtOH	250s(4.24),330(4.31), 364s(4.14)	44-2781-86
$C_{25}H_{21}NS$			
1H-Pyrrolizine, 2,3-dihydro-6,7-diphenyl-5-(phenylthio)-	n.s.g.	204(4.55),246(4.48), 272s(--)	83-0065-86
Pyrrolo[2,1-a]isoquinolin-1(5H)-one, 6,10b-dihydro-10b-methyl-2,3-diphenyl-	MeCN	255(4.12),423(4.12)	118-0899-86
$C_{25}H_{21}N_7O_3$			
Benzamide, N-[2-[(2,6-dicyano-4-nitrophenyl)azo]-5-(diethylamino)phenyl]-	CHCl$_3$	260(4.23),315s(3.76), 430s(3.52),585s(4.63), 622(4.75)	48-0497-86
$C_{25}H_{21}O$			
Pyrylium, 2,6-bis(4-methylphenyl)-4-phenyl-, chloride	EtOH	365(4.44)	86-0274-86
$C_{25}H_{22}BrNS$			
2H-Thiopyran-2-amine, 4-(4-bromophenyl)-N,N-dimethyl-2,6-diphenyl-	dioxan	260(4.42),347(3.73)	48-0567-86
$C_{25}H_{22}Br_2NTe$			
4H-Tellurin, 1,1-dibromo-4-[4-(dimethylimino)-2,5-cyclohexadien-1-ylidene]-1,1-dihydro-2,6-diphenyl-, tetrafluoroborate	CH$_2$Cl$_2$	522(4.26)	157-2250-86
$C_{25}H_{22}ClNS$			
2H-Thiopyran-2-amine, 4-(4-chlorophenyl)-N,N-dimethyl-2,6-diphenyl-	dioxan	258(4.40),347(3.72)	48-0567-86
$C_{25}H_{22}F_9N_4$			
1H-Benzimidazolium, 2-[2-[(1,3-dihydro-1,3-dimethyl-2H-benzimidazol-2-ylidene)methyl]-3,3,4,4,5,5,6,6,6-	n.s.g.	500(4.22)	104-0144-86

Compound	Solvent	λ_{max}(log ε)	Ref.
nonafluoro-1-hexenyl]-1,3-dimethyl-, perchlorate (cont.)			104-0144-86
$C_{25}H_{22}NO$			
1-Benzopyrylium, 4-[2-[4-(dimethylamino)phenyl]ethenyl]-2-phenyl-, perchlorate	CH_2Cl_2	690(5.06)	103-0703-86
	MeCN	676(4.96)	103-0703-86
$C_{25}H_{22}NTe$			
Tellurinium, 4-[4-(dimethylamino)phenyl]-2,6-diphenyl-, tetrafluoroborate	CH_2Cl_2	655(4.82)	157-2250-86
$C_{25}H_{22}N_2$			
1H-Indole, 3,3'-(phenylmethylene)-bis[2-methyl-	EtOH	210(4.68),228(4.73), 284(4.12),292(4.07)	2-1204-86
	EtOH-$HClO_4$	212(4.04),239(4.13), 263(4.17),310(3.86)	2-1204-86
$C_{25}H_{22}N_2O$			
Phenol, 2-[bis(2-methyl-1H-indol-3-yl)methyl]-	EtOH	209(4.58),228(4.69), 282(4.14),291(4.06)	2-1204-86
	EtOH-$HClO_4$	210(4.05),218(4.03), 267(3.85),315(3.65)	2-1204-86
$C_{25}H_{22}N_2O_2$			
Benzo[h]quinolin-2(1H)-one, 3,4,5,6-tetrahydro-3-(4-methoxyphenyl)-4-(3-pyridinyl)-	EtOH	223(4.02),290(4.18)	4-1789-86
$C_{25}H_{22}N_2O_9$			
2,4(1H,3H)-Pyrimidinedione, 1-(2-O-acetyl-3,5-di-O-benzoyl-β-D-xylofuranosyl)-	EtOH	235(4.41),261(4.06)	87-0203-86
$C_{25}H_{22}N_4O_2$			
[1,2,4]Triazolo[5,1-a][2,7]naphthyridine-6-carboxylic acid, 5,9-dihydro-9-methyl-2,5-diphenyl-, ethyl ester	EtOH	213(4.36),253(4.37), 365(3.97)	128-0027-86
$C_{25}H_{22}OS$			
2H-Thiopyran, 2-methoxy-4-(4-methylphenyl)-2,6-diphenyl-	dioxan	255(4.36),330(3.93)	48-0373-86
$C_{25}H_{22}O_2S$			
2H-Thiopyran, 2-methoxy-4-(4-methoxyphenyl)-2,6-diphenyl-	dioxan	258(4.33),314(3.96)	48-0373-86
$C_{25}H_{22}O_3$			
1,4-Methanonaphthalene-5,8,9-trione, 1,4,4a,6,7,8a-hexahydro-1,4-dimethyl-2,3-diphenyl-, (1α,4α,4aα,8aα)-	MeCN	223(4.06),260(3.82)	12-1537-86
2H-2,4,7a-Methenoindeno[7,1-bc]furan-3,5-dione, dihydro-2a,4-dimethyl-2,8-diphenyl-	$CHCl_3$	240(3.78),277(3.61)	12-1537-86
2,5,9-Trioxabicyclo[4.2.1]non-7-ene, 1-methyl-6,7,8-triphenyl-	ether	225(4.19),252(4.03)	78-1345-86
$C_{25}H_{22}O_4$			
4H-1-Benzopyran-4-one, 5,7-dihydroxy-2,3-dimethyl-6,8-bis(phenylmethyl)-	MeOH	210(4.40),254(4.22), 298(3.67)	2-0862-86

Compound	Solvent	λ_{max}(log ϵ)	Ref.
$C_{25}H_{22}O_9$			
Cadensin D	MeOH	242(4.51),253(4.50), 317(4.18),363s(3.77)	100-0095-86
	MeOH-NaOMe	255(--),275s(--), 319(--),360s(--)	100-0095-86
$C_{25}H_{23}NOS$			
2H,6H-[1,3]Thiazino[2,3-a]isoquinoline, 7,11b-dihydro-2-methoxy-3,4-diphenyl-	MeCN	277(3.86),295(3.92)	118-0899-86
$C_{25}H_{23}NO_2$			
2,5-Cyclohexadiene-1,4-dione, 2-[[3-(5H-dibenzo[a,d]cyclohepten-5-yl)-propyl]amino]-5-methyl-	CH_2Cl_2	240s(4.20),281(4.36), 480(3.40)	106-0029-86
2,5-Cyclohexadiene-1,4-dione, 2-[[3-(10,11-dihydro-5H-dibenzo[a,d]cyclo-hepten-5-ylidene)propyl]amino]-5-methyl-	CH_2Cl_2	238s(4.30),272s(4.10), 478(3.32)	106-0029-86
2,5-Cyclohexadiene-1,4-dione, 2-[[3-(10,11-dihydro-5H-dibenzo[a,d]cyclo-hepten-5-ylidene)propyl]amino]-6-methyl-	CH_2Cl_2	275(3.99),477(3.32)	106-0029-86
2,5-Cyclohexadiene-1,4-dione, 2-[(9,10-ethanoanthracene-9(10H)-ylmethyl)methylamino]-5-methyl-	CH_2Cl_2	250(4.04),264s(3.90), 272(3.87),280s(3.80), 485(3.47)	106-0029-86
$C_{25}H_{23}NO_4$			
Propanedioic acid, [(2,3-dihydro-6,7-diphenyl-1H-pyrrolizin-5-yl)methyl-ene]-, dimethyl ester	$CHCl_3$	243(4.36),311s(--), 375(4.39)	83-0234-86
$C_{25}H_{23}NS$			
2H-Thiopyran-2-amine, N,N-dimethyl-2,4,6-triphenyl-	dioxan	251(4.34),349(3.79)	48-0567-86
$C_{25}H_{23}N_3$			
Propanedinitrile, (2-ethyl-7,8-di-hydro-1-methyl-3,5-diphenyl-4(1H)-azocinylidene)-	MeCN	240s(--),310(4.01), 456(4.23)	118-0908-86
$C_{25}H_{23}N_3O$			
2H-1,2,3-Triazole, 4,5-diphenyl-2-[2-(4-pentenyloxy)phenyl]-	EtOH	265(c.4.26)	33-0927-86
$C_{25}H_{23}N_3O_{10}$			
1-Azabicyclo[3.2.0]hept-2-ene-2-carb-oxylic acid, 3-methyl-6-[1-[[(4-nitrophenyl)methoxycarbonyl]oxy]-ethyl]-7-oxo-, (4-nitrophenyl)methyl ester	EtOH	267(4.36)	39-0973-86C
$C_{25}H_{23}N_5$			
1H-Pyrazolo[3,4-b]quinoxalin-7-amine, N,N-diethyl-1,3-diphenyl-	BuOAc	429(4.13),461(3.99)	48-0342-86
$C_{25}H_{23}N_7O_3S$			
Thioguanine-elliptinium adduct	50% HOAc	320(4.59),386(3.86), 450(3.52)	87-1350-86
$C_{25}H_{24}ClN_2$			
2-Propynylium, 3-(4-chlorophenyl)-1,1-	CH_2Cl_2	498(4.53),695(4.95)	138-0329-86

Compound	Solvent	λ_{max}(log ε)	Ref.
bis[4-(dimethylamino)phenyl]-, perchlorate (cont.)			138-0329-86
$C_{25}H_{24}N_2OS$			
Benzeneacetonitrile, α-[(phenylmethyl)(phenylthio)methyl]amino-2-(2-propynyloxy)-	EtOH	253(3.95),340(3.08)	24-0813-86
$C_{25}H_{24}N_2O_3$			
2H-Indol-3-amine, 2-(ethoxyphenylmethylene)-7-methoxy-N-(2-methoxyphenyl)-	MeCN	262(4.61),326(3.84), 340(3.73),400(3.51)	24-2289-86
$C_{25}H_{24}N_3O_2$			
2-Propynylium, 1,1-bis[4-(dimethylamino)phenyl]-3-(4-nitrophenyl)-, perchlorate	CH_2Cl_2	494(4.48),713(4.91)	138-0329-86
$C_{25}H_{24}O_2$			
1,4-Methanonaphthalene-5,9(1H)-dione, 4,4a,6,7,8,8a-hexahydro-1,4-dimethyl-2,3-diphenyl-, endo	MeCN	225(4.18),262(3.94)	12-1537-86
exo	MeCN	218(4.30),255(4.10)	12-1537-86
2-Oxapentacyclo[$6.4.0.0^{1,4}.0^{3,7}.0^{5,9}$]-dodecan-6-one, 5,7-dimethyl-3,4-diphenyl-	CHCl	277(3.69)	12-1537-86
$C_{25}H_{24}O_4S$			
15-Thiaestra-1,3,5(10),8-tetraen-17-one, 3-methoxy-16-(phenylmethylene)-, S,S-dioxide, (14β)-	EtOH	275(4.20),304(4.04), 324(4.07)	104-0107-86
$C_{25}H_{24}O_5$			
4H,8H-Benzo[1,2-b:3,4-b']dipyran-4-one, 5-hydroxy-6-(3-hydroxy-3-methyl-1-butenyl)-8,8-dimethyl-2-phenyl-, (E)- (fulvinervin C)	MeOH	263(4.45),277s(4.38), 301(4.40),356(3.79)	102-1507-86
$C_{25}H_{24}O_8$			
Rehmaglutin A 8-acetate 6,7-dibenzoate	MeOH	229(4.32)	94-1399-86
$C_{25}H_{24}O_{10}$			
6H-Benz[b]indeno[1,2-d]furan-1,3,8,9-tetrol, 5a-ethoxy-5a,10b-dihydro-, tetraacetate	MeOH	277(3.74)	163-0569-86
$C_{25}H_{25}BN_2O_6$			
Boron, [1,7-bis[4-(dimethylamino)phenyl]-1,6-heptatriene-3,5-dionato-O,O'][ethanedioato(2-)-O,O']-	CH_2Cl_2	646(5.13)	97-0399-86
$C_{25}H_{25}BrN_6O_5$			
Benzamide, N-[2-[(2-bromo-4,6-dinitrophenyl)azo]-5-(dipropylamino)phenyl]-	$CHCl_3$	403(3.72),571(4.62)	48-0497-86
$C_{25}H_{25}Cl_2N_3O_2$			
2,5-Cyclohexadiene-1,4-dione, 2,5-dichloro-3-[[3-(10,11-dihydro-5H-dibenzo[b,f]azepin-5-yl)propyl]-methylamino]-	CH_2Cl_2	250s(4.30),275s(4.10), 392(4.11),553(2.69)	106-0832-86

Compound	Solvent	$\lambda_{max}(\log \epsilon)$	Ref.
$C_{25}H_{25}F_3N_3O$			
1H-Indolium, 1-[5-(2,3-dihydro-1H-indol-1-yl)-3-[2-[(trifluoroacetyl)amino]ethyl]-2,4-pentadienylidene]-2,3-dihydro-, chloride	MeOH	522s(4.92),557(5.13)	33-1588-86
$C_{25}H_{25}N$			
Benzenemethanamine, N-(2,2-dimethyl-4,4-diphenyl-3-butenylidene)-	CH_2Cl_2	254(4.31)	150-0631-86M
2H-Benzo[a]quinolizine, 1,3,4,6,7,11b-hexahydro-2,3-diphenyl-	MeCN	265(3.64)	118-0908-86
5H-Indolo[3,2,1-de]acridine, 6,7,7a,8-9,10,11,12-octahydro-8-phenyl-	MeOH	318(4.40)	103-0038-86
$C_{25}H_{25}NO_4$			
Propanedioic acid, [(2,3-dihydro-6,7-diphenyl1H-pyrrolizin-5-yl)methyl]-, dimethyl ester	$CHCl_3$	243(4.43),350s(--), 360(2.79)	83-0234-86
$C_{25}H_{25}NO_5S$			
Benzenesulfonamide, N-[2-(3,8-dihydro-5-methoxy-3-oxophenanthro[4,5-bcd]-furan-9b(3aH)-yl)ethyl]-N,4-dimethyl-, (3aR-cis)-	EtOH	248(3.92)	44-2594-86
$C_{25}H_{25}N_2$			
2-Propynylium, 1,1-bis[4-(dimethylamino)phenyl]-3-phenyl-, perchlorate	CH_2Cl_2	493(4.56),688(5.00)	138-0329-86
$C_{25}H_{25}N_3O_3$			
4H-Imidazo[1,2-a]pyrrolo[3,4-b]indole-3,5,6(2H)-trione, 2-cyclohexyl-1,11b-dihydro-11b-(phenylmethyl)-	EtOH	264(3.88),286s(3.76), 306s(3.50)	118-0731-86
$C_{25}H_{25}N_3O_7$			
1,3-Diazabicyclo[4.2.0]octane-2-carboxylic acid, 4-(2-methoxy-2-oxoethylidene)-8-oxo-7-[(phenoxyacetyl)-amino]-, phenylmethyl ester	EtOH	281(4.21)	39-1077-86C
$C_{25}H_{25}N_5O$			
4(3H)-Quinazolinone, 2-[4-(dimethylamino)phenyl]-3-[[[4-(dimethylamino)phenyl]methylene]amino]-	MeOH	222(3.53),261(3.63), 277(3.68),352(3.79)	18-1575-86
$C_{25}H_{25}N_5O_6$			
L-Glutamic acid, N-[4-[[(2-amino-1,4-dihydro-4-oxo-6-quinazolinyl)methyl]-2-propynylamino]-2-methylbenzoyl]-	pH 13	229.5(4.69),277(4.43)	87-0468-86
$C_{25}H_{25}N_5O_8S$			
1H,4aH-[1,2,4]Triazolo[1',2':1,2]diazeto[3,4-d]azepine-4a,9-dicarboxylic acid, 8-amino-2,3,5,6,7,9a-hexahydro-7-[(4-methylphenyl)sulfonyl]-1,3-dioxo-2-phenyl-, dimethyl ester	$CHCl_3$	228(4.21),288(4.33)	24-2114-86
$C_{25}H_{25}N_7O_7$			
Propanamide, N-[5-[bis(2-acetoxyethyl)-	$CHCl_3$	304(3.77),314s(3.67),	48-0497-86

Compound	Solvent	λ_{max}(log ε)	Ref.
amino]-2-[(2,6-dicyano-4-nitrophenyl)azo]phenyl]- (cont.)		417(3.47),578s(4.64), 608(4.74)	48-0497-86
$C_{25}H_{25}N_7O_8$			
Acetamide, N-[5-[bis(2-acetoxyethyl)-amino]-2-[(2,6-dicyano-4-nitrophenyl)azo]-4-methoxyphenyl]-	$CHCl_3$	267(4.04),289s(3.94), 321(3.55),363(3.47), 473s(3.69),604s(4.60), 649(4.86)	48-0497-86
$C_{25}H_{26}N_2OS$			
2H,11bH-Pyrrolo[2',1':3,4]pyrazino-[2',1':3,4]pyrazino[2,1-b][1,3]-thiazine, 2-ethoxy-6,7-dihydro-11b-methyl-3,4-diphenyl-	MeCN	222(4.36),290(3.90)	118-0899-86
$C_{25}H_{26}N_2O_2$			
2,5-Cyclohexadiene-1,4-dione, 2-[[3-(10,11-dihydro-5H-dibenzo[b,f]azepin-5-yl)propyl]methylamino]-5-methyl-	CH_2Cl_2	253(4.22),270(4.20), 275(4.20),497(3.55)	106-0029-86
2,5-Cyclohexadiene-1,4-dione, 2-[[3-(10,11-dihydro-5H-dibenzo[b,f]azepin-5-yl)propyl]methylamino]-6-methyl-	CH_2Cl_2	252s(4.20),275s(4.10), 498(3.50)	106-0029-86
$C_{25}H_{26}N_2O_2S$			
Benzenesulfonamide, N-[3-(diphenylamino)-1-(1-methylethyl)-2-propenylidene]-4-methyl-, (?,E)-	EtOH	222s(4.21),265s(3.79), 365(4.54)	161-0539-86
$C_{25}H_{26}N_2O_3$			
Ergoline-8-carboxylic acid, 1-benzoyl-9,10-didehydro-2,3-dihydro-6-methyl-, ethyl ester	EtOH	254(4.51),307(3.80)	94-0442-86
$C_{25}H_{26}N_2O_4$			
1H-[3]Benzazecino[9,8,7-abc]carbazole-4-carboxylic acid, 8,9-dimethoxy-2,3,4,5,6,10-hexahydro-	MeOH	262(4.73),283(4.70), 298(4.44),333(3.86), 353(3.70),371(3.72)	12-0001-86
$C_{25}H_{26}N_2O_8$			
Butanedioic acid, mono[2-[[3,4-dihydro-2,4-dioxo-5-[[3-(phenylmethoxy)-phenyl]methyl]-1(2H)-pyrimidinyl]-methoxy]ethyl] ester, monosodium salt	pH 1	267(3.92)	4-1651-86
	pH 11	268(3.86)	4-1651-86
$C_{25}H_{26}N_4$			
2,3'-Bipyridinium, 1',1"'-(1,5-pentanediyl)bis-, dibromide	pH 8.2	210(4.40),229(4.38), 276(4.28)	64-0239-86B
$C_{25}H_{26}O_4$			
1-Propanone, 1-[3-(diphenylmethyl)-2,4,6-trimethoxyphenyl]-	MeOH	258(4.59),281(4.48), 296(4.64)	2-0478-86
$C_{25}H_{26}O_6$			
1,3-Azulenedicarboxylic acid, 6-ethoxy-2-(4-methoxyphenyl)-, diethyl ester	CH_2Cl_2	275.5(4.33),323(4.72), 366(4.14),455.5(2.99)	24-2272-86
2H,6H-Benzo[1,2-b:5,4-b']dipyran-6-one, 7-(2,4-dihydroxyphenyl)-7,8-dihydro-5-hydroxy-2,2-dimethyl-10-(3-methyl-2-butenyl)- (2,3-	MeOH	275(4.72)	100-0670-86

Compound	Solvent	λ_{max}(log ϵ)	Ref.
dihydroauriculatin) (cont.)			100-0670-86
$C_{25}H_{27}Cl_3O_5$			
Napyradiomycin C2	MeOH	206(4.22),271(4.23), 340(3.84)	158-0487-86
	MeOH-HCl	208(4.22),272(4.25), 340(3.85)	158-0487-86
	MeOH-NaOH	209(4.49),259(3.98), 311(4.18),404(4.05)	158-0487-86
$C_{25}H_{27}NO_3Se_2$			
Benzene, 1,1'-[[(4-nitrophenoxy)methylene]bis(seleno)]bis[2,4,6-trimethyl-	hexane	215(4.44),248s(4.31), 300(4.22)	104-0092-86
$C_{25}H_{27}NO_6$			
2,5-Pyrrolidinedione, 1-(4-methoxyphenyl)-3-(1-methylethylidene)-4-[1-(3,4,5-trimethoxyphenyl)ethylidene]-, (E)-	toluene	340(3.89)	39-0315-86C
$C_{25}H_{27}NO_8$			
2-Butenedioic acid, 2-[2,3,3a,4,5,6-hexahydro-3-(2-methoxy-2-oxoethylidene)-3a,11-dimethyl-2-oxofuro-[3,2-j]carbazol-6a(11H)-yl]-, dimethyl ester	n.s.g.	213(4.34),242(4.20), 283(3.42)	150-2782-86M
$C_{25}H_{27}N_5O$			
Benzoic acid, 2-[[[4-(dimethylamino)-phenyl]methylene]amino]-, [[4-(dimethylamino)phenyl]methylene]hydrazide	MeOH	205(3.14),220(3.16), 265(3.42),347(3.28)	18-1575-86
$C_{25}H_{27}N_5O_9$			
L-erythro-Pentos-2-ulose, 1-(acetylphenylhydrazone) 2-[(4-nitrophenyl)-hydrazone], 3,4,5-triacetate	MeOH	277(3.92),404(4.63)	136-0316-86G
$C_{25}H_{27}N_7O_4S$			
1H-Imidazole-1-carbothioic acid, O-[4-(6-amino-9H-purin-9-yl)dihydro-2,2-dimethyl-6-[(phenylmethoxy)-methyl]-4H-cyclopenta-1,3-dioxol-5-yl] ester, (3aα,4α,5β,6α,6aα)-(±)-	MeOH	263(4.20)	44-1287-86
$C_{25}H_{28}$			
Pyrene, 1,3,6-tris(1-methylethyl)-	n.s.g.	246.9(4.7),281.5(4.7), 355.3(4.6)	1-0665-86
$C_{25}H_{28}Cl_2O_5$			
Napyradiomycin B2	MeOH	206(4.11),220s(4.01), 256(4.16),280s(3.96), 312(3.86),362(3.92), 400s(3.53)	158-0487-86
	MeOH-HCl	208(4.04),220s(4.01), 311(3.86),357(3.91)	158-0487-86
	MeOH-NaOH	209(4.53),241(4.10), 272(3.97),309(3.87), 393(4.21)	158-0487-86

Compound	Solvent	λ_{max}(log ϵ)	Ref.
Napyradiomycin Cl	MeOH	205(4.22),274(4.23), 320s(3.78),343(3.82)	158-0487-86
	MeOH-HCl	206(4.20),275(4.25), 345(3.84)	158-0487-86
	MeOH-NaOH	208(4.54),311(4.28), 397(3.99)	158-0487-86
$C_{25}H_{28}CrO_3Si_2$			
Chromium, tricarbonyl[[(10,11,12,13,14-η)-tricyclo[8.2.2.2^{4,7}]hexadeca-2,4,6,8,10,12,13,15-octaene-2,9-diyl]bis[trimethylsilane]]-	CH_2Cl_2	345(3.97)	88-2353-86
$C_{25}H_{28}N_2O$			
1H-Benzimidazole, 2-[5-(1,1-dimethylethoxy)bicyclo[4.4.1]undeca-2,4,6-8,10-pentaen-2-yl)-1,5,6-trimethyl-, (±)-	CH_2Cl_2	264(4.59),348(4.26)	33-0315-86
$C_{25}H_{28}N_2O_9S$			
α-D-Glucopyranose, 2-deoxy-2-[[(1-naphthalenylamino)thioxomethyl]-amino]-, 1,3,4,6-tetraacetate	EtOH	243(4.19)	136-0049-86I
$C_{25}H_{28}N_4$			
2,3'-Bipyridinium, 1',1"'-(1,3-propanediyl)bis[1-methyl-, tetraperchlorate	pH 2.3	206(4.49),271(4.44)	64-0239-86B
4,4'-Bipyridinium, 1,1"-(1,3-propanediyl)bis[1'-methyl-	n.s.g.	261(4.64)	35-3380-86
diradical dication	n.s.g.	534(4.36)	35-3380-86
radical trication	n.s.g.	560s(3.83),620(4.04)	35-3380-86
$C_{25}H_{28}N_6O$			
1H-Indole, 3,3'-[1-ethyl-1,2,5,6-tetrahydro-3-(4-morpholinyl)-1,2,4-triazine-5,6-diyl]bis-, hydriodide	EtOH	222(4.90),282(4.15), 290(4.09)	103-1242-86
$C_{25}H_{28}OSe_2$			
Benzene, 1,1'-[(phenoxymethylene)-bis(seleno)]bis[2,4,6-trimethyl-	hexane	213(4.43),250s(4.27), 299(3.60)	104-0092-86
$C_{25}H_{28}O_3$			
3,5,7,9,11,13,15,17,19,21-Tetracosadecaenoic acid, 23-oxo-, methyl ester	MeOH	290(4.32),440(4.87), 463s(4.85)	5-0741-86
$C_{25}H_{28}O_5$			
9H-Xanthen-9-one, 1-hydroxy-3,7-dimethoxy-2,8-bis(3-methyl-2-butenyl)-	MeOH	242(4.38),264(4.37), 309(4.11),365(3.54)	102-1957-86
$C_{25}H_{28}O_6$			
4H-1-Benzopyran-4-one, 2-[3,4-dihydroxy-2,5-bis(3-methyl-2-butenyl)-phenyl]-2,3-dihydro-5,7-dihydroxy-(sigmoidin A)	MeOH	288(4.08)	39-0033-86C
	MeOH-NaOMe	323(4.32)	39-0033-86C
	MeOH-NaOAc	325(4.27)	39-0033-86C
	MeOH-$AlCl_3$	309(4.19)	39-0033-86C
	+ HCl	309(4.19)	39-0033-86C
4H-1-Benzopyran-4-one, 2-(2,3,4,7,8,9-hexahydro-2,2,9,9-tetramethylbenzo-[2,1-b:3,4-b']dipyran-5-yl)-2,3-dihydro-5,7-dihydroxy-, (S)-	MeOH	229(3.91),289(3.85)	39-0033-86C

Compound	Solvent	λ_{max}(log ϵ)	Ref.
Bicyclo[6.2.2]dodeca-8,10,11-triene-9,10-dicarboxylic acid, 4-(benzoyloxy)-, diethyl ester	EtOH	229(4.34),267s(3.56), 280s(3.38),325(3.05)	24-0297-86
$C_{25}H_{29}BrCl_2O_5$			
Napyradiomycin B3	MeOH	204(4.13),251(4.19), 270s(4.02),298(3.97), 363(3.91),400s(3.59)	158-0487-86
	MeOH-HCl	205(4.09),252(4.27), 273s(4.02),305(3.85), 358(3.88)	158-0487-86
	MeOH-NaOH	207(4.53),245s(3.97), 261(4.03),296(4.15), 383(4.09)	158-0487-86
$C_{25}H_{29}Cl_3O_5$			
Napyradiomycin B1	MeOH	204(4.03),252(4.20), 270s(4.00),300(3.88), 360(3.86),400s(3.29)	158-0487-86
	MeOH-HCl	204(3.99),251(4.24), 270s(4.02),305(3.81), 357(3.85)	158-0487-86
	MeOH-NaOH	206(4.53),245s(3.94), 260(4.00),296(4.13), 383(4.08)	158-0487-86
$C_{25}H_{29}NO_2$			
2H-[1,3]Oxazino[2,3-j]quinoline, 6,7,8,8a,9,10,11,12-octahydro-2-methoxy-3,4-diphenyl-	MeCN	230(4.07),289(3.89)	118-0899-86
$C_{25}H_{29}NO_4$			
9-Acridinecarboxylic acid, 1,2,3,4,5-6,7,8,9,10-decahydro-3,3,6,6,9-pentamethyl-1,8-dioxo-10-phenyl-	EtOH	248(4.39),373(3.97)	2-0347-86
$C_{25}H_{30}$			
Azulene, 6-(4-methylphenyl)-2-octyl-	CH_2Cl_2	238.5(--),304(4.84), 378(4.11),569(2.49), 603.5(2.43)	24-2272-86
$C_{25}H_{30}Cl_2O_5$			
2H-Naphtho[2,3-b]pyran-5,10-dione, 3,4a-dichloro-10a-(3,7-dimethyl-2,6-octadienyl)-3,4,4a,10a-tetrahydro-6,8-dihydroxy-2,2-dimethyl-(napyradiomycin A)	MeOH	205(4.25),251(4.14), 270s(4.04),295s(3.87), 360(3.80),400s(3.24)	158-0487-86
	MeOH-HCl	205(4.23),250(4.18), 270s(4.07),359(3.81)	158-0487-86
	MeOH-NaOH	208(4.54),263(3.99), 298(4.18),385(4.00)	158-0487-86
sodium salt	MeOH	204(4.41),254(4.27), 265s(4.23),298(4.31), 365s(4.08),377(4.12), 400s(4.00)	158-0494-86
$C_{25}H_{30}N_2O_2$			
1,3-Cyclohexanedione, 2-[bis[4-(dimethylamino)phenyl]methylene]-5,5-dimethyl-	MeOH	285(3.99),577(4.76)	24-3276-86
	acetone	497(4.53)	24-3276-86
	$CHCl_3$	276(4.33),526(4.59)	24-3276-86
$C_{25}H_{30}N_3$			
Crystal violet (chloride)	MeCN	310(4.5),590(5.0)	35-0128-86

Compound	Solvent	λ_{max}(log ε)	Ref.
$C_{25}H_{30}N_4OS_2$			
Propanamide, N-2-benzothiazolyl-3-(2-benzothiazolylbutylamino)-N-butyl-	pH 1	210(4.57),260(4.32), 281(4.13),289(4.13)	103-0795-86
	EtOH	224(4.54),278(4.37), 301(4.18)	103-0795-86
$C_{25}H_{30}O_4$			
1,3-Benzenediol, 4-[3-[8-hydroxy-2,2-dimethyl-5-(3-methyl-2-butenyl)-2H-1-benzopyran-6-yl]propyl]- (kazinol M)	EtOH	228(4.52),266s(3.87), 277(4.06),286s(4.00), 325(3.21)	94-2448-86
$C_{25}H_{31}Cl_2FO_4S$			
Androsta-1,4-diene-17-carbothioic acid, 9,11-dichloro-6-fluoro-16-methyl-3-oxo-17-(1-oxopropoxy)-, S-methyl ester, (6α,11β,16β,17α)-	MeOH	237(4.28)	44-2315-86
$C_{25}H_{31}NOSi$			
Bicyclo[4.4.1]undeca-1,3,5,7,9-pentaene-2-carboxamide, N,N-diethyl-3-phenyl-9-(trimethylsilyl)-	MeOH	275(4.67),315(3.99)	33-1851-86
$C_{25}H_{31}NO_4$			
1,3-Azulenedicarboxylic acid, 6-cyano-2-octyl-, diethyl ester	CH_2Cl_2	245.5(4.68),308.5(4.88), 316.5(4.96),366.5(3.93), 566(2.47)	24-2272-86
$C_{25}H_{31}NO_5$			
1,3-Azulenedicarboxylic acid, 6-cyano-2-(octyloxy)-, diethyl ester	CH_2Cl_2	246.5(4.76),318(5.14), 379(4.22),395.5(4.07), 526(2.30)	24-2272-86
$C_{25}H_{31}N_3O$			
2(1H)-Pyridinone, 4,5-bis(diethylamino)-1,3-diphenyl-	MeCN	200(4.42),242(4.24), 326(4.02)	24-1315-86
$C_{25}H_{31}N_3O_5$			
3H-Phenoxazine-1,9-dicarboxamide, N,N'-diethyl-7-methoxy-4,6-dimethyl-3-oxo-, acetone adduct hydrate	EtOH	227(4.62),253s(4.28), 438(3.60),475(3.60)	4-0329-86
$C_{25}H_{32}$			
Benzene, [[2,2,3,3-tetramethyl-1-[(pentamethylcyclopropyl)ethynyl]cyclopropyl]ethynyl]-	MeOH	240(4.25),252(4.24)	138-1075-86
$C_{25}H_{32}ClN_5OS$			
Chlorimpiphenine, dihydrochloride	MeOH	214.5(4.338),257(4.573), 311(3.682)	106-0571-86
$C_{25}H_{32}F_2O_5S$			
Androsta-1,4-diene-17-carbothioic acid, 17-acetoxy-6,9-difluoro-11-hydroxy-16-methyl-3-oxo-, S-ethyl ester, (6α,11β,16α,17α)-	MeOH	238(4.29)	44-2315-86
Androsta-1,4-diene-17-carbothioic acid, 6,9-difluoro-11-hydroxy-16,17-[(1-methylethylidene)bis(oxy)]-3-oxo-, S-ethyl ester, (6α,11β,16α-17α)-	MeOH	239(4.30)	44-2315-86

Compound	Solvent	λ_{max}(log ϵ)	Ref.
Androsta-1,4-diene-17-carbothioic acid, 6,9-difluoro-11-hydroxy-16-methyl-3-oxo-17-(1-oxopropoxy)-, S-methyl ester, (6α,11β,16α,17α)-	MeOH	238(4.29)	44-2315-86
$C_{25}H_{32}N_6O_6$			
Benzamide, 4-(1,1-dimethylethyl)-N-[6-(4-morpholinyl)-9-β-D-ribofuranosyl-9H-purin-2-yl]-	pH 7	254(4.45),280s(4.32)	1-0806-86
	pH 2	270(--)	1-0806-86
	pH 13	255(--),280(--)	1-0806-86
$C_{25}H_{32}O_4$			
1,2-Benzenediol, 5-[3-(2,4-dihydroxyphenyl)propyl]-3,4-bis(3-methyl-2-butenyl)- (kazinol F)	EtOH	282(3.91),286s(3.89)	94-1968-86
2(5H)-Furanone, 4-hydroxy-3-methyl-5-[6-methyl-2-methylene-9-(4,5,6,7-tetrahydro-7-methyl-7-benzofuranyl)-6,8-nonadienyl]- (hippospongin)	EtOH	241(4.40)	88-2113-86
$C_{25}H_{32}O_8$			
Eucannabinolide 18,19-O-acetonide	MeOH	205(4.25)	102-0745-86
$C_{25}H_{32}O_{15}$			
6,9-Epoxy-1H-pyrano[3,4-d]oxepin-4-carboxylic acid, 4a,5,6,8,9,9a-hexahydro-1-[(2,3,4,6-tetraacetyl-β-D-glucopyranosyl)oxy]-, methyl ester, (1S,4aS,6R,9S,9aS)-	ether	235(4.04)	5-1413-86
2H-Pyran-2-one, 6-[2,4-bis(β-D-glucopyranosyloxy)-6-methylphenyl]-4-methoxy-	MeOH	205(4.55),224(4.16), 232s(4.10),297(4.00)	100-0800-86
$C_{25}H_{33}FO_5S$			
Androsta-1,4-diene-17-carbothioic acid, 9-fluoro-11-hydroxy-16-methyl-3-oxo-17-(1-oxopropoxy)-, S-methyl ester, (11β,16β,17α)-	MeOH	239(4.28)	44-2315-86
$C_{25}H_{33}NOSi$			
Bicyclo[4.4.1]undeca-4,6,8,10-tetraene-2-carboxamide, N,N-diethyl-3-phenyl-10-(trimethylsilyl)-	MeOH	260(3.75)	33-1597-86
$C_{25}H_{34}N_4$			
7H-1,2,4-Triazolo[4,3-a]perimidine, 10-tridecyl-	dioxan	225(4.29),252(3.85), 286(3.61),297(3.66), 353(3.80)	48-0237-86
$C_{25}H_{34}OSe_2$			
Benzene, 1,1'-[[(cyclohexyloxy)methylene]bis(seleno)]bis[2,4,6-trimethyl-	hexane	215(4.47),243(4.35), 299(3.66)	104-0092-86
$C_{25}H_{34}O_4$			
Elysione	MeOH	255(4.20)	88-2559-86
$C_{25}H_{34}O_6$			
Card-20(22)-enolide, 3-acetoxy-14-hydroxy-16-oxo-, (3β,5β,14β,17α)-	MeOH	216(3.98)	87-0997-86
(3β,5β,14β,17β)-	MeOH	216(4.00)	87-0997-86

Compound	Solvent	λ max (log ε)	Ref.
$C_{25}H_{34}O_7$			
Card-20(22)-enolide, 3,16-bis(formyloxy)-14-hydroxy- (gitoxigenin 3β,16β-diformate)	MeOH	218(4.19)	87-0997-86
$C_{25}H_{34}O_9$			
Renillafoulin B	MeOH	204(4.10),219(3.97)	44-4450-86
$C_{25}H_{34}O_{13}$			
8-Epideoxyloganin tetraacetate, [10-$^2H_3$]-	MeOH	236(4.02)	102-2309-86
$C_{25}H_{35}ClN_4O_6$			
Glycine, N-[N-N-[1-(4-chlorobenzoyl)-L-prolyl]-L-valyl]-2-methylalanyl]-, ethyl ester	MeOH	225(4.11)	23-0277-86
$C_{25}H_{35}N_5O_8$			
4H-Pyrrolo[2,3-d]pyrimidin-4-one, 2-amino-5-[[(3,6a-dihydro-2,2-dimethyl-4H-cyclopenta-1,3-dioxol-4-yl)-amino]methyl]-3-(methoxymethyl)-7-[2,3-O-(1-methylethylidene)-β-D-ribofuranosyl]-, [3aR-(3aα,4α,6aα)]-	MeOH	264(4.00),292(3.85)	78-0207-86
$C_{25}H_{36}BrNO_4$			
Androst-4-en-3-one, 17-acetoxy-4-bromo-2-(4-morpholinyl)-, (2α,17β)-	EtOH	261(4.07)	118-0461-86
(2β,17α)-	EtOH	263.5(4.14)	118-0461-86
$C_{25}H_{36}O_2Si$			
4H-Cyclopropa[3,4]cyclohept[1,2-e]inden-4-one, 1,1a,5,6,7,8,9,9a-octahydro-1,1,2-trimethyl-7-methylene-3-[(trimethylsilyl)oxy]-, (1aR-cis)-	EtOH	206(4.25),228(4.31), 272(4.04),321(3.43)	35-3040-86
6H-Cyclopropa[3,4]cyclohept[1,2-e]inden-6-one, 1,1a,4,5,7,8,9,9a-octahydro-1,1,2-trimethyl-7-methylene-3-[(trimethylsilyl)oxy]-, (1aR-cis)-	EtOH	200(4.03),225(4.29), 269(3.81),317(3.32)	35-3040-86
$C_{25}H_{36}O_4$			
4H-Pyran-4-one, 2-(10-hydroxy-3,7,9,11-tetramethyl-2,5,7,11-tridecatetraenyl)-6-methoxy-3,5-dimethyl- (actinopyrone A)	MeOH	239(4.53)	158-0038-86
$C_{25}H_{36}O_4S$			
Androst-5-en-17-one, 3-[[[(1,1-dimethylethyl)thio]oxoacetyl]oxy]-, (3β)-	EtOH	270(3.62)	39-1603-86C
$C_{25}H_{36}O_5$			
Chol-4-en-24-oic acid, 7α-(formyloxy)-3-oxo-	EtOH	248(4.23)	94-1929-86
$C_{25}H_{36}O_6$			
Card-20(22)-enolide, 3β-acetoxy-14,16-dihydroxy- (gitoxigenin 3β-acetate)	MeOH	220(4.15)	87-0997-86
Card-20(22)-enolide, 16β-acetoxy-3,14-dihydroxy-	MeOH	219(4.11)	87-0997-86

Compound	Solvent	λ_{max}(log ϵ)	Ref.
Pseudopterosin A	MeOH	230(4.05),278(3.31), 283(3.34)	44-5140-86
	MeOH-KOH	247(--),285(--), 295(--)	44-5140-86
$C_{25}H_{37}NO_2S_2$			
2(3H)-Thiazolethione, 3-[(1-oxohexadecyl)oxy]-4-phenyl-	CH_2Cl_2	242(3.88),324(4.04)	39-0039-86C
$C_{25}H_{38}OSi$			
Silane, triethyl[(1a,4,5,6,7,8,9,9a-octahydro-1,1,2-trimethyl-7-methylene-1H-cyclopropa[3,4]cyclohept[1,2-e]inden-3-yl)oxy]-, (1aR-cis)-	EtOH	211(4.38)	35-3040-86
$C_{25}H_{39}NO_2$			
Cycloxobuxoviricine	MeOH	265(3.59)	100-0106-86 +164-0663-86
$C_{25}H_{39}NO_4Si$			
Morphinan-7-one, 3,4,5,6,8,14-hexadehydro-4-[[(1,1-dimethylethyl)dimethylsilyl]oxy]-11,12-dihydro-3,6-dimethoxy-17-methyl-, (±)-	hexane	247(4.2)	89-1025-86
$C_{25}H_{41}NO_4$			
2-Azabicyclo[2.2.0]hepta-2,5-diene-1,6-dicarboxylic acid, 3,4,5-tris-(1,1-dimethylethyl)-, 6-(1,1-dimethylethyl) 1-ethyl ester	hexane	250(3.54)	22-0239-86
2-Azabicyclo[2.2.0]hepta-2,5-diene-3,4-dicarboxylic acid, 1,5,6-tris-(1,1-dimethylethyl)-, 4-(1,1-dimethylethyl) 3-ethyl ester	hexane	236(3.66)	22-0239-86
2,3-Pyridinedicarboxylic acid, 4,5,6-tris(1,1-dimethylethyl)-, 3-(1,1-dimethylethyl) 2-ethyl ester	hexane	233(4.05),265s(3.59), 315(3.32)	22-0239-86
$C_{25}H_{41}N_5$			
4,5,6-Pyrimidinetriamine, N^5-[3,7-dimethyl-9-(1,2,6-trimethyl-2-cyclohexen-1-yl)-2,6-nonadienyl]-N^4-methyl-	MeOH	272(3.92)	18-2495-86
$C_{25}H_{42}O_5Si$			
Labda-12,14-dien-11-one, 7β-acetoxy-8-hydroxy-5β-(trimethylsilyloxy)-, (12E)-	MeCN	270(4.34)	78-5949-86
(12Z)-	MeCN	272(4.11)	78-5949-86
$C_{25}H_{44}O_4$			
1,2-Dioxin-3-ethanol, 3,6-dihydro-6-methoxy-6-(14,16-octadecadienyl)-	MeOH	227(4.46)	44-4260-86
$C_{25}H_{45}NO_2Si_2$			
Silane, (3-nitrobicyclo[4.1.0]hepta-1,3,5-trien-7-ylidene)bis[tris(1-methylethyl)-	CH_2Cl_2	295(4.10),363(3.74)	33-1623-86

Compound	Solvent	$\lambda_{max}(\log \epsilon)$	Ref.
$C_{25}H_{46}Si_2$			
Silane, bicyclo[4.1.0]hepta-1,3,5-trien-7-ylidenebis[tris(1-methylethyl)-	CH_2Cl_2	287(3.80)	33-1623-86
$C_{25}H_{47}NSi_2$			
Bicyclo[4.1.0]hepta-1,3,5-trien-3-amine, 7,7-bis[tris(1-methylethyl)-silyl]-	CH_2Cl_2	274(3.92)	33-1623-86
$C_{25}H_{49}NSi_2$			
Benzenamine, 3-[tris(1-methylethyl)-silyl]-4-[[tris(1-methylethyl)-silyl]methyl]-	CH_2Cl_2	269(3.81)	33-1623-86

Compound	Solvent	λ_{max}(log ϵ)	Ref.
$C_{26}H_{12}$			
Naphtho[2',3':7,8]cyclodeca[1,2,3,4-def]biphenylene, 7,8,15,16-tetradehydro-	C_6H_{12}	233(4.54),241(4.73), 284(4.83),293(4.91), 303(5.27),321(4.17), 331(4.07),351(3.77), 371(3.71),389(3.77), 406(3.78),412(3.77)	44-1088-86
$C_{26}H_{12}N_4O_8$			
Benz[5,6]indolo[2,1-a]isoquinoline-8,13-dione, 14-(2,4,6-trinitrophenyl)-	DMF	462(3.7853)	2-1126-86
$C_{26}H_{13}N_3O_6$			
Benz[5,6]indolo[2,1-a]isoquinoline-8,13-dione, 14-(2,4-dinitrophenyl)-	DMF	468(3.8976)	2-1126-86
Benz[5,6]indolo[1,2-a]quinoline-8,13-dione, 7-(2,4-dinitrophenyl)-	DMF	472(3.9294)	2-1126-86
$C_{26}H_{14}BeO_4$			
Beryllium, bis(1H-phenalene-1,9(9aH)-dionato-O,O')-, (T-4)-	propylene carbonate	243(4.73),260s(4.30), 330s(4.05),347(4.32), 365(4.72),400(3.82), 415s(3.92),425(4.14), 439(4.06),450(4.32)	78-6293-86
$C_{26}H_{14}O_2$			
6,15-Hexacenedione	$CHCl_3$	328(4.3),342(4.4), 371(3.9),407(3.8), 456(3.8)	44-3762-86
$C_{26}H_{15}NO_3$			
7H-Dibenzo[a,h]carbazole-7,12(13H)-dione, 5-hydroxy-13-phenyl-	EtOH	280(4.74),333(3.74), 502(3.88)	104-0748-86
$C_{26}H_{15}N_3O_4$			
Benz[5,6]indolo[2,1-a]isoquinoline-8,13-dione, 14-(4-amino-2-nitrophenyl)-	DMF	530(3.6532)	2-1126-86
$C_{26}H_{15}N_3O_6$			
Benzo[f]pyrido[1,2-a]indole-2-carboxylic acid, 12-(2-cyano-4-nitrophenyl)-6,11-dihydro-6,11-dioxo-	DMF	488(3.8388)	2-1126-86
$C_{26}H_{16}$			
5H-Dibenzo[a,d]cycloheptene, 5-(1H-cyclopropa[b]naphthalen-1-ylidene)-	C_6H_{12}	336(3.82),414(4.29), 473s(3.21)	88-5159-86
	MeCN	340s(3.60),411(4.39), 468s(3.36)	88-5159-86
15H-Dibenzo[a,g]cyclotridecene, 15-(2,4-cyclopentadien-1-ylidene)-5,6,7,8-tetradehydro-, (E,E)-	THF	280s(4.51),294(4.62), 322(4.45),379(4.46), 401s(4.31)	18-1723-86
relative absorbance given	MeCN	277s(0.75),291(1.00), 320(0.71),376(0.70), 397s(0.51)	18-1723-86
$C_{26}H_{16}BrN_5O$			
1,1,2,2-Ethanetetracarbonitrile, 1-[1-(4-bromophenyl)-4,5,6,7-tetra-	MeOH	248(4.41),286(4.19), 468(4.45)	103-0172-86

Compound	Solvent	λ_{max}(log ϵ)	Ref.
hydro-4-oxo-2-phenyl-1H-indol-5-yl]- (cont.)			103-0172-86
$C_{26}H_{16}O_4$			
1,4:8,11-Diepoxypentacene-6,13-dione, 1,2,3,4,8,9,10,11-octahydro-2,3,9,10-tetrakis(methylene)-	dioxan	271(4.80),332(3.69)	44-4160-86
$C_{26}H_{16}S_2$			
1,6-Dithiapyrene, 3,8-diphenyl-	toluene	417(4.09),456(3.92)	35-3460-86
$C_{26}H_{16}S_6$			
2,2':5',2"-Terthiophene, 5,5"'-(1,2-ethenediyl)bis-, (E)-	$CHCl_3$	324(4.13),460(4.84)	142-2639-86
$C_{26}H_{17}BO_5$			
Boron, (1,3-diphenyl-1,3-propanedionato-O,O')[1-hydroxy-2-naphthalenecarboxylato(2-)-O^1, O^2]-, (T-4)-	CH_2Cl_2	287(4.26),296s(4.11), 370(4.62),381s(4.54)	48-0755-86
$C_{26}H_{17}BrN_2OS$			
1,3,4-Oxadiazole, 2-[4-[2-[5-(4-bromophenyl)-2-thienyl]ethenyl]phenyl]-5-phenyl-	toluene	299(4.44),377(4.38)	103-1028-86
$C_{26}H_{17}ClN_2OS$			
1,3,4-Oxadiazole, 2-[4-[2-[5-(4-chlorophenyl)-2-thienyl]ethenyl]phenyl]-5-phenyl-	toluene	298(4.37),381(4.32)	103-1028-86
$C_{26}H_{17}N_3$			
3,5-Pyridinedicarbonitrile, 2-[1,1'-biphenyl]-4-yl-6-methyl-4-phenyl-	MeCN	313(4.49)	73-1061-86
$C_{26}H_{17}N_3O_2$			
7H-2,2a,8-Triazabenzo[cd]cyclopent[g]-azulene-7,9(8H)-dione, 8-methyl-1,10-diphenyl-	$CHCl_3$	320(3.95),335(3.94), 372(4.18),389(4.33), 410(4.30),500(2.78), 540(2.87),585(2.81), 640(2.57)	94-3588-86
$C_{26}H_{17}N_3O_3S$			
1,3,4-Oxadiazole, 2-[4-[2-[5-(4-nitrophenyl)-2-thienyl]ethenyl]phenyl]-5-phenyl-, (E)-	toluene	415(4.54)	103-1028-86
$C_{26}H_{17}N_3O_4$			
9,10-Anthracenedione, 5-nitro-1,4-bis(phenylamino)-	EtOH	624(4.16),661(4.19)	104-0547-86
$C_{26}H_{17}N_5O$			
1,1,2,2-Ethanetetracarbonitrile, 1-(4,5,6,7-tetrahydro-4-oxo-1,2-diphenyl-1H-indol-5-yl)-	MeOH	248(4.44),278(4.23), 474(4.02)	103-0172-86
$C_{26}H_{18}$			
2,2'-Bi-9H-fluorene	dioxan	323(4.76)	46-2666-86
	EtOH	320(--)	46-2666-86
Dibenzo[a,g]cyclooctadecene, 19,20,21,22-tetradehydro-,	isooctane	269s(4.35),284s(4.50), 323s(4.85),342(4.93),	18-2209-86

Compound	Solvent	λmax(log ε)	Ref.
(cont.)		360s(4.52),417(4.17), 435(4.16),460s(3.95)	18-2209-86
$C_{26}H_{18}ClO$			
5H-Indeno[1,2-b]naphtho[2,1-e]pyrylium, 7-(4-chlorophenyl)-6,8-dihydro-, perchlorate	MeCN	226(4.46),230s(4.45), 286(4.20),352(4.15), 442(4.58)	22-0600-86
$C_{26}H_{18}F_{10}N_4Pt$			
Platinum, [N,N'-bis(pentafluorophenyl)-1,2-ethanediaminato(2-)-N,N']-bis(4-methylpyridine)-, (SP-4-2)-	EtOH	354(3.70)	12-2013-86
$C_{26}H_{18}N_2$			
8,8'-Bibenzo[b]quinolizinium	EtOH	308(3.6),370(3.5)	97-0333-86
$C_{26}H_{18}N_2OS$			
1,3,4-Oxadiazole, 2-phenyl-5-[4-[2-(5-phenyl-2-thienyl)ethenyl]phenyl]-, (E)-	toluene	297(4.43),393(4.29)	103-1028-86
$C_{26}H_{18}N_2O_3$			
m-Azoxybenzophenone	EtOH	253(4.40),320s(4.00)	23-2076-86
$C_{26}H_{18}N_2O_3S$			
2,4-Cyclohexadien-1-one, 6-[(2,3-dihydro-1H-isoquino[4,3,2-kl]phenothiazin-4-yl)methylene]-4-nitro-	EtOH	230(4.52),260(4.51), 310(4.28),375(3.88), 400(3.85)	135-1208-86
$C_{26}H_{18}N_4$			
Propanedinitrile, (5,6-dihydro-9,10-diphenyl-8H-pyrido[1,2-a]pyrrolo-[2,1-c]pyrazin-8-ylidene)-	MeCN	275(4.11),337(4.27), 458(4.15)	118-0908-86
$C_{26}H_{18}N_4O_2$			
Benzo[de][1,2,3]triazolo[4,5-g]isoquinoline-4,6(5H,9H)-dione, 5,9-bis(4-methylphenyl)-	DMF	362(4.17)	2-0496-86
2(1H)-Pteridinone, 3,4-dihydro-4-(2-oxo-2-phenylethylidene)-6,7-diphenyl-	pH 3	228(4.54),310(4.28), 422(4.45),440s(4.46)	128-0199-86
	pH 11.6	237(4.49),324(4.30), 463(4.47),486(4.48)	128-0199-86
	MeOH	228(4.23),310(3.96), 422(4.23),440s(4.16)	128-0199-86
$C_{26}H_{18}N_4O_2S_2$			
4,10-Epithio-4H,6H,10H-isothiazolo-[3,4-d][1,2,4]triazolo[1,2-a]pyridazine-6,8(7H)-dione, 7-methyl-3,4,10-triphenyl-	C_6H_{12}	221(4.42),241s(4.21), 247(4.07),252(4.01), 258(4.09),265s(4.17), 274(4.19)	5-1796-86
$C_{26}H_{18}N_4O_3$			
Benzo[de][1,2,3]triazolo[4,5-g]isoquinoline-4,6(5H,9H)-dione, 9-(4-methoxyphenyl)-5-(4-methylphenyl)-	DMF	358(4.22)	2-0496-86
$C_{26}H_{18}N_4O_7$			
Acridinium, 10-methyl-4-phenyl-, picrate	MeOH	206.4(4.825),263.1(4.541), 364.6(4.412)	150-2367-86M

Compound	Solvent	$\lambda_{max}(\log \epsilon)$	Ref.
$C_{26}H_{18}O_2$			
6,10:13,17-Dimethenocyclotetradeca[b]-naphthalene-11,12-dione, 19,20-dimethyl-	EtOH	221(4.45),242(4.14), 264(4.00),320(4.56), 333(4.80),360(3.96), 420(3.08)	77-0444-86
2,4-Pentadienal, 5,5'-(1,3-butadiyne-1,4-diyldi-2,1-phenylene)bis-, (all-E)-	THF	227(4.39),269(4.52), 283s(4.49),300(4.63), 327(4.63),355s(4.48)	18-2209-86
$C_{26}H_{18}O_4$			
7H-Furo[3,2-g][1]benzopyran-7-one, 2-benzoyl-3,5-dimethyl-6-phenyl-	$CHCl_3$	290(4.23),355(4.23)	18-1299-86
7H-Furo[3,2-g][1]benzopyran-7-one, 2-benzoyl-5,9-dimethyl-3-phenyl-	$CHCl_3$	290(4.45),358(4.50)	18-1299-86
$C_{26}H_{18}O_5$			
7H-Furo[3,2-g][1]benzopyran-7-one, 2-(4-hydroxybenzoyl)-3,5-dimethyl-6-phenyl-	$CHCl_3$	290(4.87),358(4.92)	18-1299-86
$C_{26}H_{19}ClN_6OS$			
Propanedinitrile, [2-[[4-chloro-2-[[4-(diethylamino)phenyl]azo]-5-thiazolyl]methylene]-2,3-dihydro-3-oxo-1H-inden-1-ylidene]-	CH_2Cl_2	700(4.83)	77-1639-86
	DMSO	720(--)	77-1639-86
$C_{26}H_{19}Cl_2N$			
Benzenemethanamine, N-[bis(4-chlorophenyl)methylene]-α-phenyl-	EtOH	209(4.59),255(4.32)	39-0799-86C
$C_{26}H_{19}NO_2$			
2H-Naphtho[1,2-b]pyran-2-one, 4-(methylphenylamino)-3-phenyl-	EtOH	273(4.39),279(4.40), 315(4.09),324(4.11), 363(4.07)	4-1067-86
$C_{26}H_{19}N_3O_5$			
9,12-Epoxy-1H-diindolo[1,2,3-fg:3',2'-1'-kl]pyrrolo[3,4-i][1,6]benzodiazocine-10-carboxylic acid, 2,3,9,10-11,12-hexahydro-10-hydroxy-9-methyl-1-oxo-, [9S-(9α,10α,12α)]- (K-252b)	H_2O	232(4.46),246(4.45), 268s(4.46),280s(4.62), 291(4.79),323(4.08), 337(4.20),353(4.11), 371(4.11)	158-1072-86
	0.01M Na_2CO_3	279(4.9),330(4.3), 350(4.3),370(4.3)	158-1066-86
$C_{26}H_{19}O$			
5H-Indeno[1,2-b]naphtho[2,1-e]pyrylium, 6,8-dihydro-7-phenyl-, perchlorate	MeCN	228(4.51),286(4.32), 344(4.16),439(4.63)	22-0600-86
$C_{26}H_{20}$			
Ethene, tetraphenyl-	$CDCl_3$	311(4.2)	77-0974-86
Phenanthrene, 9,10-dihydro-2,7-diphenyl-	EtOH	318(4.63)	46-2666-86
	dioxan	321(4.63)	46-2666-86
$C_{26}H_{20}Br_2OS$			
2-Propanone, 1-[2,6-bis(4-bromophenyl)-4-phenyl-2H-thiopyran-2-yl]-	dioxan	264(4.42),350(3.82)	48-0573-86
$C_{26}H_{20}Cl_2OS$			
2-Propanone, 1-[2,6-bis(4-chlorophenyl)-4-phenyl-2H-thiopyran-2-yl]-	dioxan	263(4.41),348(3.80)	48-0573-86

Compound	Solvent	$\lambda_{max}(\log \epsilon)$	Ref.
$C_{26}H_{20}F_8N_4Pt$			
Platinum, [N,N'-bis(2,3,5,6-tetrafluoro-4-methylphenyl)-1,2-ethanediaminato(2-)-N,N']bis(pyridine)-	EtOH	355(3.64)	12-2013-86
Platinum, [N,N'-bis(2,3,5,6-tetrafluorophenyl)-1,2-ethanediaminato(2-)-N,N']bis(4-methylpyridine)-	EtOH	359(3.72)	12-2013-86
$C_{26}H_{20}N$			
Benz[g]isoquinolinium, 2-methyl-5,10-diphenyl-, iodide	$CHCl_3$	244.3(4.43),269.0(4.56), 449.2(3.83)	83-0886-86
$C_{26}H_{20}N_2O_3$			
2,4-Cyclohexadien-1-one, 4-nitro-6-[(5,7,8,9-tetrahydro-5-phenyl-10-phenanthridinyl)methylene]-	EtOH	240(4.71),310(4.40), 375(4.13),400(4.08), 515(3.60)	135-1208-86
$C_{26}H_{20}N_2O_8$			
Pyrrolo[1,2-a]quinoline-3-carboxylic acid, 1-[5-(methoxycarbonyl)-2-methyl-3-furanyl]-, ethyl ester	MeOH	220(2.20),256(2.21), 315(1.68),433(1.84)	73-0412-86
$C_{26}H_{20}N_4$			
Propanedinitrile, (5,6,9,10-tetrahydro-9,10-diphenyl-8H-pyrido[1,2-a]-pyrrolo[2,1-c]pyrazin-8-ylidene)-	MeCN	275(4.18),307(4.30)	118-0908-86
$C_{26}H_{20}O$			
Ethanone, tetraphenyl-	n.s.g.	251(4.06),329(2.48)	101-0021-86C
$C_{26}H_{20}O_2$			
1H-Naphtho[2,3-c]pyran, 1-methoxy-5,10-diphenyl-	$CHCl_3$	262(4.26),306(4.18), 318(--)	83-0886-86
$C_{26}H_{20}O_3$			
1H-Naphtho[2,3-c]pyran-4(3H)-one, 1-methoxy-5,10-diphenyl-	$CHCl_3$	259(4.68),312(--), 359(3.58)	83-0886-86
$C_{26}H_{20}O_{11}$			
2H-Pyran-2,5(6H)-dione, 3-acetoxy-4-(4-acetoxyphenyl)-6-[(3,4-diacetoxyphenyl)methylene]-	MeOH	219(4.32),279(3.85), 359(3.50)	5-0177-86
$C_{26}H_{21}ClOS$			
2-Propanone, 1-[4-(4-chlorophenyl)-2,6-diphenyl-2H-thiopyran-2-yl]-	dioxan	263(4.43),350(3.71)	48-0573-86
$C_{26}H_{21}N$			
Benzenemethanamine, N-(diphenylmethylene)-α-phenyl-	EtOH	209(4.70),250(4.34)	39-0799-86C
Benzenemethanamine, α-(diphenylmethylene)-N-phenyl-	EtOH	215(4.37),258(4.24), 353(4.15)	22-0781-86
$C_{26}H_{21}NO_2$			
Benzo[h]quinolin-2(1H)-one, 5,6-dihydro-4-(4-methoxyphenyl)-3-phenyl-	EtOH	236(4.02),260(3.70), 364(3.52)	4-1789-86
$C_{26}H_{21}NO_4$			
3,6-Methanonaphtho[2,3-b]oxirene-8-carbonitrile, 1a,2,2a,3,6,6a,7,7a-octahydro-8-hydroxy-3,6-dimethyl-	MeCN	218(4.24)	12-1537-86

Compound	Solvent	λ_{max}(log ϵ)	Ref.
2,7-dioxo-4,5-diphenyl- (cont.)			12-1537-86
$C_{26}H_{21}N_3$			
Benzenamine, 2-(5,12-dihydro-5,12-dimethylindolo[3,2-a]carbazol-6-yl)-	EtOH	242(4.49),287(4.30), 355(3.59),370(3.79)	78-5019-86
Benzenamine, 2-(5,12-dihydro-7-methylindolo[3,2-a]carbazol-6-yl)-N-methyl-	EtOH	243(4.44),260(4.39), 287(4.33),338(3.61), 353(3.72)	78-5019-86
1H-Indene-4,6-dicarbonitrile, 5-amino-2,3-dihydro-7-(4-methylphenyl)-3-[(4-methylphenyl)methylene]-	EtOH	411(4.27)	64-1471-86B
2,3':2',3"-Ter-1H-indole, 1",2"-dimethyl-	EtOH	290(4.06)	78-5019-86
$C_{26}H_{21}O_2$			
Pyrylium, 2-[2-(4-methoxyphenyl)ethenyl]-4,6-diphenyl-, tetrafluoroborate	CH_2Cl_2	355(4.53),530(4.43)	104-1786-86
$C_{26}H_{22}$			
1,3,5,7,13,15,17-Cyclononadecaheptaene-9,11-diyne, 19-(2,4-cyclopentadien-1-ylidene)-8,13-dimethyl-	THF	303(4.54),371(4.68), 460s(4.32)	18-1723-86
relative absorbance	MeCN	302(0.73),369(1.00), 458s(0.41)	18-1723-86
p-Quaterphenyl, 2,2"'-dimethyl-	EtOH	278(4.55)	46-2661-86
p-Quaterphenyl, 3,3"'-dimethyl-	EtOH	298(4.68)	46-2661-86
	dioxan	301(4.67)	46-2661-86
$C_{26}H_{22}Br_2N_4O_6$			
Benzoic acid, 4-bromo-, 5'-ester with 4-bromo-N-[4-cyano-1-[2,3-O-(1-methylidene)-β-D-ribofuranosyl]-1H-imidazol-5-yl]benzamide	MeOH	244(4.56)	23-2109-86
$C_{26}H_{22}Cl_2N_4O_5$			
1H-Pyrazolo[3,4-d]pyrimidine, 4,6-dichloro-1-[2-deoxy-3,5-bis-O-(4-methylbenzoyl)-β-D-erythro-pentofuranosyl]-	MeOH	240(4.55),270(3.90)	5-1213-86
$C_{26}H_{22}Cl_2N_4O_6$			
Benzoic acid, 4-chloro-5'-ester with 4-chloro-N-[4-cyano-1-[2,3-O-(1-methylethylidene)-β-D-ribofuranosyl]-1H-imidazol-5-yl]benzamide	MeOH	239(4.51)	23-2109-86
$C_{26}H_{22}Fe_2$			
Benzene, 1,3-diferrocenyl-	MeOH	280(4.5),475(3.1)	18-3589-86
Benzene, 1,4-diferrocenyl-	MeOH	310(4.0),350s(3.7), 480(3.0)	18-3589-86
$C_{26}H_{22}N_2$			
2,4,6,12,14,16-Cycloheptadecahexaene-8,10-diyn-1-one, 7,12-dimethyl-, 2,4,6-cycloheptatrien-1-ylidenehydrazone, (E,E,Z,Z,E,E)-	THF	222(4.37),264s(4.42), 281s(4.72),298(4.92), 306(4.93),422(3.86)	18-2509-86
relative absorbance given	CF_3COOH	264(0.33),308s(0.58), 326(1.00),358(0.67), 437(0.68),671(0.03)	18-2509-86

Compound	Solvent	$\lambda_{max}(\log \epsilon)$	Ref.
$C_{26}H_{22}N_2O$			
Ethanone, 2-phenoxy-1,2-diphenyl-, phenylhydrazone, (E)-	MeOH	285(4.1),314s(4.00)	48-0181-86
$C_{26}H_{22}N_2S_2$			
Dicyclopenta[a,e]dicyclopropa[c,g]-cyclooctene-2,3-dicarbonitrile, 1,5-bis[(1,1-dimethylethyl)thio]-	CH_2Cl_2	268s(4.45),277(4.49), 310s(4.21),344(4.59), 365(4.37),413(4.25), 520(2.88)	35-7032-86
$C_{26}H_{22}N_6O$			
2,4(1H,3H)-Pteridinedione, 1,3-dimethyl-6,7-diphenyl-, 4-(phenylhydrazone)	pH 0-13	264(4.39),340(3.78), 464(4.10),535(3.75)	128-0199-86
	$CHCl_3$	267(4.49),346(3.93), 496(4.22)	128-0199-86
$C_{26}H_{22}OS$			
2-Propanone, 1-(2,4,6-triphenyl-2H-thiopyran-2-yl)-	dioxan	260(4.38),345(3.72)	48-0573-86
$C_{26}H_{22}O_2$			
p-Quaterphenyl, 2,2"'-dimethoxy-	EtOH	301(4.56)	46-2661-86
	dioxan	303(4.55)	46-2661-86
p-Quaterphenyl, 2',3"-dimethoxy-	EtOH	309(4.46)	46-2661-86
p-Quaterphenyl, 3,3"'-dimethoxy-	dioxan	304(4.65)	46-2661-86
$C_{26}H_{22}O_3$			
1,4-Methanonaphthalene-5,8,9-trione, 1,4,4a,8a-tetrahydro-1,4,6-trimethyl-2,3-diphenyl-, (1α,4α,4aα,8aα)-	MeCN	228(4.39),273s(3.81)	12-1537-86
1H-Naphtho[2,3-c]pyran-4-ol, 3,4-dihydro-1-methoxy-5,10-diphenyl-, cis	$CHCl_3$	238(4.81),294(4.07)	83-0886-86
trans	$CHCl_3$	238(4.79),295(4.03)	83-0886-86
2-Oxapentacyclo[$6.4.0.0^{1,4}.0^{3,7}.0^{5,9}$]-dodec-11-ene-6,10-dione, 5,7,12-trimethyl-3,4-diphenyl-	MeCN	218(4.11),261(3.70)	12-1537-86
Pentacyclo[$5.4.0.0^{2,6}.0^{3,10}.0^{5,9}$]undecane-4,8,11-trione, 1,3,5-trimethyl-2,6-diphenyl-	MeCN	219(4.19)	12-1537-86
$C_{26}H_{22}O_3S$			
2-Propanone, 1-(2,4,6-triphenyl-2H-thiopyran-2-yl)-, S,S-dioxide	dioxan	241(4.35),324(3.76)	48-0573-86
$C_{26}H_{22}O_4$			
1,4:8,11-Diepoxypentacene-6,13-dione, 1,2,3,4,4a,6a,7,8,9,10,11,12,12a,14a-tetradecahydro-2,3,9,19-tetrakis-(methylene)-	dioxan	242(4.45),262(4.16)	44-4160-86
Ethanedione, 1,1'-(2,3,5,6-tetramethyl-1,4-phenylene)bis[2-phenyl-	dioxan	268(5.15),411(3.41)	104-1117-86
$C_{26}H_{22}O_6$			
9(10H)-Anthracenone, 10-acetoxy-10-[(4-acetoxy-3-methoxyphenyl)methyl]-	EtOH	275(4.22)	104-0158-86
$C_{26}H_{22}PS_2$			
Phosphonium, triphenyl[[(phenylthioxomethyl)thio]methyl]-, iodide	CH_2Cl_2	308(4.37),483(2.34)	18-1403-86

Compound	Solvent	λ_{max}(log ε)	Ref.
$C_{26}H_{23}BrN_2O_8$			
Uridine, 3-benzoyl-5-bromo-2',3'-O-(1-methylethylidene)-, 5'-benzoate	MeOH	230(4.29),252(4.27), 280s(4.01)	94-4585-86
$C_{26}H_{23}N$			
Benz[g]isoquinoline, 1,2,3,4-tetrahydro-2-methyl-5,10-diphenyl-	$CHCl_3$	248.0(4.18),298.0(4.04)	83-0886-86
$C_{26}H_{23}NO$			
Benzo[h]quinolin-2(1H)-one, 3,4,5,6-tetrahydro-4-(4-methylphenyl)-3-phenyl-	EtOH	227(4.15),265(3.52), 314(3.58)	4-1789-86
Cycloprop[hi]indolizin-2(1H)-one, dihydro-1,6b,6c-triphenyl-, (1α,6aα,6bα,6cα)-	MeCN	264(4.06),276s(--)	118-0899-86
1(5H)-Indolizinone, 6,7,8,8a-tetrahydro-2,3,8a-triphenyl-	MeCN	223(4.24),276(4.08), 348(3.95)	118-0899-86
$C_{26}H_{23}NOS$			
1H-Pyrrolizin-1-one, 5,6,7,7a-tetrahydro-7a-[4-(methylthio)phenyl]-2,3-diphenyl-	MeCN	258(4.48),323(3.75), 371(3.76)	118-0899-86
$C_{26}H_{23}NO_2$			
Benzo[h]quinolin-2(1H)-one, 3,4,5,6-tetrahydro-3-(4-methoxyphenyl)-4-phenyl-	EtOH	224(4.44),295(3.69)	4-1789-86
Benzo[h]quinolin-2(1H)-one, 3,4,5,6-tetrahydro-4-(4-methoxyphenyl)-3-phenyl-	EtOH	228(4.43),251s(3.92), 292(3.79)	4-1789-86
1H-Pyrrolizin-1-one, 5,6,7,7a-tetrahydro-7a-(4-methoxyphenyl)-2,3-diphenyl-	MeCN	234(4.36),324(3.76), 368(3.80)	118-0899-86
$C_{26}H_{23}NO_4S$			
Benzenesulfonamide, 4-methyl-N-(4-methyl-2-oxo-3-phenyl-2H-1-benzopyran-7-yl)-N-2-propenyl-	MeOH	235(4.31),294(4.21), 320(4.31)	151-0069-86A
	$CHCl_3$	295(3.99),323(4.20)	151-0069-86A
Pyrano[2,3-a]indol-2(7H)-one, 8,9-dihydro-4,8-dimethyl-7-[(4-methylphenyl)sulfonyl]-3-phenyl-	MeOH	235(4.29),250(4.17), 328(4.33)	151-0069-86A
	$CHCl_3$	256(3.91),330(4.24)	151-0069-86A
$C_{26}H_{23}NS$			
Cycloprop[hi]indolizine-2(1H)-thione, hexahydro-1,6b,6c-triphenyl-, (1α,6aα,6bα,6cα)-	MeCN	269s(--),294(4.08), 333s(--)	118-0899-86
$C_{26}H_{23}N_3O_5$			
5H-Indolo[2,3-a]pyrrolo[3,4-c]carbazol-5-one, 13-(6-deoxy-α-L-mannopyranosyl)-6,7,12,13-tetrahydro-	MeOH	300(5.0),325f(4.3), 367(4.3)	158-1066-86
$C_{26}H_{23}N_5O_7$			
9H-Purin-6-amine, 9-(2-O-acetyl-3,5-di-O-benzoyl-β-D-xylofuranosyl)-	EtOH	230(4.40),258(4.11)	87-0203-86
$C_{26}H_{24}ClN_5O_5$			
1H-Pyrazolo[3,4-d]pyrimidin-6-amine, 4-chloro-1-[2-deoxy-3,5-bis-O-(4-methylbenzoyl)-β-D-erythro-pentofuranosyl]-	MeOH	233(4.75),272(4.62), 283(3.63),305(3.81)	33-1602-86

Compound	Solvent	λ_{max}(log ε)	Ref.
2H-Pyrazolo[3,4-d]pyrimidin-6-amine, 4-chloro-2-[2-deoxy-3,5-bis-O-(4-methylbenzoyl)-β-D-erythro-pentofuranosyl]-	MeOH	230(4.62),278(3.90), 327(3.69)	33-1602-86
$C_{26}H_{24}NO_2$			
1-Benzopyrylium, 4-[2-[4-(dimethylamino)phenyl]ethenyl]-2-(2-methoxyphenyl)-, perchlorate	CH_2Cl_2	686(5.05)	103-0703-86
	MeCN	664(4.99)	103-0703-86
1-Benzopyrylium, 4-[2-[4-(dimethylamino)phenyl]ethenyl]-2-(4-methoxyphenyl)-, perchlorate	CH_2Cl_2	682(5.17)	103-0703-86
	MeCN	666(5.01)	103-0703-86
1-Benzopyrylium, 4-[2-[4-(dimethylamino)phenyl]ethenyl]-7-methoxy-2-phenyl-, perchlorate	CH_2Cl_2	668(5.04)	103-0703-86
	MeCN	646(4.94)	103-0703-86
$C_{26}H_{24}NS$			
5H-Pyrrolo[2,1-a]isoquinolinium, 6,10b-dihydro-10b-methyl-1-(methylthio)-2,3-diphenyl-, iodide	MeCN	246(4.27),288(3.82), 367(4.02)	118-0899-86
$C_{26}H_{24}N_2O$			
1H-Pyrrole-2-carboxaldehyde, 1-[4-(dimethylamino)phenyl]-5-methyl-3,4-diphenyl-	ether	262(4.50),301(4.27)	78-1355-86
$C_{26}H_{24}N_2O_2$			
Pyridine, 1,2-dihydro-2-(1-methyl-1-nitroethyl)-1,4,6-triphenyl-	DMSO	355(4.00)	44-2481-86
Pyridine, 1,4-dihydro-4-(1-methyl-1-nitroethyl)-1,2,6-triphenyl-	DMSO	331(3.49)	44-2481-86
$C_{26}H_{24}N_2O_4$			
Pyrazino[1,2-a:4,3-a']diindole-6,7,13,14-tetrone, 3,10-bis(1,1-dimethylethyl)-	benzene	290(4.36),431(4.04)	138-1393-86
$C_{26}H_{24}N_2O_5$			
1H-Pyrrole-3-carbonitrile, 1-[2-deoxy-3,5-bis-O-(4-methylbenzoyl)-β-D-erythro-pentofuranosyl]-	EtOH	238(4.60)	78-5869-86
$C_{26}H_{24}N_2O_8$			
[2,2'-Bi-1H-indole]-5,5',6,6'-tetrol, 1,1'-dimethyl-, tetraacetate	$CHCl_3$	302(4.31)	78-2083-86
[2,4'-Bi-1H-indole]-5,5',6,6'-tetrol, 1,1'-dimethyl-, tetraacetate	$CHCl_3$	303(4.20)	78-2083-86
Uridine, 3-benzoyl-2',3'-O-(1-methylethylidene)-, 5'-benzoate	MeOH	236(4.32),251(4.35)	94-4585-86
$C_{26}H_{24}O_5$			
2H-Furo[2,3-h]-1-benzopyran, 3,4-dihydro-5-methoxy-2,8-bis(4-methoxyphenyl)-	dioxan	221(4.38),247(4.01), 256(4.00),306s(4.40), 319(4.47),334(4.29)	94-0595-86
$C_{26}H_{24}O_6$			
1(2H)-Naphthalenone, 4-ethoxy-2-(4-ethoxy-5-hydroxy-7-methyl-1-oxo-2(1H)-naphthalenylidene)-5-hydroxy-7-methyl-	$CHCl_3$	297(4.48),326s(4.30), 671s(4.53),698(4.55)	39-0675-86C

Compound	Solvent	λ_{max}(log ϵ)	Ref.
$C_{26}H_{24}O_9$			
Sigmoidin C triacetate	EtOH	284(3.95)	39-0033-86C
$C_{26}H_{25}NO$			
Azacyclotetracosa-3,5,7,9,15,17,19,21-23-nonaene-11,13-diyn-2-one, 3,10,15-trimethyl-, (Z,E,Z,Z,E,E,E,E,E)-	THF	238(4.41),265s(4.39), 327s(4.75),355(4.93), 492s(4.31)	39-0933-86C
2,4,6,8,10,16,18,20,22-Cyclotricosa-nonaene-12,14-diyn-1-one, 11,16,23-trimethyl-, (Z,E,E,Z,Z,E,E,E,E,E)-	THF	260(4.16),276s(4.11), 320s(4.52),362(4.92), 456s(3.65)	39-0933-86C
$C_{26}H_{25}NOS$			
2H,6H-[1,3]Thiazino[2,3-a]isoquinoline, 7,11b-dihydro-2-methoxy-11b-methyl-3,4-diphenyl-	MeCN	274(3.83),295(3.90)	118-0899-86
2H-Thiopyran-2-amine, 4-(4-methoxy-phenyl)-N,N-dimethyl-2,6-diphenyl-	dioxan	239(4.33),254s(4.32), 340(3.65)	48-0567-86
$C_{26}H_{25}NO_2$			
2,5-Cyclohexadiene-1,4-dione, 2-[[3-(5H-dibenzo[a,d]cyclohepten-5-yl)-propyl]methylamino]-5-methyl-	CH_2Cl_2	240s(4.40),287(4.30), 497(3.52)	106-0029-86
2,5-Cyclohexadiene-1,4-dione, 2-[[3-(5H-dibenzo[a,d]cyclohepten-5-yl)-propyl]methylamino]-6-methyl-	CH_2Cl_2	240s(4.40),288(4.24), 498(3.48)	106-0029-86
2,5-Cyclohexadiene-1,4-dione, 2-[[3-(10,11-dihydro-5H-dibenzo[a,d]-cyclohepten-5-ylidene)propyl]-methylamino]-5-methyl-	CH_2Cl_2	245s(4.30),278s(4.00), 497(3.51)	106-0029-86
2,5-Cyclohexadiene-1,4-dione, 2-[[3-(10,11-dihydro-5H-dibenzo[a,d]-cyclohepten-5-ylidene)propyl]-methylamino]-6-methyl-	CH_2Cl_2	275s(3.90),497(3.52)	106-0029-86
2,5-Cyclohexadiene-1,4-dione, 2-[[3-(9,10-ethanoanthracen-9(10H)-yl)-propyl]amino]-5-methyl-	CH_2Cl_2	266s(4.00),273(4.11), 480(3.44)	106-0029-86
2,5-Cyclohexadiene-1,4-dione, 2-[[3-(9,10-ethanoanthracen-9(10H)-yl)-propyl]amino]-6-methyl-	CH_2Cl_2	265s(3.90),273(3.96), 280s(3.90),478(3.5)	106-0029-86
2,5-Cyclohexadiene-1,4-dione, 5,6-di-methyl-2-[[3-(5H-dibenzo[a,d]cyclo-hepten-5-yl)propyl]amino]-	CH_2Cl_2	240s(4.20),291(4.31), 479(3.05)	83-0421-86
$C_{26}H_{25}NS$			
2H-Thiopyran-2-amine, N,N-dimethyl-4-(4-methylphenyl)-2,6-diphenyl-	dioxan	256(4.37),345(3.69)	48-0567-86
$C_{26}H_{25}N_3O_{11}$			
Tunichrome B-1	MeOH	210(4.83),245s(--), 285s(--),340(4.29)	100-0193-86
$C_{26}H_{25}N_5O_5S$			
1H-Pyrazolo[3,4-d]pyrimidine-4(5H)-thione, 6-amino-1-[2-deoxy-3,5-bis-O-(4-methylbenzoyl)-β-D-erythro-pentofuranosyl]-	MeOH	238(4.67),272s(3.97), 336(4.34)	33-1602-86
$C_{26}H_{25}N_7O_5$			
Benzamide, N-[2-[(2-cyano-4,6-dinitro-phenyl)azo]-5-(dipropylamino)phenyl]-	$CHCl_3$	290(4.09),616(4.70)	48-0497-86

Compound	Solvent	λ_{max}(log ε)	Ref.
$C_{26}H_{26}$			
1,3,5-Methenocyclopenta[cd]pentalene, 6b,6'-b-(1,3-butadiyne-1,4-diyl)-bis[decahydro-	hexane	223(2.51),234(2.67), 247(2.71),260(2.51)	78-1763-86
$C_{26}H_{26}N_2$			
Benzenamine, 4-(2,5-dimethyl-3,4-di-phenyl-1H-pyrrol-1-yl)-N,N-dimethyl-	ether	266(4.59)	78-1355-86
$C_{26}H_{26}N_2O$			
1H-Pyrrole-2-methanol, 1-[4-(dimethyl-amino)phenyl]-5-methyl-3,4-diphenyl-	ether	267(4.61)	78-1355-86
2H-Pyrrol-2-one, 1-[4-(dimethylamino)-phenyl]-3,5-dimethyl-3,4-diphenyl-1,3-dihydro-	ether	267(4.44),290s(4.27)	78-1355-86
$C_{26}H_{26}N_2OS_2$			
1-Triphenodithiazinol, 2,4-bis(1,1-di-methylethyl)-	MeOH-10% $CHCl_3$	286(4.44),324(4.35), 528s(4.43),564(4.52)	39-2233-86C
	+ HCl	283(4.40),338(4.42), 620(4.25),670(4.40), 735(4.32)	39-2233-86C
$C_{26}H_{26}N_2O_{10}$			
1,3(2H,4H)-Isoquinolinedione, 4-(2,3-dihydro-5,6,7-trimethoxy-2-methyl-1,3-dioxo-4(1H)-isoquinolinylidene)-5,6,7-trimethoxy-2-methyl-, (E)-	$CHCl_3$	395(4.16)	139-0337-86C
$C_{26}H_{26}N_4P_2$			
1,5-Diaza-3,7-diphosphacyclooctane, P,P'-diphenyl-N,N'-dipyridinyl-	MeCN	253(4.33),307(4.11)	65-0362-86
$C_{26}H_{26}O_2Si$			
Silane, [(1,1-dimethylethyl)dioxy]-1-naphthalenyldiphenyl-	MeCN	205(4.9),267(3.72), 275(3.86),286(3.94), 296(3.79)	116-0968-86
$C_{26}H_{26}O_4$			
1,3-Azulenedicarboxylic acid, 2-meth-yl-6-[2-(4-methylphenyl)ethenyl]-, diethyl ester, (E)-	CH_2Cl_2	246(4.56),357.5(4.66), 408.5(4.62),517.5(3.00)	24-2272-86
$C_{26}H_{26}O_5$			
1,3-Azulenedicarboxylic acid, 6-eth-oxy-2-[2-(4-methylphenyl)ethenyl]-, ethyl methyl ester, (E)-	CH_2Cl_2	342.5(4.79),411(4.32)	24-2272-86
$C_{26}H_{26}O_6$			
2H,8H-Benzo[1,2-b:3,4-b']dipyran-2-one, 4-hydroxy-3-(4-hydroxyphenyl)-5-methoxy-8,8-dimethyl-6-(3-methyl-2-butenyl)- (scandenin)	MeOH	237(4.63),286(4.20), 340(4.21)	102-2425-86
$C_{26}H_{26}O_8$			
4H-1-Benzopyran-4-one, 7-acetoxy-3-[2-acetoxy-4,6-dimethoxy-3-(3-methyl-2-butenyl)phenyl]-	EtOH	246s(4.41),294(3.86), 302(3.86)	94-2369-86
4H-1-Benzopyran-4-one, 7-acetoxy-3-[6-acetoxy-2,4-dimethoxy-3-(3-methyl-2-butenyl)phenyl]-	EtOH	256s(4.16),282.5(4.03), 301.5s(3.87)	94-2369-86

Compound	Solvent	$\lambda_{max}(\log \epsilon)$	Ref.
$C_{26}H_{27}ClFNO$			
Ethanamine, 2-[4-[2-chloro-1-(4-fluorophenyl)-2-phenylethenyl]phenoxy]-N,N-diethyl- (fluoroclomiphene)	EtOH	236(4.16),286(3.66), 298s(3.60)	13-0073-86B
$C_{26}H_{27}N$			
Benzeneethanamine, N-(2,2-dimethyl-4,4-diphenyl-3-butenylidene)-	EtOH	249(4.14)	150-0631-86M
Benzenemethanamine, N-(2,2-dimethyl-4,4-diphenyl-3-butenylidene)-α-methyl-	EtOH	249(4.14)	150-0631-86M
$C_{26}H_{27}NO$			
Azacyclodocosa-1,3,5,7,9,15,17,19,21-nonaene-11,13-diyne, 2-ethoxy-10,15,22-trimethyl-	THF	243(4.51),273s(4.59), 292s(4.62),353(4.87), 453s(3.92)	39-0933-86C
5H-Indolo[3,2,1-de]acridine, 6,7,7a,8-9,10,11,12-octahydro-8-(4-methoxyphenyl)-	MeOH	318(4.40)	103-0038-86
$C_{26}H_{27}N_2O$			
2-Propynylium, 1,1-bis[4-(dimethylamino)phenyl]-3-(4-methoxyphenyl)-, perchlorate	CH_2Cl_2	530(4.63),680(5.01)	138-0329-86
$C_{26}H_{28}$			
Azulene, 2-octyl-6-(phenylethynyl)-	CH_2Cl_2	301.5(4.80),379(4.40), 395(4.60),600.5(2.57), 647.5(2.48)	24-2272-86
$C_{26}H_{28}BrNO_2$			
Phenol, 4-[2-bromo-1-[4-[2-(diethylamino)ethoxy]phenyl]-2-phenylethenyl]-	EtOH	310(4.00)	87-2511-86
$C_{26}H_{28}FNO$			
Ethanamine, N,N-diethyl-2-[4-[1-(4-fluorophenyl)-2-phenylethenyl]phenoxy]-	EtOH	243(4.03),304(4.08)	13-0073-86B
$C_{26}H_{28}F_2N_4O_{13}$			
α-D-Glucofuranose, 1,2-O-(1-methylethylidene)-3,5-bis(5-fluoro-3,4-dihydro-2,4-dioxo-1(2H)-pyrimidinepropanoate) 6-(2-propenoate)	MeOH	270.5(4.25)	47-2059-86
$C_{26}H_{28}N_2$			
Benzenamine, 4-(2,3-dihydro-3,5-diphenyl-1H-pyrrol-1-yl)-N,N-dimethyl-	ether	261(4.13),317(4.24)	78-1355-86
Pyrazine, 2,5-bis(1-methylethyl)-3,6-bis(2-phenylethenyl)-, (E,E)-	EtOH	226(4.01),235s(3.94), 243s(3.82),293(4.25), 302s(4.22)	4-1481-86
$C_{26}H_{28}N_2O_2$			
2,5-Cyclohexadiene-1,4-dione, 2-[[3-(10,11-dihydro-5H-dibenzo-[b,f]azepin-5-yl)propyl]methylamino]-3,5-dimethyl-	CH_2Cl_2	238(4.20),270s(4.1), 300s(3.85),545(3.32)	83-0421-86
2,5-Cyclohexadiene-1,4-dione, 2-[[3-(10,11-dihydro-5H-dibenzo-	CH_2Cl_2	243(4.51),270s(4.1), 300s(3.7),543(3.50)	83-0421-86

Compound	Solvent	$\lambda_{max}(\log \epsilon)$	Ref.
[b,f]azepin-5-yl)propyl]methylamino]-3,6-dimethyl- (cont.)			83-0421-86
2,5-Cyclohexadiene-1,4-dione, 2-[[3-(10,11-dihydro-5H-dibenzo[b,f]azepin-5-yl)propyl]methylamino]-5,6-dimethyl-	CH_2Cl_2	252s(4.20),284(4.12), 493(3.47)	83-0421-86
Ethanedione, bis(2,3,6,7-tetrahydro-1H,5H-benzo[ij]quinolizin-9-yl)-	EtOH	263(3.95),365s(--), 390(4.64)	48-0253-86
$C_{26}H_{28}N_2O_4$			
Phenol, 4-[1-[4-[2-(diethylamino)ethoxy]phenyl]-2-nitro-2-phenylethenyl]-	EtOH	288(4.14)	87-2511-86
$C_{26}H_{28}N_2O_7S_2$			
Uridine, 5'-O-(methoxymethyl)-2',3'-O-(1-methylethylidene)-5,6-bis(phenylthio)-	MeOH	244(4.17),276s(4.01), 310s(3.76)	78-4187-86
$C_{26}H_{28}N_2O_{10}$			
[4,4'-Biisoquinoline]-1,1',3,3'(2H-2'H,4H,4'H)-tetrone, 5,5',6,6',7,7'-hexamethoxy-2,2'-dimethyl-, (R*,R*)-(±)-	$CHCl_3$	315(3.78)	139-0337-86C
$C_{26}H_{28}N_6O_5S$			
Benzamide, 4-(1,1-dimethylethyl)-N-[6-(2-pyridinylthio)-9-β-D-ribofuranosyl-9H-purin-2-yl]-	pH 7	252(4.40),300(4.33)	1-0806-86
	pH 2	254(--),274(--), 330(--)	1-0806-86
	pH 13	248(--),297(--)	1-0806-86
$C_{26}H_{28}N_6O_6$			
Benzamide, 4-(1,1-dimethylethyl)-N-[6-(4-oxo-1(4H)-pyridinyl)-9-β-D-ribofuranosyl-9H-purin-2-yl]-	pH 7	285(4.25),320(4.06)	1-0806-86
	pH 2	273(--),320(--)	1-0806-86
	pH 13	297(--)	1-0806-86
Benzamide, 4-(1,1-dimethylethyl)-N-[6-(3-pyridinyloxy)-9-β-D-ribofuranosyl-9H-purin-2-yl]-	pH 7	241(4.32),274(4.41)	1-0806-86
	pH 2	244(--),276(--)	1-0806-86
	pH 13	235(--),272(--)	1-0806-86
Benzamide, 4-(1,1-dimethylethyl)-N-[6-(4-pyridinyloxy)-9-β-D-ribofuranosyl-9H-purin-2-yl]-	pH 7	242(4.32),277(4.40)	1-0806-86
	pH 2	253(--),278(--)	1-0806-86
	pH 13	238(--),280(--)	1-0806-86
$C_{26}H_{28}N_8O_4S$			
Morpholine, 4,4'-[1,2,4-thiadiazole-3,5-diylbis[1-oxo-2-(phenylhydrazono)-2,1-ethanediyl]]bis-	DMF	200(4.01),403(4.41)	48-0741-86
$C_{26}H_{28}O_4$			
2H-1-Benzopyran-2-one, 4-hydroxy-3-[[6-methyl-3-(1-methylethyl)-2-oxocyclohexyl]phenylmethyl]-	MeOH	282(4.09),308(4.10)	2-0102-86
$C_{26}H_{28}O_6$			
24-Norchola-1,5,20,22-tetraene-3,7,11-trione, 14,15:21,23-diepoxy-6-hydroxy-4,4,8-trimethyl-, (13α,14β-15β,17α)-	n.s.g.	217(3.90),280(4.00)	100-0056-86
$C_{26}H_{28}O_8$			
Obacunoic acid, 4,5-didehydro-7-deoxo-4-deoxy-7,19-dihydroxy-6-oxo-, δ-lactone, (7β)-	EtOH	213(4.83),249(4.55)	142-3417-86
	EtOH-NaOH	223(4.61),353(3.89)	142-3417-86

Compound	Solvent	λ_{max}(log ϵ)	Ref.
3,6-Phenanthrenedione, 5,8-diacetoxy-7-(2-acetoxypropyl)-4,4a-dihydro-1,2,4a-trimethyl-, [R-(R*,R*)]-	ether	277(4.11),282(4.11), 315s(3.24),440(3.97)	33-0972-86
$C_{26}H_{28}O_9$			
1,4,6(5H)-Phenanthrenetrione, 3,9,10-triacetoxy-4b,8a,9,10-tetrahydro-4b,7,8-trimethyl-2-(2-propenyl)-	ether	244(3.99),265s(3.78)	33-0972-86
$C_{26}H_{28}O_{11}$			
4H-1-Benzopyran-4-one, 5,6,7,8-tetra-acetoxy-5,6,7,8-tetrahydro-2-[2-(4-methoxyphenyl)ethyl]-, [5S-(5α,6β,7β,8α)]-	EtOH	210(4.27),213(4.23), 247(4.08)	94-3033-86
$C_{26}H_{29}ClN_2O$			
Benzenamine, 4-[2-chloro-1-[4-[2-(di-ethylamino)ethoxy]phenyl]-2-phenyl-ethenyl]-	EtOH	230(4.26),250(4.18), 295(3.97),345(3.69)	13-0073-86B
$C_{26}H_{29}NO_2$			
Phenol, 4-[1-[4-[2-(diethylamino)eth-oxy]phenyl]-2-phenylethenyl]-	EtOH	314(4.28)	87-2511-86
$C_{26}H_{29}N_2O_3$			
1-Benzopyrylium, 7-(diethylamino)-2-[7-(diethylamino)-2-oxo-2H-1-benzopyran-3-yl]-, perchlorate	CH_2Cl_2	667(5.07)	3-0384-86
$C_{26}H_{29}N_2O_6P$			
Phosphoric acid, 4-(6-amino-1,3-dioxo-1H-benz[de]isoquinolin-2(3H)-yl)-phenyl dibutyl ester	EtOH	430(4.18)	65-1105-86
$C_{26}H_{30}ClN_3O_7$			
L-Glutamic acid, N-[N-[N-(4-chloro-benzoyl)glycyl]-2-methylalanyl]-, 1-methyl 5-(phenylmethyl) ester	MeOH	236(4.19)	23-0277-86
$C_{26}H_{30}N_2O_2$			
Diazene, [4-(hexyloxy)phenyl](methoxy-diphenylmethyl)-	CH_2Cl_2	251.5(4.42),311.0(4.27), 411.0(2.51)	107-0741-86
$C_{26}H_{30}N_4$			
2,3'-Bipyridinium, 1',1"'-(1,4-butane-diyl)bis[1-methyl-, tetraperchlorate	pH 2.3	206(4.47),271(4.42)	64-0239-86B
4,4'-Bipyridinium, 1,1"-(1,4-butane-diyl)bis[1'-methyl-	n.s.g.	260(4.63)	35-3380-86
diradical dication	n.s.g.	536(4.36)	35-3380-86
monoradical trication	n.s.g.	560s(3.88),600(4.00)	35-3380-86
$C_{26}H_{30}N_6$			
1H-Indole, 3,3'-[1,2,5,6-tetrahydro-1-methyl-3-(1-pyrrolidinyl)-1,2,4-triazine-5,6-diyl]bis[2-methyl-, monohydriodide	EtOH	225(4.96),283(5.13), 290(5.08)	103-1242-86
$C_{26}H_{30}N_6O$			
1H-Indole, 3,3'-[1,2,5,6-tetrahydro-1-methyl-3-(4-morpholinyl)-1,2,4-triazine-5,6-diyl]bis[2-methyl-,	EtOH	225(4.95),283(4.16), 290(4.04)	103-1242-86

Compound	Solvent	λ max (log ε)	Ref.
monohydriodide (cont.)			103-1242-86
$C_{26}H_{30}N_6O_8$			
Carbamic acid, [[[4-(aminocarbonyl)-1-β-D-ribofuranosyl-1H-imidazol-5-yl]-amino][[(4-methoxyphenyl)methyl]-amino]methylene]-, phenylmethyl ester	pH 1	250(4.00)	44-1277-86
	pH 11	273(4.00)	44-1277-86
	MeOH	276(4.04)	44-1277-86
$C_{26}H_{30}O$			
Ethanone, 2-(2-octyl-6-azulenyl)-1-phenyl-	CH_2Cl_2	239(4.43),281(4.88), 290.5(4.99),336.5(3.74), 352(3.79),402(2.03), 560(2.47),598.5(2.40)	24-2272-86
$C_{26}H_{30}O_5$			
24-Norchola-5,14,20,22-tetraene-3,7,11-trione, 21,23-epoxy-6-hydroxy-4,4,8-trimethyl-, (13α,17α)-	n.s.g.	215(4.00),280(4.00)	100-0056-86
$C_{26}H_{30}O_6$			
24-Norchola-5,20,22-triene-3,7,11-trione, 14,15:21,23-diepoxy-6-hydroxy-4,4,8-trimethyl-, (13α,14β,15β,17α)-	n.s.g.	217(3.90),280(4.00)	100-0056-86
$C_{26}H_{30}O_7$			
Kushenol N	MeOH	290(4.23),331(3.73)	95-0022-86
$C_{26}H_{30}O_9$			
Bufa-20,22-dienolide, 1,3,5-[ethylidynetris(oxy)]-11,14-dihydroxy-12,19-dioxo-, [1β(R),3β,5β,11α]-	MeOH	298(3.74)	33-0359-86
3-Furanmethanol, α-(4-acetoxy-3-methoxyphenyl)-4-[(4-acetoxy-3-methoxyphenyl)methyl]tetrahydro-, acetate	MeOH	223(-),274.3(4.18), 278.3(4.00)	39-1181-86C
$C_{26}H_{30}O_{11}$			
Fibleucinoside	MeOH	207(3.97),229(3.79)	102-0905-86
$C_{26}H_{30}O_{12}$			
Fibraurinoside	MeOH	212(3.94),230(3.83)	102-0905-86
$C_{26}H_{31}NO_2$			
2H-[1,3]Oxazino[2,3-j]quinoline, 2-ethoxy-6,7,8,8a,9,10,11,12-octahydro-3,4-diphenyl-	MeCN	235(4.09),292(3.90)	118-0899-86
$C_{26}H_{31}N_3O_5$			
Capparisine	MeOH	218(2.71),280(2.90), 308s(--)	64-1033-86B
$C_{26}H_{32}N_2O_6$			
Erichsonine, N,O,O-triacetyl-	EtOH	238(4.19),318(4.35)	102-0965-86
$C_{26}H_{32}N_4O_4$			
2,5-Cyclohexadiene-1,4-dione, 2,5-bis[[4-[(dimethylamino)methyl]-phenyl]amino]-3,6-dimethoxy-	MeOH	276(4.51),406(4.24)	87-1792-86
$C_{26}H_{32}N_8O_{15}P_2$			
Adenosine 5'-(trihydrogen triphosphate),	pH 1	257(4.20)	87-0268-86

Compound	Solvent	λ max (log ε)	Ref.
5'→5'-ester with 5-β-D-ribofuranosyl-1H-pyrazole-3-carboxamide (cont.)	pH 7 and 11	257(4.20)	87-0268-86
$C_{26}H_{32}O_2$			
2-Naphthalenecarboxylic acid, 3,7,11-trimethyl-2,6,10-dodecatrienyl ester, (E,E)-	MeCN	236(4.77)	130-0046-86
$C_{26}H_{32}O_3$			
Estra-1,3,5(10)-trien-17β-ol, 4-methoxy-3-(phenylmethoxy)-	EtOH	206(4.49),230s(3.94), 275(3.22),283(3.15)	94-2231-86
Estra-1,3,5(10)-trien-17β-ol, 4-(phenylmethoxy)-3-methoxy-	EtOH	209(4.54),275(3.23), 283(3.20)	94-2231-86
24-Norchola-1,14,20,22-tetraene-3,7-dione, 21,23-epoxy-4,4,8-trimethyl-, (5α,13α,17α)-	EtOH	218(4.08),257s(3.56)	100-0583-86
$C_{26}H_{32}O_4$			
2H-1-Benzopyran-8-ol, 6-[3-(4-hydroxy-2-methoxyphenyl)propyl]-2,2-dimethyl-5-(3-methyl-2-butenyl)- (kazinol N)	EtOH	231(4.47),266s(3.84), 276(4.04),286s(3.97), 325(3.21)	94-2448-86
$C_{26}H_{32}O_6$			
24-Norchola-1,5,20,22-tetraen-7-one, 14,15:21,23-diepoxy-3,6,11-trihydroxy-4,4,8-trimethyl-, (3β,11α-13α,14β,15β,17α)-	n.s.g.	280(4.04)	100-0056-86
	+ NaOH	326(3.70)	100-0056-86
$C_{26}H_{32}O_6S$			
Estra-1,3,5(10)-trien-17β-ol, 3-methoxy-4-(phenylmethoxy)-, hydrogen sulfate, potassium salt	EtOH	210(4.46),230s(3.90), 276(3.24),283(3.22)	94-2231-86
Estra-1,3,5(10)-trien-17β-ol, 4-methoxy-3-(phenylmethoxy)-, hydrogen sulfate, potassium salt	EtOH	212(4.42),227s(4.00), 274(3.20),282(3.15)	94-2231-86
$C_{26}H_{32}O_7$			
Bufa-20,22-dienolide, 1,3,5-[ethylidynetris(oxy)]-14-hydroxy-19-oxo-, [1β(R),3β,5β)] (bersaldegenin 1,3,5-orthoacetate)	MeOH	298(3.75)	33-0359-86
$C_{26}H_{32}O_8$			
Phenanthro[3,2-b]furan-1,2,3,4a(2H)-tetrol, 1,3,4,5,6,11b-hexahydro-4,4,7,11b-tetramethyl-, 1,2,3-triacetate	EtOH	251(4.06),282(3.40), 292(3.43)	100-0913-86
$C_{26}H_{32}O_9$			
1-Butanone, 1-[3-[(3-acetyl-2,4-dihydroxy-6-methoxy-5-methylphenyl)methyl]-2,4,6-trihydroxy-5-(2-hydroxy-3-methyl-3-butenyl)phenyl]-, (±)- (mallotolerin)	EtOH	230(4.20),287(4.12), 320(4.00)	100-0298-86
$C_{26}H_{32}O_{12}$			
Loganin, 7-O-p-coumaroyl-	MeOH	239(3.7),281(2.8)	102-1907-86
Shinjulactone M triacetate	EtOH	238(3.94)	18-1638-86

Compound	Solvent	λ_{max}(log ε)	Ref.
$C_{26}H_{32}O_{14}$			
Mulberroside A	EtOH	218(4.59),235s(4.42), 290(4.43),300(4.43), 325(4.56),336s(4.54)	100-0218-86
$C_{26}H_{33}ClO_9$			
Ptilosarcen-12-acetate	EtOH	218(3.69)	78-6565-86
$C_{26}H_{33}ClO_{10}$			
Benzo[4,5]cyclodeca[1,2-b]furan-2(1H)-one, 8,11,13-triacetoxy-4-chloro-3a,4,5,8,8a,11,12,12a,13,13a-decahydro-12,13a-dihydroxy-1,8a,12-trimethyl-5-methylene- (minabein-4)	MeOH	222(3.62)	20-0835-86
$C_{26}H_{33}NOS$			
2H,6H-Pyrido[2,1-b][1,3]oxazine, 2-[(1,1-dimethylethyl)thio]-9a-ethyl-7,8,9,9a-tetrahydro-3,4-diphenyl-	MeCN	228(4.13),292(3.97)	118-0899-86
$C_{26}H_{33}NO_4$			
1,3-Azulenedicarboxylic acid, 6-(cyanomethyl)-2-octyl-, diethyl ester	CH_2Cl_2	239(4.51),270(4.30), 312.5(4.83),350(3.88), 379(3.74),507.5(2.57)	24-2272-86
$C_{26}H_{34}N_6O_5$			
Benzamide, 4-(1,1-dimethylethyl)-N-[6-(1-piperidinyl)-9-β-D-ribofuranosyl-9H-purin-2-yl]-	pH 7	254(4.46),280(4.32)	1-0806-86
	pH 2	270(--)	1-0806-86
	pH 13	255(--),280(--)	1-0806-86
$C_{26}H_{34}O_4$			
1,2-Benzenediol, 5-[3-(4-hydroxy-2-methoxyphenyl)propyl]-3,4-bis(3-methyl-2-butenyl)- (kazinol J)	EtOH	282(4.04),288s(4.00)	94-2448-86
A-Homo-24-nor-4-oxachola-1,14,20,22-tetraen-3-one, 21,23-epoxy-7-hydroxy-4,4,8-trimethyl-, (5α,7α,13α,17α)-(7-deacetylproceranone)	EtOH	213(4.20)	100-0583-86
$C_{26}H_{34}O_5$			
Ouabanginone	EtOH	217(4.26)	100-0583-86
$C_{26}H_{34}O_8$			
Bufa-20,22-dienolide, 1-acetoxy-3,5,14-trihydroxy-19-oxo-, (1β,3β,5β)- (bersaldegenin 1-acetate)	MeOH	298(3.71)	33-0359-86
Bufa-20,22-dienolide, 3-acetoxy-1,5,14-trihydroxy-19-oxo-, (1β,3β,5β)- (bersaldegenin 3-acetate)	MeOH	298(3.72)	33-0359-86
Phenanthro[3,2-b]furan-1,2,3,4a(2H)-tetrol, 1,3,4,5,6,6a,7,11,11a,11b-decahydro-4,4,11b-trimethyl-7-methylene-, 1,2,3-triacetate	EtOH	238(3.91)	100-0913-86
$C_{26}H_{34}O_{16}$			
2H-Pyran-5-carboxylic acid, 4-(1,3-dioxolan-2-ylmethyl)-3-formyl-3,4-dihydro-2-[(2,3,4,6-tetra-O-acetyl-β-D-glucopyranosyl)oxy]-, methyl ester, [2S-(2α,3β,4β)]-	ether	227(3.98)	5-1413-86

Compound	Solvent	λ max (log ε)	Ref.
$C_{26}H_{36}N_4O$			
Propanedinitrile, [[(5α,17β)-17-hydroxy-3-(1-pyrrolidinyl)androst-2-en-2-yl]imino]- (hydrate)	EtOH	365(3.91)	33-0793-86
$C_{26}H_{36}O_5$			
1,3-Azulenedicarboxylic acid, 6-ethoxy-2-octyl-, diethyl ester	CH_2Cl_2	275.5(4.31),324(4.82), 365.5(4.06),453.5(2.90)	24-2272-86
$C_{26}H_{36}O_7$			
Bufa-20,22-dienolide, 3-acetoxy-1,5,14-trihydroxy-, (1β,3β,5β)-	MeOH	298(3.74)	33-0359-86
Card-20(22)-enolide, 3-acetoxy-16-(formyloxy)-14-hydroxy-	MeOH	218.5(4.16)	87-0997-86
Card-20(22)-enolide, 16-acetoxy-3-(formyloxy)-14-hydroxy-, (3β,5β,16β)-	MeOH	218(4.15)	87-0997-86
$C_{26}H_{36}O_9$			
Caesalpine F	EtOH	233(3.20)	100-0913-86
Renillafoulin C	MeOH	202(4.11),220(3.86)	44-4450-86
$C_{26}H_{37}BrN_4O_6$			
Alanine, N-[N-[N-[1-(4-bromobenzoyl)-L-prolyl]-L-valyl]-2-methylalanyl]-2-methyl-, methyl ester	MeOH	229(4.11)	23-0277-86
$C_{26}H_{38}O_2Si$			
6H-Cyclopropa[3,4]cyclohept[1,2-e]inden-6-one, 1,1a,4,5,7,8,9,9a-octahydro-1,1,2,5-tetramethyl-7-methylene-3-[(triethylsilyl)oxy]-, [1aR-(1aα,5α,9aα)]-	EtOH	225(4.37),269(3.90), 316(3.40)	35-3040-86
[1aR-(1aα,5β,9aα)]-	EtOH	225(4.26),270(3.80), 318(3.31)	35-3040-86
$C_{26}H_{38}O_4$			
Kaur-16-en-18-oic acid, 7-[(3-methyl-1-oxo-2-butenyl)oxy]-, methyl ester, (E)- (foetidin)	EtOH	216(4.15)	100-0334-86
(Z)-	EtOH	215(4.01)	100-0334-86
4H-Pyran-4-one, 3-ethyl-6-(10-hydroxy-3,7,9,11-tetramethyl-2,5,7,11-tridecatetraenyl)-2-methoxy-5-methyl-	MeOH	239(4.54)	158-0038-86
$C_{26}H_{38}O_5$			
5α-Pregn-9(11)-en-12-one, 3β-acetoxy-17,20α-dihydroxy-, 17,20α-acetonide	EtOH	238(4.0)	13-0381-86A
$C_{26}H_{38}O_6$			
β-D-Xylopyranoside, 2,3,7,8,9,9a-hexahydro-5-methoxy-3,6,9-trimethyl-7-(2-methyl-1-propenyl)-1H-phenalen-4-yl [3S-(3α,7β,9α,9aα)]-	MeOH	226(4.12),268(3.15)	44-5140-86
$C_{26}H_{38}O_8$			
18-Nor-16-oxaandrostane-8-carboxaldehyde, 6-acetoxy-15-hydroxy-4,4-dimethyl-17-oxo-7-(1-oxobutoxy)-, cyclic 8,15-hemiacetal	EtOH	230(2.40)	12-1629-86
18-Nor-16-oxaandrostane-8-carboxaldehyde, 7-acetoxy-15-hydroxy-4,4-di-	EtOH	223(2.48)	12-1629-86

Compound	Solvent	$\lambda_{max}(\log \epsilon)$	Ref.
methyl-17-oxo-6-(1-oxobutoxy)-, cyclic 8,15-hemiacetal (cont.)			12-1629-86
$C_{26}H_{38}PtSi_2$			
Platinum, [(1,2,5,6-η)-1,5-cycloocta- diene]bis[4-(trimethylsilyl)phenyl]-	CH_2Cl_2	270(3.82),278(3.73), 306(3.53)	101-0027-86N
$C_{26}H_{39}BrN_2O_3$			
Androst-4-en-3-one, 17-acetoxy-4-bromo- 2-(4-methyl-1-piperazinyl)-, (2β,17β)-	EtOH	263.0(4.15)	118-0461-86
$C_{26}H_{39}NO$			
Cyclobuxotriene, (+)-	MeOH	324(3.96)	78-5747-86
$C_{26}H_{40}$			
Azulene, 2,6-dioctyl-	CH_2Cl_2	231(4.16),282.5(4.84), 291(4.94),336(3.63), 350.5(3.74),365(3.37), 552(2.44),579.5(2.39), 618(2.20)	24-2272-86
$C_{26}H_{40}N_2O_3$			
Chol-5-ene-3,24-diol, 24-(diazo- acetate), (3β)-	EtOH	247(4.00)	94-0931-86
$C_{26}H_{40}N_4Se$			
Cyclopropenylium, 1,1'-selenobis[2,3- di-1-piperidinyl-, diperchlorate	n.s.g.	272(4.18)	89-0183-86
$C_{26}H_{40}N_5$			
Agelasine A, chloride	MeOH	272(3.95)	18-2495-86
Agelasine B, chloride	MeOH	272(3.92)	18-2495-86
Agelasine C, chloride	MeOH	272(3.92)	18-2495-86
Agelasine E, chloride	MeOH	272(3.99)	18-2495-86
$C_{26}H_{40}O_2$			
25-Ketovitamin D_3	ether	266(4.26)	44-1269-86
$C_{26}H_{40}O_2Si_2$			
Cyclodisiloxane, 2,4-bis(1,1-dimethyl- ethyl)-2,4-bis(2,4,6-trimethyl- phenyl)-	C_6H_{12}	204(4.62),220(4.20)	157-0531-86
$C_{26}H_{41}N_5O$			
Formamide, N-[4-amino-6-(methylamino)- 5-pyrimidinyl]-N-[3,7-dimethyl- 9-(1,2,6-trimethyl-2-cyclohexen- 1-yl)-2,6-nonadienyl]-, [1R- [1α(2E,6E),6β]]-	pH 2	270(3.96)	18-2495-86
	MeOH	260(3.64)	18-2495-86
$C_{26}H_{42}CuN_4O_{14}$			
Copper, bis[3-C-[(hydroxyamino)imino]- methyl]-1,2:5,6-bis-O-(1-methyleth- ylidene)-α-D-glucofuranosato]-, (SP- 4-1)-	EtOH	220(4.27),290(3.46), 570(1.81)	159-0631-86
$C_{26}H_{42}N$			
Quinolinium, 1-hexadecyl-2-methyl-, bromide	EtOH	303(3.84),309(3.80), 315(3.97)	4-0209-86
Quinolinium, 1-hexadecyl-4-methyl-, bromide	EtOH	235(4.62),314(3.89)	4-0209-86

Compound	Solvent	λ_{max}(log ϵ)	Ref.
$C_{26}H_{42}O_3$			
2-Oxacholest-4-en-3-one, 1-hydroxy-, (1α)-	MeOH	227(4.15)	78-5693-86
$C_{26}H_{43}NO_4$			
1,3-Cyclopentadiene-1-carboxylic acid, 2,4,5-tris(1,1-dimethylethyl)-3-[(2-ethoxy-2-oxoethylidene)amino]-, 1,1-dimethylethyl ester	hexane	246(3.83),317(3.59), 393(3.77)	24-2159-86
$C_{26}H_{44}O_5$			
1,2-Dioxin-3-acetic acid, 3,6-dihydro-6-methoxy-6-(14,16-octadecadienyl)-, methyl ester (xestin A)	MeOH	227(4.46)	44-4260-86
Xestin B	MeOH	227(4.46)	44-4260-86
$C_{26}H_{45}NO_2$			
1,4-Cyclopentadiene-1-carboxylic acid, 2,4,5-tris(1,1-dimethylethyl)-3-[(1,1-dimethylethyl)imino]-, 1,1-dimethylethyl ester	hexane	252(4.04),391(2.73)	24-2159-86
$C_{26}H_{46}O_2S$			
1-Propanol, 3-(hexadecylthio)-2-(phenylmethoxy)-	MeOH	205(4.00)	32-0025-86
$C_{26}H_{50}O_3S$			
Acetic acid, [(1,1-dimethylethyl)thio]-oxo-, 1,1-dimethyloctadecyl ester	EtOH	267(3.83)	39-1603-86C
$C_{26}H_{60}GeN_4Si_2$			
1H-Tetraazagermole, 4,5-dihydro-5,5-dimethyl-1,4-bis[tris(1,1-dimethylethyl)silyl]-	C_6H_{12}	255(4.59)	24-2980-86

Compound	Solvent	λmax(log ε)	Ref.
$C_{27}Cl_{18}O$			
1H-Inden-1-one, 2,3,7-trichloro-4,5,6-tris(pentachlorophenyl)-	$CHCl_3$	238(4.80),246s(4.77), 268s(4.59),355(3.35), 370(3.33),408s(3.05)	44-1100-86
$C_{27}H_9Cl_{14}O_3S$			
Methyl, bis(pentachlorophenyl)[2,3,5,6-tetrachloro-4-[[[(4-methylphenyl)sulfonyl]oxy]methyl]phenyl]-	C_6H_{12}	222(4.99),291(3.87), 336s(3.86),366s(4.27), 386(4.55),512(3.02), 563(3.00)	44-2472-86
$C_{27}H_{10}Cl_{14}O_3S$			
Benzenemethanol, 4-[bis(pentachlorophenyl)methyl]-2,3,5,6-tetrachloro-, 4-methylbenzenesulfonate	C_6H_{12}	222(4.09),253s(4.49), 296(3.28),325(3.33)	44-2472-86
$C_{27}H_{12}O$			
4H-Acenaphtho[6,7,8,1-pqra]perylen-4-one	EtOH	240(3.94),245(3.95), 288(4.00),364(3.30), 384(3.50),399(3.49), 407(3.56),445(2.81)	24-3521-86
$C_{27}H_{13}Cl_{14}O_4$			
Methyl, bis(pentachlorophenyl)[2,3,5,6-tetrachloro-4-[3-ethoxy-2-(ethoxycarbonyl)-3-oxopropyl]phenyl]-	C_6H_{12}	223(4.97),287(3.83), 366s(4.23),383(4.55), 510(3.03),563(3.02)	44-2472-86
$C_{27}H_{13}N_3O_4$			
Benzonitrile, 2-(8,13-dihydro-8,13-dioxobenzo[5,6]indolo[2,1-a]isoquinolin-14-yl)-5-nitro-	DMF	454(3.6990)	2-1126-86
$C_{27}H_{13}O_4$			
[2,3'-Bi-1H-indene]-1,3(2H)-dione, 1'-(1,3-dihydro-1,3-dioxo-2H-inden-2-ylidene)-, ion(1-)-, N-[3-(dimethylamino)-2-propenylidene]-N-methylmethanaminium	EtOH	268(4.71),312(4.73), 422(4.33),670(4.32), 720(4.59),785(4.54)	70-1446-86
$C_{27}H_{14}Cl_{14}NO_3$			
Methyl, bis(pentachlorophenyl)[2,3,5,6-tetrachloro-4-[3-[(2-ethoxy-1-methyl-2-oxoethyl)amino]-3-oxopropyl]-phenyl]-, (S)-	$CHCl_3$	285(3.83),335s(3.86), 365s(4.32),383(4.60), 475s(3.06),507(3.08), 560(3.08)	44-2472-86
$C_{27}H_{14}Cl_{14}O_4$			
Propanedioic acid, [[4-[bis(pentachlorophenyl)methyl]-2,3,5,6-tetrachlorophenyl]methyl]-, diethyl ester	C_6H_{12}	222(5.08),292(3.08), 302(3.10)	44-2472-86
$C_{27}H_{15}N_3O_7$			
Methanone, [2,3-bis(5-nitro-2-furanyl)pyrrolo[1,2-a]quinolin-3-yl]-phenyl-	MeOH	221(3.41),274(3.72), 412(3.36)	73-0412-86
$C_{27}H_{17}N_3O$			
2,2'-Biquinoline, 4-(5-phenyl-2-oxazolyl)-	EtOH	260(4.70),330(4.39)	103-0229-86
$C_{27}H_{17}N_5O_3S$			
Benzo[de]]1,2,3]triazolo[4,5-g]iso-	DMF	378(4.42)	2-0496-86

Compound	Solvent	λ_{max}(log ε)	Ref.
quinoline-4,6(5H,9H)-dione, 9-(6-methoxy-2-benzothiazolyl)-5-(4-methylphenyl)- (cont.)			2-0496-86
$C_{27}H_{18}BrNOS$			
Oxazole, 2-[4-[2-[5-(4-bromophenyl)-2-thienyl]ethenyl]phenyl]-5-phenyl-	toluene	317(4.35),394(4.60)	103-1028-86
$C_{27}H_{18}ClNOS$			
Oxazole, 2-[4-[2-[5-(4-chlorophenyl)-2-thienyl]ethenyl]phenyl]-5-phenyl-	toluene	320(4.36),394(4.57)	103-1028-86
$C_{27}H_{18}ClN_3O_2S$			
2H,5H-[1]Benzothiopyrano[3,4-e]-1,3-oxazin-5-one, 9-chloro-3-phenyl-2,4-di-3-pyridinyl-	MeOH	238(3.9),332(3.9)	42-0323-86
$C_{27}H_{18}N_2OS$			
Pyrrolo[3,4-a]phenothiazin-4(2H)-one, 2-methyl-1,3-diphenyl-	MeOH	257(4.67),324(4.09), 377s(3.94),486(4.20)	4-1267-86
$C_{27}H_{18}N_2O_3S$			
Oxazole, 2-[4-[2-[5-(4-nitrophenyl)-2-thienyl]ethenyl]phenyl]-5-phenyl-, (E)-	toluene	421(4.67)	103-1028-86
$C_{27}H_{19}NOS$			
Oxazole, 5-phenyl-2-[4-[2-(5-phenyl-2-thienyl)ethenyl]phenyl]-, (E)-	toluene	315(4.40),382(4.48)	103-1028-86
$C_{27}H_{19}N_3O_2S$			
2H,5H-[1]Benzothiopyrano[3,4-e]-1,3-oxazin-5-one, 3,4-dihydro-3-phenyl-2,4-di-3-pyridinyl-	MeOH	231(3.7),331(3.9)	42-0323-86
$C_{27}H_{19}N_5O$			
1,1,2,2-Ethenetetracarbonitrile, 1-[4,5,6,7-tetrahydro-1-(2-methylphenyl)-4-oxo-2-phenyl-1H-indol-5-yl]-	MeOH	245(4.32),282(4.35), 474(4.44)	103-0172-86
1,1,2,2-Ethenetetracarbonitrile, 1-[4,5,6,7-tetrahydro-1-(3-methylphenyl)-4-oxo-2-phenyl-1H-indol-5-yl]-	MeOH	250(4.60),278(4.38), 475(4.52)	103-0172-86
1,1,2,2-Ethenetetracarbonitrile, 1-[4,5,6,7-tetrahydro-1-(4-methylphenyl)-4-oxo-2-phenyl-1H-indol-5-yl]-	MeOH	217(4.60),250(4.60), 472(4.50)	103-0172-86
$C_{27}H_{20}$			
5,11[1',2']-Benzeno-5H-cyclohepta[b]naphthalene, 5a,11-dihydro-5-phenyl-	EtOH	253(3.66)	18-1095-86
$C_{27}H_{20}ClO$			
Dibenzo[c,h]xanthylium, 7-(4-chlorophenyl)-5,6,8,9-tetrahydro-, perchlorate	MeCN	228(4.54),295(4.24), 318s(4.09),340(3.91), 448(4.50)	22-0600-86
$C_{27}H_{20}Cl_2N_2O$			
Phenol, 2,4-dichloro-6-[3-(1,2-dihydro-	$CHCl_3$	225(3.038),295(2.567),	65-0301-86

Compound	Solvent	$\lambda_{max}(\log \epsilon)$	Ref.
5-acenaphthylenyl)-4,5-dihydro-1-phenyl-1H-pyrazol-5-yl]- (cont.)		377(2.605)	65-0301-86
$C_{27}H_{20}NOPS$			
5(4H)-Oxazolethione, 2-phenyl-4-(triphenylphosphoranylidene)-	CH_2Cl_2	357(4.11)	65-1323-86
$C_{27}H_{20}NO_2P$			
5(4H)-Oxazolone, 2-phenyl-4-(triphenylphosphoranylidene)-	CH_2Cl_2	338(4.23)	65-1323-86
$C_{27}H_{20}NO_3$			
Dibenz[c,h]xanthylium, 5,6,8,9-tetrahydro-7-(4-nitrophenyl)-, perchlorate	MeCN	228(4.53),250(4.29), 300(4.38),318s(4.26), 450(4.43)	22-0600-86
$C_{27}H_{20}N_2$			
Pyrrolo[3,2-a]carbazole, 3,10-dihydro-3-(3-methylphenyl)-2-phenyl-	MeOH	200(4.22),228s(4.12), 248(4.28),253(4.28), 290(4.19),350s(3.33), 432(4.06)	103-0172-86
Pyrrolo[3,2-a]carbazole, 3,10-dihydro-3-(4-methylphenyl)-2-phenyl-	MeOH	210(4.33),228s(4.19), 256(4.40),290(4.28), 351s(3.37),440(3.08)	103-0172-86
$C_{27}H_{20}N_2O_2$			
Spiro[9H-fluorene-9,3'(3'aH)-pyrazolo-[1,5-a]pyridine]-2'-carboxylic acid, 5'-phenyl-, methyl ester	CH_2Cl_2	406(3.96)	89-0572-86
$C_{27}H_{20}N_2O_2S$			
1,3,4-Oxadiazole, 2-[4-[2-[5-(4-methoxyphenyl)-2-thienyl]ethenyl]phenyl]-5-phenyl-, (E)-	toluene	301(4.48),383(4.33)	103-1028-86
$C_{27}H_{20}N_2O_9S$			
Benzoic acid, 3-(5-formyl-2-hydroxyphenoxy)-4-methoxy-, 1,2,3,4,5a,6-hexahydro-2-methyl-1,4-dioxo-3-thioxooxepino[3',4':4,5]pyrrolo[1,2-a]-pyrazil-6-yl ester, (5aS-cis)-(aurantioemestrin)	EtOH	227s(4.53),263(4.45), 287s(4.36),340(4.25), 418(4.01)	77-1495-86
$C_{27}H_{20}N_2O_{10}$			
Dethiosecoemestrin	MeOH	230(4.26),259(4.17), 286s(4.00),365(3.83)	94-2411-86
$C_{27}H_{20}N_6O_5$			
Methanone, [2,4-dihydroxy-3,5-bis(5-methoxy-2H-benzotriazol-2-yl)phenyl]phenyl-	$CHCl_3$	249(4.45),285(4.48), 340(4.57),350s(4.30)	49-0805-86
$C_{27}H_{20}O_5$			
7H-Furo[3,2-g][1]benzopyran-7-one, 2-(4-methoxybenzoyl)-3,5-dimethyl-6-phenyl-	$CHCl_3$	290(4.48),355(4.50)	18-1299-86
$C_{27}H_{21}BO_4$			
Boron, [1,3-bis(4-methylphenyl)-1,3-propanedionato-O,O'][1,8-naphthalenediolato(2-)-O,O']-, (T-4)-	DMF	318s(4.24),338s(4.37), 344s(4.36),364s(4.42), 379(4.44),397s(4.28),	48-0755-86

Compound	Solvent	λ max (log ε)	Ref.
(cont.)		446s(2.91)	48-0755-86
Boron, [1,3-bis(4-methylphenyl)-1,3-propanedionato-O,O'][2,3-naphthalenediolato(2-)-O,O']-. (T-4)-	DMF	401(4.49),455(2.68)	48-0755-86
$C_{27}H_{21}BO_6$			
Boron, [1,3-bis(4-methoxyphenyl)-1,3-propanedionato][2,3-naphthalenediolato(2-)-O,O']-, (T-4)-	DMF	312s(3.92),319s(4.04), 338(4.25),407s(4.67), 425(4.78),484s(2.56), 517s(2.34)	48-0755-86
$C_{27}H_{21}BrO$			
Acetophenone, 2-(4-bromophenyl)-3-methyl-4,6-diphenyl-	MeCN	242(4.63)	48-0359-86
Acetophenone, 6-(4-bromophenyl)-3-methyl-2,4-diphenyl-	MeCN	244(4.68)	48-0359-86
$C_{27}H_{21}BrO_4$			
Benzoic acid, 4-bromo-, 1,4-dihydro-5,7,8-trimethyl-1,4-dioxo-2-(1-propenyl)-3-phenanthrenyl ester (relative absorbance)	ether	239(1.0),294s(0.41), 302(0.43),379(0.06)	33-1395-86
$C_{27}H_{21}Br_3O_6$			
5H-Tribenzo[a,d,g]cyclononene-2,7,12-triol, 3,8,13-tribromo-10,15-dihydro-, triacetate	dioxan	210(4.62),227(4.54), 275(3.48),284(3.53)	152-0017-86
$C_{27}H_{21}ClN_2O_2$			
Benzoic acid, 4-chloro-, (2-phenoxy-1,2-diphenylethylidene)hydrazide, (E)-	MeOH	264(4.25)	48-0181-86
$C_{27}H_{21}ClO$			
Acetophenone, 2-(4-chlorophenyl)-3-methyl-4,6-diphenyl-	MeCN	242(4.57)	48-0359-86
Acetophenone, 6-(4-chlorophenyl)-3-methyl-2,4-diphenyl-	MeCN	243(4.62)	48-0359-86
$C_{27}H_{21}FO$			
Acetophenone, 2-(4-fluorophenyl)-3-methyl-4,6-diphenyl-	MeCN	240(4.55)	48-0359-86
Acetophenone, 6-(4-fluorophenyl)-3-methyl-2,4-diphenyl-	MeCN	239(4.58)	48-0359-86
$C_{27}H_{21}N_2O$			
Pyridinium, 4-(5-[1,1'-biphenyl]-4-yl-2-oxazolyl)-1-(phenylmethyl)-, chloride	H_2O	395(4.33)	135-0244-86
	EtOH	285(4.31),405(4.39)	135-0244-86
4-methylbenzenesulfonate	H_2O	390(4.42)	135-0244-86
	EtOH	285(4.37),395(4.50)	135-0244-86
	5% HOAc	390(4.45)	135-0244-86
	50% HOAc	395(4.47)	135-0244-86
$C_{27}H_{21}N_3O_2$			
9,10-Anthracenedione, 1-(methylamino)-4,5-bis(phenylamino)-	$CHCl_3$	628(4.33),676(4.40)	104-0547-86
9,10-Anthracenedione, 5-(methylamino)-1,4-bis(phenylamino)-	$CHCl_3$	628(4.32),668(4.36)	104-0547-86
1H-Pyrazole, 3-(1,2-dihydro-5-acenaphthalenyl)-4,5-dihydro-5-(4-nitro-	$CHCl_3$	230(2.631),290(2.269), 370(2.141)	65-0301-86

Compound	Solvent	λ_{max}(log ε)	Ref.
phenyl)-1-phenyl- (cont.)			65-0301-86
$C_{27}H_{21}N_3O_5$			
9,12-Epoxy-1H-diindolo[1,2,3-fg:3',2'-1'-kl]pyrrolo[3,4-i][1,6]benzodiazocine-10-carboxylic acid, 2,3,9,10,11-12-hexahydro-10-hydroxy-9-methyl-1-oxo-, methyl ester (K-252a)	MeOH	298(5.0),330(4.5), 350(4.4),366(4.4)	158-1066-86
	MeOH	226(4.62),248(4.61), 264s(4.64),280s(4.83), 290(5.01),320s(4.28), 333(4.42),350(4.28), 367(4.32)	158-1072-86
$C_{27}H_{21}O$			
Dibenzo[c,h]xanthylium, 5,6,8,9-tetrahydro-7-phenyl-, perchlorate	MeCN	228(4.60),292(4.28), 315(4.16),444(4.50)	22-0600-86
$C_{27}H_{22}$			
9H-Fluorene, 2,7-bis(3-methylphenyl)-	EtOH	323(4.69)	46-2666-86
$C_{27}H_{22}N_2$			
1H-Pyrazole, 3-(1,2-dihydro-5-acenaphthylenyl)-4,5-dihydro-1,5-diphenyl-	$CHCl_3$	225(2.641),290(2.251), 375(1.926)	65-0301-86
$C_{27}H_{22}N_2O$			
Phenol, 2-[3-(1,2-dihydro-5-acenaphthylenyl)-4,5-dihydro-1-phenyl-1H-pyrazol-5-yl]-	$CHCl_3$	230(2.728),285(2.226), 355(1.940)	65-0301-86
$C_{27}H_{22}N_2O_2$			
2-Butene-1,4-dione, 2-(4,5-dimethyl-2-phenyl-1H-imidazol-1-yl)-1,4-diphenyl-, (E)-	MeOH	215(4.36),263(4.65)	44-3420-86
(Z)-	MeOH	218(4.44),272(4.66)	44-3420-86
Methanone, (5,6-dihydro-2,3-dimethylimidazo[2,1-a]isoquinoline-5,6-diyl)bis[phenyl-	MeOH	223(4.44),256(4.60)	44-3420-86
$C_{27}H_{22}N_2O_2S$			
Methanesulfonamide, N-[4-(9-acridinyl)phenyl]-N-(phenylmethyl)-	MeOH	207(4.72),259(5.04), 358(4.39)	150-3425-86M
$C_{27}H_{22}N_2O_3$			
2,4-Cyclohexadien-1-one, 4-nitro-6-[[5,7,8,9-tetrahydro-5-(phenylmethyl)-10-phenanthridinyl]methylene]-	EtOH	243(4.59),300(4.16), 330(4.19),370(4.22), 410(4.28),500(4.15)	135-1208-86
	acetone	590(--)	135-1208-86
	$CHCl_3$	650(--)	135-1208-86
	DMF	580(--)	135-1208-86
$C_{27}H_{22}N_2O_{10}S_2$			
Emestrin	MeOH	230(4.44),262(4.19), 278(3.96)	39-0109-86C
$C_{27}H_{22}O$			
Acetophenone, 3-methyl-2,4,6-triphenyl-	MeCN	239(4.56)	48-0359-86
$C_{27}H_{22}O_4$			
Ethanone, 1-[5-(diphenylmethyl)-2,3,4-trihydroxyphenyl]-2-phenyl-	MeOH	262(4.44),296(4.6)	2-0649-86
$C_{27}H_{22}O_5$			
4H-1-Benzopyran-4-one, 7-acetoxy-3-	MeOH	216(4.53),244(4.10),	2-0755-86

Compound	Solvent	λmax(log ε)	Ref.
3-acetyl-6-(diphenylmethyl)-2-methyl- (cont.)		292(3.53)	2-0755-86
4H-1-Benzopyran-4-one, 7-acetoxy-3-acetyl-8-(diphenylmethyl)-2-methyl-	MeOH	214(4.47),244(4.03),290(3.44)	2-0755-86
2H,5H-Pyrano[3,2-c][1]benzopyran-5-one, 2-(acetoxymethyl)-3,4-dihydro-2,4-diphenyl-	EtOH	270(4.03),281(4.04),305(3.94),318s(--)	32-0501-86
$C_{27}H_{23}NO$			
Benzenemethanamine, N-methyl-α-(phenoxyphenylmethylene)-N-phenyl-, (E)-	EtOH	209(4.38),256(4.48),356(3.95)	22-0781-86
(Z)-	EtOH	213(4.42),262(4.29),358(3.99)	22-0781-86
$C_{27}H_{23}N_3$			
Benzenamine, 2-(5,12-dihydro-5,12-dimethylindolo[3,2-a]carbazol-6-yl)-N-methyl-	EtOH	242(4.38),270(4.33),284(4.28),338(3.59),353(3.73)	78-5019-86
2,3':2',3"-Ter-1H-indole, 1,1',1"-trimethyl-	EtOH	220(4.49),297(4.19)	78-5019-86
$C_{27}H_{23}O_3$			
Pyrylium, 2-[2-(3,4-dimethoxyphenyl)-ethenyl]-4,6-diphenyl-, tetrafluoroborate	CH_2Cl_2	360(4.62),560(4.54)	104-1786-86
$C_{27}H_{24}N_6O$			
2,4(1H,3H)-Pteridinedione, 1,3-dimethyl-6,7-diphenyl-, 4-(methylphenylhydrazone)	pH 0.7	228(4.55),282(4.28),387(4.21)	128-0199-86
	pH 5-13	227(4.45),260(4.43),360(4.08),490(3.69)	128-0199-86
	MeOH	224(4.46),260(4.42),360(4.07),496(3.71)	128-0199-86
$C_{27}H_{24}OPS_2$			
Phosphonium, [[[(4-methoxyphenyl)thioxomethyl]thio]methyl]triphenyl-, iodide	CH_2Cl_2	360(4.36),483(2.53)	18-1403-86
$C_{27}H_{24}OS$			
2-Propanone, 1-[4-(4-methylphenyl)-2,6-diphenyl-2H-thiopyran-2-yl]-	dioxan	262(4.38),349(3.65)	48-0573-86
$C_{27}H_{24}O_2S$			
2-Propanone, 1-[4-(4-methoxyphenyl)-2,6-diphenyl-2H-thiopyran-2-yl]-	dioxan	266(4.36),346(3.68)	48-0573-86
$C_{27}H_{24}O_3$			
1,4-Methanonaphthalene-5,8,9-trione, 1,4,4a,8a-tetrahydro-1,4,6,7-tetramethyl-2,3-diphenyl-, (1α,4α,4aα,8aα)-	MeCN	229(4.31),251(4.33)	12-1537-86
2-Oxapentacyclo[6.4.0.0^{1,4}.0^{3,7}.0^{5,9}]-dodec-11-ene-6,10-dione, 5,7,11,12-tetramethyl-3,4-diphenyl-	MeCN	220(4.15),268(3.76)	12-1537-86
Pentacyclo[5.4.0.0^{2,6}.0^{3,10}.0^{5,9}]undecane-4,8,11-trione	MeCN	217(4.10)	12-1537-86
$C_{27}H_{24}PS_2$			
Phosphonium, [[[(4-methylphenyl)thioxomethyl]thio]methyl]triphenyl-,	CH_2Cl_2	321(4.38),487(2.44)	18-1403-86

Compound	Solvent	λ_{max}(log ϵ)	Ref.
iodide (cont.)			18-1403-86
$C_{27}H_{25}ClN_4O_6$			
1H-Pyrazolo[3,4-d]pyrimidine, 6-chloro-1-[2-deoxy-3,5-bis-O-(4-methylbenzoyl)-β-D-erythro-pentofuranosyl]-4-methoxy-	MeOH	241(4.63)	5-1213-86
$C_{27}H_{25}NO$			
Pyrrolo[2,1-a]isoquinolin-1(5H)-one, 6b,10b-dihydro-10b-(1-methylethyl)-2,3-diphenyl-	MeCN	238(4.09),274(4.08), 335(3.84)	118-0899-86
$C_{27}H_{25}NOS$			
Cycloprop[hi]indolizinium, hexahydro-6c-(4-methoxyphenyl)-1,6b-diphenyl-	MeCN	264(4.31)	118-0899-86
1(5H)-Indolizinium, 6,7,8,8a-tetrahydro-8a-[4-(methylthio)phenyl]-2,3-diphenyl-	MeCN	231s(--),276(4.27), 351(3.93)	118-0899-86
Morpholine, 4-(2,4,6-triphenyl-2H-thiopyran-2-yl)-	dioxan	253(4.36),345(3.74)	48-0567-86
$C_{27}H_{25}NO_2S$			
1H-Pyrrolizine, 6-(3,4-dimethoxyphenyl)-2,3-dihydro-7-phenyl-5-(phenylthio)-	n.s.g.	211(4.91),240(3.90), 276s(--)	83-0065-86
$C_{27}H_{25}NO_5$			
8H-Dibenzo[a,g]quinolizin-8-one, 5,6-dihydro-2,3,9-trimethoxy-13-(3-methoxyphenyl)-	MeOH	250s(4.24),326s(4.15), 366(4.22),376(4.16)	44-2781-86
$C_{27}H_{25}NS$			
Pyrrolo[2,1-a]isoquinoline-1(5H)-thione, 6,10b-dihydro-10b-(1-methylethyl)-2,3-diphenyl-	MeCN	255(4.10),423(4.10)	118-0899-86
$C_{27}H_{25}NS_2$			
Cycloprop[hi]indolizine-2(1H)-thione, hexahydro-6c-[4-(methylthio)phenyl]-1,6b-diphenyl-	MeCN	266(4.34),290(4.17), 335s(--)	118-0899-86
$C_{27}H_{25}N_3$			
Propanedinitrile, (5,6,7,7a,8,9,10,11-octahydro-2,3-diphenyl-1H-pyrrolo[2,1-j]quinolin-1-ylidene)-	MeCN	243(3.76),382(4.58), 397(4.59)	118-0908-86
$C_{27}H_{25}N_3O_4$			
Spiro[2H-1-benzopyran-2,2'-[2H]indole]-1'(3H)-propanamide, 3',5'-dimethyl-6-nitro-N-phenyl-	CH_2Cl_2	340(2.99)	18-1853-86
after uv irradiation	CH_2Cl_2	345(4.00),567(4.08)	18-1853-86
$C_{27}H_{25}N_7O_3$			
Benzamide, N-[2-[(2,6-dicyano-4-nitrophenyl)azo]-5-(dipropylamino)phenyl]-	$CHCl_3$	322(3.81),424(3.64), 588s(4.65),641(4.77)	48-0497-86
$C_{27}H_{26}ClFN_2OS$			
1-Propanone, 3-[4-(2-chloro-7-fluoro-10,11-dihydrodibenzo[b,f]thiepin-10-yl)-1-piperazinyl]-, dihydrochloride	MeOH	241(4.30),270s(3.94)	73-2598-86

Compound	Solvent	λ_{max}(log ε)	Ref.
$C_{27}H_{26}ClN_3$			
Quinoline, 6-chloro-4-phenyl-2-[4-(2-phenylethyl)-1-piperazinyl]-	MeOH	221(4.58),257(4.59), 265(3.81)	83-0338-86
$C_{27}H_{26}NO_3$			
1-Benzopyrylium, 4-[2-[4-(dimethylamino)phenyl]ethenyl]-6-methoxy-2-(4-methoxyphenyl)-, perchlorate	CH_2Cl_2	684(5.13)	103-0703-86
	MeCN	666(4.97)	103-0703-86
1-Benzopyrylium, 4-[2-[4-(dimethylamino)phenyl]ethenyl]-7-methoxy-2-(4-methoxyphenyl)-, perchlorate	CH_2Cl_2	666(5.18)	103-0703-86
	MeCN	645(4.95)	103-0703-86
$C_{27}H_{26}N_2O_4$			
6H-[1,4]Diazepino[1,2-a:4,3-a']diindole-6,8,14,15(7H)-tetrone, 3,11-bis(1,1-dimethylethyl)-	benzene	296(4.36),431(3.72)	138-1393-86
$C_{27}H_{26}N_3O_4P$			
5-Pyrimidinecarboxylic acid, 1,2,3,4-tetrahydro-1,3-dimethyl-2,4-dioxo-6-[(triphenylphosphoranylidene)amino]-, ethyl ester	$CHCl_3$	268(4.11),275(4.15), 296(4.30)	44-0149-86
$C_{27}H_{26}N_4O_8$			
1,2,3,4-Butanetetrol, 1-(1-phenyl-1H-pyrazolo[3,4-b]quinoxalin-3-yl)-, tetraacetate (arabino)	MeOH	233s(4.3),267(4.7), 335(4.1),408(3.6)	159-0049-86
lyxo isomer	MeOH	233s(4.4),265(4.7), 333(4.1),408(3.6)	159-0049-86
$C_{27}H_{26}N_4O_{10}P$			
5'-Uridylic acid, 2',3'-O-(6,9-dihydro-2,5,11-trimethyl-9-oxo-10H-pyrido[4,3-b]carbazolium-10-ylidene)-, hexafluorophosphate	50% HOAc	276(4.73),302(4.60), 314(4.61),373(4.24), 450(3.60)	87-1350-86
$C_{27}H_{26}N_6O_8S$			
1H,4aH-[1,2,4]Triazolo[1',2':1,2]-[1,2]diazeto[3,4-d]azepine-4a,9-dicarboxylic acid, 2,3,5,6,7,9a-hexahydro-2-methyl-7-[(4-methylphenyl)sulfonyl]-1,3-dioxo-8-[(phenylcarbonimidoyl)amino]-, dimethyl ester	$CHCl_3$	232(4.16),258(4.13), 292(4.08)	24-2114-86
$C_{27}H_{27}NOS$			
2H,6H-[1,3]Thiazino[2,3-a]isoquinoline, 2-ethoxy-7,11b-dihydro-11b-methyl-3,4-diphenyl-	MeCN	273(3.80),293(3.88)	118-0899-86
$C_{27}H_{27}NO_2$			
2,5-Cyclohexadiene-1,4-dione, 2-[[3-(9,10-dihydro-9,10-ethanoanthracen-9(10H)-yl)propyl]methylamino]-5-methyl-	CH_2Cl_2	266s(4.00),273(4.02), 280s(4.00),496(3.54)	106-0029-86
2,5-Cyclohexadiene-1,4-dione, 2-[[3-(9,10-dihydro-9,10-ethanoanthracen-9(10H)-yl)propyl]methylamino]-6-methyl-	CH_2Cl_2	266(3.89),272(3.93), 280s(3.90),495(3.55)	106-0029-86
2,5-Cyclohexadiene-1,4-dione, 2-[[3-(5H-dibenzo[a,d]cyclohepten-5-yl)-	CH_2Cl_2	240s(4.40),295(4.33), 494(3.44)	83-0421-86

Compound	Solvent	λ_{max}(log ϵ)	Ref.
propyl]methylamino]-5,6-dimethyl- (cont.)			83-0421-86
$C_{27}H_{27}NO_7$			
Benzanthrin pseudoaglycone	MeOH	235(4.64),264(4.10), 326(4.19),450(3.97)	158-1515-86
$C_{27}H_{27}NS$			
2H-Thiopyran-2-amine, N,N,3,5-tetramethyl-2,4,6-triphenyl-	dioxan	231(4.27),339(3.61)	48-0567-86
$C_{27}H_{27}N_5O_6$			
1H-Pyrazolo[3,4-d]pyrimidin-6-amine, 1-[2-deoxy-3,5-bis-O-(4-methylbenzoyl)-β-D-erythro-pentofuranosyl]-4-methoxy-	MeOH	225(4.56),241(4.58), 274(4.04)	33-1602-86
2H-Pyrazolo[3,4-d]pyrimidin-6-amine, 2-[2-deoxy-3,5-bis-O-(4-methylbenzoyl)-β-D-erythro-pentofuranosyl]-4-methoxy-	MeOH	224(4.46),240(4.53), 283(3.94),295(3.93)	33-1602-86
$C_{27}H_{27}N_5O_9P$			
5'-Cytidylic acid, 2',3'-O-(6,9-dihydro-2,5,11-trimethyl-9-oxo-10H-pyrido[4,3-b]carbazolium-10-ylidene)-, hexafluorophosphate	50% HOAc	277(4.65),314(4.33), 375(4.05)	87-1350-86
$C_{27}H_{28}ClN_3O$			
Piperazine, 1-[1-[(2-benzoyl-4-chlorophenyl)imino]ethyl]-4-(2-phenylethyl)-	MeOH	222(4.38),252(4.46), 358(3.45)	83-0338-86
$C_{27}H_{28}CrO_5$			
Chromium, tricarbonyl[(1,10,11,12,13-14-η)-5,15-dibutoxytricyclo[8.2.2-$2^{4,7}$]hexadeca-2,4,6,8,10,12,13,15-octaene]	CH_2Cl_2	335(3.72),406(3.48)	88-2353-86
Chromium, tricarbonyl[(4,5,6,7,15,16-η)-5,15-dibutoxytricyclo[8.2.2.$2^{4,7}$]-hexadeca-2,4,6,8,10,12,13,15-octaene]-	CH_2Cl_2	337(4.01)	88-2353-86
$C_{27}H_{28}N_2O_2$			
Benzeneacetonitrile, α-[[4-(2-diethylamino)ethoxy]phenyl](4-hydroxyphenyl)methylene]-	EtOH	343(4.16)	87-2511-86
$C_{27}H_{28}N_4O_6$			
2,4(1H,3H)-Pyrimidinedione, 5,5'-methylenebis[1-(2-hydroxyethyl)-6-methyl-3-phenyl-	EtOH	281(4.25)	118-0857-86
$C_{27}H_{28}O_{12}$			
4H-1-Benzopyran-4-one, 5,6,7,8-tetraacetoxy-2-[2-(2-acetoxyphenyl)ethyl]-5,6,7,8-tetrahydro-, [5S-(5α,6β,7α,8β)]-	EtOH	204(4.18),250(3.98)	94-3033-86
$C_{27}H_{29}BrO_4$			
1-Phenanthrenecarboxylic acid, 7-ethenyl-4b,5,6,7,8,8a,9,10-octahydro-3-hydroxy-4b,7-dimethyl-, 2-(4-bromo-	MeOH	256(4.42)	94-1015-86

Compound	Solvent	λ_{max}(log ϵ)	Ref.
phenyl)-2-oxoethyl ester (cont.)			94-1015-86
$C_{27}H_{29}NO_4$			
6H-Dibenzo[a,g]quinolizine, 5,8,13,13a-tetrahydro-2,3,9,10-tetramethoxy-8-phenyl-, cis-(±)-	MeOH	240(3.61),280(3.59)	78-2075-86
trans-(±)-	MeOH	240(3.63),281(3.60)	78-2075-86
$C_{27}H_{29}NO_5$			
α-D-Ribofuranose, 3-deoxy-3-(hydroxyamino)-1,2-O-(1-methylethylidene)-5-O-(triphenylmethyl)-	EtOH	212(3.66)	33-1132-86
$C_{27}H_{29}NO_{10}$			
Daunorubicin	pH 7.4	482(3.96)	161-0881-86
$C_{27}H_{30}N_2O_2$			
2,5-Cyclohexadiene-1,4-dione, 2-[[3-(10,11-dihydro-5H-dibenzo[b,f]azepin-5-yl)propyl]methylamino]-3,5,6-trimethyl-	CH_2Cl_2	239(4.22),270s(4.11), 300s(3.8),532(3.22)	83-0421-86
$C_{27}H_{30}N_2O_3$			
1-Piperidinecarboxylic acid, 3-ethyl-4-[3-oxo-3-(4-quinolinyl)propyl]-, phenylmethyl ester, cis-(±)-	MeOH	232(4.07),238(4.03), 305(3.57),316(3.59)	77-0573-86
$C_{27}H_{30}N_3$			
2-Propynylium, 1,1,3-tris[4-(dimethylamino)phenyl]-, perchlorate	CH_2Cl_2	663(5.12)	138-0329-86
$C_{27}H_{30}N_6O_6$			
Benzamide, 4-(1,1-dimethylethyl)-N-[6-[(6-methyl-3-pyridinyl)oxy]-9-β-D-ribofuranosyl-9H-purin-2-yl]-	pH 7	243(4.25),277(4.42)	1-0806-86
	pH 2	243(--),277(--)	1-0806-86
	pH 13	238(--),275(--)	1-0806-86
$C_{27}H_{30}N_6O_7S_2$			
L-Proline, 1-[[2-[[[2-[4-amino-4-oxo-1-[[(phenylmethoxy)carbonyl]amino]butyl]-4-thiazolyl]carbonyl]amino]methyl]-4-thiazolyl]carbonyl]-, methyl ester	MeOH	239(3.90)	44-4580-86
$C_{27}H_{30}O_{13}$			
Kushenol O	MeOH	258(4.50),305(4.12)	95-0022-86
	MeOH-NaOAc	258(4.50),306(4.17)	95-0022-86
Sophoroflavone A	MeOH	259s(4.03),313s(4.27), 328(4.28)	100-0645-86
	MeOH-NaOMe	254(4.67),326(4.23), 392(4.39)	100-0645-86
	MeOH-NaOAc	270(4.43),308(4.31), 378(4.34)	100-0645-86
	MeOH-$AlCl_3$	258s(4.03),311(4.27), 329(4.28),390(3.53)	100-0645-86
$C_{27}H_{30}O_{16}$			
4H-1-Benzopyran-4-one, 3-[(6-O-β-D-glucopyranosyl-β-D-glucopyranosyl)oxy]-5,7-dihydroxy-2-(4-hydroxyphenyl)-	MeOH	270(4.65),350(4.60)	95-0982-86
	MeOH-NaOMe	283(--),406(--)	95-0982-86
	MeOH-NaOAc	276(--),370(--)	95-0982-86

Compound	Solvent	λ_{max}(log ϵ)	Ref.
$C_{27}H_{30}O_{17}$			
4H-1-Benzopyran-4-one, 3-[(2-O-D-apio-β-D-furanosyl-β-D-glucopyranosyl)-oxy]-2-(3,4-dihydroxyphenyl)-5,7-dihydroxy-6-methoxy-	MeOH	256(4.24),267s(4.18), 295(3.90),350(4.25)	102-0231-86
	MeOH-NaOMe	269(4.31),335(4.01), 401(3.33)	102-0231-86
	MeOH-NaOAc	270(4.31),330s(4.02), 382(4.21)	102-0231-86
	MeOH-$AlCl_3$	277(4.35),305s(3.90), 348(3.75),434(4.35)	102-0231-86
	MeOH-$AlCl_3$-HCl	270(4.26),300s(3.94), 383(4.22),410s(4.18)	102-0231-86
	MeOH-NaOAc-H_3BO_3	261(4.32),373(4.26)	102-0231-86
$C_{27}H_{32}CrO_5$			
Chromium, tricarbonyl[(1,10,11,12,13-14-η)-5,15-dibutoxytricyclo[8.2.2-$2^{4,7}$]hexadeca-4,6,10,12,13,15-hexa-ene]-	CH_2Cl_2	336(4.05)	88-2353-86
Chromium, tricarbonyl[(4,5,6,7,15,16-η)-5,15-dibutoxytricyclo[8.2.2.$2^{4,7}$]-hexadeca-4,6,10,12,13,15-hexaene]-	CH_2Cl_2	339(3.75)	88-2353-86
$C_{27}H_{32}CrO_5Si_2$			
Chromium, tricarbonyl[[(1,10,11,12,13-14-η)-5,15-dimethoxytricyclo[8.2.2-$2^{4,7}$]hexadeca-2,4,8,10,12,13,15-hexa-deca-2,4,6,8,10,12,13,15-octaene-2,9-diyl]bis[trimethylsilane]]-	CH_2Cl_2	339(4.08)	88-2353-86
$C_{27}H_{32}N_4$			
2,3'-Bipyridinium, 1',1'''-(1,5-pentane-diyl)bis[1-methyl-, tetraperchlorate	pH 2.3	206(4.48),271(4.42)	64-0239-86B
$C_{27}H_{32}N_6O$			
1H-Indole, 3,3'-[1-ethyl-1,2,5,6-tetra-hydro-3-(4-morpholinyl)-1,2,4-tria-zine-5,6-diyl]bis[2-methyl-, mono-tetrafluoroborate	EtOH	225(4.88),285(4.15), 290(4.11)	103-1242-86
$C_{27}H_{32}O_3$			
Phenol, 2-[1-[3-(1,1-dimethylethyl)-2-hydroxy-5-methylphenyl]ethyl]-6-[1-(2-hydroxyphenyl)ethyl]-, (R*,S*)-(-)-	EtOH	279(3.84)	116-0509-86
$C_{27}H_{32}O_4$			
1,3-Benzenediol, 4-(1,1-dimethyl-2-propenyl)-6-[3-[7-hydroxy-4-(3-methyl-2-butenyl)-5-benzofuran-yl]propyl]- (kazinol L)	EtOH	248(3.58),259(3.55), 285(3.50)	94-2448-86
$C_{27}H_{32}O_6$			
1(2H)-Anthracenone, 3-acetoxy-6-(3,7-dimethyl-2,6-octadienyl)oxy]-3,4-dihydro-8,9-dihydroxy-3-methyl-, (E)-(+)- (acetylvismione D)	$CHCl_3$	278(4.68),320(3.87), 332(3.79),400(4.09)	102-0766-86
	MeOH	274(--),319(--), 334(--),396(--)	102-0766-86
	MeOH-$AlCl_3$	269(--),290(--), 352(--),455(--)	102-0766-86
	MeOH-NaOAc	274(--),319(--), 334(--),396(--)	102-0766-86

Compound	Solvent	$\lambda_{max}(\log \epsilon)$	Ref.
$C_{27}H_{32}O_8$			
Isomargosinolide	MeOH	220(3.85)	78-4849-86
Margosinolide	MeOH	223(3.85)	78-4849-86
$C_{27}H_{32}O_{10}$			
3-Furanmethanol, 2-(4-acetoxy-3,5-dimethoxyphenyl)-4-[(4-acetoxy-3-methoxyphenyl)methyl]tetrahydro-, acetate, (2α,3β,4β)-(±)-	MeOH	223(4.23),274(3.61), 282(3.58)	100-0706-86
$C_{27}H_{33}BrO_5$			
Benzoic acid, 4-bromo-, 4-[2-(2,5-dihydro-5-oxo-3-furanyl)ethyl]decahydro-4,5,7a,7b-tetramethyl-naphth-[1,2-b]oxiren-7-yl ester, [1aS-(1aα,3aα,4α,5α,7β,7aα,7bα)]-	EtOH	243(4.24)	78-3461-86
$C_{27}H_{33}N_3O_4$			
5H,14H-Pyrrolo[1",2":4',5']pyrazino-[1',2':1,6]pyrido[3,4-b]indole-5,14-dione, 1,2,3,5a,6,11,12,14a-octahydro-6-hydroxy-9-methoxy-11-(3-methyl-2-butenyl)-12-(2-methyl-1-propenyl)-, (±)- (12-deoxy-12-epifumitremorgin B)	MeOH	208(4.56),224(4.61), 264(3.79),272(3.81), 294(3.87)	88-2391-86
$C_{27}H_{33}N_3O_5$			
Fumitremorgin B	MeOH	203(4.56),225(4.56), 275(3.90),295(3.94)	88-6361-86
isomer	MeOH	205(4.58),222(4.60), 273(3.86),296(3.95)	88-6361-86
$C_{27}H_{33}N_5O_{16}S$			
β-D-Ribofuranosylamine, N-[1-(2,3,5-tri-O-acetyl-β-D-ribofuranosyl)-1H-pyrazino[2,3-c][1,2,6]thiadiazin-4-yl]-, S,S-dioxide	MeOH	248(4.42),290s(3.93), 340(4.05)	5-1872-86
$C_{27}H_{34}N_2O_6$			
Pentanedioic acid, 3-hydroxy-3-[(8)-6'-methoxycinchonan-2'-yl]-, dimethyl ester	MeOH-HCl	258(4.40),323(3.59), 345(3.63)	78-3711-86
$C_{27}H_{34}O_{12}$			
2-Oxoborapetoside B	MeOH	213(3.84),244(3.55)	94-2868-86
$C_{27}H_{35}ClO_9$			
Ptilosarcen-12-propanoate	EtOH	218(3.74)	78-6565-86
$C_{27}H_{35}FO_7$			
Androsta-1,4-diene-17-carboxylic acid, 9-fluoro-11-hydroxy-17-(1-oxopropoxy)-16-methyl-3-oxo-, anhydride with propanoic acid, (11β,16β,17α)-	MeOH	239(4.21)	44-2315-86
$C_{27}H_{35}N_3O_5$			
Acetamide, 2,2'-[(1-acetyl-4-piperidinylidene)bis[(2,6-dimethyl-4,1-phenylene)oxy]]bis-	MeOH	197(4.90),230s(4.15), 268(2.99),275(2.90)	35-2273-86

Compound	Solvent	λ_{max}(log ϵ)	Ref.
$C_{27}H_{36}O_6$			
2H-1-Benzopyran-2-one, 6-(3,7-dimethyl-2,6-octadienyl)-5,7-dihydroxy-4-(1-hydroxypropyl)-8-(2-methyl-1-oxobutyl)- (surangin C)	MeOH	227(4.01),256(4.75), 333(4.51)	102-0555-86
$C_{27}H_{36}O_{11}$			
Tinophylloloside	MeOH	210(4.03)	102-0905-86
$C_{27}H_{36}O_{15}$			
2H-Pyran-5-carboxylic acid, 4-(1,3-dioxolan-2-ylmethyl)-3-ethenyl-3,4-dihydro-2-[(2,3,4,6-tetraacetyl-β-D-glucopyranosyl)oxy]-, methyl ester	ether	229(4.02)	5-1413-86
2H-Pyran-5-carboxylic acid, 4-(1,3-dioxolan-2-ylmethyl)-3-ethylidene-3,4-dihydro-2-[(2,3,4,6-tetraacetyl-β-D-glucopyranosyl)oxy]-, methyl ester, [2S-(2α,3E,4β)]-	ether	233(4.02)	5-1413-86
$C_{27}H_{37}F_2O_9P$			
Androsta-1,4-diene-17-carboxylic acid, 17-acetoxy-6,9-difluoro-11-hydroxy-16-methyl-3-oxo-, anhydride with diethyl phosphate, (6α,11β,16α,17α)-	MeOH	238(4.21)	44-2315-86
$C_{27}H_{37}NO_6S$			
17,18,19,20-Tetranorleucotriene E_4, 16-phenoxy-, dimethyl ester, (±)-	EtOH	268(4.52),274(4.62), 284(4.47)	104-1457-86
$C_{27}H_{38}N_4$			
7H-[1,2,4]Triazolo[4,3-a]perimidine, 10-pentadecyl-	dioxan	227(3.88),250(3.54), 267(3.26),298(3.32), 352(3.49)	48-0237-86
$C_{27}H_{38}N_4O_{12}$			
1H-Purine-2,6-dione, 3,9-dihydro-1,3-dimethyl-9-[2,3,4-tri-O-acetyl-6-O-(6-acetoxyhexyl)-β-D-galactopyranosyl]-	$CHCl_3$	279(3.87)	136-0256-86K
$C_{27}H_{38}O_3$			
Benzaldehyde, 2,4-dihydroxy-3-(3,7,11-15-tetramethyl-2,6,10,14-hexadecatetraenyl)-, (E,E,E)-	EtOH	240s(4.12),287(4.34), 325s(3.82)	64-0645-86B
$C_{27}H_{38}O_3S_2$			
Acetic acid, [[2-hydroxy-3-[(phenylmethyl)thio]-4,6,8,11-heptadecatetraenyl]thio]-, methyl ester, (±)-	EtOH	272(4.56),284(4.65), 295(4.58)	104-1875-86
$C_{27}H_{38}O_4$			
Naphtho[2,3-e][1,3]dioxin-5,10-dione, 4-heptyl-2-octyl-	EtOH	251(4.32),277(4.15), 339(3.38),381(3.04)	39-0659-86C
$C_{27}H_{38}O_5$			
Amentol	EtOH	220(4.02),290(3.52)	78-6015-86
Cystoseirol B	MeOH	230(3.94),296(3.60)	44-2707-86
Cystoseirol C	MeOH	229(3.94),296(3.62)	44-2707-86
Cystoseirol D	MeOH	229(3.95),293(3.33)	44-2707-86
Cystoseirol E	MeOH	229(3.94),296(3.35)	44-2707-86

Compound	Solvent	λ_{max}(log ϵ)	Ref.
$C_{27}H_{38}O_7$			
Card-20(22)-enolide, 3,16-diacetoxy-14-hydroxy-, (3β,5β,16β)- (gitoxigenin 3,16-diacetate)	MeOH	219(4.14)	87-0997-86
Pseudopterosin B	MeOH	233(3.78),274(3.15), 285(3.20)	44-5140-86
	MeOH-KOH	250(--),288(--), 294(--)	44-5140-86
Pseudopterosin C	MeOH	229(3.98),275(3.18), 282(3.23)	44-5140-86
	MeOH-KOH	245(--),285(--), 295(--)	44-5140-86
$C_{27}H_{38}O_8$			
Card-20(22)-enolide, 3-acetoxy-14-hydroxy-16-[(methoxycarbonyl)oxy]-, (3β,5β,14β,16β,17β)-	MeOH	218(4.16)	87-0997-86
Card-20(22)-enolide, 14-acetoxy-16-hydroxy-3-[(methoxycarbonyl)oxy]-	MeOH	220.5(4.18)	87-0997-86
Gerardiasterone	MeOH	243(4.10)	77-0040-86
Slov-3-enolide, 8α,11α-diangeloyloxy-10β-acetoxy-	EtOH	215(4.2)	102-1747-86
$C_{27}H_{38}O_9$			
Card-20(22)-enolide, 14-hydroxy-3,16-bis[(methoxycarbonyl)oxy]-, (3β,5β,16β)-	MeOH	218.5(4.17)	87-0997-86
$C_{27}H_{38}O_{16}$			
2H-Pyran-5-carboxylic acid, 4-(1,3-dioxolan-2-ylmethyl)-3,4-dihydro-3-(1-hydroxyethyl)-2-[(2,3,4,6-tetraacetyl-β-D-glucopyranosyl)oxy]-, methyl ester	ether	231(4.04)	5-1413-86
$C_{27}H_{38}O_{17}$			
2H-Pyran-5-carboxylic acid, 3-(1,2-dihydroxyethyl)-4-(1,3-dioxolan-2-ylmethyl)-3,4-dihydro-2-[(2,3,4,6-tetraacetyl-β-D-glucopyranosyl)-oxy]-, methyl ester, [2S-[2α-3β(R*),4β]]-	ether	232(4.11)	5-1413-86
$C_{27}H_{39}N_3O_2$			
Teleocidin B-4, de-N-methyl-	MeOH	230(4.55),281(3.92)	94-4883-86
$C_{27}H_{40}O$			
Cholesta-1,4,6-trien-3-one	EtOH	300(4.13)	18-3702-86
$C_{27}H_{40}OS_2$			
Benzene, 1-methyl-4-[[1-[(4-methylphenyl)sulfinyl]tridecyl]thio]-	dioxan	253(4.1)	88-6381-86
$C_{27}H_{40}O_2$			
Cholesta-1,4-diene-3,15-dione	EtOH	242(4.14)	39-1797-86C
Cholesta-1,4-diene-3,24-dione	EtOH	243(4.15)	39-1797-86C
$C_{27}H_{40}O_4$			
2H-Pyran-2-one, 3-[2-[decahydro-2,4a-dimethyl-5-(4-methyl-3-pentenyl)-6-oxo-2-naphthalenyl]ethyl]-4-hydroxy-	MeOH	231(4.20),300(3.87)	88-2121-86

Compound	Solvent	$\lambda_{max}(\log \epsilon)$	Ref.
5,6-dimethyl-, [2R-(2α,4aα,5α,8aβ)]-(cont.)			88-2121-86
$C_{27}H_{40}O_7$			
Dihydropseudopterosin C	MeOH	226(3.98),244(3.32), 251(3.36),256(3.38), 263(3.32),272(3.15), 284(3.23)	44-5140-86
$C_{27}H_{40}O_{14}$			
Furcatoside C, 2',3'-di-O-acetyl-	MeOH	209(3.57)	102-1227-86
$C_{27}H_{41}BrN_2O_4$			
Androst-4-en-3-one, 17-acetoxy-4-bromo-2-[4-(2-hydroxyethyl)-1-piperazinyl]-, (2β,17β)-	EtOH	262.0(4.13)	118-0461-86
$C_{27}H_{41}N_2$			
Pyridinium, 4-[2-[4-(dimethylamino)-phenyl]ethenyl]-1-dodecyl-, chloride	EtOH	482(4.60)	4-0209-86
perchlorate	EtOH	482(4.65)	4-0209-86
$C_{27}H_{42}N_2O_3$			
1H-Indole-2,4-dione, 3-[[4,4-dimethyl-2-[(2-methylpropyl)amino]-6-oxo-1-cyclohexen-1-yl]-3,5,6,7-tetrahydro-3,6,6-trimethyl-1-(2-methylpropyl)-	EtOH	287(4.46),345(4.26)	2-0347-86
$C_{27}H_{42}O$			
Cholesta-4,6-dien-3-one	EtOH	284(4.46)	18-3702-86
Cholesta-1,5,7-trien-3β-ol	EtOH	282(4.04)	18-3702-86
5,9-Cyclo-9,10-secocholesta-1(10),2-dien-4-one, (9ξ)	EtOH	323(3.86)	88-4505-86
19-Norcholesta-1,3,5(10)-trien-4-ol, 1-methyl-	EtOH	285(3.30)	88-4505-86
20,29,30-Trinorlup-18-en-21-one	EtOH	228(4.25)	73-0621-86
$C_{27}H_{42}O_2$			
Cholesta-1,4-dien-3-one, 2-hydroxy-	MeOH	254(4.14)	78-5693-86
Cholest-4-ene-3,12-dione	EtOH	240(4.20)	39-1797-86C
Cholest-4-ene-3,15-dione	EtOH	242(4.09)	39-1797-86C
Cholest-4-ene-3,16-dione	EtOH	242(4.03)	39-1797-86C
Cholest-4-ene-3,24-dione	EtOH	243(4.16)	39-1797-86C
Sarcodictyenone	EtOH	203(4.40),229(3.93)	33-1581-86
$C_{27}H_{42}O_3$			
Cholest-2-en-1-one, 6,7-epoxy-5-hydroxy-, (5α,6α,7α)-	EtOH	221(3.98)	39-0321-86C
$C_{27}H_{42}O_4$			
Dihydropycnophorin	MeOH	291(3.87)	88-2121-86
$C_{27}H_{42}O_6$			
Pinnasterol, 14α-hydroxy-	EtOH	256(3.88)	102-1305-86
$C_{27}H_{42}O_7$			
Raspailyne A triacetate	MeOH	274(4.27),288(4.19)	77-0077-86
$C_{27}H_{43}NO$			
6-Aza-B-homocholesta-2,4-dien-7-one	EtOH	273(3.89)	39-2123-86C

Compound	Solvent	$\lambda_{max}(\log \epsilon)$	Ref.
Cholesta-2,4-dien-6-one, oxime, (6E)-	EtOH	300(3.88)	39-2123-86C
$C_{27}H_{43}NO_3$			
Peiminine	EtOH	287(1.88)	102-2008-86
$C_{27}H_{43}N_3O$			
9,10-Secocholesta-5,7,10(19)-trien-25-ol, 3-azido-, (3α,5Z,7E)-	EtOH	267(4.30)	130-0134-86
$C_{27}H_{43}N_3O_2$			
9,10-Secocholesta-5,7,10(19)-triene-1,25-diol, 3-azido-, (1α,3α,5Z,7E)-	EtOH	301(4.34)	130-0134-86
$C_{27}H_{44}O$			
Cholesta-4,6-dien-3β-ol	EtOH	239(4.40)	18-3702-86
$C_{27}H_{44}OSi$			
Silane, (1,1-dimethylethyl)dimethyl-[[3,3,5,5-tetramethyl-4-(3-methyl-2-phenyl-2-butenylidene)cyclohexyl]-oxy]-, (±)-	C_6H_{12}	248(3.91)	44-2361-86
$C_{27}H_{44}O_2$			
9,10-Secocholesta-5,7,10(19)-triene-3,25-diol, (3β,5E,7E)-	EtOH	273(4.33)	44-4819-86
(3β,5Z,7E)-	EtOH	262(4.28)	44-4819-86
$C_{27}H_{46}O_3Si$			
5-Tetradecen-4-ol, 1-[[(1,1-dimethylethyl)dimethylsilyl]oxy]-, [R-(Z)]-	EtOH	230(4.18)	163-0187-86

Compound	Solvent	λ_{max}(log ε)	Ref.
$C_{28}Cl_{18}$			
Indeno[2,1-a]indene, 1,2,3,4,6,7,8,9-octachloro-5,10-bis(pentachlorophenyl)-	C_6H_{12}	217(5.15),244s(4.73), 300(4.77),312(4.89), 397s(3.49),423(3.90), 448(4.21),480(4.23)	44-1100-86
$C_{28}Cl_{20}$			
1H-Indene, 4,5,6,7-tetrachloro-1-[chloro(pentachlorophenyl)methylene]-2,3-bis(pentachlorophenyl)-	C_6H_{12}	275(4.59),360(3.88)	44-1100-86
Naphthalene, 1,2,3,4,5-pentachloro-6,7,8-tris(pentachlorophenyl)-	C_6H_{12}	271(4.73),322s(4.00), 335(4.03)	44-1100-86
$C_{28}Cl_{22}$			
Naphthalene, 1,1,2,5,6,7,8-heptachloro-1,2-dihydro-2,3,4-tris(pentachlorophenyl)-	C_6H_{12}	218(4.87),229(4.86), 280s(4.02),307s(3.59)	44-1100-86
$C_{28}H_{12}O_3$			
Naphtho[1',2',3',4'-1,12]perylo[2,3-c]furan-5,7-dione	5% KOH	249(4.06),297(4.18), 309(4.31),377(3.56), 397(3.73),421(3.79)	24-3521-86
$C_{28}H_{12}O_7$			
Dianthra[1,2-b:2',1'-d]furan-5,10,15,20-tetrone, 6,9-dihydroxy-	CH_2Cl_2	266(4.63),296(4.40), 335(4.01),437(4.33), 490s(4.05)	5-0839-86
$C_{28}H_{14}S_2$			
Pentaleno[1,2,3-kl:4,5,6-k'l']bis-thioxanthene	CH_2Cl_2	254(4.79),302(4.61), 328(4.30),342(4.39), 580(4.34),642(4.50)	138-0829-86
$C_{28}H_{16}CuN_2O_6$			
Copper, bis(1-amino-4-hydroxy-9,10-anthracenedionato-O^4,O^{10})-	DMF	295(3.44),497s(3.32), 535(3.39),570(3.32), 595s(3.02)	111-0385-86
$C_{28}H_{16}N_4O_2$			
Propanedinitrile, [4-cyano-4-(4-nitrophenyl)-3,5-diphenyl-2,5-cyclohexadien-1-ylidene]-	$CHCl_3$	355(4.39)	83-0898-86
$C_{28}H_{16}O_5$			
Ethanedione, 1,1'-(2,8-dibenzofurandiyl)bis[2-phenyl-	dioxan	257(5.66),285s(5.34), 378s(2.78)	104-1117-86
$C_{28}H_{16}O_8S_3$			
Thieno[2',3',4',5'-6,7]perylo[1,12-cde]-1,2-dithiin-4,5,9,10-tetracarboxylic acid, tetramethyl ester	KOH	253(4.37),260(4.44), 283(4.53),293(4.54), 372(3.89),389(4.31), 412(4.66)	104-0943-86
$C_{28}H_{17}NO_4$			
7H-Dibenzo[a,h]carbazole-7,12(13H)-dione, 5-acetoxy-13-phenyl-	$CHCl_3$	274(4.93),322(3.84), 432(4.01)	104-0748-86
Ethanedione, 1,1'-(9H-carbazole-3,6-diyl)bis[2-phenyl-	dioxan	298(4.30),358(4.30)	104-1117-86
$C_{28}H_{17}N_3$			
Propanedinitrile, (4-cyano-3,4,5-tri-	$CHCl_3$	350(4.42)	83-0898-86

Compound	Solvent	$\lambda_{max}(\log \epsilon)$	Ref.
phenyl-2,5-cyclohexadien-1-ylidene)- (cont.)			83-0898-86
Propanedinitrile, (2,3-diphenyl-4H-benzo[a]quinolizin-4-ylidene)-	MeCN	337(4.16),387(3.76), 490(4.13)	118-0908-86
$C_{28}H_{18}Cl_2O_4$			
9,10-Anthracenedione, 1,4-dichloro-5,8-bis(phenylmethoxy)-	MeOH	246(4.42),410(3.65)	4-1491-86
$C_{28}H_{18}O_4$			
Ethanedione, 1,1'-[1,1'-biphenyl]-4,4'-diylbis[2-phenyl-	dioxan	254(5.27),309(5.49), 385s(3.00)	104-1117-86
$C_{28}H_{18}O_5$			
Ethanedione, 1,1'-(oxydi-4,1-phenylene)bis[2-phenyl-	dioxan	255s(5.19),296(5.44), 375s(2.80)	104-1117-86
$C_{28}H_{19}N_3$			
Propanedinitrile, (6,7-dihydro-2,3-diphenyl-4H-benzo[a]quinolizin-4-ylidene)-	MeCN	288(4.18),343(3.98), 436(4.02)	118-0908-86
$C_{28}H_{19}N_3O$			
Acetamide, 2-cyano-2-(4-cyano-3,4,5-triphenyl-2,5-cyclohexadien-1-ylidene)-	$CHCl_3$	345(4.40)	83-0898-86
$C_{28}H_{19}N_5OS$			
Ethenetricarbonitrile, [4-(1,1-dimethylethyl)-6-phenyl[1,3,4]oxadiazino-[6,5,4-kl]phenothiazin-2-yl]-, trans	$CHCl_3$	271(4.49),317s(4.52), 330(4.58),345s(4.48), 564(4.68)	39-2233-86C
[1,4]Oxazino[2,3,4-kl]phenothiazine-1,1,2,2-tetracarbonitrile, 4-(1,1-dimethylethyl)-6-phenyl-	$CHCl_3$	264(4.48),283s(4.13), 458(4.22)	39-2233-86C
$C_{28}H_{19}N_5O_4$			
Propanenitrile, 3,3'-[[4-(6,11-dihydro-6,11-dioxobenzo[f]pyrido[1,2-a]indol-12-yl)-3-nitrophenyl]-imino]bis-	DMF	516(4.000)	2-1126-86
$C_{28}H_{20}$			
Dibenzo[a,g]cycloeicosene, 21,22,23,24-tetradehydro-, (E,E,E,E,Z)-	isooctane	293s(4.67),320s(5.03), 333(5.08),413s(4.37), 435(4.10),462s(4.29), 498s(4.16)	18-2209-86
$C_{28}H_{20}F_{11}N_2$			
Quinolinium, 2-[3,3,4,4,5,5,6,6,7,7,7-undecafluoro-2-[(1-methyl-2(1H)-quinolinylidene)methyl]-1-heptenyl]-1-methyl-, perchlorate	n.s.g.	645(4.73)	104-0144-86
$C_{28}H_{20}N_2OS$			
Pyrrolo[3,4-a]phenothiazin-4(2H)-one, 2,5-dimethyl-1,3-diphenyl-	MeOH	259(4.62),327(4.11), 374s(3.90),477(4.22)	4-1267-86
$C_{28}H_{20}N_2O_3$			
Spiro[9H-fluorene-9,3'(3'aH)-pyrazolo[1,5-a]pyridine]-2'-carboxylic	CH_2Cl_2	365(3.80),414(3.81)	89-0572-86

Compound	Solvent	λ_{max}(log ϵ)	Ref.
acid, 5'-benzoyl-, methyl ester (cont.)			89-0572-86
$C_{28}H_{20}N_4OS_3$			
4-Thiazolidinone, 3-ethyl-2-thioxo-5-(2,3,8-triphenyl-6H-thiazolo[3,4-b][1,2,4]triazin-6-ylidene)-	n.s.g.	458(4.54),645(3.60)	103-1373-86
$C_{28}H_{20}O$			
2-Propyn-1-one, 3-(1,3-dimethyl-1H-cyclopenta[l]phenanthren-1-yl)-1-phenyl-	EtOH	240(4.57),263(4.56), 322(3.89)	39-1471-86C
2-Propyn-1-one, 3-(1,3-dimethyl-1H-cyclopenta[l]phenanthren-2-yl)-1-phenyl-	EtOH	243(4.19),308(3.79), 405(3.87)	39-1471-86C
$C_{28}H_{20}O_2$			
2,4,6-Heptatrienal, 7-[2-[4-[2-(5-oxo-1,3-pentadienyl)phenyl]-1,3-butadiynyl]phenyl]-, (all-E)-	THF	225(4.42),240s(4.36), 272(4.45),284s(4.48), 300s(4.56),310s(4.64), 340(4.66)	18-2209-86
$C_{28}H_{20}O_4$			
2H-1-Benzopyran-2-one, 6-(diphenylmethyl)-4,7-dihydroxy-3-phenyl-	MeOH	214(4.2),319(3.6)	2-0623-86
2H-1-Benzopyran-2-one, 8-(diphenylmethyl)-4,7-dihydroxy-3-phenyl-	MeOH	227(4.0),298(4.1)	2-0623-86
4H-1-Benzopyran-4-one, 6-(diphenylmethyl)-7,8-dihydroxy-3-phenyl-	MeOH	240(4.52),270(4.44), 292(4.32)	2-0649-86
$C_{28}H_{21}NOS$			
Oxazole, 2-[4-[2-[5-(4-methylphenyl)-2-thienyl]ethenyl]phenyl]-5-phenyl-	toluene	314(4.42),383(4.46)	103-1028-86
$C_{28}H_{21}NO_2S$			
Oxazole, 2-[4-[2-[5-(4-methoxyphenyl)-2-thienyl]ethenyl]phenyl]-5-phenyl-	toluene	315(4.46),387(4.43)	103-1028-86
$C_{28}H_{21}N_3$			
Propanedinitrile, (2,3,6,7-tetrahydro-2,3-diphenyl-4H-benzo[a]quinolizin-4-ylidene)-	MeCN	228(4.20),268(4.29), 335(4.07)	118-0908-86
$C_{28}H_{21}N_5$			
3H-Pyrrol-3-one, 2,5-diphenyl-4-(phenylazo)-, phenylhydrazone	EtOH	200(4.12),235s(3.95), 295(4.19),480(3.98)	103-0499-86
$C_{28}H_{22}$			
Anthracene, 1,8-bis(2-methylphenyl)-	C_6H_{12}	243(4.68),249(3.94), 254(5.09),261(5.13), 332(3.51),349(3.82), 366(4.00),386(3.95)	44-0921-86
Anthracene, 1,8-bis(3-methylphenyl)-	C_6H_{12}	257s(4.91),263(5.18), 340(3.62),358(3.87), 375(4.03),395(3.97)	44-0921-86
2,2'-Biphenanthrene, 9,9',10,10'-tetrahydro-	EtOH	322(4.69)	46-2666-86
	dioxan	324(4.69)	46-2666-86
$C_{28}H_{22}BrNO_2$			
1,4-Naphthalenedione, 2-bromo-3-[[3-(5H-dibenzo[a,d]cyclohepten-5-yl)propyl]amino]-	CH_2Cl_2	240s(4.3),278(4.48), 290s(4.4),340s(3.3), 472(3.45)	83-0791-86

Compound	Solvent	λ_{max}(log ε)	Ref.
1,4-Naphthalenedione, 2-bromo-3-[(9,10-dihydro-9,10-ethanoanthracen-9(10H)-ylmethyl)methylamino]-	CH_2Cl_2	245(4.15),251(4.15), 267s(4.2),274(4.13), 280s(4.2),335s(3.65), 505(3.45)	83-0791-86
$C_{28}H_{22}ClNO_2$			
1,4-Naphthalenedione, 2-chloro-3-[(9,10-dihydro-9,10-ethanoanthracen-9-ylmethyl)methylamino]-	CH_2Cl_2	244(4.15),250(4.17), 265s(4.1),273(4.11), 280s(4.1),330s(3.6), 503(3.45)	83-0791-86
$C_{28}H_{22}ClN_7O_2S$			
Acetamide, N-[2-[[4-chloro-5-[[1-(dicyanomethylene)-1,3-dihydro-3-oxo-2H-inden-2-ylidene]methyl]-2-thiazolyl]azo]-5-(diethylamino)phenyl]-	CH_2Cl_2	710(4.87)	77-1639-86
	DMSO	730(--)	77-1639-86
$C_{28}H_{22}ClO$			
5H-Benzo[6,7]cyclohepta[1,2-b]naphtho[2,1-e]pyrylium, 7-(4-chlorophenyl)-6,8,9,10-tetrahydro-, perchlorate	MeCN	224(4.36),292(4.16), 340(3.85),426(4.35)	22-0600-86
$C_{28}H_{22}ClO_2$			
Dibenzo[c,h]xanthylium, 7-(4-chlorophenyl)-5,6,8,9-tetrahydro-3-methoxy-, perchlorate	MeCN	228(4.59),252(4.48), 312(4.35),350s(4.04), 478(4.57)	22-0600-86
$C_{28}H_{22}NO_4$			
Dibenzo[c,h]xanthylium, 5,6,8,9-tetrahydro-3-methoxy-7-(4-nitrophenyl)-, perchlorate	MeCN	228(4.61),252(4.57), 315(4.38),344(4.05), 488(4.40)	22-0600-86
$C_{28}H_{22}N_2O$			
Furo[2,3-f][4,7]phenanthroline, 2,3-dihydro-2,3-bis(2-methylphenyl)-	EtOH	239(4.44),253(4.43), 279(4.25),288(4.20), 321(3.93)	39-0415-86C
$C_{28}H_{22}N_2OS$			
4H-Isoindol-4-one, 2,7-dihydro-7-[(2-mercaptophenyl)imino]-2,5-dimethyl-1,3-diphenyl-	MeOH	258(4.60),367(4.23), 489(3.14)	4-1267-86
$C_{28}H_{22}N_2OS_2$			
1-Triphenodithiazinol, 2-(1,1-dimethyl)-4-phenyl-	MeOH-20% $CHCl_3$	278(4.31),328s(3.93), 564(4.20)	39-2233-86C
	+ HCl	281(4.46),335(4.34), 615s(4.15),670(4.30), 734(4.24)	39-2233-86C
$C_{28}H_{22}N_2O_2$			
3-Pyridinecarboxylic acid, 6-[1,1'-biphenyl]-4-yl-5-cyano-2-methyl-4-phenyl-, ethyl ester	MeCN	293(4.53)	73-1061-86
$C_{28}H_{22}N_2O_4$			
Spiro[anthracene-9(10H),1'(10'bH)-pyrrolo[2,1-a]phthalazine]-2',3'-dicarboxylic acid, dimethyl ester	CH_2Cl_2	478(--)	89-0572-86
	CH_2Cl_2	378(3.97)(photochromic form)	89-0572-86

Compound	Solvent	λ_{max}(log ϵ)	Ref.
$C_{28}H_{22}N_6O_6$			
Benzoic acid, 4-cyano-, 5'-ester with 4-cyano-N-[4-cyano-1-[2,3-O-(1-methylidene)-β-D-ribofuranosyl]-1H-imidazol-5-yl]benzamide	MeOH	247(4.58)	23-2109-86
$C_{28}H_{22}N_8S_2$			
4H-Pyrazole-3,5-diamine, N,N,N',N'-tetramethyl-4-(4-phenyl-1,3-dithiol-2-ylidene)-, conjugate monoacid, salt with 2,2'-(2,5-cyclohexadiene-1,4-diylidene)bis(propanedinitrile)	MeCN	227(4.37),280(3.75), 394(4.90),421(4.59), 665(3.59),680(3.70), 698(3.67),727(4.01), 743(4.23),761(4.15), 820(4.34),845(4.52)	78-0849-86
$C_{28}H_{22}O$			
2-Propen-1-one, 3-(1,3-dimethyl-1H-cyclopenta[l]phenanthren-1-yl)-1-phenyl-, (E)-	EtOH	245s(4.78),262(4.83), 322(4.22)	39-1471-86C
2-Propen-1-one, 3-(1,3-dimethyl-1H-cyclopenta[l]phenanthren-2-yl)-1-phenyl-, (E)-	EtOH	246s(3.85),256(3.88), 302(3.49),429(3.36)	39-1471-86C
$C_{28}H_{22}O_3$			
Cyclohepta[de]naphthalen-7(8H)-one, 8,8-bis(phenylmethoxy)-	$CHCl_3$	320(3.95),336s(3.91)	39-1965-86C
$C_{28}H_{22}O_4$			
1,4-Naphthalenedione, 2,2'-(1,2-ethanediyl)bis[3-(1-methylethenyl)-	EtOH	249(4.57),257s(4.47), 332(3.76)	39-2217-86C
$C_{28}H_{22}O_6$			
1,4,8,11-Pentacenetetrone, 6,13-dimethoxy-2,3,9,10-tetramethyl-	CH_2Cl_2	262(4.71),324(4.43), 339(4.64),429(3.51), 459(3.68),485(3.89), 524(3.96)	44-4160-86
$C_{28}H_{22}O_6S_2$			
Benzenesulfonic acid, 2,2'-([1,1'-biphenyl]-4,4'-diyldi-2,1-ethenediyl)-bis-, disodium salt	EtOH	349(4.79)	61-0439-86
$C_{28}H_{23}ClN_2O_3S$			
5-Thia-1-azabicyclo[4.2.0]oct-2-ene-2-carboxylic acid, 3-(chloromethyl)-8-oxo-7-[(phenylmethylene)amino]-, diphenylmethyl ester, (6R-trans)-	EtOH	215(4.44),258(4.41)	158-1092-86
$C_{28}H_{23}NO_2$			
1,4-Naphthalenedione, 2-[[3-(10,11-dihydro-5H-dibenzo[a,d]cyclohepten-5-ylidene)propyl]amino]-	CH_2Cl_2	240s(4.5),263(4.41), 270(4.44),295s(4.0), 330s(3.5),445(3.49)	83-0607-86
1,4-Naphthalenedione, 2-[(9,10-ethanoanthracen-9(10H)-ylmethyl)methylamino]-	CH_2Cl_2	238s(4.2),245s(4.1), 278(4.22),330s(3.5), 458(3.48)	83-0607-86
$C_{28}H_{23}NO_3$			
1H-Pyrrolizine-5-acetic acid, 2,3-dihydro-α-oxo-6,7-diphenyl-, phenylmethyl ester	MeOH	204(4.67),231(4.32), 328(4.22)	83-0500-86

Compound	Solvent	$\lambda_{max}(\log \epsilon)$	Ref.
$C_{28}H_{23}N_3O_3$			
Benzenepropanenitrile, α,α'-[[5-(dimethylamino)-2,4-furandiyl]dimethylidyne]bis[4-methyl-β-oxo-	MeOH	270(2.51),392(2.36), 524(2.86),676(2.08)	73-0879-86
$C_{28}H_{23}N_4S_2$			
Thiazolo[3,4-b][1,2,4]triazin-5-ium, 6-[(3-ethyl-2(3H)-benzothiazolylidene)methyl]-2-methyl-3,8-diphenyl-, perchlorate	n.s.g.	449(4.69),510(3.75)	103-1373-86
Thiazolo[3,4-b][1,2,4]triazin-5-ium, 6-[(3-ethyl-2(3H)-benzothiazolylidene)methyl]-3-methyl-2,8-diphenyl-, perchlorate	n.s.g.	448(4.72),522(3.79)	103-1373-86
$C_{28}H_{23}N_5O_9S$			
Acridinium, 9-(N-methyl-4-methanesulfonamidophenyl)-10-methyl-, picrate	MeOH	207(4.68),262(4.98), 362(4.53)	150-3425-86
$C_{28}H_{23}O$			
5H-Benzo[6,7]cyclohepta[1,2-b]naphtho[2,1-e]pyrylium, 6,8,9,10-tetrahydro-7-phenyl-, perchlorate	MeCN	230(4.54),294(4.18), 422(4.21)	22-0600-86
Dibenzo[c,h]xanthylium, 5,6,8,9-tetrahydro-7-(4-methylphenyl)-, perchlorate	MeCN	228(4.64),295(4.22), 316(4.18),445(4.42)	22-0600-86
$C_{28}H_{23}O_2$			
Dibenzo[c,h]xanthylium, 5,6,8,9-tetrahydro-3-methoxy-7-phenyl-, perchlorate	MeCN	228(4.50),250s(4.35), 310(4.30),350s(3.98), 475(4.51)	22-0600-86
Dibenzo[c,h]xanthylium, 5,6,8,9-tetrahydro-7-(4-methoxyphenyl)-, perchlorate	MeCN	228(4.66),294(4.16), 318(4.14),444(4.46)	22-0600-86
Pyrylium, 2-[4-(4-methoxyphenyl)-1,3-butadienyl]-4,6-diphenyl-, tetrafluoroborate	CH_2Cl_2	375(4.65),580(4.67)	104-1786-86
$C_{28}H_{23}O_3$			
5H-Indeno[1,2-b]naphtho[2,1-e]pyrylium, 7-(3,4-dimethoxyphenyl)-6,8-dihydro-, perchlorate	MeCN	230(4.60),284(4.29), 350(3.95),440(4.48)	22-0600-86
$C_{28}H_{24}$			
Azulene, 2,6-bis[2-(4-methylphenyl)-ethenyl]-, (E,E)-	CH_2Cl_2	245.5(4.18),327.5(4.74), 461.5(4.76),575.5(2.93), 608.5(2.92)	24-2272-86
Phenanthrene, 9,10-dihydro-2,7-bis(3-methylphenyl)-	EtOH	318(4.64)	46-2666-86
1,6:7,12:13,18:19,24-Tetramethano[24]-annulene	hexane	263(4.96),338(4.35), 415(3.58)	89-0367-86
$C_{28}H_{24}Cl_2N_2$			
1,2-Ethenediamine, 1,2-bis(4-chlorophenyl)-N,N'-dimethyl-N,N'-diphenyl-, (E)-	EtOH	262(4.45),402(4.02)	22-0781-86
(Z)-	EtOH	242(4.20),276(4.23), 392(3.97)	22-0781-86
$C_{28}H_{24}N_2O$			
2H-Oxazolo[5,4-b]indole, 1,3a,4,8b-	EtOH	214(4.34),247(4.26),	22-0781-86

Compound	Solvent	$\lambda_{max}(\log \epsilon)$	Ref.
tetrahydro-4-methyl-1,3a,8b-triphenyl- (cont.)		300(3.69)	22-0781-86
$C_{28}H_{24}N_2O_6$			
6,2'-O-Cyclouridine, 5'-O-(triphenylmethyl)-	MeOH	250(4.11)	94-3623-86
$C_{28}H_{24}N_4O_4$			
Picrasidine H	EtOH	228(4.29),240(4.31), 276s(3.99),286(4.06), 334s(3.33),346(3.61), 368(3.46)	94-2090-86
$C_{28}H_{24}O$			
Acetophenone, 3-ethyl-2,4,6-triphenyl-	MeCN	238(4.57)	48-0359-86
Acetophenone, 3-methyl-2-(4-methylphenyl)-4,6-diphenyl-	MeCN	241(4.58)	48-0359-86
Acetophenone, 3-methyl-6-(4-methylphenyl)-2,4-diphenyl-	MeCN	243(4.56)	48-0359-86
$C_{28}H_{24}O_3$			
Ethanone, 1-[2-hydroxy-4-(phenylmethoxy)-3-(phenylmethyl)phenyl]-2-phenyl-	MeOH	230(4.5),292(4.5)	2-0646-86
$C_{28}H_{24}O_6$			
4H-1-Benzopyran-4-one, 5,7-diacetoxy-6-(diphenylmethyl)-2,3-dimethyl-	MeOH	264(4.58),303(4.63)	2-0478-86
4H-1-Benzopyran-4-one, 5,7-diacetoxy-8-(diphenylmethyl)-2,3-dimethyl-	MeOH	266(4.58),304(4.64)	2-0478-86
Cnetin F	MeOH	226(4.00),286(3.48)	20-0737-86
$C_{28}H_{24}O_8$			
5H-Dibenzo[a,d]cycloheptene-2,3,6,8-tetrol, 11-(3,4-dihydroxyphenyl)-5-[(3,5-dihydroxyphenyl)methyl]-10,11-dihydro- (cassigarol A)	MeOH	284(4.03)	94-4418-86
1(2H)-Naphthalenone, 7-acetyl-2-(7-acetyl-8-hydroxy-4-methoxy-6-methyl-1-oxo-2(1H)-naphthalenylidene)-8-hydroxy-4-methoxy-6-methyl-	$CHCl_3$	695(4.49)	64-0377-86B
$C_{28}H_{25}BrN_2O_2$			
1,4-Naphthalenedione, 2-bromo-3-[[3-(10,11-dihydro-5H-dibenzo-[b,f]azepin-5-yl)propyl]methylamino]-	CH_2Cl_2	243(4.33),251(4.31), 279(4.36),340s(3.5), 504(3.59)	83-0791-86
$C_{28}H_{25}ClN_2O_2$			
1,4-Naphthalenedione, 2-chloro-3-[[3-(10,11-dihydro-5H-dibenzo-[b,f]azepin-5-yl)propyl]methylamino]-	CH_2Cl_2	243s(4.2),250(4.28), 280(4.26),340s(3.4), 503(3.52)	83-0791-86
$C_{28}H_{25}N_3$			
1H-Indole, 6,7-dihydro-6,6-dimethyl-1,2-diphenyl-4-(phenylazo)-	MeOH	250(3.97),296(4.04), 402(4.03)	103-0172-86
$C_{28}H_{26}$			
1,1':6',1":6",1"'-Quater-1,3,5-cycloheptatriene	hexane	241(4.57),299(4.32), 340(4.13),394(4.02)	89-0367-86
p-Quaterphenyl, 2',3,3",3"'-tetramethyl-	EtOH	277(4.55)	46-2661-86

Compound	Solvent	$\lambda_{max}(\log \epsilon)$	Ref.
p-Quaterphenyl, 3,3',2",3"'-tetramethyl-	EtOH	266(4.61)	46-2661-86
	dioxan	269(4.62)	46-2661-86
$C_{28}H_{26}Br_2N_2P_2$			
1,5,3,7-Diazadiphosphocin, 1,5-bis(4-bromophenyl)octahydro-3,7-diphenyl-	MeCN	258(4.77),283(4.55)	65-0362-86
cobalt complex	MeCN	417(3.30),583(2.91)	65-0362-86
nickel complex	MeCN	401(3.08),511(3.21)	65-0362-86
$C_{28}H_{26}Br_2N_4S_2$			
6H-1,3-Thiazine, 4-(4-bromophenyl)-6-[4-(4-bromophenyl)-2-(1-pyrrolidinyl)-6H-1,3-thiazin-6-ylidene]-2-(1-pyrrolidinyl)-	$CHCl_3$	280s(4.39),312(4.36), 525(4.06)	118-0916-86
$C_{28}H_{26}Cl_6N_2O_{10}$			
[5,5'-Bi-1,3-dioxolo[4,5-g]isoquinoline]-6,6'(5H,5'H)-dicarboxylic acid, 7,7',8,8'-tetrahydro-4,4'-dimethoxy-, bis(2,2,2-trichloroethyl) ester (same for isomers a and b)	MeCN	220(4.2),260(3.1), 284(3.5)	83-1122-86
$C_{28}H_{26}F_2N_2P_2$			
1,5,3,7-Diazadiphosphocin, 1,5-bis(3-fluorophenyl)octahydro-3,7-diphenyl-	MeCN	256(4.59),294(4.29)	65-0362-86
cobalt complex	MeCN	381(3.43),575(2.85)	65-0362-86
nickel complex	MeCN	425(2.96),494(3.32)	65-0362-86
$C_{28}H_{26}N_2$			
1,2-Ethenediamine, N,N'-dimethyl-N,N',1,2-tetraphenyl-, (E)-	EtOH	217(4.11),259(4.54), 393(4.04)	22-0781-86
(Z)-	EtOH	219(4.16),263(4.37), 382(4.07)	22-0781-86
$C_{28}H_{26}N_2OS_2$			
Furo[2,3-f][4,7]phenanthroline, 2,3-dihydro-2,3-bis(2,4,5-trimethyl-3-thienyl)-, cis	EtOH	238(4.72),252(4.70), 278(4.35),287(4.29), 319(4.07)	39-0415-86C
trans	EtOH	239(4.66),253(4.68), 279(4.30),288(4.28), 321(4.01)	39-0415-86C
$C_{28}H_{26}N_8$			
Pyrazinecarbonitrile, 4,4'-(1,4-butanediyl)bis[3,4-dihydro-3-imino-6-methyl-5-phenyl-	EtOH	438(4.13)	33-1025-86
$C_{28}H_{26}OS$			
2-Propanone, 1-[2,6-bis(4-methylphenyl)-4-phenyl-2H-thiopyran-2-yl]-	dioxan	262(4.38),349(3.75)	48-0573-86
$C_{28}H_{26}O_2$			
[9,9'-Bianthracene]-10,10'-diol, 1,1',2,2',3,3',4,4'-octahydro-	MeOH	238(4.55),249s(4.23), 260(3.21)	39-1789-86C
	MeOH-KOH	253(4.74),341(3.96)	39-1789-86C
$C_{28}H_{26}O_4$			
p-Quaterphenyl, 2",3,3',3"'-tetramethoxy-	EtOH	302(4.46)	46-2661-86

Compound	Solvent	λ_{max}(log ϵ)	Ref.
$C_{28}H_{26}O_6$			
1,6:2,5-Dimethanonaphthalene-4,8,9,10-tetrone, 3-[(1,4-dihydro-3-methyl-1,4-dioxo-2-naphthalenyl)methyl]-octahydro-2,4a,6,8a-tetramethyl-, (1α,2α,3α,4aβ,5α,6α,8aβ)-(±)-	EtOH	246(4.32),264(4.18), 328(3.59)	78-3663-86
1,4,6-Ethanylylidenenaphthalene-3,5,8,10(2H)-tetrone, 7-[(1,4-dihydro-3-methyl-1,4-dioxo-2-naphthalenyl)methyl]hexahydro-1,4a,6,8a-tetramethyl-, (1α,4α,4aβ,6α,7β,8aβ-9S*)-(±)-	EtOH	250(4.46),258s(4.37), 331(3.57)	78-3663-86
$C_{28}H_{26}O_7$			
2H-Furo[2,3-h]-1-benzopyran-3-ol, 3,4-dihydro-5-methoxy-2,8-bis(4-methoxyphenyl)-, acetate, trans	dioxan	221s(4.31),242(3.91), 251(3.91),300s(4.30), 314(4.56),330(4.24)	94-0595-86
$C_{28}H_{26}O_8$			
[2,2'-Binaphthalene]-1,1',5,5'-tetrol, 4,4'-dimethoxy-7,7'-dimethyl-, 1,1'-diacetate	EtOH	248(4.25),261s(4.07), 283s(3.58),298s(3.59), 313(3.68),328(3.69), 342(3.74)	39-0675-86C
$C_{28}H_{26}Si_2$			
Silane, (benzo[3,4]cyclobuta[1,2-a]biphenylene-5,6-diyldi-2,1-ethynediyl)-bis[trimethyl-	C_6H_{12}	235(4.53),247(4.53), 280(4.70),293(4.82), 305(4.51),314(4.37), 320(4.37),331(4.62), 398(3.53),422(3.58), 452(3.51)	35-3150-86
$C_{28}H_{27}NOS$			
2-Propanone, 1-[4-[4-(dimethylamino)-phenyl]-2,6-diphenyl-2H-thiopyran-2-yl]-	dioxan	262(4.36),282s(4.32), 329(4.12)	48-0573-86
$C_{28}H_{27}NO_2S$			
Cycloprop[hi]indolizine-2(1H)-thione, 6c-(2,5-dimethoxyphenyl)-1,6b-diphenyl-1,2,4,5,6,6a,6b,6c-octahydro-, (1α,6aα,6bα,6cα)-	MeCN	269s(--),296(4.18), 341s(--)	118-0899-86
$C_{28}H_{27}NO_3$			
1(5H)-Indolizinone, 8a-(2,5-dimethoxyphenyl)-6,7,8,8a-tetrahydro-2,3-diphenyl-	MeCN	223s(--),285(4.12), 346(3.86)	118-0899-86
$C_{28}H_{27}N_3O_7$			
1H-Pyrazole-4-carboxylic acid, 3-(cyanomethyl)-1-[2-deoxy-3,5-bis-O-(4-methylbenzoyl)-β-D-erythro-pentofuranosyl]-, methyl ester	MeOH	237(4.54)	4-0059-86
1H-Pyrazole-4-carboxylic acid, 5-(cyanomethyl)-1-[2-deoxy-3,5-bis-O-(4-methylbenzoyl)-β-D-erythro-pentofuranosyl]-, methyl ester	MeOH	235(4.74)	4-0059-86
$C_{28}H_{27}N_5O_4$			
6H-Purin-6-one, 1,9-dihydro-9-[(2-hydroxyethoxy)methyl]-2-[[(4-methoxy-	MeOH	260(4.19),277(4.15)	87-1384-86

Compound	Solvent	$\lambda_{max}(\log \epsilon)$	Ref.
phenyl)diphenylmethyl]amino]- (cont.)			87-1384-86
$C_{28}H_{27}N_5O_9$			
Azauridine-elliptinium adduct	H_2O	274(4.49),312(4.21), 370(3.95),450(3.20)	87-1350-86
$C_{28}H_{27}N_5O_9S$			
1H-Pyrazino[2,3-c][1,2,6]thiadiazin-4-amine, 6,7-diphenyl-1-(2,3,5-tri-O-acetyl-β-D-ribofuranosyl)-, 2,2-dioxide	MeOH	220(4.33),278(4.26), 367(4.03)	5-1872-86
$C_{28}H_{27}N_7O_8P$			
5'-Adenylic acid, 2',3'-O-(6,9-dihydro-2,5,11-trimethyl-9-oxo-10H-pyrido-[3,4-b]carbazolium-10-ylidene)-, hexafluorophosphate	50% HOAc	274(4.41),299(4.22), 313(4.20),370(3.90)	87-1350-86
$C_{28}H_{27}N_7O_9P$			
5'-Guanylic acid, 2',3'-O-(6,9-dihydro-2,5,11-trimethyl-9-oxo-10H-pyrido-[4,3-b]carbazolium-10-ylidene)-, hexafluorophosphate, (S)-	50% HOAc	274(4.58),295(4.39), 312(4.32),366(4.04), 450(3.40)	87-1350-86
$C_{28}H_{28}ClGaN_4$			
Gallium, chloro[2,3,7,8,12,13,17,18-octamethyl-21H,23H-porphinato(2-)-$N^{21},N^{22},N^{23},N^{24}$]-, (SP-5-12)-	MeOH	375(4.20),395(5.07), 489(3.08),529(3.64), 567(3.75)	90-1157-86
$C_{28}H_{28}N_2$			
Benzenamine, 4-(2-ethenyl-5-ethyl-3,4-diphenyl-1H-pyrrol-1-yl)-N,N-dimethyl-	ether	265(4.66),295s(4.28)	78-1355-86
$C_{28}H_{28}N_2O_5$			
2,4(1H,3H)-Pyrimidinedione, 1-[(2-hydroxyethoxy)methyl]-3-[[4-(phenylmethoxy)phenyl]methyl]-5-(phenylmethyl)-	EtOH	266(4.13)	4-1651-86
$C_{28}H_{28}N_2O_6$			
2,4(1H,3H)-Pyrimidinedione, 1-[(2-hydroxyethoxy)methyl]-5-[[3-[[4-(phenylmethoxy)phenyl]methoxy]phenyl]-methyl]-	pH 1	267(4.03)	4-1651-86
	pH 11	267(4.03)	4-1651-86
$C_{28}H_{28}N_2P_2$			
1,5,3,7-Diazadiphosphocin, octahydro-1,3,5,7-tetraphenyl-	MeCN	258(4.42),292(4.25)	65-0362-86
cobalt complex	MeCN	417(3.41),572(2.91)	65-0362-86
nickel complex	MeCN	415(3.24),509(3.32)	65-0362-86
$C_{28}H_{28}N_2S_2$			
Propanethioamide, N,N-diethyl-2-(2,4,4-trimethyl-5(4H)-thiazol-5-ylidene)-	CH_2Cl_2	252(4.37),292s(3.91), 326s(3.15),380s(2.30)	33-0174-86
2-Propanethione, 1-(diethylamino)-1-(2,4,4-triphenyl-5(4H)-thiazol-ylidene)-	CH_2Cl_2	230(4.19),250s(4.14), 328(4.02),456(4.13)	33-0174-86
$C_{28}H_{28}N_4O_2S_2$			
Morpholine, 4-[6-[2-(4-morpholinyl)-	$CHCl_3$	269(4.57),320s(4.01),	118-0916-86

Compound	Solvent	λmax(log ε)	Ref.
4-phenyl-6H-1,3-thiazin-6-ylidene]-4-phenyl-6H-1,3-thiazin-2-yl]- (cont.)		502(4.08)	118-0916-86
$C_{28}H_{28}N_4O_7$			
1H-Pyrazolo]3,4-d]pyrimidine, 1-[2-deoxy-3,5-bis-O-(4-methylbenzoyl)-β-D-erythro-pentofuranosyl]-4,6-dimethoxy-	MeOH	240(4.58)	5-1213-86
$C_{28}H_{28}N_4O_8$			
Benzoic acid, 4-methoxy-, 5'-ester with N-[4-cyano-1-[2,3-O-(1-methylethylidene)-β-D-ribofuranosyl]-1H-imidazol-5-yl]-4-methoxy-	MeOH	259(4.54)	23-2109-86
$C_{28}H_{28}N_4S_2$			
6H-1,3-Thiazine, 4-phenyl-6-[4-phenyl-2-(1-pyrrolidinyl)-6H-1,3-thiazin-6-ylidene]-2-(1-pyrrolidinyl)-	$CHCl_3$	257(4.48),320s(3.97), 518(3.98)	118-0916-86
$C_{28}H_{28}O_{10}$			
Sigmoidin B tetraacetate	MeOH	287(4.04)	39-0033-86C
$C_{28}H_{28}O_{12}$			
2-Propenoic acid, 3-(4-hydroxyphenyl)-, 2'-ester with 6-[2-(β-D-glucopyranosyloxy)-4-hydroxy-6-methylphenyl]-4-methoxy-2H-pyran-2-one, (E)-	MeOH	205(--),224s(4.33), 298(4.39),308(4.40)	100-0800-86
$C_{28}H_{29}NOS$			
2H,6H-[1,3]Thiazino[2,3-a]isoquinoline, 7,11b-dihydro-2-methoxy-11b-(1-methylethyl)-3,4-diphenyl-	MeCN	273(3.79),298(3.89)	118-0899-86
$C_{28}H_{29}NS_2$			
2H,6H-[1,3]Thiazino[2,3-a]isoquinoline, 2-[(1,1-dimethylethyl)thio]-7,11b-dihydro-3,4-diphenyl-	MeCN	275s(--),290(3.65)	118-0899-86
$C_{28}H_{29}N_5O_5S$			
Benzamide, N-[9-[2,3-O-(1-methylethylidene)-5-O-[(4-methylphenyl)sulfonyl]-β-D-allofuranosyl]-9H-purin-6-yl]-	pH 7	256(4.42),296(4.34)	1-0806-86
	pH 2	258(--),292(--)	1-0806-86
	pH 13	250(--),296(--)	1-0806-86
$C_{28}H_{30}N_2$			
Benzenamine, 4-(2,5-diethyl-3,4-diphenyl-1H-pyrrol-1-yl)-N,N-dimethyl-	ether	265(4.58)	78-1355-86
$C_{28}H_{30}N_2O$			
2H-Pyrrol-2-one, 1-[4-(dimethylamino)phenyl]-3,5-diethyl-1,3-dihydro-3,4-diphenyl-	ether	267(4.39),295s(4.10)	78-1355-86
$C_{28}H_{30}N_2O_{10}$			
3-Isoxazolidineacetic acid, 5-(hydroxyphenylamino)-4-[[(4-methoxy-4-oxo-2-butenyl)oxy]carbonyl]-2-phenyl-, 4-methoxy-4-oxo-2-butenyl ester	EtOH	223s(4.53),319(4.82)	44-3125-86

Compound	Solvent	$\lambda_{max}(\log \epsilon)$	Ref.
$C_{28}H_{30}N_6O_8$			
Benzoic acid, 4-[[2-(acetylamino)-4-[bis(2-acetoxyethyl)amino]-5-methoxyphenyl]azo]-3,5-dicyano-, ethyl ester	$CHCl_3$	293(3.82),318(3.79), 344s(3.35),365s(3.33), 580s(4.57),612(4.67)	48-0497-86
$C_{28}H_{30}O_7$			
24-Norchola-1,5,20,22-tetraene-3,7,11-trione, 6-acetoxy-14,15:21,23-diepoxy-4,4,8-trimethyl-	n.s.g.	218(4.00),240(3.90)	100-0056-86
$C_{28}H_{30}O_{11}$			
Yadanzioside M	EtOH	231(4.20),255(3.95)	18-3541-86
$C_{28}H_{31}BrN_4O_9$			
D-lyxo-Hexos-2-ulose, 1-(acetylphenylhydrazone) 2-[(4-bromophenyl)hydrazone], 3,4,5,6-tetraacetate	MeOH	248(4.36),270s(3.96), 380(4.39)	136-0316-86G
$C_{28}H_{31}BrO_7$			
Benzoic acid, 4-bromo-, 4,4a,6,6a,7,7a-9,10,10b,12,13,13a-dodecahydro-6,6,10b,11-tetramethyl-3,9-dioxo-1H,3H-pyrano[4',3':3,3a]isobenzofuro[4,5,6-ij][2]benzopyran-10-yl ester (meroliminol p-bromobenzoate)	n.s.g.	247(4.11)	138-1337-86
$C_{28}H_{31}ClN_2O_2$			
Acetamide, N-[4-[2-chloro-1-[4-[2-(diethylamino)ethoxy]phenyl]-2-phenylethenyl]phenyl]-	EtOH	258(4.28),305(4.16)	13-0073-86B
$C_{28}H_{31}N_3O_4$			
1,3-Dioxane-4,6-dione, 5-[5-[[bis[4-(dimethylamino)phenyl]methylene]amino]-2,4-pentadienylidene]-2,2-dimethyl-	acetone	544(4.45)	24-3276-86
$C_{28}H_{31}N_5O_3$			
2,4,6(1H,3H,5H)-Pyrimidinetrione, 5-[5-[[bis[4-(dimethylamino)phenyl]methylene]amino]-2,4-pentadienylidene]-1,3-dimethyl-	acetone	545(4.72)	24-3276-86
	$CHCl_3$	323(4.18),574(4.70)	24-3276-86
$C_{28}H_{31}N_5O_5S$			
Benzamide, 4-(1,1-dimethylethyl)-N-[6-[(4-methylphenyl)thio]-9-β-D-ribofuranosyl-9H-purin-2-yl]-	pH 7	256(4.43),296(4.40)	1-0806-86
	pH 2	260(--),290(--)	1-0806-86
	pH 13	245(--),298(--)	1-0806-86
$C_{28}H_{31}N_5O_7$			
Saframycin Yd-2	MeOH	268(4.23)	158-1639-86
$C_{28}H_{31}N_5O_{11}$			
D-lyxo-Hexos-2-ulose, 1-(acetylphenylhydrazone) 2-[(4-nitrophenyl)hydrazone, 3,4,5,6-tetraacetate	MeOH	225(4.23),280(3.88), 402(4.55)	136-0316-86G
$C_{28}H_{32}ClN_3O_8S$			
Prochlorperazine dimaleate	MeOH	258(4.579),313(3.655)	106-0571-86

Compound	Solvent	λ_{max}(log ε)	Ref.
$C_{28}H_{32}N_2$			
Benzenamine, 4-(3,5-diethyl-2,3-dihydro-3,4-diphenyl-1H-pyrrol-1-yl)-N,N-dimethyl-	ether	260(4.18),310(4.19)	78-1355-86
Pyrazine, 2,5-bis(2-methylpropyl)-3,6-bis(2-phenylethenyl)-, (E,E)-	EtOH	225(3.69),237s(3.60), 245s(3.47),295(3.96), 305s(3.93)	4-1481-86
$C_{28}H_{32}N_2O_2$			
Acetamide, N-[4-[1-[4-[2-(diethylamino)ethoxy]phenyl]-2-phenylethenyl]phenyl]-	EtOH	267(4.33),315(4.18)	13-0073-86B
$C_{28}H_{32}N_2O_8$			
2-Propenoic acid, 3-(3-nitrophenyl)-, 1,10-decanediyl ester, (E,E)-	MeCN	261(4.74)	18-1379-86
2-Propenoic acid, 3-(4-nitrophenyl)-, 1,10-decanediyl ester, (E,E)-	MeCN	302(4.58)	18-1379-86
$C_{28}H_{32}N_2O_{10}$			
[5,5'-Bi-1,3-dioxolo[4,5-g]isoquinoline]-6,6'(5H,5'H)-dicarboxylic acid, 7,7',8,8'-tetrahydro-4,4'-dimethoxy-, diethyl ester (same spectrum for isomers a and b)	MeOH	218(4.5),260(3.2), 283(3.5)	83-1122-86
$C_{28}H_{32}N_2O_{16}$			
Ethanedioic acid, bis[2-[[2-(2-ethoxyethoxy)ethoxy]carbonyl]-4-nitrophenyl ester	MeCN	221(4.58),296(4.16)	96-0209-86
Ethanedioic acid, bis[4-[[2-(2-ethoxyethoxy)ethoxy]carbonyl]-2-nitrophenyl ester	MeCN	232(4.59),340(3.59)	96-0209-86
$C_{28}H_{32}O_4Si$			
Silane, bis[(1,1-dimethylethyl)dioxy]-di-1-naphthalenyl-	MeCN	277(4.14),296(4.11), 300(4.02),310(4.38), 314(4.23),318(4.03)	116-0968-86
$C_{28}H_{32}O_5$			
Benzoic acid, 2-[[3-[7-(2-butoxyphenyl)-1,3,5-heptatrienyl]oxiranyl]-methoxy]-, methyl ester, [2α,3β-(1E,3E,5Z)]-	n.s.g.	274(4.46),283(4.51), 294(4.41)	104-1597-86
$C_{28}H_{32}O_7$			
Meroliminol benzoate	n.s.g.	230(3.99)	138-1337-86
24-Norchola-1,14,20(22)-triene-21,23-dioic acid, 7-acetoxy-4,4,8-trimethyl-3,6-dioxo-, cyclic anhydride, (5α,7α,13α,17α)-	MeOH	220(4.39)	39-1021-86C
$C_{28}H_{32}O_9$			
3(4H)-Dibenzofuranone, 4a,5a,9a,9b-tetrahydro-1,4,4a,6,8-pentahydroxy-4,5a,7,9b-tetramethyl-2,9-bis(1-oxo-2,4-hexadienyl)- (bisvertinolone)	MeOH	231(4.27),271(4.30), 305s(4.26),362(4.50)	78-3157-86
	MeOH-NaOH	240s(4.27),275(4.49), 394(4.23)	78-3157-86
$C_{28}H_{32}O_{10}$			
1H-Indene-1,5-diol, 2-[2-(4-acetoxy-3-methoxyphenyl)-1-(acetoxymethyl)-	MeOH	204(--),217(--), 232(4.08),280(3.98)	39-1181-86C

Compound	Solvent	$\lambda_{max}(\log \epsilon)$	Ref.
ethyl]-2,3-dihydro-6-methoxy-, diacetate (4λ,2ε) (cont.)			39-1181-86C
$C_{28}H_{32}O_{16}$			
Rhamnocitrin 3-O-β-D-galactopyranoside 4'-O-β-D-glucopyranoside	MeOH	268(4.3),328(3.6)	100-0151-86
	MeOH-NaOMe	284(--),318s(--), 390(--)	100-0151-86
	MeOH-NaOAc	253(--),272(--), 283(--),320s(--), 360s(--)	100-0151-86
	MeOH-AlCl$_3$	276(--),297s(--), 345(--),397(--)	100-0151-86
$C_{28}H_{32}O_{18}$			
Petuletin 3-O-β-D-gentiobioside	MeOH	257(4.31),270s(4.25), 300s(3.97),355(4.33)	102-0231-86
	MeOH-NaOMe	270(4.17),337(3.87), 409(4.22)	102-0231-86
	MeOH-NaOAc	270(4.16),330s(3.84), 390(4.09)	102-0231-86
	MeOH-AlCl$_3$	276(4.43),310s(3.89), 336(3.81),437(4.44)	102-0231-86
	+ HCl	270(4.30),300(3.92), 390(4.28),400s(4.28)	102-0231-86
$C_{28}H_{33}NO_2$			
Phenol, 4-[1-[4-[2-(diethylamino)ethoxy]phenyl]-2-phenyl-1-butenyl]-	EtOH	287(4.15)	87-2511-86
$C_{28}H_{33}N_2O$			
Pyrrolidinium, 1-[1-[3-ethoxy-3-phenyl-1-(1-pyrrolidinyl)-2-propenylidene]-1,3-dihydro-2H-inden-2-ylidene]-, tetrafluoroborate	MeCN	228(3.92),303(4.20), 485(3.41)	48-0314-86
$C_{28}H_{34}$			
Pyrene, 2,4,7,10-tetrakis(1-methylethyl)-	n.s.g.	251.8(5.0),282.1(4.7), 345.0(4.6)	1-0665-86
$C_{28}H_{34}N_3O_2$			
1H-Indolium, 1-[5-(2,3-dihydro-1H-indol-1-yl)-3-[2-[[(1,1-dimethylethoxy)carbonyl]amino]ethyl]-2,4-pentadienylidene]-2,3-dihydro-, perchlorate	MeOH	246(3.76),273(3.85), 350(3.48),522(4.90), 554(5.10)	33-1588-86
$C_{28}H_{34}O_6$			
Sigmoidin A trimethyl ether	EtOH	228s(4.07),288(3.92)	39-0033-86C
$C_{28}H_{34}O_7$			
24-Norchola-1,5,20,22-tetraen-7-one, 6-acetoxy-14,15:21,23-diepoxy-3,11-dihydroxy-4,4,8-trimethyl-, (3β,11α,13α,14β,15β,17α)-	n.s.g.	218(4.00),240(3.90)	100-0056-86
$C_{28}H_{34}O_8$			
Deacetylnimbin	MeOH	225(3.96)	100-1068-86
2,4-Hexadien-1-one, 1,1'-(4a,5a,6,7-9a,9b-hexahydro-2,4,5a,6,9-pentahydroxy-3,4a,6,9a-tetramethyl-1,8-di-	MeOH	233(4.09),273(4.25), 301(4.22),316s(4.20), 400(4.42)	78-3157-86

Compound	Solvent	$\lambda_{max}(\log \epsilon)$	Ref.
benzofurandiyl)bis- (bisvertinol) (cont.)	MeOH-NaOH	234(4.18),273(4.26), 301s(4.18),357(4.34), 418(4.33)	78-3157-86
	+ HCl	224(4.03),273(4.24), 301(4.24),400(4.49)	78-3157-86
$C_{28}H_{34}O_{10}$			
Deacetylnimbinolide	MeOH	208(3.94)	100-1068-86
iso-	MeOH	210(3.99),285(3.17)	100-1068-86
$C_{28}H_{34}O_{11}$			
Yadanzioside K	EtOH	222(4.23),252(3.95)	18-3541-86
$C_{28}H_{35}ClN_4Zn$			
Tetramethyl corphinate zinc chloride complex	EtOH	263(4.06),294s(4.58), 302(4.68),343.5s(4.60), 358(4.69),384s(3.80), 411.5(3.65),531(3.99)	35-7875-86
$C_{28}H_{35}NOS$			
2H-[1,3]Oxazino[2,3-j]quinoline, 2-[(1,1-dimethylethyl)thio]-6,7,8,8a,9,10,11,12-octahydro-3,4-diphenyl-	MeCN	229(4.15),285(3.86)	118-0899-86
$C_{28}H_{35}NO_6$			
Androst-5-ene-17-carboxylic acid, 3-acetoxy-, 4-nitrophenyl ester, (3β,17β)-	EtOH	217s(3.89),270(3.98)	128-0297-86
$C_{28}H_{35}N_5O_5$			
Piperazine, 1-[5-[[3-(1,1-dimethyleth-yl)-2-phenyl-5-oxazolidinyl]methoxy]-1,6-dihydro-1-methyl-6-oxo-4-pyrida-zinyl]-4-(2-furanylcarbonyl)-	EtOH	242(4.3),310(3.2)	83-0060-86
$C_{28}H_{36}N_4O_2$			
2,5-Cyclohexadiene-1,4-dione, 2,5-bis(dimethylamino)-3,6-bis[4-[(di-methylamino)methyl]phenyl]-	MeOH	305(4.10),370(4.24)	87-1792-86
$C_{28}H_{36}O_2$			
Retinoic acid, 1-phenylethyl ester, (-)	MeOH	363(4.82)	39-0635-86B
$C_{28}H_{36}O_4$			
5-Tetradecene-1,4-diol, dibenzoate, (4R,5Z)-	EtOH	230(4.40)	163-0187-86
$C_{28}H_{36}O_7$			
Ergost-2-en-26-oic acid, 5,6:24,25-di-epoxy-20,22-dihydroxy-1,4-dioxo-, δ-lactone, (5β,6β,24S,25S)-	EtOH	225(3.93)	105-0560-86
Isonimocinolide	MeOH	232(3.49)	39-1021-86C
Nimocinolide	MeOH	230(3.55)	39-1021-86C
$C_{28}H_{36}O_8$			
Bisvertinol, dihydro-	MeOH	228(3.96),270(4.19), 304(4.06),387(4.30)	78-3157-86
	MeOH-NaOH	233(4.00),274(4.20), 305(4.11),406(4.29)	78-3157-86
	+ HCl	224(3.94),270(4.20),	78-3157-86

Compound	Solvent	$\lambda_{max}(\log \epsilon)$	Ref.
(cont.)		305(4.06),379(4.22)	78-3157-86
Isobisvertinol, dihydro-	MeOH	250s(3.92),300(4.20), 312(4.20),354(4.26)	78-3157-86
	MeOH-NaOH	240s(3.88),300(4.24), 312s(4.24),342(4.29)	78-3157-86
	+ HCl	252s(3.99),300(4.13), 312(4.14),370(4.27)	78-3157-86
$C_{28}H_{36}O_{11}$			
Picras-2-en-21-oic acid, 15-[(3,4-dimethyl-1-oxo-2-pentenyl)oxy]-13,20-epoxy-2,11,12-trihydroxy-1,16-dioxo-, methyl ester, [11β,12α,15β(E)]-	EtOH	220(4.20),267(3.78)	88-0593-86
	EtOH-base	314(3.60)	88-0593-86
Yadanzioside N	EtOH	222(4.13),255(3.71)	18-3541-86
$C_{28}H_{37}Cl_2NO_6$			
Geldanamycin, 4,5-dichloro-7-de[(aminocarbonyl)oxy]-6,7-didehydro-6,17-didemethoxy-4,5-dihydro-15-methoxy-11-O-methyl-	MeOH	271(4.37)	158-0415-86
$C_{28}H_{38}CuN_2O_2$			
4,4'-Dihexyl-N,N'-disalicylidene-ethylenediamine, copper complex	$CHCl_3$	379f(4.03),565(2.63)	138-1145-86
$C_{28}H_{38}N_2NiO_2$			
4,4'-Dihexyl-N,N'-disalicylidene-ethylenediamine, nickel complex	$CHCl_3$	348(3.97),422(3.86), 554(2.23)	138-1145-86
$C_{28}H_{38}N_4O_7$			
Benzoic acid, 2-hydroxy-6-methyl-, [5-[(3-acetylphenyl)amino]-4-amino-3-[[(dimethylamino)carbonyl]amino]-3-ethyl-1,2-dihydroxy-2-methylcyclopentyl]methyl ester, (7-deoxypactamycin)	n.s.g.	209(4.64),236(4.64), 264s(4.11),310(3.58), 355(3.39)	158-1779-86
$C_{28}H_{38}N_4O_8$			
Benzoic acid, 2-hydroxy-6-methyl-, [5-[(3-acetylphenyl)amino]-4-amino-3-[[(dimethylamino)carbonyl]amino]-3-(1-hydroxyethyl)-2-methylcyclopentyl]methyl ester (pactamycin)	n.s.g.	239(--),264s(--), 313(--),356(--)	158-1779-86
$C_{28}H_{38}N_4O_9$			
Benzoic acid, 2-hydroxy-6-methyl-, [4-amino-3-[[(dimethylamino)carbonyl]amino]-1,2-dihydroxy-5-[[3-(hydroxyacetyl)phenyl]amino]-3-(1-hydroxyethyl)-2-methylcyclopentyl]-methyl ester	n.s.g.	210(4.60),238(4.61), 265s(4.02),312(3.60), 355(3.35)	158-1779-86
$C_{28}H_{38}O_5$			
Jaborosalactone L	EtOH	224(4.51)	102-1765-86
$C_{28}H_{38}O_6$			
1,3-Azulenedicarboxylic acid, 6-(2-ethoxy-2-oxoethyl)-2-octyl-, ethyl ester	CH_2Cl_2	237.5(4.51),314(4.82), 350.5(3.94),376.5(3.74), 501(2.65)	24-2272-86
Chol-8-en-24-oic acid, 4,4,14-tri-	EtOH	251.5(3.35)	94-3695-86

Compound	Solvent	λ_{max}(log ε)	Ref.
methyl-3,7,11,15-tetraoxo-, methyl ester, (5α)- (methyl lucidenate F) (cont.)			94-3695-86
$C_{28}H_{38}O_7$			
Ergost-2-en-26-oic acid, 5,6:24,25-di-epoxy-4,20,22-trihydroxy-1-oxo-, δ-lactone (4β,5β,6β,24S,25S)-	EtOH	203(4.21)	105-0560-86
$C_{28}H_{40}$			
s-Indacene, 1,3,5,7-tetrakis(1,1-di-methylethyl)-	hexane	224(4.10),234s(4.03), 278s(4.23),284s(4.37), 289s(4.48),294(4.61), 301(4.66),307(4.77), 330s(3.45),340(3.48), 451s(3.41),486s(3.83), 500s(3.96),505(4.01), 519s(4.11),527s(4.26), 539s(4.51),545(4.64)	89-0630-86
s-Indacene, 3,5,7-tris(1,1-dimethyl-ethyl)-1,7-dihydro-1-methyl-1-(1-methylethenyl)-	hexane	229s(4.02),237(4.26), 244(4.49),253(4.50), 268s(3.50),278s(3.46), 283s(3.43),296(3.38), 301s(3.29),308(3.40), 313s(3.14),321(3.20)	89-0630-86
$C_{28}H_{40}CuO_4$			
Copper, bis[3,5-bis(1,1-dimethylethyl)-1,2-benzenediolato(2-)-O,O']-, (SP-4-1)-	THF	510(2.5),713(2.90)	35-1903-86
$C_{28}H_{40}N_2O_2$			
Phenol, 4-[2-(2,5-dihydro-2,5,5-tri-methyl-4-phenyl-1H-imidazol-2-yl)-ethyl]-2,6-bis(1,1-dimethylethyl)-, N-oxide	EtOH	284(3.96)	103-0052-86
$C_{28}H_{40}N_5$			
Agelasine D (chloride)	MeOH	272(3.96)	18-2495-86
$C_{28}H_{40}O_4$			
1,2,4-Benzenetriol, 3-(3,7,11,15-tetramethyl-2,6,10,14-hexadeca-tetraenyl)-, 4-acetate, (E,E,E)-(suillin)	MeOH	275(4.15)	64-0645-86B
$C_{28}H_{40}O_4S$			
Pregnan-20-one, 3-[[(4-methylphenyl)-sulfonyl]oxy]-, (3β,5α)-	EtOH	220(4.15)	39-1797-86C
$C_{28}H_{40}O_5$			
Amentacea	EtOH	222(4.02),287(3.80)	78-6015-86
Strictaepoxide	EtOH	220(4.09),289(3.56)	78-6015-86
$C_{28}H_{40}O_6$			
Chol-8-en-24-oic acid, 7-hydroxy-4,4,14-trimethyl-3,11,15-trioxo-, methyl ester, (5α,7β)- (methyl lucidenate A)	EtOH	254(3.51)	94-3695-86
Ergosta-2,24-dien-26-oic acid, 6,7-ep-oxy-5,20,22-trihydroxy-1-oxo-, δ-	EtOH	226(3.996)	105-0300-86

Compound	Solvent	λ_{max}(log ε)	Ref.
lactone, (5α,6α,7α,22R,25R)- (ixocarpanolide) (cont.)			105-0300-86
$C_{28}H_{40}O_7$			
Ergost-2-en-26-oic acid, 6,7-epoxy-5,14,20,22-tetrahydroxy-1-oxo-, δ-lactone, (5α,6α,7α,25R)-	EtOH	225(4.02)	105-0560-86
$C_{28}H_{40}O_8$			
1,2,3,4,11,12,13,14-Cycloeicosaneoctone, 5,5,10,10,15,15,20,20-octamethyl-	CH_2Cl_2	385(2.17),525(2.22)	89-0449-86
Ergost-2-en-26-oic acid, 5,6-epoxy-4,16,20,22,23-pentahydroxy-1-oxo-, γ-lactone, (4β,5β,6β,16β,22R,23R-25R)- (ixocarpalactone A)	EtOH	216(4.03)	105-0300-86
$C_{28}H_{40}O_9$			
18-Nor-16-oxaandrostane-8-carboxaldehyde, 7-acetoxy-15-hydroxy-4,4-dimethyl-17-oxo-6-(1-oxobutoxy)-, cyclic 8,15-(acetyl acetal)	EtOH	220(2.48)	12-1629-86
18-Nor-16-oxaandrostane-8-carboxaldehyde, 7,15-diacetoxy-4,4-dimethyl-17-oxo-6-(1-oxobutoxy)-, (5α,6α,7α-13α,15β)-	EtOH	220(2.54)	12-1629-86
$C_{28}H_{41}NO_6S$			
7,9,11-Tridecatrienoic acid, 6-[(2-amino-3-methoxy-3-oxopropyl)thio]-13-(2-butoxyphenyl)-5-hydroxy-, methyl ester, (±)-	EtOH	274(4.49),283(4.58), 290(4.50)	104-1457-86
$C_{28}H_{42}$			
s-Indacene, 1,3,5,7-tetrakis(1,1-dimethylethyl)-1,7-dihydro-	hexane	223s(4.06),230s(4.19), 236(4.43),245(4.61), 253(4.55),286(3.82), 290(3.82),296(3.84), 301s(3.78),307s(3.68), 312s(3.53),321(3.17)	89-0630-86
$C_{28}H_{42}O_2$			
Cholesta-1,4-diene-3,14-dione, 25-methyl-	EtOH	245(4.13)	39-1805-86C
$C_{28}H_{42}O_3$			
Ergosta-7,22-diene-3,6-dione, 5-hydroxy-, (5 ,22E)-	MeOH	250(3.92)	94-3465-86
$C_{28}H_{42}O_4$			
Pycnophorin product from methylation with diazomethane	MeOH	255(3.92)	88-2121-86
$C_{28}H_{42}O_7$			
5α-Chol-8-en-24-oic acid, 7β,15α-dihydroxy-4β-(hydroxymethyl)-4α,14α-dimethyl-3,11-dioxo-, methyl ester (methyl lucidenate G)	MeOH	254(3.90)	163-0809-86
$C_{28}H_{42}O_8$			
Octahydrobisvertinol	MeOH	234(3.72),276s(3.94),	78-3157-86

Compound	Solvent	λmax(log ε)	Ref.
(cont.)		305(4.09),334s(4.02)	78-3157-87
	MeOH-NaOH	236(3.69),299(4.13), 340s(3.99)	78-3157-86
$C_{28}H_{44}N_2$			
Cyanamide, (cholest-4-en-3-ylidene)-	EtOH	281(4.40)	118-1055-86
$C_{28}H_{44}N_2O_6SSi_2$			
Thymidine, 6-(phenylthio)-3',5'-O-[1,1,3,3-tetrakis(1-methylethyl)-1,3-disiloxanediyl]-	MeOH	245(3.92),278(3.94)	78-4187-86
$C_{28}H_{44}O$			
Cholesta-1,4-dien-3-one, 25-methyl-	EtOH	245(4.15)	39-1805-86C
$C_{28}H_{44}O_2$			
19-Norcholesta-1(10),5-dien-3-ol, acetate, (3β)-	EtOH	243(4.57)	2-0175-86
Retinoic acid, 1-methylheptyl ester, (+)-	MeOH	360(4.66)	39-0635-86B
$C_{28}H_{44}O_3$			
Ergosta-7,22-dien-6-one, 3,5-dihydroxy-, (3β,5α,22E)-	MeOH	247(4.07)	94-3465-86
$C_{28}H_{46}N_8O_{10}S_2$			
Ethylenediammonium, N,N'-bis[2-(3-methyl-3-nitrosoureido)ethyl]-N,N,N',N'-tetramethyl-, ditosylate	H_2O	197(4.60),223(4.48), 395(2.20)	70-0555-86
$C_{28}H_{48}N_2O$			
Buximinol G	MeOH	237(3.78),245(3.77), 253(3.62)	164-0663-86
$C_{28}H_{54}Cu_2N_8$			
Copper, [μ-[11,11'-[1,4-phenylenebis(methylene)]bis[1,4,7,11-tetraazacyclotetradecane]-N^1,N^4,N^7,N^{11}-$N^{1'},N^{4'},N^{7'},N^{11'}$]]di-, tetraperchlorate	n.s.g.	553(2.56)	33-0053-86
$C_{28}H_{54}N_8Ni_2$			
Nickel, [μ-[11,11'-[1,4-phenylenebis(methylene)]bis[1,4,7,11-tetraazacyclotetradecane]-N^1,N^4,N^7,N^{11}-$N^{1'},N^{4'},N^{7'},N^{11'}$]]di-, tetraperchlorate	n.s.g.	475(1.97)	33-0053-86

Compound	Solvent	$\lambda_{max}(\log \epsilon)$	Ref.
$C_{29}H_{16}Cl_5NO_2S$			
2H,5H-[1]Benzothiopyrano[3,4-e]-1,3-oxazin-5-one, 9-chloro-2,4-bis(2,4-dichlorophenyl)-3,4-dihydro-3-phenyl-	MeOH	237(4.4),320(4.5)	42-0323-86
$C_{29}H_{17}Cl_4NO_2S$			
2H,5H-[1]Benzothiopyrano[3,4-e]-1,3-oxazin-5-one, 2,4-bis(2,4-dichlorophenyl)-3,4-dihydro-N-phenyl-	MeOH	237(4.4),321(4.5)	42-0323-86
$C_{29}H_{17}Cl_{14}NO_5$			
Propanedioic acid, (acetylamino)[[4-[bis(pentachlorophenyl)methyl]-2,3,5,6-tetrachlorophenyl]methyl]-, diethyl ester	C_6H_{12}	222(5.05),293(3.15), 304(3.18)	44-2472-86
$C_{29}H_{18}Cl_2N_2$			
Quinoline, 4,7-dichloro-3-[3-methyl-4-(2-naphthalenyl)-1-isoquinolinyl]-	pH 1	216(4.98),324(4.42)	2-1072-86
	EtOH	219(4.92),236(4.79), 280(4.26),329(4.10)	2-1072-86
$C_{29}H_{18}Cl_3NO_2S$			
2H,5H-[1]Benzothiopyrano[3,4-e]-1,3-oxazin-5-one, 9-chloro-2,4-bis(4-chlorophenyl)-3,4-dihydro-3-phenyl-	MeOH	237(4.4),320(4.5)	42-0323-86
$C_{29}H_{18}N_4O_2$			
1H-Benz[de]isoquinoline-1,3(2H)-dione, 2-(4-methylphenyl)-6-(2H-naphtho[1,2-d]triazol-2-yl)-	DMF	368(4.26)	2-0496-86
Benzo[de][1,2,3]triazolo[4,5-g]isoquinoline-4,6(5H,9H)-dione, 5-(4-methylphenyl)-9-(1-naphthalenyl)-	DMF	388(4.46)	2-0496-86
$C_{29}H_{18}O_2$			
Spiro[1H-cyclohepta[c]furan-1,15'-[15H]dibenzo[a,g]cyclotridecen]-3(8aH)-one, 5',6',7',8'-tetradehydro-, (E,E)-	EtOH	223(4.81),259(4.52), 273(4.50),321(4.18), 341(4.16)	18-1713-86
$C_{29}H_{18}O_3$			
7H-Benz[de]anthracen-7-one, 6,8-diphenoxy-	benzene	362(3.99),392(4.11), 407s(4.09)	104-0566-86
7H-Benz[de]anthracen-7-one, 6,11-diphenoxy-	benzene	346s(3.80),359(3.99), 396(4.06),411s(4.01)	104-0566-86
7H-Benz[de]anthracen-7-one, 8,11-diphenoxy-	benzene	356(3.73),410(4.14)	104-0566-86
$C_{29}H_{19}BO_4$			
Boron, [1,2-benzenediolato(2-)-O,O']-(1,3-di-2-naphthalenyl-1,3-propanedionato-O,O')-, (T-4)-	CH_2Cl_2	407(4.53),426(4.59)	48-0755-86
$C_{29}H_{19}Cl_2NO_2S$			
2H,5H-[1]Benzothiopyrano[3,4-e]-1,3-oxazin-5-one, 2,4-bis(4-chlorophenyl)-3,4-dihydro-3-phenyl-	MeOH	238(4.4),324(3.8)	42-0323-86
$C_{29}H_{19}NO$			
9H-Naphtho[2,1-a]quinolizin-8-one, 9,10-diphenyl-	MeCN	297(4.25),335(3.69), 402(4.10)	118-0908-86

Compound	Solvent	$\lambda_{max}(\log \epsilon)$	Ref.
$C_{29}H_{19}N_3O_4$			
Acetic acid, cyano[4-cyano-4-(4-nitrophenyl)-3,5-diphenyl-2,5-cyclohexadien-1-ylidene]-, methyl ester	$CHCl_3$	345(4.40)	83-0898-86
$C_{29}H_{20}$			
Cyclopropene, 1,2,3-triphenyl-3-(phenylethynyl)-	EtOH	224(4.42),231(4.45), 254(4.28),290(4.28), 304(4.39),322(4.27)	104-1599-86
Dibenzo[a,f]dibenzo[2,3:4,5]pentaleno[1,6-cd]pentalene, 4b,8b-12b,16b-tetrahydro- (fenestrindan)	heptane	261.5(3.38),267.0(3.62), 273.5(3.70)	35-8107-86
$C_{29}H_{20}ClNO_2S$			
2H,5H-[1]Benzothiopyrano[3,4-e]-1,3-oxazin-5-one, 9-chloro-3,4-dihydro-2,3,4-triphenyl-	MeOH	238(4.3),329(3.6)	42-0323-86
$C_{29}H_{20}N_2O_2$			
Acetic acid, cyano(4-cyano-3,4,5-triphenyl-2,5-cyclohexadien-1-ylidene)-, methyl ester	$CHCl_3$	345(4.44)	83-0898-86
2-Butene-1,4-dione, 1,4-diphenyl-2-(2-phenyl-1H-benzimidazol-1-yl)-, (E)-	MeOH	220(4.54),237(4.49), 271(4.52)	44-3420-86
(Z)-	MeOH	219(4.58),239(4.50), 280(4.56)	44-3420-86
Methanone, (5,6-dihydrobenzimidazo[2,1-a]isoquinoline-5,6-diyl)bis[phenyl-	MeOH	220(4.60),260(4.48)	44-3420-86
Methanone, (1,3-diphenyl-1H-pyrazole-4,5-diyl)bis[phenyl-	MeOH	255(4.60)	44-1972-86
$C_{29}H_{20}O_4$			
Bicyclo[2.2.1]hepta-2,5-diene-2,3-dicarboxylic acid, di-1-naphthalenyl ester	MeCN	280(4.16)	32-0281-86
Bicyclo[2.2.1]hepta-2,5-diene-2,3-dicarboxylic acid, di-2-naphthalenyl ester	MeCN	265(4.16)	32-0281-86
Tetracyclo[3.2.0.0^{2,7}.0^{4,6}]heptane-1,5-dicarboxylic acid, di-1-naphthalenyl ester	MeCN	265(4.15)	32-0281-86
Tetracyclo[3.2.0.0^{2,7}.0^{4,6}]heptane-1,5-dicarboxylic acid, di-2-naphthalenyl ester	MeCN	267(4.04)	32-0381-86
$C_{29}H_{20}O_7$			
4H-1-Benzopyran-4-one, 2-(2,2-diphenyl-1,3-benzodioxol-5-yl)-5,6-dihydroxy-7-methoxy-	EtOH	245(4.30),286(4.25), 342(4.44)	94-0030-86
	EtOH-$AlCl_3$	252(4.25),296(4.26), 364(4.47)	94-0030-86
$C_{29}H_{21}BO_5$			
Boron, (1,3-diphenyl-1,3-propanedionato-O,O')(α-hydroxy-α-phenylbenzeneacetato(2-)]-, (T-4)-	MeCN	253(4.08),269(4.32), 381s(4.29)	48-0755-86
$C_{29}H_{21}NO_2S$			
2H,5H-[1]Benzothiopyrano]3,4-e]-1,3-oxazin-5-one, 3,4-dihydro-2,3,4-triphenyl-	MeOH	233(4.8),318(4.1)	42-0323-86

Compound	Solvent	λ_{max}(log ε)	Ref.
$C_{29}H_{21}N_3$			
Propanedinitrile, (6,7-dihydro-1-methyl-2,3-diphenyl-4H-benzo[a]quinolizin-4-ylidene)-	MeCN	287(4.07),332(4.04), 426(4.06)	118-0908-86
$C_{29}H_{21}N_4OP$			
3a,6,7(7aH)-Benzofurantricarbonitrile, 2,3-dihydro-4-[(triphenylphosphoranylidene)amino]-, cis	$CHCl_3$	370(3.13),381(3.19), 433(3.70)	24-2723-86
$C_{29}H_{22}ClNO$			
Benzo[h]quinolin-2(1H)-one, 3-(4-chlorophenyl)-3,4,5,6-tetrahydro-4-(1-naphthalenyl)-	EtOH	223(4.68),268(4.10)	4-1789-86
$C_{29}H_{22}N$			
Pyridinium, 1,2,4,6-tetraphenyl-perchlorate	DMSO	310(4.64)	44-2481-86
	EtOH	311(4.6)	96-0441-86
	$CHCl_3$	315(4.3),412(2.9)	62-1107-86
$C_{29}H_{22}N_2O_2$			
Benzaldehyde, (1-benzoyl-3-oxo-3-phenyl-1-propenyl)phenylhydrazone, (E,?)-	MeOH	263(4.52),374(4.53)	44-1972-86
5(4H)-Oxazolone, 4-[(2,3-dihydro-6,7-diphenyl-1H-pyrrolizin-5-yl)methylene]-2-phenyl-	MeOH	203(4.71),244(4.36), 462(4.71)	83-0354-86
$C_{29}H_{22}N_2O_3$			
1H-Benzimidazole-1-acetic acid, α-(2-phenoxy-2-phenylethenyl)-2-phenyl-, (Z)-	MeOH	253(4.69)	44-3420-86
$C_{29}H_{22}N_2O_9$			
Emestrin, 3,11a-de(epidithio)-3,7',11-11a-tetradehydro-7',11-dideoxy-, 4'-acetate	EtOH	239(4.31),263(4.25), 294(4.00),360(4.19)	39-0109-86C
$C_{29}H_{22}N_2O_{11}$			
Dethiosecoemestrin, O-acetyl-	EtOH	222s(4.55),254(4.50), 287s(4.00),367(4.09)	94-2411-86
$C_{29}H_{22}O_3$			
4H-1-Benzopyran-4-one, 7-hydroxy-3-phenyl-6,8-bis(phenylmethyl)-	MeOH	266(4.0),302(3.98), 416(3.81)	2-0646-86
1,4-Methanoanthracene-9,10,11-trione, 1,4,4a,9a-tetrahydro-1,4-dimethyl-2,3-diphenyl-, (1α,4α,4aα,9aα)-	MeCN	219(3.92),250(3.56), 310(2.70)	12-1537-86
$C_{29}H_{22}O_4$			
4H-1-Benzopyran-4-one, 5,7-dihydroxy-6,8-bis(phenylmethyl)-3-phenyl-	MeOH	204(4.47),262(4.17), 290(3.79)	2-0166-86
4H-1-Benzopyran-4-one, 7,8-dihydroxy-2-methyl-3-phenyl-6-(diphenylmethyl)-	MeOH	265(4.82),300(4.62)	2-0649-86
$C_{29}H_{23}BrO_2$			
Benzophenone, 3-acetyl-4'-bromo-2,5-dimethyl-4,6-diphenyl-	MeCN	261(4.38)	48-0359-86
Benzophenone, 3-acetyl-4-(4-bromophenyl)-2,5-dimethyl-6-phenyl-	MeCN	243(4.50)	48-0359-86
Benzophenone, 3-acetyl-6-(4-bromo-	MeCN	244(4.47)	48-0359-86

Compound	Solvent	$\lambda_{max}(\log \epsilon)$	Ref.
phenyl)-2,5-dimethyl-4-phenyl- (cont.)			48-0359-86
$C_{29}H_{23}ClO_2$			
Benzophenone, 3-acetyl-4'-chloro-2,5-dimethyl-4,6-diphenyl-	MeCN	254(4.39)	48-0359-86
Benzophenone, 3-acetyl-4-(4-chlorophenyl)-2,5-dimethyl-6-phenyl-	MeCN	241(4.47)	48-0359-86
Benzophenone, 3-acetyl-6-(4-chlorophenyl)-2,5-dimethyl-4-phenyl-	MeCN	244(4.46)	48-0359-86
$C_{29}H_{23}FO_2$			
Benzophenone, 3-acetyl-4'-fluoro-2,5-dimethyl-4,6-diphenyl-	MeCN	244(4.38)	48-0359-86
Benzophenone, 3-acetyl-4-(4-fluorophenyl)-2,5-dimethyl-6-phenyl-	MeCN	242(4.40)	48-0359-86
$C_{29}H_{23}NO$			
Benzo[f]pyrrolo[2,1-a]isoquinolin-10(6H)-one, 5,10a-dihydro-10a-methyl-8,9-diphenyl-	MeCN	235s(--),270(4.26), 354(3.84)	118-0899-86
Benzo[h]quinolin-2(1H)-one, 3,4,5,6-tetrahydro-4-(1-naphthalenyl)-3-phenyl-	EtOH	227(4.27),272(3.81), 325(3.70)	4-1789-86
Benzo[h]quinolin-2(1H)-one, 3,4,5,6-tetrahydro-4-(2-naphthalenyl)-3-phenyl-	EtOH	223(4.40),268(4.32), 310s(3.90)	4-1789-86
$C_{29}H_{23}NO_2$			
Benzenemethanol, α-[phenyl[(1-phenylethylidene)amino]methylene]-, benzoate	CH_2Cl_2	237(4.37),283(4.06)	39-2021-86C
1H-Indene-1,3(2H)-dione, 2-[4-(dimethylamino)phenyl]-4,7-diphenyl-	EtOH-NH_3	310(4.30),373(3.94), 490(3.11)	70-0582-86
1(3H)-Isobenzofuranone, 3-[[4-(dimethylamino)phenyl]methylene]-4,7-diphenyl-	EtOH	262(4.57),427(4.34)	70-0582-86
	EtOH-HCl	270(4.5),360(4.2)	70-0582-86
Methanone, (2,5-dihydro-2-methyl-2,4,5-triphenyl-5-oxazolyl)phenyl-	CH_2Cl_2	251(4.24)	39-0623-86C
$C_{29}H_{23}NO_4$			
Benzophenone, 3-acetyl-2,5-dimethyl-6-(4-nitrophenyl)-4-phenyl-	MeCN	254(4.46)	48-0359-86
Isoquinolinium, 2-(9(10H)-anthracenylidene)-3-methoxy-1-(methoxycarbonyl)-3-oxopropylide	CH_2Cl_2	574(3.95)	89-0572-86
Spiro[anthracene-9(10H),1'(10'bH)-pyrrolo[2,1-a]isoquinoline]-2',3'-dicarboxylic acid, dimethyl ester	CH_2Cl_2	383(3.95)	89-0572-86
$C_{29}H_{23}NS$			
Benzo[f]pyrrolo[2,1-a]isoquinoline-10(6H)-thione, 5,10a-dihydro-10a-methyl-8,9-diphenyl-	MeCN	227(4.92),272s(--), 424(4.10)	118-0899-86
Pyridine, 1,2-dihydro-1,4,6-triphenyl-2-(phenylthio)-	DMSO	340(4.13)	44-2481-86
Pyridine, 1,4-dihydro-1,2,6-triphenyl-4-(phenylthio)-	DMSO	335(3.88)	44-2481-86
$C_{29}H_{23}N_3$			
Propanedinitrile, (7,8-dihydro-1-methyl-2,3,5-triphenyl-4(1H)-azocinylidene)-	MeCN	236(4.34),327(3.79), 483(4.29)	118-0908-86

Compound	Solvent	$\lambda_{max}(\log \epsilon)$	Ref.
Propanedinitrile, (2,3,6,7-tetrahydro-1-methyl-2,3-diphenyl-4H-benzo[a]quinolizin-4-ylidene)-	MeCN	330(3.90)	118-0908-86
$C_{29}H_{23}N_5$			
1H-Pyrrole, 1-methyl-2,5-diphenyl-3,4-bis(phenylazo)-	EtOH	208(4.54),220s(4.33), 294(4.46),335(4.27)	103-0499-86
$C_{29}H_{24}BrNO_2$			
1,4-Naphthalenedione, 2-bromo-3-[[3-(5H-dibenzo[a,d]cyclohepten-5-yl)propyl]methylamino]-	CH_2Cl_2	240s(5.3),250s(4.1), 287(4.43),350s(3.3), 505(3.58)	83-0791-86
1,4-Naphthalenedione, 2-bromo-3-[[3-(10,11-dihydro-5H-dibenzo[a,d]cyclohepten-5-ylidene)propyl]methylamino]-	CH_2Cl_2	243(4.37),250s(4.3), 284(4.18),340s(3.5), 505(3.59)	83-0791-86
$C_{29}H_{24}ClNO_2$			
1,4-Naphthalenedione, 2-chloro-3-[[3-(5H-dibenzo[a,d]cyclohepten-5-yl)propyl]methylamino]-	CH_2Cl_2	240s(4.3),248s(4.1), 288(4.38),345s(3.3), 504(3.54)	83-0791-86
1,4-Naphthalenedione, 2-chloro-3-[[3-(10,11-dihydro-5H-dibenzo[a,d]cyclohepten-5-ylidene)propyl]methylamino]-	CH_2Cl_2	243(4.31),248s(4.3), 285(4.13),340s(3.4), 500(3.52)	83-0791-86
1,4-Naphthalenedione, 2-chloro-3-[[3-(9,10-ethanoanthracen-9(10H)-yl)propyl]amino]-	CH_2Cl_2	235s(4.2),242s(4.1), 267s(4.3),275(4.43), 300s(3.9),333(3.32), 466(3.51)	83-0791-86
$C_{29}H_{24}NO_5$			
Dibenzo[c,h]xanthylium, 5,6,8,9-tetrahydro-3,11-dimethoxy-7-(4-nitrophenyl)-, perchlorate	MeCN	264(4.57),316(4.35), 367(4.09),510(4.49)	22-0600-86
$C_{29}H_{24}N_3O_5P$			
1H-Pyrrolo[3,4-c]pyridine-7-carboxylic acid, 2,4,5,6-tetrahydro-2,5-dimethyl-1,4,6-trioxo-3-[(triphenylphosphoranylidene)amino]-, methyl ester	$CHCl_3$	366(4.31),463(4.31)	44-0149-86
$C_{29}H_{24}N_4O_5$			
Picrasidine R	EtOH	234(4.22),254(3.97), 276(4.11),310s(3.27), 372(3.68)	94-2090-86
$C_{29}H_{24}O_2$			
Benzophenone, 3-acetyl-2,5-dimethyl-4,6-diphenyl-	MeCN	240(4.41)	48-0359-86
$C_{29}H_{24}O_3$			
2-Propen-1-one, 1-[3-(diphenylmethyl)-2-hydroxy-4-methoxyphenyl]-3-phenyl-, (E)-	MeOH	208(4.63),286(4.11), 320(4.08)	2-0755-86
2-Propen-1-one, 1-[5-(diphenylmethyl)-2-hydroxy-4-methoxyphenyl]-3-phenyl-, (E)-	MeOH	210(4.72),270(4.10), 314(4.24)	2-0755-86
2-Propen-1-one, 1-[2-hydroxy-4-(phenylmethyl)-3-(phenylmethyl)phenyl]-3-phenyl-	MeOH	210(4.7),268(4.2), 334(4.4)	2-0259-86

Compound	Solvent	λ_{max}(log ϵ)	Ref.
2-Propen-1-one, 1-[2-hydroxy-4-(phenylmethoxy)-5-(phenylmethyl)phenyl]-3-phenyl-	MeOH	210(4.7),268(4.2), 334(4.4)	2-0259-86
$C_{29}H_{24}O_9$			
[2,2'-Binaphthalene]-1,4-dione, 1',5,5'-triacetoxy-4'-methoxy-7,7'-dimethyl-	EtOH	248(4.59),258s(4.48), 304(4.03),331s(3.82), 357s(3.62),430s(3.10)	39-0675-86C
$C_{29}H_{25}BrO_3$			
2,4-Octadiene-1,7-dione, 1-(4-bromophenyl)-6-(1-hydroxyethylidene)-4-methyl-3,5-diphenyl-, (Z,Z,E)-	dioxan	274(4.39),285(4.39), 344s(3.73)	48-0359-86
2,4-Octadiene-1,7-dione, 5-(4-bromophenyl)-6-(1-hydroxyethylidene)-4-methyl-1,3-diphenyl-, (Z,Z,E)-	dioxan	263(4.39),278(4.38), 344s(3.64)	48-0359-86
$C_{29}H_{25}ClO_3$			
2,4-Octadiene-1,7-dione, 1-(4-chlorophenyl)-6-(1-hydroxyethylidene)-4-methyl-3,5-diphenyl-, (Z,Z,E)-	dioxan	267(4.38),282(4.38), 346s(3.68)	48-0359-86
2,4-Octadiene-1,7-dione, 5-(4-chlorophenyl)-6-(1-hydroxyethylidene)-4-methyl-1,3-diphenyl-, (Z,Z,E)-	dioxan	263(4.37),276(4.37), 348s(3.59)	48-0359-86
$C_{29}H_{25}FO_3$			
2,4-Octadiene-1,7-dione, 1-(4-fluorophenyl)-6-(1-hydroxyethylidene)-4-methyl-3,5-diphenyl-, (Z,Z,E)-	dioxan	259(4.33),283s(4.31), 350s(3.60)	48-0359-86
2,4-Octadiene-1,7-dione, 5-(4-fluorophenyl)-6-(1-hydroxyethylidene)-4-methyl-1,3-diphenyl-, (Z,Z,E)-	dioxan	261(4.33),282(4.34), 349s(3.63)	48-0359-86
$C_{29}H_{25}NO_2$			
1,4-Naphthalenedione, 2-[[3-(5H-dibenzo[a,d]cyclohepten-5-yl)propyl]-methylamino]-	CH_2Cl_2	280(4.51),290s(4.42), 340s(3.40),465(3.64)	83-0607-86
1,4-Naphthalenedione, 2-[[3-(10,11-dihydro-5H-dibenzo[a,d]cyclohepten-5-ylidene)propyl]methylamino]-	CH_2Cl_2	239(4.42),245s(4.40), 279(4.30),330s(3.40), 464(3.63)	83-0607-86
3-Penten-2-one, 3-(5,6-dihydro-2,3-diphenylpyrrolo[2,1-a]isoquinolin-1-yl)-4-hydroxy-, (Z)-	MeCN	310(4.45)	118-0906-86
$C_{29}H_{25}NO_5$			
2,4-Octadiene-1,7-dione, 6-(1-hydroxyethylidene)-4-methyl-3-(4-nitrophenyl)-1,5-diphenyl-, (Z,Z,E)-	dioxan	263s(4.34),292(4.43), 347s(3.83)	48-0359-86
$C_{29}H_{25}N_5O_2$			
9H-Purin-6-amine, 9-[1-[[1-[(triphenylmethoxy)methyl]ethenyl]oxy]ethenyl]-	EtOH	259(4.11)	87-2445-86
$C_{29}H_{25}N_5O_3$			
9H-Purin-6-one, 9-(2,5-dihydro-2-furanyl)-1,9-dihydro-2-[[(4-methoxyphenyl)diphenylmethyl]amino]-, (±)-	MeOH	234(4.21),261(4.20), 276s(4.15)	23-1885-86
$C_{29}H_{25}O_2$			
Dibenzo[c,h]xanthylium, 5,6,8,9-tetra-	MeCN	228(4.60),250(4.49),	22-0600-86

Compound	Solvent	λ_{max}(log ϵ)	Ref.
hydro-3-methoxy-7-(4-methylphenyl)-, perchlorate (cont.)		312(4.27),372(3.98), 478(4.45)	22-0600-86
$C_{29}H_{25}O_3$			
Dibenzo[c,h]xanthylium, 7-(3,4-dimethoxyphenyl)-5,6,8,9-tetrahydro-, perchlorate	MeCN	228(4.69),292(4.27), 316(4.21),445(4.38)	22-0600-86
Dibenzo[c,h]xanthylium, 5,6,8,9-tetrahydro-3,11-dimethoxy-7-phenyl-, perchlorate	MeCN	230(4.60),316(4.29), 368(4.09),504(4.59)	22-0600-86
Dibenzo[c,h]xanthylium, 5,6,8,9-tetrahydro-3-methoxy-7-(4-methoxyphenyl)-, perchlorate	MeCN	224(4.63),252(4.53), 310(4.27),479(4.53)	22-0600-86
Pyrylium, 2-[4-(3,4-dimethoxyphenyl)-1,3-butadienyl]-4,6-diphenyl-, tetrafluoroborate	CH_2Cl_2	380(4.72),600(4.68)	104-1786-86
$C_{29}H_{26}ClN_5O_3$			
6H-Purin-6-one, 9-(3-chlorotetrahydro-2-furanyl)-1,9-dihydro-2-[[(4-methoxyphenyl)diphenylmethyl]amino]-, trans-(±)-	MeOH	261(4.17),277s(4.12)	23-1885-86
$C_{29}H_{26}F_3NO_{11}$			
9,10-Anthracenedione, 2-[[[3,4,6-tri-O-acetyl-2-deoxy-2-[(trifluoroacetyl)amino]-β-D-glucopyranosyl]oxy]-methyl]-	EtOH	255(4.08),274(3.57), 325(3.11)	87-1709-86
$C_{29}H_{26}F_3NO_{12}$			
9,10-Anthracenedione, 1-hydroxy-2-[[[3,4,6-tri-O-acetyl-2-deoxy-2-[(trifluoroacetyl)amino]-β-D-glucopyranosyl]oxy]methyl]-	EtOH	258(4.45),266(4.28), 405(3.77)	87-1709-86
$C_{29}H_{26}F_3NO_{13}$			
9,10-Anthracenedione, 1,4-dihydroxy-2-[[[3,4,6-tri-O-acetyl-2-deoxy-2-[(trifluoroacetyl)amino]-β-D-glucopyranosyl]oxy]methyl]-	EtOH	246(4.11),257(4.04), 285(3.58),477(3.72), 510(3.48)	87-1709-86
9,10-Anthracenedione, 1,8-dihydroxy-2-[[[3,4,6-tri-O-acetyl-2-deoxy-2-[(trifluoroacetyl)amino]-β-D-glucopyranosyl]oxy]methyl]-	EtOH	270(4.23),485(3.28)	87-1709-86
9,10-Anthracenedione, 1,8-dihydroxy-3-[[[3,4,6-tri-O-acetyl-2-deoxy-2-[(trifluoroacetyl)amino]-β-D-glucopyranosyl]oxy]methyl]-	EtOH	269(4.11),493(3.28)	87-1709-86
$C_{29}H_{26}NO$			
Dibenzo[c,h]xanthylium, 7-[4-(dimethylamino)phenyl-5,6,8,9-tetrahydro-, perchlorate	MeCN	258(4.36),296(4.36), 382(4.03),444(3.68)	22-0600-86
$C_{29}H_{26}N_2O_3S$			
Formamide, N-[2-(2,3-dihydro-6,7-diphenyl-1H-pyrrolizin-5-yl)-1-[(4-methylphenyl)sulfonyl]ethenyl]-	MeOH	203(4.69),234(4.49), 361(4.35)	83-0354-86
$C_{29}H_{26}N_4O_3$			
Picrasidine F, hydrochloride	EtOH	252(4.29),282s(3.60),	94-3228-86

Compound	Solvent	λ_{max}(log ε)	Ref.
Picrasidine F hydrochloride (cont.)		310(3.63),347(3.42), 372(3.20)	94-3228-86
	EtOH-NaOH	242(4.08),278(4.24), 334(3.57),402(3.03)	94-3228-86
$C_{29}H_{26}O_3$			
Ethanone, 1-[2-hydroxy-4-(phenylmethoxy)-3,5-bis(phenylmethyl)phenyl]-	MeOH	229(4.7),276(3.7)	2-0259-86
2,4-Octadiene-1,7-dione, 6-(1-hydroxyethylidene)-4-methyl-1,3,5-triphenyl-, (Z,Z,E)-	dioxan	260(4.34),285(4.33), 340s(3.78)	48-0359-86
1-Propanone, 1-[2,4-dihydroxy-3,5-bis(phenylmethyl)phenyl]-3-phenyl-	MeOH	220(4.9),282(4.2)	2-0259-86
$C_{29}H_{26}O_4$			
Ethanone, 1-[2-hydroxy-3,4-bis(phenylmethoxy)-5-(phenylmethyl)phenyl]-	MeOH	217(4.9),280(4.2)	2-0259-86
$C_{29}H_{27}N$			
Naphtho[2,1-a]quinolizine, 5,8,9,10-11,11a-hexahydro-9,10-diphenyl-	MeCN	272(3.84),281(3.82)	118-0908-86
$C_{29}H_{27}N_3O_8$			
2(1H)-Pyrimidinone, 1-[[2-(benzoyloxy)ethoxy]methyl]-4-methoxy-5-[[3-(4-nitrophenyl)methoxy]phenyl]methyl]-	EtOH	273(4.22)	4-1651-86
$C_{29}H_{27}N_5O_3$			
9H-Purin-6-amine, 9-[6-[(triphenylmethoxy)methyl]-1,4-dioxan-2-yl]-, (2R-cis)-	EtOH	213(4.50),259(4.15)	87-2445-86
$C_{29}H_{27}N_5O_5$			
6H-Purin-6-one, 1,9-dihydro-2-[[(4-methoxyphenyl)diphenylmethyl]amino]-9-(tetrahydro-3,4-dihydroxy-2-furanyl)-, (2α,3β,4β)-(±)-	MeOH	260(4.05),278s(4.01)	23-1885-86
$C_{29}H_{28}N_2O_2$			
Pyridine, 1,2-dihydro-2-(nitrocyclohexyl)-1,4,6-triphenyl-	DMSO	337(4.05)	44-2481-86
Pyridine, 1,4-dihydro-4-(nitrocyclohexyl)-1,2,6-triphenyl-	DMSO	329(3.95)	44-2481-86
$C_{29}H_{28}N_2O_6$			
2(1H)-Pyrimidinone, 1-[[2-(benzoyloxy)ethoxy]methyl]-4-methoxy-5-[[3-(phenylmethoxy)phenyl]methyl]-	EtOH	271(4.03)	4-1651-86
$C_{29}H_{28}N_2O_{10}S_2$			
Emestrin, 3,11a-de(epidithio)-3,11a-bis(methylthio)-	MeOH	259(4.11),287(3.67)	39-0109-86C
$C_{29}H_{28}N_4O_9$			
Uridine-elliptinium adduct	H_2O	275(4.55),313(4.32), 368(4.05),450(3.20)	87-1350-86
$C_{29}H_{28}O_3$			
1,4-Methanonaphthalene-5,8,9-trione, 6-(1,1-dimethylethyl)-1,4,4a,8a-	MeCN	231(4.34),275s(3.75)	12-1537-86

Compound	Solvent	λ max (log ε)	Ref.
hydro-1,4-dimethyl-2,3-diphenyl-, endo (cont.)			12-1537-86
exo	MeCN	228(4.45)	12-1537-86
2-Oxapentacyclo[6.4.0.0^{1,4}.0^{3,7}.0^{5,9}]-dodec-11-ene-6,10-dione, 11-(1,1-dimethylethyl)-5,7-dimethyl-3,4-diphenyl-	MeCN	220(4.30),261(3.91)	12-1537-86
2-Oxapentacyclo[6.4.0.0^{1,4}.0^{3,7}.0^{5,9}]-dodec-11-ene-6,10-dione, 12-(1,1-dimethylethyl)-5,7-dimethyl-3,4-diphenyl-	MeCN	221(4.30),255(3.84)	12-1537-86
$C_{29}H_{29}N_5O_4$			
9H-Purine-9-ethanol, 6-amino-β-[1-(hydroxymethyl)-2-(triphenylmethoxy)-ethoxy]-, [S-(R*,S*)]-	EtOH	212(4.51),260(4.14)	87-2445-86
$C_{29}H_{29}N_5O_5$			
6H-Purin-6-one, 1,9-dihydro-9-[2-hydroxy-1-(2-hydroxyethoxy)ethyl]-2-[[(4-methoxyphenyl)diphenylmethyl]amino]-, (±)-	MeOH	235(4.23),261(4.20), 277s(4.16)	23-1885-86
$C_{29}H_{29}N_5O_6$			
Adenosine, N,N-bis(4-methylbenzoyl)-2',3'-O-(1-methylethylidene)-	MeOH	260(4.44),275s(4.41)	23-2109-86
8,5'-O-Cycloadenosine, 7,8-dihydro-N^6,7-bis(4-methylbenzoyl)-2',3'-O-(1-methylethylidene)-	MeOH	240(4.46),285(4.18)	23-2109-86
$C_{29}H_{29}N_5O_8$			
8,5'-O-Cycloadenosine, N^6,7-bis(4-methoxybenzoyl)-7,8-dihydro-2',3'-O-(1-methylethylidene)-	MeOH	266(4.59)	23-2109-86
Cytidine-elliptinium adduct	H_2O	275(4.54),313(4.20), 372(3.95),460(3.20)	87-1350-86
$C_{29}H_{30}N_2O_6$			
2(1H)-Pyrimidinone, 1-[(2-hydroxyethoxy)methyl]-4-methoxy-5-[[3-[[4-(phenylmethoxy)phenyl]methoxy]-phenyl]methyl]-	EtOH	275(3.89),281(3.90)	4-1651-86
$C_{29}H_{30}N_4O_{10}S$			
1-Azabicyclo[3.2.0]hept-2-ene-2-carboxylic acid, 6-(1-hydroxyethyl)-3-[[3-[[[(4-nitrophenyl)methoxy]-carbonyl]amino]cyclopentyl]thio]-7-oxo-, [(4-nitrophenyl)methyl] ester	EtOH	267(4.36),324(4.18)	23-2184-86
$C_{29}H_{31}NS_2$			
2H,6H-[1,3]Thiazino[2,3-a]isoquinoline, 2-[(1,1-dimethylethyl)thio]-7,11b-dihydro-11b-methyl-3,4-diphenyl-	MeCN	274s(--),288(3.85)	118-0899-86
$C_{29}H_{32}Cl_4$			
Spiro[cyclopropane-1,17'-tricyclo-[10.4.1.1^{4,9}]octadeca-2,4,6,8,12,14-16-heptaene], 2,2,8',16'-tetra-	C_6H_{12}	217s(4.26),252s(4.19), 294s(3.93),415(4.03)	44-2214-86

Compound	Solvent	λ_{max}(log ϵ)	Ref.
chloro-6',14'-bis(1,1-dimethylethyl)-18'-methylene- (cont.)			44-2214-86
$C_{29}H_{32}O_{11}$			
Aloeresin D	MeOH	226(4.53),242s(4.31), 252(4.24),300(4.47)	102-2219-86
Roritoxin D	MeOH	253(4.98)	44-2906-86
$C_{29}H_{32}O_{12}$			
Roritoxin C	MeOH	254(4.97)	44-2906-86
$C_{29}H_{33}NO_{12}$			
Doxorubicin, N-(2-hydroxyethyl)-, hydrochloride	MeOH	234(4.57),252(4.41), 289(3.94),478(4.08), 530(3.83)	87-2120-86
$C_{29}H_{33}N_5O_7$			
Saframycin Y-3	MeOH	268(4.27),340s(--)	158-1639-86
$C_{29}H_{34}N_4O$			
1H-Pyrazolium, 4-[[4-(dimethylamino)-phenyl][4-(trimethylammino)phenyl]-methylene]-4,5-dihydro-2,3-dimethyl-5-oxo-1-phenyl-, diperchlorate	MeCN	540(4.09)	104-0727-86
$C_{29}H_{34}O_9$			
2H,3H-Cyclopenta[d']naphtho[1,8-bc-2,3-b']difuran-6-acetic acid, 8-(5-acetoxy-2,5-dihydro-2-oxo-3-furanyl)-2a,5a,6,6a,8,9,9a,10a,10b-10c-decahydro-2a,5a,6,7-tetramethyl-3-oxo-, methyl ester	MeOH	222(3.80)	78-4849-86
2H,3H-Cyclopenta[d']naphtho[1,8-bc-2,3-b']difuran-6-acetic acid, 8-(2-acetoxy-2,5-dihydro-5-oxo-3-furanyl)-2a,5a,6,6a,8,9,9a,10a,10b-10c-decahydro-2a,5a,6a,7-tetramethyl-3-oxo-, methyl ester	MeOH	225(3.78)	78-4849-86
$C_{29}H_{34}O_{10}$			
Roritoxin A (7',8'-didehydroverti-sporin)	MeOH	231(4.35),260(4.7)	44-2906-86
$C_{29}H_{34}O_{11}$			
Roritoxin B (7',8'-didehydro-2',3'-epoxy-2',3'-dihydro-	MeOH	254(4.99)	44-2906-86
Youngiaside B	MeOH	277(3.35)	94-3769-86
Youngiaside C	MeOH	277(3.23)	94-3769-86
Youngiaside D	MeOH	277(3.32)	94-3769-86
$C_{29}H_{34}O_{16}$			
Tricin 7-O-rutinoside	MeOH	241(4.34),262(4.24), 346(4.43)	94-0082-86
	MeOH-NaOAc	241(--),262(--), 346(--)	94-0082-86
	MeOH-$AlCl_3$	268(--),304(--), 397(--)	94-0082-86
$C_{29}H_{34}O_{18}$			
Spinacetin 3-O-β-gentiobioside	MeOH	254(4.28),270s(4.21), 300s(3.97),350(4.33)	102-0231-86

Compound	Solvent	$\lambda_{max}(\log \epsilon)$	Ref.
(cont.)	MeOH-NaOMe	270(4.33),334(4.05), 415(4.48)	102-0231-86
	MeOH-NaOAc	272(4.35),322(3.79), 390(4.27)	102-0231-86
	MeOH-NaOAc-H_3BO_3	255(4.27),268(4.25), 355(4.29)	102-0231-86
	MeOH-$AlCl_3$	267(4.29),280s(4.25), 305s(3.95),382(4.33), 400s(4.30)	102-0231-86
	MeOH-$AlCl_3$-HCl	267(4.28),280s(4.24), 300s(3.97),382(4.32), 410s(4.25)	102-0231-86
$C_{29}H_{35}Cl_3$			
Spiro[cyclopropane-1,17'-tricyclo-[10.4.1.1^{4,8}]octadeca[2,4,6,8(18)-12,14,16]heptaene], 2,2,16'-tri-chloro-6',14'-bis(1,1-dimethylethyl)-18'-methyl-	C_6H_{12}	216(4.38),230s(4.35), 254(4.28),277s(4.04), 336(3.67)	44-2214-86
$C_{29}H_{36}F_3NO_4$			
Oxiranebutanoic acid, 3-[14-[4-[(tri-fluoroacetyl)amino]phenyl]-1,3,5,8-tetradecatetraenyl]-, methyl ester	CH_2Cl_2	270.5(4.61),282(4.69), 293(4.53)	88-6193-86
$C_{29}H_{36}O_5$			
Vouacapen-5α-ol, 6β-cinnamoyl-7β-hy-droxy-	EtOH	210s(4.24),220(4.47), 225(4.39),280(4.49)	102-0167-86
$C_{29}H_{36}O_6$			
Butanoic acid, 3-hydroxy-3-methyl-, 1,2,3,4,4a,5,6,7-octahydro-1,1,4a-trimethyl-6-methylene-5-[[(2-oxo-2H-1-benzopyran-7-yl)oxy]methyl]-2-naphthalenyl ester	MeOH	209(4.3),260s(3.2), 320(4.1)	102-2417-86
$C_{29}H_{36}O_7$			
Meroliminol 4-methylbenzoate	n.s.g.	241(4.12)	138-1337-86
$C_{29}H_{36}O_{10}$			
2-Naphthalenecarboxylic acid, 3,4-di-acetoxy-8-(1,7-diacetoxy-6-methyl-2-heptyl)-5-methyl-, methyl ester	MeOH	252(4.36),293(3.94), 305(3.87),346(3.67)	102-1377-86
$C_{29}H_{36}O_{11}$			
Yadanzioside O aglycone	EtOH	222(4.17),278(3.85)	18-3541-86
$C_{29}H_{36}O_{16}$			
β-D-Glucopyranoside, 3,4-dihydroxy-β-(phenylethyl)-O-β-D-glucopyran-osyl-(1→5)-4-O-caffeoyl-	MeOH	218(4.11),247(3.83), 291(3.94),333(4.08)	102-1633-86
$C_{29}H_{37}N_3O_6$			
Calcimycin	n.s.g.	278(4.26),378(3.91)	100-0035-86
$C_{29}H_{38}F_3N_3O_2S$			
Fluphenazine enanthate	MeOH	256(4.536),313(3.602)	106-0571-86
$C_{29}H_{38}O_5$			
1,3-Azulenedicarboxylic acid, 6-(3-hy-droxy-3-methyl-1-butynyl)-2-octyl-,	CH_2Cl_2	231.5(4.36),268.5(4.14), 328.5(4.93),372(4.25),	24-2272-86

Compound	Solvent	$\lambda_{max}(\log \epsilon)$	Ref.
diethyl ester (cont.)		528.5(2.66)	24-2272-86
$C_{29}H_{39}ClN_2O_8$			
Geldanamycin, 6-chloro-6,17-didemethoxy-15-methoxy-11-O-methyl-, (15R)-	MeOH	232(4.27)	158-0415-86
$C_{29}H_{39}PSi_2$			
Phosphorane, [bis(trimethylsilyl)methylene](diphenylmethylene)(2,4,6-trimethylphenyl)-	C_6H_{12}	258(4.20),319(4.01), 479(4.08)	24-1977-86
$C_{29}H_{40}ClNO_8$			
Geldanamycin, 19-chloro-7-O-de(aminocarbonyl)-17-demethoxy-15-methoxy-11-O-methyl-, (15R)-	MeOH	251(4.20)	158-0415-86
$C_{29}H_{40}N_2O_2$			
Pregn-5-en-3-ol, 21-[3-(3H-diazirin-3-yl)phenoxy]-20-methyl-, (3β,20S)-	EtOH	363(2.53)	94-0931-86
$C_{29}H_{40}N_8O_6S_2$			
Cyclo[L-Val-L-Leu-L-Pro-(RS)-(Gln)-Thz(Gly)Thz]	MeOH	250(3.89)	44-4586-86
Cyclo[L-Pro-L-Leu-L-Val-(RS)-(Gln)-Thz(Gly)Thz], form A	MeOH	244(3.92)	44-4580-86
form B (dolostatin 3 isomers)	MeOH	242(3.91)	44-4580-86
$C_{29}H_{40}O_7$			
Cholesta-4,7-diene-3,6,22-trione, 2-acetoxy-14,20-dihydroxy-	EtOH	262(3.52)	102-1305-86
$C_{29}H_{40}O_{14}$			
Ebuloside tetraacetate	MeOH	205(3.33)	33-0156-86
$C_{29}H_{41}FO_5S$			
Androsta-1,4-diene-17-carbothioic acid, 17-acetoxy-9-fluoro-11-hydroxy-16-methyl-3-oxo-, S-hexyl ester, (11β,16β,17α)-	MeOH	240(4.28)	44-2315-86
Androsta-1,4-diene-17-carboxylic acid, 9-fluoro-17-[1-(hexylthio)-1-hydroxyethoxy]-11-hydroxy-16-methyl-3-oxo-, γ-lactone, (11β,16β,17α)-	MeOH	237(4.21)	44-2315-86
$C_{29}H_{41}NO_8$			
Geldanamycin, 7-O-de(aminocarbonyl)-17-demethoxy-15-methoxy-11-O-methyl-, (15R)- (7-decarbamoylherbimycin A)	MeOH	267(4.34)	158-0415-86
$C_{29}H_{41}NO_9$			
Geldanamycin, 7-O-de(aminocarbonyl)-17-demethoxy-8,9-epoxy-8,9-dihydro-15-methoxy-11-O-methyl-, (15R)-	MeOH	266(4.27)	158-0415-86
$C_{29}H_{41}N_2S$			
Benzothiazolium, 2-[2-[4-(dimethylamino)phenyl]ethenyl]-3-dodecyl-, iodide	EtOH	530(4.84)	4-0209-86

Compound	Solvent	λ_{max}(log ε)	Ref.
$C_{29}H_{41}N_7O_8S_2$			
L-Proline, 1-[[2-[[[[2-[4-amino-1-[[2-[[(1,1-dimethylethoxy)carbonyl]amino]-4-oxobutyl]-4-thiazolyl]carbonyl]amino]methyl]-4-thiazolyl]carbonyl]-, methyl ester	MeOH	240(3.89)	44-4580-86
$C_{29}H_{42}N_4$			
7H-[1,2,4]Triazolo[4,3-a]perimidine, 10-heptadecyl-	dioxan	226(4.19),252(3.75), 287(3.54),298(3.59), 353(3.73)	48-0237-86
$C_{29}H_{42}O_8$			
Withaferin A, 2,3-dihydro-12β-hydroxy-3-methoxy-	MeOH	217(3.98)	102-1613-86
$C_{29}H_{42}O_{13}$			
Rehmaionoside C pentaacetate	MeOH	228(4.22)	94-2294-86
$C_{29}H_{43}Br_2SeTe$			
4H-Tellurin, 4-[3-[2,6-bis(1,1-dimethylethyl)seleninium-4-yl]-2-propenylidene]-1,1-dibromo-2,6-bis(1,1-dimethylethyl)-1,1-dihydro-, tetrafluoroborate	CH_2Cl_2	543(4.74)	157-2250-86
$C_{29}H_{43}Br_2Te_2$			
4H-Tellurin, 4-[3-[2,6-bis(1,1-dimethylethyl)tellurinium-4-yl]-2-propenylidene]-1,1-dibromo-2,6-bis(1,1-dimethylethyl)-1,1-dihydro-, tetrafluoroborate	CH_2Cl_2	565(4.81)	157-2250-86
$C_{29}H_{43}NO_{16}$			
Paulomenol A	EtOH	241(3.93),318(3.95)	44-2493-86
$C_{29}H_{43}N_3O_2$			
6H-Benzo[g][1,4]diazonino[7,6,5-cd]indol-6-one, 13-(1,1-dimethylethyl)-10-ethenyl-1,3,4,5,7,8,10,11,12,13-decahydro-4-(methoxymethyl)-8,10-dimethyl-7-(1-methylethyl)- (olivoretin E)	MeOH	234(4.51),291(3.93), 300s(3.91)	94-4883-86
$C_{29}H_{43}SeTe$			
Seleninium, 4-[3-[2,6-bis(1,1-dimethylethyl)-4H-tellurin-4-ylidene]-1-propenyl]-2,6-bis(1,1-dimethylethyl)-, tetrafluoroborate	CH_2Cl_2	786(5.45)	157-2250-86
$C_{29}H_{43}Te_2$			
Tellurinium, 4-[3-[2,6-bis(1,1-dimethylethyl)-4H-tellurin-4-ylidene]-1-propenyl]-2,6-bis(1,1-dimethylethyl)-, tetrafluoroborate	CH_2Cl_2	830(5.52)	157-2250-86
$C_{29}H_{44}$			
s-Indacene, 1,3,5,7-tetrakis(1,1-dimethylethyl)-1,7-dihydro-1-methyl-	hexane	230s(4.17),237(4.43), 245(4.65),254(4.64), 274(3.64),284(3.63), 296(3.60),302s(3.53),	89-0630-86

Compound	Solvent	λ_{max}(log ϵ)	Ref.
(cont.)		308(3.56),313s(3.38), 322(3.27)	89-0630-86
s-Indacene, 1,3,5,7-tetrakis(1,1-dimethylethyl)-3a,4-dihydro-4-methyl-	hexane	235(3.95),241(3.95), 286s(3.20),386(4.30), 471s(2.76)	89-0630-86
$C_{29}H_{44}N_2O_5$			
Androst-4-en-3-one, 17-acetoxy-2,6-di-4-morpholinyl-, (2α,6β,17β)-	EtOH	239(4.07)	107-0195-86
$C_{29}H_{44}O$			
s-Indacene, 1,3,5,7-tetrakis(1,1-dimethylethyl)-1,7-dihydro-1-methoxy-	hexane	227s(4.07),233s(4.23), 240(4.44),248(4.67), 257(4.68),281(3.38), 288s(3.37),300(3.27), 312(3.20),326(3.03)	89-0630-86
$C_{29}H_{44}O_5S$			
16,24-Cyclo-13,17-secochola-16,20(22)-23-triene-3,23,24-triol, 14,17-epoxy-23-[(2-hydroxyethyl)thio]-4,4,8-trimethyl-, (3β,5α,13α)-	MeOH	218(3.95),241(3.94), 292(3.72),397(3.51)	87-1851-86
$C_{29}H_{44}O_7$			
Cholesta-4,7-dien-6-one, 3-acetoxy-2,14,20,22-tetrahydroxy- (3-O-acetyl-14α-pinnasterol)	EtOH	255(4.00)	102-1305-86
Pinnasterol, 14α-hydroxyacetyl-	EtOH	257(3.99)	102-1305-86
22-epi-	EtOH	253(3.93)	102-1305-86
$C_{29}H_{44}O_8$			
Card-20(22)-enolide, 3-[(β-L-rhamnopyranosyl)oxy]-14-hydroxy-, (3β,5β,14β,17β)-	MeOH	217(4.22)	87-1945-86
Cholesta-4,7-dien-6-one, 2-acetoxy-3,9,14,20,22-pentahydroxy-, (2β,3α,22R)-	EtOH	232(3.82)	102-1305-86
$C_{29}H_{44}O_9$			
Card-20(22)-enolide, 3-[(α-D-mannopyranosyl)oxy]-14-hydroxy-, (3β,5β,14β,17β)-	MeOH	217(4.18)	87-1945-86
L-isomer	MeOH	217(4.20)	87-1945-86
Card-20(22)-enolide, 3-[(β-D-mannopyranosyl)oxy]-14-hydroxy-	MeOH	217(4.20)	87-1945-86
L-isomer	MeOH	217(4.18)	87-1945-86
$C_{29}H_{46}O$			
Cyclonervilasterol	EtOH	210(3.72)	94-3183-86
24-epi-	EtOH	211(3.76)	94-3183-86
$C_{29}H_{46}O_6$			
2-Deoxy-α-ecdysone, 3-acetate	EtOH	245(4.00)	105-0409-86
$C_{29}H_{47}NO$			
6-Aza-B-homostigmasta-2,4-dien-7-one	EtOH	273(3.89)	39-2123-86C
$C_{29}H_{47}N_5O_6SSi_2$			
1H-Imidazole-4-carboxamide, 5-[[(benzoylamino)thioxomethyl]amino]-1-	MeOH	248(4.39),276s(--)	87-0203-86

Compound	Solvent	λ_{max}(log ε)	Ref.
[3,5-bis-O-[(1,1-dimethylethyl)dimethylsilyl]-α-D-xylofuranosyl]-, (cont.)			87-0203-86
$C_{29}H_{48}O$			
Cyclonervilasterol, dihydro-	EtOH	211(3.75)	94-3183-86
24-epi-	EtOH	211(3.72)	94-3183-86
$C_{29}H_{50}B_2Br_2F_4N_8O_4Rh_2$			
Rhodium, dibromobis[difluoro[[2,2'-(1,3-propanediyldinitrilo)bis[3-pentanone]dioximato](2-)-O,O']-borato(1-)-N, N',N",N'"]-μ-1,3-propanediyldi-, trans,trans	acetone	372(3.98)	157-0218-86
$C_{29}H_{50}O_3$			
2H-1-Benzopyran-6(5H)-one, 3,4-dihydro-5-hydroxy-2,5,7,8-tetramethyl-2-(4,8,12-trimethyltridecyl)-, (2R,5R,4'R,8'R)-	THF	335(3.15),375(3.09)	44-1435-86
(2R,5S,4'R,8'R)-	THF	335(3.11),375(3.09)	44-1435-86
5H-1-Benzopyran-5-one, 2,3,4,6-tetrahydro-6-hydroxy-2,6,7,8-tetramethyl-2-(4,8,12-trimethyltridecyl)-, [2R,6R,4'R,8'R]-	THF	245(3.70),318(3.48)	44-1435-86
[2R,6S,4'R,8'R]-	THF	242s(3.51),318(3.54)	44-1435-86

Compound	Solvent	λ_{max}(log ε)	Ref.
$C_{30}F_{20}O_6$			
2,5-Cyclohexadiene-1,4-dione, 2,3,5,6-tetrakis(pentafluorophenoxy)-	EtOH	211(3.59),288(3.09), 416(1.76)	104-2313-86
$C_{30}H_{12}O_6$			
5,7,9,14,16,18-Heptacenehexone	dioxan	222(4.54),258(4.83), 307(4.32),430(3.75)	44-4160-86
$C_{30}H_{14}N_6$			
Propanedinitrile, 2,2'-(2,3-diphenyl-benzo[g]quinoxaline-5,10-diylidene)-bis-	MeCN	216(4.54),227(4.54), 277s(4.20),318(4.35), 370(4.49)	138-0715-86
$C_{30}H_{15}N_3O_4$			
3H-Phenoxazin-3-one, 4-(6-tripheno-dioxazinyl)-	$CHCl_3$	472(4.25),506(4.32)	4-1003-86
$C_{30}H_{16}O_2$			
7,16-Heptacenedione	$CHCl_3$	337(4.4),347(4.6), 452(4.0),463(4.0)	44-3762-86
$C_{30}H_{16}O_7$			
Dianthra[1,2-b:2',1'-d]furan-5,10,15,20-tetrone, 6,9-di-methoxy-	CH_2Cl_2	264(4.69),288s(4.52), 323s(3.92),414(4.34), 445s(4.20)	5-0839-86
$C_{30}H_{16}O_8$			
Hypericin	EtOH	308(4.33),550(--), 590(4.60)	41-0311-86
triplet	EtOH	510(4.49),630(4.12)	41-0311-86
$C_{30}H_{18}N_2O_4$			
3H-Indol-3-one, 1-benzoyl-2-(1-benzoyl-1,3-dihydro-3-oxo-2H-indol-2-yli-dene)-1,2-dihydro-, (E)-	benzene	577(3.90)	44-2529-86
$C_{30}H_{18}O_4$			
Ethanedione, 1,1'-(1,2-ethynediyldi-4,1-phenylene)bis[2-phenyl-	dioxan	328(4.078),348(4.097), 400s(2.76)	104-1117-86
$C_{30}H_{18}O_5$			
7,16-Epoxyheptacene-5,9,14,18-tetrone, 5a,6,7,16,17,17a-hexahydro-	dioxan	222(4.30),267(4.27), 305(3.41)	44-4160-86
$C_{30}H_{18}O_9$			
Cyclopenta[2,1-a:3,4-d']dianthracene-5,8,9,14,18(8aH)-pentone, 5c,15-di-hydro-1,8a,10,17-tetrahydroxy-12-methyl-	EtOH	254(4.35),283s(4.04), 358(3.71),438(3.92), 460s(3.86)	39-0255-86C
$C_{30}H_{18}S_2$			
Phenanthro[1,10-bc:8,9-b'c']bisthio-pyran, 2,11-diphenyl-	CH_2Cl_2	274(4.57),299s(--), 373(3.94),462(4.43)	35-3460-86
$C_{30}H_{19}N_3O_2$			
Benzonitrile, 4-(4,5-dibenzoyl-1-phen-yl-1H-pyrazol-3-yl)-	MeOH	251(4.65)	44-1972-86
$C_{30}H_{20}Br_2O_2S$			
2-Propen-1-one, 1,1'-(thiodi-4,1-phen-ylene)bis[3-(4-bromophenyl)-	93% H_2SO_4	357(3.9),397(4.44), 448(4.65),522(4.85)	104-0055-86

Compound	Solvent	λ_{max}(log ε)	Ref.
$C_{30}H_{20}Br_2O_3$			
2-Propen-1-one, 1,1'-(oxydi-4,1-phenylene)bis[3-(4-bromophenyl)-	93% H_2SO_4	350(4.41),475(4.95)	104-0055-86
$C_{30}H_{20}ClNOS$			
2(1H)-Pyridinethione, 3-benzoyl-6-(4-chlorophenyl)-1,4-diphenyl-	EtOH	250(4.53),285(4.39), 410(3.92)	118-0952-86
2(1H)-Pyridinethione, 3-(4-chlorobenzoyl)-1,4,6-triphenyl-	EtOH	260(4.50),283(4.35), 407(3.85)	118-0952-86
$C_{30}H_{20}ClN_3$			
Propanedinitrile, [4-(4-chlorophenyl)-4-cyano-3,5-bis(4-methylphenyl)-2,5-cyclohexadien-1-ylidene]-	$CHCl_3$	355(4.38)	83-0898-86
$C_{30}H_{20}N_2$			
3-Pyridinecarbonitrile, 2-[1,1'-biphenyl]-4-yl-4,6-diphenyl-	MeCN	280(4.65)	73-1061-86
3-Pyridinecarbonitrile, 6-[1,1'-biphenyl]-4-yl-2,4-diphenyl-	MeCN	325(4.59)	73-1061-86
$C_{30}H_{20}N_2O_2S$			
Oxazole, 5-phenyl-2-[5-[2-[4-(5-phenyl-2-oxazolyl)phenyl]ethenyl]-2-thienyl]-, (E)-	toluene	288(4.23),414(4.75)	103-1028-86
$C_{30}H_{20}N_2O_2Se_2$			
Benzenamine, N,N'-[diselenobis(3,2-benzofurandiylmethylidyne)]bis-	EtOH	239s(4.42),318(4.60), 349s(4.55)	103-0608-86
$C_{30}H_{20}N_2O_6S$			
2-Propen-1-one, 1,1'-(thiodi-4,1-phenylene)bis[3-(4-nitrophenyl)- (graph maxima)	93% H_2SO_4	339(4.03),376(4.61) 444(4.13),521(4.64)	104-0055-86
$C_{30}H_{20}N_2O_7$			
2-Propen-1-one, 1,1'-(oxydi-4,1-phenylene)bis[3-(4-nitrophenyl)-	93% H_2SO_4	358(4.46),442(4.66)	104-0055-86
$C_{30}H_{20}N_4O_2$			
Propanedinitrile, [4-cyano-3,5-bis(4-methylphenyl)-4-(4-nitrophenyl)-2,5-cyclohexadien-1-ylidene]-	$CHCl_3$	365(4.37)	83-0898-86
$C_{30}H_{20}N_4O_5$			
2,4-Imidazolidinedione, 5,5'-(2,8-dibenzofurandiyl)bis[5-phenyl-	dioxan	290(5.02)	104-1117-86
$C_{30}H_{20}N_4O_7$			
5H-Imidazo[2,1-b][1,3]oxazine-2,3-dicarbonitrile, 5,6-bis(benzoyloxy)-7-[(benzoyloxy)methyl]-6,7-dihydro-, [5R-(5α,6α,7β)]-	MeOH	230(4.64),266(4.42)	88-0323-86
$C_{30}H_{20}O$			
Ethanone, 1,2-di-9-anthracenyl-	C_6H_{12}	250(5.3),260(5.2), 350(3.9),367(4.1), 385(4.1)	44-2956-86
$C_{30}H_{20}OS_2$			
Bisdibenzo[2,3:6,7]thiepino[4,5-b-	MeOH	258(4.74),276s(4.58),	73-0141-86

Compound	Solvent	$\lambda_{max}(\log \epsilon)$	Ref.
4',5'-d]furan, 1,9-dimethyl- (cont.)		320(4.49),335(4.42)	73-0141-86
$C_{30}H_{20}O_8$			
5H-Indeno[1,2-a]anthracene-5,9,13-trione, 8a-(1,3-dihydro-4-hydroxy-6-methyl-3-oxo-1-isobenzofuranyl)-8,8a,12,12a-tetrahydro-4,6-dihydroxy-	EtOH	282s(4.07),291(4.14), 442(4.04),457(4.01), 522(3.32),535(3.28)	39-0255-86C
$C_{30}H_{20}O_9$			
Benzo[e]naphtho[2',3':5,6]fluoreno-[1,9a-b]oxepin-5,10,15,19-tetrone, 5c,8,8a,16-tetrahydro-1,8,11,18-tetrahydroxy-13-methyl-	EtOH	252(4.56),291(4.38), 443(4.37),462s(4.36)	39-0255-86C
$C_{30}H_{21}NO$			
Pyrrolo[2,1-a]isoquinolin-1(10bH)-one, 2,3,10b-triphenyl-	MeCN	278(4.26),283(4.26), 416(3.89)	118-0899-86
$C_{30}H_{21}N_3$			
Benzenamine, 2-(5,12-dihydro-7-phenyl-indolo[3,2-a]carbazol-6-yl)-	EtOH	237(4.70),280(4.58), 336(4.11),353(4.16)	78-5019-86
2,3':2',3"-Ter-1H-indole, 2"-phenyl-	EtOH	303(4.16)	78-5019-86
$C_{30}H_{21}N_3O_2$			
Benzonitrile, 4-[[(1-benzoyl-3-oxo-3-phenyl-1-propenyl)phenylhydrazono)methyl]-, (E,?)-	MeOH	259(4.43),384(4.69)	44-1972-86
$C_{30}H_{21}N_5O_4S$			
2H-Naphtho[1,2-d]triazole-5-sulfonamide, 2-[2,3-dihydro-2-(4-methylphenyl)-1,3-dioxo-1H-benz[de]isoquinolin-6-yl]-N-methyl-	DMF	372(4.36)	2-0496-86
$C_{30}H_{22}$			
Dibenzo[a,g]cyclododocosene, 23,24,25,26-tetradehydro-, (E,E,E,E,E,E,Z)-	isooctane	278s(4.45),297(4.63), 355(5.05),420s(4.10), 445(4.19),464(4.18), 501s(3.91)	18-2209-86
Lepidopterene	C_6H_{12}	253(3.23),257(3.23), 264(3.28),272(3.30)	44-3693-86
$C_{30}H_{22}BO_4$			
Boron(1+), bis(1,3-diphenyl-1,3-propanedionato-O,O')-, (T-4)-, perchlorate	CH_2Cl_2	392(4.66)	97-0399-86
$C_{30}H_{22}BO_8$			
Boron(1+), bis(1,7-di-2-furanyl-1,6-heptadiene-3,5-dionato-O^3,O^5)-, (T-4)-, perchlorate	CH_2Cl_2	529(4.96)	97-0330-86
$C_{30}H_{22}Br_2N_2O_8$			
1,2,4,5-Benzenetetrol, 3,6-bis(5-bromo-1H-indol-5-yl)-, tetraacetate	EtOH	204(4.70),224(4.82), 290(4.33),300(4.34)	5-1765-86
$C_{30}H_{22}Cl_2O$			
Methanone, [4,5-bis(4-chlorophenyl)-2-phenyl-1-cyclopenten-1-yl]phenyl]-	C_6H_{12}	246(4.36),352(2.45)	138-1631-86

Compound	Solvent	λ_{max}(log ϵ)	Ref.
$C_{30}H_{22}N_2$			
4-Pyridinecarbonitrile, 1,4-dihydro-1,2,4,6-tetraphenyl-	DMSO	327(4.16)	44-2481-86
$C_{30}H_{22}N_2O_2$			
Acetic acid, cyano(4-cyano-3,4,5-triphenyl-2,5-cyclohexadien-1-ylidene)-, ethyl ester	$CHCl_3$	345(4.44)	83-0898-86
Methanone, [3-(4-methylphenyl)-1-phenyl-1H-pyrazole-4,5-diyl]bis[phenyl-	MeOH	253(4.72)	44-1972-86
Phenol, 2,2'-(5,6,11,12-tetrahydroindolo[3,2-b]carbazole-6,12-diyl)-	EtOH	210(4.76),228(4.84), 281(4.37),291(4.18)	2-1204-86
bis-	EtOH-$HClO_4$	215(4.62),275(4.32), 325(4.08)	2-1204-86
$C_{30}H_{22}N_2O_3$			
2,4-Cyclohexadien-1-one, 4-nitro-6-[[5,7,8,9-tetrahydro-5-(1-naphthalenyl)-10-phenanthridinyl]-methylene]-	EtOH	240(4.48),290(4.18), 310(4.08),335(4.06), 405(4.00),525(3.81)	135-1208-86
2,4-Cyclohexadien-1-one, 4-nitro-6-[(2,3,4,6-tetrahydro-6-phenyl-benzo[a]phenanthridinyl)methylene]-	EtOH	240(4.74+),288(4.56), 305(4.48),370(4.30), 412(4.48),510(3.65)	135-1208-86
Methanone, [3-(4-methoxyphenyl)-1-phenyl-1H-pyrazole-4,5-diyl]bis-[phenyl-	MeOH	256(4.55)	44-1972-86
$C_{30}H_{22}N_2O_6$			
Propanedioic acid, [4-cyano-4-(4-nitrophenyl)-3,5-diphenyl-2,5-cyclohexadien-1-ylidene]-, dimethyl ester	$CHCl_3$	310(4.30)	83-0898-86
$C_{30}H_{22}N_2S_2$			
Benzothiazole, 2,2'-(2,4-diphenyl-1,3-cyclobutanediyl)bis-	$CDCl_3$	230(4.52),236s(4.43), 258(4.30),264s(4.28), 269s(4.21),285(3.77), 295(3.57)	24-3109-86
$C_{30}H_{22}N_4O_2$			
2,4-Imidazolidinedione, 5,5'-[1,1'-biphenyl]-4,4'-diylbis[5-phenyl-	dioxan	265(4.67)	104-1117-86
1H-Indole, 3-[[3-(3-methyl-7-nitro-1H-indol-2-yl)phenyl]methyl]-7-nitro-2-phenyl-	EtOH	250(4.51),272(4.41), 281(4.42),300(4.26), 323(3.59),350(4.03), 378(4.21),400(4.12), 450(3.41)	64-0768-86B
1H-Indole, 3-[[4-(3-methyl-7-nitro-1H-indol-2-yl)phenyl]methyl]-7-nitro-2-phenyl-	EtOH	250(4.56),269(4.50), 284(4.52),300(4.40), 325(3.92),350(4.18), 373(4.30),400(4.15)	64-0768-86B
$C_{30}H_{22}N_4O_5$			
2,4-Imidazolidinedione, 5,5'-(oxy-di-4,1-phenylene)bis[5-phenyl-	dioxan	240(4.72)	104-1117-86
$C_{30}H_{22}N_4O_7$			
2-Butenedinitrile, 2-amino-3-[(2,3,5-tri-O-benzoyl-D-ribofuranosylidene)amino]-	MeOH	230(4.60),324(4.20)	88-0323-86
1H-Imidazole-4,5-dicarbonitrile,	MeOH	230(4.62),260(4.05)	88-0323-86

Compound	Solvent	λ_{max}(log ε)	Ref.
2-[1,2,4-tris(benzoyloxy)-3-hydroxybutyl]-, [1S-(1R*,2S*,3S*)]- (cont.)			88-0323-86
$C_{30}H_{22}O$			
9-Anthraceneethanol, α-9-anthracenyl-	C_6H_{12}	250(5.3),260(5.3), 335(3.7),351(4.0), 370(4.3),390(4.3)	44-2956-86
$C_{30}H_{22}O_2$			
2,4,6-Heptatrienal, 7,7'-(1,3-butadiyne-1,4-diyldi-2,1-phenylene)bis-, (all-E)-	THF	241(4.53),253s(4.38), 283(4.52),296s(4.55), 321(4.74),359(4.82), 383s(4.78)	18-2209-86
$C_{30}H_{22}O_2S$			
2-Propen-1-one, 1,1'-(thiodi-4,1-phenylene)bis[3-phenyl- (middle maxima are graphical)	93% H_2SO_4	345(4.02),385(4.41), 439(4.57),513(4.82)	104-0055-86
$C_{30}H_{22}O_2Te$			
2-Propen-1-one, 3,3'-tellurobis[1,3-diphenyl-, (Z,Z)-	CH_2Cl_2	452(4.36)	44-1692-86
$C_{30}H_{22}O_3$			
8H-Anthra[9,1-bc]benz[e]oxepin-8,14(12bH)-dione, 12-(2,3-dimethylphenyl)-1-methyl-	C_6H_{12}	230(4.50),248(4.39), 283s(4.10)	44-0921-86
2-Propen-1-one, 1,1'-(oxydi-4,1-phenylene)bis[3-phenyl- (middle maximum is graphical)	93% H_2SO_4	355(4.16),420(4.57), 465(4.90)	104-0055-86
$C_{30}H_{22}O_8$			
1,4,8,11-Pentacenetetrone, 6,13-diacetoxy-2,3,9,10-tetramethyl-	CH_2Cl_2	256(4.70),300(4.29), 321(4.30),335(4.52), 411(3.47),434(3.45), 463(3.41),493(3.43)	44-4160-86
$C_{30}H_{22}O_9$			
Methanone, [3,4-dihydro-5-hydroxy-2,8-bis(4-hydroxyphenyl)-2H-furo[2,3-h]-1-benzopyran-9-yl](2,4,6-trihydroxyphenyl)- (daphnodorin A)	EtOH	209(4.61),305(4.53)	94-0595-86
	dioxan	217(4.58),307(4.58)	94-0595-86
Rubellin A	EtOH	247(4.40),288(4.04), 417s(3.91),440(4.06), 466s(3.99)	39-0255-86C
$C_{30}H_{22}O_{10}$			
Rubellin B (7λ,6ε)	EtOH	252(4.25),288(3.94), 465s(3.93),483(3.99), 495(3.91),517(3.82), 530(?)	39-0255-86C
4H-1-Benzopyran-4-one, 8-[5-(3,4-dihydro-5,7-dihydroxy-4-oxo-2H-1-benzopyran-2-yl)-2-hydroxyphenyl]-2,3-dihydro-5,7-dihydroxy-2-(4-hydroxyphenyl)- (chamaejasmine)	MeOH	296(4.54)	94-3249-86
isochamaejasmine	MeOH	296(4.48)	94-3249-86
Methanone, [3,4-dihydro-3,5-dihydroxy-2,8-bis(4-hydroxyphenyl)-2H-furo-[2,3-h]-1-benzopyran-9-yl](2,4,6-trihydroxyphenyl)-, (daphnodorin B)	dioxan	215(4.57),301(4.57)	94-0595-86

Compound	Solvent	λmax(log ε)	Ref.
$C_{30}H_{23}NO$			
Pyrrolo[2,1-a]isoquinolin-1(5H)-one, 6,10b-dihydro-2,3,10b-triphenyl-	MeCN	235(4.20),274(4.09), 360(3.87)	118-0899-86
$C_{30}H_{23}NO_2$			
4H-1,3-Oxazin-4-one, 2-(diphenylmethyl)-2,3-dihydro-3-(4-methylphenyl)-6-phenyl-	EtOH	217(4.49),284(4.42)	104-0204-86
$C_{30}H_{23}NO_4$			
Propanedioic acid, (4-cyano-3,4,5-triphenyl-2,5-cyclohexadien-1-ylidene)-, dimethyl ester	$CHCl_3$	317(4.29)	83-0898-86
$C_{30}H_{23}NO_9$			
Cervinomycin A_2, O-methyl-	$CHCl_3$	248.0(4.63),317.3(4.44)	35-6088-86
$C_{30}H_{23}NS$			
Pyrrolo[2,1-a]isoquinoline-1(5H)-thione, 6,10b-dihydro-2,3,10b-triphenyl-	MeCN	255(4.17),435(4.20)	118-0899-86
$C_{30}H_{23}N_3O_3S$			
Benzenesulfonic acid, 2-[2-[4'-[2-(2-phenyl-2H-1,2,3-triazol-4-yl)ethenyl][1,1'-biphenyl]-4-yl]ethenyl]-, sodium salt	EtOH	349(4.85)	61-0439-86
$C_{30}H_{23}N_3O_8$			
6-Oxa-1,4-diazaspiro[4.4]nona-1,3-diene-2-carbonitrile, 8,9-bis(benzoyloxy)-7-[(benzoyloxy)methyl]-3-methoxy-, [5R-(5α,7β,8α,9α)]-	MeOH	229(4.61)	88-0323-86
$C_{30}H_{24}Cl_2N_6O_4$			
Isoxazolo[3,4-d]pyridazin-7(6H)-one, 3,3'-[2,4-bis(4-chlorophenyl)-1,3-cyclobutanediyl]bis[4,6-dimethyl-, (1α,2α,3β,4β)-	n.s.g.	236(3.98),306(3.98), 366(4.27)	142-3467-86
$C_{30}H_{24}N_2$			
2,4,10,12-Cyclotridecatetraen-6,8-diyn-1-one, 5,10-dimethyl-, (5,10-dimethyl-2,4,10,12-cyclotridecatetraen-6,8-diyn-1-ylidene)hydrazone, (E,E,E,E,Z,Z,Z,Z)-	THF	255(4.52),264s(4.51), 290(4.42),420(4.45)	18-2509-86
	CF_3COOH	271(0.78),289(0.79), 445(1.00),580s(0.19) (rel. absorbance)	18-2509-86
$C_{30}H_{24}N_2O_2$			
Benzaldehyde, 4-methyl-, (1-benzoyl-3-oxo-3-phenyl-1-propenyl)phenylhydrazone, (E,?)-	MeOH	254(4.40),378(4.61)	44-1972-86
$C_{30}H_{24}N_2O_3$			
Benzaldehyde, 4-methoxy-, (1-benzoyl-3-oxo-3-phenyl-1-propenyl)phenylhydrazone, (E,?)-	MeOH	258(4.38),385(4.62)	44-1972-86
1H-Benzimidazole-1-acetic acid, α-(2-phenoxy-2-phenylethenyl)-2-phenyl-, methyl ester, (Z)-	MeOH	248(4.71)	44-3420-86
$C_{30}H_{24}N_4$			
Benzonitrile, 4,4'-[1,2-bis(methyl-	EtOH	286(4.49),460(4.01)	22-0781-86

Compound	Solvent	λ_{max}(log ε)	Ref.
phenylamino)-1,2-ethenediyl]-bis-, (E)- (cont.)			22-0781-86
(Z)-	EtOH	236(4.27),274(4.01), 340(3.69),446(3.62)	22-0781-86
$C_{30}H_{24}N_6O_7S$			
1,2,4-Triazolo[3,4-f][1,2,4]triazine-8(5H)-thione, 6-amino-3-(2,3,5-tri-O-benzoyl-β-D-ribofuranosyl)-	pH 1	240(4.48),330s(3.90)	87-2231-86
	pH 7	232(4.58),334(3.90)	87-2231-86
	pH 11	232(3.59),334(3.93)	87-2231-86
$C_{30}H_{24}N_6O_8$			
1,2,4-Triazolo[3,4-f][1,2,4]triazin-8(5H)-one, 6-amino-3-(2,3,5-tri-O-benzoyl-β-D-ribofuranosyl)-	pH 1	226(4.59),270(4.05), 282(3.95)	87-2231-86
	pH 7	226(4.69),270(3.81)	87-2231-86
	pH 11	224(4.70),270(3.77)	87-2231-86
$C_{30}H_{24}O$			
Methanone, phenyl(2,4,5-triphenyl-2-cyclopenten-1-yl)-, (1α,4α,5β)-	C_6H_{12}	247(4.32),351(2.45)	138-1631-86
$C_{30}H_{24}O_2$			
9,10-Anthracenedione, 1,8-bis(2,3-dimethylphenyl)-	C_6H_{12}	258(4.71),332(3.62)	44-0921-86
2-Propen-1-one, 1-(3,4-dihydro-2,2,4-triphenyl-2H-1-benzopyran-3-yl)-, cis	EtOH	227(3.99),254(3.90)	4-1195-86
$C_{30}H_{24}O_4$			
2H-1-Benzopyran-2-one, 8-(diphenylmethyl)-4,7-dimethoxy-3-phenyl-	MeOH	209(4.7),313(3.9)	2-0623-86
4H-1-Benzopyran-4-one, 6-(diphenylmethyl)-7,8-dimethoxy-3-phenyl-	MeOH	263(4.57),296(4.15)	2-0649-86
$C_{30}H_{24}O_4S$			
1,4-Epoxynaphthalene-2-carboxylic acid, 1,2,3,4-tetrahydro-1,4-diphenyl-3-(phenylsulfinyl)-, methyl ester, (1α,2α,3α,4α)-	EtOH	258(4.54),263(4.53), 270s(4.01)	12-0575-86
(1α,2α,3β,4α)-	EtOH	211(4.46),253(3.67)	12-0575-86
(1α,2β,3α,4α)-, isomer C	EtOH	243(3.77)	12-0575-86
(1α,2β,3α,4α)-, isomer D	EtOH	213(4.49),246(4.01)	12-0575-86
(1α,2β,3β,4α)-	EtOH	211(4.58),260(3.52)	12-0575-86
$C_{30}H_{24}O_5$			
4H-1-Benzopyran-4-one, 5,7-dihydroxy-3-methoxy-2-phenyl-6,8-bis(phenylmethyl)-	MeOH	225(4.36),323(4.37), 361(3.97)	2-0166-86
$C_{30}H_{24}O_7$			
4H-1-Benzopyran-4-one, 2-[3,4-bis(phenylmethoxy)phenyl]-5,6-dihydroxy-7-methoxy-	EtOH	286(4.30),340(4.44)	94-0030-86
	EtOH-AlCl3	296(4.33),364(4.47)	94-0030-86
$C_{30}H_{25}NO$			
Pyridine, 1,2-dihydro-2-methoxy-1,2,4,6-tetraphenyl-	DMSO	354(4.28)	44-2481-86
$C_{30}H_{25}N_3O$			
Propanedinitrile, [7,8-dihydro-2-(2-methoxy-5-methylphenyl)-3,5-diphenyl-4(1H)-azocinylidene]-	MeCN	268(4.31),324(4.15), 480(4.31)	118-0906-86

Compound	Solvent	λ max(log ε)	Ref.
Propanedinitrile, [7,8-dihydro-2-(4-methoxy-3-methylphenyl)-3,5-diphenyl-4(1H)-azocinylidene]-	MeCN	235(4.37),292(4.08), 487(4.29)	118-0908-86
Propanedinitrile, [2-(4-ethoxyphenyl)-7,8-dihydro-3,5-diphenyl-4(1H)-azocinylidene]-	MeCN	235(4.38),290(4.09), 486(4.37)	118-0906-86
$C_{30}H_{25}N_3O_2$			
Propanedinitrile, [2-(2,4-dimethoxyphenyl)-7,8-dihydro-3,5-diphenyl-4(1H)-azocinylidene]-	MeCN	235(4.28),318(3.92), 483(4.27)	118-0908-86
Propanedinitrile, [2-(2,5-dimethoxyphenyl)-7,8-dihydro-3,5-diphenyl-4(1H)-azocinylidene]-	MeCN	240(4.33),333(3.76), 480(4.30)	118-0908-86
$C_{30}H_{25}N_3O_4$			
Cyclopenta[4,5]pyrimido[6,1-a]isoquinoline-3-carboxamide, 4,5,7,8-tetrahydro-10,11-dimethoxy-5-oxo-N,4-diphenyl-	CH_2Cl_2	245(4.45),292(4.27), 372(4.45),383(4.43)	24-2553-86
$C_{30}H_{25}N_3O_8$			
2(1H)-Pyrimidinone, 4-amino-1-(2,3,5-tri-O-benzoyl-β-D-xylofuranosyl)-	EtOH	232(3.95),272(3.61)	87-0203-86
$C_{30}H_{25}O_2$			
Pyrylium, 2-[6-(4-methoxyphenyl)-1,3,5-hexatrienyl]-4,6-diphenyl-, tetrafluoroborate	CH_2Cl_2	380(4.53),630(4.64)	104-1786-86
$C_{30}H_{26}$			
Anthracene, 1,8-bis(2,3-dimethylphenyl)-	C_6H_{12}	253s(4.89),260(5.06), 332(3.52),348(3.82), 366(4.01),386(3.95)	44-0921-86
$C_{30}H_{26}BrNO_2$			
1,4-Naphthalenedione, 2-bromo-3-[[3-(9,10-ethanoanthracen-9(10H)-yl)propyl]methylamino]-	CH_2Cl_2	240(4.15),248s(4.1), 273(4.17),284(4.20), 330s(3.6),504(3.57)	83-0791-86
$C_{30}H_{26}ClNO_2$			
1,4-Naphthalenedione, 2-chloro-3-[[3-(9,10-ethanoanthracen-9(10H)-yl)propyl]methylamino]-	CH_2Cl_2	243(4.10),249(4.11), 257s(4.1),265s(4.1), 273s(4.1),285(4.18), 330s(3.5),501s(3.59)	83-0791-86
$C_{30}H_{26}ClN_7O_2S$			
Acetamide, N-[2-[[4-chloro-5-[[1-(dicyanomethylene)-1,3-dihydro-3-oxo-2H-inden-2-ylidene]methyl]-2-thiazolyl]azo]-5-(dipropylamino)phenyl]-	CH_2Cl_2	716(4.87)	77-1639-86
	DMSO	740(--)	77-1639-86
$C_{30}H_{26}N_3O_5P$			
1H-Pyrrolo[3,4-c]pyridine-7-carboxylic acid, 2,4,5,6-tetrahydro-2,5-dimethyl-1,4,6-trioxo-3-[(triphenylphosphoranylidene)amino]-, ethyl ester	$CHCl_3$	363(4.24),463(4.44)	44-0149-86
$C_{30}H_{26}N_6O_4$			
Isoxazolo[3,4-d]pyridazin-7(6H)-one,	n.s.g.	242s(3.05),302(4.12),	142-3467-86

Compound	Solvent	λ_{max}(log ε)	Ref.
3,3'-(2,4-diphenyl-1,3-cyclobutane-diyl)bis[4,6-dimethyl- (cont.)		364(4.27)	142-3467-86
$C_{30}H_{26}N_8O_4$			
1H-Purine-2,6-dione, 8,8'-(9,10-phen-anthrenediyl)bis[3,7-dihydro-1,3,7-trimethyl-	CH_2Cl_2	251s(4.59),256(4.66), 266s(4.47),288s(4.29), 300s(4.14),330s(3.60), 351s(3.17)	24-1525-86
$C_{30}H_{26}O$			
9(10H)-Anthracenone, 1,8-bis(2,3-di-methylphenyl)-	C_6H_{12}	260(4.43),304s(4.02)	44-0921-86
$C_{30}H_{26}O_2$			
Benzophenone, 3-acetyl-2,5-dimethyl-4-(4-methylphenyl)-6-phenyl-	MeCN	240(4.46)	48-0359-86
Benzophenone, 3-acetyl-2,5-dimethyl-6-(4-methylphenyl)-4-phenyl-	MeCN	244(4.46)	48-0359-86
Benzophenone, 3-acetyl-5-ethyl-2-meth-yl-4,6-diphenyl-	MeCN	240(4.38)	48-0359-86
Benzophenone, 3-acetyl-2,4',5-trimeth-yl-4,6-diphenyl-	MeCN	257(4.35)	48-0359-86
1,1'-Biphenyl, 4,4'-bis[2-(2-methoxy-phenyl)ethenyl]-	EtOH	356(--)	61-0439-86
$C_{30}H_{26}O_3$			
Benzophenone, 3-acetyl-4'-methoxy-2,5-dimethyl-4,6-diphenyl-	MeCN	286(4.33)	48-0359-86
2-Propen-1-one, 1-[2-hydroxy-4-meth-oxy-3,5-bis(phenylmethyl)phenyl]-3-phenyl-	MeOH	210(4.7),262(3.9), 326(4.0)	2-0259-86
$C_{30}H_{26}O_4$			
Anthracene, 9,10-dimethoxy-1,4-bis(phenylmethoxy)-	$CHCl_3$	248(4.55),268(4.53), 364(3.66),380(3.85), 403(3.71),422(3.60)	5-0839-86
$C_{30}H_{26}O_5$			
9(10H)-Anthracenone, 10,10-bis[(4-hy-droxy-3-methoxyphenyl)methyl]-	EtOH	278(4.24)	104-0158-86
$C_{30}H_{26}O_{11}$			
7H-Furo[3,2-g][1]benzopyran-7-one, 9,9'-[oxybis(2,1-ethanediyloxy-2,1-ethanediyloxy]bis-	1% DMSO	302.5(4.30)	149-0391-86A
$C_{30}H_{26}O_{12}$			
Echinacin	MeOH	221s(3.50),270(3.42), 320(3.58)	25-0713-86
Echinaticin	MeOH	222s(3.50),269(3.42), 317(3.85)	25-0713-86
$C_{30}H_{27}BrN_4O_6$			
Benzoic acid, 4-methyl-, diester with ethyl N-[5-bromo-3,4-dicyano-1-(2-deoxy-β-D-erythro-pentofuranosyl)-1H-pyrrol-2-yl]methanimidate	EtOH	238(4.59),270(4.13)	78-5869-86
$C_{30}H_{27}NO_2$			
Formamide, N-[2-oxo-2-phenyl-1,1-bis(phenylmethyl)ethyl]-N-benzyl-	MeOH	223(4.05),245(3.97)	44-3374-86

Compound	Solvent	λ_{max}(log ε)	Ref.
1,4-Naphthalenedione, 2-[[3-(9,10-ethanoanthracen-9(10H)-yl)propyl]methylamino]-	CH_2Cl_2	238s(4.2),254s(4.2), 278(4.35),330s(3.4), 464(3.64)	83-0607-86
$C_{30}H_{27}NO_5$			
7H-Furo[3,2-g][1]benzopyran-7-one, 2-[4-[2-(dimethylamino)ethoxy]benzoyl]-3,5-dimethyl-6-phenyl-	$CHCl_3$	292(4.28),358(4.36)	18-1299-86
$C_{30}H_{27}NO_6S$			
Benzo[b]thiophen-3(2H)-one, 4-methoxy-2-(phenylmethylene)-7-(4,5,6,7-tetrahydro-10-methoxy-5-methyl-3H-furo-[4,3,2-fg][1,3]benzazocin-6-yl]-, 1,1-dioxide, [R-(Z)]-	EtOH	246(4.18),258(4.03), 288(3.81),310(3.79)	78-0591-86
$C_{30}H_{27}N_2O$			
3H-Indolizinium, 3-[[4-(dimethylamino)-phenyl]methylene]-2-hydroxy-1-methyl-5,7-diphenyl-, chloride	80% MeOH	500(4.73)	4-1315-86
	+ NaOH	495(3.88)	4-1315-86
$C_{30}H_{27}N_3O_9$			
Benzamide, N,N',N"-(2,6,10-trioxo-1,5,9-trioxacyclododecane-3,7,11-triyl)tris-, [3S-(3R*,7R*,11R*)]-	EtOH	226(4.54)	35-7609-86
	CH_2Cl_2	223(4.50)	35-7609-86
	MeCN	223(4.33)	35-7609-86
$C_{30}H_{27}O_3$			
5H-Benzo[6,7]cyclohepta[1,2-b]naphtho[2,1-e]pyrylium, 6-(3,4-dimethoxyphenyl)-6,8,9,10-tetrahydro-, perchlorate	MeCN	285s(4.30),360(4.04), 444(3.93)	22-0600-86
$C_{30}H_{27}O_4$			
Dibenzo[c,h]xanthylium, 7-(3,4-dimethoxyphenyl)-5,6,8,9-tetrahydro-3-methoxy-, perchlorate	MeCN	229(4.63),250(4.51), 306(4.29),475(4.45)	22-0600-86
Dibenzo[c,h]xanthylium, 5,6,8,9-tetrahydro-3,11-dimethoxy-7-(4-methoxyphenyl)-, perchlorate	MeCN	229(4.68),262(4.80), 314s(4.28),500(4.67)	22-0600-86
$C_{30}H_{28}$			
5,9,13,17,21-Hexacosapentaene-3,7,11-15,19,23-hexayne, 2,2,25,25-tetramethyl-	hexane	385(4.90),402s(4.82), 418s(4.70)	35-4685-86
$C_{30}H_{28}ClN_7O_4S_2$			
Acetamide, N-[2-[[4-chloro-5-[[3-(dicyanomethylene)benzo[b]thien-2(3H)-ylidene]methyl]-2-thiazolyl]azo]-4-methoxy-5-[(1-methylpentyl)amino]-phenyl]-, S,S-dioxide	CH_2Cl_2	778(4.92)	77-1639-86
	DMSO	806(--)	77-1639-86
$C_{30}H_{28}NO_2$			
Dibenzo[c,h]xanthylium, 7-[4-(dimethylamino)phenyl]-5,6,8,9-tetrahydro-3-methoxy-, perchlorate	MeCN	256(4.42),302(4.21), 412(4.19),475(3.72)	22-0600-86
$C_{30}H_{28}N_3O_6P$			
2-Butenedioic acid, 2-[1,2,3,4-tetrahydro-1,3-dimethyl-2,4-dioxo-6-[(triphenylphosphoranylidene)amino-5-	$CHCl_3$	342(3.99)	44-0149-86

Compound	Solvent	λ_{max}(log ε)	Ref.
pyrimidinyl]-, dimethyl ester (cont.)			44-0149-86
3,4-Pyridinedicarboxylic acid, 1,2,5,6-tetrahydro-1-methyl-5-[(methylamino)[(triphenylphosphoranylidene)amino]methylene]-2,6-dioxo-, dimethyl ester	$CHCl_3$	352(4.57)	44-0149-86
$C_{30}H_{28}O_3$			
2,4-Octadiene-1,7-dione, 4-ethyl-6-(1-hydroxyethylidene)-1,3,5-triphenyl-, (Z,Z,E)-	dioxan	260(4.36),285s(4.33), 346s(3.66)	48-0359-86
2,4-Octadiene-1,7-dione, 6-(1-hydroxyethylidene)-4-methyl-1-(4-methylphenyl)-3,5-diphenyl-, (Z,Z,E)-	dioxan	256s(4.29),286(4.37), 350s(3.60)	48-0359-86
2,4-Octadiene-1,7-dione, 6-(1-hydroxyethylidene)-4-methyl-5-(4-methylphenyl)-1,3-diphenyl-, (Z,Z,E)-	dioxan	258(4.38),283s(4.33), 344s(3.62)	48-0359-86
1-Propanone, 1-[2-hydroxy-4-methoxy-3,5-bis(phenylmethyl)phenyl]-3-phenyl-	MeOH	222(4.0),270(3.8)	2-0259-86
$C_{30}H_{28}O_4$			
Ethanone, 1-[5-(diphenylmethyl)-2,3,4-trimethoxyphenyl]-2-phenyl-	MeOH	258(4.42),281(4.28), 296(4.18)	2-0649-86
$C_{30}H_{28}O_6$			
1,4-Naphthalenedione, 2,2'-(3,7-dimethyl-6-octenylidene)bis[3-hydroxy-	EtOH	272(4.62),324(3.59), 421(3.40)	39-0659-86C
$C_{30}H_{28}O_9$			
[2,2'-Binaphthalene]-1,1',5,5'-tetrol, 4,4'-dimethoxy-7,7'-dimethyl-, 1,1',5-triacetate	EtOH	248(4.34),262s(4.16), 282s(3.70),308(3.70), 324s(3.67),337s(3.64)	39-0675-86C
$C_{30}H_{28}Si_2$			
Silane, bisbenzo[3,4]cyclobuta[1,2-a:1',2'-c]biphenylene-2,3-diylbis[trimethyl-	THF	247(3.60),253(3.80), 259(4.00),300(4.20), 316(4.13),364(3.20), 377(3.20),458(3.50), 488(3.64)	35-2481-86
$C_{30}H_{29}ClN_4O_2$			
Phenol, 4-[[7-chloro-3-(3,4-dihydro-3-methyl-1-isoquinolinyl)-4-quinolinyl]amino]-2-(4-morpholinylmethyl)-	EtOH	220(4.50),350(3.81)	2-1072-86
$C_{30}H_{29}NO_5S$			
[1,4]Oxazino[2,3,4-kl]phenothiazine-1,2-dicarboxylic acid, 4-(1,1-dimethylethyl)-6-phenyl-, diethyl ester	$CHCl_3$	276(4.31),304s(3.95), 404(3.53)	39-2233-86C
$C_{30}H_{29}NO_6S$			
4,11-Etheno-5,10b-(iminoethano)-4H-,10bH-benzo[4,5]cycloprop[1,7a]indeno[1,2-b]thiophen-1(2H)-one, 5,6,11,11a-tetrahydro-10-hydroxy-9,12-dimethoxy-16-methyl-2-(phenyl-	EtOH	215(4.29),237(3.82), 289(3.10),295(2.83), 340(2.23),350(2.15)	78-0591-86

Compound	Solvent	λ max (log ε)	Ref.
methylene)-, 3,3-dioxide, [4S-(2Z-3aR*,4α,4aR*,5β,10bβ,11α,11aα)]- (cont.)			78-0591-86
4a,8-Etheno-4,13-methano-4aH-benzofuro[3,2-e]thieno[3,2-h]isoquinolin-7(6H)-one, 1,2,3,4,4b,7a,8,8a-octahydro-8,10-dimethoxy-3-methyl-6-(phenylmethylene)-, 5,5-dioxide, [4R-(4α,4aα,6bβ,6Z,7aβ,8α,8aβ,13α,13bR*)]-	EtOH	210(4.32),292(3.89), 333(4.05)	78-0591-86
$C_{30}H_{29}N_7O_7$			
Adenosine-ellipticinium adduct	H_2O	272(4.56),315(4.13), 370(3.85)	87-1350-86
$C_{30}H_{29}N_7O_8$			
Guanosine-elliptinium adduct	H_2O	275(4.52),315(4.31), 372(3.97)	87-1350-86
$C_{30}H_{30}N_2O_2$			
1,2-Ethenediamine, 1,2-bis(4-methoxyphenyl)-N,N'-dimethyl-N,N'-diphenyl-, (E)-	EtOH	216(4.08),261(4.34), 287(4.34),381(4.06)	22-0781-86
(Z)-	EtOH	217(4.10),278(4.38), 376(4.11)	22-0781-86
$C_{30}H_{31}ClN_4O$			
Phenol, 4-[[7-chloro-3-(3,4-dihydro-3-methyl-1-isoquinolinyl)-4-quinolinyl]amino]-2-[(diethylamino)-methyl]-	EtOH	220(4.59),350(3.88)	2-1072-86
$C_{30}H_{31}N_3O_5S_2$			
Acridium, 10-amino-9-(N-methyl-4-methanesulfonamidophenyl)-, 2,4,6-trimethylbenzenesulfonate	MeOH	205(4.95),262(4.97), 362(4.29)	150-3425-86M
$C_{30}H_{31}N_5O_6$			
6H-Purin-6-one, 1,9-dihydro-9-[[2-hydroxy-1,1-bis(hydroxymethyl)ethoxy]-methyl]-2-[[(4-methoxyphenyl)diphenylmethyl]amino]-	MeOH	260(4.17),277(4.12)	87-1384-86
$C_{30}H_{31}N_7O_7S$			
Thioguanoside-elliptinium adduct	50% HOAc	323(4.63),386(3.90), 450(3.57)	87-1350-86
$C_{30}H_{32}Cl_6$			
Dispiro[cyclopropane-1,17'-tricyclo-[10.4.1.1^{4,9}]octadeca[2,4,6,8,12-14,16]heptaene-18',1"-cyclopropane], 2,2,2",2",5',13'-hexachloro-7',15'-bis(1,1-dimethylethyl)-	C_6H_{12}	218(4.39),272(4.31), 350s(3.88)	44-2214-86
$C_{30}H_{32}N_2P_2$			
1,5,3,7-Diazadiphosphocine, octahydro-1,5-bis(4-methylphenyl)-3,7-diphenyl-	MeCN	253(4.66),290(4.42)	65-0362-86
cobalt complex	MeCN	408(3.41),562(2.92)	65-0362-86
nickel complex	MeCN	426(3.08),498(3.41)	65-0362-86
1,5,3,7-Diazadiphosphocine, octahydro-3,7-diphenyl-1,5-bis(phenylmethyl)-	MeCN	247(4.04),276(3.86)	65-0362-86

Compound	Solvent	λ_{max}(log ε)	Ref.
cobalt complex	MeCN	403(3.31),566(2.58)	65-0362-86
nickel complex	MeCN	363(3.28),486(3.26)	65-0362-86
$C_{30}H_{32}N_2S_2$			
Butanethioamide, N,N-diethyl-3-methyl-2-(2,4,4-triphenyl-5(4H)-thiazolylidene)-, (E)-	CH_2Cl_2	253(4.41),292s(3.96), 332s(3.00),386s(2.30)	33-0174-86
$C_{30}H_{32}N_2S_3$			
2,5-Cyclopentadiene-1,2-dicarbonitrile, 4-[2-[5-[2,3-bis[(1,1-dimethylethyl)-thio]-2-cyclopropen-1-ylidene]-1,3-cyclopentadien-1-yl]-3-[(1,1-dimethylethyl)thio]-2-cyclopropen-1-ylidene]-	CH_2Cl_2	246(4.55),260s(4.49), 344(4.60),432(4.35), 448s(4.32)	35-7032-86
$C_{30}H_{32}N_4O_2S_2$			
Morpholine, 4-[4-(4-methylphenyl)-6-[4-(4-methylphenyl)-2-(4-morpholinyl)-6H-1,3-thiazin-6-ylidene]-6H-1,3-thiazin-2-yl]-	benzene	283(4.68),333s(4.45), 513(4.52)	118-0916-86
$C_{30}H_{32}N_4O_8$			
Saframycin Ad-1	MeOH	268(4.30)	158-1639-86
$C_{30}H_{32}N_4O_{10}S$			
1-Azabicyclo[3.2.0]hept-2-ene-2-carboxylic acid, 6-(1-hydroxyethyl)-3-[[4-[[[(4-nitrophenyl)methoxy]carbonyl]amino]cyclohexyl]thio]-7-oxo-, (4-nitrophenyl)methyl ester, [5R-[2(cis),5α,6α(R*)]]-	EtOH	267(4.37),323(4.18)	23-2184-86
$C_{30}H_{32}N_4S_2$			
6H-1,3-Thiazine, 4-(4-methylphenyl)-6-[4-(4-methylphenyl)-2-(1-pyrrolidinyl)-6H-1,3-thiazin-6-ylidene]-2-(1-pyrrolidinyl)-	$CHCl_3$	277(4.94),330s(3.98), 514(4.03)	118-0916-86
$C_{30}H_{32}O_7$			
Broussoflavonol D	EtOH	206(4.78),222s(4.58), 263(4.35),283s(4.15), 304s(3.90),350(3.90)	94-1987-86
	EtOH-$AlCl_3$	206(4.78),224s(4.54), 269(4.41),310s(3.76), 405(3.90)	94-1987-86
$C_{30}H_{32}O_{12}$			
2-Propenoic acid, 3-(4-methoxyphenyl)-, 2'-ester with 6-[2-(β-D-glucopyranosyloxy)-4-methoxy-6-methylphenyl]-4-methoxy-2H-pyran-2-one	MeOH	205(4.60),224s(4.43), 308(4.45)	100-0800-86
$C_{30}H_{32}O_{15}$			
Elloramycin B	MeOH	286(4.49),386(3.90), 406(3.94)	158-0856-86
	MeOH-NaOH	253(4.46),386(3.90), 440(3.94)	158-0856-86
$C_{30}H_{33}N_3O_{10}$			
Azinomycin A	MeOH	217(4.72),248s(--),	158-1527-86

Compound	Solvent	λ_{max}(log ε)	Ref.
Azinomycin A (cont.)		344(3.70)	158-1527-86
$C_{30}H_{33}N_5O_{12}$			
Acetic acid, 1-(4-nitrophenyl)hydrazide, hydrazone with D-lyxo-hexos-2-ulose 1-(acetylphenylhydrazone), 3,4,5,6-tetraacetate	MeOH	288(4.35),349(3.86)	136-0316-86G
$C_{30}H_{34}$			
Benzene, [[2,2,3,3-tetramethyl-1-[(2,2,3,3-tetramethyl-1-phenylcyclopropyl]rthynyl]cyclopropyl]ethynyl]-	MeOH	242(4.20),253(4.21)	138-1075-86
$C_{30}H_{34}Cl_6$			
Dispiro[cyclopropane-1,17'-tricyclo[10.4.1.1^{4,9}]octadeca[4,6,8,12,14-16]hexaene-18',1"-cyclopropyl], 2,2,2",2",5',13'-hexachloro-7',15'-bis(1,1-dimethylethyl)-	C_6H_{12}	213(4.32),237(4.33), 283(3.70)	44-2214-86
$C_{30}H_{34}N_2O_{11}$			
Daunorubicin, N-(2-amino-2-methyl-1-oxoethyl)-	pH 7	480(3.96)	161-0881-86
Daunorubicin, N-(3-amino-1-oxopropyl)-	pH 7.4	480(3.86)	161-0881-86
$C_{30}H_{34}O_7$			
4H-1-Benzopyran-4-one, 2-[6-(1,1-dimethyl-2-propenyl)-3,4-dihydroxy-2-(3-methyl-2-butenyl)phenyl]-3,5,7-trihydroxy-8-(3-methyl-2-butenyl)- (broussoflavonol C)	EtOH	212(4.81),261(4.44), 294s(3.98),306(4.01), 352(4.03)	94-1987-86
	EtOH-$AlCl_3$	212(4.82),269(4.56), 310(3.96),333s(3.81), 405(4.08)	94-1987-86
	EtOH-$AlCl_3$-HCl	269(4.43),314(4.08), 354s(3.94),394s(3.83)	94-1987-86
$C_{30}H_{34}O_8$			
24-Norchola-1,5,20,22-tetraene-3,7-dione, 6,11-diacetoxy-14,15:21,23-diepoxy-4,4,8-trimethyl-, (11α,13α-14β,15β,17α)-	n.s.g.	217(4.00),239(3.90)	100-0056-86
$C_{30}H_{35}N_3O_7$			
Capparisine diacetate	MeOH	204(3.12),277(3.81), 310s(--)	64-1033-86B
$C_{30}H_{35}N_5O_7$			
Saframycin Yd-1	MeOH	269(4.26)	158-1639-86
$C_{30}H_{36}N_2O_{18}$			
Ethanedioic acid, bis[2-[[2-[2-(2-methoxyethyl)ethoxy]ethoxy]carbonyl]-4-nitrophenyl] ester	MeCN	220(4.59)	96-0209-86
$C_{30}H_{36}N_5O_6PS$			
9H-Purin-6-amine, 9-[6-deoxy-5-S-(1,1-dimethylethyl)-6-(diphenoxyphosphinyl)-2,3-O-(1-methylethylidene)-5-thio-β-D-allofuranosyl]-	MeOH	261(4.17)	87-1030-86

Compound	Solvent	λ_{max}(log ϵ)	Ref.
$C_{30}H_{36}O_4$			
4H-1-Benzopyran-4-one, 2,3-dihydro-7-hydroxy-2-[4-hydroxy-3,5-bis(3-methyl-2-butenyl)phenyl]-8-(3-methyl-2-butenyl)-, (S)- (kazinol P)	EtOH	211(4.61),228s(4.29), 240s(4.17),282(3.93)	142-2141-86
Kazinol H	EtOH	229(4.60),265s(3.87), 277(4.05),286(4.05), 320(3.28)	94-1968-86
$C_{30}H_{36}O_8$			
24-Norchola-1,5,20,22-tetraen-7-one, 6,11-diacetoxy-14,15:21,23-diepoxy-3-hydroxy-4,4,8-trimethyl-, (3β,11α,13α,14β,15β,17α)-	n.s.g.	218(4.00),240(3.90)	100-0056-86
$C_{30}H_{36}O_{10}$			
Orbicuside A	MeOH	298(3.39)	39-1633-86C
$C_{30}H_{36}O_{11}$			
Affinoside I	MeOH	260(4.04)	94-2774-86
Orbicuside C	MeOH	298(3.39)	39-1633-86C
$C_{30}H_{36}O_{12}$			
Ixerin U	MeOH	244(4.07),299s(4.15), 330(4.27)	94-4170-86
$C_{30}H_{38}BrClO_6$			
Prosta-5,7,14-trien-1-oic acid, 12-acetoxy-9-[(4-bromobenzoyl)oxy]-10-chloro-, methyl ester, (5E,7Z,9α,10β,14Z)-	EtOH	248(4.32)	88-0223-86
(5E,7Z,9β,10β,14Z)-	EtOH	247(4.32)	88-0223-86
$C_{30}H_{38}O_3$			
Phenol, 2-[1-[3,5-bis(1,1-dimethylethyl)-2-hydroxyphenyl]ethyl]-6-[1-(2-hydroxyphenyl)ethyl]-, (R*,S*)-(-)-	EtOH	279(3.82)	116-0509-86
$C_{30}H_{38}O_4$			
1,2-Benzenediol, 5-[6-(1,1-dimethyl-2-propenyl)-3,4-dihydro-7-hydroxy-2H-1-benzopyran-2-yl]-3,4-bis(3-methyl-2-butenyl)-, (S)- (kazinol E)	EtOH	224(4.92),232s(4.36), 288(3.91)	94-1968-86
1,3-Benzenediol, 4-(1,1-dimethyl-2-propenyl)-6-[3-[8-hydroxy-2,2-dimethyl-5-(3-methyl-2-butenyl)-2H-1-benzopyran-6-yl]propyl]-(kazinol K)	EtOH	226(4.45),266s(4.00), 277(4.18),285(4.16), 324(3.34)	94-1968-86
$C_{30}H_{38}O_4S_2$			
Acetic acid, [[10-(2-butoxyphenyl)-2-hydroxy-3-[(phenylmethyl)thio]-4,6,8-decatrienyl]thio]-, methyl ester	n.s.g.	272(4.59),284(4.65), 295(4.56)	104-1598-86
	EtOH	272(4.56),284(4.65), 295(4.58)	104-1875-86
$C_{30}H_{38}O_{10}$			
Orbicuside B	MeOH	298(3.69)	39-1633-86C
$C_{30}H_{39}Br_2NO_{10}$			
Geldanamycin, 9,19-dibromo-7,8-[carbo-	MeOH	268(4.26)	158-0415-86

Compound	Solvent	$\lambda_{max}(\log \epsilon)$	Ref.
nylbis(oxy)]-7-de[(aminocarbonyl-oxy]-17-demethoxy-8,9-dihydro-15-methoxy-11-O-methyl- (cont.)			158-0415-86
$C_{30}H_{40}$			
Dicyclopenta[a,e]pentalene, 1,3,5,7-tetrakis(1,1-dimethylethyl)-	hexane	226(4.10),249s(3.92), 349(4.58),399s(3.68), 518(3.25),808s(1.57)	89-0466-86
$C_{30}H_{40}Fe$			
Ferrocene, 1,1':2,2':3,3':4,4':5,5'-penta-1,4-butanediyl-	THF	403(1.95)	35-1333-86
$C_{30}H_{40}N_2O_4$			
Lophocine, (±)-	MeOH	230(4.55),275(3.83), 287(3.88),338(3.88)	100-0745-86
$C_{30}H_{40}O_4$			
1,2-Benzenediol, 5-[3-[5-(1,1-dimethyl-2-propenyl)-2,4-dihydroxyphenyl]-propyl]-3,4-bis(3-methyl-2-butenyl)-(kazinol C)	EtOH	218(4.59),286(3.97)	94-1968-86
2H-1-Benzopyran-8-ol, 6-[3-(2,3-dihydro-6-hydroxy-2,3,3-trimethyl-5-benzofuranyl)propyl]-3,4-dihydro-2,2-dimethyl-5-(3-methyl-2-butenyl)-(kazinol D)	EtOH	221(4.67),230s(4.07), 283(3.91)	94-1968-86
Hexadecanoic acid, (2-oxo-2H-naphtho-[1,2-b]pyran-4-yl)methyl ester	EtOH	265(4.41),275(4.49), 292(3.78),304(3.87), 317(3.86),354(3.70)	94-0390-86
$C_{30}H_{40}O_7$			
Pulcherralpin	EtOH	209(4.14),220(4.27), 226(4.23),280(4.50)	100-0561-86
$C_{30}H_{40}O_8$			
Methyl lucidenate D2	EtOH	252.5(3.96)	94-4018-86
$C_{30}H_{40}O_{10}$			
Affinoside O	MeOH	219(4.27),262(3.88)	94-2774-86
1,4,7,10,13-Benzopentaoxacyclopentadecin, 15,15'-(1,2-ethenediyl)bis-[2,3,5,6,8,9,11,12-octahydro-, (E)-	MeCN	333(4.55)	1-0545-86
(Z)-	MeCN	298(4.10)	1-0545-86
$C_{30}H_{41}$			
Dicyclopenta[a,e]pentalen-1-ylium, 1,3,5,7-tetrakis(1,1-dimethylethyl)-1,5-dihydro-, (deloc-1,2,3,3a,3b,4-4a,7a,7b,8,8a)-	CH_2Cl_2	265s(3.83),308(4.06), 369(4.26),435s(3.84), 470s(3.50),547s(2.97), 572s(3.06),622s(3.33), 663(3.43),722s(3.24), 862(2.54),974(2.23)	89-0466-86
$C_{30}H_{41}BrN_2O_9$			
Geldanamycin, 19-bromo-17-demethoxy-15-methoxy-11-O-methyl- (19-bromo-herbimycin A)	MeOH	258(4.27)	158-0415-86
$C_{30}H_{41}BrN_2O_{10}$			
Geldanamycin, 19-bromo-17-demethoxy-8,9-epoxy-8,9-dihydro-15-methoxy-	MeOH	262(4.30)	158-0415-86

Compound	Solvent	$\lambda_{max}(\log \epsilon)$	Ref.
11-O-methyl-, (15R)- (cont.)	MeOH	262(4.30)	158-0415-86
$C_{30}H_{41}NO_{10}$			
Geldanamycin, 7,8-[carbonylbis(oxy)]-7-de(aminocarbonyl)oxy]-17-demethoxy-8,9-dihydro-15-methoxy-11-O-methyl-	MeOH	265(4.24)	158-0415-86
$C_{30}H_{41}N_3O_5$			
Androst-5-ene-17-carboxamide, 3-acetoxy-N-[(2,3-dihydro-6-methyl-7-oxo-7H-oxazolo[3,2-a]pyrimidin-2-yl)-methyl]-, [3β,17β(R)]-	n.s.g.	228(3.98),262(4.00)	128-0297-86
(S)-	n.s.g.	229(3.84),260(3.92)	128-0297-86
2,4(1H,3H)-Pyrimidinedione, 1-[[2-(3β,17β)-3-acetoxyandrost-5-en-17-yl]-4,5-dihydro-5-oxazolyl]methyl-5-methyl-	n.s.g.	269(4.02)	128-0297-86
(R)-	n.s.g.	269(3.98)	128-0297-86
(S)-	n.s.g.	269(4.00)	128-0297-86
$C_{30}H_{41}N_5O_8S_2Si$			
2',3'-Secoadenosine, 5'-O-(tert-butyldimethylsilyl)-2',3'-di-O-p-toluenesulfonyl-	EtOH	215(4.40),226(4.39), 262(4.15)	87-2445-86
$C_{30}H_{42}$			
Dicyclopenta[a,e]pentalene, 1,3,5,7-tetrakis(1,1-dimethylethyl)-2,3-dihydro-	hexane	275s(3.97),327(4.68), 337s(4.63),435(3.33), 497s(3.04)	89-0466-86
Dicyclopenta[a,e]pentalene, 1,3,5,7-tetrakis(1,1-dimethylethyl)-4,8-dihydro-	hexane	255(3.36),264(3.37), 305s(3.79),320s(4.14), 333(4.46),350(4.68), 368(4.64),474(3.24)	89-0466-86
$C_{30}H_{42}N_2O_3S$			
Neothiobinupharidine, 11-oxide, (7S,11R,13S)-	EtOH	end absorption	102-2227-86
$C_{30}H_{42}N_2O_4S$			
Neothiobinupharidine, 6-hydroxy-, 11-oxide, (6 ,7S,11S,13S)-	EtOH	end absorption	102-2227-86
$C_{30}H_{42}N_8O_{14}S_2$			
Glycine, 2,2'-[(2,6-dimethyl-1,7-dioxo-1H,7H-pyrazolo[1,2-a]pyrazole-3,5-diyl)bis(methylene)]bis[L-γ-glutamyl-L-cysteinyl-	3% MeCN-pH 7.3	234(--),393(3.65)	35-4527-86
$C_{30}H_{42}O_2Si_4$			
Benzo[a]benzo[3,4]cyclobuta[1,2-e]cyclooctene-6,11-dione, 2,3,8,9-tetrakis(trimethylsilyl)-	isooctane	274(4.84),356(4.22), 374(4.24)	77-0281-86
Bicyclo[4.2.0]octa-1,3,5-trien-7-one, 8-[2-[2-ethynyl-4,5-bis(trimethylsilyl)phenyl]-2-oxoethylidene]-3,4-bis(trimethylsilyl)-, (E)-	isooctane	221(4.67),248(4.69), 286(4.34),299(4.36), 338(4.32),348(4.30)	77-0281-86
(Z)-	isooctane	219(4.64),248(4.64), 286(4.30),301(4.33), 339(4.34),352(4.33)	77-0281-86

Compound	Solvent	$\lambda_{max}(\log \epsilon)$	Ref.
$C_{30}H_{42}O_8$			
Methyl lucidenate E2	EtOH	256(3.75)	94-4018-86
$C_{30}H_{44}$			
Dicyclopenta[a,e]pentalene, 1,3,5,7-tetrakis(1,1-dimethylethyl)-2,3,4,8-tetrahydro-	hexane	244s(3.63),252(3.70), 260(3.69),288s(3.62), 302s(3.71),358s(4.46), 369(4.53),382s(4.46), 425s(3.28)	89-0466-86
Dicyclopenta[a,e]pentalene, 1,3,5,7-tetrakis(1,1-dimethylethyl)-4,4a,8,8a-tetrahydro-	hexane	227(4.09),276(3.44), 392(4.19)	89-0466-86
$C_{30}H_{44}O_2$			
5α-Lanosta-7,9(11),24-trien-26-al, 3-oxo-, (24E)- (ganoderal A)	EtOH	235(4.17)	94-3025-86
$C_{30}H_{44}O_3$			
5α-Lanosta-7,9(11),24-trien-26-oic acid, 3-oxo-, (24E)- (ganoderic acid S)	EtOH	226(4.14),235(4.19), 243(4.15),252(3.98)	94-3025-86
$C_{30}H_{44}O_4$			
Lanosta-7,24-dien-26-oic acid, 3,23-dioxo-	EtOH	239(4.24)	105-0613-86
$C_{30}H_{44}O_6$			
Lanosta-8,24-dien-26-oic acid, 15,23-dihydroxy-, (15α,24E)- (ganolucidic acid D)	EtOH	257(3.89)	163-0809-86
$C_{30}H_{44}O_9$			
Card-20(22)-enolide, 3-[(2,6-dideoxy-β-D-ribo-hexopyranosyl)oxy]-16-(formyloxy)-14-hydroxy-, (3β,5β,16β)-	MeOH	218(4.22)	87-0997-86
$C_{30}H_{45}NO_4S$			
1,3-Cyclopentadiene-1-carboxylic acid, 2,3,4-tris(1,1-dimethylethyl)-5-[[[(4-methylphenyl)sulfonyl]methyl]imino]-, 1,1-dimethylethyl ester	hexane	221(4.34),243s(4.15), 344(2.34),418(2.38)	24-2159-86
$C_{30}H_{46}N_2O_2$			
Cyanamide, [(3β)-3-acetoxycholest-5-en-7-ylidene]-	EtOH	275(4.28)	118-1055-86
$C_{30}H_{46}N_2O_9$			
Geldanamycin, 17-demethoxy-2,3,4,5-tetrahydro-15-methoxy-11-O-methyl-, (15R)- (2,3,4,5-tetrahydroherbimycin A)	MeOH	275(4.16)	158-0415-86
$C_{30}H_{46}O_2$			
5α-Lanosta-7,9(11),24-trien-3-one, 26-hydroxy-, (24E)- (ganoderol A)	EtOH	236(4.13),243(4.18), 252(4.00)	94-3025-86
	MeOH	236(4.17),243(4.23), 251(4.06)	100-0621-86
$C_{30}H_{46}O_3$			
Bauer-7-en-28-oic acid, 3-oxo-	EtOH	209(3.7)	102-1685-86

Compound	Solvent	λ_{max}(log ε)	Ref.
Friedel-1(10)-en-4β-ol-2,3-dione	n.s.g.	278(3.93)	2-0721-86
4-Norfriedel-1(10)-en-3β-ol-2-one, 3α-acetyl-	n.s.g.	237(4.08)	2-0721-86
$C_{30}H_{46}O_4$			
5α-Lanosta-7,9(11),24-trien-3-one, 15α,26,27-trihydroxy- (ganoderiol B)	EtOH	237(3.68),245(3.73), 253(3.45)	163-2887-86
$C_{30}H_{47}N_3O_4$			
Androst-4-en-3-one, 17-acetoxy-2-(4-methyl-1-piperazinyl)-6-(4-morpholinyl)-, (2α,6β,17β)-	EtOH	236.5(4.13)	118-0461-86
Androst-4-en-3-one, 17-acetoxy-6-(4-methyl-1-piperazinyl)-2-(4-morpholinyl)-, (2α)-	EtOH	237.5(4.13)	118-0461-86
(2β)-	EtOH	239.5(4.03)	118-0461-86
$C_{30}H_{48}O$			
Cyclohomonervilasterol	EtOH	211(3.75)	94-3244-86
5,9,13,17-Nonadecatetraenal, 6,10,14-18-tetramethyl-2-(4-methyl-4-hexenylidene)-, (E,E,E,E,Z)-	EtOH	232(4.08)	70-1851-86
$C_{30}H_{48}O_2$			
5α-Lanosta-7,9(11),24-triene-3,26-diol, (24E)- (ganoderol B or ganodermadiol)	EtOH	237(4.10),244(4.17), 252(3.98)	94-3025-86
	MeOH	236(4.08),243(4.12), 251(3.95)	100-0621-86
$C_{30}H_{48}O_3$			
Bauer-7-en-28-oic acid, 3β-hydroxy-	MeOH	208(3.84),240(2.78)	102-1685-86
Ganodermanondiol	MeOH	235(4.00),243(4.06), 251(3.89)	100-1122-86
5α-Lanosta-7,9(11),24-triene-3β,26,27-triol	EtOH	237(3.68),245(3.73), 253(3.56)	163-2887-86
$C_{30}H_{48}O_4$			
Ganodermanontriol	MeOH	236(4.06),243(4.11), 251(3.95)	100-1122-86
$C_{30}H_{48}O_5$			
5α-Cholestane-3,6-dione, 9α-hydroxy-24ξ-methyl-25-acetoxy-	n.s.g.	203.3(4.0)	13-0233-86B
$C_{30}H_{50}N_2O_2$			
Cyanamide, [(3β,5α)-3-(methoxymethoxy)cholestan-6-ylidene]-	EtOH	222s(--)	118-1055-86
$C_{30}H_{50}O_2$			
19-Norlanosta-1(10),5-diene-3,16-diol, 9-methyl-, (3β,9β,16β)-	MeOH	235(4.10),241.5(4.15), 251(3.94)	94-0088-86
$C_{30}H_{50}O_3$			
Ganodermanondiol, dihydro-	MeOH	236(3.97),243(4.03), 251(3.86)	100-1122-86
$C_{30}H_{50}O_4$			
5α-Lanosta-7,9(11)-diene-3β,24,25,26-tetrol (ganoderiol A)	EtOH	237(3.91),244(3.97), 253(3.82)	163-2887-86

Compound	Solvent	λ_{max}(log ε)	Ref.
$C_{30}H_{52}O_5$			
Lanost-9(11)-ene-3,16,24,25,28-pentol, (3β,4β,16β,24S)-	EtOH	204(3.69)	105-0288-86
$C_{30}H_{54}O_8$			
Anhydroarjunolic acid, methyl ester	EtOH	242(4.30),251(4.40), 262(4.20)	2-0113-86
$C_{30}H_{56}N_2O_4Si_2$			
1-Pyrrolidinecarboxylic acid, 3-[2-[[(1,1-dimethylethyl)dimethylsilyl]-oxy]ethyl]-2-[[[(1,1-dimethylethyl)-dimethylsilyl]oxy]methyl]-4-(6-methyl-3-pyridinyl)-, 1,1-dimethylethyl ester, [2S-(2α,3β,4β)]-	EtOH	268(3.66)	88-0607-86
$C_{30}H_{58}N_2O_2Si_6$			
Cyclodisiloxane-2,4-diamine, 2,4-bis(2,4,6-trimethylphenyl)-N,N,N',N'-tetrakis(trimethylsilyl)-, cis	C_6H_{12}	202(5.03),229(4.30)	157-0531-86
$C_{30}H_{66}Si_3$			
Cyclotrisilane, hexakis(2,2-dimethyl-propyl)-	n.s.g.	310(2.59)	77-1768-86

Compound	Solvent	$\lambda_{max}(\log \epsilon)$	Ref.
$C_{31}H_{14}O$			
15H-Dibenzo[fg,ij]indeno[2,1,7,6,5-pqrst]pentaphen-15-one	dioxan	245(3.80),253(3.74), 291(3.76),303(3.74), 346(3.46),371(3.12), 400(2.33),420(2.09), 458(2.10)	24-3521-86
$C_{31}H_{16}$			
15H-Dibenzo[fg,ij]indeno[2,1,7,6,5-pqrst]pentaphene	dioxan	248(3.61),292(3.90), 304(2.98),322(3.43), 336(3.39),350(3.51), 360(3.31),366(3.31), 372(3.36),382(2.99)	24-3521-86
$C_{31}H_{19}N_3$			
3,5-Pyridinedicarbonitrile, 2-[1,1'-biphenyl]-4-yl-4,6-diphenyl-	MeCN	285(4.51)	73-1061-86
$C_{31}H_{19}N_3O_2$			
7H-2,2a,8-Triazabenzo[cd]cyclopent[g]-azulene-7,9(8H)-dione, 1,8,10-tri-phenyl-	$CHCl_3$	320(3.88),370s(3.88), 390(3.95),410(3.91), 507s(2.44),547(2.51), 650(2.06)	94-3588-86
$C_{31}H_{20}N_2O_2$			
2-Butene-1,4-dione, 2-(1H-phenanthro-[9,10-d]imidazol-1-yl)-1,4-diphenyl-	MeOH	226(4.51),270(4.62)	44-3420-86
$C_{31}H_{20}N_4O_2S_2$			
4,10-Epithio-4H,6H,10H-isothiazolo-[3,4-d][1,2,4]triazolo[1,2-a]pyri-dazine-6,8(7H)-dione, 3,4,7,10-tetraphenyl-	C_6H_{12}	222(4.47),240(3.95), 245(3.70),253(3.65), 258(3.85),265(3.99), 268s(3.98)	5-1796-86
$C_{31}H_{21}Cl_2NOS$			
Methanone, [1,4-bis(4-chlorophenyl)-1,2-dihydro-5-methyl-6-phenyl-2-thioxo-3-pyridinyl]phenyl-	EtOH	250(4.40),294(4.14), 404(3.85)	118-0952-86
$C_{31}H_{22}Br_2O_2$			
2-Propen-1-one, 1,1'-(methylenedi-4,1-phenylene)bis[3-(4-bromophenyl)-	93% H_2SO_4	340(4.32),467(4.99)	104-0055-86
$C_{31}H_{22}ClNOS$			
Methanone, [1-(4-chlorophenyl)-1,2-di-hydro-5-methyl-4,6-diphenyl-2-thi-oxo-3-pyridinyl]phenyl-	EtOH	248(4.38),294(4.10), 402(3.84)	118-0952-86
Methanone, [4-(4-chlorophenyl)-1,2-di-hydro-5-methyl-1,6-diphenyl-2-thi-oxo-3-pyridinyl]phenyl-	EtOH	249(4.37),294(4.12), 402(3.84)	118-0952-86
$C_{31}H_{22}ClNO_2S$			
Methanone, (4-chlorophenyl)[1,2-dihy-dro-1-(4-methoxyphenyl)-4,6-diphen-yl-2-thioxo-3-pyridinyl]-	EtOH	257(4.56),285(4.39), 407(3.89)	118-0952-86
Methanone, [6-(4-chlorophenyl)-1,2-di-hydro-1-(4-methoxyphenyl)-4-phenyl-2-thioxo-3-pyridinyl]phenyl-	EtOH	253(4.58),280(4.45), 407(3.95)	118-0952-86
$C_{31}H_{22}N_2O_2$			
2-Butene-1,4-dione, 2-(7a,7b-dihydro-	MeOH	239(4.57),268(4.60),	44-3420-86

Compound	Solvent	λ_{max}(log ε)	Ref.
1H-phenanthro[9,10-d]imidazol-1-yl)-1,4-diphenyl- (cont.)		312(4.24)	44-3420-86
2-Butene-1,4-dione, 2-(4,5-diphenyl-1H-imidazol-1-yl)-1,4-diphenyl-, (E)-	MeOH	225(4.56),261(4.58)	44-3420-86
(Z)-	MeOH	228(4.58),270(4.60)	44-3420-86
$C_{31}H_{22}N_2O_4$			
Benzoic acid, 4-(4,5-dibenzoyl-1-phenyl-1H-pyrazol-3-yl)-, methyl ester	MeOH	254(4.49)	44-1972-86
Methanone, [3-(4-acetoxyphenyl)-1-phenyl-1H-pyrazole-4,5-diyl]bis[phenyl-	MeOH	252(4.60)	44-1972-86
$C_{31}H_{22}N_2O_6$			
2-Propen-1-one, 1,1'-(methylenedi-4,1-phenylene)bis[3-(4-nitrophenyl)-, (E,E)-	93% H_2SO_4	330(4.43),420(4.81)	104-0055-86
$C_{31}H_{22}O_8$			
4H-1-Benzopyran-4-one, 6-acetoxy-2-(2,2-diphenyl-1,3-benzodioxol-5-yl)-5-hydroxy-7-methoxy-	EtOH	276(4.15),285s(4.13), 342(4.43)	94-0030-86
	EtOH-$AlCl_3$	255(4.26),292(4.21), 361(4.42)	94-0030-86
$C_{31}H_{23}NOS$			
Methanone, (1,2-dihydro-5-methyl-1,4,6-triphenyl-2-thioxo-3-pyridinyl)phenyl-	EtOH	248(4.44),295(4.17), 399(3.93)	118-0952-86
$C_{31}H_{23}NO_{10}$			
Cervinomycin A_2, monoacetyl-	$CHCl_3$	244(4.49),274(4.33), 307(4.41),374(3.96)	35-6088-86
$C_{31}H_{23}N_3O_4$			
Acetic acid, cyano[4-cyano-3,5-bis(4-methylphenyl)-4-(4-nitrophenyl)-2,5-cyclohexadien-1-ylidene]-, methyl ester	$CHCl_3$	345(4.40)	83-0898-86
$C_{31}H_{24}$			
Spiro[cyclopropane-1,9'-[9H]fluorene], 2,3-bis(2-phenylethenyl)-, (+)-	n.s.g.	213(4.86),271(4.78)	88-4493-86
$C_{31}H_{24}ClNO_2S$			
2H,5H-[1]Benzothiopyrano[3,4-e]-1,3-oxazin-5-one, 9-chloro-3,4-dihydro-2,4-bis(4-methylphenyl)-3-phenyl-	MeOH	231(4.6),313(4.0)	42-0323-86
$C_{31}H_{24}ClN_5O_7$			
7H-Purin-2-amine, 6-chloro-7-(2,3,5-tri-O-benzoyl-β-D-xylofuranosyl)-	EtOH	228(--),273s(--), 323(--)	87-0203-86
9H-Purin-2-amine, 6-chloro-9-(2,3,5-tri-O-benzoyl-β-D-xylofuranosyl)-	EtOH	224(--),298s(--), 309(--)	87-0203-86
$C_{31}H_{24}ClN_5O_7S$			
D-Ribitol, 1,4-anhydro-1-C-[6-chloro-8-(methylthio)-1,2,4-triazolo[3,4-f][1,2,4]triazin-3-yl]-, 2,3,5-tribenzoate, (S)-	pH 1	240(4.09),312(3.76)	87-2231-86
	pH 7	242(4.11),315(3.76)	87-2231-86
	pH 11	245(3.89),325(3.59)	87-2231-86

Compound	Solvent	$\lambda_{max}(\log \epsilon)$	Ref.
$C_{31}H_{24}N$			
Methylium, [4-(diphenylamino)phenyl]-diphenyl-	HOAc-CF_3COOH	524(4.23)	64-1045-86B
$C_{31}H_{24}N_2O_3$			
2,4-Cyclohexadien-1-one, 4-nitro-6-[[2,3,4,6-tetrahydro-6-(phenylmethyl)benzo[a]phenanthridinyl]-methylene]-	EtOH	240(4.74+),300(4.20), 335(4.26),370(4.34), 410(4.42),500(4.35)	135-1208-86
2,4-Cyclohexadien-1-one, 4-nitro-6-[[5,7,8,9-tetrahydro-5-(phenylmethyl)benzo[c]phenanthridin-10-yl]methylene]-	EtOH	235(4.72),285(4.64), 310(4.38),335(4.35), 410(4.06),505(3.70)	135-1208-86
$C_{31}H_{24}N_2O_4$			
Benzaldehyde, 4-acetoxy-, 1-[(1-benzoyl-3-oxo-3-phenyl-1-propenyl)-phenylhydrazone]-, (E,?)-	MeOH	252(4.63),374(4.62)	44-1972-86
Benzoic acid, 4-[[(1-benzoyl-3-oxo-3-phenyl-1-propenyl)phenylhydrazono]-methyl]-, methyl ester, (E,?)-	MeOH	246(4.48),384(4.65)	44-1972-86
$C_{31}H_{24}O_2$			
2-Propen-1-one, 1,1'-(methylenedi-4,1-phenylene)bis[3-phenyl-	93% H_2SO_4	329(4.17),417(4.53), 455(4.83)	104-0055-86
$C_{31}H_{24}O_4$			
4H-1-Benzopyran-4-one, 7-acetoxy-3-phenyl-6,8-bis(phenylmethyl)-	MeOH	268(4.5),311(4.18)	2-0646-86
$C_{31}H_{24}O_{10}$			
Sikokianin A	MeOH	298(4.51)	94-3631-86
Sikokianin B	MeOH	298(4.49)	94-3631-86
$C_{31}H_{24}O_{11}$			
1H-Anthra[2,3-c]pyran-8-acetic acid, 12-hydroxy-9-[3-hydroxy-3-(2-hydroxyphenyl)-1-oxo-2-propenyl]-3-(hydroxymethyl)-10,11-dimethoxy-1-oxo-, (Z)-	EtOH-1% DMF	297(4.72),355(4.38), 400(4.18)	78-0727-86
$C_{31}H_{25}NO$			
Pyrrolo[2,1-a]isoquinolin-1(5H)-one, 6,10b-dihydro-2,3-diphenyl-10b-(phenylmethyl)-	MeCN	234(4.14),273(4.06), 357(3.84)	118-0899-86
$C_{31}H_{25}NO_2$			
4H-1,3-Oxazin-4-one, 2-(diphenylmethylene)-2,3-dihydro-3,6-bis(4-methylphenyl)-	EtOH	217(4.48),287(4.44)	104-0204-86
$C_{31}H_{25}NO_2S$			
2H,5H-[1]Benzothiopyrano[3,4-e]-1,3-oxazin-5-one, 3,4-dihydro-2,4-bis-(4-methylphenyl)-3-phenyl-	MeOH	235(4.5),329(4.7)	42-0323-86
$C_{31}H_{25}NO_9$			
[1]Benzopyrano[2',3':6,7]naphth[2,1-g]oxazolo[3,2-b]isoquinoline-8,14,15,17-tetrone, 1,2,3a,4-tetrahydro-11,12,16-trimethoxy-	$CHCl_3$	252.5(4.54),327.2(4.33)	35-6088-86

Compound	Solvent	λ_{max}(log ε)	Ref.
3a,5-dimethyl-, (-)- (C,O-dimethylcervinomycin A_2) (cont.)			35-6088-86
$C_{31}H_{25}NS$			
Pyrrolo[2,1-a]isoquinoline-1(5H)-thione, 6,10b-dihydro-2,3-diphenyl-10b-(phenylmethyl)-	MeCN	255(4.06),428(4.04)	118-0899-86
$C_{31}H_{25}N_5O_7S$			
6H-Purine-6-thione, 2-amino-1,7-dihydro-7-(2,3,5-tri-O-benzoyl-β-D-xylofuranosyl)-	EtOH	230(4.68),268s(--), 355(4.22)	87-0203-86
6H-Purine-6-thione, 2-amino-1,9-dihydro-9-(2,3,5-tri-O-benzoyl-β-D-xylofuranosyl)-	EtOH	230(4.66),274(3.93), 345(4.41)	87-0203-86
D-Ribitol, 1,4-anhydro-1-C-[8-(methylthio)-1,2,4-triazolo[3,4-f][1,2,4]-triazin-3-yl]-, 2,3,5-tribenzoate, (S)-	MeOH	240(4.08),310(3.72)	87-2231-86
$C_{31}H_{25}N_5O_9S$			
1H-Pyrazino[2,3-c][1,2,6]thiadiazin-4-amine, 1-(2,3,5-tri-O-benzoyl-β-D-ribofuranosyl)-, 2,2-dioxide	MeOH	231(4.86),273s(4.04), 281s(3.99),335(3.89)	5-1872-86
8H-Pyrazino[2,3-c][1,2,6]thiadiazin-4-amine, 8-(2,3,5-tri-O-benzoyl-β-D-ribofuranosyl)-, 2,2-dioxide	MeOH	232(4.48),260s(3.92), 282s(3.41),412(3.65)	5-1872-86
$C_{31}H_{26}Br_2NO$			
3H-Indolium, 2-[3-[4,6-bis(4-bromophenyl)-2H-pyran-2-ylidene]-1-propenyl-1,3,3-trimethyl-, tetrafluoroborate	CH_2Cl_2	670(4.75),710(4.62)	103-1295-86
$C_{31}H_{26}Cl_2NO$			
3H-Indolium, 2-[3-[4,6-bis(4-chlorophenyl)-2H-pyran-2-ylidene]-1-propenyl-1,3,3-trimethyl-, tetrafluoroborate	CH_2Cl_2	665(4.74),700(4.61)	103-1295-86
$C_{31}H_{26}F_2NO$			
3H-Indolium, 2-[3-[4,6-bis(4-fluorophenyl)-2H-pyran-2-ylidene]-1-propenyl-1,3,3-trimethyl-, tetrafluoroborate	CH_2Cl_2	658(4.75),700(4.64)	103-1295-86
$C_{31}H_{26}NS$			
5H-Pyrrolo[2,1-a]isoquinolinium, 6,10b-dihydro-1-(methylthio)-2,3,10b-triphenyl-, iodide	MeCN	247(4.22),292(3.84), 376(3.91)	118-0899-86
6H-[1,3]Thiazino[2,3-a]isoquinolinium, 7,11b-dihydro-3,4-diphenyl-11b-(phenylmethyl)-, tetrafluoroborate	MeCN	247(4.06),274s(--), 313s(--),416(3.59)	118-0899-86
$C_{31}H_{26}N_2O_9$			
2,4(1H,3H)-Pyrimidinedione, 5-methyl-1-(2,3,5-tri-O-benzoyl-β-D-xylofuranosyl)-	EtOH	232(4.57),266(3.98)	87-0203-86
$C_{31}H_{26}N_2S_2$			
Acridine, 9,9'-(1,5-pentanediyl)-	$CHCl_3$	259(5.30),350(4.01),	4-1141-86

Compound	Solvent	λ$_{max}$(log ε)	Ref.
bis(thio)]bis- (cont.)		365(4.26),376(4.11), 395(4.01)	4-1141-86
$C_{31}H_{26}N_3O_2PS$			
Benzo[b]thiophene-3a(7aH)-carboxylic acid, 6,7-dicyano-2,3-dihydro-4-[(triphenylphosphoranylidene)amino]-, ethyl ester, cis	$CHCl_3$	278(3.40),287(3.36), 441(3.90)	24-2723-86
$C_{31}H_{26}N_3O_3P$			
3a(7aH)-Benzofurancarboxylic acid, 6,7-dicyano-2,3-dihydro-4-[(triphenylphosphoranylidene)amino]-, ethyl ester, cis	$CHCl_3$	262(3.18),272(3.14), 433(4.20)	24-2723-86
$C_{31}H_{26}N_3O_5$			
3H-Indolium, 2-[3-[4,6-bis(4-nitrophenyl)-2H-pyran-2-ylidene]-1-propenyl]-1,3,3-trimethyl-, tetrafluoroborate	CH_2Cl_2	645(4.58),690(4.54)	103-1295-86
$C_{31}H_{26}N_4O_5$			
Cyclopenta[4,5]pyrimido[6,1-a]isoquinoline-1,3-dicarboxamide, 4,5,7,8-tetrahydro-10,11-dimethoxy-5-oxo-N^3,4-diphenyl-	CH_2Cl_2	255(4.45),283(4.30), 388(4.42)	24-2553-86
$C_{31}H_{26}N_6O_7S$			
D-Ribitol, 1-C-[6-amino-8-(methylthio)-1,2,4-triazolo[3,4-f][1,2,4]triazin-3-yl]-1,4-anhydro-, 2,3,5-tribenzoate, (S)-	MeOH	227(4.65),270(4.14)	87-2231-86
$C_{31}H_{26}O_4$			
4H-1-Benzopyran-4-one, 6-(diphenylmethyl)-7,8-dimethoxy-2-methyl-3-phenyl-	MeOH	260(4.22),298(4.6)	2-0649-86
$C_{31}H_{26}O_7$			
4H-1-Benzopyran-4-one, 2-[3,4-bis(phenylmethoxy)phenyl]-5-hydroxy-6,7-dimethoxy-	EtOH	277(4.25),340(4.43)	87-2256-86
	EtOH-$AlCl_3$	258(4.17),292(4.29), 358(4.41)	87-2256-86
4H-1-Benzopyran-4-one, 2-[3,4-bis(phenylmethoxy)phenyl]-5-hydroxy-7,8-dimethoxy-	EtOH	254(4.27),276(4.31), 337(4.30)	87-2256-86
	EtOH-$AlCl_3$	261(4.23),285(4.29), 302(4.22),348(4.33), 404(4.05)	87-2256-86
4H-1-Benzopyran-4-one, 2-[3,4-bis(phenylmethoxy)phenyl]-6-hydroxy-5,7-dimethoxy-	EtOH	278(4.26),331(4.46)	94-0030-86
$C_{31}H_{26}O_{10}$			
1H,9H-Anthra[2,3-c:7,6-c']dipyran-1,9-dione, 8,11-dihydro-14-hydroxy-11-[2-hydroxy-2-(2-hydroxyphenyl)-ethyl]-3-(hydroxymethyl)-12,13-dimethoxy-	EtOH-1% DMF	300(4.96),375(3.90), 395(4.17)	78-0727-86
$C_{31}H_{27}NOS$			
2H,6H-[1,3]Thiazino[2,3-a]isoquinoline, 7,11b-dihydro-2-methoxy-	MeCN	276(3.81),297(3.84)	118-0899-86

Compound	Solvent	$\lambda_{max}(\log \epsilon)$	Ref.
3,4,11b-triphenyl- (cont.)			118-0899-86
$C_{31}H_{27}NO_4$			
1,3-Dioxane-4,6-dione, 2,2-dimethyl-5-(2,3,6,7-tetrahydro-2,3-diphenyl-4H-benzo[a]quinolizin-4-ylidene)-	MeCN	255(4.40),315(4.16), 413(3.46)	118-0908-86
$C_{31}H_{27}N_3O$			
Propanedinitrile, [7,8-dihydro-2-(4-methoxy-3-methylphenyl)-1-methyl-3,5-diphenyl-4(1H)-azocinylidene]-	MeCN	244(4.36),323(3.88), 488(4.30)	118-0908-86
$C_{31}H_{27}N_3O_2$			
Benzaldehyde, 4-(dimethylamino)-, (1-benzoyl-3-oxo-3-phenyl-1-propenyl)-phenylhydrazone, (E,?)-	MeOH	253(4.44),414(4.63)	44-1972-86
Propanedinitrile, [2-(2,5-dimethoxyphenyl)-7,8-dihydro-1-methyl-3,5-diphenyl-4(1H)-azocinylidene]-	MeCN	243(4.33),338(3.78), 483(4.35)	118-0908-86
$C_{31}H_{27}O_3$			
Pyrylium, 2-[6-(3,4-dimethoxyphenyl)-1,3,5-hexatrienyl]-4,6-diphenyl-, tetrafluoroborate	CH_2Cl_2	380(4.48),640(4.53)	104-1786-86
$C_{31}H_{28}$			
Anthracene, 1,8-bis(2,3-dimethylphenyl)-9-methyl-	C_6H_{12}	262s(4.80),270(4.94), 350(3.44),370(3.74), 390(3.93),410(3.85)	44-0921-86
$C_{31}H_{28}ClN_7O_3S$			
Acetamide, N-[2-[[4-chloro-5-[[1-(dicyanomethylene)-1,3-dihydro-3-oxo-2H-inden-2-ylidene]methylene]-2-thiazolyl]azo]-4-methoxy-5-[(1-methylpentyl)amino]phenyl]-	CH_2Cl_2	750(4.92)	77-1639-86
	DMSO	799(--)	77-1639-86
$C_{31}H_{28}NO$			
3H-Indolium, 2-[3-(4,6-diphenyl-2H-pyran-2-ylidene)-1-propenyl]-1,3,3-trimethyl-, tetrafluoroborate	CH_2Cl_2	654(4.71),682(4.63)	103-1295-86
$C_{31}H_{28}O_2$			
1-Azulenecarboxylic acid, 2,6-bis-[2-(4-methylphenyl)ethenyl]-, ethyl ester, (E,E)-	CH_2Cl_2	253.5(4.20),349.5(4.66), 461.5(4.74),575.5(3.31)	24-2272-86
$C_{31}H_{28}P$			
Phosphonium, [(6,8-dimethyl-4-azulenyl)methyl]triphenyl-, tetrafluoroborate	MeOH	227(4.57),251(4.49), 279s(4.43),290(4.58), 309s(4.00),341(3.52), 356(3.58),566(2.71), 600s(2.65),657s(2.23)	5-1222-86
$C_{31}H_{29}NO_5$			
7H-Furo[3,2-g][1]benzopyran-7-one, 2-[4-[2-(dimethylamino)-1-methylethoxy]benzoyl]-3,5-dimethyl-6-phenyl-	$CHCl_3$	292(4.74),355(4.77)	18-1299-86

Compound	Solvent	λ_{max}(log ε)	Ref.
$C_{31}H_{29}NO_6$			
1H-Pyrrole-2,5-dione, 3-[2,3-O-(1-methylethylidene)-β-D-ribofuranosyl)-1-(triphenylmethyl)-	EtOH	220(4.12),233(4.11), 331(2.98)	44-0495-86
$C_{31}H_{30}ClOP_2RhS_6$			
Rhodium, carbonylchloro[tris(5-methyl-2-thienyl)phosphine-P]-, (SP-4-3)-	CH_2Cl_2	248(4.57),276(4.53), 366(3.45)	70-1919-86
$C_{31}H_{30}O$			
1-Pentanone, 3,3-dimethyl-1,4,5,5-tetraphenyl-	EtOH	255(3.99),270(3.43), 280(2.86)	44-4604-86
4-Penten-1-ol, 3,3-dimethyl-1,1,5,5-tetraphenyl-	EtOH	260(4.34),270(4.20), 280(4.09),290(3.95), 320(2.65)	44-4604-86
$C_{31}H_{31}ClN_4O$			
Phenol, 4-[[7-chloro-3-(3,4-dihydro-3-methyl-1-isoquinolinyl)-4-quinolinyl]amino]-2-(1-piperidinylmethyl)-	EtOH	220(4.71),350(4.01)	2-1072-86
$C_{31}H_{31}N_5O_7$			
Adenosine, N,N-bis(4-methylbenzoyl)-2',3'-O-(1-methylethylidene)-, 5'-acetate	MeOH	260(4.46),275s(4.43)	23-2109-86
$C_{31}H_{32}O_{16}$			
Pentanedioic acid, 3-hydroxy-3-methyl-, mono[5-carboxy-2,3-bis[[3-(3,4-dihydroxyphenyl)-1-oxo-2-propenyl]-oxy]-5-hydroxycyclohexyl] ester	MeOH	216(4.42),242(4.25), 329(4.49)	94-1419-86
$C_{31}H_{33}NO_9$			
1H-Pyrrole-3,4-dicarboxylic acid, 1-[2-deoxy-3,5-bis-O-(4-methylbenzoyl)-β-D-erythro-pentofuranosyl]-, diethyl ester	EtOH	240(4.63)	78-5869-86
$C_{31}H_{33}N_3O_2S$			
2,6-Diazabicyclo[2.2.2]oct-7-ene-3,5-dione, 8-(diethylamino)-1-(ethylthio)-7-methyl-	MeCN	221(4.55),257s(4.20), 358(3.84)	24-1315-86
$C_{31}H_{33}N_3O_{11}$			
Azinomycin B	MeOH	217(4.82),250(4.48), 290s(--),340(3.93)	158-1527-86
$C_{31}H_{34}N_4O_{10}S$			
1-Azabicyclo[3.2.0]hept-2-ene-2-carboxylic acid, 6-(1-hydroxyethyl)-3-[[4-[[[[(4-nitrophenyl)methoxy]-carbonyl]amino]methyl]cyclohexyl]-thio]-7-oxo-, (4-nitrophenyl)methyl ester, [5R-[3(cis),5α,6α(R*)]]-	n.s.g.	268(4.32),323(4.11)	23-2184-86
$C_{31}H_{34}O_{15}$			
Elloramycin C	MeOH	238(4.41),286(4.61), 391(4.06),410(4.07)	158-0856-86
	MeOH-NaOH	252(4.53),424s(4.08), 440(4.11)	158-0856-86

Compound	Solvent	λ_{max}(log ϵ)	Ref.
Elloramycin D	MeOH	240(4.40),286(4.62), 391(4.07),408(4.08)	158-0856-86
	MeOH-NaOH	251(4.57),424s(4.10), 440(4.14)	158-0856-86
$C_{31}H_{35}N_5O_8$			
Acetylsaframycin Y3	MeOH	264(4.26),340s(--)	158-1639-86
$C_{31}H_{36}$			
Benzene, 1,3,5-tris(1-methylethyl)-2-[2-[3-(2-phenylethenyl)phenyl]-ethenyl]-, (E,E)-	$C_6H_{11}Me$	285(4.59),307s(4.51), 323s(4.23)	35-0841-86
$C_{31}H_{36}N_2OS$			
5α-Androst-16-eno[17,16-b]pyridine-5'-carbonitrile, 6'-[(2-oxo-2-phenylethyl)thio]-	EtOH	217(4.29),248(4.08), 269(4.13),319(3.86), 410(3.01)	70-1504-86
Methanone, (10-amino-2,3,4,4a,4b,5,6-6a,12,12a,12b,13,14,14a-tetradeca-hydro-4a,6a-dimethyl-1H-naphth[2',1'-4,5]indeno[1,2-b]thieno[3,2-e]pyri-din-9-yl)phenyl-	EtOH	257(3.69),280(3.76), 300(3.96),320(4.06), 330(4.04),412(3.64)	70-1504-86
$C_{31}H_{36}N_2O_{11}$			
Daunorubicin, N-(4-amino-1-oxobutyl)-	pH 7	480(3.90)	161-0881-86
$C_{31}H_{36}N_4O_2$			
Ceridimine	EtOH	236(4.29),278s(4.09), 287(4.14),295(4.11)	77-0001-86
$C_{31}H_{37}ClN_4O$			
Ethanamine, 2-[4-[2-chloro-2-phenyl-1-[4-(1-piperidinylazo)phenyl]eth-enyl]phenoxy]-N,N-diethyl-	EtOH	229(4.24),300s(4.23), 345(4.28)	13-0073-86B
$C_{31}H_{38}O_4$			
1,3-Azulenedicarboxylic acid, 6-(4-methylphenyl)-2-octyl-, diethyl ester	CH_2Cl_2	238.5(4.50),269(4.18), 332.5(4.80),363(4.31), 504(2.79)	24-2272-86
$C_{31}H_{38}O_{11}$			
Tyledoside A	MeOH	297(3.74)	39-0429-86C
$C_{31}H_{40}$			
Pyrene, 1,3,4,6,8-pentakis(1-methyl-ethyl)-	n.s.g.	252.1(4.7),289.9(4.7), 367.6(4.6)	1-0665-86
Pyrene, 1,3,5,7,9-pentakis(1-methyl-ethyl)-	n.s.g.	252.1(4.9),287.7(4.8), 358.5(4.7)	1-0665-86
$C_{31}H_{40}O_{10}$			
Bufa-20,22-dienolide, 7,8-epoxy-11,14-dihydroxy-2,3-[(tetrahydro-4-methoxy-6-methyl-3-oxo-2H-pyran-4,2-diyl)bis(oxy)]- (tyledoside B)	MeOH	298(3.70)	39-0429-86C
$C_{31}H_{40}O_{11}$			
Bufa-20,22-dienolide, 7,8-epoxy-11,14-dihydroxy-12-oxo-2,3-[(tetra-hydro-3-hydroxy-4-methoxy-6-methyl-2H-pyran-4,2-diyl)bis(oxy)]- (tyle-doside D)	MeOH	298(3.69)	39-0429-86C

Compound	Solvent	λ_{max}(log ϵ)	Ref.
$C_{31}H_{41}N_3O_6$			
1H-Pyrrole-2-carboxylic acid, 3-(2-acetoxyethyl)-5-[[(2-acetoxyethyl)-5-[(3,5-dimethyl-2H-pyrrol-2-ylidene)methyl]-4-methyl-1H-pyrrol-2-yl]methyl]-4-methyl-, 1,1-dimethylethyl ester, monohydrobromide, (Z)-	CH_2Cl_2	484(4.94)	44-4660-86
$C_{31}H_{42}O_5$			
Milbemycin β_3, (+)-	EtOH	246(4.37)	44-4840-86
$C_{31}H_{42}O_7$			
Cystoseirol B acetate	MeOH	228(3.96),287(3.23)	44-2707-86
Methyl ganoderate D	EtOH	251(3.38)	94-4018-86
Methyl ganoderate E	EtOH	251.5(3.93)	94-3695-86
Methyl ganoderate F	EtOH	251(3.9)	102-1189-86
$C_{31}H_{42}O_{10}$			
Tyledoside F	MeOH	298(3.69)	39-0429-86C
Tyledoside G	MeOH	298(3.67)	39-0429-86C
$C_{31}H_{43}NO_4$			
X-14,547A	n.s.g.	244(4.51),291(4.21)	100-0035-86
$C_{31}H_{44}N_2O_2$			
Chol-5-en-3-ol, 24-[3-(3H-diazirin-3-yl)phenoxy]-, (3β)-	EtOH	363(2.53)	94-0931-86
$C_{31}H_{44}O_5$			
Mediterraneol A	MeOH	215(4.28),237s(--), 289(3.45)	44-1115-86
Mediterraneol B	MeOH	215(4.27),257s(--), 289(3.46)	44-1115-86
$C_{31}H_{44}O_6$			
Lanost-8-en-26-oic acid, 3,11,15,23-tetraoxo-, methyl ester	EtOH	254(3.59)	94-4030-86
$C_{31}H_{44}O_7$			
Methyl ganoderate C	EtOH	250(3.9)	102-1189-86
Methyl ganoderate I	EtOH	254.5(3.86)	94-3695-86
$C_{31}H_{44}O_8$			
Lanost-8-en-26-oic acid, 3,12-dihydroxy-7,11,15,23-tetraoxo-, methyl ester, (3β,12β)-	EtOH	253.5(3.43)	94-4018-86
$C_{31}H_{44}O_{18}$			
Argylioside	MeOH	234(3.9)	102-0946-86
$C_{31}H_{44}O_{19}$			
Radiatoside	MeOH	234(3.9)	100-0519-86
$C_{31}H_{46}Cl_3N_7O_{11}$			
HV-toxin M	MeOH	244(4.26)	163-2689-86
$C_{31}H_{46}N_2O$			
A(1)-Norolean-1-eno[1,2-b]pyrazine, 19,28-epoxy-, (18α,19β)-	C_6H_{12}	273(3.93),279(3.85), 327(3.04)	73-0118-86

Compound	Solvent	λ_{max}(log ε)	Ref.
$C_{31}H_{46}O_2$			
1,4-Naphthalenedione, 2-methyl-3-(3,7,11,15-tetramethyl-2-hexadecenyl) (vitamin K_1)	MeOH	244.5(4.25),248.5(4.26), 263(4.19),271(4.21), 330(3.49)	160-0177-86
	EtOH	244.5(4.25),248.5(4.26), 263(4.20),271(4.21), 330(3.49)	160-0177-86
$C_{31}H_{46}O_4$			
Lanosta-7,24-dien-26-oic acid, 3,23-dioxo-, methyl ester	EtOH	239s(3.94)	105-0613-86
$C_{31}H_{46}O_5$			
Cholest-8-ene-3,7,11-trione, 24-acetoxy-4,14-dimethyl-, (4α,5α)-	EtOH	269(3.86)	18-1767-86
$C_{31}H_{46}O_6$			
Methyl ganolucidate A	EtOH	256.5(4.00)	94-4030-86
$C_{31}H_{46}O_7$			
Methyl ganoderate A	EtOH	252.5(3.92)	94-3695-86
Methyl ganoderate B	EtOH	254.5(4.01)	94-3695-86
Methyl ganoderate K	EtOH	273(3.68)	94-3695-86
Methyl ganolucidate B	EtOH	254.5(3.49)	94-3695-86
$C_{31}H_{46}O_8$			
Methyl ganoderate G	EtOH	252.5(4.01)	94-4018-86
$C_{31}H_{48}O_2$			
1,4-Naphthalenedione, 2-methyl-3-(3,7,11,15-tetramethylhexadecyl)-	MeOH	244.5(4.69),333(3.66)	160-0177-86
	EtOH	244.5(4.68)	160-0177-86
18α-Olean-2-ene-2-carboxaldehyde, 19β,28-epoxy-	C_6H_{12}	228(4.16)	73-2869-86
$C_{31}H_{48}O_3$			
Bauer-7-en-28-oic acid, 3-oxo-, methyl ester	EtOH	209(3.54)	102-1685-86
$C_{31}H_{48}O_5$			
Holotoxinogenin, 25-methoxy-	EtOH	none above 210 nm	100-0809-86
$C_{31}H_{48}O_6$			
Methyl ganolucidate B	EtOH	257.5(3.73)	94-4030-80
$C_{31}H_{48}O_7$			
Methyl ganoderate C2	EtOH	253(4.01)	94-3695-86
Methyl ganoderate E	EtOH	250(3.9)	102-1189-86
Lanost-8-en-26-oic acid, 3,7,15-trihydroxy-11,23-dioxo-, methyl ester, (3β,7β,11α)-	EtOH	255(3.43)	94-3695-86
$C_{31}H_{48}O_8$			
5α-Lanost-8-en-26-oic acid, 3β,7β,15α,20-tetrahydroxy-11,23-dioxo-, methyl ester (methyl ganoderate L)	MeOH	256(3.83)	163-0809-86
$C_{31}H_{49}N_2$			
Pyridinium, 4-[2-[4-(dimethylamino)-phenyl]ethenyl]-1-hexadecyl-, chloride	EtOH	482(4.62)	4-0209-86

Compound	Solvent	λ max (log ε)	Ref.
iodide (cont.)	EtOH	482(4.65)	4-0209-86
nitrate	EtOH	482(4.65)	4-0209-86
$C_{31}H_{49}N_3O_5$			
Androst-4-en-3-one, 17-acetoxy-2-[4-(2-hydroxyethyl)-1-piperazinyl]-6-(4-morpholinyl)-, (2α,6β,17β)-	EtOH	236.5(4.15)	118-0461-86
Androst-4-en-3-one, 17-acetoxy-6-[4-(2-hydroxyethyl)-1-piperazinyl]-2-(4-morpholinyl)-, (2α,6β,17β)-	EtOH	237.5(4.12)	118-0461-86
$C_{31}H_{50}BrNO_2$			
Cholest-4-en-3-one, 4-bromo-2-(4-morpholinyl)-, (2β)-	EtOH	267.0(4.16)	118-0461-86
$C_{31}H_{50}N_4O_3$			
Androst-4-en-3-one, 17-acetoxy-2,6-bis(4-methyl-1-piperazinyl)-, (2α,6β,17β)-	EtOH	238(4.09)	107-0195-86
$C_{31}H_{50}O$			
Neocyclonervilasterol	EtOH	209(3.74)	94-3244-86
$C_{31}H_{50}O_3$			
Cholest-8-en-7-one, 3-acetoxy-4,14-dimethyl-, (3β,4α,5α)-	EtOH	252(4.0)	18-1767-86
$C_{31}H_{52}O_3$			
19-Norlanosta-1(10),5-diene-3,16-diol, 25-methoxy-9-methyl-, (3β,9β,16β)-	MeOH	235(4.09),242(4.13), 251(3.96)	94-0088-86
2-Propenoic acid, 3-(4-hydroxyphenyl)-, docosyl ester, (E)-	MeOH	228(4.11),299s(--), 312(4.38)	64-0164-86C
	MeOH-KOH	240(4.02),311(4.04), 358(4.56)	64-0164-86C
(Z)-	MeOH	225s(--),296s(--), 309(3.99)	64-0164-86C
	MeOH-KOH	309s(--),357(4.14)	64-0164-86C
$C_{31}H_{52}O_3S$			
Acetic acid, oxo[(2,4,6-trimethylphenyl)thio]-, 1,1-dimethyloctadecyl ester	EtOH	236(4.24),276(3.58)	39-1603-86C
$C_{31}H_{54}BrN_4Zn$			
Zinc(1+), bromo(2-heptadecyl-12-methyl-3,7,11,17-tetraazabicyclo[11.3.1]heptadeca-1(17),2,11,13,15-pentaene-N^3,N^7,N^{11},N^{17})-, perchlorate	MeCN	218(4.31),298(3.55)	35-2388-86

Compound	Solvent	λ_{max}(log ϵ)	Ref.
$C_{32}H_6Cl_{20}$			
Benzene, 1,4-bis[bis(2,3,5,6-tetrachlorophenyl)methyl]-2,3,5,6-tetrachloro-	$CHCl_3$	284s(3.46),291(3.50),301(3.60)	118-0064-86
$C_{32}H_{12}MgN_{12}O_8$			
Magnesium, [2,9,16,23-tetranitro-29H,31H-phthalocyaninato(2-)-N^{29},N^{30},N^{31},N^{32}]-, (SP-4-1)-	90% DMF	675(5.45)	149-0125-86B
$C_{32}H_{16}MgN_8$			
Magnesium, [29H,31H-phthalocyaninato(2-)-N^{29},N^{30},N^{31},N^{32}]-, (SP-4-1)-	90% DMF	674(5.26)	149-0125-86B
$C_{32}H_{18}F_2N_6O_2S_2$			
Imidazo[5,1-a]phthalazin-3(2H)-one, 1,1'-dithiobis[2-(3-fluorophenyl)-	EtOH	250(4.92),268(4.98),328(4.63)	150-1901-86M
Imidazo[5,1-a]phthalazin-3(2H)-one, 1,1'-dithiobis[2-(4-fluorophenyl)-	EtOH	206(4.67),224(4.50),250(4.20),268(4.35),344(3.52),444(3.21)	150-1901-86M
$C_{32}H_{19}N_3$			
Propanedinitrile, (9,10-diphenyl-8H-naphtho[2,1-a]quinolizin-8-ylidene)-	MeCN	347(4.04),362(4.17),494(4.23)	118-0908-86
$C_{32}H_{20}Br_2N_2S_2$			
6H-1,3-Thiazine, 4-(4-bromophenyl)-6-[4-(4-bromophenyl)-2-phenyl-6H-1,3-thiazin-6-ylidene]-2-phenyl-	$CHCl_3$	286(4.66),378s(4.25),567(3.98)	118-0916-86
$C_{23}H_{20}N_2O_2$			
[2,2'-Binaphthalene]-1,1'(4H,4'H)-dione, 4,4'-bis(phenylimino)-	EtOH	248(4.54),296(4.42),465(3.86)	104-0748-86
7H-Dibenzo[a,h]carbazole-7,12(13H)-dione, 13-phenyl-5-(phenylamino)-	EtOH	272(4.74),286(4.70),340(3.89),454(3.66),531(3.71)	104-0748-86
$C_{23}H_{20}N_4O_4S_2$			
6H-1,3-Thiazine, 4-(4-nitrophenyl)-6-[4-(4-nitrophenyl)-2-phenyl-6H-1,3-thiazin-6-ylidene]-2-phenyl-	$CHCl_3$	286(4.00),418(3.64),581(3.72)	118-0916-86
$C_{32}H_{20}N_6O_2S_2$			
Imidazo[5,1-a]phthalazin-3(2H)-one, 1,1'-dithiobis[2-phenyl-	EtOH	267(4.42),348(3.91),367(3.88),438(3.68)	150-1901-86M
$C_{32}H_{20}N_{10}O_8S_2$			
4-Thiazolamine, 2,2'-(1,2-ethanediylidene)bis[2,5-dihydro-N-(4-nitrophenyl)-5-[(4-nitrophenyl)imino]-, (E,E,?,?)-	n.s.g.	404(4.24),587s(4.59),629(4.69)	48-0805-86
$C_{32}H_{21}N_3$			
Propanedinitrile, (5,6-dihydro-9,10-diphenyl-8H-naphtho[2,1-a]quinolizin-8-ylidene)-	MeCN	318(4.25),340(4.12),443(4.06)	118-0908-86
$C_{32}H_{22}BrN_3O_2$			
Ethanone, 1-(4-bromophenyl)-2-[5,6-diphenyl-4-(phenylimino)furo[2,3-d]pyrimidin-1(4H)-yl]-, HBr salt	EtOH	263(4.45),350(4.42)	11-0337-86

Compound	Solvent	λ_{max}(log ε)	Ref.
$C_{32}H_{22}N_2O_4$			
Anthracene, 9-(2-nitroethenyl)-, dimer 3	CH_2Cl_2	360(3.8),380(3.9), 400(3.9)	44-3223-86
$C_{32}H_{22}N_2S_2$			
6H-1,3-Thiazine, 6-(2,4-diphenyl-6H-1,3-thiazin-6-ylidene)-2,4-diphenyl-	$CHCl_3$	278(4.74),366(4.33), 550(4.07)	118-0916-86
$C_{32}H_{22}N_4O_4$			
2,4-Imidazolidinedione, 5,5'-(1,2-ethynediyldi-4,1-phenylene)bis[5-phenyl-	dioxan	302(4.65),320(4.51)	104-1117-86
$C_{32}H_{22}O_8$			
[2,2'-Bianthracene]-1,1',4,4'-tetrone, 9,9',10,10'-tetramethoxy-	CH_2Cl_2	242(5.01),288(4.51), 299s(4.46),438(4.15)	5-0839-86
[2,2'-Bianthracene]-9,9',10,10'-tetrone, 1,1',4,4'-tetramethoxy-	CH_2Cl_2	228(4.53),256(4.72), 320(3.68),405(4.16)	5-0839-86
$C_{32}H_{22}O_{12}$			
[8,8'-Bi-2H-furo[3,2-b]naphtho[2,3-d]-pyran]-2,2',6,6',11,11'-hexone, 3,3',3a,3'a,5,5',11b,11'b-octahydro-10,10'-dihydroxy-5,5'-dimethyl- (crisamicin A)	MeCN	215(4.46),231(4.57), 267(4.52),435(4.07)	158-0345-86
$C_{32}H_{23}NO_6$			
Dibenzo[a,h]carbazole-5,7,12-triol, 13-phenyl-, triacetate	EtOH	284(4.89),337(4.14), 354s(3.84),366s(3.71), 387(3.78),407(3.87)	104-0748-86
$C_{32}H_{23}N_3$			
Propanedinitrile, (5,6,9,10-tetrahydro-9,10-diphenyl-8H-naphtho-[2,1-a]quinolizin-8-ylidene)-	MeCN	295(4.27),307(4.23), 330(4.00)	118-0908-86
$C_{32}H_{23}N_3O_2$			
9,10-Anthracenedione, 1,4,5-tris(phenylamino)-	EtOH	619(4.16),663(4.22)	104-0547-86
$C_{32}H_{24}BrOP$			
Ethanone, 2-bromo-1-(1,2-dihydro-3-acenaphthylenyl)-2-(triphenylphosphoranylidene)-	$CHCl_3$	270(4.362),307(3.850), 318(3.860),389(3.704)	65-0301-86
Ethanone, 2-bromo-1-(1,2-dihydro-5-acenaphthylenyl)-2-(triphenylphosphoranylidene)-	$CHCl_3$	266(4.408),273(4.393), 382(4.189)	65-0301-86
$C_{32}H_{24}ClNO_2S$			
Methanone, [4-(4-chlorophenyl)-1,2-dihydro-1-(4-methoxyphenyl)-5-methyl-6-phenyl-2-thioxo-3-pyridinyl]phenyl-	EtOH	248(4.45),289(4.21), 402(3.88)	118-0952-86
$C_{32}H_{24}N_2O_2Se_2$			
Benzenamine, N,N'-[diselenobis(3,2-benzofurandiylmethylidyne)]bis[2-methyl-	CH_2Cl_2	250s(4.30),310(4.47), 360s(4.29)	103-0608-86
Benzenamine, N,N'-[diselenobis(3,2-benzofurandiylmethylidyne)]bis[4-methyl-	EtOH	244(4.33),316(4.55), 363(4.49)	103-0608-86

Compound	Solvent	λ_{max}(log ε)	Ref.
$C_{32}H_{24}N_2O_4Se_2$			
Benzenamine, N,N'-[diselenobis(3,2-benzofurandiylmethylidyne)]bis[4-methoxy-	EtOH	250s(4.29),315(4.47), 382(4.47)	103-0608-86
$C_{32}H_{24}N_6S_2$			
4-Thiazolamine, 2,2'-(1,2-ethanediylidene)bis[2,5-dihydro-N-phenyl-5-(phenylimino)-	n.s.g.	463s(3.97),578s(4.65), 615(4.75)	48-0805-86
$C_{32}H_{24}O_2$			
Naphth[2,3-a]azulene, 5,12-bis(phenylmethoxy)-	$CHCl_3$	272(4.39),288(4.46), 3-H(4.72)[sic], 323(4.66),352(4.04), 373(4.01),398(3.93), 423(3.99),450(3.85), 495s(3.94),535(4.09), 586(4.18),643(4.16), 717(3.98)	118-0686-86
$C_{32}H_{24}O_6$			
2H-1-Benzopyran-2-one, 4,7-diacetoxy-6-(diphenylmethyl)-3-phenyl-	MeOH	225(4.52),243(4.32), 254(4.3),260(4.3)	2-0623-86
2H-1-Benzopyran-4-one, 4,7-diacetoxy-8-(diphenylmethyl)-3-phenyl-	MeOH	224(4.2),270(3.9)	2-0623-86
4H-1-Benzopyran-4-one, 7,8-diacetoxy-6-(diphenylmethyl)-3-phenyl-	MeOH	263(4.70),298(4.26), 310(4.22)	2-0649-86
$C_{32}H_{24}O_{12}$			
Thermorubin	EtOH-1% DMF	297(4.74),330(4.71), 430(4.71)	78-0727-86
$C_{32}H_{25}NO_2S$			
Methanone, [1,2-dihydro-1-(4-methoxyphenyl)-5-methyl-4,6-diphenyl-2-thioxo-3-pyridinyl]phenyl-	EtOH	247(4.36),293(4.11), 399(3.82)	118-0952-86
Methanone, [1,2-dihydro-4-(4-methoxyphenyl)-5-methyl-1,6-diphenyl-2-thioxo-3-pyridinyl]phenyl-	EtOH	248(4.33),296(4.25), 398(3.86)	118-0952-86
$C_{32}H_{25}NO_{11}$			
1H-Anthra[2,3-c]pyran-8-acetic acid, 12-hydroxy-9-[3-hydroxy-3-(2-hydroxyphenyl)-1-oxo-2-propenyl]-10,11-dimethoxy-3-[(methylamino)carbonyl]-1-oxo-, (Z)-	EtOH-1% DMF	298(4.79),327(4.76), 425(4.23)	78-0727-86
Naphth[2,3-g]isoquinoline-8-acetic acid, 3-carboxy-12-hydroxy-9-[3-hydroxy-3-(2-hydroxyphenyl)-1-oxo-2-propenyl]-10,11-dimethoxy-2-methyl-1-oxo-, (Z)-	EtOH-1% DMF	305(4.77),330s(4.65), 418(4.31)	78-0727-86
$C_{32}H_{25}OP$			
Ethanone, 1-(1,2-dihydro-3-acenaphthylenyl)-2-(triphenylphosphoranylidene)-	$CHCl_3$	243(4.563),302(3.947), 341(4.012)	65-0301-86
Ethanone, 1-(1,2-dihydro-5-acenaphthylenyl)-2-(triphenylphosphoranylidene)-	$CHCl_3$	330(4.0745)	65-0301-86

Compound	Solvent	λ_{max}(log ε)	Ref.
$C_{32}H_{26}N_2$			
2,4,6,12,14-Cyclopentadecapentaene-8,10-diyn-1-one, 7,12-dimethyl-, (5,10-dimethyl-2,4,10,12-cyclotridecatetraene-6,8-diyn-1-ylidene)hydrazone (relative absorbance)	CF_3COOH	258s(0.64),274(0.76), 293(0.69),346s(0.73), 382(1.00),593(0.95), 608s(0.92)	18-2509-86
$C_{32}H_{26}N_2O_6$			
Propanedioic acid, [4-cyano-3,5-bis(4-methylphenyl)-4-(4-nitrophenyl)-2,5-cyclohexadien-1-ylidene]-, dimethyl ester	$CHCl_3$	318(4.32)	83-0898-86
$C_{32}H_{26}N_4O_9$			
4H-Pyrazolo[3,4-d]pyrimidin-4-one, 1,5-dihydro-3-methoxy-1-(2,3,5-tri-O-benzoyl-β-D-ribofuranosyl)-	pH 1 and 7	226(4.68),260s(4.38)	4-1869-86
	pH 11	265(4.23)	4-1869-86
$C_{32}H_{26}OP$			
Phosphonium, [2-(1,2-dihydro-3-acenaphthylenyl)-2-oxoethyl]triphenyl-, bromide	$CHCl_3$	267(4.305),285(4.170), 288(3.596),301(3.677), 313(3.478),345(3.367), 366(3.392),385(3.253)	65-0301-86
Phosphonium, [2-(1,2-dihydro-5-acenaphthylenyl)-2-oxoethyl]triphenyl-, bromide	$CHCl_3$	263(4.455),300(3.588), 357(4.111),363(3.772)	65-0301-86
$C_{32}H_{26}O_4S$			
2-Propen-1-one, 1,1'-(thiodi-4,1-phenylene)bis[3-(4-methoxyphenyl)- (two middle maxima are graphical)	93% H_2SO_4	364(4.02),403(4.54), 444(4.22),508(4.79)	104-0055-86
$C_{32}H_{26}O_5$			
2-Propen-1-one, 1,1'-(oxydi-4,1-phenylene)bis[3-(4-methoxyphenyl)-	93% H_2SO_4	347(4.49),436(4.48), 482(4.95)	104-0055-86
$C_{32}H_{26}O_6$			
9,10-Anthracenediol, 1,4-bis(phenylmethoxy)-, diacetate	$CHCl_3$	276(4.50),360(3.92), 377(4.20),400(4.08), 422(3.97)	5-0839-86
$C_{32}H_{26}O_8$			
4H-1-Benzopyran-4-one, 6-acetoxy-2-[3,4-bis(phenylmethoxy)-5-hydroxy-7-methoxy-	EtOH	251s(4.26),271(4.22), 341(4.41)	94-0030-86
	EtOH-$AlCl_3$	259(4.20),282(4.24), 355(4.38)	94-0030-86
$C_{32}H_{26}O_{14}$			
[8,8'-Bi-1H-naphtho[2,3-c]pyran]-3,3'-diacetic acid, 3,3',4,4',5,5',10,10'-octahydro-4,4',6,6'-tetrahydroxy-1,1'-dimethyl-5,5',10,10'-tetraoxo-, [1S-[1α,3β,4β,8(1'R*,3'R*,4'R*)]-	MeCN	229(4.42),265(4.35), 430(3.90)	158-0345-86
$C_{32}H_{27}NO_4$			
Propanedioic acid, (4-cyano-3,4,5-triphenyl-2,5-cyclohexadien-1-ylidene)-, diethyl ester	$CHCl_3$	312(4.25)	83-0898-86
$C_{32}H_{27}NO_6$			
1,3-Azulenedicarboxylic acid, 6-[2-(4-	CH_2Cl_2	260.5(4.34),352.5(4.64),	24-2272-86

Compound	Solvent	$\lambda_{max}(\log \epsilon)$	Ref.
methylphenyl)ethenyl]-2-[2-(4-nitrophenyl)ethenyl]-, ethyl methyl ester, (E,E)- (cont.)		447.5(4.77)	24-2272-86
$C_{32}H_{27}NO_9Se$			
4-Selenazolecarboxylic acid, 2-(2,3,5-tri-O-benzoyl-β-D-ribofuranosyl)-, ethyl ester	MeOH	229(4.60),259(3.86)	4-0155-86
$C_{32}H_{27}N_3O$			
2,3-Diazabicyclo[3.2.0]hepta-3,6-diene, 2-benzoyl-6-[4-(dimethylamino)phenyl]-1,4-diphenyl-	ether	308(3.2)	151-0311-86D
$C_{32}H_{27}N_5O_6$			
Cyclopenta[4,5]pyrimido[6,1-a]isoquinoline-3-carboximidic acid, 1-(aminocarbonyl)-4,5,7,8-tetrahydro-10,11-dimethoxy-5-oxo-N,4-diphenyl-, anhydride with carbamic acid	CH_2Cl_2	255(4.37),305(4.22), 391(4.44)	24-2553-86
$C_{32}H_{27}N_5O_8$			
1H-Pyrazolo[3,4-d]pyrimidin-4-amine, 3-methoxy-1-(2,3,5-tri-O-benzoyl-β-D-ribofuranosyl)-	pH 1	224(4.70);260(3.83)	4-1869-86
	pH 7	230(4.49),269(4.23)	4-1869-86
	pH 11	225(4.61),268(4.20)	4-1869-86
$C_{32}H_{27}O_2$			
Pyrylium, 2-[8-(4-methoxyphenyl)-1,3,5,7-octatetraenyl]-4,6-diphenyl-, tetrafluoroborate	CH_2Cl_2	380(4.68),670(4.89)	104-1786-86
$C_{32}H_{28}ClN_3O_4$			
Anthra[1,9-cd]pyrazol-6(2H)-one, 5-chloro-2-[2-[(2-hydroxyethyl)amino]ethyl]-7,10-bis(phenylmethoxy)-	n.s.g.	242(4.41),273(4.05), 329(3.76),442(3.98)	4-1491-86
$C_{32}H_{28}N_2S_2$			
Acridine, 9,9'-[1,6-hexanediylbis(thio)]bis-	$CHCl_3$	258(5.29),349(4.04), 365(4.28),378(4.11), 398(4.02)	4-1141-86
$C_{32}H_{28}N_3O_3P$			
3a(7aH)-Benzofurancarboxylic acid, 6,7-dicyano-2,3-dihydro-2-methyl-4-[(triphenylphosphoranylidene)amino]-, ethyl ester, (2α,3aβ,7aβ)-	$CHCl_3$	265(3.22),275(3.19), 435(4.29)	24-2723-86
$C_{32}H_{28}O$			
Methanone, [4,5-bis(4-methylphenyl)-2-phenyl-1-cyclopenten-1-yl]phenyl-	C_6H_{12}	247(4.34),351(2.43)	138-1631-86
$C_{32}H_{28}O_2$			
9-Anthracenol, 1,8-bis(2,3-dimethylphenyl)-, acetate, cis	C_6H_{12}	257s(4.81),264(4.98), 326s(3.15),345(3.50), 358(3.79),376(3.99), 397(3.93)	44-0921-86
trans	C_6H_{12}	258s(4.91),264(5.07), 327s(3.20),347(3.54), 360(3.85),378(4.04), 400(3.98)	44-0921-86
2,4,10,12,15,17,23,25-Cyclohexacosa-	THF	253(4.47),291(4.50)	18-2209-86

Compound	Solvent	λ_{max}(log ϵ)	Ref.
octaene-6,8,19,21-tetrayne-1,14-dione, 2,5,10,13,18,23-hexamethyl-, (E,E,E,E,Z,Z,Z,Z)- (cont.)		329s(4.64),350(4.68), 403(4.20)	18-2209-86
$C_{32}H_{28}O_3$			
Methanone, [4,5-bis(4-methoxyphenyl)-2-phenyl-1-cyclopenten-1-yl]phenyl-	C_6H_{12}	246(4.46),351(2.56)	138-1631-86
Methanone, (4-methoxyphenyl)[2-(4-methoxyphenyl)-4,5-diphenyl-1-cyclopenten-1-yl]-	C_6H_{12}	266(4.42)	138-1631-86
$C_{32}H_{28}O_6$			
2-Butene-1,4-dione, 1,2,3,4-tetrakis(4-methoxyphenyl)-	EtOH	228(4.38),290(4.47)	24-2220-86
$C_{32}H_{29}NO_5$			
7H-Furo[3,2-g][1]benzopyran-7-one, 3,5-dimethyl-6-phenyl-2-[4-[2-(1-pyrrolidinyl)ethoxy]benzoyl]-	MeOH	290(4.30),358(4.36)	18-1299-86
$C_{32}H_{29}NO_6$			
7H-Furo[3,2-g][1]benzopyran-7-one, 3,5-dimethyl-2-[4-[2-(4-morpholinyl)ethoxy]benzoyl]-6-phenyl-	$CHCl_3$	288(4.33),355(4.91)	18-1299-86
$C_{32}H_{30}$			
1,1'-Biphenyl, 4,4'-bis[2-(2,5-dimethylphenyl)ethenyl]-	EtOH	348(4.83)	61-0439-86
$C_{32}H_{30}N_2$			
Benzeneacetonitrile, α-(octahydro-1,10-diphenyl-1H-1,8-ethanocyclopropa[i]quinolin-9-ylidene)-	MeCN	260(4.06),267(4.05), 274(3.99)	118-0908-86
$C_{32}H_{30}N_2O_4$			
2,5-Cyclohexadiene-1,4-dione, 2,5-dihydroxy-3,6-bis[5-(3-methyl-2-buten-yl)-1H-indol-3-yl]- (cochliodinol)	EtOH	214(4.77),280(4.51), 470(3.66)	5-1765-86
$C_{32}H_{30}O_2$			
3,3'-Bidibenzofuran, 9,9'-bis(2-methylpropyl)-	dioxan	336(4.72)	46-2666-86
3,5,8,10,16,18,21,23-Hexacosaoctaene-1,12,14,15-tetrayne-7,20-dione, 3,8,11,16,19,24-hexamethyl-, (E,E,E,E,Z,Z,Z,Z)-	THF	246s(4.38),258s(4.43), 284(4.43),302s(4.44), 353(4.55),385(4.54), 420s(4.30)	18-2209-86
$C_{32}H_{30}O_{10}$			
[2,2'-Binaphthalene]-1,1',5,5'-tetrol, 4,4'-dimethoxy-7,7'-dimethyl-, tetraacetate	EtOH	248(4.70),262s(4.49), 295s(4.11),305(4.11), 318s(4.03),335(3.96), 360s(2.94)	39-0675-86C
$C_{32}H_{31}NO_2S$			
2,4-Pentanedione, 3-[6,7,8,8a-tetrahydro-8a-[4-(methylthio)phenyl]-1,2-diphenyl-3(3H)-indolizinylidene]-	MeCN	270(4.32)	118-0908-86
$C_{32}H_{31}NO_5$			
7H-Furo[3,2-g][1]benzopyran-7-one,	$CHCl_3$	290(4.30),355(4.36)	18-1299-86

Compound	Solvent	λ_{max}(log ε)	Ref.
2-[4-[2-(diethylamino)ethoxy]-benzoyl]-3,5-dimethyl-6-phenyl- (cont.)			18-1299-86
$C_{32}H_{32}N_2O_9$			
Butanedioic acid, mono[2-[[3,4-dihydro-2,4-dioxo-5-[[3-[[4-(phenylmethoxy)-phenyl]methoxy]phenyl]methyl]-1(2H)-pyrimidinyl]methoxy]ethyl] ester	pH 11 in EtOH	267(4.04)	4-1651-86
sodium salt	EtOH	268(4.05)	4-1651-86
$C_{32}H_{32}N_2O_{12}$			
4',5"-Anhydro-3'-deamino-3'-(3"-cyano-5"-hydroxy-4"-morpholinyl)doxorubicin	MeOH	233(4.61),252(4.44), 289(3.98),478(4.11), 495(4.11),530(3.85)	87-1225-86
$C_{32}H_{32}N_3O_6P$			
2-Butenedioic acid, 2-[1,2,3,4-tetra-hydro-1,3-dimethyl-2,4-dioxo-6-[(triphenylphosphoranylidene)-amino]-5-pyrimidinyl]-, diethyl ester, (E)-	$CHCl_3$	348(3.52)	44-0149-86
3,4-Pyridinedicarboxylic acid, 1,2,5,6-tetrahydro-1-methyl-5-[(methylamino)-[(triphenylphosphoranylidene)amino]-methylene]-2,6-dioxo-, diethyl ester	$CHCl_3$	353(4.56)	44-0149-86
$C_{32}H_{32}N_4O_6S$			
Methanimidic acid, N-[3,4-dicyano-1-[2-deoxy-3,5-bis-O-(4-methyl-benzoyl)-β-D-erythro-pentofurano-syl]-5-(ethylthio)-1H-pyrrol-2-yl]-, ethyl ester	EtOH	236(4.52),281(4.05)	78-5869-86
$C_{32}H_{32}O$			
1-Pentene, 1-methoxy-1,4,5,5-tetra-phenyl-3,3-dimethyl-, (E)-	EtOH	250(3.70),270(3.42), 290(3.04)	44-4604-86
(Z)-	EtOH	260(3.95),270(3.85)	44-4604-86
4-Pentene, 1-methoxy-1,1,5,5-tetra-phenyl-3,3-dimethyl-	EtOH	255(4.35),270(3.99), 290(3.85)	44-4604-86
$C_{32}H_{33}N_5O_5$			
Adenoside, 5'-O-[bis(4-methoxyphenyl)-phenylmethyl]-2'-deoxy-N-methyl-	MeOH	236(4.42),271(4.16)	33-1034-86
$C_{32}H_{34}$			
p-Quaterphenyl, 3,3"'-bis(1,1-dimeth-ylethyl)-	EtOH	297(4.67)	46-2661-86
$C_{32}H_{34}N_4S_2$			
3-Pyridinecarbonitrile, 2,2'-dithio-bis[6-tricyclo[3.3.1.1^{3,7}]dec-1-yl-	EtOH	218(4.56),257(4.28), 298(4.05)	70-0131-86
$C_{32}H_{34}N_4Se_2$			
3-Pyridinecarbonitrile, 2,2'-diseleno-bis[6-tricyclo[3.3.1.1^{3,7}]dec-1-yl-	EtOH	224(4.34),266(4.08), 303(3.77)	70-0376-86
$C_{32}H_{34}O_7$			
Meroliminol 1-naphthoate	n.s.g.	238(4.71)	138-1337-86
$C_{32}H_{34}O_{15}$			
2-Naphthacenecarboxylic acid, 3-[(6-	MeOH	242s(4.41),285(4.61),	158-0856-86

Compound	Solvent	λ_{max}(log ϵ)	Ref.
deoxy-2,3,4-tri-O-methyl-α-L-manno-pyranosyl)oxy]-6,6a,7,10,10a,11-hexahydro-6a,12-dihydroxy-8,10a-dimethoxy-1-methyl-6,7,10,11-tetra-oxo-, methyl ester, (6aR-cis)-, (elloramycin E) (cont.)		388(4.06),405(4.08)	158-0856-86
	MeOH-NaOH	256(4.58),312s(3.90), 440(4.09)	158-0856-86
$C_{32}H_{35}N_5O_2$			
1H-Pyrazolium, 4-[(2,3-dihydro-1,5-di-methyl-3-oxo-2-phenyl-1H-pyrazol-4-yl)[4-(trimethylammonio)phenyl]-methylene]-4,5-dihydro-2,3-dimethyl-5-oxo-1-phenyl-, diperchlorate	MeCN	495(3.69)	104-0727-86
$C_{32}H_{36}N_2O_7$			
α-D-Allofuranose, 3-C-[amino[(triphen-ylmethoxy)imino]methyl]-1,2:5,6-bis-O-(1-methylethylidene)-, (Z)-	EtOH	203(4.44),290(3.00)	159-0631-86
α-D-Glucofuranose, 3-C-[amino[(tri-phenylmethoxy)imino]methyl]-1,2:5,6-bis-O-(1-methylethylidene)-, (Z)-	EtOH	203(4.75)	159-0631-86
$C_{32}H_{36}O_4$			
1,3-Azulenedicarboxylic acid, 2-octyl-6-(phenylethynyl)-, diethyl ester	CH_2Cl_2	242.5(4.53),347.5(4.77), 368.5(4.51),399.5(4.55), 532(2.77)	24-2272-86
$C_{32}H_{36}O_{10}$			
2-Oxabicyclo[2.2.1]heptane-1-carbox-ylic acid, 4,7,7-trimethyl-3-oxo-, 2-(2,4,6,10b-tetrahydro-7-hydroxy-9,10-dimethoxy-3,10b-dimethyl-4,6-dioxo-1H-benz[4,5]indeno[7,1-bc]-pyran-8-yl)phenyl ester	ether	250s(3.72),288s(3.76), 294s(3.76),299s(3.78), 303.7(3.78),312.1(3.80), 318(3.79),348.4(4.02), 350.8(4.02),358(4.02), 390s(3.95)	33-1513-86
$C_{32}H_{36}O_{16}$			
Elloramycin F	MeOH	286(4.31),412(3.91), 439(3.78)	158-0856-86
	MeOH-NaOH	254(4.32),448(3.72), 510s(3.45)	158-0856-86
$C_{32}H_{38}N_4$			
21,22,23,24-Tetraazapentacyclo[16.2-$1.1^{2,5}.1^{8,11}.1^{12,15}$]tetracosa-1,3,5,7,9,11(23),12,14,16,18(21),19-undecaene, 4,9,14,19-tetrapropyl-	toluene	372(5.14),565(4.59), 604(4.54),637(4.70)	149-0555-86B
$C_{32}H_{38}N_6O_{10}$			
[1,4,7,10,13]Pentaoxacyclopentadec-ino[2,3-g]quinoxaline, 2,2]-azo-bis[7,8,10,11,13,14,16,17-octa-hydro-, (E)-	$CHCl_3$	240(4.12),286(4.32), 424(4.44)	24-3870-86
$C_{32}H_{38}O_{12}$			
Deacetylisonimbinolide diacetate	MeOH	210(3.71)	100-1068-86
Deacetylnimbinolide diacetate	MeOH	210(3.67)	100-1068-86
$C_{32}H_{38}O_{13}$			
Lustromycin	MeCN	277(3.79),350(3.67)	158-1205-86

Compound	Solvent	$\lambda_{max}(\log \epsilon)$	Ref.
$C_{32}H_{38}O_{17}$			
Sucrose, 3,6'-diferuloyl-	EtOH	237(4.21),300s(4.40), 327(4.58)	102-2897-86
$C_{32}H_{40}O_{8}S_{8}$			
Cyclobuta[1,2-d:3,4-d']bis[1,3]dithiole-3a,6a(3bH,6bH)-dicarboxylic acid, 2,5-bis[4-(butoxycarbonyl)-1,3-dithiol-2-ylidene]-, dibutyl ester	dioxan	302s(4.51),314(4.59), 390(3.97)	104-0367-86
4,7,12,15,17,18,19,20-Octathiaheptacyclo[12.2.1.1^{2,5}.1^{6,9}.1^{10,13}.0^{3,6}-0^{8,11}]eicosa-5,13-diene-1,3,8,10-tetracarboxylic acid, tetrabutyl ester	dioxan	240s(4.21),275(4.47), 370(2.13)	104-0367-86
$C_{32}H_{40}O_{9}$			
Isonimocinolide diacetate	MeOH	233(3.57)	39-1021-86C
Nimocinolide diacetate	MeOH	225(3.62)	39-1021-86C
$C_{32}H_{40}O_{12}$			
Bufa-20,22-dienolide, 2,3-[(4-acetoxytetrahydro-3-hydroxy-6-methyl-2H-pyran-3,2-diyl]bis(oxy)]-7,8-epoxy-11,14-dihydroxy-12-oxo-, (tyledoside C)	MeOH	298(3.69)	39-0429-86C
$C_{32}H_{41}N_{5}O_{4}$			
Discarin I	MeOH	227.5($\underline{3.9}$),270($\underline{3.6}$), 278($\underline{3.6}$),289($\underline{3.5}$)	64-1180-86B
$C_{32}H_{43}BrN_{2}O_{10}$			
Geldanamycin, 22-acetyl-19-bromo-17-demethoxy-15-methoxy-11-O-methyl-, (15R)- (N-acetyl-19-bromoherbimycin A)	MeOH	272(4.26)	158-0415-86
$C_{32}H_{43}BrN_{2}O_{11}$			
Geldanamycin, 22-acetyl-19-bromo-17-demethoxy-8,9-epoxy-8,9-dihydro-15-methoxy-11-O-methyl-, (15R)-	MeOH	260(4.22)	158-0415-86
$C_{32}H_{44}F_{3}N_{3}O_{2}S$			
Fluphenazine decanoate	MeOH	261(4.538),313(3.589)	106-0571-86
$C_{32}H_{44}N_{2}O_{10}$			
Herbimycin A, N-acetyl-	MeOH	269(4.32)	158-0415-86
$C_{32}H_{44}O_{5}$			
Mibemycin β_3, 5-O-methyl-	EtOH	246(4.38)	44-4840-86
$C_{32}H_{44}O_{8}$			
Cucurbitacin E	EtOH	231(4.09),267(3.92)	163-2831-86
$C_{32}H_{46}N_{2}O_{2}$			
18α-Olean-2-eno[2,3-b]pyrazin-28-oic acid, 19β-hydroxy-, γ-lactone	C_6H_{12}	270(3.94),276(3.89), 315(2.96),321(3.01)	73-0118-86
$C_{32}H_{46}O_{3}Se$			
Benzoic acid, 4-[[2,4,6-tris(1-methylethyl)phenyl]seleninyl]-, 5-methyl-2-(1-methylethyl)cyclohexyl ester, (-)-	MeOH	220(4.35),240(4.35), 275(4.01)	138-0161-86

Compound	Solvent	$\lambda_{max}(\log \epsilon)$	Ref.
$C_{32}H_{46}O_5$			
Abiesonic acid, dimethyl ester	EtOH	238(2.54)	105-0548-86
$C_{32}H_{46}O_6$			
Benzoic acid, 2-[9-[4-hydroxy-8,9-dimethyl-1,7-dioxaspiro[5.5]undec-2-yl]-1,5,7-trimethyl-1,3,7-nonatrienyl]-4-methoxy-5-methyl-, [2R-[2α(1E,3E,5R*,7E),4β,6β(8R*,9S*)]]-	EtOH	242(4.38),270s(4.11)	44-4840-86
$C_{32}H_{46}O_7$			
2,10,12,16,18-Nonadecapentaenoic acid, 19-(3,6-dihydro-3-methyl-6-oxo-2H-pyran-2-yl)-6-hydroxy-9-(hydroxymethyl)-3,5,7,11,15,17-hexamethyl-8-oxo- (PD-124,895)	MeOH	234(4.64)	158-1651-86
$C_{32}H_{46}O_8$			
19-Norlanosta-5,23-diene-3,11,22-trione, 25-acetoxy-2,16,20-trihydroxy-9-methyl-, (2β,9β,10α,16α,23E)-	EtOH	228(4.04)	163-2831-86
$C_{32}H_{46}O_9$			
Lanost-8-en-26-oic acid, 12-acetoxy-3,7-dihydroxy-11,15,23-trioxo-, (3β,7β,12β)- (ganoderic acid K)	EtOH	251.5(4.14)	94-3025-86
$C_{32}H_{48}N_2O$			
18α-Olean-1-eno[1,2-b]pyrazine, 19β,28-epoxy-	C_6H_{12}	268(3.97),273(3.89), 313(2.88)	73-0118-86
18α-Olean-2-eno[2,3-b]pyrazine, 19β,28-epoxy-	C_6H_{12}	270(3.94),277(3.89), 315(2.97),320(3.01), 326(2.92)	73-0118-86
$C_{32}H_{48}N_2O_2$			
18α-Olean-2-eno[2,3-b]pyrazine, 19β,28-epoxy-, 1'-oxide	C_6H_{12}	229(4.53),234(4.54), 272(4.20)	73-0118-86
$C_{32}H_{48}O_4$			
1,3-Azulenedicarboxylic acid, 2,6-dioctyl-, diethyl ester	CH_2Cl_2	236.5(4.48),275.5(4.34), 305(4.67),314.5(4.81), 351(4.00),376(3.73), 487.5(2.75)	24-2272-86
$C_{32}H_{48}O_5$			
Ganoderic acid S	n.s.g.	231(4.2),240(4.2), 248(4.0)	94-2282-86
$C_{32}H_{48}O_8$			
1,2,3,4,13,14,15,16-Cyclotetracosaneoctone, 5,5,12,12,17,17,24,24-octamethyl-	CH_2Cl_2	380(1.98),522(2.00)	89-0449-86
$C_{32}H_{50}O_2$			
Benzene, 1,1'-[1,4-butanediylbis(oxy)]-bis[2,6-bis(1,1-dimethylethyl)-	EtOH	220(4.23),268(2.81)	104-1079-86
$C_{32}H_{50}O_4$			
Ganodermanondiol monoacetate	MeOH	237(3.91),243(3.97), 251(3.81)	100-1122-86

Compound	Solvent	$\lambda_{max}(\log \epsilon)$	Ref.
$C_{32}H_{52}O$			
19-Nor-9,10-secoergosta-3,5,7,22-tetraen-1-ol, 4-pentyl-, (1α,5E,7E,22E)-	EtOH	278(4.55),288(4.64), 301(4.51)	88-4903-86
$C_{32}H_{52}S_2$			
1,3-Dithiane, 2-cholesta-3,5-dien-3-yl-2-methyl-	EtOH	240(4.33),247(4.36), 256s(4.16)	39-0261-86C
$C_{32}H_{54}N_2O$			
2H-Pyrrol-2-one, 5-[(3,5-dimethyl-4-octadecylpyrrol-2-yl)methylene]-4-ethyl-1,5-dihydro-3-methyl-	$CHCl_3$	409(4.51)	49-0849-86
$C_{32}H_{54}O_4$			
2-Propenoic acid, 3-(4-hydroxy-3-methoxyphenyl)-, docosyl ester, (E)-	EtOH	218(4.15),236(4.09), 299s(4.15),326(4.28)	64-0164-86C
	EtOH-KOH	253(--),310(--), 376(4.47)	64-0164-86C
$C_{32}H_{60}O_2Si_2$			
4,6-Dioxa-3,8-disiladecane, 5-(2,6-dimethyl-1,5-heptadienyl)-7-(2,6-dimethyl-2,5-heptadienylidene)-2,2,3,3,8,8,9,9-octamethyl-	pentane	233s(3.88)	33-1378-86
$C_{32}H_{88}Si_{12}$			
Cyclotetrasilane, octakis[(trimethylsilyl)methyl]-	C_6H_{12}	300s(3.28)	138-1643-86

Compound	Solvent	λ_{max}(log ϵ)	Ref.
$C_{33}H_{21}NO_2$			
1H-Perylo[1,12-efg]isoindole-1,3(2H)-dione, 2-(2,4,6-trimethylphenyl)-	$CHCl_3$	421(3.28),449(3.61), 478(3.79)	24-2373-86
$C_{33}H_{22}N_2OS$			
Pyrrolo[3,4-a]phenothiazin-4(2H)-one, 2-methyl-1,3,5-triphenyl-	MeOH	259(4.54),329(4.01), 377s(3.80),487(4.08)	4-1267-86
$C_{33}H_{23}Cl_2NO$			
Ethanone, 2,2-bis(4-chlorophenyl)-2-[(diphenylmethylene)amino]-1-phenyl-	EtOH	207(4.69),235(4.36), 251(4.31)	39-0799-86C
Methanone, [2-[[[bis(4-chlorophenyl)-methylene]amino]phenylmethyl]phen-yl]phenyl-	CH_2Cl_2	231(4.73),251(4.67)	39-0799-86C
Methanone, [2-[[[bis(4-chlorophenyl)-methyl]imino]phenylmethyl]-phenyl]phenyl-	CH_2Cl_2	256(4.69)	39-0799-86C
$C_{33}H_{23}N_3$			
Propanedinitrile, (5,6-dihydro-11-methyl-9,10-diphenyl-8H-naphtho-[2,1-a]quinolizin-8-ylidene)-	MeCN	316(4.27),332(4.23), 432(4.15)	118-0908-86
$C_{33}H_{24}N_4$			
2,3':2',3":2",3"'-Quater-1H-indole, 2"'-methyl-	EtOH	295(3.94)	78-5019-86
$C_{33}H_{24}O$			
2-Propen-1-one, 3-phenyl-1-(5'-phen-yl[1,1':3',1"-terphenyl]-2'-yl)-	n.s.g.	247(4.68),299(4.42)	48-0573-86
$C_{33}H_{24}O_{12}$			
1H-Anthra[2,3-c]pyran-8-acetic acid, 12-hydroxy-9-[(4-hydroxy-2H-1-benzo-pyran-3-yl)carbonyl]-10,11-dimeth-oxy-3-(methoxycarbonyl)-1-oxo-	EtOH-1% DMF	296(4.78),328(4.76), 425(4.25)	78-0727-86
$C_{23}H_{25}Br_3N_4O_7$			
Benzoic acid, 4-bromo-, 5'-ester with 4-bromo-N-(4-bromobenzoyl)-N-[4-cya-no-1-[2,3-O-(1-methylethylidene)-β-D-ribofuranosyl]-1H-imidazol-5-yl]-benzamide	MeOH	248(4.55)	23-2109-86
$C_{33}H_{25}NO$			
Ethanone, 2-[(diphenylmethylene)amino]-1,2,2-triphenyl-	EtOH	222(4.12),252(4.12)	39-0799-86C +39-2021-86C
Methanone, [2-[[(diphenylmethylene)-amino]phenylmethyl]phenyl]phenyl-	EtOH	208(4.59),248(4.39)	39-0799-86C
Methanone, [2-[[(diphenylmethyl)-imino]phenylmethyl]phenyl]phenyl-	EtOH	210(4.64),249(4.41)	39-0799-86C
$C_{33}H_{25}N_3$			
Propanedinitrile, (5,6,9,10-tetrahy-dro-11-methyl-9,10-diphenyl-8H-naphtho[2,1-a]quinolizin-8-ylidene)-	MeCN	263(3.84),294(4.10), 330(3.85)	118-0908-86
$C_{33}H_{25}N_3O_2$			
9,10-Anthracenedione, 4-[(4-methyl-phenyl)amino]-1,5-bis(phenylamino)-	EtOH	635(4.30),675(4.36)	104-0547-86

Compound	Solvent	$\lambda_{max}(\log \epsilon)$	Ref.
$C_{33}H_{25}N_3O_3$			
9,10-Anthracenedione, 4-[(4-methoxyphenyl)amino]-1,5-bis(phenylamino)-	EtOH	635(4.27),675(4.33)	104-0547-86
$C_{33}H_{25}N_4S_2$			
Thiazolo[3,4-b][1,2,4]triazin-5-ium, 6-[(3-ethyl-2(3H)-benzothiazolylidene)methyl]-2,3,8-triphenyl-, perchlorate	n.s.g.	454(4.71),555(3.87)	103-1373-86
$C_{33}H_{25}N_5O_5S$			
Morpholine, 1-[[2-[2,3-dihydro-2-(4-methylphenyl)-1,3-dioxo-1H-benz[de]isoquinolin-6-yl]-2H-naphtho[1,2-d]triazol-4-yl]sulfonyl]-	DMF	386(4.50)	2-0496-86
$C_{33}H_{26}O_6$			
4H-1-Benzopyran-4-one, 7,8-diacetoxy-6-(diphenylmethyl)-2-methyl-3-phenyl-	MeOH	262(4.1),298(4.39)	2-0649-86
$C_{33}H_{26}O_9$			
Benzo[e]naphtho[2',3':5,6]fluoreno-[1,9a-b]oxepin-5,10,15,19-tetrone, 5c,8,8a,16-tetrahydro-8-hydroxy-1,11,18-trimethoxy-13-methyl-	EtOH	280s(3.99),316(3.60), 408(3.77)	39-0255-86C
$C_{33}H_{26}O_{12}$			
1,3,5-Cyclohexanetrione, 2-hydroxy-4-[(3-hydroxy-5-[3-(4-hydroxyphenyl)-1-oxo-2-propenyl]-3-methyl-2,4,6-trioxocyclohexylidene]methyl]-6-[3-(4-hydroxyphenyl)-1-oxo-2-propenyl]-2-methyl-	EtOH	385(4.23),524(4.47)	138-0495-86
$C_{33}H_{27}N_3O_9$			
1H-Pyrazole-4-carboxylic acid, 3-(cyanomethyl)-1-(2,3,5-tri-O-benzoyl-β-D-ribofuranosyl)-, methyl ester	MeOH	227(4.66)	4-0059-86
$C_{33}H_{28}N_2O_4$			
1,2-Dehydro-2-nortelobine	MeOH	220(4.45),265(4.06), 294(3.73),335(3.57)	164-0663-86
$C_{33}H_{28}N_2O_{13}S_2$			
Emestrine triacetate	MeOH	257(4.21),263s(4.21), 288s(3.77),330(3.20)	39-0109-86C
$C_{33}H_{28}N_4O_9$			
4H-Pyrazolo[3,4-d]pyrimidin-4-one, 3-ethoxy-1,5-dihydro-1-(2,3,5-tri-O-benzoyl-β-D-ribofuranosyl)-	pH 1 and 7	230(4.35),273s(3.88)	4-1869-86
	pH 11	230(4.39),273s(3.79)	4-1869-86
$C_{33}H_{28}O_4$			
2-Propen-1-one, 1,1'-(methylenedi-4,1-phenylene)bis[3-(4-methoxyphenyl)-	93% H_2SO_4	334(3.24),442(4.49), 480(4.99)	104-0055-86
$C_{33}H_{29}N_5O_9S$			
1H-Pyrazino[2,3-c][1,2,6]thiadiazin-4-amine, 6,7-dimethyl-1-(2,3,5-tri-O-benzoyl-β-D-ribofuranosyl)-, 2,2-dioxide	MeOH	230(4.94),256s(4.49), 281s(4.08),338(4.13)	5-1872-86

Compound	Solvent	λ_{max}(log ϵ)	Ref.
$C_{33}H_{29}O_3$			
Pyrylium, 2-[8-(3,4-dimethoxyphenyl)-1,3,5,7-octatetraenyl]-4,6-diphenyl-, tetrafluoroborate	$CHCl_3$	380(4.89),675(3.78)	104-1786-86
$C_{33}H_{30}BrN_5O_4$			
7H-Pyrrolo[2,3-d]pyrimidine-5-carbonitrile, 4-amino-6-bromo-7-[2,3,5-tris-O-(phenylmethyl)-β-D-arabinofuranosyl]-	EtOH	288(4.29)	78-5869-86
$C_{33}H_{30}N_2O_{11}$			
2-Anthraceneacetic acid, 6-carboxy-5-hydroxy-3-[3-hydroxy-3-(2-hydroxyphenyl)-1-oxo-2-propenyl]-4,10-dimethoxy-7-[3-(methylamino)-2-(methylimino)-3-oxopropyl]-	EtOH-1% DMF	277(4.71),297(4.70), 355(4.20),385(4.14)	78-0727-86
$C_{33}H_{30}N_3O_3P$			
3a(7aH)-Benzofurancarboxylic acid, 6,7-dicyano-2,3-dihydro-2,2-dimethyl-4-[(triphenylphosphoranylidene)amino]-, ethyl ester, cis	$CHCl_3$	266(3.23),274(3.20), 438(4.36)	24-2723-86
$C_{33}H_{30}O_4$			
1,3-Azulenedicarboxylic acid, 2,6-bis[2-(4-methylphenyl)ethenyl]-, ethyl methyl ester, (E,E)-	CH_2Cl_2	231.5(4.52),354.5(4.66), 453.5(4.71)	24-2272-86
$C_{33}H_{30}O_{11}$			
[2,2'-Binaphthalene]-1',4,4',5',6-pentol, 5-methoxy-2,7'-dimethyl-, pentaacetate	MeOH	230(4.90),292(4.12), 315s(3.78),330(3.64)	39-0675-86C
[2,2'-Binaphthalene]-1,1',4,5,5'-pentol, 4'-methoxy-7,7'-dimethyl-, pentaacetate	EtOH	251(4.63),279s(4.08), 291(4.05),302(3.98), 332(3.68)	39-0675-86C
$C_{33}H_{31}NO$			
Benzonitrile, 4-[1-(diphenylmethyl)-4-methoxy-2,2-dimethyl-4-phenyl-3-butenyl]-, (E)-	EtOH	250(4.18),270(3.90), 290(3.38)	44-4604-86
(Z)-	EtOH	250(4.17),270(3.88), 290(3.32)	44-4604-86
Benzonitrile, 4-(1-methoxy-3,3-dimethyl-1,5,5-triphenyl-4-pentenyl)-	EtOH	255(4.31),290(3.85)	44-4604-86
$C_{33}H_{31}NO_2$			
1,3-Cyclohexanedione, 5,5-dimethyl-2-(2,3,6,7-tetrahydro-2,3-diphenyl-4H-benzo[a]quinolizin-4-ylidene)-	MeCN	265(4.44),315(4.10), 420(3.40)	118-0908-86
$C_{33}H_{31}NO_5$			
7H-Furo[3,2-g][1]benzopyran-7-one, 3,5-dimethyl-6-phenyl-2-[4-[2-(1-piperidinyl)ethoxy]benzoyl]-	$CHCl_3$	290(4.42),355(4.47)	18-1299-86
$C_{33}H_{31}N_5O_4$			
7H-Pyrrolo[2,3-d]pyrimidine-5-carbonitrile, 4-amino-7-[2,3,5-tris-O-(phenylmethyl)-β-D-arabinofuranosyl]-	EtOH	231(4.03),280(4.21)	78-5869-86

Compound	Solvent	$\lambda_{max}(\log \epsilon)$	Ref.
$C_{33}H_{32}$			
Benzene, 1,1'-[3-(2,2-diphenylethenyl)-2-methyl-2-(1-methylethyl)cycloprop-ylidene]bis-, cis	EtOH	222s(4.36),270(4.26)	44-0196-86
trans	EtOH	220s(4.35),267(4.23)	44-0196-86
Benzene, 1,1',1",1"'-[3-methyl-3-(1-methylethyl)-1,4-pentadiene-1,5-diylidene)tetrakis-	EtOH	250(4.38),289(3.36), 337(<0)	44-0196-86
Benzene, 1,1',1",1"'-(2,3,3,4-tetra-methyl-1,4-pentadiene-1,5-diyli-dene)tetrakis-	EtOH	228s(4.32),268s(3.64), 289(2.34),337(<0)	44-0196-86
$C_{33}H_{32}FeNO_5$			
Pyridinium, 3,5-bis(ethoxycarbonyl)-1-(4-ferrocenylphenyl)-4-(2-furan-yl)-2,6-dimethyl-, perchlorate	MeCN	207(4.6),248(4.1), 287(3.9),357(4.4), 472(3.8)	103-0886-86
$C_{33}H_{32}NO$			
3H-Indolium, 2-[3-[4,6-bis(4-methyl-phenyl)-2H-pyran-2-ylidene]-1-pro-penyl]-1,3,3-trimethyl-, tetra-fluoroborate	CH_2Cl_2	665(4.76),700(4.71)	103-1295-86
$C_{33}H_{32}NO_3$			
3H-Indolium, 2-[3-[4,6-bis(4-methoxy-phenyl)-2H-pyran-2-ylidene]-1-prop-enyl]-1,3,3-trimethyl-, tetra-fluoroborate	CH_2Cl_2	664(4.78),700(4.79)	103-1295-86
$C_{33}H_{32}N_4O_2$			
$13^2,17^3$-Cyclopheophorbide enol	n.s.g.	287(3.63),359(3.68), 426(3.68),452(3.67), 626(3.60),686(3.65)	88-2177-86
$C_{33}H_{33}FeNO_5$			
Ferrocene, [4-[3,5-bis(ethoxycarbonyl)-4-(2-furanyl)-2,6-dimethyl-1(4H)-pyridinyl]phenyl]-	MeCN	212(4.54),247(4.4), 286(4.27),345(4.1), 468(3.11)	103-0886-86
$C_{33}H_{33}N_5O_5$			
7H-Pyrrolo[2,3-d]pyrimidine-5-carbox-amide, 4-amino-7-[2,3,5-tris-O-(phenylmethyl)-β-D-arabino-furanosyl]-	pH 1	269(3.70)	78-5869-86
	pH 7	269(3.71)	78-5869-86
	pH 11	272(3.67)	78-5869-86
$C_{33}H_{34}Cl_2N_4O_4$			
21H,23H-Porphine-2-propanoic acid, 18-carboxy-8,13-bis(2-chloroethyl)-3,7,12,17-tetramethyl-, α-methyl ester	CH_2Cl_2	404(5.28),506(4.04), 544(4.07),574(3.85), 634(3.30)	44-0657-86
$C_{33}H_{34}O_2$			
1-Pentene, 1-methoxy-1-(4-methoxy-phenyl)-3,3-dimethyl-4,4,5-tri-phenyl-, (E)-	EtOH	250(4.02),270(3.84)	44-4604-86
(Z)-	EtOH	250(4.08),260(3.98), 280(3.84)	44-4604-86
1-Pentene, 1-methoxy-4-(4-methoxy-phenyl)-3,3-dimethyl-1,5,5-tri-phenyl-, (E)-	EtOH	250(3.79),270(3.72), 290(3.28)	44-4604-86
(Z)-	EtOH	250(3.98),270(3.91), 290(3.63)	44-4604-86

Compound	Solvent	λ_{max}(log ε)	Ref.
$C_{33}H_{35}ClN_2O_6S$			
Aspidospermidine-2,3-diol, 1-[(4-methoxyphenyl)sulfonyl]-, 3-(3-chlorobenzoate), (3α)-(±)-	MeCN	230(4.24),254(3.93)	78-3215-86
$C_{33}H_{35}NO_4$			
1,3-Azulenedicarboxylic acid, 6-[(4-cyanophenyl)ethynyl]-2-octyl-, diethyl ester	CH_2Cl_2	234.5(4.42),344(4.84), 372(4.52),400(4.60), 540.5(2.74)	24-2272-86
$C_{33}H_{35}N_3OS_2$			
2-Propenethioamide, 3-(3-benzoyl-5-phenyl-4-isothiazolyl)-2-(diethylamino)-N,N-diethyl-3-phenyl- (87%) (13% isomer 12b)	MeOH	250(4.51),353(3.97)	5-1796-86
$C_{33}H_{36}N_4O_6$			
Bilirubin photoisomer	MeOH-NH_3	437(4.52)	77-0249-86
$C_{33}H_{36}N_4O_8$			
1H-Purine-2,6-dione, 9-[2-O-benzoyl-6-O-[6-(benzoyloxy)hexyl]-3,4-dideoxy-β-D-erythro-hex-3-enopyranosyl]-3,9-dihydro-1,3-dimethyl-	$CHCl_3$	277(3.96)	136-0256-86K
$C_{33}H_{36}O_{10}$			
Sigmoidin A tetraacetate	MeOH	260(4.04)	39-0033-86C
$C_{33}H_{37}NO_4$			
1,3-Azulenedicarboxylic acid, 6-[2-(4-cyanophenyl)ethenyl]-2-octyl-, diethyl ester, (E)-	CH_2Cl_2	245.5(4.67),352.5(4.97), 399.5(4.78),414(4.70), 539.5(3.04)	24-2272-86
(Z)-	CH_2Cl_2	237(4.43),268(4.29), 273.5(4.29),338(4.55), 368(4.28),400(4.11), 528.5(2.55)	24-2272-86
$C_{33}H_{38}N_4O_6$			
21H-Biline-8,12-diacetic acid, 3,18-diethyl-1,19,22,24-tetrahydro-2,7,13,17-tetramethyl-1,19-dioxo-, dimethyl ester	$CHCl_3$	360(4.37),640(3.71)	78-4137-86
	acid	366(4.44),673(4.28)	78-4137-86
$C_{33}H_{38}N_4O_{10}$			
1H-Purine-2,6-dione, 9-[2-O-benzoyl-6-O-[6-(benzoyloxy)hexyl]-β-D-galactopyranosyl]-3,9-dihydro-1,3-dimethyl-	$CHCl_3$	277(3.90)	136-0256-86K
$C_{33}H_{39}NO_{10}$			
10,25b-Etheno-1,5-methano-25bH-benzofuro[3,2-e][1,4,7,10,13]benzopentaoxacyclopentadecino[15,16-h]isoquinoline-11,25-diol, 1,2,3,4,9a,10,13-14,16,17,19,20,22,23-tetradecahydro-8,10-dimethoxy-2-methyl-, [1R-(1α,4aS*,5α,9aα,10β,25bβ)]-	MeCN	308(3.83)	77-0268-86
$C_{33}H_{39}P$			
Phosphine, (3,3-diphenyl-1,2-propadienylidene)[2,4,6-tris(1,1-dimethyl-	hexane	238(4.43),268(4.30), 305(3.85),396(4.29),	89-1003-86

Compound	Solvent	$\lambda_{max}(\log \epsilon)$	Ref.
ethyl)phenyl]- (cont.)		412(4.31)	89-1003-86
$C_{33}H_{40}O_{19}$			
Robinin	EtOH	268(4.30),356(4.21)	105-0601-86
$C_{33}H_{40}O_{22}$			
Patuletin 3-O-β-D-glucopyranosyl-(1→6)-[β-D-apiofuranosyl-(1→2)]-β-D-glucopyranoside	MeOH	256(4.25),270s(4.19),294(3.92),351(4.28)	102-0231-86
	MeOH-NaOMe	269(4.32),333(3.92),400(4.36)	102-0231-86
	MeOH-NaOAc	270(4.33),330(4.03),380(4.23)	102-0231-86
	+ H_3BO_3	261(4.34),374(4.28)	102-0231-86
	MeOH-$AlCl_3$	268(4.37),310s(3.86),348(3.80),438(4.38)	102-0231-86
	+ HCl	270(4.26),300s(3.93),385(4.23),410s(4.20)	102-0231-86
$C_{33}H_{44}CrO_5Si_2$			
Chromium, tricarbonyl[[(1,10,11,12,13-14-η)-5,15-dibutoxytricyclo[8.2.2-$2^{4,7}$]hexadeca-2,4,6,8,10,12,13,15-octaene-2,9-diyl]bis[(trimethylsilane]]-	CH_2Cl_2	338(3.99)	88-2353-86
$C_{33}H_{44}O_9$			
Lanost-8-en-26-oic acid, 12-acetoxy-3,7,11,15,23-pentaoxo-, methyl ester, (12β)- (methyl ganoderate F)	EtOH	251.5(3.91)	94-4018-86
$C_{33}H_{44}O_{10}$			
Pseudopterosin C triacetate	MeOH	229(4.15),269(3.11)	44-5140-86
$C_{33}H_{46}O_8$			
Lanost-8-en-26-oic acid, 7-acetoxy-3,11,15,23-tetraoxo-, methyl ester, (7β,25R)- (methyl O-acetylganoderate C)	EtOH	251(3.9)	102-1189-86
$C_{33}H_{46}O_9$			
Lanost-8-en-26-oic acid, 12-acetoxy-3-hydroxy-7,11,15,23-tetraoxo-, methyl ester, (3β,12β)- (methyl ganoderate H)	EtOH	256(3.87)	94-4018-86
$C_{33}H_{48}CrO_5Si_2$			
Chromium, tricarbonyl[[(4,5,6,7,15,16-η)-11,13-dibutoxytricyclo[8.2.2.$2^{4,7}$]-hexadeca-4,6,10,12,13,15-hexaene-5,15-diyl]bis[trimethylsilane]]-	CH_2Cl_2	340(3.97)	88-2353-86
$C_{33}H_{48}N_2O$			
Benzamide, N-[(3β,5α,20S)-20-(dimethylamino)-4,4,14-trimethyl-B(9a)-homo-19-norpregna-9,11(9a)-dien-3-yl]-, (+)- (buxabenzamidienine)	MeOH	228s(4.37),238(4.46),245(4.49),255s(4.31)	78-5747-86
$C_{33}H_{48}N_2O_3$			
Benzamide, N-[(3β,4β,5α,16α,20S)-20-(dimethylamino)-16-hydroxy-4-(hydroxymethyl)-4,14-dimethyl-B(9a)-	MeOH	228(4.07)	78-5747-86

Compound	Solvent	$\lambda_{max}(\log \epsilon)$	Ref.
homo-19-norpregna-1(10),9(11)-dien-3-yl]-, (-)- (buxanoldine) (cont.)			78-5747-86
$C_{33}H_{48}O_9$			
Card-20(22)-enolide, 3-[[2,6-dideoxy-3,4-O-(1-methylethylidene)-β-D-ribo-hexopyranosyl]oxy]-16-(formyloxy)-14-hydroxy-, (3β,5β,16β)-	MeOH	218(4.21)	87-0997-86
$C_{33}H_{50}N_4O_5$			
Piperazine, 1,1'-[(2α,6β,17β)-17-acetoxy-3-oxoandrost-4-ene-2,6-diyl]-bis[4-acetyl-	EtOH	235(4.10)	107-0195-86
$C_{33}H_{51}NO_5$			
6-Aza-B-homostigmasta-2,4-dien-7-one, 22,23-diacetoxy-, (22S,23S)-	EtOH	273(3.89)	39-2123-86C
$C_{33}H_{52}N_2O_4$			
1H-Pyrrole-3-propanoic acid, 5-[(3-ethyl-1,5-dihydro-4-methyl-5-oxo-2H-pyrrol-2-ylidene)methyl]-2-formyl-4-methyl-, hexadecyl ester, (Z)-	$CHCl_3$	400(4.45),420(4.40)	49-1081-86
$C_{33}H_{54}N_4O_5$			
Androst-4-en-3-one, 17-acetoxy-2,6-bis[4-(2-hydroxyethyl)-1-piperazinyl]-, (2α,6β,17β)-	EtOH	239(4.04)	107-0195-86
$C_{33}H_{54}O_{10}$			
Blechnoside A	EtOH	243(4.02)	102-1301-86
Blechnoside B	EtOH	243(4.03)	102-1301-86
Sileneoside F	EtOH	245(4.0)	105-0297-86

Compound	Solvent	λ_{max}(log ϵ)	Ref.
$C_{34}H_{16}N_4$			
Propanedinitrile, [10-[10-(dicyano-methylene)-9,10-anthracenylidene]-9(10H)-anthracenylidene]-	DMF	423(4.22)	88-2411-86
	DMSO	427(4.12)	88-2411-86
$C_{34}H_{20}N_4O_2$			
Pyrazino[2,3-g]quinoxaline-5,10-dione, 2,3,7,8-tetraphenyl-	MeCN	218s(4.68),223(4.69), 251(4.34),255s(4.34), 262s(4.29),275(4.29), 363(4.48)	138-0715-86
$C_{34}H_{22}N_4O_2S_2$			
Imidazo[5,1-a]isoquinolin-3(2H)-one, 1,1'-dithiobis[2-phenyl-	EtOH	266(4.12),366(3.77), 480(3.46)	150-1901-86M
$C_{34}H_{22}N_4O_8$			
Benzenemethanol, α-[[(diphenylmethyl-ene)amino](4-nitrophenyl)methylene]-4-nitro-, 4-nitrobenzoate	CH_2Cl_2	267(4.46),376(4.14)	39-2021-86C
$C_{34}H_{22}N_8$			
Benzene, 1,3-bis[diazo[3-(diazophen-ylmethyl)phenyl]methyl]-	n.s.g.	249(4.91),518(2.65)	35-2147-86
$C_{34}H_{22}O_9$			
8aH-Benzo[3,4][2]benzopyrano[1,8-bc]-[1]benzopyran-1,4,11-triol, 8a-(2,4-dihydroxyphenyl)-6-(6-hydroxy-2-benzofuranyl)-2-methyl-, (+)-, (mulberrofuran P)	EtOH	210(4.72),284(4.17), 315s(4.33),336(4.53), 351(4.61),370(4.56)	142-1381-86
$C_{34}H_{22}O_{10}$			
Mulberrofuran M	EtOH	208(4.71),225s(4.42), 261(4.30),299s(4.26), 323s(4.44),335(4.53), 350s(4.45)	142-1251-86
$C_{34}H_{23}NO_2$			
1H-Indene-1,3(2H)-dione, 2-(6,7-di-hydro-2,3-diphenyl-4H-benzo[a]-quinolizin-4-ylidene)-	MeCN	243(4.60),321(4.25), 440(4.00)	118-0908-86
$C_{34}H_{23}N_3$			
Propanedinitrile, (6,7-dihydro-1,2,3-triphenyl-4H-benzo[a]quinolizin-4-ylidene)-	MeCN	305(4.05),338(4.09), 430(3.98)	118-0908-86
$C_{34}H_{24}BCl_2O_4$			
Boron(1+), bis[5-(4-chlorophenyl)-1-phenyl-4-pentene-1,3-dionato-O,O']-, (T-4)-, perchlorate	CH_2Cl_2	448(5.07)	97-0399-86
$C_{34}H_{24}BrNO_2$			
Benzoic acid, 3-bromo-, 2-[(diphenyl-methylene)amino]-1,2-diphenylethen-yl ester	CH_2Cl_2	238(4.53),255(4.42), 285(4.34),302(4.24)	39-0623-86C
Methanone, (3-bromophenyl)(2,5-dihy-dro-2,2,4,5-tetraphenyl-5-oxazolyl)-	CH_2Cl_2	233(4.43),251(4.51)	39-0623-86C

Compound	Solvent	λ_{max}(log ϵ)	Ref.
$C_{34}H_{24}ClNO_2$			
Methanone, [4-(4-chlorophenyl)-2,5-dihydro-2,2,5-triphenyl-5-oxazolyl]-phenyl-	CH_2Cl_2	232(4.35),257(4.56)	39-0623-86C
$C_{34}H_{24}N_2O_2$			
[2,2'-Binaphthalene]-1,1'(4H,4'H)-dione, 4,4'-bis[(4-methylphenyl)-imino]-	EtOH	248(4.56),298(4.40), 480(3.95)	104-0748-86
7H-Dibenzo[a,h]carbazole-7,12(13H)-dione, 13-(4-methylphenyl)-5-[(4-methylphenyl)amino]-	EtOH	272(4.45),277(4.41), 342(3.90),441(3.57), 560(3.73)	104-0748-86
$C_{34}H_{24}N_2O_4$			
Benzenemethanol, α-[[(diphenylmethylene)amino](4-nitrophenyl)methylene]-, benzoate, (E)-	CH_2Cl_2	234(4.74),258(4.72), 276(4.68),371(4.30)	39-2021-86C
$C_{34}H_{24}O_{10}$			
Mulberrofuran Q	EtOH	222s(4.43),280(4.16), 284s(4.15),321(4.45), 335s(4.32)	142-1807-86
$C_{34}H_{24}S_2$			
2H-Thiopyran, 2-(4,6-diphenyl-2H-thiopyran-2-ylidene)-4,6-diphenyl-	$CHCl_3$	277(4.38),316s(4.19), 376(3.86),558(3.93)	118-0916-86
$C_{34}H_{25}B$			
1H-Borole, 1,2,3,4,5-pentaphenyl-pyridine adduct	toluene	567(--)	35-0379-86
	heptane	209(4.072),231s(3.850), 259s(3.659),332s(3.193)	35-0379-86
$C_{34}H_{25}F_9O_5S$			
Benzeneacetic acid, 4-methyl-α-phenyl-, 1-(4-methylphenyl)-2-[[(nonafluorobutyl)sulfonyl]oxy]-2-phenylethenyl ester, (Z)-	EtOH	227(4.55),273(4.16)	24-2220-86
$C_{34}H_{25}F_9O_7S$			
Benzeneacetic acid, 4-methoxy-α-phenyl-, 1-(4-methoxyphenyl)-2-[[(nonafluorobutyl)sulfonyl]oxy]-2-phenylethenyl ester, (Z)-	EtOH	208(4.53),229(4.50), 293(4.15)	24-2220-86
$C_{34}H_{25}NO_2$			
Benzenemethanol, α-[[(diphenylmethylene)amino]phenylmethylene]-, benzoate, (E)-	CH_2Cl_2	254(4.33),270(4.00)	39-2021-86C
1H-Indene-1,3(2H)-dione, 2-(2,3,6,7-tetrahydro-2,3-diphenyl-4H-benzo-[a]quinolizin-4-ylidene)-	MeCN	277(4.30),315(3.95), 400(4.35)	118-0908-86
Methanone, (2,5-dihydro-2,2,4,5-tetraphenyl-5-oxazolyl)phenyl-	CH_2Cl_2	234(4.24),254(4.37)	39-0623-86C
$C_{34}H_{25}N_3$			
Propanedinitrile, (2,3,6,7-tetrahydro-1,2,3-triphenyl-4H-benzo[a]quinolizin-4-ylidene)-	MeCN	280(4.11),338(4.33)	118-0908-86
$C_{34}H_{26}Br_3N_5O_7$			
Adenosine, N,N-bis(4-bromobenzoyl)-	MeOH	242(4.83),275s(4.27)	23-2109-86

Compound	Solvent	$\lambda_{max}(\log \epsilon)$	Ref.
2',3'-O-(1-methylethylidene)-, 5'-(4-bromobenzoate) (cont.)			23-2109-86
Benzamide, 4-bromo-N-(4-bromobenzoyl)-N-[7-(4-bromobenzoyl)-3a,4,6a,7,13-13a-hexahydro-2,2-dimethyl-4,13-epoxy-5H-1,3-dioxolo[5,6][1,3]oxazino-[3,2-e]purin-8-yl]-, [3aR-(3aα,4β-6aβ,13β,13aα)]-	MeOH	254(4.62),315s(3.83)	23-2109-86
$C_{34}H_{26}Cl_3N_5O_7$			
Adenosine, N,N-bis(4-chlorobenzoyl)-2',3'-O-(1-methylethylidene)-, 5'-(4-chlorobenzoate)	MeOH	237(4.91),275(4.62)	23-2109-86
Benzamide, 4-chloro-N-(4-chlorobenzoyl)-N-[7-(4-chlorobenzoyl)-3a,4-6a,7,13,13a-hexahydro-2,2-dimethyl-4,13-epoxy-5H-1,3-dioxolo[5,6][1,3]-oxazocino[3,2-e]purin-8-yl]-, [3aR-(3aα,4β,6aβ,13β,13aα)]-	MeOH	248(4.60),310(3.90)	23-2109-86
$C_{34}H_{26}N_2O_2S_2$			
6H-1,3-Thiazine, 4-(4-methoxyphenyl)-6-[4-(4-methoxyphenyl)-2-phenyl-6H-1,3-thiazin-6-ylidene]-2-phenyl-	$CHCl_3$	287(4.59),345(4.38), 549(3.95)	118-0916-86
$C_{34}H_{26}N_2O_6$			
4,9-Epoxyazeto[1,2-a]benz[f]indole-3a(2H)-carboxylic acid, 1,4,9,9a-10,10a-hexahydro-2-oxo-4,9-diphenyl-, (4-nitrophenyl)methyl ester, (3aα,4α,9α,9aα,10aβ)-(±)-	MeOH	266(4.15)	39-0973-86C
$C_{34}H_{26}N_2S_2$			
6H-1,3-Thiazine, 4-(4-methylphenyl)-6-[4-(4-methylphenyl)-2-phenyl-6H-1,3-thiazin-6-ylidene]-2-phenyl-	$CHCl_3$	282(4.66),352s(4.31), 546(4.01)	118-0916-86
$C_{34}H_{26}N_2S_4$			
6H-1,3-Thiazine, 4-phenyl-2-[(phenylmethyl)thio]-6-[4-phenyl-2-[(phenylmethyl)thio]-6H-1,3-thiazin-6-ylidene]-	$CHCl_3$	276(4.50),336(4.23), 513(3.95)	118-0916-86
$C_{34}H_{26}N_4$			
2,3':2',3":2",3"'-Quater-1H-indole, 1"',2"'-dimethyl-	EtOH	298(3.94)	78-5019-86
$C_{34}H_{26}O_2$			
Benzophenone, 3-benzoyl-2,5-dimethyl-4,6-diphenyl-	MeCN	248(4.54)	48-0359-86
$C_{34}H_{26}O_6$			
[5,5'-Bi-4H-1-benzopyran]-4,4'-dione, 6,8'-dihydroxy-2,2'-bis(2-phenylethyl)-	MeOH	234(4.63)	94-4889-86
$C_{34}H_{26}O_{10}$			
Kuwanon Z	EtOH	208(4.75),283(4.29), 304(4.27),327(4.38)	142-2603-86

Compound	Solvent	$\lambda_{max}(\log \epsilon)$	Ref.
$C_{34}H_{26}O_{12}$			
[8,8'-Bi-1H-naphtho[2,3-c]pyran]-3,3'-diacetic acid, 5,5',10,10'-tetrahydro-6,6'-dihydroxy-1,1'-dimethyl-5,5',10,10'-tetraoxo-, dimethyl ester, [S-(R*,R*)]-	MeCN	229(4.56),274(4.40), 338(4.16),426(4.20)	158-0345-86
$C_{34}H_{26}O_{23}$			
Coriariin F	MeOH	221(4.84),261(4.53)	94-4533-86
$C_{34}H_{27}N_5O_4S$			
Piperidine, 1-[[2-[2,3-dihydro-2-(4-methylphenyl)-1,3-dioxo-1H-benz[de]isoquinolin-6-yl]-2H-naphtho[1,2-d]triazol-5-yl]sulfonyl]-	DMF	384(4.43)	2-0496-86
$C_{34}H_{27}N_5O_9S$			
Acridinium, 9-(N-benzyl-4-methanesulfonamidophenyl)-10-methyl-, methosulfate, picrate	MeOH	205(4.73),261(4.94), 362(4.53)	150-3425-86M
$C_{34}H_{28}N_2$			
2,4,6,12,14-Cyclopentadecapentaene-8,10-diyn-1-one, 7,12-dimethyl-, 7,12-dimethyl-2,4,6,12,14-cyclopentadecapentaen-8,10-diyn-1-ylidene)hydrazone (rel. absorbances)	CF_3COOH	258(0.37),277s(0.28), 366(1.00),375s(0.99), 429(0.36),529s(0.20), 598(0.82),634s(0.54)	18-2509-86
$C_{34}H_{28}N_2O_6$			
Siddiquine	MeOH	232(4.70),267(4.39), 355(3.68)	164-0663-86
$C_{34}H_{28}O_3$			
Ethanone, 1-[3,5-bis(diphenylmethyl)-2,4-dihydroxyphenyl]-	MeOH	202(4.61),278(3.65)	2-0755-86
2,4-Heptadiene-1,7-dione, 6-(1-hydroxyethylidene)-4-methyl-1,3,5,7-tetraphenyl-, (Z,Z,E)-	dioxan	250(4.35),275s(4.32), 347s(3.85)	48-0359-86
$C_{34}H_{28}O_4$			
1-Cyclopropene-1-carboxylic acid, 3-[2-(methoxycarbonyl)-1,3-diphenylcyclopropyl]-2,3-diphenyl-, methyl ester	n.s.g.	216(4.20),310(3.20)	104-2213-86
$C_{34}H_{29}O_2$			
Pyrylium, 2-[10-(4-methoxyphenyl)-1,3,5,7,9-decapentaenyl]-4,6-diphenyl-, tetrafluoroborate	CH_2Cl_2	380(4.56),700(4.78)	104-1786-86
$C_{34}H_{30}BO_8$			
Boron(1+), bis[1,3-bis(4-methoxyphenyl)-1,3-propanedionato]-, (T-4)-, perchlorate	CH_2Cl_2	446(5.03)	97-0399-86
$C_{34}H_{30}N_2O_6$			
Cocsupendine	MeOH	208(4.58),212(4.54), 278(4.35),337(3.25)	164-0663-86
Dehydrokohatine	MeOH	235(4.46),254(4.29), 287(3.47)	164-0663-86

Compound	Solvent	λ_{max}(log ε)	Ref.
$C_{34}H_{30}N_2O_7$			
1,2-Dehydrokohatine 2β-N-oxide	MeOH	225(4.54),278s(3.96), 335(3.52)	164-0663-86
$C_{34}H_{30}N_4O_8$			
Butanedioic acid, 2,3-bis[[4-(phenylazo)benzoyl]oxy]-, diethyl ester	EtOH	228(4.33),325(4.67), 448(3.15)	5-1956-86
$C_{34}H_{30}N_4O_8S_2$			
1H-Pyrazolo[3,4-d]pyrimidine, 3-methoxy-4,6-bis(methylthio)-1-(2,3,5-tri-O-benzoyl-β-D-ribofuranosyl)-	MeOH	230(4.81),258(4.64), 295(4.18)	4-1869-86
$C_{34}H_{38}N_4O_9S$			
4H-Pyrazolo[3,4-d]pyrimidin-4-one, 3-ethoxy-1,5-dihydro-6-(methylthio)-1-(2,3,5-tri-O-benzoyl-β-D-ribofuranosyl)-	pH 1	238(4.51),280(4.21)	4-1869-86
	pH 7	241(4.43),280(4.18)	4-1869-86
	pH 11	232(4.40),263(3.97)	4-1869-86
$C_{34}H_{38}N_4O_{11}S$			
4H-Pyrazolo[3,4-d]pyrimidin-4-one, 3-ethoxy-1,5-dihydro-6-(methylsulfonyl)-1-(2,3,5-tri-O-benzoyl-β-D-ribofuranosyl)-	pH 1	222(4.69),270(4.29)	4-1869-86
	pH 7	220(4.62),270(3.96)	4-1869-86
	pH 11	216(4.69),270(4.02)	4-1869-86
$C_{34}H_{30}O_2Si$			
Silane, [(1,1-dimethylethyl)dioxy]tri-2-naphthalenyl-	MeCN	269.5(4.21),277.5(4.20), 308(3.23),315(3.15), 323(3.22)	116-0968-86
$C_{34}H_{30}O_3$			
Cyclohexanone, 2,6-bis[[2-(phenylmethoxy)phenyl]methylene]-, (E,E)-	MeOH	208(4.38),234(3.81), 255(3.72),307(3.73), 328(3.74)	78-5637-86
$C_{34}H_{30}O_7$			
9(10H)-Anthracenone, 10,10-bis[(4-acetoxy-3-methoxyphenyl)methyl]-	EtOH	274(4.24)	104-0158-86
$C_{34}H_{30}O_8$			
4H-1-Benzopyran-4-one, 5,6,7,8-tetrahydro-6,7,8-trihydroxy-5-[(4-oxo-2-(2-phenylethyl)-4H-1-benzopyran-6-yl]oxy]-2-(2-phenylethyl)-, [5S-(5α,6β,7β,8α)]-	MeOH	241(4.55)	94-4889-86
$C_{34}H_{30}O_9$			
Kuwanon Y	EtOH	218(4.60),285(4.40), 300s(4.34),326(4.44)	142-2603-86
	EtOH-AlCl$_3$	218(4.60),286(4.39), 300(4.37),326(4.43)	142-2603-86
$C_{34}H_{32}ClN_3O_3$			
Anthra[1,9-cd]pyrazol-6(2H)-one, 5-chloro-2-[2-(diethylamino)ethyl]-7,10-bis(phenylmethoxy)-	n.s.g.	241(4.43),275(4.09), 330(3.79),445(4.01)	4-1491-86
$C_{34}H_{32}F_3N_2$			
1H-Benz[e]indolium, 2-[2-[(1,3-dihydro-1,1,3-trimethyl-2H-benz[e]indol-2-ylidene)methyl]-1-propenyl-, perchlorate	n.s.g.	592(4.61)	104-0144-86

Compound	Solvent	λmax(log ε)	Ref.
$C_{34}H_{32}N_2O_2S$			
2-Propen-1-one, 1,1'-(thiodi-4,1-phenylene)bis[3-[4-(dimethylamino)-phenyl]-	93% H_2SO_4	330(3.94),373(4.54), 435(4.18),512(4.64)	104-0055-86
$C_{34}H_{32}N_2O_3$			
2-Propen-1-one, 1,1'-(oxydi-4,1-phenylene)bis[3-[4-(dimethylamino)-phenyl]-	93% H_2SO_4	336(4.61),426(4.76)	104-0055-86
$C_{34}H_{32}N_2O_6$			
Apateline, 5-hydroxy-	MeOH	210(4.32),234(4.13), 274(3.31)	164-0663-86
$C_{34}H_{33}ClN_4O_3$			
Anthra[1,9-cd]pyrazol-6(2H)-one, 5-chloro-2-[2-[2-[(dimethylamino)-ethyl]amino]ethyl]-7,10-bis(phenyl-methoxy)-	n.s.g.	243(4.33),278(3.91), 330(3.77),443(4.05)	4-1491-86
$C_{34}H_{33}NO_5$			
7H-Furo[3,2-g][1]benzopyran-7-one, 3,5-dimethyl-6-phenyl-2-[4-[3-(1-piperidinyl)propoxy]benzoyl]-	MeOH	292(4.44),258(4.53)	18-1299-86
$C_{34}H_{34}N_2O_2S_2$			
Spiro[[1,4]oxazino[2,3,4-kl]phenothiazine-2(1H),4'(10'H)-[1H]phenothiazin]-1'-one, 2',4-bis(1,1-dimethyl-ethyl)-6-methyl-	$CHCl_3$	267(4.50),297s(4.03), 320s(3.72),415(3.37)	39-2233-86C
$C_{34}H_{34}N_4O$			
1,4-Benzenediamine, N'-[[1-[4-(dimethylamino)phenyl]-5-methyl-3,4-diphenyl-1H-pyrrol-2-yl]methylene]-N,N-dimethyl-, N'-oxide	ether	266(4.54),372(4.11)	78-1355-86
1H-Pyrrole-2-carboxamide, N,1-bis[4-(dimethylamino)phenyl]-5-methyl-3,4-diphenyl-	$CHCl_3$	265(4.57),312(4.37)	78-1355-86
$C_{34}H_{34}O_8$			
[6,6'-Biazulene]-1,1',3,3'-tetracarboxylic acid, 2,2'-dimethyl-, tetraethyl ester	CH_2Cl_2	231.5(5.31),342.5(4.92), 375.5(4.72),399(4.36), 512(3.09)	24-2272-86
7H-Furo[3,2-g][1]benzopyran-7-one, 9,9'-[1,2-dodecanediylbis(oxy)]bis-	1% DMSO	306(4.40)	149-0391-86A
$C_{34}H_{34}O_{10}$			
[6,6'-Biazulene]-1,1',3,3'-tetracarboxylic acid, 2,2'-dimethoxy-, tetraethyl ester	CH_2Cl_2	246.5(4.74),344(4.95), 405.5(4.46),491(3.17)	24-2272-86
[2,2'-Binaphthalene]-1,1',8,8'-tetrol, 4,4'-diethoxy-6,6'-dimethyl-, tetraacetate	EtOH	285(4.40),307(4.10), 320s(4.06),335(4.01)	39-0675-86C
$C_{34}H_{35}F_3N_2O_{14}$			
Adriamycin, N-(trifluoroacetyl)-14-O-(N-acetylalanyl)-	MeOH	232(4.63),250(4.47), 480(4.04),497(4.05), 532(3.95)	87-1273-86

Compound	Solvent	$\lambda_{max}(\log \epsilon)$	Ref.
$C_{34}H_{35}N_2OS$			
Benzoxazolium, 3-ethyl-2-[[7-[(3-ethyl-2(3H)-benzothiazolylidene)methyl]-4,4a,5,6,10,10a-hexahydro-2(3H)-anthracenylidene]methyl]-, iodide	DMSO	840(5.00)	24-3102-86
$C_{34}H_{35}N_2O_2$			
Benzoxazolium, 3-ethyl-2-[[7-[(3-ethyl-2(3H)-benzoxazolylidene)methyl]-4,4a,5,6,10,10a-hexahydro-2(3H)-anthracenylidene]methyl]-, iodide	DMSO	793(5.13)	24-3102-86
$C_{34}H_{35}N_2SSe$			
Benzothiazolium, 3-ethyl-2-[[7-[(3-ethyl-2(3H)-benzoselenazolylidene)methyl]-4,4a,5,6,10,10a-hexahydro-2(3H)-anthracenylidene]methyl]-, iodide	DMSO	863(4.79)	24-3102-86
$C_{34}H_{35}N_2S_2$			
Benzothiazolium, 3-ethyl-2-[[7-[(3-ethyl-2(3H)-benzothiazolylidene)methyl]-4,4a,5,6,10,10a-hexahydro-2(3H)-anthracenylidene]methyl]-, iodide	DMSO	855(5.19)	24-3102-86
$C_{34}H_{35}N_2Se_2$			
Benzoselenazolium, 3-ethyl-2-[[7-(3-ethyl-2(3H)-benzoselenazolylidene)-methyl]-4,4a,5,6,10,10a-hexahydro-2(3H)-anthracenylidene]methyl]-, iodide	DMSO	872(5.15)	24-3102-86
$C_{34}H_{35}N_5O_6$			
Adenosine, 5'-O-[bis(4-methoxyphenyl)-phenylmethyl]-2',3'-O-(1-methylethylidene)-	MeOH	234(4.40),257(4.20)	23-2109-86
$C_{34}H_{38}N_2O_8$			
1,3(2H,4H)-Isoquinolinedione, 2-cyclohexyl-4-(2-cyclohexyl-2,3-dihydro-5,7-dimethoxy-1,3-dioxo-4(1H)-isoquinolinylidene)-5,7-dimethoxy-, (E)-	$CHCl_3$	403(4.18)	139-0337-86C
$C_{34}H_{38}N_4O_4$			
21H,23H-Porphine-2,18-dicarboxylic acid, 7,13-diethyl-3,8,12,17-tetramethyl-, diethyl ester	$CHCl_3$	413(4.99),513(4.12), 549(3.97),584(3.82), 638(3.70)	12-0399-86
$C_{34}H_{38}O_{12}$			
Tyledoside C	MeOH	298(3.40)	39-1633-86C
$C_{34}H_{38}O_{17}$			
Aloenin B	MeOH	205(4.75),224s(4.48), 298(4.44),308(4.45)	100-0800-86
$C_{34}H_{39}Cl_2N_3O_4$			
1H-Pyrrole-2-carboxylic acid, 4-(2-chloroethyl)-5-[[4-(2-chloroethyl)-5-[[4-(3-methoxy-3-oxopropyl)-3,5-dimethyl-2H-pyrrol-2-ylidene]methyl]-3-methyl-1H-pyrrol-2-yl]methyl]-3-methyl-, phenylmethyl ester, HBr salt, (Z)-	CH_2Cl_2	492(4.91)	44-4667-86

Compound	Solvent	λ_{max}(log ϵ)	Ref.
$C_{34}H_{40}N_2O_8$			
[4,4'-Biisoquinoline]-1,1',3,3'(2H-2'H,4H,4'H)-tetrone, 2,2'-dicyclohexyl-5,5',7,7'-tetramethoxy-, (R*,R*)-(±)-	$CHCl_3$	332(3.78)	139-0337-86C
$C_{34}H_{40}N_8O_2$			
Pyridinium, 1,1',1",1"'-(3,6-dioxo-1,4-cyclohexadiene-1,2,4,5-tetrayl)tetrakis[4-(dimethylamino)-, salt with trifluoromethanesulfonic acid (1:4)	CH_2Cl_2-MeCN	450.9(3.94)	89-0917-86
$C_{34}H_{40}O_{18}$			
Sucrose, 4-acetyl-3,6'-diferuloyl-	EtOH	237(4.45),300s(4.45), 328(4.62)	102-2897-86
Sucrose, 4'-acetyl-3,6'-diferuloyl-	EtOH	237(4.37),300s(4.40), 328(4.59)	102-2897-86
Sucrose, 6-acetyl-3,6'-diferuloyl-	EtOH	235(4.27),300s(4.33), 327(4.50)	102-2897-86
$C_{34}H_{41}N_5O_2$			
2H-Pyrrol-2-one, 5,5'-[(3,4-dimethyl-1H-pyrrole-2,5-diyl)bis[methylene-(3,4-dimethyl-1H-pyrrole-5,2-diyl)-methylidyne]]bis[1,5-dihydro-3,4-dimethyl-, (Z,Z)-	$CHCl_3$	384(4.83)	49-0057-86
$C_{34}H_{42}N_4O_4$			
21H-Biline-2-propanoic acid, 7,13,17-triethyl-1,19,22,24-tetrahydro-3,8,12,18-tetramethyl-1,19-dioxo-, ethyl ester	$CHCl_3$	368(4.71),644(4.17)	78-4137-86
	acid	366(4.77),664(4.59)	78-4137-86
$C_{34}H_{42}N_8O_2$			
Pyridinium, 1,1',1",1"'-(3,6-dihydroxy-1,2,4,5-benzenetetrayl)tetrakis[4-(dimethylamino)-, salt with trifluoromethanesulfonic acid (1:4)	MeCN	397.5(4.00)	89-0917-86
$C_{34}H_{44}N_2O_{16}S$			
Paulomycinone A	MeOH	232(4.18),264(4.27), 440(3.32)	44-2493-86
$C_{34}H_{44}O_{11}$			
14β,15β-Epoxymeliacin-1,5-diene-3-O-α-L-rhamnopyranoside, 6-acetoxy-11α-hydroxy-7-oxo-	n.s.g.	217(4.00),239(3.90)	100-0056-86
$C_{34}H_{45}NO_7$			
Rhizoxin, 2,3:11,12-dideepoxy-2,3-11,12-tetradehydro-17-O-demethyl-	MeOH	215(4.45),232(4.42), 238(4.41),297(4.63), 308(4.74),323(4.61)	94-1387-86
WF 1360C	MeOH	296(4.59),308(4.68), 323(4.56)	158-0762-86
$C_{34}H_{45}NO_8$			
Rhizoxin, 2,3-deepoxy-2,3-didehydro-17-O-demethyl-	MeOH	297(4.59),308(4.70), 323(4.56)	94-1387-86
WF 1360B	MeOH	296(4.63),308(4.72), 323(4.61)	158-0762-86

Compound	Solvent	λ_{max}(log ε)	Ref.
$C_{34}H_{45}NO_9$			
Rhizoxin, 17-O-demethyl- (norrhizoxin)	MeOH	297(4.51),309(4.61), 323(4.48)	94-1387-86
$C_{34}H_{46}N_4O_5$			
9,10-Secocholesta-5,7,10(19)-triene-3,25-diol, 3-(2-azido-4-nitrobenzoate), (3β,5Z,7E)-	EtOH	255(4.30),340(3.20)	130-0134-86
9,10-Secocholesta-5,7,10(19)-triene-3,25-diol, 3-(4-azido-2-nitrobenzoate), (3β,5Z,7E)-	EtOH	268(4.36)	130-0134-86
$C_{34}H_{46}O_{12}$			
D-Homo-24-nor-17-oxachola-1,20,22-trien-16-one, 7-acetoxy-14,15:21,23-diepoxy-3-(β-D-glucopyranosyloxy)-4,4,8-trimethyl-, (3β,5α,7α,13α-14β,15β,17aα)-	EtOH	258(1.8)	2-1087-86
$C_{34}H_{46}O_{16}$			
Yadanzioside N	EtOH	222(4.11),255(3.70)	88-0593-86
Yadanzioside P	EtOH	222(4.22),252(3.94)	94-4447-86
$C_{34}H_{47}N_3O_3$			
9,10-Secocholesta-5,7,10(19)-triene-3,25-diol, 3-(3-azidobenzoate), (3β,5Z,7E)-	EtOH	254(4.32)	130-0134-86
9,10-Secocholesta-5,7,10(19)-triene-3,25-diol, 3-(4-azidobenzoate)-, (3β,5Z,7E)-	EtOH	271(4.42)	130-0134-86
$C_{34}H_{48}O_8$			
Cholest-7-en-6-one, 22-(benzoyloxy)-2,3,14,20,25-pentahydroxy-	EtOH	235(4.36)	105-0071-86
$C_{34}H_{48}O_{12}$			
1,4,7,10,13,16-Benzohexaoxacyclooctadecin, 18,18'-(1,2-ethenediyl)-bis[2,3,5,6,8,9,11,12,14,15-decahydro-, (E)-	MeCN	335(4.56)	1-0545-86
(Z)-	MeCN	301(4.11)	1-0545-86
$C_{34}H_{50}O_6$			
Ganoderic acid R	n.s.g.	224(4.0),234(4.0), 242(4.0),250(3.8)	94-2282-86
$C_{34}H_{52}N_4$			
Propanedinitrile, [[(5α,17β)-17-octyl-3-(1-pyrrolidinyl)androst-2-en-2-yl]imino]-	EtOH	438(4.61)	33-0793-86
$C_{34}H_{52}O_4S$			
Cholestan-6-one, 3-[[(4-methylphenyl)-sulfonyl]oxy]-, (3β,5α)-	EtOH	220(4.14)	39-1797-86C
Cholestan-7-one, 3-[[(4-methylphenyl)-sulfonyl]oxy]-, (3β,5α)-	EtOH	220(4.14)	39-1797-86C
Cholestan-15-one, 3-[[(4-methylphenyl)-sulfonyl]oxy]-, (3β,5α)-	EtOH	220(4.15)	39-1797-86C
Cholestan-16-one, 3-[[(4-methylphenyl)-sulfonyl]oxy]-, (3β,5α)-	EtOH	220(4.14)	39-1797-86C

Compound	Solvent	$\lambda_{max}(\log \epsilon)$	Ref.
Cholestan-24-one, 3-[[(4-methylphenyl)-sulfonyl]oxy]-, (3β,5α)-	EtOH	220(4.14)	39-1797-86C
$C_{34}H_{52}O_6$			
Ganodermanontriol diacetate	MeOH	236(4.23),243(4.28), 251(4.12)	100-1122-86
$C_{34}H_{54}O_6$			
Anhydrophomenolactone	MeOH	207(4.10),233(4.06)	102-0531-86
$C_{34}H_{54}O_8$			
Lasalocide A	n.s.g.	248(3.83),318(3.63)	100-0035-86
$C_{34}H_{55}NO_{11}$			
Tylonolide, 5-O-[3,6-dideoxy-3-(dimethylamino)-β-D-glucopyranosyl]-, 23-propanoate	EtOH	280(4.31)	158-1108-86
$C_{34}H_{56}O_7$			
Phomenolactone	MeOH	233(4.04)	102-0531-86

Compound	Solvent	$\lambda_{max}(\log \epsilon)$	Ref.
$C_{35}H_{21}NO_2$			
9H-Benzo[6,7]perylo[1,12-efg]isoindole-9,11(10H)-dione, 10-(2,4,6-trimethylphenyl)-	$CHCl_3$	417(3.68),442(3.76), 470(3.89)	24-2373-86
$C_{35}H_{22}N_6O_6S$			
2H-Naphtho[1,2-d]triazole-5-sulfonamide, 2-[2,3-dihydro-2-(4-methylphenyl)-1,3-dioxo- 1H-benz[de]isoquinolin-6-yl]-N-(4-nitrophenyl)-	DMF	398(4.10)	2-0496-86
$C_{35}H_{25}Br_2Te_2$			
4H-Tellurin, 1,1-dibromo-4-[(2,6-diphenyltellurinium-4-yl)methylene]-1,1-dihydro-2,6-diphenyl-, tetrafluoroborate	CH_2Cl_2	525(4.45)	157-2250-86
$C_{35}H_{25}Cl_2Te_2$			
4H-Tellurin, 1,1-dichloro-4-[(2,6-diphenyltellurinium-4-yl)methylene]-1,1-dihydro-2,6-diphenyl-, tetrafluoroborate	CH_2Cl_2	497(4.41)	157-2250-86
$C_{35}H_{25}Te_2$			
Tellurinium, 4-[(2,6-dimethyl-4H-tellurin-4-ylidene)methyl]-2,6-diphenyl-, tetrafluoroborate	CH_2Cl_2	759(5.20)	157-2250-86
$C_{35}H_{27}NO_2$			
Methanone, [2,5-dihydro-4-(4-methylphenyl)-2,2,5-triphenyl-5-oxazolyl]-phenyl-	CH_2Cl_2	233(4.80),242(4.60), 281(3.82)	39-0623-86C
$C_{35}H_{27}NO_3$			
Benzenemethanol, α-[[(diphenylmethylene)amino](4-methoxyphenyl)methylene]-, benzoate	CH_2Cl_2	255(4.55),258(4.52), 282(4.46),386(3.58)	39-2021-86C
Methanone, [2,5-dihydro-4-(4-methoxyphenyl)-2,2,5-triphenyl-5-oxazolyl]-phenyl-	CH_2Cl_2	230(4.35),265(4.43), 280(4.38)	39-0623-86C
$C_{35}H_{29}N_2O$			
1H-Indolizinium, 1-[[4-(dimethylamino)-phenyl]methylene]-2,3-dihydro-2-oxo-3,5,7-triphenyl-, chloride	80% MeOH	460(3.82)	4-1315-86
	MeOH-HCl	532(4.67)	4-1315-86
	MeOH-NaOH	435(4.00)	4-1315-86
$C_{35}H_{30}N_2O_6$			
Siddiquamine	MeOH	232(4.63),267(4.32), 355(3.63)	164-0663-86
Stephasubimine	MeOH	242(4.64),281(3.91), 323(3.70)	100-0424-86
	MeOH-acid	264(4.57),307(4.02), 362(3.92),368(3.91)	100-0424-86
$C_{35}H_{30}N_4O_{10}$			
Imidazo[4,5-d][1,2]diazepine-6(1H)-carboxylic acid, 2,3-dihydro-2-oxo-1-(2,3,5-tri-O-benzoyl-β-D-ribofuranosyl)-, ethyl ester	MeOH	266(3.79),335(3.04)	78-1511-86
Imidazo[4,5-d][1,2]diazepine-6(1H)-carboxylic acid, 2,3-dihydro-2-oxo-	MeOH	268(3.98),349(3.00)	78-1511-86

Compound	Solvent	$\lambda_{max}(\log \epsilon)$	Ref.
3-(2,3,5-tri-O-benzoyl-β-D-ribo-furanosyl)-, ethyl ester (cont.)			78-1511-86
$C_{35}H_{30}O_3$			
Ethanone, 1-[2-hydroxy-4-(phenylmeth-oxy)-3,5-bis(phenylmethyl)phenyl]-2-phenyl-	MeOH	220(4.6),282(4.41)	2-0646-86
$C_{35}H_{30}O_4$			
1-Propanone, 1-[3,5-bis(diphenylmeth-yl)-2,4,6-trihydroxyphenyl]-	MeOH	220(4.42),286(4.78), 296(4.75)	2-0478-86
$C_{35}H_{30}O_{13}$			
[8,8'-Bi-1H-naphtho[2,3-c]pyran]-3,3'-diacetic acid, 3,4,5,5',10,10'-hexa-hydro-6,6'-dihydroxy-4-methoxy-1,1'-dimethyl-5,5',10,10'-tetraoxo-, di-methyl ester, [1S-[1α,3β,4β,8(R*)]-	MeCN	229(4.58),268(4.46), 320s(--),427(4.18)	158-0345-86
$C_{35}H_{32}F_5N_2$			
1H-Benz[e]indolium, 2-[2-[(1,3-dihy-dro-1,1,3-trimethyl-2H-benz[e]ind-ol-2-ylidene)methyl]-3,3,4,4,4-pentafluoro-1-butenyl]-1,1,3-tri-methyl-, perchlorate	n.s.g.	578(4.41)	104-0144-86
$C_{35}H_{32}N_2O_6$			
Dehydrokohatine, 12-O-methyl-	MeOH	274(4.14),283(4.12), 305(4.08)	164-0663-86
1,2-Dehydronorkohatine, O,O-dimethyl-	MeOH	215(4.64),226(4.56), 331(3.62)	164-0663-86
Norstephasubine, (+)-	MeOH	240(4.53),286(3.62), 338(3.42)	100-0424-86
	MeOH-acid	235(4.30),264(4.36), 321(3.32),368(3.40), 375(3.39)	100-0424-86
$C_{35}H_{32}O_9$			
2H,13H-Furo[2,3-b:5,4-h']bis[1]benzo-pyran-13-one, 3,4,7a,13a-tetrahydro-5,10,12-trimethoxy-2,7a-bis(4-meth-oxyphenyl)-	dioxan	230(4.77),282(4.29)	94-0595-86
$C_{35}H_{32}O_{14}$			
[8,8'-Bi-1H-naphtho[2,3-c]pyran-3,3'-diacetic acid, 3,3',4,4',5,5',10,10'-octahydro-4,4',6-trihydroxy-6'-meth-oxy-1,1'-dimethyl-5,5',10,10'-tetra-oxo-, dimethyl ester, [1S-[1α,3β-4β,8(1'R*,3'R*,4'R*)]]-	MeCN	228(4.53),265(4.47), 420(4.00)	158-0345-86
$C_{35}H_{33}FeN_2O_6$			
Pyridinium, 3,5-bis(ethoxycarbonyl)-1-(4-ferrocenylphenyl)-2,6-dimethyl-4-(3-nitrophenyl)-, perchlorate	MeCN	207(4.8),251(4.5), 275(4.7),335s(3.8), 467(2.8)	103-0886-86
Pyridinium, 3,5-bis(ethoxycarbonyl)-1-(4-ferrocenylphenyl)-2,6-dimethyl-4-(4-nitrophenyl)-, perchlorate	MeCN	210(4.9),252s(4.5), 286(4.7),472(2.8)	103-0886-86
$C_{35}H_{33}NO$			
Benzenemethanol, 4-[bis(4-methylphen-	HOAc	391(4.37),550(4.73)	64-1045-86B

Compound	Solvent	λmax(log ε)	Ref.
yl)amino]-α,α-bis(4-methylphenyl)- (cont.)	$MeSO_3H$	391(4.26),480s(4.35), 553(4.64)	64-1045-86B
	CF_3COOH	393(4.50),551(4.77)	64-1045-86B
	$CHCl_2COOH$	401(4.37),565(4.62)	64-1045-86B
$C_{35}H_{34}FeN_2O_6$			
Ferrocene, [4-[3,5-bis(ethoxycarbonyl)-2,6-dimethyl-4-(2-nitrophenyl)-1(4H)-pyridinyl]phenyl]-	MeCN	212(4.65),247(4.53), 287(4.35),330s(4.0), 465(3.17)	103-0886-86
Ferrocene, [4-[3,5-bis(ethoxycarbonyl)-2,6-dimethyl-4-(3-nitrophenyl)-1(4H)-pyridinyl]phenyl]-	MeCN	211(4.66),248(4.62), 276(4.49),340(3.98), 460(3.0)	103-0886-86
Ferrocene, [4-[3,5-bis(ethoxycarbonyl)-2,6-dimethyl-4-(4-nitrophenyl)-1(4H)-pyridinyl]phenyl]-	MeCN	210(4.83),217(4.49), 288(4.41),330s(3.5), 472(2.04)	103-0886-86
$C_{35}H_{34}N_2O_2$			
2-Propen-1-one, 1,1'-(methylenedi-4,1-phenylene)bis[3-[4-(dimethylamino)-phenyl]-	93% H_2SO_4	328(4.24),415(4.79)	104-0055-86
$C_{35}H_{34}N_2O_5$			
Telobine, 5-hydroxy-	MeOH	237(4.30),267(3.93), 352(3.25)	164-0663-86
$C_{35}H_{34}N_2O_6$			
Kohatine, 12-O-methyl-	MeOH	233(4.35),235(4.35), 274(3.52),288(3.51)	164-0663-86
$C_{35}H_{34}N_2O_{13}S_2$			
Emestrin, 3,11a-deepidithio-3,11a-bis(methylthio)-, 4',7',11-triacetate	EtOH	210(4.76),261(4.30), 291(3.73)	39-0109-86C
$C_{35}H_{34}O_{11}$			
Benzaldehyde, 2-[1-[2-(3,4-dimethoxybenzoyl)-4,5-dimethoxyphenyl]-2-(5-methoxy-3,6-dioxo-1,4-cyclohexadien-1-yl)ethyl]-3,5-dimethoxy-	MeOH	208.5(4.77),227(4.55), 272(4.32),304(4.08), 397(2.80)	94-4418-86
$C_{35}H_{35}NS_2$			
2H,6H-[1,3]Thiazino[2,3-a]isoquinoline, 2-[(1,1-dimethylethyl)thio]-7,11b-dihydro-3,4-diphenyl-11b-(phenylmethyl)-	MeCN	276s(--),290(3.87)	118-0899-86
$C_{35}H_{36}$			
Benzene, 1,1'-[[2,2-bis(1-methylethyl)-3,3-diphenylcyclopropyl]ethenylidene)bis-	EtOH	246(4.34)	44-0196-86
1,4-Pentadiene, 3,3-bis(1-methylethyl)-1,1,5,5-tetraphenyl-	EtOH	246(4.34),289(3.20), 337(<0)	44-0196-86
$C_{35}H_{36}NO_3$			
3H-Indolium, 2-[3-[4,6-bis(4-ethoxyphenyl)-2H-pyran-2-ylidene]-1-propenyl]-1,3,3-trimethyl-, tetrafluoroborate	CH_2Cl_2	664(4.78),700(4.81)	103-1295-86
$C_{35}H_{38}N_4O_6$			
21H,23H-Porphine-3,7,13-tripropanoic acid, 2,8,12-trimethyl-, trimethyl ester	$CHCl_3$	323s(4.19),346s(4.46), 371s(4.82),399(5.32)	35-6449-86

Compound	Solvent	λ_{max}(log ε)	Ref.
$C_{35}H_{39}N_3O_9$			
Spiro[2H-1-benzopyran-2,2'-[2H]indole]-1'(3'H)-propanamide, 3',3'-dimethyl-6-nitro-N-(2,3,5,6,8,9-11,12-octahydro-1,4,7,10,13-benzopentaoxacyclopentadecin-14-yl)-	CH_2Cl_2	340(4.01)	18-1853-86
after uv irradiation	CH_2Cl_2	345(4.02),567(4.04)	18-1853-86
$C_{35}H_{40}Cl_2N_6$			
a,c-Biladiene, 4,6-bis(2-chloroethyl)-1,8-bis(2-cyanoethyl)-1',2,3,5,7,8'-hexamethyl-, dihydrobromide	CH_2Cl_2	450(4.74),524(4.73)	k50-3333-86M
$C_{35}H_{40}CuN_4O_4$			
Copper, [dimethyl 7,8-dihydro-3,7,7-8,12,13,17-heptamethyl-21H,23H-porphine-2,18-dipropanoato(2-)-N^{21},N^{22}-N^{23},N^{24}]-, (SP-4-2)-	$CHCl_3$	410(5.11),500(3.75), 631(4.65)	94-2428-86
$C_{35}H_{41}NO_5S$			
Sulfilimine, N-[(11β,16α)-21-acetoxy-11-hydroxy-3,20-dioxopregn-4-en-16-yl]-S,S-diphenyl-	EtOH	233(4.44),290s(3.39)	70-1072-86
$C_{35}H_{42}N_2O_9$			
Benzanthrin A	MeOH	230(4.67),260(4.38), 304(4.41),428(3.91)	158-1515-86
	MeOH-HCl	230(--),260(--), 304(--),428(--)	158-1515-86
	MeOH-NaOH	230(--),245(--), 290(--),510(--)	158-1515-86
Benzanthrin B	MeOH	230(4.69),260(4.40), 304(4.43),426(3.92)	158-1515-86
	MeOH-HCl	230(--),260(--), 304(--),426(--)	158-1515-86
	MeOH-NaOH	230(--),243(--), 289(--),510(--)	158-1515-86
$C_{35}H_{42}N_4O_6$			
21H-Biline-8,12-dipropanoic acid, 3,18-diethyl-1,19,22,24-tetrahydro-2,7,13,17-tetramethyl-1,19-dioxo-, dimethyl ester	$CHCl_3$	367(4.60),628(4.04)	78-4137-86
	acid	366(4.60),660(4.56)	78-4137-86
$C_{35}H_{42}O_{10}$			
Daphnodorin B pentamethyl ether	dioxan	218(4.58),303(4.55)	94-0595-86
$C_{35}H_{45}NO_3$			
Benzoic acid, 4-cyano-, 3-methyl-4-oxo-6-(3,7,11,15-tetramethyl-2,6,10-14-hexadecatetraenyl)-2-cyclohexen-1-yl ester, [1α,6α(2E,6E,10E)]-(sarcodictyenone p-cyanobenzoate)	EtOH	203.8(4.56),240(4.42)	33-1581-86
$C_{35}H_{45}N_5$			
21H,23H-5-Azaporphine, 2,3,7,8,12,13-17,18-octaethyl-	CH_2Cl_2	376(5.12),505(3.91), 536(4.42),558(3.98), 610(4.43)	39-0001-86C
$C_{35}H_{46}O_{20}$			
β-D-Glucopyranoside, 3,4-dihydroxy-	MeOH	218(4.28),247(3.94),	102-1633-86

Compound	Solvent	λ_{max}(log ε)	Ref.
β-phenylethyl-O-β-D-glucopyranosyl-(1→3)-O-α-L-rhamnopyranoosyl)-(1→6)-4-caffeoyl- (cont.)		290(4.07),330(4.20)	102-1633-86
$C_{35}H_{47}NO_7$			
Rhizoxin, 2,3:11,12-dideepoxy-2,3,11,12-tetradehydro-	MeOH	216(4.41),235(4.39), 238(4.39),297(4.59), 309(4.70),323(4.56)	94-1387-86
$C_{35}H_{47}NO_8$			
Rhizoxin, 2,3-deepoxy-2,3-didehydro-	MeOH	297(4.66),308(4.81), 323(4.67)	94-1387-86
WF 1360F	MeOH	296(4.55),309(4.65), 323(4.52)	158-0762-86
$C_{35}H_{47}NO_9$			
Rhizoxin	MeOH	295(4.63),308(4.73), 325(4.59)	94-1387-86
WF 1360	MeOH	296(4.61),308(4.72), 324(4.58)	158-0762-86
$C_{35}H_{47}N_3O_8$			
1H-Pyrrole-2-carboxylic acid, 3-(2-acetoxyethyl)-5-[[3-(2-acetoxyethyl)-5-[[4-(3-methoxy-3-oxopropyl)-3,5-dimethyl-2H-pyrrol-2-ylidene]-methyl-4-methyl-1H-pyrrol-2-yl]-methyl]-4-methyl-, 1,1-dimethylethyl ester, (Z)-, hydrobromide	CH_2Cl_2-TFA	484(4.95)	44-0657-86
$C_{35}H_{48}N_4O_2$			
Bilirubin, octaethyl-	MeOH	395(4.76),430(4.78)	64-0437-86C
	$CHCl_3$	278(--),395(--)	64-0437-86C
$C_{35}H_{48}O_4$			
1,3-Benzenediol, 4-(1,1-dimethyl-2-propenyl]-6-[3-[4-hydroxy-2,3-bis(3-methyl-2-butenyl)-5-[(3-methyl-2-butenyl)oxy]phenyl]propyl]- (kazinol G)	EtOH	221s(4.9),285(3.88)	94-1968-86
$C_{35}H_{48}O_{18}$			
Yadanzioside O	EtOH	224(4.18),254(3.90)	88-0593-86
$C_{35}H_{49}NO_{10}$			
Rhizoxin, 11,12-deepoxy-10,11-didehydro-9,10-dihydro-9,12-dihydroxy-	MeOH	298(4.63),310(4.75), 325(4.62)	158-0424-86
$C_{35}H_{50}N_2O_3$			
Benzamide, N-[(3β,5α,16α,20S)-16-acetoxy-20-(dimethylamino)-4,4,14-trimethyl-B(9a)-homo-19-norpregna-9(11),9a-dien-3-yl]- (buxabenzamidienine, 16α-acetoxy)	MeOH	228s(4.31),237(4.34), 245(4.35),254(4.15)	78-5747-86
$C_{35}H_{50}N_2O_4$			
Benzamide, N-[(3β,4β,5α,16α,20S)-16-acetoxy-20-(dimethylamino)-4-formyl-4,14-dimethyl-9,19-cyclo-19-norpregn-1(10)-en-3-yl]- (buxanaldinine)	MeOH	228(4.18)	78-5747-86

Compound	Solvent	λ_{max}(log ε)	Ref.
Benzamide, N-[(3β,4α,5α,16α,20S)-4-(acetoxymethyl)-20-(dimethylamino)-16-hydroxy-4,14-dimethyl-B(9a)-homo-19-norpregna-9(11),9a-dien-3-yl]-	$CHCl_3$	238(4.13),245(4.29), 253(4.17),268(3.92), 279(3.89),290(3.87)	39-0919-86C
$C_{35}H_{50}N_4O_2$			
Bilirubin, 2,3-dihydro-2,3,7,8,12,13-17,18-octaethyl-	MeOH	275(--),400(4.48), 425s(--)	64-0437-86C
	$CHCl_3$	300(--),392(--)	64-0437-86C
$C_{35}H_{51}N_3O_2$			
2-Phthalazinecarboxamide, 1,2,3,4,5,6-7,8-octahydro-7-hydroxy-1-[[octahydro-7a-methyl-1-(1,4,5-trimethyl-2-hexenyl)-4H-inden-4-ylidene)methyl]-N-phenyl-	EtOH	243(4.39),275(3.09)	44-4819-86
$C_{35}H_{52}N_8O_9S_2$			
L-Proline, 1-[N-[[2-[[[[2-[4-amino-1-[[2-[[(1,1-dimethylethoxy)carbonyl]amino]-4-methyl-1-oxopentyl]-amino]-4-oxobutyl]-4-yjiazolyl]-carbonyl]amino]methyl]-4-thiazolyl]-carbonyl]-L-valyl]-, methyl ester, [S-(R*,R*)]-	MeOH	240(3.89)	44-4580-86
$C_{35}H_{54}O_7$			
Stigmast-4-en-6-one, 3,22,23-triacetoxy-, (3β,22S,23S)-	EtOH	238(3.83)	39-2123-86C
$C_{35}H_{55}NO_{12}$			
Tylonolide, 5-O-[2,4-di-O-acetyl-3,6-dideoxy-3-(dimethylamino)-β-D-glucopyranosyl]-	EtOH	283(4.35)	158-1108-86
$C_{35}H_{58}N_2O_3$			
Cholest-4-en-3-one, 2,6-di-4-morpholinyl-	EtOH	241(4.10)	107-0195-86
isomer	EtOH	239(4.09)	107-0195-86

Compound	Solvent	$\lambda_{max}(\log \epsilon)$	Ref.
$C_{36}H_{14}F_{18}N_4Pt$			
Platinum, [N,N'-bis(2,2',3,3',4',5,5'-6,6'-nonafluoro[1,1'-biphenyl]-4-yl)-1,2-ethanediaminato(2-)-N,N']-bis(pyridine)-, (SP-4-2)-	EtOH	349(4.11),390s(--)	12-2013-86
$C_{36}H_{18}N_4O_4$			
6,6'-Bitriphenodioxazine	$CHCl_3$	472(4.78),506(4.86)	4-1003-86
$C_{36}H_{24}Br_2N_2O_4$			
2,5-Cyclohexadiene-1,4-dione, 2,5-bis(5-bromo-1H-indol-3-yl)-3,6-bis(phenylmethoxy)-	MeCN	230(4.69),273(4.43), 455(3.74)	5-1765-86
$C_{36}H_{24}MgN_8O_4$			
Magnesium, [2,9,16,23-tetramethoxy-29H,31H-phthalocyaninato(2-)-N^{29},N^{30},N^{31},N^{32}]-, (SP-4-1)-	90% DMF	685(5.24)	149-0125-86B
$C_{36}H_{24}N_2$			
3-Pyridinecarbonitrile, 2-[1,1'-biphenyl]-4-yl-4,5,6-triphenyl-	MeCN	286(4.52)	73-1061-86
3-Pyridinecarbonitrile, 2,6-bis([1,1'-biphenyl]-4-yl)-4-phenyl-	MeCN	300(4.61)	73-1061-86
3-Pyridinecarbonitrile, 2,4-diphenyl-6-[1,1':4',1"-terphenyl]-4-yl-	MeCN	335(4.70)	73-1061-86
$C_{36}H_{25}Cl_6N_5O_8$			
Cyclopenta[4,5]pyrimido[6,1-a]isoquinoline-3-carboximidic acid, 4,5,7,8-tetrahydro-10,11-dimethoxy-5-oxo-N,4-diphenyl-1-[[(trichloroacetyl)-amino]carbonyl]-, anhydride with trichloroacetylcarbamic acid	CH_2Cl_2	255(4.41),310(4.32), 401(4.40)	24-2553-86
$C_{36}H_{26}N_4O_2S_2$			
Imidazo[5,1-a]isoquinolin-3(2H)-one, 1,1'-dithiobis[5-methyl-2-phenyl-	EtOH	232(4.89),266(4.69), 374(3.84),490(3.51)	150-1901-86M
$C_{36}H_{26}O_7$			
4H-1-Benzopyran-4-one, 2-(2,2-diphenyl-1,3-benzodioxol-5-yl)-6-hydroxy-7-methoxy-5-(phenylmethoxy)-	EtOH	242(4.40),279(4.22), 333(4.50)	94-0030-86
$C_{36}H_{28}FN_3O_8S_2$			
1,2,4-Triazin-3(2H)-one, 6-[[(2-fluorophenyl)methyl]thio]-4,5-dihydro-5-thioxo-2-(2,3,5-tri-O-benzoyl-β-D-ribofuranosyl)-	pH 13	338(3.06)	80-0881-86
	EtOH	330(4.01)	80-0881-86
1,2,4-Triazin-3(2H)-one, 6-[[(3-fluorophenyl)methyl]thio]-4,5-dihydro-5-thioxo-2-(2,3,5-tri-O-benzoyl-β-D-ribofuranosyl)-	pH 13	338(3.04)	80-0881-86
	EtOH	330(4.01)	80-0881-86
1,2,4-Triazin-3(2H)-one, 6-[[(4-fluorophenyl)methyl]thio]-4,5-dihydro-5-thioxo-2-(2,3,5-tri-O-benzoyl-β-D-ribofuranosyl)-	pH 13	338(3.03)	80-0881-86
	EtOH	330(4.01)	80-0881-86
$C_{36}H_{28}N_2O_6$			
Benzoic acid, 4-methoxy-, 2-[(diphenylmethylene)amino]-1-(4-methoxy-	CH_2Cl_2	268(4.58),384(3.99)	39-2021-86C

Compound	Solvent	$\lambda_{max}(\log \epsilon)$	Ref.
phenyl)-2-(4-nitrophenyl)ethenyl ester, (E)- (cont.)			39-2021-86C
$C_{36}H_{28}O_3$			
2H-1-Benzopyran-2-one, 3,6-bis(diphenylmethyl)-7-hydroxy-4-methyl-	MeOH	209(4.24),329(3.65)	2-0623-86
2H-1-Benzopyran-2-one, 3,8-bis(diphenylmethyl)-7-hydroxy-4-methyl-	MeOH	213(4.39),329(3.94)	2-0623-86
4H-1-Benzopyran-4-one, 6,8-bis(diphenylmethyl)-7-hydroxy-2-methyl-	MeOH	194(4.60),250(4.08), 300(3.77)	2-0755-86
$C_{36}H_{29}NO_2S$			
1H-Indene-1,3(2H)-dione, 2-(6,7,8,8a-tetrahydro-8a-[4-(methylthio)phenyl]-2,3-diphenyl-1(5H)-indolizinylidene]-	MeCN	255(4.38),418(4.26)	118-0908-86
$C_{36}H_{29}NO_4$			
Benzoic acid, 4-methoxy-, 2-[(diphenylmethylene)amino]-1-(4-methoxyphenyl)-2-phenylethenyl ester, (E)-	CH_2Cl_2	260(4.63),285(4.41), 370(3.60)	39-2021-86C
$C_{36}H_{30}BO_6$			
Boron(1+), bis[5-(4-methoxyphenyl)-1-phenyl-4-pentene-1,3-dionato-O^1,O^3]-, (T-4)-, perchlorate	CH_2Cl_2	488(5.04)	97-0399-86
$C_{36}H_{30}O_3$			
2-Propen-1-one, 1-[2-hydroxy-4-(phenylmethoxy)-3,5-bis(phenylmethyl)-phenyl]-3-phenyl-	MeOH	220(5.0),254(4.7), 326(4.9)	2-0259-86
$C_{36}H_{30}O_4$			
2-Propen-1-one, 1-[2-hydroxy-3,4-bis(phenylmethoxy)-5-(phenylmethyl)-phenyl]-3-phenyl-	MeOH	210(4.6),260(4.2), 337(4.5)	2-0259-86
$C_{36}H_{30}O_8$			
4H-1-Benzopyran-4-one, 7-(benzoyloxy)-3-[6-(benzoyloxy)-2,4-dimethoxy-3-(3-methyl-2-butenyl)phenyl]-	EtOH	284s(3.98),293s(3.90), 303s(3.86)	94-2369-86
$C_{36}H_{31}CrO_2P$			
Chromium, dicarbonyl[(1,2,3,4,4a,10a-η)-9,10-dihydro-4,5-dimethylphenanthrene](triphenylphosphine)-	MeOH	240s(4.43),260s(4.26), 342(3.96)	49-1423-86
$C_{36}H_{32}N_6O_4S_2$			
4-Thiazolamine, 2,2'-(1,2-ethanediylidene)bis[2,5-dihydro-N-(4-methoxyphenyl)-5-[(4-methoxyphenyl)imino]-, (E,E,?,?)-	Me_4urea	481s(4.13),608(4.63), 641(4.76)	48-0805-86
$C_{36}H_{32}N_6S_2$			
4-Thiazolamine, 2,2'-(1,2-ethanediylidene)bis[2,5-dihydro-N-(4-methylphenyl)-5-[(4-methylphenyl)imino]-, (E,E,?,?)-	Me_4urea	471s(4.03),589s(4.70), 625(4.79)	48-0805-86
$C_{36}H_{34}N_2O_6$			
Caryolivine, (-)-	MeOH	212(4.67),254(4.53), 286(3.88),336(3.51)	94-1148-86

Compound	Solvent	λ_{max}(log ε)	Ref.
Stephasubine, (+)-	MeOH	240(4.56),287(3.61), 337(3.43)	100-0424-86
	MeOH-acid	235(4.35),264(4.41), 290s(3.69),321(3.39), 368(3.48),374(3.48)	100-0424-86
$C_{36}H_{34}N_2O_7$			
2(1H)-Pyrimidinone, 1-[[2-(benzoyl-oxy)ethoxy]methyl]-4-methoxy-5-[[3-[[4-(phenylmethoxy)phenyl]methoxy]-phenyl]methyl]-	EtOH	274(3.93)	4-1651-86
Secolucidine	EtOH	222(4.63),254s(4.32), 284(4.12)	100-0854-86
	EtOH-NaOH	220(4.66),292(3.98), 346(4.10)	100-0854-86
$C_{36}H_{34}O_8$			
5,8,11-Metheno-1H-benzo[a]fluorene-1,4,6,7,10-pentone, 9-[(1,4-dihydro-3-methyl-1,4-dioxo-2-naphthalenyl)-methyl]-4a,5,6a,6b,8,9,10a,11,11a,11b-decahydro-11-hydroxy-2,4a,5,6b,9,11a-hexamethyl-, (4aα,5β,6aβ,6bβ,8β,9β-10aβ,11α,11aβ,11bα)-(±)-	EtOH	245(4.50),261(4.25), 330(3.59)	78-3663-86
$C_{36}H_{34}O_9$			
Methanone, [3,4-dihydro-5-methoxy-2,8-bis(4-methoxyphenyl)-2H-furo[2,3-h]-1-benzopyran-9-yl](2,4,6-trimethoxy-phenyl)-, (-)-	dioxan	220(4.66),283(4.36)	94-0595-86
$C_{36}H_{35}N_5O_6S$			
2',3'-Secoadenosine, 2'-O-[(4-methyl-benzene)sulfonyl]-5'-O-(triphenyl-methyl)-	EtOH	213(4.54),261(4.10)	87-2445-86
2',3'-Secoadenosine, 3'-O-[(4-methyl-benzene)sulfonyl]-5'-O-(triphenyl-methyl)-	EtOH	212(4.60),260(4.14)	87-2445-86
$C_{36}H_{36}Cl_2N_6O_3$			
21H,23H-Porphine-2-propanoic acid, 8,13-bis(2-chloroethyl)-18-(1H-imidazol-2-ylcarbonyl)-3,7,12,17-tetramethyl-, methyl ester	CH_2Cl_2	406(5.34),508(4.09), 546(4.21),574(3.99), 632(3.29)	44-0657-86
$C_{36}H_{36}FeNO_5$			
Pyridinium, 3,5-bis(ethoxycarbonyl)-1-(4-ferrocenylphenyl)-4-(4-meth-oxyphenyl)-2,6-dimethyl-, per-chlorate	MeCN	210(4.7),250s(4.5), 287(4.3),356(4.1), 480(3.4)	103-0886-86
$C_{36}H_{36}N_2O_5$			
Dinklacorine	MeOH	203(4.32),235(4.11), 290(3.41)	83-0841-86
Tiliacorine	MeOH	204(4.44),235(4.27), 290(3.61)	83-0126-86
Tiliacorinine	MeOH	204(4.55),235(4.32), 288(3.68)	83-0126-86
Yanangcorinine	MeOH	205(4.57),335(4.27), 290(3.62)	83-0126-86

Compound	Solvent	λ_{max}(log ϵ)	Ref.
$C_{36}H_{36}N_2O_6$			
Thalsivasine, (+)-	MeOH	234s(4.39),281(4.05), 313(3.78)	100-0488-86
	MeOH-acid	239(4.36),285(3.99), 308(3.69),354(3.82)	100-0488-86
Tilianangin	MeOH	205(4.28),233(4.05), 290(3.42)	83-0872-86
Yanangine	MeOH	205(4.50),233(4.27), 290(3.62)	83-0841-86
$C_{36}H_{36}N_4$			
Dipyrrolo[3,4-b:3',4'-b'][1,3]diazeto[1,2-a:3,4-a']diindole, 1,2,3,3a-9,10,11,11a-octahydro-2,10-dimethyl-3a,11a-bis(phenylmethyl)-	EtOH	261(4.43),311(3.79)	118-0731-86
	EtOH-HCl	245(4.34),296(3.74)	118-0731-86
$C_{36}H_{36}N_4O_5$			
21H,23H-Porphine-2,18-dipropanoic acid, 7,12-diethenyl-3,8,13,17-tetramethyl-β^2-oxo-, dimethyl ester	CH_2Cl_2	412(5.25),514(4.04), 556(4.15),582(3.95), 640(3.21)	44-0657-86
$C_{36}H_{37}F_3N_2O_{14}$			
Adriamycin, N-(trifluoroacetyl)-14-O-(N-acetylprolyl)-	MeOH	232(4.60),250(4.43), 285(4.03),480(4.02), 494(4.02),528(3.78)	87-1273-86
$C_{36}H_{37}F_3N_2O_{16}$			
Adriamycin, N-(trifluoroacetyl)-14-O-(N-acetyl-5-glutamyl)-	MeOH	232(4.60),250(4.44), 290(4.11),480(3.98), 494(3.99),528(3.76)	87-1273-86
$C_{36}H_{37}FeNO_5$			
Ferrocene, [4-[3,5-bis(ethoxycarbonyl)-4-(4-methoxyphenyl)-2,6-dimethyl-1(4H)-pyridinyl]phenyl]-	MeCN	212(4.8),250(4.7), 287(4.6),350(4.02), 462(2.3)	103-0886-86
$C_{36}H_{37}N_5O_6$			
21H,23H-Porphine-2,8,12-tripropanoic acid, 17-cyano-3,7,13-trimethyl-, trimethyl ester	$CHCl_3$	333s(4.33),358s(4.37), 382s(4.77),407(5.33), 417s(5.07)	35-6449-86
at 295°K	CH_2Cl_2	624(4.15)	35-6449-86
at 199°K	CH_2Cl_2	621(4.28)	35-6449-86
$C_{36}H_{37}N_5O_7$			
Benzoic acid, 4-methyl-, ester with N,N'-[6-[[2,3-O-(1-methylethylidene)-β-D-ribofuranosyl]amino]-4,5-pyrimidinediyl]bis[4-methylbenzamide]	MeOH	240(4.60)	23-2109-86
$C_{36}H_{38}Cl_2N_4O_5$			
21H,23H-Porphine-2,18-dipropanoic acid, 7,12-bis(2-chloroethyl)-3,8,13,17-tetramethyl-β^2-oxo-, dimethyl ester	CH_2Cl_2	408(5.38),508(4.18), 546(4.24),576(4.05), 632(3.01)	44-0657-86
$C_{36}H_{38}F_3N_3O_{15}$			
Adriamycin, N-(trifluoroacetyl)-14-O-(N-acetylglutaminyl)-	MeOH	233(4.59),250(4.42), 286(3.99),480(4.06), 495(4.06),529(3.84)	87-1273-86

Compound	Solvent	λ_{max}(log ε)	Ref.
$C_{36}H_{38}N_2O_6$			
2-Norlimacine, (-)-	MeOH	211(4.91),238(4.46), 282(3.96)	94-1148-86
2'-Northalictine, (-)-	MeOH	237s(4.33),284(3.90)	100-0488-86
2'-Northaliphylline, (+)-	MeOH	237s(4.16),285(3.83)	100-0488-86
$C_{36}H_{38}N_4$			
21H,23H-Porphine, 5,10,15,20-tetrakis(2-methyl-1-propenyl)-	benzene	423(5.43),494s(3.53), 523(4.03),561(3.99), 600(3.56),661(3.46)	77-1725-86
$C_{36}H_{38}N_4O_4$			
Protoporphyrin IX dimethyl ester, [1-^{13}C]-	CH_2Cl_2	406(5.16),506(4.13), 540(4.04),576(3.80), 630(3.65),654(2.70)	44-4667-86
[3-^{13}C]-	CH_2Cl_2	404(5.16),506(4.11), 540(4.00),576(3.78), 630(3.60),654(2.70)	44-4667-86
[5-^{13}C]-	CH_2Cl_2	404(5.16),504(4.11), 540(3.99),576(3.76), 630(3.60),654(2.95)	44-4667-86
[8-^{13}C]-	CH_2Cl_2	408(5.15),506(4.14), 540(4.02),576(3.79), 630(3.64),654(2.85)	44-4667-86
Porphyrin, 6,7-bis[2-(methoxycarbonyl)ethyl]-1,4,5,8-tetramethyl-2,3-divinyl-	CH_2Cl_2	406(5.16),504(4.20), 540(4.11),576(3.88), 630(3.78)	44-0666-86
Porphyrin, 6,7-bis[2-(methoxycarbonyl)ethyl]-2,3,5,8-tetramethyl-1,4-divinyl-	CH_2Cl_2	402(5.15),504(4.19), 540(4.10),576(3.87), 630(3.75)	44-0666-86
$C_{36}H_{38}N_4O_7$			
21H,23H-Porphine-2,8,12-tripropanoic acid, 17-formyl-3,7,13-trimethyl-, trimethyl ester	$CHCl_3$	340s(4.40),366s(4.49), 392s(4.74),414(5.22), 427s(4.96)	35-6449-86
at 295°K	CH_2Cl_2	634(4.34)	35-6449-86
at 199°K	CH_2Cl_2	633(4.38)	35-6449-86
$C_{36}H_{39}F_3N_2O_{14}$			
Adriamycin, N-(trifluoroacetyl)-14-O-(N-acetylvalyl)-	MeOH	232(4.63),250(4.47), 287(4.04),480(4.09), 495(4.10),530(3.90)	87-1273-86
$C_{36}H_{39}F_3N_2O_{14}S$			
Adriamycin, N-(trifluoroacetyl)-14-O-(N-acetylmethionyl)-	MeOH	232(4.66),250(4.50), 287(4.07),480(4.11), 496(4.13),531(4.01)	87-1273-86
$C_{36}H_{39}NO_8$			
1H-Pyrrole-3,4-dicarboxylic acid, 1-[2,3,5-O-tris-(phenylmethyl)-β-D-arabinofuranosyl]-, diethyl ester	EtOH	250(3.87)	78-5869-86
$C_{36}H_{39}N_5O_7$			
21H,23H-Porphine-2,8,12-tripropanoic acid, 17-[(hydroxyimino)methyl]-, trimethyl ester, (E)-	$CHCl_3$	337s(4.35),359s(4.53), 384s(4.87),409(5.24)	35-6449-86
(Z)-	$CHCl_3$	332s(4.36),357s(4.57), 386s(4.92),409(5.31)	35-6449-86

Compound	Solvent	$\lambda_{max}(\log \epsilon)$	Ref.
$C_{36}H_{40}Cl_2N_4O_4$			
Porphyrin, 1,4-bis(2-chloroethyl-6,7-bis[2-(methoxycarbonyl)ethyl]-2,3,5,8-tetramethyl-	CH_2Cl_2	402(5.07),498(4.19), 532(4.00),568(3.84), 622(3.67)	44-0666-86
Porphyrin, 2,3-bis(2-chloroethyl)-6,7-bis[2-(methoxycarbonyl)ethyl]-1,4,5,8-tetramethyl-	CH_2Cl_2	400(5.18),502(4.15), 534(3.96),574(3.78), 622(3.43)	44-0666-86
$C_{36}H_{40}N_2O_6$			
Berbamunine	EtOH	208(4.81),230s(4.42), 228(3.95)	100-0854-86
	EtOH-NaOH	222(4.86),250s(4.56), 300(4.37)	100-0854-86
2'-Norpisopowiardine	EtOH	209(4.30),230s(4.18), 289(3.73)	100-1018-86
$C_{36}H_{40}N_4O_6$			
21H,23H-Porphine-2-propanoic acid, 7,12-bis(2-acetoxyethyl)-3,8,13,17-tetramethyl-, methyl ester	CH_2Cl_2	400(5.26),496(4.11), 530(3.93),566(3.53), 620(2.91)	44-4660-86
$C_{36}H_{42}ClN_3O_6$			
1H-Pyrrole-3-propanoic acid, 5-[[3-(2-chloroethyl)-4,5-dimethyl-2H-pyrrol-2-ylidene]methyl]-2-[[3-(3-methoxy-3-oxopropyl)-4-methyl-5-[(phenylmethoxy)carbonyl]-1H-pyrrol-2-yl]-methyl]-4-methyl-, methyl ester, hydrobromide	CH_2Cl_2	496(4.92)	44-4667-86
1H-Pyrrole-3-propanoic acid, 5-[[4-(2-chloroethyl)-3,5-dimethyl-2H-pyrrol-2-ylidene]methyl]-2-[[3-(3-methoxy-3-oxopropyl)-4-methyl-5-[(phenylmethoxy)carbonyl]-1H-pyrrol-2-yl]-methyl-4-methyl-, methyl ester, hydrobromide	CH_2Cl_2	492(4.92)	44-4667-86
$C_{36}H_{42}N_2O_{10}$			
1,3(2H,4H)-Isoquinolinedione, 2-cyclohexyl-4-(2-cyclohexyl-2,3-dihydro-5,6,7-trimethoxy-1,3-dioxo-4(1H)-isoquinolinylidene)-5,6,7-trimethoxy-, (E)-	$CHCl_3$	397(4.19)	139-0337-86C
$C_{36}H_{42}N_4O_5$			
2-Mesoporphyrinone dimethyl ester	CH_2Cl_2	407(5.22),508(3.94), 547(4.09),585(3.76), 642(4.54)	44-2134-86
4-Mesoporphyrinone dimethyl ester	CH_2Cl_2	407(5.24),547(4.08), 585(3.78),642(4.52)	44-2134-86
6-Mesoporphyrinone dimethyl ester	CH_2Cl_2	407(5.25),508(4.00), 547(4.08),585(3.77), 642(4.52)	44-2134-86
7-Mesoporphyrinone dimethyl ester	CH_2Cl_2	407(5.24),508(4.00), 547(4.08),585(3.78), 642(4.52)	44-2134-86
$C_{36}H_{42}N_4O_6$			
2,3-Mesoporphyrinedione dimethyl ester	CH_2Cl_2	417(5.12),435(5.01), 592(3.97),622(4.26)	44-2134-86

Compound	Solvent	λ_{max}(log ε)	Ref.
2,4-Mesoporphyrindione dimethyl ester (7,12-diethyl-3,7,12,17-tetramethyl-8,13-dioxoporphinedipropanoic acid dimethyl ester)	CH_2Cl_2	402(4.87),417(4.97), 438(4.99),544(3.98), 584(4.20),592(4.18), 638(4.23)	44-2134-86
2,6-Mesoporphyrinedione dimethyl ester	CH_2Cl_2	401(5.21),411(5.27), 486(3.76),514(3.91), 556(4.03),622(3.83), 652(3.86),685(4.98)	44-2134-86
4,5-Mesoporphyrinedione dimethyl ester	CH_2Cl_2	417(5.13),436(5.00), 592(3.98),623(4.28)	44-2134-86
4,6-Mesoporphyrinedione dimethyl ester	CH_2Cl_2	402(4.86),417(4.95), 437(4.96),544(3.96), 583(4.18),592(4.16), 637(4.20)	44-2134-86
$C_{36}H_{42}N_4O_{10}$			
1H-Purine-2,6-dione, 9-[2-O-benzoyl-6-O-[6-(benzoyloxy)hexyl]-3,4-O-(1-methylethylidene)-β-D-galactopyranosyl]-3,9-dihydro-1,3-dimethyl-	$CHCl_3$	277(4.14)	136-0256-86K
$C_{36}H_{42}O_6$			
3-Butyn-2-ol, 4,4',4",4"',4"",4""'-(1,2,3,4,5,6-benzenehexayl)hexakis[2-methyl-	MeOH	273s(4.88),289(5.13), 297s(4.83),309(4.85)	89-0268-86
$C_{36}H_{42}O_{19}$			
Sucrose, 3',6-diacetyl-3,6'-diferuloyl-	EtOH	234(4.57),300s(4.65), 327(4.81)	102-2897-86
Sucrose, 4,6-diacetyl-3,6'-diferuloyl-	EtOH	235(4.35),300s(4.58), 327(4.54)	102-2897-86
$C_{36}H_{43}ClN_4$			
21H,23H-Porphine, 2-(2-chloroethenyl)-3,7,8,12,13,17,18-heptaethyl-, (E)-	$CHCl_3$	404(5.18),503(4.01), 541(4.05),571(3.79), 627(3.54)	150-0452-86S
	+ CF_3COOH	407.5(5.37),553(4.08), 598(3.84)	150-0452-86S
$C_{36}H_{43}N_5O_3S$			
21H-Bilin-1(10H)-one, 19-[(1,5-dihydro-3,4-dimethyl-5-oxo-2H-pyrrol-2-ylidene)methyl]-22,23-dihydro-10-[(2-hydroxyethyl)thio]-2,3,7,8-12,13,17,18-octamethyl-, [19(Z)]-	$CHCl_3$	318(4.58),394(4.51), 554(4.43)	49-0057-86
$C_{36}H_{44}Br_2N_4Ru$			
Ruthenium, dibromo[2,3,7,8,12,13,17,18-octaethyl-21H,23H-porphinato(2-)-N^{21},N^{22},N^{23},N^{24}]-, (OC-6-12)-	$CHCl_3$	360(4.71),398(4.91), 505(4.22),535(4.12)	77-0787-86
$C_{36}H_{44}ClGaN_4$			
Gallium, chloro[2,3,7,8,12,13,17,18-octaethyl-21H,23H-porphinato(2-)-N^{21},N^{22},N^{23},N^{24}]-, (SP-5-12)-	MeOH	377(4.16),397(5.00), 489(2.50),531(3.52), 568(3.66)	90-1157-86
$C_{36}H_{44}N_2O_{10}$			
[4,4'-Biisoquinoline]-1,1',3,3'(2H-2'H,4H,4'H)-tetrone, 2,2'-dicyclohexyl-5,5',6,6',7,7'-hexamethoxy-, (R*,R*)-(±)-	$CHCl_3$	317(3.49)	139-0337-86C

Compound	Solvent	$\lambda_{max}(\log \epsilon)$	Ref.
$C_{36}H_{44}N_4O$			
Manzamine A	MeOH	219(4.36),236(4.27), 280(4.03),290s(3.99), 346(3.72),357(3.75)	35-6404-86
$C_{36}H_{44}N_4O_6$			
21H-Biline-2-propanoic acid, 7-(2-ethoxy-2-oxoethyl)-13,17-diethyl-1,19,22,24-tetrahydro-3,8,12,18-tetramethyl-1,19-dioxo-, ethyl ester	$CHCl_3$	366(4.80),613(4.35)	78-4137-86
	acid	366(4.63),660(4.64)	78-4137-86
21H-Biline-2-propanoic acid, 18-(2-ethoxy-2-oxoethyl)-7,13-diethyl-1,19,22,24-tetrahydro-3,8,12,17-tetramethyl-1,19-dioxo-, ethyl ester	$CHCl_3$	372(4.42),651(3.59)	78-4137-86
	acid	368(4.43),665(4.21)	78-4137-86
$C_{36}H_{44}N_4Rh$			
Rhodium(1+), [2,3,7,8,12,13,17,18-octaethyl-21H,23H-porphinato(2-)-N^{21},N^{22},N^{23},N^{24}]-, (SP-4-1)-, perchlorate	CH_2Cl_2	396(5.00),516(4.20), 549(4.54)	157-0168-86
tetrafluoroborate	CH_2Cl_2	396(5.26),514(4.20), 548(4.51)	157-0168-86
$C_{36}H_{44}N_4Zn$			
Zinc, [2,3,7,8,12,13,17,18-octaethyl-21H,23H-porphinato(2-)-N^{21},N^{22},N^{23}-N^{24}]-, (SP-4-1)-	benzene	403(5.55),530(4.23), 568(4.49)	35-1155-86
cyanide adduct	benzene	425(5.62),554(4.33), 578(4.28)	35-1155-86
pyridine adduct	benzene	425(5.62),554(4.33), 590(4.01)	35-1155-86
$C_{36}H_{44}O_2Si_2$			
Cyclodisiloxane, tetrakis(2,4,6-trimethylphenyl)-	C_6H_{12}	208(5.06),234(4.40)	157-0531-86
$C_{36}H_{44}O_{10}$			
Daphnodorin B hexamethyl ether	dioxan	218(4.37),288(4.03)	94-0595-86
$C_{36}H_{45}BrN_4O_6$			
Jasplakinolide	MeOH	281(3.73),290(3.61)	88-2797-86
$C_{36}H_{45}IrN_4$			
Iridium, hydro[2,3,7,8,12,13,17,18-octaethyl-21H,23H-porphinato(2-)-N^{21},N^{22},N^{23},N^{24}]-, (SP-5-31)-	benzene	392(4.95),501(3.97), 531(4.41)	35-3659-86
$C_{36}H_{46}$			
6,6'-Biazulene, 2,2'-dioctyl-	CH_2Cl_2	253.5(4.46),317(5.21), 394.5(4.39),581(2.80)	24-2272-86
$C_{36}H_{46}N_2O_4$			
3H-Indol-3-one, 2-(1-acetyl-1,3-dihydro-3-oxo-2H-indol-2-ylidene)-1,2-dihydro-1-(1-oxooctadecyl)-, (E)-	benzene	418(3.73),564(3.82)	44-2529-86
	butyl stearate	440(3.62),567(3.85)	44-2529-86
$C_{36}H_{46}N_4$			
21H,23H-Porphine, 2,3,7,8,12,13,17,18-octaethyl-	$CHCl_3$	330s(4.26),352s(4.54), 376s(4.94),399(5.23)	35-6449-86

Compound	Solvent	$\lambda_{max}(\log \epsilon)$	Ref.
$C_{36}H_{48}N_4O_2$			
Biliverdin, octaethyl-O-methyl-	CH_2Cl_2	369(4.58),620s(4.00), 369(4.09)	39-0001-86C
$C_{36}H_{48}O_{14}$			
Rehmaionoside pentaacetate benzoate, A-	MeOH	228(4.15)	94-2294-86
B-	MeOH	229(4.11)	94-2294-86
$C_{36}H_{50}N_2O$			
18α-Olean-2-eno[2,3-b]quinoxaline, 19β,28-epoxy-	C_6H_{12}	232(4.42),236(4.53), 240(4.48),310(3.92), 321(3.99)	73-0118-86
$C_{36}H_{52}O_8$			
Lanosta-7,9(11),24-trien-26-oic acid, 3,15,22-triacetoxy- (ganoderic acid T)	n.s.g.	216(4.0),225(4.0), 234(4.0),242(4.0)	94-2282-86
Oleana-11,13(18)-diene-23,28-dioic acid, 2α,3β-diacetoxy-, dimethyl ester	EtOH	244(4.30),252(4.40), 261(4.22)	2-0113-86
Oleana-12,18-diene-23,28-dioic acid, 2,3-diacetoxy-, dimethyl ester	EtOH	242(3.90)	2-0113-86
$C_{36}H_{52}O_9$			
Olean-12-ene-23,28-dioic acid, 2,3-diacetoxy-19-oxo-, dimethyl ester, (2α,3β,4β)-	EtOH	292(1.95)	2-0113-86
Olean-13(18)-ene-23,28-dioic acid, 2,3-diacetoxy-19-oxo-, dimethyl ester, (2α,3β)-	EtOH	253(3.85)	2-0113-86
$C_{36}H_{54}O_6$			
5α-Lanosta-7,9(11),24-triene-3,26,27-triol, triacetate	MeOH	236(3.89),243(3.94), 251(3.78)	100-0621-86
$C_{36}H_{54}Si_6$			
Silane, (1,2,3,4,5,6-benzenehexayl-hexa-2,1-ethynediyl)hexakis[tri-methyl-	hexane	286s(4.76),291(4.86), 302(5.18),315s(4.87), 321(4.95)	89-0268-86
$C_{36}H_{55}NO_2S$			
3-Thiazoline, 2,4-dimethyl-2-[2-(3,5-di-tert-butyl-4-hydroxyphenyl)ethyl]-5-(3,5-di-tert-butyl-4-hydroxybenzyl)-	EtOH	278(3.60)	103-0052-86
$C_{36}H_{58}P_2$			
Diphosphene, bis[2,4,6-tris(1,1-dimethylethyl)phenyl]-, cis	heptane	343(3.87),452(3.09)	139-0091-86A
trans	heptane	347(3.89),463(3.13)	139-0091-86A
$C_{36}H_{61}N_3O_2$			
Cholest-4-en-3-one, 6-(4-methyl-1-piperazinyl)-2-(4-morpholinyl)-, (2α,6β)-	EtOH	239.0(4.14)	118-0461-86
$C_{36}H_{90}Si_9$			
Cyclotrisilane, hexakis(triethyl-silyl)-	n.s.g.	335s(3.11)	77-1768-86

Compound	Solvent	λ_{max}(log ε)	Ref.
$C_{37}H_{20}$			
Dibenzo[fg,ij]phenanthro[9,10,1,2,3-pqrst]pentaphene, 9-methyl-	dioxan	224(5.02),237(4.96), 289(4.50),302(4.75), 320(4.77),330(4.99), 362(4.59),373(4.16), 405(3.04),413(2.96), 428(2.75)	78-1127-86
$C_{37}H_{22}N_4O_2$			
Benzonitrile, 4,4'-[5-(4-cyanobenzoyl)-2,5-dihydro-2,2-diphenyl-4,5-oxazolediyl]bis-	CH_2Cl_2	233(4.10),248(4.21), 256(4.15)	39-0623-86C
$C_{37}H_{23}Br_4O_2$			
Pyrylium, 2-[3-[4,6-bis(4-bromophenyl)-2H-pyran-2-ylidene]-1-propenyl]-4,6-bis(4-bromophenyl)-, tetrafluoroborate	CH_2Cl_2	745(4.75),820(4.88)	103-1295-86
$C_{37}H_{23}Cl_4O_2$			
Pyrylium, 2-[3-[4,6-bis(4-chlorophenyl)-2H-pyran-2-ylidene]-1-propenyl]-4,6-bis(4-chlorophenyl)-, tetrafluoroborate	CH_2Cl_2	740(4.75),815(4.88)	103-1295-86
$C_{37}H_{23}F_4O_2$			
Pyrylium, 2-[3-[4,6-bis(4-fluorophenyl)-2H-pyran-2-ylidene]-1-propenyl]-4,6-bis(4-fluorophenyl)-, tetrafluoroborate	CH_2Cl_2	730(4.77),810(4.92)	103-1295-86
$C_{37}H_{23}N_3$			
3,5-Pyridinedicarbonitrile, 2,6-bis([1,1'-biphenyl]-4-yl)-4-phenyl-	MeCN	315(4.73)	73-1061-86
$C_{37}H_{23}N_4O_{10}$			
Pyrylium, 2-[3-[4,6-bis(4-nitrophenyl)-2H-pyran-2-ylidene]-1-propenyl]-4,6-bis(4-nitrophenyl)-, tetrafluoroborate	CH_2Cl_2	760(4.62),830(4.59)	103-1295-86
$C_{37}H_{24}N_2O_2$			
2-Butene-1,4-dione, 1,4-diphenyl-2-(2-phenyl-1H-phenanthro[9,10-d]imidazol-1-yl)-, (E)-	MeOH	223(4.49),254(4.74)	44-3420-86
$C_{37}H_{25}ClFNOS$			
Methanone, [4-(4-chlorophenyl)-1-(4-fluorophenyl)-1,2-dihydro-6-phenyl-5-(phenylmethyl)-2-thioxo-3-pyridinyl]phenyl-	EtOH	248(4.36),296(4.15), 400(3.85)	118-0952-86
$C_{37}H_{26}ClNOS$			
Methanone, [1-(4-chlorophenyl)-1,2-dihydro-4,6-diphenyl-5-(phenylmethyl)-2-thioxo-3-pyridinyl]phenyl-	EtOH	248(4.33),297(4.19), 400(3.88)	118-0952-86
$C_{37}H_{26}N_2O$			
1H,3H-[1,3,5]Oxadiazino[3,4-a]quinoline, 1,3-bis(diphenylmethylene)-	EtOH	312(4.42),453(4.11)	97-0100-86

Compound	Solvent	λ_{max}(log ε)	Ref.
$C_{37}H_{26}N_2O_2$			
2-Butene-1,4-dione, 2-(7a,7b-dihydro-2-phenyl-1H-phenanthro[9,10-d]imidazol-1-yl)-1,4-diphenyl-, [1(E)-7aα,7bβ]-	MeOH	236(4.47),290(4.20), 314(4.28)	44-3420-86
2-Butene-1,4-dione, 2-(2,4,5-triphenyl-1H-imidazol-1-yl)-, (E)-	MeOH	223(4.68),270(4.66)	44-3420-86
(Z)-	MeOH	220(4.74),278(4.72)	44-3420-86
Methanone, (5,6-dihydro-2,3-diphenylimidazo[2,1-a]isoquinoline-5,6-diyl)bis[phenyl-, trans	MeOH	220(4.16),249(4.17)	44-3420-86
$C_{37}H_{26}N_8O_7$			
Benzamide, 4-cyano-N-(4-cyanobenzoyl)-N-[7-(4-cyanobenzoyl)-3a,4,6a,6,13-13a-hexahydro-2,2-dimethyl-4,13-epoxy-5H-1,3-dioxolo[5,6][1,3]oxazocino[3,2-e]purin-8-yl]-, [3aR-(3aα,4β,6aβ,13β,13aα)]-	MeOH	243(4.68)	23-2109-86
$C_{37}H_{27}NOS$			
Methanone, [1,2-dihydro-1,4,6-triphenyl-5-(phenylmethyl)-2-thioxo-3-pyridinyl]phenyl-	EtOH	248(4.40),296(4.15), 398(3.87)	118-0952-86
$C_{37}H_{27}O_2$			
Pyrylium, 2-[3-(4,6-diphenyl-2H-pyran-2-ylidene)-1-propenyl]-4,6-diphenyl-, tetrafluoroborate	CH_2Cl_2	730(4.77),800(4.93)	103-1295-86
$C_{37}H_{29}N_2$			
Benzenamine, 4-[[4-(diphenylamino)-phenyl]methyl]-N,N-diphenyl-, cation	HOAc-10% CF_3COOH	412(3.87),657(5.01)	64-1045-86B
$C_{37}H_{30}O_4$			
4H-1-Benzopyran-4-one, 6,8-bis(diphenylmethyl)-5,7-dihydroxy-2,3-dimethyl-	MeOH	266(4.42),304(4.42)	2-0478-86
$C_{37}H_{30}O_7$			
4H-1-Benzopyran-4-one, 2-[3,4-bis(phenylmethoxy)phenyl]-6-hydroxy-7-methoxy-5-(phenylmethoxy]-	EtOH	279(4.28),330(4.48)	94-0030-86
$C_{37}H_{31}NO_4$			
1H-Indene-1,3(2H)-dione, 2-[8a-(2,5-dimethoxyphenyl)-6,7,8,8a-tetrahydro-2,3-diphenyl-1(5H)-indolizinylidene]-	MeCN	276(4.38),413(4.14)	118-0908-86
$C_{37}H_{31}NO_5$			
Benzoic acid, 4-methoxy-, 2-[(diphenylmethylene)amino]-1,2-bis(4-methoxyphenyl)ethenyl ester	CH_2Cl_2	261(4.75),293(4.43), 379(3.69)	39-2021-86C
Methanone, [2,5-dihydro-4,5-bis(4-methoxyphenyl)-2,2-diphenyl-5-oxazolyl](4-methoxyphenyl)-	CH_2Cl_2	230(4.54),289(4.65)	39-0623-86C
$C_{37}H_{31}N_4O_4PS$			
3aH-Indole-3a-carboxylic acid, 6,7-dicyano-1,2,3,7a-tetrahydro-1-[(4-	$CHCl_3$	438(3.82)	24-2723-86

Compound	Solvent	$\lambda_{max}(\log \epsilon)$	Ref.
methylphenyl)sulfonyl]-4-[(triphenylphosphoranylidene)amino]- (cont.)			24-2723-86
$C_{37}H_{32}N_4O_6$			
Acetamide, N-phenyl-2-[3,4,6,7-tetrahydro-9,10-dimethoxy-3-phenyl-1-[2-[[(phenylamino)carbonyl]oxy]ethenyl]-2H-pyrimido[6,1-a]isoquinolin-2-ylidene]-	CH_2Cl_2	272(3.98),302(3.91), 366(4.05)	24-2553-86
$C_{37}H_{34}O_{12}$			
Benzoic acid, 4-methoxy-, 5,8-diacetoxy-5,6,7,8-tetrahydro-4-oxo-2-(2-phenylethyl)-4H-1-benzopyran-6,7-diyl ester, [5S-(5α,6β,7α,8β)]-	MeOH	210(4.75),258(4.74)	94-2766-86
[5S-(5α,6β,7β,8α)]-	EtOH	209(4.65),259(4.67)	94-2766-86
$C_{37}H_{35}Br_2NO_{10}$			
Phenol, 2,2'-[3,5-bis(2-bromo-4,5-dimethoxyphenyl)-2,4-pyridinediyl]bis[5,6-dimethoxy-	$CHCl_3$	238(4.66),293(4.46), 315(4.40)	142-1943-86
$C_{37}H_{35}N_5O_7$			
Adenosine, N,N-bis(4-methylbenzoyl)-2',3'-O-(1-methylethylidene)-, 5'-(4-methylbenzoate)	MeOH	243(4.54),275s(4.43)	23-2109-86
Benzamide, N-[3a,4,6a,7,13,13a-hexahydro-2,2-dimethyl-7-(4-methylbenzoyl)-4,13-epoxy-5H-1,3-dioxolo[5,6]-[1,3]oxazocino[3,2-e]purin-8-yl]-4-methyl-N-(4-methylbenzoyl)-	MeOH	253(4.67),310s(4.10)	23-2109-86
$C_{37}H_{35}N_5O_{10}$			
Adenosine, N,N-bis(4-methoxybenzoyl)-2',3'-O-(1-methylethylidene)-, 5'-(4-methoxybenzoate)	MeOH	260s(4.59),272(4.61), 290(4.58)	23-2109-86
8,5'-O-Cycloadenosine, 7,8-dihydro-2',3'-O-isopropylidene-N^6,N^6,7-tri-p-anisoyl-	MeOH	278(4.65)	23-2109-86
$C_{37}H_{37}FeNOP$			
Iron(1+), carbonyl(η^5-2,4-cyclopentadien-1-yl)[1,3,3-trimethyl-4-(4-methylphenyl)-2-azetidinylidene]-(triphenylphosphine)-, tetrafluoroborate	CH_2Cl_2	236(4.20),264(4.00), 322s(3.43)	88-3811-86
$C_{37}H_{37}NO_{10}$			
Phenol, 2,2'-[3,5-bis(3,4-dimethoxyphenyl)-2,4-pyridinediyl]bis[5,6-dimethoxy-	$CHCl_3$	238(4.66),293(4.46), 315(4.40)	142-1943-86
$C_{37}H_{38}N_2O_6$			
Medelline	EtOH	220(4.38),250s(3.98), 282(3.82)	142-0607-86
	EtOH-NaOH	222(4.58),300(3.95)	142-0607-86
$C_{37}H_{38}N_2O_7$			
Oxandrine	EtOH	209(4.73),282(4.02)	23-1390-86
	EtOH-NaOH	224(4.92),296(4.30)	23-1390-86

Compound	Solvent	λmax (log ε)	Ref.
Pseudoxandrine	EtOH	208(4.78),282(3.94)	23-1390-86
	EtOH-NaOH	222(4.96),300(4.24)	23-1390-86
Thalmiculimine, (-)-	MeOH	274(3.97)	100-0488-86
	MeOH-acid	237s(4.40),284(3.93)	100-0488-86
$C_{37}H_{40}$			
Benzene, 1,1'-[(2,2-dimethyl-3,3-diphenylcyclopropyl)ethenylidene]-bis[2,4,6-trimethyl-	EtOH	242s(4.31),260(4.29), 264s(4.27)	44-0196-86
Benzene, 1,1'-(3,3-dimethyl-5,5-diphenyl-1,4-pentadienylidene)-bis[2,4,6-trimethyl-	EtOH	242(4.46),256s(4.26), 289(3.36),337(0)	44-0196-86
$C_{37}H_{40}N_2O_6$			
Gyroamericine, (-)-	MeOH	241(4.30),282(3.84)	100-0101-86
Gyrocarpusine, (+)-	MeOH	234(4.42),283(3.85)	100-0101-86
Gyrocarpine, (-)-	MeOH	239s(4.34),284(3.87)	100-0101-86
Homoaromoline	EtOH	222(4.59),282(3.96)	100-0854-86
	EtOH-NaOH	223(4.90),291(3.90)	100-0854-86
$C_{37}H_{40}N_2O_7$			
Berbamine 2'-β-N-oxide, (+)-	MeOH	234s(4.55),282(4.04)	100-0538-86
2'-Northalmiculine, (-)-	MeOH	236s(4.33),281(3.63)	100-0488-86
Oxandrine, dihydro-	EtOH	210(4.99),285(4.02)	23-1390-86
	EtOH-NaOH	220(5.00),292(4.09)	23-1390-86
Thalisopidine	MeOH	234s(4.25),285(3.86)	102-0935-86
Thalmine, 5-hydroxy-, (-)-	MeOH	236s(4.43),281(3.78)	100-0488-86
$C_{37}H_{40}N_4O_6$			
21H,23H-Porphine-3,13,17-tripropanoic acid, 8-ethenyl-2,12,18-trimethyl-, trimethyl ester	$CHCl_3$	330s(4.22),360s(4.60), 373s(4.80),406(5.24)	35-6449-86
at 295°K	CH_2Cl_2	628(4.56)	35-6449-86
at 199°K	CH_2Cl_2	626(4.60)	35-6449-86
$C_{37}H_{40}N_4O_7$			
Acetamide, N-[4,7-dihydro-3-(methoxymethyl)-5-methyl-4-oxo-7-[2,3,5-tris-O-(phenylmethyl)-α-D-arabinofuranosyl]-3H-pyrrolo[2,3-d]pyrimidin-2-yl]-	MeCN	306(3.98)	18-1915-86
β-	MeCN	306(3.98)	18-1915-86
21H,23H-Porphine-3,13,17-tripropanoic acid, 8-acetyl-2,12,18-trimethyl-, trimethyl ester	$CHCl_3$	335s(4.44),366s(4.55), 388s(4.80),413(5.32), 424s(5.07)	35-6449-86
at 295°K	CH_2Cl_2	633(4.34)	35-6449-86
at 199°K	CH_2Cl_2	630(4.28)	35-6449-86
$C_{37}H_{40}N_4O_8$			
21H,23H-Porphine-2-propanoic acid, 8,13-bis(2-acetoxyethyl)-18-carboxy-3,7,12,17-tetramethyl-, α-methyl ester	CH_2Cl_2	404(5.26),508(3.91), 548(4.05),576(3.83), 634(3.12)	44-0657-86
$C_{37}H_{41}F_3N_2O_{14}$			
Adriamycin, N-(trifluoroacetyl)-14-O-(N-acetylleucyl)-	MeOH	232(4.64),250(4.46), 287(4.02),480(4.11), 494(4.11),529(3.87)	87-1273-86
$C_{37}H_{41}N_2S$			
Benzothiazolium, 3-ethyl-2-[[7-[(1-	DMSO	864(5.05)	24-3102-86

Compound	Solvent	λmax(log ε)	Ref.
ethyl-1,3-dihydro-3,3-dimethyl-2H-indol-2-ylidene)methyl]-4,4a,5,6-10,10a-hexahydro-2(3H)-anthracen-ylidene]methyl-, iodide (cont.)			24-3102-86
$C_{37}H_{42}N_2O_6$			
Pisopowamine	EtOH	208(4.69),228s(4.38), 288(3.92)	100-1018-86
Pisopowiardine	MeOH	209(4.75),226s(4.55), 288(4.15)	100-1018-86
Popisopine	EtOH	208(4.65),228s(4.37), 286(3.92)	100-1018-86
$C_{37}H_{42}N_2O_9$			
Popisonine	EtOH	208(4.66),229s(4.44), 286(3.93)	100-1018-86
$C_{37}H_{44}N_4O_8$			
21H-Biline-2-propanoic acid, 7,12-bis(2-acetoxyethyl)-1-carboxy-10,23-dihydro-3,8,13,17,19-pentamethyl-, α-methyl ester, dihydrobromide	CH_2Cl_2-CF_3COOH	444(4.90),516(4.35)	44-4660-86
$C_{37}H_{44}O_{13}$			
Urdamycin B	MeOH	268(4.43),403(3.59)	158-1657-86
	MeOH-NaOH	248(4.15),266(4.24), 315s(--),505(3.56)	158-1657-86
$C_{37}H_{46}Cl_2N_4O_4$			
a,c-Biladiene, 1,8-bis(2-chloroethyl)-4,5-bis[2-(methoxycarbonyl)ethyl]-1',2,3,6,7,8'-hexamethyl-, dihydrobromide	CH_2Cl_2	448(4.76),524(5.05)	44-0666-86
a,c-Biladiene, 2,7-bis(2-chloroethyl)-4,5-bis[2-(methoxycarbonyl)ethyl]-1,1',3,3,6,8'-hexamethyl-, dihydrobromide	CH_2Cl_2	448(4.74),524(5.05)	44-0666-86
a,c-Biladiene, 2,8-bis(2-chloroethyl)-bis[2-(methoxycarbonyl)ethyl]-hexamethyl-, dihydrobromide, [1-^{13}C]-	CH_2Cl_2	450(4.97),524(5.10)	44-4667-86
[7-^{13}C]-	CH_2Cl_2	450(4.95),524(5.09)	44-4667-86
a,c-Biladiene, 4,6-bis(2-chloroethyl)-bis[2-(methoxycarbonyl)ethyl]-hexamethyl-, dihydrobromide, [2-^{13}C]-	CH_2Cl_2	448(4.87),524(5.06)	44-4667-86
[7-^{13}C]-	CH_2Cl_2	450(4.83),524(5.08)	44-4667-86
$C_{37}H_{47}IrN_4$			
Iridium, methyl[2,3,7,8,12,13,17,18-octaethyl-21H,23H-porphinato(2-)-N^{21},N^{22},N^{23},N^{24}]-, (SP-5-31)-	benzene	338(4.30),391(4.98), 500s(--),531(4.34)	35-3659-86
$C_{37}H_{52}O_6$			
Avenestergenin B-2	EtOH	229(4.08),257s(2.93), 273(2.72),281(2.56)	39-1905-86C
$C_{37}H_{52}O_7$			
Avenestergenin A-2	EtOH	228(4.09),254s(3.23), 274(2.91),281(2.85)	39-1905-86C
$C_{37}H_{53}NO_9$			
Rhizoxin, 2,3:5b,7-dideepoxy-2,3-dide-	MeOH	296(4.63),309(4.73),	158-0762-86

Compound	Solvent	λ_{max}(log ε)	Ref.
hydro-5b-ethoxy-7-hydroxy- (cont.)		323(4.60)	158-0762-86
$C_{37}H_{53}NO_{10}$			
L-681,217	pH 1	232(4.71),285s(4.43), 299(4.52),310s(4.46)	158-1361-86
	pH 6	234(4.72),278s(4.48), 280(4.59),300s(4.51)	158-1361-86
	pH 13	230(4.71),278s(4.49), 288(4.55),300s(4.29)	158-1361-86
	MeOH	234(4.72),278s(4.50), 288(4.61),298s(4.55)	158-1361-86
$C_{37}H_{54}O_2$			
5,9,13,17-Nonadecatetraenal, 6,10,14-18-tetramethyl-2-[4-[(phenylmethoxy)methyl]-4-hexenylidene]-, (all-E)-	EtOH	200(4.63),233(4.01)	70-1851-86
$C_{37}H_{54}O_8$			
Oleana-11,13(18)-dien-28-oic acid, 2α,3β,23-triacetoxy-, methyl ester	EtOH	244(4.28),252(4.40), 261(4.20)	2-0113-86
$C_{37}H_{54}O_9$			
Diaulusterol A triacetate	MeOH	264(4.11)	23-1527-86
Olean-12-en-28-oic acid, 2α,3β,23-triacetoxy-19-oxo-, methyl ester	EtOH	292(2.04)	2-0113-86
$C_{37}H_{54}O_{10}$			
Lanost-8-en-26-oic acid, 3,7,15-triacetoxy-11,23-dioxo-, methyl ester, (3β,7β,15α,25R)-	EtOH	249(4.0)	102-1189-86
$C_{37}H_{56}O_{15}$			
Aldgamycin F, 8-deoxy- (aldgamycin G)	MeOH	218(4.35),241(4.09)	158-1776-86
$C_{37}H_{56}O_{16}$			
Aldgamycin F	MeOH	218(4.32),240(4.09)	158-1776-86
$C_{37}H_{57}NO_{13}$			
Tylonolide, 5-O-[2,4-di-O-acetyl-3,6-dideoxy-3-(dimethylamino)-β-D-glucopyranosyl]-, 23-acetate	EtOH	279(4.41)	158-1108-86
$C_{37}H_{60}N_2O_5$			
1H-Pyrrole-3-propanoic acid, 2-[(1,1-dimethylethoxy)carbonyl]-5-[(3-ethyl-1,5-dihydro-4-methyl-5-oxo-2H-pyrrol-2-ylidene)methyl]-4-methyl-, hexadecyl ester, (Z)-	$CHCl_3$	383(4.44)	49-1081-86
$C_{37}H_{64}N_4O$			
Cholest-4-en-3-one, 2,6-bis(4-methyl-1-piperazinyl)-, (2α,6β)-	EtOH	239(4.01)	107-0195-86
isomer	EtOH	241(4.03)	107-0195-86
$C_{37}H_{64}O_{10}$			
Misakinolide A	MeOH	220(3.70)	138-1499-86

Compound	Solvent	λ_{max}(log ε)	Ref.
$C_{38}H_{24}$			
5,16[1',2']:8,13[1",2"]-Dibenzeno-hexacene, 5,8,13,16-tetrahydro-	MeCN	248(5.21),261(4.91), 269(4.92),313(4.08), 320(3.79),328(3.84), 340(3.56),356(3.51), 376(3.26)	78-1641-86
$C_{38}H_{24}F_{18}O_7S_2$			
1-Butanesulfonic acid, 1,1,2,2,3,3,4-4,4-nonafluoro-, oxybis[2-(4-methyl-phenyl)-1-phenyl-2,1-ethenediyl] ester, (Z,Z)-	EtOH	227(4.62),283(4.34)	24-2220-86
$C_{38}H_{24}F_{18}O_9S_2$			
1-Butanesulfonic acid, 1,1,2,2,3,3,4-4,4-nonafluoro-, oxybis[2-(4-meth-oxyphenyl)-1-phenyl-2,1-ethenediyl]-ester, (Z,Z)-	EtOH	203.5(4.71),230.5(4.57), 292.5(4.35)	24-2220-86
$C_{38}H_{24}O_8$			
6,21:10,17-Diepoxynonacene-1,4,8,12-15,19-hexone, 4a,5,6,10,11,11a,15a-16,17,21,22,22a-dodecahydro-	dioxan	277(4.63),335(3.63)	44-4160-86
$C_{38}H_{25}NO_2$			
1H-Indene-1,3(2H)-dione, 2-(5,6-dihy-dro-9,10-diphenyl-8H-naphtho[2,1-a]quinolizin-8-ylidene)-	MeCN	290(4.47),365(4.10), 453(3.97)	118-0906-86
$C_{38}H_{26}BCl_4O_4$			
Boron(1+), bis[1,7-bis(4-chlorophen-yl)-1,6-heptadiene-3,5-dionato-O,O']-, (T-4)-, perchlorate	CH_2Cl_2	488(5.18)	97-0399-86
$C_{38}H_{27}NO_2$			
1H-Indene-1,3(2H)-dione, 2-(5,6,9,10-tetrahydro-9,10-diphenyl-8H-naph-tho[2,1-a]quinolizin-8-ylidene)-	MeCN	278(4.41),317(4.15), 397(4.24)	118-0906-86
$C_{38}H_{27}N_5O_4$			
1H-Indole, 2,3-bis[1H-indol-3-yl-(3-nitrophenyl)methyl]-	EtOH	220(4.74),272(4.18)	2-1204-86
	$EtOH-HClO_4$	204(4.52),218(4.49), 262(4.26)	2-1204-86
$C_{38}H_{28}ClNO_2S$			
Methanone, [4-(4-chlorophenyl)-1,2-dihydro-1-(4-methoxyphenyl)-6-phen-yl-5-(phenylmethyl)-2-thioxo-3-pyridinyl]phenyl-	EtOH	247(4.38),296(4.16), 401(3.81)	118-0952-86
$C_{38}H_{28}N_2O_2Se$			
1H-Pyrrolizin-1-one, 2,2'-seleno-bis[2,3-dihydro-6,7-diphenyl-	$CHCl_3$	256(4.53),470.05(4.06)	83-0749-86
$C_{38}H_{28}N_2O_8$			
Bipowinone	MeOH	237(4.56),248s(4.49), 278s(4.24),290s(4.18), 300s(4.11),410(3.97), 474(3.92),506s(3.93)	100-1028-86
	MeOH-HCl	209(--),249(--), 265s(--),285s(--),	100-1028-86

Compound	Solvent	λ_{max}(log ϵ)	Ref.
Bipowinine (cont.)		300s(--),410s(--), 433(--),496(--)	100-1028-86
$C_{38}H_{28}O_{13}$			
Benzo[e]naphtho[2',3':5,6]fluoreno-[1,9a-b]oxepin-5,10,15,19-tetrone, 1,8,11,18-tetraacetoxy-5c,8,8a,16-tetrahydro-13-methyl-, (5cα,8β,8aα-15aα)-	EtOH	254(4.49),320(3.83), 445(3.48)	39-0255-86C
$C_{38}H_{29}NO_2S$			
Methanone, [1,2-dihydro-1-(4-methoxy-phenyl)-4,6-diphenyl-5-(phenylmeth-yl)-2-thioxo-3-pyridinyl]phenyl-	EtOH	246(4.37),296(4.17), 400(3.82)	118-0952-86
$C_{38}H_{30}$			
Benzene, 1,2,4,5-tetrakis(2-phenyl-ethenyl)-, (all-E)-	toluene	334(4.67)	24-1716-86
$C_{38}H_{30}BO_4$			
Boron(1+), bis(1,7-diphenyl-1,6-hepta-diene-3,5-dionato-O,O')-, (T-4)-, perchlorate	CH_2Cl_2	480(5.10)	97-0399-86
$C_{38}H_{30}BO_8$			
Boron(1+), bis[1,7-bis(2-hydroxyphen-yl)-1,6-heptadiene-3,5-dionato-O^3,O^5]-, (T-4)-, perchlorate	CH_2Cl_2	505(4.83)	97-0330-86
Boron(1+), bis[1,7-bis(4-hydroxyphen-yl)-1,6-heptadiene-3,5-dionato-O^3,O^5]-, (T-4)-, perchlorate	CH_2Cl_2	521(4.66)	97-0330-86
$C_{38}H_{31}NO_2$			
1H-Perylo[1,12-efg]isoindole-1,3(2H)-dione, 2-[2,5-bis(1,1-dimethylethyl)-	$CHCl_3$	428(3.29),455(3.68), 482(3.83)	24-2373-86
$C_{38}H_{32}$			
5,9,13,17,21,25,29-Tetratriacontahep-taene-3,7,11,15,19,23,27,31-octa-yne, 2,2,33,33-tetramethyl-	benzene	414(4.96),436s(4.87), 454s(4.69)	35-4685-86
$C_{38}H_{32}O_4$			
4H-1-Benzopyran-4-one, 6,8-bis(diphen-ylmethyl)-5-hydroxy-7-methoxy-2,3-dimethyl-	MeOH	264(4.72),298(4.24), 299(4.34)	2-0478-86
$C_{38}H_{32}O_6$			
1(2H)-Naphthalenone, 4-methoxy-2-[4-methoxy-6-methyl-1-oxo-8-(phenyl-methoxy)-2(1H)-naphthalenylidene]-6-methyl-8-(phenylmethoxy)-	$CHCl_3$	252(4.72),322(4.44), 332s(4.43),387s(3.79), 585(4.33)	5-1669-86
$C_{38}H_{33}NO_{10}$			
Scleroderris Blue	$CHCl_3$	268(4.48),368(4.36), 405s(--),608(4.26)	23-1585-86
$C_{38}H_{35}F_3N_2O_{14}$			
Adriamycin, N-(trifluoroacetyl)-14-O-(N-benzoylglycyl)-	MeOH	232(4.70),248(4.48), 287(4.04),480(4.08), 494(4.08),529(3.86)	87-1273-86

Compound	Solvent	λ_{max}(log ϵ)	Ref.
$C_{38}H_{36}BN_2O_4$			
Boron(1+), bis[5-[4-(dimethylamino)-phenyl]-1-phenyl-4-pentene-1,3-di-onato-O,O']-, (T-4)-, perchlorate	CH_2Cl_2	612(5.12)	97-0399-86
$C_{38}H_{36}N_2O_6$			
Thalmiculatimine, (+)-	MeOH	280(4.02)	100-0488-86
	MeOH-acid	237s(4.38),285(4.01)	100-0488-86
$C_{38}H_{36}N_2O_8$			
Bipowine	MeOH	212(4.14),266(4.51), 335(3.87),388(3.66)	100-1028-86
	MeOH-HCl	208(--),262(--), 272s(--),293(--), 340(--),358(--), 377(--)	100-1028-86
	MeOH-NaOH	228s(--),263(--), 293(--),354(--), 384(--),400(--), 485(--)	100-1028-86
$C_{38}H_{36}N_4O_2$			
2,3-Piperazinedicarbonitrile, 1,4-bis(phenylmethyl)-2,3-bis[2-(2-propenyloxy)phenyl]-, trans	EtOH	274(3.68)	24-0813-86
$C_{38}H_{36}N_4O_4$			
2-Propenoic acid, 3,3'-(10,11,22,23-tetrahydro-6H,18H-6,9:18,21-dieth-eno-5,24:17,12-dimethenodibenzo-[f,p][1,5,11,15]tetraazacycloei-cosine-7,19-diyl)bis-, dimethyl ester	CH_2Cl_2	237(4.91),287(4.18), 372(4.57)	44-2995-86
$C_{38}H_{36}N_5O_8PS$			
1H,4aH-[1,2,4]Triazolo[1',2':1,2][1,2]-diazeto[3,4-d]azepine-4a,9-dicarbox-ylic acid, 2,3,5,6,7,9a-hexahydro-2-methyl-7-[(4-methylphenyl)sulfonyl]-1,3-dioxo-8-[(triphenylphosphoran-ylidene)amino]-, dimethyl ester	$CHCl_3$	248(4.19),313(4.20)	24-2114-86
$C_{38}H_{36}O_9$			
5,8,11-Metheno-1H-benzo[a]fluorene-1,4,6,7,10-pentone, 11-acetoxy-9-[(1,4-dihydro-3-methyl-1,4-dioxo-2-naphthalenyl)methyl]-4a,5,6a,6b-8,9,10a,11,11a,11b-decahydro-2,4a,5,6b,9,11a-hexamethyl-, (4aα,5β,6aβ,6bβ,8β,9β,10aβ,11α-11aβ,11bα)-(±)-	EtOH	244(4.43),260(4.16), 328(3.51)	78-3663-86
$C_{38}H_{37}FeNOP$			
Iron(1+), carbonyl(η^5-2,4-cyclopenta-dien-1-yl)[1,3,3-trimethyl-4-(2-phenylethenyl)-2-azetidinylidene]-(triphenylphosphine)-, tetrafluoro-borate	CH_2Cl_2	236(4.16),257(4.18), 327(3.60)	88-3811-86
$C_{38}H_{40}N_2O_7$			
Oxandrinine	EtOH	208(4.50),282(3.64)	23-1390-86
	EtOH-NaOH	216(4.78),290(4.00)	23-1390-86

Compound	Solvent	λ_{max}(log ϵ)	Ref.
Pseudoxandrinine	EtOH	208(4.44),224s(4.34), 280(3.80)	23-1390-86
	EtOH-NaOH	220(4.90),296(4.33)	23-1390-86
$C_{38}H_{40}N_4O_2$			
[18,19'-Bieburnamenine]-14,14'(15H-15'H)-dione, 18,19-didehydro-, (3α,16α)-(3'α,16'α,19'S)-	EtOH-HCl	202(4.69),240(4.60), 265(4.24),300(3.97)	88-2775-86
$C_{38}H_{40}N_4O_6$			
1H-Purine-2,6-dione, 9-[3,4-dihydro-3-oxo-6-[[[6-(triphenylmethoxy)hexyl]oxy]methyl]-2H-pyran-2-yl]-3,6-dihydro-1,3-dimethyl-	$CHCl_3$	279(3.93)	136-0256-86K
$C_{38}H_{42}N_2O_6$			
Gyrolidine, (-)-	MeOH	261(4.23),282(4.12)	100-0101-86
$C_{38}H_{42}N_2O_7$			
Thaligosinine	MeOH	235s(4.44),283(3.93)	102-0935-86
Thalmiculine, (-)-	MeOH	235s(4.28),281(3.58)	100-0488-86
$C_{38}H_{42}N_4O_6$			
1H-Purine-2,6-dione, 9-[3,4-dideoxy-6-O-[6-(triphenylmethoxy)hexyl]-β-D-erythro-hex-3-enopyranosyl]-3,9-dihydro-1,3-dimethyl-	$CHCl_3$	279(3.95)	136-0256-86K
$C_{38}H_{42}N_4O_8$			
21H,23H-Porphine-3,13,17-tripropanoic acid, 8-(ethoxycarbonyl)-2,12,18-trimethyl-, trimethyl ester	$CHCl_3$	330s(4.36),355s(4.46), 385s(4.82),409(5.33), 420s(5.07)	35-6449-86
$C_{38}H_{44}N_2O_4$			
Benzo[1,2-c:4,5-c']dipyrrole-1,3,5,7-(2H,6H)-tetrone, 2,6-bis[2,5-bis-(1,1-dimethylethyl)phenyl]-	$CHCl_3$	308(3.50),316(3.50)	24-2373-86
$C_{38}H_{44}N_2O_6$			
Pisopowetine	EtOH	210(4.74),232s(4.54), 284(4.19)	100-1018-86
Pisopowiarine	MeOH	209(4.58),226s(4.36), 288(3.91)	100-1018-86
Popisine	EtOH	208(4.64),228s(4.36), 285(3.91)	100-1018-86
$C_{38}H_{44}N_4O_6$			
21H,23H-Porphine-2,8,18-tripropanoic acid, 15-ethyl-3,7,12,17-tetramethyl-, trimethyl ester	$CHCl_3$	407(5.33),504(4.17), 539(3.72),573(3.82), 627(3.26)	12-0419-86
$C_{38}H_{44}O_{20}$			
Sucrose, 3',4,6-triacetyl-3,6'-diferuloyl-	EtOH	234(4.44),300s(4.52), 327(4.67)	102-2897-86
$C_{38}H_{45}N_5O_5$			
21H,23H-Porphine-2,18-dipropanoic acid, 7-[2-(dimethylamino)-2-oxoethyl)-8-ethylidene-7,8-dihydro-3,7,12,17-tetramethyl-, dimethyl ester, (Z)-(±)-	$CHCl_3$	399(5.176),496(3.99), 503(3.95),532(4.03), 602(3.56),629(3.48), 659(4.61)	89-0458-86

Compound	Solvent	λ_{max}(log ε)	Ref.
$C_{38}H_{47}N_5O_5$			
21H,23H-Porphine-2,18-dipropanoic acid, 7-[2-(dimethylamino)-2-oxoethyl]-8-ethyl-7,8-dihydro-3,7,12,17-tetramethyl-, dimethyl ester, cis-(±)-	$CHCl_3$	392(5.195),489(4.025), 496(4.041),522(3.185), 532(3.047),592(3.505), 616(3.466),645(4.583)	89-0458-86
$C_{38}H_{49}N_5O_4S_2$			
2H-Pyrrol-2-one, 5,5'-[(3,4-dimethyl-1H-pyrrole-2,5-diyl)bis[[[(2-hydroxyethyl)thio]methylene](3,4-dimethyl-1H-pyrrole-5,2-diyl)methylidyne]]-bis[1,5-dihydro-3,4-dimethyl-	$CHCl_3$	394(4.81)	49-0057-86
$C_{38}H_{49}N_5O_6Si_2$			
Benzamide, N-benzoyl-N-[6,7,7a,7b-13a,14a-hexahydro-9,9,11,11-tetrakis(1-methylethyl)-13H-[1,3,5,2,4]-trioxadisilocino[6",7":4',5']furo-[3',2':5,6]pyrido[1,2-e]purin-4-yl]-, [7aS-(7aα,7bβ,13aα,14aα)-	MeOH	249(4.38),278(4.28)	94-0015-86
$C_{38}H_{50}N_2O_4$			
Phthalazino[2,3-b]phthalazine-5,14-dione, 9-acetoxy-7,8,9,10,11,12-hexahydro-7-[[octahydro-7a-methyl-1-(1,4-5-trimethyl-2-hexenyl)-4H-inden-4-ylidene]methylene]-	EtOH	238(4.38),312(4.05)	44-4819-86
$C_{38}H_{50}N_4Ru$			
Ruthenium, dimethyl[2,3,7,8,12,13,17-18-octaethyl-21H,23H-porphinato(2-)-N^{21},N^{22},N^{23},N^{24}]-, (OC-6-12)-	CH_2Cl_2	334(4.80),377(5.04), 543(4.12)	77-0787-86
$C_{38}H_{50}N_8O_2$			
Pyridinium, 1,1',1",1"'-(3,6-diethoxy-1,2,4,5-benzenetetrayl)tetrakis-[4-(dimethylamino)-, salt with trifluoromethanesulfonic acid (1:4)	MeCN	393(3.99)	89-0917-86
$C_{38}H_{53}N_3$			
Postsecamidine	MeOH	224(2.16),284(1.45), 290s(1.38)	164-0663-86
$C_{38}H_{54}N_4O_{14}$			
L-Alaninamide, N-[(2,3,5,6,8,9,11,12-octahydro-1,4,7,10,13-benzopentaoxacyclopentadecin-15-yl)carbonyl]-L-alanyl-L-alanyl-N-(2,3,5,6,8,9,11,12-octahydro-1,4,7,10,13-benzopentaoxacyclopentadecin-15-yl)-	n.s.g.	207(4.77),257(4.37), 288(4.02)	126-0205-86B
$C_{38}H_{55}NO_6$			
Avenestergenin B-1	EtOH	225(4.37),255(3.84), 357(3.67)	39-1905-86C
$C_{38}H_{55}NO_7$			
Avenestergenin A-1	EtOH	224(4.36),255(3.81), 359(3.66)	39-1905-86C
$C_{38}H_{56}O_8$			
Oleana-11,13(18)-diene-2,3,23,28-	EtOH	244(4.30),251(4.40),	2-0113-86

Compound	Solvent	λ max (log ε)	Ref.
tetrol, tetraacetate, (2α,3β)- (cont.)		262(4.22)	2-0113-86
$C_{38}H_{59}AsO_2$			
Benzenemethanol, 2,4,6-tris(1,1-dimethylethyl)-α-[[2,4,6-tris(1,1-dimethylethyl)benzoyl]arsinidene]-, (Z)-	heptane	231s(4.21),269s(3.58), 382s(3.76),422(3.90)	88-1771-86
$C_{38}H_{59}O_2P$			
Benzenemethanol, 2,4,6-tris(1,1-dimethylethyl)-α-[[2,4,6-tris(1,1-dimethylethyl)benzoyl]phosphinidene]-, (Z)-	heptane	232s(4.23),370(4.01), 387s(3.97),400s(3.91)	88-1771-86
$C_{38}H_{60}O_2S_2$			
Benzo[1,2-b:4,5-b']dithiophene-4,8-dione, 2-(1,5,9,13,17,21-hexamethyldocosyl)-	$CHCl_3$	250(3.93),297(3.87), 346(3.49)	77-0733-86
$C_{38}H_{61}NO_{12}$			
Tylonolide, 5-O-[2-O-acetyl-3,6-dideoxy-3-(dimethylamino)-4-O-(3-methyl-1-oxobutyl)-β-D-glucopyranosyl]-	EtOH	283(4.32)	158-1108-86
$C_{38}H_{62}F_6P_2Pt$			
Platinum, bis(tributylphosphine)-bis[4-(trifluoromethyl)phenyl]-, cis	n.s.g.	257(3.99),303(2.81)	64-1281-86B
trans	n.s.g.	249(4.23),264(3.99), 301(2.81)	64-1281-86B
$C_{38}H_{66}O_4$			
2-Propenoic acid, 3-(3-hydroxy-4-methoxyphenyl)-, octacosyl ester, (E)-	EtOH	235(4.03),325(4.06)	102-0757-86
$C_{38}H_{68}O_2P_2Pt$			
Platinum, bis(4-methoxyphenyl)bis(tributylphosphine)-, cis	CH_2Cl_2	234(4.23),264(3.78)	64-1281-86B
trans	CH_2Cl_2	243(4.16),258(3.93)	64-1281-86B
$C_{38}H_{68}O_2Si_2$			
4,6-Dioxa-3,8-disiladecane, 2,2,3,3-8,8,9,9-octamethyl-5-[1-methyl-2-(2,6,6-trimethyl-2-cyclohexen-1-yl)ethenyl]-7-[1-methyl-2-(2,6,6-trimethyl-2-cyclohexen-1-ylidene)-ethylidene]-	pentane	205(4.50),245(4.17)	33-1378-86
isomer B	pentane	206(4.38),244(4.13)	33-1378-86
$C_{38}H_{68}P_2Pt$			
Platinum, bis(4-methylphenyl)bis(tributylphosphine)-	CH_2Cl_2	234(4.38),271(4.29), 306(2.76)	64-1281-86B

Compound	Solvent	$\lambda_{max}(\log \epsilon)$	Ref.
$C_{39}H_{25}NO_2S$			
Methanone, (3,4,7-triphenyl-2,1-benzisothiazole-5,6-diyl)bis[phenyl-	C_6H_{12}	229(4.60),235s(4.53), 240(4.38),246(4.20), 252(4.17),258(4.33), 264(4.45),276s(4.35), 358(4.07)	5-1796-86
$C_{39}H_{25}NS_2$			
Thieno[3,4-f]-2,1-benzisothiazole-2-S^{IV}, 3,4,5,7,8-pentaphenyl-	C_6H_{12}	221s(4.55),230(4.58), 235s(4.52),241(4.37), 246(4.23),252(4.21), 258(4.32),264(4.43), 298s(4.39),352s(3.88), 390s(3.66),648(4.07)	24-3158-86
$C_{39}H_{29}NO_2$			
Benzenamine, 4,5-dibenzoyl-3,6-diphenyl-2-(phenylmethyl)-	C_6H_{12}	221(4.62),235(4.50), 241(4.32),246(4.08), 252(4.00),258(4.23), 264(4.30),306s(3.80)	5-1796-86
$C_{39}H_{29}NO_2S$			
2,1-Benzisothiazole-5,6-dimethanol, α,α',3,4,7-pentaphenyl-	C_6H_{12}	221(4.57),235s(4.05), 265(4.07),305s(3.91), 319(3.99),346(4.00)	5-1796-86
$C_{39}H_{40}N_2O_7$			
Medelline, 12'-O-acetyl-	EtOH	222(4.60),283(3.90)	142-0607-86
$C_{39}H_{42}N_2O_7$			
Pseudoxandrine, O,O-dimethyl-	EtOH	210(4.70),286(3.95)	23-1390-86
$C_{39}H_{44}ClNO_9$			
Naphthomycin A, 2-demethyl- (naphthoquinomycin C)	MeOH	232(4.34),305(4.25), 380s(3.55),580(1.98)	158-0157-86
	MeOH-HCl	227(4.35),283(4.21), 300(4.23)	158-0157-86
	MeOH-NaOH	233(4.34),303(4.25), 400s(3.57),575(2.85)	158-0157-86
$C_{39}H_{44}N_2O_7$			
Thalifaramine, (+)-	MeOH	228(4.75),270s(4.35), 280(4.42),308(4.04)	100-0494-86
$C_{39}H_{44}N_2O_8$			
Thalidasine, 5-hydroxy-, (-)-	MeOH	237s(4.20),282(3.54)	100-0488-86
Thalifaricine, (+)-	MeOH	224(4.74),274s(4.31), 283(4.40),295s(4.31), 310s(4.17)	100-0494-86
	MeOH-base	282s(4.19),315(4.29)	100-0494-86
$C_{39}H_{45}N_2O_6$			
Isotetrandrine, 2'-N-methyl-, (iodide)	EtOH	284(3.91)	105-0235-86
$C_{39}H_{46}N_2O_6$			
Thaligrisine, 7,7'-dimethyl-	EtOH	230(4.42),280(3.96)	100-0854-86
$C_{39}H_{46}N_2O_{10}$			
Benzeneacetamide, N-[2-(3,4-dimethoxyphenyl)-2-methoxyethyl]-3-[4-[2-[[2-(3,4-dimethoxyphenyl)-2-methoxy-	MeOH	203(5.09),228(4.58), 277(4.02)	83-0950-86

Compound	Solvent	λ max (log ε)	Ref.
ethyl]amino]-2-oxoethyl]phenoxy]-4-methoxy- (cont.)			83-0950-86
$C_{39}H_{46}N_4O_6$			
Porphyrin-5-propanoic acid, 3,7-bis(ethoxycarbonyl)-12,18-diethyl-2,8,13,17-tetramethyl-, ethyl ester	$CHCl_3$	408(5.03),507(4.15),542(3.74),377(3.75),631(3.32)	12-0399-86
$C_{39}H_{46}N_4O_{10}$			
Coprobiliverdin I, tetramethyl ester	$CHCl_3$	368(4.80),649(4.23)	44-3001-86
	$CHCl_3$-HCl	366(4.84),664(4.65)	44-3001-86
Coprobiliverdin IIIα, tetramethyl ester	$CHCl_3$	369(4.83),644(4.23)	44-3001-86
	$CHCl_3$-HCl	365(4.87),659(4.76)	44-3001-86
Coprobiliverdin IIIβ, tetramethyl ester	$CHCl_3$	366(4.70),648(4.11)	44-3001-86
	$CHCl_3$-HCl	365(4.88),664(4.59)	44-3001-86
Coprobiliverdin IIIγ, tetramethyl ester	$CHCl_3$	367(4.77),642(4.17)	44-3001-86
	$CHCl_3$-HCl	373(4.93),662(4.66)	44-3001-86
Coprobiliverdin IIIδ, tetramethyl ester	$CHCl_3$	370(4.86),650(4.33)	44-3001-86
	$CHCl_3$-HCl	366(4.92),664(4.70)	44-3001-86
$C_{39}H_{56}O_2$			
β-Amyrin cinnamate	MeOH	202(3.71),216(4.20),222(4.16),280(3.93)	102-0274-86
$C_{39}H_{57}NO_{11}$			
Tylonolide, 5-O-[3,6-dideoxy-3-(dimethylamino)-β-D-glucopyranosyl]-, 23-benzeneacetate	EtOH	281(4.47)	158-1108-86
$C_{39}H_{64}N_4O_3$			
Piperazine, 1,1'-[(2α,6β)-3-oxocholest-4-ene-2,6-diyl]bis[4-acetyl-	EtOH	240(4.09)	107-0195-86
$C_{39}H_{66}O_2S_2$			
Benzo[b]thiophene-4,7-dione, 6-(3,7-11,15,19,23-hexamethyltetracosyl)-5-methyl- (relative absorbances)	hexane	239(1.0),273(0.6),280(0.6),320(0.4),456(0.1)	63-0191-86
$C_{39}H_{68}N_4O_3$			
Cholest-4-en-3-one, 2,6-bis[4-(2-hydroxyethyl)-1-piperazinyl]-, (2α,6β)-	EtOH	239(4.15)	107-0195-86

Compound	Solvent	λ_{max}(log ε)	Ref.
$C_{40}H_4Cl_{28}$			
Methyl, [1,2-ethanediylbis(2,3,5,6-tetrachloro-4,1-phenylene]bis[bis-(pentachlorophenyl)-	$CHCl_3$	283(4.08),370(4.59), 385(4.87),510(3.39), 562(3.39)	44-2472-86
$C_{40}H_{20}N_8$			
Propanedinitrile, 2,2'-(2,3,7,8-tetraphenylpyrazino[2,3-g]quinoxaline-5,10-diylidene)bis-	MeCN	216(4.52),239(4.59), 295(4.30),398(4.56), 435s(4.37)	138-0715-86
radical anion, tetraethylammonium salt	MeCN	249(4.29),263(4.81), 375s(4.88),392(4.95), 430s(4.59),580s(4.02), 632(4.12),702(4.63)	138-0715-86
$C_{40}H_{24}$			
Dibenzo[a,o]perylene, 7,16-diphenyl-	CH_2Cl_2	300(4.5),390(3.3), 595(4.5)	61-0813-86
$C_{40}H_{24}N_6O_2S_2$			
Imidazo[5,1-a]phthalazin-3(2H)-one, 1,1'-dithiobis[2-(1-naphthalenyl)-	EtOH	220(4.28),276(3.64), 350(3.40),440(3.04)	150-1901-86M
$C_{40}H_{24}O_2$			
7H-7,11b-Epidioxydibenzo[a,o]perylene, 7,16-diphenyl-	CH_2Cl_2	270(4.7),300(3.8), 315(3.8),429f(4.3)	61-0813-86
$C_{40}H_{28}N_2$			
1H-Indole, 5-(2,3-diphenyl-3H-indol-3-yl)-2,3-diphenyl-	MeOH	257(4.59),313(4.50)	39-2305-86C
$C_{40}H_{28}O_2$			
[1,1'-Biphenyl]-2-acetic acid, α-phenyl-, 9H-fluoren-9-ylidenephenylmethyl ester	MeOH	252(4.5),262(4.5), 282(4.1),304(4.1), 316(4.1)	64-0772-86B
$C_{40}H_{31}NO_2$			
3,4-Coronenedicarboximide, N-[2,5-bis(1,1-dimethylethyl)phenyl]-	$CHCl_3$	400(3.69),420(3.76), 444(3.84),473.5(3.97)	24-2373-86
$C_{40}H_{32}O_5$			
4H-1-Benzopyran-4-one, 7-acetoxy-3-acetyl-6,8-bis(diphenylmethyl)-2-methyl-	MeOH	212(4.59),244(4.12), 292(3.54)	2-0755-86
$C_{40}H_{32}O_{14}$			
Rubellin A pentaacetate	EtOH	253(4.57),274s(4.24), 352(3.85)	39-0255-86C
$C_{40}H_{33}N_5O_7S_2$			
Adenosine, N,N-dibenzoyl-5'-S-methyl-5'-C-(methylthio)-5'-thio-, 2',3'-dibenzoate	MeOH	230(4.68),272s(4.41)	44-1258-86
$C_{40}H_{38}N_5O_{10}PS$			
1H,5H-[1,2,4]Triazolo[1',2':1,2][1,2]-diazeto[3,4-d]azepin-4a,9,9a-tricarboxylic acid, 2,3,6,7-tetrahydro-2-methyl-7-[(4-methylphenyl)sulfonyl]-1,3-dioxo-8-[(triphenylphosphoranylidene)amino]-, trimethyl ester	$CHCl_3$	255(4.17),315(4.26)	24-2114-86

Compound	Solvent	λmax(log ε)	Ref.
$C_{40}H_{38}O_8$			
2-Propen-1-one, 1-[3-[6-[2,4-dihydroxy-3-(3-methyl-2-butenyl)benzoyl]-5-(4-hydroxyphenyl)-3-methyl-2-cyclohexen-1-yl]-2,4-dihydroxyphenyl]-3-(4-hydroxyphenyl)-, [1R-[1α(E)-5β,6α]]- (kuwanon V)	EtOH	226s(4.70),296(4.37), 370(4.56)	94-2471-86
	EtOH-AlCl₃	226s(3.86),296(3.47), 370(3.59)	94-2471-86
$C_{40}H_{38}O_9$			
Kuwanon Q	EtOH	222s(4.37),266s(3.94), 300(4.06),389(4.26)	94-2471-86
	EtOH-AlCl₃	222s(4.38),266s(3.94), 300(4.06),389(4.24)	94-2471-86
Kuwanon R	EtOH	224s(4.47),299(4.22), 370(4.35)	94-2471-86
	EtOH-AlCl₃	224s(4.98),299(4.22), 370(4.35)	94-2471-86
$C_{40}H_{38}O_{10}$			
2-Propen-1-one, 1-[3-[6-[2,4-dihydroxy-3-(3-methyl-2-butenyl)benzoyl]-5-(2,4-dihydroxyphenyl-3-methyl-2-cyclohexen-1-yl]-2,4-dihydroxyphenyl)-3-(2,4-dihydroxyphenyl)-(kuwanon J)	EtOH	223s(4.56),264s(4.03), 298(4.30),390(4.47)	94-2471-86
	EtOH-AlCl₃	223s(4.58),266s(4.04), 301(4.30),392(4.44), 460s(3.65)	94-2471-86
$C_{40}H_{39}F_3N_2O_{14}$			
Adriamycin, N-(trifluoroacetyl)-14-O-(N-acetylphenylalanyl)-	MeOH	232(4.64),250(4.48), 287(4.08),480(4.02), 497(4.06),532(3.97)	87-1273-86
$C_{40}H_{40}Cl_2N_4O_4$			
21H,23H-Porphine-2-propanoic acid, 8,13-bis(2-chloroethyl)-3,7,12,17-tetramethyl-18-[(phenylmethoxy)-carbonyl]-, methyl ester	CH_2Cl_2	406(5.46),510(4.21), 548(4.29),576(4.05), 634(3.43)	44-0657-86
$C_{40}H_{40}N_2O_8$			
1,2,4,5-Benzenetetrol, 3,6-bis[5-(3-methyl-2-butenyl)-1H-indol-3-yl]-, tetraacetate	EtOH	223(4.93),274(4.47), 308(4.48)	5-1765-86
Bisdehydronorglaucine	EtOH	208s(4.21),215(4.22), 265(4.45),273(4.44), 340(4.14),390s(3.41)	100-1028-86
	EtOH-HCl	208(--),218s(--), 268s(--),274(--), 284s(--),330(--), 350s(--),368s(--), 390s(--)	100-1028-86
$C_{40}H_{40}N_2O_{16}S_2$			
15,16-Diazahexacyclo[8.4.1.1⁴,⁷.0²,⁹-0³,⁸.0¹¹,¹⁴]hexadeca-5,12-diene-2,5,6,9,12,13-hexacarboxylic acid, hexamethyl ester	MeCN	229(4.56),253s(3.82), 258s(3.68),263s(3.61), 274s(3.42)	24-0616-86
$C_{40}H_{42}N_4O_6$			
21H,23H-Porphine-2-propanoic acid, 8,13-bis(2-hydroxyethyl)-3,7,12,17-tetramethyl-18-[(phenylmethoxy)carbonyl]-, methyl ester	CH_2Cl_2	406(5.38),508(4.15), 548(--),574(4.03), 634(3.42)	44-0657-86

Compound	Solvent	$\lambda_{max}(\log \epsilon)$	Ref.
$C_{40}H_{44}N_2O_9$			
Vilaportine	MeOH	227s(4.57),257s(4.30), 320(4.45),392(3.67)	23-0123-86
$C_{40}H_{44}N_4O_8$			
21H,23H-Porphine-2-propanoic acid, 7,12-bis(2-acetoxyethyl)-18-(3-methoxy-3-oxo-2-propenyl)-3,8,13,17-tetramethyl-, methyl ester	CH_2Cl_2	412(5.22),508(4.30), 548(4.38),576(4.26)	44-4660-86
$C_{40}H_{44}N_4O_{10}$			
21H,23H-Porphine-2,8,17-tripropanoic acid, 15,17-bis(methoxycarbonyl)-3,7,13,18-tetramethyl-, trimethyl ester	$CHCl_3$	407(5.29),508.5(4.09), 543(4.01),576(3.84), 628.5(3.54)	12-0399-86
$C_{40}H_{45}N_5O_9$			
L-Aspartic acid, N-[3-(2,17-diethenyl)-1,19,23,24-tetrahydro-12-(3-methoxy-3-oxopropyl)-3,7,13,18-tetramethyl-1,19-dioxo-21H-bilin-8-yl]-1-oxopropyl]-, dimethyl ester	benzene	380(4.68),655(4.15)	49-1205-86
	EtOH	376(4.69),664(4.12)	49-1205-86
	$CHCl_3$	378(4.70),660(4.13)	49-1205-86
isomer	benzene	380(4.68),655(4.15)	49-1205-86
	EtOH	376(4.68),665(4.12)	49-1205-86
	$CHCl_3$	378(4.70),658(4.13)	49-1205-86
$C_{40}H_{46}ClNO_9$			
Naphthomycin A	MeOH	231(4.59),275(4.52)	158-0316-86
$C_{40}H_{46}N_2O_7$			
Thalifaronine, (+)-	MeOH	227(4.75),268s(4.30), 280(4.38),304s(4.08)	100-0494-86
$C_{40}H_{46}N_2O_8$			
Thalibulamine, (+)- (5λ,4ε)	MeOH	225(4.79),270s(--), 281(4.42),301(4.31), 314(4.24)	100-0494-86
Thalidasine, 5-methoxy-, (-)-	MeOH	237s(4.40),281(3.67)	100-0488-86
Thalifalandine	MeOH	205(4.47),225s(4.37), 285(4.03),308s(3.89)	142-2731-86
	MeOH-NaOH	204(4.37),226s(4.30), 290(3.83),313(3.95), 325(3.95)	142-2731-86
Thalifarazine, (+)-	MeOH	228(4.70),270s(4.29), 283(4.38),297s(4.25), 310s(4.02)	100-0494-86
Thalifaretine, (+)-	MeOH	222(4.76),272s(4.24), 283(4.39),293s(4.27), 310s(4.04)	100-0494-86
Thalifaroline, (+)-	MeOH	225(4.69),275s(4.28), 285(4.37),300s(4.24), 310s(4.16)	100-0494-86
	MeOH-base	284s(4.15),292s(4.17), 325(4.34)	100-0494-86
$C_{40}H_{46}N_4O_3$			
Hazuntiphylline	EtOH and EtOH-base	255(4.27),304(3.83)	100-0321-86
$C_{40}H_{47}NO_9S$			
Naphthoquinomycin B	MeOH	232(4.31),306(4.19),	158-0157-86

Compound	Solvent	λmax(log ε)	Ref.
Naphthoquinomycin B (cont.)		580(2.74)	158-0157-86
	MeOH-HCl	233(4.28),295(4.15)	158-0157-86
	MeOH-NaOH	232(4.30),308(4.19), 410s(3.46),580(2.90)	158-0157-86
$C_{40}H_{47}NO_{10}$			
Naphthomycin A, 30-dechloro-2-demethyl-30-methoxy- (naphthoquinomycin A)	MeOH	233(4.31),302(4.20)	158-0157-86
	MeOH-HCl	232(4.30),276(4.18), 298(4.17)	158-0157-86
	MeOH-NaOH	233(4.32),303(4.23), 400s(3.54)	158-0157-86
$C_{40}H_{48}N_2O_6$			
Pisopowine	EtOH	209(4.68),228s(4.49), 288(4.03)	100-1018-86
$C_{40}H_{48}O_{16}$			
Cinerubin X	MeOH	258(4.37),290(3.89), 493(4.15),505(4.04), 525(3.96)	158-1178-86
$C_{40}H_{52}O_4$			
Astaxanthin	EtOH	486(5.13)	33-0012-86
	solid	490(--)	33-0012-86
$C_{40}H_{56}O$			
ε,ε-Carotene, 4,5-epoxy-4,5-dihydro-, (4S,5R,6R,6'R)-	hexane	263(4.54),392s(4.69), 412(4.99),463(5.18), 465(5.19)	33-0816-86
$C_{40}H_{56}O_2$			
ε,ε-Carotene, 4,5:4',5'-diepoxy-4,4',5,5'-tetrahydro-, (±)-	hexane	262(4.55),390s(4.68), 411(4.99),434(5.18), 464(5.19)	33-0816-86
isomer 17	hexane	262(4.56),390s(4.68), 411(4.98),434(5.17), 464(5.18)	33-0816-86
meso	hexane	262(4.46),390s(4.58), 411(4.88),434(5.08), 464(5.08)	33-0816-86
isomer 16	hexane	262(4.40),390s(4.52), 411(4.81),434(5.00), 464(5.01)	33-0816-86
ε,ε-Carotene-3,3'-diol (zeaxanthin)	EtOH	461(5.11)	33-0012-86
	solid	435(--)	33-0012-86
Lycopene diepoxide, (all-E)-	ether	293(4.69),442(5.04), 468(5.21),499(5.17)	33-0106-86
(2R,2'R,5Z)-	ether	293(4.65),442(5.04), 468(5.23),499(5.19)	33-0106-86
(2R,2'R,7Z)-	ether	294(4.52),444(5.03), 467(5.16),497(5.09)	33-0106-86
$C_{40}H_{56}O_4$			
Violaxanthin	benzene	426(4.95),453(5.13), 483(5.12)	102-0195-86
	+ acid	387(--),410(--), 436(--)	102-0195-86
	EtOH	417(--),441(--), 471(--)	102-0195-86
9,9'-di-cis	benzene	417(4.91),442(5.09), 472(5.06)	102-0195-86

Compound	Solvent	λmax(log ε)	Ref.
(cont.)	benzene-acid	387(--),408(--), 435(--)	102-0195-86
	EtOH	409(--),432(--), 461(--)	102-0195-86
Violaxanthin, 9,13-di-cis	benzene	426(4.95),453(5.13), 483(5.12)	102-0195-86
	+ acid	387(--),410(--), 436(--)	102-0195-86
	EtOH	417(--),441(--), 471(--)	102-0195-86
Violaxanthin, 9,13'-di-cis	benzene	416(4.84),440(4.95), 470(4.92)	102-0195-86
	+ acid	385(--),407(--), 432(--)	102-0195-86
	EtOH	407(--),430(--), 459(--)	102-0195-86
Violaxanthin, 9,15-di-cis	benzene	419(4.85),443(4.98), 474(4.91)	102-0195-86
	+ acid	386(--),408(--), 434(--)	102-0195-86
	EtOH	411(--),434(--), 462(--)	102-0195-86
$C_{40}H_{60}O_4$			
ε,ε-Carotene-4,4',5,5'-tetrol, 4,4',5,5'-tetrahydro-, (±)-	EtOH	263(4.54),394s(4.65), 414(4.94),438(5.12), 467(5.11)	33-0816-86
(4R,4'R,5R,5'R,6R,6'R)-	EtOH	262(4.58),394s(4.64), 414(4.93),438(5.11), 467(5.11)	33-0816-86
ψ,ψ-Carotene-1,1',2,2'-tetrol, 1,1',2,2'-tetrahydro-, all-E-	CH_2Cl_2	296(4.61),452(5.01), 478(5.19),511(5.13)	33-0106-86
7Z-	CH_2Cl_2	297(4.48),454(5.00), 477(5.13),508(5.05)	33-0106-86
7Z,7'Z-	CH_2Cl_2	297(4.38),454(5.00), 475(5.09),503(4.98)	33-0106-86
$C_{40}H_{60}Sn_2$			
Stannane, dicyclopenta[a,e]dicyclopropa[c,g]cyclooctene-1,5-diylbis[tributyl-	CH_2Cl_2	240(4.76),313(4.75), 369(4,60),460s(3.21)	35-7032-86
$C_{40}H_{63}NO_{13}$			
Tylonolide, 5-O-[2-O-acetyl-3,6-dideoxy-3-(dimethylamino)-4-O-(3-methyl-1-oxobutyl)-β-D-glucopyranosyl]-, 23-acetate	EtOH	280(4.19)	158-1108-86
Tylosin, 23-de[(6-deoxy-2,3-di-O-methyl-β-D-allopyranosyl)oxy]-, 2A-acetate	EtOH	284(4.32)	158-1724-86
$C_{40}H_{65}NO_{14}$			
Tylosin, 23-O-de(6-deoxy-2,3-di-O-methyl-β-D-allopyranosyl)-, 2A-acetate	EtOH	283(4.28)	158-1108-86
$C_{40}H_{68}O_5$			
2-Propenoic acid, 3-(3-acetoxy-4-methoxyphenyl)-, octacosyl ester, (E)-	EtOH	234(3.99),325(4.04)	102-0757-86

Compound	Solvent	λ_{max}(log ε)	Ref.
$C_{40}H_{73}N_2$			
1H-Benzimidazolium, 1,3-dihexadecyl-2-methyl-, bromide	EtOH	255(3.75),270(3.88), 278(3.91)	4-0209-86
$C_{40}H_{88}Si_4$			
Cyclotetrasilane, octakis(2,2-dimethylpropyl)-	C_6H_{12}	286(2.64)	138-1643-86
$C_{41}H_{30}O_4$			
2H-1-Benzopyran-2-one, 6,8-bis(diphenylmethyl)-4,7-dihydroxy-3-phenyl-	MeOH	228(4.1),299(3.2)	2-0623-86
$C_{41}H_{30}O_{27}$			
Geraniinic acid	MeOH	216(4.84),271(4.48)	94-4075-86
$C_{41}H_{31}S_2$			
Thiopyrylium, 4-[7-(2,6-diphenyl-4H-thiopyran-4-ylidene)-1,3,5-heptatrienyl]-2,6-diphenyl-, perchlorate	CH_2Cl_2	1020(5.47)	104-0150-86
	$C_2H_4Cl_2$	1021(5.39)	104-0150-86
	o-$C_6H_4Cl_2$	1040(5.43)	104-0150-86
	EtOH	1000(5.26)	104-0150-86
	acetone	1000(5.16)	104-0150-86
	MeCN	995(5.22)	104-0150-86
	DMF	1010(5.11)	104-0150-86
	C_6H_5CN	1030(5.24)	104-0150-86
	$C_6H_5NO_2$	1035(5.29)	104-0150-86
	pyridine	1030(5.23)	104-0150-86
	DMSO	1020(5.09)	104-0150-86
$C_{41}H_{32}NO_2P$			
2,5-Pyrrolidinedione, 1-(triphenylmethyl)-3-(triphenylphosphoranylidene)-	$CHCl_3$	258(3.85),305(3.45)	44-0495-86
$C_{41}H_{34}O_6$			
4H-1-Benzopyran-4-one, 5,7-diacetoxy-6,8-bis(diphenylmethyl)-2,3-dimethyl-	MeOH	240(5.03),306(4.35), 333(4.50)	2-0478-86
$C_{41}H_{35}O_2$			
Pyrylium, 2-[3-[4,6-bis(4-methylphenyl)-2H-pyran-2-ylidene]-1-propenyl]-4,6-bis(4-methylphenyl)-, tetrafluoroborate	CH_2Cl_2	730(4.77),810(4.95)	103-1295-86
$C_{41}H_{35}O_6$			
Pyrylium, 2-[3-[4,6-bis(4-methoxyphenyl)-2H-pyran-2-ylidene]-1-propenyl]-4,6-bis(4-methoxyphenyl)-, tetrafluoroborate	CH_2Cl_2	746(4.70),824(4.92)	103-1295-86
$C_{41}H_{39}N_2O_{10}PS$			
1H-Indole-3a,5,7,7a-tetracarboxylic acid, 2,3-dihydro-1-[(4-methylphenyl)sulfonyl]-4-[(triphenylphosphoranylidene)amino]-, tetramethyl ester	$CHCl_3$	293(3.94),388(4.33)	24-2114-86
$C_{41}H_{40}Cl_2N_4O_2$			
2,5-Cyclohexadiene-1,4-dione, 2,5-dichloro-3-[[3-(10,11-dihydro-5H-dibenzo[b,f]azepin-5-yl)propyl]-	CH_2Cl_2	254(4.43),277s(2.30), 395(4.10),540s(2.70)	106-0832-86

Compound	Solvent	λ_{max}(log ε)	Ref.
amino]-6-[[3-(10,11-dihydro-5H-dibenzo[b,f]azepin-5-yl)propyl]methylamino]- (cont.)			106-0832-86
$C_{41}H_{42}N_2O_9$			
Oxandrine, O,O-diacetyl-	EtOH	224(4.62),282(3.69)	23-1390-86
Pseudoxandrine, O,O-diacetyl-	EtOH	212(4.75),281(3.92)	23-1390-86
Thalicarpine, 2'-N-oxide, (+)-	MeOH	215(4.36),280(3.98), 300s(3.86)	23-0123-86
$C_{41}H_{48}N_6O_{10}$			
L-Serine, N,N'-[(3,18-diethenyl-1,19,22,24-tetrahydro-2,7,13,17-tetramethyl-1,19-dioxo-21H-biline-8,12-diyl)bis(1-oxo-3,1-propanediyl)]bis-	benzene	381(4.69),656(4.17)	49-1205-86
	EtOH	378(4.72),662(4.17)	49-1205-86
	$CHCl_3$	378(4.67),656(4.11)	49-1205-86
$C_{41}H_{49}NO_9S$			
Naphthomycin A, 30-dechloro-30-(methylthio)-	MeOH	230(4.76),275(4.52)	158-0316-86
$C_{41}H_{49}N_5O_7$			
21H-Biline-8-propanoic acid, 2,17-diethenyl-1,19,23,24-tetrahydro-12-[3-[[1-(methoxycarbonyl)-2-methylpropyl]methylamino]-3-oxopropyl]-3,7,13,18-tetramethyl-1,19-dioxo-, methyl ester, (S)-	benzene	381(4.72),657(4.20)	49-1205-86
	EtOH	376(4.73),663(4.18)	49-1205-86
	$CHCl_3$	380(4.72),657(4.15)	49-1205-86
$C_{41}H_{50}N_4O_2$			
20-Phorbinecarboxylic acid, 3,4,20,21-tetradehydro-3,4,8,9,13,14,19-heptaethyl-18-ethylidene-18,19-dihydro-, ethyl ester, (Z)-	CH_2Cl_2	438(5.02),510(3.98), 540(4.05),583(4.19), 653(3.96),715(4.63)	44-1347-86
$C_{41}H_{50}N_4O_2Zn$			
Zinc, [ethyl 3,4,20,21-tetradehydro-3,4,8,9,13,14,18,19-octaethyl-18,19-dihydro-20-phorbinecarboxylato(2-)-N^{23},N^{24},N^{25},N^{26}]-, (SP-4-2)-	CH_2Cl_2	413(5.29),435(5.34), 535(4.15),578(4.28), 618(4.46),663(4.93)	44-1347-86
$C_{41}H_{50}N_6O_{10}$			
RA-III	EtOH	276(3.36),281(3.26)	94-3762-86
RA-IV	EtOH	276(3.42),284(3.30)	94-3762-86
$C_{41}H_{52}N_4NiO_2$			
Nickel, [ethyl 3,4-didehydro-3,4,8,9-13,14,18,19-octaethyl-18,19-dihydro-20-phorbinecarboxylato(2-)-N^{23},N^{24},N^{25},N^{26}]-, (SP-4-2)-	CH_2Cl_2	405(5.17),498(4.04), 533(3.94),588(4.29), 630(4.81)	44-1347-86
$C_{41}H_{52}N_4O_2$			
20-Phorbinecarboxylic acid, 3,4,20,21-tetradehydro-3,4,8,9,13,14,18,19-octaethyl-18,19-dihydro-, ethyl ester	CH_2Cl_2	433(4.95),453(4.95), 503(4.16),530(4.09), 568(4.28),648(4.03), 695(4.63)	44-1347-86
$C_{41}H_{52}N_4O_2Zn$			
Zinc, [ethyl 3,4-didehydro-3,4,8,9,13-14,18,19-octaethyl-18,19-dihydro-20-phorbinecarboxylato(2-)-	CH_2Cl_2	408(5.16),515(4.00), 545(3.73),590(4.20), 633(4.77)	44-1347-86

Compound	Solvent	λ_{max}(log ε)	Ref.
N^{23},N^{24},N^{25},N^{26}]-, (SP-4-2)- (cont.)			44-1347-86
$C_{41}H_{52}N_4O_4$			
21H,23H-Porphine-2-acetic acid, 2,3,7,8,12,13,17,18-octaethyl-20-formyl-2,3-dihydro-α-oxo-, ethyl ester	CH_2Cl_2	413(5.00),510(3.97), 545(3.88),640(3.77), 690(4.53)	44-1347-86
$C_{41}H_{53}NO_{13}$			
Rhodilunacin A	MeOH	235(4.52),253(4.48), 294(3.89),464(4.04), 480(4.11),493(4.18), 513(4.08),527(4.08)	158-0430-86
	MeOH-HCl	234(4.54),253(4.46), 293(3.90),464(4.08), 480(4.11),492(4.18), 513(4.08),526(4.04)	158-0430-86
	MeOH-NaOH	240(4.56),296(3.88), 493(3.97),540(4.11), 584(3.97)	158-0430-86
$C_{41}H_{53}NO_{14}$			
Rhodilunacin B	MeOH	235(4.52),253(4.48), 294(3.88),465(4.08), 480(4.11),493(4.18), 513(4.04),528(4.04)	158-0430-86
	MeOH-HCl	234(4.56),253(4.46), 294(3.89),465(4.08), 480(4.11),495(4.18), 513(4.04),528(4.04)	158-0430-86
	MeOH-NaOH	240(4.58),295(3.88), 495(3.96),543(4.11), 548(3.98)	158-0430-86
$C_{41}H_{54}N_4O_2$			
20-Phorbinecarboxylic acid, 3,4-didehydro-3,4,8,9,13,14,18,19-octaethyl-18,19-dihydro-, ethyl ester	CH_2Cl_2	403(5.06),500(4.37), 535(3.75),558(3.63), 610(3.93),660(4.60)	44-1347-86
$C_{41}H_{55}N_5O_4Si_2$			
Guanosine, 2'-deoxy-3',5'-bis-O-[(1,1-dimethylethyl)dimethylsilyl]-N-(triphenylmethyl)-	EtOH	261(4.14),274s(4.09)	152-0643-86
$C_{41}H_{56}N_4O_7$			
Docosahexaenoic acid, 6-[[5,6-dihydro-5-oxo-6-(1,2,3,6-tetrahydro-1,3-dimethyl-2,6-dioxo-9H-purin-9-yl)-2H-pyran-2-yl]methoxy]hexyl ester, (2S-cis)-	$CHCl_3$	278(3.84)	136-0256-86K
$C_{41}H_{62}O_{16}$			
Bryostatin-9 hydrolysis product	MeOH	229(4.56)	100-0661-86
$C_{41}H_{67}NO_{14}$			
Tylosin, 23-O-de(6-deoxy-2,3-di-O-methyl-β-D-allopyranosyl)-23-O-(1-oxopropyl)-	EtOH	280(4.25)	158-1108-86
$C_{41}H_{71}N_6O_6PSi_3$			
1H-[1,3,5,2]Oxadiazaphosphorino[3,4-a]-	EtOH	262(4.29),344(4.16),	44-3559-86

Compound	Solvent	λmax(log ε)	Ref.
purin-9(6H)-one, 1-[bis(1-methylethyl)amino]-3-phenyl-6-[2,3,5-tris-O-[(1,1-dimethylethyl)dimethylsilyl]-β-D-ribofuranosyl]- (cont.)		356(4.15),377s(3.94)	44-3559-86
$C_{41}H_{72}O_9$			
Ionomycin (calcium salt)	n.s.g.	200(4.25),300(4.33)	100-0035-86
$C_{42}H_{26}$			
5,18[1',2']:7,16[1",2"]-Dibenzenoheptacene, 5,7,16,18-tetrahydro-	C_6H_{12}	261(3.57),278(4.37), 286(4.33),295(4.30), 320(3.49),337(3.68), 355(3.73),374(3.56)	78-1641-86
$C_{42}H_{28}N_2$			
3-Pyridinecarbonitrile, 2-[1,1'-biphenyl]-4-yl-4-phenyl-6-[1,1':4',1"-terphenyl]-4-yl-	MeCN	332(4.69)	73-1061-86
$C_{42}H_{28}N_2O_2$			
Pyridine, 4,4'-[2,2'-bifuran]-5,5'-diylbis[2,6-diphenyl-	MeCN	372(4.61)	103-1158-86
$C_{42}H_{28}N_4O_2S_2$			
5,11-Epithio-5H,7H,11H-isothiazolo-[3,4-g][1,2,4]triazolo[1,2-b]phthalazine-7,9(8H)-dione, 8-methyl-3,4,5,11,12-pentaphenyl-	CH_2Cl_2	233(4.60),270s(4.32), 332(4.08),345s(4.06)	24-3158-86
$C_{42}H_{32}O_9$			
Copalliferol B	EtOH	283(3.43)	102-1498-86
$C_{42}H_{33}Br_4NO_3$			
9H-Xanthene-9-methanamine, 2,4,5,6-tetrabromo-3,6-dimethoxy- ,9-diphenyl-N,N-bis(phenylmethyl)-	EtOH	270(4.21),300s(3.49)	39-0671-86C
$C_{42}H_{34}O_3$			
2-Propen-1-one, 1-[3,5-bis(diphenylmethyl)-2-hydroxy-4-methoxyphenyl]-3-phenyl-	EtOH	206(4.58),286(4.58), 326(3.93)	2-0755-86
$C_{42}H_{34}O_{15}$			
Methanone, [5-acetoxy-2,8-bis(4-acetoxyphenyl)-3,4-dihydro-2H-furo[2,3-h]-1-benzopyran-9-yl](2,4,6-triacetoxyphenyl)-	dioxan	216(4.72),287(4.46)	94-0595-86
$C_{42}H_{34}O_{16}$			
Rubellin B hexaacetate	EtOH	250(4.50),350(3.82)	39-0255-86C
$C_{42}H_{35}N_5O_7S$			
Adenosine, N,N-dibenzoyl-4',5'-didehydro-5'-S-(2-methylpropyl)-5'-thio-, (Z)-	MeOH	233(4.68),250s(4.61), 270s(4.44)	44-1258-86
$C_{42}H_{36}N_2$			
2,4,6,8,14,16,18-Cyclononadecaheptaene-10,12-diyn-1-one, 9,14-dimethyl-, (9,14-dimethyl-2,4,6,8,14,16,18-	CF_3COOH	275s(0.93),285(1.00), 306(0.93),423(0.90), 447s(0.96),464s(0.74),	18-2509-86

Compound	Solvent	$\lambda_{max}(\log \epsilon)$	Ref.
cyclononadecaheptaene-10,12-diyn-1-ylidene)hydrazone (relative absorbance given) (cont.)		628(0.56),701s(0.47)	18-2509-86
$C_{42}H_{36}N_4O_2$			
2,36:3,6:27,30:31,33-Tetraetheno-12,16:17,21-dimetheno-7,26,1,15-18,32-benzodioxatetraazacyclotetratriacontine, 8,9,10,11,22,23,24,25-octahydro-, cuprous complex	CH_2Cl_2	417(3.52)	88-0865-86
Ru(bipyridinyl)$_2$ complex	DMSO	458(4.12)	88-0865-86
$C_{42}H_{38}BO_4$			
Boron(1+), bis[1,7-bis(4-methylphenyl)-1,6-heptadiene-3,5-dionato-O,O']-, (T-4)-, perchlorate	CH_2Cl_2	496(5.03)	97-0330-86
$C_{42}H_{38}BO_8$			
Boron(1+), bis[1,7-bis(4-methoxyphenyl)-1,6-heptadiene-3,5-dionato-O^3,O^5]-, (T-4)-, perchlorate	CH_2Cl_2	536(5.25)	97-0399-86
$C_{42}H_{42}Cl_2N_4O_2$			
2,5-Cyclohexadiene-1,4-dione, 2,5-dichloro-3,6-bis[3-(10,11-dihydro-5H-dibenzo[b,f]azepin-5-yl)propyl]-methylamino]-	CH_2Cl_2	244(4.39),255s(4.40), 275s(4.30),432(3.94), 570s(2.60)	106-0832-86
$C_{42}H_{42}N_4O_8$			
21H,23H-Porphine-2,8-dipropanoic acid, 13,17-bis(methoxycarbonyl)-3,7,12,18-tetramethyl-15-phenyl-, methyl ester	$CHCl_3$	410(5.25),508(4.19), 541(3.80),580(3.81), 632(3.41)	12-0399-86
$C_{42}H_{45}N_5O_6$			
Nummularine O	MeOH	270(3.54),318(3.36)	102-2690-86
$C_{42}H_{48}$			
Tricyclo[3.3.1.1^{3,7}]decane, 1,3,5,7-tetrakis(5,5-dimethyl-1,3-hexadiynyl)-	hexane	217(3.42),230(3.48), 241(3.46),255(3.29)	78-1763-86
$C_{42}H_{48}N_4$			
1,17-Metheno-5,8:10,13-dinitrilo-8H-dipyrrolo[1,2-a:2',1'-n][1,15]benzodiazacycloheptadecine, 2,3,6,7,11-12,15,16-octaethyl-, monoperchlorate	CH_2Cl_2	410(4.89),542(3.83), 576(3.94),624(3.40)	77-0767-86
$C_{42}H_{50}N_4O_8$			
21H,23H-Porphine-2,8,18-tripropanoic acid, 13-(ethoxycarbonyl)-3,7,12,17-tetramethyl-15-propyl-, trimethyl ester	$CHCl_3$	409(5.35),509(4.13), 545.5(3.82),577(3.84), 628(3.17)	12-0419-86
$C_{42}H_{52}N_4O_8$			
Vincamine N-oxide dimer	EtOH-HCl	203(4.52),224(4.82), 276(4.19)	88-2775-86
$C_{42}H_{54}N_2O_2$			
Ethanaminium, 2,2'-[[1,1'-biphenyl]-4,4'-diylbis(2,1-ethenediyl-2,1-phenyleneoxy)]bis[N,N-diethyl-N-	EtOH	353(4.83)	61-0439-86

Compound	Solvent	λ$_{max}$(log ε)	Ref.
methyl-, bis(methyl sulfate) (cont.)			61-0439-86
$C_{42}H_{57}N_5O_5Si_2$			
Guanosine, 2'-deoxy-3',5'-bis-O-[(1,1-dimethylethyl)dimethylsilyl]-N-[(4-methoxyphenyl)diphenylmethyl]-	EtOH	262(4.15),275(4.12)	152-0643-86
$C_{42}H_{60}Si_6$			
Silane, bisbenzo[3,4]cyclobuta[1,2-a-1',2'-c]biphenylene-2,3,6,7,10,11-hexaylhexakis[trimethyl-	hexane	242(4.72),251(4.79), 291(4.90),299(5.09), 314(5.33),347(4.47), 364(4.33),376s(4.21), 394s(4.06)	35-3150-86
$C_{42}H_{64}Cl_3NO_{15}$			
Tylosin, 23-O-de(6-deoxy-2,3-di-O-methyl-β-D-allopyranosyl)-23-O-(trichloroacetyl)-, 2A-acetate	EtOH	279(4.38)	158-1724-86
$C_{42}H_{64}O_4$			
Abietic acid dimer	EtOH	251(4.33),256(4.33)	77-1038-86
$C_{42}H_{70}O_{11}$			
Salinomycin	n.s.g.	285(2.03)	100-0035-86
$C_{43}H_{31}N_3O_4S_2$			
4,8-Biimino-4H,8H-thieno[3,4-f]-2,1-benzisothiazole-9,10-dicarboxylic acid, 3,4,5,7,8-pentaphenyl-, dimethyl ester	MeOH	265(4.41)	24-3158-86
$C_{43}H_{31}S_2$			
1H-Benzo[c]thiolium, 3-[3-(3,3-diphenylbenzo[c]thien-1(3H)-ylidene)-1-propenyl]-1,1-diphenyl-, perchlorate	CH_2Cl_2	431(4.11),611s(4.50), 660.5(5.20)	104-0791-86
	MeCN	415(4.19),597s(4.60), 645(4.88)	104-0791-86
$C_{43}H_{33}N_2$			
Methylium, bis[4-(diphenylamino)-phenyl]phenyl-	HOAc-CF_3COOH	461(4.29),663(4.91)	64-1045-86B
$C_{43}H_{33}OS_2$			
Thiopyrylium, 4-[2-[5-[(2,6-diphenyl-4H-thiopyran-4-ylidene)ethylidene]-5,6-dihydro-2H-pyran-3-yl]ethenyl]-2,6-diphenyl-, perchlorate	CH_2Cl_2	1035(5.39)	104-0150-86
$C_{43}H_{34}O_3$			
2H-1-Benzopyran-2-one, 6,8-bis(diphenylmethyl)-4,7-dimethoxy-3-phenyl-	MeOH	208(5.00),319(3.6)	2-0623-86
$C_{43}H_{38}N_5O_8PS$			
1H,4aH-[1,2,4]Triazolo[1',2':1,2][1,2]-diazeto[3,4-d]azepine-4a,9-dicarboxylic acid, 2,3,5,6,7,9a-hexahydro-7-[(4-methylphenyl)sulfonyl]-1,3-dioxo-2-phenyl-8-[(triphenylphosphoranylidene)amino]-, dimethyl ester	$CHCl_3$	251(4.33),313(4.32)	24-2114-86
$C_{43}H_{43}N_2O_{10}PS$			
1H-Indole-3a,5,7,7a-tetracarboxylic	$CHCl_3$	294(3.70),389(4.24)	24-2114-86

Compound	Solvent	λ_{max}(log ε)	Ref.
acid, 2,3-dihydro-1-[(4-methylphenyl)sulfonyl]-4-[(triphenylphosphoranylidene)amino]-, 7,7a-diethyl 3a,5-dimethyl ester (cont.)			24-2114-86
$C_{43}H_{47}N_5O_6$			
Jubanine B	MeOH	270(3.54),318(3.36)	102-2690-86
$C_{43}H_{49}NO_{18}$			
Mayteine	EtOH	233(4.30),269(3.79)	102-1776-86
	1% HCl	235(4.20),269(3.92)	102-1776-86
$C_{43}H_{50}N_4O_6$			
Vincarubine	n.s.g.	273(3.21),340(3.29), 480(2.40)	88-5413-86
$C_{43}H_{50}N_4O_{10}$			
21H,23H-Porphine-2,5,12,18-tetrapropanoic acid, 7-(ethoxycarbonyl)-3,8,13,17-tetramethyl-, tetramethyl ester	$CHCl_3$	406(5.10),507(3.86), 534(3.98),570(3.98), 621(3.02)	12-0399-86
$C_{43}H_{56}O_{17}$			
Urdamycin A	EtOH and EtOH-HCl	319(3.65),426(3.74), 440s(3.73)	150-0104-86S +158-1657-86
	EtOH-NaOH	325(3.91),404(3.23), 580(3.73)	150-0104-86S
$C_{43}H_{58}O_{18}$			
Urdamycin F	MeOH	218(4.49),250(3.89), 278(3.77),432(3.63), 460s(--)	158-1657-86
	MeOH-NaOH	226(4.41),277(3.82), 565(3.65)	158-1657-86
$C_{43}H_{61}NO_{13}$			
Tylonolide, 5-O-[2,4-di-O-acetyl-3,6-dideoxy-3-(dimethylamino)-β-D-glucopyranosyl]-, 23-benzeneacetate	EtOH	280(4.45)	158-1108-86
$C_{43}H_{61}NO_{14}$			
Tylonolide, 5-O-[2,4-di-O-acetyl-3,6-dideoxy-3-(dimethylamino)-β-D-glucopyranosyl]-, 23-(phenoxyacetate)	EtOH	212(4.04),276(4.32)	158-1108-86
$C_{43}H_{64}O_{14}$			
Noboritomycin A (sodium salt)	n.s.g.	244(3.61),300(3.56)	100-0035-86
$C_{43}H_{64}O_{17}$			
Bryostatin-9	MeOH	229(4.56)	100-0661-86
$C_{43}H_{66}ClOP_2RhS_6Si_6$			
Rhodium, carbonylchlorobis[tris[5-(trimethylsilyl)-2-thienyl]phosphine-P]-, (SP-4-3)-	CH_2Cl_2	256(4.67),278s(4.60), 365(3.46)	70-1919-86
$C_{43}H_{69}NO_{13}$			
Tylonolide, 5-O-[2-O-acetyl-3,6-dideoxy-3-(dimethylamino)-4-O-(3-methyl-1-oxobutyl)-β-D-glucopyranosyl]-, 23-(3-methylbutanoate)	EtOH	279(4.42)	158-1108-86

Compound	Solvent	$\lambda_{max}(\log \epsilon)$	Ref.
$C_{43}H_{69}NO_{14}$			
Tylosin, 23-O-de(6-deoxy-2,3-di-O-methyl-β-D-allopyranosyl)-23-O-(1-oxopropyl)-, 2A-acetate	EtOH	284(4.38)	158-1724-86
$C_{43}H_{69}NO_{15}$			
Tylosin, 23-O-de(6-deoxy-2,3-di-O-methyl-β-D-glucopyranosyl)-23-O-(1-oxopropyl)-, 2A-acetate	EtOH	280(4.33)	158-1108-86
$C_{44}H_{24}AuN_4O_{12}S_4$			
Aurate(3-)-, [[4,4',4",4"'-(21H,23H-porphine-5,10,15,20-tetrayl)tetrakis[benzenesulfonato]](6-)-N^{21},N^{22},N^{23},N^{24}]-, (SP-4-1)-, chloride, tetrasodium salt	H_2O	406(5.57),483(3.62), 522(4.22)	152-0213-86
$C_{44}H_{26}Br_4N_4$			
21H,23H-Porphine, 5,10,15,20-tetrakis(2-bromophenyl)-	pyridine	422(5.99),515(4.69), 548(4.26),592(4.33), 648(4.09)	103-0629-86
21H,23H-Porphine, 5,10,15,20-tetrakis(3-bromophenyl)-	pyridine	421(5.74),515(4.39), 549(4.24),541(3.95), 650(3.76)	103-0397-86 +103-0629-86
21H,23H-Porphine, 5,10,15,20-tetrakis(4-bromophenyl)-	pyridine	422(5.69),517(4.30), 553(3.97),595(3.77), 652(3.74)	103-0397-86 +103-0629-86
$C_{44}H_{26}Cl_4N_4$			
21H,23H-Porphine, 5,10,15,20-tetrakis(2-chlorophenyl)-	pyridine	420(5.20),514(4.38), 546(3.74),590(3.89), 656(3.54)	103-0629-86
21H,23H-Porphine, 5,10,15,20-tetrakis(3-chlorophenyl)-	pyridine	421(5.65),515(4.29), 550(3.85),592(3.76), 648(3.52)	103-0397-86 +103-0629-86
21H,23H-Porphine, 5,10,15,20-tetrakis(4-chlorophenyl)-	pyridine	422(5.46),517(4.34), 551(4.10),593(3.68), 651(3.65)	103-0397-86
$C_{44}H_{26}F_4N_4$			
21H,23H-Porphine, 5,10,15,20-tetrakis(2-fluorophenyl)-	pyridine	419(5.93),512(4.33), 546(3.81),593(3.89), 653(3.62)	103-0629-86
21H,23H-Porphine, 5,10,15,20-tetrakis(3-fluorophenyl)-	pyridine	421(5.37),518(4.37), 552(3.92),591(3.86), 647(3.56)	103-0397-86 +103-0629-86
21H,23H-Porphine, 5,10,15,20-tetrakis(4-fluorophenyl)-	pyridine	421(5.39),517(4.30), 552(3.94),591(3.85), 647(3.62)	103-0397-86 +103-0629-86
$C_{44}H_{26}I_4N_4$			
21H,23H-Porphine, 5,10,15,20-tetrakis(2-iodophenyl)-	pyridine	427(5.76),520(4.46), 555(3.91),597(3.98), 659(3.58)	103-0629-86
21H,23H-Porphine, 5,10,15,20-tetrakis(3-iodophenyl)-	pyridine	423(5.66),518(4.33), 552(3.96),593(3.81), 653(3.72)	103-0629-86
21H,23H-Porphine, 5,10,15,20-tetrakis(4-iodophenyl)-	pyridine	423(5.70),517(4.35), 551(4.07),590(3.84), 649(3.82)	103-0397-86 +103-0629-86

Compound	Solvent	λ_{max}(log ε)	Ref.
$C_{44}H_{28}Br_2N_4Sn$			
Tin, dibromo[5,10,15,20-tetraphenyl-21H,23H-porphinato(2-)-N^{21},N^{22},N^{23}-N^{24}]-, (OC-6-12)-	$CHCl_3$	429(5.79),525(3.54), 565(4.30),605(4.16)	90-1957-86
$C_{44}H_{28}ClGaN_4$			
Gallium, chloro[5,10,15,20-tetraphenylporphinato(2-)-N^{21},N^{22},N^{23},N^{24}]-, (SP-5-12)-	MeOH	395(3.65),414(4.80), 509(2.50),548(3.36), 587(2.90)	90-1157-86
$C_{44}H_{28}CoN_4$			
Cobalt, [5,10,15,20-tetraphenyl-21H,23H-porphinato(2-)-N^{21},N^{22},N^{23}-N^{24}]-, (SP-4-1)-	benzene	413(5.43),529(4.22)	44-0602-86
$C_{44}H_{28}FGaN_4$			
Gallium, fluoro[5,10,15,20-tetraphenylporphinato(2-)-N^{21},N^{22},N^{23},N^{24}]-, (SP-5-12)-	MeOH	394(4.01),414(5.16), 509(3.04),548(3.69), 587(3.23)	90-1157-86
$C_{44}H_{28}N_4Sn$			
Tin, [5,10,15,20-tetraphenyl-21H,23H-porphinato(2-)-N^{21},N^{22},N^{23},N^{24}]-, (SP-4-1)-, diperchlorate	$CHCl_3$	424(5.72),513(3.49), 552(4.31),591(3.93)	90-1957-86
$C_{44}H_{28}N_4Zn$			
Zinc, [5,10,15,20-tetraphenyl-21H,23H-porphinato(2-)-N^{21},N^{22},N^{23},N^{24}]-	benzene	421(5.76),559(4.38), 587(3.59)	35-1155-86
	benzene	549.8(4.35)	94-1865-86
	EtOH	557.0(4.33)	94-1865-86
	$CHCl_3$	547.9(4.29)	94-1865-86
cyanide adduct	benzene	440(5.84),578(4.26), 620(4.28)	35-1155-86
pyridine adduct	benzene	430(5.85),563(4.36), 601(4.02)	35-1155-86
$C_{44}H_{30}N_4$			
21H,23H-Porphine, 5,10,15,20-tetraphenyl-	$CHCl_3$	416(5.68),514(4.27), 548(3.92),588(3.77), 645(3.74)	110-1023-86
	pyridine	420(5.73),515(4.27), 549(3.96),591(3.71), 646(3.71)	103-0629-86
$C_{44}H_{30}N_4O_4$			
21H,23H-Porphine, 3,7,12,17-tetrakis(3-hydroxyphenyl)-	pyridine	423(5.65),517(4.26), 552(3.87),593(3.72), 649(3.57)	103-0397-86
21H,23H-Porphine, 3,7,12,17-tetrakis(4-hydroxyphenyl)-	pyridine	426(5.64),521(4.16), 561(4.11),598(3.67), 656(3.81)	103-0397-86
$C_{44}H_{30}O_4$			
[9,9'(10H,10'H)-Bi-9,10[1',2']benzenoanthracene]-2,2'-dicarboxylic acid, dimethyl ester, anti	$CHCl_3$	239.3(4.59),264.8(4.18), 292.1(3.48)	44-0995-86
gauche	$CHCl_3$	238.8(4.55),263.1(4.16), 293.6(3.51)	44-0995-86
$C_{44}H_{32}N_8Zn$			
Zinc, [[2,2',2'',2'''-(21H,23H-porphine-	DMSO	438(5.10),565(--),	126-2535-86

Compound	Solvent	λ_{max}(log ε)	Ref.
5,10,15,20-tetrayl)tetrakis[benzen-aminato]](2-)-N^{21},N^{22},N^{23},N^{24}]-, (SP-4-1)- (cont.)		614(--)	126-2535-86
$C_{44}H_{32}O_2P_2$			
Phosphine oxide, [1,1'-binaphthalene]-2,2'-diylbis[diphenyl-, (R)-	EtOH	229(5.00),273(4.11), 288(4.08),298s(4.04), 316s(3.53),332(3.56)	44-0629-86
$C_{44}H_{32}P_2$			
Phosphine, [1,1'-binaphthalene]-2,2'-diylbis[diphenyl-	EtOH	222(5.00),235s(4.96)	44-0629-86
$C_{44}H_{33}N$			
Methylium, [(phenylimino)di-4,1-phenylene]bis[diphenyl-	HOAc-CF_3COOH	528(4.54)	64-1045-86B
dication	CF_3COOH	421(4.64),498(4.37), 675(5.09)	64-1045-86B
$C_{44}H_{34}ClO_2$			
Pyrylium, 2-[2-[2-chloro-3-[(4,6-diphenyl-2H-pyran-2-ylidene)ethylidene]-1-cyclohexen-1-yl]ethenyl]-4,6-diphenyl-, perchlorate	CH_2Cl_2	1072(4.78)	103-0250-86
$C_{44}H_{35}ClO_3$			
2,4,6-Heptatrien-1-one, 7-[2-chloro-3-[(4,6-diphenyl-2H-pyran-2-ylidene]-ethylidene]-1-cyclohexen-1-yl]-5-hydroxy-1,3-diphenyl-	CH_2Cl_2	520(3.61)	103-0250-86
$C_{44}H_{35}S_2$			
Thiopyrylium, 4-[2-[3-[(2,6-diphenyl-4H-thiopyran-4-ylidene)ethylidene]-1-cyclohexen-1-yl]ethenyl]-2,6-diphenyl-, perchlorate	CH_2Cl_2	1034(5.39)	104-0150-86
$C_{44}H_{36}AuN_8$			
Gold(5+), [[4,4',4",4"'-(21H,23H-porphine-5,10,15,20-tetrayl)tetrakis[1-methylpyridiniumato]](2-)-N^{21},N^{22}-N^{23},N^{24}]-, pentachloride	H_2O	406(5.44),481(3.61), 523(4.18),558(3.78)	152-0213-86
$C_{44}H_{36}O_{17}$			
Methanone, [3,5-diacetoxy-2,8-bis(4-acetoxyphenyl)-3,4-dihydro-2H-benzo-[2,3-h]-1-benzopyran-9-yl](2,4,6-triacetoxyphenyl)-, trans	dioxan	216(4.79),287(4.56)	94-0595-86
$C_{44}H_{38}N_8$			
Pyridinium, 3,3',3",3"'-(21H,23H-porphine-5,10,15,20-tetrayl)tetrakis[1-methyl-	n.s.g.	417(5.44)	149-0717-86B
palladium complex	n.s.g.	416(5.34)	149-0717-86B
$C_{44}H_{40}N_6O_8S_2$			
Benzoic acid, 4,4'-[1,2-ethanediylidenebis[[4-[[4-(ethoxycarbonyl)-phenyl]amino]-2,5-thiazolylidene]-nitrilo]]bis-. diethyl ester, (E,E,?,?)-	Me_4urea	451s(4.16),524s(4.49), 576(4.78),619(4.80)	48-0805-86

Compound	Solvent	$\lambda_{max}(\log \epsilon)$	Ref.
$C_{44}H_{46}N_2O_{11}$			
5,12-Naphthacenedione, 8-acetyl-10-[[3-[[2-[bis(phenylmethyl)amino]-1-oxopropyl]amino]-2,3,6-trideoxy-α-L-lyxo-hexopyranosyl]oxy]-7,8,9,10-tetrahydro-6,8,11-trihydroxy-1-methyl-	$CHCl_3$	495(4.92)	161-0881-86
5,12-Naphthacenedione, 8-acetyl-10-[[3-[[3-[bis(phenylmethyl)amino]-1-oxopropyl]amino]-2,3,6-trideoxy-α-L-lyxo-hexopyranosyl]oxy]-7,8,9,10-tetrahydro-6,8,11-trihydroxy-1-methyl- (daunorubicin, N-1-oxo-3-dibenzylaminopropyl-)	$CHCl_3$	495(4.10)	161-0881-86
$C_{44}H_{46}N_4O_8$			
21H,23H-Porphine-2-propanoic acid, 8,13-bis(2-acetoxyethyl)-3,7,12,17-tetramethyl-18-[(phenylmethoxy)-carbonyl]-, methyl ester	CH_2Cl_2	406(5.36),508(4.09), 548(4.20),574(3.96), 634(3.32)	44-0657-86
$C_{44}H_{48}N_4O_8$			
23H,25H-Benzo[b]porphine-9,13-dipropanoic acid, 18-ethenyl-3,4-bis(ethoxycarbonyl)-2,4a-dihydro-4a,8,14,19-tetramethyl-, dimethyl ester	CH_2Cl_2	403(4.94),498(3.75), 505(3.74),534(3.80), 606(3.26),635(3.28), 663(4.25)	44-1094-86
cis rearrangement product	CH_2Cl_2	359(4.26),424(4.66), 435(4.64),508(3.56), 576(3.88),625(3.45), 688(4.13)	44-1094-86
trans rearrangement product	CH_2Cl_2	354(4.30),417(5.57), 432(4.57),500(3.51), 552(3.70),577(4.24), 625(3.53),688(4.20)	44-1094-86
$C_{44}H_{50}N_4O_8$			
21H-Biline-2-propanoic acid, 8,13-bis(2-acetoxyethyl)-10,23-dihydro-1,3,7,14,17-pentamethyl-18-[(phenylmethoxy)carbonyl]-, methyl ester, dihydrobromide	CH_2Cl_2-TFA	442(4.13),516(4.47)	44-0657-86
$C_{44}H_{52}N_2O_{12}S$			
Naphthomycin A, 30-[[2-(acetylamino)-2-carboxyethyl)thio]-30-dechloro-2-demethyl-, (4E,6Z)- (diastovaricin II)	MeOH	210(4.40),229(4.43), 290(4.32),470(2.72)	158-0311-86
$C_{44}H_{52}N_{10}O_{16}S_2$			
1H,4aH-[1,2,4]Triazolo[1',2':1,2]diazeto[3,4-d]azepine-4a,9-dicarboxylic acid, 2,2'-(1,6-hexanediyl)-bis[8-amino-2,3,5,6,7,9a-hexahydro-7-[(4-methylphenyl)sulfonyl]-1,3-dioxo-, dimethyl ester	$CHCl_3$	237(4.45),288(4.62)	24-2114-86
$C_{44}H_{54}N_4$			
21H,23H-Porphine, 5,10,15,20-tetrakis(2-ethyl-1-butenyl)-	benzene	421(5.43),491s(3.51), 521(4.03),559(3.99), 596(3.57),657(3.40)	77-1725-86

Compound	Solvent	λ_{max}(log ε)	Ref.
$C_{44}H_{56}N_4NiS_4$			
Nickel, bis[1,2-bis[4-(diethylamino)-phenyl]-1,2-ethenedithiolato(2-)-S,S']-, (SP-4-1)-	$C_2H_4Cl_2$	1188(4.46)	48-0253-86
$C_{44}H_{56}Ru_2$			
Bis(η^6-hexamethylbenzene)(η^4,η^6-[2]-(1,2,4,5-cyclophenediruthenium bis-(tetrafluoroborate)	MeOH	484.5(2.97)	35-7010-86
	MeCN	487.0(2.87)	35-7010-86
	CH_2Cl_2	506.0(2.69)	35-7010-86
$C_{44}H_{58}O_{17}S$			
Urdamycin E	MeOH	310(3.80),460(3.65)	158-1657-86
	MeOH-NaOH	225(4.21),315(3.70), 480(3.72)	158-1657-86
$C_{44}H_{64}O_{16}$			
Card-20(22)-enolide, 16-acetoxy-3-[(O-3,4-O-carbonyl-2,6-dideoxy-β-D-ribo-hexopyranosyl-(1→4)-O-2,6-dideoxy-β-D-ribo-hexopyranosyl)-(1→4)-2,6-dideoxy-β-D-ribo-hexopyranosyl)oxy]-14-hydroxy-, (3β,5β,14β,16β,17β)-	n.s.g.	218(4.16)	87-0997-86
$C_{44}H_{64}Si_4$			
Cyclotetrasilane, 1,2,3,4-tetrakis-(2,2-dimethylpropyl)-1,2,3,4-tetraphenyl-	C_6H_{12}	230s(4.63),264s(4.35), 313s(3.28)	18-3314-86
$C_{44}H_{66}O_{16}$			
Bryostatin A	MeOH	228(4.56)	164-0415-86
$C_{44}H_{72}O_{11}$			
Rutamycin	MeOH	225(4.54),230(--)	44-4264-86
$C_{44}H_{76}O_2P_2Si_2$			
3,8-Dioxa-5,6-diphospha-2,9-disila-deca-4,6-diene, 2,2,9,9-tetramethyl-4,7-bis[2,4,6-tris(1,1-dimethylethyl)phenyl]-, (Z,Z)-	$CHCl_3$	370(4.31)	88-0171-86
$C_{44}H_{80}P_2Pt$			
Platinum, bis[4-(1,1-dimethylethyl)-phenyl]bis(tributylphosphine)-, cis	CH_2Cl_2	266(3.75),307(2.77)	64-1281-86B
trans	CH_2Cl_2	240(4.39),267(3.87), 304(2.93)	64-1281-86B
$C_{45}H_{25}NO_2$			
6H-Diindeno[1',2',3':3,4;1",2",3"-9,10]perylo[1,12-efg]isoindole-6,8(7H)-dione, 7-(2,4,6-trimethyl-phenyl)-	$CHCl_3$	458(4.41),490(4.12), 530(3.95)	24-2373-86
$C_{45}H_{29}N_3O$			
3H-Pyrrolo[2,1,5-de]quinolizin-3-one, 2-phenyl-4,5-bis(2-phenyl-3-indolizinyl)-	EtOH	247(4.66),319(4.01), 448(3.96),530(4.04)	95-1098-86
$C_{45}H_{30}N_3$			
Cyclopropenylium, tris(2-phenyl-3-indolizinyl)-, perchlorate	EtOH	247(4.87),270s(4.66), 320(4.38),460s(4.62), 490(4.84)	95-1098-86

Compound	Solvent	λ_{max}(log ϵ)	Ref.
$C_{45}H_{32}N_4O_{12}S_4$			
Benzenesulfonic acid, 4,4',4",4"'-(21-methyl-21H,23H-porphine-5,10,15,20-tetrayl)tetrakis-	pH 2	450(5.6),650(5.2)	74-0033-86A
$C_{45}H_{34}O_6$			
2H-1-Benzopyran-2-one, 4,7-diacetoxy-6,8-bis(diphenylmethyl)-3-phenyl-	MeOH	278(3.9)	2-0623-86
$C_{45}H_{35}N_3O_4S_2$			
4,8-Biimino-4H,8H-thieno[3,4-f]-2,1-benzisothiazole-9,10-dicarboxylic acid, 3,4,5,7,8-pentaphenyl-, diethyl ester	C_6H_{12}	222(4.56),241(4.30), 246(4.16),252(4.10), 258(4.26),265(4.41), 269s(4.40)	24-3158-86
$C_{45}H_{35}O_2$			
5H-Cyclopenta[b]pyrylium, 7-[5-(5,6-dihydro-2,4-diphenylcyclopenta[b]-pyran-7-yl)-2,4-pentadienylidene]-6,7-dihydro-2,4-diphenyl-, perchlorate	CH_2Cl_2	956(5.04),1080(5.37)	104-0150-86
5H-Cyclopenta[c]pyrylium, 5-[5-(6,7-dihydro-1,3-diphenylcyclopenta[c]-pyran-5-yl)-2,4-pentadienylidene]-6,7-dihydro-1,3-diphenyl-, perchlorate	CH_2Cl_2	980(5.49)	104-0150-86
$C_{45}H_{35}S_2$			
5H-Cyclopenta[c]thiopyrylium, 5-[5-(6,7-dihydro-1,3-diphenylcyclopenta[c]thiopyran-5-yl)-2,4-pentadien-1-ylidene]-6,7-dihydro-1,3-diphenyl-, perchlorate	CH_2Cl_2	1052(5.44)	104-0150-86
$C_{45}H_{37}S_2$			
Thiopyrylium, 4-[2-[3-[(2,6-diphenyl-4H-thiopyran-4-ylidene)ethylidene]-2-methyl-1-cyclohexen-1-yl]ethenyl]-2,6-diphenyl-, perchlorate	CH_2Cl_2	1045(5.40)	104-0150-86
$C_{45}H_{40}N_5O_{10}PS$			
1H,5H-[1,2,4]Triazolo[1',2':1,2][1,2]-diazeto[3,4-d]azepine-4a,9,9a-tricarboxylic acid, 2,3,6,7-tetrahydro-7-[(4-methylphenyl)sulfonyl]-1,3-dioxo-2-phenyl-8-[(triphenylphosphoranylidene)amino]-, trimethyl ester	$CHCl_3$	248(2.27),315(4.31)	24-2114-86
$C_{45}H_{43}O_6$			
Pyrylium, 2-[3-[4,6-bis(4-ethoxyphenyl)-2H-pyran-2-ylidene]-1-propenyl]-4,6-bis(4-ethoxyphenyl)-, tetrafluoroborate	CH_2Cl_2	750(4.71),830(4.93)	103-1295-86
$C_{45}H_{44}O_{12}$			
Sanggenon E	EtOH	218s(4.93),288(4.60), 302(4.57)	142-2285-86
	EtOH-$AlCl_3$	220s(4.93),288(4.58), 307(4.58)	142-2285-86

Compound	Solvent	λ$_{max}$(log ε)	Ref.
Sanggenon P (stereoisomer of E)	EtOH	218s(4.47),236s(4.29), 287(4.21),296(4.19)	142-2285-86
	EtOH-$AlCl_3$	218s(4.39),237s(4.16), 291(4.12),300(4.14)	142-2285-86
$C_{45}H_{48}N_2O_{11}$			
5,12-Naphthacenedione, 8-acetyl-10-[[3-[[4-[bis(phenylmethyl)amino]-1-oxobutyl]amino]-2,3,6-trideoxy-α-L-lyxo-hexopyranosyl]oxy]-7,8,9,10-tetrahydro-6,8,11-trihydroxy-1-methoxy-, (8S-cis)-	$CHCl_3$	495(4.14)	161-0881-86
$C_{45}H_{52}N_6O_8$			
L-Proline, 1,1'-[(3,18-diethenyl-1,19,22,24-tetrahydro-2,7,13,17-tetramethyl-1,19-dioxo-21H-biline-8,12-diyl)bis(1-oxo-3,1-propanediyl)]bis-, dimethyl ester	benzene	382(4.68),652(4.18)	49-1205-86
	EtOH	378(4.70),664(4.14)	49-1205-86
	$CHCl_3$	382(4.67),657(4.11)	49-1205-86
$C_{45}H_{52}N_6O_{12}$			
L-Aspartic acid, N,N'-[(3,18-diethenyl-1,19,22,24-tetrahydro-2,7,13,17-tetramethyl-1,19-dioxo-21H-biline-8,12-diyl)bis(1-oxo-3,1-propanediyl)]bis-, tetramethyl ester	benzene	381(4.70),653(4.15)	49-1205-86
	EtOH	377(4.70),659(4.12)	49-1205-86
	$CHCl_3$	379(4.70),658(4.10)	49-1205-86
$C_{45}H_{54}N_4O_7$			
Aspidospermidine-3-carboxylic acid, 15-[(16S)-10-acetoxy-17-methoxy-17-oxo-2,4(1H)-cyclo-4,5-secoakuammilan-11-yl]-2,3-didehydro-16-methoxy-1-methyl-, methyl ester, (5α,12β,19α)-	n.s.g.	259(3.11),347(3.46)	88-5413-86
$C_{45}H_{64}O_{14}$			
Olean-12-ene-24,28-dioic acid, 2-acetoxy-3-[(2,3-di-O-acetyl-4-deoxy-6-methyl-α-L-threo-hex-4-enopyranosyl)oxy]-, dimethyl ester, (2β,3β,4α)-	MeOH	211(2.87),232(2.85)	20-0207-86
$C_{45}H_{66}O_{18}$			
Bryostatin-9 acetate	MeOH	228(4.55)	100-0661-86
$C_{45}H_{72}O_{12}$			
Oligomycin B	MeOH	218(4.50),224(4.51), 233(4.47),242(4.25), 285(2.36)	44-4264-86
$C_{45}H_{73}NO_{12}$			
Scytophycin B	EtOH	264(4.45)	44-5300-86
$C_{45}H_{74}O_{10}$			
Oligomycin C	MeOH	219(4.51),224(4.53), 231(4.47),239(4.23), 275(2.16)	44-4264-86
$C_{45}H_{74}O_{11}$			
Oligomycin A	MeOH	220(4.53),225(4.57), 232(4.52),242(4.27), 275(2.11)	44-4264-86

Compound	Solvent	λ_{max}(log ϵ)	Ref.
$C_{45}H_{75}NO_{11}$			
Scytophycin C	MeOH	262(4.68)	44-5300-86
$C_{45}H_{75}NO_{12}$			
Scytophycin A	EtOH	264(4.54)	44-5300-86
Scytophycin D	MeOH	264(4.67)	44-5300-86
Scytophycin E	MeOH	265(4.59)	44-5300-86
$C_{46}H_{28}$			
5,20[1',2']:10,15[1",2"]-Dibenzeno-octacene, 5,10,15,20-tetrahydro-	CH_2Cl_2	276(3.86),284(3.85), 298(4.13),310(4.42), 364(3.48),382(3.61), 392(3.55),404(3.83), 412(3.75),438(3.93), 450(3.83),466(3.83)	78-1641-86
$C_{46}H_{32}N_4$			
1H-1,3,5-Triazino[1,2-a]quinoline, 1,3-bis(diphenylmethylene)-2,3-dihydro-2-(2-quinolinyl)-	EtOH	293(4.44),432(4.06)	97-0100-86
$C_{46}H_{40}N_4O_2$			
6H,13H-Pyrazino[1,2-a:4,5-a']diindole-7,14-diamine, 6,13-diethoxy-N,N'-6,13-tetraphenyl-	MeCN	240s(4.72)	24-2289-86
$C_{46}H_{44}N_2$			
1H-Indole, 3,3'-(1,4-phenylenedi-2,1-ethenediyl)bis[1-butyl-2-phenyl-	dioxan	396(4.64)	103-1262-86
$C_{46}H_{45}N_5O_7S_2$			
Adenosine, N,N-dibenzoyl-5'-S-(2-methylpropyl)-5'-C-[(2-methylpropyl)-thio]-5'-thio-, 2',3'-dibenzoate	MeOH	231(4.63),272(4.34)	44-1258-86
$C_{46}H_{46}N_4O_8S$			
23H,25H-Benzo[b]porphine-9,13-dipropanoic acid, 18-ethenyl-2,4a-dihydro-4-(methoxycarbonyl)-4a,8,14,19-tetramethyl-3-(phenylsulfonyl)-, dimethyl ester (or isomer)(23)	CH_2Cl_2	402(5.21),488(4.09), 506s(4.06),534(3.95), 605(3.61),662(4.57)	44-1094-86
24	CH_2Cl_2	404(5.23),500(4.09), 507s(4.09),536(3.99), 605(3.65),632s(3.53), 662(4.58)	44-1094-86
27	CH_2Cl_2	348(4.68),432(4.87), 552s(4.01),579(4.18), 623(3.90),683(4.47)	44-1094-86
28	CH_2Cl_2	353(4.66),422(4.87), 579(4.22),622(3.92), 682(4.50)	44-1094-86
31	CH_2Cl_2	353(4.71),395s(4.89), 557s(4.00),583(4.15), 626(3.92),687(4.52)	44-1094-86
32	CH_2Cl_2	350(4.60),396s(4.63), 396s(4.63),436(4.72), 468s(4.52),557s(3.92), 583(4.06),627(3.83), 687(4.43)	44-1094-86

Compound	Solvent	λ_{max}(log ϵ)	Ref.
$C_{46}H_{50}BN_4O_4$			
Boron(1+), bis[1,7-bis[4-(dimethyl-amino)phenyl]-1,6-heptadiene-3,5-dionato-O,O']-, (T-4)-, perchlorate	CH_2Cl_2	658(5.39)	97-0399-86
$C_{46}H_{52}N_4$			
Dipyrrolo[3,4-b:3',4'-b'][1,3]diazeto-[1,2-a:3,4-a']diindole, 2,10-dicyclo-hexyl-1,2,3,3a,9,10,11,11a-octahydro-3a,11a-bis(phenylmethyl)-	EtOH	257(4.28),302(3.70)	118-0731-86
	EtOH-HCl	244(4.02),290(3.61)	118-0731-86
diastereoisomer	EtOH	264(4.33),317(3.70)	118-0731-86
	EtOH-HCl	248(4,29),300(3.72)	118-0731-86
$C_{46}H_{64}N_4O_{13}$			
Ulapualide A	MeOH	246(4.53)	35-0846-86
$C_{46}H_{68}N_4O_4$			
21H-Biline-8-propanoic acid, 3-ethyl-1,17,18,19,22,24-hexahydro-2,7,12-13,17,17-hexamethyl-1,19-dioxo-, hexadecyl ester	$CHCl_3$	356(4.59),580(4.20)	49-1081-86
$C_{46}H_{69}NO_{13}$			
Tylosin, 23-de[(6-deoxy-2,3-di-O-meth-yl-β-D-allopyranosyl)oxy]-, 4^B-benzeneacetate	EtOH	278(4.21)(end abs.)	158-1724-86
$C_{46}H_{69}NO_{14}$			
Tylosin, 23-O-de(6-deoxy-2,3-di-O-methyl-β-D-allopyranosyl)- 23-O-(phenylacetyl)-	EtOH	281(4.31)	158-1108-86
$C_{46}H_{70}O_{16}$			
Bryostatin B	MeOH	227(4.56)	164-0415-86
$C_{46}H_{73}NO_{15}$			
Tylosin, 23-de[(6-deoxy-2,3-di-O-meth-yl-β-D-allopyranosyl)oxy]-, 2^A-acet-ate 3,4^B-dipropanoate	EtOH	282(4.39)	158-1724-86
$C_{46}H_{73}NO_{16}$			
Tylosin, 23-O-de(6-deoxy-2,3-di-O-meth-yl-β-D-allopyranosyl)-23-O-(1-oxo-propyl)-, 2^A-acetate 4^B-propanoate	EtOH	280(4.34)	158-1724-86
$C_{47}H_{30}N_4O_2S_2$			
5,11-Epithio-5H,7H,11H-isothiazolo-[3,4-g][1,2,4]triazolo[1,2-b]phthal-azine-7,9(8H)-dione, 3,4,5,8,11,12-hexaphenyl-	C_6H_{12}	220(4.41),265(4.20), 322(4.38)	24-3158-86
$C_{47}H_{34}O_{32}$			
Elaeocarpusin	MeOH	223(4.89),282(4.50)	94-4075-86 +142-1841-86
$C_{47}H_{39}O_2$			
1-Benzopyrylium, 8-[5-(6,7-dihydro-2,4-diphenyl-5H-1-benzopyran-8-yl)-2,4-pentadienylidene]-5,6,7,8-tetra-hydro-2,4-diphenyl-, perchlorate	CH_2Cl_2	934(4.92),1050(5.21)	104-0150-86

Compound	Solvent	λ max (log ε)	Ref.
2-Benzopyrylium, 5-[5-(7,8-dihydro-1,3-diphenyl-6H-2-benzopyran-5-yl)-2,4-pentadienylidene]-5,6,7,8-tetrahydro-1,3-diphenyl-, perchlorate	CH_2Cl_2	942(5.41)	104-0150-86
$C_{47}H_{39}S_2$			
1H-Benzo[c]thiolium, 3-[3-[3,3-bis(4-methylphenyl)benzo[c]thien-1(3H)-ylidene]-1-propenyl]-1,1-bis(4-methylphenyl)-, perchlorate	CH_2Cl_2	439(4.10),613s(4.52), 662(5.22)	104-0791-86
	MeCN	429(3.99),602s(4.34), 650(4.94)	104-0791-86
$C_{47}H_{48}N_4O_8S$			
23H,25H-Benzo[b]porphine-9,13-dipropanoic acid, 18-ethenyl-4-(ethoxycarbonyl)-2,4a-dihydro-4a,8,14,19-tetramethyl-3-(phenylsulfonyl)-, dimethyl ester	CH_2Cl_2	350(4.67),397s(4.79), 420(4.85),551s(3.99), 579(4.16),625(3.86), 684(4.42)	44-1094-86
isomer 29	CH_2Cl_2	352(4.67),396s(4.70), 435(4.81),466s(4.61), 554s(3.99),583(4.15), 626(3.91),686(4.53)	44-1094-86
isomer 30	CH_2Cl_2	354(4.61),396(4.64), 417(4.77),431(4.79), 471s(4.43),555s(3.95), 583(4.13),625(3.89), 686(4.55)	44-1094-86
$C_{47}H_{54}N_4O_{18}$			
Urobiliverdin I, octamethyl ester	$CHCl_3$	372(4.76),633(4.19)	44-3001-86
	+ HCl	370(4.71),674(4.65)	44-3001-86
Urobiliverdin IIIα, octamethyl ester	$CHCl_3$	371(4.77),625(4.21)	44-3001-86
	+ HCl	370(4.77),662(4.77)	44-3001-86
Urobiliverdin IIIβ, octamethyl ester	$CHCl_3$	372(4.72),654(4.14)	44-3001-86
	+ HCl	371(4.74),667(4.55), 706(4.61),718(4.62)	44-3001-86
Urobiliverdin IIIγ, octamethyl ester	$CHCl_3$	372(4.78),637(4.18)	44-3001-86
	+ HCl	371(4.91),666(4.67), 674(4.66)	44-3001-86
Urobiliverdin IIIδ, octamethyl ester	$CHCl_3$	372(4.62),633(4.07)	44-3001-86
	+ HCl	371(4.63),664(4.57), 669(4.56),719(4.46)	44-3001-86
$C_{47}H_{60}N_4O$			
4,8,12-Tetradecatrien-1-one, 5,9,13-trimethyl-1-(7,12,17-triethyl-3,8,13,18-tetramethyl-21H,23H-porphin-2-yl)-, (E,E)-	$CHCl_3$	406.5(5.26),508(4.00), 547(4.13),574(3.94), 630(3.08)	39-1565-86C
$C_{47}H_{60}N_6O_8$			
L-Alanine, N,N'-[(3,18-diethenyl-1,19,22,24-tetrahydro-2,7,13,17-tetramethyl-1,19-dioxo-21H-biline-8,12-diyl)bis(1-oxo-3,1-propanediyl)]bis-, bis(1,1-dimethylethyl) ester	benzene	380(4.68),640(4.18)	49-1205-86
	EtOH	377(4.70),664(4.13)	49-1205-86
	$CHCl_3$	379(4.68),652(4.08)	49-1205-86
L-Alanine, N,N'-[(3,18-diethenyl-1,19,22,24-tetrahydro-2,7,13,17-tetramethyl-1,19-dioxo-21H-biline-8,12-diyl)bis(1-oxo-3,1-propanediyl)]bis[N-methyl-, dimethyl ester	benzene	381(4.71),657(4.18)	49-1205-86
	EtOH	377(4.72),661(4.16)	49-1205-86
	$CHCl_3$	379(4.74),658(4.17)	49-1205-86

Compound	Solvent	$\lambda_{max}(\log \epsilon)$	Ref.
$C_{47}H_{62}N_2OS$			
Cholestan-3-ol, (triphenylmethyl)diazinecarbothioate, (3β,5α)-	C_6H_{12}	255(3.87)	78-2329-86
$C_{47}H_{62}N_4O$			
21H,23H-Porphine-2-methanol, 7,12,17-triethyl-3,8,13,18-tetramethyl-α-(4,8,12-trimethyltrideca-3,7,11-trienyl)-, (E,E)-(±)-	$CHCl_3$	399(5.24),497(4.16), 531(4.03),567(3.86), 620(3.08)	39-1565-86C
$C_{47}H_{70}O_{16}$			
Desfontainic acid	MeOH	236(3.15)	102-1907-86
$C_{47}H_{78}O_{13}$			
Lenoremycin (sodium salt)	n.s.g.	235(4.15)	100-0035-86
$C_{47}H_{78}O_{14}$			
Dianemycin	n.s.g.	232(4.18)	100-0035-86
$C_{48}H_{26}$			
8,8'-Binaphtho[1,2,3,4-def]chrysene	benzene	297(4.70),309(4.74), 353(3.99),377(4.18), 398(4.27)	24-1683-86
$C_{48}H_{30}N_4$			
3-Pyridinecarbonitrile, 6,6'-[1,1'-biphenyl]-4,4'-diylbis[2,4-diphenyl-	MeCN	348(4.82)	73-1061-86
$C_{48}H_{30}N_4O_8$			
21H,23H-Porphine, 5,10,15,20-tetrakis(1,3-benzodioxol-5-yl)-	pyridine	428(5.47),520(4.12), 557(3.90),594(3.63), 652(3.57)	103-0629-86
$C_{48}H_{30}O_{12}$			
Anthra[2,3-j]heptaphene-6,7,14,15,22-23-hexone, 5,8,13,16,21,24-hexamethoxy-	CH_2Cl_2	240(4.95),304(4.79), 442(4.35)	5-0839-86
$C_{48}H_{34}O_{31}$			
Coriariin B	MeOH	222(4.96),277(4.62)	94-4092-86
$C_{48}H_{35}N_3O$			
3H-Pyrrolo[2,1,5-de]quinolizin-3-one, 7-methyl-4,5-bis(7-methyl-2-phenyl-3-indolizinyl)-2-phenyl-	EtOH	250(4.64),316s(3.95), 452(3.94),534(4.02)	95-1098-86
$C_{48}H_{36}$			
[2.2.2.2.2.2]Orthoparacyclophene, (all-E)-	C_6H_{12}	341(4.85)	88-1063-86
(all-Z)-	C_6H_{12}	312(4.66)	88-1063-86
$C_{48}H_{36}CoN_4$			
Cobalt, [5,10,15,20-tetrakis(2-methylphenyl)-21H,23H-porphinato(2-)-N^{21},N^{22},N^{23},N^{24}]-, (SP-4-1)-	benzene	412(5.40),528(4.19)	44-0602-86
Cobalt, [5,10,15,20-tetrakis(4-methylphenyl)-21H,23H-porphinato(2-)-N^{21},N^{22},N^{23},N^{24}]-, (SP-4-1)-	benzene	415(5.43),530(4.22)	44-06012-86
$C_{48}H_{36}N_3$			
Cyclopropenylium, tris(7-methyl-	EtOH	249(4.88),320s(4.22),	95-1098-86

Compound	Solvent	λ_{max}(log ϵ)	Ref.
2-phenyl-3-indolizinyl)-, perchlorate (cont.)		468s(4.66),496(4.80)	95-1098-86
$C_{48}H_{38}N_4$			
21H,23H-Porphine, 5,10,15,20-tetrakis(2-methylphenyl)-	pyridine	419(5.64),514(4.29), 547(3.76),592(3.73), 648(3.54)	103-0629-86
21H,23H-Porphine, 5,10,15,20-tetrakis(3-methylphenyl)-	pyridine	421(5.64),516(4.21), 550(3.89),592(3.67), 648(3.71)	103-0629-86
21H,23H-Porphine, 5,10,15,20-tetrakis(4-methylphenyl)-	pyridine	422(5.68),517(4.25), 554(4.01),595(3.72), 652(3.80)	103-0629-86
$C_{48}H_{38}N_4O_4$			
21H,23H-Porphine, 5,10,15,20-tetrakis(2-methoxyphenyl)-	pyridine	420(5.62),513(4.27), 546(3.79),590(3.73), 648(3.53)	103-0629-86
21H,23H-Porphine, 5,10,15,20-tetrakis(3-methoxyphenyl)-	pyridine	422(5.66),518(4.28), 549(3.86),591(3.74), 649(3.73)	103-0629-86
21H,23H-Porphone, 5,10,15,20-tetrakis(4-methoxyphenyl)-	pyridine	424(5.65),518(4.16), 556(4.04),595(3.66), 651(3.79)	103-0629-86
$C_{48}H_{39}O_2$			
5H-Cyclopenta[c]pyrylium, 5-[[3-[(6,7-dihydro-1,3-diphenylcyclopenta[c]-pyran-5-yl)methylene]-1-cyclohexen-1-yl]methylene]-6,7-dihydro-1,3-diphenyl-, perchlorate	CH_2Cl_2	1010(5.35)	104-0150-86
$C_{48}H_{40}O_2P_2$			
Phosphine oxide, [1,1'-binaphthalene]-2,2'-diylbis[bis(4-methylphenyl)-	EtOH	232(5.04),275(4.08), 287(4.08),299s(4.00), 316s(3.54),332(3.53)	44-0629-86
$C_{48}H_{40}P_2$			
Phosphine, [1,1'-binaphthalene]-2,2'-diylbis[bis(4-methylphenyl)-	EtOH	223(4.98),236s(4.90)	44-0629-86
$C_{48}H_{41}N_5O_4$			
6H-Purin-6-one, 1,9-dihydro-9-[6-[(triphenylmethoxy)methyl]-1,4-dioxan-2-yl]-2-(triphenylmethyl)-amino]-, (2R-cis)-	EtOH	206(4.88),260(4.20)	87-2445-86
$C_{48}H_{41}N_5O_5$			
Guanosine, N-(triphenylmethyl)-5'-O-(triphenylmethyl)-	MeOH	260(4.19),273s(4.09)	87-2445-86
$C_{48}H_{41}O_2$			
Pyrylium, 4-[3-[3-[3-(2,6-diphenyl-4H-pyran-4-ylidene)-1-propenyl]-5,5-dimethyl-2-cyclohexen-1-ylidene]-1-propenyl]-2,6-diphenyl-, perchlorate	CH_2Cl_2	1032(5.51)	104-0150-86
$C_{48}H_{41}S_2$			
Thiopyrylium, 4-[3-[3-[3-(2,6-diphenyl-4H-pyran-4-ylidene)-1-propenyl]-	CH_2Cl_2	1115(5.50)	104-0150-86

Compound	Solvent	λ_{max}(log ε)	Ref.
5,5-dimethyl-2-cyclohexen-1-ylidene]-1-propenyl]-2,6-diphenyl-, perchlorate (cont.)			104-0150-86
$C_{48}H_{41}Se_2$			
Seleninium, 4-[3-[3-[3-(2,6-diphenyl-4H-selenin-4-ylidene)-1-propenyl]-5,5-dimethyl-2-cyclohexen-1-ylidene]-1-propenyl]-2,6-diphenyl-, perchlorate	CH_2Cl_2	1142(5.45)	104-0150-86
$C_{48}H_{43}N_5O_5$			
6H-Purin-6-one, 1,9-dihydro-9-[2-hydroxy-1-[1-(hydroxymethyl)-2-(triphenylmethoxy)ethoxy]ethyl]-2-[(triphenylmethyl)amino]-, [S-(R*,S*)]-	MeOH	261(4.15)	87-2445-86
$C_{48}H_{48}AlClN_8$			
Aluminum, chloro[2,9,16,23-tetrakis(1,1-dimethylethyl)-29H,31H-phthalocyaninato(2-)-N^{29},N^{30},N^{31}-N^{32}]-, (SP-5-12)-	toluene	340(4.72),355s(4.36), 573s(3.63),623(4.44), 660(4.34),696(5.11)	49-0475-86
	$C_6H_5NO_2$	578s(3.50),623(4.33), 659(4.33),692(5.10)	49-0475-86
	DMSO	348(4.62),360(4.67), 575s(3.69),616(4.35), 655(4.27),768.5(5.09)	49-0475-86
	1-$C_{10}H_7Cl$	362s(4.51),585s(3.69), 631(4.33),671(4.26), 702(5.10)	49-0475-86
$C_{48}H_{48}CuN_8$			
Copper, [2,9,16,23-tetrakis(1,1-dimethylethyl)-29H,31H-phthalocyaninato(2-)-N^{29},N^{30},N^{31},N^{32}]-, (SP-4-1)-	toluene	338s(4.55),346(4.89), 550s(3.82),609(4.65), 646(4.58),677(5.40)	49-0475-86
	$C_6H_5NO_2$	570s(3.84),614(4.59), 654(4.56),683(5.34)	49-0475-86
	DMSO	336(4.80),346(4.81), 560s(4.15),608(4.63), 648(4.63),675(5.35)	49-0475-86
	1-$C_{10}H_7Cl$	565s(4.07),614(4.63), 653(4.57),684(5.37)	49-0475-86
$C_{48}H_{48}N_8OV$			
Vanadium, oxo[2,9,16,23-tetrakis(1,1-dimethylethyl)-29H,31H-phthalocyaninato(2-)-N^{29},N^{30},N^{31},N^{32}]-, (SP-5-12)-	toluene	346(5.04),365s(4.52), 580s(3.81),628(4.69), 670(4.69),698(5.39)	49-0475-86
	DMSO	350(5.01),362(4.97), 580s(3.98),624(4.68), 660(4.66),692(5.36)	49-0475-86
	1-$C_{10}H_7Cl$	634(4.34),678(4.31), 707(5.36)	97-0216B-86
	H_2SO_4	738(--),795s(--), 834(--)	97-0216B-86
$C_{48}H_{52}N_6O_4$			
Ethanediamide, N,N'-bis[2-cyclohexyl-2,3,4,8b-tetrahydro-3-oxo-8b-(phenylmethyl)pyrrolo[3,4-b]indol-3a(1H)-yl]-	EtOH	242.5s(4.23),294(3.70)	118-0731-86

Compound	Solvent	$\lambda_{max}(\log \epsilon)$	Ref.
$C_{48}H_{54}N_4Ru$			
Ruthenium, [2,3,7,8,12,13,17,18-octa-ethyl-21H,23H-porphinato(2-)-N^{21},N^{22},N^{23},N^{24}]diphenyl-	CH_2Cl_2	377(4.82),516(4.27)	77-0787-86
$C_{48}H_{62}O_8$			
[6,6'-Biazulene]-1,1',3,3'-tetracarb-oxylic acid, 2,2'-dioctyl-, tetra-ethyl ester	CH_2Cl_2	244(4.73),343(4.94), 378(4.76),523.5(3.12)	24-2272-86
$C_{48}H_{70}ClNO_{15}$			
Tylosin, 23-O-[(4-chlorophenyl)acetyl]-23-O-de(6-deoxy-2,3-di-O-methyl-β-D-allopyranosyl)-, 2^A-acetate	EtOH	219(4.20),280(4.28)	158-1108-86
$C_{48}H_{70}N_2O_{17}$			
Tylosin, 23-O-[(4-nitrophenyl)acetyl]-23-O-de(6-deoxy-2,3-di-O-methyl-β-D-allopyranosyl)-, 2^A-acetate	EtOH	277(4.49)	158-1108-86
$C_{48}H_{71}NO_{14}$			
Tylosin, 23-de[(6-deoxy-2,3-di-O-meth-yl-β-D-allopyranosyl)oxy]-, 2^A-acetate 4^A-benzeneacetate	EtOH	279(4.24)	158-1724-86
$C_{48}H_{71}NO_{15}$			
Tylosin, 23-O-de(6-deoxy-2,3-di-O-meth-yl-β-D-allopyranosyl)-23-)-(phenyl-acetyl)-, 2^A-acetate	EtOH	204(4.15),280(4.31)	158-1724-86
$C_{48}H_{71}N_5O_{14}$			
Kabiramide C	MeOH	245(4.42)	35-0847-86
$C_{48}H_{74}O_{19}$			
Tenacissoside A	MeOH	216(4.06)	102-2861-86
$C_{49}H_{31}N_3$			
3,5-Pyridinedicarbonitrile, 4-phenyl-2,6-bis([1,1':4',1"-terphenyl]-4-yl)-	MeCN	332(4.84)	73-1061-86
$C_{49}H_{35}N_5O$			
2,4-Pentadienal, 5-[(5,10,15,20-tetra-phenyl-21H,23H-porphin-2-yl)amino]-, (E,E)-	$CHCl_3$	419(5.23),471(4.86), 530(4.49),573(4.26), 599(4.20),655(3.45)	39-0575-86
$C_{49}H_{37}S_2$			
Thiopyrylium, 4-[2-[3-[(2,6-diphenyl-4H-thiopyran-4-ylidene)ethylidene]-2-phenyl-1-cyclopenten-1-yl]ethen-yl]-2,6-diphenyl-, perchlorate	EtOH	1054(5.32)	104-0150-86
	acetone	1050(5.20)	104-0150-86
	CH_2Cl_2	1064(5.52)	104-0150-86
	$C_6H_4Cl_2$	1092(5.53)	104-0150-86
	$C_2H_4Cl_2$	1070(5.51)	104-0150-86
	MeCN	1045(5.26)	104-0150-86
	pyridine	1084(5.26)	104-0150-86
	$PhNO_2$	1088(5.39)	104-0150-86
	PhCN	1082(5.42)	104-0150-86
	DMF	1068(5.14)	104-0150-86
	DMSO	1076(5.13)	104-0150-86
$C_{49}H_{38}N_4O_3$			
Acetic acid, [2-[10,15,20-tris(4-	$CHCl_3$	420(5.67),516(4.22),	35-0618-86

Compound	Solvent	λ_{max}(log ϵ)	Ref.
methylphenyl)-21H,23H-porphin-5-yl]phenoxy]- (cont.)		551(3.89),590(3.71), 644(3.66)	35-0618-86
$C_{49}H_{41}O_3$			
1-Benzopyrylium, 8-[[5-[(6,7-dihydro-2,4-diphenyl-5H-1-benzopyran-8-yl)-methylene]-5,6-dihydro-2H-pyran-3-yl]methylene]-5,6,7,8-tetrahydro-2,4-diphenyl-, perchlorate	CH_2Cl_2	964(4.85),1082(5.11)	104-0150-86
$C_{49}H_{43}N$			
Benzenamine, 4-methyl-N,N-bis[[4-(4-methylphenyl)methyl]phenyl]-, dication	CF_3COOH	447(5.05),507(4.69), 681(5.15)	64-1045-86B
$C_{49}H_{43}O_2$			
1-Benzopyrylium, 8-[5-(6,7-dihydro-2-(4-methylphenyl)-4-phenyl-5H-1-benzopyran-8-yl]-2,4-pentadienylidene]-5,6,7,8-tetrahydro-2-(4-methylphenyl)-4-phenyl-, perchlorate	CH_2Cl_2	940(4.89),1054(5.20)	104-0150-86
$C_{49}H_{44}N_8O_5$			
6H-Purin-6-one, 9-[[2-azido-1-[[(4-methoxyphenyl)diphenylmethoxy]-methyl]ethoxy]methyl]-1,9-dihydro-2-[[(4-methoxyphenyl)diphenyl-methyl]amino]-, (±)-	MeOH	260(4.17),276(4.14)	87-1384-86
$C_{49}H_{45}N_5O_5$			
6H-Purin-6-one, 1,9-dihydro-9-[2-hydroxy-1-[1-(methoxymethyl)-1-(triphenylmethoxy)ethoxy]ethyl]-2-[(triphenylmethyl)amino]-, [S-(R*,S*)]-	EtOH	212(4.68),261(4.16)	87-2445-86
$C_{49}H_{45}N_5O_6$			
6H-Purin-6-one, 1,9-dihydro-9-[[2-hydroxy-1-[[(4-methoxyphenyl)diphenyl-methoxy]methyl]ethoxy]methyl]-2-[[(4-methoxyphenyl)diphenylmethyl]-amino]-, (±)-	MeOH	260(4.08),279(4.11)	87-0671-86 +87-1384-86
$C_{49}H_{46}N_6O_5$			
6H-Purin-6-one, 9-[[2-amino-1-[[(4-methoxyphenyl)diphenylmethoxy]methyl]ethoxy]methyl]-1,9-dihydro-2-[[(4-methoxyphenyl)diphenylmethyl]-amino]-, (±)-	MeOH	232(4.48),260(4.21), 276(4.17)	87-1384-86
$C_{49}H_{58}N_8O_{10}$			
Glycine, 1,1'-[(3,18-diethenyl-1,19-22,24-tetrahydro-2,7,13,17-tetra-methyl-1,19-dioxo-21H-biline-8,12-diyl)bis(1-oxo-3,1-propanediyl)]-bis[L-prolyl-, dimethyl ester	benzene	380(4.73),658(4.19)	49-1205-86
	EtOH	377(4.74),663(4.17)	49-1205-86
	$CHCl_3$	378(4.75),660(4.14)	49-1205-86
$C_{49}H_{60}N_4O_{14}$			
Chromoxymycin	H_2O	276(4.56),340s(3.87)	158-0012-86
	+ HCl	272(4.57),330s(3.88)	158-0012-86
	+ NaOH	243(4.41),278(4.61), 410(3.46)	158-0012-86

Compound	Solvent	$\lambda_{max}(\log \epsilon)$	Ref.
$C_{49}H_{62}N_4O_3$			
21H,23H-Porphine-2-propanoic acid, 7,12,17-triethyl-3,8,13,18-tetramethyl-β-oxo-α-(3,7,11-trimethyl-2,6,10-dodecatrienyl)-, methyl ester, (E,E)-	$CHCl_3$	408(5.25),507(4.09), 548(4.23),573(4.10), 625(3.48)	39-1565-86C
$C_{49}H_{66}ClOP_2RhS_6$			
Rhodium, carbonylchlorobis[tris-[5-(1,1-dimethylethyl)-2-thienyl]-phosphine-P]-, (SP-4-3)-	CH_2Cl_2	249(4.62),276(4.61), 366(3.43)	70-1919-86
$C_{49}H_{71}N_4OOsP$			
Osmium, carbonyl[2,3,7,8,12,13,17,18-octaethyl-21H,23H-porphinato(2-)-N^{21},N^{22},N^{23},N^{24}](tributylphosphine)-, (OC-6-42)-	CH_2Cl_2	257(4.74),331(4.57), 348(4.59),405(5.51), 516(4.24),541(4.24)	77-0386-86
$C_{49}H_{77}NO_{17}$			
Tylosin, 23-O-de(6-deoxy-2,3-di-O-methyl-β-D-allopyranosyl)-23-O-(1-oxopropyl)-, 2^A-acetate $3,4^B$-dipropanoate	EtOH	278(4.34)	158-1724-86
$C_{49}H_{77}N_2OS$			
Benzoxazolium, 3-hexadecyl-2-[3-(3-hexadecyl-2(3H)-benzothiazolylidene)-1-propenyl]-, perchlorate	EtOH	524(5.14)	4-0209-86
$C_{49}H_{77}N_2O_2$			
Benzoxazolium, 3-hexadecyl-2-[3-(3-hexadecyl-2(3H)-benzoxazolylidene)-1-propenyl]-, iodide	EtOH-iso-PrOH	487(5.15)	4-0209-86
$C_{49}H_{80}N_8O_{15}$			
Azinothricin, compd. with acetone	dioxan	225s(4.28)	158-0017-86
$C_{50}H_{32}N_4$			
1,17-Metheno-5,8:10,13-dinitrilo-8H-dipyrrolo[1,2-a:2',1'-n][1,15]benzodiazacycloheptadecine, 4,9,14,24-tetraphenyl-, perchlorate	CH_2Cl_2	442(4.96),566(3.88), 604(4.07),656(3.79)	77-0767-86
$C_{50}H_{34}N_4$			
3-Pyridinecarbonitrile, 6,6'-[1,1'-biphenyl]-4,4'-diylbis[2-(4-methylphenyl)-4-phenyl-	MeCN	346(4.67)	73-1061-86
3-Pyridinecarbonitrile, 6,6'-[1,1'-biphenyl]-4,4'-diylbis[4-(4-methylphenyl)-2-phenyl-	MeCN	345(4.91)	73-1061-86
$C_{50}H_{35}NO_2$			
6H-Diindeno[1',2',3':3,4;1",2",3"-9,10]perylo[1,12-efg]isoindole-6,8(7H)-dione, 7-[2,5-bis(1,1-dimethylethyl)phenyl]-	$CHCl_3$	433(4.33),458.5(4.44), 485(4.21),517(3.86)	24-2373-86
$C_{50}H_{38}N_4O_8$			
Phthalazinium, 2,2'-(7,14-dihydro-5,12-pentacenediylidene)bis[3-methoxy-1-(methoxycarbonyl)-3-	CH_2Cl_2	481(--)	89-0572-86

Compound	Solvent	λ_{max}(log ε)	Ref.
oxopropylide] (cont.)			89-0572-86
photochromic form	CH_2Cl_2	378(4.20)	89-0572-86
$C_{50}H_{39}S_2$			
Thiopyrylium, 4-[2-[3-[(2,6-diphenyl-4H-thiopyran-4-ylidene)ethylidene]-2-phenyl-1-cyclohexen-1-yl]ethenyl]-2,6-diphenyl-, perchlorate	EtOH	1030(5.32)	104-0150-86
	acetone	1025(5.19)	104-0150-86
	CH_2Cl_2	1040(5.48)	104-0150-86
	$C_6H_4Cl_2$	1062(5.49)	104-0150-86
	$C_2H_4Cl_2$	1042(5.44)	104-0150-86
	MeCN	1020(5.23)	104-0150-86
	PhCN	1050(5.41)	104-0150-86
	$PhNO_2$	1060(5.38)	104-0150-86
	pyridine	1055(5.26)	104-0150-86
	DMF	1038(5.14)	104-0150-86
	DMSO	1050(5.11)	104-0150-86
$C_{50}H_{41}N_5O_4$			
6H-Purin-6-amine, 9-[5-[(4-methoxyphenyl)diphenylmethoxy]methyl]-2-furanyl]-N-[(4-methoxyphenyl)diphenylmethyl]-	EtOH	254(4.29),268(4.32)	11-0225-86
$C_{50}H_{43}O_2$			
1-Benzopyrylium, 8-[[3-[(6,7-dihydro-2,4-diphenyl-5H-1-benzopyran-8-yl)methylene]-1-cyclohexen-1-ylmethylene]-5,6,7,8-tetrahydro-2,4-diphenyl-, perchlorate	CH_2Cl_2	966(4.89),1088(5.06)	104-0150-86
$C_{50}H_{44}DN_5O_5$			
Adenosine-2'-C-d, 2'-deoxy-N-[(4-methoxyphenyl)diphenylmethyl]-5'-O-[(4-methoxyphenyl)diphenylmethyl]-	EtOH	275(4.38)	118-0196-86
Adenosine-3'-C-d, 3'-deoxy-N-[(4-methoxyphenyl)diphenylmethyl]-5'-O-[(4-methoxyphenyl)diphenylmethyl]-	EtOH	275(4.40)	118-0196-86
$C_{50}H_{45}N_5O_5$			
Adenine, 6-N-(4-methoxyphenyldiphenylmethyl)-9-(5'-O-4-methoxyphenyldiphenylmethyl)-2'-deoxy-β-D-threo-pentofuranosyl)-	EtOH	275(4.42)	118-0196-86
Adenine, 6-N-(4-methoxyphenyldiphenylmethyl)-9-(5'-O-4-methoxyphenyldiphenylmethyl)-3'-deoxy-β-D-threo-pentofuranosyl)-	EtOH	265(4.37)	118-0196-86
$C_{50}H_{47}N_5O_8$			
Adenosine, 5'-O-[bis(4-methoxyphenyl)phenylmethyl]-N,N-bis(4-methylbenzoyl)-2',3'-O-(1-methylethylidene)-	MeOH	234(4.53),260(4.47), 275s(4.43)	23-2109-86
$C_{50}H_{52}$			
1,4-Ethanonaphthalene, 6,6',6",6"'-(1,2-ethenediylidene)tetrakis-[1,2,3,4-tetrahydro-	$CDCl_3$	295(3.9),324(4.0)	77-0974-86
$C_{50}H_{64}O_4$			
2-Cyclohexen-1-one, 3,3'-(3,7,11,16-20,24-hexamethyl-1,3,5,7,9,11,13,15-17,19,21,23,25-hexacosatridecaene-	EtOH	527(5.28)	33-0012-86
	THF	352(4.59),527(5.28)	33-0012-86
	solid	525(--)	33-0012-86

Compound	Solvent	λmax(log ε)	Ref.
1,26-diyl)bis[6-hydroxy-2,4,4-trimethyl-, [S-[R*,R*-(all-E)]]- (cont.)			33-0012-86
$C_{50}H_{68}O_2$			
3-Cyclohexen-1-ol, 4,4'-(3,7,11,16,20-24-hexamethyl-1,3,5,7,9,11,13,15,17-19,21,23,25-hexacosatridecaene-1,26-diyl)bis[3,5,5-trimethyl-, [R-[R*,R*-(all-E)]]-	EtOH	518(5.26)	33-0012-86
	THF	327(4.66),472(5.12), 511(5.26),564(5.19)	33-0012-86
	solid	530(--)	33-0012-86
$C_{50}H_{81}N_4O_{15}P$			
Calyculin A	n.s.g.	230(4.28),342(4.08)	35-2780-86
$C_{50}H_{90}N_2Si_4$			
Diazene, bis[7,7-bis[tris(1-methylethyl)silyl]bicyclo[4.1.0]hepta-1,3,5-trien-3-yl]-	CH_2Cl_2	258(3.39),402(4.30)	33-1623-86
$C_{50}H_{91}N_3Si_4$			
Bicyclo[4.1.0]hepta-1,3,5-trien-3-amine, 4-[[7,7-bis[tris(1-methylethyl)-silyl]bicyclo[4.1.0]hepta-1,3,5-trien-3-yl]azo]-7,7-bis[tris(1-methylethyl)silyl]-	CH_2Cl_2	278(4.32),493(4.29)	33-1623-86
$C_{51}H_{36}N_4$			
21H,23H-Porphine, 5,10,15,20-tetraphenyl-21-(phenylmethyl)-	n.s.g.	434(5.59),533(4.11), 573(4.34),615(3.85), 675(3.82)	88-3521-86
$C_{51}H_{36}N_4O$			
21H,23H-Porphine, 21-(4-methoxyphenyl)-5,10,15,20-tetraphenyl-, pentachloroantimony adduct	CH_2Cl_2	448(5.26),602(3.96), 664(4.26)	77-0767-86
$C_{51}H_{43}N_5O_4$			
9H-Purin-6-amine, 9-[5-[[(4-methoxyphenyl)diphenylmethoxy]methyl]-2-furanyl]-N-[(4-methoxyphenyl)phenylmethyl]-8-methyl-	EtOH	274(4.40),290s(3.95)	11-0225-86
$C_{51}H_{45}O_2$			
1-Benzopyrylium, 8-[[3-[(6,7-dihydro-2,4-diphenyl-5H-1-benzopyran-8-yl)-methylene]-2-methyl-1-cyclohexen-1-yl]methylene]-5,6,7,8-tetrahydro-2,4-diphenyl-, perchlorate	CH_2Cl_2	999(4.76),1100(4.91)	104-0150-86
$C_{51}H_{47}N_3O_{14}S_3$			
Benzenesulfonamide, 4-methyl-N,N-bis-[2-[[(4-methylphenyl)sulfonyl][2-[(7-oxo-7H-furo[3,2-g][1]benzopyran-9-yl)oxy]ethyl]amino]ethyl]-	1% DMSO	311(3.92)	149-0391-86A
$C_{51}H_{50}O_9$			
Copalliferol B, nona-O-methyl-	EtOH	273(3.79),290(3.60)	102-1498-86
$C_{51}H_{68}N_6O_8$			
L-Valine, N,N'-[(3,18-diethenyl-1,19,22,24-tetrahydro-2,7,13,17-	benzene	381(4.70),643(4.18)	49-1205-86
	EtOH	377(4.71),660(4.14)	49-1205-86

Compound	Solvent	λ_{max}(log ϵ)	Ref.
tetramethyl-1,19-dioxo-21H-biline-8,12-diyl)bis(1-oxo-3,1-propanediyl)]bis-, bis(1,1-dimethylethyl) ester (cont.)	$CHCl_3$	378(4.72),652(4.12)	49-1205-86
$C_{51}H_{74}N_4O_{16}$			
Ulapualide B	MeOH	246(4.52)	35-0846-86
$C_{51}H_{78}O_{17}$			
Avermectin A $_{2a}$, 3α-acetoxy-5-dehydro-3-hydroxy-	MeOH	236(4.47),243(4.50), 251s(4.31)	44-3058-86
$C_{51}H_{78}O_{19}$			
Tenacissoside B	MeOH	218(4.32)	102-2861-86
$C_{51}H_{79}N_2S$			
Quinolinium, 1-hexadecyl-4-[3-(3-hexadecyl-2(3H)-benzothiazolylidene)-1-propenyl]-, perchlorate	EtOH	633(5.14)	4-0209-86
$C_{51}H_{80}O_{19}$			
Tenacissoside D	MeOH	221(4.07)	102-2861-86
$C_{51}H_{81}N_2S_2$			
Benzothiazolium, 3-hexadecyl-2-[2-[(3-hexadecyl-2(3H)-benzothiazolylidene)methyl]-1-butenyl]-, perchlorate	MeOH	551(4.96)	4-0209-86
$C_{51}H_{93}N_3O_{14}$			
Demalonylcopiamycin	EtOH	222(3.24),320s(2.53)	142-3351-86
$C_{52}H_{34}Cl_5FeN_4O_2$			
Ferrate(1-), chloro[3-hydroxy-5,5-dimethyl-2-[5,10,15,20-tetrakis(4-chlorophenyl)-21H,23H-porphin-21-yl]-2-cyclohexen-1-onato-N^{21},N^{22}-N^{23},N^{24},O^3]-, (OC-6-25)-	benzene	457(5.02),524(4.34), 578(4.04),652(3.75), 726(3.81)	35-5598-86
$C_{52}H_{38}ClFeN_4O_2$			
Iron, chloro[3-hydroxy-5,5-dimethyl-2-(5,10,15,20-tetraphenyl-21H,23H-porphin-21-yl)-2-cyclohexen-1-onato(2-)-N^{21},N^{22},N^{23},N^{24},O^3]-	benzene	454(4.92),524(4.26), 580(3.94),655(3.60), 736(3.69)	35-5598-86
$C_{52}H_{38}N_4$			
3-Pyridinecarbonitrile, 6,6'-[1,1'-biphenyl]-4,4'-diylbis[2,4-bis(4-methylphenyl)-	MeCN	347(4.84)	73-1061-86
$C_{52}H_{40}N_2O_8$			
Isoquinolinium, 2,2'-(7,14-dihydro-5,12-pentacenediylidene)bis[3-methoxy-1-(methoxycarbonyl)-3-oxopropylide]	CH_2Cl_2	577(4.11)	89-0572-86
photochromic product	CH_2Cl_2	396(4.15)	89-0572-86
$C_{52}H_{41}N_7O_7$			
Benzamide, N-benzoyl-N-[9-[3,4-bis-(benzoyloxy)-5-(1,3-diphenyl-2-imidazolidinyl)tetrahydro-2-furanyl]-9H-purin-6-yl]-, [2R-(2α,3β,4β,5α)]-	MeOH	239(4.74),249(4.72), 280s(4.38)	44-1258-86

Compound	Solvent	λ_{max}(log ε)	Ref.
$C_{52}H_{44}ClFeN_4$			
Iron, chloro[5,10,15,20-tetrakis(3,5-dimethylphenyl)-21H,23H-porphinato(2-)-N[21],N[22],N[23],N[24]]-	$CHCl_3$	378(4.76),420(5.04), 512(4.16),576(3.56), 663(3.46),695(3.54)	64-1265-86B
$C_{52}H_{44}N_4Zn$			
Zinc, [5,10,15,20-tetrakis(2,6-dimethylphenyl)-21H,23H-porphinato(2-)-N[21],N[22],N[23],N[24]]-, (SP-4-1)-	CH_2Cl_2	391(4.68),418(5.81), 483s(--),510(3.51), 548(4.36),584(3.30)	35-4433-86
$C_{52}H_{46}N_4$			
21H,23H-Porphine, 5,10,15,20-tetrakis(2,3-dimethylphenyl)-	pyridine	420(5.61),515(4.30), 547(3.91),592(3.81), 657(3.84)	103-0629-86
21H,23H-Porphine, 5,10,15,20-tetrakis(2,4-dimethylphenyl)-	pyridine	422(5.55),516(4.29), 548(3.88),592(3.79), 650(3.77)	103-0629-86
21H,23H-Porphine, 5,10,15,20-tetrakis(2,5-dimethylphenyl)-	pyridine	420(5.63),516(4.30), 548(3.84),592(3.81), 651(3.77)	103-0629-86
21H,23H-Porphine, 5,10,15,20-tetrakis(2,6-dimethylphenyl)-	pyridine	420(5.43),514(4.18), 546(3.76),591(3.73), 649(3.75)	103-0629-86
21H,23H-Porphine, 5,10,15,20-tetrakis(3,4-dimethylphenyl)-	pyridine	424(5.76),518(4.30), 555(4.06),595(3.82), 651(3.80)	103-0629-86
21H,23H-Porphine, 5,10,15,20-tetrakis(3,5-dimethylphenyl)-	pyridine	423(5.74),518(4.29), 553(4.01),594(3.81), 650(3.80)	103-0629-86
	$CHCl_3$	402s(--),420(5.74), 485s(--),516(4.30), 551(3.95),590(3.76), 647(3.68)	64-1265-86B
$C_{52}H_{46}N_4O_8$			
21H,23H-Porphine, 5,10,15,20-tetrakis(2,3-dimethoxyphenyl)-	pyridine	420(5.68),513(4.33), 545(3.77),590(3.77), 653(3.78)	103-0629-86
21H,23H-Porphine, 5,10,15,20-tetrakis(3,4-dimethoxyphenyl)-	pyridine	428(5.58),520(4.22), 558(4.05),595(3.72), 653(3.80)	103-0629-86
$C_{52}H_{46}N_4O_{12}S_4$			
Benzenesulfonic acid, 3,3',3",3"'-(21H,23H-porphine-5,10,15,20-tetrayl)tetrakis[2,4-dimethyl-, tetrasodium salt	H_2O	413(5.80),482s(3.49), 513(4.30),550(3.68), 580(3.84),633(3.34)	35-4433-86
$C_{52}H_{48}N_8O_8Zn$			
Zinc, [2,9,16,23-tetrakis(1,1-dimethylethyl)-6,13,20,27-tetranitro-29H,31H-tetrabenzo[b,g,l,q]porphinato(2-)-N[29],N[30],N[31],N[32]]-, (SP-4-1)-	$CHCl_3$	412s(4.80),436(5.41), 596(4.17),646(5.08)	103-0960-86
$C_{52}H_{49}N_7O_6Zn$			
Zinc, [2,9,16,23-tetrakis(1,1-dimethylethyl)-6,13,20-trinitro-29H,31H-tetrabenzo[b,g,l,q]porphinato(2-)-N[29],N[30],N[31],N[32]]-, (SP-4-2)-	$CHCl_3$	410s(4.81),434(5.40), 590(4.18),644(5.08)	103-0960-86

Compound	Solvent	λ_{max}(log ε)	Ref.
$C_{52}H_{50}N_8O_8$			
29H,31H-Tetrabenzo[b,g,l,q]porphine, 2,9,16,23-tetrakis(1,1-dimethylethyl)-6,13,20,27-tetranitro-	$CHCl_3$	400s(4.89),428(5.22), 582(4.24),618(4.83), 624(4.82),632(4.803), 688(4.75)	103-0960-86
$C_{52}H_{55}ClCoN_9O_4$			
Cobalt(4+), (1-butanamine)chloro-[4,4',4",4"'-(21H,23H-porphine-5,10,15,20-tetrayl)tetrakis[1-(2-hydroxyethyl)pyridiniumato]](2-)-N^{21},N^{22},N^{23},N^{24}]-, tetrachloride (OC-6-42)-	n.s.g.	436(5.06),548(4.03), 585s(3.64)	103-0167-86
$C_{52}H_{56}N_4NiS_4$			
Nickel, bis[1,2-bis(2,3,6,7-tetrahydro-1H,5H-benzo[ij]quinolizin-9-yl)-1,2-ethenedithiolato(2-)-S,S']-, (SP-4-1)-	$C_2H_4Cl_2$	1298(4.68)	48-0253-86
$C_{52}H_{60}O_8$			
Oleana-11,13(18)-dien-28-oic acid, 2,3,23-tris(benzoyloxy)-, methyl ester, (2α,3β,4α)-	EtOH	230(4.51),244(4.28), 251(4.40),260(4.23)	2-0113-86
(2α,3β,4β)-	EtOH	230(4.48),244(4.30), 252(4.40),261(4.23)	2-0113-86
$C_{52}H_{70}N_4$			
21H,23H-Porphine, 5,10,15,20-tetrakis[3-methyl-2-(1-methylethyl)-1-butenyl]-	benzene	421(5.44),488(3.60), 520(4.05),556(3.99), 596(3.56),657(3.48)	77-1725-86
$C_{52}H_{76}CoN_4$			
Cobalt, [2,3,7,8,12,13,17,18-octabutyl-21H,23H-porphinato(2-)-N^{21},N^{22},N^{23},N^{24}]-, (SP-4-1)-	$CHCl_3$	394(5.47),521(4.19), 554(4.50)	18-1259-86
$C_{52}H_{76}CuN_4$			
Copper, [2,3,7,8,12,13,17,18-octabutyl-21H,23H-porphinato(2-)-N^{21},N^{22},N^{23},N^{24}]-, (SP-4-1)-	$CHCl_3$	401(5.59),527(4.20), 564(4.48)	18-1259-86
$C_{52}H_{76}N_4Ni$			
Nickel, [2,3,7,8,12,13,17,18-octabutyl-21H,23H-porphinato(2-)-N^{21},N^{22},N^{23},N^{24}]-, (SP-4-1)-	$CHCl_3$	395(5.30),519(4.08), 554 (4.53)	18-1259-86
$C_{52}H_{78}N_2O_4$			
3H-Indol-3-one, 2-[1,3-dihydro-3-oxo-1-(1-oxooctadecyl)-2H-indol-2-ylidene]-1,2-dihydro-1-(1-oxooctadecyl)-	benzene	412(3.60),569(3.86)	44-2529-86
	butyl stearate	425(3.68),560(3.80)	44-2529-86
$C_{52}H_{78}O_{22}$			
Desfontainoside	MeOH	236(3.10)	102-1907-86
$C_{53}H_{49}ClCoN_9O_4$			
Cobalt(4+), chloro[[4,4',4",4"'-(21H-23H-porphine-5,10,15,20-tetrayl)-tetrakis[1-(2-hydroxyethyl)pyridiniumato]](2-)-N^{21},N^{22},N^{23},N^{24}](pyridine)-, tetrachloride, (OC-6-32)-	$CHCl_3$-MeOH	440(5.06),562(4.00), 601(3.60)	103-0167-86

Compound	Solvent	λ_{max}(log ε)	Ref.
$C_{53}H_{55}ClCoN_9O_4$			
Cobalt(4+), chloro(piperidine)[[(4,4'-4",4"'-(21H,23H-porphine-5,10,15,20-tetrayl)tetrakis[1-(2-hydroxyethyl)-pyridiniumato]](2-)-N^{21},N^{22},$N^{23}N^{24}$]-, tetrachloride, (OC-6-32)-	$CHCl_3$-MeOH	439(4.96),552(4.02), 591s(3.65)	103-0167-86
$C_{53}H_{61}FeN_6O_5S_2$			
Iron(1+), bis[1-[(isocyanomethyl)sulfonyl]-4-methylbenzene][2,3,7,8,12-13,17,18-octaethyl-21H-5-oxaporphinato-N^{21},N^{22},N^{23},N^{24}]-, (OC-6-34)-, tetrafluoroborate	CH_2Cl_2-5% pyridine	371(4.73),405s(4.56), 491(4.00),525(4.04), 590s(4.11),631(4.65)	39-0001-86C
$C_{53}H_{61}FeN_7O_4S_2$			
Iron, bis[1-[(isocyanomethyl)sulfonyl]-4-methylbenzene][2,3,7,8,12,13-17,18-octaethyl-21H,23H-5-azaporphinato(2-)-N^{21},N^{22},N^{23},N^{24}]-, (OC-6-24)	CH_2Cl_2	364(5.06),490(4.12), 555(4.21)	39-0001-86C
$C_{53}H_{76}O_{19}$			
Tenacissoside C	MeOH	225(4.36),274(3.00), 282(3.00)	102-2861-86
$C_{53}H_{78}O_{19}$			
Tenacissoside E	MeOH	231(4.13),275(3.07), 282(3.00)	102-2861-86
$C_{53}H_{81}N_2$			
Quinolinium, 1-hexadecyl-2-[3-(1-hexadecyl-2(1H)-quinolinylidene)-1-propenyl]-, bromide	EtOH	607(5.06)	4-0209-86
Quinolinium, 1-hexadecyl-4-[3-(1-hexadecyl-4(1H)-quinolinylidene)-1-propenyl]-, bromide	isoPrOH	711(5.35)	4-0209-86
$C_{53}H_{82}O_{24}S$			
Neothyonidioside (sodium salt)	EtOH	none above 210nm	100-0809-86
$C_{54}H_{40}N_2O_2P_2S_2$			
Phosphonium, [dithiobis(2-phenyl-5,4-oxazolediyl)]bis[triphenyl-, diperchlorate	CH_2Cl_2	330(4.08)	65-1323-86
$C_{54}H_{57}ClCoN_9O_4$			
Cobalt(4+), chloro(cyclohexanamine)-[[4,4',4",4"'-(21H,23H-porphine-5,10,15,20-tetrayl)tetrakis[1-(2-hydroxyethyl)pyridiniumato]](2-)-N^{21},N^{22},N^{23},N^{24}]-, tetrachloride, (OC-6-42)-	$CHCl_3$-MeOH	436(5.06),548(4.01), 587s(3.59)	103-0167-86
$C_{54}H_{66}Ge_3$			
Trigermirane, hexakis(2,4,6-trimethylphenyl)-	C_6H_{12}	268(4.69),310s(--)	88-3251-86
$C_{54}H_{70}N_4O_7$			
Dispiro[piperidine-4,2'-[7,13,23,29]-tetraoxa[10,26]diazapentacyclo[28.2-	MeOH	198(5.18),228s(4.39), 268(3.38),275s(3.29)	35-2273-86

Compound	Solvent	$\lambda_{max}(\log \epsilon)$	Ref.
2.2^{3,6}.2^{14,17}.2^{19,22}]tetraconta-[3,5,14,16,19,21,30,32,33,35,37,39]-dodecaene-18',4"-piperidine]-9',27'-dione, 1-acetyl-1"-ethyl-5',15',21'-31',34',35',38',39'-octamethyl- (cont.)			35-2273-86
$C_{54}H_{72}N_6$			
Cyclohexatriaconta[1,2-b:13,14-b':25-26-b"]triquinoxaline, 6,7,8,9,10,11-12,13,14,15,22,23,24,25,26,27,28,29-30,31,38,39,40,41,42,43,44,45,46,47-triacontahydro-	n.s.g.	241(4.92),319(4.42)	44-3257-86
$C_{54}H_{80}O_{20}$			
Oleanan-29-al, 21-(benzoyloxy)-12,13-epoxy-3-[(O-β-D-glucopyranosyl-(1→-2)-O-(β-D-glucopyranosyl-(1→4)-β-L-arabinopyranosyl)oxy]-16-hydroxy-(avenacin B-2)	EtOH	228(4.16),274(3.11), 281(3.10)	39-1905-86C +102-2069-86
$C_{54}H_{80}O_{21}$			
Avenacin A-2	EtOH	223(4.17),274(3.08), 281(3.04)	39-1905-86C +102-2069-86
$C_{55}H_{38}N_6O$			
1,2-Benzenedicarbonitrile, 4-[4-[10-15,20-tris(4-methylphenyl)-21H,23H-porphin-5-yl]phenoxy]-	toluene	420(5.57),515(4.28), 550(4.08),592(3.90), 648(3.85)	77-1239-86
$C_{55}H_{42}N_3$			
Methylium, tris[4-(diphenylamino)-phenyl]-	HOAc-5% CF_3COOH	409(3.27),627(4.95)	64-1045-86B
$C_{55}H_{45}O_2$			
1-Benzopyrylium, 8-[[3-[(6,7-dihydro-2,4-diphenyl-5H-1-benzopyran-8-yl)-methylene]-2-phenyl-1-cyclopenten-1-yl]methylene]-5,6,7,8-tetrahydro-2,4-diphenyl-, perchlorate	CH_2Cl_2	1005(4.86),1136(5.10)	104-0150-86
2-Benzopyrylium, 5-[[3-[(7,8-dihydro-1,3-diphenyl-6H-2-benzopyran-5-yl)-methylene]-2-phenyl-1-cyclopenten-1-yl]methylene]-5,6,7,8-tetrahydro-1,3-diphenyl-, perchlorate	CH_2Cl_2	1000(5.35)	104-0150-86
$C_{55}H_{49}N_5O_7S$			
2',3'-Secoguanosine, N^2,O$^{5'}$-bis(tri-phenylmethyl)-2'-O-[(4-methylphen-yl)sulfonyl]-	EtOH	211(4.73),262(4.16)	87-2445-86
2',3'-Secoguanosine, N^2,O$^{5'}$-bis(tri-phenylmethyl)-3'-O-[(4-methylphen-yl)sulfonyl]-	EtOH	206(4.94),261(4.18)	87-2445-86
$C_{55}H_{70}N_8O_{10}$			
Glycine, 1,1'-[(3,18-diethenyl-1,19,22,24-tetrahydro-2,7,13,17-tetramethyl-1,19-dioxo-21H-biline-8,12-diyl)bis(1-oxo-3,1-propanediyl)]-bis[L-prolyl-, bis(1,1-dimethylethyl) ester	benzene	382(4.72),655(4.20)	49-1205-86
	EtOH	377(4.74),664(4.15)	49-1205-86
	$CHCl_3$	380(4.73),657(4.13)	49-1205-86

Compound	Solvent	λ_{max}(log ϵ)	Ref.
$C_{55}H_{83}NO_{20}$			
Avenacin B-1	EtOH	223(4.37),255(3.90), 356(3.72)	39-1905-86C +102-2069-86
$C_{55}H_{83}NO_{21}$			
Avenacin A-1	EtOH	223(4.40),255(3.90), 357(3.74)	39-1905-86C +102-2069-86
$C_{56}H_{32}MgN_8$			
Magnesium, [2,9,16,23-tetraphenyl-29H,31H-phthalocyaninato(2-)-N^{29},N^{30},N^{31},N^{32}]-, (SP-4-1)-	90% DMF	694(5.27)	149-0125-86B
$C_{56}H_{32}N_8O_8Zn$			
Zinc, [[4,4',4",4"'-(29H,31H-phthalocyanine-2,9,16,23-tetrayl)tetrakis[phenolato]](2-)-N^{29},N^{30},N^{31},N^{32}]-, (SP-4-1)-	DMSO	355(--),610(--), 678(5.07)	126-2535-86
$C_{56}H_{34}$			
7,16[2',3']-Anthraceno-5,18[1',2']-9,14[1",2"]-dibenzenoheptacene, 5,7,9,14,16,18-hexahydro-	CH_2Cl_2	293(5.68),324(3.68), 357(4.02),377(3.86)	35-6675-86
5,22[1',2']:7,20[1",2"]:11,16[1"',2"']-Tribenzenononacene, 5,7,11,16,20,22-hexahydro-	C_6H_{12}	238(4.55),248(4.57), 260(4.80),266(4.82), 277(4.80),286(4.90), 319(3.68),335(3.80), 352(3.84),371(3.75)	78-1641-86
$C_{56}H_{46}ClFeN_4O_3$			
Iron, chloro-[3-hydroxy-5,5-dimethyl-2-[5,10,15,20-tetrakis(4-methylphenyl)-21H,23H-porphin-21-yl]-2-cyclohexen-1-onato(2-)-N^{21},N^{22},N^{23}-N^{24},O^3]-, (OC-6-25)-	benzene	458(4.92),526(4.26), 581(3.98),654(3.73), 736(3.69)	35-6598-86
$C_{56}H_{46}N_4$			
21H,23H-Porphine, 5,10,15,20-tetrakis(2-phenyl-1-propenyl)-	benzene	429(5.46),497s(3.57), 528(4.01),570(4.18), 600s(3.72),663(3.61)	77-1725-86
$C_{56}H_{47}O_2$			
1-Benzopyrylium, 8-[[3-[(6,7-dihydro-2,4-diphenyl-5H-1-benzopyran-8-yl)methylene]-2-phenyl-1-cyclohexen-1-yl]methylene]-5,6,7,8-tetrahydro-2,4-diphenyl-, perchlorate	CH_2Cl_2	990(4.58),1100(4.76)	104-0150-86
2-Benzopyrylium, 5-[[3-[(7,8-dihydro-1,3-diphenyl-6H-2-benzopyran-5-yl)methylene]-2-phenyl-1-cyclohexen-1-yl]methylene]5,6,7,8-tetrahydro-1,3-diphenyl-, perchlorate	CH_2Cl_2	980(5.23)	104-0150-86
$C_{56}H_{49}FeN_6O_2$			
Iron(1+), [2,2,5,5-tetramethyl-3-[[2-[10,15,20-tris(4-methylphenyl)-21H,23H-porphin-5-yl]benzoyl]amino]-1-pyrrolidinyloxyato(2-)-N^{21},N^{22}-N^{23},N^{24}]-, chloride, (SP-4-2)-	$CHCl_3$	382(4.77),418(5.01), 511(4.13),575(3.59), 660s(3.44),694(3.50)	35-0618-86

Compound	Solvent	$\lambda_{max}(\log \epsilon)$	Ref.
$C_{56}H_{51}N_5O_8S$			
6H-Purin-6-one, 1,9-dihydro-9-[[1-[[(4-methoxyphenyl)diphenylmethoxy]-methyl]-2-[[(4-methylphenyl)sulfonyl]oxy]ethoxy]methyl]-2-[[(4-methoxyphenyl)diphenylmethyl]amino]-, (±)-	MeOH	260(4.21),273(4.18)	87-1384-86
$C_{56}H_{52}ClFeN_4$			
Iron, chloro[5,10,15,20-tetrakis(2,4,6-trimethylphenyl)-21H,23H-porphinato(2-)-N^{21},N^{22},N^{23},N^{24}]-, (SP-5-12)-	$CHCl_3$	376(4.77),420(5.06), 512(4.18),577(3.60), 665(3.49),696(3.52)	64-1265-86B
$C_{56}H_{53}FeN_4O$			
Iron, hydroxy[5,10,15,20-tetrakis(2,4,6-trimethylphenyl)-21H,23H-porphinato(2-)-N^{21},N^{22},N^{23},N^{24}]-	$CHCl_3$	338(4.40),420(4.98), 580(3.79),630s(--)	64-1265-86B
$C_{56}H_{54}N_4$			
21H,23H-Porphine, 5,10,15,20-tetrakis(2,4,6-trimethylphenyl)-	$CHCl_3$	402s(--),419(5.73), 481s(--),514(4.32), 548(3.79),590(3.79), 645(3.43)	64-1265-86B
$C_{56}H_{56}P_4Rh$			
Rhodium(1+), bis[(1,2-dimethyl-1,2-ethanediyl)bis[diphenylphosphine]-P,P']-, chloride	MeOH	237(4.65),265(4.47), 313(3.85),345s(3.66), 410(3.62)	23-0051-86
	$MeNO_2$	410(3.61)	23-0051-86
$C_{56}H_{58}N_4O_2$			
2-Naphthalenol, 1,1'-(2,3,7,8,12,13-17,18-octaethyl-21H,23H-porphine-5,15-diyl)bis-, cis	CH_2Cl_2	413(5.33),510(4.26), 545(3.89),576(3.89), 628(3.55)	88-6365-86
trans	CH_2Cl_2	413(5.32),510(4.25), 545(3.90),576(3.90), 628(3.53)	88-6365-86
$C_{56}H_{77}NO_{16}$			
Tylosin, 23-O-de(6-deoxy-2,3-di-O-methyl-β-D-allopyranosyl)-23-O-(phenylacetyl)-, 2^A-acetate 4^B-propanoate	EtOH	282(4.27)	158-1724-86
$C_{56}H_{86}O_{19}$			
Hebevinoside I	MeOH	end absorption	94-0088-86
$C_{57}H_{42}N$			
Methylium, (nitrilotri-4,1-phenylene)-tris[diphenyl-	HOAc-CF_3COOH	534(4.56)	64-1045-86B
dication	45% HOAc-CF_3COOH	424(4.50),515(4.47), 676(4.81)	64-1045-86B
trication	CF_3COOH	426(4.35),641(5.26)	64-1045-86B
$C_{57}H_{45}N_5O_{16}S$			
β-D-Ribofuranosylamine, N-[1-(2,3,5-tri-O-benzoyl-β-D-ribofuranosyl)-1H-Pyrazino[2,3-c][1,2,6]thiadiazin-4-yl]-, 2,3,5-tribenzoate, S,S-dioxide	MeOH	230(5.04),273s(4.17), 281s(4.09),340(3.86)	5-1872-86

Compound	Solvent	λ_{max}(log ε)	Ref.
$C_{57}H_{51}FeN_6O_2$			
Iron(1+), [2,2,5,5-tetramethyl-3-[[[2-[10,15,20-tris(4-methylphenyl)-21H,23H-porphin-5-yl]benzoyl]amino]-methyl]-1-pyrrolidinyloxyato(2-)-N^{21},N^{22},N^{23},N^{24}]-, chloride, (SP-4-2)-	$CHCl_3$	350s(4.60),380(4.75), 419(5.03),510(4.15), 575(3.61),655(3.46), 692(3.54)	35-0618-86
$C_{57}H_{55}FeN_4O$			
Iron, methoxy[5,10,15,20-tetrakis(2,4,6-trimethylphenyl)-21H,23H-porphinato(2-)-N^{21},N^{22},N^{23},N^{24}]-, (SP-5-12)-	toluene	330(4.38),420(5.02), 582(3.76),630s(--)	64-1265-86B
$C_{57}H_{56}N_4NiO_3$			
Nickel, [2,3,7,8,12,13,17,18-octaethyl-10,20-bis(2-hydroxy-1-naphthalenyl)-21H,23H-porphine-5-carboxaldehydato(2-)-N^{21},N^{22},N^{23},N^{24}]-, [SP-4-2-(R*,R*)]-	CH_2Cl_2	442(4.99),554(3.69), 671(4.09)	88-6365-86
$C_{57}H_{57}N_3O_{22}S_3$			
Benzo[3,4]cyclobuta[1,2-b]cyclobuta-[h]biphenylene-3,10:4,9:5,8-triimine-1,2,3a,4a,6,7,8b,9b-octacarboxylic acid, 2a,3,3b,4,4b,5,8,8a,9,9a-10,10a-dodecahydro-11,12,13-tris[(4-methylphenyl)sulfonyl]-, octamethyl ester, (2aα,3β,3aα,3bβ,4α,4aβ,4bα-5β,8β,8aα,8bβ,9α,9aβ,9bα,10β,10aα)-	MeCN	231(4.68),253s(4.01), 258s(3.83),264s(3.73), 274(3.56)	24-0616-86
$C_{57}H_{60}N_4O_3$			
21H,23H-Porphine-5-methanol, 2,3,7,8-12,13,17,18-octaethyl-10,20-bis(2-hydroxy-1-naphthalenyl)-	CH_2Cl_2	429(5.25),554(4.10), 592(4.21)	88-6365-86
$C_{57}H_{60}N_6O_8$			
L-Proline, 1,1'-[(3,18-diethenyl-1,19,22,24-tetrahydro-2,7,13,17-tetramethyl-1,19-dioxo-21H-biline-8,12-diyl)bis(1-oxo-3,1-propanediyl)]bis-, bis(phenylmethyl) ester	benzene	381(4.67),655(4.18)	49-1205-86
	EtOH	377(4.69),665(4.14)	49-1205-86
	$CHCl_3$	381(4.67),660(4.13)	49-1205-86
$C_{58}H_{42}Fe$			
Ferrocene, 1,1',2,2',3,3',4,4'-octaphenyl-	CH_2Cl_2	262(4.72),340(4.15), 502(2.72)	157-1116-86
Ferrocenium, 1,1',2,2',3,3',4,4'-octaphenyl-, hexafluorophosphate	CH_2Cl_2	248(4.65),372(4.18), 495(3.70),750(2.22)	157-1116-86
$C_{58}H_{53}FeN_6O_2$			
Iron(1+), [2,2,6,6-tetramethyl-4-[[[2-[10,15,20-tris(4-methylphenyl)-21H,23H-porphin-5-yl]benzoyl]amino]-methyl]-1-piperidinyloxyato(2-)-N^{21},N^{22},N^{23},N^{24}]-, chloride, (SP-4-2)-	$CHCl_3$	354s(4.57),381(4.71), 420(4.99),511(4.15), 578(3.66),655(3.53), 694(3.57)	35-0618-86
$C_{58}H_{53}FeN_6O_3$			
Iron(1+), [[2,2,6,6-tetramethyl-4-[[[2-[10,15,20-tris(4-methylphenyl)-21H,23H-porphin-5-yl]phenoxy]acetyl]-amino]-1-piperidinyloxyato(2-)-N^{21},N^{22},N^{23},N^{24}]-, chloride, (SP-4-2)-	$CHCl_3$	380(4.78),419(5.03), 511(4.16),576(3.55), 654(3.47),692(3.55)	35-0618-86

Compound	Solvent	$\lambda_{max}(\log \epsilon)$	Ref.
$C_{58}H_{64}N_{10}O_{14}$			
Saframycin Y2b	MeOH	268(4.66)	158-1639-86
$C_{58}H_{88}N_{10}O_{33}S_2$			
β-Cyclodextrin, 6A,6B-bis-S-[3-(6-amino-9H-purin-9-yl)propyl]-6A,6B-dithio-	pH 7.0	262(4.33)	44-3931-86
isomer b	pH 7.0	262(4.39)	44-3931-86
isomer c	pH 7.0	262(4.43)	44-3931-86
$C_{58}H_{89}N_7O_{35}S_2$			
β-Cyclodextrin, 6A-S-[3-(6-amino-9H-purin-9-yl)propyl]-6B-S-[3-(3,4-dihydro-5-methyl-2,4-dioxo-1(2H)-pyrimidinyl)propyl]-6A,6B-dithio-	pH 7.0	265(4.23)	44-3931-86
$C_{58}H_{90}N_4O_{37}S_2$			
β-Cyclodextrin, 6A,6B-bis-S-[(3,4-dihydro-5-methyl-2,4-dioxo-1(2H)-pyrimidinyl)propyl]-6A,6B-dithio-	pH 7.0	273(4.27)	44-3931-86
isomer b	pH 7.0	273(4.28)	44-3931-86
isomer c	pH 7.0	273(4.26)	44-3931-86
$C_{59}H_{55}FeN_6O_3$			
Iron(1+), [2,2,6,6-tetramethyl-4-[[1-oxo-3-[2-[10,15,20-tris(4-methylphenyl)-21H,23H-porphin-5-yl]phenoxy]propyl]amino]-1-piperidinyloxyato(2-)-N21,N22,N23,N24]-, chloride, (SP-4-2)-	$CHCl_3$	350s(4.53),383(4.74), 418(4.98),512(4.10), 579(3.50),657(3.39), 694(3.45)	35-0618-86
$C_{59}H_{68}N_6O_8$			
L-Phenylalanine, N,N'-[(3,18-diethenyl-1,19,22,24-tetrahydro-2,7,13,17-tetramethyl-1,19-dioxo-21H-biline-8,12-diyl)bis[1-oxo-3,1-propanediyl)]bis-, bis(1,1-dimethylethyl) ester	benzene	380(4.68),650(4.15)	49-1205-86
	EtOH	377(4.70),662(4.10)	49-1205-86
	$CHCl_3$	379(4.68),655(4.10)	49-1205-86
$C_{60}H_{38}N_4$			
3-Pyridinecarbonitrile, 6,6'-[1,1'-biphenyl]-4,4'-diylbis[2-[1,1'-biphenyl]-4-yl-4-phenyl-	MeCN	348(4.73)	73-1061-86
$C_{60}H_{44}Cl_4FeN_4O_4$			
Ferrate(1-), [[2,2'-[5,10,15,20-tetrakis(4-chlorophenyl)-21H,23H-porphine-21,23-diyl]bis[3-hydroxy-5,5-dimethyl-2-cyclohexen-1-onato]](2-)-N21,N22,N23,N24,O3,O3']-, (OC-6-12)-	benzene	467(5.04),562(4.09), 620(4.30),750(3.71)	35-5598-86
$C_{60}H_{48}CoN_8O_{16}$			
Cobalt, [tetrakis[2-[(2-methyl-1-oxo-2-propenyl)oxy]ethyl]-29H,31H-phthalocyaninetetracarboxylato(2-)-N29,N30,N31,N32]-	DMF	289(4.72),338(4.71), 613(4.52),677(5.04)	126-0585-86B
$C_{60}H_{48}FeN_4O_4$			
Iron, [[2,2'-(5,10,15,20-tetraphenyl-21H,23H-porphine-21,23-diyl)bis[3-hydroxy-5,5-dimethyl-2-cyclohexen-1-onato]](2-)-N21,N22,N23,N24,O3,O3']-	benzene	464(5.06),562(4.04), 615(4.21),746(3.64)	35-5598-86

Compound	Solvent	λmax(log ε)	Ref.
$C_{60}H_{50}O_{18}$			
Copalliferol B nonaacetate	EtOH	284(3.99),313(3.82)	102-1498-86
$C_{60}H_{64}N_8O_4Zn$			
Zinc, [[N,N',N",N"'-[2,9,16,23-tetrakis(1,1-dimethylethyl)-29H,31H-phthalocyanine-6,13,20,27-tetrayl]-tetrakis[acetamidato]](2-)-N^{29},N^{30},N^{31},N^{32}]-, (SP-4-1)-	$CHCl_3$	410(4.95),436(5.65), 594(4.22),648(5.17)	103-0960-86
$C_{60}H_{64}O_2P_2$			
Phosphine oxide, [1,1'-binaphthalene]-2,2'-diylbis[bis[4-(1,1-dimethylethyl)phenyl]-, (S)-(-)-	EtOH	233(5.11),273s(4.08), 287(4.08),300s(4.00), 316s(3.56),332(3.54)	44-0629-86
$C_{60}H_{64}P_2$			
Phosphine, [1,1'-binaphthalene]-2,2'-diylbis[bis[4-(1,1-dimethylethyl)-phenyl-, (S)-(-)-	EtOH	221(5.10),237s(5.00)	44-0629-86
$C_{60}H_{68}N_{10}O_{14}$			
Saframycin Y2b-d	MeOH	269(4.64)	158-1639-86
$C_{60}H_{107}N_{11}O_{13}$			
Leucinostatin C	EtOH	220s(4.24),225(4.42)	111-0527-86
$C_{61}H_{50}N_4O_{17}$			
Imidazo[4,5-d][1,2]diazepine-6(1H)-carboxylic acid, 2,3-dihydro-2-oxo-1,3-bis(2,3,5-tri-O-benzoyl-β-D-ribofuranosyl)-, ethyl ester	MeOH	274(4.26),352(3.20)	78-1511-86
$C_{61}H_{59}FeN_6O_3$			
Iron(1+), [2,2,6,6-tetramethyl-4-[[1-oxo-5-[2-[10,15,20-tris(4-methylphenyl)-21H,23H-porphin-5-yl]phenoxy]pentyl]amino]-1-piperidinyloxyato(2-)-N^{21},N^{22},N^{23},N^{24}]-, chloride, (SP-4-2)-	$CHCl_3$	351s(4.54),385(4.76), 418(4.99),512(4.12), 579(3.51),660(3.39)	35-0618-86
$C_{61}H_{88}N_{18}O_{21}S_2$			
Peplomycin	MeOH	290(4.11)	35-7089-86
phototransformed product	MeOH	293(3.78)	35-7089-86
$C_{62}H_{38}$			
28,33[1',2']-Benzeno-7,16[2',3']-anthraceno-5,18[1',2']:9,14[1",2"]-dibenzenoheptacene	CH_2Cl_2	273(4.35),279(4.42), 289(4.49),297(4.60)	35-6675-86
$C_{62}H_{42}N_4$			
3-Pyridinecarbonitrile, 6,6'-[1,1'-biphenyl]-4,4'-diylbis[2-[1,1'-biphenyl]-4-yl-4-(4-methylphenyl)-	MeCN	348(4.80)	73-1061-86
$C_{62}H_{54}N_5O_8P$			
Phosphonic acid, [3-[[1,6-dihydro-2-[[(4-methoxyphenyl)diphenylmethyl]amino]-6-oxo-9H-purin-9-yl]methoxy]-4-[(4-methoxyphenyl)diphenylmethoxy]-1-butenyl]-, diphenyl ester, (E)-(±)-	MeOH	232(4.48),261(4.22)	87-0671-86

Compound	Solvent	λ_{max}(log ϵ)	Ref.
$C_{62}H_{55}N_5O_9S_2$			
2',3'-Secoguanosine, N^2,$O^{5'}$-bis(triphenylmethyl)-2',3'-bis-O-[(4-methylphenyl)sulfonyl]-	EtOH	210(4.79),262(4.18)	87-2445-86
$C_{62}H_{56}N_5O_8P$			
Phosphonic acid, [3-[[1,6-dihydro-2-[[(4-methoxyphenyl)diphenylmethyl]amino]-6-oxo-9H-purin-9-yl]methoxy]-4-[(4-methoxyphenyl)diphenylmethoxy]butyl]-, diphenyl ester, (±)-	MeOH	232(4.48),261(4.22)	87-0671-86
$C_{62}H_{86}N_{12}O_{16}$			
Actinomycin D	MeOH	238(4.51),426(4.31), 441(4.33)	158-0721-86
$C_{62}H_{109}N_8OZn_2$			
Zinc(3+), bis(2-heptadecyl-12-methyl-3,7,11,17-tetraazabicyclo[11.3.1]heptadeca-1(17),2,11,13,15-pentaene-N^3,N^7,N^{11},N^{17})-μ-hydroxydi-, triperchlorate	MeCN	198(4.34),282(3.71)	35-2388-86
$C_{63}H_{54}N$			
Methylium, (nitrilotri-4,1-phenylene)-tris[bis(4-methylphenyl)-	HOAc-1% CF_3COOH	553(4.65)	64-1045-86B
dication	HOAc-30% CF_3COOH	453(4.95),536(4.50), 671(5.04)	64-1045-86B
trication	CF_3COOH	456(5.36),639(5.41)	64-1045-86B
$C_{63}H_{00}N_{12}O_{16}$			
Actinomycin C_2	MeOH	236s(4.59),425(4.38), 442(4.40)	158-0721-86
$C_{63}H_{102}N_4O_2$			
21H-Biline-1,19-dione, 3,17-diethyl-23,24-dihydro-2,7,13,18-tetramethyl-8,12-dioctadecyl-	$CHCl_3$	366(4.14),618(3.66)	49-0849-86
$C_{64}H_{32}N_8OV$			
Vanadium, oxo[45H,47H-tetraanthra-[2,3-b:2',3'-g:2",3"-1:2"',3"'-q]-porphyrazinato(2-)-N^{45},N^{46},N^{47},N^{48}]-, (SP-5-12)-	1-$C_{10}H_7Cl$	380(5.13),405(4.67), 536(4.46),605s(4.39), 780s(4.43),836(4.78), 874(4.83),932(5.32)	49-0475-86
	$C_6H_5NO_2$	870(4.86),934(5.36)	49-0475-86
	DMSO	844(--),890(--)	49-0475-86
$C_{64}H_{56}AlClN_8$			
Aluminum, chloro[2,11,20,29-tetrakis(1,1-dimethylethyl)-37H,39H-tetranaphtho[2,3-b:2',3'-g:2",3"-1:2"',3"'-q]porphyrazinato(2-)-N^{37},N^{38},N^{39},N^{40}]-, (SP-5-12)-	toluene	335(5.16),385s(4.83), 625s(4.58),702(4.99), 739(4.97),779(5.25)	49-0475-86
	1-$C_{10}H_7Cl$	365s(5.09),410s(4.75), 450s(4.58),635s(4.02), 725(4.87),790(5.00), 817(5.22)	49-0475-86
	$C_6H_5NO_2$	640s(3.96),711(4.75), 751(4.80),792(5.21)	49-0475-86
	DMSO	340(4.85),366s(4.82), 380s(4.66),440s(4.30), 635s(3.90),697(4.50), 748(4.53),783(5.21)	49-0475-86

Compound	Solvent	λmax(log ε)	Ref.
$C_{64}H_{56}CuN_8$			
Copper, [2,11,20,29-tetrakis(1,1-dimethylethyl)-37H,39H-tetranaphtho[2,3-b:2',3'-g:2",3"-1:2"',3"'-q]porphyrazinato(2-)-N^{37},N^{38},N^{39},N^{40}]-, (SP-4-1)-	toluene	330(5.01),380s(4.68), 410s(4.62),630s(4.26), 687(4.72),730(4.80), 770(5.25)	49-0475-86
	1-$C_{10}H_7Cl$	360s(4.75),418s(4.32), 420s(4.68),635s(4.23), 693(4.67),740(4.82), 781(5.18)	49-0475-86
	$C_6H_5NO_2$	635s(4.22),696(4.75), 743(4.80),781(5.27)	49-0475-86
	DMSO	330(5.05),360s(4.56), 410s(4.25),630s(4.18), 694(4.66),740s(4.78), 769(5.18)	49-0475-86
$C_{64}H_{56}FeN_4O_4$			
Iron, [[2,2'-[5,10,15,20-tetrakis(4-methylphenyl)-21H,23H-porphine-21,23-diyl]bis[3-hydroxy-5,5-dimethyl-2-cyclohexen-1-onato]](2-)-N^{21},N^{22},$N^{23}N^{24}$,O^3,$O^{3'}$]-, (OC-6-12)-	benzene	466(5.06),562(4.00), 622(4.25),754(3.72)	35-5598-86
isomer	benzene	450(4.86),543(4.35), 583s(--)	35-5598-86
$C_{64}H_{56}N_8OV$			
Vanadium, oxo[2,11,20,29-tetrakis(1,1-dimethylethyl)-37H,39H-tetranaphtho[2,3-b:2',3'-g:2",3"-1:2"',3"'-q]porphyrazinato(2-)-N^{37},N^{38},N^{39},N^{40}]-, (SP-5-12)-	toluene	410s(4.38),444(4.35), 665s(3.85),717(4.56), 764(4.53),807(5.40)	49-0475-86
	1-$C_{10}H_7Cl$	418s(4.64),450(4.43), 670s(4.02),727(4.68), 777(4.55),821(5.29)	49-0475-86 +97-0216B-86
	$C_6H_5NO_2$	452(4.35),665s(4.12), 727(4.68),777(4.65), 821(5.46)	49-0475-86
	DMSO	390s(4.65),440(4.43), 650s(4.06),712(4.65), 760(4.61),801(5.38)	49-0475-86
	H_2SO_4	685(--),970(--), 1060s(--)	97-0216B-86
$C_{64}H_{62}ClFeN_4O_2$			
Iron, chloro[3-hydroxy-5,5-dimethyl-2-[5,10,15,20-tetrakis(2,4,6-trimethylphenyl)-21H,23H-porphin-21-yl]-2-cyclohexen-1-onato(2-)-N^{21},N^{22},N^{23},N^{24},O^3]-, (OC-6-25)	benzene	464(5.06),518(4.10), 574(3.88),638(3.56), 696s(--)	35-6598-86
$C_{64}H_{64}N_8O_4Zn$			
Zinc, [[N,N',N",N"'-(21H,23H-porphine-5,10,15,20-tetrayltetra-2,1-phenylene)tetrakis[2,2-dimethylpropanamidato]](2-)-N^{21},N^{22},$N^{23}N^{24}$]-, α,α,α,α-	benzene	547.4(4.32)	94-1865-86
	EtOH	556.4(4.32)	94-1865-86
	$CHCl_3$	584.4(4.26)	94-1865-86
α,α,α,β-	benzene	549.2(4.33)	94-1865-86
	EtOH	557.6(4.34)	94-1865-86
	$CHCl_3$	549.0(4.32)	94-1865-86
α,α,β,β-	benzene	548.4(4.31)	94-1865-86
	EtOH	557.8(4.32)	94-1865-86
	$CHCl_3$	547.5(4.33)	94-1865-86

Compound	Solvent	$\lambda_{max}(\log \epsilon)$	Ref.
α,β,α,β-	benzene	549.2(4.26)	94-1865-86
	EtOH	559.2(4.30)	94-1865-86
	$CHCl_3$	555.1(4.27)	94-1865-86
$C_{64}H_{90}N_{12}O_{16}$			
Actinomycin C_3	MeOH	236s(4.59),425(4.36), 442(4.39)	158-0721-86
$C_{65}H_{51}Cl_2N_7O_{18}$			
Degluco-teicoplanin benzyl ester, hydrochloride	MeOH	282(4.07)	158-1430-86
$C_{65}H_{102}N_4O_8$			
21H-Biline-8,12-dipropanoic acid, 3,17-diethyl-1,19,22,24-tetrahydro-2,7,13,18-tetramethyl-1,19-dioxo-, dihexadecyl ester	$CHCl_3$	365(4.96),635(4.48)	49-1081-86
$C_{65}H_{106}N_6O_6$			
Bilirubin IXα, bis(tetrabutylammonium) salt, (4Z,15Z)-	MeOH	440(4.52)	112-0311-86
	$CHCl_3$	448(4.58)	112-0311-86
Bilirubin XIIIα, bis(tetrabutylammonium) salt	MeOH	424(4.61)	112-0311-86
	$CHCl_3$	449(4.71)	112-0311-86
$C_{68}H_{46}N_{12}O_4$			
Phenol, 4,4',4",4"'-[21H,23H-porphine-5,10,15,20-tetrayltetrakis(2,1-phenyleneazo)]tetrakis-	pyridine	436(5.36),526(4.38), 561(4.06),599(3.95), 656(3.84)	103-0397-86
Phenol, 4,4',4",4"'-[21H,23H-porphine-5,10,15,20-tetrayltetrakis(3,1-phenyleneazo)]tetrakis-	pyridine	428(5.72),517(4.48), 553(4.15),590(4.00), 647(3.91)	103-0397-86
Phenol, 4,4',4",4"'-[21H,23H-porphine-5,10,15,20-tetrayltetrakis(4,1-phenyleneazo)]tetrakis-	pyridine	435(5.54),521(4.39), 559(4.35),595(3.96), 650(4.00)	103-0397-86
$C_{68}H_{50}O_{44}$			
Coriariin E	MeOH	221(5.11),261(4.80)	94-4533-86
$C_{68}H_{111}NO_{30}S$			
Notonesomycin A, disodium salt	MeOH	310(3.99)	88-1603-86
$C_{69}H_{46}CuN_{11}O_7P_3$			
Copper, [2,2,4,4,6,6-hexahydro-2,2,4,4,6-pentaphenoxy-6-[4-[(29H-31H-phthalocyanin-2-yloxy)methyl]-phenoxy]-1,3,5,2,4,6-triazatriphosphorinato(2-)-N^{29},N^{30},N^{31},N^{32}]-, (SP-4-2)-	benzene	666(5.32)	116-1495-86
polymer	benzene	668(5.08)	116-1495-86
$C_{70}H_{64}$			
1,3,5-Methenocyclopenta[cd]pentalene, 6b,6'b,6"b,6"'b-(tricyclo[3.3.1.$1^{3,7}$]-decane-1,3,5,7-tetrayltetra-1,3-butadiyne-4,1-diyl)tetrakis[decahydro-	hexane	223(3.45),234(3.50), 247(3.49),261(3.23)	78-1763-86
$C_{72}H_{60}Ge_5$			
Pentagermane, dodecaphenyl-	n.s.g.	228.5(4.8),293(4.6)	101-0057-86S

Compound	Solvent	λ_{max}(log ϵ)	Ref.
$C_{72}H_{78}N_4O_8$			
Benzoic acid, 4,4',4",4"'-(21H,23H-porphine-5,10,15,20-tetrayl)tetrakis-, tetrahexyl ester	CH_2Cl_2	400s(4.95),419(5.68), 450(3.98),514(4.29), 549(3.92),588(3.71), 644(3.52)	33-0849-86
$C_{72}H_{88}Ir_2N_8$			
Iridium, bis[2,3,7,8,12,13,17,18-octaethyl-21H,23H-porphinato(2-)-N^{21},N^{22},N^{23},N^{24}]di- (Ir-Ir)-	benzene	350(4.75),385(4.72), 532(4.01)	77-1653-86
$C_{72}H_{88}N_8Os_2$			
Osmium(II,III), bis[2,3,7,8,12,13,17,18-octaethyl-21H,23H-porphinato(2-)-N^{21},N^{22},N^{23},N^{24}]di-, bis(hexachloroantimonate)	CH_2Cl_2	358(4.58),392(4.45), 500(3.83),596(3.67)	35-2916-86
bis(hexafluorophosphate)	CH_2Cl_2	328(4.71),358(4.83), 488(3.93)	35-2916-86
bis(tetrafluoroborate)	CH_2Cl_2	329(4.77),358(4.90), 488(3.97)	35-2916-86
Osmium(III,III), bis[2,3,7,8,12,13,17-18-octaethyl-21H,23H-porphinato(2-)-N^{21},N^{22},N^{23},N^{24}]di-, bis(hexachloroantimonate)	CH_2Cl_2	334s(--),355(5.05), 482s(--),537s(--)	35-2916-86
bis(hexafluorophosphate)	CH_2Cl_2	331(4.81),364(4.75), 448s(--),481(3.97)	35-2916-86
bis(tetrafluoroborate)	CH_2Cl_2	333(4.93),358(4.91), 449s(--),481(4.09)	35-2916-86
$C_{72}H_{88}N_8Ru_2$			
Ruthenium(II,III), bis[2,3,7,8,12,13-17,18-octaethyl-21H,23H-porphinato(2-)-N^{21},N^{22},N^{23},N^{24}]di-, bis(hexachloroantimonate)	CH_2Cl_2	360s(5.12),504(4.14), 530(4.09)	35-2916-86
bis(hexafluorophosphate)	CH_2Cl_2	345(5.10),384s(--), 474(4.01),504s(--), 546(3.89),568(3.89), 642(3.72)	35-2916-86
bis(tetrafluoroborate)	CH_2Cl_2	346(5.26),384s(--), 476(4.14),504s(--), 546(4.02),572(4.02), 644(3.87)	35-2916-86
Ruthenium(III,III), bis[2,3,7,8,12,13-17,18-octaethyl-21H,23H-porphinato(2-)-N^{21},N^{22},N^{23},N^{24}]di-, bis(hexafluorophosphate)	CH_2Cl_2	347(5.20),388s(--), 486(4.05),514(4.05), 560(4.01)	35-2916-86
bis(tetrafluoroborate)	CH_2Cl_2	347(5.21),388s(--), 486(4.06),513(4.07), 560(4.02)	35-2916-86
$C_{72}H_{110}N_{12}O_{20}$			
Plipastatin A 1 (potassium salt)	MeOH	276(3.28)	158-0737-86
$C_{73}H_{112}N_{12}O_{20}$			
Plipastatin A 2 (potassium salt)	MeOH	276(3.25)	158-0737-86
$C_{74}H_{114}N_{12}O_{20}$			
Plipastatin B 1 (potassium salt)	MeOH	276(3.23)	158-0737-86
$C_{75}H_{58}CuN_{11}O_7P_3$			
Copper, [2-[4-[[(9,10,16,17,23,24-	benzene	604(4.58),668(5.26)	116-1495-86

Compound	Solvent	λ_{max}(log ε)	Ref.
hexamethyl-29H,31H-phthalocyanin-2-yl)oxy]methyl]phenoxy]-2,2,4,4,6,6-hexahydro-2,4,4,6,6-pentaphenoxy-1,3,5,2,4,6-triazatriphosphorinato(2-)-N^{29},N^{30},N^{31},N^{32}]-, (SP-4-2)-(cont.)			116-1495-86
polymer	benzene	666(5.23)	116-1495-86
$C_{75}H_{116}N_{12}O_{20}$			
Plipastatin B 2 (potassium salt)	MeOH	276(3.26)	158-0737-86
$C_{76}H_{54}N_4$			
21H,23H-Porphine, 5,10,15,20-tetrakis(2,2-diphenylethenyl)-	benzene	448(5.48),513s(3.89), 548(4.11),593(4.39), 686(4.00)	77-1725-86
$C_{76}H_{64}N_2O_{20}Zn$			
Green pigment from fungus roesleria hypogea	CH_2Cl_2	240s(4.87),268(4.95), 340s(4.69),368(4.85), 406(4.80),614(4.85)	5-0305-86
$C_{76}H_{92}ClFeN_4$			
Iron, chloro[5,10,15,20-tetrakis[3,5-bis(1,1-dimethylethyl)phenyl]-21H,23H-porphinato(2-)-N^{21},$N^{22}N^{23}$-N^{24}]-, (SP-5-12)-	CH_2Cl_2	382(4.81),420(5.09), 511(4.18),578(3.53), 664s(--),698(3.57)	64-1265-86B
$C_{76}H_{94}N_4$			
21H,23H-Porphine, 5,10,15,20-tetrakis[3,5-bis(1,1-dimethylethyl)-phenyl]-	$CHCl_3$	405s(--),423(5.75), 487(3.60),519(4.28), 554(4.03),592(3.76), 647(3.77)	64-1265-86B
$C_{76}H_{94}N_4O_4$			
Phenol, 4,4',4",4"'-(21H,23H-porphine-5,10,15,20-tetrayl)tetrakis[2,6-bis(1,1-dimethylethyl)-	pyridine	430(5.86),525(4.35), 564(4.39),599(3.94), 658(4.17)	103-0629-86
$C_{77}H_{71}BrCoN_4O_8$			
Benzoic acid, 4,4',4",4"'-(21H,23H-porphine-5,10,15,20-tetrayl)tetrakis-, tetrahexyl ester, bromo(pyridine) complex	CH_2Cl_2	322(4.43),441(5.24), 520s(3.71),556(4.12), 595(3.82)	33-0849-86
	pyridine	338(4.33),438(5.44), 552(4.15),590s(3.81)	33-0849-86
$C_{78}H_{54}N_3O_{12}P_3$			
1,3,5,2,4,6-Triazatriphosphorin, 2,2,4,4,6,6-hexakis(4-benzoylphenoxy)-2,2,4,4,6,6-hexahydro-	CH_2Cl_2	258(4.51),338(2.53)	116-0574-86
$C_{78}H_{96}O_{42}$			
γ-Cyclodextrin, 6^A,6^D(or 6^E)-bis(9-anthracenecarboxylate)	10% $C_2H_6O_2$	260(4.1),367(3.9)	138-1865-86
$C_{79}H_{78}N_4O_3$			
21H,23H-Porphine, 2,3,7,8,12,13,17,18-octaethyl-10-(phenylmethoxy)-5,15-bis[2-(phenylmethoxymethyl)-1-naphthalenyl]-	hexane-1% isoPrOH	235(5.02),283(4.33), 324(4.11),428(4.97), 552(3.87),592(4.04)	88-6365-86

Compound	Solvent	λ_{max}(log ε)	Ref.
$C_{79}H_{98}N_9O_{16}PSi_3$			
Adenosine, N^4-benzoyl-2'-O-[(1,1-dimethylethyl)dimethylsilyl]-5'-O-(monomethoxytrityl)cytidylyl]-3'-[O^P-2-(4-nitrophenyl)ethyl]→-5']-N^6-benzoyl-2',3'-bis-O-[(1,1-dimethylethyl)dimethylsilyl]-	MeOH	230(4.66),270(4.62), 315s(4.03)	33-1768-86
$C_{81}H_{105}N_5O_9$			
Trispiro[4,14,20,30,35,45-hexaoxa-1,17-diazaoctacyclo[15.15.15.$2^{5,8}$-$2^{10,13}$.$2^{21,24}$.$2^{26,29}$.$2^{36,39}$.$2^{41,44}$]-nonapentaconta-5,7,10,12,21,23,26,28-36,38,41,43,48,50,52,54,56,58-octadecaene-9,4'"25,4":40,4"'-trispiperidine]-2,16-dione, 1'-acetyl-1",1"'-diethyl-6,12,22,28,37,43,48,51,52-55,56,59-dodecamethyl-	MeOH	199(5.30),232s(4.36), 269(3.51),277s(3.46)	35-2273-86
$C_{81}H_{107}N_5O_8$			
Trispiro[4,14,20,30,35,45-hexaoxa-1,17-diazaoctacyclo[15.15.15.$2^{5,8}$-$2^{10,13}$.$2^{21,24}$.$2^{26,29}$.$2^{36,39}$.$2^{41,44}$]-nonapentaconta-5,7,10,12,21,23,26,28-36,38,41,43,48,50,52,54,56,58-octadecaene-9,4':25,4":40,4"'-trispiperidine]-2,16-dione, 1',1",1"'-triethyl-6,12,22,28,37,43,48,51,52,55,56-59-dodecamethyl-	MeOH	198(5.26),232s(4.38), 269(3.48),278s(3.42)	35-2273-86
$C_{81}H_{111}N_5O_6$			
Trispiro[4,14,20,30,35,45-hexaoxa-1,17-diazaoctacyclo[15.15.15.$2^{5,8}$-$2^{10,13}$.$2^{21,24}$.$2^{26,29}$.$2^{36,39}$.$2^{41,44}$]-nonapentaconta-5,7,10,12,21,23,26,28-36,38,41,43,48,50,52,54,56,58-octadecaene-9,4':25,4":40,4"'-trispiperidine], 1',1",1"'-triethyl-6,12,22-28,37,43,48,51,52,55,56,59-dodecamethyl-	MeOH	197(5.29),269(3.54), 278(3.48)	35-2273-86
$C_{81}H_{143}N_4$			
1H-Benzimidazolium, 2-[3-(1,3-dihexadecyl-2H-benzimidazol-2-ylidene)-1-propenyl]-1,3-dihexadecyl-, bromide	EtOH	504(5.11)	4-0209-86
$C_{82}H_{52}$			
9,10[1',2']-Benzenoanthracene, 2,2',2",2"'-(1,2-ethenediylidene)-tetrakis[9,10-dihydro-	$CDCl_3$	267s(4.43),273(4.46), 280(4.49),340(4.26)	77-0974-86
$C_{82}H_{58}O_{52}$			
Coriariin A	MeOH	224(5.20),280(4.92)	94-4092-86
$C_{82}H_{58}O_{53}$			
Coriariin D	MeOH	220(5.15),274(4.82)	94-4533-86
$C_{84}H_{140}CoN_4$			
Cobalt, [2,3,7,8,12,13,17,18-octaethyl-21H,23H-porphinato(2-)-N^{21},N^{22},N^{23},N^{24}]-, (SP-4-1)-	$CHCl_3$	395(5.30),522(4.04), 555(4.33)	18-1259-86

Compound	Solvent	λ_{max}(log ε)	Ref.
$C_{84}H_{140}CuN_4$			
Copper, [2,3,7,8,12,13,17,18-octaethyl-21H,23H-porphinato(2-)-N^{21},N^{22}-N^{23},N^{24}]-, (SP-4-1)-	$CHCl_3$	401(5.61),527(4.23), 564(4.51)	18-1259-86
$C_{84}H_{140}N_4Ni$			
Nickel, [2,3,7,8,12,13,17,18-octaethyl-21H,23H-porphinato(2-)-N^{21},N^{22}-N^{23},N^{24}]-, (SP-4-1)-	$CHCl_3$	396(5.32),519(4.11), 555(4.51)	18-1259-86
$C_{86}H_{94}Cl_2N_8O_{22}P_2Si_2$			
3'-Cytidylic acid, N-benzoyl-2'-O-[(1,1-dimethylethyl)dimethylsilyl]-5'-O-[(4-methoxyphenyl)diphenylmethyl]-P-[2-(4-nitrophenyl)ethyl]cytidylyl-(3'→5')-N-benzoyl-2'-O-[(1,1-dimethylethyl)dimethylsilyl]-, 2,5-dichlorophenyl 2-(4-nitrophenyl)ethyl ester	MeOH	230s(4.68),262(4.81), 300s(4.36)	33-1768-86
$C_{86}H_{106}N_6O_{22}P_2Si_2$			
Cytidine, N^4-benzoyl-2'-O-[(1,1-dimethyl)dimethylsilyl]-5'-O-(monomethoxytrityl)cytidylyl-[3'-[O^P-[2-(4-nitrophenyl)ethyl]→5']-N -benzoyl-2'-O-[(1,1-dimethylethyl)dimethylsilyl]-, 3'-[2-(4-nitrophenyl)ethyl] triethylammonium phosphate	MeOH	230(4.55),260(4.73), 300(4.35)	33-1768-86
$C_{87}H_{94}Cl_2N_{10}O_{21}P_2Si_2$			
3'-Adenylic acid, N-benzoyl-2'-O-[(1,1-dimethylethyl)dimethylsilyl]-5'-O-[(4-methoxyphenyl)diphenylmethyl]-P-[2-(4-nitrophenyl)ethylcytidylyl-(3'→5')-N-benzoyl-2'-O-[(1,1-dimethyl)dimethylsilyl]-, 2,5-dichlorophenyl 2-(4-nitrophenyl)ethyl ester	MeOH	228s(4.74),268(4.73), 315s(4.09)	33-1768-86
3'-Cytidylic acid, N-benzoyl-2'-O-[(1-1-dimethylethyl)dimethylsilyl]-5'-O-[(4-methoxyphenyl)diphenylmethyl]-P-[2-(4-nitrophenyl)ethyl]adenylyl-(3'→5')-N-benzoyl-2'-O-[(1,1-dimethylethyl)dimethylsilyl]-, 2,5-dichlorophenyl 2-(4-nitrophenyl)ethyl ester	MeOH	230s(4.75),268(4.78), 315s(4.03)	33-1768-86
$C_{89}H_{62}O_{57}$			
Coriariin C	MeOH	220(5.24),277(4.93)	94-4533-86
$C_{89}H_{119}N_{13}O_{25}P_2Si_4$			
Adenosine, N-benzoyl-2'-O-[(1,1-dimethylethyl)dimethylsilyl]-P-[2-(4-nitrophenyl)ethyl]cytidylyl-(3'→5')-N-benzoyl-2'-O-[(1,1-dimethylethyl)-dimethylsilyl]-P-[2-(4-nitrophenyl)-ethyl]cytidylyl-(3'→5')-N-benzoyl-2',3'-bis-O-[(1,1-dimethylethyl)dimethylsilyl]-	MeOH	230s(4.67),262(4.71), 317s(3.99)	33-1768-86
$C_{91}H_{72}N_{12}OZn_2$			
Zinc, μ-[2,9,16-tris(1,1-dimethyleth-	toluene	349(4.87),424(5.60),	77-1239-86

Compound	Solvent	$\lambda_{max}(\log \epsilon)$	Ref.
yl)-23-[4-[10,15,20-tris(4-methyl-phenyl)-21H,23H-porphin-5-yl]phen-oxy]-29H,31H-phthalocyaninato(4-)-$N^{21},N^{22},N^{23},N^{24}:N^{29},N^{30},N^{31},N^{32}$]]di- (cont.)		357(4.30),607(4.53), 673(5.28)	77-1239-86
$C_{93}H_{94}CuN_{11}O_{19}P_3$			
Copper, [2-[4-[[[9,10,16,17,23,24-hexa-kis[(2-methoxyethoxy)methyl]-29H,31H-phthalocyanin-2-yl]oxy]methyl]phen-oxy]-2,2,4,4,6,6-hexahydro-2,4,4,6,6-pentaphenoxy-1,3,5,2,4,6-triazatri-phosphorinato(2-)-$N^{29},N^{30},N^{31},N^{32}$]-, (SP-4-2)-	benzene	681(4.23)	116-1495-86
$C_{94}H_{90}Co_2N_{16}O_7$			
Cobalt, [μ-[[2,2'-oxybis[9,16,23-tris(2,2-dimethylpropoxy)-29H,31H-phthalocyaninato]](4-)-N^{29},N^{30},N^{31}-$N^{32}:N^{29'},N^{30'},N^{31'},N^{32'}$]]di-	o-$C_6H_4Cl_2$	300(4.94),330(4.93), 650(4.83),680(4.93)	118-0406-86
$C_{94}H_{94}N_{16}O_7$			
29H,31H-Phthalocyanine, 2,2'-oxybis-[9,16,23-tris(2,2-dimethylpropoxy)-	CH_2Cl_2	336(5.19),388(5.04), 616s(5.20),640(5.25), 666(5.23),696(5.13)	118-0406-86
$C_{104}H_{88}Fe_2N_8O$			
Iron, μ-oxobis[5,10,15,20-tetra-kis(3,5-dimethylphenyl)-21H,23H-porphinato(2-)-$N^{21},N^{22},N^{23},N^{24}$]di-	$CHCl_3$	410(5.33),571(4.23), 611(3.88)	64-1265-86B
$C_{109}H_{135}N_{13}O_{26}P_2Si_4$			
Adenosine, N-benzoyl-2'-O-[(1,1-di-methylethyl)dimethylsilyl]-5'-O-[(4-methoxyphenyl)diphenylmethyl]-P-[2-(4-nitrophenyl)ethyl]cytidyl-yl-(3'→5')-N-benzoyl-2'-O-[(1,1-di-methylethyl)dimethylsilyl]-P-[2-(4-nitrophenyl)ethyl]cytidylyl-(3'→5')-N-benzoyl-2',3'-bis-O-[(1,1-dimethyl-ethyl)dimethylsilyl]-	MeOH	230(4.80),262(4.81), 317s(4.09)	33-1768-86
$C_{111}H_{82}CuN_{11}O_{13}P_3$			
Copper, [2-[4-[[[9,10,16,17,23,24-hexakis(phenoxymethyl)-29H,31H-phthalocyanin-2-yl]oxy]methyl]-phenoxy]-2,2,4,4,6,6-hexahydro-2,4,4,6,6-pentaphenoxy-1,3,5,2,4,6-triazatriphosphorinato(2-)-N^{29},N^{30}-N^{31},N^{32}]-, (SP-4-2)-	benzene	332(4.26),604(4.15), 670(5.45)	116-1495-86
$C_{112}H_{64}AlClN_8$			
Aluminum, chloro[5,11,16,22,27,33,38 44-octaphenyl-45H,47H-tetraanthra-[2,3-b:2',3'-g:2",3"-l:2"',3"'-q]-porphyrazinato(2-)-$N^{45},N^{46},N^{47},N^{48}$]-, (SP-5-12)-	toluene	365(--),415s(--), 540(--),600s(--), 725s(--),870(5.21), 900(5.26),933(5.29)	49-0475-86
	1-$C_{10}H_7Cl$	365(5.19),415s(4.83), 560(4.67),610(4.53), 725s(4.57),868(4.94), 919(5.09),959(5.31)	49-0475-86

Compound	Solvent	λmax(log ε)	Ref.
(cont.)	$C_6H_5NO_2$	540(4.49),610s(4.36), 730s(4.26),816(4.69), 872(4.84),920(5.35)	49-0475-86
	DMSO	365(5.20),410s(4.82), 550(4.46),610s(4.27), 725s(4.26),808(4.62), 872(4.69),921(5.31)	49-0475-86
$C_{112}H_{64}CuN_8$ Copper, [5,11,16,22,27,33,38,44-octaphenyl-45H,47H-tetraanthra[2,3-b-2',3'-g:2",3"-l:2"',3"'-q]porphyrazinato(2-)-N^{45},N^{46},N^{47},N^{48}]-, (SP-4-1)-	toluene	340(5.12),405s(4.72), 500s(4.37),590(4.12), 745(4.16),775(4.65), 834(4.74),878(5.28)	49-0475-86
	1-$C_{10}H_7Cl$	362(5.10),405s(4.70), 520(4.52),598(4.42), 725s(4.06),789(4.65), 840(4.69),894(5.30)	49-0475-86
	$C_6H_5NO_2$	530(4.27),600s(4.01), 730(4.00),794(4.50), 850(4.53),900(5.22)	49-0475-86
	DMSO	350s(5.16),400(4.87), 510(4.40),590(4.15), 713s(4.16),776(4.57), 832s(4.60),878(5.30)	49-0475-86
$C_{112}H_{64}N_8OV$ Vanadium, [5,11,16,22,27,33,38,44-octaphenyl-45H,47H-tetraanthra-[2,3-b:2',3'-g:2",3"-l:2"',3"'-q]-porphyrazinato(2-)-N^{45},N^{46},N^{47},N^{48}]-oxo-, (SP-5-12)-	toluene	358(5.02),375s(4.90), 415(4.43),540(4.34), 600(4.09),768s)4.31), 824(4.50),882(4.73), 935(5.34)	49-0475-86
	1-$C_{10}H_7Cl$	360(5.12),410s(4.65), 555(4.52),620s(4.35), 795s(4.38),840(4.65), 895(4.86),955(5.30)	49-0475-86
	$C_6H_5NO_2$	565(4.48),620s(4.21), 796(4.35),842(4.64), 898(4.90),956(5.33)	49-0475-86
	DMSO	367(5.13),410s(4.66), 540(4.41),590(4.17), 738s(4.17),808(4.67), 880s(4.83),918(5.29)	49-0475-86
$C_{118}H_{131}Cl_2N_{16}O_{30}P_3Si_3$ 3'-Adenylic acid, N-benzoyl-2'-O-[(1,1-dimethylethyl)dimethylsilyl]-5'-O-[(4-methoxyphenyl)diphenylmethyl]-P-[2-(4-nitrophenyl)ethyl]cytidylyl-(3'→5')-N-benzoyl-2'-O-[(1,1-dimethylethyl)dimethylsilyl]-P-[2-(4-nitrophenyl)ethyl]adenylyl-(3'→5')-N-benzoyl-2'-O-[(1,1-dimethylethyl)-dimethylsilyl]-, 2,5-dichlorophenyl 2-(4-nitrophenyl)ethyl ester	MeOH	228s(4.92),270(4.94), 315s(4.05)	33-1768-86
$C_{120}H_{112}CuN_8$ Copper, [2,13,24,35,48,57,64,71-octakis(1,1-dimethylethyl)-5,11,16,22-27,33,38,44-octahydro-5,44[1',2']-11,16[1",2"]:22,27[1"',2"']:33,38-[1"",2""]-tetrabenzeno-7,42:20,29-	pentane	628(5.04),670(4.96)	65-0341-86
	$CHCl_3$	284(4.88),314(4.88), 342(4.95),488(4.68), 606(4.83),642(4.83), 676(5.58)	65-0341-86

Compound	Solvent	λ_{max}(log ε)	Ref.
diimino-9,18:40,31-dinitrilotetra-anthra[2,3-c:2',3'-h:2",3"-m:2"',3"'-r][1,6,11,16]tetraazacycloeicosin-ato(2-)-N^{51},N^{52},N^{59},N^{66}]-, (SP-4-1)- (cont.)			65-0341-86
$C_{120}H_{156}N_{19}O_{34}P_3Si_5$			
Adenosine, N-benzoyl-2'-O-[(1,1-dimethylethyl)dimethylsilyl]-P-[2-(4-nitrophenyl)ethyl]adenylyl-(3'→5')-N-benzoyl-2'-O-[(1,1-dimethylethyl)dimethylsilyl]-P-2-[(4-nitrophenyl)ethyl]-adenylyl-(3'→5')-N-benzoyl-2'-O-[(1,1-dimethylethyl)dimethylsilyl]-P-[2-(4-nitrophenyl)ethyl]cytidylyl-(3'→5')-N-benzoyl-2',3'-bis-O-[(1,1-dimethylethyl)dimethylsilyl]-	MeOH	233s(4.84),267(4.95), 315s(4.20)	33-1768-86
$C_{124}H_{192}N_8O_4Zn$			
Zinc, [[N,N',N",N"'-[2,9,16,23-tetrakis(1,1-dimethylethyl)-29H,31H-phthalocyanine-6,13,20,27-tetrayl]-tetrakis[octadecanamidato]](2-)-N^{29},N^{30},N^{31},N^{32}]-, (SP-4-1)-	$CHCl_3$	414(4.88),422(5.61), 596(4.33),650(5.18)	103-0960-86
$C_{136}H_{130}N_{24}O_{10}$			
29H,31H-Phthalocyanine, 2,16-bis(2,2-dimethylpropoxy)-9,23-bis[[9,16,23-tris(2,2-dimethylpropoxy)-29H,31H-phthalocyanin-2-yl]oxy]-	CH_2Cl_2	340(5.16),388(4.89), 614s(4.89),642(5.02), 672(5.18),702(5.16)	118-0406-86
$C_{140}H_{172}N_{19}O_{35}P_3Si_5$			
Adenosine, N-benzoyl-2',3'-bis-O-[(1,1-dimethylethyl)dimethylsilyl]-P-[2-(4-nitrophenyl)ethyl]adenylyl-(5'→3')-N-benzoyl-2'-O-[(1,1-dimethylethyl)dimethylsilyl]-P-[2-(4-nitrophenyl)ethyl]cytidylyl-(5'→3')-N-benzoyl-2'-O-[(1,1-dimethylethyl)-dimethylsilyl]-P-[2-(4-nitrophenyl)-ethyl]cytidylyl-(5'→3')-N-benzoyl-2'-O-[(1,1-dimethylethyl)dimethylsilyl]-5'-O-[(4-methoxyphenyl)diphenylmethyl]-	MeOH	232s(4.89),267(4.88), 315s(4.18)	33-1768-86
$C_{197}H_{246}N_{28}O_{54}P_5Si_7$			
Adenosine, N^4-benzoyl-2'-O-[(1,1-dimethylethyl)dimethylsilyl]-5'-O-(monomethoxytrityl)cytidylyl-[3'-[O^P-[2-(4-nitrophenyl)ethyl]→5']-N^6-benzoyl-2'-O-[(1,1-dimethylethyl)dimethylsilyl]adenylyl-[3'-[O^P-[2-(4-nitrophenyl)ethyl]→5']-N^4-benzoyl-2'-O-[(1,1-dimethylethyl)dimethylsilyl]cytidylyl-[3'-[O^P-[2-(4-nitrophenyl)ethyl]→-5']-N^6-benzoyl-2'-O-[(1,1-dimethylethyl)-dimethylsilyl]-cytidylyl-[3'-[O^P-[2-(4-nitrophenyl)ethyl]→5']-N^6-benzoyl-2',3'-bis-O-[(1,1-dimethylethyl)dimethylsilyl]-	MeOH	235(5.00),266(5.16), 315s(4.50)	33-1768-86

1- -86, Acta Chem. Scand. B, 40 (1986)
0210 L. Eberson and B. Larsson
0545 G. Lindsten et al.
0555 P. Keil et al.
0625 I. Nilsson et al.
0665 A. Berg et al.
0806 X.-X. Zhou et al.

2- -86, Indian J. Chem. Sect. B, 25 (1986)
0015 S.B. Maiti et al.
0102 V. Khanna et al.
0113 G.S.R. Subba Rao et al.
0137 A. Chandrasekhar et al.
0166 A.C. Jain et al.
0175 P. Sengupta et al.
0180 R. Gulge
0212 C. Paparao and V. Sundaramurthy
0215 P.M. Pillai and P. Ramabhadran
0228 N. Narayanan and T.R. Balasubramanian
0233 M.G. Gogte et al.
0256 R.S. Mali et al.
0259 A.C. Jain et al.
0275 P.B. Talukdar et al.
0304 A. Banerji and N.C. Goomer
0319 C.L. Gajurel
0347 N. Viswanathan et al.
0360 V.S. Ekkundi and G.K. Trivedi
0377 K.J.R. Prasad et al.
0478 A.C. Jain et al.
0481 A.C. Jain and N.K. Nayyar
0496 S.B. Lokhande and D.W. Rangnekar
0525 M. Goswami and N. Borthakur
0616 M.I. Younes and S.A. Metwally
0619 B.S. Rani and M. Darbarwar
0623 A.C. Jain et al.
0638 S.B. Lokhande and D.W. Rangnekar
0646 A.C. Jain et al.
0649 A.C. Jain and P.K. Bambah
0675 S.B. Thakur et al.
0721 B.P. Pradhan et al.
0755 A.C. Jain et al.
0760 S.T. Vijayaraghaven and T.R. Balasubramanian
0779 S.S. Hanmantgad et al.
0799 G.B.V. Subramanian and A. Mehrotra
0860 D.R. Samant and N.A. Kudav
0862 A.C. Jain et al.
0870 S.S. Kumari and R.S.R.K.M. Rao
0887 D.N. Chatterjee and J. Guha
0901 P.M. Pillai and P. Ramabhadran
0960 P.M. Pillai and P. Ramabhadran
0981 M.M. Kulkarni et al.
1057 P.V. Tagdiwala and D.W. Rangnekar
1072 M. Das et al.
1087 M. Saxena and S.K. Srivastava
1126 N.R. Ayyangar et al.
1133 M. Sarkar et al.
1167 A. Patra et al.
1200 S.K. Chatterjee et al.
1204 J. Banerji et al.
1247 T.P. Veluchamy et al.

2- -86A, Indian J. Chem. Sect. A, 25 (1986)
0127 H. Ranganathan et al.
0178 M. Krishnamurthy and S.K. Dogra
0509 T. Mukherjee et al.
0513 P. Phaniraj et al.
0517 R.V. Subba Rao et al.
0923 A. Akelah et al.
1092 H.K. Sinha and S.K. Dogra

3- -86, Anal. Chem., 58 (1986)
0192 K.S. Patel and K.H. Lieser
0289 Z. Shishan and Z. Mei-yi
0384 E. Brecht
0455 S.M. Han and N. Purdie
0588 M.H. Jones and J.T. Woodcock
0631 M.C. Gosnell and H.A. Mottola
0639 G. Norwitz et al.
1845 M.H. Jones and J.T. Woodcock
2366 S.T. Colgan et al.
2504 S.A. Margolis and B. Coxon
3184 J. Kohn et al.

4- -86, J. Heterocyclic Chem., 23 (1986)
0001 J.I. Degraw et al.
0033 C.C. Tzeng et al.
0059 P.K. Gupta et al.
0135 N.R. El-Rayyes and A. Al-Jawhary
0153 Ji-Wang et al.
0155 P.D. Cook and D.J. McNamara
0209 E. Barni et al.
0245 R.A. Nugent and M. Murphy
0271 J.L. Kelley and H.J. Schaeffer
0329 R.K. Sehgal et al.
0333 K. Imafuku and H. Matsuura
0349 C.K. Chu et al.
0369 G.A. Brine et al.
0401 J. Reiter et al.
0449 G. Menozzi et al.
0455 G. Menozzi et al.
0505 R. Beyerle-Pfnur et al.
0567 J. Becher et al.
0589 S. Nan'ya et al.
0629 O.W. Lever, Jr. et al.
0679 J.I. Andres and M.T. Garcia-Lopez
0685 T. Hirota et al.
0689 R.P. Thummel et al.
0705 A.L. Weis et al.
0727 F.S. Al-Hajjar and S.S. Sabri
0737 K.G. Grozinger and K.D. Onan
0747 M.J. Tanga and E.J. Reist
0825 H. Feuer et al.
0849 R.W. Middleton et al.
0921 E. Pedroso et al.
0963 S.K. Arora et al.
1003 A. Bolognese et al.
1067 A. Bargagna et al.
1091 D.J. Anderson and A.J. Taylor
1141 P. Poulallion et al.
1189 J.L. Kelley and E.W. McLean
1195 E.A. Adegoke et al.
1207 K. Ito et al.
1263 D.W. Combs et al.
1265 K. Liu and B. Shih
1267 S. Nan'ya et al.
1295 A. Horvath and I. Hermecz
1303 A. Walser et al.
1315 A.R. Katritzky et al.
1347 T. Hirota et al.
1401 L. DeNapoli

1423 J.E. Johnson et al.
1451 N.F. Eweiss et al.
1475 H.M. Gilow et al.
1481 Y. Akita et al.
1491 H.D.H. Showalter et al.
1515 E.W. Badger and W.H. Moos
1535 K. Tanaka et al.
1551 E. Lee-Ruff and Y.-S. Chung
1625 E. Campaigne and R.F. Weddleton
1641 T. Troll and K. Schmid
1645 D. Chiarino and A.M. Contri
1651 S.-H. Chu et al.
1685 T. Hirota et al.
1697 S. Nan'ya et al.
1709 T.A. Riley et al.
1715 T. Hirota et al.
1721 D. Chergui et al.
1747 G. Alberghina et al.
1757 W.-Y. Ren et al.
1777 C.K. Chu et al.
1781 A. Alberola et al.
1789 N.R. El-Rayyes et al.
1805 S.T. Ross et al.
1869 J.D. Anderson et al.
1897 E. Mohasci

5- -86, Ann. Chem. Liebigs (1986)
0142 A. Aumiller and S. Hünig
0177 H.-J. Lohrisch et al.
0195 H.-J. Lohrisch et al.
0269 T.P. Andersen et al.
0305 O. Bachmann et al.
0479 H.J. Bestmann et al.
0533 M. Mittelbach and H. Junek
0741 A. Kleversaat et al.
0751 H.A. Staab et al.
0765 H.-J. Hasselbach et al.
0839 H. Laatsch
0947 S.P. Spyroudis
0957 H. Renz and E. Schlimme
1003 N.R. Guirguis et al.
1012 K.E. Andersen and E.B. Pedersen
1098 R. Askani and M. Wieduwilt
1104 R. Askani and M. Wieduwilt
1109 J. Becher, H. Nissen and K.S. Varma
1127 W.H. Gundel and S. Bohnert
1179 H.-J. Gais et al.
1213 F. Seela and S. Menkhoff
1222 S. Hünig et al.
1241 J.A. Farrera et al.
1398 N. Krause et al.
1413 L.F. Tietze et al.
1506 K. Krohn and W. Priyano
1540 P. Molina et al.
1621 U. Pindur et al.
1655 H. Laatsch
1669 H. Laatsch
1705 M. Böhme et al.
1728 S.V. Nielsen and E.B. Pedersen
1749 H. Beere, R. Russe and W. Seelen
1765 U. Hörcher et al.
1796 H. Gotthardt and F.R. Böhm
1847 H. Laatsch and H. Ludwig-Köhn
1872 P. Goya et al.
1891 H. Quast et al.
1956 G. Konrad and H. Musso
1968 G. Stoll et al.
1990 J. Frank et al.
2065 P.H. Gotz, J.W. Bats and H. Fritz
2142 O. Hofer et al.

7- -86, Ann. chim.(Rome), 76 (1986)
0473 G. Seu and L. Mura

9- -86, Appl. Spectroscopy, 40 (1986)
0556 R. Hilal et al.
1058 M. Garcia-Vargas et al.

11- -86, Chem. Scripta, 26 (1986)
0013 H. Bazin and J. Chattopadhyaya
0135 T. Pathak and J. Chattopadhyaya
0173 F. Seela et al.
0225 J.-M. Vial and J. Chattopadhyaya
0325 J.M. Castresana et al.
0331 S.V. Nielsen and E.B. Pedersen
0337 F. Johannsen et al.
0347 F. Johannsen et al.
0379 S.K. El-Sadany et al.
0607 P. Goya et al.

12- -86, Australian J. Chem., 39 (1986)
0001 J.B. Bremner and K.N. Winzenberg
0015 D. St.C. Black, N. Kumar and L.C.H. Wong
0031 W.L.F. Armarego et al.
0059 M.J. Gray et al.
0089 L.K. Dyall
0103 B.F. Bowden et al.
0123 B.F. Bowden et al.
0271 W.M.P. Johnson et al.
0281 R.K. Norris and A. Wright
0399 P.S. Clezy et al.
0419 P.S. Clezy et al.
0433 R. Urech
0487 R.C. Cambie et al.
0575 K. Picker et al.
0591 R.J. Goodridge et al.
0635 W.M. Best and D. Wege
0647 W.M. Best and D. Wege
0687 D.D. Ridley and G.W. Simpson
0817 R.M. Carman and J.M.A.M. Van Dongen
0821 I.K. Boddy et al.
0893 J.B. Bremner et al.
1101 G.A. Lawrance et al.
1231 D. St.C. Black et al.
1401 J.G. Hall and J.A. Reiss
1525 D.A. House et al.
1537 J.M. Coxon et al.
1629 P. Karuso and W.C. Taylor
1643 P. Karuso et al.
1655 B. Raguse and D.D. Ridley
1747 A.J. Liepa and T.C. Morton
1833 T.W. Hambley et al.
1843 R.M. Carman et al.
2013 D.P. Buxton et al.
2067 H. Greenland et al.
2075 I.K. Boddy et al.
2121 M.P. Hartshorn et al.

13- -86A, Steroids, 47 (1986)
0381 C.R. Engel et al.

0431 S.-q. Dang

13- -86B, Steroids, 48 (1986)
0047 V.S. Salvi et al.
0073 A. Gazit et al.
0233 J. Su et al.
0347 N. Numazawa et al.
0439 R.J. Park et al.

18- -86, Bull. Chem. Soc. Japan, 59 (1986)
0073 F. Ishibashi et al.
0083 T. Sone et al.
0511 A. Mori et al.
0569 N. Kanamaru and J. Tanaka
0661 M. Ochi et al.
0843 N. Sano and J. Tanaka
0961 T. Moriya
1095 K. Saito and H. Ishihara
1125 H. Takeshita et al.
1181 S. Ohta
1235 F.F. Abd et al.
1245 R.C. Phadke and D.W. Rangnekar
1259 C.Q. Ye et al.
1299 Y. Geetanjali et al.
1379 M. Kuzuya et al.
1403 M. Ishida et al.
1481 Y. Kudo et al.
1501 H. Kamogawa and M. Yamada
1515 M. Sato et al.
1575 P.P. Reddy et al.
1601 N. Kamigata et al.
1626 A. Sugimoto et al.
1638 Y. Niimi et al.
1709 K. Tsukahara et al.
1713 T. Asao et al.
1723 J. Ojima et al.
1767 M. Naori et al.
1791 J. Ojima et al.
1801 J. Ojima et al.
1915 K. Okamoto et al.
1953 H. Sasaki et al.
2032 M. Kodaka and T. Fukaya
2179 N. Kamigata et al.
2185 K. Inami and T. Shiba
2209 J. Ojima et al.
2317 Y. Ohta et al.
2495 H. Wu et al.
2509 J. Ojima et al.
2811 T. Ogino and K. Awano
2899 K. Mukai et al.
2947 K. Yamauchi et al.
2953 M. Suzuki et al.
3169 M. Kobayashi et al.
3257 E.J. Shin, B.S. An and S.C. Shim
3314 H. Matsumoto et al.
3320 K. Fujimori et al.
3326 K. Yamane et al.
3393 K. Murakami et al.
3535 M. Nakatani et al.
3541 T. Sakaki et al.
3589 T. Akiyama et al.
3611 M. Sato et al.
3702 Y. Tachibana et al.
3871 T. Tokumitsu
3983 S. Yamaguchi et al.

19- -86, Bull. Polish Acad. Sci., 34 (19
0183 R. Dabrowski and K. Kenig
0313 G. Groszek et al.

20- -86, Bull. soc. chim. Belges, 95 (19
0631 C.A.T. Fall et al.
0699 W.A. Ayer and C. Dufresne
0737 A.P. Lins et al.
0815 J.C. Coll et al.
0835 M.B. Ksebati and F.J. Schmitz
0895 W. Eng and G.D. Prestwich
1107 A. Maquestiau et al.

22- -86, Bull. soc. chim. France (1986)
0239 J. Fink and M. Regitz
0600 S. Tripathi et al.
0659 R. Martin et al.
0781 M. Cariou et al.
0817 S. Berrada and P. Metzner
0876 T.D. Spawn and D.J. Burton

23- -86, Can. J. Chem., 64 (1986)
0051 C.G. Young et al.
0067 P.S. Surdhar et al.
0123 M.C. Chalandre et al.
0180 E. Piers et al.
0237 C.O. Bender et al.
0265 S. Brownstein et al.
0277 W.D. Jamieson and A. Taylor
0514 P. Kulanthairel and M.H. Benn
0837 J.-S.-Liu et al.
0845 S.G. Bertolotti et al.
0950 N. Burlinson et al.
1116 T. Okuyama et al.
1308 S.V. Pathre et al.
1312 M.V. Laycock et al.
1390 D. Cortes et al.
1491 J.D. Arunziata et al.
1527 D.E. Williams et al.
1560 H. Tanaka et al.
1567 R.D. Singer et al.
1585 W.A. Ayer et al.
1764 A. Fisher et al.
1885 D.P.C. McGee and J.C. Martin
1969 N.R. Ayyangar et al.
2064 M. Cygler et al.
2076 P. Wan and K. Yates
2109 K. Anzai et al.
2115 E.Buncel and I. Onyido
2184 Y.Ueda and V. Vinet
2382 G.W. Bushnell et al.

24- -86, Chem. Ber., 119 (1986)
0050 G. Kossmehl and H.-C. Frohberg
0162 J. Goerdeler et al.
0182 K. Köhler et al.
0297 J.L. Jessen et al.
0387 W. Sucrow and H. Wolter
0589 H. Prinzbach et al.
0616 H. Prinzbach et al.
0794 F.-G. Klärner, V. Glock and H. Figge
0813 A. Padwa et al.
0844 R. Neidlein and R. Leidholdt
0950 M. Christl et al.

1016 H. Quast and H.-L. Fuchsbauer
1070 H. Wamhoff and H.-A. Thiemig
1105 H. Hopf et al.
1111 G. Maier et al.
1244 U.H. Brinker and I. Fleischauer
1315 H. Gotthardt et al.
1525 G. Kaupp and E. Ringer
1569 M. Fuchs et al.
1627 V. Sachweh and H. Langhals
1683 R. Bunte et al.
1716 H. Meier et al.
1801 H. Quast et al.
1836 H. Hopf and M. Psiorz
1911 W. Bernlöhr et al.
1977 R. Appel et al.
2094 H. Gotthardt and M. Oppermann
2104 F.-J. Kaiser et al.
2114 H. Wamhoff et al.
2159 J. Fink and M. Regitz
2208 Z. Huang and L. Tzai
2220 W. Lorenz and G. Maas
2272 D. Balschukat and E.V. Dehmlow
2289 R. Aumann and H. Heinen
2308 H. Gotthardt and H.-J. Kinzelmann
2317 H. Gotthardt et al.
2339 K. Sarma et al.
2373 H. Langhals and S. Grundner
2531 R. Sustmann and J. Lau
2553 W. Lösel and K.-H. Pook
2631 J. Daub et al.
2698 J. Hunger et al.
2723 H. Wamhoff et al.
2859 T. Gillman and K. Hartke
2889 R. Zeiger et al.
2956 E.V. Dehmlov et al.
2963 L. Somogyi
2980 N. Wiberg and C.-K. Kim
3102 G. Heilig and W. Lüttke
3109 G. Kaupp et al.
3158 H. Gotthardt and F.-R. Bohm
3198 P. Beimling and G. Kossmehl
3213 K. Karaghiosoff et al.
3276 B. Feith et al.
3316 E. Tauer et al.
3442 G. Sedelmeier et al.
3487 U. Kuckländer and B. Schneider
3521 R. Bunte et al.
3862 R. Neidlein and L. Tadesse
3870 A. Gül et al.

25- -86, Chem. and Ind.(London), (1986)
0107 B. Robinson and M.I. Khan
0246 P. Bhattacharyya et al.
0669 P. Kumar et al.
0713 K.N. Singh et al.
0823 B. Dinda and S. Saha
0827 F. Gomez-Garibay et al.

27- -86, Chimia, 40 (1986)
0074 G. Calzaferri

28- -86, Compt. rend., 303 (1986)
1179 A. Rossi et al.

30- -86, Doklady Akad. Nauk S.S.S.R., 286-291 (1986)
0055 I.V. Berezin et al.
0105 N.S. Kozlov et al.
0176 L.Y. Ukhin et al.
0387 M.G. Zimin et al.

32- -86, Gazz. chim. ital., 116 (1986)
0025 M. Nali et al.
0067 A. Bianco et al.
0115 F. Orsini et al.
0281 G.G. Aloisi et al.
0285 P. Bravo et al.
0291 P. Gariboldi et al.
0303 D. Monti et al.
0405 P.G. Tsoungas
0407 M. d'Ischia and G. Prota
0431 M. Nicoletti et al.
0441 P. Bravo et al.
0485 G. Jommi et al.
0491 P. Bravo et al.
0501 P. Bravo et al.
0533 F. Ciardelli et al.
0653 C. Galli et al.
0705 G. Galiazzo et al.

33- -86, Helv. Chim. Acta, 69 (1986)
0012 A. Milon et al.
0053 R. Schneider et al.
0071 R. Gleiter et al.
0091 R. Sakai et al.
0106 H. Meier et al.
0156 G.A. Gross et al.
0174 C. Jenny and H. Heimgartner
0184 B. Scholl et al.
0315 R. Neidlein and G. Hartz
0359 H. Wagner et al.
0374 C. Jenny and H. Heimgartner
0419 C. Jenny and H. Heimgartner
0456 P. Buss et al.
0469 G. Quinkert et al.
0555 A. O'Sullivan, B. Frei and O. Jeger
0560 M. Keller et al.
0659 G.G.G. Manzardo et al.
0692 E.P. Müller
0704 A. Heckel and W. Pfleiderer
0708 A. Heckel and W. Pfleiderer
0726 G. Guella et al.
0734 G. Snatzke et al.
0761 J.L. Birbaum and P. Vogel
0793 M. Lang et al.
0816 P. Uebelhart et al.
0849 D. Ammann et al.
0865 R. Gabioud and P. Vogel
0927 H. Meier and H. Heimgartner
0972 P. Rüedi
1017 M. Süsse et al.
1025 M. Lang et al.
1034 A. Guy et al.
1095 A. Heckel and W. Pfleiderer
1113 G.A. Gross and O. Sticher
1132 J.M.J. Tronchet et al.
1137 H. Felber et al.
1224 H. Moser et al.
1263 R. Neidlein and W. Wirth
1287 J.-L. Metral et al.
1310 G. Burnier et al.

1378 M.E. Scheller et al.
1395 A.C. Alder et al.
1513 J.M. Künzle et al.
1521 T. Eichenberger and H. Balli
1546 P.Müller and D. Rodriguez
1581 M.D'Ambrosio et al.
1588 I. Parikh et al.
1597 R. Neidlein et al.
1602 F. Seela and H. Steker
1623 R. Neidlein and D. Christen
1644 A. Escher et al.
1734 V. Bilinski et al.
1768 M. Ichiba et al.
1833 W.Keller-Schierlein
1851 R. Neidlein and W. Wirth
1872 R. Gleiter et al.
1936 B. Scholl and H.-J. Hansen
2048 C.W. Jefford et al.
2087 H. Hilpert et al.

34- -86, J. Chem. Eng. Data, 31 (1986)
0369 N.R. El-Rayyes et al.

35- -86, J. Am. Chem. Soc., 108 (1986)
0017 W. Leupin et al.
0099 M.A. Kesselmayer and R.S. Sheridan
0128 Y.M.A. Naguib et al.
0134 R.A. Moss et al.
0343 R.S. Macomber and T.C. Hemling
0379 J.J. Eisch et al.
0490 V. Kumar et al.
0512 L.A. Paquette et al.
0618 L. Fielding et al.
0806 R.L. Danheiser et al.
0841 Y. Ito et al.
0846 J.A. Roesener and P.J. Scheuer
0847 S. Matsunaga et al.
0936 D.T. Sawyer et al.
0948 R.T. Edidin and J.R. Norton
1006 P.-H. Bong et al.
1155 A. Ceulemans et al.
1234 Y.L. Chow and G.E. Buono-Core
1333 M. Hisatome et al.
1338 H.R. Pfaendler et al.
1589 H. Ikezawa et al.
1617 T. Miyashi et al.
1903 J.S. Thompson and J.C. Calabrese
2023 R.S. Phillips and L.A. Cohen
2147 Y. Teki et al.
2273 F. Diederich et al.
2372 C.F. Bernasconi et al.
2388 S.H. Gellman et al.
2416 R.A. McClelland et al.
2479 T.N. Sorrell and A.S. Borovik
2481 M. Hirthammer and K.P.C. Vollhardt
2691 S.M. Reddy et al.
2773 Y. Sugihara et al.
2780 Y. Kato et al.
2916 J.P. Collman et al.
2985 E. Vedejs et al.
3005 F.D. Lewis et al.
3016 F.D. Lewis et al.
3040 A.B. Smith, III et al.
3088 D.J. Brand and J. Fisher
3145 S. Bantia and R.M. Pollack
3150 R. Diercks and K.P.C. Vollhardt
3186 S.A. Asher et al.
3304 J. Vites et al.
3380 S.J. Atherton et al.
3443 G. Mehta et al.
3460 K. Nakasuji et al.
3659 M.D. Farnos et al.
3731 H. Jendralla et al.
3739 L.A. Paquette et al.
3907 A.M. Halpern et al.
4105 G. Neumann and K. Müllen
4111 D.P. Kjell and R.S. Sheridan
4115 G. Büchi et al.
4126 B. Zielinska et al.
4149 M. Demuth et al.
4237 H.H. Wasserman et al.
4433 M.F. Zipplies et al.
4459 J.S. Miller et al.
4484 D.A. Lightner et al.
4498 A. Thompson et al.
4527 A.E. Radkowsky and E.M. Kosower
4532 A.E. Radkowsky et al.
4556 W. Adam et al.
4561 A.C. Chan and D.I. Schuster
4614 A. Albeck et al.
4644 C.-M. Che and W.K. Cheng
4676 T. Shono et al.
4685 F. Wudl and S.P. Bitler
4856 T.T. Wenzel and R.G. Bergman
4881 J. Oren and B. Fuchs
4920 T. Gotoh et al.
4953 K.J. Shea et al.
5109 L.B. Clark
5158 A. Hamilton et al.
5264 R.G. Zimmermann et al.
5503 S.F. Nelsen et al.
5527 B. Reimann et al.
5543 K. Llamas et al.
5589 R.W. Spencer et al.
5598 J.-P. Battioni et al.
5949 B. Halton et al.
5964 F.D. Lewis et al.
5972 M.M. Al-Arab and G.A. Hamilton
6039 M. Jako et al.
6074 G.R. Newkome et al.
6088 S. Omura et al.
6276 H.E. Zimmerman and R.D. Solomon
6404 R. Sakai et al.
6440 M. Sheves et al.
6449 R.A. Goldbeck et al.
6579 F. Brisse et al.
6675 A. Bashir-Hashemi et al.
6680 D.M. Roll et al.
6713 D.L. Boger and C.E. Brotherton
6739 A. Padwa et al.
7010 R.H. Voegeli et al.
7016 S. Gronert and A. Streitwieser,Jr.
7032 T. Sugimoto et al.
7089 T. Morii et al.
7575 J.B. Lambert et al.
7609 A. Shanzer et al.
7693 C.F. Wilcox, Jr. et al.
7744 C.F. Bernasconi and A. Kamavarioti
7858 M. Chang et al.
7875 G. Müller et al.
7926 S.F. Nelsen et al.
8002 M.P. Groziak et al.

8088 K.J. Stone et al.
8107 D. Kuck and H. Bogge
8283 V. Nair et al.

36- -86, J. Pharm. Sci., 75 (1986)
0612 N. Shoji et al.
0784 C. Rossi et al.
1188 N. Shoji et al.

39- -86B, J. Chem. Soc., Perkin II (1986)
0043 K. Ohkata et al.
0065 D.J. Palling et al.
0105 I. Pallagi and P. Dvortsak
0117 A. Castro et al.
0123 G. Hallas et al.
0359 F. Billes and A. Toth
0525 F. Seela et al.
0537 I.D. Cunningham and A.F. Hegarty
0635 A.K. Singh
0727 Y.L. Chow and J.S. Polo
0931 J. Riand et al.
0961 M.A. Pericas et al.
0973 S.-F. Tan et al.
1003 S. Steenken et al.
1115 H. Kothandaraman and N. Arumugasamy
1147 T. Hoshi et al.
1217 R. Matsushima and K. Sakai
1247 M. Krishnamurthy and S.K. Dogra
1391 J.L. Mokrosz and M.H. Paluchowska
1439 A. Albini et al.
1561 W.M. Odijk and G.-J. Koomer
1623 Y. Sakaino et al.
1875 Y.S. Chong and H.H. Huang
1917 M. Krishnamurthy et al.
1941 S.-F. Tan et al.
1987 T. Kitagawa et al.

39- -86C, J. Chem. Soc., Perkin I (1986)
0001 S. Saito and H. Itano
0033 Z.T. Fomum et al.
0039 D.H.R. Barton et al.
0083 A.L. Weis and F. Frolow
0091 D. Armesto et al.
0109 H. Seya et al.
0145 A.C. Gibbard et al.
0157 C. Bleasdale and D.W. Jones
0169 N. Haider and G. Heinisch
0215 A.C. Jain and A. Mehta
0255 A. Arnone et al.
0261 J. Blumbach et al.
0291 G. Mehta et al.
0315 P.J. Darcy et al.
0321 E. Glotter and M. Zviely
0381 G. Barbaro et al.
0415 D.N. Nicolaides et al.
0421 D.F. Corbett
0429 P.S. Steyn et al.
0483 R.S. Gairns et al.
0491 R.S. Gairns et al.
0515 G.M. Brooke and J.A.K.J. Ferguson
0525 A. Arnone et al.
0551 A.G. Gonzalez et al.
0575 J.A.S. Cavaleiro et al.
0585 M. Burton and D.J. Robins
0595 M. Nitta and H. Miyano
0619 R.L. Beddoes et al.
0623 D. Armesto et al.
0659 K. Bock et al.
0671 K. Phillips and G. Read
0675 B.C. Maiti and O.C. Musgrave
0701 A. Matsuo et al.
0711 J. Clark et al.
0729 K. Satake et al.
0741 C.D. Burt et al.
0753 M. Balogh et al.
0789 S.T. Saengchantara and T.W. Wallace
0799 D. Armesto et al.
0885 M. Kimura et al.
0905 M. Ito et al.
0919 Atta-ur-Rahman and M.I. Choudhary
0933 J. Ojima et al.
0973 J.H. Bateson et al.
1021 S. Siddiqui et al.
1043 M. Cushman et al.
1051 M. Cushman et al.
1077 C.L. Branch amd M.J. Pearson
1145 J.F. Grove and P.B. Hitchcock
1147 T. Nishio et al.
1157 D. Moderhack and A. Lembcke
1181 J.B. Hanuman et al.
1203 C.W. Spangler et al.
1235 I.D. Cunningham et al.
1243 A.K. Chakraborti et al.
1267 J.G. Buchanan et al.
1297 J. Herscovici et al.
1303 P.A. Brown and P.R. Jenkins
1317 M.M. Blight and J.F. Grove
1329 D.N. Harcourt et al.
1363 G. Appendino et al.
1471 M.J. Collett et al.
1479 P.J. Battye and D.W. Jones
1507 V.L. Bell et al.
1551 P. Petchnaree et al.
1565 A.R. Battersby et al.
1599 D. Crescente et al.
1603 D.H.R. Barton and D. Crich
1627 A. Takuwa et al.
1633 P.S. Steyn et al.
1651 D.W.M. Benzies et al.
1691 A.C. Pratt and Q. Abdul-Majid
1759 M. Sakamoto et al.
1769 R.E. Banks et al.
1781 T. Polonski and Z. Dauter
1789 H. Greenland et al.
1797 D.H.R. Barton et al.
1805 D.H.R. Barton et al.
1809 M. Furber et al.
1825 S. Shinkai et al.
1905 M.J. Begley et al.
1923 P. Cano et al.
1965 J. Tsunetsuga et al.
2005 M. Cavazza et al.
2011 H. Greenland et al.
2021 D. Armesto et al.
2051 R. Zelnik et al.
2075 K. Smith et al.
2123 M. Anastasia et al.
2127 D.R. Gedge et al.
2213 B. Hanlon et al.
2217 M.F. Aldersley et al.

2233 N.E. Mackenzie et al.
2253 M. Hanaoka et al.
2295 B.C. Uff et al.
2305 E. Balogh-Hergovich and G. Speier

41- -86, J. chim. phys., 83 (1986)
0311 P. Jardon et al.
0577 A. Brembilla et al.
0589 E. Casassas and T. Visa

42- -86, J. Indian Chem. Soc., 63(1986)
0035 D.K. Palit et al.
0323 B.S. Reddy and M. Darbarwar
0420 D. Sen et al.
0448 K.M. Biswas and H. Mallik
0449 A. Banerji and S. Jana
0600 K.R.S. Reddy et al.
1060 S. Uma Rani and M. Darbarwar

44- -86, J. Org. Chem., 51 (1986)
0057 E. Fahy et al.
0075 P.M. Chouinard and P.A. Bartlett
0088 A. Albini et al.
0149 H. Wamhoff et al.
0196 H.E. Zimmerman and D.N. Schissel
0247 H. Aoyama et al.
0251 K. Yamamura et al.
0256 R.W. Curley, Jr. and C.J. Ticoras
0268 R.L. Shone et al.
0287 S.-I. Kang and J.L. Kice
0295 S.-I. Kang and J.L. Kice
0314 T. Katada et al.
0495 A.G.M. Barrett et al.
0522 E.B. Skibo
0589 G. Gottarelli et al.
0602 K. Maruyama and H. Tamiaki
0629 H. Takaya et al.
0657 K.M. Smith and D.A. Goff
0666 K.M. Smith et al.
0755 X. Gao et al.
0773 M.P. Bosch et al.
0921 H.O. House et al.
0945 D.J. Anderson
0950 L. Maggiora and M.P. Mertes
0995 L.H. Schwartz et al.
1050 O.L. Acevedo et al.
1058 T.L. Cupps et al.
1065 M.P. Groziak et al.
1088 C.F. Wilcox, Jr. and K.A. Weber
1094 V.S. Pangka et al.
1100 M. Ballester et al.
1115 C. Francisco et al.
1253 C.-Q. Han et al.
1258 G.W. Craig et al.
1269 J.L. Mascarenas et al.
1277 M.P. Groziak and L.B. Townsend
1282 G.R. Pettit and P.S. Nelson
1287 G.V.B. Madhaven and J.C. Martin
1324 A.D. Allen et al.
1347 A.R. Morgan and N.C. Tertel
1374 J.H. Rigby and F.J. Burkhardt
1407 R.G. Harvey et al.
1413 M. Ballester et al.
1419 B. Gregoire et al.
1435 S. Matsumoto et al.
1440 E.L. Clennan et al.
1490 C.-K. Sha et al.
1498 D.G. Melillo et al.
1692 M.R. Detty et al.
1719 W. Lwowski et al.
1773 S.K. Balani et al.
1866 A.L. Schroll and G. Barany
1967 C.V. Kumar et al.
1972 S. Pratapan et al.
2134 C.K. Chang and W. Wu
2168 R.A. Moss et al.
2184 E.W. Thomas
2214 T. Yamato et al.
2257 A. Nishinaga et al.
2315 D.J. Kertesz and M. Marx
2343 E. Wenkert et al.
2361 S.M. Reddy et al.
2385 R. Gabioud and P. Vogel
2408 H.O. House et al.
2428 J.E. Rice et al.
2445 S. Kumar and N.L. Agarwal
2461 D.H. Bohn et al.
2472 M. Ballester et al.
2481 A.R. Katritzky et al.
2493 P.F. Wiley et al.
2529 S. Ganapathy et al.
2550 M.E. Jason et al.
2594 J.E. Toth and P.L. Fuchs
2601 T.F. Molinski and D.J. Faulkner
2707 C. Francisco et al.
2736 C.B. Rao et al.
2781 C. Saa et al.
2784 M.J. McManus et al.
2863 J.K. Gawronski and H.M. Walborsky
2906 B.B. Jarvis and C.S. Yatawara
2956 H.-D. Becker et al.
2961 A.G. Anderson, Jr. et al.
2969 M. Bruch et al.
2988 B. Podanyi et al.
2995 E. Wenkert et al.
3001 A. Valasinas et al.
3058 H. Mrozik et al.
3116 R.B. Gammill and S.A. Nash
3123 T. Sheradsky and R. Moshenberg
3125 A. Padwa and G.S.K. Wong
3223 H.-D. Becker et al.
3228 D.P. Matthews et al.
3232 W.J. Gottstein
3244 G. Sabbioni et al.
3257 W.H. Mandeville and G.M. White-sides
3364 T. Ichiba and T. Higa
3374 J. Ackrell et al.
3407 S.K. Dubey and S. Kumar
3413 E. Perrone et al.
3420 R. Barik et al.
3453 S.P. Spyroudis
3468 R.I. Pacut and E. Kariv-Miller
3502 H. Lee and R.G. Harvey
3542 S.V. Copaja et al.
3551 J.L. Foc, C.H. Chen and H.R. Luss
3559 M.J. Damha and K.K. Ogilvie
3693 L.A. Deardurff and D.M.Camaioni
3751 R.H. Smith, Jr. et al.
3762 J.G. Smith et al.
3811 G. Balina et al.
3931 K. Nagai et al.

3973 R. Rodrigo et al.
3978 R.G. Carlson et al.
4075 M. Caron
4087 J.C. Gilbert and B.K. Blackburn
4107 Z. Rappoport and A. Gazit
4112 Z. Rappoport and A. Gazit
4160 A.D. Thomas and L.L. Miller
4196 D. Belotti et al.
4254 A.G.M. Barrett et al.
4260 E. Quinoa et al.
4264 G.T. Carter
4272 P.W. Jeffs et al.
4385 G. Doddi and G. Ercolani
4417 K. Minamoto et al.
4424 A. Feigenbaum et al.
4432 H. Garcia et al.
4450 P.A. Keifer et al.
4477 T. Akasaka et al.
4568 B.W. Sullivan et al.
4580 G.R. Pettit and C.W. Holzapfel
4586 G.R. Pettit and C.W. Holzapfel
4590 R.C. Kelly et al.
4604 H.E. Zimmerman and J.M. Nuss
4644 R.F.X. Klein and V. Horak
4660 K.M. Smith et al.
4667 K.M. Smith et al.
4681 H. Morrison et al.
4784 C.-H. Lee et al.
4792 P.S. Engel and G.A. Bodager
4813 G. Buchi and J.C. Leung
4819 D.R. Andrews et al.
4840 A.G.M. Barrett et al.
4934 S. Yamada et al.
5001 J. Szmuszkovicz et al.
5002 J. Szmuszkovicz et al.
5051 M.G. Steinmetz et al.
5134 B.W. Sullivan et al.
5136 N.K. Gulavita et al.
5140 S.A. Look et al.
5148 I. Saito et al.
5198 R.M. Coates and S.J. Firsan
5300 M. Ishibashi et al.
5447 M.F. Boehm and G.D. Prestwich
5476 N.B. Perry et al.

46- -86, J. Phys. Chem., 90 (1986)
0328 R.G. Barr and T.J. Pinnavaia
0603 G. Vanermen et al.
0607 R.W. Carr, Jr. et al.
1179 S. Monti and L. Flamigni
1455 A. Mordzinski and W. Kuhnle
1541 A.K. Wisor and L. Czuchajowski
1921 Y. Yamamoto et al.
2340 E.D. Schmid et al.
2469 Y. Kawanishi et al.
2589 T. Handa et al.
2635 R.D. Kenner et al.
2651 C. Lenoble and R.S. Becker
2661 M. Rinke et al.
2666 M. Rinke et al.
2670 X. Xie et al.
5654 H. Abdel-Halim et al.
6068 J. Riefkohl et al.
6266 A.I. Ononye et al.
6314 A. Maciejewski et al.
6464 N.C. Rothman et al.

47- -86, J. Polymer Sci., Chem. Ed., 24 (1986)
0119 Y. Inaki et al.
0133 R.J.G. Roth et al.
0241 V. Rasmuswami and D.A. Tirrell
1726 D.T. Longone and D.T. Glatzhofer
1839 E. Isohe et al.
1879 C.E. Hoyle and K.-J. Kim
1943 R.V. Todesco and R. Basheer
2021 Y.S. Gal, H.N. Cho and S.K. Choi
2059 T. Ouchi et al.
3201 Y. Inaki et al.

48- -86, J. prakt. Chem., 328 (1986)
0081 W. Ortmann et al.
0089 R. Gase et al.
0113 C. Haessner and H. Mustroph
0149 J. Gaca et al.
0181 H.-J. Timpe et al.
0237 U. Burkhardt and S. Johne
0253 W. Freyer
0261 H. Heberer and H. Matschiner
0314 R. Spitzner et al.
0342 L. Hennig et al.
0359 T. Zimmermann and G.W. Fischer
0373 T. Zimmermann and G.W. Fischer
0459 K. Gewald, U. Hain and M. Schmidt
0497 W. Thiel et al.
0522 N.A. Shams et al.
0539 H.-H. Ruttinger et al.
0567 T. Zimmermann and G.W. Fischer
0573 T. Zimmermann and G.W. Fischer
0635 M. Süsse and S. Johne
0695 C. Lindig and K.R.H. Repke
0741 K. Gewald, O. Calderon and U. Hain
0755 H. Hartmann
0777 R. Mahrwald et al.
0805 R. Beckert and J. Fabian
0812 W. Thiel, J. Fabian and B. Friehe
0906 U. Burkhardt and S. Johne

49- -86, Monatsh., 117 (1986)
0057 H. Falk and H. Flödl
0185 J.M. Ribo and X. Serra
0475 W. Freyer and L. Minh
0631 R. Merce et al.
0805 S. Fu, S. Li and O. Vogl
0849 H. Falk et al.
1081 H. Falk and A. Hinterberger
1205 D. Krois and H. Lehner
1339 U. Jordis
1423 K. Schlogl et al.

54- -86, Rec. trav. chim., 105 (1986)
0054 T.S.V. Buys et al.
0111 N.P. van Vliet et al.
0156 L. Rodenburg et al.
0188 T.S.V. Buys et al.
0360 J.J.G.S. van Es et al.
0386 H. Cerfontain and J.A.J. Geenevasen

56- -86, Polish. J. Chem., 60 (1986)
0065 M. Janczewski and J. Pekalska
0143 N.A. Shams et al.
0163 T. Widernik and T. Skarzynska-Klentak

0311 B. Osipowicz et al.
0315 P. Kus
0429 W. Baran et al.
0441 W. Baran et al.
0541 J. Maslowska and J. Jaroszynska
0599 A. Lapucha and E. Wyrzykiewicz
0605 M. Stobiecki
0615 Z. Szulc and J.Młahowski
0767 E.E. Pasternak et al.
0797 K. Wojciechowski and J. Szadowski
0825 M. Paluchowska and J. Bojarski
0831 R. Gawinecki
0837 W. Cisowski

59- -86, Spectrochim. Acta, 42A (1986)
0405 S.R. Salman et al.
0669 S. Sawamura et al.
0701 M.P. Martinez-Martinez et al.
0771 G.R. Ramos et al.
0793 M. Krishnamurthy and S.K. Dogra
0929 K. Inuzuka and A. Fujimoto
1127 G. Kohler and K. Rotkiewicz
1217 B.H. Bhowmik and A. Bhattacharyya
1289 E. Spinner
1393 K.C. Mehdi

60- -86, J. Chem. Soc., Faraday Trans. I (1986)
0281 R.A. Schoonheydt et al.
0291 P. Molyneaux and S. Vekavakuyan-ondha
0359 P. Maruthamuthu et al.
1307 S. Mishima and T. Nakajima
2123 R.L. Schiller et al.
2233 R.J. Clarke et al.
3141 K. Hamada et al.

61- -86, Ber. Bunsen Gesell. Phys. Chem., 90 (1986)
0439 M. Rinke and H. Gusten
0813 R. Schmidt
0891 A. Castro et al.

62- -86, Z. physik. Chem.(Wiesbaden), 267 (1986)
1107 S.M.M. Elshafie

63- -86, Biol. Chem.Hoppe-Seyler, 367 (1986)
0191 S. Thurl et al.
0757 K. Schafer et al.

64- -86B, Z. Naturforsch., 41b (1986)
0093 R. Neidlein et al.
0223 M.W. Haenel et al.
0239 S.K. Singh and L.A. Summers
0265 C.O. Meese and H. Güsten
0377 H. Laatsch
0645 E. Jägers et al.
0768 R.E. Balsells et al.
0772 W. Rühl et al.
1033 V.U. Ahmad et al.
1045 D. Hellwinkel et al.
1151 J. Daub et al.
1161 U. Hildebrand and H. Budzikiewicz
1180 P. Hennig et al.
1265 H. Volz et al.
1281 H.A. Brune et al.
1471 J. Mirek and P. Milart
1587 W.I. Owtscharenko et al.

64- -86C, Z. Naturforsch., 41c (1986)
0164 H. Achenbach et al.
0187 A. Baumert et al.
0437 W. Kufer et al.
0677 A. Stoessl and J.B. Stothers
0681 M. Iinuma et al.
0695 O. Spring et al.
1119 M. Martinez V. et al.

65- -86, Zhur. Obshchei Khim., 56 (1965) (English translation pagination)
0179 R.N. Khelevin
0301 M.I. Shevchuk et al.
0341 T.A. Shatskaya et al.
0362 R.G. Khairullina et al.
0365 L.E. Protosova et al.
0391 P.V. Tarasenko et al.
0613 V.F. Pozdnev
0727 G.I. Validuda et al.
0758 A.A. Vyazgina et al.
1020 V.M. Tsentovskii et al.
1105 G.A. Mezentseva et al.
1167 A.L. Nikonov et al.
1323 V.S. Brovarets et al.
1397 E.R. Luchkevich et al.
1553 V.A. Krutikov et al.
1848 M.G. Voronkov et al.
1891 M.A. Dyadchenko et al.
2088 E.K. Rukhadze et al.
2109 E.V. Borisov
2114 L.V. Shmelev et al.
2372 A.D. Sinitsta et al.
2389 V.I. Krutikov et al.
2391 V.I. Krutikov et al.
2461 V.V. Khoroshilova et al.

67- -86, J. Structural Chem., 27 (1986)
0062 A.I. Ermakov et al.
0225 A.I. Ermakov et al.

69- -86, Biochemistry, 25 (1986)
1186 L. Davenport et al.
1799 J.W. Walker et al.
2374 O. Bangcharoenpaurpong et al.
3286 S. Ghisla et al.
3290 M. Shahbaz et al.
3752 S.-H. Park et al.
4699 C.M. Cook et al.
4762 D.C. Pike et al.
5571 J.M. Konopke et al.
5602 C. D'Silva et al.
5774 T.P. King, S.W. Zhao and T. Lam
7118 G. Vanderkooi and A.B. Adade
7288 D.F. Covey et al.
7423 E. Kanaya and H. Yamagawa
8095 V. Massey et al.
8103 V. Massey et al.

70- -86, Izvest. Akad. Nauk S.S.S.R., 35 (1986)
0120 Z.G. Aliev et al.
0131 V.P. Litvinov et al.

0363 G.N. Bondarev et al.
0376 E.E. Apenova et al.
0446 G.F. Bannikov et al.
0458 Z.A. Krasnaya
0555 A.A. Belyaev and L.B. Radina
0582 V.I. Nikulin and L.M. Pisarenko
0733 Y.A. Ershov et al.
0804 V.M. Karpov et al.
0832 A.M. Turuta et al.
0871B V.P. Mamaev et al.
0994 A.N. Mirskova et al.
1058 O.P. Petrenko et al.
1064 I.S. Salikhov et al.
1072 A.V. Kamernitskii et al.
1140 A.I. Nikolaev et al.
1251 V.N. Nesterov et al.
1391 I.K. Korobeinicheva et al.
1446 Z.A. Krasnaya et al.
1470 Y.L. Gol'dfarb et al.
1504 Y.A. Sharanin et al.
1677 A.K. Filippova et al.
1685 E.F. Kolchina et al.
1690 N.A. Polezhaeva et al.
1720 A.M. Turuta et al.
1741 A.V. Sultanov and S.B. Savvin
1851 N.Y. Grigor'eva et al.
1883 V.M. Karpov et al.
1919 Y.L. Gol'dfarb et al.
2105 L.I. Leont'eva et al.
2267 S.G. Bitiev et al.
2409 O.M. Nefedov et al.
2511 L.A. Myshkina et al.
2522 B.A. Trofimov et al.

73- -86, Coll. Czech. Chem. Comm., 51 (1986)
0106 E. Kral'ovicova et al.
0118 J. Sejbal et al.
0145 I. Cervena et al.
0156 M. Protiva et al.
0215 M.P. Nemeryuk et al.
0412 J. Stetinova et al.
0459 A. Holy et al.
0573 T. Gracza et al.
0621 V. Krecek et al.
0698 M. Protiva et al.
0879 T. Gracza et al.
0937 V. Kmonicek et al.
1061 S. Marchalin et al.
1127 J. Hrabovsky et al.
1140 J. Krepelka et al.
1311 H. Hrebabecky
1316 A.M. Moiseenkov and B.A. Czekis
1450 D. Berkes and J. Kovac
1455 E. Kralovicova et al.
1512 M.K. Spassova et al.
1665 P. Kuzmic et al.
1678 M. Uher et al.
1685 E. Kral'ovicova et al.
1692 S. Radl and V. Zikan
1731 J. Hajicek and J. Trojanek
1743 J. Slavik and L. Slavikova
2002 P. Kutschy et al.
2034 Z. Polivka et al.
2158 L. Fisera et al.
2167 L. Fisera et al.
2186 J. Stetinova et al.
2232 E. Taborska et al.
2598 M. Protiva et al.
2817 L. Stibranyi et al.
2826 L. Stibranyi et al.
2839 P. Kutschy et al.
2848 K. Sindelar et al.
2869 J. Klinot et al.

74- -86A, Mikrochim. Acta I (1986)
0027 E. Kavlentis
0033 S. Funahashi et al.
0097 P. Tarin et al.

74- -86C, Mikrochim. Acta III (1986)
0251 E. Kavlentis
0321 M.T.M. Zaki et al.
0417 S.S. Rahim et al.

77- -86, J. Chem. Soc. Chem. Comm. (1986)
0001 G. Baudouin et al.
0019 S. Yoneda et al.
0040 A. Guerriero et al.
0077 G. Guella et al.
0249 A.F. McDonagh et al.
0268 K. Hayakawa et al.
0281 H. Mestdagh and K.P.C. Vollhardt
0310 C.-K. Sha and C.-P. Tsou
0386 C.M. Che and W.C. Chung
0444 A. Gazit
0573 M. Ihara et al.
0574 Y. Nambu et al.
0733 A. Trincone et al.
0767 H.J. Callot et al.
0787 C. Sishta et al.
0935 D.D. Rowan, M.B. Hunt and D.L. Gaynor
0964 B.F. Bonini et al.
0974 J. Nakayama et al.
0981 K. Iguchi et al.
1019 D.W.M. Benzies and R.A. Jones
1038 B. Gigante et al.
1076 S. Matsumoto et al.
1077 M. Takasugi et al.
1083 S.C. Rawle et al.
1086 U. Dressler et al.
1110 E. Kimura et al.
1130 Y. Sugihara et al.
1162 P. Carter et al.
1188 J. Kurita et al.
1239 S. Gaspard et al.
1247 Y.M. Goo and Y.B. Lee
1257 B. Miller and J. Baghdadchi
1322 E. Kimura et al.
1385 P.A.J. Link et al.
1472 K. Kikuchi et al.
1480 M. Cavazza and F. Pietra
1481 B.K. Cassels et al.
1489 Y. Yamashita et al.
1495 N. Kawahara et al.
1516 Y. Okamoto et al.
1518 P. Berno et al.
1616 A. Vogler and H. Kunkely
1639 K.A. Bello and J. Griffiths
1645 A. Lannikonis et al.
1653 K.J. Del Rossi and B.B. Wayland
1725 L. Hevesi et al.

1756 M. Ladlow et al.
1768 H. Matsumoto et al.

78- 086, Tetrahedron, 42 (1986)
0137 Z. Kilic and N. Gunduz
0199 T. Kondo et al.
0207 T. Kondo et al.
0245 F.S. Abbott et al.
0305 M. Mizutani et al.
0409 M. Largeron et al.
0529 J.W. Ann et al.
0591 G.A. Tolstikov et al.
0655 T.-L. Chan et al.
0699 J. Alvarez-Builla et al.
0727 M. Turconi et al.
0747 M. Poje and L. Sokolic-Maravic
0835 H. Bader and H.-U. Reissig
0839 R. Gompper and R. Guggenberger
0849 R. Gompper and R. Guggenberger
0961 A. Ohno et al.
1127 W. Hendel et al.
1139 H. Singh et al.
1285 M. Feldhues and H.J. Schafer
1315 E. Monnier et al.
1345 J. Rigaudy et al.
1355 J. Rigaudy et al.
1511 A. Frankowski
1581 G.E. Renzoni et al.
1585 M. Christl et al.
1641 H. Hart et al.
1655 H. Hopf et al.
1693 A. Krebs et al.
1721 H. Butenschon and A. De Meijere
1745 I. Murata et al.
1763 K. Naemura et al.
1797 W.-D. Fessner and H. Prinzbach
1805 H. Quast et al.
1823 L.T. Scott and M.H. Hashemi
1831 E. Carceller et al.
1851 Y. Tobe et al.
1903 A.J.H. Klunder et al.
1971 H. Tanaka et al.
2075 R.S. Mali et al.
2083 M.G. Corradini et al.
2221 W. Eberbach and J. Roser
2329 D.H.R. Barton et al.
2377 J. De Mendoza et al.
2417 A.S.R. Anjaneyulu et al.
2575 J.E. Baldwin et al.
2685 G. Costerousse and G. Teutsch
2757 R. Uma et al.
2771 C.O. Okafor
3157 L.S. Trifonov et al.
3215 P. Magnus et al.
3303 G.A. Morris et al.
3311 A.F. Thomas and C. Perret
3461 S. Manabe and C. Nishino
3491 H. Cervantes et al.
3547 H. Herzog et al.
3579 B.M. Gianetti et al.
3587 B.M. Gianetti et al.
3655 M. Carda et al.
3663 F.M. Dean et al.
3683 T. Jagodzinski et al.
3711 W.A. Laurie et al.
3767 B. Simoneau and P. Brassard
3781 H. Kigoshi et al.
3807 T. Akasaka et al.
3943 J.E. Baldwin et al.
4137 J. Awruch and B. Frydman
4187 H. Tanaka et al.
4197 H. Nakamura et al.
4493 R. Capasso et al.
4563 J. De Pascual Teresa et al.
4849 S. Siddiqui et al.
5019 V. Bocchi and G. Palla
5081 E. Malamidou-Xenikaki and D.N. Nicolaides
5281 K. Mori and H. Kisida
5321 T. Cuvigny et al.
5341 H. Wingert et al.
5369 L. Mayol et al.
5415 A. Messmer et al.
5427 T. Pathak et al.
5637 S. Antus et al.
5693 A.A. Frimer et al.
5747 M.I. Choudhary et al.
5869 R. Ramasamy et al.
5949 J. Scherkenbeck et al.
5969 U. Hildebrand et al.
6015 V. Amico et al.
6047 G. Costello and J.E. Saxton
6175 A. Okamoto et al.
6195 Y. Kawamura et al.
6293 R.C. Haddon et al.
6443 Y. Oshima et al.
6487 A. De Meijere and D. Kaufmann
6545 R.J. Capon et al.
6555 O.E. Jensen and A. Senning
6565 R.L. Hendrickson and J.N. Cardellina, II
6615 A.K. Banerjee et al.
6635 K. Krohn and U. Muller
6713 T.M. Cresp and D. Wege
6719 J.P. Brennan and J.E. Saxton

80- -86, Revue Roumaine Chim., 31 (1986)
0137 M.H. Terdic et al.
0273 H.-H. Glatt et al.
0503 M. Banciu et al.
0541 N.A. Shams et al.
0661 V. Gorduza et al.
0761 D.P. Khurana et al.
0857 G. Ivan and M. Giurginga
0869 R. Kumari et al.
0881 C. Cristescu
1019 M. Banciu et al.

83- -86, Arch. Pharm., 319 (1986)
0025 J. Reisch and S.M.E. Aly
0043 V. Ashauer et al.
0052 H.-J. Kallmayer and K. Seyfang
0060 M.C. Gomez Gil et al.
0065 G. Dannhardt and L. Steindl
0079 J. Dusemund and T. Schurreit
0081 K.-A. Kovar and M. Teutsch
0108 K.Görlitzer and K. Michele
0120 K. Görlitzer and D. Hölscher
0126 P. Pachaly et al.
0154 G. Kiefer et al.
0188 K.-C. Liu and M.-K. Hu
0231 G. Dannhardt and L. Steindl

0234 G. Dannhardt and L. Steindl
0242 F. Eiden and G. Patzelt
0266 K. Seifert et al.
0332 K. Görlitzer and D. Hölscher
0338 F. Eiden and K. Berndl
0347 F. Eiden and K. Berndl
0354 G. Dannhardt and L. Steindl
0421 H.-J. Kallmayer and C. Tappe
0465 I. Körfer and A.W. Frahm
0500 G. Dannhardt and L. Steindl
0607 H.-J. Kallmayer and C. Tappe
0690 G. Seitz and R. Mohr
0694 D.U. Lee and W. Wiegrebe
0720 J. Reisch et al.
0735 G. Dannhardt et al.
0749 G. Dannhardt and L. Steindl
0791 H.-J. Kallmayer and C. Tappe
0798 G. Seitz et al.
0841 P. Pachaly and T.J. Tan
0872 P. Pachaly and T.J. Tan
0886 F. Eiden and B. Wünsch
0898 R. Kühn and H.-H. Otto
0910 H. Möhrle and B. Grimm
0947 H. Mertens and R. Troschütz
0950 J. Knabe and B. Hanke
1018 H. Möhrle and B. Grimm
1122 Y. Okamoto et al.

86- -86, Talanta, 33 (1986)
0170 F. Belal et al.
0209 M. Garcia-Vargus et al.
0274 K. Nakashima et al.
0288 Y. Akama et al.
0375 M. Villarreal et al.
0627 M.C. Mochon et al.

87- -86, J. Med. Chem., 29 (1986)
0079 Y.F. Shealy et al.
0125 C. Muller-Uri et al.
0138 T.A. Krenitsky et al.
0203 G. Gosselin et al.
0213 E. De Clerq et al.
0224 L. Brehn et al.
0261 I. Sircar et al.
0268 C.R. Petrie, III et al.
0318 F. Kappler et al.
0468 T.R. Jones et al.
0483 Y.F. Shealy et al.
0671 E.J. Prisbe et al.
0681 T. Lin et al.
0709 T.-L. Su et al.
0809 W.L. Mitchell et al.
0842 J.D. Karkas et al.
0862 T.-S. Lin et al.
0997 T. Hashimoto et al.
1030 F. Kappler et al.
1052 R.D. Elliott et al.
1080 J.R. Piper et al.
1099 D.J. Blythin et al.
1114 T.R. Jones et al.
1225 E.M. Acton et al.
1237 P.D. Devanesan and A.M. Bobst
1273 M. Israel et al.
1311 L. Strekowski et al.
1346 M.J. Cho et al.
1350 G. Pratviel et al.
1384 J.C. Martin et al.
1461 R.A. Johnson et al.
1544 C.C. Cheng et al.
1596 R.H. Bocker and F.P. Guengerich
1683 R.A. Olsson et al.
1709 H.N. Abramson et al.
1720 Y.F. Shealy et al.
1737 P.J. Harvison et al.
1792 A.E. Mathew et al.
1851 E.T. O'Brien et al.
1864 B.S. Iyengar et al.
1945 H. Rathore et al.
1996 J.A. Levine et al.
2034 D.A. Berry et al.
2088 S. Thaisrivongs et al.
2117 C.A. Caperelli and J. Conigliaro
2120 E.M. Acton et al.
2231 K. Ramasamy et al.
2256 T. Horie et al.
2445 E.J. Prisbe
2511 P.C. Ruenitz et al.

88- -86, Tetrahedron Letters (1986)
0143 Y. Sato et al.
0159 R. Gompper and P. Guggenberger
0167 G. Opitz et al.
0171 G. Märkl and H. Sejpka
0223 H. Nagaoka et al.
0255 M.F. Aldersley et al.
0323 J.P. Ferris and B. Devadas
0403 G. Pattenden et al.
0485 A. Beck et al.
0559 K. Dornberger et al.
0593 T. Sakaki et al.
0607 K. Konno et al.
0691 R. Gompper et al.
0699 R.L. Snowden and M. Wüst
0703 R.L. Snowden and M. Wüst
0865 J.C. Chambron and J.P. Savvage
0981 K. Mori and M. Kato
1063 M. Sundahl et al.
1161 A. Isogai et al.
1191 J. Kobayashi et al.
1269 G. Fischer et al.
1603 T. Sasaki et al.
1625 V.A. Reznikov and L.B. Volodarsky
1627 B. Alcaide et al.
1665 K. Hafner and G.L. Knaup
1669 K. Hafner et al.
1673 K. Hafner and G.L. Knaup
1771 G. Markl and H. Sejpka
1929 Y. Fukazawa et al.
2011 H. Miyamoto et al.
2015 I. Uchida et al.
2111 B. Maurer et al.
2113 J. Kobayashi et al.
2121 T. Sassa et al.
2169 A.D. Ryabov et al.
2177 P. Karuso et al.
2257 C.O. Dietrich-Buchecker et al.
2353 M. Stobbe et al.
2391 S. Nakatsuka et al.
2411 S. Yamaguchi et al.
2439 W. Naengchomnong et al.
2501 S.H. Goh and A.R.M. Ali
2559 R.C. Dawe and J.L.C. Wright

2591 G. Mignani et al.
2621 M. Guyot and M. Meyer
2775 I. Moldvai et al.
2797 P. Crews et al.
2907 Y. Fujise et al.
3005 T. Nakazawa et al.
3045 J. Moursounidis and D. Wege
3177 S.A. Fedoreyev et al.
3251 W. Ando and T. Tsumuraya
3283 G. Lidgren et al.
3399 S. Nakatsuka et al.
3449 Y.L. Chen et al.
3521 D.K. Lavallee et al.
3811 A.G.M. Barrett et al.
3873 N. Morita et al.
3935 E. Ghera et al.
4077 J.M. Boente et al.
4129 D.E. Pereira and N.J. Leonard
4281 K. Kirschke et al.
4335 T. Yoshioka et al.
4489 H. Kawata et al.
4493 K. Okada et al.
4505 T. Koga and Y. Nogami
4585 S. Sakai et al.
4595 S. Kato et al.
4741 K. Yoshima et al.
4903 M.J. Calverley
5003 J. Cornelisse et al.
5159 B. Halton et al.
5189 J. Shin, V.J. Paul and W. Fenical
5339 D. Lenoir et al.
5367 F.A. Neugebauer and H. Fischer
5413 B. Proksa et al.
5459 L. Rodriguez-Hahn et al.
5463 K. Nickisch et al.
5515 K. Takahashi et al.
5535 J.M. Boente et al.
5621 Y. Fukazawa et al.
5653 M. Aga et al.
5675 W. Naengchomnong et al.
5699 J. Heinicke
5923 D.T. Glatzhofer and D.T. Longone
5967 C.M. Dicken et al.
6001 J.H. Bateson et al.
6049 T.R. Kelly et al.
6051 A.S. Kende and K. Koch
6189 H. Hauptmann et al.
6193 P. Perrin et al.
6217 M. Nakagawa et al.
6221 N. Tanaka et al.
6225 T. Kumagai et al.
6361 S. Nakatsuka et al.
6365 H. Ogoshi et al.
6381 K. Ogura et al.
6385 I. Saito et al.

89- -86, Angew. Chem.(Intl. Ed.), 25 (1986)
0079 N. Wiberg et al.
0086 K. Tatsumi et al.
0171 M. Stobbe et al.
0183 R. Allmann et al.
0187 K. Beck and S. Hunig
0255 W. Sander
0257 E. Vogel et al.
0268 R. Diercks et al.
0339 K. Krohn and W. Priyono
0340 H. Irngartinger and W. Götzmann
0367 R. Mynott et al.
0369 Y. Tobe et al.
0374 H. Fischer et al.
0449 R. Gleiter and G. Krenrich
0458 F.-P. Montforts and G. Zimmermann
0466 B. Stowasser and K. Hafner
0572 H. Dürr et al.
0578 T. Loerzer et al.
0630 K. Hafner et al.
0632 K. Hafner and Y. Kuhn
0633 K. Hafner et al.
0635 K. Yamamoto et al.
0720 E. Vogel et al.
0723 E. Vogel et al.
0818 W. Adam et al.
0819 G. Maier and H.P. Reisenauer
0828 G. Kaupp and J.A. Dohle
0842 U.-J. Vogelbacher et al.
0906 R. Gleiter et al.
0917 R. Weiss et al.
0919 P. Jutzi et al.
0999 R. Gleiter et al.
1000 E. Vogel et al.
1003 G. Märkl et al.
1018 M. Kummer et al.
1023 K. Elbl et al.
1025 W. Ludwig and H.J. Schäfer
1116 R. Kramme et al.
1117 M. Demuth et al.
1131 M. Franck-Neumann et al.
1132 H. Gstach et al.

90- -86, Polyhedron, 5 (1986)
0877 L.Tusek-Bozic and D. Sevdic
0983 V.M. Leovac et al.
1157 A. Coutsolelos et al.
1957 D.P. Arnold

94- -86, Chem. Pharm. Bull., 34 (1986)
0001 M. Nagai et al.
0007 S. Kondo et al.
0015 H. Usui and T. Ueda
0030 T. Horie et al.
0077 E. Yamanaka et al.
0082 Y. Hirai et al.
0088 H. Fujimoto et al.
0390 K. Ito and J. Maruyama
0406 T. Tomimori et al.
0423 T. Sano and T. Ueda
0442 T. Kurihara et al.
0471 T. Maruyama et al.
0516 K. Yamagata et al.
0564 Y. Takagawa et al.
0595 K. Baba et al.
0664 M. Sako et al.
0669 T. Fujii et al.
0858 T. Oda, Y. Yamaguchi and Y. Sato
0931 T. Terasawa et al.
0944 H. Hayatsu et al.
1015 N. Tanaka et al.
1094 T. Fujii et al.
1148 M. Lavault et al.
1299 A. Kusai and S. Ueda
1319 K. Yamakawa et al.
1333 S. Inoshiri et al.

1387 S. Iwasaki et al.
1399 I. Kitagawa et al.
1419 M. Nishizawa and Y. Fujimoto
1468 M. Ono et al.
1479 H. Ogura et al.
1505 T. Naito et al.
1518 H. Usui and T. Ueda
1561 H. Takagi et al.
1573 A. Matsuda and T. Ueda
1656 M. Iinuma et al.
1667 T. Tanaka et al.
1794 K. Kikugawa and Y. Sugimura
1809 T. Kinoshita et al.
1821 T. Fujii et al.
1840 H. Kanazawa et al.
1865 K. Anzai et al.
1871 T. Ishida et al.
1929 T. Iida et al.
1961 H. Usui, A. Matsuda and T. Ueda
1968 J. Ikuta et al.
1987 T. Fukai et al.
2037 N.J. Leonard, T. Fujii and T. Saito
2056 H. Yamaguchi et al.
2090 K. Koike and T. Ohmoto
2298 M. Iinuma et al.
2231 K. Watanabe et al.
2282 M. Hirotani et al.
2290 K. Yazaki et al.
2294 M. Yoshikawa et al.
2369 M. Tsukayama et al.
2411 H. Seya et al.
2428 M. Sato et al.
2443 S. Kobayashi et al.
2448 S. Kato et al.
2471 J. Ikuta et al.
2487 M. Noji et al.
2506 T. Saitoh et al.
2609 M. Muraoka et al.
2743 C. Zhengxiong et al.
2766 Y. Shimada et al.
2774 F. Abe et al.
2786 T. Kurihara et al.
2810 M. Watanabe et al.
2868 N. Fukuda et al.
3025 A. Morigiwa et al.
3033 Y. Shimada et al.
3067 T. Kinoshita et al.
3097 M. Kubo et al.
3130 S. Tsukamoto et al.
3183 T. Kikuchi et al.
3202 G. Goto et al.
3228 K. Koike et al.
3244 S. Kadota et al.
3249 M. Niwa et al.
3465 G. Kusano et al.
3588 Y. Miki et al.
3623 T. Maruyama et al.
3631 M. Niwa et al.
3635 T. Kato et al.
3658 C. Kaneko et al.
3695 T. Kikuchi et al.
3713 E. Yamanaka et al.
3727 N. Tanaka et al.
3762 H. Itokawa et al.
3769 S. Adegawa et al.
3866 T. Ohta et al.
3922 H. Furukawa et al.
4012 M. Kuroyanagi et al.
4018 T. Kikuchi et al.
4030 T. Kikuchi et al.
4050 S. Kobayashi et al.
4075 T. Okuda et al.
4092 T. Hatano et al.
4170 M. Seto et al.
4346 M. Ito et al.
4418 K. Baba et al.
4447 T. Sakaki et al.
4533 T. Hatano et al.
4554 K. Yokoi et al.
4585 H. Inoue et al.
4641 I. Kitagawa et al.
4682 H. Itokawa et al.
4883 S. Sakai et al.
4889 K. Iwagoe et al.
4892 M. Maeda et al.
4916 S. Ohta et al.
4939 C. Iwata et al.
4947 C. Iwata et al.
4984 Y. Tagawa et al.
5140 S. Kyosaka et al.
5157 M. Haruna et al.

95- -86, J. Pharm. Soc. Japan (1986) (Yakugaku Zasshi)
0022 L.J. Wu et al.
0982 T. Murakami et al.
0989 H. Wada et al.
1002 Y. Fujimura et al.
1098 Y. Matsuda et al.

96- -86, The Analyst, 111 (1986)
0057 Y.K. Agrawal and V.J. Bhatt
0209 K. Imai et al.
0411 R.J.H. Clark
0429 M.A.B. Lopez
0441 J.A. Ortuno
0923 A.E. Gassim et al.
0983 B.B. Ibraheem and W.A. Bashir
1051 M.B. Fleury et al.

97- -86, Z. Chemie, 26 (1986)
0094 W. Schroth and H. Kluge
0100 J. Bodeker and A. Kockritz
0132 W. Schroth et al.
0134B J. Slouka and K. Nalepa
0140 U. Abram et al.
0165 E. Schmitz and G. Lutze
0209 K. Gustav
0216B W. Freyer
0330 H. Hartmann et al.
0333 S. Helm et al.
0378B J. Kuthan et al.
0399 H. Ilge and H. Hartmann
0437 M. v. Janta-Lipinski et al.

98- -86, J. Agr. Food Chem., 34 (1986)
1005 C.E. Ward et al.

99- -86, Theor. Exptl. Chem., 22 (1986)
0219 A.P. Popov et al.
0409 G.G. Dyadyusha et al.

100- -86, J. Natural Products, 49 (1986)
0035 J.W. Westley
0048 S.-C. Kuo et al.
0056 S.D. Srivastava
0090 A. Evidente
0095 M.L. Cardona et al.
0101 M.-C. Chalandre et al.
0106 Atta-Ur-Rahman et al.
0111 D.M.X. Donnelly et al.
0122 V. Guay and P. Brassard
0145 M. D'Agostino et al.
0151 J.A. Marco et al.
0159 S.A. Ross et al.
0168 A. Evidente
0180 V. Tantishaiyakul et al.
0193 R.C. Bruening et al.
0218 K. Hirakura et al.
0236 C. Tringali et al.
0253 N. Ruangrungsi et al.
0298 M. Arisawa et al.
0303 B. Charles et al.
0313 Y. Oshima et al.
0318 A. Reyes et al.
0321 A.-M. Bui et al.
0334 M. Pinar and M.P. Galan
0336 A. Ulubelen and S. Oksuz
0343 J.U.M. Rao et al.
0360 M. Hoeneisen and H. Becker
0398 M. Shamma and M. Rahimizadeh
0424 A. Patra et al.
0428 N. Fukamiya et al.
0452 S. Michel et al.
0483 Y. Schun et al.
0488 S.F. Hussein et al.
0494 S.F. Hussein et al.
0514 A.J. Chulia et al.
0519 A. Bianco et al.
0538 S.F. Hussain
0561 C.-T. Che et al.
0583 J.F. Ayafor et al.
0588 M. Matsui and Y. Yamamura
0593 A. Evidente et al.
0602 A. Cave et al.
0608 A. Guerriero et al.
0614 B.S. Joshi et al.
0621 M. Arisawa et al.
0645 Y. Shirataki et al.
0661 G.R. Pettit et al.
0670 R.B. Taylor et al.
0677 A. San Feliciano et al.
0690 P.P. Diaz et al.
0692 A. Ulubelen et al.
0695 T.M. Zennie et al.
0706 C.-Y. Duh et al.
0729 V. Lakshmi and F.J. Schmitz
0745 M.D. Menachery et al.
0800 G. Speranza et al.
0806 Y. Schun and G.A. Cordell
0809 M.B. Zurita et al.
0823 L. Mayol et al.
0845 A. San Feliciano et al.
0854 D. Cortes et al.
0866 M.E. Stack et al.
0878 D. Cortes et al.
0901 M.H. Abu Zarga
0913 K.D. Pascoe et al.
0922 A. Urzua and L. Mendoza
0932 R. Promsattha et al.
0995 G.R. Pettit et al.
1010 N.Crespi-Perellino et al.
1018 A. Jossang et al.
1028 A. Jossang et al.
1061 P. Satyanarayana et al.
1068 S. Siddiqui et al.
1078 S. Rasamizaty et al.
1091 S. Mitaku et al.
1122 A. Fujita et al.
1130 C. Kan-Fan et al.
1173 V.M. Hernandez et al.

101- -86C, J. Organomet. Chem., 300 (1986)
0021 A.G. Brook

101- -86D, J. Organomet. Chem., 301 (1986)
0061 M. Franck-Neumann et al.
0173 N.M. Loim et al.

101- -86N, J. Organomet. Chem., 311 (1986)
0027 H.A. Brune et al.
0387 A.A. Watson et al.

101- -86O, J. Organomet. Chem., 312 (1986)
0097 V. Guerchais and D. Astruc

101- -86Q, J. Organomet. Chem., 314 (1986)
0001 F.K. Mitley et al.

101- -86S, J. Organomet. Chem., 316 (1986)
0057 S. Roller and M. Drager

102- -86, Phytochemistry, 25 (1986)
0061 E. Tlomak et al.
0077 B. Tal and D.J. Robeson
0125 A. Evidente et al.
0167 D.D. McPherson et al.
0171 M.C. de la Torre et al.
0185 J. de Pascual Teresa et al.
0195 P. Molnar et al.
0213 S.C.M. Dias et al.
0231 M. Aritomi et al.
0253 J.R. Sierra et al.
0266 A. Michavila et al.
0272 M.C. Garcia-Alvarez et al.
0274 J. Bhattacharyya and C.B. Barros
0281 H. Chiji et al.
0429 I.H. Bowen and Y.N. Patel
0449 M. Nakatani et al.
0453 S. Huneck et al.
0467 A.G. Ober et al.
0471 A. Arnone et al.
0487 L.P. Badheka et al.
0503 T.J. Nagem and J.C. da Silveira
0505 M. Grande et al.
0517 T. Furata et al.
0525 A. Evidente et al.
0527 J.L. Corbin et al.
0529 L. Hansen and P.M. Boll
0531 M. Devys et al.
0555 M.M. Mahandru and V.K. Ravindran
0557 R. Chakrabarti, B.Das and J. Banerji
0649 A.R. Hayman et al.

0695 L. Quijano et al.
0703 J. de Pascual Teresa et al.
0711 J. de Pascual Teresa et al.
0745 A.L. Perez et al.
0755 F. Simoes et al.
0757 Z.T. Fomum et al.
0764 J. Bhattacharyya and V.R. de Carvalho
0766 B. Botta et al.
0767 H.A. Khan et al.
0865 P. Gariboldi et al.
0877 A.G. Ober et al.
0905 H. Itokawa et al.
0909 S.B. Mahato and B.C. Pal
0935 S. Al-Khalil and P.L. Schiff, Jr.
0946 A. Bianco et al.
0965 K. Mahmood et al.
0969 P. Forgacs et al.
0972 I.R.C. Bick et al.
1093 M. Cojocaru et al.
1097 S. Ghosal et al.
1103 G.L. Boyer and J.A.D. Zeevaart
1189 M. Hirotani and T. Furuya
1217 B. Botta et al.
1224 D.H.G. Crout et al.
1227 T. Iwagawa and T. Hase
1240 F.S. Knudsen et al.
1255 A.A.L. Mesquita et al.
1257 M. Iinuma et al.
1301 A. Suksamrarn et al.
1305 A. Fukuzawa et al.
1351 Y. Shiobara et al.
1365 J. de Pascual Teresa et al.
1377 P.G. Forster et al.
1393 A. San-Martin et al.
1411 N.K. Satti et al.
1427 B. Burke and M. Nair
1453 J.M. Llabres et al.
1472 M.-A. Baute et al.
1484 B. Esquirel et al.
1498 Y.A. Geewananda et al.
1505 L. Labbiento et al.
1507 G. Venkataratnam et al.
1509 K. Mahmood et al.
1613 J. Fuska et al.
1633 Y. Shoyama et al.
1673 M. Miski and T.J. Mabry
1677 J.B. Hanuman et al.
1685 D. Meksuriyen et al.
1691 M.O.F. Goulart et al.
1701 E. Morelli et al.
1711 F. Delle Monache et al.
1727 K.W. Biswas and H. Mallik
1743 J.D. Hernandez et al.
1747 G. Appendino et al.
1757 A. San Feliciano et al.
1765 D. Lavie et al.
1776 J.R. de Sousa et al.
1781 Atta-Ur-Rahman et al.
1783 C. Kan et al.
1865 N. Hirai et al.
1903 J.T. Etse and P.G. Waterman
1907 P.J. Houghton et al.
1931 R. Pereda-Miranda et al.
1935 A. Michavila et al.
1949 U. Evcim et al.
1953 X. Yanez R. et al.
1957 H.R.W. Dharmaratne
1975 S. Ghosal et al.
1992 F.S. El-Feruly et al.
1999 K.C. Chan and H.T. Toh
2000 I. Murakoshi et al.
2008 M. Zhi-Da et al.
2069 W.M.L. Crombie and L. Crombie
2093 W.M. Kamil and P.M. Dewick
2155 R.V. Gerard et al.
2175 V. Gambaro et al.
2209 J. Borges del Castillo et al.
2219 G. Speranza et al.
2223 A. Rathore et al.
2227 A. Iwanow et al.
2241 I.A. Al-Meshal et al.
2245 M. Rahimizadeh et al.
2289 I.J.O. Jondiko
2309 S. Uesato et al.
2381 B. Esquivel et al.
2395 P.P. Diaz D. and A.M.P. de Diaz
2399 A.A. Ali et al.
2403 G. Sariyar and M. Shamma
2412 G. Morales B. et al.
2417 H.M.G. Al-Hazimi
2425 M. Garcia et al.
2543 Y. Asakawa et al.
2567 Y. Oshima et al.
2609 M.N. Lopes et al.
2613 D.A. Dias et al.
2625 A.V. Vyas and N.B. Mulchandani
2637 J.M. Llabres et al.
2681 T.J. Nagem and J.C. Da Silveira
2690 V.B. Pandey et al.
2692 A. Ulubelen et al.
2693 D. Cortes et al.
2739 A. Evidente et al.
2803 S.-I. Ayabe et al.
2821 H. Abdullahi et al.
2857 G. Savona et al.
2861 S. Miyakawa et al.
2897 H. Shimomura et al.
2910 P.J. Houghton and I.M. Said

103- -86, <u>Khim. Geterosikl. Soedin.</u>, <u>22</u> (1986)

0038 T.V. Moskovkina and M.N. Tilichenko
0052 T.F. Titova et al.
0057 B.E. Zaitsev et al.
0076 T.V. Sycheva et al.
0085 Y.A. Maurin'sh et al.
0167 V.N. Madakyan et al.
0172 K. Dagher et al.
0176 B.L. Moldarev et al.
0180 Y.M. Yutilov and K.M. Khabarov
0192 N.S. Prostakov et al.
0200 N.S. Prostakov et al.
0225 L.A. Tolochko et al.
0229 B.M. Krasovitskii et al.
0250 L.S. Shishkanova et al.
0257 L.N. Astakhova et al.
0268 L.N. Grigor'eva et al.
0287 N.V. Belova et al.
0318 V.N. Charushin et al.
0370 L.P. Glushko et al.

0397 A.S. Semeikin et al.
0401 B.A. Vigante et al.
0417 T.I. Reznikova et al.
0456 K.V. Anan'eva and N.K. Rozhkov
0499 A.N. Grinev et al.
0503 I.Y. Kvitko
0532 N.B. Chernyshova and V.N. Shibuev
0538 V.D. Orlov et al.
0543 V.L. Rusinov et al.
0564 K.V. Ahan'eva and N.K. Rozhkova
0567 Y.M. Shafran et al.
0608 V.P. Litvinov et al.
0629 A.S. Semeikin et al.
0680 L.M. Gol'denberg and R.N. Lyubovskaya
0703 I.M. Gavrilyuk et al.
0710 N.S. Prostakov et al.
0713 Y.B. Vysotskii et al.
0740 Y.M. Shafran et al.
0759 N.S. Kozlov et al.
0774 O.A. Zagulyaeva et al.
0795 V.A. Saprykina et al.
0800 L.G. Levkovskaya et al.
0844 D.S. Yufit et al.
0856 N.V. Dulepova et al.
0886 A.K. Sheinkman et al.
0918 N.N. Kolos et al.
0927 V.D. Orlov et al.
0943 V.G. Kul'nevich et al.
0960 V.N. Kopranenkov et al.
0980 A.T. Soldatenkov et al.
0990 E.L. Khanina et al.
1001 L.M. Yun et al.
1022 B.M. Krasovitskii et al.
1026 B.M. Krasovitskii et al.
1028 L.S. Afanasiudi et al.
1072 G.D. Krapivin et al.
1097 V.A. Lepikhin et al.
1118 V.N. Charushin et al.
1139 T. Yagodzinski et al.
1158 V.M. Feigel'man et al.
1186 Y.N. Kreitsberga et al.
1229 N.S. Kozlov et al.
1232 G.G. Moskalenko et al.
1242 S.G. Alekseev et al.
1250 T.L. Pilicheva et al.
1262 E.A. Andreeshchev et al.
1267 R.R. Dubure et al.
1276 A.A. Ladatko et al.
1295 I.I. Boiko et al.
1311 D.A. Partsvaniya et al.
1322 E.A. Zvesdina et al.
1336 A.D. Shutalev et al.
1342 G.V. Shishkin and V.N. Sil'nikov
1345 G.P. Tokmakov et al.
1363 L.M. Gornostaev and G.F. Zeibert
1373 A.D. Kachkovskii et al.

104- -86, Zhur. Organ. Khim., 22 (1986)
0055 V.P. Chuev et al.
0092 I.I. Lapkin et al.
0107 G.A. Tolstikov et al.
0124 B.L. Litvin et al.
0129 M.V. Gorelik et al.
0138 V.V. Litvak et al.
0144 L.M. Yagupol'skii et al.
0150 A.A. Ishchenko et al.
0158 S.M. Shevchenko et al.
0177 A.F. Shivanyuk et al.
0204 Y.S. Andreichikov et al.
0207 A.A. Nekhoroshev et al.
0230 Y.T. Abramenko et al.
0258 S.M. Makin et al.
0353 A.M. Sipyagin and V.G. Kartsev
0367 Y.N. Kreitsberga et al.
0379 M.V. Vasil'eva et al.
0420 B.A. Trofimov and A.N. Vavilova
0483 N.B. Kuplet-skaya et al.
0495 V.V. Nurgatin et al.
0503 V.P. Rybalkin et al.
0513 K.A. Petriashvili et al.
0547 M.V. Gorelik et al.
0563 V.F. Anikin et al.
0566 Y.E. Gerasimenko et al.
0582 N.P. Smirnova et al.
0599 V.A. Vasin et al.
0644 M.P. Sokolov and G.E. Avakyan
0647 D.A. Oparin et al.
0720 E.V. Malykhin et al.
0727 V.V. Sinev and A.V. Aleksandr
0748 L.V. Ektova et al.
0791 D.A. Oparin
0826 M.B. Taraban et al.
0894 N.V. Toropin et al.
0943 T.I. Solomentseva et al.
0947 M.V. Gorelik and V.I. Lomzakova
0957 A.I. Pavlyuchenko et al.
0973 A.F. Shivanyuk et al.
1079 A.P. Krysin
1114 N.B. Kuplet-skaya and L.A. Kazitsyna
1117 T.I. Savchenko and A.K. Yatsimirskii
1138 N.N. Alekhina et al.
1159 V.S. Chesnovskii and V.A. Shigalevskii
1192 L.M. Gornostaev and G.F. Zeibert
1302 A.V. Metelitsa et al.
1342 V.V. Mezheritskii et al.
1352 S.K. Sarkar et al.
1371 V.V. Kravchenko et al.
1375 A.I. Pavlyuchenko et al.
1389 M.I. Komendantov et al.
1407 S.P. Titova et al.
1457 G.A. Tolstikov et al.
1547 N.V. Kondratenko et al.
1554 V.B. Muratov and A.P. Kharchevnikov
1560 B.L. Litvin and V.T. Kolesnikov
1566 B.E. Zaitsev et al.
1589 N.G. Luk'yanenko et al.
1597 G.A. Tolstikov et al.
1598 G.A. Tolstikov et al.
1599 I.N. Domnin et al.
1602 V.A. Vinokurov et al.
1658 S.P. Korshunov et al.
1762 Y.A. Sharanin et al.
1786 S.M. Makin et al.
1805 G.A. Tolstikov et al.
1806 S.A. Ismailov
1814 T.A. Chibisova et al.
1865 V.A. Gailite et al.

1875 A.G. Tolstikov et al.
1943 I.G. Il'ina et al.
1968 A. Degutene et al.
1979 V.F. Anikin et al.
2085 V.V. Kravchenko et al.
2108 V.P. Rybalkin et al.
2117 L.M. Gornostaev et al.
2121 A.V. El'tsov et al.
2126 I.I. Potapochkina et al.
2131 Y.N. Kreitsberga and O.Y. Neiland
2163 A.B. Tomchin et al.
2173 A.B. Tomchin and I.M. Krylova
2186 V.S. Velezheva et al.
2213 R.R. Kostikov et al.
2230 M.A. Ikrina et al.
2313 A.A. Bogachev et al.
2329 V.F. Anikin et al.

105- -86, Khim. Prirodn. Soedin., 22 (1986)
0071 Z. Saatov et al.
0078 V.I. Akhmedzhanova et al.
0107 E.K. Batirov et al.
0112 I.M. Saibaeva et al.
0228 S. Narantuyaa et al.
0230 M.T. Ikramov et al.
0234 V.V. Melik-Guseinov and V.A. Mnatskanyan
0235 A. Karimov and K.L. Lutfullin
0236 I.A. Israilov et al.
0267 S. Narantuyaa et al.
0288 T.V. Ganenko et al.
0297 E. Saatov et al.
0300 N.D. Abdullaev et al.
0409 Z. Saatov et al.
0536 V.I. Bol'shakova et al.
0548 V.A. Raldugin et al.
0560 O.E. Vasina et al.
0601 V. Akhmedzhanova
0603 S.S. Yusupova et al.
0608 S.M. Adekenov et al.
0609 S.V. Serkerov and A.I. Aleskerova
0613 V.I. Roshcin et al.
0618 E.N. Zhukovich et al.
0684 I.A. Bessonova and S.Y. Yunusov
0732 G.B. Oganesyan et al.

106- -86, Die Pharmazie, 41 (1986)
0029 H.-J. Kallmayer and C. Tappe
0096 S. Leistner et al.
0571 J. Kracmar et al.
0703 B. Proksa et al.
0832 H.-J. Kallmayer and C. Tappe

107- -86, Synthetic Comm., 16 (1986)
0195 T. Koga et al.
0627 R.W. Curley, Jr. and C.J. Ticoras
0741 H. Gstach and J.G. Schantl
0957 P. Zhang and L. Li
1177 A. Zasada-Parzynska et al.
1469 J. Cooper et al.

108- -86, Israel J. Chem., 27 (1986)
0033 S. Schatzmiller et al.
0081 R.H. Baur and W. Pfleiderer

109- -86, Doklady Akad. Nauk S.S.S.R., Phys. Chem. Sect., 286-291 (1986)
0607 O.K. Bazil' et al.
1028 A.M. Sergeev et al.

110- -86, Russian J. Phys. Chem., 60 (1986)
0195 A.N. Petrov and Y. Nikozlov
0205 S.N. Shtykov et al.
0379 I.L. Belaits et al.
0857 N.O. Mohedlov-Petrosyan and V.F. Mindrina
1000 N.Y. Vasil'eva et al.
1023 A.I. V'yugin et al.
1666 G.A. Val'kova et al.

111- -86, European J. Med. Chem., 21 (1986)
0111 J.M.J. Tronchet et al.
0143 L. Campayo and P. Navarro
0385 M.N. Bakola-Christianopoulou
0527 C.G. Casinovi et al.

112- -86, Spectroscopy Letters, 19 (1986)
0033 B. Vidal et al.
0265 A. Darry-Henaut and B. Vidal
0311 D.A. Lightner et al.
0321 I.M. Khasawneh and J.D. Winefordner
0517 J. Rawlings and C.A.L. Mahaffy
0669 R. Salim and A.H. Laila
1049 M.P. Crozet et al.
1099 A. Darry-Henaut et al.

116- -86, Macromolecules, 19 (1986)
0196 R.V. Todesco and P.V. Kamat
0291 R. Roussel et al.
0509 G. Casiraghi et al.
0574 M. Gleria et al.
0968 S. Hayase et al.
1495 H.R. Allcock and T.X. Neenaz
2390 R.V. Todesco et al.
2397 K. Clausen and K. Brunfeldt
2472 H. Yamamoto
2932 M.R. Gomez-Anton et al.

117- -86, Org. Preps. Procedures Intl., 18 (1986)
0007 Y.S. Kulkarni and B.B. Snider
0209 F.V. Bright et al.

118- -86, Synthesis (1986)
0064 M. Ballester et al.
0071 D.R. Sliskovic, M. Siegel and Y. Lin
0190 L.-F. Tietze et al.
0196 A. Nyilas and J. Chattopadhyaya
0212 C. Boga et al.
0216 G. Seitz and K. Kohler
0239 G. Zvilichovsky and M. David
0406 S. Greenberg et al.
0430 M. d'Iscia and G. Prota
0450 V. Nair and D.A. Young
0461 T. Koga et al.
0473 A. Citteriu and L. Filippini
0484 N. Fukada et al.
0565 L. Fisera et al.

0620 W. Hinz et al.
0686 U. Seitz and J. Daub
0731 P.L. Southwick and D.S. Sullivan, III
0857 T. Kinoshita et al.
0860 R.C. Phadke and D.W. Rangnekar
0899 T. Eicher and D. Krause
0908 T. Eicher and W. Freihoff
0916 W. Schroth et al.
0942 K. Krohn and W. Baltus
0952 J. Becher et al.
0968 A. Sawluk and J. Voss
1027 K.T. Potts and J.M. Kane
1055 R. Carrau et al.

121- -86, J. Macromol. Sci., Chem. A23 (1986)
0355 K.-D. Ahn et al.
1233 M. Akashi et al.

126- -86, Makromol. Chem., 187 (1986)
0763 T. Koyama et al.
1573 E. de B. LoboFilho et al.
2369 S. Perlier et al.
2535 D. Wohrle and G. Krawczyk
2841 C. Bonnans-Plaisance and G. Levesque

126- -86B, Makromol. Chem., Rapid Comm., 7 (1986)
0205 M. Berthet et al.
0585 H. Itol et al.

128- -86, Croatica Chem. Acta, 59 (1986)
0027 A.R. Katritzky et al.
0183 R.C. Boruah et al.
0199 H. Lutz and W. Pfleiderer
0297 V. Skaric et al.
0491 V. Turjak-Zebic et al.

130- -86, Bioorg. Chem., 14 (1986)
0017 J.B. Noar et al.
0046 V.J. Davisson et al.
0103 I.A. McDonald et al.
0119 S. Shinkai et al.
0134 A. Kutner et al.

131- -86B, J. Mol. Structure, 136 (1986)
0275 K. Gustav and C. Seydenschwanz

131- -86C, J. Mol. Structure, 137 (1986)
0055 G. Buemi and F. Zuarello

131- -86E, J. Mol. Structure, 139 (1986)
0013 W. Nowak et al.

131- -86F, J. Mol. Structure, 140 (1986)
0353 L. Nyulaszi and T. Vezpremi

131- -86H, J. Mol. Structure, 142 (1986)
0167 I. Petrov and B. Simonovska
0197 E. Martin et al.

131- -86I, J. Mol. Structure, 143 (1986)
0345 J. Tortajada et al.
0375 L. Chmurzynski et al.
0509 E.S. Romero et al.
0513 E. Pastor et al.
0517 G. Galan et al.

131- -86N, J. Mol. Structure, 148 (1986)
0119 J. Singh et al.

135- -86, J. Appl. Spectr. U.S.S.R., 44-45 (1986)
0244 V.I. Alekseeva et al.
0967 N.I. Kunavich et al.
1208 B.M. Gutsulyak et al.

136- -86A, Carbohydrate Res., 146 (1986)
0107 J.C. Goodwin et al.

136- -86B, Carbohydrate Res., 147 (1986)
0237 F.J. Lopez Aparicio et al.

136- -86D, Carbohydrate Res., 149 (1986)
0329 A. Gomez-Sanchez et al.

136- -86G, Carbohydrate Res., 152 (1986)
0143 R.W. Myers and Y.C. Lee
0283 F.J. Lopez Herrera and M.S.P. Gonzalez
0316 L. Somogyi

136- -86H, Carbohydrate Res., 153 (1986)
0271 H.S. El Khadem and J. Kawai
0318 J.F. Mota et al.

136- -86I, Carbohydrate Res., 154 (1986)
0049 M.A. Gonzalez et al.

136- -86K, Carbohydrate Res., 156 (1986)
0256 T. Halmos et al.

136- -86M, Carbohydrate Res., 158 (1986)
0037 H. Ogura et al.
0053 M.A. Gonzalez et al.

137- -86, Finnish Chem. Letters (1986)
0119 R. Petrola et al.
0177 R. Petrola and R. Larja
0185 R. Petrola

138- -86, Chemistry Letters (1986)
0161 T. Shimizu and M. Kobayashi
0177 T. Nakajima et al.
0225 H. Matsunaga and T.M. Suzuki
0329 S. Nakatsuji et al.
0495 H. Obara et al.
0715 Y. Yamashita et al.
0781 K. Inoue et al.
0829 K. Nakasuji et al.
0907 M. Fujiwara et al.
0969 K. Nakasuji et al.
1021 K. Fujimori et al.
1059 K. Yoshida et al.
1075 S.C. Shim and T.S. Lee
1129 Y. Shizuri et al.
1137 E. Kimura et al.
1145 K. Miyamura and Y. Goshi
1217 Y. Tobe et al.
1279 Y. Yamashita et al.
1337 M. Nakatani et al.

1393 J. Setsune et al.
1399 K. Hashimoto et al.
1499 R. Sakai, T. Higa and Y. Kashman
1519 T. Kuwamura et al.
1537 M. Kajitani et al.
1585 A. Ohsaki et al.
1631 T. Ueda and Y. Otsuji
1643 H. Matsumoto et al.
1781 H. Sakurai et al.
1789 K. Iguchi et al.
1865 F. Moriwaki et al.
1925 M. Nitta and N. Kanomata
2025 A. Sekiguchi and W. Ando
2039 S. Kuroda et al.
2045 T. Nakazawa et al.
2139 T. Muramatsu et al.

139- -86A, P and S and Related Elements, 26 (1986)
0091 A.M. Caminade et al.

139- -86C, P and S and Related Elements, 28 (1986)
0337 I.H. Jo and D.-Y. Oh

140- -86, J. Anal. Chem. S.S.S.R., 41 (1986)
0140 V.Y. Veselov et al.
0239 L.N. Bazhenova et al.
0439 I.A. Zheltukhin et al.
0502 G.M. Pisichenko et al.
0584 G.K. Budnikov and O.Y. Kargina
0653 A.T. Pilipenko et al.
0957 A.T. Pilipenko et al.
1223 N.O. Mchedlov-Petrosyan

142- -86, Heterocycles, 24 (1986)
0001 S. Han-dong et al.
0009 T. Yamazaki et al.
0013 F.J. Perez and H.M. Niemeyer
0041 T.S. Wu et al.
0143 T. Hirota et al.
0335 H.R. Bravo and H.M. Niemeyer
0345 T. Fujii et al.
0355 C. Parkanyi et al.
0365 D.M. Perrine and J. Kagan
0437 J. Kunitomo and Y. Miyata
0579 K. Saito and Y. Horie
0601 M. Nagai and S. Nagumo
0607 D. Cortes et al.
0617 T. Tatsuoka et al.
0625 Y. Kosugi et al.
0679 M.V. Fernandez et al.
0687 J.M. Bobbitt et al.
0703 Atta-Ur-Rahman et al.
0771 T. Hirota et al.
0777 N. Langlois et al.
0785 A. Ohta et al.
0923 Y. Fukuda et al.
0967 J. Dusemund and E. Kroger
1119 T. Hirota et al.
1227 T. Gozler et al.
1251 Y. Hano et al.
1291 K. Saito and H. Ishihara
1381 Y. Hano and T. Nomura
1565 T. Sugimoto et al.
1567 M. Bert et al.
1807 Y. Hano, H. Tsubura and T. Nomura
1815 M.G. Knize and J.S. Felton
1831 K. Saito
1841 T. Okuda et al.
1893 S. Munavalli et al.
1943 M.J. Villa et al.
1973 H. Hodjat-Kashani et al.
2089 G. Ma et al.
2133 T. Tatsuoka et al.
2141 S. Kato et al.
2195 D. Sowmithran and K.J.R. Prasad
2233 S. Julia and C. Martinez-Martorell
2247 Z. Huang and Z. Liu
2261 N. Jayasuriya and J. Kagan
2285 Y. Hano et al.
2293 K. Hirota et al.
2299 J. Becher et al.
2403 H.B. Mereyala
2603 Y. Hano et al.
2639 J. Nakayama et al.
2731 L. Lin et al.
2749 M. Sindler-Kulyk et al.
2777 M. Ju-ichi et al.
2781 L. Castedo et al.
2901 N. Sayasuriya and J. Kagan
3023 Y.H. Kim and H.R. Kim
3071 Y. Tominaga et al.
3079 K.-K. Chan et al.
3097 J. Castells et al.
3181 G. Cusmano et al.
3351 T. Fukai et al.
3359 J.M. Boente et al.
3417 J. Banerji et al.
3451 P. Goya and A. Martinez
3467 S. Chimichi et al.

145- -86, Arzneimittel. Forsch., 36 (1986)
0784 G. Cleve et al.

149- -86A, Photochem. Photobiol., 43 (1986)
0121 R.D. Birdal and J.A. Katzenellenbogen
0221 H. Fujita and H. Kakishima
0291 K. Freedman et al.
0391 K.E. Yokobata et al.

149- -86B, Photochem. Photobiol., 44 (1986)
0001 M. Palumbo et al.
0075 T. Hamanake et al.
0125 H. Ohtani et al.
0159 A. Jankowski and P. Dobryszycki
0441 C. Evans et al.
0551 O.S. Wolfbeis et al.
0555 P.F. Aramendia et al.
0571 M. Krishnamurthy and S.K. Dogra
0689 B. Pilas et al.
0717 D. Praseuth et al.

150- -86M, J. Chem. Research (M), (1986)
0277 M. Loriga and G. Paglietti
0343 A.F. Casy et al.
0514 R.M. Letcher and J.S.M. Wai
0631 D. Armesto et al.
0670 R. Rydzkowski et al.
0718 M. Numazawa et al.

0735 P.R. Boshoff and G.W. Perold
0901 C. Kashima et al.
0920 T.Y. Chan and M.P. Sammes
1348 M. Dorta de Marquez et al.
1555 A.M. Lithgow et al.
1901 B.C. Uff et al.
2049 S. Kanoktanaporn et al.
2367 R.M. Acheson et al.
2384 M.N. Khan
2471 G. Paglietti and M. Loriga
2492 A.M. Mehta et al.
2625 T. Nishiwaki and T. Fukui
2731 G. Rapi et al.
2782 R.M. Letcher and J.S.M. Wai
3060 V. Bertini et al.
3101 S. Radzki et al.
3137 A. Echavarren et al.
3162 A. Echavarren et al.
3333 K.M. Smith et al.
3425 R.M. Acheson and D.H. Birtwistle
3514 A. Mylona et al.
3601 F.R. Ahmed and T.P. Toube

150- -86S, J. Chem. Research (S), (1986)
0024 V. Gold and W.N. Wassef
0028 S. Ghosal et al.
0088 C. Kashima et al.
0102 S.A. Gadir et al.
0104 A. Zeeck et al.
0110 L. Forlani and M. Sintoni
0112 S. Ghosal et al.
0116 T.V. Rao et al.
0196 J. Herzig et al.
0222 S.A. Gadir et al.
0312 S. Ghosal and S.K. Singh
0382 Y. Yano et al.
0433 A. Mylona et al.
0452 L.R. Milgrom

151- -86A, Photochem., 32 (1986)
0069 A.R. Reddy et al.
0105 W.H. Laarhoven and T.J.H.M. Cuppen
0331 K. Bhattacharyya et al.
0341 K. Kikuchi et al.
0369 S.C. Shim et al.

151- -86B, Photochem., 33 (1986)
0297 W.H. Laarhoven et al.

151- -86C, Photochem., 34 (1986)
0055 R.V. Subba Rao et al.
0209 P. Phaniraj et al.

151- -86D, Photochem., 35 (1986)
0177 G. Gennari et al.
0311 W. Abraham et al.
0353 K. Meier and H. Zweifel

152- -86, Nouveau J. Chim., 10 (1986)
0017 J. Canceill and A. Collet
0165 F. Arnaud-Neu et al.
0213 T. Shimidzu et al.
0399 C. Roussel and A. Djafri
0643 A.F. Maggio et al.

154- -86, J. Chem. Ecology, 12 (1986)
0561 M. Chang and D.G. Lynn
1205 K.L. Stevens

155- -86, J. Fluorine Chem., 34 (1986)
0201 M.J. Mokrosz

156- -86A, Photobiochem. Photobiophys., 11 (1986)
0139 A. Chow et al.

157- -86, Organometallics, 5 (1986)
0128 T.A. Blinka and R. West
0168 Y. Aoyama et al.
0218 J.P. Collman et al.
0531 M.J. Michalczyk et al.
1116 M.P. Castellani et al.
1395 C.G. Marcellus et al.
1449 H. tom Dieck et al.
2250 M.R. Detty and H.R. Luss
2500 P. Lei and P. Vogel

158- -86, J. Antibiotics, 39 (1986)
0001 T. Okabe et al.
0012 Y. Hori et al.
0017 H. Maehr et al.
0026 L.A. Dolak et al.
0038 K. Yano et al.
0121 W.J. Wheeler et al.
0149 A. Hirota et al.
0157 J. Mochizuki et al.
0164 H. Kachi et al.
0184 M. Yamato et al.
0198 M. Okamoto et al.
0309 S. Omura et al.
0311 M. Hotta et al.
0316 T. Okabe et al.
0345 D. Ling et al.
0415 K. Shibata et al.
0424 S. Iwasaki et al.
0430 M. Li and Y.-L. Chen
0469 D.L. Kern et al.
0487 K. Shiomi et al.
0494 K. Shiomi et al.
0601 H. Imai et al.
0624 K. Takahashi et al.
0629 D. Rodphaya et al.
0721 D.G. Martin et al.
0727 S. Kondo et al.
0737 H. Umezawa et al.
0762 S. Kiyoto et al.
0780 J. Itoh et al.
0800 T.A. Smitka et al.
0856 H.-P. Fiedler et al.
1013 K. Isshiki et al.
1016 H. Nakayama et al.
1047 H. Yamaguchi et al.
1066 S. Nakanishi et al.
1072 T. Yasuzawa et al.
1092 T. Naito et al.
1108 H.A. Kirst et al.
1178 M. Nakagawa et al.
1205 H. Tomoda et al.
1211 S. Omura et al.

1288 M. Kitahara et al.
1343 K. Tsuzaki et al.
1361 A.J. Kempf et al.
1430 A. Malabarba et al.
1495 A. Kawashima et al.
1515 R.R. Rasmussen et al.
1527 K. Nagaoka et al.
1551 I. Kawamoto et al.
1623 N. Shimada et al.
1639 K. Yazawa et al.
1651 T.R. Hurley et al.
1657 H. Drautz et al.
1685 S. Sano et al.
1724 H.A. Kirst et al.
1760 U. Fauth et al.
1776 S. Mizobuchi et al.
1779 K. Dobashi et al.

159- -86, J. Carbohydrate Chem., 5 (1986)
0049 M.A.E. Sallam and S.M.E.A. Megid
0631 J.M.J. Tronchet et al.

160- -86, Anal. Letters, 19 (1986)
0025 Y.K. Agrawal and K.H. Desai
0177 S.E. Hamilton et al.
0713 G.V. Rathaiah and M.C. Eshwar
0979 A.A. Gallo and F.H. Walters
1533 F. Belal et al.
1853 B. Gallo et al.

161- -86, Il Farmaco, 41 (1986)
0499 G. Cristalli et al.
0539 G. Romussi et al.
0666 G. Auzzi et al.
0881 M. Dzieduszycka et al.

163- -86, Agr. Biol. Chem., 50 (1986)
0143 M. Yoshida et al.
0187 Y. Nishida et al.
0239 M. Awata et al.
0263 T. Washino et al.
0565 T. Washino et al.
0569 S. Takeuchi et al.
0801 Y. Kono et al.
0809 T. Nishitoba et al.
0813 Y. Nishida et al.
0997 T. Okuno et al.
1139 K. Akasaka et al.
1355 E. Matsumura et al.
1589 S. Takahashi et al.
1597 Y. Kono et al.
1649 Y. Kimura et al.
1655 A. Okamoto et al.
1669 H. Kachi and T. Sassa
2003 N. Morooka et al.
2243 S. Nishikawa et al.
2315 G. Saito et al.
2427 K. Konishi et al.
2655 H. Ohashi et al.
2667 H.G. Cutler et al.
2689 Y. Kono et al.
2693 M. Murata et al.
2697 M. Nishii et al.
2831 R. Nishida et al.
2887 H. Sato et al.
2943 H.G. Cutler et al.
3053 M.G. Nair et al.
3083 J. Ueda et al.
3179 K. Yamazawa et al.
3209 S. Hayashi and K. Mori

164- -86, Pure and Applied Chem., 58 (1986)
0007 M. Oda
0197 J.L. Morris and C.W. Rees
0415 G.R. Pettit et al.
0663 Atta-Ur-Rahman
0719 F. Dergwini et al.

165- -86, Acta Phys.Chem.Univ. Seged., 32 (1986)
0017 J. Csaszar